The chemistry of the
metal–carbon bond
Volume 1

THE CHEMISTRY OF FUNCTIONAL GROUPS

A series of advanced treatises under the general editorship of Professor Saul Patai

The chemistry of alkenes (2 volumes)
The chemistry of the carbonyl group (2 volumes)
The chemistry of the ether linkage
The chemistry of the amino group
The chemistry of the nitro and nitroso groups (2 parts)
The chemistry of carboxylic acids and esters
The chemistry of the carbon—nitrogen double bond
The chemistry of amides
The chemistry of the cyano group
The chemistry of the hydroxyl group (2 parts)
The chemistry of the azido group
The chemistry of the acyl halides
The chemistry of the carbon—halogen bond (2 parts)
The chemistry of the quinonoid compounds (2 parts)
The chemistry of the thiol group (2 parts)
The chemistry of the hydrazo, azo and azoxy groups (2 parts)
The chemistry of the amidines and imidates
The chemistry of cyanates and their thio derivatives (2 parts)
The chemistry of diazonium and diazo groups (2 parts)
The chemistry of the carbon—carbon triple bond (2 parts)
Supplement A: The chemistry of double-bonded functional groups (2 parts)
Supplement B: The chemistry of acid derivatives (2 parts)
The chemistry of ketenes, allenes and related compounds (2 parts)
The chemistry of the sulphonium group (2 parts)
Supplement E: The chemistry of ethers, crown ethers, hydroxyl groups and their sulphur analogues
Supplement F: The chemistry of amino, nitroso and nitro compounds and their derivatives
Supplement C: The chemistry of triple-bonded functional groups (2 parts)
The chemistry of the metal—carbon bond Volume 1

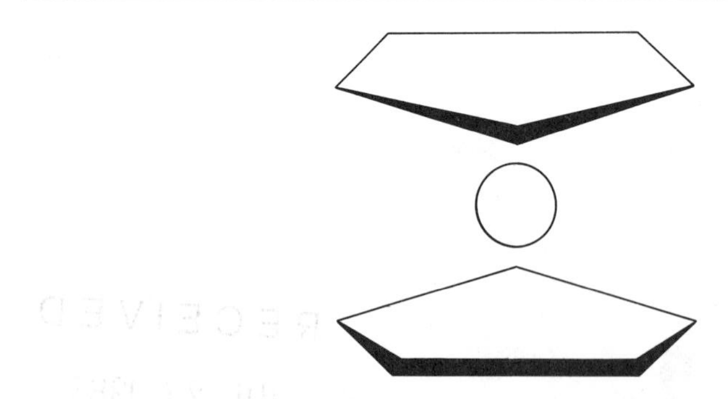

The chemistry of the metal–carbon bond

Volume 1

The structure, preparation, thermochemistry and characterization of organometallic compounds

Edited by

FRANK R. HARTLEY

The Royal Military College of Science
Shrivenham, England

and

SAUL PATAI

The Hebrew University
Jerusalem, Israel

1982

JOHN WILEY & SONS

CHICHESTER – NEW YORK – BRISBANE – TORONTO – SINGAPORE

An Interscience ® Publication

British Library Cataloguing in Publication Data:

The chemistry of the metal–carbon bond.
  Vol. I – (The chemistry of functional groups)
  1. Organometallic compounds
  I. Hartley, Frank R.
  II. Patai, Saul
  547′.05   QD411

ISBN 0 471 10058 7

Typeset by Preface Ltd., Salisbury, Wiltshire, and printed in the United States of America by Vail-Ballou Press., Binghamton, N.Y.

Volume 1—Contributing authors

T. R. Crompton	"Beechcroft", Whittingham Lane, Goosnargh, Near Preston, Lancashire, UK
Julian A. Davies	Department of Chemistry, University of Toledo, Toledo, Ohio 43606, USA
William J. Evans	Department of Chemistry, University of Chicago, 5735 South Ellis Avenue, Chicago, Illinois 60637, USA
Helmut Fischer	Anorganisch-Chemisches Institut der Technischen Universität München, Lichtenbergstrasse 4, D-8046 Garching, German Federal Republic
Glenn S. Lewandos	Department of Chemistry, Central Michigan University, Mount Pleasant, Michigan 48859, USA
George Marr	Department of Physical Sciences, The Polytechnic, Wolverhampton WV1 1LY, UK
Michael J. McGlinchey	Department of Chemistry, McMaster University, 1280 Main Street West, Hamilton, Ontario, L8S 4M1, Canada
Geoffrey Pilcher	Department of Chemistry, University of Manchester, Manchester M13 9PL, UK
Paul Powell	Department of Chemistry, Royal Holloway College, Egham Hill, Surrey TW20 0EX, UK
Richard J. Puddephatt	Department of Chemistry, University of Western Ontario, London, Ontario N6A 5B7, Canada
Bernard W. Rockett	Department of Physical Sciences, The Polytechnic, Wolverhampton WV1 1LY, UK
Alan D. Redhouse	Department of Chemistry and Applied Chemistry, University of Salford, Salford M5 4WT, UK
Ulrich Schubert	Anorganisch-Chemisches Institut der Technischen Universität München, Lichtenbergstrasse 4, D-8046 Garching, German Federal Republic
Henry A. Skinner	Department of Chemistry, University of Manchester, Manchester M13 9PL, UK
Trevor Spalding	Department of Chemistry, University College, Cork, Ireland
Michael J. Taylor	Department of Chemistry, University of Auckland, Auckland 1, New Zealand
Lothar Weber	Fachbereich Chemie, Universität Essen, Universitätsstrasse 5–7, D-4300 Essen 1, German Federal Republic

Foreword

The Chemistry of the Metal—Carbon Bond will be a multi-volume work within the well established series of books covering *The Chemistry of Functional Groups*. It aims to cover the chemistry of the metal—carbon bond as a whole, but lays emphasis on the carbon end. It should therefore be of particular interest to the organic chemist. The general plan of the material will be the same as in previous volumes with the exception that, because of the large amount of material involved, this will be a multi-volume work. This present volume is concerned with:

(a) Structure and thermochemistry of organometallic compounds.
(b) The preparation of organometallic compounds.
(c) The analysis and spectroscopic characterization of organometallic compounds.

Chapters on the theoretical understanding of the metal—carbon bond and the preparation of organometallic compounds of the main group metals have not yet been completed. They will be included in the second volume, which will also cover metal—carbon bond cleavage reactions. Later volumes will be concerned with the use of organometallic compounds for the formation of new carbon—carbon, carbon—hydrogen and other carbon—element bonds. In classifying organometallic compounds we have used Cotton's hapto-nomenclature (η-) to indicate the number of carbon atoms directly linked to a single metal atom.

In common with other volumes in *The Chemistry of the Functional Groups* series, the emphasis is laid on the functional group treated and on the effects which it exerts on the chemical and physical properties, primarily in the immediate vicinity of the group in question, and secondarily on the behaviour of the whole molecule. The coverage is restricted in that material included in easily and generally available secondary or tertiary sources, such as Chemical Reviews and various 'Advances' and 'Progress' series as well as textbooks (i.e. in books which are usually found in the chemical libraries of universities and research institutes) is not, as a rule, repeated in detail, unless it is necessary for the balanced treatment of the subject. Therefore each of the authors has been asked *not* to give an encyclopaedic coverage of his subject, but to concentrate on the most important recent developments and mainly on material that has not be adequately covered by reviews or other secondary sources by the time of writing of the chapter, and to address himself to a reader who is assumed to be at a fairly advanced postgraduate level. With these restrictions, it is realised that no plan can be devised for a volume that would give a *complete* coverage of the subject with *no* overlap between the chapters, while at the same time preserving the readability of the text. The Editors set themselves the goal of attaining *reasonable* coverage with *moderate* overlap, with a minimum of cross-references between the chapters of each volume. In this manner sufficient freedom is given to each author to produce readable quasi-monographic chapters. Such a plan necessarily means that the breadth, depth and thought-provoking nature of each chapter will differ with the views and inclinations of the author.

The publication of the Functional Group Series would never have started without the support of many people. Foremost among these is Dr Arnold Weissberger, whose reassurance and trust encouraged the start of the task and who continues to help and advise. This volume would never have reached fruition without Mrs Trembath's help with typing and the efficient and patient cooperation of several staff members of the Publisher whose code of ethics does not allow us to thank them by name. Many of our colleagues in England, Israel and elsewhere gave great help in solving many problems, especially Professor Z. Rappoport. Finally, that the project ever reached completion is due to the essential support and partnership of our wives and families.

Shrivenham, England — FRANK HARTLEY
Jerusalem, Israel — SAUL PATAI

Contents

List of Abbreviations Used

A	appearance potential
acac	acetylacetone
ac	acrylonitrile
all	allyl
An	actinide metal
ap	*o*-allylphenyldiphenylphosphine
appe	$Ph_2AsCH_2CH_2PPh_2$
bipy	2,2′-bipyridyl
cdt	*E*,*E*,*E*-1,5,9-cyclododecatriene
cht	cycloheptatriene
CI	chemical ionization
1,5-cod	1,5-cyclooctadiene
cot	cyclooctatetraene
Cp	η^5-cyclopentadienyl
C.P.	cross polarization
Cy	cyclohexyl
dba	dibenzylideneacetone
dccd	dicyclohexylcarbodiimide
def	diethyl fumarate
dem	diethyl maleate
dme	1,2-dimethoxyethane
dmf	dimethyl formamide
dmfm	dimethyl fumarate
dmg	dimethyl glyoxime
dmm	dimethyl maleate
dmpe	bis(1,2-dimethylphosphino)ethane
dmso	dimethyl sulphoxide
dpm	dipivaloylmethanato
dppb	bis(1,4-diphenylphosphino)butane
dppe	bis(1,2-diphenylphosphino)ethane
dppm	bis(1,1-diphenylphosphino)methane
dppp	bis(1,3-diphenylphosphino)propane
EI	electron impact
Fc	ferrocene
FD	field desorption
FI	Field ionization
FID	flame-ionization detector in Chapter 21 free induction decay in Chapter 22
fmn	fumaronitrile
fod	$F_3C(CF_2)_2COCH{=}C(O)C(CH_3)_3$

Fp	$Fe(\eta^5\text{-}C_5H_5)(CO)_2$
FT	Fourier transform
hfacac	hexafluoroacetylacetonato
hmdb	hexamethyl(Dewar)benzene
hmpt	hexamethylphosphorotriamide
I	ionization potential
ICR	ion cyclotron resonance
INDOR	inter-nuclear double resonance
LCAO	linear combination of atomic orbitals
Ln	lanthanide metal
M	metal
M	parent molecule (in Chapter 22)
ma	maleic anhydride
MAS	magic angle spinning
nbd	norbornadiene
oA	*o*-allylphenyldimethylarsine
phen	*o*-phenanthroline
ppm	parts per million
pz	pyrazolyl
sp	*o*-styryldiphenylphosphine
SPT	selective population transfer
tba	tribenzylideneacetylacetone
tcne	tetracyanoethylene
thf	tetrahydrofuran
tmeda	tetramethylethylenediamine
tms	tetramethylsilane
tond	1,3,5,7-tetramethyl-2,6,9-trioxobicyclo[3.3.1]nona-3,7-diene
un	olefin or acetylene
X	halide

The Chemistry of the Metal–Carbon Bond
Edited by F. R. Hartley and S. Patai

CHAPTER **1**

Structure of organometallic compounds

A. D. REDHOUSE
Department of Chemistry and Applied Chemistry, University of Salford, Salford M5 4WT, UK

I. INTRODUCTION

The increasing interest in the synthesis and chemistry of organometallic compounds during the period 1950–65 was accompanied by the desire to know more about the structures of these complexes. It was fortunate that this desire came at a time of great advance in X-ray structural analysis. Many organometallic compounds were ideally suited to the application of the 'heavy atom' method of phase determination and this, together with the very rapid improvement in computing facilities, enabled the structural features of these complexes to be revealed. More recent improvements in data collection and handling, and also in methods of structure solution, have led to a continual expansion in the number of structure determinations of organometallic compounds. If one adds to these the structural work carried out using electron and neutron diffraction and also the various spectroscopic techniques, the amount of structural information available to the organometallic chemist is truly overwhelming. To survey the whole of structural organometallic chemistry would be a task of Herculean proportions and perhaps as pointless as many of those classical labours. It has therefore been necessary to restrict the aims of this chapter.

A. The Aims of the Chapter

The Patai series of textbooks is directed towards the research worker in organic chemistry, who need not be an organometallic chemist, and is almost certainly not a crystallographer. The aims of this chapter on the structures of organometallic compounds have been determined with this in mind and are as follows:

(a) to survey the main types of organometallic structure, keeping this within manageable bounds;
(b) to provide details of the structural effect on the organic group of being bound to a metal atom rather than detailing metal geometry or overall molecular structure;
(c) to provide a starting point from which the interested organic chemist may quickly increase his knowledge of the structural chemistry of any of the main types of organometallic complexes.

To try to achieve these aims the following guidelines have been used.

(i) The structures described are mainly solid-state structures determined using X-ray and, more rarely, neutron diffraction methods.

(ii) The structures have been classified in terms of the number of carbon atoms in the ligand which can be considered to be bound to the metal atom.
(iii) In order to reduce the number of structures described for each class, the reader is referred wherever possible to sources in the literature which contain comprehensive lists of structure determinations relevant to that class.
(iv) Many of the structures reported in the chapter are recent examples of types originally reported a number of years ago. This is because reference to the most recent structural papers will frequently lead the reader back through the literature, by way of similar complexes, down to the prototype. Thus, what the examples quoted have lost in terms of primacy it is hoped that they have gained as starting points for literature searches and also, probably, in accuracy of structural parameters.

B. Sources of Structural Data on Crystalline Organometallic Complexes Obtained by Diffraction Methods

In addition to the normal journals reporting structural data for organometallic complexes, there are several sources which provide comprehensive listings of such data on an annual basis. They vary in content and also in method of access. Thus some provide considerable structural detail and others simply a bibliography; some are available in book form and others as computer files. A useful general guide to these sources is given in an article by M. R. Truter[1]. The two most comprehensive general sources are:

(i) *Structure Reports.* An annual survey of all published crystal structures carried out under the auspices of the International Union of Crystallography.
(ii) *The Cambridge Crystallographic Data Files.* Computer-based files containing bibliographic and structural details of all organic and organometallic structure determinations from 1935 onwards. The files are available in a growing number of countries and are compiled by the Cambridge Crystallographic Data Centre, Lensfield Road, Cambridge, UK. The bibliographic files are published in book form under the title *Molecular Structures and Dimensions*.

Two annual compilations of structural data for organometallic complexes only are published in:

(a) *Organometallic Chemistry*, Specialist Periodical Report, Royal Society of Chemistry, London, UK.
(b) *Journal of Organometallic Chemistry*. Annual Surveys.

C. Structural Results from X-ray Diffraction Data

The bond lengths and angles quoted in this chapter are followed by their estimated standard deviations in parentheses, where these are available. However, it should be noted that the standard deviations are derived from parameters obtained by least-squares procedures which assume that all of the data errors are random and that no errors reside in the theoretical model. These assumptions are not strictly valid and care should therefore be exercized in the interpretation of the significance of small differences in bond lengths and angles.

II. σ-BONDED (MONOHAPTO) LIGANDS

Complexes involving metal—carbon σ-bonds can be classified as follows:

Class A: complexes where the organic ligand can be described as an anionic σ-donor(alkyls, aryls, acyls, etc.).

Class B: complexes where the organic ligand can be described as a neutral σ-donor/π-acceptor(carbenes, carbynes).

These two broad classes will be subdivided into structural groups where the complexes contain (i) terminal metal—carbon bonds, (ii) bridging metal—carbon bonds. The structures of complexes containing only π-acceptor ligands, such as carbonyls and isonitriles, are beyond the scope of this chapter.

A. Class A(i) Structures

1. Alkyl, alkenyl, alkynyl, and aryl complexes

Alkyl compounds involving most metals have been prepared. Those containing the most electropositive metals are in general ionic[2] and the bonding and structures of the other alkyls are normally those which would be expected. Indeed, the main structural interest frequently centres on the metal atom rather than the organic ligand to the extent that such complexes are a rich source of unusual stereochemistries, oxidation states and multiple metal—metal bonds.

Thus, in $[\{(Me_3Si)_2CH\}_3Cr]$ the chromium is three-coordinate (**1**) and 0.32 Å out of the plane defined by the ligated carbon atoms[3]. The Cr—C distance is 2.07(1) Å. The complex $[Cr(CO)_5(Ge\{CH(SiMe_3)_2\}]$ contains Ge(II) in a planar three-coordinate environment (**2**)[4]. The Ge—C distances are 1.984(7) and 1.989(6)

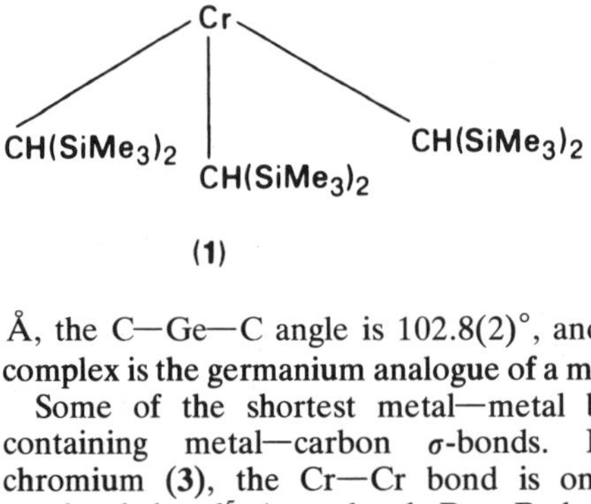

Cr
CH(SiMe$_3$)$_2$ CH(SiMe$_3$)$_2$ CH(SiMe$_3$)$_2$

(**1**)

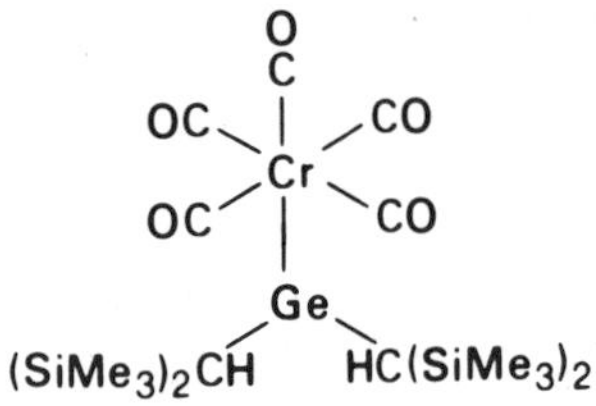

O C OC CO Cr OC CO Ge (SiMe$_3$)$_2$CH HC(SiMe$_3$)$_2$

(**2**)

Å, the C—Ge—C angle is 102.8(2)°, and the Ge—Cr distance is 2.406(1) Å. This complex is the germanium analogue of a metal—carbene complex (see Section II.B.1).

Some of the shortest metal—metal bonds have been reported in complexes containing metal—carbon σ-bonds. In tetrakis(2-methoxy-5-methylphenyl)dichromium (**3**), the Cr—Cr bond is only 1.828(2) Å and represents a Cr—Cr quadruple bond[5]. A quadruple Re—Re bond is found in the octamethyldirhenate(III) ion (**4**); the $ReMe_4$ units are eclipsed with Re—Re and Re—C distances of 2.198(1) and 2.19(1) Å, respectively[6].

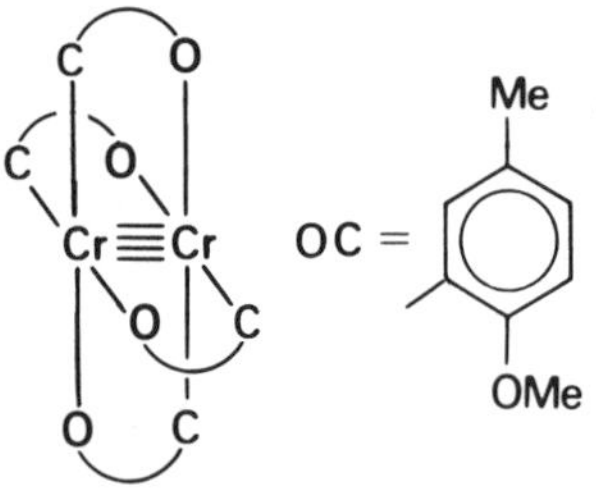

C O C O Cr≡Cr O C O C   OC = (Me, OMe-substituted ring)

(**3**)

$[Me_4Re{\equiv}ReMe_4]^{2-}$

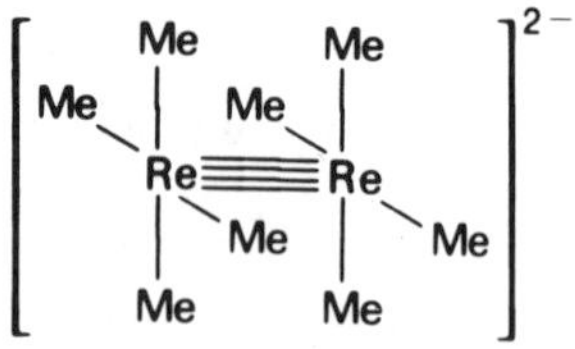

(**4**)

The length of the metal—carbon bond in metal alkyls, in general, compares favourably with calculated values using appropriate covalent radii. For example, the Re—C(sp^3) distances in $[(\eta^5\text{-}C_5H_5)(CO)_2HRe(CH_2Ph)]$[7] and $[ReMe_2(\eta^4\text{-}C_5H_5Me)(\eta^5\text{-}C_5H_5)]$[8] are 2.29(1) Å and 2.232(26), 2.251(56) Å, respectively, and are close to the calculated Re(II)—C(sp^3) distance of 2.28 Å. However, care must be taken in using calculated bond distances owing to the uncertainty in the values of the radii for many metal atoms.

When the metal—carbon bond length in alkyl complexes deviates from that expected, it is normally because of steric factors. Thus, in $[(C_5H_5)_4Sn]$ the lengthening of the Sn—C bond to 2.23–2.29(3) Å from 2.17 Å (calculated) is ascribed to steric hindrance[10]. The tin coordination geometry is a slightly distorted tetrahedron described by the σ-bonded C_5H_5 groups (**5**).

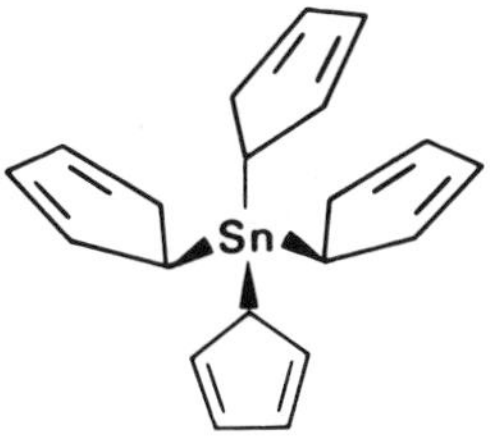

Sn

(**5**)

In metal aryls, alkenyls and alkynyls the metal—carbon distance should be shorter than that observed in the corresponding alkyls because of the smaller carbon atom radius and the possibility of multiple bond character in such a bond. Sometimes this difference is not detected, as in $[SnPh_3(CH_2I)]$, where the Sn—C(aryl) and Sn—C(alkyl) distances of 2.133(6) and 2.134(6) Å are not significantly different[11]. Similarly, in $[Cr\{C(Ph){=}CMe_2\}_4]$ the Cr—C bonds are equal within experimental error, averaging 2.03(2) Å[12], and this is in the range 2.01–2.07(3) Å found in the complex $[Cr(CH_2CMe_2Ph)_4]$[13]. This observation is taken to indicate that there is no significant multiple bond character in the metal—carbon bond in such complexes.

Evidence supporting an identical proposal for transition metal—alkynyl complexes is found in the structure of *trans*-chlorobis(diethylphenylphosphine)(phenylethynyl)-platinum(II), where the observed Pt—C(sp) distance of 1.98(2) Å is similar to those reported for a range of Pt(II)—alkynyl complexes[14] and agrees with the predicted value of 2.00 Å calculated using the covalent radii of Pt(II) and C(sp). The C≡C bond distance in the complex is 1.18(3) Å.

However, considerable shortening of the transition metal—C(aryl) bond has been reported. Thus, in tris(η^1-phenyl)bis(diethylphenylphosphine)rhenium the shortening of the Re—C(sp^2) bond to 2.024–2.029(11) Å from the calculated 2.22 Å value is believed to indicate significant metal $d\pi$–phenyl $p\pi^*$ back donation[15]. An interesting feature of this molecule is that all of the phenyl groups lie in the equatorial plane of the trigonal bipyramidal ligand arrangement, and are nearly coplanar (**6**). Complete planarity is restricted by the close approach of the *ortho*-H atoms.

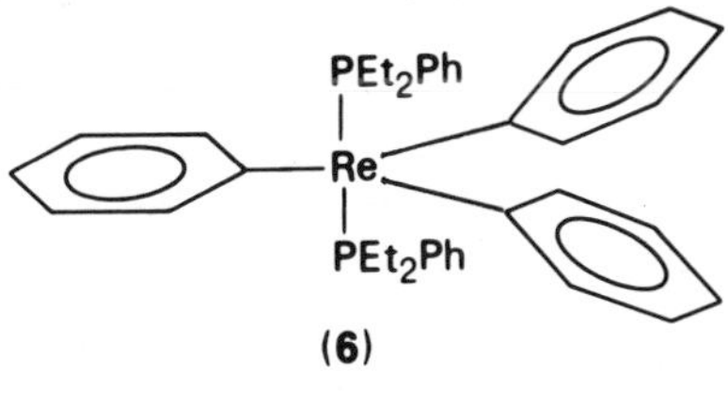

PEt_2Ph
Re
PEt_2Ph

(**6**)

2. *Acyl complexes*

Multiple bonding in the metal—carbon bond in acyl complexes is expected because of the possible interaction between the metal $d\pi$ orbitals and the $p\pi^*$ orbitals of the acyl group. Shortening of the M—C(acyl) bond, which could be in part related to such an interaction, is observed in (maleonitriledithiolato)bis(triethylphosphine)-propanoylrhodium(III)[16], where the Rh—C(acyl) distance of 2.002(7) Å is significantly shorter than most Rh(III)—C σ-bonds, which range from 2.05 to 2.26 Å[17].

In their report of the structure of $[(C_5H_4C_6H_6CO)Fe_2(CO)_5]$ (**7**), Churchill and Chang tabulate known Fe—C(σ) bond distances for iron alkyls (2.06–2.11 Å), alkenyls (1.987–1.996 Å), and alkynyls (1.906–1.920 Å)[18]. They compare these values with the observed Fe—C(acyl) distance [1.9596(30) Å] in the complex and conclude that this bond must have some multiple bond character.

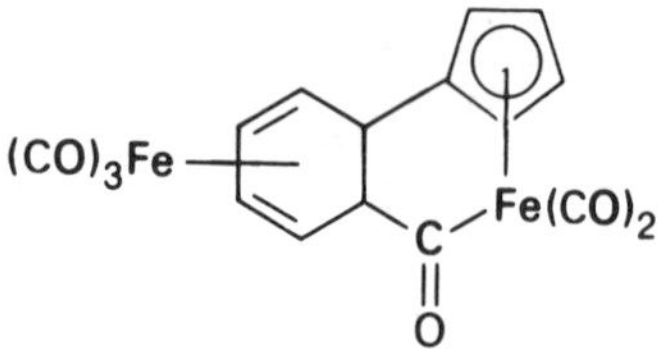

$(CO)_3Fe$ $Fe(CO)_2$ C O

(**7**)

The significant differences observed in the M—C(acyl) distances in different complexes of the same transition metal are presumably related to the extent of the metal—ligand π-interaction. Thus, the Fe—C(acyl) distances in the metallo-β-diketonate complex $[(\eta^5\text{-}C_5H_5)(OC)Fe(MeCO)(i\text{-}C_3H_7CO)]BF_2$ are 1.859(2) and 1.867(2) Å[19], approximately 0.1 Å shorter than those observed in Churchill and Chang's complex. Again, the Re—C(acyl) distance in *cis*-$[(CO)_4Re(COMe)(NH_2Ph)]$ is 2.211(6) Å[20], whereas in the formyl complex $[(\eta^5\text{-}C_5H_5)Re(PPh_3)(NO)(CHO)]$ it is 2.055(10) Å[21]. The acyl C—O distances reported for the latter two complexes of 1.214(7) and 1.220(12) Å, respectively, are not significantly different.

The substituent on the acyl carbon does not appear to affect the metal—carbon bond length unduly. In $[(MeCO)(PhCO)(CO)_4Mn]^-$, where two different acyl groups are attached to the same metal, the Mn—C(acetyl) and Mn—C(benzoyl) distances of 2.045(11) Å and 2.091(11) Å, respectively, are not significantly different[22], and this is also the case for the C—O distances of 1.203(13) Å (acetyl) and 1.218(13) Å (benzoyl). The acyl groups are in the *cis*-positions (**8**), with both oxygen atoms

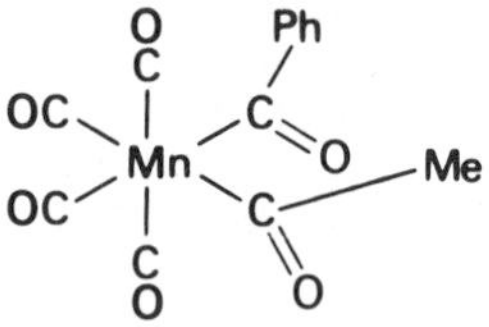

O C Ph OC C Mn O Me OC C C O O

(**8**)

orientated in the same direction with respect to the plane defined by the Mn and the two acyl carbon atoms. The planes of the acetyl and benzoyl groups are twisted out of the equatorial plane by 73.9° and 66.1°, respectively, the oxygen atoms being tilted away from each other.

The impetus for the determination of the structures of many metal—acyl complexes has been provided by the long-standing interest in the 'carbonyl insertion reaction', which is of fundamental importance in many processes involving homogeneous catalysis. Calderazzo[23] and Berke and Hoffmann[24], although primarily concerned with the mechanistic and theoretical aspects of carbonyl insertion, provide useful sources of structural information on metal—acyl complexes.

3. *Ylide complexes*

The structures of complexes containing ligands derived from ylide systems have been determined. Phosphorus ylides, $R_3P^+—C^-R_2$, provide the most frequently encountered ligands of this type[25,26].

In the carbonyl—ylide complex $[(CO)_3Ni\{CH(Me)—P(C_6H_{11})_3\}]$ **(9)** the Ni—C(ylide) and P—C(ylide) are 2.09(6) and 1.74(5), respectively[27]. This complex shows the ylide ligand in a terminal bonding mode, whereas in the dimer $[\{Me_2P(CH_2)_2\}_2Ni]_2$ the structure determination[28] reveals the use of the ylide as a chelating and a bridging ligand **(10)**. The Ni—C (chelating) and P—C (chelating) distances are 2.031(3) and 1.736(3) Å, respectively, the equivalent distances for the bridging ylide being 1.978(3) and 1.754(3) Å.

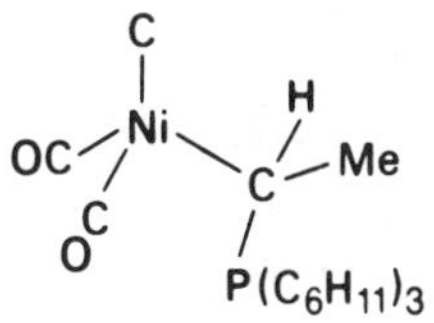

(9)

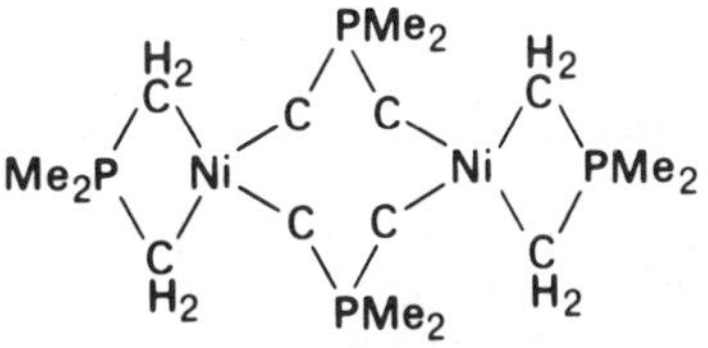

(10)

The use of ylide ligands as bridges between transition metals has been exploited by Cotton and his colleagues in their studies of multiple metal—metal bonds. Thus, the Cr—Cr and Mo—Mo distances of 1.895(3) and 2.082(2) Å in the ylide complexes $[M_2\{(CH_2)_2PMe_2\}_4]$, where M = Cr or Mo, correspond to metal—metal quadruple bonds **(11)**. However, it should be noted that the metal—C(ylide) distances observed

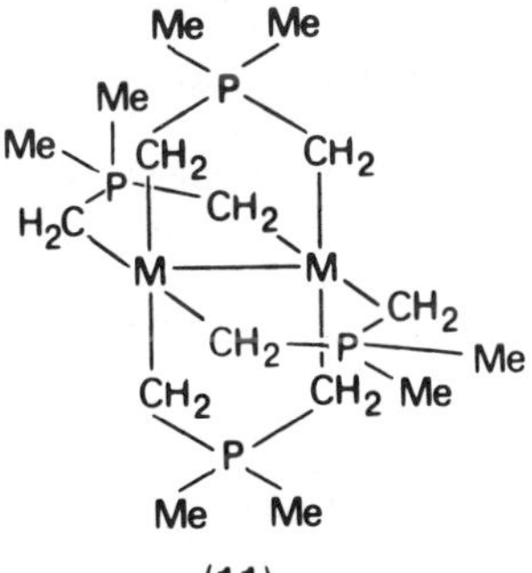

(11)

in these complexes are not significantly different from the metal—alkyl distances observed in the corresponding methyl complexes $[M_2Me_8]^{4-}$, for example the relevant average Cr—C distances are 2.22(1) Å in the ylide complex and 2.199(13) Å in the alkyl complex. Nevertheless, consideration of the structural parameters for a number of complexes containing the $[Me_2P(CH_2)_2]^-$ ligand[29] shows that the P—CH_2 bond is always shorter than the P—CH_3 bond and that the CH_2—P—CH_2 angle has expanded to 114°. These observations are consistent with the retention of ylidic character by the ligand.

B. Class A(ii) Structures (Complexes with Bridging Ligands)

The alkyls of lithium, beryllium and aluminium are associated and contain bridging alkyl groups. Thus, in dimeric trimethylaluminium the two tetrahedrally coordinated aluminium atoms are linked by two methyl bridges (**12**)[30]. The Al—C(bridge) bond

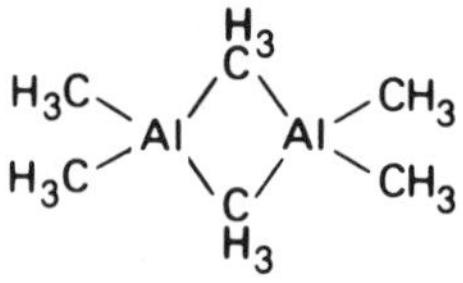

(**12**)

(2.15 Å) is longer than the Al—C(terminal) bond (1.96 Å) and the Al—C (bridge)—Al angle is 74.7°. This pattern of metal—carbon distances and very acute bridging angles is typical of this type of complex. Another feature observed in Me_6Al_2 which is found in similar complexes is the short metal—metal distance which at 2.60 Å is only slightly longer than the sum of the covalent radii.

In the mixed metal complex $[(\eta^5\text{-}C_5H_5)_2Y(\mu\text{-}Me_2)AlMe_2]$ (**13**) the same pattern of metal—carbon bonds is found [Al—C(terminal) 1.90–1.97(3) Å, Al—C(bridge) 2.08–2.11(2) Å] and the Y—Al distance of 3.056 Å indicates some metal—metal interaction[31]. The Y—C—Al angle has increased to 91°, presumably reflecting the disparity in size between the two metal atoms.

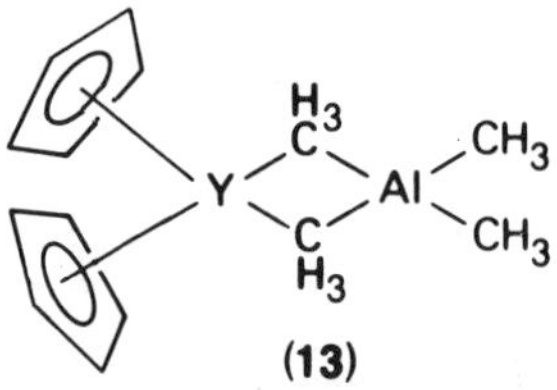

(**13**)

Alkyllithium complexes are tetrameric and hexameric in the solid state. In [(methyllithium)$_2$(*NNN'N'*-tetramethylethylenediamine)] the diamine links together $(LiMe)_4$ groups[32]. These methyllithium tetramers (**14**) consist of a tetrahedral cluster of lithium atoms [Li—Li 2.561–2.571(6) Å] and μ_3-Me groups on each tetrahedral face [Li—C 2.234–2.279(6) Å]. Bridged structures similar to those described above are found for alkenyl and aryl complexes of aluminium and lithium.

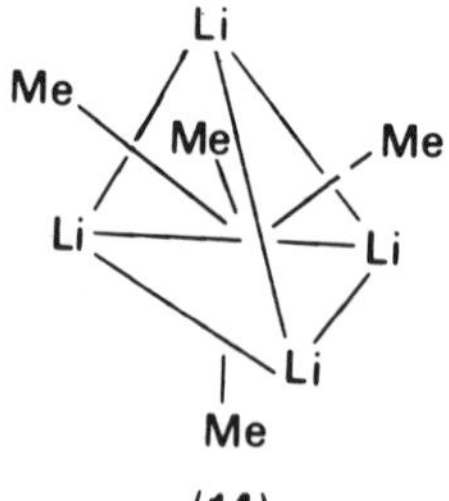

(**14**)

Binuclear transition metal complexes containing bridging alkyl groups have been reported. In bis (μ-methyl-1,3-dimethyl-η^3-allylnickel)[33] (**15**) and di-μ-trimethylsilylmethylbis[(trimethylphosphine)(trimethylsilylmethyl)chromium(II)][34] (**16**) the complexes are folded across the vector joining the two bridging carbon atoms.

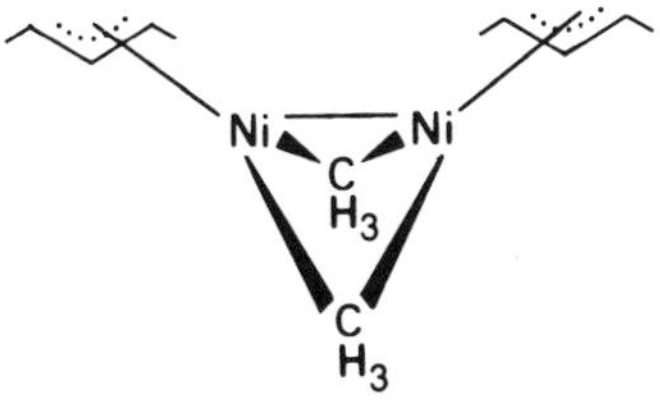

Ni Ni CH$_3$ CH$_3$

(**15**)

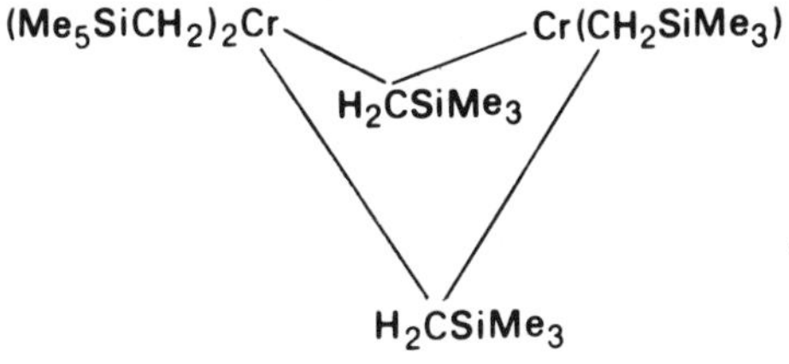

$(Me_5SiCH_2)_2Cr$ $Cr(CH_2SiMe_3)$ H_2CSiMe_3 H_2CSiMe_3

(**16**)

Both structures show evidence of asymmetric bridging [Ni—μC 2.045–2.067 Å, Cr—μC 2.181–2.269(3) Å], its more pronounced nature in the chromium complex being due to the close contact of one of the hydrogens on the μ-carbon with the chromium atom. Both complexes contain metal—metal bonds [Ni—Ni 2.371 Å, Cr—Cr 2.1007(5) Å] and in the chromium compound the metal—C(sp^3, terminal) distance is shorter [2.123–2.140(3) Å] than the metal—C(sp^3, bridging) distance, as would be expected.

C. Class B(i) Structures

1. Carbene (alkylidene) complexes

The stabiliza ion of carbenes by bonding to transition metals was first achieved by Fischer and Maasbol[35] and confirmed by the structure determination of pentacarbonyl[methoxy(phenyl)carbene]chromium[36] (**17**). The main structural features of

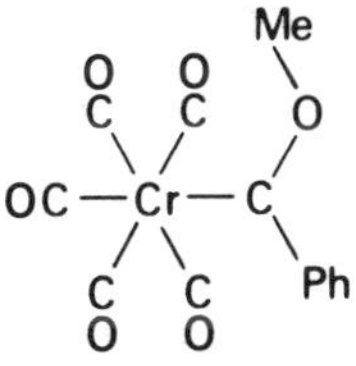

Me O O C C O OC—Cr—C C C Ph O O

(**17**)

this molecule, which are also typical of carbene complexes with heteroatom substituents, are as follows:

(i) The carbenoid carbon atom is coplanar with the metal atom and the carbene substituent atoms [O and C (phenyl)].
(ii) The metal—C(carbene) bond [2.04(3) Å] is shorter than that calculated for a metal—carbon(sp^2) single bond (Cr—C 2.21 Å), but appreciably longer than the metal—carbonyl bonds (Cr—CO 1.88 Å, average).
(iii) The C(carbene)—X(heteroatom) bond length [Cr—O 1.33(2) Å] is shorter than the corresponding C—X single bond length (Cr—O 1.46 Å), indicating substantial multiple bond character.

These features imply that the main stabilizing influence on the carbene is the interaction between the carbon p_z orbital and lone-pair electrons associated with the heteroatom.

Complexes where the carbenoid carbon interacts with two heteroatoms have been extensively studied by Lappert and his colleagues[37]. They found, in general, no evidence for very short metal—carbene bonds, thus indicating minimal multiple bond character. Indeed, in the carbene complex *trans*-dichlorotetrakis-(1,3-diethylimidazolidin-2-ylidene)ruthenium(II)[38] (**18**) the Ru—C(carbene) distances

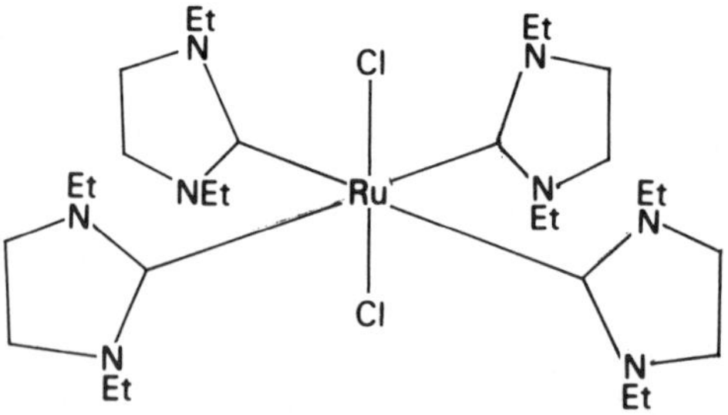

(**18**)

average 2.105(5) Å, which compares well with the Ru—C(naphthyl) single bond length of 2.16(1) Å found in hydridobis[1,2-bis(dimethylphosphino)ethane]-naphthylruthenium(II)[39]. However, multiple bond character in the carbene—heteroatom linkage is indicated by the short C—N distance of 1.349(11) Å. The carbene ligands are in a 'propeller-like' arrangement in the equatorial plane.

Although a number of structure determinations of carbene complexes without heteroatom substituents have been reported [e.g. pentacarbonyl(2,3-diphenylcyclopropenylidene)chromium[40] and (η^5-cyclopentadienyl)(phenylbenzoylcarbene)dicarbonylmanganese[41]], and have revealed metal—carbene distances compatible with significant multiple bond character, the first conclusive metal—carbon double bond was reported with the structure determination of $[(\eta^5\text{-}C_5H_5)_2Ta(CH_2)(CH_3)]$[42,43] (**19**). The CH_2 plane is perpendicular to the plane defined by Ta, C(Me), and

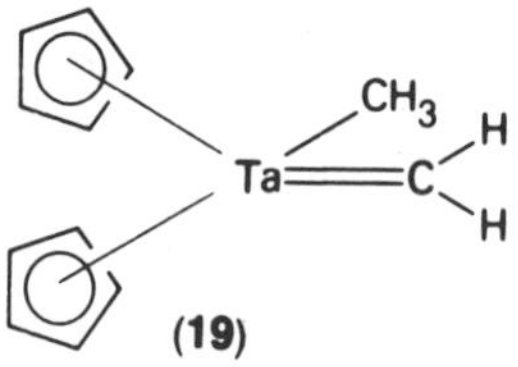

(**19**)

C(carbene). The H—C—H angle is 107(9)° and the Ta—CH_2 system is planar. The orientation of the CH_2 group allows the interaction of the p_z orbital on the methylene carbon with the appropriate π-orbitals of the tantalum, resulting in a Ta—C distance of 2.026(10) Å, which is considerably shorter than the Ta—C(methyl) single bond of 2.246(12) Å.

In complexes containing the group Ta=CHR (R ≠ H), considerable obtuseness of the Ta—C—C angle has been observed. Thus, in mesitylbis(neopentylidene)-bis(trimethylphosphine)tantalum(V) (**20**), the Ta—$C_{(\alpha)}$—$C_{(\beta)}$ angles at formally sp^2 hybridized carbon atoms are 154.0(6)° and 168.9(6)°, respectively[44]. This obtuseness has been noted before in $[(\eta^5\text{-}C_5H_5)_2TaCl(CHMe_3)]$, 150.4(5)°[45], and in $[W(O)(CHCMe_3)(PEt_3)Cl_2]$, 140.6(11)°[46]. The apparent flexibility of the M—$C_{(\alpha)}$—$C_{(\beta)}$ angle in such complexes would appear to be related to the ease of α-hydrogen abstraction in them and also to their role in olefin metathesis reactions.

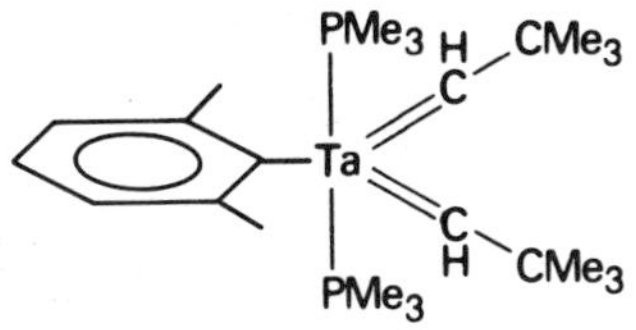

(20)

In complexes containing the Ta=C bond the carbene carbon is nucleophilic in character and the compounds have been compared to the main group ylides of phosphorus and antimony[43]. In contrast, the carbene carbon in the heteroatom-substituted carbene complexes of Fischer is electrophilic[47].

2. *Carbyne (alkylidyne) complexes*

By analogy with carbene complexes where a formal metal—carbon double bond is postulated, Fischer named compounds containing formal metal—carbon triple bonds as carbyne complexes[47]. The basic structural features associated with such complexes are seen in the structure of *trans*-(iodo)tetracarbonyl(methylcarbyne)chromium[48] (**21**).

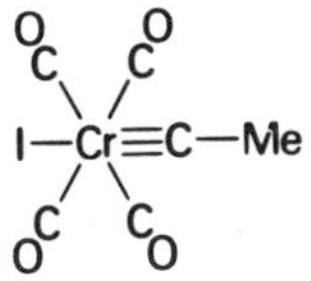

(21)

(i) The three-atom group consisting of metal, carbyne carbon, and carbyne substituent is linear.
(ii) The metal—C(carbyne) distance [Cr—C 1.69(1) Å] is significantly shorter than the corresponding metal—C(sp^2), metal—carbene, and metal—carbonyl distances, which in this case are 2.17[47], 2.04[36], and 1.946(9) Å[48], respectively.
(iii) The nature of the carbyne substituent determines the C(carbyne)—X(substituent) bond length and can also affect the metal—carbyne distance. Thus, in the methylcarbyne complex the C(carbyne)—C(methyl) distance of 1.49(2) Å is in agreement with a normal C(*sp*)—C(sp^3) bond length (1.46 Å). However, in the cationic carbyne complex $[(CO)_5CrCNEt_2]^+$ the short C(carbyne)—N distance [1.282(2) Å] together with the comparatively long metal—carbyne bond [1.782(1) Å] indicates delocalization of the positive charge over the complete Cr—C—N system[49]. This proposal is strengthened by the occurrence of a long Cr—CO bond [1.975(2) Å] *trans* to the carbyne ligand.

The tungsten carbyne complex $[W(CCMe_3)(CHCMe_3)(CH_2CMe_3)(Me_2PCH_2CH_2PMe_2)]$ is interesting because it allows the comparison of alkyl, carbene and carbyne ligands (all related by α-hydrogen abstraction) at the same metal centre[50]. The W—C(sp^3), W=C(sp^2), and W≡C(*sp*) distances are 2.258(9), 1.942(9), and 1.785(8) Å, respectively. The corresponding W—Cα—Cβ angles of 124.53(69)°, 150.44(67)°, and 175.34(69)° again reflect the flexibility of the system. The tungsten coordination geometry is a distorted square-based pyramid (**22**).

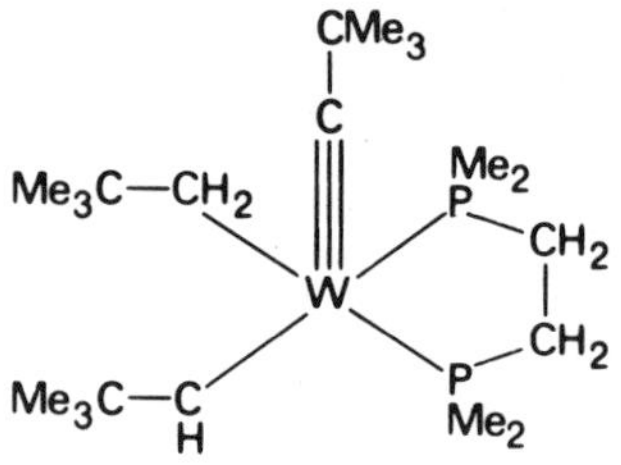

(22)

D. Class B(II) Structures (Complexes with Bridging Ligands)

Carbene and carbyne systems can bridge two metal atoms[51]. Thus, in (μ-methylene)bis[(η^5-cyclopentadienyl)carbonylrhodium] **(23)** the μ-carbon atom has tetrahedral geometry and the Rh—C(μ) distances are 2.045(4) and 2.029(4) Å[52], and the Rh—C—Rh angle of 81.7(1)° is similar to the acute values observed in many organometallic complexes. The C_5H_5 groups are arranged in a *trans* configuration with respect to the Rh—Rh bond.

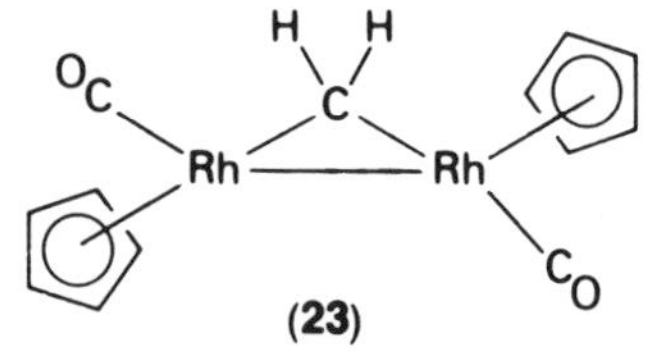

(23)

Similar methylene bridged binuclear transition metal complexes whose structures have been determined include [(μ-CH_2){(η^5-MeC_5H_4)$Mn(CO)_2$}$_2$][53], [(μ-CH_2)-{$Fe(CO)_4$}$_2$][54], [(μ-CH_2)(μ-CO){Co(η^5-Me_5C_5)}$_2$][55], and [(μ-CH_2)$_3${$Ru_2(PMe_3)_6$}][56], all of which contain metal—metal bonds. μ-Methylene complexes have been described as dimetallacyclopropanes[57].

The structure of the carbyne complex [(μ-CMe)(μ-CO){(η^5-C_5H_5)(CO)Ru}$_2$] shows the C(carbyne) atom to have a trigonal geometry **(24)**. The Ru—C(carbyne) distances

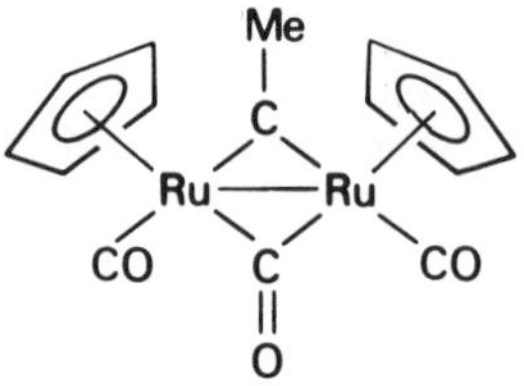

(24)

are 1.933 and 1.941(5) Å, and the Ru—C(carbyne)—Ru angle is 89.0(2)°. The C_5H_5 groups are in a *cis* arrangement with respect to the Ru—Ru bond. This complex, together with the analogous μ-methylcarbene and μ-vinylidene compounds, was prepared and characterized during investigations into the reactions of ethyne at a di-metal centre[58]. On passing from the μ-carbyne to the μ-vinylidene complex the Ru—C(μ) distance increases [2.029(7) Å], as would be expected, and consequently the angle at the μ-C(vinylidene) atom decreases [83.2(3)°]. The structure of the μ-methylcarbene complex has not been determined but is believed to be similar to that of the corresponding iron complex[59] in which the Fe—C(carbene)—Fe angle is even more acute [78.8(1)°].

III. DIHAPTO LIGANDS

A. Transition Metal—Alkene Complexes

There have been numerous structure determinations of η^2-alkene—metal complexes, many of which have been undertaken because of the role of such compounds in chemical processes of considerable industrial importance. The simplest η^2-alkene ligand, and one of the most studied, is ethene.

1. η^2-ethene complexes

The structures of $[PtCl_3(C_2H_4)]^-$ [60] **(25)** and $[Pt(Ph_3P)_2(C_2H_4)]$[61] **(26)** show features associated with the complexed ethene that are representative of those found for more complicated alkene ligands and are listed below.

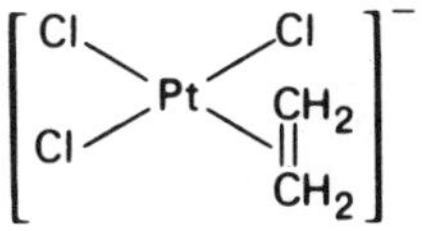

(25) 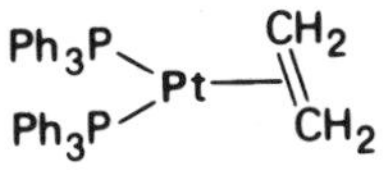 **(26)**

(a) The C—C bond distance in the complexed alkene is significantly longer than in the free alkene [1.375(4) Å in **25**, 1.43(1) Å in **26**, and 1.337(2) in ethene].

(b) The planar alkene becomes non-planar on coordination, the substituents bending away from the metal. This distortion of the ligand is very evident in some substituted ethene complexes. Thus, in $[Pt\{C_2(CN)_4\}(PPh_3)_2]$[62] **(27)** the angle between the planes defined by $C_{(1)}$, $C_{(2)}$, $C_{(3)}$, $C_{(4)}$, and $C_{(5)}$, $C_{(3)}$, $C_{(2)}$, $C_{(6)}$ is 141.3°, and in $[(C_2Cl_4)Pt(PPh_3)_2]$[63] the corresponding angle is 132(2)°. It is noteworthy that in both these complexes the C=C bond is longer than usual, 1.49(5) Å in **27** and an exceptional 1.62(3) Å in $[(C_2Cl_4)Pt(PPh_3)_2]$. This very long bond length is probably suspect[64].

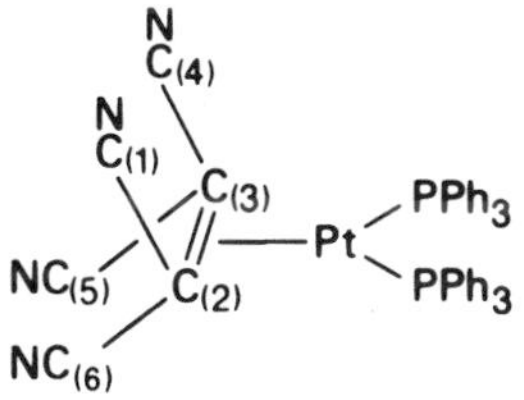

(27)

(c) The alkene carbon atoms are essentially equidistant from the metal atom [Pt—C 2.128, 2.135(3) Å in **25**, Pt—C 2.106(8), 2.116(9) Å in **26**].

(d) the alignment of the alkene double bond with respect to the metal coordination geometry can be generalized as follows:
   (1) trigonal planar geometry; the C=C bond is in the plane as in **26**.
   (2) square planar geometry; the C=C bond is perpendicular to the plane as in **25**.
   (3) trigonal bipyramidal geometry; the C=C bond lies in the equatorial plane as in $[\{Me(H)NNC(Me)C(Me)NN(H)Me\}PtCl_2(C_2H_4)]$[65] **(28)**. it is pertinent to note here that this complex contains Pt(II) in an unusual coordination geometry. However, the C—C and Pt—C distances of 1.46(2) Å and 2.073(12) Å, respectively, compare well with those reported for alkene—platinum(0) complexes[66].

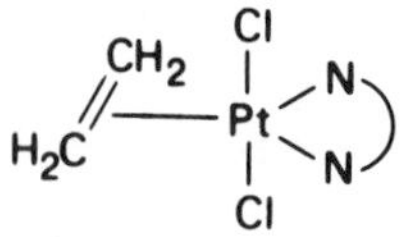

(28)

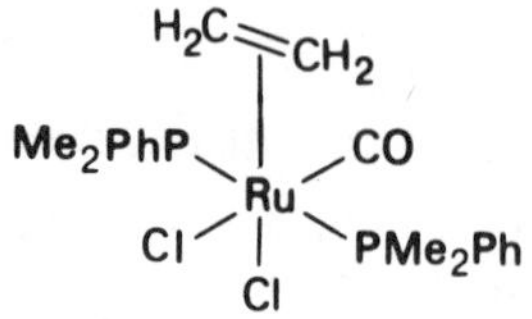

(29)

(4) octahedral geometry; the C=C bond is aligned parallel to one axis as in $[(Me_2PhP)_2Cl_2(CO)Ru(C_2H_4)]$[64] (**29**).

A detailed discussion of the features listed above, together with a list of structural parameters for a number η^2-alkene and alkyne—transition metal complexes, is given in a recent review[67].

It is possible for more than one ethene molecule to be bound to the same metal atom and complexes of this type include $[(C_2H_4)_2Rh(acac)]$[68] (**30**) and $[(C_2H_4)_2NiCH_3]^{-}$[69] (**31**). In **30** the two ethene molecules are perpendicular to the rhodium coordination plane and the C=C distance is 1.41(3) Å, whereas in **31** they lie in the nickel coordination plane and the shorter C—C distances (1.36, 1.38 Å) appear to indicate weaker alkene—metal bonding.

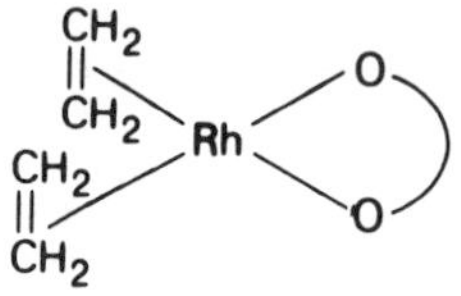

(30)

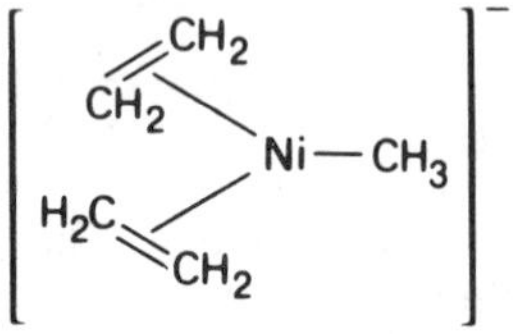

(31)

Rosch and Hoffmann[70] have discussed the possible structure that could be adopted by metal complexes with η^2-ethene as the only ligand, but as yet crystal structures of such compounds do not appear to have been reported. However, the all-η^2-alkene ligand complexes $[Pt(C_2H_4)_2(C_2F_4)]$ and [Pt(norbornene)$_3$] have been studied; the alkene ligands all lie in the trigonal plane[70a].

2. *Alkenes as chelating ligands*

The use of alkenes as chelating ligands in organometallic chemistry is well documented. Thus, in bis(cycloocta-1,5-diene)nickel (**32**) both double bonds in the

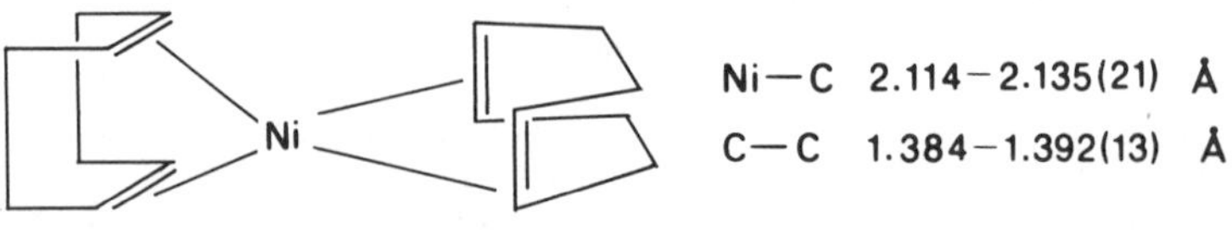

Ni—C 2.114–2.135(21) Å
C—C 1.384–1.392(13) Å

(32)

diene are coordinated to the nickel, giving the metal a distorted tetrahedral arrangement[71]. Similarly in the Dewar-benzene complex $[(Me_6C_6)Cr(CO)_4]$[72] the double bonds of the organic ligand occupy *cis*-positions in the Cr coordination octahedron (**33**).

An example of an alkene acting as a tridentate chelating ligand is found in the complex *trans,trans,trans*-1,5,9-cyclododecatrienenickel[73]. The nickel has trigonal

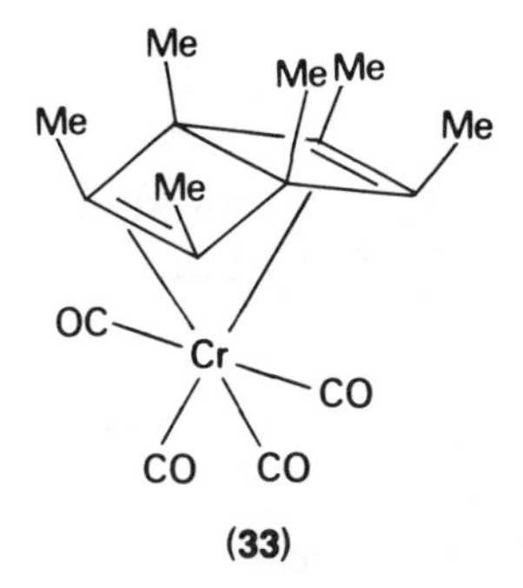

(33)

planar geometry with respect to the centres of the double bonds (**34**) and these bonds are arranged in a propeller-like fashion (**35**).

Ni—C 2.021–2.027 Å
C=C 1.370–1.374 Å

(34)

(35)

The structural features associated with the C=C bonds in the chelate complexes are, in general, similar to those described for the ethene complexes unless steric factors involving the rest of the organic ligand force changes. Thus, in (but-2-enyl-1-methylallylether)pentane-2,4-dionatorhodium(I) (**36**) the double bonds are perpendicular to the rhodium coordination plane, as would be expected[74].

O⌒O = pentane-2,4-dionato
Rh—C 2.108–2.135(6) Å
C=C 1.400, 1.368(9) Å

(36)

However, in chloro[hex-3-ene-1,6-diylbis(di-*t*-butylphosphine)]rhodium(I) (**37**) the coordinated double bond is at an angle of 74° to the rhodium coordination plane because of the conformation of the chelate ring[75]. An interesting feature concerning **37** is that it is prepared from $RhCl_3$ and the fully saturated diphosphine. Therefore, it is presumed that the steric requirement that the P atoms be in *trans* positions forces the

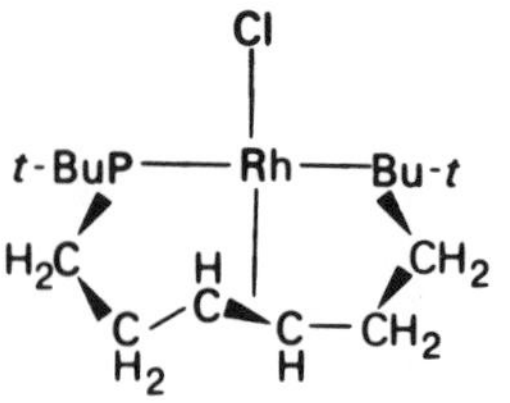

Rh—C 2.139–2.160(21) Å
C=C 1.37(3) Å

(37)

short methylene chain to interact with the metal and reductive elimination occurs to produce the η^2-bonded alkene.

Alkene complexes of Rh(I) are commonly associated with hydrogenation reactions and much structural work has been undertaken in order to gain an insight into the mechanisms involved in these reactions. In the square planar complex **[{3-(diphenylphosphino)propyl}(3-butenyl)phenylphosphinechlororhodium(I)]**[76] **(38)**, the Rh—C distances of 2.203–2.208(6) Å are long compared with the values observed in the two Rh(I) complexes described above, which indicates the weaker metal—alkene bonding. It is suggested that the influence of the phosphorus atom *trans* to the double bond may cause the alkene to dissociate easily in solution or activate it sufficiently to react with molecular hydrogen[76].

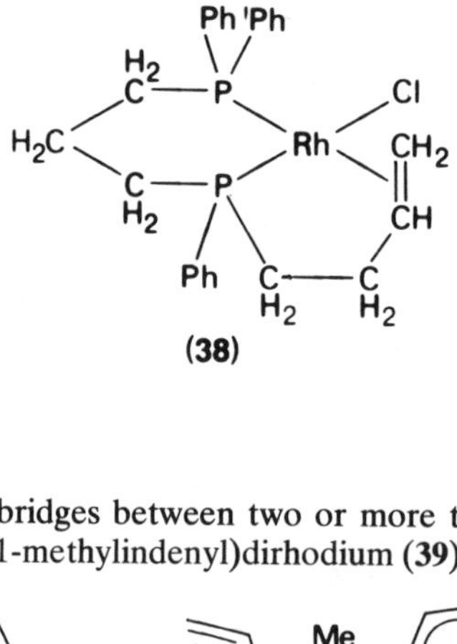

(38)

3. μ-η^2-Alkene ligands

Alkenes can function as bridges between two or more transition metal atoms. In μ-but-2-ene-μ-ethenebis(η^5-1-methylindenyl)dirhodium **(39)** the alkene is σ-bonded to

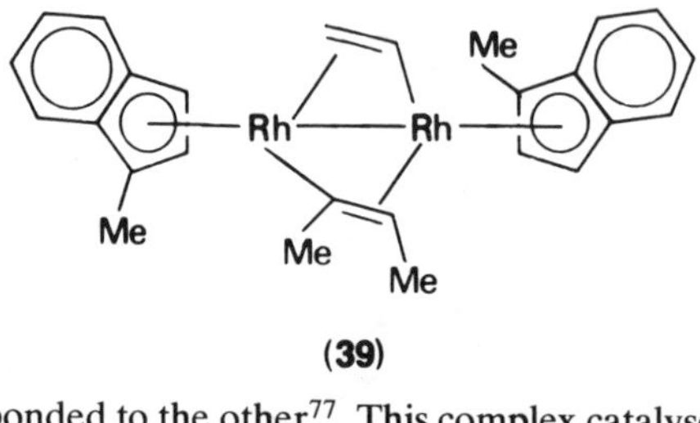

(39)

one rhodium and η^2-bonded to the other[77]. This complex catalyses the trimerization of alkynes, e.g. 2,2-dimethylbut-1-yne to 1,2,4-tri-*t*-butylbenzene and the unsubstituted indenyl analogue reacts with carbon monoxide to give the αβ-unsaturated ketone **40**.

The bridging ligand in the cluster complex [(*s-trans*-C_4H_6)$(CO)_{10}Os_3$] is a diene **(41)**. The C=C distances are not equal [1.32(4), 1.45(4) Å] but the author[78] notes that the inner atoms of the C_4 chain are not well determined. However, this is an

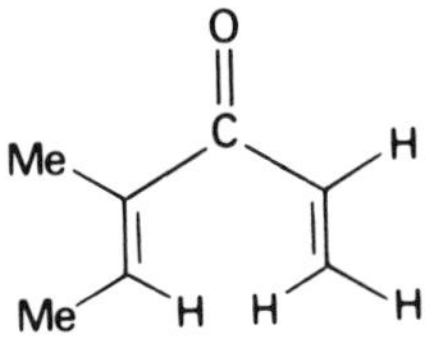

(40)

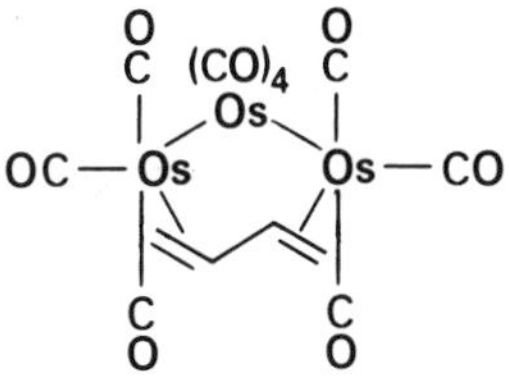

(41)

unusual mode of bonding for buta-1,3-diene, which is more normally encountered in organometallic complexes as the *cis*-isomer acting as an η^4-ligand to a single metal centre.

B. η^2-Ligands Containing Heteroatoms

The complex $[(\eta^5\text{-}C_5H_5)Mo(CO)_2(\eta^2\text{-MeCNPh})]$ contains an η^2-CN system (**42**). The Mo—C and Mo—N distances are 2.106(5) and 2.143(4) Å, respectively[79], and the C═N distance is 1.233(6) Å.

The linear carbon disulphide molecule is bent on coordination. Thus, in $[(\eta^2\text{-}CS_2)(CO)_2(PMe_3)(PPh_3)Fe]$ (**43**) the S—C—S angle is 138.9(1)° and also the

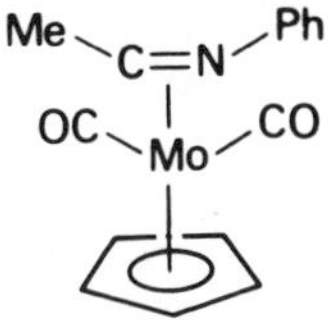

(42)

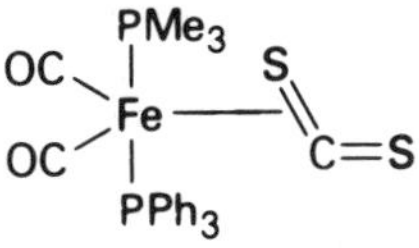

(43)

C—S distances of 1.676(7) and 1.615(8) Å are both longer than those observed in free carbon disulphide[80]. Similar distortions of the CO_2 molecule have been observed when it is η^2-bonded to transition metals[81].

Interaction of acetyl chloride and $[Ti(C_5H_5)_2(CO)_2]$ affords a complex (**44**) where the acyl group acts as an η^2-ligand. The Ti—C and Ti—O distances are 2.07(2) and

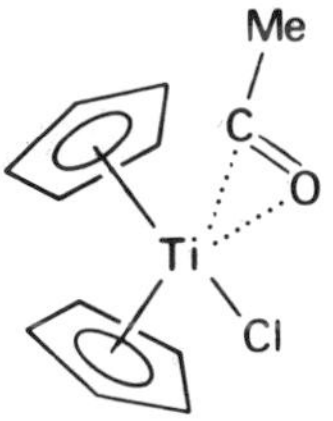

(44)

2.194(14) Å, respectively[82]. This structure provides an explanation for the low CO stretching frequency ($\nu_{CO} \approx 1600\ cm^{-1}$) observed for the complex. Low values of ν_{CO} are normally found for acyl derivatives of the early members of the transition series[83].

Diphenylketene can act both as an η^2-CO ligand as in $[(C_5H_5)_2Ti(PhCCO)]_2$[84] (**45**), and as an η^2-CC ligand as in $[(C_5H_5)(CO)_2Mn(PhCCO)]$[85] (**46**). In both complexes the ketene is no longer linear the C—C—O angle being 128.8° in the titanium complex and 145° in the manganese complex.

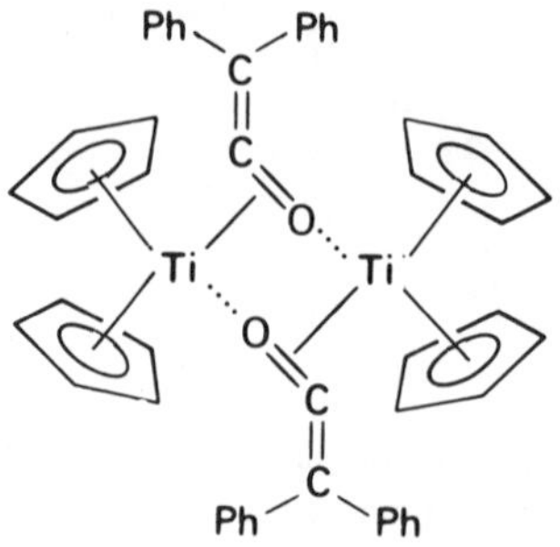

Ti—O 2.037(2), 2.099(3) Å
C=C 1.357(4) Å, C=O 1.311(4) Å

(45)

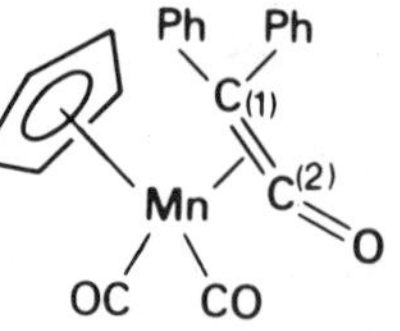

Mn—C(1)	2.17 Å
Mn—C(2)	1.96 Å
C=C	1.35 Å
C=O	1.21 Å

(46)

C. Transition Metal – Alkyne Complexes

The structural features possessed by η^2-alkene ligands are shared to a large extent by their alkyne counterparts. However, the bending away from the metal of the ligand substituents is much more marked in the alkyne systems where the ligand assumes the geometry associated with a *cis*-alkene. In the trigonal planar platinum complex $[(Ph_3P)_2Pt(CF_3CCCF_3)]$[86] **(47)** the deviation of the alkyne from linearity may be judged from the average C≡C—C angle, which is 140.8°. The alkyne makes an angle of 3.6(4)° with the coordination plane of the platinum and the triple bond has lengthened to 1.255(9) Å in comparison with the value of 1.204(2) Å for an uncoordinated alkyne. In bis(diphenylethyne)platinum[87] the platinum has an essentially tetrahedral configuration **(48)**; the angle between the two (Pt—C—C) planes is 82°. The C≡C and Pt—C distances are 1.291(5) Å and 2.021, 2.022(5) Å and the Ph—C—C angle is 153°.

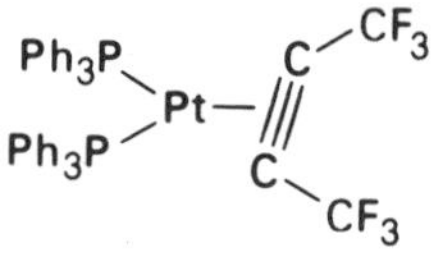

(47) 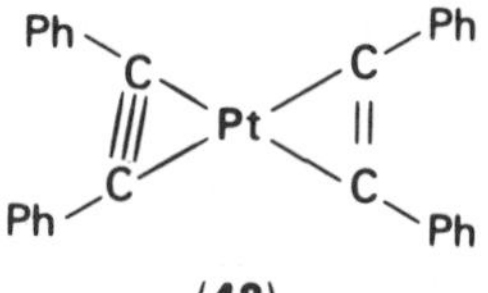 **(48)**

In general, the metal—carbon bonds in the alkyne complexes are about 0.07 Å shorter than in the related alkene compounds. The carbon atoms are essentially equidistant from the metal atom unless steric requirements dictate otherwise. Thus, in $[(C_5H_5)_2Ti(CO)(Ph_2C_2)]$[88] **(49)** steric factors result in Ti—C distances of 2.107(7) and 2.230(7) Å, respectively. The C≡C bond length is 1.285(10) Å and the average C—C—Ph angle 142.3°.

The complex $[(\eta^5\text{-}C_5H_5)(\eta^2\text{-}C_6H_4)Me_2Ta]$[89] contains benzyne complexed to a single metal atom **(50)**. The Ta—C(benzyne) distances are 2.059 and 2.091(4) Å, which are significantly shorter than the Ta—C(methyl) distances of 2.169 and 2.181(6) Å. The benzyne system is perpendicular to the cyclopentadienyl ring and bisects the Me—Ta—Me angle. The C—C distances in the C_6 ring alternate as follows: $C_{(1)}$—$C_{(6)}$, $C_{(2)}$—$C_{(3)}$, $C_{(4)}$—$C_{(5)}$ are 1.364(5), 1.362(6) and 1.375(6) Å, respectively,

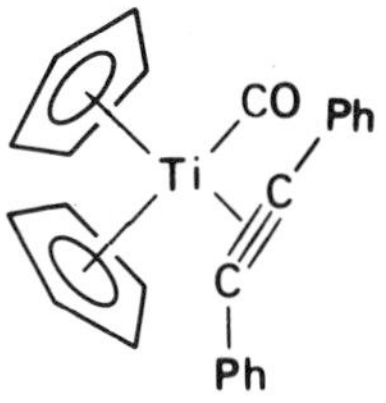

(49) 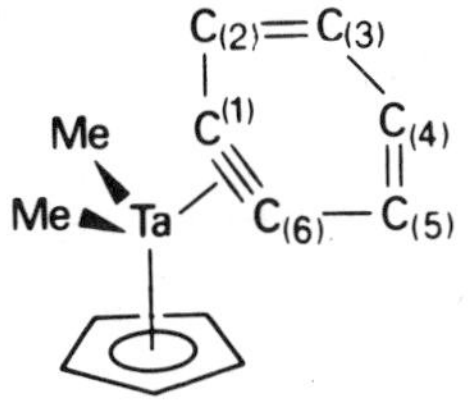 (50)

whereas $C_{(1)}—C_{(2)}$, $C_{(3)}—C_{(4)}$, and $C_{(5)}—C_{(6)}$ are 1.410(5), 1.403(6) and 1.408(6) Å, respectively.

Although alkynes coordinated to one metal centre are generally regarded as 2-electron donors, the extra π-bond allows the possibility of 4-electron donation. Thus, in the first structure determination of a mononuclear complex containing coordinated unsubstituted ethyne, $[(Et_2NCS_2)_2(CO)(C_2H_2)W]$ (**51**), the C≡C bond is aligned parallel with the W—CO axis of the pseudo-octahedral coordination polyhedron. The authors[90] believe this to be consistent with donation from both of the filled ethyne $1\pi_u$ bonding orbitals to metal orbitals.

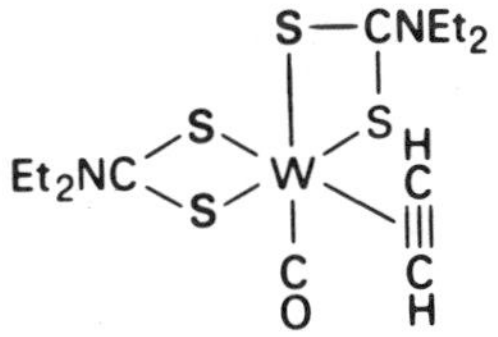

(51)

Another example of an alkyne apparently donating more than two electrons to one metal atom is found in the novel tungsten complex $[(Ph_2C_2)_3W(CO)]$[91]. This molecule has essentially C_{3v} symmetry with the alkyne groups inclined to the nearly linear W—CO axis by an average of 13.4° (**52**). The C—C—Ph angle is 139.6° and the triple bond has lengthened to 1.30 Å, giving the ligand a geometry approaching that of *cis*-stilbene. In order for the tungsten to have an 18-electron structure the alkyne groups must donate ten electrons.

The commonest structural mode for an alkyne to adopt when donating four electrons is that of a bridging ligand as shown in $[(\mu\text{-}Ph_2C_2)\{(CO)_3CoNi(C_5H_5)\}]$[92] (**53**). In this complex the C≡C bond has lengthened considerably [1.337(5) Å],

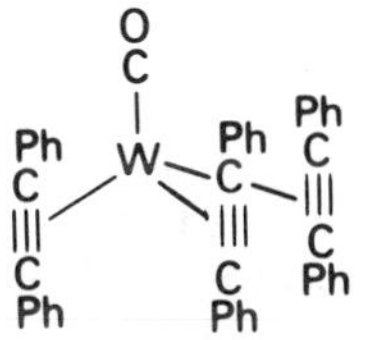

(52) 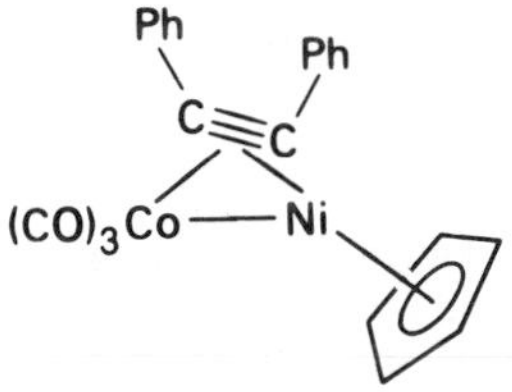 (53)

reflecting the involvement of both the alkyne π-bonds. The M—C distances lie in the range 1.904–1.917(4) Å (M = Ni) and 1.963–1.989(4) Å (M = Co), and the C—C—Ph angle is 144.2(3)°.

IV. TRIHAPTO LIGANDS

A. η^3-Allyl Ligands

The first three-dimensional X-ray structure determination of an η^3-allyl complex,

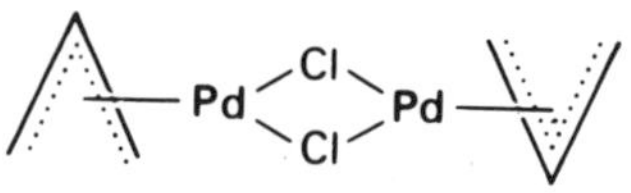

(54)

$[(\eta^3\text{-}C_3H_5)PdCl]_2$ **(54)**, was published in 1965 and revealed some of the typical structural features associated with the η^3-allyl ligand:

(1) The C—C distances in the allyl group are not significantly different [1.357(15), 1.395(15) Å].
(2) The C—C—C angle is close to 120° [119.8(9)°].
(3) The metal—carbon distances are equal, within experimental error. It should be noted that the central carbon atom frequently appears nearer to the metal atom than the two terminal carbons [Pd—C(central) 2.108(9) Å; Pd—C(terminal) 2.121, 2.123(7) Å], although these differences are not always significant.
(4) The dihedral angle between the C_3 plane and the metal—chloride coordination plane is not 90° (111.5°).

The above details show the allyl ligand to be symmetrically bound to the metal atom. However, there are a number of complexes where this is not so and the asymmetry leads to differences between the metal—C(terminal) distances and also between the two C—C distances. This asymmetry occurs when there are differences in the electronic character of the other ligands in the complex. A good example of this is seen when the structures of $[(\eta^5\text{-}C_5H_5)(\eta^3\text{-}C_3H_5)(CO)_2Mo]$ **(55)** and

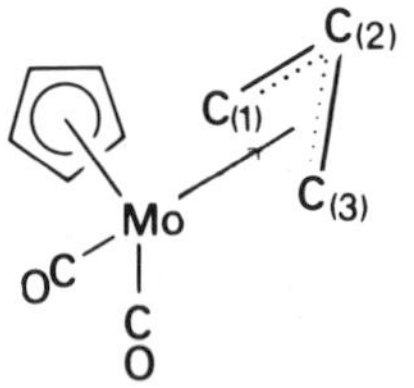

Mo—C(1) 2.359(3) Å, C(1)—C(2) 1.380(4) Å

Mo—C(2) 2.359(3) Å, C(2)—C(3) 1.380(4) Å

Mo—C(2) 2.236(4) Å

(55)

$[(\eta^5\text{-}C_5H_5)(\eta^3\text{-}C_3H_5)(NO)(I)Mo]$ **(56)** are compared[95]. In **55** the η^3-allyl group is symmetrical, whereas in **56** a pronounced asymmetry is observed, the origin of which can be ascribed to the different π-accepting ability of the nitrosyl and iodide ligands. Note also that **55** has the allyl group in the *exo* conformation, whereas in **56** it is *endo*.

Structural differences of the type just described have been used to rationalize the stereoselectivity observed during the reactions of some η^3-allyl complexes with nucleophiles[95–97]. In some cases the asymmetry of the allyl system is so marked that authors have represented the metal—allyl linkage in terms of a σ–π arrangement **(57)**[96].

Mo—$C_{(1)}$ 2.433(3) Å, $C_{(1)}$—$C_{(2)}$ 1.366(5) Å

(56) Mo—$C_{(3)}$ 2.294(3) Å, $C_{(2)}$—$C_{(3)}$ 1.418(5) Å **(57)**

Mo—$C_{(2)}$ 2.357(2) Å

An attempt to bring some order into the structural chemistry of η^3-allyl complexes has been made by Ibers and his colleagues[98]. In a study of the correlations between geometric structure and the number of metal *d* electrons they 'describe the geometry of the M—(η^3-C_3H_5) linkage by three parameters which are independent of the geometry and number of other ligands present' (Figure 1). This study, based upon a large sample of η^3-C_3H_5 structures found that:

(a) the bow angle β was almost constant (*ca.* 90°), indicating nearly symmetrical π-bonding of the allyl group;
(b) there was a strong correlation between α and D (the larger D, the larger the value of α). It would seem that the allyl geometry alters, in order to maximize the overlap of the orbitals responsible for the M—allyl bonding, with the varying size of the metal atom;
(c) there was a correlation between the average C—C distance and D. The larger the value of D the shorter the C—C distance. The d^8 complexes tend to have smaller D values than d^4 complexes and also their C—C distances are shorter. This is taken to indicate that transference of electrons from the metal *d* orbitals into π-antibonding allyl orbitals is of greater importance than some previous workers[99] have believed.

A similar study has been carried out by Putnik and his colleagues[100].

In substituted η^3-allyl ligands, it is observed that the substituents are, in general, bent out of the allyl plane. Thus, in bis[(η^3-2-methylallyl)palladium chloride] the methyl group is 0.29 Å out of plane[101]. Usually the substituent is bent towards the metal, although this is not exclusively the case, and in bis(η^3-2-methylallyl)-bis(trimethylphosphite)ruthenium[102] the methyl groups are bent away from the ruthenium atom.

The structures of a number of complexes containing only η^3-allyl ligands have been determined. Bis (η^3-2-methylallyl)nickel is a sandwich complex **(58)** adopting a *trans* arrangement with the methyl groups bent towards the metal and 12° out of plane[103]. It is interesting that frequently bis(η^3-allyl) complexes of Ni, Pd, and Pt are proposed as

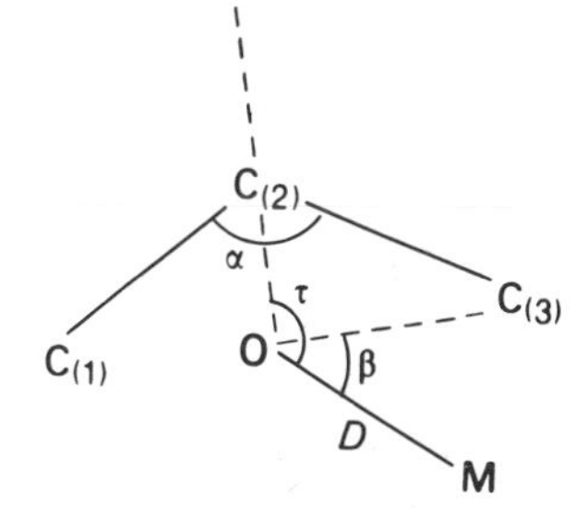

FIGURE 1. Geometry of the metal–η^3-allyl linkage. O = Centre of mass of the allyl group; D = distance from the metal atom M to the centre of mass of C_3H_5; α = $C_{(1)}$—$C_{(2)}$—$C_{(3)}$ angle; τ = tilt angle, i.e. angle between the vector $\overrightarrow{OM}$ and $\overrightarrow{OC_{(2)}}$; β = bowing angle, i.e. angle between the vector $\overrightarrow{OM}$ and the vector parallel to $\overrightarrow{C_{(1)}C_{(3)}}$ passing through O.

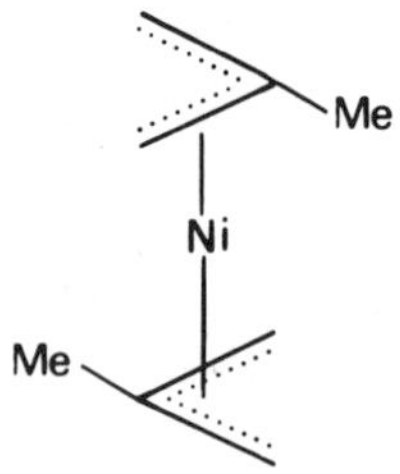

Me

Ni

Me

(58)

intermediates in metal-catalysed cyclo-oligomerization and cycloco-oligomerization reactions involving 1,3-dienes[104].

In tetra(η^3-allyl)dirhenium (**59**) the allyl groups present a staggered arrangement when viewed along the Re—Re axis[105]. A completely different structure is observed for the analogous molybdenum complex (**60**), where two allyl groups act as symmetrical bridging ligands, their planes being parallel with the Mo—Mo axis[106].

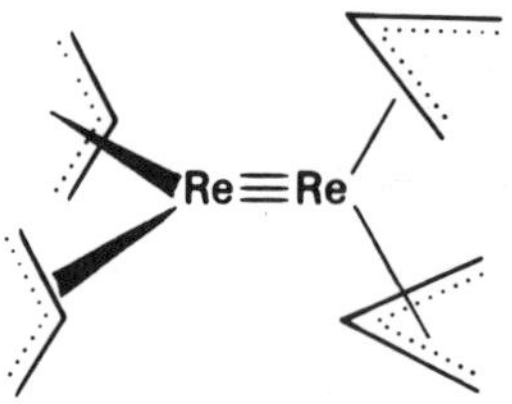

Re≡Re

(59)

Mo≡Mo

(60)

The use of η^3-allyl groups to bridge two metal centres is well documented. Thus, in $[Pd_3(\mu\text{-}\{\eta^3\text{-}C_3(p\text{-}MeOC_6H_4)_2Ph\}_2)(acac)_2]$ the allyl ligands bridge the palladium atoms in the bent Pd_3 chain[107] (**61**). The terminal carbon atoms of the C_3 groups are closer to the terminal Pd atoms [2.01(2) Å] than to the central Pd atom [2.12(2), 2.13(2) Å], and the central carbon atoms are at longer distances from the metal atoms [Pd(terminal)—C 2.55(2) Å, Pd(central)—C 2.31(2) Å]. However, in $[Ni_2(CO)_2(\mu\text{-}C_3Cl_3)(\mu\text{-}Cl)]_2$ the η^3-allyl groups are orthogonal to the planar Ni_4Cl_2 ring (**62**). The Ni—C(terminal) distances lie in the range 1.920–1.978(8) Å and for Ni—C (central) the range is 2.254–2.328(8) Å[108].

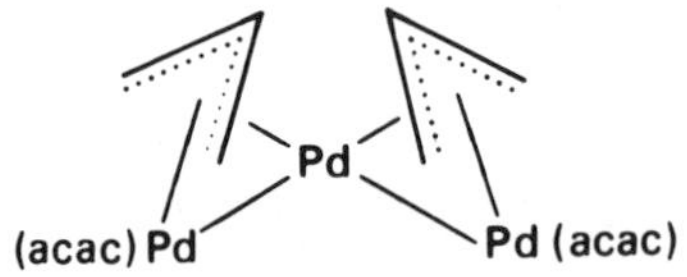

Pd

(acac) Pd

Pd (acac)

(allyl substituents omitted)

(61)

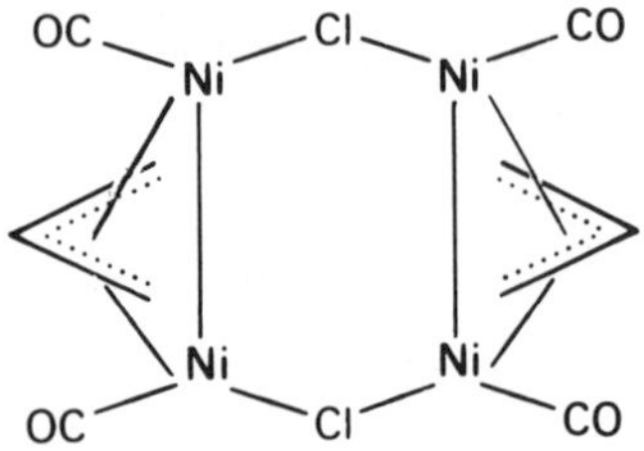

OC Ni Cl Ni CO

OC Ni Cl Ni CO

(allyl substituents omitted)

(62)

B. Cyclic η^3-Ligands

Many cyclic organic molecules are coordinated to transition metals through an η^3-allyl fragment. Thus, in the sandwich complex bis(η^3-cyclooctatrienyl)nickel (**63**)

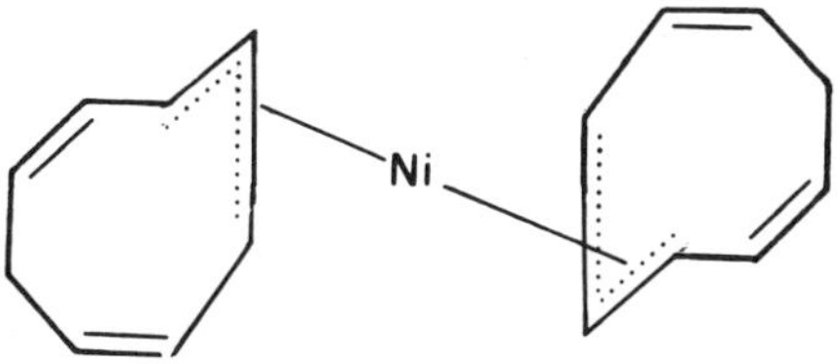

(63) 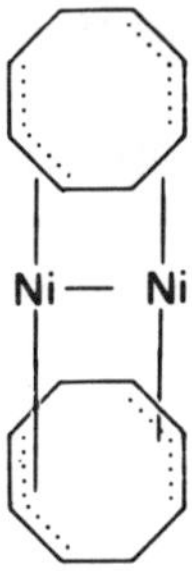 (64)

the two allyl fragments are mutually *trans*[104] and in bis(cyclooctatetraene)dinickel (**64**) both metal atoms are sandwiched between the C_8H_8 rings, which each use two η^3-allyl fragments for metal—carbon bonding[109].

Bis(η-cyclopentadienyl)dicarbonyltungsten (**65**) provides an example of the normally η^5 and planar C_5H_5 ligand acting in a trihapto manner with corresponding

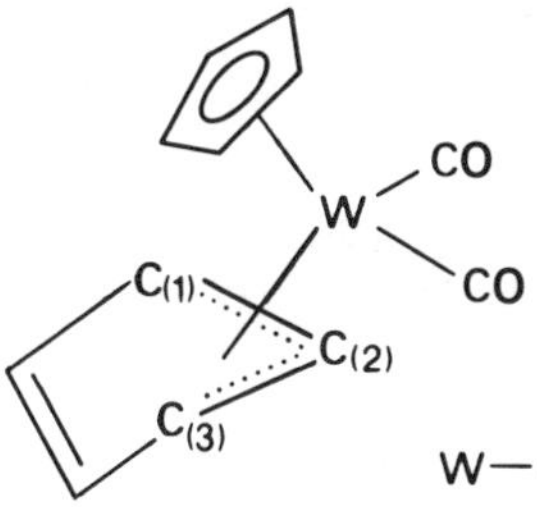

W—$C_{(1)}$,$C_{(3)}$ 2.40–2.45(2) Å

W—$C_{(2)}$ 2.28–2.32(2) Å

(65)

loss of planarity[110]. The angle between the olefinic and allylic fragments of the η^3-C_5H_5 ligand is 20°. This mode of bonding for the C_5H_5 group is necessary in order to give the metal an 18-electron configuration. It is noteworthy that the tungsten achieves this configuration by distortion of the C_5H_5 ring rather than by eliminating one of the carbonyl ligands.

A neutron diffraction study of the $[Fe\{P(OMe)_3\}_3(\eta^3\text{-}C_8H_{13})]^+$ ion (**66**) revealed that the η^3-bonding of the C_8H_{13} ring allows one of the aliphatic hydrogen atoms to enter into a strong interaction with the iron atom [H···Fe 1.879(9) Å][111]. C—H bond activation of this type has been considered to be of importance in many homogeneously catalysed chemical reactions.

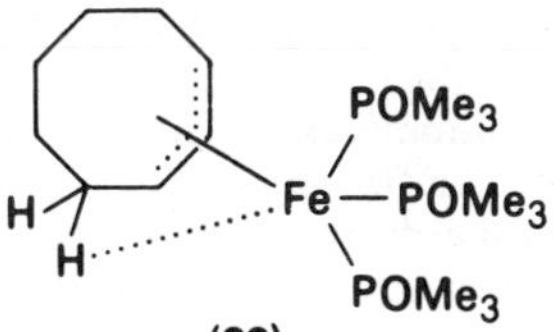

(66)

C. η^3-Ligands Containing Heteroatoms

Structural data for these systems are scarce. An η^3-CCN system is reported in $[(\eta^5\text{-}C_5H_5)Mo(CO)(I)\{C(NMe_2)C(Me)N(Me)\}]$[112] **(67)** and bridging η^3-xanthato groups are found in $[Mo_2(S_2COEt)_4I_2]$[113] **(68)**.

NMe 1.332 Å
Me CMe
I
OC CNMe$_2$ 1.455 Å

Mo—N 2.257 Å
Mo—C(Me) 2.354 Å
Mo—C(NMe$_2$) 2.027 Å

(67)

I S COEt S Mo S COEt S EtOC S Mo S S EtOC I

(68)

V. TETRAHAPTO LIGANDS

A. η^4-*cis*-1,3-Diene Complexes

There has been considerable discussion on the most appropriate bonding model for η^4-*cis*-1,3-diene—metal complexes[114–117]. The main structural features reported for such complexes are consistent with a model intermediate between **(69)** and **(70)** and are as follows:

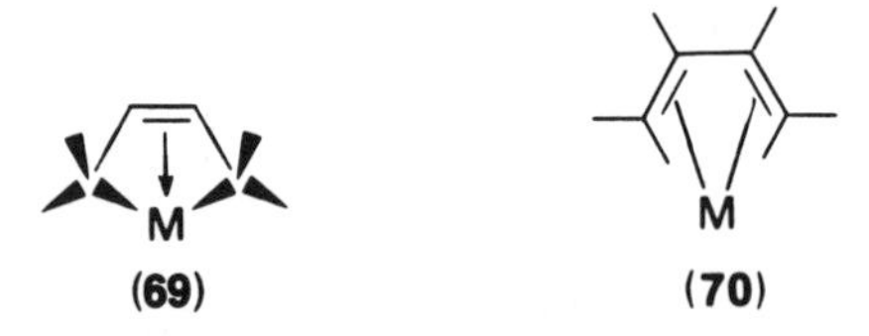

(69) **(70)**

(1) The inner carbon atoms of the diene ligand are equidistant from the metal atom and significantly nearer to it than the two terminal carbon atoms which are also equidistant from the metal.
(2) The length of the inner C—C bond tends to be shorter than the average of the two outer C—C bonds. This difference between the two types of bond has been quoted to lie between 0.015 and 0.021 Å[118,119]. However, this trend is not always observed and it is perhaps better to compare the C—C distances of the complexed diene with those of the free diene [C—C (outer) 1.341(2) Å, C—C(inner) 1.463(3) Å][164]. A reasonable generalization is that on coordination the C—C(inner) bond becomes significantly shorter than that observed in the free diene and the outer bonds become significantly longer.
(3) The internal C—C—C angles close up [values in the range 114–121°, compared with 122.9(5)° in the uncoordinated diene]. These angles are sometimes affected by the stereochemical requirements of the rest of the molecule of which the η^4-*cis*-1,3-diene may be only a fragment.
(4) The four carbon atoms of the diene are coplanar.
(5) The substituents on the outer carbon atoms tend to be twisted out of the C_4 plane. Immirzi designated these substituents *syn* and *anti*[120] **(71)** and noted that the *anti*

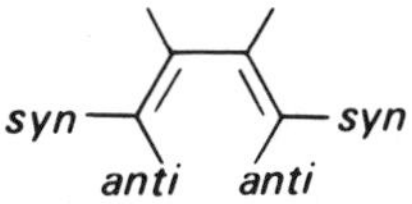

(**71**)

substituents are twisted away from the metal atom whereas the *syn* substituents (together with the substituents on the inner carbon atoms) are twisted towards the metal.

Compilations of structural data for η^4-*cis*-1,3-diene complexes and attempts to rationalize the data in terms of various bonding models can be found in references 118, 119 and 121.

The structure of (η^4-buta-1,3-diene)(tricarbonyl)iron is a classic example of the type of complex under discussion (**72**). Unfortunately, the X-ray data were such that the

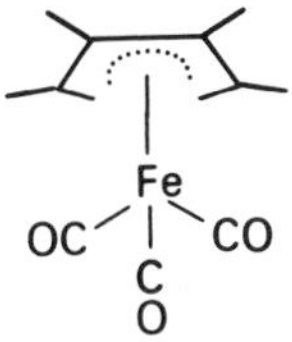

(**72**)

positions of the H atoms could not be determined[114]. Immirzi[120] has suggested, from the structure analyses of *syn*- and *anti*-substituted *cis*-buta-1,3-diene(tricarbonyl)iron complexes, that the structural parameters for the parent compound should be C—C(outer) 1.416 Å, C—C(inner) 1.404 Å, and that the *anti*-H atom is displaced 30° from the diene plane away from the iron atom, whereas the *syn*-H atom is displaced by 20° towards the iron. The inner H atoms are displaced 6° towards the iron.

A similar arrangement of substituents is found in bis(buta-1,3-diene)-carbonylmanganese[122] (**73**), where the *anti*-H atoms lie 0.72 Å above the C_4 plane and away from the metal, *syn*-H atoms lie in the plane, and the inner H atoms lie 0.24 Å below the plane towards the metal. The C—C distances are 1.39(1) Å(outer) and 1.46(1) Å(inner), and the Mn—C distances are 2.15(1) Å(outer) and 2.06(1) Å(inner).

Tetrahapto bonding of dimethylvinylketene to a tricarbonyl iron moiety results in the distortion of the linear ketene system (**74**). The C=C=O angle is reduced from 180° to 136.5(6)° and the Fe—C(O) distance is 1.929(6) Å. This iron—carbon bond length is considerably shorter than the other three, Fe—C(outer) 2.235(5) Å, Fe—C (inner) 2.090(5), 2.109(6) Å[123].

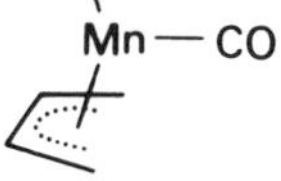

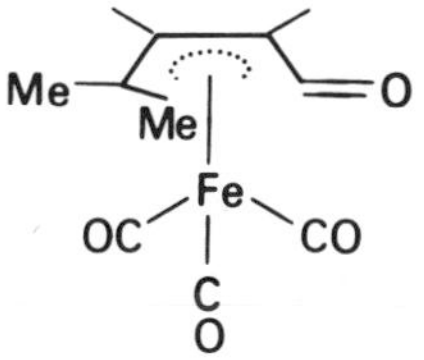

C—C (outer) 1.416, 1.442(9) Å

C—C (inner) 1.407(8) Å

(**73**) (**74**)

B. Cyclic η^4-Ligands

1. Cyclic ligands excluding cyclobutadiene

When cyclic ligands are bound to a transition metal by a tetrahapto *cis*-1,3-diene fragment, the out-of-plane twisting of the outer carbon substituents, noted with the acyclic ligands, results in the folding of the cyclic molecule. Thus, in $[\{\eta^4\text{-}C_6(CF_3)_6\}(\eta^5\text{-}C_5H_5)Rh]$ **(75)** the benzene ring is bent along the 1,4-axis, the dihedral angle between the two planes being 47.9°[124].

The η^4-diene fragment in (η^4-cyclooctatetraene)(η^6-hexamethylbenzene)ruthenium **(76)** is accurately planar with the two H atoms on the outer carbon atoms bent towards

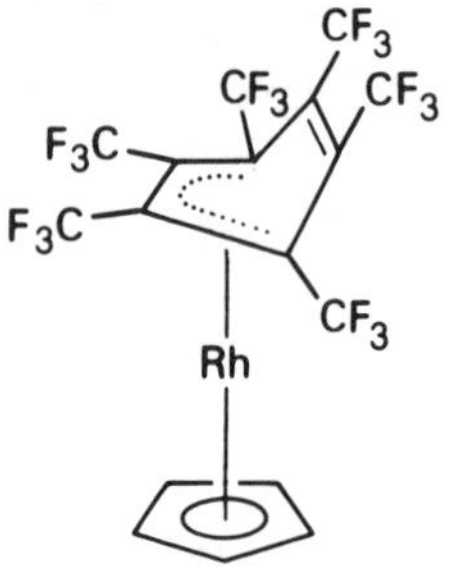

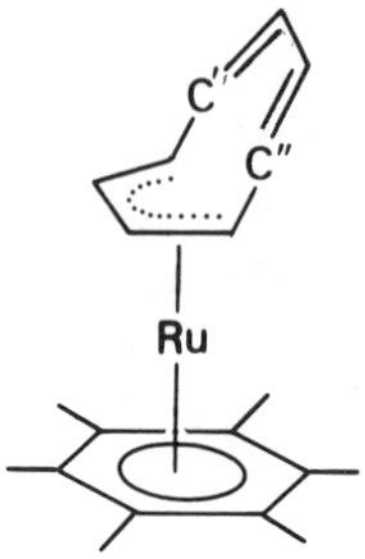

Ru—C (inner) 2.115, 2.126(10) Å
Ru—C (outer) 2.232, 2.234(10) Å
C—C (outer) 1.413 (16), 1.423(15) Å
C—C (inner) 1.400(16) Å

(75) **(76)**

the ruthenium (*ca.* 0.4 Å out of plane). Atom C′ and C″ are out of plane by *ca.* 0.8 Å, resulting in the folding of the C_8H_8 ring[125].

The structures of some other transition metal—η^4-diene complexes with carbo-, hetero-, and metallocyclic ligands are illustrated in **77**[115], **78**[126], **79**[127] and **80**[128].

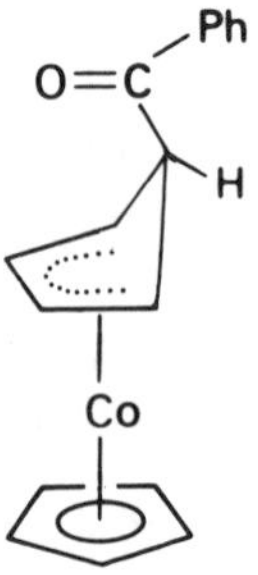

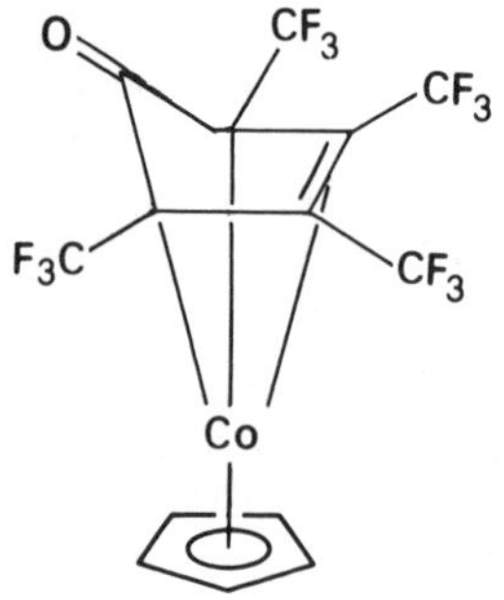

(77) **(78)**

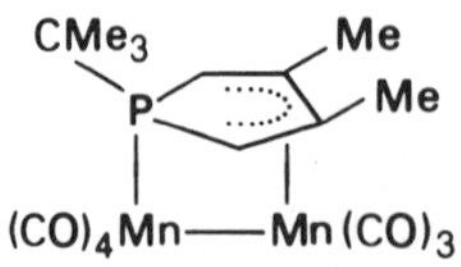

(79)

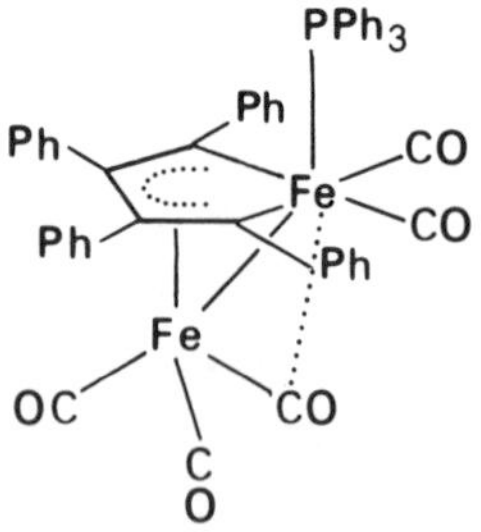

2. η^4–Cyclobutadiene complexes

Although the structures of several substituted cyclobutadiene—metal complexes have been known for some time[129,130], very few data are available for the unsubstituted η^4-cyclobutadiene ligand. The structure of (η^4-cyclobutadiene)(1,2-bis-(diphenylphosphino)ethane-*PP'*)carbonyliron has been determined[131] and shows the coordination geometry of the iron to be tetrahedral if the centroid of the ring is taken as one apex (**81**). The C_4 ring is planar and the H atoms appear to lie out of

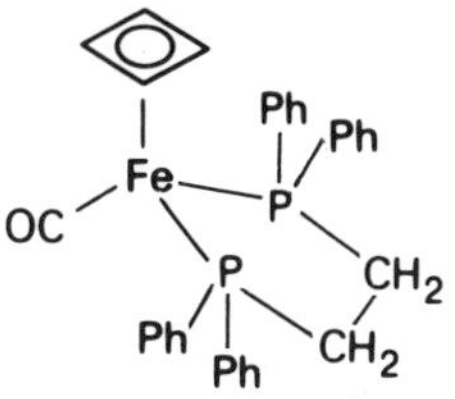

C—C	1.438–1.453 (5) Å
Fe—C	2.023–2.040 (4) Å
∠CCC	89.4–90.5 (3) Å

the plane and away from the metal. (This is the reverse of the H atom distortion observed in η^5-C_5H_5 complexes; see Section VI.A.)

Davis and Riley[131] analysed the structural results for cyclobutadiene complexes of type $[(C_4R_4)ML_3]$ and showed that the structures exhibit C_4 orientations over the entire range between the two idealized arrangements (**82**), thus indicating a low barrier to rotation about the metal—C_4 axis.

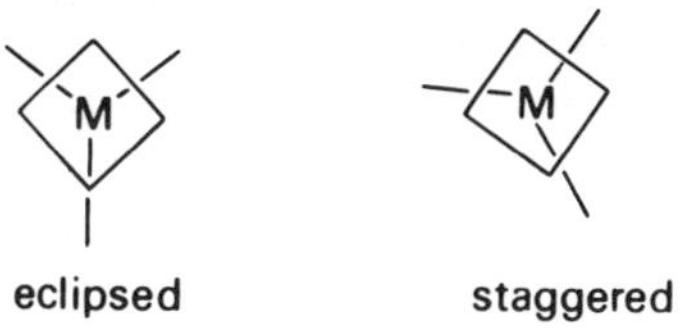

eclipsed staggered

(**82**)

C. Trimethylenemethane Complexes

The structure of (phenyltrimethylenemethane)tricarbonyliron reveals the iron to be positioned directly below the central carbon atom of the trimethylenemethane ligand, which in turn adopts a staggered arrangement with respect to the $Fe(CO)_3$ group (**83**)[132]. The central carbon atom lies above the plane defined by the three methylene

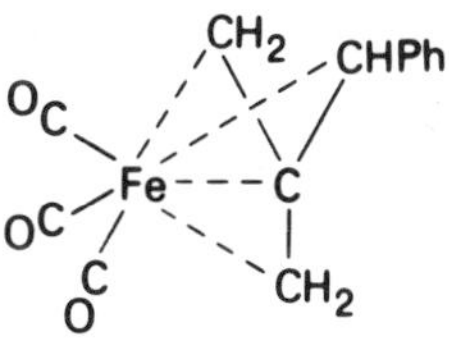

(**83**)

carbons (away from the iron atom) by 0.315 Å. The C—C distances in the η^4-ligand are in the range 1.405–1.436(13) Å, and the internal C—C—C angles lie between 114.4° and 116.2(9)°. Two distinct Fe—C distances are observed, firstly that to the central carbon atom [1.932(10) Å] and secondly those to the methylene carbons [2.098—2.162(11) Å].

Cyclic molecules are known to coordinate to transition metals via an η^4-trimethylenemethane fragment. Thus, in (η^4-1,6,7,8-heptafulvene)tricarbonyliron (**84**) the carbonyls and the trimethylenemethane moiety again adopt a mutually staggered arrangement[133] and the molecular parameters are similar to those observed in the phenyltrimethylenemethane complex described above.

Fe
OC CO CO

(**84**)

VI. PENTAHAPTO LIGANDS

A. Cyclopentadienyl Complexes

The main structural features associated with the η^5-cyclopentadienyl ligand are:

(1) The carbon atoms of the η^5-C_5H_5 group are accurately coplanar and equidistant from the metal atom.

(2) The C—C bonds in the complexed C_5H_5 are, in general, of equal length and longer than those observed in uncoordinated benzene. Thus the electron diffraction data for a number of $(\eta^5\text{-}C_5H_5)_2M$ complexes lead to C—C distances in the range 1.423–1.440(2) Å[134], whereas the C—C distances in benzene have been reported as 1.397(3) Å (electron diffraction data)[135], 1.377 Å (X-ray data at −3°C)[136], and 1.390 Å (neutron data at −55°C)[137]. It is pertinent to note here that some workers have reported distortion of the C_5 ring from regular pentagonal geometry. Thus, in (η^5-pentachlorocyclopentadienyl)(η^2:η^2-cycloctadiene)-rhodium the C_5 ring has two short C—C bonds [1.399(6) Å] and three long bonds [1.436(7) Å], and the ring is folded about the $C_{(1)}$—$C_{(2)}$ direction by 4.6° (**85**)[138]. The most conclusive evidence for C_5 ring distortion was presented by Byers and Dahl[139] in their structure analysis of $[(\eta^5\text{-}C_5Me_5)(CO)_2Co]$. The C—C distances shown in (**86**) led the authors to describe the distortion as a move from η^5 to η^3:η^2 geometry and they noted a slight bending of the ring about the terminal carbon atoms of the 'allyl system'.

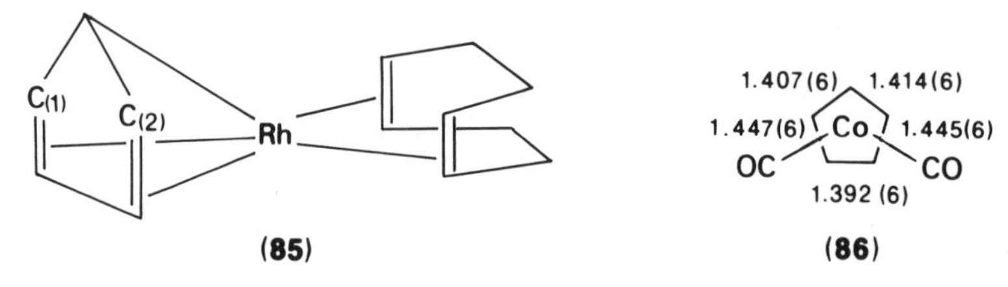

(**85**) (**86**)

(3) The substituents on the aromatic framework are bent out of the ring plane. This feature is also observed for the η^6-benzene, η^7-cycloheptatrienyl, and η^8-cyclooctatetraene ligands. At present there seems to be no consistent rationalization for the conflicting evidence concerning these deviations from planarity. Hydrogen atom substituents are usually displaced towards the metal as

in ferrocene[140], ferrocenedicarboxylic acid[140], (η^6-benzene)tricarbonylchromium[141], η^7-cycloheptatrienyl[η^6-phenyl(triphenyl)borato]molybdenum[142], and η^8-cyclooctatetraene(tetrahydrofuran)dichlorozirconium[143]. However, larger substituents are normally displaced away from the metal, as in *sym*-octamethylferrocene[144] and (η^6-phenyldimethylphosphine)tris(dimethylphenylphosphine)molybdenum[145].

1. $[M(C_5H_5)_x]$ *complexes*

The best known member of this class is ferrocene or bis(η^5-cyclopentadienyl)iron. The two C_5 rings in ferrocene are planar and are also parallel to each other. It was originally thought, from crystallographic considerations, that the two rings were staggered with respect to one another (**87**)[146]. However, recent work by Seiler and Dunitz, who carried out a re-determination of the crystal structure of ferrocene[147,148], has called into question the assignment of a staggered arrangement for the two rings.

(**87**)

Ferrocene undergoes a phase transition at 164 K, and in the low-temperature (triclinic) phase Seiler and Dunitz found that the two rings are neither staggered nor eclipsed but are closer to the latter conformation, the rings being mutually rotated by about 9° from the eclipsed orientation. The high-temperature (monoclinic) phase is disordered and can be described in terms of the average superposition of the four nearly eclipsed molecules which are present in the low temperature unit cell.

It would appear, therefore, that the rings in ferrocene in the solid state are nearer to the eclipsed arrangement observed for the gas-phase molecule using electron diffraction techniques[134]. Ferrocene thus joins ruthenocene[149] and osmocene[150], both of which have eclipsed rings in the solid state. A neutron diffraction study[140] of ferrocene at 173 and 298 K resulted in conclusions similar to those arrived at by Seiler and Dunitz.

In contrast to ferrocene, no phase change is observed when nickelocene is cooled from room temperature to 101 K. The structure analysis of $[(C_5H_5)_2Ni]$ shows essentially the same structure at both temperatures (i.e. a staggered arrangement). At 101 K the average Ni—C and C—C distances (corrected for vibrational motion) are 2.185(4) are 1.423(6) Å, respectively, and the H atoms are displaced out of the ring plane towards the nickel by 0.028 Å[151]. There is no reason to believe that the staggered arrangement is observed because of random disorder over two eclipsed molecules as in ferrocene, and it is suggested that the energy barrier between the two conformations for nickelocene is so small that the arrangement is almost exclusively determined by packing forces, which in the monoclinic crystal structure would favour the centrosymmetric staggered conformation.

The work of Seiler and Dunitz raises the question of whether further structural reinvestigation of metallocenes isomorphous with ferrocene (e.g. cobaltocene: staggered arrangement[152]) would be advisable. One particularly interesting re-investigation would be that of beryllocene whose crystal structure at room

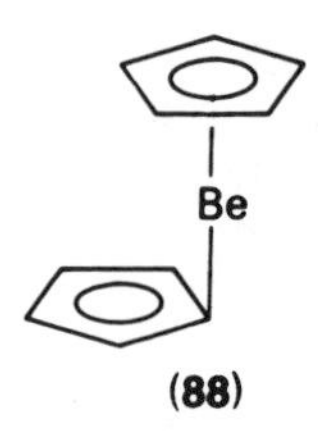

(88)

temperature[153] and 153 K[154] is disordered and has been interpreted in terms of a 'slip-sandwich' with η^5 and η^1-C_5H_5 groups **(88)**. However, the room temperature crystals are isomorphous with ferrocene and it seems likely that the proposed average structure is very different from the actual structure of the individual molecules.

Manganocene is unlike the other first transition series biscyclopentadienyl complexes in that the structure analyses[155–157] indicate essentially ionic bonding. The dark brown anti-ferromagnetic modification of manganocene consists of a polymeric zig-zag chain of manganese atoms linked by C_5H_5 bridges. Each manganese atom is bound to a η^5-C_5H_5 group and also interacts with one μ-C_5H_5 via a η^3-system and with a second μ-C_5H_5 via an η^2-system **(89)**. This structure is similar to that observed in the solid state for $[(C_5H_5)_2Pb]$[158], although in this case the μ-C_5H_5 groups are η^5-bonded to the lead atoms **(90)**.

Mn—C 2.40–2.43(4) Å

(89) **(90)**

The alkali metal cyclopentadienides are ionic in nature, as are the biscyclopentadienyl complexes of strontium and barium. Monocyclopentadienylindium and the corresponding thallium compound are polymeric with ionic μ-(η^5: η^5-C_5H_5) groups[159]. Biscyclopentadienylcalcium is also polymeic with η^5, η^3, and η^1-C_5H_5 groups, and is considered to be predominantly ionic in character[160], although there is still some debate as to the exact nature of the bonding[161,162].

(91)

$[(C_5H_5)_xM]$ complexes with $x > 2$ are generally found with the lanthanide and actinide elements, although the structures of some such complexes for the heavier d-block elements have been determined. Thus the zirconium atom in $[(C_5H_5)_4Zr]$ is coordinated to three η^5- and one η^1-cyclopentadienyl ligands **(91)**[163].

2. $[(C_5H_5)_xML_y]$ complexes

Numerous mono-, bis-, and tris-η^5-cyclopentadienyl—metal complexes containing in addition σ-bonded, π-bonded and π-acceptor ligands have been prepared and structurally characterized. Many examples of the structures of such complexes have already been given in this chapter under other headings. The structural features involving the C_5H_5 group are as listed at the opening of this section.

Lauher and Hoffmann[83] discussed the structure and chemistry associated with $[(C_5H_5)_2ML_y]$ complexes and made reference to the structures of many such compounds. It should be noted that in these biscyclopentadienyl complexes the two C_5H_5 rings are not parallel but are inclined to each other **(92)**.

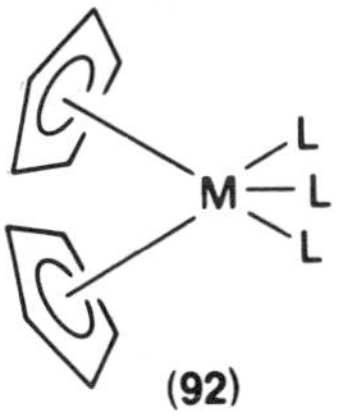

M L L L

(92)

B. Cyclohexadienyl Complexes

Pentahapto coordination of cyclic dienyls other than cyclopentadienyl usually leads to some distortion of the ligand geometry. Thus, the uncomplexed cyclohexadienyl ligand is either planar or only moderately distorted[165], whereas in its metal complexes five of the ring carbon atoms remain approximately planar whilst the sixth moves significantly out of plane away from the metal to give an envelope conformation **(93)**. The dihedral angle between the two planes intersecting along the fold line XY is normally between 40° and 50°, being 42.8° in $[(C_6H_7)Mn(CO)_3]$[166], 41° in $[\{C_6H_6CH(COOEt)_2)\}Mn(CO)_3]$[167], and 42.9° in $[Rh(\eta^5\text{-}C_5Me_4Et)\{\eta^5\text{-}C_6H_6P(O)\text{-}(OMe)_2\}]^{+}$[168]. The larger angle of 58° found in 1-5-η-*exo*-1-acetyl-2,4,6-tris(trifluoromethyl)cyclohexadienyl(η^5-cyclopentadienyl)iron **(94)** is presumably due to steric hindrance[169].

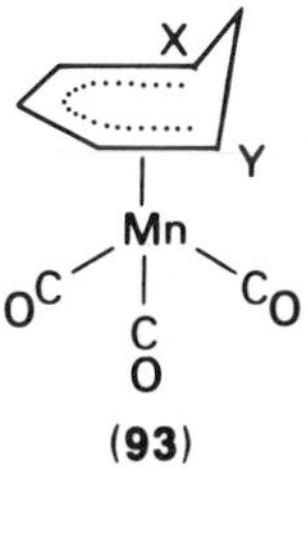

X Y Mn OC CO C O

(93)

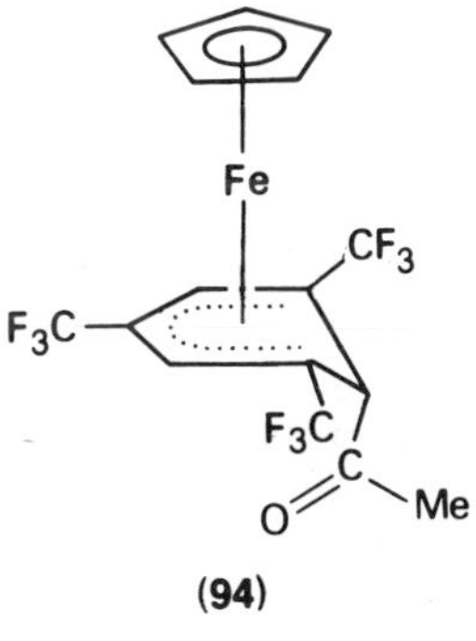

Fe CF3 F3C F3C C O Me

(94)

The C—C distances between the ring carbon atoms bound to the metal are not significantly different from one another {1.388—1.415(12) Å in [(C_6H_7)Mn(CO)$_3$], 1.408—1.419(5) Å in **94**}, indicating the delocalization of electrons in this portion of the ring, whereas the C—C distances in the bent back portion are normal single bond lengths [1.511(13) Å in [(C_6H_7Mn(CO)$_3$], 1.510, 1.518(5) Å in **94**]. The interior angle at the saturated carbon is significantly smaller than the ideal tetrahedral angle {104.1(8)° [(C_6H_7)Mn(CO)$_3$], 98.7(3)° in **94**}.

The metal—carbon distance in η^5-cyclohexadienyl complexes show a pattern in which the metal atom is significantly nearer to the central atom of the delocalized set than to the two terminal carbon atoms {Mn—C(central) 2.141(9) Å, Mn—C(terminal) 2.219(7) Å, in [C_6H_7)Mn(Co)$_3$]; Fe—C(central) 2.042(4) Å, Fe—C(terminal) 2.060, 2.069(3) Å in **94**}.

C. Heterocyclic Ligands

There are many examples of heterocycles acting as η^5-ligands, for example the anion of 1-*t*-butyl-3-methyl-2-phenyl-Δ^3-1,2-azaboroline can act as a $C_5H_5^-$ analogue (**95**)[170].

Triple[171] and tetra-decker sandwich complexes have been synthesized using heterocyclic ligands. Thus, bis(η^5-cyclopentadienyliron-μ-η^5-thiadiborolene)iron is a tetradecker sandwich molecule (**96**) and the distances between the iron atoms and the best planes through the ligands are 1.67 Å (to C_5H_5), and 1.63, 1.64 Å (to C_2B_2S), respectively[172].

(**95**)

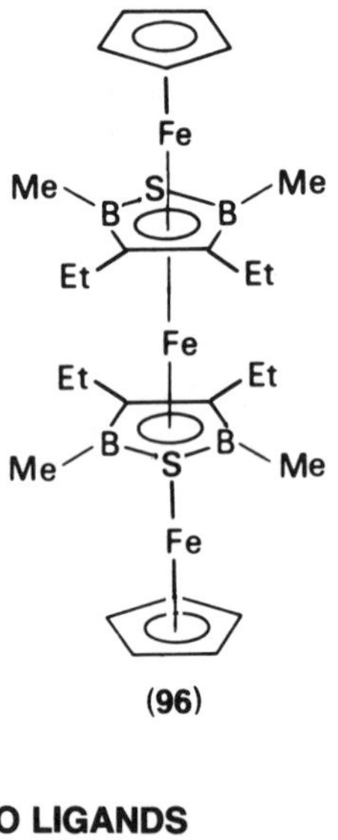

(**96**)

VII. HEXA-, HEPTA-, AND OCTAHAPTO LIGANDS

The structural features associated with η^6, η^7, and η^8-cyclic ligands are similar to those described for η^5-systems. Thus, the metal-bonded cabon atoms are coplanar and equidistant from the metal atom, their substituents are bent out of the plane, and the C—C distances are equivalent and slightly longer than those found in free benzene. There are exceptions to these generalizations and structures illustrating these, together with more typical examples, are described below.

A. η^6-Ligands

Bis(η^6-benzene)chromium has an eclipsed sandwich structure (**97**). The rings are planar and parallel to each other and all the Cr—C distances are equivalent. The possibility of distortion of the benzene to give C—C bonds of alternating length has been the subject of some debate. The consensus of opinion is that the C—C bond lengths are equal within experimental error[173].

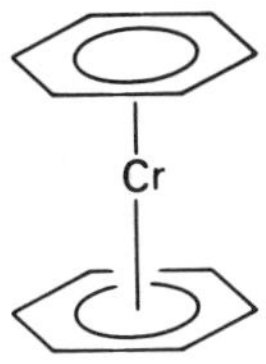

(**97**)

However, in an X-ray and neutron diffraction study of (η^6-benzene)tricarbonylchromium, Rees and Coppens[141] found that the C—C bonds in the staggered molecule (**98**) alternate in length. Those bonds *cis* to CO groups are longer (1.423 Å average) than those *trans* to CO groups (1.406 Å average). The ring is accurately planar, the hydrogen atoms are bent out of the ring plane by *ca.* 0.03 Å, bending towards the metal, and the Cr—C(benzene) distances range from 2.223 to 2.243 Å. Alternation of the C—C bond lengths in the benzene ring is reported in the similar molecule (η^6-hexamethylbenzene)tricarbonylmolybdenum[174] (**99**). The three

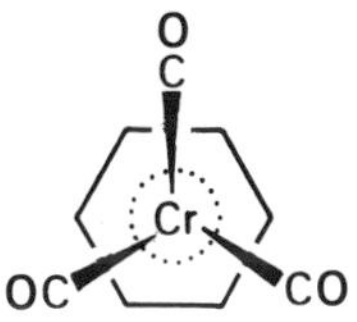

(**98**) 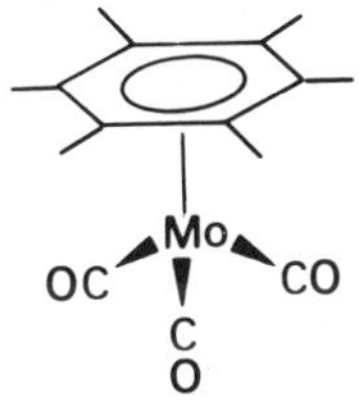 (**99**)

long bonds average 1.441(9) Å and the three short ones average 1.405(5) Å. The ring is planar and the Mo—C(benzene) distances lie in the range 2.386–2.397(5) Å. However, the methyl substituents are displaced out of the ring plane and away from the metal atom, in contrast to the substituents in $[(C_6H_6)Cr(CO)_3]$. This reversal of direction of displacement is ascribed to steric factors.

There is no indication of alternation of the C—C bond distances in the C_6 ring in (η^6-mesitylene)(η^2-maleic anhydride)dicarbonylchromium, where the bond lengths lie in the range 1.393–1.428(10 Å[175]. Likewise, there is no significant difference between the C—C bond lengths in the η^6-ring in the anion $[Mo(CO)_3\{\eta^6\text{-}C_6H_5B(C_6H_5)_3\}]^-$ (**100**). However, the lengthening of the C—C distance on complexing is illustrated by comparing the average C—C bond length in the ligated ring [1.415(6) Å] with those observed for the other three rings [1.396(6), 1.394(6), and 1.394(7) Å][142].

The benzene ring does not always remain planar on η^6-coordination Thus, in $[(\eta^6\text{-}C_6H_5Me)Ni(C_6F_5)_2]$ the η^6-ring has a small boat-type distortion (**101**), two carbon atoms being *ca.* 0.05 Å out of plane and away from the metal[176]. This results in two distinct types of metal—carbon (η^6) bond length, i.e. four in the range 2.164–2.171(5) Å and two in the range 2.232–2.251(7) Å. Similar boat distortions of

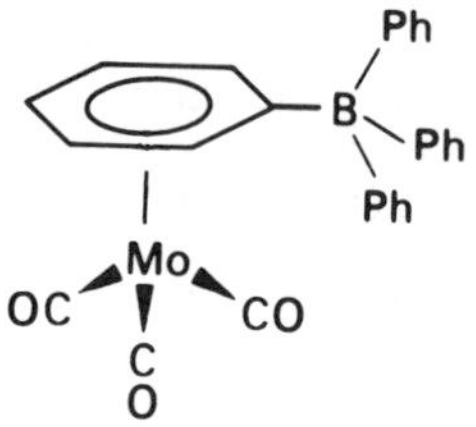

(100) 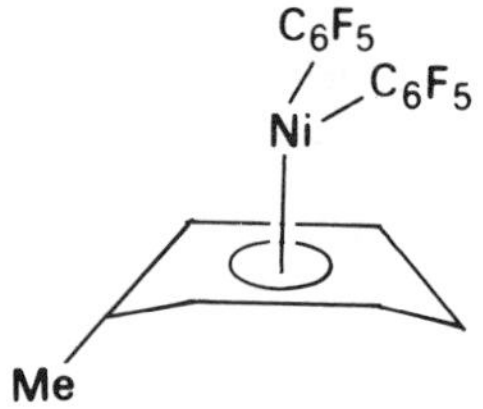 (101) 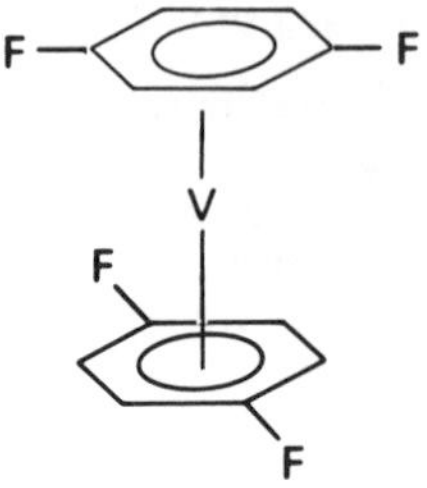 (102)

the η^6-ring have been observed in $[(\eta^6\text{-}C_6H_5Me)Mo(\mu\text{-}SMe)_3Mo(\eta^6\text{-}C_6H_5Me)]^{2+}$ [177] and in $[(\eta^6\text{-}C_6H_4F_2)_2V]$[178]. In the vanadium complex the two rings are parallel, with the carbon atoms nearly eclipsed **(102)**, and the fluorine-substituted carbon atoms are displaced by 0.06 Å from the mean plane of the ligand.

When the η^6-ligand is a polycyclic molecule, coordination can result in a slight distortion of the ring system. Thus, in the (η^6-phenanthrene)tricarbonylchromium: 1,3,5-trinitrobenzene adduct **(103)** the angle between the plane of the η^6-ring and the mean plane of the non-bonded part of the phenanthrene is 4.7° [179]. The mean C—C bond length for the complexed side ring is 1.418 Å and the equivalent value for the other two rings is 1.398 Å.

In (η^6-fluorenyl)(η^5-cyclopentadienyl)iron the fluorenyl group is bound to the iron atom not through the C_5 ring but through one of the C_6 rings **(104)**[180]. Five carbon

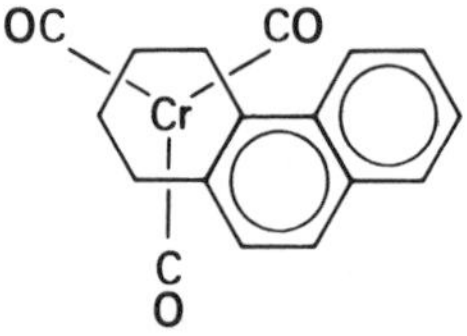

(103) 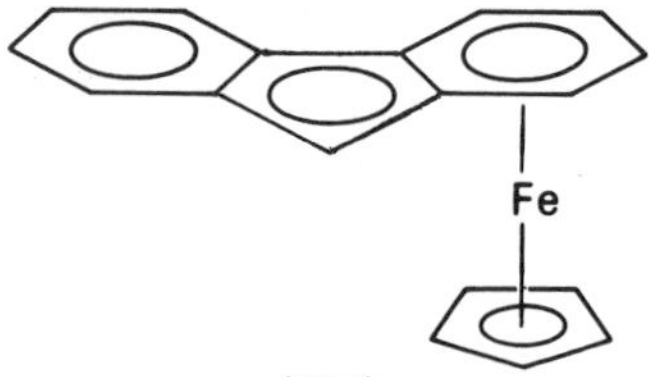 (104)

atoms of the η^6-ring are coplanar but the sixth (that bound to the unique atom of the C_5 ring) is diplaced by 0.15 Å out of the plane and away from the metal atom. This stereochemistry is reminiscent of that observed in (η^5-cyclohexadienyl)—metal complexes (see Section VI.B), as is the pattern of iron—carbon bond lengths [Fe—C(central) 2.039(5) Å, and Fe—C(terminal) 2.122, 2.152(5) Å]. However, the sixth carbon atom is close enough to the iron to be considered to be bonded [2.316(5) Å], whereas in cyclohexadienyl complexes the equivalent atom lies at distances >2.7 Å from the metal. Perhaps a better structural model for this complex is a zwitterionic formulation with the positive charge on the metal atom and the negative charge centred primarily on the unique atom of the C_5 ring, a description which is in keeping with the carbanion character of the latter atom.

Cyclohepta-1,3,5-triene can act as an η^6-ligand on coordination to a transition metal. In (η^5-methylcyclopentadienyl)(η^6-7-*exo*-phenylcyclohepta-1,3,5-triene)-manganese the six atoms of the triene fragment are coplanar and the Mn—C distances are in the range 2.06–2.15(2) Å, the larger values corresponding to the distances between the Mn atom and the terminal atoms of the triene system[181]. The final atom of the C_7 ring is too far away to be considered to be bonded to the Mn atom **(105)**. Bonding the triene to the metal in this manner results in a flattening of the conjugated system. Thus the dihedral angle between the planes defined by atoms 1,2,5,6 and

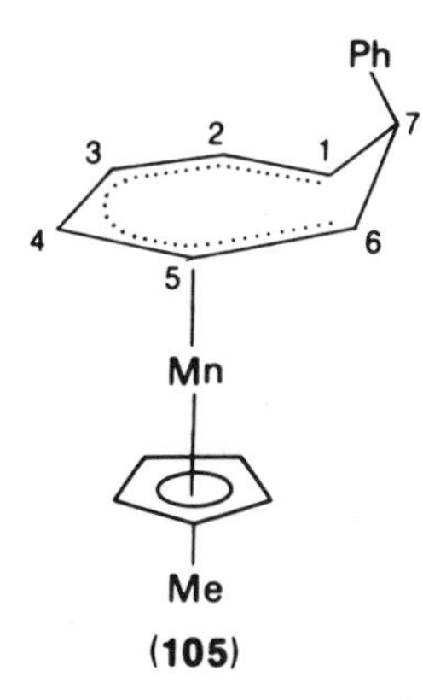

(105)

2,3,4,5 is 8°, a significant decrease from the 23.2° found in the non complexed *p*-bromophenacyl ester of 5,5-dimethylcyclohepta-1,3,5-triene-1-carboxylic acid[182].

B. η^7-Ligands

Complexes of molybdenum containing the η^7-cycloheptatrienyl ligand are well documented[183–185] and, as expected, the C_7 ring is planar which, together with the equality of the C—C distances, is taken to indicate delocalization of the ring π-electrons. The rings in $[(\eta^7\text{-}C_7H_7)Mo(\mu\text{-}Cl)_2(\mu\text{-}OH)Mo(\eta^7\text{-}C_7H_7)]^+$ (**106**) can be

(106)

described as essentially regular heptagons, the C—C—C angles vary from 125 to 132(1)° and the C—C distances lie in the range 1.39–1.43(2) Å[183]. Similarly, in (η^7-cycloheptatrienyl)(η^6-tetraphenylborato)molybdenum the interior angles of the planar C_7 ring range from 128.4 to 129.0(4)° and the C—C distances lie between 1.401 and 1.413(7) Å[142]. In these two complexes the Mo—C(η^7) distances are in the range 2.218–2.297(13) Å and 2.263–2.286(5) Å respectively.

C. η^8-Ligands

Cyclooctatetraene, which can act as an η^2-, η^3- and η^4-ligand when coordinated to transition metals, can also be used as an η^8-ligand. Bis(η^8-cyclooctatetraene)uranium (uranocene) is a sandwich complex in which the two rings are parallel and eclipsed, and each is planar (**107**)[186]. The C—C distances range from 1.337 to 1.439(23) Å and the carbon atoms are equidistant from the uranium atom [2.635–2.675(11) Å]. The interior ring angles lie between 133.3 and 136.7(12)°, compared with 135° for the interior angle of a regular octagon.

In the two substituted uranocenes bis(1,3,5,7-tetramethylcyclooctatetraene)-uranium[187] and the phenyl analogue[188], the C_8 rings are nearly eclipsed. The substituents in the methyl complex are found in nearly eclipsed and nearly staggered arrangements in the two crystallographically independent molecules and are bent out

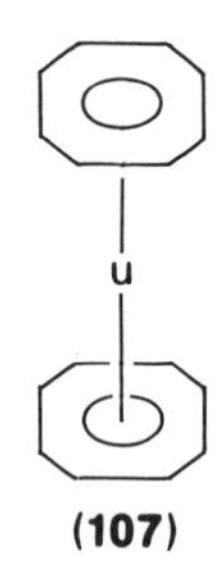

(107)

of the C_8 plane towards the uranium atom. In the phenyl complex only the staggered form is found, with the phenyl groups tilted at an average angle of 42° to the C_8 plane.

The uranium atom is centred accurately on the cyclooctatetraene fragment of the bicyclic ligand bicyclo[6.2.0]deca-1,3,5,7-tetraene in the sandwich complex $[(C_{10}H_{10})_2U]$[189], and the C_4 ring is set at an angle of 6.8° to the plane of the C_8 ring.

Hydrogen atoms in the η^8-C_8H_8 ligand are bent out of the ring plane and towards the metal atom giving the 'umbrella' arrangement observed in several cycloctatetraene complexes, e.g. $[(\eta^8\text{-}C_8H_8)(C_4H_8O)ZrCl_2]$[190] and $[(\eta^8\text{-}C_8H_8)(C_4H_8O)TiCl]_2$[191].

In the sandwich anion $[(\eta^8\text{-}C_8H_8)_2Nd]^-$ the C—C distances for the two rings average 1.40(2) and 1.42(1) Å, respectively[192]. However, these rings are not parallel, the angle between the two ring planes being 7.28°. This results in two sets of Nd–C distances: the average value for ring 1 is 2.68(1) Å and that for ring 2 is 2.79(1) Å. This deviation from the normal arrangement for $[(\eta^8\text{-}C_8H_8)_2M]$ complexes is thought to arise from an interaction of one of the rings with the neodymium atom of the cation $[(\eta^8\text{-}C_8H_8)Nd(C_4H_8O)_2]^+$ (**108**).

THF THF

Nd Nd⁺

(108)

VIII. REFERENCES

1. M. R. Truter, *Molecular Structure by Diffraction Methods*, Specialist Periodical Report, Royal Society of Chemistry, **6**, 93 (1978).
2. E. Weiss and H. Koster, *Chem. Ber.*, **110**, 717 (1977).
3. G. K. Barker, M. F. Lappert, and J. A. K. Howard, *J. Chem. Soc., Dalton Trans.*, 734 (1978).
4. M. F. Lappert, S. J. Miles, P. P. Power, A. J. Carty, and N. J. Taylor, *J. Chem. Soc. Chem. Commun.*, 458 (1977).
5. F. A. Cotton, S. A. Koch, and M. Millar, *Inorg. Chem.*, **17**, 2084 (1978).
6. F. A. Cotton, L. D. Gage, K. Mertis, L. W. Shire, and G. Wilkinson, *J. Am. Chem. Soc.*, **98**, 6922 (1976).
7. E. O. Fischer and A. Frank, *Chem. Ber.*, **111**, 3740 (1978).
8. N. W. Alcock, *J. Chem. Soc., A*, 2001 (1967).
9. N. I. Gapotchenko, N. W. Alekseev, N. E. Kolobova, K. N. Anisimov, I. A. Ronova, and A. A. Johansson, *J. Organomet. Chem.*, **35**, 319 (1972).
10. V. I. Kulishov, N. G. Bokii, A. F. Prikhotko, and Yu. T. Struchkov, *J. Struct. Chem.*, **16**, 231 (1975).

11. P. G. Harrison and K. Molloy, *J. Organomet. Chem.*, **152**, 53 (1978).
12. C. J. Cardin, D. J. Cardin, and A. Roy, *J. Chem. Soc., Chem. Commun.*, 899 (1978).
13. V. Gramlich and K. Pfefferkorn, *J. Organomet. Chem.*, **61**, 247 (1973).
14. C. J. Cardin, D. J. Cardin, M. F. Lappert, and K. W. Muir, *J. Chem. Soc., Dalton Trans.*, 46 (1978).
15. W. E. Carroll and R. Bau, *J. Chem. Soc., Chem. Commun.*, 825 (1978).
16. C. H. Cheng and R. Eisenberg, *Inorg. Chem.*, **18**, 1418 (1978).
17. J. P. Collman, P. A. Christian, S. Current, P. Denisevich, T. R. Halbert, E. R. Schmitton, and K. O. Hodgson, *Inorg. Chem.*, **15**, 223 (1976).
18. M. R. Churchill and S. W. Y. Chang, *Inorg. Chem.*, **14**, 1680 (1975).
19. P. G. Lenhert, C. M. Lukehart, and L. T. Warfield, *Inorg. Chem.*, **19**, 2343 (1980).
20. C. M. Lukehart and J. V. Zeile, *J. Organomet. Chem.*, **140**, 309 (1977).
21. W. K. Wong, W. Tam, C. E. Strouse and J. A. Gladysz, *J. Chem. Soc., Chem. Commun.*, 530 (1979).
22. C. P. Casey and C. A. Bunnell, *J. Am. Chem. Soc.*, **98**, 436 (1976).
23. F. Calderazzo, *Angew. Chem., Int. Ed. Engl.*, **16**, 299 (1977).
24. H. Berke and R. Hoffmann, *J. Am. Chem. Soc.*, **100**, 7224 (1978).
25. H. Schmidbaur, *Acc. Chem. Res.*, **8**, 62 (1975).
26. H. Schmidbaur, *Pure Appl. Chem.*, **50**, 19 (1978).
27. C. Kruger, *Angew. Chem. Int. Ed. Engl.*, **11**, 387 (1972).
28. D. J. Brauer, C. Kruger, P. J. Roberts, and Y. H. Tsay, *Chem. Ber.*, **107**, 3706 (1974).
29. F. A. Cotton, B. E. Hanson, W. H. Ilslay, and G. W. Rice, *Inorg. Chem.*, **18**, 2713 (1979).
30. R. G. Vranka and E. L. Amma, *J. Am. Chem. Soc.*, **89**, 3121 (1967).
31. G. R. Scollary, *Aust. J. Chem.*, **31**, 411 (1978).
32. H. Koster, D. Thoennes, and E. Weiss, *J. Organomet. Chem.*, **160**, 1 (1978).
33. C. Kruger, J. Sekutowski, H. Berke, and R. Hoffmann, *Z. Naturforsch. B, Anorg. Chem., Org. Chem.*, **33B**, 1110 (1978).
34. M. B. Hursthouse, K. M. A. Malik and K. D. Sales, *J. Chem. Soc., Dalton Trans.*, 1314 (1978).
35. E. O. Fischer and A. Maasbol, *Angew. Chem. Int. Ed. Engl.*, **3**, 580 (1964).
36. O. S. Mills, and A. D. Redhouse, *J. Chem. Soc., A*, 642 (1968).
37. M. F. Lappert, R. W. McCabe, J. J. MacQuilty, P. L. Pye, and P. I. Riley, *J. Chem. Soc., Dalton Trans.*, 90 (1980).
38. P. B. Hitchcock, M. F. Lappert, and P. L. Pye, *J. Chem. Soc., Dalton Trans.*, 826 (1978).
39. U. A. Gregory, S. D. Ibekwe, B. T. Kilbourn, and D. R. Russell, *J. Chem. Soc., A*, 1118 (1971).
40. G. Huttner, S. Schelle, and O. S. Mills, *Angew. Chem. Int. Ed. Engl.*, **8**, 515 (1969).
41. A. D. Redhouse, *J. Organomet. Chem.*, **99**, C29 (1975).
42. L. J. Guggenberger and R. R. Schrock, *J Am. Chem. Soc.*, **97**, 6578 (1975).
43. R. R. Schrock, *Acc. Chem. Res.*, **12**, 98 (1979).
44. M. R. Churchill and W. J. Youngs, *Inorg. Chem.*, **18**, 1930 (1979).
45. M. R. Churchill and F. J. Hollander, *Inorg. Chem.*, **17**, 1957 (1978).
46. J. H. Wengrovius, R. R. Schrock, M. R. Churchill, J. R. Missert, and W. J. Youngs, *J. Am. Chem. Soc.*, **102**, 4515 (1980).
47. E. O. Fischer, *Adv. Organomet. Chem.*, **14**, 1, (1976).
48. G. Huttner, H. Lorenz, and W. Gartzke, *Angew. Chem. Int. Ed. Engl.*, **13**, 609 (1974).
49. U. Schubert, E. O. Fischer, and D. Wittmann, *Angew. Chem. Int. Ed. Engl.*, **19**, 643 (1980).
50. M. R. Churchill and W. J. Youngs, *Inorg. Chem.*, **18**, 2454 (1979).
51. W. A. Herrmann, *Angew. Chem. Int. Edn. Engl.*, **17**, 800 (1978).
52. W. A. Herrmann, C. Kruger, R. Goddard, and I. Bernal, *J. Organomet. Chem.*, **140**, 73 (1977).
53. M. Creswick, I. Bernal, and W. A. Herrmann, *J. Organomet. Chem.*, **172**, C39 (1979).
54. C. E. Summer, P. E. Riley, R. E. Davis and R. Pettit, *J. Am. Chem. Soc.*, **102**, 1752 (1980).
55. T. R. Halbert, M. E. Leonowicz, and D. J. Maydonovitch, *J. Am. Chem. Soc.*, **102**, 5101 (1980).

56. M. B. Hursthouse, R. A. Jones, K. M. A. Malik, and G. Wilkinson, *J. Am. Chem. Soc.*, **101**, 4128 (1979).
57. P. Hoffmann, *Angew. Chem. Int. Ed. Engl.*, **18**, 554 (1979).
58. D. L. Davis, A. F. Dyke, A. Endesfelder, S. A. R. Knox, P. J. Naish, A. G. Orpen, D. Plaas, and G. E. Taylor, *J. Organomet. Chem.*, **198**, C43 (1980).
59. A. F. Dyke, S. A. R. Knox, P. J. Naish, and A. G. Orpen, *J. Chem. Soc., Chem. Commun.*, 441 (1980).
60. R. A. Love, T. F. Koetzle, G. J. B Williams, and L. C. Andrews, *Inorg. Chem.*, **14**, 2653 (1975).
61. P. T. Cheng, and S. C. Nyburg, *Can. J. Chem.*, **50**, 912 (1972).
62. G. Bombieri, E. Forsellini, C. Panattoni, R. Graziani, and G. Bandoli, *J. Chem. Soc., A*, 1313 (1970).
63. J. N. Francis, A. McAdam, and J. A. Ibers, *J. Organomet. Chem.*, **29**, 131 (1971).
64. L. D. Brown, C. F. J. Barnard, J. A. Daniels, R. J. Mawby and J. A. Ibers, *Inorg. Chem.*, **17**, 2932 (1978).
65. L. Maresca, G. Natile, M. Calligaris, M. Delise, and L. Randaccio, *J. Chem. Soc., Dalton Trans.*, 2386 (1976).
66. J. M. Baraban and J. A. McGinnety, *J. Am. Chem. Soc.*, **97**, 4232 (1975).
67. S. D. Ittel and J. A. Ibers, *Adv. Organomet. Chem.*, **14**, 33 (1976).
68. J. A. Evans and D. R. Russell, *J. Chem. Soc., Chem. Commun.*, 197 (1971).
69. K. Jonas, K. R. Porschke, C. Kruger, and Y. H. Tsay, *Angew. Chem. Int. Ed. Engl.*, **15**, 621 (1976).
70. N. Rosch and R. Hoffmann, *Inorg. Chem.*, **13**, 2656 (1974).
70a. M. Green, J. A. K. Howard, J. L. Spencer, and F. G. A. Stone, *J. Chem. Soc., Chem. Commun.*, 449 (1975).
71. H. Dierks and H. Dietrich, *Z. Kristallogr.*, **122**, 1 (1965).
72. G. Huttner and O. S. Mills, *J. Organomet. Chem.*, **29**, 275 (1971).
73. D. J. Brauer and C. Kruger, *J. Organomet. Chem.*, **44**, 397 (1972).
74. R. Grigg, B. Kongakathip, and T. J. King, *J. Chem. Soc., Dalton Trans.*, 333 (1978).
75. R. Mason and G. R. Scollary, *Aust. J. Chem.*, **31**, 781 (1978).
76. K. D. Tau, D. W. Meek, T. Sorrell and J. A. Ibers, *Inorg. Chem.*, **17**, 3454 (1978).
77. P. Caddy, M. Green, L. E. Smart, and N. White, *J. Chem. Soc., Chem. Commun.*, 839 (1978).
78. C. G. Pierpont, *Inorg. Chem.*, **17**, 1976 (1978).
79. R. D. Adams and D. F. Chodosh, *Inorg. Chem.*, **17**, 41 (1978).
80. H. Le Bozec, P. H. Dixneuf, A. J. Carty, and N. J. Taylor, *Inorg. Chem.*, **17**, 2568 (1978).
81. M. Aresta, C. F. Nobile, V. G. Albano, E. Forni, and M. Manassero, *J. Chem. Soc., Chem. Commun.*, 636 (1975).
82. G. Fachinetti, C. Floriani, and H. Stoeckli-Evans, *J. Chem. Soc., Dalton Trans.*, 2297 (1977).
83. J. W. Lauher and R. Hoffmann, *J. Am. Chem. Soc.*, **98**, 1729 (1976).
84. G. Fachinetti, C. Biron, C. Floriani, A. C. Villa, and C. Guastini, *Inorg. Chem.*, **17**, 2995 (1978).
85. A. D. Redhouse and W. A. Herrmann, *Angew. Chem. Int. Ed. Engl.*, **15**, 615 (1976).
86. B. W. Davies and N. C. Payne, *Inorg. Chem.*, **13**, 1848 (1974).
87. N. M. Boaz, M. Green, D. M. Grove, J. A. K. Howard, J. L. Spencer, and F. G. A. Stone, *J. Chem. Soc., Dalton Trans.*, 2170 (1980).
88. G. Fachinetti, C. Floriani, F. Marchetti and M. Mellini, *J. Chem. Soc., Dalton Trans.*, 1398 (1978).
89. M. R. Churchill and W. J. Youngs, *Inorg. Chem.*, **18**, 1697 (1979).
90. L. Ricard, R. Weiss, W. E. Newton, G. J. J. Chen, and J. W. McDonald, *J. Am. Chem. Soc.*, **100**, 1318 (1978).
91. R. M. Laine, R. E. Moriarty, and R. Bau, *J. Am. Chem. Soc.*, **94**, 1402 (1972).
92. B. H. Freeland, J. E. Hux, N. C. Payne, and K. G. Tyers, *Inorg. Chem.*, **19**, 693 (1980).
93. W. E. Oberhansli and L. F. Dahl, *J. Organomet. Chem.*, **3**, 43 (1965).
94. A. E. Smith, *Acta Crystallogr.*, **18**, 331 (1965).
95. J. W. Faller, D. F. Chodosh and D. Katahiri, *J. Organomet. Chem.*, **187**, 227 (1980).

96. T. J. Greenhough, P. Legzdius, D. T. Martin and J. Trotter, *Inorg. Chem.*, **18**, 3268 (1979).
97. R. D. Adams, D. F. Chodosh, J. W. Faller, and A. M. Rosan, *J. Am. Chem. Soc.*, **101**, 2570 (1979).
98. J. A. Kaduk, A. T. Poulos, and J. A. Ibers, *J. Organomet. Chem.*, **127**, 245 (1977).
99. S. F. A. Kettle and R. Mason, *J. Organomet. Chem.* **5**, 573 (1966).
100. C. F. Putnik, J. J. Welter, G. D. Stucky, M. J. D'Aniello, B. A. Sosinsky, J. F. Kirner, and E. L. Muetterties, *J. Am. Chem. Soc.*, **100**, 4107 (1978).
101. R. Mason and A. G. Wheeler, *J. Chem. Soc., A*, 2549 (1968).
102. R. A. Marsh, J. Howard and P. Woodward, *J. Chem. Soc., Dalton Trans.*, 778 (1973).
103. R. Uttech and H. Dietrich, *Z. Kristallogr.*, **122**, 60 (1965).
104. B. Henc, P. W. Jolly, R. Salz, G. Wilke, R. Benn, E. G. Hoffmann, R. Mynott, G. Schroth, K. Seevogel, J. C. Sekutowski, and C. Kruger, *J. Organomet. Chem.*, **191**, 425 (1980).
105. F. A. Cotton and M. W. Extine, *J. Am. Chem. Soc.*, **100**, 3788 (1978).
106. F. A. Cotton and J. R. Pipal, *J. Am. Chem. Soc.*, **93**, 5441 (1971).
107. A. Keasey, P. M. Bailey and P. M. Maitlis, *J. Chem. Soc., Chem. Commun.*, 178 (1977).
108. P. D. Frisch, R. G. Posey and G. P. Khare, *Inorg. Chem.*, **17**, 402 (1978).
109. D. J. Brauer and C. Kruger, *J. Organomet. Chem.*, **122**, 265 (1976).
110. G. Huttner, H. Brintzinger, L. G. Bell, P. Friedrich, V. Bejenke, and D. Neugebauer, *J. Organomet. Chem.*, **145**, 329 (1978).
111. J. M. Williams, R. K. Brown, A. J. Schultz, G. D. Stucky and S. D. Ittel, *J. Am. Chem. Soc.*, **100**, 7407 (1978).
112. R. D. Adams and D. F. Chodosh, *J. Am. Chem. Soc.*, **99**, 6544 (1977).
113. F. A. Cotton, M. W. Extine, and R. H. Niswander, *Inorg. Chem.*, **17**, 692 (1978).
114. O. S. Mills and G. Robinson, *Acta Crystallogr.*, **16**, 758 (1963).
115. M. R. Churchill, *J. Organomet. Chem.*, **4**, 258 (1965).
116. M. R. Churchill and R. Mason, *Adv. Organomet. Chem.*, **5**, 93 (1967).
117. R. Mason, *Nature*, **217**, 543 (1968).
118. F. H. Herbstein and M. G. Reisner, *Acta Crystallogr. Sect. B*, **33B**, 3304 (1977).
119. F. A. Cotton, V. W. Day, B. A. Frenz, K. I. Hardcastle, and J. M. Troup *J. Am. Chem. Soc.*, **95**, 4522 (1973).
120. A. Immirzi, *J. Organomet. Chem.*, **76**, 65 (1974).
121. M. R. Churchill and P. H. Bird, *Inorg. Chem.*, **8**, 1941 (1969).
122. G. Hunter, D. Neugebauer, and A. Razavi, *Angew. Chem. Int. Ed. Engl.*, **14**, 352 (1975).
123. P. Binger, B. Cetinkaya, and C. Kruger, *J. Organomet. Chem.*, **159**, 63 (1978).
124. M. R. Churchill and R. Mason, *Proc. Roy. Soc. London, Ser. A*, **292**, 61 (1966).
125. M. A. Bennett, T. W. Matheson, G. B. Robertson, A. K. Smith, and P. A. Tucker, *Inorg. Chem.*, **19**, 1014 (1980).
126. M. Gerloch and R. Mason, *Proc. Roy. Soc. London, Ser. A*, **279**, 170 (1964).
127. J. Rosalky, B. Metz, F. Mathey, and R. Weiss, *Inorg. Chem.*, **16**, 3307 (1977).
128. L. J. Todd, J. P. Hickey, J. R. Wilkinson, J. C. Huffman, and K. Folting, *J. Organomet. Chem.*, **112**, 167 (1976).
129. J. D. Dunitz, H. C. Mez, O. S. Mills, and H. M. M. Shearer, *Helv. Chim. Acta*, **45**, 647 (1962).
130. R. P. Dodge and V. Schomaker, *Acta Crystallogr.*, **18**, 614 (1965).
131. R. E. Davis and P. E. Riley, *Inorg. Chem.*, **19**, 674 (1980).
132. M. R. Churchill and K. Gold, *Inorg. Chem.*, **8**, 401 (1969).
133. M. R. Churchill and B. G. Deboer, *Inorg. Chem.*, **12**, 525 (1973).
134. A. Haaland, *Topics Curr. Chem.*, **53**, 1, (1975).
135. K. Kimura and M. Kuber, *J. Chem. Phys.*, **32**, 1776 (1960).
136. E. G. Cox, D. W. J. Cruickshank, and J. A. S. Smith, *Proc. Roy. Soc. London, Ser. A*, **247**, 1 (1958).
137. G. E. Bacon, N. A. Curry, and S. A. Wilson, *Proc. Roy. Soc. London, Ser. A*, **279**, 98, (1964).
138. V. W. Day, K. J. Reimer, and A. Shaver, *J. Chem. Soc., Chem. Commun.*, 403 (1975).
139. L. R. Byers and L. F. Dahl, *Inorg. Chem.*, **19**, 277 (1980).
140. F. Takusagawa and T. F. Koetzle, *Acta Crystallogr., Sect. B*, **35B**, 1074 (1979).

141. B. Rees and P. Coppens, *Acta Crystallogr., Sect. B*, **29B**, 2516 (1973).
142. M. B. Hossain and D. van der Helm, *Inorg. Chem.*, **17**, 2893 (1978).
143. D. J. Brauer and C. Kruger, *Inorg. Chem.*, **14**, 3053 (1975).
144. Yu. T. Struchkov, V. G. Andrianov, T. N. Sal'nikova, I. R. Lyatifov, and R. B. Materikova, *J. Organomet. Chem.*, **145**, 213 (1978).
145. R. Mason, K. M. Thomas, and G. A. Heath, *J. Organomet. Chem.*, **90**, 195 (1975).
146. J. D. Dunitz, L. E. Orgel, and A. Rich, *Acta Crystallogr.*, **9**, 373 (1956).
147. P. Seiler and J. D. Dunitz, *Acta Crystallogr., Sect. B*, **35B**, 1068 (1979).
148. P. Seiler and J. D. Dunitz, *Acta Crystallogr., Sect. B*, **35B**, 2020 (1979).
149. G. L. Hardgrove and D. H. Templeton, *Acta Crystallogr.*, **12**, 28 (1959).
150. F. Jellinek, *Z. Naturforsch. B: Anorg. Chem., Org. Chem.*, **14B**, 737 (1959).
151. P. Seiler and J. D. Dunitz, *Acta Crystallogr., Sect. B*, **36B**, 2255 (1980).
152. W. Bunder and E. Weiss, *J. Organomet. Chem.*, **92**, 65 (1975).
153. C. Wang, T. Y. Lee, T. J. Lee, T. W. Chang, and C. S. Liu, *Inorg. Nucl. Chem. Lett.*, **9**, 667 (1973).
154. C. Wang, T. Y. Lee, K. J. Chao, and S. Lee, *Acta Crystallogr., Sect. B*, **28B**, 1662 (1972).
155. E. Weiss and E. O. Fischer, *Z. Naturforsch. B: Anorg. Chem., Org. Chem.*, **10B**, 58 (1955).
156. P. Coppens, *Nature*, **204**, 1298 (1964).
157. W. Bunder and E. Weiss, *Z. Naturforsch B: Anorg. Chem., Org. Chem.*, **33B**, 1235 (1978).
158. C. Panattoni, G. Bombieri, and U. Croatto, *Acta Crystallogr.*, **21**, 823 (1966).
159. E. Frasson, F. Menegus and C. Panattoni, *Nature*, **199**, 1087 (1963).
160. R. Zerger and G. Stucky, *J. Organomet. Chem.*, **80**, 7 (1974).
161. W. Bunder and E. Weiss, *J. Organomet. Chem.*, **92**, 1 (1975).
162. V. T. Aleksanyan, I. A. Garbuzova, V. V. Garvrilenko, and L. I. Zakharbin, *J. Organomet. Chem.*, **129**, 139 (1977).
163. R. D. Rogers, R. V. Bynum, and J. L. Atwood, *J. Am. Chem. Soc.*, **100**, 5238 (1978).
164. J. K. Kuchitsu, T. Fukuyama, and Y. Morino, *J. Mol. Struct.*, **1**, 463, (1968).
165. R. Hoffmann and P. Hoffmann, *J. Am. Chem. Soc.*, **98**, 598 (1976).
166. M. R. Churchill and F. R. Scholer, *Inorg. Chem.*, **8**, 1950 (1969).
167. A. Mawby, P. J. C. Walker, and R. J. Mawby, *J. Organomet. Chem.*, **55**, C39 (1973).
168. N. A. Bailey, E. H. Blunt, G. Fairhurst, and C. White, *J. Chem. Soc., Dalton Trans.*, 829 (1980).
169. M. Bottrill, M. Green, E. O'Brien, L. E. Smart, and P. Woodward, *J. Chem. Soc., Dalton Trans.*, 292 (1980).
170. J. Schulze and G. Schmid, *Angew. Chem. Int. Ed. Engl*, **19**, 54 (1980).
171. W. Siebert, W. Rothermel, C. Bohle, C. Kruger, Y. H. Tsay, D. J. Brauer, *Angew. Chem. Int. Ed. Engl.*, **18**, 949 (1979).
172. W. Siebert, C. Bohle and C. Kruger, *Angew. Chem. Int. Ed. Engl.*, **19**, 746 (1980).
173. E. Keulen and F. Jellinek, *J. Organomet. Chem.*, **5**, 490 (1966).
174. D. E. Koshland, S. E. Myers, and J. P. Chesick, *Acta Crystallogr., Sect. B*, **33B**, 2013 (1977).
175. Yu. T. Struchkov, V. G. Andrianov, V. N. Setkina, N. K. Baranetskaya, V. I. Losilkina, and D. N. Kursanov, *J. Organomet. Chem.*, **182**, 213 (1979).
176. L. J. Radonovich, K. J. Klabunde, C. B. Behrens, D. P. McCollor, and B. B. Anderson, *Inorg. Chem.*, **19**, 1221 (1980).
177. W. E. Silverthorn, C. Couldwell, and K. Prout, *J. Chem. Soc., Chem. Commun.*, 1009 (1978).
178. L. J. Radonovich, E. C. Zuerner, H. F. Efner, and K. J. Klabunde, *Inorg. Chem.*, **15**, 2976 (1976).
179. R. Lal De, J. Seyerl, L. Zsolnai, and G. Huttner, *J. Organomet. Chem.*, **175**, 185 (1979).
180. J. W. Johnson and P. M. Treichel, *J. Am. Chem. Soc.*, **99**, 1427 (1977).
181. J. A. D. Jeffreys and J. Macfie, *J. Chem. Soc., Dalton Trans.*, 144 (1978).
182. R. E. Davis and A. Tulinsky, *J. Am. Chem. Soc.*, **88**, 4583 (1966).
183. C. Couldwell and K. Prout, *Acta Crystallogr., Sect. B*, **34B**, 2439 (1978).
184. A. J. Welch, *Inorg. Chim. Acta*, **24**, 97 (1977).
185. M. Green, H. P. Kirsch, F. G. A. Stone, and A. J. Welch, *J. Chem. Soc., Dalton Trans.*, 1755 (1977).

186. A. Avdeef, K. N. Raymond, K. O. Hodgson, and A. Zalkin, *Inorg. Chem.*, **11**, 1083 (1972).
187. K. O. Hodgson and K. N. Raymond, *Inorg. Chem.*, **12**, 458 (1973).
188. L. K. Templeton, D. H. Templeton, and R. Walker, *Inorg. Chem.*, **15**, 3000 (1976).
189. A. Zalkin, D. H. Templeton, S. R. Berryhill, and W. D. Lube, *Inorg. Chem.*, **18**, 2287 (1979).
190. D. J. Brauer and C. Kruger, *Inorg. Chem.*, **14**, 3053 (1975).
191. H. R. van der Wal, F. Overzet, H. O. van Oven, J. L. De Boer, H. J. De Liefde Meijer, and F. Jellinek, *J. Organomet. Chem.*, **92**, 329 (1975).
192. C. W. De Kock, S. R. Ely, T. E. Hopkins, and M. A. Brault, *Inorg. Chem.*, **17**, 625 (1978).

The Chemistry of the Metal–Carbon Bond
Edited by F. R. Hartley and S. Patai

CHAPTER **2**

Thermochemistry of organometallic compounds

G. PILCHER and H. A. SKINNER
Department of Chemistry, University of Manchester, Manchester M13 9PL, UK

I. INTRODUCTION

The main interest in organometallic thermochemistry in recent years has been on compounds involving transition metals, which exhibit great variety in the types of chemical binding between the metal and ligand. Although one definition of organometallic compounds restricts these to compounds containing metal–carbon bonds, the thermochemical interest is somewhat wider. This review will refer to some compounds which do not contain metal–carbon bonding but which are of relevance in organometallic thermochemistry.

Recent reviews of this topic were made by Skinner[1] in 1964, Cox and Pilcher[2] in 1970, Pilcher[3] in 1975, and Pedley and Rylance[4] in 1977. In this chapter, the literature is surveyed to the end of 1979; all thermal data are presented in SI units.

II. EXPERIMENTAL METHODS

In this section we consider the more important experimental methods that have been used to obtain the data presented in Table 1.

A. Enthalpies of Combustion

Static-bomb calorimetry is the most widely used method for determining enthalpies of formation of organic compounds containing C,H,O, and N by measurement of their energies of combustion in oxygen; rotating-bomb calorimetry is preferred and is more reliable for organic compounds containing S and the halogens. Both types of bomb calorimetry have been applied to organometallic compounds.

1. Static-bomb calorimetry

Early static-bomb measurements on organometallic compounds often gave results which were misleading, but this method has become more successful in recent years with improved techniques of chemical analysis, and the use of 'combustion aids'. The main problems are: (a) to contain the substance within the bomb when it is volatile, or reacts spontaneously with oxygen; (b) to achieve complete combustion of both the organic ligand, and of the metal; and (c) to form products which can be thermodynamically defined. Careful quantitative analysis of the combustion products (in particular the amount of carbon dioxide) can establish completeness of combustion. X-ray powder photographs of the oxides produced can yield information on the state of the final products. The use of combustion aids can significantly improve the completeness of the oxidation process, but this depends on the metal, and in some cases the static-bomb method is not satisfactory, and other methods are used.

The static-bomb combustion method has been shown to be satisfactory for the following classes of compound.

a. Dialkylzinc compounds. The static-bomb results of Long and Norrish[5] for $ZnMe_2$(l) and $ZnEt_2$(l) agree to within the limits of experimental error with values derived from enthalpies of hydrolysis by Carson *et al.*[6].

b. Alkyl- and arylmercury compounds. The results of the study by Jones *et al.*[7] in 1935 are doubtful, but in 1952 Carson *et al.*[8] made reliable measurements on $HgMe_2$(l) and $HgEt_2$(l). The amount of compound burned was determined from the carbon dioxide produced and most of the mercury in the products was elemental, only a small correction being required for the HgO formed. The enthalpies of formation agreed with those from reaction calorimetry by Hartley *et al.*[9]. For $HgPh_2$(c) there is excellent agreement between the static-bomb measurements by Carson and Wilmshurst[10] and Fairbrother and Skinner[11], and the enthalpies of formation from reaction calorimetry by Chernick *et al.*[12].

c. Organoboron compounds. Close agreement on the enthalpies of formation of BMe_3(l) and BEt_3(l) have been obtained by Long and Norrish[5] and Johnson *et al.*[13], although the combustion conditions differed greatly and the solid products were B_2O_3(am) and H_3BO_3(c) for the two investigations, respectively. Gal'chenko and co-workers have studied several boron alkyls[14] and some carboranes[15], and studies on enthalpies of hydroborination of alkenes[16] yielded enthalpies of formation in agreement with combustion values.

d. Alkyl- and aryltin compounds. Static-bomb combustions in 45 atm of oxygen by Davies *et al.*[17] yielded a consistent set of enthalpy of formation data when tested against bond-energy schemes. A small thermal correction was required for the incomplete combustion of the tin, and the amount of carbon dioxide was measured to demonstrate the completeness of combustion of the organic part of the molecule. Other static-bomb measurements on tin compounds reported in the literature, in which the amount of carbon dioxide was not determined, are unsatisfactory.

e. Transition metal organometallic compounds. The problems in making successful static-bomb measurements on these compounds are severe, and only in the last decade has sufficient progress been made to yield reliable results.

Static-bomb combustion of metal carbonyls can be successful when done in dry oxygen because there will be no complication from hydration of the oxide formed. Barnes *et al.*[18] measured the energies of combustion of $Mo(CO)_6$(c) and $W(CO)_6$(c) in this manner; the amount of carbon dioxide was determined and the oxides MoO_3 and WO_3 easily dissolved to determine trace amounts of unburned metal. The derived enthalpies of formation agreed with those from enthalpies of decomposition (see later). Similar measurements on $Cr(CO)_6$(c)[19] were less successful owing to the difficulty of analysis of the mixture of Cr_2O_3 and Cr. The enthalpy of formation of $Ni(CO)_4$(l) from static-bomb measurements by Fischer *et al.*[20] is in agreement with that derived from equilibrium studies of the thermal decomposition by Sykes and Townshend[21]. Cotton *et al.*[22] have used this technique for $Fe(CO)_5$(l).

Rabinovich and co-workers, in a recent series of papers, appear to have solved many of the difficulties in the static-bomb combustion of organometallic compounds. Their method is to burn a relatively small amount of compound with a large amount of auxiliary material such as polyethylene or benzoic acid. Although a price is paid in the precision of measurement (because the compound contributes a smaller fraction than usual to the measured enthalpy), the advantages are that the auxiliary material promotes completeness of combustion. Moreover, because the oxide product is formed at high temperature, it will be fused and is unlikely to hydrate, e.g. Tel'noi *et al.*[23] burned bisbenzene chromium in heavy polyethylene ampoules, and product analyses (CO_2) indicated complete combustion of the benzene and of the chromium metal. The derived value, $\Delta H_f^0\{[Cr(C_6H_6)_2],c\} = (+146 \pm 8.4)$ kJ/mol is in excellent agreement with the value derived from the enthalpy of decomposition[24], $(+142.3 \pm 8.4)$ kJ/mol; earlier static-bomb combustion values were discordant, being $(+89.1 \pm 32.0)$ kJ/mol[25] and $(+213.0 \pm 12.0)$ kJ/mol[26].

For compounds which react spontaneously with oxygen, such as the dicyclopentadienyl derivatives $[Ti(C_5H_5)_2]$(c) and $[V(C_5H_5)_2]$(c), Tel'noi *et al.*[27] protected the sample before combustion by coating it with paraffin wax. This served additionally as the auxiliary combustion aid. An analysis for complete combustion by determining carbon dioxide was made, and there was no evidence of incomplete combustion of the metals.

The static-bomb combustion method may be satisfactory but has not been proved to be so far the following:

f. Alkylphosphorus, alkylarsenic, alkylantimony and alkylbismuth compounds. The enthalpies of formation of these compounds were all obtained by static-bomb combustion calorimetry, in which the experimental problems resemble those encountered with organoboron compounds. There are, however, no independent investigations by other methods, and insufficient data to test for internal consistency so that we cannot accept the values for these compounds as established.

The static-bomb combustion method has been shown to be unsatisfactory for the following classes of compound.

g. Alkylaluminium compounds. Values from static-bomb calorimetry for these compounds are discrepant, e.g. for $AlEt_2H(l)$ the energy of combustion obtained by Shaulov *et al.*[28] is 103 kJ/mol less than that by Pawlenko[29]; incomplete combustion may have been responsible for difficulties with this compound. Apparently reliable reaction calorimetry studies of enthalpies of hydrolysis gave enthalpies of formation differing from those obtained by combustion, for $AlEt_3(l)$ by 30 kJ/mol[30] and for $AlMe_3(l)$ by 80 kJ/mol[31]. Shmyreva *et al.*[32] measured both the energy of combustion and the enthalpy of hydrolysis in dilute hydrochloric acid of $Al(n\text{-}Pr_3)(l)$, but the enthalpies of formation from the two methods differed by 46 kJ/mol. The magnitude of the discrepancies makes it difficult to accept that any of the published enthalpies of formation of aluminium alkyls are reliable.

h. Organosilicon compounds. In 1964, Good *et al.*[33] developed a rotating-bomb technique for organosilicon compounds; this work demonstrated that the static-bomb technique for these compounds is completely unsatisfactory and that all static-bomb results should be rejected.

i. Alkylgermanium compounds. The problem with these compounds lies in the state of the germanium oxide produced, and most authors, on the evidence of X-ray powder photographs, have taken this product to be crystalline, in either the hexagonal or tetragonal forms. It appears, however, that it is largely amorphous and the enthalpies of transition between the three forms are fairly large[34]:

$$GeO_2(am) \longrightarrow GeO_2(c, hex) \quad \Delta H = -15.7 \text{ kJ/mol} \tag{1}$$

$$GeO_2(c, hex) \longrightarrow GeO_2(c, tet) \quad \Delta H = -25.4 \text{ kJ/mol} \tag{2}$$

This uncertainty regarding the state of the oxide product necessarily enters the enthalpy of formation of the alkylgermanium. Although the static-bomb combustion of germanium metal with benzoic acid auxiliary probably produced an amorphous oxide similar to that from the combustion of alkylgermanium compounds, the rotating-bomb technique for germanium compounds avoids any argument and is clearly preferable.

j. Alkyllead compounds. Good and Scott[35] compared the results from static and rotating-bomb techniques, and showed that the former method was inadequate for these compounds. The differences in the enthalpies of formation were −123.0 kJ/mol for $PbMe_3(g)$ and +166.6 kJ/mol for $PbEt_3(l)$, despite the care taken in the analysis of the combustion products from the static-bomb experiments.

2. *Rotating-bomb calorimetry*

The rotating bomb was originally developed for combustion of organic compounds containing sulphur, since by rotation of the bomb and its contents after combustion the sulphuric acid produced was of uniform concentration. This method has since been applied to several organometallic compounds, but the chemical problems associated with the combustion reaction are specific to each element, and must be resolved individually. With properly designed comparison experiments, the rotating bomb method is superior to conventional combustion techniques and has been successfully applied in the following examples.

a. Organosilicon compounds. Good *et al.*[33] applied the rotating-bomb calorimeter to hexamethyldisiloxane: the compound, mixed with $\alpha\alpha\alpha$-trifluorotoluene, was burned in oxygen in a bomb containing water. After combustion and rotation a homogeneous solution of hexafluorosilicic acid was produced according to the reaction

$$5PhCF_3(l) + (Me_3Si)_2O(l) + 49.5O_2(g) + 406H_2O(l) \longrightarrow 41CO_2(g) + 2\{(H_2SiF_6 + 1.5HF)(212H_2O)\}(l) \quad (3)$$

The enthalpy of formation of the final solution in the bomb was determined by burning silicon mixed with poly(vinylidene fluoride) to produce a final solution of the same composition. Hence the enthalpy of formation of hexamethyldisiloxane was determined with respect to elemental silicon. Iseard *et al.*[36] have described an aneroid rotating-bomb calorimeter, and used a similar procedure, with trifluoromethylbenzoic as combustion aid; measurements were made on $SiEt_4(l)$ and $Si_2Et_6(l)$. Steele[37] has measured the enthalpy of formation of $SiPh_4(c)$ by rotating-bomb calorimetry.

Hajiev and Agarunov[38] studied the combustion of chlorosilicon alkyls by rotating-bomb calorimetry. A solution of hydrazinium chloride was placed in the bomb to reduce any chlorine produced on combustion to chloride ion. The silicon on combustion was transformed into a homogeneous colloidal solution of amorphous silica. The enthalpies of formation of $Me_2SiCl_2(l)$ and $Me_3SiCl(l)$ so derived agree with those obtained from enthalpies of hydrolysis[39].

b. Alkyl- and arylgermanium compounds. The rotating-bomb technique for these compounds was introduced by Bills and Cotton[40], who studied $GeEt_4(l)$. An aqueous HF solution placed in the bomb dissolved the GeO_2 formed on combustion. Comparison experiments were made by burning benzoic acid and at the same time dissolving GeO_2 (c,hex.) in the acid to produce a final solution of the same composition as that formed in the $GeEt_4(l)$ combustion. The enthalpy of formation of $GeEt_4(l)$ was determined with respect to GeO_2(c,hex.) from the enthalpy of the reaction

$$GeEt_4(l) + 14O_2(g) \longrightarrow GeO_2(c, hex.) + 8CO_2(g) + 10H_2O(l) \quad (4)$$

Carson and co-workers used an aneroid rotating-bomb calorimeter charged with KOH solution to determine the enthalpies of formation of $GePh_4(c)$[41], $Ge(CH_2Ph)_4(c)$ and $Ge_2Ph_6(c)$[42]. The KOH solution dissolved both the GeO_2 and the CO_2 formed.

C. Alkyl- and aryllead compounds. Good *et al.*[43] determined the enthalpy of formation of $PbEt_4(l)$ by rotating-bomb calorimetry. A solution of nitric acid containing some arsenious oxide was placed in the bomb and, after combustion and rotation, the nitric acid dissolved the metallic lead and the lead oxides whilst the arsenious oxide ensured that the dissolved lead was in the Pb^{2+} oxidation state. For the comparison experiments, hydrocarbon oil was burned in the presence of crystalline lead nitrate to produce a final solution of the same composition as in the normal combustion experiments. The enthalpy of formation of $PbEt_4(l)$ was determined with respect to the enthalpy of formation of lead nitrate according to the reaction

$$PbEt_4(l) + 13.5O_2(g) + 2(HNO_3 \text{ in } 30H_2O)(l) \longrightarrow 8CO_2(g) + 11H_2O(l) + Pb(NO_3)_2(c) \quad (5)$$

Using this method, Good *et al.*[44] determined the enthalpy of formation of $PbMe_4(l)$ and Carson *et al.*[45] determined that for $PbPh_4(c)$.

d. Organoarsenic and -bismuth compounds. Mortimer and Sellers[46] used the rotating-bomb calorimeter to determine the enthalpy of formation of $AsPh_3(c)$. Aqueous NaOH was placed in the bomb to dissolve the CO_2 and the arsenious oxide formed as products of combustion (a small amount of sodium arsenate was also formed). The enthalpy of solution of the CO_2 was determined in comparison experiments in which benzoic acid was burned with the same initial solution in the bomb. The enthalpy of formation was obtained from the enthalpy of the reaction

$$AsPh_3(c) + 22.5O_2(g) + 43(NaOH, 12.93H_2O)(l) \longrightarrow 18CO_2(g) + (40NaOH, Na_3AsO_3, 565H_2O)(l) \quad (6)$$

This method should be applicable to other arsenic compounds. Steel[47] has applied a similar rotating-bomb method for determination of the enthalpy of formation of $BiPh_3(c)$.

e. Organoselenium compounds. Combustion of selenium and its compounds produces selenium dioxide which is soluble in water. Barnes and Mortimer[48] have studied several selenium compounds by rotating-bomb calorimetry; typical is the example of $SePh_2(l)$:

$$SePh_2(l) + 15.5O_2(g) + 396H_2O(l) = 12CO_2(g) + (SeO_2.401H_2O)(l) \quad (7)$$

Comparison experiments were made by burning benzoic acid with a solution of selenium dioxide in the bomb: the enthalpy of solution of SeO_2 was measured in separate experiments.

f. *Transition metal organometallic compounds*. The only published work on a rotating-bomb study of a compound of this type is that for dimanganesedecacarbonyl by Good *et al.*[49]. A procedure similar to that for the lead alkyls was used, except that hydrogen peroxide was added to the nitric acid solvent to ensure solution of the manganese as Mn^{2+}, and a small thermal correction was required because of the catalytic decomposition of hydrogen peroxide by Mn^{2+}. The bomb was charged at a lower than usual oxygen pressure (*ca.* 5 atm), to assist the solution of the manganese oxides by ensuring that they were not fused, but there was incomplete combustion of the organic part of the molecule requiring a thermal correction for the carbon formed. For the comparison experiments, hydrocarbon oil was burned in the presence of manganous nitrate in solution; the enthalpy of formation of $Mn_2(CO)_{10}$ was determined from the enthalpy of the reaction

$$Mn_2(CO)_{10}(c) + 6O_2(g) + 4(HNO_3{\cdot}16H_2O)(l) \longrightarrow 10CO_2(g) + 45.4H_2O(l) + 2(Mn(NO_3)_2{\cdot}10.3H_2O)(l) \quad (8)$$

B. Enthalpies of Reaction

The commonly used reactions of hydrolysis and halogenation of organic thermochemistry have been extensively applied with organometallic compounds, but also there are some specific reactions for this class of compound.

1. Enthalpies of hydrolysis

For many organometallic compounds, the measurement of the enthalpy of hydrolysis is straightforward. It is preferable that the hydrolysis results in a homogeneous solution rather than a precipitate, which may introduce uncertainties due to surface adsorption: e.g. Carson *et al.*[50] preferred the results of the hydrolysis of $CdMe_2(l)$ in acid, even though the acidic and neutral hydrolysis gave results which agreed to within the limits of experimental error. The two reactions were

$$CdMe_2(l) + 2H_2O(l) \longrightarrow Cd(OH)_2(c, \text{precipitate}) + 2CH_4(g) \quad (9)$$

$$CdMe_2(l) + (H_2SO_4 \cdot 100H_2O)(l) \longrightarrow (CdSO_4 \cdot 100H_2O)(l) + 2CH_4(g) \quad (10)$$

For some compounds, the reaction with liquid water can be violent, even explosive, and Fowell and Mortimer devised a method to control the rate of reaction by bubbling nitrogen saturated with water vapour through the reactive liquids. This method was applied to Li-*n*-Bu(l)[51] and to $AlEt_3(l)$[31].

An interesting example of hydrolysis in acidic solution was of dicyclopentadienylmagnesium by Hull *et al.*[52], and the reverse of hydrolysis was used by Hull and Turnbull[53] to measure the enthalpy of formation of cyclopentadienylthallium from the enthalpy of reaction in aqueous solution:

$$(TlOH \cdot 2050H_2O)(l) + c\text{-}C_5H_6(g) = Tl(C_5H_5)(c) + 2051H_2O(l) \quad (11)$$

Oxidative hydrolysis has proved to be a useful method for organometallic compounds in which the metal is in a low oxidation state. Adedeji *et al.*[54] found that hexakis (dimethylamido)ditungsten, $[W_2(NMe_2)_6](c)$, is rapidly hydrolysed in water, yielding a black powdered precipitate, but in the presence of a suitable oxidizing agent this black precipitate is gradually oxidized to a precipitate of yellow tungstic acid. The oxidizing agent used was potassium dichromate in sulphuric acid and the enthalpy of the following reaction (ΔH_1) was measured:

$$[W_2(NMe_2)_6](c) + (Cr_2O_7^{2-} + 14H^+ + H_2O)(\text{soln.}) \longrightarrow$$
$$2H_2WO_4(\text{ppt./soln.}) + (2Cr^{3+} + 6NMe_2H_2^+)(\text{soln}) \quad (12)$$

From separate experiments on the oxidation of hydrazine by the same acid dichromate solution, ΔH_2,

$$1.5N_2H_4(l) + (Cr_2O_7^{2-} + 8H^+ + 6NMe_2H_2^+)(\text{soln.}) \longrightarrow$$
$$1.5N_2(g) + (2Cr^{3+} + 7H_2O + 6NMe_2H_2^+)(\text{soln.}) \quad (13)$$

and the enthalpy of solution of $Me_2NH(l)$ in the same solvent, ΔH_3,

$$6NMe_2H(l) + 6H^+(\text{soln.}) = 6(NMe_2H_2^+)(\text{soln.}) \quad (14)$$

are obtained and then the combination, $\Delta H_1 - \Delta H_2 - \Delta H_3$, yields the enthalpy of reaction

$$[W_2(NMe_2)_6](c) + 1.5N_2(g) + 8H_2O(l) \longrightarrow$$
$$2H_2WO_4(\text{ppt./soln.}) + 1.5N_2H_4(l) + 6NMe_2H(l) \quad (15)$$

from which $\Delta H_f^0([W_2(NMe_2)_6],c)$ was derived without reference to the enthalpies of formation of dichromate or chromic ions in solution. Adedeji *et al.*[54] also studied $[Mo_2(NMe_2)_6](c)$ by this method.

Oxidative hydrolysis was used by Cavell *et al.*[55] to determine the enthalpies of formation of several tetra-μ-acetato derivatives, including those of dimolybdenum(II) and dichromium(II). The oxidizing agent was iron(III) chloride in acidic solution, and the thermochemical reaction is represented by

$$[Mo_2(OAc)_4](c) + 8FeCl_3(c) + 8H_2O(l) + 4NaCl(c) \longrightarrow 2Na_2MoO_4(c) + 8FeCl_2(c) + 12HCl(\text{in } 7.97H_2O)(l) + 4AcOH(l) \quad (16)$$

The enthalpy of reaction was determined by measuring the enthalpy of solution of each reactant and product successively in the calorimetric solvent, so that the final solution from dissolution of all the reactants was of the same composition as the corresponding solution from the dissolution of the products.

2. *Enthalpies of halogenation*

Enthalpies of halogenation are obviously important for determining the enthalpies of formation of halogeno-organometallic compounds; e.g. Pedley *et al.*[56] measured the enthalpies of several reactions, such as

$$SnMe_4(l) + Br_2(g) = Me_3SnBr(l) + MeBr(g) \quad (17)$$

Nitrogen carried bromine vapour (from a weighted evaporator containing liquid bromine) into a reaction vessel which contained the liquid tin alkyl within the calorimeter, and the amount of reaction was determined from the weight loss of the bromine evaporator. From the value of $\Delta H_f^\circ(Me_3SnBr(l))$ so derived, the enthalpy of formation of hexamethyldistannane was calculated from the enthalpy of the reaction

$$Me_3SnSnMe_3(l) + Br_2(l) = 2Me_3SnBr(l) \quad (18)$$

Cartner *et al.*[57] have determined the enthalpy of formation of dicobaltoctacarbonyl from the enthalpy of reaction with bromine in carbon tetrachloride:

$$[Co_2(CO)_8](c) + (2Br_2{\cdot}n\,CCl_4)(l) = 2CoBr_2(c) + 8CO(g) + n\,CCl_4(l) \quad (19)$$

The derived-enthalpy of formation agreed, within the limits of experimental error. with the value obtained independently from thermal decomposition studies.

3. *Enthalpies of redistribution reactions*

Single-centre redistribution reactions, defined by Skinner[58] as those reactions in which the chemical bonds change in relative position but not in number or formal character, have been useful in the organometallic field, notably for organomercury and organotin compounds. For example, the enthalpy of the reaction

$$Et_2Hg(l) + HgCl_2(c) = 2EtHgCl(c) \quad (20)$$

was derived from the enthalpy of reaction in solution and the appropriate enthalpies of solution to obtain the enthalpy of formation of EtHgCl(c)[59]. The enthalpies of formation of alkylmercury halides are all derived from reactions of this type.

Nash *et al.*[60] studied the enthalpies of redistribution reactions of $SnCl_4(l)$ and $SnMe_4(l)$: on mixing these liquids the whole range of redistribution products, $MeSnCl_3$, Me_2SnCl_2, and Me_3SnCl, can be obtained, the relative amounts of each in the final mixture depending on the composition of the initial reaction mixture. The composition of the final mixture was determined by GLC, and by investigating over a wide range of composition the enthalpies of formation of all the chlorotinmethyls were obtained.

4. Enthalpies of hydroborination

Bennett and Skinner[16] measured the enthalpies of hydroborination of several alkenes to produce alkyls according to

$$6RCH{=}CH_2(l) + B_2H_6(g) \longrightarrow 2(RCH_2CH_2)_3B(l) \qquad (21)$$

Gaseous diborane was admitted to a reaction vessel which contained a solution of the alkene in diglyme, through a sintered-glass disk covered with mercury. The enthalpies of formation of the boron alkyls obtained in this way are in excellent agreement with those obtained from static-bomb combustion experiments by Gal'chenko and Zaugol'nikova[14].

5. Enthalpies of thermal decomposition

Many organometallic compounds thermally decompose at elevated temperatures to comparatively simple and stable products which can be thermodynamically defined. Measurement of the enthalpy of decomposition provides a route to the enthalpy of formation of the organometallic compound, but this method requires either a calorimeter capable of operation at elevated temperatures or a room-temperature calorimeter which contains a small high-temperature region, a 'hot-zone'.

Connor *et al.*[24,61,62] have measured the enthalpies of decomposition of several transition metal carbonyls and related compounds using the Calvet high-temperature microcalorimeter. Two types of reaction, exemplified by reference to dimanganesedecacarbonyl, were studied: (i) thermal decomposition:

$$[Mn_2(CO)_{10}](c) = 2Mn(c) + 10CO(g) \qquad (22)$$

and (ii) thermal decomposition in the presence of gaseous iodine:

$$[Mn_2(CO)_{10}](c) + 2I_2(g) = 2MnI_2(c) + 10CO(g) \qquad (23)$$

More recently, Zafarani-Moattar[63] used bromine to replace iodine in reactions of this type. One advantage is that the bromination reaction can be used at lower temperatures. For the thermal decomposition, *ca.* 3 mg of the sample at 25°C, enclosed in a glass capillary, were dropped into a glass reaction vessel filled with argon in the microcalorimeter at 241°C. The rapid decomposition gave a manganese mirror on the walls of the glass reaction vessel, and the enthalpy of reaction is subject to possible error because of the exothermic adsorption of carbon monoxide on the thin metal film. The iodination enthalpies do not suffer from this defect, and for $[Mn_2(CO)_{10}](c)$ the largest thermal decomposition enthalpies were *ca.* 120 kJ/mol less than that required to conform to the iodination value. Thermal corrections were required to convert the observed enthalpies of reaction to the value at 298 K, and to derive the enthalpy of formation of the carbonyl, the enthalpy of formation of the transition metal iodide, which could be non-stoichiometric, had to be estimated. The iodination results are preferred and believed to be reliable to ± 10 kJ/mol. The compounds examined by this method include simple and polynuclear carbonyls, arene derivatives (e.g. benzenechromiumtricarbonyl), and 'sandwich' complexes including dibenzenechromium, dibenzenemolybdenum, and ditoluenetungsten.

The Calvet high-temperature microcalorimeter has also proved useful in measuring the enthalpies of sublimation of organometallic compounds, by dropping *ca.* 3 mg of the compound into the reaction vessel in the microcalorimeter at a temperature below that for decomposition, and then evaporating the solid by applying a vacuum. The procedure was calibrated by dropping in substances of known enthalpies of sublimation[64], and the results appear to be reliable to ±4kJ/mol.

The thermal decomposition enthalpies of the hexacarbonyls of chromium, molybdenum and tungsten were measured by Pittam and co-workers[18,19] using a hot-zone calorimeter. For these measurements *ca.* 2.5 g of the carbonyl were dropped into a reaction vessel heated at *ca.* 400°C by a hydrogen-in oxygen flame. The carbon monoxide left the calorimeter through a heat-exchange spiral, was then oxidized to carbon dioxide, and the amount of this determined the amount of carbonyl decomposed. The enthalpy of decomposition contributed *ca.* 10% of the total heat produced, the remainder being generated by the hydrogen flame (the latter was determined precisely from the mass of water formed). Adsorption of carbon monoxide on the metal surface was not significant in these macro-scale experiments because of the small surface to volume ratio of the metal produced. For $[Mo(CO)_6]$ and $[W(CO)_6]$ the results were in agreement with static-bomb measurements, and for $[Mo(CO)_6]$ were in good agreement with the microcalorimeter iodination study[61].

Differential scanning calorimetry has been used to study the enthalpies of dissociation of metal–ligand bonds in a number of transition metal complexes. One advantage of this method is that only a few milligrams of sample are needed for each determination, and the extent of decomposition is derived from the decrease in weight of the sample. The temperature of the calorimeter is raised linearly with respect to time until decomposition occurs and is completed; thus the observed enthalpies of decomposition for a series of compounds will refer to different temperatures and for proper comparison, heat capacity corrections are needed to convert the values to ΔH (298 K).

Investigations using differential scanning calorimetry were initiated about 15 years ago, since when the technique has been widely applied.

Mortimer and co-workers[65–69] studied the thermal decomposition of complexes of the general formula MX_2L_n, where M is a first-row transition metal, X is a halogen and the ligand L is pyridine, one of the methylpyridines, quinoline, pyrazine, pyrimidine, or aniline. The decompositions followed the equation

$$MX_2L_n(c) = MX_2(c) + nL(g) \tag{24}$$

and the results were interpreted in terms of σ- and π-bonding components and of steric interactions.

Another type of thermal decomposition, studied by Ashcroft *et al.*[70], was the decarbonylation of *trans*-bis(triphenylphosphine)chlorobenzoylplatinium(I), according to the equation

$$[\mathrm{Pt(Cl)(PPh_3)_2(COPh)}](c) \longrightarrow [\mathrm{Pt(Cl)(PPh_3)_2(Ph)}](c) + CO(g) \tag{25}$$

and the results were interpreted to give 180.7 kJ/mol for the dissociation energy of the Pt—C bond in the grouping Pt—COPh.

Mortimer *et al.*[71] have also studied thermal decompositions of the type

$$[IrX(CO)(PPh_3)_2A](c) \longrightarrow [IrX(CO)(PPh_3)_2](c) + A(g) \tag{26}$$

where X is a halogen and A is either $F_2C{=}CF_2$ or $CF_3C{\equiv}CCF_3$. The results suggested that the strength of the iridium–olefin bonding depends on the combination of the σ(olefin $\rightarrow$ metal) and the π(metal$\rightarrow$olefin) components of the bond.

III. ENTHALPIES OF FORMATION AND ATOMIZATION OF ORGANOMETALLIC COMPOUNDS

A compilation of enthalpies of formation of organometallic compounds is given in Table 1. The metals are ordered according to the Groups of the Periodic Table. Values are given at 298.15 K in kJ/mol for the enthalpy of formation in the condensed state (crystal or liquid), the enthalpy of sublimation or evaporation, the enthalpy of formation in the gaseous state and the enthalpy of atomization, ΔH_a. Estimated values are given in parentheses. The enthalpies of atomization were calculated using the enthalpies of formation of the gaseous atoms listed in Table 3.

IV. STRENGTHS OF CHEMICAL BONDS IN ORGANOMETALLIC COMPOUNDS

A. Definitions of Thermochemical Bond Strengths

It is necessary to distinguish between the commonly used terms bond dissociation energy, mean bond dissociation energy, bond energy, and bond enthalpy contribution.

For a diatomic molecule, the spectroscopist distinguishes between the dissociation energy, D_e, measured from the minimum of the potential energy curve of the molecule, and D_0, determined from the lowest energy level of the molecule. D_e and D_0 differ by the zero-point energy of the molecule. The spectroscopic D_0 corresponds to ΔU_0^0 for the process

$$\text{AB(g, 0 K)} \longrightarrow \text{A(g, 0 K)} + \text{B(g, 0 K)} \tag{27}$$

where all the species are in their ground states at the absolute zero temperature, and is correctly called a *bond dissociation energy*. The thermochemist measures standard enthalpies of formation, ΔH_f^0, usually at 298.15 K, and for the process

$$\text{AB(g, 298 K)} \longrightarrow \text{A(g, 298 K)} + \text{B(g, 298 K)} \tag{28}$$

thermochemical data can provide the enthalpy of dissociation:

$$\Delta H_r^0(298\text{ K}) = \Delta H_f^0(\text{A, g, 298 K}) + \Delta H_f^0(\text{B, g, 298 K}) - \Delta H_f^0(\text{AB, g, 298 K}) \tag{29}$$

ΔH_r^0(298 K) is a *bond dissociation enthalpy*, and will normally differ from D_0 and from D_{298}^0, where

$$D_{298}^0 = \Delta U_r^0(298) = \Delta H_r^0(298) - RT \tag{30}$$

For a gaseous polyatomic molecule MX_n (where X is an atom), the *enthalpy of atomization*, ΔH_{atom}^0, is given by

$$\Delta H_{\text{atom}}^0 = \Delta H_f^0(\text{M, g}) + n\Delta H_f^0(\text{X, g}) - \Delta H_f^0(\text{MX}_n\text{, g}) \tag{31}$$

where the normal thermochemical convention is followed of not specifying temperature for reactions at 298.15 K. In so far as structural evidence confirms equivalence of the (M—X) bonds in MX_n, the quantity $\Delta H_{\text{atom}}^0/n$ measures the mean

TABLE 1. Enthalpies of formation of organometallic compounds

Compound	State	ΔH_f^0 (c/l) (kJ/mol)	ΔH^0 (sub/vap) (kJ/mol)	ΔH_f^0 (g) (kJ/mol)	ΔH_a^0 (kJ/mol)	Ref.
Group IA						
EtLi	c	−58.7 ± 5.4	116.7 ± 0.8	58.0 ± 5.5	2626.0	72
n-BuLi	l	−132.2 ± 3.3	107.1 ± 2.9	−25.1 ± 4.4	5014.5	51
MeNHLi	c	−95.9 ± 1.7				73
HC_2Na	c	96.7 ± 1.1				74
C_2Na_2	c	20.1 ± 1.7				74
HC_2Cs	c	78.8 ± 1.1				75
Group IIA						
$(C_5H_5)_2Mg$	c	62.5 ± 8.1	68.2 ± 1.3	130.7 ± 8.2	9363.1	52
MeMgI (in ether)		−287.4 ± 2.5				1
Group IIB						
Me_2Zn	l	20.8 ± 1.2	29.5 ± 0.4	50.3 ± 1.3	2821.5	6
Et_2Zn	l	16.7 ± 6.3	40.2 ± 2.1	56.9 ± 6.6	5120.2	6
$(n\text{-}Pr)_2Zn$	l	−59.3 ± 23.0	45.6 ± 2.5	−13.7 ± 23.1	7496.1	5
$(n\text{-}Bu)_2Zn$	l	−105.9 ± 23.5	54.4 ± 3.3	−51.5 ± 23.7	9839.3	5
Me_2Cd	l	67.8 ± 1.3	37.9 ± 0.1	105.7 ± 1.3	2745.6	50
Et_2Cd	l	59.3 ± 1.8	46.0 ± 2.1	105.3 ± 2.8	5051.4	6
Me_2Hg	l	59.5 ± 0.5	34.6 ± 0.8	94.1 ± 0.9	2708.6	9
Et_2Hg	l	27.6 ± 0.8	44.8 ± 1.7	72.4 ± 1.9	5035.7	59
$(n\text{-}Pr)_2Hg$	l	−19.5 ± 5.4	55.2 ± 1.3	35.7 ± 5.6	7377.7	76
$(i\text{-}Pr)_2Hg$	l	−13.3 ± 4.2	53.6 ± 1.7	40.3 ± 4.5	7373.1	76
Ph_2Hg	c	278.6 ± 3.1	112.8 ± 0.8	391.4 ± 3.2	10 450.0	11
MeHgCl	c	−116.2 ± 2.4	64.9 ± 1.7	−51.3 ± 2.9	1604.7	9
MeHgBr	c	−86.1 ± 2.6	67.8 ± 1.7	−18.3 ± 3.1	1562.2	9
MeHgI	c	−42.9 ± 0.9	65.3 ± 1.7	22.4 ± 1.9	1516.4	9
EtHgCl	c	−142.2 ± 2.8	76.1 ± 2.9	−66.1 ± 4.0	2772.1	59

EtHgBr	c	−108.8 ± 3.1	76.6 ± 2.9	−32.2 ± 4.2	2728.8	59
EtHgI	c	−66.6 ± 2.8	79.5 ± 2.9	12.9 ± 4.0	2678.6	59
(*n*-Pr)HgCl	c	−166.2 ± 3.8				76
(*i*-Pr)HgCl	c	−165.9 ± 3.7				76
(*n*-Pr)HgBr	c	−133.1 ± 3.8				76
(*i*-Pr)HgBr	c	−136.4 ± 3.6				76
(*n*-Pr)HgI	c	−88.9 ± 3.2				76
(*i*-Pr)HgI	c	−92.0 ± 3.0				76
PhHgCl	c	−0.7 ± 2.8				12
PhHgBr	c	36.2 ± 2.7				59
PhHgI	c	75.7 ± 1.6				59
Group III						
Me_3B	l	−143.0 ± 10.7	20.2 ± 0.1	−122.8 ± 10.7	4794.9	13
Et_3B	l	−185.7 ± 5.6	36.8 ± 0.4	−148.9 ± 5.6	8278.9	77
n-Bu—B<(CH_2—CH_2)(CH_2—CH_2) (ring)	l	−194.5 ± 4.3				78
$(n\text{-}Pr)_3B$	l	−278.5 ± 13.0	41.8 ± 1.3	−236.7 ± 13.1	11 824.7	79
$(i\text{-}Pr)_3B$	l	−293.9 ± 11.4	41.8 ± 1.3	−252.1 ± 11.5	11 840.1	79
$(n\text{-}Bu)_3B$	l	−352.5 ± 4.2	(61.9 ± 2.1)	−290.6 ± 4.7	15 336.6	78
$(i\text{-}Bu)_3B$	l	−338.1 ± 3.6	(59.8 ± 2.1)	−278.3 ± 4.2	15 324.3	77
$(s\text{-}Bu)_3B$	l	−305.6 ± 25.3	(60.7 ± 2.1)	−244.9 ± 25.4	15 290.9	80
$[Me_2CH(CH_2)_2]_3B$	l	−453.7 ± 7.0	(72.0 ± 2.5)	−381.7 ± 7.4	18 822.8	79
Ph_3B	c	48.4 ± 7.8	81.6 ± 2.1	130.0 ± 8.1	16 600.1	81
$(cyclo\text{-}C_6H_{11})_3B$	c	−481.7 ± 10.2	(81.6 ± 4.2)	−400.1 ± 11.0	21 054.2	81
$(n\text{-}C_6H_{13})_3B$	l	−486.1 ± 2.4	(88.7 ± 2.9)	−397.4 ± 3.8	22 359.5	16
$(n\text{-}C_7H_{15})_3B$	l	−558.8 ± 1.8	(102.1 ± 2.9)	−456.7 ± 3.4	25 876.8	16
$(n\text{-}C_8H_{17})_3B$	l	−630.7 ± 3.6	(115.5 ± 2.9)	−515.2 ± 4.6	29 393.3	16
$(s\text{-}C_8H_{17})_3B$	l	−631.7 ± 9.0	(113.0 ± 4.2)	−518.7 ± 9.9	29 396.8	16
o-Carborane, $C_2H_{12}B_{10}$	c	−171.0 ± 7.1	65.4 ± 1.0	−105.6 ± 7.2	9754.9	82
m-Carborane, $C_2H_{12}B_{10}$	c	−241.7 ± 8.1	58.5 ± 1.0	−183.2 ± 8.2	9832.5	82
p-Carborane, $C_2H_{12}B_{10}$	c	−312.0 ± 8.1	61.3 ± 1.0	−250.7 ± 8.2	9900.0	82
Methyl-*o*-carborane, $C_3H_{14}B_{10}$	c	−244.5 ± 7.5	63.8 ± 0.6	−180.7 ± 7.5	10 982.7	82
Dimethyl-*o*-carborane, $C_4H_{16}B_{10}$	c	−280.7 ± 11.0	65.3 ± 0.7	−215.4 ± 11.0	12 170.1	82
1-Hexyl-*o*-carborane, $C_8H_{24}B_{10}$	c	−452.3 ± 12.0	86.2 ± 1.4	−366.1 ± 12.1	16 931.5	83
MeO—B<(CH_2—CH_2)(CH_2—CH_2) (ring)	l	−390.8 ± 2.6				84

TABLE 1. *continued*

Compound	State	ΔH_f^0 (c/l) (kJ/mol)	ΔH^0 (sub/vap) (kJ/mol)	ΔH_f^0 (g) (kJ/mol)	ΔH_a^0 (kJ/mol)	Ref.
$(n\text{-Bu})_2BOH$	l	-615.5 ± 3.6	(62.8 ± 8.4)	-552.7 ± 9.1	11 237.2	85
$(n\text{-Bu})_2BOMe$	l	-552.2 ± 3.6				84
$(n\text{-Bu})_2B(On\text{-Bu})$	l	-649.2 ± 4.2				86
$PhB(OH)_2$	c	-720.2 ± 1.9				87
$n\text{-BuB}(On\text{-Bu})_2$	l	-937.9 ± 10.2				88
$[(n\text{-Bu})_2B]_2O$	l	-891.2 ± 5.2				89
$(Ph_2B)_2O$	c	-184.8 ± 8.4				87
$(PhBO)_3$	c	-1261.3 ± 8.7				87
Hydroxymethyl-*o*-carborane, $C_3H_{14}OB_{10}$	c	-341.6 ± 8.8	77.0 ± 1.3	-264.6 ± 8.9	11 315.8	82
Hydroxymethyl-*m*-carborane, $C_3H_{14}OB_{10}$	c	-439.0 ± 7.8	78.3 ± 1.3	-360.7 ± 7.9	11 411.9	82
Hydroxymethyl-*p*-carborane, $C_3H_{14}OB_{10}$	c	-454.5 ± 6.0	83.9 ± 1.3	-370.6 ± 6.1	11 421.8	82
o-Carboranecarboxylic acid, $C_3H_{12}O_2B_{10}$	c	-542.8 ± 8.8	97.0 ± 1.7	-445.8 ± 9.0	11 310.2	90
m-Carboranecarboxylic acid, $C_3H_{12}O_2B_{10}$	c	-659.9 ± 8.8	97.7 ± 0.7	-562.2 ± 9.0	11 426.6	90
p-Carboranecarboxylic acid, $C_3H_{12}O_2B_{10}$	c	-627.3 ± 13.0	96.3 ± 0.7	-531.0 ± 13.0	11 395.4	90
Me_2BNHMe	c	-138.3 ± 2.1	56.9 ± 0.8	-81.4 ± 2.2	5444.1	91
Et_2BNHEt	l	-192.3 ± 2.0	60.7 ± 0.8	-131.6 ± 2.2	8952.3	92
$(n\text{-Bu})_2BNH_2$	l	-398.6 ± 3.6				93
$(n\text{-Bu})_2BNH(n\text{-Bu})$	l	-452.4 ± 5.3				93
$CH_2{-}S{>}BPh$ / $CH_2{-}S$ (cyclic)	l	-135.4 ± 2.8				94
$CH_2{<}(CH_2{-}S)(CH_2{-}S){>}BPh$ (cyclic)	l	-154.6 ± 3.2				94
$(n\text{-Bu})_2BCl$	l	-419.0 ± 3.7	50.2 ± 1.3	-368.8 ± 3.9	10 707.5	85
$(n\text{-Bu})_2BBr$	l	-357.5 ± 3.2	52.3 ± 1.3	-305.2 ± 3.5	10 634.4	85
$(n\text{-Bu})_2BI$	l	-283.8 ± 2.8	54.4 ± 2.5	-229.4 ± 3.8	10 553.5	85
Ph_2BCl	l	-134.9 ± 3.7	41.4 ± 2.1	-93.5 ± 4.3	11 554.8	81
Ph_2BBr	l	-69.7 ± 3.7	60.3 ± 2.1	-9.4 ± 4.3	11 461.3	81
$PhBCl_2$	l	-299.9 ± 2.1	33.9 ± 0.8	-266.0 ± 2.2	6458.6	81
$PhBBr_2$	l	-173.2 ± 2.0	43.9 ± 2.1	-129.3 ± 2.9	6303.0	81
Me_3Al	l	-144.4 ± 11.1	63.2 ± 1.7	-81.2 ± 11.2	4522.9	30
Et_2AlH	l	-204.3 ± 6.7	57.7 ± 2.1	-146.5 ± 7.1	5740.9	29
$(n\text{-Pr})_2AlH$	l	-243.2 ± 3.5				29

Et_3Al	l	−236.8 ± 3.1	73.2 ± 2.1	−163.6 ± 3.7	8063.3	29
$(n\text{-}Bu)_2AlH$	l	−283.0 ± 5.6				29
$(i\text{-}Bu)_2AlH$	l	−289.4 ± 4.4	42.3 ± 2.1	−247.1 ± 4.9	10 452.2	29
$(n\text{-}Pr)_3Al$	l	−281.6 ± 15.0	42.5 ± 1.2	−239.1 ± 15.1	11 596.8	29
$(n\text{-}Bu)_3Al$	l	−372.4 ± 5.7				29
$(i\text{-}Bu)_3Al$	l	−388.3 ± 7.7				29
Et_2AlOEt	l	−523.1 ± 5.5	48.6 ± 1.0	−474.5 ± 5.6	8623.4	95
$Et_2AlO(n\text{-}Pr)$	l	−653.9 ± 14.2	50.9 ± 1.0	−603.0 ± 14.3	9904.6	95
$Et_2AlO(t\text{-}Bu)$	l	−773.8 ± 17.2				95
Tris(pentan-2,4-dionato)Al, $Al(acac)_3$	c	−1793.3 ± 2.0	121.7 ± 4.2	−1671.6 ± 4.7	18 824.4	96
Et_2AlCl	l	−414.9 ± 2.5				97
Et_2AlBr	l	−364.1 ± 2.6				97
Et_2AlI	l	−286.9 ± 2.5				97
$EtAlCl_2$	l	−553.5 ± 5.1				97
$EtAlBr_2$	l	−451.7 ± 5.2				97
$EtAlI_2$	l	−300.7 ± 5.1				97
$Et_3Al_2Cl_3$	l	−970.6 ± 5.1				97
$Et_3Al_2Br_3$	l	−818.3 ± 5.2				97
$(C_5H_5)_3Sc$	c	−13.6 ± 4.6	97.1 ± 3.5	83.5 ± 5.8	14 318.2	98
$(C_5H_5)_3Y$	c	−45.2 ± 4.6	111.7 ± 3.5	66.5 ± 5.8	14 378.3	98
$(C_5H_5)_3La$	c	35.7 ± 5.1	114.6 ± 4.0	150.3 ± 6.5	14 300.8	98
$(C_5H_5)_3Pr$	c	−28.2 ± 8.7	125.5 ± 3.0	97.3 ± 9.2	14 295.6	98
$(C_5H_5)_3Tm$	c	−49.5 ± 5.1	111.3 ± 3.5	61.8 ± 6.2	14 205.6	98
$(C_5H_5)_3Yb$	c	29.3 ± 5.1	108.8 ± 3.5	138.1 ± 6.2	14 033.9	98
Me_3Ga	l	−74.7 ± 6.2	33.1 ± 0.8	−41.6 ± 6.3	4442.3	99
Et_3Ga	l	−99.9 ± 8.1	38.5 ± 0.4	−61.4 ± 8.2	7900.3	100
$(n\text{-}Bu)_3Ga$	l	−284.6 ± 5.0				100
$(i\text{-}Bu)_3Ga$	l	−293.0 ± 5.0				100
Tris(pentan-2,4-dionato)gallium, $Ga(acac)_3$	c	−1476.0 ± 4.5	(121.7 ± 4.2)	−1354.3 ± 6.0	18 466.1	96
$MeGaI_2$	c	−212.6 ± 7.6				101
Me_3In	c	124.8 ± 6.5	48.5 ± 2.5	173.4 ± 6.9	4181.3	102
Tris(pentan-2,4-dionato)indium, $In(acac)_3$	c	−1405.7 ± 3.6	(121.7 ± 4.2)	−1284.0 ± 5.5	18 349.8	96
$(C_5H_5)Tl$	c	97.7 ± 4.3				53

TABLE 1. *continued*

Compound	State	ΔH_f^0 (c/l) (kJ/mol)	ΔH^0 (sub/vap) (kJ/mol)	ΔH_f^0 (g) (kJ/mol)	ΔH_a^0 (kJ/mol)	Ref.
Group IVA						
$[(C_5H_5)_2Ti]$	c	−70.7 ± 9.5	58.5 ± 8.0	−12.2 ± 12.4	9829.6	27
$[(C_5H_5)_2TiMe_2]$	c	54.4 ± 8.4	79.5 ± 8.4	133.9 ± 11.8	12 424.8	103
$[(C_5H_5)_2TiPh_2]$	c	294.4 ± 8.8	88.0 ± 8.0	382.4 ± 11.9	20 215.0	197
$[(C_5H_5)_2Ti(CH_2Ph)_2]$	c	195.8 ± 5.0	83.7 ± 8.4	279.5 ± 9.8	22 623.3	103
$[(C_5H_5)_2Ti(OCOPh)_2]$	c	−775.2 ± 8.1	(112.0 ± 8.0)	−663.2 ± 11.4	23 690.7	165
$[(C_5H_5)_2TiCl_2]$	c	−384.8 ± 8.7	118.8 ± 2.1	−266.0 ± 8.9	10 326.0	103
$[(C_5H_5)TiCl_3]$	c	−610.7 ± 7.9	104.6 ± 8.4	−506.1 ± 11.5	2275.4	103
$[(C_5H_5)_2Ti(OCOCF_3)_2]$	c	−2219.0 ± 8.0	108.0 ± 8.0	−2111.0 ± 11.3	16 268.1	165
$[(C_5H_5)_2ZrMe_2]$	c	−44.4 ± 2.1	81.2 ± 2.1	36.8 ± 3.0	12 659.6	105
$[(C_5H_5)_2ZrPh_2]$	c	275.7 ± 10.9	(92.0 ± 4.2)	367.7 ± 11.7	20 367.4	105
$[(C_5H_5)_2ZrCl_2]$	c	−538.1 ± 2.9	105.0 ± 2.1	−433.1 ± 3.6	10 630.8	105
$[(C_5H_5)_2HfCl_2]$	c	−536.0 ± 2.5	106.7 ± 2.1	−429.3 ± 3.3	10 637.8	105
Group IVB						
Et_2SiH_2	l	−212.5 ± 5.8	30.0 ± 0.4	−182.5 ± 5.8	6115.2	106
Me_4Si	l	−271.5 ± 9.9	26.2 ± 0.4	−245.4 ± 9.9	6178.1	106
$H_2C\langle^{CH_2}_{CH_2}\rangle SiMe_2$	l	−172.9 ± 11.2	34.7 ± 2.1	−138.2 ± 11.4	6787.6	106
$^{CH_2-CH_2}_{CH_2-CH_2}\rangle SiMe_2$	l	−219.3 ± 12.2	37.7 ± 2.1	−181.6 ± 12.4	7983.6	106
Et_3SiH	l	−237.3 ± 15.2	36.4 ± 1.3	−200.9 ± 15.2	8438.9	106
Et_4Si	l	−305.2 ± 15.2	39.7 ± 2.1	−265.4 ± 15.3	10 808.8	106
Ph_4Si	c	184.9 ± 5.9	156.9 ± 1.7	341.8 ± 6.1	21 668.3	37
$Me_2Si\langle^{CH_2}_{CH_2}\rangle SiMe_2$	l	−342.3 ± 13.6	41.0 ± 2.1	−301.3 ± 13.8	8989.3	106
$(Me_3Si)_2$	l	−400.5 ± 8.5	37.4 ± 0.4	−363.1 ± 8.6	9487.1	106
$(Me_2Si(SiMe_3)_2$	l	−516.4 ± 17.1	46.0 ± 0.8	−470.4 ± 17.2	12 785.8	106
$(Me_3Si(SiMe_2)_2SiMe_3$	l	−620.7 ± 23.5	52.3 ± 1.7	−568.4 ± 23.5	16 075.1	106
$Si(SiMe_3)_4$	c	−643.0 ± 33.6	83.7 ± 20.9	−559.3 ± 39.6	19 257.3	106

Me_3SiOH	l	−545.1 ± 2.7	45.6 ± 1.7	−499.5 ± 3.2	5528.7	109
$(Me_3Si)_2O$	l	−814.5 ± 5.4	37.2 ± 1.7	−777.2 ± 5.7	10 150.4	33
Octamethyltrisiloxane, $Me_8Si_3O_2$	l	−1420.4 ± 11.7	39.7 ± 2.1	−1380.7 ± 11.9	14 194.4	106
Decamethyltetrasiloxane, $Me_{10}Si_4O_3$	l	−1982.7 ± 22.5	48.1 ± 2.1	−1934.6 ± 22.6	18 188.8	106
Dodecamethylpentasiloxane, $Me_{12}Si_5O_4$	l	−2621.0 ± 23.3	53.1 ± 2.1	−2567.9 ± 23.4	22 262.6	106
$Me_3SiNHMe$	l	−263.1 ± 2.8	36.0 ± 2.1	−227.1 ± 3.5	6850.5	39
Me_3SiNMe_2	l	−279.5 ± 2.8	31.8 ± 1.7	−247.7 ± 3.2	8023.7	39
Me_3SiNEt_2	l	−367.1 ± 2.2				107
$(Me_3Si)_2NH$	l	−518.0 ± 5.5	41.4 ± 2.1	−476.6 ± 5.8	10 291.3	108
$(Me_3Si)_2NMe$	l	−487.2 ± 5.5	38.9 ± 2.1	−448.3 ± 5.9	11 415.7	39
$(Me_3Si)_3N$	l	−724.7 ± 8.4	54.4 ± 8.4	−670.3 ± 11.9	14 829.0	39
$(Me_3SiS(n\text{-}Bu)$	l	−381.4 ± 3.0	40.6 ± 2.1	−340.9 ± 3.6	10 008.6	39
$MeSiCl_3$	l	−602.8 ± 3.4	31.0 ± 2.1	−571.8 ± 4.0	2756.4	109
$MeSiHCl_2$	l	−430.0 ± 3.5	28.0 ± 1.7	−402.0 ± 3.8	2683.3	38
Me_2SiHCl	l	−321.1 ± 2.7	28.5 ± 2.1	−292.6 ± 3.4	3823.2	38
Me_2SiCl_2	l	−482.6 ± 2.3	34.3 ± 1.7	−448.3 ± 2.8	3882.2	38
Me_3SiCl	l	−383.9 ± 2.7	30.1 ± 1.7	−353.8 ± 3.2	5037.1	39
Me_3SiBr	l	−325.8 ± 2.7	32.6 ± 2.1	−293.2 ± 3.5	4967.1	39
Ph_2SiCl_2	l	−278.2 ± 4.2	69.5 ± 4.2	−208.7 ± 5.9	11 681.3	110
Me_4Ge	l	−98.3 ± 9.4	27.6 ± 2.1	−70.7 ± 9.6	5930.4	111
Et_4Ge	l	−206.4 ± 7.5	44.8 ± 1.3	−161.6 ± 7.6	10 632.0	40
$(n\text{-}Pr)_4Ge$	l	−288.3 ± 4.9	61.5 ± 4.2	−226.8 ± 6.5	15 307.8	112
Ph_4Ge	c	288.6 ± 23.6	156.9 ± 4.2	445.5 ± 23.9	21 491.6	41
$(PhCH_2)_4Ge$	c	223.3 ± 11.9	168.6 ± 8.4	391.9 ± 14.6	26 155.9	42
$(Et_3Ge)_2$	l	−372.9 ± 11.9	62.8 ± 2.1	−310.2 ± 12.1	16 204.2	113
$(Ph_3Ge)_2$	c	453.7 ± 14.2	209.2 ± 4.2	662.9 ± 14.8	32 431.2	42
$Et_3GeOO(t\text{-}Bu)$	l	−486.2 ± 7.0	43.5 ± 4.2	−442.7 ± 8.1	13 716.7	114
$(Et_3Ge)_2O$	l	−611.3 ± 11.9	58.6 ± 4.2	−552.7 ± 12.6	16 695.9	114
Octamethyltetragermoxane, $(Me_2GeO)_4$	c	−1514.5 ± 25.9	68.2 ± 4.2	−1446.3 ± 26.2	14 916.3	115
Octaethyltetragermoxane, $(Et_2GeO)_4$	c	−1519.3 ± 27.6				116
Et_3GeNEt_2	l	−342.5 ± 6.1	46.0 ± 4.8	−296.5 ± 7.8	13 762.9	117
$(Et_3Ge)_2Hg$	l	−101.0 ± 9.5	62.8 ± 4.2	−38.2 ± 10.4	15 993.6	118
$[(i\text{-}Pr)_3Ge]_2Hg$	l	−273.4 ± 9.7	54.4 ± 4.2	−219.0 ± 10.6	23 090.4	118
Me_4Sn	l	−52.3 ± 1.7	33.1 ± 1.3	−19.2 ± 2.1	5803.1	17
$Me_3Sn(CH{=}CH_2)$	l	54.4 ± 13.3	37.2 ± 2.1	91.7 ± 13.4	6408.9	119

TABLE 1. *continued*

Compound	State	ΔH_f^0 (c/l) (kJ/mol)	ΔH^0 (sub/vap) (kJ/mol)	ΔH_f^0 (g) (kJ/mol)	ΔH_a^0 (kJ/mol)	Ref.
Me_3SnEt	l	-67.2 ± 2.4	37.7 ± 1.7	-29.5 ± 3.0	6966.1	17
$Me_3Sn(i\text{-}Pr)$	l	-87.4 ± 4.3	40.6 ± 2.1	-46.8 ± 4.8	8136.0	120
$Me_3Sn(t\text{-}Bu)$	l	-121.1 ± 4.5	54.0 ± 4.2	-67.1 ± 6.2	9309.0	120
$(CH_2{=}CH)_4Sn$	l	300.8 ± 7.6				121
$Et_3Sn(CH{=}CH_2)$	l	36.2 ± 3.2				121
Et_4Sn	l	-95.9 ± 2.5	51.0 ± 2.1	-44.9 ± 3.3	10 439.5	17
Me_3SnPh	l	60.8 ± 3.1	52.3 ± 4.2	113.1 ± 5.2	9690.1	119
Me_3SnCH_2Ph	l	26.3 ± 3.9	56.5 ± 4.2	82.8 ± 5.7	10 873.1	119
$(CH_2{=}CHCH_2)_4Sn$	l	-170.2 ± 7.3				121
$(n\text{-}Pr)_4Sn$	l	-211.3 ± 5.3	66.9 ± 2.1	-144.4 ± 5.7	15 149.6	17
$(i\text{-}Pr)_4Sn$	l	-188.0 ± 5.6	64.9 ± 4.2	-123.1 ± 7.0	15 128.3	120
$(n\text{-}Bu)_4Sn$	l	-302.1 ± 3.7	82.8 ± 2.1	-219.2 ± 4.2	19 835.1	17
$(i\text{-}Bu)_4Sn$	l	-330.9 ± 6.3				121
Ph_4Sn	c	411.6 ± 3.7	161.1 ± 4.2	572.7 ± 5.6	21 288.6	122
$(cyclo\text{-}C_6H_{11})_4Sn$	c	-364.7 ± 28.6				121
$(Me_3Sn)_2$	l	-77.1 ± 7.3	50.2 ± 4.2	-26.9 ± 8.4	8853.3	56
$(Et_3Sn)_2$	l	-217.4 ± 8.6	62.8 ± 4.2	-154.6 ± 9.5	15 897.0	123
$(Ph_3Sn)_2$	c	661.4 ± 15.6	188.3 ± 4.2	849.7 ± 16.2	32 091.0	124
$Me_3SnOCOPh$	c	-491.7 ± 8.5				123
$Et_3SnOO(t\text{-}Bu)$	l	-421.3 ± 16.0	48.8 ± 2.1	-372.5 ± 16.1	13 570.7	114
$Et_3SnOCOPh$	c	-575.8 ± 4.5				123
$Et_3SnOOC(Me)_2Ph$	l	-285.6 ± 8.6	56.5 ± 2.1	-229.1 ± 8.9	17 446.7	114
$(Et_3Sn)_2O$	l	-427.0 ± 12.7	52.3 ± 2.1	-374.7 ± 12.9	16 366.3	114
Et_3SnNEt_2	l	-210.3 ± 4.4	50.2 ± 4.2	-160.1 ± 6.1	13 550.7	117
$MeSnCl_3$	l	-443.1 ± 7.6				60
Me_2SnCl_2	l	-330.3 ± 7.7				60
Me_3SnCl	l	-213.0 ± 7.1				60
Me_3SnBr	l	-185.5 ± 3.5	47.3 ± 4.2	-138.2 ± 5.5	4663.3	56
Me_3SnI	l	-130.6 ± 3.9	48.1 ± 4.2	-82.5 ± 5.7	4602.5	56
$(n\text{-}Bu)_3SnBr$	l	-356.2 ± 1.3	83.7 ± 12.6	-272.5 ± 12.6	15 171.6	119
Ph_2SnBr_2	c	-19.5 ± 6.8				119
Ph_3SnBr	c	188.0 ± 7.7				119

Me_4Pb	l	98.1 ± 4.4	38.1 ± 0.4	136.2 ± 4.4	5541.7	44
Et_4Pb	l	53.1 ± 5.0	56.9 ± 2.5	109.5 ± 5.3	10 179.1	43
Ph_4Pb	c	515.3 ± 15.4	194.6 ± 6.3	709.9 ± 16.7	21 045.4	45
Ph_3PbPr	c	271.5 ± 18.0	134.7 ± 3.3	406.2 ± 18.3	16 070.9	125
Ph_2PbBr_2	c	36.0 ± 17.6	141.8 ± 0.8	177.8 ± 17.6	11 021.2	125
Ph_3PbI	c	326.8 ± 15.5	130.1 ± 0.4	456.9 ± 15.5	16 015.1	125
Ph_2PbI_2	c	152.7 ± 15.5	(138.0 ± 4.2)	290.7 ± 16.1	10 898.1	125
Group VA						
$[(C_5H_5)_2V]$	c	144.9 ± 8.8	58.6 ± 4.2	203.5 ± 9.7	9657.8	27
$[(C_6H_6)_2V]$	c	37.2 ± 15.3	(70.0 ± 10.0)	107.2 ± 18.3	11 623.4	126
$[Me_5Ta]$	l	169.8 ± 26.0	(42.0 ± 4.2)	211.8 ± 26.3	7428.2	127
$[(Me_2N)_5Ta]$	c	−362.2 ± 11.0	(89.0 ± 10.0)	−237.2 ± 14.9	17 093.9	54
Group VB						
Me_3P	l	−129.1 ± 4.8	28.0 ± 2.1	−101.1 ± 5.3	4529.6	128
Et_3P	l	−89.2 ± 12.6	39.7 ± 2.1	−49.4 ± 12.8	7539.5	121
$(n\text{-}Bu)_3P$	l	−123.9 ± 33.8				129
9-Phenyl-9-phosphafluorene, $C_{18}H_{13}P$	c	183.9 ± 17.0				130
Ph_3P	c	218.0 ± 10.7	96.2 ± 8.4	314.2 ± 13.6	16 172.4	131
Pentaphenylphosphole, $C_{34}H_{25}P$	c	388.9 ± 28.8				130
Me_3PO	c	−484.6 ± 6.4	50.2 ± 4.2	−434.4 ± 7.6	5112.1	132
$(n\text{-}Bu)_3PO$	c	−468.1 ± 32.7				133
Ph_3PO	c	−67.2 ± 25.3				131
$MePO(OH)_2$	c	−1053.2 ± 25.2	48.1 ± 4.2	−1005.1 ± 25.6	3875.8	134
$EtPO(OH)_2$	c	−1060.0 ± 14.7	50.6 ± 4.2	−1009.4 ± 15.2	5032.8	134
$MePO(OEt)_2$	l	−1029.3 ± 25.4	56.5 ± 4.2	−967.4 ± 25.7	8448.8	135
$EtPO(Oi\text{-}Pr)_2$	l	−1330.1 ± 9.4	60.7 ± 4.2	−1069.4 ± 10.3	12 008.8	135
$Me_3P{=}NEt$	l	−156.3 ± 6.8	61.5 ± 4.2	−94.8 ± 8.0	7519.3	132
$Ph_3P{=}NEt$	c	110.9 ± 2.2	75.3 ± 8.4	186.2 ± 8.7	19 296.4	132
Hexamethylcyclophosphazatriene, $C_6H_{18}N_3P_3$	c	−538.1 ± 23.0				136
$MePO(NHPh)_2$	c	−358.7 ± 25.2				135
$EtPO(NHPh)_2$	c	−386.2 ± 12.7				135
$EtPCl_2$	l	−311.3 ± 14.6				134
$MePOCl_2$	c	−618.5 ± 25.2	62.3 ± 4.2	−556.2 ± 25.5	2735.1	135
$EtPOCl_2$	l	−627.2 ± 13.3	42.7 ± 4.2	−584.5 ± 13.9	3916.1	135

TABLE 1. *continued*

Compound	State	ΔH_f^0 (c/l) (kJ/mol)	ΔH^0 (sub/vap) (kJ/mol)	ΔH_f^0 (g) (kJ/mol)	ΔH_a^0 (kJ/mol)	Ref.
Me_3As	l	−16.2 ± 10.1	28.9 ± 1.3	12.6 ± 10.1	4388.4	137
Et_3As	l	13.1 ± 16.8	43.1 ± 4.2	56.2 ± 17.3	7802.8	121
Ph_3As	c	309.8 ± 6.7	98.3 ± 8.4	408.1 ± 10.7	16 051.0	46
Me_3Sb	l	0.8 ± 25.2	31.4 ± 1.3	32.1 ± 25.2	4343.5	138
Et_3Sb	l	5.2 ± 10.7	43.5 ± 4.2	48.7 ± 11.5	7784.9	121
Ph_3Sb	c	329.2 ± 17.0	106.3 ± 8.4	435.4 ± 19.0	15 998.3	139
Me_3Bi	l	158.4 ± 14.2	36.0 ± 1.3	194.4 ± 14.3	4124.7	140
Et_3Bi	l	169.7 ± 16.8	46.0 ± 4.2	215.7 ± 17.3	7561.4	121
Ph_3Bi	c	489.7 ± 5.2	110.9 ± 8.4	600.6 ± 9.9	15 776.6	47
Group VIA						
$[(C_5H_5)_2Cr]$	c	178.1 ± 2.6	62.8 ± 4.2	240.9 ± 5.0	9502.4	141
$[(C_6H_6)_2Cr]$	c	141.4 ± 4.9	78.2 ± 6.3	219.6 ± 8.0	11 393.0	24
$[(EtPh)_2Cr]$	l	59.4 ± 4.0	75.3 ± 8.4	134.7 ± 9.3	16 088.6	142
$[(1,3,5\text{-}Me_3\text{-}C_6H_3)_2Cr]$	c	−40.5 ± 12.0	104.0 ± 1.0	63.5 ± 12.0	18 465.2	143
$[(1,2,4\text{-}Me_3\text{-}C_6H_3)_2Cr]$	c	−37.4 ± 16.9				142
Bis(naphthalene)chromium, $[(C_{10}H_8)_2Cr]$	c	301.7 ± 4.0	(105.0 ± 10.0)	406.7 ± 10.8	17 811.3	143
$[(1,2\text{-}Et_2\text{-}C_6H_4)_2Cr]$	l	−55.6 ± 7.6	75.3 ± 8.4	19.7 ± 11.4	20 814.3	142
Isopropylbenzene-1,2-diisopropylbenzene chromium, $[(i\text{-}PrPh)(1,2\text{-}(i\text{-}Pr)_2C_6H_4)Cr]$	l	−132.4 ± 7.5	100.4 ± 8.4	−32.0 ± 11.3	20 866.0	142
$[(1,2\text{-}(i\text{-}Pr)_2C_6H_4)_2Cr]$	l	−204.4 ± 8.2	100.4 ± 8.4	−104.0 ± 11.7	25 548.7	142
Bis(hexamethylbenzene)chromium, $[(Me_6C_6)_2Cr]$	c	−207.0 ± 11.0	119.0 ± 4.0	−88.0 ± 11.7	25 532.7	143
$[(C_6H_5)Cr(CO)_3]$	c	−441.2 ± 8.4	91.2 ± 4.2	−350.0 ± 9.4	9252.1	64
Cycloheptatnenechromiumtricarbonyl, $[(c\text{-}C_7H_8)Cr(CO)_3]$	c	−307.8 ± 8.5	87.9 ± 4.0	−219.9 ± 9.4	10 274.7	64
$[(MePh)Cr(CO)_3]$	c	−476.1 ± 4.2	94.6 ± 4.2	−381.5 ± 6.0	10 436.3	64
1,3,5-Trimethylbenzenechromiumtricarbonyl, $[(C_9H_{12})Cr(CO)_3]$	c	−547.7 ± 8.5	108.4 ± 4.2	−466.3 ± 9.5	12 826.5	64
$[(Me_6C_6)Cr(CO)_3]$	c	−671.0 ± 12.0	123.0 ± 4.0	−548.0 ± 12.6	16 366.2	144
Naphthalenechromiumtricarbonyl, $[(C_{10}H_8)Cr(CO)_3]$	c	−365.0 ± 7.0	107.0 ± 3.0	−258.0 ± 7.6	12 462.8	143

Norbornadienechromiumtetracarbonyl, $[(nor\text{-}C_7H_8)Cr(CO)_4]$	c	−400.0 ± 12.0	89.0 ± 4.0	−311.0 ± 12.6	11 331.7	144
Methylbenzoatechromiumtricarbonyl, $[(PhCO_2Me)Cr(CO)_3]$	c	−769.0 ± 8.0	114.0 ± 5.0	−655.0 ± 9.4	11 924.8	145
$[Cr(CO)_6]$	c	−979.7 ± 2.0	72.0 ± 4.2	−907.7 ± 4.7	7099.3	19
Tris(2,4-pentandionato)chromium, $[Cr(acac)_3]$	c	−1554.0 ± 6.2	123.0 ± 3.0	−1431.0 ± 6.9	18 650.7	146
Bis(pyridine)chromiumtetracarbonyl, $[Py_2Cr(CO)_4]$	c	−515.0 ± 20.0	(117.0 ± 10.0)	−398.0 ± 22.4	14 950.0	147
$[(C_6H_6)_2CrCl]$	c	−43.9 ± 12.7				23
Chlorobenzenechromiumtricarbonyl, $[(PhCl)Cr(CO)_3]$	c	−466.3 ± 4.3	102.5 ± 4.2	−363.8 ± 6.0	9169.2	64
$[(C_6H_6)_2CrBr]$	c	−3.6 ± 8.6				23
$[(C_6H_6)_2CrI]$	c	49.7 ± 8.7				24
$[(MePh)_2CrI]$	c	−69.4 ± 4.7				23
Bis(*m*-xylene)chromium iodide, $[C_{16}H_{20}CrI]$	c	−114.3 ± 8.7				23
Bis(mesitylene)chromium iodide, $[C_{18}H_{24}CrI]$	c	−188.5 ± 8.8				23
Bis(diphenyl)chromium iodide, $[(PhPh)_2CrI]$	c	151.8 ± 9.0				23
Tetra-μ-acetatodichromium(II), $[(MeCO_2)_4Cr_2]$	c	−2297.5 ± 6.6	313.8 ± 27.0	−1983.7 ± 27.8	13 119.6	55
$[(MeCO_2)_4Cr_2].2H_2O$	c	−2875.4 ± 6.7				55
Tetra(6-methyl-2-hydroxypyridyl)-dichromium(II), $[(Mhp)_4Cr_2]$	c	−948.2 ± 9.0	150.0 ± 4.0	−798.2 ± 9.8	26 910.1	150
$[(C_5H_5)_2MoH_2]$	c	210.3 ± 5.8	92.5 ± 2.1	302.8 ± 6.2	10 138.0	148
$[(C_6H_6)_2Mo]$	c	307.1 ± 18.6	94.6 ± 8.4	401.7 ± 20.4	11 472.4	126
$[(C_5H_5)_2MoMe_2]$	c	283.8 ± 3.8	70.4 ± 4.2	354.2 ± 5.8	12 391.9	198
Cycloheptatrienemolybdenumtricarbonyl, $[(c\text{-}C_7H_8)Mo(CO)_3]$	c	−297.0 ± 6.0	88.0 ± 4.0	−209.0 ± 7.2	10 525.3	144
1,3,5-Trimethylbenzenemolybdenumtricarbonyl, $[(C_9H_{12})Mo(CO)_3]$	c	−533.0 ± 12.0	(109.0 ± 6.0)	−424.0 ± 13.4	13 045.7	144
Hexamethylbenzenemolybdenumtricarbonyl, $[(Me_6C_6)Mo(CO)_3]$	c	−630.9 ± 8.5	(123.0 ± 4.0)	−507.9 ± 9.4	16 587.6	144
Norbornadienemolybdenumtricarbonyl, $[(nor\text{-}C_7H_8)Mo(CO)_3]$	c	−427.6 ± 10.0	92.0 ± 4.0	−335.6 ± 10.8	11 617.8	144
$[(C_5H_5)_2Mo(OCOPh)_2]$	c	−486.2 ± 3.4	(94.0 ± 10.0)	−392.2 ± 10.6	23 607.1	165
Dioxobis(pentane-2,4-dionato)molybdenum(VI), $[MoO_2(acac)_2]$	c	−1342.1 ± 4.5				151
$[Mo(acac)_3]$	c	−1327.0 ± 6.0	[125.0 ± 5.0]	−120.2 ± 7.8	18 683.2	151
$[Mo(CO)_6]$	c	−989.1 ± 1.7	73.8 ± 1.0	−915.3 ± 2.0	7368.4	18

TABLE 1. *continued*

Compound	State	ΔH_f^0 (c/l) (kJ/mol)	ΔH^0 (sub/vap) (kJ/mol)	ΔH_f^0 (g) (kJ/mol)	ΔH_a^0 (kJ/mol)	Ref.
$[Mo(NMe_2)_4]$	c	59.0 ± 5.0	72.4 ± 6.0	131.4 ± 7.8	13 382.8	54
$[(MeCN)_3Mo(CO)_3]$	c	−410.0 ± 12.0	96.0 ± 10.0	−314.0 ± 15.6	11 549.7	147
Tris(pyridine)molybdenumtricarbonyl, $[Py_3Mo(CO)_3]$	c	−275.0 ± 12.0	142.0 ± 10.0	−133.0 ± 15.6	19 126.7	147
$[(C_5H_5)_2Mo(S\text{-}n\text{-}Pr)_2]$	c	4.6 ± 5.3	(90.0 ± 10.0)	94.6 ± 11.3	17 816.2	166
$[(C_5H_5)_2Mo(S\text{-i-}Pr)_2]$	c	57.1 ± 5.7	(90.0 ± 10.0)	147.1 ± 11.5	17 763.7	166
$[(C_5H_5)_2Mo(S\text{-}n\text{-}Bu)_2]$	c	14.0 ± 5.7	(92.0 ± 10.0)	106.0 ± 11.5	20 110.1	166
$[(C_5H_5)_2Mo(S\text{-}t\text{-}Bu)_2]$	c	6.9 ± 4.4	(92.0 ± 10.0)	98.9 ± 10.9	20 117.2	166
$[(C_5H_5)_2MoCl_2]$	c	−95.8 ± 2.5	(100.4 ± 4.2)	4.6 ± 4.9	10 242.8	152
$[(C_5H_5)_2MoBr_2]$	c	8.4 ± 18.3	(100.4 ± 4.2)	108.8 ± 18.8	10 119.7	199
$[(C_5H_5)_2MoI_2]$	c	69.8 ± 7.8	(100.4 ± 4.2)	170.2 ± 8.9	10 048.1	200
$[(C_5H_5)_2Mo(OCOCF_3)_2]$	c	−1952.0 ± 3.9	(90.0 ± 10.0)	−1862.0 ± 10.7	16 206.5	165
$[Mo_2(O\text{-}iPr)_6]$	c	−1661.8 ± 9.0	(113.0 ± 10.0)	−1548.8 ± 13.5	25 668.6	153
$[Mo_2(O\text{-}iPr)_8]$	c	−2292.5 ± 10.0	(137.0 ± 15.0)	−2155.5 ± 18.0	34 873.1	153
Tetra-μ-acetatodimolybdenum(II), $[(MeCO_2)_4Mo_2]$	c	−1970.7 ± 8.4	165.0 ± 8.4	−1805.7 ± 11.9	13 464.6	55
Di-μ-acetatobis(pentane-2,4-dionato)-dimolybdenum(II), $[(MeCO_2)_2Mo_2(acac)_2]$	c	−1808.4 ± 8.9	163.0 ± 5.0	−1645.5 ± 10.2	19 348.3	55
$[Mo_2(NMe_2)_6]$	c	17.2 ± 10.0	(111.0 ± 8.0)	128.2 ± 12.8	20 472.1	54
Tetra(6-methyl-2-hydroxypyridyl)-dimolybdenum(II), $[(Mhp)_4Mo_2]$	c	−754.0 ± 9.0	157.0 ± 3.0	−597.0 ± 9.5	27 232.7	150
Di(6-methyl-2-hydroxypyridyl)diacetato-dimolybdenum(II), $[(Mhp)_2Mo_2(MeCO_2)_2]$	c	−1366.8 ± 12.0	161.0 ± 4.0	−1205.8 ± 12.6	20 353.1	150
Tetra-μ-acetatomolybdenum(II)chromium(II), $[(MeCo_2)_4MoCr]$	c	−2113.9 ± 6.4	(165.0 ± 8.4)	−1948.9 ± 10.6	13 346.3	55
$[WMe_6]$	l	738.7 ± 34.0	(33.0 ± 10.0)	771.7 ± 35.4	8312.2	127
$[(C_5H_5)_2WH_2]$	c	214.8 ± 5.0	96.2 ± 2.1	311.0 ± 5.4	10 331.6	148
$[(C_5H_5)_2WMe_2]$	c	284.8 ± 3.5	(74.6 ± 4.2)	359.4 ± 5.5	12 588.5	198
Cycloheptatrienetungstentricarbonyl, $[(c\text{-}C_7H_8)W(CO)_3]$	c	−236.0 ± 6.0	92.0 ± 4.0	−144.0 ± 7.2	8512.1	144
1,3,5-Trimethylbenzenetungstentricarbonyl, $[(C_9H_{12})W(CO)_3]$	c	−477.4 ± 12.0	(111.0 ± 6.0)	−366.4 ± 13.4	13 189.9	144

$[(C_5H_5)_2W(OCOPh)_2]$	c	−448.9 ± 3.5	(98.0 ± 10.0)	−350.9 ± 10.6	23 767.6	165
$[W(CO)_6]$	c	−960.2 ± 2.9	76.4 ± 1.3	−883.8 ± 3.2	7538.7	18
$[W(NMe)_2)_6]$	c	178.9 ± 12.0	89.1 ± 7.0	268.0 ± 13.9	19 876.0	54
$[(CH_3CN)_3W(CO)_3]$	c	−405.0 ± 12.0	(100.0 ± 10.0)	−305.0 ± 15.6	11 742.5	147
Tris(pyridine)tungstentricarbonyl, $[(Py_3W(CO)_3]$	c	−250.0 ± 12.0	(146.0 ± 10.0)	−104.0 ± 15.6	19 299.5	147
$[(C_5H_5)_2WCl_2]$	c	−71.1 ± 2.5	(104.6 ± 4.2)	33.5 ± 4.9	10 415.7	152
$[(C_5H_5)_2WBr_2]$	c	6.9 ± 17.7	(104.6 ± 4.2)	111.5 ± 18.2	10 318.8	199
$[(C_5H_5)_2WI_2]$	c	57.8 ± 7.6	(104.6 ± 4.2)	162.4 ± 8.7	10 257.7	200
$[(C_5H_5)_2W(OCOCF_3)_2]$	c	−1914.5 ± 3.8	(94.0 ± 10.0)	−1820.5 ± 10.7	16 366.8	165
$[W_2(NMe_2)_6]$	c	19.2 ± 9.0	113.3 ± 6.0	132.5 ± 10.8	20 871.4	54
Group VIB						
CSe_2	l	218.7 ± 20.4	37.2 ± 0.8	255.9 ± 20.4	874.2	154
Et_2Se	l	−96.4 ± 3.4	38.9 ± 4.2	−57.5 ± 5.4	5310.9	155
Ph_2Se	l	222.8 ± 4.5	63.6 ± 2.5	286.4 ± 5.2	10 700.3	48
$(PhCH_2)_2Se$	c	2.2 ± 21.0				155
Group VIIA						
$[(C_5H_5)_2Mn]$	c	201.0 ± 2.8	75.7 ± 1.7	276.7 ± 3.3	9349.1	27
$[(C_5H_5)Mn(CO)_3]$	c	−528.1 ± 6.8	52.4 ± 3.1	−475.7 ± 7.5	8325.5	156
$[MeMn(CO)_5]$	c	−790.7 ± 8.5	60.3 ± 1.0	−730.4 ± 8.6	7209.4	157
Tris(pentan-2,4-dionato)manganese, $[Mn(acac)_3]$	c	−1373.2 ± 6.2	77.8 ± 0.8	−1295.3 ± 6.3	18 397.5	158
$[Mn(CO)_5Cl]$	c	−1009.4 ± 4.3	58.6 ± 8.4	−950.8 ± 9.4	6180.4	61
$[Mn(CO)_5Br]$	c	−970.1 ± 4.3	58.6 ± 8.4	−911.5 ± 9.4	6131.7	61
$[Mn_2(CO)_{10}]$	c	−1676.9 ± 5.9	80.3 ± 4.2	−1596.6 ± 7.2	11 813.2	49
$[MeRe(CO)_5]$	c	−828.3 ± 8.5	65.3 ± 1.0	−763.0 ± 8.6	7745.3	157
$[Re_2(CO)_{10}]$	c	−1651.2 ± 12.8	93.3 ± 4.2	−1557.9 ± 13.5	12 781.1	157
Group VIII						
$[(C_5H_5)_2Fe]$	c	168.2 ± 2.6	73.6 ± 0.4	241.8 ± 2.6	9521.2	141
$[Fe(CO)_5]$	l	−765.1 ± 6.6	40.2 ± 0.8	−724.9 ± 6.7	10 487.9	22
Bis(1,3-butadiene)ironcarbonyl, $[(C_4H_6)_2Fe(CO)]$	c	−24.6 ± 8.4	76.1 ± 4.2	51.5 ± 9.4	9680.0	159
Bis(1,3-cyclohexadiene)ironcarbonyl, $[(c\text{-}C_6H_8)_2Fe(CO)]$	c	−82.0 ± 8.4	95.0 ± 4.2	13.0 ± 9.4	13 457.2	159
1,3-Butadieneirontricarbonyl, $[(C_4H_6)Fe(CO)_3]$	l	−407.8 ± 8.4	49.0 ± 4.2	−358.8 ± 9.4	7847.3	159
Cyclooctatetraeneirontricarbonyl, $[(c\text{-}C_8H_8)Fe(CO)_3]$	c	−237.0 ± 12.0	87.0 ± 4.0	−150.0 ± 12.6	10 941.2	160
Ethyleneirontetracarbonyl, $[(C_2H_4)Fe(CO)_4]$	l	−582.3 ± 9.6	41.8 ± 4.0	−540.5 ± 10.4	7125.5	159

TABLE 1. *continued*

Compound	State	ΔH_f^0 (c/l) (kJ/mol)	ΔH^0 (sub/vap) (kJ/mol)	ΔH_f^0 (g) (kJ/mol)	ΔH_a^0 (kJ/mol)	Ref.
Tris(pentane-2,4-dionato)iron(III), $[Fe(acac)_3]$	c	−1308.9 ± 4.6	65.3 ± 3.3	−1243.6 ± 5.7	18 483.0	161
Allylirontricarbonyl iodide, $[(C_3H_5)Fe(CO)_3I]$	c	−428.0 ± 10.0	84.5 ± 4.0	−343.5 ± 10.8	7004.1	160
$[Fe(CO)_4I_2]$	c	−722.0 ± 8.0	86.0 ± 4.0	−636.0 ± 8.9	5129.2	160
$[Fe_2(CO)_9]$	c	−1409.0 ± 8.5	75.3 ± 21.0	−1333.7 ± 22.7	10 858.9	62
$[Fe_3(CO)_{12}]$	c	−1849.3 ± 16.8	96.0 ± 21.0	−1753.3 ± 26.9	14 592.3	62
$[(C_5H_5)_2Co]$	c	236.6 ± 2.5	70.3 ± 4.2	306.9 ± 4.9	9464.9	141
$[Co(CO)_4H]$	g			−569.2 ± 2.2	5075.7	162
$[Co_2(CO)_8]$	c	−1250.6 ± 5.0	65.2 ± 3.3	−1185.4 ± 6.0	9762.3	163
Chloromethylidynetricobaltenneacarbonyl, $[Co_3(CCl)(CO)_9]$	c	−1186.2 ± 10.0	117.6 ± 2.5	−1068.6 ± 10.3	11 874.4	163
Bromomethylidynetricobaltenneacarbonyl, $[Co_3(CBr)(CO)_9]$	c	−1189.5 ± 9.2	99.6 ± 1.7	−1089.9 ± 9.4	11 886.3	163
$[Co_4(CO)_{12}]$	c	−1845.1 ± 12.7	96.2 ± 4.2	−1748.9 ± 13.3	15 039.4	62
$[(C_6H_6)Co_4(CO)_9]$	c	−1313.0 ± 12.0	(117.0 ± 21.0)	−1196.0 ± 24.2	17 197.0	144
1,3,5-Trimethylbenzenetetracobaltenneacarbonyl, $[(C_9H_{12})Co_4(CO)_9]$	c	−1444.0 ± 12.0	(134.0 ± 21.0)	−1310.0 ± 24.2	20 769.0	144
Hexamethylbenzenetetracobaltenneacarbonyl, $[(Me_6C_6)Co_4(CO)_9]$	c	−1555.0 ± 17.0	(148.0 ± 25.0)	−1407.0 ± 30.2	24 324.0	144
$[(C_5H_5)_2Ni]$	c	284.8 ± 4.6	72.4 ± 1.3	357.2 ± 4.8	9419.6	141
$[Ni(CO)_4]$	l	−626.4 ± 4.3	27.6 ± 1.3	−598.8 ± 4.5	4892.3	20
$[Ru_3(CO)_{12}]$	c	−1920.4 ± 17.1	(100.0 ± 20.0)	−1820.4 ± 26.3	15 331.1	62
$[Rh_4(CO)_{12}]$	c	−1824.2 ± 12.8	(100.0 ± 20.0)	−1724.2 ± 23.7	15 540.3	62
$[Rh_6(CO)_{16}]$	c	−2417.9 ± 16.9	117.2 ± 20.0	−2300.7 ± 26.2	21 093.1	164
$[Os_3(CO)_{12}]$	c	−1748.9 ± 16.8	104.6 ± 20.0	−1644.3 ± 26.1	15 581.6	62
$[Ir_4(CO)_{12}]$	c	−1820.0 ± 16.8	104.6 ± 20.0	−1715.4 ± 26.1	15 966.7	62
$[(C_5H_5)PtMe_3]$	c	168.6 ± 2.1	77.8 ± 2.0	246.4 ± 2.9	9104.7	103
$[(C_5H_5)_2PtMe_2]$	c	130.5 ± 7.1	83.7 ± 3.5	214.2 ± 7.9	12 439.5	103

TABLE 2. Mean bond dissociation energies and enthalpies for some simple species

Molecule	Bond (M—X)	ΔH^0_{atom} (kJ/mol)	T (K)	$\bar{D}$(M—X) (kJ/mol)	D(M—X) − RT (kJ/mol)
H_2O	H—O	917.76	0	458.9	458.9
		926.98	298.15	463.5	461.0
H_3N	H—N	1157.87	0	386.0	386.0
		1172.61	298.15	390.9	388.4
$H_3C\cdot$	H—C	1209.60	0	403.2	403.2
		1224.36	298.15	408.1	405.6
H_4C	H—C	1641.95	0	410.5	410.5
		1663.26	298.15	415.8	413.3
Me_2O	Me—O	725.77[a]	298.15	362.9	360.4
Me_3N	Me—N	935.30[a]	298.15	311.8	309.3
Me_4C	Me—C	1469.27[a]	298.15	367.3	364.8

[a] $\Delta H^0_{disrupt}$ (kJ/mol).

bond dissociation enthalpy, $\bar{D}$(M—X). Correspondingly from the *energy of atomization*

$$\Delta U^0_{atom} = \Delta H^0_{atom} - nRT \tag{32}$$

the quantity $\Delta U^0_{atom}/n$ measures the *mean bond dissociation* energy, equal to $\bar{D}$(M—X) − RT.

For a polyatomic molecule, MR_n, where R is a radical, the *enthalpy of the disruption reaction*

$$MR_n(g) \longrightarrow M(g) + nR(g) \tag{33}$$

is given by

$$\Delta H^0_{disrupt} = \Delta H^0_f(M, g) + n\Delta H^0_f(R, g) - \Delta H^0_f(MR_n, g) \tag{34}$$

and the quantity $\Delta H^0_{disrupt}/n$ may properly be termed the *mean bond disruption enthalpy*, symbolized by $\bar{D}$(M—R). It is a measurable quantity provided that the enthalpy of formation of the radical R is known. Some well established examples are listed in Table 2, together with mean bond dissociation energies and enthalpies at 0 and 298.15 K for some simple species.

When ΔH^0_f(r, g) is not known, it remains possible to evaluate ΔH^0_{atom} and thence to apportion this total enthalpy amongst the constituent bonds of the MR_n molecule. Values obtained in this way have been referred to variously as 'bond energies', 'bond energy terms', and more recently 'bond enthalpy contributions'. The last terminology is preferred here and symbolized $\bar{E}$(M—X), to distinguish from bond disruption enthalpies, $\bar{D}$(M—X). It is important to stress at the outset that $\bar{E}$(M—X) values depend on the *distribution rules of the scheme* adopted in apportioning ΔH^0_{atom} amongst the bonds present in MR_n. To illustrate this point, consider the molecule Me_4C, for which the mean bond disruption enthalpy, $\bar{D}$(C—Me) is experimentally determined ($\bar{D}$ = 367.3 kJ/mol at 298.15 K), and ΔH^0_{atom}(298.15 K) = 6367.0 kJ/mol. The starting point

$$6367.0 = 4\bar{E}(C—C) + 12\bar{E}(C—H) \tag{35}$$

is insufficient in itself to proceed further. The distribution scheme of Laidler[167] distinguishes between C—H bonds according to their description as primary, $(C—H)_p$, secondary, $(C—H)_s$, or tertiary, $(C—H)_t$. Applied to the alkanes Me_4C and C_2H_6 at 298 K, the Laidler scheme gives

$$\Delta H^0_{atom}(CMe_4) = 6367.0 = 4\bar{E}(C—C) + 12\bar{E}(C—H)_p \quad (36)$$

$$\Delta H^0_{atom}(C_2H_6) = 2825.3 = \bar{E}(C—C) + 6\bar{E}(C—H)_p \quad (37)$$

leading to $\bar{E}(C—C) = 358.2$ kJ/mol and $\bar{E}(C—H)_p = 411.2$ kJ/mol. The Laidler bond enthalpy contribution, $\bar{E}(C—C)$ in alkanes compares closely with $\bar{E}(C—C)$ in diamond (evaluated from $\Delta H^0_{atom}/2 = 357.4$ kJ/mol).

The larger difference between $\bar{E}(C—C)$ and $\bar{D}(Me—C)$ in Me_4C arises from the difference in meaning of $\bar{D}$ and $\bar{E}$ *values.* $\bar{D}$ values, and individual bond dissociation enthalpies, such as

$$D(Me_3C—Me) = 354.5 \text{ kJ/mol}$$
$$D(Me—Me) = 376.6 \text{ kJ/mol}$$
$$D(CH_3CH_2—H) = 410.9 \text{ kJ/mol}$$

incorporate the *enthalpies of reorganization* of the radicals as they become individual entities on disruption, $\bar{E}$ values, in contrast, derive from ΔH^0_{atom} and are in no way associated with radical reorganization enthalpies. When there is a substantial radical reorganization, there are necessarily large differencies between $\bar{D}$ and the corresponding $\bar{E}$; azomethane is a notable example. The disruption process

$$Me—N{=}N—Me(g) \longrightarrow 2Me\, g) + N_2(g) \quad (38)$$

has $\Delta H^0_{disrupt} = 144$ kJ/mol, corresponding to $D(Me—N) = 72$ kJ/mol. This may be compared with 312 kJ/mol for $\bar{D}(Me—N)$ in Me_3N (Table 2), and with $\bar{E}(C—N) = 302.5$ kJ/mol (Laidler scheme) in the same molecule. There is no reason to expect a large difference between $\bar{E}(C—N)$ in Me_2N_2 and Me_3N, but the large differences in $\bar{D}(Me—N)$ reflect the change $—N{=}N \rightarrow N{\equiv}N$ on dissociation of azomethane. $\bar{D}$ values are particularly relevant in considerations of the thermal stability and reactivity of the parent molecule, whereas $\bar{E}$ values correlate more closely with the equilibrium molecular structure.

For molecules in which the central atom is bonded to dissimilar atoms or ligands (e.g. MX_mY_n or $MR_mR^*_n$), the distribution of ΔH^0_{atom} or $\Delta H^0_{disrupt}$ amongst the bonds becomes a matter of choice. For the disruption

$$MR_mR^*_n(g) \longrightarrow M(g) + mR(g) + nR^*(g) \quad (39)$$

$$\Delta H^0_{disrupt} = m\bar{D}(M—R) + n\bar{D}(M—R^*) \quad (40)$$

and a common assumption is that $\bar{D}(M—R)$ is transferable from MR_{m+n} without change. Transferability in general would require that a redistribution reaction,

$$MR_{m+n}(g) + MR^*_{m+n}(g) \longrightarrow MR_mR^*_n(g) + MR_nR^*_m(g) \quad (41)$$

be thermoneutral. This is rarely the case, as will be examined later, but this redistribution enthalpy is often small.

B. Metal–Carbon Mean Bond Disruption Enthalpies

1. *Metal alkyl and aryl derivatives*

For molecules of the general type MR_n (where R is an alkyl or aryl radical), mean bond dissociation enthalpies may be evaluated based on available values for $\Delta H^0_f(M,g)$

TABLE 3. Enthalpies of formation of gaseous atoms in their ground states at 298.15 K

Atom	ΔH_f^0 (g) (kJ/mol)	Atom	ΔH_f^0 (g) (kJ/mol)	Atom	ΔH_f^0 (g) (kJ/mol)
H	218.00 ± 0.01	Co	425.1	Cd	110.0 ± 0.4
Li	160.7 ± 1.7	Ni	430.1	In	243 ± 8
Be	324 ± 5	Cu	337.6 ± 1.2	Sn	301.2 ± 1.7
B	560 ± 12	Zn	130.42 ± 0.20	Sb	264 ± 8
C	716.67 ± 0.44	Ga	288.7	Te	193 ± 8
N	472.68 ± 0.40	Ge	377 ± 13	I	106.76 ± 0.04
O	249.17 ± 0.10	As	289 ± 13	Cs	78.2 ± 1.3
F	79.39 ± 0.30	Se	206.7 ± 4.2	Ba	177.8
Na	107.9 ± 0.4	Br	111.86 ± 0.12	Hf	619.2
Mg	147.1 ± 0.8	Rb	81.6 ± 4.2	Ta	786.6 ± 4.0
Al	329.7 ± 4.0	Sr	143.6 ± 4.2	W	859.9 ± 4.6
Si	450 ± 8	Y	424.7 ± 0.8	Re	783 ± 8
P	316.5 ± 1.0	La	431.0 ± 0.4	Os	783 ± 8
S	276.98 ± 0.25	Pr	372.8 ± 1.3	Ir	665 ± 8
Cl	121.30 ± 0.01	Tm	247.3 ± 0.8	Pt	565.7 ± 4.2
K	89.1 ± 0.8	Yb	151.9 ± 0.4	Au	369.4 ± 3.8
Ca	177.8 ± 0.8	Zr	608.4 ± 1.7	Hg	61.38 ± 0.04
Sc	381.6 ± 1.3	Nb	724 ± 8	Tl	179.9 ± 4.2
Ti	470.7	Mo	658.1 ± 2.1	Pb	195.20 ± 0.80
V	514.6	Ru	640 ± 8	Bi	207.1 ± 4.2
Cr	396.6 ± 4.2	Rh	556.5 ± 4.2	Th	598 ± 6
Mn	279.1	Pd	380.7 ± 4.2	U	481 ± 13
Fe	416.3 ± 4.2	Ag	284.9 ± 0.8		

TABLE 4. Mean bond disruption enthalpies in metal methyl, ethyl and phenyl derivatives at 298.15 K

Molecule	$\bar{D}$(M—R) (kJ/mol)	Molecule	$\bar{D}$(M—R) (kJ/mol)	Molecule	$\bar{D}$(M—R) (kJ/mol)
$ZnMe_2$	186.4	$ZnEt_2$	145.0		
$CdMe_2$	148.5	$CdEt_2$	110.6		
$HgMe_2$	130.0	$HgEt_2$	102.7	$HgPh_2$	160.1
BMe_3	373.9	BEt_3	344.5	BPh_3	468.4
$AlMe_3$	283.3	$AlEt_3$	272.6		
$GaMe_3$	256.4	$GaEt_3$	224.9		
$InMe_3$	169.4				
CMe_4	367.3	CEt_4	345.3	CPh_4	404.8
$SiMe_4$	320.2	$SiEt_4$	287.1	$SiPh_4$	352.2
$GeMe_4$	258.2	$GeEt_4$	242.9	$GePh_4$	308.0
$SnMe_4$	226.4	$SnEt_4$	194.7	$SnPh_4$	257.2
$PbMe_4$	161.1	$PbEt_4$	129.6	$PbPh_4$	196.4
NMe_3	311.8	NEt_3	296.7	NPh_3	373.7
PMe_3	285.5	PEt_3	230.2	PPh_3	325.9
$AsMe_3$	238.4	$AsEt_3$	185.8	$AsPh_3$	285.4
$SbMe_3$	223.5	$SbEt_3$	179.8	$SbPh_3$	267.8
$BiMe_3$	150.5	$BiEt_3$	105.3	$BiPh_3$	193.9
OMe_2	362.9	OEt_2	358.7	OPh_2	423.8
SMe_2	303.6	SEt_2	293.5	SPh_2	348.0
		$SeEt_2$	240.3	$SePh_2$	285.3
$TaMe_5$	261.3				
WMe_6	161.0				

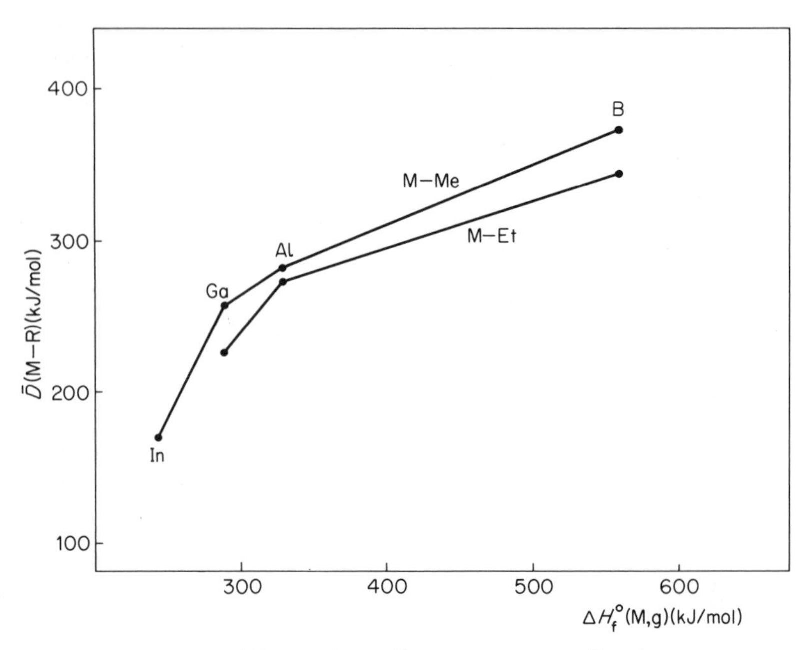

FIGURE 1. Plots of $\bar{D}$(M—Me) and $\bar{D}$(M—Et) *versus* ΔH_f^0(M,g) for B, Al, Ga, and In.

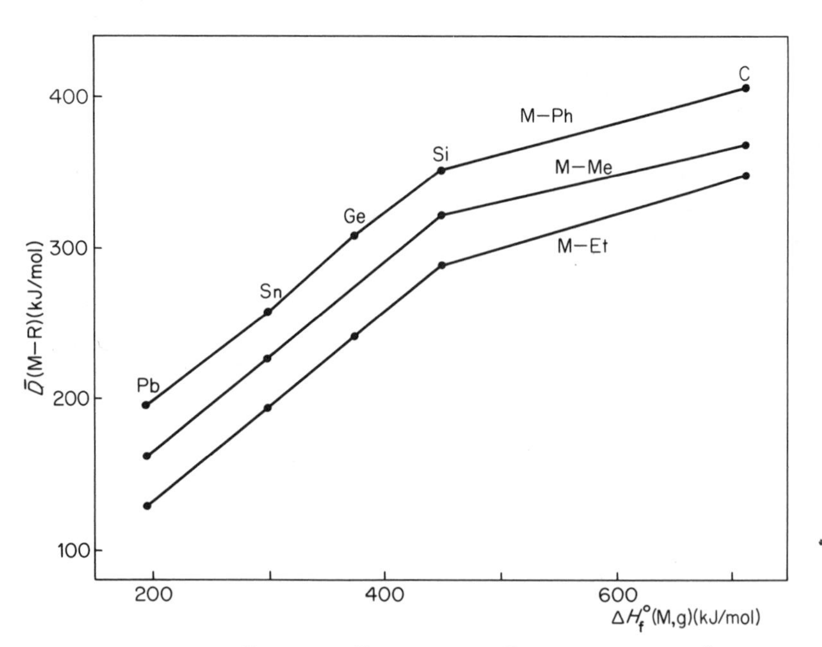

FIGURE 2. Plots of $\bar{D}$(M—Me), $\bar{D}$(M—Et), and $\bar{D}$(M—Ph) *versus* ΔH_f^0(M,g) for C, Si, Ge, Sn, and Pb.

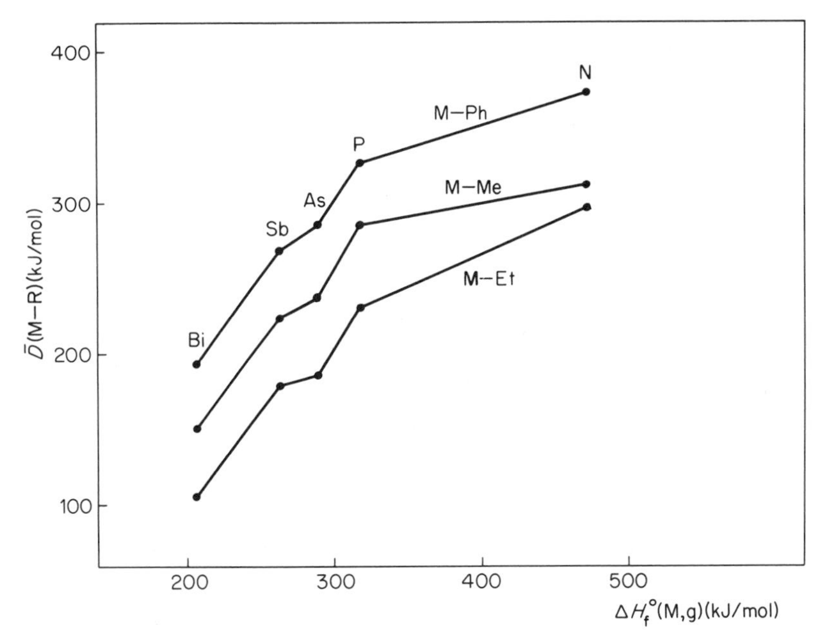

FIGURE 3. Plots of $\bar{D}$(M—Me), $\bar{D}$(M—Et), and $\bar{D}$(M—Ph) *versus* ΔH_f^0(M,g) for N, P, As, Sb, and Bi.

and ΔH_f^0(R,g). Enthalpies of formation of the gaseous atoms in their ground states are listed in Table 3, taken mainly from data compilations by CODATA[168] and by Pilcher[3], except for Mo[169] and W[170]. The following enthalpies of formation of gaseous radicals were taken from reference 3 where the original references are cited: ΔH_f^0(R,g) in kJ/mol at 298.15 K: R = Me, $+146.3 \pm 0.6$; R = Et, $+108.2 \pm 4.3$; R = Ph, $+325.1 \pm 4.3$. Table 4 lists mean bond disruption enthalpies in metal methyl, ethyl, and phenyl derivatives.

Although more data are now available, Skinner's[1] conclusions of 1964 remain valid; (i) $\bar{D}$(M—Me) > $\bar{D}$(M—Et) by *ca.* 20 kJ/mol; (ii) $\bar{D}$(M—Ph) >$\bar{D}$(M—Me) by *ca.* 40 kJ/mol; and (iii) $\bar{D}$(M—R) falls progressively as M descends a particular B group (this trend is reversed for A group elements). When $\bar{D}$(M—R) is plotted against ΔH_f^0(M, g), the A and B group elements show similar behaviour, $\bar{D}$(M—R) increasing with increasing ΔH_f^0(M,g) as shown in Figures 1, 2 and 3.

2. *Metal cyclopentadienyls*

To calculate $\bar{D}$(M—Cp) in a metal cyclopentadienyl, ΔH_f^0(Cp,g) = $+264.4 \pm 9.0$ kJ/mol is used, recently reported by De Frees *et al.*[171] from ion-cyclotron resonance studies. The mean bond disruption enthalpies are listed in Table 5, and are plotted against ΔH_f^0(M,g) in Figure 4.

It should be noted that:

(i) the ΔH_f^0 values for the rare-earth compounds, [MCp_3], were obtained from measurements of ΔU_c^0 by static-bomb combustion calorimetry, and it is yet to be established that this method is reliable for compounds of these metals;

TABLE 5. Mean bond disruption enthalpies in metal cyclopentadienyls at 298.15 K

Molecule	$\bar{D}$(M—Cp) (kJ/mol)	Molecule	$\bar{D}$(M—Cp) (kJ/mol)
[$MgCp_2$]	272.6	[$TiCp_2$]	505.9
[$ScCp_3$]	363.8	[VCp_2]	420.0
[YCp_3]	385.9	[$CrCp_2$]	340.3
[$LaCp_3$]	337.0	[$MnCp_2$]	265.6
[$PrCp_3$]	343.0	[$FeCp_2$]	351.7
[$TmCp_3$]	326.1	[$CoCp_2$]	323.5
[$YbCp_3$]	339.7	[$NiCp_2$]	300.9

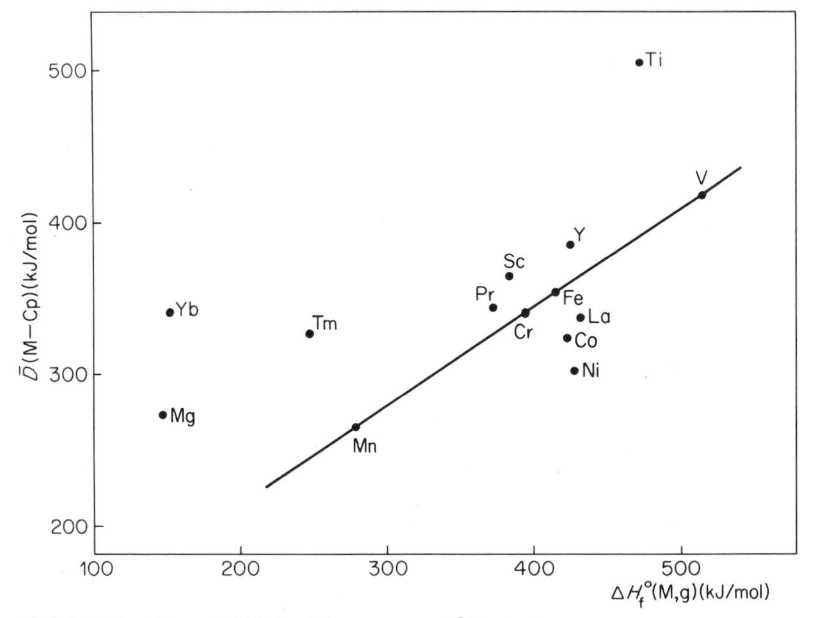

FIGURE 4. Plot of $\bar{D}$(M—Cp) *versus* ΔH_f^0(M,g) for metal cyclopentadienyl derivatives.

(ii) the values of $\bar{D}$(Co—Cp) and D(Ni—Cp) lie below the line drawn, but this is to be expected in terms of the occupancy of anti-bonding orbitals in cobaltocene (by one electron) and nickelocene (by two electrons);

(iii) the point for [$TiCp_2$] (which lies well above the line through [$MnCp_2$], [$CrCp_2$], [$FeCp_2$] and [VCp_2]) is questionable, as solid '[$TiCp_2$]' is of unusual structure[172], and it is not established that the vapour is to be classified with the other transition metallocenes included in Figure 4.

3. *Metal arene 'sandwich' compounds*

From thermochemical studies, ΔH_f^0(g) values are now available for several 'sandwich' compounds, MAr_2 (Ar = arene), leading to the disruption enthalpies,

$$MAr_2(g) \longrightarrow M(g) + 2Ar(g) \tag{42}$$

and the mean bond disruption enthalpies, $\bar{D}$(M—Ar), listed in Table 6.

TABLE 6. Mean bond disruption enthalpies in metal arenes at 298.15 K

Molecule	$\Delta H_{disrupt}$ (kJ/mol)	$\bar{D}$(M—Ar) (kJ/mol)
[Cr(C_6H_6)$_2$]	330 ± 9	165 ± 5
[Cr(PhEt)$_2$]	314 ± 9	157 ± 5
[Cr(mesitylene)$_2$]	302 ± 12	151 ± 6
[Cr(C_6Me_6)$_2$]	311 ± 12	155 ± 7
[Cr(naphthalene)$_2$]	290 ± 12	145 ± 6
[Mo(C_6H_6)$_2$]	494 ± 11	247 ± 6
[W(PhMe)$_2$]	608 ± 12	304 ± 6

4. Metal cyclopentadienyl derivatives

The enthalpies of formation[173] of $TiCl_2$(g) (−238.5 ± 12.0) kJ/mol) and $TiCl_3$(g) (−541 ± 5.0 kJ/mol), coupled with the reported ΔH_f^0(g) values (Table 1) for [Cp_2TiCl_2] and [$CpTiCl_3$], lead directly to

$$[Cp_2TiCl_2](g) \longrightarrow 2Cp(g) + TiCl_2(g);\ \Delta H_{disrupt} = 556 \pm 23\ \text{kJ/mol} \quad (43)$$

$$[CpTiCl_3](g) \longrightarrow Cp(g) + TiCl_3(g);\ \Delta H_{disrupt} = 229 \pm 16\ \text{kJ/mol} \quad (44)$$

and to the mean bond disruption enthalpies, $\bar{D}$(Cp—$TiCl_2$) = 278 ± 12 kJ/mol and D(Cp—$TiCl_3$) = 229 ± 16 kJ/mol. These values are decidely less than for $\bar{D}$(Cp—Ti) from the reported ΔH_f^0($TiCp_2$,g), but as previously remarked, there are reasons to doubt the validity of the latter. The linear plot in Figure 4 would place $\bar{D}$(Ti—Cp) at *ca.* 383 kJ/mol, which remains, however, much larger than $\bar{D}$(Cp—$TiCl_2$) or D(Cp—$TiCl_3$).

The Ti—Cl bond length in [Cp_2TiCl_2], 2.24 + 0.01 Å, is slightly longer[174] than in $TiCl_4$, 2.21 ± 0.03 Å, and so it is reasonable on this basis to allocate a bond-enthalpy contribution to each Ti—Cl bond in [Cp_2TiCl_2] of magnitude similar to $\bar{E}$(Ti—Cl) in $TiCl_4$ for which ΔH_{atom} = 1722 ± 5 and $\bar{E}$(Ti—Cl) = 430.5 ± 1.3 kJ/mol. In [Cp_2TiCl_2], ΔH_{atom} = 10 329 ± 10 kJ/mol, and the assignment $\bar{E}$(Ti—Cl) ⩽ 430 kJ/mol leaves the residue of 9469 ± 10 kJ/mol for the Cp_2Ti part of the molecule. The same starting point, $\bar{E}$(Ti—Cl) ⩽ 430 kJ/mol applied to [$CpTiCl_3$] for which ΔH_{atom} = 6017 ± 13 kJ/mol, leaves >4727 ± 13 kJ/mol for the Cp—Ti atomization enthalpy. These 'in-molecule' ΔH_a values correspond to 'in-molecule' ΔH_f^0(g) values of <351 ± 10 kJ/mol for Cp_2Ti in [Cp_2TiCl_2] and <420 ± 13 kJ/mol for CpTi in [$CpTiCl_3$], leading to $\bar{D}$(Cp—Ti) ⩾326 ± 8 kJ/mol in [Cp_2TiCl_2] and D(Cp—Ti) ⩾318 ± 16 kJ/mol in [$CpTiCl_3$]. $\bar{D}$

TABLE 7. 'In-molecule' $\bar{D}$(Cp—M) values in [Cp_2MCl_2] (M = Ti, Zr, Hf, Mo, W) at 298.15 K

Molecule	ΔH_f^0 (g) (kJ/mole)	ΔH_{atom}^0 (kJ/mole)	$\bar{E}$(M—Cl) (kJ/mole)	Origin	ΔH_f^0 (MCp_2) (kJ/mole)	$\bar{D}$(M—Cp) (kJ/mole)
[Cp_2TiCl_2]	−266 ± 9	10326 ± 10	<430 ± 1	$TiCl_4$	<351 ± 10	>328 ± 8
[Cp_2ZrCl_2]	−433 ± 4	10631 ± 6	<489 ± 2	$ZrCl_4$	<302 ± 8	>418 ± 8
[Cp_2HfCl_2]	−429 ± 3	10638 ± 5	<496 ± 1	$HfCl_4$	<306 ± 8	>421 ± 8
[Cp_2MoCl_2]	5 ± 5	10243 ± 6	<304 ± 7	$MoCl_6$	<370 ± 15	>409 ± 12
[Cp_2WCl_2]	34 ± 5	10416 ± 6	<347 ± 1	WCl_6	<485 ± 8	>450 ± 10

TABLE 8. Derived $\bar{D}$(M—X) values in $[Cp_2MX_2]$ (X = H, Br, I, Me, Ph, $PhCH_2$) at 298.15 K

Molecule	ΔH_f^0 (g) (kJ/mol)	ΔH_f^0 (Cp_2M) (kJ/mol)	ΔH_f^0 (X) (kJ/mol)	$\bar{D}$(M—X) (kJ/mol)
$[Cp_2TiMe_2]$	134 ± 12	351 ± 10	146	255 ± 8
$[Cp_2TiPh_2]$	155 ± 17	351 ± 10	325	423 ± 12
$[Cp_2Ti(CH_2Ph)_2]$	280 ± 10	351 ± 10	179	215 ± 15
$[Cp_2ZrMe_2]$	37 ± 3	302 ± 8	146	278 ± 8
$[Cp_2ZrPh_2]$	368 ± 12	302 ± 8	325	292 ± 8
$[Cp_2MoH_2]$	303 ± 6	370 ± 15	218	252 ± 8
$[Cp_2MoBr_2]$	109 ± 19	370 ± 15	112	242 ± 12
$[Cp_2MoI_2]$	170 ± 9	370 ± 15	107	207 ± 9
$[Cp_2MoMe_2]$	354 ± 6	370 ± 15	146	154 ± 8
$[Cp_2WH_2]$	311 ± 6	485 ± 8	218	305 ± 5
$[Cp_2WBr_2]$	112 ± 18	485 ± 8	112	298 ± 14
$[Cp_2WI_2]$	162 ± 9	485 ± 8	107	268 ± 6
$[Cp_2WMe_2]$	359 ± 6	485 ± 8	146	209 ± 5

values obtained in this way allow for the reorganization of the Cp ligand on disruption, but not for changes in $TiCl_2$ or TCl_3 as they become free radicals.

'In molecule' values were evaluated for Cp—Mo bonds in like manner by Tel'noi and Rabinovich[103], from $\Delta H_f^0[Cp_2MoCl_2](g)$ by assigning $\bar{E}$(Mo—Cl) in $MoCl_6$ to the Mo—Cl bonds in the cyclopentadienyl complex. These and other related results are summarized in Tables 7 and 8.

5. *Transition metal carbonyls*

For a mononuclear metal carbonyl, the mean bond disruption enthalpy, $\bar{D}$(M—CO), is readily derived from the enthalpy of disruption

$$[M(CO)_n](g) \longrightarrow M(g) + n\,CO(g) \quad (45)$$

given $\Delta H_f^0(CO,g) = -110.5 \pm 0.2$ kJ/mol[168]. The known values are listed in Table 9.

Connor *et al.*[62] transferred these mean bond disruption enthalpies to *polynuclear* transition-metal carbonyls, but pointed out the need to distinguish between the two types of metal–carbonyl bonding, the *terminal* type, $\bar{T}$, [(M—CO) as in $M(CO)_n$], and the *bridging* type, (M—CO—M), signified as *B*. Each metal–metal bond was assigned a bond enthalpy contribution $\bar{M}$, and it was assumed that for any particular metal, $\bar{T}$, $\bar{B}$

TABLE 9. $\bar{D}$(M—CO) values in mononuclear transition metal carbonyls at 298.15 K

Molecule	$\bar{D}$(M—CO) (kJ/mol)	Ref.
$[Cr(CO)_6]$	106.9	
$[\cdot Mn(CO)_5]$	97.5	175
$[Fe(CO)_5]$	117.7	
$[\cdot Co(CO)_4]$	135.9	176
$[Ni(CO)_4]$	146.7	
$[Mo(CO)_6]$	151.7	
$[W(CO)_6]$	180.1	
$[\cdot Re(CO)_5]$	180.7	177

TABLE 10. Mean bond disruption enthalpies in polynuclear transition metal carbonyls at 298.15 K

Molecule	$\Delta H^0_{\text{disrupt}}$ (kJ/mol)	Bonds	
$[Mn_2(CO)_{10}]$	1049.8	$10\bar{T} + \bar{M}$	$\bar{M} = 74.8$
$[Fe_2(CO)_9]$	1171.8	$6\bar{T} + 6\bar{B} + \bar{M}$	$\bar{B} = 64.1$
$[Fe_3(CO)_{12}]$	1676.2	$10\bar{T} + 4\bar{B} + 3\bar{M}$	$\bar{M} = 80.9$
$[Co_2(CO)_8]$	1151.6	$6\bar{T} + 4\bar{B} + \bar{M}$	$\bar{B} = 62.1$
$[Co_4(CO)_{12}]$	2123.3	$9\bar{T} + 6\bar{B} + 6\bar{M}$	$\bar{M} = 87.8$
$[Ru_3(CO)_{12}]$	2415.0	$12\bar{T} + 3\bar{M}$	$\bar{T}^* = 172.0$; $\bar{M}^* = 117.0$
$[Rh_4(CO)_{12}]$	2624.2	$9\bar{T} + 6\bar{B} + 6\bar{M}$	$\bar{T}^* = 163.2$; $\bar{M}^* = 111.0$
$[Rh_6(CO)_{16}]$	3871.7		
$[Re_2(CO)_{10}]$	2017.7	$10\bar{T} + \bar{M}$	$\bar{M} = 210.7$
$[Os_3(CO)_{12}]$	2665.5	$12\bar{T} + 3\bar{M}$	$\bar{T}^* = 189.9$; $\bar{M}^* = 129.0$
$[Ir_4(CO)_{12}]$	3050.6	$12\bar{T} + 6\bar{M}$	$\bar{T}^* = 189.7$; $\bar{M}^* = 129.0$

and $\bar{M}$ were constant in all the carbonyl derivatives of that metal, and also that the structure in the crystalline state is retained in the gaseous state. The polynuclear carbonyls to which this treatment was applied are listed in Table 10. It was found that for the iron and cobalt carbonyls, $\bar{B} \approx 0.5\bar{T}$ and $\bar{M} \approx 0.68$T, and by applying these approximate relations to other carbonyls it was possible to derive additional $\bar{T}$ and $\bar{M}$ values indicated in Table 10 by asterisks.

The assumption that the enthalpies of disruption of $[Fe(CO)_5]$, $[Fe_2(CO)_9]$, and $[Fe_3(CO)_{12}]$ can be divided into $\bar{T}$, $\bar{B}$, and $\bar{M}$ contributions allowed the evaluation of the three unknowns from the three items of experimental data. The $\bar{M}$ value so obtained is an average Fe—Fe bond disruption enthalpy from *four* Fe—Fe bonds of

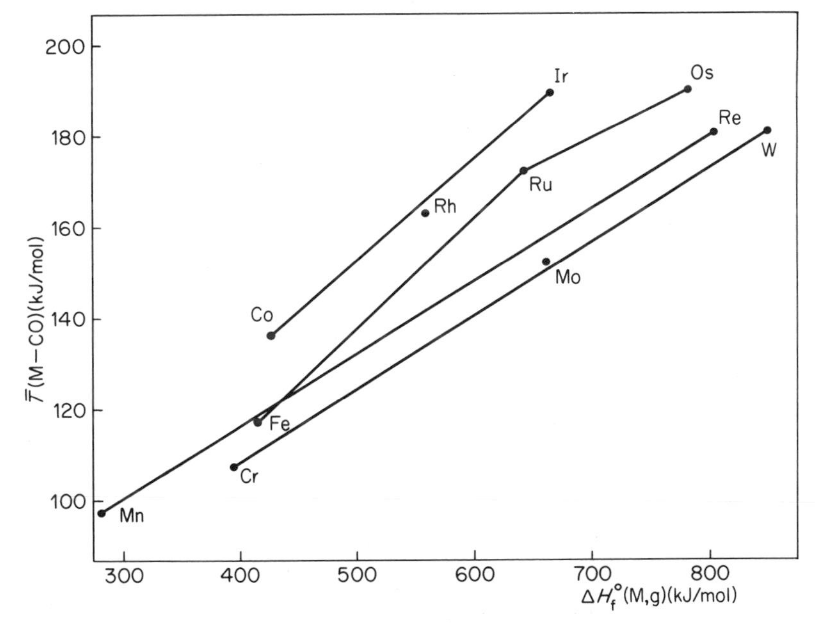

FIGURE 5. Plots of $\bar{T}$(M—CO) *versus* ΔH^0_f(M,g) for transition metal carbonyls.

different lengths {2.46 Å in $[Fe_2(CO)_9]$ and 2.56 Å and 2.67 Å (twice) in $[Fe_3(CO)_{12}]$}. Some of these Fe—Fe bonds are associated with bridging carbonyls, e.g. in $[Fe_2(CO)_9]$ there are *three* CO molecules bridging the Fe—Fe bond, whereas in $[Fe_3(CO)_{12}]$, the shorter Fe—Fe bond has *two* bridging CO molecules and the two longer Fe—Fe bonds have none. Other transition-metal carbonyls were similarly treated and the derived $\overline{T}$ values are plotted against ΔH_f^0(m,g) in Figure 5. The curves for each group are approximately parallel.

6. *Transition metal carbonyl derivatives*

In so far as $\overline{T}$ and $\overline{B}$ bond disruption enthalpies are constant and transferable, a single-centre redistribution reaction in the gaseous state of the type

$$MA_n(g) + MB_m(g) \longrightarrow 2MA_{n/2}B_{m/2}(g) \qquad (46)$$

should have zero enthalpy of reaction. A selection of actual enthalpies of redistribution are listed in Table 11.

The non-zero enthalpies of redistribution are not large, so that enthalpies of formation based on the transferability of bond enthalpy contributions are unlikely to be seriously in error.

The redistribution of the chromium arenes and chromium hexacarbonyl becomes more exothermic with increasing methyl substitution of the benzene ring, but there is no significant strengthening of the arene–chromium bonding in bis(arene)chromiums with increasing methyl substitution in the arene, as indicated by the following mean bond disruption enthalpies [$\overline{D}$(Cr—arene): $[Cr(benzene)_2]$, 171.4, $[Cr(mesitylene)_2]$, 175.7, $[Cr(hexamethylbenzene)_2]$, 155.5, and $[Cr(napthalene)_2]$, 145.4 kJ/mol.

The exothermic redistribution reactions with $[Cr(CO)_6]$ are more probably due to stronger bonding in the Cr—CO bonds of $[(arene)Cr(CO)_3]$ relative to those in the hexacarbonyl, reflected by the shorter Cr—CO bond lengths in the arene–carbonyl complexes[143].

TABLE 11. Enthalpies of redistribution in the gaseous state at 298.15 K

Reaction[a]			Enthalpy of redistribution (kJ/mol)
$\frac{1}{2}HgMe_2 + \frac{1}{2}HgCl_2$	$\rightleftharpoons$	MeHgCl	-28.5 ± 3.6
$\frac{1}{2}HgMe_2 + \frac{1}{2}HgBr_2$	$\rightleftharpoons$	MeHgBr	-21.8 ± 3.8
$\frac{1}{2}HgMe_2 + \frac{1}{2}HgI_2$	$\rightleftharpoons$	MeHgI	-17.2 ± 3.4
$\frac{1}{2}HgPh_2 + \frac{1}{2}HgCl_2$	$\rightleftharpoons$	PhHgCl	-23.4 ± 4.3
$\frac{1}{2}SiMe_4 + \frac{1}{2}SiCl_4$	$\rightleftharpoons$	Me_2SiCl_2	-8.0 ± 5.8
$\frac{1}{2}SnMe_4 + \frac{1}{2}SnCl_4$	$\rightleftharpoons$	Me_2SnCl_2	-29.3 ± 8.1
$\frac{1}{2}[Cr(CO)_6] + \frac{1}{2}[Cr(C_6H_6)_2]$	$\rightleftharpoons$	$[(C_6H_6)Cr(CO)_3]$	-6.0 ± 10.5
$\frac{1}{2}[Cr(CO)_6] + \frac{1}{2}[Cr(mes)_2]$	$\rightleftharpoons$	$[(mes)Cr(CO)_3]$	-44.2 ± 11.5
$\frac{1}{2}[Cr(CO)_6] + \frac{1}{2}[Cr(hmb)_2]$	$\rightleftharpoons$	$[(hmb)Cr(CO)_3]$	-50.2 ± 14.1
$\frac{1}{2}[Cr(CO)_6] + \frac{1}{2}[Cr(nap)_2]$	$\rightleftharpoons$	$[(nap)Cr(CO)_3]$	-7.5 ± 9.6

[a]mes = 1,3,5-trimethylbenzene; hmb = hexamethylbenzene; nap = naphthalene.

C. Metal–Metal Bond Enthalpy Contributions

1. *Metal–metal single bonds*

For compounds of the type $R_nM—MR_n$ (R = alkyl or aryl radical), the enthalpy of the disruption

$$M_2R_{2n}(g) \longrightarrow 2M(g) + 2n\,R(g) \qquad (47)$$

TABLE 12. $E(M—M)$ values in polynuclear metal alkyls and aryls at 298.15 K

Molecule	$\bar{D}(M—R)$ (kJ/mol)	$E(M—M)$ (kJ/mol)	$D(M_2)$ (kJ/mol)	$E(M—M)^*$ (kJ/mol)
$Me_3Si—SiMe_3$	320.15	220.0	309.6	225.0
$Me_3Si—SiMe_2—SiMe_3$		214.8		
$Me_3Si—(SiMe_2)_2—SiMe_3$		210.0		
$(Me_3Si)_4Si$		180.7		
$Ph_3Ge—GePh_3$	308.0	193.7	272.0	188.5
$Me_3Sn—SnMe_3$	226.4	148.7	191.6	150.6
$Ph_3Sn—SnPh_3$	257.2	160.1		

can be thermochemically evaluated and allocated to the M—R and M—M bonds according to

$$\Delta H_{\text{disrupt}} = 2n\bar{D}(M—R) + E(M—M) \tag{48}$$

The common assumption is made that $\bar{D}(M—R)$, determined from the disruption enthalpy of MR_{n+1}, is transferable to M_2R_{2n} and to higher homologues. Some examples using this procedure are given in Table 12.

The values of $E(M—M)$ compare reasonably with $E(M—M)^*$, obtained from the enthalpies of sublimation of the crystalline elements (diamond structures), $E(M—M)^* = \frac{1}{2}\Delta H$ (sub.), and are in all cases less than the dissociation energies, $D(M_2)$, of the diatomic molecules in the gas phase. The approximate constancy and trends in the $E(M—M)$ values for the silicon methyls suggests that the Allen bond-energy scheme[178] could be satisfactory when data are sufficient to apply this scheme to these compounds.

The same values for $E(M—M)$ can be obtained, starting from the enthalpies of atomization, ΔH_{atom}, of M_2R_{2n} and MR_{n+1}, and the assumption that the sum of bond enthalpy contributions $[3E(C—H) + E(M—C)]$ in MR_{n+1} remain the same in M_2R_{2n}. For example,

$$\Delta H_{\text{atom}}(SiMe_4) = 12E(C—H) + 4E(Si—C) = 6178.1 \text{ kJ/mol} \tag{49}$$

and

$$\Delta H_{\text{atom}}(Si_2Me_6) = 18E(C—H) + 6E(Si—C) + E(Si—Si) = 9487.1 \text{ kJ/mol} \tag{50}$$

so that

$$\Delta H_{\text{atom}}(Si_2Me_6) - \tfrac{3}{2}\Delta H_{\text{atom}}(SiMe_4) = E(Si—Si) = 220.0 \text{ kJ/mol} \tag{51}$$

which is identical with the value given in Table 12.

The bond enthalpy contribution $E(Si—Si)$ is substantially less than the bond dissociation enthalpy, $D(Me_3Si—SiMe_3)$, in Si_2Me_6. From kinetic studies on the iodination of Me_3SiH, Doncaster and Walsh[179] found that $D(Me_3Si—H) = 378 \pm 6$ kJ/mol, from which (with estimated $\Delta H_f^0(Me_3SiH,g) \approx -186$ kJ/mol), $\Delta H_f^0(Me_3Si) \approx 26$ kJ/mol and $D(Me_3Si—SiMe_3) \approx 311$ kJ/mol. Similar studies on the kinetics of iodination of SiH_4[180] led to $D(H_3Si—H) \approx 376 \pm 8$ kJ/mol, which corresponds to $D(H_3Si—SiH_3) = 305 \pm 11$ kJ/mol compared with $E(Si—Si) = 98$ kJ/mol in this molecule.

The bond enthalpy contributions $E(M—M)$ in the poly-nuclear transition metal carbonyls may be similarly derived and compared with $E(M—M)^*$ values in the

TABLE 13. Comparison of $E(M—M)$ values in transition metal carbonyls with $E(M—M)^*$ values at 298.15 K

Bond	$E(M—M)$ (kJ/mol)	$E(M—M)^*$ (kJ/mol)
Re—Re	210.7	130.4
Fe—Fe	80.9	69.4
Ru—Ru	117.0	106.7
Os—Os	129.0	130.4
Co—Co	87.8	70.9
Rh—Rh	110.0	92.8
Ir—Ir	129.0	110.9

crystalline metals, obtained by dividing the enthalpy of sublimation of the metal by half of the coordination number of each atom in the lattice. The 'metallic bond' in a transition metal is electron deficient with respect to a normal covalent bond and might be expected to be longer and weaker in consequence. Metallic $E(M—M)^*$ values from the enthalpies of sublimation are compared with $E(M—M)$ bond enthalpy contributions in the transition-metal carbonyls in Table 13.

The reasonable agreement makes the 'metallic bond' enthalpy derived from the enthalpy of sublimation of the metal a useful first approximation for an unknown metal–metal bond enthalpy contribution in other polynuclear transition metal carbonyls.

2. *Metal–metal multiple bonds*

Multiple bonds between transition metal atoms have become well known[181] and many compounds containing such bonds have been characterized structurally. There are difficulties in analysing available thermochemical data for these compounds, particularly when attempting to make realistic estimates of the strengths of the metal–metal bonds.

The problems have been highlighted in the study of the hexakis(dimethylamido) derivatives of dimolybdenum(III) and ditungsten(III)[54], where the metal–metal bond lengths suggest triple bonding, i.e. these compounds are represented by $[(Me_2N)_3M{\equiv}M(NMe_2)_3]$, where M = Mo or W. The enthalpies of disruption,

$$[M_2(NMe_2)_6](g) \longrightarrow 2M(g) + 6(\cdot NMe_2)(g) \tag{52}$$

can be calculated, given that $\Delta H_f^0(\cdot NMe_2, g) = 123.4 \pm 4.2$ kJ/mol[182], and

$$\Delta H_{disrupt} = 6\bar{D}(M—NMe_2) + E(M{\equiv}M) \tag{53}$$

The only mononuclear dimethylamido derivatives of Mo and W which can be used to derive the $\bar{D}(M—NMe_2)$ values are:

$$[Mo(NMe_2)_4];\ \bar{D}(Mo—NMe_2) = 255.4 \text{ kJ/mol};$$
$$[W(NMe_2)_6];\ \bar{D}(W—NMe_2) = 222.1 \text{ kJ/mol}.$$

When these values are transferred to the $[M_2(NMe_2)_6]$ molecules, the derived $E(M{\equiv}M)$ values seem unreasonable; moreover, $\bar{D}(Mo—NMe_2) > \bar{D}(W—NMe_2)$, not in accord with the general trend of increasing $\bar{D}(M—X)$ with increasing $\Delta H_f^0(M,g)$. It is known that $\bar{D}(Mo—X)$ and $\bar{D}(W—X)$ (X = halogen) in MoX_n and WX_n depend on n, the number of attached atoms or the formal oxidation number of the central atom.

TABLE 14. $\bar{D}$(M—NMe$_2$) and corresponding E(M≡M) values for various formal oxidation numbers of Mo and W at 298.15 K

Molecule	$\bar{D}$(Mo—NMe$_2$) (kJ/mol)	E(Mo≡Mo) (kJ/mol)	$\bar{D}$(W—NMe$_2$) (kJ/mol)	E(W≡W) (kJ/mol)
$[M(NMe_2)_3]$	288	200	331	340
$[M(NMe_2)_4]$	255[a]	396	295	558
$[M(NMe_2)_5]$	223	592	258	775
$[M(NMe_2)_6]$	190	788	222[a]	995

[a]Experimental value.

If it be assumed that the trend of $\bar{D}$(Mo—NMe$_2$) and $\bar{D}$(W—NMe$_2$) in $[Mo(NMe_2)_n]$ and $[W(NMe_2)_n]$ with n parallels that in MoX$_n$ and WX$_n$, then the values given in Table 14[183] are derived.

This approach gives no clear answer to the question of the thermochemical strengths of these metal–metal triple bonds. The structure of $[W(NMe_2)_6]$[184] shows evidence of considerable strain in this molecule: the C—N—C angles in the ligand are compressed to 103°, compared with 110° in $[W_2(NMe_2)_6]$; also r(W—N) in $[W(NMe_2)_6]$ is 2.02 Å compared with 1.98 Å in $[W_2(NMe_2)_6]$. Transfer of $\bar{D}$(W—NMe$_2$) derived from the strained $[W(NMe_2)_6]$ to the relatively unstrained $[W_2(NMe_2)_6]$ could lead to a spuriously high value of E(W≡W). The structure of $[Mo(NMe_2)_4]$[185] does not suggest strain and the transfer in this case may be more acceptable. the problem of transferability shows up well in this study because $\bar{D}$(M—NMe$_2$) enters six-fold into the derived value of E(M≡M).

A different approach was used by Cavell *et al.*[55] in considering the metal–metal quadruple bonds in the μ-tetraacetates of dimolybdenum(II) and dichromium(II). For the disruption of the following molecules:

$$[Mo_2(O_2CMe)_4](g) \longrightarrow 2Mo(g) + 4(O_2CMe, g) \quad (54)$$

$$[Mo_2(CO_2CMe)_2(acac)_2](g) \longrightarrow 2Mo(g) + 2(O_2CMe, g) + 2(acac, g) \quad (55)$$

$$[Mo(acac)_3](g) \longrightarrow Mo(g) + 3(acac, g) \quad (56)$$

it was assumed that

(i) E(Mo≡Mo) has the same value in $[Mo_2(O_2CMe)_4]$ and in $[Mo_2(O_2CMe)_2(acac)_2]$; r(Mo—Mo) in these molecules in the crystalline state is 2.093 and 2.129 Å[186], respectively;
(ii) $\bar{D}$(Mo—O) for the acetato ligands is the same in the two compounds, and in each molecule the dimensions within the bridge acetato groups are not significantly different;
(iii) $\bar{D}$(Mo—O) for the pentane-2,4-dionato groups are the same in $[Mo_2(O_2CMe)_2(acac)_2]$ and in $[Mo(acac)_3]$.

Application of these assumptions gives

$$\begin{aligned} E(\text{Mo}\equiv\text{Mo}) &= \tfrac{2}{3}\Delta H_f^0(\text{Mo, g}) + \Delta H_f^0([\text{Mo}_2(\text{O}_2\text{CMe})_4], \text{g}) \\ &\quad -2\Delta H_f^0([\text{Mo}_2(\text{O}_2\text{CMe})_2(\text{acac})_2], \text{g}) + \tfrac{4}{3}\Delta H_f^0([\text{Mo(acac)}_3], \text{g}) \\ &= 321 \text{ kJ/mol} \end{aligned} \quad (57)$$

Although E(Mo≡Mo) so derived is unambiguous, the value does depend on the assumptions made and, of these, the third is the most critical and may be incorrect. In effect, assumption (iii) is that $\bar{D}$(Mo—O) for the pentane-2,4-dionato group is independent of the oxidation state of Mo and, in view of the variation in the mean bond disruption enthalpy illustrated in Table 14, this assumption is a weak point in the derivation.

By making the further reasonable assumption that

$$\bar{D}(\text{Mo—O})_{\text{acac}} - \bar{D}(\text{Cr—O})_{\text{acac}} = \bar{D}(\text{Mo—O})_{\text{Ac}} - \bar{D}(\text{Cr—O})_{\text{Ac}} \quad (58)$$

the following expressions were derived:

$$\begin{aligned} E(\text{Cr}{\equiv}\text{Mo}) - E(\text{Mo}{\equiv}\text{MO}) &= \Delta H_f^0(\text{Cr, g}) - \Delta H_f^0(\text{Mo, g}) \\ &\quad - \Delta H_f^0([\text{CrMo}(O_2\text{CMe})_4], \text{g}) \\ &\quad + \Delta H_f^0([\text{Mo}_2(O_2\text{CMe})_4], \text{g}) + 33.6 \\ &= -84.7 \text{ kJ/mol} \end{aligned} \quad (59)$$

$$\begin{aligned} E(\text{Cr}{\equiv}\text{Cr}) - E(\text{Mo}{\equiv}\text{MO}) &= 2\Delta H_f^0(\text{Cr, g}) - 2\Delta H_f^0(\text{Mo, g}) \\ &\quad - \Delta H_f^0([\text{Cr}_2(O_2\text{CMe})_4], \text{g}) \\ &\quad + \Delta H_f^0([\text{Mo}_2(O_2\text{CMe})_4], \text{g}) + 67.2 \\ &= -277.8 \text{ kJ/mol} \end{aligned} \quad (60)$$

so that

$$E(\text{Mo}{\equiv}\text{Cr}) = 236.3 \text{ kJ/mol}$$
$$E(\text{Cr}{\equiv}\text{Cr}) = 43.2 \text{ kJ/mol}$$

The small value for E(Cr≡Cr) may reflect the fact that r(Cr—Cr) in [$Cr_2(O_2CMe)_4$] is one of the longest (Cr≡Cr) bonds, but more probably this value is low because of the assumptions made in the derivation. Although the differences between E(Mo≡Mo), E(Mo≡Cr) and E(Cr≡Cr) may be realistic, the absolute values must be tentative.

D. Bond Energy Schemes for Organometallic Compounds

For any compound, the enthalpy of atomization may be equated to the total chemical binding energy, so that

$$\Delta H_a = \Sigma \text{ bond energies} + \text{stabilization energy—strain energy} \quad (61)$$

It is clear that bond energies can only be derived from ΔH_a values for compounds in which it is reasonable to assume that exceptional stabilization or strain are absent. The simplest assumption is that the energy of a bond between two particular atoms is constant and transferable between molecules containing the same type of bond. In the late 1950s, however, three schemes of slightly greater complexity were proposed, the Laidler scheme[167], the Group scheme[87], and the Allen scheme[178]. It has been demonstrated that these schemes are equivalent[2], so that if the parameters are chosen in accord with the equivalence relations, the three schemes will produce identical results. Essentially, the reason for this equivalence is the assumption that the energy of a bond of a particular type is constant *provided that the nearest neighbours of the bond are the same*. As the schemes are in effect identical, it is only necessary to apply one, and here we select the Laidler scheme.

1. *The Laidler scheme*

Laidler parameters for metal alkyl and aryl derivatives are listed in Table 15. The parameters previously derived for hydrocarbons[2] are listed first. The Laidler scheme

TABLE 15. Laidler bond energy parameters (kJ/mol) for metal alkyl and aryl derivatives at 298.15 K

Parameter	Value	Parameter	Value	Parameter	Value
$E(\text{C—C})$	358.46	$E(\text{C=C})$	556.50	$E(\text{C}_b\text{—C}_b)$	499.44
$E(\text{C—H})_p$	411.26	$E(\text{C}_d\text{—H})_2$	424.20	$E(\text{C}_b\text{—H})$	421.41
$E(\text{C—H})_s$	407.40	$E(\text{C}_d\text{—H})_1$	421.41	$E(\text{C}_b\text{—C})$	372.81
$E(\text{C—H})_t$	404.30	$E(\text{C}_d\text{—C})$	378.05		
$E(\text{C—H})_p^{M}$	411.26	for all metals			
$E(\text{C—Zn})$	176.97	$E(\text{C—Si})$	310.75	$E(\text{C—As})$	229.02
$E(\text{C—H})_s^{Zn}$	395.45	$E(\text{C—H})_s^{Si}$	399.61	$E(\text{C—H})_s^{As}$	389.83
$E(\text{C—Cd})$	139.02	$E(\text{C—Ge})$	248.82	$E(\text{C—Sb})$	214.05
$E(\text{C—H})_s^{Cd}$	397.22	$E(\text{C—H})_s^{Ge}$	406.32	$E(\text{C—H})_s^{Sb}$	394.34
$E(\text{C—Hg})$	120.52	$E(\text{C—Sn})$	217.00	$E(\text{C—Bi})$	141.12
$E(\text{C—H})_s^{Hg}$	402.55	$E(\text{C—H})_s^{Sn}$	401.64	$E(\text{C—H})_s^{Bi}$	393.55
$E(\text{C—H})_t^{Hg}$	381.55	$E(\text{C—H})_t^{Sn}$	381.39		
$E(\text{C}_b\text{—Hg})$	121.31	$E(\text{C}_d\text{—Sn})$	230.25		
		$E(\text{C}_b\text{—Sn})$	234.07		
$E(\text{C—B})$	364.52	$E(\text{Sn—Sn})$	148.62		
$E(\text{C—H})_s^{B}$	401.50				
$E(\text{C—H})_t^{B}$	397.70	$E(\text{C—Pb})$	151.65		
		$E(\text{C—H})_s^{Pb}$	400.44		
$E(\text{C—Al})$	273.85				
$E(\text{C—H})_s^{Al}$	410.84	$E(\text{C—Ta})$	251.86		
$E(\text{C—Ga})$	246.99	$E(\text{C—P})$	276.09		
$E(\text{C—H})_s^{Ga}$	397.10	$E(\text{C—H})_s^{P}$	322.42		
$E(\text{C—In})$	159.99				

for alkanes has one C—C bond energy and three C—H bond energies, $E(\text{C—H})_p$, $E(\text{C—H})_s$, and $E(\text{C—H})_t$ for primary, secondary and tertiary C—H bonds, respectively. For olefins additional parameters are employed: $E(\text{C=C})$ for the carbon–carbon double bond, $E(\text{C}_d\text{—H})_2$ for the C—H bond in $=\text{C}\begin{smallmatrix}\diagup\text{H}\\ \diagdown\text{H}\end{smallmatrix}$ and $E(\text{C}_d\text{—H})_1$ for the C—H bond in $=\text{C}\begin{smallmatrix}\diagup\text{H}\\ \diagdown\text{C}\end{smallmatrix}$, where the C—C single bond involving the doubly bound carbon atom C_d has the parameter $E(\text{C}_d\text{—C})$. For benzene derivatives we follow the procedure due to Cox[188] by devising parameters which include the π-delocalization energy; thus for benzene

$$\Delta H_a = 6E(\text{C}_b\text{—C}_b) + 6E(\text{C}_b\text{—H}) \tag{62}$$

and $E(\text{C}_b\text{—H})$ has been taken as equal to $E(\text{C}_d\text{—H})_1$. Metal alkyls can be regarded as substituted alkanes, for which there are two equivalent procedures for applying the Laidler scheme:

(a) The bond energy $E(\text{M—C})$ can be taken as constant and the values for $E(\text{C—H})$ involving the carbon atom of M—C depend on the degree of substitution of that carbon atom, thus giving rise to the bond energies, $E(\text{C—H})_p^M$, $E(\text{C—H})_s^M$, and $E(\text{C—H})_t^M$.

(b) An alternative approach is that the C—H bonds have the same bond energies as in alkanes but that $E(\text{M—C})$ depends on the degree of substitution of the carbon atom, giving rise to $E(\text{M—C})_p$, $E(\text{M—C})_s$, $E(\text{M—C})_t$, and $E(\text{M—C})_q$.

Clearly, procedures (a) and (b) are equivalent, as they contain the same number of parameters. Procedure (a) will be followed here as it is more in keeping with the spirit of Laidler's original scheme. For all metals, $E(C-H)_p^M$ will be taken as equal to $E(C-H)_p$ in alkanes. Table 15 concentrates primarily on metal—carbon bonds. Laidler parameters for other types of bond in organometallic compounds, e.g. M—O, M—N or M—halogen, if required, can be derived from the appropriate enthalpies of atomization given in Table 1 together with the relevant parameters from Table 15. Care must be taken, however, to derive bond energies from ΔH_a values which are considered reliable and from molecules for which steric strain should be absent.

As an illustration, the Laidler parameters are applied to tin compounds in Table 16: tin compounds were chosen because the experimental data are reliable.

TABLE 16. Calculated and observed ΔH_a values for tin compounds at 298.15 K

Molecule	ΔH_a (calc.) (kJ/mol)	ΔH_a (obs.) (kJ/mol)	Difference (kJ/mol)
Me_4Sn	5803.1	5803.1	0.0
Me_3SnEt	6964.9	6966.1	−1.2
$Me_3Sn(i\text{-}Pr)$	8135.2	8136.0	−0.8
$Me_3Sn(t\text{-}Bu)$	9346.1	9309.0	37.1
Et_4Sn	10 450.1	10 439.5	10.6
$(n\text{-}Pr)_4Sn$	15 143.1	15 149.6	−6.5
$(n\text{-}Bu)_4Sn$	19 836.2	19 835.1	1.1
$Me_3(Sn(CH{=}CH_2)$	6408.9	6408.9	0.0
Me_3SnPh	9686.3	9690.1	−3.8
Me_3SnCH_2Ph	10 849.1	10 873.1	−24.0
Ph_4Sn	21 335.8	21 288.6	47.2
$(Me_3Sn)_2$	8853.3	8853.3	0.0
$(Ph_3Sn)_2$	32 152.3	32 091.0	61.3

It can be seen that for those compounds in which steric strain is expected, i.e. $Me_3Sn(t\text{-}Bu)$, Ph_4Sn, and $(Ph_3Sn)_2$, the observed ΔH_a is less than the calculated value. For this reason, $E(C_b-M)$ values from MPh_4 compounds were not derived and are not listed in Table 15. The large negative deviation for Me_3SnCH_2Ph may arise because the secondary C—H bonds are subject to both the influence of the Sn atom and the Ph group, and allowance cannot be made for this.

It appears that steric strain is present in $(Ph_3Sn)_2$ relative to $(Me_3Sn)_2$, whereas in Table 12 it can be seen that the bond enthalpy contribution $E(Sn-Sn)$ is larger in $(Ph_3Sn)_2$ than in $(Me_3Sn)_2$. The reason for this apparent contradiction is that in deriving $E(Sn-Sn)$ from $\Delta H_{disrupt}$ of $(Ph_3Sn)_2$, $\bar{D}(Sn-Ph)$ was taken from $\Delta H_{disrupt}$ of $SnPh_4$, which exhibits steric strain: whereas the bond energy term $E(Sn-Sn)$ in Table 15 was derived from $\Delta H_2(Me_3Sn)_2$ and the value of $E(C_b-Sn)$ derived from $\Delta H_a(Me_3SnPh)$, which is less likely to be affected by steric hindrance than $SnPh_4$.

This example highlights the danger of simply considering values of mean bond disruption enthalpies, bond enthalpy contributions, or bond energy terms as measures of bond strength. It is necessary to take into account the assumptions made in the derivations and the structures of the molecules and radicals involved.

2. *Bond energy differences*

A recent series of studies by Mortimer and co-workers concentrated on deriving differences in bond dissociation enthalpies for the attachment of various ligands to

TABLE 17. Enthalpies of reaction for determining differences in bond dissociation enthalpies at 298.15 K

Reaction	ΔH (kJ/mol)	Ref.
1. $[Pt(PPh_3)_2(CH_2{=}CH_2)](c) + C(CN)_2{=}C(CN)_2(g) \rightarrow [Pt(PPh_3)_2\{C(CN)_2{=}C(CN)_2\}](c) + C_2H_4(g)$	$\Delta H_1 = -155.8 \pm 8.0$	189
2. $[Pt(PPh_3)_2(CH_2{=}CH_2)](c) + PhC{\equiv}CPh(g) \rightarrow [Pt(PPh_3)_2(PhC{\equiv}CPh)](c) + C_2H_4(g)$	$\Delta H_2 = -82 \pm 12$	190
3. $[Pt(PPh_3)_2(CH_2{=}CH_2)](c) + CS_2(g) \rightarrow [Pt(PPh_3)_2(CS_2)](c) + C_2H_4(g)$	$\Delta H_3 = -44.0 \pm 2.2$	191
4. $[Pt(PPh_3)_2(CH_2{=}CH_2)](c) + I_2(g) \rightarrow$ *trans*-$[Pt(PPh_3)_2I_2](c) + C_2H_4(g)$	$\Delta H_4 = -176.6 \pm 5.4$	192
5. $[Pt(PPh_3)_2(CH_2{=}CH_2)](c) + CH_2ICH_2I(g) \rightarrow$ *cis*-$[Pt(PPh_3)_2I_2(c) + 2C_2H_4(g)$	$\Delta H_5 = -107.8 \pm 6.0$	192
6. $[Pt(PPh_3)_2(CH_2{=}CH_2)](c) + CH_3I(g) \rightarrow$ *cis*-$[Pt(PPh_3)_2(CH_3I)](c) + C_2H_4(g)$	$\Delta H_6 = -78.9 \pm 2.0$	192
7. $[Pt(PPh_3)_2(PhC{\equiv}CPh)](c) + HCl(g) \rightarrow [Pt(PPh_3)_2Cl(PhC{\equiv}CPh)](c)$	$\Delta H_7 = -90.2 \pm 6.0$	193
8. $[Pt(PPh_3)_2(PhC{\equiv}CPh)](c) + 2HCl(g) \rightarrow$ *cis*-$[Pt(PPh_3)_2Cl_2](c)$ + *trans*-$CHPh{=}CHPh$	$\Delta H_8 = -139.0 \pm 16.0$	193

platinum, rather than primarily attempting to determine the absolute values. From this work it is possible to see how the strength of bonding depends on the nature of the ligand. A selection of the results obtained by reaction calorimetric studies is given in Table 17.

The general arguments used to derive differences in bond dissociation enthalpies can be illustrated by considering the first reaction in Table 17, for which it would be desirable to know ΔH_r^0 for all the reactants and products in the gaseous phase. It is reasonable, however, to assume that the enthalpies of sublimation of the two crystalline complexes will not be very different, hence the observed ΔH_r^0 should be close to the gaseous value. The enthalpy of reaction can be equated to the difference, D(Pt–ethylene) – D(Pt–tetracyanoethylene), but changes in the olefinic bonds should also be considered. On dissociation the C=C bond in the ethylene molecule shortens and thus presumably makes an exothermic contribution to ΔH_r^0, whereas there will be an endothermic contribution arising from the lengthening of the C=C bond in $(CN)_2C{=}C(CN)_2$ when it bonds to platinum.

By applying arguments of this type to the enthalpies of reaction listed in Table 17, Mortimer and co-workers deduced the following differences in bond dissociation enthalpies:

$$
\begin{aligned}
D(\text{Pt—tetracyanoethylene}) - D(\text{Pt—ethylene}) &= 156 \text{ kJ/mol} \\
D(\text{Pt—tolane}) - D(\text{Pt—ethylene}) &= 82 \text{ kJ/mol} \\
D(\text{Pt—CS}_2) - D(\text{Pt—ethylene}) &= 44 \text{ kJ/mol} \\
D(\text{Pt—CH}_3) - D(\text{Pt—I}) &= 6 \text{ kJ/mol} \\
D(\text{Pt—tolane}) - D(\text{Pt—Cl}) &= 1 \text{ kJ/mol}
\end{aligned}
$$

It should be noted that these differences in bond dissociation enthalpies are derived directly from the enthalpies of reaction and that it has not been necessarry to derive the enthalpies of formation of the compounds involved.

3. *Bond energy–bond length relations*

That the bond energy and bond length of a given bond are interrelated has been generally accepted for some time, e.g. C—C bond lengths diminish with increasing multiplicity [r(C—C)diamond 1.544 Å, r(C⩦C)benzene 1.39 Å, r(C=C) ethylene 1.33 Å, r(C≡C)acetylene 1.20 Å] and the C—C bond energy increases with increasing bond order. Dewar and Schmeising[194] postulated that

$$r_i = (1/b_i)\{a_i \log[a_i + (a_i^2 - E_i^2)^{1/2}] - a_i \log E_i - (a_i^2 - E_i^2)^{1/2}\} \qquad (63)$$

where a_i and b_i are constant for a given bond type, but when applied to C—C and C—H bonds, the values of E_i are very sensitive to changes in r_i, especially for C—H bonds. The most precise structural data for benzene by Langseth and Stoicheff[195], r(C—C) = 1.397 ± 0.001 Å, r(C—H) = 1.084 ± 0.005 Å when substituted into the appropriate bond energy–bond length equations, the uncertainty in the bond lengths gives rise to an uncertainty in the sum of the bond energies of ±45 kJ mol^{-1}, and possibly for this reason this approach has not been widely applied to organic compounds.

Attempts have been made recently[150,153] to analyse the bond enthalpy contributions of metal–oxygen and other metal–ligands in organometallic complexes of Cr and Mo, from which bond enthalpy–bond length curves for Mo—O and Cr—O bonds have been derived. Cavell *et al.*[153] considered the bonding in [Mo_2(O—*i*-Pr)$_8$] in this way starting from a provisional plot of E(Mo—Mo) against r(Mo—Mo). This served to

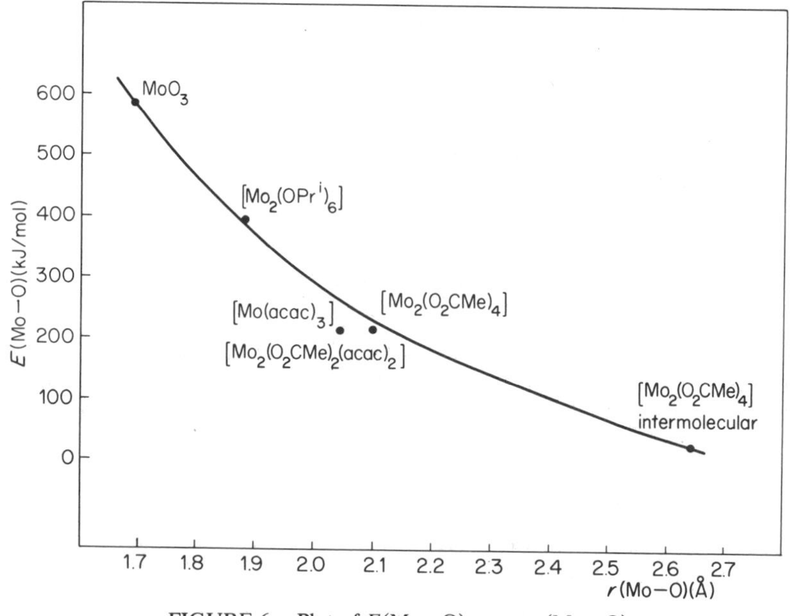

FIGURE 6. Plot of E(Mo—O) *versus* r(Mo—O).

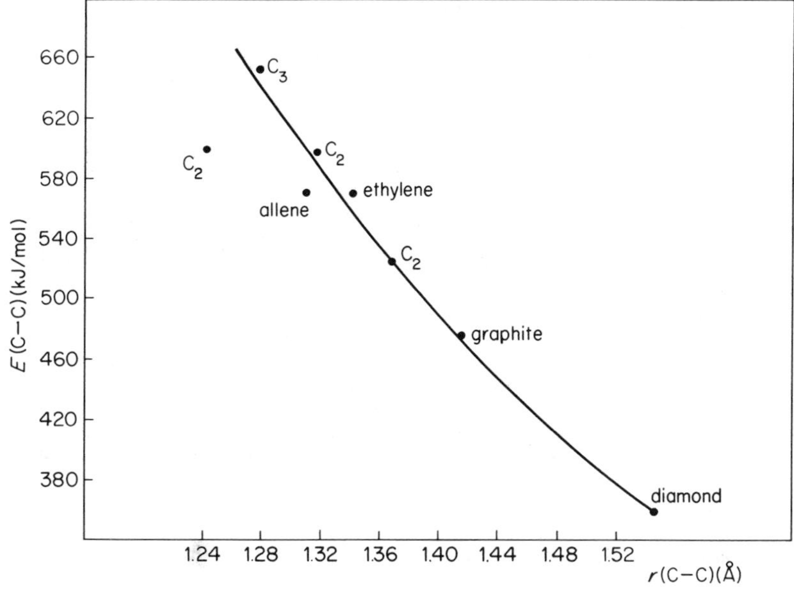

FIGURE 7. Plot of E(C—C) *versus* r(C—C).

obtain Mo—O bond enthalpy contributions in complexes of dimolybdenum, e.g. $[Mo_2(O_2CMe)_4]$, and to combine these with other data for Mo—O bonds from MoO_3, $[Mo_2(O—i\text{-}Pr)_6]$ and $[Mo(acac)_3]$. The resultant smoothed energy–length curve is shown in Figure 6: the sources of the data points are indicated on the figure.

The molecular structure of $[Mo_2(O—i\text{-}Pr)_8]$ in the crystal is known[196]:

i-Pr
O
i-Pr—O 1.884 O—i-Pr
1.976
i-Pr—O 1.872 Mo ═ Mo—O—i-Pr   r (Mo—Mo) 2.52
1.958 2.111
i-Pr—O O O—i-Pr
i-Pr

and the bond energies from Figure 6 corresponding to the five Mo—O lengths are 378, 365, 300, 285, and 202 kJ/mol. The Mo—Mo energy–length plot indicates E(Mo—Mo) = 230 kJ/mol. For the disruption reaction

$$[Mo_2(O—i\text{Pr})_8] \longrightarrow 2Mo + 8(O—i\text{Pr}) \qquad (64)$$

summation of these bond energies gives 3290 kJ/mol compared with the experimental ΔH^0 (disruption) of 3257 kJ/mol.

Similar plots have been given for Cr—Cr and Cr—O bonds[150], and these have recently been used with an energy–length plot for C—C bonds, shown in Figure 7, in an attempt to construct Cr—C and Cr—N energy–length curves.

The available thermochemical data are insufficient at present to progress very far, but the approach is a promising one for future development.

V. REFERENCES

1. H. A. Skinner, *Adv. Organomet. Chem.*, **2**, 49 (1964).
2. J. D. Cox and G. Pilcher, *Thermochemistry of Organic and Organometallic Compounds*, Academic Press, London, 1970.
3. G. Pilcher, *Thermochemistry and Thermodynamics*, Phys. Chem. Series 2, Vol. 10, Int. Rev. Sci., Butterworths, London, 1975, Chapter 2.
4. J. B. Pedley and J. Rylance, *Sussex–NPL Computer Analysed Thermochemical Data, Organic and Organometallic Compounds*, University of Sussex, Brighton, 1977.
5. L. H. Long and R. G. W. Norrish, *Phil. Trans. Roy. Soc. (London)*, **A241**, 587 (149).
6. A. S. Carson, K. Hartley, and H. A. Skinner, *Trans. Faraday Soc.*, **45**, 1159 (1949).
7. W. J. Jones, D. P. Evans, T. Gulwell, and D. C. Griffiths, *J. Chem. Soc.*, 39 (1935).
8. A. S. Carson, E. M. Carson, and B. R. Wilmshurst, *Nature (London)*, **170**, 320 (1952).
9. K. Hartley, H. O. Pritchard, and H. A. Skinner, *Trans. Faraday. Soc.*, **46**, 1019 (1950).
10. A. S. Carson and B. R. Wilmshurst, *J. Chem. Thermodyn.*, **3**, 251 (1971).
11. D. M. Fairbrother and H. A. Skinner, *Trans. Faraday. Soc.*, **52**, 956 (1956).
12. C. L. Chernick, H. A. Skinner, and I. Wadsö, *Trans. Faraday. Soc.*, **52**, 1088 (1956).
13. W. H. Johnson, M. V. Kilday, and E. J. Prosen, *J. Res. Nat. Bur. Stand.*, **65A**, 215 (1961).
14. G. L. Gal'chenko and N. S. Zaugol'nikova, *Russ. J. Phys. Chem.*, **41**, 538, 1174 (1967).
15. G. L. Gal'chenko, L. N. Martynovskaya, and V. I. Stanko, *Russ. J. Gen. Chem.*, **40**, 2397 (1970).
16. J. E. Bennett and H. A. Skinner, *J. Chem. Soc.*, 2472 (1961).
17. J. V. Davies, A. E. Pope, and H. A. Skinner, *Trans. Faraday. Soc.*, **59**, 2233 (1963).
18. D. S. Barnes, D. A. Pittam, G. Pilcher, H. A. Skinner and Y. Virmani, *J. Less-Common Met.*, **36**, 177; **38**, 53 (1974).
19. D. A. Pittam, G. Pilcher, D. S. Barnes, H. A. Skinner, and D. Todd, *J. Less-Common Met.*, **42**, 217 (1975).
20. A. K. Fischer, F. A. Cotton, and G. Wilkinson, *J. Amer. Chem. Soc.*, **79**, 2044 (1957).
21. K. W. Sykes and S. C. Townshend, *J. Chem. Soc.*, 2528 (1955).
22. F. A. Cotton, A. K. Fischer, and G. Wilkinson, *J. Am. Chem. Soc.*, **81**, 800, (1959).
23. V. I. Tel'noi, I. B. Rabinovich, B. G. Gribov, A. S. Pashinkin, B. A. Salamatin, and V. I. Chernova, *Russ. J. Phys. Chem.*, **46**, 465 (1972).
24. J. A. Connor, H. A. Skinner, and Y. Virmani, *J. Chem. Soc. Faraday* Trans. I, **69**, 1218 (1973).
25. A. K. Fischer, F. A. Cotton, and G. Wilkinson, *J. Phys. Chem.*, **63**, 154 (1959).
26. E. O. Fischer and S. Schreiner, *Chem. Ber.*, **91**, 2213 (1958).
27. V. I. Tel'noi, I. B. Rabinovich, V. N. Latyaeva, and A. N. Lineva, *Dokl. Akad. Nauk SSSR*, **197**, 353 (1971).
28. Yu. Kh. Shaulov, G. O. Shmyreva, and V. S. Tubyanskaya, *Russ. J. Phys. Chem.*, **39**, 51 (1965).
29. S. Pawlenko, *Chem. Ber.*, **100**, 3951 (1967).
30. C. T. Mortimer and P. Sellers, *J. Chem. Soc.*, 1978 (1963).
31. P. A. Fowell, *PhD Thesis*, University of Manchester, 1961.
32. G. O. Shmyreva, G. B. Sakharovskaya, A. F. Popov, N. N. Korneev, and A. A. Smolyaninova, *Russ. J. Phys. Chem.*, **45**, 260 (1971).
33. W. D. Good, J. L. Lacina, B. L. De Prater, and J. P. McCullough, *J. Phys. Chem.*, **68**, 579 (1964).
34. P. Cross, C. Haymann, and J. T. Bingham, *Trans. Faraday. Soc.*, **62**, 2338 (1966).
35. W. D. Good and D. W. Scott, *Experimental Thermochemistry*, Vol. 2, Interscience, New York, 1962, Chapter 4.
36. B. S. Iseard, J. B. Pedley, and J. A. Treverton, *J. Chem. Soc. A*, 3095 (1971).
37. W. V. Steele, *J. Chem. Thermodyn.*, **10**, 445 (1978).
38. S. N. Hajiev and M. J. Agarunov, *J. Organomet. Chem.*, **11**, 415 (1968).
39. J. C. Baldwin, M. F. Lappert, J. B. Pedley, and J. A. Treverton, *J. Chem. Soc. A*, 1980 (1967).
40. J. L. Bills and F. A. Cotton, *J. Phys. Chem.*, **68**, 806 (1964).
41. G. P. Adams, A. S. Carson and P. G. Laye, *Trans. Faraday Soc.*, **65**, 113 (1969).
42. A. S. Carson, E. M. Carson, P. G. Laye, J. A. Spencer, and W. V. Steele, *Trans. Faraday Soc.*, **66**, 2459 (1970).

43. W. D. Good, D. W. Scott, and G. Waddington, *J. Phys. Chem.*, **60**, 1090 (1956).
44. W. D. Good, D. W. Scott, J. L. Lacina, and J. P. McCullough, *J. Phys. Chem.*, **63**, 1139 (1959).
45. A. S. Carson, P. G. Laye, J. A. Spencer, and W. V. Steele, *J. Chem. Thermodyn.*, **4**, 783 (1972).
46. C. T. Mortimer and P. Sellers, *J. Chem. Soc.*, 1965 (1964).
47. W. V. Steele, *J. Chem. Thermodyn.*, **11**, 187 (1979).
48. D. S. Barnes and C. T. Mortimer, *J. Chem. Thermodyn.*, **5**, 371 (1963).
49. W. D. Good, D. M. Fairbrother, and G. Waddington, *J. Phys. Chem.*, **62**, 853 (1958).
50. A. S. Carson, K. Hartley, and H. A. Skinner *Proc. R. Soc. (London)*, **A195**, 500 (1949).
51. P. A. Fowell and C. T. Mortimer, *J. Chem. Soc.*, 3793 (1961).
52. H. S. Hull, A. F. Reid, and A. G. Turnbull, *Inorg. Chem.*, **6**, 805 (1967).
53. H. S. Hull and A. G. Turnbull, *Inorg. Chem.*, **6**, 2020 (1967).
54. F. A. Adedeji, K. J. Cavell, S. Cavell, J. A. Connor, G. Pilcher, H. A. Skinner, and M. T. Zafarani-Moattar, *J. Chem. Soc. Faraday Trans. I*, **75**, 603 (1979).
55. K. J. Cavell, C. D. Garner, G. Pilcher, and S. Parkes, *J. Chem. Soc. Dalton Trans.*, 1914 (1979).
56. J. B. Pedley, H. A. Skinner, and C. L. Chernick, *Trans. Faraday Soc.*, **53**, 1612 (1957).
57. A. Cartner, B. Robinson, and P. J. Gardner, *J. Chem. Soc. Chem. Commun.*, 317 (1973).
58. H. A. Skinner, *Rec. Trav. Chim.*, **73**, 991 (1954).
59. K. Hartley, H. O. Pritchard, and H. A. Skinner, *Trans. Faraday Soc.*, **47**, 254 (1951).
60. G. A. Nash, H. A. Skinner, and W. F. Stack, *Trans. Faraday Soc.*, **61**, 640 (1965).
61. J. A. Connor, H. A. Skinner, and Y. Virmani, *J. Chem. Soc., Faraday Trans. I*, **68**, 1754 (1972).
62. J. A. Connor, H. A. Skinner, and Y. Virmani, *Faraday Symposium 8, High Temperature Chemistry*, 18 (1974).
63. M. T. Zafarani-Moattar, *PhD Thesis*, University of Manchester, 1979.
64. F. A. Adedeji, D. L. S. Brown, J. A. Connor, M. Leung, I. M. Paz-Andrade, and H. A. Skinner, *J. Organomet. Chem.*, **97**, 221 (1975).
65. G. Beech, C. T. Mortimer, and E. G. Tyler, *J. Chem. Soc. A*, 925 (1967).
66. G. Beech, S. J. Ashcroft, and C. T. Mortimer, *J. Chem. Soc. A*, 929 (1967).
67. G. Beech, C. T. Mortimer, and E. G. Tyler, *J. Chem. Soc. A*, 1111 (1967).
68. G. Beech and C. T. Mortimer, *J. Chem. Soc. A*, 1115 (1967).
69. G. Beech, C. T. Mortimer, and E. G. Tyler, *J. Chem. Soc. A*, 512 (1969).
70. S. J. Ashcroft, G. Beech, and A. Maddock, *Conference de Thermodynamique Chimique*, Soc. Chim. France, Bordeaux, 1972.
71. C. T. Mortimer, J. L. McNaughton, J. Burgess, M. J. Hacker, R. D. W. Kemmitt, M. I. Bruce, G. Shaw, and F. G. A. Stone, *J. Organomet. Chem.*, **47**, 439 (1973).
72. Yu. A. Lebedev, E. A. Miroschnichenko, and A. M. Chaikin, *Dokl. Akad. Nauk SSSR*, **145**, 751 (1962).
73. R. Juza and E. Hillenbrand, *Z. Anorg. Chem.*, **273**, 297 (1953).
74. G. K. Johnson, E. H. van Deventer, J. P. Ackerman, W. N. Hubbard, D. W. Osborne, and H. L. Flotow, *J. Chem. Thermodyn.* **5**, 57 (1973).
75. M. Ader and W. N. Hubbard, *J. Chem. Thermodyn.*, **5**, 607 (1973).
76. C. T. Mortimer, H. O. Pritchard, and H. A. Skinner, *Trans. Faraday Soc.*, **48**, 220 (1952).
77. A. E. Pope and H. A. Skinner, *J. Chem. Soc.*, 3704 (1963).
78. G. L. Gal'chenko, N. S. Zaugol'nikova, S. M. Skuratov, L. S. Vasil'ev, Yu. N. Bubnov, and B. M. Mikhailov, *Dokl. Akad. Nauk. SSSR*, **169**, 715 (1966).
79. G. L. Gal'chenko and R. M. Varushchenko, *Zh. Fiz. Khim.*, **37**, 1355 (1963).
80. E. A. Haseley, A. B. Garrett, and H. H. Sisler, *J. Phys. Chem.*, **60**, 1136 (1956).
81. A. Finch, P. J. Gardner, E. J. Pearn, and G. B. Watts, *Trans. Faraday Soc.*, **63**, 1880 (1967).
82. G. L. Gal'chenko, B. K. Lavlova, and V. I. Stanko, USSR Calorimetry Conference, Tiblisi, April 1973; E. A. Miroshnichenko, B. K. Lavlova, Yu. A. Lebedev, V. I. Stanko, and G. L. Gal'chenko, *Termokhim. Org. Svedin*, **5**, 3 (1976).
83. G. L. Gal'chenko, V. K. Pavlovich, and V. N. Siryatskaya, *Termokhim. Org. Svedin*, **7**, 38 (1978).
84. G. L. Gal'chenko, N. S. Zaugolnikova, S. M. Skuratov, L. S. Vasil'ev, A. Ya. Bezmanov, and B. M. Mikhailov, *Dokl. Akad. Nauk. SSSR*, **166**, 12 (1966).

85. H. A. Skinner and T. F. S. Tees, *J. Chem. Soc.*, 3378 (1953).
86. G. L. Gal'chenko, R. M. Varuschchenko, Yu. N. Bubnov, and B. M. Mikhailov, *Zh. Obsch. Khim.*, **32**, 284, (1962).
87. A. Finch and P. J. Gardner, *Trans. Faraday Soc.*, **62**, 3314 (1966).
88. G. L. Gal'chenko, R. M. Varushchenko, Yu. N Bubnov, and B. M. Mikhailov, *Zh. Obsch. Khim.*, **32**, 2373 (1962).
89. G. L. Gal'chenko, M. M. Ammar, S. M. Skuratov, Yu. N. Bubnov, and B. M. Mikhailov, *Vestn. Mosk. Univ., Ser. II, Khim.*, **20**, 3 (1965).
90. G. L. Gal'chenko, L. N. Martynovskaya, V. I. Stanko, and A. I. Klimova, *Dokl. Akad. Nauk. SSSR*, **193**, 483 (1970).
91. W. D. Good and M. Mansson, *J. Phys. Chem.*, **70**, 97 (1966).
92. N. K. Smith and W. D. Good, *J. Chem. Eng. Data*, **12**, 570 (1967).
93. G. L. Gal'chenko, M. M. Ammar, S. M. Skuratov, Yu. N. Bubnov, and B. M. Mikhailov, *Vestn. Mosk. Univ., Ser. II, Khim.*, **20**, 10 (1965).
94. A. Finch, P. J. Gardner, and E. J. Pearn, *Trans. Faraday. Soc.*, **62**, 1072 (1966).
95. G. O. Shmyreva, R. M. Golosova, G. Sakharovskaya, A. F. Popov, N. N. Korneev, and A. A. Smolyanmova, *Russ. J. Phys. Chem.*, **48**, 447 (1974).
96. K. J. Cavell and G. Pilcher, *J. Chem. Soc., Faraday Trans. I*, **73**, 1590 (1977).
97. M. B. Smith, *J. Organomet. Chem.*, **76**, 171 (1974).
98. C. G. Devyatykh, I. B. Rabinovich, V. I. Tel'noi, G. K. Borisov, and L. F. Zyazina, *Dokl. Akad. Nauk. SSSR*, **217**, 673 (1974).
99. L. H. Long and J. F. Sackman, *Trans. Faraday Soc.*, **54**, 1797 (1958).
100. G. M. Kol'yakova, I. B. Rabinovich, and E. N. Zorina, *Dokl. Akad. Nauk. SSSR.*, **209**, 245 (1973).
101. P. A. Fowell and C. T. Mortimer, *J. Chem. Soc.*, 3734 (1958).
102. W. D. Clarke and S. J. W. Price, *Can. J. Chem.*, **46**, 1633 (1968).
103. V. I. Tel'noi and I. B. Rabinovich, *Usp. Khim.*, **46**, 1337 (1977).
104. V. I. Tel'noi, I. B. Rabinovich, V. D. Tikhonov, V. N. Latyaeva, L. I. Vyshinskaya, and G. A. Razuvaev, *Dokl. Akad. Nauk. SSSR*, **174**, 1374 (1969).
105. K. V. Kir'yanov, V. I. Tel'noi, G. A. Vasil'eva, and I. B. Rabinovich, *Dokl. Akad. Nauk. SSSR*, **231**, 1021 (1976).
106. J. B. Pedley and B. S. Iseard, *CATCH Tables, Silicon Compounds*, University of Sussex, Brighton, 1972.
107. A. P. Claydon and C. T. Mortimer, *J. Chem. Soc.*, 3212 (1962).
108. A. E. Beezer and C. T. Mortimer, *J. Chem. Soc. A*, 514 (1966).
109. M. J. Agarunov and S. N. Hadjiev, *Dokl. Akad. Nauk. SSSR*, **185**, 221 (1969).
110. M. A. Ring, H. E. O'Neal, A. H. Kadhim, and F. Jappe, *J. Organomet. Chem.*, **5**, 124 (1966).
111. Yu. Kh. Shaulov, A. K. Federov, and V. G. Genchel, *Russ. J. Phys. Chem.*, **43**, 744 (1969).
112. A. E. Pope and H. A. Skinner, *Trans. Faraday Soc.*, **60**, 1404 (1964).
113. I. B. Rabinovich, V. I. Tel'noi, N. V. Karyakin, and G. A. Razuvaev, *Dokl. Akad. Nauk. SSSR*, **149**, 217 (1963).
114. I. B. Rabinovich, E. G. Kiparisov, and Yu. A. Aleksandrov, *Dokl. Akad. Nauk. SSSR*, **200**, 842 (1971).
115. E. A. Volchkova, D. D. Smol'yanimova, V. G. Genchel, K. Lapatkina, and Yu. Kh. Shaulov, *Russ. J. Phys. Chem.*, **46**, 1053 (1972).
116. V. G. Genchel, A. I. Toporkova, Yu. Kh. Shaulov, and D. D. Smol'yanimova, *Russ. J. Phys. Chem.*, **48**, 1085 (1974).
117. G. M. Kol'yakova, I. B. Rabinovich, and N. S. Vyazankin, *Dokl. Akad. Nauk. SSSR*, **200**, 735 (1971).
118. G. M. Kol'yakova, I. B. Rabinovich, E. N. Gladyshev, and N. S. Vyazankin, *Dokl. Akad. Nauk SSSR*, **204**, 419 (1972).
119. J. B. Pedley and H. A. Skinner, *Trans. Faraday Soc.*, **55**, 544 (1959).
120. D. J. Coleman and H. A. Skinner, *Trans. Faraday Soc.*, **62**, 1721 (1966).
121. W. F. Lautsch, A. Tröber, W. Zimmer, L. Mehner, W. Linck, H. M. Lehmann, H. Brandenberger, H. Korner, H.-J. Metschker, K. Wagner, and R. Kaden, *Z. Chem.*, **3**, 415 (1963).
122. G. P. Adams, A. S. Carson, and P. G. Laye, *J. Chem. Thermodyn.*, **1**, 393 (1969).

123. V. I. Tel'noi and I. B. Rabinovich, *Russ. J. Phys. Chem.*, **40**, 842 (1966).
124. A. S. Carson, P. G. Laye, J. A. Spencer, and W. V. Steele, *J. Chem. Thermodyn.*, **2**, 659 (1970).
125. R. S. Butler, A. S. Carson, P. G. Laye, and W. V. Steele, *J. Chem. Thermodyn.*, **8**, 1153 (1976).
126. E. O. Fischer and A. Reckziegel, *Chem. Ber.*, **94**, 2204 (1961).
127. F. A. Adedeji, J. A. Connor, H. A. Skinner, L. Galyer, and G. Wilkinson, *J. Chem. Soc., Chem. Commun.*, 159 (1976).
128. L. H. Long and J. F. Sackman, *Trans. Faraday Soc.*, **563**, 1606 (1957).
129. C. L. Chernick and H. A. Skinner, *J. Chem. Soc.*, 1401 (1956).
130. A. F. Bedford, D. M. Heinekey, I. T. Millar, and C. T. Mortimer, *J. Chem. Soc.*, 2932 (1962).
131. A. F. Bedford and C. T. Mortimer, *J. Chem. Soc.*, 1622 (1960).
132. A. P. Claydon, P. A. Fowell, and C. T. Mortimer, *J. Chem. Soc.*, 3284 (1960).
133. A. D. Starostin, A. V. Nikolaev, and Yu. A. Afanas'ev, *Izvest. Akad. Nauk. SSSR, Ser. Khim.*, 1255 (1966).
134. E. Neale and L. T. D. Williams, *J. Chem. Soc.*, 2485 (1955).
135. E. Neale, L. T. D. Williams, and V. T. Moores, *J. Chem. Soc.*, 422 (1956).
136. A. F. Bedford and C. T. Mortimer, *J. Chem. Soc.*, 4649 (1960).
137. L. H. Long and J. F. Sackman, *Trans. Faraday Soc.*, **52**, 1201 (1956).
138. L. H. Long and J. F. Sackman, *Trans. Faraday Soc.*, **51**, 1062 (1955).
139. K.-H. Birr, *Z. Anorg. Allg. Chem.*, **306**, 21 (1960).
140. L. H. Long and J. F. Sackman, *Trans. Faraday Soc.*, **50**, 1177 (1954).
141. V. I. Tel'noi, K. V. Kirynov, V. I. Ermolaev, and I. B. Rabinovich, *Dokl. Akad. Nauk. SSSR*, **220**, 1088 (1975).
142. V. I. Tel'noi, I. B. Rabinovich, and V. A. Umilin, *Dokl. Akad. Nauk. SSSR*, **209**, 197 (1973).
143. J. A. Connor, J. A. Martinho-Simoes, H. A. Skinner, and M. T. Zarfarani-Moattar, *J. Organomet. Chem.*, **179**, 331 (1979).
144. D. L. S. Brown, J. A. Connor C. P. Demain, M. L. Leung, J. A. Martinho-Simoes, H. A. Skinner, and M. T. Zafarani-Moattar, *J. Organomet. Chem.*, **142**, 321 (1977).
145. M. T. Zafarani-Moattar, J. A. Connor, and H. A. Skinner, *J. Organomet. Chem.*, in press
146. J. O. Hill and R. J. Irving, *J. Chem. Soc. A*, 1413 (1967).
147. F. A. Adedeji, J. A. Connor, C. P. Demain, J. A. Martinho-Simoes, H. A. Skinner, and M. T. Zafarani-Moattar, *J. Organomet. Chem.*, **149**, 333 (1978).
148. J. C. G. Calado, A. R. Dias, J. A. Martinho-Simoes, and M. A. V. Ribeiro da Silva, *J. Organomet. Chem.*, **142**, 321 (1979).
149. J. C. G. Calado, A. R. Dias, J. A. Martinho-Simoes and M. A. V. Ribeiro da Silva, *J. Chem. Soc., Chem. Commun.*, **737** (1978).
150. K. J. Cavell, C. D. Garner, J. A. Martinho-Simoes, G. Pilcher, H. Al-Samman, H. A. Skinner, G. Al-Takhin, I. B. Walton, and M. T. Zafarani-Moattar, *J. Chem. Soc. Faraday Trans. I*, **77**, 2927 (1981).
151. G. Pilcher, K. J. Cavell, C. D. Garner, and S. Parkes, *J. Chem. Soc., Dalton Trans.*, 1311 (1978).
152. V. I. Tel'noi, I. B. Rabinovich, K. V. Kir'yanov, and A. S. Smirnov, *Dokl. Akad. Nauk. SSSR*, **231**, 733 (1976).
153. K. J. Cavell, J A. Connor, G. Pilcher, M. A. V. Ribeiro da Silva, M. D. M. C. Ribeiro da Silva, H. A. Skinner, Y. Virmani, and M. T. Zafarani-Moattar, *J. Chem. Soc., Faraday Trans. I.*, **77**, 1585 (1981).
154. G. Gattow and M. Dräger, *Z. Anorg. Allg. Chem.*, **343**, 232 (1966).
155. H. Merten and H. Schlüter, *Chem. Ber.*, **69**, 1364 (1936).
156. E. V. Evstigneeva and G. O. Shmyreva, *Russ. J. Phys. Chem.*, **39**, 529 (1965).
157. D. L. S. Brown, J. A. Connor, and H. A. Skinner, *J. Organomet. Chem.*, **81**, 403 (1974).
158. J. O. Hill and R. J. Irving, *J. Chem. Soc. A*, 3116 (1968).
159. D. L. S. Brown, J. A. Connor, M. L. Leung, M. I. Paz-Andrade, and H. A. Skinner, *J. Organomet. Chem.*, **110**, 79 (1976).
160. J. A. Connor, C. P. Demain, H. A. Skinner, and M. T. Zafarani-Moattar, *J. Organomet. Chem.*, **170**, 117 (1979).

161. J. O. Hill and R. J. Irving, *J. Chem. Soc. A.*, 1052 (1968).
162. Yu. E. Bronshstein, V. Yu. Gankin, D. P. Krinkin, and D. M. Rudkovskii, *Russ. J. Phys. Chem.*, **40**, 802 (1966).
163. P. J. Gardner, A. Cartner, R. G. Cunningham, and B. H. Robinson, *J. Chem. Soc., Dalton* Trans., 2582 (1978)
164. D. L. S. Brown, J. A. Connor, and H. A. Skinner, *J. Chem. Soc.*, Faraday Trans. **I, 71**, 699 (1975).
165. J. C. G. Calado, A. R. Dias, M. S. Salema, and J. A. Martinho-Simoes, *J. Chem. Soc., Dalton Trans.*, 1174 (1981).
166. A. R. Dias, J. A. Martinho-Simoes, and C. Teixeira, *J. Chem. Soc., Dalton Trans.*, 1178 (1981).
167. K. J. Laidler, *Can. J. Chem.*, **34**, 626 (1956).
168. CODATA, *J. Chem. Thermodyn.*, **10**, 903 (1978).
169. I. Dellien, F. M. Hall, and L. G. Hepler, *Chem. Rev.*, **76**, 283 (1976).
170. E. R. Plante and A. B. Sessoms, *J. Res. Nat. Bur. Stand.*, **77A**, 237 (1973).
171. D. J. De Frees, R. T. McIver, and W. J. Hehre, *J. Am. Chem. Soc.*, **102**, 3334 (1980).
172. L. J. Guggenberger and F. N. Tebbe, *J. Am. Chem. Soc.*, **98**, 4137 (1976).
173. V. P. Glusko and B. A. Medvedev, Thermal Constants of Chemical Compounds, Vol. 7, *Akad. Nauk SSSR*, Moscow, 1974.
174. I. A. Ponova and N. B. Alekseev, *Dokl. Akad. Nauk. SSSR*, **174**, 614 (1967).
175. D. R. Bidinosti and N. S. McIntyre, *Can. J. Chem.*, **45**, 641 (1967).
176. D. R. Bidinosti and N. S. McIntyre, *J. Chem. Soc., Chem. Commun.*, 1 (1967).
177. H. J. Svec and G. A. Junk, *J. Chem. Soc. A*, 2102 (1970).
178. T. L. Allen, *J. Chem. Phys.*, **31**, 1039 (1959).
179. A. M. Doncaster and R. Walsh, *J. Chem. Soc., Faraday Trans. I*, **75**, 1126 (1979).
180. A. M. Doncaster and R. Walsh, *J. Chem. Soc. Chem. Commun.*, 904 (1979).
181. F. A. Cotton, *Chem. Soc. Rev.*, **4**, 27 (1975).
182. L. V. Gurvich, G. V. Karachevtsev, V. N. Kondratiev, Yu. A. Lebedev, V. A. Medvedev, V. K. Potapov, and Yu. S. Hodiev, *Energies of Dissociation of Chemical Bonds*, Akad. Nauk SSSR, Moscow, 1974.
183. J. A. Connor, G. Pilcher, H. A. Skinner, M. H. Chisholm, and F. A. Cotton, *J. Am. Chem. Soc.*, **100**, 7738 (1978).
184. D. C. Bradley, M. H. Chisholm, and M. W. Extine, *Inorg. Chem.*, **16**, 1791 (1977).
185. M. H. Chisholm, F. A. Cotton, and M. W. Extine, *Inorg. Chem.*, **17**, 1329 (1978).
186. C. D. Garner, I. B. Walton, S. Parkes, and W. Clegg, *Inorg. Chim. Acta*, **31**, L451 (1978).
187. S. W. Benson and J. H. Buss, *J. Chem. Phys.*, **29**, 546 (1958).
188. J. D. Cox, *Tetrahedron*, **19**, 1175 (1963).
189. A. Evans, C. T. Mortimer, and R. J. Puddephat, *J. Organomet. Chem.*, **72**, 295 (1974).
190. A. Evans, C. T. Mortimer, and R. J. Puddephat, *J. Organomet. Chem.*, **85**, 101 (1975).
191. C. T. Mortimer, M. P. Wilkinson, and R. J. Puddephat, *J. Organomet. Chem.*, **102**, C43 (1975).
192. C. T. Mortimer, M. P. Wilkinson, and R. J. Puddephat, *J. Organomet. Chem.*, **65**, 265 (1979).
193. A. Evans, C. T. Mortimer, and R. J. Puddephat, *J. Organomet. Chem.*, **96**, C58 (1975).
194. M. J. S. Dewar and H. N. Schmeising, *Tetrahedron*, **11**, 96 (1960).
195. A. Langseth and B. P. Stoicheff, *Can. J. Phys.*, **34**, 350 (1956).
196. M. H. Chisholm, F. A. Cotton, M. W. Extine, and W. W. Reichert, *Inorg. Chem.*, **17**, 2944 (1978).
197. A. R. Dias, M. S. Salema and J. A. Martinho-Simoes (unpublished results).
198. J. C. G. Calado, A. R. Dias, M. E. Minas da Piedade and J. A. Martinho-Simoes, *Rev. Port. Quim.*, in press.
199. J. C. G. Calado, A. R. Dias and J. A. Martinho-Simoes, *J. Organomet. Chem.*, **195**, 203 (1980).
200. J. C. G. Calado, A. R. Dias, J. A. Martinho-Simoes and M. A. V. Ribeiro da Silva, *Rev. Port. Quim.*, **21**, 129 (1979).

The Chemistry of the Metal–Carbon Bond
Edited by F. R. Hartley and S. Patai

CHAPTER **3**

Synthesis of ylide complexes

LOTHAR WEBER
Fachbereich Chemie, Universität Essen, Universitätstrasse 5–7, D-4300 Essen 1, Federal Republic of Germany

---

I. INTRODUCTION

Classical organic ylides are neutral molecules possessing a carbanionic function in an α-position to an onium centre. The onium function can be represented by all the Group Va elements, as well as by sulphur, selenium, tellurium, and iodine. The most common method of organic ylide generation consists in proton abstraction from the corresponding onium salts $(R_3{}^3E—CR^1R^2H)^+X^-$. This chapter is mainly concerned with ylide compounds in which the negative charge, adjacent to an onium centre, is located at a carbon atom. In Section **IV** base-stabilized germylenes and stannylenes, which are formally analogous to classical ylides, will be given brief consideration. Metal complexes of compounds such as $R_3\overset{+}{N}—\overset{-}{\ddot{N}}R$, $R_3P{=}NR$, $R_2S{=}NR$, etc., with the isoelectronic imino group in an α-position to the onium centre are omitted because they do not contain new metal–carbon bonds.

In non-stabilized ylides the negative charge achieves stabilization only by the inductive effect and the π-acceptor capacity of the adjacent onium atoms. The extreme of this interaction is shown in the ylene resonance formula **B** in Figure 1. In contrast, compounds in which the negative charge is further delocalized by anion-stabilizing carbonyl, cyano, sulphonyl, nitro, or vinyl substituents are regarded as stabilized ylides. The chemical and physico-chemical properties of ylidic compounds can in most cases be described with aid of the zwitterionic resonance formula **A** (Figure 1). In

$$R^1R^2\overset{-}{\ddot{C}}—\overset{+}{E}R^3_n \longleftrightarrow R^1R^2C{=}ER^3_n$$

(**A**) (ylide) (**B**) (ylene)

FIGURE 1. Resonance structures of simple ylides.

this chapter the ylene formula is also used.

Although phosphorus[1] and sulphur[2] ylides have been known as extremely useful reagents in organic synthesis for about 30 years, their inorganic and organometallic chemistry were neglected for a long time. Only within the last decade has the coordination chemistry of ylides developed to a wide and rapidly expanding field of research which has attracted a great deal of attention[3]. Almost all metals form a wide variety of novel types of complexes containing the ylide ligand, in which the latter functions as an extremely powerful electron donor. In these complexes the metal–carbon bonds generally exhibit surprising thermal and chemical stability and thus differ from other organometallics to a significant extent. Some ylide complexes serve as excellent catalysts[4–9] and others may be useful in pharmacology[10].

Figure 2 illustrates the positions of an ylidic species $R^1R^2C{=}ER^3_{n-1}(CR^4_2H)$ to which metals (M) can be attached.

The simplest way of bonding the ylide to a metal can be best described in terms of the donation of the lone-pair electrons of the ylide carbon atom to the metal atom. This coordination is accompanied by a change in the hybridization of the ylide carbon atom from sp^2 to sp^3 in the resulting metal–carbon σ-bond.

Ylides with anion-stabilizing substituents such as R^1 = vinyl or $R^1R^2 = —(CH{=}CH)_2—$ can be best regarded as allylide or cyclopentadienide anions which experience intramolecular stabilization by the adjacent onium centre. As η^3- and η^5-ligands they are capable of forming organometallic π-complexes (Figure 3). In both σ- and π-complexes the metal is linked to the ylide at the same position as is a proton in the corresponding onium salts. Therefore, the previous complexes (**IIA** in Figure 2) are described as metal-substituted onium salts in the case of cationic com-

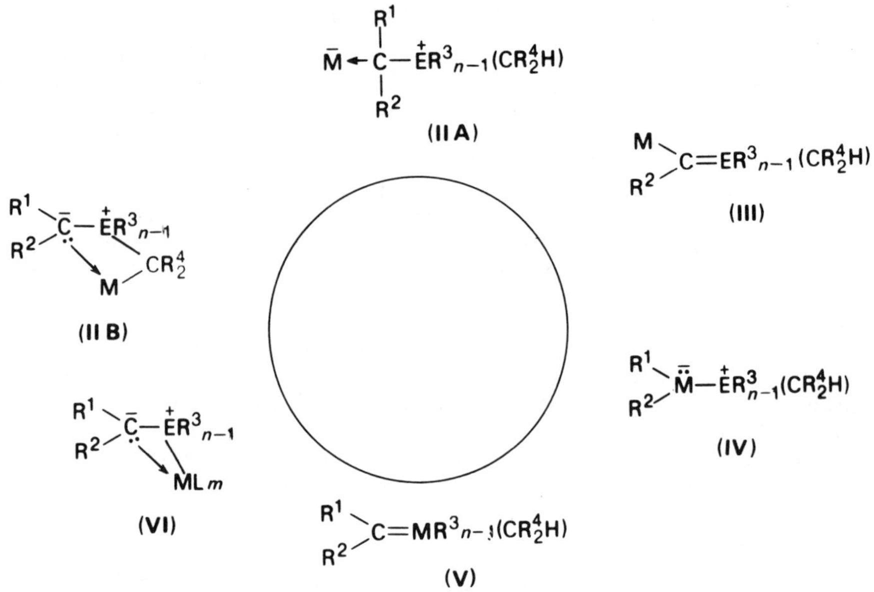

FIGURE 2. An illustration of the different sites a metal atom can occupy in an ylidic molecule. The Roman numerals refer to the respective sections of this chapter.

FIGURE 3. π-Interactions of vinyl- and cyclopentadiene-substituted ylides and metal atoms.

pounds such as **1**[11], or as onium-metallates in the case of neutral betain-like species such as **2**[12].

$$[(\eta^5\text{-Cp})(CO)_2Fe{-}CH_2\overset{+}{P}Ph_3]^+BF_4^- \quad [(CO)_5\overset{-}{Cr}{-}CH_2\overset{+}{P}Ph_3]$$
$$\textbf{(1)} \qquad\qquad\qquad\qquad \textbf{(2)}$$

Complexes such as **III** in Figure 2 are obtained if one or both substituents R^1 and R^2 at the carbanionic centre are replaced by organometallic fragments. Metal-substituted ylides are known for main-group metals as well as for transition metals.

In the so-called classical ylides the carbon atom always carries the anionic function. If the carbon atom is substituted by its higher homologues, especially by metals such as germanium and tin, a novel class of ylides is established in the base-stabilized germylenes and stannylenes (Figure 4). The metal now takes the position of the carbanionic centre (**IV** in Figure 2). The formal analogy of such ylides to the classical ones finds additional experimental support in the trapping of the bivalent species CX_2[13–16], GeX_2[17], and SnX_2[18] with tertiary phosphines.

The replacement of phosphorus or sulphur in classical ylides by main-group or transition metals leads to ylides in which the onium centre is formed by metal atoms. Metal–onium ylides (**V** in Figure 2) are known for arsenic, antimony, bismuth, and

$R^1R^2\ddot{\bar{G}}e-\overset{+}{E}R^3_n$ $R^1R^2\ddot{\bar{S}}n-\overset{+}{E}R^3_n$

FIGURE 4. Base-stabilized germylenes and stannylenes as germanium and tin ylides.

tellurium. Most surprisingly, the recently discovered niobium and tantalum alkylidenes also exhibit ylidic properties. In sharp contrast to Fischer's carbene complexes, the alkylidene carbon atom in compounds such as $R^1R^2R^3Ta{=}CHR$ displays nucleophilic character.

Another type of coordination compound containing an ylidic ligand is obtained by displacing one of the organic substituents at the onium centre by an organometallic group (**VI** in Figure 2). As the carbanionic centre in such ylides is a powerful donor, it may undergo an additional coordinative interaction to form a three-membered ring (**3**)[19a,b] or a metallocycle with a bridging ylidic unit (**4**)[2]. This is illustrated in Figure 5. Experimental results have confirmed the analogies between this kind of ylide complex and those of type **IIA**.

$H_2\bar{C}-\overset{+}{P}Me_2$ / $Co(PMe_3)_3$
(**3**)

$H_2\bar{C}-\overset{+}{P}Me_2$ / $(Me_3P)_2Co-Co(PMe_3)_2$ / PMe_2
(**4**)

FIGURE 5. Ylides with organometallic substituents at the onium centre.

The alkylation of a functionalized methyl complex leads to metal–ylide species of type **IIA** whereas complexes such as **3** and **4** result from coordination reactions at the donor function of the alkyl ligand (Scheme 1).

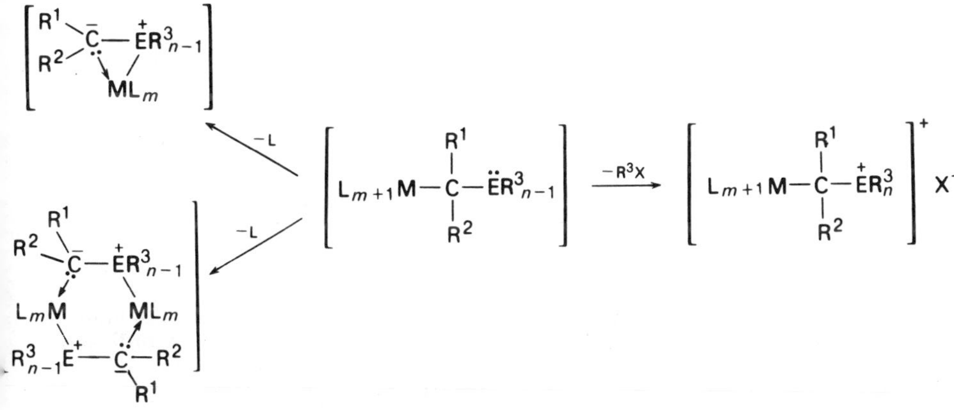

$[L_{m+1}M-C(R^1)(R^2)-\ddot{E}R^3_{n-1}] \xrightarrow{-R^3X} [L_{m+1}M-C(R^1)(R^2)-\overset{+}{E}R^3_n]^+ X^-$

$\xrightarrow{-L}$ $[R^1R^2\bar{C}-\overset{+}{E}R^3_{n-1}$, $ML_m]$ (three-membered ring)

$\xrightarrow{-L}$ $[R^1R^2\bar{C}-\overset{+}{E}R^3_{n-1}$, L_mM, ML_m, $R^3_{n-1}\overset{+}{E}-\bar{C}(R^1)-R^2]$ (six-membered ring)

SCHEME 1

Proton abstraction from the organic substituents of the onium centre by strongly basic organometallics with subsequent metallation leads to another possibility of ylide–metal interaction (Scheme 2). The nucleophilicity of the carbanionic carbon and

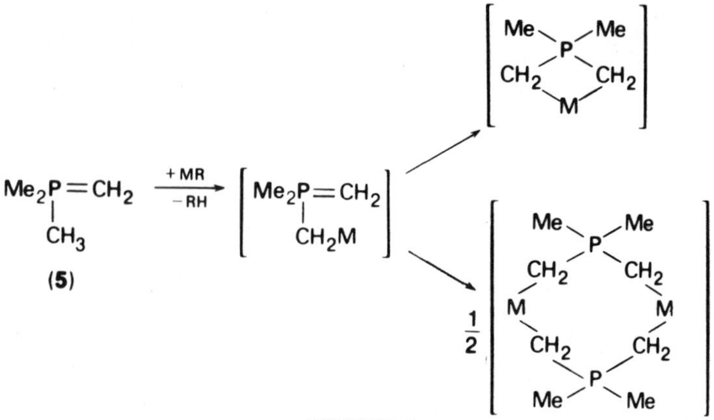

$Me_2P{=}CH_2$ (with CH_3) **(5)** $\xrightarrow[-RH]{+MR}$ $[Me_2P{=}CH_2$ (with CH_2M)$]$ → $[Me_2P(CH_2)_2M]$ or $\frac{1}{2}[Me_2P(CH_2M CH_2)_2PMe_2]$

SCHEME 2

the electrophilicity of the metal generally result in stabilization of the so generated species by chelation[21], or dimerization to eight-membered heterocycles[22] (see also Section II.B).

Three-fold lithiated ylides of the type $(LiCH_2)_3ECH_2$ (E = N[23], P[24]), accessible by multiple metallation of the free ylides or the corresponding onium salts, are described as reactive intermediates.

II. ONIUM-METALLATES AND METAL-SUBSTITUTED ONIUM SALTS

A. Non-stabilized Ylides as Terminal, Monodentate Ligands

Complexes with monodentate ylides as terminal ligands are accessible via a series of synthetic routes as illustrated in Figure 6, in which the dotted lines indicate the formation of which respective bond eventually leads to the new ylide complex.

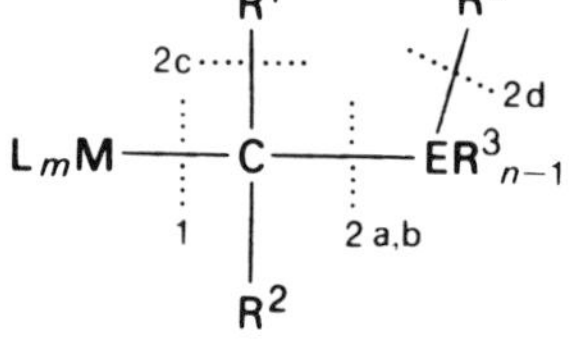

L_mM—C(R^1)(R^2)—ER^3_{n-1}(R^3); bonds labelled 1, 2a,b, 2c, 2d

FIGURE 6. An illustration of synthetic approaches to ylide complexes. Numbers 1 and 2 refer to subheadings in Section II.A.

A simple synthesis of metal–ylide complexes involves the attack of a free intact organic ylide on a metal complex fragment with the formation of bond 1. In indirect metal–ylide complex syntheses, the ylide ligand is constructed in the coordination sphere of the metal. As the existence of the free ylide is of no importance, these procedures are useful for stabilizing unstable ylides by complex formation. Addition of the donor molecules, ER^3_n, to the electrophilic carbene carbon atom of carbene–metal complexes leads, with the formation of bond 2a, to ylide complexes. The displacement of good leaving groups X in compounds of the general type $[L_mMCH_2X]$ by similar donor systems also affords bond formation between the heteroatom and the ylidic carbon (bond 2b). The bond 2c of the species under discussion results from the addition of lithium alkyls, LiR^1, to the ylidic centre in cationic, metal-substituted ylides $[L_mM{=}C(R){-}ER^3_n]^+$. Alkylating the heteroatom of aminomethyl, phosphinomethyl, and thiomethoxymethyl complexes leads to the formation of bond 2d in ylide complexes.

1. Direct syntheses

a. Addition reactions with coordinatively unsaturated metal compounds. Coordinatively unsaturated complexes or metal alkyls containing electron-deficient bonds react with phosphorus ylides to give stable betain-like adducts (equation 1). In the tetra-

$$\underset{(\mathbf{5})}{Me_3PCH_2} + MMe_3 \longrightarrow \underset{(\mathbf{6})}{[Me_3\overset{+}{P}CH_2\overset{-}{M}Me_3]} \qquad (1)$$

$M = Al$[25], Ga, In, Tl[26]

hedral molecule **6**, four metal–carbon σ-bonds exist with surprisingly thermal stability. Generally, the thermal stability of metal–carbon σ-bonds in ylide complexes far exceeds the stability of simple metal alkyl or aryl species; this stability is only comparable to that of corresponding bonds in stable perfluoroalkyl complexes. The stabilization is attributed to the inductive effect and the *d*-orbital acceptor capability of the onium centre in an α-position to the metal–carbon bond. As a result of this, $[Me_3PCH_2TlMe_3]$ decomposes at temperatures higher than 100°C, whereas the conventional tetraalkylthallates TlR_4^- are known as thermolabile systems.

Simple and highly reactive adducts are obtained from the reaction of $Me_3PCHSiMe_3$ (**7**) with zinc and cadmium alkyls (equations 2–4)[22]. The size of the metals and the bulk of the ligands determine the coordination number of **8**, **9** and **10**. Lithium salt-

$$[Cd(CH_2SiMe_3)_2] \xrightarrow{+7} \left[(Me_3SiCH_2)_2Cd{-}CH\begin{matrix} PMe_3 \\ SiMe_3 \end{matrix}\right] \qquad (2)$$

(**8**)

$$[CdMe_2] \xrightarrow{+7} \left[Me_2Cd\left(CH\begin{matrix} PMe_3 \\ SiMe_3 \end{matrix}\right)_2\right] \qquad (3)$$

(**9**)

$$[ZnEt_2] \xrightarrow{+7} \left[Et_2Zn{-}CH\begin{matrix} PMe_3 \\ SiMe_3 \end{matrix}\right] \qquad (4)$$

(**10**)

bearing ylides, synthesized according to equation 5, are best described as lithium–ylide complexes[25].

$$(Me_4P)Cl + n\text{-}C_4H_9Li \longrightarrow Me_3PCH_2{\cdot}LiCl + n\text{-}C_4H_{10} \qquad (5)$$

b. Ligand substitution reactions. Most of the complexes under discussion result from the thermal or photochemical substitution of ligands such as carbonmonoxide, phosphines, olefins, ethers and halides by a suitable ylide. Ylide ligands usually occupy one or more coordination sites in the standard coordination polyhedrons. Thermally stable linear complexes are known with one or two ylide ligands (**11**[27] and **12**[28]). In contrast to the extremely sensitive dialkyl aurate(I) complexes $(RAuR')^-$, complex **11** melts at 119–121°C without decomposition.

With respect to the marked thermolability of dialkyl cuprate(I) complexes $(RCuR')^-$, the stabilization of the CuC_2 moiety by two onium centres is remarkable. Complexes similar to **12**[28] but with triarylphosphonium ylides have also been intensively studied[29,30]

$$[MeAuPMe_3] + \mathbf{5} \longrightarrow PMe_3 + \underset{(\mathbf{11})}{[MeAuCH_2PMe_3]} \qquad (6)$$

$$[Me_3PMCl] + \mathbf{27} \longrightarrow \underset{(\mathbf{12})}{\left[\begin{matrix} Me_3Si \\ Me_3P \end{matrix}\!\!>\!CH{-}M{-}CH\!<\!\!\begin{matrix} PMe_3 \\ SiMe_3 \end{matrix}\right]} Cl \qquad (7)$$

12	a	b	c
M	Cu	Ag	Au
m.p.(°C)	124	141	138–142

The carbonyl ligands of nickel tetracarbonyl are kinetically labile and easily susceptible to substitution (equation 8)[31]. The ylidic ligands in **13** function as powerful donors

$$[Ni(CO)_4] + R^1_3PCHR^2 \longrightarrow \underset{(\mathbf{13})}{[R^1_3PCH(R^2){-}Ni(CO)_3]} + CO \qquad (8)$$

without considerable acceptor properties. The electron density thus accumulated at the nickel is to some extent removed by the remaining three carbonyl ligands via π back-bonding. From i.r. spectroscopic studies it can be deduced that the ylides in complexes **13** behave as stronger donor ligands than $P(n\text{-}Bu)_3$. Proton n.m.r. spectra and the X-ray structure analysis of $[(CO)_3NiCH(Me)P(C_6H_{11})_3]$ confirm the σ-coordination of an sp^3-hybridized ylide carbon atom to the nickel. As a result of π back-bonding the nickel carbonyl carbon bonds are reinforced so that further CO displacement is inhibited (equation 9).

$$[(CO)_3NiCH_2PPh_3] + Ph_3PCH_2 \not\longrightarrow [(CO)_2Ni(CH_2PPh_3)_2] + CO \qquad (9)$$

The reaction of the double ylide **14** with $Ni(CO)_4$ produces **15**, which exhibits a σ-bond between the metal and the central ylide carbon atom.

$$[Ni(CO)_4] + \underset{(\mathbf{14})}{Ph_3P{=}C{=}PPh_3} \longrightarrow \underset{(\mathbf{15})}{[(CO)_3Ni{-}C(PPh_3)_2]} + CO \qquad (10)$$

The neutral tetrahedral complex **9** with two terminal ylide ligands has been mentioned above. Similar ionic compounds involving tetravalent tin centres are known[32].

$$\underset{(\mathbf{16})}{2\,Ph_3PCH_2} + 2\,Me_2SnX_2 \longrightarrow [(Ph_3PCH_2)_2SnMe_2](Me_2SnX_4) \xrightarrow{(NH_4)[Cr(SCN)_4(NH_3)_2]} \underset{(\mathbf{17})}{[(Ph_3PCH_2)_2SnMe_2][Cr(SCN)_4(NH_3)_2]_2} \qquad (11)$$

Metals with d^8-electron configuration such as Ni(II), Pd(II), Pt(II), and Au(III) give rise to well investigated ylide complexes of square-planar configuration (equations 12–14)[33,34].

$$[(Me_3P)_2NiMe_2] + \mathbf{5} \longrightarrow [Me_3P{-}Ni(Me)_2{-}CH_2PMe_3] + PMe_3 \quad (12)$$

(18)

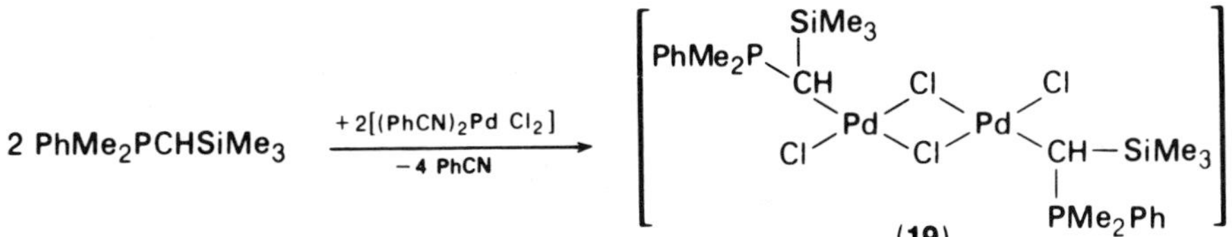

$$2\ PhMe_2PCHSiMe_3 \xrightarrow[-4\ PhCN]{+2[(PhCN)_2PdCl_2]} [(PhMe_2P)(Me_3Si)CH(Cl)Pd(\mu\text{-}Cl)_2Pd(Cl)CH(SiMe_3)(PMe_2Ph)] \quad (13)$$

(19)

$$\mathbf{19} + 2\ L{-}L \xrightarrow[-2\ AgCl]{+2\ AgPF_6} 2[(L{-}L)(Cl)Pd{-}CH(SiMe_3)(PMe_2Ph)]^+\ PF_6^- \quad (14)$$

(20)

20	**a**	**b**
L—L	1,5-cyclooctadiene (cod)	norbornadiene (nbd)

To elucidate the mode of coordination of the ylide ligand in **20a** a molecular structure analysis was undertaken[35] (Figure 7). The ylide alkyl bond of **20a** exerts a strong *trans*-influence on the opposite cyclooctadiene olefin bond. Lengths to the cod carbon atoms C_1' and C_2' are significantly longer than lengths to C_5' and C_6', the latter being of similar length to those in $[(cod)PdCl_2]$ (2.20 Å). Olefinic carbon–carbon distances also reflect the difference in bond strength to the metal, with the $C_5'{-}C_6'$ bond lengthened to 1.370 Å compared with the $C_1'{-}C_2'$ distance, which is essentially that of a free olefin.

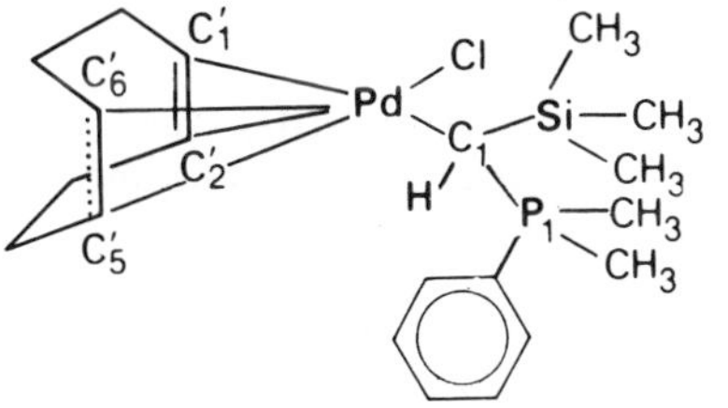

	Bond length (Å)
$C_1'{-}C_2'$	= 1.327
$C_5'{-}C_6'$	= 1.370
$Pd{-}C_1$	= 2.097
$Pd{-}C_1'$	= 2.317
$Pd{-}C_2'$	= 2.306
$Pd{-}C_5'$	= 2.178
$Pd{-}C_6'$	= 2.198

FIGURE 7. Molecular structure of and selected bond distances in **20a**.

The displacement of bidentate olefin ligands generally affords a mild and convenient access to bis-ylide complexes (equation 15)[36]. Neither methane nor ethane is observed

$$2R_2MePCH_2 + [(cod)PtMe_2] \longrightarrow cis\text{-}[(R_2MePCH_2)_2PtMe_2] + cod \quad (15)$$

(21)

(a) R = Me
(b) R = *t*-Bu

during the synthesis of **21**, whereas the reaction of $(Me_3Pt)^+PF_6^-$ with **14** is accompanied by spontaneous generation of methane[37]. The reactions illustrated in Schemes 3 and 4 further emphasize the exceptional thermal and chemical stability of metal–carbon σ-bonds in the compounds under discussion[38].

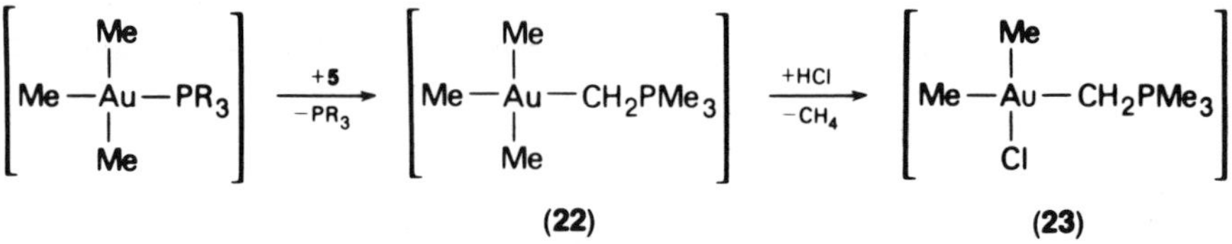

$$[Me_3Au{-}PR_3] \xrightarrow[-PR_3]{+\mathbf{5}} [Me_3Au{-}CH_2PMe_3] \xrightarrow[-CH_4]{+HCl} [Me_2ClAu{-}CH_2PMe_3]$$

(**22**) (**23**)

SCHEME 3

$$[Me_2AuBr]_2 + \mathbf{45} \longrightarrow 2\,[Me_2Au(CH_2PMe_3)_2]^+\,Br^- \quad (\mathbf{24})$$

$$\mathbf{24} \xrightarrow[-CH_4]{+HCl} [MeClAu(CH_2PMe_3)_2]^+\,Br^- \quad (\mathbf{25})$$

$$\mathbf{24} \xrightarrow{\Delta,\ T > 205\ ^\circ C} [Me_3PCH_2{-}Au{-}CH_2PMe_3]^+\,Br^- \quad (\mathbf{26})$$

$$\mathbf{26} \xrightarrow{+X_2} [Me_3PCH_2{-}AuX_2{-}CH_2PMe_3]^+\,Br^- \quad (\mathbf{27})$$

27	a	b
X	Br	I

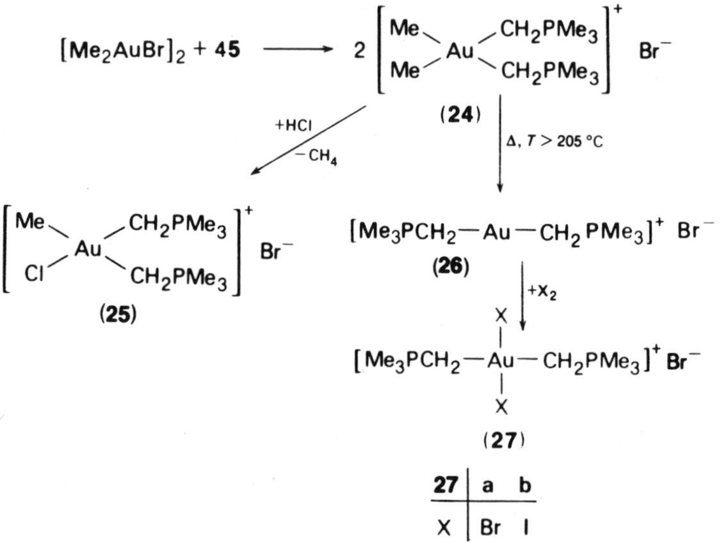

SCHEME 4

Compound **22**, a representative of a tetraalkylaurate(III) compound, does not decompose at temperatures below 185°C, whereas AuR_4^- anions are not capable of existence at ambient temperature. Treatment of **22** with gaseous hydrogen chloride leads only to the scission of a gold–methyl bond, whilst the ylide carbon–metal bond remains unaffected. A similar behaviour is encountered in the reaction of **24** with HCl. When heated to temperatures higher than 205°C, **24** is readily converted to **26** with loss of ethane. In aqueous solution, **24** resists hydrolysis. The carbon–gold σ-bonds in **26** are also resistant to oxidizing reagents such as bromine and iodine.

The reaction of $[V(CO)_6]$ with Ph_3PCH_2 produces ionic **28** via disproportionation. In this complex four ylide ligands occupy the coordination sphere of the V^{2+} ion[39].

$$3[V(CO)_6] + 4Ph_3PCH_2\ (\mathbf{16}) \longrightarrow [V(CH_2PPh_3)_4][V(CO)_6]_2\ (\mathbf{28}) + 6CO \qquad (16)$$

If the η^5-C_5H_5 moiety is regarded as a tridendate ligand, the coordination number 5 is established in a series of cyclopentadienyl–nickel–ylide complexes[40]. In the formation of **29**, a cyclopentadienyl ligand serves as the leaving group.

$$[(\eta^5\text{-Cp})_2\text{Ni}] \xrightarrow[\text{(2) NaBPh}_4]{\text{(1) 216}} [(\eta^5\text{-Cp})\text{Ni}(\text{CH}_2\text{PPh}_3)_2]^+ \ \text{BPh}_4^- \quad (17)$$

(29)

In octahedral complexes monodentate ylides are capable of occupying one, two, or three coordination sites. Octahedral **30** is one of the few ylide complexes derived from ruthenium[41] (equation 18).

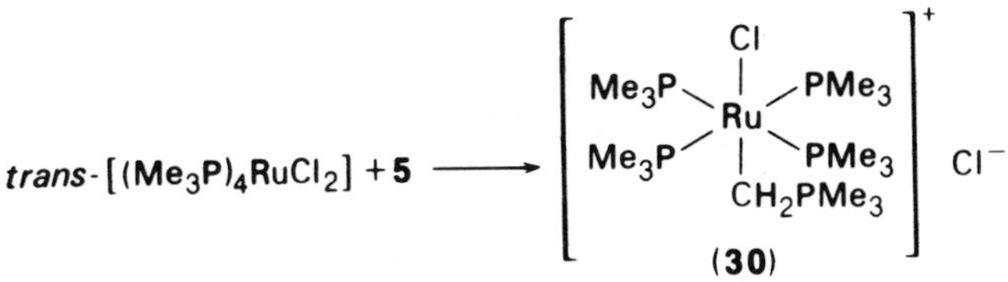

$$trans\text{-}[(\text{Me}_3\text{P})_4\text{RuCl}_2] + \mathbf{5} \longrightarrow [\text{RuCl}(\text{PMe}_3)_4(\text{CH}_2\text{PMe}_3)]^+ \ \text{Cl}^- \quad (18)$$

(30)

When synthesizing metal carbonyl ylide complexes of chromium, molybdenum, and tungsten, it is of advantage to employ organometallic precursors with labile ligands in order to avoid side-reactions. The strongly basic ylides especially favour addition to the electrophilic carbon atom of the carbonyl ligand in hexacarbonyls with subsequent transylidations. Section VII provides further coverage. When the electrophilic feature is sufficiently decreased, as in **31**, the reaction proceeds smoothly and displacement becomes the major course of reaction (Scheme 5)[42].

$$(\text{NEt}_4)[\text{Cr(CO)}_5\text{Br}] + \text{Ph}_{3-n}\text{Me}_n\text{PCH}_2 \xrightarrow[-(\text{NEt}_4)\text{Br}]{\text{thf}} [(\text{CO})_5\text{CrCH}_2\text{PPh}_{3-n}\text{Me}_n]$$

(31) → **(2)**, **(32)**

(31) $\xrightarrow[-(\text{NEt}_4)\text{Br}]{+\text{Ph}_{3-n}\text{Me}_n\text{PCHPh}}$ $[(\text{CO})_5\text{CrCH(Ph)PPh}_{3-n}\text{Me}_n]$ **(33)**

$n = 0$	1	2	3
2	**32b**	**32c**	**32d**
33a	**33b**	**33c**	**33d**

SCHEME 5

The analysis of the ^{31}P n.m.r. data leads to the conclusion that a d_π–d_π interaction exists between the highly charged $(\text{CO})_5\text{Cr}$ moiety and the phosphonium centre which, presumably for steric reasons, increases with increasing methyl substitution. This effect is especially evident in **33** in which the steric requirements of the ylidic phenyl group account for a decrease of the P—C—Cr angle. Similar interactions are also postulated for other ylide complexes[43–46]. The preparation of **33c,d** is accompanied by the production of **34**, the formation of which is probably due to base-catalysed rearrangements of the free ylides with subsequent complex formation.

$$[\{(\text{PhCH}_2)\text{Ph}_{(2-m)}\text{Me}_m\text{PCH}_2\}\text{Cr(CO)}_5]$$

(34)

$m = 1$	2
34 a	**b**

Phosphorus ylide complexes of the type [(arene)M(CO)$_2$(ylide)] (arene = $\text{C}_5\text{H}_4\text{Me}$, M = Mn; arene = C_6H_6, M = Cr) can be generated by treating the corresponding photochemically prepared thf complexes with the appropriate ylides[43].

Photochemically synthesized [(thf)M(CO)$_5$] solutions have been utilized for the coordination of the less basic sulphur ylides **35** and **37**[47,48].

$$[(thf)M(CO)_5] + \underset{(\mathbf{35})}{Me_2S(O)CH_2} \xrightarrow[20°C]{thf} \underset{(\mathbf{36})}{[\{Me_2S(O)CH_2\}M(CO)_5]} + thf \quad (19)$$

M	Cr	Mo	W
36	**a**	**b**	**c**

$$[(thf)M(CO)_5] + \underset{(\mathbf{37})}{(Me_2N)(Me)S(O)CH_2} \xrightarrow[20°C]{thf} \underset{(\mathbf{38})}{[\{(Me_2N)(Me)S(O)CH_2\}M(CO)_5]} + thf \quad (20)$$

M	Cr	Mo
38	**a**	**b**

The analogous tungsten species **38c**, however, can be prepared only by direct photolysis of excess of $[W(CO)_6]$ in a light petroleum solution of **37** (equation 21).

$$\mathbf{37} + [W(CO)_6] \xrightarrow[h\nu]{\text{light petroleum}} \underset{(\mathbf{38c})}{[\{(Me_2N)(Me)S(O)CH_2\}W(CO)_5]} + CO \quad (21)$$

On the other hand, the photochemical reactions of $M(CO)_6$ with excess of **35** in diethyl ether produce the disubstituted *cis*-complexes **39** (equation 22)[49].

$$[M(CO)_6] + 2\mathbf{35} \xrightarrow[h\nu]{Et_2O} \underset{(\mathbf{39})}{cis\text{-}[\{Me_2S(O)CH_2\}_2M(CO)_4]} + 2CO \quad (22)$$

M	Cr	Mo	W
39	**a**	**b**	**c**

The displacement of norbornadiene (nbd) and cycloheptatriene (cht) ligands provides a convenient and mild method for the synthesis of di- and trisubstituted ylide carbonyl complexes[49,50] (equations 23 and 24).

$$[(nbd)M(CO)_4] + 2\mathbf{35} \xrightarrow[r.t.]{Et_2O} \mathbf{39a\text{–}c} + nbd \quad (23)$$

$$[(cht)M(CO)_3] + 3\mathbf{35} \xrightarrow[r.t.]{Et_2O} \underset{\mathbf{40a\text{–}c}}{fac\text{-}[\{Me_2S(O)CH_2\}_3M(CO)_3]} + cht \quad (24)$$

For steric reasons, the yellow chromium derivative **40a** exhibits a considerably lower thermal stability than **40b** and **40c**[51]. The more bulky ylide **37** only permits the generation of the molybdenum complexes **41** and **42** under otherwise similar reaction conditions. It is generally observed that increasing substitution of carbonyls for ylides results in a decrease in the chemical and thermal stability of the product.

$$[\{(Me_2N)(Me)S(O)CH_2\}_nMo(CO)_{6-n}]$$

n	
2	**41**
3	**42**

The tetrahalides of germanium, tin, titanium, and zirconium are reported to be starting materials for adducts with ylides of octahedral geometry and of even higher coordination numbers[52,53].

c. Oxidative additions of halogenated onium salts to metals. The reaction of $[Pt(PPh_3)_4]$ with CH_2ClI to form ionic **43** is believed to involve the oxidative addition of a chlorinated phosphonium intermediate to the coordinatively unsaturated and electron-rich $[Pt(PPh_3)_2]$ which is known to be present in benzene solutions of $[Pt(PPh_3)_4]$ (Scheme 6)[54].

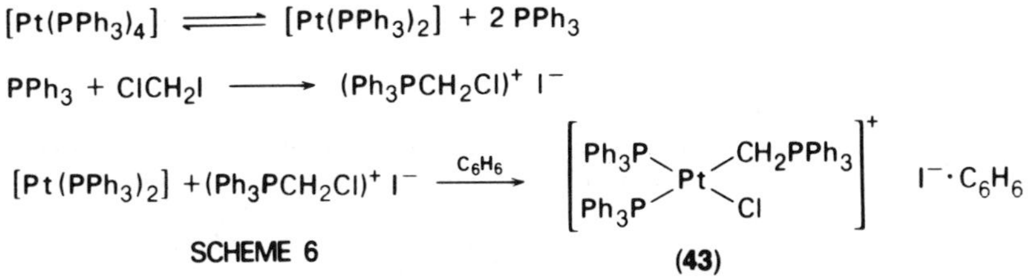

$$[Pt(PPh_3)_4] \rightleftharpoons [Pt(PPh_3)_2] + 2\,PPh_3$$

$$PPh_3 + ClCH_2I \longrightarrow (Ph_3PCH_2Cl)^+\,I^-$$

$$[Pt(PPh_3)_2] + (Ph_3PCH_2Cl)^+\,I^- \xrightarrow{C_6H_6} \left[(Ph_3P)_2Pt(CH_2PPh_3)(Cl)\right]^+ I^-\cdot C_6H_6$$

(**43**)

SCHEME 6

SCHEME 6

An alternative mechanism is based on the oxidative addition of the halomethane to the electron-rich platinum species with subsequent rearrangement as is postulated for the formation of $[(Me_3PCH_2)_2CoCl_2]$ in Section II.A.2b[55].

Halogenated phosphonium salts, available from the reaction of polyhalomethanes with phosphines, are dehalogenated by treatment with zinc, cadmium and mercury to generate ylide complexes of varying stability[56–58].

$$PPh_3 + CXY_3 + Zn \longrightarrow [Ph_3PCXY(ZnY)]^+Y^- \rightleftharpoons Ph_3PCXY + ZnY_2 \qquad X = F,\ Y = Cl \quad (25)$$

$$(Ph_3PCHFI)I + Zn(Cu) \xrightarrow[0^\circ C]{dmf} [Ph_3PCHF(ZnI)]^+I^- \rightleftharpoons Ph_3PCHF + ZnI_2 \quad (26)$$

$$[(Me_2N)_3PCFCl_2]Cl + Zn(Cu) \xrightarrow[60^\circ C]{thf} [(Me_2N)_3PCFCl(ZnCl)]^+Cl^- \ (\mathbf{44a}) \rightleftharpoons (Me_2N)_3PCFCl + ZnCl_2 \quad (27)$$

Organic carbonyls, when exposed to these crude reaction mixtures, undergo Wittig reactions, thus supporting the postulated dissociation processes as shown in equations 25–27 (see also Section VIII). In contrast to the free ylide, thf solutions of **44a** exhibit remarkable stability (>10 h at 60°C, >30 days at 20°C). The stability of the mercury derivative **44c** in benzonitrile solution of 60°C far exceeds those of the zinc and cadmium homologues. Despite the fact that compounds **44a–c** resist isolation and therefore complete analytical characterization, the observation of metal–fluorine coupling constants in the ^{19}F n.m.r. spectra of **44b** and **44c** proves the presence of

$$[(Me_2N)_3PCFCl_2]^+Cl^- + M \xrightarrow[60^\circ C]{PhCN} [(Me_2N)_3PCFCl(MCl)]^+Cl^- \ (\mathbf{44}) \quad (28)$$

M	Zn	Cd	Hg
44	**a**	**b**	**c**

covalent metal–carbon σ-bonds. The metal atoms in **44a** and **44b** are readily exchanged by mercury, and **44d**, possessing a complex anion of low nucleophilic power, is an isolated and analytically characterized compound. The exact molecular structures of **44a–d**, however, have yet to be established.

$$\mathbf{44a} + HgCl_2 \longrightarrow \underset{(\mathbf{44d})}{[(Me_2N)_3PCFCl(HgCl)]^+[ZnCl_3]^-} \quad (29)$$

2. *Indirect syntheses*

a. Addition of donor molecules to carbene complexes. In some cases ylides have been synthesized by the trapping reactions of carbenes with phosphines[13–16]. Transition metal–carbene complexes are converted to ylide complexes by addition of Group Va bases to the sp^2-hybridized carbene carbon atom. Although this synthetic method is independent of the existence of free, stable ylides, the unavailability of many carbene complexes poses a limitation for the application of these reagents.

Whereas primary and secondary amines react with alkoxycarbene complexes of the Fischer type to give aminocarbene complexes with elimination of alcohol[59], tertiary amines such as 1-azabicyclo[2.2.2]octane (**45**) or 1,4-diazabicyclo[2.2.2]octane (**46**) are added to the electrophilic carbene centre to generate nitrogen ylide complexes[60,61] (Scheme 7).

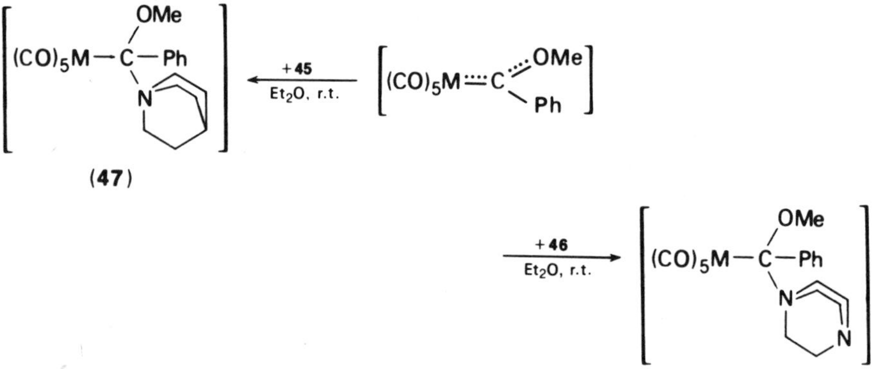

(CO)5M—C(OMe)(Ph)—N(quinuclidine) **(47)** ← +**45**, Et2O, r.t. — [(CO)5M=C(OMe)Ph] — +**46**, Et2O, r.t. → (CO)5M—C(OMe)(Ph)—N(DABCO) **(48)**

M	Cr	W
47, 48	**a**	**b**

SCHEME 7

The ylide complexes **47** and **48** are postulated to be important as model compounds for the first step in the aminolysis of alkoxy-substituted carbene complexes. As a result of kinetic investigations, it was concluded that aminolysis proceeds by reversible nucleophilic attack of the amine at the carbene carbon atom which produces such tetravalent carbon species[62]. The phosphorus ylide complexes **49a** and **49b** also resemble the intermediates of the reaction under discussion. In contrast to the homologous amines, the reaction of $HPMe_2$ with carbene complexes ceases with the intermediate ylide complex without eliminating methanol and generating phosphinocarbene complexes (equation 30)[63,64]. The thermolability of **49a** and **49b** permits manipulation of these compounds only at temperatures below −40°C. In acetone solution, **49a** is converted to **50** by initial scission of the carbene–chromium

$$\left[(CO)_5M\overset{\cdots}{=}C\begin{matrix}OMe\\R\end{matrix}\right] + HPMe_2 \xrightarrow[T]{pentane} \left[(CO)_5M-C(OMe)(R)(HPMe_2)\right] \quad (30)$$

(49)

49	M	R	T(°C)
a	Cr	Ph	−50
b	W	Me	−78

$$\left[(CO)_5Cr-C(OMe)(Ph)(HPMe_2)\right] \xrightarrow[8\,h,\ 20°C]{acetone} \left[(CO)_5Cr\overset{\cdots}{=}C\begin{matrix}OMe\\Ph\end{matrix}\right] + \left[(CO)_5Cr-PMe_2-CH(OMe)(Ph)\right] \quad (31)$$

(49a) (50)

bond and subsequent rearrangement (equation 31). This result is also in contrast with the reactivity of carbene complexes towards Me_2NH (see Chapter 4, Section III.A).

The same synthetic method has been applied successively to the preparation of ylide complexes from the reaction of alkoxycarbene-, alkylthiocarbene-, and alkylselenocarbene-chromium and -tungsten complexes with phosphines, e.g. PMe_3, PEt_3, and $P(n\text{-}Bu)_3$[64,65]. Diarylcarbene tungsten pentacarbonyls have also proved to be suitable precursors for ylide complexes[66].

$$[(CO)_5W{=}CR^1R^2] + PMe_3 \xrightarrow[-78°C]{Et_2O} [(CO)_5WCR^1R^2(PMe_3)] \quad (32)$$

(51)

51	a	b	c	d	e	f
R_1	Ph	2-C_4H_3S	2-C_4H_3O	2-C_4H_3O	2-C_4H_3S	2-C_4H_3O
R_2	Ph	Ph	Ph	2-C_4H_3S	2-C_4H_3S	2-C_4H_3O

With the preparation of **52**, this synthetic principle was extented to a Group VIIB metal[67].

$$\left[\eta^5\text{-}Cp(CO)_2Re{=}C\begin{matrix}Me\\Ph\end{matrix}\right] + PMe_3 \longrightarrow \left[\eta^5\text{-}Cp(CO)_2Re-C(Me)(Ph)(PMe_3)\right] \quad (33)$$

(52)

Aminocarbene complexes resist addition of donors, which emphasizes the importance of a sufficient electron deficiency at the coordinated carbene carbon atom required for donor attack.

The ylide complexes under discussion gain further importance as intermediates in carbonyl displacement reactions of carbene complexes carried out at elevated temperatures with phosphines. Kinetic investigations show that the initial adduct formation of pentacarbonylmethoxymethylcarbenechromium and -tungsten with a series of phosphines represents an equilibrium reaction strongly dependent on the choice of temperature. For the same carbene ligand, low temperature, polar solvents, decreased steric bulk, and increased basicity of the phosphines, and also an increase in the size of the central metal with a concomitant decrease in steric hindrance, favour the generation of ylide complexes[68].

The presence of stable and isolable carbene complexes is not crucial for the successful application of this ylide complex synthesis. On the contrary, amine and phosphine additions are a useful probe for the existence of highly reactive metal–carbene species (equations 34[69] and 35[70]).

$$[\eta^5\text{-CpRe(NO)(PPh}_3\text{)CH}_3] + (\text{Ph}_3\overset{+}{\text{C}})\text{X}^- \xrightarrow[-70^\circ\text{C}]{\text{CH}_2\text{Cl}_2} [\eta^5\text{-CpRe(NO)(PPh}_3\text{)(CH}_2\text{)]}^+\text{X}^-$$

$$\xrightarrow{+\text{PR}_3} [\eta^5\text{-Cp(PPh}_3\text{)(NO)ReCH}_2\text{PR}_3]^+\text{X}^- \ \textbf{(53)} \qquad \xrightarrow{+\text{NC}_5\text{H}_5} [\eta^5\text{-Cp(PPh}_3\text{)(NO)ReCH}_2\text{NC}_5\text{H}_5]^+\text{X}^- \ \textbf{(54)} \tag{34}$$

53	**a**	**b**
R	*n*-Bu	Ph

$$[(\eta^5\text{-Cp})(\text{CO})_3\text{WCH}_2\text{OMe}] \xrightarrow[-\text{Me}_3\text{SiOMe},\ -35^\circ\text{C}]{+\text{Me}_3\text{SiO}_3\text{SCF}_3} [\eta^5\text{-Cp(CO)}_3\text{W(CH}_2\text{)]}^+\text{SO}_3\text{CF}_3^-$$

$$\xrightarrow{+\text{NEt}_3,\ -35^\circ\text{C}} [\eta^5\text{-Cp(CO)}_3\text{WCH}_2\text{NEt}_3]^+\text{SO}_3\text{CF}_3^- \ \textbf{(55)} \tag{35}$$

Reversible α-elimination of hydrogen from a methyl–tungsten compound to give a tungsten–methylene hydride derivative has been proposed as a key step in the preparation of ylide complexes **57** (Scheme 8)[71].

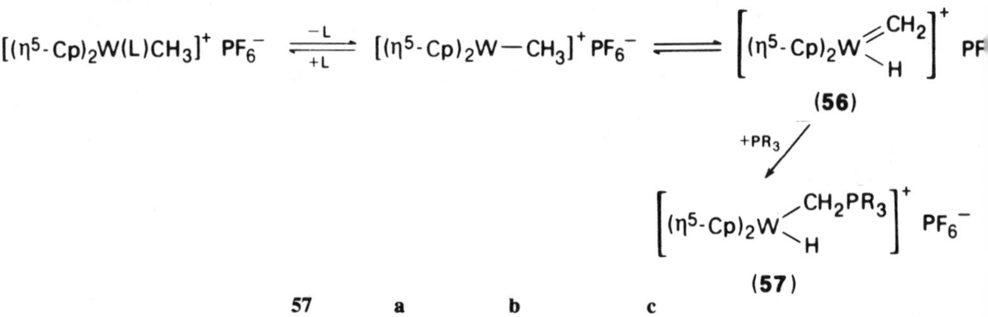

$$[(\eta^5\text{-Cp})_2\text{W(L)CH}_3]^+\ \text{PF}_6^- \underset{+\text{L}}{\overset{-\text{L}}{\rightleftharpoons}} [(\eta^5\text{-Cp})_2\text{W}-\text{CH}_3]^+\ \text{PF}_6^- \rightleftharpoons \left[(\eta^5\text{-Cp})_2\text{W}{<}^{\text{CH}_2}_{\text{H}}\right]^+ \text{PF}_6^- \ \textbf{(56)}$$

$$\xrightarrow{+\text{PR}_3} \left[(\eta^5\text{-Cp})_2\text{W}{<}^{\text{CH}_2\text{PR}_3}_{\text{H}}\right]^+ \text{PF}_6^- \ \textbf{(57)}$$

57	**a**	**b**	**c**
PR_3	PMe_3	PMe_2Ph	$PMePh_2$

SCHEME 8

The formation of **57** is facilitated by increasing phosphine basicity: $Ph_2PMe < PhPMe_2 < PMe_3$. Rearrangement to phosphinemethyl–tungsten compounds is achieved by prolonged heating of acetone solutions of **57**[72].

The rearrangement of platinacyclobutanes to pyridium ylide complexes proceeds from treatment with pyridine bases in warm benzene (equations 36–38)[73–75].

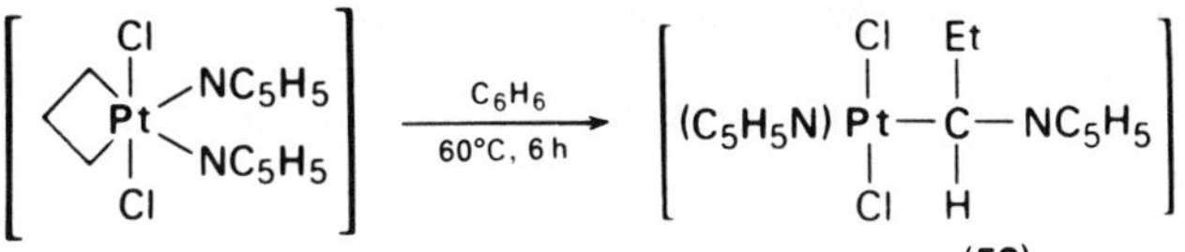

$$[\text{Pt(cyclobutane)Cl}_2(NC_5H_5)_2] \xrightarrow[60°C,\ 6\ h]{C_6H_6} [(C_5H_5N)PtCl_2\text{—}C(Et)(H)\text{—}NC_5H_5] \quad (\mathbf{58}) \qquad (36)$$

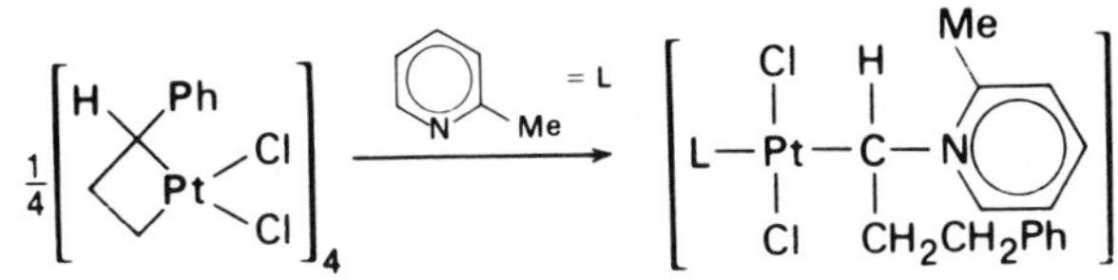

$$\tfrac{1}{4}[\text{Pt(CH}_2\text{CHPhCH}_2)\text{Cl}_2]_4 \xrightarrow{\text{2-methylpyridine} = L} [L\text{—}PtCl_2\text{—}C(H)(CH_2CH_2Ph)\text{—}N(2\text{-Me-pyridinium})] \quad (\mathbf{59}) \qquad (37)$$

$$\tfrac{1}{4}[\text{Pt(CH}_2\text{CMe}_2\text{CH}_2)\text{Cl}_2]_4 \xrightarrow{C_5H_5N} [(C_5H_5N)PtCl_2\text{—}C(H)(CH_2CHMe_2)\text{—}NC_5H_5] \quad (\mathbf{60}) \qquad (38)$$

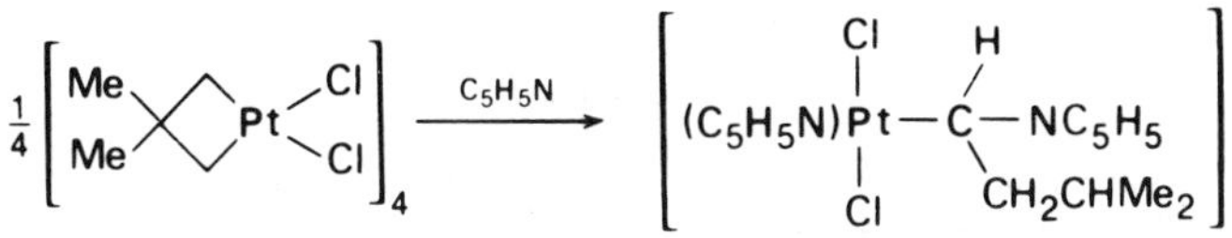

Experiments with partially deuterated platinacyclobutanes lead to the conclusion that reversible α-elimination again produces a reactive carbene species, which is subsequently intercepted by the pyridine base (Scheme 9)[75].

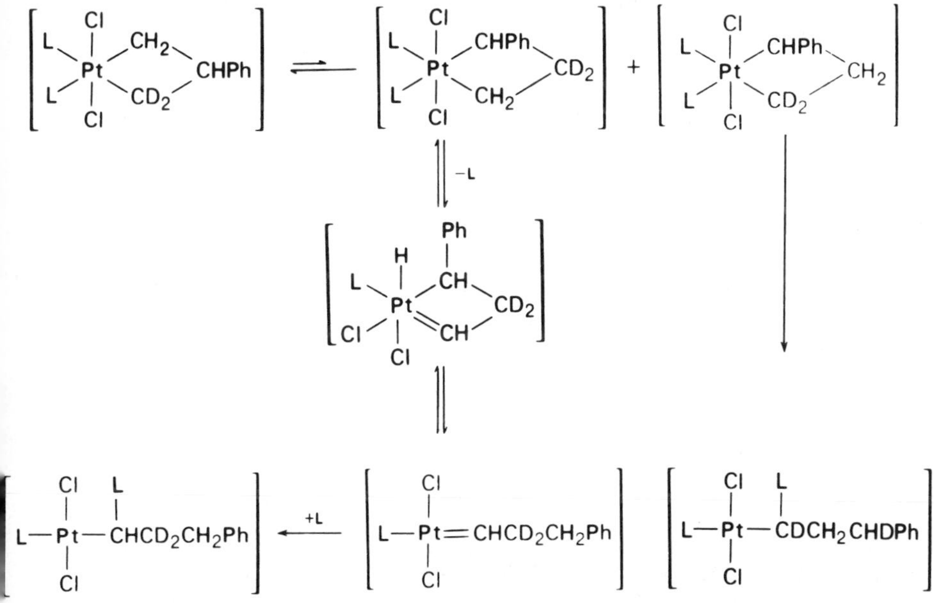

$$[L_2Cl_2Pt(CH_2CHPhCD_2)] \rightleftharpoons [L_2Cl_2Pt(CHPhCD_2CH_2)] + [L_2Cl_2Pt(CHPhCH_2CD_2)]$$

$$[L_2Cl_2Pt(CHPhCD_2CH_2)] \underset{}{\overset{-L}{\rightleftharpoons}} [LCl_2(H)Pt(CHPh)CD_2(=CH)] \rightleftharpoons [L\text{—}PtCl_2{=}CHCD_2CH_2Ph] \xrightarrow{+L} [L\text{—}PtCl_2\text{—}CH(L)CD_2CH_2Ph]$$

$$[L_2Cl_2Pt(CHPhCH_2CD_2)] \longrightarrow [L\text{—}PtCl_2\text{—}CD(L)CH_2CHDPh]$$

SCHEME 9

The synthesis of vinylene phosphorane complexes **(61)** proceeds via phosphine addition to the electrophilic carbon atom of an intermediate vinylene–iron complex[76] (Scheme 10). The reverse of this method offers a possible route to novel carbene

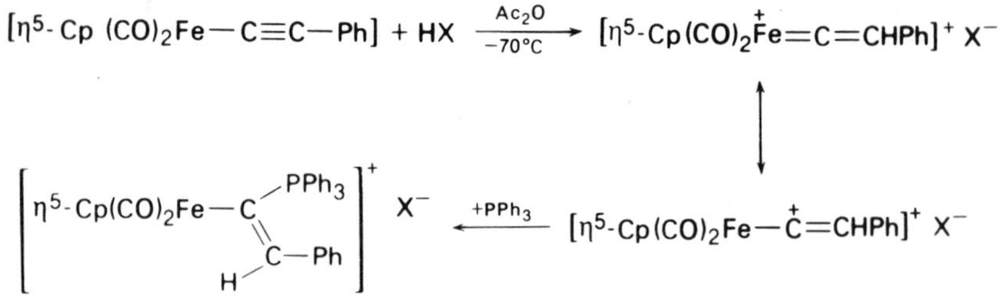

$$[\eta^5\text{-Cp}(CO)_2Fe{-}C{\equiv}C{-}Ph] + HX \xrightarrow[-70°C]{Ac_2O} [\eta^5\text{-Cp}(CO)_2\overset{+}{Fe}{=}C{=}CHPh]^+ X^-$$

$$\updownarrow$$

$$\left[\eta^5\text{-Cp}(CO)_2Fe{-}C(PPh_3){=}C(H){-}Ph\right]^+ X^- \xleftarrow{+PPh_3} [\eta^5\text{-Cp}(CO)_2Fe{-}\overset{+}{C}{=}CHPh]^+ X^-$$

(61)

61	**a**	**b**
X^-	ClO_4^-	$\bar{B}F_4^-$

SCHEME 10

complexes via organometallic ylide species[77]. This method also provides excellent access to alkylidene complexes of tantalum[78] and zirconium[79].

$$[(thf)Cr(CO)_5] \xrightarrow[-thf]{Ph_3PC(SPh)_2} \underset{(62)}{[(CO)_5CrC(SPh)_2PPh_3]} \xrightarrow[-Ph_3PS]{+\frac{1}{6}S_6} \underset{(63)}{[(CO)_5Cr{=}C(SPh)_2]} \quad (39)$$

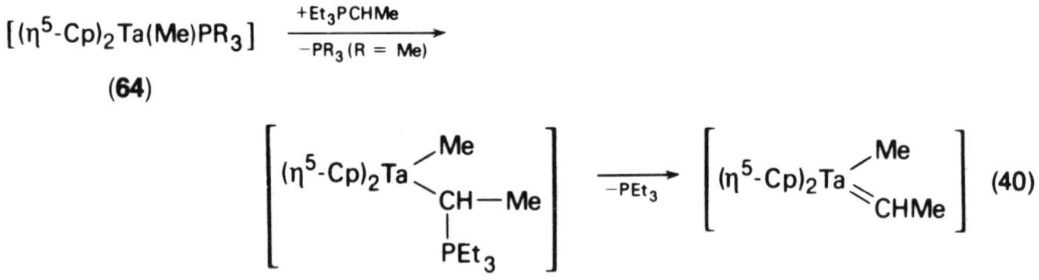

$$\underset{(64)}{[(\eta^5\text{-Cp})_2Ta(Me)PR_3]} \xrightarrow[-PR_3\ (R\,=\,Me)]{+Et_3PCHMe} \underset{(65)}{\left[(\eta^5\text{-Cp})_2Ta(Me){-}CH(Me){-}PEt_3\right]} \xrightarrow{-PEt_3} \underset{(66)}{\left[(\eta^5\text{-Cp})_2Ta(Me){=}CHMe\right]} \quad (40)$$

The labile phosphine ligand in **64** is readily displaced by the ylide, generating **65**, which is stabilized by loss of PEt_3 to the ethylidene complex **66**. Ethylidene complexes such as **66** cannot be prepared successfully via the routes usually employed for the generation of tantalum–alkylidene compounds. Analogously, treatment of **64** ($R = Me_2PPh$) with Me_3PCHPh affords **67**[78].

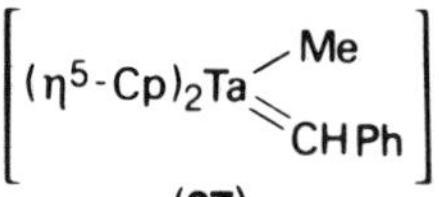

$$\underset{(67)}{\left[(\eta^5\text{-Cp})_2Ta(Me){=}CHPh\right]}$$

Attempts to isolate the first zirconium–carbene species **69**[79], which is isoelectronic with $[(\eta^5\text{-Cp})_2Ta(Me)(CH_2)]$, failed. Nevertheless, the presence of this compound generated from unstable **68** in about 40% yield can be confirmed by ^1H n.m.r. studies (equation 41).

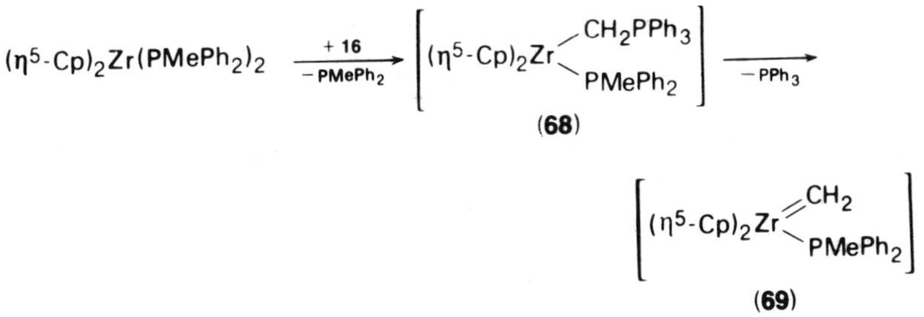

$$(\eta^5\text{-}Cp)_2Zr(PMePh_2)_2 \xrightarrow[-PMePh_2]{+16} \left[(\eta^5\text{-}Cp)_2Zr\begin{matrix}CH_2PPh_3\\PMePh_2\end{matrix}\right] \xrightarrow[-PPh_3]{}$$

(68)

$$\left[(\eta^5\text{-}Cp)_2Zr\begin{matrix}=CH_2\\PMePh_2\end{matrix}\right] \quad (41)$$

(69)

b. Displacement reactions on functionalized metal–alkyl complexes. Another method of indirect ylide complex formation is accomplished by nucleophilic displacements with metal alkyl complexes with a good leaving group X (X = dmso, halide) at the carbon atom adjacent to the metal.

A characteristic feature of the organic chemistry of sulphur ylides is the capability of sulphanes to serve as reasonably good leaving groups[2]. In accordance with this, nucleophilic attack at the sulphur ylide ligand in **36a** by phosphines and arsines results in the formation of other ylide complexes concomitant with the loss of dmso (equation 42)[80]. **70** possesses the otherwise unstable $(MeO)_3PCH_2$ ligand.

$$[(CO)_5CrCH_2S(O)Me_2] + L \xrightarrow[60^\circ C]{C_6H_6} [(CO)_5CrCH_2L] + Me_2SO \quad (42)$$

(36a) → **(2), (32b,c), (70), (71)**

L	PPh_3	$PMePh_2$	PMe_2Ph	$P(OMe)_3$	$AsPh_3$
	2	**32b**	**32c**	**70**	**71**

The displacement of halide ions in the M—CH_2X unit by tertiary amines and phosphines has been studied. The addition of NMe_3 and PPh_3 to ethereal solutions of oligomeric $[(ClCH_2)_2Zn]_n$, pre-generated by treatment of $ZnCl_2$ with diazomethane, leads to the ylide complexes **72** and **73**[81] (Scheme 11).

$$[(Ph_3PCH_2)_2ZnCl_2] \xleftarrow[Et_2O]{PPh_3} \frac{1}{n}[Zn(CH_2Cl)_2]_n \xrightarrow[Et_2O]{NMe_3} [(Me_3NCH_2)_2ZnCl_2]$$

(73) **(72)**

SCHEME 11

Analogous transformations have been reported with mercury compounds[81]. Attempts to synthesize $[(t\text{-}BuCH_2)_2SCH_2]_2ZnI_2$ resulted in the isolation of saline $[(t\text{-}BuCH_2)_2SMe][(t\text{-}BuCH_2)_2SCH_2ZnI_3]$ in which the sulphur ylide is ligated to a tetrahedral zincate anion[82].

Equation 43 describes the preparation of germanylated phosphonium salts[83].

$$Me_3GeCH_2Cl + PMe_3 \longrightarrow (Me_3PCH_2GeMe_3)Cl \quad (43)$$

(74)

$$3[(Me_3P)_4Co] + 2R^1R^2CCl_2 \xrightarrow[-10^\circ C]{pentane} [(Me_3PCR^1R^2)_2CoCl_2] + 2[(Me_3P)_3CoCl] + 4PMe_3 \quad (44)$$

(75)

75	**a**	**b**	**c**	**d**
R^1	H	D	H	H
R^2	H	D	Me	$SiMe_3$

An ylide synthesis at the cobalt centre has been shown to be feasible with $(Me_3P)_4Co$ and geminal organodihalides[55]. This synthesis is believed to be initiated by the oxidative addition of the dihalide to two cobalt complex molecules. The postulated intermediate **76** undergoes rearrangement to the cobalt(I) ylide complex **77**, which eventually disproportionates to **75a** and $(Me_3P)_4Co$ (Scheme 12). All attempts to obtain **77** independently from $[(Me_3P)_3CoCl]$ and Me_3PCH_2 failed and produced the same disproportionated materials.

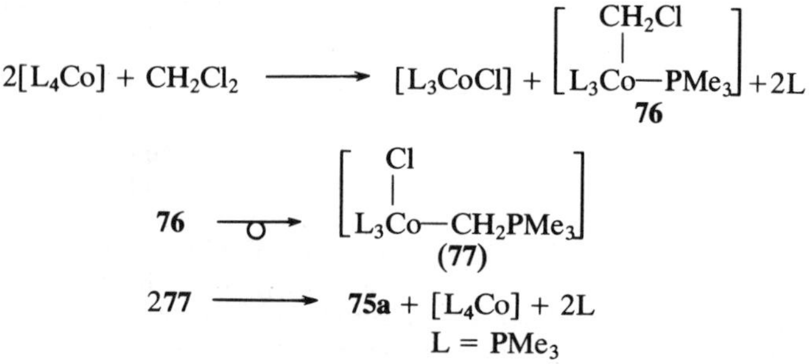

$$2[L_4Co] + CH_2Cl_2 \longrightarrow [L_3CoCl] + \left[L_3Co(CH_2Cl){-}PMe_3\right] + 2L \quad (\mathbf{76})$$

$$\mathbf{76} \longrightarrow \left[L_3Co(Cl){-}CH_2PMe_3\right] \quad (\mathbf{77})$$

$$2\,\mathbf{77} \longrightarrow \mathbf{75a} + [L_4Co] + 2L$$

$L = PMe_3$

SCHEME 12

c. Addition of nucleophiles and electrophiles to metal-substituted ylides. The ylidic carbon atom in the transition-metal substituted cationic ylide **78** exhibits electrophilic character and thus allows the addition of lithium organyls to yield the onium metallate **52**[67]. (equation 45). In comparison with this, the organotin-substituted ylide **79**, with the usual nucleophilicity at the tervalent carbon atom, is susceptible to alkylation with metal iodide[84].

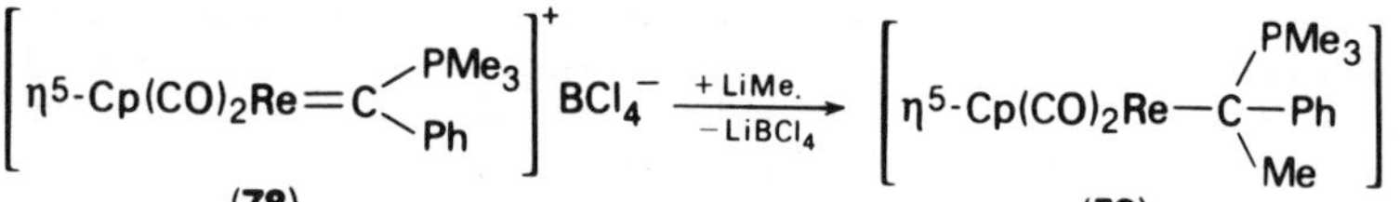

$$\left[\eta^5\text{-}Cp(CO)_2Re{=}C(PMe_3)(Ph)\right]^+ BCl_4^- \xrightarrow[-LiBCl_4]{+\,LiMe} \left[\eta^5\text{-}Cp(CO)_2Re{-}C(PMe_3)(Ph)(Me)\right] \quad (45)$$

(**78**) (**52**)

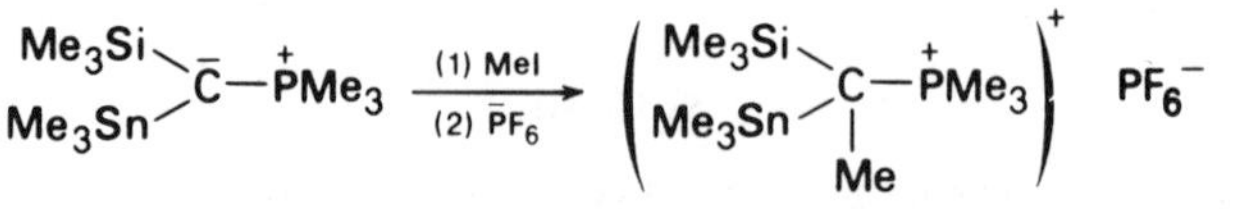

$$(Me_3Si)(Me_3Sn)\bar{C}{-}\overset{+}{P}Me_3 \xrightarrow[(2)\ \bar{P}F_6]{(1)\ MeI} \left((Me_3Si)(Me_3Sn)C(Me){-}\overset{+}{P}Me_3\right)^+ PF_6^- \quad (46)$$

(**79**) (**80**)

d. Alkylation reactions on η^1-aminomethyl and η^1-thiomethoxymethyl complexes. Compounds of the type $[L_nMCH_2SMe]$ or $[L_nMCH_2NMe_2]$ can be both protonated and alkylated at the basic heteroatom of the alkyl ligand, thus producing an ylide complex by the creation of an onium centre (equations 47–50)[85–88].

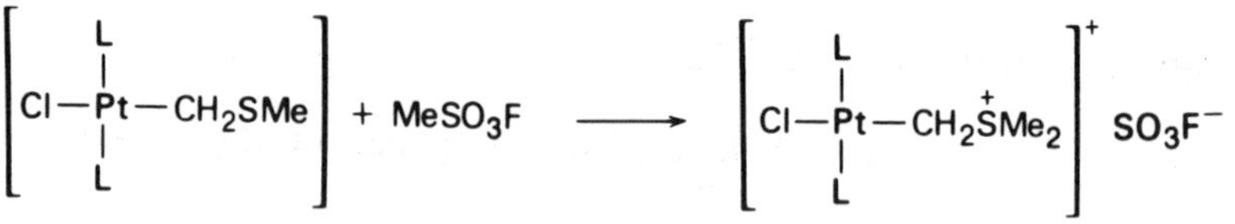

$$\left[Cl{-}PtL_2{-}CH_2SMe\right] + MeSO_3F \longrightarrow \left[Cl{-}PtL_2{-}CH_2\overset{+}{S}Me_2\right]^+ SO_3F^- \quad (47)$$

(**81**) (**82**)

82	a	b
L	PPh_3	$PMePh_2$

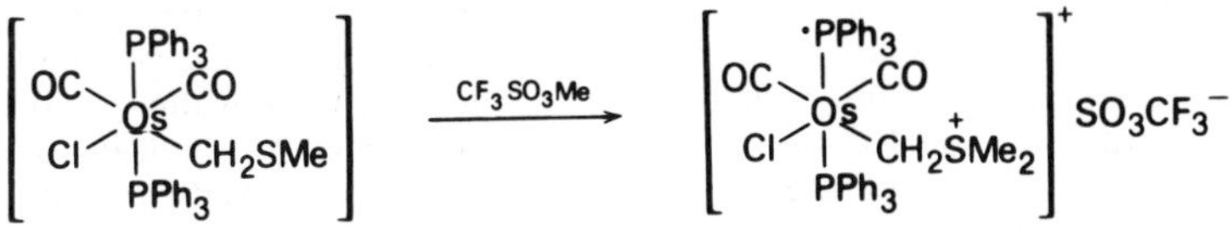

$$\left[\mathrm{OC(Cl)Os(PPh_3)_2(CO)CH_2SMe}\right] \xrightarrow{\mathrm{CF_3SO_3Me}} \left[\mathrm{OC(Cl)Os(PPh_3)_2(CO)CH_2\overset{+}{S}Me_2}\right]^+ \mathrm{SO_3CF_3^-} \quad (48)$$

(83)

$$[\eta^5\text{-}\mathrm{Cp(Ph_3P)NiCH_2NMe_2}] \xrightarrow{\mathrm{MeI}} [\eta^5\text{-}\mathrm{Cp(Ph_3P)NiCH_2\overset{+}{N}Me_3}]^+\ \mathrm{I^-} \quad (49)$$

(84)

$$[\eta^5\text{-}\mathrm{Cp(CO)_2FeCH_2\overset{+}{N}HMe_2}]^+\ \mathrm{Cl^-} \xleftarrow{\mathrm{HCl}} [\eta^5\text{-}\mathrm{Cp(CO)_2FeCH_2NMe_2}] \quad (50)$$

(85)

$$[\eta^5\text{-}\mathrm{Cp(CO)_2FeCH_2NMe_2}] \xrightarrow{\mathrm{MeI}} [\eta^5\text{-}\mathrm{Cp(CO)_2FeCH_2NMe_3}]^+\ \mathrm{I^-}$$

(86)

In contrast to phosphorus ylides, simple nitrogen ylides are not isolable and can apparently only be generated as lithium halide-stabilized species. Me_2SCH_2 is a thermally very sensitive compound whereas free Me_2NHCH_2 cannot be generated in the usual fashion from the quaternary salt. Thus, indirect syntheses such as protonation and derivatization of substituted alkyl complexes could develop into a further general method for generating metal–ylide complexes, especially when the free ylides are inherently unstable.

B. Non-stabilized Double Ylides as Ligands

Double ylides possess two carbanionic donor sites adjacent to one or two onium centres. In the simplest double ylide **87**, the two ylidic functions interact with the same phosphonium moiety (Scheme 13). Double ylides of the shown type may occur as bridging groups (as in Figure 8, **A**) or as chelating species (as in Figure 8, **B**).

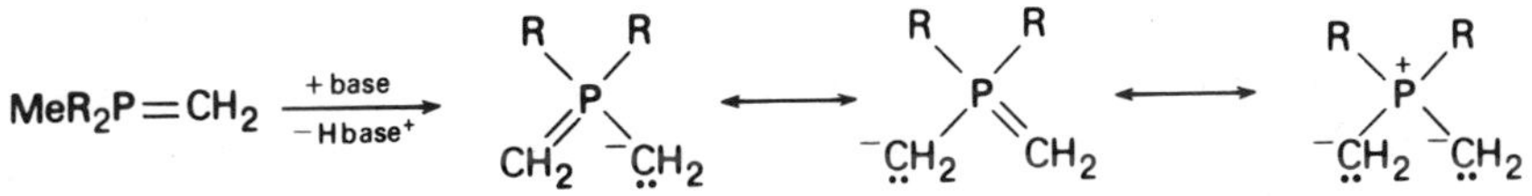

$$\mathrm{MeR_2P{=}CH_2} \xrightarrow[-\mathrm{Hbase^+}]{+\mathrm{base}} \mathrm{R_2P({=}CH_2)(\ddot{C}H_2^-)} \longleftrightarrow \mathrm{R_2P(\ddot{C}H_2^-)({=}CH_2)} \longleftrightarrow \mathrm{R_2\overset{+}{P}(\ddot{C}H_2^-)_2}$$

(87)

R = alkyl, aryl

SCHEME 13

$$\mathrm{R_2P(CM)_2}\ (\mathbf{A}) \qquad \mathrm{R_2P(C)_2M}\ (\mathbf{B})$$

FIGURE 8. Bonding interactions between **87** and metal atoms.

1. *Ylide-bridged complexes*

In complexes of type **A** (Figure 8), the two metal atoms can be linked by two or four bridges. Their metal–carbon σ-bonds are characterized by the usual high thermal and chemical stability, inherent in ylide complexes. Further, it is remarkable that ligands such as **87** keep the metal atoms of binuclear complexes in close proximity, thereby favouring not only weak intermetallic interactions, but also providing the fundamental requirements that stabilize metal–metal multiple bonds. The indispensable proton abstraction accompanying the coordination of **87** is performed in most cases with excess of ylide via transylidation (Scheme 14).

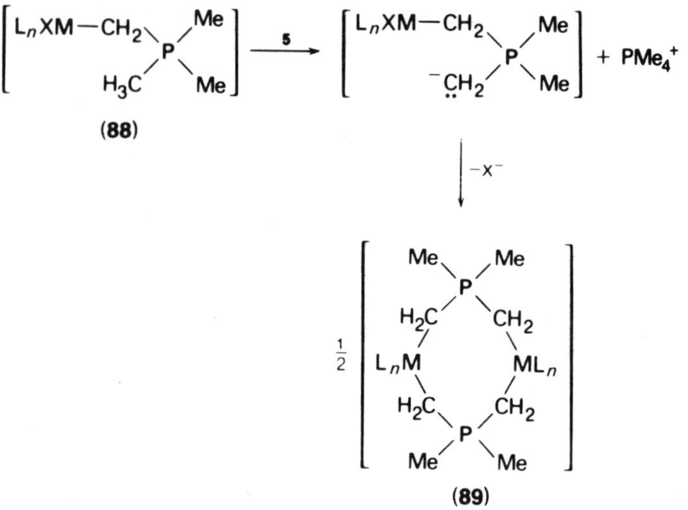

$$[L_nXM{-}CH_2P(Me)_2CH_3] \ \textbf{(88)} \xrightarrow{\textbf{5}} [L_nXM{-}CH_2P(Me)_2\ddot{C}H_2^-] + PMe_4^+ \xrightarrow{-X^-} \tfrac{1}{2}[L_nM(CH_2PMe_2CH_2)_2ML_n] \ \textbf{(89)}$$

SCHEME 14

The bridging ylide ligand **87** originates from a species which is already terminally ligated to the metal. The alternative deprotonation at the methylene bridge in **88**, which is realized with Group IVA and VA metals, does not occur with Group IIA and IIB metals or with the d^8—d^{10} electron systems at the end of the transition period[3] (see also Section III.A.1).

Organometallic starting materials **90** with strongly basic alkyl or aryl ligands react with Me_3PCH_2 to generate ligated **87**, presumably via intramolecular proton abstraction (Scheme 15).

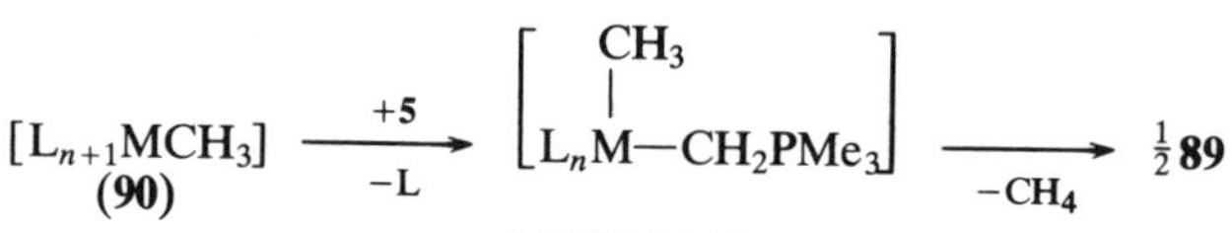

$$[L_{n+1}MCH_3] \ \textbf{(90)} \xrightarrow[-L]{+5} [L_nM(CH_3){-}CH_2PMe_3] \xrightarrow[-CH_4]{} \tfrac{1}{2}\textbf{89}$$

SCHEME 15

The direct displacement of ligands, providing a total of three electrons, by the lithium or magnesium complexes of **87** affords another useful synthetic tool for the compounds under discussion (Scheme 16).

The synthetic strategies developed here are also valid for the preparation of chelate complexes **B** depicted in Figure 8 (see Section II.B.2).

The treatment of phosphine halide complexes of the coinage metals with excess of ylide allows the isolation of $M_2P_2C_4$ eight-membered rings with alternating onium and metallate centres[28,89,90] (equation 51).

$$5 \xrightarrow[-C_4H_{10}]{+\text{LiBu}} \left[\begin{array}{c} Me \quad Me \\ P \\ CH_2 \quad CH_2 \\ Li \end{array} \right] \xrightarrow[-\text{LiX}, -L]{+L_{n+1}MX} \tfrac{1}{2}\,\mathbf{89}$$

SCHEME 16

$$\frac{2}{n}[(Me_3P)MCl]_n + \mathbf{45} \xrightarrow[-2\,(Me_4P)Cl]{-2\,PMe_3} \left[Me_2P \begin{array}{c} CH_2-M-CH_2 \\ \vdots \\ CH_2-M-CH_2 \end{array} PMe_2 \right] \quad (51)$$

(**91**)

91	**a**	**b**	**c**
M	Cu	Ag	Au

$$\left[Et_2P \begin{array}{c} CH_2-Au-CH_2 \\ \vdots \\ CH_2-Au-CH_2 \end{array} PEt_2 \right]$$

(**92**)

Molecular structural analyses of **91a**[91] and **92**[92] confirm the presence of tetrahedral phosphorus atoms and approximately linear C_2M arrangements in the ring. The relatively short metal–metal distances (Cu—Cu = 2.843 Å and Au—Au = 3.023 Å) indicate weak transannular metal–metal bonds. Metallocycles originating from arsonium ylides[92,93] as well as double spirocyclic complexes such as **93**[94] and **94**[95] have also been the subjects of investigation (equation 52).

$$\frac{2}{n}[(Me_3P)MCl]_n \xrightarrow[-2[(CH_2)_xPMe_2]Cl,\ -2\,PMe_3]{+4(CH_2)_xP(Me)CH_2} \left[(CH_2)_x P \begin{array}{c} CH_2-M-CH_2 \\ CH_2-M-CH_2 \end{array} P (CH_2)_x \right] \quad (52)$$

(**93**) (**94**)

M	Cu	Ag	Au
$x = 4$			**93**
$x = 5$	**94a**	**94b**	**94c**

By coordination to gold atoms in **93** the otherwise unstable monomeric λ^5-phospholane $(CH_2)_4P(Me)CH_2$ achieves stabilization.

Eight-membered heterocycles have been reported for gallium, indium and thallium (equation 53)[96]. The synthesis of **95** is accompanied by polymer formation; however, monomeric species cannot be detected. Strongly covalent thallium–carbon σ-bonds in **95c** are indicated by the observation of coupling constants such as $^2J(TlCH_3) = 305.5$ Hz. and $^2J(TlCH_2) = 152$ Hz in the ^1H n.m.r. spectrum. Magnesium, zinc, and cadmium alkyls have also been found to undergo reactions of this type with phosphorus ylides. With these reagents, the higher acceptor capacity of the metals leads to the formation of the novel organometallic polymers **96**[22]. Mercury ylide complexes, on the other hand, always exhibit low molecular or ionic characteristics[97,98].

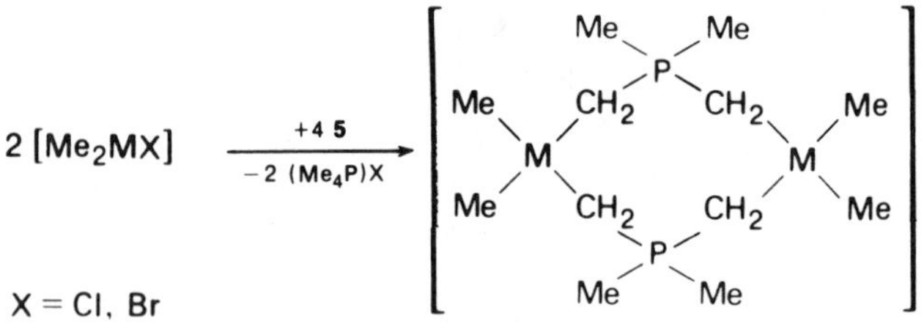

$$2\,[Me_2MX] \xrightarrow[-2\,(Me_4P)X]{+4\ \mathbf{5}} [Me_2M(CH_2)_2PMe_2]_2 \quad (53)$$

X = Cl, Br

(95)

95	a	b	c
M	Ga	In	Tl

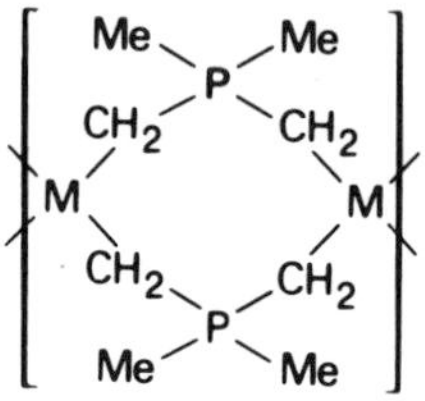

(96)

96	a	b	c
M	Mg	Zn	Cd

As shown below, both the carbanionic centres of **87** can be accommodated at the acceptor sites of gold, rhodium, nickel, palladium and platinum, resulting in complexes of square-planar configuration. The close vicinity of the gold atoms in **91c** and **92** provides the conditions for the transannular oxidative addition of electrophiles such as halogens, methyliodide, and disulphides with the generation of metal–metal multiple bonds[99–101] (equation 54).

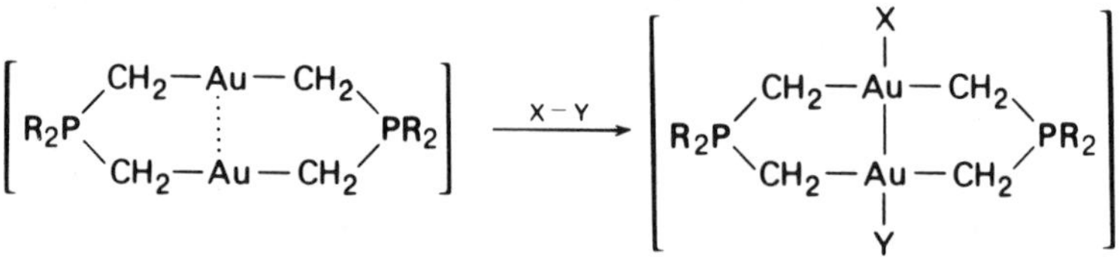

$$[R_2P(CH_2\text{–}Au\text{–}CH_2)_2PR_2] \xrightarrow{X-Y} [R_2P(CH_2)_2Au(X)\text{–}Au(Y)(CH_2)_2PR_2] \quad (54)$$

(91c), (92) (97), (98)

X—Y	Cl_2	Br_2	I_2	MeI	$(S_2CNEt_2)_2$
R = Me	**97a**	**97b**	**97c**	**97d**	
R = Et	**98a**	**98b**	**98c**		**98e**

The molecular structure of **98a** was elucidated by X-ray analysis[100] (Figure 9). The gold atoms, surrounded by ligands of approximately square-planar configuration, are arranged in the same plane as the chlorine and phosphorus atoms from which the carbon atoms C^1 and C^2 deviate slightly. The gold–gold bond distance of 2.597 Å is the shortest reported to date, and can be compared with the corresponding bond lengths in the metal of 2.884 Å, and with those in gold clusters, which range from 2.68 to 2.98 Å. The metal carries a formal oxidation state of +II in a linear

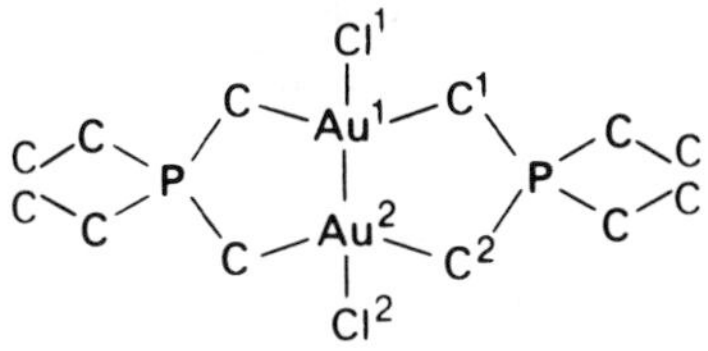

FIGURE 9. Molecular structure of **98a** (hydrogen atoms have been omitted).

X—Au—Au—X moiety. Thus, **98a** represents the first example of a compound containing a multiple bond between two Group IB metals.

The metal–metal linkage in **97a** and **97b**, unlike that in **97c**, can be cleaved by treatment with another equivalent of halogen.

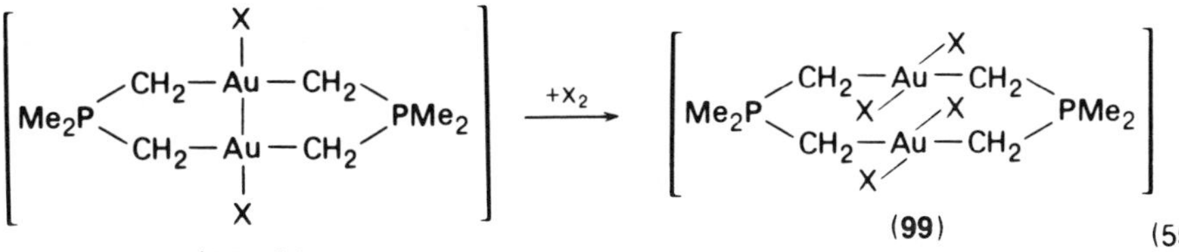

(**97a, b**) $\xrightarrow{+X_2}$ (**99**) (55)

99	a	b
X	Cl	Br

The methylation of **99b** with methyllithium yields the complex **101** instead of the expected tetramethyl derivative **100**. Severe steric hindrance of the gold substituents in **100**, as may also be true for the hypothetical tetraiodo complex **99c**, provides a rationale for this result (equation 56)[99].

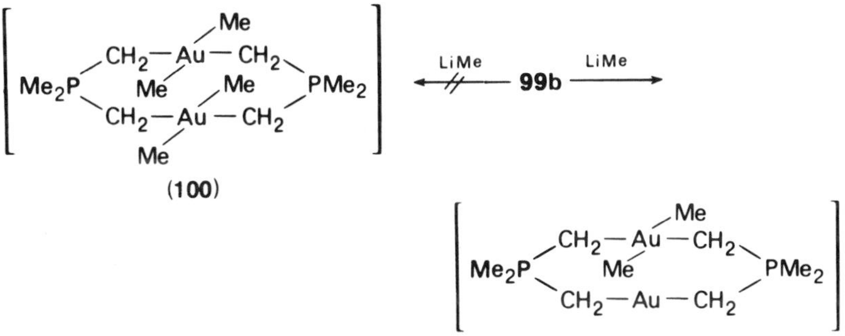

(**100**) $\xleftarrow{LiMe}$ (not formed) **99b** $\xrightarrow{LiMe}$ (**101**) (56)

In **102**, heterocycles containing gold serve as building blocks for a polymeric chain (equation 57)[102].

In the eight-membered heterocycle **105**, nickel atoms, coordinated in a square-planar environment, are incorporated as metallate centres. Whereas in pentane solution the reaction of **103** with stoichiometric amounts of ylide leads to the salt-like **104**, the synthesis of **105** is accomplished by employing **103** and excess of ylide in ether. Catalytic amounts of ylide also effect the conversion of **104** to **105**[33] (equations 58 and 59).

$$2_n Me_2P(CH_2)_6PMe_2(=CH_2)_2 \xrightarrow[-2_n PMe_3, -n[Me_3P(CH_2)_6PMe_3]Cl_2]{+2_n[Me_3PAuCl]} [(CH_2)_6P(Me)(CH_2{-}Au{-}CH_2)_2P(Me)CH_2]_n \quad (57)$$

(102)

$$[(Me_3P)_2NiCl(Me)] + 2\ \mathbf{5} \xrightarrow[-PMe_3]{pentane} [(Me)(Me_3P)Ni(CH_2PMe_3)_2]^+ Cl^- \quad (58)$$

(103) (104)

$$2\ \mathbf{103} + 4\ \mathbf{5} \xrightarrow[-2\ PMe_3,\ -2(Me_4P)Cl]{Et_2O} [(Me)(Me_3P)Ni(CH_2)_2P(Me)_2(CH_2)_2Ni(PMe_3)(Me)\ \text{with second } P(Me)_2 \text{ bridge}] \quad (59)$$

(105)

The metal–metal quadruple bonds of **107** between two nuclei, chromium(II) and molybdenum(II), are excellently stabilized by four bridging ylides. The preparation of **107** involves a stoichiometric amount of phosphonium salt which is deprotonated by the methyl ligands of the precursor **106**[103,104] (equation 60).

$$[Li_4M_2Me_8]\cdot 4\ thf \xrightarrow[-8\ CH_4,\ -4\ LiCl]{+4\ (Me_4P)Cl} [M_2\{(CH_2)_2PMe_2\}_4] + 4\ thf \quad (60)$$

(106) (107)

107	a	b
M	Cr	Mo
m.p.(°C)	165–167 (dec.)	256–275 (dec.)

The thermodynamic and kinetic stability of **107**, which possess eight metal–carbon σ-bonds per molecule, is attributed to the inductive effect of the phosphonium centres and to the existence of four bridges embedding and protecting the multiple bond. X-ray analyses reveal extremely short metal–metal distances, the chromium–chromium bond length of 1.895 Å being one of the shortest between two chromium nuclei. Further, it is noteworthy that the phosphorus–methylene carbon distances (1.75 Å) are always significantly shorter than the phosphorus–methyl carbon bond lengths (1.84 Å), which suggests that at least some of the ylidic character of the PCH_2 linkage is maintained after complex formation[105].

In the organotitanium compound **108**, two octahedrally coordinated metal atoms are held together by bridging ylide ligands[106] (equation 61). Here again phosphorus–

$$2\,[(MeO)_3TiCl] \xrightarrow[-2(Me_4P)Cl]{+4\ \mathbf{5}} \left[(MeO)_2Ti(\mu\text{-}OMe)_2Ti(OMe)_2(\mu\text{-}CH_2PMe_2CH_2)_2 \right] \quad (61)$$

(108)

methylene carbon distances of 1.71 and 1.75 Å in addition to the long titanium–carbon σ-bonds (2.30 Å) confirm the inheritance of some ylide nature in the bridging unit.

The two anionic functions at the central carbon atom of hexamethylcarbodiphosphorane are coordinated to the acceptor sites of two gold atoms in **109**, thus illustrating the bridging properties of this ambidentate double ylide[107] (equation 62).

$$Me_3P{=}C{=}PMe_3 \xrightarrow[-2\,PMe_3]{+2[MeAuPMe_3]} \left[(Me_3P)_2C(AuMe)_2 \right] \quad (62)$$

(109)

2. *Chelates*

As mentioned earlier, the double ylide **87** also possesses chelating properties (see **B** in Figure 8). With R = Me, however, complications are often encountered owing to the presence of monomer–oligomer equilibria or to the appearance of different oligomers. The employment of ylides with two bulky *t*-butyl substituents at the phosphorus atom enhances the selectivity of the reaction and favours only the formation of monomeric chelates. Thus, the carbonyl rhodium derivative **111** exists in solution as a strongly temperature-dependent equilibrium mixture of monomeric and dimeric species[4,108]. The same reaction sequence involving $t\text{-}Bu_2CH_3PCH_2$ instead of Me_3PCH_2 yields monomeric **112** and **113**, respectively, as the sole organometallic products[109] (Scheme 17).

In contrast to **110**, the homologous iridium compound can only be obtained as mononuclear chelate complex[110]. The nickel complex **114** is converted to the salt-like **115** by treatment with Me_3PCH_2 in a pentane slurry. In thf as solvent the ionic intermediate cannot be detected, and transylidation proceeds to give a mixture of the two binuclear isomers **116a** and **116b** (equations 63 and 64)[111].

One of the isomers, **116a**, obtained in platelets by slow crystallization, was subjected to a crystal structure analysis and its constitution fully established[46] (Figure 10). This isomer contains the ylidic ligand **87** (R = Me) both in the bridging position **A** and in

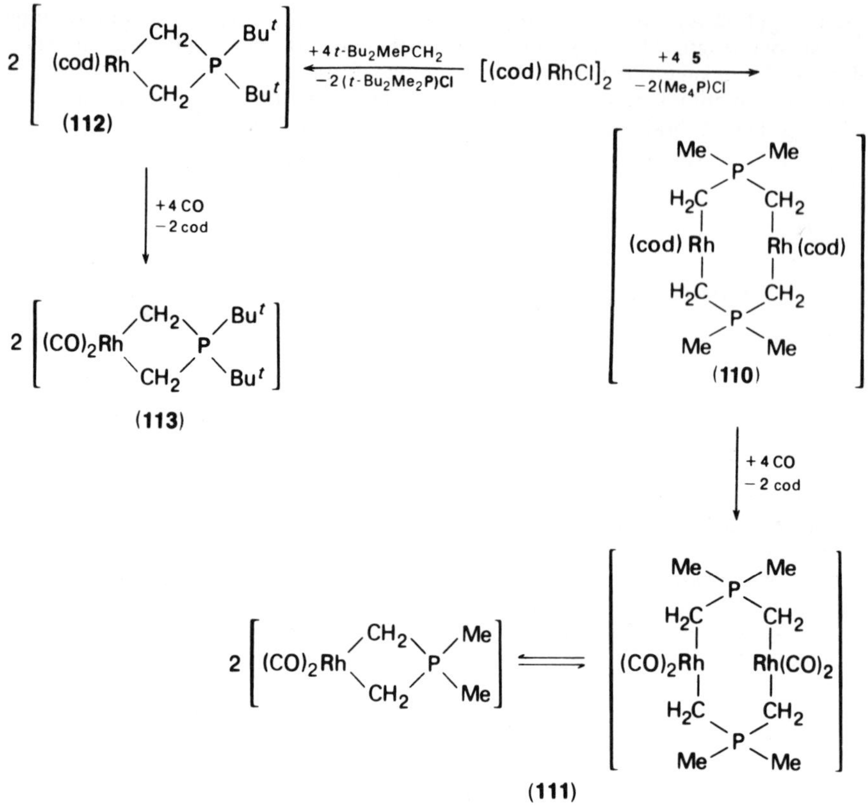

$$2\left[(cod)Rh(CH_2)_2PBu^t_2\right] \ (\mathbf{112}) \xleftarrow[-2(t\text{-}Bu_2Me_2P)Cl]{+4\ t\text{-}Bu_2MePCH_2} [(cod)RhCl]_2 \xrightarrow[-2(Me_4P)Cl]{+4\ \mathbf{5}} \left[(cod)Rh(CH_2PMe_2CH_2)_2Rh(cod)\right] \ (\mathbf{110})$$

$$\mathbf{112} \xrightarrow[-2\,cod]{+4\,CO} 2\left[(CO)_2Rh(CH_2)_2PBu^t_2\right] \ (\mathbf{113})$$

$$\mathbf{110} \xrightarrow[-2\,cod]{+4\,CO} 2\left[(CO)_2Rh(CH_2)_2PMe_2\right] \rightleftharpoons \left[(CO)_2Rh(CH_2PMe_2CH_2)_2Rh(CO)_2\right] \ (\mathbf{111})$$

SCHEME 17

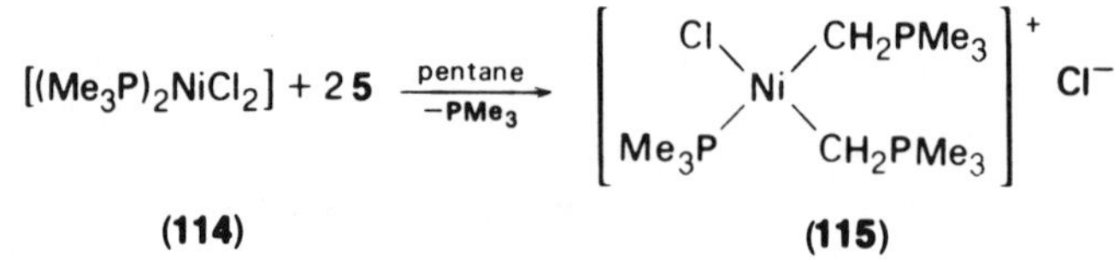

$$[(Me_3P)_2NiCl_2] + 2\,\mathbf{5} \xrightarrow[-PMe_3]{pentane} \left[(Cl)(Me_3P)Ni(CH_2PMe_3)_2\right]^+ Cl^- \quad (63)$$

(**114**) (**115**)

$$2\ \mathbf{114} + 8\ \mathbf{5} \xrightarrow{thf} 4\,(Me_4P)Cl + \mathbf{116\,a,b} + 4\,PMe_3 \quad (64)$$

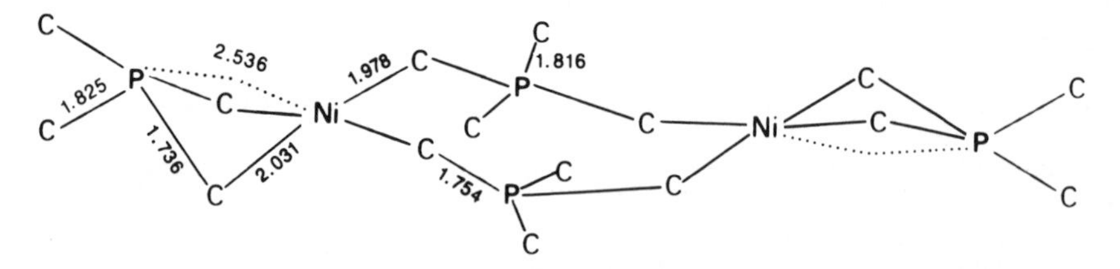

FIGURE 10. Molecular structure of and selected bond distances in **116a**.

the chelating position **B**, as shown in Figure 8. The phosphorus atom of each chelating ligand is displaced 0.79 Å from the plane defined by the four carbon atoms around each nickel atom. The dihedral angle of 43.9° between this plane and the one defined by the methylene carbon atoms and the phosphorus for each chelating ligand gives rise to a weak nickel–phosphorus bonding interaction (Ni—P = 2.536 Å). The phosphorus–methylene bond lengths in each chelate part of the molecule clearly indicate slight multiple bond character. The nickel–phosphorus interaction is best described in terms of the resonance structures **A–D** (Figure 11) with predominance of **A** and **C**. Therefore, the geometry of the ligand resembles the well known array of π-allyl groups with the restriction of a minor interaction between the metal and the phosphorus in the heteroatomic system.

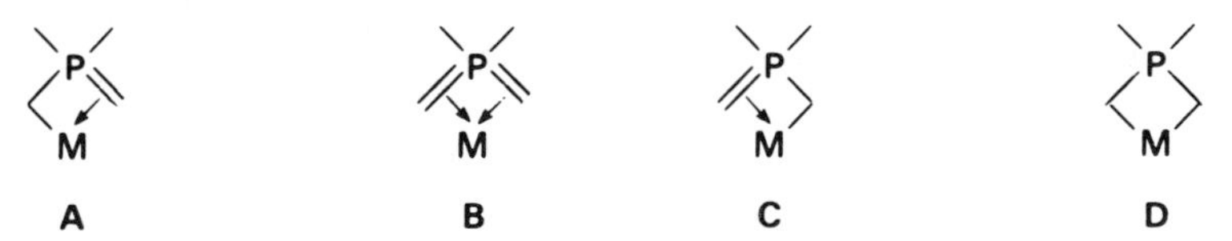

FIGURE 11. Various resonance structures for the description of the nickel chelate interaction in **116a**.

Rapid cooling of the mixture resulting from the reaction of **114** and Me_3PCH_2, or rapid removal of solvent leads to the isolation of yellow **116b**, the cage structure of which is postulated on analytical and spectroscopic grounds. In the gas phase, the existence of monomer **116c** is also evident.

$$[Me_2P(CH_2)_2Ni(CH_2)_2PMe_2]$$

(**116b**) (**116c**)

In contrast to the above, the utilization of bulky t-Bu_2MePCH_2 furnishes only the specific generation of bischelate **117**. The analogous palladium and platinum derivatives of **117** have also been reported[112].

$$\mathbf{114} \xrightarrow[-2(t\text{-}Bu_2Me_2P)Cl]{+4\,t\text{-}Bu_2MePCH_2} [Bu^t_2P(CH_2)_2Ni(CH_2)_2PBu^t_2] \qquad (65)$$

(**117**)

The complications encountered in the production of complexes derived from **87** (R = Me) are also circumvented when two alkyl substituents and the phosphorus atom of the ylidic substrate are incorporated into a phospholane or phosphorinane system. Although the low steric requirements of these ligands suggest a straightforward synthetic route to oligomers, most surprisingly exclusive monomer formation is observed with nickel[112] and rhodium[109] (equations 66 and 67). The unexpected generation of **118** is attributed to a small CPC angle in the heterocycle concomitant with opening of the ylidic CPC angle, which in turn may allow a more pronounced pseudo-allylic interaction between ligand and metal atom.

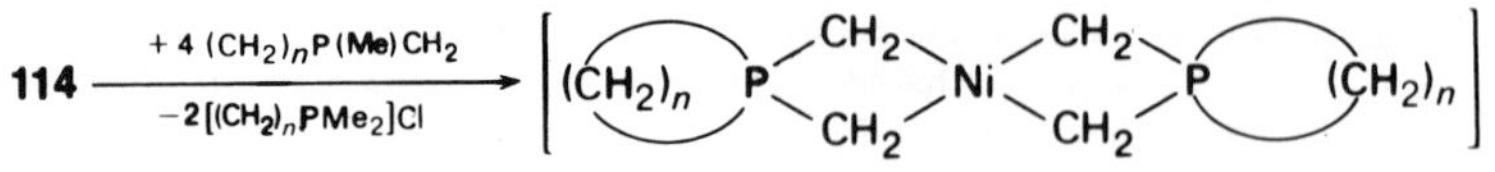

$$114 \xrightarrow[-2[(CH_2)_nPMe_2]Cl]{+4\,(CH_2)_nP(Me)CH_2} \left[(CH_2)_n\,P\begin{smallmatrix}CH_2\\ CH_2\end{smallmatrix}Ni\begin{smallmatrix}CH_2\\ CH_2\end{smallmatrix}P\,(CH_2)_n\right] \quad (66)$$

(118)

118	a	b
n	4	5

$$[(cod)RhCl]_2 \xrightarrow[-2[(CH_2)_5PMe_2]Cl]{+4\,(CH_2)_5P(Me)CH_2} 2\left[(cod)Rh\begin{smallmatrix}CH_2\\ CH_2\end{smallmatrix}P(CH_2)_5\right] \quad (67)$$

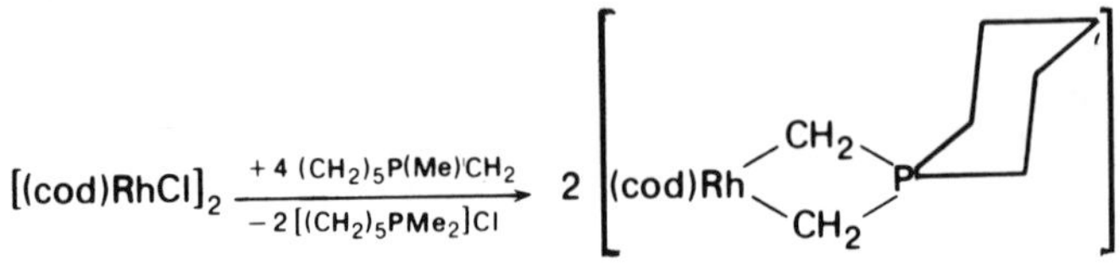

(119)

With palladium in the form of $[(Me_3P)_2PdCl_2]$, excess of Me_3PCH_2 in benzene solution affords a mixture of the ionic chelate **120** and the dimer **121** (Scheme 18)[113].

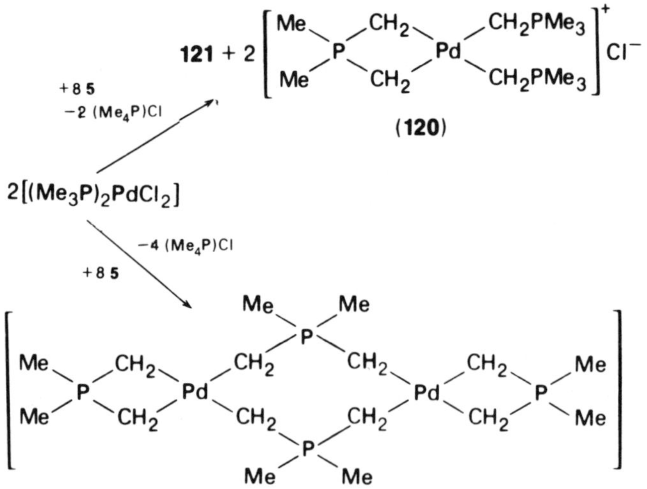

$$2[(Me_3P)_2PdCl_2] \xrightarrow[-2\,(Me_4P)Cl]{+8\,5} 121 + 2\left[Me_2P\begin{smallmatrix}CH_2\\ CH_2\end{smallmatrix}Pd\begin{smallmatrix}CH_2PMe_3\\ CH_2PMe_3\end{smallmatrix}\right]^+ Cl^-$$

(120)

$$2[(Me_3P)_2PdCl_2] \xrightarrow[+8\,5]{-4\,(Me_4P)Cl} \left[Me_2P\begin{smallmatrix}CH_2\\ CH_2\end{smallmatrix}Pd\begin{smallmatrix}CH_2-PMe_2-CH_2\\ CH_2-PMe_2-CH_2\end{smallmatrix}Pd\begin{smallmatrix}CH_2\\ CH_2\end{smallmatrix}PMe_2\right]$$

(121)

SCHEME 18

Obviously, the low solubility of **120** in benzene prevents its subsequent transformation to **121**. Treatment of $[(Me_3P)_2PdCl_2]$ with 1-methyl-1-methylene-λ^5-phospholane and -phosphorinane proceeds with exclusive transylidation to the structurally fascinating dimers **122** and **123**, respectively. This result can be attributed to the minimal steric demands of the ligands and altered solubilities (Scheme 19)[113].

The introduction of the chelate ligands into quasi-tetrahedral complexes of scandium, vanadium, and titanium is achieved by employment of pre-lithiated ylides[5] (Scheme 20).

The ylide in **128**[114] functions as a chelating ligand occupying one octahedral edge. However, compared with the nickel compound **116a**, there are significant stereochemical differences in the configuration of the four-membered ring. The

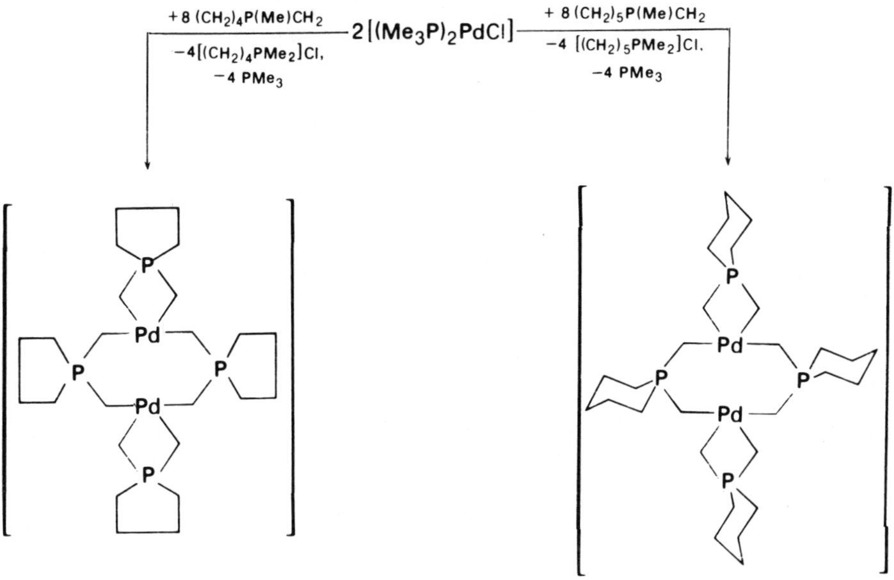

+8 $(CH_2)_4P(Me)CH_2$ / $-4[(CH_2)_4PMe_2]Cl$, $-4\ PMe_3$ ← $2[(Me_3P)_2PdCl]$ → +8 $(CH_2)_5P(Me)CH_2$ / $-4\ [(CH_2)_5PMe_2]Cl$, $-4\ PMe_3$

(**122**) (**123**)

SCHEME 19

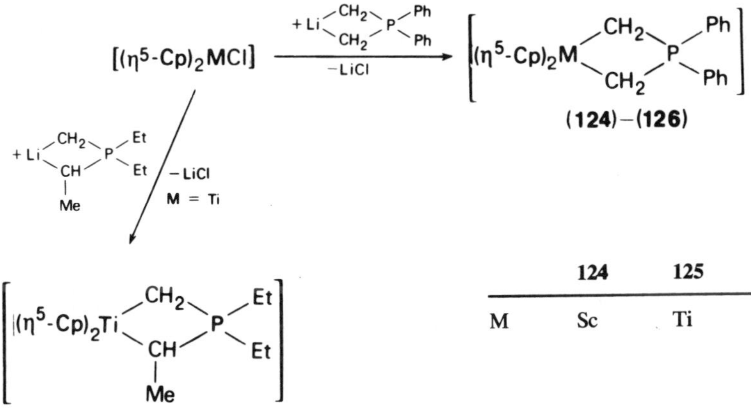

$[(\eta^5\text{-Cp})_2MCl]$ + $Li(CH_2)_2PPh_2$, $-LiCl$ → $[(\eta^5\text{-Cp})_2M(CH_2)_2PPh_2]$ (**124**)–(**126**)

$[(\eta^5\text{-Cp})_2MCl]$ + $Li[CH_2CH(Me)PEt_2]$, $-LiCl$, M = Ti → $[(\eta^5\text{-Cp})_2Ti(CH_2)(CH(Me))PEt_2]$ (**127**)

	124	125	126
M	Sc	Ti	V

SCHEME 20

planarity of the CoC_2P ring suggests more regular σ-type bonding for the cobalt(III) species[46] (equation 68).

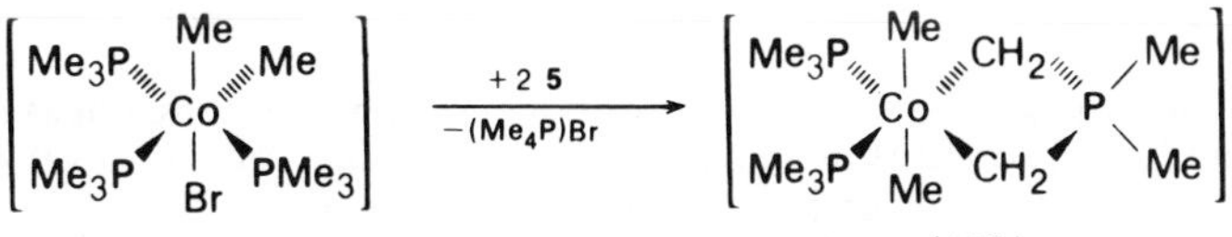

$$[(Me_3P)_3Co(Me)_2Br] \xrightarrow[-(Me_4P)Br]{+2\ \mathbf{5}} [(Me_3P)_2Co(Me)_2(CH_2)_2PMe_2] \quad (68)$$

(**128**)

In the distorted octahedral chromium derivative **130**, the double ylide **87** appears as the sole ligand comprising all six chromium–carbon σ-bonds. The mobile carbanionic ligands of precursor **129** behave as bases during the generation of **87** from the

$$Li_3[CrPh_6] + 3(Me_4P)Cl \xrightarrow[-3\,LiCl,\ -6\,C_6H_6]{thf} [Cr\{(CH_2)_2PMe_2\}_3] \qquad (69)$$

(**129**) (**130**)

phosphonium salt[115a]. This reaction can be described in terms of various dissociation and deprotonation processes which occur either simultaneously or consecutively. Intermediate *trans*-$[PhCr(CH_2PMe_3)\{(CH_2)_2PMe_2\}_2]$ was isolated in low yields. Because of the mutually *trans*-configuration of the phenyl and terminal ylide ligands, subsequent benzene release from this intermediate leading to the product **130** is hindered[115b]. Regardless of its extreme sensitivity towards oxidation and hydrolysis and despite the presence of six covalent carbon–metal σ-bonds, **130** exhibits remarkably thermal stability. It melts in the range 59–68°C and can be sublimed at 10^{-3} torr and 30°C.

The lanthanide compounds **132**, analogous to **130**, were synthesized by treatment of the salt-like ylide complexes **131** with *n*-butyllithium. Transylidation reactions, however, failed[116] (equation 70).

$$LnCl_3 \xrightarrow{+3\,5} \underset{(\mathbf{131})}{[Ln(CH_2PMe_3)_3]Cl_3} \xrightarrow[-3C_4H_{10},\ -3LiCl]{+3LiBu} \underset{(\mathbf{132})}{Ln[(CH_2)_2PMe_2]_3} \qquad (70)$$

Ln	La	Pr	Nd	Sm	Ho	Er	Lu
131, 132	**a**	**b**	**c**	**d**	**e**	**f**	**g**

Chelates **133**[115a] and **134**[37] represent a related class of complexes which, in addition to the ylidic centres, possess *ortho*-metallated aromatic rings at the onium functions (equations 71 and 72).

$$[Ph_3Cr(thf)_3] + 3\,Ph_3PCH_2 \xrightarrow{-3\,thf,\ -3\,C_6H_6} [Cr(CH_2PPh_2C_6H_4)_3] \qquad (71)$$

(**133**)

A crystalline sodium complex (**136**) was prepared from the phosphine-substituted ylide **135** and $NaNH_2$ or NaH in an ether solvent (equation 73). Its novel structure, elucidated by X-ray crystallography, is shown in Figure 12[117]. The ylidic ligand is attached to the metal via a P→Na donor bond and via a sodium–benzylide

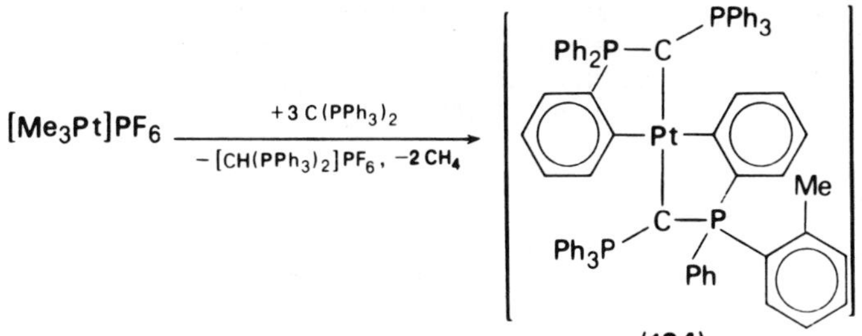

$$[Me_3Pt]PF_6 \xrightarrow[-[CH(PPh_3)_2]PF_6,\ -2CH_4]{+3\,C(PPh_3)_2} \text{(134)} \qquad (72)$$

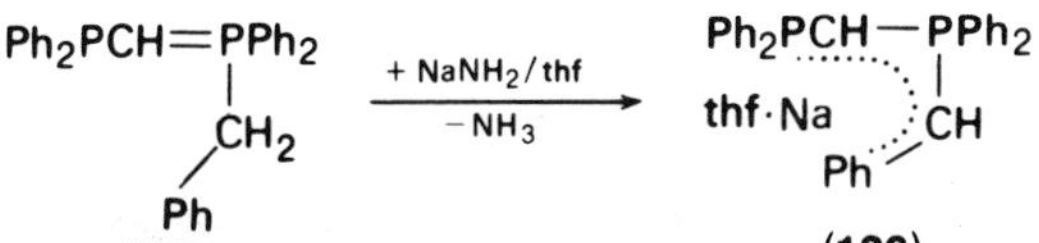

(134)

$$\underset{\mathbf{(135)}}{Ph_2PCH{=}PPh_2(CH_2Ph)} \xrightarrow[-NH_3]{+\,NaNH_2/thf} \underset{\mathbf{(136)}}{\text{thf}\cdot Na[Ph_2PCH{-}PPh_2{-}CHPh]} \qquad (73)$$

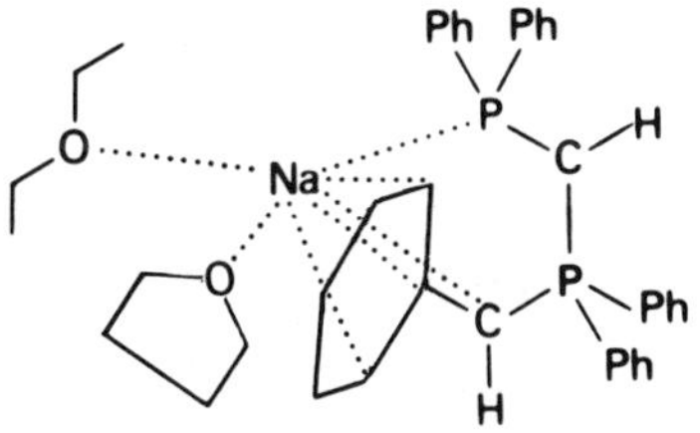

FIGURE 12. Molecular structure of **136**.

interaction. The latter appears to result from electrostatic forces between Na^+, the negatively charged benzylide, and *ortho*-carbon atoms, which is in accordance with the classical charge distribution in a benzylide anion.

Double ylide bridges with additional chelating functions are characteristic of the uranium binuclear complex **137** (Figure 13). In this organometallic species, the uranium atoms of formal oxidation state +IV utilize the unusual coordination number 9[118] (equation 74).

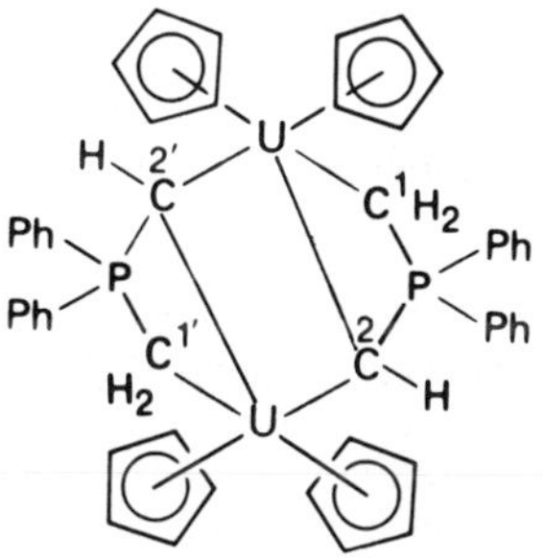

FIGURE 13. Molecular structure of **137**.

$$[Cp_3UCl] + 2Li(CH_2)_2PPh_2 \xrightarrow[-50^\circ C]{Et_2O} \tfrac{1}{2}\mathbf{137} + LiCl + Ph_2MePCH_2 + LiCp \qquad (74)$$

In contrast to numerous reports on the ligating properties of systems such as **87** with phosphonium centres, there exists only one well characterized and low-molecular-weight transition metal compound in which a sulphur ylide accomplishes chelating functions[119] (Scheme 21).

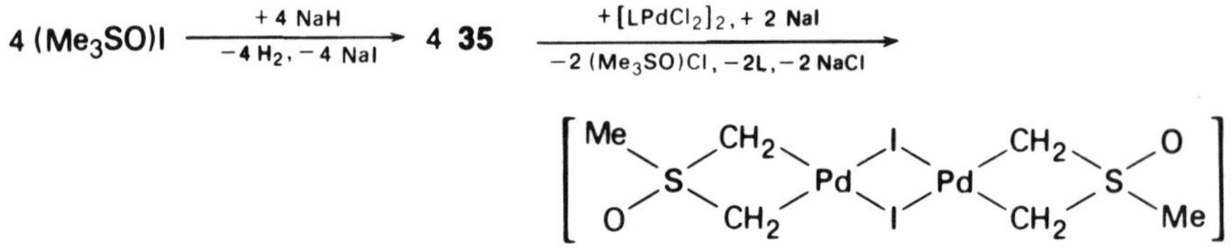

$$4\ (Me_3SO)I \xrightarrow[-4\ H_2,\ -4\ NaI]{+4\ NaH} 4\ \mathbf{35} \xrightarrow[-2\ (Me_3SO)Cl,\ -2L,\ -2\ NaCl]{+[LPdCl_2]_2,\ +2\ NaI}$$

$$\left[Me(O)S(CH_2)_2Pd(\mu\text{-}I)_2Pd(CH_2)_2S(O)Me \right] \quad \mathbf{(138)}$$

L = styrene

SCHEME 21

Chelating and bridging $R(O)S(CH_2)_2$ moieties are also assumed to be present in coordination polymers[98] as well as in some magnesium derivatives[120].

Both carbanionic centres of double ylide **140**, each stabilized by its own onium function, are accommodated at the available acceptor sites of zinc and cadmium alkyls[121] (equation 75).

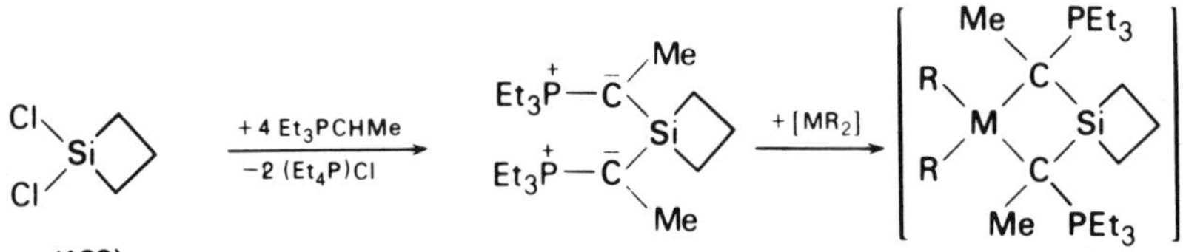

$$Cl_2Si(CH_2)_3\ \mathbf{(139)} \xrightarrow[-2\ (Et_4P)Cl]{+4\ Et_3PCHMe} (Et_3\overset{+}{P}-\overset{-}{C}Me)_2Si(CH_2)_3\ \mathbf{(140)} \xrightarrow{+[MR_2]} \left[R_2M(C(Me)PEt_3)_2Si(CH_2)_3 \right]\ \mathbf{(141)},\ \mathbf{(142)} \qquad (75)$$

	141	**142**
M	Zn	Cd
R	Et	Me

Proton abstraction from the isoelectronic onium salts **143a–c** affords the cumulated double ylide **144a**, the conjugated double ylide **144b** and the monofunctional ylide **144c**, respectively (Scheme 22).

In alkyl-substituted carbodiphosphoranes such as **144a**, the ylidic function is not strictly limited to the central carbon atom as it is in the hexaphenyl derivative **14**, but can, under appropriate conditions, be prototropically transferred to the periphery of the molecule. This behaviour turns out to be of eminent importance for complex formation with metals (Scheme 23). Similar prototropic phenomena are also observed with **144b**.

Repeated deprotonations at the peripheral methyl groups convert **144a–c** into a series of isoelectronic bidentate ligand anions (**145a–c**) (Scheme 22). Using their peripheral donor sites, these anions operate as extremely powerful ligand systems in main group and transition metal chelates. Their generation from **143** or **144** in a metal coordination sphere follows the same principles which have been applied to the synthesis of ligated **87**.

With **145**, Group IIIA metals generate tetrahedral complexes, the metal chelate moiety of which appears to be isostructural, irrespective of the nature of the central ligand atom Y (Scheme 24)[122–124].

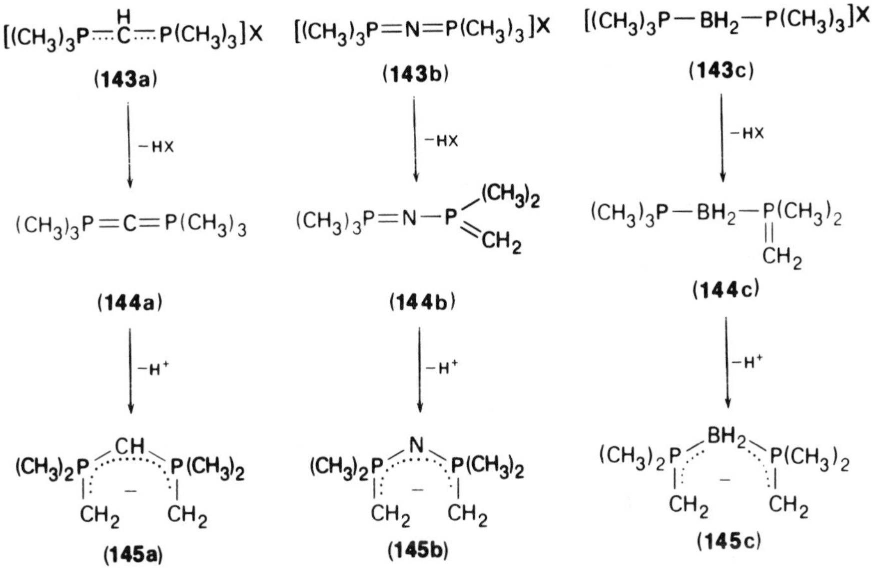

$[(CH_3)_3P\!\cdots\!C(H)\!\cdots\!P(CH_3)_3]X$ (143a) $[(CH_3)_3P{=}N{=}P(CH_3)_3]X$ (143b) $[(CH_3)_3P{-}BH_2{-}P(CH_3)_3]X$ (143c)

−HX

$(CH_3)_3P{=}C{=}P(CH_3)_3$ (144a) $(CH_3)_3P{=}N{-}P(CH_3)_2{=}CH_2$ (144b) $(CH_3)_3P{-}BH_2{-}P(CH_3)_2{=}CH_2$ (144c)

$-H^+$

(145a) (145b) (145c)

SCHEME 22

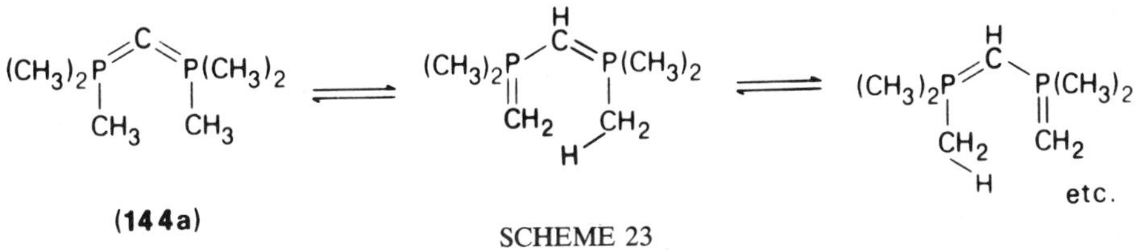

(144a) etc.

SCHEME 23

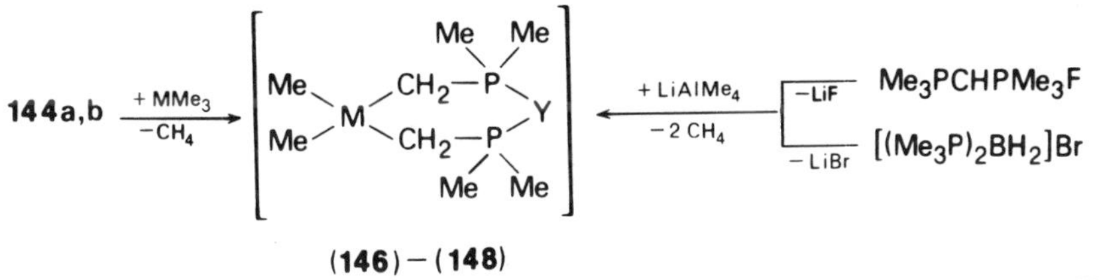

144a,b $\xrightarrow[-CH_4]{+MMe_3}$ $[Me_2M(CH_2PMe_2)_2Y]$ (146)–(148) $\xleftarrow[-2\,CH_4]{+LiAlMe_4}$ $Me_3PCHPMe_3F$ (−LiF); $[(Me_3P)_2BH_2]Br$ (−LiBr)

Y	CH	N	BH_2
M = Al	**146a**		**146c**
M = Ga	**147a**	**147b**	
M = Tl		**148b**	

SCHEME 24

With these chelating ligands, Group IIB metals lead to tetrahedral complexes in which all the metal acceptor sites are occupied[122–126] (Scheme 25).

In contrast to their zinc and cadmium analogues, dialkylmercury compounds are extremely weak acceptor molecules with non-basic alkyl groups. Despite their inherent inability to form simple tri- or tetraalkylmercurates(II), such as HgR_3^- or

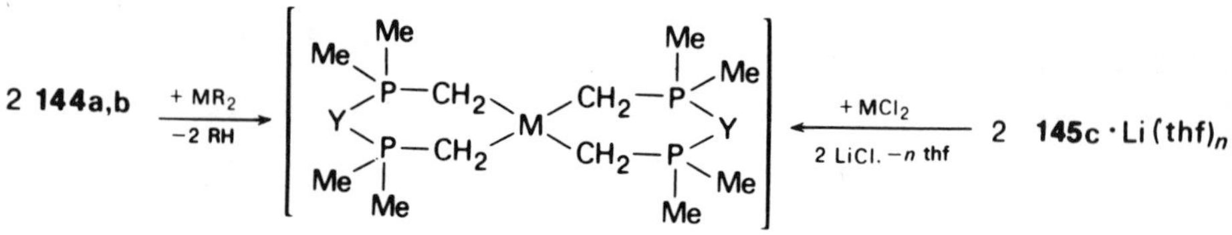

2 **144a,b** $\xrightarrow[-2\ RH]{+\ MR_2}$ **(149)**, **(150)** $\xleftarrow[2\ LiCl,\ -n\ thf]{+\ MCl_2}$ 2 **145c**·Li(thf)$_n$

	M	R	Y
149a	Zn	Et	CH
149b	Zn	Et	N
149c	Zn		BH_2

	M	R	Y
150a	Cd	Me	CH
150b	Cd	Me	N
150c	Cd		BH_2

SCHEME 25

HgR_4^-, using **144a** the synthesis of the first species (**151**) with a tetraalkylmercurate structure has been accomplished[127] (equation 76).

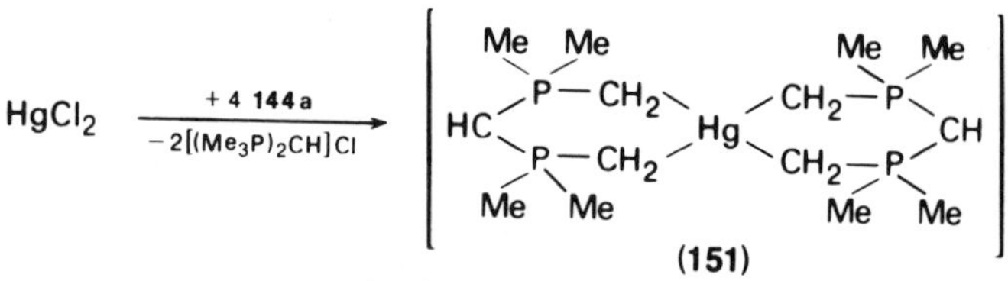

$$HgCl_2 \xrightarrow[-2[(Me_3P)_2CH]Cl]{+4\ \mathbf{144a}} \mathbf{(151)} \qquad (76)$$

The excellent ligating capacity of these chelating systems certainly also accounts for the successful preparation of the first low-molecular-weight magnesium ylide complex containing four metal–carbon σ-bonds[123].

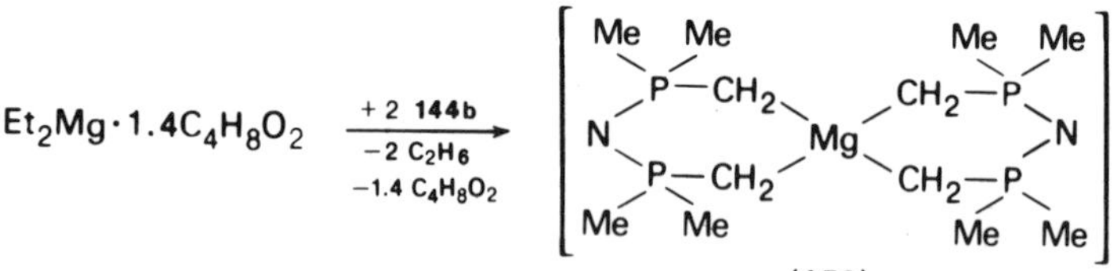

$$Et_2Mg\cdot 1.4C_4H_8O_2 \xrightarrow[-2\ C_2H_6,\ -1.4\ C_4H_8O_2]{+2\ \mathbf{144b}} \mathbf{(152)} \qquad (77)$$

The same structural principle has been found in a number of square-planar complexes (equations 78[122,126] and 79[109,110]) (Scheme 26[122,125,126]). A pentacoordinate iridium derivative of **145b** has also been reported[110].

In all the complexes under discussion, the ligand is attached to the metal centre by covalent σ-bonds. Molecular structural analyses of **153a,b**[128], **156b**[126] and **156c**[125] indicate metal atoms in square-planar environments as members of largely isostructural heterocycles which maintain strainless chair conformations. In **153a** the P—CH bonds of the planar PCHP moiety appear significantly shorter than the P—CH_2 bond distances. These results, in addition to the high field resonance of the CH protons in the n.m.r. experiment, confirm the strong ylidic character of the bridge, illustrated by resonance structures **A–C** in Figure 14.

Figure 15 shows the molecular structure and selected data for **153a**[128].

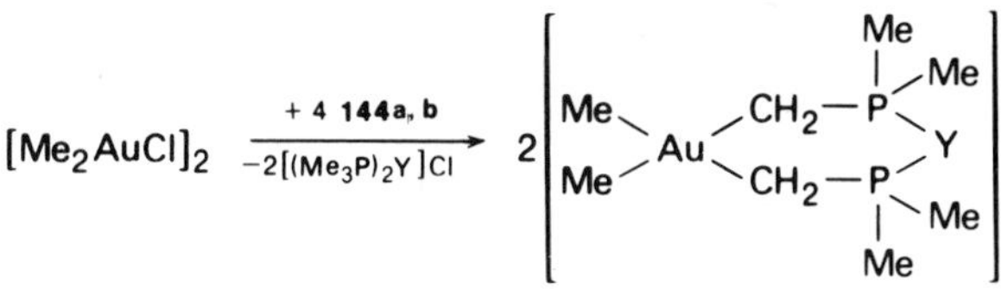

$$[Me_2AuCl]_2 \xrightarrow[-2[(Me_3P)_2Y]Cl]{+4\ \mathbf{144a,b}} 2\left[Me_2Au(CH_2PMe_2)_2Y\right] \quad (78)$$

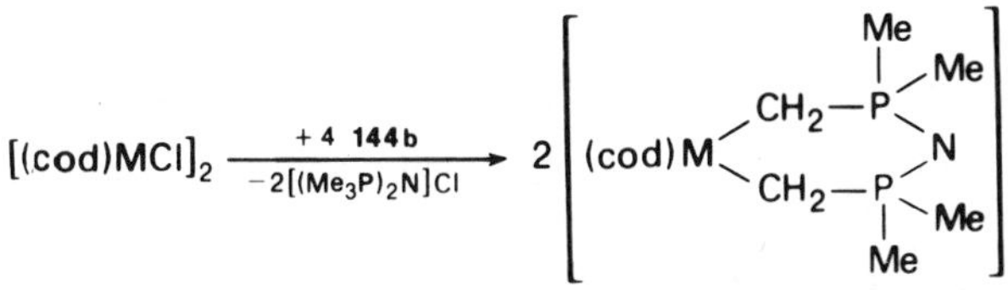

(153)

Y	CH	N
153	a	b

$$[(cod)MCl]_2 \xrightarrow[-2[(Me_3P)_2N]Cl]{+4\ \mathbf{144b}} 2\left[(cod)M(CH_2PMe_2)_2N\right] \quad (79)$$

(154), **(155)**

M	Rh	Ir
	154	**155**

$$[(Me_3P)_2MCl_2] \xrightarrow[-2\,PMe_3]{+4\ \mathbf{144a,b},\ -2[(Me_3P)_2Y]Cl \text{ or } +2\ \mathbf{145c}\cdot Li(thf)_n,\ -2\,LiCl,\ -n\,thf} \left[Y(Me_2PCH_2)_2M(CH_2PMe_2)_2Y\right]$$

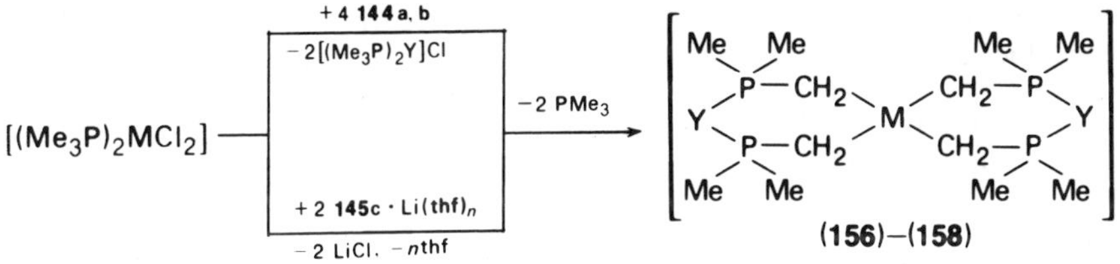

(156)–**(158)**

Y	CH	N	BH_2
M = Ni	**156a**	**156b**	**156c**
M = Pd	**157a**		**157c**
M = Pt	**158a**	**158b**	**158c**

SCHEME 26

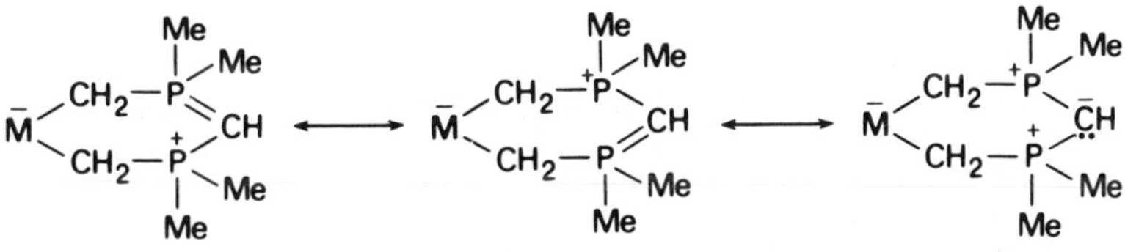

$$\bar{M}(CH_2)_2(PMe_2)(\overset{+}{P}Me_2)CH \ (\mathbf{A}) \longleftrightarrow (\mathbf{B}) \longleftrightarrow (\mathbf{C})$$

A B C

FIGURE 14. Resonance structures of ligating **145a**.

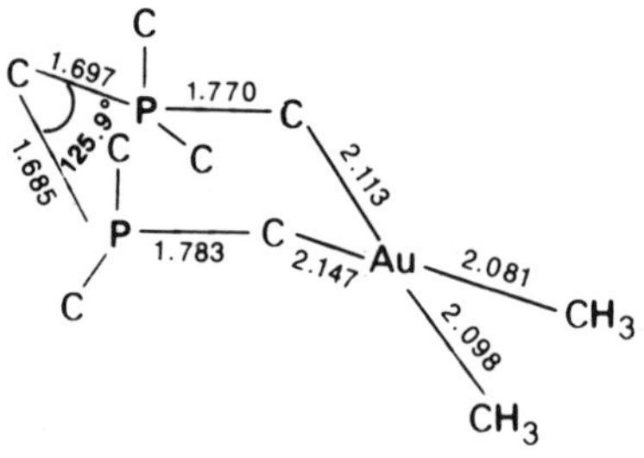

FIGURE 15. Molecular structure and selected data for **153a**[128].

C. Carbonyl-stabilized Ylides as Terminal, Monodentate Ligands

Ylides with acyl, aroyl, alkoxy carbonyl, and cyano substituents at the carbanionic centre experience additional stabilization from efficient delocalization of the negative charge (Figure 16).

Employment of the carbonyl group for charge delocalization shown in resonance structure **B** in Figure 16, is supported by the i.r. spectra of this class of ylides. The bands due to carbonyl stretching vibrations have considerable longwave shifts compared with those in the spectra of the corresponding onium salts [e.g. $Ph_3PCHCOPh$: $\nu(CO) = 1523\ cm^{-1}$; $(Ph_3PCH_2COPh)Br$: $\nu(CO) = 1656\ cm^{-1}$][129]. The ylidic carbon centre and the enolate oxygen site of such carbonyl-stabilized ylides can be utilized for coordination to metal atoms in a monodentate fashion. The first mode of bonding enhances the weight of resonance structure **A** in Figure 16. As a result, the carbonyl band appears between those of the free ylide and the onium salt in the i.r. spectra of the metal complex. On the other hand, in metal–enolate coordinated species, resonance structure **B** is stabilized, and hence carbonyl bands close to those of the free ylides are exhibited.

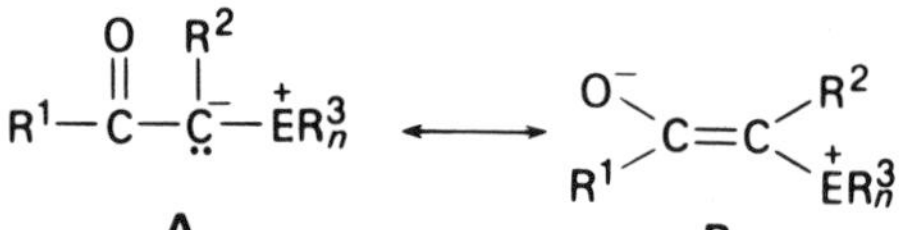

FIGURE 16. Resonance structures of carbonyl-stabilized ylides.

1. Metal–carbon coordination

The bonding interactions between carbonyl-stabilized ylides derived from phosphorus, arsenic, sulphur, and nitrogen, and dihalides of palladium and platinum have been well studied, and it has been shown that the ylidic carbon represents the effective donor centre of the ligand. In order to form the metal–carbon σ-bond, the carbon atom is rehybridized from sp^2 to sp^3. Palladium(II) mainly forms monomeric square-planar complexes with *trans*-ylide ligands (Figure 17).

Such complexes are readily obtained from the reaction of free ylides with palladium dihalides in the presence of donor solvents such as acetonitrile or dimethyl sulphoxide, or from labile ligand-bearing palladium precursors (equation 80).

Analogous complexes with $Ph_3PC(R^1)COR^2$ (R^1 = H, PhCO, Me; R^2 = OMe, OEt, Ph), Ph_3PCHCN, $Bu_3PCHCOPh$, $Ph_3AsCHCOPh$, and $Ph_3AsCHCO_2Me$ are

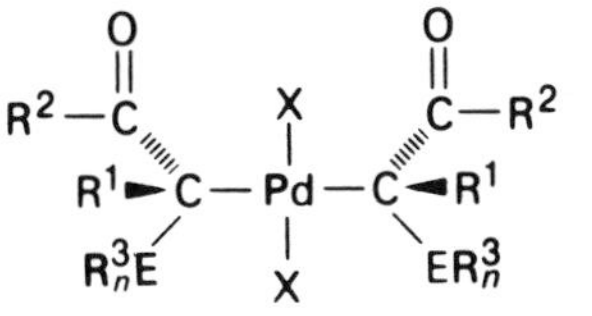

FIGURE 17. General constitution of palladium complexes containing carbonyl-stabilized ylide ligands.

$$PdCl_2 + 2PhCOCHPPh_3 \xrightarrow{CH_3CN} \textit{trans-}[(Ph_3PCHCOPh)_2PdCl_2] \quad \textbf{(159)} \qquad (80)$$

$$2R^1COCHER^2_n + [(PhCN)_2PdCl_2] \xrightarrow{-2PhCN} \textit{trans-}[(R^2_nECHCOR^1)_2PdCl_2] \qquad (81)$$

accessible in a similar way[129] (equation 81). The substituent R^1 at the carbonyl function is mainly a methyl or an aryl group, whereas the ER^2_n moiety is more varied since it can originate from aryl and alkyl phosphines, triphenylarsine, pyridine, and diorganyl sulphanes[130,131]. The procedure exemplified by Scheme 27 involves deprotonation of the onium salts of halopalladates, leading directly to the complexes under discussion. One major advantage of this approach arises from the fact that the syntheses of sensitive free ylides, which often require special precautions, are circumvented[132].

$$2(PhCOCH_2SMe_2)Br \xrightarrow[-2NaBr]{+Na_2[PdBr_4]/MeOH} [(PhCOCH_2SMe_2)_2PdBr_4] \quad \textbf{(160)}$$

$$\textbf{160} \xrightarrow[-2HOAc,\ -2NaBr]{+2NaOAc/MeOH} [(PhCOCHSMe_2)_2PdBr_2] \quad \textbf{(161)}$$

SCHEME 27

The application of this reaction to the complex phosphonium salts **162** makes feasible the synthesis of binuclear halide bridged systems **163**[133] (Scheme 28).

$$2(R_3PCH_2Z)Cl \xrightarrow[-4\ NaCl]{+\ 2\ Na_2[PdCl_4]/MeOH} (R_3PCH_2Z)_2[Pd_2Cl_6] \quad \textbf{(162)}$$

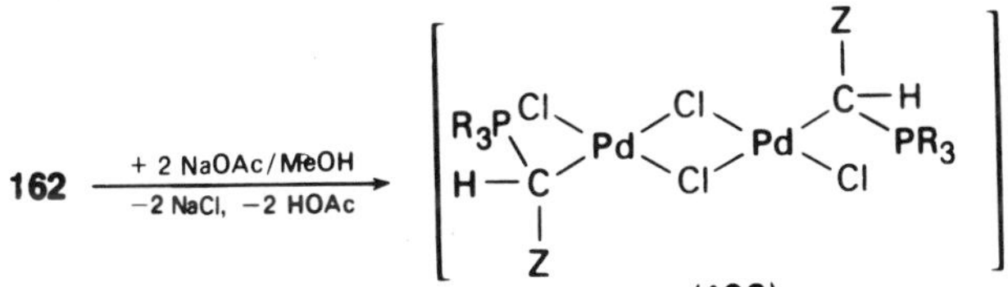

$$\textbf{162} \xrightarrow[-2\ NaCl,\ -2\ HOAc]{+\ 2\ NaOAc/MeOH} \textbf{(163)}$$

$PR_3 = PPh_3,\ PMePh_2;\ Z = COMe,\ CO_2Et,\ CONH_2,\ CN$

SCHEME 28

The cleavage of the halide bridges in **164** by ylide ligands produces the mixed substituted compounds **165**, for which *cis–trans* isomerism is observed depending on the respective ylide and the state of aggregation. In solution, complexes **165** isomerize utilizing dissociation–association mechanisms[134].

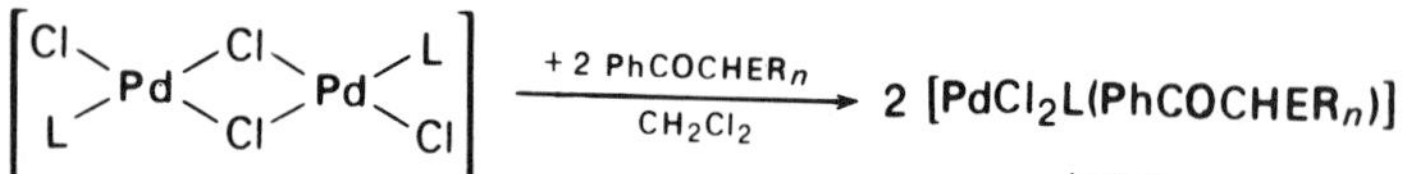

$$[LClPd(\mu\text{-}Cl)_2PdClL] \; (164) \xrightarrow[CH_2Cl_2]{+2\,PhCOCHER_n} 2\,[PdCl_2L(PhCOCHER_n)] \; (165) \quad (82)$$

$L = PMe_3, PMe_2Ph;\; ER_n = PMePh_2, AsPh_3, SMe_2$

Information about the nature of the palladium ylide bond was provided by an X-ray diffraction analysis of **166**, which was shown to possess *trans*-ylide ligands attached through sp^3-hybridized carbon atoms to the metal centre (Figure 18). The consequent asymmetry at the ylidic carbon atoms leads to additional isomers, but only the *d,l-trans*-configuration is realized in the solid state[130].

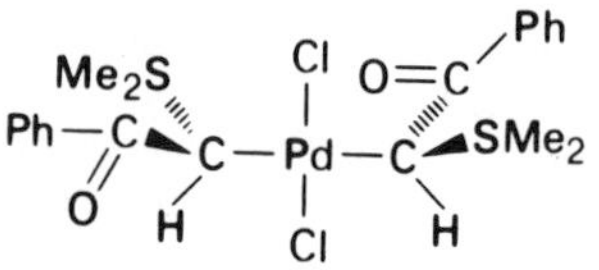

FIGURE 18. Molecular structure of **166**.

The interconversion of *d,l*- and *meso*-configurations[130] as well as ligand-exchange reactions with different ylides reflect the existence of dissociation–association equilibria in solutions of **166**, in addition to the kinetic lability of its ligands[131].

$$\mathbf{166} + 2\,p\text{-}MeOC_6H_4COCHSMe_2 \rightleftharpoons$$

$$2\,PhCOCHSMe_2 + [(p\text{-}MeOC_6H_4COCHSMe_2)_2PdCl_2] + \text{mixed complexes} \quad (83)$$

Similar ligand-exchange experiments employing phosphonium, arsonium, and pyridinium ylides have shown that the stability of the respective complexes in solution is determined by both the nature of the heteroatom and the substituents at the ylidic carbon atom. With comparable ylide substitution, the complex stability, depending on the onium centre, varies as follows: $S > P$; $As \gg P$; $N_{(pyridine)} \geqslant P$. The application of these synthetic methods to platinum chemistry led to a series of structurally analogous derivatives which have also been carefully studied[129,134–136].

In the most competent work concerning mercury complexes of carbonyl-stabilized ylides[137–142], a dimeric halide bridged structure (**167**) is assigned to this kind of mercurial[129]. The ylides examined in that study were derived from PPh_3, PBu_3, $AsPh_3$, SMe_2, and pyridine.

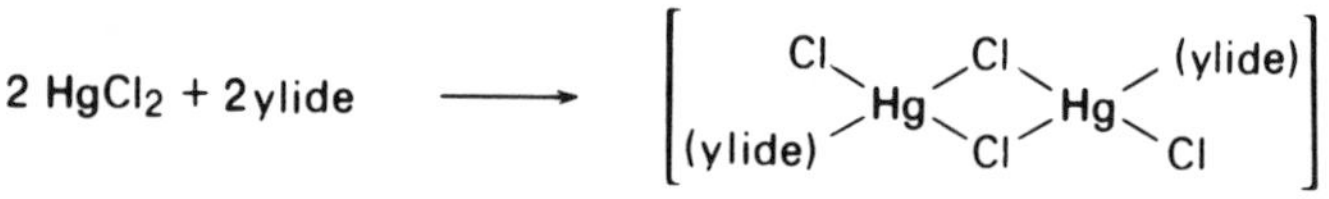

$$2\,HgCl_2 + 2\,\text{ylide} \longrightarrow [(\text{ylide})ClHg(\mu\text{-}Cl)_2HgCl(\text{ylide})] \; (167) \quad (84)$$

Cobalt(III) forms an octahedral complex (**168**), which is of interest because of its strong Co—C σ-bond which is comparable to that in cobalt alkyls[143].

2. *Metal–oxygen coordination*

Whereas for all metal complexes of carbonyl-stabilized ylides, so far discussed, metal–carbon σ-bonds are structurally characteristic, tin and lead allow in addition the synthesis of metal–oxygen bonded species. This mode of attachment is realized in the

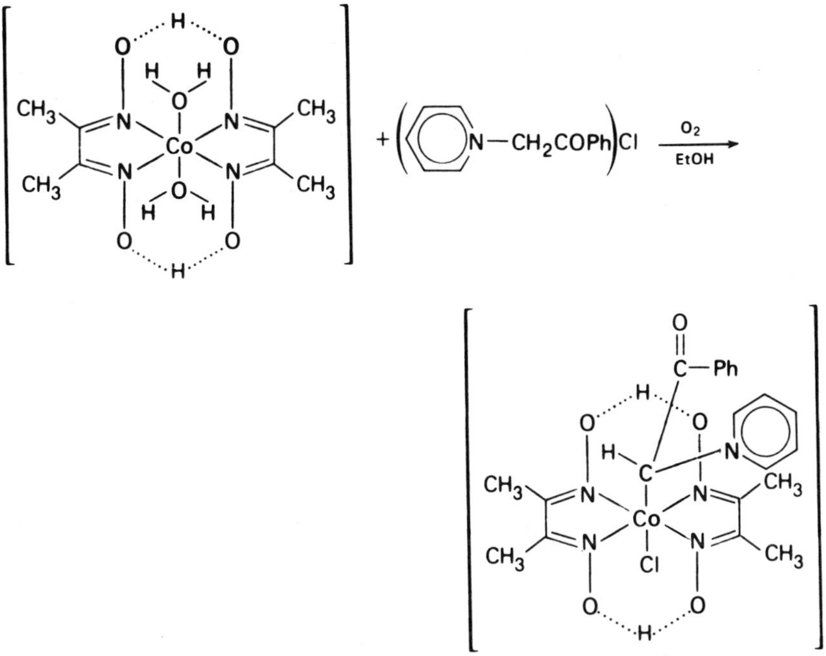

(85)

(168)

1:1 adducts of Me_3MCl (M = Sn, Pb) and ylides such as $Ph_3PCHCOR$ (R = Me, Ph, OMe), $Ph_3PC(Me)CO_2Me$, and $Ph_3AsCHCOPh$[144], where carbonyl bands in the range 1465–1510 cm^{-1} in the i.r. spectra imply a metal–enolate structure. The carbonyl stretching mode of **169** (1665 cm^{-1}), however, seems to be consistent with a metal–carbon interaction[144]. The X-ray diffraction analysis of **170** proves the tin–oxygen interaction beyond any doubt[145].

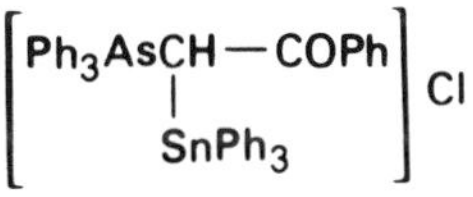

(169)

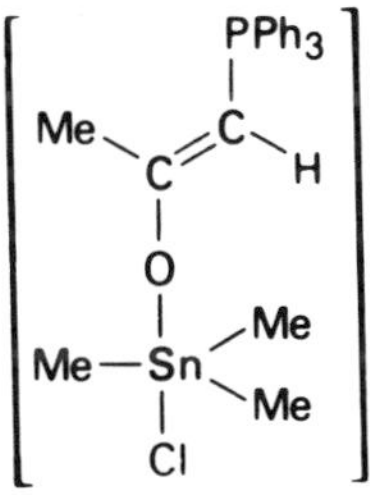

(170)

D. Carbonyl-stabilized Ylides as Chelating Ligands

Carbonyl-stabilized ylides should be capable of functioning as chelating ligands via the ylidic carbon and the oxygen atom. In the chelates examined to date, however, the ylide ligands contain additional substituents exhibiting donor properties and thus provide a second coordination site apart from carbon or oxygen.

1. Picolinyl methylide ligands

With ylides **171**, substituted at the carbanionic function by a picolinyl group, coordination to palladium and platinum dihalides proceeds via the ylidic carbon and the pyridine nitrogen atom irrespective of the nature of the onium centre[146] (equation 86).

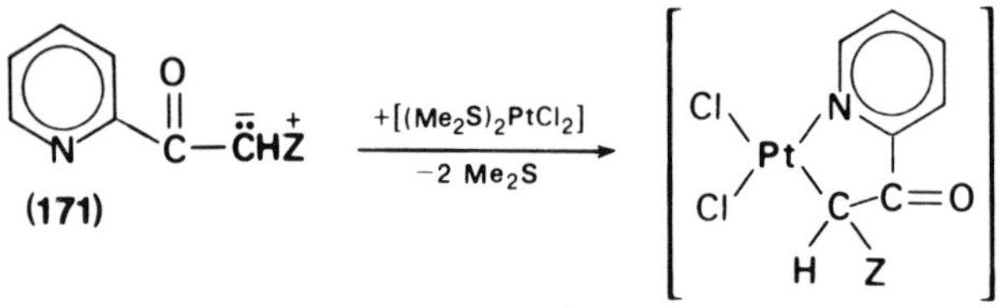

$$\text{(171)} \xrightarrow[-2\,Me_2S]{+[(Me_2S)_2PtCl_2]} \text{(172)} \qquad (86)$$

171, 172	a	b	c
Z	$4\text{-}MeC_5H_4N$	Me_2S	PPh_3

Similar bonding interactions operate in the analogous palladium derivatives as well as in the salt-like bis-chelates **173** and in the carbonyl–tungsten compound **174**[147].

$[Pt\{C_5H_4NC(O)CHZ\}_2](ClO_4)_2$

(**173**)

173	a	b
Z	$4\text{-}MeC_5H_4N$	$3,5\text{-}Me_2C_5H_3N$

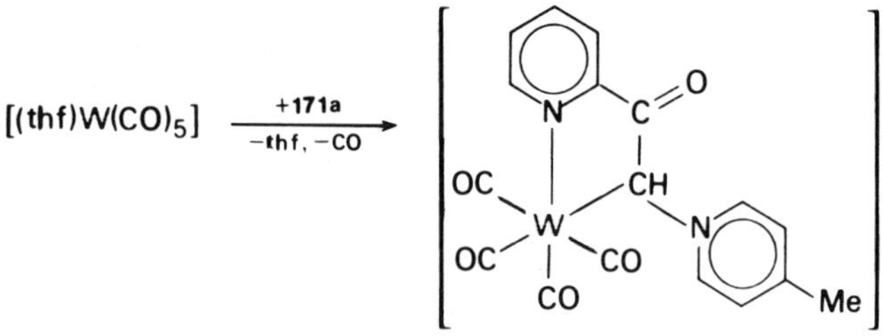

$$[(thf)W(CO)_5] \xrightarrow[-thf,\,-CO]{+171a} \text{(174)} \qquad (87)$$

The structural alternative is realized in **175** where the pyridine nitrogen atom and the enolate function occupy the two available acceptor sites at one edge of the octahedron[147].

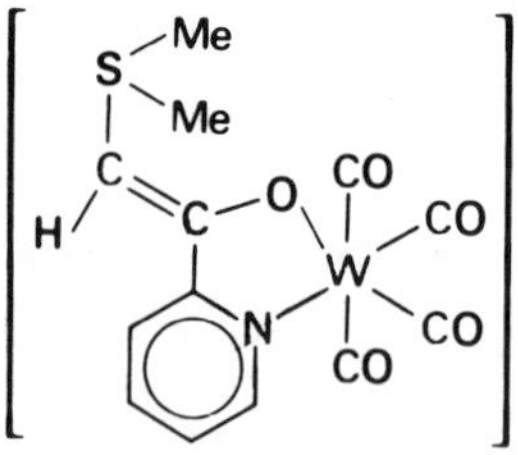

(**175**)

2. Acyl methylene ω-phosphinoalkyldiphenylphosphoranes

Bidentate ligands such as **176** have been utilized in a series of palladium and platinum dihalide chelates with the carbanionic function and the trivalent phosphorus atom as the sole donor centres[148].

$Ph_2P(CH_2)_n\overset{+}{P}Ph_2\overset{-}{C}HCOR$
(176)

R	Ph	Me	OMe
$n = 1$	**a**	**b**	**c**
$n = 2$	**d**	**e**	

$$\mathbf{176} + [(PhCN)_2PdCl_2] \longrightarrow [(\mathbf{176})PdCl_2] + 2PhCN \quad (88)$$
(177)

$$\mathbf{176} + [(Me_2S)_2PtCl_2] \longrightarrow [(\mathbf{176})PtCl_2] + 2Me_2S \quad (89)$$
(178)

Substitution studies of **177** and **178** with pyridine led to the conclusion that the ylidic moiety in the chelates is more tightly bound to the metal atoms than for example the monodentate $Ph_3PCHCOR$ in $[(Ph_3PCHCOR)_2MCl_2]$. The chelate ligands of **178** do not undergo displacement reactions with pyridine at all, whereas with palladium the five-membered ring chelates **177a–c** appear to be kinetically more stable than the six-membered ring species **177d–e**. On top of this, the reluctance towards pyridine attack is influenced by the substituents R at the carbonyl group in the same way as is the basicity of the free ylides: Ph > Me > OMe[148]. The mode of bonding in **177** and **178** was deduced mainly from i.r spectroscopic observations, and molecular structural analysis of **177d** (Figure 19) fully confirms these assumptions[149]. The different palladium–chlorine bond distances in **177d** imply that the *trans*-influence of the phosphine group is superior to that of the ylidic carbon atom.

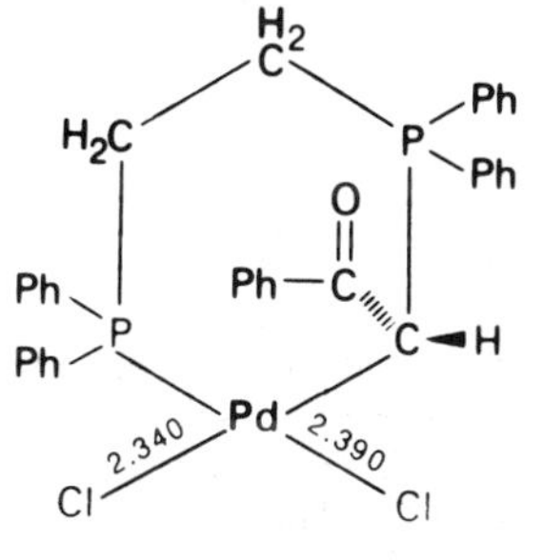

FIGURE 19. Molecular structure of **177d**.

A chloride ligand in **177** and **178** is labilized to such an extent that replacement by monodentate neutral donor molecules readily takes place. Both chloride ligands may also be exchanged for other halide and pseudo-halide ions (Scheme 29)[150].

$$[(\mathbf{176})MCl_2] + NaBPh_4 + L \xrightarrow[\text{acetone}]{CH_2Cl_2/} [(\mathbf{176})MClL]BPh_4 + NaCl$$
(179)

$[(\mathbf{176})MCl_2] \xrightarrow[-2M'Cl]{+2M'X} [(\mathbf{176})MX_2]$
(180)

M = Pd, Pt; M′ = alkali metal

L = PPh_3, $P(C_6H_{11})_3$, $P(OMe)_3$, $PMePh_2$, $AsPh_3$, $SbPh_3$ $4\text{-}MeC_5H_4N$, $3,5\text{-}Me_2C_5H_3N$

X^- = Br^-, I^-, SCN^-

SCHEME 29

3. Others

The oxygen atom of the oxosulphonium group in **181** inherits sufficient donor capability to displace the adjacent diethylthioether with the generation of chelate **182** (equation 90). The alteration in the coordination sphere of the metal atom is reflected in the longwave shift of the SO stretching mode in the i.r. spectra[151].

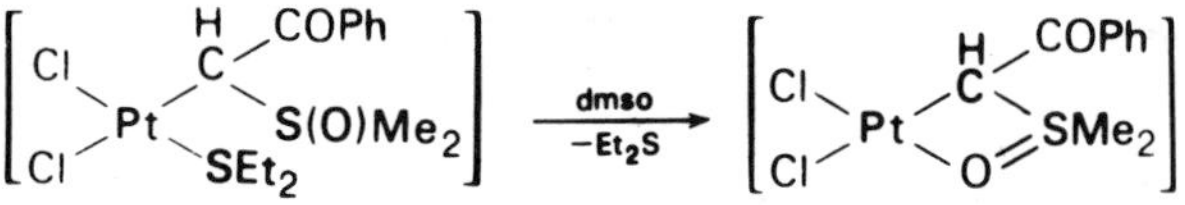

$$[Cl_2Pt\{CH(COPh)S(O)Me_2\}(SEt_2)] \xrightarrow[-Et_2S]{dmso} [Cl_2Pt\{CH(COPh)SMe_2O\}] \quad (90)$$

(**181**) (**182**)

	$\nu(CO)(cm^{-1})$	$\nu(SO)(cm^{-1})$
181	1650	1230
182	1660	1140

In the chelate complexes discussed so far, the free carbonyl-stabilized ylides used as ligands are stable compounds which do not rearrange while being attached to the metal atom. In contrast, the binuclear *ortho*-metallated complex **183** was isolated from the reaction of $PdCl_2$ with $Ph_3PCHCOPh$ in the presence of sodium acetate in boiling methanol[152] (equation 91).

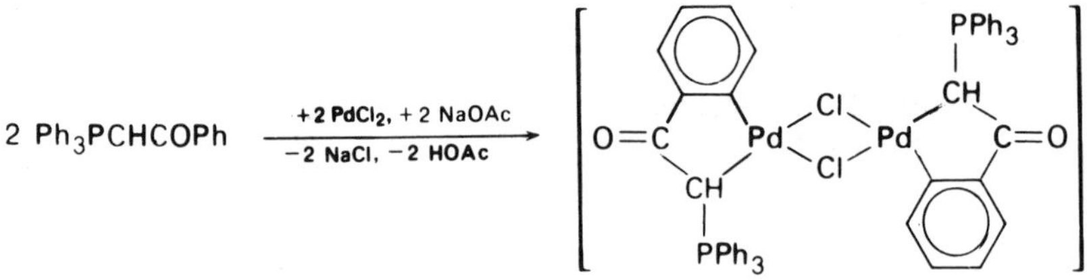

$$2\ Ph_3PCHCOPh \xrightarrow[-2\ NaCl,\ -2\ HOAc]{+2\ PdCl_2,\ +2\ NaOAc} \quad (91)$$

(**183**)

The ligand of the unusual molecule **185** is generated from the ylide **184** and a carbonyl ligand from iron. The formyl proton originates from one of the *o*-phenyl ring positions of the phosphonium moiety. This occurs in the coordination sphere of the iron atoms[153] (equation 92). In **185** (Figure 20) the $(CO)_3Fe—Fe(CO)_3$ unit is doubly bridged by the $C(CHO)P(Ph)_2(C_6H_4)$ ligand which bonds from the original ylide carbon atom and the *ortho*-carbon atom of one phenyl ring. The *ortho*-dimetallated

$$\underset{(\mathbf{184})}{Ph_3PC(SnMe_3)_2} \xrightarrow[-[(Me_3Sn)_2Fe(CO)_4],\ -CO]{+[Fe_3(CO)_{12}]} \underset{(\mathbf{185})}{[Fe_2(CO)_6\{C(CHO)P(Ph)_2(C_6H_4)\}]} \quad (92)$$

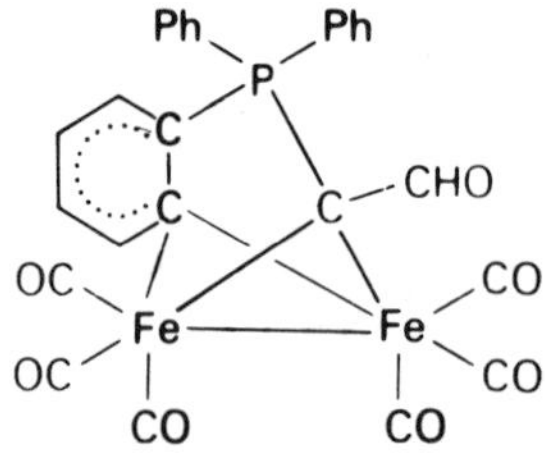

FIGURE 20. Molecular structure of **185**.

phenyl ring is similar to that of a Meisenheimer complex found in nucleophilic aromatic substituion reactions. **185** can also be considered as a chelate complex where each donor site is attached to two metal centres simultaneously.

Phosphine addition to the metallocyclic carbene species **186a,b** results in the formation of complexes containing chelating ylide systems **(187a,b)** (Scheme 30)[154].

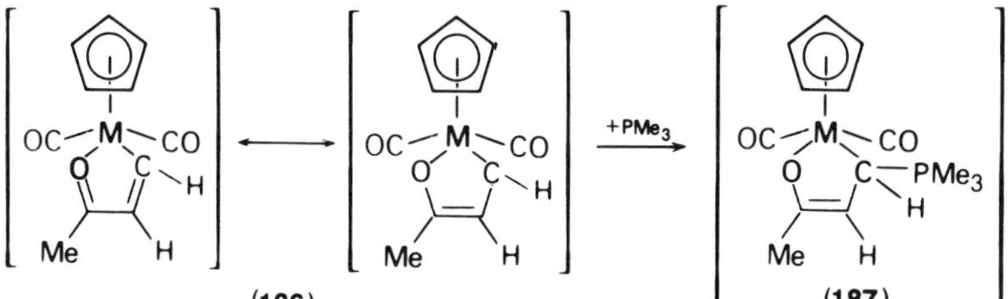

186, 187	a	b
M	Mo	W

SCHEME 30

E. Alkenyl-stabilized Ylides (Onium Allylides) as Ligands

Ylidic centres are also remarkably stabilized by resonance interactions with open-chained or cyclic olefins. Vinyl substituents effect conversion of the ylidic carbon to an integral constituent of an onium-stabilized allyl anion which can be employed in coordination chemistry as η^1- and η^3-ligands.

1. η^1-Coordination

η^1-Phosphonium allylide pentacarbonyl complexes of chromium and molybdenum are accessible from the corresponding hexacarbonyls by photochemical as well as thermal routes (equation 93)[155].

$$Ph_3PCHR + [M(CO)_6] \xrightarrow[50\text{–}60^\circ C \text{ or } h\nu]{\text{light petroleum}} [Ph_3P\underset{\displaystyle R}{\underset{|}{C}}H\text{—}M(CO)_5] + CO \qquad (93)$$

(188), **(189)**

R	—CH=CH$_2$	—CH=CHMe	CH=CHPh
M = Cr		**188b**	
M = Mo	**189a**	**189b**	**189c**

2. η^3-Coordination

When phosphonium allylides and metal hexacarbonyls are allowed to react under more forcing thermal conditions, the η^3-phosphonium allylide complexes **190** and **191**[155,156] are obtained (equation 94).

An X-ray analysis showed **190a** to be a distorted octahedral molecule in which the molybdenum atom is attached only to the planar allyl unit of the ligand (Figure 21)[157].

A similar mode of interaction is encountered in a variety of palladium complexes of alkenyl-substitued ylides. Base treatment of the corresponding onium halopalladates

$$Ph_3PCH{-}CR{=}CH_2 + [M(CO)_6] \xrightarrow{\Delta} [(Ph_3PCHCR{=}CH_2)M(CO)_4] \quad \textbf{(190), (191)} \qquad (94)$$

R	H	Me
M = Mo	**190a**	**190b**
M = W	**191a**	

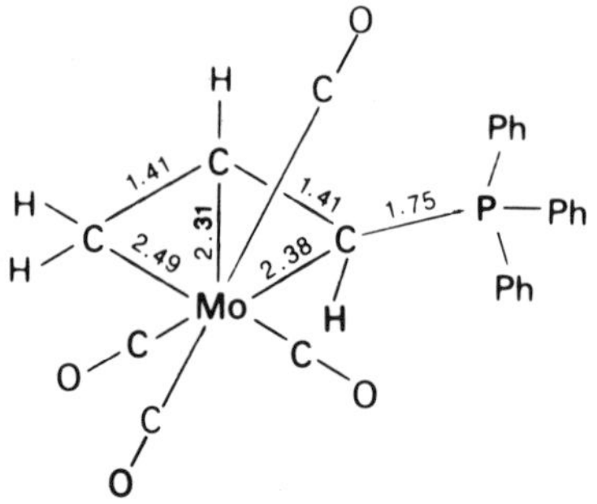

FIGURE 21. Molecular structure of **190a**.

constitutes a useful synthesis of compounds such as **192**[132] and **193**[158]. The circumvention of the free ylide in the reaction, as shown in Scheme 31, turns out to be crucial, as free $C_3H_5SMe_2$ suffers rearrangement to $MeS(CH_2)_2CH{=}CH_2$.

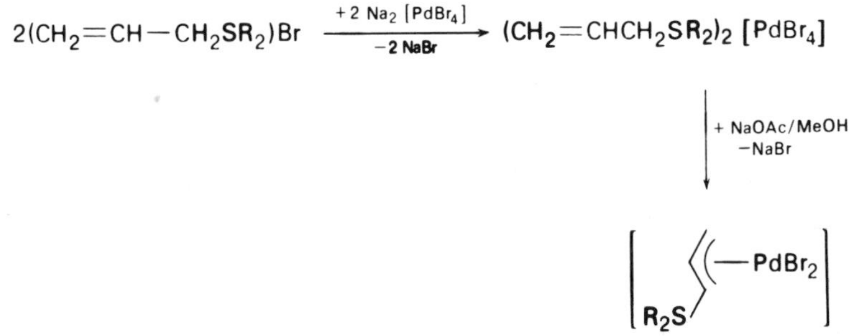

$$2(CH_2{=}CH{-}CH_2SR_2)Br \xrightarrow[-2\,NaBr]{+2\,Na_2[PdBr_4]} (CH_2{=}CHCH_2SR_2)_2[PdBr_4]$$

$$\downarrow + NaOAc/MeOH,\ -NaBr$$

$$[R_2S\text{-allyl-}PdBr_2] \quad \textbf{(192)}$$

R_2	Me_2	$-(CH_2)_4-$
192	**a**	**b**

SCHEME 31

The addition of **193** to polymeric **194** leads to the neutral binuclear complexes **195**[159] (equation 95).

The interaction of **193** with an equimolar amount of silver ions, followed by the addition of donor ligands, yielded the cationic species **197** in a reaction sequence implying halide bridge cleavage of intermediate binuclear **196** (Scheme 32)[160]. The employment of twice the molar amount of silver salt in the presence of neutral

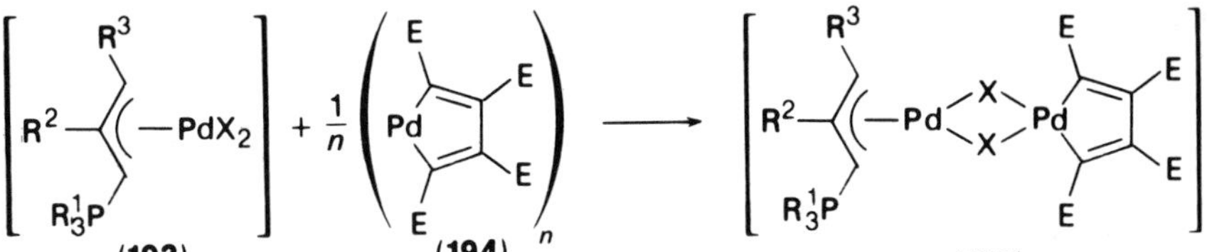

$$[R^2C(CHR^3)(CHPR^1_3)PdX_2] \text{ (193)} + \frac{1}{n}(\text{Pd}C_4E_4)_n \text{ (194)} \longrightarrow [R^2C(CHR^3)(CHPR^1_3)Pd(\mu\text{-}X)_2PdC_4E_4] \text{ (195)} \quad (95)$$

R^1_3P = PPh_3, PMe_2Ph; X = Cl, Br; E = CO_2Me; R^2, R^3 = H, Me

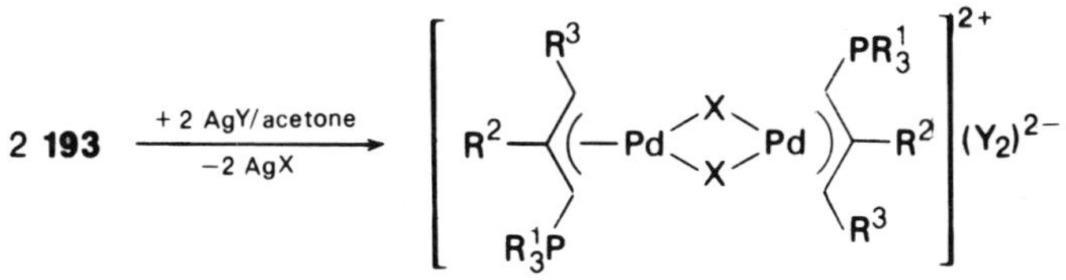

$$2\ \mathbf{193} \xrightarrow[-2\ AgX]{+2\ AgY/acetone} [R^2C(CHR^3)(CHPR^1_3)Pd(\mu\text{-}X)_2Pd(CHPR^1_3)(CHR^3)CR^2]^{2+}(Y_2)^{2-} \text{ (196)}$$

$$\mathbf{196} \xrightarrow{+2\ PPh_3} 2[R^2C(CHR^3)(CHPR^1_3)Pd(X)(PPh_3)]^+ Y^- \text{ (197)}$$

R^1 = Ph, Et; R^2 = H, Me

R^3 = H, Me; X = Cl, Br

Y^- = PF_6^-, BF_4^-, $CF_3SO_3^-$

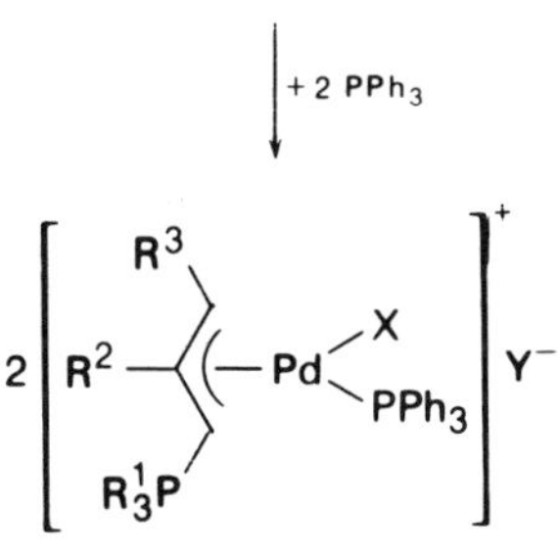

SCHEME 32

bidentate ligands led to the conversion of **193** into the bivalent complex cations **198**[161,162] (equation 96). Apparently the *cis–trans* ratio of **198d** is temperature dependent, which is in contrast to the behaviour of $[(\eta^3\text{-}C_3H_5)_2Pd]$. The stability of

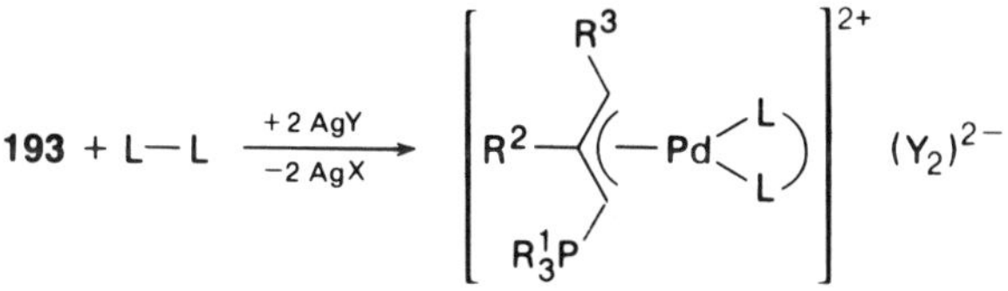

$$\mathbf{193} + L\text{—}L \xrightarrow[-2\ AgX]{+2\ AgY} [R^2C(CHR^3)(CHPR^1_3)Pd(L\text{—}L)]^{2+}(Y_2)^{2-} \text{ (198)} \quad (96)$$

198	**a**	**b**	**c**	**d**
L—L	bipyridine	cod	nbd	$R^3CH{=}C(R^2)CHPR^1_3$

bis-allylide complexes such as **198d**, effected by the onium centres, merits consideration because in contrast to the related $[(\eta^3\text{-}C_3H_5)_2Pd]$, they do not suffer facile decomposition with ligand coupling[162].

F. Cyclic Ylides as Ligands

1. Cyclopentadienylide complexes

According to calculations, the electronic ground state of the sulphonium and phosphonium cyclopentadienylides **199** and **200** can be understood in terms of resonance structures **A** and **B** with contributions of 80% from the ylide structure **A** and 20% from the ylene structure **B** (Figure 22)[163].By analogy with the cyclopentadienyl ligand, η^1, η^3, and η^5-coordinated **199** and **200** can be predicted.

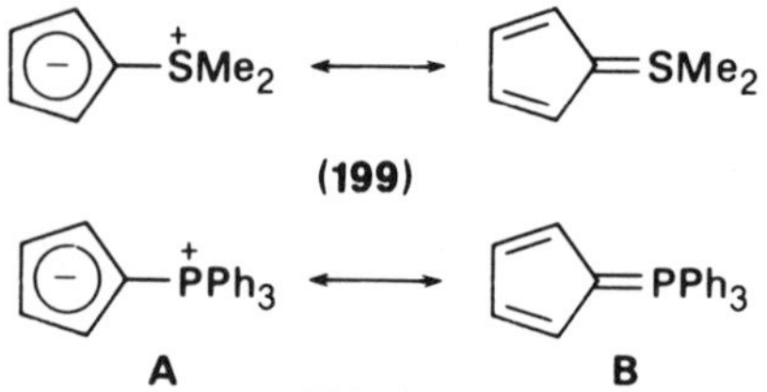

$\overset{+}{S}Me_2$ ⟷ $=SMe_2$

(199)

$\overset{+}{P}Ph_3$ ⟷ $=PPh_3$

A B

(200)

FIGURE 22. Resonance structures of simple cyclopentadienylides.

a. η^1-Coordination. This mode of combination is encountered in the adducts of **200** and mercury dihalides. As is evident from an X-ray analysis, a σ-bond exists between the metal atom and the $C_{(3)}$ position of the ring in crystalline **201c**. ^1H and ^{13}C n.m.r. data, however, are consistent with the presence of fluxional molecules in solution[164] (equation 97).

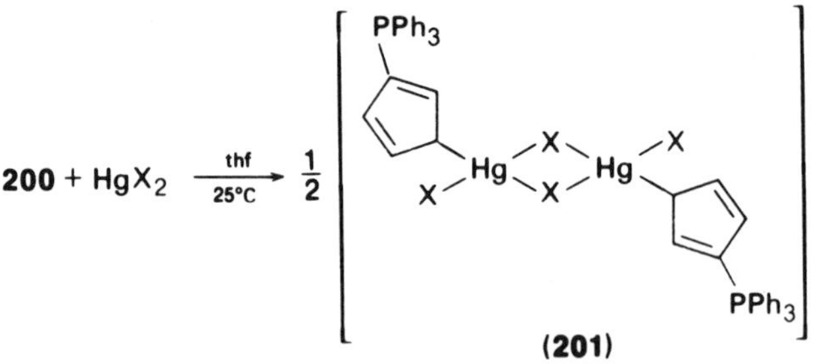

$$\mathbf{200} + HgX_2 \xrightarrow[25°C]{thf} \frac{1}{2}\,[\mathbf{201}] \qquad (97)$$

(201)

201	a	b	c
X	Cl	Br	I

b. η^3-Coordination. Treatment of polymer **194** with **200** results in the formation of a complex (**202**) which has allylic coordination of the ligand (equation 98). In **202**, bond lengths $C_1—C_5$ and $C_2—C_3$ of the ylide ring appear to indicate almost single bond

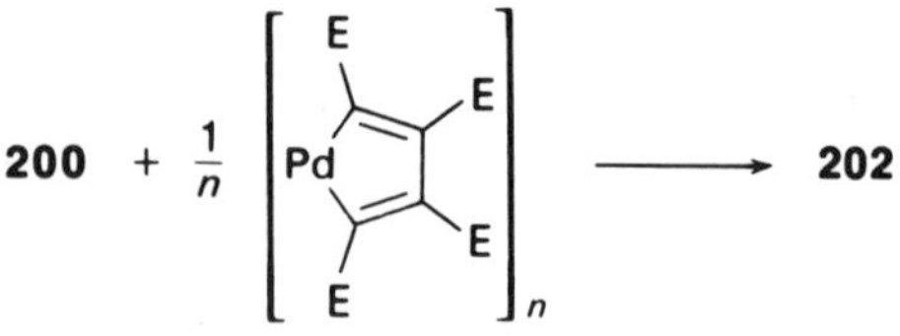

$$\mathbf{200} + \frac{1}{n}\,[\mathbf{194}]_n \longrightarrow \mathbf{202} \qquad (98)$$

(194)

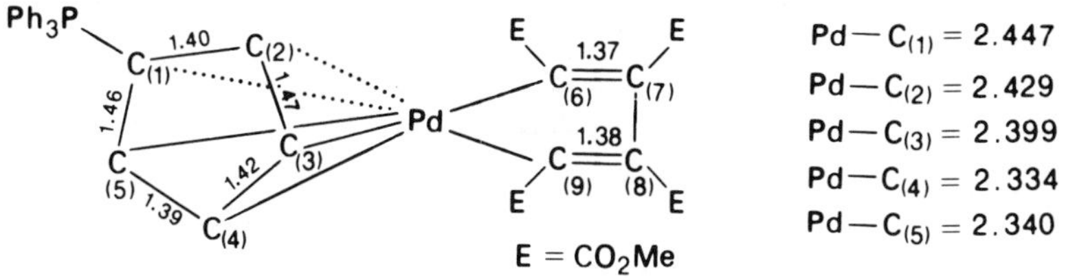

FIGURE 23. Molecular structure of **202** with selected bond lengths in Å.

character. The allylic interaction between the palladium atom and the ring atoms $C_3—C_4—C_5$ is reflected in significantly shorter bond distances between these carbon atoms and the metal centre. Intramolecular repulsions between the phosphonium phenyl rings and the palladocycle methyl groups restrict the Pd atom to the lower region of the ylide ring, thus precluding a more regular η^5-coordination (Figure 23)[165].

c. *η^5-Coordination*. The attachment of a metal complex fragment to all five carbon atoms represents the mode of bonding most frequently realized in organometallic cyclopentadienylide compounds. Silver ion-induced halide abstraction in **193**, and interception of the coordinatively unsaturated intermediate by **200**, provided a synthetic approach to a series of stable palladium species (**203**) (equation 99)[162]. The acceptor sites at the palladium atom, occupied by the allylide in **203**, may also be utilized by other four-electron donors such as nbd, cod[161], and the η^3-allyl group[162].

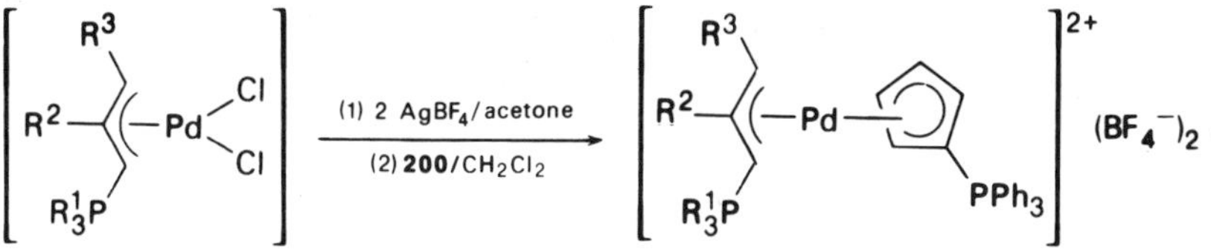

Neutral as well as cationic cyclopentadienylide complexes, including carbonyl–metal derivatives of Group VIB[45,166,167a,b], VIIB, and VIIIB metals where the ring also manifests the properties of an η^5-ligand, are obtained by labile ligand displacement reactions in ether solvents. The latter constitutes a very important general process used for (6π-ring) $M(CO)_3$ generation. Ionic complexes **206**[168,169], **207**[169], and **208**[8] were prepared analogously.

The presence of the onium functions accounts for the remarkable increase in the stability and retarded reactivity of **204** and **205** compared with the related $[(\eta^5\text{-Cp})M(CO)_3]^-$ anions. X-ray structural analyses of **205a**[170] and **208**[8] show that the

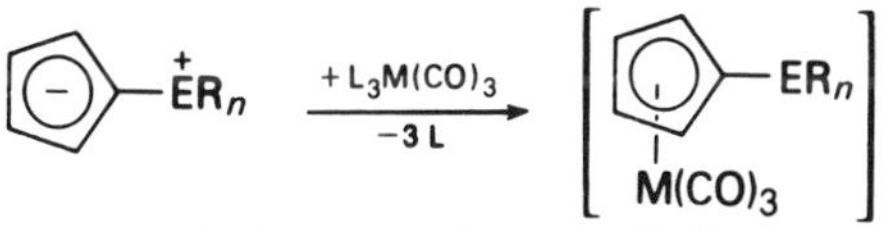

M	Cr	Mo	W
ER_n = PPh_3	**204a**	**204b**	**204c**
ER_n = SMe_2	**205a**	**205b**	**205c**

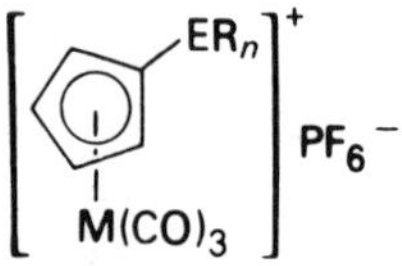

$[(\eta^5\text{-}C_5H_4ER_n)M(CO)_3]^+ PF_6^-$

M	Mn	Re
$ER_n = PPh_3$	**206a**	**206b**
$ER_n = SMe_2$	**207a**	**207b**

(206), **(207)**

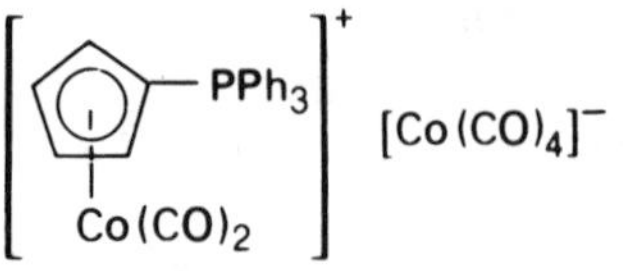

$[(\eta^5\text{-}C_5H_4PPh_3)Co(CO)_2]^+ [Co(CO)_4]^-$

(208)

combination of the ylides with the transition metal atoms does not result in a drastic perturbation of the ylide structure itself, but only in an enhancement of the contribution of resonance structure **A** (Figure 22) to the electronic ground state of the η^5-bonded ring. η^5-Coordination effects a significant balance of the bond distances as well as the electron density distribution in the ring[163]. Differences in the reactivities at the α- and β-positions in the free ylides are also compensated by complexation as is inferred from H/D exchange experiments with **199**, **200**, and the corresponding $M(CO)_3$ derivatives[170]. A bonding interaction between the chromium and sulphur atoms in **205** can be excluded. Complexes **204a–c** display Lewis base behaviour towards $AlMe_3$, protons, halogens, nitrosyl, and diazonium ions as well as towards the halides of mercury, cadmium, boron, gallium, and indium[45,171–173]. Whereas $AlMe_3$ is reversibly added to the carbonyl oxygens, the other Lewis acids exclusively attack the metal centre.

Ionic complexes of **200** and the halides of Group IVB and VIB elements[174] and the arsonium ylide compound **209**[45] have also been reported.

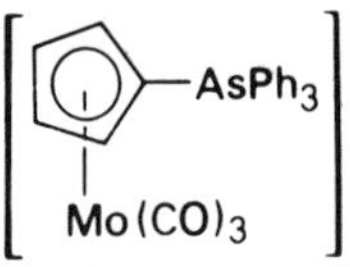

$[(\eta^5\text{-}C_5H_4AsPh_3)Mo(CO)_3]$

(209)

2. *Phosphabenzene and thiabenzene complexes*

λ^5-Phosphabenzenes **210**[175] and also λ^4-and λ^6-thiabenzenes **211**[176] and **212**[177] differ from cyclopentadienylides, because their onium units are incorporated as integral constituents in the ring systems. In these ylides, the negative charge achieves stabilization both from delocalization over the five ring carbon atoms and the onium function. All three ring systems exhibit ligand properties towards tri-carbonyl metal fragments of chromium, molybdenum, and tungsten.

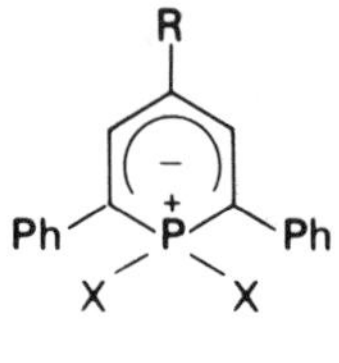

R; Ph, P⁺, Ph; X, X — **(210)**

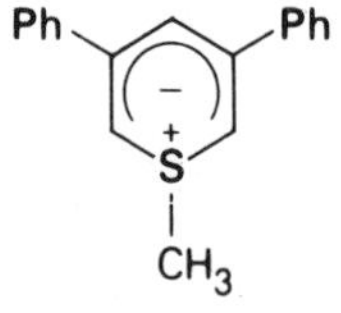

Ph, Ph; S⁺; CH_3 — **(211)**

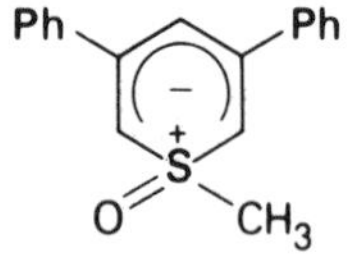

Ph, Ph; S⁺; O, CH_3 — **(212)**

a. λ^5-Phosphabenzene complexes. When the above mentioned procedure used to generate cyclopentadienylide organometallics was applied to **210**, a series of λ^5-phosphabenzene complexes (**213**) were obtained (equation 101)[178]. **210** suffers loss of planarity when coordinated to the $Cr(CO)_3$ moiety. In **213a**, the phosphorus and C_4

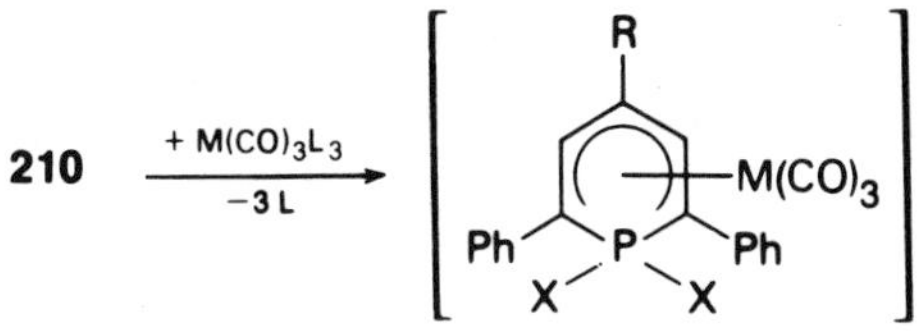

$$210 \xrightarrow[-3\,L]{+\,M(CO)_3L_3} [\text{(213)}] \quad (101)$$

213	a	b	c	d	e	f	g	h	i
X	OMe	OMe	OMe	OMe	OMe	OMe	OMe	F	F
R	Ph	Ph	Ph	Me	CH_2Ph	*t*-Bu	phthalimido	Ph	*t*-Bu
M	Cr	Mo	W	Cr	Cr	Cr	Cr	Cr	Cr

carbon atom deviate from the plane defined by the remaining ring atoms, whereas in **213i** only the heteroatom, at the face opposite to the metal, is located above the planar arrangement of the ring carbon atoms. These results, obtained by X-ray analyses, clearly exclude any bonding interaction between chromium and the onium centre[179].

b. λ^4-Thiabenzene complexes. λ^4-Thiabenzenes such as **211** have so far resisted isolation owing to the inherent instability generally associated with these species[176,180]. However, the decomposition of **211** in dmso solution is retarded sufficiently to allow for the interception and fixation of this species at the $M(CO)_3$ template. To achieve the synthesis of **215**, the thiinium salt **214**, an appropriate precursor of **211**, is subjected to deprotonation in the presence of metal complexes, the labile ligands of which are readily displaced by the heterocycle[181a,b] (equation 102).

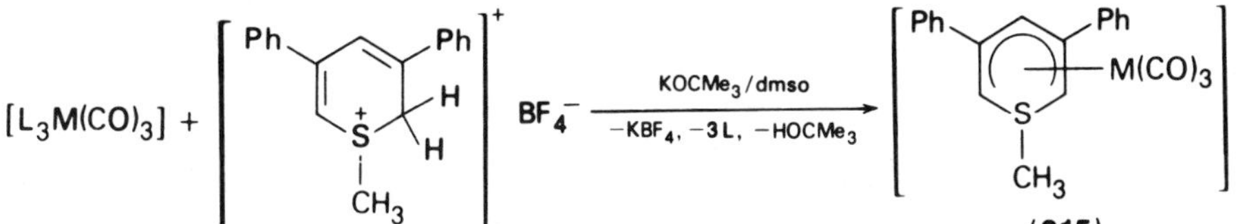

$$[L_3M(CO)_3] + [\text{(214)}]^+ BF_4^- \xrightarrow[-KBF_4,\ -3\,L,\ -HOCMe_3]{KOCMe_3/dmso} [\text{(215)}] \quad (102)$$

L_3 = cht, $3\,CH_3CN$

215	a	b	c
M	Cr	Mo	W

The combination of the heterocycle and the chromium complex fragment is achieved by utilization of the five carbon atoms of the ring alone, which are arranged in a coplanar array. The position of the sulphur atom is at a distance of 0.75 Å above this plane, and the chromium–sulphur bond length of 2.88 Å undoubtedly proves the absence of any bond interaction between these two atoms (Figure 24).

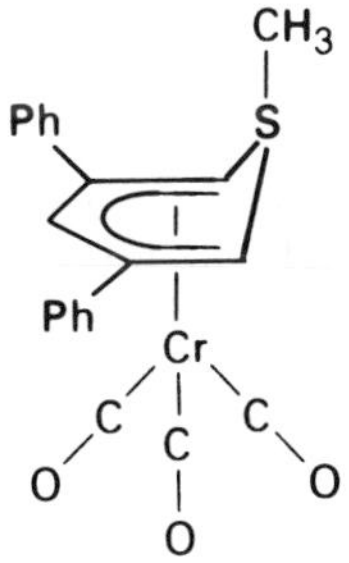

FIGURE 24. Molecular structure of **215a**.

c. λ^6-Thiabenzene complexes. Molecular structural analysis identified **212** as a tilted ring molecule in which the sulphur atom deviates by 0.46 Å from the plane of the five carbon atoms[182]. During complex formation with **212** the occurrence of isomers, inferred from the different possible orientations of the S(O)Me group to the $M(CO)_3$ moiety, can be predicted. This phenomenon, however, can only be detected in the case of chromium (equation 103). In contrast, a *syn* orientation of the ligand, similar to that

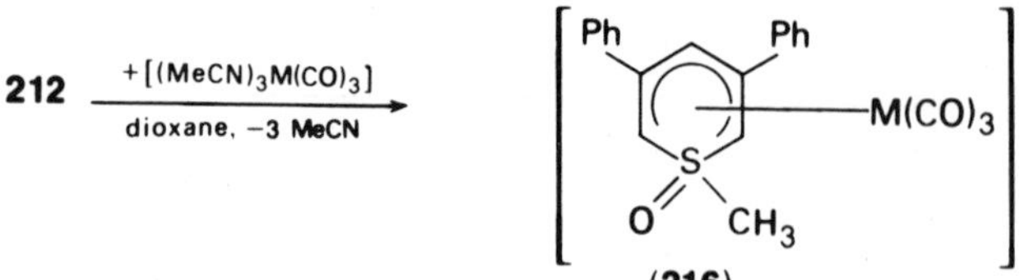

$$\mathbf{212} \xrightarrow[\text{dioxane, } -3\,\text{MeCN}]{+[(MeCN)_3M(CO)_3]} [\text{Ph}_2\text{C}_5\text{H}_3\text{S(O)CH}_3\text{–}M(CO)_3] \quad (103)$$

(216)

216	a	b	c	d
M	Cr	Cr	Mo	W

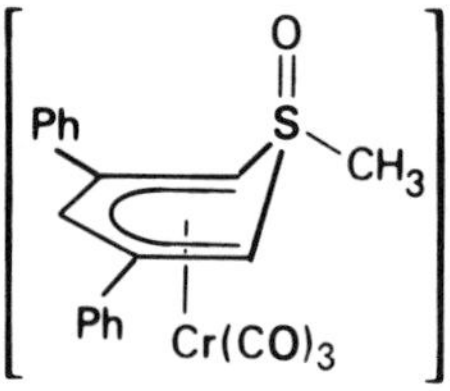

(216a) (*syn*)

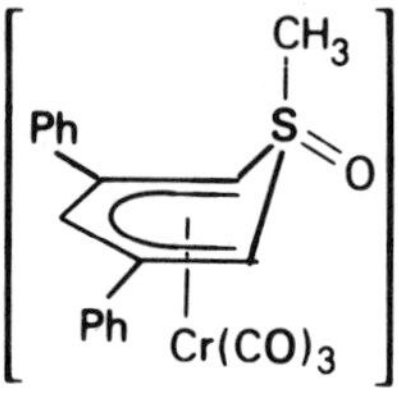

(216b) (*anti*)

in free **212**, is encountered in the molybdenum and tungsten derivatives **216c,d**, with chromium *syn*- and *anti*-isomers are produced in a 10:3 ratio[183]. The different ring geometry in the two isomers is reflected in both their spectroscopic and chemical properties. Complexes **216** contain a series of reactive centres which are susceptible to further chemical transformations, as depicted in Scheme 33[184].

III. METAL-SUBSTITUTED YLIDES

This section is concerned with syntheses and characteristics of ylidic compounds substituted at the carbanionic centre by main group or transition metal atoms. A more detailed account of the chemistry of silicon-substituted ylides is not given here, owing to the purely non-metallic nature of this element.

A. Syntheses via Metallation of Simple Ylides

1. *Deprotonation of metallated onium salts*

Proton abstractions from onium salts containing d^8–d^{10} transition metals and d^0 group IA–IIIA metals generally proceeds at one of the alkyl groups attached to the phosphorus, followed by ring closure through intercomplexation (see Section II.B). On the other hand, it is apparent that d^0 Group IVA and VA metals exert a considerable stabilizing effect on ylidic carbanions concomitant with a reduction in ylide basicity. In accordance with this, the acidity of the methylene protons in the corresponding onium salts such as **74** is sufficiently enhanced so that an ylidic centre is regenerated at that position upon base treatment. In addition, the lack of a suitable

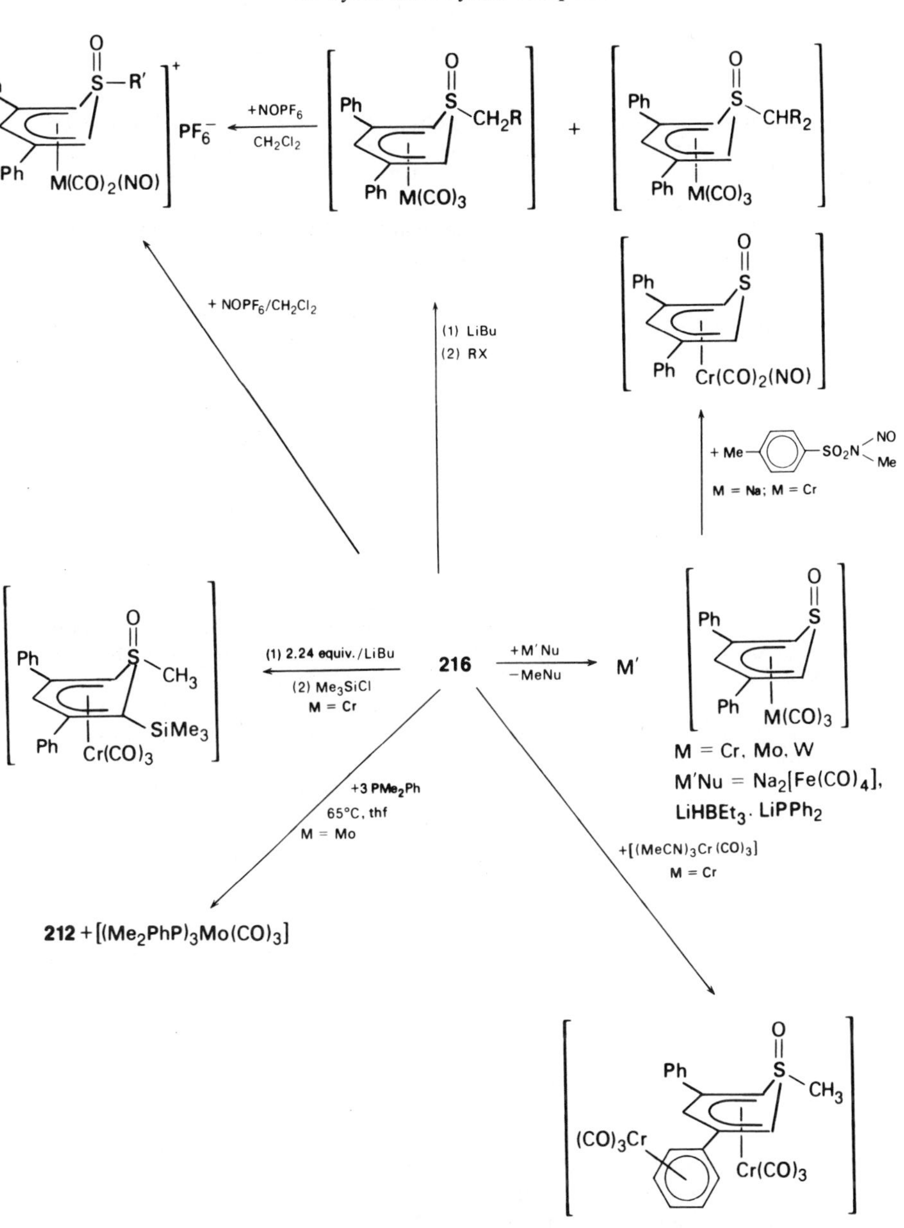
+NOPF$_6$
CH$_2$Cl$_2$
PF$_6^-$
M(CO)$_2$(NO)
M(CO)$_3$
CH$_2$R
CHR$_2$
Cr(CO)$_2$(NO)
+ NOPF$_6$/CH$_2$Cl$_2$
(1) LiBu
(2) RX
+ Me–C$_6$H$_4$–SO$_2$N(NO)Me
M = Na; M = Cr
(1) 2.24 equiv./LiBu
(2) Me$_3$SiCl
M = Cr
216
+M′Nu
−MeNu
M′
SiMe$_3$
Cr(CO)$_3$
M = Cr, Mo, W
M′Nu = Na$_2$[Fe(CO)$_4$],
LiHBEt$_3$, LiPPh$_2$
+3 PMe$_2$Ph
65°C, thf
M = Mo
+[(MeCN)$_3$Cr(CO)$_3$]
M = Cr
212 + [(Me$_2$PhP)$_3$Mo(CO)$_3$]
(CO)$_3$Cr

SCHEME 33

$$(\mathrm{Me_3\overset{+}{P}CH_2GeMe_3})^+\mathrm{Cl}^- \xrightarrow[-\mathrm{C_4H_{10}},\ -\mathrm{LiCl}]{+\mathrm{LiBu}} \mathrm{Me_3\overset{+}{P}{-}\overset{-}{C}HGeMe_3} \qquad (104)$$

(74) → **(217)**

acceptor site with the Group IVA elements obviously renders unfavourable a reaction pathway similar to those discussed in Section II.B[83] (equation 104).

The synthesis of metallated ylides such as **217** generally does not require isolated onium salts as precursors, but may also be accomplished by the reaction of simple ylides with 0.5 equiv of organometallic halides. This procedure involves the transylidation of *in situ* generated metallated onium salts (equation 105)[185].

$$\mathrm{Me_3PCHSiMe_3} \xrightarrow{+\mathrm{Me_3MCl}} \left(\mathrm{Me_3\overset{+}{P}CH}\begin{matrix}\mathrm{MMe_3}\\ \mathrm{SiMe_3}\end{matrix}\right)\mathrm{Cl}^- \xrightarrow[-(\mathrm{Me_3SiCH_2PMe_3})\mathrm{Cl}]{+7} \mathrm{Me_3\overset{+}{P}{-}\overset{-}{C}}\begin{matrix}\mathrm{MMe_3}\\ \mathrm{SiMe_3}\end{matrix} \qquad (105)$$

(218) → **(79)**, **(219)**

M	Ge	Sn
	219	**79**

Transylidation occurs between ylides and onium salts only when the competing ylides differ significantly in basicity or when there are large differences in the lattice energies of the onium salts. Another requirement for the successful application of this process concerns the stability of the metallated onium salts, which should not equilibrate with their starting materials. Under appropriate conditions the subsequent transylidations proceed at rates which preclude the detection of salt-like species such as **218** even in the presence of excess of Me_3MCl. In some cases the application of suitable stoichiometries afforded double metallated ylides[185,186] (equation 106).

$$3\mathrm{R}_n\mathrm{ECH_2} + 2\mathrm{Me_3MCl} \longrightarrow 2(\mathrm{R}_n\mathrm{ECH_3})\mathrm{Cl} + \mathrm{R}_n\mathrm{EC(MMe_3)_2} \qquad (106)$$

(184), **(220)**, **(221)**

	R_nE	M
220a	$(Me_2N)(Me)S(O)$	Ge
220b	$(Me_2N)(Me)S(O)$	Sn
221	PPh_3	Ge
184	PPh_3	Sn

Often the chemistry of d^0-configurated tetravalent titanium parallels that of the d^0 elements of Group IVA. Thus, the preparation and structures of the dititanacyclobutane (**222**) and the silicon derivative (**223**) are comparable[53].

The reactions between $[Cp_2MCl_2]$ (M = Ti, Zr) and 2 equiv of Me_3PCH_2 lead to ionic complexes (**224**) with terminal ylide ligands. They cannot be converted into metal-substituted ylides on base treatment (Scheme 34).

In contrast, employment of the more bulky ylide Ph_3PCH_2 obviously inhibits the formation of analogous ionic products. The proposed intermediates $[Cp_2MCl_2(CH_2PPh_3)]$ (M = Zr, Hf) are transylidated by a second molecule of Ph_3PCH_2 and thus generate transition metal-substituted ylides **225**[187].

Mercury atoms are generally not capable of acidifying adjacent hydrogen atoms in mercurated onium salts, but like other d^{10}-systems they tend to do the reverse.

$$2\,[(Me_2N)_2TiCl_2] \xrightarrow{+4\,\mathbf{5}} 2\,[(Me_3PCH_2)_2Ti(NMe_2)_2]Cl_2 \xrightarrow[-4(Me_4P)Cl]{+2\,\mathbf{5}} \mathbf{222} \quad (107)$$

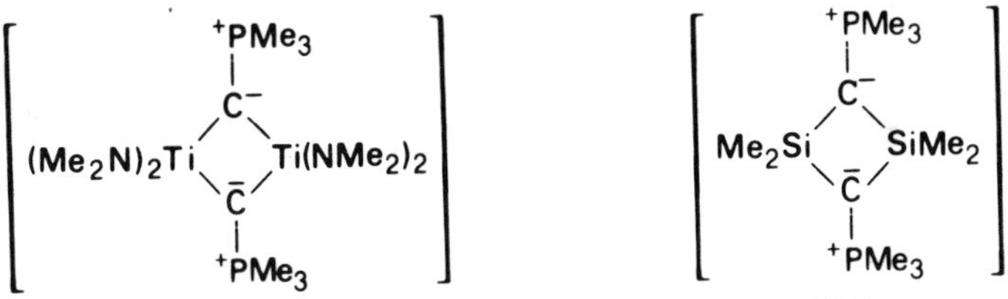

(222) (223)

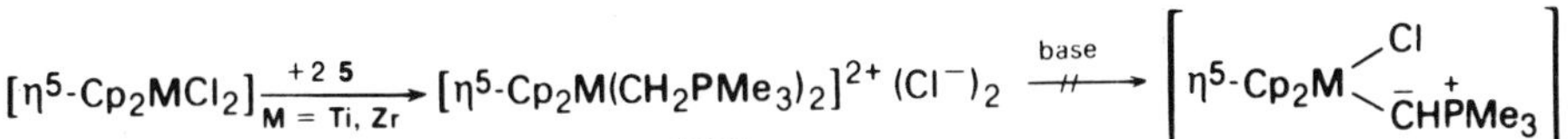

$$[\eta^5\text{-}Cp_2MCl_2] \xrightarrow[M = Ti, Zr]{+2\,\mathbf{5}} [\eta^5\text{-}Cp_2M(CH_2PMe_3)_2]^{2+}(Cl^-)_2 \text{ (224)} \xrightarrow{\text{base}} [\eta^5\text{-}Cp_2M(Cl)\bar{C}H\overset{+}{P}Me_3]$$

$$\xrightarrow[-(Ph_3PMe)Cl]{+2\,\mathbf{16}} \ (M = Zr, Hf) \quad [\eta^5\text{-}Cp_2M(Cl)\bar{C}H\overset{+}{P}Ph_3] \text{ (225)}$$

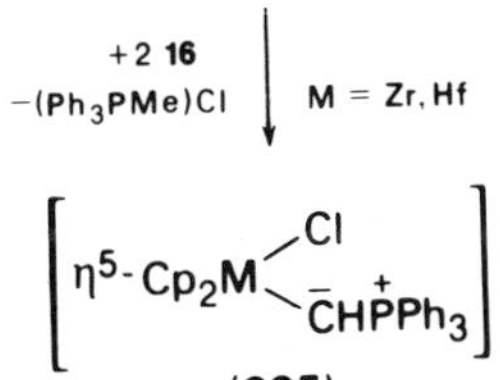

M	Ti	Zr	Hf
224	**a**	**b**	
225		**b**	**c**

SCHEME 34

Therefore, neither transylidation nor treatment with lithium alkyls can achieve the conversion of ionic mercurials of Me_3PCH_2 into mercurated ylides or heterocycles[97]. Mercurated ylides, however, are accessible when anion-stabilizing substituents at the ylidic carbon compensate for the retarding influence of the metal (equation 108).

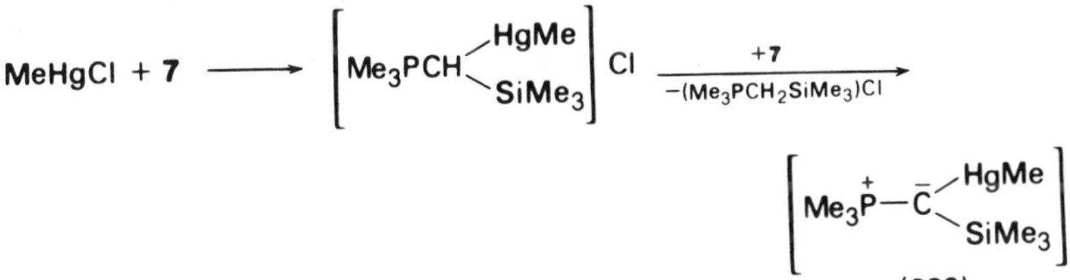

$$MeHgCl + \mathbf{7} \longrightarrow [Me_3PCH(HgMe)(SiMe_3)]Cl \xrightarrow[-(Me_3PCH_2SiMe_3)Cl]{+\mathbf{7}} [Me_3\overset{+}{P}\text{—}\bar{C}(HgMe)(SiMe_3)] \text{ (226)} \quad (108)$$

The successful syntheses of **227** and **228**[137–139,142] depend strongly on the electron-withdrawing capacity of the benzoyl, cyano, or carbomethoxy substituents and on the increased electronegative effects of a triphenylphosphonium group compared with the trimethylphosphonium analogue (equations 109 and 110).

At this point the problem concerning the electronic and structural features of transition metal complex frameworks required for stabilization of adjacent ylidic centres merits discussion. In contrast with the behaviour of electron-rich d^8–d^{10} transition metals, d^0 titanium, zirconium, and hafnium in some cases stabilize an anionic ylide function. The synthesis of compounds **124–128** and **130** containing tervalent metals with d^0, d^1, d^2, d^3, and d^6 configurations, respectively, clearly illustrates that low numbers of d electrons can hardly be the sole factor responsible for ylide generation, as might be expected. It is well known that the electron density at the

$$\left[\begin{array}{c}\text{Ph}_3\text{P—CH(COPh)—Hg}(\mu\text{-Cl})_2\text{Hg—CH(COPh)PPh}_3 \\ \text{(Cl on each Hg)}\end{array}\right] \xrightarrow[\substack{-2\ \text{MeOH/DMF} \\ -2\ \text{NaCl}}]{2\ \text{NaOMe}} 2\left[\text{Ph}_3\overset{+}{\text{P}}\text{—}\overset{-}{\text{C}}(\text{HgCl})(\text{COPh})\right] \quad \mathbf{(227)} \qquad (109)$$

$$\text{Ph}_3\text{PCHR} + \text{Hg(OAc)}_2 \longrightarrow [(\text{Ph}_3\text{PCHR})\text{Hg(OAc)}_2] \xrightarrow[\text{(2) NH}_3]{\text{(1) NaOMe/MeOH}} [\{\text{Ph}_3\text{PC(R)}\}_2\text{Hg}] \quad \mathbf{(228)} \qquad (110)$$

R = CN, CO_2Me

metal atom is extensively governed by the nature of the ligands present, and this is also obvious for a number of d^6-configurated systems that have been exposed to reaction with ylides. Thus, when $[(Me_3P)_3Me_2CoBr]$ was allowed to combine with Me_3PCH_2, chelate formation took preference over the generation of metallated ylides. On the other hand, iron complex **1** and the chromium derivative **2** were reluctant to undergo proton abstraction at all, whereas five-electron withdrawing carbonyl ligands in combination with formal positive charges at manganese(I) and rhenium(I) in **229** constitute a situation suitable for transylidation (equation 111)[12]. Furthermore, steric hindrance may play an important role in these reactions[187]. Clearly, further experimentation is necessary in this area.

$$[(\text{CO})_5\text{MBr}] \xrightarrow{+16} [(\text{CO})_5\text{MCH}_2\text{PPh}_3]\text{Br} \; \mathbf{(229)} \xrightarrow[-(\text{Ph}_3\text{PMe})\text{Br}]{+16} [(\text{CO})_5\text{M—}\overset{-}{\text{C}}\text{H}\overset{+}{\text{P}}\text{Ph}_3] \; \mathbf{(230)} \qquad (111)$$

230	**a**	**b**
M	Mn	Re

An interesting alternative to the reaction of ylides with organometallic halides arose in the heterolytic cleavage of metal–metal bonds in species such as **231** by ylides (Scheme 35)[188–193].

$$[\text{L}_m\text{M}^1\text{—M}^2\text{R}_n] \; \mathbf{(231)} \xrightarrow{+5} [\text{R}_n\text{M}^2\text{CH}_2\text{PMe}_3]^+[\text{M}^1\text{L}_m]^- \; \mathbf{(232)} \xrightarrow[-(\text{Me}_4\text{P})(\text{M}^1\text{L}_m)]{+5} [\text{R}_2\text{M}^2\overset{-}{\text{C}}\text{H—}\overset{+}{\text{P}}\text{Me}_3] \; \mathbf{(7)}, \mathbf{(233)}\text{–}\mathbf{(236)}$$

L_mM^1 = η^5-Cp(CO)$_3$Cr; η^5-Cp(CO)$_3$Mo; η^5-Cp(CO)$_3$W; η^5-Cp(CO)$_2$Fe; (CO)$_4$Co

	7, 233	**234**	**235**	**236**
R_nM^2	SiR'_3	$SnMe_3$	$AsMe_2$	$SbMe_2$

SCHEME 35

Compared with organometallic halides, the reactivity of **231** towards ylides is considerably diminished. Thus, cleavage of the metal–metal linkage becomes the slowest step in the reaction sequence, followed by very rapid transylidation of *in situ* generated **232**. The retarded reactivities of **231** also rationalize the omission of double metallations and disproportionations generally associated with the utilization of organometallic halides. Monostannylated ylides such as **234** are especially suitable targets for this approach, as other synthetic techniques failed to produce clean, isolable samples. The applicability of the above transformation seems to depend on the stability of the heterometallic bond and on the basicity and the stereochemistry of the ylidic centre. Ylide activity decreases in the order $Me_3PCH_2 > Et_3PCHMe > Me_3PCHSiMe_3 > Me_2S(O)CH_2 > Me_3PC(SiMe_3)_2$[188].

2. *Metallation of ylidic anions*

Treatment of anionic double ylides such as **87** with equimolar amounts of organometallic halide R^2Cl is also applicable to the formation of metal-substituted ylides[83,97,194].

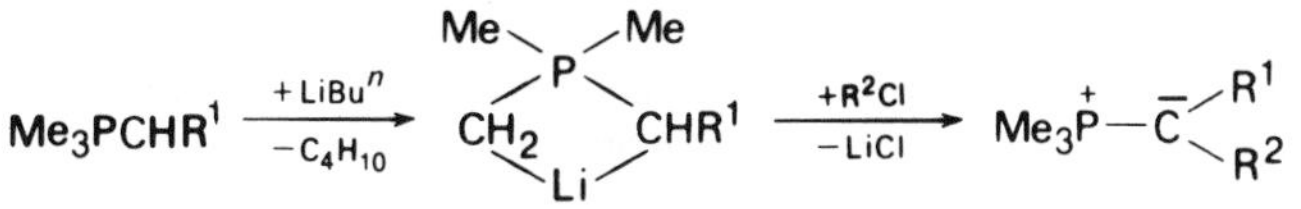

$$Me_3PCHR^1 \xrightarrow[-C_4H_{10}]{+LiBu^n} Me_2P(CH_2)(CHR^1)Li \xrightarrow[-LiCl]{+R^2Cl} Me_3\overset{+}{P}-\overset{-}{C}R^1R^2 \qquad (112)$$

	226	219	79	236
R^1	$SiMe_3$	$SiMe_3$	$SiMe_3$	H
R^2	HgMe	$GeMe_3$	$SnMe_3$	$SbMe_2$

This mode of synthesis exhibits interesting phenomena of prototropy and rearrangements leading to products in which the substituents are exclusively situated at the ylide carbon centre[21,194]. Multiply lithiated ylides constitute precursors to other multiply metallated species[24] (Scheme 36).

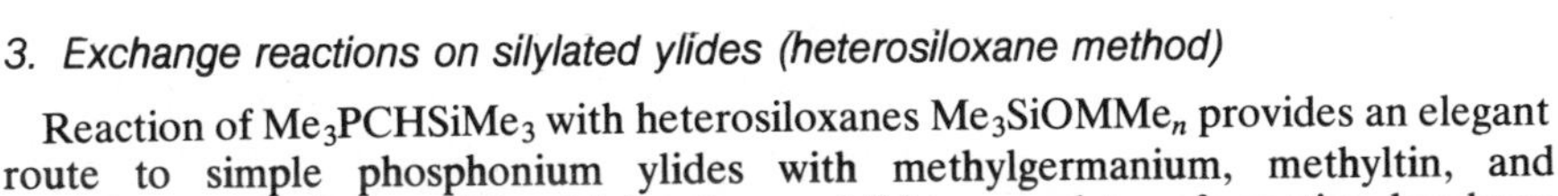

$$\underset{(238)}{Me_2\overset{+}{P}(CH_2GeMe_3)-\overset{-}{C}(GeMe_3)_2} \xleftarrow[(2)\ +3\ Me_3GeCl,\ -3\ LiCl]{(1)\ +3\ LiR,\ -3\ RH} \mathbf{5} \xrightarrow[(2)\ +2\ Me_3GeCl,\ -2\ LiCl]{(1)\ +2\ LiR,\ -2\ RH} \underset{(237)}{Me_3\overset{+}{P}-\overset{-}{C}(GeMe_3)_2}$$

SCHEME 36

3. *Exchange reactions on silylated ylides (heterosiloxane method)*

Reaction of $Me_3PCHSiMe_3$ with heterosiloxanes $Me_3SiOMMe_n$ provides an elegant route to simple phosphonium ylides with methylgermanium, methyltin, and methyllead substituents. The driving force of this unusual transformation has been attributed to the marked propensity for disiloxane formation[83] (equation 113). The generation of **217** and **235** proceeded smoothly, whereas **234**, which under these conditions underwent facile disproportionation to Me_3PCH_2 and **239**, was only

$$Me_3PCHSiMe_3 + Me_3SiOMMe_n \longrightarrow (Me_3Si)_2O + \underset{(217),\ (234),\ (235)}{Me_3PCHMMe_n} \qquad (113)$$

	217	234	235
MMe_n	$GeMe_3$	$SnMe_3$	$AsMe_2$

$$2Me_3PCHSiMe_3 + 2Me_3SiOMMe_3 \longrightarrow 2(Me_3Si)_2O + Me_3PC(MMe_3)_2 + \mathbf{5}$$

(239), (240) (114)

	239	240
M	Sn	Pb

detected spectroscopically[83]. The analogous lead derivative was found to be even more sensitive[24] (equation 114).

The susceptibility to disproportionation exhibited by the higher homologues contrast sharply with the behaviour of $Me_3PCHSiMe_3$, which is accessible via comproportionation according to equation 115. Silicon and, to some extent,

$$Me_3PCH_2 + Me_3PC(SiMe_3)_2 \longrightarrow 2Me_3PCHSiMe_3 \qquad (115)$$

germanium atoms stabilize ylidic carbanions, whereas this function is destabilized and thus enhanced in reactivity by tin- and lead-containing substituents. The increased acceptor capacities of tetracoordinated tin and lead species, compared with silicon and germanium analogues, facilitate the nucleophilic attack of the ylide, inducing disproportionation[24,83]. The reactivity of metallated ylides is evidently also governed by the nature of the onium centre. In accordance with this, monogermylated derivatives of sulphonium and arsonium ylides such as **241** and **242** undergo rapid disproportionation (equations 116[186] and 117[24]). Similar behaviour was encountered with the ylides $Me_3PCHMMe_2$ (M = P, As, Sb)[83].

$$2\ Me(O)(Me_2N)\overset{+}{S}\overline{C}HGeMe_3\ (\mathbf{241}) \longrightarrow Me(O)(Me_2N)\overset{+}{S}\overline{C}H_2\ (\mathbf{37}) + Me(O)(Me_2N)\overset{+}{S}\overline{C}(GeMe_3)_2\ (\mathbf{220}a) \qquad (116)$$

$$2\ Me_3\overset{+}{As}\overline{C}HSiMe_3 \xrightarrow[-2\ (Me_3Si)_2O]{+2\ Me_3SiOGeMe_3} 2\ Me_3\overset{+}{As}\overline{C}HGeMe_3\ (\mathbf{242}) \longrightarrow Me_3\overset{+}{As}\overline{C}(GeMe_3)_2\ (\mathbf{243}) + Me_3\overset{+}{As}\overline{C}H_2 \qquad (117)$$

B. Phosphine Addition to Carbyne Complexes

Phosphine addition converts triply coordinated carbene carbon atoms of carbene complexes (see Chapter 5) into tetravalent carbon atoms of the corresponding ylide complexes. In the case of transition metal carbyne complexes (Chapter 7), phosphine addition to the dihedral carbyne carbon atom generates transition metal-substituted ylides possessing three substituents at the ylide centre as usual.

$$[X(CO)_4Cr{\equiv}CR]\ (\mathbf{244}) + PMe_3 \xrightarrow{-60°C} \left[X(CO)_4Cr{=}C\begin{matrix}PMe_3\\R\end{matrix}\right]\ (\mathbf{245}) \qquad (118)$$

R	Me	CH_2Ph	Ph	p-MeC_6H_4	2,4,6-$Me_3C_6H_2$	$SiPh_3$
X = Cl				**d**		
X = Br	**a**	**b**	**c**	**e**	**g**	**h**
X = I				**f**		

Extremely thermolabile carbonylchromium ligated ylides such as **245** were obtained from carbyne complexes **244** and PMe_3 at low temperatures (equation 118)[195,196].

Chromium compounds such as **245** and **246** were decomposed by excess of PMe_3, as illustrated in equation 119. Intermediate **247**, containing a semi-ylidic ligand, resisted isolation[197]. In contrast, the analogous molybdenum and tungsten carbyne complexes incorporate two PMe_3 molecules, the second of which displaces a carbonyl ligand[198,199] (equation 120).

$$[(CO)_5Cr{\equiv}CNEt_2]^+ BF_4^- \xrightarrow{+2PMe_3}$$

(246)

$$\left[(CO)_5Cr{-}C(PMe_3)_2{-}NEt_2\right]^+ BF_4^- \xrightarrow[-(CO)_5CrPMe_3]{+PMe_3} [Et_2NC(PMe_3)_2]^+ BF_4^- \quad (119)$$

(247)

$$[X(CO)_4M{\equiv}CR] + 2PMe_3 \xrightarrow{-60°C} \left[X(CO)_3(PMe_3)M{=}C{<}^{PMe_3}_{R}\right] + CO \quad (120)$$

(248), (249)

248: M = Mo, X = Cl, R = *p*-MeC_6H_4

249: M = W; R	*p*-MeC_6H_4	$SiPh_3$	Ph
X = Cl	**249a**		
X = Br	**249b**	**249c**	**249d**

The thermal stabilities of phosphine complexes **248** and **249** far exceed those of ylides **245**. At elevated temperatures repeated carbonyl displacement seems to be possible with **249b,d** producing **250b,d**, which are very stable at ambient temperatures[198,199] (equation 121).

$$Br(CO)_3(PMe_3)W{=}C{<}^{PMe_3}_{R} \xrightarrow[-CO]{+PMe_3/30°C} \left[Br(CO)_2(PMe_3)_2W{=}C{<}^{PMe_3}_{R}\right] \quad (121)$$

(249 b,d) **(250 b,d)**

250	b	d
R	*p*-MeC_6H_4	Ph

A pronounced degree of thermolability is also inherent in the cationic arene-substituted chromium species **251**[67] (equation 122).

$$[Ar(CO)_2M{\equiv}CPh]^+ BCl_4^- \xrightarrow[-50°C]{+PMe_3} \left[Ar(CO)_2M{=}C{<}^{Ph}_{PMe_3}\right]^+ BCl_4^- \quad (122)$$

(78), (251), (252)

	251a	251b	251c	252	78
M	Cr	Cr	Cr	Mn	Re
Ar	C_6H_6	$1,4Me_2C_6H_4$	$1,3,5Me_3C_6H_3$	$C_5H_5^-$	$C_5H_5^-$

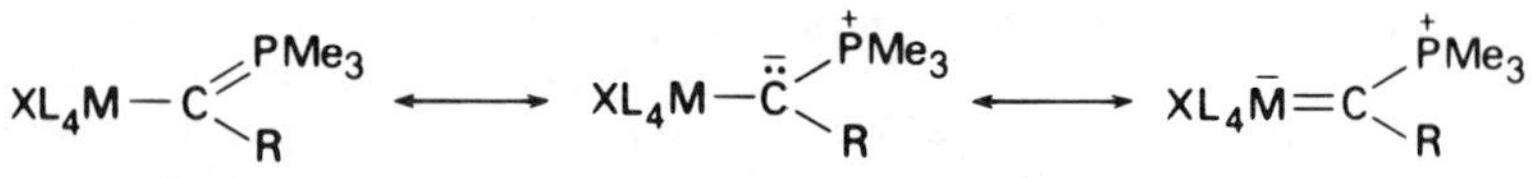

$$XL_4M-C(=PMe_3)(R) \longleftrightarrow XL_4M-\ddot{\bar{C}}(\overset{+}{P}Me_3)(R) \longleftrightarrow XL_4\bar{M}=C(\overset{+}{P}Me_3)(R)$$

A B C

FIGURE 25. Resonance structures of transition metal-substituted ylides.

In **78** the short bond distance between the rhenium atom and the ylide carbon atom (1.97 Å)[200], indicating considerable double bond order, favours resonance structure **C** for the description of the mode of coordination in transition metal-substituted ylides (Figure 25). **78**, possessing an electrophilic site at the ylide carbon atom, is susceptible to nucleophilic addition with LiMe (equation 45) and PMe_3[67] (equation 123). Unlike

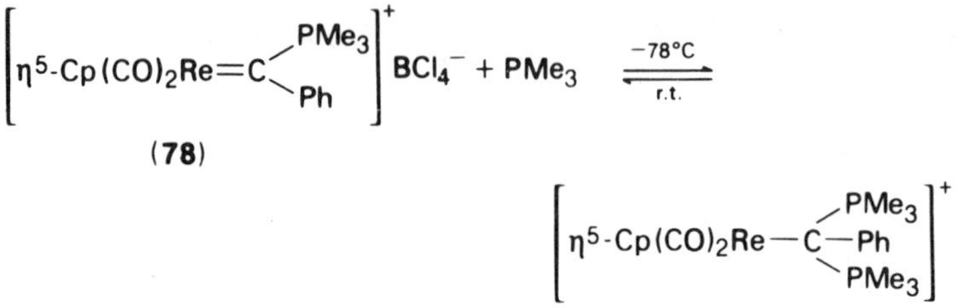

$$\left[\eta^5\text{-Cp(CO)}_2\text{Re}=\text{C}(\text{PMe}_3)(\text{Ph})\right]^+ \text{BCl}_4^- + \text{PMe}_3 \underset{\text{r.t.}}{\overset{-78^\circ\text{C}}{\rightleftharpoons}} \left[\eta^5\text{-Cp(CO)}_2\text{Re}-\text{C}(\text{PMe}_3)_2\text{Ph}\right]^+ \text{BCl}_4^- \quad (123)$$

(**78**) (**253**)

the chromium intermediate **247**, which decomposes via metal–carbon bond fission with expulsion of the semi-ylidic ligand, **253** regenerates **78** with loss of phosphine at room temperature. A different mode of reaction is encountered when phosphine is added to the tungsten carbyne derivative **254**, containing the complex anion $[Re(CO)_5]^-$ in the *trans*-position instead of a halide ligand. The availability of acceptor sites at the adjacent rhenium atom in **255** obviously induces the ylidic ligand to function as a bridge in which it donates three electrons[201].

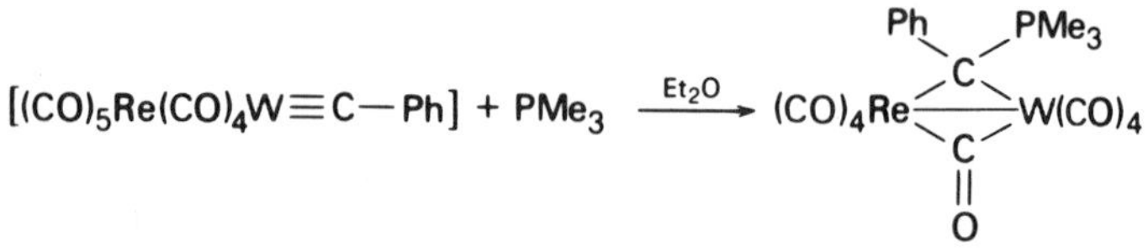

$$[(\text{CO})_5\text{Re(CO)}_4\text{W}\equiv\text{C}-\text{Ph}] + \text{PMe}_3 \xrightarrow{\text{Et}_2\text{O}} (\text{CO})_4\text{Re}\,[\mu\text{-C(Ph)(PMe}_3)]\,[\mu\text{-C=O}]\,\text{W(CO)}_4 \quad (124)$$

(**254**) (**255**)

Complexes **137** and **185** may also be considered as compounds in which carbon atoms of transition metal-substituted ylides are utilized as bridges.

The unavailability of many carbyne complexes poses a severe limitation on the general application of these reagents. Moreover, the competing formation of transition metal-substituted ketenes, which are the sole products of the reaction between $[Cp(CO)_2M\equiv CR]$ (M = Mo, W) and PMe_3[202], also limits the scope of this method.

In this connection, the addition of phosphines such as $P(OMe)_3$, $P(OEt)_3$, and $P(OMe)Ph_2$ to the σ–π-acetylide diironhexacarbonyl derivative **256** is worth mentioning. During the course of the reaction, the acetylide is transformed to a two-carbon, three-electron ligand bridging the metal atoms. The mode of coordination is best described in terms of the resonance structures **257a** and **257b**[203] (equation 125). Structure **257a** features a metal-substituted ylide which is attached to the second iron atom through a carbene function in a side-chain.

$$\left[(CO)_3Fe(\mu\text{-}PPh_2)(\mu\text{-}C{\equiv}CPh)Fe(CO)_3\right] \xrightarrow{+P(OEt)_3} \left[(EtO)_3P{=}C{-}\ddot{C}(Ph)\ (CO)_3Fe(\mu\text{-}PPh_2)Fe(CO)_3\right] \quad (125)$$

(256) (257a)

$$\longleftrightarrow \left[(EtO)_3P{-}C{=}C{-}Ph\ (CO)_3Fe(\mu\text{-}PPh_2)Fe(CO)_3\right]$$

(257b)

IV. METALS AS ANIONIC CENTRES IN YLIDES

A. Base-stablized Germylenes and Stannylenes

1. Complexation of metal dihalides

In contrast to carbenes, the homologous germylenes, with the exception of dialkylated and diarylated derivatives, represent a class of compounds already stable under normal conditions. For a long period information about them was restricted to coordination polymers[204]. Like dihalocarbenes, germanium dihalides (dihalogermylenes) form adducts with tertiary phosphines to give products which are at least formally analogous to dihalomethylenephosphoranes[17,18,205].

$$GeI_2 + PR_3 \xrightarrow[\text{xylene}]{\Delta} R_3\overset{+}{P}{-}\overset{-}{\ddot{G}}eI_2 \quad (126)$$

(258)

258	a	b	c	d	e	f
R_3P	PPh_3	$PMePh_2$	$PEtPh_2$	PPh_2Pr^i	PBu^nPh_2	PBu^n_3

$$GeCl_2{\cdot}C_4H_8O_2 + PR_3 \xrightarrow{C_6H_6} R_3\overset{+}{P}{-}\overset{-}{\ddot{G}}eCl_2 + C_4H_8O_2 \quad (127)$$

(259) (260)

260	a	b	c
R	Ph	Et	*t*-Bu

Whereas in classical ylides triply coordinated carbon atoms represent the donor functions, the negative charges in α-positions to the onium centres of **258** and **260** are located on triply coordinated germanium atoms. Owing to large differences in the stabilities of divalent carbon and germanium, differences in the chemical properties of the two classes of ylides are not unexpected. Their dissociation phenomena in solution[18,206] or behaviour on heating[17] lead one to the conclusion that base-stabilized germylenes and their tin analogues (equation 128)[206] are best described as metal coordination compounds.

$$SnX_2 + PR_3 \longrightarrow R_3\overset{+}{P}\overset{-}{\ddot{S}}nX_2 \quad (128)$$

(261)

R	*t*-Bu	NMe_2
X = Cl	**261a**	**261b**
X = Br	**261c**	**261d**

2. *α-Elimination reactions on trihalogermanes*

Compounds such as **260** were prepared from dialkylphosphinotrihalogermanes under very mild conditions, which was rationalized as being due to a marked tendency of formation of the germanium phosphorus linkage in the adduct[18,207].

$$t\text{-}Bu_2PGeX_3 + PR_3 \xrightarrow[20°C]{C_6H_6} t\text{-}Bu_2PX + \underset{\mathbf{(260)}}{R_3PGeX_2} \qquad (129)$$

R	Ph	*t*-Bu	*n*-Bu
X = Cl	**260a**	**260c**	**260d**

α-Elimination of Me_3SiCl from $Cl_3Ge—SiMe_3$ is believed to be a key step in the synthesis of monomeric germylene complexes of heteroaromatic nitrogen bases such as **263** (Scheme 37)[208,209].

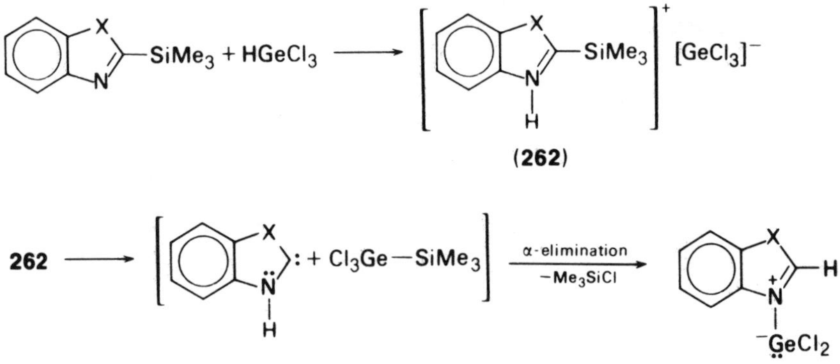

(262)

(263)

262, 263	a	b
X	S	NMe

SCHEME 37

The decomposition of intermediate **262b** proceeds spontaneously in solution at room temperature, whereas **262a** can be isolated as an ionic solid which is converted to **263a** at 91°C. To explain the surprisingly short bond distance between the germanium and nitrogen atoms (2.092 Å) in **263a**, it is necessary to invoke interactions between the π-system of the heteroaromatic ring and vacant *d*-orbitals of germanium. This interaction appears to be crucial for the stabilization of germanium–nitrogen ylides, which is consistent with the fruitless attempts to obtain $GeCl_2$ complexes of tertiary aliphatic and aromatic amines of the general type NR_3. Nevertheless, the dative bond between a donor molecule and MX_2 in base-stabilized germylenes and stannylenes is kinetically labile, thus facilitating ligand-exchange reactions[18,209].

B. Transition Metal Complexes with Ylides of Germanium and Tin

Base-stabilized germylenes and stannylenes are capable of functioning as ligands in transition metal complexes by utilizing the free electron pair at the main group metal. In contrast to classical ylides, the tin and germanium species exhibit excellent π-acceptor properties due to vacant *d*-orbitals at the metal atoms.

1. Direct syntheses

A direct access to germanium and tin ylide complexes is available from the photochemical displacement of carbonyl ligands by base-stabilized germylenes and stannylenes. **263a** and the thf adduct of $GeCl_2$, the latter of which may be pre-generated in thf from **259** or $GeCl_3^-$, have been used as germanium ylide ligands, whereas anhydrous tin dihalides represent a convenient source of tetrahydrofuranodihalostannide ligands (Scheme 38[210] and equation 130[211,212]).

$$\underset{\mathbf{(264)}}{[(CO)_5M\{GeCl_2(thf)\}]} \xleftarrow[h\nu]{CsGeCl_3/thf} [M(CO)_6] \xrightarrow[h\nu]{SnX_2/thf} \underset{\mathbf{(265)}}{[(CO)_5M\{SnX_2(thf)\}]}$$

264	**a**	**b**	**c**
M	Cr	Mo	W

X	Cl	Br	I
M = Cr	**265a**	**d**	**g**
M = Mo	**b**	**e**	**h**
M = W	**c**	**f**	**i**

SCHEME 38

$$[L_n(CO)_3M] \xrightarrow[h\nu]{\mathbf{263a},\ thf} \underset{\mathbf{(266)}}{[L_n(CO)_2M\{GeCl_2(benzothiazole)\}]} + CO \qquad (130)$$

266	**a**	**b**	**c**	**d**	**e**	**f**
L_nM	$Cr(CO)_3$	$Mo(CO)_3$	$W(CO)_3$	$Fe(CO)_2$	$(C_6H_6)Cr$	$(1,3,5\text{-}Me_3C_6H_3)Cr$

266	**g**	**h**
L_nM	C_6Me_6Cr	(MeCp)Mn

2. Indirect syntheses

a. Addition of base to germylene complexes. Addition of base to germylene complexes such as **267** and **268**[213] parallels the synthesis of classical ylide complexes from carbene complexes and donor molecules (Section II.A.2a) (equation 131)[214,215].

$$\underset{\mathbf{(267),\ (268)}}{[(CO)_5CrGeR_2]} + B \longrightarrow \underset{\mathbf{(269),\ (270)}}{[(CO)_5Cr{-}GeR_2(B)]} \qquad (131)$$

(In the product, B is drawn bonded below Ge.)

	R
267	2,4,6-$Me_3C_6H_2S$
268	mesityl

B	NMe_3	C_5H_5N	PPh_3
R = 2,4,6-$Me_3C_6H_2S$	**269a**	**269b**	**269c**
R = mesityl		**270**	

Steric crowding at the germanium atom in **268** presumably precludes the addition of bulky bases such as PPh_3 and benzothiazole as well as thf and diethyl ether.

b. Base-exchange reactions on complexed ylides. The nucleophilic displacement of donor molecules constituting onium centres of ylide complexes (see Section II.2.b) is paralleled in the complex chemistry of tin ylides. The isolation of **271** shows that tin–phosphorus coordination clearly takes preference over the respective tin–oxygen interaction[216–218] (equation 132).

The coordination of dihalogermylenes and stannylenes to transition metals leads to a considerable enhancement in the Lewis acidity of the coordinated ligands compared

$$[(CO)_5M\{M'Cl_2(thf)\}] \xrightarrow[-thf]{+PR_3} [(CO)_5M\{M'Cl_2(PR_3)\}] \quad (132)$$

(271)

M = Cr, W; M' = Ge, Sn; PR_3 = PPh_3, $P(t\text{-Bu})_3$

with the free species. Owing to extensive dissociation, free Ph_3PSnCl_2 is obviously non-existent, whereas the corresponding pentacarbonyl chromium complex **271** appears to be an isolable compound[217].

The occurrence of the exchange of substituents between base-stabilized germanium dichloride and mercaptosilanes yielding base-free low-molecular-weight germylenes **272**[219] suggested the applicability of this method as a procedure for base-free germylene complex formation.

$$GeCl_2{-}B + 2Me_3SiSR \longrightarrow Ge(SR)_2 + 2Me_3SiCl + B \quad (133)$$

(259), (363) → **(272)**

R = Ph, $PhCH_2$, n-Bu

As expected, the transformation of **264a** into germylene complexes **267** and **273** is accompanied by removal of the weakly basic thf; however, the more basic benzothiazole of **266a** remains attached to the germanium throughout the reaction[213,215] (Scheme 39).

$$\left[(CO)_5Cr{-}Ge(SR)_2 \leftarrow \text{N-benzothiazole}\right] \; \mathbf{(274)} \xleftarrow[-2\,Me_3SiCl]{+2\,Me_3SiSR} [(CO)_5Cr{-}GeCl_2{-}B] \; \mathbf{(266a)}, \mathbf{(264a)} \xrightarrow[-2\,Me_3SiCl]{+2\,Me_3SiSR} [(CO)_5Cr{-}Ge(SR)_2] + thf \; \mathbf{(267)}, \mathbf{(273)}$$

274	**a**	**b**	**c**	**d**	**e**
R	Me	Et	Ph	$PhCH_2$	mesityl

	267	**273**
R	mesityl	Me

SCHEME 39

The excellent π-acceptor capacities of germanium and tin ylide ligands are illustrated in the i.r. spectra of their metal–carbonyl complexes and in a remarkably short Mo—Ge bond length (2.521 Å), deduced from the X-ray structural determination of **266b**[211].

c. Dehalogenation reactions of halogenated germanes and stannanes by metal carbonylates. The reaction of sodium transition metal carbonylates with geminal dihalo- and trihaloalkanes constitutes a valuable route to carbene complexes. The generation of base-stabilized germylene and stannylene complexes, inevitable in thf as a solvent, from the reaction of dihalodialkylgermanes and -stannanes or the tetrahalides of tin and germanium with organometallic nucleophiles undoubtedly resembles this carbene complex synthesis (Scheme 40)[220,221].

The attachment of the pyridine molecule to the tin atom in **275** was confirmed by an X-ray diffraction analysis[222] (Figure 26).

On the other hand, dechlorination of $R_2M'Cl_2$ with $Na_2[Fe(CO)_4]$ produces binuclear complexes with germylene and stannylene bridges which, however, suffer facile and reversible cleavage with Lewis bases[223] (equation 134).

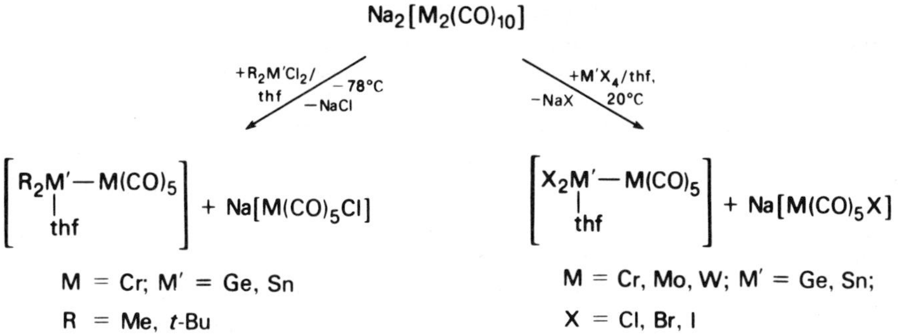

$$
\begin{array}{c}
Na_2[M_2(CO)_{10}] \\
\swarrow \; {}^{+R_2M'Cl_2/thf,\ -78^{\circ}C}_{-NaCl} \qquad\qquad {}^{+M'X_4/thf,\ 20^{\circ}C}_{-NaX} \; \searrow
\end{array}
$$

$$
\left[\begin{array}{c} R_2M'{-}M(CO)_5 \\ | \\ thf \end{array}\right] + Na[M(CO)_5Cl] \qquad\qquad \left[\begin{array}{c} X_2M'{-}M(CO)_5 \\ | \\ thf \end{array}\right] + Na[M(CO)_5X]
$$

M = Cr; M′ = Ge, Sn
R = Me, *t*-Bu

M = Cr, Mo, W; M′ = Ge, Sn;
X = Cl, Br, I

SCHEME 40

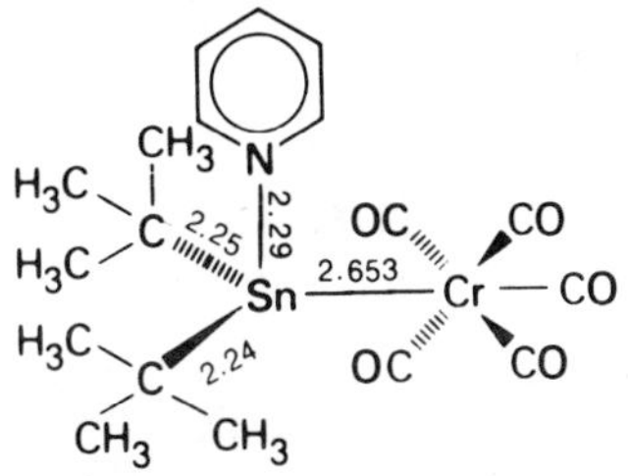

FIGURE 26. Molecular structure of **275** with selected bond distances in Å.

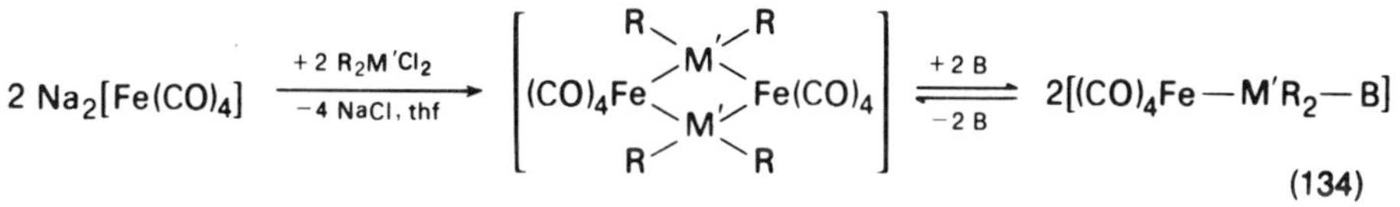

$$
2\,Na_2[Fe(CO)_4] \xrightarrow[-4\,NaCl,\ thf]{+2\,R_2M'Cl_2} \left[(CO)_4Fe\begin{array}{c} R_2 \\ M' \\ M' \\ R_2 \end{array}Fe(CO)_4\right] \underset{-2\,B}{\overset{+2\,B}{\rightleftharpoons}} 2[(CO)_4Fe{-}M'R_2{-}B] \qquad (134)
$$

M′ = Ge, Sn, Pb; R = Me, Ph, *t*-Bu, *n*-Bu;
B = thf, C_5H_5N, MeCN, Et_2O, dmf, acetone

Attempts to isolate the iron-containing ylide species proved to be successful only for $[(CO)_4Fe\{GeMe_2(NC_5H_5)\}]$ and $[(CO)_4Fe\{Sn(t\text{-}Bu)_2(NC_5H_5)\}]$, whereas when they were not isolated their presence was claimed from solution i.r. data. Removal of the solvent results in complete formation of the dimers. The susceptibility to fission of the bridges is enhanced by the basicity of the solvent, and in thf as solvent ylide complex formation is impeded with increasing atomic weight of the bridging atoms.

V. METALONIUM YLIDES

Semi-metals and metals of main Groups V and VI are known to form ylides containing the metal atom in the centre of the onium function. Replacement of the phosphorus atom of classical ylides by the higher homologues such as arsenic, antimony, and bismuth leads to successive loss of stability of the corresponding ylides with increasing atomic weight of the metals. Although simple non-stabilized arsonium ylides such as Me_3AsCH_2[224] and Ph_3AsCH_2[225] appear to be isolable compounds, their thermolability is evident when compared with Me_3PCH_2 and Ph_3PCH_2. Related stibonium ylides have not been isolated. However, the introduction of stabilizing trimethylsilyl substituents into the molecule allows the preparation of **276**, which is further stabilized by isomerization to **277**[226] (equation 135). Resonance-stabilized ylide **279** exhibits

$$(Me_3SiCH_2)_4SbCl \xrightarrow[-(Et_4P)Cl]{+Et_3PCHMe} [(Me_3SiCH_2)_3\overset{+}{Sb}-\overset{-}{C}HSiMe_3] \ (\mathbf{276}) \xrightarrow{\Delta} [Me(Me_3SiCH_2)_2\overset{+}{Sb}-\overset{-}{C}(SiMe_3)_2] \ (\mathbf{277}) \quad (135)$$

sufficient stability at ambient temperature, whereas slow decomposition of the bismuth derivative **280** is inevitable under these conditions (Scheme 41)[227].

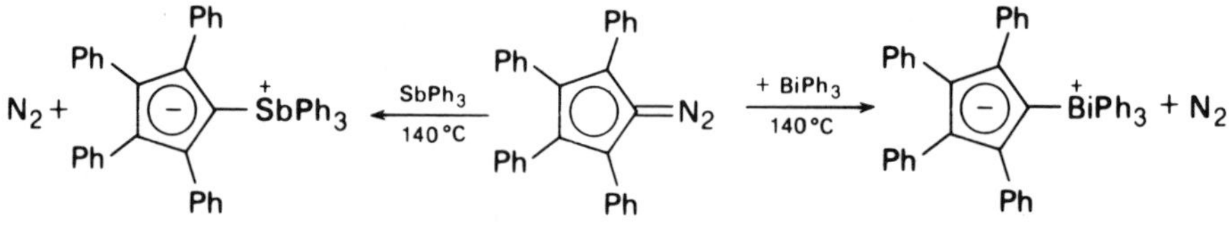

SCHEME 41

Similar observations have also been made for selenium and tellurium ylides, which are capable of existence only when extensively resonance stabilized, and even then most of these species suffer decomposition above 0°C[227–234].

Concerning synthesis and reactivity toward electrophiles, alkylides of niobium and tantalum bear a close resemblance to phosphorus ylides (Scheme 42)[235].

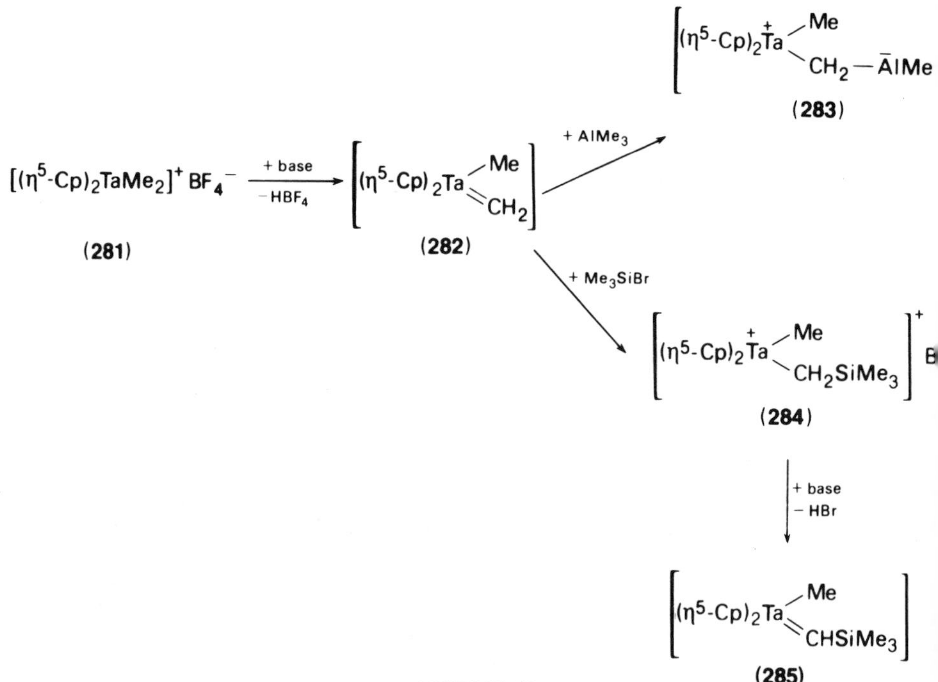

SCHEME 42

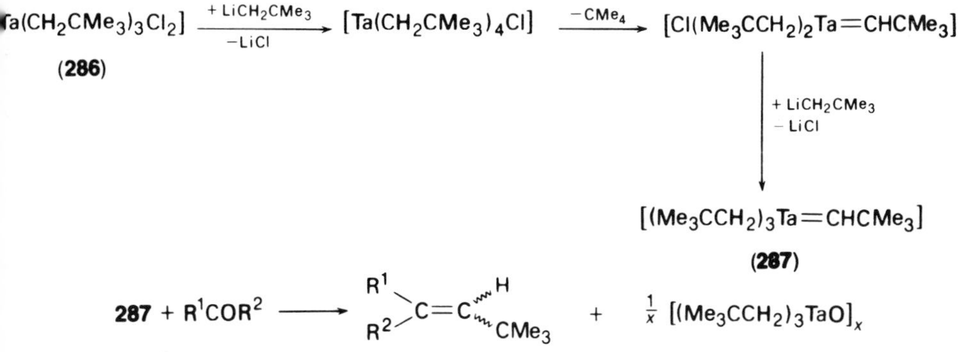

$$[Ta(CH_2CMe_3)_3Cl_2] \xrightarrow[-LiCl]{+LiCH_2CMe_3} [Ta(CH_2CMe_3)_4Cl] \xrightarrow{-CMe_4} [Cl(Me_3CCH_2)_2Ta{=}CHCMe_3]$$

(286)

$$[Cl(Me_3CCH_2)_2Ta{=}CHCMe_3] \xrightarrow[-LiCl]{+LiCH_2CMe_3} [(Me_3CCH_2)_3Ta{=}CHCMe_3]$$

(287)

$$\mathbf{287} + R^1COR^2 \longrightarrow R^1R^2C{=}C(H)CMe_3 + \frac{1}{x}[(Me_3CCH_2)_3TaO]_x$$

R^1	Me	Me	H	Me	H	H
R^2	Me	3-MeC_6H_4	Ph	OEt	OEt	NMe_2

SCHEME 43

Proton abstraction with bases such as Me_3PCH_2 or $LiN(SiMe_3)_2$ converts the tantalonium salt **281** to the tantalonium ylide **282**. The nucleophilic methylene carbon atom of **282** permits entry into a series of reactions with electrophiles such as $AlMe_3$ or Me_3SiBr, thus providing the ylide complex **283** or the tantalonium salt **284**, respectively. Dehydrohalogenation of **284** leads to the silylated ylide **285** in a reaction sequence that is known for phosphorus ylide chemistry. Transition metal ylide **287**, generated from **286** and 2 equiv of $LiCH_2CMe_3$[236], was utilized in the conversion of a series of organic carbonyl compounds into olefins[237] (Scheme 43).

In contrast to the Wittig reaction, esters, amides, and even CO_2 prove to be susceptible to this olefination procedure. It is assumed that the first step of the reaction consists in adduct formation between the metal atom and the carbonyl oxygen atom. This complexation proceeds efficeintly with 14-electron systems where vacant acceptor sites are readily available. On the other hand, reaction of the 18-electron complex **282** with carbonyls appears to be sluggish. At this point the very interesting transition metal ylide chemistry of niobium and tungsten is not discussed further. The very competent review by Schrock is recommended to the interested reader[238].

Resonance structure **B** of the iridium complex **288** also possesses the features of an iridium(III) ylide[239] (Figure 27).

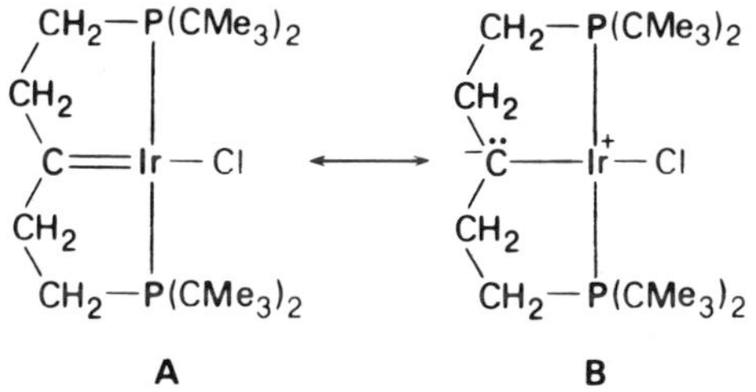

A: $CH_2CH_2P(CMe_3)_2$ / $CH_2CH_2P(CMe_3)_2$ chelate with $C{=}Ir{-}Cl$ $\longleftrightarrow$ B: same chelate with $^-\ddot{C}{-}\overset{+}{Ir}{-}Cl$

A **B**

FIGURE 27. Resonance structures of **288**.

VI. YLIDE COMPLEXES WITH METAL-SUBSTITUTED ONIUM CENTRES

The compounds discussed in this section are not real ylides with respect to the classical ones. Their relationship, however, is revealed if the transition metal atom attached to the heteroatom is regarded as a normal substituent (Figure 28).

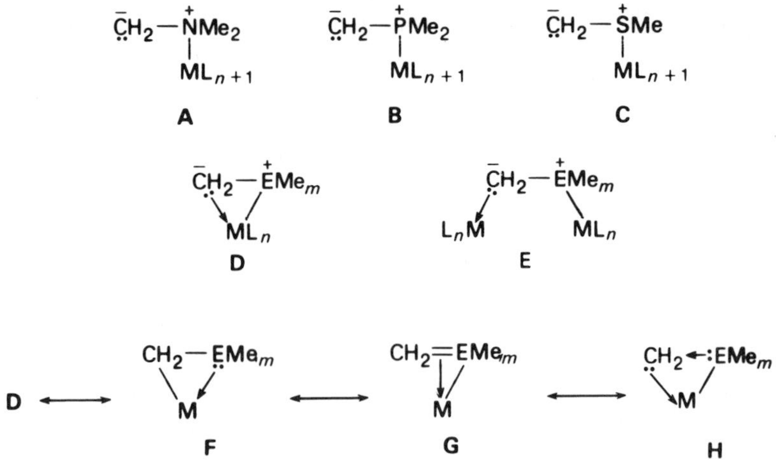

$$\overset{-}{\ddot{C}}H_2-\overset{+}{N}Me_2(ML_{n+1}) \quad \textbf{A} \qquad \overset{-}{\ddot{C}}H_2-\overset{+}{P}Me_2(ML_{n+1}) \quad \textbf{B} \qquad \overset{-}{\ddot{C}}H_2-\overset{+}{S}Me(ML_{n+1}) \quad \textbf{C}$$

$$\overset{-}{C}H_2-\overset{+}{E}Me_m \text{ bridged by } ML_n \quad \textbf{D} \qquad L_nM \leftarrow \overset{-}{C}H_2-\overset{+}{E}Me_m \rightarrow ML_n \quad \textbf{E}$$

$$\textbf{D} \longleftrightarrow \underset{\textbf{F}}{CH_2-\ddot{E}Me_m / M} \longleftrightarrow \underset{\textbf{G}}{CH_2=EMe_m / M} \longleftrightarrow \underset{\textbf{H}}{\ddot{C}H_2 \leftarrow :EMe_m / M}$$

FIGURE 28. Descriptions of the mode of bonding of ylides and ylide metal complexes with metal-substituted onium centres.

Thus, heteroatoms in species **A–C** in Figure 28 formally represent onium centres of ylides. Owing to its pronounced donor capacity, the anionic function is always accommodated at the acceptor site of a metal, thus leading either to the formation of three-membered ring units **D** or to complexes with ligand bridges such as **E**. The mode of bonding in the three-membered ring **D** is further illustrated by resonance structures **F–H**.

A. Ylide Complexes Containing Three-membered Metallocycles

1. Syntheses involving neutral ligands

The dimethylphosphinomethyl(hydrido)iron complex **289** results from the reduction of $[(Me_3P)_2FeCl_2]$ with magnesium in the presence of excess of phosphine (equation 136). The generation of the ligand can be understood in terms of the oxidative

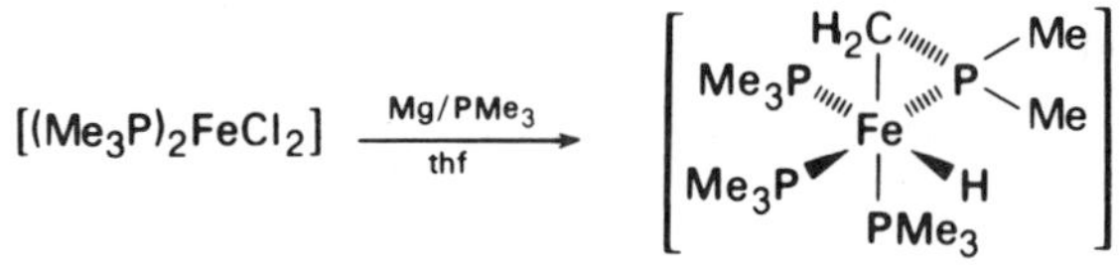

$$[(Me_3P)_2FeCl_2] \xrightarrow[\text{thf}]{\text{Mg/PMe}_3} [(Me_3P)_3Fe(H)(\eta^2\text{-}CH_2PMe_2)] \quad \textbf{(289)} \tag{136}$$

addition of PMe_3 to the coordinatively unsaturated and electron-rich reduced metal centre. The reactivity of **289** toward π-acceptor ligands stabilizing low oxidation states indicates the presence of **290**, which equilibrates with **289** in a temperature- and solvent-dependent manner[19a] (equation 137).

$$\textbf{289} \rightleftharpoons \underset{\textbf{(290)}}{Fe(PMe_3)_4} \tag{137}$$

The same structural array is encountered in the ruthenium species **292**, where the phosphinomethyl moiety is believed to originate from a PMe_3 ligand deprotonated by the strongly basic $Li(CH_2)_2PMe_2$ (**87**)[41] (equation 138).

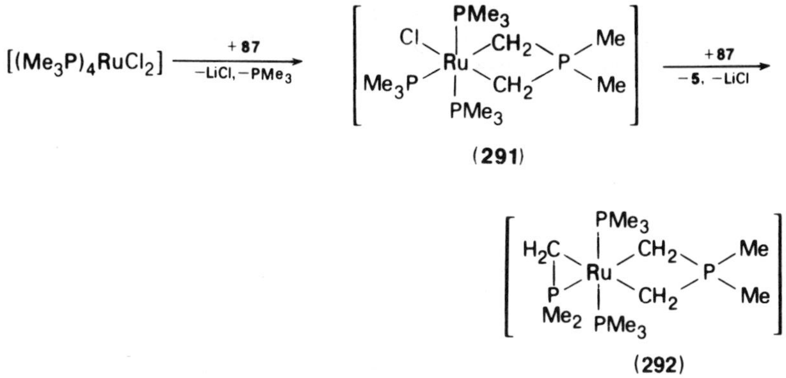

$$[(Me_3P)_4RuCl_2] \xrightarrow[-LiCl,\ -PMe_3]{+\mathbf{87}} [(Cl)(Me_3P)(PMe_3)_2Ru(CH_2)_2PMe_2]\ (\mathbf{291}) \xrightarrow[-\mathbf{5},\ -LiCl]{+\mathbf{87}} [(Me_2PCH_2)(PMe_3)_2Ru(CH_2)_2PMe_2]\ (\mathbf{292}) \qquad (138)$$

2. *Syntheses with nucleophilic methyl compounds*

This synthetic approach involves the treatment of transition metal halide complexes with metallated aminomethyl[240] and phosphinomethyl[19b] compounds, respectively.

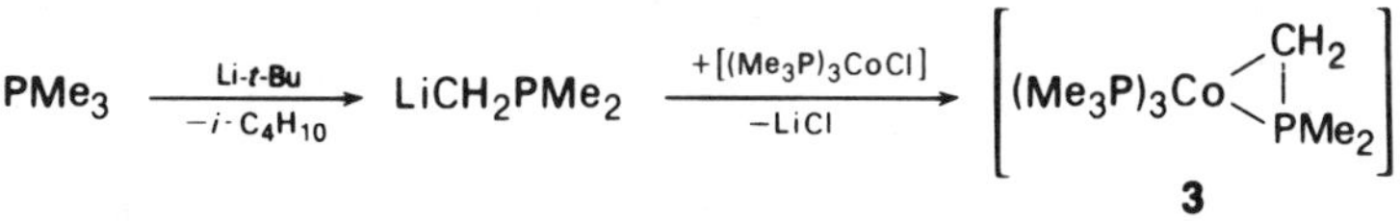

$$PMe_3 \xrightarrow[-i\text{-}C_4H_{10}]{Li\text{-}t\text{-}Bu} LiCH_2PMe_2 \xrightarrow[-LiCl]{+[(Me_3P)_3CoCl]} [(Me_3P)_3Co(CH_2PMe_2)]\ \mathbf{3} \qquad (139)$$

The utilization of aminomethylstannanes **293** has led to the successful development of a new synthesis of η^2-aminomethyl complexes (Scheme 44)[240].

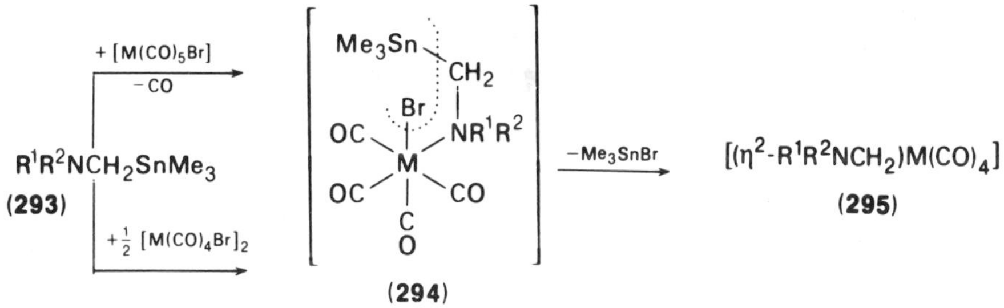

$$R^1R^2NCH_2SnMe_3\ (\mathbf{293}) \xrightarrow[-CO]{+[M(CO)_5Br]} \text{ or } \xrightarrow{+\frac{1}{2}[M(CO)_4Br]_2} [(OC)_4MBr(NR^1R^2CH_2SnMe_3)]\ (\mathbf{294}) \xrightarrow{-Me_3SnBr} [(\eta^2\text{-}R^1R^2NCH_2)M(CO)_4]\ (\mathbf{295})$$

M = Mn, Re; R^1,R^2 = Me, Et, *n*-Pr, *n*-Bu, C_6H_{11}, Ph, H, $-(CH_2)_2-$, $-(CH_2)_5-$

SCHEME 44

By analogy with the synthesis of η^3-allyl complexes from $M(CO)_5Br$ and allylstannanes, the formation of **295** is proposed to proceed via an addition–elimination mechanism. This necessitates invoking of the non-isolable intermediate **294** produced by CO-substitution in $Mn(CO)_5Br$ or bridge cleavage in $[Mn(CO)_4Br]_2$. **294** eventually collapses to the final products with release of Me_3SnBr. Fruitless attempts to detect an η^1-aminomethyl complex species are consistent with the proposed mechanism.

3. *Syntheses with electrophilic methyl compounds*

Nucleophilic displacement of chloride in $MeSCH_2Cl$ by organometallics constitutes a convenient synthesis of a series of complexes containing the η^2-thiomethoxymethyl ligand. The compounds with η^1-coordinated thiomethoxymethyl ligands, obtained as primary products, are readily transformed into the desired complex **296** by the thermally or photochemically induced expulsion of a carbonyl ligand[241] (equation 140).

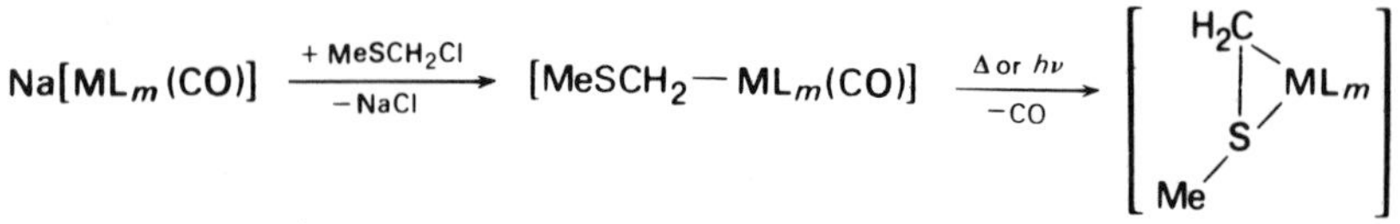

$$\mathrm{Na[ML_m(CO)]} \xrightarrow[-\mathrm{NaCl}]{+\mathrm{MeSCH_2Cl}} \mathrm{[MeSCH_2{-}ML_m(CO)]} \xrightarrow[-\mathrm{CO}]{\Delta \text{ or } h\nu} \left[\mathrm{H_2C(\eta^2\text{-}S(Me))ML_m}\right] \quad \mathbf{(296)} \qquad (140)$$

296	**a**	**b**	**c**
ML_m	η^5-$Cp(CO)_2Mo$	η^5-$Cp(CO)_2W$	$(OC)_4Mn$

The application of the same procedure to metal carbonylates and methylene iminium salts provides an alternative route to η^2-aminomethyl complexes[88] (equation 141).

$$\mathrm{Na[ML_m(CO)]} \xrightarrow[-\mathrm{NaX}]{+(\mathrm{Me_2NCH_2})X} \mathrm{[(Me_2NCH_2)ML_m(CO)]} \xrightarrow[-\mathrm{CO}]{\Delta \text{ or } h\nu} \mathrm{[(\eta^2\text{-}Me_2NCH_2)ML_m]} \quad \mathbf{(297)} \qquad (141)$$

297	**a**	**b**
ML_m	η^5-$Cp(CO)_2Mo$	η^5-$Cp(CO)Fe$

The oxidative addition of $MeSCH_2Cl$ or methylene iminium salts to zero-valent metal complexes of Mo, W, Fe, Ni, Pd, and Pt opens a general pathway to the coordination compounds under discussion[87,242] (Scheme 45).

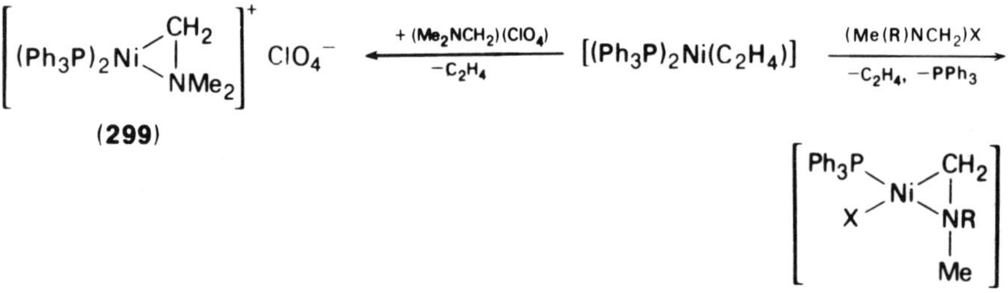

$$\left[\mathrm{(Ph_3P)_2Ni(\eta^2\text{-}CH_2NMe_2)}\right]^+ \mathrm{ClO_4^-} \ \mathbf{(299)} \xleftarrow[-\mathrm{C_2H_4}]{+(\mathrm{Me_2NCH_2})(\mathrm{ClO_4})} \mathrm{[(Ph_3P)_2Ni(C_2H_4)]} \xrightarrow[-\mathrm{C_2H_4},\ -\mathrm{PPh_3}]{(\mathrm{Me(R)NCH_2})X} \left[\mathrm{(Ph_3P)XNi(\eta^2\text{-}CH_2NRMe)}\right] \ \mathbf{(298)}$$

X	Cl	Br	I
R = Me	**298**	**b**	**c**
R = Et	**d**		

SCHEME 45

The stable ring arrangement of the nickel complexes is manifested by the occurrence of diastereotopic methylene protons in the n.m.r. spectrum of **298d** and by the molecular structural analysis of **298a**[87].

Complex **300**, synthesized by oxidative addition of $MeSCH_2Cl$ to $[Pd(PPh_3)_4]$, undergoes facile dissociation in solution, as was concluded from its osmometric molecular weight determination and n.m.r. data[243] (Scheme 46).

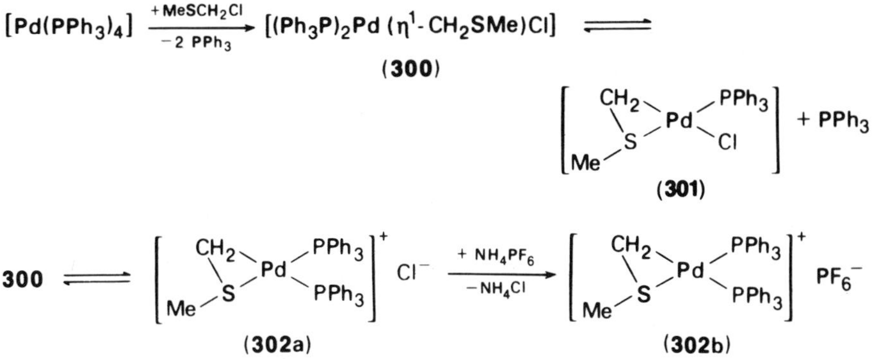

$$[Pd(PPh_3)_4] \xrightarrow[-2\,PPh_3]{+MeSCH_2Cl} [(Ph_3P)_2Pd(\eta^1\text{-}CH_2SMe)Cl] \rightleftharpoons$$

(300)

$$\left[\begin{array}{c} CH_2 \\ \backslash \quad Pd(PPh_3)(Cl) \\ Me{-}S \end{array} \right] + PPh_3$$

(301)

$$\mathbf{300} \rightleftharpoons \left[(CH_2\text{-}S(Me))Pd(PPh_3)_2 \right]^+ Cl^- \xrightarrow[-NH_4Cl]{+NH_4PF_6} \left[(CH_2\text{-}S(Me))Pd(PPh_3)_2 \right]^+ PF_6^-$$

(302a) **(302b)**

SCHEME 46

The phosphine ligands and the chloride ions are engaged in the dissociation processes of **300** leading to ionic **302a**, from which the chloride is easily replaced by the PF_6^- ion upon treatment with NH_4PF_6. Repeated recrystallization of **300** effects removal of a phosphine, accompanied by ring closure to give **301**. N.m.r. spectroscopy and X-ray analysis provide evidence for the identity of the palladothia-cyclopropanes[244].

B. Ylide Complexes Containing Bridging Phosphinomethyl and Thiomethoxymethyl Groups

1. Syntheses from neutral ligands

The reaction of nucleophilic **303** with $[(Me_3P)_3CoCl]$ did not yield the expected dinuclear $[(Me_3P)_3CoN_2Co(PMe_3)_3]$ **(304)**, but rather the five-membered heterocyclic **4**[20] (Scheme 47). The Co_2P_2C-ring of **4** almost acquires a chair conformation. Its

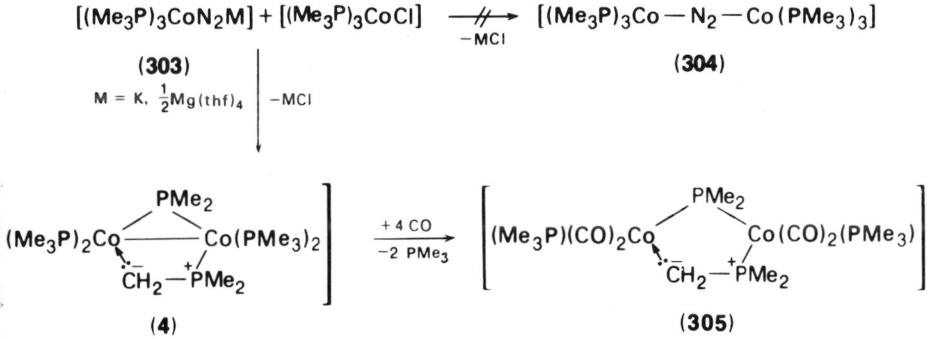

$$[(Me_3P)_3CoN_2M] + [(Me_3P)_3CoCl] \xrightarrow[-MCl]{\not\to} [(Me_3P)_3Co{-}N_2{-}Co(PMe_3)_3]$$

(303) **(304)**

$M = K, \tfrac{1}{2}Mg(thf)_4$; $-MCl$ (down arrow)

$$\left[(Me_3P)_2Co(\mu\text{-}PMe_2)(\mu\text{-}\bar{C}H_2{-}\overset{+}{P}Me_2)Co(PMe_3)_2 \right] \xrightarrow[-2\,PMe_3]{+4\,CO} \left[(Me_3P)(CO)_2Co(\mu\text{-}PMe_2)(\mu\text{-}\bar{C}H_2{-}\overset{+}{P}Me_2)Co(CO)_2(PMe_3) \right]$$

(4) **(305)**

SCHEME 47

remarkably short PCH_2 bond length (1.707 Å), of the order of P—C distances expected for stabilized ylides, and the respective bond angle (88.0°), led to the postulation of an sp^2-hybridized methylene carbon atom. The cobalt–cobalt interaction in **4** is disrupted by the addition of carbonyl ligands, producing another heterocycle with a bridging CH_2PMe_2 unit[20].

2. Syntheses with electrophilic methylene compounds

In contrast to the behaviour of **300**, the analogous platinum complex **81** is a η^1-thiomethoxymethyl ligand containing compound which resists phosphine dissociation[85]. When compared with simple alkyl complexes such as *trans*-

$$[Pt(PPh_3)_4] + MeSCH_2Cl \xrightarrow[-2PPh_3]{} trans\text{-}[(\eta^1\text{-}MeSCH_2)(PPh_3)_2PtCl] \quad (142)$$
(81)

$[(Ph_3P)_2(Me)PtCl]$, it is seen that the sulphur atom in the alkyl ligand of **81** effects considerable activation of the complex. Hydrogen peroxide affords oxidative cleavage to give Ph_3PO and a coordinatively unsaturated fragment which undergoes dimerization to **306** (equation 143). Reaction of $[Pt(AsPh_3)_4]$ and $MeSCH_2Cl$ leads directly to a dimeric product structurally resembling **306**[85].

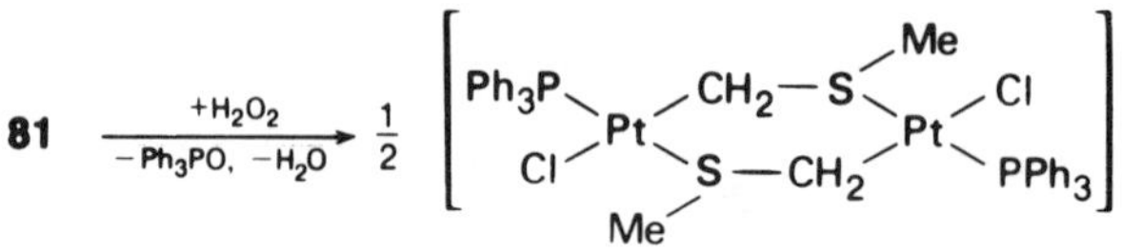

$$\mathbf{81} \xrightarrow[-Ph_3PO,\ -H_2O]{+H_2O_2} \frac{1}{2}\left[\begin{matrix} Ph_3P & & CH_2-S(Me) & & Cl \\ & Pt & & Pt & \\ Cl & & (Me)S-CH_2 & & PPh_3 \end{matrix}\right] \quad (143)$$

(306)

C. Others

As outlined in Scheme 48, the phosphonium centre of $Ph_3PCHCOPh$ oxidatively adds to zero-valent nickel atoms, generating a chelate complex which also possesses the feature of an ylide substituted at the onium centre by a metal atom. Further, the enolate oxygen is accommodated in the coordination sphere of the nickel so that the product provides an example of a case where five-membered ring formation takes preference over three-membered ring formation (Scheme 48)[6].

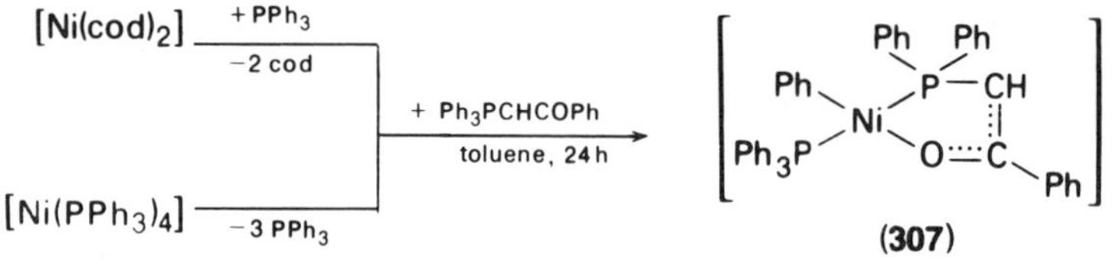

$[Ni(cod)_2]$ ($+PPh_3$, -2 cod) or $[Ni(PPh_3)_4]$ ($-3\ PPh_3$) $\xrightarrow[\text{toluene, 24h}]{+\ Ph_3PCHCOPh}$ [(Ph)(Ph_3P)Ni(–PPh_2–CH⋯C(Ph)⋯O–)] **(307)**

SCHEME 48

VII. REACTIONS OF YLIDES WITH COORDINATED LIGANDS

This section deals with reactions between free ylides and ligated molecules within the coordination sphere of a metal. In this connection, bond formation between ylide and metal centre is generally not observed.

A. Organic Carbonyl Functions in Olefinic Ligands

Wittig reactions of phosphorus ylides and organic carbonyl groups attached to coordinated species gain special importance when the free ligand cannot be employed owing to its inherent instability. Thus, iron complexes of formylated cyclobutadienes such as **308** and **310** are typical precursors providing a route to a wide variety of derivatized novel cyclobutadiene ligand systems[245–249] (equations 144, 145).

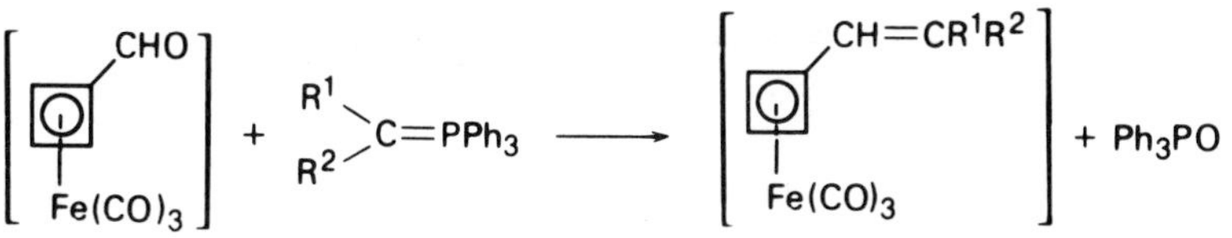

(308) (309) (144)

309	a	b	c	d	e	f	g
R^1	H	H	Me	H	H	Br	Br
R^2	H	Ph	Me	Me	CO_2Et	Br	CO_2Me

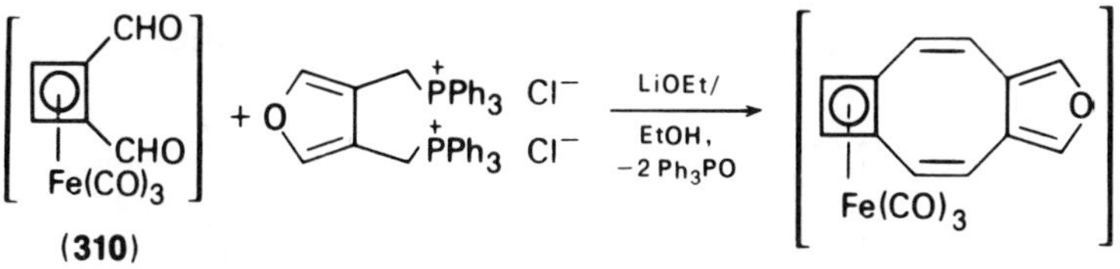

(310) (145)

Free norbornadien-7-one spontaneously decomposes to benzene and carbon monoxide. The stability of the related tricarbonyliron complex, however, allows it to react with weakly basic ylides of phosphorus and sulphur at the ketone function. Transformations of **311** with Ph_3PCH_2 or Me_2SCH_2 were prevented by the strongly basic conditions used for *in situ* ylide generation[250] (Scheme 49).

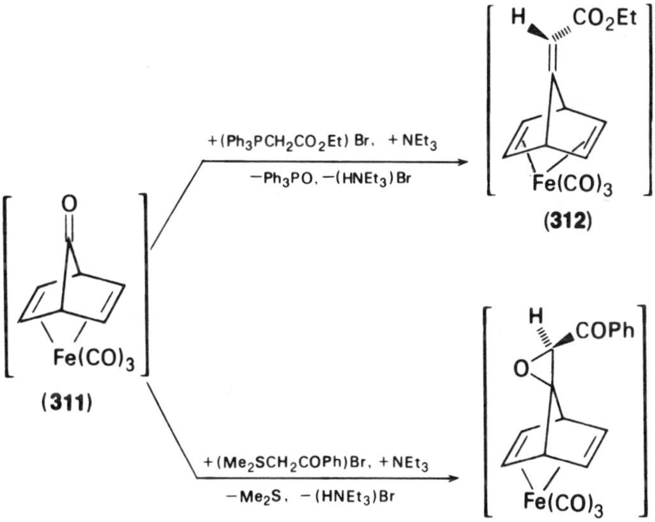

(311) (312) (313)

SCHEME 49

Attempts to synthesize regiospecifically either **314** or **315** from the reaction of $[Fe_3(CO)_{12}]$ and the free polyolefin invariably led to inseparable mixtures of both isomers. Each species, however, is obtainable cleanly by means of carbonyl olefination, as illustrated in Scheme 50[251].

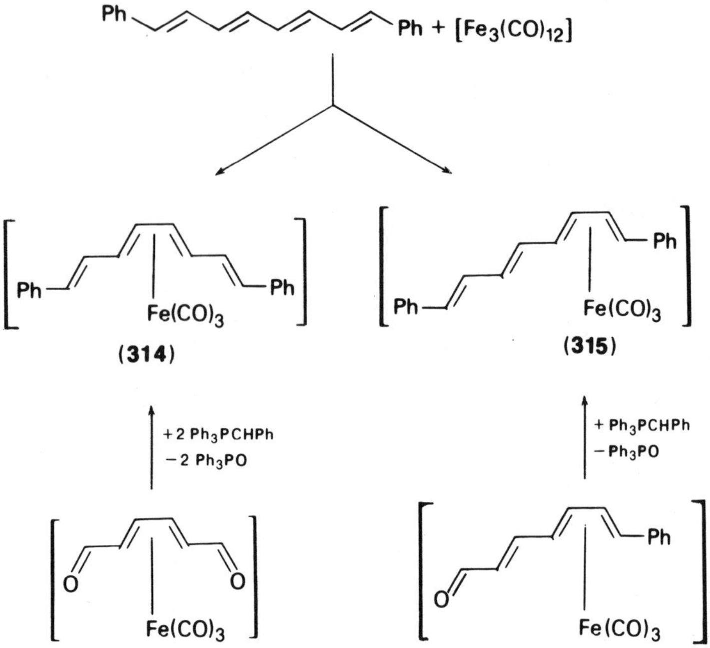

SCHEME 50

B. Carbon Monoxide Ligands

1. Transformations resembling the Wittig reaction

The conversion of coordinated CO into vinylidene complexes by simple Wittig reagents has not been observed so far. Nevertheless, transformations formally analogous to the Wittig reaction have been described in a very few cases in which the double ylide **14** was employed (equation 146).

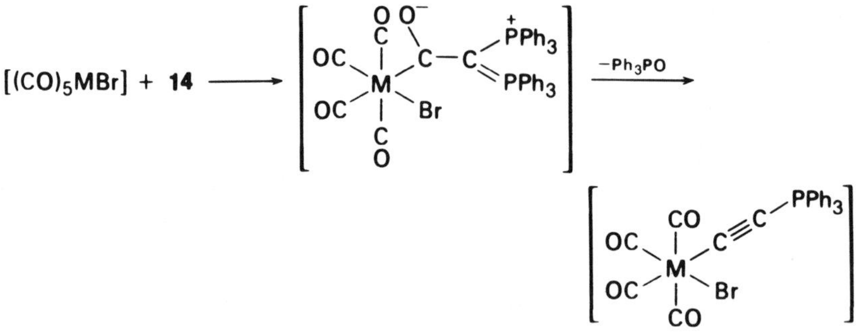

(146)

316	a	b
M	Mn	Re

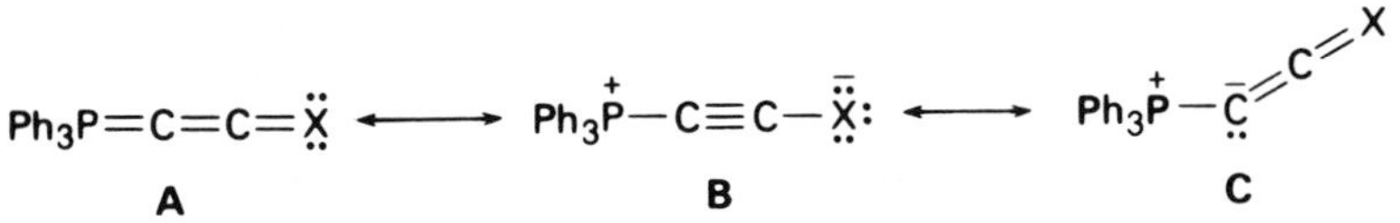

$$\mathrm{Ph_3P{=}C{=}C{=}\ddot{X}} \longleftrightarrow \mathrm{Ph_3\overset{+}{P}{-}C{\equiv}C{-}\overset{-}{\ddot{X}}:} \longleftrightarrow \mathrm{Ph_3\overset{+}{P}{-}\overset{-}{\ddot{C}}{=}C{=}X}$$

A B C

FIGURE 29. Resonance structures of cumulated ylides (X = O, S) and complex **316**.

316a contains a bent MnC_2P array containing a carbon–carbon triple bond[252]. The bond angles in **316a** are comparable to those found in the cumulated ylide $Ph_3P{=}C{=}C{=}S$, which can be described by the resonance structures **A–C** in Figure 29.

For **316a** (X = $Mn(CO)_4Br$), structure **B** appears to be preponderant. U.v. irridiation of a mixture of **14** and $[W(CO)_6]$ furnished small amounts of the tungsten derivative **317**[44] (equation 147).

$$\underset{}{[W(CO)_6]} + \underset{\mathbf{(14)}}{(Ph_3P)_2C} \xrightarrow[\text{thf}]{h\nu} \underset{\mathbf{(317)}}{[(CO)_5W{-}C{\equiv}C{-}PPh_3]} + Ph_3PO + \ldots \qquad (147)$$

2. *Metal acylate formation*

Kinetically labile carbonyl ligands are subject to facile displacement by ylides (Section II.A.1). As an alternative pathway, carbonyl ligands possessing sufficient electrophilicity at the carbon atom allow nucleophilic addition of an ylide at this position. The methylene protons in the primary adduct experience additional acidification by the adjacent carbonyl group, causing subsequent transylidation[12,253,254] (equation 148).

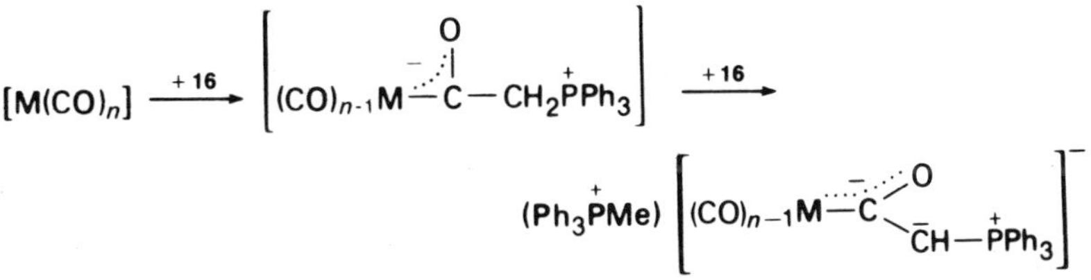

$$[M(CO)_n] \xrightarrow{+16} \left[(CO)_{n-1}\overset{-}{M}{-}\overset{O}{\overset{\vdots}{C}}{-}CH_2\overset{+}{P}Ph_3\right] \xrightarrow{+16} (Ph_3\overset{+}{P}Me)\left[(CO)_{n-1}M{\overset{-}{\cdots}}C{\genfrac{}{}{0pt}{}{\cdots O}{\overset{-}{C}H{-}\overset{+}{P}Ph_3}}\right]^- \qquad (148)$$

(318)

M	Fe	Cr	W
$n = 5$	**318a**		
$n = 6$		**318b**	**318c**

The anion in **318**, as an ambident base, is alkylated at the acylate oxygen atom by hard alkylating reagents[12,253,254].

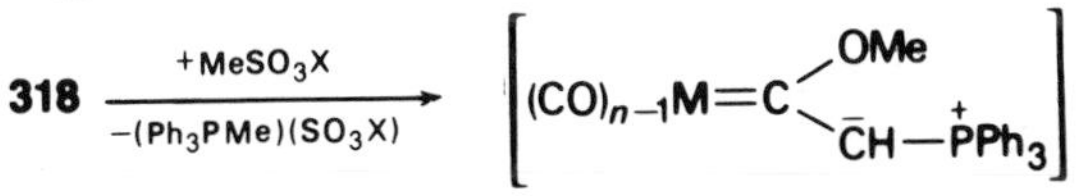

$$\mathbf{318} \xrightarrow[-(Ph_3PMe)(SO_3X)]{+MeSO_3X} \left[(CO)_{n-1}M{=}C{\genfrac{}{}{0pt}{}{OMe}{\overset{-}{C}H{-}\overset{+}{P}Ph_3}}\right] \qquad (149)$$

(319)

M	Fe	Cr	W
$n = 5$	**319a**		
$n = 6$		**319b**	**319c**

If one regards the $M(CO)_{n-1}$ fragment to be analogous to oxygen, complexes **319** may be considered as organometallic derivatives of $Ph_3P{=}CH{-}CO_2Me$. A different situation is encountered in the reaction of $[(MeCp)Mn(CO)_3]$ with Me_3PCH_2 and Et_3PCHMe. **321**, which is prepared from $[(MeCp)Mn(CO)_3]$ and 2 equiv of Me_3PCH_2 in pentane, dissolves in thf, benzene, or large amounts of pentane and subsequently undergoes a reverse reaction with complete regeneration of the starting materials[255] (Scheme 51).

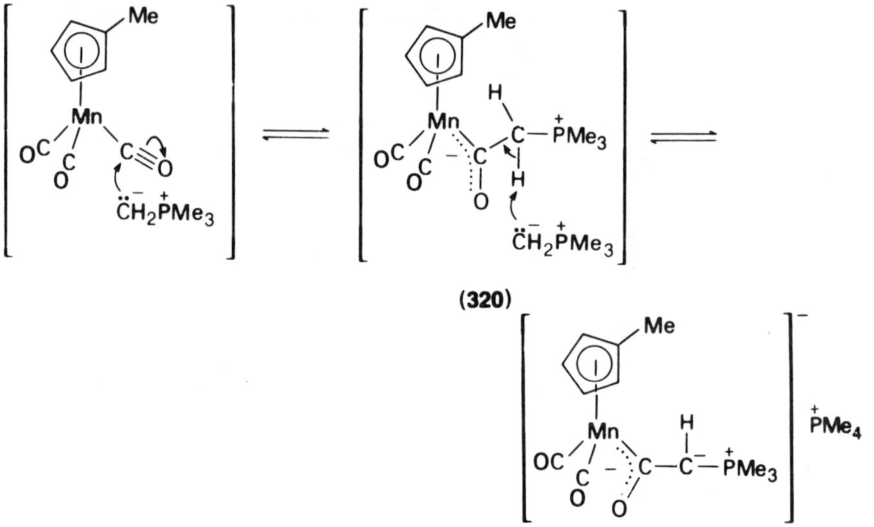

(**320**)

(**321**)

SCHEME 51

Apparently, addition of the ylide to the carbonyl carbon atom and the subsequent proton abstraction at the methylene bridge in adduct **320** by a second molecule of ylide are completely reversible processes. The enhanced σ-donor/π-acceptor ratio of the cyclopentadienyl ligand compared with carbon monoxide may account for this phenomenon, which is unknown with binary transition metal carbonyls. The ligand properties of the ring bring about (a) a low electrophilicity of the carbonyl carbon atoms in concert with low stability of the primary adduct **320** and (b) high electron density at the manganese atom, thus increasing the basicity of the ylide function of the metal acylate **321** to such an extent that deprotonation of the counterion PMe_4^+ appears to be feasible.

Exhaustive methylation of **321** with excess of $MeSO_3F$ yields the salt-like **322**, which is readily convertible to the phosphonium ylide–carbene complex **323** by equivalent amounts of free ylide (equation 150).

Resonance structures **A** and **B** (Figure 30) help to give a suitable description of the mode of bonding in **319** and in **323**, the latter of which has been investigated by X-ray structural analysis[256].

The sulphur ylide $Me_2S(O)CH_2$ (**35**) exhibits less basicity than, for example, Me_3PCH_2 or Ph_3PCH_2, which is in agreement with i.r. studies of the carbonyl stretching vibration of complexes such as (ylide)$M(CO)_5$. Thus, it was impossible to add **35** to a carbonyl unit in the hexacarbonyls of chromium and tungsten. However, the increased electrophilicity inherent in carbonyl ligands of $Fe(CO)_5$ and

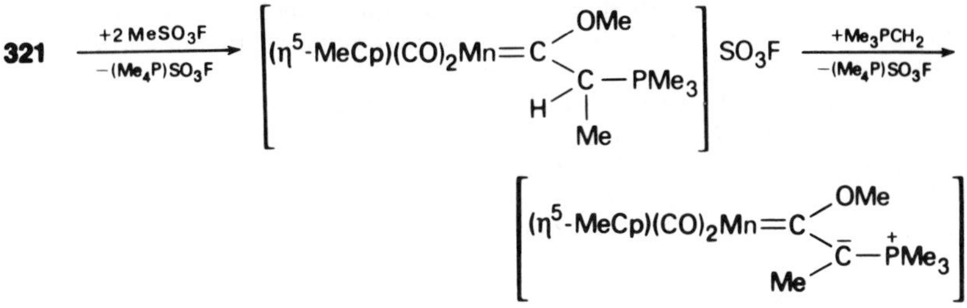

$$\mathbf{321} \xrightarrow[-(Me_4P)SO_3F]{+2\,MeSO_3F} \left[(\eta^5\text{-MeCp})(CO)_2Mn{=}C(OMe){-}CH(Me){-}\overset{+}{P}Me_3\right]SO_3F \xrightarrow[-(Me_4P)SO_3F]{+Me_3PCH_2}$$

$$\left[(\eta^5\text{-MeCp})(CO)_2Mn{=}C(OMe){-}\overset{-}{C}(Me){-}\overset{+}{P}Me_3\right] \quad (150)$$

(**323**)

$$L_nM{=}C(OMe){-}\ddot{\overset{-}{C}}(R^1){-}\overset{+}{P}R^2_3 \longleftrightarrow L_n\ddot{\overset{-}{M}}{-}C(OMe){=}C(R^1){-}\overset{+}{P}R^2_3$$

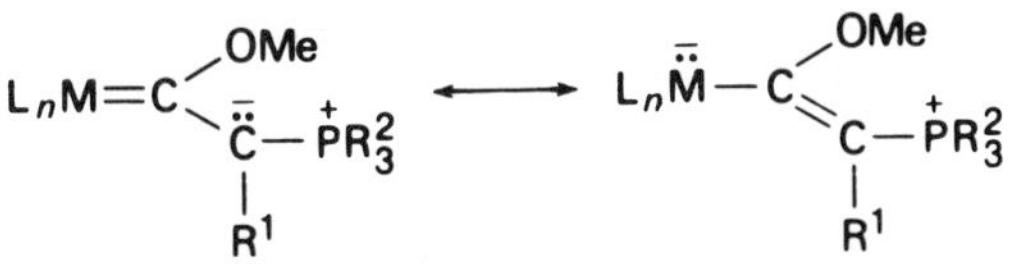

A **B**

FIGURE 30. Resonance structures of carbene–ylide complexes such as **319** and **323**.

$[(Cp)Fe(CO)_3]^+$ provides enough reactivity for the performance of the desired reaction[257a,b] (Scheme 52).

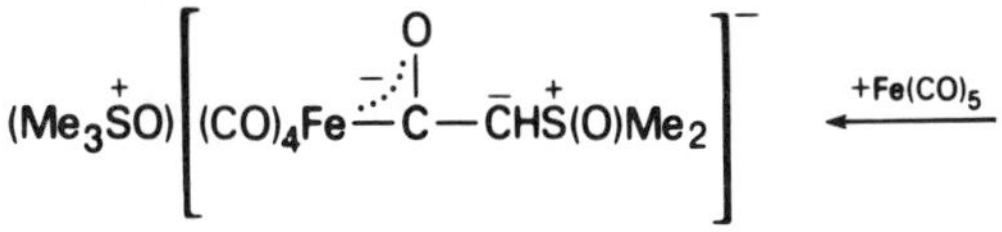

$$(Me_3\overset{+}{S}O)\left[(CO)_4\overset{-}{Fe}{\cdots}C(O){-}\overset{-}{C}H\overset{+}{S}(O)Me_2\right]^- \xleftarrow{+Fe(CO)_5}$$

(**324**)

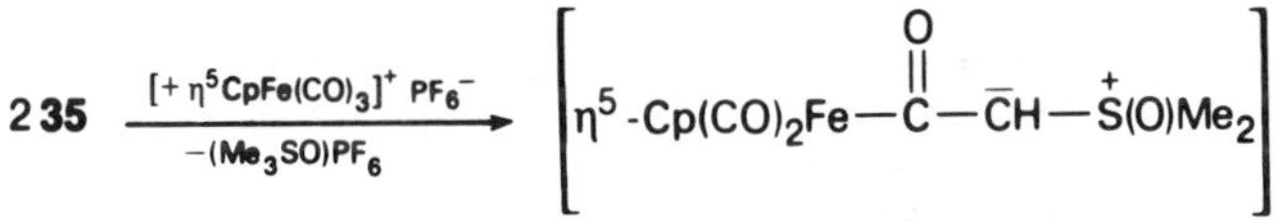

$$2\,\mathbf{35} \xrightarrow[-(Me_3SO)PF_6]{[+\eta^5CpFe(CO)_3]^+\,PF_6^-} \left[\eta^5\text{-Cp}(CO)_2Fe{-}C({=}O){-}\overset{-}{C}H{-}\overset{+}{S}(O)Me_2\right]$$

(**325**)

SCHEME 52

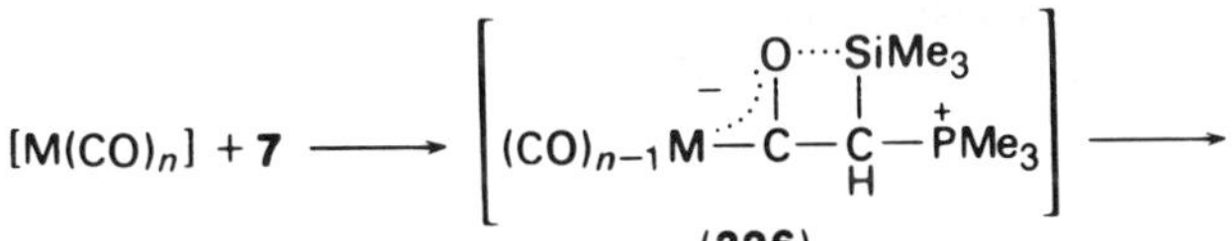

$$[M(CO)_n] + \mathbf{7} \longrightarrow \left[(CO)_{n-1}\overset{-}{M}{\cdots}C(O{\cdots}SiMe_3){-}C(H)(SiMe_3){-}\overset{+}{P}Me_3\right] \longrightarrow$$

(**326**)

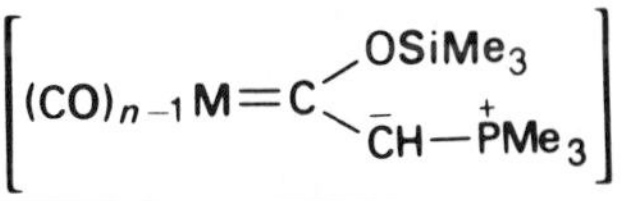

$$\left[(CO)_{n-1}M{=}C(OSiMe_3){-}\overset{-}{C}H{-}\overset{+}{P}Me_3\right] \quad (151)$$

(**327**)

M	E	Cr	Mo	W
$n = 5$	**327a**			
$n = 6$		**327b**	**327c**	**327d**

Neutral **325** may be described either as a sulphuranylidene derivative of $[(Cp)(CO)_2FeCH_3]$ or as a metal-substituted carbonyl-stabilized sulphur ylide such as $RC(O)CHS(O)Me_2$.

Reaction of $Me_3PCHSiMe_3$ with $[Fe(CO)_5]$ and Group VIB metal carbonyls appears also to be initiated by the attack of the ylide at the carbonyl function. However, in this case, stabilization of adduct **326** does not proceed via transylidation, but is achieved by a 1,3-silyl shift, thus affording carbene complexes which are functionalized in the side-chain by an ylidic moiety[258] (equation 151).

C. Carbene Ligands

The electrophilic carbene carbon atom of carbene–metal complexes is another suitable target for ylide attack, which furnishes in this case the initial step of a novel enol-ether synthesis[259] (Scheme 53).

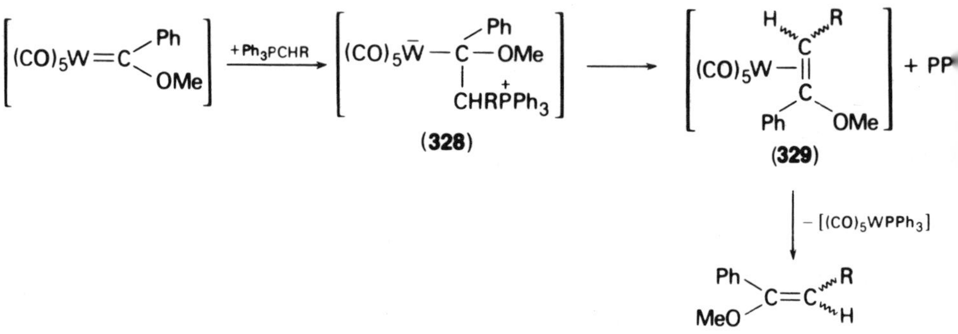

$$[(CO)_5W{=}C(Ph)(OMe)] \xrightarrow{+Ph_3PCHR} [(CO)_5\bar{W}{-}C(Ph)(OMe){-}CHR\overset{+}{P}Ph_3] \; (\mathbf{328}) \longrightarrow [(CO)_5W{-}(\eta^2\text{-}HRC{=}C(Ph)(OMe))] \; (\mathbf{329}) + PPh_3$$

$$(\mathbf{329}) \xrightarrow{-[(CO)_5WPPh_3]} Ph(MeO)C{=}CRH$$

SCHEME 53

The betain-like intermediate **328** collapses to the enol-ether complex **329** and free PPh_3, followed by formation of the final products. The proposed mechanism of this transformation is supported by the observation of $[(CO)_5W\{P(C_6H_4Me\text{-}p)_3\}]$ when the reaction is run in the presence of tri-*p*-tolylphosphine. The generation of enol-ethers proceeds smoothly at ambient temperatures with the ylides Ph_3PCH_2 and Ph_3PCHMe. The employment of Ph_3PCHPh necessitates heating at 60°C, whereas Ph_3PCMe_2 as well as carbonyl-stabilized ylides, appear to be unreactive. Attempts to expand this reaction to alkylmethoxy–carbene complexes failed owing to the abstraction of a proton from the carbon adjacent to the carbene carbon atom (equation 152).

$$[(CO)_5W{=}C(OMe)(CH_3)] + Ph_3PCH_2 \; (\mathbf{16}) \longrightarrow (Ph_3PMe)^+ [(CO)_5W{=}C(OMe)(CH_2^-)] \; (\mathbf{330}) \qquad (152)$$

The occurrence of small amounts of the *trans*-stilbene complex **334** in the synthesis of the ylide complex **332** is assumed to result from a nucleophilic attack of the ylide at carbene complex **333**. In solution **333** apparently equilibrates with **332**[43] (Scheme 54).

A novel ring expansion takes place when pyridinium ylides are allowed to react with carbene complexes **335**[260,261] (equation 153).

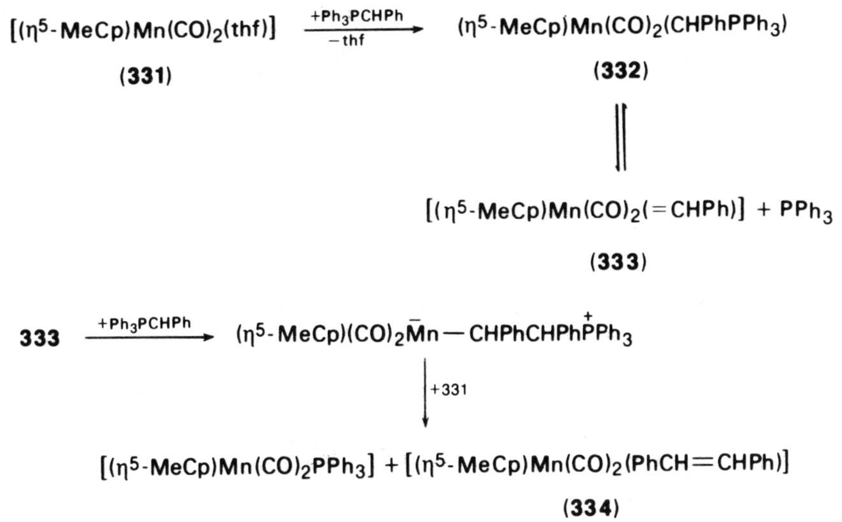

$$[(\eta^5\text{-MeCp})Mn(CO)_2(thf)] \xrightarrow[-thf]{+Ph_3PCHPh} (\eta^5\text{-MeCp})Mn(CO)_2(CHPhPPh_3)$$

(**331**) (**332**)

$$\Updownarrow$$

$$[(\eta^5\text{-MeCp})Mn(CO)_2(=CHPh)] + PPh_3$$

(**333**)

$$\mathbf{333} \xrightarrow{+Ph_3PCHPh} (\eta^5\text{-MeCp})(CO)_2\bar{M}n{-}CHPhCHPh\overset{+}{P}Ph_3$$

$$\downarrow +331$$

$$[(\eta^5\text{-MeCp})Mn(CO)_2PPh_3] + [(\eta^5\text{-MeCp})Mn(CO)_2(PhCH{=}CHPh)]$$

(**334**)

SCHEME 54

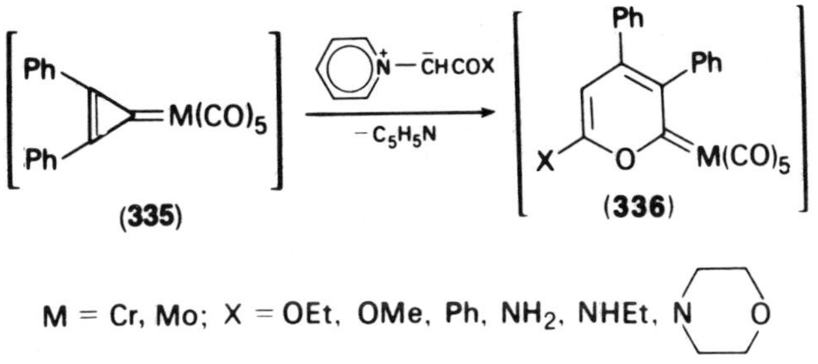

$$\left[\text{Ph}_2\text{C}_3{=}M(CO)_5\right] \xrightarrow[-C_5H_5N]{C_5H_5\overset{+}{N}{-}\bar{C}HCOX} \left[\text{pyranylidene: 4-Ph, 3-Ph, 6-X, } {=}M(CO)_5\right] \quad (153)$$

(**335**) (**336**)

M = Cr, Mo; X = OEt, OMe, Ph, NH_2, NHEt, N(morpholino)O

D. Olefinic Ligands

1. *Addition reactions*

Ph_3PCH_2 cleanly displaces the olefinic ligand of $[(\eta^5\text{-Cp})(CO)_2Fe(C_2H_4)]^+ BF_4^-$ (**337**), but the phosphine-containing derivative **338**, in a competing reaction, also adds to the coordinated ethylene[262] (equation 154).

$$[(\eta^5\text{-Cp})(CO)(Ph_3P)Fe(\pi\text{-}C_2H_4)]^+ BF_4^- \xrightarrow{+Ph_3PCH_2\cdot LiBr}$$

(**338**) $[(\eta^5\text{-Cp})(CO)(Ph_3P)Fe\text{-}(CH_2)_3\text{-}PPh_3]^+ BF_4^-\cdot LiBr$

(**339**)

$$+ [\eta^5\text{-Cp})(CO)(Ph_3P)FeCH_2PPh_3]^+ BF_4^-\cdot LiBr \quad (154)$$

The addition of less basic **340** to **337** yielded the phosphonium salt **341**, which, upon treatment with alkali, was converted into the organometallic ylide **342**. The Wittig reaction of **342** with aldehydes was used for the production of highly functionalized iron complexes such as **343** (Scheme 55)[263].

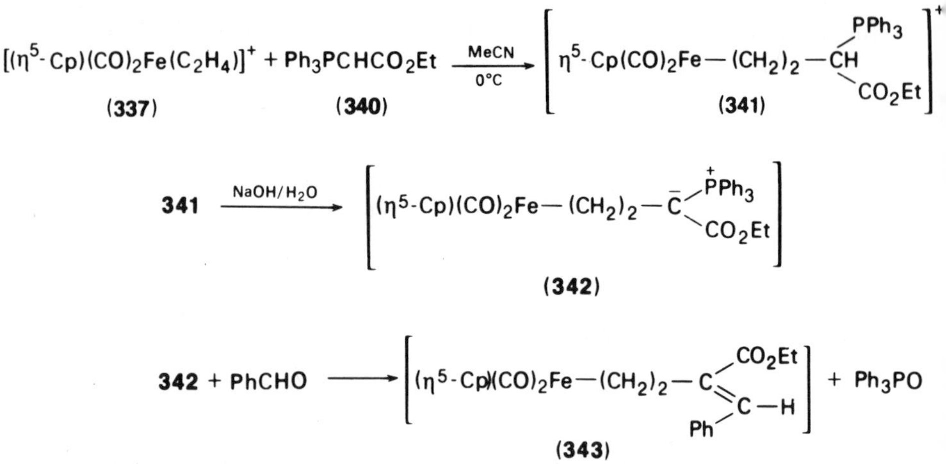

$$[(\eta^5\text{-Cp})(CO)_2Fe(C_2H_4)]^+ + Ph_3PCHCO_2Et \xrightarrow[0^\circ C]{MeCN} [\eta^5\text{-Cp}(CO)_2Fe{-}(CH_2)_2{-}CH(PPh_3)(CO_2Et)]^+$$

(337) (340) (341)

$$\mathbf{341} \xrightarrow{NaOH/H_2O} [(\eta^5\text{-Cp})(CO)_2Fe{-}(CH_2)_2{-}\bar{C}(\overset{+}{P}Ph_3)(CO_2Et)]$$

(342)

$$\mathbf{342} + PhCHO \longrightarrow [(\eta^5\text{-Cp})(CO)_2Fe{-}(CH_2)_2{-}C(CO_2Et){=}C(H)Ph] + Ph_3PO$$

(343)

SCHEME 55

2. *Stevens rearrangement*

Nickel–olefin complexes such as [(cod)$_2$Ni], all-*trans*-1,5,9-cyclododecatriene-nickel [(cdt)Ni], or [(Ph$_3$P)$_2$Ni(C$_2$H$_4$)] induce the Stevens rearrangement of arylphosphonium ylides (equation 155).

$$[Ni(cod)_2] + 4Ph_3PCH_2 \xrightarrow{70^\circ C} 2cod + [(Ph_2PCH_2Ph)_4Ni] \qquad (155)$$

Monosubstitution at the carbanionic centre is tolerated, although the yields of the resulting phosphines decrease with increasing alkyl substitution. Ph$_3$PCMe$_2$ and trialkylphosphoranes resist this isomerization. A possible mechanism to explain the course of the reaction invokes the *ortho*-metallation of a coordinated ylide as an initial step, followed by the fission of the aryl–phosphorus linkage. The subsequent formation of the final product is then preceded by reductive elimination of the benzylphosphine[31] (Scheme 56).

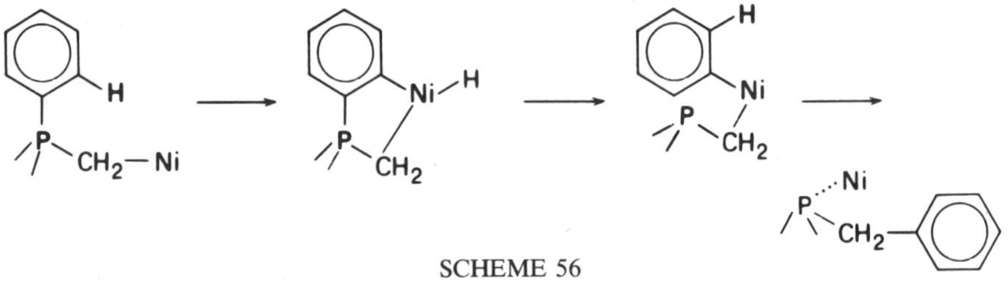

SCHEME 56

E. Organocyanide Ligands

Phosphonium ylides are also capable of nucleophilically attacking the Lewis acidic carbon atom of coordinated organocyanides. The occurrence of prototropy following the addition yields *cis–trans* mixtures of ketimine complexes which possess ylide

$$[(CO)_5Cr{-}N{\equiv}C{-}R] \xrightarrow[Et_2O/0°C]{+Ph_3PCH_2} \left[(CO)_5Cr{-}\overset{H}{N}{=}C\begin{smallmatrix}\bar{C}H\overset{+}{P}Ph_3\\ R\end{smallmatrix}\right] \quad (156)$$

(344)

functions in the side-chain[264,265] (equation 156). The success of this procedure is determined by the nature of the substituent R. Whereas nitriles containing substituents such as Me, Et, *i*-Pr, CH_2OMe, Ph, and *p*-$R^1C_6H_4$ are smoothly converted to **344**, the alternative reaction, displacement of the nitrile, is observed with *t*-BuCN, $PhCH_2CN$, *p*-$O_2NC_6H_4CH_2CN$, and 2,6-$Me_2C_6H_3CN$. The use of this approach to ketimine complex synthesis can also be applied to manganese chemistry[266].

F. Halophosphine Ligands

Bimolecular substitution of halide ions in diphenylchlorophosphine–metal complexes has been accomplished by ylides. Upon subsequent transylidation, complexes **345** were generated where the ligand remains attached to the metal atoms through the trivalent phosphorus atom[267] (equation 157).

$$[(CO)_nMPPh_2Cl] \xrightarrow[-(Ph_3PCH_2R)Cl]{+2Ph_3PCHR} [(CO)_nMPPh_2\bar{C}(R)\overset{+}{P}Ph_3] \quad (157)$$

(345)

R	H	Me
M(*n*) = Cr(5)	**345a**	**345b**
M(*n*) = Fe(4)	**345c**	

G. Hydrido Complexes

The interaction of hydrido complexes **346** with ylides of phosphorus and sulphur can be described in terms of acid–base reactions yielding ionic onium metallates **347**[268] (equation 158). In the case of reactions with weakly basic ylides such as

$$\underset{(346)}{[HM(CO)_3(Cp)]} + R^1_3PCR^2R^3 \longrightarrow \underset{(347)}{(R^1_3PCHR^2R^3)^+[M(CO)_3(Cp)]^-} \quad (158)$$

R^1 = Me, Et, Ph; R^2, R^3 = H, Me, $SiMe_3$;
M = Cr, Mo, W

$Me_2S(O)CHSiMe_3$, proton abstraction is followed by cleavage of the Me_3Si group, thus providing desilylated onium salts such as $(Me_3SO)^+$ $[M(CO)_3Cp]^-$. Even silyl-substituted phosphorus ylides suffer desilylation when exposed to reaction with $[HCr(CO)_3(Cp)]$, which is the most acidic carbonyl hydride of the Group VIB metals.

VIII. YLIDE COMPLEXES IN SYNTHETIC CHEMISTRY

A. Wittig Reactions

A series of ylide complexes of Group IIB metals are capable of performing carbonyl olefination reactions with aldehydes and ketones[56–58]. The complexes generated according to equations 25–27 contain labile halomethylene phosphorane ligands, the treatment of which with organic carbonyls provides a route to a class of halogenated

$$(Me_2N)_3PCFCl \cdot ZnCl_2 \rightleftharpoons (Me_2N)_3PCFCl + ZnCl \xrightarrow{R^1R^2C{=}O} R^1R^2C{=}CFCl + (Me_2N)_3PO \qquad (159)$$

$R^1 = CF_3, CF_2Cl, CF_2H, C_2F_5$;
R^2 = OMe, OEt, OPr-*i*

olefins. Activated esters are transformed into *cis–trans* mixtures of the respective vinyl ethers when subjected to reaction with **44a**[58] (equation 159).

The use of modified ylides containing Me_2N substituents at the phosphorus atom is of advantage in comparison with the utilization of the corresponding triphenylphosphorane derivatives mainly because of the ready availability of the halogenated precursors, high yields, and a facile work-up procedure (no Ph_3PO!). Additionally, the zinc reagent does not require the strongly basic reaction conditions that are necessary for the performance of classical Wittig reactions. Dissociation processes are also believed to be responsible for the successful performance of Wittig reactions with $[(PhCOCHPPh_3)HgCl_2]_2$[137–142] or $[\{(Me_2N)_3PCFCl\}HgCl_2]$[58]. Kinetically labile ylide complexes are presumably the reactive species in a halo-olefin synthesis involving carbonyls, PPh_3, and $PhHgCX_2Br$ (X = halogen)[269]. Carbonyl olefination is also reported to be feasible with chelates $[(Cp)_2M(CH_2)_2PPh_2]$ (**124–126**)[5].

B. Cyclopropanations

Sulphur ylides have gained use as suitable cyclopropanating reagents for activated olefins in transformations initiated by Michael addition to α,β-unsaturated carbonyl species[2]. However, a few sulphur ylide complexes of copper and iron exhibit the tendency to transfer methylene groups to non-activated alkenes. Reactive carbene complexes such as $[(Cp)(CO)_2Fe{=}CH_2]^+$, derived from stable or transient sulphur ylide complexes by release of thioether, have been postulated as intermediates[270] (equation 160). This method works for the cyclopropanation of 1-heptene,

$$Ph_2SCH_2 + [Cu(acac)_2] \xrightarrow[\text{thf, 20°C}]{>C=C<} >C\overset{CH_2}{—}C< + Ph_2S + \cdots \qquad (160)$$

isobutyl vinyl ether, *cis*- and *trans*-2-octene, cyclohexene, and 3-methylcyclohexene with moderate yields[270]. The cyclopropanation of cyclohexene by $PhCOCHSMe_2$ in the presence of $CuSO_4$ proceeds in only 5% yield[271] (equation 161). Significantly better results are obtained when **348** and olefins are allowed to react in boiling dioxane (equation 162).

$$PhCOCHSMe_2 + \text{cyclohexene} \xrightarrow{CuSO_4} \text{7-(H)(COPh)-bicyclo[4.1.0]heptane} + \cdots \qquad (161)$$

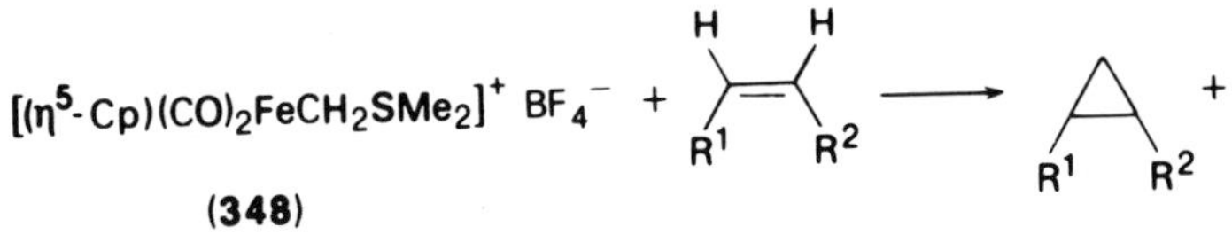

$$[(\eta^5\text{-Cp})(CO)_2FeCH_2SMe_2]^+ BF_4^- + R^1HC{=}CHR^2 \longrightarrow \text{cyclopropane}(R^1,R^2) + \cdots \quad (162)$$

(348)

This apparently stereospecific cyclopropanation suffers complications from competing reactions which consume the organometallic precursor. Thus, to achieve satisfactory conversions, the use of excess **348** is required. The synthetic potential of this method was ascertained from reactions with olefins such as cyclooctene, *n*-dec-1-ene, *cis*-*n*-dec-5-ene and *trans*-*n*-dec-5-ene[272]. The harsh reaction conditions can be avoided by utilizing **349** instead of **348** as cyclopropanating reagent[272].

$$[(\eta^5\text{-Cp})(CO)_2FeCH_2S(Me)Ph]^+ BF_4^-$$
(349)

Compared with Me_2S, phenyl methyl thioether shows a greater propensity of serving as a leaving group and facilitates smooth accomplishment of the reaction at ambient temperature[272]. 7,7′-Spirobinorcarane **(351)** was isolated in poor yields as a product from the thermolysis of ylide complex **350** in the presence of cyclohexene. The mechanism outlined in Scheme 57 was postulated to account for these results[273].

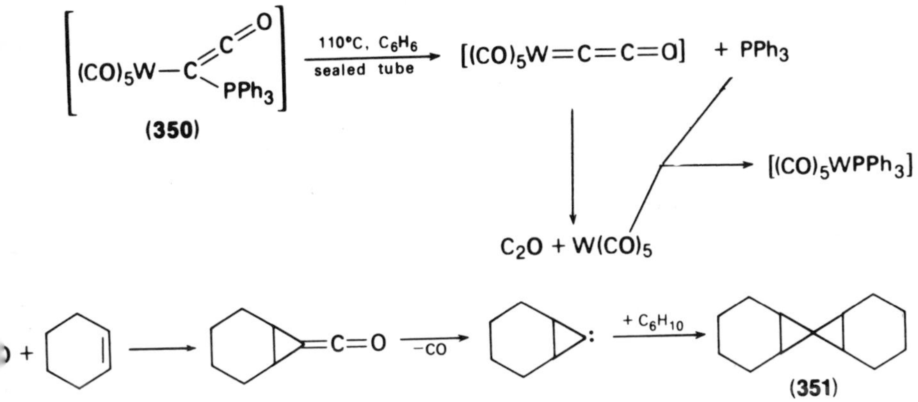

$$[(CO)_5W{-}C(=C{=}O)PPh_3] \ \textbf{(350)} \xrightarrow[\text{sealed tube}]{110\,°C,\ C_6H_6} [(CO)_5W{=}C{=}C{=}O] + PPh_3 \longrightarrow [(CO)_5WPPh_3]$$

$$[(CO)_5W{=}C{=}C{=}O] \longrightarrow C_2O + W(CO)_5$$

$$C_2O + C_6H_{10} \longrightarrow C_6H_{10}{>}C{=}C{=}O \xrightarrow{-CO} C_6H_{10}{>}C{:} \xrightarrow{+C_6H_{10}} \textbf{(351)}$$

SCHEME 57

C. Catalysis

Some transition metal ylide complexes have been found useful as catalysts in various processes. Thus, **307** accelerates the polymerization of ethylene in hexane solution at room temperature to give linear crystalline polyethylene, whereas at elevated temperatures and pressures of about 50 bar it favours the generation of linear olefins[6].

The formation of crystalline polyethylene and polypropylene is also achieved by a mixture of Ph_3PCH_2, LiI, and $TiCl_4$. The ionic cobalt complex **208** trimerizes alkynes such as tolane and oct-4-yne to the corresponding benzene derivatives[8]. The dimeric rhodium complex **110** catalyses the hydrogenation of olefins such as hexe-1-ne, hexe-2-ne, and cyclohexene, but not that of methyl acrylate. The catalyst requires an induction period necessary for the hydrogenation of ligated cod in order to provide vacant acceptor sites at the metal. When hydrogenation has been completed, the catalyst decomposes unless stabilizing ligands are added[4].

IX. REFERENCES

1. A. W. Johnson, *Ylid Chemistry*, Academic Press, New York, 1966
2. B. M. Trost and L. M. Melvin, Jr., *Sulfur Ylides, Emerging Synthetic Intermediates*, Academic Press, New York, 1975.
3. H. Schmidbaur, *Acc. Chem. Res.*, **8**, 62 (1975); *Pure Appl. Chem.*, **50**, 19 (1978).
4. R. P. Grey and L. R. Anderson, *Inorg. Chem.*, **16**., 3187 (1977).
5. L. E. Manzer, *Inorg. Chem.*, **15**, 2567 (1976).
6. W. Keim, F. H. Kowaldt, R. Goddard, and C. Krüger, *Angew. Chem.*, **90**, 493 (1978); *Angew. Chem. Int. Ed. Engl.*, **17**, **466** (**1978**).
7. Shell Internationale Research Maatschappij N.V., *Neth. Pat. Appl.*, 7018, 381 (1971); *Chem. Int. Ed. Engl.*, **17**, 686 (1978).
8. N. L. Holy, N. C. Baenziger, and R. M. Flynn, *Angew. Chem.*, **90**, 732 (1978); *Angew. Chem. Int. Ed. Engl.*, **17**, 686 (1978).
9. A. Mendel, *U.S. Pat.*, 2 998 416 (1960); *Chem. Abstr.*, **56**, 587g (1962).
10. H. Schmidbaur, J. R. Mandl, A. Wohlleben-Hammer, and A. Fügner, *Z. Naturforsch.*, **33B**, 1325 (1978).
11. L. Knoll, *Z. Naturforsch.*, **32B**, 1268 (1977).
12. W. C. Kaska, D. K. Mitchell, R. F. Reichelderfer, and W. D. Korte, *J. Am. Chem. Soc.*, **96**, 2847 (1974).

13a. A. J. Speziale, G. J. Marco, and K. W. Ratts, *J. Am. Chem. Soc.*, **82**, 1260 (1960).

13b. A. J. Speziale and K. W. Ratts, *J. Am. Chem. Soc.*, **84**, 854 (1962).

14. D. Seyferth, S. O. Grim, and T. O. Read, *J. Am. Chem. Soc.*, **82**, 1510 (1960).
15. D. M. Lemal and E. H. Banitt, *Tetrahedron Lett.*, 245 (1964).
16. D. Seebach, *Angew. Chem.*, **79**, 469 (1967); *Angew. Chem. Int. Ed. Engl.*, **6**, 443 (1967).
17. R. B. King, *Inorg. Chem.*, **2**, 199 (1963).
18. W. W. du Mont, B. Neudert, G. Rudolph, and H. Schumann, *Angew. Chem.*, **88**, 303 (1976); *Angew. Chem. Int. Ed. Engl.*, **15**, 308 (1976).

19a. H. H. Karsch, H. F. Klein, and H. Schmidbaur, *Angew. Chem.*, **87**, 630 (1975); *Angew. Chem. Int. Ed. Engl.*, **14**, 637 (1975); *Chem. Ber.*, **110**, 2200 (1977).

19b. H. H. Karsch and H. Schmidbaur, *Z. Naturforsch.*, **32B**, 726 (1977).

20. H. F. Klein, J. Wenninger, and U. Schubert, *Z. Naturforsch.*, **34B**, 1391 (1979).
21. H. Schmidbaur and W. Tronich, *Chem. Ber.*, **101**, 3556 (1968).
22. H. Schmidbaur and J. Eberlein, *Z. Anorg. Allg. Chem.*, **434**, 145 (1977).
23. G. Wittig and M. Rieber, *Justus Liebigs Ann. Chem.*, **562**, 177 (1949).
24. H. Schmidbaur, J. Eberlein, and W. Richter, *Chem. Ber.*, **110**, 677 (1977).
25. H. Schmidbaur and W. Tronich, *Chem. Ber.*, **101**, 595 (1968).
26. H. Schmidbaur, H. F. Füller and F. H. Köhler, *J. Organomet. Chem.*, **99**, 353 (1975).
27. H. Schmidbaur and R. Franke, *Chem. Ber.*, **108**, 1321 (1975).
28. H. Schmidbauer, J. Adlkofer, and M. Heimann, *Chem. Ber.*, **107**, 3697 (1974).
29. Y. Yamamoto and H. Schmidbauer, *J. Organomet. Chem.*, **96**, 133 (1975).
30. Y. Yamamoto and Z. Kanda, *Bull. Chem. Soc. Jap.*, **52**, 2560 (1979).
31. F. Heydenreich, A. Mollbach, G. Wilke, H. Dreeskamp, E. G. Hoffmann, G. Schroth, K. Seevogel, and W. Stempfle, *Isr. J. Chem.*, **10**, 293 (1972).
32. D. Seyferth and S. O. Grim, *J. Am. Chem. Soc.*, **83**, 1610 (1961).
33. H. H. Karsch, H. F. Klein, and H. Schmidbaur, *Chem. Ber.*, **107**, 93 (1974).
34. K. Itoh, M. Fukui, and Y. Ishii, *J. Organomet. Chem.*, **129**, 259 (1977).
35. R. M. Buchanan and C. G. Pierpont, *Inorg. Chem.*, **18**, 3608 (1979).
36. G. Blaschke, H. Schmidbaur, and W. C. Kaska, *J. Organomet. Chem.*, **182**, 251 (1979).
37. J. C. Baldwin and W. C. Kaska, *Inorg. Chem.*, **18**, 687 (1979).
38. H. Schmidbaur and R. Franke, *Inorg. Chim. Acta*, **13**, 79 (1975).
39. W. Hieber, E. Winter, and E. Schubert, *Chem. Ber.*, **95**, 3070 (1962).
40. B. L. Booth and K. G. Smith, *J. Organomet. Chem.*, **178**, 361 (1979).
41. H. Schmidbaur and G. Blaschke, *Z. Naturforsch.*, **35B**, 584 (1980).
42. L. Knoll, *J. Organomet. Chem.*, **182**, 77 (1979).
43. L. Knoll, *J. Organomet. Chem.*, **193**, 47 (1980).
44. W. C. Kaska, D. K. Mitchell, and R. F. Reichelderfer, *J. Organomet. Chem.*, **47**, 391 (1973).
45. D. Cashman and F. J. Lalor, *J. Organomet. Chem.*, **32**, 351 (1971).

46. D. J. Brauer, C. Krüger, P. J. Roberts, and Y.-H. Tsay, *Chem. Ber.*, **107**, 3706 (1974).
47. L. Weber, *J. Organomet. Chem.*, **105**, C9 (1976).
48. L. Weber, *J. Organomet. Chem.*, **142**, 309 (1977).
49. L. Weber, *Z. Naturforsch.*, **31B**, 780 (1976).
50. H. Bock and H. tom Dieck, *Z. Naturforsch.*, **21B**, 739 (1966).
51. L. Weber, unpublished results.
52. V. L., Shelepina, O. A. Osipov, and O. E. Shelepin, *Zh. Vses. Khim. Obshchest.*, **14**, 586 (1969); *Chem. Abstr.*, **72**, 37299 (1970).
53. H. Schmidbaur, W. Scharf, and H. J. Füller, *Z. Naturforsch.*, **32B**, 853 (1977).
54. J. R. Moss and J. C. Spiers, *J. Organomet. Chem.*, **182**, C20 (1979).
55. H. F. Klein and R. Hammer, *Angew. Chem.*, **88**, 61 (1976); *Angew. Chem. Int. Ed. Engl.*, **16**, 42 (1976).
56. M. J. van Hamme and D. J. Burton, *J. Fluorine Chem.*, **10**, 131 (1977).
57. D. J. Burton and P. E. Greenlimb, *J. Fluorine Chem.*, **3**, 447 (1973/74); *J. Org. Chem.*, **40**, 2796 (1975).
58. M. J. van Hamme and D. J. Burton, *J. Organomet. Chem.*, **169**, 123 (1979).
59. U. Klabunde and E. O. Fischer, *J. Am. Chem. Soc.*, **89**, 7147 (1967).
60. F. R. Kreissl, E. O. Fischer, C. G. Kreiter, and K. Weiss, *Angew. Chem.*, **85**, 617 (1973); *Angew. Chem. Int. Ed. Engl.*, **12**, 563 (1973).
61. F. R. Kreissl and E. O. Fischer, *Chem. Ber.*, **107**, 183 (1974).
62. H. Werner, E. O. Fischer, B. Heckl, and C. G. Kreiter, *J. Organomet. Chem.*, **28**, 367 (1971).
63. F. R. Kreissl, C. G. Kreiter, and E. O. Fischer, *Angew. Chem.*, **84**, 679 (1972); *Angew. Chem. Int. Ed. Engl.*, **11**, 643 (1972).
64. F. R. Kreissl, E. O. Fischer, C. G. Kreiter, and H. Fischer, *Chem. Ber.*, **106**, 1262 (1973).
65. E. O. Fischer, G. Kreis, F. R. Kreissl, C. G. Kreiter, and J. Müller, *Chem. Ber.*, **106**, 3910 (1973).
66. F. R. Kreissl and W. Held, *Chem. Ber.*, **110**, 799 (1977).
67. F. R. Kreissl, P. Stückler, and E. W. Meinecke, *Chem. Ber.*, **110**, 3040 (1977) and references cited therein.
68. H. Fischer, E. O. Fischer, C. G. Kreiter, and H. Werner, *Chem. Ber.*, **107**, 2459 (1974).
69. W.-K. Wong, W. Tam, and J. A. Gladysz, *J. Am. Chem. Soc.*, **101**, 5440 (1979).
70. W. Beck, K. Schloter, and H. Ernst, Paper presented at the IXth International Conference on Organometallic Chemistry, Dijon, 1979, Abstr. C53.
71. N. J. Cooper and M. L. H. Green, *J. Chem. Soc., Chem. Commun.*, 208 and 761 (1974); *J. Chem. Soc., Dalton Trans.*, 1121 (1979).
72. M. Canestrari and M. L. H. Green, *J. Chem. Soc., Chem. Commun.*, 913 (1979).
73. D. M. Adams, J. Chatt, R. G. Guy, and N. Sheppard, *J. Chem. Soc.*, 738 (1961).
74. N. A. Bailey, R. D. Gillard, M. Keeton, R. Mason, and D. R. Russell, *J. Chem. Soc., Chem. Commun.*, 396 (1966); M. Keeton, R. Mason, and D. R. Russell, *J. Organomet. Chem.*, **33**, 247 (1971); R. D. Gillard, M. Keeton, R. Mason, M. F. Pilbrow, and D. R. Russell *J. Organomet. Chem.*, **33**, 259 (1971).
75. R. J. Al-essa and R. J. Puddephatt, *J. Chem. Soc., Chem. Commun.*, 45 (1980).
76. N. Y. Kolobova, V. V. Skripkin, G. G. Alexandrov, and Y. T. Struchkov, *J. Organomet. Chem.*, **169**, 293 (1979).
77. E. Lindner, *J. Organomet. Chem.*, **94**, 229 (1975).
78. P. R. Sharp and R. R. Schrock, *J. Organomet. Chem.*, **171**, 43 (1979).
79. J. Schwartz and K. I. Gell, *J. Organomet. Chem.*, **184**, C1 (1980).
80. L. Weber, *J. Organomet. Chem.*, **131**, 49 (1977).
81. G. Wittig and K. Schwarzenbach, *Justus Liebigs Ann. Chem.*, **650**, 1 (1961).
82. B. T. Kilbourn and D. Felix, *J. Chem. Soc. A*, 163 (1969).
83. H. Schmidbaur and W. Tronich, *Chem. Ber.*, **101**, 3545 (1968).
84. D. R. Mathiason and N. E. Miller, *Inorg. Chem.*, **7**, 709 (1968).
85. G. Yoshida, H. Kurosawa, and R. Okawara, *J. Organomet. Chem.*, **131**, 309 (1977).
86. T. J. Collins and W. R. Roper, *J. Chem. Soc., Chem. Commun.*, 901 (1977); *J. Organomet. Chem.*, **159**, 73 (1978).
87. D. J. Sepelak, C. G. Piermont, E. K. Barefield, J. T. Budz, and C. A. Poffenberger, *J. Am. Chem. Soc.*, **98**, 6178 (1976).
88. E. K. Barefield and D. J. Sepelak, *J. Am. Chem. Soc.*, **101**, 6542 (1979).

89. H. Schmidbaur, J. Adlkofer, and W. Buchner, *Angew. Chem.*, **85**, 448 (1973); *Angew. Chem. Int. Ed. Engl.*, **12**, 415 (1973).
90. H. Schmidbaur and R. Franke, *Angew. Chem.*, **85**, 449, (1973); *Angew. Chem. Int. Ed. Engl.*, **12**, 416 (1973).
91. G. Nardin, L. Randaccio, and E. Zangrando, *J. Organomet. Chem.*, **74**, C23 (1974).
92. H. Schmidbaur, J. E. Mandl, W. Richter, V. Bejenke, A. Frank, and G. Huttner, *Chem. Ber.*, **110.**, 2236 (1977).
93. H. Schmidbaur and W. Richter, *Chem. Ber.*, **108.**, 2656 (1975).
94. H. Schmidbaur, H.-P. Scherm, and U. Schubert, *Chem. Ber.*, **111**, 764 (1978).
95. H. Schmidbaur and H.-P. Scherm, *Chem. Ber.*, **110**, 1576 (1977).
96. H. Schmidbaur and H. J. Füller, *Chem. Ber.*, **107**, 3674 (1974).
97. H. Schmidbaur and K. H. Räthlein, *Chem. Ber.*, **107**, 102 (1974).
98. H. Schmidbaur and W. Richter, *Z. Anorg. Allg. Chem.*, **429**, 222 (1977).
99. H. Schmidbaur and R. Franke, *Inorg. Chim. Acta*, **13**, 85 (1975).
100. H. Schmidbaur, J. R. Mandl, A. Frank, and G. Huttner, *Chem. Ber.*, **109**, 466 (1976).
101. H. Schmidbaur and J. R. Mandl, *Naturwissenschaften*, **63**, 585 (1976).
102. H. Schmidbaur and H.-P. Scherm, *Z. Naturforsch.*, **34B**, 1347 (1979).
103. E. Kurras, U. Rosenthal, H. Mennenga, G. Oehme, and G. Engelhardt, *Z. Chem.*, **14**, 160 (1974).
104. E. Kurras, H. Mennenga, G. Oehme, U. Rosenthal, and G. Engelhardt, *J. Organomet. Chem.*, **84**, C13 (1975).
105. F. A. Cotton, B. E. Hanson, W. H. Ilsley, and G. W. Rice, *Inorg. Chem.*, **18**, 2713 (1979).
106. W. Scharf, D. Neugebauer, U. Schubert, and H. Schmidbaur, *Angew. Chem.*, **90**, 628 (1978); *Angew Chem. Int. Ed. Engl.*, **17**, 601 (1978).
107. H. Schmidbaur and O. Gasser, *Angew. Chem.*, **88**, 542 (1976); *Angew. Chem. Int. Ed. Engl.*, **15**, 502 (1976).
108. R. L. Lapinski, H. Yue, and R. A. Grey, *J. Organomet. Chem.*, **174**, 213 (1979).
109. H. Schmidbaur, G. Blaschke, H. J. Füller, and H.-P. Scherm, *J. Organomet. Chem.*, **160**, 41 (1978).
110. T. E. Fraser, H. J. Füller, and H. Schmidbaur, *Z. Naturforsch.*, **34B**, 1218 (1979).
111. H. H. Karsch and H. Schmidbaur, *Chem. Ber.*, **107**, 3684 (1974).
112. H. Schmidbaur, G. Blaschke, and H.-P. Scherm, *Chem. Ber.*, **112**, 3311 (1979).
113. H. Schmidbaur and H.-P. Scherm, *Chem. Ber.*, **111**, 797 (1978).
114. H. H. Karsch, H. F. Klein, C. G. Kreiter, and H. Schmidbaur, *Chem. Ber.*, **107**, 3692 (1974).
115a. E. Kurras, U. Rosenthal, H. Mennenga, and G. Oehme, *Angew. Chem.*, **85**, 913 (1973); *Angew. Chem. Int. Ed. Engl.*, **12**, 854 (1973).
115b. E. Kurras and U. Rosenthal, *J. Organomet. Chem.*, **160**, 35 (1978).
116. H. Schumann and S. Hohmann, *Chem. Ztg.*, **100**, 336 (1976).
117. H. Schmidbaur, U. Deschler, B. Zimmer-Gasser, D. Neugebauer, and U. Schubert, *Chem. Ber.*, **113**, 902 (1980).
118. R. E. Cramer, R. B. Maynard, and J. W. Gilje, *J. Am. Chem. Soc.*, **100** 5562 (1978).
119. P. Bravo, G. Fronza, and G. Ticozzi, *J. Organomet. Chem.*, **118**, C78 (1976).
120. O. I. Kolodyazhnyi, *Zh. Obshch. Khim.*, **45**, 704 (1975).
121. H. Schmidbaur and W. Wolf, *Chem. Ber.*, **108**, 2851 (1975).
122. H. Schmidbaur, O. Gasser, C. Krüger, and J. C. Sekutowski, *Chem. Ber.*, **110**, 3517 (1977).
123. H. Schmidbaur and H. J. Füller, *Angew. Chem.*, **88**, 541 (1976); *Angew. Chem. Int. Ed. Engl.*, **15**, 501 (1976); *Chem. Ber.*, **110**, 3528 (1977).
124. H. Schmidbaur, G. Müller, U. Schubert, and O. Orama, *Angew. Chem.*, **90**, 126 (1978); *Angew. Chem. Int. Ed. Engl.*, **17**, 126 (1978).
125. G. Müller, U. Schubert, O. Orama, and H. Schmidbaur, *Chem. Ber.*, **112**, 3302 (1979).
126. H. Schmidbaur, H. J. Füller, V. Bejenke, A. Frank, and G. Huttner, *Chem. Ber.*, **110**, 3536 (1977).
127. H. Schmidbaur, O. Gasser, T. E. Fraser, and E. A. V. Ebsworth, *J. Chem. Soc., Chem. Commun.*, 334 (1977).
128. C. Krüger, J. C. Sekutowski, R. Goddard, H. J. Füller, O. Gasser, and H. Schmidbaur, *Isr. J. Chem.*, **15**, 149 (1976/77).

129. E. T. Weleski, Jr., J. L. Silver, M. D. Jansson, and J. L. Burmeister, *J. Organomet. Chem.*, **102**, 365 (1975).
130. P. Bravo, G. Fronza, G. Gaudiano, and C. Ticozzi, *Gazz. Chim. Ital.*, **103**, 623 (1973); *J. Organomet. Chem.*, **74**, 143 (1974).
131. P. Bravo, G. Fronza, and C. Ticozzi, *J. Organomet. Chem.*, **111**, 361 (1976).
132. H. Nishiyama, *J. Organomet. Chem.*, **165**, 407 (1979).
133. H. Nishiyama, K. Itoh, and Y. Ishii, *J. Organomet. Chem.*, **87** 129 (1975).
134. H. Koezuka, G. Matsubayashi, and T. Tanaka, *Inorg. Chem.*, **15**, 417 (1976).
135. P. A. Arnup and M. C. Baird, *Inorg. Nucl. Chem. Lett.*, **5**, 65 (1969).
136. H. Koezuka, G. Matsubayashi, and T. Tanaka, *Inorg. Chem.*, **13**, 443 (1974).
137. N. A. Nesmeyanov, V. M. Novikov, and O. A. Reutov, *Izv. Akad. Nauk SSSR, Ser. Khim.*, 772 (1964); *Chem. Abstr.*, **61**, 3143e (1964).
138. N. A. Nesmeyanov and V. M. Novikov, *Dokl. Akad. Nauk SSSR*, **162**, 350 (1965); *Chem. Abstr.*, **63**, 5671f (1965).
139. N. A. Nesmeyanov, V. M. Novikov, and O. A. Reutov, *J. Organomet. Chem.*, **4**, 202 (1965).
140. N. A. Nesmeyanov, V. M. Novikov, and O. A. Reutov, *Zh. Org. Khim.*, **2**, 942 (1966); *Chem. Abstr.*, **65**, 15420h (1966).
141. N. A. Nesmeyanov, S. T. Berman, L. D. Ashkinadze, L. A. Kazitna, and O. A. Reutov, *Zh. Org. Khim.*, **4**, 1685 (1968); *Chem. Abstr.*, **70**, 19469v (1969).
142. N. A. Nesmeyanov, A. V. Kalinin, and O. A. Reutov, *Dokl. Akad. Nauk SSSR*, **195**, 98 (1970); *Chem. Abstr.*, **74**, 100185k (1971).
143. T. Saito, *Bull. Chem. Soc. Jap.*, **51**, 169 (1978).
144. J. Buckle and P. G. Harrison, *J. Organomet. Chem.*, **49**, C17 (1973).
145. J. Buckle, P. G. Harrison, T. J. King, and J. A. Richards, *J. Chem. Soc., Chem. Commun.*, 1104 (1972).
146. G. Matsubayashi, Y. Kondo, T. Tanaka, S. Nishigaki, and K. Nakatsu, *Chem. Lett.*, 375 (1979).
147. G. Matsubayashi, I. Kawafume, T. Tanaka, S. Nishigaki, and K. Nakatsu, *J. Organomet. Chem.*, **187**, 113 (1980).
148. Y. Oosawa, T. Miyamoto, T. Saito, and Y. Sasaki, *Chem. Lett.*, 33 (1975).
Y. Oosawa, T. Saito, and Y. Sasaki, *Chem. Lett.*, 1259 (1975).
Y. Oosawa, H. Urabe, T. Saito, and Y. Sasaki, *J. Organomet. Chem.*, **122**, 113 (1976).
149. H. Takahashi, Y. Oosawa, A. Kolobayashi, T. Saito, and Y. Sasaki, *Chem. Lett.*, 15 (1976); *Bull. Chem. Soc. Jap.*, **50**, 1771 (1977).
150. M. Kato, H. Urabe, Y. Oosawa, T. Saito, and Y. Sasaki, *Chem. Lett.*, 51 (1976).
N. Sugita, T. Miyamoto, and Y. Sasaki, *Chem. Lett.*, 659 (1976).
M. Kato, H. Urabe, Y. Oosawa, T. Saito, and Y. Sasaki, *J. Organomet. Chem.*, **121**, 81 (1976).
151. M. Seno and S. Tsuchiya, *J. Chem. Soc., Dalton Trans.*, 751 (1977).
152. S. A. Dias, A. W. Downs, and W. R. McWhinnie, *J. Chem. Soc., Dalton Trans.*, 162 (1975).
153. M. R. Churchill, F. J. Rotella, E. W. Abel, and St. A. Mucklejohn, *J. Am. Chem. Soc.*, **99**, 5820 (1977); M. R. Churchill and F. R. Rotella, *Inorg. Chem.*, **17**, 2614 (1978).
154. H. G. Alt, J. A. Schwärzle, and F. R. Kreissl, *J. Organomet. Chem.*, **152**, C57 (1978).
155. K. A. Ostoja Starzewski, H. tom Dieck, K. D. Franz, and F. Hohmann, *J. Organomet. Chem.*, **42**, C35 (1972).
156. A. Greco, *J. Organomet. Chem.*, **43**, 351 (1972).
157. I. W. Bassi and R. Scordamaglia, *J. Organomet. Chem.*, **51**, 273 (1973).
158. K. Itoh, H. Nishiyama, T. Oshnishi, and Y. Ishii, *J. Organomet. Chem.*, **76**, 401 (1974).
159. K. Itoh, H. Nishiyama, and Y. Ishii, Paper presented at the VIIth International Conference on Organometallic Chemistry, Venice, 1975, Abstr. 243.
160. M. F. Hirai, M. Miyasaka, K. Itoh, and Y. Ishii, *J. Organomet. Chem.*, **160**, 25 (1978).
161. M. F. Hirai,, M. Miyasaka, K. Itoh, and Y. Ishii, *J. Organomet. Chem.*, **165**, 391 (1979).
162. M. F. Hirai, M. Miyasaka, K. Itoh, and Y. Ishii, *J. Chem. Soc., Dalton Trans.*, 1200 (1979).
163. V. N. Setkina, A. Zh. Zhakaeva, G. A. Panosyan, V. I. Zdanovitch, P. V. Petrovskii, and D. N. Kursanov, *J. Organomet. Chem.*, **129**, 361 (1977).

164. N. L. Holy, N. C. Baenziger, R. M. Flynn, and D. C. Swenson, *J. Am. Chem. Soc.*, **98**, 7823 (1976).
165. C. G. Pierpont, H. H. Downs, K. Itoh, N. Nishiyama, and Y. Ishii, *J. Organomet. Chem.*, **124**, 93 (1976).
166. E. W. Abel, A. Singh, and G. Wilkinson, *Chem. Ind. (London)*, 1067 (1959).
167a. V. I. Zdanovitch, A. Zh. Zhakaeva, V. N. Setkina, and D. N. Kursanov, *J. Organomet. Chem.*, **64**, C25 (1974).
167b. V. N. Setkina, V. I. Zdanovitch, A. Zh. Zhakaeva, Yu. S. Nekrasov, N. I. Vasyukova, and D. N. Kursanov, *Kokl. Akad. Nauk SSSR*, **219**, 1137 (1974); *Chem. Abstr.*, **82**, 125460z (1975).
168. A. Nesmeyanov, N. E. Kolobova, V. I. Zdanovitch, and A. Zh. Zhakaeva, *J. Organomet. Chem.*, **107**, 319 (1976).
169. V. I. Zdanovitch, N. E. Kolobova, N. I. Vasyukova, Yu. S. Nekrasov, G. A. Panosyan, P. V. Petrovskii, and A. Zh. Zhakaeva, *J. Organomet. Chem.*, **148**, 63 (1978).
170. V. G. Andrianov, Yu. T. Struchkov, V. N. Setkina, A. Zh. Zhakaeva, and V. I. Zdanovitch, *J. Organomet. Chem.*, **140**, 169 (1977).
171. J. C. Kotz and C. D. Turnipseed, *J. Chem. Soc., Chem. Commun.*, 41 (1970).
172. J. C. Kotz and D. G. Pedrotty, *J. Organomet. Chem.*, **22**, 425 (1970).
173. D. Cashman and F. J. Lalor, *J. Organomet. Chem.*, **24**, C29 (1970).
174. N. L. Holy, T. E. Nalesnik and L. T. Warfield, *Inorg. Nucl. Chem. Lett.*, **13**, 523 and 569 (1977).
175. K. Dimroth, *Top. Curr. Chem.*, **38**, 1 (1973).
176. A. G. Hortmann, R. L. Harris, and J. A. Miles, *J. Am. Chem. Soc.*, **96**, 6119 (1974).
177. A. G. Hortmann and R. L. Harris, *J. Am. Chem. Soc.*, **93**, 2471 (1971).
178. M. Lückoff and K. Dimroth, *Angew. Chem.*, **88**, 543 (1976); *Angew. Chem. Int. Ed. Engl.*, **15**, 503 (1976).
179. T. Debaerdemaeker, *Angew. Chem.*, **88**, 544 (1976); *Angew. Chem. Int. Ed. Engl.*, **15**, 504 (1976); *Acta Crystallogr.*, **B35**, 1686 (1979).
180. B. E. Maryanoff, J. Stackhouse, G. H. Senkler, Jr., and K. Mislow, *J. Am. Chem. Soc.*, **97**, 2718 (1975).
181a. L. Weber, *Angew. Chem.*, **93**, 304 (1981); *Angew. Chem. Int. Ed.*, **20**, 297 (1981).
181b. L. Weber and R. Boese, *Chem. Ber.*, **115**, 1775 (1982).
182. R. Boese, personal communication.
183. L. Weber, C. Krüger, and Y.-H. Tsay, *Chem. Ber.*, **111**, 1709 (1978).
184. L. Weber, *Chem. Ber.*, **112**, 99, 3828 (1979); **114**, 1 (1981); L. Weber, D. Vehreschild-Yzermann, C. Krüger, and G. Wollmershäuser, *Z. Naturforsch.*, **36B**, 198 (1981).
185. H. Schmidbaur and W. Tronich, *Chem. Ber.*, **100**, 1032 (1967).
186. H. Schmidbaur and G. Kammel, *Chem. Ber.*, **104**, 3252 (1971).
187. J. C. Baldwin, N. L. Keder, C. E. Strouse, and W. C. Kaska, *Z. Naturforsch.*, **35B**, 1289 (1980).
188. W. Malisch, *J. Organomet. Chem.*, **61**, C15 (1973).
189. W. Malisch, *J. Organomet. Chem.*, **77**, C15 (1974).
190. W. Malisch, *J. Organomet. Chem.*, **82**, 185 (1974).
191. W. Malisch and P. Panster, *Chem. Ber.*, **108**, 2554 (1975).
192. W. Malisch, M. Kuhn, W. Albert, and H. Rössner, *Chem. Ber.*, **113**, 3318 (1980).
193. W. Malisch and P. Panster, *J. Organomet. Chem.*, **99**, 421 (1975).
194. H. Schmidbaur and W. Malisch, *Chem. Ber.*, **102**, 83 (1969).
195. F. R. Kreissl, *J. Organomet. Chem.*, **99**, 305 (1975).
196. F. R. Kreissl, W. Uedelhoven, and G. Kreis, *Chem. Ber.*, **111**, 3283 (1978).
197. F. R. Kreissl, K. Eberl, and W. Kleine, *Chem. Ber.*, **111**, 2451 (1978).
198. F. R. Kreissl, W. Uedelhoven, and A. Ruhs, *J. Organomet. Chem.*, **113**, C55 (1976).
199. E. O. Fischer, A. Ruhs, and F. R. Kreissl, *Chem. Ber.*, **110**, 805 (1977).
200. F. R. Kreissl and P. Friedrich, *Angew. Chem.*, **89**, 553 (1977); *Angew. Chem. Int. Ed. Engl.*, **16**, 543 (1977).
201. F. R. Kreissl, P. Friedrich, T. L. Lindner, and G. Huttner, *Angew. Chem.*, **89**, 325 (1977); *Angew. Chem. Int. Ed. Engl.*, **16**, 314 (1977).
202. W. Uedelhoven, K. Eberl, and F. R. Kreissl, *Chem. Ber.*, **112**, 3376 (1979).

203. Y. S. Wong, H. N. Paik, P. C. Chieh, and A. J. Carty, *J. Chem. Soc., Chem. Commun.*, 309 (1975).
204. S. Satgé, P. Massol, and P. Rivière, *J. Organomet. Chem.*, **56**, 1 (1973).
205. W. W. duMont and G. Rudolph, *Chem. Ber.*, **109**, 3419 (1976).
206. W. W. duMont and B. Neudert, *Z. Anorg. Allg. Chem.*, **441**, 86 (1978).
207. W. W. duMont and H. Schumann, *J. Organomet. Chem.*, **85**, 245 (1975).
208. P. Jutzi, H. J. Hoffmann, D. J. Brauer, and C. Krüger, *Angew. Chem.*, **85**, 1116 (1973); *Angew. Chem. Int. Ed. Engl.*, **12**, 1002 (1973).
209. P. Jutzi, H. J. Hoffmann, and K. H. Wyes, *J. Organomet. Chem.*, **81**, 341 (1974).
210. D. Uhlig, H. Behrens, and E. Lindner, *Z. Anorg. Allg. Chem.*, **401**, 233 (1973).
211. P. Jutzi and H. J. Hoffmann, *Chem. Ber.*, **107**, 3616 (1974).
212. P. Jutzi and W. Steiner, *Chem. Ber.*, **109**, 3473 (1976).
213. P. Jutzi and W. Steiner, *Angew. Chem.*, **88**, 720 (1976); *Angew. Chem. Int. Ed. Engl.*, **15**, 684 (1976).
214. P. Jutzi and W. Steiner, *Angew. Chem.*, **89**, 675 (1977); *Angew. Chem. Int. Ed. Engl.*, **16**, 639 (1977).
215. P. Jutzi, W. Steiner, E. König, G. Huttner, A. Frank, and U. Schubert, *Chem. Ber.*, **111**, 606 (1978).
216. W. W. duMont, *J. Organomet. Chem.*, **153**, C11 (1978).
217. W. W. duMont and B. Neudert, *Chem. Ber.*, **111**, 2267 (1978), and references cited therein.
218. G. Rudolph and W. W. duMont, Paper presented at the XXIth International Conference on Coordination Chemistry, Toulouse, 1980, Abstr. 157.
219. P. Jutzi and W. Steiner, *Chem. Ber.*, **109**, 1575 (1976).
220. T. J. Marks, *J. Am. Chem. Soc.*, **93**, 7090 (1971).
221. H. Behrens, M. Moll, and E. Sixtus, *Z. Naturforsch.*, **32B**, 1105 (1977).
222. M. D. Brice and F. A. Cotton, *J. Am. Chem. Soc.*, **95**, 4529 (1973).
223. T. J. Marks and A. R. Newman, *J. Am. Chem. Soc.*, **95**, 769 (1973).
224. H. Schmidbaur and W. Tronich, *Inorg. Chem.*, **7**, 168 (1968).
225. Y. Yamamoto and H. Schmidbaur, *J. Chem. Soc., Chem. Commun.*, 668 (1975).
226. H. Schmidbaur and G. Hasslberger, *Chem. Ber.*, **111**, 2702 (1978).
227. B. H. Freeman, D. Lloyd, and M. I. C. Singer, *Tetrahedron*, **28**, 343 (1972).
228. N. N. Magdasieva and R. Kandgetcyan, *Zh. Org. Khim.*, **7**, 2228 (1971).
229. N. N. Magdasieva, R. Kandgetcyan, and A. A. Ibragimov, *J. Organomet. Chem.*, **42**, 399 (1972).
230. A. Tamagaki,and K. Sakaki, *Chem. Lett.*, 503 (1975).
231. I. D. Sadekov, A. I. Usachev, A. A. Maksimenko, and V. I. Minkin, *Zh. Obshch. Khim.*, **45**, 2563 (1975).
232. N. Ya. Derkach, N. P. Tishchenko, and V. G. Volushchuk, *Zh. Org. Khim.*, **14**, 958 (1978).
233. V. V. Semenov, L. G. Melnikova, S. A. Shevelev, and A. A. Fainzilberg, *Izv. Akad. Nauk SSSR, Ser. Khim.*, 138 (1980).
234. S. Tamagaki, R. Akatsuka, and S. Kozuka, *Bull. Chem. Soc. Jap.*, **53**, 817 (1980).
235. R. R. Schrock and R. P. Sharp, *J. Am. Chem. Soc.*, **100**, 2389 (1978).
236. R. R. Schrock and J. D. Fellmann, *J. Am. Chem. Soc.*, **100**, 3359 (1978).
237. R. R. Schrock, *J. Am. Chem. Soc.*, **98**, 5399 (1976).
238. R. R. Schrock, *Acc. Chem. Res.*, **12**, 98 (1979).
239. H. D. Empsall, E. M. Hyde, R. Markham, W. S. McDonald, N. C. Norton, B. L. Shaw, and B. Weeks, *J. Chem. Soc., Chem. Commun.*, 589 (1977).
240a. E. W. Abel, R. J. Rowley, R. Mason, and K. M. Thomas, *J. Chem. Soc., Chem. Commun.*, 72 (1974).
240b. E. W. Abel and R. J. Rowley, *J. Chem. Soc., Dalton Trans.*, 1096 (1975).
241. R. B. King and M. B. Bisnette, *J. Am. Chem. Soc.*, **86**, 1267 (1964); *Inorg. Chem.*, **4**, 475 and 486 (1965).
242. C. W. Fong and G. Wilkinson, *J. Chem. Soc., Dalton Trans.*, 1100 (1975).
243. G. Yoshida, H. Kurosawa, and R. Okawara, *J. Organomet. Chem.*, **113**, 85 (1976).
244. K. Miki, Y. Kai, N. Yasuoka, and N. Kasai, *J. Organomet. Chem.*, **135**, 53 (1977).
245. E. R. Biehl and P. C. Reeves, *Synthesis*, 883 (1974).

246. G. Berens, F. Kaplan, R. Rimerman, B. W. Roberts, and A. Wissner, *J. Am. Chem. Soc.*, **97**, 7076 (1975).
247. E. E. Nunn, *Aust. J. Chem.*, **29**, 2549 (1976).
248. M. B. Stringer and D. Wege, *Tetrahedron Lett.*, 65 (1977).
249. F. A. Kaplan and B. W. Bryan, *J. Am. Chem. Soc.*, **99**, 513 (1977).
250. J. M. Landesberg and J. Sieczkowski, *J. Am. Chem. Soc.*, **93**, 972 (1971).
251. H. W. Whitlock, Jr., C. Reich, and W. D. Woessner, *J. Am. Chem. Soc.*, **93**, 2483 (1971).
252. S. Z. Goldberg, E. N. Duesler, and K. N. Raymond, *J. Chem. Soc., Chem. Commun.*, 826 (1971); *Inorg. Chem.*, **11**, 1397 (1972).
253. W. C. Kaska and C. S. Creaser, *Transition Met. Chem.*, **3**, 360 (1978).
254. L. Knoll, *Chem. Ber.*, **111**, 814 (1978).
255. H. Blau and W. Malisch, *Angew. Chem.*, **92**, 1063 (1980); *Angew. Chem. Int. Ed.*, **19**, 1020 (1980).
256. W. Malisch, H. Blau, and U. Schubert, *Angew. Chem.*, **92**, 1065 (1980); *Angew. Chem. Int. Ed.*, **19**, 1020 (1980).
257a. L. Weber, *J. Organomet. Chem.*, **122**, 69 (1976).
257b. L. Weber, unpublished results.
258. W. Malisch, H. Blau, and S. Voran, *Angew. Chem.*, **90**, 827 (1978); *Angew. Chem. Int. Ed. Engl.*, **17**, 780 (1978).
259. C. P. Casey and T. J. Burkhardt, *J. Am. Chem. Soc.*, **94**, 6543 (1972).
260. C. W. Rees and E.v. Angerer, *J. Chem. Soc., Chem. Commun.*, 420 (1972).
261. T. L. Gilchrist, R. Livingston, C. W. Rees, and E.v. Angerer, *J. Chem. Soc., Perkin Trans. I*, 2535 (1973).
262. D. L. Reger and E. C. Culbertson, *J. Organomet. Chem.*, **131**, 297 (1977).
263. P. Lennon, A. M. Rosan, and M. Rosenblum, *J. Am. Chem. Soc.*, **99**, 8426 (1977).
264. L. Knoll, *J. Organomet. Chem.*, **155**, C63 (1978).
265. L. Knoll and H. Wolff, *Chem. Ber.*, **112**, 2709 (1979).
266. L. Knoll, *J. Organomet. Chem.*, **186**, C42 (1980).
267. L. Knoll, *Z. Naturforsch.*, **33B**, 396 (1978).
268. W. Malisch, *Angew. Chem.*, **85**, 228 (1973); *Angew. Chem. Int. Ed. Engl.*, **12**, 235 (1973).
269. D. Seyferth, H. D. Simmons, Jr., and G. Singh, *J. Organomet. Chem.*, **3**, 337 (1965).
270. T. Cohen, G. Herman, T. M. Chapman, and D. Kuhn, *J. Am. Chem. Soc.*, **96**, 5627 (1974).
271. B. M. Trost, *J. Am. Chem. Soc.*, **89**, 138 (1967).
272. S. Brandt and P. Helquist, *J. Am. Chem. Soc.*, **101**, 6473 (1979); Paper presented at the IXth International Conference on Organometallic Chemistry, Dijon, 1979, Abstr. B4.
273. H. Berke and E. Lindner, *Angew. Chem.*, **85**, 668 (1973); *Angew. Chem. Int. Ed. Engl.*, **12**, 667 (1973); *Chem. Ber.*, **107**, 1360 (1974).

The Chemistry of the Metal–Carbon Bond
Edited by F. R. Hartley and S. Patai

CHAPTER **4**

Synthesis of transition metal–carbene complexes

HELMUT FISCHER

Anorganisch-Chemisches Institut der Technischen Universität München, Lichtenbergstrasse 4, D-8046 Garching, German Federal Republic

I. INTRODUCTION

Since the first planned synthesis and characterization of a stable transition metal–carbene complex by Fischer and Maasböl[1] in 1964, this field of organometallic chemistry has expanded at a rapid rate. Stable carbene complexes had already been prepared before[2–4] but were not recognized as such. Only as late as 1970 were they assigned correct structures[5,6]. The current interest in carbene complexes focuses mainly on three areas: (a) synthesis, structure, bonding properties, and reactivity, (b) their use as model compounds for the study of catalytic processes (e.g. olefin metathesis), and (c) their potential use as intermediates and as starting compounds for the synthesis of organic and organometallic compounds. A whole series of new preparative routes have been developed during the last 15 years and many hundreds of carbene complexes have been isolated, characterized, and studied. Several reviews have appeared, covering the whole area[7–10] or parts of it[11–16].

This chapter deals with the different routes employed in the preparation of transition metal–carbene complexes. Owing to the limited space available, some of the great number of complexes and synthetic methods already available have necessarily been omitted. Especially emphasized are synthetic routes with a wide range of applicability. Excluded are complexes (1) in which the carbene carbon atom is not

$$L_nM{=}C\genfrac{}{}{0pt}{}{\diagup X}{\diagdown Y}$$

(1)

essentially sp^2-hybridized, e.g. the carbene carbon atom and the three substituents (M, X, Y) in **1** do not lie within one plane, (2) with 'bridging carbene ligands' and (3) with a coordinated, sp^2-hybridized carbon atom that is formally double bonded to one of its two nonmetal substituents (e.g. acyl– or imino–metal complexes), and (4) prepared by more 'exotic' procedures.

The preparation of carbene complexes can be divided into roughly three different strategies: (a) synthesis from non-carbene complex precursors, (b) synthesis from carbene complexes by modification of the carbene ligand, and (c) synthesis from carbene complexes by modification of the metal–ligand framework.

II. SYNTHESIS FROM NON-CARBENE–METAL COMPLEX PRECURSORS

A. Nucleophilic Attack on Metal Carbonyls

In spite of the variety of methods available, the original preparation by Fischer and Maasböl is probably still the most useful and general procedure for the direct synthesis of carbene complexes from non-carbene complex precursors. It involves attack of a nucleophile R^- in LiR (R = Me, Ph, etc.) at the carbon atom of a coordinated carbon monoxide in a metal–carbonyl complex to give the anionic acyllithium salt **2**. This complex can be converted into and isolated as the corresponding stable tetraalkylammonium salt (equation 1). Acidification of **2** gives the hydroxycarbene complex **3**, which decomposes rapidly in solution at room temperature [by a 1,2-hydrogen shift to form benzaldehyde (for R = Ph) as demonstrated for iron–carbene complexes[17,18]]; compound **3**, was not isolated in an analytically pure form until 1973[19]. Reaction of **3** with diazomethane finally yields the methoxycarbene complex **4**[1].

Carbene complexes such as **4** are thermally stable. On the other hand, they possess a strongly electrophilic centre at the carbene carbon atom and simultaneously a nucleophilic centre at the heteroatom (e.g. O in **4**), allowing a variety of reactions.

$$[(CO)_6Cr] + LiR \longrightarrow \left[(CO)_5Cr-C\begin{matrix}{=}O\\ {\diagdown}R\end{matrix}\right]^- Li^+ \xrightarrow{H^+} \left[(CO)_5Cr{\cdots}C\begin{matrix}{\cdots}OH\\ {\diagdown}R\end{matrix}\right]$$

(2) (3)

$$\textbf{(2)} \xrightarrow[\text{or } MeOSO_2F]{+Me_3O^+\bar{B}F_4} \left[(CO)_5Cr{\cdots}C\begin{matrix}{\cdots}OMe\\ {\diagdown}R\end{matrix}\right] \xleftarrow{+CH_2N_2} \textbf{(3)} \quad (1)$$

(4)

By substituting $PhN_2^+\bar{B}F_4$ for CH_2N_2, the corresponding phenoxycarbene complex[20] can be isolated in very low yield (equation 2).

$$[(CO)_6Cr] \xrightarrow[(2)\ PhN_2^+\bar{B}F_4]{(1)\ LiPh} \left[(CO)_5Cr{\cdots}C\begin{matrix}{\cdots}OPh\\ {\diagdown}Ph\end{matrix}\right] + \cdots \quad (2)$$

In 1967 Aumann and Fischer reported that better yields of **4** (>80%) can be obtained by direct alkylation of the acyllithium salt (**2**) with trimethyloxonium tetrafluoroborate[21]. Casey *et al.*[22] later introduced methyl fluorosulphonate, $MeOSO_2F$. In addition, **2** and other related metal acyl compounds also react with triethyloxonium tetrafluoroborate[23], acyl halides[24–26], trimethylchlorsilane[27,28], and dichlorodicyclopentadienyltitanium[29,30] to give the corresponding ethoxy-, acyloxy-, trialkylsiloxy-, and titanoxy-substituted carbene complexes, respectively, e.g.

$$[(CO)_6Cr] \xrightarrow[(2)\ Cp_2TiCl_2]{(1)\ LiMe} (CO)_5Cr{\cdots}C\begin{matrix}{\cdots}OTiCp_2Cl\\ {\diagdown}Me\end{matrix} + \cdots \quad (3)$$

Among the different nucleophiles used are e.g. the anions in alkyl-[1], aryl-[1,31], vinyl-[32,33], phenylacetylenyl-[34], furyl-[32], thienyl-[32], dialkylamido-[35] (equation 4), diphenylmethyleneamido-[36], triorganylsilyl-[37,38] (equation 5), and ferrocenyllithium[39], as well as 2-lithiodithiane[40], potassium ethoxide[41] and benzylmagnesium chloride[42]. Generally, Grignard compounds react much more slowly than organolithium reagents[42] and alkoxides give extremely low yields.

$$[(CO)_6Cr] \xrightarrow[(2)\ Et_3O^+\bar{B}F_4]{(1)\ LiNR_2} \left[(CO)_5Cr{\cdots}C\begin{matrix}{\cdots}OEt\\ {\cdots}NR_2\end{matrix}\right] \quad (4)$$

$$[(CO)_6W] \xrightarrow[(2)\ MeOSO_2F]{(1)\ LiSiR_3} \left[(CO)_5W{\cdots}C\begin{matrix}{\cdots}OMe\\ {\diagdown}SiR_3\end{matrix}\right] \quad (5)$$

A great number of binary or substituted metal carbonyls have been employed as precursors for carbene complexes, e.g. $[(CO)_6M]$ (M = Cr, Mo, W), $[(CO)_{10}M_2]$ (M = Mn, Tc, Re)[7,8], $[(CO)_5Fe]$[7,8], $[(CO)_4Ni]$[43], $[(CO)_3NOCo]$[43], $[(CO)_2(NO)_2Fe]$[43], $[(CO)_5WXPh_3]$ (X = P, As, Sb)[7,8], $[(CO)_5MnGeR_3]$[44], $[(CO)_2(PPh_3)_2Cl(N_2)Re]$[45], $[(\eta^5\text{-}C_5H_5)(CO)_3Mn]$[7], $[(\eta^6\text{-}C_6H_6)(CO)_3Cr]$[46] and $[(CO)_5Cr\{C(NMeCH_2)_2\}]$[47] (equation 6).

Monosubstituted metal hexacarbonyls tend to give the corresponding *cis*-substituted carbene complexes[48,49], which, on heating in solution, isomerize to yield a solution containing both *cis*- and *trans*-isomers (equation 7)[50]. Both isomers can be isolated[51]. The equilibrium ratio of *cis*- to *trans*-isomers varies depending on the steric

$$\left[(CO)_5Cr{=}C\langle N(Me)CH_2CH_2N(Me)\rangle\right] \xrightarrow[(2)\ MeOSO_2F]{(1)\ LiMe} cis\text{-}\left[(CO)_4Cr(=C(OMe)Me)(=C\langle N(Me)CH_2CH_2N(Me)\rangle)\right] \quad (6)$$

(5) (6)

$$[Ph_3P(CO)_5Cr] \xrightarrow[(2)\ Me_3O^+\ \bar{B}F_4]{(1)\ LiMe} cis\text{-}\left[Ph_3P(CO)_4Cr{=}C(OMe)Me\right] \rightleftharpoons trans\text{-}\left[Ph_3P(CO)_4Cr{=}C(OMe)Me\right] \quad (7)$$

requirements of the carbene and the phosphine ligands, on the central metal (Cr, W), and on the solvent used[50].

Treatment of trimethylgermyl(pentacarbonyl)manganese with methyllithium and subsequent reaction with dilute hydrochloric acid results in the evolution of methane and the formation of the cyclic carbene complex **7**, probably via the hydroxycarbene complex as an intermediate[44] (equation 8). Compound **7** is in equilibrium with its dimer (**8**)[52].

$$[Me_3Ge(CO)_5Mn] \xrightarrow{LiMe} cis\text{-}\left[Me_3Ge(CO)_4Mn{=}C(OLi)Me\right] \xrightarrow{HCl/H_2O} cis\text{-}\left[(CO)_4Mn{=}C(Me){-}O{-}GeMe_2\right] \ (\mathbf{7}) \rightleftharpoons \left[(CO)_4Mn(C(Me){=}O{-}GeMe_2)_2Mn(CO)_4\right] \ (\mathbf{8}) \quad (8)$$

$$[(CO)_nM] + [(Me_2N)_4Ti] \longrightarrow \left[(CO)_{n-1}M{=}C(NMe_2)OTi(NMe_2)_3\right] \quad (9)$$

M = Cr, W ($n = 6$); M = Fe ($n = 5$)

An interesting variation of the stepwise synthesis of carbene ligands outlined in equation 1 is the addition of a metal amide across the carbon–oxygen bond (equation 9)[53]. A similar reaction of tris(dimethylamino)aluminium with tetracarbonylnickel (equation 10)[54] or pentacarbonyliron[55] yields binuclear species such as **9**.

$$2\,[(CO)_4Ni] + 2\,Al(NMe_2)_3 \longrightarrow \left[(CO)_3Ni{=}C(NMe_2){-}O{-}Al(NMe_2)(\mu\text{-}NMe_2)_2Al(NMe_2){-}O{-}C(NMe_2){=}Ni(CO)_3\right] \quad (10)$$

(**9**)

Dimethylaminotin compounds were also found to add across the carbon—oxygen bond of pentacarbonyliron according to equation 9[56,57]. Zirconoxycarbene complexes can be prepared by addition of $[(C_5Me_5)_2ZrH_2]$ to carbonyl ligands of transition metal complexes[58], e.g.

$$[Cp_2MCO] + [H_2(C_5Me_5)_2Zr] \longrightarrow \left[Cp_2M{=}C\begin{matrix}H\\OZr(H)(C_5Me_5)_2\end{matrix}\right] \quad (11)$$

M = Cr, Mo, W

Because of the highly electrophilic character of the carbene carbon in Fischer-type complexes such as **4**, a nucleophilic anion in excess does not attack **4** at a carbon monoxide ligand but instead at the carbene carbon atom. Therefore, biscarbene complexes of type **4** could not, in general, be prepared by this route (the carbene atom in **5** is less electrophilic in nature and thus formation of **6** is possible). Another exception is the formation of the biscarbene complex **10** (yield 1.5%) in the very complex reaction of $[(CO)_6M]$ (M = Cr, W) with lithium dimethylphosphide and triethyloxonium tetrafluoroborate[59] (equation 12).

$$[(CO)_6M] \xrightarrow[(2)\ Et_3O^+\ \bar{B}F_4]{(1)\ LiPMe_2} cis\text{-}\left[(CO)_4M\left(=C\begin{matrix}OEt\\PMe_2\end{matrix}\right)_2\right] \quad (12)$$

(**10**)

M = Cr, W

B. Nucleophilic Attack on Isocyanide Complexes

Like metal carbonyls, isonitrile complexes can also react with nucleophiles. Alcohols add to complexed isonitrile ligands to give alkoxy(amino)carbene complexes[60,61] (equation 13). Mainly Pd(II) and Pt(II) compounds were employed.

$$cis\text{-}[Cl_2(PEt_3)PtCNR] + HOR' \longrightarrow cis\text{-}\left[Cl_2(PEt_3)Pt{=}C\begin{matrix}NHR\\OR'\end{matrix}\right] \quad (13)$$

R = Me, Ph; R′ = Me, Et, *i*-Pr

However, complexes from Au(I) and Ni(II) have also been obtained. Using isonitriles with a β- or γ-hydroxy group, cyclic carbene complexes can be obtained[62] (equation 14). Similarly, thiols or primary and secondary amines can function as nucleophiles to produce amino(thio)-[63] (equation 15) or bisaminocarbene complexes[64] (equation 16), respectively.

$$\mathrm{Pd^{2+} + 4\ CN(CH_2)_nOH \longrightarrow \left[Pd\left\{ \overset{\cdots}{=} C \begin{matrix} \cdot\cdot N(H) \\ \cdot\cdot O \end{matrix} (CH_2)_n \right\}_4\right]^{2+}} \quad (14)$$

$n = 2, 3$

$$trans\text{-}[\mathrm{(PR_3)_2Pt(CNEt)_2}] + \mathrm{HSCH_2Ph} \longrightarrow trans\text{-}\left[\mathrm{(PR_3)_2Pt(CNEt)}\left(\overset{\cdots}{=}\mathrm{C}\begin{matrix}\mathrm{NHEt} \\ \mathrm{SCH_2Ph}\end{matrix}\right)\right] \quad (15)$$

$$cis\text{-}[\mathrm{X_2LPdCNC_6H_4Y}(p)] + \mathrm{H_2NC_6H_4Z}(p) \longrightarrow cis\text{-}\left[\mathrm{X_2LPd}\overset{\cdots}{=}\mathrm{C}\begin{matrix}\mathrm{NHC_6H_4Y}(p) \\ \mathrm{NHC_6H_4Z}(p)\end{matrix}\right] \quad (16)$$

$\mathrm{X = Cl, Br;\ L = PPh_3, AsPh_3;\ Y = OMe, Me, H, NO_2;}$

$\mathrm{Z = OMe, Me, H, Cl, NO_2}$

A kinetic investigation of the latter reaction[64] showed that electron-withdrawing groups Y and electron-donating groups Z increase the reaction rate. Further, the chloro complexes react faster than the bromo complexes. These observations are consistent with an electrophilic attack of the amine at the isonitrile carbon atom in the rate-determining reaction step.

The first non-chelated biscarbene complex was prepared in 1967 from mercury acetate, methylisonitrile and secondary amines[65], presumably via the intermediate formation of a mercury–isonitrile compound (equation 17). Likewise, carbene

$$\mathrm{Hg(OAc)_2 \xrightarrow[+2\ HNR_2]{+2\ CNMe} \left[\begin{matrix}\mathrm{R_2N} \\ \mathrm{MeHN}\end{matrix}C\overset{\cdots}{=}Hg\overset{\cdots}{=}C\begin{matrix}\mathrm{NHMe} \\ \mathrm{NR_2}\end{matrix}\right](OAc)_2} \quad (17)$$

complexes of Au(I)[66,67], Rh(I)[68], Rh(III)[69], and Ni(II)[70,71] have been synthesized (equation 18). A remarkably stable tetracarbene complex was obtained by the reaction of $[(\mathrm{MeNC})_4\mathrm{M}]^{2+}$ (M = Pd, Pt) with methylamine[72] (equation 19). An X-ray crystallographic analysis of the platinum compound[73] confirmed that (a) the nitrogen substituents of the carbene ligands are in the amphi-configuration shown in equation

$$\left[\begin{matrix}\mathrm{(CF_2)_2} \\ | \\ \mathrm{(CF_2)_2}\end{matrix}\mathrm{Ni(CNR)_2}\right] \xrightarrow{+\mathrm{HNMe_2}} \left[\begin{matrix}\mathrm{(CF_2)_2} \\ | \\ \mathrm{(CF_2)_2}\end{matrix}\mathrm{Ni(CNR)}\left(\overset{\cdots}{=}\mathrm{C}\begin{matrix}\mathrm{NHR} \\ \mathrm{NMe_2}\end{matrix}\right)\right] \quad (18)$$

$$[\mathrm{(MeNC)_4M}]^{2+} \xrightarrow{\mathrm{H_2NMe}} \left[\mathrm{M}\left(\overset{\cdots}{=}\mathrm{C}\begin{matrix}\mathrm{N(Me)-H} \\ \mathrm{N(H)-Me}\end{matrix}\right)_4\right]^{2+} \quad (19)$$

M = Pd, Pt

19 and (b) the carbene planes form angles between 77° and 82° with the Pt-$(C_{Carb})_4$ plane. Such an arrangement is believed to protect the metal from any attack and may thus account for the overall stability of the complex.

Reaction of $[(MeNC)_6M]^{2+}$ (M = Fe, Ru, Os) with amines yields different products depending on the metal and on the amine (equation 20). Whereas the iron complex

$$[(MeNC)_6Fe]^{2+}\ (\mathbf{11}) \xrightarrow{+NH_3} [(MeNC)_5Fe{=}C(NH_2)(NHMe)]^{2+}\ (\mathbf{12})$$

$$[(MeNC)_6Fe]^{2+}\ (\mathbf{11}) \xrightarrow{+H_2NMe} [(MeNC)_4Fe\{C(=NMeH)\}_2NMe]^{2+}\ (\mathbf{13}) \qquad (20)$$

(**11**) adds methylamine to form the cyclic biscarbene compound **13**[74], with the ruthenium analogue of **11** and amines no complex such as **13** is obtained but, instead, the analogue of **12** and, on prolonged heating with excess of ethylamine, the open *cis*-biscarbene complex **16** can be isolated[74] (equation 21). On warming solutions of **15**, rearrangement takes place, probably via intramolecular cyclization to form **17**[75].

$$[(MeNC)_6Ru]^{2+}\ (\mathbf{14}) \xrightarrow{H_2NEt} [(MeNC)_5Ru{=}C(NHMe)(NHEt)]^{2+}\ (\mathbf{15})$$

$$(\mathbf{15}) \rightleftharpoons [(MeNC)_4(EtNC)Ru{=}C(NHMe)_2]^{2+}\ (\mathbf{17})$$

$$(\mathbf{15}) \longrightarrow cis\text{-}[(MeNC)_4Ru\{{=}C(NHMe)(NHEt)\}_2]^{2+}\ (\mathbf{16}) \qquad (21)$$

Reaction of the osmium analogue of **11** with methylamine finally gives mixtures of bis- (type **16**) and meridional-triscarbene complexes[76]. Compound **14** also reacts with hydrazine to form a cyclic carbene complex[74] (equation 22). This type of reaction was first performed in 1915 with tetrakis(methylisonitrile)platinum(II) and hydrazine[2,3] but the resulting product **19** was assigned the wrong structure (equation 23). Compound **19** can be protonated[77] and, additionally, the same reaction can be carried out with the palladium analogue of **18**[77,78] as well as with monosubstituted hydrazines and hydroxylamine[72]. The correct structures of these compounds, which are best described as 'resonance stabilized' carbene complexes, were finally established by

$$[(MeNC)_6Ru]^{2+} \xrightarrow{H_2NNH_2} \left[(MeNC)_4Ru\text{(C(NHMe)NH–NHC(NHMe))}\right]^{2+} \quad (22)$$

(**14**)

$$[(MeNC)_4Pt]^{2+} \xrightarrow{N_2H_4} \left[\text{MeHN}{=}\text{C}{-}\text{N}{-}\text{N(H)}{-}\text{C}{=}\text{NHMe}, \text{Pt(CNMe)}_2\right]^{+} \underset{^{-}OH}{\overset{HBF_4}{\rightleftharpoons}}$$

(**18**) (**19**)

$$\left[\text{MeHN}{=}\text{C}{-}\text{N(H)}{-}\text{N(H)}{-}\text{C}{=}\text{NHMe}, \text{Pt(CNMe)}_2\right]^{2+} \quad (23)$$

(**20**)

X-ray crystallographic analyses of **19**[79] and its Pd analogue[80]. Recently, an aminocarbene complex of molybdenum formed by reaction of pentaisonitrile-(nitrosyl)molybdenum(0) with amines was prepared[81] (equation 24). Earlier attempts with chromium(0) and molybdenum(0) carbonyls, however, failed[82].

$$[(NO)Mo(CNR)_5]^{+}\ I^{-} \xrightarrow{H_2NR'} \left[(NO)(RNC)_4Mo{=}C\begin{smallmatrix}NHR\\NHR'\end{smallmatrix}\right]^{+} I^{-} \quad (24)$$

C. Nucleophilic Attack on Carbyne Complexes

A relatively new method for the preparation of carbene complexes is the nucleophilic addition to the carbyne carbon atom of cationic carbyne complexes (see Chapter 5). Although the latter first have to be synthesized from carbene complexes, this route offers some advantages. Thus compounds become available which are inaccessible via any other synthetic route. Two types of cationic carbyne precursors have been used: (a) $[(Ar)(CO)_2M{\equiv}CR]^{+}$ with Ar = η^5-Cp or MeC_5H_4 (M = Mn, Re) and Ar = η^6-benzene (M = Cr) and (b) pentacarbonyl(dialkylaminocarbyne)-chromium. Dicarbonyl(cyclopentadienyl)phenylcarbynemanganese tetrafluoroborate (**21**) adds thiocyanate to form an isothiocyanato(phenyl)carbene complex (**22**) (equation 25)[83].

$$[Cp(CO)_2Mn{\equiv}CPh]^{+} + SCN^{-} \longrightarrow \left[Cp(CO)_2Mn{=}C\begin{smallmatrix}NCS\\Ph\end{smallmatrix}\right] \quad (25)$$

(**21**) (**22**)

Similar reactions can be carried out with cyanate[84] and cyanide[83] as nucleophiles and with the rhenium analogue of **21**. Likewise, the sodium salts of long-chain alcohols[85] and methyllithium[86,87] or cyclopentadienyllithium[84] (equation 26) were employed.

$$[(\eta^5\text{-MeC}_5\text{H}_4)(\text{CO})_2\text{Mn}{\equiv}\text{CPh}]^+ + \text{C}_5\text{H}_5^- \longrightarrow \left[(\eta^5\text{-MeC}_5\text{H}_4)(\text{CO})_2\text{Mn}{=}\text{C}\begin{matrix}\text{C}_5\text{H}_5\\ \text{Ph}\end{matrix}\right] \quad (26)$$

Addition of neutral nucleophiles such as trimethylphosphine[88] or isonitrile[89] yields cationic carbene complexes (equation 27 and 28). An X-ray crystallographic analysis[90] indicated a Re=C_{Carb} double bond and a C_{Carb}—P single bond supporting the formulation of **23** as a carbene-ylide. On thermolysis of **24** (R = *t*-Bu) the butyl group

$$[\text{Cp(CO)}_2\text{Re}{\equiv}\text{CPh}]^+ + \text{PMe}_3 \xrightarrow{-50°\text{C}} \left[\text{Cp(CO)}_2\text{Re}{=}\text{C}\begin{matrix}\text{Ph}\\ \text{PMe}_3\end{matrix}\right]^+ \quad (27)$$

(23)

$$[\text{Cp(CO)}_2\text{Mn}{\equiv}\text{CPh}]^+ + \text{CNR} \longrightarrow \left[\text{Cp(CO)}_2\text{Mn}{\equiv}\text{C}\begin{matrix}\text{CNR}\\ \text{Ph}\end{matrix}\right]^+ \quad \textbf{(24)}$$

$$\xrightarrow[\Delta]{\text{R} = t\text{-Bu}} \left[\text{Cp(CO)}_2\text{Mn}{\equiv}\text{C}\begin{matrix}\text{CN}\\ \text{Ph}\end{matrix}\right] \quad 48\% \quad (28)$$

is split off and the resulting neutral cyano(phenyl)carbene complex can be isolated[89]. The first dimethylcarbene complex was obtained by reaction of the dicarbonyl(cyclopentadienyl)methylcarbynemanganese cation with methyllithium[86] (equation 29).

$$[\text{Cp(CO)}_2\text{Mn}{\equiv}\text{CMe}]^+ + \text{LiMe} \longrightarrow \left[\text{Cp(CO)}_2\text{Mn}{=}\text{C}\begin{matrix}\text{Me}\\ \text{Me}\end{matrix}\right] + \text{Li}^+ \quad (29)$$

From substituted benzene(dicarbonyl)phenylcarbynechromium tetrafluoroborate and dimethylamine or ammonia a series of carbene complexes were synthesized[91] (equation 30).

$$[(\eta^6\text{-C}_6\text{H}_3\text{R}_3)(\text{CO})_2\text{Cr}{\equiv}\text{CPh}]^+ + \text{HNR}_2 \longrightarrow \left[(\eta^6\text{-C}_6\text{H}_3\text{R}_3)(\text{CO})_2\text{Cr}{\equiv}\text{C}\begin{matrix}\text{NR}_2\\ \text{Ph}\end{matrix}\right] + \text{H}^+ \quad (30)$$

Secondary carbene complexes are accessible via reduction of the metal–carbon triple bond of cationic carbyne complexes. This was shown with the rhenium analogue of **21**[92] (equation 31) and by reduction of dicarbonyl(methylcyclopentadienyl)diethyl-

$$[\text{Cp(CO)}_2\text{Re}{\equiv}\text{CPh}]^+ \xrightarrow{\text{Et}_2\text{AlH}} \left[\text{Cp(CO)}_2\text{Re}{=}\text{C}\begin{matrix}\text{Ph}\\ \text{H}\end{matrix}\right] \quad (31)$$

aminocarbynemanganese tetrafluoroborate with aluminium lithium hydride[93]. Pentacarbonyl(dialkylaminocarbyne)chromium tetrafluoroborate turned out to be excellent precursors for the preparation of interesting and previously inaccessible carbene complexes. By reaction of boron trifluoride with pentacarbonyl-[dialkylamino(ethoxy)carbene]chromium at −100°C **24** are synthesized[93,94] (equation 32). The cation **24** (R = Me) can also be produced via the reaction

$$\left[(CO)_5Cr\cdots C\begin{matrix}NR_2\\OEt\end{matrix}\right] \xrightarrow[-100\ ^\circ C]{BF_3} [(CO)_5Cr\equiv CNR_2]^+\ \bar{B}F_4 \qquad (32)$$

(24)

$NR_2 = NMe_2$, NEt_2, **(25)**, NC_5H_{10}

of pentacarbonyl[chloro(dimethylamino)carbene]chromium with silver salts[95]. The reaction in equation 32 using boron trichloride instead of boron trifluoride yields **24** with tetrachloroborate as counter ion. Whereas for R = Et the tetrafluoroborate salt is relatively stable and can be handled at room temperature in the crystalline state for short periods of time, the analogous tetrachloroborate salt forms pentacarbonyl[chloro(diethylamino)carbene]chromium (**26**)[93] spontaneously above −25°C (equation 33). The complex **26** can also be obtained by reaction of the tetrafluoroborate salt of **25** with chloride[93].

$$[(CO)_5Cr\equiv CNEt_2]^+\bar{B}Cl_4 \xrightarrow{-25\ ^\circ C} \left[(CO)_5Cr\cdots C\begin{matrix}NEt_2\\Cl\end{matrix}\right] + BCl_3 \qquad (33)$$

(25) **(26)**

Hartshorn and Lappert[95] found that the dimethylanalogue of **25** easily adds dimethylamide or cyanide to form pentacarbonyl[bis(dimethylamino)carbene]- and -[cyano(dimethylamino)carbene]chromium, respectively. They further noticed that in the dimethyl analogue of **26** synthesized by a different route (see Section II.G) the chlorine substituent can be replaced with cyanide[95] (equation 34). In addition to

$$\left[(CO)_5Cr\cdots C\begin{matrix}NMe_2\\Cl\end{matrix}\right] + K^+CN^- \longrightarrow \left[(CO)_5Cr\cdots C\begin{matrix}NMe_2\\CN\end{matrix}\right] + K^+Cl^- \qquad (34)$$

chloride, several other nucleophiles were also employed in the reaction with the cation **25**. Thus **25** adds, e.g. fluoride[96], bromide[97], iodide[97], thiocyanate[97], and cyanate[97] (equation 35). In **30** and **31** NCO and NCS are bonded to the carbene carbon atom via

$$[(CO)_5Cr\equiv CNEt_2]^+ + X^- \longrightarrow \left[(CO)_5Cr\cdots C\begin{matrix}NEt_2\\X\end{matrix}\right] \qquad (35)$$

X = F(**27**), Cl(**26**), Br(**28**), I(**29**), NCO(**30**), NCS(**31**)

nitrogen. The first carbene complex with a main group metal bonded to the carbene carbon atom was also prepared by this route[98] (equation 36). The compounds **26**, **28**, **29**, and **32** show an unusual feature. In solution the carbene complexes spontaneously

$$[(CO)_5Cr\equiv CNEt_2]^+ + KSnPh_3 \longrightarrow \left[(CO)_5Cr\cdots C\begin{matrix}NEt_2\\SnPh_3\end{matrix}\right] + K^+ \qquad (36)$$

(32)

$$\left[(CO)_5Cr\cdots C\begin{smallmatrix}\cdots NEt_2\\ \diagdown X\end{smallmatrix}\right] \longrightarrow \textit{trans}\text{-}[X(CO)_4Cr\equiv CNEt_2] + CO \quad \underset{\textbf{(33)}}{} \qquad (37)$$

rearrange with loss of one carbon monoxide, forming *trans*-tetracarbonyl(diethylaminocarbyne)halogeno(or triphenylstannyl)chromium[97,99,100] (equation 37). The rate of the reaction strongly depends on the migrating group X[99]. Whereas for X = I the reaction takes place above −60°C, for X = Br a temperature of −35°C and for X = Cl and $SnPh_3$ + 35°C is required. Finally the fluorine compound **27** is stable up to 100°C, above which decomposition occurs[99,101]. However, for X = CN, NCS[100], NCO, or $SiPh_3$[103] no rearrangement could be observed. The compounds **28**, **29**, and **32** rearrange even in the crystalline state, forming **33**; **26** gives pentacarbonyl-(ethylisonitrile)chromium in 28% yield. Kinetic investigations[99,100,102] have shown (a) that the migration of X from C to Cr follows a first-order rate law and (b) neither the presence of free carbon monoxide nor of an excess of X^- nor radical initiators or radical inhibitors have any significant influence on the rate of the reaction. Further, the presence of PPh_3 does not influence the reaction rate but does lead to a different product[103] (equation 38). Compound **34** is also accessible via reaction of **33**

$$\left[(CO)_5Cr\cdots C\begin{smallmatrix}\cdots NEt_2\\ \diagdown Cl\end{smallmatrix}\right] + PPh_3 \longrightarrow \underset{\textbf{(34)}}{\textit{mer}\text{-}[Cl(PPh_3)(CO)_3Cr\equiv CNEt_2]} + 2\ CO \qquad (38)$$

(X = Cl) with PPh_3[104]. These results indicate an intramolecular mechanism, although chlorine in **26** may be exchanged with fluorine and, with bromide in large excess, also with bromine to form **27** and **28**, respectively[103]. The attempt to synthesize the *trans*-PPh_3-substituted analogues of the complexes **26–29** via addition of halide to *trans*-tetracarbonyl(diethylaminocarbyne)triphenylphosphinechromium tetrafluoroborate failed. Surprisingly, **34** and its analogues (with PPh_3 now *cis* to the carbyne ligand) were isolated, but no carbene complex[104] (equation 39).

$$\textit{trans}\text{-}[(PPh_3)(CO)_4Cr\equiv CNEt_2]^+ + X^- \;\not\longrightarrow\; \textit{trans}\text{-}\left[(PPh_3)(CO)_4Cr\cdots C\begin{smallmatrix}\cdots NEt_2\\ \diagdown X\end{smallmatrix}\right]$$

$$\textit{trans}\text{-}[(PPh_3)(CO)_4Cr\equiv CNEt_2]^+ + X^- \longrightarrow \textit{mer}\text{-}[X(PPh_3)(CO)_3Cr\equiv CNEt_2] \qquad (39)$$

D. Alkylation of Acyl Complexes

The second step in the classical Fischer two-step synthesis of carbene complexes involves alkylation of the anionic acylmetallate. The same type of reaction can be performed with stable neutral acyl complexes, producing cationic carbene complexes. This was demonstrated as early as 1967 by Schöllkopf and Gerhart with alkylation of dicarbamoyl mercury(II)[105] (equation 40). Subsequently, several other carbene complexes of iron[106,107], ruthenium[106], and molybdenum[106] were obtained by this

$$[(R_2NCO)_2Hg] \xrightarrow{+2Me_3O^+\bar{B}F_4} \left[Hg\left(\cdots C\begin{smallmatrix}\cdots OMe\\ \cdots NR_2\end{smallmatrix}\right)_2\right](BF_4)_2 + 2\ Me_2O \qquad (40)$$

method (equation 41). Compound **35** can also be protonated at the acyl oxygen to form the hydroxycarbene complex **36** which, on reaction with diazomethane, yields **37**.

$$[Cp(CO)PPh_3FeCOMe] \underset{^-OH}{\overset{H^+}{\rightleftharpoons}} \left[Cp(CO)PPh_3Fe \cdots C \begin{matrix} OH \\ Me \end{matrix}\right]^+$$

(35) (36)

$$\xrightarrow{CH_2N_2} \quad (36) \rightarrow (37); \quad (35) \underset{LiMe}{\overset{Me_3O^+ \bar{B}F_4}{\rightleftharpoons}} (37) \qquad (41)$$

$$\left[Cp(CO)PPh_3Fe \cdots C \begin{matrix} OMe \\ Me \end{matrix}\right]^+$$

(37)

Treatment of **37** with methyllithium[106] or sodium iodide[108] reverses the alkylation and **35** is regenerated. Similarly, **36** is easily deprotonated. Water converts **36** back to **35**. Similar procedures were applied to prepare aryl(methoxy)carbene and bis(methoxy)carbene complexes of nickel(II)[71,109] (equation 42).

$$trans\text{-}\left[C_6Cl_5(PMe_2Ph)Ni—C \begin{matrix} O \\ R \end{matrix}\right] \xrightarrow{MeOSO_2F} trans\text{-}\left[C_6Cl_5(PMe_2Ph)Ni \cdots C \begin{matrix} OMe \\ R \end{matrix}\right]^+ \qquad (42)$$

R = OMe, *p*-Tol

A special variation of *O*-alkylation of acyl complexes is intramolecular alkylation resulting in cyclic carbene complexes. This has been performed several times, mainly with manganese compounds. The cyclization can be initiated by silver salts as in equation 43[110]. Whether the intramolecular alkylation takes place or not depends to a

$$\left[(CO)_5Mn—C \begin{matrix} O \\ O(CH_2)_2Cl \end{matrix}\right] \xrightarrow{Ag^+} \left[(CO)_5Mn \cdots C \begin{matrix} O \\ O \end{matrix} \right]^+ \qquad (43)$$

(38)

great extent on the nucleophilicity of the acyl oxygen. Thus, the complex $(CO)_5MnC(O)(CH_2)_3Cl$, related to **38**, does not react with Ag^+ according to equation 43[110]; the *cis*-phosphite- and *cis*-phosphine-substituted compounds, however, do react[111] (equation 44). Intramolecular cyclization can also be initiated by strong

$$cis\text{-}\left[L(CO)_4Mn—C \begin{matrix} O \\ (CH_2)_3Cl \end{matrix}\right] + Ag^+ \underset{L = P(OMe)_3}{\overset{L = CO}{\longrightarrow}} cis\text{-}\left[L(CO)_4Mn—C \begin{matrix} O \\ \end{matrix}\right]^+ + AgCl \qquad (44)$$

nucleophiles X (I^- [111], $[(CO)_5Mn]^-$ [111]) (equation 45). Obviously, exchange of one carbon monoxide ligand with X^- increases the nucleophilic character of the acyl oxygen so much that nucleophilic attack of the oxygen at the ω-carbon atom becomes possible. In **39** the α-hydrogen atoms are slightly acidic and thus base-catalysed H/D exchange is observed[111]. This type of carbene complex synthesis can also be carried out with pentacarbonyl(ω-halogenalkyl)manganese as precursors by making use of the

$$\left[(CO)_5Mn{-}C\overset{\displaystyle O}{\underset{\displaystyle (CH_2)_3Cl}{\lessgtr}}\right] + X^- \xrightarrow[-CO]{} cis\text{-}\left[X(CO)_4Mn{-}C\overset{O}{\frown}\right] \quad (45)$$

(**39**)

$X = I^-, [(CO)_5Mn]^-$

well known fact that nucleophiles react with pentacarbonyl(alkyl)manganese to form acyl compounds which may function as intermediates for carbene complex synthesis[112,113] (equation 46). By this method non-cyclic carbene complexes can also be prepared (equation 47).

$$[(CO)_5Mn]^- + Br(CH_2)_3Br \xrightarrow[-Br^-]{} [(CO)_5MnCH_2CH_2CH_2Br] \xrightarrow{[(CO)_5Mn]^-} \left[\begin{matrix}(CO)_4Mn{-}C({=}O)(CH_2)_3Br \\ | \\ (CO)_5Mn\end{matrix}\right] \xrightarrow[-Br^-]{} \left[\begin{matrix}(CO)_4Mn{\cdots}C(O\frown) \\ | \\ (CO)_5Mn\end{matrix}\right] \quad (46)$$

$$[(CO)_5MnMe] + X^- \longrightarrow$$

(**40**)

$$cis\text{-}\left[X(CO)_4Mn{-}C\overset{\displaystyle O}{\underset{\displaystyle Me}{\lessgtr}}\right]^- \xrightarrow{Me_3O^+\bar{B}F_4} cis\text{-}\left[X(CO)_4Mn{\cdots}C\overset{\displaystyle OMe}{\underset{\displaystyle Me}{\lessgtr}}\right] \quad (47)$$

(**41**) (**42**)

$X = [(CO)_5Mn]^-$ [114], $GeCl_3$ [115]

The corresponding reaction of **40** with X = $[(CO)_5Re]^-$, however, did not produce the manganese carbene complex (**42**) but rather the rearranged rhenium complex **43**[116] (equation 48). Reaction of **40** with X = Br or X = I yields a product analogous to **41**, which can subsequently be protonated with phosphoric acid but not alkylated with oxonium salts[117].

$$[(CO)_5MnMe] + [(CO)_5Re]^- \longrightarrow cis\text{-}\left[\begin{matrix}(CO)_4Re{\cdots}C\overset{\displaystyle OMe}{\underset{\displaystyle Me}{\lessgtr}} \\ | \\ (CO)_5Mn\end{matrix}\right] \quad (48)$$

(**43**)

In addition to manganese, molybdenum[118,119], iron[111], and ruthenium[111] compounds were also employed in the synthesis of carbene complexes via intramolecular cyclization. Closely related to *O*-alkylation of acyl complexes is *S*-alkylation of thioacyl compounds[120,121], e.g.

$$trans\text{-}\left[Cl(PPh_3)_2Pt{-}C\overset{\displaystyle S}{\underset{\displaystyle Y}{\lessgtr}}\right] \xrightarrow{MeOSO_2F} trans\text{-}\left[Cl(PPh_3)_2Pt{\cdots}C\overset{\displaystyle SMe}{\underset{\displaystyle Y}{\lessgtr}}\right]^+ \quad (49)$$

Y = OMe, SEt, NMe_2

E. Electrophilic Addition to Coordinated Imidoyls

The nitrogen atom of imidoyl ligands may also be attacked electrophilically. It may be protonated reversibly as in equation 50[122]:

$$trans\text{-}\left[CO(PPh_3)_2(OAc)Ru{-}C{\overset{\textstyle =NR}{\underset{\textstyle H}{}}}\right] \underset{^-OH}{\overset{H^+}{\rightleftharpoons}} trans\text{-}\left[CO(PPh_3)_2(OAc)Ru{\cdots}C{\overset{\textstyle \cdots NHR}{\underset{\textstyle H}{}}}\right]^+ \qquad (50)$$

or it may be alkylated with methyl iodide[123] (equation 51) or methyl sulphate. Several metal complex systems have been employed as precursors, e.g. systems of nickel[71,124].

$$\left[(CO)(PPh_3)_2(CNR)(OAc)Ru{-}C{\overset{\textstyle =NR}{\underset{\textstyle H}{}}}\right] \xrightarrow{MeI} \left[CO(PPh_3)_2(CNR)(OAc)\,Ru{\cdots}C{\overset{\textstyle \cdots NRMe}{\underset{\textstyle H}{}}}\right]^+ \qquad (51)$$

F. From Acetylene or Acetylide Complexes

Chisholm and Clark reported[125] that platinum(II) complexes react with monosubstituted acetylenes and alcohol in the presence of silver salts to give alkoxycarbene complexes (equation 52). The originally proposed mechanism –

$$trans\text{-}[Me(PMe_2Ph)_2PtCl] \xrightarrow[AgPF_6]{HC\equiv CR/MeOH} trans\text{-}\left[Me(PMe_2Ph)_2Pt{\cdots}C{\overset{\textstyle \cdots OMe}{\underset{\textstyle CH_2R}{}}}\right]^+ \qquad (52)$$

$$trans\text{-}[Me(PMe_2Ph)_2PtCl] \xrightarrow[AgPF_6]{HC\equiv C(CH_2)_2OH} trans\text{-}\left[Me(PMe_2Ph)_2Pt{\cdots}C\langle\text{cyclic }O(CH_2)_3\rangle\right]^+$$

exchange of chloride with acetylene, subsequent nucleophilic addition of alcohol to the coordinated acetylene to form a vinyl ether complex, which then rearranges via a hydride shift to give the carbene complex – could not be confirmed because vinyl ether complexes do not react under these reaction conditions. Cationic vinylidene complexes were therefore postulated as reaction intermediates[126]. Several carbene complexes of iridium(II)[127] and platinum(II)[126,128,129] were obtained by a similar procedure. By reaction of acetylide complexes with alcohol and acid, carbene complexes of nickel(II)[130] and platinum(II)[131,132] were synthesized (equation 53).

$$trans\text{-}[(C_6Cl_5)(PMe_2Ph)_2Ni(C\equiv CH)] \xrightarrow[ROH]{HClO_4} trans\text{-}\left[(C_6Cl_5)(PMe_2Ph)_2Ni{\cdots}C{\overset{\textstyle \cdots OR}{\underset{\textstyle Me}{}}}\right]ClO_4 \qquad (53)$$

G. From Salt-like Precursors and Oxidative Addition

The reaction of anionic metal complexes with some organic salts or neutral compounds with highly ionic bonds was used for the preparation of carbene complexes. In 1968 Öfele[133] reported the formation of pentacarbonyl(2,3-diphenylcyclopropenylidene)chromium (**45**) from dichloro-2,3-diphenylcyclopropene (**44**) and sodium pentacarbonylchromate via oxidative addition (equation 54). The

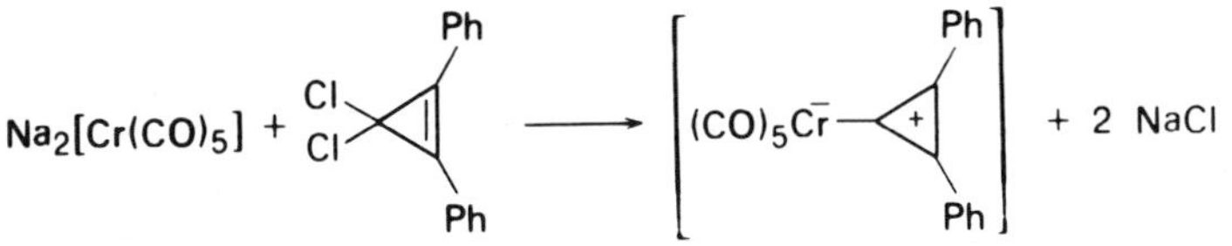

$$Na_2[Cr(CO)_5] + \text{(44)} \longrightarrow \text{(45)} + 2\ NaCl \qquad (54)$$

(**44**) (**45**)

reaction of **44** with metallic palladium yields the dimeric species **46**, which on heating in pyridine forms **47**[134] (equation 55). These carbene compounds (**45**,

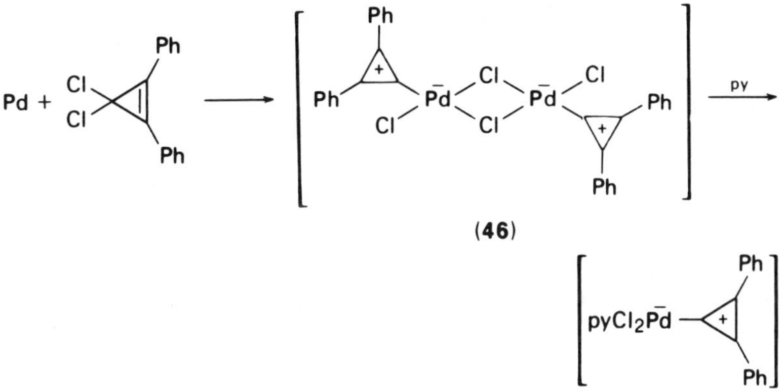

(**46**)

(55)

(**47**)

46, and **47**) are thermally very stable, probably owing to the aromaticity of the 'diphenylcyclopropenium cation'. A marked stability is also characteristic of complexes such as **48**, which can be prepared from imidazolium salts and carbonylhydridometal anions with elimination of molecular hydrogen[135,136] (equation 56). On heating, **48** disproportionates to yield a mixture of the hexacarbonyl and *cis*-tetracarbonyl(biscarbene) complexes[137]. Irradiation converts the latter into the corresponding *trans*-isomers, which isomerize thermally back to the *cis*-isomer[138].

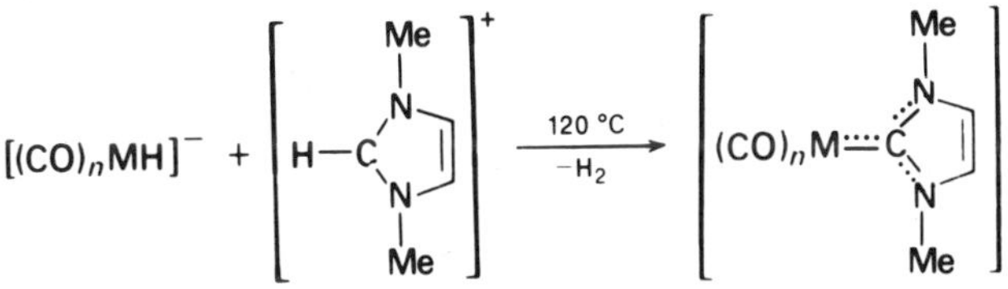

$$[(CO)_nMH]^- + [\text{imidazolium}]^+ \xrightarrow[-H_2]{120\ °C} [(CO)_nM\text{-carbene}] \qquad (56)$$

(**48**) $n = 5$; M = Cr, Mo, W

(**49**) $n = 4$; M = Fe

In addition to **44**, other imidazolium salts[139–141] and pyrazolium, triazolium, tetrazolium[138], and thiazolium salts[142,143] have been employed in the preparation of cyclic carbene complexes via oxidative addition. Among the different metal complexes and metal salts used as precursors are compounds of mercury[144], nickel[145], palladium[145], platinum[145], iron[143], rhodium[142], iridium[142], ruthenium[143], osmium[143], and manganese[142].

Related to **48** are carbene complexes of ruthenium obtained by acid-catalysed rearrangement of the *N*-bonded imidazole compound[146] (equation 57). Similar

$$\left[(NH_3)_5Ru{-}N\text{(imidazole)}N{-}Me\right]^{2+} \xrightarrow{H^+} \left[H_2O(NH_3)_4Ru{\cdots}C\text{(N(H), N(Me) ring)}\right]^{2+} \quad (57)$$

complexes from xanthine have also been synthesized[147]. Another synthetic route developed in recent years is oxidative addition of immonium salts, derived from dimethyl(methylene)ammonium halides (**50**), to metal complexes. Thus, by reaction of **50**[148,149], **51**[148] or **52**[149,150] with either neutral metal complexes (equation 58) or

$$\left[Me_2N{=}CR^1R^2\right]^+ X^-$$

(**50**): $R^1 = R^2 = H$; $X = I, Cl$

(**51**): $R^1 = H$; $R^2 = SMe$; $X = Br$

(**52**): $R^1 = H$; $R^2 = Cl$; $X = Cl$

(**53**): $R^1 = Cl$; $R^2 = NMe_2$; $X = Cl$

(**54**): $R^1 = R^2 = Cl$; $X = Cl$

$$trans\text{-}[Cl(PPh_3)_2IrN_2] + [Me_2N{=}CHCl]^+Cl^- \xrightarrow{-N_2} \left[Cl_3(PPh_3)_2Ir{\cdots}C(NMe_2)H\right] \quad (58)$$

$$[Cp(CO)_3V]^{2-} + [Me_2N{=}CHCl]^+Cl^- \longrightarrow \left[Cp(CO)_3V{\cdots}C(NMe_2)H\right] + 2\,Cl^- \quad (59)$$

carbonyl metallates (equation 59) a series of secondary carbene complexes of Cr, Fe[148,149], Mo, W, Re, Co, Ru, Ir, Pt[149], Rh[151], and even V[149] has been prepared. Although the yields are sometimes relatively low, this method is of special importance because of its wide applicability.

Whereas compounds **50–52** function as precursors for the formation of the secondary carbene ligand, the reactions of **54** with e.g. pentacarbonyl chromate and of **53** with diironnonacarbonyl result in the formation of chloro(dimethylamino)-[152] and bis(dimethylamino)carbene complexes[153], respectively (equations 60 and 61). The related reaction of imidoyl chloride (**55**) and hydrochloric acid with binuclear

$$[(CO)_5Cr]^{2-} + [Me_2N{=}CCl_2]^+Cl^- \longrightarrow \left[(CO)_5Cr{\cdots}C(NMe_2)Cl\right] + \cdots \quad (60)$$

$$3[Fe_2(CO)_9] + [Me_2N{=}C(Cl)NMe_2]^+Cl^- \longrightarrow \left[(CO)_4Fe{\cdots}C(NMe_2)NMe_2\right] + \cdots \quad (61)$$

$$[PPh_3(CO)ClRh]_2 + [MeN{=}C(Ph)Cl] + HCl \longrightarrow \left[Cl_3(CO)PPh_3Rh{-}C\begin{matrix}NHMe\\Ph\end{matrix}\right] + \cdots \quad (62)$$

(55) (56)

$[PPh_3(CO)ClRh]_2$ affords the rhodium(III) carbene complex **56**[154] (equation 62). In the complete absence of hydrogen chloride, however, $[(CO)_2ClRh]_2$ and **55** furnished a rhodium(III) carbene chelate complex (**57**), the chelate bridge of which can be cleaved with tertiary phosphines[155] (equation 63).

$$1/2[(CO)_2ClRh]_2 + 2\ MeN{=}C(Ph)Cl \longrightarrow \left[Cl_3(CO)Rh\begin{matrix}C(Ph){\cdots}NMe\\|\\N(Me){=}CPh\end{matrix}\right]$$

(57)

$$\xrightarrow[-CO]{+PMe_2Ph} \left[Cl_3(PMe_2Ph)_2Rh{\cdots}C(Ph){\cdots}N(Me){-}C(Ph){=}NMe\right] \quad (63)$$

H. From Electron-rich Olefins

Closely related in stability to the compounds **48** and **49** (see Section II.G) are complexes obtained from electron-rich olefins such as **59** and metal complexes via a bridge-cleaving reaction of binuclear (equation 64) and/or a ligand substitution reaction of mononuclear metal complexes (equation 65). Thus, reaction of the chlorine-bridged platinum(II) dimer **58** with **59** yields the *trans*-dichloro(triethylphosphine)platinum carbene complex **60**, which can be isomerized to the thermodynamically more stable *cis*-compound[156].

Similar rhodium–carbene complexes are isolable intermediates in the metathesis of two different electron-rich olefins[157]. Ligand-exchange reactions turned out to be

$$[PEt_3Cl_2Pt]_2 + \left(\begin{matrix}Ph & & Ph\\ N & & N\\ & C{=}C & \\ N & & N\\ Ph & & Ph\end{matrix}\right) \longrightarrow trans\text{-}\left[PEt_3Cl_2Pt{\cdots}C\begin{matrix}N(Ph)\\N(Ph)\end{matrix}\right] \quad (64)$$

(58) (59) (60)

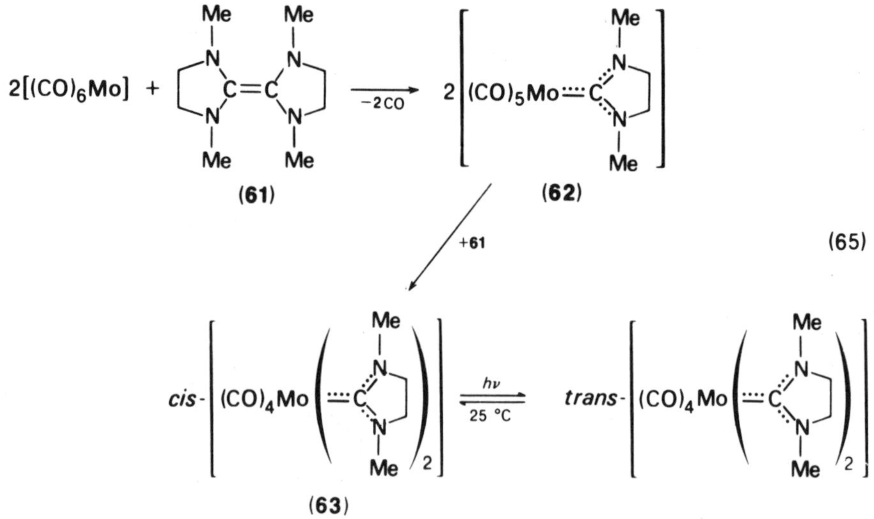

$$2[(CO)_6Mo] + \mathbf{61} \xrightarrow{-2CO} 2[(CO)_5Mo{=}C(NMeCH_2)_2]\ (\mathbf{62}) \xrightarrow{+\mathbf{61}} cis\text{-}[(CO)_4Mo(C(NMeCH_2)_2)_2]\ (\mathbf{63}) \underset{25\,^{\circ}C}{\overset{h\nu}{\rightleftharpoons}} trans\text{-}[(CO)_4Mo(C(NMeCH_2)_2)_2] \qquad (65)$$

very convenient routes for the synthesis of such cyclic carbene complexes. Hexacarbonylmolybdenum reacts with **61** (the *N*-methyl analogue of **59**) to form either pentacarbonyl(monocarbene)- or *cis*-tetracarbonyl(biscarbene)molybdenum, depending on the reaction conditions[158]. The chemical behaviour of **48** (see Section II.G) is paralleled in **62**. Like **48**, **62** disproportionates above 100°C to hexacarbonylmolybdenum and **63**, which upon irradiation isomerizes reversibly to its *trans*-isomer[158]. A great number of different mono-, bis-, tris-, and even tetrakis-carbene complexes were synthesized from various metal complexes, e.g. from Cr[47], Mo[158], W[159], Mn, Fe, Ru, Ni, Co[160], Rh, Ir[161], Os[162], and Au[163]. In addition to carbon monoxide, other groups e.g. phosphines, isonitriles, norbornadiene, and halides, were replaced by cyclic carbene ligands resulting from electron-rich olefins. Further, other olefins such as **64**, **65** and the *N*-Et-, *N*-*p*-Tol- or *N*-CH_2Ph analogue of **59** were

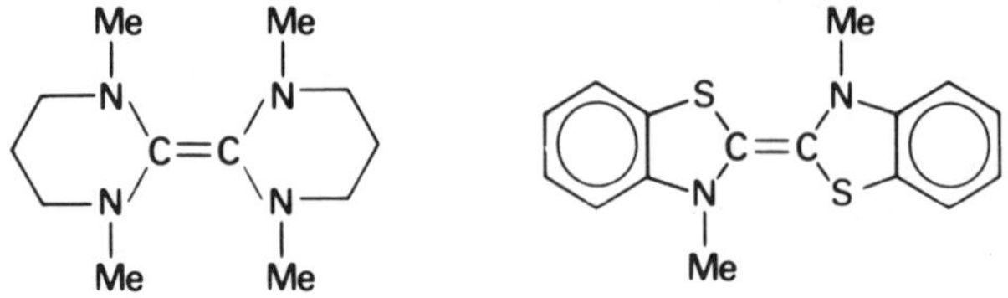

(**64**) (**65**)

employed. The olefins, however, show significant differences in reactivity, which decreases in the order **61** > **65** > **64**. This is also illustrated in the reaction of hexacarbonylmolybdenum with **59**, **61**, and **64**. Thus, whereas $(CO)_6Mo$ and **61** yield mono- and biscarbene complexes (**62** and **63**), only a monocarbene complex is obtained from $(CO)_6Mo$ and **64**. Under similar conditions, however, **59** does not react any more. From tungsten a triscarbene complex was prepared[159] and by reaction of dichlorobis(triphenylphosphine)ruthenium(II) with **61** the first neutral tetracarbene complex was synthesized[164] (equation 66). On the other hand, **66** reacts with **59** with elimination of PPh_3 and HCl, forming a five-coordinated carbene complex containing an *ortho*-metallated *N*-aryl carbene ligand[165] (equation 67). From **61** and

$$[Cl_2(PPh_3)_3Ru] + \mathbf{61} \longrightarrow trans\text{-}\left[Cl_2Ru\left(\overset{\cdot\cdot\cdot}{=}C\langle N(Me)CH_2CH_2N(Me)\rangle\right)_4\right] + \cdots \quad (66)$$

(**66**)

$$[Cl_2(PPh_3)_3Ru] + \mathbf{59} \xrightarrow[-PPh_3,\,-HCl]{} \left[Cl(PPh_3)_2Ru\overset{\cdot\cdot\cdot}{=}C\langle N(C_6H_4)CH_2CH_2N(Ph)\rangle\right] \quad (67)$$

$$[(CO)_4Mo(nbd)] + \mathbf{61} \longrightarrow cis\text{-}\left[(CO)_4Mo\ldots\right] \xrightarrow{140\ °C} \mathbf{63} \quad (68)$$

(**67**)

tetracarbonyl(norbornadiene)molybdenum the *NN'*-bonded isomer of **63**, **67**, can be isolated[166], which can be transformed thermally into **63** (equation 68).

The conditions for the rearrangement of **67** into **63** are, however, more vigorous than those for the direct synthesis of **63**. Compound **67** can therefore be excluded as an intermediate in the preparation of **63** according to equation 65. Thus, a mechanistic scheme was proposed which involves replacement of one complex ligand by the electron-rich olefin. The olefin, initially *N*-bonded, rearranges in a fast second step to a *C*-bonded species, which may then fragment to form the carbene–metal complex with expulsion of a resonance-stabilized carbene fragment. The latter may then react with another metal centre[47].

A comparison between this type of cyclic carbene complex and Fischer-type complexes (e.g. **4**) shows some significant differences. The carbene ligand in **4** has both a nucleophilic and an electrophilic centre and may thus react with acids, bases, electrophiles, and nucleophiles to give new compounds. The heterocyclic carbene ligand (e.g. in **62**), on the other hand, does not show the same high reactivity. On the contrary, reactions of these complexes are almost exclusively restricted to the metal–ligand framework. Thus, whereas mixed biscarbene complexes can be prepared from the chromium analogue of **62**, organyllithium and $MeOSO_2F$ (see equation 6), the reverse procedure [reaction of **4** (R = Me) with **61**] causes displacement of the methyl(methoxy)carbene ligand[47].

I. From Diazoalkane Precursors

Other reactive molecules were also used as precursors for the synthesis of carbene complexes. Diazo compounds react with the dicarbonyl(methylcyclopentadienyl) manganese–tetrahydrofuran complex to give carbene complexes[167–169] (equation 69). Upon irradiation, **69** rearranges intramolecularly to form the corresponding π-diphenylketene complex, which may also be obtained from **70** via high-pressure

$$[(MeCp)(CO)_2Mn(thf)] + N_2{=}C\begin{matrix}R^1\\R^2\end{matrix} \longrightarrow \left[(MeCp)(CO)_2Mn{=}C\begin{matrix}R^1\\R^2\end{matrix}\right] \quad (69)$$

(**68**)

(**69**): $R^1 = Ph$, $R^2 = C(O)Ph$

(**70**): $R^1 = R^2 = Ph$

(**71**): $R^1 = p\text{-}C_6H_4NO_2$, $R^2 = COOEt$

carbonylation[170]. Reaction of **68** or its cyclopentadienyl analogue with diazocyclopentadiene yields an oligomeric biscarbene complex[171] with high thermal stability (equation 70). A bis(alkylthio)carbene complex was synthesized from pentacarbonyl(methylnitrile)chromium or -tungsten and the sodium salt of the tosylhydrazone of a dithiocarbonate as carbenoid precursor[172] (equation 71).

$$[Cp(CO)_2Mn(thf)] + N_2{=}C_5H_4 \xrightarrow{-N_2} \frac{1}{2}\left[Cp(CO)_2Mn{=}C_{10}H_8{=}Mn(CO)_2Cp\right] \quad (70)$$

$$[(CO)_5M(NCMe)] + Na[TosNNC(SEt)_2] \longrightarrow \left[(CO)_5M{\cdots}C\begin{matrix}SEt\\SEt\end{matrix}\right] + \cdots \quad (71)$$

J. Alkylidene Complexes

The first example of a secondary alkylidene complex (**77**) was obtained by Schrock in 1974[173] by reaction of $[(Me_3CCH_2)_3TaCl_2]$ (**72**) with Me_3CCH_2Li (**73**) (equation 72). The rate-determining step is believed to be the formation of thermally unstable **74**, which can also be prepared from **77** and HCl at -78°C. **74** reacts very rapidly with **73** compared with the rate at which **72** reacts. It was proposed that **77** is formed from **74**

$$\underset{(\mathbf{72})}{[Cl_2(Me_3CCH_2)_3Ta]} \xrightarrow[-LiCl]{Me_3CCH_2Li\ (\mathbf{73})} \underset{(\mathbf{74})}{[(Me_3CCH_2)_4ClTa]}$$

$$(\mathbf{74}) \xrightarrow{-CMe_4} \underset{(\mathbf{75})}{[(Me_3CCH_2)_2ClTa{=}CHCMe_3]} \qquad (\mathbf{74}) \xrightarrow{+\mathbf{73}} \underset{(\mathbf{76})}{[(Me_3CCH_2)_5Ta]} \quad (72)$$

$$(\mathbf{75}) \xrightarrow{+\mathbf{73}} \underset{(\mathbf{77})}{\left[(Me_3CCH_2)_3Ta{=}C\begin{matrix}H\\CMe_3\end{matrix}\right]} \xleftarrow{-CMe_4} (\mathbf{76})$$

by two different rapid paths: (a) via **75** and (b) directly from **74** by formal dehydrohalogenation possibly via short-lived **76**[174]. In the fast hydrogen abstraction step a relatively more nucleophilic axial alkyl α-carbon atom of the trigonal bipyramidal complex removes a relatively more acidic proton from an equatorial alkyl α-carbon atom. Steric crowding about the metal–alkyl precursor is believed to be an important factor in determining when α-hydrogen abstraction occurs. The analogous niobium compound has also been prepared[175] but, in general, niobium–alkylidene complexes appear to be less stable than their tantalum analogues.

Hydrogen abstraction may also be induced by substitution of a halide ligand with a cyclopentadienyl ligand in neopentyl and benzyl complexes. The neopentyl compound **78** reacts with CpTl to give the monocyclopentadienyl complex **79** which can be isolated for X = Cl[176]. The compound **79** decomposes smoothly to form neopentane and the neopentylidene complex **80**[176]. For X = Br the intermediate **79** could not be

$$[X_3(Me_3CCH_2)_2Ta] \xrightarrow{CpTl} [X_2Cp(Me_3CCH_2)_2Ta] \xrightarrow{-CMe_4} [X_2CpTa{=}C(CMe_3)H] \qquad (73)$$

(**78**) (**79**) (**80**)

isolated but it could be detected by ^{1}H n.m.r. spectroscopy. A tetragonal pyramidal structure is assigned to **79**, for which only the *trans*-form could be observed. In a solution of the neopentylbenzyl complex $[X_2Cp(Me_3CCH_2)(PhCH_2)Ta]$ (**81**), both isomers are present and the equilibrium constants for the *cis/trans* interconversion could be measured. Like **79**, compound **81** (X = Br) decomposes in solution at room temperature to give **80** (X = Br). It was proposed that the electrophilic d^0 metal in **79** interacts with the C—H$_\alpha$ electron pair in one neopentyl ligand, thus facilitating abstraction of H$_\alpha$ by a second neopentyl ligand. Schrock and co-workers showed that (a) α-abstraction in these compounds is intramolecular, (b) the rate of the reaction for X = Br is 200–400 times faster than for X = Cl and varies with solvent, and (c) the rate of decomposition (**79** → **80**) of the C_5H_5 complex is 5×10^3 times that of the corresponding C_5Me_5 analogue. Furthermore, a deuterium isotope effect of $k_H/k_D \approx 6$ was observed. It was concluded that only the *cis*-isomer reacts with α-hydrogen abstraction, whereas the *trans*-form is inert. Similarly, benzylidene complexes were prepared[177,178] (equation 74) and finally the first transition metal methylene complex

$$[Cl_2(PhCH_2)_3Ta] \xrightarrow{2\ CpTl} [Cp_2(PhCH_2)Ta{=}C(Ph)H] \qquad (74)$$

was obtained by treatment of Cp_2Me_3Ta with trityl tetrafluoroborate and deprotonation of the resulting salt with a phosphorane[179,180] (equation 75).

$$[Cp_2Me_3Ta] \xrightarrow{Ph_3C^+\bar{B}F_4} [Cp_2Me_2Ta]^+\bar{B}F_4 \xrightarrow{Ph_3P{=}CH_2} [Cp_2MeTa{=}CH_2] \qquad (75)$$

For the preparation of alkylidene complexes containing β-hydrogen atoms another technique has to be employed, as β-hydrogen atoms are lost more readily than α-hydrogen atoms. Alkylidene transfer via replacement of the PMe_3 ligand in **82** by $Et_3P{=}CHMe$ and subsequent loss of PEt_3 from the intermediate gave **83**[181] (equation 76). Bisalkylidene complexes have been prepared by addition of PMe_3 to solutions of **77**[182] (equation 77) and also by reaction of Me_3CCH_2Li with the alkylidyne complex **85**[182] (equation 78). X-ray analysis showed **84** to be a distorted trigonal bipyramid with both PMe_3 ligands in the axial positions. The Ta$=$C$_\alpha$—C$_\beta$ angles in the equatorial plane are unusually large (154.0° and 168.9°) for sp^2-hybridized carbon atoms[183].

$$[Cp_2PMe_3TaMe] \xrightarrow[-PMe_3,\ -PEt_3]{Et_3P=CHMe} [Cp_2MeTa=C(H)(Me)] \quad (76)$$

(**82**) (**83**)

$$[(Me_3CCH_2)_3Ta=C(H)(CMe_3)] \xrightarrow[-CMe_4]{2PMe_3} [(Me_3CCH_2)(PMe_3)_2Ta(=C(H)(CMe_3))_2] \quad (77)$$

(**84**)

$$[Cl(C_5Me_5)(PMe_3)_2Ta\equiv CCMe_3] \xrightarrow{Me_3CCH_2Li} [(C_5Me_5)(PMe_3)Ta(=C(H)(CMe_3))_2] \quad (78)$$

(**85**)

Another alkylidyne complex (**86**), obtained from WCl_6 and Me_3CCH_2Li, reacts at 100°C rapidly and quantitatively with PMe_3 to give the unusual alkylalkylidenealkylidyne (or alkylcarbenecarbyne according to the Fischer nomenclature) complex **87**[184] (equation 79). Deprotonation of the cationic alkyl

$$(Me_3CCH_2)_3W\equiv CCMe_3 \xrightarrow{Me_3P} [(Me_3P)_2(Me_3CCH_2)W(=C(H)(CMe_3))(\equiv CCMe_3)] \quad (79)$$

(**86**) (**87**)

metal(V) complexes **88** with $[(Me_3Si)_2N]Li$ also yields alkylidene complexes[185] (equation 80), as well as the reaction of Me_3SiCH_2Li with the bis(trimethylsilyl)amide complex **89**[186] (equation 81). Instead of the expected methylene complex a

$$[Cp_2(Me_3SiCH_2)_2M]^+ \xrightarrow{[(Me_3Si)_2N]Li} [Cp_2(Me_3SiCH_2)M=C(H)(SiMe_3)] \quad (80)$$

(**88**)

M = Nb, Ta

$$[Cl_3\{(Me_3Si)_2N\}_2Ta] \xrightarrow{Me_3SiCH_2Li} [\{(Me_3Si)_2N\}_2(Me_3SiCH_2)Ta=C(H)(SiMe_3)] \quad (81)$$

(**89**)

benzylidene complex was obtained by reaction of Me_3P with the mesityl (mes) complex **90** via γ-hydrogen abstraction[187] (equation 82).

$$[X_3(mes)MeTa] \xrightarrow{Me_3P} [X_3(Me_3P)_2Ta=C(H)(C_6H_3\text{-}3,5\text{-}Me_2)] \quad (82)$$

(**90**)

X = Cl, Br

In most cases the alkylidene complexes are surprisingly stable. In contrast to the electrophilic nature of the carbene carbon atom in all other carbene complexes the alkylidene carbon atom exhibits nucleophilic behaviour.

K. Miscellaneous Methods

Abstraction of atoms or groups bonded to the α-carbon atom of alkyl complexes has been employed several times in the synthesis of carbene complexes. Thus, a non-heteroatom-stabilized iron–carbene complex was obtained via hydride abstraction with triphenylmethyl hexafluorophosphate from **91**[188,189] (equation 83).

$$[Cp(CO)_2Fe(H)C_8H_7] + Ph_3C^+\bar{P}F_6 \xrightarrow{-Ph_3CH} [Cp(CO)_2Fe{=}C_8H_6]^+ \bar{P}F_6 \quad (83)$$

(91)

Hydride abstraction was also used in the preparation of other iron–carbene complexes[190], e.g. equation 84 and also in the generation of the cationic carbene

$$[Cp(CO)LFe{-}CH_2OR] + Ph_3C^+\bar{B}F_4 \longrightarrow$$

$$L = CO,\ PPh_3;\quad R = Me,\ Et$$

$$[Cp(CO)LFe{\cdots}C(OR)H]^+\ \bar{B}F_4 + Ph_3CH \quad (84)$$

complex **92** from [Cp(NO)PPh$_3$ReMe] and a triphenylmethyl salt. Compound **92** could not be isolated but could be observed spectroscopically and was trapped with PPh$_3$[191] (equation 85). Closely related to this route is alkoxide abstraction. By this

$$[Cp(NO)PPh_3ReMe] \xrightarrow{Ph_3C^+} [Cp(NO)PPh_3Re{=}CH_2] \xrightarrow{PPh_3}$$

(92)

$$[Cp(CO)PPh_3Re{-}CH_2{-}PPh_3]^+ \quad (85)$$

method the cationic dicarbonyl(cyclopentadienyl)phenylcarbene complex was synthesized[192] (equation 86). The corresponding cationic methylene complex

$$[Cp(CO)_2Fe{-}CH(Ph){-}OMe] \underset{CF_3COOH}{\overset{Ph_3C^+\bar{P}F_6}{\rightleftharpoons}} [Cp(CO)_2Fe{=}C(H)Ph]^+ \quad (86)$$

[Cp(CO)$_2$FeCH$_2$]$^+$ has been postulated several times as a reaction intermediate[192–194] formed on acid treatment of the alkoxymethyl compound, but has never been isolated because of its instability. The comparable cation [Cp(Ph$_2$PC$_2$H$_4$PPh$_2$)FeCH$_2$]$^+$, however, could be generated by the same method from the ethoxymethyl compound and characterized spectroscopically[195]. Another related complex was obtained from an alkenyl precursor. Trimethyloxonium tetrafluoroborate reacts with **93** to form **94**[196] (equation 87). Fluoride abstraction was employed in the reaction of [Cp(CO)$_2$LMoCF$_3$] (L = CO, PPh$_3$) with SbF$_5$. Again, the resulting difluorcarbene complex could not be isolated but could be observed by ^{1}H n.m.r. spectroscopy[197]

$$[Cp(Ph_2PC_2H_4PPh_2)FeCH{=}CMe_2] \xrightarrow{Me_3O^+\bar{B}F_4} \left[Cp(Ph_2PC_2H_4PPh_2)Fe{=}C\begin{matrix}H\\CMe_3\end{matrix}\right]^+ \bar{B}F_4 \quad (87)$$

(**93**) (**94**)

Dichlorcarbene complexes, however, can be obtained in an analytically pure form: reaction of **95** with $Hg(CCl_3)_2$ proceeds to give **96** in 80% yield[198] (equation 88).

$$[(PPh_3)_3Cl(CO)OsH] \xrightarrow{Hg(CCl_3)_2} [(PPh_3)_2Cl_2(CO)Os{=}CCl_2] \quad (88)$$

(**95**) (**96**)

A different approach to the synthesis of dichlorcarbene complexes embraces the reaction of 5,10,15,20-tetraphenylporphinatoiron(II), [(tpp)Fe(II)], with carbon tetrachloride in the presence of an excess of reducing agent[199]. The structure of the resulting dichlorcarbene complex (which is believed to be formed via the [(tpp)Fe(III)Cl] complex and CCl_3 radicals) was confirmed by X-ray analysis[200]. This route was extended to the synthesis of related tetraphenylporphinato(carbene)iron

$$[(tpp)Fe(II)] + CCl_3R \xrightarrow[-2Cl^-]{+2e} \left[(tpp)Fe{=}C\begin{matrix}Cl\\Cl\end{matrix}\right] \quad (89)$$

$R = Cl, CN, COOEt, SCH_2Ph$

complexes[201,202]. Hydride abstraction was also used in the preparation of iron- and tungsten–carbene complexes[203,204], e.g.

$$\left[(CO)_5W{-}C_7H_7\right]^- \xrightarrow{Ph_3C^+} (CO)_5W{=}C_7H_6 \quad (90)$$

A series of thiocarbene complexes were synthesized from thiocarbonyl, CS_2, thiocarbamoyl or thioester complexes (see equation 49). Cyclopentadienyl(triphenylphosphine)thiocarbonylrhodium and -iridium react with methyl iodide to form the methyl(thiomethyl)carbene compound **97**[205] (equation 91). The proposed

$$[Cp(PPh_3)M(CS)] \xrightarrow{+2\,MeI} \left[Cp(PPh_3)IM{\cdots}C\begin{matrix}SMe\\Me\end{matrix}\right]^+ I^- \quad (91)$$

$M = Rh, Ir$ (**97**)

mechanism involves oxidative addition of MeI to the metal followed by rearrangement via methyl transfer to the thiocarbonyl carbon atom and coordination of the iodide to the metal. This thioacyl intermediate then undergoes electrophilic attack at the sulphur with formation of **97**. Another iridium complex, $[(PPh_3)_2I_2(C_6F_5)Ir\{C(SMe)Me\}]$, was prepared similarly[206].

Cyclic dithiocarbene complexes are formed in the reactions of CS_2 complexes with 1,2-dibromethane[207] or substituted acetylenes[208] (equations 92 and 93). The addition of the acetylene to **98** is reversible, as the carbene complex **99** isomerizes to the heterometallocycle **100**, leading to an equilibrium mixture of **99** and **100**. For this transformation (**99** → **100**) a mechanism was proposed involving retrocycloaddition

$$[(PPh_3)_2(CO)(RNC)OsCS_2] \xrightarrow[(2)\ NaClO_4]{(1)\ BrC_2H_4Br} \left[Br(PPh_3)_2(CO)(RNC)Os{\cdots}C\overset{S}{\underset{S}{\langle\rangle}}\right]^+ ClO_4^- \quad (92)$$

$$[(PMe_2Ph)_2(CO)_2FeCS_2]\ (\mathbf{98}) + MeOOCC{\equiv}CCOOMe \rightleftharpoons [(PMe_2Ph)_2(CO)_2Fe{\cdots}C(SC(COOMe){=}C(COOMe)S)]\ (\mathbf{99}) \quad (93)$$

$$\rightleftharpoons [(PMe_2Ph)_2(CO)_2Fe\text{–}C(=S)\text{–}S\text{–}C(COOMe){=}C(COOMe)\text{–}Fe]\ (\mathbf{100})$$

(**99** → **98**) followed by 1,3-dipolar addition (**98** → **100**)[208]. In addition to **99**, other iron–[209] and manganese–carbene complexes[210] were prepared by the same route.

Non-cyclic dithiocarbene complexes (see also equation 49 and Section II.I) were prepared by trapping $:C(SPh)_2$ on pentacarbonylchromium and -tungsten moieties[211]:

$$[(CO)_5M(thf)] + :C(SPh)_2 \longrightarrow [(CO)_5M{\cdots}C(SPh)_2] \quad (94)$$

$M = Cr, W$; $:C(SPh)_2$ generated from $LiC(SPh)_3$ ($-LiSPh$)

Desulphuration[212–214] was also used in the preparation of carbene complexes, e.g. equations 95[213] and 96[214]. Formal desulphuration with concomitant ligand transfer

$$[Cp(SnPh_3)COFeCS] \xrightarrow{H_2NC_2H_4NH_2} [Cp(SnPh_3)COFe{\cdots}C(NHC_2H_4NH)] + H_2S \quad (95)$$

$$[Fe_2(CO)_9] + S{=}C(NH_2)_2 \longrightarrow [(CO)_4Fe{\cdots}C(NH_2)_2] + \cdots \quad (96)$$

was observed in the photochemical reaction of pentacarbonyl iron with **101**[215] (equation 97). Transfer of the carbene ligand from carbene complexes to metal carbonyls can be induced by irradiation[216] (equation 98) or thermally[217] (equation 99). For other transfer reactions see also Sections II.G and II.H.

$$[(CO)_5Fe] + \left[(CO)_5CrS{=}C\left\langle{}^{O}_{O}\right\rangle\right] \longrightarrow \left[(CO)_4Fe{\cdots}C\left\langle{}^{O}_{O}\right\rangle\right] + \cdots \quad (97)$$

(**101**)

$$\left[Cp(NO)COMo{\cdots}C\left\langle{}^{R}_{Ph}\right.\right] + [Fe(CO)_5] \xrightarrow{h\nu}$$

R = OMe, OEt, NMe_2

$$[Cp(NO)(CO)_2Mo] + \left[(CO)_4Fe{\cdots}C\left\langle{}^{R}_{Ph}\right.\right] \quad (98)$$

$$\left[(CO)_5Cr{\cdots}C\left\langle{}^{O}\right.\right] + [(CO)_6W] \underset{K_{eq} = 3}{\overset{140\ °C}{\rightleftharpoons}} [(CO)_6Cr] + \left[(CO)_5W{\cdots}C\left\langle{}^{O}\right.\right] \quad (99)$$

In addition to the methods already mentioned, several other routes for the synthesis of carbene complexes from non-carbene precursors were developed in recent years, including the following:

(a) reaction of metal carbonyl anions, e.g. **102**, with a carbodiimide and subsequent protonation of the adduct **103**[218] (equation 100);

$$[(CO)_5Cr]^{2-} + RN{=}C{=}NR \longrightarrow [(CO)_5Cr{-}C(NR)_2]^{2-} \xrightarrow{H^+}$$

(**102**) (**103**)

$$\left[(CO)_5Cr{\cdots}C\left\langle{}^{NHR}_{NHR}\right.\right] \quad (100)$$

(b) reaction of silyl-substituted ylides with metal carbonyls[219] (equation 101);

$$[(CO)_nM] + Me_3P{=}C(H)SiMe_3 \longrightarrow \left[(CO)_{n-1}M{\cdots}C\left\langle{}^{OSiMe_3}_{CH=PMe_3}\right.\right] \quad (101)$$

M = Cr, Mo, W($n = 6$) M = Fe($n = 5$)

(c) addition of an ynamine to a metal carbonyl hydride and subsequent alkylation[220] (equation 102);

$$[Cp(CO)_3WH] \xrightarrow{MeC\equiv CNEt_2} \left[Cp(CO)_2W\left\langle{}^{C(NEt_2)}_{C(=O)}\right\rangle\right] \xrightarrow{Me_3O^+\bar{B}F_4} \left[Cp(CO)_2W\left\langle{}^{C{\cdots}NEt_2}_{C{\cdots}OMe}\right\rangle\right]^+ \quad (102)$$

(d) addition of cyclohexene oxide to an anionic iron complex and subsequent alkylation[221] (equation 103);

$$[Cp(CO)_2Fe]^- \xrightarrow[(2)\ Me_3O^+\bar{B}F_4]{(1)\ O{=}C_6H_8}$$

$$\left[Cp(CO)Fe{=}C(OMe)O\text{-}C_6H_{10}\right] \xrightarrow{HBF_4/CO} \left[Cp(CO)_2Fe{\cdots}C(OC_6H_{11})(OMe)\right] \quad (103)$$

(e) thermally induced reversible dehydrogenation[222] (equation 104);

$$\left[HC(CH_2CH_2PBu_2)_2Ir(H)Cl\right] \underset{H_2}{\overset{170\ °C/1\ torr}{\rightleftharpoons}} C(CH_2CH_2PBu_2)_2{=}Ir{-}Cl \quad (104)$$

(f) deprotonation of sulphide complexes containing an active α-hydrogen atom by double consecutive carbonylation; alkylation affords neutral (sometimes *fac*-trisubstituted) chelates[223] (equation 105).

$$[(CO)_5MS(CH_2R^1)R^2] \xrightarrow[(2)\ L \quad (3)\ Et_3O^+]{(1)\ BuLi} \left[(CO)_3LM{\cdots}C(OEt){-}C(OEt){=}C(R^1){-}S(R_2){\to}M\right] \quad (105)$$

M = Cr, W, L = CO, CNBu-*t*, $P(OMe)_3$

III. SYNTHESIS BY MODIFICATION OF THE CARBENE LIGAND

In addition to the synthesis of carbene complexes from non-carbene complex precursors, several routes for the modification of the carbene ligand were developed, thus extending the variety of preparative methods. Three different reactive centres in non-cyclic mainly Fischer-type carbene complexes (but also in some cyclic complexes such as **39**) can be used for alterations of the carbene ligand: (a) the carbene carbon atom is highly electrophilic and is therefore susceptible to nucleophilic attack, (b) the heteroatom bonded to the carbene carbon atom is nucleophilic and may therefore react with electrophiles, and (c) the hydrogen atom alpha to the carbene carbon atom is acidic and may be split off as a proton by bases, generating an anion which may be used in further reactions.

A. Nucleophilic Attack at the Carbene Carbon Atom

Although Mulliken population analysis[224,225] and ESCA measurements[226] indicate that the carbon atoms of the carbonyl ligands of Fischer-type pentacarbonyl(carbene) complexes carry a greater positive charge than the carbene carbon atom, experimental results show that nucleophiles add to carbene complexes such as **4** at the carbene carbon and not at one of the carbonyl carbon atoms, e.g. in equation 106[227–229]. This was explained in terms of frontier orbital control of these reactions as the lowest unoccupied molecular orbital (LUMO) in **4** is energetically and spatially localized on

$$\left[(CO)_5M\overset{\cdots}{=}C\begin{smallmatrix}R\\ \\ OMe\end{smallmatrix}\right] + PR_3 \rightleftharpoons \left[(CO)_5-\underset{OMe}{\overset{PR_3}{C}}-R\right] \qquad (106)$$

M = Cr (**4**), W (**104**)

the carbene carbon atom[225]. The magnitude of the electrophilicity depends to a great extent on the substituents bonded to the carbene carbon atom. Whereas phosphines react with e.g. alkoxy- and alkylthiocarbene complexes of type **4** to form ylide complexes, e.g. **104**, the same reaction with aminocarbene complexes failed. For some phosphines adduct formation (equation 106) was shown to be reversible and the equilibrium constants were measured[229,230]. Sterically fixed tertiary amines (e.g. 1-azabicyclo[2.2.2]octane) may also add to the carbene carbon atom to form the analogues of **104**[231,232]. With ammonia, primary and secondary amines, however, instead of an adduct the substitution product is isolated in high yield, e.g. **105** in equation 107[233–235].

$$\left[(CO)_5Cr\overset{\cdots}{=}C\begin{smallmatrix}R\\ OMe\end{smallmatrix}\right] + NH_2R \longrightarrow \left[(CO)_5Cr\overset{\cdots}{=}C\begin{smallmatrix}R\\ NHR\end{smallmatrix}\right] + MeOH \qquad (107)$$

(**105**)

A detailed kinetic investigation[236,237] showed that this aminolysis is a reaction with a negative Arrhenius activation energy. The rate of reaction in non-polar solvents (e.g. hexane) is first order in carbene complex and third order in amine, and in dioxane it is first order in carbene complex and second order in amine. A mechanism was proposed involving nucleophilic attack of an amine (the nucleophilicity of which is increased by hydrogen bonding to another amine or dioxane, respectively) at the carbene carbon atom and labilization of the carbon–oxygen bond via hydrogen bonding of another amine with the oxygen atom at the carbene carbon. The elimination of methanol from the intermediate is considered to be the rate-determining step.

The aminolysis reaction is a fairly general type of carbene complex reaction. A whole series of amines including amino acid esters[238,239] and alkoxycarbene complexes, both cationic cyclic and non-cyclic[240], were employed. With secondary amines, however, steric factors become important. The reaction of **4** with the sterically hindered diisopropylamine did not yield the expected methyl(diisopropylamino)-carbene but instead via propylene elimination the methyl(monoisopropylamino)-carbene complex[241].

An interesting competition is observed in the reaction of dimethylamine with pentacarbonyl[ethoxy(phenylacetylenyl)carbene]chromium and -tungsten. At room temperature, with the amine in excess, aminolysis and conjugate addition of one molecule of amine to the triple bond occur and the carbene complex **109** is isolated[34]. At -115°C the aminolysis product **108** and at -20°C the conjugate addition product **107** are formed[242]. At -78°C, however, a mixture of **107** and **108** (3:2) is obtained. Compound **108**, but not **107**, can finally be converted into **109** at room temperature.

These results can be understood when considering the unusual negative Arrhenius activation energy observed for the aminolysis reaction and a positive $\Delta G^{\neq}$ value which is to be expected for the conjugate addition reaction. Thus, with decreasing temperature the rate for the conjugate addition reaction decreases. At the same time the rate for the aminolysis reaction increases, dominating at -115°C. With increasing temperature the rate of aminolysis decreases and that for conjugate addition increases

$$\left[(CO)_5Cr\overset{\cdots}{=}C\begin{matrix}\cdot\cdot OEt\\ \diagdown C\equiv CPh\end{matrix}\right] \textbf{(106)} \xrightarrow{HNMe_2,\ -20\ ^\circ C} \left[(CO)_5Cr\overset{\cdots}{=}C\begin{matrix}\cdot\cdot OEt\\ \diagdown CH=C(Ph)NMe_2\end{matrix}\right] \textbf{(107)}$$

$$\textbf{(106)} \xrightarrow{HNMe_2,\ -115\ ^\circ C} \left[(CO)_5Cr\overset{\cdots}{=}C\begin{matrix}\cdot\cdot NMe_2\\ \diagdown C\equiv CPh\end{matrix}\right] \textbf{(108)} \xrightarrow{+HNMe_2} \left[(CO)_5Cr\overset{\cdots}{=}C\begin{matrix}\cdot\cdot NMe_2\\ \diagdown CH=C(Ph)NMe_2\end{matrix}\right] \textbf{(109)} \qquad (108)$$

more and more in proportion. This explains the formation of a mixture of **107** and **108** at −78°C. At −20°C conjugate addition dominates. The failure to convert **107** into **109** at room temperature or even at −50°C may be due to a strongly diminished electron deficiency of the carbene carbon atom caused by the higher electron-donating properties of the enamine moiety in relation to the acetylenic group.

Closely related to aminolysis is the reaction of alkoxycarbene complexes with thiols[233,243] (e.g. equation 109) or with thiolates and subsequently with acids to form

$$\left[(CO)_5Cr\overset{\cdots}{=}C\begin{matrix}\cdot\cdot OMe\\ \diagdown Ph\end{matrix}\right] + HSMe \longrightarrow \left[(CO)_5Cr\overset{\cdots}{=}C\begin{matrix}\cdot\cdot SMe\\ \diagdown Ph\end{matrix}\right] + HOMe \qquad (109)$$

thiocarbene complexes[244]. Methylselenol reacts in a similar manner to give the corresponding selenocarbene complex[245]. The more acidic phenylselenol, however, yielded a rearranged complex in which the ligand was bonded via selenium to the metal[246]. In some cases alcohols[132] or enolate anions[247] were used as nucleophiles (equation 111).

$$\left[(CO)_5Cr\overset{\cdots}{=}C\begin{matrix}\cdot\cdot OMe\\ \diagdown Me\end{matrix}\right] + HSePh \longrightarrow \left[(CO)_5Cr-Se\begin{matrix}\diagup CH(Me)OMe\\ \diagdown Ph\end{matrix}\right] \qquad (110)$$

$$\left[(CO)_5W\overset{\cdots}{=}C\begin{matrix}\cdot\cdot OMe\\ \diagdown Ph\end{matrix}\right] \xrightarrow[(2)\ H^+]{(1)\ >C=C(O^-)Ph} \left[(CO)_5W\overset{\cdots}{=}C\begin{matrix}\cdot\cdot O-C(=CMe_2)Ph\\ \diagdown Ph\end{matrix}\right] \qquad (111)$$

Taking advantage of the weaker carbene carbon–oxygen bond in acetoxycarbene in relation to alkoxycarbene complexes, aminolysis and thiolysis reactions can also be performed with acetoxycarbene complexes[25]. By this route phenoxycarbene complexes are easily accessible, usually in relatively high yields (up to 92%)[25,248], e.g. equation 112. The reaction of a siloxycarbene complex with methanol, on the other hand, gave a hydroxycarbene complex (91%)[28] (equation 113).

$$\left[(CO)_5M\cdots C\begin{matrix}\cdot OCOMe\\ \diagdown R\end{matrix}\right] \xrightarrow{NaOPh} \left[(CO)_5M\cdots C\begin{matrix}\cdot OPh\\ \diagdown R\end{matrix}\right] \quad (112)$$

M = Cr, W; R = Ph, furyl

$$\left[(CO)_5W\cdots C\begin{matrix}\cdot OSiMe_3\\ \diagdown Th\end{matrix}\right] \xrightarrow[pentane]{MeOH,\ -20\,^{\circ}C} \left[(CO)_5W\cdots C\begin{matrix}\cdot OH\\ \diagdown Th\end{matrix}\right] \quad (113)$$

Th = thienyl

Other nucleophiles have been used. Reaction of pentacarbonyl[methyl(methoxy)-carbene]chromium (**110**) with benzophenoneimine gave **111**[249] (equation 114). Similar imino-substituted carbene complexes were isolated from **110** and benzaldoxime[249] and from pentacarbonyl[methoxy(phenyl)carbene]chromium and 1-aminoethanol[250] (equation 115).

$$\underset{(\mathbf{110})}{\left[(CO)_5Cr\cdots C\begin{matrix}\cdot OMe\\ \diagdown Me\end{matrix}\right]} + HN{=}CPh_2 \longrightarrow \underset{(\mathbf{111})}{\left[(CO)_5Cr\cdots C\begin{matrix}\cdot N{=}CPh_2\\ \diagdown Me\end{matrix}\right]}\ (21\%) \quad (114)$$

$$\left[(CO)_5Cr\cdots C\begin{matrix}\cdot OMe\\ \diagdown Ph\end{matrix}\right] + H_2NCH_2CH_2OH \longrightarrow \left[(CO)_5Cr\cdots C\begin{matrix}\cdot N{=}CHMe\\ \diagdown Ph\end{matrix}\right]\ (5.7\%) \quad (115)$$

In the reaction of **110** with substituted hydrazines the hydrazinocarbene complex could not be isolated, but a nitrile complex was obtained as the product of further rearrangement and nitrogen–nitrogen bond cleavage[251]. The palladium compound **112**, however, yielded an amino(hydrazino)carbene complex (**113**)[252] and, in a similar reaction with benzalhydrazone, complex **114**[252] was isolated (equation 116).

$$\underset{(\mathbf{112})}{\left[Cl_2(RNC)Pd\cdots C\begin{matrix}\cdot NHR'\\ \cdot OEt\end{matrix}\right]} \xrightarrow{H_2NNHPh} \underset{(\mathbf{113})}{\left[Cl_2(RNC)Pd\cdots C\begin{matrix}\cdot NHR\\ \cdot NHNHPh\end{matrix}\right]}\ (51\%)$$

$$\underset{(\mathbf{112})}{\left[Cl_2(RNC)Pd\cdots C\begin{matrix}\cdot NHR'\\ \cdot OEt\end{matrix}\right]} \xrightarrow{H_2NN{=}CHPh} \underset{(\mathbf{114})}{\left[Cl_2(RNC)Pd\cdots C\begin{matrix}\cdot NHR'\\ \cdot NHN{=}CHPh\end{matrix}\right]} \quad (116)$$

Non-heteroatom-stabilized carbene complexes may be obtained via a two-step reaction first reported by Casey and Burkhardt in 1973[253]. Nucleophilic attack of phenyllithium to a Fischer-type carbene complex, e.g. **115** at $-78°C$, gives the adduct **116**[253] which, in solution, decomposes rapidly at room temperature but may be isolated at low temperatures[254]. Treatment of **116** at $-78°C$ with acids such as HCl[253,255] or SiO_2[256] induces loss of methanol and the diphenylcarbene complex **117** is formed (equation 117). This procedure was later extended to phenyl-, furyl-, thienyl-, and

$$[(CO)_5W{=}C(OMe)Ph] \; (\mathbf{115}) + LiPh \longrightarrow [(CO)_5W{-}C(Ph)_2(OMe)]^- \, Li^+ \; (\mathbf{116}) \xrightarrow{H^+} [(CO)_5W{=}CPh_2] \; (\mathbf{117}) \qquad (117)$$

pyrrolyl(methoxy)carbene complexes of chromium and tungsten as precursors and to other organolithium compounds including substituted phenyl-, furyl-, thienyl-, and pyrrolyllithium[256].

Attempts to synthesize alkylphenylcarbene complexes by a similar method employing alkyllithium failed, as the resulting alkyl(phenyl)carbenetungsten complex immediately rearranges with 1,2-hydrogen migration to form an olefin complex[257] (equation 118). However, methyl(phenyl)carbene complexes of manganese and rhenium[86,87] and a dimethylcarbene manganese complex[86] were prepared by another route via addition of methyl anions to cationic carbyne complexes (see Section II.C).

$$[(CO)_5W{=}C(OMe)Ph] \xrightarrow[(2)\ SiO_2/\text{pentane}]{(1)\ RCH_2Li} [(CO)_5W \leftarrow \| (HRC{=}CHPh)] \qquad (118)$$

In pentacarbonyl[methoxy(styryl)carbene]chromium (**118**), phenyllithium adds to the carbon–carbon double bond to give a conjugate addition product (**119**) which may be obtained in higher yield from **118** and lithium diphenylcuprate (equation 119). The

$$[(CO)_5Cr{=}C(OMe)CH{=}C(H)Ph] \; (\mathbf{118}) \xrightarrow[(2)\ HCl]{(1)\ LiPh} [(CO)_5Cr{=}C(OMe)CH_2CHPh_2] \; (\mathbf{119}) \qquad (119)$$

phenyl(styryl)carbene complex could not be isolated[258]. A similar reaction of **115** using the anion in potassium triisopropoxyborohydride as nucleophile led to the isolation of the tungstate complex **120**[259], which, on reaction with trifluoroacetic acid at $-78°C$, gave **121**[260,261], the first phenylhydrogencarbene complex. Because of its instability **121** could not be obtained in analytically pure form; it decomposes at −56°C with a half-life of 24 min, but gives isolable phosphine adducts analogous to **104** (see equation 106) with tributylphosphine (equation 120).

$$[(CO)_5W\!\equiv\!C(OMe)Ph] \xrightarrow[(2)\ [NEt_4]Br]{(1)\ K[HB(O\text{-}Pr)_3]} [(CO)_5W{-}C(H)(OMe)(Ph)]^- NEt_4^+ \ \textbf{(120)}$$

$$\textbf{(120)} \xrightarrow[-78\,°C]{CF_3COOH} [(CO)_5W{=}C(H)Ph] \ \textbf{(121)} \xrightarrow{P(n\text{-}Bu)_3} [(CO)_5W{-}C(P(n\text{-}Bu)_3)(H)(Ph)] \qquad (120)$$

Chloride in pentacarbonyl[chloro(dimethylamino)carbene]chromium was exchanged with cyanide probably via nucleophilic attack of cyanide at the carbene carbon atom in the first reaction step (see equation 34).

B. Electrophilic Attack at the Carbene Ligand

Electrophilic attack by Group III halides at the heteroatom of the carbene ligand results in the formation of carbyne complexes[262]. Using boron trifluoride and pentacarbonyl[dialkylamino(ethoxv)carbene]chromium (equation 32)[93,94] or boron trichloride and complexes of the general type [(Ar)(CO)$_2$M{C(Y)OMe}] (Ar = η^6-C_6H_6, M = Cr, Y = Ph; or Ar = η^5-C_5H_5, M = Mn, Re, Y = Me, Ph)[86,87,91] cationic carbyne complexes are obtained which can, in turn, again function as precursors for further preparation of carbene complexes (see Section II.C).

The formation of pentacarbonyl[chloro(diethylamino)carbene]chromium (**26**) from **122** and boron trichloride[263] (equation 121) probably also proceeds via the cationic

$$[(CO)_5Cr\!\equiv\!C(NEt_2)(OEt)] \ \textbf{(122)} + BCl_3 \longrightarrow [(CO)_5Cr\!\equiv\!C(NEt_2)Cl] \ \textbf{(26)} + \cdots \qquad (121)$$

species **25** as the tetrachloroborate salt of **25** reacts above $-25°C$ to form **26**[93]. Further, the cation **25** has been obtained by reaction of **26** with boron trichloride. From *cis*-bromo(tetracarbonyl)[hydroxy(methyl)carbene]manganese and boron tribromide the unusual carbene complex **123** was prepared[14] (equation 122).

$$cis\text{-}\left[Br(CO)_4Mn{\cdots}C(OH)Me\right] + BBr_3 \xrightarrow[-HBr]{} \left[(CO)_4Mn{\cdots}C(Me){\cdots}O{-}BBr_2{\cdots}Br{-}Mn\right] \quad (122)$$

(123)

Closely related complexes were obtained by a different route via addition of $AlBr_3$ to pentacarbonyl(methyl)manganese (equation 123) or tricarbonyl(cyclopentadienyl)methylmanganese[264].

$$[(CO)_5MnMe] + AlBr_3 \longrightarrow \left[(CO)_4Mn{\cdots}C(Me){\cdots}O{-}AlBr_2{\cdots}Br{-}Mn\right] \quad (123)$$

C. Reactions of Carbene Anions

Alkoxy(alkyl)carbene complexes are remarkably acidic. The hydrogen atoms attached to the α-carbon atom undergo rapid hydrogen/deuterium exchange in alkaline deuteromethanol[265]. This is explained by the existence of an intermediate carbene anion. These carbene anions can be generated stoichiometrically from carbene complexes by reaction with sodium methanolate or butyllithium[266]. For example, with pentacarbonyl[methoxy(methyl)carbene]chromium (**110**) this anion (**124**) has been isolated as the bis(triphenylphosphine)iminium salt[267] (equation 124).

$$\underset{(110)}{\left[(CO)_5Cr{\cdots}C(OMe)Me\right]} \xrightarrow[-78°C]{BuLi} \underset{(124)}{\left[(CO)_5Cr{\cdots}C(OMe)\bar{C}H_2\right] \longleftrightarrow \left[(CO)_5\bar{C}r{-}C(OMe){=}CH_2\right]} \quad (124)$$

It was demonstrated that **124** is not measurably protonated by methanol and that **110** has an acidity in tetrahydrofuran comparable to that of *p*-cyanophenol ($pK_a = 8$ in water)[267]. By addition of DCl in excess to solutions of **124** the d_1 analogue of **110** could be recovered in 90% yield and with 90% monodeuteration[267].

Casey and co-workers have used the moderate reactivity of these carbene anions towards carbon nucleophiles including epoxides, aldehydes, α-bromoesters, and α,β-unsaturated carbonyl compounds to prepare a series of carbene complexes inaccessible by other synthetic routes. Reaction of the anion **124** with, for example, epoxides yields cyclic carbene complexes[268] (equation 125) and with ethylene sulphide the

$$[(CO)_5Cr{=}C(OMe)Me] \xrightarrow[(2)\ \text{epoxide (R)}]{(1)\ BuLi} [(CO)_5Cr{=}C(\text{2-oxacyclopentylidene, 5-R})] \tag{125}$$

(**125**) R = H

(**126**) R = Me

analogous thiocarbene complexes[269]. The nucleophilic attack by the carbene anion occurs at the least hindered carbon of propyleneoxide (R = Me).

The carbene anion generated from **126** reacts with benzyl iodide to give the alkylated carbene complexes[9]. Methyl fluorosulphonate or trimethyloxonium tetrafluoroborate were used to methylate carbene anions[266] (equation 126). Acetyl chloride

$$[(CO)_5W{=}C(OMe)Me] \xrightarrow[(2)\ Me_3O^+\ \bar{B}F_4]{(1)\ BuLi} [(CO)_5W{=}C(OMe)Et] \tag{126}$$

also reacts with metal carbene anions[269] (equation 127); however, when the initial acylated product contains an enolizable hydrogen an enol ester is isolated[266] (equation 128). A similar reaction of acetyl chloride with the carbene anion generated from

$$[(CO)_5Cr{=}C(\text{2-oxacyclopentylidene, 3-Me})] \xrightarrow[(2)\ AcCl]{(1)\ BuLi} [(CO)_5Cr{=}C(\text{2-oxacyclopentylidene, 3-Me, 3-COMe})] \tag{127}$$

$$[(CO)_5W{=}C(OMe)Me] \xrightarrow[(2)\ AcCl]{(1)\ BuLi} [(CO)_5W{=}C(OMe)CH{=}C(Me)OAc] \tag{128}$$

cis-(2-oxacyclopentylidene)nonacarbonylmanganese and butyllithium resulted in the formation of a mononuclear manganese–vinylcarbene complex (**127**)[270] (equation 129).

$$[(CO)_5Mn{-}Mn(CO)_4{=}C(\text{2-oxacyclopentylidene})] \xrightarrow[(2)\ AcCl]{(1)\ BuLi} [(CO)_4Mn{-}O{-}C(Me){=}C\text{(ring)}{-}C{=}Mn] \tag{129}$$

(**127**)

Reaction of the carbene anion from **125** with acetaldehyde yielded a vinylcarbene complex[271] (equation 130); with formaldehyde, on the other hand, a dimeric product

$$[(CO)_5Cr{=}C(\text{2-oxacyclopentylidene})] \xrightarrow[(2)\ MeCHO]{(1)\ BuLi} [(CO)_5Cr{=}C(\text{2-oxacyclopentylidene, 3-}{=}CHMe)] \tag{130}$$

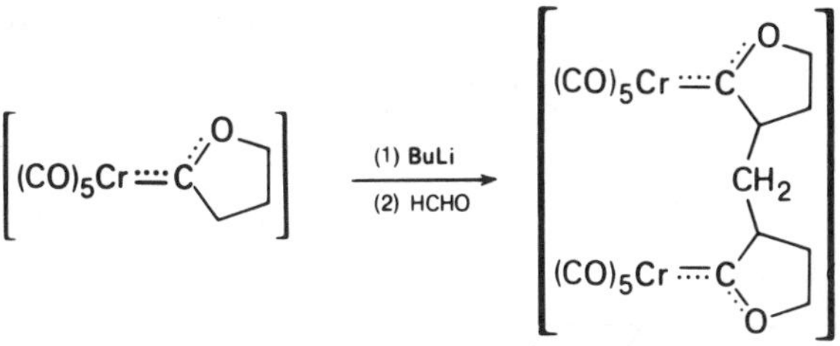

(131)

(**128**)

(**128**)[271] was obtained (equation 131). The monomeric species **129** was finally obtained by inverse addition of the carbene anion to an excess of a formaldehyde equivalent, chloromethoxymethane, and subsequent treatment of the solution with basic alumina[271] (equation 132).

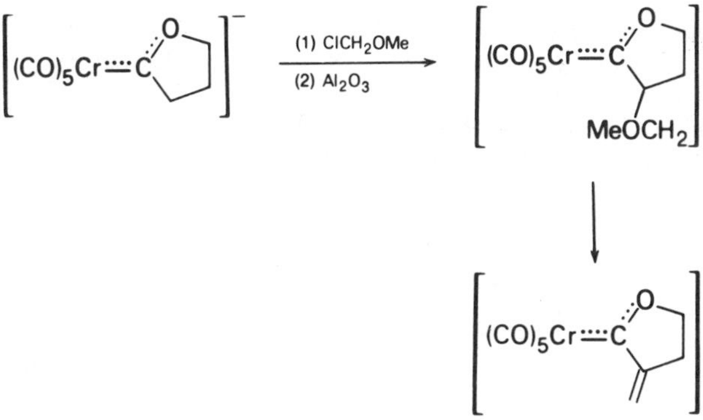

(132)

(**129**)

Sometimes problems arise from dialkylation as a side-reaction in the alkylation of carbene anions, e.g. with α-bromo esters or chloromethoxymethane. Conjugate addition of carbene anions to α,β-unsaturated esters and ketones has also been investigated[269]. An α-bromocarbene complex was obtained as the product of bromination of a carbene anion[269] (equation 133).

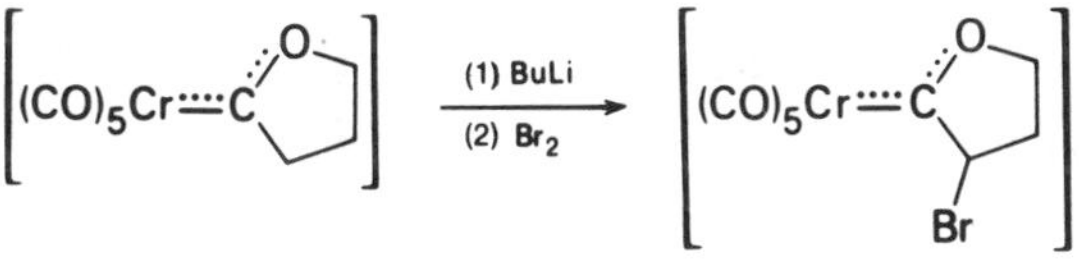

(133)

D. Miscellaneous

Vinylcarbene complexes such as pentacarbonyl[isobutenyl(methoxy)carbene]-chromium (**130**) or pentacarbonyl[methoxy(styryl)carbene]chromium (**118**) add to a number of enolate anions to form new carbene complexes[247]. Thus, **130** reacts with the lithium enolate of cyclopentanone or the potassium enolate of isobutyrophenone to give **131** and **132**, respectively (equation 134). Similar products are obtained using (**118**) instead of **130** and from the reaction of **118** with the sodium enolate of dimethyl malonate. The reactions of **130** or **118** with the lithium enolate of acetone, however, did not give the analogous carbene complexes, but organic products derived from

$$[(CO)_5Cr{=}C(OMe)CH{=}CMe_2]\ (130) \xrightarrow{(1)\ \text{cyclopentenolate};\ (2)\ H^+} (131)$$

$$[(CO)_5Cr{=}C(OMe)CH{=}CMe_2]\ (130) \xrightarrow{(1)\ Me_2C{=}C(O^-)Ph;\ (2)\ H^+} (132) \qquad (134)$$

addition to the carbene carbon atom. For conjugate addition of dimethylamine to ethoxy(phenylacetylenyl)carbene complexes (equation 108) and of phenyllithium to **118** (equation 119), see Section III.A.

An unusual addition was observed in the reaction of pentacarbonyl[amino-(organyl)carbene]chromium with 1-diethylaminopropyne. Instead of the conjugate addition product, compound **133** was isolated[272] (equation 135).

$$[(CO)_5Cr{=}C(NH_2)R] + MeC{\equiv}CNEt_2 \longrightarrow [(CO)_5Cr{=}C(R)N{=}C(Et)NEt_2]\ (133) \qquad (135)$$

R = Me, Ph

Hydroxycarbene complexes react readily with dicyclohexylcarbodiimide (**134**) to form different products depending on (a) the central metal atom and (b) whether the hydroxy(phenyl)- or the hydroxy(methyl)carbene complex is used. Whereas the reaction of pentacarbonyl[hydroxy(phenyl)carbene]chromium with **134** gave a carbene complex anhydride via intermolecular elimination of water[273] (equation 136), from the reaction of the tungsten complexes and **134** under identical conditions bimetallic carbene–carbyne complexes (**135**) were obtained[273,274] (equation 137). Mixtures of

$$2[(CO)_5Cr{=}C(OH)Ph] + C_6H_{11}NCNC_6H_{11}\ (134) \longrightarrow [(CO)_5Cr{=}C(Ph){-}O{-}C(Ph){=}Cr(CO)_5] \qquad (136)$$

$$2[(CO)_5W{=}C(OH)R] + 134 \longrightarrow [(CO)_5W{=}C(R){-}O{-}W(CO)_4({\equiv}CR)]\ (135) \qquad (137)$$

R = Me, Ph

the hydroxy(phenyl)carbene complexes of chromium and tungsten finally yielded with **134** the chromium carbene–tungsten carbyne compound **136** (55%) (equation 138). These results may be explained by a stronger W—C_{Carb} bond in relation to the

$$\left[(CO)_5Cr\overset{\cdots}{=}C\begin{matrix}OH\\C_6H_4Me\text{-}p\end{matrix}\right] + \left[(CO)_5W\overset{\cdots}{=}C\begin{matrix}OH\\Ph\end{matrix}\right] \xrightarrow{+134} \left[(CO)_5Cr\overset{\cdots}{=}C\begin{matrix}O-W(CO)_4(\equiv CPh)\\C_6H_4Me\text{-}p\end{matrix}\right] \quad (138)$$

(136)

Cr—C_{Carb} bond and simultaneously a weaker C_{Carb}—O bond in the tungsten compound compared with the chromium compound. Thus, C—O bond rupture in an initially formed carbene complex–diimide adduct is facilitated for the tungsten complex, resulting in the formation of a cationic carbyne intermediate which may react further with another molecule of hydroxycarbenetungsten or -chromium to yield **135** or **136**, respectively. On the other hand, in the case of the chromium complex (equation 136) the carbodiimide adduct is already attacked by another molecule of hydroxycarbene complex to give the anhydride.

Employing pentacarbonyl[hydroxy(methyl)carbene]chromium instead of the corresponding phenylcarbene complex in the reaction with **134**, a mixture of **137** (27%) and **138** (47%) is isolated. The latter is probably formed via 2 + 2 cycloaddition of the carbodiimide and a ketene complex intermediate[275] (equation 139).

$$\left[(CO)_5Cr\overset{\cdots}{=}C\begin{matrix}Me\\OH\end{matrix}\right] \xrightarrow{+134} \left[(CO)_5Cr\overset{\cdots}{=}C\begin{matrix}Me\\N(C_6H_{11})-CONHC_6H_{11}\end{matrix}\right] + [(CO)_5Cr=C=CH_2] \xrightarrow{+134} \left[(CO)_5Cr\overset{\cdots}{=}C\langle CH_2, N(C_6H_{11})\rangle C=NC_6H_{11}\right] \quad (139)$$

(137) **(138)**

IV. SYNTHESIS BY MODIFICATION OF THE METAL–LIGAND FRAMEWORK

A great number of reactions involving modification of the metal–ligand framework have been reported in the literature and only a few can be quoted here as illustrative examples. Modification of the metal–ligand framework can be grouped together in: (a) reactions involving the metal carbene—ligand bond (e.g. insertion, metathesis), (b) substitution of non-carbene ligands and, (c) oxidative addition to the metal.

A. Reactions Involving the Metal Carbene–Ligand Bond

Pentacarbonyl(carbene) complexes of chromium, molybdenum, and tungsten react with ynamines via insertion into the metal–carbene bond and redistribution of the π-electrons to give aminocarbene complexes[276,277] (equation 140). The reaction is stereoselective, yielding predominantly the E-configurated insertion product. From the

$$\left[(CO)_5M\!\cdots\!C\!\begin{smallmatrix}\cdots OMe\\ \diagdown R\end{smallmatrix}\right] + R'C{\equiv}CNEt_2 \longrightarrow \left[(CO)_5M\!\cdots\!C(NEt_2)\!-\!C(R')\!=\!C(R)OMe\right] \quad (140)$$

M = Cr, Mo, W; R = Me, Ph; R′ = H, Me

results of a kinetic investigation an associative stepwise mechanism was deduced. A nucleophilic attack of the ynamine at the carbene carbon atom in the first reaction step is followed by the formation of a metallocycle. Ring opening finally leads to the aminocarbene complex[278].

Other ynamines and other carbene complexes have also been employed[278–281]. Ethoxyacetylene reacts with different carbene complexes to give similar products in which ethoxyacetylene is inserted into the carbene–metal bond[261] (equation 141).

$$\left[(CO)_5W{=}C(Ph)R\right] + HC{\equiv}COEt \longrightarrow \left[(CO)_5W\!\cdots\!C(OEt)CH{=}C(Ph)R\right] \quad (141)$$

R = H, Ph, OMe

Insertion with concomitant carbon monoxide substitution was found in the reaction of bis(diethylamino)acetylene with pentacarbonyl[methoxy(phenyl)carbene]chromium and -tungsten[282] (equation 142).

$$\left[(CO)_5M\!\cdots\!C(OMe)Ph\right] + Et_2NC{\equiv}CNEt_2 \longrightarrow cis\text{-}\left[(CO)_4M\!\cdots\!C(NEt_2)\!-\!C(=C(Ph)OMe)\!-\!NEt_2\ (\text{ring})\right] \quad (142)$$

M = Cr, W

In addition to amino and alkoxyacetylenes, other molecules containing a polar triple bond can insert into the carbene–metal bond, e.g. dimethylcyanamide[283,284] (equation 143) or isonitriles[285,286] (equation 144). In the latter reaction a non-carbene complex

$$\left[(CO)_5M{=}C(R^1)R^2\right] + Me_2NC{\equiv}N \longrightarrow \left[(CO)_5M\!\cdots\!C(NMe_2)\!\cdots\!N{=}C(R^1)R^2\right] \quad (143)$$

M = Cr, W; R^1 = Me, Ph; R^2 = Ph, OMe

$$\left[(CO)_5Cr\!\cdots\!C(OMe)Me\right] + C{\equiv}NC_6H_{11} \longrightarrow [(CO)_5Cr\{H_{11}C_6N{=}C{=}C(Me)OMe\}] \quad (144)$$

(139)

(139) is formed, which can be used as precursor in the synthesis of new carbene complexes[285,286] (equation 145). The carbene complex **140**[287] produced in the reaction

$$139 \xrightarrow{HCl} \left[(CO)_5Cr{\cdots}C\begin{matrix}NHC_6H_{11}\\C(O)Me\end{matrix}\right]$$

$$139 \xrightarrow{(PhCO)_2O_2} \left[(CO)_5Cr{\cdots}C\begin{matrix}NHC_6H_{11}\\C(OMe){=}CH_2\end{matrix}\right] \qquad (145)$$

$$139 \xrightarrow{MeOH} \left[(CO)_5Cr{\cdots}C\begin{matrix}NHC_6H_{11}\\CMe(OMe)_2\end{matrix}\right]$$

of **117** with 1-methoxycyclopentene can also be regarded as the product of a formal insertion of the carbon–carbon double bond into the carbene–metal bond (equation 146). 2-Ethoxynorbornene and **117** gave a similar complex. Six- and eight-membered ring enol ethers, however, failed to react[288]. On the other hand, reaction of the

$$\left[(CO)_5W{=}C\begin{matrix}Ph\\Ph\end{matrix}\right] + \text{1-ethoxycyclopentene} \longrightarrow \left[(CO)_5W{\cdots}C\begin{matrix}OEt\\(CH_2)_3CH{=}CPh_2\end{matrix}\right] \qquad (146)$$

(117) (140)

non-cyclic ethyl vinyl ether with **117** gave a cyclopropane derivative, 1-ethoxy-2,2-diphenylcyclopropane, and hexacarbonyltungsten as the only metal complex isolated[289], whereas from the reaction of **117** with 1-methoxy-1-phenylethylene the carbene complex **115** was obtained in 24% yield[289] (equation 147).

$$117 \xrightarrow{CH_2=CHOEt} \text{1-ethoxy-2,2-diphenylcyclopropane } (65\%) + Ph_2C{=}CH_2\ (11\%) + [(CO)_6W]\ (45\%)$$

$$117 \xrightarrow{CH_2=C(OMe)Ph} \left[(CO)_5W{\cdots}C\begin{matrix}OMe\\Ph\end{matrix}\right] + CH_2{=}CPh_2 \qquad (147)$$

(115)

A formal carbonyl insertion was observed in the reaction of pentacarbonyl-[methoxy(phenyl)carbene]chromium with sodium alkoxide. A neutral carbene complex was obtained on subsequent alkylation with oxonium salts[290,291] (equation 148).

The reaction of $[(CO)_6M]$ (M = Cr, W) with $LiCH(SR^1)R^2$ solutions does not yield simple carbene complexes on alkylation but rather, with further carbonyl insertion, heterometallocyclic chelates[40] (equation 149).

Alkylidene complexes react with nitriles via insertion of the carbon–nitrogen triple

$$[(CO)_5Cr{\cdots}C(OMe)Ph] + NaOMe \longrightarrow [(CO)_5Cr{-}C(=O)CPh(OMe)_2]^- Na^+ \xrightarrow{Me_3O^+ \bar{B}F_4} [(CO)_5Cr{\cdots}C(OMe)CPh(OMe)_2] \quad (14\%) \qquad (148)$$

$$[(CO)_6M] \xrightarrow[(2)\ Et_3O^+ \bar{B}F_4]{(1)\ LiCH(SR^1)R^2} [(CO)_4M{\cdots}C(OEt){-}C(OH){=}CR^2{-}S(R^1){\rightarrow}M] + [(CO)_4M{\cdots}C(OEt){-}C(OEt){\cdots}CR^2{-}S(R^1){\rightarrow}M] \qquad (149)$$

M = Cr, W; R^1 = Me, Ph; R^2 = Ph, SPh

bond into the metal–carbon bond. The reaction product, however, is not an alkylidene or carbene complex (as in equation 143) but an imido complex[174] owing to the reversed polarization of the metal–alkylidene carbon bond in relation to carbene complexes (equation 150).

$$[(Me_3CCH_2)_3Ta{=}C(CMe_3)H] + MeCN \longrightarrow [(Me_3CCH_2)_3Ta{=}N{-}C(Me){=}C(CMe_3)H] \qquad (150)$$

B. Ligand Substitution Reactions

Ligand substitution reactions are known for many types of carbene complexes. For Fischer-type carbonylcarbene complexes the displacement of carbon monoxide by phosphines has been particularly well studied[292,293] (e.g. equation 151). Small amounts of the pentacarbonyl(phosphine) and *trans*-tetracarbonyl(bisphosphine) complexes are formed in this reaction (equation 151), depending on the reaction conditions. A series of alkoxy(organyl)carbene complexes of chromium, molybdenum, and tungsten and of phosphines as well as phosphites[292], arsines, and stibines[294] were employed. Phosphorus trihalides[295] and phosphine[296,297], however, do not react with **110** via

$$\underset{(\mathbf{110})}{[(CO)_5Cr{\cdots}C(OMe)Me]} + PBu_3 \xrightarrow{-CO} \textit{cis-} + \textit{trans-}\underset{(\mathbf{141})}{[(CO)_4PBu_3Cr{\cdots}C(OMe)Me]} \qquad (151)$$

displacement of one carbon monoxide, but rather by substitution of the carbene ligand.

Comprehensive kinetic and mechanistic investigations of the carbene complex/tertiary phosphine system showed that the formation of the substitution products such as **141** followed an additive rate law of first and second order: $-d[\mathbf{110}]/dt = k_1[\mathbf{110}] + k_2[\mathbf{110}][PR_3]$. The initial reaction product is the *cis*-isomer; however, an equilibrium mixture of *cis*- and *trans*-isomers is obtained since equilibration is faster than the substitution reaction. The first-order term indicates a rate-limiting dissociation of CO from the complex as observed in the ligand substitution reactions of many carbonyl complexes. The second-order term, which is negligible for some phosphines, suggests an associative mechanism: nucleophilic addition of a phosphine to the carbene carbon atom yields an adduct which is in equilibrium with starting materials. On dissociation of one carbon monoxide (probably a *cis*-CO) this adduct gives a coordinatively unsaturated intermediate which is likely to react further by migration of the phosphine, bonded to the carbene carbon atom, to the metal. Capture of an external phosphine by the intermediate, however, cannot be totally excluded[229,298,299].

The isomerization reaction was also studied and an intramolecular mechanism was proposed for the rearrangement[300]. Whereas thermally induced displacement of CO by PR_3 yields an equilibrium mixture of the *cis*- and *trans*-isomers, irradiation of solutions containing **110** and phosphine gives only the *cis*-form[299]. The similar photochemical reaction of pentacarbonyl[chloro(dimethylamino)carbene]chromium with triethylphosphine also produced the *cis*-isomer[95] (equation 152). In contrast, the

$$[(CO)_5Cr{=}C(NMe_2)Cl] + PEt_3 \xrightarrow{-CO} cis\text{-}[(CO)_4PEt_3Cr{=}C(NMe_2)Cl] \quad (152)$$

thermal substitution of CO by PPh_3 in **142**[143] or **143**[216] was found to give the *trans*-isomers (equation 153 and 154). In compounds such as **141** only one carbon

$$[(CO)_4Os{=}\overline{C{-}S{-}CH{=}C(Me){-}N}(Me)] + PPh_3 \longrightarrow trans\text{-}[(CO)_3PPh_3Os{=}\overline{C{-}S{-}CH{=}C(Me){-}N}(Me)] + CO \quad (153)$$

(142)

$$[(CO)_4Fe{=}C(OEt)C_6Cl_5] + PPh_3 \longrightarrow trans\text{-}[(CO)_3PPh_3Fe{=}C(OEt)C_6Cl_5] + CO \quad (154)$$

monoxide ligand could be displaced by PR_3 [a second molecule of PR_3 displaces the carbene ligand, forming *trans*-tetracarbonyl(bisphosphine) complexes]. In carbonylcarbene complexes containing a cyclic bisaminocarbene ligand, however, two

$$\left[(CO)_5W\overset{\dots}{=}C\begin{matrix}N(Et)\\N(Et)\end{matrix}\right] \xrightarrow[-CO]{P(OMe)_3} fac\text{-}\left[(CO)_3[P(OMe)_3]_2W\overset{\dots}{=}C\begin{matrix}N(Et)\\N(Et)\end{matrix}\right] \quad (155)$$

CO ligands could be substituted for $P(OR)_3$[159], e.g. equation 155. In **144** finally all carbon monoxide ligands were found to be displaced by a chelating bisphosphine molecule[301] (equation 156). The reverse process, exchange of phosphine for carbon monoxide, was also observed for some rhodium(I) complexes[302], e.g. equation 157.

$$\left[(MeCp)(CO)_2Mn\overset{\dots}{=}C\begin{matrix}OMe\\Me\end{matrix}\right] + MeN(PF_2)_2 \xrightarrow[-2\,CO]{h\nu}$$

$$\left[(MeCp)MeN\begin{matrix}PF_2\\PF_2\end{matrix}Mn\overset{\dots}{=}C\begin{matrix}OMe\\Me\end{matrix}\right] \quad (156)$$

(144)

$$\left[(PPh_3)_2ClRh\overset{\dots}{=}C\begin{matrix}N(Me)\\N(Me)\end{matrix}\right] + CO \xrightarrow{-PPh_3} \left[CO(PPh_3)ClRh\overset{\dots}{=}C\begin{matrix}N(Me)\\N(Me)\end{matrix}\right] \quad (157)$$

A multitude of other ligand substitution reactions have been reported, especially for carbene complexes containing cyclic carbene ligands. Some examples are shown in the equations 158[303], 159[304], and 160[305]. Intramolecular substitution reactions have also been observed (see equations 142 and 161[306]).

$$trans\text{-}[Cl_2PEt_3PtL] \xrightarrow{LiBr} trans\text{-}[Br_2PEt_3PtL]$$

$$trans\text{-}[Cl_2PEt_3PtL] \xrightarrow{LiMe} trans\text{-}[Me_2PEt_3PtL] \quad (158)$$

$$trans\text{-}[Cl_2PEt_3PtL] \xrightarrow{Et_3SiH} trans\text{-}[Cl(H)PEt_3PtL]$$

$$L = \overset{\dots}{=}C\begin{matrix}NMe\\NMe\end{matrix}$$

$$\textit{trans}\text{-}[Cl_2RuL_4] \begin{cases} \xrightarrow{CO} \textit{trans}\text{-}[Cl(CO)RuL_4]^+Cl^- \\ \xrightarrow{PF_3} \textit{trans}\text{-}[Cl(PF_3)RuL_4]^+Cl^- \\ \xrightarrow{CO,\,py} \textit{trans}\text{-}[Cl(CO)py_2RuL_2]^+Cl^- \\ \xrightarrow{P(OMe)_3} [Cl[P(OMe)_3]_3RuL_2]^+Cl^- \end{cases} \quad (159)$$

L = :C(N(Me)CH$_2$CH$_2$N(Me)) (cyclic)

$$\textit{cis}\text{-}[Cl_2(CNR)Pt{-}C(OEt)(NHR)] \xrightarrow{Na[CpMo(CO)_3]} [Cp(CO)_3Mo{-}Pt(CNR)(=C(OEt)NHR){-}Mo(CO)_3Cp] \quad (160)$$

$$[Cl_2(PPh_3)Pd{=}C(NHPh)(NH(CH_2)_2NH_2)] \xrightarrow{AgBF_4} [Cl_2(PPh_3)Pd{=}C(NHPh)(NH(CH_2)_2NH_2{-}\kappa N)]^+ \bar{B}F_4 \quad (161)$$

C. Oxidative Addition Reactions

Oxidative addition to carbene complexes has also been used for the modification of already existing carbene complexes. Employing this method, gold(III) complexes can be synthesized from gold(I) compounds[307] (equation 162), and similarly platinum(IV)

$$[Au(=C(NHPh)_2)_2]^+ ClO_4^- \xrightarrow{I_2} [I_2Au(=C(NHPh)_2)_2]^+ ClO_4^- \quad (162)$$

from platinum(II) compounds[308]. The reaction of **145** with iodine gave a seven-coordinate species[159] (equation 163). In addition, methyl iodide has also been used in oxidative addition reactions[240] (equation 164). A high electron density at the metal in **146** seems to be decisive for the reaction, since the analogous complex in which Me is replaced by Cl failed to react. Several oxidative addition reactions were performed with compound **147**[142], e.g. equation 165.

$$[(CO)_5W{=}C(NMe)CH_2CH_2(NMe)] + I_2 \xrightarrow{-CO} [I_2(CO)_4W{=}C(NMe)CH_2CH_2(NMe)] \quad (163)$$

(145)

$$trans\text{-}[Me(PMe_2Ph)Pt{=}C(OCH_2CH_2CH_2)]^+ \xrightarrow{MeI} [I(Me)_2(PMe_2Ph)Pt{=}C(OCH_2CH_2CH_2)]^+ \quad (164)$$

(146)

$$trans\text{-}[CO(PPh_3)_2Ir{=}C(SCH{=}C(Me)NMe)]^+ \xrightarrow{H_2} trans\text{-}[(H)_2CO(PPh_3)_2Ir{=}C(SCH{=}C(Me)NMe)]^+$$

$$trans\text{-}[CO(PPh_3)_2Ir{=}C(SCH{=}C(Me)NMe)]^+ \xrightarrow{HCl} trans\text{-}[H(Cl)CO(PPh_3)_2Ir{=}C(SCH{=}C(Me)NMe)]^+ \quad (165)$$

(147)

V. OXIDATION OF CARBENE COMPLEXES

One-electron oxidation of carbene complexes using silver salts as oxidants was employed to generate paramagnetic carbene complexes, e.g. equation 166. In addition

$$[(CO)_3Fe(=C(NMe)CH_2CH_2(NMe))_2] \xrightarrow{AgBF_4/thf} [(CO)_3Fe(=C(NMe)CH_2CH_2(NMe))_2]^+ \ \bar{B}F_4 \quad (166)$$

(148)

to **148**, other paramagnetic carbene—iron(I) salts which are stable at room temperature have also been prepared. Infrared and e.s.r. measurements indicate that the odd electron is substantially metal-centred. The carbene ligand contributes to stability by its strong Fe—C bond and delocalization of the positive charge[309]. A stable paramagnetic chromium(I)–carbene complex has also been synthesized and characterized[310].

VI. REFERENCES

1. E. O. Fischer and A. Maasböl, *Angew. Chem.*, **76**, 645 (1964); *Angew. Chem. Int. Ed. Engl.*, **3**, 580 (1964).
2. L. Chugaev and M. Skanavy-Grigorizeva, *J. Russ. Chem. Soc.*, **47**, 776 (1915).
3. L. Chugaev, M. Skanavy-Grigorizeva, and A. Posnjak, *Z. Anorg. Chem.*, **148**, 37 (1925).

4. R. B. King, *J. Am. Chem. Soc.*, **85**, 1922 (1963).
5. G. Rouschias and B. L. Shaw, *J. Chem. Soc., Chem. Commun.*, 183 (1970).
6. C. P. Casey, *J. Chem. Soc., Chem. Commun.*, 1220 (1970).
7. D. L. Cardin, B. Cetinkaya, and M. F. Lappert, *Chem. Rev.*, **72**, 545 (1972).
8. F. A. Cotton and C. M. Lukehart, *Progr. Inorg. Chem,.*, **16**, 487 (1972).
9. C. P. Casey, in *Transition Metal Organometallics in Organic Synthesis*, (Ed. H. Alper), Vol. 1, Academic Press, New York, 1976, pp. 189–233.
10. F. J. Brown, *Progr. Inorg. Chem.*, **27**, 1 (1980).
11. E. O. Fischer, *Pure Appl. Chem.*, **24**, 407 (1970); **30**, 353 (1972).
12. C. G. Kreiter and E. O. Fischer, in *XXIIIrd International Congress of Pure and Applied Chemistry*, Vol. 6, Butterworths, London, 1971, pp. 151–168.
13. E. O. Fischer, U. Schubert, and H. Fischer, *Pure Appl. Chem.*, **50**, 857 (1978).
14. E. O. Fischer, *Angew. Chem.*, **86**, 651 (1974); *Adv. Organomet. Chem.*, **14**, 1 (1976).
15. M. F. Lappert, *J. Organomet. Chem.*, **100**, 139 (1975).
16. R. R. Schrock, *Acc. Chem. Res.*, **12**, 98 (1979).
17. M. Ryang, I. Rhee, and S. Tsutsumi, *Bull. Chem. Soc. Jap.*, **37**, 341 (1964).
18. E. O. Fischer and V. Kiener, *J. Organomet. Chem.*, **23**, 215 (1970).
19. E. O. Fischer, G. Kreis, and F. R. Kreissl, *J. Organomet. Chem.*, **56**, C37 (1973).
20. E. O. Fischer and W. Kalbfus, *J. Organomet. Chem.*, **46**, C15 (1972).
21. R. Aumann and E. O. Fischer, *Angew. Chem.*, **79**, 900 (1967); *Angew. Chem. Int. Ed. Engl.*, **6**, 879 (1967).
22. C. P. Casey, C. R. Cyr, and R. A. Boggs, *Synth. Inorg. Met. -Org. Chem.*, **3**, 249 (1973).
23. E. O. Fischer and A. Maasböl, *J. Organomet. Chem.*, **12**, P15 (1968).
24. J. A. Connor and E. M. Jones, *J. Chem. Soc., Chem. Commun.*, 570 (1971).
25. J. A. Connor and E. M. Jones, *J. Chem. Soc. A*, 3368 (1971).
26. E. O. Fischer, T. Selmayr, and F. R. Kreissl, *Chem. Ber.*, **110**, 2947 (1977).
27. E. Moser and E. O. Fischer, *J. Organomet. Chem.*, **12**, P1 (1968).
28. E. O. Fischer, T. Selmayr, F. R. Kreissl, and U. Schubert, *Chem. Ber.*, **110**, 2574 (1977).
29. E. O. Fischer and S. Fontana, *J. Organomet. Chem.*, **40**, 159 (1972).
30. H. G. Raubenheimer and E. O. Fischer, *J. Organomet. Chem.*, **91**, C23 (1975).
31. E. O. Fischer, H.-J. Kollmeier, C. G. Kreiter, J. Müller, and R. D. Fischer, *J. Organomet. Chem.*, **22**, C39 (1970).
32. J. A. Connor and E. M. Jones, *J. Chem. Soc. A*, 1974 (1971).
33. J. W. Wilson and E. O. Fischer, *J. Organomet. Chem.*, **57**, C63 (1973).
34. E. O. Fischer and F. R. Kreissl, *J. Organomet. Chem.*, **35**, C47 (1972).
35. E. O. Fischer and H.-J. Kollmeier, *Angew. Chem.*, **82**, 325 (1970); *Angew. Chem. Int. Ed. Engl.*, **9**, 309 (1970).
36. M. J. Doyle, M. F. Lappert, G. M. McLaughlin, and J. McMeeking, *J. Chem. Soc. Dalton Trans.*, 1494 (1974).
37. E. O. Fischer, H. Hollfelder, F. R. Kreissl, and W. Üdelhoven, *J. Organomet. Chem.*, **113**, C31 (1976).
38. E. O. Fischer, H. Hollfelder, P. Friedrich, F. R. Kreissl, and G. Huttner, *Chem. Ber.*, **110**, 3467 (1977).
39. G. A. Moser, E. O. Fischer, and M. D. Rausch, *J. Organomet. Chem.*, **27**, 379 (1971).
40. H. G. Raubenheimer, S. Lotz, H. W. Viljoen, and A. A. Chalmers, *J. Organomet. Chem.*, **152**, 73 (1978).
41. E. O. Fischer, K. Scherzer, and F. R. Kreissl, *J. Organomet. Chem.*, **118**, C33 (1976).
42. D. J. Darensbourg and M. Y. Darensbourg, *Inorg. Chim. Acta*, **5**, 247 (1971).
43. E. O. Fischer, F. R. Kreissl, E. Winkler, and C. G. Kreiter, *Chem. Ber.*, **105**, 588 (1972).
44. M. J. Webb, R. P. Stewart, Jr., and W. A. G. Graham, *J. Organomet. Chem.*, **59**, C21 (1973).
45. J. Chatt, G. J. Leigh, C. J. Pickett, and D. R. Stanley, *J. Organomet. Chem.*, **184**, C64 (1980).
46. E. O. Fischer, P. Stückler, H.-J. Beck, and F. R. Kreissl, *Chem. Ber.*, **109**, 3089 (1976).
47. P. B. Hitchcock, M. F. Lappert, and P. L. Pye, *J. Chem. Soc., Dalton Trans.*, 2160 (1977).
48. E. O. Fischer and R. Aumann, *Chem. Ber.*, **102**, 1495 (1969).
49. C. L. Hyde and D. J. Darensbourg, *Inorg. Chim. Acta*, **7**, 145 (1973).
50. H. Fischer and E. O. Fischer, *Chem. Ber.*, **107**, 673 (1974).

51. E. O. Fischer and H. Fischer, *Chem. Ber.*, **107**, 657 (1974).
52. M. J. Webb, M. J. Bennett, L. Y. Y. Chan, and W. A. G. Graham, *J. Am. Chem. Soc.*, **96**, 5931 (1974).
53. W. Petz, *J. Organomet. Chem.*, **72**, 369 (1974).
54. W. Petz, *J. Organomet. Chem.*, **55**, C42 (1973).
55. W. Petz and G. Schmid, *Angew. Chem.*, **84**, 997 (1972); *Angew. Chem. Int. Ed. Engl.*, **11**, 934 (1972).
56. W. Petz and A. Jonas, *J. Organomet. Chem.*, **120**, 423 (1976).
57. W. Petz, *J. Organomet. Chem.*, **165**, 199 (1979).
58. P. T. Wolczanski, R. S. Threlkel, and J. E. Bercaw, *J. Am. Chem. Soc.*, **101**, 218 (1979).
59. E. O. Fischer, F. R. Kreissl, C. G. Kreiter, and E. W. Meineke, *Chem. Ber.*, **105**, 2558 (1972).
60. E. M. Badley, J. Chatt, R. L. Richards, and G. A. Sim, *J. Chem. Soc., Chem. Commun.*, 1322 (1969).
61. E. M. Badley, J. Chatt, and R. L. Richards, *J. Chem. Soc. A*, 21 (1971).
62. K. Bartel and W. P. Fehlhammer, *Angew. Chem.*, **86**, 588 (1974); *Angew. Chem. Int. Ed. Engl.*, **13**, 600 (1974).
63. H. C. Clark and L. E. Manzer, *Inorg. Chem.*, **11**, 503 (1972).
64. B. Crociani, T. Boschi, M. Nicolini, and U. Belluco, *Inorg. Chem.*, **11**, 1292 (1972).
65. U. Schöllkopf and F. Gerhart, *Angew. Chem.*, **79**, 990 (1967); *Angew. Chem. Int. Ed. Engl.*, **6**, 970 (1967).
66. F. Bonati and G. Minghetti, *Gazz. Chim. Ital.*, **103**, 373 (1973).
67. J. E. Parks and A. L. Balch, *J. Organomet. Chem.*, **57**, C103 (1973).
68. P. R. Branson and M. Green, *J. Chem. Soc., Dalton Trans.*, 1303 (1972).
69. P. R. Branson, R. A. Cable, M. Green, and M. K. Lloyd, *J. Chem. Soc., Dalton Trans.*, 12 (1976).
70. C. H. Davies, C. H. Game, M. Green, and F. G. A. Stone, *J. Chem. Soc., Dalton Trans.*, 357 (1974).
71. M. Wada, S. Kanai, R. Maeda, M. Kinoshita, and K. Oguro, *Inorg. Chem.*, **18**, 417 (1979).
72. J. S. Miller and A. L. Balch, *Inor. Chem.*, **11**, 2069 (1972).
73. S. Z. Goldberg, R. Eisenberg, and J. S. Miller, *Inorg. Chem.*, **16**, 1502 (1977).
74. D. J. Doonan and A. L. Balch, *Inorg. Chem.*, **13**, 921 (1974).
75. D. J. Doonan and A. L. Balch, *J. Am. Chem. Soc.*, **95**, 4769 (1973).
76. J. Chatt, R. L. Richards, and G. H. D. Royston, *J. Chem. Soc., Dalton Trans.*, 1433 (1973).
77. G. Rouschias and B. L. Shaw, *J. Chem. Soc. A*, 2097 (1971).
78. A. Burke, A. L. Balch, and J. H. Enemark, *J. Am. Chem. Soc.*, **92**, 2555 (1970).
79. W. M. Butler, J. H. Enemark, J. Parks, and A. L. Balch, *Inorg. Chem.*, **12**, 451 (1973).
80. W. M. Butler and J. H. Enemark, *Inorg. Chem.*, **10**, 2416 (1971).
81. J. A. McCleverty and J. Williams, *Transition Met. Chem.*, **1**, 288 (1976).
82. J. A. Connor, E. M. Jones, G. K. McEwen, M. K. Lloyd, and J. A. McCleverty, *J. Chem. Soc., Dalton Trans.*, 1246 (1972).
83. E. O. Fischer, P. Stückler, and F. R. Kreissl, *J. Organomet. Chem.*, **129**, 197 (1977).
84. E. O. Fischer and G. Besl, *Z. Naturforsch.*, **34B**, 1186 (1979).
85. E. O. Fischer, E. W. Meineke, and F. R. Kreissl, *Chem. Ber.*, **110**, 1140 (1977).
86. E. O. Fischer, R. L. Clough, G. Besl, and F. R. Kreissl, *Angew. Chem.*, **88**, 584 (1976); *Angew. Chem. Int. Ed. Engl.*, **15**, 543 (1976).
87. E. O. Fischer, R. L. Clough, and P. Stückler, *J. Organomet. Chem.*, **120**, C6 (1976).
88. F. R. Kreissl, P. Stückler, and E. W. Meineke, *Chem. Ber.*, **110**, 3040 (1977).
89. E. O. Fischer, W. Schambeck, and F. R. Kreissl, *J. Organomet. Chem.*, **169**, C27 (1979).
90. F. R. Kreissl and P. Friedrich, *Angew. Chem.*, **89**, 553 (1977); *Angew. Chem. Int. Ed. Engl.*, **16**, 543 (1977).
91. E. O. Fischer, P. Stückler, H.-J. Beck, and F. R. Kreissl, *Chem. Ber.*, **109**, 3089 (1976).
92. E. O. Fischer and A. Frank, *Chem. Ber.*, **111**, 3740 (1978).
93. A. Motsch, *Dissertation*, Tech. Univ. München, 1980.
94. E. O. Fischer, W. Kleine, G. Kreis, and F. R. Kreissl, *Chem. Ber.*, **111**, 3542 (1978).
95. A. J. Hartshorn and M. F. Lappert, *J. Chem. Soc., Chem. Commun.*, 761 (1976).
96. E. O. Fischer, W. Kleine, and F. R. Kreissl, *Angew. Chem.*, **88**, 646 (1976); *Angew. Chem. Int. Ed. Engl.*, **15**, 616 (1976).

97. E. O. Fischer, W. Kleine, F. R. Kreissl, H. Fischer, P. Friedrich, and G. Huttner, *J. Organomet. Chem.*, **128**, C49 (1977).
98. E. O. Fischer, R. B. A. Pardy, and U. Schubert, *J. Organomet. Chem.*, **181**, 37 (1979).
99. H. Fischer, A. Motsch, and W. Kleine, *Angew. Chem.*, **90**, 914 (1978); *Angew. Chem. Int. Ed. Engl.*, **17**, 842 (1978).
100. E. O. Fischer, H. Fischer, U. Schubert, and R. B. A. Pardy, *Angew. Chem.*, **91**, 929 (1979); *Angew. Chem. Int. Ed. Engl.*, **18**, 871 (1979).
101. H. Fischer and A. Motsch, unpublished results.
102. H. Fischer, *J. Organomet. Chem.*, **195**, 55 (1980).
103. H. Fischer, unpublished results.
104. H. Fischer, A. Motsch, U. Schubert, and D. Neugebauer, *Angew. Chem.*, **93**, 483 (1981); *Angew. Chem. Int. Ed. Engl.*, **20**, 463 (1981).
105. U. Schöllkopf and F. Gerhart, *Angew. Chem.*, **79**, 578 (1967); *Angew. Chem. Int. Ed. Engl.*, **6**, 560 (1967).
106. M. L. H. Green, L. C. Mitchard, and M. G. Swanwick, *J. Chem. Soc. A*, 794 (1971).
107. H. Felkin, B. Meunier, C. Pascard, and T. Prange, *J. Organomet. Chem.*, **135**, 361 (1977).
108. A. Davison and D. L. Reger, *J. Am. Chem. Soc.*, **94**, 9237 (1972).
109. M. Wada, N. Asada, and K. Oguro, *Inorg. Chem.*, **17**, 2353 (1978).
110. D. H. Bowen, M. Green, D. M. Grove, J. R. Moss, and F. G. A. Stone, *J. Chem. Soc., Dalton Trans.*, 1189 (1974).
111. C. H. Game, M. Green, J. R. Moss, and F. G. A. Stone, *J. Chem. Soc., Dalton Trans.*, 351 (1974).
112. R. B. King, *J. Am. Chem. Soc.*, **85**, 1922 (1963).
113. C. P. Casey, *J. Chem. Soc., Chem. Commun.*, 1220 (1970).
114. C. P. Casey and R. L. Anderson, *J. Am. Chem. Soc.*, **93**, 3554 (1971).
115. W. K. Dean and W. A. G. Graham, *J. Organomet. Chem.*, **120**, 73 (1976).
116. C. P. Casey, C. R. Cyr, R. L. Anderson, and D. F. Marten, *J. Am. Chem. Soc.*, **97**, 3053 (1975).
117. J. R. Moss, M. Green, and F. G. A. Stone, *J. Chem. Soc., Dalton Trans.*, 975 (1973).
118. F. A. Cotton and C. M. Lukehart, *J. Am. Chem. Soc.*, **93**, 2672 (1971).
119. T. Kruck and L. Liebig, *Chem. Ber.*, **106**, 1055 (1973).
120. E. D. Dobrzynski and R. J. Angelici, *Inorg. Chem.*, **14**, 1513 (1975).
121. F. B. McCormick and R. J. Angelici, *Inorg. Chem.*, **18**, 1231 (1979).
122. D. F. Christian, G. R. Clark, W. R. Roper, J. M. Waters, and K. R. Whittle, *J. Chem. Soc., Chem. Commun.*, 458 (1972).
123. D. F. Christian, G. R. Clark, and W. R. Roper, *J. Organomet. Chem.*, **81**, C7 (1974).
124. Y. Yamamoto and H. Yamazaki, *Bull. Chem. Soc. Jap.*, **48**, 3691 (1975).
125. M. H. Chisholm and H. C. Clark, *Inorg. Chem.*, **10**, 1711 (1971).
126. M. H. Chisholm, and H. C. Clark, *J. Am. Chem. Soc.*, **94**, 1532 (1972).
127. H. C. Clark and L. E. Manzer, *J. Organomet. Chem.*, **47**, C17 (1973).
128. T. G. Attig and H. C. Clark, *Can. J. Chem.*, **53**, 3466 (1975).
129. G. K. Anderson, R. J. Cross, L. Manojlović-Muir, K. W. Muir, and R. A. Wales, *J. Chem. Soc., Dalton Trans.*, 684 (1979).
130. K. Oguro, M. Wada, and R. Okawara, *J. Chem. Soc., Chem. Commun.*, 899 (1975).
131. R. A. Bell and M. H. Chisholm, *J. Chem. Soc., Chem. Commun.*, 818 (1974).
132. R. A. Bell, M. H. Chisholm, D. A. Couch, and L. A. Rankel, *Inorg. Chem.*, **16**, 677 (1977).
133. K. Öfele, *Angew. Chem.*, **80**, 1032 (1968); *Angew. Chem. Int. Ed. Engl.*, **7**, 950 (1968).
134. K. Öfele, *J. Organomet. Chem.*, **22**, C9 (1970).
135. K. Öfele, *J. Organomet. Chem.*, **12**, P42 (1968).
136. K. Öfele, *Angew. Chem.*, **81**, 936 (1969); *Angew. Chem. Int. Ed. Engl.*, **8**, 916 (1969).
137. C. G. Kreiter, K. Öfele, and G. W. Wieser, *Chem. Ber.*, **109**, 1749 (1976).
138. K. Öfele, E. Roos, and M. Herberhold, *Z. Naturforsch.*, **31B**, 1070 (1976).
139. R. Gompper and E. Bartmann, *Angew. Chem.*, **90**, 490 (1978); *Angew. Chem. Int. Ed. Engl.*, **17**, 456 (1978).
140. R. Weiss and C. Priesner, *Angew. Chem.*, **90**, 491 (1978); *Angew. Chem. Int. Ed. Engl.*, **17**, 457 (1978).
141. Y. Kawada and W. M. Jones, *J. Organomet. Chem.*, **192**, 87 (1980).
142. P. J. Fraser, W. R. Roper, and F. G. A. Stone, *J. Chem. Soc., Dalton Trans.*, 760 (1974).

143. M. Green, F. G. A. Stone, and M. Underhill, *J. Chem. Soc., Dalton Trans.*, 939 (1975).
144. H.-W. Wanzlick and H.-J. Schönherr, *Angew. Chem.*, **80**, 154 (1968); *Angew. Chem. Int. Ed. Engl.*, **7**, 141 (1968).
145. P. J. Fraser, W. R. Roper, and F. G. A. Stone, *J. Chem. Soc., Dalton Trans.*, 102 (1974).
146. R. J. Sundberg, R. F. Bryan, I. F. Taylor, Jr., and H. Taube, *J. Am. Chem. Soc.*, **96**, 381 (1974).
147. M. J. Clarke and H. Taube, *J. Am. Chem. Soc.*, **97**, 1397 (1975).
148. C. W. Fong and G. Wilkinson, *J. Chem. Soc., Dalton Trans.*, 1100 (1975).
149. A. J. Hartshorn, M. F. Lappert, and K. Turner, *J. Chem. Soc., Dalton Trans.*, 348 (1978).
150. B. Cetinkaya, M. F. Lappert, and K. Turner, *J. Chem. Soc., Chem. Commun.*, 851 (1972).
151. B. Cetinkaya, M. F. Lappert, G. M. McLaughlin, and K. Turner, *J. Chem. Soc., Dalton Trans.*, 1591 (1974).
152. A. J. Hartshorn, M. F. Lappert, and K. Turner, *J. Chem. Soc., Chem. Commun.*, 929 (1975).
153. W. Petz, *Angew. Chem.*, **87**, 288 (1975); *Angew. Chem. Int. Ed. Engl.*, **14**, 367 (1975).
154. M. F. Lappert and A. J. Oliver, *J. Chem. Soc., Dalton Trans.*, 65 (1974).
155. P. B. Hitchcock, M. F. Lappert, G. M. McLaughlin, and A. J. Oliver, *J. Chem. Soc., Dalton Trans.*, 68 (1974).
156. D. J. Cardin, B. Cetinkaya, M. F. Lappert, L. Manojlović-Muir, and K. W. Muir, *J. Chem. Soc., Chem. Commun.*, 400 (1971).
157. D. J. Cardin, M. J. Doyle, and M. F. Lappert, *J. Chem. Soc., Chem. Commun.*, 927 (1972).
158. M. F. Lappert, P. L. Pye, and G. M. McLaughlin, *J. Chem. Soc., Dalton Trans.*, 1272 (1977).
159. M. F. Lappert and P. L. Pye, *J. Chem. Soc., Dalton Trans.*, 1283 (1977).
160. M. F. Lappert and P. L. Pye, *J. Chem. Soc., Dalton Trans.*, 2172 (1977).
161. B. Cetinkaya, P. Dixneuf, and M. F. Lappert, *J. Chem. Soc., Dalton Trans.*, 1827 (1974).
162. M. F. Lappert and P. L. Pye, *J. Chem. Soc., Dalton Trans.*, 837 (1978).
163. B. Cetinkaya, P. Dixneuf, and M. F. Lappert, *J. Chem. Soc., Chem. Commun.*, 206 (1973).
164. P. B. Hitchcock, M. F. Lappert, and P. L. Pye, *J. Chem. Soc., Chem. Commun.*, 644 (1976).
165. P. B. Hitchcock, M. F. Lappert, P. L. Pye, and S. Thomas, *J. Chem. Soc., Dalton Trans.*, 1929 (1979).
166. B. Cetinkaya, P. B. Hitchcock, M. F. Lappert, and P. L. Pye, *J. Chem. Soc., Chem. Commun.*, 683 (1975).
167. W. A. Herrmann, *Angew. Chem.*, **86**, 556 (1974); *Angew. Chem. Int. Ed. Engl.*, **13**, 599 (1974).
168. W. A. Herrmann, *Chem. Ber.*, **108**, 486 (1975).
169. W. A. Herrmann, *Chem. Ber.*, **108**, 3412 (1975).
170. W. A. Herrmann and J. Plank, *Angew. Chem.*, **90**, 555 (1978); *Angew. Chem. Int. Ed. Engl.*, **17**, 525 (1978).
171. W. A. Herrmann, J. Plank, M. L. Ziegler, and K. Weidenhammer, *Angew. Chem.*, **90**, 817 (1978); *Angew. Chem. Int. Ed. Engl.*, **17**, 777 (1978).
172. M. F. Lappert and D. B. Shaw, *J. Chem. Soc., Chem. Commun.*, 146 (1978).
173. R. R. Schrock, *J. Am. Chem. Soc.*, **96**, 6796 (1974).
174. R. R. Schrock and J. D. Fellmann, *J. Am. Chem. Soc.*, **100**, 3359 (1978).
175. R. R. Schrock, *J. Am. Chem. Soc.*, **98**, 5399 (1976).
176. C. D. Wood, S. J. McLain, and R. R. Schrock, *J. Am. Chem. Soc.*, **101**, 3210 (1979).
177. R. R. Schrock, L. W. Messerle, C. D. Wood, and L. J. Guggenberger, *J. Am. Chem. Soc.*, **100**, 3793 (1978).
178. S. J. McLain, C. D. Wood, L. W. Messerle, R. R. Schrock, F. J. Hollander, W. J. Youngs, and M. R. Churchill, *J. Am. Chem. Soc.*, **100**, 5962 (1978).
179. R. R. Schrock, *J. Am. Chem. Soc.*, **97**, 6577 (1975).
180. R. R. Schrock and P. R. Sharp, *J. Am. Chem. Soc.*, **100**, 2389 (1978).
181. P. R. Sharp and R. R. Schrock, *J. Organomet. Chem.*, **171**, 43 (1979).
182. J. D. Fellmann, G. A. Rupprecht, C. D. Wood, and R. R. Schrock, *J. Am. Chem. Soc.*, **100**, 5964 (1978).
183. M. R. Churchill and W. J. Youngs, *Inorg. Chem.*, **18**, 1930 (1979).
184. D. N. Clark and R. R. Schrock, *J. Am. Chem. Soc.*, **100**, 6774 (1978).
185. M. F. Lappert and C. R. C. Milne, *J. Chem. Soc., Chem. Commun.*, 925 (1978).
186. R. A. Andersen, *Inorg. Chem.*, **18**, 3622 (1979).

187. P. R. Sharp, D. Astruc, and R. R. Schrock, *J. Organomet. Chem.*, **182**, 477 (1979).
188. A. Sanders, L. Cohen, W. P. Giering, D. Kenedy, and C. V. Magatti, *J. Am. Chem. Soc.*, **95**, 5430 (1973).
189. A. Sanders, T. Bauch, C. V. Magatti, C. Lorenc, and W. P. Giering, *J. Organomet. Chem.*, **107**, 359 (1976).
190. A. R. Cutler, *J. Am. Chem. Soc.*, **101**, 604 (1979).
191. W.-K. Wong, W. Tam, and J. A. Gladysz, *J. Am. Chem. Soc.*, **101**, 5440 (1979).
192. M. Brookhart and G. O. Nelson, *J. Am. Chem. Soc.*, **99**, 6099 (1977).
193. P. W. Jolly and R. Pettit, *J. Am. Chem. Soc.*, **88**, 5044 (1966).
194. M. L. H. Green, M. Ishaq, and R. N. Whiteley, *J. Chem. Soc. A*, 1508 (1967).
195. M. Brookhart, J. R. Tucker, T. C. Flood and J. Jensen, *J. Am. Chem. Soc.*, **102**, 1203 (1980).
196. A. Davison and J. P. Selegue, *J. Am. Chem. Soc.*, **102**, 2455 (1980).
197. D. L. Reger and M. D. Dukes, *J. Organomet. Chem.*, **153**, 67 (1978).
198. G. R. Clark, K. Marsden, W. R. Roper, and L. J. Wright, *J. Am. Chem. Soc.*, **102**, 1206 (1980).
199. D. Mansuy, M. Lange, J.-C. Chottard, P. Guerin, P. Morliere, D. Brault, and M. Rougee, *J. Chem. Soc., Chem. Commun.*, 648 (1977).
200. D. Mansuy, M. Lange, J.-C. Chottard, J. F. Bartoli, B. Chevrier, and R. Weiss, *Angew. Chem.*, **90**, 828 (1978); *Angew. Chem. Int. Ed. Engl.*, **17**, 781 (1978).
201. D. Mansuy, P. Guerin, and J.-C. Chottard, *J. Organomet. Chem.*, **171**, 195 (1979).
202. J.-P. Battioni, D. Mansuy, and J.-C. Chottard, *Inorg. Chem.*, **19**, 791 (1980).
203. N. T. Allison, Y. Kawada, and W. M. Jones, *J. Am. Chem. Soc.*, **100**, 5224 (1978).
204. P. E. Riley, R. E. Davis, N. T. Allison, and W. M. Jones, *J. Am. Chem. Soc.*, **102**, 2458 (1980).
205. F. Faraone, G. Tresoldi, and G. A. Loprete, *J. Chem. Soc., Dalton Trans.*, 933 (1979).
206. G. Tresoldi, F. Faraone, and P. Piraino, *J. Chem. Soc., Dalton Trans.*, 1053 (1979).
207. T. J. Collins, K. R. Grundy, W. R. Roper, and S. F. Wong, *J. Organomet. Chem.*, **107**, C37 (1976).
208. H. Le Bozec, A. Gorgues, and P. Dixneuf, *J. Chem. Soc., Chem. Commun.*, 573 (1978).
209. H. Le Bozec, A. Gorgues, and P. Dixneuf, *J. Am. Chem. Soc.*, **100**, 3946 (1978).
210. J. Y. Le Marouille, C. Lelay, A. Benoit, D. Grandjean, D. Touchard, H. Le Bozec, and P. Dixneuf, *J. Organomet. Chem.*, **191**, 133 (1980).
211. H. G. Raubenheimer and H. E. Swanepoel, *J. Organomet. Chem.*, **141**, C21 (1977).
212. J. Daub, U. Erhardt, J. Kappler, and V. Trautz, *J. Organomet. Chem.*, **69**, 423 (1974).
213. M. H. Quick and R. J. Angelici, *J. Organomet. Chem.*, **160**, 231 (1978).
214. W. Petz, *J. Organomet. Chem.*, **172**, 415 (1979).
215. J. Daub and J. Kappler, *J. Organomet. Chem.*, **80**, C5 (1974).
216. E. O. Fischer, H.-J. Beck, C. G. Kreiter, J. Lynch, J. Müller, and E. O. Winkler, *Chem. Ber.*, **105**, 162 (1972).
217. C. P. Casey and R. L. Anderson, *J. Chem. Soc., Chem. Commun.*, 895 (1975).
218. W. P. Fehlhammer, A. Mayr, and M. Ritter, *Angew. Chem.*, **89**, 660 (1977); *Angew. Chem. Int. Ed. Engl.*, **16**, 641 (1977).
219. W. Malisch, H. Blau, and S. Voran, *Angew. Chem.*, **90**, 827 (1978); *Angew. Chem. Int. Ed. Engl.*, **17**, 780 (1978).
220. W. Beck, H. Brix, and F. H. Köhler, *J. Organomet. Chem.*, **121**, 211 (1976).
221. P. Klemarczyk, T. Price, W. Priester, and M. Rosenblum, *J. Organomet. Chem.*, **139**, C25 (1977).
222. H. D. Empsall, E. M. Hyde, R. Markham, W. S. McDonald, M. C. Norton, B. L. Shaw, and B. Weeks, *J. Chem. Soc., Chem. Commun.*, 589 (1977).
223. H. G. Raubenheimer, S. Lotz, H. E. Swanepoel, H. W. Viljoen, and J. C. Rautenbach, *J. Chem. Soc., Dalton Trans.*, 1701 (1979).
224. T. F. Block and R. F. Fenske, *J. Organomet. Chem.*, **139**, 235 (1977).
225. T. F. Block, R. F. Fenske, and C. P. Casey, *J. Am. Chem. Soc.*, **98**, 441 (1976).
226. W. B. Perry, T. F. Schaaf, W. L. Jolly, L. J. Todd, and D. L. Cronin, *Inorg. Chem.*, **13**, 2038 (1974).
227. F. R. Kreissl, C. G. Kreiter, and E. O. Fischer, *Angew. Chem.*, **84**, 679 (1972); *Angew. Chem. Int. Ed. Engl.*, **11**, 643 (1972).
228. F. R. Kreissl, E. O. Fischer, C. G. Kreiter, and H. Fischer, *Chem. Ber.*, **106**, 1262 (1973).

229. H. Fischer, E. O. Fischer, C. G. Kreiter, and H. Werner, *Chem. Ber.*, **107**, 2459 (1974).
230. H. Fischer, *J. Organomet. Chem.*, **170**, 309 (1979).
231. F. R. Kreissl, E. O. Fischer, C. G. Kreiter, and K. Weiss, *Angew. Chem.*, **85**, 617 (1973); *Angew. Chem. Int. Ed. Engl.*, **12**, 563 (1973).
232. F. R. Kreissl and E. O. Fischer, *Chem. Ber.*, **107**, 183 (1974).
233. U. Klabunde and E. O. Fischer, *J. Am. Chem. Soc.*, **89**, 7141 (1967).
234. J. A. Connor and E. O. Fischer, *J. Chem. Soc. A*, 578 (1969).
235. E. O. Fischer and H.-J. Kollmeier, *Chem. Ber.*, **104**, 1339 (1971).
236. B. Heckl, H. Werner, and E. O. Fischer, *Angew. Chem.*, **80**, 847 (1968); *Angew. Chem. Int. Ed. Engl.*, **7**, 817 (1968).
237. H. Werner, E. O. Fischer, B. Heckl, and C. G. Kreiter, *J. Organomet. Chem.*, **28**, 367 (1971).
238. K. Weiss and E. O. Fischer, *Chem. Ber.*, **106**, 1277 (1973).
239. K. Weiss and E. O. Fischer, *Chem. Ber.*, **109**, 1868 (1976).
240. M. H. Chisholm, H. C. Clark, W. S. Johns, J. E. H. Ward, and K. Yasufuku, *Inorg. Chem.*, **14**, 900 (1975).
241. J. A. Connor and E. O. Fischer, *J. Chem. Soc., Chem. Commun.*, 1024 (1967).
242. E. O. Fischer and H.-J. Kalder, *J. Organomet. Chem.*, **131**, 57 (1977).
243. E. O. Fischer, M. Leupold, C. G. Kreiter, and J. Müller, *Chem. Ber.*, **105**, 150 (1972).
244. C. T. Lam, C. V. Senoff, and J. E. H. Ward, *J. Organomet. Chem.*, **70**, 273 (1974).
245. E. O. Fischer, G. Kreis, F. R. Kreissl, C. G. Kreiter, and J. Müller, *Chem. Ber.*, **106**, 3910 (1973).
246. E. O. Fischer and V. Kiener, *Angew. Chem.*, **79**, 982 (1967); *Angew. Chem. Int. Ed. Engl.*, **6**, 961 (1967).
247. C. P. Casey and W. R. Brunsvold, *Inorg. Chem.*, **16**, 391 (1977).
248. E. O. Fischer, T. Selmayr, and F. R. Kreissl, *Monatsh. Chem.*, **108**, 759 (1977).
249. L. Knauss and E. O. Fischer, *Chem. Ber.*, **103**, 3744 (1970).
250. L. Knauss and E. O. Fischer, *J. Organomet. Chem.*, **31**, C68 (1971).
251. E. O. Fischer and R. Aumann, *Chem. Ber.*, **101**, 963 (1968).
252. Y. Ito, T. Hirao, and T. Saegusa, *J. Organomet. Chem.*, **131**, 121 (1977).
253. C. P. Casey and T. J. Burkhardt, *J. Am. Chem. Soc.*, **95**, 5833 (1973).
254. E. O. Fischer, W. Held, and F. R. Kreissl, *Chem. Ber.*, **110**, 3842 (1977).
255. C. P. Casey, T. J. Burkhardt, C. A. Bunnell, and J. C. Calabrese, *J. Am. Chem. Soc.*, **99**, 2127 (1977).
256. E. O. Fischer, W. Held, F. R. Kreissl, A. Frank, and G. Huttner, *Chem. Ber.*, **110**, 656 (1977).
257. E. O. Fischer and W. Held, *J. Organomet. Chem.*, **112**, C59 (1976).
258. C. P. Casey and W. R. Brunsvold, *J. Organomet. Chem.*, **77**, 345 (1974).
259. C. P. Casey, S. W. Polichnowski, H. E. Tuinstra, L. D. Albin, and J. C. Calabrese, *Inorg. Chem.*, **17**, 3045 (1978).
260. C. P. Casey and S. W. Polichnowski, *J. Am. Chem. Soc.*, **99**, 6097 (1977).
261. C. P. Casey, S. W. Polichnowski, A. J. Shusterman, and C. R. Jones, *J. Am. Chem. Soc.*, **101**, 7282 (1979).
262. E. O. Fischer and U. Schubert, *J. Organomet. Chem.*, **100**, 59 (1975).
263. E. O. Fischer, W. Kleine, and F. R. Kreissl, *J. Organomet. Chem.*, **107**, C23 (1976).
264. S. B. Butts, E. M. Holt, S. H. Strauss, N. W. Alcock, R. E. Stimson, and D. F. Shriver, *J. Am. Chem. Soc.*, **101**, 5864 (1979).
265. C. G. Kreiter, *Angew. Chem.*, **80**, 402 (1968); *Angew. Chem. Int. Ed. Engl.*, **7**, 390 (1968).
266. C. P. Casey, R. A. Boggs, and R. L. Anderson, *J. Am. Chem. Soc.*, **94**, 8947 (1972).
267. C. P. Casey and R. L. Anderson, *J. Am. Chem. Soc.*, **96**, 1230 (1974).
268. C. P. Casey and R. L. Anderson, *J. Organomet. Chem.*, **73**, C28 (1974).
269. C. P. Casey, W. R. Brunsvold, and D. M. Scheck, *Inorg. Chem.*, **16**, 3059 (1977).
270. C. P. Casey, R. A. Boggs, D. F. Marten, and J. C. Calabrese, *J. Chem. Soc., Chem. Commun.*, 243 (1973).
271. C. P. Casey and W. R. Brunsvold, *J. Organomet. Chem.*, **102**, 175 (1975).
272. K. H. Dötz, *J. Organomet. Chem.*, **118**, C13 (1976).
273. E. O. Fischer, K. Weiss, and C. G. Kreiter, *Chem. Ber.*, **107**, 3554 (1974).
274. E. O. Fischer and K. Weiss, *Chem. Ber.*, **109**, 1128 (1976).

275. K. Weiss, E. O. Fischer, and J. Müller, *Chem. Ber.*, **107**, 3548 (1974).
276. K. H. Dötz and C. G. Kreiter, *J. Organomet. Chem.*, **99**, 309 (1975).
277. K. H. Dötz, *Chem. Ber.*, **110**, 78 (1977).
278. H. Fischer and K. H. Dötz, *Chem. Ber.*, **113**, 193 (1980).
279. K. H. Dötz and I. Pruskil, *Chem. Ber.*, **111**, 2059 (1978).
280. K. H. Dötz and I. Pruskil, *J. Organomet. Chem.*, **132**, 115 (1977).
281. K. H. Dötz, B. Fügen-Köster, and D. Neugebauer, *J. Organomet. Chem.*, **182**, 489 (1979).
282. K. H. Dötz and C. G. Kreiter, *Chem. Ber.*, **109**, 2026 (1976).
283. H. Fischer and U. Schubert, *Angew. Chem.*, **93**, 482 (1981); *Angew. Chem. Int. Ed. Engl.*, **20**, 461 (1981).
284. H. Fischer, *J. Organomet. Chem.*, **197**, 303 (1980).
285. R. Aumann and E. O. Fischer, *Chem. Ber.*, **101**, 954 (1968).
286. C. G. Kreiter and R. Aumann, *Chem. Ber.*, **111**, 1223 (1978).
287. J. Levisalles, H. Rudler, and D. Villemin, *J. Organomet. Chem.*, **146**, 259 (1978).
288. J. Levisalles, H. Rudler, D. Villemin, J. Daran, Y. Jeannin, and L. Martin, *J. Organomet. Chem.*, **155**, C1 (1978).
289. C. P. Casey and T. J. Burkhardt, *J. Am. Chem. Soc.*, **96**, 7808 (1974).
290. U. Schubert and E. O. Fischer, *Justus Liebigs Ann. Chem.*, 393 (1975).
291. E. O. Fischer, U. Schubert, W. Kalbfus, and C. G. Kreiter, *Z. Anorg. Allg. Chem.*, **416**, 135 (1975).
292. H. Werner, V. Kiener, and H. Rascher, *Angew. Chem.*, **79**, 1021 (1967); *Angew. Chem. Int. Ed. Engl.*, **6**, 1001 (1967).
293. H. Werner and H. Rascher, *Inorg. Chim. Acta*, **2**, 181 (1968).
294. E. O. Fischer and K. Richter, *Chem. Ber.*, **109**, 1140 (1976).
295. E. O. Fischer and L. Knauss, *Chem. Ber.*, **102**, 223 (1969).
296. E. O. Fischer, E. Louis, and W. Bathelt, *J. Organomet. Chem.*, **20**, 147 (1969).
297. E. O. Fischer, E. Louis, W. Bathelt, E. Moser, and J. Müller, *J. Organomet. Chem.*, **14**, P9 (1968).
298. H. Werner and H. Rascher, *Helv. Chim. Acta*, **51**, 1765 (1968).
299. H. Fischer, E. O. Fischer, and F. R. Kreissl, *J. Organomet. Chem.*, **64**, C41 (1974).
300. H. Fischer, E. O. Fischer, and H. Werner, *J. Organomet. Chem.*, **73**, 331 (1974).
301. E. O. Fischer and G. Besl, *J. Organomet. Chem.*, **157**, C33 (1978).
302. D. J. Cardin, M. J. Doyle, and M. F. Lappert, *J. Organomet. Chem.*, **65**, C13 (1974).
303. B. Cetinkaya, E. Cetinkaya, and M. F. Lappert, *J. Chem. Soc., Dalton Trans.*, 906 (1973).
304. P. B. Hitchcock, M. F. Lappert, and P. L. Pye, *J. Chem. Soc., Dalton Trans.*, 826 (1978).
305. P. Braunstein, E. Keller, and H. Vahrenkamp, *J. Organomet. Chem.*, **165**, 233 (1979).
306. R. Zanella, T. Boschi, B. Crociani, and U. Belluco, *J. Organomet. Chem.*, **71**, 135 (1974).
307. G. Minghetti and F. Bonati, *J. Organomet. Chem.*, **54**, C62 (1973).
308. A. L. Balch, *J. Organomet. Chem.*, **37**, C19 (1972).
309. M. F. Lappert, J. J. MacQuitty, and P. L. Pye, *J. Chem. Soc., Chem. Commun.*, 411 (1977).
310. M. F. Lappert, R. W. McCabe, J. J. MacQuitty, P. L. Pye, and P. I. Riley, *J. Chem. Soc. Dalton*, 90 (1980).

The Chemistry of the Metal–Carbon Bond
Edited by F. R. Hartley and S. Patai

CHAPTER **5**

Syntheses of transition metal–carbyne complexes

ULRICH SCHUBERT
Anorganisch-Chemisches Institut der Technischen Universität München, Lichtenbergstrasse 4, D-8046 Garching, German Federal Republic

Carbyne complexes containing a terminally bonded *C*-alkyl or *C*-aryl ligand were first synthesized by Fischer *et al.*[1] in 1973 by reaction of methoxy(organyl)carbene complexes with boron trihalides (see Section I.A). These complexes were shown both spectroscopically and by X-ray structural analyses to have a metal–carbon triple bond. Following Fischer's nomenclature[2,3], an organometallic compound is called a 'carbyne complex', if (1) there is a terminally bonded ligand CR, (2) the carbon atom of this ligand, which is bound to the metal, is essentially *sp*-hybridized, and (3) the metal–carbon bond has a bond order of three (or at least greater than two). This definition will be followed in this chapter. Thus, for example, the syntheses of complexes which contain CR groups bridging two or three metals will not be considered, nor will Lewis acid adducts to carbonyl or thiocarbonyl complexes.

Review articles on the early developments of the chemistry of carbyne complexes have been published[2–4].

I. SYNTHESES FROM CARBENE COMPLEXES

In all syntheses of carbyne complexes known up to the present time (July 1980), organometallic compounds are used as precursors. This means that an already existing metal–carbon bond of lower bond order is transformed into a triple bond. Carbene complexes have proved to be a valuable synthetic source in this respect. Complexes with either electrophilic[3] or nucleophilic[5] carbene ligands can be used, although, of course, different types of reactions must be applied.

A. Reaction of Carbene Complexes with Lewis Acids

1. Preparation of neutral carbyne complexes

When solutions of pentacarbonylalkoxy(organyl)carbene complexes of chromium, molybdenum, or tungsten at low temperatures are treated with one of the trichlorides, tribromides or triiodides of boron, aluminium, or gallium, the alkoxy substituent and a CO ligand are split off and *trans*-halogeno(tetracarbonyl)organylcarbyne complexes are formed (equation 1). The nature of the substituent R does not affect the type of

$$[(OC)_5M{=}C(O\text{-alkyl})R] + \text{'}M'X_3\text{'} \longrightarrow [X{-}M(CO)_4{\equiv}C{-}R] + CO + \cdots \qquad (1)$$

$M = Cr, Mo, W;\; M' = B, Al, Ga;\; X = Cl, Br, I$

products. This reaction is therefore a general route for the preparation of neutral carbyne complexes with a wide variety of substituents R; R may be an alkyl[6,7] or aryl group (even chloro- and amino-substituted ones)[6,8], a vinylic[9] or acetylenic[10] group, a metallocene[11,12,13,14], an amino group[15], or a silyl group[16]. Although in most cases boron trihalides can be used as Lewis acids for these reactions, the graduated reactivity of the halides of aluminium or gallium sometimes offers preparative advantages[8,11,13,17].

Instead of alkoxy groups, other substituents at the carbene carbon atom, such as amino[18], thio[19], siloxy[20], or acetoxy[21] groups, can act as leaving groups in the reaction with Lewis acids. Since carbene complexes bearing these substituents either are less easily prepared than alkoxycarbene complexes or are derived from them, there are generally no advantages in using them for the preparation of carbyne complexes.

Reaction of *cis*-phosphine(tetracarbonyl) carbene complexes leads to the formation of *mer*-halogeno(phosphine)tricarbonylcarbyne complexes for the incoming halogeno ligand still enters the position *trans* to the carbyne (equation 2). The same is true for *cis*-arsine-and *cis*-stibine-substituted carbene complexes[22].

A slightly different reaction is observed when BF_3 is used as the Lewis acid. In this case, no *trans*-fluorocarbyne complexes are formed, but a BF_4 group enters the *trans*-position at the metal[23–25] (equation 3).

Dicyclohexylcarbodiimid (DCCD) is widely used for condensation reactions. On reaction with hydroxycarbene complexes of chromium, the corresponding symmetric

$$\left[\begin{array}{c} OC \quad YMe_3 \\ OC-Cr{\cdots}C(OMe)(Me) \\ OC \quad CO \end{array}\right] + BX_3 \longrightarrow \left[\begin{array}{c} OC \quad YMe_3 \\ X-Cr\equiv C-Me \\ OC \quad CO \end{array}\right] + CO + \cdots \quad (2)$$

X=Cl, Br, I; Y=P, As, Sb

$$\left[\begin{array}{c} OC \quad L \\ OC-M{\cdots}C(OMe)(R) \\ OC \quad CO \end{array}\right] + BF_3 \longrightarrow \left[\begin{array}{c} OC \quad L \\ BF_4-M\equiv C-Me \\ OC \quad CO \end{array}\right] + CO + \cdots \quad (3)$$

M=Cr, W; L=CO, PMe$_3$; R = alkyl, aryl

anhydrides $[(CO)_5CrC(R)—O—C(R)Cr(CO)_5]$ are obtained[26]. From the analogous hydroxycarbene complexes of tungsten, however, carbyne complexes are formed (equation 4). These complexes have the same elemental composition as the anhydrides; a metal acylate moiety is ligated *trans* to the carbyne ligand, however[27]. The mechanism of this reaction is believed to be similar to that of the syntheses of carbyne complexes with Group III halides (see Section I.A.3).

$$\left[(CO)_5M{\cdots}C(OH)(R)\right] + \left[(CO)_5W{\cdots}C(OH)(R')\right] + dccd \longrightarrow$$

$$\left[\begin{array}{c} OC \quad CO \\ (CO)_5M{\cdots}C(R){\cdots}O-W\equiv C-R' \\ OC \quad CO \end{array}\right] + CO + \cdots \quad (4)$$

M=Cr, W; R, R'= alkyl, aryl

2. *Preparation of cationic carbyne complexes*

If *trans*-substituted carbene complexes containing ligands which have a greater σ-donor/π-acceptor ratio than CO (e.g. PR_3, AsR_3, SbR_3[28], or a π-aromatic system[29–31]) are reacted analogously to equations 1 and 3, neutral carbyne complexes are no longer formed. These ligands remain in the product complex. Only the heteroatom containing substituent is removed from the carbene carbon and cationic carbyne complexes are formed (equation 5). The use of the former reaction (equation 5) is limited by the fact that *trans*-substituted carbene complexes are difficult to

$$\left[\begin{array}{c} OC \quad CO \\ Me_3Y-Cr{\cdots}C(OMe)(Me) \\ OC \quad CO \end{array}\right] + BX_3 \longrightarrow \left[\begin{array}{c} OC \quad CO \\ Me_3Y-Cr\equiv C-Me \\ OC \quad CO \end{array}\right] BX_4 + \cdots \quad (5)$$

Y=P, As, Sb; X=F, Cl, Br

separate from their *cis*-isomers. The latter reaction (equation 6) seems to be more general (R = aryl[29,30] or silyl[31] groups). The preparation of the carbene complex precursors, however, is limited to a smaller variety of R groups compared with

$$\left[(\pi\text{—}C_nH_n)(CO)_2M\text{:::}C\begin{matrix}\text{OMe}\\ \text{R}\end{matrix}\right] + BX_3 \longrightarrow [(\pi\text{—}C_nH_n)(CO)_2M\equiv CR]BX_4 + \cdots \quad (6)$$

$M{=}Cr$: $n=6$[29]; Mn, Re: $n=5$[30,31]; $X = Cl, F$

pentacarbonylcarbene complexes. A unique type of cationic carbyne complex is obtained either when pentacarbonyldialkylamino(ethoxy)carbenechromium is reacted with boron trihalides[15,32] (equation 7), or when pentacarbonyldimethylamino(chloro)-carbenechromium is treated with silver salts[32] (equation 8). Up to the present, this

$$\left[(CO)_5Cr\text{:::}C\begin{matrix}NR_2\\ OEt\end{matrix}\right] + BX_3 \longrightarrow [(CO)_5Cr\equiv C\text{—}NR_2]BX_4 + \cdots \quad (7)$$

$X = Cl, F$

$$\left[(CO)_5Cr\text{:::}C\begin{matrix}NMe_2\\ Cl\end{matrix}\right] + AgQ \longrightarrow [(CO)_5Cr\equiv C\text{—}NMe_2]Q + \ldots \quad (8)$$

$Q = BF_4, PF_6\ ClO_4$

is the only carbyne complex with a *trans*-CO ligand. X-ray structural analysis[33] shows that the Cr—C bond of the *trans*-CO ligand is labilized, despite some delocalization of the positive charge to the nitrogen atom. Therefore, it is probable that complexes of this type with simple alkyl or aryl substituents at the carbyne carbon atom cannot be isolated.

3. *Reaction mechanism*

The mechanism of the reaction of carbene complexes with Lewis acids according to equations 1–8 involves initial addition of the Lewis acid to the heteroatom bound to the carbene carbon. The heteroatom-containing substituent is thus transformed to a better leaving group and splits off, e.g. equation 9. In the remaining cationic carbyne

$$\left[L\text{—}M(CO)_4\text{:::}C\begin{matrix}OMe\\ R\end{matrix}\right] + BX_3 \xrightarrow{-OMe} \left[L\text{—}M(CO)_4\equiv CR\right]^+ \quad (9)$$

complex, the positive charge is localized principally on the metal atom. This increased positive charge at the metal decreases back-bonding to the ligands, particularly to the *trans*-ligand L. Weakening of the back-bonding does not greatly affect ligands L with a high σ-donor/π-acceptor ratio; therefore, cationic carbyne complexes with those ligands can be isolated. However, if the σ-donor/π-acceptor ratio falls below a certain limit, the bond is broken and the ligand is replaced by a more suitable one, e.g. equation 10. Only the aminocarbyne ligand offers a possibility for charge delocalization into the carbyne ligand, reducing the positive charge at the metal[33] (equation 11). Therefore, cationic pentacarbonylaminocarbyne complexes can be isolated. Whether the steps in reactions 9 and 10 are concerted or not cannot be decided in most cases. In the reaction of pentacarbonyldiethylamino(ethoxy)carbene-chromium with boron trichloride, pentacarbonyldiethylamino(chloro)carbene-

$$\text{L=CO:}\quad \left[\begin{array}{c} \text{OC}\quad\text{CO} \\ \text{L—M}{\equiv}\text{CR} \\ \text{OC}\quad\text{CO} \end{array}\right]^+ \xrightarrow[+X^-]{-L} \left[\begin{array}{c} \text{OC}\quad\text{CO} \\ \text{X—M}{\equiv}\text{CR} \\ \text{OC}\quad\text{CO} \end{array}\right] \tag{10}$$

$$[(CO)_5\overset{+}{Cr}{\equiv}C{-}NR_2] \longleftrightarrow [(CO)_5Cr{=}C{=}\overset{+}{N}R_2] \tag{11}$$

chromium can be isolated[34]; this rearranges to give *trans*-chloro(tetracarbonyl)diethylaminocarbynechromium (see Section I.B). The intermediacy of halo(organyl)carbene complexes in the reaction of carbene complexes other than amino-substituted ones is still in question.

B. Rearrangement of Carbene Complexes

A small number of pentacarbonyl carbene complexes of chromium rearrange spontaneously by loss of a CO ligand to give *trans*-substituted carbyne complexes.

$$\left[(CO)_5Cr\cdots C\begin{smallmatrix} NR_2 \\ X \end{smallmatrix}\right] \longrightarrow trans\text{-}[X(CO)_4Cr{\equiv}C{-}NR_2] + CO \tag{12}$$

$X = Cl, Br, SnPh_3$

Whether this reaction occurs or not depends on the nature of the group X. Up to now, it has been observed only for aminocarbyne complexes of chromium with X = Cl, Br[35], and $SnPh_3$[36], but not with X = F[35], CN, SCN[36], or $SiPh_3$[37]. Kinetic investigations[35,38] have shown that the rearrangements follow a first-order rate law and that free CO does not influence the reaction rates. Which properties of the group X are required for this intramolecular reaction is at present unknown. The intermediate occurrence of halocarbene complexes has been postulated for the reaction of hydroxycarbene complex anhydrides with tetraalkylammonium halides. The reaction

$$[(CO)_5Cr\cdots C(R)\cdots]_2O + [NR'_4]X \longrightarrow trans\text{-}X(CO)_4Cr{\equiv}CR + [NR'_4][(CO)_5CrC(O)R] + CO \tag{13}$$

R = aryl; R′ = alkyl

depends strongly on the electronic properties of the group R. The p-$CF_3C_6H_4$-substituted carbene complex is cleaved by both bromide and iodide, and the phenyl-substituted one only by iodide. However, no reaction occurs with the p-tolyl-substituted complex or between any of these complexes and tetraalkylammonium chlorides[39]. The first step in these reactions may involve cleavage of one carbene carbon–oxygen bond to give a metal acylate and a halocarbene complex; the latter would then rearrange according to equation 12.

Another carbyne complex synthesis, probably involving the rearrangement of an intermediate carbene complex (which was observed ir spectroscopically), is the reaction of a metal acylate with dibromotriphenylphosphane[40] (equation 14).

$$\left[(CO)_5W\cdots C\begin{smallmatrix} OLi \\ Ph \end{smallmatrix}\right] + Br_2PPh_3 \xrightarrow{-LiBr} \left[(CO)_5W\cdots C\begin{smallmatrix} O{-}P(Br)Ph_3 \\ Ph \end{smallmatrix}\right] \longrightarrow trans\text{-}[Br(CO)_4W{\equiv}CPh] + CO + Ph_3PO \tag{14}$$

C. Deprotonation of Carbene Complexes

A variety of carbene (alkylidene) complexes of the early transition metals were prepared by Schrock *et al.*[5]; these complexes exhibit different chemical behaviour to that of the transition metal–carbonylcarbene complexes discussed so far. Hydrogen atoms attached to the carbene carbon of these complexes can be abstracted by bases to give carbyne complexes. When trimethylphosphine is added to a solution of $[(\eta^5\text{-}C_5R_5)TaCl(CHR)(CH_2R)]$, the carbyne complex $[(\eta^5\text{-}C_5R_5)(PMe_3)_2TaCl(CR)]$ is formed (equation 15)[41]. It is believed that PMe_3 accelerates abstraction of the

$$[(\eta^5\text{-}C_5R_5)TaCl(=CHR')(CH_2R')] + 2\,PMe_3 \longrightarrow R'CH_3 + [(\eta^5\text{-}C_5R_5)TaCl(PMe_3)_2(\equiv CR')] \quad (15)$$

$$R = H,\ R' = CMe_3;\ R = Me,\ R' = CMe_3,\ Ph$$

α-hydrogen on the carbene ligand by the alkyl ligand. Similarly, $[(\eta^5\text{-}C_5H_5)Cl_2Ta = CHCMe_3]$ can be deprotonated by $Ph_3P = CH_2$ in the presence of an excess of PMe_3 to give $[(\eta^5\text{-}C_5H_5)(PMe_3)_2TaCl(CCMe_3)]$[41].

When $[(\eta^5\text{-}C_5H_5)TaCl(CH_2Ph)_3]$ is reacted with PMe_3, the complex $[(\eta^5\text{-}C_5H_5)\text{-}TaCl(PMe_3)(CH_2Ph)_3]$ is formed, which on heating also yields the tantalum carbyne complex[41] (equation 16). Whether this reaction proceeds via a carbene complex has

$$[(\eta^5\text{-}C_5H_5)TaCl(CH_2Ph)_3] + 2\,PMe_3 \longrightarrow 2\,CH_3Ph + [(\eta^5\text{-}C_5H_5)ClTa(PMe_3)_2(\equiv CPh)] \quad (16)$$

not been proved. However, since carbene complexes of tantalum have been prepared by deprotonating the α-carbon atom of an alkyl ligand, this assumption seems reasonable.

This reaction principle was extended to Group VI transition metals. If either WCl_6 or $MoCl_5$ is reacted with 6 or 5 mol of $LiCH_2CMe_3$, the readily dissociable complexes $[(Me_3CCH_2)_3MCCMe_3]_2$ (M = W, Mo) are formed. Heating the tungsten complex with trimethylphosphine or bis(trimethylphosphino)ethane gives an unique type of complex which contains alkyl, carbene, and carbyne ligands within the same molecule[42] (equation 17).

$$WCl_6 + 6LiCH_2CMe_3 \longrightarrow \tfrac{1}{2}[(Me_3CCH_2)_3WCCMe_3]_2 \xrightarrow{+(PP)} [(PP)W(CH_2CMe_3)(CHCMe_3)(CCMe_3)] \quad (17)$$

$$(PP) = 2PMe_3,\ Me_2P\text{-}CH_2CH_2\text{-}PMe_2$$

II. MODIFICATION OF CARBYNE COMPLEXES

A. Modification of the Carbyne Ligand

In principle, unsaturated or functional groups within the carbyne ligand offer the possibility of synthesizing new types of carbyne complexes by making use of the reactivity of these groups. Up to now, however, only one such example is known. If *trans*-halogeno(tetracarbonyl)phenylacetylenylcarbynetungsten is treated with

$$trans\text{-}[X(CO)_4W{\equiv}C{-}C{\equiv}C{-}Ph] + HNMe_2 \longrightarrow trans\text{-}[X(CO)_4W{\equiv}C{-}CH{=}C(NMe_2)Ph] \quad (18)$$

X = Cl, Br, I

dimethylamine, the amine adds to the organic triple bond (equation 18). In the carbyne complex formed, the carbyne carbon is substituted by an enamine moiety[10].

B. Substitution of Ligands

In a given series of carbyne complexes the ligands *cis* or *trans* to the carbyne moiety strongly influence the stability of these complexes. For example, the stability of *trans*-$X(CO)_4MCR$ (M = Cr, Mo, W) increases from X = Cl to X = I and X = carbonylmetallate (equation 19). Further, *cis*-phosphine- or similarly substituted

$$trans\text{-}[X(CO)_4M{\equiv}CR] + Y^- \longrightarrow trans\text{-}[Y(CO)_4M{\equiv}CR] + X^- \quad (19)$$

X = Cl, Br. M = Cr, Mo, W: Y = Br, I, $(CO)_5Re$, $(CO)_5Mn$.
M = W: Y = $(\eta^5\text{-}C_5H_5)(CO)_3Mo$, $(\eta^5\text{-}C_5H_5)(CO)_3W$

carbyne complexes are more stable than *cis*-CO-substituted complexes. If, for stability or any other reason, substituted carbyne complexes are desired, the preparation of the most accessible carbyne complex and subsequent substitution of ligands is often the method of choice. In some cases it is the only method of preparation available. It has been shown that a variety of such ligand-exchange reactions can be performed without affecting the carbyne moiety.

Although *trans*-iodo-substituted carbyne complexes can be synthesized according to equation 1, the iodides required for this reaction are less easily handled than the chlorides and bromides. Therefore, it is better to prepare the *trans*-chloro- or *trans*-bromo-substituted complexes and then to exchange the chloro or bromo ligand for iodide by treatment with LiI[9,11,16,43] (in the same way, the chloro ligand can be exchanged for bromide). Analogously, the halide can be substituted by a carbonyl metallate, yielding carbyne complexes with metal–metal bonds[16,43] (equation 12).

When *trans*-halogeno(tetracarbonyl)carbyne complexes are reacted with NaC_5H_5[16,44] (or lithium indenide[16]), not only the halide, but also two CO ligands are replaced by the π-aromatic ligand. Better leaving groups than halides are

$$[Br{-}W(CO)_4{\equiv}CR] + NaC_5H_5 \longrightarrow [(\eta^5\text{-}C_5H_5)W(CO)_2{\equiv}CR] + 2\,CO + NaBr \quad (20)$$

trans-$[(CO)_5M{\cdots}C(R)]$ (M = Cr, W)[27] (see equation 4) and BF_4[23,25] groups (see equation 3). BF_4 is the most labile ligand and can be substituted by neutral or anionic groups, yielding cationic or neutral carbyne complexes (equation 21). The *trans* ligand

$$trans\text{-}[X(CO)_4M{\equiv}CR] \xleftarrow[-BF_4]{+X^-,\ L=CO} [BF_4{-}M(CO)_3L{\equiv}CR] \xrightarrow{+Y} [Y{-}M(CO)_3L{\equiv}CR]\,BF_4 \quad (21)$$

M = Cr, W; L = CO, PMe_3; X^- = SCN^-, CN^-[25]; Y = H_2O, PPh_3[23,25]

cannot be replaced as easily in cationic carbyne complexes (see Section I.A.2) as in neutral complexes (see Section I.A.3). So far, only pentacarbonylaminocarbyne complexes (equation 11) are accessible to such substitutions. On reaction of $[(CO)_5CrCNMe_2]BF_4$ with triethylphosphine ($h\nu$) or tetrabutylammonium iodide the *trans*-CO ligand is split off and *trans*-$[PEt_3(CO)_4CrCNMe_2]BF_4$ or *trans*-$[I(CO)_4CrCNMe_2]$ is formed[32].

In neutral *trans*-halogeno(tetracarbonyl)carbyne complexes, one or two CO ligands are easily substituted when solutions of these complexes are treated with donor molecules (L) such as pyridine, *o*-phenanthroline, YPh_3 (Y = P, As, Sb), *t*-butylisonitrile, or triphenyl phosphite at temperatures between +20 and −20 °C[45] (equation 22). Bis(diphenylarsino)methane enters the complex as a monodentate

$$[Br{-}M(CO)_4{\equiv}CPh] + L \xrightarrow{-CO} [Br{-}M(CO)_3(L){\equiv}CPh] \xrightarrow[+L]{-CO} [Br{-}M(CO)_2(L)_2{\equiv}CPh] \text{ (L, L cis to each other, trans to CO) or } [Br{-}M(CO)_2(L)_2{\equiv}CPh] \text{ (L trans to L)} \quad (22)$$

$$M = Cr, W$$

ligand[46]. Depending on the nature of M and L, only monosubstituted complexes, or only disubstituted complexes, or both can be isolated. When disubstituted complexes are formed, the two L ligands are in *trans*-positions for L = $P(OPh)_3$, whereas they are in *cis*-positions for all other L.

The kinetics of the monosubstitution reaction was studied for a variety of triaryl-substituted phosphines, arsines, and stibines[47]. It was shown that dissociation of a CO ligand is the rate-determining step, and that the rate of reaction does not depend on the nature or concentration of L. Under a high CO pressure this substitution is partially reversible.

III. SYNTHESES FROM THIOCARBONYL AND ISONITRILE COMPLEXES

Thiocarbonyl and isonitrile complexes can be protonated, alkylated, arylated, or acylated at sulphur or nitrogen to give CSR or CNR_2 ligands. However, this method for the preparation of carbyne complexes is usually restricted to those complexes which are sufficiently electron rich to allow electrophilic attack at the heteroatom. No example is known of a similar reaction at the oxygen atom of a CO ligand.

A. Syntheses from Thiocarbonyl Complexes

When *cis*-$[W(CO)(CS)(dppe)_2]$ [dppe = bis(diphenylphosphino)ethane] or *trans*-$[I(CO)_4WCS]NBu_4$ are reacted with methyl fluorosulphonate, triethyloxonium tetrafluoroborate, or anhydrides of acids, an alkyl or acyl cation adds to the sulphur atom and cationic or neutral carbyne complexes are obtained[48] (equation 23).

$[W(CO)_5CS]$, $[W(CO)_4(CS)(PPh_3)]$, and $[W(CO)_3(CS)dppe]$ are not electron rich enough to give similar carbyne complexes. Protonation of $W(CO)(CS)(dppe)_2$ with trifluoromethylsulphonic acid does not yield a carbyne complex but a hydride

$$[LL'_4WCS] + R^+ \longrightarrow [LL'_4W{\equiv}CSR] \qquad (23)$$

$L = CO$, L'_2 = dppe, R^+ = Me, Et;
$L = I^-$, L' = CO, R^+ = Me, C(O)Me, $C(O)CF_3$

complex. Surprisingly, $[(\eta^5\text{-}C_5H_5)W(PPh_3)(CO)(CS)I]$ was found to react with phenyllithium to give the carbyne complex $[(\eta^5\text{-}C_5H_5)W(PPh_3)(CO)(CSPh)]$[49]. A direct nucleophilic attack of Ph^- on the sulphur seems unlikely and a charge-transfer–radical mechanism is suggested.

B. Syntheses from Isonitrile Complexes

When *trans*-[ReCl(CNMe)(dppe)$_2$] is treated with $[Et_2OH]BF_4$, the carbyne complex *trans*-[ReCl(dppe)$_2$(CNHMe)]BF_4 is formed[50]. Extensive experimental and mechanistic details are available for protonation of *trans*-[M(CNR)$_2$(dppe)$_2$]-(M = Mo, W) with strong acids such as HBF_4, H_2SO_4, $HBPh_4$, HCl, HBr, HSO_3F, and HSO_3Cl (Scheme 1). Depending on the solvent used and the rate of addition of acid, a hydrido, hydridocarbyne[51], monocarbyne, or dicarbyne[52] complex is formed. A study of the interconversion of these complexes showed the initial site of the reaction to be the isonitrile nitrogen[53].

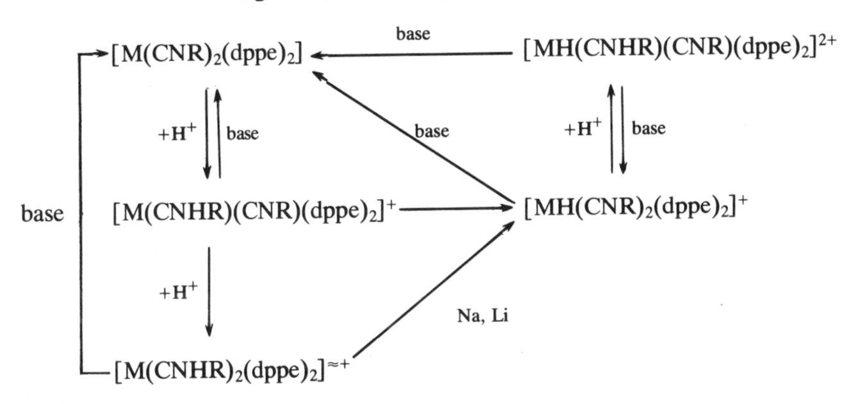

R = Me, t.Bu, M = Mo, W

SCHEME 1

As judged from their spectra, the dicarbyne complexes should be regarded as bis(iminocarbene) compounds, with double rather than triple metal–carbon bonds and their positive charges located at both nitrogen atoms[52]. These complexes can be further protonated by HBF_4 to give $[M(CNH_2Me)(CNHMe)(dppe)_2](BF_4)_3$[53].

Trans-[M(CNR)$_2$(dppe)$_2$] complexes also can be alkylated using $MeSO_3F$, Me_2SO_4, or $[Et_3O]BF_4$. In contrast to the protonation reaction, no dialkylated species could be observed. The initially formed complexes *trans*[M(CNRMe)(CNR)(dppe)$_2$]X isomerize in CH_2Cl_2 solution to give the *cis*-complexes.

IV. MISCELLANEOUS SYNTHESES

In benzene solution and also in the solid state, the vinyl complex $[(\eta^5\text{-}C_5H_5)[P(OMe)_3]_3Mo{-}CH{=}CH{-}{}^tBu]$ undergoes an unusual rearrangement to form the carbyne complex $[(\eta^5\text{-}C_5H_5)[P(OMe)_3]_2Mo{\equiv}C{-}CH_2{-}{}^tBu]$ with loss of

$P(OMe)_3$[55]. The rearrangement is suppressed by the presence of free trimethyl phosphite, suggesting the requirement of a vacant coordination site for the H shift to occur. When $[Sn(TPP)Cl_2]$ is heated with $[Re_2(CO)_{10}]$ at 160 °C for 240 h, the complex $[Sn(tpp)\{C\equiv Re(CO)_3\}_2]$ is formed (tpp = 5,10,15,20-tetraphenylporphyrinato)[56].

V. REFERENCES

1. E. O. Fischer, G. Kreis, C. G. Kreiter, J. Müller, G. Huttner, and H. Lorenz, *Angew. Chem.*, **85**, 618 (1973); *Angew. Chem. Int. Ed. Engl.*, **12**, 564 (1973).
2. E. O. Fischer and U. Schubert, *J. Organomet Chem.*, **100**, 59 (1975).
3. E. O. Fischer, *Angew. Chem.*, **86**, 651 (1974); *Adv. Organomet. Chem.*, **14**, 1 (1976).
4. E. O. Fischer, U. Schubert, and H. Fischer, *Pure Appl. Chem.*, **50**, 857 (1978).
5. R. R. Schrock, *Acc. Chem. Res.* **12**, 98 (1979).
6. E. O. Fischer and G. Kreis, *Chem. Ber.*, **109**, 1673 (1976).
7. S. Fontana, O. Orama, E. O. Fischer, U. Schubert, and F. R. Kreissl, *J. Organomet. Chem.*, **149**, C57 (1978).
8. E. O. Fischer, A. Schwanzer, H. Fischer, D. Neugebauer, and G. Huttner, *Chem. Ber.*, **110**, 53 (1977).
9. E. O. Fischer, W. R. Wagner, F. R. Kreissl, and D. Neugebauer, *Chem. Ber.*, **112**, 1320 (1979).
10. E. O. Fischer, H. J. Kalder, and F. H. Köhler, *J. Organomet. Chem.*, **81**, C23 (1974).
11. E. O. Fischer, M. Schluge, J. O. Besenhard, P. Friedrich, G. Huttner, and F. R. Kreissl, *Chem. Ber.*, **111**, 3530 (1978).
12. E. O. Fischer, V. N. Postnov, and F. R. Kreissl, *J. Organomet. Chem.*, **127**, C19 (1977).
13. E. O. Fischer, F. J. Gammel, J. O. Besenhard, A. Frank, and D. Neugebauer, *J. Organomet. Chem.*, **191**, 261 (1980).
14. E. O. Fischer, F. J. Gammel, and D. Neugebauer, *Chem. Ber.*, **113**, 1010 (1980).
15. E. O. Fischer, W. Kleine, G. Kreis, and F. R. Kreissl, *Chem. Ber.*, **111**, 3542 (1978).
16. E. O. Fischer, H. Hollfelder, and F. R. Kreissl, *Chem. Ber.*, **112**, 2177 (1979).
17. E. O. Fischer, S. Walz, and W. R. Wagner, *J. Organomet. Chem.*, **134**, C37 (1977).
18. K. Weiss and E. O. Fischer, *Chem. Ber.*, **109**, 1868 (1976).
19. A. Schwanzer, *Dissertation*, Tech. Univ. München, 1976.
20. E. O. Fischer, T. Selmayr, F. R. Kreissl, and U. Schubert, *Chem. Ber.*, **110**, 2574 (1977).
21. E. O. Fischer and T. Selmayr, *Z. Naturforsch.*, **32B**, 105 (1977).
22. E. O. Fischer and K. Richter, *Chem. Ber.*, **109**, 2547 (1976).
23. K. Richter, E. O. Fischer, and C. G. Kreiter, *J. Organomet Chem.*, **122**, 187 (1976).
24. E. O. Fischer and F. J. Gammel, *Z. Naturforsch*, **34B**, 1183 (1979).
25. E. O. Fischer, S. Walz, A. Ruhs, and F. R. Kreissl, *Chem. Ber.*, **111**, 2765 (1978).
26. E. O. Fischer, K. Weiss, and C. G. Kreiter, *Chem. Ber.*, **107**, 3554 (1974).
27. E. O. Fischer and K. Weiss, *Chem. Ber.*, **109**, 1128 (1976).
28. E. O. Fischer and K. Richter, *Chem. Ber.*, **109**, 3079 (1976).
29. E. O. Fischer, P. Stückler, H.-J. Beck, and F. R. Kreissl, *Chem. Ber.*, **109**, 3089 (1976).
30. E. O. Fischer, E. W. Meineke, and F. R. Kreissl, *Chem. Ber.*, **110**, 1140 (1977).
31. E. O. Fischer, P. Rustemeyer, and D. Neugebauer, *Z. Naturforsch.*, **35B**, 1083 (1980).
32. A. J. Hartshorn and M. F. Lappert, *J. Chem. Soc., Chem. Commun.*, 761 (1976).
33. U. Schubert, D. Neugebauer, P. Hofmann, B. E. R. Schilling, H. Fischer, and A. Motsch, *Chem. Ber.*, **114**, 3349 (1981).
34. E. O. Fischer, W. Kleine, and F. R. Kreissl, *J. Organomet. Chem.*, **107**, C23 (1976).
35. H. Fischer, A. Motsch, and W. Kleine, *Angew. Chem.*, **90**, 914 (1978), *Angew. Chem. Int. Ed. Engl.*, **17**, 842 (1978).
36. E. O. Fischer, H. Fischer, U. Schubert, and R. B. A. Pardy, *Angew. Chem.*, **91**, 929 (1979), *Angew. Chem. Int. Ed. Engl.*, **18**, 872 (1979).
37. H. Fischer, unpublished results.
38. H. Fischer, *J. Oganomet. Chem.*, **195**, 55 (1980).
39. K. Weiss and E. O. Fischer, *Chem. Ber.*, **109**, 1120 (1976).
40. H. Fischer and E. O. Fischer, *J. Organomet. Chem.*, **69**, C1 (1974).

41. S. J. McLain, C. D. Wood, L. W. Messerle, R. R. Schrock, F. J. Hollander, W. J. Youngs, and M. R. Churchill, *J. Am. Chem. Soc.*, **100**, 5962 (1978).
42. D. N. Clark and R. R. Schrock, *J. Am. Chem. Soc.*, **100**, 6774 (1978).
43. E. O. Fischer, T. L. Lindner, F. R. Kreissl, and P. Braunstein, *Chem. Ber.*, **110**, 3139 (1977).
44. E. O. Fischer, T. Lindner, G. Huttner, P. Friedrich, F. R. Kreissl, and J. O. Besenhard, *Chem. Ber.*, **110**, 3397 (1977).
45. E. O. Fischer, A. Ruhs, and F. R. Kreissl, *Chem. Ber.*, **110**, 805 (1977).
46. E. O. Fischer and A. Ruhs, *Chem. Ber.*, **111**, 2774 (1978).
47. H. Fischer and A. Ruhs, *J. Organomet. Chem.*, **170**, 181 (1979).
48. B. D. Dombek and R. J. Angelici, *J. Am. Chem. Soc.*, **97**, 1261 (1975).
49. W. W. Greaves, R. J. Angelici, B. J. Helland, R. Klima, and R. A. Jacobson, *J. Am. Chem. Soc.*, **101**, 7618 (1979).
50. A. J. L. Pombeiro, R. L. Richards, and J. R. Dilworth, *J. Organomet. Chem.*, **175**, C17 (1979).
51. J. Chatt, A. J. L. Pombeiro, and R. L. Richards, *J. Chem. Soc., Dalton Trans.*, 1585 (1979).
52. J. Chatt, A. J. L. Pombeiro, and R. L. Richards, *J. Chem. Soc., Dalton Trans.*, 492 (1980).
53. A. J. L. Pombeiro and R. L. Richards, *Transition Met. Chem.*, **5**, 55 (1980).
54. J. Chatt, A. J. L. Pombeiro, and R. L. Richards, *J: Organomet. Chem.*, **184**, 357 (1980).
55. M. Bottrill and M. Green, *J. Am. Chem. Soc.*, **99**, 5795 (1977).
56. I. Noda, S. Kato, M. Mizuta, N. Yasuoka, and N. Kasai, *Angew. Chem.*, **91**, 85 (1979), *Angew. Chem. Int. Ed. Engl.*, **18**, 83 (1979).

The Chemistry of the Metal–Carbon Bond
Edited by F. R. Hartley and S. Patai

CHAPTER **6**

Synthesis of transition metal alkyl and aryl complexes

R. J. PUDDEPHATT
Department of Chemistry, University of Western Ontario, London, Ontario, N6A 5B7, Canada

I. INTRODUCTION

Progress in the synthesis of alkyl and aryl derivatives of transition metals has occurred in a stepwise manner. Early attempts to prepare such complexes, analogous to Frankland's $ZnEt_2$, were largely unsuccessful. A few thermally stable compounds such as $[PtIMe_3]_4$ and $[AuBrEt_2]_2$ were isolated and their coordination chemistry was studied in detail, but more usually only organic decomposition products and transition metals or reduced metal species were obtained[1].

In the next step forward, it was found that thermally stable alkyls could be prepared if other 'stabilizing' ligands were present. Typical stabilizing ligands were η^5-C_5H_5, CO, or PR_3, and examples of such complexes are $[TiPh_2(\eta^5\text{-}C_5H_5)_2]$, $[MnMe(CO)_5]$, and *cis*-$[PtEt_2(PMe_3)_2]$[2,3]. The role of the 'stabilizing' ligands was not correctly interpreted at that time, but a great number of complexes were prepared and many of the characteristic reactions and general preparative methods were established.

More recently, emphasis has been placed on the synthesis of binary complexes such as $ReMe_6$, WMe_6, and $TaMe_5$, compounds containing metal—metal bonds such as $[Mo_2(CH_2SiMe_3)_6]$, and compounds containing metallacyclic rings such as $[\overline{Ti(CH_2CMe_2C}H_2)(\eta^5\text{-}C_5H_5)_2]$. Much of this research has been stimulated by the realization that the M—C bond is generally strong in the thermodynamic sense, and that the successful synthesis of a transition metal alkyl derivative often relies on blocking decomposition pathways having low activation energies. The β-elimination mechanism of decomposition is often particularly facile, and the use of bulky alkyl groups having no β-hydrogen atom (such as Me_3CCH_2—, Me_3SiCH_2—, $(Me_3Si)_2CH$— and Me_2PhCH_2—) or in which β-elimination would give an intermediate of high energy (such as norbornyl, adamantyl, or butane-1,4-diyl groups) has led to the synthesis of a very large number of simple alkyl—transition metal complexes[4–6]. However, facile decomposition by α-elimination, reductive elimination, or other mechanisms may still occur and frustrate logical synthesis.

A further area of interest has been synthesis by unusual methods, for example, by the reduction of a carbonyl ligand progressively to formyl, hydroxymethyl, and methyl ligands. These more recent advances will be emphasized in this account of the synthesis of alkyl and aryl derivatives of transition metals. The review cannot be comprehensive and emphasis will be placed on synthetic methods, illustrated by selected examples. As a further restriction, syntheses in which subsidiary ligands are displaced to give new alkyl derivatives, but in which the alkyl metal bond is not affected, will be omitted.

II. SYNTHETIC METHODS

A. Synthesis by Alkyl Transfer Reactions

This is the most widely used method for the synthesis of alkyl and aryl derivatives of transition metals. The most frequently used reagents are alkyllithium or Grignard reagents but many other alkyl—metal compounds are useful, notably dialkylzinc, dialkylmercury, or tetra-alkyltin derivatives, with which milder alkylating agents are required.

1. Methyl derivatives

Tetramethyltitanium can be prepared at $-78\,^{\circ}C$ under an inert atmosphere by reaction of $TiCl_4$ with methyllithium in diethyl ether[7,8] (equation 1). The product and ether co-distil from the reaction mixture under vacuum at $-30\,^{\circ}C$ and decomposition

$$[TiCl_4] + 4\,MeLi \longrightarrow [TiMe_4] + 4\,LiCl \quad (1)$$

occurs at around room temperature. However, the pure yellow $[TiMe_4]$ decomposes at about $-70°C$ and is very air sensitive. Tetramethylzirconium can be prepared in a similar way[8].

With excess of methyllithium, 'ate' complexes can be prepared. These are ionic in solution and examples include $Li[TiMe_5]\cdot 2(dioxane)$, $Li_2[ZrMe_6]$, and $Li[TiMe_4R]L$, where R = Ph, Bz and L = N- or O-donor ligand[9–11].

Examples of all species $[TiX_nMe_{4-n}]$ are known, where X = halide, alkoxide, or dialkylamide. They are usually prepared using mild alkylating agents such as dimethylzinc, trimethylaluminium, or tetramethyllead with TiX_4. For example, $TiCl_4$ with Me_3Al gives $[TiCl_3Me]$ and $TiBr_4$ with Me_2Zn gives $[TiBr_3Me]$[2]. Alternatively, stoichiometric reactions with methyllithium can give the same complexes. The products become progressively less stable, and hence more difficult to isolate, as more methyl groups are attached to titanium and thermal stability can be increased by addition of ligands such as 2,2'-bipyridine[12–15]. Dimethylzinc reacts with $ZrCl_4$ to give $[ZrCl_3Me]$ or $[ZrCl_2Me_2]$[16].

Complexes of the Group IVA metals in lower oxidation states have proved difficult to prepare. For example, $[TiCl_3(thf)_3]$ in ether reacts with LiMe to give a solution thought to contain $[TiMe_3]$, whereas in pyridine solution a blue complex $[TiCl_2Me(py)_3]$ is formed that can be isolated[17,18].

Probably the most closely studied methyl derivatives of Ti, Zr, and Hf are the cyclopentadienyls $[MMe_2(\eta^5\text{-}C_5H_5)_2]$ and $[MClMe(\eta^5\text{-}C_5H_5)_2]$. These can be prepared according to equation 2. These complexes decompose slowly at room

$$[TiCl_2(\eta^5\text{—}C_5H_5)_2] + 2\,MeLi \xrightarrow[-78°]{Et_2O} [TiMe_2(\eta^5\text{-}C_5H_5)_2] + 2\,LiCl$$

$$\xrightarrow[\;-CH_4\;]{HCl} [TiClMe(\eta^5\text{-}C_5H_5)_2] \quad (2)$$

temperature and (when M = Ti) can be handled in air, and are therefore much easier to prepare and study than the binary methyl derivatives. Derivatives of Ti, Zr, and Hf are known, as well as related η^5-indenyl and η^5-pentamethylcyclopentadienyl complexes[19–23].

Complexes containing one $\eta^5\text{-}C_5H_5$ ring, such as $[TiMe_3(\eta^5\text{-}C_5H_5)]$, and compounds with the metal in a lower oxidation state are less stable thermally and more sensitive to oxygen. They may be prepared in solution by reaction of methyllithium or MeMgX with the corresponding cyclopentadienyltitanium halide. With excess of Grignard reagent the anionic $[TiMe_2(\eta^5\text{-}C_5H_5)_2]^-$ appears to be formed[24,25].

The best studied alkyl derivatives of the Group V elements are those of niobium and tantalum. Partially methylated derivatives were prepared by reaction of the pentachlorides with dimethylzinc[26–29] (equation 3 and 4). The compounds $NbMeCl_4$ and $TaMeCl_4$ can be prepared in reasonable purity by the redistribution reaction 5. The monomethyl derivatives of niobium and tantalum can also be prepared using the very mild methylating agents $HgMe_2$ or $SnMe_4$ with the corresponding pentahalide. They readily form 6- and 7-coordinate complexes on reaction with donor ligands[27].

$$NbCl_5 + ZnMe_2 \longrightarrow ZnCl_2 + [NbMe_2Cl_3] \quad (3)$$

$$TaCl_5 + 1.5ZnMe_2 \longrightarrow 1.5ZnCl_2 + [TaMe_3Cl_2] \quad (4)$$

$$[NbMe_2Cl_3] + [NbCl_5] \longrightarrow 2[NbMeCl_4] \quad (5)$$

Pentamethyltantalum is a volatile solid formed by further methylation of $[TaMe_3Cl_2]$ by methyllithium. It decomposes only slowly in solution at room temperature, but the corresponding $[NbMe_5]$, which apparently can be formed at low temperature by similar methylation of $[NbMe_2Cl_3]$, decomposes at around $-20°C$ and has not been fully characterized[29]. Further methylation to 'ate' complexes such as $Li_2[TaMe_7]$ may also occur. The methyltantalum complexes are particularly significant as precursors of methylene complexes of tantalum such as $[Ta(=CH_2)Me(\eta^5\text{-}C_5H_5)_2]$, the chemistry of which has proved to be exceptionally interesting.[30]

$$[CrCl_3(thf)_3] + Me_3Al \longrightarrow [CrMeCl_2(thf)_3] \quad (6)$$

$$[WCl_6] + 6Me_3Al \longrightarrow [WMe_6] + 6AlMe_2Cl \quad (7)$$

Alkylaluminum compounds are used to prepare alkyl derivatives of chromium and tungsten[31,32] (equations 6 and 7). Hexamethyltungsten is a volatile solid, and synthesis by the above method requires trace amounts of oxygen to be present. It is potentially explosive. Methyllithium is able to form 'ate' complexes with the Group VI elements. For example, reaction with $CrCl_3$ gives $[Li(1,4\text{-dioxane})]_3[CrMe_6]$ which contains octahedral chromium(III) centres, and $CrCl_2$ gives $[Li(thf)]_4[Cr_2Me_8]$, which contains a Cr—Cr quadruple bond[33,34]. The corresponding anion $[Mo_2Me_8]^{4-}$ is also known[35]. Metal—metal multiple bonds are a feature of the organometallic chemistry of molybdenum and tungsten, and triple bonds are present in the complexes $[M_2Me_2(NR_2)_4]$[36,37] (equation 8).

$$[Mo_2Cl_2(NMe_2)_4] + 2MeLi \longrightarrow [Mo_2Me_2(NMe_2)_4] \quad (8)$$

Rhenium(III) also forms quadruple bonds and the methyl derivative $[Li(thf)]_2[Re_2Me_8]$ is a good example. It can be prepared by reaction of $[ReCl_5]$ with methyllithium. Wilkinson has characterized many other rhenium—methyl derivatives and a particularly interesting sequence of reactions is shown in equations 9–11, illustrating the high methylating power of methyllithium. Attempts to prepare $ReMe_7$ have been unsuccessful. Complexes with methyl groups bridging between rhenium atoms can also be prepared (equation 12). A structurally characterized example is $[Re_3Me_9(PEt_2Ph)_2]$, in which each rhenium is bound to two terminal methyl groups and two μ_2-methyl groups. This chemistry has been reviewed briefly[38].

$$[ReOCl_4] + MgMeBr \longrightarrow [ReOMe_4] \quad (9)$$

$$[ReOMe_4] + AlMe_3 \longrightarrow [ReMe_6] \quad (10)$$

$$[ReMe_6] + 2LiMe \longrightarrow Li_2[ReMe_8] \quad (11)$$

$$[Re_3Cl_9] + MgMe_2 \longrightarrow \{Re_3Me_9\}_n \xrightarrow{L} [Re_3Me_9L_{2\text{ or }3}] \quad (12)$$

Manganese(II) forms the polymeric $[(MnMe_2)_n]$ by reaction of MnI_2 with 2 mol of methyllithium, and with excess of methyllithium the 'ate' complex $[Li(Me_2NCH_2CH_2NMe_2)]_2[MnMe_4]$ is formed. These complexes have not yet been structurally characterized[39].

With the Group VIII elements methyllithium is again sufficiently reactive to give 'ate' complexes and some examples include $[FeMe_4]^{2-}$, $[CoMe_4]^{2-}$, $[NiMe_4]^{2-}$, and $[PtMe_6]^{2-}$ [40–42]. However, the more important derivatives contain neutral ligands in addition to methyl groups. These are too numerous to treat in detail and only representative syntheses will be given. Methyllithium is still the preferred methylating agent and reacts, for example, with $[RhCl(PPh_3)_3]$ to give $[RhMe(PPh_3)_3]$. The similar cobalt complex is also known, and both decompose at room temperature by loss of methane[43]. This method was also used by Chatt and Shaw and by Coates to prepare methylplatinum(II) and methylpalladium(II) complexes such as *cis*-$[PtMe_2(PEt^3)_2]$ and $[PdMe_2$(2,2′-bipyridine)]. Several reviews of this work have been published[44–46]. Whereas methyllithium reacts with *cis*-$[PtCl_2(PEt_3)_2]$ to give *cis*-$[PtMe_2(PEt_3)_2]$, methylmagnesium chloride only gives the monomethyl derivative *trans*-$[PtClMe(PEt_3)_2]$.

Another useful synthetic method involves the reaction of $[AlMe_2OMe]$ with a metal acetylacetonate and neutral ligand. The methyl—metal complex is the only insoluble product and so is easily isolated. This method has been used to prepare complexes such as $[NiMe_2$(2,2′-bipyridine)] and $[FeMe_2$(2,2′-bipyridine$)_2]$ as well as many methylcobalt complexes which cannot easily be prepared by other methods[47–49].

2. *Alkyl derivatives with β-hydrogen atoms*

The synthetic procedures for the synthesis of ethyl, propyl, butyl, and similar derivatives are essentially identical with those described for the methyl derivatives. The chief difference is that, when β-hydrogen atoms are present, thermal decomposition often occurs easily and hence the syntheses are much less likely to yield stable products.

The only derivatives known which do not contain ‘stabilizing’ ligands are $[Cr(^tBu)_4]$, $[Cr(Pr^i)_4]$ and salts of $[M(Bu^t)_4]^-$, where M = Sm, Er, or Yb. The remarkable thermal stability of the chromium species is thought to be due to a cog-wheel effect whereby the interlocking *t*-butyl or *i*-propyl groups are rigidly held in a conformation in which β-elimination is not possible. The complex $[Cr(Bu^t)_4]$ is prepared from $[Cr(OBu^t)_4]$[17] and Bu^tLi[50–52].

A very stable series of complexes, MR_4, can be prepared from bridgehead ligands, R = 1-norbornyl or 1-adamantyl. These are unable to β-eliminate owing to the very unfavourable conformational strain which would be necessary. The compounds are usually prepared using the corresponding organolithium reagent, and a suitable metal halide or alkoxide. Examples include [Ti(norbornyl$)_4$] and [Cr(adamantyl$)_4$], prepared from $[Cr(OBu^t)_4]$ and the adamantyl Grignard reagent[53–55].

3. *Alkyl derivatives without β-hydrogen atoms*

These now constitute an increasingly important class of compounds. The absence of β-hydrogen atoms and the bulk of many of these groups together lead to remarkable thermal stability of many such compounds. Typical alkyl groups in this class are $PhCH_2$—, Me_3CCH_2—, Me_3SiCH_2—, $(Me_3Si)_2CH$—, $PhMe_3CCH_2$—, and adamantyl-CH_2—. Some syntheses are illustrated in equations 13–15[55–58]. With the

$$[TiCl_4] + 4PhCH_2MgCl \longrightarrow [Ti(CH_2Ph)_4] + 4MgCl_2 \quad (13)$$

$$TaCl_5 + 1.5Zn(CH_2CMe_3)_2 \longrightarrow TaCl_2(CH_2CMe_3)_3 + 1.5ZnCl_2 \quad (14)$$

$$V(OPr^i)_4 + 4Li(CH_2\text{-adamantyl}) \longrightarrow V(CH_2\text{-adamantyl})_4 + 4LiOPr^i \quad (15)$$

very bulky $(Me_3Si)_2CH$— group only three alkyl groups can be easily accommodated in many cases, typical derivatives being TiR_3, VR_3, and CrR_3. Even when the synthesis involves a titanium(IV) precursor only TiR_3 is formed[59].

Reduction or α-elimination may be observed when attempts to put more than four of the smaller neopentyl or trimethylsilylmethyl groups around a metal are made. For example, reaction of Me_3SiCH_2MgCl with $[WCl_6]$ gives the dimer $[W_2(CH_2SiMe_3)_6]$, while Me_3CCH_2Li with $[TaCl_2(CH_2CMe_3)_3]$ gives the alkylidene derivative $[(Me_3CCH_2)_3Ta=CHCMe_3]$[60,61].

4. *Aryl derivatives*

Aryl derivatives are most commonly prepared by transmetallation reactions using aryllithium or Grignard reagents. Phenyl derivatives without stabilizing ligands are typically less stable towards thermolysis than neopentyl and related derivatives but considerably more stable than ethyl derivatives. For example [$TiPh_4$] can be prepared in the usual way at low temperature and can be isolated, but decomposes rapidly at room temperature[62].

Mesityl derivatives may have higher thermal stability owing to increased bulk or the absence of β-hydrogen atoms. Thus MR_4 with R = mesityl and M = Ti, V, or Cr are all thermally stable. Under suitable conditions mesityllithium and chromium halides may give $LiCrR_4 \cdot 4thf$, $LiCrR_3 \cdot dioxan \cdot Et_2O$ or $Li_2CrR_4 \cdot thf$, all of which can be isolated[63,64].

Another useful technique is to use an aryl ligand with a donor substituent in the *ortho*-position, so that chelation can occur and hence stabilize the compounds. For example, [$AuCl(AsPh_3)$] with Li-*o*-$C_6H_4CH_2NMe_2$ gives [Au-*o*-$C_6H_4CH_2NMe_2]_n$, which exists as an oligomer, and many related complexes can be prepared by similar methods[65,66].

5. *Metallacyclic compounds*

Metallacycles are often prepared from a metal halide and a difunctional organolithium or Grignard reagent. For example, *cis*[$PtCl_2(PPh_3)_2$] with $Li(CH_2)_nLi$ gives [$Pt(CH_2)_n(PPh_3)_2$], with n = 4–6[67,68]. The reaction is not successful when n = 3, perhaps because the product would be strained. These compounds are resistant to β-elimination but, in some cases, metallacyclopentanes decompose easily by carbon—carbon bond cleavage to give ethylene. Examples in which this mode of decomposition is sufficiently facile to make handling of the metallacycle difficult include [$Ti(CH_2)_4(\eta^5\text{-}C_5H_5)_2$] and [$Ni(CH_2)_4(PPh_3)_2$]. The syntheses are then carried out at low temperature[69,70].

More stable derivatives may be obtained when the β-carbon or β-hydride elimination is more difficult. For example, metallacycles formed from the Grignard reagent *o*-$C_6H_4(CH_2MgCl)_2$ have high thermal stability, an example being [$\overline{Zr(CH_2\text{-}o\text{-}C_6H_4\text{-}CH_2})(\eta^5\text{-}C_5H_5)_2$], as have derivatives prepared from the dilithium reagent LiCPh=CPh—CPh=CPhLi[71].

B. Synthesis by Oxidative Addition

1. *Oxidative addition of alkyl halides*

Oxidative addition has been used principally in synthesis from complexes having the d^8 or d^{10} electron configuration at the metal centre. One of the first such reactions involved oxidative addition of methyl iodide (equation 16). Here a d^8 square-planar

$$\left[\begin{array}{c} PEt_3 \\ | \\ I{-}Pt{-}Me \\ | \\ PEt_3 \end{array}\right] + MeI \longrightarrow \left[\begin{array}{c} Me\ \ PEt_3 \\ |\ / \\ I{-}Pt{-}Me \\ /\ | \\ Et_3P\ \ \ I \end{array}\right] \qquad (16)$$

platinum(II) complex is converted into a d^6 octahedral platinum(IV) complex, with formation of new methylplatinum and iodoplatinum bonds. Iridium(I) complexes are even more reactive, and most alkyliridium complexes known have been prepared by this method[72] (equation 17).

$$[\text{Cl}-\text{Ir}(\text{PPh}_3)(\text{PPh}_3)-\text{CO}] + \text{MeI} \longrightarrow [\text{Cl}-\text{Ir}(\text{Me})(\text{I})(\text{PPh}_3)(\text{PPh}_3)-\text{CO}] \qquad (17)$$

Complexes with d^{10} configurations are typically converted into square-planar d^8 complexes, often with ligand displacement[73,74] (equation 18). The reactions occur more readily for third-row elements (e.g. $Ir^I > Rh^I > Co^I$) and complexes in lower oxidation states are more reactive towards oxidative addition (e.g. $Ir^I > Pt^{II}$). The reactivity is also greatly influenced by the donor power and steric bulk of ancillary ligands attached to the metal, and these factors may all be significant in synthetic procedures. Alkyl iodides are more reactive than bromides or chlorides. In general, methyl and benzyl halides give straightforward oxidative addition reactions and the S_N2 mechanism has been established in many of these reactions[75]. However, methyl halides may react by radical mechanisms in some cases[74].

$$[\text{Pt}(\text{PPh}_3)_4] + \text{MeI} \longrightarrow \textit{trans}\text{-}[\text{PtIMe}(\text{PPh}_3)_2] + 2\text{PPh}_3 \qquad (18)$$

Many other alkyl halides appear to add oxidatively to metal complexes by free-radical mechanisms. One consequence of such mechanisms is the loss of optical activity on reaction of chiral alkyl halides. An example of such racemization is the oxidative addition of chiral $CH_3CH(CO_2Et)Br$ to square-planar iridium(I) complexes[76]. Another consequence of radical mechanisms is that side-reactions, for example giving coupling of alkyl groups, may occur in competition with oxidative addition. This, of course, decreases the usefulness as a synthetic method. Finally, rearrangement of intermediate alkyl radicals may occur, for example cyclopropylmethyl to but-3-enyl or hex-5-enyl to cyclopentylmethyl, and the metal derivative of the rearranged alkyl group will then be formed[77].

The complexes $[Rh(CN)_4]^{3-}$ and the macrocyclic rhodium(I) complex shown in equation 19 are particularly reactive in oxidative addition, and many primary and

$$[\text{Rh macrocycle (BF}_2\text{, O, N)}] + \text{BuCl} \longrightarrow [\text{Rh macrocycle (BF}_2\text{, O, N)}(\text{Bu})(\text{Cl})] \qquad (19)$$

secondary alkyl halides give normal adducts. In addition α,ω-dihalides react to give binuclear derivatives[78] (equation 20). In some cases binuclear rhodium(I) complexes

$$2\,\text{Rh} + \text{I}(\text{CH}_2)_n\text{I} \longrightarrow \text{I}-\text{Rh}-(\text{CH}_2)_n-\text{Rh}-\text{I} \qquad (20)$$

may be induced to undergo binuclear oxidative addition giving alkylrhodium(II) complexes with an Rh—Rh bond. Alkyl derivatives of other metals in unusual oxidation states have been prepared by similar methods[79] (equation 21).

$$[Rh(\mu\text{-}CNCH_2CH_2CH_2NC)_4Rh]Cl_2 + MeI \longrightarrow [MeRh(\mu\text{-}CNCH_2CH_2CH_2NC)_4RhI]Cl_2 \quad (21)$$

μ-Methylene complexes can be prepared by binuclear oxidative addition of dihalogenomethanes in some cases[80] (equation 22). When the transition metal complex is coordinatively saturated, simple halide displacement from alkyl halides occurs[81] (equation 23).

$$[Pd_2(\mu\text{-}Ph_2PCH_2PPh_2)_3] + CH_2Br_2 \longrightarrow [Pd_2Br_2(\mu\text{-}CH_2)(\mu\text{-}Ph_2PCH_2PPh_2)_2] \quad (22)$$

$$[Fe(CO)_2(\eta^5\text{-}C_5H_5)]^- + CH_3I \longrightarrow [CH_3Fe(CO)_2(\eta^5\text{-}C_5H_5)] \quad (23)$$

Many anionic complexes behave in a similar way and a reactivity series $[Fe(CO)_4]^{2-}$ > $[Fe(CO)_2(\eta^5\text{-}C_5H_5)]^-$ > $[Ni(CO)(\eta^5\text{-}C_5H_5)]^-$ > $[Re(CO)_5]^-$ > $[W(CO)_3(\eta^5\text{-}C_5H_5)^-$ > $[Mn(CO)_5]^-$ > $[Mo(CO)_3(\eta^5\text{-}C_5H_5)^-$ > $[Cr(CO)_3(\eta^5\text{-}C_5H_5)^-$ > $[Co(CO)_4]^-$ has been drawn up. The rates of reaction with a given alkyl halide vary over many orders of magnitude down this series[82–85]. The method has been very widely used in synthesis. It is not necessary for the initial complex to be anionic, although anionic complexes are most reactive (equation 24).

$$[Rh(CO)(PMe_2Ph)(\eta^5\text{-}C_5H_5)] + MeI \longrightarrow [MeRh(CO)(PMe_2Ph)(\eta^5\text{-}C_5H_5)]I \quad (24)$$

When carbonyl ligands are present, rearrangement to acyl derivatives by migration of the alkyl group on to a coordinated carbonyl ligand can be a complicating factor in synthesis[86].

Aryl halides are much less reactive to oxidative addition than are simple alkyl halides, but they do react with nickel(0) or palladium(0) complexes ($L = PPh_3$)[87,88] (equation 25). Vinyl halides behave similarly and intermediate alkene complexes have been isolated in some cases[89].

$$[PdL_4] + C_6H_5I \longrightarrow trans\text{-}[PdIPhL_2] + 2L \quad (25)$$

In the above examples the metal is oxidized by two units in the oxidative addition. A few examples are known in which one-electron oxidation occurs, although these are seldom used in synthesis since the separation of two products is necessary. The reactions have been studied in greatest detail for cobalt(II) and chromium(II) complexes[90,91] (equation 26).

$$2[Co(CN)_5]^{3-} + MeBr \longrightarrow [CoMe(CN)_5]^{3-} + [CoBr(CN)_5]^{3-} \quad (26)$$

2. *Oxidative addition of C—H bonds*

This reaction occurs much less easily than oxidative addition of alkyl halides but, since it is significant in the activation of alkanes, it has been studied in some depth.

Perhaps the best studied systems are based on $[Ir(dmpe)_2]^+$ and $[Fe(dmpe)_2]$. The iridium complex reacts with acetonitrile to give *trans*-$[IrH(CH_2CN)(dmpe)_2]^+$ but fails to react with arenes[92]. The iron complex is more reactive and reacts reversibly with naphthalene to give *cis*-$[FeH(naphthyl)(dmpe)_2]$, and with many other arenes to give analogous products. A C—H bond of acetonitrile, acetone, or ethyl acetate can be added also[93]. The complex $[Ru(dmpe)_2]$ undergoes oxidative addition of a C—H bond from a methylphosphorus group of the dmpe ligand giving a binuclear ruthenium(II) complex[94].

Another very reactive system is formed by photolysis of $[MH_2(\eta^5\text{-}C_5H_5)_2]$, where M = Mo or W, which give hydrogen and, almost certainly, $[M(\eta^5\text{-}C_5H_5)_2]$. This latter will then undergo oxidative addition involving C—H bonds of arenes or alcohols, or undergo self-reaction[95]. Zirconocene is similarly reactive and zirconium will insert into a C—H bond of toluene[96]. These C—H bond activations are thought to occur, at least initially, by *cis* oxidative addition and complexes have now been prepared in which the C—H bond is still present but interacts strongly with a transition metal centre, one example being the $[Fe(\eta^3\text{-cyclooctenyl})\{P(OMe)_3\}_3]^+$ cation[97]. A particularly interesting example is found in the cluster $[Os_3HMe(CO)_{10}]$, which is reversibly converted into $[Os_3H_2(\mu\text{-}CH_2)(CO)_{10}]$ in solution. In the methyl derivative a C—H bond bridges between osmium centres[98].

In many cases cyclometallated derivatives are formed by intramolecular oxidative addition[99] (equation 27). Metallacyclobutane derivatives can be prepared in a similar way (equation 28). In many cases the intermediate alkyl is not detected but undergoes rapid oxidative addition[100] (equation 29). These reactions give rise to very interesting alkyl—transition metal derivatives but they are limited to metal systems which are very reactive towards oxidative addition. It has not yet been possible to prepare metal alkyls from simple alkanes, although catalytic activation of C—H bonds in alkanes has been achieved[92].

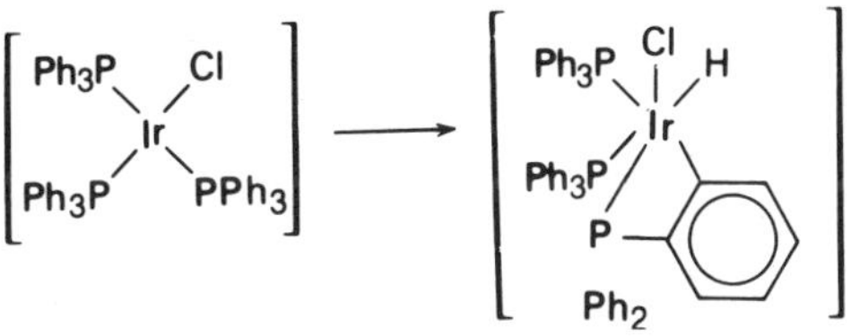

$$[IrCl(PPh_3)_3] \longrightarrow [IrHCl(PPh_3)_2(C_6H_4PPh_2)] \quad (27)$$

$$[Ir(PMe_3)_3(CH_2SiMe_3)] \longrightarrow [IrH(PMe_3)_3(CH_2SiMe_2CH_2)] \quad (28)$$

$$[Ir(PMe_3)_4]^+ + [CH_2COCH_3]^- \longrightarrow [IrH(PMe_3)_3(CH_2C(=O)CH_2)] \quad (29)$$

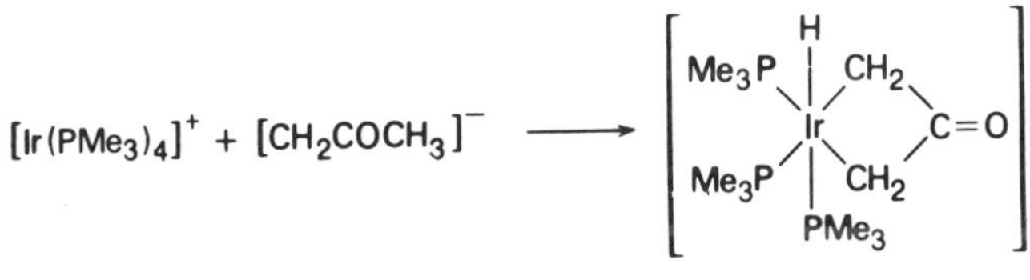

3. *Oxidative addition of C—C bonds*

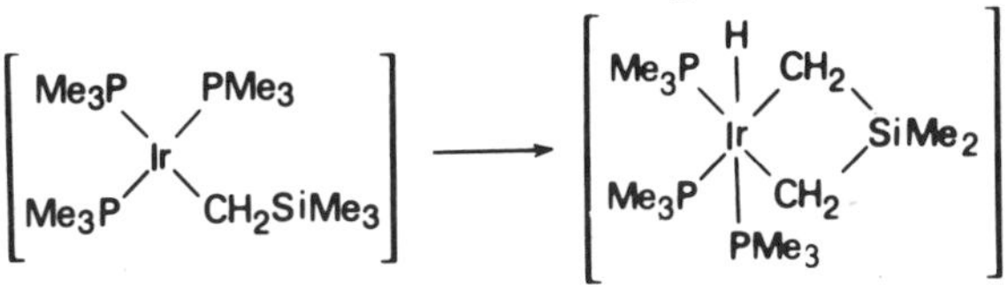

These reactions are more difficult again than C—H oxidative additions. There are reports of oxidative additions of the C—CN bond, for example of benzonitrile[92,101], but the most important reactions involve strained carbocyclic compounds.

Tetracyanocyclobutane or cyclobutenedione derivatives react with platinum(0) complexes to give platinacyclopentanes[102,103] (equations 30 and 31). However, there

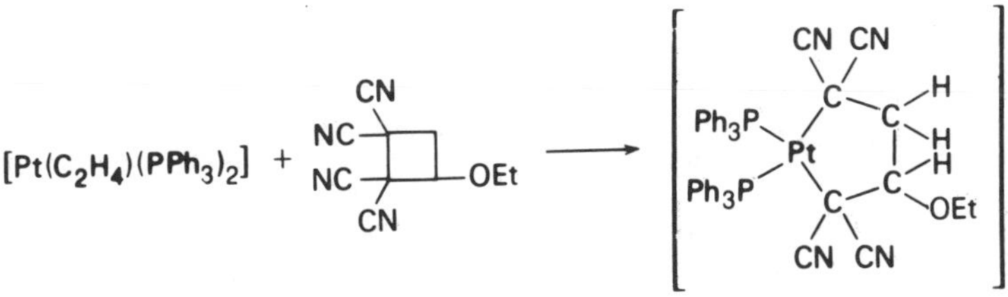

$$[Pt(C_2H_4)(PPh_3)_2] + (NC)_2C_4H_3(CN)_2(OEt) \longrightarrow [(Ph_3P)_2Pt\{C(CN)_2C(CN)_2CH_2CH(OEt)\}] \quad (30)$$

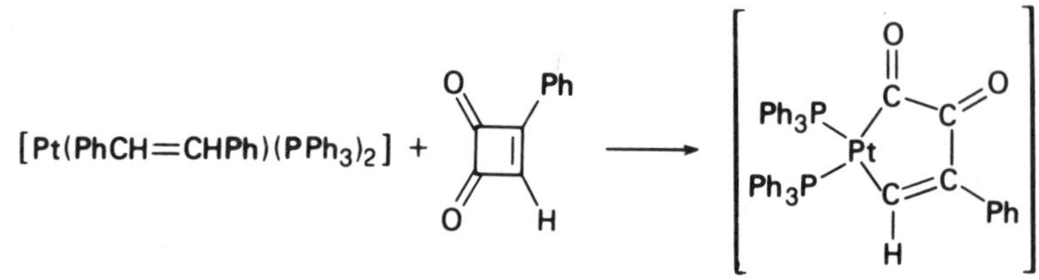

$$[Pt(PhCH{=}CHPh)(PPh_3)_2] + \text{(3-phenylcyclobutenedione)} \longrightarrow [\text{platinacycle}] \qquad (31)$$

are no examples of oxidative addition of cyclobutanes without functional group substituents. Cyclopropanes are much more reactive and many platinacyclobutanes have been prepared in this way[104,105] (equation 32). A ferracyclobutane derivative has been prepared in a similar way[106] (equation 33).

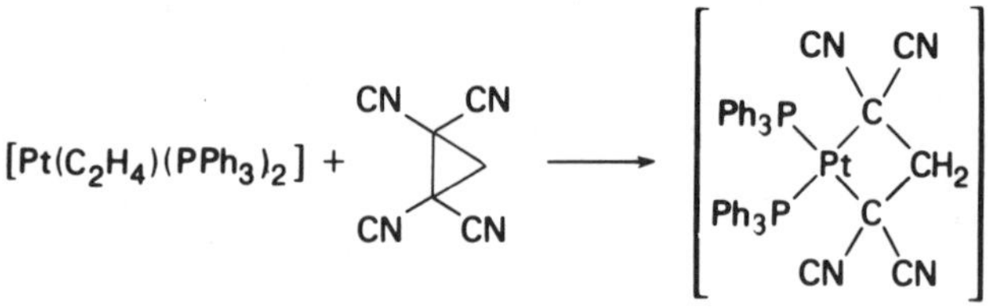

$$[Pt(C_2H_4)(PPh_3)_2] + \text{1,1,2,2-tetracyanocyclopropane} \longrightarrow [(Ph_3P)_2Pt\{C(CN)_2CH_2C(CN)_2\}] \qquad (32)$$

$$\text{(dibenzo-fused cyclopropane)} + [Fe_2(CO)_9] \longrightarrow [\text{ferracyclobutane, } Fe(CO)_4] \qquad (33)$$

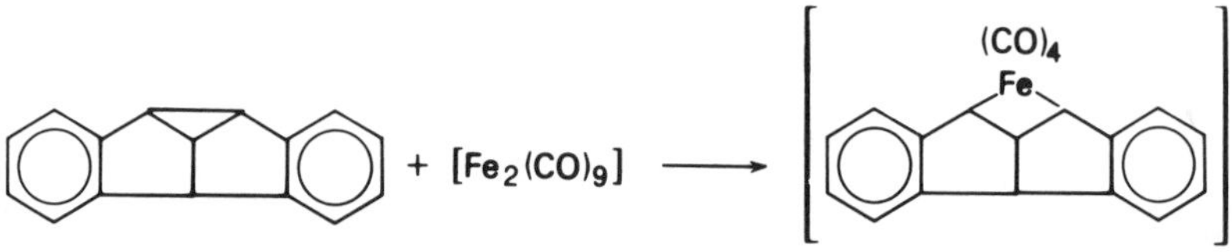

Oxidative addition to platinum(II) also occurs easily and cyclopropane itself will form platinum(IV) metallacyclobutanes[107] (equation 34). Arylcyclopropanes have been shown to react at a substituted C—C bond, but the products undergo facile isomerization to give mixtures of isomers[108].

$$[\{PtCl_2(C_2H_4)\}_2] + \triangleright \longrightarrow [\{PtCl_2(CH_2CH_2CH_2)\}_n] \qquad (34)$$

4. Reactions with metal atoms

Metal atoms, obtained by high-temperature vacuum vaporization techniques, are highly reactive and will undergo oxidative addition reactions[109–111] (equations 35 and 36). This technique is described in more detail in Chapter 13.

$$Pd(0) + C_6F_5Br \longrightarrow [\{PdBr(C_6F_5)\}_n] \xrightarrow{PPh_3} trans\text{-}[PdBr(C_6F_5)(PPh_3)_2] \qquad (35)$$

$$Pd(0) + CF_3I \longrightarrow [\{PdICF_3\}_n] \qquad (36)$$

With benzyl halides, the initial product is $[\{PdX(\eta^3\text{-}CH_2C_6H_5)\}_2]$ and η^1-benzyls are formed by addition of phosphine ligands[112]. The reactions appear to occur by free-radical mechanisms, and typical coupling products may also be formed[113]. Interesting π-arene complexes may be formed when the reactions are carried out in arene solvents.

These reactions require sophisticated equipment and attempts have been made to generate reactive metals, with chemical properties similar to those of metal atoms, by simpler methods. Two promising techniques in this regard involve reduction of transition metal halides with potassium to give reactive slurries of Ni, Pd, or Pt or electrochemical oxidation of a nickel or palladium anode in the presence of alkyl halides and stabilizing ligands. Either method gives useful synthesis of compounds $[MXRL_2]$ with M = Ni or Pd[114,115].

C. Synthesis by Cyclometallation

These reactions may be closely related to some of the intramolecular oxidative additions involving C—H bonds described above, but cyclometallations may occur by other mechanisms and they are of sufficient importance to justify a section of their own.

Key discoveries were those of Kleiman and Cope and co-workers, who found the cyclometallation of azobenzene and of *NN*-dimethylbenzylamine[116–118] (equations 37 and 38). In the latter case, and with similar compounds of platinum, the products are

$$[Ni(\eta^5\text{-}C_5H_5)_2] + PhN{=}NPh \longrightarrow [(\eta^5\text{-}C_5H_5)Ni(C_6H_4N{=}NPh)] + C_5H_6 \quad (37)$$

$$PdCl_2 + Me_2NCH_2Ph \longrightarrow 1/2\,[Pd(C_6H_4CH_2NMe_2)Cl]_2 + HCl \quad (38)$$

usually characterized by addition of ligands or by substitution of chloride by acetylacetonate or other bidentate ligands so as to give more soluble monomeric complexes[118,119].

The above examples illustrate many of the common features of the general cyclometallation or *ortho*-metallation reactions. Most commonly, the donor atom (nitrogen) and the substituted carbon atom form a five-membered chelate ring. Steric hindrance appears to promote cyclometallation. Thus, *NN*-dimethylbenzylamine will undergo facile cyclometallation, but benzylamine or *N*-methylbenzylamine will not and form simple complexes with palladium(II) chloride[118,119]. Several reviews on this topic have been published[120–122], and the number of ligands which can be cyclometallated in this way is very large. Further examples of nitrogen donor ligands which are readily cyclometallated include 2-phenylpyridine, benzo[*h*]quinoline, 8-methylquinoline, oximes and hydrazones of aryl ketones, and many more.[123–128].

With tertiary phosphine ligands, the presence of very bulky substituents such as tertiary butyl groups is often desirable or necessary in order to induce cyclometallation reactions (equation 39). The smaller methyl substituents are less effective than in the amine complexes discussed above[129]. In the above example, metallation of a methyl group occurs but metallation or aryl groups occurs very readily when such groups are present[130] (equation 40). Metallation of the tertiary butyl group to give a chelate with

$$2\,{}^tBu_2PCH_2CMe_3 + 2\,PdCl_2 \longrightarrow [\{{}^tBu_2PCH_2CMe_2CH_2\}Pd(\mu\text{-}Cl)]_2 + 2\,HCl \quad (39)$$

$$1,3\text{-}C_6H_4(CH_2P{}^tBu_2)_2 \xrightarrow{PdCl_2} [PdCl\{C_6H_3(CH_2P{}^tBu_2)_2\}] + HCl \quad (40)$$

a four-membered ring occurs in some cases when formation of the more favourable five-membered ring is not possible[131] (equation 41). This example also illustrates the use of a good 'leaving group' such as hydride or alkyl ligands in enhancing cyclometallation. As a more direct illustration, $[RhCl(PPh_3)_3]$ is not easily cyclometallated, but $[RhMe(PPh_3)_3]$ and especially $[Rh(CH_2SiMe_3)(PPh_3)_3]$ undergo cyclometallation very readily[131] (equation 42). In reactions of this kind, triphenyl phosphite, which gives a five-membered ring on *ortho*-metallation, is more readily metallated than triphenylphosphine which must form a four-membered ring[132].

$$trans\text{-}[PtH_2(P{}^tBu_3)_2] \xrightarrow{-H_2} [\{{}^tBu_2PCMe_2CH_2\}Pt(\mu\text{-}H)_2Pt\{CH_2CMe_2P{}^tBu_2\}] \quad (41)$$

$$[RhMe(PPh_3)_3] \longrightarrow CH_4 + [(Ph_3P)_2Rh\{C_6H_4PPh_2\}] \quad (42)$$

Sulphur donor ligands such a benzyl thioethers and thiobenzophenones are also metallated in many cases[133,134] (equations 43 and 44).

$$PhCH_2SMe + [MeMn(CO)_5] \xrightarrow{-CO} [\{C_6H_4CH_2S(Me)\}Mn(CO)_4] + CH_4 \quad (43)$$

$$\left[(RC_6H_4)_2C{=}S{-}Re(CO)_4{-}Re(CO)_5\right] \longrightarrow \left[\text{cyclometallated } Re(CO)_4 \text{ complex}\right] + [HRe(CO)_5] \qquad (44)$$

The selectivity in cyclometallation may differ with different metals. A particularly elegant example is seen in the metallation of *m*-fluoroazobenzene. The major products, shown in equations 45 and 46, differ when manganese or palladium complexes are used[122,135].

$$[MnMe(CO)_5] + PhN{=}N{-}C_6H_4F\text{-}m \xrightarrow[-CO]{-CH_4} [\text{cyclometallated } Mn(CO)_4 \text{ complex}] \qquad (45)$$

$$PdCl_2 + PhN{=}N{-}C_6H_4F\text{-}m \xrightarrow{-HCl} [\text{cyclometallated } PdCl \text{ complex}]_2 \qquad (46)$$

These presumably arise as a result of different mechanisms of metallation. It seems that manganese(I) acts as a nucleophile and palladium(II) as an electrophile in such reactions, but detailed mechanisms are difficult to elucidate in many cases. It is often assumed that, when the metal acts as a nucleophile, the reaction occurs by an oxidative addition–reductive elimination sequence at the metal centre. The metallation then occurs as shown for oxidative addition of C—H bonds, discussed above. When the metal acts as an electrophile, a mechanism akin to electrophilic aromatic substitution may operate. In the absence of a donor site on the arene, such electrophilic aromatic substitution occurs easily with gold(III) halides and gives a useful synthesis of arylgold(III) complexes[136] (equation 47).

$$Au_2Cl_6 + C_6H_6 \longrightarrow \tfrac{1}{2}[(AuCl_2Ph)_2] + H[AuCl_4] \qquad (47)$$

D. Synthesis of Metallacycles from Alkenes and Alkynes

The most common form of this synthetic method may be represented schematically by equation 48. The first examples of these reactions involved the use of fluorinated

$$M + 2\,C_2H_4 \rightleftharpoons M(C_2H_4)_2 \rightleftharpoons MC_4H_8 \qquad (48)$$

alkenes[137,138] (equations 49 and 50). Often intermediate alkene complexes can be isolated[139] (equation 51). It is necessary that there are vacant coordination sites present, or that such sites can readily be created by ligand dissociation, in order for the second alkene to coordinate and then form the metallacyclopentane[139].

$$[Fe(CO)_5] + 2\,C_2F_4 \xrightarrow{-CO} [(CO)_4Fe(CF_2CF_2CF_2CF_2)] \qquad (49)$$

$$[Co(CO)_2(\eta^5\text{-}C_5H_5)] + 2\,C_2F_4 \longrightarrow [(\eta^5\text{-}C_5H_5)(CO)Co(CF_2CF_2CF_2CF_2)] \qquad (50)$$

$$[Ni(PPh_3)_4] + C_2F_4 \longrightarrow [(Ph_3P)_2Ni(CF_2CF_2)] \xrightarrow{C_2F_4} [(Ph_3P)_2Ni(CF_2CF_2CF_2CF_2)] \qquad (51)$$

Fluorinated alkenes are not necessary and metallacyclopentanes have been prepared from the strained alkene norbornadiene (nbd) or even from ethylene itself[140,141] (equations 52 and 53). Reaction 53 is mechanistically much more complex

$$\text{nbd} + [Ir(cod)Cl]_2 \longrightarrow [Cl(nbd)Ir(C_7H_8)_2] \qquad (52)$$

$$[(Ph_3P)_2Ni(C_2H_4)] + C_2H_4 \rightleftharpoons [(Ph_3P)_2Ni(CH_2CH_2CH_2CH_2)] \qquad (53)$$

than is represented in this equation[141]. Many functionally substituted alkenes will behave similarly[142,143] (equations 54 and 55). Such reactions are often selective although, in principle, many isomers may be formed. When electronegative substituents are present on the alkene these usually adopt a position α to the

$$[Ni(cod)(bipy)] + \text{(cyclopropene)} \longrightarrow [(bipy)Ni(\ldots)] \qquad (54)$$

$$[Ni(cod)_2] + \text{(methylenecyclopropane)} \longrightarrow [\text{(cod)Ni metallacycle}] \xrightarrow{\text{bipy}} [\text{(bipy)Ni metallacycle}] \quad (55)$$

metallacyclopentane (equation 56). The major isomer has the substituents in mutually *trans*-positions[144]. In this case the dissociation of CO needed to create a vacant site for the incoming alkene is achieved photochemically. In other cases, displacement of weakly bound N_2 ligands or loss of hydride ligands by hydrogenation of excess alkene may give a similar result[145] (equation 57).

$$[Fe(CO)_4(CH_2{=}CHCO_2Me)] + CH_2{=}CHCO_2Me \longrightarrow [(CO)_4Fe(CH(CO_2Me)CH_2CH_2CH(CO_2Me))] \quad (56)$$

$$[\{(\eta^5\text{-}C_5Me_5)_2ZrN_2\}_2(\mu\text{-}N_2)] + C_2H_4 \xrightarrow{-N_2} [(\eta^5\text{-}C_5Me_5)_2Zr(CH_2)_4] \xleftarrow{-C_2H_6} [(\eta^5\text{-}C_5Me)_2ZrH_2] + C_2H_4 \quad (57)$$

These reactions are not limited to simple alkenes, and allene, 1,3-dienes and α,ω-dienes can also give metallacycles in this way[146–148] (equations 58a–c). In these

$$[\{(\eta^5\text{-}C_5Me_5)_2ZrN_2\}_2(\mu\text{-}N_2)] + CH_2{=}C{=}CH_2 \longrightarrow [(\eta^5\text{-}C_5Me_5)_2Zr\text{(3,4-dimethylenemetallacyclopentane)}] \quad (58a)$$

$$[(Ph_3P)_2Ni] + CH_2{=}C{=}CH_2 \longrightarrow [(Ph_3P)_2Ni\text{(3,4-dimethylenemetallacyclopentane)}] \quad (58b)$$

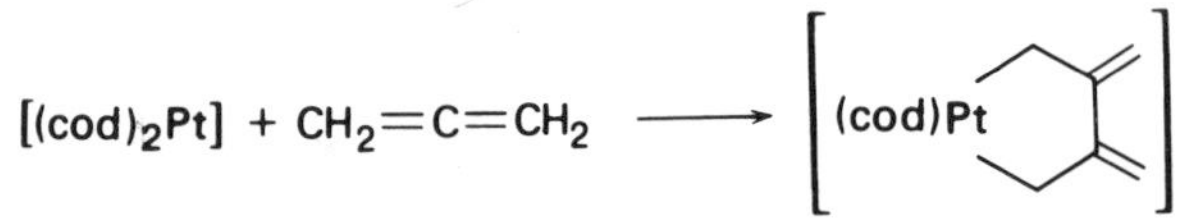

$$[(cod)_2Pt] + CH_2{=}C{=}CH_2 \longrightarrow [(cod)Pt\ \ldots] \qquad (58c)$$

examples, it is again shown that the reactions are selective, although it is difficult to predict *a priori* which isomer will be formed. 1,3-Dienes may be coupled in a similar manner[148] (equation 59). However, 1,4-addition of the metal across a single diene molecule may also occur[148] (equation 60).

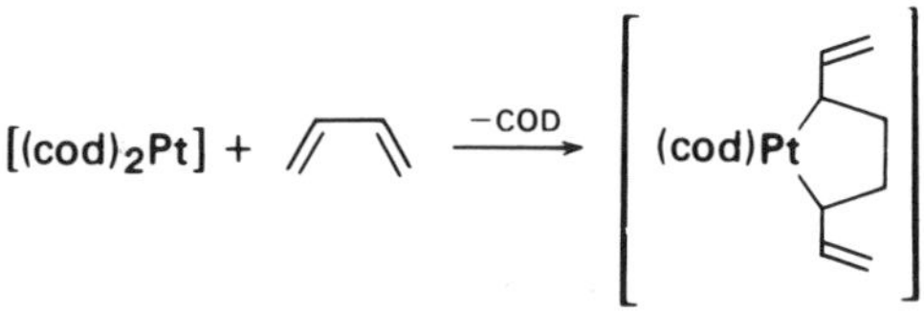

$$[(cod)_2Pt] + \ldots \xrightarrow{-COD} [(cod)Pt\ \ldots] \qquad (59)$$

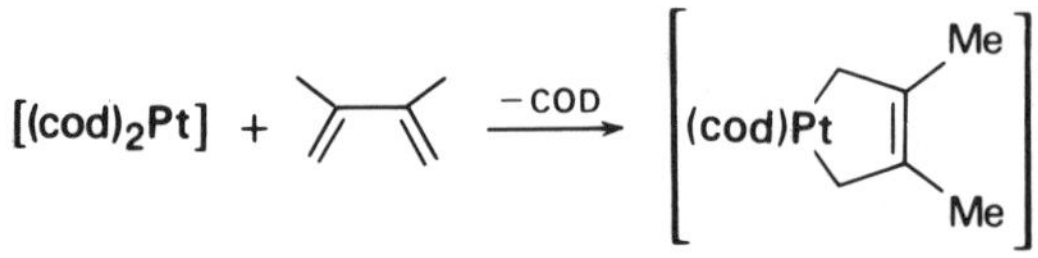

$$[(cod)_2Pt] + \ldots \xrightarrow{-COD} [(cod)Pt\ \ldots(Me)(Me)] \qquad (60)$$

Nickelahydrindane complexes may be prepared by coupling of octa-1,7-diene[149] (equation 61).

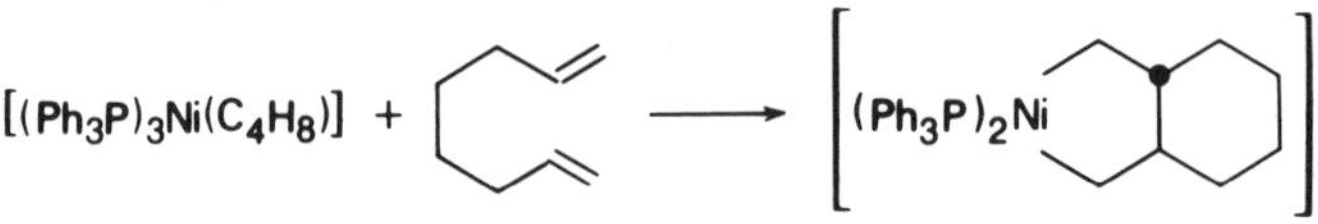

$$[(Ph_3P)_3Ni(C_4H_8)] + \ldots \longrightarrow [(Ph_3P)_2Ni\ \ldots] \qquad (61)$$

Alkynes can be coupled in a similar way to yield metallacyclopentadienes or metalloles (equations 62 and 63). Again, electronegative substituents on the alkyne often facilitate isolation of products[150,151]. Again, alkyne complexes are almost

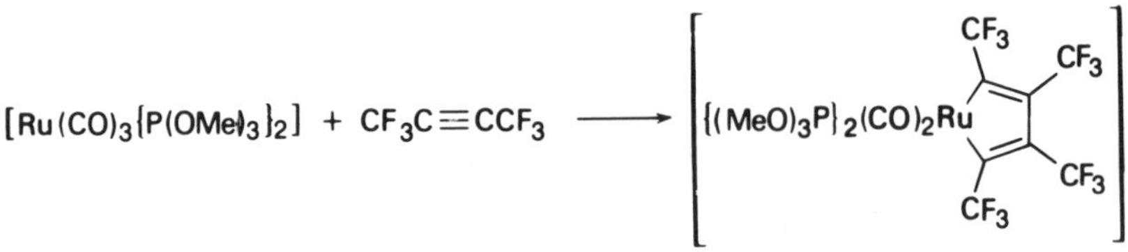

$$[Ru(CO)_3\{P(OMe)_3\}_2] + CF_3C{\equiv}CCF_3 \longrightarrow [\{(MeO)_3P\}_2(CO)_2Ru\ C_4(CF_3)_4] \qquad (62)$$

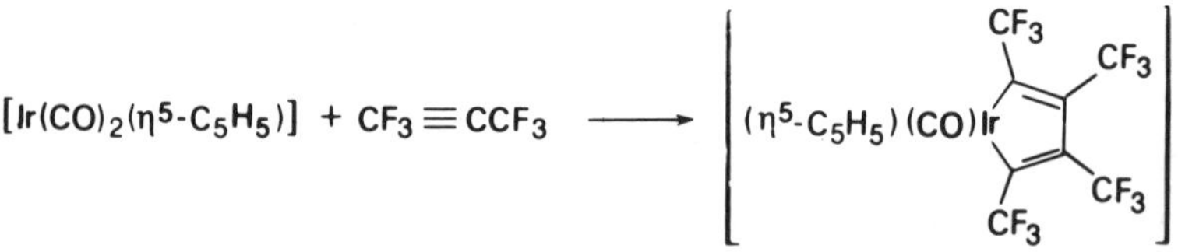

$$[Ir(CO)_2(\eta^5\text{-}C_5H_5)] + CF_3{\equiv}CCF_3 \longrightarrow [(\eta^5\text{-}C_5H_5)(CO)Ir\ C_4(CF_3)_4] \qquad (63)$$

certainly intermediates in such reactions and the synthesis may proceed in a stepwise fashion ($R = CO_2Me$)[152,153] (equation 64). In one case a ligand-free metallacyclopentadiene has been isolated and its chemistry has been thoroughly studied (equation 65; dba = dibenzylideneacetone, $R = CO_2Me$)[154]. This, and the cobalt complexes in reaction 66, are catalysts for the trimerization of alkynes or of

$$[IrCl(N_2)(PPh_3)_2] + RC{\equiv}CR \longrightarrow [Cl(PPh_3)_2Ir(\eta^2\text{-}RC{\equiv}CR)] \xrightarrow{RC\equiv CR} [Cl(PPh_3)_2IrC_4R_4] \quad (64)$$

$$[Pd_2(dba)_3] + RC{\equiv}CR \longrightarrow [PdC_4R_4]_n \quad (65)$$

$$[(\eta^5\text{-}C_5H_5)Co(PPh_3)_2] + MeC{\equiv}CMe \longrightarrow (\eta^5\text{-}C_5H_5)(PPh_3)CoC_4Me_4 \quad (66)$$

co-trimerization of alkenes and alkynes to benzene or cyclohexadiene derivatives, respectively, and the syntheses and reactions have therefore been the subject of great interest[155–158].

The formation of metallacyclopentadienes has been used to trap reactive intermediates, as in the examples in equations 67 and 68[159,160]. A somewhat different

$$[(\eta^5\text{-}C_5H_5)_2TiPh_2] \xrightarrow{h\nu} \text{"}(\eta^5\text{-}C_5H_5)_2Ti\text{"} \xrightarrow{PhC\equiv CPh} [(\eta^5\text{-}C_5H_5)_2TiC_4Ph_4] \quad (67)$$

$$[FeCl_2(PMe_3)_2] \xrightarrow[PMe_3]{Na/Hg} \text{"}Fe(PMe_3)_4\text{"} \xrightarrow{MeC\equiv CMe} [(Me_3P)_4FeC_4Me_4] \quad (68)$$

route is believed to be followed in the synthesis of zirconiacyclopentadienes[161] (equation 69). Further insertion of alkynes may occur to yield metallacycloheptatrienes and even larger rings in some cases[162–164] (equation 70). Coupling of an alkene and an alkyne to give a metallacyclopentene can also occur[155]

$$[(\eta^5\text{-}C_5H_5)_2ZrHMe] \xrightarrow{MeC\equiv CMe} CH_4 + \left[(\eta^5\text{-}C_5H_5)_2Zr(MeC\equiv CMe)\right] \xrightarrow{MeC\equiv CMe} \left[(\eta^5\text{-}C_5H_5)_2\overline{ZrC(Me)=C(Me)C(Me)=C}(Me)\right] \quad (69)$$

$$[NiHCl(PCy_3)_2] + CF_3C\equiv CCF_3 \longrightarrow \left[Cy_3P\overline{NiC(CF_3)=C(CF_3)C(CF_3)=C(CF_3)C(CF_3)=C}(CF_3)\right] \quad (70)$$

(equation 71). In one case the coupling has been shown to occur within the coordination sphere of the metal[165] (equation 72).

$$\left[(\eta^5\text{-}C_5H_5)(Ph_3P)Co(PhC\equiv CCO_2Me)\right] + (MeO_2C)HC=CH(CO_2Me) \longrightarrow \left[(\eta^5\text{-}C_5H_5)(Ph_3P)\overline{CoCH(CO_2Me)CH(CO_2Me)C(CO_2Me)=C}(Ph)\right] \quad (71)$$

$$[Rh(C_2H_4)(CF_3C\equiv CCF_3)(Me_3CCOCHCOCMe_3)] + 2C_5H_5N \longrightarrow [\overline{Rh\{CH_2CH_2C(CF_3)=C}(CF_3)\}(Me_3CCOCHCOCMe_3)(C_5H_5N)_2] \quad (72)$$

In principle, coupling of an alkene with a metal–carbene complex should yield a metallacyclobutane but, although such reactions probably play an important role in catalysis, few syntheses using this method have been reported. A classic example is shown in equation 73[166]. More commonly, the metallacyclobutanes decompose and metallacyclopentanes can be formed from the coupling of alkenes on the residual metal complex fragment[164] (equation 74).

$$[(\eta^5\text{-}C_5H_5)_2TiCH_2AlMe_2Cl] + Me_2C{=}CH_2 \xrightarrow[-AlMe_2Cl]{py} \left[(\eta^5\text{-}C_5H_5)_2Ti\square\right] \quad (73)$$

$$[(\eta^5\text{-}C_5H_5)Cl_2Ta{=}CH^tBu] + C_2H_4 \longrightarrow \left[(\eta^5\text{-}C_5H_5)Cl_2Ta\square{}^tBu\right]$$

$$\downarrow \quad (74)$$

$$\left[(\eta^5\text{-}C_5H_5)Cl_2Ta\bigcirc\right] \xleftarrow{2\,C_2H_4} \text{"}(\eta^5\text{-}C_5H_5)Cl_2Ta\text{"} + {}^tBuCH_2CH{=}CH_2$$

E. Synthesis Involving Insertion Reactions

1. Reactions of alkenes and alkynes

The best studied reactions involve insertion of alkenes and alkynes into metal–hydride bonds. These reactions are involved in many catalytic reactions and are also frequently used in the synthesis of alkyl and vinyl derivatives of transition metals. One of the first reactions to be studied is shown in equation 75[168]. The reaction is

$$trans\text{-}[PtHCl(PEt_3)_2] + C_2H_4 \rightleftharpoons trans\text{-}[PtEtCl(PEt_3)_2] \quad (75)$$

reversible and a great amount of work, both experimental and theoretical, has established that this and related reactions of hydridoplatinum(II) complexes occur by a dissociative process, with the key insertion step involving rearrangement of an intermediate *cis*-$[PtH(C_2H_4)(PEt_3)_2]^+$ [169,170]. In general, for such insertion reactions to occur the alkene must first be able to coordinate to the metal, and insertion can be induced by ligand addition[171] (equation 76).

$$[(\eta^5\text{-}C_5H_5)_2MoH(C_2H_4)]^+ \xrightarrow{PPh_3} [(\eta^5\text{-}C_5H_5)_2MoEt(PPh_3)]^+ \quad (76)$$

It should be borne in mind that the insertion reaction is easily reversed (β-elimination) and that multiple insertion/β-elimination sequences can lead to unexpected products. An important example is found in the hydrozirconation of alkenes by $[(\eta^5\text{-}C_5H_5)_2ZrHCl]$, which is very useful in organic synthesis. Both terminal and internal alkenes react to give alkylzirconium compounds in which zirconium is always attached to the least hindered terminal carbon atom. Zirconium is able to migrate to the end of the carbon chain by the above mechanism[172].

This reversibility is not observed when the alkene has electronegative substituents, as illustrated by the irreversible reactions 77 and 78[169,173]. Other electronegative

$$trans\text{-}[RhH(CO)(PPh_3)_2] + C_2F_4 \longrightarrow trans\text{-}[Rh(CF_2CF_2H)(CO)(PPh_3)_2] \quad (77)$$

$$trans\text{-}[PtHCl(PEt_3)_2] + C_2F_4 \longrightarrow trans\text{-}[Pt(CF_2CF_2H)Cl(PEt_3)_2] \quad (78)$$

substituents frequently used include —CN and —CO_2R, and the thermal stability of products usually increases with increasing number of such substituents[174,175].

$$[\eta^5\text{-}C_5H_5)(Ph_3P)_2RuH] + C(CN)_2{=}C(CF_3)_2 \longrightarrow [(\eta^5\text{-}C_5H_5)(Ph_3P)_2RuC(CN)_2C(CF_3)_2H] \quad (79)$$

More complex alkenes such as 1,3-dienes or methylenecyclopropane derivatives often give 1,2-addition products in the usual way with metal hydrides, but further reactions may then occur and lead to complex products[176,177] (equation 80).

trans-$[PtH(Me_2CO)(PPh_3)_2]^+$ + (methylenecyclopropane-1,2-di-CO_2Me) ⟶ [ring product: MeO_2C, Me, PPh_3, Pt, PPh_3, C=O, MeO]$^+$ (80)

Alkynes most typically give *cis*-insertion into metal—hydride bonds, as expected for the mechanism outlined above[169,178] (equations 81 and 82). Similar reactions frequently occur with hydrido derivatives of metal clusters, but the resulting

$$trans\text{-}[PtHCl(PEt_3)_2] + CF_3C{\equiv}CCF_3 \longrightarrow [Cl{-}Pt(PEt_3)_2{-}C(CF_3){=}C(H)CF_3] \quad (81)$$

$$[(\eta^5\text{-}C_5H_5)_2ReH] + MeO_2CC{\equiv}CCO_2Me \longrightarrow [(\eta^5\text{-}C_5H_5)_2Re\{cis\text{-}C(CO_2Me){=}C(CO_2Me)H\}] \quad (82)$$

vinyl group then typically becomes bound to more than one metal centre. For example, reaction of $CF_3C{\equiv}CCF_3$ with $[Os_3H_2(CO)_{10}]$ gives $[Os_3H\{C(CF_3){=}C(CF_3)H\}(CO)_{10}]$ with the vinyl group bound to three osmium atoms, (**1**)[179,180].

Os, H–C, CF3, CF3, C, Os, Os

(**1**)

It is certainly not safe to predict *cis*-insertion and the alternative *trans*-insertion is frequently observed[181–183] (equations 83–85). The mechanism by which the *trans* products are formed has been the subject of debate, and there is not yet a clear answer. In some cases it is thought that *cis*-addition occurs, followed by isomerization to the *trans*-isomer, while in other cases the *trans*-isomer is thought to be formed directly by free-radical or dipolar mechanisms or a mechanism in which a twisted vinyl group is formed initially[181,184–186]. The stereochemistry cannot be predicted in advance.

$$[(\eta^5\text{-}C_5H_5)_2MoH_2] + CF_3C{\equiv}CCF_3 \longrightarrow \left[(\eta^5\text{-}C_5H_5)_2Mo(H)C(CF_3){=}C(CF_3)H\right] \quad (83)$$

$$[IrH(CO)(PPh_3)_3] + PhC{\equiv}CPh \longrightarrow [Ir(trans\text{-}CPh{=}CHPh)(CO)(PPh_3)_2] \quad (84)$$

$$[MnH(CO)_5] + MeO_2CC{\equiv}CCO_2Me \longrightarrow [Mn\{trans\text{-}C(CO_2Me){=}CH(CO_2Me)\}(CO)_5] \quad (85)$$

Insertion of alkenes and alkynes into metal—carbon bonds occurs less readily than insertion into metal—hydride bonds. Thus, ethylene is not known to give a simple insertion with any metal alkyl derivatives, although there is indirect evidence for this reaction in several cases[187]. However, alkenes and alkynes with electronegative substituents will insert into many metal—carbon bonds[188–191] (equations 86–89).

$$[(\eta^5\text{-}C_5H_5)Fe(CO)_2Et] + C_2(CN)_4 \longrightarrow [(\eta^5\text{-}C_5H_5)Fe(CO)_2C(CN)_2C(CN)_2Et] \quad (86)$$

$$[(\eta^5\text{-}C_5H_5)Mo(CO)_3Me] + C_2(CN)_4 \longrightarrow [(\eta^5\text{-}C_5H_5)Mo(CO)_3C(CN)_2C(CN)_2Me] \quad (87)$$

$$[PtClMe(PMe_2Ph)_2] + C_2F_4 \longrightarrow [PtCl(CF_2CF_2Me)(PMe_2Ph)_2] \quad (88)$$

$$[AuMe(PPh_3)] + C_2F_4 \longrightarrow [Au(CF_2CF_2Me)(PPh_3)] \quad (89)$$

Allyl derivatives may react with tetracyanoethylene to give cyclopentyl derivatives[192] (equation 90).

$$[(\eta^5\text{-}C_5H_5)(CO)_2FeCH_2CH{=}CH_2] + C_2(CN)_4 \longrightarrow (\eta^5\text{-}C_5H_5)(CO)_2Fe\text{-}C_5H_5(CN)_4 \quad (90)$$

Alkynes are more reactive and insertion can occur without electro-negative substituents[193,194] (equations 91 and 92). Again *cis*-insertion is most commonly observed and, when the product has *trans*-stereochemistry, there is evidence in at least

$$[(\eta^5\text{-}C_5H_5)_2TiMe_2] + PhC{\equiv}CPh \xrightarrow{h\nu} \left[(\eta^5\text{-}C_5H_5)_2Ti(Me)C(Ph){=}C(Ph)Me\right] \quad (91)$$

$$[NiPhBr(PPh_3)_2] + MeC{\equiv}CMe \longrightarrow \left[BrNi(PPh_3)_2C(Me){=}C(Me)Ph\right] \quad (92)$$

one case that this is formed by isomerization of the initially formed *cis*-insertion product[195] (equation 93). Most synthesis still use alkynes with electronegative substituents, as illustrated by the example in equation 94[196]. In addition to the *cis*-stereochemistry at the vinyl group, it has also been shown in one case that a chiral alkyliron complex undergoes insertion with retention of stereochemistry at the chiral carbon, as expected for an intramolecular alkyl migration[197].

$$[(acac)(Ph_3P)NiMe] + PhC{\equiv}CPh \longrightarrow [(acac)(Ph_3P)Ni(Z\text{- or }E\text{-}C(Ph){=}CPhMe)] \quad (93)$$

$$[(CO)_5MnMe] + CF_3C{\equiv}CCF_3 \longrightarrow [(CO)_5Mn{-}C(CF_3){=}C(Me)CF_3] \quad (94)$$

There are many cases in which more than one alkyne may insert and very complex products may result. One relatively simple case is illustrated in equation 95[196]. Similarly, metallacyclic ring expansions may occur[198] (equation 96).

$$[(\eta^5\text{-}C_5H_5)(CO)_3MoCH_2CH{=}CH_2] \xrightarrow[-2\,CO]{CF_3C{\equiv}CCF_3} \quad (95)$$

$$\xrightarrow[-Ph_3P]{PhCCPh} \quad (96)$$

Insertion of alkenes and alkynes into metal—halogen bonds often occurs[199,200]. The alkyne insertions usually give *trans*-products, and the reactions have been used to prepare fluoroalkylsilver compounds (equations 97–99). Insertion into

$$CF_2{=}CF_2 + AgF \longrightarrow [CF_3CF_2Ag]_n \quad (97)$$

$$CF_2{=}C{=}CF_2 + AgF \longrightarrow [CF_3C(Ag){=}CF_2]_n \quad (98)$$

$$CF_3C{\equiv}CCF_3 + AgF \longrightarrow [(F)(CF_3)C{=}C(CF_3)(Ag)]_n \quad (99)$$

metal—halogen bonds may occur more easily than insertion into metal—carbon bonds[201] (equation 100). Insertion into metal—halide bonds may give *cis*-products and further insertions may then occur. Excellent examples are found in reactions of

$$[PtIMe(bipy)] + MeO_2CCCCO_2Me \longrightarrow [PtMe\{C(CO_2Me){=}C(CO_2Me)I\}(bipy)] \quad (100)$$

alkynes with palladium halides, where the final product often depends on the bulk of the substituents, R, on the alkyne[202] (equation 101).

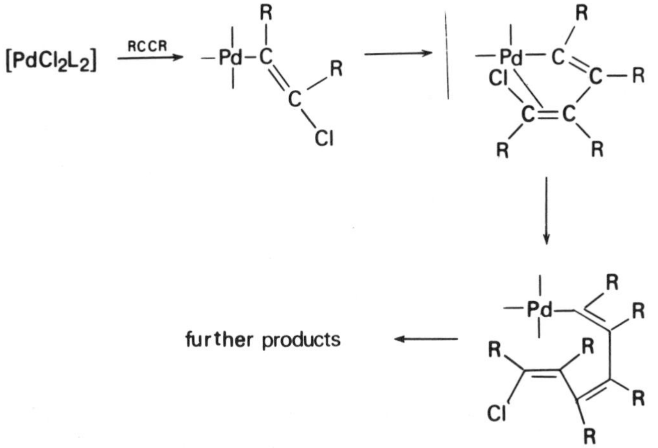
[PdCl2L2]
RCCR
further products

(101)

Many examples are known in which bridging alkyl or vinyl derivatives are formed by insertion of alkenes or alkynes into metal—metal bonds[203–205] (equations 102 and 103). A metal—bond may also be formed in such reactions[206] (equation 104).

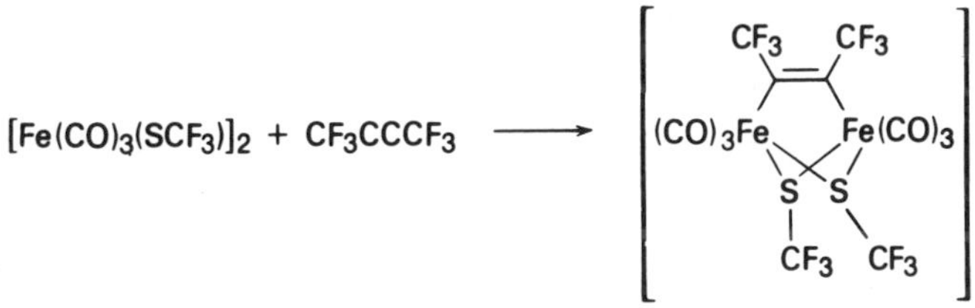
[Fe(CO)3(SCF3)]2 + CF3CCCF3
(CO)3Fe
Fe(CO)3

(102)

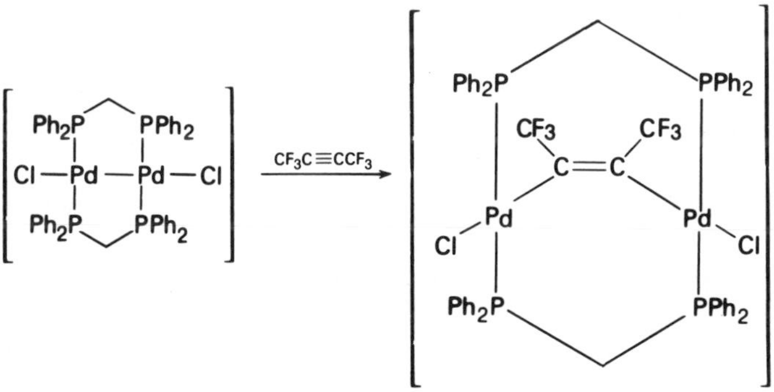
CF3C≡CCF3

(103)

$$[(\eta^5\text{-}C_5H_5)Rh(CO)_2] + CF_3C{\equiv}CCF_3 \longrightarrow [(\eta^5\text{-}C_5H_5)(CO)Rh\text{—}Rh(CO)(\eta^5\text{-}C_5H_5)(\mu\text{-}CF_3C{=}CCF_3)] \quad (104)$$

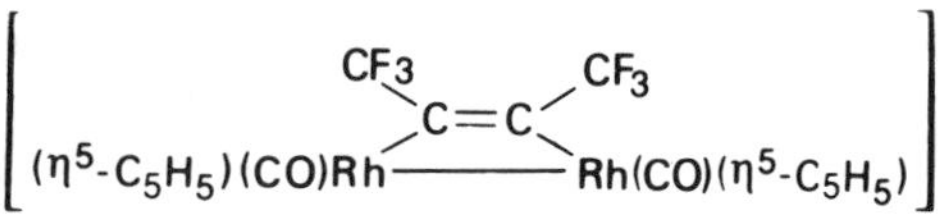

2. Reactions of carbon monoxide, isocyanides, and carbene precursors

Carbon monoxide and alkyl isocyanide insertion reactions may be represented by the general equations 105 and 106. Acylmetal and iminoalkylmetal derivatives are

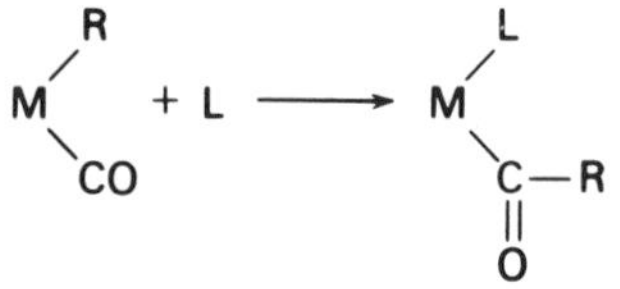

$$M(R)(CO) + L \longrightarrow M(L)(C(=O)R) \quad (105)$$

$$M(R)(CNR') + L \longrightarrow M(L)(C(=NR')R) \quad (106)$$

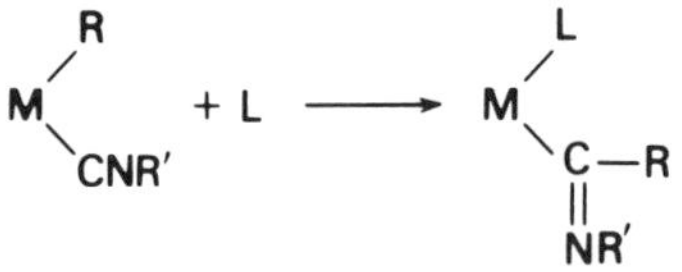

therefore formed. These topics have been thoroughly reviewed[207–209], and the mechanisms are understood well. Specific examples are shown in equations 107 and 108, and the reader is referred to the above reviews for more comprehensive coverage[210].

$$[(CO)_5MnMe] + CO \longrightarrow [(CO)_5MnCOMe] \quad (107)$$

$$[PdIMe({}^tBuNC)_2] \longrightarrow [\{Pd(\mu\text{-}I)(C(=N^tBu)Me)(CN^tBu)\}_2] \quad (108)$$

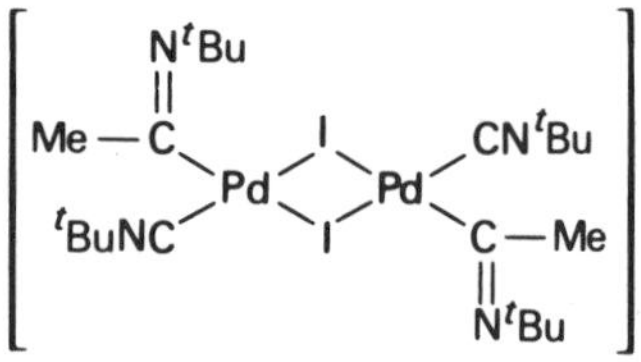

Insertion into metal—carbon bonds occurs readily in many cases, but insertion into metal—hydride bonds is generally not possible. Thus the reactions essentially convert one metal—carbon σ-bond into another. These reactions may also occur during attempted insertion of alkenes and alkynes into metal—carbon σ-bonds of alkyl(carbonyl)metal derivatives. An example is given in equation 109 and many others are known[196].

$$[(\eta^5\text{-}C_5H_5)(CO)_2FeMe] \xrightarrow[-CO]{CF_3CCCF_3} [(\eta^5\text{-}C_5H_5)Fe\{C(CF_3){=}C(CF_3)C(CF_3){=}C(CF_3)C(Me){=}O\}] \quad (109)$$

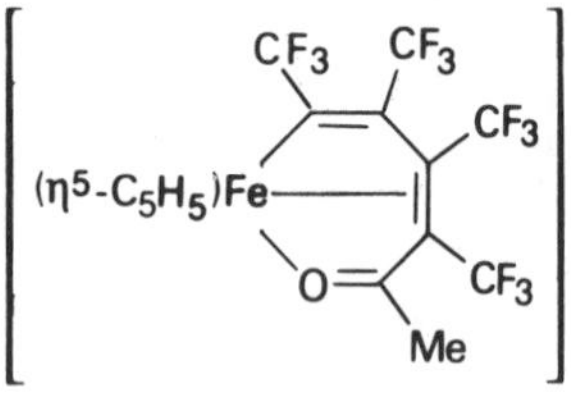

With electron-deficient oxophilic transition metal alkyls the acyl group may be η^2-bonded to the metal[211] (equation 110).

$$[(\eta^5\text{-}C_5H_5)_2ZrMe_2] + CO \longrightarrow \left[(\eta^5\text{-}C_5H_5)_2Zr(Me)(\eta^2\text{-}OCMe)\right] \quad (110)$$

The insertion of methylene also gives 1,1-addition analogous to the insertion of CO and isocyanides. Generally diazomethane is used as the source of methylene, and it is unlikely that free methylene is involved in any of the reactions. The detailed mechanisms of methylene insertions are not yet known but the overall effect is shown in reaction 111 for insertion into a metal—hydride bond[212]. Insertion into metal—chlorine bonds gives a route to chloromethyl derivatives[213] (equation 112).

$$[(\eta^5\text{-}C_5H_5)(CO)_3MoH] + CH_2N_2 \longrightarrow [(\eta^5\text{-}C_5H_5)(CO)_3MoMe] + N_2 \quad (111)$$

$$trans\text{-}[IrCl(CO)(PPh_3)_2] + CH_2N_2 \xrightarrow{-N_2} trans\text{-}[Ir(CH_2Cl)(CO)(PPh_3)_2] \quad (112)$$

There has been particular interest in the synthesis of μ-methylene complexes from diazomethane, and the topic has been reviewed[214]. When mononuclear complexes are used a metal—metal bond accompanies the formation of the μ-methylene group[215–217].

$$(\eta^5\text{-}C_5H_5)(CO)_2Mn(thf) \xrightarrow{CH_2N_2} \left[(\eta^5\text{-}C_5H_5)(CO)_2Mn\overset{CH_2}{\text{—}}Mn(CO)_2(\eta^5\text{-}C_5H_5)\right] \quad (113)$$

$$[(\eta^5\text{-}C_5H_5)(CO)_2Co] \xrightarrow{N_2CHCO_2Et} \left[(\eta^5\text{-}C_5H_5)(CO)Co\overset{C(H)(CO_2Et)}{\text{—}}Co(CO)(\eta^5\text{-}C_5H_5)\right]$$

$$[(\eta^5\text{-}C_5H_5)(CO)_2Co] \xrightarrow{N_2CPh_2} \left[(\eta^5\text{-}C_5H_5)Co\overset{CPh_2}{\underset{C=O}{=}}Co(\eta^5\text{-}C_5H_5)\right] \quad (114)$$

When the precursor complex contains a metal—metal bond this may be broken[218] (equation 115). When the precursor contains a metal—metal double bond, this may be converted to a single bond[219] (equation 116). Elimination of CO then recreates the double bond and further methylene addition can occur[220] (equation 117). A particularly interesting synthesis involves diazomethane addition to the cluster $[Os_3H_2(CO)_{10}]$[221] (equation 118).

Addition of diphenyldiazomethane to a metal—metal triple bond occurs and the *N*-bonded intermediate can be isolated (equation 19). In the product the CPh_2 ligand

$$[Pt_2Cl_2(\mu\text{-}Ph_2PCH_2PPh_2)_2] + CH_2N_2 \longrightarrow [Pt_2Cl_2(\mu\text{-}CH_2)(\mu\text{-}Ph_2PCH_2PPh_2)_2] \quad (115)$$

$$[(\eta^5\text{-}C_5Me_5)Rh(\mu\text{-}CO)_2Rh(\eta^5\text{-}C_5Me_5)] \xrightarrow{N_2CHMe} [(\eta^5\text{-}C_5Me_5)(CO)Rh(\mu\text{-}CHMe)Rh(CO)(\eta^5\text{-}C_5Me_5)] \quad (116)$$

$$[(\eta^5\text{-}C_5Me_5)(CO)Rh(\mu\text{-}CPh_2)Rh(CO)(\eta^5\text{-}C_5Me_5)] \xrightarrow{-CO} [(\eta^5\text{-}C_5Me_5)Rh(\mu\text{-}CPh_2)(\mu\text{-}CO)Rh(\eta^5\text{-}C_5Me_5)] \xrightarrow{CH_2N_2} [(\eta^5\text{-}C_5Me_5)Rh(\mu\text{-}CPh_2)(\mu\text{-}CH_2)(\mu\text{-}CO)Rh(\eta^5\text{-}C_5Me_5)] \quad (117)$$

$$[Os_3H_2(CO)_{10}] \xrightarrow{CH_2N_2} [Os_3(\mu\text{-}H)(\mu\text{-}Me)(CO)_{10}] \rightleftharpoons [Os_3(\mu\text{-}H)_2(\mu\text{-}CH_2)(CO)_{10}] \quad (118)$$

$$[(\eta^5\text{-}C_5H_5)_2(CO)_4Mo_2](Mo{\equiv}Mo) \xrightarrow{Ph_2CN_2} [(\eta^5\text{-}C_5H_5)_2(CO)_4Mo_2(\mu\text{-}N_2CPh_2)] \xrightarrow{heat} [(\eta^5\text{-}C_5H_5)_2(CO)_4Mo_2(\mu\text{-}CPh_2)] \quad (119)$$

acts as a four-electron ligand[222], and the Mo—Mo bond order is reduced from three to one. A related synthetic method is to react a carbene complex with coordinatively unsaturated metal complex, according to the general equation 120[223]. This method has

$$M{=}CR_2 + M' \longrightarrow M(\mu\text{-}CR_2)M' \quad (120)$$

proved to be very useful and examples of complexes prepared in this way are $[(CO)_5W\{\mu\text{-}CPh(OMe)\}Pt(PMe_3)_2]$ and $[(\eta^5\text{-}C_5H_5)(CO)_2Mn\{\mu\text{-}CPh(OMe)\}\text{-}Ni(PMe_3)_2]$.

F. Synthesis Involving Elimination Reactions

These reactions are seldom used in synthesis and so will be reviewed only briefly. The reaction is the opposite of the insertion reaction but is more limited in scope. For example, while insertion of alkenes into alkylmetal bonds occurs as described in Section II.E, the reverse reaction cannot be used to prepare alkylmetal compounds. However, carbon monoxide insertion is reversible and CO elimination is sometimes used to prepare metal alkyls[208] (equation 121). Some examples of this reaction are

$$L\text{-}M\text{-}C({=}O)R \xrightarrow[-L]{heat} R\text{-}M\text{-}CO \quad (121)$$

given in equations 122–124[224–226]. The mechanism is well established and occurs with retention of stereochemistry at a chiral migrating alkyl group but inversion at a chiral metal centre[196,227].

$$[(\eta^5\text{-}C_5H_5)_2ZrMe(COMe)] \xrightarrow{-CO} [(\eta^5\text{-}C_5H_5)_2ZrMe_2] \quad (122)$$

$$[(CO)_5MnCOCH_2CH_2CO_2Me] \xrightarrow{-CO} [(CO)_5MnCH_2CH_2CO_2Me] \quad (123)$$

$$[Rh(COMe)Cl_2(PPh_3)_2] \overset{-CO}{\rightleftharpoons} [RhMeCl_2(CO)(PPh_3)_2] \quad (124)$$

Another useful elimination involves loss of N_2 from aryldiazonium complexes, as illustrated in equation 25[228].

$$[O_2N\text{-}C_6H_4\text{-}N{=}N\text{-}PtCl(PEt_3)_2] \xrightarrow{-N_2} [O_2N\text{-}C_6H_4\text{-}PtCl(PEt_3)_2] \quad (125)$$

Derivatives which cannot be prepared by the conventional Grignard reagent can be prepared in this way, but the method is more commonly used in the synthesis of main group metal aryls. A related method involves use of diaryliodonium reagents[229] (equation 126).

$$[(\eta^5\text{-}C_5H_5)(CO)_2Fe]^- + Ph_2I^+ \xrightarrow{-PhI} [(\eta^5\text{-}C_5H_5)(CO)_2FePh] \qquad (126)$$

Elimination of CO_2 from metal carboxylates or SO_2 from metal sulphinates can also give alkyl or aryl derivatives. The method is especially useful for synthesis of pentafluorophenyl complexes[230] (equations 127 and 128).

$$[Ni(O_2CC_6F_5)_2(bipy)] \xrightarrow[-CO_2]{heat} [Ni(C_5F_5)_2(bipy)] \qquad (127)$$

$$[Au(O_2CC_6F_5)(PPh_3)] \xrightarrow[-CO_2]{heat} [AuC_6F_5(PPh_3)] \qquad (128)$$

G. Synthesis by Attack on Coordinated Ligands

1. Synthesis from alkene and alkyne complexes

Nucleophilic attack on coordinated alkenes is a useful synthetic method, and is represented by the general equation 129.

$$X^- + \|\text{—}M \longrightarrow X\text{—}CH_2CH_2\text{—}M^- \qquad (129)$$

There is an obvious resemblance to the alkene insertion reaction, but there is an important difference. Nucleophilic attack occurs on the face of the alkene remote from the metal and hence the stereochemistry is different. This effect is seen clearly in reaction 130[231].

$$\left[(\eta^5\text{-}C_5H_5)(Ph_3P)Pd\text{—}(CHD{=}CHD)\right]^+ + CH_3O^- \longrightarrow \left[(\eta^5\text{-}C_5H_5)(Ph_3P)Pd\text{—}CHD\text{—}CHD\text{—}OMe\right] \qquad (130)$$

Many of the earliest examples of these reactions involved nucleophilic attack on diene complexes of palladium and platinum (equation 131; M = Pd or Pt)[232]. Again, it is clear that *exo* attack by the nucleophile occurs. The nucleophile X^- may be alkoxide, primary amine, ammonia, azide, acetate, acetylacetonate, malonate, and many more. The diene may be 1,5-cyclooctadiene, *endo*-dicyclopentadiene, norbornadiene, 1,5-hexadiene, and others[233].

Hydroxide attack on coordinated alkenes is important in catalytic processes (e.g. the Wacker process). Several model studies have shown *exo* attack to occur, but there is still some doubt about the stereochemistry in the catalytic reactions. One example of

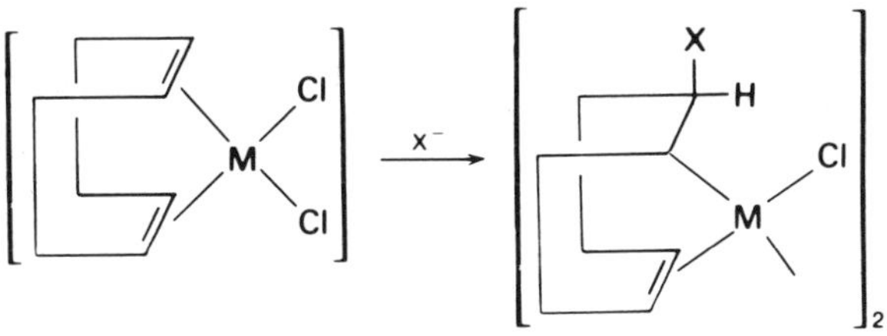

(131)

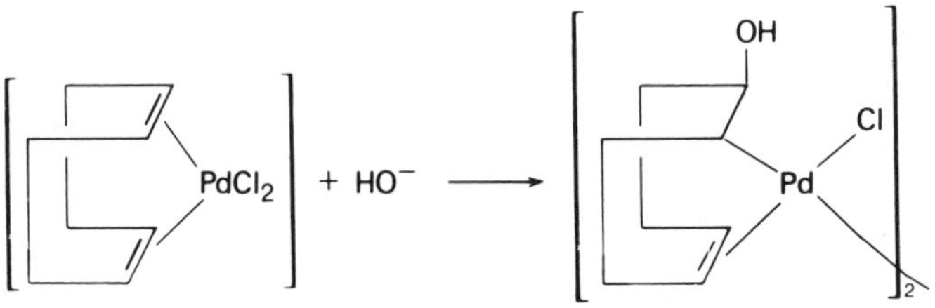

(132)

exo attack is shown in equation 132[234–236]. Where simple alkene complexes are attacked, the products are often difficult to isolate but there are many good examples of such syntheses, e.g. equation 133[231]. When the nucleophile is neutral rather than

$$[(\eta^5\text{-}C_5H_5)(Ph_3P)Pd(C_2H_4)]^+ + MeO^- \longrightarrow [(\eta^5\text{-}C_5H_5)(Ph_3P)PdCH_2CH_2OMe] \quad (133)$$

anionic, dipolar complexes may be formed[237–239] (equations 134 and 135). The reactions may be promoted by the presence of an electrophilic alkene[240] (equation 136).

$$cis\text{-}[PtCl_2(C_2H_4)(py)] + py \longrightarrow cis\text{-}[Pt^-Cl_2(CH_2CH_2py^+)(py)] \quad (134)$$

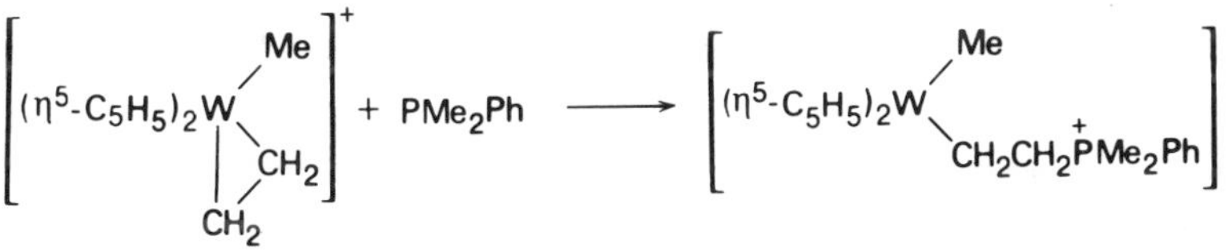

$$[(\eta^5\text{-}C_5H_5)_2W(Me)(CH_2{=}CH_2)]^+ + PMe_2Ph \longrightarrow [(\eta^5\text{-}C_5H_5)_2W(Me)(CH_2CH_2\overset{+}{P}Me_2Ph)] \quad (135)$$

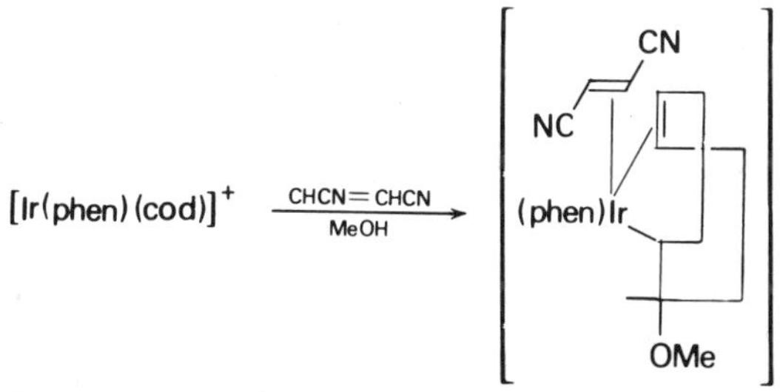

$$[Ir(phen)(cod)]^+ \xrightarrow[\text{MeOH}]{\text{CHCN=CHCN}} (136)$$

(phen)Ir … NC, CN, OMe

The presence of a donor atom may stabilize products as in reactions of coordinate alkenylamines[241–244] (equation 136). An attacking amine ligand may also coordinate and give a chelate, the strained four-membered ring being more robust with platinum than palladium in the ring $\overline{MCH_2CH_2NMe_2}$ resulting from attack of dimethylamine on coordinated ethylene[245,246].

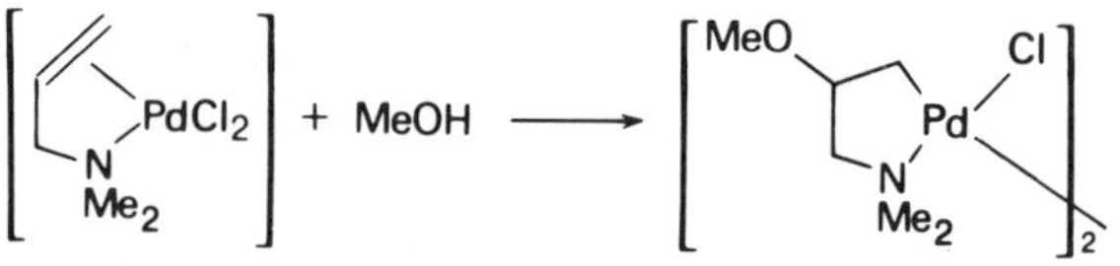

$$\left[\text{(CH}_2\text{=CHCH}_2\text{CH}_2\text{NMe}_2\text{)PdCl}_2\right] + \text{MeOH} \longrightarrow \left[\text{MeOCH(CH}_2\text{)CH}_2\text{CH}_2\text{NMe}_2\text{Pd}(\mu\text{-Cl})\right]_2 \quad (137)$$

Another versatile system involves nucleophilic attack on alkene complexes of iron[247] (equation 137). The nucleophile may be a carbanion, amine phosphine, thiol, or phosphine ylide[247,248].

$$[(\eta^5\text{-}C_5H_5)(CO)_2Fe(C_2H_4)]^+ + CH_2NO_2^- \longrightarrow [(\eta^5\text{-}C_5H_5)(CO)_2FeCH_2CH_2CH_2NO_2] \quad (137)$$

Coordinated allene derivatives also undergo nucleophilic attack to give vinylmetal derivatives[249] (equation 138). Finally, binuclear metal alkyls can be prepared by using a metal-centred nucleophile[250] (equation 139).

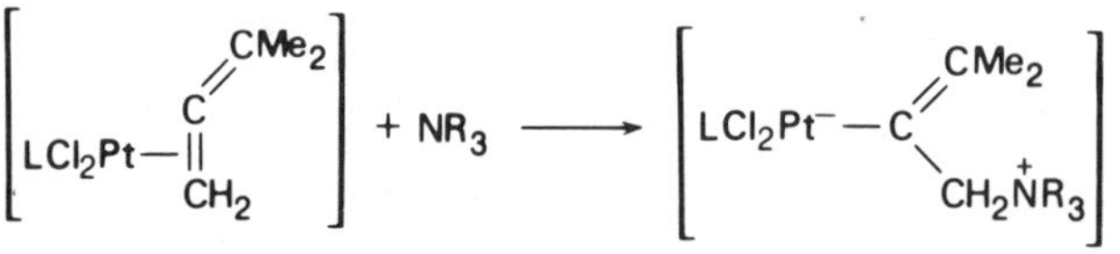

$$\left[LCl_2Pt\text{—}\|(CH_2{=}C{=}CMe_2)\right] + NR_3 \longrightarrow \left[LCl_2Pt^-\text{—}C(=CMe_2)CH_2\overset{+}{N}R_3\right] \quad (138)$$

$$[(\eta^5\text{-}C_5H_5)(CO)_3W(C_2H_4)]^+ + [Re(CO)_5]^- \longrightarrow [(\eta^5\text{-}C_5H_5)(CO)_3WCH_2CH_2Re(CO_5] \quad (139)$$

2. Synthesis from carbene complexes

The earliest examples of these reactions were discovered by Fischer[251] (equation 140). Trialkylphosphines are the most widely used reagents, but the nature of the

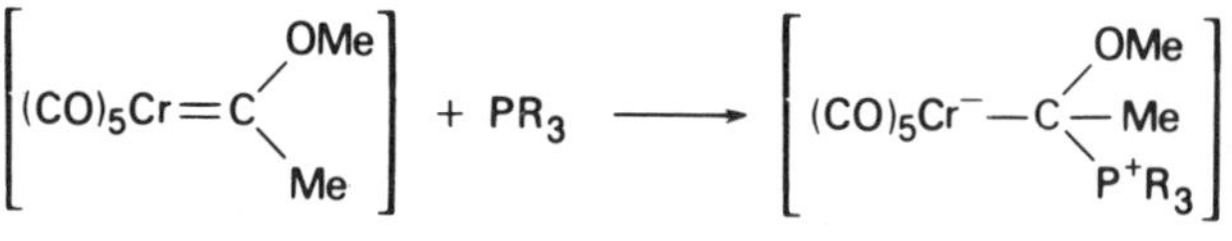

$$\left[(CO)_5Cr{=}C(OMe)Me\right] + PR_3 \longrightarrow \left[(CO)_5Cr^-\text{—}C(OMe)(Me)P^+R_3\right] \quad (140)$$

initial carbene complex can vary widely[252,253] (equations 141 and 142). In some cases substitution of one nucleophile by another can occur (equation 143). Attack by anionic nucleophiles also occurs readily[255] (equations 144 and 145). The formation of ylide complexes is sometimes even used to trap shortlived carbene complexes[256,257] (equations 146 and 147).

$$[(CO)_5WCPh_2] + PMe_3 \longrightarrow [(CO)_5\bar{W}\text{—}CPh_2\text{—}\overset{+}{P}Me_3] \quad (141)$$

$$[(CO)_5WCHPh] + PPh_3 \longrightarrow [(CO)_5\bar{W}\text{—}CHPh\text{—}\overset{+}{P}Ph_3] \quad (142)$$

$$[(CO)_5\bar{Cr}CH_2\text{—}\overset{+}{S}OMe_2] + Ph_3As \longrightarrow [(CO)_5\bar{Cr}CH_2\overset{+}{A}sPh_3] + Me_2SO \quad (143)$$

$$[(\eta^5\text{-}C_5H_5)(NO)(PPh_3)ReCH_2]^+ + PhLi \longrightarrow (\eta^5\text{-}C_5H_5)(NO)(PPh_3)ReCH_2Ph \quad (144)$$

$$(CO)_5Cr{=}C(OMe)Ph + RLi \longrightarrow Li[(OC)_5Cr\text{—}C(OMe)PhR] \quad (145)$$

$$\text{"(py)Cl}_2\text{Pt=CHCH}_2\text{CHMe}_2\text{"} + \text{py} \longrightarrow [\text{(py)Cl}_2\text{Pt}^-\text{CH(py}^+\text{)CH}_2\text{CHMe}_2] \quad (146)$$

$$[(\eta^5\text{-C}_5\text{H}_5)_2\text{W}(=\text{CH}_2)(\text{H})]^+ + \text{PMe}_2\text{Ph} \longrightarrow [(\eta^5\text{-C}_5\text{H}_5)_2\text{W}(\text{CH}_2\text{PMe}_2\text{Ph})(\text{H})]^+ \quad (147)$$

3. Synthesis from η^3-allyl complexes

Normally, nucleophiles attack a terminal carbon a η^3-allyl complexes to give alkene complexes but, in one case, attack at the central carbon occurs to give a metallacyclobutane (equation 148). Further examples of such syntheses may be expected[258].

$$[(\eta^5\text{-C}_5\text{H}_5)_2\text{W}(\eta^3\text{-CH}_2\text{CHCH}_2)]^+ \xrightarrow{D^-} [(\eta^5\text{-C}_5\text{H}_5)_2\text{W}(\text{CH}_2\text{CHDCH}_2)]$$

$$[(\eta^5\text{-C}_5\text{H}_5)_2\text{W}(\eta^3\text{-CH}_2\text{CHCH}_2)]^+ \xrightarrow{\text{MeLi}} [(\eta^5\text{-C}_5\text{H}_5)_2\text{W}(\text{CH}_2\text{CHMeCH}_2)] \quad (148)$$

4. Synthesis by reaction of carbonyl ligands

The best studied reactions of carbonyl ligands are those with carbanions to give carbene complexes and with base to give MCO_2H groups. The latter reactions are favoured in cationic carbonyl complexes and the products often eliminate CO_2 to give the metal hydride[259–262] (equation 149).

$$[\text{Cl-Pt(PEt}_3)_2\text{-CO}]^+ + \text{OH}^- \rightleftharpoons [\text{Cl-Pt(PEt}_3)_2\text{-CO}_2\text{H}] \xrightarrow{-\text{CO}_2} [\text{Cl-Pt(PEt}_3)_2\text{-H}] \quad (149)$$

More recently interest has centred on reduction of carbonyl ligands. This is generally difficult although acyl ligands can be reduced to alkyl ligands; for example, BH_3·thf will reduce $[(\eta^5\text{-}C_5H_5)(CO)_2FeCOMe]$ to $[(\eta^5\text{-}C_5H_5)(CO)_2FeEt]$[263]. However, by careful choice of reducing agents and reaction conditions, it has been possible to reduce a carbonyl ligand progressively to formyl, hydroxymethyl, and methyl ligands[262,264–268].

$$[(\eta^5\text{-C}_5\text{H}_5)\text{Re(CO)}_2\text{(NO)}]^+ \xrightarrow{\text{H}^-} [(\eta^5\text{-C}_5\text{H}_5)\text{Re(CO)(NO)(CHO)}]$$

$$\xrightarrow{\text{H}^-} [(\eta^5\text{-C}_5\text{H}_5)\text{Re(CO)(NO)(CH}_2\text{OH)}] \xrightarrow{\text{H}^-} [(\eta^5\text{-C}_5\text{H}_5)\text{Re(CO)(NO)(CH}_3)] \quad (150)$$

Several other examples are now known in which reduction of carbonyl ligands to formyl or methyl ligands occurs; some examples are shown in equations 151–153[269–272]. In one case synthesis of a μ-CH_2 group from CO has been

$$[(\eta^5\text{-}C_5H_5)(Ph_3P)W(CO)_3] \xrightarrow{NaBH_4} [(\eta^5\text{-}C_5H_5)(Ph_3P)W(CO)_2CH_3] \quad (151)$$

$$cis\text{-}[IrH(CHO)(PMe_3)_4]^+ \xrightarrow{BH_3} cis\text{-}[IrHMe(PMe_3)_4]^+ \quad (152)$$

$$[Re_2(CO)_{10}] \xrightarrow{H^-} [Re_2(CO)_9(CHO)]^- \quad (153)$$

$$[Os_3(CO)_{12}] \xrightarrow{H^-} [Os(CO)_{11}CHO]^- \xrightarrow{H^+} [Os_3(CO)_{11}(\mu\text{-}CH_2)] \quad (154)$$

achieved[273,274]. In one case rupture of a C—O bond of a formyl complex can be induced by addition of a tertiary phosphine ligand[275,276] (equation 155). In another

$$[\{(\eta^5\text{-}C_5Me_5)Cl_2Ta\}_2(H)_2] \xrightarrow{CO} [\{(\eta^5\text{-}C_5Me_5)Cl_2Ta\}_2(H)(CHO)]$$

$$\downarrow PMe_3$$

$$[\{(\eta^5\text{-}C_5Me_5)Cl_2Ta\}_2(H)(\mu\text{-}CHPMe_3)(\mu\text{-}O)] \quad (155)$$

example, intermolecular reduction of an acyl group gives a binuclear alkyl complex[277] (equation 156). Clearly the method may yield many exotic alkyl derivatives, although systematic planned syntheses by this route are not yet possible.

$$(\eta^5\text{-}C_5H_5)_2ZrMe_2] \xrightarrow{CO} [(\eta^5\text{-}C_5H_5)_2Zr(Me)(\eta^2\text{-}OCMe)]$$

$$\swarrow [(\eta^5\text{-}C_5H_5)_2WH_2]$$

$$[(\eta^5\text{-}C_5H_5)_2Zr(Me)\text{—}O\text{—}CHMe\text{—}WH(\eta^5\text{-}C_5H_5)_2] \quad (156)$$

H. Synthesis from Acidic Hydrocarbons

Alkynes are sufficiently acidic to react with many alkyl or hydrido metal complexes to give alkynylmetal complexes by the general reactions 157 and 158. An example is

$$M\text{—}R + R'C{\equiv}CH \longrightarrow RH + M\text{—}C{\equiv}CR' \quad (157)$$

$$M\text{—}R + R'C{\equiv}CH \longrightarrow H_2 + M\text{—}C{\equiv}CR' \quad (158)$$

shown in equation 159[278]. More useful syntheses are based on reactions of acidic hydrocarbons with metal hydroxides. Such syntheses are only effective for 'class b'

$$[PtMe_2(PMe_2Ph)_2] + CF_3C{\equiv}CH \xrightarrow{-CH_4} [PtMe(C{\equiv}CCF_3)(PMe_2Ph)_2] \xrightarrow[CF_3C{\equiv}CH]{-CH_4} [Pt(C{\equiv}CCF_3)_2(PMe_2(Ph)_2] \quad (159)$$

metal ions with a very low affinity for oxygen donor ligands. Platinum hydroxides are the best studied systems[279,280] (equations 160 and 161). It is possible to generate the

$$[Pt(Ph_2PCH_2CH_2PPh_2)(Me)(OH)] \xrightarrow{\text{acetone}} [Pt(Ph_2PCH_2CH_2PPh_2)(Me)(CH_2COMe)]$$
$$[Pt(Ph_2PCH_2CH_2PPh_2)(Me)(OH)] \xrightarrow{CH_3CN} [Pt(Ph_2PCH_2CH_2PPh_2)(Me)(CH_2CN)] \quad (160)$$

$$[Ph{-}Pt(PPh_3)_2{-}OH] \xrightarrow{PhCOMe} [Ph{-}Pt(PPh_3)_2{-}CH_2COPh]$$
$$[Ph{-}Pt(PPh_3)_2{-}OH] \xrightarrow{CH_3NO_2} [Ph{-}Pt(PPh_3)_2{-}CH_2NO_2] \quad (161)$$

metal hydroxide *in situ* from the more easily prepared metal chloride[281,282] (equations 162 and 163). In this region of the Periodic Table, many ligands which are normally oxygen donors may act as carbon donors, with acetylacetonates such as $[Au\{CH(COMe)_2\}(PPh_3)]$ having been best studied[283]. In some cases, both isomers may be formed (**2** and **3**)[284]. Substitution at the methyl group of the acetylacetonato ligand is also possible[285] (equation 164).

$$cis\text{-}[PtCl_2(PEt_3)_2] \xrightarrow[PhCCH]{Ag_2O} trans\text{-}[Pt(C{\equiv}CPh)_2(PEt_3)_2]$$
$$cis\text{-}[PtCl_2(PEt_3)_2] \xrightarrow[CH_3NO_2]{Ag_2O} cis\text{-}[PtCl(CH_2NO_2)(PEt_3)_2] \quad (162)$$

$$[Ph_3PAuCH{=}CH_2] \xrightarrow{KMnO_4} [Ph_3PAuOH] \xrightarrow{acetone} [Ph_3PAuCH_2COMe] \quad (163)$$

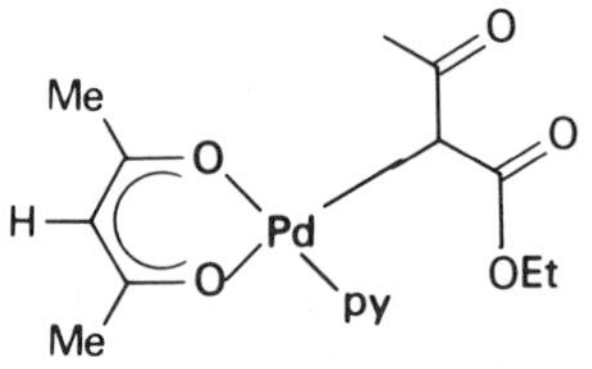

Me O O H Pd O OEt py Me

(2)

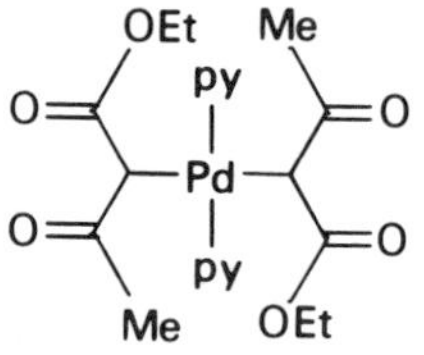

OEt Me O py O Pd O py O Me OEt

(3)

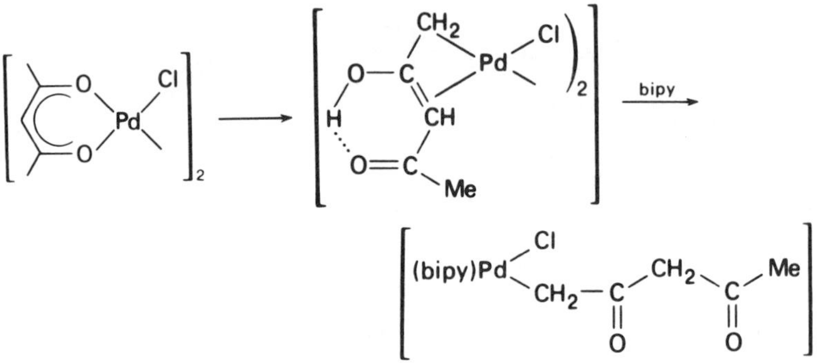

O Cl Pd O 2 → CH_2 Cl O C Pd H CH O C Me 2 $\xrightarrow{bipy}$

$$\left[(bipy)Pd(Cl)CH_2-C(=O)-CH_2-C(=O)-Me\right] \quad (164)$$

I. Miscellaneous Synthetic Methods

The transfer of alkyl groups to a metal from dialkylmercury(II) derivatives is a common synthetic method in main group organometallic chemistry but is rarely applicable for transition elements. The method is useful for synthesis of some ytterbium derivatives[286,287] (equations 165 and 166). A similar method is based on

$$[Hg(CCPh)_2] + Yb \longrightarrow [Yb(CCPh)_2] + Hg \quad (165)$$

$$[Hg(C_6F_5)_2] + Yb \xrightarrow{thf} [Yb(C_6F_5)_2(thf)_4] + Hg \quad (166)$$

$$[AuCl(PPh_3)] + (C_6F_5)_2TlBr \longrightarrow cis\text{-}[Au(C_6F_5)_2Cl(PPh_3)] + TlBr \quad (167)$$

transfer of aryl groups from thallium(III)[288] (equation 167). Clearly, in both of the above methods the oxidation state of the metal is increased by two. Transfer of aryl groups is also possible from arylhydrazine derivatives, but there is not necessarily a change in oxidation state of the metal in this case. The mechanisms are ill-understood, but an example is shown in equation 168.

$$2Et_4N[AuCl_4] + PhNHNH_2HCl \longrightarrow Et_4N[PhAuCl_3] + Et_4N[AuCl_2] + N_2 + 4HCl \quad (168)$$

The method is useful for the synthesis of functionally substituted derivatives. Thus, 4-nitrophenylhydrazine gives $Et_4N[4\text{-}NO_2C_6H_4AuCl_3]$ in the above reaction[289].

III. REFERENCES

1. F. A. Cotton, *Chem. Rev.*, **55**, 551 (1955).
2. M. L. H. Green, *Organometallic Compounds, Vol. II, The Transition Elements*, Methuen, London, 1968, Ch. 7.
3. G. W. Parshall and J. J. Mrowca, *Adv. Organomet. Chem.*, **7**, 157 (1968).
4. F. A. Cotton and G. Wilkinson, *Advanced Inorganic Chemistry*, 4th ed. Wiley, New York, 1980, pp. 1119–1140.
5. P. J. Davidson, M. F. Lappert, and R. Pearce, *Chem. Rev.*, **76**, 219 (1976).
6. R. R. Schrock and G. W. Parshall, *Chem. Rev.*, **76**, 243 (1976).
7. K. Clauss and C. Beermann, *Angew. Chem.*, **71**, 627 (1959).
8. H. J. Berthold and G. Groh, *Z. Anorg. Allg. Chem.*, **319**, 239 (1963); Angew. Chem., **78**, 495 (1966).
9. H. Rau and J. Müller, *Z. Anorg. Chem.*, **415**, 225 (1975).
10. K. H. Thiele, K. Milowski, P. Zdunnek, J. Müller, and H. Rau, *Z. Chem.*, **12**, 186 (1972).
11. J. Müller, H. Rau, P. Zdunnek, and K.-H. Thiele, *Z. Anorg. Allg. Chem.*, **401**, 113 (1973).
12. H. Burger and H.-J. Neese, *J. Organomet. Chem.*, **20**, 129 (1969).
13. K. Clauss, *Justus Liebigs Ann. Chem.*, **711**, 19 (1968).
14. G. W. A. Fowles, D. A. Rice, and J. D. Wilkins, *J. Chem. Soc. A.*, 1920 (1971).
15. R. J. H. Clark and M. Coles, *J. Chem. Soc., Chem. Commun.*, 1587 (1971).
16. K. H. Thiele, *Pure Appl. Chem.*, **30**, 575 (1972).
17. W. Schafer and K. H. Thiele, *Z. Anorg. Allg. Chem.*, **381**, 205 (1971).
18. K. Matsuzaki and T. Yasukawa, *J. Organomet. Chem.*, **10**, P9 (1967).
19. P. C. Wailes, H. Weigold, and A. P. Bell, *J. Organomet. Chem.*, **34**, 155 (1972).
20. E. Samuel and M. D. Rausch, *J. Am. Chem. Soc.*, **95**, 6263 (1973).
21. J. E. Bercaw, R. H. Marvich, L. G. Bell, and H. H. Brintzinger, *J. Am. Chem. Soc.*, **94**, 1219 (1972).
22. J. A. Waters and G. A. Mortimer, *J. Organomet. Chem.*, **22**, 417 (1970).
23. J. L. Atwood, W. E. Hunter, D. C. Hrncir, E. Samuel, H. Alt, and M. D. Rausch, *Inorg. Chem.*, **14**, 1757 (1975).
24. M. L. H. Green and C. R. Lucas, *J. Organomet. Chem.*, **73**, 259 (1974).
25. H. H. Brintzinger, *J. Am. Chem. Soc.*, **89**, 6871 (1967).
26. G. L. Juvinall, *J. Am. Chem. Soc.*, **86**, 4202 (1964).
27. G. W. A. Fowles, D. A. Rice, and J. D. Wilkins, *J. Chem. Soc., Dalton Trans.*, 961 (1973); 2313 (1972).
28. C. Santini-Scampucci and J. G. Riess, *J. Chem. Soc., Dalton Trans.*, 2436 (1973).
29. R. R. Schrock and P. Meakin, *J. Am. Chem. Soc.*, **96**, 5288 (1974).
30. R. R. Schrock, *Acc. Chem. Res.*, **12**, 98 (1979).
31. K. Nishimura, H. Kuribayashi, A. Yamamoto, and S. Ikeda, *J. Organomet. Chem.*, **37**, 317 (1972).
32. A. J. Shortland and G. Wilkinson, *J. Chem. Soc., Dalton Trans.*, 872 (1973); L. Galyer, K. Mertis, and G. Wilkinson, *J. Organomet. Chem.*, **85**, C37 (1975).
33. E. Kurras and J. Otto, *J. Organomet. Chem.*, **4**, 114 (1965).
34. E. Kurras and J. Zimmerman, *J. Organomet. Chem.*, **7**, 348 (1967); J. Krausse, G. Marx, and G. Schödl, *J. Organomet. Chem.*, **21**, 159 (1970).
35. F. A. Cotton, *Acc. Chem. Res.*, **11**, 225 (1978); F. A. Cotton, J. M. Troup, T. R. Webb, D. H. Williamson, and G. Wilkinson, *J. Am. Chem. Soc.*, **96**, 3824 (1974).
36. M. H. Chisholm and F. A. Cotton, *Acc. Chem. Res.*, **11**, 356 (1978).
37. M. H. Chisholm, D. A. Haitko, and C. A. Murillo, *J. Am. Chem. Soc.*, **100**, 6261 (1978).
38. F. A. Cotton and G. Wilkinson, *Advanced Inorganic Chemistry*, 4th ed., Wiley, New York, 1980, pp. 900–901.
39. M. Tamura and J. K. Kochi, *J. Organomet. Chem.*, **29**, 111 (1971).
40. H. J. Spiegl, G. Groh, and H. J. Berthold, *Z. Anorg. Chem.*, **398**, 225 (1973).
41. R. Taube and G. Honymus, *Angew. Chem. Int. Ed. Engl.*, **14**, 261 (1975).
42. G. W. Rice and R. S. Tobias, *J. Chem. Soc. Chem. Commun.*, 994 (1975).
43. L. S. Pu and A. Yamamoto, *J. Chem. Soc., Chem. Commun.*, 4 (1975).
44. F. R. Hartley, *The Chemistry of Platinum and Palladium*, Applied Science, London, 1973.

45. P. M. Maitlis, *The Organic Chemistry of Palladium*, Vol. 1, Academic Press, New York, 1971.
46. U. Belluco, *Organometallic and Co-ordination Chemistry of Platinum*, Academic Press, New York, 1974.
47. G. Wilke and J. Herrman, *Angew. Chem. Int. Ed. Engl.*, **5**, 581 (1966).
48. A. Yamamoto, K. Morifuji, S. Ikeda, T. Saito, Y. Uchida, and A. Misono, *J. Am. Chem. Soc.*, **90**, 1878 (1968).
49. T. Yamamoto, M. Bundo, and A. Yamamoto, *Chem. Lett.*, 833 (1977).
50. W. Kruse, *J. Organomet. Chem.*, **42**, C39 (1972).
51. A. L. Wayda and W. J. Evans, *J. Am. Chem. Soc.*, **100**, 7119 (1978).
52. J. Müller and W. Holzinger, *Angew. Chem. Int. Ed. Engl.*, **14**, 760 (1975).
53. B. K. Bower and H. G. Tennent, *J. Am. Chem. Soc.*, **94**, 2512 (1972).
54. R. M. G. Roberts, *J. Organomet. Chem.*, **63**, 159 (1973).
55. M. Bochmann, G. Wilkinson, and G. B. Young, *J. Chem. Soc., Dalton Trans*, 1879 (1980).
56. R. Tabacchi and A. Jacot-Guillarmod, *Helv. Chim. Acta*, **53**, 1977 (1970).
57. V. Malatesta, K. U. Ingold, and R. R. Schrock, *J. Organomet. Chem.*, **152**, C53 (1978).
58. E. C. Guzman, G. Wilkinson, J. L. Atwood, R. D. Rogers, W. E. Hunter, and M. J. Zaworotko, *J. Chem. Soc., Chem. Commun.*, 465 (1978).
59. G. K. Barker, M. F. Lappert, and J. A. K. Howard, *J. Chem. Soc. Dalton, Trans.* 734 (1978).
60. G. Wilkinson, *Pure Appl. Chem.*, **30**, 627 (1972).
61. R. R. Schrock, *J. Am. Chem. Soc.*, **96**, 6796 (1974).
62. V. N. Latyaeva, G. A. Razuvaev, A. V. Malysheva, and G. A. Kilyakova, *J. Organomet. Chem.*, **2**, 388 (1964).
63. W. Seidel and G. Kreisel, *Z. Anorg. Chem.*, **426**, 150 (1976).
64. K. Schmiedeknecht, *J. Organomet. Chem.*, **133**, 187 (1977).
65. G. van Koten and J. G. Noltes, *J. Organomet. Chem.*, **80**, C56 (1974).
66. L. E. Manzer, *J. Am. Chem. Soc.*, **99**, 276 (1977).
67. J. X. McDermott, J. F. White, and G. M. Whitesides, *J. Am. Chem. Soc.*, **98**, 6521 (1976).
68. M. P. Brown, A. Hollings, K. J. Houston, R. J. Puddephatt, and M. Rashidi, *J. Chem. Soc., Dalton Trans.*, 786 (1976).
69. J. X. McDermott, M. E. Wilson, and G. M. Whitesides, *J. Am. Chem. Soc.*, **98**, 6529 (1976).
70. R. H. Grubbs, A. Miyashita, M. Liu, and P. Burk, *J. Am. Chem. Soc.*, **100**, 2418 (1978).
71. M. F. Lappert, T. R. Martin, J. L. Atwood, and W. E. Hunter, *J. Chem. Soc. Chem. Commun.*, 476 (1980).
72. P. B. Chock and J. Halpern, *J. Am. Chem. Soc.*, **88**, 3511 (1966).
73. A. V. Kramer and J. A. Osborn, *J. Am. Chem. Soc.*, **96**, 7832 (1974).
74. M. F. Lappert and P. W. Lednor, *Adv. Organomet. Chem.*, **14**, 345 (1976).
75. Y. Becker and J. K. Stille, *J. Am. Chem. Soc.*, **100**, 838 (1979).
76. J. S. Bradley, D. E. Connor, D. E. Dolphin, J. A. Labinger, and J. A. Osborn, *J. Am. Chem. Soc.*, **94**, 4043 (1972).
77. P. J. Krusic, P. J. Fagan, and J. San Filippo Jr., *J. Am. Chem. Soc.*, **99**, 2515 (1977).
78. J. P. Collman and M. R. MacLaury, *J. Am. Chem. Soc.*, **96**, 3019 (1974).
79. N. S. Lewis, K. R. Mann, J. G. Gordon, and H. B. Gray, *J. Am. Chem. Soc.*, **98**, 7461 (1976).
80. A. L. Balch, unpublished work.
81. R. B. King, *Acc. Chem. Res.*, **3**, 417 (1970); *J. Organomet. Chem.*, **100**, 111 (1975).
82. R. E. Dessy, R. L. Pohl, and R. B. King, *J. Am. Chem. Soc.*, **88**, 5121 (1966).
83. R. G. Pearson, H. Sobel, and J. Songstad, *J. Am. Chem. Soc.*, **90**, 319 (1968).
84. J. P. Collman, *Acc. Chem. Res.*, **8**, 342 (1975).
85. J. P. Collman, R. G. Finke, J. N. Cawse, and J. I. Brauman, *J. Am. Chem. Soc.*, **99**, 2515 (1977).
86. A. J. Hart-Davis and W. A. G. Graham, *Inorg. Chem.*, **9**, 2658 (1978).
87. P. Fitton and E. A. Rick, *J. Organomet. Chem.*, **28**, 287 (1971).
88. T. T. Tsou and J. K. Kochi, *J. Am. Chem. Soc.*, **101**, 6319 (1979).
89. W. J. Bland and R. D. W. Kemmitt, *J. Chem. Soc. A*, 127 (1968).
90. J. Halpern and P. Phelan, *J. Am. Chem. Soc.*, **94**, 1181 (1972).

91. R. S. Nohr and J. H. Espenson, *J. Am. Chem. Soc.*, **97**, 3392 (1975).
92. G. W. Parshall, T. Herskovitz, F. N. Tebbe, A. D. English, and J. V. Zeile, in *Fundamental Research in Homogeneous Catalysis*, (ed. M. Tsutsui), Vol. 3, Plenum Press, New York, 1979.
93. C. A. Toman, S. D. Ittel, A. D. English, and J. P. Jesson, *J. Am. Chem. Soc.*, **101**, 1742 (1979).
94. F. A. Cotton, D. L. Hunter, and B. A. Frenz, *Inorg. Chim. Acta*, **15**, 155 (1975).
95. L. Farrugia and M. L. H. Green, *J. Chem. Soc., Chem. Commun.*, 416 (1975); C. Gianotti and M. L. H. Green, *J. Chem. Soc., Chem. Commun.*, 1114 (1972).
96. K. I. Grell and J. Schwartz, *J. Chem. Soc., Chem. Commun.*, 244 (1979).
97. S. D. Ittel, F. A. Van Catledge, C. A. Tolman, and J. P. Jesson, *J. Am Chem. Soc.*, **100**, 1317 (1978).
98. R. B. Calvert and J. R. Shapley, *J. Am. Chem. Soc.*, **100**, 7726 (1978).
99. G. W. Parshall, *Acc. Chem. Res.*, **3**, 139 (1970); **8**, 113 (1975).
100. T. H. Tulip and D. L. Thorn, *J. Am. Chem. Soc.*, **103**, 2448 (1981).
101. G. Favero, A. Morvillo, and A. Turco, *J. Organomet. Chem.*, **162**, 99 (1978); P. Rigo and A. Turco, *Co-ord. Chem. Rev.*, **13**, 133 (1974); J. L. Bernmeister and L. M. Edwards, *J. Chem. Soc. A*, 1663 (1971).
102. R. Ros, M. Lenarda, N. B. Pahor, M. Calligaris, P. Delise, L. Randaccio, and M. Graziani, *J. Chem. Soc.*, Dalton Trans., 1937 (1976).
103. E. R. Hamner, R. D. W. Kemmitt, and M. A. R. Smith, *J. Chem. Soc.*, Chem. Commun., 841 (1974).
104. M. Graziani, M. Lenarda, R. Ros, and U. Belluco, *Co-ord. Chem. Rev.*, **16**, 35 (1975).
105. J. Rajaram and J. A. Ibers, *J. Am. Chem. Soc.*, **100**, 829 (1978).
106. J. L. Flippen, *Inorg. Chem.*, **13**, 1054 (1974).
107. F. J. McQuillin and K. G. Powell, *J. Chem. Soc., Dalton Trans.*, 2123 (1972).
108. R. J. Puddephatt, *Co-ord. Chem. Rev.*, **33**, 149 (1980).
109. M. J. McGlinchey and P. S. Skell, in *Cryochemistry* (ed. M. Moskovits and G. A. Ozin), Wiley, New York, 1976.
110. K. J. Klabunde, *Ann. N. Y. Acad. Sci.*, **295**, 83 (1977).
111. P. L. Timms and T. W. Turney, *Adv. Organomet. Chem.*, **15**, 53 (1977).
112. J. S. Roberts and K. J. Klabunde, *J. Am. Chem. Soc.*, **99**, 2509 (1977).
113. K. J. Klabunde and J. S. Roberts, *J. Organomet. Chem.*, **137**, 113 (1977).
114. R. D. Rieke, *Acc. Chem. Res.*, **10**, 301 (1977).
115. J. J. Habeeb and D. G. Tuck, *J. Organomet. Chem.*, **139**, C17 (1977).
116. J. Kleiman and M. Dubeck, *J. Am. Chem. Soc.*, **85**, 1544 (1963).
117. A. C. Cope and R. W. Siekman, *J. Am. Chem. Soc.*, **87**, 3272 (1965).
118. A. C. Cope and E. C. Friedrich, *J. Am. Chem. Soc.*, **90**, 909 (1968).
119. B. N. Cockburn, D. V. Howe, T. Keating, B. F. G. Johnson, and J. Lewis, *J. Chem. Soc. Dalton Trans.*, 404 (1973).
120. G. W. Parshall, *Acc. Chem. Res.*, **3**, 139 (1970); **8**, 113 (1975).
121. J. Dehand and M. Pfeffer, *Co-ord. Chem. Rev*., **18**, 327 (1976).
122. M. I. Bruce, *Angew. Chem. Int. Ed. Engl.*, **16**, 73 (1977).
123. M. I. Bruce, B. L. Goodall, and F. G. A. Stone, *J. Organomet. Chem.*, **60**, 343 (1973).
124. A. Kasahara, *Bull. Chem. Soc. Jap.*, **41**, 1272 (1968).
125. B. A. Grigor and A. J. Nillson, *J. Organomet. chem.*, **132**, 439 (1977).
126. M. Nonoyama, *J. Organomet. Chem.*, **154**, 169 (1978).
127. A. G. Constable, W. S. McDonald, L. C. Hawkins, and B. L. Shaw, *J. Chem. Soc., Commun.*, 1061 (1978).
128. A. J. Deeming and I. P. Rothwell, *J. Chem. Soc., Dalton*, 1497 (1978). A. J. Deeming, I. P. Rothwell, M. B. Hursthouse, and L. New, *J. Chem. Soc., Dalton Trans.*, 1490 (1978).
129. R. Mason, M. Textor, N. Al-Salem, and B. L. Shaw, *J. Chem. Soc., Chem. Commun.*, 292 (1976).
130. C. J. Moulton and B. L. Shaw, *J. Chem. Soc. Dalton Trans*., 1020 (1976).
131. H. C. Clark, A. B. Goel, R. G. Goel, and W. O. Ogini, *J. Organomet. Chem.*, **157**, C16 (1978); C. S. Cundy, M. F. Lappert, and R. Pearce, *J. Organomet. Chem.*, **59**, 161 (1973).
132. M. Y. Darensbourg, D. J. Darensbourg, and D. Drew, *J. Organomet. Chem.*, **73**, C25 (1974).

133. R. L. Bennett, M. I. Bruce, I. Matsuda, R. J. Doedens, R. G. Little, and J. T. Veal, *J. Organomet. Chem.*, **67**, C72 (1974).
134. H. Alper, *Inorg. Chem.*, **15**, 962 (1976).
135. M. I. Bruce, R. C. F. Gardner, B. L. Goodall, F. G. A. Stone, R. J. Doedens, and J. A. Moreland, *J. Chem. Soc., Chem. Commun.*, 185 (1974).
136. P. W. J. de Graaf, J. Boersma, and G. J. M. van der Kerk, *J. Organomet. Chem.*, **105**, 399 (1976).
137. T. A. Manuel, S. L. Stafford, and F. G. A. Stone, *J. Am. Chem. Soc.*, **83**, 249 (1961).
138. H. H. Hoehn, L. Pratt, K. R. Watterson, and G. Wilkinson, *J. Chem. Soc.*, 2738 (1961).
139. F. G. A. Stone, *J. Organomet. Chem.*, **100**, 257 (1975).
140. A. R. Fraser, P. H. Bird, S. A. Bezman, J. R. Shapley, R. White, and J. A. Osborn, *J. Am. Chem. Soc.*, **95**, 597 (1973).
141. R. H. Grubbs and A. Miyashita, *J. Am. Chem. Soc.*, **100**, 7416 (1978); **100**, 7418 (1978).
142. P. Binger and M. J. Doyle, *J. Organomet. Chem.*, **162**, 195 (1978).
143. P. Binger, *Angew. Chem., Int. Ed. Engl.*, **11**, 309 (1972).
144. F. W. Grevels, D. Schulz, and E. A. Koerner von Gustorf, *Angew. Chem.*, **86**, 558 (1974); C. Krüger and Y.-H. Tsay, *Cryst. Struct. Commun.*, **5**, 215 (1976).
145. J. M. Manriquez, D. R. McAlister, R. D. Sanner, and J. E. Bercaw, *J. Am. Chem. Soc.*, **100**, 2716 (1978).
146. J. R. Schmidt and M. Duggan, *Inorg. Chem.*, **20**, 318 (1981).
147. P. W. Jolly and C. Krüger, *J. Organomet. Chem.*, **165**, C39 (1979).
148. G. K. Barker, M. Green, J. A. K. Howard, J. L. Spencer, and F. G. A. Stone, *J. Chem. Soc., Dalton Trans.*, 1839 (1979).
149. R. H. Grubbs and A. Miyashita, *J. Organomet. Chem.*, **161**, 371 (1978).
150. R. Burt, M. Cooke, and M. Green, *J. Chem. Soc. A*, 2975 and 2981 (1970).
151. P. A. Corrigan and R. S. Dickson, *Aust. J. Chem.*, **32**, 2147 (1979).
152. J. P. Collman, J. W. Kang, W. F. Little, and M. F. Sullivan, *Inorg. Chem.*, **7**, 1298 (1968).
153. J. P. Collman and J. W. Kang, *J. Am. Chem. Soc.*, **89**, 844 (1967).
154. H. Suzuki, K. Itoh, Y. Ishii, K. Simon, and J. A. Ibers, *J. Am. Chem. Soc.*, **98**, 8494 (1976).
155. Y. Wakatsuki, K. Aoki, and H. Yamazaki, *J. Am. Chem. Soc.*, **96**, 5284 (1974).
156. Y. Wakatsuki and H. Yamazaki, *J. Organomet. Chem.*, **139**, 169 (1977).
157. D. R. McAllister, J. E. Bercaw, and R. G. Bergman, *J. Am. Chem. Soc.*, **99**, 1666 (1977).
158. W. S. Lee and H. H. Brintzinger, J. Organomet. Chem., **127**, 93 (1977).
159. M. D. Rausch, W. H. Boon, and E. A. Mintz, *J. Organomet. Chem.*, **160**, 81 (1978).
160. T. V. Harris, J. W. Rathke, and E. L. Muetterties, *J. Am. Chem. Soc.*, **100**, 6966 (1978).
161. K. I. Gell and J. Schwartz, *J. Organomet. Chem.*, **153**, C15 (1978).
162. H. C. Clark and A. Shaver, *Can. J. Chem.*, **53**, 3462 (1975).
163. J. Browning, M. Green, B. R. Penfold, J. L. Spencer, and F. G. A. Stone, *J. Chem. Soc., Dalton Trans.*, 97 (1974).
164. S. A. R. Knox, R. F. D. Stansfield, F. G. A. Stone, M. J. Winter, and P. Woodward *J. Chem. Soc., Chem. Commun.*, 221 (1978).
165. C. E. Dean, R. D. W. Kemmitt, D. R. Russell, and M. D. Schilling, *J. Organomet. Chem.*, **187**, C1 (1980).
166. T. R. Howard, J. B. Lee, and R. H. Grubbs, *J. Am. Chem. Soc.*, **102**, 6878 (1980).
167. S. J. McLain and R. R. Schrock, *J. Am. Chem. Soc.*, **100**, 1315 (1978). S. J. McLain, C. D. Wood, and R. R. Schrock, *J. Am. Chem. Soc.*, **101**, 4558 (1979).
168. J. Chatt and B. L. Shaw, *J. Chem. Soc.*, 5075 (1962).
169. H. C. Clark, *J. Organomet. Chem.*, **200**, 63 (1980); H. C. Clark, C. R. Jablonski, and C. S. Wong, *Inorg. Chem.*, **14**, 1332 (1975); H. C. Clark, C. R. Jablonski, J. Halpern, A. Mantorani, and T. A. Weil, *Inorg. Chem.*, **13**, 1541 (1971).
170. D. L. Thorn and R. Hoffmann, *J. Am. Chem. Soc.*, **100**, 2079 (1978).
171. F. W. S. Benfield and M. L. H. Green, *J. Chem. Soc., Chem. Commun.*, 1324 (1974).
172. D. W. Hart and J. Schwartz, *J. Am. Chem. Soc.*, **96**, 8115 (1974).
173. G. Yagupsky, C. K. Brown, and G. Wilkinson, *J. Chem. Soc., Chem. Commun.*, 1244 (1969).
174. T. Blackmore, M. I Bruce, and F. G. A. Stone, *J. Chem. Soc., Dalton Trans.*, 106 (1974).
175. J. R. Moss and W. A. G. Graham, *J. Chem. Soc., Dalton Trans*, 89 (1977).
176. T. G. Attig, R. J. Ziegler, and C. P. Brock, *Inorg. Chem.*, **19**, 2315 (1980).

177. C. A. Bertelo and J. Schwartz, *J. Am. Chem. Soc.*, **98**, 262 (1976).
178. M. Dubeck and R. A. Scheel, *Inorg. Chem.*, **3**, 1757 (1964).
179. A. G. Orpen, D. Pippard, G. M. Sheldrick, and K. D. Rouse, *Acta Crystallogr.*, **B34**, 2466 (1978).
180. M. Laing, P. Sommerville, Z. Dawoodi, M. J. Mays, and P. J. Wheatley, *J. Chem. Soc., Chem. Commun.*, 1035 (1978).
181. S. Otsuka and A. Nakamura, *Adv. Organomet. Chem.*, **14**, 245 (1976).
182. R. A. Sanchez-Delgado and G. Wilkinson, *J. Chem. Soc., Dalton Trans.*, 804 (1977).
183. B. L. Booth and R. G. Hargreaves, *J. Chem. Soc. A*, 2766 (1969).
184. H. C. Clark and C. S. Wong, *J. Am. Chem. Soc.*, **99**, 7073 (1977).
185. R. L. Sweany and J. Halpern, *J. Am. Chem. Soc.*, **99**, 8335 (1977).
186. J. Schwartz, D. W. Hart, and J. L. Holden, *J. Am. Chem. Soc.*, **94**, 9269 (1972).
187. E. R. Evitt and R. G. Bergman, *J. Am. Chem. Soc.*, **101**, 3973 (1979).
188. S. R. Su and A. Wojcicki, *Inorg. Chem.*, **14**, 89 (1975).
189. M. R. Churchill and S. W. Y. Chang, *Inorg. Chem.*, **14**, 98 (1975).
190. H. C. Clark and R. J. Puddephatt, *Inorg. Chem.*, **10**, 18 (1971).
191. C. M. Mitchell and F. G. A. Stone, *J. Chem. Soc., Dalton Trans.*, 102 (1972).
192. S. R. Su and A. Wojcicki, *Inorg. Chim. Acta*, **8**, 55 (1974).
193. W. H. Boon and M. D. Rausch, *J. Chem. Soc., Chem. Commun.*, 397 (1977).
194. D. R. Tremont and R. G. Bergman, *J. Organomet. Chem.*, **140**, C12 (1977).
195. J. M. Huggins and R. G. Bergman, *J. Am. Chem. Soc.*, **101**, 4410 (1979).
196. J. L. Davidson, M. Green, F. G. A. Stone, and A. J. Welch, *J. Chem. Soc., Dalton Trans.*, 2044 (1976).
197. P. L. Bock, D. J. Boschetto, J. R. Rasmussen, J. P. Demers, and G. M. Whitesides, *J. Am. Chem. Soc.*, **96**, 2814 (1974).
198. J. C. Ricci and J. A. Ibers, *J. Organomet. Chem.*, **27**, 261 (1971).
199. R. E. Banks, R. N. Hazeldine, D. R. Taylor, and G. Webb, *Tetrahedron Lett.*, 5215 (1970).
200. W. T. Miller, R. H. Snider, and R. J. Hummel, *J. Am. Chem. Soc.*, **91**, 6532 (1969).
201. N. Chaudhury and R. J. Puddephatt, *Can. J. Chem.*, **57**, 2549 (1979).
202. P. M. Maitlis, *J. Organomet. Chem.*, **200**, 161 (1980).
203. J. L. Davidson, W. Harrison, D. W. A. Sharp, and G. A. Sim, *J. Organomet. Chem.*, **46**, C47 (1972).
204. A. L. Balch, C. L. Lee, C. H. Lindsay, and M. M. Olmstead, *J. Organomet. Chem.*, **177**, C22 (1979).
205. D. M. Roundhill and G. Wilkinson, *J. Chem. Soc. A*, 506 (1968).
206. R. S. Dickson and H. P. Kirsch, *Aust. J. Chem.*, **25**, 2535 (1972).
207. P. M. Treichel, *Adv. Organomet. Chem.*, **11**, 21 (1973).
208. A. Wojcicki, *Adv. Organomet. Chem.*, **11**, 87 (1973).
209. Y. Yamamoto and H. Yamazaki, *Co-ord. Chem. Rev.*, **8**, 225 (1972).
210. S. Otsuka, A. Nakamura, and T. Yoshida, *J. Am. Chem. Soc.*, **91**, 7196 (1969).
211. G. Fachinetti, G. Fochi, and C. Floriani, *J. Chem. Soc., Dalton Trans.*, 1946 (1977); G. Erker and F. Rosenfeldt, *Angew. Chem. Int. Ed. Engl.*, **17**, 605 (1978).
212. T. S. Piper and G. Wilkinson, *J. Inorg. Nucl. Chem.*, **3**, 104 (1956).
213. F. D. Mango and I. Dvoretzky, *J. Am. Chem. Soc.*, **88**, 1654 (1966).
214. W. A. Herrmann, *Angew. Chem. Int. Ed. Engl.*, **17**, 800 (1978).
215. W. A. Herrmann, B. Reiter, and H. Biersack, *J. Organomet. Chem.*, **97**, 245 (1975).
216. W. A. Herrmann, *Chem. Ber.*, **111**, 1077 (1978).
217. W. A. Herrmann and I. Schweizer, *Z. Naturforsch.*, **B33**, 911 (1978).
218. M. P. Brown, J. R. Fisher, R. J. Puddephatt, and K. R. Seddon, *Inorg. Chem.*, **18**, 2808 (1979).
219. A. D. Clauss, P. A. Dimas and J. R. Shapley, *J. Organomet. Chem.*, **201**, C31 (1980).
220. C. Bauer and W. A. Herrmann, *J. Organomet. Chem.*, **209**, C13 (1981).
221. R. B. Calvert, J. R. Shapley, A. J. Schulz, J. M. Williams, S. L. Suib, and G. D. Stucky, *J. Am. Chem. Soc.*, **100**, 6240 (1978).
222. L. Messerle and M. D. Curtis, *J. Am. Chem. Soc.*, **102**, 7789 (1980).
223. T. V. Ashworth, M. J. Chetcuti, L. J. Farrugia, J. A. K. Howard, J. C. Jeffery, R. Mills, G. N. Pain, F. G. A. Stone, and P. Woodward, in *Reactivity of Metal—Metal Bonds* (ed. M. H. Chisholm) ACS Symposium Series, No. 155, American Chemical Society, Washington, D.C., 1981, p. 299.

224. G. Fachinetti, C. Floriani, F. Marchetti, and S. Merlino, *J. Chem. Soc., Chem. Commun.*, 522 (1976).
225. C. P. Casey, W. R. Brunsvold, and J. Koch, *Inorg. Chem.*, **15**, 1991 (1976).
226. D. Egglestone and M. C. Baird, *J. Organomet. Chem.*, **113**, C25 (1976).
227. P. Reich-Rohrwig and A. Wojcicki, *Inorg. Chem.*, **13**, 2457 (1974).
228. G. W. Parshall, *J. Am. Chem. Soc.*, **87**, 2133 (1965).
229. A. N. Nesmeyanov, *Adv. Organomet. Chem.*, **10**, 1 (1972).
230. P. G. Cookson and G. B. Deacon, *J. Organomet. Chem.*, **33**, C38 (1971).
231. T. Majima and H. Kurosawa, *J. Chem. Soc., Chem. Commun.*, 610 (1977).
232. J. Chatt, L. M. Vallarino, and L. M. Venanzi, *J. Chem. Soc.*, 3413 (1957).
233. P. M. Maitlis, *The Organic Chemistry of Palladium*, Academic Press, New York, 1971.
234. J. K. Stille and D. E. James, *J. Organomet. Chem.*, **108**, 401 (1976).
235. J. E. Bäckvall, B. Akermark, and S. O. Ljunggren, *J. Chem. Soc., Chem. Commun.*, 264 (1977).
236. J. K. Stille and R. Divakaruni, *J. Am. Chem. Soc.*, **100**, 1303 (1978).
237. G. Natile, L. Maresca, and L. Cattalini, *J. Chem. Soc., Dalton Trans.*, 651 (1977).
238. I. M. Al-Najjar and M. Green, *J. Chem. Soc., Chem. Commun.*, 926 (1977).
239. N. J. Cooper and M. L. H. Green, *J. Chem. Soc., Chem. Commun.*, 761 (1974).
240. N. Bresciani-Pahor, M. Calligaris, G. Nardin, and P. Delise, *J. Chem. Soc.*, Dalton Trans., 762 (1976).
241. A. C. Cope, J. M. Kliegman, and E. C. Friederich, *J. Am. Chem. Soc.*, **89**, 287 (1967).
242. R. A. Holton and R. A. Kjonaas, *J. Am. Chem. Soc.*, **99**, 4177 (1977).
243. R. McCrindle, E. Alyea, G. Ferguson, S. A. Dias, A. J. McAlees, and M. Parvez, *J. Chem. Soc., Dalton Trans.*, 137 (1980).
244. J. K. Stille, D. E. James, and L. F. Hines, *J. Am. Chem. Soc.*, **95**, 5062 (1973).
245. D. Medema, R. van Helden, and C. F. Kohll, *Inorg. Chim. Acta*, **3**, 255 (1969).
246. I. M. Al-Najjar, M. Green, and J. K. K. Sarhan, *Inorg. Chim. Acta*, **39**, L213 (1980).
247. P. Lennon, M. Madhavaro, A. Rosan, and M. Rosenblum, *J. Am. Chem. Soc.*, **99**, 2823 (1977).
248. P. Lennon, A. Rosan, and M. Rosenblum, *J. Am. Chem. Soc.*, **99**, 8426 (1977).
249. A. DeRenzi, B. DiBlasio, A. Panunuzo, C. Pedone, and A. Vitagliano, *J. Chem. Soc., Dalton Trans.*, 1392 (1978).
250. W. Beck and B. Olgemoeller, *J. Organomet. Chem.*, **127**, C45 (1977).
251. E. O. Fischer, *Adv. Organomet. Chem.*, **14**, 1 (1976).
252. F. R. Kreissl and W. Held, *Chem. Ber.*, **110**, 799 (1977).
253. C. P. Casey and S. W. Polichnowski, *J. Am. Chem. Soc.*, **99**, 6097 (1977).
254. L. Weber, *J. Organomet. Chem.*, **142**, 309 (1977).
255. E. O. Fischer and S. Riedmüller, *Chem. Ber.*, **109**, 3358 (1976).
256. R. J. Al-Essa and R. J. Puddephatt, *J. Chem. Soc., Chem. Commun.*, 45 (1980).
257. N. J. Cooper and M. L. H. Green, *J. Chem. Soc., Dalton Trans.*, 1121 (1979).
258. M. Ephritikhine, B. R. Francis, M. L. H. Green, R. E. Mackenzie, and M. J. Smith, *J. Chem. Soc. Dalton Trans.*, 1131 (1977).
259. M. Catellani and J. Halpern, *Inorg. Chem.*, **19**, 566 (1980).
260. H. C. Clark and W. J. Jacobs, *Inorg. Chem.*, **9**, 1229 (1970).
261. A. J. Deeming and B. L. Shaw, *J. Chem. Soc. A*, 443 (1969).
262. C. P. Casey, M. A. Andrews, and J. E. Rinz, *J. Am. Chem. Soc.*, **101**, 741 (1979).
263. J. A. van Doorn, C. Masters, and H. C. Volger, *J. Organomet. Chem.*, **105**, 245 (1976).
264. R. P. Stewart, N. Okamota, and W. A. G. Graham, *J. Organomet. Chem*, **42**, C32 (1972).
265. C. P. Casey, M. A. Andrews, and D. R. McAlister, *J. Am. Chem. Soc.*, **101**, 3371 (1979).
266. W. Tam, W. K. Wong and J. A. Gladysz, *J. Am. Chem. Soc.*, **101**, 1589 (1979).
267. W. K. Wong, W. Tam, C. E. Strouse, and J. A. Gladysz, *J. Chem. Soc., Chem. Commun.*, 530 (1979).
268. J. R. Sweet and W. A. G. Graham, *J. Organomet. Chem.*, **173**, C9 (1979).
269. P. M. Treichel and R. L. Shubkin, *Inorg. Chem.*, **6**, 1328 (1968).
270. D. L. Thorn, *J. Am. Chem. Soc.*, **102**, 7109 (1980).
271. S. R. Winter, G. W. Cornett, and E. A. Thompson, *J. Organomet. Chem.*, **133**, 339 (1977).
272. C. P. Casey and S. M. Neumann, *J. Am. Chem. Soc.*, **100**, 2544 (1978).
273. G. R. Steinmetz and G. L. Geoffroy, *J. Am. Chem. Soc.*, **103**, 1278 (1981).

274. R. L. Pruett, R. C. Schoening, J. L. Vidal, and R. A. Fiato, *J. Organomet. Chem.*, **182**, C57 (1979).
275. P. Belmonte, R. R. Schrock, M. R. Churchill, and W. J. Youngs, *J. Am. Chem. Soc.*, **102**, 2858 (1980).
276. M. R. Churchill and W. J. Youngs, *Inorg. Chem.*, **20**, 382 (1981).
277. J. A. Marsella and K. G. Caulton, *J. Am. Chem. Soc.*, **102**, 1747 (1980).
278. T. G. Appleton H. C. Clark, and R. J. Puddephatt, *Inorg. Chem.*, **11**, 2074 (1972).
279. T. G. Appleton and M. A. Bennett, *Inorg. Chem.*, **17**, 738 (1978).
280. T. Yoshida, T. Okano, and S. Otsuka, *J. Chem. Soc. Dalton Trans.*, 993 (1976).
281. M. A. Cairns, K. R. Dixon, and M. A. R. Smith, *J. Organomet. Chem.*, **135**, C33 (1977).
282. A. N. Nesmeyanov, K. I. Grandberg, E. I. Smyslova, and E. G. Perevalova, *Izv. Akad. Nauk SSSR, Ser. Khim.*, 2872 (1974).
283. D. Gibson, *Co-ord Chem. Rev.*, **4**, 225 (1969).
284. S. Okeya and S. Kawaguchi, *Inorg. Chem.*, **16**, 1730 (1977).
285. Z. Kanda, Y. Nakamura, and S. Kawaguchi, *Inorg. Chem.*, **17**, 910 (1978).
286. G. B. Deacon and A. J. Coplick, *J. Organomet. Chem.*, **146**, C43 (1978).
287. G. B. Deacon and D. G. Vince, *J. Organomet. Chem.*, **112**, C1 (1976).
288. R. Uson, P. Royo, and A. Laguna, *J. Organomet. Chem.*, **69**, 361 (1974).
289. P. Braunstein and R. J. H. Clark, *Inorg. Chem.*, **13**, 2224 (1974).

The Chemistry of the Metal–Carbon Bond
Edited by F. R. Hartley and S. Patai

CHAPTER 7

Synthesis of olefin and acetylene complexes of the transition metals

GLENN S. LEWANDOS
Department of Chemistry, Central Michigan University, Mount Pleasant, Michigan 48859, U.S.A.

I. INTRODUCTION

The field of transition metal—olefin and —acetylene η^2-complex synthesis has been well covered by review articles through the 1960s and early 1970s. A few of the more comprehensive reviews, with years through which the literature is covered (in parentheses), are listed below.

Fischer and Werner (1964) surveyed complexes of di- and oligo-olefinic ligands[1]. They covered general methods of preparation, followed by detailed surveys by type of η^2-organic ligand. King (1965) published a monograph on general techniques of organometallic synthesis, containing selected examples with experimental procedures[2]. Hübel (1966) covered the synthesis of metal carbonyls with acetylene ligands[3]. Green (1967) reviewed η^2-complexes by type of ligand with one chapter each on olefinic and acetylenic ligands[4]. A survey of transition metals coordinated to olefins was published by Quinn and Tsai (1968)[5]. The syntheses of acetylene η^2-complexes of nickel, platinum, and palladium were briefly described by Nelson and Jonassen (1969)[6]. Shaw and Tucker (1969) categorized syntheses by reaction type and identity of metal[7]. Herberhold (1970) surveyed exhaustively general methods of preparation and transition metal complexes known up to 1970[8]. Syntheses of both metal—olefin and —acetylene complexes were covered by Kemmitt (1970)[9]. Black *et al.* (1976) reviewed briefly common synthetic approaches with selected examples[10]. Comprehensive reviews in the areas of cobalt—acetylene (1973)[11] and iron—olefin (1976)[12] syntheses have appeared.

Because of the number of works in this area through the literature of 1970, this chapter will concentrate on more recent advances. Where necessary, for reasons of completeness or historical importance, earlier works are cited.

The coverage of this chapter is limited to non-conjugated monoenes, polyenes, monoynes, and polyynes η^2-coordinated to transition metals. Complexes have been excluded in which no discrete olefin or acetylene ligand can be identified, such as **1**. Polymetallic compounds MM′ (olefin or acetylene), where M $\neq$ M′, are not covered.

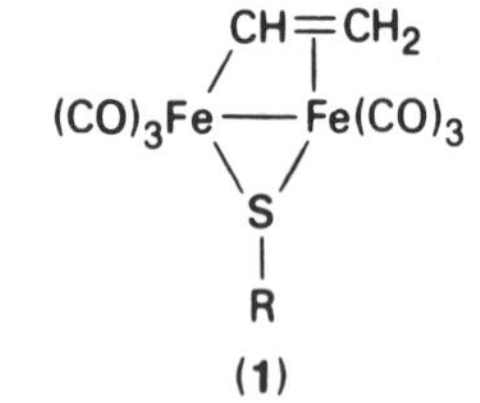

CH=CH$_2$
(CO)$_3$Fe—Fe(CO)$_3$
S
R

(1)

Reports from the patent literature have not been surveyed, since patents are rarely sufficiently detailed to allow reproduction of the reported results. In addition, syntheses of compounds unstable below −80°C or isolated in trace amounts are not included.

A limitation to accuracy is that not all molecular structures presented in this review have been verified by diffraction methods. Some of them represent those suggested only by the authors indicated.

II. GENERAL PREPARATIVE METHODS

A. Addition to Coordinatively Unsaturated Complexes

Perhaps the most facile syntheses are those involving coordinatively unsaturated transition metal complexes. With such complexes, activation of the complex is usually unnecessary. In general, product complexes are formed by direct reaction of a transition metal salt with olefin or acetylene (equation 1). Either the salt is dissolved in the liquid hydrocarbon or a stream of gaseous hydrocarbon is directed through a solution of the salt[13].

$$MX_n + (un)_m \longrightarrow [MX_n(un)_m] \qquad (1)$$

Olefin (monoene and polyene) complexes of silver(I) salts can be prepared in which the anions are $[NO_3]$[14,15], $[ClO_4]$[14,16], $[BF_4]$[17,18], [hfacac][19], [β-diketonato][20], and $[CF_3SO_3]$[21,22]. Corresponding copper(I) complexes are common when the anion is [halide][23], $[ClO_4]$[23], $[CH_3CO_2]$[24], $[CF_3CO_2]$[25,26] $[BF_4]$[27], $[AlCl_4]$[28], and $[CF_3SO_3]$[29]. The η^2-complexes are of variable stability and composition. Copper complexes with anions of low coordination ability often disproportionate to Cu(0) and Cu(II). In general, the most stable complexes of Ag(I) and Cu(I) with olefins and acetylenes are formed when the anion is $[CF_3SO_3]$[21,29]. The $CuCF_3SO_3$ complexes are prepared by the two-step method (Scheme 1) and show no tendency to disproportionate[30]. The

$$Cu_2O + C_6H_6 + 2CF_3SO_3H \longrightarrow [Cu_2(C_6H_6)](CF_3SO_3)_2$$

$$[Cu_2(C_6H_6)][(CF_3SO_3)]_2 + 2un \longrightarrow 2[Cu(un)](CF_3SO_3)$$

SCHEME 1

$AgCF_3SO_3$ η^2-complexes of terminal acetylenes are so stable that they may be isolated without contamination by the corresponding silver acetylide (Scheme 2)[22]. Analogous olefin complexes of $AuCF_3SO_3$ have been reported[31].

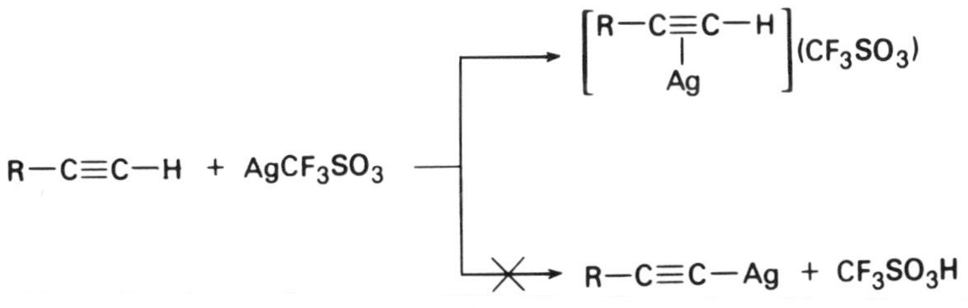

$$R{-}C{\equiv}C{-}H + AgCF_3SO_3 \longrightarrow \left[\begin{array}{c} R{-}C{\equiv}C{-}H \\ | \\ Ag \end{array}\right](CF_3SO_3)$$

$$R{-}C{\equiv}C{-}H + AgCF_3SO_3 \not\longrightarrow R{-}C{\equiv}C{-}Ag + CF_3SO_3H$$

SCHEME 2

The stoichiometry of these complexes is dependent on the ligand, medium, and anion. The most complete series studied is the 1,5-cod complexes of Cu(I)[24]. Complexes of the type $[Cu(1,5\text{-cod})_2X]$ result when the poorly coordinating anions

$[ClO_4]$, $[BF_4]$, and $[CF_3SO_3]$ are present. More strongly coordinating anions such as [Cl] and [Br] give a stoichiometry of [Cu(1,5-cod)X]. With anions $[CF_3CO_2]$ and $[CH_3CO_2]$, products of the type $[Cu_2(1,5\text{-cod})X_2]$ result, in which the anions may be bridging.

The η^2-complexes of Cu(I), Ag(I), and Au(I) readily dissociate in solution. The loss of the olefinic ligand can be prevented if it is incorporated into a ligand bearing a more strongly coordinating arsenic or phosphorus atom[32].

While Cu(η^2-3-phenylpropylene) and Cu(propylene) readily dissociate in solution, CuX (X = Cl, Br, I) reacts with the bidentate *o*-allylphenyldiphenylphosphine (ap) to form **2**[33]. The isomorphous phosphine complexes are dimeric in chloroform and

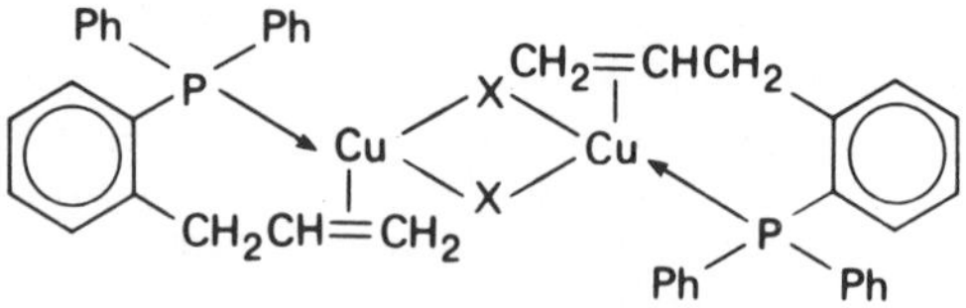

(**2**)

non-conducting. They show absorption in the i.r. of only η^2-coordinated carbon—carbon double bonds. With AgX (X = Cl, Br, I) the same ligand forms 2:1 adducts $[Ag(ap)_2X]$, which are monomeric but contain three coordinate Ag(I) and uncoordinated olefinic ligands. The analogous product is obtained with AuBr[32]. Thus, copper(I) bidentate monoolefin complexes appear more stable than those of Ag(I) and Au(I).

The available information indicates that phosphine ligands allow better olefin coordination than arsine ligands. As an example, $PtCl_2$ and NaNCS form complexes with ap and *o*-allylphenyldimethylarsine (oA)[34]. The resulting $[Pt(ap)(NCS)_2]$ (**3**) exhibits η^2-coordination of the olefinic ligand whereas $[Pt(oA)_2(NCS)_2]$ (**4**) has coordination through arsenic only.

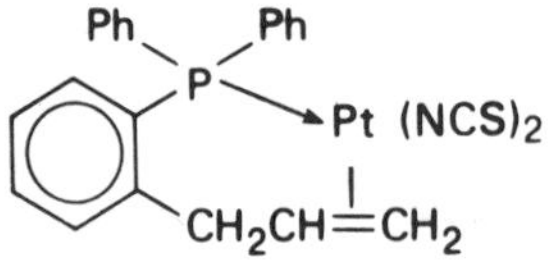

(**3**)

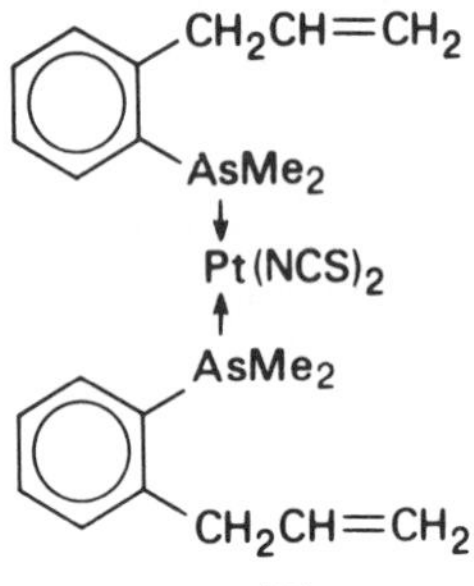

(**4**)

The ability of chelating ligands to form η^2-complexes depends on the structure of the ligand and the identity of the coordinating heteroatom. Aromatic ligands of the type (**5**) and (**6**) form more stable chelate complexes than aliphatic ligands $R_2Y(CH_2)_xCH{=}CH_2$[32]. In platinum complexes, arsine chelating ligands forming complexes with six-membered rings are more stable than those forming five-membered rings (relative stability **6** > **5**)[32].

Examples of olefin or acetylene addition to coordinatively unsaturated complexes are found in most of the transition metal groups. Vaska's complex, *trans*-$[Ir(PPh_3)_2(CO)Cl]$, undergoes olefin addition to yield $[Ir(PPh_3)_2(CO)Cl$-

$o\text{-}C_6H_4(YR_2)(CH{=}CH_2)$ **(5)** $o\text{-}C_6H_4(YR_2)(CH_2CH{=}CH_2)$ **(6)** Y = As, P

(olefin)][35,36,37]. Since olefin addition is reversible, it is desirable to use an excess of olefin. Complex stability is maximized by the presence of electron-withdrawing substituents on the olefin.

Similarly, other tetracoordinate complexes of iridium undergo addition of olefins and acetylenes (equations 2 and 3)[38]. In both examples, dienes (buta-1,3-diene, 1,5-cod, nbd) are dihapto.

$$[\mathrm{Ir(chel)(1,5\text{-}cod)}]\mathrm{X} + \mathrm{un} \longrightarrow [\mathrm{Ir(chel)(1,5\text{-}cod)(un)}]\mathrm{X} \qquad (2)$$

chel = bipy, phen; un = monoene, diene, acetylene

$$[\mathrm{Ir(chel)(CO)_2}]\mathrm{X} + \mathrm{un} \longrightarrow [\mathrm{Ir(chel)(CO)_2(un)}]\mathrm{X} \qquad (3)$$

chel = bipy, phen; un = monoene, diene

The complexes [Ir(chel)(diene)](PF_6) (diene = 1,5-cod, nbd) may add olefins and acetylenes with only five-coordinate complexes of maleic anhydride and fumaronitrile isolable (equation 4)[39]. The norbornadiene complex reacts more rapidly to give the more stable adducts.

$$[\mathrm{Ir(chel)(diene)}](\mathrm{PF_6}) + \mathrm{un} \longrightarrow [\mathrm{Ir(chel)(diene)(un)}](\mathrm{PF_6}) \qquad (4)$$

chel = bipy, phen

Activated olefins such as diethyl maleate and diethyl fumarate add to [V(Cp)$_2$] to yield the paramagnetic [V(Cp)$_2$(olefin)][40]. Olefins without strongly electron-withdrawing groups (*trans*-stilbene, norbornadiene) are unreactive.

The complex [MoO(S_2CNEt_2)$_2$(RC≡CR)] can be prepared by the direct reaction of RC≡CR (R=Ph, CO_2Me) with the coordinatively unsaturated [MoO(S_2CNEt_2)$_2$][41,42]. The phenyl-substituted complex is the first example of a Mo—acetylene η^2-complex without electron-withdrawing substituents on the acetylene[42].

Olefin complexes of iron are easily prepared from commercially available [Fe(Cp)(CO)$_2$]$_2$ (equation 5)[43]. The method is general for monoenes of the type

$$[\mathrm{Fe(Cp)(CO)_2}]_2 + \mathrm{olefin} + \mathrm{Ph_3CBF_4} \longrightarrow [\mathrm{Fp(olefin)}](\mathrm{BF_4}) \qquad (5)$$

Fp = Fe(Cp)(CO)$_2$

hept-1-ene, oct-1- and -4-ene, cycloheptene and cyclooctene. The presence of an oxidizing agent, Ph_3CBF_4, is necessary for splitting of the substrate dimer to form presumably the coordinatively unsaturated [Fe(Cp)(CO)$_2$] cation. The method is unsuccessful with norbornadiene, but if excess of Ph_3CBF_4 is quenched with cycloheptatriene prior to addition of norbornadiene, the corresponding η^2-norbornadiene complex is obtained in 30% yield. Other methods for the preparation of olefin complexes of this type require prior availability of the cation [fp(isobutylene)] (see Section II.D.2) or the olefin epoxide (see Section II.F.4).

The ligand hydrotris(pyrazolylborate), [HB(pz)$_3$], stabilizes labile, five-coordinate complexes of platinum[44–46]. The starting complex, [PtMe{HB(pz)$_3$}], is probably a polymer with bridging pyrazolyl groups. In CH_2Cl_2, this coordinatively unsaturated complex undergoes addition of olefins ($F_2C{=}CF_2$, ma, dmm, dmfm, $H_2C{=}CHCN$, $PhHC{=}CHCHO$, $H_2C{=}CHCO_2Me$, $H_2C{=}CMeCN$, $Me_2C{=}C{=}CMe_2$, $Me_2C{=}C{=}CH_2$) and acetylenes ($CF_3C{\equiv}CCF_3$, $MeO_2CC{\equiv}CCO_2Me$) (equation 6).

$$[PtMe\{HB(pz)_3\}] + un \longrightarrow [PtMe\{HB(pz)_3\}(un)] \qquad (6)$$

B. Anion Displacement

The earliest important example of anion displacement was the synthesis of Zeise's salt, $K[PtCl_3(H_2C{=}CH_2)]\cdot H_2O$[8], which is prepared by passing ethylene into an acidic aqueous K_2PtCl_4 solution[47]. The reaction and its modifications require several days and high pressure, and often yield impure products. However, the process has been useful in the synthesis of higher platinum—olefin complexes by displacement of ethylene from Zeise's salt (see Section II.D.1). Recent modification of the original procedure involves the use of $SnCl_2$ as a catalyst (equation 7)[47,48]. Gaseous olefin

$$K_2PtCl_4 + H_2C{=}CH_2 + HCl/H_2O \xrightarrow{SnCl_2} K[PtCl_3(H_2C{=}CH_2)]\cdot H_2O \qquad (7)$$

pressures may be 1 atm. Reaction times are shortened to 2–4 h and the yield is 86%. An additional modification makes use of incorporation of dimethyl sulphoxide (dmso) as a ligand[49]. Initially, halide is displaced by dmso (equation 8). Olefin ($H_2C{=}CH_2$, $CH_3CH{=}CH_2$, but-1-ene, *cis*- or *trans*-but-2-ene) can be bubbled through an acetone solution of **7** for 3 h to yield *cis*-[Pt(dmso)Cl_2(olefin)].

$$K_2PtCl_4 + H_2C{=}CH_2 + DMSO + HCl \longrightarrow \underset{(7)}{cis\text{-}[Pt(DMSO)Cl_2(H_2C{=}CH_2)]} \qquad (8)$$

An interesting η^2-complex results from the reaction of Na_2PtCl_4 with 5-methylenecycloheptene (equation 9)[50]. The product is a chelate complex in which two perpendicular π-bonds are coordinated to platinum.

$$Na_2PtCl_4 + \text{5-methylenecycloheptene} \xrightarrow{MeOH} [(C_8H_{12})PtCl_2] \qquad (9)$$

Unsaturated alcohols (prop-2-enol, but-2-enol, but-3-en-2-ol) yield $K[PtCl_3(olefin)]$ η^2-complexes[51]. The use of a stoichiometric excess of the unsaturated alcohol leads to dehydration with formation of chelating ethers (**8**). Similar intermolecular

O
Pt
Cl_2

association of the OH functional groups in acetylenic alcohols (**9**) gives bisacetylene η^2-complexes of platinum[52]. With alcohols such as $MeEtC(OH)C{\equiv}CC(OH)EtMe$, $Et_2C(OH)C{\equiv}CC(OH)Et_2$, $Me(n\text{-}Pr)C(OH)C{\equiv}CC(OH)(n\text{-}Pr)Me$, Me(*n*-

(9)

$C_5H_{11})C(OH)C{\equiv}CC(OH)(n\text{-}C_5H_{11})Me$, and $Et(n\text{-}Bu)C(OH)C{\equiv}CC(OH)(n\text{-}Bu)Et$, intermolecular hydrogen bonding is proposed on the basis of infrared spectral analysis.

Platinum alkyl complexes of olefins and acetylenes can be prepared when stabilized by phosphine and arsine ligands (equation 10)[53,54]. When $L = PMe_2Ph$ and

$$trans\text{-}[PtMeL_2Cl] + AgPF_6 + un \longrightarrow trans\text{-}[PtMeL_2(un)](PF_6) \quad (10)$$

un = $MeC{\equiv}CMe$, $PhC{\equiv}CMe$, $MeC{\equiv}CEt$, $EtC{\equiv}CEt$, $H_2C{=}CH_2$, $MeCH{=}CH_2$, allyl alcohol, buta-1,2-diene, or vinyl ethers, the corresponding η^2-complexes are obtained. When $L = AsMe_3$, the $PhC{\equiv}CPh$ complex may be prepared. Syntheses with $AsMe_3$ are unsuccessful with lower molecular weight acetylenes.

A unique anion displacement process involves the reaction of $[Pd(OAc)_2]$ with diphenylacetylene (equation 11)[55]. The origin of the η^5-C_5Ph_5 ligands is $PhC{\equiv}CPh$.

$$[Pd(OAc)_2] + PhC{\equiv}CPh \xrightarrow{MeOH} (Ph_5C_5)Pd(\mu\text{-}PhC{\equiv}CPh)Pd(C_5Ph_5) \quad (11)$$

Each is formed from two molecules of $PhC{\equiv}CPh$ and half of another $PhC{\equiv}CPh$ molecule. The other half of the acetylene yields an *ortho* ester by-product $PhC(OAc)_3$. The mechanistic pathway remains to be elucidated.

The acetylene complexes of Group VIB metals are available in high yield from the halides MX_5 (M = Mo, X = Cl and M = W, X = Br) (equation 12)[56]. The paramagnetic products (**10**) are isolated in yields of 77–97%.

$$[MX_5] + RC{\equiv}CR \xrightarrow[C_6H_6]{CCl_4\ \text{or}} [MX_4(RC{\equiv}CR)] \quad (12)$$

R=Me, Ph; product (**10**)

The classic method for preparation of olefin η^2-complexes of rhodium involves displacement of ethylene from the complex **11** by higher olefins (see Section II.D.1). This precursor (**11**) is obtained by displacement of chloride from $RhCl_3$ (equation 13)[57]. Alternatively, the higher olefin complexes may be prepared directly (equations

$$RhCl_3 \cdot H_2O + H_2C{=}CH_2 \xrightarrow[H_2O]{MeOH/} [(C_2H_4)_2Rh(\mu\text{-}Cl)_2Rh(C_2H_4)_2] \quad (13)$$

(**11**)

$$RhCl_3 + \text{hexa-1,5-diyne} \xrightarrow[H_2O]{EtOH/} [RhCl(\eta^4\text{-hexa-1,5-diyne})]_2 \quad (14)$$

$$RhCl_3 + \text{carvone (R or S)} \xrightarrow[H_2O]{EtOH/} \text{R- or S-}[\ldots]_2 \quad (15)$$

(12)

14 and 15)[58,59]. The chiral product **(12)** contains a free carbonyl group that potentially can be used for derivatization or polymer linking.

Similarly, $IrCl_3$ undergoes anion displacement by 1,5-cod and cyclooctene in refluxing EtOH and *i*-PrOH, respectively (equations 16 and 17)[60].

$$2IrCl_3 + \text{1,5-cod} \xrightarrow{EtOH} [IrCl(\eta^4\text{-1,5-cod})]_2 \quad (16)$$

$$2IrCl_3 + \text{cyclooctene} \xrightarrow{i\text{-PrOH}} [IrCl(\text{cyclooctene})_2]_2 \quad (17)$$

An important starting material in iridium chemistry is $[IrCl(1,5\text{-cod})]_2$ **(13)**. Many reagents will split the chloro bridge. It is available via reaction 18[61].

$$[IrH_2Cl_6] + \text{1,5-cod} \xrightarrow[(2)\ \text{aq. NaOAc}]{(1)\ i\text{-PrOH}, \Delta} \quad (18)$$

(13)

The osmium complex $[Os(PR_3)_2(CO)(NO)Cl]$ may be converted into acetylene complexes[62], first by reaction with $AgPF_6$ to yield an uncharacterized intermediate (equation 19). The intermediate **14** reacts with a variety of acetylenes (equation 20).

$$[Os(PR_3)_2(CO)(NO)Cl] + AgPF_6 \xrightarrow{CH_2Cl_2/\text{acetone}} [Os(PR_3)_2(CO)(NO)](PF_6)$$

or

$$[Os(PR_3)_2(CO)(NO)\text{acetone}](PF_6)$$

(14) (not isolated)

(19)

$$\mathbf{14} + \text{un} \longrightarrow [Os(PR_3)_2(CO)(NO)\text{un}](PF_6) \quad (20)$$

R = Cy, un = HC≡CH
R = Ph, un = HC≡CH
R = Ph, un = PhC≡CH
R = Ph, un = PhC≡CPh

C. Replacement of Carbonyl Ligands

Because of the variety of carbonyl complexes of the transition metals, displacement of carbon monoxide has been widely used to generate coordinatively unsaturated intermediates that subsequently will coordinate olefins and acetylenes (Scheme 3).

$$[M(CO)_n] \xrightarrow[-m(CO)]{\Delta} [M(CO)_{n-m}] \xrightarrow{un} [M(CO)_{n-m}(un)]$$

SCHEME 3

Carbonyl ligands may be activated either thermally or photochemically. This section will deal first with thermal syntheses of various transition metal η^2-complexes, followed by photochemically assisted preparations.

1. Thermal

A number of chelating diene complexes of the Group VIB metal carbonyls are known. For example, norbornadiene reacts with $[Cr(CO)_6]$[63] and $[Mo(CO)_6]$[64] in refluxing hydrocarbon solvent to yield $[M(nbd)(CO)_4]$ (M = Cr, Mo). An experimental problem often encountered in such reactions is the amount of unreacted $[M(CO)_6]$ that sublimes from the reaction flask. This problem may be partially alleviated by use of the so-called 'Strohmeier apparatus' (Figure 1)[65], which allows condensing solvent

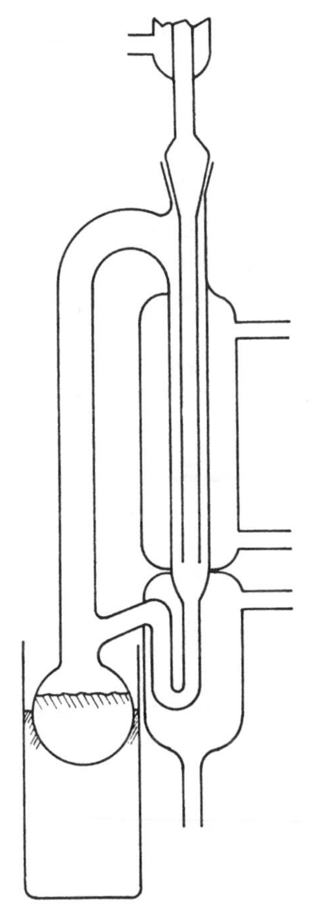

FIGURE 1. The 'Strohmeier apparatus' for metal carbonyl reactions[65]. Reproduced by permission of Verlag Chemie GmbH, Weinheim, Germany.

to return sublimed metal carbonyl to the reaction flask. Fluorine-substituted norbornadienes undergo the same reaction[66].

Reactions of acetylenes with metal carbonyls are often complicated by formation of oligomerization and/or carbonyl insertion products. Selectivity toward η^2-complex formation can be enhanced by use of low temperatures with the concomitant disadvantage of long reaction times. An example is the reaction of $[M(Cp)(CO)_3X]$ with disubstituted acetylenes (equation 21)[67,68]. Additional recent preparations of acetylene complexes of molybdenum are shown in equations 22[69] and 23[70].

$$[M(Cp)(CO)_3X] + RC{\equiv}CR \xrightarrow[8\text{–}100\ h]{<50°C} [M(Cp)X(RC{\equiv}CR)_2] \quad (21)$$

M = Mo, W; X = Cl, Br, I; R = Me, CF_3, CO_2Me, Ph

$$[Mo_2(Cp)_2(CO)_6] + HC{\equiv}CH \xrightarrow{C_6H_6} [Mo_2(Cp)_2(CO)_4(\mu\text{-}HC{\equiv}CH)] \quad (22)$$

$$[Mo(CO)_2\{S_2P(i\text{-}C_3H_7)_2\}_2] + RC{\equiv}CR' \longrightarrow [Mo(CO)(RC{\equiv}CR')\{S_2P(i\text{-}C_3H_7)_2\}_2] \quad (23)$$

R = R' = H
R = R' = Ph
R = R' = CO_2Me
R = Me, R' = H
R = Ph, R' = H
R = Ph, R' = Me
R = H, R' = CO_2Me

To date, the only simple dihapto olefin or acetylene complex of Group IVB has been the diphenylacetylene complex of titanium formed by carbonyl displacement (equation 24)[71]. Olefin interaction with these metals leads to Ziegler polymerization.

$$[Ti(Cp)_2(CO)_2] + PhC{\equiv}CPh \xrightarrow{\text{heptane}} [Ti(Cp)_2(CO)(PhC{\equiv}CPh)] \quad (24)$$

A relatively few olefin and acetylene complexes of Group VB metals are known. Those isolated are often stabilized by chelating ligands such as **15** formed from $[V(CO)_6]$ and ap[72].

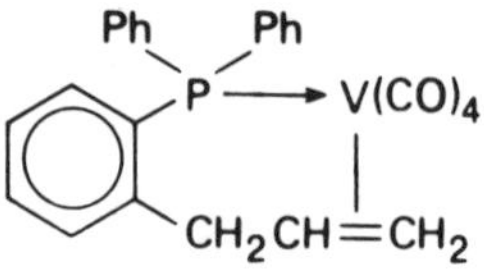

Ph Ph
P ⟶ V(CO)4
CH2CH=CH2

(15)

Activated and unactivated olefin complexes of manganese are most commonly prepared from $[Mn(Cp)(CO)_3]$ (equation 25)[73]. In a two-step sequence, thf is used to displace one CO ligand, presumably forming the intermediate $[Mn(Cp)(CO)_2(thf)]$, followed by addition of olefin. The products are very air sensitive.

$$[Mn(Cp)(CO)_3] \xrightarrow[RR'C{=}CH_2,\ \Delta]{thf,\ \Delta} [Mn(Cp)(CO)_2(RR'C{=}CH_2)] \quad (25)$$

R = R' = H
R = R' = CH_3O
R = CH_3O, R' = H

Iron—olefin complexes of formula [Fe(olefin)$(CO)_4$] can be prepared by thermal reaction of the olefin with [$Fe_2(CO)_9$] in benzene at moderate temperatures (20–60°C). While the reaction does not formally involve carbonyl replacement, but instead a bridge splitting (equation 26), it is considered in detail in this section. With activated olefins (ma, maleic acid, fumaric acid, PhCH=CHCO$_2$Me) the corresponding complexes Fe(olefin)$(CO)_4$ are obtained in good yield[74-76].

$$[Fe_2(CO)_9] \xrightarrow{\Delta} [Fe(CO)_5] + [Fe(CO)_4] \qquad (26)$$

Olefins with strained rings form $Fe(CO)_4$ derivatives. The extent to which ring opening occurs depends upon the reaction conditions. For example, [$Fe_2(CO)_9$] forms a dihapto η^2-complex with substituted methylenecyclopropanes at room temperature (equation 27)[77]. There is no equilibration of *cis*- and *trans*-isomers under the reaction

[$Fe_2(CO)_9$] + (methylenecyclopropane with CO$_2$Me, CO$_2$Me) —hexane, 25°C, 4 h→ [(complex with CO$_2$Me, CO$_2$Me) Fe(CO)$_4$] (27)

conditions. However, at 40°C in hexane, products of ring opening are obtained. In contrast, the substituted methylenecyclopropane **16** reacts with $Fe_2(CO)_9$ in ether at room temperature over a longer period to give $Fe(CO)_3$ and $Fe(CO)_4$ complexes derived from ring-opened products (equation 28)[78]. The tricyclooctadiene **17** gives a

[$Fe_2(CO)_9$] + (CH$_2$OH, CH$_2$OH) (**16**) —Et$_2$O, 20 °C, 16 h→ [O, O, Fe(CO)$_4$] + [O, O, Fe(CO)$_3$] (28)

dihapto complex with [$Fe_2CO)_9$] at room temperature, although it undergoes ring opening in refluxing hexane (equation 29)[79]. Hexafluoro-'Dewar'-benzene forms

[$Fe_2(CO)_9$] + (**17**) —hexane, 25°C→ [Fe(CO)$_4$] (29)

mono- and bis-$Fe(CO)_4$ η^2-complexes at low temperatures (equation 30)[80], although it may be converted to hexafluorobenzene at higher temperatures.

A few iron and ruthenium complexes from thermal reaction of acetylenes with [$M_3(CO)_{12}$] (M = Fe, Ru) are known[81,82], but when substituents on the acetylene are not bulky acetylene oligomerization and CO insertion are facile.

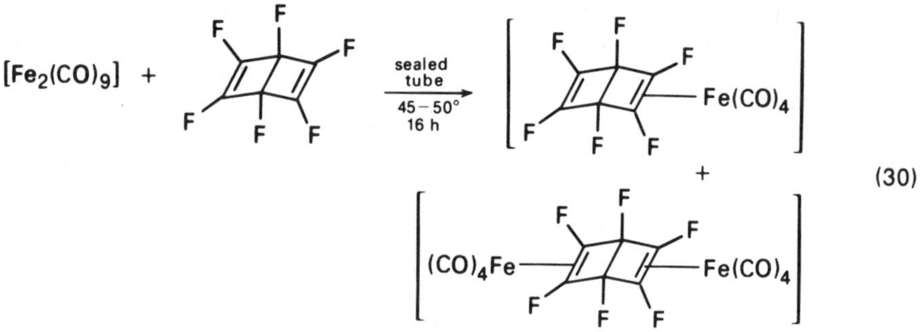

$[Fe_2(CO)_9]$ + (hexafluoro-Dewar benzene) $\xrightarrow[45-50°, 16 h]{sealed tube}$ [$Fe(CO)_4$ complex] + [$(CO)_4Fe$ … $Fe(CO)_4$ complex] (30)

Numerous complexes of the type $[Co_2(acetylene)(CO)_6]$ (**18**) have been obtained by thermal reaction of $[Co_2(CO)_8]$ with the appropriate acetylene in light petroleum[11]. A variety of substituents are tolerated on the acetylene RC≡CR in which R = H, Et, CH_2OH, Ph[83], CH_2Cl, CO_2H[84], Cl, Br, and I[85]. Unsymmetrical, cyclic, and terminal acetylenes are easily made[83,84].

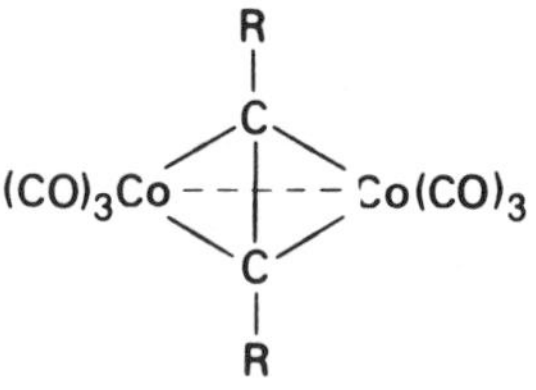

(**18**)

With some macrocyclic diynes cobalt promotes intramolecular transannular cyclizations to give tricyclic cobalt—cyclobutadiene π-complexes[86]; however, reaction of 1,7-cyclododecadiyne with $Co_2(CO)_8$ in light petroleum forms complexes **19** and **20**[87]. The stoichiometry of the products depends on the initial $[Co_2(CO)_8]$ to alkadiyne ratio.

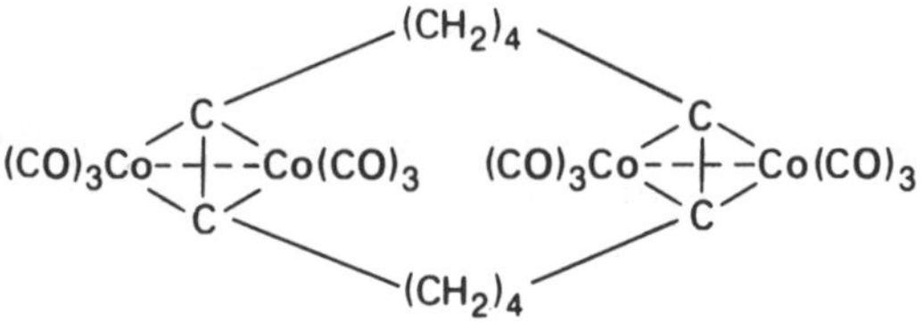

(**19**)

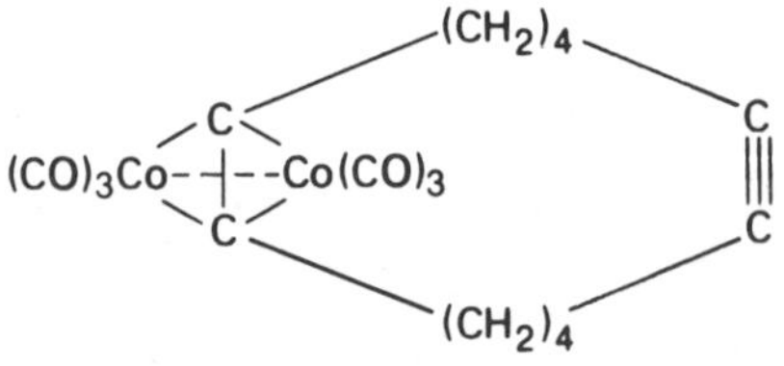

(**20**)

Comparatively few olefin complexes of cobalt are known, although chelating diene complexes have been reported by displacement of CO from $[Co_2(CO)_8]$ without cleavage of the $Co(CO)_2Co$ bridge (equation 31)[88].

$$[Co_2(CO)_8] + nbd \xrightarrow[\Delta]{\text{light petroleum}} [Co_2(nbd)_2(CO)_4] \tag{31}$$

Rhodium carbonyl complexes promote cyclotrimerization of acetylene under thermal conditions[89], but the rhodium carbonyl cluster complex $[Rh_6(CO)_{16}]$ yields olefin complexes with a variety of stoichiometries (equation 32)[90]. Also, allene can

$$[Rh_6(CO)_{16}] + nbd \xrightarrow[\Delta]{\text{methylcyclohexane}} [Rh_6(nbd)_x(CO)_y] \tag{32}$$

x	y
1	14
2	12
3	10

displace carbonyl groups from $[Rh(acac)(CO)_2]$ under mild thermal conditions (equation 33)[91].

$$[Rh(acac)(CO)_2] + H_2C{=}C{=}CH_2 \xrightarrow[30\,^\circ C]{C_6H_6} \left[\begin{array}{c} H_2C{=}C{=}CH_2 \\ | \quad\quad | \\ \text{acac—Rh} \quad \text{Rh—acac} \\ | \quad\quad | \\ CO \quad CO \end{array}\right] \tag{33}$$

The dimer $[Ni(Cp)(CO)]_2$ readily loses CO at elevated temperatures to lead to acetylene complexes of $[Ni(Cp)]_2$. The $[Ni(Cp)]_4$ and $[Ni(Cp)]_2$ complexes analogous to **19** and **20**, respectively, have been obtained via this route[87,92]. A number of nickel—acetylene (terminal and internal) complexes can be synthesized by direct reaction of $[Ni(Cp)(CO)]_2$ with the acetylene in refluxing toluene (equation 34)[93–95].

$$[Ni(Cp)(CO)]_2 + RC{\equiv}CR' \xrightarrow[\Delta]{PhMe} \left[\begin{array}{c} R' \\ | \\ C \\ (Cp)Ni \text{- - -} \text{+} \text{- - -} Ni(Cp) \\ C \\ | \\ R \end{array}\right] \tag{34}$$

$R = R' = Et$ $\quad$ $R = H,\ R' = n\text{-Bu}$

$R = R' = Ph$ $\quad$ $R = H,\ R' = Ph$

$R = R' = CO_2Me$ $\quad$ $R = Me,\ R' = Ph$

$R = H,\ R' = Me$ $\quad$ $R = t\text{-Bu},\ R' = PPh_2$

2. *Photochemical*

The field of light-initiated reactions of metal carbonyls has been extensively reviewed[96–99]. These reviews have concentrated mainly on photoreactions, rather than on the preparative aspects. The remainder of this survey on replacement of carbonyl groups will attempt to provide an overview of photochemical syntheses of representative complexes. Readers are referred to other sources for a comprehensive survey of photochemical equipment and techniques[100].

Group VIB metal carbonyls are susceptible to light-initiated activation of carbonyl ligands. After loss of one or more CO ligands, entering groups such as olefins or acetylenes give the corresponding η^2-complexes.

Molybdenum and tungsten hexacarbonyls react with low-molecular-weight olefins (ethylene, propylene, *cis*-but-2-ene) in the presence of light to give complexes of the type $[M(olefin)(CO)_5]$ and $[M(olefin)_2(CO)_4]$[101]. The products are too unstable for isolation. However, with π-ligands such as aromatic rings, carbonyl substitution gives monoolefin and -acetylene η^2-complexes (equations 35[102], 36[103], and 37[104]). Other

$$[Mo(mesitylene)(CO)_3] + H_2C{=}CH_2 \xrightarrow{h\nu} [Mo(mesitylene)(CO)_2(H_2C{=}CH_2)] \quad (35)$$

$$[Cr(C_6Ph_6)(CO)_3] + un \xrightarrow[thf]{h\nu} [Cr(C_6Ph_6)(CO)_2un] \quad (36)$$

un = ma, cyclopentene, PhC≡CPh, maleic acid, fumaric acid, endic acid

$$[Cr(C_6Me_6)(CO)_3] + RC{\equiv}CR \xrightarrow[thf]{h\nu} [Cr(C_6Me_6)(CO)_2(RC{\equiv}CR)]$$

R = Ph, *p*-$MeOC_6H_4$ (37)

stabilizing ligands may be cyclopentadienyl (equation 38)[105], $AsMe_3$ (equation 39)[106], or PMe_3 (equation 40)[106]. Similarly, if enough carbonyl ligands are displaced, a complex such as $[Mo(PhC{\equiv}CPh)_3(CO)]$ can be prepared from photoreaction of $[Mo(CO)_6]$ with PhC≡CPh[107]. The synthesis is not general and cyclooligomerization has been reported with other acetylenes.

$$[W(Cp)(CO)_3Me] + H_2C{=}CH_2 \xrightarrow{h\nu} [W(Cp)(CO)_2(H_2C{=}CH_2)Me] \quad (38)$$

$$[W(CO)_5(AsMe_3)] + un \xrightarrow[thf]{h\nu} cis\text{-}[W(CO)_4(AsMe_3)un] \quad (39)$$

un = dmm or dmfm

$$[W(CO)_5(PMe_3)] + un \xrightarrow[thf]{h\nu} cis\text{-}[W(CO)_4(PMe_3)un] \quad (40)$$

un = $H_2C{=}CH_2$, dmm, dmfm

Olefin complexes of manganese, $[Mn(Cp)(CO)_2olefin]$, are prepared easily by photolysis of $[Mn(Cp)(CO)_3]$ in hexane[108,109]. Appropriate olefins are $H_2C{=}CH_2$, $MeHC{=}CH_2$, pent-1-ene, cyclopentene, cycloheptene, *cis*-cyclooctene, and norbornene. The analogous acetylene complexes are prepared by the same method in ether (equation 41)[110]. In order to minimize photoreaction of the free acetylene, it is added to the reaction mixture only after initial irradiation of $[Mn(Cp)(CO)_3]$.

$$[Mn(Cp)(CO)_3] + HC{\equiv}CCO_2Me \xrightarrow[ether,\ -50^\circ C]{h\nu} [Mn(Cp)(CO)_2(HC{\equiv}CCO_2Me)] \quad (41)$$

Photochemical syntheses of iron complexes are possible from $[Fe(CO)_5]$, $[Fe_2(CO)_9]$ and $[Fe_3(CO)_{12}]$. A number of fluoroolefin complexes have been reported recently (equation 42)[111]. If photolyses are carried out at low temperatures (−50°C),

$$[Fe_2(CO)_9] + \text{olefin} \xrightarrow[\text{light petroleum}]{h\nu} [Fe(\text{olefin})(CO)_4] \quad (42)$$

olefin = $H_2C{=}CHF$, $H_2C{=}CF_2$, $F_2C{=}CHF$, $F_2C{=}CHBr$, $F_2C{=}CHCl$, $CF_3FC{=}CFCF_3$, $CF_3HC{=}CHCF_3$, $(CF_3)_2C{=}C(CF_3)_2$, $Cl_2C{=}CCl_2$

iron complexes of olefinic substrates containing strained rings may be isolated. Upon irradiation with the parent olefin, $[Fe(CO)_5]$ yields $Fe(CO)_4$ complexes **(21)**[112], **(22)**[113], and **(23)**[114].

$Fe(CO)_4$ $Fe(CO)_4$ $Fe(CO)_4$

(21) 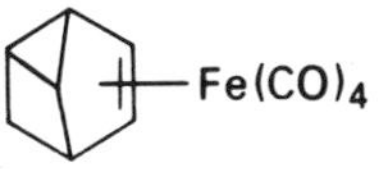 **(22)**  **(23)**

It is possible to activate selectively a carbonyl group photochemically without dissociation of other π-coordinated ligands from the metal. As an example, irradiation of $[Fe(\text{cyclobutadiene})(CO)_3]$ in the presence of dimethyl maleate or dimethyl fumarate gives the corresponding $[Fe(\text{cyclobutadiene})(CO)_2\text{olefin}]$[115]. Since dissociation of the iron—cyclobutadiene bond is irreversible, ligand exchange is specific to the carbonyl group.

Simple acetylenic complexes of iron are stable only when the acetylene is disubstituted with very bulky groups. A synthesis reported with experimental detail is the preparation of $[Fe(Me_3SiC{\equiv}CSiMe_3)(CO)_4]$ from $Me_3SiC{\equiv}CSiMe_3$ and $[Fe_3(CO)_{12}]$ in benzene[116].

Carbonyl displacement from ruthenium carbonyl complexes by photolysis gives isolable olefin complexes when they are stabilized by phosphine ligands and fluorine substituents on the olefin (equation 43)[117].

$$[Ru(CO)_3L_2] + \text{olefin} \xrightarrow[\text{hexane}]{h\nu} [Ru(CO)_2L_2\text{olefin}] \quad (43)$$

L = $P(OMe_3)_3$, $P(OEt_3)_3$, PMe_2Ph

olefin = $F_2C{=}CF_2$, $ClFC{=}CF_2$, $FHC{=}CF_2$, $CF_3FC{=}CF_2$, hexafluorocyclobutene

D. Displacement of η^2-Coordinated Ligands

Often the displacement of an η^2-coordinated ligand by another unsaturated ligand (olefin or acetylene) is of limited success. One reason is that a mixture of η^2-complexes may result. Syntheses of this type are facilitated if the ligand being replaced is sufficiently volatile to be removed at room temperature or under reduced pressure.

1. *Ethylene displacement*

Ethylene is easily displaced from a variety of platinum complexes, including Zeise's salt (equations 44 and 45)[118,119], the platinum dimer $[PtCl_2(H_2C{=}CH_2)]_2$ (equation

$$K[PtCl_3(H_2C{=}CH_2)] + \text{styrene} \rightleftharpoons K[PtCl_3(\text{styrene})] \quad (44)$$

$$Na[PtCl_3(H_2C{=}CH_2)] + \text{norbornadiene} \rightleftharpoons [PtCl_2(\eta^2\text{-norbornadiene})] \quad (45)$$

$$[PtCl_2(H_2C{=}CH_2)]_2 + \text{pent-2-ene} \rightleftharpoons [PtCl_2(\text{pent-2-ene})]_2 \quad (46)$$

$$[Pt(PPh_3)_2(H_2C{=}CH_2)] + PhC{\equiv}CH \rightleftharpoons [Pt(PPh_3)_2(PhC{\equiv}CH)] \quad (47)$$

46)[118], and $[Pt(PPh_3)_2(H_2C{=}CH_2)]$ (equation 47)[120]. The reversible nature of the above reactions is overcome by use of a vacuum to remove the liberated ethylene.

The $[Pt(PPh_3)_2(\text{ethylene})]$ complex has served as a good reagent for the preparation of η^2-complexes with thermally sensitive ligands. The strained olefin $\Delta^{1,4}$-bicyclo[2.2.0]hexene (**24**) can be stored as a dihapto ligand in a platinum phosphine complex at $-78°C$ (equation 48)[121].

$$[Pt(PPh_3)_2(H_2C{=}CH_2)] + \underset{\textbf{(24)}}{\square\!\square} \xrightarrow{PhMe} [Pt(PPh_3)_2(\square\!\square)] \quad (48)$$

In addition, the precursor to the first vinyl alcohol η^2-complex was prepared by ethylene displacement (equation 49)[122]. The platinum complex of η^2-coordinated vinyl alcohol is formed upon hydrolysis of **25**.

$$[Pt(acac)Cl(H_2C{=}CH_2)] + H_2C{=}CHOSiMe_3 \xrightarrow{C_6H_6} \underset{\textbf{(25)}}{[Pt(acac)Cl(H_2C{=}CHOSiMe_3)]} \quad (49)$$

Although there are many known platinum—olefin complexes which are stabilized by coordinating ligands such as tertiary phosphines or acac, the unique $[Pt(\text{ethylene})_3]$ may undergo ethylene displacement to yield a variety of products without such ligands, particularly with olefins bearing electron-withdrawing groups (equation 50)[123].

$$[Pt(H_2C{=}CH_2)_3] + \text{olefin} \xrightarrow{Et_2O} [Pt(H_2C{=}CH_2)(\text{olefin})_2] + [Pt(\text{olefin})_3] \quad (50)$$

olefin = dmm, dmfm, def

Bimetallic platinum—acetylene complexes can be prepared from reaction of $[Pt(PPh_3)_2(\text{ethylene})]$ with $[Pt(PhC{\equiv}CPh)_2]$ in diethyl ether[124]. The product (**26**) contains a bridging acetylene moiety.

Ph
|
C
$(Ph_3P)_2Pt$ - - | - - $Pt(PPh_3)_2$
C
|
Ph

(**26**)

The replacement of ethylene is not limited to the preparation of higher olefin complexes of platinum. The methylenecyclopropanes **27** and **28** can form complexes with platinum (equation 51), rhodium (equation 52), and iridium (equation 53)[125].

Chiral rhodium complexes result from reaction of tond (**29**) with the dimer $[RhCl(\text{ethylene})_2]_2$[126]. The resulting enantiomeric pair (**30**) may be resolved by treatment with a chiral amine (equation 54), followed by fractional crystallization. The

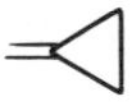

CO₂Me

CO₂Me

(27) 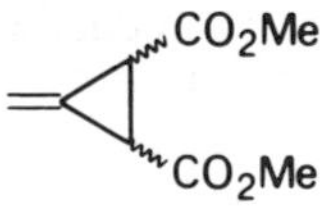 **(28)**

$$[Pt(PPh_3)_2(H_2C{=}CH_2)] + \text{olefin} \xrightarrow{C_6H_6} [Pt(PPh_3)_2\text{olefin}] \qquad (51)$$

olefin = **27**, **28**

$$[Rh(acac)(H_2C{=}CH_2)_2] + \text{olefin} \xrightarrow{Et_2O} [Rh(acac)(\text{olefin})_2] \qquad (52)$$

olefin = **27**, **28**

$$[Ir(acac)(H_2C{=}CH_2)_2] + \text{olefin} \xrightarrow{Et_2O} [Ir(acac)(\text{olefin})_2] \qquad (53)$$

olefin = **27**, **28**

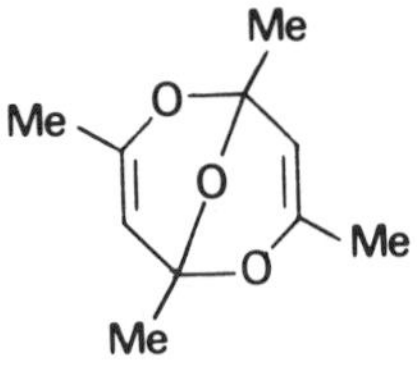

Me O O O Me Me Me

(29)

$$[RhCl(tond)]_2 + (s)\text{-}\alpha\text{-methylbenzenemethanamine} \longrightarrow [RhCl(s\text{-Am})(tond)] \qquad (54)$$

(30) (*s*-AM)

dimer can also yield rhodium η^2-complexes of the terpenes limonene **(31)**, α-phellandrene **(32)**, and carvone **(33)** (equation 55)[127].

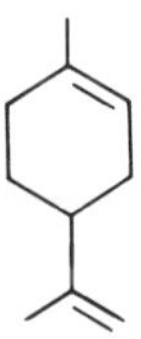

CHMe₂ O

(31) 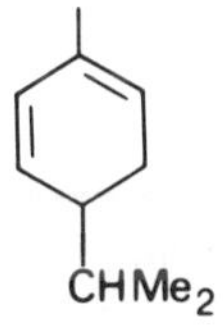 **(32)** 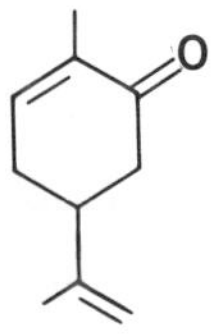 **(33)**

$$[RhCl(H_2C{=}CH_2)]_2 + \text{terpene} \longrightarrow [RhCl(\text{terpene})]_2 \qquad (55)$$

terpene = **31, 32, 33**

2. *Isobutylene displacement*

A variety of cationic iron complexes can be prepared by reaction of [Fp(isobutylene)](BF_4) (see Section II.F.1) with the appropriate olefin (equation 56)[128,129]. The reaction is complete in approximately 10 min and may be monitored by

$$[Fp(\text{isobutylene})](BF_4) + \text{olefin} \xrightarrow{CH_2Cl_2} [Fp(\text{olefin})](BF_4) \qquad (56)$$

olefin = $H_2C{=}CH_2$, cycloheptene, cyclooctene, cyclohexa-1,3-diene, cyclohexa-1,4-diene, norbornadiene

evolution of isobutylene. The product complexes are dihapto, and thus the [Fp] cation may be used as a protecting group for one double bond of a polyene (Scheme 4)[130].

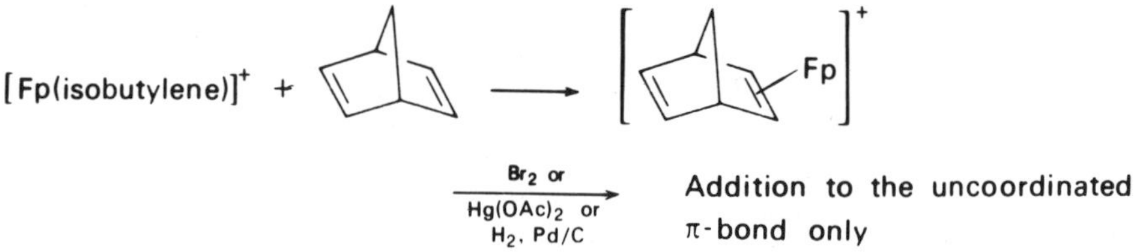

[Fp(isobutylene)]$^+$ + (norbornadiene) ⟶ [(norbornadiene)Fp]$^+$

$\xrightarrow[\text{Hg(OAc)}_2 \text{ or } \text{H}_2\text{, Pd/C}]{\text{Br}_2 \text{ or}}$ Addition to the uncoordinated π-bond only

SCHEME 4

In addition, the [Fp] group selectively coordinates carbon—carbon double bonds (Scheme 5).

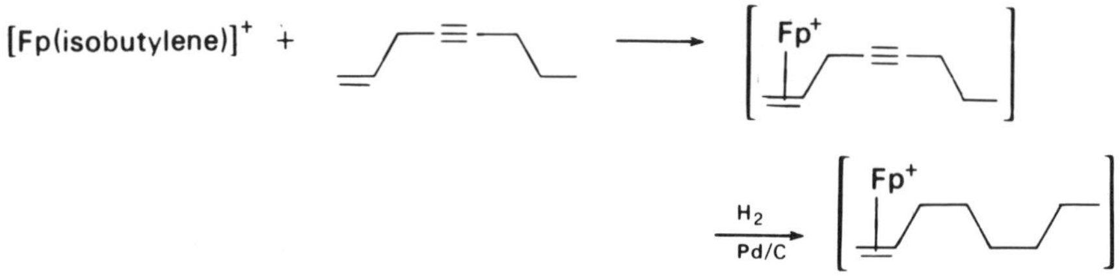

[Fp(isobutylene)]$^+$ + (enyne) ⟶ [Fp$^+$ (enyne)]

$\xrightarrow[\text{Pd/C}]{\text{H}_2}$ [Fp$^+$ (alkene)]

SCHEME 5

3. *Cyclooctene displacement*

Cyclooctene has found limited application as a leaving group. A variety of η^2-complexes have been prepared from [Cr(Cp)(CO)(NO)cyclooctene] (equation 57)[131,132].

$$[\text{Cr(Cp)(CO)(NO)cyclooctene}] + \text{un} \xrightarrow{\text{C}_6\text{H}_6} [\text{Cr(Cp)(CO)(NO)un}] \quad (57)$$

un = $H_2C{=}CH_2$, HC≡CH, ma

The cyclooctene complexes $[MCl(cyclooctene)_2]_2$ (M = Rh, Ir) undergo cyclooctene replacement to yield olefin[133,134], allene[134], and acetylene[135] complexes (equations 58–60). The syntheses (equations 58 and 59) are of interest because of the

$$[\text{IrCl(cyclooctene)}_2]_2 + \text{H}_2\text{C}{=}\text{CH}_2 \longrightarrow [\text{IrCl(H}_2\text{C}{=}\text{CH}_2)_2]_2 \quad (58)$$

$$[\text{IrCl(cyclooctene)}_2]_2 + \text{PPh}_3 + \text{un} \xrightarrow{\text{C}_6\text{H}_6} [\text{Ir(PPh}_3)_2\text{Cl(un)}] \quad (59)$$

un = ethylene, allene

$$[\text{RhCl(cyclooctene)}_2]_2 + \text{PCy}_3 + \text{RC}{\equiv}\text{CR} \xrightarrow{\text{C}_6\text{H}_6} [\text{Rh(PCy}_3)_2\text{Cl(RC}{\equiv}\text{CR)}] \quad (60)$$

R = H, Et, Ph

marked instability of iridium(I) and its ready conversion to iridium(III). In contrast to rhodium, relatively few olefin complexes of iridium without strongly π-accepting ligands such as CO or chelating alkenes are known.

4. *Cycloocta-1,5-diene displacement*

Together with ethylene, 1,5-cod is one of the most widely used ligands for easy displacement. It undergoes displacement by olefins and acetylenes from a number of metals.

The $[Ni(1,5\text{-cod})_2]$ complex may react with complete displacement of 1,5-cod (equation 61)[136,137,138].

$$[Ni(1,5\text{-cod})_2] + 2PPh_3 + PhHC{=}CHPh \xrightarrow{\text{pentane}} [Ni(PPh_3)_2(PhHC{=}CHPh)] \quad (61)$$

A two step synthesis starting from $[Ni(1,5\text{-cod})_2]$ involves the formation of a coordinatively unsaturated intermediate isonitrile complex (equation 62), which then

$$[Ni(1,5\text{-cod})_2] + 2t\text{-BuN}{\equiv}C \xrightarrow{Et_2O} [Ni(t\text{-BuN}{\equiv}C)_2] \quad (62)$$

$$[Ni(t\text{-BuN}{\equiv}C)_2] + un \xrightarrow{Et_2O} [Ni(t\text{-BuN}{\equiv}C)(un)] \quad (63)$$

un = (NC)HC=CH(CN), PhC≡CPh

adds olefins and acetylenes (equation 63)[137]. In many cases, molecules of solvent remain coordinated to nickel in the products (equation 64)[136], or only partial replacement of 1,5-cod is observed (equation 65)[124].

$$[Ni(1,5\text{-cod})_2] + 1,5\text{-cod} + \text{olefin} \xrightarrow{CH_3CN} [Ni(CH_3CN)(un)_2] \quad (64)$$

un = dmm, dmfm

$$[Ni(1,5\text{-cod})_2] + Me_3SiC{\equiv}CSiMe_3 \longrightarrow [(1,5\text{-cod})Ni(\mu\text{-}Me_3SiC{\equiv}CSiMe_3)Ni(1,5\text{-cod})] \quad (65)$$

Whereas the complex $[Ni(1,5\text{-cod})_2]$ has been particularly valuable in both the synthesis of additional nickel compounds and catalytic applications[139], the similar chemistry of platinum and palladium has been slower to develop. Now that a pathway exists to $[M(1,5\text{-cod})_2]$ (M = Pt, Pd) in good yields (see Section II.G.6), numerous olefin and acetylene complexes are becoming available. With these complexes, ethylene, *trans*-cyclooctene, and norbornadiene give the corresponding complexes $M(un)_3$ (M = Pt, Pd and un = ethylene, *trans*-cyclooctene, nbd)[140,141]. They appear to be thermodynamically more stable than similar nickel complexes. Olefins with electron-withdrawing substituents do not displace 1,5-cod as completely (equation 66)[142]. Acetylenes undergo analogous reactions (equations 67 and 68)[143,144], and in the latter reaction product yield is sensitive to the reactant complex to acetylene ratio.

$$[Pt(1,5\text{-cod})_2] + \text{olefin} \xrightarrow{Et_2O} [Pt(1,5\text{-cod})(\text{olefin})] \quad (66)$$

olefin = dmfm, defm, dem

$$[Pt(1,5\text{-cod})_2] + PhC{\equiv}CPh \xrightarrow{\text{light petroleum}} [Pt(PhC{\equiv}CPh)_2] \quad (67)$$

$$[Pt(1,5\text{-cod})_2] + CF_3C{\equiv}CCF_3 \xrightarrow{Et_2O} [Pt(\eta^4\text{-}1,5\text{-cod})(CF_3C{\equiv}CCF_3)] \quad (68)$$

The ruthenium complex $[Ru(1,5\text{-cod})(CO)_3]$ readily loses 1,5-cod to form diene complexes (equation 69)[145].

$$[Ru(1,5\text{-cod})(CO)_3] + \text{norbornadiene} \xrightarrow{C_6H_6} [(\text{norbornadiene})Ru(CO)_3] \quad (69)$$

5. *Norbornadiene displacement*

Since norbornadiene is relatively involatile and reasonably effective in coordination of transition metals, it may be replaced only when the entering ligand is even more strongly coordinating or is used in stoichiometric excess. Thus, the chelating phosphine—monoolefin complex **34** may be obtained (equation 70)[146].

$$[M(NBD)(CO)_4] + sp \xrightarrow{C_6H_6} \mathbf{(34)} \quad (70)$$

M = Cr, Mo;

sp = *o*-styryldiphenylphosphine

(34) (Ph_2P and vinyl of *o*-styryl group chelated to $M(CO)_4$)

6. *Miscellaneous π-ligand displacements*

The trihapto ligands *E,E,E*-1,5,9-cyclododecatriene and benzene are labile enough in selected transition metal systems to be replaced by low-molecular-weight olefins[147] and acetylenes[28] (equations 71 and 72).

$$[Ni(cdt)] + H_2C{=}CH_2 \xrightarrow[0^\circ C]{Et_2O} [Ni(H_2C{=}CH_2)_3] \quad (71)$$

$$[CuL_2](AlCl_4) + RC{\equiv}CH \xrightarrow{-10^\circ C} [Cu(RC{\equiv}CH)](AlCl_4) \quad (72)$$

L = C_6H_6, C_6H_5Me; R = H, *t*-Bu, Ph

E. Displacement of Miscellaneous Ligands

1. *Tertiary phosphine displacement*

Tertiary phosphines are commonly used as stabilizing ligands in transition metal complexes. However, it is possible in some cases for olefins and acetylenes to displace one or more phosphines with resulting formation of η^2-complexes.

Whereas iron carbonyl complexes of olefins and acetylenes are legion, few iron phosphine complexes have been reported. The PF_3 ligands in $[Fe(PF_3)_5]$ undergo

$$[Fe(PF_3)_5] + \text{olefin} \xrightarrow[h\nu]{\text{ether}} [Fe(PF_3)_4\text{olefin}] \tag{73}$$

olefin = MeHC=CHCN, PhHC=CH_2, H_2C=$CHCO_2Me$

substitution upon photolysis (equation 73)[148]. The resulting PF_3 complexes are more stable than the equivalent more common [Fe(olefin)$(CO)_4$] complexes. The displaced PF_3 is a gas, and thus product formation is irreversible. This is not the case with all phosphine ligands.

Phosphine—hydride complexes of ruthenium are well known homogeneous hydrogenation catalysts. However, in some cases they yield stable olefin η^2-complexes. The reaction of norbornadiene with $[Ru(PPh_3)_3ClH]$ gives $[Ru(PPh_3)_2ClH(nbd)]$, which is one of the rare examples of a hydrido—olefin complex[149]. Similarly, ethylene and styrene react with $[Ru(PPh_3)_4H_2]$ to yield $[Ru(PPh_3)_3\text{olefin}]$ (olefin = ethylene, styrene)[150]. In this reaction 1 mol of ethane or ethylbenzene, respectively, is obtained per mole of starting ruthenium complex.

Simple olefin and acetylene complexes are obtained on displacement of PPh_3 from $[Ru(CO)_2(PPh_3)_3]$ (equation 74)[151]. This pathway was the source of the first ethylene complex of ruthenium.

$$[Ru(CO)_2(PPh_3)_3] + \text{un} \xrightarrow{\text{MeCN}} [Ru(CO)_2(PPh_3)_2\text{un}] \tag{74}$$

un = H_2C=CH_2, PhC≡CPh

Cobalt—acetylene complexes $[Co(Cp)(PPh_3)(RC{\equiv}CR)]$ (R = Ph, CO_2Me) are prepared in good yield from $[Co(Cp)(PPh_3)_2]\cdot C_6H_6$[152,153].

The ethylene complex $[RhCl(PPh_3)_2(H_2C{=}CH_2)]$ is unstable with respect to ethylene dissociation. However, if the synthesis is attempted in solvent saturated with ethylene and in an ethylene atmosphere, it may be isolated (equation 75)[154].

$$[RhCl(PPh_3)_3] + H_2C{=}CH_2 \xrightarrow{CHCl_3} [RhCl(PPh_3)_2(H_2C{=}CH_2)] \tag{75}$$

Acetylene complexes of rhodium in large variety may be obtained by reaction of the acetylene with $[Rh_2(PF_3)_8]$ (equation 76)[155]. In method A, the reactants are heated in

$$[Rh_2(PF_3)_8] + RC{\equiv}CR' \longrightarrow [Rh_2(PF_3)_6(\mu\text{-}RC{\equiv}CR')] \tag{76}$$

Method A	*Method B*
R = R′ = Ph	R = R′ = H
R = R′ = CF_3	R = R′ = Me
R = Ph, R′ = Me	R = *n*-Pr, R′ = Me
R = Ph, R′ = Et	R = *n*-Bu, R′ = H
R = Ph, R′ = CO_2Me	R = *t*-Bu, R′ = H
	R = Ph, R′ = H

refluxing hexane. In method B, the acetylene is condensed with $Rh_2(PF_3)_8$ in a vacuum system, and the resulting mixture is heated at 40–90°C. Yields are consistently higher with method A. The products have structures analogous to those of $[Co_2(CO)_6(\mu\text{-}RC{\equiv}CR)]$[156].

Olefin[157] and acetylene[157,158] complexes of iridium are obtainable by PPh_3 displacement (equations 77 and 78).

$$[Ir(CO)H(PPh_3)_3] + \text{un} \xrightarrow{C_6H_6} [Ir(CO)H(PPh_3)_2(\text{un})] \tag{77}$$

un = dmfm, fmn, dmm, tcne, maleic acid, cinnamic acid, cinnamonitrile, fumaric acid, $CF_3C{\equiv}CCF_3$

$$[IrCl(PPh_3)_3] + PhC{\equiv}CPh \xrightarrow{C_6H_6} [IrCl(PPh_3)_2(PhC{\equiv}CPh)] \qquad (78)$$

Platinum(0) and palladium(0) complexes of the type $[M(PPh_3)_2(un)]$ (M = Pt, Pd) are prepared from $M(PPh_3)_4$. Activated olefins such as $F_2C{=}CF_2$[159], $Cl_2C{=}CCl_2$[159,160], tcne[161], ma[161], def[161], and $CF_3FC{=}CFCF_3$[162] are reactive, but unactivated monoolefins do not appear capable of replacing two PPh_3 moieties. Acetylenes give products of the type $[Pt(PPh_3)_2(RC{\equiv}CR')]$, (R = R′ = Ph, R = R′ = 2-py, R = R′ = 2-(6-Me-py))[163].

2. *Nitrile displacement*

Benzonitrile and acetonitrile coordinated through the non-bonding electrons on nitrogen are displaced by π-ligands. Olefin complexes of palladium(II) may be prepared by reaction of the olefin $[H_2C{=}CH_2$, $Me_2C{=}CH_2$, $PhHC{=}CH_2$, $H_2C{=}CH(CH_2)_2HC{=}CH_2]$ with $[Pd(PhCN)_2Cl_2]$[58,164]. The $[Pd(PhCN)_2Cl_2]$ starting complex is prepared by reaction of $PdCl_2$ in PhCN. The pathway cannot be extended to platinum(II) complexes.

Acetonitrile complexes of the Group VI metals have been useful intermediates in development of the chemistry of those metals. Three CO ligands of $[M(CO)_6]$(M = Cr, Mo, W) may be replaced upon refluxing in CH_3CN[165,166]. The resulting $[M(CO)_3(CH_3CN)_3]$ is mixed with the liquid olefin or acetylene or mixed with the solid olefin or acetylene in benzene. It should be noted that in most cases some carbonyl ligands are displaced along with CH_3CN ligands (equations 79 and 80)[167–170]. In the reaction of $MeSC{\equiv}CSMe$ (equation 80) there is no evidence of sulphur coordination[169]. Monosubstituted acetylenes do not give characterizable products[168].

$$[W(CO)_3(CH_3CN)_3] + RC{\equiv}CR \longrightarrow [W(CO)(RC{\equiv}CR)_3] \qquad (79)$$

R = Ph, Et, SMe

$$[W(CO)_3(CH_3CN)_3] + CF_3C{\equiv}CCF_3 \xrightarrow{100^{\circ}C} [W(CH_3CN)(CF_3C{\equiv}CCF_3)_3] \qquad (80)$$

An interesting reaction is the formation of $[M(\eta^4\text{-diene})(CO)_4]$ complexes from $[M(CO)_3(CH_3CN)_3]$ (equations 81 and 82)[170–172]. The mechanism of $M(CO)_4$ formation from the $M(CO)_3$ moiety has not been determined.

$$[M(CO)_3(CH_3CN)_3] + \text{hmdb} \xrightarrow{\text{dioxane}} [(\text{hexamethyl-Dewar-benzene})M(CO)_4] \qquad (81)$$

M = Cr, Mo, W

Me Me Me Me Me Me M (CO)₄

$$[W(CO)_3(CH_3CN)_3] + 1{,}5\text{-cod} \xrightarrow{\text{hexane}} [W(1{,}5\text{-cod})(CO)_4] \qquad (82)$$

3. *Benzylideneacetone derivative displacement*

Until the recent synthesis of $[Pd(1{,}5\text{-cod})_2]$ (see Section II.G.6), there have been few routes to palladium—olefin complexes. One of the more general methods, however, is displacement of the chelating acetone derivatives dibenzylideneacetone (dba) and tribenzylideneacetylacetone (tba) from $[Pd_2(dba)_3]$[173] and $[Pd_3(tba)_2(bipy)_3]$[174], respectively (equations 83 and 84).

$$[Pd_2(dba)_3] + L + \text{olefin} \xrightarrow{\text{acetone}} [PdL(\text{olefin})] \quad (83)$$

olefin = ma, dmm, dmfm, acrylonitrile; L = bipy, *o*-phen

$$[Pd_3(tba)_2(bipy)_3] + \text{olefin} \xrightarrow{\text{acetone}} [Pd(bipy)(\text{olefin})] \quad (84)$$

olefin = ma, dmm, dmfm

4. *Solvent displacement*

Often solvent ligands coordinated to a transition metal may be displaced. In some cases solvent coordination is so weak that the complex may be considered coordinatively unsaturated, depending on the position of equilibrium (equation 85).

$$[ML_m(\text{solvent})] \rightleftharpoons [ML_m] + \text{solvent} \quad (85)$$

Ethers form metastable complexes with many metals but can be replaced to give the more thermodynamically stable olefin[175] and acetylene[176] complexes (equations 86 and 87). More recently, acetone has been displaced from the coordination sphere of

$$[Fp(thf)](BF_4) + \text{olefin} \xrightarrow{CH_2Cl_2} [Fp(\eta^2\text{-olefin})](BF_4) \quad (86)$$

olefin = $H_2C{=}CH_2$, cyclohexene, cycloheptene, cyclohexa-1,4-diene, norbornadiene

$$[Mn(Cp)(CO)_2(thf)] + PhC{\equiv}CPh \longrightarrow [Mn(Cp)(CO)_2(PhC{\equiv}CPh)] \quad (87)$$

octahedral ruthenium complexes by a variety of unsaturated hydrocarbons (Scheme 6)[177].

$$[Ru(NH_3)_5H_2O](PF_6)_2 + \text{acetone} \longrightarrow [Ru(NH_3)_5\text{acetone}](PF_6)_2$$

$$\xrightarrow{\text{un}} [Ru(NH_3)_5\text{un}](PF_6)_2$$

un = cyclohexene, norbornene, $HC{\equiv}CH$, hex-1-yne, hex-3-yne, oct-1-yne

SCHEME 6

5. *Silane displacement*

The chelating silicon atoms in **35** are reported to be replaced by $PhC{\equiv}CH$ to give $[Pt(PPh_3)_2(PhC{\equiv}CH)]$ in 50% yield[178].

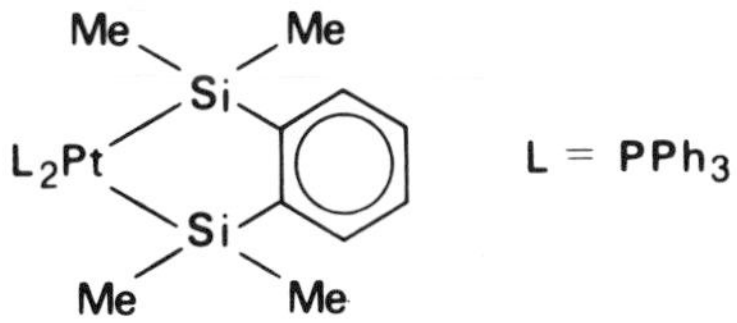

Me Me Si L_2Pt Si Me Me L = PPh_3

(35)

6. Hydrogen displacement

Elimination of molecular hydrogen has been common to preparation of several complexes of Group VB metals. Niobium and tantalum complexes of olefins are obtained from reaction of the hydrocarbon with the metal trihydrides (equation 88)[179].

$$[M(Cp)_2H_3] + H_2C{=}CH_2 \longrightarrow [M(Cp)_2(H_2C{=}CH_2)(C_2H_5)] \qquad (88)$$

M = Nb, Ta

If the metal trihydride–ethylene reaction is allowed to proceed in a limited amount of ethylene, the hydrido—ethylene complex is isolated. The first acetylene complexes of tantalum[180] and niobium[181] were prepared by a similar route (equations 89 and 90).

$$[Ta(CpMe)_2H_3] + C_5H_5I + RC{\equiv}CR' \xrightarrow{\text{dioxane}} [Ta(CpMe)_2I(RC{\equiv}CR')] \qquad (89)$$

R = R′ = *n*-Pr
R = Me, R′ = *i*-Pr
R = Me, R′ = *n*-Bu
R = Me, R′ = *t*-Bu

$$[Nb(Cp)_2H_3] + RC{\equiv}CR' \xrightarrow{C_6H_6} [Nb(Cp)_2H(RC{\equiv}CR')] \qquad (90)$$

R = R′ = Me
R = R′ = *n*-Pr
R = Me, R′ = *n*-Pr
R = Me, R′ = *i*-Pr

The dihydride complex of molybdenum $[Mo(Cp)_2H_2]$ reacts with PhC≡CPh to give the diamagnetic complex $[Mo(Cp)_2(PhC{\equiv}CPh)]$[182].

The osmium—dihydride cluster complex $[Os_3(CO)_{10}H_2]$ loses hydrogen in the presence of ethylene and cyclooctene to form $[Os_3(\text{cyclooctene})_2(CO)_{10}]$[183]. The cyclooctene is replaceable by other ligands so that $[Os_3(\text{cyclooctene})_2(CO)_{10}]$ is a convenient source of the $Os_3(CO)_{10}$ group. This trimetallic cluster has been of interest as a model for studying hydrocarbon chemisorption on metallic surfaces.

Often hydrogen displacement syntheses lead to some hydrogenation of the entering unsaturated hydrocarbon. The ruthenium—dihydride complex yields olefin η^2-complexes with ethylene and propylene (equation 91)[184]. For each mole of product

$$[Ru(PPh_3)_4H_2] + RHC{=}CH_2 \xrightarrow{C_6H_6} \underset{\mathbf{(36)}}{[Ru(PPh_3)_3(RHC{=}CH_2)} + RCH_2CH_3 \qquad (91)$$

R = H, Me

one equivalent of hydrogenation product is formed. Styrene does not form an η^2-complex by this method, but the ethylene complex **(36)** in styrene undergoes ligand

$$[Ir(P\text{-}i\text{-}Pr_3)_2H_5] + H_2C{=}CH_2 \xrightarrow{C_6H_6} \left[\begin{array}{c} P(i\text{-}Pr)_2 \\ / \quad \backslash \\ CH_2 \quad Ir(P\text{-}i\text{-}Pr_3)(H_2C{=}CH_2)_2 \\ \backslash \quad / \\ CHMe \end{array}\right] + CH_3CH_3 \qquad (92)$$

exchange to give the styrene analogue of **36**. If the synthesis (equation 91) is attempted with an excess of hydrogen, alkanes are the only hydrocarbon products with no formation of **36**. Similar chemistry is observed with the iridium complex $[Ir(P\text{-}i\text{-}Pr_3)_2H_5]$ (equation 92)[185].

7. *Oxygen displacement*

An ethylene complex of platinum has been reported from the displacement of oxygen[186]. Oxygen is bubbled through a benzene solution of $[Pt(PPh_3)_4]$ and the resulting isolable adduct $[Pt(PPh_3)_2O_2]$ is dissolved in ethanol. Ethylene is bubbled through the solution concurrent with addition of $NaBH_4$ to give $[Pt(PPh_3)_2(H_2C{=}CH_2)]$. The generality of this method is unexplored.

8. *Methane displacement*

Cationic olefin and acetylene complexes of cobalt can be prepared by displacement of methane (equation 93)[187]. The proposed mechanism involves a reductive elimination (Scheme 7).

$$[Co\{P(OMe)_3\}_4Me] + H^+ + \text{un} \longrightarrow [Co\{P(OMe)_3\}_4(\text{un})]^+ \qquad (93)$$

un = $H_2C{=}CH_2$, $MeHC{=}CH_2$, hex-1-ene, $PhC{\equiv}CPh$

$$[CoL_4Me] + H^+ \longrightarrow [CoL_4MeH]^+ \longrightarrow [CoL_4]^+ + CH_4$$

L = $P(OMe)_3$

SCHEME 7

F. Ligand Rearrangements

1. η^1-*Allyl*

A number of metal—η^1-allyl complexes undergo protonation with mineral acids in hydrocarbon solvents to form η^2-complexes (equation 94). The product salts usually

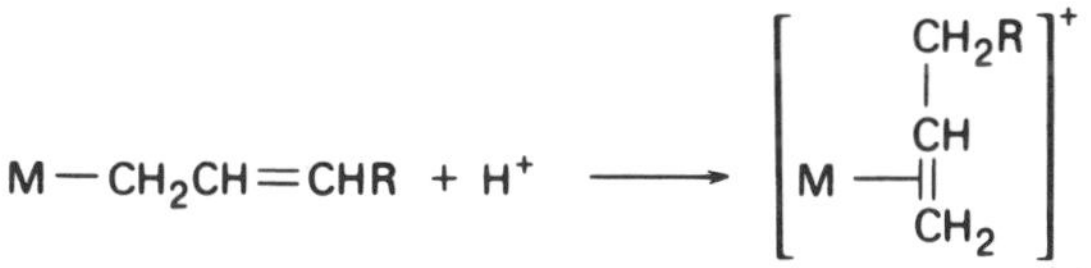

$$M{-}CH_2CH{=}CHR + H^+ \longrightarrow \left[M \leftarrow \begin{matrix} CH_2R \\ | \\ CH \\ \| \\ CH_2 \end{matrix} \right]^+ \qquad (94)$$

precipitate immediately from solution. Obviously ethylene complexes are not available by this route, but the cationic propylene complexes $[Mn(\text{propylene})(CO)_5]$[188], [Fp(propylene)][189], $[W(Cp)(CO)_3(\text{propylene})]$[190], and $[Mo(Cp)(CO)_3(\text{propylene})]$[191] are. Higher olefin η^2-complexes may be synthesized by the same general method[128,189]. Under similar conditions, iron—$CH_2C{\equiv}CR$ complexes are known to rearrange to η^2-allene complexes (equation 95)[192]. This acid-promoted rearrangement is reported to be unsuccessful with the complexes $M{-}CH_2C{\equiv}CPh$, where M = $Mn(CO)_5$, $Mo(Cp)(CO)_3$, and $W(Cp)(CO)_3$.

$$[Fp({-}CH_2C{\equiv}CR)] + HBF_4/\text{acetic anhydride} \longrightarrow [Fp(\eta^2{-}CH_2{=}C{=}CHR)](BF_4) \qquad (95)$$

R = Me, Ph

2. η^1-Alkyl

Another general method for generation of cationic metal—olefin η^2-complexes is hydride abstraction from metal—η^1-alkyl complexes (equation 96). To be successful,

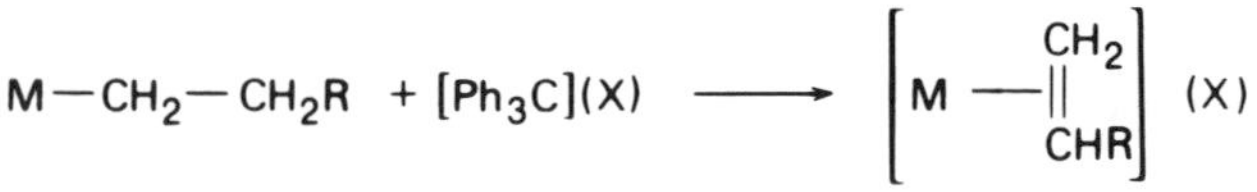

$$M{-}CH_2{-}CH_2R \; + [Ph_3C](X) \longrightarrow \left[M \longrightarrow \begin{matrix} CH_2 \\ \| \\ CHR \end{matrix} \right] (X) \qquad (96)$$

there must be a hydrogen atom on the β-carbon atom of the η^1-complex. Triphenyl—carbenium salts are the usual hydride abstraction agents in ethereal solvents. By this method the salts $[Mo(Cp)(CO)_3(ethylene)](BF_4)$[191], $[Mn(CO)_5$-(ethylene)](BF_4)[181], [Fp(ethylene)](PF_6)[193], and $[Ru(Cp)(CO)_2(ethylene)]$ (BF_4)[189] have been isolated. Although not many examples exist, higher olefin complexes such as [Fp(propylene)](BF_4)[188] are available by this path.

An ingenious adaption of the hydride abstraction process led to isolation of the unique bis-iron cation complex of cyclobutadiene (Scheme 8)[194,195].

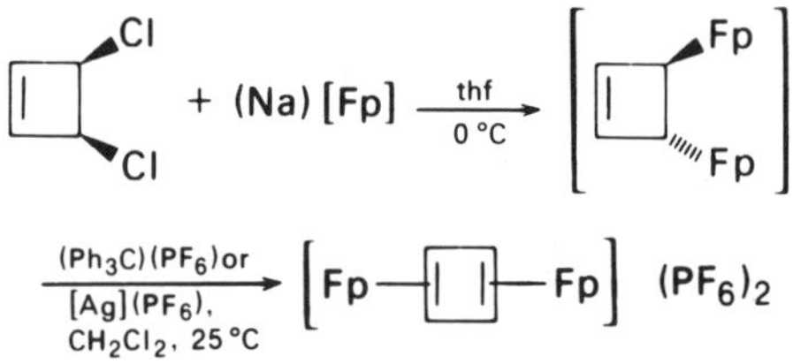

(dichlorocyclobutene) + (Na) [Fp] —thf, 0 °C→ [(cyclobutene)(Fp)(Fp)]

—$(Ph_3C)(PF_6)$ or $[Ag](PF_6)$, CH_2Cl_2, 25 °C→ [Fp—(cyclobutadiene)—Fp] $(PF_6)_2$

SCHEME 8

3. Carbenes

Metal—carbene complexes (see Chapter 5) may undergo conversion to olefin and acetylene complexes. Lithium—alkyl reagents, followed by chromatography of the reaction products on SiO_2, convert $[W(carbene)(CO)_5]$ into $[W(olefin)(CO)_5]$ products by the reaction sequence in Scheme 9[196].

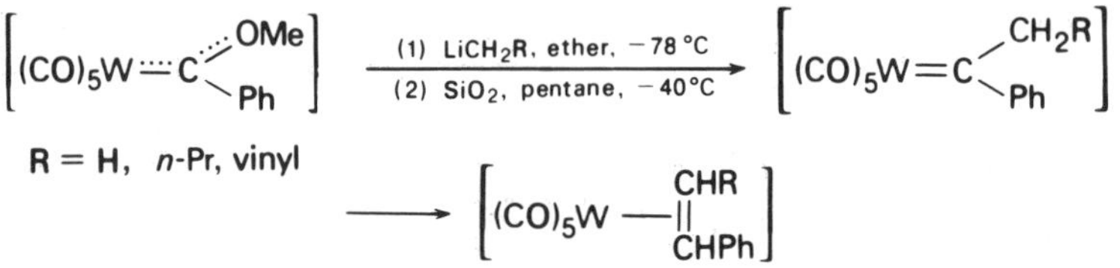

$[(CO)_5W{=}C(OMe)Ph]$ —(1) $LiCH_2R$, ether, −78 °C; (2) SiO_2, pentane, −40 °C→ $[(CO)_5W{=}C(CH_2R)Ph]$

R = H, *n*-Pr, vinyl

⟶ $\left[(CO)_5W \longrightarrow \begin{matrix} CHR \\ \| \\ CHPh \end{matrix}\right]$

SCHEME 9

A recent example involving an intermediate metal—carbene complex that is converted to an acetylene η^2-complex is outlined in Scheme 10[197]. Intermediates **37** and **38** have been confirmed as present in low concentration throughout the reaction. In addition, **38** with R = Me or Et, when heated to 45°C, produces **39**, among other products.

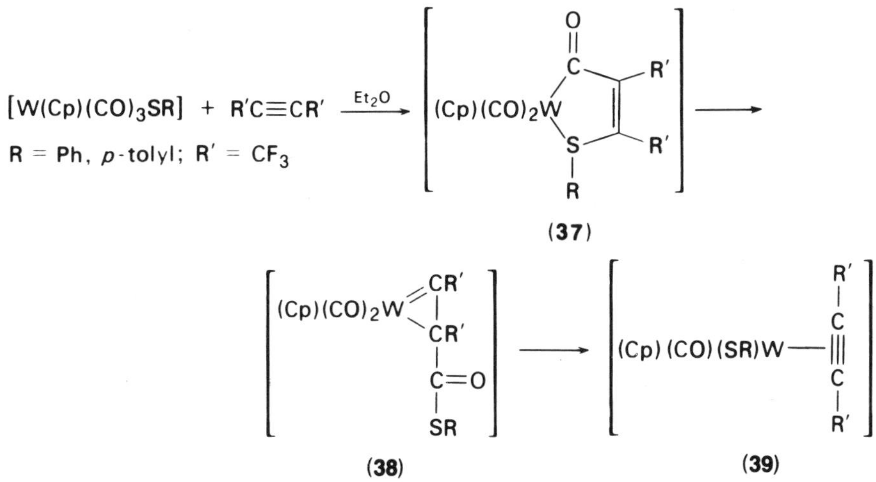

$[W(Cp)(CO)_3SR] + R'C{\equiv}CR' \xrightarrow{Et_2O}$ [(Cp)(CO)$_2$W–C(=O)–C(R')=C(R')–S(R)] (37) ⟶

R = Ph, p-tolyl; R′ = CF_3

[(Cp)(CO)$_2$W(CR′=CR′–C(=O)SR)] (38) ⟶ [(Cp)(CO)(SR)W—(R′C≡CR′)] (39)

SCHEME 10

4. Epoxides

Appropriately substituted cyclic and acyclic epoxides are susceptible to nucleophilic attack by the [Fp] anion. The reaction has led to a variety of complexes of the type [Fp(olefin)](BF_4) (Scheme 11)[129,198]. The method is limited by the availability of

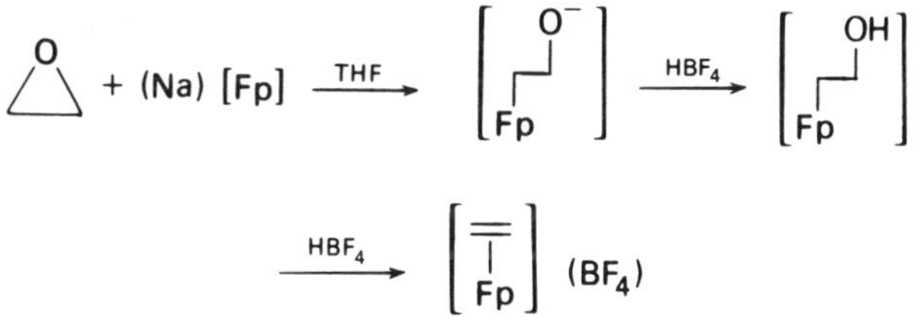

Epoxide + (Na) [Fp] $\xrightarrow{THF}$ [Fp–CH$_2$CH$_2$–O$^-$] $\xrightarrow{HBF_4}$ [Fp–CH$_2$CH$_2$–OH] $\xrightarrow{HBF_4}$ [Fp(CH$_2$=CH$_2$)] (BF_4)

SCHEME 11

epoxides but appears tolerant of a variety of functional groups. Yields are excellent (Table 1). η^2-Coordination of [Fp] cation to an olefin activates the carbon—carbon double bond to nucleophilic attack as evidenced by its susceptibility to Michael condensation[199].

TABLE 1. Preparation of [Fp (olefin)] complexes from epoxides[198]. Reproduced with permission from Giering *et al.*, *J. Am. Chem. Soc.*, **94**, 7170. Copyright 1972 by the American Chemical Society

Epoxide	Yield of Fp(olefin) complex (%)	Epoxide	Yield of Fp(olefin) complex (%)
Ethylene oxide	90	*trans*-Stilbene oxide	83
Propylene oxide	91	*cis*-Stilbene oxide	82
But-1-ene oxide	91	Cyclohexene oxide	66
cis-But-2-ene oxide	64	Butadiene monooxide	91
trans-But-2-ene oxide	50	Acrolein oxide	90
Styrene oxide	62	*trans*-Ethyl crotonate oxide	96
		4-Vinylcyclohexene dioxide	50

5. *Keto–enol tautomerism*

A final example of ligand rearrangement is the keto–enol tautomerism leading to isolation of η^2-vinyl alcohol complexes of platinum[200]. The platinum complex **40** results from reaction of $[Pt(acac)Cl(H_2C{=}CH_2)]$ with acetaldehyde, followed by treatment with base, then acid (Scheme 12)[201,202]. Similarly, the complex $[Pt(acac)Cl(\eta^2\text{-}CH_2{=}C(OH)CH_3)]$ can be prepared.

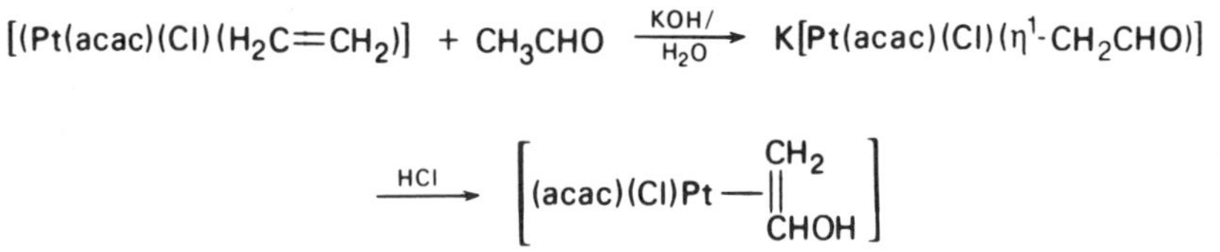

$$[(Pt(acac)(Cl)(H_2C{=}CH_2)] + CH_3CHO \xrightarrow[H_2O]{KOH/} K[Pt(acac)(Cl)(\eta^1\text{-}CH_2CHO)]$$

$$\xrightarrow{HCl} \left[(acac)(Cl)Pt - \begin{matrix} CH_2 \\ \| \\ CHOH \end{matrix}\right]$$

(40)

SCHEME 12

G. Reductive Olefination and Acetylenation

Metals in positive oxidation states (typically metal halides) can be reduced by a variety of metals, metal alkyls, and hydrides. In the presence of olefin or acetylene, the metal—olefin or metal—acetylene η^2-complex may form. Although the technique is generally applicable, it is not widely used.

1. *Sodium*

Sodium amalgam reacts with chlorides of molybdenum[203,204], tungsten[203], and niobium[205] to give olefin and acetylene complexes (equations 97 and 98). The molybdenum and tungsten syntheses with gaseous hydrocarbons require pressures from 1 to 150 atm.

$$[M(Cp)_2Cl_2] + Na/Hg + un \xrightarrow[\text{hexane}]{\text{PhMe or}} [M(Cp)_2un] \tag{97}$$

M = Mo;
un = $H_2C{=}CH_2$
$HC{\equiv}CH$
$MeC{\equiv}CH$
$CF_3C{\equiv}CCF_3$
$PhC{\equiv}CPh$

M = W;
un = $H_2C{=}CH_2$
$MeC{\equiv}CMe$

$$[Nb(Cp)_2Cl_2] + Na/Hg + un \xrightarrow{\text{PhMe}} [Nb(Cp)_2Cl(un)] \tag{98}$$

un = cyclopropene, $MeC{\equiv}CH$, $CF_3C{\equiv}CCF_3$

Cobalt(II) chloride is reduced by sodium in the presence of 1,5-cod, resulting in formation of $[Co(\eta^3\text{-}C_8H_{13})(\eta^4\text{-}C_8H_{12})]$[206].

2. *Manganese powder*

Manganese powder reacts with cobalt(II) halides (halide = Cl, Br, I) to give products of reductive olefination (equation 99)[207].

$$CoX_2 + RO_2CHC{=}CHCO_2R \xrightarrow[R'CN]{Mn} [Co(R'CN)_2(RO_2CHC{=}CHCO_2R)_2] \qquad (99)$$

X = Cl, Br, I
R = Et, R' = Me
R = *n*-Pr, R' = Me
R = *n*-Bu, R' = Et

3. *Aluminium trihalides*

Aluminium trichloride is often used to remove halide ligands irreversibly. Ethylene and $Mo(Cp)(CO)_3Cl$ thus are converted into the $[Mo(Cp)(CO)_3(H_2C{=}CH_2)]^+$ cation[208]. Ethylene[209] and 1,5-cod[210] react with $[M(CO)_5Cl]$ (M = Mn, Re) in the presence of $AlCl_3$ to yield $[M(CO)_5(H_2C{=}CH_2)]^+$ and $[M(CO)_4(1,5\text{-cod})]^+$ cations, respectively. When very high pressures of ethylene (210 atm) are used, products of halide and carbon monoxide displacement ($[Re(CO)_4(H_2C{=}CH_2)_2]$) are obtained[211]. The iron complex [Fp](X) (X = Cl, Br) with AlX_3 and olefin (olefin = $H_2C{=}CH_2$[208], cyclohexene[212], octadec-1-ene[212]) yields the cationic olefin complexes $[Fp(olefin)]^+$.

4. *Aluminium trialkyls*

A large number of nickel(0) complexes of olefins have been prepared from $[Ni(acac)_2]$. For example, reaction of $[Ni(acac)_2]$ with PR_3 (R = *o*-tolyl) in the presence of $AlEt_3$ leads to reductive olefination (equation 100)[213]. The product (**41**)

$$[Ni(acac)_2] + PR_3 + AlEt_3 + H_2C{=}CH_2 \xrightarrow{PhMe} \underset{\mathbf{(41)}}{[Ni(PR_3)_2(H_2C{=}CH_2)]} \qquad (100)$$

R = *o*-tolyl

has proved a useful catalyst for the reaction of dienes. The 1,5-cod complex of nickel is prepared in high yield from $[Ni(acac)_2]$, 1,5-cod, and $AlEt_3$[214].

Reductive olefinations of metal(acac) complexes have been reported for iron (equation 101)[215], molybdenum (equation 102)[216], and palladium (equation 103)[217]. In equations 101 and 102, the organoaluminium reagent is the source of the ethylene ligand.

$$[Fe(acac)_3] + dppe + AlEt_2(OEt) \xrightarrow{Et_2O} [Fe(dppe)_2(H_2C{=}CH_2)] \qquad (101)$$

$$[Mo(acac)_3] + dppe + AlEt_3 \longrightarrow [Mo(dppe)_2(H_2C{=}CH_2)] \qquad (102)$$

$$[Pd(acac)_2] + PR_3 + AlEt_2(OEt) + H_2C{=}CH_2 \xrightarrow{Et_2O} [Pd(PR_3)_2(H_2C{=}CH_2)] \qquad (103)$$

R = Cy, Ph, *o*-tolyl

5. *Grignard reagents*

Reduction with Grignard reagents is a general method of organometallic synthesis. The magnesium alkyl reagent Mg(*i*-Pr)Br reacts with $[Co(Cp)(PPh_3)I_2]$ in the presence of $PhC{\equiv}CPh$ to form $[Co(Cp)(PPh_3)(PhC{\equiv}CPh)]$[218]. A Grignard synthesis of $[Pt(1,5\text{-cod})_2]$ has been reported (Scheme 13), but the yields are low[219].

$$[PtCl_2(1,5\text{-cod})] + Mg(i\text{-Pr})Br \xrightarrow[-50°C]{MeOH}$$

$$[Pt(i\text{-Pr})_2(1,5\text{-cod})] \xrightarrow[1,5\text{-cod}]{h\nu} [Pt(1,5\text{-cod})_2]$$

SCHEME 13

It is possible for the alkyl group of a Grignard reagent to serve as a precursor to an olefinic ligand. When the alkyl group bears a β-hydrogen atom, a series of tantalum η^2-complexes can be prepared (equation 104)[220]. In some examples, mixtures of isomeric olefin complexes are obtained.

$$[Ta(Cp)_2Cl_2] + MgRCl \longrightarrow [Ta(Cp)_2H(\text{olefin})] \quad (104)$$

when R =	olefin =
η-Pr,	propylene
i-Pr,	propylene
n-Bu,	but-1-ene
s-Bu,	but-1-ene
n-pentyl,	pent-1-ene
cyclopentyl,	cyclopentene

6. *Lithium salts*

Development of the organometallic chemistry of platinum and palladium similar to that afforded by $[Ni(1,5\text{-cod})_2]$ has been slow to develop owing to the lack of mono- or diolefin complexes in reasonable amounts. The recent development of synthetic pathways to $[Pt(1,5\text{-cod})_2]$ and $[Pd(1,5\text{-cod})_2]$ by reductive olefination (Scheme 14)[140,141] should offer progress in this area. When M = Pd the preparation requires an

$$K_2MCl_4 + 1,5\text{-cod} \xrightarrow{HOAc} [MCl_2(1,5\text{-cod})]$$

M = Pt, Pd

$$[MCl_2(1,5\text{-cod})] + Li_2cot + 1,5\text{-cod} \xrightarrow{Et_2O} [M(1,5\text{-cod})_2]$$

SCHEME 14

atmosphere of propylene. The $[M(1,5\text{-cod})_2]$ complexes are reported to be more stable to oxidation and thermal decomposition than the corresponding nickel complex. They undergo facile 1,5-cod ligand displacement (see Section II.D.4) to yield $[M(H_2C{=}CH_2)_3]$[140,141], $[M(\text{norbornene})_3]$[140], and $[M(\textit{trans}\text{-cyclooctene})_3]$[140].

7. *Hydrazine*

Mention should be made of the reduction of metal halides with hydride reducing agents. Hydrazine hydrate reacts with $[Pt(PPh_3)_2Cl_2]$ in ethanol to give olefin[221], acetylene[222], and cycloalkyne[223] η^2-complexes (equation 105).

$$[Pt(PPh_3)_2Cl_2] + H_2NNH_2{\cdot}H_2O + un \xrightarrow{EtOH} [Pt(PPh_3)_2un] \quad (105)$$

un = PhHC=CHPh, 4,4′-dinitrostilbene, acenaphthene, HC≡CCH(OH)Me, HC≡CCH$_2$OMe, HC≡CCH$_2$OH, HC≡CCH(OH)Ph, *n*-PrC≡C*n*-Pr, PhC≡CC(OH)Me$_2$, cyclooctyne

8. Phosphites

A less common method of reductive olefination involves use of phosphites $P(OR)_3$ (R = alkyl or aryl) with CuX_2 (X = Cl, Br, ClO_4, BF_4) in alcoholic media[224]. In the presence of norbornadiene and 1,5-cod, the corresponding complexes $[Cu_2(nbd)X_2]$ and [Cu(1,5-cod)X] are obtained. The extent of carbon—carbon π-bond coordination (dihapto or tetrahapto) depends on the identity of the gegenion X.

H. Electrochemical

Electrolysis has been used in selected cases for the synthesis of organometallic η^2-complexes[225]. In general, the reduction of a mixture of a metal salt and olefin yields a metal(0)—olefin η^2-complex (Scheme 15). The polarographic half-wave reduction

$$M^+ + e^- \longrightarrow M(0) \xrightarrow{>C=C<} M—\|(C=C)$$

SCHEME 15

potentials for several salts and potential ligands (Table 2) predict that the salt cations would be reduced preferentially to olefin or acetylene[226].

Suitable solvents for electrolysis must be of adequate polarity for good conductivity and must have a high enough reduction potential to resist reduction. Diglyme, dimethoxyethane, thf, pyridine, and dmf are preferable. Additional experimental parameters have been completely reviewed[226].

Because of the tendency to form complex anions, organometallic η^2-complexes of copper(I) halides are unstable (equation 106). However, the complex

$$2[Cu(un)X] \rightleftharpoons 2un + [Cu][CuX_2] \qquad (106)$$

$[Cu(1,5\text{-cod})_2](ClO_4)$ has been prepared by electrolysis at copper electrodes of a solution of $[Cu](ClO_4)_2$ and 1,5-cod[23]. The suggested half-reactions are shown in Scheme 16. A synthesis of the corresponding tetrafluoroborate complex is reported by the same procedure[27].

TABLE 2. Polarographic half-wave reduction potentials in thf[a] [226]

Substrate	Half-wave reduction potential (V *vs.* SCE)
Ni^{II} acac	−1.57
Co^{II} acac	−1.98
Fe^{III} acac	(1) −0.82
	(2) −2.25
trans-Stilbene	(1) −2.25
	(2) −2.64
Diphenylacetylene	−2.41
Styrene	−2.69
Phenylacetylene	−2.71
Cycloocta-1,5-diene	−2.95

[a]Dropping mercury electrode, tetrabutylammonium bromide as electrolyte.

Cathode:

$$Cu^{2+} + 2(1,5\text{-cod}) + e^- + ClO_4^- \longrightarrow [Cu(1,5\text{-cod})_2](ClO_4)$$

Anode:

$$Cu + 2(1,5\text{-cod}) - e^- + ClO_4^- \longrightarrow [Cu(1,5\text{-cod})_2](ClO_4)$$

SCHEME 16

Similarly, $[Ni^{II}(acac)_2]$ can be reduced in the presence of 1,5-cod without metal deposition. In pyridine, thf, or dimethoxyethane, yields of $[Ni(1,5\text{-cod})_2]$ up to 96% are obtained[227].

Reduction of $[Co^{II}(acac)_2]$ with 1,5-cod in pyridine or a pyridine/alcohol mixture gives $[Co(1,5\text{-cod})(\eta^3\text{-cyclooctenyl})]$ in yields of 27–40% (equation 107)[228].

$$Co^{II} + 2\ 1,5\text{-cod} + 2e^- \longrightarrow [Co(\eta^3\text{-cyclooctenyl})(1,5\text{-cod})] \qquad (107)$$

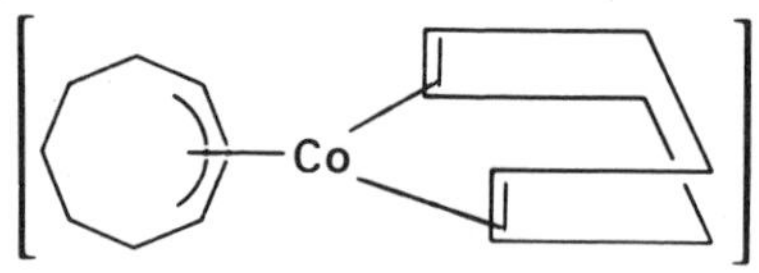

III. GENERAL PROBLEMS AND LIMITATIONS

The following is a brief summary of some of the major pitfalls and problems associated with application of the synthetic methods covered in this chapter.

Olefinic and acetylenic substrates may undergo a variety of reactions in addition to η^2-complexation. Transition metals are known to promote (often catalytically) hydrogenation, hydrogen transfer isomerizations, carbon transfer isomerizations, dimerization, oligomerization, polymerization, metathesis, carbonylation, and decarbonylation[229,230].

In photochemically initiated syntheses, the olefin or acetylene may absorb in the u.v.–visible region and undergo photoreaction. Often this problem can be overcome by carrying out the photolysis in a weakly coordinating solvent (e.g. thf), which will occupy the vacated coordination sites on the metal. Subsequent to irradiation the more strongly coordinating olefin or acetylene is added to the reaction solution, with resulting displacement of weakly coordinated solvent ligand[110].

In many organometallic syntheses reported in the literature, yields are either not reported or not optimized. In the latter case, with adjustment of reaction variables, one may obtain products in yields greater than reported. When dealing with air-sensitive complexes particular attention should be given to the purity of inert gases used to blanket the reaction mixture.

At times, dihapto olefins and acetylenes η^2-coordinated to transition metals are not as stable as polyene or polyyne complexes of the same metal. This is because of the greater entropic stabilization to be had from coordination of chelating ligands.

With terminal acetylenes the danger of the formation of explosive heavy metal acetylides is often present. In such cases examination of the literature to determine ligands, solvents, and reaction conditions that favour η^2-coordination rather than acetylide formation is recommended[22].

IV. REFERENCES

1. E. O. Fischer and H. Werner, *Metal Pi-Complexes*, Vol. I, Elsevier, Amsterdam, 1966.
2. R. B. King, in *Organometallic Syntheses* (Eds. J. J. Eisch and R. B. King), Vol. I, Academic Press, New York, 1965.

3. W. Hübel in *Organic Syntheses via Metal Carbonyls* (Eds. I. Wender and P. Pino), Vol. I, Interscience, New York, 1968, pp. 273–342.
4. M. L. H. Green, in *Organometallic Compounds* (Eds. G. E. Coates, M. L. H. Green and K. Wade), 3rd ed., Vol. II, Methuen, London, 1968, pp. 7–38 and 288–311.
5. H. W. Quinn and J. H. Tsai, *Adv. Inorg. Radiochem.*, **12**, 217 (1969).
6. J. H. Nelson and H. B. Jonassen, *Coord. Chem. Rev.*, **6**, 27 (1971).
7. B. L. Shaw and N. I. Tucker, in *Comprehensive Inorganic Chemistry* (Eds. J. C. Bailar, H. J. Emeleus, R. Nyholm and A. F. Trotman-Dickenson), Pergamon Press, Oxford, 1973, pp. 819–828.
8. M. Herberhold, *Metal Pi-Complexes*, Vol. II, Elsevier, Amsterdam, 1972.
9. R. D. W. Kemmitt, in *MTP International Review of Science* (Eds. H. J. Emeleus and M. J. Mays), Vol. VI, Part 2, Butterworth, London, 1972, pp. 227–272.
10. D. St. C. Black, W. R. Jackson, and J. M. Swan, in *Comprehensive Organic Chemistry* (Eds. D. Barton, W. D. Ollis, and D. N. Jones), Vol. III, Pergamon Press, Oxford, 1979, pp. 1156–1163.
11. R. S. Dickson and P. J. Fraser, *Adv. Organomet. Chem.*, **12**, 323 (1974).
12. R. B. King, in *The Organic Chemistry of Iron* (Eds. E. A. Koerner Von Gustorf, F.-W. Grevels and I. Fischler), Vol. I, Academic Press, New York, 1978, pp. 397–464.
13. C. D. M. Beverwijk, G. J. M. Van Der Kerk, A. J. Leusink, and J. G. Noltes, *Organomet. Chem. Rev. A*, **5**, 215 (1970).
14. A. E. Comyns and H. J. Lucas, *J. Am. Chem. Soc.*, **79**, 4339 (1957).
15. A. C. Cope, D. C. McLean, and N. A. Nelson, *J. Am. Chem. Soc.*, **77**, 1628 (1955).
16. G. Bressan, R. Broggi, M. P. Lachi, and A. L. Segre, *J. Organomet. Chem.*, **9**, 355 (1967).
17. J. Solodar and J. P. Petrovich, *Inorg. Chem.*, **10**, 395 (1971).
18. A. Albinati, S. V. Meille, and G. Carturan, *J. Organomet. Chem.*, **182**, 269 (1979).
19. W. Partenheimer and E. H. Johnson, *Inorg. Chem.*, **11**, 2840 (1972).
20. W. Partenheimer and E. H. Johnson, *Inorg. Syn.*, **16**, 117 (1976).
21. G. S. Lewandos, D. K. Gregston, and F. R. Nelson, *J. Organomet. Chem.*, **118**, 363 (1976).
22. J. P. Ginnebaugh, J. W. Maki, and G. S. Lewandos, *J. Organomet. Chem.*, **190**, 403 (1980).
23. S. E. Manahan, *Inorg. Chem.*, **5**, 2063 (1966).
24. D. A. Edwards and R. Richards, *J. Organomet. Chem.*, **86**, 407 (1975).
25. M. B. Dines, *Inorg. Chem.*, **11**, 2949 (1972).
26. D. L. Reger and M. D. Dukes, *J. Organomet. Chem.*, **113**, 173 (1976).
27. S. E. Manahan, *Inorg. Nucl. Chem. Lett.*, **3**, 383 (1967).
28. G. A. Chukhadzhyan, G. A. Gevorkyan, and V. P. Kukolev, *Zh. Obshch. Khim.*, **46**, 909 (1976).
29. R. G. Salomon and J. K. Kochi, *J. Am. Chem. Soc.*, **95**, 1889 (1973).
30. R. G. Salomon and J. K. Kochi, *J. Chem. Soc., Chem. Commun.*, 559 (1972).
31. S. Komiya and J. K. Kochi, *J. Organomet. Chem.*, **135**, 65 (1977).
32. D. I. Hall, J. H. Ling, and R. S. Nyholm, in *Structure and Bonding* (Eds. J. Dunitz, P. Hemmerich, J. A. Ibers, C. K. Jørgensen, J. B. Neilands, D. Reinen, and R. J. P. Williams), Vol. 15, Springer-Verlag, New York, 1973, pp. 3–43.
33. M. A. Bennett, W. R. Kneen, and R. S. Nyholm, *Inorg. Chem.*, **7**, 552 (1968).
34. M. A. Bennett, W. R. Kneen, and R. S. Nyholm, *Inorg. Chem.*, **7**, 556 (1968).
35. W. H. Baddley, *J. Am. Chem. Soc.*, **88**, 4545 (1966).
36. J. P. Collman and J. W. Kang, *J. Am. Chem. Soc.*, **89**, 844 (1967).
37. W. H. Baddley, *J. Am. Chem. Soc.*, **90**, 3705 (1968).
38. G. Mestroni and A. Camus, *Inorg. Nucl. Chem. Lett.*, **9**, 261 (1973).
39. G. Mestroni, A. Camus, and G. Zassinovich, *J. Organomet. Chem.*, **73**, 119 (1974).
40. G. Fachinetti, S. Del Niro, and C. Floriani, *J. Chem. Soc., Dalton Trans.*, 1046 (1976).
41. P. W. Scheneider, D. C. Bravard, J. W. McDonald, and W. E. Newton, *J. Am. Chem. Soc.*, **94**, 8640 (1972).
42. E. A. Maatta and R. A. D. Wentworth, *Inorg. Chem.*, **18**, 524 (1979).
43. P. F. Boyle and K. M. Nicholas, *J. Organomet. Chem.*, **114**, 307 (1976).
44. H. C. Clark and L. E. Manzer, *J. Chem. Soc., Chem. Commun.*, 870 (1973).
45. H. C. Clark and L. E. Manzer, *J. Am. Chem. Soc.*, **95**, 3812 (1973).
46. H. C. Clark and L. E. Manzer, *Inorg. Chem.*, **13**, 1996 (1974).
47. R. Cramer, *Inorg. Chem.*, **4**, 445 (1965).

48. P. B. Chock, J. Halpern, and F. E. Paulik, *Inorg. Syn.*, **14**, 90 (1973).
49. H. Boucher and B. Bosnich, *Inorg. Chem.*, **16**, 717 (1977).
50. C. B. Anderson and J. T. Michalowski, *J. Chem. Soc., Chem. Commun.*, 459 (1972).
51. J. Hubert and T. Theophanides, *J. Organomet. Chem.*, **93**, 265 (1975).
52. F. D. Rochon and T. Theophanides, *Can. J. Chem.*, **50**, 1325 (1972).
53. M. H. Chisholm and H. C. Clark, *Inorg. Chem.*, **10**, 2557 (1971).
54. M. H. Chisholm and H. C. Clark, *Inorg. Chem.*, **12**, 991 (1973).
55. T. R. Jack, C. J. May, J. Powell, *J. Am. Chem. Soc.*, **99**, 4707 (1977).
56. A. Greco, F. Pirinoli, G. Dall'Asta, *J. Organomet. Chem.*, **60**, 115 (1973).
57. R. Cramer, *Inorg. Syn.*, **15**, 14 (1974).
58. I. A. Zakharova, L. A. Leites, and V. A. Aleksanyan, *J. Organomet. Chem.*, **72**, 283 (1974).
59. V. Schurig, *J. Organomet. Chem.*, **74**, 457 (1974).
60. J. L. Herde, J. C. Lambert, and C. V. Senoff, *Inorg. Syn.*, **15**, 18 (1974).
61. R. H. Crabtree and G. E. Morris, *J. Organomet. Chem.*, **135**, 395 (1977).
62. J. A. Segal and B. F. G. Johnson, *J. Chem. Soc., Dalton Trans.*, 1990 (1975).
63. M. A. Bennett, L. Pratt, and G. Wilkinson, *J. Chem. Soc.*, 2037 (1961).
64. R. Pettit, *J. Am. Chem. Soc.*, **81**, 1266 (1959).
65. W. Strohmeier, *Chem. Ber.*, **94**, 2490 (1961).
66. A. R. L. Bursics, E. Bursics-Szekeres, and F. G. A. Stone, *J. Fluorine Chem.*, **7**, 619 (1976).
67. J. L. Davidson, M. Green, D. W. A. Sharp, F. G. A. Stone, and A. J. Welch, *J. Chem. Soc., Chem. Commun.*, 706 (1974).
68. J. L. Davidson and D. W. A. Sharp, *J. Chem. Soc., Dalton Trans.*, 2531 (1975).
69. W. I. Bailey, J. M. Collins, and F. A. Cotton, *J. Organomet. Chem.*, **135**, C53 (1977).
70. W. E. Newton, J. L. Corbin, and J. W. McDonald, *Inorg. Syn.*, **18**, 53 (1978).
71. G. Fachinetti, C. Floriani, F. Marchetti, and M. Mellini, *J. Chem. Soc., Dalton Trans.*, 1398 (1978).
72. L. V. Interrante and G. V. Nelson, *J. Organomet. Chem.*, **25**, 153 (1970).
73. M. Herberhold, G. O. Wiedersatz, and C. G. Kreiter, *J. Organomet. Chem.*, **104**, 209 (1976).
74. E. Weiss, K. Stark, J. E. Lancaster, and H. D. Murdoch, *Helv. Chim. Acta*, **46**, 288 (1963).
75. A. J. P. Domingos, J. A. S. Howell, B. F. G. Johnson, and J. Lewis, *Inorg. Syn.*, **16**, 103 (1976).
76. M. Bigorgne, *J. Organomet. Chem.*, **127**, 55 (1977).
77. T. H. Whitesides and R. W. Slaven, *J. Organomet. Chem.*, **67**, 99 (1974).
78. B. M. Chisnall, M. Green, R. P. Hughes, and A. J. Welch, *J. Chem. Soc., Dalton Trans.*, 1899 (1976).
79. W. Slegeir, R. Case, J. S. McKennis, and R. Pettit, *J. Am. Chem. Soc.*, **96**, 287 (1974).
80. D. J. Cook, M. Green, N. Mayne, and F. G. A. Stone, *J. Chem. Soc. (A)*, 1771 (1968).
81. K. Nicholas, L. S. Bray, R. E. Davis, and R. Pettit, *J. Chem. Soc., Chem. Commun.*, 608 (1971).
82. C. J. Sears and F. G. A. Stone, *J. Organomet. Chem.*, **11**, 644 (1968).
83. H. Greenfield, H. W. Sternberg, R. A. Friedel, J. H. Wotiz, R. Markby, and I. Wender, *J. Am. Chem. Soc.*, **78**, 120 (1956).
84. G. Cetini, O. Gambino, R. Rossetti, and E. Sappa, *J. Organomet. Chem.*, **8**, 149 (1967).
85. G. Varadi and G. Palyi, *Inorg. Chim. Acta*, **20**, L33 (1976).
86. R. B. King and A. Efraty, *J. Am. Chem. Soc.*, **94**, 3021 (1972).
87. R. B. King, I. Haiduc, and A. Efraty, *J. Organomet. Chem.*, **47**, 145 (1973).
88. P. McArdle and A. R. Manning, *J. Chem. Soc. (A)*, 2123 (1970).
89. Y. Iwashita and F. Tamura, *Bull. Chem. Soc. Jap.*, **43**, 1517 (1970).
90. J. A. J. Jarvis and R. Whyman, *J. Chem. Soc., Chem. Commun.*, 562 (1975).
91. A. Borrini and G. Ingrosso, *J. Organomet. Chem.*, **132**, 275 (1977).
92. R. B. King, *J. Indian Chem. Soc.*, **54**, 169 (1977).
93. J. F. Tilney-Bassett, *J. Chem. Soc.*, 577 (1961).
94. A. J. Carty, N. H. Paik, and T. W. Ng, *J. Organomet. Chem.*, **74**, 279 (1974).
95. E. W. Randall, E. Rosenberg, L. Milone, R. Rossetti, and P. L. Stanghellini, *J. Organomet. Chem.*, **64**, 271 (1974).
96. E. Koerner von Gustorf and F.-W. Grevels, *Fortschr. Chem. Forsch.*, **13**, 366 (1969).

97. M. Wrighton, *Chem. Rev.*, **74**, 401 (1974).
98. A. Vogler, in *Concepts of Inorganic Photochemistry* (Eds. A. W. Adamson and P. D. Fleischauer), Wiley-Interscience, New York, 1975, pp. 269–298.
99. G. L. Geoffroy and M. S. Wrighton, *Organometallic Photochemistry*, Academic Press, New York, 1979.
100. J. G. Calvert and J. N. Pitts, *Photochemistry*, Wiley, New York, 1966, pp. 686–814.
101. I. Stolz, G. R. Dobson, and R. K. Sheline, *Inorg. Chem.*, **2**, 1264 (1963).
102. E. O. Fischer and P. Kuzel, *Z. Naturforsch.*, **166**, 475 (1961).
103. R. J. Angelici and L. Busetto, *Inorg. Chem.*, **7**, 1935 (1968).
104. N. G. Connelly and A. G. Johnson, *J. Organomet. Chem.*, **77**, 341 (1974).
105. H. G. Alt, J. A. Schwarzle, and C. G. Kreiter, *J. Organomet. Chem.*, **153**, C7 (1978).
106. U. Koemm, C. G. Kreiter, and H. Strack, *J. Organomet. Chem.*, **148**, 179 (1978).
107. W. Strohmeier and D. von Hobe, *Z. Naturforsch, B*, **19**, 959 (1964).
108. R. J. Angelici and W. Loewen, *Inorg. Chem.*, **6**, 682 (1967).
109. I. S. Butler, N. J. Coville, and A. E. Fenster, *Inorg. Syn.*, **16**, 53 (1976).
110. H. Berke, *Angew. Chem., Int. Ed. Engl.*, **15**, 624 (1976).
111. R. Fields, G. L. Godwin, and R. N. Haszeldine, *J. Chem. Soc., Dalton Trans.*, 1867 (1975).
112. R. Aumann, *J. Organomet. Chem.*, **66**, C6 (1974).
113. R. Aumann, H. Wormann, and C. Kruger, *Angew. Chem., Int. Ed. Engl.*, **15**, 609 (1976).
114. R. Aumann, *J. Am. Chem. Soc.*, **96**, 2631 (1974).
115. P. C. Reeves, J. Henery, and R. Pettit, *J. Am. Chem. Soc.*, **91**, 5888 (1969)
116. K. H. Pannell and G. M. Crawford, *J. Coord. Chem.*, **2**, 251 (1973).
117. R. Burt, M. Cooke, and M. Green, *J. Chem. Soc. (A)*, 2975 (1970).
118. J. S. Anderson, *J. Chem. Soc.*, 1042 (1936).
119. R. A. Alexander, N. C. Baenziger, C. Carpenter, and J. R. Doyle, *J. Am. Chem. Soc.*, **82**, 535 (1960).
120. J. P. Birk, J. Halpern, and A. L. Pickard, *J. Am. Chem. Soc.*, **90**, 4491 (1968).
121. M. E. Jason, J. A. McGinnety, and K. B. Wiberg, *J. Am. Chem. Soc.*, **96**, 6531 (1974).
122. M. Tsutsui, M. Ori, and J. Francis, *J. Am. Chem. Soc.*, **94**, 1414 (1972).
123. M. T. Chicote, M. Green, J. L. Spencer, F. G. A. Stone, and J. Vicente, *J. Chem. Soc., Dalton Trans.*, 536 (1979).
124. N. M. Boag, M. Green, J. A. K. Howard, J. L. Spencer, R. F. D. Stansfield, F. G. A. Stone, M. D. O. Thomas, J. Vicente, and P. Woodward, *J. Chem. Soc., Chem. Commun.*, 930 (1977).
125. M. Green, J. A. K. Howard, R. P. Hughes, S. C. Kellett, and P. Woodward, *J. Chem. Soc., Dalton Trans.*, 2007 (1975).
126. A. De Rinzi, A. Panunzi, L. Paolillo, and A. Vitagliano, *J. Organomet. Chem.*, **124**, 221 (1977).
127. B. F. G. Johnson, J. Lewis, and D. J. Yarrow, *J. Chem. Soc., Dalton Trans.*, 1054 (1974).
128. W. P. Giering and M. Rosenblum, *J. Chem. Soc., Chem. Commun.*, 441 (1971).
129. M. Rosenblum, *Acc. Chem. Res.*, **7**, 122 (1974).
130. K. M. Nicholas, *J. Am. Chem. Soc.*, **97**, 3254 (1975).
131. M. Herberhold and H. Alt, *J. Organomet. Chem.*, **42**, 407 (1972).
132. M. Herberhold, H. Alt, and C. G. Kreiter, *J. Organomet. Chem.*, **42**, 413 (1972).
133. A. L. Onderdelinden and A. van der Ent, *Inorg. Chim. Acta*, **6**, 420 (1972).
134. A. van der Ent and A. L. Onderdelinden, *Inorg. Chim. Acta*, **7**, 203 (1973).
135. H. L. M. van Gaal and J. P. J. Verlaan, *J. Organomet. Chem.*, **133**, 93 (1977).
136. F. Guerrieri and G. Salerno, *J. Organomet. Chem.*, **114**, 339 (1976).
137. S. D. Ittel, *Inorg. Syn.*, **17**, 117 (1977).
138. C. S. Cundy, M. Green, and F. G. A. Stone, *J. Chem. Soc. (A)*, 1647 (1970).
139. P. W. Jolly and G. Wilke, *The Organic Chemistry of Nickel*, Vol. 1, Academic Press, New York, 1974, and Vol. 2, Academic Press, New York, 1975.
140. M. Green, J. A. K. Howard, J. L. Spencer, and F. G. A. Stone, *J. Chem. Soc., Dalton Trans.*, 271 (1977).
141. J. L. Spencer, *Inorg. Syn.*, **19**, 213 (1979).
142. M. T. Chicote, M. Green, J. L. Spencer, and F. G. A. Stone, *J. Organomet. Chem.*, **137**, C8 (1977).

143. M. Green, D. M. Grove, J. A. K. Howard, J. L. Spencer, and F. G. A. Stone, *J. Chem. Soc., Chem. Commun.*, 759 (1976).
144. L. E. Smart, J. Browning, M. Green, A. Laguna, J. L. Spencer, and F. G. A. Stone, *J. Chem. Soc., Dalton Trans.*, 1777 (1977).
145. A. J. P. Domingos, B. F. G. Johnson, and J. Lewis, *J. Organomet. Chem.*, **49**, C33 (1973).
146. M. A. Bennett, R. S. Nyholm, and J. D. Saxby, *J. Organomet. Chem.*, **10**, 301 (1967).
147. K. Fischer, K. Jonas, and G. Wilke, *Angew. Chem., Int. Ed. Engl.*, **12**, 565 (1973).
148. T. Kruck and L. Knoll, *Chem. Ber.*, **106**, 3578 (1973).
149. P. S. Hallman, B. R. McGarvey, and G. Wilkinson, *J. Chem. Soc. (A)*, 3143 (1968).
150. S. Komiya, A. Yamamoto, and S. Ikeda, *J. Organomet. Chem.*, **42**, C65 (1972).
151. B. E. Cavit, K. R. Grundy, and W. R. Roper, *J. Chem. Soc., Chem. Commun.*, 60 (1972).
152. K. Yasufuku, Y. Kuramitsu, and H. Yamazaki, *Tetrahedron Lett.*, 4549 (1974).
153. K. Yasufuku and H. Yamazaki, *J. Organomet. Chem.*, **121**, 405 (1976).
154. J. A. Osborn, F. H. Jardine, J. F. Young, and G. Wilkinson, *J. Chem. Soc. (A)*, 1711 (1966).
155. M. A. Bennett, R. N. Johnson, and T. W. Turney, *Inorg. Chem.*, **15**, 90 (1976).
156. M. A. Bennett, R. N. Johnson, G. B. Robertson, T. W. Turney, and P. O. Whimp, *J. Am. Chem. Soc.*, **94**, 6540 (1972).
157. W. H. Baddley and M. Shirley-Frazer, *J. Am. Chem. Soc.*, **91**, 3661 (1969).
158. M. A. Bennett, R. Charles, and P. J. Fraser, *Aust. J. Chem.*, **30**, 1213 (1977).
159. W. J. Bland and R. D. W. Kemmitt, *J. Chem. Soc. (A)*, 1278 (1968).
160. W. J. Bland and R. D. W. Kemmitt, *Nature*, **211**, 963 (1966).
161. P. Fitton and J. E. McKeon, *J. Chem. Soc., Chem. Commun.*, 4 (1968).
162. D. M. Roundhill and G. Wilkinson, *J. Chem. Soc. (A)*, 506 (1968).
163. G. R. Newcombe, G. L. McClure, S. F. Watkins, B. Gayle, R. E. Taylor, and R. Musselman, *J. Org. Chem.*, **40**, 3759 (1975).
164. M. S. Kharasch, R. C. Seyler, and F. R. Mayo, *J. Am. Chem. Soc.*, **60**, 882 (1938).
165. D. P. Tate, W. R. Knipple, and J. M. Augl, *Inorg. Chem.*, **1**, 433 (1962).
166. B. L. Ross, J. G. Grasselli, W. M. Ritchey, and H. D. Kaesz, *Inorg. Chem.*, **2**, 1023 (1963).
167. D. B. Tate and J. M. Augl, *J. Am. Chem. Soc.*, **85**, 2174 (1963).
168. D. P. Tate, J. M. Augl, W. M. Ritchey, B. L. Ross, and J. G. Grasselli, *J. Am. Chem. Soc.*, **86**, 3261 (1964).
169. J. A. Connor and G. A. Hudson, *J. Organomet. Chem.*, **160**, 159 (1978).
170. R. B. King and A. Fronzaglia, *Inorg. Chem.*, **5**, 1837 (1966).
171. E. O. Fischer, C. G. Kreiter, and W. Berngruber, *Angew. Chem. Int. Ed. Engl.*, **6**, 634 (1967).
172. E. O. Fischer, W. Berngruber, and C. G. Kreiter, *Chem. Ber.*, **101**, 824 (1968).
173. T. Ito, S. Hasegawa, Y. Takahashi, and Y. Ishii, *J. Organomet. Chem.*, **73**, 401 (1974).
174. Y. Ishii, S. Hasegawa, S. Kimura, and K. Itoh, *J. Organomet Chem.*, **73**, 411 (1974).
175. D. L. Reger and C. Coleman, *J. Organomet. Chem.*, **131**, 153 (1977).
176. A. N. Nesmeyanov, N. E. Kolobava, A. B. Antonova, and K. N. Anisimov, *Dokl. Chem., Engl. Ed.*, **220**, 12 (1975).
177. B. P. Sullivan, J. A. Baumann, T. J. Meyer, D. J. Salmon, H. Lehmann, and A. Ludi, *J. Am. Chem. Soc.*, **99**, 7368 (1977).
178. C. Eaborn, T. N. Metham, and A. Pidcock, *J. Organomet. Chem.*, **63**, 107 (1973).
179. F. N. Tebbe and G. W. Parshall, *J. Am. Chem. Soc.*, **93**, 3793 (1971).
180. J. A. Labinger, J. Schwartz and J. M. Townsend, *J. Am. Chem. Soc.*, **96**, 4009 (1974).
181. J. A. Labinger and J. Schwartz, *J. Am. Chem. Soc.*, **97**, 1596 (1975).
182. S. Otsuka, A. Nakamura, and H. Minamida, *J. Chem. Soc., Chem. Commun.*, 1148 (1969).
183. M. Tachikawa and J. R. Shapley, *J. Organomet. Chem.*, **124**, C19 (1977).
184. S. Komiya and A. Yamamoto, *Bull. Chem. Soc. Jap.*, **49**, 2553 (1976).
185. M. G. Clerici, S. DiGioacchino, F. Maspero, E. Perrotti, and A. Zanobi, *J. Organomet. Chem.*, **84**, 379 (1975).
186. C. D. Cook and G. S. Jauhal, *J. Am. Chem. Soc.*, **90**, 1464 (1968).
187. E. L. Muetterties and P. L. Watson, *J. Am. Chem. Soc.*, **98**, 4665 (1976).
188. M. L. H. Green and P. L. I. Nagy, *J. Organomet. Chem.*, **1**, 58 (1963).
189. J. W. Faller and B. V. Johnson, *J. Organomet. Chem.*, **88**, 101 (1975).
190. M. L. H. Green and A. N. Stear, *J. Organomet. Chem.*, **1**, 230 (1963).

191. M. Cousins and M. L. H. Green, *J. Chem. Soc.*, 889 (1963).
192. D. W. Lichtenberg and A. Wojcicki, *J. Organomet. Chem.*, **94**, 311 (1975).
193. W. H. Knoth, *Inorg. Chem.*, **14**, 1566 (1975).
194. A. Sanders and W. P. Giering, *J. Am. Chem. Soc.*, **96**, 5247 (1974).
195. A. Sanders and W. P. Giering, *J. Organomet. Chem.*, **104**, 49 (1976).
196. E. O. Fischer and W. Held, *J. Organomet. Chem.*, **112**, C59 (1976).
197. J. L. Davidson, *J. Chem. Soc., Chem. Commun.*, 597 (1979).
198. W. P. Giering, M. Rosenblum, and J. Tancrede, *J. Am. Chem. Soc.*, **94**, 7170 (1972).
199. A. Rosan and M. Rosenblum, *J. Org. Chem.*, **40**, 3621 (1975).
200. J. Frances, M. Ishaq, and M. Tsutsui, in *Organotransition–Metal Chemistry* (Eds. Y. Ishii and M. Tsutsui), Plenum, New York, 1975, pp. 57–64.
201. J. Hillis, J. Francis, M. Ori, and M. Tsutsui, *J. Am. Chem. Soc.*, **96**, 4800 (1974).
202. M. Tsutsui and A. Courtney, *Adv. Organomet. Chem.*, **16**, 241 (1977).
203. J. L. Thomas, *J. Am. Chem. Soc.*, **95**, 1838 (1973).
204. J. L. Thomas, *Inorg. Chem.*, **17**, 1507 (1978).
205. S. Fredericks and J. L. Thomas, *J. Am. Chem. Soc.*, **100**, 350 (1978).
206. S. Otsuka and M. Rossi, *J. Chem. Soc. (A)*, 2630 (1968).
207. G. Agnes, I. W. Bassi, C. Benedicenti, R. Intrito, M. Calcaterra, and C. Santini, *J. Organomet. Chem.*, **129**, 401 (1977).
208. E. O. Fischer and K. Fichtel, *Chem. Ber.*, **94**, 1200 (1961).
209. E. O. Fischer and K. Öfele, *Angew. Chem.*, **73**, 581 (1961).
210. P. J. Harris, S. A. R. Knox, R. J. McKinney, and F. G. A. Stone, *J. Organomet. Chem.*, **148**, 327 (1978).
211. E. O. Fischer and K. Öfele, *Angew. Chem., Int. Ed. Engl.*, **1**, 52 (1962).
212. E. O. Fischer and K. Öfele, *Chem. Ber.*, **95**, 2063 (1962).
213. W. C. Seidel and L. W. Gosser, *Inorg. Syn.*, **15**, 9 (1974).
214. R. A. Schunn, *Inorg. Syn.*, **15**, 5 (1974).
215. G. Hata, H. Kondo, and A. Miyake, *J. Am. Chem. Soc.*, **90**, 2278 (1968).
216. T. Ito, T. Kokubo, T. Yamamoto, A. Yamamoto, and S. Ikeda, *J. Chem. Soc., Dalton Trans.*, 1783 (1974).
217. A. Visser, R. Van Der Linde, and R. O. Dejongh, *Inorg. Syn.*, **16**, 127 (1976).
218. H. Yamazaki and N. Hagihara, *J. Organomet. Chem.*, **7**, P22 (1967).
219. J. Müller and P. Göser, *Angew. Chem., Int. Ed. Engl.*, **6**, 364 (1967).
220. A. H. Klazinga and J. H. Teuben, *J. Organomet. Chem.*, **157**, 413 (1978).
221. J. Chatt, B. L. Shaw, and A. A. Williams, *J. Chem. Soc.*, 3269 (1962).
222. H. D. Empsall, B. L. Shaw, and A. J. Stringer, *J. Chem. Soc., Dalton Trans.*, 185 (1976).
223. G. Wittig and S. Fischer, *Chem. Ber.*, **105**, 3542 (1972).
224. B. W. Cook, R. G. J. Miller, and P. F. Todd, *J. Organomet. Chem.*, **19**, 421 (1969).
225. H. Lehmkuhl, in *Organic Electrochemistry* (Ed. M. M. Baizer), Marcel Dekker, New York, 1973, p. 621.
226. H. Lehmkuhl, *Synthesis*, 377 (1973).
227. H. Lehmkuhl and W. Leuchte, *J. Organomet. Chem.*, **23**, C30 (1970).
228. W. Leuchte, *Dissertation*, Universität Bochum (1971).
229. M. L. H. Green, in *Organometallic Compounds* (Eds. G. E. Coates, M. L. H. Green, and K. Wade), 3rd ed., Vol. II, Methuen, London, 1968, pp. 312–348.
230. D. St. C. Black, W. R. Jackson, and J. M. Swan, in *Comprehensive Organic Chemistry* (Eds. D. Barton, W. D. Ollis, and D. N. Jones), Vol. III, Pergamon Press, Oxford, 1979, pp. 1178–1323.

The Chemistry of the Metal–Carbon Bond
Edited by F. R. Hartley and S. Patai

CHAPTER **8**

Synthesis of η^3-allyl complexes

P. POWELL
Department of Chemistry, Royal Holloway College, Egham Hill, Surrey TW20 0EX, UK

I. INTRODUCTION

Methods for the synthesis of η^3-allyl complexes have been discussed in several reviews[1–6], which cover comprehensively early work in this area. More recent research is described in texts which treat the organometallic chemistry of particular elements including titanium, zirconium, and hafnium[7], iron[8], nickel[9], palladium[10,11] and platinum[11]. Additional review references are given at relevant points in this chapter. Only those synthetic methods which lead either to the introduction of an η^3-allyl group or the conversion of an organic ligand already present into such a substituent are discussed here. Reactions which involve other coordinated groups but which leave the allyl ligand unchanged are in general not considered.

II. ALLYL GROUP TRANSFER

A. Allyl Grignard Reagents

The reaction between transition metal halides and allyl Grignard reagents provides a general route to binary allyl derivatives, $[M(all)_n]$[4]. The preparation of allylmagnesium halides requires more care than is necessary for alkyl Grignard reagents, on account of ready Wurtz coupling giving hexa-1,5-dienes[12]. This side reaction is minimized by slow addition of the allyl halide to magnesium, and by cooling to -10°C throughout. The synthesis of bis(η^3-allyl)nickel from anhydrous nickel(II) bromide and allylmagnesium chloride is typical. Binary allyls of the transition metals are generally extremely air sensitive, of low thermal stability, and labile towards displacement of allyl groups. Loss of ligands from bis(η^3-allyl)nickel leads to 'naked nickel', which can act as a matrix for the cyclooligomerization of buta-1,3-diene to cyclododecatrienes[4]. In conjunction with co-catalysts such as phosphines or phosphites, it and other nickel(0) complexes are versatile catalysts for diene and olefin reactions[13–15]. These properties are shared by other transition metal allyls[16–18].

Many binary allyls, including the tetraallyls $[M(all)_4]$ (M = Zr, Hf, Th, U[19], Mo, and W), the triallyls $[M(all)_3]$ (M = V, Cr[20], Fe, Co[16], Rh[21], Ir[22]), and the bisallyls $[M(all)_2]$ (M = Ni, Pd[23] and Pt[24]) have been prepared by the Grignard method. Chromium and molybdenum in addition afford the dinuclear complexes $Cr_2(all)_4$ and $Mo_2(all)_4$, respectively[25].

The Grignard method has also been widely used to introduce allyl groups into complexes bearing other ligands. In the reaction of allylmagnesium halides with $[Ti(C_5H_5)_2Cl_2]$, the Grignard reagent reduces titanium to the +3 oxidation state prior to transfer of an allyl group[26] (equation 1).

$$[Ti(C_5H_5)_2Cl_2] \xrightarrow{RMgX} [Ti(C_5H_5)(\eta^3\text{-}R)] \ (R = C_3H_5,\ C_4H_7,\ C_5H_9) \quad (1)$$

The analogous zirconium complex, however, is not reduced, and affords cream air-sensitive crystals, which probably have the formula $[Zr(\eta^1\text{-}C_3H_5)(\eta^3\text{-}C_3H_5)(C_5H_5)_2]$[27]. Similar reactions of $[M(C_5H_5)_2Cl_2]$, (M = V[28], Nb[29], and Ta[30]) yield the reduced products $[M(C_5H_5)_2(C_3H_5)]$. The infrared spectrum of the vanadium compound and its paramagnetism ($\mu_{eff} = 2.7$ BM) suggest η^1-bonding of the allyl group, while the diamagnetic niobium and tantalum complexes are probably η^3-bonded.

Treatment of titanium(III) chloride with the dianion of cyclooctatetraene in tetrahydrofuran gives **1**, which with allylmagnesium halides yields very air- and water-sensitive derivatives, **2**[31] (equation 2). It again proved possible to prepare the

$$TiCl_3 + (K^+)_2(C_8H_8^{2-}) \longrightarrow \underset{(\mathbf{1})}{[Ti(\eta^8\text{-}C_8H_8)Cl\cdot thf]_2} \xrightarrow{\text{allMgX}}$$

$$[Ti(\eta^8\text{-}C_8H_8)(all)thf] \xrightarrow[\text{pentane}]{\text{wash,}} \underset{(\mathbf{2})}{[Ti(\eta^8\text{-}C_8H_8)(\eta^3\text{-}all)]} \quad (2)$$

zirconium(IV)[32,34] and hafnium(IV)[33,34] complexes **3**, although not the corresponding titanium(IV) derivative (equation 3).

$$M(OR)_4 + AlEt_3 \xrightarrow{C_8H_8} M(C_8H_8)_2 \xrightarrow{HCl} [M(C_8H_8)Cl_2] \xrightarrow{C_3H_5MgX}$$

$$\underset{(\mathbf{3})}{[M(\eta^8\text{-}C_8H_8)(\eta^3\text{-}C_3H_5)_2]} \quad (3)$$

Some other allyl complexes which have been obtained by the Grignard method include [Ru(cod)(all)$_2$][35], [Ru(PPh$_3$)$_2$(all)][36], [Rh(cod)all][37], and [Mo(arene)(all)$_2$][38], where cod = cycloocta-1,5-diene.

B. Other Main Group Allyl Derivatives

Allyllithium reagents are prepared by a circuitous route involving transmetallation from allyltin precursors, which in turn are obtained through the Grignard reagent. Their use in synthesis is therefore inconvenient. Allylzinc halides have also been used little, if at all, in transition metal chemistry, although their preparation has been well researched[39]. Allylmercury halides, however, have proved useful for transfer of allyl groups to the platinum metals[40]. Although allylmercury iodide can be prepared from mercury and allyl iodide, and the chloride and bromide from the Grignard reagent and mercury(II) halide, Nesmeyanov *et al.* found it convenient to obtain the reagents from readily available allylpalladium halides (see Section VIII) by reaction with metallic mercury[41]:

$$[Pd(C_3H_5)Cl]_2 + 2Hg \longrightarrow 2Pd + 2C_3H_5HgCl$$

Allylmercury halides are easily handled, being stable to air and also to hydrolysis under neutral conditions. The reactions in equations 4–7 illustrate their application[40,42].

Abel and Moorhouse made a detailed study of the transfer of allyl, cyclopentadienyl, indenyl, and related groups from tin to transition metals[43]. Various allyltin reagents, in particular allyltrimethyl tin, were reacted with carbonyl halides and organocarbonyl halides (equations 8–11).

In the preparation of η^3-allyl metal carbonyls by other methods (see Sections III and IV), decarbonylation of an intermediate η^1-allyl is often required. This step is sometimes difficult with formally 6-coordinate η^1-allyls and may proceed in low yield. The successful preparation of η^3-allyl complexes under milder conditions than those

$$[Pd(C_3H_5)Cl]_2 \underset{Na_2PdCl_4}{\overset{Hg}{\rightleftharpoons}} Hg(C_3H_5)Cl \xrightarrow{Na_3RhCl_6} [Rh(C_3H_5)Cl_2]_2 \quad (4)$$

$$Hg(C_3H_5)Cl \xrightarrow[TlC_5H_5]{RhCl_3 \cdot 3\,H_2O} [RhCl(C_3H_5)(C_5H_5)] \quad (5)$$

$$Hg(C_3H_5)Cl \xrightarrow[(ii)\ S_2O_3^{2-}]{(i)\ [C_5H_5IrI_2]} [IrI(C_3H_5)(C_5H_5)] \quad (6)$$

$$Hg(C_3H_5)Cl \xrightarrow{[Ru(C_6H_6)Cl_2]_2} [RuCl(C_3H_5)(C_6H_6)] \quad (7)$$

$$CH_2{=}CH{-}CH_2SnMe_3 + [Mn(CO)_5Br] \xrightarrow[\text{reflux (4 h)}]{\text{thf}} \underset{(82\%)}{[Mn(\eta^3\text{-}C_3H_5)(CO)_4]} \quad (8)$$

$$CH_2{=}CH{-}CH_2SnMe_3 + [Fe(CO)_4I_2] \xrightarrow[25^\circ C]{Et_2O,\ 2\ h} \underset{(81\%)}{[Fe(\eta^3\text{-}C_3H_5)(CO)_3I]} \quad (9)$$

$$CH_2{=}CH{-}CH_2SnMe_3 + [Ru(\eta^3\text{-}C_3H_5)(CO)_3Br] \xrightarrow[\text{reflux}]{\text{thf, 4 h}} \underset{(83\%)}{[Ru(\eta^3\text{-}C_3H_5)_2(CO)_2]} \quad (10)$$

$$CH_2{=}CH{-}CH_2SnMe_3 + [Mo(C_5H_5)(CO)_3Cl] \xrightarrow[25^\circ C]{\text{thf, 120 h}} \underset{(60\%)}{[Mo(\eta^3\text{-}C_3H_5)(\eta^5\text{-}C_5H_5)(CO)_2]} \quad (11)$$

normally needed for the decarbonylation led the authors to suggest that η^1-allyls are not intermediates in these reactions. It seems, however, that there may be cases where their intermediacy does occur.

Surprisingly, the organotin method has not been widely employed for the synthesis of allyls, although it has proved its worth in making pentaalkylcyclopentadienyl complexes. $[RuCl(\eta^3\text{-}C_3H_5)(\eta^6\text{-}C_6H_6)]$ was obtained from $[RuCl_2(C_6H_6)]_2$ and tetraallyltin[44] and the conversion of $[W(C_5H_5)(NO)I_2]_2$ into $[W(C_5H_5)(NO)(\eta^3\text{-}C_3H_5)I]$ has also been reported[45].

C. From Other Transition Elements[46,47]

η^3-Allyl complexes have also been prepared by transfer of allyl groups from one transition metal to another. In most cases η^3-allylpalladium halides (see Section VIII) have been used as starting materials. Thus, $[Fe_2(CO)_9]$ reacts with $[Pd(\eta^3\text{-}C_3H_5)X]_2$ to yield allyltricarbonyliron halides (equation 12).

$$[(\eta^3\text{-}C_3H_5)Pd(\mu\text{-}X)_2Pd(\eta^3\text{-}C_3H_5)] \xrightarrow{[Fe_2(CO)_9]} [Fe(\eta^3\text{-}C_3H_5)(CO)_3X] + Pd + [Fe(CO)_5] + Fe \quad (12)$$

III. REACTIONS OF ANIONIC TRANSITION METAL COMPLEXES WITH ALLYL HALIDES

A. Anionic Metal Carbonyls

The attack of anionic metal carbonyls on allyl halides provides an important route to transition metal allyl complexes. An η^1-allyl derivative is formed initially[2]. If the formal coordination number of the metal in this intermediate is 5 (d^8), e.g. $[Co(\eta^1\text{-}C_3H_5)(CO)_4]$, it loses carbon monoxide spontaneously to give the η^3-allyl complex. When the formal coordination number is 6 (d^6) or 7 (d^4), however, the CO is much less labile and heating or irradiation is usually required to displace it (see also Section IV.A).

Thus, the tetracarbonylcobaltate anion yields η^3-allyltricarbonylcobalt at room temperature. The isoelectronic anion $[Fe(CO)_3(NO)]^-$ also gives η^3-allyl derivatives directly[48]. One mole of CO is easily displaced by phosphines from $[Co(\eta^3\text{-all})(CO)_3]$ **4** or from $[Fe(\eta^3\text{-all})(CO)_2(NO)]$ **5**. The resulting phosphine complexes are more stable thermally and to oxidation than the parent compounds. A dissociative mechanism has been suggested for substitution of CO in **4**[49]. A variety of pathways are indicated for **5** depending on the nature of the allyl substituent[50,51]. The salts $K[CoL_4]$ (L = a trialkyl phosphite), which are prepared by reaction of $[CoHL_4]$ with the strongly basic potassium hydride, yield η^3-allyl and η^3-methylallyl derivatives, $[Co(\eta^3\text{-all})L_3]$, with the appropriate allyl halide[52].

One of the first reactions of this type to be discovered was that between $[Mn(CO)_5]^-$ and allyl chloride, following the method used for preparing methylpentacarbonylmanganese. Decacarbonyldimanganese is reduced in tetrahydrofuran by 1% sodium amalgam and the solution of the anion treated with the allyl halide. η^1-Allylpentacarbonylmanganese **6** was isolated in high yield and decomposed to **7** by heating to 80°C[2] (equation 13).

$$[Mn_2(CO)_{10}] \xrightarrow[\text{thf } 25^\circ C/15\,h]{Na/Hg} [Mn(CO)_5]^- \xrightarrow{H_2C{=}CHCH_2Cl}$$

$$\underset{(\mathbf{6})}{[H_2C{=}CHCH_2Mn(CO)_5]} \xrightarrow[80^\circ C/12\,h\ (\text{or } h\nu)]{\Delta(-CO)} \underset{\substack{89\% \\ (\mathbf{7})}}{[(\eta^3\text{-}C_3H_5)Mn(CO)_4]} \qquad (13)$$

A similar reaction sequence starting from $NaRe(CO)_5$ and allyl chloride has been described[53].

The preparation and structures of η^3-allyl vanadium complexes have been investigated in some detail[54–56]. The appropriate salt and allyl halide are mixed in ether at room temperature and irradiated with ultraviolet light to convert the η^1-allyl formed initially into the η^3-allyl complex (equation 14). Complexes $[V(\eta^3\text{-allyl})(CO)_3(L{-}L')]$ where L—L′ is a bidentate ligand such as $Ph_2PCH_2CH_2PPh_2$, $Ph_2PCH_2CH_2AsPh_2$, or $(o\text{-}Me_2As)_2C_6H_4$ were also prepared.

Reductive cleavage of dinuclear cyclopentadienylmetal carbonyls, $[M(C_5H_5)(CO)_3]_2$ (M = Mo, W) or $[M(C_5H_5)(CO)_2]_2$ (M = Fe, Ru), with sodium amalgam yields strongly nucleophilic anions which react with allyl halides. The

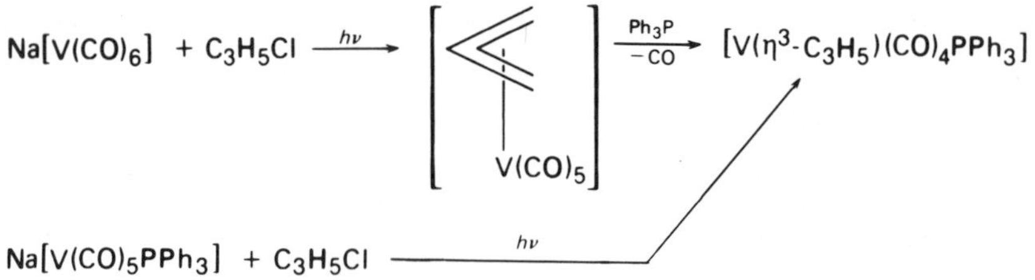

$$\mathrm{Na[V(CO)_6]} + \mathrm{C_3H_5Cl} \xrightarrow{h\nu} [(\eta^1\text{-}\mathrm{C_3H_5})\mathrm{V(CO)_5}] \xrightarrow[-\mathrm{CO}]{\mathrm{Ph_3P}} [\mathrm{V}(\eta^3\text{-}\mathrm{C_3H_5})(\mathrm{CO})_4\mathrm{PPh_3}] \qquad (14)$$

$$\mathrm{Na[V(CO)_5PPh_3]} + \mathrm{C_3H_5Cl} \xrightarrow{h\nu} [\mathrm{V}(\eta^3\text{-}\mathrm{C_3H_5})(\mathrm{CO})_4\mathrm{PPh_3}]$$

η^1-complexes formed initially are decomposed photolytically to give η^3-allyls[2,57]. The conformation of some of these compounds have been studied in solution by n.m.r. spectroscopy[58]. Irradiation of $[\mathrm{M}(\eta^1\text{-}\mathrm{CH_2Ph})(\mathrm{C_5H_5})(\mathrm{CO})_3]$ (M = Mo, W), affords the interesting η^3-benzyl complexes **8**[59], which show fluxional behaviour[60].

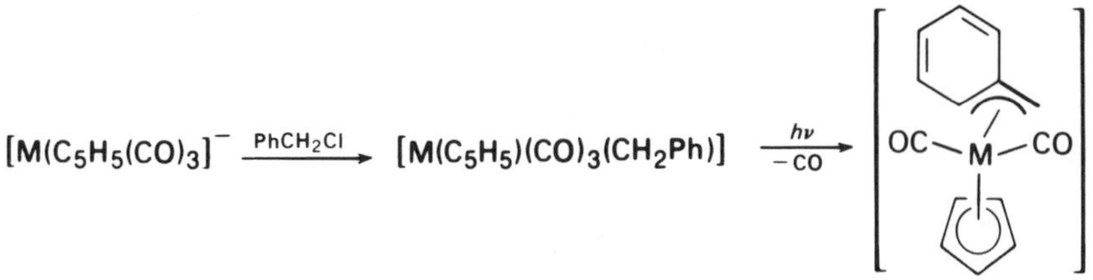

$$[\mathrm{M(C_5H_5(CO)_3}]^- \xrightarrow{\mathrm{PhCH_2Cl}} [\mathrm{M(C_5H_5)(CO)_3(CH_2Ph)}] \xrightarrow[-\mathrm{CO}]{h\nu} [\mathrm{M}(\eta^3\text{-}\mathrm{CH_2C_6H_5})(\mathrm{CO})_2(\eta^5\text{-}\mathrm{C_5H_5})] \qquad (15)$$

(**8**)

Anions $[\mathrm{M(CO)_5X}]^-$ (M = Mo, W), which are derived from the hexacarbonyl and tetraethylammonium halides, form triply halogeno-bridged anions **9** with allyl halides[61]. Under different conditions a mononuclear anion of tungsten **10** was isolated[62].

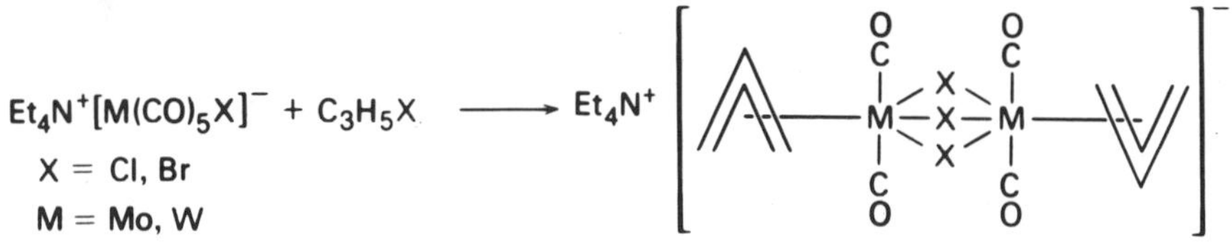

$$\mathrm{Et_4N^+[M(CO)_5X]^-} + \mathrm{C_3H_5X} \longrightarrow \mathrm{Et_4N^+}[(\eta^3\text{-}\mathrm{C_3H_5})(\mathrm{CO})_2\mathrm{M}(\mu\text{-}\mathrm{X})_3\mathrm{M}(\mathrm{CO})_2(\eta^3\text{-}\mathrm{C_3H_5})]^-$$

X = Cl, Br

M = Mo, W

(**9**) (16)

$$\mathrm{Et_4N^+[W(CO)_5Cl]^-} + \mathrm{C_3H_5Cl} \xrightarrow[\mathrm{CH_3CN}]{\text{benzene/}} \text{red solution} \xrightarrow{\mathrm{PPh_3}} \mathrm{Et_4N^+}[\mathrm{W}(\eta^3\text{-}\mathrm{C_3H_5})(\mathrm{CO})_2\mathrm{Cl_2}(\mathrm{PPh_3})]^-$$

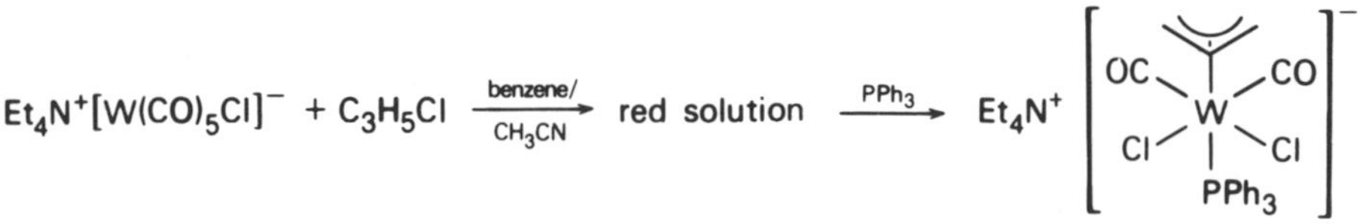

(**10**) (17)

Use of phase-transfer catalysis in the synthesis of η^3-allyl complexes leads to improved yields, and in some cases obviates the need for photolysis of an intermediate η^1-allyl[63]. For example, **7** was obtained from $\mathrm{Mn(CO)_5Cl}$ and allyl bromide in benzene or dichloromethane, using aqueous sodium hydroxide and benzyltriethylammonium chloride as catalyst. A 95% yield of $[\mathrm{Mo}(\eta^3\text{-}\mathrm{C_3H_5})(\eta^5\text{-}\mathrm{C_5H_5})(\mathrm{CO})_2]$ from $[\mathrm{Mo(C_5H_5)(CO)_3Cl}]$ was also reported.

Although sodium amalgam is the usual reagent for converting carbonyls into their anions, it can have the disadvantage of contaminating the product with small amounts of by-products which contain metal—mercury bonds. Other reagents which have been employed for generating carbonylate anions include lithium triethylborohydride in tetrahydrofuran[64] and potassium hydride in tetrahydrofuran or hexamethylphosphoramide[65].

B. Reactions of Complexes With M—Li, M—Mg, and M—Al Bonds

Some unusual inorganic Grignard, lithium, and organoaluminium reagents have been isolated from the reactions of $[M(C_5H_5)_2H_2]$ (M = Mo, W) with a main group organometallic reagent[66–71]. These compounds are strong nucleophiles and react *inter alia* with allyl halides. One example is given in equation 18.

$$[(C_5H_5)_2WH_2] \xrightarrow{n\text{-BuLi}} [\{(C_5H_5)_2W(H)Li\}_4] \xrightarrow{\text{allCl}} [(C_5H_5)_2W(CH_2CH{=}CH_2)_2] \xrightarrow{H^+} [(C_5H_5)_2W^{+}(\eta^3\text{-}C_3H_5)] + C_3H_6 \quad (18)$$

IV. ADDITION OF TRANSITION METAL HYDRIDES AND ALKYLS TO 1,3-DIENES

A. Direct Addition

The addition of transition metal hydrides to 1,3-dienes has been reviewed recently[72]. Two general mechanisms for these additions might be considered. In the first [path (a)], initial dissociation of a ligand L from the hydride is followed by coordination of olefin[73] (equation 19). Alternatively, [path (b)], 1,4-addition to the diene without prior coordination could occur, followed by loss of L[74] (equation 20).

$$\text{(a)}\quad HM_n \underset{+L}{\overset{-L}{\rightleftharpoons}} HML_{n-1} \xrightarrow{\text{diene}} H{-}ML_{n-1}(\text{diene}) \longrightarrow (\eta^3\text{-allyl})ML_{n-1} \quad (19)$$

$$\text{(b)}\quad H{-}ML_n + \text{diene} \longrightarrow H_3C{-}C{=}C{-}C(ML_n) \rightleftharpoons (\eta^3\text{-allyl})ML_{n-1} + L \quad (20)$$

TABLE 1. Hydrides which add to 1,3-dienes[a]

Labile ligand present: Stable η^1-intermediate not observed.

Square-planar: (4-coordinate) (d^8)

$[PtH(NO_3)(PPh_3)_2]$; $[PtH(PR_3)_2(acetone)]^+$; $[PtH(ClO_4)(PPh_3)_2]$

5-coordinate: (d^8)
$[CoH(CO)_4]$; $[NiH\{P(OR)_3\}_4]^+$; $[FeH(NO)(CO)_3]$; $[CoH(CO)_{4-n}(PBu_3)_n]$; $[CoH(N_2)(PPh_3)_3]$; $[CoH(PF_3)_4]$; $[RhH(CO)(PPh_3)_3]$; $[RhH(PF_3)_4]$; $[RhH(PPh_3)_4]$; $[IrH(CO)_2(PPh_3)_2]$

No labile ligand present: Stable η^1-intermediate observed.

6-coordinate: (d^6)
$[MnH(CO)_5]$; $[FeH(CO)_4SiCl_3]$; $[FeH(CO)_2(C_5H_5)]$; $[CoH(CN)_5]^{3-}$

[a]For relevant literature see ref. 72.

If path (b) operates, the η^1-allyl will be observed only if it is sufficient inert to ligand dissociation. Path (a) will be favoured if one ligand in the hydride itself is labile.

Somer hydrides which have been shown to add to 1,3-dienes are given in Table 1. They are classified into two main groups. In the first group, either the hydride itself, or the η^1-allyl complex which would be formed by addition to the diene, contains a ligand which is labile to dissociation. In these cases an intermediate η^1-allyl is observed only transiently, if at all. The second group consists of compounds which do not possess a labile ligand, so that an isolable η^1-allyl complex results[75]. This η^1-allyl is converted into the η^3-allyl by heating or, often more cleanly, by photolysis.

The 5-coordinate 'labile' hydrides normally yield the η^3-allyl complex directly. Tolman studied the addition of $[NiH\{P(OEt)_3\}_4]^+$ to a wide range of dienes[76]. He suggested that path (a) is followed, that is, dissociation of one phosphite ligand occurs to give a 16-electron coordinatively unsaturated intermediate, to which the diene then adds. Butadiene yields a mixture of *anti*- and *syn*-η^3-1-methylallyl products in the ratio 88:12, which is converted into the thermodynamically controlled mixture (5:95) by heating to 70°C. The preferential formation of the *anti*-isomer arises, it is suggested, by synchronous addition of the hydride $[NiHL_3]^+$ to the diene in a *cisoid* configuration (equation 21). If the diene adds in a *transoid* configuration, the *syn*-η^3-1-methylallyl complex results.

$[NiHL_4]^+$ ⇌ (−L / +L) $[NiHL_3]^+$ → $[H–NiL_3 \cdot \text{diene}]^+$ → $[H_3C\ \text{allyl}\ NiL_3]^+$; transoid: $[HNiL_3 \cdot \text{diene}]^+$ → $[H_3C\ \text{allyl}\ L_3Ni]^+$ (21)

Hydridotetracarbonylcobalt adds to 1,3-butadiene at 25°C to give a mixture of *syn* (35%) and *anti* (65%) isomers. The *anti*-isomer isomerizes to the thermodynamically more stable *syn*-form on heating. The additions of $[CoH(CO)_4]$ to many other dienes including isoprene, 2,3-dimethylbutadiene, cyclohexa-1,3-diene, and penta-1,3- and 1,4-dienes have also been studied on account of their relevance to industrially important hydroformylation reactions[74]. The usual mode of addition to isoprene yields an η^3-1,1-dimethylallyl derivative (as with $[CoH(CO)_4]$). $[RhH(PF_3)_4]$[77] and probably

$[CoH(PF_3)_4]$[78,79], however, afford η^3-1,2-dimethylallyl complexes. These may arise by isomerization of a 1,1-dimethylallyl complex, which is formed initially, through an intermediate diene metal hydride (equation 22).

$$RhH(PF_3)_4 \xrightarrow{\text{isoprene}} [(\eta^3\text{-}CH_2C(CH_3)C(H)CH_3)Rh(PF_3)_3] \rightleftharpoons [(\eta^4\text{-diene})RhH(PF_3)_3] \rightleftharpoons [(\eta^3\text{-}CH_2C(CH_3)CHCH_3)Rh(PF_3)_3] \qquad (22)$$

The square-planar platinum hydrides $[PtHYL_2]$ ($Y = NO_3$ or ClO_4) and $[PtHQ(PPh_3)_2]^+PF_6^-$ (Q = acetone, methanol) all possess one ligand which is a good leaving group in the presence of an incoming diene[80–82]. On treatment with buta-1,3-diene a cationic η^3-1-methylallyl derivative is formed (equation 23). Allene gives the corresponding η^3-allyl complex (equation 24).

$$trans\text{-}[PtH(NO_3)(PEt_3)_2] + CH_2{=}CHCH{=}CH_2 \xrightarrow[PF_6^-]{\text{then}} [(\eta^3\text{-}C_3H_4CH_3)Pt(PEt_3)_2]^+ PF_6^- \qquad (23)$$

$$trans\text{-}[PtH(acetone)(PPh_3)_2]^+ PF_6^- + H_2C{=}C{=}CH_2 \longrightarrow [(\eta^3\text{-}C_3H_5)Pt(PPh_3)_2]^+ PF_6^- \qquad (24)$$

The ligands in the octahedral hydrides (R = H) $[MnR(CO)_5]$, $[FeR(C_5H_5)(CO)_2]$, $[FeR(CO)_4SiCl_3]$, and $[CoR(CN)_5]^{3-}$ or in their derivatives (R = η^1-allyl) are much less labile to substitution than those in the 4- or 5-coordinate complexes discussed above. When they react with dienes, therefore, an η^1-allyl intermediate is observed, and can commonly be isolated. Thus, the initial product from the reaction between buta-1,3-diene and $[MnH(CO)_5]$ is an 85:15 mixture of *cis*- **11** and *trans*- **12** geometrical isomers (equation 25).

Thermal or photolytic decomposition of the mixture yields the *syn*-1-methylallyl tetracarbonyl, **13**. A synchronous 1,4-addition of hydride to *cis*-butadiene is proposed to account for the major product **11**. This is supported by deuterium labelling studies of the reaction with isoprene[83,84] (equation 26).

The substitution of a carbonyl ligand in the octahedral complexes $[MX(CO)_5]$ (M = Mn, Re) (X = Cl, Br, $GePh_3$, etc.) proceeds by a dissociative mechanism in which a carbonyl ligand *cis* to the substituent X is lost[85,86]. The formation of **13** from **11** and **12** may take place similarly. In the presence of $[Ir(dppe)_2Cl]$, a strong CO acceptor, the η^1-complex is converted into the η^3-allyl even at room temperature. Under these conditions the reaction proceeds with retention of configuration of the 1-methylallyl ligands: the product contains the same ratio of *anti* to *syn*-methyl groups (85:15) as the *cis* to *trans* ratio in the precursor[84].

When the hydride $[CoH(CN)_5]^{3-}$ is treated with buta-1,3-diene in aqueous media the η^1-1-methylallyl derivative $[Co(CN)_5\text{-}CH_2CH{=}CHCH_3]^{3-}$ is produced. An equilibrium between this and η^3-complex $[Co(CN)_4(\eta^3\text{-}CH_2CHCHCH_3)]^{2-}$ is gradually set up and can be reversed by addition of cyanide. The η^3-allyl compound

$[MnH(CO)_5]$ + butadiene → $[(OC)_5Mn{-}CH_2CH{=}CHCH_3]$ (11) *cis-* 85% + *trans*-isomer 15% (12)

(11) —$h\nu$ or Δ, $-CO$→ $[Mn(\eta^3\text{-}C_4H_7)(CO)_4]$ (13)

(11) —$[Ir(dppe)_2Cl]$→ $[Mn(\eta^3\text{-}C_4H_7)(CO)_4]$ 85%; (12) —$[Ir(dppe)_2Cl]$→ $[Mn(\eta^3\text{-}C_4H_7)(CO)_4]$ 15% (25)

$[MnH(CO)_5] + D_2C{=}C(CD_3)CH{=}CH_2$ → $(D_3C)(D_2C{=})C{-}CH{=}CH_2 \cdots H{-}Mn(CO)_5$ → $(D_3C)(D_2CH)C{=}CH{-}CH_2Mn(CO)_5$ —$[Ir(dppe)_2Cl]$, $-CO$→ $[Mn(\eta^3\text{-}CH_2CHC(CD_3)(CD_2H))(CO)_4]$ (26)

could not be isolated pure from this system but it has been obtained from $[Co(\eta^3\text{-}C_4H_7)(CO)_3]$ by the method shown in equation 27[87].

$$[Co(\eta^3\text{-}C_4H_7)(CO)_3] + I_2 + 4KCN \longrightarrow K_2[Co(\eta^3\text{-}C_4H_7)(CN)_4] + 3CO + 2KI \quad (27)$$

Cobalt cyanides are catalysts for the monohydrogenation of 1,3-dienes to alkenes[88].

The reaction of $[VH(CO)_6]$ with butadiene at 20°C under ultraviolet irradiation provides an alternative method of preparation of $[V(\eta^3\text{-}CH_2CHCHCH_3)(CO)_5]$ (see Section IV.A)[54].

Maitlis and his group, in the course of their detailed work on pentamethylcyclopentadienyl complexes of rhodium and iridium[89], studied the reactions of $[M(C_5Me_5)Cl_2]_2$ (**14**) (M = Rh, Ir) with olefins and dienes in the presence of an alcohol and a base, usually sodium carbonate. Two reactions were identified, the first leading to η^3-allyl complexes, the second to η^4-diene derivatives. Thus, **16** was obtained from buta-1,3-diene while cyclohexa-1,3-diene gave $[M(C_5Me_5)(\eta^4\text{-}C_6H_8)]$. The reaction proceeds via a bridged chlorohydrido complex (**15**), which could be prepared by reduction of **14** (equation 28).

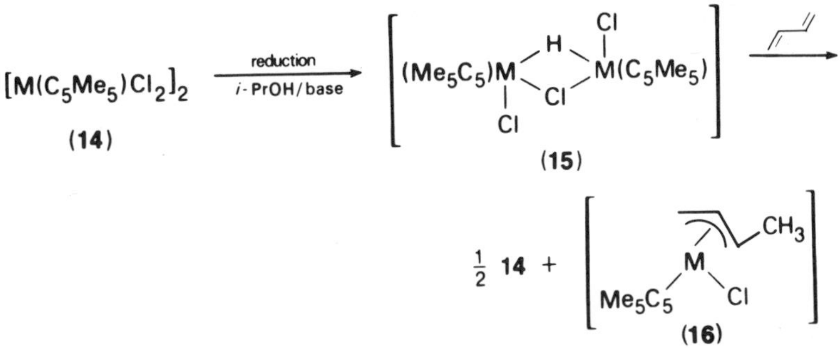

$$[M(C_5Me_5)Cl_2]_2 \ (\mathbf{14}) \xrightarrow[i\text{-PrOH/base}]{\text{reduction}} [(Me_5C_5)MCl(\mu\text{-H})(\mu\text{-Cl})MCl(C_5Me_5)] \ (\mathbf{15}) \longrightarrow \tfrac{1}{2}\,\mathbf{14} + [(\eta^3\text{-}CH_2CHCHCH_3)M(C_5Me_5)Cl] \ (\mathbf{16}) \qquad (28)$$

Some of the η^3-allylic compounds eliminate hydrogen chloride, either spontaneously or under forcing conditions. Those with 1-*anti*-alkyl substituents lose HCl so rapidly that they can barely be detected even in the absence of base. The mechanism shown in equation 29 is suggested[90]. This is essentially the reverse of the mechanism proposed for addition of hydrogen chloride to η^4-diene complexes.

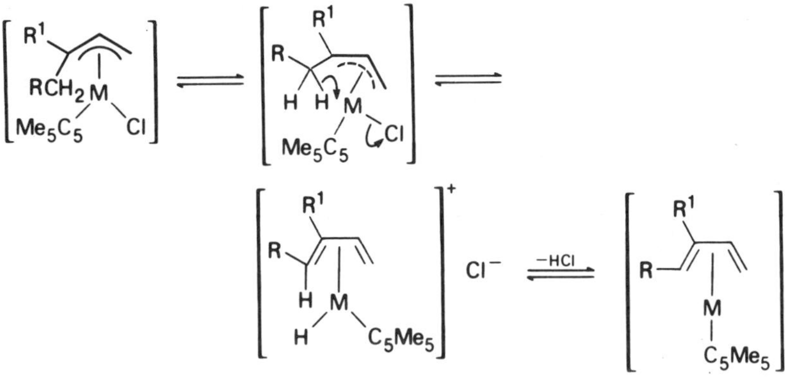

$$[(R^1,RCH_2\text{-allyl})M(C_5Me_5)Cl] \rightleftharpoons [\ldots] \rightleftharpoons [(R^1,R\text{-diene})MH_2(C_5Me_5)]^+Cl^- \underset{}{\overset{-HCl}{\rightleftharpoons}} [(R^1,R\text{-diene})M(C_5Me_5)] \qquad (29)$$

A study of several dienes showed that isomers with *syn*-1- or -3-alkyl substituents were always obtained. It was concluded that the diene is in the *transoid* form when it reacts with the hydride, in contrast to additions by $[NiH\{P(OEt)_3\}_4]^+$ or $[MnH(CO)_5]$. On steric grounds a 1,2-addition of the rhodium hydride to the diene to give the *syn*-isomer directly is proposed (equation 30).

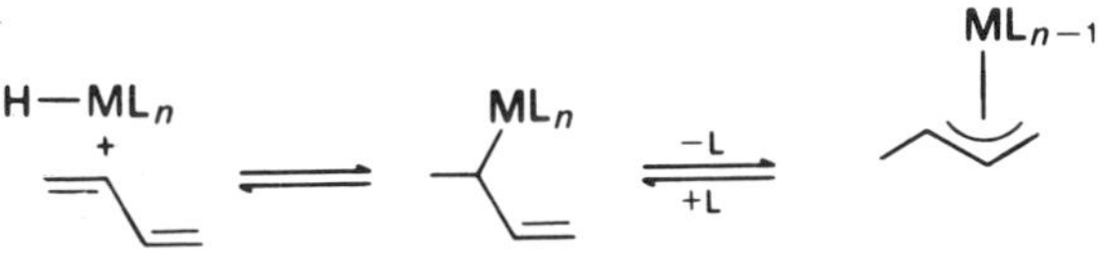

$$H{-}ML_n + \text{diene} \rightleftharpoons (\eta^1\text{-allyl})ML_n \underset{+L}{\overset{-L}{\rightleftharpoons}} (\eta^3\text{-allyl})ML_{n-1} \qquad (30)$$

B. The Isopropylmagnesium Bromide Method

When a halogeno complex of a transition metal is treated with isopropylmagnesium bromide at low temperature, an η^1-isopropyl complex is probably formed initially. This is susceptible to β-elimination, yielding propene and a metal hydride[91]. In principle the hydride can then add to a diene to give an allyl complex.

Treatment of $[Ti(C_5H_5)_2Cl_2]$ with two molar proportions of *i*-PrMgBr and one of diene yields **17**, possibly as shown in equation 31[92,93]. Other examples of its use include

$$[Ti(C_5H_5)_2Cl] \xrightarrow{i\text{-PrMgBr}} [Ti(C_5H_5)_2(\eta^1\text{-Pr-}i)] \longrightarrow$$

$$[Ti(C_5H_5)_2H] \longrightarrow [(C_5H_5)_2Ti(\eta^3\text{-}1,3\text{-dimethylallyl})] \quad (\mathbf{17}) \qquad (31)$$

the synthesis of $[Nb(C_5H_5)_2(\eta^3\text{-}C_8H_9)]$ from $[Nb(C_5H_5)_2Cl_2]$ and cyclooctatetraene[94], and in a series of rhodium complexes[95] (equation 32).

$$[Rh(cod)Cl]_2 + \text{cyclohexa-1,3-diene} \xrightarrow[h\nu]{i\text{-PrMgBr}} [(cod)Rh\text{—}(\eta^3\text{-cyclohexenyl})] \qquad (32)$$

Analogous η^3-cycloheptenyl and -cyclooctenyl complexes were obtained from cycloheptadiene and cyclooctadiene and butadiene and isoprene yielded the *syn*-1-methylallyl and 1,2-dimethylallyl derivatives, respectively[96]. Irradiation of the reaction mixture was found to improve yields.

The isopropylmagnesium bromide method has been extensively used, especially by Müller and coworkers[97,98] and by the Gröningen group[99], for the synthesis of compounds containing hydrocarbon ligands. It is by no means restricted to the preparation of η^3-allyl complexes (see, for example, Chapter 7, Section II.G.5).

C. Additions of Alkyls and Aryls

There are a few examples of the insertion of 1,3-dienes into metal carbon η^1-bonds. These are covered in a recent review[72]. The η^1-alkyl and aryl complexes of manganese, iron, and cobalt carbonyls are considered to react via the acyl derivative[100]. For example, phenylpentacarbonylmanganese adds to *trans*-1,3-pentadiene to yield **18**. On heating, **18** is converted into the novel oxopentadienyl complex **19**[101].

$$[PhMn(CO)_5] + H_3C\text{—CH=CH—CH=CH}_2 \longrightarrow$$

$$[H_3C\text{—}CH_2\text{—}(\eta^3\text{-allyl})\text{—}COPh]Mn(CO)_4 \ (\mathbf{18}) \longrightarrow [(\eta^5\text{-Ph—C(O)—CH—CH—CH—}CH_2CH_3)Mn(CO)_3] \ (\mathbf{19}) \qquad (33)$$

Insertions of dienes into metal—η^3-allyl bonds are considered in Section VIII.C, and addition of metal—carbon bonds to allenes in Section VIII.D.

V. ELECTROPHILIC ATTACK ON COORDINATED DIENES

A. Protonation

The protonation of η^4-diene complexes should provide a synthetic route to η^3-allyl derivatives, but it suffers from the disadvantage that a reduction of 2 in the formal electron number of the metal occurs if a non-coordinating acid is used for the protonation. The resulting 16-electron complex therefore may be labile and not readily isolated. Proton addition to dienetricarbonyliron complexes has been studied by n.m.r spectroscopy, but conflicting and confusing results have been reported. The position has now apparently been clarified[102].

Hydrogen chloride adds to butadienetricarbonyliron (**20**) at room temperature to give covalent chloro(*syn*-1-methylallyl)tricarbonyliron[103]. Whitesides and Arhart showed that addition of DCl to 1-phenyl-3-methylbutadienetricarbonyliron (**21**) at $-78°C$ occurs stereospecifically to give **22**, in which the deuterium is entirely in the *anti* substituent[104].

Ph—(diene)—$Fe(CO)_3$ (**21**) $\xrightarrow[-78°C]{DCl}$ Ph—(allyl, CH_2D)—$Fe(CO)_3Cl$ (**22**) (34)

It is likely, therefore, that in the reaction of **20** the *anti*-1-methylallyl derivatives is formed initially but then rearranges to the *syn*-isomer.

With acids containing weakly coordinating anions the situation is more complicated. Trifluoroacetic acid gives a covalent trifluoroacetate in solution, but after addition of HBF_4 in acetic anhydride the allyltetracarbonyliron salt **24** could be isolated[105]. This species must arise by capture of CO by the 16-electron allyltricarbonyliron cation. Better yields of **24** are obtained by conducting the reaction under an atmosphere of carbon monoxide[106]. The allyltetracarbonyliron tetrafluoroborates are yellow crystalline salts, fairly stable thermally and in dry air.

In CF_3COOD the terminal protons of the complexed diene are rapidly and completely exchanged. Studies of H/D exchange in (η^4-cyclohexadiene)tricarbonyliron[104] and (η^4-cyclohexadiene)cyclopentadienylrhodium and -iridium[107] in CF_3COOD suggest that protonation occurs stereospecifically *endo*. A metal hydride intermediate is likely to be involved. Such intermediates have been observed by n.m.r. measurements at $-80°C$ in the system HSO_3F/SO_2 and at 0°C in CF_3COOH/HBF_4[102,108].

Some reactions of **20** with acids are summarised in equation 35.

In the carbonyl series, it has not proved possible to isolate directly protonated species such as **23** with weakly coordinating anions. Recently, however, Ittel and co-workers found that the phosphite complexes $[Fe(diene)\{P(OR)_3\}_3]$ are very readily protonated even by such weak acids as the ammonium ion[109]. Whereas $[Fe(diene)(CO)_3]$ requires HBF_4 in acetic anhydride ($pK_a \approx -6$), $[Fe(diene)\{P(OMe)_3\}]$ is protonated by ammonium hexafluorophosphate in methanol ($pK_a \approx 9$) and $[Fe(diene)(PMe_3)_3]$ reacts even with methanol itself ($pK_a \approx 16$)[110]. The resulting salts $[Fe(\eta^3\text{-allyl})\{P(OR)_3\}_3]^+BF_4^-$ can be isolated as air-stable crystalline materials[109,111].

The apparent 16-electron configuration of the iron atom is only nominal. N.m.r. studies reveal a bonding interaction between iron and a hydrogen atom attached to

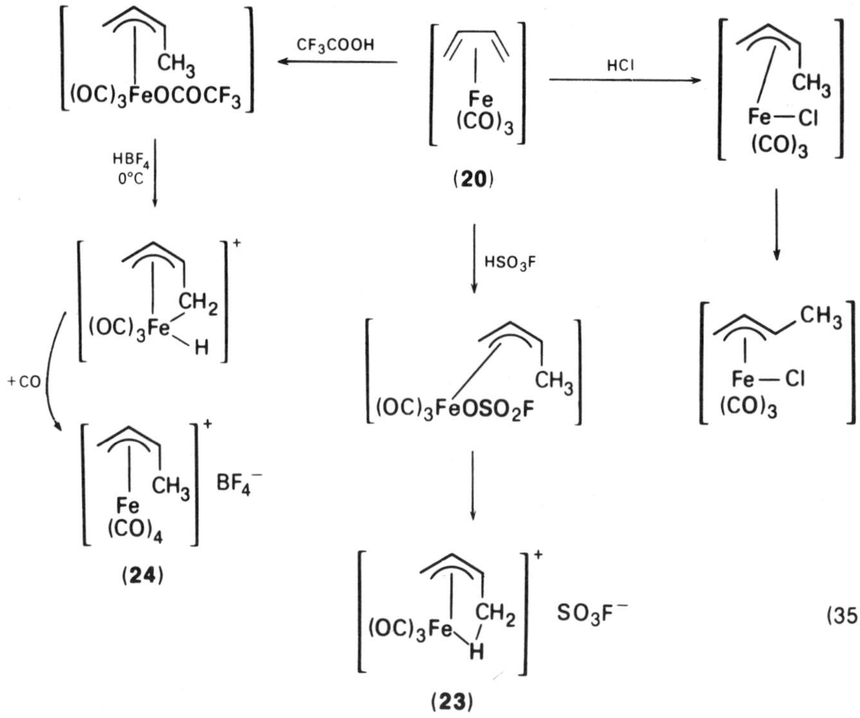

(20) (24) (23) (35)

carbon adjacent to the allylic group. This interaction has been confirmed by a neutron diffraction study of the complex $[Fe(\eta^3\text{-}C_8H_{13})\{P(OMe)_3\}_3]^+BF_4^-$ **25** which is prepared by protonation of the η^4-1,3-cyclooctadiene derivative[112].

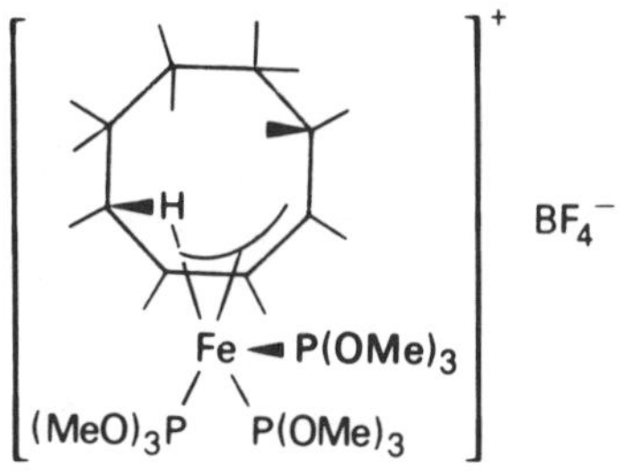

(25)

Protonation of η^4-dienonetricarbonyliron complexes occurs at oxygen to give η^5-hydroxypentadienyltricarbonyliron cations via intermediate *trans*-ions[113–115]. The complexes **26** are protonated at carbon affording η^3-allyl derivatives (**27**). The infrared spectrum of **27** indicates that the acyl CO group is coordinated to the metal. In this way an 18-electron configuration is achieved in the presence of a weakly coordinating anion, PF_6^- [116].

$$[\text{Ph}-\text{allyl}-\text{CH}_2\text{C(=O)Ph}\;\text{RhX}(C_5H_5)] \xleftarrow[(X = Cl, Br)]{HX} [\text{Ph}-\text{diene}-\text{C(=O)Ph}\;\text{Rh}(C_5H_5)]\;(\mathbf{26}) \xrightarrow{HPF_6} [\text{Ph}-\text{allyl}-\text{CH}-\text{C(Ph)=O}\rightarrow\text{Rh}(C_5H_5)]^+ PF_6^-\;(\mathbf{27}) \quad (36)$$

B. Friedel–Crafts Acylation

Friedel–Crafts acylation of dienetricarbonyliron complexes with $RCOCl/AlCl_3$ proceeds via an intermediate, **28**, which has a structure analogous to that of **27**[117,118]. This shows that, as with protonation, attack of the RCO^+ electrophile occurs *endo* to the organic ligand[119]. Hydrolysis of **28** gives mainly the *anti*-acetyl isomer **29**.

$$[\text{diene}\,Fe(CO)_3] \xrightarrow[AlCl_3]{MeCOCl} [\text{allyl}\,Fe(CO)_3\leftarrow O{=}C(Me)]^+ AlCl_4^-\;(\mathbf{28}) \xrightarrow{H_2O} [\text{diene-COMe}\,Fe(CO)_3]\;(\mathbf{29}) \quad (37)$$

Acylation of tri- or tetracarbonyliron complexes of unsaturated aldehydes or ketones leads to similar products (**30**)[120].

$$[RC(O)CH{=}CHR'\,Fe(CO)_4] \xrightarrow{MeCO^+\cdot BF_4^-} [R(OC(O)Me)\text{-allyl-}R'\,Fe(CO)_3]^+ BF_4^-\;(\mathbf{30}) \xleftarrow{MeCO^+\cdot BF_4^-} [RC(O)CH{=}CHR'\,Fe(CO)_3] \quad (38)$$

C. Addition of Fluoroolefins

The addition of fluoroolefins to η^4-diene complexes has been studied in detail[121–125]. A typical example is the reaction which occurs when butadienetricarbonyliron and tetrafluoroethene in hexane are irradiated by ultraviolet light (equation 39)[121]. Similar reactions have been observed between $[Fe(diene)(CO)_3]$ (diene = cyclobutadienes,

$$\left[\text{Fe}(CO)_3 \text{ diene}\right] \xrightarrow[h\nu]{C_2F_4} \left[\text{Fe}(CO)_3\text{-}CH_2CF_2CF_2 \text{ allyl}\right] \quad (39)$$

C_4R_4 where R = H or Me, isoprene, 2,3-dimethylbutadiene, *trans*-1,3-pentadiene) and a variety of substrates including $CF_2{=}CF_2$, $CF_3CF{=}CF_2$, $(CF_3)_2CO$, $CF_2{=}CFH$ and $CF_2{=}CFCl$[122–125]. In common with the addition of other electrophiles such as H^+ or CH_3CO^+, *endo* attack of the fluoroolefin is observed[124].

VI. OXIDATIVE ADDITION OF ALLYL HALIDES

A. Introduction

The oxidative addition reaction

$$[L_nM] + XY \longrightarrow [L_nMXY] \quad (40)$$

requires (a) non-bonding electron density on the metal M, (b) two vacant coordination sites on the complex to permit formation of bonds X and Y, and (c) a metal with its oxidation states separated by two units. The addition of an allyl halide can yield initially an η^1-allyl complex, from which a ligand must be lost for conversion into the η^3-allyl derivative. Thus the 18-electron d^{10} complex bis(cyclooctadiene)platinum(0) affords **31**, from which **32** can be obtained by treatment with $AgPF_6$[126].

$$[Pt(cod)_2] + C_3H_5X \longrightarrow \underset{\textbf{(31)}}{[PtX(\eta^1\text{-}C_3H_5)(cod)]} \xrightarrow{AgPF_6} \underset{\textbf{(32)}}{[Pt(\eta^3\text{-}C_3H_5)(cod)]PF_6^-} \quad (41)$$

Oxidative addition to an 18-electron complex requires prior dissociation of a 2-electron ligand. The d^{10} platinum(0) compounds $[Pt(PR_3)_4]$ provide good illustrations (see Section VI.D.2).

B. Reactions of Allyl Halides With Metal Complexes

The addition of allyl halides to iron carbonyls provided one of the earliest routes to η^3-allyl complexes[8]. Both thermal[127] and photochemical methods have been described. These reactions proceed by addition to the coordinatively unsaturated species $Fe(CO)_4$ to yield η^3-allyltricarbonyliron halides $[FeX(\eta^3\text{-}C_3H_5)(CO)_3]$ (X = Cl, Br, I). These compounds consist of a mixture of two isomers in solution, the equilibria between which have been studied by n.m.r. spectroscopy[128].

The claim that $Fe_2(CO)_9$ reacts with disilanes to give η^3-1-silapropenyl complexes (**33**) has recently been shown to be incorrect.[129] The products are in fact η^2-vinylsilane complexes[129a].

Pentakis(trimethyl phosphite)iron, unlike $Fe(CO)_5$, yields ionic $[Fe(\eta^3\text{-}C_3H_5)\{P(OMe)_3\}_4]^+Br^-$ on reaction with allyl bromide[130]. $[Fe(PF_3)_5]$, however, gives $[FeX(\eta^3\text{-}C_3H_5)(PF_3)_3]$ (X = Br, I) on irradiation with the appropriate halide in solution in diethyl ether[131]. $[Ru_3(CO)_{12}]$ and allyl bromide afford the complex $[RuBr(\eta^3\text{-}C_3H_5)(CO)_3]$ in 97% yield on heating at 60–70°C in isooctane[132].

$$[Fe_2(CO)_9] + RMe_2Si{-}SiMe_2CH{=}CH_2 \longrightarrow [Fe(CO)_4(\eta^2\text{-}H_2C{=}CH{-}SiMe_2R)]$$

R = Me, vinyl (42)

$$\left[HC(\text{—SiMe}_2)(\text{—CH}_2)\text{—Fe(CO)}_3\text{SiMe}_2R \right]$$ **(33)**

Reduction of $[FeI(\eta^3\text{-}C_3H_5)(CO)_3]$ yields allyltricarbonyliron, which exists as a monomer/dimer equilibrium mixture in solution[133]. The monomer is paramagnetic and its e.s.r. spectrum has been studied between −90 and +40°C[134]. The enthalpy of dissociation of the dimer is 55.0 kJ mol in pentane. An X-ray diffraction study shows a very long Fe—Fe bond (314 pm)[135].

A dark blue–green monomeric 17-electron complex, $[Fe(\eta^3\text{-}C_8H_{13})\{P(OMe)_3\}_3]$, is similarly obtained by reduction of **25**[136].

Oxidative addition of allyl halides to tetracarbonylnickel yields allylnickel halides (equation 43), which are important catalysts for olefin reactions. This aspect of their chemistry has been reviewed[9,14,15,137], and their application in organic synthesis is covered by a chapter in 'Organic Reactions' which includes detailed procedures for their handling and isolation[138].

$$[Ni(CO)_4] + C_3H_5X \longrightarrow [(\eta^3\text{-}C_3H_5)Ni(\mu\text{-}X)_2Ni(\eta^3\text{-}C_3H_5)] \quad (43)$$

Cyclopentadienyldicarbonylcobalt (**34**) is very susceptible to addition of allyl halides. Depending on the halide and the solvent, either covalent $[Co(all)(C_5H_5)X]$ or ionic $[Co(all)(C_5H_5)(CO)]^+X^-$ is obtained, or a mixture of both[139–141]. The bromides allBr (all = C_3H_5, 1-MeC_3H_4, 2-MeC_3H_4) in a non-polar solvent yield ionic species preferentially, which can conveniently be isolated as air-stable hexafluorophosphates. Allyl iodides in tetrahydrofuran or in light petroleum give a mixture. Phosphine-substituted allyl complexes are obtained as shown in equation 44[193]. Similar reactions with $[Co(C_5Me_4Et)(CO)_2]$ have been reported[142].

$$[Co(C_5H_5)(CO)_2] \xrightarrow{PR_3} [Co(C_5H_5)(PR_3)(CO)] \xrightarrow[\text{light petroleum}]{C_3H_5Br} [Co(C_5H_5)(PR_3)(\eta^3\text{-}C_3H_5)]^+ Br^- \quad (44)$$

(34)

Dicobalt octacarbonyl is not usually a useful starting material for the synthesis of allyls in one step. Initial conversion into the anion (see Section III.A) or the hydride (see Section IV.A) is usually necessary. Triphenylcyclopropenium tetrafluoroborate, however, gives an η^3-cyclopropenium complex directly (equation 45)[143].

The addition of allyl halides to the hexacarbonyls of chromium, molybdenum and tungsten has been studied in acetonitrile solution. $[M(CH_3CN)_3(CO)_3]$ is probably formed initially. No addition product could be isolated for M = Cr, but with M = Mo or W high yields of $[M(all)(CH_3CN)_2(CO)_2X]$ were obtained[144]. In the absence of any structure determination by X-ray crystallography these complexes have usually been written with the allyl and halide ligands mutually *trans*. Recent work suggests that the

$$Ph_3C_3^+ BF_4^- + [Co_2(CO)_8] \longrightarrow [(\eta^3\text{-}C_3Ph_3)Co(CO)_3] \quad (45)$$

correct formulation may be *cis*[62]. The acetonitrile ligands are readily displaced by bidentate ligands such as 2,2′-bipyridyl, *o*-phenanthroline or $Ph_2PCH_2CH_2PPh_2$[145]. Tri-*n*-butylphosphine displaces allyl bromide, however, to give Friedel's complex **(35)**[146], in which the acetonitrile ligands are labile, being replaced even by butadiene or cyclooctatetraene[147].

$$[Mo(CO)_6] \xrightarrow{CH_3CN} [Mo(CH_3CN)_3(CO)_3] \xrightarrow{C_3H_5Br} [Mo(CH_3CN)_2(C_3H_5)(CO)_2Br]$$

$$[Mo(CH_3CN)_2(C_3H_5)(CO)_2Br] \xrightarrow{Bu_3P} [Mo(CH_3CN)_2(PBu_3)_2(CO)_2] \quad \mathbf{35}$$

$$[Mo(CH_3CN)_2(C_3H_5)(CO)_2Br] \xrightarrow{bipy} [Mo(bipy)(C_3H_5)(CO)_2Br]$$

$$[Mo(CO)_6] \xrightarrow{bipy} [Mo(bipy)(CO)_4] \xrightarrow[\text{(ref. 148)}]{C_3H_5Br} [Mo(bipy)(C_3H_5)(CO)_2Br] \quad (46)$$

$$[Mo(bipy)(C_3H_5)(CO)_2Br] \xrightarrow[\text{(ref. 149)}]{AgBF_4/L = NH_3, py, PPh_3} [Mo(bipy)(C_3H_5)(CO)_2L]^+ BF_4^-$$

No allyl complexes were isolated from the reaction of $[Mo(CO)_6]$ and allyl halides in cyclohexane. When $[W(CO)_6]$ was irradiated in this solvent for 100 h, and then left for several weeks, $[W_2(CO)_6(C_3H_5)Cl_3]$ (44%) or $[W(CO)_4(\eta^3\text{-}C_3H_5)X]$ (X = Br, 64%; X = I, 67% were isolated. Further reaction with 2,2′-bipyridyl gave $[W(CO)_2(\eta^3\text{-}C_3H_5)(bipy)X]$, and with triphenylphosphine or pyridine, $[W(CO)_2(\eta^3\text{-}C_3H_5)L_2X]$[150].

Various cationic η^3-allyl carbonyl nitrosyl derivatives of molybdenum and tungsten **(37)** have been obtained from neutral precursors **(36)** by treatment with allyl bromides in the presence of silver salts (equation 47)[151,152]. A more general route is from **38** and

$$[Mo(C_5H_5)(CO)_2NO] \xrightarrow{RC_3H_4Br/AgPF_6} [Mo(C_5H_5)(C_3H_4R)(CO)(NO)]^+PF_6^- \; \mathbf{(37)}$$

$$\underset{\mathbf{(38)}}{[Mo(C_5H_5)(CO)_2(C_3H_4R)]} \xrightarrow{NO^+PF_6^-} \mathbf{(37)} \quad (47)$$

$NO^+PF_6^-$ [153]. Complexes **37** and **38** show conformational equilibria in solution. The stereochemistry has been established from an n.m.r. study of the indenyl complexes. The indenyl ligand adopts a preferred conformation in which the C_6 ring lies above the allyl ligand (equation 48)[154].

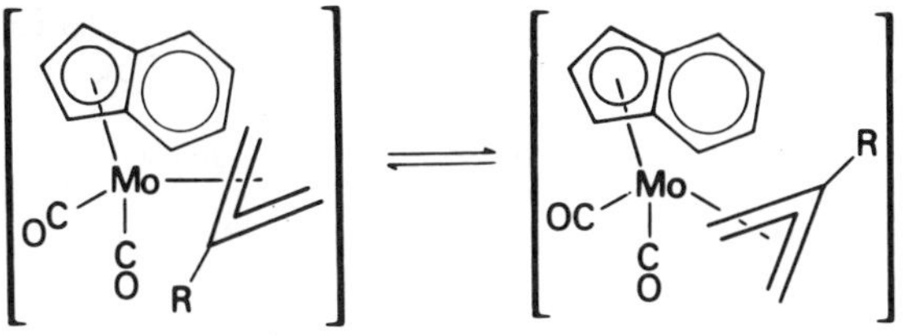

(48)

endo ⇌ *exo*

C. Direct Reaction Between Allyl Halides and Metals

The direct reaction between allyl halides and metals could, in principle, provide a route to allyl metal halides. Fischer and Bürger found that allyl bromide and finely divided metallic palladium afforded $[Pd(\eta^3\text{-}C_3H_5)Br]_2$ in 56% isolated yield[155]. Allyl chloride and allyl alcohol did not appear to react, while allyl iodide gave a red solution which decomposed to a palladium mirror. There was no reaction, however, between nickel or platinum and allyl halides.

Combination of bulk transition metals with organic halides may be disfavoured thermodynamically, especially if the organometallic products are endothermic. There is also likely to be an appreciable kinetic barrier to their interaction which could necessitate heating to temperatures at which the products would decompose. The metal atom technique is thus useful for preparing gram amounts of complexes which are thermally labile or otherwise difficult to obtain. The metal is vapourized under vacuum either thermally or with an electron gun and co-condensed with the reactant(s) on the walls of a vessel cooled to $-196^\circ C$. On allowing the condensate to warm up gradually, the metal atoms, which have a high energy content, combine with the other reactant(s). Difficulties lie not so much in effecting a reaction, but in working up and isolating products which may be thermally labile. The method can be scaled up to produce up to 50 g of organometallic compound in a single run. The technique is described more fully in Chapter 13 and in recent reviews[156,157].

Nickel, palladium, or platinum atoms, co-condensed with allyl halides, afford $[Ni(\eta^3\text{-}C_3H_5)X]_2$, $[Pd(\eta^3\text{-}C_3H_5)X]_2$ (X = Cl, Br), and $[Pt(\eta^3\text{-}C_3H_5)Cl]_4$, respectively[156]. Possibly of more interest is the synthesis of η^3-benzyl complexes of palladium (**39**) which are not otherwise accessible[158,159]. Co-condensation of PF_3 and propene with cobalt or iron atoms affords $[M(\eta^3\text{-}C_3H_5)(PF_3)_3]$ (M = Co, Fe)[156].

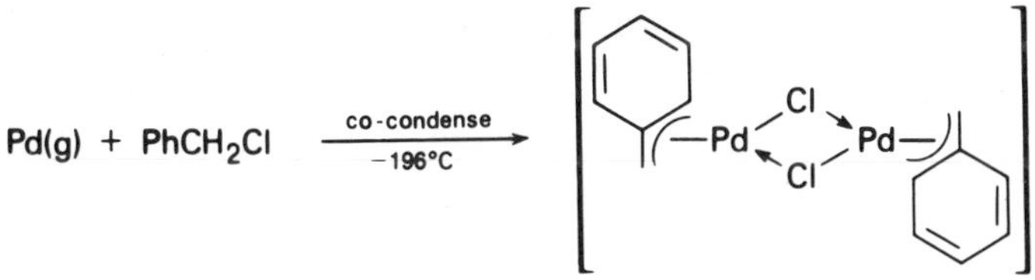

$$Pd(g) + PhCH_2Cl \xrightarrow[-196^\circ C]{\text{co-condense}} [(\eta^3\text{-}C_6H_5CH_2)Pd(\mu\text{-}Cl)_2Pd(\eta^3\text{-}CH_2C_6H_5)] \qquad (49)$$

(39)

Another approach to the production of highly reactive metals is to reduce a salt with an alkali metal in an organic solvent such as 1,2-dimethoxyethane, preferably in the presence of an electron carrier such as naphthalene[160]. Solutions of allylnickel halides have been prepared in this way[161].

Reduction of metal salts either electrolytically[162] or with zinc in the presence of suitable ligands has also been used successfully. Allylnickel halides result from conjugated dienes[163]. Cobalt(II) chloride yields complexes of the type **41**.

$$NiX_2 + CH_2{=}CH{-}CH{=}CH_2 \xrightarrow{Zn} [Ni(\eta^3\text{-}C_4H_7)X]_2 \quad \textbf{(40)} \qquad (50)$$

$$NiCl_2 + Ph_3P + CH_2{=}CH{-}CH{=}CH_2 \xrightarrow{Zn} [(\eta^3\text{-}CH_3C_3H_4){-}Ni(PPh_3)]ZnCl_3 \xrightarrow{2\,py} [(\eta^3\text{-}CH_3C_3H_4)Ni(Cl)(PPh_3)] \qquad (51)$$

$$CoCl_2 + PPh_3 + CH_2{=}CH{-}CH{=}CH_2 \xrightarrow{Zn} [(\eta^3\text{-}CH_3C_3H_4)Co(\eta^4\text{-}C_4H_6)(PPh_3)] \quad \textbf{(41)} \qquad (52)$$

The *anti*-butenyl isomer **41** is formed initially and is stable at room temperature. It is converted into the *syn* isomer by treatment with ligands such as pyridine or triphenylphosphine[164].

D. Other Oxidative Additions

1. To d^{10} complexes of Ni(0), Pd(0), and Pt(0)

Bis(cycloocta-1,5-diene)nickel reacts with allyl halides with displacement of both the chelating olefin groups to yield allylnickel halides (**40**). This provides a convenient method of preparing these complexes which avoids the use of the highly toxic tetracarbonylnickel[138]. [Ni(cod)$_2$] can itself be obtained by reduction of nickel salts in the presence of cycloocta-1,5-diene by triethylaluminium[165] or more easily by manganese powder[166]. Bis(cycloocta-1,5-diene)platinum[167,168] adds allyl to afford η^1-allyl halides, which form η^3-allyl complexes on treatment with silver ion[126], pyridine[169], or TlC$_2$H$_5$[170] (equation 53).

$$Pt(cod)_2 + allX \longrightarrow PtX(\eta^1\text{-}all)(cod) \begin{cases} \xrightarrow{AgPF_6} [Pt(\eta^3\text{-}all)(cod)]^+PF_6^- \\ \xrightarrow{py} [PtX(\eta^3\text{-}all)py] \\ \xrightarrow{TlC_5H_5} [Pt(\eta^3\text{-}all)(C_5H_5)] \end{cases} \qquad (53)$$

Oxidative addition to coordinatively saturated d^{10} complexes ML_4 requires prior generation of a vacant coordination site. Dissociation of one or two ligands in solution is well established[171]. It is aided by increased σ-donation from ligand to metal and, very significantly, by increased size of the ligand[172]. Thus there is little dissociation for L = CO, PF_3, or $P(OMe)_3$, although kinetic studies show that ligand exchange and substitution occur by a dissociative mechanism. $[Pt(PPh_3)_4]$, however, is completely dissociated in solution to $[Pt(PPh_3)_3]$, but the equilibrium constant for

$$[Pt(PPh_3)_3] \rightleftharpoons [Pt(PPh_3)_2]] + PPh_3 \qquad (54)$$

is only about 10^{-4} in benzene. With very bulky phosphines such as $P(i\text{-}Pr)_3$ or $P(t\text{-}Bu)_3$ monomeric species, PtL_2 can be isolated, although as expected their reactivity in oxidative addition is low.

An excess of allyl chloride reacts with $[Pd(PPh_3)_4]$ to give $[Pd(\eta^3\text{-}C_3H_5)(PPh_3)Cl]$[173]. $[Pt(PPh_3)_3]$ or $[Pt(AsPh_3)_4]$ affords ionic products (equation 55)[174,175]. This contrasts with a report that 2-methylallyl chloride forms only $[Pt(PPh_3)_2Cl_2]$. Kurosawa, however, found that η^3-allylplatinum complexes result even from the oxidative addition of 2-butenylamine or 2-buten-1-ol to $[Pt(C_2H_4)(PPh_3)_2]$[176], or of allyl acetate to $[Pt(PPh_3)_3]$[177].

$$[Pt(PPh_3)_3] + C_3H_5 \longrightarrow [Pt(PPh_3)_2(\eta^3\text{-}C_3H_5)]^+X^- \qquad (55)$$

Dibenzylidene acetone complexes of Pd(0) and Pt(0) are useful precursors to η^3-allyl and related derivatives. They are obtained from dibenzylidene acetone (dba) (1,5-diphenylpenta-1,4-dien-3-one) and Na_2PdCl_4 or K_2PtCl_4 in methanol in the presence of sodium acetate[178,179]. Species $[Pd_2(dba)_3]\cdot$ solvent (solvent = $CHCl_3$, CH_2Cl_2) have been established by X-ray diffraction. They react with allyl halides as shown in equations 56–58. Metallacyclobutenyl complexes result from their reactions with triarylcyclopropenium salts[183].

$$[Pd_2(dba)_3]\cdot\text{solvent} + 2RCH{=}CR'CH_2X \xrightarrow[\text{(ref 180)}]{\text{benzene, }25^\circ C} [R'\text{-}(\eta^3\text{-allyl})\text{-}Pd(\mu\text{-}X)]_2 + 3\,dba \qquad (56)$$

$$[Pd_2(dba)_3] + 2PhCH{=}CRCH{=}CH{=}CHCH_2Cl \xrightarrow{\text{(ref. 181)}} [PhCH{=}CR\text{-}(\eta^3\text{-allyl})Pd(\mu\text{-}Cl)]_2 + 3\,dba \qquad (57)$$

$$1/2\,[Pd_2(dba)_3] + nPh_3CCl \xrightarrow[\frac{3n}{2}\,dba]{\text{(ref. 182)}} [Pd(CPh_3)Cl]_n \xrightarrow{\text{Tlacac}} [(\eta^3\text{-}C_6H_5{=}CPh_2)Pd(acac)] \qquad (58)$$

2. To d^8 complexes

The basis for a good understanding of the chemistry of allylrhodium complexes was laid by Powell and Shaw[184]. Allyl chloride reacts with $[Rh(CO)_2Cl]_2$ in aqueous methanol at pH 7 to yield the chloro-bridged dimer **42**. A detailed mechanism for this reaction was proposed.

$$[Rh(CO)_2Cl]_2 + 4H_2O + 6C_3H_5Cl \longrightarrow [(\eta^3\text{-}C_3H_5)_2Rh(\mu\text{-}Cl)_2Rh(\eta^3\text{-}C_3H_5)_2] + 4CO_2 + 2C_3H_6 + 6HCl \quad (59)$$

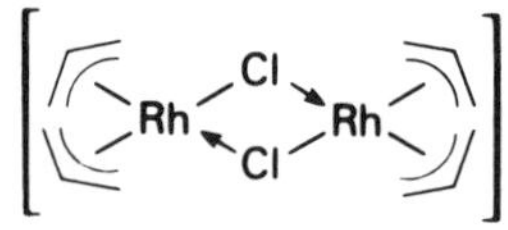

(**42**)

Some further transformations of **42** are shown in Scheme 1. They include some which are relevant to other sections of this Chapter.

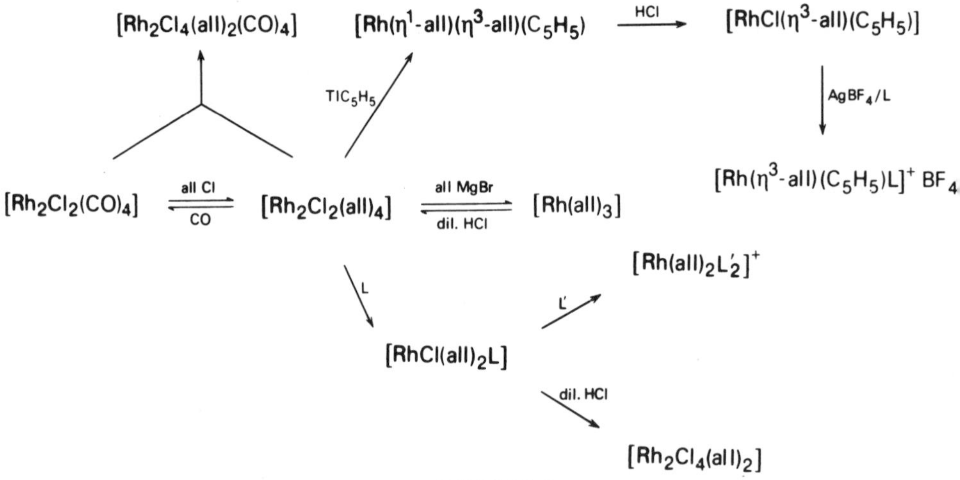

$[Rh_2Cl_2(CO)_4]$ ⇌ (all Cl / CO) $[Rh_2Cl_2(all)_4]$

$[Rh_2Cl_2(all)_4]$ → $[Rh_2Cl_4(all)_2(CO)_4]$

$[Rh_2Cl_2(all)_4]$ → (TlC_5H_5) $[Rh(\eta^1\text{-all})(\eta^3\text{-all})(C_5H_5)]$ → (HCl) $[RhCl(\eta^3\text{-all})(C_5H_5)]$ → ($AgBF_4$/L) $[Rh(\eta^3\text{-all})(C_5H_5)L]^+BF_4^-$

$[Rh_2Cl_2(all)_4]$ ⇌ (all MgBr / dil. HCl) $[Rh(all)_3]$

$[Rh_2Cl_2(all)_4]$ → (L) $[RhCl(all)_2L]$ → (L′) $[Rh(all)_2L'_2]^+$

$[RhCl(all)_2L]$ → (dil. HCl) $[Rh_2Cl_4(all)_2]$

SCHEME 1

Isocyanide complexes of Rh(I) and Ir(I) react with allyl halides as shown in equations 60 and 61[185].

$$[Rh(RNC)_4]^+PF_6^- + C_3H_5X \longrightarrow [Rh(RNC)_4(\eta^1\text{-}C_3H_5X)]^+PF_6^- \quad (60)$$

$$[Ir(t\text{-}BuNC)_4]^+PF_6^- + 2MeC_3H_4X \xrightarrow{PF_6^-} [Ir(t\text{-}BuNC)_4(\eta^3\text{-}2\text{-}MeC_3H_4)]^{2+}(PF_6^-)_2 \quad (61)$$

The square-planar complexes $[IrCl(CO)L_2]$ (**43**) are converted into octahedral d^6 derivatives with allyl halides[186]. In benzene *cis*-addition occurs, which is very unusual, as alkyl halides normally add *trans* to Vaska's compound and its analogues[187]. Recrystallization of **44** from ethanol yields isomer **45**, and treatment with $NaPF_6$ in methanol leads to the η^3-allyl derivative **46**.

The initial oxidative addition could go through an η^3-allyl in benzene but not in methanol, where normal *trans*-addition occurs. Wilkinson's complex, $[Rh(PPh_3)_3Cl]$,

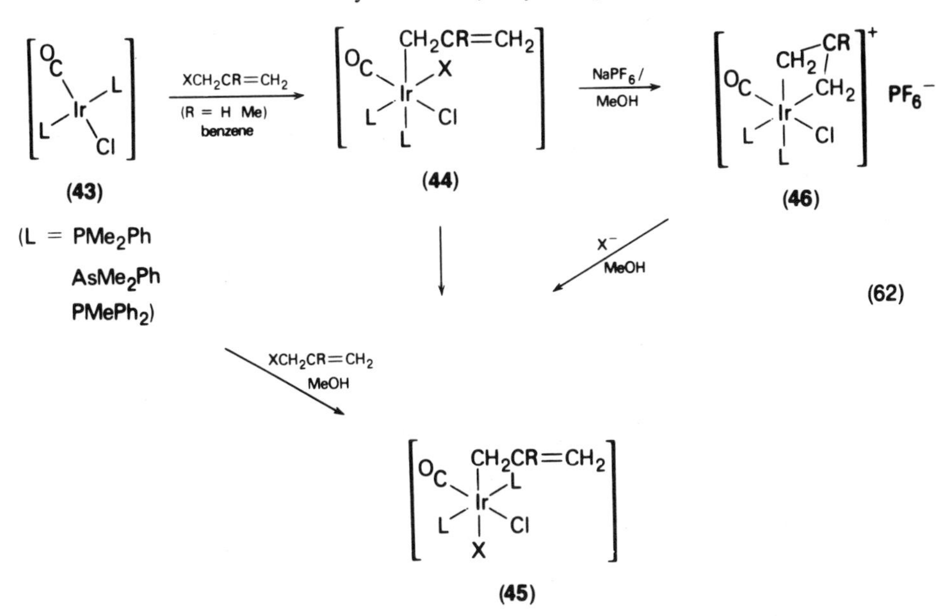

and the triphenylarsine and -stibine analogues react with allyl halides to afford $[RhL_2(allyl)Cl_2]$. These derivatives are fluxional in solution; exchange of *syn* and *anti* substituents occurs by an η^3–η^1–η^3 mechanism[174].

3. *To d^6 complexes*

Bis(arene)molybdenum complexes, prepared by the reducing Friedel–Crafts procedure or by metal atom synthesis, are electron rich and hence readily undergo oxidative addition with allyl halides[188]. The chlorine-bridged products **47** catalyse the 1,2-polymerization of butadiene and the conversion of propyne into a mixture of a polymer and 1,3,5- and 1,2,4-trimethylbenzenes[189]. Some reactions are illustrated in Scheme 2.

Nucleophiles attack the cations **48** at the butadiene ligand (see Section VII.A). For further details of the chemistry of these molybdenum complexes, the original literature should be consulted[190].

VII. NUCLEOPHILIC ATTACK ON COORDINATED DIENES

A. 1,3-Dienes

Nucleophilic addition to a coordinated 1,3-diene yields an η^3-allyl complex. Thus, reduction of **49** affords a mixture of *syn*- **(50)** and *anti*- **(51)** 1-methylallyl isomers[191].

Similarly, reduction of $[Mo(\eta^5\text{-indenyl})(\eta^4\text{-}C_4H_6)(CO)_2]^+$ with sodium borohydride yields the *anti*-η^3-1-methylallyl complex[192]. Other examples are quoted in a recent review, which includes three simple rules to enable the site of attack to be predicted in cases where more than one possibility exists[193].

Tetraphenylcyclobutadiene palladium and platinum complexes **(52)** are attacked by nucleophiles such as ethoxide ion[194,195] to yield *exo*-substituted cyclobutenyl derivatives **(53)**.

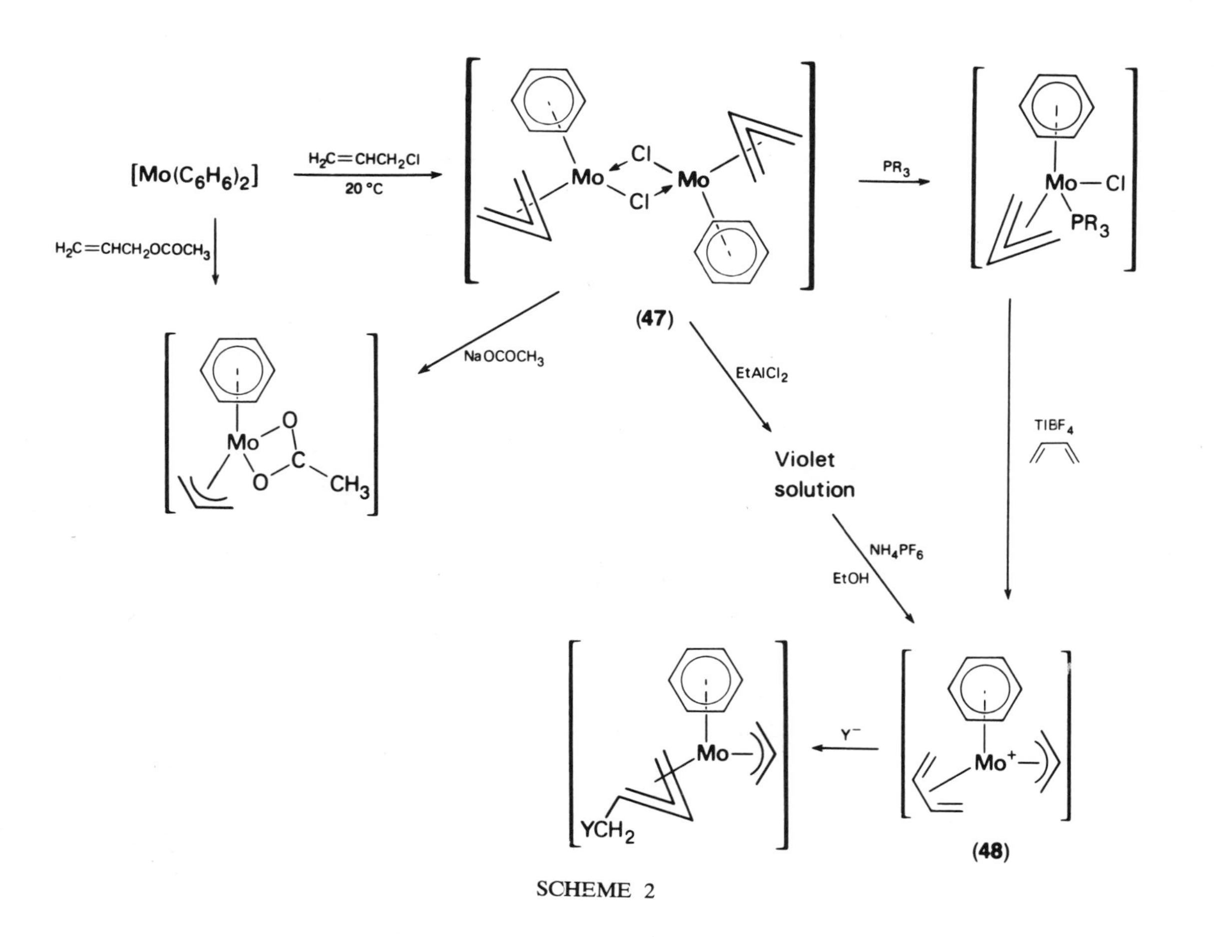

[Mo(C6H6)2]
H2C=CHCH2Cl
20 °C
Mo
Cl
Cl
Mo
PR3
Mo—Cl
PR3
H2C=CHCH2OCOCH3
(47)
NaOCOCH3
EtAlCl2
Mo
O
O
C
CH3
TlBF4
Violet
solution
NH4PF6
EtOH
Mo
YCH2
Y−
Mo+
(48)

SCHEME 2

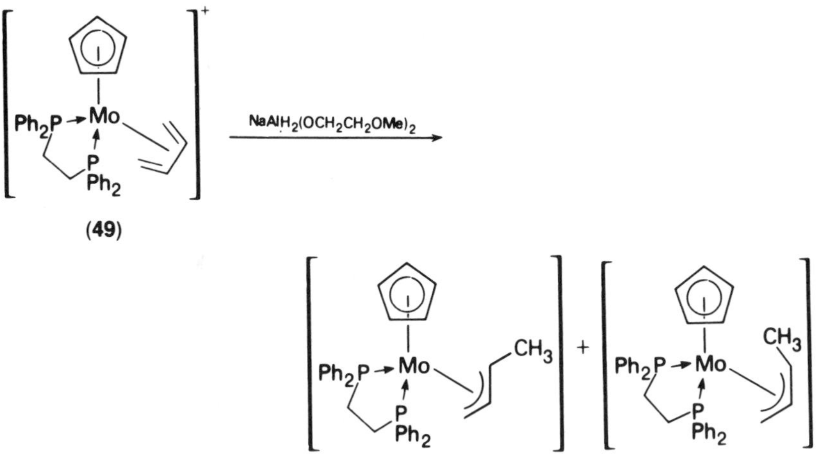

$[Mo(C_5H_5)(Ph_2PCH_2CH_2PPh_2)(allyl)]^+$ **(49)** $\xrightarrow{NaAlH_2(OCH_2CH_2OMe)_2}$ **(50)** + **(51)** (63)

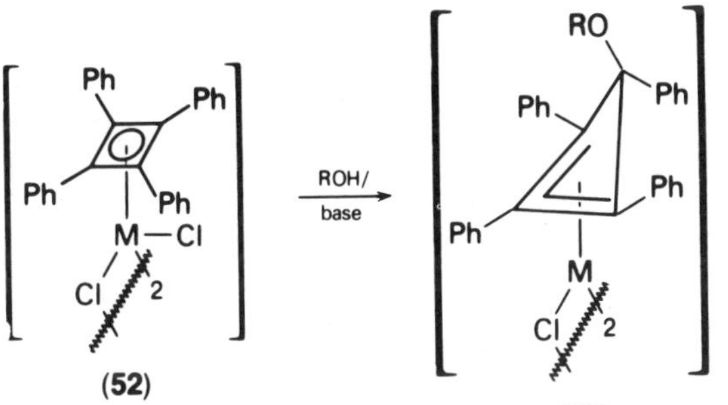

(52) $\xrightarrow{ROH/base}$ **(53)** (M = Pd, Pt) (64)

The *endo*-ethoxy isomer (**55**) of the palladium complex can be prepared from $PdCl_2$ and diphenylacetylene in ethanol. A butadienyl complex (**54**) is an intermediate[196,197]. Alkenes can insert into the Pd—butadienyl bond of **54** to yield η^3-allyl derivatives[198].

B. Alkoxycarbonylation and Related Reactions

Treatment of η^1-propargyl complexes of transition metal carbonyls with alcohols yields 2-carboxyalkylallyl derivatives (**57**). Mercaptans give the corresponding thio compounds[199–210]. The propargyl derivatives **56** are very air and light sensitive, especially when R′ = H. Improved stability is achieved when one carbonyl ligand is substituted by a phosphine or phosphite. The products $[M(C_5H_5)(CO)L(allyl)]$ show conformational equilibria in solution[200]. The mechanism in equation 67 has been proposed for the reactions, which takes into account the observed acid catalysis. The first step involves the formation of a cationic allene complex, followed by nucleophilic attack of alcohol or thiol on a coordinated carbonyl ligand. The —COXR group finally transfers to the central carbon atom of the allene. Some related reactions which lead to η^3-allylcyclopentenone complexes have been described by the same research group[202–205].

$$PdCl_2 + PhC{\equiv}CPh \longrightarrow [PhC{\equiv}CPh \rightarrow PdCl_2] \xrightarrow{OR^-} [RO(Ph)C{=}C(Ph)PdCl]_2 \xrightarrow{PhC{\equiv}CPh} (54) \rightleftharpoons (55) \quad (65)$$

(54)

(55)

$$[\{M\}(CO)(CH_2{-}C{\equiv}CR')] \xrightarrow[25°C\ (1-7d)]{RXH/thf} [\{M\}{-}(\eta^3\text{-}CH_2C(R')CH{-})C(=O)XR] \quad (66)$$

(56) (57)

$\{M\} = Mn(CO)_4, C_5H_5Mo(CO)_2, C_5H_5W(CO)_2$; $RXH = H_2O$, ROH

$$Cp(CO)_2LMo{-}CH_2{-}C{\equiv}CR' \xrightarrow{slow} [Cp(CO)LMo(C{\equiv}O)(CH_2{=}C{-}CHR')]^+ \xrightarrow{-H^+} [Cp(CO)LMo(C(=O)XR)(\eta^3\text{-}allyl)] \longrightarrow [Cp(CO)LMo{-}(\eta^3\text{-}CH_2C(R')CH{-})C(=O){-}XR] \quad (67)$$

5-Bromopenta-1,2-dienes, on treatment with carbonyl anions, yield air-sensitive η^1-complexes, which rearrange in tetrahydrofuran into the products 58.

$$[\{M\}(CO)]^- + Br(CH_2)_2C(R){=}C{=}CH(R') \xrightarrow[r.t.]{thf} [\{M\}(CO)(CH_2)_2C(R){=}C{=}CHR'] \longrightarrow \quad (68)$$

(58)

The insertion of alkynes into Mo—alkyl and W—alkyl bonds to give, *inter alia*, η^3-bonded lactone complexes have been described[206].

Treatment of 1,4-dichlorobut-2-yne with $[Fe(C_5H_5)(CO)_2]^-$, $[Fp(CO)_2]$, in tetrahydrofuran yields $[(CO)_2FpCH_2{-}C{\equiv}C{-}CH_2Fp(CO)_2]$ in 66% yield. If the reaction is carried out in the presence of methanol, however, the product is $[Fp\{\eta^3\text{-}CH_2CH{-}C(CH_3)(COOCH_3)\}(CO)_2]$ (40%)[207].

VIII. REACTIONS OF METAL SALTS WITH UNSATURATED COMPOUNDS

In this section, various reactions of transition metal salts with unsaturated organic compounds are described. Palladium in particular shows a strong tendency to afford η^3-allyl complexes from such reactions. As detailed accounts of this area of chemistry have been published[10,11], the emphasis here will be on recent developments and on modifications of the original methods which lead to improved yields. Where similar reactions are known for other metals, notably platinum, these will be mentioned.

A. Allyl Alcohols, Halides, and Amines

Allyl alcohol, on warming with palladium(II) chloride in the presence of dilute hydrochloric acid, yields the dimeric complex $[Pd(\eta^3\text{-}C_3H_5)Cl]_2$ (**59**). The bromide is obtained similarly. Allyl chloride can with advantage replace the alcohol. Hüttel obtained a 50% yield of **59** from allyl chloride and $PdCl_2$ in 50% acetic acid at 30–60°C[207a]. Yields are improved by carrying out the reaction in the presence of a reducing agent[208]. Carbon monoxide[209], tin(II) chloride, ethene, carbon monoxide an amine, titanium(III) chloride, and even metal powders such as iron, zinc, or copper have been used[210].

The corresponding platinum complexes have not been so widely studied. Lukas and Blom obtained $[Pt(\eta^3\text{-}C_3H_5)Cl]_4$ in 90% yield from K_2PtCl_4/LiCl and allyl chloride in tetrahydrofuran, using tin(II) chloride as reducing agent[210]. The 2-methylallyl complex was prepared in 85% yield by heating K_2PtCl_4 and 2-methylallyl chloride under reflux with sodium acetate in aqueous ethanol[211].

An ingenious method was developed by Kurosawa involving the decarboxylation of allyloxycarbonyl complexes (equation 69)[212]. Decarboxylation can be effected either by refluxing in benzene or at room temperature by treatment with $AgClO_4$ (on grounds of safety, $AgBF_4$ or $AgPF_6$ would seem preferable). Even allylamines have

$$[PtClL_2(CO)]^+ + R^3CH{=}CR^2CHR^1OH \xrightarrow{Na_2CO_3} [PtCl(COOR^1CH{-}CR^2{=}CHR^3)L_2] \xrightarrow{AgClO_4} \left[L_2Pt(\eta^3\text{-}CHR^1CR^2CHR^3) \right]^+ ClO_4^- + CO_2 \quad (69)$$

$L = PPh_3, PMePh_2$

$R^1 = R^2 = R^3 = H$

$R^1 = Me;\ R^2 = R^3 = H$

$R^2 = Me;\ R^1 = R^3 = H$

been used to prepare η^3-allyl platinum complexes, in spite of the rather poor leaving group, ammonia, which must separate (equation 70)[213].

$$[PtH(PPh_3)_2L]ClO_4 \xrightarrow[CH_2Cl_2(-L)]{H_2C{=}CHCH_2NH_2} [Pt(\eta^3\text{-}C_3H_5)(PPh_3)_2]^+ ClO_4^- + NH_3 \quad (70)$$

$L = Ph_3As, CO$

B. Alkenes

The reactions of alkenes with palladium(II) chloride were first investigated by Hüttel and Christ, who obtained between 20% and 90% yields of allylpalladium chlorides from suitable olefins by heating in 50% acetic acid. The reaction often did not proceed cleanly; organic carbonyl compounds were obtained as by-products[214]. Volger modified this procedure by using a lower temperature (85°C), and by substituting Na_2PdCl_4 for the sparingly soluble $PdCl_2$ and glacial acetic acid for the aqueous solvent[215]. No allyl complexes were obtained unless sodium acetate was added, but then high yields (60–95%) resulted. The method was restricted, however, to olefins which yield complexes with 2-alkyl or 2-aryl substituents. It fails in other cases for various reasons, such as the formation of olefin acetates or the oxidation of the olefin with deposition of palladium metal.

Volger proposed that the first step is the rapid formation of an η^2-alkene complex, which then eliminates hydrogen chloride in the rate-determining step (equation 71).

$$(CH_3)_2C{=}CH_2 + [PdCl_4^{2-}] \underset{k_{-1}\ \text{fast}}{\overset{k_1}{\rightleftharpoons}} [(\eta^2\text{-}(CH_3)_2C{=}CH_2)PdCl_3]^- \rightleftharpoons [(\eta^3\text{-}C_3H_4CH_3)PdHCl_2] \underset{+HCl}{\overset{-HCl}{\rightleftharpoons}} [(\eta^3\text{-}C_3H_4CH_3)PdCl]_2 \quad (71)$$

The function of the base, sodium acetate, is therefore to assist in the removal of HCl. Similarly, the use of the basic solvent dimethylformamide[215a] or sodium carbonate in dichloromethane at room temperature leads to improvements over the original Hüttel and Christ procedure[216]. Even so, the reaction often proved capricious. Cycloalkenes in particular gave poor yields. Trost *et al.* devised a procedure which, when carried out under carefully controlled conditions, is successful with most olefins, including cycloalkenes[217]. A solution of $PdCl_2$, NaCl, sodium acetate, and copper(II) acetate is prepared at 90–95°C in glacial acetic acid, then cooled to 60°C, olefin is added and the temperature is kept at 60°C for 2–4 h. The inclusion of copper(II) salts overcomes the problem of reduction to palladium(0) by the olefin. The products obtained in the presence of copper are approximately those expected for kinetic control; the slower deprotonation by base yields product compositions which are roughly thermodynamically controlled. Many reports have appeared of η^3-allylpalladium halides derived from terpenes and steroids[218]. Most of these preparations have not included copper(II), so it may be possible to obtain improved yields using Trost's procedure[217a].

Monosubstituted terminal olefins fail or give poor yields using the methods of Hüttel and Christ[214], Volger[215] or Conti[215a]. In a recent modification the olefin is first irradiated with tertiary butyl hypochlorite followed by treatment with $PdCl_4^{2-}$ and carbon monoxide in aqueous methanol (equation 72)[219].

$$H_2C{=}CHCH_2R \xrightarrow[t\text{-BuOCl}]{h\nu} \left\{ \begin{matrix} H_2C{=}CHCHClR \\ H_2CClCH{=}CHR \end{matrix} \right\} \xrginalxrightarrow[CO]{[PdCl_4]^{2-}} [(\eta^3\text{-}CH_2CHCHR)PdCl]_2 \quad (72)$$

R = *n*-Bu; 84%

A specialized reaction in which an alkyne and an alkene are linked to form an η^3-allylpalladium chloride has been described by Avram and co-workers (equation 73)[220].

$$PhC{\equiv}CAr + RR'C{=}CH_2 \xrightarrow{[Pd(PhCN)_2Cl_2]} [(\eta^3\text{-}ClC(Ar)CCH_2\text{-}CHRR')PdCl]_2 \quad (73)$$

(26–81%)

Linear alkenes and cycloalkenes react with trimeric palladium(II) acetate in glacial acetic acid to yield acetato bridged complexes **60**. Exclusion of oxygen is important[221].

$$[Pd_3(OAc)_6] + RCH_2CH{=}CH_2 \xrightarrow[25\,^\circ C]{AcOH} [(\eta^3\text{-}C_3H_4R)Pd(\mu\text{-}OAc)_2Pd(\mu\text{-}OAc)_2Pd(\eta^3\text{-}C_3H_4R)] \quad (74)$$

(60)

C. 1,3-Dienes

1. Palladium salts

Palladium(II) complexes show a strong tendency to form η^3-allyl complexes with conjugated dienes. When $[Pd(PhCN)_2Cl_2]$ is treated with buta-1,3-diene in benzene at room temperature the dimeric chloro-bridged compound **61** (X = Cl) is obtained. An η^2-butadiene complex (**62**) was isolated by ligand exchange with $[Pd(pent\text{-}1\text{-}ene)Cl_2]_2$ at −40°C, which on warming to 25°C was quantitatively converted into **61**. This suggests that the initial step is coordination of one double bond, followed by addition of Pd—Cl across it[222].

$[Pd(PhCN)_2Cl_2]$ → (**61**) ← 25 °C ← (**62**) ← −40 °C ← $[Pd(olefin)Cl_2]_2$ (75)

In methanol, Na_2PdCl_4 and buta-1,3-diene give **61** (X = OMe). Other dienes and alcohols react similarly[223]. The reactions of isoprene[224] and of cyclohexa-1,3-diene[225] have been studied kinetically, to shed further light on their mechanisms. In acetic acid at 100°C in the presence of lithium acetate palladium chloride affords **61** (X = OAc). Yields are improved by addition of copper(II) acetate, which reoxidizes any palladium produced in side-reactions[226]. Decomposition to palladium is also minimized by working at room temperature. Under these conditions 2,3-dimethylbutadiene gives a mixture of *syn* and *anti* isomers[227]. The diamination of 1,3-dienes proceeds through similar intermediates[228].

The addition of dienes to palladium chloride occurs stereo-specifically[229]. Moreover, the ready solvolysis of the 4-chloro derivatives **61** (X = Cl) by nucleophilic media such as methanol proceeds with retention of configuration at C_4, and inversion at palladium. A carbonium ion intermediate has been suggested.

Divinyl carbinols react with palladium (II) chloride in methanol to give η^3-allyls (**62**) with two —CH_2OMe substituents[230].

$CH_2CHCR(OH)CH{=}CH_2 + [PdCl_4]^{2-}$ —MeOH→ $[MeOCH_2(\eta^3\text{-}C_3H_2R)CH_2OMe\,PdCl]_2$ **62** (76)

Direct formation of η^3-allyls from dienes and metal halides is uncommon outside palladium chemistry. Rhodium(III) chloride, however, reacts with buta-1,3-diene in methanol to form the unusual complex **63**, in which a bridging butadiene ligand is present[231]. Similar complexes catalyse the codimerization of ethylene and buta-1,3-diene[232].

$$\text{(63)} \xrightarrow[\text{CH}_2\text{Cl}_2]{\text{TlC}_5\text{H}_5} [(\eta^3\text{-MeC}_3\text{H}_4)\text{Rh}(\eta^5\text{-C}_5\text{H}_5)] \qquad (77)$$

(63)

2. *Insertion into allyl—palladium bonds*

Buta-1,3-diene reacts with η^3-allyl(2,4-pentadionato)palladium by insertion into the allyl—Pd bond (equation 78). On account of their relevance to catalytic oligomerization

$$\text{C}_4\text{H}_6 + [\text{R}\text{-}(\eta^3\text{-C}_3\text{H}_4)\text{—Pd(acac)}] \longrightarrow [\text{H}_2\text{C}{=}\text{CRCH}_2\text{CH}_2(\eta^3\text{-C}_3\text{H}_4)\text{—Pd(acac)}] \qquad (78)$$

and polymerization of 1,3-dienes, reactions of this type aroused much interest[17]. A study of the effects of changing the allyl group, the diene, and ancillary ligands showed that the rate of insertion decreases in the order 2-chloroallyl > 2-methylallyl ≫ 1-methylallyl; OAc > acac > Cl > Br > NCS > I; butadiene > isoprene > *trans*-penta-1,3-diene ≫ 2,5-dimethylhexa-2,4-diene. 1,2-Dienes also insert[233]. A mechanism which apparently accounts for all the stereochemical and kinetic observations has been proposed[234].

D. 1,2-Dienes

Allene (propa-1,2-diene) and its substituted derivatives form a few η^2-complexes with transition metals[235]. Tetraphenyl-[235] and tetramethylallene[236] yield such complexes with iron carbonyls. Electrophilic addition of H^+ or CH_3CO^+ to **64** (R = Me) affords η^3-allyltetracarbonyls (**65**), which are deprotonated by acetone to η^4-diene complexes (**66**)[237].

$$\text{R}_2\text{C}{=}\text{C}{=}\text{CR}_2 \xrightarrow[\text{or [Fe}_2\text{(CO)}_9]]{[\text{Fe(CO)}_5]} [\text{R}_2\text{C}{=}\text{C}{=}\text{CR}_2\ \text{Fe(CO)}_4]\ \textbf{(64)} \xrightarrow[\text{Ac}_2\text{O}]{\text{HBF}_4} [(\eta^3\text{-R}_2\text{C}\text{C(}\text{CR}_2\text{)}...)\text{Fe(CO)}_4]^+\text{BF}_4^-\ \textbf{(65)} \xrightarrow{-\text{H}^+} \textbf{(66)}\ (\eta^4\text{-diene})\text{Fe(CO)}_3 \qquad (79)$$

Tetramethylallene displaces ethene from $[Rh(C_2H_4)_2acac]$ to give $[Rh(\eta^2\text{-}Me_2C{=}C{=}CMe_2)_2(acac)]$[238]. Similar complexes of platinum, rhodium, and iridium, including some containing allene monomer itself, are known[239]. Allene, however, is very susceptible to oligomerization or polymerization in the presence of transition metal complexes. Often mixtures of products are obtained. Many of the resulting compounds contain η^3-allyl ligands. Thus, $[Pd(PhCN)_2Cl_2]$ affords **67**, **68**, or **69**, depending on the reaction conditions[240].

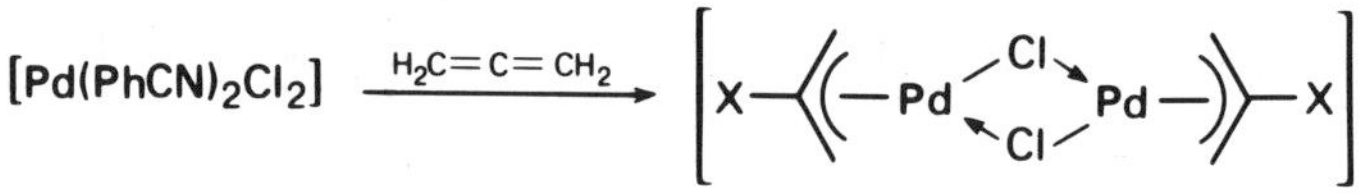
[Pd(PhCN)2Cl2]
H2C=C=CH2
X
Pd
Cl
Cl
Pd
X

(80)

(**67**) X = Y = Cl; (**68**) X = Y = —C(=CH$_2$)CH$_2$Cl;
(**69**) X = Cl, Y = C(=CH$_2$)CH$_2$Cl; (**70**) X = Y = OAc

In the presence of acetate ion, **70** is formed[241]. Palladium acetate in benzene gives **71** in 18% yield, in which the organic ligand is an allene trimer. Higher yields (94%) are obtained from η^3-allylpalladium acetate and allene (equation 81)[242].

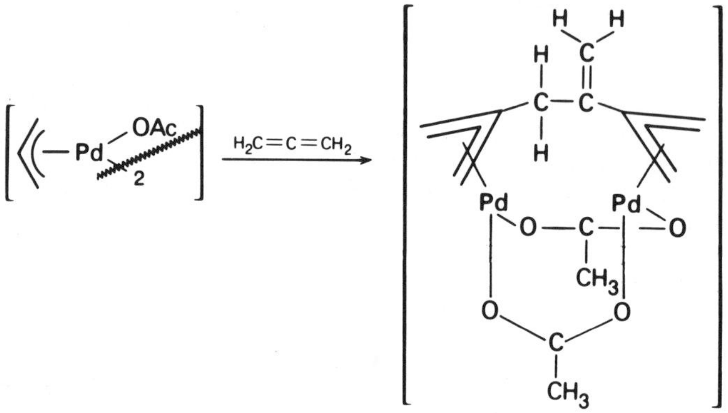
Pd
OAc
2
H2C=C=CH2
H H
H C
C—C
H
Pd
Pd
O—C—
O
CH3
O
O
C
CH3

(81)

(**71**)

Triirondodecacarbonyl (1 mol) reacts with allene (3 mol) at 80–95°C in hydrocarbon solvents under pressure to give a biallyl complex. With 4–6 mol of allene a small proportion of **72**, $[Fe_2(CO)_6(C_9H_{12})]$, is obtained in addition. This isomerizes first to **73** and then further to **74** on refluxing in toluene[243]. The structures of **73** and **74** have been determined by X-ray crystallography[244].

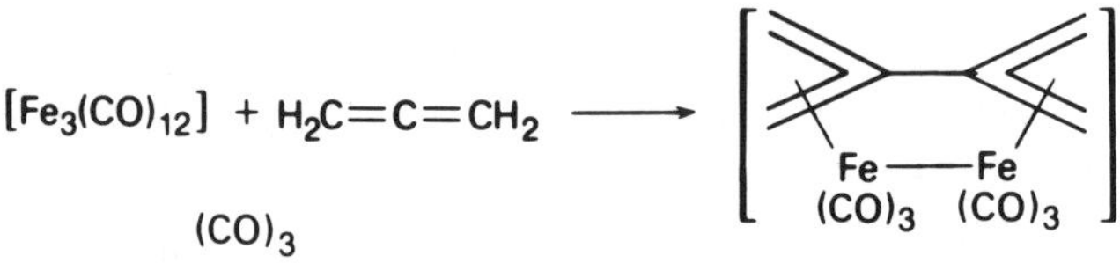
[Fe3(CO)12] + H2C=C=CH2
Fe
Fe
(CO)3
(CO)3

(82)

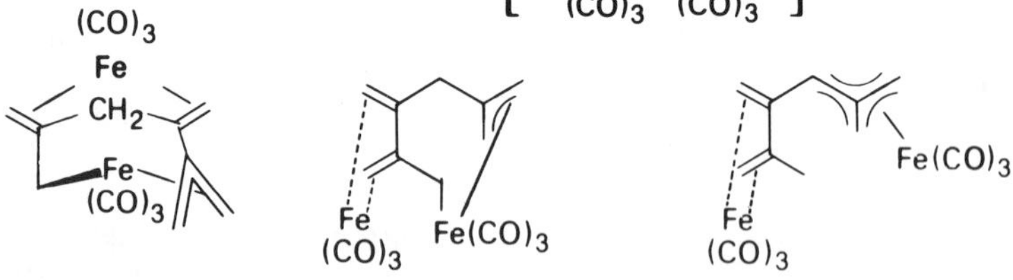
(CO)3
Fe
CH2
Fe
(CO)3
Fe
(CO)3
Fe(CO)3
Fe(CO)3
Fe
(CO)3

(**72**) (**73**) (**74**)

The cyclooligomerization of allene to 1,2,4,6,9-pentamethylenecyclodecane is catalysed by zerovalent nickel compounds. Addition of three equivalents of allene at $-70°C$ to bis(cyclooctadiene)nickel gives a red solution which contains $Ni(C_9H_{12})$. This has been stabilized and isolated as the triphenylphosphine complex **75**[245]. Reactions of allene with $[Rh(C_2H_4)_2acac)]$[246] and with $[Rh(CO)_2Cl]_2$[247] have also been studied. At $-35°C$ the former yields a complex of an allene pentamer (**76**), which is thought to arise via an intermediate rhodocyclopentane.

Ni—PPh$_3$

(**75**)

$[Rh(C_2H_4)_2(acac)] + H_2C{=}C{=}CH_2 \longrightarrow$

[$H_2C{=}C{=}CH_2$, CH_2, Rh(acac), CH_2] $\xrightarrow{H_2C=C=CH_2}$ [Rh(acac)] (83)

(**76**)

E. Cyclopropenes and Cyclopropanes

Palladium chloride, usually in the form of $[Pd(PhCN)Cl_2]_2$ or $[Pd(C_2H_4)_2Cl]_2$, yields η^3-allyl complexes with substituted cyclopropenes, methylenecyclopropanes and vinylcyclopropanes[6]. The initial step in these reactions is probably coordination of the olefin to the metal. Complexes of rhodium(I), iridium(I), and platinum(II) with η^2-bonded methylenecyclopropanes have been characterized[248]. 1,2-Chloropalladation of the double bond then occurs, followed by opening of the cyclopropane ring[249,250].

R^1, R^2, R^3 $\longrightarrow$ [—PdCl$_2$, R^1, R^2, R^3] $\longrightarrow$ [Pd, Cl, Cl, R^1, R^2, R^3] $\longrightarrow$

[Cl, —Pd, Cl, R^1, CR^2R^3H]$_2$ (84)

2,2-Diphenylmethylenecyclopropane, however, yields **77**, possibly via a trimethylenemethane complex[251]. Triphenylcyclopropene with $[(PhCN)_2PdCl_2]$

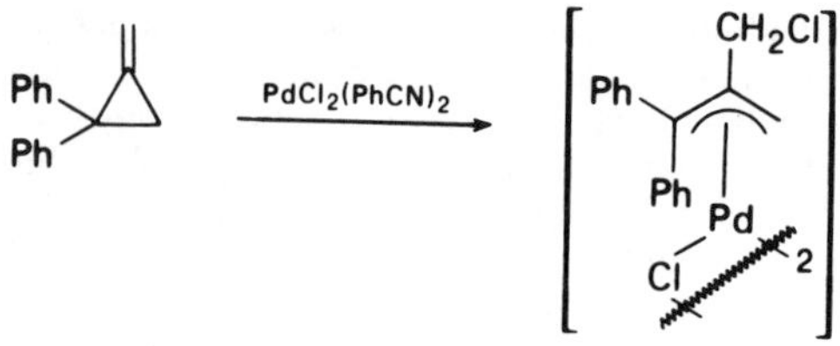

(85)

(77)

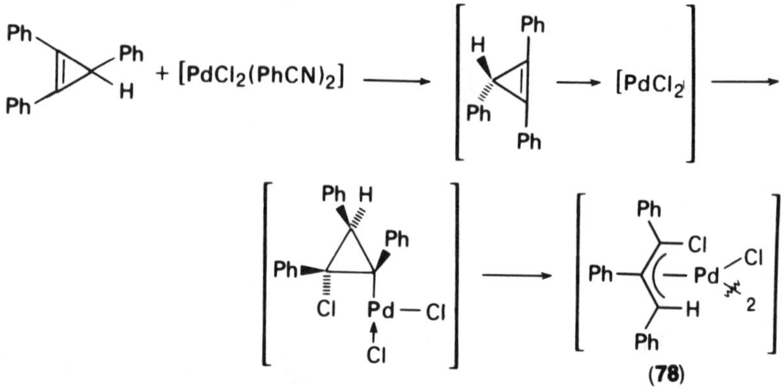

(86)

(78)

affords **78**[252]. The postulated rearrangement in the final step is supported by the observation that when chloride is removed from the coordination sphere of a cyclopropylplatinum complex[79], an η^3-allyl results (equation 87)[253].

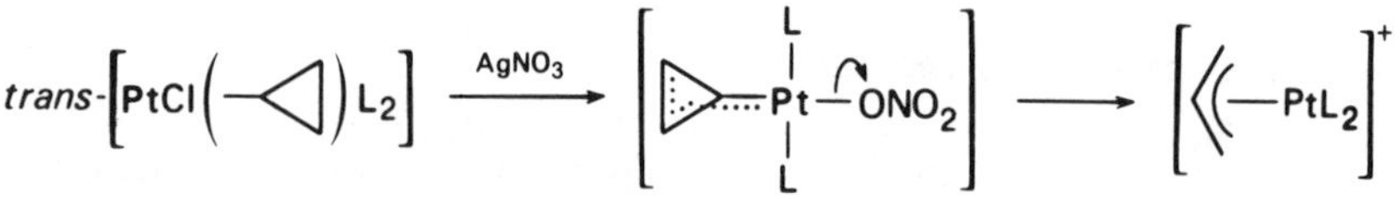

(87)

Although simple vinylcyclopropanes react with $[Fe_2(CO)_9]$ to give η^4-diene complexes, the η^3-allyl-η^1-alkyl intermediate can be stabilised if C_4 is the bridgehead of a polycyclic system[8]. This is illustrated by semi-bullvalene (**81**).

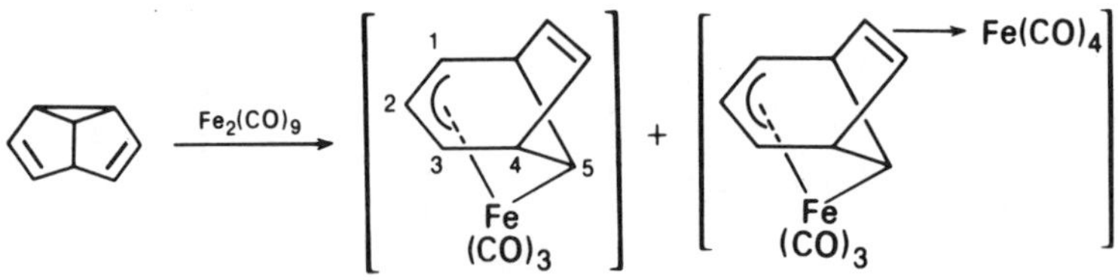

(88)

(81) (82) (83)

Certain cyclopropanes can also yield η^3-allyl complexes. When *trans*$[IrCl(N_2)(PPh_3)_2]$ reacts with neat phenylcyclopropane $[IrClH(\eta^3\text{-}C_3H_4Ph)(PPh_3)_2]$, an η^3-allylmetal hydride of unusual stability results[254]. The first step in the reaction is thought to be insertion of the metal into the cyclopropane ring to form a metallocyclobutane.

The reactions described in this section do not provide general methods for the synthesis of allyl derivatives, with the possible exception of palladium. Even here the relative inaccessibility of starting materials imposes a severe disadvantage.

IX. BINUCLEAR COMPLEXES OF PALLADIUM AND PLATINUM

An interesting series of complexes was discovered in 1975 by Felkin and co-workers from the reaction of $[PdBr(C_5H_5)P]$ with magnesium in tetrahydrofuran[255].

$[(i\text{-Pr})_3P\text{—Pd(Cp)—Br}]$ $\xrightarrow{\text{0.5 mol Mg in thf}}$ $[(i\text{-Pr})_3P\text{—Pd—Pd—P}(i\text{-Pr})_3]$ (Cp and Br bridging) $\xrightarrow{\text{LiAlH}_4}$ **(84)**

$[\text{L—Pd—Pd—L}]$ (Cp and Br bridging) **(85)** $\xrightarrow{\text{allMgBr}}$ $[\text{L—Pd—Pd—L}]$ (two allyl bridging) **(86)** (89)

The product (**84**), which was obtained in 80% yield, formed dark red crystals, stable in air. The cyclopentadienyl group bridges the two palladium atoms. Reduction of **84** with lithium aluminium hydride gave **85**, which on treatment with allylmagnesium bromide yielded **86** with two η^3-allyl groups bridging the Pd—Pd bond[256].

Other routes to such sandwich complexes with Pd—Pd or Pt—Pt bonds have been explored by Werner and his group[257]. Treatment of **87** with a phosphine or phosphite

$[\text{CpPd(2-methylallyl)}]$ **(87)** + L $\xrightarrow{-50\,°\text{C}}$ $[\text{CpPd(L)CH}_2\text{—C(Me)=CH}_2]$ isolated L = P(i-Pr)$_3$ $\xrightarrow{\text{L}}$

$[\text{L—Pd—Pd—L}]$ (Cp and 2-methylallyl bridging) **(88)** (90)

$$[Pd(2\text{-}MeC_3H_4)_2] + PdL_2 \xrightarrow[PCy_3]{L = P(i\text{-}Pr)_3,} \left[L\text{—}Pd\text{—}Pd\text{—}L \right] \quad (91)$$

(**89**)

ligand at $-50°C$ yields a range of complexes of type 88[258]. With bulky phosphines such as $P(i\text{-}Pr)_3$ or PCy_3, complexes PdL_2 can be isolated. These react with bis(2-methylallyl)palladium to give the complexes **89**[259,260].

X. REFERENCES

1. R. G. Guy and B. L. Shaw, *Adv. Inorg. Chem. Radiochem.*, **4**, 78 (1962).
2. M. L. H. Green and P. L. I. Nagy, *Adv. Organomet. Chem.*, **2**, 325 (1964).
3. M. L. H. Green, *Organometallic Compounds*, 2nd ed., Vol. 2, Methuen, London, 1968.
4. G. Wilke, B. Bogdanovic, P. Hardt, P. Heimbach, W. Keim, M. Kröner, W. Oberkirch, K. Tanaka, E. Steinrücke, D. Walter, and H. Zimmermann, *Angew. Chem. Int. Ed. Engl.*, **5**, 151 (1966).
5. M. I. Lobach, B. D. Babitskii, and V. A. Kormer, *Russ. Chem. Rev.*, **36**, 1158 (1967).
6. J. Powell, *MTP International Review of Science, Inorganic Chemistry*, Series 1, Volume 6, Butterworths, London, 1972, pp. 273–308.
7. P. C. Wailes, R. S. P. Coutts, and H. Weigold, *Organometallic Chemistry of Titanium, Zirconium and Hafnium*, Academic Press, New York, 1974.
8. R. B. King, in *The Organic Chemistry of Iron*, (Eds. E. A. Koerner von Gustorf, F.-W. Grevels, and I. Fischler), Vol. 1, Academic Press, New York, 1978, pp. 463–523.
9. P. W. Jolly and G. Wilke, *The Organic Chemistry of Nickel*, Academic Press, New York, Vol. 1, 1974, Vol. 2, 1975.
10. P. M. Maitlis, *The Organic Chemistry of Palladium*, Vols. 1 and 2, Academic Press, New York, 1971.
11. F. R. Hartley, *Chemistry of Platinum and Palladium*, Applied Science, Barking, 1973.
12. S. O'Brien, M. Fishwick, B. McDermott, M. G. H. Wallbridge, and G. A. Wright, *Inorg. Syn.*, **13**, 74 (1972).
13. P. Heimbach, P. W. Jolly, and G. Wilke, *Adv. Organomet. Chem.*, **8**, 29 (1970).
14. G. P. Chiusoli and G. Salermo, *Adv. Organomet. Chem.*, **17**, 195 (1979), and references cited therein.
15. R. Baker, *Chem. Rev.*, **73**, 487 (1973).
16. H. Bönnemann, C. Grard, W. Kopp, W. Pump, K. Tanata, and G. Wilke, *Angew. Chem. Int. Ed. Engl.*, **12**, 964 (1973).
17. J. Tsuji, *Adv. Organomet. Chem.*, **17**, 141 (1979).
18. H. L. Clarke, *J. Organomet. Chem.*, **80**, 155 (1974).
19. T. J. Marks, *Prog. Inorg. Chem.*, **25**, 223 (1979).
20. S. O'Brien, M. Fishwick, B. McDermott, M. G. H. Wallbridge, and G. A. Wright, *Inorg. Syn.*, **13**, 73 (1972).
21. J. Powell and B. L. Shaw, *J. Chem. Soc. A*, 583 (1968).
22. P. Chini and S. Martinengo, *Inorg. Chem.*, **6**, 837 (1967).
23. J. L. Becconsall, B. E. Job, and S. O'Brien, *J. Chem. Soc. A*, 423 (1967).
24. A. M. Lazutkin and A. J. Lazutkina, *Zh. Obshch. Khim.*, **47**, 1398 (1977).
25. F. A. Cotton, *Chem. Soc. Rev.*, **4**, 27 (1975).
26. H. A. Martin and F. Jellinek, *J. Organomet. Chem.*, **8**, 115 (1967).
27. H. A. Martin, P. J. Lemaire, and F. Jellinek, *J. Organomet. Chem.*, **14**, 149 (1968).

28. F. W. Siegert and H. J. de Liefde Meijer, *J. Organomet. Chem.*, **15**, 131 (1968).
29. F.W. Siegert and H. J. de Liefde Meijer, *J. Organomet. Chem.*, **23**, 177 (1970).
30. A. van Baalen, C. J. Groenenboom, and H. J. de Liefde Meijer, *J. Organomet. Chem.*, **74**, 245 (1974).
31. H. K. Hofstee, C. J. Groenenboom, H. O. van Ofen, and H. J. de Liefde Meijer *J. Organomet. Chem.*, **85**, 193 (1975).
32. H. J. Kablitz and G. Wilke, *J. Organomet. Chem.*, **51**, 241 (1973).
33. H. K. Kablitz, R. Kallweit, and G. Wilke, *J. Organomet. Chem.*, **44**, C49 (1972).
34. E. G. Hoffmann, R. Kallweit, G. Schroth, K. Seevogel, W. Stempfle, and G. Wilke, *J. Organomet. Chem.*, **97**, 183 (1975).
35. M. Cooke, R. J. Goodfellow, M. Green, and G. Parker, *J. Chem. Soc. A*, 16 (1971).
36. A. E. Smith, *Inorg. Chem.*, **11**, 2306 (1972).
37. A. J. Sivak and E. L. Muetterties, *J. Am. Chem. Soc.*, **101**, 4878 (1979).
38. M. L. H. Green, L. C. Mitchard, and W. E. Silverthorn, *J. Chem. Soc., Dalton Trans.*, 1952 (1974).
39. M. Gaudemar, *Bull. Soc. Chim. Fr.*, 974 (1962).
40. A. N. Nesmeyanov and A. Z. Rubezhov, *J. Organomet. Chem.*, **164**, 259 (1979).
41. A. N. Nesmeyanov, A. Z. Rubezhov, L. A. Leites, and S. P. Gubin, *J. Organomet. Chem*, 187 (1968).
42. A. S. Ivanov and A. Z. Rubezhov, *Izv. Akad. Nauk SSSR, Ser. Khim.*, 1349 (1979).
43. E. E. Abel and S. Moorhouse, *J. Chem. Soc., Dalton Trans.*, 1706 (1973).
44. R. A. Zelonka and M. C. Baird, *J. Organomet. Chem.*, **44**, 383 (1972).
45. T. J. Greenhough, P. Legzdins, D. T. Martin, and J. Trotter, *Inorg. Chem.*, **18**, 3268 (1979).
46. A. Z. Rubezhov and S. P. Gubin, *Adv. Organomet. Chem.*, **10**, 347 (1972).
47. A. Efraty, *J. Organomet. Chem.*, **57**, 1 (1973).
48. F. M. Chaudhari, G. R. Knox, and P. L. Pauson, *J. Chem. Soc. C*, 2255 (1967).
49. R. F. Heck, *J. Am. Chem. Soc.*, **85**, 655 (1963).
50. G. Cardaci and A. Foffani, *J. Chem. Soc., Dalton Trans.*, 1808 (1974).
51. G. Cardaci, *J. Chem. Soc., Dalton Trans.*, 2452 (1974).
52. E. L. Muetterties and F. J. Hirsekorn, *J. Am. Chem. Soc.*, **96**, 7920 (1974).
53. B. J. Brisdon, D. A. Edwards, and J. A. White, *J. Organomet. Chem.*, **175**, 113 (1979).
54. M. Schneider and E. Weiss, *J. Organomet. Chem.*, **73**, C7 (1974).
55. J. E. Ellis and R. A. Faltynek, *J. Organomet. Chem.*, **93**, 205 (1975).
56. U. Franke and E. Weiss, *J. Organomet. Chem.*, **172**, 341 (1979); **193**, 329 (1980).
57. R. B. King and M. Ishaq, *Inorg. Chim. Acta*, **4**, 258 (1970).
58. J. W. Faller, B. V. Johnson, and T. P. Dryja, *J. Organomet. Chem.*, **65**, 395 (1974).
59. R. B. King and A. Fronzglia, *J. Am. Chem. Soc.*, **88**, 709 (1966).
60. F. A. Cotton and T. J. Marks, *J. Am. Chem. Soc.*, **91**, 1339 (1969).
61. H. D. Murdoch, *J. Organomet. Chem.*, **4**, 119 (1965); H. D. Murdoch and R. Henzi, *J. Organomet. Chem.*, **5**, 552 (1966).
62. M. Boyer, J. C. Daran and Y. Jeannin, *J. Organomet. Chem.*, **190**, 177 (1980).
63. D. H. Gibson, W.-L. Hsu and D.-S. Lin, *J. Organomet. Chem.*, **172**, C7 (1979).
64. J. A. Gladysz, G. M. Williams, W. Tam, and D. L. Johnson, *J. Organomet. Chem.*, **140**, C1 (1977).
65. K. Inkrott, R. Goetze, and S. G. Shore, *J. Organomet. Chem.*, **154**, 337 (1978).
66. F. W. Benfield, B. R. Francis, M. L. H. Green, N. T. Luong-Thi, G. Moser, J. S. Poland and D. M. Roe, *J. Less-Common Met.*, **36**, 187 (1974).
67. B. R. Francis, M. L. H. Green, L. Luong-Thi, and G. A. Moser, *J. Chem. Soc., Dalton Trans.*, 1339 (1976).
68. M. L. H. Green, T. Luong-Thi, G. A. Moser, I. Packer, F. Pettit, and D. M. Roe, *J. Chem. Soc., Dalton Trans.*, 1988 (1976).
69. M. L. H. Green, R. E. Mackenzie, and J. S. Poland, *J. Chem. Soc., Dalton Trans.*, 1993 (1976).
70. M. Ephritikhine, B. R. Francis, M. L. H. Green, R. E. Mackenzie, and M. J. Smith, *J. Chem. Soc., Dalton Trans.*, 1131 (1977).
71. S. G. Davies and M. L. H. Green, *J. Chem. Soc., Dalton Trans.*, 1510 (1978).
72. M. I. Lobach and V. A. Korner, *Russ. Chem. Rev.*, **48**, 758 (1979).

73. D. M. Roundhill, *Adv. Organomet. Chem.*, **13**, 273 (1975).
74. R. F. Heck, *Organotransition Metal Chemistry*, Academic Press, New York, 1974).
75. J. W. Connelly and C. D. Hoff, *J. Organomet. Chem.*, **160**, 467 (1978).
76. C. A. Tolman, *J. Amer. Chem. Soc.*, **92**, 6785 (1970).
77. J. F. Nixon, B. Wilkins, and D. A. Clement, *J. Chem. Soc., Dalton Trans.*, 1993 (1974).
78. M. A. Cairns and J. F. Nixon, *J. Chem. Soc., Dalton Trans.*, 2001 (1974).
79. Th. Kruck, I. P. Kuhnau, and G. Sylvester, *Z. Naturforsch.*, **28B**, 28 (1973).
80. A. J. Deeming, B. F. G. Johnson, and J. Lewis, *J. Chem. Soc., Chem. Commun.*, 598 (1970); *J. Chem. Soc., Dalton Trans.*, 1848 (1973).
81. H. C. Clark and H. Kurosawa, *J. Organomet. Chem.*, **36**, 399 (1972).
82. H. C. Clark and H. Kurosawa, *J. Chem. Soc., Chem. Commun.*, 957 (1971).
83. N. N. Druz, V. I. Klepikova, M. I. Lobach, and V. A. Kormer, *J. Organomet. Chem.*, **162**, 343 (1978).
84. V. A. Kormer, M. I. Lobach, and N. N. Druz, *Dokl. Akad. Nauk SSSR*, **246**, 1372 (1979).
85. G. R. Dobson, *Acc. Chem. Res.* **9**, 300 (1976).
86. J. D. Atwood and T. L. Brown, *J. Am. Chem. Soc.*, **98**, 3160 (1976).
87. J. L. Dineen and P. L. Pauson, *J. Organomet. Chem.*, **71**, 87 (1974).
88. J. Kwaitek, J. L. Mador, and J. K. Seyler, *Adv. Chem. Ser.*, **37**, 201 (1963).
89. P. M. Maitlis, *Acc. Chem. Res.*, **11**, 301 (1978).
90. H. B. Lee and P. M. Maitlis, *J. Chem. Soc., Dalton Trans.*, 2316 (1975).
91. P. S. Braterman and R. J. Cross, *Chem. Soc. Rev.*, **2**, 271 (1973).
92. H. A. Martin and F. Jellinek, *J. Organomet. Chem.*, **6**, 293 (1966).
93. H. A. Martin and F. Jellinek, *J. Organomet. Chem.*, **12**, 149 (1968).
94. A. Westerhof and H. J. de Liefde Meijer, *J. Organomet. Chem.*, **139**, 71 (1977).
95. J. Müller, H.-O. Stühler and W. Goll, *Chem. Ber.*, **108**, 1074 (1975).
96. H.-O. Stühler and J. Müller, *Chem. Ber.*, **112**, 1359 (1979).
97. J. Müller, H. Menig, G. Huttner, and A. Frank, *J. Organomet. Chem.*, **185**, 251 (1980).
98. J. Müller and W. Goll, *J. Organomet. Chem.*, **71**, 257 (1974).
99. H. O. van Oven, C. J. Groenenboom, and H. J. de Liefde Meijer, *J. Organomet. Chem.*, **81**, 379 (1974).
100. R. F. Heck, *Adv. Organomet. Chem.*, **4**, 343 (1966).
101. M. Green and R. I. Hancock, *J. Chem. Soc. A*, 109 (1968).
102. M. Brookhart, T. H. Whitesides, and J. M. Crockett, *Inorg. Chem.*, **15**, 1550 (1976), and references cited therein.
103. D. H. Gibson and D. K. Erwin, *J. Organomet. Chem.*, **86**, C31 (1975).
104. T. H. Whitesides and R. W. Arhart, *J. Am. Chem. Soc.*, **93**, 5296 (1971).
105. D. H. Gibson and R. L. Vonnahme, *J. Am. Chem. Soc.*, **94**, 5090 (1972).
106. T. H. Whitesides, R. W. Arhart, and R. W. Slaven, *J. Am. Chem. Soc.*, **95**, 5792 (1973).
107. B. F. G. Johnson, J. Lewis, and D. J. Yarrow, *J. Chem. Soc., Dalton Trans.*, 2084 (1972).
108. T. H. Whitesides and R. W. Arhart, *Inorg. Chem.*, **14**, 209 (1975).
109. S. D. Ittel, F. A. Van-Catledge, and J. P. Jesson, *J. Am. Chem. Soc.*, **101**, 6905 (1979).
110. T. V. Harris, J. W. Rathke, and E. L. Muetterties, *J. Am. Chem. Soc.*, **100**, 6966 (1978).
111. S. D. Ittel, F. A. Van-Catledge, C. A. Tolman, and J. P. Jesson, *J. Am. Chem. Soc.*, **100**, 1317 (1978).
112. J. M. Williams, R. K. Brown, A. J. Schultz, S. D. Stucky, and S. D. Ittel, *J. Am. Chem. Soc.*, **100**, 7407 (1978).
113. N. A. Clinton and C. P. Lillya, *J. Am. Chem. Soc.*, **92**, 3065 (1970).
114. C. P. Lillya and R. A. Sahatjian, *J. Organomet. Chem.*, **32**, 371 (1971).
115. M. Brookhart and D. L. Harris, *J. Organomet. Chem.*, **42**, 441 (1972).
116. P. Powell, *J. Organomet. Chem.*, **206**, 229 (1981).
117. E. O. Greaves, G. R. Knox, and P. L. Pauson, *J. Chem. Soc., Chem. Commun.*, 1124 (1969).
118. A. D. U. Hardy and G. A. Sim, *J. Chem. Soc., Dalton Trans.*, 2305 (1972).
119. E. O. Greaves, G. R. Knox, P. L. Pauson, S. Toma, G. A. Sim, and D. I. Woodhouse, *J. Chem. Soc., Chem. Commun.*, 257 (1974).
120. A. N. Nesmeyanov, L. V. Rybin, N. T., Gubenko, M. I. Rybinskaya, and P. V. Petrovskii, *J. Organomet. Chem.*, **71**, 271 (1974).

121. A. Bond, M. Green, B. Lewis, and S. F. W. Lowrie, *J. Chem. Soc., Chem. Commun.*, 1230 (1971).
122. A. Bond and M. Green, *J. Chem. Soc., Dalton Trans.*, 763 (1972).
123. A. Bond, B. Lewis, and M. Green, *J. Chem. Soc., Dalton Trans.*, 1109 (1975).
124. M. Green, B. Lewis, J. J. Daly, and F. Sanz, *J. Chem. Soc., Dalton Trans.*, 1118 (1975).
125. M. Green and B. Lewis, *J. Chem. Soc., Dalton Trans.*, 1137 (1975).
126. N. M. Boag, M. Green, J. L. Spencer, and F. G. A. Stone, *J. Chem. Soc., Dalton Trans.*, 1200 (1980).
127. R. B. King, *Organometallic Syntheses*, Vol. 1, Academic Press, New York, 1965, p. 176.
128. J. W. Faller, *Adv. Organomet. Chem.*, **16**, 211 (1977).
129. H. Sakurai, Y. Kamiyama, and Y. Nakadoira, *J. Organomet. Chem.*, **184**, 13 (1980); *J. Am. Chem. Soc.*, **98**, 7453 (1976).
129a. P. Radnia and J. S. McKennis, *J. Am. Chem. Soc.*, **102**, 6349 (1980).
130. E. L. Muetterties and J. W. Rathke, *J. Chem. Soc., Chem. Commun.*, 850 (1974).
131. T. Kruck and L. Knoll, *Z. Naturforsch.*, **26B**, 34 (1973).
132. G. Sbrana, G. Braca, F. Piacenti, and P. Pino, *J. Organomet. Chem.*, **13**, 240 (1968).
133. H. D. Murdoch and E. A. C. Lucken, *Helv. Chim. Acta*, **47**, 1517, (1964).
134. E. L. Muetterties, B. A. Sosinsky, and K. I. Zamarev, *J. Am. Chem. Soc.*, **97**, 5299 (1975).
135. C. F. Putrick, J. J. Welter, G. D. Stucky, M. J. D'Aniello, B. A. Sosinsky, J. F. Kirner, and E. L. Muetterties, *J. Am. Chem. Soc.*, **100**, 4107 (1978).
136. S. D. Ittel, P. J. Krusie, and P. Meakin, *J. Am. Chem. Soc.*, **100**, 3264 (1978).
137. B. Bogdanovic, *Adv. Organomet. Chem.*, **17**, 105 (1979).
138. M. F. Semmelhack, *Org. React.*, **19**, 115 (1972).
139. E. O. Fischer and R. D. Fischer, *Z. Naturforsch*, **16B**, 475 (1961).
140. R. F. Heck, *J. Org. Chem.*, **28**, 604 (1963).
141. T. Aviles and M. L. H. Green, *J. Chem. Soc., Dalton Trans.*, 1116 (1979).
142. M. L. H. Green and R. B. A. Pardy, *J. Chem. Soc., Dalton Trans.*, 355 (1979).
143. T. Chiang, R. C. Kerber, D. S. Kimball, and J. W. Lauher, *Inorg. Chem.*, **18**, 1687 (1979).
144. R. G. Hayter, *J. Organomet. Chem.*, **13**, P1 (1968).
145. B. J. Brisdon, *J. Organomet. Chem.*, **125**, 225 (1977).
146. H. Friedel, I. W. Renk, and H. tom Dieck, *J. Organomet. Chem.*, **26**, 247 (1971).
147. H. tom Dieck and F. Hohmann, *J. Less-Common Met.*, **54**, 221 (1977).
148. C. G. Hull and M. H. B. Stiddard, *J. Organomet. Chem.*, **9**, 519 (1967).
149. P. Powell, *J. Organomet. Chem.*, **129**, 175 (1977).
150. C. E. Holloway, J. D. Kelly, and M. H. B. Stiddard, *J. Chem. Soc. A*, 931 (1969).
151. N. A. Bailey, W. G. Kita, J. A. McCleverty, A. J. Murray, B. E. Mann, and N. W. J. Walker, *J. Chem. Soc., Chem. Commun.*, 592 (1974).
152. J. A. McCleverty and A. J. Murray, *Transition Met. Chem.*, **4**, 273 (1979).
153. J. W. Faller and A. M. Rosan, *J. Am. Chem. Soc.*, **98**, 3388 (1976).
154. J. W. Faller, C. C. Chen, M. J. Mattina, and A. Jakubowski, *J. Organomet. Chem.*, **52**, 361 (1973).
155. E. O. Fischer and G. Bürger, *Z. Naturforsch.*, **16B**, 702 (1961).
156. P. L. Timms and T. W. Turney, *Adv. Organomet. Chem.*, **15**, 53 (1977), and references cited therein.
157. J. R. Blackborow and D. Young, *Metal Vapour Synthesis in Organometallic Chemistry*, Springer Verlag, Berlin, 1979.
158. K. J. Klabunde, *Acc. Chem. Res.*, **8**, 393 (1975).
159. J. S. Roberts and K. J. Klabunde, *J. Am. Chem. Soc.*, **99**, 2509 (1977).
160. R. D. Rieke, *Acc. Chem. Res.*, **10**, 301 (1977).
161. R. D. Rieke, A. V. Kavaliunas, L. D. Rhyme, and D. J. J. Fraser, *J. Am. Chem. Soc.*, **101**, 246 (1979).
162. W. Schaefer, H. J. Kerrines, and U. Langbein, *Z. Anorg. Allg. Chem.*, **406**, 101 (1974).
163. G. Vitulli, P. Pertici, C. Agami, and L. Porri, *J. Organomet. Chem.*, **84**, 399 (1975).
164. G. Vitulli, L. Porri, and A. L. Segre, *J. Chem. Soc. A*, 3246 (1971).
165. R. A. Schunn, *Inorg. Synth.*, **15**, 5 (1974).
166. F. Guerrieri and G. Salerno, *J. Organomet. Chem.*, **114**, 339 (1976).
167. J. L. Spencer, *Inorg. Syn.*, **19**, 213 (1978).
168. G. E. Herberich and B. Hessner, *Z. Naturforsch*, **34B**, 638 (1979).

169. N. M. Boag, M. Green, J. L. Spencer, and F. G. A. Stone, *J. Chem. Soc., Dalton Trans.*, 1208 (1980).
170. N. M. Boag, M. Green, J. L. Spencer, and F. G. A. Stone, *J. Chem. Soc., Dalton Trans.*, 1220 (1980).
171. P. M. Maitlis, *The Organic Chemistry of Palladium*, Vol. 1, Academic Press, New York, 1971, pp. 15–26.
172. C. A. Tolman, *Chem. Rev.*, **77**, 313 (1977).
173. J. Powell and B. L. Shaw, *J. Chem. Soc. A*, 774 (1968).
174. H. C. Volger and K. Vrieze, *J. Organomet. Chem.*, **9**, 527 and 537 (1967).
175. H. Kurosawa and G. Yoshida, *J. Organomet. Chem.*, **120**, 297 (1976).
176. H. Kurosawa, *J. Organomet. Chem.*, **112**, 369 (1976).
177. H. Kurosawa, *J. Chem. Soc., Dalton Trans.*, 939 (1979).
178. Y. Takahashi, T. Ito, S. Sakai, and Y. Ishii, *J. Chem. Soc., Chem. Commun.*, 1065 (1970).
179. K. Moseley and P. M. Maitlis, *J. Chem. Soc., Chem. Commun.*, 982 and 1604 (1971).
180. T. S. Ito, S. Hasegawa, Y. Takahashi, and Y. Ishii, *J. Organomet. Chem.*, **73**, 401 (1974).
181. T. Ukai, H. Kawazura, Y. Ishii, J. J. Bonnet, and J. A. Ibers, *J. Organomet. Chem.*, **65**, 253 (1974).
182. A. Sonoda, B. E. Mann, and P. M. Maitlis, *J. Chem. Soc., Chem. Commun.*, 108 (1975).
183. A. Keasey and P. M. Maitlis, *J. Chem. Soc., Dalton Trans.*, 1830 (1978).
184. J. Powell and B. L. Shaw, *J. Chem. Soc. A*, 583 (1968).
185. J. W. Dart, M. K. Lloyd, R. Mason, and J. A. McCleverty, *J. Chem. Soc., Dalton Trans.*, 2039 and 2046 (1973).
186. A. J. Deeming and B. L. Shaw, *J. Chem. Soc. A*, 1562 (1969).
187. R. G. Pearson and A. T. Poulos, *Inorg. Chim. Acta*, **34**, 67 (1979).
188. M. L. H. Green and W. E. Silverthorn, *J. Chem. Soc., Dalton Trans.*, 301 (1973); *J. Chem. Soc., Chem. Commun.*, 577 and 1619 (1971).
189. M. L. H. Green and J. Knight, *J. Chem. Soc., Dalton Trans.*, 311 (1974).
190. M. L. H. Green, L. C. Mitchard, and W. E. Silverthom, *J. Chem. Soc., Dalton Trans.*, 1403 (1973).
191. M. L. H. Green, J. Knight, and J. A. Segal, *J. Chem. Soc., Dalton Trans.*, 2189 (1977).
192. M. Bottrill and M. Green, *J. Chem. Soc., Dalton Trans.*, 2365 (1977).
193. S. G. Davies, M. L. H. Green, and D. M. P. Minogs, *Tetrahedron*, **34**, 3047 (1978).
194. D. F. Pollock and P. M. Maitlis, *J. Organomet. Chem.*, **26**, 407 (1971).
195. F. Canziani, P. Chini, A. Quartz, and A. DiMartino, *J. Organomet. Chem.*, **26**, 285 (1971).
196. S. H. Taylor and P. M. Maitlis, *J. Am. Chem. Soc.*, **100**, 4700 (1978).
197. T. R. Jack, C. J. May, and J. Powell, *J. Am. Chem. Soc.*, **100**, 5057 (1978).
198. C. J. May and J. Powell, *J. Organomet. Chem.*, **184**, 385 (1980).
199. C. Charrier, J. Collin, J. Y. Mérour, and J. L. Roustan, *J. Organomet. Chem.*, **162**, 57 (1978).
200. J. Collin, J. L. Roustan, and P. Cadiot, *J. Organomet. Chem.*, **162**, 67 (1978).
201. J. Collin, C. Charrier, M. J. Pouet, P. Cadiot, and J. L. Roustan, *J. Organomet. Chem.*, **168**, 321 (1979).
202. J. L. Roustan, J. Y. Mérour, C. Charrier, J. Bénaïm, and P. Cadick, *J. Organomet. Chem.*, **168**, 61 (1979).
203. J. Y. Mérour, J. L. Roustan, C. Charrier, J. Bénaïm, J. Collin, and P. Cadiot, *J. Organomet. Chem.*, **168**, 337 (1979).
204. J. L. Roustan, J. Y. Mérour, C. Charrier, J. Bénaïm, and P. Cadiot, *J. Organomet. Chem.*, **169**, 39 (1979).
205. J. Collin, J. L. Roustan, and P. Cadiot, *J. Organomet. Chem.*, **169**, 53 (1979).
206. M. Bottrill and M. Green, *J. Chem. Soc., Dalton Trans.*, 820 (1979).
207. T. E. Bauch and W. P. Giering, *J. Organomet. Chem.*, **114**, 165 (1976).
207a. R. Hüttel, J. Kratzer and M. Bechter, *Chem. Ber.*, **94**, 766 (1961).
208. F. R. Hartley and S. R. Jones, *J. Organomet. Chem.*, **66**, 465 (1974), and references cited therein.
209. Y. Tatsuno, T. Yoshida, and S. Otsuka, *Inorg. Syn.*, **19**, 220 (1979).
210. J. Lukas and J. E. Blom, *J. Organomet. Chem.*, **26**, C25 (1971).
211. D. J. Mabbott, B. E. Mann, and P. M. Maitlis, *J. Chem. Soc., Dalton Trans.*, 294 (1977).

212. H. Kurosawa, *Inorg. Chem.*, **14**, 2148 (1975).
213. H. Kurosawa, *Inorg. Chem.*, **15**, 120 (1976).
214. R. Hüttel and H. Christ, *Chem. Ber.*, **96**, 3101 (1963), and references cited therein.
215. H. C. Volger, *Rec. Trav. Chim.*, **87**, 225 (1969).
215a. J. Morelli, R. Ugo, F. Canti and M. Donati, *J. Chem. Soc. Chem. Commun.*, 801 (1967).
216. J. Lukas, *Inorg. Syn.* **15**, 75 (1974).
217. B. M. Trost, P. E. Strege, L. Weber, T. J. Fullerton, and T. Dietsche, *J. Am. Chem. Soc.*, **100**, 3407 (1978).
217a. B. M. Trost, *Acc. Chem. Res.*, **13**, 385 (1980).
218. J. Y. Satoh and C. A. Horiuchi, *Bull. Chem. Soc. Japan,* **52**, 2653 (1979), and references cited therein.
219. R. C. Larock and J. P. Burkhart, *Synth. Commun.*, **9**, 659 (1979).
220. I. G. Dinulescu, S. Staicu, F. Chiraleu, and M. Avram, *J. Organomet. Chem.*, **140**, 91 (1977).
221. R. G. Brown, R. V. Chaudhari, and J. M. Davidson, *J. Chem. Soc., Dalton Trans.*, 176 (1977).
222. M. Donati and F. Conti, *Tetrahedron Lett.*, 1219 (1966).
223. S. D. Robinson and B. L. Shaw, *J. Chem. Soc.*, 4806 (1963).
224. R. Pietropaolo, F. Faraone, D. Pietropaolo, and P. Pirairo, *J. Organomet. Chem.*, **64**, 403 (1974).
225. R. Pietropaolo, E. Rotondo, F. Faraone, and D. Pietropaolo, *J. Organomet. Chem.*, **60**, 197 (1973).
226. J. M. Rowe and D. A. White, *J. Chem. Soc. A*, 1451 (1967).
227. O. G. Levanda, G. Yu. Pek, and I. I. Moiseev, *Zh. Org. Khim.*, **7**, 217 (1971).
228. B. Akermark, J. E. Backvall, A. Lowenberg, and K. Zetterberg, *J. Organomet. Chem.*, **166**, C33 (1979).
229. J. Lukas, P. W. N. M. van Leewen, H. C. Volger, and A. P. Kouwenhoven, *J. Organomet. Chem.*, **47**, 153 (1973).
230. K. Tsukiyama, Y. Takahashi, S. Sakai, and Y. Ishii, *J. Chem. Soc. A*, 3112 (1971).
231. J. Powell and B. L. Shaw, *J. Chem. Soc. A*, 597 (1968).
232. A. C. L. Su, *Adv. Organomet. Chem.*, **17**, 269 (1979).
233. D. Medema and R. van Helden, *Rec. Trav. Chim.*, **90**, 304 (1971).
234. R. P. Hughes and J. Powell, *J. Am. Chem. Soc.*, **94**, 7723 (1972).
235. F. L. Bowden and R. Giles, *Coord. Chem. Rev.*, **20**, 81 (1976).
236. A. Nakamura, P.-J. Kim and N. Hagihara, *J. Organomet. Chem.*, **3**, 7 (1965); R. Ben-Shoshan and R. Pettit, *J. Am. Chem. Soc.*, **89**, 2231 (1967).
237. D. H. Gibson, R. I. Vonnahme, and J. E. McKiernan, *J. Chem. Soc., Chem. Commun.*, 720 (1971).
238. P. Racanelli, G. Pantini, A. Immerzi, G. Allegra, and L. Porri, *J., Chem. Soc., Chem Commun.*, 361 (1969).
239. J. A. Osborn, *J. Chem. Soc., Chem. Commun.*, 1231 (1968).
240. M. S. Lupin, J. Powell, and B. L. Shaw, *J. Chem. Soc. A*, 1688 (1966); R. G. Schultz, *Tetrahedron*, **20**, 2809 (1964).
241. T. Susuki and J. Tsuji, *Bull. Chem. Soc. Jap.*, **41**, 1954 (1968).
242. T. Okamoto, *Bull. Chem. Soc. Jap.*, **44**, 1353 (1971).
243. S. Otsuka, A. Nakamura, and K. Tani, *J. Chem. Soc. A*, 154 (1971).
244. N. Yasuda, Y. Kai, N. Yasuoka, and N. Kasai, *J. Chem. Soc. Chem. Commun.*, 157 (1972).
245. S. Otsuka, A. Nakamura, S. Veda, and K. Tani, *J. Chem. Soc., Chem Commun.*, 963 (1971).
246. G. Ingrosso, L. Porri, G. Pantini, and P. Racinelli, *J. Organomet. Chem.*, **84**, 75 (1975).
247. G. Ingrosso, P. Gronchi, and L. Porri, *J. Organomet. Chem.*, **86**, C20 (1975).
248. M. Green, J. A. K. Howard, R. P. Hughes, S. C. Kellett, and P. Woodward, *J. Chem. Soc., Dalton Trans.*, 2007 (1975).
249. M. Green and R. P. Hughes, *J. Chem. Soc., Dalton Trans.*, 1880 (1976).
250. R. P. Hughes, D. E. Hunton, and K. Schumann, *J. Organomet. Chem.*, **169**, C37 (1979).
251. B. K. Dallas and R. P. Hughes, *J. Organomet. Chem.*, **184**, C67 (1980), and references cited therein.

252. P. Mushak and M. A. Battiste, *J. Organomet. Chem.*, **17**, P46 (1969); M. A. Battiste, L. E. Friedrich, and R. A. Fiato, *Tetrahedron Lett.*, 45 (1975).
253. R. L. Phillips and R. J. Puddephatt, *J. Chem. Soc., Dalton Trans.*, 1732 (1978).
254. T. H. Tulip and J. A. Ibers, *J. Am. Chem. Soc.*, **101**, 4201 (1979).
255. A. Ducruix, H. Felkin, C. Pascard, and G. K. Turner, *J. Chem. Soc., Chem. Commun.*, 615. (1975).
256. H. Felkin and G. K. Turner, *J. Organomet. Chem.*, **129**, 429 (1977).
257. H. Werner, *Angew. Chem. Int. Ed. Engl.*, **16**, 1 (1977).
258. H. Werner, A. Kühn, D. J. Tune, C. Kruger, D. J. Brauer, J. C. Sekutowski, and Y. H. Tsay, *Chem. Ber.*, **110**, 1763 (1977).
259. A. Kühn and H. Werner, *J. Organomet. Chem.*, **179**, 421 (1979).
260. H. Werner and A. Kühn, *J. Organomet. Chem.*, **179**, 439 (1979).

The Chemistry of the Metal–Carbon Bond
Edited by F. R. Hartley & S. Patai

CHAPTER **9**

The synthesis of η^4-butadiene and cyclobutadiene complexes

G. MARR and B. W. ROCKETT
Department of Physical Sciences, The Polytechnic, Wolverhampton WY1 1LY, UK

I. η^4-BUTA-1,3-DIENE COMPLEXES

A. Introduction

Iron shows a high chemical affinity for compounds containing the buta-1,3-diene group and (η^4-buta-1,3-diene)tricarbonyliron complexes dominate the chemistry of this ligand[1]. In general terms, three synthetic approaches to these compounds are available. The first route involves the reaction between an unsaturated hydrocarbon ligand such as a butadiene, an α,β-unsaturated ketone, a methylenecyclopropane, a vinylcyclopropane, a cyclopropene, an allyl compound or a cyclobutene, together with a transition metal carbonyl, carbonyl derivative, or salt. The second route brings together metal atoms, the vapour of the hydrocarbon ligand, and either carbon monoxide or a similar ligand, and these are co-condensed at low temperature and pressure. The third route uses a preformed (η-hydrocarbon)transition metal complex such as an η^2-alkene, an η^3-allyl, an η^4-trimethylenemethane, or an η^5-dienyl complex and converts this, with the aid of a suitable reagent, to the desired η^4-butadiene compound.

B. From Buta-1,3-dienes and Transition Metal Carbonyls, Carbonyl Derivatives or Transition Metal Salts

The reaction between a buta-1,3-diene and a transition metal carbonyl provides the most important source of η^4-buta-1,3-diene complexes and large numbers of these complexes have been prepared in this way. The parent complex, (η^4-buta-1,3-diene)tricarbonyliron (**1**) was first prepared in 1930 by heating butadiene with pentacarbonyliron in a pressure tube[2]. Essentially the same procedure is still employed;

$$\text{butadiene} + [Fe(CO)_5] \xrightarrow[\text{autoclave}]{140°C} [(\eta^4\text{-}C_4H_6)Fe(CO)_3] \quad (1)$$

(**1**)

the two reagents are introduced into a rocking autoclave and heated at 140°C for 12 h. The product is obtained in 42% yield as an orange–yellow oily liquid by vacuum distillation and it crystallizes from light petroleum on cooling to give a solid, m.p. 19°C[2–4]. This preparation is exceptional in requiring a pressure vessel, as most 1,3-diene ligands are sufficiently involatile to undergo reaction at atmospheric pressure. The latter procedure has the advantage that the carbon monoxide liberated in the displacement can escape and the equilibrium in the reaction (equation 2) is displaced to the right. The

$$2\ \text{diene} + [M(CO)_mCO \rightleftharpoons [(\text{diene})_aM(CO)_b] + (m - b) \qquad (2)$$

reaction is usually carried out at elevated temperatures using hydrocarbon solvents. Many complexes have been prepared in this way and some typical examples are discussed. Diphenylbuta-1,4-diene is heated with dodecacarbonyltriiron in benzene at the reflux temperature for 6 h to give the tricarbonyliron complex (**2**)[5,6]. Substituted tricarbonyl(η^4-penta-1,3-diene)iron complexes (**3**) are obtained by heating dodecacar-

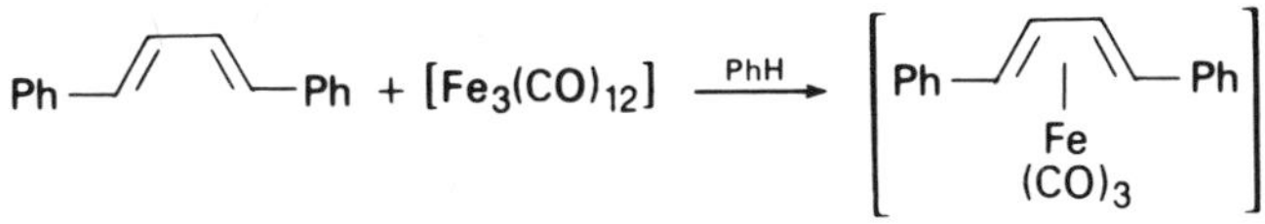

Ph—CH=CH—CH=CH—Ph + $[Fe_3(CO)_{12}]$ —PhH→ [Ph—diene—Ph, Fe$(CO)_3$] (3)

(**2**)

bonyltriiron with the appropriate penta-1,3-diene in benzene under nitrogen until the green colour of the original metal carbonyl is discharged; this takes about 10 h. Complexes prepared in this way include the ester, amide, ketone, aldehyde, and nitrile (**3**; X = CO_2Et, $CONH_2$, COMe, CHO, CN, respectively), as yellow or orange oils or

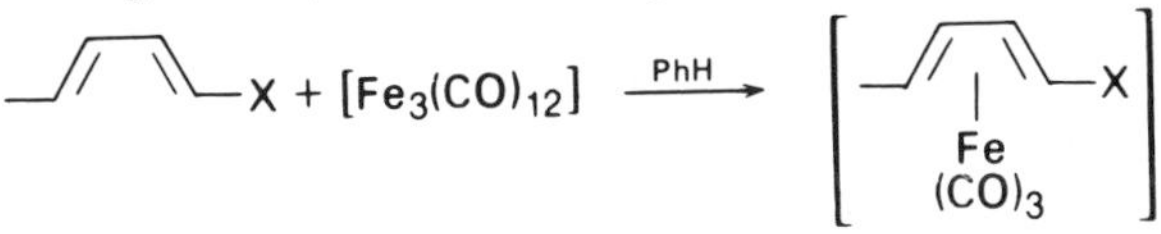

diene—X + $[Fe_3(CO)_{12}]$ —PhH→ [diene—X, Fe$(CO)_3$] (4)

(**3**)

X = CO_2Et, $CONH_2$, COMe, CHO, CN

low-melting solids[7]. A series of mono-, di-, and tri-methylbutadieneirontricarbonyls may be prepared by heating enneacarbonyldiiron with the butadiene in anhydrous ether at 40–45 °C until the iron carbonyl disappears, which takes about 1.5 h. The same complexes are also formed by irradiation of the butadiene with pentacarbonyliron at 0 °C in the absence of air. Compounds obtained in this way include **4** (R = H, Me), **5** (R = H, Me), **6** (R = H, Me), and **7** (R = H, Me)[8]. Siloxybutadienes combine with dodecacarbonyltriiron on heating to give tricarbonyliron complexes[9].

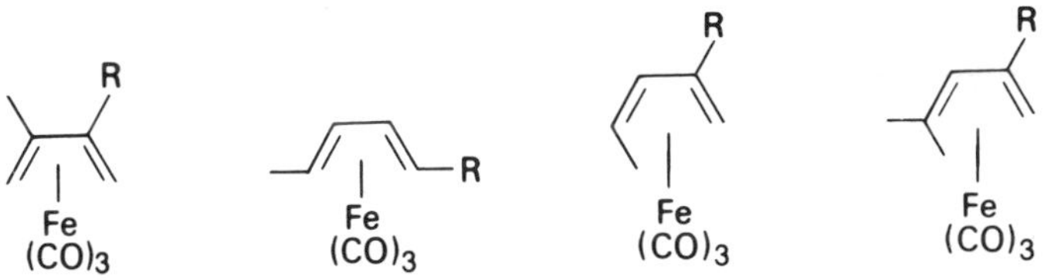

R, Fe$(CO)_3$ (**4**) R, Fe$(CO)_3$ (**5**) R, Fe$(CO)_3$ (**6**) R, Fe$(CO)_3$ (**7**)

In these reactions and in many related ones, pentacarbonyliron is the preferred reagent for use in photochemically induced processes, whereas enneacarbonyldiiron and dodecacarbonyltriiron are usually employed in thermally induced processes. In some cases, irradiation of the metal carbonyl and diene gives a different product or product mixture from that obtained by heating. Thus, irradiation of isoprene with pentacarbonyliron gives tricarbonyl(η^4-isoprene)iron (**8**) in low yield, whereas heating under pressure gives a mixture of the isoprene complex (**8**) and the α-terpene complex (**9**), the ligand in this last complex being formed by dimerization of isoprene[7].

The aromatic nucleus in naphthalene contributes a pair of electrons to bonding with iron in tricarbonyl (η^4-1-vinylnaphthalene)iron (**10**), which may be regarded as an analogue of the butadiene complex (**1**). The naphthalene ring is sufficiently activated

$[Fe(CO)_3]$ (8) $\xleftarrow[h\nu]{[Fe(CO)_5]}$ $\xrightarrow[\text{heat, autoclave}]{[Fe(CO)_5]}$ (8) + $[Fe(CO)_3]$ (9) (5)

$+ [Fe_3(CO)_{12}] \longrightarrow$ $[\ —Fe(CO)_3]$ (10) (6)

by the vinyl group to allow formation of the complex (**10**) to occur on heating 1-vinylnaphthalene with dodecacarbonyltriiron in hexane at 90°C for 16 h. The product is obtained in 31% yield[10]. In a number of other cases extended conjugation of an aromatic residue permits the π-electrons to take part in bonding to iron[10,11]. Hexa-1,3,5-triene, with three conjugated double bonds, forms a tricarbonyliron complex in the same way on heating with enneacarbonyldiiron[12].

The reaction between 1,2-dimethylenecyclobutane and dodecarbonyltriiron in benzene at reflux temperature gives the expected tricarbonyl(η^4-1,2-dimethylenecyclobutane)iron (**11**) in 24% yield, together with a bis(η^3-allyl)dicarbonyliron complex (**12**)

$+ [Fe_3(CO)_{12}] \xrightarrow[\text{heat}]{PhH}$ $[Fe(CO)_3]$ (11) + $[Fe(CO)_2]$ (12) (7)

in similar yield (25%)[13]. This reaction illustrates another important feature found in the reactions between dienes and iron carbonyls, namely the tendency for a mixture of products to be obtained, often in low yields. Ligand dimerization[7,13] and rearrangement by hydrogen migration[7,14] are frequently encountered but are not necessarily disadvantageous since the desired product may be obtained by the controlled application of such transformations.

It is also possible for more than one molecule of the ligand to be bound to iron, as occurs on irradiation of butadiene and pentacarbonyliron in benzene or pentane to form bis(η^4-butadiene)carbonyliron (**13**)[15].

$+ [Fe(CO)_5] \xrightarrow[h\nu]{PhH}$ $[FeCO]$ (13) (8)

Similar (η^4-diene)$_2$FeCO complexes are formed from isoprene, 2,3-dimethylbutadiene and methyl sorbate by the same reaction[15]. Alternatively, iron(III) chloride combines with butadiene, isoprene and penta-1,3-diene in the presence of carbon monoxide to form the corresponding bis(η^4-diene)iron carbonyl complexes[16]. Further examples include the formation of the *exo*- and *endo*-bicyclooctane complexes (**14**) and the corresponding hexacarbonyldiiron complexes from the free ligand and enneacarbonyldiiron[17]; other dimethylidene bicycloheptanes, -heptenes, -octanes, and -octenes combine with the same carbonyl in the same way[18,19] or with pentacarbonyliron on ultraviolet irradiation[19,20]. The (η^4-butadiene)iron complex (**15**) and related species may be prepared from the ligand and pentacarbonyliron or enneacarbonyldiiron[21] and tricarbonyl(η^4-homobarrelene)iron is obtained from homobarrelene and pentacarbonyliron[22]. Pentamethyldisilane combines with enneacarbonyldiiron to form $[Me_2Si{=}FeH(SiMe_3)(CO)_3]$, which transfers tricarbonyliron to 1,4-diphenylbutadiene to give the (η^4-diphenylbutadiene)iron complex (**2**)[23].

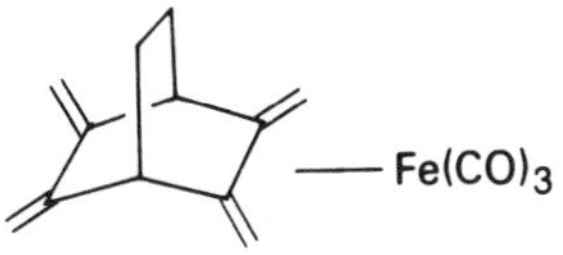

(**14**) 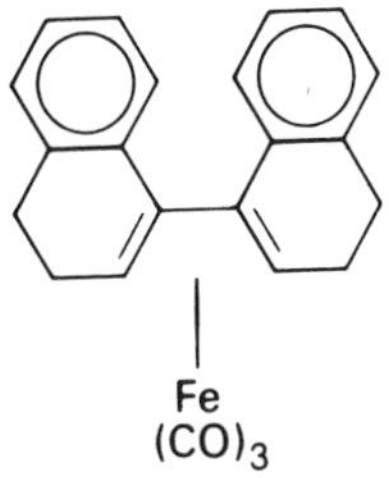 (**15**)

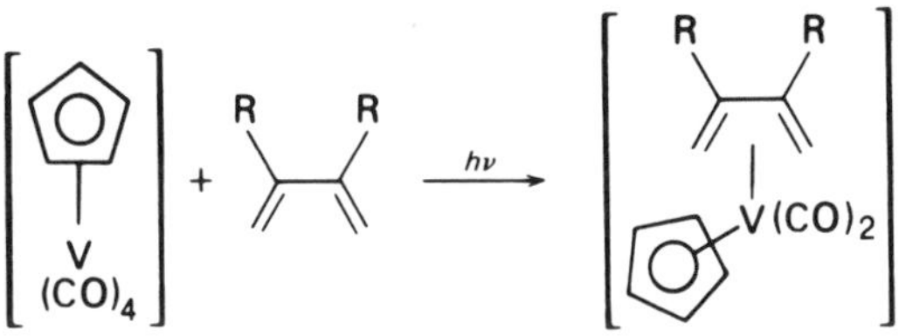

(**16**) (**17**) (9)

R = H, Me

In addition to iron, several other transition metals may form η^4-buta-1,3-diene complexes. Irradiation of tetracarbonyl(η^5-cyclopentadienyl)vanadium (**16**) with butadiene or 2,3-dimethylbutadiene gives the mixed (η^4-buta-1,3-diene)(η^5-cyclopentadienyl)vanadium complex (**17**; R = H, Me)[21]. The molybdenum (**18**)[21], manganese (**19**)[21], cobalt (**20**; R = H, Me)[24,25], (**21**)[26], and rhodium (**22**)[27] complexes may be prepared in the same way and isolated as crystalline solids by vacuum sublimation or chromatography. (η^4-Buta-1,3-diene)tricarbonylruthenium is not, however, obtained by direct reaction between butadiene and dodecacarbonyltriruthenium[28], but is formed by heating tricarbonyl(η^4-cycloocta-1,5-diene)ruthenium with butadiene in benzene in a sealed tube at 100°C[29].

α,β-Unsaturated ketones and α,β-unsaturated imines may behave as 1,3-diene ligands to transition metals such as iron. Thus, heating benzylideneacetone (**23**) with diiron enneacarbonyl in toluene at 60°C gives the η^4-heterodiene complex (**24**) in 32% yield[30]. The same procedure may be used to form iron tricarbonyl complexes of chalcone (PhCH=CHCOPh), dypnone (PhCMe=CHCOPh), and 2,6-bis(benzylidene)cyclohexanone (**25**)[30]. It has been suggested that the benzylideneacetone complex (**24**) may be used as a convenient source of η^4-diene com-

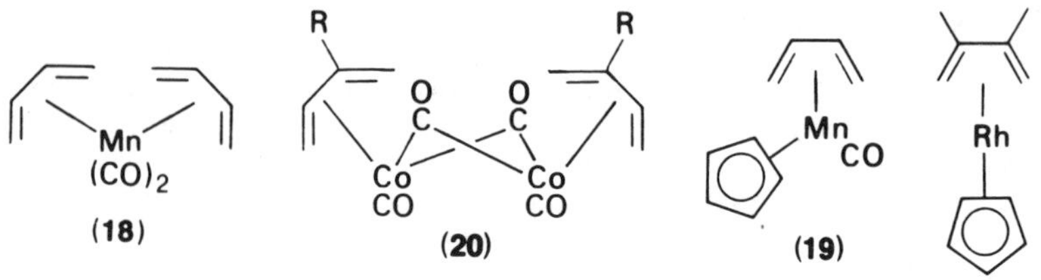

R R

Mn (CO)$_2$ (18)

O C O C Co CO Co CO (20)

Mn CO (19)

Rh (22)

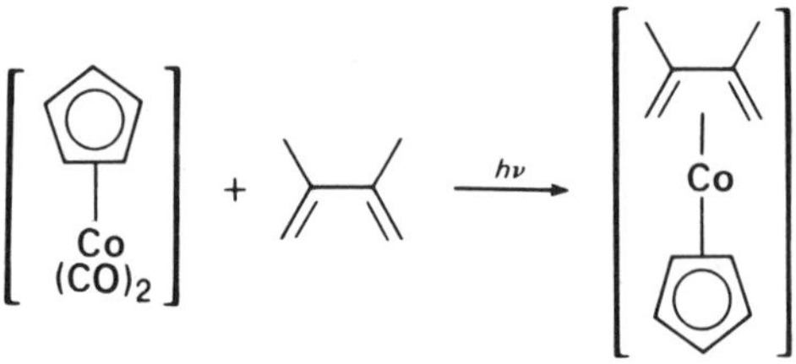

(10)

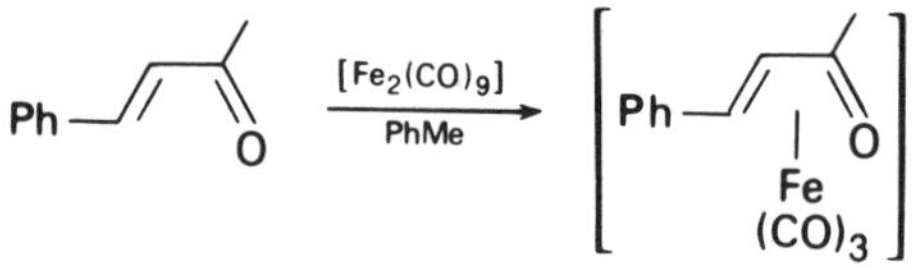

Co (CO)$_2$ + $\xrightarrow{h\nu}$ Co

(21)

Ph O $\xrightarrow[\text{PhMe}]{[Fe_2(CO)_9]}$ Ph O Fe (CO)$_3$ (11)

(23) (24)

Ph

Ph O Fe (CO)$_3$

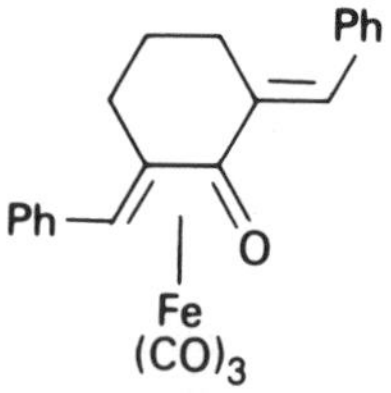

(25)

plexes since the heterodiene ligand is displaced by buta-1,3-dienes under mild conditions to give (η^4-buta-1,3-diene)tricarbonyliron derivatives[30]. In addition to α,β-unsaturated ketones, other heterodiene systems, namely C=C—C=N—[31], —N=C—C=N—[32], and —N=N—N=N—[33], form iron tricarbonyl complexes. Details of some typical preparations which involve direct combination of the ligand with an iron carbonyl are collected in Table 1.

C. From Buta-1,3-dienes and Transition Metal Carbonyls with Rearrangement or Modification of the Ligand

It has frequently been observed that the reaction between a buta-1,3-diene and a metal carbonyl gives an η^4-1,3-diene complex in which the identity of the ligand has changed on complexation[14,34]. Butadienes containing *cis*-alkyl substituents are subject to rearrangement; thus, *cis*-penta-1,3-diene **(26)** on heating with pentacarbonyliron

TABLE 1. Reagents and conditions used in the synthesis of (η^4-Buta-1,3-diene)tricarbonyliron complexes by direct combination between the ligand and an iron carbonyl

Carbonyl	Conditions	Product	Reference
$[Fe(CO)_5]$	140°C, autoclave	$[(CH_2{=}CHCH{=}CH_2)Fe(CO)_3]$	2, 3, 4
$[Fe_3(CO)_{12}]$	100°C, trimethylpentane	$[(CH_2{=}CHCH{=}CHMe)Fe(CO)_3]$	7
$[Fe(CO)_5]$	Irradiation	$[(CH_2{=}CMeCH{=}CH_2)Fe(CO)_3]$	7
$[Fe_3(CO)_{12}]$	80°C, benzene	$[(MeCH{=}CHCH{=}CHCOMe)Fe(CO)_3]$	7
$[Fe_3(CO)_{12}]$	80°C, benzene	$[(MeCH{=}CHCH{=}CHCHO)Fe(CO)_3]$	7
$[Fe_3(CO)_{12}]$	80°C, benzene	$[(MeCH{=}CHCH{=}CHCO_2Et)Fe(CO)_3]$	7
$[Fe_3(CO)_{12}]$	80°C, benzene	$[(MeCH{=}CHCH{=}CHCONH_2)Fe(CO)_3]$	7
$[Fe_3(CO)_{12}]$	80°C, benzene	$[(MeCH{=}CHCH{=}CHCN)Fe(CO)_3]$	7
$[Fe_3(CO)_{12}]$	80°C, benzene	$[(PhCH{=}CHCH{=}CHPh)Fe(CO)_3]$	5, 6
$[Fe_2(CO)_9]$	45°C, ether	$[(CH_2{=}CMeCR{=}CH_2)Fe(CO)_3]$**(4)**	8
$[Fe(CO)_5]$	Irradiation	$[(CH_2{=}CMeCR{=}CH_2)Fe(CO)_3]$**(4)**	8
$[Fe_2(CO)_9]$	45°C, ether	$[(MeCH{=}CHCH{=}CHR)Fe(CO)_3]$**(5)**	8
$[Fe(CO)_5]$	Irradiation	$[(MeCH{=}CHCH{=}CHR)Fe(CO)_3]$**(5)**	8
$[Fe_2(CO)_9]$	45°C, ether	$[(MeCH{=}CHCR{=}CH_2)Fe(CO)_3]$**(6)**	8
$[Fe(CO)_5]$	Irradiation	$[(MeCH{=}CHCR{=}CH_2)Fe(CO)_3]$**(6)**	8
$[Fe_2(CO)_9]$	45°C, ether	$[(Me_2C{=}CHCR{=}CH_2)Fe(CO)_3]$**(7)**	8
$[Fe(CO)_5]$	Irradiation	$[(Me_2C{=}CHCR{=}CH_2)Fe(CO)_3]$**(7)**	8
$[Fe_3(CO)_{12}]$	80°C. benzene	$[(PhCH{=}CHCH{=}CHPh)Fe(CO)_3]$	5
$[Fe_3(CO)_{12}]$	80°C, benzene	$[(4\text{-}CH_2{=}CHC_6H_4CH{=}CH_2)Fe_2(CO)_6]$	5
$[Fe_3(CO)_{12}]$	90°C, hexane	$[(1\text{-}CH_2{=}CHC_{10}H_7)Fe(CO)_3]$**(10)**	10
$[Fe_3(CO)_{12}]$	80°C, benzene	$[(C_6H_8)Fe(CO)_3]$**(11)**	13
$[Fe_2(CO)_9]$	60°C, toluene	$[(PhCH{=}CHCOMe)Fe(CO)_3]$**(24)**	30
$[Fe_2(CO)_9]$	80°C, toluene	$[(PhCH{=}CHCOPh)Fe(CO)_3]$	30
$[Fe_2(CO)_9]$	60°C, toluene	$[(PhCMe{=}CHCOPh)Fe(CO)_3]$	30
$[Fe_2(CO)_9]$ or $[Fe_3(CO)_{12}]$	40°C, benzene	$[(PhCH{=}CHCH{=}NPh)Fe(CO)_3]$	31
$[Fe_2(CO)_9]$	40°C, benzene	$[(MeCH{=}CHCH{=}NC_4H_9)Fe(CO)_3]$	31

(26) —[Fe(CO)5]→ [(27) Fe(CO)3] ←[Fe(CO)5] or [Fe3(CO)12]— (12)

(27)

gives tricarbonyl(η^4-*trans*-penta-1,3-diene)iron (**27**) by geometrical inversion of the ligand[14]. The same product is obtained by heating the non-conjugated diene, penta-1,4-diene, with either iron pentacarbonyl or triiron dodecacarbonyl[7]. 4-Methylpenta-1,3-diene (**28**) combines with pentacarbonyliron to give the *trans*-2-methylpenta-1,3-diene product (**31**)[34]. Mechanistic studies suggest that the product is determined by kinetic rather than thermodynamic factors. The reaction appears to proceed through the iron tetracarbonyl intermediate (**29**), which loses CO and undergoes hydrogen transfer to iron forming an allyl complex (**30**). A reversed metal—carbon hydrogen transfer to the other terminal carbon atom then gives the product (**31**)[35].

(28) —[Fe(CO)5]→ [Fe(CO)4] (29) → [Fe(CO)3H] (30) → [Fe(CO)3] (31) (13)

2,5-Dimethylhexa-2,4-diene (**32**) rearranges on heating with pentacarbonyliron to form tricarbonyl-(η^4-*trans*-2,5-dimethylhexa-2,4-diene)iron (**33**)[14]. However, as mentioned previously, *cis*-alkyl-substituted butadienes are attacked by iron carbonyls without ligand rearrangement under the appropriate conditions, as exemplified by the formation of the compounds **6** and **7**.

(32) —[Fe(CO)5]→ [Fe(CO)3] (33) (14)

2-Chlorobutadiene behaves normally towards enneacarbonyldiiron, giving the corresponding tricarbonyliron complex (**34**) together with polymeric products[36,37]. However, while 2,3-dichlorobutadiene gives the expected product (**35**) with dodecacarbonyltriiron[36], it undergoes insertion of an iron carbonyl fragment into the carbon—chlorine bond with enneacarbonyldiiron. This is followed by CO insertion and coupling to give the hexacarbonyldiiron product (**36**) or by reaction with by-product hydrogen chloride to give the 2-chlorobutadiene complex (**34**)[36] (Scheme 1).

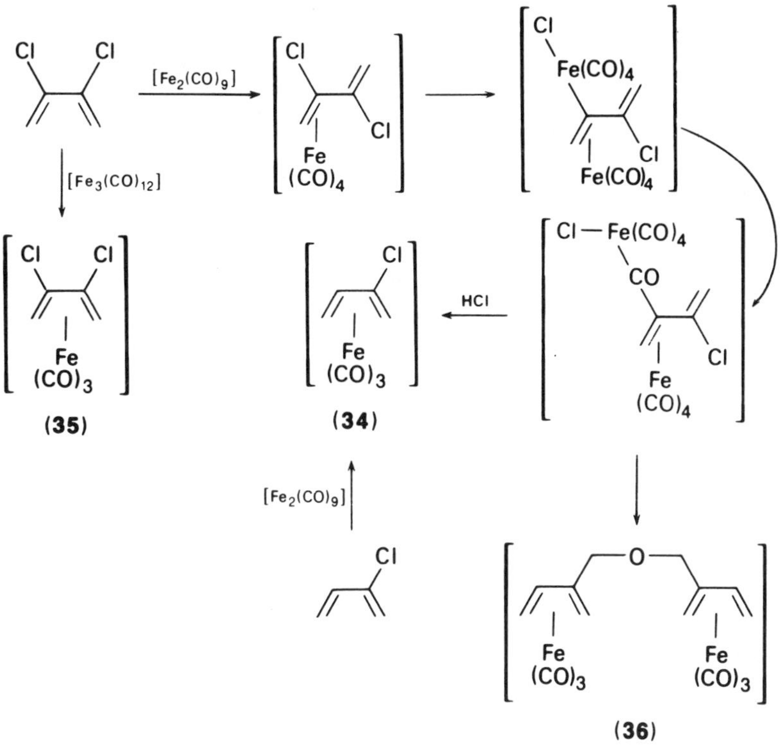

Cl Cl

$[Fe_2(CO)_9]$

Cl Cl Fe $(CO)_4$

Cl $Fe(CO)_4$ Cl $Fe(CO)_4$

$[Fe_3(CO)_{12}]$

Cl Cl Fe $(CO)_3$

(**35**)

Cl Fe $(CO)_3$

(**34**)

HCl

Cl — $Fe(CO)_4$ CO Cl Fe $(CO)_4$

$[Fe_2(CO)_9]$

Cl

O

Fe $(CO)_3$ Fe $(CO)_3$

(**36**)

SCHEME 1

Br Fe $(CO)_3$

(**37**)

Br $Fe(CO)_4$ Fe $(CO)_3$

HBr

Fe $(CO)_3$

(**38**)

$[Fe_2(CO)_9]$

Br

O

Fe $(CO)_3$ Fe $(CO)_3$

(**40**)

Fe $(CO)_3$ Fe $(CO)_3$

(**39**)

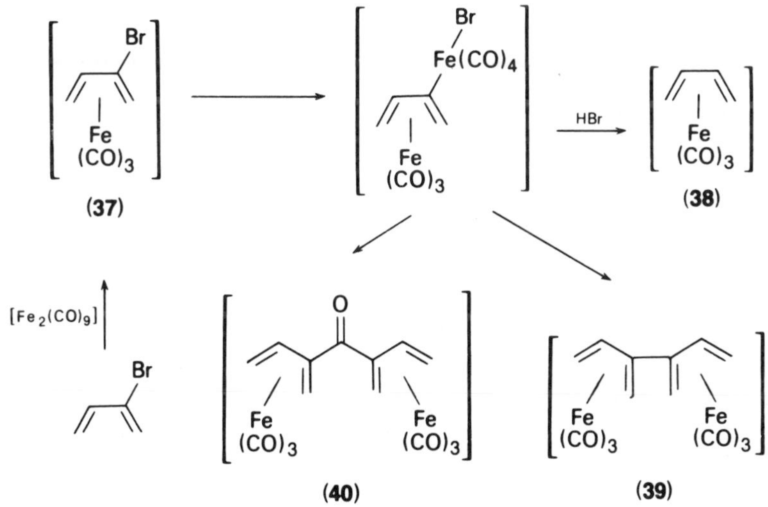

SCHEME 2

Several diene—iron complexes are isolated from the reaction of 2-bromobutadiene with enneacarbonyldiiron in boiling hexane including the expected 2-bromobutadiene complex (**37**), (η^4-butadiene)tricarbonyliron (**38**) and the binuclear complexes **39** and **40**, as shown in Scheme 2. Additional products are formed under different reaction conditions and the mechanism appears to involve insertion of a tricarbonyliron group into the carbon—bromine bond followed by one of several subsequent pathways[37]. The reaction between 2-bromobutadiene and dodecacarbonyltriiron in boiling benzene gives only the complexes **37** and **38** and provides a more satisfactory route to the bromo compound (**37**)[37]. Debromination also features in the reaction of α,α'-dibromo-*o*-xylylene with enneacarbonyldiiron in ether, which leads to tricarbonyl(η^4-*o*-xylene)iron (**41**) in low yield[38].

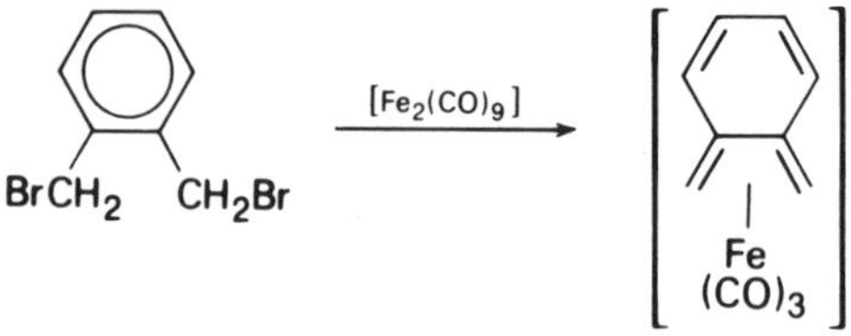

BrCH$_2$ CH$_2$Br $\xrightarrow{[Fe_2(CO)_9]}$ [Fe(CO)$_3$ complex] (15)

(**41**)

D. Cleavage of Strained Rings with Metal Carbonyls

1. *From vinylcyclopropanes and metal carbonyls*

Vinylcyclopropanes may undergo ring opening in the presence of metal carbonyls to give η^4-butadiene complexes. Irradiation of vinylcyclopropane with pentacarbonyliron in ether at -50 °C gives tetracarbonyl(η^2-vinylcyclopropane)iron (**42**) as the major product, which, on heating to 50 °C with benzene in a sealed tube, forms a 3:1 mixture of the isomeric (η^4-penta-1,3-diene)iron complexes **43** and **44**[39]. While the initial low-

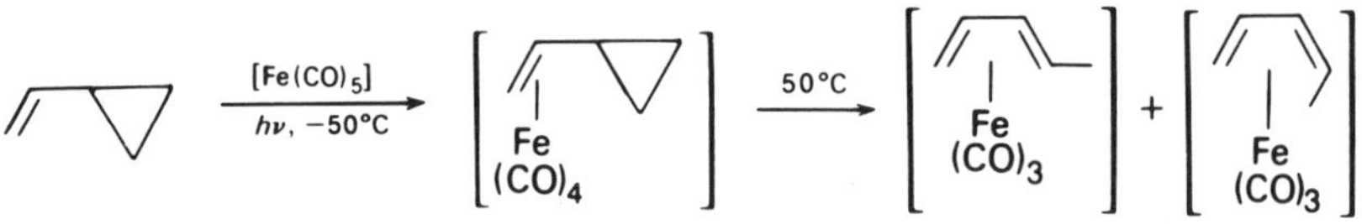

$\xrightarrow[h\nu,\ -50°C]{[Fe(CO)_5]}$ [Fe(CO)$_4$] $\xrightarrow{50°C}$ [Fe(CO)$_3$] + [Fe(CO)$_3$] (16)

(**42**) (**43**) (**44**)

temperature reaction to form the intermediate (**42**) is of mechanistic significance, it is unnecessary for the effective synthesis of the products **43** and **44**. Phenyl-[40], 4-chlorophenyl-[40], and 4-methoxyphenyl-substituted vinylcyclopropanes[41] undergo the same reaction with pentacarbonyliron to form the corresponding tricarbonyliron complexes (**45**; X = H, Cl, OMe) in yields of up to 61%.

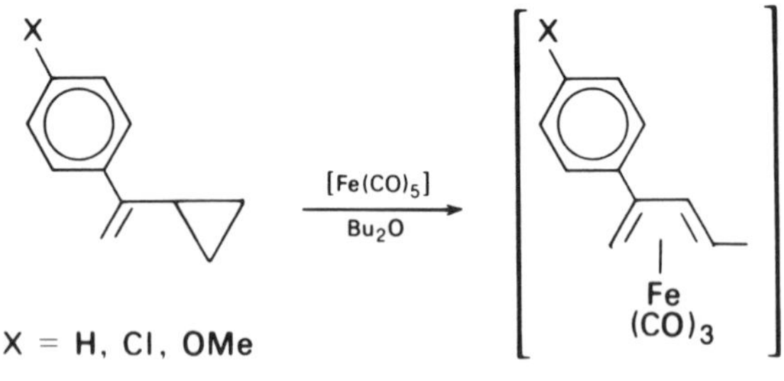

X $\xrightarrow[Bu_2O]{[Fe(CO)_5]}$ [X, Fe(CO)$_3$] (17)

X = H, Cl, OMe

(**45**)

The expected product (**46**) is obtained by heating 1,1-dicyclopropylethylene with pentacarbonyliron for 5 h[40]; however, on prolonged heating the dienone complex (**47**) is the dominant product. It is found that this product (**47**) is not obtained by heating

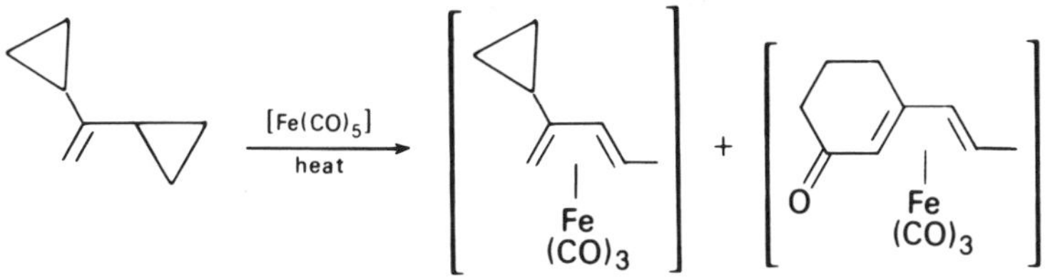
[Fe(CO)5]
heat
Fe
(CO)3
+
O
Fe
(CO)3

(**46**) (**47**)

(18)

the diene complex (**46**) with pentacarbonyliron[40]. In an extension of the last ring-expansion process, the methylenecyclohexane (**48**), on irradiation with pentacarbonyliron, yields the tricarbonyliron complex (**50**) as the principal product together with the 7-keto-octalin (**49**) which, on subsequent irradiation with pentacarbonyliron, gives the dienone complex (**51**) in low yield[42]. Recently, it has been

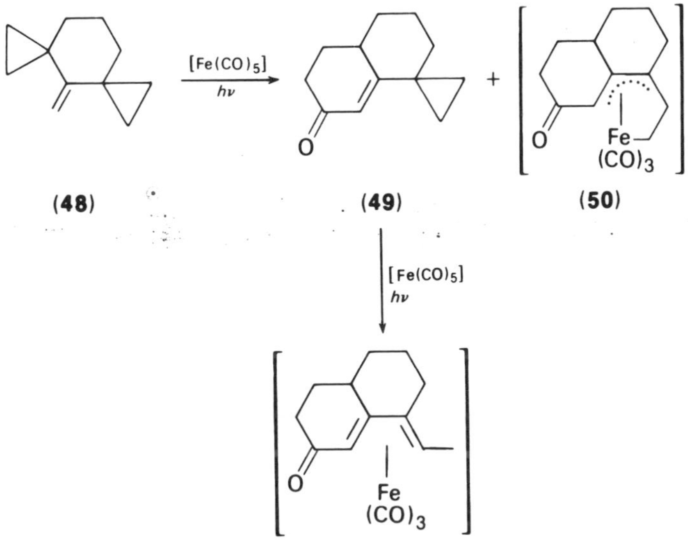
[Fe(CO)5]
hν
O
+
O
Fe
(CO)3
[Fe(CO)5]
hν
O
Fe
(CO)3

(**48**) (**49**) (**50**)

(19)

(**51**)

shown that photolysis of the methylenespiroalkanes (**52**; $n = 0, 1, 2$) with pentacarbonyliron gives a mixture of the two (η^4-butadiene)iron complexes (**53** and **54**) together with a bis(η^3-allyl)diiron complex of a branched triene[43].

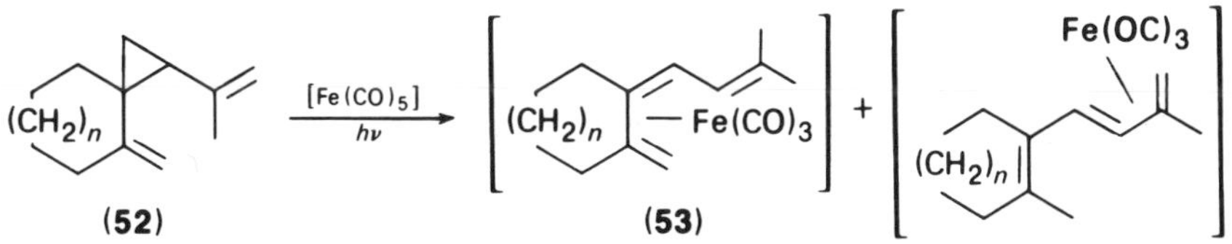
(CH2)n
[Fe(CO)5]
hν
(CH2)n
Fe(CO)3
+
Fe(OC)3
(CH2)n

(**52**) (**53**) (**54**) (20)

$n = 0,1,2$

2. From methylenecyclopropanes and metal carbonyls

Methylenecyclopropanes undergo ring opening in the presence of metal carbonyls to give η^4-diene complexes in reactions which parallel those observed for vinylcyclopropanes. The *trans*-isomer of Feist's ester **(55)** combines with enneacarbonyldiiron in hexane at room temperature to give the tetracarbonyliron complex **(56)** in 88% yield with retention of the *trans*-configuration. Warming this intermediate with enneacarbonyldiiron at 40 °C afforded the (η^4-*syn*-diene)iron complex **(57)** in 78% yield. The same transformation is achieved by thermolysis in toluene[44].

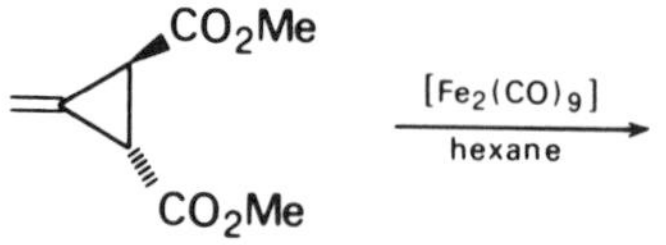

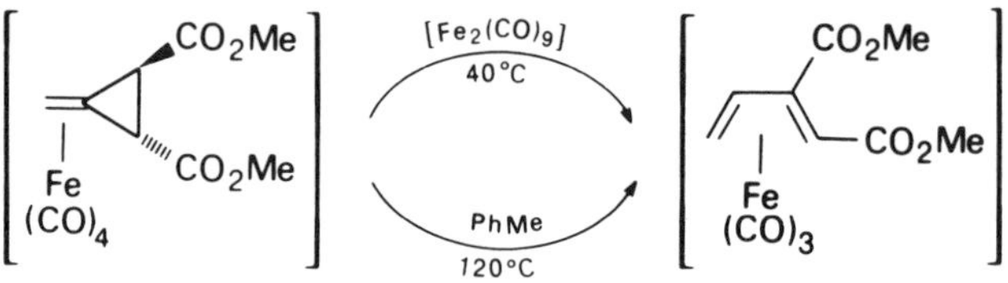

(21)

A similar sequence of reactions may be carried out with the *cis*-isomer of Feist's ester **(58)** to form the (η^4-*anti*-diene)iron complex **(59)**. In this case the second step of the sequence may involve any one of three alternatives: (i) warming with enneacarbonyldiiron to give an 88% yield of product, (ii) irradiation in hexane to give a 62% yield, and (iii) thermolysis in toluene to give 49% yield[44]. Formation of the two η^4-butadiene complexes **(57** and **59)** is highly stereospecific and the mechanism appears to involve two metal centres[44].

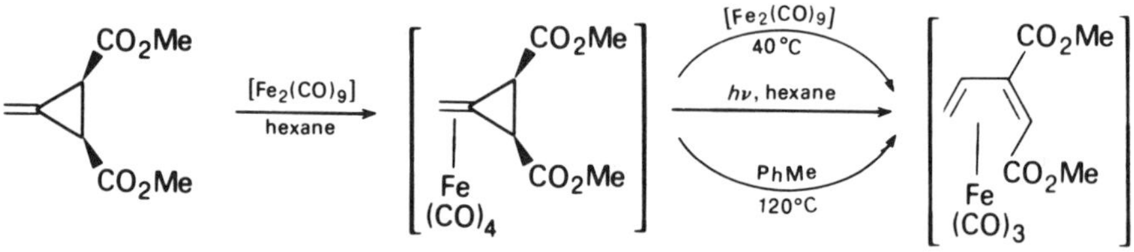

(22)

trans-2,3-Bis(hydroxymethyl)methylenecyclopropane **(60)** combines with an excess of enneacarbonyldiiron in ether at room temperature under an atmosphere of carbon monoxide to yield the tricarbonyl(η^4-3-methylene-4-vinyldihydrofuranone)iron complex **(62)** (48%), together with the tetracarbonyliron complex **(63)** (30%)[45,46] (Scheme 3). Each of these intermediates is converted to tricarbonyl(η^4-1,3-diene)iron products on heating in diethyl ether for several hours. The tricarbonyliron complex **(62)** undergoes regiospecific rearrangement to give a 1:2.3 mixture of the isomeric tricarbonyliron complexes **(64** and **65)**, while the tetracarbonyliron intermediate **(63)** also undergoes regiospecific rearrangement under the same conditions to form a third

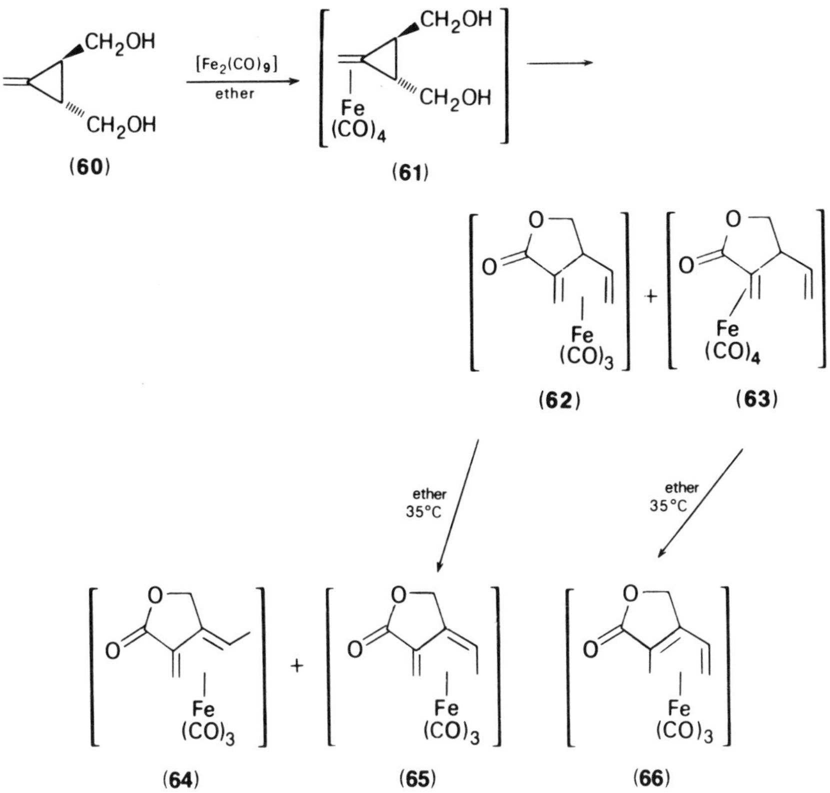

(60) (61) (62) (63) (64) (65) (66)

SCHEME 3

isomeric tricarbonyl(η^4-1,3-diene)iron complex (**66**)[47]. The *cis*-isomer of the methylenecyclopropane (**60**) undergoes the same reaction with enneacarbonyldiiron to form a mixture of the complexes **62** and **63**[46]. Mechanistic studies indicate that the first formed intermediate in the reactions is the tetracarbonyliron species (**61**), a sequence of steps lead to formation of the lactone ring and thence the complexes **62** and **63**[46]. These, in turn, undergo rearrangement by way of a 'σ-allyl metal hydride' mechanism[47].

In contrast to the previous reactions, transition metal-promoted cleavage of the strained ring in *syn*- and *anti*-2,2-dimethylallylidenecyclopropane (**67**) gives η^4-trimethylenemethane complexes (**68** and **69**) as products[47]. However, the ligand (**67**) contains a buta-1,3-diene residue and the tricarbonyl(η^4-allylidenecyclopropane)iron complex (**70**) is also formed in the reaction[47]. The overall yield of the three complexes **68**, **69** and **70** is 32%. Separate experiments with *syn*- and *anti*-isomers of the ligand (**67**) show that the η^4-butadiene complex (**70**) is derived exclusively from the *anti*-form (**67**) and can be obtained in 30% yield together with the η^4-trimethylenemethane complex (**68**) in 15% yield[47].

Related to the cleavage of methylenecyclopropanes to form butadiene ligands is the cleavage of 1,3,3-trimethylcyclopropene in the presence of dodecacarbonyltriiron to give the η^4-vinylketene complex (**71**; R = Me) in 5% yield. In addition to ring opening, a molecule of carbon monoxide is incorporated into the ligand[48]. 3,3-

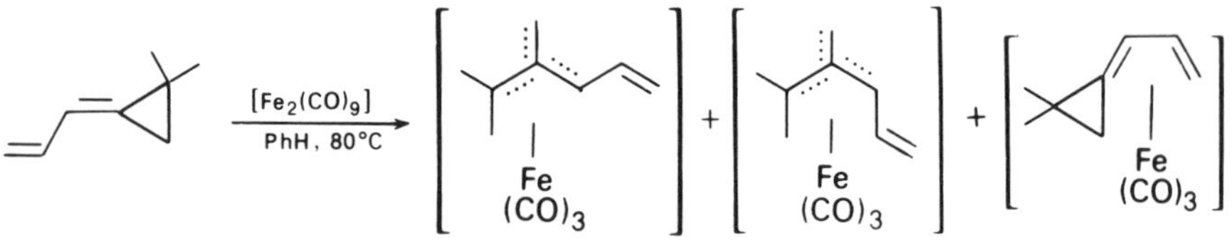

$[Fe_2(CO)_9]$, PhH, 80°C

(67) (68) (69) (70) (23)

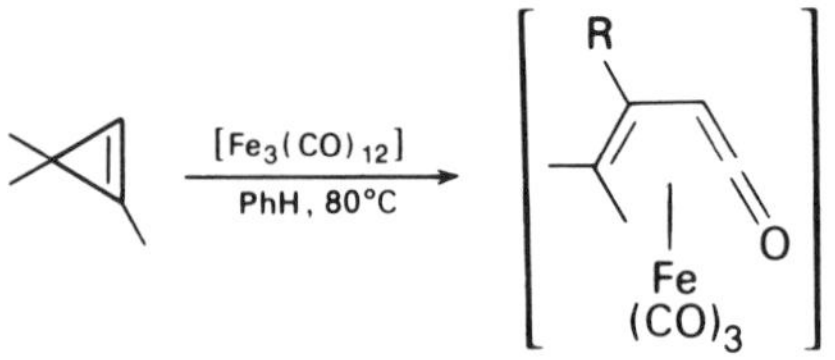

$[Fe_3(CO)_{12}]$, PhH, 80°C

(71) (24)

Dimethylcyclopropene undergoes a similar reaction with enneacarbonyldiiron to form the dienone complex (**71**; R = H)[49]. The same cyclopropene combines with dicarbonyl-(η^5-cyclopentadienyl)(tetrahydrofuran)manganese to give the (η^4-dienone)manganese complex (**72**)[49].

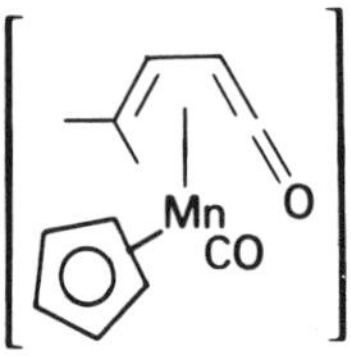

(72)

3. *From thiacyclobutenes and metal carbonyls*

The thiacyclobutenes (**73**; $R^1 = R^2$ = H, R^1 = Me, R^2 = Et) combine readily with enneacarbonyldiiron on heating or with pentacarbonyliron on photolysis to form the binuclear η^4-thioacrolein complexes (**74**; $R^1 = R^2$ = H, R^1 = Me, R^2 = Et) in 30% yield[50]. The parent complex (**74**; $R^1 = R^2$ = H) may be converted to the mononuclear species (**75**) with triphenylphosphine and to the vinylsulphine complex (**76**) with hydrogen peroxide in acetic acid[50] (Scheme 4).

E. From Allyl Alcohols and Metal Carbonyls

Allyl alcohols may undergo dehydration in the presence of iron carbonyls to give η^4-1,3-diene complexes and allyl halides undergo similar reactions with these carbonyls. The allyl alcohols (**77**; R^1 = ferrocenyl, R^2 = H; R^1 = H, R^2 = ferrocenyl) are heated with either enneacarbonyldiiron or dodecacarbonyltriiron and copper sulphate in benzene to form the ferrocenylbutadiene complexes (**78**; R^1 = ferrocenyl, R^2 = H; R^1 = H, R^2 = ferrocenyl) in yields of 10 and 18%, respectively. Ferrocenyl ketones are formed as the major reaction products[51]. The reaction mechanism has been explained by a study of the dehydration of the vinyl alcohol (**79**) with copper sulphate to the butadiene (**80**), which then gives the tricarbonyliron complex (**81**) with dodecacarbonyltriiron[52].

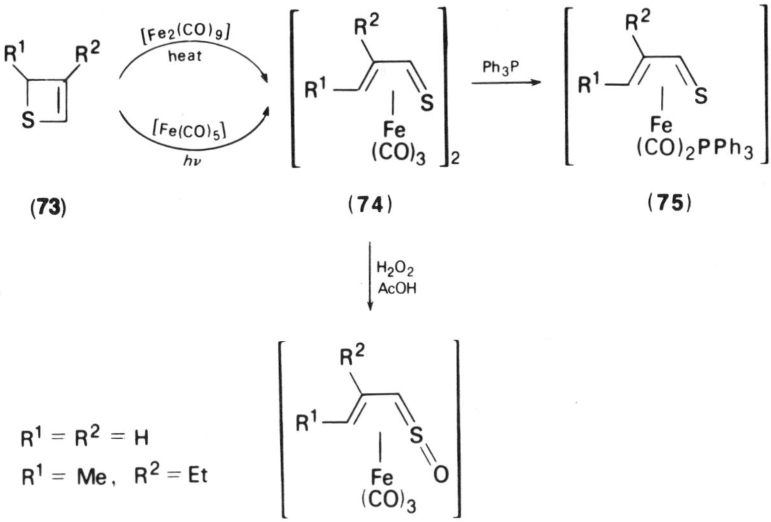

SCHEME 4

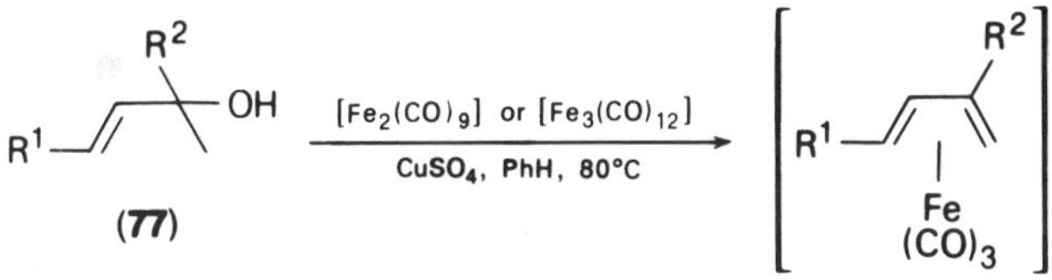

(25)

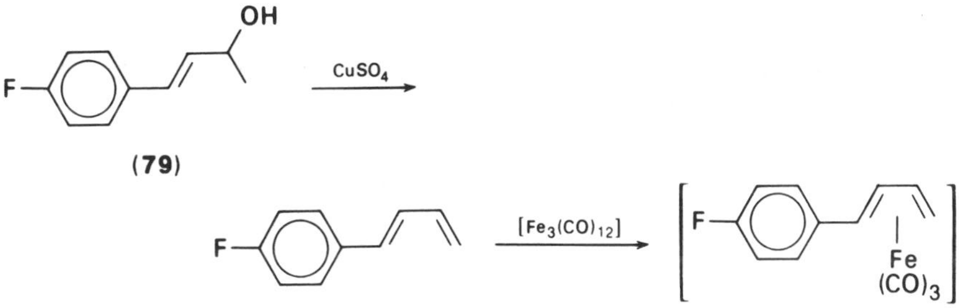

(26)

Penta-1,4-dien-3-ol combines with pentacarbonyliron on heating for several hours in light petroleum to form the bis[tricarbonyl(η^4-pentadiene)iron] complex (**82**) in low yield as a mixture of two diasterioisomers[53].

Allyl halides may also be used as precursors for buta-1,3-diene complexes, although the reactions involved are different from those based on allyl alcohols. 2-Methoxyallyl chloride combines with enneacarbonyldiiron on heating at 40°C in benzene to give tricarbonyl(η^4-3-methoxybuta-1,3-dienone)iron (**84**) in 10% yield[54]. The reaction may involve the tetracarbonyliron chloride intermediate **83**, which undergoes carbonyl

$Fe(CO)_5$; light petroleum, 140°C → (82) (27)

$[Fe_2(CO)_9]$; PhH, 40°C → (83) → (84) (28)

insertion and loss of chloride to form the product (**84**). The corresponding 2-methoxyallyl bromide and iodide give η^3-allyl complexes of iron rather than butadiene complexes under the same conditions[54].

F. From Butadienes, Transition Metal Atoms and Carbon Monoxide

Metal atoms, formed by thermal vaporization of metals under vacuum (10^{-3} torr), travel by a line-of-sight path to the cold walls of the vacuum chamber and co-condense with either an excess of the butadiene vapour or into a solution of the ligand[55] (see Chapter 13). Chromium atoms may be co-condensed with butadiene at −196°C and carbon monoxide added to form the tetracarbonylchromium complex (**85**) in 4% yield[57,58]. Replacing carbon monoxide with trifluorophosphine gives the tetrakis(trifluorophosphine)chromium complex (**86**) in 3% yield[59]. The reaction between molyb-

(86) ← (i) $Cr_{(g)}$ −196°C, (ii) PF_3; (i) $Cr_{(g)}$ −196°C, (ii) CO → (85) (29)

denum vapour or tungsten vapour and butadiene follows a different path and yields the tris(η^4-butadiene)metal complexes (**87**; M = Mo, W) in 50–60% yield[60]. Cocon-

$M_{(g)}$ −196°C → (87) (30)

M = Mo, W

densation of manganese atoms with butadiene followed by the addition of carbon monoxide gives a very low yield of bis(η^4-butadiene)carbonylmanganese (**88**)[61].

Iron atoms combine with butadiene at −196°C to afford a red–brown complex (**89**) which decomposes on warming to −5°C. However, when the complex is warmed in

(i) $Mn_{(g)}$ −196°C; (ii) CO → Mn CO **(88)** (31)

$Fe_{(g)}$ −196°C → Fe **(89)** —L→ Fe–L **(90)** (32)

L = CO, PF_3, $P(OMe)_3$, *t*-BuNC

the presence of a suitable ligand such as carbon monoxide or trifluorophosphine, then bis(η^4-butadiene)iron complexes [**90**; L = CO, PF_3, $P(OMe)_3$, *t*-BuNC] are obtained[57,58,61]. Thus the product (**90**; L = CO) is obtained in 31% yield overall[57] and yields are usually in the range 20–50%[55]. Recently, complexes of the type [(η^4-butadiene)FeL_3] and [(η^4-butadiene)$_2$FeL], where L = a phosphorus ligand such as $P(OMe)_3$ have been formed by a metal atom evaporation technique[56]. Iron complexes containing a single butadiene ligand are not usually formed in these reactions, but co-condensation of iron and toluene gives the unstable bis(η^6-arene)iron intermediate (**91**), which undergoes ligand displacement on condensation of butadiene to give (η^4-butadiene)(η^6-toluene)iron (**92**) in 10–20% yield[61]. Styrene behaves as a four-electron

$Fe_{(g)}$ −196°C → Fe **(91)** —(butadiene) −196°C→ Fe **(92)** (33)

donor when it is co-condensed with iron atoms at −196°C and the resulting mixture is allowed to warm under an atmosphere of carbon monoxide. The products obtained are tricarbonyl(η^4-styrene)iron (**93**) in 10% yield, tetracarbonyl(η^2-styrene)iron (**94**) and pentacarbonyliron[62].

(i) $Fe_{(g)}$ −196°C; (ii) CO → Fe $(CO)_3$ **(93)** + Fe $(CO)_4$ **(94)** (34)

A complex reaction takes place between cobalt atoms and butadiene, but in the presence of a hydrogen donor such as *t*-butane the cobalt hydride **95** may be isolated[60].

$Co_{(g)}$ − 196 °C, *t*-BuH (35)

(95)

The (η^4-isoprene)cobalt complex **(96)** may be obtained by sequential reaction of the metal atoms with cyclopentadiene and then isoprene[60], whereas bis(η^4-penta-1,3-diene)cobalt hydride **(97)** is formed from penta-1,4-diene by isomerization in a complex reaction which also yields polymers of the original diene and olefins formed from it[60].

(i) $Co_{(g)}$ − 196°C, (ii) isoprene (36)

(96)

$Co_{(g)}$, − 196°C (37)

(97)

G. From (η-Hydrocarbon)transition Metal Complexes

1. (η^5-Dienium)transition metal complexes

η^5-Dienium, η^4-trimethylenemethane, η^3-allyl, and η^2-olefin transition metal carbonyls may all act as sources for (η^4-buta-1,3-diene)metal complexes, although the usefulness of such compounds is restricted in some cases where the starting material may itself be derived from an η^4-diene. A convenient route to tricarbonyl(η^4-*syn*-1-vinylbutadiene)iron **(100)** involves dehydration of the alcohol complex **(98)** to the (η^5-dienium)iron cation **(99)** with fluoroboric acid and then treatment with alumina in

HBF_4, Ac_2O; BF_4^-; Al_2O_3, CH_2Cl_2 (38)

(98) (99) (100)

methylene chloride for 1 h at room temperature to give the product in 30% yield[63]. Prolonged treatment with alumina (60 h) gives binuclear complexes[63]. An alternative route from the dienium complex **(99)** requires slow distillation of the triethylamine adduct under reduced pressure to remove the volatile product **(100)** as it is formed[64]. The methods appear to provide a general route to triene complexes[64]. The action of perchloric acid on the (η^4-*trans,trans*-hexa-2,4-dien-1-ol)iron complex **(101)**, itself formed from the free ligand and pentacarbonyliron, gives the (η^5-*syn*-1-methylpentadienyl)iron cation (**102**) in quantitative yield, which is selectively attacked by water and alcohols to give one diastereoisomer of the complex **103** (R = H, Me, Et)

(**101**) —$HClO_4$, Ac_2O, 0 °C→ (**102**) ClO_4^- —ROH→ (**103**) (39)

R = H, Me, Et

in a yield of about 90%[65,66]. Primary amines attack the (η^5-*syn*-1,5-dimethylpentadienyl)iron cation (**104**) to give either *cis,trans*-secondary amine (**105**) or *trans,trans*-secondary amine (**106**) products, depending on the basicity of the amine. Thus the strongly basic (pK_b 3–6) amines ethylamine, isopropylamine, and methylbenzylamine give exclusively the *cis,trans* products (**105**; R = Et, *i*-Pr, CHMePh), whereas weakly basic (pK_b 10–13) amines such as *p*-bromoaniline and *p*-nitroaniline give exclusively and quantitatively the *trans,trans* products (**106**; R = C_6H_4Br, $C_6H_4NO_2$)[67]. The tricarbonyl(η^5-*syn*-pentadienyl)iron cation may also be used as a source of (η^4-butadiene—amine)iron complexes by treatment with primary amines; however, binuclear products are formed in addition to the expected mononuclear complexes and separation is not always easily achieved[68].

(**105**) ←RNH_2— (**104**) BF_4^- —RNH_2→ (**106**) (40)

R = Et, *i*-Pr, CHMePh R = C_6H_4Br, $C_6H_4NO_2$

2. (η^4-Trimethylenemethane)-, (η^3-allyl)-, and (η^2-olefin)-transition metal complexes

The binuclear trimethylenemethane complex (**108**), obtained in 7% yield from 2,3-bis(bromomethyl)buta-1,3-diene (**107**) and enneacarbonyldiiron in hexane, rearranges in 85% sulphuric acid to a 1:1 mixture of bis-diene complexes (**109** and **110**) in 81% yield[69]. (η^3-Allyl)iron complexes may be converted under suitable conditions to (η^4-buta-1,3-diene)iron complexes.

Tricarbonyl(η^4-α-pyrone)iron (**111**) is formed in 85% yield by heating α-pyrone with enneacarbonyldiiron in benzene for 1 h[70]. This intermediate contains a highly reactive ester group and undergoes cleavage with bases to give (η^3-allyl)iron anions, as shown in Scheme 5. Thus, attack by methoxide ion leads to the allyl anion (**112**), which

(107) (108) (109) (110) (41)

combines with acetic anhydride to give the (η^4-butadiene)iron complex (**113**) in almost quantitative yield[70]. Cleavage with methyllithium and phenyllithium gives the allyl anions (**114**; R = Me, Ph) and treatment with acetic anhydride forms the *cis,trans*-acetoxy ketones (**115**; R = Me, Ph), which may be isomerized thermally to *trans,trans* compounds (**116**; R = Me, Ph) on heating in benzene[70]. Reduction of the α-pyrone complex (**111**) with lithium aluminium hydride gives the *cis,trans*-aldehyde (**118**), apparently by way of the allyldialdehyde anion (**117**). The *cis,trans* product readily isomerizes to the *trans,trans* form[70]. Benzoylation of (η^3-allyl)tricarbonyliron using benzoyl chloride gives the (η^4-enone)iron complex (**119**) as a minor product[71]. In a related reaction, the η^3-allyl lactam (**120**) is converted to the tricarbonyliron complex (**121**) on heating in methanol[72]. When mild conditions are used in the reaction between dienes or enones and iron carbonyls then (η^2-olefin)iron complexes may be obtained. These are usually converted smoothly to η^4-diene or η^4-enone complexes by using more vigorous conditions. Thus, chalcone and 2′-hydroxychalcone (**122**; R = H, OH, respectively) give the tetracarbonyl complexes (**123**; R = H, OH) on heating with enneacarbonyldiiron at 40°C. On raising the reaction temperature to 70–80°C the corresponding tricarbonyl complexes are obtained (**124**; R = H, OH)[73]. η^4-Enone complexes may be converted back to η^2-enone complexes on treatment with basic phosphine ligands and this reaction may be used to provide a route to (η^4-enone)iron dicarbonyl phosphine complexes. The η^4-chalcone and η^4-benzilideneacetone complexes (**125**; R = Ph, Me) are treated with a phosphite or phosphine PX_3, where X = OMe, Ph, to give the η^2-enone intermediates (**126**; R = Me, Ph) and these undergo thermolysis to form the η^4-enone products (**127**; R = Me, Ph)[74,75].

The enone ligand in (η^4-benzylideneacetone)tricarbonyliron is displaced easily by butadiene groups and it is used as a convenient intermediate in the formation of (η^4-butadiene)iron complexes. Thus, 2,3-dimethylbutadiene, *trans,trans*-hexa-2,4-diene, and *trans,trans*-hexa-2,4-dienal are heated with (η^4-benzylideneacetone)tricarbonyliron to form the corresponding (η^4-butadiene)tricarbonyliron complexes in yields of 54–96%[76].

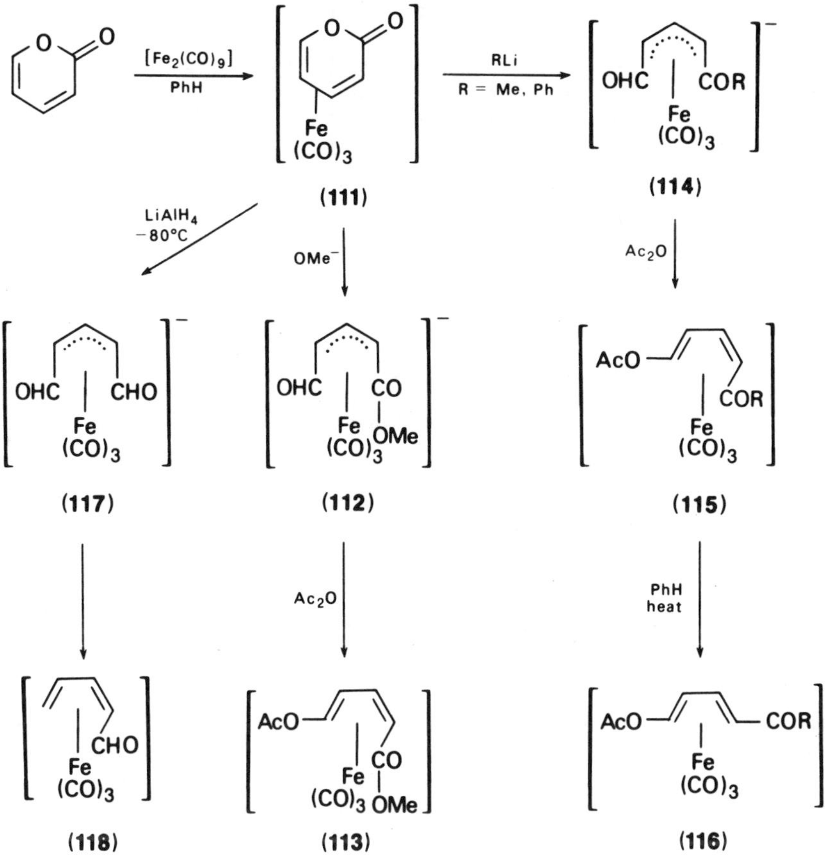

(111) (114) (117) (112) (115) (118) (113) (116)

SCHEME 5

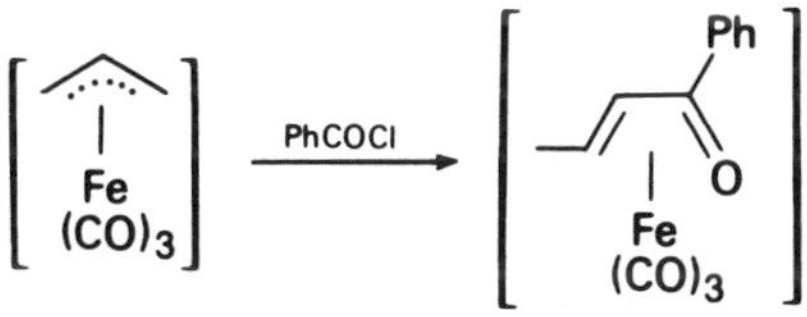

(42)

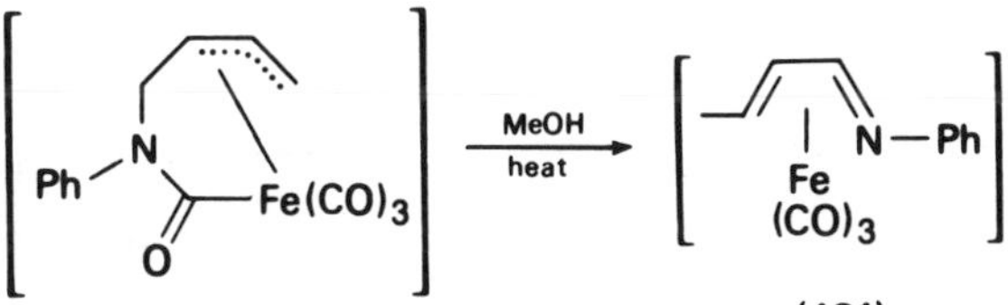

(119)

(43)

(120) (121)

(122) → $[Fe_2(CO)_9]$, PhMe, 40°C → (123) → PhMe, 80°C → (124) (44)

R = H, OH

(125) → PX_3 → (126) → heat → (127) (45)

R = Me, Ph
X = OMe, Ph

II. η^4-TRIMETHYLENEMETHANE COMPLEXES

A. Introduction

The hypothetical trimethylenemethane radical may be stabilized by complex formation with the appropriate transition metals and two important routes are available for the preparation of these complexes. Allyl halides, substituted allyl halides, and related compounds combine with enneacarbonyldiiron to form η^4-trimethylenemethene complexes, whereas methylenecyclopropanes undergo ring opening in the presence of the same carbonyl to give η^4-vinyltrimethylenemethane complexes. In a third and less important route, (η^3-allyl)tricarbonyliron halides may eliminate the halogen or hydrogen halide with the formation of η^4-trimethylenemethane complexes.

B. From Allyl Halides and Enneacarbonyldiiron

The parent complex tricarbonyl(η^4-trimethylenemethane)iron (**129**) is obtained in 66% yield as a pale yellow solid, melting at 28–29°C, by stirring 3-chloro-(2-chloromethyl)propene (**128**) with enneacarbonyldiiron in ether at room temperature for 24 h[77,78]. When $Na_2[Fe(CO)_4]$ is used instead of enneacarbonyldiiron as the reagent, the same product (**129**) is obtained but only in low yield[78]. Substituted

$$CH_2{=}C(CH_2Cl)_2 + [Fe_2(CO)_9] \longrightarrow [Fe(CO)_3] \quad (46)$$

(128) (129)

η^4-trimethylenemethane complexes (**130**) are accessible from the appropriately substituted allyl halide. Methyl, phenyl, and carboalkoxy complexes[78] may be prepared in this way (see Table 2). 2,3-Di(bromomethyl)butadiene (**131**) may be regarded as a diallyl halide and a precursor of the tetramethyleneethane diradical. Thus, the reaction between the dibromide (**131**) and enneacarbonyldiiron gives the bis(tricarbonyliron) complex of the dimer of tetramethyleneethane (**132**). A small proportion of the isomeric product (**133**) is also obtained[79]. The η^4-trimethylenemethane complex (**134**)

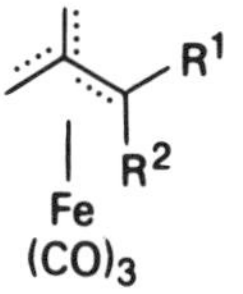

(**130**) 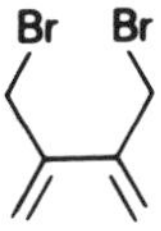 (**131**) 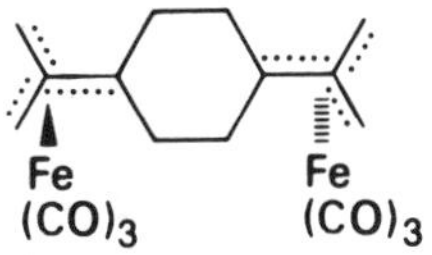 (**132**)

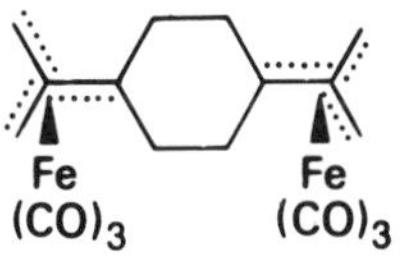

(**133**) 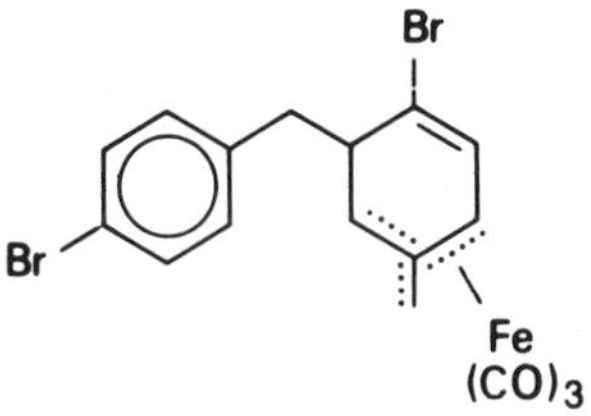 (**134**)

is formed from *p*-bromobenzyl bromide and enneacarbonyldiiron by coupling of two ligand molecules[80]. Coupling of ligand molecules also occurs when the tribromide (**135**) is treated with the same carbonyl to form the bis(tricarbonyliron) complex (**136**)[78]. This product is obtained from 2-bromomethylbutadiene using

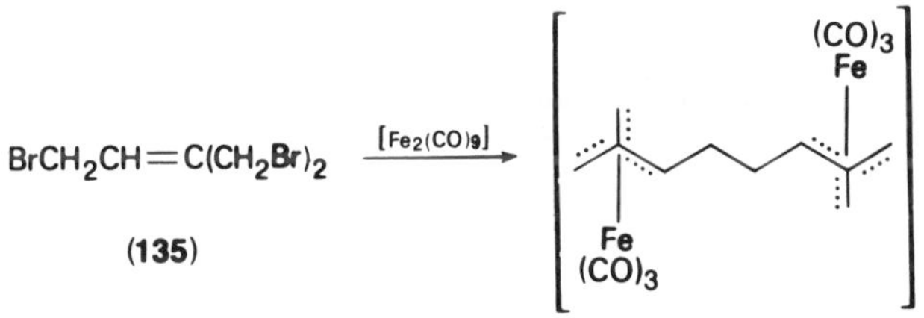

$$BrCH_2CH{=}C(CH_2Br)_2 \xrightarrow{[Fe_2(CO)_9]} \quad (47)$$

(**135**) (**136**)

enneacarbonyldiiron as the reagent[78]. Further details of preparations utilizing allyl halides are collected in Table 2.

A related route involves the treatment of 5-bromopenta-1,2-diene (**137**) with sodium tetracarbonylferrate to form the cyclopentenyl anion (**138**), which may be trapped by trimethylchlorosilane as the neutral η^4-trimethylenemethane complex (**139**)[81]. The formation of α,β-unsaturated ketones in the reaction of allene and an alkyl bromide with sodium tetracarbonylferrate proceeds by insertion of allene into an iron—acyl bond and protonation to form an η^4-trimethylenemethane intermediate (**140**; R = H), which may be characterized as the trimethylsilyl derivative (**140**); R = $SiMe_3$) on the addition of trimethylchlorosilane[82].

TABLE 2. Products, yields and conditions used in the preparation of trimethylenemethane complexes from allyl halides and diiron enneacarbonyl

Allyl halide	Product	Conditions	Yield (%)	Reference
$CH_2{=}C(CH_2Cl)_2$	**129**	Et_2O, room temp., 24 h	66	77, 78
$CH_2{=}C(Me)CH_2Cl$	**129**	Cyclohexane, room temp.	<10	78
$PhCH{=}C(Me)CH_2Cl$	**130**; $R^1 = Ph$, $R^2 = H$	Hexane, heat	36	78
$Me_2C{=}C(Me)CH_2Cl$	**130**; $R^1 = R^2 = Me$	—	—	78
$MeOCOCH{=}C(Me)CH_2Cl$	**130**; $R^1 = CO_2Me$, $R^2 = H$	—	—	78
$(EtOCO)_2C{=}C(Me)CH_2Cl$	**130**; $R^1 = R^2 = CO_2Et$	—	—	78
$CH_2{=}C(CH_2Br)C(CH_2Br){=}CH_2$	**132**	—	7.5	79
$CH_2{=}C(CH_2Br)C(CH_2Br){=}CH_2$	**133**	—	2	79
4-Br·C_6H_4·CH_2Br	**134**	Hexane, 45°C	<10	80
$BrCH_2CH{=}C(CH_2Br)_2$	**136**	—	<10	78
$CH_2{=}C(CH_2Br)CH{=}CH_2$	**136**	—	—	78

$Br(CH_2)_2CH{=}C{=}CH_2 + Na_2[Fe(CO)_4] \longrightarrow$

(137)

(138) —Me_3SiCl→ (139); (138) —AcOH→; (139) —AcOH→ (48)

Fe (CO)3 R = H, SiMe3

(140)

C. From Methylenecyclopropanes and Enneacarbonyldiiron

The strained three-membered ring in methylenecyclopropanes opens under mild conditions in the presence of enneacarbonyldiiron to form trimethylenemethane complexes. Thus, 2-vinylmethylenecyclopropane is attacked by enneacarbonyldiiron in benzene at 40°C to form the vinyltrimethylenemethane complex **(141)** as a green oil in 43% yield[83]. In a similar reaction, a mixture of the *anti*- and *syn*-isomers of 2,2-dimethylallylidenecyclopropane **(142)** combines with the same reagent in benzene at

(142a) (142b)

$[Fe_2(CO)_9]$, PhH, 40°C (49)

(141)

reflux temperature to give the trisubstituted trimethylenemethane complex **(143)** in 19% yield, together with a small proportion of the isomeric product **(144)**[84]. Under the same conditions, the *anti*-isomer **(142a)** gives the complex **(144)** as the only η^4-trimethylenemethane product, whereas the *syn*-isomer **(142b)** gives a mixture of the two complexes **(143** and **144)**, with the former as the dominant product[84].

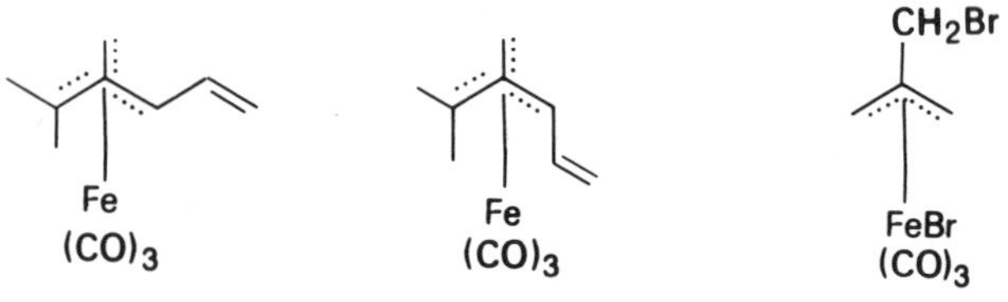

Fe (CO)3 (143) Fe (CO)3 (144) CH2Br FeBr (CO)3 (145)

D. From η^3-Allyl Complexes

Halogen or hydrogen halide may be eliminated from (η^3-allyl)tricarbonylirion complexes to leave the corresponding η^4-trimethylenemethane complexes. Stirring (η^3-2-bromomethylallyl)tricarbonyliron bromide (**145**) with enneacarbonyldiiron in ether at room temperature gives tricarbonyl(η^4-trimethylenemethane)iron (**129**) in 90% yield[78]. The same dibromide (**145**) gives the same product (**129**) on heating in cyclohexane but in low yield[78]. The action of heat on tricarbonyl(η^3-2-methylallyl)iron chloride (**146**) also forms this complex (**129**) in 39% yield[78].

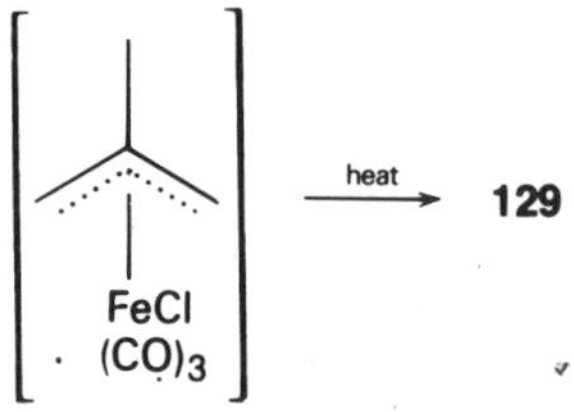

FeCl (CO)3 —heat→ 129 (50)

(146)

III. η^4-CYCLOBUTADIENE COMPLEXES

A. Introduction

Stabilization of the hitherto elusive cyclobutadiene by complex formation with a transition metal was postulated in 1956[85]. Since that time, many routes to (η^4-cyclobutadiene)transition metal complexes have been developed. These routes may be conveniently, though arbitrarily, grouped into methods which use a reagent with a four-membered carbocyclic ring and those which close the ligand ring during the course of the reaction. Among the former methods are routes employing tetrahalocyclobutanes, cyclobutenes, η^3-cyclobutenyl complexes, methylene cyclobutenes, and η^4-cyclobutadiene complexes. Included in the latter methods are those utilizing photo-α-pyrone, acetylenes, heterocycles, and cyclooctatetraenes as cyclobutadiene precursors.

B. From Cyclobutenes and Cyclobutanes with Transition Metal Carbonyls

Historically, methods based on dihalocyclobutenes have assumed importance[86,93], and they still provide useful routes to a number of η^4-cyclobutadiene complexes.

The parent complex (**148**) is formed in 40% yield as a pale yellow solid, m.p. 26°C, by stirring a suspension of enneacarbonyldiiron in pentane at 30°C for 2 h with

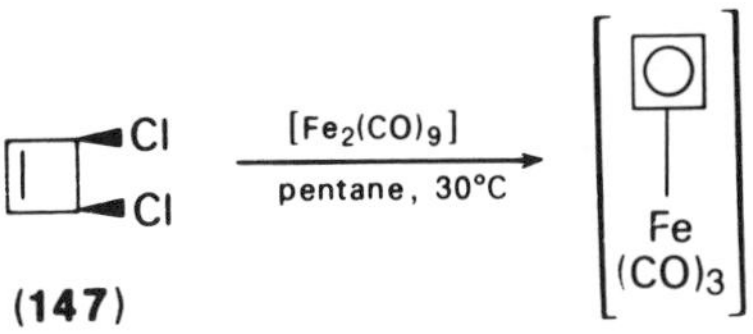

(51)

(147) (148)

cis-3,4-dichlorocyclobutene (**147**)[87]. The (η^4-cyclobutadiene)cobalt complex (**151**) is obtained from the same dichloride (**147**) by treatment with a two-fold excess of sodium tetracarbonylcobaltate in thf to give hexacarbonyl(η^4-cyclobutadiene)dicobalt (**149**), which, on reaction with iodine, forms dicarbonyl(η^4-cyclobutadiene)cobalt iodide (**150**). This iodide is heated with cyclopentadiene and triethylamine in thf to give the product (**151**) as a yellow crystalline solid, m.p. 88–89°C[88]. Mono-[89], di-[87,91], tri-[91], and

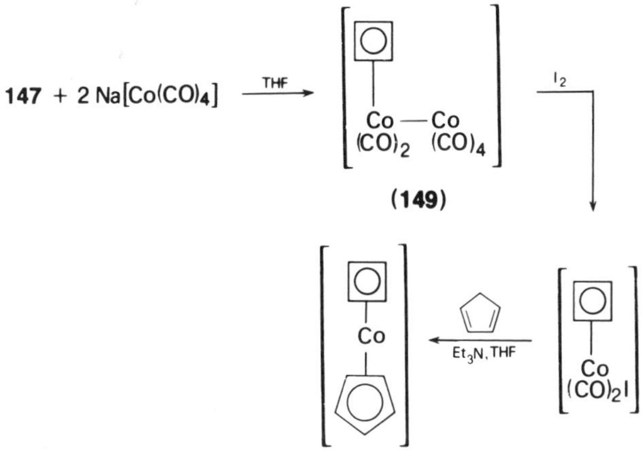

(52)

(149) (151) (150)

tetra-substituted[91–94] η^4-cyclobutadiene complexes may be prepared from similar starting materials. The molybdenum and tungsten tetracarbonyls (**154**; M = Mo, W) are formed by treatment of the dihalide (**147**) with hexacarbonylmolybdenum or -tungsten and sodium amalgam, which generates the reactive dianion $[M(CO)_5]^{2-}$ [92]. The dihalide (**147**) is obtained from cyclooctatetraene in a sequence of reactions involving chlorination, addition of acetylene dicarboxylic ester, and thermolysis of the adduct (Scheme 6). Hexacarbonylmolybdenum and -tungsten attack the dihalide (**147**) in the presence of sodium amalgam to form the tetracarbonyl complexes (**152**; M = Mo, W). The ruthenium tricarbonyl complex (**153**) and the parent iron tricarbonyl (**148**) are obtained in similar reactions[92]. This route has been exploited also for the preparation of tetracarbonyl(η^4-tetramethylcyclobutadiene)molybdenum and -tungsten (**154**; M = Mo, W), respectively[92], together with the chromium tricarbonyl (**155**)[92] and the iron tricarbonyl (**165**)[92].

The η^4-bromocyclobutadiene complex (**156**; X = Br) is prepared from the *trans*-tribromide (**157**) and enneacarbonyldiiron[89] (Scheme 7). A similar reaction with the trichlorocyclobutene (**158**) gives the corresponding chlorocyclobutadiene (**156**; X = Cl)[90], whereas the mixed bromodichlorocyclobutene (**159**) gives the bis[tricarbonyl(η^4-cyclobutadiene)iron] complex (**160**) rather than a mononuclear

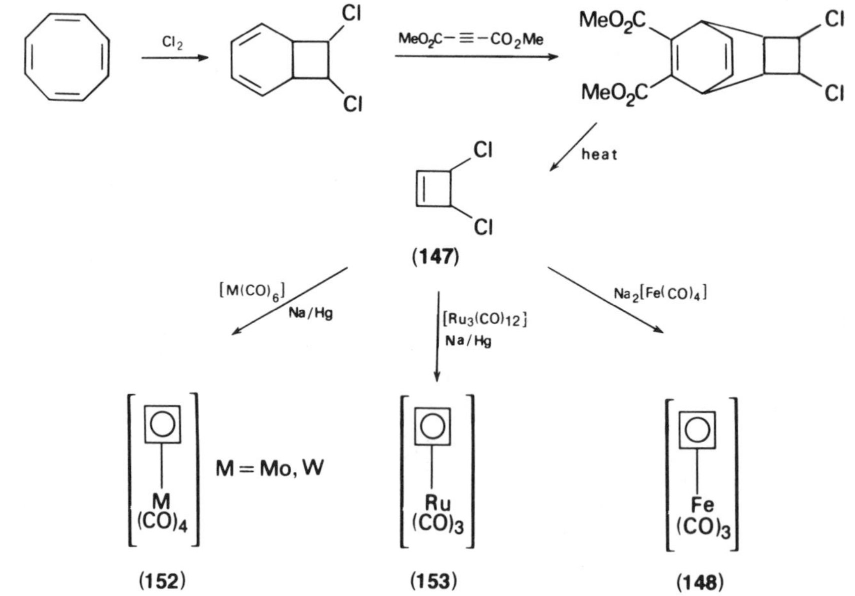

Cl_2 Cl Cl

$MeO_2C-\equiv-CO_2Me$

MeO_2C MeO_2C Cl Cl

heat

Cl Cl

(147)

$[M(CO)_6]$ Na/Hg

$[Ru_3(CO)_{12}]$ Na/Hg

$Na_2[Fe(CO)_4]$

M $(CO)_4$ M = Mo, W

Ru $(CO)_3$

Fe $(CO)_3$

(152) (153) (148)

SCHEME 6

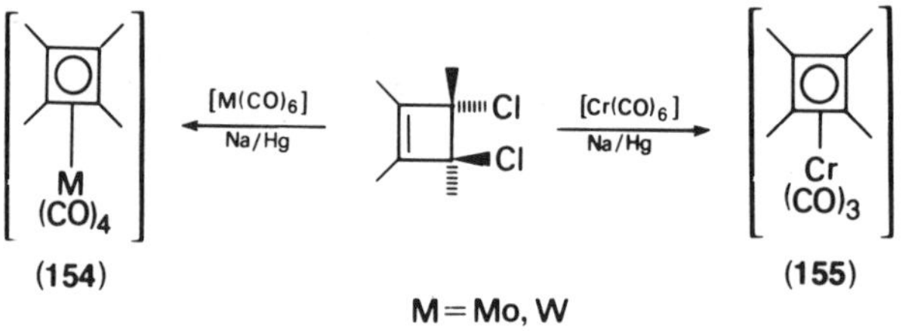

M $(CO)_4$

$[M(CO)_6]$ Na/Hg

Cl Cl

$[Cr(CO)_6]$ Na/Hg

Cr $(CO)_3$

(53)

(154) (155)

M = Mo, W

Br Br Br

$[Fe_2(CO)_9]$

X Fe $(CO)_3$

$[Fe_2(CO)_9]$

Cl Cl Cl

(157) (156) (158)

$[Fe_2(CO)_9]$

Cl Cl Br

$[Fe_2(CO)_9]$

Fe $(CO)_3$ Fe $(CO)_3$

(159)

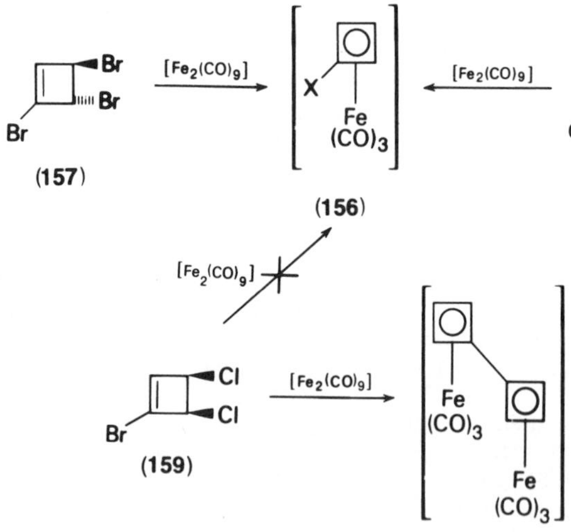

SCHEME 7 (160)

[$Fe_2(CO)_9$] (54)

(161) (162)

[$Fe_2(CO)_9$] (55)

(163) (164)

bromocyclobutadiene (**156**; X = Br)[89]. The benzocyclobutene (**161**; X = Br, I) is dehalogenated by enneacarbonyldiiron to form the benzocyclobutadiene complex (**162**)[87] and the related complex (**164**) is obtained in the same way from the dichloride (**163**). The dimethyl- and trimethylcyclobutadiene iron complexes (**166** and **167**) are prepared by the same route[91].

(165) (166) (167)

(168) (169)

The influence of stereochemical factors in the syntheses employing 3,4-dichlorodimethylcyclobutenes, derived from dimethylhexa-1,5-diynes, has been explored. The 3,4-dimethyl isomer (**169**) combined with enneacarbonyldiiron to give the cyclobutadiene complex (**166**) in 10% yield. The 1,2-dimethyl isomer gave a 7% yield of the same product whereas the 2,3-dimethyl isomer gave the product in only 5% yield[95]. Reagents, conditions, and products for some typical reactions are collected in Table 3. Although dihalocyclobutenes have been used extensively in the preparation of cyclobutadiene complexes, the method is limited by the availability of the appropriate dihalocyclobutenes, many of which are difficult to prepare. These halides also show a strong tendency to undergo side-reactions with dehalogenating agents, which frequently lead to cyclobutadiene dimers[96].

The *cis*-3,4-carbonyldioxycyclobutenes (**170**; $R^1 = R^2$ = H, Me; R^1 = Bu, CH_2OMe, R^2 = H; R^1 = Me, $R^2 = CH_2OMe$) prepared from an appropriate

TABLE 3. Reagents and conditions used for the preparation of η^4-cyclobutadiene complexes from halocyclobutenes

Halocyclobutene	Reagent(s)	Conditions	Product	Yield (%)	Reference
cis-2,3-$Cl_2C_4H_4$	$[Fe_2(CO)_9]$	Pentane, 30°C	**148**	40	87
cis-2.3-$Cl_2C_4H_4$	$Na_2[Fe(CO)_4]$	—	**148**	—	92
cis-2,3-$Cl_2C_4H_4$	$[Mo(CO)_6]$, Na/Hg	Thf	**152**; M=Mo	35	92
cis-2,3-$Cl_2C_4H_4$	$[W(CO)_6]$, Na/Hg	Thf	**152**; M=W	1	92
cis-2,3-$Cl_2C_4H_4$	$[Ru_3(CO)_{12}]$, Na/Hg	Thf	**153**	—	92
trans-1,3,4-$Br_3C_4H_3$	$[Fe_2(CO)_9]$		**156**; X=Br		89
cis-1-Br-3,4-$Cl_2C_4H_3$	$[Fe_2(CO)_9]$		**160**		89
3,4-Cl_2-$Me_2C_4H_2$	$[Fe_2(CO)_9]$	Thf, 60°C	**166**	16	91
161; X = Br, I	$[Fe_2(CO_9]$	Pentane, 30°C	**162**	—	87
163	$[Fe_2(CO)_9]$		**164**		
3,4-$Cl_2Me_3C_4H$	$[Fe_2(CO_9]$	Thf, 60°C	**167**	23	91
3,4-$Cl_2Me_4C_4$	$[Fe_2(CO)_9]$	Thf, 60°C	**165**	26	91
3,4-$Cl_2Me_4C_4$	$Na_2[Fe(CO)_4]$	Thf	**165**	—	92
3,4-$Cl_2Me_4C_4$	$[Cr(CO)_6]$, Na/Hg	Thf	**155**	—	92
3,4-$Cl_2Me_4C_4$	$[Mo(CO)_6]$, Na/Hg	Thf	**154**; M=Mo	—	92
3,4-$Cl_2Me_4C_4$	$[W(CO)_6]$, Na/Hg	Thf	**154**; M=W	—	92
3,4-$Cl_2Me_4C_4$	$[Ni(CO)_4]$	PhH, 78°C	**168**	90	86, 93
3,4-$Cl_2Me_4C_4$	$NiBr_2$	Li, $C_{10}H_8$, thf	**168**	>95	94

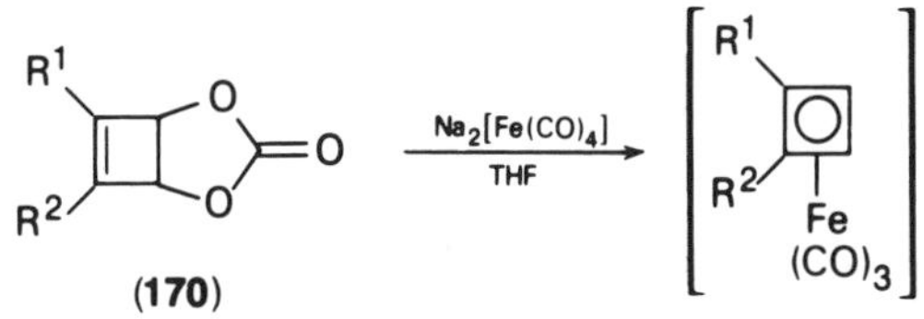

$$\xrightarrow[\text{THF}]{Na_2[Fe(CO)_4]} \quad (56)$$

(170) (171)

acetylene and vinylenecarbonate by photolysis, are converted to substituted (η^4-cyclobutadiene)iron complexes (**171**) with sodium tetracarbonylferrate(−II) or enneacarbonyldiiron. Yields are in the range 25–50%[97–99]. The same route is used in the stereospecific preparation of tricarbonyl-(A-di-nor-17β-acetoxy-1,5(10)-estradiene)iron (**172**) from the appropriate *cis*-3,4-carbonyldioxycyclobutene precursor[100].

In a few cases, cyclobutadiene complexes may be obtained from tetrahalocyclobutanes by methods analogous to those used for dihalocyclobutenes. The dichlorodibromocyclobutane (**173**; X = Br) is treated with enneacarbonyldiiron to form the

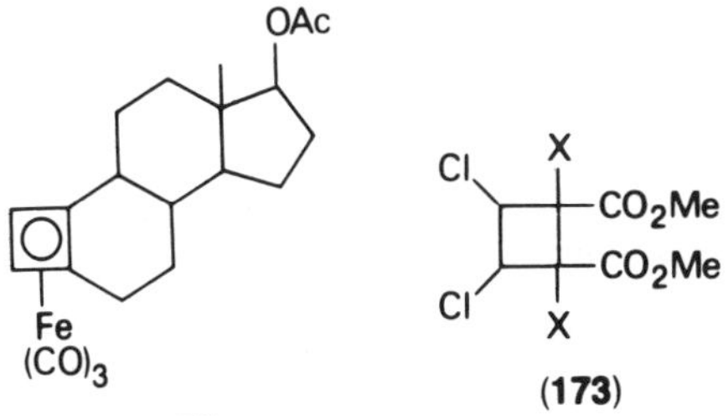

(172) (173)

diester (**174**)[101], and the tetrachlorocyclobutane (**173**; X = Cl) is dehalogenated by zinc and acetic acid in the presence of enneacarbonyldiiron to give the same product (**174**)[102]. Sodium tetracarbonylferrate(−II) may replace the iron carbonyl in the final step[103].

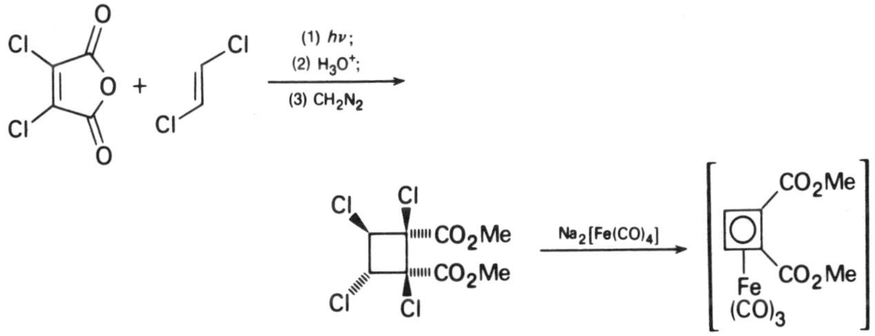

$$\xrightarrow[(3)\ CH_2N_2]{(1)\ h\nu;\ (2)\ H_3O^+;} \qquad \xrightarrow{Na_2[Fe(CO)_4]} \quad (57)$$

(174)

[η^4-Tetra(carbomethoxy)cyclobutadiene]tetracarbonylmolybdenum (**178**) is obtained as a minor product from the tetrachlorocyclobutane (**177**) and hexacarbonylmolybdenum. The tetrahalide (**177**) is in turn obtained by irradiation of dichloromaleic anhydride (**175**) to form the dimer (**176**) and subsequent treatment with diazomethane[104].

(175) (176)

(177) (178)

(58)

C. From (η^3-Cyclobutenyl)transition Metal Complexes

Products from the reaction between acetylenes and palladium(II) chloride are sensitive to the nature of the substituents on the acetylene and the conditions used[96]. Sodium or ammonium tetrachloropalladate(II) combines with diphenylacetylene in aqueous ethanol to form the η^3-alkoxycyclobutenyl complex (**179**) in 92% yield and this intermediate is converted to the (η^4-cyclobutadiene)palladium complex (**180**; X = Cl) in 76% yield[105,106]. The corresponding bromide and iodide complexes (**180**; X = Br, I) are formed direct from the chloride (**179**) by treatment with anhydrous hydrogen bromide in chloroform and aqueous hydriodic acid in chloroform; the yields are 82% and 50%, respectively[104].

$PhC\equiv CPh + Na_2[PdCl_4] \xrightarrow{EtOH}$ (179) $\xrightarrow{HX}$ (180)

(59)

The benzonitrile complex of palladium chloride, $[(PhCN)_2PdCl_2]$, may be used in place of the simple chloride but the yields are lower[107]. It is important to note that the reaction is reversible and treatment of the η^4-cyclobutadiene complex (**180**) with alkoxide ion under mild conditions gives a cyclobutenyl complex isomeric with (**179**) but with the alkoxide groups in the *exo*-position with respect to the metal[105,108]. This is exemplified by the interconversion of the (η^3-tetraphenylcyclobutenyl)nickel and -palladium complexes (**181**; M = Ni, Pd; R = H, alkyl) with the η^4-cyclobutadiene cation (**182**)[109,110]. The di-*t*-butyldiphenylcyclobutadiene complex (**183**) is prepared by a similar route[111].

Mechanistic investigations of these reactions indicate that cyclobutenyl complexes and thence cyclobutadiene complexes are formed only when the acetylene carries bulky substituents such as phenyl or *t*-butyl[112]. The mechanism appears to involve

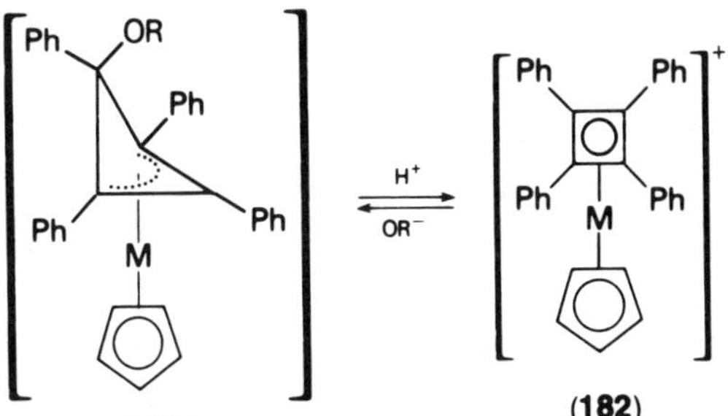

(181) (182) (60)

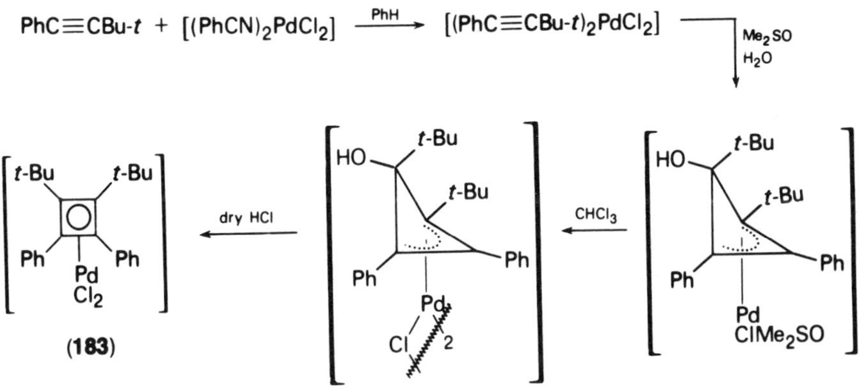

$$PhC{\equiv}CBu\text{-}t + [(PhCN)_2PdCl_2] \xrightarrow{PhH} [(PhC{\equiv}CBu\text{-}t)_2PdCl_2] \xrightarrow{Me_2SO,\ H_2O}$$

(183) (61)

initial formation of a π-complex followed by *cis*-'insertion' of the acetylene into the Pd—X bond to form a σ-vinyl complex (a slow step for X = Cl but fast for X = OR), then a new π-complex is formed and finally *cis*-'insertion' of the acetylene into the Pd—vinyl bond occurs[113].

D. From Methylenecyclobutenes and Metal Carbonyls

Olefin isomerization of an *exo*-methylenecyclobutene will give a cyclobutadiene and this reaction may be exploited to prepare η^4-cyclobutadiene complexes. Thus, a route to tricarbonyl(η^4-1,2-dimethylcyclobutadiene)iron (**185**) from the methylenecyclobutene compound (**184**) is available[114]. The bicycloheptatriene (**186**) combines with

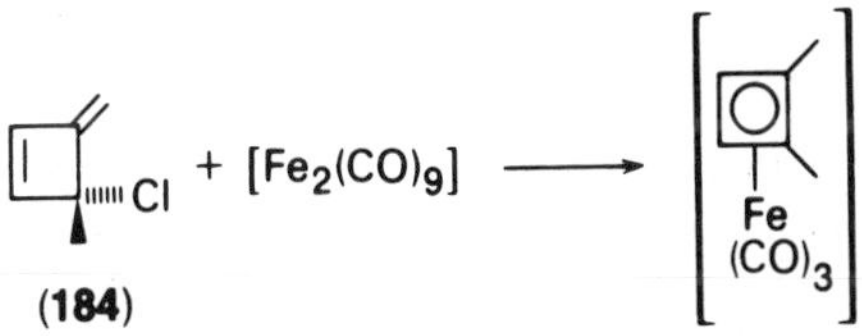

$$(\mathbf{184}) + [Fe_2(CO)_9] \longrightarrow (\mathbf{185}) \qquad (62)$$

dodecacarbonyltriiron in boiling hexane to form the cycloalkenylcyclobutadiene complex (**187**) in 41% yield by hydrogen migration[115].

$$\text{(186)} + [Fe_3(CO)_{12}] \xrightarrow[69\ ^\circ C]{hexane} [\text{(187)}\ Fe(CO)_3] \quad (63)$$

(186) (187)

E. From (η^4-Cyclobutadiene)transition Metal Complexes by Ligand Transfer

η^4-Tetraphenylcyclobutadiene and η^4-tetramethylcyclobutadiene complexes of nickel and palladium are readily accessible and are used extensively to prepare cyclobutadiene complexes of other metals by ligand exchange and ligand transfer[113]. Metal carbonyls and η^5-cyclopentadienylmetal carbonyls are frequently used as reagents. In a typical reaction (η^4-tetraphenylcyclobutadiene)palladium dibromide dimer (**188**; X = Br) is heated with an excess of pentacarbonyliron in xylene to give the tricarbonyliron complex (**189**) in 88% yield[116]. A number of these reactions are shown in Scheme 8 and details of reaction conditions and yields are given in Table 4. The reactions are usually carried out in an aromatic solvent (benzene, chlorobenzene, xylene) at reflux temperatures. Yields are very variable, iron, cobalt, and nickel complexes being formed in quantitative or near quantitative yields whereas vanadium, molybdenum and tungsten complexes are formed in low yields.

The reaction mechanism appears to consist of two steps: (i) decomposition of the palladium complex (**189**) by the transition metal carbonyl reagent (or low-valent complex) and (ii) displacement of two or more carbonyl groups by the cyclobutadiene ligand[113].

This approach may be an over-simplification and the steps may be merged into a sequence of intramolecular rearrangements involving a cyclobutadiene ring bonded simultaneously to two transition metals[128]. Reactions which compete with cyclobutadiene ligand transfer lead to the formation of metallic palladium, which is almost always obtained, and derivatives of cyclobutadiene such as octaphenylcyclooctatetraene and tetraphenylcyclopentadienone[113]. Reagents, conditions, products, and yields for some typical reactions are collected in Table 4.

F. From Acetylenes and Transition Metal Carbonyls

Acetylenes undergo a number of reactions, including cyclodimerization, in the presence of transition metal complexes and transition metal carbonyls. These reactions have been extensively investigated and exploited to provide routes to η^4-cyclobutadiene complexes[129,130]. The reactions of macrocylic alkadiynes with transition metal carbonyl derivatives afford η^4-cyclobutadiene complexes and may be regarded as a special case of the more general cyclodimerization process[131,133].

Disubstituted acetylenes, in particular diphenylacetylene, are the preferred reagents in reactions with metal carbonyls and related complexes leading to η^4-tetrasubstituted cyclobutadiene products. Thus, titanium(III) chloride combines with diphenylacetylene and cyclooctatetraene in the presence of isopropylmagnesium bromide in ether to give the green (η^4-tetraphenylcyclobutadiene)titanium complex (**190**) in 11% yield[134].

Irradiation of tetracarbonyl(η^5-cyclopentadienyl)vanadium with diphenylacetylene gives the acetylene complex (**191**), which undergoes thermal addition of a second acetylene to form the (η^4-cyclobutadiene)vanadium complex (**193**), presumably through the intermediate metallocycle (**192**)[117]. A similar sequence of reactions is used to obtain the mixed (η^4-cyclobutadiene)(η^2-acetylene)niobium complex (**194**) starting

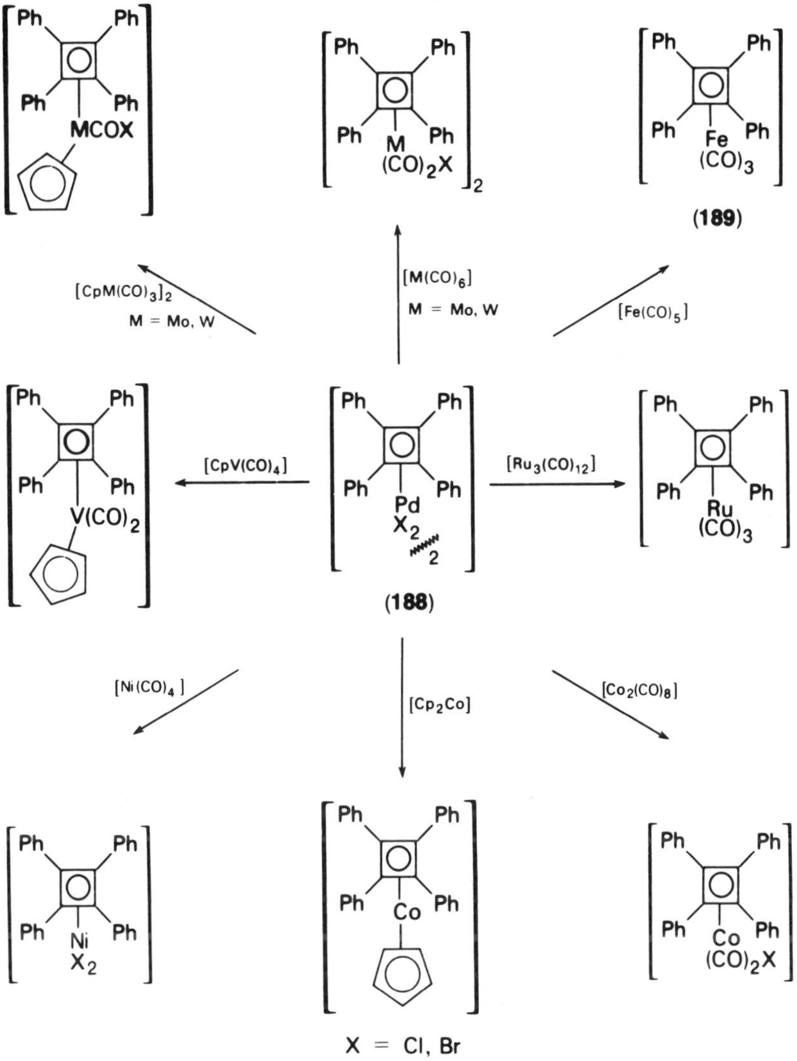

X = Cl, Br

SCHEME 8

from tetracarbonyl(η^5-cyclopentadienyl)niobium and diphenylacetylene in the molar ratio 1:2[135,136]. The correct structure **(194)** was assigned by X-ray crystallography shortly after the synthesis was reported[137].

The reaction between hexacarbonylmolybdenum and diphenylacetylene gives four different (η^4-cyclobutadiene)molybdenum complexes of which the two dicarbonyl complexes **(195** and **196)** are important[138]. The diacetylene molybdenum complex **(197)** is obtained by a related reaction using diphenylacetylene and tricarbonyl diglymemolybdenum[138].

The direct combination of acetylenes and pentacarbonyliron gives (η^4-cyclobutadiene)iron complexes. The high-temperature reaction using diphenylacetylene gives

TABLE 4. Reaction conditions and yields for η^4-cyclobutadiene ligand transfer reactions

Cyclobutadiene complex	Reagent	Conditions	Product	Yield	Reference
$[(\eta^4\text{-}Ph_4C_4)PdBr_2]_2$	$[(\eta^5\text{-}Cp)V(CO)_4]$	Toluene, reflux, 10 min	$[(\eta^4\text{-}Ph_4C_4)CpV(CO)_2]$	15	117
$[(\eta^4\text{-}Ph_4C_4)PdCl_2]_2$	$[(\eta^5\text{-}Cp)Mo(CO)_3]_2$	Benzene, 78°C, 50 h	$[(\eta^4\text{-}Ph_4C_4)CpMoCOCl]$	55	118
$[(\eta^4\text{-}Ph_4C_4)PdBr_2]_2$	$[(\eta^5\text{-}Cp)W(CO)_3]_2$	Benzene, 78°C, 50 h	$[(\eta^4\text{-}Ph_4C_4)CpWCOCl]$	0.7	118
$[(\eta^4\text{-}Ph_4C_4)PdBr_2]_2$	$[Mo(CO)_6]$	Benzene, 78°C, 67 h	$[(\eta^4\text{-}Ph_4C_4)Mo(CO)_2Br]_2$	18	119
$[(\eta^4\text{-}Ph_4C_4)PdBr_2]_2$	$[Fe(CO)_5]$	Xylene, 140°C, 20 min	$[(\eta^4\text{-}Ph_4C_4)Fe(CO)_3]$	88	116
$[\{\eta^4(4\text{-}ClC_6H_4)_4C_4\}PdBr_2]_2$	$[Fe(CO)_5]$	Benzene, 78°C, 10 min	$[\{\eta^4(4\text{-}ClC_6H_4)_4C_4\}Fe(CO)_3]$	46	120
$[(\eta^4\text{-}t\text{-}Bu_2Ph_2C_4)PdCl_2]_2$	$[Fe(CO)_5]$	Benzene, 78°C, 4 h	$[(\eta^4\text{-}Bu^t_2Ph_2C_4)Fe(CO)_3]$	65	121
$[(\eta^4\text{-}Ph_4C_4)PdBr_2]_2$	$[Ru_3(CO)_{12}]$	Chlorobenzene, 132°C, 12 h	$[(\eta^4\text{-}Ph_4C_4)Ru(CO)_3]$	42	122, 123
$[(\eta^4\text{-}Ph_4C_4)PdCl_2]_2$	$[Co_2(CO)_8]$	CH_2Cl_2, room temp., 45 h	$[(\eta^4\text{-}Ph_4C_4)Co(CO)_2Cl]$	65	124
$[(\eta^4\text{-}Ph_4C_4)PdBr_2]_2$	$[(\eta^5\text{-}Cp)_2Co]$	Xylene, 140°C, 2.5 h	$[(\eta^4\text{-}Ph_4C_4)CoCp]$	12	116
$[(\eta^4\text{-}Ph_4C_4)PdBr_2]_2$	$[Ni(CO)_4]$	Benzene, 78°C, 2.5 h	$[(\eta^4\text{-}Ph_4C_4)NiBr_2]_2$	47	116
$[(\eta^4\text{-}Ph_4C_4)PdBr_2]_2$	$[(n\text{-}Bu_3P)NiBr_2]$	Chlorobenzene, 132°C, 2 h	$[(\eta^4\text{-}Ph_4C_4)NiBr_2]_2$	90	125
$[\{\eta^4(4\text{-}MeC_6H_4)_4C_4\}PdBr_2]_2$	$[(n\text{-}Bu_3P)NiBr_2]$	Chlorobenzene, 132°C, 2 h	$[\{\eta^4(4\text{-}MeC_6H_4)_4C_4\}NiBr_2]_2$	73	125
$[(\eta^4\text{-}Ph_4C_4)NiCl_2]_2$	$[Fe(CO)_5]$	Benzene, 78°C 2 h	$[(\eta^4\text{-}Ph_4C_4)Fe(CO)_3]$	90	122
$[(\eta^4\text{-}Me_4C_4)NiCl_2]_2$	$[Fe_3(CO)_{12}]$	Thf, 66°C 20 min	$[(\eta^4\text{-}Me_4C_4)Fe(CO)_3]$	35	126
$[(\eta^4\text{-}Me_4C_4)NiI_2]_2$	$[Co_2(CO)_8]$	Thf, 25°C 14 h	$[(\eta^4\text{-}Me_4C_4)Co(CO)_2I]$	100	127

$TiCl_3$ + PhC≡CPh + [cyclooctatetraene] $\xrightarrow[Et_2O]{i\text{-}PrMgBr}$ [(Ph₄C₄)Ti(C₈H₈)] (64)

Ph Ph Ph Ph Ti

(190)

[CpV(CO)₄] $\xrightarrow[h\nu]{PhC\equiv CPh}$ [CpV(CO)₂(PhC≡CPh)] $\xrightarrow[110\ °C]{PhC\equiv CPh}$

(191)

[CpV(CO)₂(C₄Ph₄)] ⟶ [(C₄Ph₄)V(CO)₂Cp] (65)

Ph Ph Ph Ph V(CO)₂

(192) (193)

[CpNb(CO)₄] $\xrightarrow[h\nu]{2\,PhC\equiv CPh}$ [CpNbCO(PhC≡CPh)₂] $\xrightarrow[80\ °C]{PhC\equiv CPh}$ [(C₄Ph₄)NbCO(Cp)(PhC≡CPh)] (66)

Nb (CO)₄ NbCO PhC CPh C C Ph Ph Ph Ph Ph Ph NbCO CPh PhC

(194)

tricarbonyl(η^4-tetraphenylcyclobutadiene)iron (**189**) in 16% yield; the major product from the reaction is the tetraphenylcyclopentadienone complex (**198**)[139]. The yield of the desired complex (**189**) is improved by carrying out the reaction in a sealed tube at 240°C[140].

Acetylene combines with pentacarbonyliron at high pressure to form tricarbonyl(η^4-cyclobutadiene)iron (**148**) together with the metallocycle (**199**), which does not appear to be an intermediate in the formation of the η^4-cyclobutadiene complex (**148**). This cyclodimerization of the parent acetylene is a reaction confined to

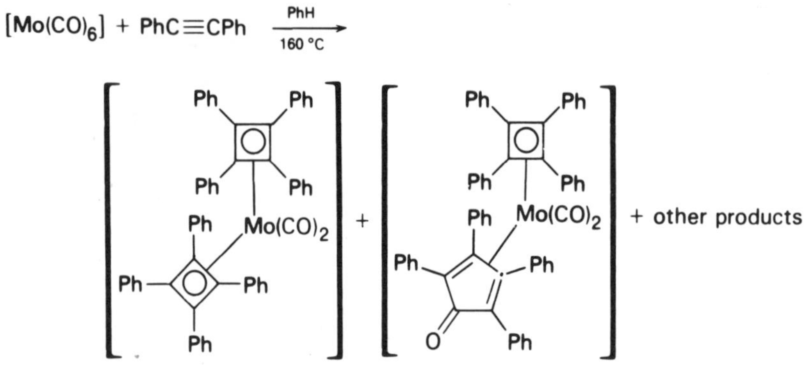

$[Mo(CO)_6] + PhC{\equiv}CPh \xrightarrow[160\,°C]{PhH}$ [(195)] + [(196)] + other products (67)

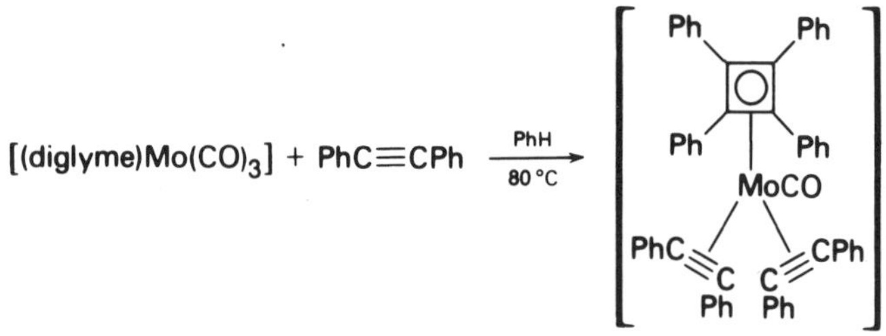

Ph Ph Ph Ph Ph $Mo(CO)_2$ Ph Ph Ph Ph

(**195**)

Ph Ph Ph Ph Ph $Mo(CO)_2$ Ph Ph O Ph

(**196**)

$[(diglyme)Mo(CO)_3] + PhC{\equiv}CPh \xrightarrow[80\,°C]{PhH}$ [(197)] (68)

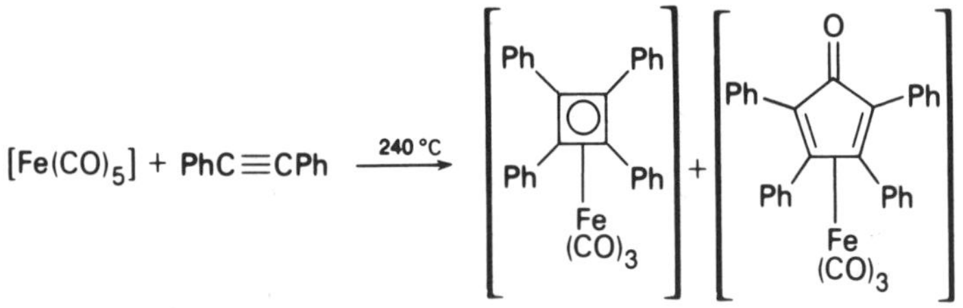

Ph Ph Ph Ph MoCO $PhC{\equiv}C$ Ph $C{\equiv}CPh$ Ph

(**197**)

$[Fe(CO)_5] + PhC{\equiv}CPh \xrightarrow{240\,°C}$ [(189)] + [(198)] (69)

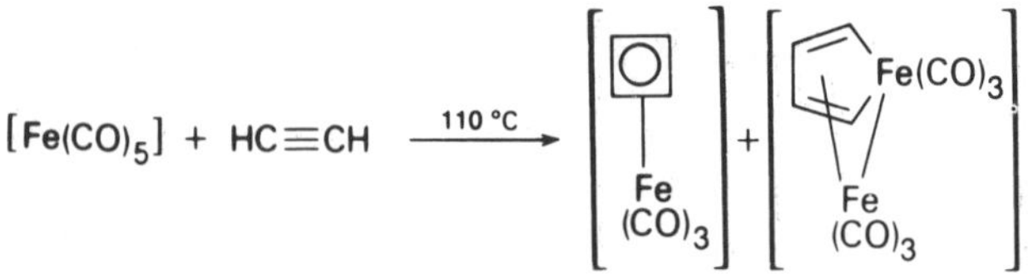

Ph Ph Ph Ph Fe $(CO)_3$

(**189**)

O Ph Ph Ph Ph Fe $(CO)_3$

(**198**)

$[Fe(CO)_5] + HC{\equiv}CH \xrightarrow{110\,°C}$ [(148)] + [(199)] (70)

Fe $(CO)_3$

(**148**)

$Fe(CO)_3$ Fe $(CO)_3$

(**199**)

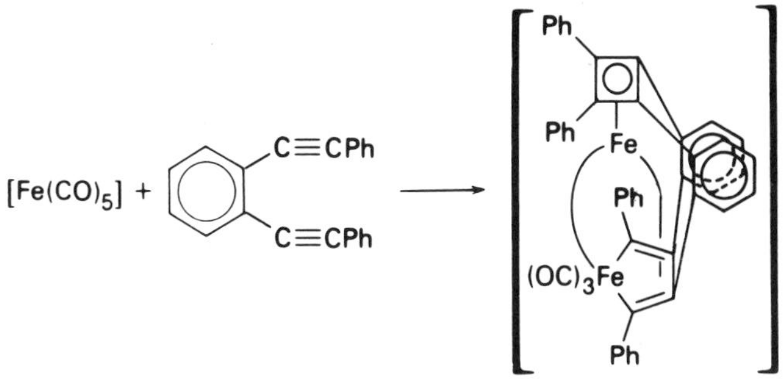

$[Fe(CO)_5]$ + o-C₆H₄(C≡CPh)₂ ⟶ [(200)] (71)

(200)

the formation of the iron complex (**148**)[141]. A mixed η^4-cyclobutadiene–metallocycle (**200**) is obtained from pentacarbonyliron and *o*-di(phenylethynyl)benzene[142]. The thiacycloheptyne (**201**) combines with palladium chloride to form the (η^4-cyclobutadiene)palladium intermediate (**202**), which is converted to the tricarbonyliron complex (**203**) with pentacarbonyliron[143].

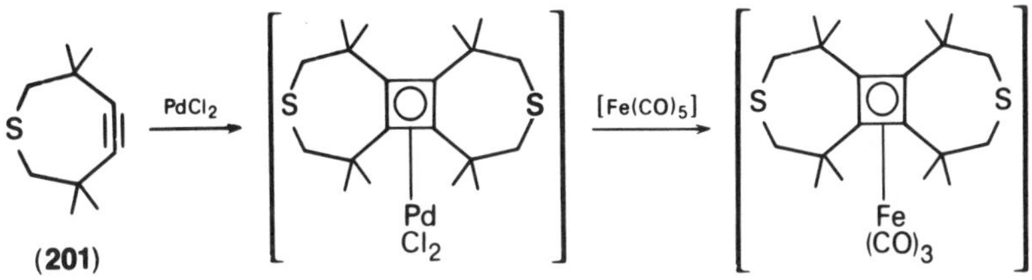

(201) $\xrightarrow{PdCl_2}$ [(202)] $\xrightarrow{[Fe(CO)_5]}$ [(203)] (72)

(201) (202) (203)

Several reactions involving acetylenes are available for the preparation of (η^4-cyclobutadiene)cobalt complexes (Scheme 9). Diphenylacetylene combines with cobaltocene[144], dicarbonyl(η^5-cyclopentadienyl)cobalt[145], and (η^4-cycloocta-1,5-diene)(η^5-cyclopentadienyl)cobalt (**204**)[146,147] to give the (η^4-tetraphenylcyclobutadiene)cobalt complex (**205**). This product is also obtained in a stepwise reaction sequence where the cobalt diiodide (**206**) is converted to the acetylene complex (**207**) by use of isopropylmagnesium bromide as a reducing agent and diphenylacetylene, followed by addition of a second molecule of diphenylacetylene to form the cobalt metallocycle (**208**), which on strong heating gives the η^4-cyclobutadiene complex (**205**) with the loss of triphenylphosphine[150,151]. These reactions are significant in that they indicate how the cyclobutadiene ligand may be formed by successive addition of two acetylene molecules to cobalt, metallocycle formation and subsequent rearrangement.

The basic synthesis has been extended to produce complexes substituted in the η^5-cyclopentadienyl ring by employing $(\eta^5\text{-}C_5H_4X)Co(CO)_2$, where X = $SiMe_3$, Ph, as the reagent[150]. An unsymmetrical acetylene may be used with cobaltocene or (η^4-cycloocta-1,5-diene)(η^5-cyclopentadienyl)cobalt to form the tetrasubstituted cyclobutadiene complexes (**209** and **210**) in yields of 38–59% (Scheme 10). The silyl and stannyl complexes (**209** and **210**; X = $SiMe_3$, $SnPh_3$) are cleaved with HCl to give the 1,2- and 1,3-diphenylcyclobutadiene complexes (**211** and **212**)[151]. In a related reaction, η^5-cyclopentadienylcobaltdicarbonyl combines with phenyl-2-thienylacetylene to form the two isomeric [η^4-diphenylbis(2-thienyl)cyclobutadiene]cobalt complexes (**209** and **210**; X = 2-thienyl)[153].

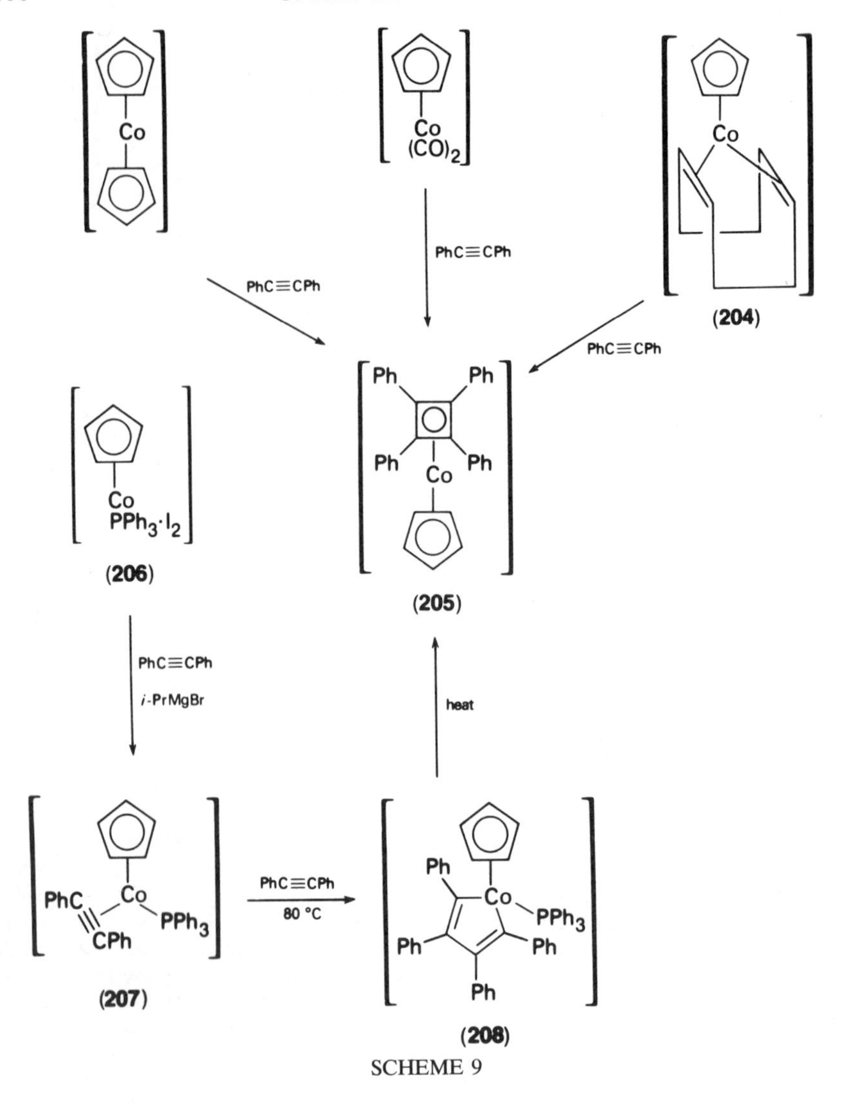

SCHEME 9

The tetrakis(trimethylsilyl)cyclobutadiene complex **(214)** is obtained in 5% yield by pyrolysis of the bridged dicobalt complex **(213)** with an excess of bis(trimethylsilyl)acetylene using xylene as solvent in an autoclave[152].

The rhodium analogue of the η^4-cyclooctadiene complex **(204)** combines with diphenylacetylene to form the rhodium complex **(215)** in low yield[154], whereas the binuclear complex **(216)** is formed in 50% yield from $[RhCl(PF_3)_2]_2$ and diphenylacetylene[155,156].

Dicarbonyldichloroplatinum combines with various disubstituted acetylenes to give mono-, di-, and trinuclear (η^4-cyclobutadiene)platinum compounds. The reaction with diphenylacetylene in ether gives first the polymeric complex **(217)**, which, on treatment with lithium bromide and lithium iodide in acetone at reflux temperature, gives

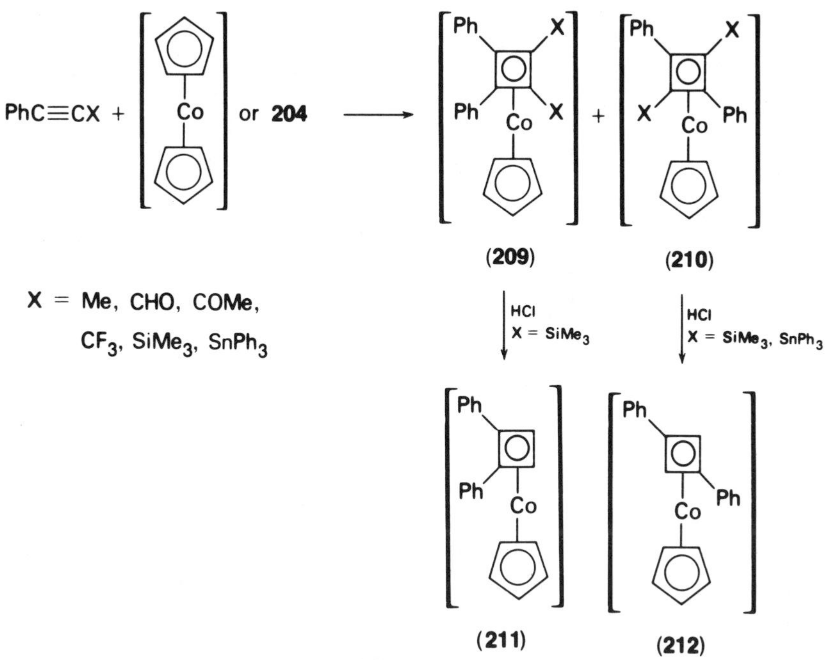

PhC≡CX + [CpCoCp] or **204** ⟶ (**209**) + (**210**)

X = Me, CHO, COMe, CF_3, $SiMe_3$, $SnPh_3$

(**209**) —HCl, X = $SiMe_3$⟶ (**211**)

(**210**) —HCl, X = $SiMe_3$, $SnPh_3$⟶ (**212**)

SCHEME 10

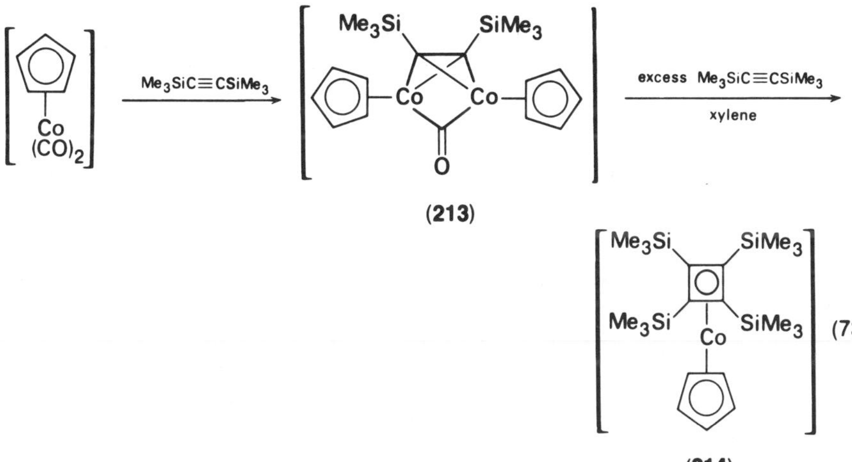

$$[CpCo(CO)_2] \xrightarrow{Me_3SiC\equiv CSiMe_3} (\mathbf{213}) \xrightarrow[\text{xylene}]{\text{excess } Me_3SiC\equiv CSiMe_3} (\mathbf{214}) \quad (73)$$

the binuclear bromide and iodide (**218**; X = Br, I)[157]. The corresponding reaction with hex-3-yne affords a mixture of products containing cyclopentadienone and quinone complexes of platinum in addition to trinuclear (**220**) and polymeric (**219**) cyclobutadiene complexes. Lewis bases such as pyridine and triphenylphosphine attack the complex **220** to form the mononuclear cyclobutadiene complex **221**[158]. A similar route is used to form (η^4-tetramethylcyclobutadiene)platinum dichloride starting from dicarbonyldichloroplatinum and but-2-yne[159].

+ PhC≡CPh ⟶ (**215**) (74)

$[Rh(PF_3)_2Cl]_2$ + PhC≡CPh ⟶ (**216**) (75)

$[Pt(CO)_2Cl_2]$ $\xrightarrow[\text{ether}]{\text{PhC≡CPh}}$ $[\ldots]_n$ (**217**) $\xrightarrow[\text{Me}_2\text{CO}]{\text{LiX}}$ (**218**) (76)

$[Pt(CO)_2Cl_2]$ $\xrightarrow{\text{EtC≡CEt}}$ $[\ldots]_n$ (**219**) + (**220**) $\xrightarrow{\text{PPh}_3}$ (**221**) (77)

Cyclobutadiene complexes of iron and cobalt are obtained by the intramolecular cyclization of macrocyclic alkadiynes in the presence of metal carbonyl derivatives. In the case of iron, the reactions give several products which vary considerably with the nature of the alkadiyne. Pentacarbonyliron combines with the alkadiyne **222** ($n = 4$, $m = 6$) in boiling toluene to form the (η^4-cyclobutadiene)iron complex (**223**; $n = 4$, $m = 6$) as the major product (26% yield)[132]. In contrast, the reaction of pentacarbonyliron with the alkadiyne **222** ($n = 4, m = 4$) gives a ferrole as the major product (31%) together with the cyclobutadiene (**223**; $n = m = 4$) (<1% yield)[132]. The complexes (**223**; $m = n = 5; m = 5, n = 6$) are obtained in only trace amounts from similar reactions using the alkadiynes (**222**; $m = n = 5$; $m = 5, n = 6$) and dodecacarbonyltriiron[133].

$(CH_2)_m$ $(CH_2)_m$ Fe $(CO)_3$ (**223**) ← $[Fe(CO)_5]$, PhMe, 111°C — $(CH_2)_m$ $(CH_2)_n$ (**222**) — $[(\eta^5\text{-}C_5H_5)Co(CO)_2]$, $n\text{-}C_8H_{18}$, 126°C → $(CH_2)_m$ $(CH_2)_n$ Co (**224**) (78)

In the case of cobalt, cyclobutadiene complexes are the major products and high yields are obtained in several reactions. Thus, dicarbonyl(η^5-cyclopentadienyl)cobalt combines with cyclododeca-1,7-diyne (**222**; $m = n = 4$), cyclotrideca-1,7-diyne (**222**; $m = 4$, $n = 5$), cyclotetradeca-1,8-diyne (**222**; $m = n = 5$), cyclotetradeca-1,7-diyne (**222**; $m = 4$, $n = 6$), and cyclopentadeca-1,8-diyne (**222**; $m = 5$, $n = 6$) in boiling octane or cyclooctane to give the respective cobalt complexes (**224**; $m = n = 4$, 85% yield; $m = 4$, $n = 5$, 75% yield; $m = n = 5$, 2% yield; $m = 4$, $n = 6$, 40% yield; and $n = 5$, $m = 6$, 52% yield)[160]. It is possible to replace (η^5-cyclopentadienyl)cobalt dicarbonyl with the cyclooctadiene complex (**204**) in the preparation of the complexes (**224**; $m = 4$, $n = 5$; $m = n = 5$)[160].

The mechanism of this reaction has attracted attention since molecular orbital calculations suggest that the concerted cyclization of bisacetylene coordinated with a single transition metal to form a (η^4-cyclobutadiene)metal complex is symmetry forbidden[161]. This approach indicates that two non-restrictive-field transition metal centres sharing opposite faces of a plane containing a bisacetylene remove the symmetry restrictions for the bisacetylene to cyclobutadiene conversion[162]. These ideas are compatible with the formation of metallocycle intermediates in a non-concerted reaction and suggest that polynuclear intermediates may be involved in the concerted process[130]. Chemical evidence in support of a stepwise process where acetylene groups are coordinated in turn and cyclize by way of a metallocyclopentadiene is provided by the conversion of the (η^5-cyclopentadienyl)cobalt complex (**206**) to the (η^4-cyclobutadiene)cobalt complex (**205**).

G. From Photo-α-pyrone and Transition Metal Carbonyls

Photolysis of α-pyrone (**225**; X = H) gives a reactive bicyclic intermediate, photo-α-pyrone (**226**; X = H)[163], which combines with metal carbonyls and carbonyl derivatives on further irradiation to give cyclobutadiene complexes (Scheme 11). The use of

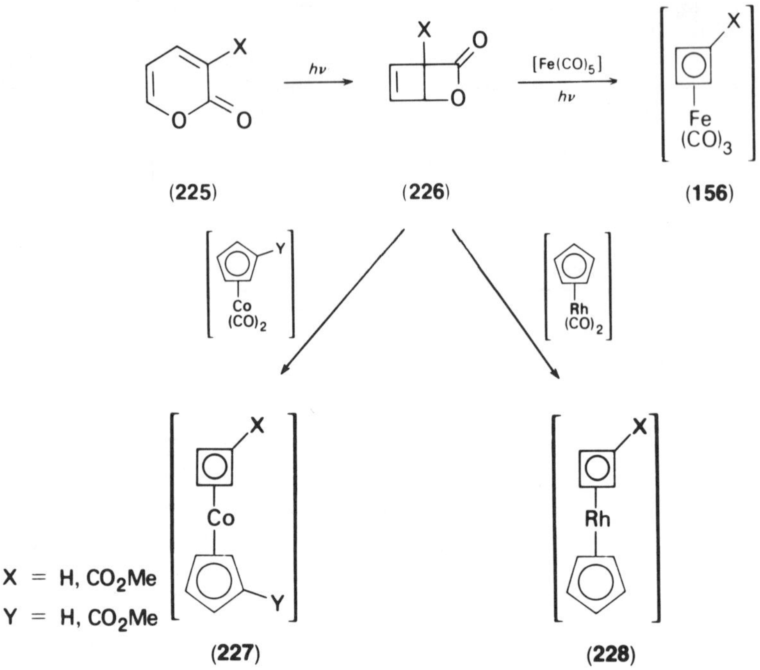

(**225**) (**226**) (**156**)

$h\nu$; $[Fe(CO)_5]$, $h\nu$

X = H, CO_2Me
Y = H, CO_2Me

(**227**) (**228**)

SCHEME 11

pentacarbonyliron in the reaction leads to tricarbonyl(η^4-cyclobutadiene)iron (**156**; X = H) in 7% yield[164], whereas the carboxymethyl derivative (**156**; X = CO_2Me) is obtained in 21% yield when the substituted α-pyrone (**225**; X = CO_2Me) in thf is used[165]. The tricarbonyliron complex of α-pyrone is also a significant product in these reactions. Both this product and the η^4-cyclobutadiene complex are labile to irradiation and thus careful control of the reaction conditions is necessary. When dicarbonyl(η^5-cyclopentadienyl)cobalt[166] or its carboxymethyl derivative[74] is irradiated with photo-α-pyrone, the (η^4-cyclobutadiene)cobalt complex (**227**; X = H, Y = H, CO_2Me) is obtained. The same route may be used to prepare the rhodium complex (**228**; X = H)[168].

H. From 1-Metallocyclopentadienes

It has been mentioned previously that metallocyclopentadienes may be intermediates in the preparation of η^4-cyclobutadiene complexes (see **192**, **208**, Section F). This method relies on the controlled formation of such an intermediate or on its use as starting material. Thus the stannole (**229**) is treated with bromine to cleave the heterocyclic ring and then with nickel bromide to form the (η^4-cyclobutadiene)nickel complex (**230**)[169]. A related reaction involves the treatment of the 1,4-dilithio-

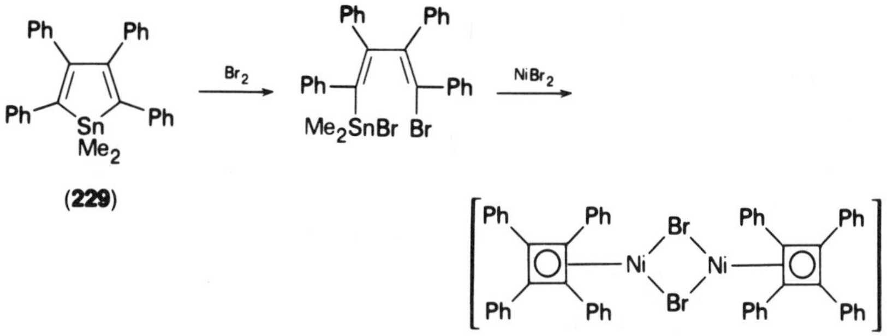

(229)

(230)

(79)

butadiene (**231**) with dibromotetracarbonyliron to give tricarbonyl(η^4-tetraphenylcyclobutadiene)iron (**189**) in 5% yield together with the corresponding ferrole[170]. Although these routes were developed early in the study of η^4-cyclobutadiene complexes, they have not attracted attention in recent years. However, an analogous reaction has been used in which tricarbonyl(η^4-1,2-diacetylbutadiene)iron is treated with hydrazine in acetic acid to give the (η^4-pyridazinocyclobutadiene)iron complex (**232**).

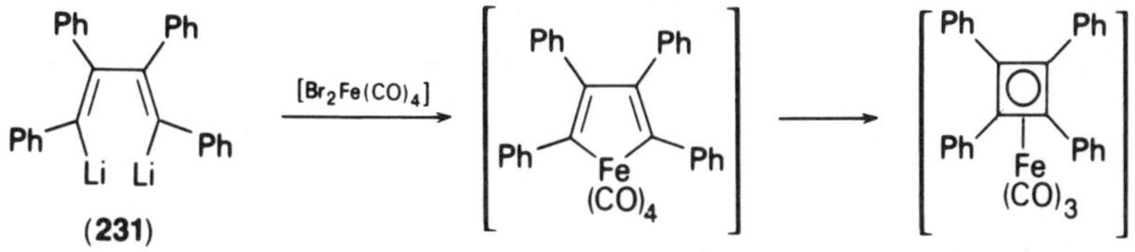

(231)

(189)

(80)

I. From Cyclooctatetraenes and Carbonylate Anions

A method which appears to be unique is based on the debromination and intramolecular cyclization of 1,4-dibromocyclooctatetraene in the presence of enneacarbonyldiiron in hexane to form the benzocyclobutadiene complex (**162**)[78]. A good yield (64%) may be obtained by using sodium tetracarbonylferrate(−II) in thf[78].

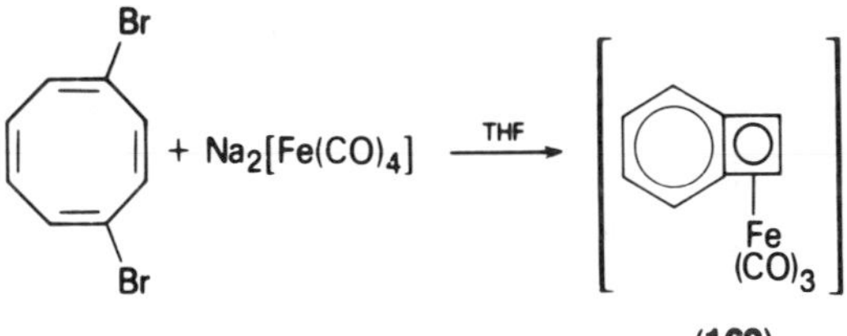

(162)

(81)

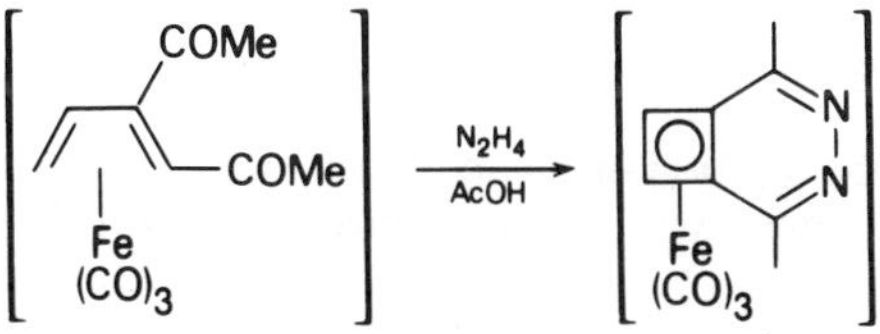

(232)

(82)

IV. η^4-CYCLO-1,3-DIENE COMPLEXES

A. Introduction

Many similarities exist between the preparative routes used for (η^4-cyclo-1,3-diene)- and (η^4-buta-1,3-diene)transition metal complexes. These are, however, outweighed by the differences observed, making it useful to discuss the two groups separately. It is also convenient to divide the survey into sections on the basis of the size of the hydrocarbon ring which is incorporated into the complex. Thus, cyclopentadiene is included with cyclopentadienone, fulvene, and five-membered heterocycles. Cyclohexadiene is grouped with cyclohexadienone, benzene and six-membered heterocycles. Cycloheptadiene, cycloheptatriene, cycloheptatrienone, and seven-membered heterocycles make up the third section, and hydrocarbons with eight-membered rings comprise the fourth section. (η^4-Cyclobutadiene)transition metal complexes are discussed separately (Section III). Iron is the most important transition metal in the chemistry of η^4-cyclo-1,3-diene complexes and receives major emphasis in this section. Complexes containing η^4-cyclodiene ligands with non-conjugated double bonds are not discussed.

B. Complexes Containing the η^4-Cyclopentadiene, η^4-Cyclopentadienone, η^4-Fulvene and η^4-Heterocyclopentadiene Ligands

Important routes to these complexes include the direct reaction between the ligand and a metal carbonyl, which may involve rearrangement of the ligand, the attack of acetylenes on metal carbonyls, and the reduction of η^5-cyclopentadienyl complexes. The first reaction is complicated by the strong tendency of (η^4-cyclopentadiene)iron complexes to undergo elimination of hydrogen and form the corresponding (η^5-cyclopentadienyl)iron complexes[173]. Cyclopentadienes and cyclopentadienones may undergo facile dimerization and further complicate the direct synthesis. Alkynes are important precursors for these complexes since the cyclopentadienone group is formed readily from two alkyne groups together with a carbonyl group abstracted from a metal carbonyl.

Cyclopentadiene combines with enneacarbonyldiiron in boiling diethyl ether to form tricarbonyl(η^4-cyclopentadiene)iron **(233)** in 27% yield as a yellow oil which freezes at $-6°C$[174]. It is purified by distillation under reduced pressure (30–35°C and 0.2 torr). Cyclopentadiene undergoes an analogous reaction on irradiation with $[Fe(PF_3)_5]$ in diethyl ether with the formation of the tris(trifluorophosphine) complex **(234)**[175].

Fe $(PF_3)_3$ ← $[Fe(PF_3)_5]$, $h\nu$ — cyclopentadiene — $[Fe_2(CO)_9]$, Et_2O, 36°C → Fe $(CO)_3$ **(83)**

(234) **(233)**

Whereas the cocondensation of metal atoms with cyclopentadiene usually affords η^5-cyclopentadienyl complexes, cobalt atoms give the (η^4-cyclopentadiene)cobalt complex **(235)**[176].

Substituents in the 5-position of cyclopentadiene reduce the tendency for hydrogen to be lost on treatment of the ligand with a metal carbonyl. Thus, acetyl(pentamethyl)-

+ $Co_{(g)}$ →(cocondense, low pressure, −196 °C) (235) (84)

cyclopentadiene combines with enneacarbonyldiiron in pentane to form the η^4-cyclopentadiene compound (**236**) in 25% yield, in addition to the η^5-cyclopentadienyl compound (**237**) in 8.5% yield. Tricarbonyl(η^5-pentamethylcyclopentadienyl)iron is the dominant product when the reaction is carried out at elevated temperatures[177,178].

COMe + $[Fe_2(CO)_9]$ →(pentane, 20 °C) COMe, Fe $(CO)_3$ (**236**) + Fe $(OC)_2$ COMe (**237**) (85)

In the same way, 5-hydroxymethyl-5-methylcyclopentadiene is attacked by enneacarbonyldiiron to give the η^4-cyclopentadiene complex (**238**) as a mixture of two isomers (**238a and 238b**) in the proportions 4:1[179]. Initial attempts to convert the spirane (**239**) to a tricarbonyliron complex with pentacarbonyliron caused rearrangement with the formation of a tetrahydroindenyl complex[180]. However, the use of enneacarbonyldiiron as reagent under mild conditions (boiling benzene for 1.5 h) gives the desired product (**240**) in 31% yield[181].

Me CH_2OH Fe $(CO)_3$ (**238**a) HOCH$_2$ Me Fe $(CO)_3$ (**238**b) (**239**) →($[Fe_2(CO)_9]$, PhH, 80 °C) Fe $(CO)_3$ (**240**) (86)

Although the reactions between iron carbonyls and fulvenes are often complex and may lead to several products, the use of mild conditions enables (η^4-fulvene)iron complexes to be obtained. Thus, 6,6-cyclopentamethylenefulvene, 6,6-diphenylfulvene, and 6,6-bis(4-chlorophenyl)fulvene each combine with enneacarbonyldiiron at ~40°C to give the tricarbonyliron compounds [**241**; $R_2 = (CH_2)_5$, R = Ph, 4-ClC_6H_4][182,183].

Pentalene and its derivatives combine directly with transition metal carbonyls to give η^4-cyclodiene complexes. The parent complex (**243**; R = H) is obtained by heating pentalene dimer (**242**) with enneacarbonyldiiron at 50°C in methylcyclohexane under carbon monoxide in a sealed tube. The yellow–brown solid product (**243**; R = H), obtained in 9% yield, is stable at room temperature[184]. The corresponding

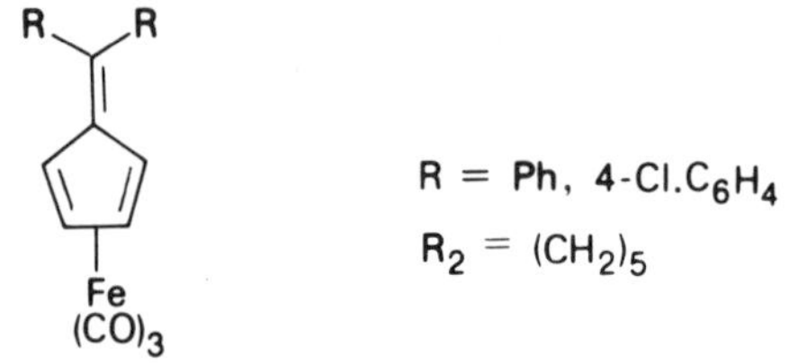

R = Ph, 4-Cl.C_6H_4

R_2 = $(CH_2)_5$

(**241**)

complex of 1,3-dimethylpentalene (**243**; R = Me) is formed in the same way in 21% yield[184]. Pentalene dimer is used in the reactions since the monomer shows a strong tendency to dimerize. However, dihydropentalenes do not show the same tendency and may be used to form pentalene complexes. 1,2-Dihydro-3-dimethylaminopentalene (**244**; R = NMe_2) is attacked by pentacarbonyliron in methylcyclohexane at 110°C to form the pentalene complex (**245**; R = NMe_2) in 11% yield[185]. In the same way, 1,2-dihydro-3-phenylpentalene gives the (1-phenylpentalene)iron complex (**245**; R = Ph) in 12% yield[186].

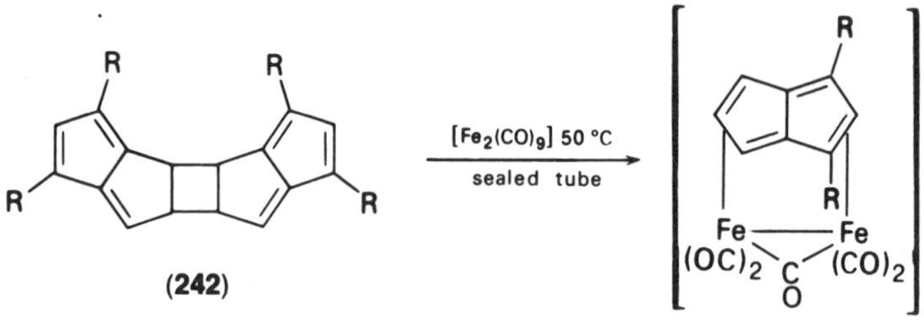

(**242**) $\xrightarrow[\text{sealed tube}]{[Fe_2(CO)_9]\ 50\ °C}$ (**243**) (87)

Many cyclopentadienones are unstable with respect to dimerization or other reactions and do not form suitable starting materials for the preparation of the corresponding transition metal complexes. However, tetraaryl- and other tetra-substituted cyclopentadienones are stable as monomers and are synthetically useful, as shown in Scheme 12. The direct reaction of tetraphenylcyclopentadienone (**246**) with pentacarbonyliron, enneacarbonyldiiron, or dodecacarbonyltriiron in boiling benzene, toluene at 100°C, or xylene at 150°C gives the (η^4-cyclopentadienone)iron complex (**247**) as a yellow, diamagnetic, air-stable solid[187,188]. The yield is almost quantitative when pentacarbonyliron is the reagent[14]. The related *p*-chlorophenylcyclopentadienone complexes (**251**; X = H, Cl) and the di-substituted cyclopentadienone complex (**252**) are formed in similar reactions with pentacarbonyliron or enneacarbonyldiiron[187].

Dodecacarbonyltriruthenium combines with tetraphenylcyclopentadienone to form the tricarbonylruthenium complex (**248**)[189], whereas hexacarbonylmolybdenum gives a bis(η^4-cyclopentadienone) complex (**249**)[187] and tetracarbonylnickel forms the complex **250**[190].

Some cyclopentadienes form colourless dimers which dissociate reversibly on heating. When a suitable metal carbonyl is introduced then the equilibrium proportion of monomer present is removed by complex formation and the method may be used in the synthesis of η^4-cyclopentadiene complexes. Tricarbonyl(η^4-2,5-dimethyl-3,4-diphenylcyclopentadienone)iron (**254**) is prepared in this way (79% yield) by heating the dimer (**253**) with pentacarbonyliron at 190°C in benzene in an autoclave[187]. Santonin undergoes rearrangement on heating with enneacarbonyldiiron in benzene at 40°C to

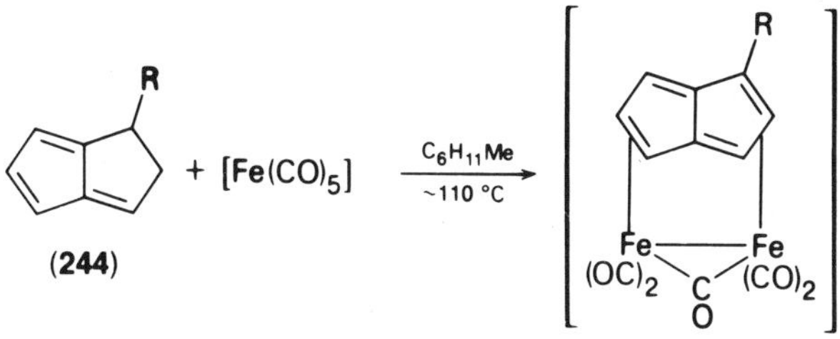
R
+ [Fe(CO)5]
C6H11Me
~110 °C
R
Fe
Fe
(OC)2
C
(CO)2
O

(244) (245) (88)

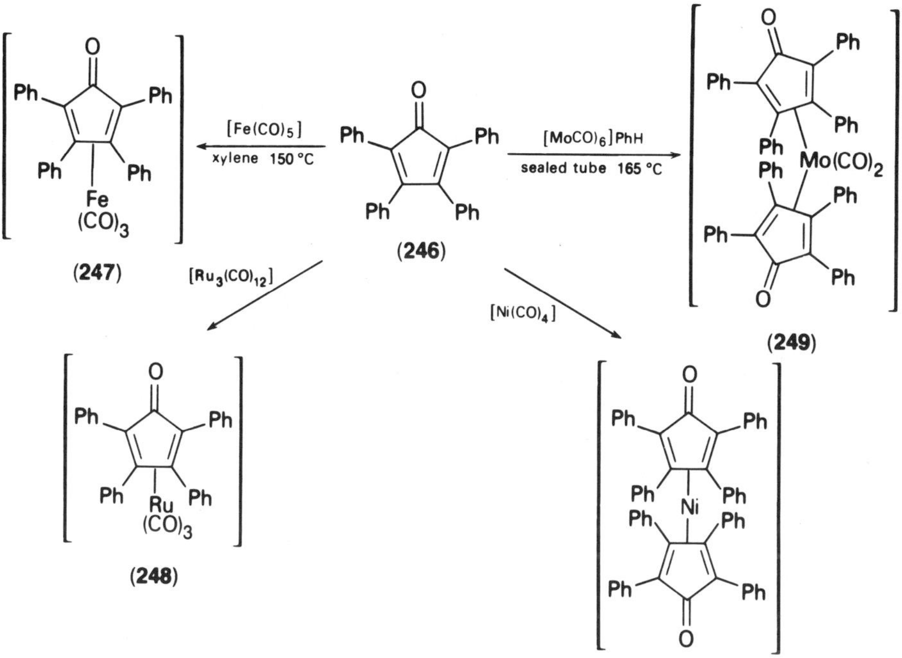
[Fe(CO)5]
xylene 150 °C
[MoCO)6]PhH
sealed tube 165 °C
[Ru3(CO)12]
[Ni(CO)4]
Mo(CO)2
Fe
(CO)3
Ru
(CO)3
Ni

(247) (246) (249) (248) (250)

SCHEME 12

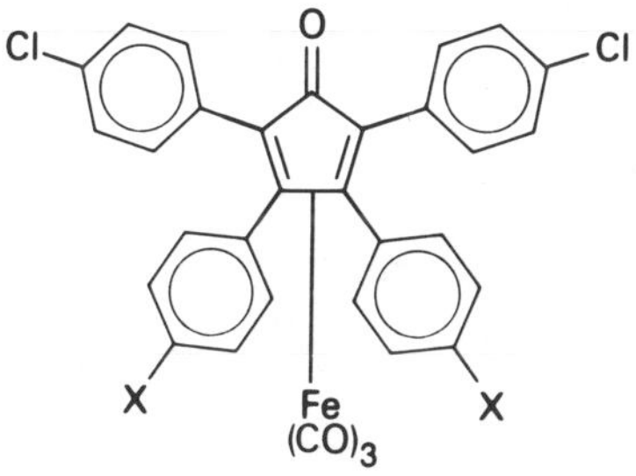
Cl
O
Cl
X
Fe
(CO)3
X

(251)

X = H, Cl

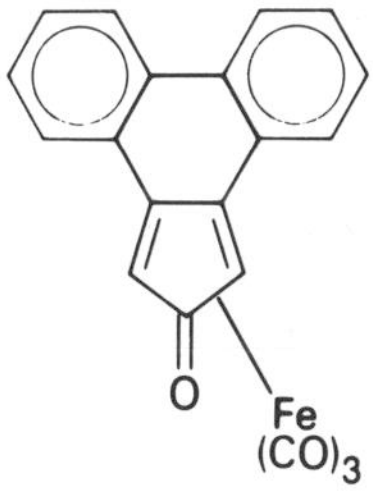
O
Fe
(CO)3

(252)

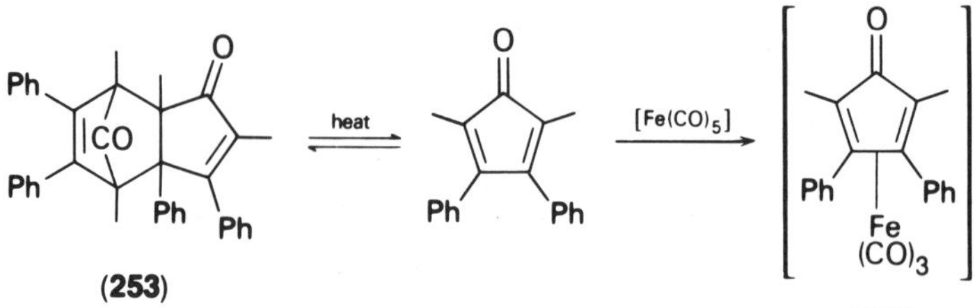

(89)

(253) (254)

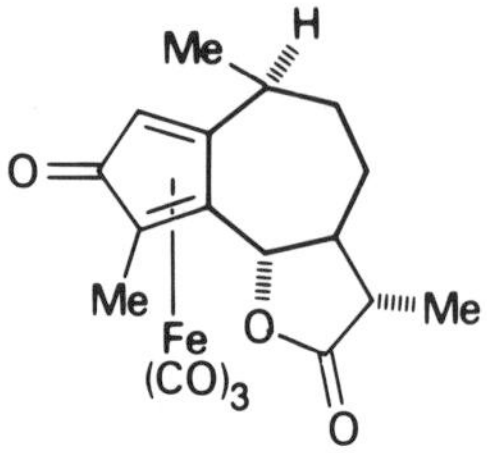

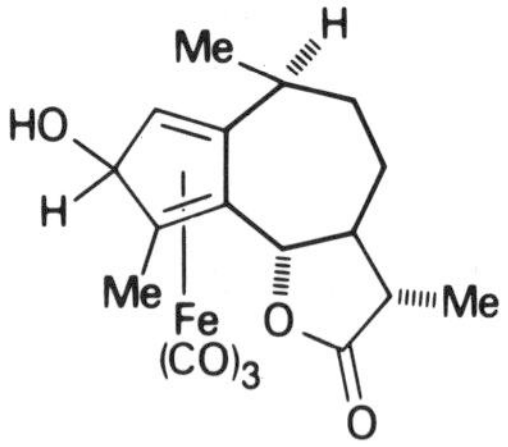

(255) (256)

form a trisubstituted cyclopentadienone complex (**255**) in addition to the tricarbonyliron complex (**256**) formed by rearrangement and reduction[191].

Free cyclopentadienone is unstable and cannot be used in the synthesis of (η^4-cyclopentadienone)metal complexes; however, the ketal (**257**) is stable and combines with enneacarbonyldiiron to form tricarbonyl(η^4-cyclopentadienone)iron (**258**) as one of the reaction products[192].

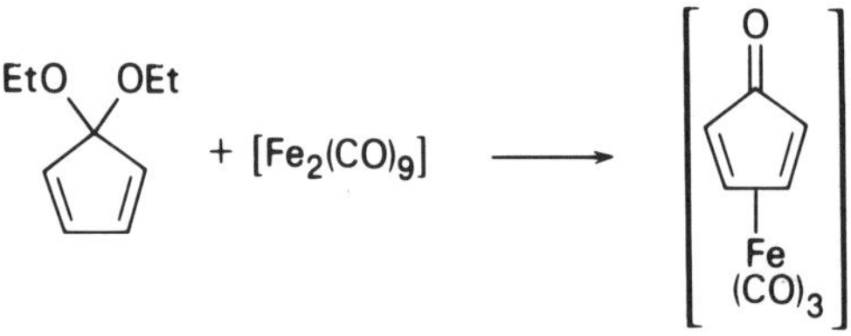

(90)

(257) (258)

Complexes containing heterocyclopentadienes, that is cyclopentadienes where carbon-5 is replaced with a heteroatom, are formed readily by direct reaction between the ligand and a transition metal carbonyl. The η^4-tetraphenylsilacyclopentadiene complex **260** is obtained from pentacarbonyliron and the free ligand (**259**)[193,194],

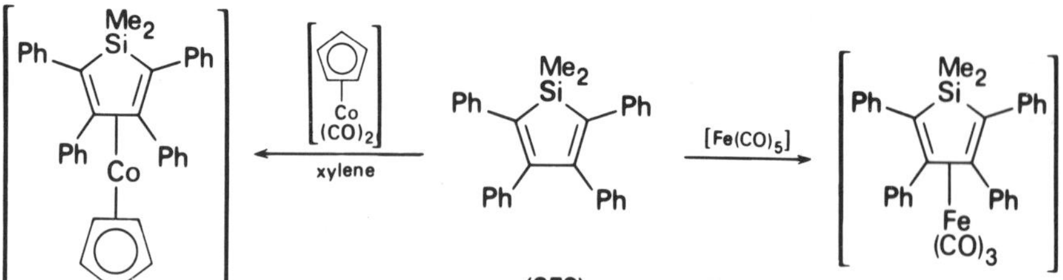

(261) (259) (260)

(91)

whereas the same reaction using any one of the neutral iron carbonyls as reagent converts the 2,5-diphenylsilacyclopentadiene **262** to the tricarbonyliron complex (**263**)[195,196]. The same silacyclopentadiene (**262**) combines with dodecacarbonyltriruthenium in toluene to form the tricarbonylruthenium complex (**264**)[197] (Scheme 13).

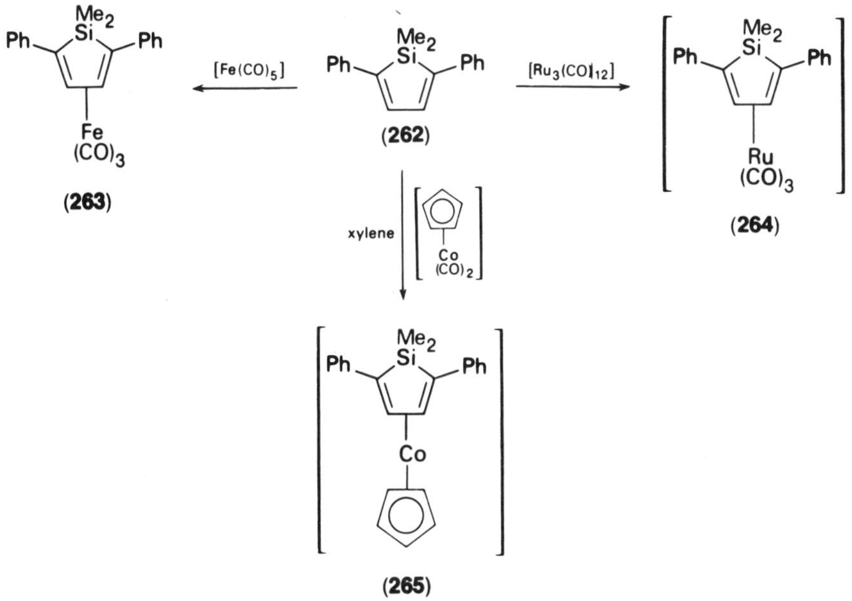

Ph Me2 Si Ph Fe (CO)3 (**263**) [Fe(CO)5] Ph Me2 Si Ph (**262**) [Ru3(CO)12] Ph Me2 Si Ph Ru (CO)3 (**264**) xylene Co (CO)2 Ph Me2 Si Ph Co (**265**)

SCHEME 13

Similar reactions may be used to form cobalt complexes of silacyclopentadiene. Dicarbonyl(η^5-cyclopentadienyl)cobalt is attacked by the silacyclopentadienes **259** and **262** to give the mixed (η^4-silacyclopentadiene)(η^5-cyclopentadienyl)-cobalt complexes (**261** and **265**, respectively)[198]. Pentaphenylphosphole (**266**) combines with dodecacarbonyltriiron in boiling isooctane to form the tricarbonyliron complex (**267**) in low yield; the major product is the tetracarbonyliron compound (**268**)[199]. In contrast, pentaphenylphosphole oxide (**269**) combines with pentacarbonyliron in a sealed tube at 150°C to give the tricarbonyliron complex (**270**) in 94% yield[199]. The arsenic analogue of the pentaphenylphosphole (**267**) is prepared in 69% yield from the free ligand and pentacarbonyliron in benzene by heating in a sealed tube at 150°C[199].

Ultraviolet irradiation of thiophen-1,1-dioxide (**271**; $R^1 = R^2 = H$) (formed *in situ* by heterogeneous debromination of 3,4-dibromotetrahydrothiophen-1,1-dioxide) in benzene solution with pentacarbonyliron gives the (η^4-thiophen-1,1-dioxide)iron complex (**272**; $R^1 = R^2 = H$) in 60% yield[200]. The same procedure may be used to prepare the 2,5-dimethyl- and 2,3,4,5-tetraphenylthiophen-1,1-dioxide complexes (**272**; $R^1 = Me$, $R^2 = H$; $R^1 = R^2 = Ph$, respectively) in yields of 90% and 50%[200]. These two complexes may also be obtained by the thermal reaction between the ligand and pentacarbonyliron in benzene using a sealed tube at 170°C. The yield of the tetraphenyl complex (**272**; $R^1 = R^2 = Ph$) is 66% using this procedure[187]. (η^4-Benzo[*b*]-thiophen-1,1-dioxide)tricarbonyliron (**274**) is formed as a minor product on irradiation of the ligand (**273**) together with pentacarbonyliron in benzene. The major product is

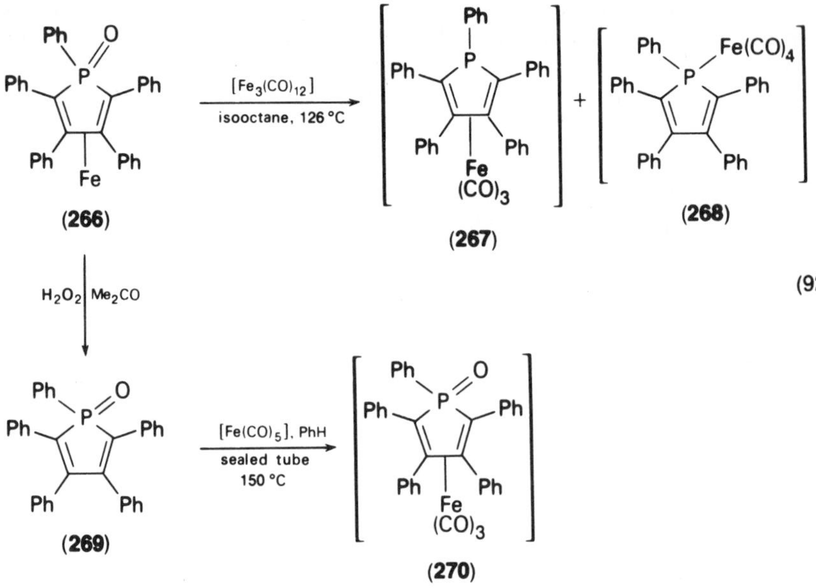

(266) (267) (268) (269) (270)

(92)

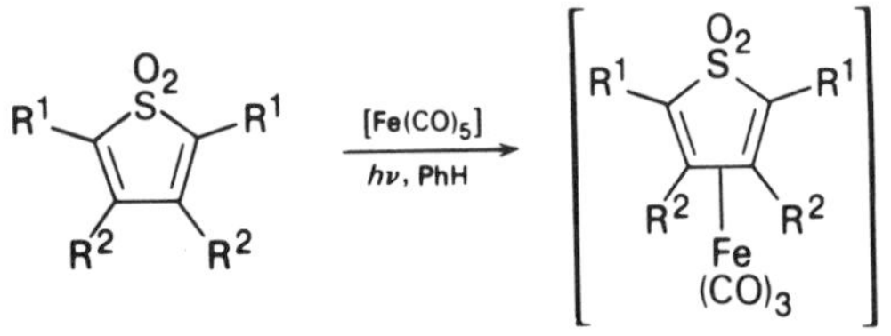

(271) (272)

(93)

the tetracarbonyliron complex (**275**)[200]. The proportion of the complex (**274**) in the product mixture is enhanced by using enneacarbonyldiiron in pentane as the solvent in the absence of light[200].

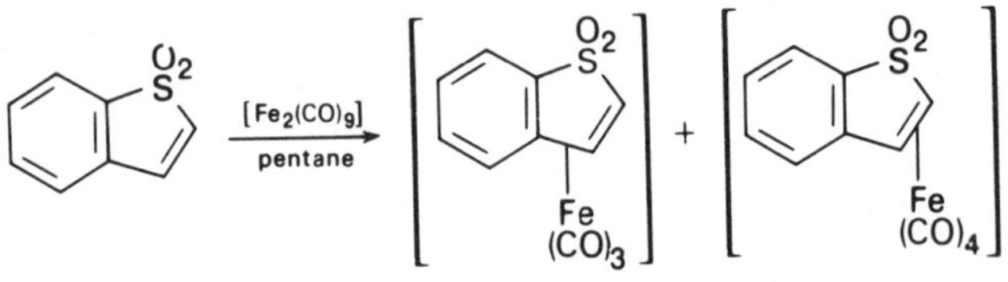

(273) (274) (275)

(94)

In contrast to thiophen-1,1-dioxide, thiophen itself combines with dodecacarbonyltriiron to give the ferrole (**276**) in 5% yield by desulphurization, rather than a thiophen complex[201].

Cyclopentadienone and ferrole complexes may be formed from alkynes and metal carbonyls. This route is widely applicable but does not appear to be used for cyclopentadienes. Tricarbonyl(η^4-cyclopentadienone)iron (**258**) may be prepared by the reaction of pentacarbonyliron with acetylene in light petroleum or benzene[202]. Substituted

$$\text{thiophene} \xrightarrow{[Fe_3(CO)_{12}]\ \text{Soxhlet}} [(C_4H_4)Fe(CO)_3Fe(CO)_3] \quad (95)$$

(**276**)

alkynes undergo similar reactions. Thus, dichloroethyne combines with enneacarbonyldiiron in diethyl ether at 25°C to give the tetrachlorocyclopentadienone complex (**277**; X = Cl) in 27% yield[203]. Tricarbonyl(η^4-tetraphenylcyclopentadienone)iron (**277**; X = Ph) is prepared in 45% yield by irradiation of diphenylethyne and pentacarbonyliron in benzene[188]. The same complex (**277**; X = Ph) may be formed under thermal conditions using dodecacarbonyltriiron. The following alkynes may also be converted to the corresponding η^4-cyclopentadienone complexes on heating at 60–100°C with enneacarbonyldiiron or dodecacarbonyltriiron in an inert solvent: $4\text{-ClC}_6H_4C{\equiv}CC_6H_4\text{-4-Cl}$, $C_6H_5C{\equiv}CMe$, $C_6H_5C{\equiv}CH$, $C_6H_5C{\equiv}CSiMe_3$, and $4\text{-BrC}_6H_4C{\equiv}CH$[204]. Hexafluorobut-2-yne (**278**) is converted to the [η^4-tetrakis(trifluoromethyl)cyclopentadienone]iron complex (**279**) in 60% yield on heating with

(**277**)

$$[(CF_3)_4C_4CO]Ru(CO)_3\ (\mathbf{280}) \xleftarrow[\text{sealed tube } 135\,°C]{[Ru_3(CO)_{12}]} F_3C{-}{\equiv}{-}CF_3\ (\mathbf{278}) \xrightarrow[\text{sealed tube } 110\,°C]{[Fe(CO)_5]} [(CF_3)_4C_4CO]Fe(CO)_3\ (\mathbf{279}) \quad (96)$$

pentacarbonyliron at 110°C under pressure[205]. The tricarbonylruthenium complex (**280**) is formed in the same way in 10% yield using dodecacarbonyltriruthenium as the reagent[189]. Pentacarbonyliron and bis(diethylamino)ethyne combine on irradiation to form the tricarbonyliron complex (**277**; X = NEt_2) in 7% yield[206]. The same complex may be obtained in 10% yield by heating the same ligand with dodecacarbonyltriiron in hexane at the reflux temperature[206]. Dicarbonyl(η^5-cyclopentadienyl)cobalt (**281**)

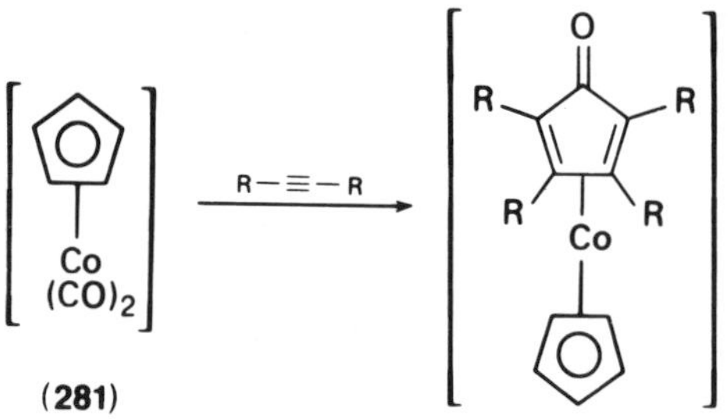

(281) → (282) (97)

$R{-}{\equiv}{-}R$

R = Me, CF$_3$, Ph

combines with alkynes to form (η^4-cyclopentadienone)cobalt complexes; typical examples are the methyl, trifluoromethyl and phenyl complexes (**282**; R = Me, CF$_3$, Ph), which may be prepared by ultraviolet irradiation of the reactants[207]. Dicarbonyl(η^5-cyclopentadienyl)rhodium (**283**) also combines with alkynes to form (η^4-cyclopentadienone)rhodium complexes. Thus, hexafluorobut-2-yne at 110°C in a sealed tube with the dicarbonyl **283** gives a mixture of the η^4-cyclopentadiene complex (**284**) in 48% yield and the η^6-benzene complex (**285**) in 45% yield[208]. Physico-

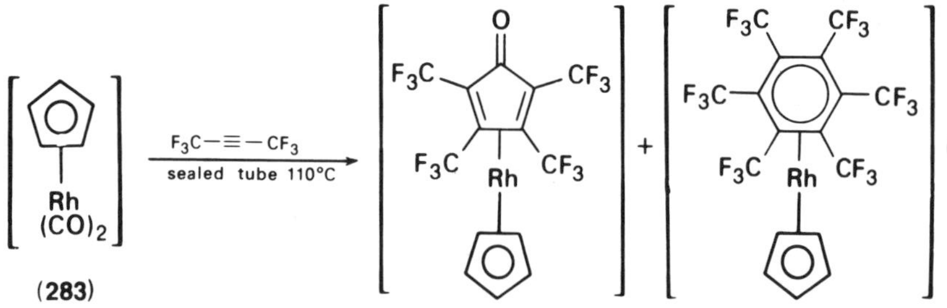

(283) → (284) + (285) (98)

$F_3C{-}{\equiv}{-}CF_3$, sealed tube 110°C

chemical evidence suggests that structures in which the cyclopentadiene group in **284** is coordinated through two σ-bonds and an η^2-bond may be important and that the metal—ligand interaction in the rhodium complex **285** may similarly receive a major contribution from a structure with two σ-bonds and two η^2-bonds[208]. The (η^4-cyclopentadienone)rhodium complex (**286**) is formed in 19% yield by heating dicarbonyl(η^5-cyclopentadienyl)rhodium (**283**) with bis(diethylamino)ethyne in octane at the reflux temperature[206].

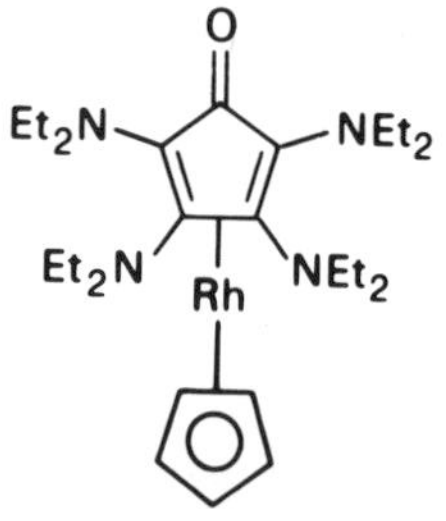

(**286**)

In addition to η^4-cyclopentadienone complexes, ferroles are formed as important products from the reaction between iron carbonyls and alkynes. The parent complex (**276**) is obtained by heating acetylene with pentacarbonyliron in ethanol under pressure or with dodecacarbonyltriiron[209–211]. The tetraphenylferrole **287** is the dominant product from the reaction of iron carbonyls with diphenylacetylene. The photochemical reaction using pentacarbonyliron in benzene gives a higher yield (42%) than the thermal reaction employing either one of the two polynuclear carbonyls[188,212]. The analogous ruthenium complex (**288**) is prepared in 47% yield from dodecacarbonyltriruthenium and diphenylacetylene in decalin at 200°C[213].

Ph—≡—Ph $\xrightarrow[\text{decalin, 200°C}]{[Ru_3(CO)_{12}]}$ (**288**)   (99)

$\xleftarrow[h\nu,\ PhH]{[Fe(CO)_5]}$ **277** + (X = Ph) + (**287**)

Disubstituted ferroles are formed when monosubstituted acetylenes are used as reagents[212]. Cyclic and acyclic diacetylenes combine with iron carbonyls to give ferroles. Thus, irradiation of the diacetylene **289** and pentacarbonyliron gives the ferrole **290** as the only product while the corresponding thermal reaction gives the same

Me—≡—≡—Me (**289**) $\xrightarrow[h\nu]{[Fe(CO)_5]}$ (**290**)   (100)

ferrole (**290**) together with two isomers[214]. Pentacarbonyliron attacks cyclododeca-1,7-diyne to form the ferrole (**291**), presumably through an (η^4-cyclobutadiene)iron intermediate[215,216]. Several other alkadiynes (**292**; $m = 4$, $n = 5$ and 6; $m = 5$, $n = 5$ and 6) take part in similar reactions on heating with pentacarbonyliron or dodecacarbonyltriiron to form ferroles (**293**) that can undergo further reaction to give η^4-cyclobutadiene or η^5-cyclopentadienyl complexes[217].

$(CH_2)_4$ $(CH_2)_4$ $\xrightarrow{[Fe(CO)_5]}$ $(CH_2)_4$ $(CH_2)_4$ Fe $(CO)_3$ Fe $(CO)_3$ (101)

(291)

$(CH_2)_m$ $(CH_2)_n$ $\xrightarrow[{[Fe_3(CO)_{12}]}]{[Fe(CO)_5] \text{ or}}$ $(CH_2)_m$ $Fe(CO)_3$ $(CH_2)_n$ Fe $(CO)_3$ (102)

(292) (293)

C. Complexes Containing the η^4-Cyclohexa-1,3-diene and η^4-Heterocyclohexa-1,3-diene Ligands

1,3-Cyclohexadiene readily forms complexes with transition metals in which the ligand acts as a four-electron donor. Important preparative routes involve combination of a metal carbonyl with the free ligand which may undergo rearrangement during complexation, or reduction of an (η^5-cyclohexadienyl)metal complex, usually by nucleophilic attack on a cationic complex. Since η^5-cyclohexadienyl complexes are frequently derived from η^4-cyclohexadiene precursors, the second route would not appear to be valuable. Its usefulness is vested in the ligand rearrangement that may occur during the $\eta^4 \rightarrow \eta^5 \rightarrow \eta^4$ transformation and in the groups that may be conveniently introduced into the ligand by this route.

The formation of η^4-cyclohexa-1,3-diene complexes is favoured by the *cisoid* configuration of the conjugated ethylenic double bonds in the ligand which is essential for complex formation[218]. Further, the complexes, once formed, do not show a strong tendency to lose hydrogen and give η^5-cyclohexadienyl or η^6-benzene complexes.

The direct reaction between cyclohexa-1,3,-diene and pentacarbonyliron in an autoclave at 135°C gives tricarbonyl(η^4-cyclohexa-1,3-diene)iron **(294)** as a yellow oil freezing at 8°C[218] (Scheme 14). Cyclohexa-1,3-diene also combines with tetracarbonyl(η^5-cyclopentadienyl)vanadium[219]; tricarbonyl(η^6-mesitylene)molybdenum[220] and tricarbonyl(η^6-mesitylene)chromium on irradiation[220]; dicarbonyl(η^5-cyclopentadienyl)cobalt on heating[221] or on irradiation[222]; dicarbonyl(η^5-cyclopentadienyl)rhodium on irradiation[222]; and dodecacarbonyltriruthenium on heating in benzene[223]. In each case an (η^4-cyclohexadiene)metal complex is formed. The vanadium, cobalt and rhodium complexes **(295**, **296**, and **297)** contain both η^4-cyclohexadiene and η^5-cyclopentadienyl ligands whereas the chromium and molybdenum complexes **(298** and **299)** contain two η^4-cyclohexadiene groups. The tricarbonylruthenium complex **(300)** is analogous to the iron complex **(294)**.

Large numbers of tricarbonyl(η^4-cyclohexa-1,3-diene)iron complexes have been prepared from the free ligand or its 1,4-isomer and an iron carbonyl. An important route uses a substituted benzene as the cyclohexadiene precursor. Sodium in liquid ammonia is used as the reducing agent (Birch reduction) and the resulting substituted cyclohexadiene is treated with pentacarbonyliron in boiling di-*n*-butyl ether to form

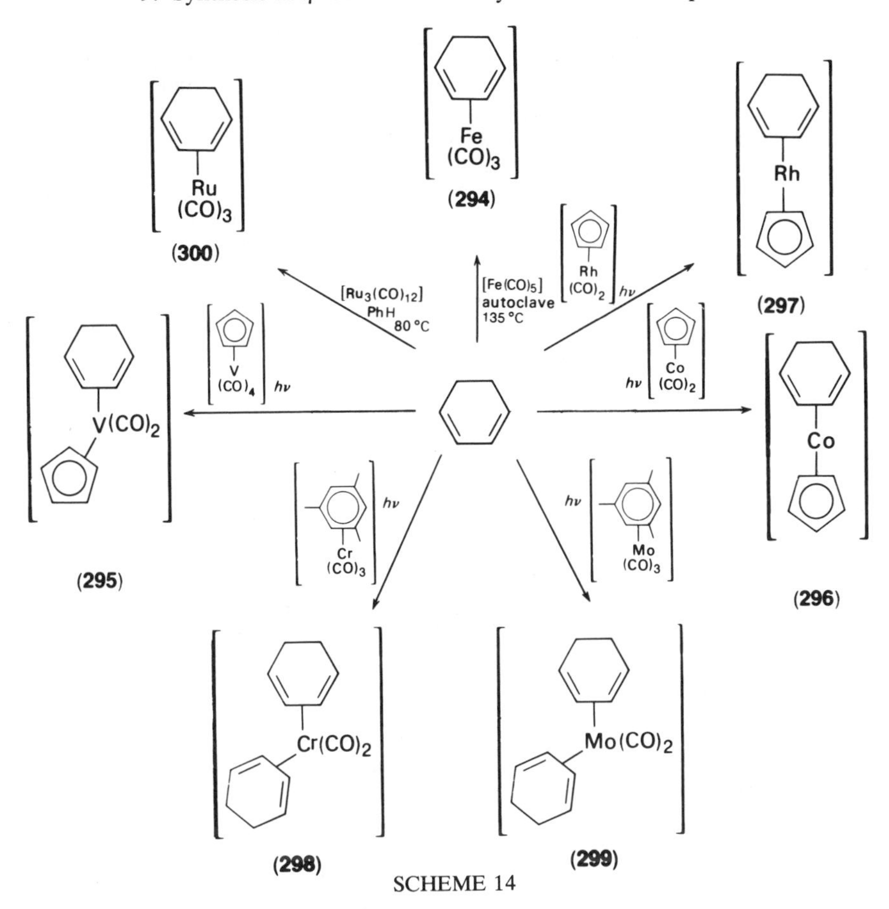

SCHEME 14

the tricarbonyliron product **(301)**. Birch reduction of a benzenoid compound usually gives a cyclohexa-1,4-rather than a cyclohexa-1,3-diene, which isomerizes on complex formation to the conjugated structure. Typical benzene compounds that may be con-

R1–(benzene)–R2 →(Na liq. NH_3) R1–(cyclohexa-1,4-diene)–R2 →($[Fe(CO)_5]$, n-Bu_2O 142 °C) [R1–(cyclohexa-1,3-diene)–R2 $Fe(CO)_3$] **(301)** (103)

$R^1 = R^2 = H$

$R^1 = OMe$, $R^2 = OMe$, Me

verted to cyclohexa-1,3-diene complexes by this route include benzene[224], toluene[224,225] *p*-xylene[224,225], *m*-xylene[225], anisole[224,225], *o*-, *m*-, and *p*-methoxytoluene[224,225], and mesitylene[226,227]. Birch reduction of benzoic acid and *o*-toluic acid followed by esterification with diazomethane to form **302** and complex formation with pentacarbonyliron gives (η^4-cyclohexa-1,3-diene)iron complexes[229]. Whereas benzoic acid leads to a mixture of two products (**303** and **304**) in 34% yield, *o*-toluic acid affords a mixture of four isomeric products in an overall yield of 12% for the last step[228]. A single product (**303**) is obtained in 38% yield from benzoic acid when the 1,4-diene is subjected to base-catalysed conjugation before esterification[228]. These reactions usually lead to mixtures of isomeric cyclohexa-1,3-diene products since the reduction of the benzene derivative often gives more than one cyclohexadiene and conversion of the cyclohexa-1,4-diene to a cyclohexa-1,3-diene complex may occur in more than one way.

Na liq. NH_3 ; CO_2H ; CH_2N_2 ; CO_2Me

(**302**)

$[Fe(CO)_5]$, *n*-Bu_2O, 142 °C

CO_2Me ; Fe $(CO)_3$; + ; Fe CO_2Me $(CO)_3$ (104)

(**303**) (**304**)

Further examples of the direct reaction of ligand with metal carbonyl are provided by the conversion of 2-chlorocyclohexa-1,3-diene to the tricarbonyliron complexes **305** and **306** with dodecacarbonyltriiron in boiling benzene[229], the preparation of the complexes **307** (n = 2, 3) from the free ligand and pentacarbonyliron[230] and the formation

Cl ; $[Fe_3(CO)_{12}]$ PhH, 80°C ; Cl ; Fe $(CO)_3$; + ; Fe Cl $(CO)_3$ (105)

(**305**) (**306**)

MeO ; $(CH_2)_nCO_2Me$; Fe $(CO)_3$; $n = 2,3$

(**307**)

of the η^4-cyclohexa-1,3-diene complex (**309**) from the vinylcyclohexene (**308**) and enneacarbonyldiiron[231]. Complexation of the diene **308** with pentacarbonyliron has been investigated under thermal and photochemical conditions. The thermal reaction

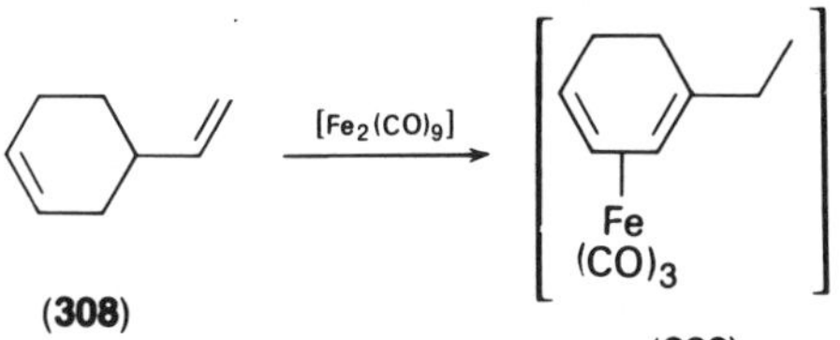

[Fe2(CO)9]

(308) (309) (106)

leads to the 1-ethylcyclohexadiene (**309**) together with the 2-ethyl isomer in the ratio 5:1. Under photochemical conditions the same two products were obtained in the proportions 1:5[232].

Heterocyclohexadiene complexes may be formed by direct reaction between the ligand and an iron carbonyl. Either *N*-carboalkoxy-1,2- or -1,4-dihydropyridine (**310** or **311**; R = Me, Et) combines with enneacarbonyldiiron to form the (η^4-*N*-carboalkoxy-1,2-dihydropyridine)iron complex (**312**; R = Me, Et)[233]. The sila- and

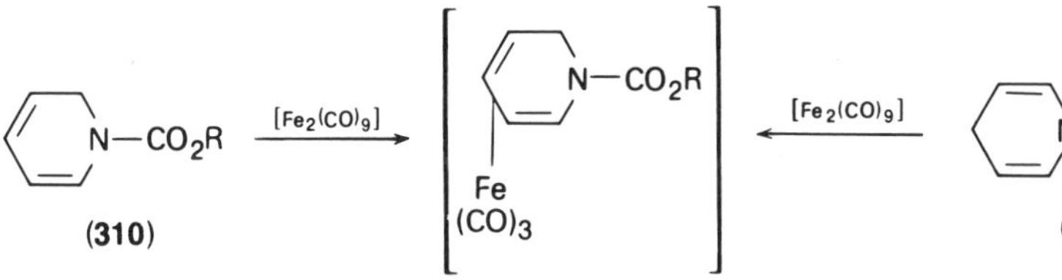

N—CO_2R [$Fe_2(CO)_9$] N—CO_2R Fe $(CO)_3$ [$Fe_2(CO)_9$] N—CO_2R (107)

(310) (312) (311)

disilacyclohexadiene complexes **313** and **314** are each obtained by heating the free ligand with pentacarbonyliron in benzene[234,235].

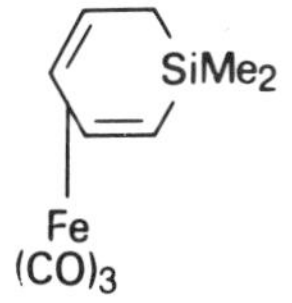

$SiMe_2$ Fe $(CO)_3$ Ph Me_2 Si $SiMe_2$ Ph Fe $(CO)_3$

(313) 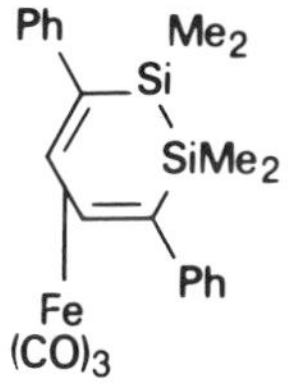 (314)

In a number of cases the ligand may undergo structural isomerization during complex formation. Thus, the cyclooctatriene **315** is attacked by enneacarbonyldiiron or dodecacarbonyltriiron to give the (η^4-bicyclooctadiene)iron complex (**316**)[236]. A similar isomerization takes place when cycloocta-1,3,5-triene combines with tetracarbonylbis(trimethylsilyl)ruthenium in boiling heptane to form the complex **317** in 15% yield[237]. Analogous reactions between bicyclononatrienes or bicyclodecatrienes and iron carbonyls may be carried out[238–240].

Closely related to these synthetic methods are those which involve the ring scission of a vinylcyclopropane in the presence of a metal carbonyl to give an η^4-cyclohexa-1,3-diene complex. Thus, bicyclo[3.1.0]hex-2-ene combines with enneacarbonyldiiron in ether to form, by way of an allyl intermediate, tricarbonyl(η^4-cyclohexa-1,3-diene)iron (**294**). Similar scission of a cyclopropane ring occurs with *cis*-bicyclo[6.1.0]nonatriene[241], spiro(cyclopenta-2,4-diene-1,7′-norcara-2′,4′-diene) (**318**)[242,243] and the 2-vinylepoxides (**319**; R = H, Me)[244].

It is interesting that the benzene ring in styrene and other vinylbenzenes may act as a

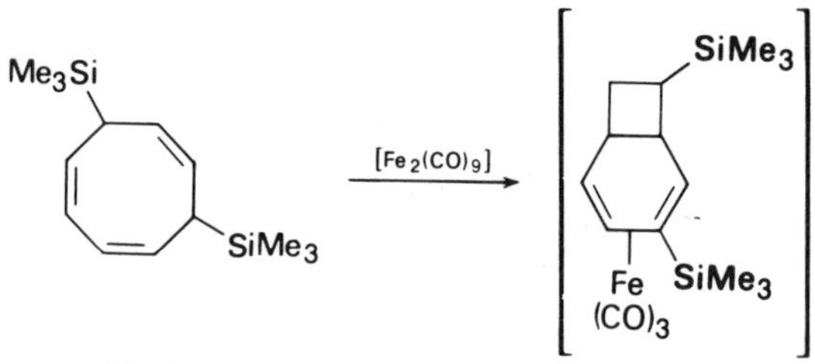

Me3Si SiMe3 $[Fe_2(CO)_9]$ SiMe3 Fe SiMe3 $(CO)_3$ (108)

(**315**) (**316**)

Ru $(CO)_3$

(**317**)

O R

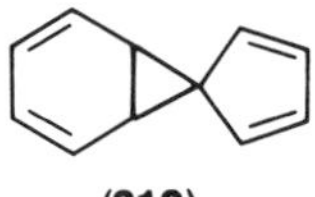

(**318**) 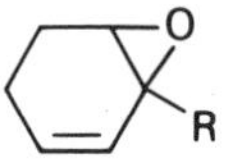 (**319**)
R = H, Me

cyclohexa-1,3-diene group by formation of the diiron complex (**320**), among other products, from the irradiation of styrene with pentacarbonyliron[245].

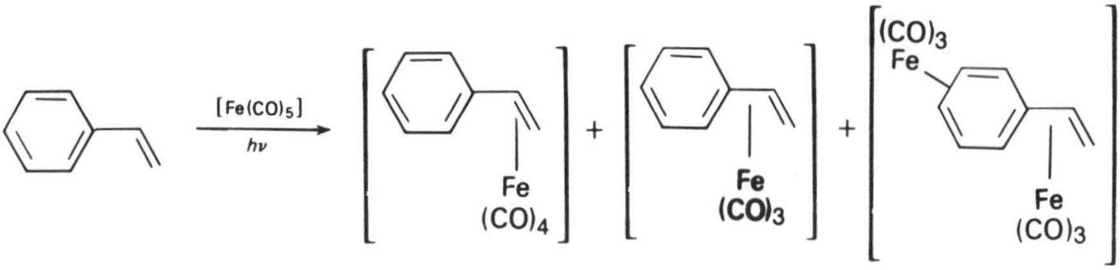

$[Fe(CO)_5]$ $h\nu$ Fe $(CO)_4$ + Fe $(CO)_3$ + $(CO)_3$ Fe Fe $(CO)_3$

(**320**)
(109)

η^5-Cyclohexadienylium transition metal cations are attacked by nucleophiles to form neutral η^4-cyclohexa-1,3-diene transition metal complexes. The nucleophile almost invariably enters the ligand at C-5 to give only one of several potential isomeric products. It is observed that the nucleophile approaches the face of the cyclohexadienyl ring remote from the pendant metal carbonyl residue and this leads to the formation of an *exo*-substituted product[246] which may, however, isomerize readily to the *endo*-isomer[246]. Reagents, reactions and products may be conveniently summarized in tabular form (Table 5).

(η^5-Cyclohexadienylium)metal complexes with substituents in the six-membered ring undergo similar addition reactions with nucleophiles. The methoxy complexes **323** [R = $(CH_2)_2CO_2Me$, $(CH_2)_3CO_2Me$, Me] combine with sodio-dimethylmalonate to form the products **324**[261,262]. Hydrolysis of the methoxy-substituted complex (**325**) gives the (η^4-cyclohexadienone)iron complex (**326**)[263].

TABLE 5. Reagents, conditions and products for the nucleophilic addition to η^5-cyclohexadienylium complexes to form η^4-cyclohexa-1,3-diene complexes

$$[(\eta^5\text{-}C_6H_7)M(CO)_3]^+ \ (\mathbf{321}) + N \longrightarrow [(\eta^4\text{-}C_6H_7Y)M(CO)_3] \ (\mathbf{322}) \qquad (110)$$

M in cation **321**	N	Conditions	Y in complex **322**	Yield	Reference
Fe	$PhSiMe_3$	MeCN or Me_2CO	Ph	High	247
Fe	$PhSnMe_3$	MeCN or Me_2CO	Ph	High	247
Fe	p-$MeOC_6H_4SiMe_3$	MeCN or Me_2CO	p-$MeOC_6H_4$	High	247
Fe	p-$MeOC_6H_4SnMe_3$	MeCN or Me_2CO	p-$MeOC_6H_4$	High	247
Fe	p-$Me_2NC_6H_4SiMe_3$	MeCN or Me_2CO	p-$Me_2NC_6H_4$	50%	247
Fe	p-$Me_2NC_6H_4SnMe_3$	MeCN or Me_2CO	p-$Me_2NC_6H_4$	50%	247
Fe	2-furyl$SiMe_3$	MeCN or Me_2CO	2-Furyl	High	247
Fe	2-thienyl$SiMe_3$	MeCN or Me_2CO	2-Thienyl	High	247
Fe	1,3-$(MeO)_2C_6H_4$	$MeNO_2$	1,3-$(MeO)_2C_6H_3$	—	248
Fe	1,3,5-$(MeO)_3C_6H_3$	$MeNO_2$	1,3,5-$(MeO)_3C_6H_2$	—	248
Fe	Pyrrole	MeCN	Pyrrolyl	—	249
Fe	Furan	MeCN	Furyl	—	249
Fe	Thiophen	MeCN	Thienyl	—	249
Fe	Indole	MeCN	3-Indolyl	—	249
Fe	NaOMe	H_2O, 0°C	OMe	57%	224

Table 5 (*continued*)

M in cation **321**	N	Conditions	Y in complex **322**	Yield	Reference
Fe	NaCN	H_2O	CN	52%	224
Fe	$NaHCO_3$	H_2O	OH	75%	224
Fe	Pyrrolidine	H_2O, 0°C	5-Pyrrolidinyl	82%	224
Fe	Morpholine	H_2O, 0°C	5-Morpholinyl	—	224
Fe	MeLi	Et_2O, −20°C	Me	39%	224
Fe	MeMgI	Et_2O, −20°C	$(\eta^4\text{-}C_6H_7)Fe(CO)_3$	15%	224
Fe	$P(OMe)_3$	—	$PO(OMe)_2$	100%	250
Fe	Hypophosphorous acid	H_2O, 65°C	P(O)H(OH)	—	250
Fe	$NaHSO_3$	H_2O	SO_3H	—	250
Fe	Na_2S	H_2O	$(\eta^4\text{-}C_6H_7S)Fe(CO)_3$	—	250
Fe	$Zn(CH_2CH{=}CH_2)_2$	Thf, 0°C	$CH_2CH{=}CH_2$	65%	251
Fe	$Cd(CH_2CH{=}CH_2)_2$	Thf, 0°C	$CH_2CH{=}CH_2$	82%	251, 254
Fe	$Et_3NH^+SiMe_3^-$	—	$SiMe_3$	—	252
Fe	$Et_3NH^+GeMe_3^-$	—	$GeMe_3$	—	252
Fe	$Et_3NH^+SnMe_3^-$	—	$SnMe_3$	—	252
Fe	Adenine	MeCN, 20°C	Adenyl	88%	253
Fe	Guanosine	$HCONMe_2$, 20°C	Guanosyl	13%	253
Fe	$LiCuMe_2$	Et_2O, 0°C	Me	80%	255
Fe	Acetylacetone	$MeNO_2$	Acetylacetonate	High	256
Ru	Acetylacetone	$MeNO_2$	Acetylacetonate	High	256
Fe	Indole	$MeNO_2$	Indolyl	High	257
Ru	Indole	$MeNO_2$	Indolyl	High	257
Fe	PPh_3	$MeNO_2$	PPh_3^+	High	258
Fe	$[(\eta^4\text{-}C_7H_7)Fe(CO)_3]^+$	—	$(\eta^4\text{-}C_7H_7)Fe(CO)_3$	—	259
Fe	$C_6H_5NMe_2$	MeCN	$p\text{-}Me_2NC_6H_4$	95%	260
Ru	$C_6H_5NMe_2$	$MeNO_2$	$p\text{-}Me_2NC_6H_4$	High	260
Os	$C_6H_5NMe_2$	Me_2CO	$p\text{-}Me_2NC_6H_4$	High	260

(111)

(323) (324)

(112)

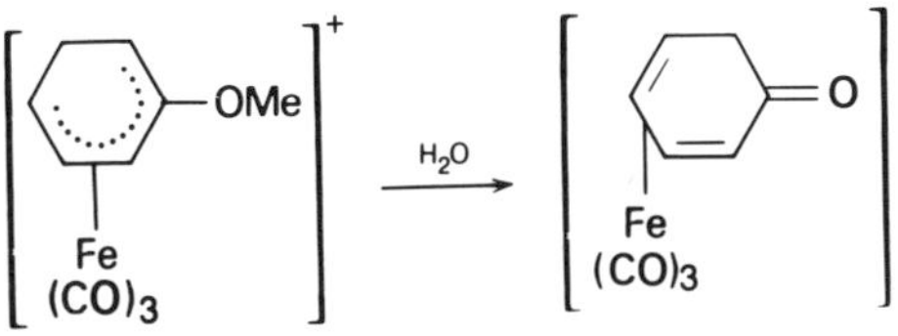

(325) (326)

D. Complexes Containing the η^4-Cyclohepta-1,3-diene Ligand

Cycloheptadiene, cycloheptatriene, and cycloheptatrienone may all behave as four-electron donor ligands to transition metals, particularly to iron. The most important method of preparation is by direct reaction between the ligand and the transition metal carbonyl, although nucleophilic attack on an η^5-cycloheptadienyl complex and reduction of a cycloheptatriene complex are also used.

Cycloheptatriene combines directly with pentacarbonyliron on heating to the reflux temperature for 21 h to give tricarbonyl(η^4-cycloheptatriene)iron (**327**) in 52% yield as

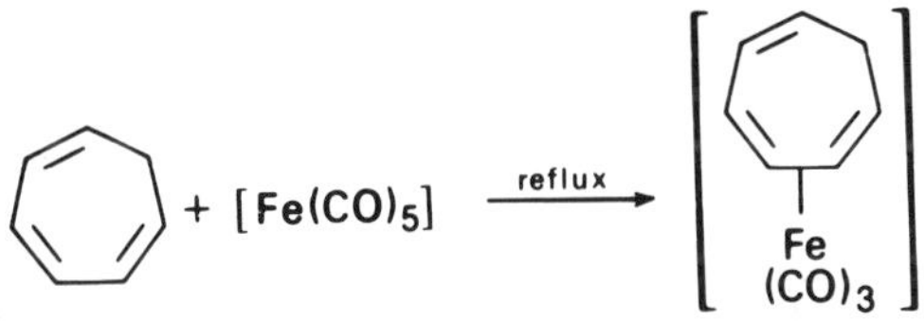

(113)

(327)

an orange–red liquid[264,265]. Substituted cycloheptatrienes undergo the same reaction and typical products are the η^4-benzo-, η^4-furano-, and η^4-thiophenocycloheptatriene complexes (**328 and 329**; X = O, S), which are obtained in yields of 23, 11, and 18%,

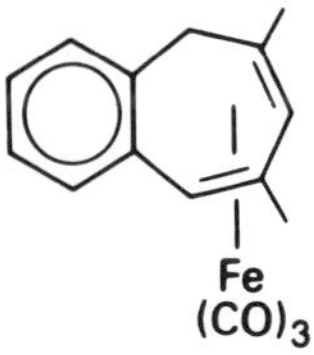

(328) 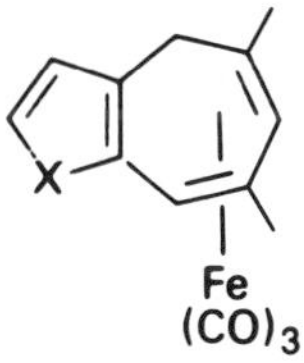 (329)

respectively[266]. The reaction between cycloheptatriene and dodecacarbonyltriruthenium is more complex; heating in hexane gives the η^4-cycloheptatriene and

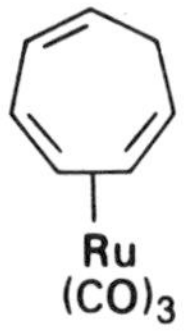

(330) 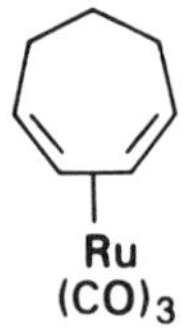 (331)

η^4-cycloheptadiene complexes (330 and 331) as minor products, the major product being the cluster compound $Ru_3(CO)_6(\eta\text{-}C_7H_7)(\eta\text{-}C_7H_9)$[267].

Cycloheptatrienone combines with dodecacarbonyltriiron[268] and with enneacarbonyldiiron[269] to form tricarbonyl(η^4-cycloheptatrienone)iron (332; R = H); the same product is obtained from acetylene and enneacarbonyldiiron under pressure[270]. Substituted tropones also combine with enneacarbonyldiiron to form, for example, the (η^4-tropone)iron complexes (332; R = Me, Ph, Cl)[271]. Heterocycloheptatrienes undergo

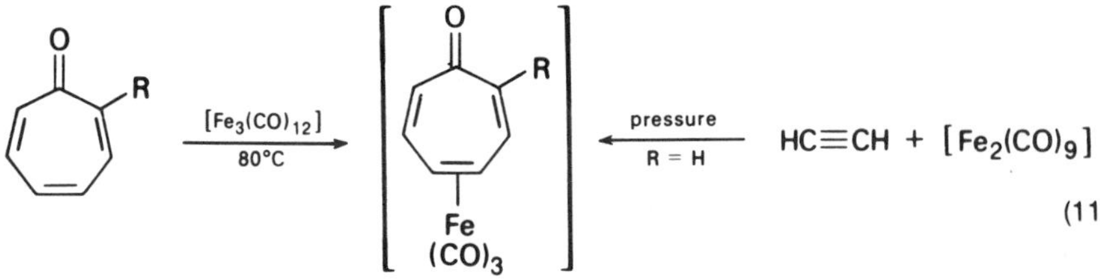

$[Fe_3(CO)_{12}]$, 80°C; pressure, R = H; $HC{\equiv}CH + [Fe_2(CO)_9]$ (114)

(332)

R = H, Me, Ph, Cl

reaction with metal carbonyls to yield complexes in which the ligand is a four-electron donor. Thus, 2,7-dimethyloxepin is converted to the complex 333, among other products, on irradiation with pentacarbonyliron[272,273] and the (η^4-diazepine)iron complexes 334 (R = COMe, CO_2Et) are also characterized[274,275].

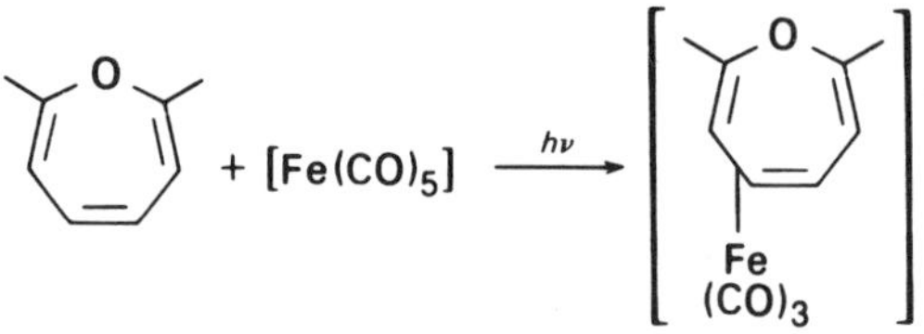

$+ [Fe(CO)_5] \xrightarrow{h\nu}$ (115)

(333)

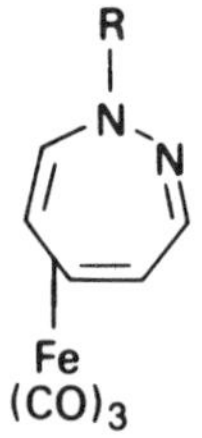

(334)

Indirect methods of preparation may be used to obtain η^4-cycloheptatriene complexes. The tricarbonyl(η^7-cycloheptatrienyl)ferrate anion is attacked by electrophiles such as Me_3SiCl and Me_3GeCl to form 7-substituted η^4-cycloheptatrienyl complexes (**335**)[274]. The same anion combines with the tricarbonyl(η^5-cyclohexadienyl)iron cation to give the corresponding neutral 7-substituted η^4-cycloheptatriene complex (Table 5)[259].

Me_3MCl; $Fe(CO)_3$; MMe_3 (116)

(**335**)

M = Si, Ge

η^4-Cycloheptadiene and η^4-cycloheptadienone complexes are available by direct reaction between the ligand and metal carbonyl, by hydrogenation of the uncomplexed olefinic bond in η^4-cycloheptatriene complexes, and by reduction of, or nucleophilic addition to, η^5-cycloheptadienyl complexes. Cycloheptadiene combines with pentacarbonyliron on heating at 160°C in methylcyclohexane to form tricarbonyl(η^4-cycloheptadiene)iron (**336**)[275]. The same product is obtained together with tricarbonyl(η^4-cycloheptatriene)iron (**327**) when cycloheptatriene is heated with pentacarbonyliron for several days[275,276]. Substituted η^4-cycloheptadiene complexes are formed in the same way; thus, the polyfluoro complexes (**337**; X = H, F) are prepared in low yield from the free ligand and dodecacarbonyltriiron[277].

+ $[Fe(CO)_5]$ —methylcyclohexane, 160°C→ $Fe(CO)_3$ (117)

(**336**)

X, F, X + $[Fe_3(CO)_{12}]$ —sealed tube, 130°C→ X, F, X, $Fe(CO)_3$ (118)

(**337**)

X = H, F

Hydrogenation of the cycloheptatriene complex **327** using Raney nickel gives the cycloheptadiene product **336**[276]; the tropone complex **332** (R = H) also undergoes hydrogenation in the presence of palladium on charcoal[270] or reduction with triethylsilane in trifluoroacetic acid[278] to form tricarbonyl(η^4-2,4-cycloheptadienone)iron (**343**; N = H). The tricarbonyl(η^5-cycloheptadienyl)iron cation (**338**) is attacked by nucleophiles to give neutral η^4-cycloheptadiene complexes. Sodium borohydride[279] as the reagent leads to the parent complex (**339**; N = H), whereas $^-$OMe[281], $^-N_3$[280], $^-$OEt[280], $^-$OPh[281], $^-$SPh[281], Me_2NH[281], and Me_3CNH_2[281] give the 5-substituted products (**339**; N = OMe, N_3, OEt, OPh, SPh, NMe_2, $NHCMe_3$, respectively). Reduction of the cation **338** as the tetrafluoroborate takes place in acetonitrile at 90°C in the absence of an added nucleophile to form the tetracarbonyldiiron complex (**340**) in 32%

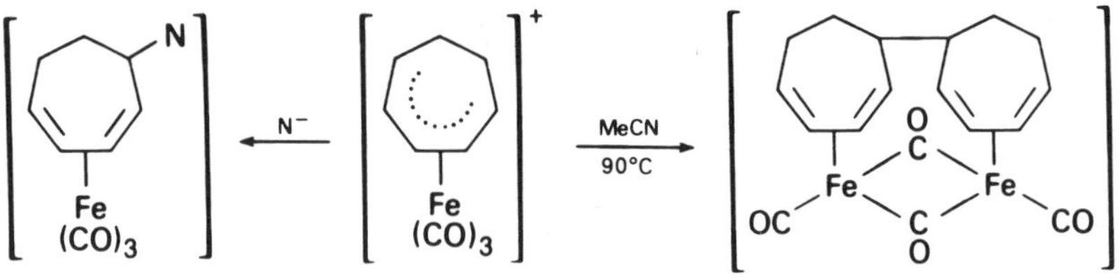

Fe (CO)3 (339) ← N⁻ — Fe (CO)3 (338)⁺ — MeCN, 90°C → OC Fe C O C O Fe CO (340)

N = H, OMe, OEt, OPh, SPh

N_3, NMe_2, $NHCMe_3$

yield[282]. Co-condensation of toluene with iron atoms at liquid air temperature and low pressure followed by treatment with cycloheptatriene gives the (η^4-cycloheptatriene)iron complex (**341**)[283].

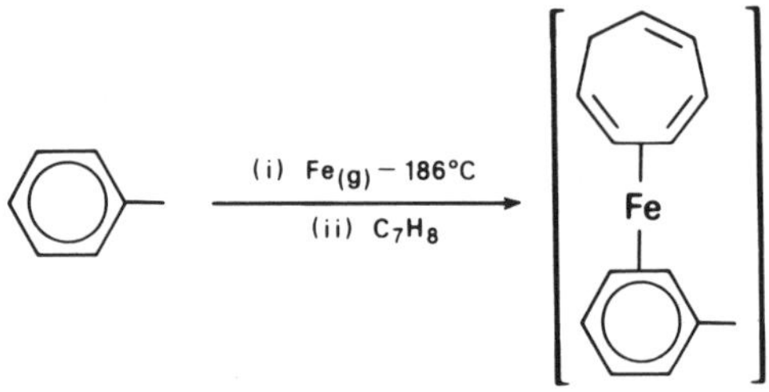

(i) $Fe_{(g)}$ − 186°C; (ii) C_7H_8 → Fe (**341**) (120)

Substituted η^4-cycloheptadienone complexes are formed by treatment of the (η^4-cycloheptadienonyl)iron cation (**342**) with nucleophiles. Thus, methanol, aniline, *t*-butylamine, and azide ion lead to the products **343** (N = OMe, NHPh, NHBu-*t* and N_3, respectively)[284]. However, borohydride and cyanide gave (η^3-cycloheptenyl)iron products (**344**; N = H, CN)[284].

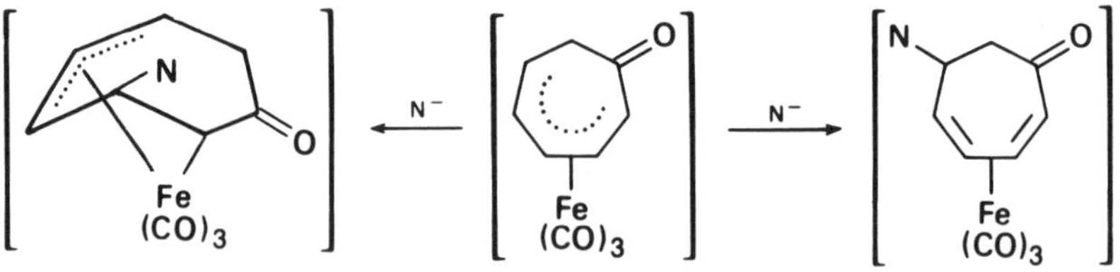

Fe (CO)3 (**344**) N = H, CN ← N⁻ — Fe (CO)3 (**342**) — N⁻ → Fe (CO)3 (**343**) N = OMe, NHPh, NHBu-*t*, N_3 (121)

E. Complexes Containing the η^4-Cycloocta-1,3-diene Ligand

Cyclooctatetraene, cyclooctatriene, and cycloocta-1,3-diene form η^4-cyclodiene complexes on treatment with transition metal carbonyls. However, the ligands show a marked tendency to undergo reactions such as ring contraction and dimerization, leading to several products in addition to the desired complexes[285]. The tricarbonyliron complex **344** is formed by irradiation of cyclooctatraene with pentacarbonyliron. When pentacarbonyliron is present in excess or when the tricarbonyliron complex **344** is irradiated with pentacarbonyliron, the binuclear complex (**345**) with *trans*-

$(CO)_3$Fe / Fe$(CO)_3$ (**345**) + **344** ← excess [$Fe(CO)_5$], $h\nu$ — cyclooctatetraene — [$Fe(CO)_5$], $h\nu$ → [Fe$(CO)_3$] (**344**) (122)

stereochemistry is obtained as an additional product[264,286–289]. The complexes **344** and **345** are also the principal products when bicyclooctatetraenyl is treated with iron carbonyls[290]. Substituted cyclooctatetraenes are attacked by iron carbonyls in the same way as the parent ligand. Thus, the (η^4-methoxycyclooctatetraene)iron complex (**346**; X = OMe) is formed from the free ligand and dodecacarbonyltriiron[291] while the trimethylsilyl-, trimethylgermyl-, and trimethylstannyl-substituted complexes (**346**; X = $SiMe_3$, $GeMe_3$, $SnMe_3$) are obtained from the appropriate ligand and enneacarbonyldiiron[292]. Binuclear species related to the hexacarbonyldiiron complex (**345**) are also obtained in these reactions[291,292].

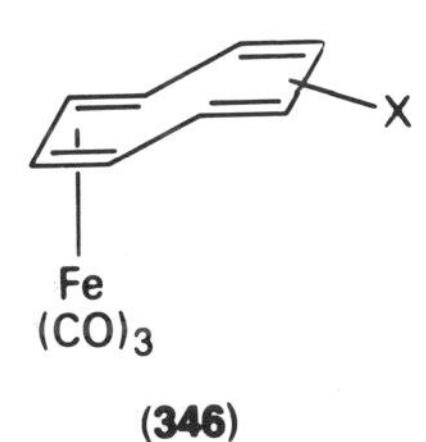

(**346**)

Cyclooctatetraene behaves as a cyclo-1,3-diene ligand towards transition metals other than iron, although the reactions may be complicated by the tendency of the ligand to oligomerize, polymerize, or become bound in the 1,2,5,6-*tetrahapto*[293,294], 1,2,3,6-*tetrahapto*[295], *hexahapto*[296] and *octahapto* forms[297].

Dodecacarbonyltriruthenium combines with cyclooctatetraene in boiling heptane to give the tricarbonylruthenium complex **347** as a minor product in low yield; the major products are the binuclear complexes [(*cis*-C_8H_8)$Ru_2(CO)_6$] and [C_8H_8)$Ru_2(CO)_5$][298]. The cyclooctatetraene dianion combines with the complex [(η^6-C_6H_6)$RuCl_2$]$_2$ to form (η^6-benzene)(η^4-cyclooctatetraene)ruthenium (**348**)[299]. The osmium complex **350** is obtained in 24% yield from the free ligand and dodecacarbonyltriosmium by way of the intermediate **349**[295]. Cobalt chloride is reduced by sodium borohydride in ethanol in the presence of cyclooctatetraene to give (η^5-cyclooctatrienyl)(η^4-cyclooctatetraene)cobalt (**351**) in 4% yield[300]. This complex (**351**) is the parent of several (η^4-cyclooctatetraene)cobalt derivatives[300].

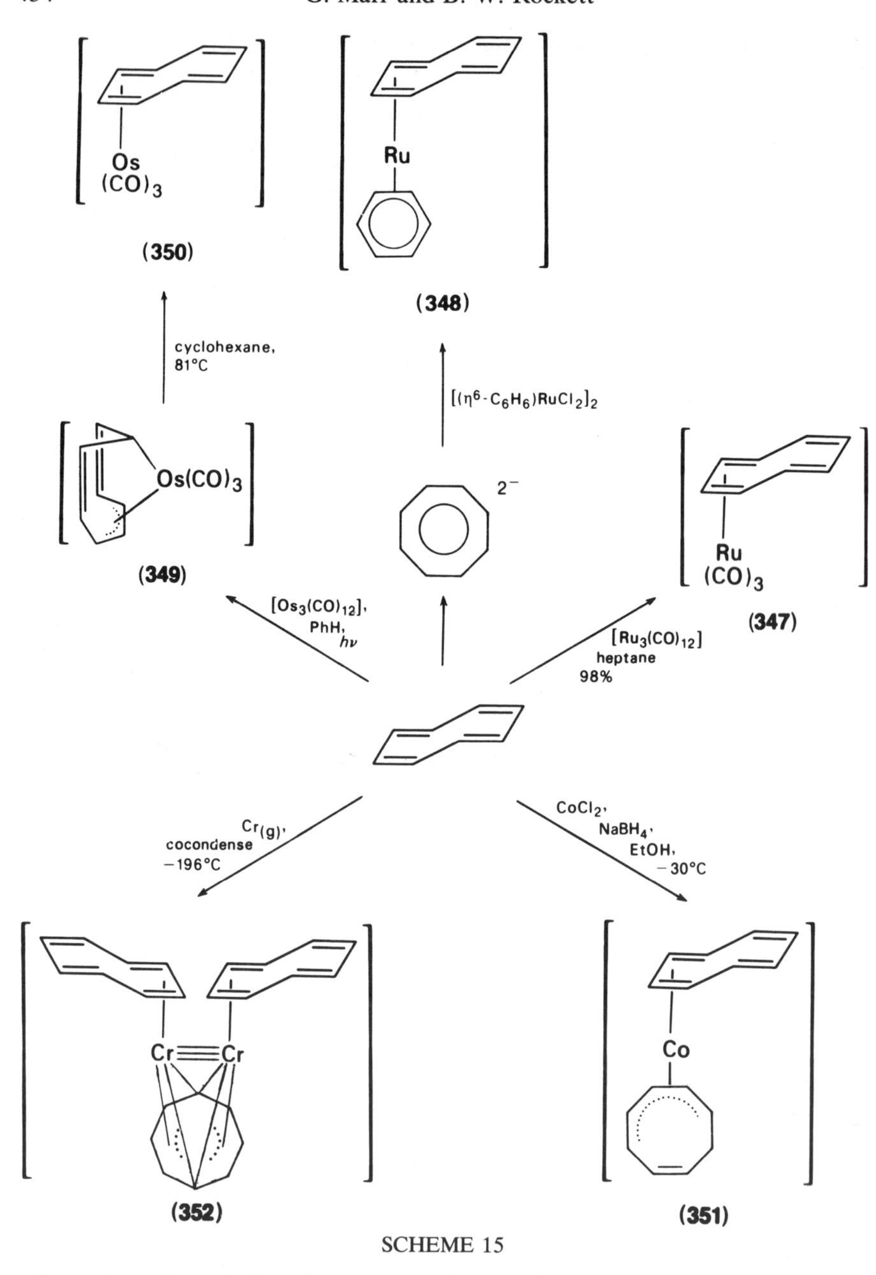

SCHEME 15

Co-condensation of chromium atoms with cyclooctatetraene and isopentane at −196°C gives the tris(cyclooctatetraene)dichromium(0) complex **352** in 43% yield[301]. The same complex is obtained by reduction of chromium(III) chloride–cyclooctatetraene mixtures with isopropylmagnesium bromide[302] (Scheme 15).

Cycloocta-1,3,5-triene combines with pentacarbonyliron on irradiation to give the

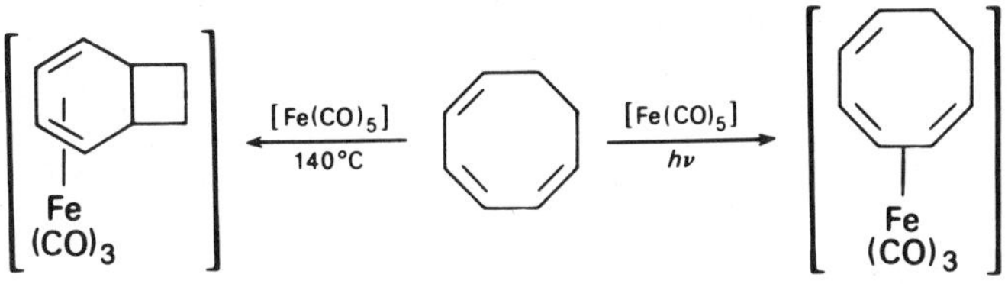

(354) ← [Fe(CO)5], 140°C — cyclooctatetraene — [Fe(CO)5], hν → (353) (123)

(η^4-cycloocta-1,3,5-triene)iron complex (**353**)[285]. This ligand is attacked by pentacarbonyliron at 140°C to yield the η^4-bicyclooctadiene complex (**354**), whereas the same reagents in benzene at the reflux temperature form a mixture of the two complexes (**353** and **354**)[303]. Irradiation of cycloocta-1,3-diene with pentacarbonyliron gives the expected complex (**355**; X = H)[304], whereas the (η^4-cycloocta-1,3-diene)iron complex **358**) is formed by heating tricarbonyl(η^4-cycloocta-1,5-diene)iron (**357**) with trimethyl

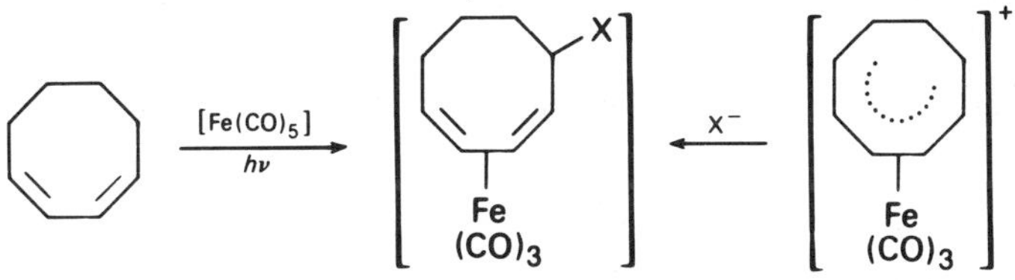

cycloocta-1,3-diene — [Fe(CO)5], hν → (355) ← X⁻ — (356) (124)

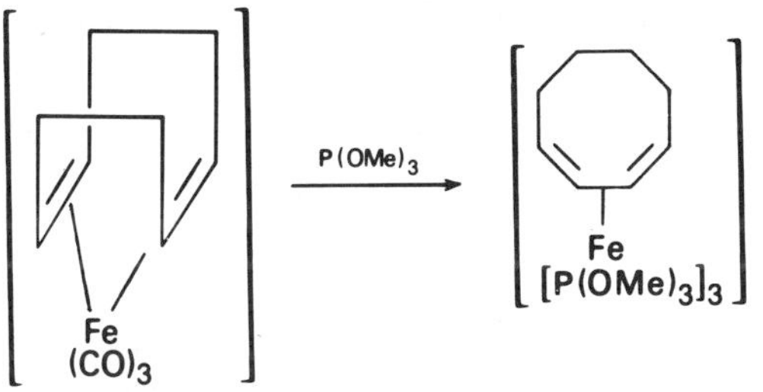

(357) — P(OMe)3 → (358) (125)

phosphite[305]. The attack of nucleophiles on (η^5-cyclooctadienyl)iron cations gives η^4-cyclooctadiene complexes; thus, methoxide ion combines with the complex **356** to give the methoxy derivative (**355**; X = OMe) as a mixture of the *exo*- and *endo*-isomers[306]. Cycloocta-1,3,5-triene attacks tricarbonyl(η^4-cycloocta-1,3,5-diene)-ruthenium in benzene to form tricarbonyl(η^4-cycloocta-1,3,5-triene)ruthenium in good yield[307].

V. REFERENCES

1. The 'Annual Surveys' contained in *J. Organomet. Chem.* provide near comprehensive coverage of (η^4-diene) transition metal complexes on a year-to-year basis. Other important reviews are: η^4-butadiene complexes, (a) R. B. King, in *Organic Chemistry of Iron,* (Ed. E. A. K. Von Gustorf, Vol. 1, Academic Press, New York, 1978, p. 525, (b) ref. 4, (c) ref. 14, (d) ref. 34, (e) ref. 55, (f) ref. 190; η^4-trimethylenemethane complexes, (g) J. M. Landesberg, in ref. la, p. 627; η^4-cyclobutadiene complexes, (h) ref. 130.
2. H. Reihlen, A. Gruhl, G. V. Hessling, and O. Pfrengle, *Justus Liebigs Ann. Chem.*, **482**, 161 (1930).

3. B. F. Hallam and P. L. Pauson, *J. Chem. Soc*., 642 (1958).
4. R. B. King, *Organometallic Syntheses,* Vol. 1, Academic Press, New York, 1965, p. 128.
5. T. A. Manuel, S. L. Stafford, and F. G. A. Stone, *J. Am. Chem. Soc*., **83**, 3597 (1961).
6. M. Cais and M. Feldkimel, *Tetrahedron Lett*., **444** (1961).
7. R. B. King, T. A. Manuel, and F. G. A. Stone, *J. Inorg. Nucl. Chem*., **16**, 233 (1961).
8. M. A. Busch and R. J. Clark, *Inorg. Chem*., **14**, 219 (1975).
9. G. A. Tolstikov, M. S. Miftakhov, Yu. B. Monakov, and S. I. Lomakina, *Zh. Obshch. Khim*., **46**, 2630 (1976).
10. T. A. Manuel, *Inorg. Chem*., **3**, 1794 (1964).
11. P. Boudjouk and S. Lin, *J. Organomet. Chem*., **155**, C13 (1978).
12. B. F. G. Johnson, J. Lewis, D. G. Parker, and S. R. Postle, *J. Chem. Soc., Dalton Trans*., 794 (1977).
13. R. B. King and C. A. Harmon, *J. Am. Chem. Soc*., **98**, 2409 (1976).
14. R. Pettit and G. F. Emerson, *Advances in Organometallic Chemistry*, Vol. 1, Academic Press, New York, 1964, p. 1.
15. E. K. Von Gustorf, J. Buchkremer, Z. Pfajfer, and F. W. Grevels, *Ger. Pat*., 2105627, 1972; *Chem. Abstr*., **77**, 62138 (1972).
16. Montedison S.p.A., *Ital. Pat*., 907326, 1972; *Chem. Abstr*., **83**, 59055 (1975).
17. A. A. Pinkerton, G. Chapuis, P. Vogel, U. Haenisch, P. Narbel, T. Boschi, and R. Roulet, *Inorg. Chim. Acta*, **35**, 197 (1979).
18. U. Steiner, H. J. Hansen, K. Bachmann, and W. Von Philipsborn, *Helv. Chim. Acta*, **60**, 643 (1977).
19. T. Boschi, P. Vogel, and R. Roulet, *J. Organomet. Chem*., **133**, C39 (1977).
20. T. Boschi, P. Narbel, P. Vogel, and R. Roulet, *J. Organomet. Chem*., **133**, C36 (1977).
21. E. O. Fischer, H. P. Kogler, and P. Kuzel, *Chem. Ber*., **93**, 3006 (1960).
22. J. Daub, J. Kappler, K. P. Krenkler, S. Schreiner, and V. Trautz, *Justus Liebigs Ann. Chem*., 1730 (1977).
23. S. Sahurai, Y. Kamiyama, and Y. Nakadaird, *Angew. Chem*., **90**, 718 (1978).
24. E. O. Fischer, P. Kuzel, and H. P. Fritz, *Z. Naturforsch*., **16B**, 138 (1961).
25. P. Kuzel, *Thesis*, Univ. Munchen, May 1972.
26. W. Frohlich, *Thesis*, Univ. Munchen, January 1961.
27. K. Bittler, *Thesis*, Tech. Hochschule, Munchen, November 1962.
28. O. Gambino, M. Valle, S. Aime, and G. A. Vaglio, *Inorg. Chim. Acta*, **8**, 71 (1974).
29. S. Ruh and W. Von Philipsborn, *J. Organomet. Chem*., **127**, C59 (1977).
30. J. A. S. Howell, B. F. C. Johnson, P. L. Josty, and J. Lewis, *J. Organomet. Chem*., **39**, 329 (1972).
31. S. Otsuka, T. Yoshida, and A. Nakamura, *Inorg. Chem*., **6**, 20 (1967).
32. H. T. Doeck and H. Bock, *J. Chem. Soc., Chem. Commun*., 678 (1968).
33. M. Dekkar and G. R. Knox, *J. Chem. Soc., Chem. Commun*., 1243 (1967).
34. J. Powell, *MTP International Review of Science, Transition Metals, Part 2; Inorganic Chemistry, Series 1*, Vol. 6, Butterworths, London, 1972, p. 309
35. S. M. Nelson and M. Sloan, *J. Chem. Soc., Chem. Commun*., 745 (1972).
36. S. M. Nelson, C. M. Regan, and M. Sloan, *J. Organomet. Chem*., **96**, 383 (1975).
37. R. N. Greene, C. H. DePuy, and T. E. Schroer, *J. Chem. Soc. C*, 3115 (1971).
38. W. R. Roth and J. D. Meier, *Tetrahedron Lett*., 2053 (1967).
39. R. Aumann, *J. Am. Chem. Soc*., **96**, 2631 (1974).
40. S. Sarel, R. Ben-Shosham, and B. Kirson, *Isr. J. Chem*., **10**, 787 (1972).
41. S. Sarel, R. Ben-Shosham, and B. Kirson, *J. Am. Chem. Soc*., **87**, 2517 (1965).
42. S. Sarel, A. Felzenstein, R. Victor, and J. Yovell, *J. Chem. Soc., Chem. Commun*., 1025 (1974).
43. S. Sarel, M. Langeheim, and I. Ringel, *J. Chem. Soc., Chem. Commun*., 73 (1979).
44. T. H. Whitesides and R. W. Slaven, *J. Organomet. Chem*., **67**, 99 (1974).
45. B. M. Chisnall, M. Green, R. P. Hughes, and A. J. Welch, *J. Chem. Soc., Dalton Trans*., 1899 (1976).
46. M. Green and R. P. Hughes, *J. Chem. Soc., Dalton Trans*., 1907 (1976).
47. W. E. Billups, L. P. Lin, and B. A. Baker, *J. Organomet. Chem*., **61**, C55 (1973).
48. R. B. King, *Inorg. Chem*., **2**, 624 (1963).
49. P. Binger, B. Catinkaya, and Kruger, *J. Organomet. Chem*., **159**, 63 (1978).

50. K. Takahashi, M. Iwanami, A. Tsai, P. L. Chang. R. L. Harlow, L. E. Harris, J. E. McCaskie, C. E. Pfluger, and D. C. Dittmer, *J. Am. Chem. Soc.*, **95**, 6113 (1973).
51. N. S. Nametkine, V. D. Tyurine, A. I. Nejhaev, V. I. Ivanov, and F. S. Bayaouova, *J. Organomet. Chem.*, **107**, 377 (1976).
52. N. S. Nametkin, A. I. Nekhaev, and V. D. Tyurin, *Izv. Akad. Nauk SSSR, Ser. Khim.*, 890 (1974).
53. R. W. Jotham, S. F. A. Kettle, D. B. Moll, and P. J. Stamper, *J. Organomet. Chem.*, **118**, 59 (1976).
54. A. E. Hill and H. M. R. Hoffmann, *J. Chem. Soc., Chem. Commun.*, 574 (1972).
55. P. L. Timms and T. W. Turney, *Advances in Organometallic Chemistry*, Vol. 13, Academic Press, New York, 1977, p. 53.
56. S. D. Ittel, F. A. Van-Catledge, and J. P. Jesson, *J. Am. Chem. Soc.*, **101**, 3874 (1979).
57. E. A. K. Von Gustorf, O. Jacnicke, O. Wolfbeis, and C. R. Eady, *Angew. Chem., Int. Ed. Engl.*, **14**, 278 (1975).
58. E. K. Von Gustorf, O. Jaenicke, and O. E. Polansky, *Angew. Chem., Int. Ed. Engl.*, **11**, 532 (1972).
59. P. S. Skell, D. L. Williams-Smith, and M. J. McGlinchey, *J. Am. Chem. Soc.*, **95**, 3337 (1973).
60. P. S. Skell and M. J. McGlinchey, *Angew. Chem., Int. Ed. Engl.*, **14**, 195 (1975).
61. D. L. Williams-Smith, L. R. Wolf, and P. S. Skell, *J. Am. Chem. Soc.*, **94**, 4042 (1972).
62. J. R. Blackborow, C. R. Eady, E. A. K. Von Gustorf, A. Scrivanti, and O. Wolfbeis, *J. Organomet. Chem.*, **111**, C3 (1976).
63. M. Anderson, A. D. H. Clague, L. P. Blaauw, and P. A. Couperus, *J. Organomet. Chem.*, **56**, 307 (1973).
64. P. McCardle and H. Sherlock, *J. Chem. Soc., Chem. Commun.*, 537 (1976).
65. J. E. Mahler and R. Pettit, *J. Am. Chem. Soc.*, **85**, 3955 (1963).
66. J. E. Mahler, D. H. Gibson, and R. Pettit, *J. Am. Chem. Soc.*, **85**, 3959 (1963).
67. G. Maglio, A. Musco, and R. Palumbo, *J. Organomet. Chem.*, **32**, 127 (1971).
68. G. Maglio and R. Palumbo, *J. Organomet. Chem.*, **76**, 367 (1974).
69. S. Sadeh and Y. Gaoni, *J. Organomet. Chem.*, **93**, C31 (1975).
70. C. H. DePuy, T. Jones, and R. L. Parton, *J. Am. Chem. Soc.*, **96**, 5602 (1974).
71. A. N. Nesmeyanov, N. P. Avakyan, G. M. Babakhina, and I. I. Kritskaya, *Izv. Akad. Nauk SSSR, Ser. Khim.*, 2133 (1975).
72. Y. Becker, A. Eisenstadt, and Y. Shovo, *Tetrahedron*, **30**, 839 (1974).
73. A. M. Brodie, B. F. G. Johnson, P. L. Josty, and J. Lewis, *J. Chem. Soc., Dalton Trans.*, 2031 (1972).
74. A. Vessieres and P. Dixneuf, *Tetrahedron Lett.*, 1499 (1974).
75. G. Cardaci and G. Concetti, *J. Orgonomet. Chem.*, **90**, 49 (1974).
76. M. Brookhart and G. O. Nelson, *J. Organomet. Chem.*, **164**, 193 (1979).
77, G. F. Emerson, K. Ehrlich, W. P. Giering, and P. C. Lauterbur, *J. Am. Chem. Soc.*, **88**, 3172 (1966).
78. G. G. Emerson and K. Ehrlich, *J. Am. Chem. Soc.*, **94**, 2464 (1972).
79. S. Sadeh and Y. Gaoni, *J. Organomet. Chem.*, **93**, C31 (1975).
80. A. N. Nesmeyanov, G. P. Zol'nikova, I. F. Lesheheva, and I. I. Kritskaya, *Izv. Akad. Nauk SSSR, Ser, Khim.*, 2388 (1974).
81. J. Y. Mevour, J. L. Roustan, C. Charrier, J. Benaim, J. Collin, and P. Cadiot, *J. Organomet. Chem.*, **168**, 337 (1979).
82. A. Guignot, P. Cadiot, and J. L. Roustan, *J. Organomet. Chem.*, **166**, 379 (1979).
83. W. E. Billups, L. P. Lin, and O. A. Gansow, *Angew. Chem., Int. Ed. Engl.*, **11**, 637 (1972).
84. W. E. Billups, L. P. Lin, and B. A. Baker, *J. Organomet. Chem.*, **61**, C55 (1973).
85. H. C. Longuet-Higgins and L. E. Orgel, *J. Chem. Soc.*, 1969 (1956).
86. R. Criegee and G. Schroder, *Angew. Chem.*, **71**, 70 (1959).
87. G. E. Emerson, L. Watts, and R. Pettit, *J. Am. Chem. Soc.*, **87**, 131 (1965).
88. R. G. Amiett and R. Pettit, *J. Am. Chem. Soc.*, **90**, 1059 (1968).
89. H. A. Brune, G. Horbeck, and U. Zahorszky, *Z. Naturforsch.*, **26B**, 222 (1971); H. A. Brune, G. Horbeck, and U. Zahorszky, *Z. Naturforsch.*, **26B**, 222 (1971); H. A. Brune, G. Horbeck, and H. Rottele, *Z. Naturforsch.*, **27B**, 505 (1972).
90. H. A. Brune, W. Eberius, and H. P. Wolff, *J. Organomet. Chem.*, **12**, 485 (1968).

91. H. A. Brune, W. Eberius, and H. P. Wolff, *J. Organomet. Chem.*, **12**, 485 (1968).
92. R. G. Amiet, P. C. Reeves, and R. Pettit, *J. Chem. Soc., Chem. Commun.*, 1208 (1967).
93. R. Criegee and G. Schroder, *Justus Liebigs Ann. Chem.*, **623**, 1 (1959).
94. H. G. Olive and S. Olive, *Angew. Chem.*, **79**, 897 (1967).
95. H. A. Brune, H. P. Wolff, W. Klein, and U. Zahorszky, *Z. Naturforsch.*, **27B**, 639 (1972); H. P. Wolff, P. Muller, and H. A. Brune, *Z. Naturforsch.*, **27B**, 915 (1972).
96. P. M. Maitlis, *Adv. Organomet. Chem.*, **4**, 95 (1966); M. P. Cava and M. J. Mitchell, *Cyclobutadiene and Related Compounds,* Interscience, New York, 1967.
97. R. H. Grubbs, *J. Am. Chem. Soc.*, **92**, 6693 (1970).
98. R. H. Grubbs, T. A. Pancoast, and R. A. Grey, *Tetrahedron Lett.*, 2425 (1974).
99. R. H. Grubbs and T. A. Pancoast, *Synth. React. Inorg. Met.-Org. Chem.*, **8**, 1 (1978).
100. F. I. Carroll, H. H. Seltzman, and F. M. Hauser, *Tetrahedron Lett.*, 4237 (1976).
101. E. K. G. Schmidt, *Chem. Ber.*, **107**, 2440 (1974).
102. A. Wissner, *Diss. Abstr. Int. B*, **32**, 2085 (1971).
103. G. Berens, F. Kaplan, R. Rimerman, B. W. Roberts, and A. Wissner, *J. Am. Chem. Soc.*, **97**, 7076 (1975).
104. H. D. Scharf and K. R. Stahlke, *Angew. Chem., Int. Ed. Engl.*, **9**, 810 (1970).
105. P. L. Maitlis and M. L. Games, *Canad. J. Chem.*, **42**, 183 (1964).
106. A.T. Blomquist and P. M. Maitlis, *J. Am. Chem. Soc.*, **84**, 2329 (1962).
107. L. Malatesta, G. Santarella, L. M. Vallarino, and F. Zingales, *Atti Accad. Naz. Lincei, Cl. Sci. Fis. Mat. Nat., Rend.*, **27**, 230 (1959).
108. L. F. Dahl and W. E. Oberhansli, *Inorg. Chem.*, **4**, 629 (1965).
109. P. M. Maitlis, A. Efraty, and M. L. Games, *J. Organomet. Chem.*, **2**, 284 (1964).
110. P. M. Maitlis, A. Efraty, and M. L. Games, *J. Am. Chem. Soc.*, **87**, 719 (1965).
111. T. Hosokawa and I. Maritani, *Tetrahedron Lett.*, 3021 (1969).
112. H. Dietl, J. Moffat, D. Wolff, and P. M. Maitlis, *J. Am. Chem. Soc.*, **90**, 5321 (1968).
113. P. M. Maitlis and K. W. Eberius, *Cyclobutadiene – Metal Complexes in Nonbenzenoid Aromatics*, Vol. 16–II, Academic Press, New York, 1971, p. 359.
114. H. A. Brune, G. Horlbeck, and H. Roettele, *Z. Naturforsch.*, **27B**, 505 (1972).
115. R. B. King and C. A. Harmon, *J. Am. Chem. Soc.*, **98**, 2409 (1976).
116. P. M. Maitlis and M. L. Games, *J. Am. Chem. Soc.*, **85**, 1887 (1963).
117. A. N. Nesmeyanov, K. N. Anisimov, N. E. Kolobova, and A. A. Pasynskii, *Dokl. Acad. Nauk SSSR*, **182**, 112 (1968).
118. P. M. Maitlis and A. Efraty, *J. Organomet. Chem.*, **4**, 172 (1965).
119. A. Efraty, *Ph.D. Dissertation*, McMaster University, Hamilton, Ontario, Canada, 1967.
120. P. M. Maitlis, D. F. Pollock, M. L. Games, and W. J. Pryde, *Can. J. Chem.*, **43**, 470 (1965).
121. M. Avram, I. G. Dinulescru, G. D. Matusu, E. Avram, and C. D. Nenitzesou, *Rev. Roum. Chem.*, **14**, 1181 (1964).
122. D. F. Pollock and P. M. Maitlis, *J. Organomet. Chem.*, **26**, 407 (1971).
123. C. T. Sears and F. G. A. Stone, *J. Organomet. Chem.*, **11**, 644 (1968).
124. A. Efraty and P. M. Maitlis, *J. Am. Chem. Soc.*, **89**, 3744 (1967).
125. D. F. Pollsak and P. M. Maitlis, *Canad. J. Chem.*, **44**, 2673 (1966).
126. R. Bruce, K. Moseley, and P. M. Maitlis, *Canad. J. Chem.*, **45**, 2011 (1967).
127. R. Bruce and P. M. Maitlis, *Canad. J. Chem.*, **45**, 2017 (1967).
128. A. Efraty, *J. Organomet. Chem.*, **57**, 1 (1973).
129. F. L. Bowden and A. B. P. Lever, *Organomet. Chem. Rev.*, **3**, 227 (1968).
130. A. Efraty, *Chem. Rev.*, **77**, 691 (1977).
131. R. B. King and A. Efraty, *J. Am. Chem. Soc.*, **92**, 6071 (1970).
132. R. B. King and I. Haiduc, *J. Am. Chem. Soc.*, **94**, 4044 (1972).
133. R. B. King, I. Haiduc, and C. W. Eavenson, *J. Am. Chem. Soc.*, **95**, 2508 (1973).
134. H. O. Van Oven, *J. Organomet. Chem.*, **53**, 309 (1973).
135. A. N. Nesmeyanov, K. N. Asimov, N. E. Kolobova, and A. A. Pasynskii, *Izv. Akad. Nauk SSSR, Ser. Khim.*, 100 (1969).
136. A. N. Nesmeyanov, A. I. Gusev, A. A. Pasynskii, K. N. Anisimov, N. E. Kolobova, and Yu. T. Struchkov, *J. Chem. Soc., Chem. Commun.*, 277 (1969).
137. A. N. Nesmeyanov, A. I. Gusev, A. A. Pasynskii, K. N. Anisimov, N. E. Kolobova, and Yu. T. Struchkov, *J. Chem. Soc., Chem. Commun.*, 739 (1969).
138. W. Hubel and R. Merenyi, *J. Organomet. Chem.*, **2**, 213 (1964).

139. W. Hubel, E. H. Braye, A. Clauss, E. Weiss, U. Kruerke, D. A. Brown, G. S. D. King, and C. Hoogzand, *J. Inorg. Nucl. Chem.*, **9**, 204 (1959).
140. L. P. Motz, J. Merritt, and R. P. Pinnell, *Synthesis,* 305 (1971).
141. R. Buehler, R. Geist, R. Muendnich, and H. Plieninger, *Tetrahedron Lett.*, 1919 (1973).
142. H. W. Whitlock and P. E. Sandvick, *J. Am. Chem. Soc.*, **88**, 4526 (1966).
143. A. Krebs, H. Kimling, and R. Kemper, *Justus Liebigs Ann. Chem.*, 431 (1978).
144. J. L. Boston, D. W. Sharpe, and G. Wilkinson, *J. Chem. Soc.*, 3488 (1962).
145. M. D. Rausch and R. A. Genetti, *J. Am. Chem. Soc.,* **89**, 5502 (1967).
146. A. Nakamura and N. Hagihara, *Bull. Chem. Soc. Jap.*, **34**, 452 (1961).
147. A. Nakamura, *Mem. Inst. Sci. Ind. Res. Osaka Univ.*, **19**, 81 (1962).
148. H. Yamazaki and N. Hagihara, *J. Organomet. Chem.*, **7**, 22 (1967).
149. H. Yamazoki and N. Hagihara, *J. Organomet. Chem.*, **21**, 431 (1970).
150. M. D. Rausch and R. A. Genetti, *J. Org. Chem.*, **35**, 3888 (1970).
151. J. F. Helling, S. C. Rennison, and A. Merigan, *J. Am. Chem. Soc.*, **89**, 7140 (1967).
152. H. Sakurai and J. Hayaski, *J. Organomet. Chem.*, **70**, 85 (1974).
153. A. Clearfield, R. Gopal, M. D. Rausch, E. F. Tokas, F. A. Higbie, and I. Bernal, *J. Organomet. Chem.*, **135**, 229 (1977).
154. G. G. Cash, J. F. Helling, M. Mathew, and G. J. Palenik, *J. Organomet. Chem.*, **50**, 227 (1973).
155. J. F. Nixon and M. Kooti, *J. Organomet. Chem.*, **104**, 231 (1976).
156. M. Kooti and J. F. Nixon, *Inorg. Nucl. Chem. Lett.*, **9**, 1031 (1973).
157. F. Canziani, P. Chini, A. Quarta, and A. Dimartin, *J. Organomet. Chem.*, **26**, 285 (1971).
158. F. Canziani and M. C. Malatesto, *J. Organomet. Chem.*, **90**, 235 (1975).
159. S. M. Kim, Yu. A. Kushnikov, L. V. Levchenko, and N. N. Paramonova, *Dokl. Vses., Konf. Khim. Atsetilena, 4th*, **2**, 272 (1972); *Chem. Abstr.*, **79**, 105384n (1973).
160. R. B. King and A. Efraty, *J. Am. Chem. Soc.*, **94**, 3021 (1972).
161. F. D. Mango and J. H. Schachtschneider, *J. Am. Chem. Soc.*, **91**, 1030 (1969).
162. F. D. Mango, *Prospects in Organotransition-metal Chemistry. 1st Seminar, University of Hawaii, Honolulu*, 1974, p. 169.
163. E. J. Carey and J. Streith, *J. Am. Chem. Soc.*, **86**, 950 (1964).
164. M. Rosenblum and C. Gatsonis, *J. Am. Chem. Soc.*, **89**, 5074 (1967).
165. J. Agar, F. Keplan, and B. W. Roberts, *J. Org. Chem.*, **39**, 3451 (1974).
166. M. Rosenblum and B. North, *J. Am. Chem. Soc.*, **90**, 1060 (1968).
167. M. Rosenblum, B. North, D. Wells, and W. P. Giering, *J. Am. Chem. Soc.*, **94**, 1239 (1972).
168. S. A. Gardner and M. D. Rausch, *J. Organomet. Chem.*, **56**, 365 (1973).
169. H. H. Freedman, *J. Am. Chem. Soc.*, **83**, 2194, 2195 (1961).
170. W. Hubel, *Conf. Current Trends Organomet. Chem., Cincinnati,* 1963.
171. I. G. Dinulescu, E. G. Georgescu, and M. Avram, *J. Organomet. Chem.*, **127**, 193 (1977).
172. P. J. Harris, J. A. K. Howard, S. A. R. Knox, R. P. Phillips, F. G. A. Stone, and P. Woodward, *J. Chem. Soc., Dalton Trans.*, 379, (1976).
173. T. S. Piper, F. A. Cotton, and G. Wilkinson, *J. Inorg. Nucl. Chem.*, **1**, 165 (1955).
174. R. C. Kochar and R. Pettit, *J. Organomet. Chem.*, **6**, 272 (1966).
175. T. Kruck and L. Knoll, *Chem. Ber.*, **105**, 3783 (1972).
176. P. L. Timms, *Adv. Inorg. Chem. Radiochem.*, **14**, 121 (1972).
177. R. B. King and A. Efraty, *J. Am. Chem. Soc.*, **93**, 4950 (1971).
178. R. B. King and A. Efraty, *J. Am. Chem. Soc.*, **94**, 3773 (1972).
179. H. Muller and G. E. Herberich, *Chem. Ber.*, **104**, 2772 (1971).
180. B. F. Hallam and P. L. Pauson, *J. Chem. Soc.*, 646 (1958).
181. G. F. Grant and P. L. Pauson, *J. Organomet. Chem.*, **9**, 553 (1967).
182. E. Weiss and W. Hubel, *Angew. Chem.*, **73**, 298 (1961); E. Weiss and W. Hubel, *Chem. Ber.*, **95**, 1186 (1962).
183. R. C. Kerber and D. J. Ehntholt, *Synthesis*, **2**, 449 (1970); a useful review of fulvene complexes.
184. W. Weidemuller and K. Hafner, *Angew. Chem., Int. Ed. Engl.*, **12**, 925 (1973).
185. D. F. Hunt and J. W. Russell, *J. Am. Chem. Soc.*, **94**, 7198 (1972).
186. D. F. Hunt and J. W. Russell, *J. Am. Chem. Soc.*, **46**, C22 (1972).
187. E. Weiss and W. Hubel, *J. Inorg. Nucl. Chem.*, **11**, 42 (1959).
188. G. N. Schrauzer, *J. Am. Chem. Soc.*, **81**, 5307 (1959).

189. M. I. Bruce and J. R. Knight, *J. Organomet. Chem.*, **12**, 411 (1968).
190. E. O. Fischer and H. Werner, *Metal π-Complexes*, Vol. 1, Elsevier, Amsterdam, 1966, p. 60.
191. H. Alper and E. C.-H. Keung, *J. Am. Chem. Soc.*, **94**, 2144 (1972).
192. A. Eisenstadt, G. Scharf, and B. Fuchs, *Tetrahedron Lett.*, 679 (1971).
193. J. C. Brunet, B. Resibois, and J. Bertrand, *Bull. Soc. Chim. Fr.*, 3424 (1969).
194. B. Resibois and J. C. Brunet, *Ann. Chim.*, **5**, 199 (1970).
195. J. C. Brunet and N. Demey, *Ann. Chim.*, **8**, 123 (1973).
196. W. Fink, *Helv. Chim. Acta,* **57**, 167 (1974).
197. K. W. Muir, R. Walker, E. W. Abel, T. Blackmore, and R. J. Whitley, *J. Chem. Soc., Chem. Commun.*, 698 (1975).
198. J. Sokurai and J. Hayashi, *J. Organomet. Chem.*, **63**, C10 (1973).
199. E. H. Braye, W. Hubel and I. Caplier, *J. Am. Chem. Soc.*, **83**, 4406 (1961).
200. R. Guilard and Y. Dusausoy, *J. Organomet. Chem.*, **77**, 393 (1974).
201. H. D. Kaesz, R. B. King, T. A. Manuel, L. D. Nichols, and F. G. A. Stone, *J. Am. Chem. Soc.*, **82**, 4749 (1960).
202. E. Weiss, W. Hubel, and R. Merenyi, *Chem. Ind. (London),* 407 (1960).
203. C. G. Krespan, *J. Org. Chem.*, **40**, 261 (1975).
204. W. Hubel, E. H. Braye, A. Claus, E. Weiss, U. Kruerke, G. A. Brown, G. S. D. King, and C. Hoogzand, *J. Inorg. Nucl. Chem.*, **9**, 204 (1959).
205. J. L. Boston, D. W. A. Sharp, and G. Wilkinson, *J. Chem. Soc.*, 3488 (1962).
206. R. B. King and C. A. Harman, *Inorg. Chem.*, **15**, 879 (1976).
207. R. Markby, H. W. Sternberg, and I. Wender, *Chem. Ind. (London)*, 1381 (1959).
208. R. S. Dickson and G. Wilkinson, *J. Chem. Soc.*, 2699 (1964).
209. M. L. H. Green, L. Pratt, and G. Wilkinson, *J. Chem. Soc.*, 989 (1960).
210. W. Hubel and E. Weiss, *Chem. Ind. (London)*, 703 (1959).
211. E. Weiss, W. Hubel, and R. Merenyi, *Chem. Ber.*, **95**, 1155 (1962).
212. W. Hubel and E. H. Braye, *J. Inorg. Nucl. Chem.*, **10**, 250 (1959).
213. C. F. Sears and F. G. A. Stone, *J. Organomet. Chem.*, **11**, 644 (1968).
214. F. R. Young, D. H. O'Brien, R. C. Pettersen, R. A. Levenson, and D. L. von Minden, *J. Organomet. Chem.*, **114**, 157 (1976).
215. H. B. Chin and R. Bau, *J. Am. Chem. Soc.*, **95**, 5068 (1973).
216. R. E. Davis, B. L. Barnett, R. G. Amiet, W. Merk, J. S. McKennis, and R. Pettit, *J. Am. Chem. Soc.*, **96**, 7108 (1974).
217. R. B. King, I. Haiduc, and C. W. Eavenson, *J. Am. Chem. Soc.*, **95**, 4311 (1973).
218. B. F. Hallam and P. L. Pauson, *J. Chem. Soc.*, (1958) 642.
219. E. O. Fischer, H. P. Kogler and P. Kuzel, *Chem. Ber.*, **93**, 3006 (1960).
220. E. O. Fischer and H. Werner, *Metal π-Complexes*, Vol. 1, Elsevier, Amsterdam, 1966, p. 66.
221. R. B. King, P. M. Treichel, and F. G. A. Stone, *J. Am. Chem. Soc.*, **83**, 3593 (1961).
222. E. O. Fischer and H. Werner, *Metal π-Complexes*, Vol. 1, Elsevier, Amsterdam, 1966, p. 67.
223. T. H. Whitesides and R. A. Budnik, *J. Chem. Soc., Chem. Commun.*, 87 (1973).
224. A. J. Birch, P. E. Cross, J. Lewis, D. A. White, and S. B. Wild, *J. Chem. Soc. A*, 332 (1968).
225. A. J. Birch and M. A. Haas, *J. Chem. Soc. C*, 2465 (1971).
226. R. B. King, T. A. Manuel, and F. G. A. Stone, *J. Inorg. Nucl. Chem.*, **16**, 233 (1961).
227. V. N. Piottukh-Peletskii, R. N. Berezina, A. I. Rezbukhin, and V. G. Shubin, *Izv. Akad. Nauk. SSSR, Ser. Khim.*, 2083 (1973); *Bull Acad. Sci., USSR, Div. Chem. Ser.*, 2027 (1973).
228. A. J. Birch and D. A. Williamson, *J. Chem. Soc., Perkin Trans. 1*, 1892 (1973).
229. N. S. Nametkin, V. D. Tyurin, M. Slupczynski, and A. I. Nekhaev, *Izv. Akad. Nauk SSSR, Ser. Khim.*, 2131 (1975).
230. A. J. Pearson, *J. Chem. Soc., Perkin Trans. 1*, 1255 (1979).
231. N. S. Nametkin, V. D. Tyurin, A. I. Nekhaev, V. I. Ivanov, and F. S. Bayaouova, *J. Organomet. Chem.*, **107**, 337 (1976).
232. P. McCardle and T. Higgins, *Inorg. Chim. Acta*, **30**, L303 (1978).
233. H. Alper, *J. Organomet. Chem.*, **96**, 95 (1975).
234. W. Frink, *Helv. Chim. Acta*, **58**, 1205 (1975).

235. N. Nakadaira, T. Kobayashi, and H. Sokurai, *J. Organomet. Chem.*, **165**, 399 (1979).
236. J. B. Davison and J. M. Bellama, *Inorg. Chim. Acta*, **14**, 263 (1975).
237. A. C. Szary, S. A. R. Knox, and F. G. A. Stone, *J. Chem. Soc., Dalton Trans.*, 662 (1974).
238. E. J. Reardon and A. Brockhart, *J. Am. Chem. Soc.*, **96**, 4311 (1974).
239. M. Cooke, J. A. K. Howard, C. R. Russ, F. G. A. Stone, and P. Woodward, *J. Organomet. Chem.*, **78**, C43 (1974).
240. F. A. Cotton and G. Deganello, *J. Organomet. Chem.*, **38**, 147 (1972).
241. R. Aumann, *J. Organomet. Chem.*, **47**, C29 (1973).
242. R. B. King, I. Haiduc, and C. W. Eavenson, *J. Am. Chem. Soc.*, **95**, 2508 (1973).
243. R. M. Moriarty, K.-N. Chen. M. R. Churchill, and S. W.-Y. Chang, *J. Am. Chem. Soc.*, **96**, 3661 (1974).
244. R. Aumann, K. Froelich, and H. Ring, *Angew. Chem.*, **86**, 309 (1974).
245. R. Victor, R. Ben-Shoshan, and S. Sarel, *J. Org. Chem.*, **37**, 1930 (1972).
246. R. B. King, in *The Organic Chemistry of Iron*, (Ed. E. A. K. von Gustorf), Vol. 1, Academic Press, New York, 1978, p. 564.
247. G. R. John, L. A. P. Kane-Maguire, and C. Eaborn, *J. Chem. Soc., Chem. Commun.*, 481 (1975).
248. C. A. Mansfield, K. M. Al-Kathumi, and L. A. P. Kane-Maguire, *J. Organomet. Chem.*, **71**, C11 (1974).
249. L. A. P. Kane-Maguire and C. A. Mansfield, *J. Chem. Soc., Chem. Commun.*, 540 (1973).
250. A. J. Birch, I. D. Jenkins, and A. J. Liepa, *Tetrahedron Lett.*, 1723 (1975).
251. A. J. Birch and A. J. Pearson, *Tetrahedron Lett.*, 2379 (1975).
252. N. S. Nametkin, V. D. Tyurin, and V. I. Ivanov, *Izv. Akad. Nauk SSSR, Ser. Khim.*, 1911 (1975).
253. F. Franke and I. D. Jenkins, *Aust. J. Chem.*, **31**, 595 (1978).
254. A. J. Birch and A. J. Pearson, *J. Chem. Soc., Perkin Trans. 1,* 954 (1976).
255. A. J. Pearson, *Aust. J. Chem.*, **29**, 1101 (1976).
256. C. A. Mansfield and L. A. P. Kane-Maguire, *J. Chem. Soc., Dalton Trans.*, 2187 (1976).
257. L. A. P. Kane-Maguire and C. A. Mansfield, *J. Chem. Soc., Dalton Trans.*, 2192 (1976).
258. G. R. John and L. A. P. Kane-Maguire, *J. Chem. Soc., Dalton Trans.*, 873 (1939).
259. M. Moll, P. Wuerstl, H. Behrens, and P. Merbach, *Z. Naturforsch.*, **33B**, 1304 (1978).
260. G. R. John and L. A. P. Kane-Maguire, *J. Chem. Soc., Dalton Trans.*, 1196 (1979).
261. A. J. Pearson, *J. Chem. Soc., Perkin Trans. 1,* 1255 (1979).
262. A. J. Pearson, *J. Chem. Soc., Perkin Trans. 1,* 2069 (1977).
263. A. J. Birch and K. B. Chamberlain, *Org. Synth.*, **57**, 107 (1977).
264. T. A. Manuel and F. G. A. Stone, *J. Am. Chem. Soc.*, **82**, 366 (1960).
265. R. Burton, M. L. H. Green, E. W. Abel, and G. Wilkinson, *Chem. Ind. (London),* 1592 (1958).
266. M. El Borai, R. Gailard, P. Fournari, Y. Dusausoy, and J. Protes, *J. Organomet. Chem.*, **148**, 285 (1978).
267. J. C. Burt, S. A. R. Knox, and F. G. A. Stone, *J. Chem. Soc., Dalton Trans.*, 731 (1975).
268. R. B. King, *Inorg. Chem.*, **2**, 807 (1963).
269. D. F. Hunt, G. C. Farrant, and G. T. Rodeheaver, *J. Organomet. Chem.*, **38**, 349 (1972).
270. E. Weiss and W. Hubel, *Chem. Ber.*, **95**, 1179 (1962).
271. A. Eisenstadt, *J. Organomet. Chem.*, **97**, 443 (1975).
272. E. O. Fischer, C. G. Kreiter, H. Ruhle, and K. E. Schwarzans, *Chem. Ber.*, **100**, 1905 (1967).
273. L. K. K. Li Shing Man and J. Takets, *J. Organomet. Chem.*, **117**, C104 (1976).
274. T. Tsuchiya and V. Snieckus, *Can. J. Chem.*, **53**, 519 (1975).
275. H. J. Dauben and D. J. Bertelli, *J. Am. Chem. Soc.*, **83**, 497 (1961).
276. R. Burton, L. Pratt and G. Wilkinson, *J. Chem. Soc.*, 594 (1961).
277. P. Dodman and J. C. Tatlow, *J. Organomet. Chem.*, **67**, 87 (1974).
278. D. F. Hunt, G. C. Farrant, and G. T. Rodeheaver, *J. Organomet. Chem.*, **38**, 349 (1972).
279. A. Davison, W. McFarlane, L. Pratt, and G. Wilkinson, *J. Chem. Soc.*, 4821 (1962).
280. D. A. Brown, S. K. Chawla, and W. K. Glass, *Inorg. Chim. Acta,* **19**, L31 (1976).
281. B. Y. Shu, E. R. Biehl, and P. C. Reeves, *Synth. Commun.*, **8**, 523, (1978).
282. B. R. Reddy and J. S. McKennis, *J. Organomet. Chem.*, **182**, C61 (1979).
283. S. D. Ittel and C. A. Tolman, *J. Organomet. Chem.*, **172**, C47 (1979).

284. R. Aumann and J. Krecht, *Chem. Ber.*, **109**, 174 (1976).
285. R. B. King, in *Organic Chemistry of Iron,* (Ed. E. A. K. von Gustorf), Vol. 1, Academic Press, New York, 1978, pp. 576–578.
286. T. A. Manuel and F. G. A. Stone, *Proc. Chem. Soc.*, 90 (1959).
287. A. Nakamura and N. Hagihara, *Bull. Chem. Soc. Jap.*, **32**, 881 (1959).
288. M. D. Rausch and G. N. Schrauzer, *Chem. Ind. (London)*, 957 (1959).
289. C. E. Keller, G. F. Emerson, and R. Pettit, *J. Am. Chem. Soc.*, **87**, 1388 (1965).
290. J. D. Edwards, J. A. K. Howard, S. A. R. Knox, V. Riera, F. G. A. Stone, and P. Woodward, *J. Chem. Soc., Dalton Trans.*, 75 (1976).
291. B. F. G. Johnson, J. Lewis, and D. Wege, *J. Chem. Soc., Dalton Trans.*, 1874 (1976).
292. M. Cooke, C. R. Russ, and F. G. Stone, *J. Chem. Soc., Dalton Trans.*, 256 (1975).
293. M. A. Bennett and J. D. Saxby, *Inorg. Chem.*, **7**, 321 (1968).
294. A. Nakamura and N. Hagihara, *Bull. Chem. Soc. Jap.*, **33**, 425 (1960).
295. M. I. Bruce, M. Cooke, M. Green, and D. J. Westlake, *J. Chem. Soc. A,* 987 (1969).
296. R. B. King and A. Fronzaglia, *Inorg. Chem.*, **5**, 1837 (1966).
297. G. Wilke and H. Brail, *Angew. Chem., Int. Ed. Engl.*, **5**, 898 (1966).
298. F. A. Cotton, A. Davison, T. J. Marks, and A. Musco, *J. Am. Chem. Soc.*, **91**, 6598 (1969).
299. M. A. Bennett, T. W. Matheson, G. B. Robertson, A. K. Smith, and P. A. Tucker, *J. Organomet. Chem.*, **121**, C18 (1976).
300. A. Greco, M. Green, and F. G. A. Stone, *J. Chem. Soc. A,* 285 (1971).
301. P. L. Timms and T. W. Turney, *J. Chem. Soc., Dalton Trans.*, 2021 (1976).
302. S. Muller, W. Holzinger, and M. Menig, *Proc. VIIIth Int. Conf. Organometallic Chem., Venice*, 1975.
303. T. A. Manuel and F. G. A. Stone, *J. Am. Chem. Soc.*, **82**, 6240 (1960).
304. A. J. Deeming, S. S. Ullah, A. J. P. Domingos, B. F. G. Johnson, and J. Lewis, *J. Chem. Soc. Dalton Trans.*, 2093 (1974).
305. A. D. English, J. P. Jesson, and C. A. Tolman, *Inorg. Chem.*, **15**, 1730 (1976).
306. G. Schiavon, C. Paradisi, and C. Boanini, *Inorg. Chim. Acta,* **14**, L5 (1975).
307. A. J. P. Domingos, B. F. G. Johnson, and J. Lewis, *J. Chem. Soc., Dalton Trans.*, 2288 (1975).

The Chemistry of the Metal—Carbon Bond
Edited by F. R. Hartley and S. Patai

CHAPTER **10**

Synthesis of complexes of η^5-bonded ligands

G. MARR and B. W. ROCKETT
Department of Physical Sciences, The Polytechnic, Wolverhampton WV1 1LY, UK

I. η^5-CYCLOPENTADIENYL COMPLEXES

A. Introduction

There are many η^5-hydrocarbon—transition metal complexes and they can be prepared by a variety of different routes. The chemistry of the η^5-cyclopentadienyl complexes and related compounds has been reviewed in recent years[1]. The 'Annual Surveys' in the *Journal of Organometallic Chemistry* provide a near comprehensive coverage of η^5-hydrocarbon—transition metal complexes on a year to year basis.

B. Reaction of the Cyclopentadienide Anion With a Transition Metal Halide or a Transition Metal Complex

The reaction of the anionic cyclopentadienide group with a transition metal salt (MCl_2) (equation 1) is the most widely used method for preparing η^5-cyclopentadienyl—metal complexes.

$$Na^+ \, C_5H_5^- + MCl_2 \longrightarrow [(\eta^5\text{-}C_5H_5)M(\eta^5\text{-}C_5H_5)] + 2\,NaCl \qquad (1)$$

Cyclopentadiene is acidic ($pK_a = 17$)[2], and on treatment with an alkali metal in a solvent with a high dielectric constant the alkali metal cyclopentadienide salt is formed readily. The solvents most frequently used are tetrahydrofuran and 1,2-dimethoxyethane but dioxane, ethanol, ammonia, and diethylamine are also suitable.

Examples of the direct reaction of a metal cyclopentadienide with a transition metal halide are given in Table 1. This reaction (equation 1) does not always proceed according to the stoichiometry shown. Ionic cyclopentadienides provide a reducing medium and, if the transition metal ion is readily reducible, reduction occurs. For example, the treatment of ruthenium(III) chloride and osmium(IV) chloride with sodium cyclopentadienide produces ruthenocene and osmocene, respectively, where the transition metal ions are formally in the +2 oxidation states (Table 1)[3,4].

TABLE 1. Synthesis of the iron group metallocenes from a metal cyclopentadienide and a transition metal salt

Transition metal salt	Metal cyclopentadienide	Product	Solvent	Yield (%)	Reference
$FeCl_2$	C_5H_5Na	$[(\eta^5\text{-}C_5H_5)_2Fe]$	Thf or $(MeOCH_2)_2$	67, 85	5, 6
$FeCl_2$	C_5H_5Na	$[(\eta^5\text{-}C_5H_5)_2Fe]$	Thf, EtOH	90	7
$FeCl_2$	C_5H_5Tl	$[(\eta^5\text{-}C_5H_5)_2Fe]$	Thf	49	8
$FeCl_3$	C_5H_5Tl	$[(\eta^5\text{-}C_5H_5)_2Fe]$	Thf	98	8
$FeCl_3$	C_5H_5MgBr	$[(\eta^5\text{-}C_5H_5)_2Fe]$	Thf	51	9
$FeCl_3$	C_5H_5MgBr	$[(\eta^5\text{-}C_5H_5)_2Fe]$	Et_2O or $(n\text{-}Bu)_2O$	60	10
				61	10
$RuCl_3$	C_5H_5Na	$[(\eta^5\text{-}C_5H_5)_2Ru]$	Thf or $(MeOCH_2)_2$	43–52	4, 11
$RuCl_3$/Ru	C_5H_5Na	$[(\eta^5\text{-}C_5H_5)_2Ru]$	$(MeOCH_2)_2$	56–69	12
$OsCl_4$	C_5H_5Na	$[(\eta^5\text{-}C_5H_5)_2Os]$	Thf or $(MeOCH_2)_2$	18–23	4, 12, 13

This method has been used extensively in the preparation of a wide range of bis(η^5-cyclopentadienyl)transition metal complexes[1]. For example, cobaltocene is prepared readily by the reaction of hexaamminecobalt(II) chloride with sodium cyclopentadienide in tetrahydrofuran[14]. Recently a modification of this method has been used to prepare substituted cobaltocenes. Treatment of all the homologues of the $(Me)_nC_5H_{6-n}$ series with sodamide in liquid ammonia produces the corresponding ions $(Me_nC_5H_{5-n})^-$. Reaction of these anions with hexaamminecobalt(II) chloride gives symmetrically substituted polymethylcobaltocenes, which are oxidized to the corresponding cobaltocenium ions and these are isolated as the hexafluorophosphate salts; for example, sodium pentamethylcyclopentadienide gives the cobaltocenium salt (**1**)[15].

Sodium cyclopentadienide will combine directly with a transition metal salt in the presence of carbon monoxide to give the η^5-cyclopentadienyl—transition metal carbonyl derivative. For example, the treatment of manganese(II) bromide with sodium cyclopentadienide and carbon monoxide under pressure produces tricarbonyl(η^5-cyclopentadienyl)manganese (**2**; M = Mn) in 40% yield[16]. Similar reactions are used to prepare the rhenium (**2**; M = Re) (16% yield)[17] and technetium (**2**; M = Tc) (86% yield)[18] analogues and tetracarbonyl(η^5-cyclopentadienyl)-vanadium[19,20].

$[C_5Me_5]^-\ Na^+ + [Co(NH_3)_6]Cl_2 \rightarrow (C_5Me_5)_2Co \xrightarrow{(1)\ \text{oxidise};\ (2)\ NH_4PF_6} [(C_5Me_5)_2Co]^+\ PF_6^-$ (**1**) (2)

Alkali metal cyclopentadienides displace carbon monoxide from metal carbonyls and this method is used to prepare η^5-cyclopentadienyl—metal complexes[21,22]. Sodium

$[C_5H_5]^-\ Na^+ + MBr_2 + 3CO \xrightarrow{\text{pressure and heat}} [(C_5H_5)M(CO)_3]$ (**2**) (3)

cyclopentadienide combines directly with carbonylmetal halides to give the complexes $[(\eta\text{-}C_5H_5)M(CO)_x]_y$ where M = Pt[23], Rh[24], Ir[25], Ru[26], Os, or Mn[27]; a typical example of this reaction is the preparation of dicarbonyl(η^5-cyclopentadienyl)rhodium **(3)**[24].

$$[Rh(CO)_2Cl]_2 + [C_5H_5]^- Na^+ \longrightarrow [(\eta^5\text{-}C_5H_5)Rh(CO)_2] \quad (4)$$

(3)

Thallium cyclopentadienide finds increasing use because it is more convenient to handle than sodium cyclopentadienide. It is prepared in 95% yield by adding cyclopentadiene to an aqueous solution of thallium sulphate and potassium hydroxide[28–31]. Thallium cyclopentadienide is an air-stable solid, a weak nucleophile, and a poor reducing agent, and hence many of the undesirable side reactions associated with the corresponding sodium salt are minimized. Thallium cyclopentadienide has been used for the preparation of η^5-cyclopentadienyl—transition metal complexes (Table 2). Most of these reactions are characterized by relatively simple experimental procedures and high yields.

A number of substituted thallium cyclopentadienides are known. These are easy to prepare but most of these compounds are less stable than thallium cyclopentadienide itself. For example, thallium methylcyclopentadienide is oxidized rapidly in air and the following order of stability has been established: Tl cyclopentadienide > Tl hydropentalenide > Tl isodicyclopentadienide > Tl methylcyclopentadienide. A number of transition metal derivatives have been prepared from these substituted and ring-annelated thallium cyclopentadienides as shown in Table 3. For example, the reaction of thallium hydropentalenide **(4)** with iron(II) chloride gives bis(η-hydropentalenyl)iron (**5a** or **5b**).

Cyclopentadiene is deprotonated by a base having $K_B > 10^{-5}$ and bases such as diethylamine, triethylamine, piperidine, ammonia, and potassium hydroxide are all

TABLE 2. Synthesis of η^5-cyclopentadienyl–transition metal complexes from thallium cyclopentadienide and a metal halide

Transition metal reagent	Product	Yield (%)	Reference
$ScCl_3$	$[(\eta^5\text{-}C_5H_5)_2ScCl]$	85	32
$TiCl_3$	$[(\eta^5\text{-}C_5H_5)_2TiCl]$	98	32
$TiCl_4$	$[(\eta^5\text{-}C_5H_5)_2TiCl_2]$	61	33
$HfCl_4$	$[(\eta^5\text{-}C_5H_5)_2HfCl_2]$	58	34
VCl_3	$[(\eta^5\text{-}C_5H_5)_2VCl]$	92	32
$CrCl_3$ and NO	$[(\eta^5\text{-}C_5H_5)Cr(NO)_2Cl]$	—	35
$[Mn(CO)_5Cl]$	$[(\eta^5\text{-}C_5H_5)Mn(CO)_3]$	93	36
$FeCl_2$	$[(\eta^5\text{-}C_5H_5)_2Fe]$	87	29
		70	33
		68	35
$FeCl_3$	$[(\eta^5\text{-}C_5H_5)_2Fe]$	96	35
$[(\eta^3\text{-}C_3H_5)Fe(CO)_3I]$	$[(\eta^5\text{-}C_5H_5)Fe(CO)_2]_2$	—	37
$[\{(CF_3)_2CF\}Fe(CO)_4I]$	$[\{(CF_3)_2CF\}Fe(CO)_2(\eta^5\text{-}C_5H_5)]$	70	38
$CoCl_2$	$[(\eta^5\text{-}C_5H_5)_2Co]$	83	33
$[RhCl(\eta^4\text{-}C_4Ph_4)]_2$	$[(\eta^5\text{-}C_5H_5)Rh(\eta^4\text{-}C_4Ph_4)]$	80	39
$NiBr_2$	$[(\eta^5\text{-}C_5H_5)_2Ni]$	26	33
$[Ni(NH_3)_6]Cl_2$	$[(\eta^5\text{-}C_5H_5)_2Ni]$	70	40

TABLE 3. Synthesis of η^5-cyclopentadienyl–transition metal complexes by the reaction of substituted or ring annelated thallium cyclopentadienides with a metal halide

Transition metal reagent	Thallium derivative	Product	Yield (%)	Reference
$ZrCl_4$	$Tl(t\text{-}BuC_5H_4)$	$[(\eta^5\text{-}t\text{-}BuC_5H_4)_2ZrCl_2]$	76	34
$[BrMn(CO)_5]$	$Tl\{C_5H_4C(CN)=C(CN)_2\}$	$[\{\eta^5\text{-}C_5H_4C(CN)=C(CN)_2\}Mn(CO)_3]$	6	41
$[(\eta^5\text{-}C_5H_5)Fe(CO)_2I]$	$Tl\{C_5H_4C(CN)=C(CN)_2\}$	$[\{\eta^5\text{-}C_5H_4C(CN)=C(CN)_2\}Fe(\eta^5\text{-}C_5H_5)]$	—	41
$FeCl_2$	$Tl(C_5D_5)$	$[(\eta^5\text{-}C_5D_5)_2Fe]$	—	42
$FeCl_2$	Tl hydropentalenide	$[(\eta^5\text{-hydropentalenyl})_2Fe]$	49	43
$CoBr_2/Br_2$	$Tl(t\text{-}BuC_5H_4)$	$[(\eta^5\text{-}t\text{-}BuC_5H_4)_2Co]^+Br_3^-$	65	44

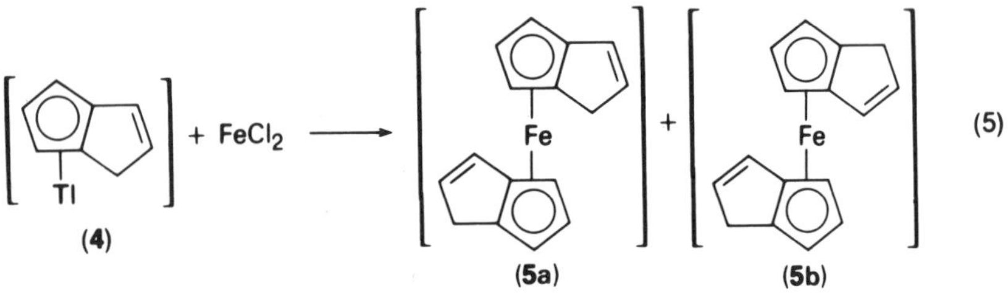

(4) + $FeCl_2$ ⟶ (5a) + (5b) (5)

used. When a solution of cyclopentadiene in diethylamine is treated with either anhydrous iron(II) or iron(III) chloride, ferrocene is obtained in high yield[45].

The amine method has also been applied successfully to the synthesis of nickelocene (equation 6)[45] and the cobaltocenium ion[46]. The reaction of the iodide **6** with cyclo-

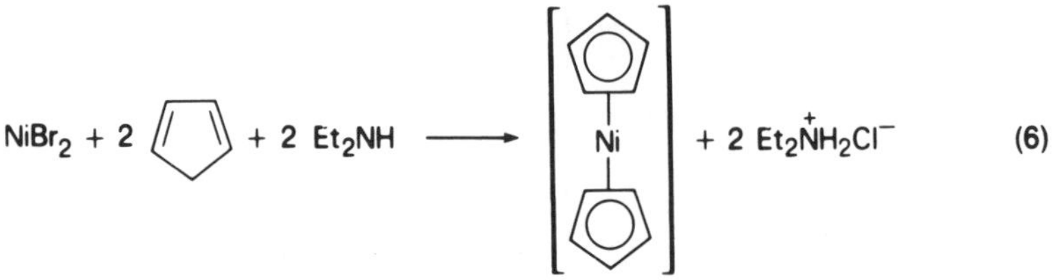

$NiBr_2$ + 2 C_5H_6 + 2 Et_2NH ⟶ $Ni(C_5H_5)_2$ + 2 $Et_2\overset{+}{N}H_2Cl^-$ (6)

pentadiene and triethylamine gives the mixed sandwich compound **7** together with dicarbonyl(η^5-cyclopentadienyl)cobalt[47]. Potassium hydroxide can be used to generate the cyclopentadienide ion[48] and a recent publication describes the use of potassium hydroxide in the presence of 18-crown-6, as a phase-transfer catalyst, for the preparation of ferrocene and a series of substituted ferrocenes[49].

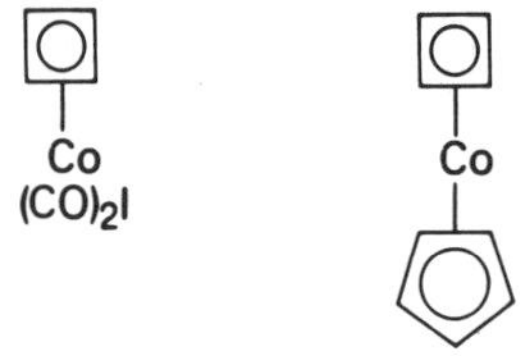

(**6**) (**7**)

C. Reaction of Cyclopentadiene With a Transition Metal or a Transition Metal Derivative

One of the most important methods developed in recent years for the synthesis of organometallic derivatives is the direct reaction of metal atoms with a suitable ligand[50,51] (see Chapter 13). In the comprehensive review by Timms and Turney[51] there is a short section on the experimental methods used in this relatively new method of synthesis. The usual procedure is to vaporize the metal under high vacuum. The metal atoms are condensed on the cooled walls of the vacuum chamber in the presence of an excess of the vapour of an organic compound or into a solution of the compound in an inert solvent. Liquid nitrogen is usually used to cool the vacuum chamber. This technique is particularly valuable for the preparation of less stable molecules but it has been applied to compounds readily prepared by more conventional methods. Cyclopentadiene and iron vapour give ferrocene, nickel vapour produces (η^5-cyclopentadienyl)(η^3-cyclopentenyl)nickel (**8**)[52], and finely divided nickel produced by an electric arc gives nickelocene (**9**)[53].

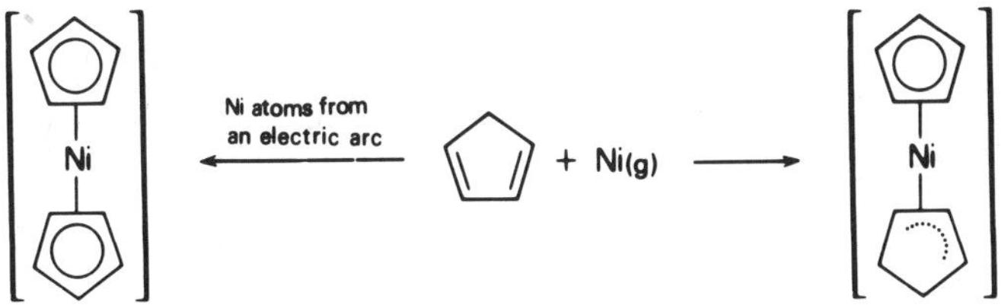
Ni
Ni atoms from
an electric arc
+ Ni(g)
Ni

(9) (8) (7)

The direct reaction between cyclopentadiene and a transition metal salt or complex is often used to prepare η^5-cyclopentadienyl derivatives. Ruthenocene is prepared in good yield by treating β-ruthenium(III) chloride with cyclopentadiene in ethanol[54]. Pentacarbonyliron and cyclopentadiene give the dimer **10**[55], which decomposes above 220°C to give ferrocene[56].

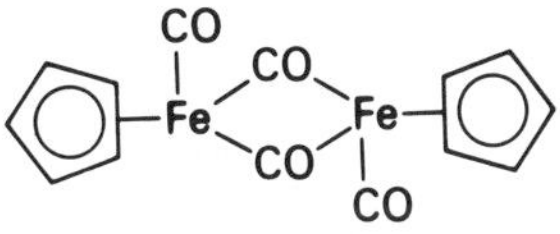
CO
CO
Fe
Fe
CO
CO

(10)

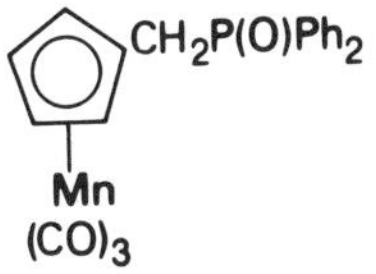
CH2P(O)Ph2
Mn
(CO)3

(11)

A substituted cyclopentadiene can be used in this reaction. For example, the treatment of diphenyl(cyclopentadienemethyl)phosphine with manganese carbonyl gives the cymantrene derivative **11** in 60% yield[57]. King and co-workers[58] reported that the reaction of pentamethylcyclopentadiene with manganese carbonyl or pentacarbonyl(methyl)manganese in boiling *n*-decane gives pentamethylcymantrene in 30% yield. In the same paper the reactions of pentamethylcyclopentadiene with some other metal carbonyls are described (Scheme 1)[58].

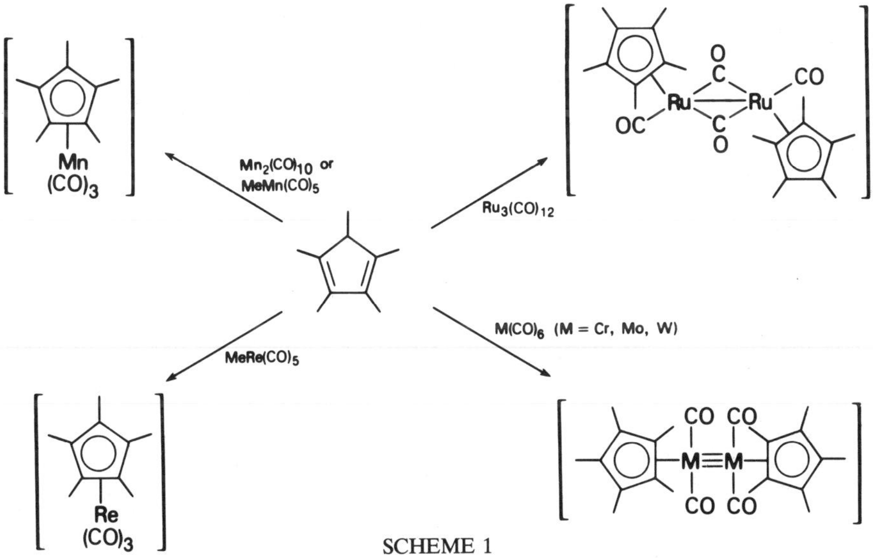
Mn
(CO)3
Mn2(CO)10 or
MeMn(CO)5
Ru3(CO)12
O
C
CO
Ru
Ru
OC
C
O
MeRe(CO)5
M(CO)6 (M = Cr, Mo, W)
Re
(CO)3
CO CO
M≡M
CO CO

SCHEME 1

If the reaction is slow or problems are encountered with the volatility of the carbonyl, complexes of the type $[L_3M(CO)_3]$ (M = Cr, Mo, W; L = MeCN, C_5H_5N, NH_3) are used to replace the metal carbonyl. The reaction of cyclopentadiene with the tris(methyl cyanide) complexes $[(MeCN)_3M(CO)_3]$(M = Cr, Mo, W) gives the corresponding metal hydrides (**12**)[59].

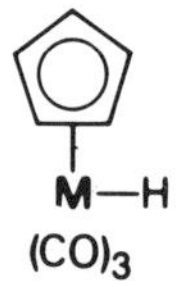

M—H
$(CO)_3$
M = Cr, Mo, W
(**12**)

D. Reaction of Fulvenes and Related Compounds With a Transition Metal or a Transition Metal Compound

The reactions of fulvenes and azulenes to give transition metal complexes have been reviewed[60]. Many of the complexes formed in these reactions have an η^5-cyclopentadienyl ring; for example, the treatment of 6-alkyl- or 6-arylpentafulvenes with hexacarbonyl-molybdenum or -tungsten gives the tricarbonyl derivatives (**13**; R = alkyl, aryl; M = Mo, W)[61].

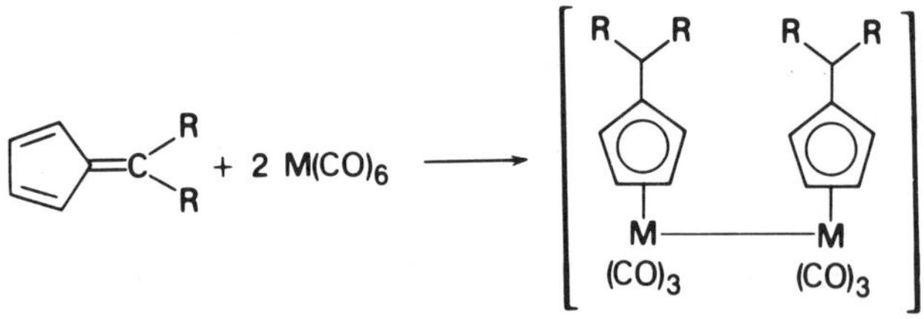

$+ 2\ M(CO)_6 \longrightarrow$ (8)

(**13**)

Several tetracarbonylvanadium complexes (**14**; R^1 = H, Me, Et, Ph; R^2 = H, Me) have been prepared by treatment of 6-alkylfulvenes with hexacarbonylvanadium[62].

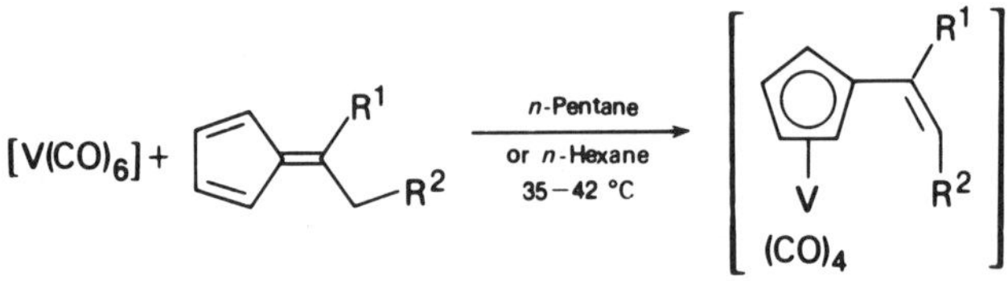

$[V(CO)_6]+$ (*n*-Pentane or *n*-Hexane, 35–42 °C) (9)

(**14**)

The direct reaction of fulvenes with metal vapour could prove a good method for the preparation of η^5-cyclopentadienyl derivatives[51]. The co-condensation of 6,6-dimethylfulvene and iron vapour at a low temperature gives the ferrocenophane (**15**) and 1,1′-diisopropylferrocene[63].

Insertion reactions of diazocyclopentadienes with pentacarbonylhalomanganese complexes are useful for preparing halogenated cymantrenes. Reaction of the pentacarbonylhalomanganese complexes $Mn(CO)_5X$ (X = Br, I) with the diazocyclopentadienes (**16**; $R^1 = R^2$ = Br; R^1 = H, R^2 = I) produces the

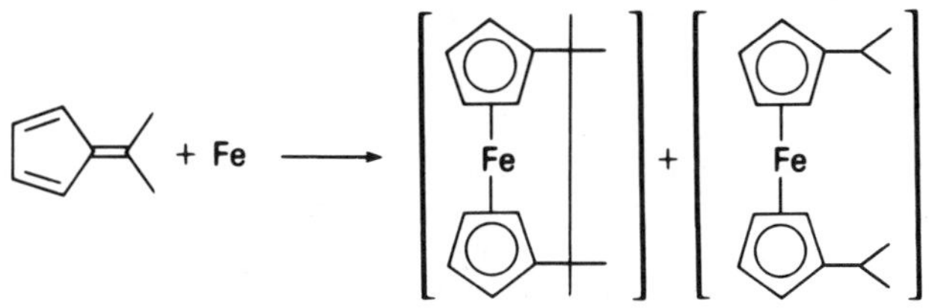

+ Fe ⟶ [Fe] + [Fe] (10)

(15)

N_2, R^2, R^2, R^1, R^1 — Br, Br, Br, Br, Br, Mn $(CO)_3$ — I, Br, Br, Br, Br, Mn $(CO)_3$ — I, I, I, Mn $(CO)_3$

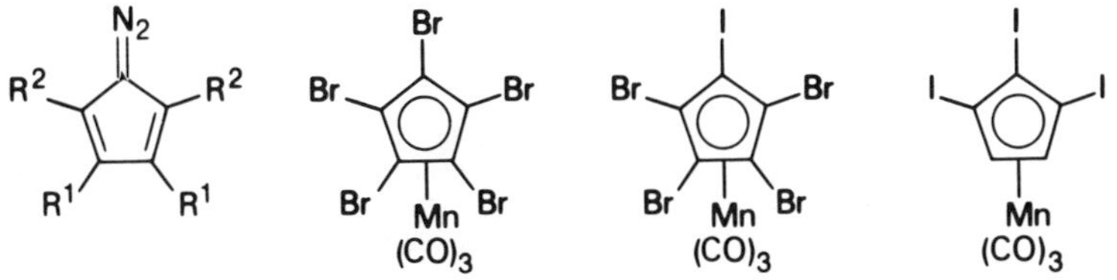

(16) (17) (18) (19)

corresponding cymantrene derivatives (**17**, **18**, and **19**) in good yields[64]. Similar reactions are used to prepare a variety of halogenated cymantrenes[65].

Reaction of the diazoindene **20** with a pentacarbonylhalomanganese complex gives the corresponding substituted cymantrene (**21**; X = Cl, Br, I)[66].

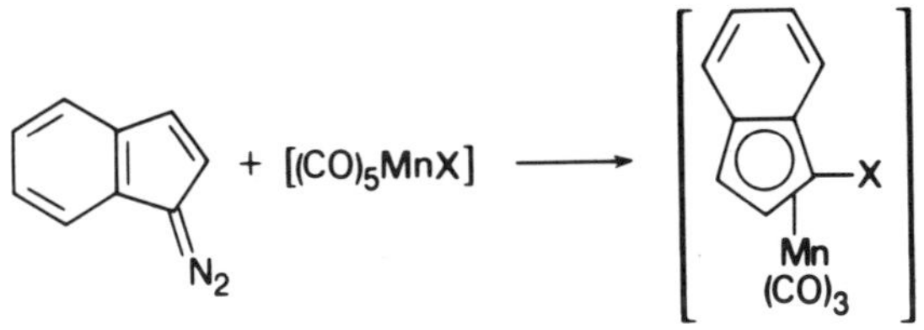

N_2 + $[(CO)_5MnX]$ ⟶ [X, Mn $(CO)_3$] (11)

(20) (21)

E. Transfer of the η^5-Cyclopentadienyl Group From One Transition Metal to Another

The transfer of the η^5-cyclopentadienyl group from one transition metal to another is of limited use as a synthetic method. It was originally used to prepare the η^5-cyclopentadienyl complexes (**22**; M = Pd, Ni)[67,68].

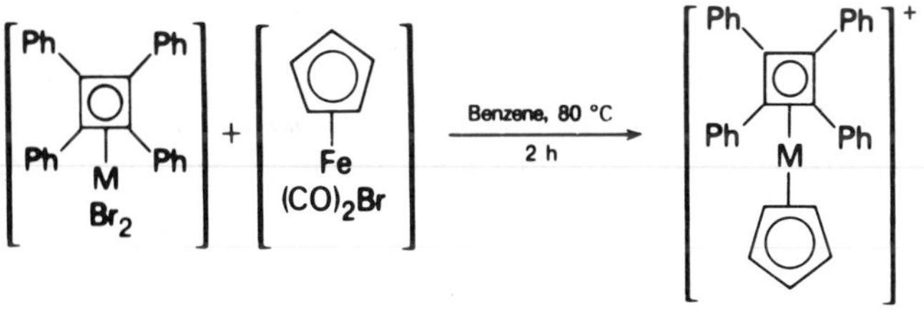

[Ph, Ph, Ph, Ph, M Br_2] + [Fe $(CO)_2Br$] —Benzene, 80 °C, 2 h⟶ [Ph, Ph, Ph, Ph, M]$^+$ (12)

(22)

The ligand transfer reaction is a good route for the synthesis of ruthenocene[69] and labelled ruthenocene and osmocene. Ferrocene or a substituted ferrocene is heated with $^{103}RuCl_3$ or $^{191}OsCl_4$ in a sealed tube to give the corresponding labelled ruthenocene or osmocene[70].

Displacement of an η^5-cyclopentadienyl ligand by carbon monoxide is a useful route to some η^5-cyclopentadienylmetalcarbonyls. For example, tricarbonyl[η^5-(triorganosilyl)cyclopentadienyl]manganese compounds are prepared by treating the (triorganosilyl)cyclopentadiene with an alkali metal, heating the resultant metal derivative with manganese(II) salts and treating the intermediate bis(η^5-cyclopentadienyl)manganese with carbon monoxide at 100–200°C and 50–200 atm[71].

F. Cycloaddition of Alkynes and Alkenes in the Presence of a Transition Metal Carbonyl

η^5-Cyclopentadienyl complexes can be formed in the reaction of alkynes[72] and alkenes with transition metal carbonyls. For example, the treatment of acetylene with manganese carbonyl gives the cymantrene derivative **23** in 40% yield[73]. The cycloaddition of [$(PhC{\equiv}CCH_2M(CO)_5$] to $MeO_2CC{\equiv}CCO_2Me$ gives the corresponding η^5-cyclopentadienyl complexes (**24**; M = Mn, Re). This reaction proceeds through the σ-bonded intermediate **25**[74].

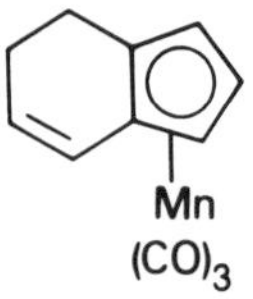

(23) 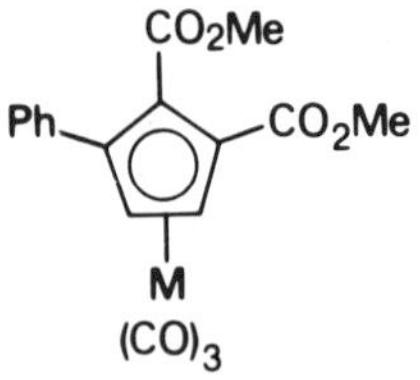 **(24)** 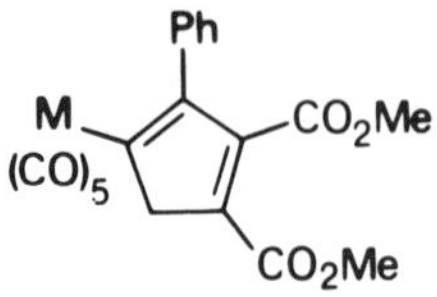 **(25)**

Treatment of butenes and pentenes with titanium(IV) chloride at 300°C under pressure gives [$(\eta^5\text{-}Me_5C_5)TiCl_3$][75]. Reaction of cyclooctatetraene with either manganese or rhenium carbonyl yields the corresponding tricarbonylmetal derivative (**26**; M = Mn, Re)[73,76].

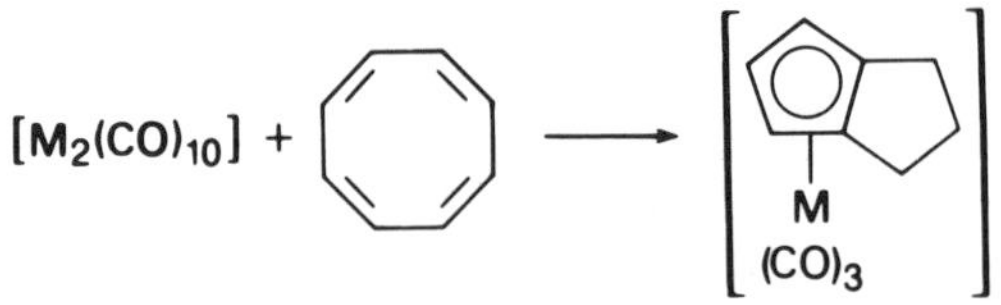

$[M_2(CO)_{10}]$ + (cyclooctatetraene) ⟶ **(26)** (13)

G. Electrochemical Preparations

In recent years the electrosynthesis of organic molecules has become increasingly important[77], but this method is seldom used for the preparation of organometallic compounds.

Cymantrene and alkyl-substituted cymantrenes are prepared by the electrolysis of an alkali metal cyclopentadienide in diethyl ether or tetrahydrofuran using a manganese electrode at an elevated temperature under a pressure of 200–500 psi of carbon monoxide[78]. Electrolysis of manganese(II) salts in the presence of cyclopentadiene, *N*-methylpyrrolidine, and nickel carbonyl at 165°C and 500 psi of carbon monoxide produces cymantrene[79].

Electrolysis of thallium cyclopentadienide in dimethylformamide with an iron anode produces ferrocene[80]. Ferrocene is prepared in a current yield of 88% by the electrolysis of a solution containing dimethylformamide, cyclopentadiene, and lithium bromide between an iron anode and a nickel cathode[81].

The electrolytic reduction of bis(pentane-2,4-dionato)cobalt(II) and cycloocta-1,5-diene in pyridine gives (η^4-cycloocta-1,5-diene)(η^3-cyclooctenyl)cobalt which, when heated in the presence of cycloocta-1,5-diene, rearranges with dehydrogenation to give the η^5-cyclopentadienyl derivative **27**[82].

[Co(AcAc)$_2$]+ (cyclooctadiene) —Electrolysis→ [Co complex] —Cyclooctadiene, 60 °C→ [Co complex]

(27) (14)

II. η^5-CYCLOHEXADIENYL COMPLEXES

A. Hydride Ion Abstraction from an (η^4-Diene)transition Metal Complex

η^5-Cyclohexadienyl complexes can be prepared by hydride abstraction, with triphenylmethylium tetrafluoroborate, from a neutral η^4-diene—transition metal complex[83,84]. For example, the treatment of the η^4-cyclohexadiene complex **28** with triphenylmethylium tetrafluoroborate gives the tricarbonyl (η^5-cyclohexadienyl)iron complex **29**[83].

[Fe(CO)$_3$ complex] + Ph_3CBF_4 —$-H^-$→ [Fe(CO)$_3$ complex]$^+$ BF_4^- + Ph_3CH (15)

(28) **(29)**

Many similar reactions involving the use of triphenylmethylium tetrafluoroborate have been reported. When a mixture of the *endo*- and *exo*-methyl isomers of the tricarbonyliron complex **30** undergoes hydride abstraction the corresponding mixture of isomers of the dienyl complex **31** is obtained[85].

MeO, Fe (CO)$_3$ MeO, Fe (CO)$_3$ $^+$

(30) **(31)**

When a reaction with a diene is carried out in the presence of aluminium chloride it is possible for hydride abstraction to occur to give the corresponding dienyl complex[86]. The 19-electron complex (η^5-cyclopentadienyl)(η^6-hexamethylbenzene)iron **(32)** undergoes hydrogen abstraction with molecular oxygen to form the corresponding (η^5-hexadienyl)iron complex **(33)**[87].

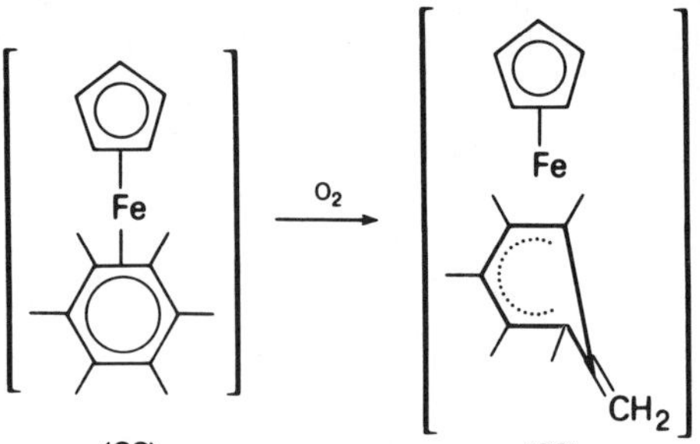

(16)

(32) **(33)**

B. Nucleophilic Attack by Hydride Ion or an Organolithium Reagent on an (η^6-Arene)transition Metal Complex

η^5-Cyclohexadienyl complexes can be prepared by nucleophilic attack of hydride ion or an organolithium reagent on an (η^6-arene)—transition metal cation. For example, the reduction of the manganese cation **34** with sodium borohydride in aqueous solution or lithium aluminium hydride in ether gives the η^5-cyclohexadienyl complex **35** (R=H) in 18% and 41% yields, respectively[88]. The substituted η^5-cyclohexadienyl complexes (**35**; R = Ph, Me) are obtained by treatment of the manganese complex **34** with either phenyl- or methyllithium. The substituent R is in the *exo*-position[89,90].

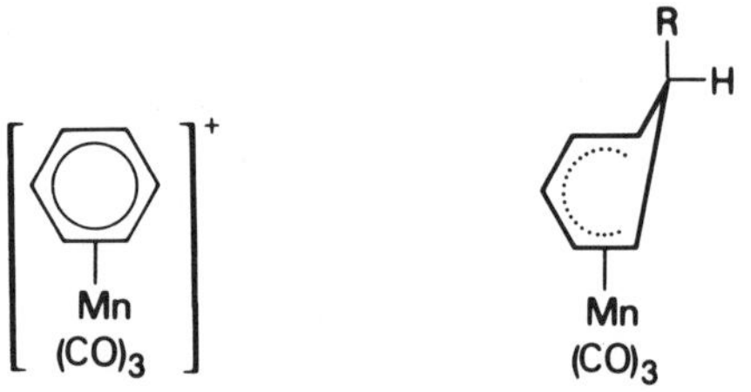

(34) **(35)**

Reduction of tricarbonyl(η^6-hexamethylbenzene)rhenium cation with lithium aluminium hydride produces the η^5-cyclohexadienyl complex **36**[91] and X-ray analysis shows that the entering hydride ion occupies the *exo*-position[92].

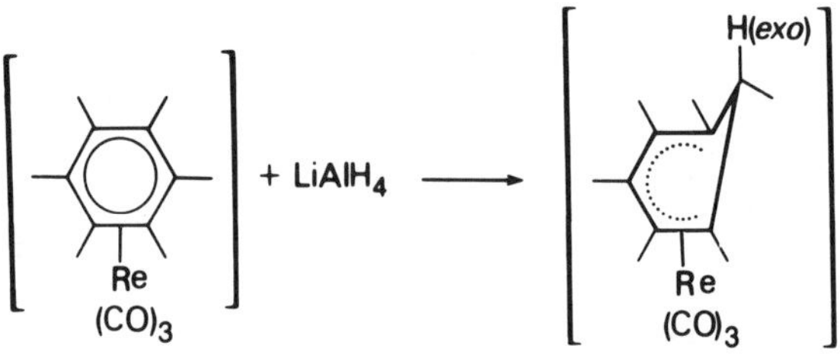

(17)

(36)

The hydride reduction route may be used to prepare η^5-cyclohexadienyl complexes of a variety of metals. For example, the reduction of the (η^6-benzene)(η^7-cycloheptatrienyl)molybdenum and (η^6-benzene)tetracarbonylvanadium cations gives the η^5-cyclohexadienyl complexes **37** and **38**, respectively[93,94]. The treatment of the dication **39** with a series of nucleophiles gives the mono adducts (**40**; N = CN, CH_2NO_2, $CHMeNO_2$, CH_2CO_2Bu-*t*)[95].

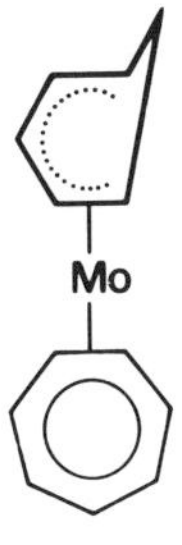

(**37**)  (**38**)

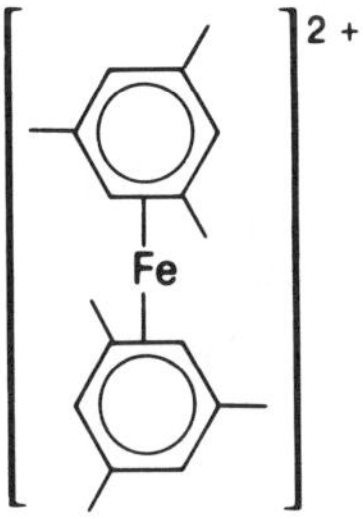

(**39**) 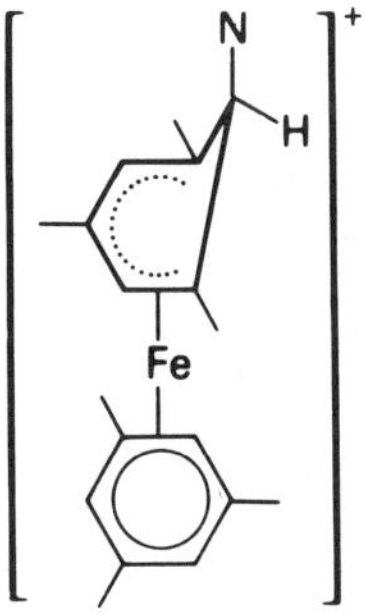 (**40**)

C. Electrochemical Reduction

Electrochemical reduction is rarely used for the synthesis of η-cyclohexadienyl complexes. One recent example is the electro-chemical reduction of (η^6-benzene)- (η^5-cyclopentadienyl)iron cations to give the corresponding (η^5-cyclohexadienyl)iron complexes (**41**; n = 1–5) through dimerization of the intermediate radical[96].

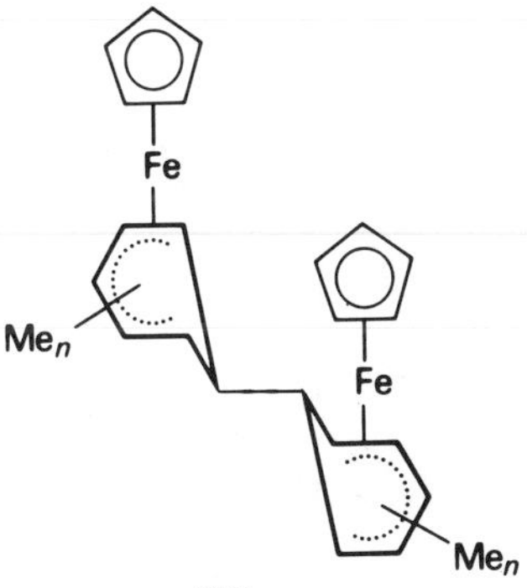

(**41**)

III. η^5-CYCLOHEPTADIENYL COMPLEXES

η^5-Cycloheptadienyl complexes are prepared by similar methods to those already described for η^5-cyclohexadienyl complexes. The tricarbonyl(η^5-cycloheptadienyl)iron cation **(42)** is prepared via protonation with a strong acid or by hydride abstraction with triphenylmethylium tetrafluoroborate in methylene chloride[97,98].

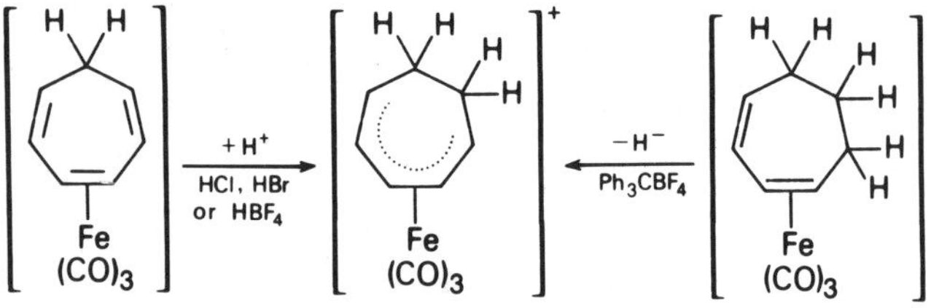
+H+
HCl, HBr
or HBF4
−H−
Ph3CBF4
Fe
(CO)3

(18)

(42)

Recent work with deuterated trifluoroacetic acid shows that deuterium addition occurs exclusively *exo* to the iron atom to give the deuterated η^5-cycloheptadienyl complex **43**[99].

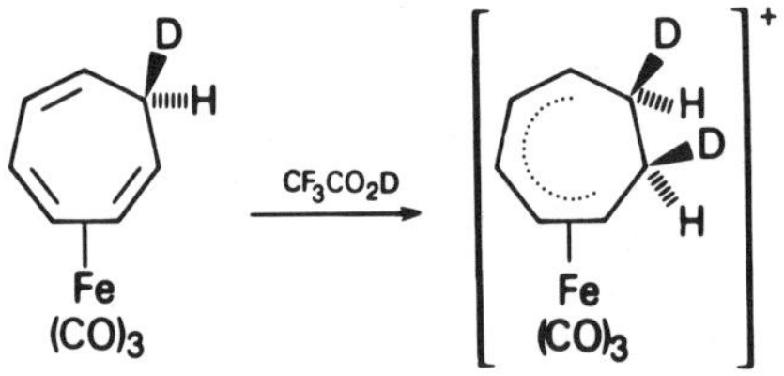
CF3CO2D
Fe
(CO)3

(43)

Tricarbonyl(η^4-tropone)iron **(44)** is protonated on the oxygen atom with trifluoroacetic acid at -78°C to form the blood-red cation **45**. On warming to 0°C the cation isomerizes to give the cycloheptadienyl complex **46**. The 7-methyltropone complex is also protonated to give the corresponding η^5-cycloheptadienyl derivative[100].

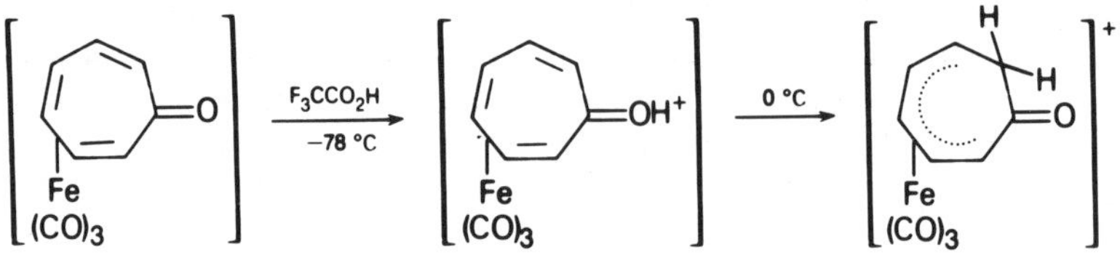
F3CCO2H
−78 °C
0 °C
Fe
(CO)3

(19)

(44) **(45)** **(46)**

(η^4-1,2-Benzocycloheptatriene)tricarbonyliron undergoes hydride abstraction to give the corresponding (η^5-benzotropylium)iron complex **(47)**[101].

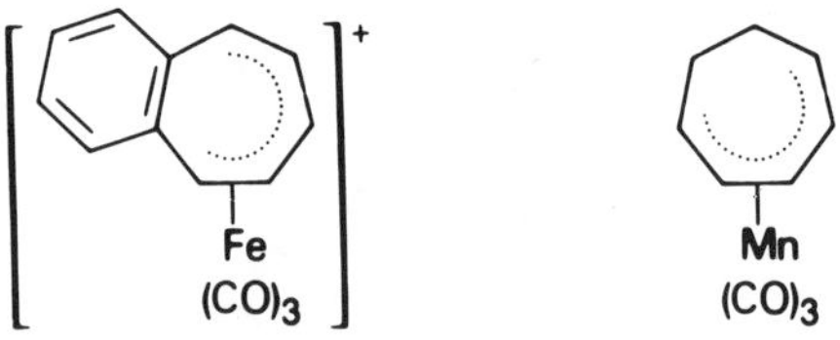
Fe
(CO)3
Mn
(CO)3

(47) **(48)**

The η^5-cycloheptadienyl complex **48** is prepared in 70% yield by direct reaction of cyclohepta-1,3-diene with manganese carbonyl[102].

IV. η^5-CYCLOOCTADIENYL COMPLEXES

η^5-Cyclooctadienyl complexes can be prepared by hydride abstraction from (η^4-cyclooctadiene)transition metal complexes. Tricarbonyl(η^4-cycloocta-1,3-diene)iron and -osmium undergo hydride abstraction with triphenylmethylium tetrafluoroborate to give the corresponding η^5-cyclooctadienyl cations (**49**: M = Fe, Os)[103].

$$[M(CO)_3(\eta^4\text{-}C_8H_{12})] \xrightarrow[Ph_3CBF_4]{-H^-} [M(CO)_3(\eta^5\text{-}C_8H_{11})]^+ \quad (20)$$

(49)

Treatment of chromium(III) chloride with isopropylmagnesium bromide in a mixture of cycloocta-1,3-diene and cycloocta-1,3,5-triene gives the chromium complexes **50** and **51**[104].

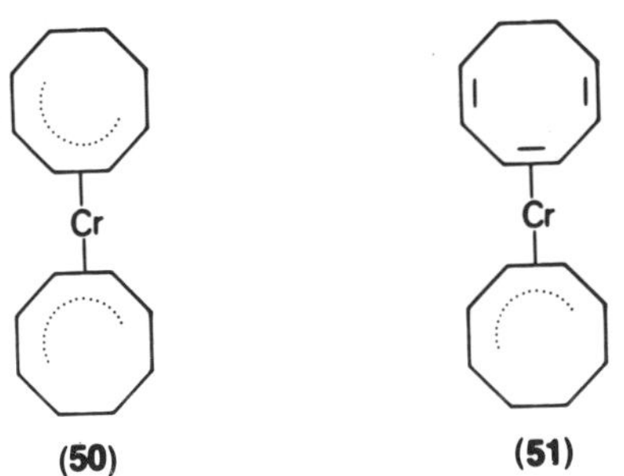

(50) **(51)**

Tricarbonyl(η^4-cyclooctatetraene)iron (**52**) protonates in a mixture of fluorosulphonic acid and fluorosulphonyl chloride at -120°C to form the dienyl cation **53**. When the solution is allowed to warm to -60° the cation rearranges to give the bicyclic derivative **54** by electrocyclic ring closure[105].

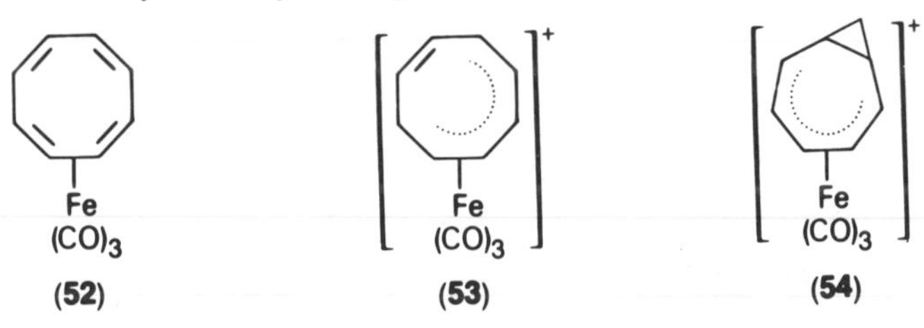

(52) **(53)** **(54)**

The formation of bicyclic carbocations is not unusual in this type of system when acidic conditions are used[106]. Treatment of the cyclooctatetraene complex **52** with acetyl chloride and aluminium chloride produces the bicyclic derivative **55** as the major product together with a low yield of tricarbonyl(η^4-acetylcyclooctatetraene)iron (**56**), whereas formylation gives the expected aldehyde (**57**) in 60% yield[107].

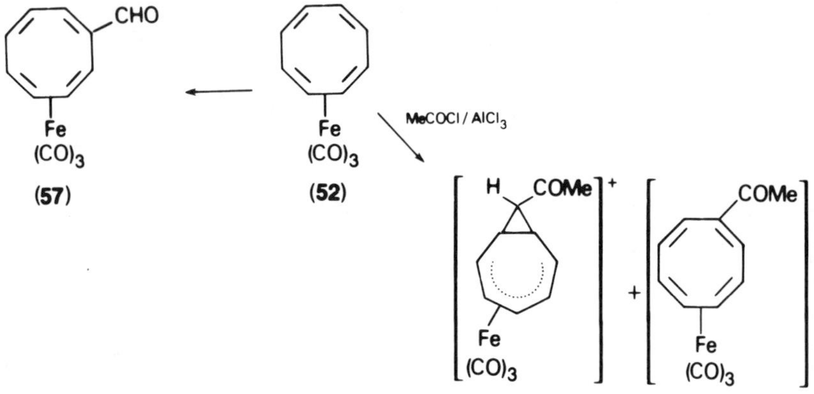

(57) (52) MeCOCl/AlCl$_3$ (55) (56)

V. η^5-PENTADIENYL COMPLEXES

A. Protonation and Dehydration of Alcohols Complexed with Transition Metals

The protonation and dehydration of the η^4-*trans*-pentadienol complexes (**58**; $R^1 = R^2 = H$; $R^1 = H$, $R^2 = Me$; $R^1 = R^2 = Me$) by treatment with a strong acid gives the corresponding η^5-dienyl carbocations (**59**) in good yields[108–110].

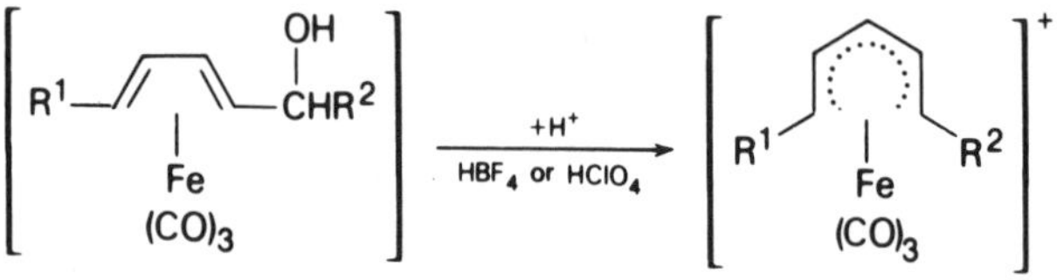

(58) $\xrightarrow[\text{HBF}_4 \text{ or HClO}_4]{+\text{H}^+}$ (59) (22)

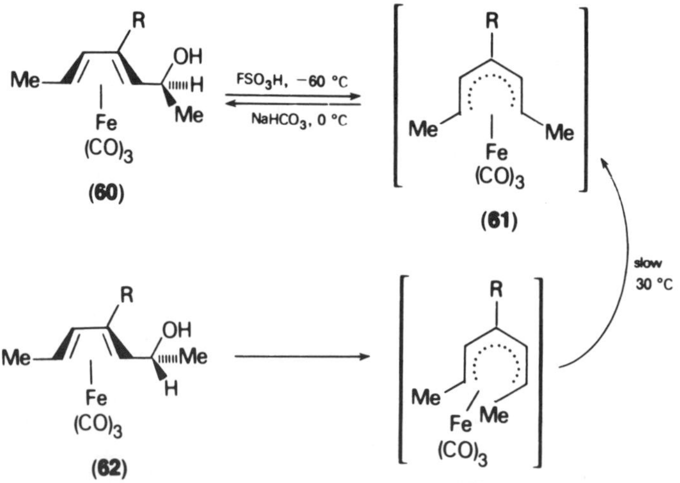

(60) $\underset{\text{NaHCO}_3,\ 0\ ^\circ\text{C}}{\overset{\text{FSO}_3\text{H},\ -60\ ^\circ\text{C}}{\rightleftharpoons}}$ (61)

(62) $\longrightarrow$ (63) $\xrightarrow{\text{slow, } 30\ ^\circ\text{C}}$ (61)

SCHEME 2

Recently the stereospecific generation and quenching of acyclic tricarbonyl(η^5-dienyl)iron cations has been investigated[111] (Scheme 2). The (η^4-ψ-*exo-trans*-dienol)iron tricarbonyl complexes (**60**; R = H, Me) give the (η^5-*syn,syn*-dienyl)iron cations (**61**; R = H, Me) with high stereospecificity in fluorosulphonic acid at −60°C, whereas the (η^4-ψ-*endo-trans*-dienol)iron complexes (**62**; R = H, Me) produce the (η^5-*syn,anti*-dienyl)iron cations (**63**; R = H, Me). The cation **63** (R = Me) is configurationally stable for several hours at room temperature. On standing at 30°C it slowly converts to the (η^5-*syn,syn*-dienyl)iron cation (**61**; R = Me).

A related reaction is the protonation of the *anti*-isomer of tricarbonyl(η^4-hexa-1,3,5-triene)iron (**64**) with fluoroboric acid which produces the tricarbonyliron complex **65** in 74% yield[112].

Fe (CO)3 —(+H⁺)→ [Me, Fe (CO)3]⁺ (23)

(**64**) (**65**)

B. Hydride Abstraction from an (η^4-Penta-1,3-diene)transition Metal Complex

Acyclic η^5-pentadienyl complexes re prepared by hydride abstraction from a neutral η^4-*cis*-penta-1,3-diene transition metal complex. Treatment of tricarbonyl-(η^4-*cis*-penta-1,3-diene)iron (**66**) with triphenylmethylium tetrafluoroborate in nitromethane produces the η^5-pentadienyl complex **67** in good yield[113].

[Me, Fe (CO)3] —($-H^-$, Ph_3CBF_4)→ [Fe (CO)3]⁺ (24)

(**66**) (**67**)

VI. REFERENCES

1. J. M. Birmingham, in *Advances in Organometallic Chemistry* (Ed. F. G. A. Stone and R. West), Vol. 2, Academic Press, New York, 1964, p. 365; M. Rosenblum, *Chemistry of the Iron Group Metallocenes, Part 1,* Wiley – Interscience, New York, 1965; G. R. Knox and W. E. Watts, in *International Review of Science, Inorganic Chemistry, Series 2* (Ed. H. J. Emeleus and M. J. Mays), Vol. 6, Butterworths, London, 1975, p. 219.
2. A. Streitwieser, *Tetrahedron Lett.*, 23 (1960).
3. M. D. Rausch, E. O. Fischer, and H. Grubert, *Chem. Ind. (London)* 756 (1958).
4. M. D. Rausch, E. O. Fischer, and H. Grubert, *J. Am. Chem. Soc.*, **82**, 76 (1960).
5. G. Wilkinson, *Org. Synth.*, **36**, 31 (1956).
6. G. Wilkinson, F. A. Cotton, and J. M. Birmingham, *J. Inorg. Nucl. Chem.*, **2**, 95 (1956); F. A. Cotton and G. Wilkinson, *Z. Naturforsch.*, **96**, 417 (1954); G. Wilkinson and J. M. Birmingham, *J. Am. Chem. Soc.*, **76**, 4281 (1954).
7. W. F. Little, R. C. Koestler, and R. Eisenthal, *J. Org. Chem.*, **25**, 1435 (1960).
8. A. N. Nesmeyanov, R. B. Materikova, and N. S. Kochetkova, *Izv. Akad. Nauk SSSR, Ser. Khim.*, 1334 (1963).

9. L. Kaplan, W. L. Kester, and J. J. Katz, *J. Am. Chem. Soc.*, **74**, 5531 (1952).
10. E. B. Sokolova, M. P. Shebanova, and V. A. Zhichkina, *Zh. Obshch. Khim.*, **30**, 2040 (1960).
11. E. O. Fischer and H. Grubert, *Chem. Ber.*, **92**, 2302 (1959).
12. D. E. Bublitz, W. E. McEwen, and J. Kleinberg, *Org. Synth.*, **41**, 96 (1961).
13. E. O. Fischer, M. D. Rausch, and H. Grubert, *Chem. Ind. (London),* 765 (1958).
14. J. F. Cordes, *Chem. Ber.*, **95**, 3084 (1962).
15. A. N. Nesmeyanov, R. B. Materikova, I. R. Lyatifov, T. Kh. Kurbanov, E. V. Leonova, and N. S. Kochetkova, *J. Organomet. Chem.*, **136**, C55 (1977).
16. T. S. Piper, F. A. Cotton, and G. Wilkinson, *J. Inorg. Nucl. Chem.*, **1**, 165 (1955).
17. R. L. Pruett and E. L. Morehouse, *Chem. Ind. (London)*, 980 (1958).
18. C. Palm, E. O. Fischer, and F. Baumgartner, *Naturwissenschaften*, **49**, 279 (1962).
19. R. B. King and F. G. A. Stone, *Inorg. Synth.*, **7**, 100 (1963).
20. R. B. King, *Organomet. Synth.*, **1**, 105 (1965).
21. R. B. King, *Adv. Organomet. Chem.*, **2**, 157 (1964).
22. R. B. King, *Inorg. Nucl. Chem. Lett.*, **5**, 905 (1969).
23. E. O. Fischer, H. Schuster-Woldan, and K. Bittler, *Z. Naturforsch.*, **18B**, 429 (1963).
24. E. O. Fischer and K. Bittler, *Z. Naturforsch.*, **16B**, 225 (1961).
25. E. O. Fischer and K. S. Brenner, *Z. Naturforsch.*, **17B**, 774 (1962).
26. E. O. Fischer and A. Vogler, *Z. Naturforsch.*, **17B**, 421 (1962).
27. J. Kozikowski, *US Pat.*, 3015668, Jan. 2 (1962); *Chem. Abstr.*, **57**, 866 (1962).
28. H. Meister, *Angew. Chem.*, **69**, 533 (1957).
29. F. A. Cotton and L. T. Reynolds, *J. Am. Chem. Soc.*, **80**, 269 (1958).
30. C. C. Hunt and J. R. Doyle, *Inorg. Nucl. Chem. Lett.*, **2**, 283 (1966).
31. G. M. Kuzyants, *Izv. Akad. Nauk SSSR, Ser. Khim.*, 1895 (1976).
32. L. E. Manzer, *J. Organomet. Chem.*, **110**, 291 (1976).
33. C. C. Hunt and J. R. Doyle, *Inorg. Nucl. Chem. Lett.*, **2**, 283 (1966).
34. T. Kobayashi, K. Funahashi, and I. Hirose (Teijin Ltd.), *Jap. Pat.*, 6911887, May 30 (1969); *Chem. Abstr.*, **71**, 112416 (1969).
35. A. N. Nesmeyanov, R. B. Materikova, and N. S. Kochetkova, *Izv. Akad. Nauk SSSR*, 1334 (1963).
36. A. N. Nesmeyanov, K. N. Anisimov, and N. E. Kolobova, *Izv. Akad. Nauk SSSR*, 2220 (1964).
37. R. B. King, *Inorg. Chem.*, **7**, 90 (1968).
38. R. B. King, R. N. Kapoor, and K. H. Pannell, *J. Organomet. Chem.*, **20**, 187 (1969).
39. J. F. Nixon and M. Kooti, *J. Organomet. Chem.*, **104**, 231 (1976).
40. A. N. Nesmeyanov, R. B. Materikova, N. S. Kochetkova, and R. V. Luk'yanova, *Izv. Akad. Nauk SSSR, Ser. Khim.*, 975 (1969).
41. M. B. Freeman, L. G. Sneddon, and J. C. Huffman, *J. Am. Chem. Soc.*, **99**, 5194 (1977).
42. G. K. Anderson, R. J. Cross, and I. G. Phillips, *J. Chem. Soc., Chem. Commun.*, 709 (1978).
43. T. J. Katz and J, J, Mrowca, *J. Am. Chem. Soc.*, **89**, 1105 (1967).
44. A. N. Nesmeyanov, R. B. Materikova, N. S. Kochetkova, and E. V. Leonova, *Dokl. Akad. Nauk SSSR,* **177**, 131 (1967).
45. G. Wilkinson, *Org. Synth.*, **36**, 31 (1956); *Org. Synth.*, Collect. Vol. IV, 476 (1963).
46. A. I. Titov, E. S. Lisitsyna, and M. A. Shemtova, *Dokl. Akad. Nauk SSSR,* **130**, 341 (1960).
47. R. G. Amiet and R. Pettit, *J. Am. Chem. Soc.*, **90**, 1059 (1968).
48. W. L. Jolly, *Inorg. Synth.*, **11**, 120 (1968).
49. M. Salisova and H. Alper, *Angew. Chem., Int. Ed. Engl.*, **18**, 792 (1979).
50. M. J. McGlinchey and P. S. Skell, *Cryochemistry,* 167 (1976).
51. P. L. Timms and T. W. Turney, *Adv. Organomet. Chem.*, **15**, 53 (1977); P. S. Skell and M. J. McGlinchey, *Angew. Chem., Int. Ed. Engl.*, **14**, 195 (1975).
52. P. L. Timms, *J. Chem. Soc., Chem. Commun.*, 1033 (1969).
53. J. L. Lang, *US Pat.*, 3880743, Apr. 29 (1975); *Chem. Abstr.*, **83**, 97557 (1975).
54. A. Z. Rubezhov, A. S. Ivanov, and A. A. Bezrukova, *Izv. Akad. Nauk SSSR, Ser. Khim.*, 1606 (1979).
55. B. F. Hallam and P. L. Pauson, *J. Chem. Soc.*, 3030 (1956); H. W. Sternberg and I. Wender, *Chem. Soc. Spec. Publ.*, **13**, 35 (1959).

56. G. Wilkinson, P. L. Pauson, and F. A. Cotton, *J. Am. Chem. Soc.*, **76**, 1970 (1954); W. F. Anzilotti and V. Weinmayr, *US Pat.*, 2791597, May 7 (1957); *Chem. Abstr.*, **51**, 15560 (1957).
57. C. Charrier and F. Mathey, *Tetrahedron Lett.*, 2407 (1978).
58. R. B. King, M. Z. Iqbal, and A. D. King, *J. Organomet. Chem.*, **171**, 53 (1979).
59. R. B. King and A. Fronzaglia, *J. Chem. Soc., Chem. Commun.*, 547 (1965); S. A. Keppie and M. F. Lappert, *J. Chem. Soc., A.* 3216 (1971).
60. R. C. Kerber and D. J. Ehntholt, *Synthesis*, **9**, 449 (1970).
61. E. W. Abel, A. Singh, and G. Wilkinson, *J. Chem. Soc.*, 1321 (1960).
62. K. Hoffmann and E. Weiss, *J. Organomet. Chem.*, **131**, 273 (1977).
63. T.-S. Tan, J. L. Fletcher, and M. J. McGlinchey, *J. Chem. Soc., Chem. Commun.*, 771 (1975).
64. W. A. Herrmann and M. Huber, *J. Organomet. Chem.*, **140**, 55 (1977).
65. K. J. Reimer, *Diss. Abstr. Int. B*, **36**, 3948 (1976); K. J. Reimer and A. Shaver, *Inorg. Chem.*, **14**, 2707 (1975).
66. W. A. Herrmann, B. Reiter, and M. Huber, *J. Organomet. Chem.*, 139 (1977) C4.
67. P. M. Maitlis, A. Efraty, and M. L. Games. *J. Am. Chem. Soc.*, **87**, 719 (1965); P. M. Maitlis, A. Efraty, and M. L. Games, *J. Organomet. Chem.*, **2**, 284 (1964).
68. P. M. Maitlis and D. F. Pollock, *J. Organomet. Chem.*, **26**, 407 (1971).
69. G. J. Gauthier, *J. Chem. Soc., Chem. Commun.*, 690 (1969).
70. M. Wenzel and D. Langheim (Schering A.-G.), *Ger. Pat.*, 2246460, Apr. 4 (1974); *Chem. Abstr.*, **81**, 13651 (1974); D. Langheim, M. Wenzel, and E. Nipper, *Chem. Ber.*, **108**, 146 (1975); E. A. Stadbauer, E. Nipper, and M. Wenzel, *J. Labelled Compd. Radiopharm.*, **13**, 491 (1977); M. Schneider, M. Wenzel, and B. Reisselman, *J. Labelled Compd. Radiopharm.*, **15**, 295 (1978).
71. E. A. Charryshev, A. S. Dmitriev, A. Ya. Yurin, G. M. Grandel, and E. P. Nikitenko, USSR Pat., 431 168, Jun. 5 (1974); *Chem. Abstr.*, **81**, 63779 (1974).
72. F. L. Bowden and A. B. P. Lever, *Organomet. Chem. Rev. A*, **3**, 227 (1968).
73. T. H. Coffield, K. G. Ihrman, and W. Burns, *J. Am. Chem. Soc.*, **82**, 4209 (1960).
74. J. P. Williams and A. Wojcicki, *Inorg. Chim. Acta*, **15**, L19 (1975).
75. H. Rohl, E. Lang, T. Goessel, and G. Roth, *Angew. Chem., Int. Ed. Engl.*, **1**, 117 (1962).
76. K. K. Joshi, R. H. B. Mais, F. Nyman, P. G. Owston, and A. M. Wood, *J. Chem. Soc., A*, 318 (1968).
77. M. Fleischmann and D. Fletcher, *Chem. Brit.*, **11**, 50 (1975).
78. A. P. Giraitis, T. H. Pearson, and R. C. Pinkerton, *US Pat.*, 2960450, Nov. 15 (1960); *Chem. Abstr.*, **55**, 8430 (1960).
79. T. H. Pearson, *US Pat.*, 2915440, Dec. 1 (1959); *Chem. Abstr.*, **54**, 17415 (1959).
80. S. Valcher, *Corsi Semin. Chem.*, 37 (1968); *Chem. Abstr.*, **72**, 17834 (1970).
81. H. Lehmkul and W. Eisenbach (Studiengesellschafte Kohle), *Ger. Pat.*, 2720165, May 18 (1978); *Chem. Abstr.*, **89**, 50600 (1978).
82. H. Lehmkuhl, W. Leuchte, and E. Janssen, *J. Organomet. Chem.*, **30**, 407 (1971).
83. E. O. Fischer and R. D. Fischer, *Angew. Chem.*, **72**, 919 (1960).
84. D. A. White, *Organomet. Chem. Rev. A*, **3**, 497 (1968).
85. A. J. Pearson, *J. Chem. Soc., Perkin Trans. 1*, 495 (1978).
86. N. S. Nametkin, A. I. Nakhaev, V. D. Tyurin, and S. P. Gubin, *Izv. Akad. Nauk SSSR, Ser. Khim.*, 676 (1975).
87. D. Astruc, E. Roman, J. R. Hamon, and P. Batail, *J. Am. Chem. Soc.*, **101**, 2240 (1979).
88. G. Winkhaus, L. Pratt, and G. Wilkinson, *J. Chem. Soc.*, 3807 (1961); G. Winkhaus, *Z. Anorg. Allg. Chem.*, **319**, 404 (1963).
89. D. Jones, L. Pratt, and G. Wilkinson, *J. Chem. Soc.*, 4458 (1962); D. Jones and G. Wilkinson, *J. Chem. Soc.*, 2479 (1964).
90. M. R. Churchill and R. Mason, *Proc. Chem. Soc.*, 112 (1963).
91. G. Winkhaus and H. Singer, *Z. Naturforsch.*, **186**, 418 (1963).
92. P. H. Bird and M. R. Churchill, *J. Chem. Soc., Chem. Commun.*, 777 (1967).
93. E. F. Ashworth, M. L. H. Green, and J. Knight, *J. Less-Common Met.*, **36**, 213 (1974).
94. F. Calderazzo, *Inorg. Chem.*, **5**, 429 (1966).
95. J. F. Helling and G. G. Cash, *J. Organomet. Chem.*, **73**, C10 (1974).
96. C. Moinet, E. Roman, and D. Astruc, *J. Organomet. Chem.*, **128**, C45 (1977).

97. H. J. Dauben and D. J. Bertelli, *J. Am. Chem. Soc.*, **83**, 497 (1961).
98. R. Burton, L. Pratt, and G. Wilkinson, *J. Chem. Soc.*, 594 (1961).
99. M. Brookhart, K. J. Karel, and L. E. Nance, *J. Organomet. Chem.*, **140**, 203 (1977).
100. M. Brookhart, C. P. Lewis, and A. Eisenstadt, *J. Organomet. Chem.*, **127**, C14 (1977); C. P. Lewis, *Diss. Abstr. Int. B,* **38**, 692 (1977).
101. D. J. Bertelli and J. M. Viebrock, *Inorg. Chem.*, **7**, 487 (1968).
102. F. Haque, J. Miller, P. L. Pauson, and J. B. Pd. Tripathi, *J. Chem. Soc., C,* 743 (1971).
103. F. A. Cotton, A. J. Deeming, P. L. Josty, S. S. Ullah, A. J. P. Domingos, B. F. G. Johnson, and J. Lewis, *J. Am. Chem. Soc.*, **93**, 4624 (1971).
104. J. Mueller, W. Holzinger, and F. H. Koehler, *Chem. Ber.*, **109**, 1222 (1976).
105. G. A. Olah, G. Liang, and Yu. Simon, *J. Org. Chem.*, **42**, 4262 (1977).
106. A. Davison, W. McFarlane, L. Pratt, and G. Wilkinson, *J. Chem. Soc.*, 4821 (1962).
107. B. F. G. Johnson, J. Lewis, and G. L. P. Randall, *J. Chem. Soc., A,* 422 (1971).
108. J. E. Mahler and R. Pettit, *J. Am. Chem. Soc.*, **84**, 1511 (1962); J. E. Mahler and R. Pettit, *J. Am. Chem. Soc.*, **85**, 3955 (1963).
109. E. O. Greaves, G. R. Knox, and P. L. Pauson, *J. Chem. Soc., Chem. Commun.*, 1124 (1969).
110. R. S. Bayoud, E. R. Biehl, and P. C. Reeves, *J. Organomet. Chem.*, **150**, 75 (1978).
111. D. G. Gresham, D. J. Kowalski, and C. P. Lillya, *J. Organomet. Chem.*, **144**, 71 (1978).
112. B. F. G. Johnson, J. Lewis, D. G. Parker, and S. R. Postle, *J. Chem. Soc., Dalton Trans.*, 794 (1977).
113. J. E. Mahler and R. Pettit, *J. Am. Chem. Soc.*, **85**, 3955 (1963).

The Chemistry of the Metal–Carbon Bond
Edited by F. R. Hartley and S. Patai

CHAPTER **11**

Synthesis of complexes of η^6-, η^7-, and η^8-bonded ligands

G. MARR and B. W. ROCKETT
Department of Physical Sciences, The Polytechnic, Wolverhampton WV1 1LY, UK

I. η^6-ARENE COMPLEXES

A. Introduction

Benzene and its derivatives are the most important η^6-arene ligands and when these are attached to the Group VIA elements (chromium, molybdenum, and tungsten) they form the most widely investigated section of (η^6-ligand)transition metal chemistry. Some heterocyclic molecules and cyclic systems containing more than six carbon atoms also behave as six-electron donors but these systems have attracted less attention. The chemistry of all these molecules has been reviewed extensively[1-3], and an up-to-date account is contained in the 'Annual Surveys' in the *Journal of Organometallic Chemistry*[4].

B. Reduction of a Metal Salt in the Presence of the Ligand

Fischer and Hafner first used this method for the synthesis of bis(η^6-benzene)chromium (**2**; M = Cr)[5]. A metal salt is reduced with aluminium powder in the presence of the arene ligand and aluminium chloride. The yields from this reaction are high if a large excess of anhydrous aluminium chloride is used. The cation **1** can be reduced with aqueous sodium dithionate, or it will undergo disporportionation in aqueous alkaline solution[6,7] to give the neutral η^6-benzene complex. This route has been used to prepare bis(η^6-cumene)chromium[8] and bis(η^6-ethylbenzene)chromium (**3**, M = Cr). In the latter preparation any impurities present are determined by mass spectral and gas chromatographic methods[9,10].

$$3MCl_3 + 2Al + AlCl_3 + 6C_6H_6 \longrightarrow 3[(\eta^6\text{-}C_6H_6)_2M]^+ AlCl_4^- \; (\mathbf{1}) \xrightarrow{Na_2S_2O_4/OH^-} [(\eta^6\text{-}C_6H_6)_2M] \; (\mathbf{2}) \qquad (1)$$

In some reactions the reduction of the metal salt is not necessary and aluminium powder is not added (Equation 2; M = Fe, Ni)[11,12]. This general method is used to prepare the η^6-arene complexes of many transition metals (equations 3[7,13,14], 4, and 5[15,16]). The reaction of molybdenum(V) chloride with ethylbenzene in the presence of aluminium chloride gives bis(η^6-ethylbenzene)molybdenum (**3**; M = Mo)[17].

$$MBr_2 + 2\text{arene} + 2AlBr_3 \longrightarrow [(\eta^6\text{-arene})_2M][AlBr_4]_2 \qquad (2)$$

$$VCl_4 + 2C_6H_6 + Al \xrightarrow{AlCl_3} [(\eta^6\text{-}C_6H_6)_2V][AlCl_4] \qquad (3)$$

$$3RuCl_3 + 6C_6H_6 + Al + 5AlCl_3 \longrightarrow 3[(\eta^6\text{-}C_6H_6)_2Ru][AlCl_4]_2 \qquad (4)$$

$$3MoCl_5 + 6C_6H_6 + 4Al \xrightarrow{AlCl_3} 2[\eta^6\text{-}C_6H_6)_2Mo][AlCl_4] + AlCl_3 \qquad (5)$$

The treatment of phenylferrocene with chromium(III) chloride in the presence of aluminium chloride and aluminium gives the mixed ferrocene—chromium complex **4**[18].

Et
M
Et

Fe Cr Fe

(3) **(4)**

C. The Co-condensation of Metal Atoms with Aromatic Hydrocarbons

One of the first reactions reported that used the metal atom technique was the synthesis of bis(η^6-benzene)chromium in 60% yield[19]. Details of the many reactions of metal atoms with aromatic hydrocarbons are included in Chapter 13 as well as in several reviews[20]. This synthetic route has proved to be particularly valuable and has provided routes to many new compounds that were inaccessible by conventional reductive procedures. The only route to bis(η^6-benzene)titanium reported is the cocondensation of titanium atoms with benzene. Bis(η^6-mesitylene)- and bis(η^6-toluene)titanium are prepared similarly in approximately 30% yield[21,22]. Bis(η^6-benzene)vanadium is formed in low yield from vanadium vapour and benzene[23].

The metal atom route has found wide application in the preparation of (η^6-arene)chromium complexes. One of the disadvantages of the reductive procedure is that in the presence of aluminium chloride isomerization of the aromatic system often occurs and a mixture of products is obtained. In the presence of chromium vapour no isomerization occurs, although the yield of bis(η^6-alkylbenzene)chromium compounds rarely exceeds 20%. For example, bis(η^6-*o*-diisopropylbenzene)chromium (**5**) is formed in 18% yield by the metal atom route[24]; some other examples are given in Table 1.

Conventional methods of preparing bis(η^6-arene)chromium complexes fail when the arene possesses a strongly electron-withdrawing substituent, but with the co-condensation method chlorobenzenes and fluorobenzenes may be used to prepare

TABLE 1. The formation of some bis(η^6-alkylbenzene)chromium complexes by co-condensation of chromium vapour with the ligand

Bis(η^6-arene)chromium complex	Yield (%)	Reference
$[(\eta^6\text{-}o\text{-}Me_2C_6H_4)_2Cr]$	25	25
$[(\eta^6\text{-}m\text{-}Me_2C_6H_4)_2Cr]$	25	25
$[(\eta^6\text{-}p\text{-}Me_2C_6H_4)_2Cr]$	29	25
$[(\eta^6\text{-}o\text{-}MeFC_6H_4)_2Cr]$	16	25, 26
$[(\eta^6\text{-}m\text{-}MeFC_6H_4)_2Cr]$	16	25, 26
$[(\eta^6\text{-}p\text{-}MeFC_6H_4)_2Cr]$	21	25, 26
$[(\eta^6\text{-}1,2,3\text{-}Me_3C_6H_3)_2Cr]$	5	25
$[(\eta^6\text{-}1,3,5\text{-}Me_3C_6H_3)_2Cr]$	13	25
$[(\eta^6\text{-}1,2,4\text{-}Me_3C_6H_3)_2Cr]$	5	25

bis(η^6-arene)chromium complexes[27]. For example, the co-condensation of chromium vapour with a mixture of pentafluorobenzene and benzene produces the air-stable chromium complex (**6**; X = H)[28,29]. A similar reaction with hexafluorobenzene and benzene gives the η^6-hexafluorobenzene complex **6** (X = F), but bis(η^6-hexafluorobenzene)chromium is not formed[30]. When chromium atoms and 1,3-bis(trifluoromethyl)benzene vapour are co-condensed on to a cold finger the [η^6-1,3-bis(trifluoromethyl)benzene] complex (**7**) is formed in 44% yield[31]. This latter preparation is included in a chapter on metal atom preparation in *Inorganic Syntheses*, where the apparatus required and the synthetic techniques are discussed[32].

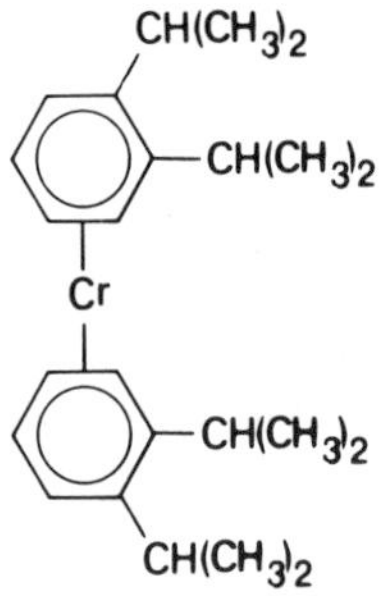

(**5**) 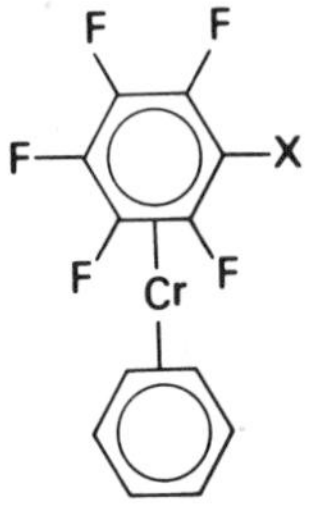 (**6**) 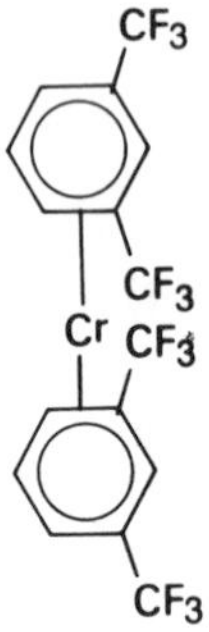 (**7**)

The metal atom technique is well suited to preparing compounds that are sensitive to oxygen and to different solvents. For example, the η^6-toluene complexes **8** (M = Co, Ni) are prepared by the triple deposition of toluene, bromopentafluorobenzene, and the metal vapour. Both of these compounds (**8**; M = Co, Ni) are air sensitive and the product is handled under nitrogen[33,34].

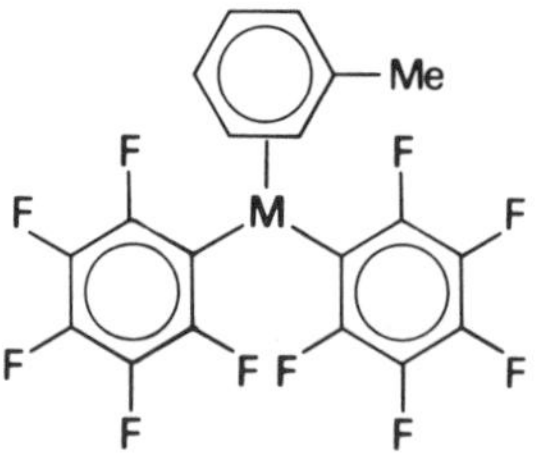

(**8**)

D. Treatment of a Metal Carbonyl with an Aromatic Hydrocarbon

The direct reaction of an aromatic hydrocarbon with a metal carbonyl often gives an (η^6-arene)transition metal complex[35]. For example, the reaction of hexacarbonylchromium with methoxybenzene, in dibutyl ether and tetrahydrofuran, gives tricarbonyl(η^6-methoxybenzene)chromium (**9**) as a yellow crystalline solid in 92% yield[36]. This type of reaction is usually carried out in an electron donor solvent (L) as its assists in the removal of the carbonyl ligand via intermediates of the type $[Cr(CO)_{6-n}L_n]$ (n = 1–3). Tetrahydrofuran is a good donor solvent but in this solvent alone the reaction is slow owing to the low boiling point (67°C). Dibutyl ether is a

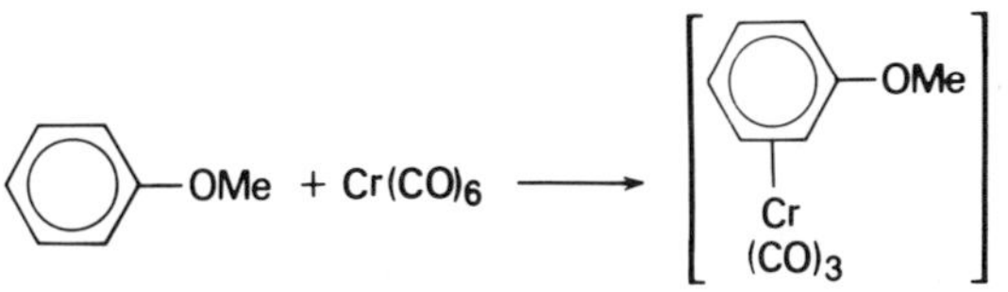

OMe + Cr(CO)6 ⟶ [OMe / Cr (CO)3] (6)

(9)

poorer electron donor but it has a higher boiling point (142°C) and a combination of dibutyl ether and tetrahydrofuran enables the reaction to proceed relatively quickly. Metal carbonyls are volatile and they tend to volatilize out of the reaction medium but in this solvent mixture the tetrahydrofuran washes back most of the chromium carbonyl. The reaction is also carried out in bis(2-methoxyethyl)ether (diglyme), but this solvent is difficult to remove at the end of the reaction because it is a good ligand and it has a high boiling point (162°C). The procedure described for tricarbonyl (η^6-methoxybenzene)chromium is also used to prepare a series of (η^6-arene)(tricarbonyl)chromium complexes (where arene = benzene, fluorobenzene, chlorobenzene, *NN*-dimethylaminobenzene, and methylbenzoate) in high yields[36]. Recently, Top and Jaouen discussed the scope, limitations, and advantages of this method and concluded that the use of a dibutyl ether–tetrahydrofuran mixture promotes the reaction between an arene and hexacarbonyl chromium. In this paper the preparation of a series of (η^6-polyaromatic)tricarbonylchromium complexes is described; for example, the reaction of biphenyl with hexacarbonylchromium gives a mixture of the tricarbonylchromium complexes **10** and **11** in 51% and 47% yields, respectively[37].

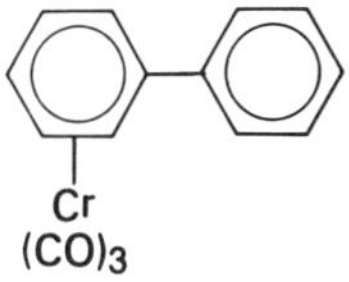

Cr (CO)3   Cr (CO)3   Cr (CO)3

(10) 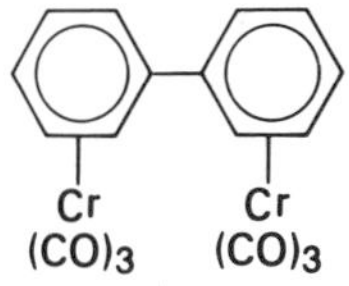   (11)

Derivatives of hexacarbonylchromium for example tricarbonyltris(pyridine)-chromium, combine with aromatic hydrocarbons to give the corresponding (η^6-arene)-tricarbonylchromium complexes[38].

The tricarbonylchromium complexes **12** ($R^1 = R^4 = R^5 = H$, $R^2 = R^3 = Me$; $R^1 = R^2 = R^3 = R^4 = Me$, $R^5 = H$; $R^1R^5 = CH_2CH_2$, $R^2 = R^3 = R^4 = H$) are prepared by the reaction of hexacarbonylchromium with the corresponding cyclophane[39]. The chromium complexes [**13**; $X = CO_2H$, $n = 0, 1, 2$; $X = Cl, Br$, $n = 2$; **14**; $R^1 = CH_2CH_2Br$, $R^2 = H$; $R^1 = Me$, $R^2 = (CH_2)_3CO_2H$, $(CH_2)_3OH$,

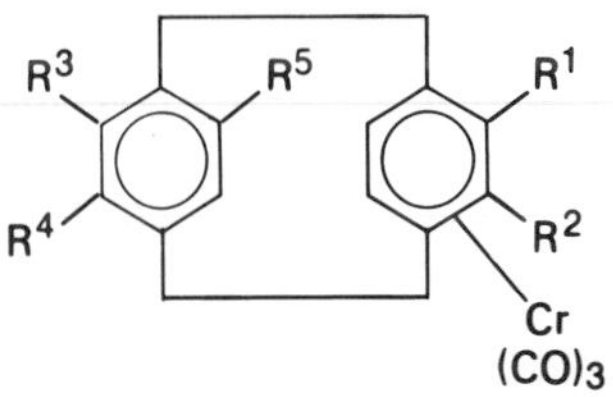

R3 R5 R1 R4 R2 Cr (CO)3   (CH2)nX Cr (CO)3   R1 R1 R2 R1 Cr (CO)3

(12) 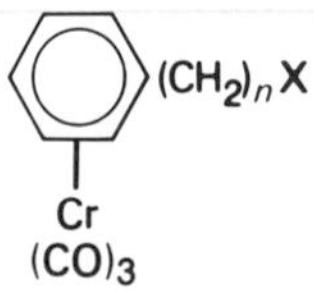   (13) 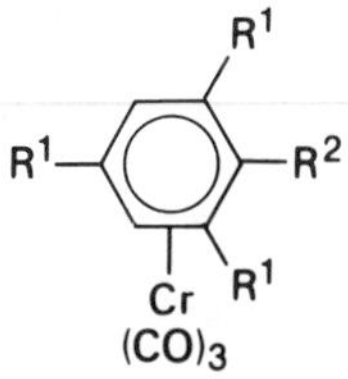   (14)

$(CH_2)_4OH$, $(CH_2)_3Br$] are formed by treatment of the appropriate ligand with triamminetricarbonylchromium or tricarbonyltris(pyridine)chromium[40]. The reactions of benzo-crown ethers with transition metal carbonyls have been investigated. Hexacarbonylchromium combines with benzo-18-crown-6 to give the tricarbonylchromium complex (**15**). Similar complexes are isolated also with benzo-15-crown-5 and methylbenzo-15-crown-5[41]. The irradiation of hexacarbonylchromium with dibenzo-18-crown-6 gives the mono- and bis(tricarbonylchromium) compounds **16** and **17**. The chromium complexes **16** and **17** exhibit a decreased ability to extract alkali metal salts into organic solvents and this is attributed to electron withdrawal from the oxygen crown by the $Cr(CO)_3$ substituents[42].

Reaction of methyl 2-amino-2-(*o*-tolyl)benzoate with hexacarbonylchromium produces the isomeric tricarbonylchromium complexes **18** and **19**. N.m.r. spectroscopy shows that both of these complexes exist as two torsional isomers. The activation energy for the interconversion of the isomers of the major product (**18**) is approximately 91 kJ/mol[43].

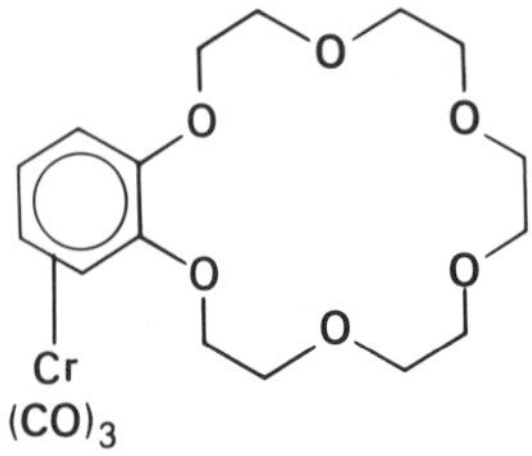

(15) 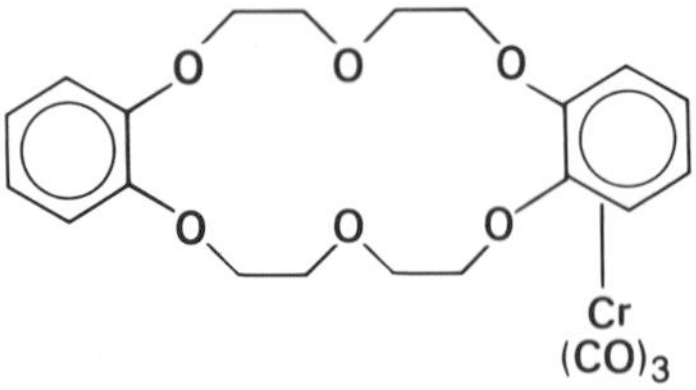 (16)

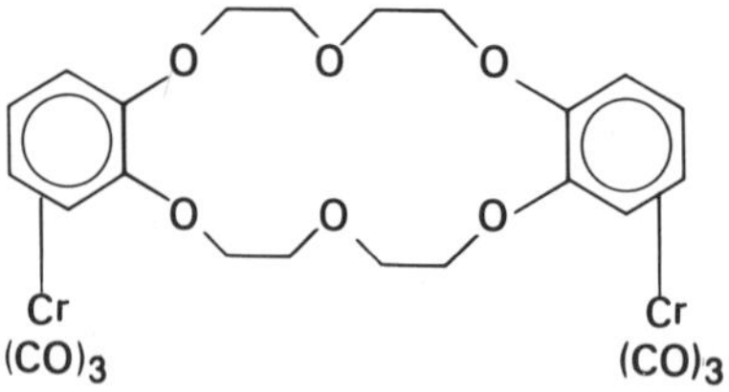

(17) 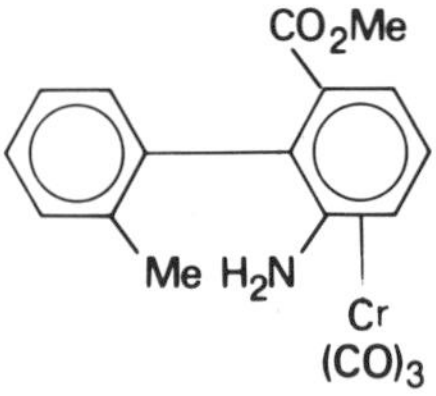 (18) 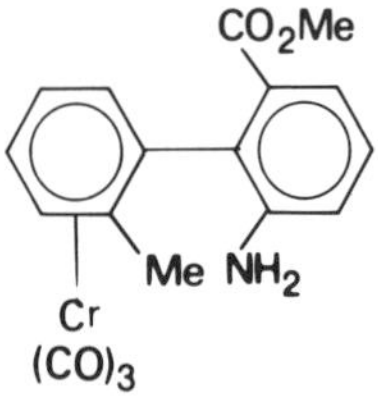 (19)

Other metal carbonyls besides hexacarbonylchromium will combine directly with certain aromatic hydrocarbons to form (η^6-arene)transition metal complexes but the chemistry of these compounds is restricted in comparison with that of chromium. Hexacarbonylmolybdenum and -tungsten behave in a similar manner to hexacarbonylchromium. However, tris(methyl cyanide)tricarbonyltungsten, which is easily prepared in high yield by heating hexacarbonyltungsten with methyl cyanide[44], is often used instead of hexacarbonyltungsten as it enables milder reaction conditions to be used with less decomposition[45]. For example, the reaction of tris(methyl cyanide)tricarbonyltungsten, -chromium, and -molybdenum with 1-methyl-3,5-diphenylthiabenzene-1-oxide produces the corresponding tricarbonyl—metal complexes (**20**; M = W, Cr, Mo)[46]. A series of (η^6-arene)tricarbonyl molybdenum complexes, which cannot be prepared by the direct reaction of hexacarbonylmolybdenum with the arene, are prepared by the reaction of tricarbonyltris(pyridine)molybdenum with the ligand in the presence of boron trifluoride etherate. This method of preparation

(20)

$$[(C_5H_5N)_3Mo(CO)_3] + \text{arene} + 3BF_3 \xrightarrow[20^\circ C]{Et_2O} [(\eta^6\text{-arene})Mo(CO)_3] + 3C_5H_5N{\cdot}BF_3 \quad (7)$$

is superior to the direct synthesis in that it gives higher yields, a shorter reaction time is required, and the reaction can be carried out at room temperature. When this method is extended to the synthesis of (η^6-arene)tricarbonyltungsten complexes only low yields of products are obtained[47].

Benzene and its methyl derivatives undergo direct reaction with hexacarbonylvanadium to give the (η^6-arene)tetracarbonyl vanadium cation **21**[48].

$$2V(CO)_6 + C_6H_6 \longrightarrow [(\eta^6\text{-}C_6H_6)V(CO)_4]^+ [V(CO)_6]^- \quad (8)$$

(21)

When bromopentacarbonylmanganese is heated in benzene in the presence of aluminium chloride it forms the (η^6-benzene)manganese complex **22**[49].

$$[Mn(CO)_5Br] + C_6H_6 \xrightarrow{AlCl_3} [(\eta^6\text{-}C_6H_6)Mn(CO)_3]^+ Br^- \quad (9)$$

(22)

E. Cyclic Trimerization of Acetylenes

The cyclic trimerization of disubstituted acetylenes in the presence of a transition metal salt can give η^6-arene complexes[50,51]. For example, diphenylmanganese is prepared *in situ* from manganese(II) chloride and phenylmagnesium bromide and on treatment with but-2-yne it gives the η^6-hexamethylbenzene complex **23** (M = Mn) in 10% yield. The corresponding cobalt complex (**23**; M = Co) is prepared by a similar route using mesitylmagnesium bromide[52]. The cyclic oligomerization of acetylenes is rarely used now as a preparative route for specific η^6-arene—transition metal complexes as the routes already described give better yields of products.

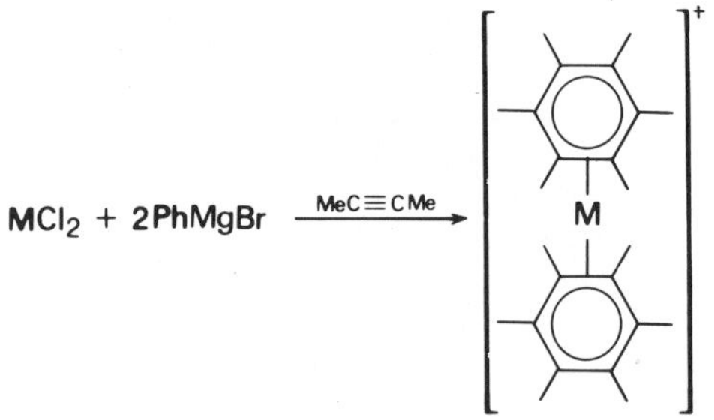

$$MCl_2 + 2PhMgBr \xrightarrow{MeC\equiv CMe} [(C_6Me_6)_2M]^+ \qquad (10)$$

(23)

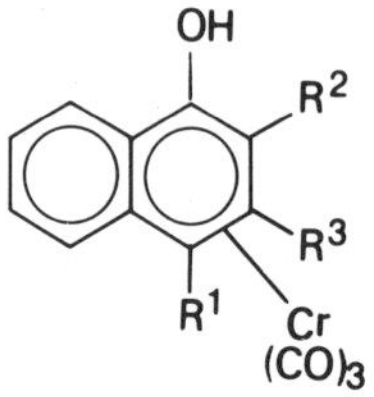

(24) 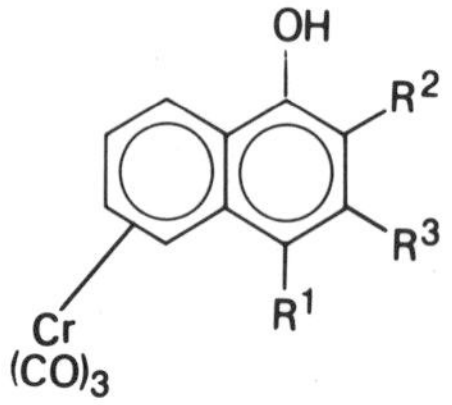 (25)

Carbenes $[PhR^1C{=}Cr(CO)_5]$undergo stereoselective reaction with acetylenes ($R^2C{\equiv}CR^3$) to give the naphthalene derivatives (**24** and **25**; R^1 = Ph, OMe; R^2 = Et, Me, Pr, Bu, Ph; R^3 = H, Me, Et, Ph, CO_2Et)[53]. Similarly, treatment of the carbenes $[(CO)_5Cr{=}C(OMe)R]$, ($R = p\text{-}MeC_6H_4$, $p\text{-}CF_3C_6H_4$, 1- and 2-naphthyl, 2- and 3-furyl, 2-thienyl, cyclopentyl) and $[(CO)_5Cr{=}CR^1R^2]$ (R^1 = Ph, $p\text{-}MeC_6H_4$, R^2 = 2-furyl, 2-naphthyl) with the acetylenes $R^3C{\equiv}CR^4$ produces the tricarbonylchromium complexes [(**26**; $R^3 = R^4$ = Ph, R^5 = Me, CF_3), (**27**; $R^3 = R^4$ = Ph, $R^6 = R^7$ = OH, OMe), (**28**; X = O, S; R^3 = Ph, Pr; R^4 = H, Ph; R^6, R^7 = OH, OMe), and (**29**; R^2 = 2-furyl, 2-naphthyl, $p\text{-}MeC_6H_4$; $R^3 = R^4$ = Ph,

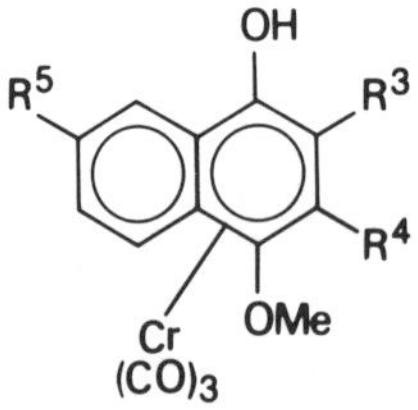

(26) 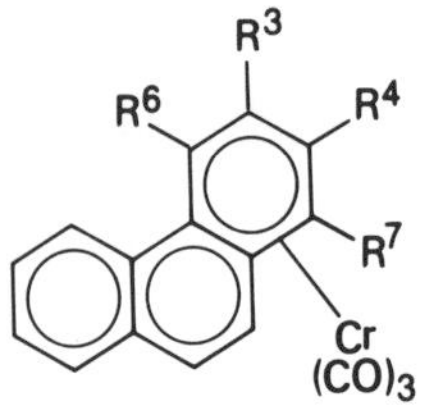 (27)

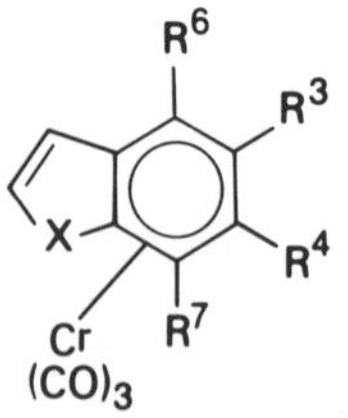

(28) 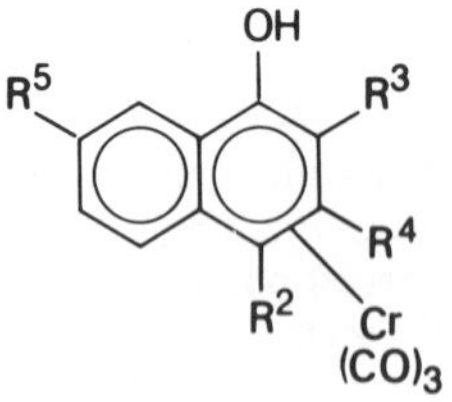 (29)

R^5 = H, Me)]. In these reactions the phenyl group in the carbenes [$(CO)_5Cr{=}CR^1R^2$] is anellated rather than the furyl and naphthyl rings[54].

F. Displacement of the η^5-Cyclopentadienyl Ligand by an η^6-Arene Ligand

One of the η^5-cyclopentadienyl ligands of ferrocene is replaced readily by benzene to give the cation **30**[55,56]. This reaction is carried out in the presence of aluminium chloride and the cation is readily isolated as a stable salt with anions such as hexafluorophosphate, tetrafluoroborate, and tetraphenylborate. The ferrocenium cation does not exchange an η^5-cyclopentadienyl ring with an arene and powdered aluminium is usually added to the reaction mixture to prevent formation of ferrocenium ion. The presence of powdered aluminium in an exchange reaction involving a halogenated arene can promote dehalogenation[57].

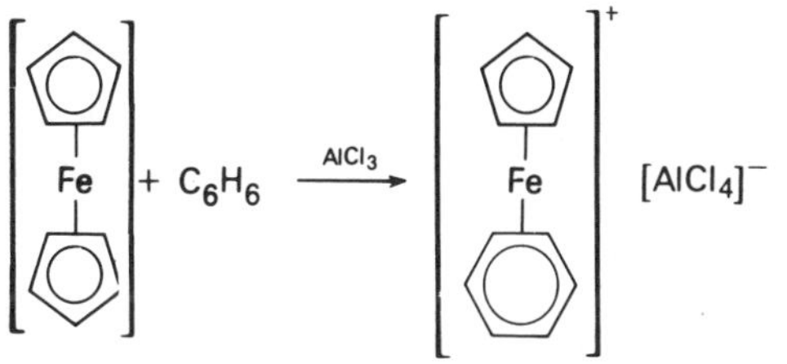

$$[(C_5H_5)_2Fe] + C_6H_6 \xrightarrow{AlCl_3} [(C_5H_5)Fe(C_6H_6)]^+ [AlCl_4]^- \quad (11)$$

(30)

Sutherland has comprehensively reviewed the literature on (η^6-arene)(η^5-cyclopentadienyl)iron cations and related systems up to the end of 1975[58]. Details of the preparation and reactivity of mixed sandwich complexes are included in the survey of (η^6-arene)transition metal chemistry by Silverthorn[59]. The hydrogenation of aromatic ligands during the η^6-arene-η^5-cyclopentadienyl ligand-exchange reaction has been surveyed[60]. It has been postulated that this hydrogenation is initiated by the abstraction of H^- by aluminium chloride to give a carbenium ion which is stabilized by a reversible intramolecular electron transfer. Successive abstractions of H by the iron atom and subsequent transfer to the ligand effect the hydrogenation[61,62]. Hydrogenation occurs in the reaction between ferrocene and pyrene in the presence of aluminium chloride–aluminium powder mixtures. Ligand exchange occurs to give the expected cation **(31)** and hydrogenated products (**32** and **33**), together with two *trans*-dications (**34** and **35**)[62].

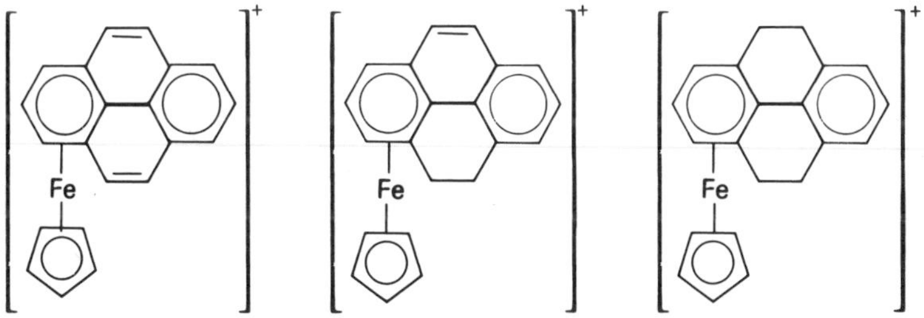

Fe Fe Fe

(31) **(32)** **(33)**

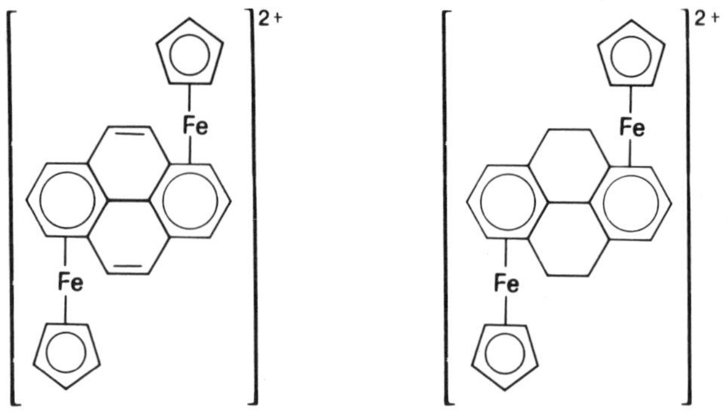

(34) (35)

The exchange reaction has been carried out with a wide variety of substituted ferrocenes and substituted arenes. Electron-donating substituents attached to ferrocene or to the arene facilitate the exchange, whereas electron removal hinders the reaction, especially when the electron-withdrawing substituent is situated in the benzene ring[56,57,63]. For example, when ethylferrocene is treated with benzene or mesitylene the predominant product (**36**; 73% yield) is that formed by exchange of

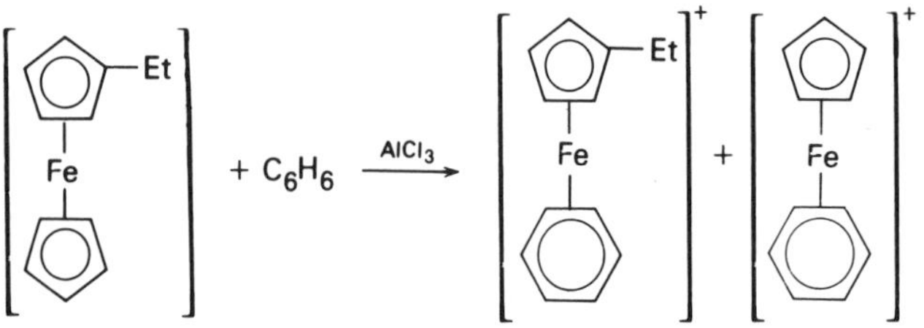

$$+ C_6H_6 \xrightarrow{AlCl_3} \tag{12}$$

(36)

the η^5-ethylcyclopentadienyl ring[56,63]. When acetylferrocene is treated with mesitylene only the unsubstituted ring exchanges to give the η^6-arene complex **37**[63].

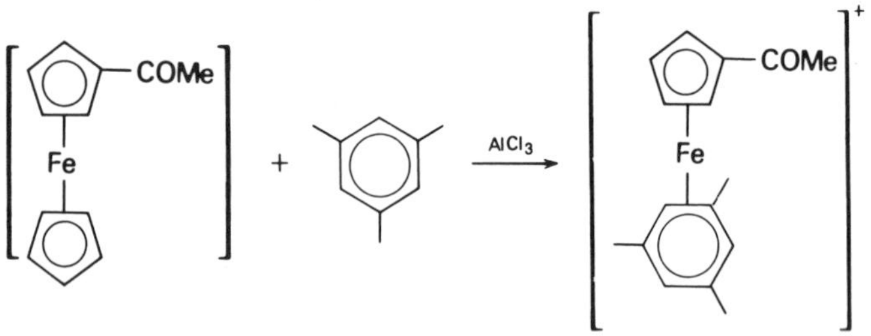

$$\xrightarrow{AlCl_3} \tag{13}$$

(37)

A recent example of an exchange reaction in the presence of aluminium chloride is that between ferrocene and aniline, which gives the (η^6-aniline)(η^5-cyclopentadienyl)-iron cation **38** (X = NH_2). This cation readily undergoes nucleophilic attack with either sodium hydroxide or sodium hydrogen sulphide to give the (η^6-phenol)- and (η^6-thiophenol)iron complexes (**38**; X = OH and SH), respectively[64].

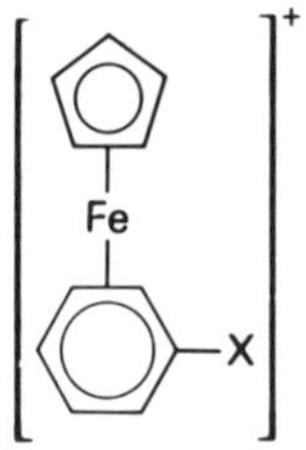

(**38**)

G. Preparation of (η^6-Heterocyclic)transition Metal Complexes

Pyridine, thiophene, borazine, and similar compounds behave as six-electron ligands with transition metals. They form complexes that are analogous to the (η^6-arene)transition metal complexes and they are prepared via similar routes.

Fischer and Ofele reported that an η^6-methylpyridine complex was formed by the pyrolysis of *N*-methylpyridinium pentacarbonyliodochromate(0)[65], but on the basis of n.m.r. evidence they have reformulated the complex as the dihydropyridine derivative **39**[66]. However, the η^6-pyridine complex **40** is formed by the co-condensation of pyridine, trifluorophosphine, and chromium vapour. The σ-pyridine complex **41** is also isolated from this reaction[67].

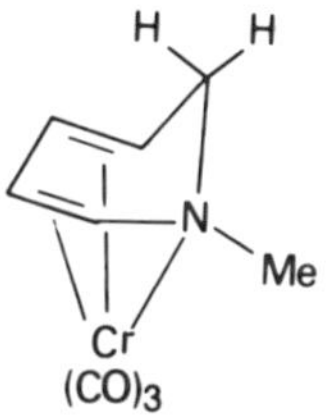

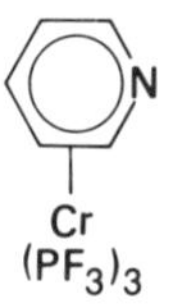

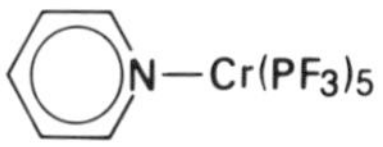

(**39**) (**40**) (**41**)

The reaction of pentamethylpyridine or 2,4,6-trimethylpyridine with hexacarbonylchromium gives the corresponding η^6-methylpyridine complexes (**42**; R = Me, H)[68]. The (η^6-2,6-dimethylpyridine)- and (η^6-2,3,5,6-tetramethylpyridine)-tricarbonylchromium complexes are prepared similarly[69].

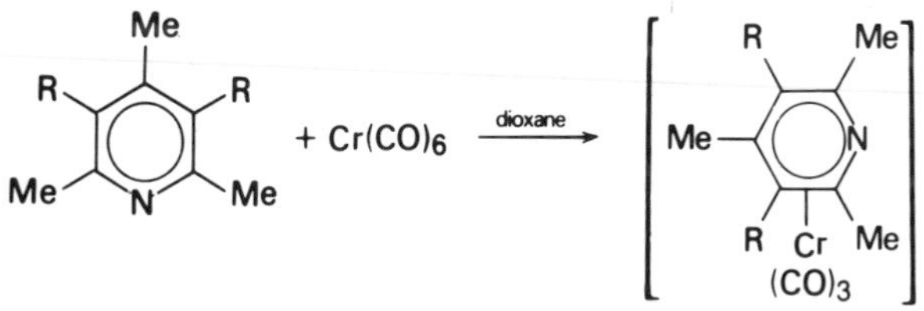

$$+ \ Cr(CO)_6 \xrightarrow{\text{dioxane}} \qquad (14)$$

(**42**)

When arsabenzene is heated with hexacarbonylmolybdenum in diglyme or treated with tricarbonyltris(pyridine)molybdenum in the presence of boron trifluoride etherate the η^6-arsabenzene complex **43** (X = As) is formed as a red air-stable solid. The corresponding stibabenzene complex (**43**; X = Sb) is easily prepared by boron trifluoride etherate catalysed ligand displacement from tricarbonyltris(pyridine)-molybdenum[70].

The co-condensation of chromium vapour with 2,6-dimethylpyridine at 77°K gives bis(η^6-2,6-dimethylpyridine)chromium (**44**). The structure of (**44**) has been confirmed

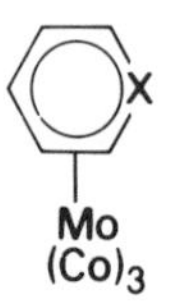

(**43**) 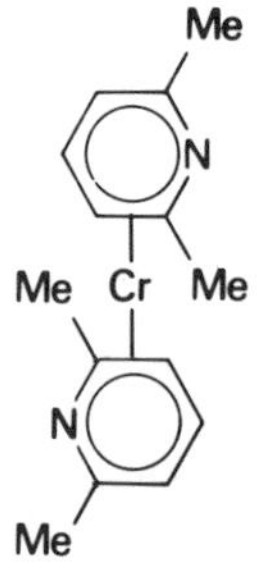 (**44**)

by X-ray analysis[71]. The reaction of 2,6-diphenylpyridine and 2,4,6-triphenylpyridine produces a series of complexes, in all of which compounds the tricarbonylchromium group is π-bonded to the phenyl groups and not to the pyridine ring[72].

The reaction of thiophen with triamminetricarbonylchromium produces the η^6-thiophen complex (**45**). Similar reactions with tricarbonyltris(methyl cyanide)-chromium also produce tricarbonyl(η^6-thiophen)chromium complexes. The use of $[(NH_3)_3Cr(CO)_3]$ and $[(MeCN)_3Cr(CO)_3]$ is preferred in these reactions as they do not sublime and they are more reactive in solution than hexacarbonylchromium[73].

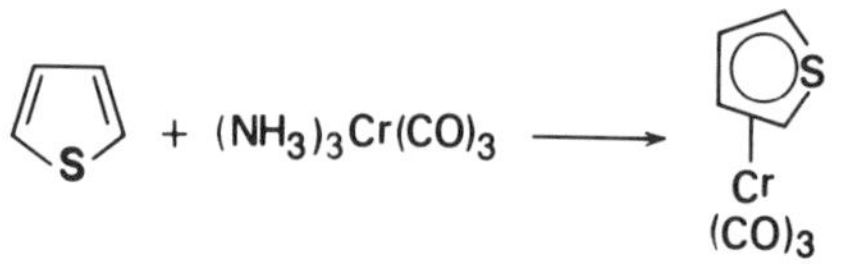

$$\text{thiophen} + (NH_3)_3Cr(CO)_3 \longrightarrow \text{(45)} \quad (15)$$

(**45**)

Treatment of the η^6-thiophene complexes, for example tricarbonyl(η^6-2-methylthiophene)chromium (**46**), with benzene results in ligand exchange to give (η^6-benzene)tricarbonylchromium[74].

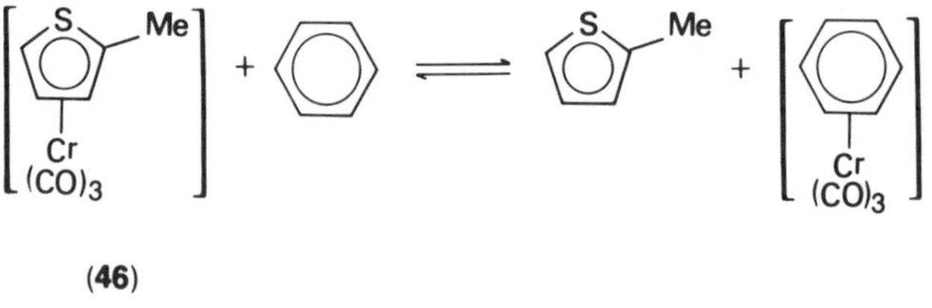

$$[\text{(46)}] + C_6H_6 \rightleftharpoons \text{2-methylthiophene} + [C_6H_6Cr(CO)_3] \quad (16)$$

(**46**)

(η^6-Borazine)tricarbonylchromium complexes (**47**; R^1 = Pr, R^2 = Me; R^1 = Me, R^2 = Pr; R^1 = *iso*-Pr, R^2 = Me; R^1 = Me, R^2 = *iso*-Pr) are prepared by the reaction of the corresponding borazine with tricarbonyltris(methyl cyanide)chromium[75].

$R^1_3B_3N_3R^2_3$ + $(MeCN)_3Cr(CO)_3$ → $[(\eta^6\text{-}R^1_3B_3N_3R^2_3)Cr(CO)_3]$ (17)

(47)

H. (η^6-Cycloheptatriene)transition Metal Complexes

η^6-Cycloheptatriene—transition metal complexes are prepared by routes similar to those used to prepare η^6-arene complexes[76].

Cycloheptatriene and its derivatives displace three carbonyl groups from the Group VI metal carbonyls to give the corresponding (η^6-cycloheptatriene)transition metal complexes (**48**; M = Cr, Mo, W)[77–80].

C_7H_8 + $[M(CO)_6]$ → $[(\eta^6\text{-}C_7H_8)M(CO)_3]$ + 3CO (18)

(**48**)

The reaction of cycloheptatriene with tricarbonyltris(methyl cyanide)molybdenum or with tricarbonyltris(pyridine)chromium and boron trifluoride in ether produces the corresponding η^6-cycloheptatriene complexes (**48**; M = Mo, Cr) in yields of 69% and 80%, respectively[45,80]. The 7-substituted cycloheptatrienes [**49**, R = Me, CH_2CO_2Et, $CH(CO_2Et)_2$, $CH_2C{\equiv}CH$, OMe] combine with hexacarbonylchromium or tricarbonyltris(pyridine)chromium and boron trifluoride to give exclusively the corresponding η^6-7-*endo*-cycloheptatriene complexes (**50**) in good yields[80].

(**49**) + $[Cr(CO)_6]$ or $[(pyridine)_3Cr(CO)_3]$ + BF_3 → $[(\eta^6\text{-}C_7H_7RH)Cr(CO)_3]$ (19)

(**50**)

Some more recent examples of this reaction are the preparations of tricarbonylchromium complexes (**51**; $R^1 = R^2$ = H, Me, Ph; R^1 = H, R^2 = Ph; R^1 = H, R^2 = Me) from the corresponding cycloheptatrienes and tricarbonyltris(methyl cyanide)chromium (Scheme 1). These η^6-cycloheptatriene

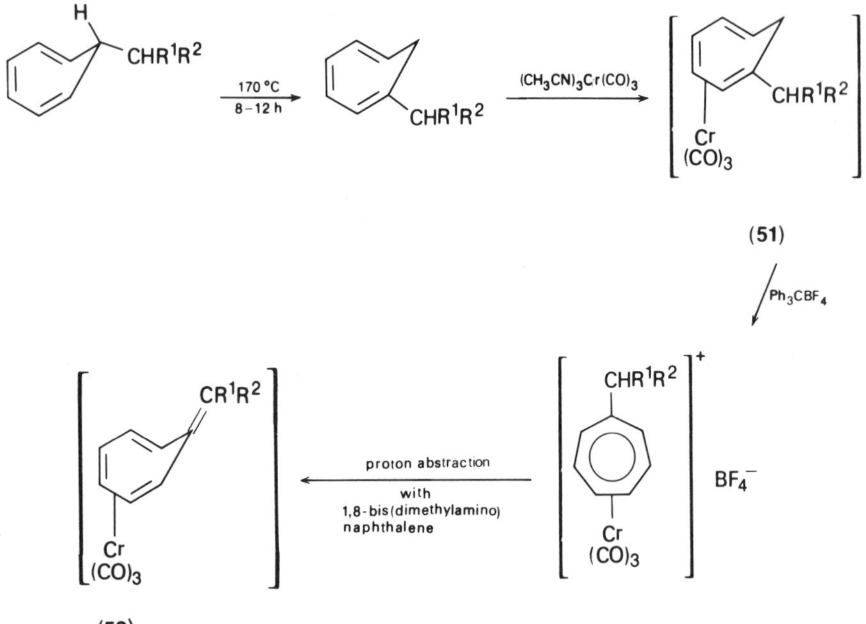
H
CHR1R2
170 °C
8–12 h
CHR1R2
(CH3CN)3Cr(CO)3
CHR1R2
Cr
(CO)3
(51)
Ph3CBF4
CHR1R2
+
Cr
(CO)3
BF4−
proton abstraction
with
1,8-bis(dimethylamino)
naphthalene
CR1R2
Cr
(CO)3
(52)

SCHEME 1

complexes are used to prepare the corresponding η^6-heptafulvene complexes (**52**)[81]. Reaction of furotropylidenes with hexacarbonylchromium produces the corresponding tricarbonylchromium complexes (**53** and **54**; R = H, Me) where the metal is bonded to the cycloheptatriene ring[82].

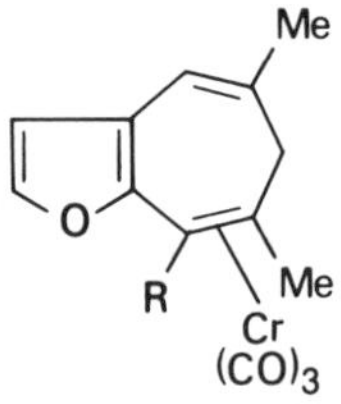
Me
O
R
Me
Cr
(CO)3

(**53**)

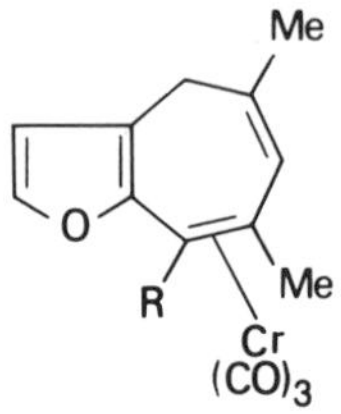
Me
O
R
Me
Cr
(CO)3

(**54**)

Substituted η^6-cycloheptatriene complexes are prepared by the reaction of the η^7-cycloheptatrienyl complex (**55**) with nucleophiles (N) where N = OMe^-, CN^-, $CH(CO_2Et)_2^-$, $CMe(CO_2Et)_2^-$, $C_5H_5^-$, $C_5H_4Bu\text{-}t^-$, SH^-[83,84]. Stereospecific addition occurs and the *exo*-isomer (**56**) is formed[85,86]. A good alternative route to the *exo*-products is by the displacement of methoxide by other anions from tricarbonyl(η^6-7-*exo*-methoxycycloheptatriene)chromium (**56**; N = OMe)[85].

Treatment of the η^7-cycloheptatrienyl complex **55** with trialkyl- or triarylphosphines gives the η^6-cycloheptatriene complexes (**57**). ^{1}H n.m.r. spectroscopy indicates that the phosphonium group occupies the *exo*-position relative to the chromium[87,88].

The co-condensation of cycloheptatriene with transition metal atoms at low temperatures is often accompanied by the extensive migration of hydrogen and the

$$[(C_7H_7)Cr(CO)_3]^+ \; (55) + N^- \longrightarrow (C_7H_7N)Cr(CO)_3 \; (56) \tag{20}$$

$$(55) \xrightarrow{PR_3} [(C_7H_7PR_3)Cr(CO)_3]^+ \; (57)$$

(55) (56) (57)

η^6-cycloheptatriene complex is not usually isolated[89–92]. However, the co-condensation of chromium vapour with cycloheptatriene in the presence of trifluorophosphine gives the chromium complex **58**[92].

$$C_7H_8 + Cr_{(g)} + 9\,PF_3 \longrightarrow [(C_7H_8)Cr(PF_3)_3] + [Cr(PF_3)_6] \tag{21}$$

(58)

I. η^6-Cyclooctatriene and η^6-Cyclooctatetraene Transition Metal Complexes

η^6-Cyclooctatriene and η^6-cyclooctatetraene transition metal complexes are prepared by routes similar to those used in the preparation of η^6-arene complexes.

Reaction of cycloocta-1,3,5-triene with hexacarbonylchromium gives the deep-red tricarbonylchromium complex **59**[93]. Many of the earlier preparations of

$$C_8H_{10} + [Cr(CO)_6] \xrightarrow{80\,°C,\,1\,h} [(C_8H_{10})Cr(CO)_3] \tag{22}$$

(59)

η^6-cyclooctatriene complexes are documented in the book by Fischer and Werner[94]. The reaction of [14]-annulene with triamminetricarbonylchromium gives the bis(tricarbonylchromium) complex **60**[95].

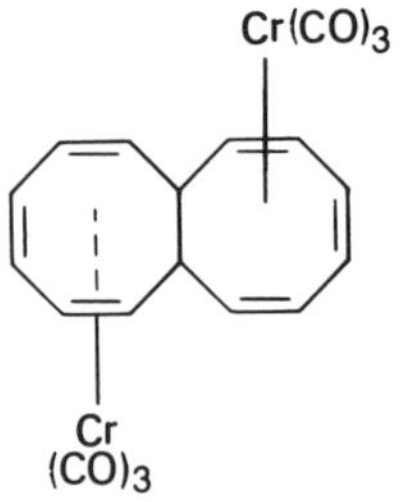

Cr(CO)3

Cr
(CO)3

(**60**)

Treatment of cycloocta-1,3,6-triene with tricarbonyltri(methyl cyanide)chromium or tricarbonyldiglymemolybdenum gives the corresponding tricarbonylmetal compounds (**61**; M = Cr, Mo) in good yields. When the molybdenum complex (**61**, M = Mo) is heated it rearranges to give the corresponding cycloocta-1,3,5-triene complex (**62**)[96].

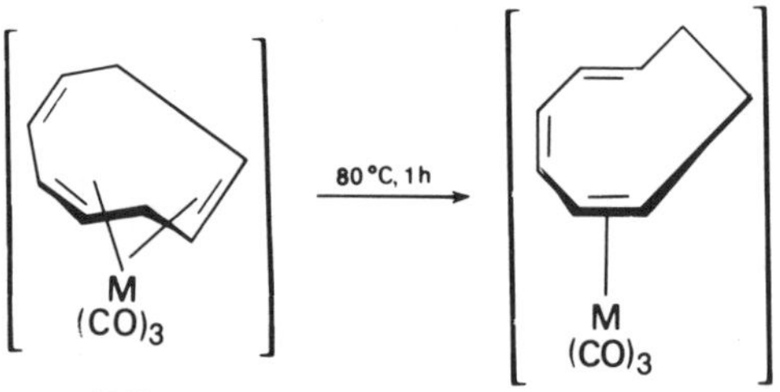

M
(CO)3

80 °C, 1 h

M
(CO)3

(23)

(**61**) (**62**)

Tricarbonyl(η^6-cyclooctatetraene)chromium (**63**) is obtained in 25% yield by stirring cyclooctatraene with triamminetricarbonylchromium in hot hexane for 65 h[97]. The reaction of tricarbonyltris(methyl cyanide)chromium with 1,3,5,7-tetramethylcyclooctatetraene produces the corresponding tricarbonylchromium complex (**64**; M = Cr) in 60% yield. The molybdenum and tungsten complexes (**64**; M = Mo, W) are prepared via similar routes[98].

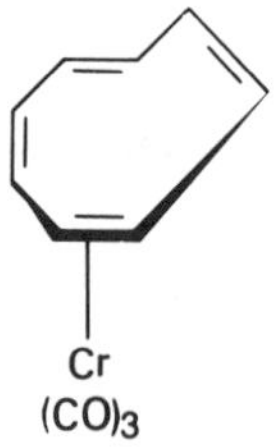

Cr
(CO)3

Me
Me
Me
Me
M
(CO)3

(**63**) 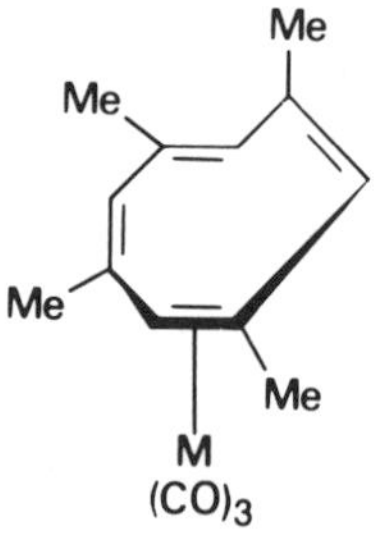 (**64**)

II. η^7-CYCLOHEPTATRIENYL COMPLEXES[99]

A. Introduction

The η^7-cycloheptatrienyl ligand is planar and (η^7-cycloheptatrienyl)transition metal complexes are formed readily by hydride abstraction from the corresponding η^6-cycloheptatriene compounds. η^7-Cycloheptatrienyl complexes are also formed by the direct reaction of cycloheptatriene with a transition metal derivative and by ring expansion of an η^6-arene ligand already attached to a metal. The chemistry of (η^7-cycloheptatrienyl)transition metal complexes has been reviewed[99]. An up-to-date account of these systems is contained in the 'Annual Surveys' in the *Journal of Organometallic Chemistry*.

B. Hydride Abstraction from an η^6-Cycloheptatriene Complex

In 1958, Dauben and Honnen[100] showed that triphenylmethylium tetrafluoroborate abstracts hydride ion from tricarbonyl(η^6-cycloheptatriene)molybdenum **(65)** to give the corresponding η^7-cycloheptatrienyl complex **66**[101]. The starting material, that is the

$$[(\eta^6\text{-}C_7H_8)Mo(CO)_3] \xrightarrow{-H^-} [(\eta^7\text{-}C_7H_7)Mo(CO)_3]^+ \quad (24)$$

(65) **(66)**

η^6-cycloheptatriene complex (**48**; M = Cr, Mo, W), is prepared by direct reaction of cycloheptatriene with the metal carbonyl $[M(CO)_6]$ (M = Cr, Mo)[102], with $[(MeCN)_3W(CO)_3]$[103], or with [(pyridine)$_3$Cr(CO)$_3$], and boron trifluoride in ether[104]. Hydride abstraction from a η^6-cycloheptatriene complex is used extensively for the preparation of η^7-cycloheptatrienyl complexes (Table 2).

The triphenylmethylium cation is often used as the hydride ion acceptor, but Connor and Rasburn[106] have reported that triethyloxonium tetrafluoroborate **(67)** can be used in this reaction and that it possesses certain advantages. The two main advantages cited are first that the reagent is easily prepared in large amounts from simple, cheap starting materials, and second that the diethyl ether and ethane produced in the reaction are volatile and hence the isolation of the η^7-cycloheptatrienyl salt is simplified[106].

$$[(\eta^6\text{-}C_7H_8)Cr(CO)_3] + \underset{(67)}{Et_3O^+BF_4^-} \longrightarrow [(\eta^7\text{-}C_7H_7)Cr(CO)_3]^+BF_4^- + Et_2O + C_2H_6 \quad (25)$$

TABLE 2. Synthesis of η^7-cycloheptatrienyl–metal complexes from the corresponding η^6-cycloheptatriene complex

Transition metal complex combining with a cycloheptatriene	η^6-Cycloheptatriene complex	η^7-Cycloheptatrienyl complex formed after hydride abstraction	Yield of η^7-cycloheptatrienyl complex (%)	Reference
$[Mo(CO)_6]$	$[(\eta^6\text{-}C_7H_8)Mo(CO)_3]$	$[(\eta^7\text{-}C_7H_7)Mo(CO)_3]^+$	98–100	100
$[Cr(CO)_6]$	$[(\eta^6\text{-}C_7H_8)Cr(CO)_3]$	$[(\eta^7\text{-}C_7H_7)Cr(CO)_3]^+$	99	101
$[Cr(CO)_6]$	$[(\eta^6\text{-}MeC_7H_7)Cr(CO)_3]$	$[(\eta^7\text{-}MeC_7H_6)Cr(CO)_3]^{+a}$	100	101
$[(\eta^5\text{-}C_5H_5)Mn(CO)_3]^b$	$[(\eta^6\text{-}C_7H_8)Mn(\eta^5\text{-}C_5H_5)]$	$[(\eta^7\text{-}C_7H_7)Mn(\eta^5\text{-}C_5H_5)]^{+c}$	92	105
$[(\eta^5\text{-}MeC_5H_4)Mn(CO)_3]^b$	$[(\eta^6\text{-}C_7H_8)Mn(\eta^5\text{-}MeC_5H_4)]$	$[(\eta^7\text{-}C_7H_7)Mn(\eta^5\text{-}MeC_5H_4)]^{+c}$	88	105
$[(MeCN)_3Cr(CO)_3]$	$[(\eta^6\text{-}C_7H_8)Cr(CO)_3]$	$[(\eta^7\text{-}C_7H_7)Cr(CO)_3]^{+d}$	91	103, 106
$[(MeCN)_3Cr(CO)_3]$	$[(\eta^6\text{-}MeO_2CC_7H_7)Cr(CO)_3]$	$[(\eta^7\text{-}MeO_2CC_7H_6)Cr(CO)_3]^+$	83	107

[a] Hydride abstraction with Ph_3CClO_4.
[b] Photochemical reaction.
[c] Hydride abstraction with Ph_3CPF_6.
[d] Hydride abstraction with Et_3OBF_4.

7-Substituted η^6-cycloheptatriene metal carbonyl complexes exist as two isomers, the *exo* (**68**) and *endo* forms (**69**). Hydride abstraction with triphenylmethylium tetrafluoroborate from the complexes **69** (M = Cr, R = OMe, OEt, CO_2Me, Me) involves stereospecific removal of the 7-*exo*-hydride ion to give the corresponding substituted η^7-cycloheptatrienyl cation **70**[107–109].

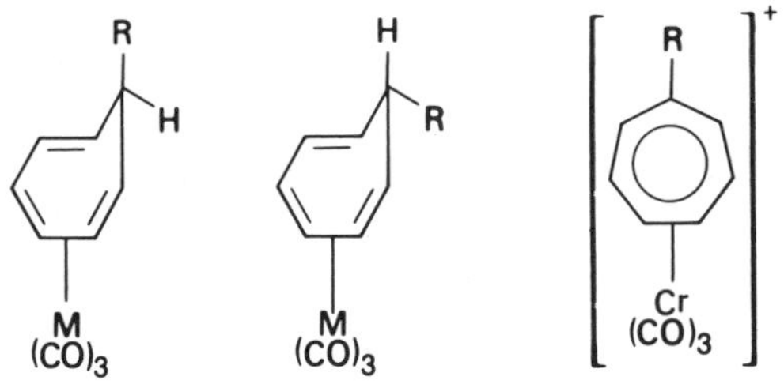

(**68**) (**69**) (**70**)

The η^7-cycloheptatrienyl cation **72** is formed on catalytic dehydrogenation of the corresponding η^6-cycloheptatriene complex (**71**)[110].

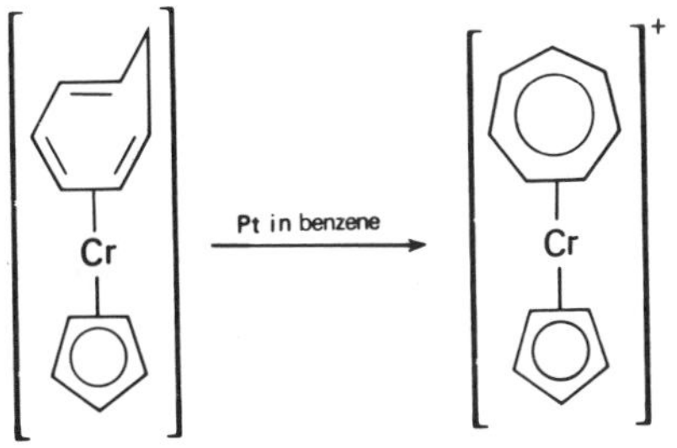

Pt in benzene (26)

(**71**) (**72**)

C. Direct Reaction of Cycloheptatriene with a Transition Metal Derivative

In some cases the direct reaction of cycloheptatriene with a transition metal derivative results in formation of an η^7-cycloheptatrienyl compound. In these reactions the expulsion of a hydride ion occurs spontaneously. The treatment of tetracarbonyl(η^5-cyclopentadienyl)vanadium or hexacarbonylvanadium with cycloheptatriene gives the η^7-cycloheptatrienyl derivatives **73** and **74** (R = H),

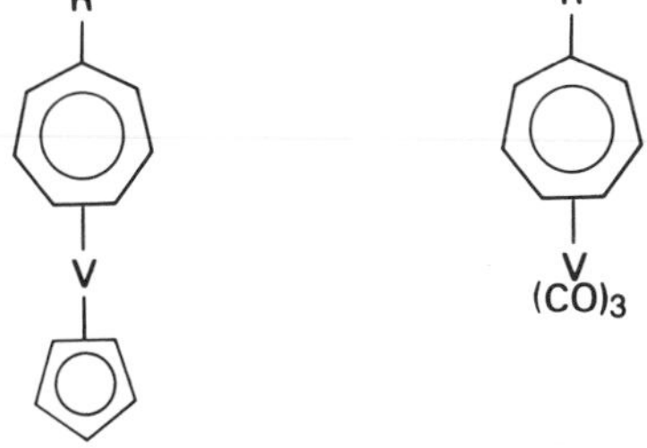

(**73**) (**74**)

respectively[111–113]. The same routes are used to prepare a series of substituted η^7-cycloheptatrienyl compounds [(**73**; R = Me, Ph, CN, CH_2CO_2Me, OMe, CO_2Me, CO_2Et) and (**74**; R = Me, Ph, OMe, OEt, OPr, CO_2Et)][114].

Ring exchange reactions can be used to prepare η^7-cycloheptatrienyl complexes. The reaction of (η^6-benzene)(η^5-cyclopentadienyl)chromium (**75**) with cycloheptatriene in the presence of aluminium chloride or with cycloheptatrienyl tetrafluoroborate produces the η^7-cycloheptatrienyl complex **76**, which can be reduced to neutral (η^7-cycloheptatrienyl)(η^5-cyclopentadienyl)chromium (**77**)[115].

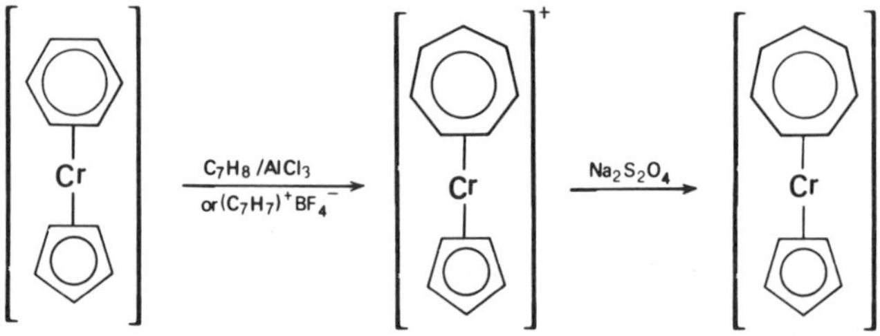

Cr; $C_7H_8/AlCl_3$ or $(C_7H_7)^+BF_4^-$; Cr; $Na_2S_2O_4$; Cr (27)

(**75**) (**76**) (**77**)

The chromium complex **77** is formed in low yield (2%) by the treatment of a mixture of chromium(III) chloride, cycloheptatriene and cyclopentadiene with isopropylmagnesium bromide in ether[116]. An improvement on this reaction is to use $CrCl_3 \cdot 3thf$ in tetrahydrofuran (thf), when the η^7-cycloheptatrienyl complex (**77**) is isolated in 15% yield. The same reaction with $MoCl_3 \cdot 3thf$ produces (η^7-cycloheptatrienyl)η^5-cyclopentadienyl)molybdenum in 13% yield[117].

Reaction of vanadium(IV) chloride with isopropylmagnesium bromide and cycloheptatriene gives bis(η^6-cycloheptatriene)vanadium, which undergoes stepwise hydride abstraction with triphenylmethylium tetrafluoroborate to give the vanadium complex **78** and then the bis(η^7-cycloheptatrienyl) complex **79**[118].

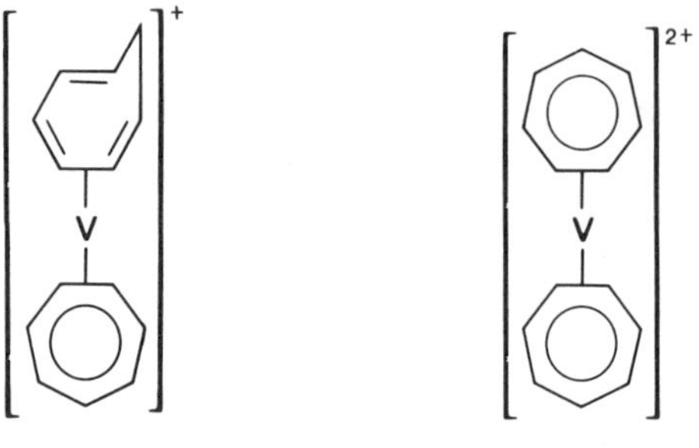

V; V

(**78**) (**79**)

A good general route for preparing non-carbonyl-containing η^7-cycloheptatrienylmolybdenum compounds is by the facile displacement of the η^6-arene ligand from the molybdenum complex **80** (Scheme 2)[119].

D. Ring Expansion of an η^6-Arene Ligand

When an attempt is made to acylate the η^6-benzene complexes **81** (M = Cr, Mn) under Friedel–Crafts conditions, ring expansion occurs and the corresponding

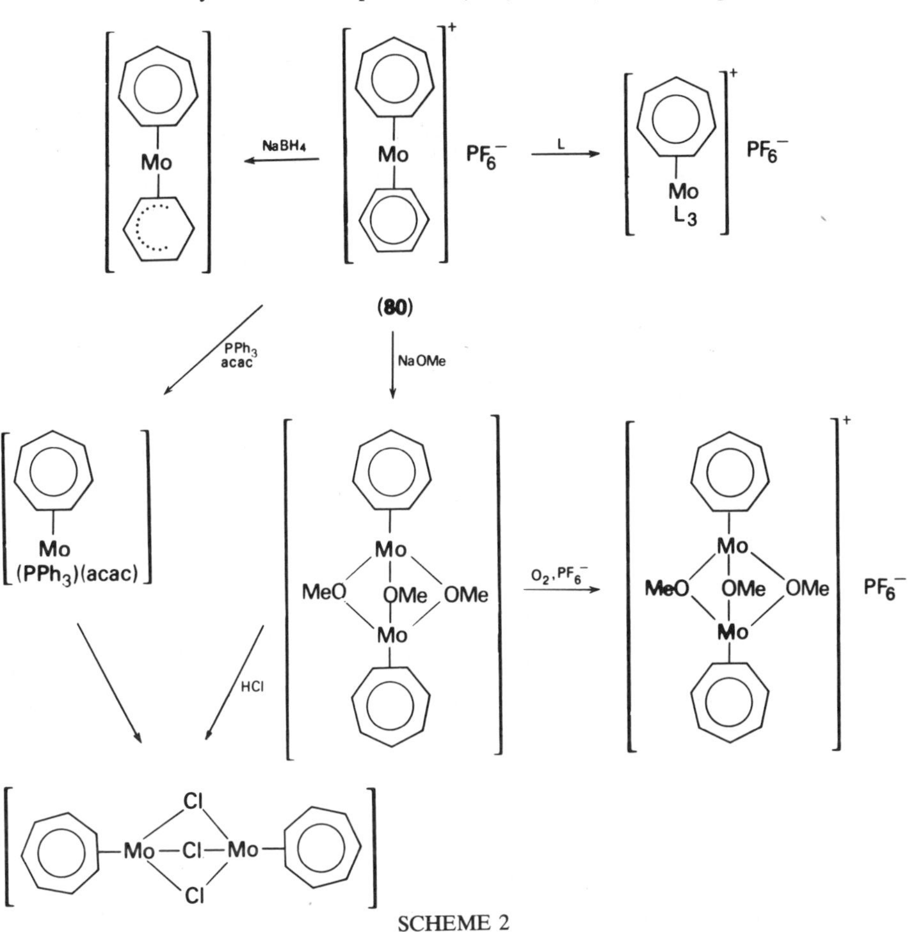

SCHEME 2

substituted η^7-cycloheptatrienyl complexes (**82**; R = Me, Ph; M = Cr, Mn) are formed[120]. The cationic η^7-cycloheptatrienyl complexes are reduced readily to the neutral complexes (**83**). The mechanism of the insertion has not been investigated in detail but it is suggested that the first step involves attack by the acyl cation on the metal followed by insertion into the η^6-benzene ring[120].

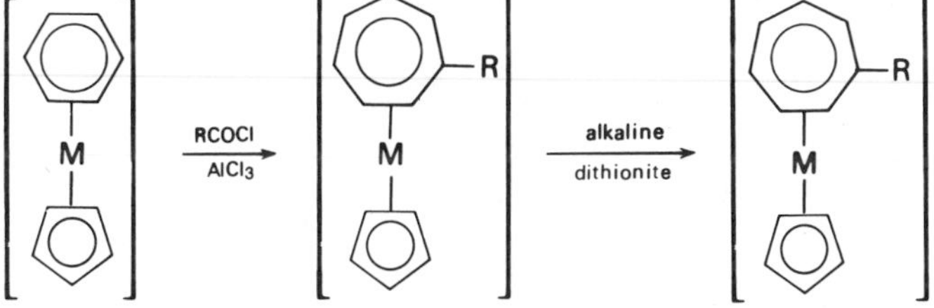

M —(RCOCl, $AlCl_3$)→ M(R)$^+$ —(alkaline dithionite)→ M(R) (28)

(**81**) (**82**) (**83**)

III. η^7-CYCLOOCTATRIENYL COMPLEXES

A. General Methods of Preparation

The chemistry of η^7-cyclooctatrienyl complexes has attracted little attention. They are readily prepared by conventional methods, for example hydride ion abstraction with triphenylmethylium tetrafluoroborate from the η^6-cyclooctatriene complexes **84** (M = Cr, Mo, W) gives the corresponding η^7-cyclooctatrienyl complexes (**85**; M = Cr, Mo, W)[121]. This reaction is in contrast to earlier unsuccessful attempts to abstract two hydride ions from the molybdenum complex (**84**; M = Mo) when an attempt was made to form a metal complex of the hypothetical $C_8H_8^{2+}$ cation[122].

Protonation of the cyclooctatetraene derivative **86** (M = Mo), where the ligand behaves as a six-electron donor[123,124], with concentrated sulphuric acid gives the molybdenum complex **85** (M = Mo)[125]. Stereospecific deuteration occurs with D_2SO_4, the electrophile D^+ attacking the uncoordinated double bond on the same side as the metal atom[125].

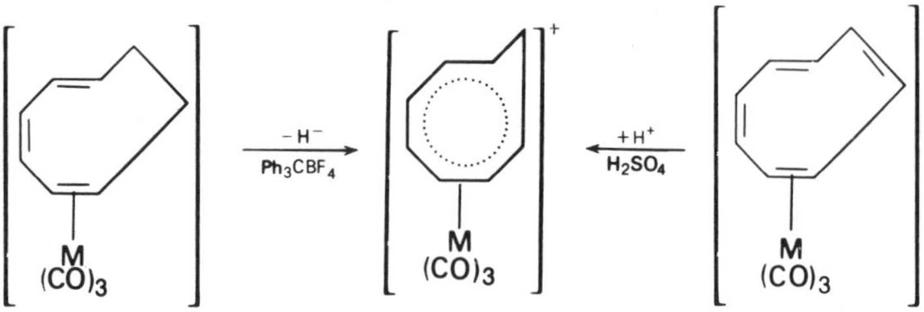

(29)

(84) (85) (86)

IV. CYCLOOCTATETRAENYL COMPLEXES

A. General Methods of Preparation[126]

Cyclooctatetraene is reduced by alkali metals or at a dropping mercury electrode to give the planar cyclooctatetraenyl anion $C_8H_8^{2-}$, which is a ten π-electron system[127,128].

The titanium complexes $[Ti(C_8H_8)_2]$ and $[Ti_2(C_8H_8)_3]$ are prepared in good yields (both 80%) by the reaction of tetra(n-butoxy)titanium with cyclooctatraene and triethylaluminium in different molar ratios. The reaction of sodium cyclooctatetraenyl with transition metal halides produces. $[Ti(C_8H_8)_2]$, $[V(C_8H_8)_2]$, $[Cr_2(C_8H_8)_3]$, $[Mo_2(C_8H_8)_3]$, $[W_2(C_8H_8)_3]$, $[Co(C_8H_8)]$, and $[Ni(C_8H_8)]$[129]. X-ray analysis of the titanium complex $[Ti_2(C_8H_8)_3]$ (**87**) confirms that this compound contains two planar eight-membered rings[130].

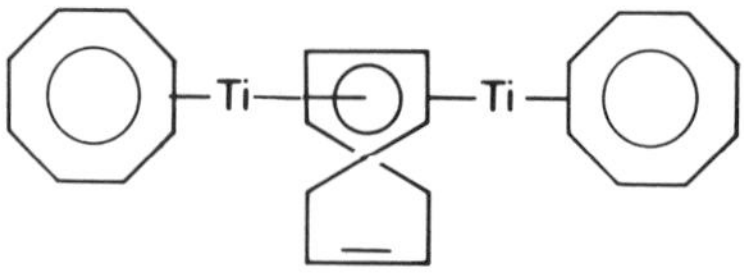

(87)

$$\text{C}_8\text{H}_8 + 2\text{K} \xrightarrow[-30^\circ\text{C}]{\text{thf}} 2\text{K}^+ [\text{RC}_8\text{H}_7]^{2+} \xrightarrow[\text{thf, 0}^\circ\text{C}]{\text{UCl}_4} [(\text{RC}_8\text{H}_7)_2\text{U}] \quad (30)$$

(**88**)

Treatment of uranium(IV) chloride with potassium cyclooctatetraenyl produces bis(η^8-cyclooctatetraenyl)uranium (uranocene) (**88**) in 80% yield[131]. X-ray analysis of uranocene shows that it has a sandwich structure with the uranium between the two planar η^8-cyclooctatetraenyl rings in D_{8h} symmetry[132,133].

Substituted uranocenes (**88**; R = ethyl, *n*-butyl, phenyl, vinyl, cyclopropyl) are prepared by the same general reaction[134,135] and this route is also suitable for the preparations of bis(η^8-alkylcyclooctatetraenyl)neptunium and -plutonium[135].

Recently, a convenient synthesis of 1,1′-di-*n*-butyluranocene (**88**; R = *n*-butyl) has been reported. Cyclooctatetraene is converted to the dianion of butylcyclooctatetraene with *n*-butyllithium and this dianion combines with uranium(IV) chloride to form the uranocene (**88**; R = *n*-butyl)[136].

V. REFERENCES

1. H. Zeiss, in *Organometallic Chemistry* (Ed. H. Zeiss), ACS Monograph 147, Reinhold, New York, 1960, p. 380; E. O. Fischer and H. P. Fritz, *Angew. Chem.*, **73**, 353 (1961); H. Zeiss, P. J. Wheatley, and H. J. S. Winkler, *Benzenoid—Metal Complexes*, Ronald Press, New York, 1966; M. A. Bennett, *Rodd's Chemistry of Carbon Compounds* (Ed. S. Coffey), Vol. 3, Elsevier, Amsterdam, 1974, p. 357.
2. R. P. A. Sneeden, *Organochromium Compounds,* Academic Press, New York, 1975.
3. W. E. Silverthorn, *Adv. Organomet. Chem.*, **13**, 47 (1975).
4. G. Marr and B. W. Rockett, *J. Organomet. Chem.*, **58**, 205 (1973); **79**, 223 (1974); **103**, 197 (1975); **126**, 227 (1977); **151**, 193 (1978); **163**, 325 (1978); **189**, 163 (1980).
5. E. O. Fischer and W. Hafner, *Z. Naturforsch.*, **106**, 665 (1955); E. O. Fischer and W. Hafner, *Z. Anorg. Chem.*, **286**, 146 (1956).
6. E. O. Fischer and J. Seeholzer, *Z. Anorg. Chem.*, **312**, 244 (1961).
7. E. O. Fischer and H. P. Koegler, *Chem. Ber.*, **90**, 250 (1957).
8. V. A. Umilin, Yu. B. Zverev, and V. K. Vanchagova, *Izv. Akad. Nauk SSSR, Ser. Khim.*, 2726 (1971); V. A. Umilin, Yu. B. Zverev, and V. K. Vanchagova, *Izv. Akad. Nauk SSSR, Ser. Khim.*, 1560 (1972).
9. Yu. B. Zverev, *Gidridy, Galedy Metalloorg. Soedin. Osoboi Chist.*, 123 (1976).
10. A. D. Zorin, V. A. Umilin, V. K. Vanchagova, and P. E. Gaivorenskii, *Zh. Obshch. Khim.*, **48**, 1092 (1978).
11. E. O. Fischer and R. Boettcher, *Chem. Ber.*, **89**, 2397 (1956).
12. H. H. Lindner and E. O. Fischer, *J. Organomet. Chem.*, **12**, P18 (1968).
13. E. O. Fischer and A. Reckziegel, *Chem. Ber.*, **94**, 2204 (1961).
14. E. O. Fischer, G. Joos, and W. Meer, *Z. Naturforsch.*, **13B**, 456 (1958).
15. E. O. Fischer and H. O. Stahl, *Chem. Ber.*, **89**, 1805 (1956).
16. E. O. Fischer, F. Scherer, and H. O. Stahl, *Chem. Ber.*, **93**, 2065 (1960).
17. L. S. Zborovskaya, T. N. Aizenshtadt, A. N. Artemov, and A. S. Emel'yanova, *Khim. Elementoorg. Soedin.*, **4**, 92 (1976).

18. A. N. Nesmeyanov, N. N. Zaitseva, L. P. Yur'eva, and R. A. Stukan, *Izv. Akad. Nauk SSSR, Ser. Khim.*, 1428 (1978).
19. P. L. Timms, *J. Chem. Soc., Chem. Commun.*, 1033 (1969).
20. P. L. Timms and T. W. Turney, *Adv. Organomet. Chem.*, **15**, 53 (1977); P. L. Timms, in *Cryochemistry* (Ed. G. A. Ozin and M. Moskovits), Wiley, New York, 1976, pp. 61–136; M. J. McGlinchey and P. S. Skell, *ibid.*, pp. 137–194; K. J. Klabunde, *Acc. Chem. Res.*, **8**, 383 (1975); Merck Symposium Proceedings, Metal Atoms in Chemical Synthesis, *Angew. Chem., Int. Ed. Engl.*, **14**, 193 and 273 (1975).
21. F. W. S. Benfield, M. L. H. Green, J. G. Ogden, and D. Young, *J. Chem. Soc., Chem. Commun.*, 866 (1973).
22. M. T. Anthony, M. L. H. Green, and D. Young, *J. Chem. Soc., Dalton Trans.*, 1419 (1975).
23. K. J. Klabunde and H. F. Efner, *Inorg. Chem.*, **14**, 789 (1975).
24. R. Middleton, J. R. Hull, S. R. Simpson, C. H. Tomlinson, and P. L. Timms, *J. Chem. Soc., Dalton Trans.*, 120 (1973).
25. V. Graves and J. J. Lagowski, *Inorg. Chem.*, **15**, 577 (1976).
26. M. J. McGlinchey and T.-S. Tan, *Can. J. Chem.*, **52**, 2439 (1974).
27. P. S. Skell, D. L. Williams-Smith, and M. J. McGlinchey, *J. Am. Chem. Soc.*, **95**, 3337 (1973).
28. A. Agarwal, M. J. McGlinchey, and T.-S. Tan, *J. Organomet. Chem.*, **141**, 85 (1977).
29. N. Hao and M. J. McGlinchey, *J. Organomet. Chem.*, **165**, 225 (1979).
30. R. Middleton, J. R. Hull, S. R. Simpson, C. H. Tomlinson, and P. L. Timms, *J. Chem. Soc., Dalton Trans.*, 120 (1973).
31. K. J. Klabunde, H. F. Efner, and T. O. Murdock, *Inorg. Synth.*, **19**, 70 (1979).
32. D. F. Shriver (Ed.), *Inorganic Syntheses*, Vol. 19, Ch. 2, *Metal Atom Syntheses*, Wiley – Interscience, New York, 1979.
33. B. B. Anderson, C. L. Behren, L. Radonovich, and K. J. Klabunde, *J. Am. Chem. Soc.*, **98**, 5390 (1976).
34. K. J. Klabunde, B. B. Anderson, and M. Bader, *Inorg. Synth.*, **19**, 72 (1979).
35. E. O. Fischer and K. Ofele, *Chem. Ber.*, **90**, 2532 (1957); E. O. Fischer, K. Ofele, H. Essler, W. Frohlich, J. P. Mortensen, and W. Semmlinger, *Chem. Ber.*, **91**, 2763 (1958); B. Nicholls and M. C. Whiting, *J. Chem. Soc.*, 551 (1959); G. Natta, R. Ercoli, F. Calderazzo, and S. Santambrogio, *Chim. Ind. (Milan)*, **40**, 1003 (1958).
36. C. A. L. Mahaffy and P. L. Pauson, *Inorg. Synth.*, **19**, 154 (1979).
37. S. Top and G. Jaouen, *J. Organomet. Chem.*, **182**, 381 (1979).
38. E. O. Fischer and K. Ofele, *Chem. Ber.*, **91**, 2395 (1958); K. Ofele, *Chem. Ber.*, **99**, 1732 (1966).
39. A. F. Mourad and H. Hopf, *Tetrahedron Lett.*, 1209 (1979).
40. A. N. Nesemeyanov, M. I. Rybinskaya and V. S. Kaganovich, *Izv. Akad. Nauk SSSR, Ser. Khim.*, 2824 (1978).
41. K. J. Odell, E. M. Hyde, B. L. Shaw and I. Sheperd, *J. Organomet. Chem.*, **168**, 103 (1979).
42. K. H. Pannell, D. H. Hambrick, and G. S. Lewandos, *J. Organomet. Chem.*, **99**, C21 (1975).
43. O. Hofer, K. Schloegl, and R. Schoelm, *Monatsh. Chem.*, **110**, 437 (1979).
44. D. P. Tate, J. M. Augl, and W. R. Knipple, *Inorg. Chem.*, **1**, 433 (1963).
45. R. B. King and A. Fronzaglia, *Inorg. Chem.*, **5**, 1837 (1966).
46. L. Weber, C. Krueger, and Y.-H. Tsay, *Chem. Ber.*, **111**, 1709 (1978).
47. A. N. Nesmeyanov, V. V. Krivykh, V. S. Kaganovich, and M. I. Rybinskaya, *J. Organomet. Chem.*, **102**, 185 (1975).
48. F. Calderazzo, *Inorg. Chem.*, **3**, 1207 (1964); F. Calderazzo, *Inorg. Chem.*, **4**, 233 (1965).
49. G. Winkhaus, L. Pratt, and G. Wilkinson, *J. Chem. Soc.*, 3807 (1961).
50. R. S. Dickson and G. Wilkinson, *Chem. Ind. (London)*, 1432 (1963).
51. E. O. Fischer and M. W. Schmidt, *Chem. Ber.*, **100**, 3782 (1967).
52. M. Tsutsui and H. Zeiss, *J. Am. Chem. Soc.*, **83**, 825 (1961).
53. K. H. Doetz and R. Doetz, *Chem. Ber.*, **110**, 1555 (1977).
54. K. H. Doetz and R. Doetz, *Chem. Ber.*, **111**, 2517 (1978).
55. A. N. Nesmeyanov, N. A. Vol'kenau, and I. N. Bolesova, *Dokl. Akad. Nauk SSSR*, **149**, 615 (1963).
56. A. N. Nesmeyanov, N. A. Vol'kenau, and I. N. Bolesova, *Tetrahedron Lett.*, 1725 (1963).

57. A. N. Nesmayanov, N. A. Vol'kenau, and I. N. Bolesova, *Dokl. Akad. Nauk SSSR*, **166**, 607 (1966).
58. R. G. Sutherland, *J. Organomet. Chem. Libr.*, **3**, 311 (1977).
59. W. E. Silverthorn, *Adv. Organomet. Chem.*, **13**, 95 (1975).
60. R. G. Sutherland, W. J. Pannekock, and C. C. Lee, *Ann. N.Y. Acad. Sci.*, **295**, 192 (1977).
61. W. J. Pannekoek, *Diss. Abstr. Int. B*, **38**, 3200 (1978).
62. C. C. Lee, B. R. Steel, K. J. Demchuk, and R. G. Sutherland, *Can. J. Chem.*, **58**, 93 (1979).
63. A. N. Nesmeyanov, N. A. Vol'kenau, and L. S. Shilovtseva, *Dokl. Akad. Nauk SSSR*, **160**, 1327 (1965).
64. J. F. Helling and W. A. Hendrickson, *J. Organomet. Chem.*, **168**, 87 (1979).
65. E. O. Fischer and K. Ofele, *Chem. Ber.*, **93**, 1156 (1960).
66. E. O. Fischer and K. Ofele, *J. Organomet. Chem.*, **8**, P5 (1967).
67. P. L. Timms, *6th Int. Conf. Organomet. Chem., Massachusetts, Abstr. Pap.*, P8 (1973); *Angew. Chem., Int. Ed. Engl.*, **14**, 273 (1975).
68. H.-G. Biedermann, K. Ofele, N. Schuhbauer, and J. Tajtelbaum, *Angew. Chem., Int. Ed. Engl.*, **14**, 639 (1975).
69. H.-G. Biedermann, K. Ofele, and J. Tajtelbaum, *Z. Naturforsch. B., Anorg. Chem. Org. Chem.*, **31B**, 321 (1976).
70. A. J. Ashe and J. C. Colburn, *J. Am. Chem. Soc.*, **99**, 8099 (1977).
71. L. H. Simona, P. E. Riley, R. E. Davis, and J. J. Lagowski, *J. Am. Chem. Soc.*, **98**, 1044 (1976).
72. J. Deberitz and H. Noth, *J. Organomet. Chem.*, **61**, 271 (1973).
73. G. A. Moser and M. D. Rausch, *Synth. React. Inorg. Met.-Org. Chem.*, **4**, 37 (1974).
74. C. Segard, C. Pommier, B. P. Roques, and G. Guiochon, *J. Organomet. Chem.*, **77**, 49 (1974).
75. M. Scotti and H. Werner, *Helv. Chim. Acta*, **57**, 1234 (1974).
76. M. A. Bennett, *Adv. Organomet. Chem.*, **4**, 353 (1966).
77. E. W. Abel, M. A. Bennett, R. Burton, and G. Wilkinson, *J. Chem. Soc.*, 4559 (1958).
78. E. W. Abel, M. A. Bennett, and G. Wilkinson, *Proc. Chem. Soc.*, 152 (1958).
79. M. A. Bennett, L. Pratt, and G. Wilkinson, *J. Chem. Soc.*, 2037 (1961).
80. P. L. Pauson, G. H. Smith, and J. H. Valentine, *J. Chem. Soc., C*, 1061 (1967).
81. J. A. S. Howell, B. F. G. Johnson, and J. Lewis, *J. Organomet. Chem.*, **42**, C54 (1972).
82. M. El Borai, R. Guilard, P. Fournari, Y. Dusausoy, and J. Protas, *Bull. Soc. Chim. Fr.*, 75 (1977).
83. J. D. Munro and P. L. Pauson, *J. Chem. Soc.*, 3475 (1961).
84. J. D. Munro and P. L. Pauson, *J. Chem. Soc.*, 3484 (1961).
85. P. L. Pauson, G. H. Smith, and J. H. Valentine, *J. Chem. Soc., C*, 1057 (1967).
86. P. E. Baikie and O. S. Mills, *J. Chem. Soc., A*, 2704 (1968).
87. P. Hackett and G. Jaouen, *Inorg. Chim. Acta*, **12**, L19 (1975).
88. D. A. Sweigart, M. Gower, and L. A. P. Kane-Maguire, *J. Organomet. Chem.*, **108**, C15 (1976).
89. J. R. Blackborow, C. R. Eady, E. A. K. von Gustorf, A. Scrivanti, and O. Wolfbeis, *J. Organomet. Chem.*, **108**, C32 (1975).
90. J. R. Blackborow, K. Hildenbrand, E. A. K. von Gustorf, A. Scrivanti, C. R. Eady, D. Entholt, and C. Kruger, *J. Chem. Soc., Chem. Commun.*, 16 (1976).
91. E. M. Van Dam, W. N. Brent, M. P. Silvon, and P. S. Skell, *J. Am. Chem. Soc.*, **97**, 465 (1975).
92. P. L. Timms and T. W. Turney, *J. Chem. Soc., Dalton Trans.*, 2021 (1976).
93. E. O. Fischer, C. Palm, and H. P. Fritz, *Chem. Ber.*, **92**, 2645 (1959).
94. E. O. Fischer and H. Werner, *Metal π-Complexes*, Vol. 1, Elsevier, Amsterdam, 1966.
95. J. Mitchell Guss and R. Mason, *J. Chem. Soc., Dalton Trans.*, 1834 (1973).
96. A. Salzer, *J. Organomet. Chem.*, **117**, 245 (1976).
97. C. G. Kreiter, A. Maasbol, F. A. L. Anet, H. D. Kaesz, and S. Winstein, *J. Am. Chem. Soc.*, **88**, 3444 (1966).
98. F. A. Cotton, J. W. Faller, and A. Musco, *J. Am. Chem. Soc.*, **90**, 1438 (1968).
99. M. A. Bennett, in *Adv. Organomet. Chem.*, **4**, 353 (1966); T. A. Stephenson, in *International Review of Science, Inorganic Chemistry, Series 1 and 2,* (Ed. H. J. Emeleus and M. J. Mays), Butterworths, London, 1972, p. 401, and 1975, p. 287.

100. H. J. Dauben and L. R. Honnen, *J. Am. Chem. Soc.*, **80**, 5570 (1958).
101. J. D. Munro and P. L. Pauson, *Proc. Chem. Soc.*, 267 (1959); J. D. Munro and P. L. Pauson, *J. Chem. Soc.*, 3475 (1961).
102. E. W. Abel, M. A. Bennett, R. Burton, and G. Wilkinson, *J. Chem. Soc.*, 4559 (1958).
103. R. B. King and A. Fronzaglia, *Inorg. Chem.*, **5**, 1837 (1966).
104. P. L. Pauson, G. H. Smith, and J. H. Valentine, *J. Chem. Soc., C*, 1061 (1967).
105. P. L. Pauson and J. A. Segal, *J. Chem. Soc., Dalton Trans.*, 2387 (1975).
106. J. A. Connor and E. J. Rasburn, *J. Organomet. Chem.*, **24**, 441 (1970).
107. P. L. Pauson and K. H. Todd, *J. Chem. Soc., C*, 2636 (1970).
108. P. L. Pauson, G. H. Smith, and J. H. Valentine, *J. Chem. Soc., C*, 1057 (1967).
109. P. L. Pauson and K. H. Todd, *J. Chem. Soc., C*, 2315 (1970).
110. E. O. Fischer and J. Mueller, *Z. Naturforsch*, **186**, 1137 (1963).
111. R. B. King and F. G. A. Stone, *J. Am. Chem. Soc.*, **81**, 5263 (1959).
112. R. P. M. Werner and S. A. Manastyrskj, *J. Am. Chem. Soc.*, **83**, 2023 (1961).
113. F. Calderazzo and P. L. Calvi, *Chem. Ind. (Milan)*, **44**, 1217 (1962).
114. J. Mueller and B. Mertschenk, *J. Organomet. Chem.*, **34**, 165 (1972).
115. E. O. Fischer and S. Breitschaft, *Angew. Chem.*, **75**, 94 (1963); E. O. Fischer and S. Breitschaft, *Angew. Chem.*, **75**, 167 (1963).
116. R. B. King and M. B. Bisnette, *Tetrahedron Lett.*, 1137 (1963); R. B. King and M. B. Bisnette, *Inorg. Chem.*, **3**, 785 (1964).
117. H. O. Van Oven, C. J. Groenenboom, and H. J. De Liefde Meijer, *J. Organomet. Chem.*, **81**, 379 (1974).
118. J. Mueller and B. Mertschenk, *J. Organomet. Chem.*, **34**., C41 (1972); J. Mueller and B. Mertschenk, *Chem. Ber.*, **105**, 3346 (1972).
119. E. F. Ashworth, M. L. H. Green and J. Knight, *J. Less-Common Metals*, **36**, 213 (1974).
120. E. O. Fischer and S. Breitschaft, *Angew. Chem., Int. Ed. Engl.*, **22**, 100 (1963); E. O. Fischer and S. Breitschaft, *Chem. Ber.*, **99**, 2213 (1966).
121. R. Aumann and S. Winstein, *Tetrahedron Lett.*, 903 (1970).
122. E. O. Fischer, C. Palm, and H. P. Fritz, *Chem. Ber.*, **92**, 2645 (1959).
123. J. S. McKechnie and I. C. Paul, *J. Am. Chem. Soc.*, **88**, 5927 (1966).
124. M. J. Bennett, F. A. Cotton, and J. Takats; *J. Am. Chem. Soc.*, **90**, 903 (1968).
125. S. Winstein, H. D. Kaesz, C. G. Kreiter, and E. C. Friedrich, *J. Am. Chem. Soc.*, **87**, 3267 (1965).
126. A. Streitweiser, *Proc. 11th Rare Earth Res. Conf.*, **1**, 278 (1974).
127. H. P. Fritz and H. Keller, *Z. Naturforsch.*, **166**, 231 (1961).
128. T. J. Katz, *J. Am. Chem. Soc.*, **82**, 3784 (1960).
129. H. Breil and G. Wilke, *Angew. Chem., Int. Ed. Engl.*, **5**, 898 (1966).
130. H. Dietrich and H. Dierks, *Angew. Chem., Int. Ed. Engl.*, **5**, 899 (1966).
131. A. Streitweiser and A. Mueller-Westerhoff, *J. Am. Chem. Soc.*, **90**, 7364 (1968); A. Streitweiser, U. Mueller-Westerhoff, C. B. Grant, and D. G. Morrell, *Inorg. Synth.*, **19**, 149 (1979).
132. A. Zalkin and K. N. Raymond, *J. Am. Chem. Soc.*, **91**, 5667 (1969).
133. A. Avdeef, K. N. Raymond, K. O. Hodgson, and A. Zalkin, *Inorg. Chem.*, **11**, 1083 (1972).
134. A. Streitweiser and C. A. Harmon, *Inorg. Chem.*, **12**, 1102 (1973).
135. D. G. Karraker, *Inorg. Chem.*, **12**, 1105 (1973).
136. J. T. Miller and C. W. DeKock, *Inorg. Chem.*, **18**, 1305 (1979).

The Chemistry of the Metal–Carbon Bond
Edited by F. R. Hartley and S. Patai

CHAPTER **12**

Synthesis of organolanthanide and organoactinide complexes

WILLIAM J. EVANS
Department of Chemistry, University of Chicago, 5735 South Ellis Avenue, Chicago, Illinois 60637, USA

I. INTRODUCTION

Although the first well characterized organometallic derivatives of the lanthanide elements were prepared in 1954[1], and the first organoactinide was described in 1956[2], the organometallic chemistry of the *f*-orbital metals received comparatively little attention during the subsequent years in which the organometallic chemistry of transition metals was extensively investigated. One major reason that little effort was devoted to the *f* elements is that their organometallic complexes are substantially more ionic than the transition metal analogues[3] and, consequently, these elements appeared to have a less extensive organometallic chemistry. Indeed, for many years organometallic *f*-element chemistry was limited primarily to stable organic anions, in contrast to the wide variety of neutral, cationic, and anionic organic species which complex with transition metals. The fact that the major part of this chapter involves cyclopentadienyl and cyclooctatetraenyl anions is a reflection of that limitation. Only recently[5–7] have major efforts been made to extend the organo-metallic chemistry of the *f* elements to include new classes of neutral organic ligands.

An equally important reason for the relative lack of emphasis on the *f* elements is that the organometallic chemistry of these metals is experimentally more difficult than the analogous transition metal chemistry. Not only are most of the organometallic complexes of these electropositive metals hydrolytically and oxidatively unstable, but even the *f* element precursors, e.g. the metal halides, are susceptible to hydrolysis and often are not commercially available in pure form. In addition, isolation and purification of the relatively ionic organo-*f*-element complexes is often more difficult since chromatographic supports usually decompose the complexes, sublimations are often very low in yield, and recrystallizations are complicated by ionic redistribution reactivity. Characterization of organo-*f*-element complexes also tends to be more difficult, since the paramagnetism of many complexes precludes n.m.r. spectroscopy, the low volatility precludes mass spectrometry, and with some systems it has even been claimed that complete elemental analysis is unobtainable[8–11].

Despite the experimental difficulties and the ionic character, the organometallic chemistry of the *f* elements is very interesting and potentially very useful. It is well recognized that each of the *f*-element series of metals has a unique combination of physical properties which makes individual elements particularly useful in special applications[12–14]. Associated with these unique physical properties may be an organometallic chemistry which is unique for the lanthanides and for the actinides, and which may develop in importance as these areas are more thoroughly examined. The discovery in 1968 of uranocene, $[(C_8H_8)_2U]$[15], stimulated interest in the *f* elements owing to their most commonly recognized special property, the existence of *f* orbitals. More recently, the interesting developments in early transition metal chemistry have stimulated activity in the *f*-element area, since the lanthanides and the actinides, in a sense, are the earliest of the transition series metals. Although a number of similarities have recently been observed between the *f* elements and the early transition metals[16,17], it is the differences due to the special properties of the *f* elements which are likely to be most important.

Evidence of increasing interest in organo-*f*-element chemistry can be found in the increasing number of recent reviews in this area[6,7,18–27]. These sources provide compendia of known complexes, physical properties, spectroscopic properties, and structural details. Such information which will not be duplicated here so that emphasis can be placed on synthesis.

II. GENERAL SYNTHETIC CONSIDERATIONS

A. Special Properties of the *f* Elements and Their Organometallic Complexes

The lanthanide and actinide elements are differentiated from other metals by the fact that their valence orbitals are *f* orbitals. Many of the properties of the organometallic complexes of the lanthanide and actinide metals are determined by the fact that the radial extension of the *f* orbitals is not nearly as large as that of comparable *d* and *p* valence orbitals. Calculations on lanthanide ions, which have $[Xe]4f^n$ electron configurations, suggest that the $4f$ orbitals do not extend significantly beyond the filled $5s^25p^6$ orbitals of the inert gas xenon core[28]. Consistent with this, the trivalent lanthanide ions have very similar chemical properties regardless of the $4f^n$ electron configuration. The $5f$ orbitals have a greater radial extension than the $4f$ orbitals and are not shielded as much by the filled $6s^26p^6$ shell, but they still are less available than comparable *d* and *p* valence orbitals[7].

As a result of the limited radial extension of the *f* orbitals and the fact that the predominant oxidation states for organometallic lanthanides and actinides are relatively high (+3 and +4), the overlap between lanthanide and actinide valence orbitals and the orbitals of organic ligands is small. Consequently, organo-*f*-element complexes tend to be more ionic than their transition metal or main group metal analogues[3,6,7], and orbital effects are much less important in determining structure and stability in these complexes. Stability generalizations analogous to the 18-electron rule, which is based on full utilization of available orbitals, are not used with *f* elements. Instead, the stability and structure of organo-*f*-element complexes are determined primarily by the optimization of electrostatic interactions[22]. Stability is also enhanced by using large, bulky ligands which can completely occupy the coordination sphere of the metal and sterically block decomposition pathways. Hence, in general, the most stable organo-*f*-element complexes are those in which large organic anions can sterically saturate the coordination sphere of the metal without forming a complex with too high a net charge on the metal.

Two other features distinguish the *f* elements from the transition and main group metals: they are large and highly electropositive. The ionic radii of the *f* elements are 0.15–0.30 Å larger than comparable radii for transition metal ions of the same charge. Consequently, coordination numbers of 9–12 are not unusual for organometallic derivatives of these elements[22]. If these coordination positions are not filled by large anionic ligands, the metals readily form adducts with any available base. These electropositive elements have a strong affinity for oxygen in any form and ether adducts are particularly common. In the absence of coordinating solvents, the complexes frequently dimerize or oligomerize by interaction of the metal with whatever electron pair is most available in the complex.

B. General Synthetic Methods

The most common method of synthesizing organolanthanide and organoactinide complexes is an ionic metathesis involving a metal halide and a stable organic anion (equations 1–3)[a]. Most successful organo-*f*-element syntheses are of this type. When

$$LnX_3 + 3MC_5H_5 \longrightarrow [(C_5H_5)_3Ln] + 3MX \quad (1)$$

$$LnX_3 + 3MR \longrightarrow [R_3Ln] + 3MX \quad (2)$$

$$AnX_4 + 2M_2C_8H_8 \longrightarrow [(C_8H_8)_2An] + 4MX \quad (3)$$

these metathesis reactions fail, it is usually because the organic anion is too small to saturate sterically the coordination sphere of the metal. In such a case, a variety of other bonding arrangements of comparably low stability become available, and a complex and inseparable mixture of products often results (e.g. see Section III.A.2). Equations 1–3 also fail when the anion is too strongly reducing and the metal halide is reduced either to a lower oxidation state or to elemental metal.

An alternative and less common method of organo-*f*-element synthesis is the acid base metathesis exemplified in equations 4–5. This method has the advantage that the

$$[(C_5H_5)_4An] + HR \longrightarrow [(C_5H_5)_3AnR] + \tfrac{1}{2}(C_5H_6)_2 \quad (4)$$

$$Li[LnR_4] + 4HR' \longrightarrow Li[LnR'_4] + 4RH \quad (5)$$

by-products are easily separated organic molecules rather than alkali halides, a feature which simplifies purification and isolation procedures considerably.

A third general synthetic method is the reaction of the elemental metal directly with a neutral precursor to a stable anion (equations 6–8). This method includes reactions

$$Ln + 2C_5H_6 \xrightarrow{NH_3(l)} [(C_5H_5)_2Ln] + H_2 \quad (6)$$

$$An + 2C_8H_8 \longrightarrow [(C_8H_8)_2An] \quad (7)$$

$$Ln + HgR_2 \longrightarrow [R_2Ln] + Hg \quad (8)$$

of the metal with readily reducible organic molecules, and also transmetallation reactions (equation 8). The metals are commonly used in some reactive form: dissolved in liquid ammonia, as amalgams, or as finely divided powders generated by decomposition of the hydride or reduction of a halide. A recent development in this area is the exploration of the chemistry of the zero-valent *f*-element metal atoms using metal vaporization techniques (Chapter 13). This approach has been used successfully with precursors of anions common in *f*-element chemistry, and with a wide range of potentially new ligands for the *f* elements.

Each of the three synthetic methods described above has many possible variations. In practice, the application of these methods has often been investigated with only a few, presumably representative, systems. This is particularly true for the lanthanides,

[a]In addition to abbreviations used throughout this book, the symbol M is used in this chapter to indicate an alkali, alkaline earth or main group metal.

where each new reaction is rarely examined with each of the 14 non-radioactive members of the series. In this chapter, only syntheses which have been described in the literature for a specific metal–ligand combination are reported. Extension of these reactions to other closely related metals and ligands should be possible within the limits described in the following section.

C. Factors Affecting Selection of Metals in Synthesis

1. Lanthanides

Owing to the limited radial extension of the $4f$ orbitals, all 15 of the lanthanide ions are chemically very similar, regardless of their $4f^n$ electron configuration. Hence, excluding the radioactive promethium, there are 14 metals to choose from at the start of any synthetic organolanthanide project. This choice can be narrowed somewhat by considering secondary chemical factors and the preferred methods for physical characterization.

For each lanthanide metal, the +3 oxidation state is the most stable, and for many years this was the only oxidation state which received much attention in organolanthanide chemistry. The two main features which differentiate the individual trivalent ions are (1) a gradually diminishing radial dimension as atomic number increases across the series, i.e. the lanthanide contraction, and (2) the accessibility of other oxidation states. Since it is easier to saturate sterically the coordination sphere of the smaller, later lanthanides, these metals are employed more frequently in synthetic studies. Indeed, for several classes of organolanthanides, only the complexes of the later lanthanides are stable enough to be isolated and characterized. Samarium is often the largest metal which readily gives pure complexes of such a series. Unfortunately, this trend in organometallic stability is just the opposite of the availability of the metals in the earth's crust. La, Ce, Pr, and Nd are the most common lanthanides, being comparable in abundance to Co or Pb, whereas Sm–Lu are less abundant. These latter elements are not 'rare', however, since they still are more abundant than Ag or Hg. The opportunity to adjust size to meet stability or perhaps reactivity requirements is one of the unique features of the lanthanide series. No other collection of metals has so many chemically similar members which differ slightly in radial dimension.

Four of the lanthanides are distinguished by the fact that they have non-trivalent oxidation states which are accessible in organometallic systems. Specifically, europium, ytterium, and samarium form divalent organometallic complexes and cerium forms tetravalent species. The availability of Ce^{4+} has little effect on organolanthanide synthesis, since strongly oxidizing conditions are rarely used. The accessibility of Eu^{2+}, Yb^{2+}, and Sm^{2+}, on the other hand, is important in trivalent reactions involving either strongly reducing anions or reducing conditions, since reduction is a viable alternative reaction path. Of these three elements, europium is the most easily reduced and samarium is the least susceptible to reduction.

Simple chemical considerations, therefore, suggest that (1) the later lanthanide metals are preferrable for initial synthetic studies in organolanthanide chemistry and (2) europium, ytterbium, and samarium should be avoided if reduction is an undesirable possibility.

The choice of a preferred lanthanide can be further narrowed by considering the methods by which the complex will be identified. Most new organolanthanides are characterized by i.r., n.m.r., and near i.r.–visible spectroscopy, by magnetic moment and molecular weight measurement, by hydrolytic decomposition, by complexometric metal analysis or complete elemental analysis, and by X-ray crystallography when suitable single crystals are available. I.r. analysis, which provides information on the organic ligand, is generally applicable to all the metals. For a series of analogous complexes of the different metals, the i.r. spectra are nearly superimposable and strongly resemble the spectrum of the alkali metal derivative of the anion. ^{1}H and ^{13}C n.m.r. spectroscopy are more powerful methods of ligand characterization, but are not generally applicable. For the trivalent ions, only La^{3+} ($4f^0$), Sm^{3+} ($4f^5$, $\mu^{298} \approx 1.6$ BM) and Lu^{3+} ($4f^{14}$) form complexes which display sharp, unshifted n.m.r. spectra. Near-i.r.–visible spectroscopy and magnetic moment measurements identify the presence of the metal and its oxidation state, and are informative for all the lanthanides except La^{3+} ($4f^0$) and Lu^{3+} ($4f^{14}$). Hydrolytic decomposition provides qualitative but not necessarily quantitative information on the organic ligand, since the decomposition may not go to completion. Complexometric metal analysis[29] is a more reliable analytical method and obviously complete elemental analysis is most definitive. All of these analytical techniques, including molecular weight measurement, are generally applicable to all of the metals[29,30]. Obviously, X-ray crystallography is the most powerful solid-state characterization method and is applicable to all systems for which single crystals can be obtained.

It is usually desirable in organometallic chemistry to choose a system that can be analysed by n.m.r. spectroscopy and hence La, Sm, and Lu would be the most preferred metals. Unfortunately, each of these has some drawbacks. Lanthanum is the largest of the metals and therefore often forms less stable complexes. Samarium is on the borderline in this respect and also has a mode of more complicated reactivity via reduction to the divalent oxidation state. Lanthanum is also less desirable since the yields of its complexes are often substantially less than those of other members of the series (Section III.A.1). Lutetium, on the other hand, is ideal with respect to yield, but it is at least an order of magnitude more expensive than other late lanthanides, such as erbium and ytterbium, and is 30–100 times more expensive than samarium and lanthanum. Both lutetium and lanthanum have the disadvantage that their complexes are colourless. Hence, a sparingly soluble organometallic complex of these metals cannot be differentiated from an alkali metal halide by-product, and an alkali metal salt of the organic anion cannot be differentiated from a lanthanum or lutetium complex by colour or by near-i.r. or n.m.r. spectroscopy.

Since lanthanum and samarium may not form the most stable and easily isolable complexes, and since colourless lutetium cannot be followed visually in the workup of a reaction, other metals are usually chosen for an initial organolanthanide synthesis. Erbium and ytterbium are the most common choices, since they are the smallest, least expensive later lanthanides which allow the synthesis to be followed visually by colour. Ytterbium is more desirable in this regard for two reasons. First, the fine structure in the $^2F_{5/2} \leftarrow {}^2F_{7/2}$ transition at *ca.* 10,000 cm^{-1} is often sensitive to the ligand environment and can be used as fingerprint identification. Second, the position of the allowed $5d \leftarrow 4f$ charge-transfer band is in the visible region, which means that different ytterbium complexes can have different colours. In contrast, the near-i.r.–visible spectrum of erbium and the other lanthanides usually contains only LaPorte-forbidden $4f$–$4f$ transitions which are unaffected by the ligand environment owing to the limited radial extensions of the $4f$ orbitals. The importance of these spectral features can be appreciated by considering the colours of the various species which can be present in a $LnCl_3$–$C_5H_5^-$ reaction mixture. The complexes $[(C_5H_5)_3Er]$, $[(C_5H_5)_2ErCl]$, $[(C_5H_5)ErCl_2(thf)_3]$, and $ErCl_3$ are all orange–pink and cannot be differentiated by optical spectroscopy. In comparison, $[(C_5H_5)_3Yb]$ is dark green, $[(C_5H_5)_2YbCl]$ is orange–red, $[C_5H_5YbCl_2(thf)_3]$ is orange and $YbCl_3$ is colourless. Ytterbium is also more desirable since its organometallic complexes are frequently more soluble than those of other lanthanides. Ytterbium is a less desirable choice, however, if reduction to a divalent state is a possible complication. In practice, erbium is frequently used initially, since it is a small, stable, inexpensive, reduction-free system which allows the lanthanide products to be distinguished from alkali metal by-products by colour. Once synthetic details have been worked out for erbium, the analogous lutetium system frequently is examined on a small scale and characterized by n.m.r. spectroscopy.

For reactions in which divalent chemistry occurs, samarium and ytterbium are frequently examined together, since at least one of these metals will provide n.m.r. information regardless of the oxidation state: Sm^{3+} is accessible to n.m.r. and Yb^{2+} ($4f^{14}$) is diamagnetic. Europium is less desirable since it is much more expensive and both Eu^{2+} and Eu^{3+} are paramagnetic.

2. *Actinides*

There is relatively less choice in selecting actinide metals for synthesis. The only members of the series which can be handled without elaborate radio-chemical facilities are thorium and uranium, and these are the metals most commonly studied. For both of these metals, the most common oxidation state in organometallic complexes is the tetravalent state, which, in many respects, is synthetically analogous to the trivalent state for the lanthanides. In other properties, thorium and uranium differ and often provide complementary information. Uranium has a relatively accessible trivalent state, whereas Th^{3+} is much more difficult to form. Th^{f+} ($5f^0$) is diamagnetic and colourless whereas U^{4+} ($5f^2$) provides shifted n.m.r. spectra and is commonly green or red. Organothorium complexes are frequently more ionic, less soluble, less polymeric, and more stable than their uranium analogues, although these differences vary, depending on the ligand environment. In practice, both metals are usually examined in organoactinide synthesis.

Specific organo-*f*-element syntheses are described in the following sections which have been arranged arbitrarily according to ligand with separate subsections for the lanthanides and for the actinides.

III. CYCLOPENTADIENYL COMPLEXES

A. Trivalent Lanthanide Complexes

1. *Tris(cyclopentadienyl)lanthanides*

a. $[(C_5H_5)_3Ln]$. The most common method of synthesizing tris(cyclopentadienyl)lanthanide complexes is the reaction of a lanthanide trihalide with an alkali or alkaline earth metal salt of the cyclopentadienide anion (equation 9).

$$LnX_3 + \frac{3}{n}M(C_5H_5)_n \longrightarrow [(C_5H_5)_3Ln] + \frac{3}{n}MX_n \qquad (9)$$

Wilkinson and Birmingham synthesized the first organolanthanides in this manner by reacting lanthanide trichlorides with NaC_5H_5 for 2–4 h in thf at reflux[1,31]. Solvated complexes, $[(C_5H_5)_3Ln(thf)]$, are formed initially[6,19,32] and can be purified and freed from thf by sublimation along a Pyrex tube at 150–260°C. Yields in excess of 70% were reported for Ln = Ce, Pr, Nd, Sm, Gd, Dy, Er, and Yb[31]. For Ln = La, however, a yield of only 25% was found, possibly owing to the low solubility of $LaCl_3$ compared to other lanthanide trichlorides[33].

Solvate-free tris(cyclopentadienyl) complexes can be made directly by using arenes or diethyl ether as solvents. Reaction times are typically 16–20 h for syntheses conducted at reflux. Tris(cyclopentadienyl) complexes of Tb, Ho, and Tm were first prepared from NaC_5H_5 in benzene and from KC_5H_5 in ether. In the ether synthesis, yields of 34, 75, and 60%, respectively, were obtained[34]. $[(C_5H_5)_3Lu]$ has been obtained from NaC_5H_5 in benzene in 66% yield[34]. Complexes of these last four metals were again purified by sublimation. The preparations of $[(C_5H_5)_3Pr]$[35] and $[(C_5H_5)_3Tm]$[36] from KC_5H_5 in benzene and $[(C_5H_5)_3Yb]$ from KC_5H_5[6] and NaC_5H_5[37] in toluene have also been reported. In the latter case, purification was achieved by toluene Soxhlet extraction of the more soluble ytterbium complex.

The tris(cyclopentadienyl) complex of the remaining non-radioactive lanthanide, europium, requires a slightly modified synthesis, since it is not as thermally stable as the other members of the series and sublimation results in decomposition. The thf adduct $[(C_5H_5)_3Eu(thf)]$ can be obtained in 66% yield from $EuCl_3$ and NaC_5H_5 in thf at room temperature[38] or from the monocyclopentadienyl complex (Section III.A.3) according to equation 10[39]. In the latter reaction, $[(C_5H_5)EuCl_2(thf)_3]$ was chosen as

$$[(C_5H_5)EuCl_2(thf)_3] + 2NaC_5H_5 \xrightarrow{\text{thf}} [(C_5H_5)_3Eu(thf)] + 2NaCl \quad (10)$$

a starting material because it is easily purified and is soluble in thf. Purification of $[(C_5H_5)_3Eu(thf)]$ can be effected by recrystallization at low temperature, and thf can be removed by heating to 70°C at 0.1 Torr[39].

Tris(cyclopentadienyl)lanthanides can also be obtained directly from mono-cyclopentadienyl species by ligand redistribution. This reaction recently has been demonstrated for gadolinium (equation 11)[32] and emphasizes the ionic nature of this class of complexes.

$$3GdCl_3 + 3NaC_5H_5 \xrightarrow{\text{thf}} 3[C_5H_5GdCl_2] \longrightarrow [(C_5H_5)_3Gd(thf)] + 2GdCl_3 + 3NaCl \quad (11)$$

Syntheses of $[(C_5H_5)_3Ln]$ which are free of organic solvents have been reported according to equation 9 using either $Mg(C_5H_5)_2$ (m.p. 177–178°C) or $Be(C_5H_5)_2$ (m.p. 59–60°C) as low-melting cyclopentadienyl precursors/solvents. CeF_3, NdF_3, and SmF_3 react with $Mg(C_5H_5)_2$ at 220–260°C over a 0.5–4-h period to form the appropriate $[(C_5H_5)_3Ln]$ complexes in 48, 39, and 55% yield, respectively[40]. Unreacted $Mg(C_5H_5)_2$ is removed by sublimation at 80–120°C, and final purification is accomplished by sublimation of the product. Interestingly, the non-trivalent precursors, CeF_4 and SmI_2, also form the trivalent complexes $[(C_5H_5)_3Ln]$ in these melt reactions (equations 12 and 13). Lanthanide trichlorides have also been successfully

$$CeF_4 + 2Mg(C_5H_5)_2 \xrightarrow{220^\circ C} \underset{15\%}{[(C_5H_5)_3Ce]} \quad (12)$$

$$SmI_2 + Mg(C_5H_5)_2 \xrightarrow{220^\circ C} \underset{25\%}{[(C_5H_5)_3Sm]} \quad (13)$$

used as precursors in reactions with $Mg(C_5H_5)_2$ in open tubes. For samarium, however, this was not a clean synthesis since, despite a 6-fold excess of $Mg(C_5H_5)_2$, some $[(C_5H_5)_2SmCl]$ still formed[35]. $Be(C_5H_5)_2$ was employed as the molten reagent for the first non-radiochemical[41] preparation of $[(C_5H_5)_3Pm]$ (equation 14)[35]. Samarium and terbium complexes were prepared similarly.

$$2PmCl_3 + 3Be(C_5H_5)_2 \xrightarrow[\text{20 h}]{70^\circ C} 2[(C_5H_5)_3Pm] + 3BeCl_2 \quad (14)$$

In summary, synthesis of $[(C_5H_5)_3Ln]$ by metathesis according to equation 9 can be accomplished with a variety of cyclopentadienyl reagents and solvent systems. Considerable variation of reaction conditions is also possible with accompanying variation in yield. A general claim of >95% yields in the preparation of *f*-element cyclopentadienyl complexes has been made for syntheses with reaction times of about 200 h[42].

Alternative syntheses of [$(C_5H_5)_3Ln$] based on C_5H_6 rather than $C_5H_5^-$ are also available. The addition of C_5H_6 to ytterbium in liquid ammonia forms [$(C_5H_5)_3Yb$] as well as [$(C_5H_5)_2Yb$][34,43]. Vaporization of Nd atoms into a C_5H_6 matrix at −196°C, followed by extraction with thf yields [$(C_5H_5)_3Nd$][44]. Neither of these methods is superior to equation 9 in convenience or yield, however.

The formation of [$(C_5H_5)_3Ln$] complexes by thermal decomposition of [$(C_5H_5)_2Ln(BH_4)$][11] (Ln = Sm, Yb) and [$(C_5H_5)_2Gd(\mu\text{-}Me)_2AlMe_2$] (Section V.B) has also been reported.

b. Substituted cyclopentadienyl complexes. In general, the synthetic methods described for C_5H_5 complexes can be extended to substituted cyclopentadienyl ring systems such as $CH_3C_5H_4$, $(Me_2CH)C_5H_4$, $C_5(CH_3)_5$, $C_5(CH_3)_4(C_2H_5)$, and indenyl (C_9H_7). Complexes of the substituted ring systems are often more soluble, more volatile, and more easily crystallized. The bulkier, substituted ligands are also more effective in stabilizing the larger, earlier lanthanides. In some cases, however, synthesis of tris(cyclopentadienyl) species may be precluded, since the rings may be too large to fit around the metal centre. Rather few substituted species have been specifically described, but it is expected that increased effort will be seen in this area in the future.

The first reported complex of a substituted lanthanide was [$(CH_3C_5H_4)_3Nd$], prepared in 70% yield by reacting $NdCl_3$ with $NaCH_3C_5H_4$ for 2 h in thf[45]. Sublimation at 200°C produces X-ray quality crystals[46]. [$(CH_3C_5H_4)_3Tm$] has been prepared from $TmCl_3$ and $KCH_3C_5H_4$ in benzene and both of these metatheses seem generally applicable. However, in the preparation of [$(CH_3C_5H_4)_3Yb$] from $YbCl_3$ and $NaCH_3C_5H_4$ in thf, there is a preference for the sterically less crowded bis species [$(CH_3C_5H_4)_2YbCl$], even in the presence of an excess of $NaCH_3C_5H_4$[47]. Separation of this mixture requires a low-yield differential sublimation.

Tris(indenyl)lanthanide complexes, [$(C_9H_7)_3Ln(thf)$], have been prepared for Ln = La, Sm, Gd, Tb, Dy, and Yb by addition of NaC_9H_7 to $LnCl_3$ in thf at low temperature followed by stirring at room temperature for 3 h to 5 days[48]. Purification was effected by recrystallization.

Tris(isopropylcyclopentadienyl) complexes of La, Pr, and Nd have been reported, but no synthetic details were given[49].

c. Base adducts, [$(C_5H_5)_3LnB$]. As previously noted, when tris(cyclopentadienyl) complexes are synthesized in thf, they are initially isolated as the base adducts [$(C_5H_5)_3LnOC_4H_8$][6,19,32]. A variety of other good σ-donor bases will also interact with [$(C_5H_5)_3Ln$]. Some of these complexes can be synthetically and spectroscopically useful, since they have enhanced solubility in non-polar solvents. In general, the base adducts are formed by adding the base to a suspension of [$(C_5H_5)_3Ln$] in a non-coordinating solvent. In cases where the base is an inexpensive liquid, [$(C_5H_5)_3Ln$] is added directly to the base.

The cyclohexylisonitrile adducts [$(C_5H_5)_3LnCNC_6H_{11}$] constitute the best characterized class of base adducts, having been examined by i.r.[50] and n.m.r. spectroscopy[51–53] and X-ray crystallography[54]. These complexes are formed[50] by adding the isonitrile to a suspension of [$(C_5H_5)_3Ln$] in benzene. The suspension quickly becomes a solution as the adduct forms and dissolves. The complexes are formed in 70–80% yield, can be further purified by sublimation at 150–160°C, and are reported to have some solubility in pentane.

A triphenylphosphine adduct of ytterbium, [$(C_5H_5)_3Yb\{P(C_6H_5)_3\}$], was prepared similarly by heating the phosphine and [$(C_5H_5)_3Yb$] in benzene for 2 h[50]. A more extensive study of the interactions of phosphines with $[(C_5H_5)_3Yb]$ has been reported using $HP(C_6H_5)_2$, $H_2PC_6H_5$, $P(C_6H_{11})_3$, $HP(C_6H_{11})_2$, $H_2PC_6H_{11}$, and $(CH_3)_2PC_6H_5$. Although base adduct formation could be detected spectroscopically, isolable analytically pure products could not be obtained[37]. $[(C_5H_5)_3Yb]$ was chosen as the lanthanide complex in this case, because this compound is soluble in arene solvents, and because organoytterbium complexes have colours which are more intense and more sensitive to a given ligand environment than complexes of the other lanthanide metals. Attempts to form a CO adduct by these methods were unsuccessful[50].

The ammonia complexes $[(C_5H_5)_3LnNH_3]$ can be prepared by adding the cyclopentadienyl complex to liquid ammonia. For Ln = Pr and Sm, the adducts are stable under vacuum at 25°C, but lose NH_3 at 100–150°[31]. $[(C_5H_5)_3YbNH_3]$, on the other hand, can be sublimed intact at 150°C[50]. It has also been reported that this ytterbium complex can be formed by displacement of thf in solution[8]. The pyridine adduct $[(C_5H_5)_3Yb(NC_5H_5)]$ has been characterized by near-i.r.–visible spectroscopy[55].

A variety of transition metal complexes have been examined in the presence of $[(C_5H_5)_3Ln]$ complexes in CH_2Cl_2 and C_6H_6, and evidence for base adduct formation presented[56–58]. Interactions of the type $[(C_5H_5)_3LnONM'L_x]$, $[(C_5H_5)_3LnOCM'L_x]$, $[(C_5H_5)_3LnXM'L_x]$, and $[(C_5H_5)_3LnW(C_5H_5)_2H_2]$, were postulated, based on i.r. and n.m.r. shifts (M' = transition metal, L = other ligands). An η^2 alkyne complex, $[(C_5H_5)_3Ln(HC{\equiv}CC_6H_5)]$, was also postulated on the basis of n.m.r. data. Additional halide-bridged adducts include $[(C_5H_5)_3YbFU(C_5H_5)_3]$[59] and $[(C_5H_5)_3LnXH]$, postulated as a precursor to $[(C_5H_5)_2LnX]$[50] (see next section).

2. *Bis(cyclopentadienyl)lanthanide halides and derivatives*

a. $[(C_5H_5)_2LnX]$. The reaction of two equivalents of a cyclopentadienyl salt with a lanthanide trichloride forms the bis(cyclopentadienyl)lanthanide chlorides (equation 15). The reaction is most commonly conducted using NaC_5H_5 in thf at room

$$LnCl_3 + \frac{2}{n} M(C_5H_5)_n \longrightarrow [(C_5H_5)_2LnCl] + \frac{2}{n} MCl_n \qquad (15)$$

temperature with a 1–2 day reaction time. The complexes are monomeric in thf[60] and can be obtained as mono-thf solvates even after sublimation[61]. The most effective means of purification is removal of thf by rotary evaporation, followed by extraction into toluene. The complexes exist as unsolvated chloro-bridged dimers in arene solvents. Since solvated thf is often persistent, it is necessary to check the degree of solvation by infrared spectroscopy (absorptions at 1010–1050 and 880–890 cm^{-1})[62] and complexometric metal analysis[29], before using these complexes as precursors in quantitative reactions. For preparative-scale syntheses, sublimation is not a desirable method of purification, since the sublimation and decomposition temperatures are sufficiently close that low yields are usually obtained.

Yields of 50–70% have been reported for Ln = Sm, Gd, Dy, Ho, Er, Tm, Yb, and Lu according to equation 15 using NaC_5H_5, but the reaction fails for the earlier lanthanides[6,60]. For Ln = Nd only an inseparable mixture of $[(C_5H_5)_2NdCl]$ and $[(C_5H_5)_3Nd]$ is obtained, and for Ln = La, Ce, and Pr only traces of $[(C_5H_5)_3Ln]$ were found. It is a general result for many classes of organolanthanides that complexes of the smaller lanthanides, usually Sm through Lu, are readily obtainable, whereas complexes of larger, early lanthanide metals cannot be synthesized and/or purified. Recently the use of the sterically bulky pentamethylcyclopentadienyl ligand has allowed the isolation of a pure bis(cyclopentadienyl)neodymium complex[63] (equation 16) and this result indicates that by suitable choice of ligand, pure

$$NdCl_3 + 2LiC_5Me_5 \xrightarrow{\text{thf}} [(C_5Me_5)_2NdCl_2][Li(thf)_2] + LiCl \qquad (16)$$

bis(cyclopentadienyl)lanthanide halides can be obtained with the early members of the series. Analytically pure crystals of the bis(pentamethylcyclopentadienyl)neodymium complex were obtained by recrystallization at 0°C from pentane. The monomethyl-substitued complexes $[(CH_3C_5H_4)_2LnCl]$, have been prepared for Ln = Gd, Er, and Yb by reaction of $LnCl_3$ with 2 equivalents of $NaCH_3C_5H_4$ in thf, but attempts with earlier lanthanides were not described[60].

An alternative synthesis of $[(C_5H_5)_2LnCl]$ is the redistribution reaction[60] shown in equation 17, which again emphasizes the ionic character of these complexes.

$$LnCl_3 + 2[(C_5H_5)_3Ln] \xrightarrow[\text{2 days}]{\text{thf}} 3[(C_5H_5)_2LnCl(thf)] \qquad (17)$$

Since such redistribution reactions are so facile with the f elements, it is often important in synthesis to use exact stoichiometric amounts of reagents. Convenient solid cyclopentadienyl reagents used in the synthesis of $[(C_5H_5)_2LnCl]$ include solvent-free NaC_5H_5[64], $Mg(C_5H_5)_2$[65,66] and the air- and moisture-stable TlC_5H_5[11].

The additional syntheses[60] of $[(C_5H_5)_2LnX]$ shown in equations 18 and 19,

$$2[(C_5H_5)_3Er] + I_2 \xrightarrow[\text{3–4 h}]{\text{thf}} \underset{58\%}{2[(C_5H_5)_2ErI(thf)]} \qquad (18)$$

$$[(C_5H_5)_3Yb] + HCl \xrightarrow{\text{thf}} \underset{70\%}{[(C_5H_5)_2YbCl(thf)]} \qquad (19)$$

although not superior to equation 15 or 17 in convenience or yield, exemplify useful approaches to bis(cyclopentadienyl)lanthanide species in general (see next section) which avoid alkali metal halide by-products. An additional reaction of this type (equation 20) is reported to form $[(C_5H_5)_2SmCl]$ in 95% yield and $[(C_5H_5)_2YbCl]$ in 84% yield[67].

$$[(C_5H_5)_3Ln] + NH_4Cl \longrightarrow [(C_5H_5)_2LnCl] + NH_3 + \tfrac{1}{2}(C_5H_6)_2 \qquad (20)$$

b. $[(C_5H_5)_2Ln(\textit{non-halide anion})]$. A number of derivatives of $[(C_5H_5)_2LnX]$ have been reported in which the halide, X, is replaced by another anion. These complexes are most often prepared by the reaction of $[(C_5H_5)_2LnCl]$ with an alkali metal salt of the substituting anion (equation 21). In some cases, this approach fails and an alternate synthesis based on the acid HZ (equation 22) can be used. Derivatives in which Z is a carbanion will be described in Section V on alkyl complexes.

$$[(C_5H_5)_2LnCl] + MZ \longrightarrow [(C_5H_5)_2LnZ] + MCl \qquad (21)$$

$$[(C_5H_5)_3Ln] + HZ \longrightarrow [(C_5H_5)_2LnZ] + \tfrac{1}{2}(C_5H_6)_2 \qquad (22)$$

Bis(cyclopentadienyl) derivatives of erbium, $[(C_5H_5)_2ErZ]$, were first obtained according to equation 21 for Z = CH_3CO_2, HCO_2, CH_3O, and NH_2 by stirring the reagents in thf for 8–65 h at room temperature followed by an additional 3–5 h at reflux. Purification was achieved by sublimation and yields of 25–55% were reported[60]. Neither the formate nor the amid derivatives of Gd and Yb were isolable by this method, however, reportedly because, upon heating the reaction mixture, thermal decomposition occurs before sublimation. The ytterbium formate derivatives $[(C_5H_5)_2YbO_2CH]$ can be obtained via equation 22 (1 h, room temperature) since the by-product $(C_5H_6)_2$ does not interfere with purification by low temperature recrystallization. The ytterbium amide derivative, $[(C_5H_5)_3YbNH_2]$, is also obtainable by an alternate route, namely the thermolysis of the base adduct $[(C_5H_5)_3YbNH_3]$ at

200–250°C followed by sublimation at 230°[43,50]. Although $[(C_5H_5)_2LnZ]$ derivatives have been obtained for Yb according to equation 21 when Z = CH_3CO_2, $C_6H_5CO_2$, CH_3O, and C_6H_5O, only $[(CH_3C_5H_4)_2GdO_2CCH_3]$ was obtained by this route with gadolinium[60]. Hence, the generality of the equations 21 and 22 should not be assumed without further investigation. The synthesis of $[(C_5H_5)_2YbZ]$ derivatives via reaction 22, using HZ = $CH_3COCH_2COCH_3$, t-BuCOCH$_2$COBu-t, and $CH_3COCH{=}C(NHPh)CH_3$, has been investigated, but only with the latter reagent was a tractable compound isolated.[68]

Solvated borohydride derivatives, $[(C_5H_5)_2LnBH_4(thf)]$, have been synthesized from $NaBH_4$ according to equation 21 for Ln = Sm, Yb, Er[11] using reaction times of 2 days in thf at room temperature. For Ln = Yb and Er, the thf can be removed at 70°C under vacuum. The complex of the larger metal, samarium, decomposes to $[(C_5H_5)_3Sm]$ upon heating, however.

Dialkylphosphido derivatives of bis(cyclopentadienyl)lanthanides can be prepared using lithium reagents (equation 23). Di-t-butylphosphido complexes were synthesized

$$[(C_5H_5)_2LnCl] + LiPR_2 \longrightarrow [(C_5H_5)_2LnPR_2] + LiCl \qquad (23)$$

for Ln = Ho and Yb by this method, but analytically pure samples could not be obtained by this approach for Ln = Sm and Gd, R = t-Bu[69] and for Ln = Yb, R = C_6H_{11}[37]. For Ln = Er, R = t-Bu, the novel reaction 24 was used. The reagents

$$[(C_5H_5)_2ErCl] + (CH_3)_3SiP(t\text{-Bu})_2 \rightarrow [(C_5H_5)_2Er\{P(t\text{-Bu})_2\}] + (CH_3)_3SiCl \qquad (24)$$

were stirred neat at 50°C and 13 Torr for 6 days. The product was separated by filtration, purified by recrystallization from benzene, and ultimately obtained in 44% yield[69].

Triaryl germanium and tin derivatives of bis(cyclopentadienyl)lanthanides can be generated from lithium precursors in thf at room temperature (equation 25). The tin analogues have been obtained with Er and Yb. Attempts to obtain similar tin and germanium complexes of the earlier lanthanides and analogous silicon derivatives of any of the lanthanides were not successful[70].

$$[(C_5H_5)_2ErCl] + LiGe(C_6H_5)_3 \longrightarrow [(C_5H_5)_2Er\{Ge(C_6H_5)_3\}] + LiCl \qquad (25)$$

An extensive class of base adducts, analogous to those of the tris(cyclopentadienyl)lanthanide complexes, is not found for the bis(cyclopentadienyl) complexes, since in the non-polar solvents normally used in adduct forming reactions, the complexes already exist as chloride bridged adducts. Since the adduct–lanthanide interactions are often weak, the dimeric chloride adduct is frequently the most stable species[56]. In thf, however, monomeric thf adducts, such as $[(C_5H_5)_2LnCl(thf)]$ and $[(C_5H_5)_2LnBH_4(thf)]$[11], predominate. The di-t-butylphosphido species, $[(C_5H_5)_2Yb\{P(t\text{-Bu})_2\}]$, are reported to be trimeric in benzene, possibily owing to intermolecular Ln—P base adduct formation[71].

3. *Cyclopentadienyl lanthanide dihalides, $[(C_5H_5)LnX_2(THF)_3]$*

The single report[72] on the synthesis of monocyclopentadienyl lanthanide dihalides contains three preparative routes (equations 26–28), all of which had previously been

$$LnCl_3 + NaC_5H_5 \xrightarrow{\text{thf}} [(C_5H_5)LnCl_2(thf)_3] + NaCl \qquad (26)$$

$$[(C_5H_5)_3Ln] + 2LnCl_3 \xrightarrow{\text{thf}} 3[(C_5H_5)LnCl_2(thf)_3] \qquad (27)$$

$$[(C_5H_5)_2LnCl] + HCl \longrightarrow [(C_5H_5)LnCl_2(thf)_3] + \tfrac{1}{2}(C_5H_6)_2 \qquad (28)$$

used to synthesize $[(C_5H_5)_2LnX]$ complexes. Equation 26 is general for Ln = Sm, Eu, Gd, Dy, Ho, Er, and Lu and provides 30–60% of product purified by recrystallization at 0°C. The complexes are non-volatile and are unstable in the absence of a coordinating base such as thf. Thf can be removed with concomitant decomposition of the complex at 40°C under high vacuum. As in the case of $[(C_5H_5)_2LnCl]$, attempts to make $[(C_5H_5)LnCl_2]$ complexes for the larger metals, La–Nd, failed. Equation 27 was reported only for Er (61% yield) and equation 28 only for Yb (30% yield). Since the ionic cyclopentadienyl lanthanide complexes are prone to redistribution and since the reaction of 1 equivalent of NaC_5H_5 with $GdCl_3$ (equation 11) has also been found to form only $[(C_5H_5)_3Gd(thf)]$[32], it is obvious that subtle differences in handling these reactions can give products which differ in how the ligands are distributed.

B. Divalent Lanthanide Complexes

The +2 oxidation state is much more accessible for Eu, Yb, and Sm than for any of the other members of the lanthanide series and, at present, divalent organometallic lanthanide chemistry is limited to these three metals. Reduction potentials for the $Ln^{3+}–Ln^{2+}$ couple are reported[73] to be −0.34 V for Eu, −1.04 V for Yb, and −1.40 V for Sm (*vs*. NHE), indicating that Eu^{2+} should be the most stable and Sm^{2+} the most reactive of these divalent species. Bis(cyclopentadienyl)lanthanide complexes can be generated by either direct reaction of the elements with C_5H_6 or by reduction of trivalent species.

The first divalent complexes of this type were formed by taking advantage of the solubility of europium and ytterbium in liquid ammonia. For europium (equation 29),

$$Eu + 3C_5H_6 \xrightarrow{NH_3(l)} [(C_5H_5)_2EuNH_3] + C_5H_8 \quad (29)$$

the by-product was identified as cyclopentene by gas chromatography. The solvating ammonia can be removed at 120–200°C under vacuum and flash sublimation at 400–420°C yields pure $[(C_5H_5)_2Eu]$ in 20% yield[34].

The analogous ytterbium reaction, on the other hand, is more complex. Despite the reducing conditions, oxidation to trivalent species such as $[(C_5H_5)_3YbNH_3]$ occurs and sublimation of the crude reaction product generates three complexes[34,43]. $[(C_5H_5)_3Yb]$ sublimes first at 150°C, the desired $[(C_5H_5)_2Yb]$ sublimes next at 170°C (11% yield), and finally at 360°C another trivalent product is obtained, which may be $[(C_5H_5)_2YbNH_2]_2$ or $[(C_5H_5)_3YbN_2H_4]$ (formulae postulated on the basis of mass spectral measurements)[43]. The greater tendency of ytterbium to oxidize to trivalent species compared to europium is expected, based on the reduction potentials.

A cleaner, direct synthesis of bis(cyclopentadienyl)ytterbium is provided by the metal vaporization method (Chapter 13). Co-condensation of Yb atoms with C_5H_6 at −196°C, vacuum removal of excess of C_5H_6, and extraction of the remaining solids with thf generates $[(C_5H_5)_2Yb(thf)]$ in 27% yield based on the metal[44].

A variety of trivalent precursors and reducing agents can be used in solution syntheses of $[(C_5H_5)_2Yb]$ in thf. Sodium and ytterbium metal were the reducing agents initially used (equations 30–32), and equation 30 was reportedly most efficient, giving

$$[(C_5H_5)_2YbCl] + Na \longrightarrow [(C_5H_5)_2Yb] + NaCl \quad (30)$$

$$3[(C_5H_5)_2YbCl] + Yb \longrightarrow 3[(C_5H_5)_2Yb] + YbCl_3 \quad (31)$$

$$[(C_5H_5)_3Yb] + Na \longrightarrow [(C_5H_5)_2Yb] + NaC_5H_5 \quad (32)$$

a 39% yield[8]. Obviously, Na(Hg), NaK alloy, $NaC_{10}H_8$, Li and K can also be used as reducing agents, and even *t*-butyllithium functions in this capacity (equation 33)[74].

$$[(C_5H_5)_2YbCl]_2 + 2t\text{-}C_4H_9Li \rightarrow 2[(C_5H_5)_2Yb] + 2LiCl + C_4H_8 + C_4H_{10} \quad (33)$$

Syntheses of $[(C_5H_5)_2Yb]$ according to equations 30 and 32 are most conveniently conducted in toluene, since in this solvent the trivalent precursor is soluble and can be easily separated from the precipitated product by filtration.

Recently, several new syntheses of bis(cyclopentadienyl)ytterbium complexes have been discovered (equations 34–36) which demonstrate how easily Yb^{2+} can be

$$[(CH_3C_5H_4)_2YbCH_3]_2 \xrightarrow[80°C]{\text{pentane/Et}_2\text{O}} [(CH_3C_5H_4)_2Yb] \quad (34)$$

$$[(CH_3C_5H_4)_2YbCH_3]_2 \xrightarrow[\text{toluene, }5°C]{h\nu} [(CH_3C_5H_4)_2Yb] \quad (35)$$

$$[(CH_3C_5H_4)_2YbCH_3]_2 + H_2 \xrightarrow[50°C]{\text{toluene}} CH_4 + [(CH_3C_5H_4)_2Yb] \quad (36)$$

generated[75]. The most efficient of these reactions is the thermolysis (equation 34), in which 50% conversion is observed after 8 h. The solvent is extremely important: in toluene the thermolytic reaction requires months at 80°C. A 25% conversion after 32 h is observed for the photolysis (equation 35). Equation 36, which apparently proceeds through a hydride intermediate, $[(CH_3C_5H_4)_2YbH]$, based on deuterolysis and CH_3I decomposition reactions, is the slowest of these reactions, requiring 2 weeks for significant conversion. Since $[(CH_3C_5H_4)_2Yb]$ can also be obtained by reduction of $[(CH_3C_5H_4)_2YbCl]_2$ with NaK in toluene, and since this chloride is a precursor to $[(CH_3C_5H_4)_2YbCH_3]_2$, equations 34–36 are not superior preparative routes. Instead, their importance lies in the fact that they provide rather facile routes to the reactive divalent species, an important component of any potential catalytic cycle based on the Yb^{2+}/Yb^{3+} couple[75].

$[(C_5H_5)_2Sm(thf)]$ was first synthesized by reduction of $[(C_5H_5)_3Sm]$ with $KC_{10}H_8$ in thf[76], a variation of equation 32. The purple, thf-insoluble complex was purified by washing with thf. $[(CH_3C_5H_4)_2Sm]$ has been obtained similarly[66].

C. Tetravalent Lanthanide Complexes

The most accessible tetravalent lanthanide ion is Ce^{4+}, a powerful oxidizing agent ($E_{1/2} = 1.44–1.77$, depending on the acid medium[73]), strong enough to convert HCl to Cl_2 in aqueous solution. The only, stable, binary halide of Ce^{4+} is CeF_4, although the double salt $(C_5H_5NH)_2[CeCl_6]$ is a stabilized chloride form of this ion. The synthesis of a large number of Ce^{4+} organometallics including $[(C_5H_5)_4Ce]$[77], $[(\text{indenyl})_4Ce]$[77], $[(\text{fluorenyl})_4Ce]$[78], $[(C_5H_5)_3CeZ]$, and $[(\text{indenyl})_2CeZ_2]$(Z = Cl[79], CN, NCO, NCS, N_3[80], CH_3, C_2H_5, C_6H_5, $CH_2C_6H_5$, COC_6H_5[81], BH_4[82], SR[83]) has been claimed using the standard metathesis reactions used for trivalent lanthanides (equations 37–39). The validity of this work has repeatedly been challenged on several counts[6,11,84] and 'these reactions should be regarded with reserve and

$$(C_5H_5NH)_2[CeCl_6] + 4NaC_5H_5 \longrightarrow [(C_5H_5)_4Ce] + 4NaCl + 2C_5H_5N\cdot HCl \quad (37)$$

$$(C_5H_5NH)_2[CeCl_6] + 3[(C_5H_5)_4Ce] \longrightarrow 4[(C_5H_5)_3CeCl] + 2C_5H_5N\cdot HCl \quad (38)$$

$$[(C_5H_5)_3CeCl] + NaBH_4 \longrightarrow [(C_5H_5)_3CeBH_4] + NaCl \quad (39)$$

reinvestigated'[84]. First it is surprising that the strongly reducing anions used in these reactions do not reduce the Ce^{4+} precursor to Ce^{3+} products. This is particularly remarkable considering some of the vigorous reaction conditions, e.g. equation 38 is reportedly run in thf 'refluxed at 80–85°C for 4–5 h'[82]. Equally remarkable are the water and acid stabilities of products such as $[(C_5H_5)_4Ce]$[77] and $[(indenyl)_2CeCl_2]$[79]. Since many of these complexes are reported to crystallize from light petroleum, it would be desirable to have X-ray structural data to confirm the reported reaction chemistry.

The synthesis of $[(C_5H_5)_3CeOCH(CH_3)_2]$ has been reported (equation 40)[85] using the benzene-soluble isopropoxide precursor $[Ce(OR)_4(ROH)](R = CH(CH_3)_2)$. The complex was isolated in 4% yield by sublimation at 150°C. Attempts to prepare $[(C_5H_5)_4Ce]$ in an analogous manner failed[85].

$$2[Ce(OR)_4(ROH)] + 3Mg(C_5H_5)_2 \longrightarrow 2[(C_5H_5)_3CeOR] + 3Mg(OR)_2 + 2ROH \quad (40)$$

D. Tetravalent Actinide Complexes

1. *Tetrakis(cyclopentadienyl)actinides*

$[(C_5H_5)_4An]$ complexes are known for An = U, Th, Pa, and Np and are synthesized according to equation 41 by the metathesis methods described previously for $[(C_5H_5)_3Ln]$ (Section III.A.1). Again, a variety of cyclopentadienyl reagents and solvents can be used in these reactions. $[(C_5H_5)_4U]$ has been obtained in 6% yield from

$$AnX_4 + \frac{4}{n} M(C_5H_5)_n \longrightarrow [(C_5H_5)_4An] + \frac{4}{n} MX_n \quad (41)$$

UCl_4 and NaC_5H_5 in benzene at room temperature[86] and in 30% yield from UF_4 and a $Mg(C_5H_5)_2$ melt at 230°[40]. $[(C_5H_5)_4U]$ is also reportedly obtainable in 99% yield[87]. Purification was originally effected by a 10-day Soxhlet extraction with pentane[86]. $[(C_5H_5)_4Th]$ has been obtained in 40% yield from $ThCl_4$ and NaC_5H_5 in thf[88] (although the use of $ThCl_4$ in thf is not recommended[89]) and in 61% yield from ThF_4 and $Mg(C_5H_5)_2$ at 200°C[40]. Purification can be achieved by sublimation at 250–290°C. $[(C_5H_5)_4Pa]$ has been obtained in 54% yield according to equation 42

$$Pa_2O_3 \xrightarrow[600°C]{Cl_2/CCl_4} \xrightarrow[65°C]{2Be(C_5H_5)_2} [(C_5H_5)_4Pa] + 2BeCl_2 \quad (42)$$

and was purified by extraction for 110 h with benzene[90]. $[(C_5H_5)_4Np]$ can be generated in 77% yield by reacting $NpCl_4$ with KC_5H_5 in benzene at reflux and extracting with benzene for 50 h[91]. Attempts to prepare the analogous plutonium complex, $[(C_5H_5)_4Pu]$, from $[PuCl_6]^{2-}$ and $Mg(C_5H_5)_2$[92] or KC_5H_5[93] and from PuF_4 and $Be(C_5H_5)_2$[87], resulted in the formation of the trivalent complex, $[(C_5H_5)_3Pu]$ (Section III.E).

$[(C_5H_5)_4U]$ can also be synthesized directly from C_5H_6 and the metal, if the metal is prepared in the appropriate reactive form. Uranium, obtained by subliming Hg from U(Hg), forms $[(C_5H_5)_4U]$ when treated with C_5H_6 at room temperature[94], but since $[(C_5H_5)_3U]$ is also formed, this is not a useful preparation of the pure complex. Finely divided uranium powder obtained by decomposition of uranium hydride, on the other hand, does not react with C_5H_6 even at 150°[95].

Fully characterized tetrakis complexes of substituted cyclopentadienyl ligands have not yet been reported, possibly because steric crowding may make this type of complex less stable. Mass spectral evidence has been presented for the indenyl complex, $[(C_9H_7(_4Th]$, formed from $ThCl_4$ and 4 equivalents of KC_9H_7, but neither this complex nor the analogous uranium species could be isolated in a pure state[96]. Syntheses of $[(C_9H_7)_4Th]$ and $[(C_9H_7)_4U]$ in yields of 60 and 65% respectively, have been claimed[97], and the synthesis of $[\{Q(C_5H_4)_2\}_2U]$, where Q = CH_2, $(CH_3)_2Si$, and $CH_2CH_2CH_2$, has been noted[98], but no experimental details were presented in either case.

2. *Tris(cyclopentadienyl)actinide halides and derivatives*

a. $[(C_5H_5)_3AnX]$. Tris(cyclopentadienyl)actinide halides are known for An = U, Th, Pa, and Np, and are most commonly synthesized according to equation 43, which

$$AnX_4 + \frac{3}{n} M(C_5H_5)_n \longrightarrow [(C_5H_5)_3AnX] + \frac{3}{n} MX_n \qquad (43)$$

is analogous to the primary synthesis of $[(C_5H_5)_2LnCl]$ (equation 15). The first organoactinide complex, $[(C_5H_5)_3UCl]$, was obtained by this method in 82% yield from UCl_4 and NaC_5H_5 in thf at reflux[2]. Syntheses using KC_5H_5 in benzene[99] and TlC_5H_5 in thf at room temperature or in benzene at reflux[100] have also been reported. In the latter case, a yield of 90% was obtained. A detailed preparative description of the TlC_5H_5 reaction using dimethoxyethane as a solvent is available and has the advantage that sublimation is not required in order to obtain an analytically pure product[89]. $[(C_5H_5)_3UCl]$ is also formed when less than 3 equivalents of $C_5H_5^-$ are added to UCl_4[101], as described in Section III.D.3. The thorium analogue, $[(C_5H_5)_3ThCl]$, can be synthesized according to equation 43 from TlC_5H_5[89] or from KC_5H_5 in $(C_2H_5)_2O$ at room temperature[4]. $[(C_5H_5)_3NpCl]$ has been prepared from $NpCl_4$ and $Be(C_5H_5)_2$ at 70°C in 45% yield[102] and from $NpCl_4$ and KC_5H_5 in thf[103].

Several alternative syntheses of $[(C_5H_5)_3AnCl]$ based on $[(C_5H_5)_4An]$ precursors are available (equations 44–46) but, as in the case of the similar syntheses of

$$[(C_5H_5)_4U] + HCl \xrightarrow[\text{r.t.}]{C_6H_6} [(C_5H_5)_3UCl] + \tfrac{1}{2}(C_5H_6)_2 \qquad (44)^{97}$$

$$[(C_5H_5)_4An] + NH_4X \longrightarrow [(C_5H_5)_3AnX] + NH_3 + \tfrac{1}{2}(C_5H_6)_2 \qquad (45)^{67}$$

$$3[(C_5H_5)_4Th] + ThCl_4 \longrightarrow 4[(C_5H_5)_3ThCl] \qquad (46)^{104}$$

$[(C_5H_5)_2LnCl]$, the more direct one-step reaction (equation 43) is usually preferable. The only reported synthesis of $[(C_5H_5)_3PaCl]$ is via equation 45, however[67].

$[(C_5H_5)_3UBr]$ and $[(C_5H_5)_3UI]$ can be prepared according to equation 43 by reaction of UBr_4 and UI_4 with KC_5H_5 in benzene[105]. $[(C_5H_5)_3UF]$ can be obtained via equation 45 using NH_4F or by equation 47 at 150–180°C in a sealed glass tube[87].

$$[(C_5H_5)_3UCl] + NaF \longrightarrow [(C_5H_5)_3UF] + NaCl \qquad (47)$$

A variety of tris(cyclopentadienyl)actinide halide complexes involving substituted cyclopentadienyls have been synthesized via the metathesis reaction 43. All six indenyl derivatives, $[(C_9H_7)_3AnX]$ (An = U, Th; X = Cl[96], Br[106], I[107]), are known and were obtained in 18–50% yield by reaction of KC_9H_7 or NaC_9H_7 with the appropriate actinide halide in thf for 3–10 days at room temperature. Purification was accomplished by Soxhlet extraction with pentane or benzene. $[(CH_3C_5H_4)_3UCl]$ has been prepared from $NaCH_3C_5H_4$ in benzene at reflux[108]. $[(C_6H_5CH_2C_5H_4)_3UCl]$, which was synthesized for crystallographic purposes, was obtained from $TlCH_2C_6H_5C_5H_4$ in dme[109].

The synthesis of a mixed tris(cyclopentadienyl) complex, $[(C_5H_5)(CH_3C_5H_4)_2UCl]$, has also been reported (equation 48)[110]. $[(C_5Me_4Et)(C_5H_5)_2UCl]$ is also known[84].

$$[(C_5H_5)UCl_3(thf)_2] + 2TlCH_3C_5H_4 \xrightarrow[16\ h]{thf} [(C_5H_5)(CH_3C_5H_4)_2UCl] + 2TlCl \quad (48)$$

b. *[(C₅H₅)₃An(non-halide ion)]*. The primary methods for replacing the halide ion in $[(C_5H_5)_3UX]$ with a new anion are metathesis reactions (equations 49 and 50), which are normally conducted in benzene or thf at room temperature for a few hours to a few days. Purification of these derivatives is accomplished by alkane or arene extraction or by sublimation. Reaction 49 has been used to prepare complexes in which Z = BH_4[100], BH_3CN[111], $B(C_6H_5)_3CN$[111], SCN[20], OR[25,112–115], and $C(CN)_3$[108], as well as a large number of alkyl and aryl derivatives which will be discussed in Section V. The synthesis of the oligomeric complexes $[(C_5H_4R)_3UNCC(CN)_2]$[108] via reaction 49 is unusual in that it can be conducted in water. Reaction 50 has been used to form mononuclear complexes, $[(C_5H_5)_3UZ]$, in which Z = NO_3, ClO_4, ReO_4, and $B(C_6H_5)_4$, and dinuclear complexes, $[\{(C_5H_5)_3U\}_2Z']$, in which Z′ = SO_4 and C_2O_4[67]. The relationship of these species to the $[(C_5H_5)_3U]^+$ ions, isolated from aqueous solutions of $[(C_5H_5)_3UCl]$ by addition of $H_4[SiW_{12}O_{40}]$, $H_2[PtCl_6]$, $K_3[Cr(NH_3)_2(SCN)_4]$, and KI_3[2], has not been determined.

$$[(C_5H_5)_3AnCl] + MZ \longrightarrow [(C_5H_5)_3AnZ] + MCl \quad (49)$$

$$[(C_5H_5)_4An] + NH_4Z \longrightarrow [(C_5H_5)_3AnZ] + NH_3 + \tfrac{1}{2}(C_5H_6)_2 \quad (50)$$

A variety of alternative syntheses for $[(C_5H_5)_3AnZ]$ are available, many of which are specific for a given Z. Most of these relatively high yield reactions have been reported for uranium only and attest to the stability of the $(C_5H_5)_3U$— unit and its capacity for diverse chemistry. $[(C_5H_5)_3UOH]$ has been prepared according to reaction 49 from $(C_5H_5)_3UF$ and NaOH in the solid state[97]. The sulphur analogue $[(C_5H_5)_3USH]$ has been obtained according to reaction 44 from $[(C_5H_5)_4U]$ and H_2S[97]. Pyrolysis of $[(C_5H_5)_3UOH]$ and $[(C_5H_5)_3USH]$ gives the bridged species $[\{(C_5H_5)_3U\}_2O]$ and $[\{(C_5H_5)_3U\}_2S]$, which also can be synthesized according to equations 51–53[87,97]. The alkoxy derivative $[(C_5H_5)_3U\{O(n\text{-}Bu)\}]$, has been obtained by the U^{3+}-assisted decomposition of thf at reflux (equations 54[4] and 55[116]) and by a more conventional route starting with a chloroalkoxyuranium complex (equation 56, R = *n*-Bu, CH_3)[4]. Alkoxy derivatives, $[(C_5H_5)_3AnOR]$, An = U, Th, can also be prepared from alkyl complexes $[(C_5H_5)_3AnR']$ by reaction with ROH (equation 57)[25]. An interesting addition reaction (equation 58) has been used to prepare $[(C_5H_5)_3UBF_4]$[97]. Mixed alkyl borohydride complexes, $[(C_5H_5)_3UBH_3R]$, can be obtained by alkylation of the parent borohydride with R_3B where R = C_2H_5 or C_6H_5 (equation 59)[111].

$$2[(C_5H_5)_3UCl] + Ag_2O \longrightarrow [\{(C_5H_5)_3U\}_2O] + 2AgCl \quad (51)$$

$$2[(C_5H_5)_3U(thf)] + \tfrac{1}{2}O_2 \longrightarrow [\{(C_5H_5)_3U\}_2O] \quad (52)$$

$$2[(C_5H_5)_3UBr] + K_2S \longrightarrow [\{(C_5H_5)_3U\}_2S] + 2KBr \quad (53)$$

$$UCl_3 + 3NaC_5H_5 + C_4H_8O \longrightarrow [(C_5H_5)_3U\{O(n\text{-}Bu)\}] \quad (54)$$

$$UCl_4 + 2KC_5H_5 + 2Na \longrightarrow [(C_5H_5)_3U\{O(n\text{-}Bu)\}] \quad (55)$$

$$UCl_4 + NaOR \longrightarrow [UCl_3OR] \xrightarrow{3NaC_5H_5} [(C_5H_5)_3UOR] + 4NaCl \quad (56)$$

$$[(C_5H_5)_3AnR'] + ROH \longrightarrow [(C_5H_5)_3AnOR] + R'H \quad (57)$$

$$[(C_5H_5)_3UF] + BF_3 \longrightarrow [(C_5H_5)_3UBF_4] \quad (58)$$

$$[(C_5H_5)_3UBH_4] + R_3B \longrightarrow [(C_5H_5)_3UBH_3R] + HBR_2 \quad (59)$$

An alternative method of synthesizing $[(C_5H_5)_3UZ]$ complexes is to add $C_5H_5^-$ to precursors rich in Z and deficient in C_5H_5. This method is less common, however, since few such starting materials exist. The homoleptic amide $[U(NEt_2)_4]$[117] is one of the few members of this rare class and can be used to prepare $[(C_5H_5)_3UNEt_2]$[118] (equation 60). The thio derivatives $[(C_5H_5)_3USR]$ (R = C_2H_5 and C_6H_5) have also

$$UCl_4 + 4LiNEt_2 \longrightarrow [U(NEt_2)_4] \xrightarrow{3C_5H_6} [(C_5H_5)_3UNEt_2] + 3HNEt_2 \quad (60)$$

been obtained from cyclopentadienyl-deficient precursors[118]. In this case (equation 61), these complexes are ligand redistribution by-products in the synthesis of $[(C_5H_5)_2U(SR)_2]$ from $[(C_5H_5)_2U(NEt_2)_2]$ (Section III.D.3). $[(C_5H_5)_3U(SBu\text{-}t)]$ is similarly a by-product (5%) of reaction 61 when it is conducted in pentane, although in thf no redistribution occurs for this thio group[118].

$$[(C_5H_5)_2U(NEt_2)_2] + 2HSR \longrightarrow [(C_5H_5)_2U(SR)_2] + [(C_5H_5)_3USR] + HNEt_2 \quad (61)$$

3. *Bis(cyclopentadienyl)actinide dihalides and derivatives*

The synthesis of $[(C_5H_5)_2UCl_2]$ was reported[119] in 1971 via the metathesis reaction normally used for mixed cyclopentadienyl halide complexes, namely the reaction of UCl_4 and a cyclopentadienyl salt in the appropriate stoichiometry (equation 62).

$$UCl_4 + 2TlC_5H_5 \xrightarrow[3\text{ h, r.t.}]{\text{dme}} [(C_5H_5)_2UCl_2] + 2TlCl \quad (62)$$

However, subsequent studies[101,109,120] indicated that the product of this reaction was not the expected $[(C_5H_5)_2UCl_2]$, but instead a mixture of $[(C_5H_5)_3UCl]$ and $[(C_5H_5)UCl_3(dme)]$[121] (equation 63). If the formally 8-coordinate[122] $[(C_5H_5)_2UCl_2]$ did form, it was apparently unstable with respect to the 10-coordinate[108,123] $[(C_5H_5)_3UCl]$ and the 8-coordinate[101] $[(C_5H_5UCl_3(dme)]$. This reaction demonstrates the importance of coordinative saturation in organo-*f*-element chemistry and the tendency of *f*-element reaction mixtures to undergo redistribution reactions to form products with higher overall coordinative saturation.

$$2UCl_4 + 4TlC_5H_5 \xrightarrow[3\text{ h, r.t.}]{\text{dme}} [(C_5H_5)_3UCl] + [(C_5H_5)UCl_3(dme)] + 4TlCl \quad (63)$$

The only report of a bis(cyclopentadienyl)thorium dihalide complex is the reaction of ThI_4 with $Mg(C_5H_5)_2$ in the melt, which produced a compound 'which appeared to be $[(C_5H_5)_2ThI_2]$'[40]. No other details were given, however, and this also may be a mixture of the tris- and monocyclopentadienyl complexes. Attempts to form $[(C_5H_5)_2NpCl_2]$ from $NpCl_4$ and 2 equivalents of KC_5H_5 formed only $[(C_5H_5)_4Np]$ and $[(C_5H_5)_3NpCl]$, identified by Mössbauer spectroscopy[103].

Although $[(C_5H_5)_2UCl_2]$ is apparently too coordinatively unsaturated to be stable, other complexes of general formula $[(\text{cyclopentadienyl})_2U(\text{anion})_2]$ can be obtained by increasing the size and/or coordination number of the ligands or by connecting the ligands with a bridging group. For example, the bis(diethylamido) complex $[(C_5H_5)_2U(NEt_2)_2]$ is stable and can be obtained from the reaction of $[U(NEt_2)_4]$ with

C_5H_6 in pentane (equation 64)[118]. As one of the only readily available bis(cyclopentadienyl)uranium(IV) complexes, $[(C_5H_5)_2U(NEt_2)_2]$ is an important precursor for other members of this class of complexes. Reactions with substrates, HZ, which are more acidic than diethylamine, provide a general route to $[(C_5H_5)_2UZ_2]$ complexes (equation 65). For R = *t*-butyl, reactions in thf form the pure bis(thio) derivative, whereas in pentane some ligand redistribution occurs to form the more highly coordinated $[(C_5H_5)_3USBu\text{-}t]$ complex. For R = Et and C_6H_4OH, the thio ligands are barely large enough to stabilize the bis(cyclopentadienyl) species, and reaction 65 results in mixtures of $[(C_5H_5)_2U(SR)_2]$ and $[(C_5H_5)_3USR]$. The reaction of $[(C_5H_5)_2U(NEt_2)_2]$ with diprotic species, such as *o*-mercaptophenol, catechol, ethane-1,2-dithiol and toluene-3,4-dithiol, forms complexes which may be bridged dimers (equation 66)[118].

$$[U(NEt_2)_4] + 2C_5H_6 \longrightarrow [(C_5H_5)_2U(NEt_2)_2] + 2HNEt_2 \qquad (64)$$

$$[(C_5H_5)_2U(NEt_2)_2] + 2RSH \longrightarrow [(C_5H_5)_2U(SR)_2] + 2HNEt_2 \qquad (65)$$

$$2[(C_5H_5)_2U(NEt_2)_2] + 2HSCH_2CH_2SH \longrightarrow [(C_5H_5)_2U(SCH_2CH_2S)]_2 + 4HNEt_2 \qquad (66)$$

The complex $[(C_5H_5)_2U(BH_4)_2]$ has been reported[124] and found to contain borohydride ligands which are tridentate. This compound is reportedly synthesized from $[UCl_2(BH_4)_4]$ generated *in situ* in thf or dme (equation 67) or from the mixture formed from UCl_4 and $2TlC_5H_5$ in dme (equation 68). In this latter case, the

$$UCl_4 + 2NaBH_4 \xrightarrow{-2NaCl} [UCl_2(BH_4)_2] \xrightarrow{2TlC_5H_5} [(C_5H_5)_2U(BH_4)_2] + 2TlCl \qquad (67)$$

$$2UCl_4 + 4TlC_5H_5 \xrightarrow{-4TlCl} [(C_5H_5)_3UCl] + [(C_5H_5)UCl_3(DME)] \xrightarrow{4NaBH_4} 2[(C_5H_5)_2U(BH_4)_2] + 4NaCl \qquad (68)$$

multidentate nature of the BH_4 ligand may drive the redistribution reactions (which presumably are occurring) to the observed product. Crystals of $[(C_5H_5)_2U(BH_4)_2]$ were obtained by sublimation at 60°C. The acetylacetonate complexes, $[(C_5H_5)_2U(acac)_2]$[125] and $[(C_5H_5)_2UCl(acac)]$, and the poly-l-pyrazolylborate complexes, $[(C_5H_5)_2U\{HB(C_3H_3N_2)_3\}_2]$ and $[(C_5H_5)_2UCl\{HB(C_3H_3N_2)_3\}]$, are also known[84].

The alternative method to stabilize '$[(C_5H_5)_2UX_2]$' complexes with respect to ligand redistribution reactions is to utilize substituted cyclopentadienyl groups, and this has proved to be one of the most important uses of substituted ring systems in *f*-element chemistry. Early attempts to make the sterically bulkier and possibly more stable indenyl derivatives $[(C_9H_7)_2UCl_2]$ and $[(C_9H_7)_2ThCl_2]$ (equation 69) did not produce pure complexes, although mass spectral evidence for these species was obtained[96]. The synthesis of $[(C_9H_7)_2UI_2]$ according to equation 69 subsequently has been reported, but only analytical data were presented[107].

$$AnX_4 + 2KC_9H_7 \xrightarrow{thf} [(C_9H_7)_2AnX_2] + 2KX \qquad (69)$$

Reactions of bridged dicyclopentadienyl ligands, $LiC_5H_4QC_5H_4Li$ [Q = CH_2, $(CH_3)_2Si$, and $CH_2CH_2CH_2$], with UCl_4 were examined in efforts to form bis(cyclopentadienyl) complexes in which the rings were tied together and thereby stabilized with respect to redistribution (equation 70)[98]. Instead of the desired monomeric product, $[\{CH_2(C_5H_4)_2\}UCl_2]$, a dimeric complex which had incorporated LiCl was formed. Although the uranium atoms in this pentane-soluble species were

$$2UCl_4 + 2CH_2(C_5H_4Li)_2 \xrightarrow[-78°C]{thf} [\{CH_2(C_5H_4)_2\}_2U_2Cl_5]Li(thf)_2 + 3LiCl \quad (70)$$

bridged by three chloride anions, and the lithium ion interacted with four chlorides, this species reacted like a monomeric dichloride in substitution reactions (equation 71). With Lewis bases such as 2,2′-bipyridyl (bipy) and 1,10-phenanthroline, monomeric base adducts, e.g. $[\{CH_2(C_5H_4)_2\}UCl_2(bipy)]$, were formed[98].

$$[\{Q(C_5H_4)_2\}_2U_2Cl_5]Li(thf)_2 + 4NaBH_4 \longrightarrow 2[\{Q(C_5H_4)_2\}U(BH_4)_2] + LiCl + 4NaCl \quad (71)$$

Although the bridged dicyclopentadienyl ligands provide access to bis(cyclopentadienyl)actinide dihalides, large-scale synthetic utilization of this approach is inconvenient since high dilution conditions are sometimes necessary in order to avoid polymerization[126]. Hence, subsequent efforts to form stable derivatives of '$(C_5H_5)_2UCl_2$' avoided these bridged systems in favour of sterically bulky cyclopentadienyl ligands. One interesting approach was the use of the carboranyl analogue of $C_5H_5^-$, namely $C_2B_9H_{11}^{2-}$ (equation 72)[127]. Ligand redistribution is not a problem in this case, since the tris- and tetrakis(carboranyl) complexes not only would be sterically crowded but also would have an unfavourably high negative charge. In fact, reactions involving excess of $C_2B_9H_{11}^{2-}$ formed only the bis(carboranyl) product[127].

$$UCl_4 + 2C_2B_9H_{11}^{2-} \longrightarrow [(C_2B_9H_{11})_2UCl_2]^{2-} + 2Cl^- \quad (72)$$

A similar result was obtained using the C_5Me_4Et ligand. Treatment of UCl_4 with 4 equivalents of $(C_5Me_4Et)Sn(n\text{-}Bu)_3$ in toluene at reflux forms only the bis(cyclopentadienyl) species (equation 73)[128]. In contrast, $C_5H_5Sn(n\text{-}Bu)_3$ reacts with UCl_4 to form $[(C_5H_5)_3UCl]$[128].

$$UCl_4 + 2(C_5Me_4Et)Sn(n\text{-}Bu)_3 \xrightarrow[120°C]{toluene} [(C_5Me_4Et)_2UCl_2] + 2(n\text{-}Bu)_3SnCl \quad (73)$$

The most widely used synthesis of bis(cyclopentadienyl) actinide dihalides involves C_5Me_5, a ligand which previously had proven to be extremely useful in early transition metal chemistry[17]. This synthesis (equation 74), like equations 72 and 73, forms the

$$AnCl_4 + \underset{excess}{C_5Me_5^-} \xrightarrow[100°C]{toluene} [(C_5Me_5)_2AnCl_2] + 2Cl^- \quad (74)$$

bis species, $[(C_5Me_5)_2AnCl_2]$, from an excess of cyclopentadienyl reagent[7]. The synthesis has been reported for both U and Th using LiC_5Me_5 and the Grignard reagent $(C_5Me_5)MgCl{\cdot}thf$[7,129]. Since the uranium derivative is less soluble than the thorium derivative (an unusual situation), it has been proposed that these species have different structures[129]. Both are reportedly monomeric in benzene, however[7]. These $[(C_5Me_5)_2AnCl_2]$ complexes should be excellent precursors and extensive derivatization is expected. These complexes have already proven to be valuable precursors to $[(C_5Me_5)_2AnR_2]$ and $[(C_5Me_5)_2AnH_2]_2$ complexes as described in Section V.

4. *Cyclopentadienyl actinide trihalides and derivatives*

In contrast to '$(C_5H_5)_2UCl_2$', some mono(cyclopentadienyl)uranium trihalides appear to be stable with respect to disproportionation when isolated as oxygen base adducts. Hence $[(C_5H_5)UCl_3dme]$[121] and $[(C_5H_5)UX_3(thf)_2](X = Cl, Br)$[110,130] can be synthesized by metathesis using one equivalent of a cyclopentadienyl salt with a uranium tetrahalide in the appropriate solvent (equation 75). When

$$UX_4 + TlC_5H_5 \xrightarrow{\text{thf}} [C_5H_5UX_3(thf)_2] + TlX \tag{75}$$

$[UCl_4(t\text{-}BuCONMe_2)_2]$ and $[UCl_4(MeCONMe_2)_{2.5}]$ are used as precursors in reaction 75, the corresponding $[(C_5H_5)UCl_3(t\text{-}BuCONMe_2)_2]$ and $[(C_5H_5)UCl_3(MeCONMe_2)_2]$ complexes are obtained in 85 and 70% yields, respectively. The analogous reactions involving $[UBr_4(t\text{-}BuCONMe_2)_2]$ and $[UBr_4(OPPh_3)_2]$ fail, however, forming mixtures which contain $[(C_5H_5)_3UBr]$. Surprisingly, $[(C_5H_5)UBr_3(t\text{-}BuCONMe_2)_2]$ and $[(C_5H_5)UBr_3(OPPh_3)_2]$ can be obtained in *ca.* 60% yield from $[(C_5H_5)UBr_3(thf)_2]$ by displacement (equation 76)[110]. Hence, not only the nature of the anionic ligands and bases, but also the method of combination of these components is important in the synthesis of monocyclopentadienyl complexes which are marginally stable with respect to disproportionation.

$$[(C_5H_5)UBr_3(thf)_2] + 2OPPh_3 \xrightarrow{\text{thf}} [(C_5H_5)UBr_3(OPPh_3)_2] + 2thf \tag{76}$$

Stable monocyclopentadienyl complexes of other actinides have not been described. The reaction of one equivalent of KC_5H_5 with $NpCl_4$ in thf was observed to form only $[(C_5H_5)_4Np]$ and $[(C_5H_5)_3NpCl]$[103].

Several substituted cyclopentadienyl actinide trihalide complexes are known, including $[(MeC_5H_4)UCl_3(thf)_2]$[101,110], $[(C_5Me_5AnCl_3(thf)_2](An = Th, U)$[27], $[(C_5Me_5)ThCl_3(dme)]$[7], $[(C_5Me_4Et)UCl_3(RCONR'_2)_2]$[84], and $[(C_9H_7)AnX_3(thf)_2](An = Th, U; X = Cl, Br)$[131]. All of these were prepared by standard metathesis routes. The indenyl derivatives can also be prepared by ligand redistribution reactions (equation 77). Triphenylphosphine oxide adducts of the indenyl species can be prepared by displacement of thf (equation 78) or from $[AnX_4(OPPh_3)_2]$ and NaC_9H_7. The $OPPh_3$ adducts are unstable in thf at 30–40°C and when 10% pentane is added to a thf solution (equation 79)[131].

$$[(C_9H_7)_3UCl] + 2UCl_4 \xrightarrow[\text{2 days}]{\text{thf}} 3[(C_9H_7)UCl_3(thf)_2] \tag{77}$$

$$[(C_9H_7)AnX_3(thf)_2] + n\,OPPh_3 \xrightarrow[<0°C]{\text{thf}} [(C_9H_7)AnX_3(thf)_{2-n}(OPPh_3)_n] \qquad n = 1, 2 \tag{78}$$

$$3[(C_9H_7)AnX_3(OPPh_3)_2] \longrightarrow [(C_9H_7)_3AnX] + 2[AnX_4(OPPh_3)_2] + 2OPPh_3 \tag{79}$$

Several examples of monocyclopentadienyl complexes in which the halides have been replaced are known. When $[U(NEt_2)_4]$ is treated with excess of C_5Me_5H, only the monocyclopentadienyl complex is formed (equation 80)[7]. Substitution of one

$$[U(NEt_2)_4] + C_5Me_5H \longrightarrow [(C_5Me_5)U(NEt_2)_3] + HNEt_2 \tag{80}$$

chloride in $[(C_5H_5)UCl_3(thf)_2]$ can be effected by a poly-1-pyrazolylborate ligand (equation 81).[130]. $[(C_5H_5)UCl_2(acac)]$, $[(C_5H_5)UCl(acac)_2]$, and $[(C_5H_5)UCl\{HB(C_3H_3N_2)_3\}_2]$ are also known[84].

$$[(C_5H_5)UCl_3(thf)_2] + K[HB(C_3H_3N_2)_3] \longrightarrow [(C_5H_5)UCl_2\{HB(C_3H_3N_2)_3\}] + KCl \tag{81}$$

E. Trivalent Actinide Complexes

1. Tris(cyclopentadienyl)actinides and derivatives

As atomic number increases across the actinide series, the trivalent oxidation state becomes relatively more stable with respect to the tetravalent state[7,13]. This trend is reflected in the trivalent cyclopentadienyl series in that the number of actinide metals for which trivalent cyclopentadienyl complexes have been reported is greater than the number of metals for which tetravalent cyclopentadienyl complexes are known. The trivalent syntheses for the actinides with higher atomic number are also chemically more straightforward. At present, $[(C_5H_5)_3An]$ complexes have been reported for An = Th, U, Np, Pu, Am, Cm, Bk, and Cf. The most common method of synthesis is metathesis involving a cyclopentadienyl salt and the actinide trihalide. For the earlier members of the series, reductive methods using tetravalent precursors are equally common.

For Pu[132], Am[133], Cm[134,135] Bk[35], and Cf[35], the solvent-free metathesis of the trihalide and $Be(C_5H_5)_2$ in the melt, conducted on a microgram scale, forms the desired complexes in good yield (equation 82). $[(C_5H_5)_3Pu]$ can also be synthesized in thf solution using either $Mg(C_5H_5)_2$ or NaC_5H_5. Yields of 75% are reported for these reactions, which require 3 h and 10 days, respectively. Using the alternative tetravalent precursor $Cs_2[PuCl_6]$, $[(C_5H_5)_3Pu]$ can be prepared in 75% yield from $Mg(C_5H_5)_2$ in thf in 2–3 min![136].

$$2AnCl_3 + 3Be(C_5H_5)_2 \xrightarrow{70°C} 2[(C_5H_5)_3An] + 3BeCl_2 \qquad (82)$$

The earlier actinide complexes, $[(C_5H_5)_3Np(thf)_3]$ and $[(C_5H_5)_3U(thf)]$, can be obtained by potassium reduction of the appropriate $[(C_5H_5)_3AnCl]$ in thf at room temperature[103]. Li(Hg) can also be used as a reductant[137]. Solvent-free $[(C_5H_5)_3U]$ can be prepared from the insoluble trivalent precursor UCl_3 by reaction with KC_5H_5 in benzene at reflux for long time periods[138] (equation 83), but similar reactions in thf are

$$UCl_3 + 3KC_5H_5 \xrightarrow[\text{reflux, 7 days}]{C_6H_6} [(C_5H_5)_3U] + 3KCl \qquad (83)$$

not so straightforward. The reaction of UCl_3 with KC_5H_5 in thf at reflux for 50 h reportedly forms $[(C_5H_5)_3U(OC_4H_8)]$[138], but the analogous reaction of UCl_3 with NaC_5H_5 in thf at reflux for 3 days generates $[(C_5H_5)_3U(OC_4H_9)]$[4], a complex formed by decomposition of thf. $[(C_5H_5)_3U(OC_4H_9)]$ is also formed when UCl_3, generated *in situ* from UCl_4 and Na, reacts with KC_5H_5 in thf at reflux over a 5-day period[115]. $[(C_5H_5)_3An(OC_4H_9)]$ complexes may also be formed when $[(C_5H_5)_3U(thf)]$ and $[(C_5H_5)_3Np(thf)_3]$ decompose upon heating under vacuum[103]. The decomposition of thf in U^{3+} reactions may be less common in future syntheses which use the recently reported soluble precursor $UCl_3(thf)$[139]. This reagent may allow lower reaction temperatures and shorter reaction times, which may help avoid undesirable side reactions.

An interesting reductive synthesis of $[(C_5H_5)_3U]$ is shown in equation 84. Potassium reduction of $[(C_5H_5)_4U]$ in benzene forms a colourless solution and what is believed to be uranium metal. Subsequent addition of $[(C_5H_5)_4U]$ and heating at reflux for 3 weeks forms $[(C_5H_5)_3U]$ in 43% yield. Purification was effected by a 3–4-week Soxhlet extraction with benzene[138].

$$[(C_5H_5)_4U] + 4K \xrightarrow{-4KC_5H_5} U \xrightarrow{3[(C_5H_5)_4U]} 4[(C_5H_5)_3U] \qquad (84)$$

The direct synthesis of $[(C_5H_5)_3U]$ from C_5H_6 and uranium metal has been described using uranium obtained by subliming Hg away from U(Hg). A 30% yield was reported for this room-temperature reaction[94].

Two syntheses of $[(C_5H_5)_3Th]$ have been reported (equations 85 and 86). However, since the product of reaction 85 is violet and has $\mu_{eff}^{298} = 0.4$ BM[140] and the photolysis product is green and has $\mu_{eff}^{298} = 2.1$ BM[141], and since no structural data are yet available, the nature of these reactions remains open to question. The possibility that metallation of a cyclopentadienyl ring has occurred in reaction 85 to form a product such as $[(C_5H_5)_3Th{-}C_5H_4Th(C_5H_5)_2]$ has been discussed[7], and is consistent with the low moment and quantitative hydrolysis studies[7]. The fact that photolysis of $[(C_5H_5)_3U(Pr\text{-}i)]$ forms $[(C_5H_5)_3U]$[7] supports equation 86 as written, but provides no guarantee that the thorium product ultimately isolated is truly $[(C_5H_5)_3Th]$.

$$[(C_5H_5)_3ThCl] + Na \xrightarrow[\text{thf}]{C_{10}H_8} [(C_5H_5)_3Th] + NaCl \tag{85}$$

$$2[(C_5H_5)_3Th(Pr\text{-}i)] \xrightarrow[C_6H_6]{h\nu} 2[(C_5H_5)_3Th] + C_3H_6 + C_3H_8 \tag{86}$$

Relatively few studies of tris(cyclopentadienyl)actinide complexes involving substituted ring systems have been reported. The tris(indenyl) species $[(C_9H_7)_3Th(thf)]$ and $[(C_9H_7)_3U(thf)]$ have been prepared for i.r. studies[142]. In contrast to the analogous $[(C_5H_5)_3An(thf)_x]$ complexes which decompose upon heating in vacuum, these indenyl thf solvates are stable to 220°C at 10^{-4} Torr[142]. The synthesis of $[(C_9H_7)_3U]$ by reaction KC_9H_7 with UCl_3 and by reduction of $[(C_9H_7)_3UCl)]$ with $LiC_6H_4CH_3$ has also been reported[143]. $[(C_9H_7)_3Np(thf)]$ has been synthesized for examination by Mössbauer spectroscopy[115] $[(CH_3C_5H_4)_3Th]$ has been prepared photolytically according to equation 86[141].

As indicated throughout this section, $[(C_5H_5)_3An]$ complexes are usually isolated as base adducts when prepared in thf. In reactions similar to those of tris(cyclopentadienyl)lanthanide complexes, this thf can be displaced by other bases to form new adducts (equation 87). The cyclohexylisonitrile complexes $[(C_5H_5)_3AnCNC_6H_{11}]$ have been prepared in this manner for An = U[138], Np[97], Pu[97], and Am[97]. *l*-Nicotine derivatives, $[(C_5H_5)_3AnN_2C_{10}H_{14}]$, have been similarly obtained for An = U[138] and Pu[87].

$$[(C_5H_5)_3U(thf)] + CNC_6H_{11} \xrightarrow{\text{pentane}} [(C_5H_5)_3U(CNC_6H_{11})] + thf \tag{87}$$

2. *Bis(cyclopentadienyl)actinide halides and derivatives*

Only a few examples of trivalent bis(cyclopentadienyl)actinide complexes are known at present. $[(C_5H_5)_2BkCl]$ can be prepared from $BkCl_3$ and $Be(C_5H_5)_2$ in the melt using a 1:1 stoichiometry[144]. $[(C_5H_5)_3Bk]$ is also formed in this reaction, however, and differential sublimation did not allow complete separation of the two products. The synthesis of $[(C_5H_5)_2ThCl]$ from $[(C_5H_5)_3Th]$ and NH_4Cl in thf has been reported, using, presumably, the violet variation of $[(C_5H_5)_3Th]$ (see Section III.E.1). The brown complex was characterized only by metal analysis[67].

The most extensive series of trivalent bis(cyclopentadienyl)actinide derivatives has been obtained using the pentamethylcyclopentadienyl ligand[145]. The parent compound of the series, $[(C_5Me_5)_2UCl]_3$, is trimeric, in contrast to the dimeric lanthanide analogues (Section III.A.2). It can be prepared in several ways from tetravalent precursors[145] (equations 88–91). The trimer forms monomeric base adducts, $[(C_5Me_5)_2UCl(base)]$, with pyridine, thf, diethyl ether, and trimethylphosphine, and

$$3[(C_5Me_5)_2UCl_2] + 3Na(Hg) \xrightarrow[-3NaCl]{toluene} \quad (88)$$

$$3[(C_5Me_5)_2UCl_2] + 3t\text{-}BuLi \xrightarrow[-3LiCl,\ organics]{ether,\ -78°C} \quad (89)$$

$$\rightarrow [(C_5Me_5)_2UCl]_3$$

$$3[(C_5Me_5)_2U(R)Cl] + \tfrac{3}{2}H_2 \xrightarrow[-3RH]{toluene,\ 25°C} \quad (90)$$

$$\tfrac{3}{4}[(C_5Me_5)_2UH_2]_2 + \tfrac{3}{2}[(C_5Me_5)_2UCl_2] \xrightarrow{-3/2\ H_2} \quad (91)$$

reacts with alkali metals salts of bulky ligands to form monomeric alkyl (Section V) and amide (equation 92) derivatives[145].

$$[(C_5Me_5)_2UCl]_3 + 3NaN(SiMe_3)_2 \longrightarrow 3[(C_5Me_5)U\{N(SiMe_3)_2\}] + 3NaCl \quad (92)$$

IV. CYCLOOCTATETRAENYL COMPLEXES

A. Trivalent Lanthanide Complexes

Following the synthesis of uranocene from UCl_4 and $K_2C_8H_8$ in 1968 (Section IV.D), the reactions of lanthanide trichlorides with $K_2C_8H_8$ were investigated[10] and found to provide straightforward routes to two types of cyclooctatetraenyl lanthanide complexes. As in the case of cyclopentadienyl metathesis reactions (Section III.A), the stoichiometry can be adjusted to provide homoleptic complexes (equation 93) or mixed cyclooctatetraenyl lanthanide chlorides (equation 94). As might be expected for these systems, the two types of cyclooctatetraenyl complexes described above can be readily interconverted (equations 95 and 96) and the bis(cyclooctatetraenyl)

$$LnCl_3 + 2K_2C_8H_8 \xrightarrow{thf} K[(C_8H_8)_2Ln] + 3KCl \quad (93)$$

$$2LnCl_3 + 2K_2C_8H_8 \xrightarrow{thf} [C_8H_8LnCl(thf)_2]_2 + 4KCl \quad (94)$$

$$[C_8H_8LnCl(thf)_2]_2 + 2K_2C_8H_8 \longrightarrow 2K[(C_8H_8)_2Ln] + 2KCl \quad (95)$$

$$K[(C_8H_8)_2Ln] + LnCl_3 \longrightarrow [C_8H_8LnCl(thf)_2]_2 + KCl \quad (96)$$

complexes, $K[(C_8H_8)_2Ln]$, are formed in reaction 94 as by-products. This side-reaction can be diminished by adding $K_2C_8H_8$ to $LnCl_3$ rather than *vice versa*. Separation of the two products is facilitated by the fact that $K[(C_8H_8)_2Ln]$ complexes are much more soluble in thf. Yields of 36–78% have been reported for equation 93 where Ln = La, Ce, Pr, Nd, Sm, Gd, and Tb[10], but the reaction fails for Ln = Eu and Yb owing to reduction to the divalent state[146]. The monochloride complexes $[C_8H_8LnCl(thf)_2]_2$ have been obtained in 30–60% yield[147] according to equation 94 for Ln = Ce, Pr, Nd, and Sm[10]. The stability of these early lanthanide derivatives should be compared with that of the cyclopentadienyl monochlorides $[(C_5H_5)_2LnCl]_2$, which are also formally 8-coordinate, but are not stable with respect to ligand redistribution for metals larger than Sm (Section II.A.2). Both $[C_8H_8LnCl(thf)_2]_2$ and crystals of thf-solvated $K[(C_8H_8)_2Ln]$ readily lose thf at atmospheric pressure. Structural characterization of the cerium derivative of the latter complex was possible only by recrystallization from diglyme, $(CH_3OCH_2CH_2)_2O$, which formed the ion-pair complex $[(diglyme)K][C_8H_8CeC_8H_8]$[148].

An alternative synthesis of $K[(C_8H_8)_2Ln]$, specific to cerium, has been reported, involving the reduction of $[(C_8H_8)_2Ce]$ (Section IV.C) with potassium[85].

A third general class of cyclooctatetraenyl-lanthanide complexes has been prepared using metal atom vaporization techniques. Co-condensation of La, Ce, Nd, and Er atoms with cycloocta-1,3,5,7-tetraene (1,3,5,7-C_8H_8) at -196°C forms coloured matrices, which, upon warm-up and Soxhlet extraction with thf, yield crystals of composition $[(C_8H_8)Ln(thf)_2][(C_8H_8)_2Ln]$ (equation 97)[149]. These complexes are similar to the $[(diglyme)K][(C_8H_8)_2Ln]$ ion pair except that the lanthanide in the cation is displaced to one edge of the bridging $C_8H_8^{2-}$ anion[149], whereas the potassium is symmetrically located with respect to the $C_8H_8^{2-}$ 'bridge'[148].

$$2Ln + 3C_8H_8 \longrightarrow [(C_8H_8)_3Ln_2] \xrightarrow{\text{thf}} [(C_8H_8)Ln(thf)_2][(C_8H_8)_2Ln] \quad (97)$$

An alternative synthesis of $[(C_8H_8)_3Ce_2]$ has been reported (equation 98)[85] which follows the procedure originally used to synthesize $[(C_8H_8)_3Ti_2]$[150]. The hydrocarbon-soluble isopropoxide $[Ce(OR)_4](R = CHMe_2)$ is the precursor in this reaction, which is conducted in cycloheptatriene as solvent. If this reaction is conducted in toluene for a longer period of time, an alkoxy-bridged aluminium bimetallic complex, $[(C_8H_8)Ce(\mu\text{-}OR)_2AlEt_2]$, is formed in 30% yield (equation 99)[151]. Using the pyridine adduct $[Ce(OR)_4(C_5H_5N)]$ as precursor and cycloheptatriene as co-solvent with toluene, an 85% yield can be obtained. The synthesis of $[(C_8H_8)_3Sm_2]$ according to equation 98 has been claimed in a patent[152].

$$2[Ce(OR)_4(ROH)] + 10AlEt_3 + 3C_8H_8 \xrightarrow[0.5\ \text{h}]{110^\circ\text{C}} [(C_8H_8)_3Ce_2] + 10Et_2AlOR + 2C_2H_6 + \text{`}8C_2H_5^{\bullet}\text{'} \quad (98)$$

$$[Ce(OR)_4(ROH)] + 4AlEt_3 + C_8H_8 \xrightarrow[24\ \text{h}]{105^\circ\text{C}} [(C_8H_8)Ce(\mu\text{-}OR)_2AlEt_2] + C_2H_6 + 3Et_2AlOR + \text{`}3C_2H_5^{\bullet}\text{'} \quad (99)$$

The synthesis of neutral, mixed cyclooctatetraenyl–cyclopentadienyl complexes $[(C_8H_8)Ln(C_5H_5)(thf)]$, has been reported according to equations 100 and 101[10,153].

$$[(C_5H_5)LnCl_2(thf)_3] + K_2C_8H_8 \longrightarrow [(C_8H_8)Ln(C_5H_5)(thf)] + 2KCl \quad (100)$$

$$[(C_8H_8)LnCl(thf)_2]_2 + 2NaC_5H_5 \longrightarrow 2[(C_8H_8)Ln(C_5H_5)(thf)] + 2NaCl \quad (101)$$

The first method was used for Ln = Sm, Ho, and Er, whereas the second route was used for Ln = Nd since $[(C_5H_5)NdCl_2(thf)_3]$ is not available (Section III.A.3). Some ligand redistribution occurs in these reactions, since $[(C_5H_5)_3Ln]$ is a by-product. Attempts to make an analogous neutral complex, $[(C_8H_8)Ln(C_9H_9)]$, from the cyclononatetraenide anion were unsuccessful, since the charge density in the ring was too diffuse to allow displacement of the chloride. This attempted reaction (equation 102) is actually the synthesis by which the $[(C_8H_8)LnCl(thf)_2]_2$ complexes were first discovered[147]. As expected, the thf in $[(C_8H_8)Ln(C_5H_5)(thf)]$ can be readily displaced by NH_3, C_5H_5N, and CNC_6H_{11} to form new base adducts[153].

$$LnCl_3 + K_2C_8H_8 + KC_9H_9 \longrightarrow [(C_8H_8)LnCl(thf)_2]_2 \quad (102)$$

Cyclooctatetraenyl complexes of praseodymium have been synthesized from *cis,cis*-cycloocta-1,5-diene (1,5-C_8H_{12}) by reaction of the diene with the product of the reduction of $PrCl_3$ with 3 equivalents of potassium[154]. Although the analytical data supported the formation of complexes such as $K[(C_8H_8)_2Pr]$ and $K[(C_8H_8)Pr(C_8H_{10})]$, the reaction was too complex to allow definitive structural characterization. The reaction did lead to the discovery of a convenient inexpensive synthesis of $K_2C_8H_8$ from 1,5-C_8H_{12}, however[155] (equation 103).

$$2K + 1,5\text{-}C_8H_{12} \xrightarrow[\text{4–5 days}]{108^\circ C} \underset{(60\% \text{ based on K})}{K_2C_8H_8} \qquad (103)$$

B. Divalent Lanthanide Complexes

The first cyclooctatetraenyl complexes of the lanthanides were the divalent species $[Eu(C_8H_8)]$ and $[Yb(C_8H_8)]$, prepared by adding 1,3,5,7-C_8H_8 to solutions of the metals in liquid ammonia[156]. The compounds precipitate from the ammonia solution are insoluble in hydrocarbons and ethers. Soluble adducts are formed in pyridine and dimethylformamide, however. $[(C_8H_8)Yb]$ has also been prepared by cocondensation of Yb and C_8H_8 at $-196^\circ C$[149].

A most unusual Ce^{2+} complex, $[K(glyme)]_2[(C_8H_8)_2Ce]$, has reportedly been formed by reduction of $[(C_8H_8)_2Ce]$ with excess of potassium in glyme ($CH_3OCH_2CH_2OCH_3$) at 60°[85]. The complex was characterized by i.r. spectroscopy and elemental analysis.

C. Tetravalent Lanthanide Complexes

Two syntheses of the tetravalent cyclooctatraenyl complex $[(C_8H_8)_2Ce]$ have been reported (equations 104[85] and 105[157]), but the physical properties of the two supposedly identical products differ considerably. In the first case, a dark red–violet complex which decomposes in alcohol to form trienes, is formed in 65% yield[85]. In the second reaction, greenish yellow crystalline material, which is stable to water and soluble in hot acid, was reported to form in 64% yield[157] (cf. Section III.C).

$$[Ce(OR)_4(ROH)] + 5AlEt_3 \xrightarrow[\text{1 h, 140°C}]{C_8H_8} [(C_8H_8)_2Ce] + 5Et_2A1OR + C_2H_6 + \text{‘}4C_2H_5\cdot\text{’} \qquad (104)$$

$$(C_5H_6N)_2[CeCl_6] + 2K_2C_8H_8 \longrightarrow [(C_8H_8)_2Ce] + 4KCl + 2C_5H_5N\cdot HCl \qquad (105)$$

D. Tetravalent Actinide Complexes

Since the synthesis of the first cyclooctatetraenyl *f*-element complex, $[U(C_8H_8)_2]$(uranocene), in 1968[15], considerable efforts have been made to extend this class of complexes to other actinide metals and to a wide variety of substituted cyclooctatetraenyl ligands. The original synthesis of uranocene by metathesis[15,158] is the most widely applicable preparative route to cyclooctatetraenyl actinide complexes in general, and recently has been described in detail in *Inorganic Synthesis*[159]. The dianion is generated from cyclooctatetraene at low temperature and subsequently treated with a thf solution of UCl_4 for at least 3 h (equations 106 and 107). The product can be precipitated with water, filtered and extracted[15] or, more simply, the thf can be

$$1,3,5,7\text{-}C_8H_8 + 2K \xrightarrow[-30°C]{thf} K_2C_8H_8 \quad (106)$$

$$UCl_4 + 2K_2C_8H_8 \xrightarrow{thf} [(C_8H_8)_2U] + 4KCl \quad (107)$$

removed under vacuum[158,159] and the product extracted directly. Soxhlet extraction with toluene for 7–9 days provides 60–80% yields. Alternatively, the product can be purified by sublimation at 140°C, but the yield drops to 35% using this method of isolation. The Th[160], Pa[161,162], and Np[163] analogues have been prepared in a similar manner.

Uranocene can be obtained from a variety of alternative uranium halide and cyclooctatetraenyl precursors via similar metathesis reactions (equations 108–115).

$$(C_5H_5NH)_2[UCl_6] + 2K_2C_8H_8 \longrightarrow [(C_8H_8)_2U] + 4KCl + 2C_5H_5N\cdot HCl \quad (108)^{163}$$

$$[(C_5H_5)_3UCl] + C_8H_7Bu^{2-} \longrightarrow [(C_8H_7Bu)_2U] \quad (109)^{164}$$

$$[(C_5H_5)_3UBu] + C_8H_7Bu^{2-} \longrightarrow [(C_8H_7Bu)_2U] \quad (110)^{164}$$

$$[(C_2B_9H_{11})_2UCl_2]^{2-} + K_2C_8H_8 \longrightarrow [(C_8H_8)_2U] \quad (111)^{126}$$

$$UBr_3 + 2K_2C_8H_8 \longrightarrow [(C_8H_8)_2U] \quad (112)^{165}$$

$$UCl_4 + [(C_8H_7Bu)_2Th] \longrightarrow [(C_8H_7Bu)_2U] \quad (113)^{164}$$

$$UCl_4 + K[Ce(C_8H_8)_2] \longrightarrow [(C_8H_8)_2U] \quad (114)^{10}$$

$$UF_4 + 2MgC_8H_8 \xrightarrow{\Delta} [(C_8H_8)_2U] + 2MgF_2 \quad (115)^{159}$$

Some of these reactions were attempted in efforts to obtain products other than uranocene, and the fact that uranocene is the preferred product emphasizes the stability of this molecule. In other cases, new synthetic approaches to bis(cyclooctatetraenyl) complexes were often tried initially with uranium because uranocene is so stable and relatively easily prepared. $[(C_8H_8)_2Pu]$ can be prepared by metathesis according to equation 108 in a reaction which is sensitive to the cation in the plutonium precursor. The reaction succeeds starting from $(Et_4N)_2[PuCl_6]$, but fails for $Cs_2[PuCl_6]$ and $(C_5H_5NH)_2[PuCl_6]$[163]. The solvent free synthesis of equation 115 has also been applied to thorium[161].

A major alternative to the synthesis of uranocene by metathesis is the direct reaction of the metal with cyclooctatetraene. Several techniques have been used to generate the metal in a reactive form. Finely divided uranium powder, formed by repeatedly decomposing and regenerating uranium hydride, reacts with cyclooctatetraene at 150°C to form uranocene in 57% yield in 2.5 h (equation 116). The Th and Pu analogues could be prepared similarly.[95] Interestingly, this reaction was found to be catalysed by mercury. A related synthesis (equation 117) involves

$$UH_3 \xrightarrow[-H_2]{\Delta} U \xrightarrow[150°C]{2C_8H_8} [(C_8H_8)_2U] \quad (116)$$

$$U(Hg) \xrightarrow{150°C} U \xrightarrow{2C_8H_8} [(C_8H_8)_2U] \quad (117)$$

formation of reactive metallic uranium by sublimation of Hg from U(Hg). Uranium generated in this way is reported to react with cyclooctatetraene at room temperature over a 4 h period to form uranocene in 70% yield[94]. Uranocene can also be generated from the 'activated uranium' formed by reduction of UCl_4 with NaK alloy in the presence of naphthalene in dimethoxyethane for 24 h. This reaction produces a 35% yield of uranocene in 2.4 h at 85°[166]. Finally, it has been reported that the product, formed when UCl_4 is treated with 4 equivalents of butyllithium and warmed to room temperature, reacts with cyclooctatetraene to form uranocene[21,164].

Metal vaporization techniques have also been used to synthesize uranocene by co-condensation of cyclooctatetraene and uranium atoms at −196°C[167,168]. This method is not preparatively preferable, however, since the relatively expensive cyclooctatetraene ligand is required in excess and since uranium is difficult to vaporize, owing to its high heat of vaporization. Although uranium can be vaporized by resistive heating[167], heating by electron gun is preferable[168]. Using this latter method, a yield of 90% (based on metal vaporized) has been claimed for uranocene[168].

A large number of substituted cyclooctatetraenyl complexes have been prepared primarily with uranium as the actinide metal. The only limitations on synthesis are the availability of the substituted cyclooctatetraene[164,169] and the possibility that the substituted dianion will reduce the tetravalent actinide. Most preparations of substituted species employ the original metathesis reaction of UCl_4 with the cyclooctatetraenyl dianion in thf. The main difference is that the dianon is frequently generated from the substituted cyclooctatetraene with potassium naphthalide rather than potassium. This soluble reducing agent allows faster reduction and decomposition/polymerization of the cyclooctatetraene is more easily avoided[170]. Recently, an alternative synthesis of alkyl-substituted uranocenes has been developed which uses cyclooctatetraene rather than substituted cyclooctatetraenes as a starting material[169]. In this reaction alkyllithium reagents are used to reduce cyclooctatetraene to form dianions of monosubstituted ring systems which can be used directly in uranocene synthesis (equation 118).

$$4RLi + 2C_8H_8 \xrightarrow{-2RH} 2Li_2C_8H_7R \xrightarrow{UCl_4} [(C_8H_7R)_2U] + 4LiCl \quad (118)$$

Over 20 uranium complexes of monosubstituted cyclooctatetraenyl dianions, $C_8H_7R^{2-}$, are now known where R = alkyl[164,171,172], aryl[171], vinyl[171], alkoxy[170], alkoxycarbonyl[164], amino[170], alkylammonio[170], silyl[173], and others[164]. $[(C_8H_7Et)_2An]$ and $[(C_8H_7Bu)_2An]$ are also known for An = Np and Pu[172]. In comparison, disubstituted cyclooctatetraenyl complexes are relatively rare. The complexes $[(C_8H_6R_2)_2U]$ are known only for R = *t*-Bu[164] and for fused-ring ligands such as $C_8H_6(CH_2)_x$ (x = 2[174], 3[164], and 4[164]). 1,3,5,7-Tetrasubstituted cyclooctatetraenyl uranocenes, $[(C_8H_4R_4)_2An]$, have been prepared for R = CH_3[175] and Ph[176]. The latter complex, $[(C_8H_4Ph_4)_2U]$, is interesting in that it is air stable, presumably because decomposition pathways are sterically blocked by the eight phenyl groups[177]. The tetramethyl-substituted complexes $[(C_8H_4Me_4)_2An]$ are also known for An = Pa[178], Np[174], and Pu[178]. For the latter two metals, the borohydride complexes $[An(BH_4)_4]$ have been used as precursors since they are more soluble than the corresponding halides (equation 119). This alternative metathesis precursor is especially valuable in the plutonium reaction since $C_8H_4Me_4^{2-}$ reduces Pu^{4+} halides to Pu^{3+}[175].

$$[An(BH_4)_4] + 2K_2C_8H_4Me_4 \longrightarrow [(C_8H_4Me_4)_2An] + 4KBH_4 \quad (119)$$

Indirect synthesis of substituted uranocenes can be accomplished via substitution reactions on the ring (equation 120)[179] and by exchange with other dianions (equation 121)[164]. Exchange with other uranocenes and with neutral cyclooctatetraene is not observed, however (equations 122 and 123[164]).

$$[(C_8H_7NMe_2)_2U] \xrightarrow{2MeI} [(C_8H_7NMe_3^+I^-)_2U] \xrightarrow{2t\text{-}BuLi} [(C_8H_7Bu\text{-}t)_2U] + 2NMe_3 + 2LiI \quad (120)$$

$$[(C_8H_7Et)_2U] + 2C_8H_7Me^{2-} \longrightarrow [(C_8H_7Me)_2U] + 2C_8H_7Et^{2-} \quad (121)$$

$$[(C_8H_7Bu)_2U] + [(C_8H_7Et)_2U] \xrightarrow[\text{diglyme}]{150^\circ\text{C, 5 h}} \text{no exchange} \quad (122)$$

$$[(C_8H_7Bu)_2U] + C_8H_8 \xrightarrow[\text{diglyme}]{\Delta} \text{no exchange} \quad (123)$$

Since uranocene is so stable, monocyclooctatetraenyl uranium complexes analogous to mixed cyclopentadienyl chloride complexes are difficult to obtain. Recently the synthesis of some 'half-sandwich' monocyclooctatetraenyl actinide complexes has been discussed[180] using An = Th, since the formation of the more ionic thorocene is not as favoured (equations 124–126). A uranium derivative can be obtained analog-

$$ThCl_4 + [(C_8H_7R)_2Th] \xrightarrow[\text{thf}]{\Delta} 2[(C_8H_7R)ThCl_2(thf)_2] \quad (124)$$

$$[(C_8H_7R)_2Th] + 2HCl \xrightarrow{\text{thf}} [(C_8H_7R)ThCl_2(thf)_2] + C_8H_9R \quad (125)$$

$$[Th(BH_4)_4(thf)_2] + K_2C_8H_7R \longrightarrow [(C_8H_7R)Th(BH_4)_2(thf)_2] + 2KBH_4 \quad (126)$$

ously (equation 127) if $K_2C_8H_8$ is added slowly. Fast addition forms uranocene. These complexes should be excellent precursors for a wide variety of monocyclooctatetraenyl derivatives.

$$UCl_4 + 2LiBH_4 \xrightarrow{\text{thf}} \xrightarrow{K_2C_8H_8} [(C_8H_8)U(BH_4)_2(thf)_2] \quad (127)$$

One mixed cyclopentadienyl cyclooctatetraenyl derivative has been synthesized by addition of $[(C_5H_5)_3UCl]$ to $K_2C_8H_8$ at low temperature (equation 128; cf. equations 109–110). The rather unstable complex loses C_8H_8 at 0°C[173].

$$2[(C_5H_5)_3UCl] + K_2C_8H_8 \longrightarrow [\{(C_5H_5)_3U\}_2(C_8H_8)] + 2KCl \quad (128)$$

E. Trivalent Actinide Complexes

Bis(cyclooctatetraenyl) complexes of trivalent actinides, $K[(C_8H_8)_2An]$, analogous to the lanthanide complexes described in Section IV.A are known for the later actinides, Np[181], Pu[181], and Am[182], a result expected based on the stability of the trivalent state for these metals. Actinide bromides and iodides were used as precursors in the metathesis synthesis (equation 129) since they are more soluble than the chlorides. For An = U, equation 129 forms uranocene rather than the trivalent product. Attempts to reduce uranocene to $M[(C_8H_8)_2U]$ with K, NaK, and $KC_{10}H_8$ were unsuccessful[181]. Both $K[(C_8H_8)_2Np]$ and $K[(C_8H_8)_2Pu]$ can be readily oxidized to $[(C_8H_8)_2An]$.

$$AnBr_3 + 2K_2C_8H_8 \longrightarrow K[(C_8H_8)_2An](thf)_2 + 3KBr \quad (129)$$

A mono-cyclooctatetraenyl neptunium complex, $[(C_8H_8)NpI(thf)_x]$, has been prepared from the reaction of equimolar amounts of NpI_3 and $K_2C_8H_8$. A two-step mechanism (equation 130) is proposed based on colour changes[183].

$$2NpI_3 + 2K_2C_8H_8 \xrightarrow{-3KI} K[(C_8H_8)_2Np] + NpI_3 \longrightarrow 2[(C_8H_8)NpI(thf)_x] + KI \quad (130)$$

V. COMPLEXES CONTAINING METAL—CARBON SIGMA BONDS

A. Homoleptic Lanthanide Complexes

The synthesis of homoleptic lanthanide complexes can be accomplished by the same general methods discussed in previous sections. Metathesis reactions with $LnCl_3$ and reactions involving the elemental metal have been used with nearly equal frequency in the synthesis of this class of complexes. Although efforts to form complexes containing lanthanide metal-carbon single bonds started as early as 1935[184], it was not until 1972[185] that complexes were obtained which were sufficiently tractable to allow crystallographic identification of a Ln–C linkage. The early attempts to make σ-bonded species included the reaction of lanthanum metal with (a) methyl radicals formed by decomposition of $[(CH_3)_4Pb]$[184], (b) diphenylmercury[186], and (c) iodobenzene[186]. The reaction of $LaCl_3$ with LiC_6H_5, $LiCH_3$, and C_2H_5MgBr was also investigated[186]. This research was hindered by the fact that the largest lanthanide, lanthanum, was used instead of a smaller metal which would have been easier to stabilize by coordinative saturation (Section II). The syntheses were also hampered by lack of effective coordinating bases such as thf and tetramethylethylenediamine (tmeda). The reaction of $LaCl_3$ with C_6H_5Li was reinvestigated in 1969 using thf as the solvent and found to form an insoluble material formulated as $Li[La(C_6H_5)_4]$ by metal analysis[9]. A similar result was obtained for praseodymium. The reaction of MeLi with $LaCl_3$ in thf was also reported, but the insoluble product, thought to be $[La(CH_3)_3(thf)_n]$, could not be positively defined[9]. However, when this metathesis approach was extended to the smallest member of the series, lutetium, and to a bulkier organic group, 2,6-dimethylphenyl, a pure, crystalline complex was obtained (equation 131), which was shown by X-ray crystallography to have four metal aryl bonds in a tetrahedral geometry[185]. This reaction clearly demonstrates the importance of saturating the coordination sphere of the metal. An anionic tetrakis species is preferentially formed in this reaction instead of a neutral tris complex. Although the isostructural ytterbium analogue was also reported, crystalline material was not obtainable by using larger metals, such as erbium, or by using unsubstituted phenyls[26].

$$LnCl_3 + 4LiC_6H_3Me_2 \xrightarrow[-78°C]{thf} [Li(THF)_4][Ln(C_6H_3Me_2)_4] + 3LiCl \tag{131}$$

Equation 131 constitutes a general method for synthesizing a variety of homoleptic lanthanide complexes providing the metal is small enough and the ligand large enough. In recent years, several classes of homoleptic lanthanides have been obtained in this way. The reaction of $LnCl_3$ with $LiCH_2SiMe_3$ is interesting because it occurs in stepwise fashion forming both the neutral tris and anionic tetrakis species[188–190] (equations 132, 133; n = 2 and 3). The neutral complexes, which have been reported for

$$LnCl_3 + 3LiCH_2SiMe_3 \longrightarrow [Ln(CH_2SiMe_3)_3(thf)_n] + 3LiCl \tag{132}$$

$$[Ln(CH_2SiMe_3)_3(thf)_n] + LiCH_2SiMe_3 \longrightarrow [Li(thf)_4][Ln(CH_2SiMe_3)_4] \tag{133}$$

Ln = Tm, Tb, Er (25–30% yield), Yb, and Lu, are soluble in hexane and stable to CH_2Cl_2. These species reportedly lose $SiMe_4$ after several days in pentane at room temperature and form insoluble polymers. A mechanism involving α-hydrogen elimination and formation of bridging $CHSiMe_3$ groups has been proposed[190] (equation 134). Reaction 132 could not be extended to Sm, Tb, and Dy owing to this proposed α-elimination. $LaCl_3$ and $NdCl_3$ reportedly did not react under the conditions used

$$[Ln(CH_2SiMe_3)_3(thf)_2] \longrightarrow [Ln(CH_2SiMe_3)(CHSiMe_3)]_n + 2thf + SiMe_4 \quad (134)$$

successfully for the later lanthanides[189]. The anionic complexes (equation 133) reported for Ln = Er, Yb, and Lu are insoluble in hexane, form oils in toluene and react vigorously with halogenated solvents. The thf which solvates the lithium can be replaced by tmeda or Et_2O. When the ether-solvated lutetium complex $[Li(Et_2O)_4][Lu(CH_2SiMe_3)_4]$ is placed in benzene, slow dissociation to the neutral $[Lu(CH_2SiMe_3)_3(Et_2O)_2]$ is observed by n.m.r. spectroscopy. Over a period of 1 week $SiMe_4$ is observed to form and the formation of anionic species such as $Li[Ln(CH_2SiMe_3)_2(CHSiMe_3)]$ and $\{[Li(thf)_2][Ln_2(CH_2SiMe_3)_2(CHSiMe_3)(CSiMe_3)]\}_n$ has been postulated[190].

The reaction of $LnCl_3$ with the bulkier trimethylsilyl reagent $LiCH(SiMe_3)_2$ takes a different course[188] (equation 135; Ln = Er, Yb) than that observed for $LiCH_2SiMe_3$ (equations 132 and 133). The hexane-soluble products, $[Er\{CH(SiMe_3)_2\}_3Cl]^-$, could not be converted to $[Er\{CH(SiMe_3)_2\}_4]^-$ by adding $LiCH(SiMe_3)_2$, but upon heating the chloro derivative in hexane, a ligand redistribution reaction occurs to form the homoleptic anion, $[Er\{CH(SiMe_3)_2\}_4]^-$, and presumably $Li[ErCl_4]$[188].

$$LnCl_3 + 3LiCH(SiMe_3)_2 \longrightarrow [Li(thf)_4][Ln\{CH(SiMe_3)_2\}_3Cl] + 2LiCl \quad (135)$$

The trimethylsilyl ligands CH_2SiMe_3 and $CH(SiMe_3)_2$ were chosen not only because they were sterically bulky, but also because they lacked hydrogen at the β-position. The β-hydrogen elimination reaction, a common mode of decomposition of transition metal alkyl complexes (equation 136), was thereby precluded. This precaution,

$$[M(CH_2CHR_2)_n] \longrightarrow CH_2{=}CR_2 + [HM(CH_2CHR_2)_{n-1}] \longrightarrow CH_3CHR_2 + [M(CH_2CHR_2)_{n-2}] \longrightarrow \text{etc} \quad (136)$$

although important in transition metal chemistry, is not necessarily a dominant consideration in lanthanide chemistry as demonstrated by the successful synthesis of the *t*-butyl complexes, $Li[Ln(t\text{-}C_4H_9)_4(thf)_n]$[191] (equation 137, Ln = Sm, Er, Yb, Lu; n = 3,4), which contain 36 β-hydrogen atoms. Not only are these complexes stable enough to be isolated with the later lanthanides, but the series can even be extended to samarium. The complexes are isolated in 50–75% yield and can be purified by recrystallization from thf/hexane. A strong coordinating base such as thf or tmeda is essential in order to avoid oils. The anionic tetrakis species is the preferred product regardless of reagent stoichiometry. Attempts to obtain the neutral tris species with 3 equivalents of $t\text{-}C_4H_9Li$ have been reported, but pure lithium-free products were not obtained[190]. The alkoxy complexes $Ln(O\text{-}t\text{-}Bu)_3$ have been used as alternative precursors[190].

$$LnCl_3 + 4t\text{-}BuLi \longrightarrow Li[Ln(t\text{-}Bu)_4(thf)_n] + 3LiCl \quad (137)$$

The least thermally stable member of this homoleptic *t*-butyl series, $Li[Sm(t\text{-}Bu)_4(thf)_4]$, decomposes at 40°C over a 16-h period to form 3.25 equivalents of 2-methylpropane[191]. The fact that equal amounts of 2-methylpropene and 2-methylpropane are *not* formed suggests that β-hydrogen elimination (equation 136) is *not* the preferred mode of decomposition. This is a clear example of the potential of the lanthanide metals to display unusual organometallic reactivity.

Isolable, homoleptic, methyl lanthanide complexes have also been obtained by the metathesis route by conducting the synthesis in the presence of tmeda[192] (equation 138, Ln = Er, Yb, Lu). The crystalline products were formed in 30–40% yield. Interestingly, the preferred product is a trianionic species containing six methyl groups. The formation of this highly charged species again emphasizes the importance of coordinative saturation in lanthanide chemistry. The reaction of CH_3Li with $ErBr_3$ is thought to form $[ErMe_3(thf)_x]$ and $[Li(thf)_4][ErMe_4]$, but the highly reactive products were not fully characterized[190].

$$LnCl_3 + 6MeLi + 3tmeda \longrightarrow [Li(tmeda)]_3[LnMe_6] + 3LiCl \quad (138)$$

Another interesting series of formally 6 coordinate alkyl lanthanide complexes, the phosphorus ylide complexes $[Ln\{(CH_2)_2PMe_2\}_3]$, has been prepared by a less direct route[193] (equations 139 and 140). These coordinatively saturated, chelated species have been reported for Ln = La, Pr, Nd, Sm, Gd, Ho, Er, and Lu.

$$LnCl_3 + Me_3P{=}CH_2 \longrightarrow [Ln(CH_2\overset{+}{P}Me_3)_3]Cl_3 \quad (139)$$

$$[Ln(CH_2PMe_3)_3]Cl_3 + 3BuLi \longrightarrow \left[Ln\left\{\begin{matrix}CH_2\\ \\CH_2\end{matrix}\right\rangle PMe_2\right]_3 + 3LiCl + 3BuH \quad (140)$$

Acid–base metathesis can also be used to form homoleptic lanthanide complexes[194] (equation 141). The *t*-butyl complexes $Li[Ln(t\text{-}Bu)_4(thf)_n]$ are particularly useful precursors in this reaction since they are formed from the readily available *t*-BuLi, they are easily purified and they are fairly reactive. The homoleptic alkynides $Li[Ln(C{\equiv}CR)_4(thf)]$ are formed in nearly quantitative yield for R = C_6H_5, *n*-Bu, and *t*-Bu and Ln = Sm, Er, and Lu. This approach provides a convenient chloride free synthesis of homoleptic complexes[194]. This route should be especially important when the formation of stable mixed alkyl chloride species is a complicating factor in homoleptic lanthanide synthesis (equation 135).

$$Li[Ln(t\text{-}Bu)_4(thf)_n] + 4HC{\equiv}CR \longrightarrow Li[Ln(C{\equiv}CR)_4(thf)] + 4t\text{-}BuH \quad (141)$$

Although the initial attempts to use elemental lanthanum metal to form organolanthanides were unsuccessful, subsequent efforts with smaller metals have allowed Ln—C bond formation. This approach has been most commonly used with those elements with the most accessible divalent states. Ytterbium and europium react with alkyl and aryl iodides in thf at −15°C to form brown solutions which exhibit Grignard behaviour[195,196]. Magnetic susceptibility measurement on the products of the reaction of MeI, EtI, C_6H_5I, *o*-MeC_6H_4I, 2,6-$Me_2C_6H_3I$, and 2,4,6-$Me_3C_6H_2I$ with ytterbium indicated that 7–17% of the metal was oxidized to Yb^{3+}, i.e. the bulk of the metal was divalent. For europium, which has a more stable divalent state, almost all the metal was divalent. Samarium was much less reactive with EtI and PhI, and it was necessary to add iodine to initiate the reaction. The products were reported to be 50% divalent and displayed near i.r.–visible spectra very similar to $[SmI_2(THF)_n]$[181]. Cerium reacts rapidly with RI to form dark solutions and appreciable amounts of precipitate. The products were observed to have a Ce:I ratio of 1:1.5, consistent with a mixture of $[R_2CeI]$ and $[RCeI_2]$ or $[R_3Ce]$ and CeI_3. Lanthanum was observed to form considerable amounts of LaI_3 when treated with RI. The yield of soluble organolanthanum iodide complexes was estimated to be only 25%. It is not surprising that no pure complexes were isolated in these studies, since there was no easy route to coordinatively saturated complexes. Although ligand redistribution could have allowed the formation of known stable anions, e.g. $[Yb(2,6\text{-}Me_2C_6H_3)_4]^-$, the only available counter cations, Yb^{3+} and Yb^{2+}, were evidently not suitable for overall stabilization of such a salt.

A cleaner synthesis of Ln—C bonds starting with elemental metals is the

$$\mathrm{Yb} + (\mathrm{C_6F_5})_2\mathrm{Hg} \xrightarrow[4\,\mathrm{h}]{\mathrm{thf}} [(\mathrm{C_6F_5})_2\mathrm{Yb(thf)_4}] + \mathrm{Hg} \tag{142}$$

transmetallation reaction[197] (equation 142). The orange crystalline pentafluorophenyl complex $[(C_6F_5)_2Yb(thf)_4]$ can be obtained in 29% yield in this manner. This complex, which is isolated by cooling a thf/alkane solution, has only moderate thermal stability, decomposing in 48 h at room temperature. Tetraflurophenyl complexes prepared similarly are less stable. Neither $(C_6Cl_5)_2Hg$ nor $(C_6H_5)_2Hg$ was observed to react with ytterbium[197]. Analogous transmetallation reactions using europium and samarium and also other organomercury compounds have been studied in solution[198], but isolable products were not obtained. The reaction of europium metal with $(C_6F_5)_2Hg$ required excess of mercury to initiate the reaction. Attempts to obtain crystals from the green solution of the bis(pentafluorophenyl) complex of this larger metal were unsuccessful. Analogous samarium reactions were more complex owing to fluoride abstraction and oxidation to the trivalent state[198].

The reaction of ytterbium metal with $Hg(C{\equiv}CPh)_2$ forms an isolable alkynide complex, $[Yb(C{\equiv}CPh)_2]$, which can also be obtained by acid–base metathesis from $[(C_6F_5)_2Yb(thf)_4]$[199] (equation 143). A 22% yield was observed in the transmetallation reaction. A 41% yield was reported for reaction 143, although the product was described as less pure than the transmetallation product.

$$[(\mathrm{C_6F_5})_2\mathrm{Yb(thf)_4}] + 2\mathrm{HC{\equiv}CPh} \longrightarrow [(\mathrm{PhC{\equiv}C})_2\mathrm{Yb}] + 2\mathrm{C_6F_5H} \tag{143}$$

Europium metal in liquid ammonia has been reported to react with propyne to form the alkynide $[Eu(C{\equiv}CCH_3)_2]$. In the case of ytterbium, the reaction does not allow the isolation of a pure compound owing to the presence of substantial amounts of amide by-products presumed to be $[Yb(NH_2)_2]$. Only traces of amide were found in the europium product[200].

Divalent alkynides can also be obtained by the metal vaporization method[201] (equation 144). Co-condensation of ytterbium vapour and hex-1-yne does not simply form the divalent insertion product, $[HYbC{\equiv}CBu]$, however, since the two purple products isolated from this reaction have alkynide to hydride ratios (determined by elemental analysis and decomposition with D_2O, CCl_4, and Me_3SiCl) which are greater than or equal to 3. Trivalent species are also formed in this synthesis as by-products (8–15%). Consistent with the presence of hydride, the alkynide hydrides are precursors to catalysts for hydrogenation of unsaturated hydrocarbons.

$$\mathrm{Yb} + \mathrm{HC{\equiv}CBu} \longrightarrow \text{`}[\mathrm{Yb_2(C{\equiv}CBu)_3H}]_n\text{'} \tag{144}$$

B. Heteroleptic Cyclopentadienyl Lanthanide Complexes

The most common method for synthesizing mixed-ligand lanthanide complexes containing cyclopentadienyl groups and σ-bonded carbon species is the metathesis reaction (equation 145), discussed earlier as a method of forming $[(C_5H_5)_2Ln(anion)]$

$$[(\mathrm{C_5H_5})_2\mathrm{LnCl}] + \mathrm{RLi} \xrightarrow[\mathrm{thf}]{-78\,^\circ\mathrm{C}} [(\mathrm{C_5H_5})_2\mathrm{LnR}] + \mathrm{LiCl} \tag{145}$$

complexes (equation 21). Lithium reagents are most commonly used in this synthesis, which is initially conducted at low temperature to minimize decomposition of thf by the lithium reagent[202,203]. The products are typically purified by arene extraction and can be isolated as either alkyl bridged dimers, e.g. $[(C_5H_5)_2YbCH_3]_2$[204], or thf solvates, e.g. $[(C_5H_5)_2Lu(t\text{-}Bu)(thf)]$[205]. Both structures have been confirmed crystallographically. Since the unsubstituted cyclopentadienyl precursors are known only for the later lanthanides (Section III.2), and since complexes of the later lanthanides are more coordinatively saturated and hence more stable and tractable, most syntheses have involved those metals. The following combinations have been used successfully with equation 145: R = CH_3, Ln = Gd, Er, Yb[206], Lu[74]; R = C_6H_5, Ln = Gd, Er, Yb[206]; R = $C{\equiv}CC_6H_5$, Ln = Gd, Er, Yb[206], R = CH_2SiMe_3, Ln = Sm, Er, Yb[190]. Yields of 40–50% are obtained. The *t*-butyl complexes $[(C_5H_5)_2Ln(t\text{-Bu})(thf)]$ can be obtained via reaction 145 for Ln = Er and Lu, but the reaction with ytterbium fails owing to reduction of $[(C_5H_5)_2YbCl]$ to $[(C_5H_5)_2Yb]$ by *t*-BuLi[74]. The preparation of $[(C_5H_5)_2Yb(t\text{-Bu})]$ from $[(C_5H_5)_2Yb(O\text{-}t\text{-Bu})]$ and *t*-BuLi has been claimed, however[190]. The monocyclopentadienyl dialkynide complex $[(C_5H_5)Ho(C{\equiv}CC_6H_5)_2]$ has been prepared according to equation 145 starting from $[(C_5H_5)HoCl_2(thf)_3]$[206]. The synthesis of the cerium(IV) derivatives $[(C_5H_5)_3CeR]$ and $[(C_9H_7)_2CeR_2]$ by similar metathesis reactions using RLi or RMgBr has been claimed for R = CH_3, C_2H_5, C_6H_5, $CH_2C_6H_5$, and C_6H_5CO[81] (see Section III.C).

A second method for the formation of $[(C_5H_5)_2LnR]$ complexes is the decomposition of bridged alkylaluminum complexes $[(C_5H_5)_2Ln(\mu\text{-}R)_2AlR_2]$[207] (equation 146) with pyridine[204] (equation 147). The bridged aluminum complexes

$$[(C_5H_5)_2LnCl]_2 + 2LiAlR_4 \xrightarrow[\text{toluene}]{0^\circ C} 2\left[(C_5H_5)_2Ln(\mu\text{-}R)_2AlR_2\right] + 2LiCl \quad (146)$$

$$[(C_5H_5)_2Ln(\mu\text{-}R)_2AlR_2] + C_5H_5N \xrightarrow[\text{toluene}]{\text{r.t.}} \tfrac{1}{2}[(C_5H_5)_2LnR]_2 + [AlR_3(C_5H_5N)] \quad (147)$$

have been prepared for R = CH_3 and Ln = Gd, Dy, Ho, Er, Tm, and Yb in 45–78% yield and for R = C_2H_5 and Ln = Ho. The pyridine decomposition has been used to generate $[(C_5H_5)_2LnCH_3]_2$ for Ln = Dy, Ho, Er, Tm, and Yb in 80% yield.

Mixed cyclopentadienyl σ-bonded complexes can also be obtained in suitable cases by acid–base metathesis starting with the tris(cyclopentadienyl) complexes (equation 148). This method has the advantage that it is halide free and that the lanthanide

$$[(C_5H_5)_3Ln] + HR \longrightarrow [(C_5H_5)_2LnR] + \tfrac{1}{2}(C_5H_6)_2 \quad (148)$$

precursors are available for the early members of the series. The synthesis of the sparingly soluble complexes $[(C_5H_5)_2NdCN]$ and $[(C_5H_5)_2YbCN]$ using this approach with HR = HCN has been reported[208]. The synthesis of $[(C_5H_5)_2Yb(C{\equiv}CR)]_x$ (R = $n\text{-}C_6H_{13}$, $n\text{-}C_4H_9$, C_6H_{11}, C_6H_5, and $C_5H_4FeC_5H_5$; $x = 2.5\text{–}3$) has been accomplished in 60–80% yield using terminal alkyne reagents. For neodymium, the latter reaction is more complex, however, and the products have not been definitively identified[209]. These alkyne reactions are unusual in that the product, C_5H_6, is a stronger acid than the precursor, $HC{\equiv}CR$, which formally loses a proton in the reaction.

A fourth synthetic method for the formation of $[(C_5H_5)_2LnR]_2$ complexes is the acid–base metathesis[194] (equation 149), which starts from $[(C_5H_5)_2LnCH_3]_2$. This synthesis, like reaction 148, has the advantage of being halide free. In addition, since CH_4 is volatile, by-product separation is particularly easy. Since CH_4 is such a weak base, the reaction is favoured as written for most HR reagents. A yield of 85% has been observed for reaction 149 for Ln = Er. $[(CH_3C_5H_4)_2YbC{\equiv}CCMe_3]_2$ can be

$$[(C_5H_5)_2LnCH_3]_2 + 2HC{\equiv}CCMe_3 \xrightarrow{\text{thf}} [(C_5H_5)_2LnC{\equiv}CCMe_3]_2 + 2CH_4 \quad (149)$$

similarly prepared[194]. The erbium product $[(C_5H_5)_2Er(C{\equiv}CCMe_3)]_2$ is interesting in that it maintains its dimeric alkynide-bridged structure even when recrystallized from thf[205]. This dimer is only sparingly soluble in toluene, in contrast to the readily soluble $[(C_5H_5)_2ErCH_3]_2$. As a result, this is a case where reaction 149 is synthetically preferable to the alternative route, the reaction of $[(C_5H_5)_2ErCl]_2$ with $LiC{\equiv}CCMe_3$ (equation 145), which requires separation of LiCl from $[(C_5H_5)_2Er(C{\equiv}CCMe_3)]_2$ by arene extraction. Both syntheses require two steps starting from $HC{\equiv}CCMe_3$ and $[(C_5H_5)_2ErCl]_2$, but it is easier to separate LiCl from the $[(C_5H_5)_2ErCH_3]_2$ intermediate than from the final alkynide product.

C. Homoleptic Actinide Complexes

In contrast to homoleptic alkyl lanthanide complexes, actinide complexes of this type are not very thermally stable and few have been isolated and definitively characterized. As a result of this apparent thermal instability, the reaction mixtures formed in these synthetic attempts are complex and are not easily analysed. This has led to disagreements concerning the actual course of these syntheses and, consequently, homoleptic actinide chemistry is one of the more controversial areas in organo-*f*-element chemistry. Most synthetic approaches have involved metathesis reactions.

The most stable neutral complex of this class is the thorium species $[(C_6H_5CH_2)_4Th]$, prepared from $[ThCl_4(THF)_3]$ and benzyllithium at −20°C (equation 150). Although the complex is stable enough to be handled at room

$$[ThCl_4(thf)_3] + 4LiCH_2C_6H_5 \xrightarrow[-20°C]{thf} [(C_6H_5CH_2)_4Th] + 4LiCl \quad (150)$$

temperature, it must be stored at temperatures below 0°C to prevent decomposition[210]. This complex may have enhanced stability compared with the other actinide alkyls discussed below because the benzyl ligands may have more than a simple σ-bonded interaction with the metal. The analogous tetrabenzylzirconium and -hafnium complexes have distorted structures in which a weak interaction between the aromatic ring and the metal has been postulated[211].

In contrast to the synthesis of other organo-*f*-element complexes, the conditions necessary for successful synthesis of $[(C_6H_5CH_2)_4Th]$ cannot be varied too greatly. The complex could not be obtained in pure form using $C_6H_5CH_2MgCl$ or $(C_6H_5CH_2)_2Mg$ and, if reaction 150 is conducted at room temperature, a different complex is obtained. The product under the latter conditions was formulated as $[(C_6H_5CH_2)_3Th(thf)]$ on the basis of metal analysis, but no evidence for Th^{3+} was found. Consequently, tetravalent formulations involving metallated phenyl rings were proposed[210]. Alternative structures in which thf is metallated also seem plausible considering the reported instability of $ThCl_4$ in thf[89].

Attempts to prepare simple uranium alkyls date back to efforts during the Manhattan project in 1941 to make complexes such as $[(CH_3)_4U]$[212]. In retrospect, four methyl groups seem insufficient to saturate the uranium coordination sphere and provide a stable complex. Attempts to form 2,6-dimethylphenyl uranium complexes related to the stable $[(2,6\text{-}Me_2C_6H_3)_4Ln]^-$ complexes (Section V.A) were also unsuccessful, however[26]. The reaction of the bulky $LiCH_2SiMe_3$ with UCl_4 was reported to form a petroleum-soluble complex[213], but isolable complexes were not obtained[214]. An attempt to form a benzyl uranium complex analogous to the thorium complex discussed above led to a magnesium-containing complex of unknown structure instead[215] (reaction 151).

$$[UCl_4(thf)_3] + 2(C_6H_5CH_2)_2Mg \xrightarrow[-40^{\circ}C]{Et_2O} [(C_6H_5CH_2)_4UMgCl_2] + MgCl_2 + 3thf \quad (151)$$

An extensive study of the metathesis reaction of UCl_4 with alkyllithium reagents has been conducted in order to understand the decomposition of alkyl actinide complexes[216] (equation 152; R = Me, *n*-Bu, *t*-Bu, *i*-Pr, 2-butenyl, neopentyl). No

$$UCl_4 + 4RLi \xrightarrow{-78^{\circ}C} [R_4U] \xrightarrow{25^{\circ}C} \text{organic products} + U \quad (152)$$

efforts were made to isolate the putative alkyl products, which were unstable at room temperature. When the alkyllithium reagent contained a β-hydrogen atom, the organic decomposition products were the appropriate alkane and alkene consistent with a β-hydrogen elimination reaction (e.g. equation 153). The amount of alkane formed

$$UCl_4 + 4t\text{-}BuLi \xrightarrow{\text{pentane}} [t\text{-}Bu_4U] \xrightarrow[96\ h]{25^{\circ}C} \underset{77\%}{Me_3CH} + \underset{22\%}{Me_2C{=}CH_2} \quad (153)$$

was often higher than expected, however, suggesting that other reactions were occurring. The possibility of an alternative decomposition route is further supported by the fact that alkyls lacking β-hydrogen atoms decompose to form the corresponding alkane (equation 154). The solvent was proposed as the source of hydrogen. Although

$$UCl_4 + 4MeLi \xrightarrow{\text{hexane}} [Me_4U] \xrightarrow[48\ h]{55^{\circ}C} \underset{98\%}{CH_4} \quad (154)$$

the intermediate actinide alkyls in all these reactions were regarded as R_4U complexes, the presence of additional halide or solvent coordination could not be excluded[216]. Since *t*-BuLi is known to reduce U^{4+} to U^{3+} [145], it is possible that U^{3+} rather than U^{4+} is involved in some of these reactions, which further complicates attempts to make mechanistic conclusions.

The presence of tetrakis(alkyl) complexes in reaction 152 was subsequently challenged in a study which claimed $[R_6U]^{2-}$ complexes were the dominant species[214] (equation 155). Anionic hexakis(alkyl) uranium complexes were claimed for

$$UCl_4 + 6RLi \xrightarrow{Et_2O} [Li(OEt_2)_4]_2[R_6U] + 4LiCl \quad (155)$$

R = CH_3, CH_2SiMe_3, C_6H_5, and *o*-$C_6H_4CH_2NMe_2$ based on Li:U ratios and hydrolytic analysis of the alkyl ligand and the coordinating base, which could be Et_2O, thf, or tmeda. Accurate C and H analyses could not be obtained for these complexes, which decompose between −20 and 35°C. As expected, the complexes of the smallest ligand, methyl, are least stable. The reaction of 4 equivalents of MeLi with UCl_4 was repeated, and a 45% yield of $[Me_6U]^{2-}$ and 50% recovery of unreacted UCl_4 were claimed. Details of this work have also been challenged[7,217].

The reaction of the pentavalent $[U_2(OEt)_{10}]$ with alkyllithium reagents forms complexes which can be stabilized by dioxane and are formulated as $Li_3[UR_8(dioxane)_3]$[214] (equation 156, R = Me, CH_2CMe_3, CH_2SiMe_3). Since the methyl complexes are the

$$[U_2(OEt)_{10}] + 16RLi \xrightarrow[-70^{\circ}C]{\text{alkane}} \xrightarrow{\text{dioxane}} 2Li_3[UR_8(C_4H_8O_2)_3] + 10LiOEt \quad (156)$$

most stable in this series, these octakis(alkyl) complexes are apparently coordinatively saturated. Attempts to make analogous octakis(alkyl) complexes of tetravalent uranium, $[R_8U]^{4-}$, were unsuccessful. The reaction of hexavalent $[U(OPr\text{-}i)_6]$ with alkyllithium, Grignard, and alkylaluminum reagents fails to generate σ-bonded species. Instead, adducts such as $[U(OPr\text{-}i)_6(LiMe)_3]$ are formed, in which each lithium may interact with two alkoxy oxygen atoms[214].

The synthesis of $[U(CF_3)_4]$ and $[U(CF_3)_6]$ by reaction of uranium halides with CF_3 radicals, generated from C_2F_6 by plasma discharge, has been described in two patent claims[218].

D. Heteroleptic Cyclopentadienyl Actinide Complexes

1. *Unsubstituted cyclopentadienyl derivatives*

Sigma-bonded actinide complexes containing cyclopentadienyl rings are much more stable than the homoleptic systems described above and, consequently, these complexes have been investigated more extensively and definitively. The stability of these heteroleptic alkyls is presumably due to the tricoordinate nature of the cyclopentadienyl rings, which allows coordinative saturation of the metal centre. Experimentally, it is found that three unsubstituted rings or two permethylated rings provided the necessary steric bulk.

The predominant member of this class of complexes, $[(C_5H_5)_3UR]$, most commonly is prepared by metathesis using $[(C_5H_5)_3UCl]$ and an organolithium reagent[219–222] (equation 157). Purification is effected by Soxhlet extraction with alkanes and arenes.

$$[(C_5H_5)_3UCl] + RLi \xrightarrow[-78°C]{\text{thf}} [(C_5H_5)_3UR] + LiCl \qquad (157)$$

Yields of 20–90% have been reported for a wide variety of R groups, including Me, *i*-Pr, *n*-Bu, *t*-Bu, CH_2CMe_3, $CH_2C_6H_5$, C_6H_5, C_6F_5, $C{\equiv}CH$, $CH{\equiv}CH_2$ and $C_5H_4FeC_5H_5$. Organosodium[222] and Grignard reagents[219] can also be used in this reaction. When the alkyllithium reagent in equation 157 is the ylide precursor $Li(CH_2)_2P(C_6H_5)_2$, the complex $[(C_5H_5)_3U\{CHPMe(C_6H_5)_2\}]$ is formed[222]. If 2 equivalents of $Li(CH_2)_2P(C_6H_5)_2$ are added to $[(C_5H_5)_3UCl]$, the uranium centre loses $C_5H_5^-$ and a dimeric bis(cyclopentadienyl) complex, $[(C_5H_5)_2U\{(\mu\text{-}CH)(CH_2)P(C_6H_5)_2\}U(C_5H_5)_2]$, results[223]. A 3:1 ratio of RLi to $[(C_5H_5)_3UCl]$ generates a monocyclopentadienyl ylide complex $[(C_5H_5)U\{(CH_2)_2P(C_6H_5)_2\}_3]$[222].

Using bifunctional organolithium reagents such as $[(LiC_5H_4)_2Fe]$tmeda and *p*-LiC_6H_4Li, complexes containing two tris(cyclopentadienyl) units can be prepared[143] (equation 158). These dinuclear species are unfortunately too insoluble in thf to allow

$$2[(C_5H_5)_3UCl] + [(LiC_5H_4)_2Fe]\text{tmeda} \longrightarrow [\{(C_5H_5)_3UC_5H_4\}_2Fe] + 2LiCl \qquad (158)$$

characterization by n.m.r. spectroscopy. An attempt to make $[(C_5H_5)_3UC{\equiv}CU(C_5H_5)_3]$ from $[(C_5H_5)_3UCl]$ and Li_2C_2 was unsuccessful. Efforts to form the diuranium precursor $[C_5H_5)_3UC{\equiv}C]^-M^+$ from $[(C_5H_5)_3UC{\equiv}CH]$ using NaH or *n*-BuLi also failed. In the latter case, the product was $[(C_5H_5)_3UBu]$[143] (equation 159). Another type of diuranium complex can be obtained using $K_2C_8H_8$ as a precursor[173] (equation 160). The exact mode of attachment of the bridging $C_8H_8^{2-}$ unit is not known, but it is interesting to note that C_8H_8 can be easily removed from this complex at 0°C.

$$[(C_5H_5)_3UC{\equiv}CH] + n\text{-BuLi} \longrightarrow [(C_5H_5)_3UBu] \qquad (159)$$

$$2[(C_5H_5)_3UCl] + K_2C_8H_8 \longrightarrow [\{(C_5H_5)_3U\}_2C_8H_8] + 2KCl \qquad (160)$$

Alkyl thorium complexes, $[(C_5H_5)_3ThR]$, can be prepared from $[(C_5H_5)_3ThCl]$ and RLi or RMgX in reactions analogous to equation 157[224]. These thorium syntheses require more precise control of reaction conditions than the analogous uranium reactions. Reduction to elemental thorium is an important side-reaction and is best controlled using low reaction temperatures and non-coordinating solvents. Yields of the cyclopentadienyl thorium alkyl products vary from 20 to 70%, although in some cases, e.g. for R = C_6H_5 and *t*-Bu, successful syntheses of pure complexes could not be achieved. Ring closure was observed when 5-hexenylmagnesium bromide was used as a precursor (equation 161).

$$[(C_5H_5)_3ThCl] + CH_2{=}CH(CH_2)_4MgBr \longrightarrow \left[(C_5H_5)_3ThCH_2{-}CH\begin{matrix} \diagup CH_2{-}CH_2 \\ \quad\quad\quad | \\ \diagdown CH_2{-}CH_2 \end{matrix}\right] + MgClBr \quad (161)$$

Cyclopentadienylneptunium alkyls, $[(C_5H_5)_3NpR]$, have been prepared for Mössbauer study, but pure complexes could not be obtained owing to reductive decomposition. $[(C_5H_5)_3Np(n\text{-}Bu)]$ and $[(C_5H_5)_3NpC_6H_5]$ were synthesized in 80–90% and 40% purity, respectively, with $[(C_5H_5)_4Np]$ and $[(C_5H_5)_3NpCl]$ constituting the major impurities[115].

Only a few cyclopentadienyl actinide complexes containing metal—carbon σ-bonds have been prepared by routes other than the ionic metathesis of equation 157. Acid–base metathesis is one alternative synthesis and has been used to generate cyanide derivatives[208] (equations 162 and 163).

$$[(C_5H_5)_4U] + HCN \longrightarrow [(C_5H_5)_3UCN] + \tfrac{1}{2}(C_5H_6)_2 \quad (162)$$

$$[(C_5H_5)_3U] + HCN \longrightarrow [(C_5H_5)_2UCN] + \tfrac{1}{2}(C_5H_6)_2 \quad (163)$$

Thermolysis of $[(C_5H_5)_3ThR]$ provides another route to complexes of this class (equation 164). In this case the σ-bonded ligand is a bridging, monohapto–pentahapto cyclopentadienyl ring[224,225]. In this reaction, intramolecular hydrogen abstraction is more facile than β-hydrogen elimination. Thermolysis of $[(C_5H_5)_3Th(i\text{-}Pr)]$ at 220°C, however, does produce 20% propene in addition to 80% propane. The analogous uranium complexes, $[(C_5H_5)_3UR]$, decompose thermally to form the alkane (RH), traces of the corresponding alkene, the dimer (R_2), and an intractable material formulated as $[(C_{5.5}H_{10.4})U]$[221].

$$2[(C_5H_5)_3ThR] \xrightarrow[\text{toluene}]{167^\circ\text{C}} [(C_5H_5)_2Th(\eta^1{:}\eta^5\text{-}C_5H_4)_2Th(C_5H_5)_2] + 2RH \quad (164)$$

With the exception of the thermolysis reaction, the $[(C_5H_5)_3AnR]$ complexes have not proved to be exceptional synthetic precursors {cf. $[(C_5Me_5)_2UR_2]$ below}. They do not react with acetone, CO_2[221] or CO[7]. For $[(C_5H_5)_3UR]$, alcoholysis generates the alkoxide (Section III.D.2), but it is formed as a mixture with the ring cleavage product[7,219,221] (equation 165). In the thorium case, the Th—R bond is more reactive to

$$2[(C_5H_5)_3UR] + 2R'OH \longrightarrow [(C_5H_5)_3UOR'] + [(C_5H_5)_2U(R)(OR')] + RH + \tfrac{1}{2}(C_5H_6)_2 \quad (165)$$

alcoholysis and $[(C_5H_5)_3ThOR']$ is exclusively formed at first[7,224] (equation 166). The $[(C_5H_5)_3UR]$ complexes have been observed by n.m.r. spectroscopy to participate in alkyl-exchange reactions with AlR_3' (equation 167), but this reaction is not useful preparatively[226].

$$[(C_5H_5)_3ThR] \xrightarrow[-RH]{R'OH} [(C_5H_5)_3ThOR'] \xrightarrow[-\frac{1}{2}(C_5H_6)_2]{R'OH} [(C_5H_5)_2Th(OR')_2] \quad (166)$$

$$[(C_5H_5)_3UR] + AlR_3' \longrightarrow [(C_5H_5)_3UR'] + AlR_2'R \quad (167)$$

2. *Substituted cyclopentadienyl derivatives*

The use of substituted cyclopentadienyl ligands has had a major impact on organoactinide chemistry in that it has allowed the isolation of stable complexes containing more than one alkyl group. These dialkyls have proved to have interesting reactivity. Since the $[R_nU]^{(n-4)-}$ complexes which have been examined are thermally unstable[214,216], whereas the $[(C_5H_5)_3AnR]$ complexes are stable[221,224], the most likely candidate for a stable uranium polyalkyl was $[(C_5H_5)_2UR_2]$. The normal precursor, $[(C_5H_5)_2UCl_2]$, is unstable with respect to disproportionation, however[101,109,120], precluding the synthesis of this dialkyl. As described in Section III.D.3, the stability of the bis(cyclopentadienyl) precursor can be circumvented by using substituted ring systems.

The first reported attempt to prepare dialkyls employed the complex ring-bridged compound $[\{Q(C_5H_4)_2UCl(\mu\text{-}Cl)\}_2(\mu\text{-}Cl)Li(thf)_2]$[98]. This complex reacted with n-BuLi and neopentyllithium to form alkyl complexes formulated as $[Q(C_5H_4)_2UR_2]$[227]. These complexes still were thermally unstable. The butyl complex decomposes at room temperature to form 60% butene and 40% butane, whereas the neopentyl complex formed neopentane exclusively. Methyl and Me_3SiCH_2 complexes were also examined. 2,2-Bipyridine complexes of these products were reported to be more stable.[98] Bis(indenyl)actinide dialkyls reportedly were prepared by sequential addition of 2 equivalents of NaC_9H_7 and 2 equivalents of MeLi (or t-BuLi) to UCl_4[228].

The most extensively examined series of bis(cyclopentadienyl)actinide dialkyls employs the pentamethylcyclopentadienyl ligand[129,227] (equation 168,

$$[(C_5Me_5)_2AnCl_2] + 2RLi \xrightarrow[-78°C]{Et_2O} [(C_5Me_5)_2AnR_2] + 2LiCl \quad (168)$$

An = U, Th, R = Me, CH_2SiMe_3, $CH_2C_6H_5$; An = Th, R = CH_2CMe_3, C_6H_5). Yields of 65–70% have been reported for the methyl derivative, which can be purified by recrystallization from toluene. These dialkyls are the most stable actinide polyalkyls known, with half-lives of 1 week and 16 h for the thorium and uranium methyl derivatives, respectively. The synthesis according to reaction 168 appears to be general, except with strongly reducing alkyllithium reagents. For example, t-butyllithium reduces $[(C_5Me_5)_2UCl_2]$ to the uranium(III) complex $[(C_5H_5)_2UCl]_3$[145] (Section III.E.2). The related reduction of $[(C_9H_7)_3UCl]$ by p-tolyllithium to form $[(C_9H_7)_3U]$ has been reported[143]. Using the dilithio reagent LiPhC=C(Ph)—C(Ph)=C(Ph)Li, the metallocyclic complex $\left[(C_5Me_5)_2U\overset{C(Ph)=C-Ph}{\underset{C(Ph)=C-Ph}{<}}\right]$ has been prepared[129]. This complex can also be obtained by the novel dimerization of PhC≡CPh by $[(C_5Me_5)_2UCl]_3$[145].

Monoalkylchloro complexes, $[(C_5Me_5)_2AnRCl]$, can be prepared by the stoichiometric reaction of $[(C_5Me_5)AnCl_2]$ with RLi[227] (An = Th, U; R = Me, CH_2SiMe_3, CH_2CMe_3, C_6H_5, $CH_2C_6H_5$) or by ligand redistribution (equation 169,

$$[(C_5Me_5)_2AnR_2] + [(C_5Me_5)_2AnCl_2] \xrightarrow{\text{toluene}} 2[(C_5Me_5)_2AnRCl] \quad (169)$$

An = Th, U; R = Me, CH_2SiMe_3). The latter reaction proceeds in high yield for R = Me and CH_2SiMe_3[129], but fails for R = CH_2CMe_3[227]. $[(C_5Me_5)_2UMeCl]$ is a monomer, whereas the thorium analogue appears to be dimeric.

Both the dialkyl- and the monoalkylchloropentamethylcyclopentadienyl actinide complexes have proved to be interesting and valuable precursors in actinide chemistry. The first molecular actinide hydrides were prepared from the dialkyls by hydrogenolysis[129,229] (equation 170). The uranium complex loses H_2 at room temperature whereas the thorium hydride is stable at 80°C. Hydrogenolysis of the alkylchloro complexes forms the trivalent complex $[(C_5Me_5)_2UCl]_3$ in the case of uranium (equation 171), and the expected $[(C_5Me_5)_2ThHCl]_2$ for thorium, which does not have an accessible trivalent state[145]. Ethene inserts into the Th—H bonds of $[(C_5Me_5)_2ThH_2]_2$ and $[(C_5Me_5)_2ThHCl]_2$ to form the corresponding ethyl complexes $[(C_5Me_5)_2ThEt_2]$ and $[(C_5Me_5)_2ThClEt]$[227].

$$2[(C_5Me_5)_2AnMe_2] + 4H_2 \xrightarrow{\text{toluene}} [(C_5Me_5)_2AnH(\mu\text{-H})]_2 + 4CH_4 \quad (170)$$

$$6[(C_5Me_5)_2URCl] + 3H_2 \xrightarrow{\text{toluene}} 2[(C_5Me_5)_2UCl]_3 + 6RH \quad (171)$$

Both the dialkyl- and alkylchloro complexes react with CO to form a variety of unusual An—O bonded species in which CO has inserted into An—C bonds. The precise product obtained is strongly dependent on the alkyl group involved. The dimethyl complexes react with CO to form a dimeric complex containing two bridging but-2-ene-2,3-diolate ligands[16] (equation 172). For $[(C_5Me_5)_2AnR_2]$, where R is the bulkier ligand CH_2SiMe_3, a monomeric complex involving a five-membered metallocyclic ring is formed (equation 173). CO inserts into the alkylchloro complex

$$2[(C_5Me_5)_2AnMe_2] + 4CO \xrightarrow{\text{toluene}} [(C_5Me_5)_2An(O\overset{\text{Me}}{\overset{|}{C}}{=}\overset{\text{Me}}{\overset{|}{C}}O)_2An(C_5Me_5)_2] \quad (172)$$

$$[(C_5Me_5)_2AnR_2] + 2CO \xrightarrow{\text{toluene}} \left[(C_5Me_5)_2An\begin{matrix}\diagup O-C-R \\ \qquad \| \\ \diagdown O-C-R\end{matrix}\right] \quad (173)$$

$[(C_5Me_5)AnCl(CH_2SiMe_3)]$ to form a non-cyclic product in which the $SiMe_3$ has migrated to the carbon of the CO[16] (equation 174). Using the neopentyl derivative

$$[(C_5Me_5)_2AnCl(CH_2SiMe_3)] + CO \longrightarrow [(C_5Me_5)_2AnCl(O-\overset{\text{SiMe}_3}{\overset{|}{C}}{=}CH_2)] \quad (174)$$

instead of the trimethylsilyl complex, the intermediate bihaptoacyl compound can be isolated[230] (equation 175). This complex rearranges at 100°C to

$$[(C_5Me_5)_2ThCl(CH_2CMe_3)] + CO \xrightarrow[0°C]{\text{toluene}} [(C_5Me_5)_2ClTh\overset{O}{\triangle}CCH_2CMe_3] \quad (175)$$

$[(C_5Me_5)_2ClThOCH{=}CH(CMe_3)]$ and reacts further with CO to form a complex which has incorporated 4 mol of CO into the starting alkyl complex[230] (equation 176). The importance of the pentamethylcyclopentadienyl ligand in allowing the development of this chemistry and the potential of this ligand system in *f*-element chemistry are clearly evident.

$$2[(C_5Me_5)_2ClTh\overset{O}{\triangle}CR] + 2CO \longrightarrow \left[(C_5Me_5)_2ClTh\begin{matrix}O{=}CR \\ | \\ O{-}C{=}\end{matrix}\right]_2 \qquad (176)$$

The pentamethylcyclopentadienyl ligand has been used to obtain a trivalent uranium alkyl, $[(C_5Me_5)_2UCH_2SiMe_3]$[145] (equation 177). This complex has the interesting property that it generates $[(C_5Me_5)_2UH_2]_2$ rather than a trivalent hydride upon hydrogenolysis.

$$[(C_5Me_5)_2UCl]_3 + 3LiCH_2SiMe_3 \longrightarrow 3[(C_5Me_5)_2UCH_2SiMe_3] + 3LiCl \qquad (177)$$

A monocyclopentadienyl actinide alkyl complex has been isolated from $[(C_5Me_5)ThCl_3(thf)_2]$ (equation 178), but the trialkyl complex could not be isolated from an analogous reaction with 3 equivalents of alkyllithium reagent[227].

$$[(C_5Me_5)ThCl_3(thf)_2] + LiCH_2SiMe_3 \xrightarrow{Et_2O} [(C_5Me_5)ThCl_2CH_2SiMe_3] + LiCl + 2thf \qquad (178)$$

As expected, tris(cyclopentadienyl)actinide alkyl complexes involving smaller substituted ring systems such as MeC_5H_4 and C_9H_7 can be obtained via metathesis from the corresponding tris(cyclopentadienyl) chlorides[114,143].

VI. MISCELLANEOUS

A. Allyl Complexes

Both σ- and π-bonded allyl complexes of the lanthanides and actinides can be prepared by standard metathesis routes. The coordination mode of the allyl group is generally determined by the size and flexibility of the auxiliary ligands. Homoleptic allyl complexes of uranium[231,232] and thorium[233] can be prepared from the allyl-Grignard at low temperature (equation 179). The complexes decompose at temperatures above

$$UCl_4 + 4C_3H_5MgCl \xrightarrow[-30°C]{Et_2O} [(C_3H_5)_4U] + 4MgCl_2 \qquad (179)$$

$-20°C$ to form propene (80%) and propane (20%) in the case of $[(C_3H_5)_4U]$[231]. The corresponding neutral tris(allyl) lanthanide complexes have not been prepared despite numerous attempts[234]. This is strange considering the relative stabilities of alkyl complexes such as $[U(CH_2SiMe_3)_6]^{2-}$ [214] and $[Ln(CH_2SiMe_3)_3(thf)_n]$[188–190] (Section V). Anionic homoleptic lanthanide allyl complexes, Li[Ln(allyl)$_4$] (Ln = Ce, Nd, Sm, Gd, Dy), have been obtained, however, by a novel reaction in which $Sn(C_3H_5)_4$ is the allyl precursor[234] (equation 180). Trihapto coordination of the allyl ligand is proposed for all of these homoleptic species.

$$\text{LnCl}_3 + 4n\text{-BuLi} + \text{Sn(C}_3\text{H}_5)_4 \xrightarrow{\text{thf}} \text{Li[Ln(C}_3\text{H}_5)_4] + \text{Sn}(n\text{-Bu})_4 + 3\text{LiCl} \qquad (180)$$

The derivative chemistry of the homoleptic uranium complexes has been investigated, and heteroleptic halide[235] and alkoxide[236] allyl complexes have been obtained (equation 181, X = Cl, Br, I; equation 182, R = Et, *i*-Pr, *t*-Bu). The alkoxide derivatives are only slightly more stable than the homoleptic allyls: decomposition occurs at room temperature.

$$[(\text{C}_3\text{H}_5)_4\text{U}] + \text{HX} \xrightarrow{\text{Et}_2\text{O}} [(\text{C}_3\text{H}_5)_3\text{UX}] + \text{C}_3\text{H}_6 \qquad (181)$$

$$[(\text{C}_3\text{H}_5)_4\text{U}] + 2\text{ROH} \xrightarrow[-30^\circ\text{C}]{\text{hexane}} [(\text{C}_3\text{H}_5)_2\text{U(OR)}_2] + 2\text{C}_3\text{H}_6 \qquad (182)$$

Cyclopentadienylallyl lanthanide and actinide complexes can also be prepared by standard metathesis routes (equation 183, Ln = Sm, Er, Ho[237]; equation 184, An = U[232], Th[233]). I.r. and n.m.r. spectroscopy indicate that these unsubstituted allyls form π-complexes, whereas an X-ray crystallographic study of $[(C_5H_5)_3U\{CH_3C(CH_2)_2\}]$ revealed a σ-bonded species[238]. The thermal stability of these complexes is comparable to that of the corresponding alkyls.

$$[(\text{C}_5\text{H}_5)_2\text{LnCl}] + \text{C}_3\text{H}_5\text{MgBr} \xrightarrow[-78^\circ\text{C}]{\text{thf}} [(\text{C}_5\text{H}_5)_2\text{LnC}_3\text{H}_5] + \text{MgBrCl} \qquad (183)$$

$$[(\text{C}_5\text{H}_5)_3\text{AnCl}] + \text{C}_3\text{H}_5\text{MgBr} \xrightarrow[-70^\circ\text{C}]{\text{Et}_2\text{O}} [(\text{C}_5\text{H}_5)_3\text{AnC}_3\text{H}_5] + \text{MgBrCl} \qquad (184)$$

B. Pyrrolyl Complexes

The pyrrolyl ligand, as a heterocyclic analogue of the cyclopentadienyl ring, can conceivably form extensive series of *f*-element complexes. The synthesis of tetrakis(2,5-dimethylpyrrolyl)uranium has been reported[239] (equation 185) and the complex has been found to be thermally stable (m.p. 98–102, dec.). Based on n.m.r. studies an idealized structure containing two monohapto and two pentahapto pyrrolyl moieties has been proposed. Rapid interconversion of the different types of rings may enhance the stability of this bis(cyclopentadienyl) uranium analogue.

$$\text{UCl}_4 + 4\text{KNC}_4\text{H}_2\text{Me}_2 \xrightarrow{\text{thf}} [(\text{NC}_4\text{H}_2\text{Me}_2)_4\text{U}] + 4\text{KCl} \qquad (185)$$

C. Complexes of Neutral Unsaturated Hydrocarbons

The only crystallographically characterized *f*-element complex of a neutral unsaturated hydrocarbon is the benzene complex $[(C_6H_6)U(AlCl_4)_3]$[240] (equation 186). The coordination environment in this complex can be most easily described as a tetrahedron in which three positions are occupied by $AlCl_4^-$ anions coordinating the metal through two bridging chlorides. The fourth position is occupied by a planar arene ring. The complex is insoluble in all common solvents and reverts to UCl_3, benzene, and $AlCl_3$(solvent) in polar solvents such as thf.

$$3\text{UCl}_4 + 9\text{AlCl}_3 + \text{Al} \xrightarrow[\text{reflux}]{\text{C}_6\text{H}_6} 3[(\text{C}_6\text{H}_6)\text{U(AlCl}_4)_3] + \text{AlCl}_3 \qquad (186)$$

The interaction of unsaturated hydrocarbons with the *f* elements has recently been probed synthetically using metal vaporization techniques[5,241,242]. Survey investigations of the metal atom technique by some of the initial workers in the field originally indicated that reactions occurred with *f* elements, but no isolable organo-*f*-element compounds were characterized[167,168,243,244]. More recently, vaporization of the lanthanide metals at 500–1600°C into frozen matrices of dienes[241], alkenes[245], and alkynes[5,242] at −196°C has been shown to generate preparative-scale amounts of several new types of organolanthanide complexes (equations 187–191). The thermally stable, air- and moisture-sensitive complexes are isolated by extraction of the matrix at room temperature with alkane, arene, or ether solvents (depending on the particular metal–ligand combination). These dark complexes differ from previously described organolanthanide complexes in their spectral and magnetic properties, thermal stabilities (both early and late lanthanides form stable complexes of the same formula), and solution behaviour (the degree of association, n, is higher in thf than in arene solvents; e.g. in equation 191, n=2 in arenes and $n > 10$ in thf).

$$\mathrm{Ln + H_2C{=}CH{-}CH{=}CH_2 \longrightarrow [Ln(C_4H_6)_3]_n \quad Ln = La, Nd, Sm, Er} \tag{187}$$

$$\mathrm{Ln + H_2C{=}C(Me){-}C(Me){=}CH_2 \longrightarrow [Ln(C_6H_{10})_2]_n \quad Ln = La, Nd, Sm, Er} \tag{188}$$

$$\mathrm{Er + CH_3CH{=}CH_2 \longrightarrow [Er(C_3H_6)_3]_n} \tag{189}$$

$$\mathrm{Ln + CH_3CH_2C{\equiv}CCH_2CH_3 \longrightarrow [LnC_6H_{10}]_n \quad Ln = Sm, Yb} \tag{190}$$

$$\mathrm{Er + CH_3CH_2C{\equiv}CCH_2CH_3 \longrightarrow [Er(C_6H_{10})_{1.5}]_n} \tag{191}$$

The reactivity of zero-valent lanthanide atoms with unsaturated hydrocarbons is apparently extensive. In addition to the species in equations 187–191, soluble products have been obtained from diphenylethyne, bis(trimethylsilyl)ethyne, cyclohexene, norbornadiene, cyclohexa-1,3-diene, cycloocta-1,5-diene, and isoprene[44], and reactions have been observed with ethene, allene, and cyclopropane[245]. There is a definite metal dependence in reactivity and at least three classes of metals can be identified. The chemistry of elements with the most accessible divalent oxidation states often differs from the other members of the series. For example, Eu and Yb polymerize dienes rather than form isolable complexes[44] (cf. equations 187 and 188), Yb and Sm react with hex-3-yne to form products with different formulae than La, Nd, or Er[47,242] (equations 189 and 190), and Sm forms a soluble complex with $\mathrm{Me_3SiC{\equiv}CSiMe_3}$, whereas Er does not[44]. Lutetium may constitute a second class of metal, since it has been found[44] to form products with hex-3-yne which are spectrally distinct from the products of other hex-3-yne reactions (equation 191). The remaining elements in the lanthanide series comprise the third class of metals with respect to reactivity in these syntheses.

VII. REFERENCES

1. G. Wilkinson and J. M. Birmingham, *J. Am. Chem. Soc.*, **76**, 6210 (1954).
2. L. T. Reynolds and G. Wilkinson, *J. Inorg. Nucl. Chem.*, **2**, 246 (1956).
3. The classic chemical evidence for this ionic character is that $(C_5H_5)_3Ln$ complexes, as well as some cyclopentadienyl actinide derivatives, react with $FeCl_2$ to form $(C_5H_5)_2Fe$ and f element chloride[1,2,4].
4. G. L. TerHaar and M. Dubeck, *Inorg. Chem.*, **3**, 1648 (1964).
5. W. J. Evans, S. C. Engerer, P. A. Piliero, and A. L. Wayda, in *Fundamental Research in Homogeneous Catalysis* (Ed. M. Tsutsui), Vol. 3, Plenum, New York, 1979, p. 941.
6. T. J. Marks, *Prog. Inorg. Chem.*, **24**, 51 (1978).
7. T. J. Marks, *Prog. Inorg. Chem.*, **25**, 223 (1979).
8. F. Calderazzo, R. Pappalardo, and S. Losi, *J. Inorg. Nucl. Chem.*, **28**, 987 (1966).
9. F. A. Hart, A. G. Massey, and M. S. Saran, *J. Organomet. Chem.*, **21**, 147 (1970).

10. K. O. Hodgson, F. Mares, D. F. Starks, and A. Streitwieser, Jr., *J. Am. Chem. Soc.*, **95**, 8650 (1973).
11. T. J. Marks and G. W. Grynkewich, *Inorg. Chem.*, **15**, 1302 (1976).
12. T. Moeller, in *Comprehensive Inorganic Chemistry* (Exec. Ed. A. F. Trotman-Dickenson), Vol. 4, Pergamon Press, Oxford, 1973, pp. 1–101.
13. A. F. Trotman-Dickenson (Exec. Ed.), *Comprehensive Inorganic Chemistry,* Vol. 5, Pergamon Press, Oxford, 1973, pp. 1–637.
14. G. J. McCarthy, J. J. Rhyne and H. B. Silber (Eds.), *Proc. Conf. Rare Earth Research, 1960–1980; Rare Earths in Modern Science and Technology*, Vol. 2, Plenum, New York, 1980.
15. A. Streitwieser, Jr., and U. Muller-Westerhoff, *J. Am. Chem. Soc.*, **90**, 7364 (1968).
16. J. M. Manriquez, P. J. Fagan, T. J. Marks, C. S. Day, and V. W. Day, *J. Am. Chem. Soc.*, **100**, 7112 (1978).
17. J. M. Manriquez, D. R. McAlister, R. D. Sanner, and J. E. Bercaw, *J. Am. Chem. Soc.*, **100**, 2716 (1978).
18. H. Gysling and M. Tsutsui, *Adv. Organomet. Chem.*, **9**, 361 (1970).
19. R. G. Hayes and J. L. Thomas, *Organomet. Chem. Rev., A*, **7**, 1 (1971).
20. B. Kanellakopulos and K. W. Bagnall, in *Lanthanides and Actinides, MTP International Review of Science, Inorganic Chemistry, Series One* (Ed. H. J. Emeleus and K. W. Bagnall), Vol. 7, Butterworths, London, 1972, pp. 299–322.
21. E. Cernia and A. Mazzei, *Inorg. Chim. Acta,* **10**, 239 (1974).
22. E. C. Baker, G. W. Halstead, and K. N. Raymond, *Struct. Bonding (Berl.),* **25**, 23 (1976).
23. M. Tsutsui, N. Ely, and R. DuBois, *Acc. Chem. Res.*, **9**, 217 (1976).
24. T. J. Marks, *Acc. Chem. Res.*, **9**, 223 (1976).
25. T. J. Marks, *Adv. Chem. Ser.*, **150**, 232 (1976).
26. S. A. Cotton, *J. Organomet. Chem. Libr.*, **3**, 189 (1977).
27. T. J. Marks and R. D. Fischer (Eds.), *Organometallics of the f-Elements*, D. Reidel Publishing Co., Dordrecht, 1979.
28. A. J. Freeman and R. E. Watson, *Phys. Rev.*, **127**, 2058 (1962).
29. G. Schwarzenbach and H. Flaschka, *Complexometric Titrations,* Methuen, London, 1969, pp. 194–197.
30. Reference 12, p. 72.
31. J. M. Birmingham and G. Wilkinson, *J. Am. Chem. Soc.*, **78**, 42 (1956).
32. R. D. Rogers, R. V. Bynum and J. L. Atwood, *J. Organomet. Chem.*, **192**, 65 (1980).
33. K. Rossmanith and C. Auer-Welsbach, *Monatsh. Chem.*, **96**, 602 (1965).
34. E. O. Fischer and H. Fischer, *J. Organomet. Chem.*, **3**, 181 (1965).
35. P. G. Laubereau and J. H. Burns, *Inorg. Chem.*, **9**, 1091 (1970).
36. R. Pappalardo, *J. Mol. Spect.*, **29**, 13 (1969).
37. G. Bielang and R. D. Fischer, *J. Organomet. Chem.*, **161**, 335 (1978).
38. S. Manastyrskyj and M. Dubeck, *Inorg. Chem.*, **3**, 1647 (1964).
39. M. Tsutsui, T. Takino, and D. Lorenz, *Z. Naturforsch.*, **21**, 1 (1966).
40. A. F. Reid and P. C. Wailes, *Inorg. Chem.*, **5**, 1213 (1966).
41. F. Baumgärtner, E. O. Fischer, and P. G. Laubereau, *Radiochim. Acta*, **7**, 188 (1967).
42. B. Kanellakopulos, reference 20, p. 303.
43. R. G. Hayes and J. L. Thomas, *Inorg. Chem.*, **8**, 2521 (1969).
44. W. J. Evans and S. C. Engerer, unpublished results.
45. L. T. Reynolds and G. Wilkinson, *J. Inorg. Nucl. Chem.*, **9**, 86 (1959).
46. J. H. Burns, W. H. Baldwin, and F. H. Fink, *Inorg. Chem.*, **13**, 1916 (1974).
47. W. J. Evans and N. Leipzig, unpublished results.
48. M. Tsutsui and H. J. Gysling, *J. Am. Chem. Soc.*, **91**, 3175 (1969).
49. P. E. Gaivoronskii, E. M. Gavrishchuk, N. P. Chernyaev, and Yu. B. Zverev, *Russ. J. Inorg. Chem.*, **23**, 1742 (1978).
50. E. O. Fischer and H. Fischer, *J. Organomet. Chem.*, **6**, 141 (1966).
51. R. von Ammon, B. Kanellakopulos, R. D. Fischer, and P. Laubereau, *Inorg. Nucl. Chem. Lett.*, **5**, 315 (1969).
52. R. von Ammon, R. D. Fischer, and B. Kanellakopulos, *Chem. Ber.*, **104**, 1072 (1971).
53. R. von Ammon, B. Kanellakopulos, R. D. Fischer and V. Formacek, *Z. Naturforsch.*, **28B**, 200 (1973).

44. J. H. Burns and W. H. Baldwin, *J. Organomet. Chem.*, **120**, 361 (1976).
45. R. D. Fischer and H. Fischer, *J. Organomet. Chem.*, **4**, 412 (1965).
46. A. E. Crease and P. Legzdins, *J. Chem. Soc., Dalton Trans.*, 1501 (1973).
47. S. Onaka and N. Furuichi, *J. Organomet. Chem.*, **173**, 77 (1979).
48. S. Onaka, *Inorg. Chem.*, **19**, 2132 (1980).
49. B. Kanellakopulos, E. Dornberger, R. von Ammon, and R. D. Fischer, *Angew. Chem., Int. Ed. Engl.*, **9**, 957 (1970).
50. R. E. Maginn, S. Manastyrskyj, and M. Dubeck, *J. Am. Chem. Soc.*, **85**, 672 (1963).
51. F. Gomez-Beltran, L. A. Oro, and F. Ibanez, *J. Inorg. Nucl. Chem.*, **37**, 1541 (1975).
52. R. J. H. Clark, J. Lewis, D. J. Machin, and R. S. Nyholm, *J. Chem. Soc.*, 379 (1963).
53. A. L. Wayda and W. J. Evans, *Inorg. Chem.*, **19**, 2190 (1980).
54. T. Aoyagi, H. M. M. Shearer, K. Wade, and G. Whitehead, *J. Organomet. Chem.*, **175**, 21 (1979).
55. R. S. P. Coutts and P. C. Wailes, *J. Organomet. Chem.*, **25**, 117 (1970).
56. W. J. Evans and H. A. Zinnen, to be published.
57. E. Dornberger, R. Klenze, and B. Kanellakopulos, *Inorg. Nucl. Chem. Lett.*, **14**, 319 (1978).
68. G. Bielang and R. D. Fischer, *Inorg. Chim. Acta*, **36**, L389 (1979).
69. H. Schumann and H. Jarosch, *Z. Anorg. Allg. Chem.*, **426**, 127 (1976).
70. H. Schumann and M. Cygon, *J. Organomet. Chem.*, **144**, C43 (1978).
71. H. Schumann, personal communication reported in reference 6.
72. S. Manastyrskyj, R. E. Maginn, and M. Dubeck, *Inorg. Chem.*, **2**, 904 (1963).
73. F. A. Cotton and G. Wilkinson, *Advanced Inorganic Chemistry,* 4th ed., Wiley, New York, 1980, Ch. 23, pp. 999–1002.
74. W. J. Evans, A. L. Wayda, W. E. Hunter and J. L. Atwood, *J. Chem. Soc., Chem. Commun.*, **292** (1981).
75. H. A. Zinnen, J. J. Pluth, and W. J. Evans, *J. Chem. Soc., Chem. Commun.*, 810 (1980).
76. G. W. Watt and E. W. Gillow, *J. Am. Chem. Soc.*, **91**, 775 (1969).
77. B. L. Kalsotra, S. P. Anand, R. K. Multani, and B. D. Jain, *J. Organomet. Chem.*, **28**, 87 (1971).
78. B. L. Kalsotra, R. K. Multani, and B. D. Jain, *J. Inorg. Nucl. Chem.*, **34**, 2679 (1972).
79. B. L. Kalsotra, R. K. Multani, and B. D. Jain, *Isr. J. Chem.*, **9**, 569 (1971).
80. B. L. Kalsotra, R. K. Multani, and B. D. Jain, *J. Inorg. Nucl. Chem.*, **34**, 2265 (1972).
81. B. L. Kalsotra, R. K. Multani, and B. D. Jain, *J. Inorg. Nucl. Chem.*, **35**, 311 (1973).
82. S. Kapur, B. L. Kalsotra, R. K. Multani, and B. D. Jain, *J. Inorg. Nucl. Chem.*, **35**, 1689 (1973).
83. S. Kapur, B. L. Kalsotra, and R. K. Multani, *J. Inorg. Nucl. Chem.*, **35**, 3966 (1973).
84. K. W. Bagnall, in reference 27, pp. 221–248.
85. A. Greco, S. Cesca, and G. Bertolini, *J. Organomet. Chem.*, **113**, 321 (1976).
86. E. O. Fischer and Y. Hristidu, *Z. Naturforsch.*, **17B**, 275 (1962).
87. F. Baumgärtner, E. Dornberger, and B. Kanellakopulos, unpublished results reported in reference 27, pp. 1–35.
88. E. O. Fischer and Y. Hristidu, *Z. Naturforsch.*, **17B**, 276 (1962).
89. T. J. Marks, A. M. Seyam and W. A. Wachter, *Inorg. Synth.*, **16**, 147 (1976).
90. F. Baumgärtner, E. O. Fischer, B. Kanellakopulos, and P. Laubereau, *Angew. Chem., Int. Ed. Engl.*, **8**, 202 (1969).
91. F. Baumgärtner, E. O. Fischer, B. Kanellakopulos, and P. Lauberau, *Angew. Chem., Int. Ed. Engl.*, **7**, 634 (1968).
92. L. R. Crisler and W. G. Eggerman, *J. Inorg. Nucl. Chem.*, **36**, 1424 (1974).
93. D. G. Karraker, personal communication reported in reference 27, p. 5.
94. C. C. Chang, N. K. Sung-Yu, C. S. Hseu, and C. T. Chang, *Inorg. Chem.*, **18**, 885 (1979).
95. D. F. Starks and A. Streitwieser, Jr., *J. Am. Chem. Soc.*, **95**, 3423 (1973).
96. P. G. Laubereau, L. Ganguly, J. H. Burns, B. M. Benjamin, J. L. Atwood, and J. Selbin, *Inorg. Chem.*, **10**, 2274 (1971).
97. B. Kanellakopulos, in reference 27, pp. 1–35.
98. C. A. Secaur, V. W. Day, R. D. Ernst, W. J. Kennelly and T. J. Marks, *J. Am. Chem. Soc.*, **98**, 3713 (1976).
99. J. H. Burns, *J. Organomet. Chem.*, **69**, 225 (1974).
100. M. L. Anderson and L. R. Crisler, *J. Organomet. Chem.*, **17**, 345 (1969).

101. R. D. Ernst, W. J. Kennelly, C. S. Day, V. W. Day, and T. J. Marks, *J. Am. Chem. So* **101**, 2656 (1979).
102. E. O. Fischer, P. Laubereau, F. Baumgärtner, and B. Kanellakopulos, *J. Organom Chem.*, **5**, 583 (1966).
103. D. G. Karraker and J. A. Stone, *Inorg. Chem.*, **11**, 1742 (1972).
104. P. Laubereau, personal communication cited in reference 20, p. 303.
105. R. D. Fischer, R. von Ammon, and B. Kanellakopulos, *J. Organomet. Chem.*, **25**, 1 (1970).
106. J. Goffart, J. Fuger, B. Gilbert, L. Hocks, and G. Duyckaerts, *Inorg. Nucl. Chem. Lett.*, **1** 569 (1975).
107. J. Goffart and G. Duyckaerts, *Inorg. Nucl. Chem. Lett.*, **14**, 15 (1978).
108. R. D. Fischer and G. R. Sienel, *Z. Anorg. Allg. Chem.*, **419**, 126 (1976).
109. J. Leong, K. O. Hodgson, and K. N. Raymond, *Inorg. Chem.*, **12**, 1329 (1973).
110. K. W. Bagnall, J. Edwards, and A. C. Tempest, *J. Chem. Soc., Dalton Trans.*, 295 (1978)
111. T. J. Marks and J. R. Kolb, *J. Am. Chem. Soc.*, **97**, 27 (1975).
112. R. von Ammon, B. Kanellakopulos, and R. D. Fischer, *Radiochim. Acta,* **11**, 162 (1969).
113. R. von Ammon, R. D. Fischer, and B. Kanellakopulos, *Chem. Ber.*, **105**, 45 (1972).
114. J. Goffart, B. Gilbert, and G. Duyckaerts, *Inorg. Nucl. Chem. Lett.*, **13**, 189 (1977).
115. D. G. Karraker, in reference 27, pp. 395–420.
116. A. B. McLaren, B. Kanellakopulos, and E. Dornberger, *Inorg. Nucl. Chem. Lett.*, **16**, 22 (1980).
117. R. G. Jones, G. Karmas, G. A. Martin, Jr., and H. Gilman, *J. Am. Chem. Soc.*, **78**, 428 (1956).
118. J. D. Jamerson and J. Takats, *J. Organomet. Chem.*, **78**, C23 (1974).
119. P. Zanella, S. Faleschini, L. Doretti, and G. Faraglia, *J. Organomet. Chem.*, **26**, 353 (1971).
120. B. Kanellakopulos, C. Aderhold, and E. Dornberger, *J. Organomet. Chem.*, **66**, 447 (1974).
121. L. Doretti, P. Zanella, G. Faraglia, and S. Faleschini, *J. Organomet. Chem.*, **43**, 339 (1972).
122. This analysis assumes the cyclopentadienyl ligand occupies 3 coordination positions.
123. C. H. Wong, T. M. Yen, and T. Y. Lee, *Acta Crystallogr., Sect. B,* **18**, 340 (1965).
124. P. Zanella, G. DePaoli, G. Bombieri, G. Zanotti, and R. Rossi, *J. Organomet. Chem.*, **142**, C21 (1977).
125. M. F. Brady and R. S. Marianelli, *8th Midwest Regional ACS Meeting, Columbia, Missouri, November 1972,* Paper No. 235.
126. T. J. Marks, NATO Advanced Study Institute, Sogesta, Italy, September 1978.
127. F. R. Fronczek, G. W. Halstead, and K. N. Raymond, *J. Am. Chem. Soc.*, **99**, 1769 (1977).
128. J. C. Green and O. Watts, *J. Organomet. Chem.*, **153**, C40 (1978).
129. J. M. Manriquez, P. J. Fagan, and T. J. Marks, *J. Am. Chem. Soc.*, **100**, 3939 (1978).
130. K. W. Bagnall and J. Edwards, *J. Organomet. Chem.*, **80**, C14 (1974).
131. J. Goffart, J. Piret-Meunier, and G. Duyckaerts, *Inorg. Nucl. Chem. Lett.*, **16**, 233 (1980).
132. F. Baumgärtner, E. O. Fischer, B. Kanellakopulos, and P. Laubereau, *Angew. Chem., Int. Ed. Engl.*, **4**, 878 (1965).
133. F. Baumgärtner, E. O. Fischer, B. Kanellakopulos, and P. Laubereau, *Angew. Chem., Int. Ed. Engl.*, **5**, 134 (1966).
134. F. Baumgärtner, E. O. Fischer, H. Billich, E. Dornberger, B. Kanellakopulos, W. Roth, and L. Stieglitz, *J. Organomet. Chem.*, **22**, C17 (1970).
135. P. G. Laubereau and J. H. Burns, *Inorg. Nucl. Chem. Lett.*, **6**, 59 (1970).
136. L. R. Crisler and W. G. Eggerman, *J. Inorg. Nucl. Chem.*, **36**, 1424 (1974).
137. Y. Hristidu, *Dissertation,* University of Munich, 1962.
138. B. Kanellakopulos, E. O. Fischer, E. Dornberger, and F. Baumgärtner, *J. Organomet. Chem.*, **24**, 507 (1970).
139. D. C. Moody and J. D. Odom, *J. Inorg. Nucl. Chem.*, **41**, 533 (1979).
140. B. Kanellakopulos, E. Dornberger, and F. Baumgärtner, *Inorg. Nucl. Chem. Lett.*, **10**, 155 (1974).
141. D. G. Kalina, T. J. Marks, and W. A. Wachter, *J. Am. Chem. Soc.*, **99**, 3877 (1977).
142. J. Goffart, in reference 27, p. 481.
143. M. Tsutsui, N. Ely, and A. Gebala, *Inorg. Chem.*, **14**, 78 (1975).
144. P. G. Laubereau, *Inorg. Nucl. Chem. Lett.*, **6**, 611 (1970).

145. J. M. Manriquez, P. J. Fagan, T. J. Marks, S. H. Vollmer, C. S. Day, and V. W. Day, *J. Am. Chem. Soc.*, **101**, 5075 (1979).
146. F. Mares, K. Hodgson, and A. Streitwieser, Jr., *J. Organomet. Chem.*, **24**, C68 (1970).
147. F. Mares, K. O. Hodgson, and A. Streitwieser Jr., *J. Organomet. Chem.*, **28**, C24 (1971).
148. K. O. Hodgson and K. N. Raymond, *Inorg. Chem.*, **11**, 3030 (1972).
149. C. W. DeKock, S. R. Ely, T. E. Hopkins, and M. A. Brault, *Inorg. Chem.*, **17**, 625 (1978).
150. H. Breil and G. Wilke, *Angew. Chem., Int. Ed. Engl.*, **5**, 898 (1966).
151. A. Greco, G. Bertolini, and S. Cesca, *Inorg. Chim. Acta*, **21**, 245 (1977).
152. G. Wilke, E. W. Mueller, M. Kroener, P. Heimbach, and H. Breil, *Ger. Pat.*, 1191375 (1965); *Chem. Abstr.*, **63**, 7045b (1965).
153. J. D. Jamerson, A. P. Masino and J. Takats, *J. Organomet. Chem.*, **65**, C33 (1974).
154. W. J. Evans, A. L. Wayda, C. W. Chang, and W. M. Cwirla, *J. Am. Chem. Soc.*, **100**, 333 (1978).
155. W. J. Evans, D. J. Wink, A. L. Wayda, and D. A. Little, *J. Org. Chem.*, **46**, 3925 (1981).
156. R. G. Hayes and J. L. Thomas, *J. Am. Chem. Soc.*, **91**, 6876 (1969).
157. B. L. Kalsotra, R. K. Multani, and B. D. Jain, *Chem. Ind. (London)*, 339 (1972).
158. A. Streitwieser, Jr., U. Muller-Westerhoff, G. Sonnichsen, F. Mares, D. G. Morrell, K. O. Hodgson, and C. A. Harmon, *J. Am. Chem. Soc.*, **95**, 8644 (1973).
159. A. Streitwieser, Jr., U. Muller-Westerhoff, F. Mares, C. B. Grant, and D. G. Morrell, *Inorg. Synth.*, **19**, 149 (1979).
160. A. Streitwieser, Jr., and N. Yoshida, *J. Am. Chem. Soc.*, **91**, 7528 (1969).
161. D. F. Starks, T. C. Parsons, A. Streitwieser, Jr., and N. Edelstein, *Inorg. Chem.*, **13**, 1307 (1974).
162. J. Goffart, J. Fuger, D. Brown, and G. Duyckaerts, *Inorg. Nucl. Chem. Lett.*, **10**, 413 (1974).
163. D. G. Karraker, J. A. Stone, E. R. Jones, Jr., and N. Edelstein, *J. Am. Chem. Soc.*, **92**, 4841 (1970).
164. A. Streitwieser, Jr., in reference 27, pp. 149–177.
165. D. G. Karraker and J. A. Stone, *J. Am. Chem. Soc.*, **96**, 6885 (1974).
166. R. D. Rieke and L. D. Rhyne, *J. Org. Chem.*, **44**, 3445 (1979).
167. V. Graves, L. H. Simons, and J. J. Lagowski, *Chem. Eng. News*, 26, Dec. 24 (1973).
168. J. R. Blackborow and D. Young, *Metal Vapor Synthesis in Organometallic Chemistry*, Springer-Verlag, Berlin, 1979, p. 180.
169. J. T. Miller and C. W. DeKock, *Inorg. Chem.*, **18**, 1305 (1979).
170. C. A. Harmon, D. P. Bauer, S. R. Berryhill, K. Hagiwara, and A. Streitwieser, Jr., *Inorg. Chem.*, **16**, 2143 (1977).
171. A. Streitwieser, Jr., and C. A. Harmon, *Inorg. Chem.*, **12**, 1102 (1973).
172. D. G. Karraker, *Inorg. Chem.*, **12**, 1105 (1973).
173. G. R. Sienel, A. W. Spiegl, and R. D. Fischer, *J. Organomet. Chem.*, **160**, 67 (1978).
174. A. Zalkin, D. H. Templeton, S. R. Berryhill, and W. D. Luke, *Inorg. Chem.*, **18**, 2287 (1979).
175. A. Streitwieser, Jr., D. Dempf, G. N. La Mar, D. G. Karraker, and N. Edelstein, *J. Am. Chem. Soc.*, **93**, 7343 (1971).
176. A. Streitwieser, Jr., and R. Walker, *J. Organomet. Chem.*, **97**, C41 (1975).
177. L. K. Templeton, D. H. Templeton, and R. Walker, *Inorg. Chem.*, **15**, 3000 (1976).
178. J. P. Solar, H. P. G. Burghard, R. H. Banks, A. Streitwieser, Jr., and D. Brown, *Inorg. Chem.*, **19**, 2186 (1980).
179. C. A. Harmon and A. Streitwieser, Jr., *J. Am. Chem. Soc.*, **94**, 8926 (1972).
180. J. P. Solar, A. Streitwieser, Jr., and N. Edelstein, *178th ACS National Meeting, Washington, D.C., September 1979*, Abstract INOR 135 (1979).
181. P. Girard, J. L. Namy, and H. B. Kagan, *J. Am. Chem. Soc.*, **102**, 2693 (1980).
182. D. G. Karraker, *J. Inorg. Nucl. Chem.*, **39**, 87 (1977).
183. D. G. Karraker and J. A. Stone, *J. Inorg. Nucl. Chem.*, **39**, 2215 (1977).
184. F. O. Rice and K. K. Rice, *The Aliphatic Free Radicals,* Johns Hopkins Press, Baltimore, 1935, p. 58.
185. S. A. Cotton, F. A. Hart, M. B. Hursthouse, and A. J. Welch, *J. Chem. Soc., Chem Commun.*, 1225 (1972).
186. H. Gilman and R. G. Jones, *J. Org. Chem.*, **10**, 505 (1945).

187. J. A. Butcher, Jr., J. Q. Chambers, and R. M. Pagni, *J. Am. Chem. Soc.*, **100**, 1012 (1978).
188. J. L. Atwood. W. E. Hunter, R. D. Rogers, J. Holton, J. McMeeking, R. Pearce, and M. F. Lappert, *J. Chem. Soc., Chem. Commun.*, 140 (1978).
189. H. Schumann and J. Müller, *J. Organomet. Chem.*, **169**, C1 (1979).
190. H. Schumann, in reference 27, pp. 81–112.
191. A. L. Wayda and W. J. Evans, *J. Am. Chem. Soc.*, **100**, 7119 (1978).
192. H. Schumann and J. Müller, *Angew. Chem., Int. Ed. Engl.*, **17**, 276 (1978).
193. H. Schumann and S. Hohmann, *Chem.-Ztg.*, **100**, 336 (1976).
194. W. J. Evans and A. L. Wayda, *J. Organomet. Chem.*, **202**, C6 (1980).
195. D. F. Evans, G. V. Fazakerley and R. F. Phillips, *J. Chem. Soc., Chem. Commun.*, 24 (1970).
196. D. F. Evans, G. V. Fazakerley, and R. F. Phillips, *J. Chem. Soc., A*, 1931 (1971).
197. G. B. Deacon, W. D. Raverty, and D. G. Vince, *J. Organomet. Chem.*, **135**, 103 (1977).
198. G. B. Deacon, A. J. Koplick, W. D. Raverty, and D. G. Vince, *J. Organomet. Chem.*, **182**, 121 (1979).
199. G. B. Deacon and A. J. Koplick, *J. Organomet. Chem.*, **146**, C43 (1978).
200. E. Murphy and G. E. Toogood, *Inorg. Nucl. Chem. Lett.*, **7**, 755 (1971).
201. W. J. Evans, S. C. Engerer, and K. M. Coleson, *J. Am. Chem. Soc.*, **103**, 6672 (1981).
202. H. Gilman and G. L. Schwebke, *J. Organomet. Chem.*, **4**, 483 (1965).
203. R. B. Bates, L. M. Kroposki, and D. E. Potter, *J. Org. Chem.*, **37**, 560 (1972).
204. J. Holton, M. F. Lappert, D. G. H. Ballard, R. Pearce, J. L. Atwood, and W. E. Hunter, *J. Chem. Soc., Dalton Trans.*, 54 (1979).
205. W. J. Evans, A. L. Wayda, W. E. Hunter, and J. L. Atwood, *J. Chem. Soc., Chem. Commun.*, 292 (1981).
206. N. M. Ely and M. Tsutsui, *Inorg. Chem.*, **14**, 2680 (1975).
207. J. Holton, M. F. Lappert, D. G. H. Ballard, R. Pearce, J. L. Atwood, and W. E. Hunter, *J. Chem. Soc., Dalton Trans.*, 45 (1979).
208. B. Kanellakopulos, E. Dornberger, and H. Billich, *J. Organomet. Chem.*, **76**, C42 (1974).
209. R. D. Fischer and G. Bielang, *J. Organomet. Chem.*, **191**, 61 (1980).
210. E. Köhler, W. Brüser, and K. H. Thiele, *J. Organomet. Chem.*, **76**, 235 (1974).
211. G. R. Davies, J. A. J. Jarvis, B. T. Kilbourn, and A. J. P. Pioli, *J. Chem. Soc., Chem. Commun.*, 1511 (1971).
212. H. Gilman, R. G. Jones, E. Bindschadler, D. Blume, G. Karmas, G. A. Martin, Jr., J. F. Nobis, J. R. Thirtle, H. L. Yale, and F. A. Yoeman, *J. Am. Chem. Soc.*, **78**, 2790 (1956).
213. G. Yagupsky, W. Mowat, A. Shortland, and G. Wilkinson, *J. Chem. Soc., Chem. Commun.*, 1369 (1970).
214. E. R. Sigurdson and G. Wilkinson, *J. Chem. Soc., Dalton Trans.*, 812 (1977).
215. K. H. Thiele, R. Opitz, and E. Köhler, *Z. Anorg. Allg. Chem.*, **435**, 45 (1977).
216. T. J. Marks and A. M. Seyam, *J. Organomet. Chem.*, **67**, 61 (1974).
217. T. J. Marks, *J. Organomet. Chem.*, **158**, 325 (1978).
218. R. J. Lagow, L. L. Gerchman, and R. A. Jacob, *U.S. Pat.*, 3954585 and 3992424; *Chem. Abstr.*, **85**, 160324s (1976) and **86**, 72887t (1976).
219. G. Brandi, M. Brunelli, G. Lugli, and A. Mazzei, *Inorg. Chim. Acta*, **7**, 319 (1973).
220. A. E. Gebala and M. Tsutsui, *J. Am. Chem. Soc.*, **95**, 91 (1973).
221. T. J. Marks, A. M. Seyam, and J. R. Kolb, *J. Am. Chem. Soc.*, **95**, 5529 (1973).
222. R. E. Cramer, J. W. Gilje, and R. B. Maynard, *178th ACS National Meeting, Washington, D.C., September 1979*, Abstract INOR 134 (1979).
223. R. E. Cramer, R. B. Maynard, and J. W. Gilje, *Inorg. Chem.*, **19**, 2564 (1980).
224. T. J. Marks and W. A. Wachter, *J. Am. Chem. Soc.*, **98**, 703 (1976).
225. E. C. Baker, K. N. Raymond, T. J. Marks, and W. A. Wachter, *J. Am. Chem. Soc.*, **96**, 7586 (1974).
226. V. K. Vasil'ev, V. N. Sokolov, and G. P. Kondratenkov, *J. Organomet. Chem.*, **142**, C7 (1977).
227. P. J. Fagan, J. M. Manriquez, and T. J. Marks, in reference 27, pp. 113–148.
228. A. M. Seyam and G. A. Eddein, *Inorg. Nucl. Chem. Lett.*, **13**, 115 (1977).
229. R. W. Broach, A. J. Schultz, J. M. Williams, G. M. Brown, J. M. Manriquez, P. J. Fagan and T. J. Marks, *Science,* **203**, 172 (1979).

230. P. J. Fagan, J. M. Manriquez, T. J. Marks, V. W. Day, S. H. Vollmer, and C. S. Day, *J. Am. Chem. Soc.*, **102**, 5393 (1980).
231. G. Lugli, W. Marconi, A. Mazzei, N. Paladino, and U. Pedretti, *Inorg. Chim. Acta*, **3**, 253 (1969).
232. M. Brunelli, G. Lugli, and G. Giacometti, *J. Magn. Reson.*, **9**, 247 (1973).
233. G. Wilke, B. Bogdanovic, P. Hardt, P. Heimbach, W. Keim, M. Kroner, W. Oberkirch, K. Tanaka, E. Steinrucke, D. Walter, and H. Zimmerman, *Ang. Chem., Int. Ed. Engl.*, **5**, 151 (1966).
234. A. Mazzei, in reference 27, pp. 379–393.
235. G. Lugli, A. Mazzei, and S. Poggio, *Makromol. Chem.*, **175**, 2021 (1974).
236. M. Brunelli, G. Perego, G. Lugli, and A. Mazzei, *J. Chem. Soc., Dalton Trans.*, 861 (1979).
237. M. Tsutsui and N. Ely, *J. Am. Chem. Soc.*, **97**, 3551 (1975).
238. G. W. Halstead, E. C. Baker, and K. N. Raymond, *J. Am. Chem. Soc.*, **97**, 3049 (1975).
239. T. J. Marks and J. R. Kolb, *J. Organomet. Chem.*, **82**, C35 (1974).
240. M. Cesari, U. Pedretti, A. Zazzetta, G. Lugli, and W. Marconi, *Inorg. Chim. Acta*, **5**, 439 (1971).
241. W. J. Evans, S. C. Engerer, and A. C. Neville, *J. Am. Chem. Soc.*, **100**, 331 (1978).
242. W. J. Evans, S. C. Engerer, P. A. Piliero, and A. L. Wayda, *J. Chem. Soc., Chem. Commun.*, 1007 (1979).
243. P. S. Skell, *Proc. Int. Cong. Pure Appl. Chem.*, **23**, 215 (1971).
244. P. S. Skell, *U.S. Govt. Rep.*, No. AFOSR-75-0200 (1975).
245. W. J. Evans, K. M. Coleson, and S. C. Engerer, *Inorg. Chem.*, **20**, 4320 (1981).

The Chemistry of the Metal–Carbon Bond
Edited by F. R. Hartley and S. Patai

CHAPTER **13**

Metal atoms in organometallic synthesis

MICHAEL J. McGLINCHEY
Department of Chemistry, McMaster University, 1280 Main Street West, Hamilton, Ontario, L8S 4M1, Canada

I. THE METAL ATOM VAPORIZATION TECHNIQUE

The use of metal vapours as synthetic reagents has, over the last decade, undergone a rapid transition from that of laboratory curiosity to a ubiquitous weapon in the synthetic chemist's armoury. The experimental techniques *per se* will not be discussed here in depth since a number of authoritative reviews are devoted to this topic[1–6]; further, the apparatus is now commercially available. Suffice it to say that atomic vapours are readily produced in a variety of ways (resistive heating[1], laser[6] or electron beam vaporization[7,8], electric arcs[9], etc.) and allowed to condense on a cold (usually $-196°C$) surface with a large excess of the co-reactant(s). This latter criterion, *viz.* large excess of substrate, minimizes the agglomeration of single atoms to bulk metal –

a process which is generally thermodynamically favoured over reaction with the substrate. Of course, the migration of metal atoms through low-temperature matrices to form clusters under controlled conditions is itself a fascinating topic[10], but is outside the scope of this review, which is aimed primarily at the macro-scale synthetic chemist.

We consider here the common organic functionalities as target moieties and examine the range of possible reactions which metal vapours can induce at that site. These can run the gamut from simple electron transfer, through σ- or π-type attachment of the site, to atom abstraction. These in turn can lead to direct syntheses of organometallics, template reactions of the organic moieties, or formation of unstable intermediates which must be investigated by further interactions with added reagents. Comparisons are drawn with the chemistry of the bulk metal, and the relative synthetic advantages or disadvantages of the co-condensation method *vis-à-vis* conventional routes are discussed.

In some cases, mechanistic details have been elucidated by matrix isolation spectroscopy in which the reaction is performed on the micro-scale at liquid helium temperature in a matrix of a noble gas[11]. This allows the direct measurement of the i.r., Raman, e.s.r., electronic, Mössbauer, or other type of spectra[2] and the progress of the reaction can be monitored as the temperature is gradually increased.

In a typical synthetic procedure, about 25 g of a ligand would be condensed on to a cold surface over a period of perhaps 1 hour while 500–1000 mg of a metal were evaporated on to that surface over the same period. At the conclusion of the co-deposition, the excess of unreacted substrate would be removed and the products sublimed or extracted, as appropriate, and purified via conventional techniques. A number of useful synthetic procedures have been gathered together in a recent volume of *Inorganic Syntheses*[12].

Recent experimental advances have not only removed the requirement that the substrate be relatively volatile but have also facilitated the evaporation of useful amounts of the catalytically important second-and third-row transition metals[13]; thus essentially all of the metals in the Periodic Table are candidates for use as reagents in organic and organometallic syntheses.

The scale of these reactions is very much a matter of availability of materials but, in principle, they could be performed on an industrially useful scale if the economics were favourable[14]. The advent of several types of commercial apparatus for metal vaporization syntheses indicates that the transition to industrial utility may not be too distant.

II. ORGANIC HALIDES

Following the pioneering studies of Polanyi on sodium vapour reactions, the organic halides were a natural early target for the co-condensation chemist. Elegant studies by Mile[15], who used a rotating cryostat to allow Na atoms to dehalogenate organic halides, led to the formation of radicals which could be studied by e.s.r. and i.r. spectroscopy, or which could themselves be reagents in subsequent chemistry (equation 1). Several reviews on spectroscopic studies of alkali metal atoms in matrices have appeared recently[2,11,16].

$$RX \xrightarrow{Na^{\bullet}} NaX + R^{\bullet} \begin{cases} \xrightarrow{O_2} R{-}O{-}O^{\bullet} \\ \xrightarrow{C_2H_4} R{-}CH_2{-}CH_2^{\bullet} \end{cases} \tag{1}$$

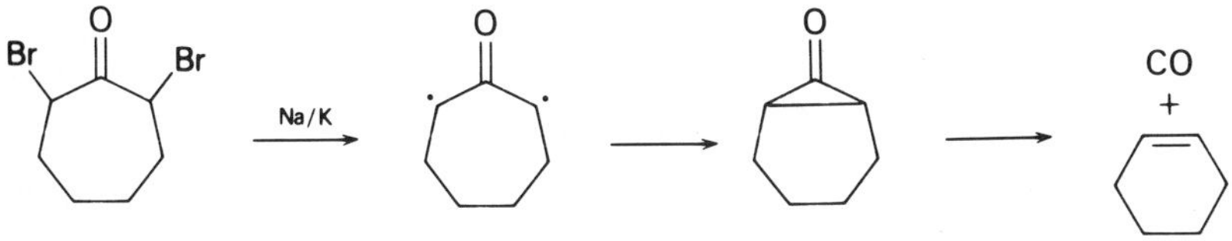

$$\text{2,7-dibromocycloheptanone} \xrightarrow{Na/K} \text{diradical} \longrightarrow \text{bicyclic ketone} \longrightarrow CO + \text{cyclohexene} \quad (2)$$

Skell and his colleagues[17] extended these ideas to the macro-scale production of short-lived species such as diradicals[18] and silenes[19]. When haloderivatives of carbon (reaction 2) and of silicon (Scheme 1) were dehalogenated by hot Na/K alloy, Skell showed that not only was it possible to trap these intermediates chemically but also, in some cases, their spin states could be elucidated. This work provided early evidence for the transitory existence of silacyclopropanes[19], which culminated in their isolation by Lambert and Seyferth 8 years later[20].

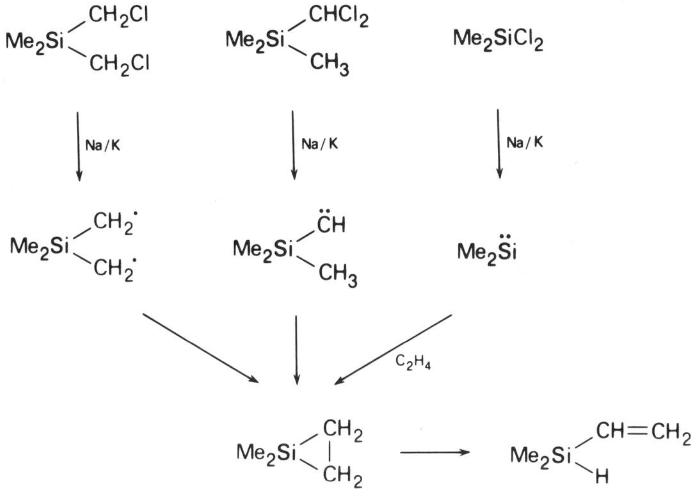

$$Me_2Si(CH_2Cl)_2 \xrightarrow{Na/K} Me_2Si(CH_2^{\bullet})_2$$
$$Me_2Si(CHCl_2)(CH_3) \xrightarrow{Na/K} Me_2Si(\ddot{C}H)(CH_3)$$
$$Me_2SiCl_2 \xrightarrow{Na/K} Me_2\ddot{Si} \xrightarrow{C_2H_4}$$
$$\longrightarrow Me_2Si\langle(CH_2)_2 \longrightarrow Me_2Si(CH{=}CH_2)(H)$$

SCHEME 1

A particularly exciting development in this area is the direct synthesis of lithiocarbons via the co-condensation of Li vapour with a deficit of halocarbon[21–23]. Thus reaction of CCl_4 and Li led to tetralithiomethane, an extremely moisture-sensitive molecule which reacts explosively with water. This approach clearly has enormous potential and some typical reactions are shown in equation 3.

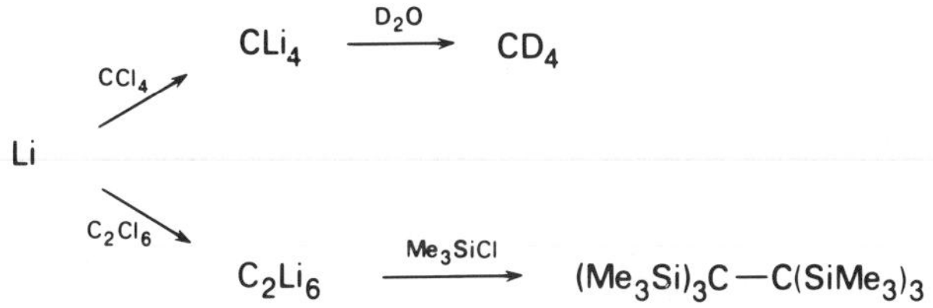

$$Li \xrightarrow{CCl_4} CLi_4 \xrightarrow{D_2O} CD_4$$
$$Li \xrightarrow{C_2Cl_6} C_2Li_6 \xrightarrow{Me_3SiCl} (Me_3Si)_3C{-}C(SiMe_3)_3 \quad (3)$$

Halocarbons can also be dehalogenated by the vapour of Cu, Ag, or Au[24,25] but the reactions may be contrasted with those of the alkali metals in that, although the radicals can undergo dimerization and disproportionation, the relative yields from these processes are highly dependent on the nature of the metal and on the ratio of metal to ligand. The increased yield of dimerization products as the ratio of substrate to metal was lowered from 100:1 to 10:1 has led to speculation as to the possible intermediacy of metal alkyls. A crucial result[26] is that while Cu or Ag atoms react with *R*(−)-*sec*-butyl chloride to give *S,S*(−)-3,4-dimethylhexane (70% optically pure) the use of Na vapour gives the optically inactive product. One is drawn to the conclusion that, in the former case, stereospecific coupling occurred on the metal whereas in the Na reaction random coupling of free radicals occurred. More compelling, perhaps, is the fact that, when perfluoroalkyl groups are used, alkylsilver derivatives are isolable (equation 4)[27]. These results may go some way to clarifying the detailed mechanism of the classic Ullman biaryl synthesis in which copper is used to couple haloarenes.

$$(CF_3)_2CFI + 2Ag \longrightarrow (CF_3)_2CFAg + AgI \qquad (4)$$

Timms has demonstrated[25] that dehalogenative coupling with copper vapour is a viable route to many inorganic molecules previously synthesized by discharge methods (equation 5)[28].

$$Cu \xrightarrow{MeBCl_2} MeBCl{-}BClMe$$
$$Cu \xrightarrow{BCl_3} B_2Cl_4\ (70\%) \qquad (5)$$
$$Cu \xrightarrow{PCl_3} P_2Cl_4$$

It is now clear that in the processes discussed for Cu, Ag, and Au vapours the organohalide need not only suffer halogen abstraction but can also take part in oxidative addition reactions. This aspect has been extensively studied notably by Girard[29] and Skell and Girard[30] in their now classic work on Mg vapour and also by Timms, Klabunde and others using Group VIII metals.

An important point to make concerning the reactions of Mg vapour is that its chemistry is crucially dependent upon its mode of generation. Thus, thermally generated Mg atoms (produced by resistive heating of a crucible) are in their electronic ground state (^1S); in contrast, when an arc is struck between two Mg electrodes the high electron flux produced by the arcing procedure leads to a great excess of electronically excited atoms over that expected from a simple Boltzmann distribution. The ^3P excited state of Mg lies 65.8 kcal above the ^1S ground state and even a low-voltage Mg arc contains between 65 and 75% of the ^3P state[29]. As one would have expected from the known chemistry of carbon atoms in different spin states[9], the singlet Mg atoms generally give insertion (oxidative addition) reactions whereas the ^3P atoms are primarily diradical in character.

When ^1S magnesium atoms are co-condensed with alkyl halides the initial product is a jet-black matrix (probably a charge-transfer complex) which, on warming, forms a non-solvated Grignard reagent. These latter products are, of course, readily hydrolysed to the corresponding alkanes but their chemistry does not always parallel that of conventional Grignards in coordinating solvents. Thus, whereas normal Grignard reagents react with ketones to give carbinols, non-solvated Grignards preferentially abstract an enolizable proton to generate the alkane and (in the case of acetone) diacetone alcohol. This latter molecule presumably arises by nucleophilic attack by acetone enolate upon another molecule of acetone. Furthermore,

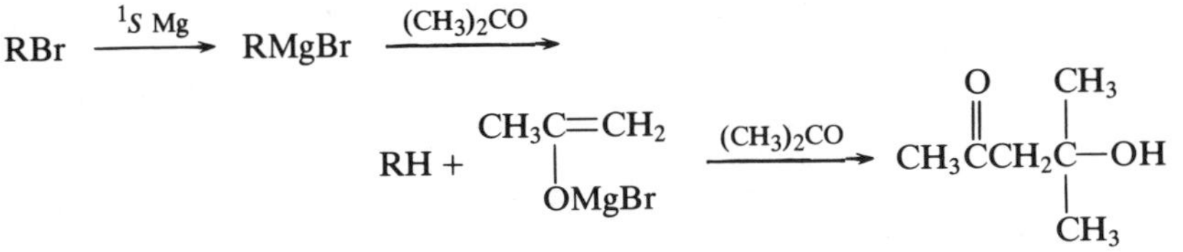

$$RBr \xrightarrow{^{1}S\ Mg} RMgBr \xrightarrow{(CH_3)_2CO} RH + CH_3C(OMgBr){=}CH_2 \xrightarrow{(CH_3)_2CO} CH_3C(=O)CH_2C(CH_3)_2OH \quad (6)$$

non-solvated Grignard reagent (which is presumably associated via halogen bridges) reacts with a non-enolizable substrate such as crotonaldehyde to give 3-penten-2-ol upon hydrolysis rather than the isovaleraldehyde which would be generated in a conventional 1,4-addition via a 6-membered transition state (equation 7). This result

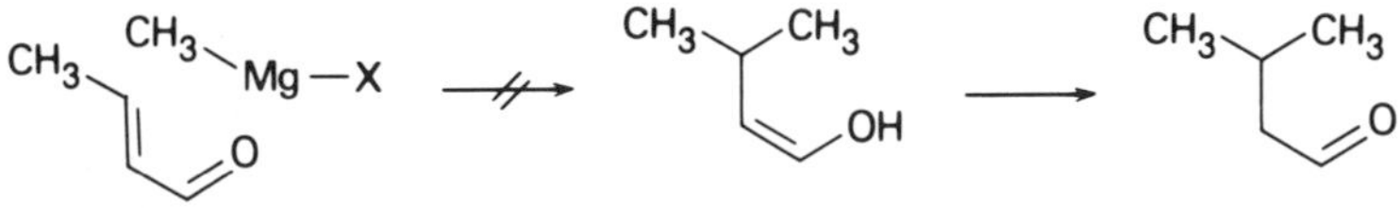

CH3 Mg—X (7)

has been rationalized[30] on the basis of a 1,2-addition involving two molecules of Grignard reagent but still proceeding via the favourable 6-membered transition state (equation 8).

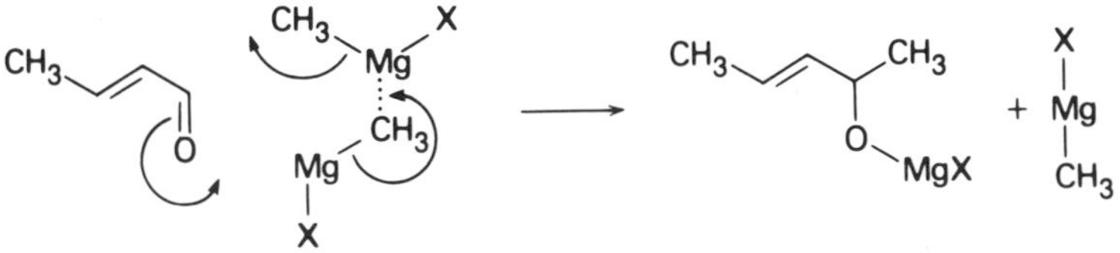

CH3 CH3 X Mg CH3 Mg X → CH3 CH3 O MgX + X–Mg–CH3 (8)

In complete contrast, arc-produced ^{3}P Mg atoms give less than 1% of non-solvated Grignard reagent and the major products result from disproportionation and dimerization reactions of the radicals produced by dehalogenation[29]. In accord with this view, 1,3-dihalides yield cyclopropanes via diradicals and geminal dihalides react with ^{3}P Mg atoms to yield carbenes (Scheme 2).

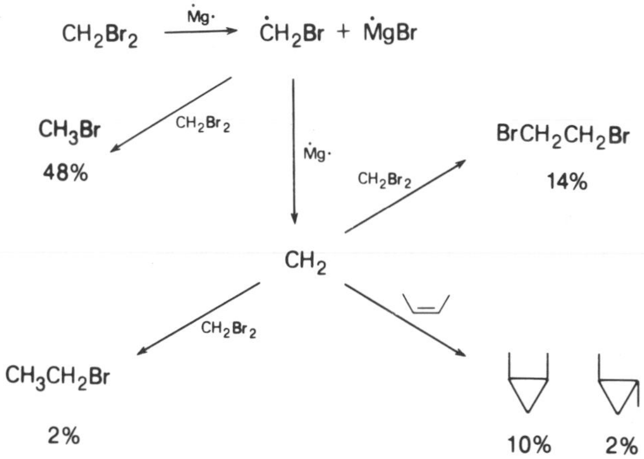

$$CH_2Br_2 \xrightarrow{Mg\cdot} \dot{C}H_2Br + \dot{M}gBr$$

$\dot{C}H_2Br \xrightarrow{CH_2Br_2} CH_3Br$ (48%)

$\dot{C}H_2Br \xrightarrow{Mg\cdot} CH_2$

$CH_2 \xrightarrow{CH_2Br_2} BrCH_2CH_2Br$ (14%)

$CH_2 \xrightarrow{CH_2Br_2} CH_3CH_2Br$ (2%)

CH_2 + cis-2-butene → cyclopropanes (10%, 2%)

SCHEME 2

Finally, ^3P Mg vapour has been utilized to debrominate *d,l*- and *meso*-2,3-dibromobutane in the presence of penta-1,3-diene (a radical trap) and the ratio of *cis*- to *trans*-2-butenes has been rationalized in terms of a bromine-bridged free-radical intermediate[31] which maintains its stereochemical integrity.

Calcium atoms have been shown[32] to undergo a reaction rare in organic chemistry, viz. insertion into C—F bonds. Typically, perfluorobut-2-ene is defluorinated to perfluorobut-2-yne (equation 9). It is clear that organocalcium intermediates are

$$CF_3CF{=}CF{-}CF_3 \xrightarrow{Ca} CF_3{-}\underset{\underset{\displaystyle Ca{-}F}{|}}{C}{=}CF{-}CF_3 \xrightarrow{warm} \underset{\displaystyle CaF_2}{\overset{\displaystyle CF_3{-}C{\equiv}C{-}CF_3}{+}} \tag{9}$$

involved since hydrolysis yields a C—H bond and also a calcium hydroxide moiety (equation 10). Klabunde *et al*[32] have tried (so far unsuccessfully) to trap the suspected tetrafluorobenzyne which would be formed by defluorination of C_6F_6. These defluorinations are unsuccessful with saturated fluorocarbons yet can give good yields with vinyl or aromatic fluorines. One might therefore postulate an initial π-type coordination of the calcium atom to the molecule with subsequent fluorine migrations leading to the observed products.

$$C_6F_6 + Ca \longrightarrow C_6F_5{-}CaF \xrightarrow{H_2O} C_6F_5H + CaF(OH) \tag{10}$$

Zn vapour is not reactive towards most alkyl halides, although perfluoroalkyl iodides do appear to undergo oxidative addition reactions producing R_fZnI which decompose readily via radical processes[33].

It is particularly interesting to see what can happen when organic halides are treated with vapours of elements which have valencies exceeding 2 or when the initial oxidative addition process leads to an intermediate which is still coordinatively unsaturated. The trivalent B and Al vapours are classifiable as falling into the former category but the radical nature of the processes precludes clean, high-yield reactions. Although such molecules as $C_6H_5BBr_2$ or the sesquihalides $R_3Al_2X_3$ are isolable from such reactions[34], the co-condensation method offers no major advantages over more conventional synthetic routes.

In contrast, the reactions of germanium vapour should, by analogy to the well studied reactions of atomic carbon[9], yield germene intermediates. The chemistry of Ge atoms should mimic that of Mg in being heavily dependent upon its mode of generation. Thermally produced Ge must be in its ^3P ground state and the initial product is thus a triplet germene. However, the ultimate fate of the germene apparently depends on the substituents since the chlorotrichloromethylgermene (From Ge and CCl_4) reacts as a triplet[35] while the trimethylsilylgermene exhibits singlet character[36].

$$CCl_4 \xrightarrow{:\dot{Ge}\cdot} Cl_3C{-}\dot{\underset{\cdot}{Ge}}{-}Cl \xrightarrow{CCl_4} \underset{\displaystyle Cl_3C\cdot}{\overset{\displaystyle Cl_3C{-}GeCl_2}{+}} \xrightarrow{CCl_4} \underset{\displaystyle C_2Cl_6}{\overset{\displaystyle Cl_3C{-}GeCl_3}{+}} \tag{11}$$

$$Me_3SiH \xrightarrow{:\dot{Ge}\cdot} Me_3Si{-}\dot{\underset{\cdot}{Ge}}{-}H \longrightarrow Me_3Si{-}\ddot{Ge}{-}H \xrightarrow{Me_3SiH} Me_3Si{-}\underset{\underset{\displaystyle SiMe_3}{|}}{GeH_2} \tag{12}$$

One might postulate (in this era of silicon—carbon double bonds) that a Ge—Si $p\pi-d\pi$ interaction as in **1** could stabilise the singlet manifold and thus govern the course of the reaction. However, few theoretical data on the spin states of germenes are yet available[37] and more studies in this area are needed.

$$Me_3Si-Ge-H \longrightarrow Me_3-Si-Ge-H$$

(**1**)

It has been shown already that organic halides can undergo oxidative addition reactions with some main-group elements but the scope of such reactions was originally delineated using square-planar transition metal complexes. Archetypal of such processes is the addition of alkyl halides to Vaska's compound, $[(Ph_3P)_2Ir(CO)Cl]$, whereby the central metal undergoes oxidation of the d^8 environment to a d^6 system; concomitantly, the coordination about the iridium increases from square planar to octahedral[38]. In similar fashion one would expect the oxidative addition of suitable addenda to zerovalent transition metals in the form of free atoms which have no steric or orbital constraints. The prototypical reaction was realized by Piper and Timms[39], who showed that η^3-allylnickel halides are efficiently produced in a *ca*. 60% yield by the co-condensation of Ni atoms with allyl chlorides or bromides. This approach avoids the toxicity hazards of $[Ni(CO)_4]$ inherent in the conventional syntheses and indeed metal vapour synthesis probably provides the best route available at the moment.

Two obvious mechanistic rationales are reasonable candidates for this process; one could visualise initial π-complexation of the nickel atom to the allyl moiety and subsequent insertion into the carbon—halogen bond, or the reverse situation, i.e. direct formation of an η^1-allylnickel halide with subsequent conversion to the η^3-complex and ultimate dimerization to the observed product. Matrix isolation spectroscopic studies[40] favour an initial π-complexation but the process of oxidative insertion is complicated by the onset of metal cluster formation in the matrix (equation 13).

$$CH_2{=}CH{-}CH_2X \xrightarrow{Ni} \left\{ \begin{array}{l} CH_2{=}CH{-}CH_2X\ (\downarrow Ni) \\ CH_2{=}CH{-}CH_2{-}Ni{-}X \end{array} \right\} \longrightarrow [(\eta^3\text{-}C_3H_5)Ni(\mu\text{-}X)]_2 \quad (13)$$

This concept has been extended to include direct syntheses of η^3-allyl halide complexes of Pd[41] and Pt[42] but, in the latter case, large-scale reactions are rendered difficult by the low volatility of platinum. The closely related η^3-benzylpalladium halide systems have been studied by variable-temperature n.m.r. spectroscopy and shown to undergo a rapid $\sigma-\pi$ rearrangement process (equation 14)[43]. This not only

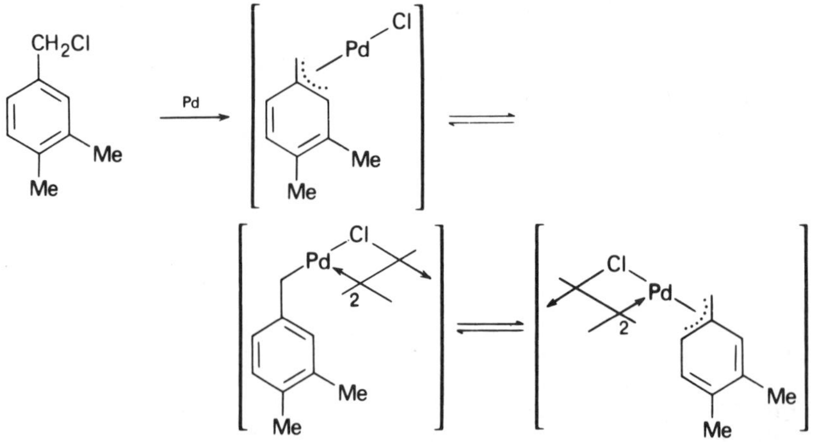

(14)

rationalizes site exchanges in the benzyl halide complex but also the isomerization of substituted benzylpalladium systems.

It is believed that the initial process at cryogenic temperatures is the formation of an arene—Pd π-bond. Thus the aromatic system is already drastically perturbed even before the oxidative addition process occurs. A similar situation requiring two oxidative additions occurs when 3,4-dichloro-1,2,3,4-tetramethylcyclobutene reacts with atomic Ni or Pd[5] to give the known η^4-cyclobutadiene complexes (equation 15).

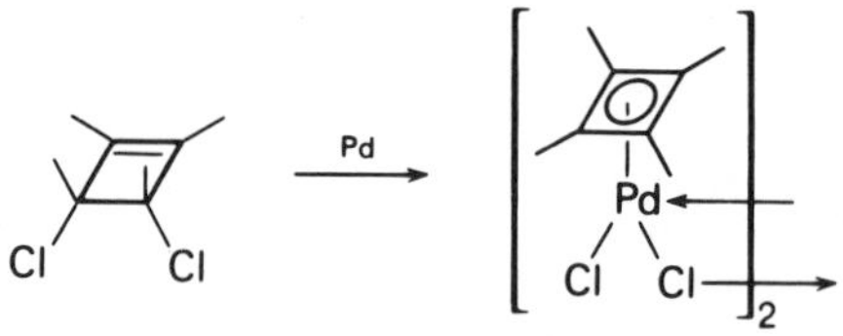

(15)

In each of the systems just discussed the coordinative unsaturation of the R—M—X system can be satisfied through coordination by the contiguous alkene moiety but, in the absence of such a donor, it is necessary to add an extra ligand to stabilize the oxidative adduct. This idea has been elegantly exploited by Klabunde and his co-workers[41,44,45], who showed that addition of phosphines at low temperature to the co-condensate of aryl halides and palladium yielded *trans*-[ArPdX$(PR_3)_2$] and the corresponding biaryl, showing that the ArPdX system was thermally extremely labile. Use of C_6F_5Br, however led to a stable, but reactive, product of empirical formula C_6F_5PdBr; it is proposed that its coordinative unsaturation may be alleviated by bromide and possible even by arene bridges.

A particularly exciting development here is the discovery that trapping of the C_6F_5NiBr species by toluene (preferred over benzene by virtue of its lower melting point) leads to production of the novel molecule [(toluene)Ni$(C_6F_5)_2$] (**2**) the first π-arene complex of Ni(II)[46]. The π-toluene is readily displaced by stronger ligands and furthermore the molecule is an active catalyst for the polymerisation of norbornadiene, trimerization of butadiene and hydrogenation of toluene. The molecule **2** is destroyed by CO, giving [Ni$(CO)_4$] and perfluorobiphenyl, and by water, which leads to pentafluorobenzene[46]. The analogous 17-electron complexes [(arene)Co$(C_6F_5)_2$] are also available[47] and undoubtedly will be the source of much novel chemistry.

$$C_6F_5Br \xrightarrow{Ni} [C_6F_5NiBr] \xrightarrow{PhMe} [(\eta^6\text{-}CH_3C_6H_5)Ni(C_6F_5)_2] \; (\mathbf{2}) \xrightarrow{L} LNi(C_6F_5)_2 \quad (16)$$

L = diene, arene, $(R_3P)_2$

In an attempt to understand better the mechanism of the formation of the oxidative adducts of Pd atoms and alkyl halides, Klabunde and Roberts[48,49] systematically varied the alkyl group and also doped the matrices with HX and with radical scavengers. The conclusions drawn were that radical chain processes were not operating but that the initial R—X—Pd complex on warming underwent rearrangement to R—Pd—X via a caged radical pair mechanism. The possibility of an S_N2 process involving Pd attack to displace the halide was shown to be very unlikely as tertiary halides reacted more efficiently than primary halides. Of course, a nucleophilic attack by a 10-electron Pd(0) atom is extremely unlikely; in fact, electrophilic behaviour is a much more reasonable expectation and the proposal of initial interaction with (and perhaps charge transfer from) the halogen is in keeping with this picture.

The isolation of many of the molecules just discussed has given impetus to the use of the metal vapour co-condensation technique for the study of transition metal-catalysed processes. Thus one can look for the existence of proposed reaction intermediates untrammelled by complicating factors such as extraneous ligands and solvents. Indeed, the η^3-benzylpalladium chloride isolated by Klabunde[41] had previously been postulated as an intermediate in the palladium acetate-catalysed reaction of benzyl chloride with methyl acrylate[50].

In like manner, the co-condensation of acyl chlorides with Ni or Pd atoms led to the production of RCO—M—Cl system which readily lost CO at low temperatures[44,45]. Subsequent trapping with phosphines yielded $[(R_3P)_2MCl_2]$, $[(R_3P)_2M(Cl)R]$, or $[(R_3P)_2M(Cl)COR]$, depending on the nature of the R group. The intermediacy of acylpalladium chlorides has long been invoked in the Pd-catalysed Rosenmund reduction of acyl chlorides to aldehydes. Hence once again a metal atom reaction provides convincing support for a mechanism operative in a conventional organometallic procedure.

III. ALKENES

A. Monoalkenes

The reactions of alkenes with a wide range of metals have been surveyed [51–53] and, whereas many form relatively unstable π-complexes, others form involatile (and sometimes polymeric) σ-complexes which can be decomposed hydrolytically. A widely used technique[51] is to add D_2O to the residue remaining after removal of the excess of unreacted ligand from the co-condensate; by this means metal—carbon σ-bonds are marked with a deuterium atom. Typically, hydrolysis of the aluminium—propene co-condensate yields a mixture of dideuterated propane and isomeric hexanes[51,54]. These products suggest the coupling of radical intermediates and the average composition may be typified by 3 (Scheme 3).

$$:\dot{Al}\cdot \xrightarrow{CH_3CH{=}CH_2} \dot{C}H_2{-}CH(CH_3){-}Al[CH_2\dot{C}H{-}CH_3]{-}CH(CH_3){-}\dot{C}H_2 \longrightarrow$$

$$[CH_3{-}CH{-}CH_2 / CH_3{-}CH{-}CH_2]Al{-}CH(CH_3){-}CH_2{-}Al[CH_2{-}CH(CH_3){-}CH_2{-}CH(CH_3)]$$

(3)

$\downarrow D_2O$ $\qquad$ $\downarrow D_2O$ $\qquad$ $\downarrow D_2O$

$$CH_3{-}CH(CH_2D){-}CH(CH_3){-}CH_2D \qquad CH_3{-}CHD{-}CH_2D \qquad DCH_2{-}CH(CH_3){-}CH_2{-}CHD{-}CH_3$$

SCHEME 3

It has been argued[51] that aluminocyclopropanes are involved and that these can couple to give hitherto unknown Al—Al bonds. Matrix e.s.r. spectroscopic studies[55] suggest that at 4°K the initial product involves π-donation from the alkene to the Al atom with some degree of back-donation into the alkene π^* level as in **4**.

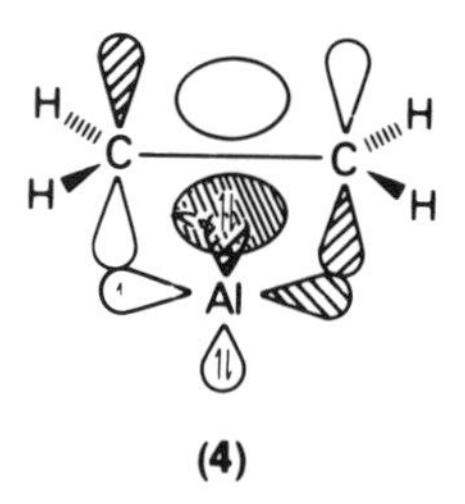

(4)

A survey of the reactions of simple alkenes with the first-row transition metals (and many of the lanthanides) yielded few isolable complexes but deuterolysis studies revealed a general trend. The early transition metals yielded most dideuterated alkanes while the later transition metals gave the original alkenes. One can rationalize this behaviour in terms of a gradual change from σ- to π-bonding character as one traverses the transition metal series from left to right[51]; this trend correlates well with the availability of metal d-orbitals for back-donation to the alkene. Although these reactions have been exploited synthetically to prepare olefin π-complexes of Ni and Pd[41], the propene—nickel systems are not just simple π-complexes.

$$Ni \xrightarrow{C_2H_4} [Ni(C_2H_4)_3] \qquad (17)$$

Thus, 1-butene was catalytically isomerized to *cis*- and *trans*-but-2-enes by nickel atoms. It was suggested[51] that these processes involved η^3-allylnickel hydrides of the

$$\text{Pd} \xrightarrow{C_7H_{10}} [(C_7H_{10})_3\text{Pd}]$$

$$\text{Pd} \xrightarrow{C_4F_8} \left[\left(\begin{matrix} CF_3-CF \\ \| \\ CF_3-CF \end{matrix}\right)_3 \text{Pd}\right] \xrightarrow[-78\,^{\circ}C]{L} \left[\begin{matrix} CF_3-CF \\ | \\ CF_3-CF \end{matrix} \text{Pd}L_2\right] \quad (18)$$

L = Et_3P or pyridine

type previously characterized by Bönneman[56]. To check this idea, Ni vapour was co-condensed in separate experiments with 3-deuteriopropene and 2-deuteriopropene. The former gave a mixture which also contained 1-deuteriopropene, whereas 2-deuteriopropene apparently remained unscrambled. Furthermore, 3-deuteriopropene led to both undeuterated and dideuterated propene, demonstrating that both inter- and intro-moleuclar exchange occurs[51]; these data are consistent with a 1,3-hydrogen shift, which would also explain migration of double bonds in longer chain alkenes.

$$\underset{\text{Ni}}{CH_2D-CH{=}CH_2} \rightleftharpoons \underset{\underset{H}{|}}{\underset{\text{Ni}}{CHD\overset{CH}{\cdots}CH_2}} \rightleftharpoons \underset{\text{Ni}}{CHD{=}CH-CH_3} \quad (19)$$

Allylmetal hydride formation is also supported by isolation of (η^3-allyl)tris(trifluorophosphine)cobalt(I) from the co-condensation of propene, PF_3, and cobalt vapour (equation 20)[26].

$$CH_3-CH{=}CH_2 \xrightarrow{Co} \left[(\eta^3\text{-}C_3H_5)\text{Co}-H\right] \xrightarrow{PF_3} \left[(\eta^3\text{-}C_3H_5)\text{Co}(PF_3)_3\right] + [HCo(PF_3)_4] \quad (20)$$

B. Dienes

The reactions of metal vapours with dienes have been intensely studied and this has led to the enrichment of an already well investigated area of organometallic chemistry. Indeed, the chemistry of the diene complexes of the later transition metals had already reached a high point with the masterful elucidation of the mechanism of the trimerisation of butadiene by 'naked nickel' catalysts. The metal vaporization route appeared to offer an alternative approach to the reductive methods pioneered by Wilke and his colleagues in Mülheim[57].

Blackborow and Young[58] have made the point that the reactions of metal atoms with unsaturated hydrocarbons may be placed into two distinct categories. We shall discuss first the cases in which the diene complexes to the metal and, upon addition of another reagent, leads to isolable products; secondly, one must consider the situation where the metal can increase its hapticity by, for example, insertion into an adjacent C—H bond. The latter circumstance frequently obtains and the resulting metal hydrides allow ready transfer of hydrogen between ligands bonded to the same metal atom. A straightforward example in the latter category is the production of (η^5-cyclopentadienyl)(η^3-cyclopentenyl)nickel(II) (**5**) from nickel vapour and cyclopentadiene (equation 21)[59].

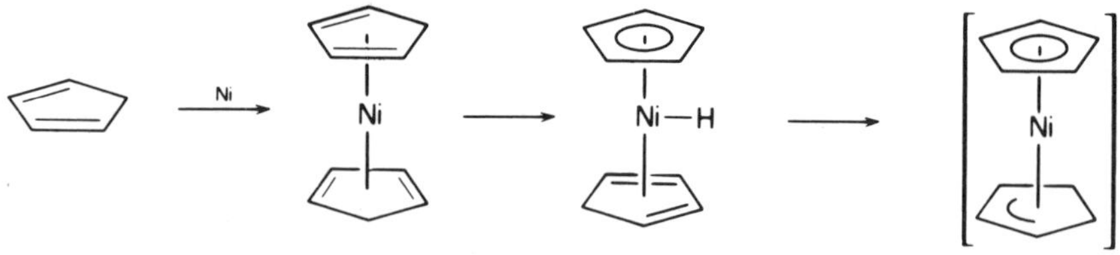

Ni

Ni

Ni—H

Ni

(5)

(21)

The simplest diene, buta-1,3-diene, has been the subject of a comprehensive study in which a series of transition metal vapours were co-condensed with the diene and the variety of other ligands including arenes, phophines, phosphites, and alkyl halides[60–62]. In some cases, aluminium vapour was also co-condensed, thus generating alkylaluminium halides *in situ*[61]. In general, the results paralleled those obtained in the more conventional reactions in that the usual butadiene dimers (cyclooctadiene, vinylcyclohexene) and trimers (isomeric dodecatrienes) are produced in varying yields dependent upon the identity of the transition metal and the added ligands. The current situation with respect to these oligomerization studies has been succinctly summarized[63].

Although the co-condensation of butadiene and many of the first-row transition metals does not yield isolable complexes directly, addition of extra ligands to the co-condensate has become a viable route to systems otherwise difficult or impossible to obtain. Scheme 4 indicates some of these reactions[52,53,64–66].

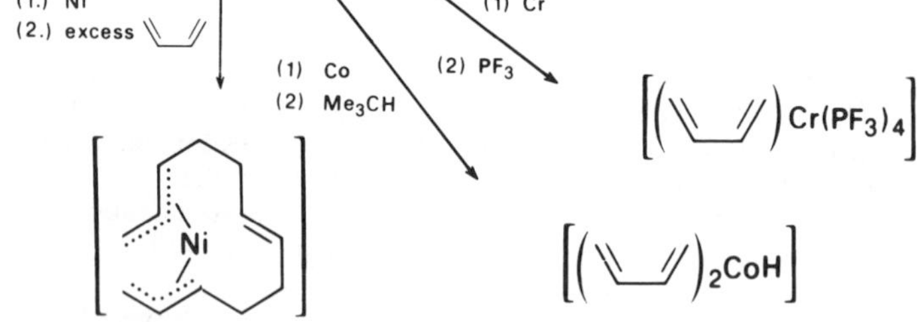

(1) Fe
(2) CO
$[(C_4H_6)_2Fe(CO)]$
(1) Fe
(2) PhMe
$[(C_4H_6)Fe(PhMe)]$
(1) Cr
(2) PF_3
$[(C_4H_6)Cr(PF_3)_4]$
(1) Co
(2) Me_3CH
$[(C_4H_6)_2CoH]$
(1.) Ni
(2.) excess C_4H_6
Ni

SCHEME 4

In contrast, molybdenum and tungsten react with butadiene to produce tris(butadiene)molybdenum and -tungsten[67]. The only comparable previously characterized system was tris(methyl vinyl ketone)tungsten, which adopts the relatively unusual trigonal prismatic structure[68]. The novel tris(butadiene)-molybdenum (**6**), is a sublimable, air-stable solid in which the butadienes are all *cisoid* and in a trigonal prismatic arrangement in a girdle around the Mo such that they are all aligned in the same direction. A series of products derived from the co-condensation of butadiene with lanthanides analyse as [(butadiene)$_2$Ln] or [(butadiene)$_3$Ln] but, as yet, no crystallographic data are available[69].

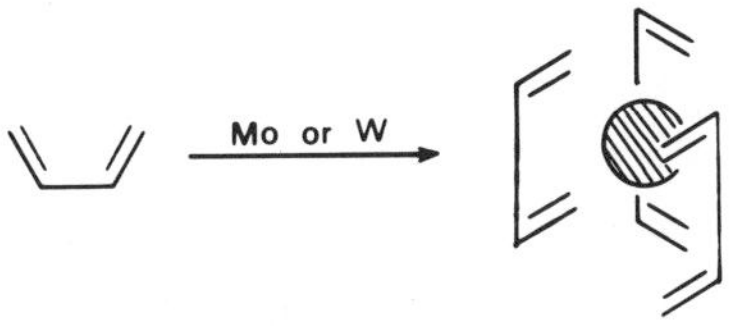

Mo or W

(22)

(**6**)

An interesting diene which has received some study is norbornadiene. It was reported in 1972 that nickel atoms effect dimerization, primarily to the *exo-trans-exo* isomer **7**[53]. Later work in which bipyridyl was added to the nickel—norbornadiene co-condensate revealed the intermediacy of the nickelacyclopentane **8**[70]; such an intermediate is closely analogous to an iridacyclopentane previously characterized crystallographically[71].

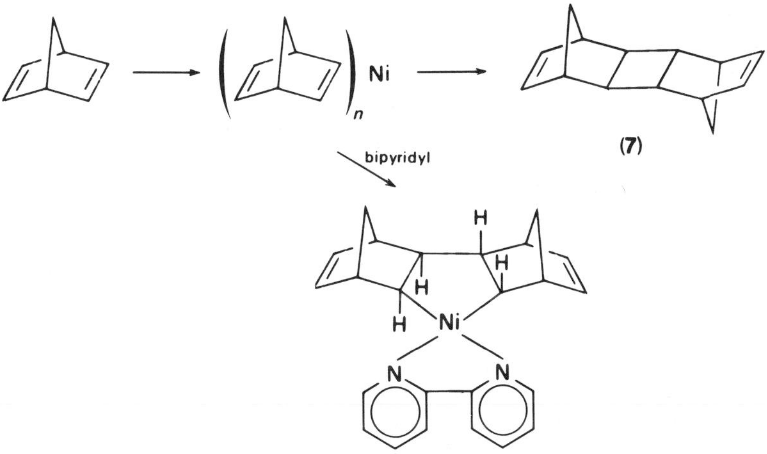

Ni

bipyridyl

(**7**)

(23)

(**8**)

The chemistry of bis(cycloocta-1,5-diene)nickel was pioneered by Wilke, Bogdanovic and their colleagues[72] and the tremendous success of 'naked nickel' chemistry prompted the search for other routes to [(cod)$_2$M] complexes. Indeed [(cod)$_2$M] systems are readily synthesizable when M is Ni[73], Pd[73], or Pt[42] but, at the moment, the scale on which these reactions are performed does not compare well with the more conventional routes to [(cod)$_2$Ni]. However, [(cod)$_2$Fe], which was first made by evaporating iron atoms into a cold 10% solution of cycloocta-1,5-diene in methylcyclohexane in a rotating reactor[74], has proved an extraordinarily versatile intermediate. In this species, which is apparently paramagnetic, the cyclooctadienes are presumably tetrahedrally disposed about the iron atom and reactions with additional ligands have led to a plethora of products such as those shown in Scheme 5[75,76].

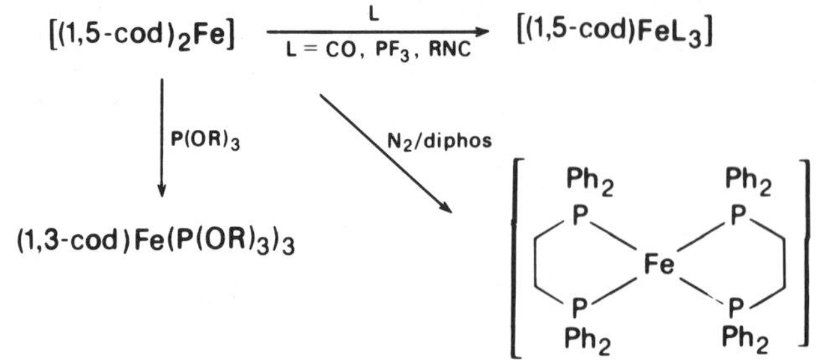

[(1,5-cod)$_2$Fe] —L; L = CO, PF$_3$, RNC→ [(1,5-cod)FeL$_3$]

[(1,5-cod)$_2$Fe] —P(OR)$_3$→ (1,3-cod)Fe(P(OR)$_3$)$_3$

[(1,5-cod)$_2$Fe] —N$_2$/diphos→ [Fe(Ph$_2$P–PPh$_2$)$_2$]

SCHEME 5

We note here the isomerization of 1,5-cod to 1,3-cod in the reaction of [(cod)$_2$Fe] with phosphites. In fact, the ability of these diene—metal co-condensates to yield products in which hydrogen migration has occurred is a widespread phenomenon. The intermediacy of η^3-allylmetal hydrides has been previously discussed in Section IIIA, and the analogous formation of η^5-cyclopentadienyl metal hydrides derived by oxidative addition of allylic C—H bonds to the coordinatively unsaturated metal atom is a logical extension of this concept. A clear example of this is provided by the observation[6] that cocondensation of either 1,3- or 1,5-cod with chromium vapour and PF$_3$ yields (η-cycloocta-1,5-dienyl)hydridotris(trifluorophosphine)chromium (**9**).

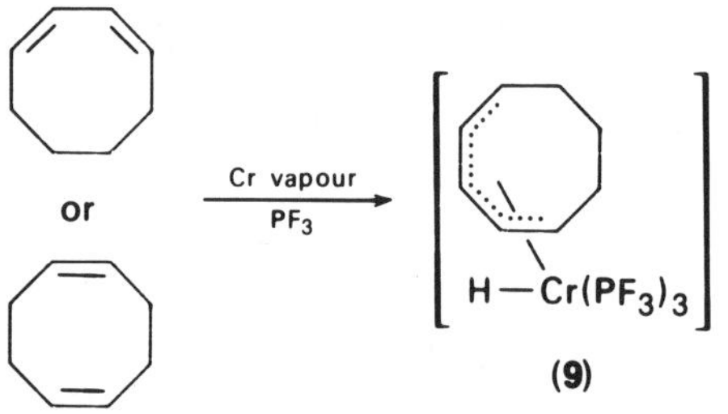

1,3-cod or 1,5-cod —Cr vapour, PF$_3$→ [(C$_8$H$_{11}$)H—Cr(PF$_3$)$_3$] (**9**) **(24)**

The equilibrium between 1,3-, 1,5-, and 1–5-η-cod complexes is very sensitive to the identity of the metals and ligands involved. Thus, treatment of the 1,3-cod—Cr or the 1,5-cod—Cr condensates with CO yields only the [(1,5-cod)Cr(CO)$_4$]complex[6]. The factors which influence particular metals in their preferences for conjugated or non-conjugated dienes have been examined[77]. Some reactions of other metals with cyclooctadienes are shown in Scheme 6[5,6,73].

The clarification of the preceding reactions in which allylic C—H bonds underwent oxidative addition to metals allowed the rationalization of an early observation that the co-condensate of cyclohexa-1,3- or -1,4-diene with any of Cr, Mn, Fe, Co, or Ni, when stirred at 0°C, catalytically produced benzene and cyclohexene (equation 24)[52,63,64,66]. It is tempting to postulate the formation of a bis(diene)—metal complex, which then undergoes hydrogen migration via a metal hydride intermediate.

Support of such a mechanism comes from the isolation of (η^6-benzene)(1–4-η-cyclohexadiene)iron when M is Fe[64]; furthermore, treatment of the cyclohexa-1,3-diene—Cr co-condensate with PF$_3$ at -196°C yielded the bis(diene)complex **10**[78].

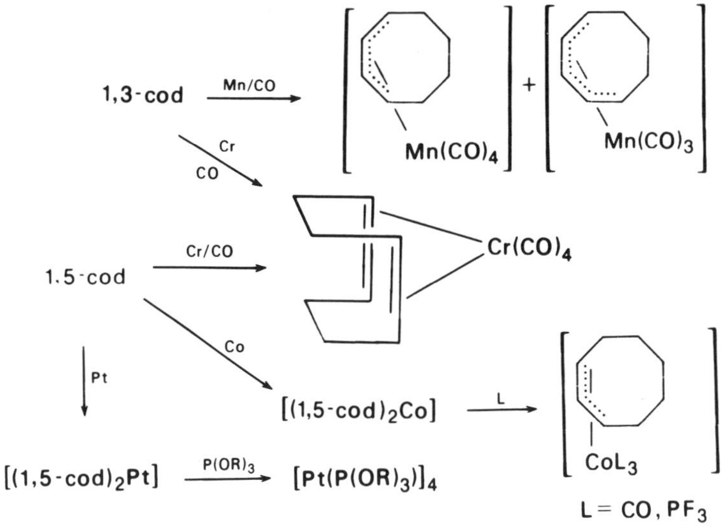

SCHEME 6

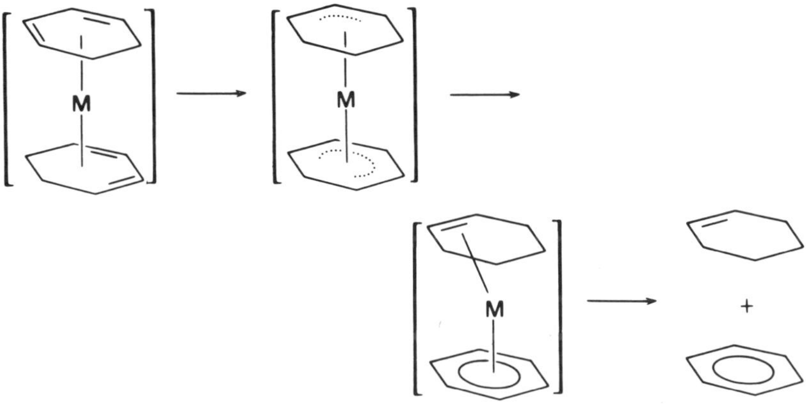

(25)

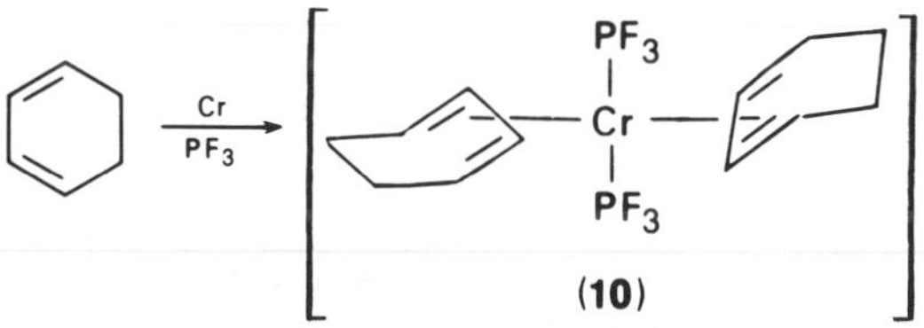

(26)

The major synthetic use of this class of reaction has been the direct synthesis of η^5-cyclopentadienyl derivatives; indeed, the synthesis of ferrocene by Timms in 1969 may be considered the genesis of organotransition metal vapour synthesis[59]. Some typical syntheses are presented in Scheme 7[1,52,59,78–80]. Of some interest here is indene, which can act as a source of either η^5-cyclopentadienyl or η^6-arene complexes (Scheme 8)[80].

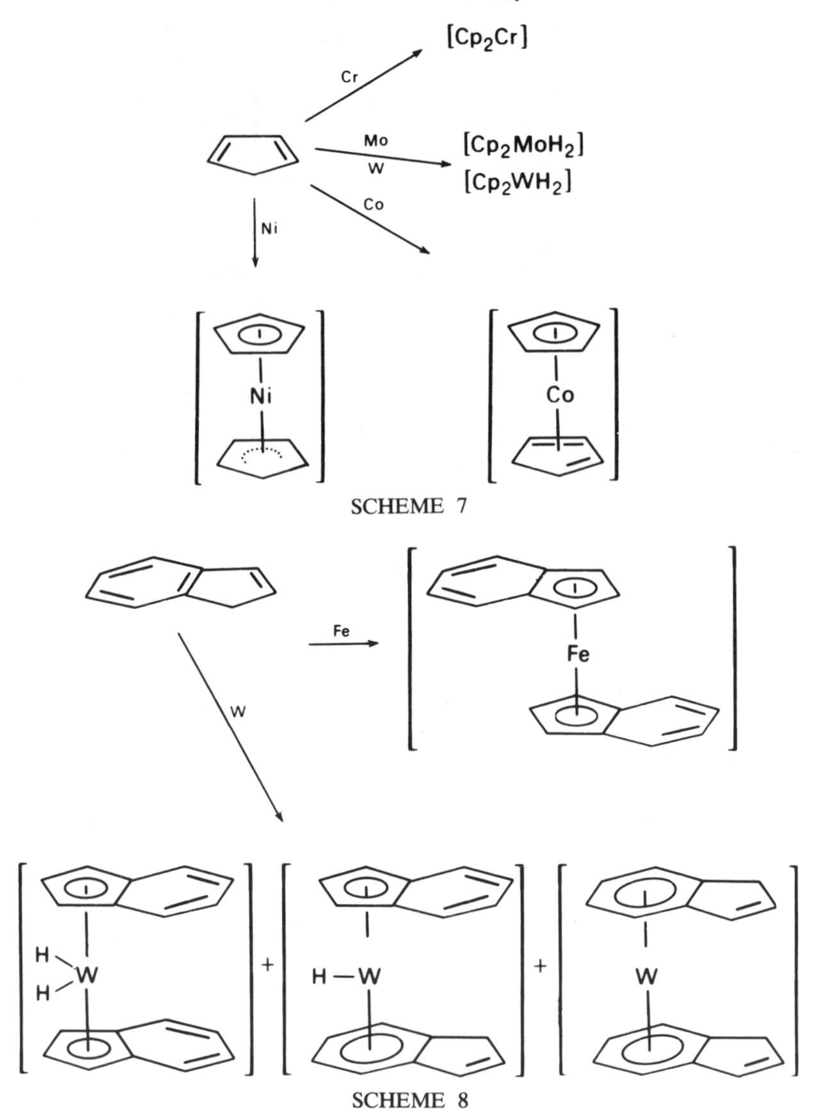

SCHEME 7

SCHEME 8

C. Polyenes

The reactions of trienes and tetraenes with metal vapours have concentrated mainly on cycloheptatriene and cyclooctatetraene. As with cyclopentadiene, cycloheptatriene has a tendency to utilize its methylene group to allow metals to increase their coördination number and thus form η^7-cycloheptatrienyl complexes. Indeed, the early transition metals reveal a proclivity for hydrogen transfer from one ring to the other presumably via an initial bis(cycloheptatriene)metal complex **(11)** (reaction 26), leading ultimately to (η^7-cycloheptatrienyl)(η^5-cycloheptadienyl)metal complexes **(12)**[80,81]. Analogous hydrogen atom transfers between rings in sandwich compounds have been

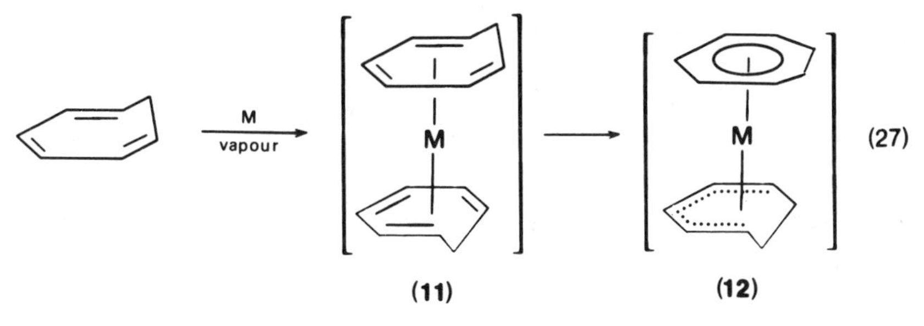

previously noted in Section III.B for both five- and six-membered rings. The later transition metals, of course, cannot accommodate a total ligand hapticity of twelve and, although there seems to be a strong tendency to yield complexes containing an η^5-cycloheptadienyl moiety, the other portion of the sandwich must match the electronic requirements of the central metal. Hence, in the iron complex **13** the other ring is of the η^5-C_7H_7 type[81,82] whereas for cobalt the η^4-C_7H_{10} complex **14** is formed[81]. Co-condensation of cycloheptatriene and trifluorophosphine with metals yields mixed Complexes as typified by **15** and **16** (Scheme 9)[52,82]. Another interesting triene is

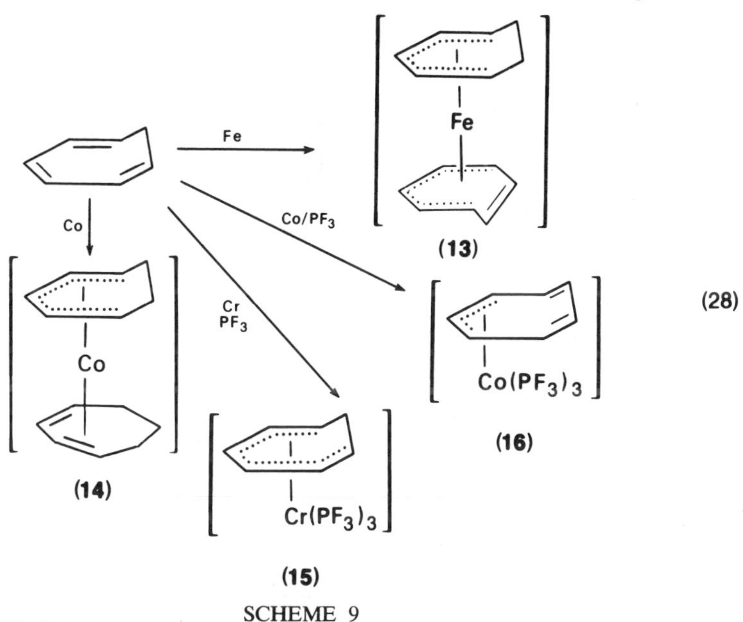

SCHEME 9

6,6-dimethylfulvene, for which it was suggested that diradical intermediates such as **17** might be viable[83]. Indeed, this is true and with iron a hydrogen transfer has been observed to give 1-isopropenyl-1-isopropylferrocene[84]. A minor product has the formula (fulvene)$_3$Fe and has been tentatively formulated as **18**[85].

In contrast, with nickel the reaction yields no organonickel species but only a 1:1 mixture of the symmetrical tetramethyltetrahydroindacenes **19** and **20**[86]. Although this reaction has the appearance of a [6+6] cycloaddition (reaction 28), the reaction

(17)

(18)

(29)

(19)

(30)

(20)

seems to be promoted exclusively by nickel atoms and so is better formulated as a coupling of the initially produced diradical to yield a bisallylnickel system (**21**) which closes in the manner previously delineated for the butadiene oligomerization process[57] The conversion of **22** to the mixture of **19** and **20** merely requires a series of 1,5-suprafacial sigmatropic shifts, but such hydrogen migrations are well documented in co-condensation processes[53].

The reactions of cyclooctatetraene (cot) with metal vapours frequently give rise to polymeric products[26,81], however, some known complexes such as [$(cot)_3Cr_2$] (**23**), [$(cot)_3Ti_2$] (**24**), and uranocene (**25**) have been obtained (Scheme 10)[81,87,88]. Iron vapour with cot alone yields no isolable complexes[81], but the inclusion of PF_3 yields the complex **26**, analogous to the known molecule [$(cot)Fe(CO)_3$][89].

The lanthanide vapours yield complexes of the type [$Ln_2(cot)_3$] and the neodymium member, which can be crystallized from tetrahydrofuran, has been studied by X-ray crystallography[90]. It has a particularly interesting structure, which may best be described as the ion pair $[(\eta\text{-cot})Nd(thf)_2]^+[(\eta\text{-cot})_2Nd]^-$, (**27**); the rings in the anion are planar but not parallel and the Nd atom is not symmetrically positioned between them but is somewhat closer to the peripheral ring (average Nd–C = 2.660 Å) than to the central ring (average Nd–C = 2.787 Å). The neodymium in the cation completes its coordination sphere by also bonding to the central cot.

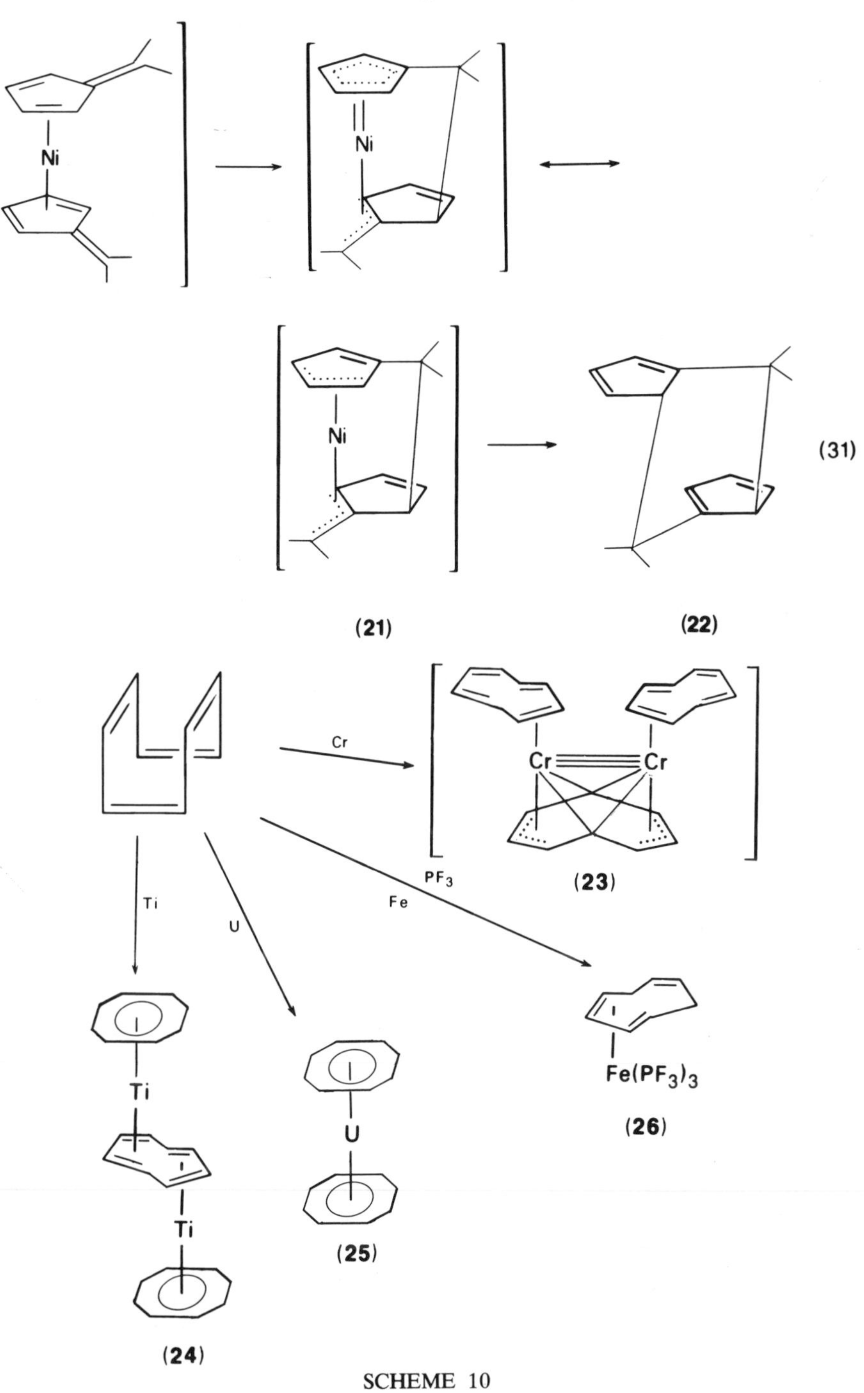

SCHEME 10

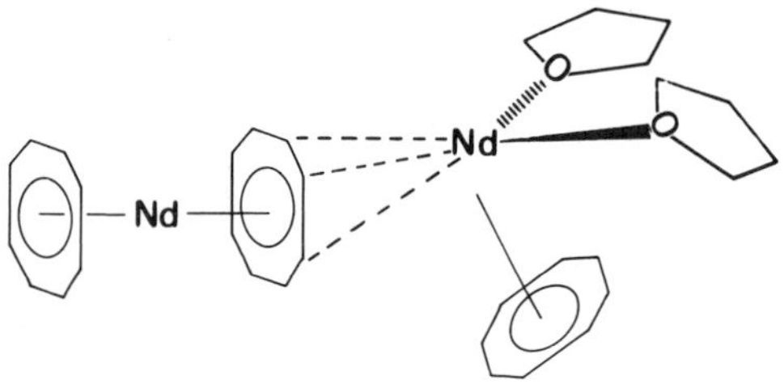

(27)

IV. ARENES

A. Chromium—Arene Complexes

The discovery of ferrocene in 1952[91] led to an avalanche of papers delineating the chemistry of the bis(cyclopentadienyl)metal sandwich compounds. In contrast, the chemistry of the bis(arene)chromium system, which had been synthesized by Hein in 1919 but whose identity remained unrecognized for more than three decades[92], seemed to be very restricted. The Fischer–Hafner synthesis[93], which utilized reducing Friedel–Crafts conditions, did not permit the incorporation of functional groups into the arenes because of side-reactions with the Lewis acid catalyst (usually aluminum trichloride) (equation 30). Thus, the chemistry of $[(C_6H_6)_2Cr]$ was limited only to

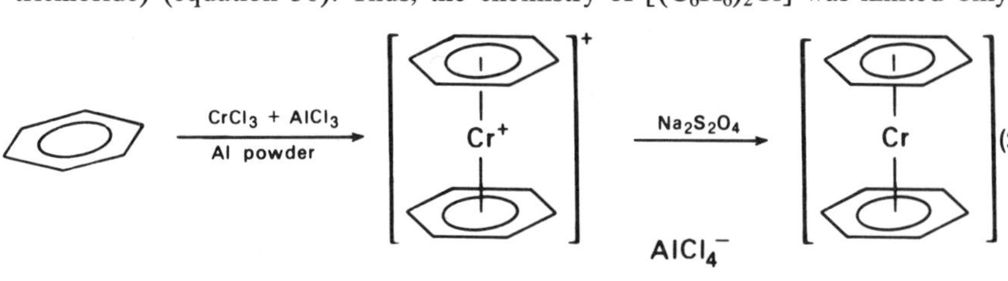

$CrCl_3 + AlCl_3$, Al powder; $Na_2S_2O_4$; Cr^+; $AlCl_4^-$; Cr (32)

some hydrogen–deuterium exchange studies[94] and some experimentally difficult metallation reactions using isoamyl sodium[95] (equation 31).

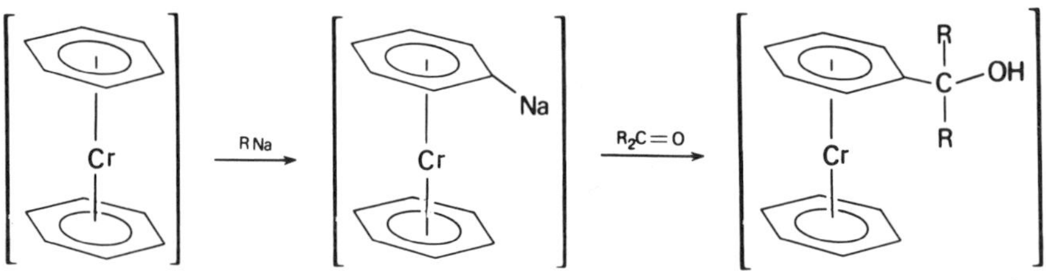

Cr; RNa; Na; $R_2C{=}O$; R, C, OH, R (33)

The situation was alleviated somewhat by use of NNN′N′-tetramethylethylenediamine to activate alkyllithiums[96]. Nevertheless, the prospect of developing a chemistry of the $[(arene)_2Cr]$ system to match that of ferrocene seemed very unlikely.

However, Timms' announcement[59] of the direct synthesis of $[(C_6H_6)_2Cr]$ via the co-condensation of chromium vapour and benzene proved to be a major turning point. It was soon shown that the restrictions which the Fischer–Hafner synthesis imposes on the availability of Ar_2Cr systems do not hold for the metal vapour synthetic route[52,97–99]. Thus, a wide range of chromium—arene (chromarene) complexes were prepared containing such functionalities as F, Cl, CH_3O, R_2N and CO_2R. This allowed the investigation of the chromarene system by a variety of spectroscopic techniques.

A particularly thorough ^{13}C n.m.r. spectroscopic study indicated that the perturbing effect of a π-complexed chromium atom on the arenes was such as to reduce the transmission of substituent effects across the ring; this was taken as indicating a loss of aromatic character and was also in accord with the ^1H n.m.r. data[100]. These studies as well as ^{19}F n.m.r. results[101] suggested that whereas electrophilic attack on the π-complexed ring was unlikely (the incoming electrophile finds the zerovalent chromium atom a more attractive target), nucleophilic substitution appeared viable, assuming that a suitable leaving group was present. These predictions soon gained experimental support as shown in reactions 32 and 33[101].

F
Cr
F
MeO⁻
−78°C
OMe
Cr
OMe
(34)

F
F
F
F
F
F
Cr
BuLi
−78°C
F
Bu
Bu
Bu
Bu
F
Cr
(35)

Furthermore, the incorporation of electronegative atoms into the arene ring(s) of the sandwich compounds enhanced the acidity of the arene protons and greatly facilitated metallation[102]. This provided another versatile route to functionalized chromarenes Scheme 11)[103].

Other spectroscopic investigations on these molecules have focused on i.r. and Raman spectra[104], e.s.r. data[105] and u.v.–visible spectra[106]. The vibrational spectra of $[(C_6F_6)Cr(C_6H_6)]$ have been interpreted[104] in terms of a migration of electron density from the C_6H_6 ring to the C_6F_6 ring via the Cr atom. Thus, the symmetrical ring-breathing modes of both rings are reduced in frequency relative to those of the free arenes; donation of π-electron density from the C_6H_6 ring reduces the C—C bond order (and lowers the stretching frequency) while the C_6F_6 accepts electron density into the π^* system. This interpretation is in accord with the chemical behaviour of the molecule, which shows enhanced acidity of the C_6H_6 protons and somewhat reduced susceptibility of the π-complexed C_6F_6 ring to nucleophilic attack relative to that of C_6F_6 itself[101]. Interestingly, the Cr—C(F) distance (2.10 Å) is noticably shorter than the Cr—C(H) distance (2.17 Å)[107] typical of the bonding of metals to fluorinated alkenes for which back-donation into the π^* orbitals seems to predominate.

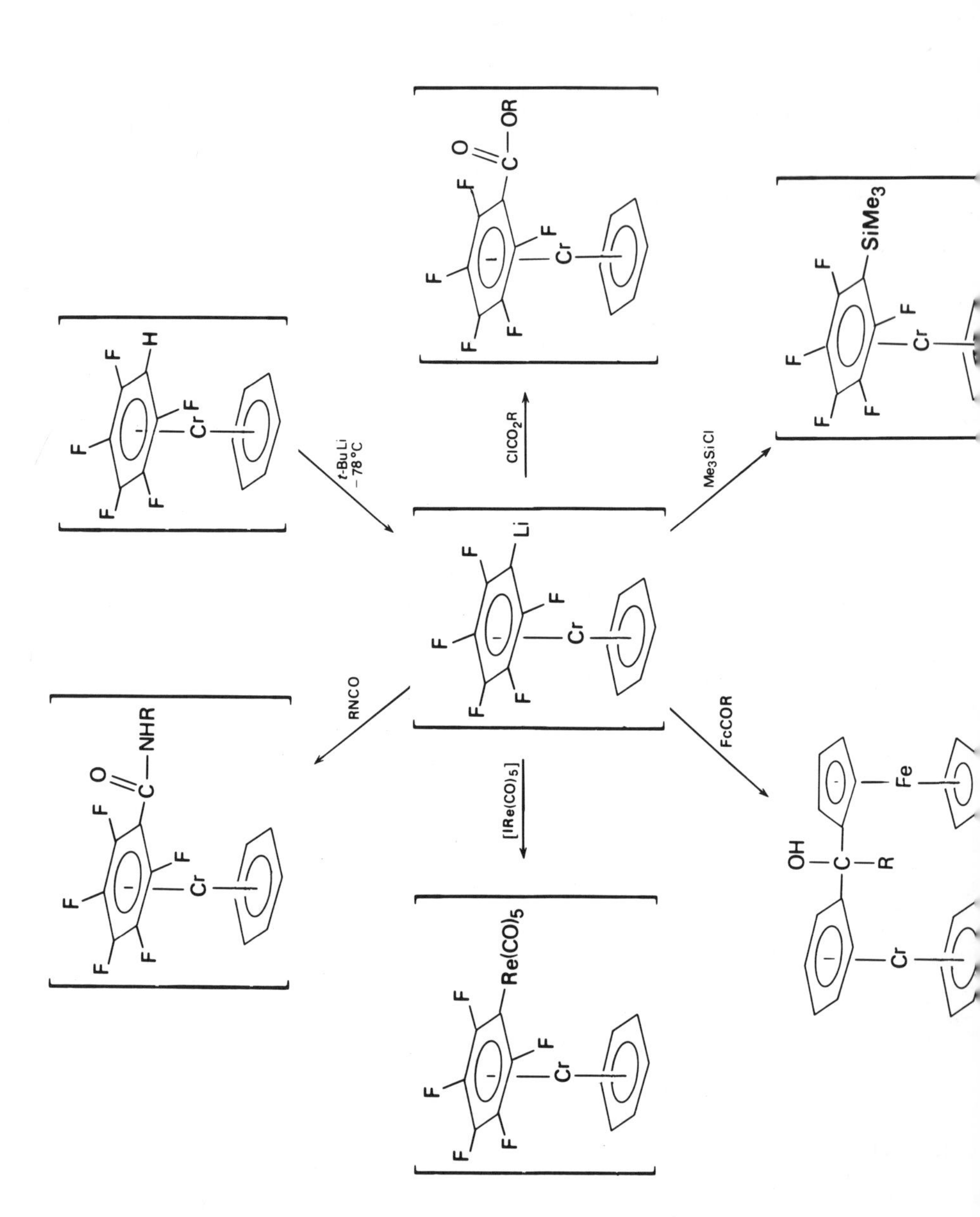
t-BuLi
−78°C
ClCO2R
Me3SiCl
RNCO
FcCOR
[IRe(CO)5]
Li
SiMe3
Re(CO)5
NHR
OR
OH
Cr
Fe

The very existence of these chromium—arene sandwich complexes depends upon a delicate balance of thermodynamic stability and kinetic lability[108]. Thus, the incorporation of electronegative moieties such as fluorine atoms or trifluoromethyl groups enhances the resistance to oxidation of the complexes and so renders them air stable. The oxidation mode of decomposition of chromarenes involves a one-electron oxidation to give the 17-electron $[Ar_2Cr]^+$ species, but this oxidation is inhibited by the presence of the electron-withdrawing groups. In contrast to the kinetic inhibition of oxidative decomposition, the thermodynamic stability of the molecules is lowered when CF_3 or similar groups are present. The 1500 cm^{-1} vibrational bond in benzene shifts to 1428 cm^{-1} in $[(C_6H_6)_2Cr]$; however, the corresponding bond in 1,4-$C_6H_4(CF_3)_2$ moves from 1529 to 1480 cm^{-1} in the corresponding chromarene[98]. The smaller complexation shift in the trifluoromethyl-substituted complex is indicative of less C—C bond weakening and therefore a weaker metal–ring interaction. Eventually one reaches the stage where the number of electron-withdrawing groups is such that the arene is insufficiently basic to donate bonding electron density to the chromium atom. Thus, although the bis(difluorobenzene)chromium complexes can be synthesized and are air stable[97], the corresponding trifluorobenzene complexes have so far defied isolation. Interestingly, although the hexafluorochromarenes with three fluorine atoms in each ring are not known, the isomeric 1,2,3,4,5,6-hexafluorochromarene $[(C_6F_6)Cr(C_6H_6)]$ is readily obtainable[101,109]. This has been rationalized[108] in terms of an internal electron buffering effect somewhat analogous to the ability of a chromium tricarbonyl group to tolerate an excess or deficit of electrons by changing the relative proportions of σ-donation to a metal and back-donation from the metal into the carbonyl π^* orbitals.

A particularly good example of the influences of kinetic inertness and thermodynamic stability is provided by Timms' experiment in which an equimolar mixture of C_6H_6, C_6H_5F, 1,4-$C_6H_4F_2$, 1,3,5-$C_6H_3F_3$, 1,2,3,5-$C_6H_2F_4$, C_6HF_5, and C_6F_6 was co-condensed with chromium vapour and PF_3[26]. The yield of $[(arene)Cr(PF_3)_3]$ was a maximum for $C_6H_4F_2$ and fell to a minimum for C_6F_6; thus, the kinetically controlled rate of complexation is favoured by increasing numbers of fluorine atoms but thermodynamic stability follows the reverse sequence. Apparently these factors are optimized in the difluoro complex, but another factor which may be important is the competing process of defluorination.

The synthetic utility of chromium vapour as a reagent for the syntheses of chromium–arene sandwich complexes is not limited solely to monoarenes. The development of the rotating metal reactor[7] in which metal atoms are sprayed into a solution of the substrate in an inert solvent, has removed the restriction on ligand volatility which the cocondensation procedure necessitated. Thus $[(naphthalene)_2Cr]$ has been prepared[110] and its chemistry investigated[111]. Because of its lower resonance energy relative to monoarenes, a naphthalene moiety is readily displaced, presumably via a tetrahapto intermediate, and this holds promise of allowing ready entry to a whole range of low valent derivatives of Cr (and also Ti, V, and Mo for which analogous naphthalene complexes are known)[111].

Cyclophanes have fascinated organometallic chemists since systematic routes to them were developed in the 1950s[112]. The $Cr(CO)_3$ derivative of 2,2-paracyclophane (**28**) was prepared[113] in the hope that subsequent loss of the three carbonyl groups would allow access to the chromium sandwich compound **29**. However, at that time the significance of the strong *trans*-effect of the carbonyl group was not fully appreciated; it is, therefore, easy to displace three mutually *cis*-carbonyl groups from $[Cr(CO)_6]$ but rather difficult to replace the other three. The molecule **29** can, however, be prepared by direct reaction of 2,2-paracyclophane with chromium vapour; the bis(2,2-paracyclophane)chromium complex **30** is also obtained. The corresponding 17-

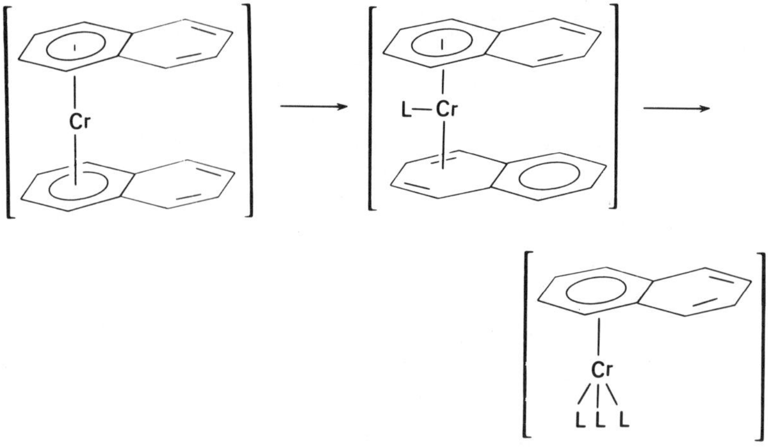

(36)

L = CO, PF_3, PMe_3, $P(OMe)_3$

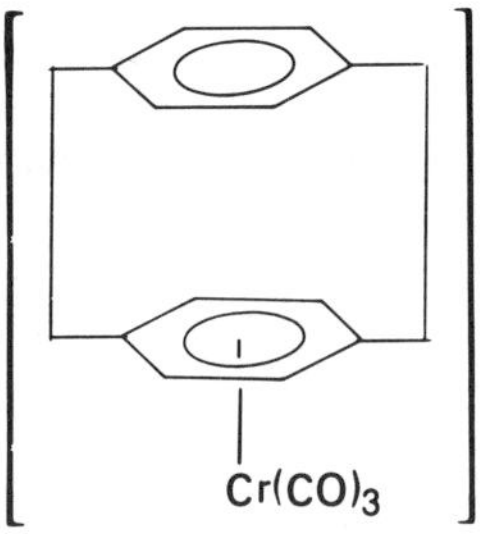

(28) 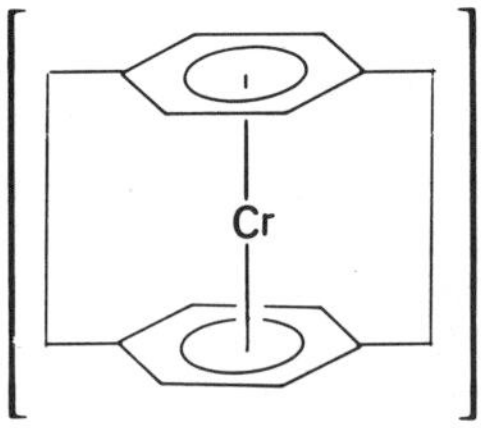 **(29)** 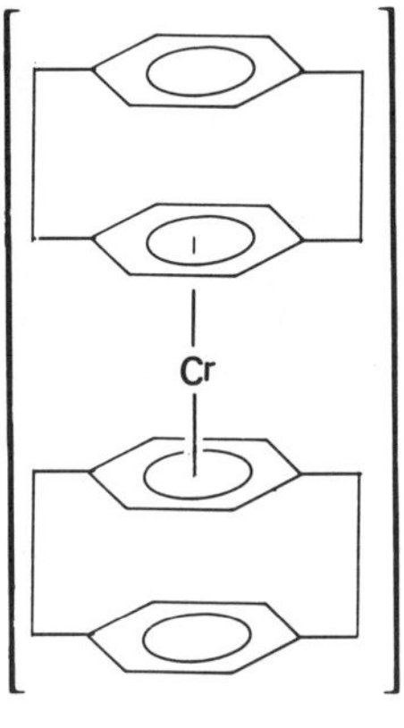 **(30)**

electron cations of **29** and **30** have been studied by e.s.r. and **29** displayed compressed-sandwich properties in that the hyperfine coupling constants for the ring protons and the ^{53}Cr site are increased and decreased, respectively, compared with the corresponding values for the monocation of **30**. This suggests a slightly more extensive $Cr(3d_{z^2}) \rightarrow$ arene $(\sigma_{a_{1g}})$ spin delocalization in **29** probably attributable to a decrease in the metal–ligand distance in **29**[105].

Other approaches to bridged chromarenes have been reported by reaction of α,ω-diarylalkanes with chromium vapour[114]. Typically, 1,4-diphenylbutane gave rise to a series of products (**31–33**).

However, the most elegant approach to bridged chromium—arene sandwiches (chromarenophanes) involved a Dieckmann cyclization of the diester **34** which itself was prepared by co-condensation with Cr vapour[115]. Clearly, this synthetic route will allow a much more controlled approach to the bridged chromarenes and may eventually lead to as extensive a chemistry as is already known for ferrocenophanes[116].

An exciting recent development[117] has been the reaction of chromium atoms with phenylsiloxanes such as Dow Corning DC510 or Silicone 704 (a tetramethyltet-

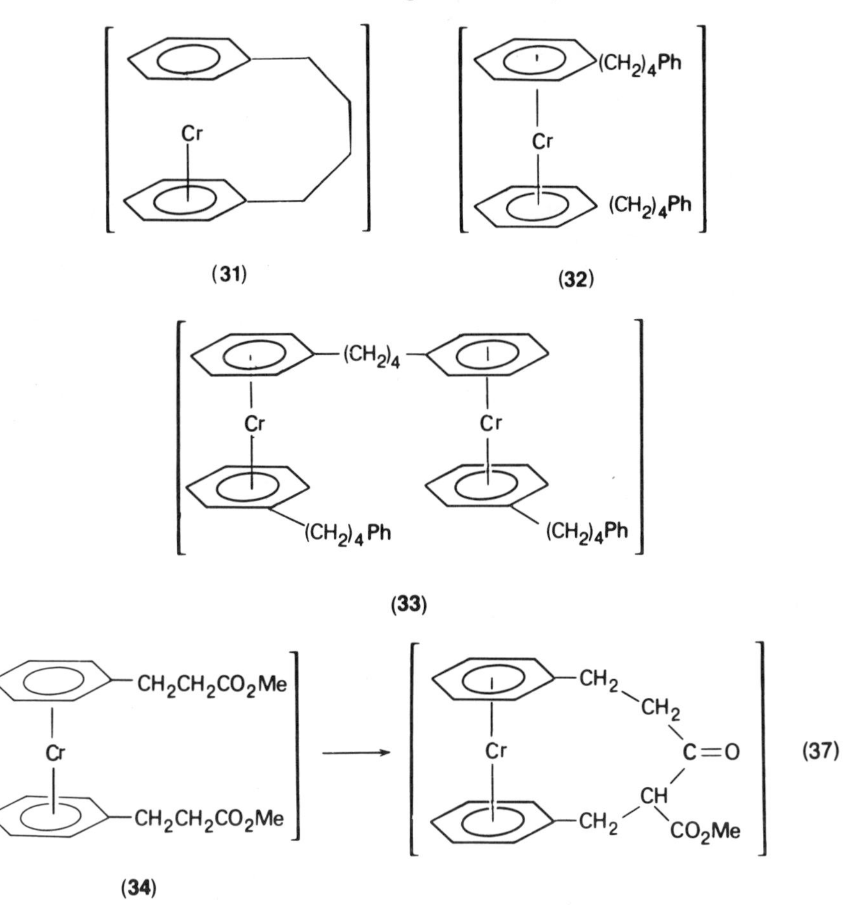

(31) (32)

(33)

(34) (37)

raphenyltrisiloxane) to give complexes in which up to 50% of the phenyl groups are coordinated. Not only can these reactions with silicone oils be carried out at 0°C, but also the products are stable up to 300°C in the absence of air. Furthermore, reaction with tetracyanoethylene gave rise to spectra typical of $[(arene)_2Cr]^+TCNE^-$ systems; hence the reactions of chromium atoms with phenylsiloxane polymers apparently yield sandwich complexes (**35**).

Si
Cr
Si

Si
$(Cr)_n$
Si

(**35**) (**36**)

These systems have been thoroughly investigated[118] and the data suggest that at low temperatures products **(36)** containing clusters of chromium atoms sandwiched between arene rings are formed. It is obvious that these reactions of metal vapours with silicone polymers will continue to attract attention for some time.

B. Arenes with Other Metals

Although the initial emphasis was on the use of chromium vapour as a synthetic reagent in the preparation of [Ar_2M] sandwich complexes, recent technological developments have allowed the use of other high-boiling metals. In particular, a variety of derivatives of molybdenum and tungsten have been prepared[13,119]. In order to vaporize synthetically useful amounts of refractory metals it is necessary to have a high power input and lasers and electron guns have been used for this purpose[2–6]. Of course, cost considerations now become a factor but, after the initial outlay, the savings in time and convenience are enormous. Typically, [$(C_6H_6)_2W$] was preparable by the Fischer–Hafner reducing Fridel–Crafts method but only in yields of 1–2%[120]. Metal vapour techniques facilitate the syntheses of multi-gram amounts in several hours. An early problem with electron gun sources was that of considerable damage to the substrate owing to electron bombardment[121]. However, reversal of the polarity of the target considerably alleviates this problem.

The availability of such techniques has now allowed development of the chemistry of the [$(arene)_2W$]system; thus, protonation of the tungsten sandwich molecule yields a bent cation **(37)**, whereas allyl chloride gives rise to the oxidative addition product **38**[13]. [$(Arene)_2Mo$] systems containing substituents such as F, Cl, CH_3O, R_2N, and CO_2CH_3 have also been isolated.

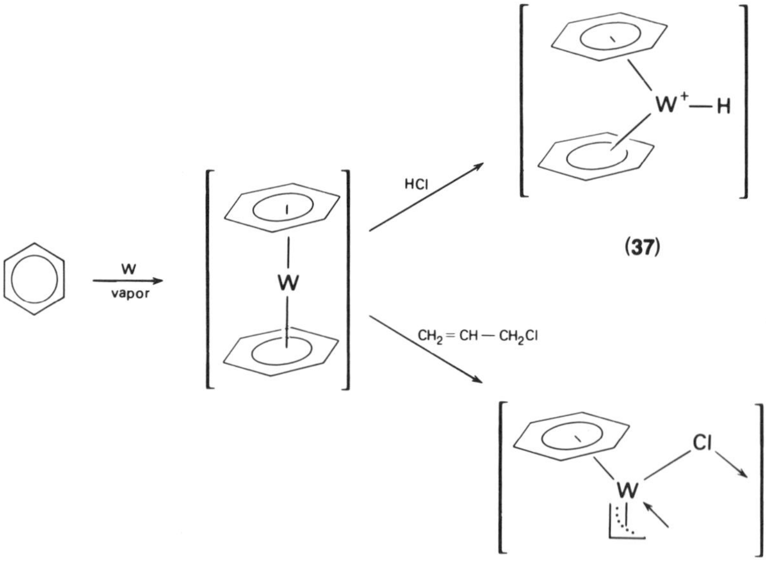

(37)

(38)

(38)

A particularly exciting development has been the synthesis of arene sandwiches of the early transition metals and the stabilization of these metals in the zero oxidation

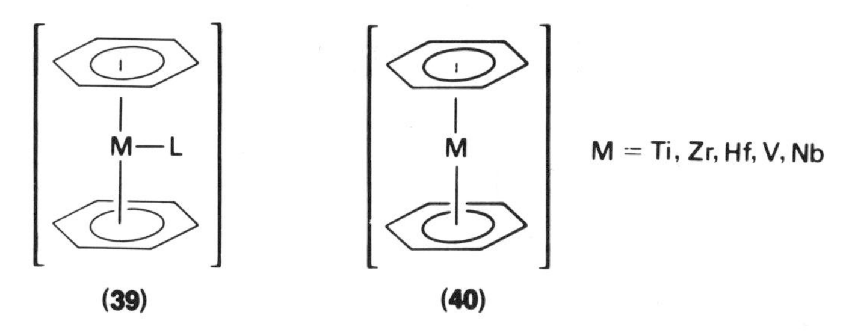

(39) **(40)**

state as in **39** or **40**[122,123]. Of course, molecules of the type [(arene)$_2$M], where M is Zr or Hf, have 16-electron configurations and it is not surprising that, when these sandwiches are prepared in the presence of an appropriate donor ligand, 18-electron complexes **(39)**, where L is trimethylphosphine, are obtainable.

An alternative route to some of these arene sandwiches of the early transition metals has been reported by Hawker *et al*[124]. They showed that evaporation of alkali metals into a solution containing a metal halide and an organic ligand can give acceptable yields of organometallics. Typically, evaporation of potassium vapour into a mixture of vanadium trichloride and toluene gave a 30% yield of [(toluene)$_2$V]. Similarly, derivatives of chromium, molybdenum, and titanium have been prepared. Naturally, vaporization of relatively volatile metals such as sodium and potassium is a much cheaper proposition than acquiring an electron gun and this dehalogenation approach appears to have a bright future.

The interesting molecule (1,4-difluorobenzene)vanadium **(41)** has been the subject of a crystallographic study which showed that the complexed *para*-difluorobenzene rings adopted a slight boat conformation with the fluorine-bearing carbon atoms bent out of the plane away from the vanadium. Furthermore, the rings were rotated relative to each other to minimize fluorine–fluorine interactions[125].

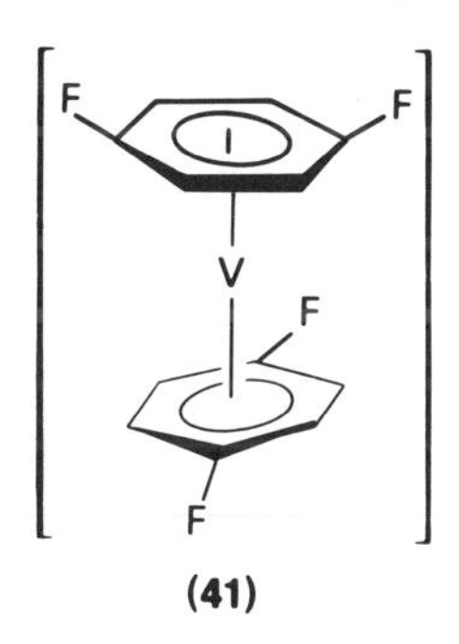

(41)

Whereas the [Ar$_2$Cr] systems have 18-electron configurations, complexes of the early transition metals, Ti, Zr, Hf, etc., do not (and thus can bond an extra ligand); in contrast, the arene sandwiches of the later transition metals are required to populate antibonding molecular orbitals. These systems circumvent this problem in numerous ways. Iron atoms and mesitylene yield ultimately the η^6—arene—η^4—cyclohexadiene complex **42**, possibly via the diarene complex **43** and η^3—benzyl—iron—hydride complex **44**[5]. Cobalt and nickel react with halopentafluorobenzenes (upon subsequent addition of toluene at low temperature) to give the novel [(toluene)M(C$_6$F$_5$)$_2$] complexes previously discussed in Section II[46]. With hexafluorobenzene, nickel yields a shock- and thermally-sensitive compound suggested to be C$_6$F$_6$Ni, which reacts vigor-

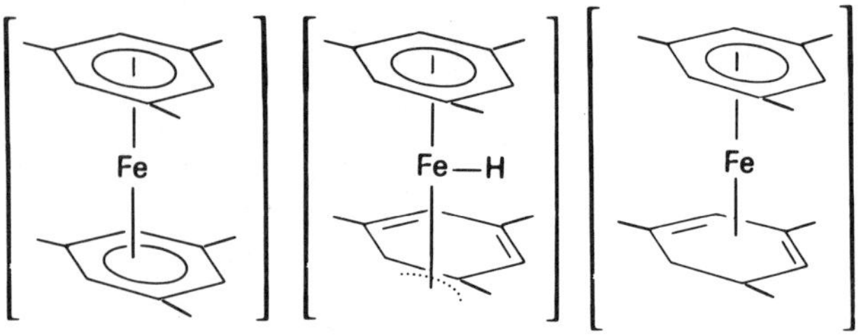

(**43**) (**44**) (**42**)

ously with carbon monoxide or trialkyl phosphites giving $[NiL_4]$ complexes and releasing C_6F_6[126].

Finally, one should mention that the metal vapour synthesis procedure has allowed the synthesis of η^6-pyridine complexes of chromium (equation 37)[26,127]. The bis(2,6-dimethylpyridine) complex **45** exists in rotameric forms in the crystal[127].

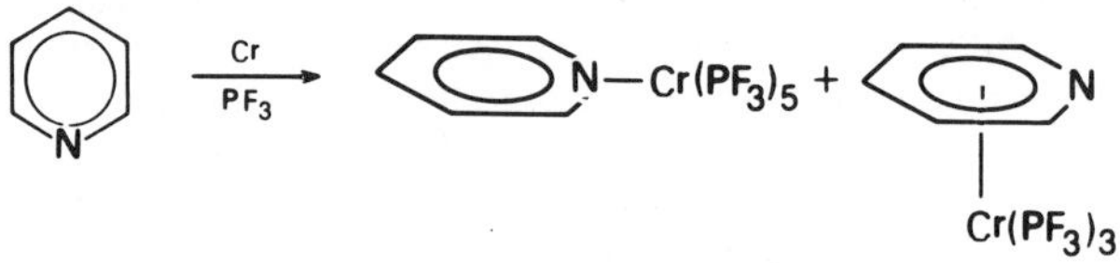

$$\text{C}_5\text{H}_5\text{N} \xrightarrow[\text{PF}_3]{\text{Cr}} \text{C}_5\text{H}_5\text{N—Cr(PF}_3)_5 + (\eta^6\text{-C}_5\text{H}_5\text{N})\text{Cr(PF}_3)_3 \quad (39)$$

$$\text{2,6-Me}_2\text{C}_5\text{H}_3\text{N} \xrightarrow[\text{vapor}]{\text{Cr}} (\textbf{45}) + (\textbf{45}) \quad (40)$$

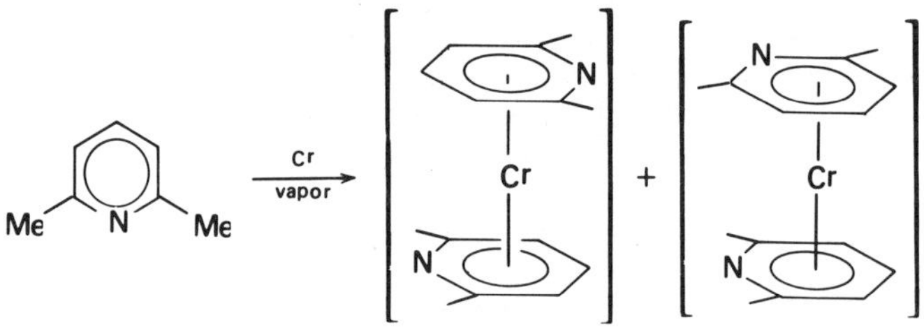

V. ALKYNES

It is well known that sodium in liquid ammonia is an excellent reagent for the controlled reduction of multiple bonds[128]. In particular, disubstituted acetylenes yield the corresponding *trans*-alkene[129]—a useful contrast to catalytic hydrogenations which yield the *cis*-isomers.

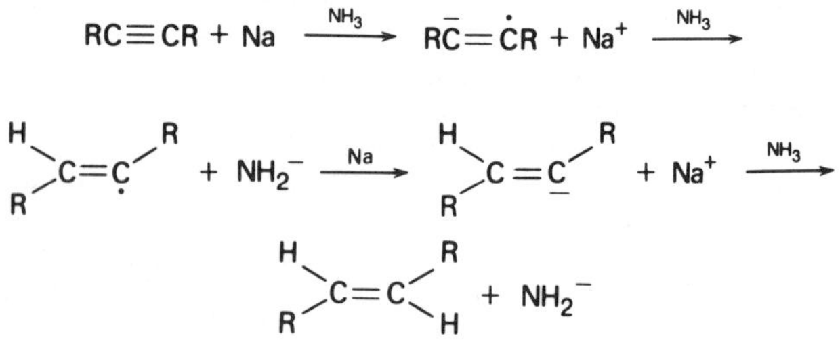

$$\text{RC}\equiv\text{CR} + \text{Na} \xrightarrow{\text{NH}_3} \text{R}\bar{\text{C}}=\dot{\text{C}}\text{R} + \text{Na}^+ \xrightarrow{\text{NH}_3}$$

$$\text{H(R)C}=\dot{\text{C}}\text{R} + \text{NH}_2^- \xrightarrow{\text{Na}} \text{H(R)C}=\bar{\text{C}}\text{R} + \text{Na}^+ \xrightarrow{\text{NH}_3}$$

$$\text{H(R)C}=\text{C(R)H} + \text{NH}_2^- \quad (41)$$

The production of the *trans*-radical on protonation of the radical ion may be considered analogous to the *trans*-addition of radicals or electrophiles to alkynes. Presumably, the further reduction of the radical to the anion and formation of the olefinic product occurs more rapidly than inversion of the vinyl radical and carbanion.

A closely analogous situation is found during the co-condensation reaction of thermal (^1S) magnesium atoms (see Section II) with ammonia containing but-2-yne, which produces a maroon-coloured matrix that, upon warming, gives an 87% yield of *trans*-but-2-ene. Girard[29] has proposed the mechanism in equation 40.

$$RC{\equiv}CR + Mg\colon \xrightarrow{NH_3} R\bar{C}{=}\dot{C}R + Mg^{\dot{+}} \xrightarrow{NH_3}$$

$$(H)(R)C{=}\dot{C}(R) + NH_2^- \xrightarrow{Mg\colon} (H)(R)C{=}\bar{C}(R) + Mg^{\dot{+}} \xrightarrow{NH_3}$$

$$(H)(R)C{=}C(R)(H) + NH_2^- \qquad (42)$$

Such a mechanism would require 0.5 mol of hydrogen to be released on hydrolysis per Mg atom reaction; also, hydrolysis of the residue should release two molecules of NH_3 for every *trans*-but-2-ene molecule produced, and both of these criteria are realized experimentally. It is not surprising that magnesium gives only one electron per atom to but-2-yne, since the first and second ionization potentials for magnesium are 7.61 and 14.96 eV, respectively. It is much easier for two atoms to give one electron each than for one atom to give both.

With ^3P, that is, arc-produced, magnesium (see Section II), the absolute yield of *trans*-olefin is low (28%) since there is a competition between the ammonia and the but-2-yne for the electrons.

In contrast to the magnesium reaction, the reaction of dialkylacetylenes with aluminium vapour yields a mixture of *cis*- and *trans*-olefins only after hydrolysis of the involatile residue. Matrix isolation studies have revealed an unexpected difference between the aluminium–ethylene and aluminium–acetylene systems. As described in Section II, the Al/C_2H_4 matrix contains a π-complex[55] but e.s.r. studies at 4 K on the Al/C_2H_2 matrix have been interpreted in terms of an aluminium—carbon σ bond and a vinyl radical with the odd electron in an orbital of essentially *p* character (**46**)[130].

$$(H)(Al)C{=}C(H)\ [\text{odd electron in } p \text{ orbital}]$$

(**46**)

The reactions of alkynes with transition metal vapours were originally reported in 1973 when it was shown that chromium atoms effected the trimerization of alkylacetylenes to mixtures of arenes (equation 41). No chromium—arene sandwich complexes were isolated and it was suggested[52] that only one arene is bonded to the metal chromium atom at any one time.

In an imaginative variation on this theme, Gladysz *et al.*[131] treated large cyclic diynes with chromium vapour and observed the formation of the novel arene—triyne derivatives (**49**). These must have arisen via the chromium complex **47**, which might

$$RC{\equiv}CH \xrightarrow{Cr} \text{1,3,5-}R_3C_6H_3 + \text{1,2,4-}R_3C_6H_3 \qquad (43)$$

have allowed a second template process leading to the hexabridged cyclophane—Cr complex **48**; however, neither this sandwich compound nor the free ligand was observable[131].

(47) (48) (49) (44)

Other less well characterized products have been obtained in reactions of alkynes with Fe, Co, and Ni[132]. However, co-condensation of cobalt vapour with PhC≡CH and a nitrile yields a mixture of arenes and pyridines (equation 43)[133].

$$PhC{\equiv}CH + RC{\equiv}N \xrightarrow[\text{vapour}]{Co} \text{Ph}_3C_6H_3 + \text{Ph}_2(R)C_5H_2N \qquad (45)$$

Some success has been achieved with hexafluorobut-2-yne (HFB), which yields complexes with Co, Ni, Pd, Pt, Cu, and Ag; all of these, however, are unstable with respect to decomposition to metal particles and the acetylene trimer hexakis(trifluoromethyl)benzene[134]. Addition of carbon monoxide to the HFB—Ni or HFB—Pd complexes before warming the matrices led to the formation of the volatile (HFB)—$M(CO)_2$ clusters which spontaneously converted to the $[M_4(HFB)_3(CO)_4]$ clusters, of which only the nickel one had previously been characterized.

VI. PHOSPHINES

Although metal vapours have been shown to function as very versatile reagents in organometallic syntheses, they have also been used extensively to prepare molecules which, although not strictly organometallic, continue to attract the attentions of organometallic chemists. Thus, reactions of metal vapours with trifluorophosphine, which may be a better π-acid than carbon monoxide yet which has the advantage of being condensable at liquid nitrogen temperature, have led to a plethora of interesting molecules[135], some of which had only been previously obtained via high-pressure routes (Scheme 12)[136].

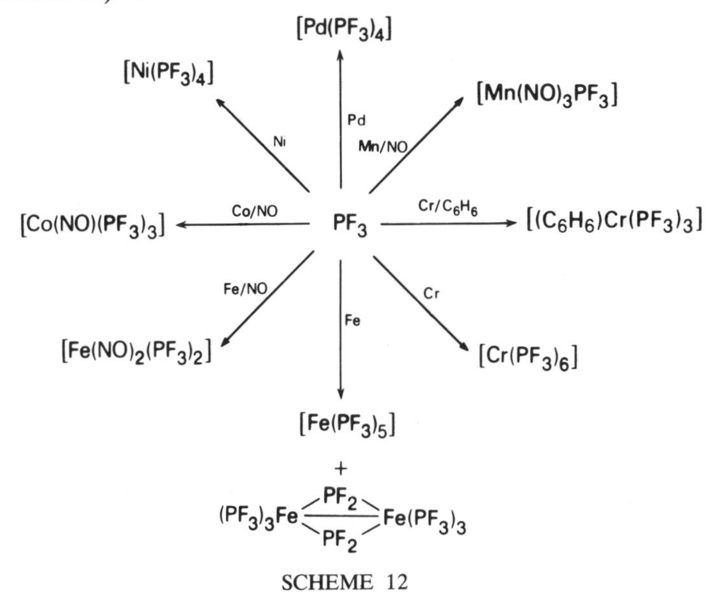

SCHEME 12

This idea has been extended to a variety of other fluorophosphines such as PF_2Cl[135], PF_2H[137], and PF_2NMe_2[138,139]. Furthermore, co-condensation of nickel atoms with an equimolar mixture of trifluorophosphine (a good π-acid) and phosphine (a notoriously weak ligand gave modest yields of $[Ni(PF_3)_n(PH_3)_{4-n}]$, where $n = 1$ or 2[135]. Other phosphines which have been co-condensed with metal vapours include trimethylphosphite[140] and trimethylphosphine[26]; a complicating feature for phosphines with alkyl groups possessing β-hydrogen atoms is the insertion of metal atoms into P—C or C—H bonds, and indeed PEt_3 has been reported to yield H_2, C_2H_4, and other products[141].

The availability of the rotating reactors which permit the use of involatile ligands[7] such as triphenylphosphine has allowed the direct syntheses of $[M(PPh_3)_4]$, where M is Ni[26] or Pd[5], and thus obviated the need for the prior preparation of $[(cod)_2M]$.

VII. MISCELLANEOUS LIGANDS

The high affinity of transition metal atoms for oxygen has been turned to synthetic advantage by several workers who have deoxygenated organic molecules with different metals. Gladysz and co-workers compared the ability of Ti, V, Cr, Co, and Ni to deoxygenate cyclohexene oxide. Vanadium and chromium were found to be the most efficient and they produced almost 3 mol of cyclohexene per gram-atom of metal[142]. Chromium was further shown to deoxygenate 2,6-dimethylpyridine-*N*-oxide, triethylphosphine oxide, dimethyl sulphoxide, and nitro- and nitrosoarenes. Although in most cases the yields were not synthetically useful, mechanistic inferences could be drawn and in the nitro- and nitrosoarene deoxygenations it was suggested that nitrene or nitrenoid intermediates were involved[143].

The vapours of Ti, V, and Cr have also been used to deoxygenate ketones and ethers[144]. The reactions of ethers with alkaline earth vapours (Ca, Sr, Ba) have been shown to yield low molecular weight alkanes, alkenes, and alkynes[145]. Typically, strontium vapour and dimethyl ether give good yields of C_2H_2 and the butenes but only small amounts of higher molecular weight products although they are detectable up to C_8. It is thought that carbide-like species are involved since addition of D_2O to the residue yields products which are extensively deuterated[146].

Transition metal vapours can also desulphurize organic systems. Thus, thiophenes are desulphurized by chromium vapour and with iron atoms the ferradiene complex **50** is obtained[147].

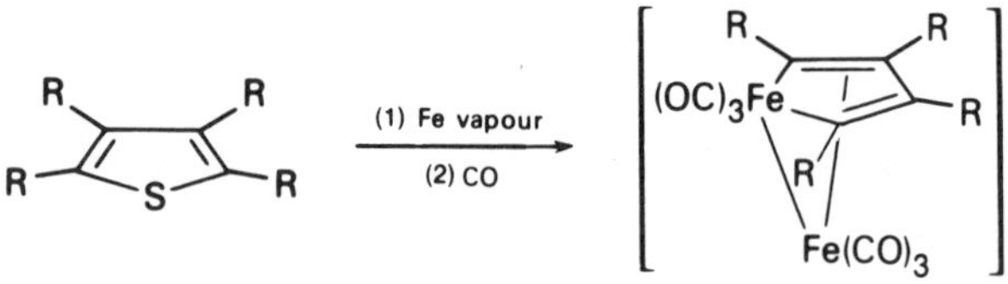

(1) Fe vapour (2) CO → $(OC)_3Fe$ ... $Fe(CO)_3$ (46)

(50)

Alkyl isonitriles have been co-condensed with metals and the complexes $[Cr(CNR)_6]$, $[Fe(CNR)_5]$, and $[Ni(CNR)_4]$ have been isolated[148]. Co-condensation of isonitriles and PF_3 with metal vapours gave mixed complexes.

Recently it has been claimed that some transition metal atoms can insert into C—H and C—C bonds at cryogenic temperatures. Thus, zirconium atoms react with neopentane[149] to give, upon deuterolysis, products clearly derivable from Zr insertion into C—H and C—C bonds; similarly, it is reported that Fe atoms can insert into the C—H bonds of methane[150]. Undoubtedly these reactions will attract the attention of the matrix isolation spectroscopists.

Finally, metal vapours have been used as synthetic reagents in the production of novel metallocarborane clusters such as **51** and **52**[26,151].

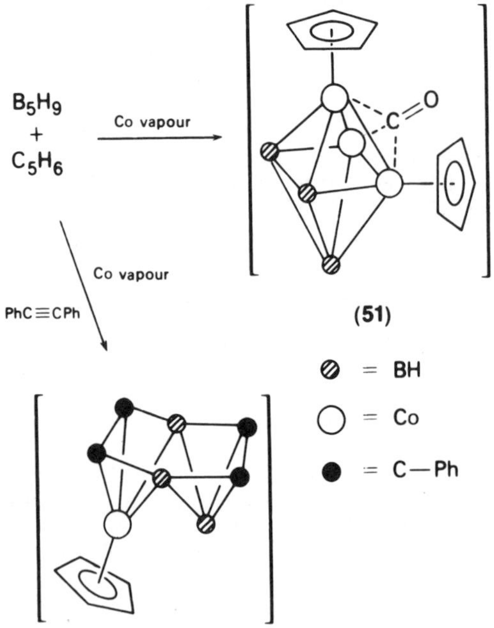

B_5H_9 + C_5H_6 —Co vapour→ **(51)** (47)

—Co vapour, PhC≡CPh→ **(52)**

⊘ = BH
○ = Co
● = C—Ph

VIII. REFERENCES

1. P. L. Timms, *Adv. Inorg. Chem. Radiochem.*, **14**, 121 (1972).
2. M. Moskovits and G. A. Ozin, *Cryochemistry*, Wiley – Interscience, New York, 1976.
3. J. R. Blackborow and D. Young, *Metal Vapour Synthesis in Organometallic Chemistry*, Springer-Verlag, Berlin, 1979.
4. K. J. Klabunde, *Chemistry of Free Atoms and Particles*, Academic Press, New York, 1980.
5. P. L. Timms and T. W. Turney, *Adv. Organomet. Chem.*, **15**, 53 (1977).
6. E. A. Koerner von Gustorf, O. Jaenicke, O Wolfbeis, and C. R. Eady, *Angew. Chem., Int. Ed. Engl.*, **14**, 278 (1975).
7. M. L. H. Green, *J. Organomet. Chem.*, **200**, 119 (1980).
8. P. S. Skell and P. W. Owen, *J. Am. Chem. Soc.*, **94**, 5434 (1972).
9. P. S. Skell, J. J. Havel, and M. J. McGlinchey, *Acc. Chem. Res.*, **6**, 97 (1973).
10. W. J. Power and G. A. Ozin, *Adv. Inorg. Chem. Radiochem.*, **23**, 79 (1980).
11. G. A. Ozin and A. Vander Voet, *Acc. Chem. Res.*, **6**, 313 (1973).
12. K. J. Klabunde, P. S. Skell, S. D. Ittel, and P. L. Timms, *Inorg. Synth.*, **19**, 59 (1979).
13. F. G. N. Cloke, M. L. H. Green, and G. E. Morris, *J. Chem. Soc., Chem. Commun.*, 72 (1978).
14. W. Reichelt, *Angew. Chem., Int. Ed. Engl.*, **14**, 218 (1975).
15. B. Mile, *Angew. Chem., Int. Ed. Engl.*, **7**, 507 (1968).
16. H. Hallam, *Vibrational Spectroscopy of Trapped Species*, Wiley, New York, 1972.
17. P. S. Skell, E. J. Goldstein, R. J. Peterson, and G. L. Tingey, *Chem. Ber.*, **100**, 1442 (1967).
18. R. G. Doerr and P. S. Skell, *J. Am. Chem. Soc.*, **89**, 4684 (1967).
19. P. S. Skell and E. J. Goldstein, *J. Am. Chem. Soc.*, **86**, 1442 (1964).
20. R. L. Lambert, Jr., and D. Seyferth, *J. Am. Chem. Soc.*, **94**, 9246 (1972).
21. C. Chung and R. J. Lagow, *J. Chem. Soc., Chem Commun.*, 1078 (1972).
22. J. A. Morrison and R. J. Lagow, *Inorg. Chem.*, **16**, 2972 (1977).
23. L. G. Sneddon and R. J. Lagow, *J. Chem. Soc., Chem. Commun.*, 302 (1975).
24. P. L. Timms, *Chem. Commun.*, 1525 (1968).
25. P. L. Timms, *J. Chem. Soc., A*, 830 (1972).
26. P. L. Timms, *Angew. Chem., Int. Ed. Engl.*, **14**, 273 (1975).
27. K. J. Klabunde, *J. Fluorine Chem.*, **7**, 95 (1976).
28. A. G. Massey, *Adv. Inorg. Chem. Radiochem.*, **10**, 1 (1967).
29. J. E. Girard, *PhD Thesis*, Pennsylvania State University, 1971.
30. P. S. Skell and J. E. Girard, *J. Am. Chem. Soc.*, **94**, 5518 (1972).
31. P. S. Skell and K. J. Shea, *Isr. J. Chem.*, **10**, 493 (1972).
32. K. J. Klabunde, J. Y. F. Low, and M. S. Key, *J. Fluorine Chem.*, **2**, 207 (1972).
33. K. J. Klabunde, M. S. Key, and J. Y. F. Low, *J. Am. Chem. Soc.*, **94**, 999 (1972).
34. W. N. Brent, *PhD Thesis*, Pennsylvania State University, 1974.
35. M. J. McGlinchey and T.-S. Tan, *Inorg. Chem.*, **14**, 1209 (1975).
36. R. T. Conlin, S. H. Lockhart, and P. P. Gaspar, *J. Chem. Soc., Chem. Commun.*, 825 (1975).
37. J.-C. Barthelat, B. S. Roch, G. Trinquier, and J. Satgé, *J. Am. Chem. Soc.*, **102**, 4080 (1980).
38. L. Vaska, *Acc. Chem. Res.*, **1**, 335 (1968).
39. M. Piper and P. L. Timms, *Chem. Commun.*, 52 (1972).
40. G. A. Ozin and W. J. Power, *Inorg. Chem.*, **17**, 2836 (1978).
41. K. J. Klabunde, *Acc. Chem. Res.*, **8**, 393 (1975).
42. P. S. Skell and J. J. Havel, *J. Am. Chem. Soc.*, **93**, 6687 (1971).
43. J. S. Roberts and K. J. Klabunde, *J. Am. Chem. Soc.*, **99**, 2509 (1977).
44. K. J. Klabunde and J. Y. F. Low, *J. Am. Chem. Soc.*, **96**, 7674 (1974).
45. K. J. Klabunde, *Angew. Chem., Int. Ed. Engl.*, **14**, 287 (1975).
46. K. J. Klabunde, B. B. Anderson, M. Bader, and L. J. Radonovich, *J. Am. Chem. Soc.*, **100**, 1313 (1978).
47. B. B. Anderson, C. Behrens, L. J. Radonovich, and K. J. Klabunde, *J. Am. Chem. Soc.*, **98**, 5390 (1976).
48. K. J. Klabunde and J. S. Roberts, *J. Organomet. Chem.*, **137**, 113 (1977).
49. J. S. Roberts, *PhD Thesis*, University of North Dakota, 1975.
50. R. F. Heck and J. P. Nolley, Jr., *J. Org. Chem.*, **37**, 2320 (1972).
51. P. S. Skell and M. J. McGlinchey, *Angew. Chem., Int. Ed. Engl.*, **14**, 195 (1975).

52. P. S. Skell, D. L. Williams-Smith, and M. J. McGlinchey, *J. Am. Chem. Soc.*, **95**, 3337 (1973).
53. P. S. Skell, J. J. Havel, D. L. Williams-Smith, and M. J. McGlinchey, *Chem. Commun.*, 1098 (1972).
54. P. S. Skell and L. R. Wolf, *J. Am. Chem. Soc.*, **94**, 7919 (1972).
55. P. Kasai and D. McLeod, Jr., *J. Am. Chem. Soc.*, **97**, 5609 (1975).
56. H. Bönneman, *Angew. Chem., Int. Ed. Engl.*, **9**, 736 (1970).
57. P. W. Jolly and G. Wilke, *The Organic Chemistry of Nickel*, Academic Press, New York, 1974.
58. Reference 3, p. 94.
59. P. L. Timms, *Chem. Commun.*, 1033 (1969).
60. V. M. Akhmedov, M. T. Anthony, M. L. H. Green, and D. Young, *J. Chem. Soc., Dalton Trans.*, 1412 (1975).
61. D. Young and M. L. H. Green, *J. Appl. Chem. Biotechnol.*, **25**, 641 (1975).
62. R. Rautenstrauch, *Angew. Chem., Int. Ed. Engl.*, **14**, 259 (1975).
63. Reference 3, pp. 94–101.
64. D. L. Williams-Smith, L. R. Wolf, and P. S. Skell, *J. Am. Chem. Soc.*, **94**, 4042 (1972).
65. E. Koerner von Gustorf, O. Jaenicke, and O. E. Polansky, *Angew. Chem., Int. Ed. Engl.*, **11**, 532 (1972).
66. J. J. Havel, *PhD Thesis,* Pennsylvania State University, 1972.
67. P. S. Skell, E. M. Van Dam, and M. P. Silvon, *J. Am. Chem. Soc.*, **96**, 626 (1974).
68. R. B. King and A. Fronzaglia, *Inorg. Chem..*, **5**, 1837 (1966).
69. W. J. Evans, S. C. Engerer, and A. C. Neville, *J. Am. Chem. Soc.*, **100**, 331 (1978).
70. J. R. Blackborow, unpublished data reported in reference 3.
71. A. R. Fraser, P. H. Bird, S. A. Bezman, J. R. Shapley, R. White, and J. A. Osborn, *J. Am. Chem. Soc.*, **95**, 597 (1973).
72. B. Bogdanovic, M. Kroener, and G. Wilke, *Justus Liebigs Ann. Chem.*, **699**, 1 (1966).
73. R. M. Atkins, R. Mackenzie, P. L. Timms, and T. W. Turney, *J. Chem. Soc., Chem. Commun.*, 764 (1975).
74. R. Mackenzie and P. L. Timms, *J. Chem. Soc., Chem. Commun.*, 650 (1974).
75. R. A. Cable, M. Green, R. E. Mackenzie, P. L. Timms, and T. W. Turney, *J. Chem. Soc., Chem. Commun.*, 270 (1976).
76. A. D. English, J. P. Jesson, and C. A. Tolman, *Inorg. Chem.*, **15**, 1730 (1976).
77. E. Elian and R. Hoffman, *Inorg. Chem.*, **14**, 1058 (1975).
78. J. R. Blackborow, R. H. Grubbs, A. Miyashita, A. Scrivanti, and E. A. Koerner von Gustorf, *J. Organomet. Chem.*, **122**, C6 (1976).
79. M. J. D'Aniello and E. K. Barefield, *J. Organomet. Chem.*, **76**, C50 (1974).
80. E. M. Van Dam, W. N. Brent, M. P. Silvon, and P. S. Skell, *J. Am. Chem. Soc.*, **97**, 467 (1975).
81. P. L. Timms and T. W. Turney, *J. Chem. Soc., Dalton Trans.*, 2021 (1976).
82. J. R. Blackborow, K. Hildebrand, E. A. Koerner von Gustorf, A. Scrivanti, C. R. Eady, D. Ehnhart, and C. Krueger, *J. Chem. Soc., Chem. Commun.*, 16 (1976).
83. T.-S. Tan, J. L. Fletcher, and M. J. McGlinchey, *J. Chem. Soc., Chem. Commun.*, 771 (1975).
84. P. Eilbracht, E. Henkes, W. Totzauer, and A. Landers, *J. Chem. Soc., Chem. Commun.*, 717 (1980).
85. T.-S. Tan, *PhD Thesis*, McMaster University, 1976.
86. Nguyen Hao, J. F. Sawyer, B. G. Sayer, and M. J. McGlinchey, *J. Am. Chem. Soc.*, **101**, 2203 (1979).
87. S. P. Kolesnikov, J. E. Dobson, and P. S. Skell, *J. Am. Chem. Soc.*, **100**, 999 (1978).
88. V. Graves, J. J. Lagowski, and L. H. Simons, unpublished data.
89. T. A. Manual and F. G. A. Stone, *J. Am. Chem. Soc.*, **82**, 366 (1960).
90. S. R. Ely, T. E. Hopkins, and C. W. DeKock, *J. Am. Chem. Soc.*, **98**, 1624 (1976).
91. G. Wilkinson, M. Rosenblum, M. C. Whiting, and R. B. Woodward, *J. Am. Chem. Soc.*, **74**, 2125 (1952).
92. H. Zeiss, M. Tsutsui, and L. Onsager, *Angew. Chem.*, **67**, 282 (1955).
93. E. O. Fischer and W. Hafner, *Z. Naturforsch.*, **10B**, 665 (1955).
94. D. N. Kursanov, V. N. Setkina, and B. G. Gribov, *J. Organomet. Chem.*, **37**, C35 (1972).

95. E. O. Fischer and H. Brunner, *Chem. Ber.*, **95**, 1999 (1962); **98**, 175 (1965).
96. C. Elschenbroich, *J. Organomet. Chem.*, **14**, 157 (1968).
97. M. J. McGlinchey and T.-S. Tan, *Can. J. Chem.*, **52**, 2439 (1974).
98. K. J. Klabunde and H. F. Efner, *Inorg. Chem.*, **14**, 789 (1975).
99. V. Graves and J. J. Lagowski, *Inorg. Chem.*, **15**, 577 (1976).
100. V. Graves and J. J. Lagowski, *J. Organomet. Chem.*, **120**, 397 (1976).
101. M. J. McGlinchey and T.-S. Tan, *J. Am. Chem. Soc.*, **98**, 2271 (1976).
102. T.-S. Tan and M. J. McGlinchey, *J. Chem. Soc., Chem. Commun.*, 155 (1976).
103. A. Agarwal, M. J. McGlinchey, and T.-S. Tan, *J. Organomet. Chem.*, **141**, 85 (1978).
104. J. D. Laposa, Nguyen Hao, B. G. Sayer, and M. J. McGlinchey, *J. Organomet. Chem.*, **195**, 193 (1980).
105. U. Zenneck, C. Elschenbroich, and R. Moeckel, *Angew Chem., Int. Ed. Engl.*, **17**, 531 (1978).
106. J. D. Laposa, Nguyen Hao, and M. J. McGlinchey, unpublished data.
107. M. J. McGlinchey, Nguyen Hao, R. Faggiani, and C. J. L. Lock, unpublished data.
108. Nguyen Hao and M. J. McGlinchey, *J. Organomet. Chem.*, **161**, 381 (1978).
109. R. Middleton, J. R. Hull, S. R. Simpson, C. H. Tomlinson, and P. L. Timms, *J. Chem. Soc., Dalton Trans.*, 120 (1973).
110. R. Moeckel and C. Elschenbroich, *Angew. Chem., Int. Ed. Engl.*, **16**, 870 (1977).
111. E. P. Kündig and P. L. Timms, *J. Chem. Soc., Dalton Trans.*, 991 (1980).
112. D. J. Cram and J. M. Cram, *Acc. Chem. Res.*, **4**, 204 (1971).
113. D. J. Cram and D. I. Wilkinson, *J. Am. Chem. Soc.*, **82**, 5721 (1960).
114. A. N. Nesmeyanov, N. N. Zaitseva, G. A. Domrachev, V. D. Zinov'ev, L. P. Yur'eva, and I. I. Tverdokhlebova, *J. Organomet. Chem.*, **121**, C52 (1976).
115. A. N. Nesmeyanov, L. P. Yur'eva, N. N. Zaitseva, and N. I. Vasyukova, *J. Organomet. Chem.*, **153**, 341 (1978).
116. G. B. Shul'pin and M. I. Rybinskaya, *Russ. Chem. Rev.*, **43**, 716 (1974).
117. C. G. Francis and P. L. Timms, *J. Chem. Soc., Dalton Trans.*, 1401 (1980).
118. C. G. Francis, G. A. Ozin, and H. Huber, *Inorg. Chem.*, **19**, 219 (1980).
119. M. P. Silvon, E. M. Van Dam, and P. S. Skell, *J. Am. Chem. Soc.*, **96**, 1945 (1974).
120. E. O. Fischer, *Inorg. Synth.*, **6**, 132 (1960).
121. P. S. Skell and P. W. Owen, *J. Am. Chem. Soc.*, **94**, 5434 (1972).
122. F. G. N. Cloke and M. L. H. Green, *J. Chem. Soc., Chem. Commun.*, 127 (1979).
123. F. G. N. Cloke, M. L. H. Green, and D. H. Price, *J. Chem. Soc., Chem. Commun.*, 431 (1978).
124. P. N. Hawker, E. P. Kündig, and P. L. Timms, *J. Chem. Soc., Chem. Commun.*, 730 (1978).
125. L. Radonovich, C. Zuerner, H. F. Efner, and K. J. Klabunde, *Inorg. Chem.*, **15**, 2976 (1976).
126. K. J. Klabunde and H. F. Efner, *J. Fluorine Chem.*, **4**, 115 (1974).
127. L. H. Simons, P. E. Riley, R. E. Davis, and J. J. Lagowski, *J. Am. Chem. Soc.*, **98**, 1044 (1976).
128. A. J. Birch, *Q. Rev. Chem. Soc.*, **4**, 69 (1950).
129. A. Farkas and L. Farkas, *Trans. Faraday Soc.*, **33**, 337 (1937).
130. P. H. Kasai, D. McLeod, Jr., and T. Watanabe, *J. Am. Chem. Soc.*, **99**, 3522 (1977).
131. J. A. Gladysz, J. G. Fulcher, S. J. Lee, and A. B. Bocarsley, *Tetrahedron Lett.*, 3421 (1977).
132. L. H. Simons and J. J. Lagowski, *J. Org. Chem.*, **43**, 3247 (1978).
133. Reference 3, p. 85.
134. K. J. Klabunde, T. Groshens, M. Brezinski, and W. Kennelly, *J. Am. Chem. Soc.*, **100**, 4437 (1978).
135. P. L. Timms, *J. Chem. Soc., A*, 2526 (1970).
136. Th. Kruck, *Angew. Chem., Int. Ed. Engl.*, **6**, 53 (1967).
137. D. Staplin and R. W. Parry, *170th Nat. Meeting Am. Chem. Soc.*, Paper INOR 117 (1975).
138. M. Chang, R. B. King, and M. G. Newton, *J. Am. Chem. Soc.*, **100**, 998 (1978).
139. R. B. King and M. Chang, *Inorg. Chem.*, **18**, 364 (1979).
140. C. A. Tolman, L. W. Yarbrough, and J. G. Verkade, *Inorg. Chem.*, **16**, 479 (1977).
141. Reference 4, p. 108.
142. J. A. Gladysz, J. G. Fulcher, and S. Togashi, *J. Org. Chem.*, **41**, 3647 (1976).
143. J. A. Gladysz, J. G. Fulcher, and S. Togashi, *Tetrahedron Lett.*, 521 (1977).

144. Reference 4, p. 55.
145. W. E. Billups, M. M. Konarski, R. H. Hauge, and J. L. Margrave, *J. Organomet. Chem.*, **194**, C22 (1980).
146. W. E. Billups, M. M. Konarski, R. H. Hauge, and J. L. Margrave, *J. Am. Chem. Soc.*, **102**, 3649 (1980).
147. T. Chivers and P. L. Timms, *J. Organomet. Chem.*, **118**, C37 (1976).
148. F. R. Scholar and D. Gladkowski, *171st Nat. Meeting Am. Chem. Soc.*, Paper INOR 133 (1976).
149. R. J. Remick, T. S. Asunta, and P. S. Skell, *J. Am. Chem. Soc.*, **101**, 1320 (1979).
150. P. H. Barrett, M. Pasternack, and R. G. Pearson, *J. Am. Chem. Soc.*, **101**, 222 (1979).
151. G. J. Zimmerman, L. W. Hall, and L. G. Sneddon, *Inorg. Chem.*, **19**, 3642 (1980).

The Chemistry of the Metal–Carbon Bond
Edited by F. R. Hartley and S. Patai

CHAPTER **14**

Analysis of organometallic compounds: determination of elements and functional groups

T. R. CROMPTON
'Beechcroft', Whittingham Lane, Goosnargh, Preston, Lancashire, UK

I. INTRODUCTION TO CHAPTERS 14–19

The analysis of organometallic compounds is considered in this and the next five chapters, each of which covers element by element a different aspect of analysis. This chapter deals with the determination of elements and functional groups. Methods are discussed for the determination of various metals and the non-metals carbon, hydrogen, oxygen, halogens, nitrogen, and sulphur, whilst the functional groups include hydride, active hydrogen, alkyl, alkoxide, amino, and thioalkoxide groups. Chapter 15 discusses various titration procedures applied to organometallic compounds, including classical titration and conductometric, potentiometric, amperometric, dielectric constant, coulometric, high-frequency, thermometric, and lumometric titration methods.

In Chapter 16 are discussed visible and ultraviolet spectroscopic methods of analysis of various types of organometallic compounds. Polarographic methods of analysis are discussed in Chapter 17. In Chapter 18 is presented a comprehensive discussion of the applications of gas chromatography to the analysis of organometallic compounds. The various types of detectors that may be used, including element-specific detectors, are discussed in detail. Chapter 19 deals with other chromatographic techniques that have been applied to organometallic compounds, including thin-layer chromatography, paper chromatography, column chromatography, and electrophoretic techniques. Table 1 lists the various sections dealing with each particular element, technique by technique.

II. ORGANOALUMINIUM COMPOUNDS

A. Determination of Aluminium

Many organoaluminium compounds are spontaneously pyrophoric upon contact with air and, as such, are extremely hazardous and should be handled with care. Even contact with traces of oxygen will alter the composition of the sample during analysis and vitiate the analytical results obtained. A sampling procedure[1] for organoaluminium compounds is illustrated in Figure 1.

To determine aluminium, the organoaluminium sample must first be decomposed by the addition of an aqueous reagent in order to provide an aqueous extract in which the aluminium is quantitatively recovered. A cooled hydrocarbon solution of the organoaluminium sample is hydrolysed by the gradual addition of aqueous hydrochloric acid in an inert atmosphere. Aluminium is quantitatively recovered in the

TABLE 1. Sections in Chapters 14–19 describing techniques applied to the detection of organometallic compounds

Element	Elements and functional groups	Titration methods	Spectroscopic methods	Polarographic techniques	Gas chromatography	Other chromatographic methods
Aluminium	14.II.A to 14.II.I	15.I.A to 15.I.G	16.I.A	—	18.I	—
Antimony	14.III.A	—	—	—	18.II	19.I.A
Arsenic	14.IV.A to 14.IV.C; 14.XXI.B	15.II	16.I.B	17.I	18.I	19.I.B; 19.II.A
Beryllium	14.V.A	—	—	—	18.IV	—
Bismuth	14.VI.A	—	16.I.C	—	—	19.I.C; 19.II.B
Boron	14.VII.A to 14.VII.D	15.III.A and 15.III.B	16.I.D; 16.II.A	17.II	18.V	19.I.D
Calcium	14.VIII.A	15.IV	—	—	—	—
Chromium	14.IX.A	—	16.I.F	—	18.VI	—
Cobalt	14.X.A	15.V	16.I.E	17.VIII	—	19.I.E; 19.II.C; 19.III.A
Copper	14.XI.A	15.VI	16.I.G	—	18.VII	19.I.F; 19.II.D
Gallium	—	—	—	—	18.VIII	
Germanium	14.XII.A and 14.XII.B	15.VII	—	—	18.IX	—
Iron	14.XIII.A; 14.XXI.B	15.VIII.A and 15.VIII.B	16.I.H	17.III	18.X	19.I.G; 19.II.E; 19.III.B; 19.IV.A
Lead	14.XIV.A to 14.XIV.D	15.IX.A to 15.IX.C	16.I.I	17.IV	18.XI	19.I.H; 19.II.F; 19.III.C; 19.IV.B

Lithium	14.XV.A	15.X.A to 15.X.E	16.II.B	—	18.XII	—
Magnesium	14.XVI.A and 14.XVI.B	15.XI.A to 15.XI.C	16.II.C	—	18.XIII	—
Manganese	14.XVII.A	—	16.I.J	17.VI	18.XIV	—
Mercury	14.XVIII.A to 14.XVIII.C	15.XII.A and 15.XII.B	15.I.K; 16.II.D	17.V	18.XV	19.I.I; 19.II.G; 19.III.D
Molybdenum	—	—	—	—	18.XVI	—
Nickel	14.XIX.A	—	16.I.L	17.VII	18.XVII	—
Palladium	14.XX	—	—	—	—	—
Phosphorus	14.XXI.A to 14.XXI.G	15.XIII.A and 15.XIII.B	16.I.M	17.IX	18.XVIII	19.I.J; 19.III.E
Platinum	14.XXII					
Potassium	14.XXIII	15.XIV	—	—	—	—
Ruthenium	—	—	—	—	18.XIX	—
Selenium	14.XXIV.A and 14.XXIV.B	15.XV	16.I.N	17.X	18.XX	19.II.H; 19.III.F
Silicon	14.XXI.B	—	—	—	—	—
Sodium	—	15.XVI	—	—	—	—
Tellurium	—	—	—	—	—	19.I.K
Thallium	14.XXV.A	—	16.I.O	—	—	19.IV.C
Tin	14.XXVI.A to 14.XXVI.E	15.XVII.A to 15.XVII.C	16.I.P; 16.II.E	17.XI	18.XXI	19.I.L; 19.II.I
Titanium	14.XXVII.A; 14.XXI.B	—	—	—	—	—
Zinc	14.XXVIII.A to 14.XXVIII.D	15.XVIII.A and 15.XVIII.B	16.I.Q	17.XII	18.XXII	19.I.M
Zirconium	14.XXIX	—	—	—	—	—

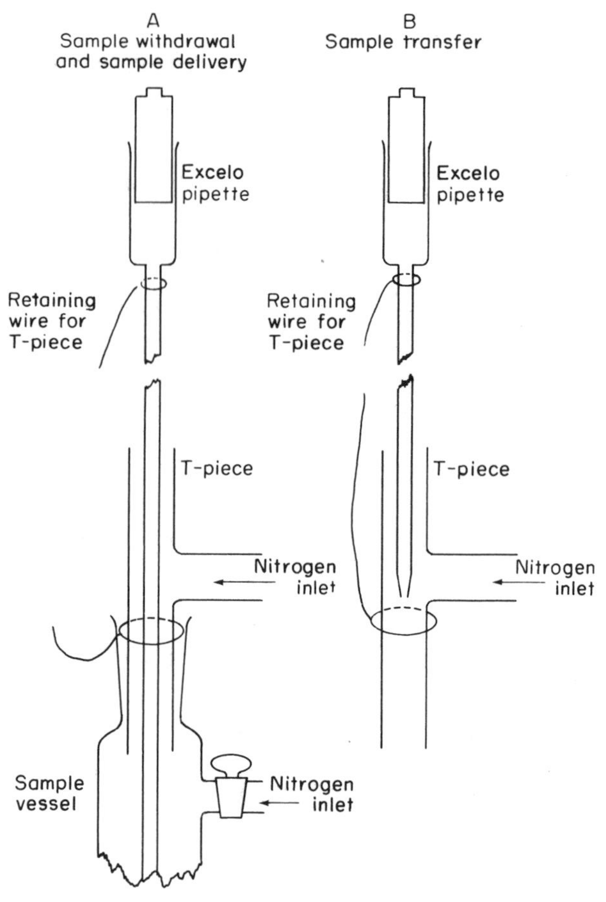

FIGURE 1. Sampling procedure for organoaluminium compounds.

aqueous extract and is then determined in this extract complexometrically using disodium EDTA. This gives satisfactory aluminium recoveries from all types of organoaluminium compounds, from the most reactive types such as neat triethylaluminium to the less reactive higher molecular weight compounds containing alkyl groups up to C_{18}[2]. Sample decomposition is performed in a specially designed flask of the type illustrated in Figure 2. In order to obtain 100% recovery of aluminium in this method, it is necessary for a 20% excess of EDTA over the amount of aluminum present to be added. For this reason a trial titration is carried out, the data obtained being used to calculate the volume of disodium EDTA to be added in the final titration. Cyclohexanediamine tetraacetate may be used as an alternative to EDTA[1,3,4].

Aluminium in organoaluminium compounds has been determined by hydrolysis with dilute nitric acid followed by addition of excess of EDTA to the aqueous extract and determination of unused EDTA by titration with standard lead solution[5]. A similar procedure involves decomposition of the sample with aqueous hydrochloric

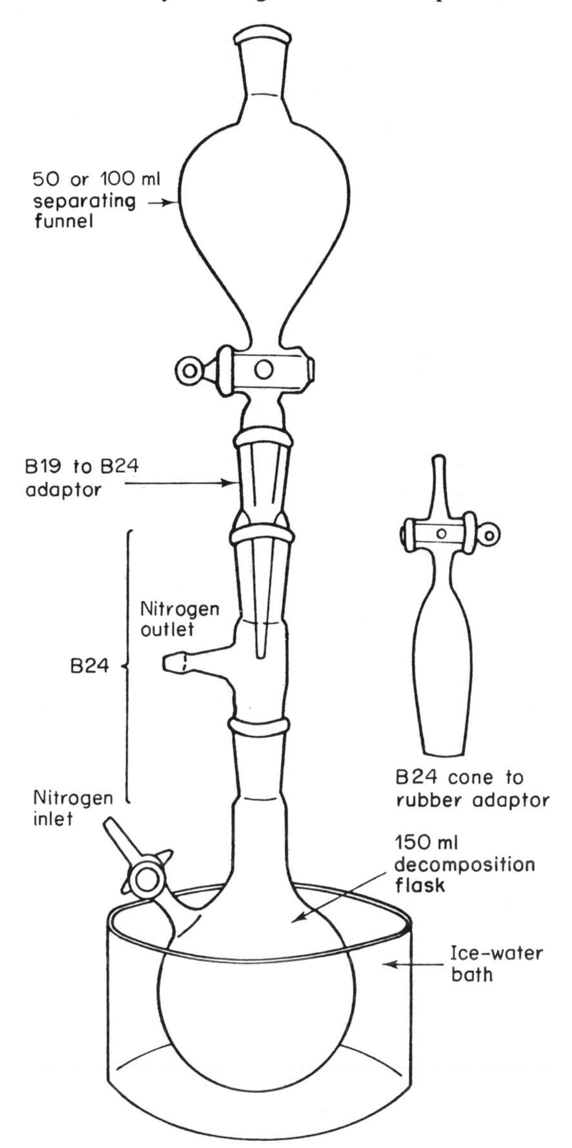

FIGURE 2. Sample decomposition flask for organo-aluminium compounds.

acid, followed by titration of the excess of EDTA with standard copper sulphate solution to the catechol violet end-point[6]. Alternatively, a xylene solution of the organoaluminium sample is transferred into a Erlenmeyer flask[7] to which is added aqueous acetone containing hydrochloric acid. An excess of 0.05 M cyclohexanediaminetetraacetic acid (CDTA) solution[1] and 200 ml of isopropanol is added and the solution heated nearly to boiling and buffered. Excess of CDTA is titrated with standard 0.05 M zinc sulphate solution[1] to the dithizone end-point. The end-point is very sharp from green to red. Excellent accuracy and precision are claimed for this method. It has been applied to very pure and to very complex mixtures without interference or matrix effects.

Organoaluminium compounds can be analysed by digestion with a nitric–sulphuric acid mixture in a sealed tube, followed by spectrophotometric determination of aluminium as the 8-hydroxyquinoline complex[8].

An amount of 20–40 mg of organoaluminium samples that also contain silicon and/or phosphorus can be fused with sodium peroxide in a bomb under an atmosphere of oxygen. After dissolving the product in sulphuric acid, excess of EDTA is added and the excess is determined by titration with standard copper sulphate solution to the catechol violet end-point[9]. An absolute error of less than 0.3% was claimed at the 14% aluminium level in the sample.

X-ray fluoroescence can be used for determining aluminium in organoaluminium compounds[10]. The use of a chromium target tube, a pulse height discriminator, and a modified sample chamber for sample cooling, together with nylon sample cells, permits the determination of aluminium over the range 0.05–10% in organoaluminium compounds.

B. Determination of Carbon and Hydrogen

Carbon, hydrogen, halogens, and aluminium in reactive or non-reactive organoaluminium compounds may be determined by burning 3–12 mg of the sample in an open capillary tube in an atmosphere of argon or nitrogen[11]. Aluminium is determined gravimetrically as alumina in an oxygen medium and carbon, hydrogen, and halogens are determined by the procedure described by Korshun *et al.*[12]. An error of less than ±0.5% is claimed with trialkylaluminium compounds. A modification to the conventional combusion method for carbon and hydrogen for the analysis of lithium and aluminium hydrides has been described[13]. Recoveries of carbon and hydrogen were 99% or better and agreed closely with results obtained by hydrolysis procedures.

C. Determination of Aluminium-bound Halogens

Aluminium-bound halogens may be determined by addition of an aqueous solution of nitric acid to a cooled hydrocarbon solution of the organoaluminium[14]. This converts aluminium-bound halogens into the halogen hydracid, which is extracted into the aqueous phase and titrated potentiometrically or directly with N/15 or N/100 silver nitrate:

$$R_2AlX + 3H_2O \longrightarrow HX + 2RH + Al(NO_3)_3 \quad \text{X = halogen} \tag{1}$$

If a low concentration of aluminium-bound bromine is to be determined, an alternative procedure is available[14] for the determination of this element in amounts down to 50 ppm with an accuracy of ±1% of the determined value. Aluminium-bound iodine interferes in the determination of bromine but chlorine does not. The sample, diluted with isooctane, is quantitatively decomposed at 50°C with aqueous sulphuric acid, converting bromine to the ionic form, which is then extracted from the organic phase with dilute aqueous sulphuric acid:

$$2R_2AlBr + 3H_2SO_4 \longrightarrow 2HBr + 4RH + Al_2(SO_4)_3 \tag{2}$$

The bromide content of the extract is then determined by a volumetric procedure in which the buffered bromide solution is treated with excess of sodium hypochlorite to oxidize bromide to bromate. Excess of hypochlorite is destroyed with sodium formate. Acidification of the test solution followed by addition of excess potassium iodide liberates an amount of iodine equivalent to the bromide content of the sample. The iodine is titrated with sodium thiosulphate. A method for the specific determination of iodine is also available[14]. It is applicable at concentrations as low as 40 ppm of iodine and also for iodine at the macro level. In the procedure for determining iodine the diluted sample is decomposed quantitatively at 50°C by the addition of hydrochloric acid. To the acid extract is added standard potassium iodate solution which converts iodide to iodine monochloride:

$$R_2AlI + 3HCl \longrightarrow HI + 2RH + AlCl_3 \quad (3)$$

$$KIO_3 + 2KI + 6HCl \longrightarrow 3KCl + 3ICl + 3H_2O \quad (4)$$

The end-point, which occurs with the complete conversion of iodide to iodine monochloride, is indicated by the disappearance of the violet iodine colour from a chloroform layer present in the titration flask. The silver nitrate titration obtained in the chlorine determination would include iodide if any were present. If the iodine analysis of the material is available, the iodine-corrected chlorine analysis may be calculated from the following equations:

$$\%\ \text{chlorine (wt./wt.) (corrected for iodine)} = \left(\frac{T_A \times f_A}{w} - \frac{2 \times T_B \times f_B}{W}\right) \times 3.5456 \quad (5)$$

$$\text{equivalent chlorine/100 g alkyl} = \left(\frac{T_A \times f_A}{w} - \frac{2 \times T_B \times f_B}{W}\right) \times 0.100\ \text{g} \quad (6)$$

where

Chlorine determination:
- T_A = titre of silver nitrate;
- f_A = normality of silver nitrate;
- w = grams of alkyl represented by the aliquot of decomposed alkyl solution employed in a silver nitrate titration.

Iodine determination:
- T_B = titre of f_B molar potassium iodate;
- f_B = molarity of potassium iodate;
- W = grams of alkyl employed per iodine determination.

N.B. These corrections can be ignored if the iodine content of the sample is less than 0.5%.

D. Determination of Aluminium-bound Alkyl Groups up to Butyl and Hydride Groups

Lower alkyl and hydride groups in organoaluminium compounds may be determined by reacting a known weight of sample at a low temperature with 2-ethylhexanol in a specially constructed nitrogen- or helium-filled gasometric system[15,16]. Upon alcoholysis, each alkyl group liberates 1 mol of an alkane gas and each hydride group liberates 1 mol of hydrogen, as follows:

$$>AlC_nH_{2n+1} + HOCH_2{-}\underset{\displaystyle C_2H_5}{\underset{|}{CH}}{-}CH_2CH_3 \longrightarrow >AlOCH_2{-}\underset{\displaystyle C_2H_5}{\underset{|}{C}}HCH_2CH_3 + C_nH_{2n+2} \tag{7}$$

$$>Al{-}H + HO{-}CH_2{-}\underset{\displaystyle C_2H_5}{\underset{|}{CH}}{-}CH_2{-}CH_3 \longrightarrow >Al{-}OCH_2{-}\underset{\displaystyle C_2H_5}{\underset{|}{C}}HCH_2CH_3 + H_2 \tag{8}$$

The alkyl and hydride contents of the samples are then calculated from the amount of gas evolved from a known weight of sample and from the composition of the gas withdrawn from the system at the end of the analysis, obtained by mass spectrometry and other methods. Gas recoveries obtained by this procedure were lower than expected from the composition of the samples analysed. Although originally attributed to partial solution of the evolved alkane–hydrogen mixture in the 2-ethylhexanol reagent[16], the principle cause is incomplete reaction of alkyl and hydride groups with the alcoholic reagent[17,18]. Thus, appreciably higher gas yields were obtained when sample decomposition was effected using a mixture of *n*-hexanol and monoethylene glycol or a mixture of water and monoethylene glycol than when 2-ethylhexanol was used. It can be seen in Table 2 (experiments 1 and 2) that a higher gas yield is obtained when a 4:1 mixture of anhydrous *n*-hexanol and anhydrous monoethylene glycol is used to decompose triethylaluminium, instead of anhydrous *n*-hexanol alone[17]. Aluminium-bound ethyl and hydride groups react very vigorously with water (reaction 9). It is not feasible to add water directly to triethylaluminium as the ensuing reaction is extremely vigorous, even when carried out at −70°C. Also, an undesirable 'fissioning' side-reaction (reaction 10), which converts alkyl groups to ethylene and

$$>AlC_2H_5 + H_2O \longrightarrow C_2H_6 + >AlOH \tag{9}$$

$$>AlC_2H_5 + H_2O \longrightarrow C_2H_4 + H_2 + >AlOH \tag{10}$$

hydrogen instead of ethane, occurs to some extent when aqueous reagents or aqueous monoethyleneglycol reagents are added directly to neat organoaluminium compounds of low molecular weight[17]. Fission does not occur, however, when anhydrous *n*-hexanol is added to neat ethylaluminium compounds (see Table 2, experiment 1), the trace of hydrogen in the gas obtained in this experiment being due to a small amount of aluminium hydride in the sample. When 20% aqueous sulphuric acid is added to the reaction product obtained in experiment 1, a further appreciable liberation of gas takes place very smoothly and, as can be seen from the results in Table 2, no fissioning occurs. Thus, by successively reacting triethylaluminium with anhydrous *n*-hexanol and then with an aqueous reagent a maximum gas yield is achieved smoothly and without the 'fissioning' side-reaction.

A one-stage procedure for determining alkyl groups up to butyl and hydride groups has been described[19]. A cyclohexane solution of the organoaluminium sample is

TABLE 2. Gas yields obtained by the use of various reagents in the alcoholysis and/or hydrolysis of triethylaluminium

Experiment No.	Alcoholysis and/or hydrolysis reagent and decomposition technique employed	Gas yield at S.T.P. (ml evolved/g sample)				
		Hydrogen	Ethane	Ethylene	Butane	Total
1	Added 3 ml of anhydrous hexanol to sample at −30°C, then slowly heated to 100°C	0.5	324	0	34	358.5
2	As above, using 3 ml of *n*-hexane (80 vol-%)–monoethylene glycol (20 vol.-%) mixture	0.5	337	0	32	369.5
3	Added 1.5 ml of hexanol to sample at −30°C, then 1 ml of 20 wt.-% aqueous sulphuric acid at −30°C. Slowly heated to 100°C	0.4	351	0	36	387.4
4	Added 1.5 ml of hexanol to sample at −30°C then heated to 100°C. Added 1 ml of 20 wt.-% aqueous sulphuric acid at 100°C	0.4	351	0	36	387.4

injected directly on to a small reagent-containing pre-column, usually containing lauric acid on a porous carrier (Sil-O-1,50–80 mesh), which is connected directly prior to the gas chromatographic column. Alkyl and hydride groups are decomposed by lauric acid as follows:

$$>AlC_2H_5 + RCOOH \longrightarrow >AlOOCR + C_2H_6 \tag{11}$$

$$>AlC_4H_9 + RCOOH \longrightarrow >AlOOCR + C_4H_{10} \tag{12}$$

$$>AlH + RCOOH \longrightarrow >AlOOCR + H_2 \tag{13}$$

The alkane gases and hydrogen are then swept on to the chromatographic column by the carrier gas and are resolved and determined. To determine the total volume of gas evolved a suitable marker compound (*n*-pentane) is added to the organoaluminium sample prior to injection into the gas chromatograph.

In a further procedure the organoaluminium sample is exposed to water vapour at sub-atmospheric pressures and the generated gases are analysed[20]. A simple, more rapid, hydrolysis method has been developed to meet the requirements of a routine quality control check and yet retain the same degree of precision and accuracy of the more time-consuming procedures[21]. A measured amount of paraffin oil is first added to a pre-evacuated small, crown-capped, thick-walled borosilicate glass bottle to serve as a diluent and reaction medium for the metal alkyl sample. The metal alkyl is then injected via a syringe, through the rubber liner of the crown cap, thoroughly mixed with the oil and immediately hydrolysed to its corresponding alkanes and hydrogen by addition of dilute hydrochloric acid at room temperature:

$$R_yAlX_{3-y} + 3HOH \longrightarrow Al(OH)_3 + yRH\uparrow + (3-y)HX \tag{14}$$

where X = Cl, Br, I, H and y = 1,2,3. With proper technique, the degree of alkene formation with this method is minimized to the same degree as in the previous methods. The major advantages claimed for this new technique over the former methods are that the hydrolysis is carried out at room temperature in a simply designed hydrolysis flask, a four-fold smaller sample size is required, and the time required to complete the hydrolysis is reduced ten-fold for the hydrolysis of any aluminium (or zinc) alkyl containing up to five carbon atoms with no decrease in accuracy or precision. Table 3 compares results obtained for a range of aluminium alkyls using both the Stauffer[20] and the rapid[21] methods. The methods agree well within experimental error. The agreement achieved using the wide variety of products demonstrates that the rapid method is a completely suitable replacement of the Stauffer method, as none of the values fell out of the expected range of values for each component. The rapid method has also been compared with the alcoholysis hydrolysis procedure[17,18]. Again, good to excellent agreement is claimed and in much less time.

A hydrolysis procedure for the determination of ethyl groups in triethylaluminium in which a correction is made for the solubility of ethane in the solvent has been described[22].

An *N*-methylaniline method for the rapid, accurate determination of aluminium-bound hydride groups in organoaluminium compounds is particularly useful in the case of organoaluminium samples which contain colloidal aluminium that cannot be removed from the sample by filtration or centrifugation[23]. Aluminium metal interferes in the alcoholysis/hydrolysis and the lauric and decomposition methods by reacting with these reagents to produce hydrogen, thereby leading to falsely high aluminium-bound hydride determinations. In this procedure (equation 15) an excess of anhydrous *N*-methylaniline is added to a known weight of the organoaluminium compound at -40°C, in a nitrogen-filled gasometric apparatus. The temperature of the reaction mixture is then slowly increased to 20°C. The volume of hydrogen evolved under these conditions is equivalent to the dialkylaluminium hydride content of the sample.

$$AlR_2H + HN(CH_3)(C_6H_5) \longrightarrow H_2 + (CH_3)(C_6H_5)N{-}AlR_2 \quad (15)$$

The method can be used to determine the concentration of hydride groups in all types of dialkylaluminium hydride-containing samples. Dialkylaluminium hydride contents between 1 and 60% may be determined by this procedure. Halogen and alkoxide substituents and higher alkyl groups do not interfere. The accuracy of individual determinations is of the order of ±5% of the determined value.

Table 4 shows results obtained by applying this method to a range of triethylaluminium–diethylaluminium hydride mixtures containing between 10 and 60% of the latter component. These results are in good agreement with dialkylaluminium hydride contents obtained by the alcoholysis–hydrolysis procedure[17] described earlier in this section. The *N*-methylaniline method takes less than 1 h, compared with 4–5 h for the alcoholysis hydrolysis procedure.

E. Determination of Higher Aluminium-bound Alkyl Groups

The gasometric alcoholysis–hydrolysis procedure[17] described earlier for the determination of alkyl groups up to butyl is not applicable to the determination of aluminium-bound alkyl groups in the C_5–C_{10} range which, upon hydrolysis, produce liquid alkanes. Hydride groups in these compounds can, of course be determined by alcoholysis–hydrolysis or lauric acid decomposition[19] or *N*-methylaniline procedures.

:ABLE 3. Comparison of hydrolysis methods using hydrolysis gas products
a) From the hydrolysis of R_3Al:

	Trimethyl-aluminium		Triethylaluminium		Triisobutyl-aluminium	
:omponent (wt.-%)	Stauffer[20]	Rapid	Stauffer[20]	Rapid	Stauffer[20]	Rapid
Aethane as Me_3Al	99.9	99.9	0.3	0.3	0.1	0.1
:thane as Et_3Al	0.0	0.0	93.5	93.5	0.2	0.2
ropane as Pr_3Al	0.0	0.0	0.4	0.4	0.2	0.2
sobutane as i-Bu_3Al	0.0	0.0	0.7	0.8	96.0	96.1
-Butane as n-Bu_3Al	0.0	0.0	4.5	4.4	0.2	0.2
sobutylene as isobutylene	0.0	0.0	0.0	0.0	2.5[a]	2.4[a]
Iydrogen as AlH_3	0.1	0.1	0.6	0.6	0.8	0.9
Jumber of runs averaged	2	3	2	6	2	6
Vt.-% Al, calculated	37.5	37.5	23.5	23.5	13.9	13.9
Vt.-% Al, found (by EDTA titration)	37.4		23.4		13.8	

Isobutylene actually exists in samples of triisobutyaluminium.

b) From the hydrolysis of R_yAlX_{3-y}:

	Diethylaluminium chloride		Diethylaluminium iodide		Ethylaluminium dichloride	
Component (mol-%)	Stauffer[20]	Rapid	Stauffer[20]	Rapid	Stauffer[20]	Rapid
Methane	0.3	0.3	0.5	0.6	0.0	0.1
Ethane	97.7	97.8	97.1	97.0	99.5	99.4
Propane	0.1	0.1	0.2	0.2	0.0	0.0
Isobutane	0.3	0.2	0.2	0.2	0.4	0.4
n-Butane	1.5	1.5	1.8	1.8	0.1	0.1
Hydrogen	0.1	0.1	0.2	0.2	0.0	0.0
Number of runs averaged	6	6	2	2	2	2
Wt.-% Al, calculated	22.2	22.2	12.5	12.5	21.3	21.3
Wt.-% Al, found	22.0		12.5		20.9	

(c) From the hydrolysis of miscellaneous metal alkyls:

	Diethylzinc		Diisobutyl-aluminium hydride		Ethylaluminium sesquichloride	
Component (mol-%)	Stauffer[20]	Rapid	Stauffer[20]	Rapid	Stauffer[20]	Rapid
Methane	0.5	0.5	0.4	0.4	0.1	0.1
Ethane	97.1	97.1	0.3	0.3	99.4	99.4
Propane	0.2	0.2	0.3	0.3	0.0	0.0
Isobutane	0.3	0.3	65.9	54.7	0.1	0.1
n-Butane	1.9	1.8	0.2	0.2	0.4	0.4
Hydrogen	0.0	0.1	32.9	33.1	0.0	0.0
Number of runs averaged	2	4	9	32	3	3
Wt.-% metal, calculated	53.1	53.1	19.0	19.0	21.8	21.8
Wt.-% metal, found	52.6		18.8		21.6	

TABLE 4. Comparison of the alcoholysis–alcoholysis procedure and the *N*-methylaniline procedure for the determination of hydride groups in triethylaluminium

Sample No.	$AlEt_2H$(% wt./wt.) *N*-methylaniline procedure	Alcoholysis–hydrolysis procedure
1	1.2	1.1, 1.2
2	2.2	1.2, 1.6
3	7.6	8.2, 8.3
4	12.4	12.3
5	13.0	13.0
6	22.6	21.3, 21.3
7	24.3	23.5
8	32.1	28.9, 31.9
9	60.7	57.5

C_5–C_{10} alkyl groups can be determined by conversion to the corresponding alkane with a proton-donating reagent[24]. The organoaluminium sample is decomposed smoothly without loss of liquid paraffins at $-60°C$, under nitrogen, by the addition of glacial acetic acid dissolved in ethylbenzene:

$$>AlR + CH_3COOH \longrightarrow >AlOOCR + RH \qquad (16)$$

The cold ethylbenzene solution is then contacted with aqueous sodium hydroxide to extract aluminium acetate and excess of acetic acid and provide an ethylbenzene solution of the C_5–C_{10} alkanes, which can be directly injected into a gas chromatographic column.

F. Determination of higher alkyl and alkoxide groups

Alkyl and alkoxide groups up to C_{20} can be determined by hydrolysis to produce the corresponding alkanes and alcohols, respectively[24]:

$$>AlR + H_2O \longrightarrow >AlOH + RH \qquad (17)$$

$$>AlOR + H_2O \longrightarrow >AlOH + ROH \qquad (18)$$

This is achieved by diluting the organoaluminium sample with light petroleum and refluxing the solution with an aqueous solution of sulphuric acid and sodium sulphate. The alkanes and/or alcohols obtained as hydrolysis products are recovered for analysis by evaporating the light petroleum extract to dryness. The mixture so obtained usually has a fairly wide carbon number range and it is convenient to carry out a detailed analysis of the individual alcohols and hydrocarbons in these mixtures by gas chromatography and of total alcohols by catalysed acetylation. Figure 3 shows a typical GLC trace obtained in the analysis of a mixture of partially oxidized C_6 to C_{16} aluminium alkyls. In the catalysed acetylation procedure for the determination of higher alcohols in the alcohol–hydrocarbon sample obtained by hydrolysis of the organoaluminium sample, a portion of the extract is reacted with a reagent consisting of acetic anhydride and a perchloric acid acetylation catalyst dissolved in pyridine[25]. Following the reaction period the mixture is diluted with aqueous pyridine and titrated with standard sodium hydroxide solution to the cresol red–thymol blue mixed indicator end-point. Alternatively, the mixture may be titrated potentiometrically

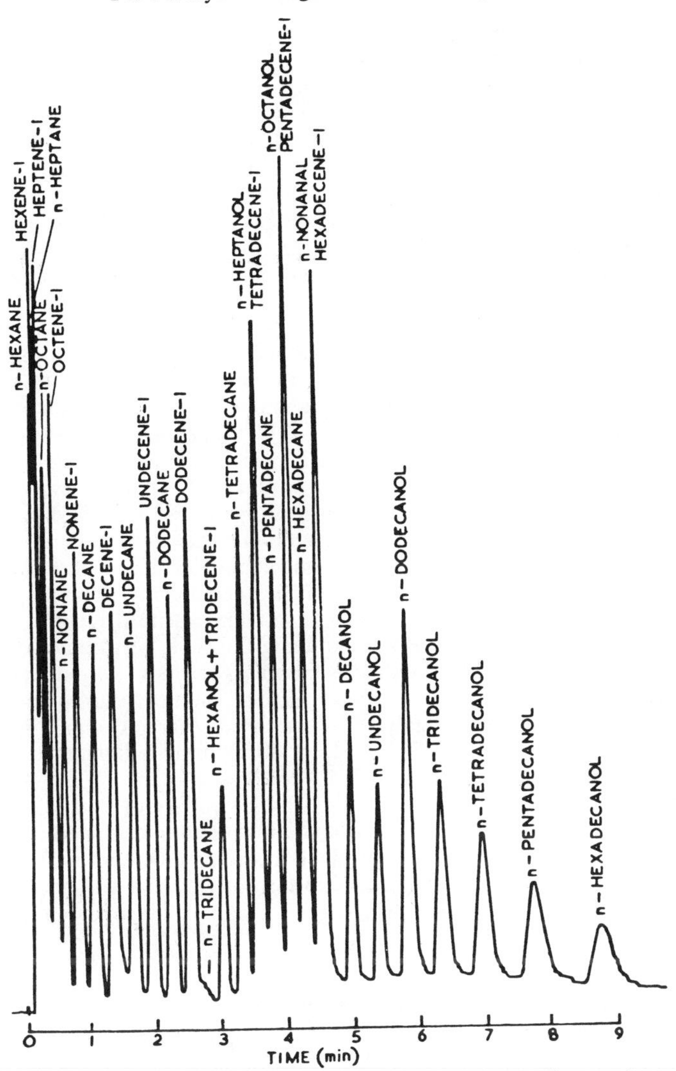

FIGURE 3. Gas chromatography, using polyethylene adipate column with temperature programming and flame-ionization detector, of a mixture of C_6 to C_{16} alkanes, alkenes, and *n*-alcohols.

using an electrode system. The analytical results are calculated from the measured consumption of acetic anhydride, after correcting for any small amount of 'free organic acidity' in the sample. The accuracy of the procedure is of the order of ±1.5% of the determined result.

G. Determination of Aluminium-bound Alkoxide Groups up to Butoxy

Trialkylaluminium or dialkylaluminium halide compounds containing alkyl groups up to butyl are often contaminated with small amounts of alkoxide groups produced by oxygen contamination during manufacture or subsequently. Alkoxide groups in reactive low molecular weight organoaluminium compounds may be determined by reacting the organoaluminium compound with dilute aqueous acid to produce the corresponding alcohols in quantitative yield[26]. Even the most reactive organoaluminium compounds could be smoothly decomposed without loss of the alcohol, when a dilute solution of glacial acetic acid dissolved in toluene was slowly added under nitrogen to the diluted sample maintained at −60°C. The subsequent addition of an aqueous solution of sodium hydroxide to this mixture dissolves the precipitated aluminium salts and extracts the alcohol quantitatively into the aqueous phase, from which it can be isolated from electrolytes by steam distillation[27] and then determined colormetrically in the distillate by the cerium(IV) ammonium nitrate method[28,29]. At the 0.1% alkoxide level the accuracy is within ±5% of the actual alkoxide content of the sample. Higher concentrations of alkoxide in the range 10–60% can be determined with an accuracy of ±1%. Ethoxide groups in triethylaluminium can be determined by crushing an ampoule containing about 0.1 g of organoaluminium compound under aqueous mixture of a potassium dichromate and sulphuric acid[30]. The mixture is refluxed and then cooled. Unconsumed potassium dichromate is determined iodimetrically. Three procedures have been described for the determination of isobutoxide groups in triisobutylaluminium[31]. In one the sample is hydrolysed under an inert gas to produce isobutyl alcohol, which is then reacted with sodium nitrite to produce isobutyl nitrite. The latter is diazotized with sulphanilic acid and coupled with 1-naphthylamine, which can be determined spectrophotometrically. The second procedure for determining isobutoxide groups utilizes the cerium(IV) ammonium nitrite reagent[31]. The error in this method is less than ±3% when determining 0.09–1% of isobutoxide groups in triisobutylaluminium. The third method is based on reaction with ethanol under an inert atmosphere[32]. The ethanol–isobutyl alcohol mixture thus obtained is centrifuged to remove alumina and analysed by gas chromatography on a column consisting of 20% tritolyl phosphate on Celite 545 at 80°C, using helium as the carrier gas. The error is claimed to be ±10%.

Ethoxydiethylaluminium may be analysed by hydrolysis to ethanol, followed by oxidation with excess of potassium dichromate in sulphuric acid medium and subsequent coulometric titration of the unconsumed dichromate with iron(II) electrogenerated from iron(II) ammonium sulphate with ferroin as visual end-point indicator[33]. The coefficient of variation for 0.04–0.16 mg of ethanol in aqueous solution is 2.4% or less.

H. Determination of Aluminium-bound Amino Groups

Aluminium-bound amino groups can be determined by procedures based on hydrolysis or ethanolysis as described below. Both of these methods are free from interference by aluminium-bound alkyl, alkoxide, hydride, halogen, or S-alkyl groups in the sample.

In the hydrolysis method for determining Al-NH_2 groups[34] a suitable weight of the neat or dilute sample is dissolved in isooctane under nitrogen and cooled to 50°C and then decomposed with dilute hydrochloric acid:

$$R_2AlNH_2 + 4HCl(aq) \longrightarrow NH_4Cl + 2RH + AlCl_3 \qquad (19)$$

The ammonium chloride is extracted with water from the isooctane phase. Steam distillation of this extract in the presence of excess of sodium hydroxide provides an amount of ammonia proportional in amount to the $AlNH_2$ content of the original organoaluminium sample. Higher concentrations of ammonia in the steam distillate are determined by a titrimetric procedure. Traces of ammonia are determined by the colorimetric indophenol blue method. This procedure is more suitable for higher molecular weight organoaluminium compounds which do not react too vigorously with water. An alternative method[35] for determining aluminium-bound amino groups utilizes ethanol as the reagent:

$$R_2AlNH_2 + C_2H_5OH \longrightarrow R_2AlOC_2H_5 + NH_3 \qquad (20)$$

This procedure is suitable for the analysis of more reactive types of organoaluminium compounds. The ammonia produced is determined by conventional spectrophotometric or titrimetric procedures.

I. Determination of Aluminium-bound Thioalkoxide Groups

SR groups containing alkyl groups up to $C_{20}H_{41}$ can be determined in amounts down to 1% in the sample by a procedure[36] involving hydrolytic decomposition of the sample at −60°C followed by argentimetric determination of the thiol produced.

III. ORGANOANTIMONY COMPOUNDS

A. Determination of Antimony

Several methods for the determination of antimony in organoantimony compounds have been described[37–41]. In one the organoantimony compound is burnt with metallic magnesium to convert antimony to magnesium antimonide[37,38]. This is then decomposed with dilute sulphuric acid to produce stibine, which is absorbed in 6 N hydrochloric acid containing sodium nitrite. The resulting hexachloroantimonic acid can be determined colorimetrically after having been converted to a blue coloured compound by reaction with methyl violet. The blue colour is extracted with toluene for spectrophotometric evaluation. Nitrogen, phosphorus, and arsenic do not interfere. An alternative procedure uses Rhodamine B (C.I. Basic Violet 10)[39]. The decomposition of the sample is best effected by the use of concentrated sulphuric acid and potassium sulphate, followed by the addition of 30% hydrogen peroxide. The solution is adjusted to 6 N with respect to hydrochloric acid and antimony is extracted with diisopropyl ether. Colour is developed by the addition to the ether phase of 0.02% Rhodamine B in 1 N hydrochloric acid. Compounds containing trivalent antimony which are unstable in oxygen or moist air may be stabilized by exposure to sulphur for 8–48 h *in vacuo* and the resulting compounds are examined by pyrolysis[41].

IV. ORGANOARSENIC COMPOUNDS

A. Determination of Arsenic

Various techniques have been applied to the determination of arsenic in organoarsenic compounds. These include oxygen flask combustion, digestion with mineral acids, and fusion with magnesium.

1. *Oxygen flask combustion*[42–50]

Since arsenic attacks platinum sample holders, many workers recommend a silica spiral, although combustion with a silica spiral is never as satisfactory as with platinum[42,43]. A modified silica holder has been described[45] which, it is claimed, is less prone to dropping the sample into the absorption solution. Use of a quartz spiral is unsatisfactory because the quartz devitrifies during even a single combustion. Some workers are of the opinion that wet combusion methods are preferable to the oxygen flask method for the determination of organic arsenic[43]. After combustion the arsenite and arsenate formed are absorbed in a suitable solution; commonly a solution of sodium hydroxide is used. Following conversion to arsenic trichloride, this is distilled into sodium hydrogen carbonate solution and is determined iodimetrically. Alternatively, the combustion products may be absorbed in dilute iodine solution in whch trivalent arsenic is quantitatively oxidized to pentavalent arsenic[44]. The arsenic is determined by the molybdoarsenate blue reaction, with hydrazine sulphate as reductant; the excess of iodine from the absorber does not interfere.

Concentrations of arsenic greater than 10 μg are measured on a filter photometer and those of less than 10 μg with a spectrophotometer. The mean error for a single determination is $\pm$0.2%. Alkaline hydrogen peroxide has been used instead of iodine solution in the oxygen flask method to oxidize arsenic to arsenate[47]. The arsenate is titrated directly with standard lead nitrate solution with 4-(2-pyridylazo)resorcinol or 8-hydroxy-7-(4-sulpho-1-napththylazo)quinoline-5-sulphonic acid as indicator. Phosphorus interferes in this method. The precision at the 99% confidence limit is within $\pm$0.67% for a 3-mg sample. A further variant of the oxygen flask method uses sodium acetate as the absorbing liquid[48]. The arsenite and arsenate so produced are precipitated with silver nitrate solution. The precipitate is dissolved in potassium nickelocyanide $[K_2Ni(CN)_4]$ solution and the displaced nickel is titrated with EDTA solution, with murexide as indicator. The average error is within $\pm$0.19% for a 3-mg sample. Halogens and phosphate interfere in the procedure.

Arsenic can be determined after oxygen combustion by precipitation as quinoline molybdoarsenate, which is then reduced with hydrazine sulphate and determined spectrophotometrically at 840 nm[49]. The absolute accuracy is within $\pm$1% for arsenic. Phosphorus interferes in this procedure. Alternatively, precipitation as ammonium uranyl arsentate and subsequent ignition under controlled conditions to triuranium octaoxide may be used[50]. Phosphate and vanadate form similar insoluble ammonium uranyl salts and the method is not applicable to organic compounds containing these elements without prior separation.

Various sample supports have been tried in the hope that they might prove more satisfactory than platinum[50]. For this purpose the stopper of the oxygen flask was

TABLE 5. Recovery of arsenic after oxygen flask combustion with various supporting materials. *o*-Arsanilic acid was the organic compound used in these recovery experiments

Support Material	Arsenic added (mg)	Arsenic recovered (mg)	Obervations
Platinum gauze	11.16	7.52	Platinum pitted and swollen. Good combustions
Glass spiral	11.02	9.98	Poor combustion, smoke and carbon deposits
	14.65	13.26	
Copper spiral	17.32	13.75	Good combustions. Cooper melts
Copper gauze	14.03	7.12	
	14.01	6.92	
Aluminium spiral	14.29	14.28	Moderate combustions. Aluminium melts.
	14.11	13.77	
Stainless-steel gauze	16.02	15.90	Good combustions
	13.92	13.53	
	14.04	13.74	

TABLE 6. Recovery of arsenic from various compounds when steel or aluminium supports were used

Support material	Arsenical compound	Arsenic calculated (mg)	Arsenic found (mg)
Aluminium spiral	Arsenious oxide	16.04	16.08
		15.50	15.60
Stainless-steel gauze	Arsenious oxide	15.76	15.88
Mild-steel gauze	Arsenious oxide	15.71	15.74
Aluminium spiral	*o*-Arsanilic acid	14.29	14.28
		14.11	13.77
		14.08	14.10
		13.87	13.70
Stainless-steel gauze	*o*-Arsanilic acid	16.02	15.90
		13.92	13.53
		14.04	13.74
Mild steel gauze	*o*-Arsanilic acid	14.49	14.49
		14.13	13.85
		15.63	15.23
Aluminium spiral	Acetarsol	11.29	11.28
		12.06	12.04
		11.48	11.04
Mild steel gauze	Acetarsol	11.20	11.18
		11.18	10.81

fitted with a glass hook, so that spirals and gauzes of various materials could be hung from it. Oxidized copper spirals were first used in the hope that the oxide film would assist combustion and prevent reduction to arsenic but, despite excellent combustions, poor results were obtained, again, presumably, because of alloy formation. The use of steel gauzes and aluminium spirals gave much improved results; there were no signs of attack on these materials (although the aluminium tended to melt) and combustion was adequate. Steel spirals gave good combustions and aluminium less so; with aluminium it was necessary to use tissue-paper rather than filter-paper and to get the combustion going well initially. After dissolution of the white sublimate of arsenic oxides, which were formed on the sides of the flask, and subsequent acidification and oxidation to arsenate with bromine water, the arsenic was determined as previously described. The results of these preliminary investigations are given in Tables 5 and 6. Further results obtained with steel and aluminium supports gave satisfactory recoveries of arsenic from *o*-arsanilic acid, arsenious oxide, and acetarsol.

2. *Digestion procedures*[51–58]

Organic arsenicals may be analysed by heating them in a Kjeldahl flask with a mixture of concentrated nitric and sulphuric acids and hydrogen peroxide[51]. After cooling, the mixture is diluted with water and an aliquot is treated with concentrated sulphuric acid and zinc. The arsine evolved passes into a solution in pyridine of silver diethyldithiocarbamate, and the molar absorptivity of the resulting solution is measured at 540 nm. Halogens and sulphur do not interfere. Arsenic in 10-mg amounts of organic compounds may be determined iodimetrically[52]. The method is suitable for the determination of all types of organic arsenic compounds, including those which give low results by the classical wet oxidation methods using sulphuric and nitric acids. The sample is weighed out into a Kjeldahl flask, concentrated sulphuric acid is added, and the solution is set aside for 5 min, warming it if necessary to dissolve any solids. Small amounts of potassium permanganate are added, warming gently after each addition until the dirty green of the solution disappears and the precipitated manganese no longer dissolves. Water and 30% hydrogen peroxide are added and the solution is heated until a clear solution is obtained. Arsenic is then determined iodimetrically.

A digestion procedure for arsenic which is not subject to interference by magnesium, calcium, strontium, barium, cobalt, nickel, zinc, manganese cadmium, copper, and halogens involves heating the sample to fuming in a Kjeldahl flask containing concentrated sulphuric acid, copper sulphate, and concentrated nitric acid[53]. After cooling, further portions of nitric acid are added and the sample is heated to fuming until colourless or pale blue. It is then diluted, heated to boiling point, and a solution of 7 g of barium nitrate in 50 ml of water is added. The solution is then cooled, filtered, and the filtrate neutralized to methyl orange with sodium hydroxide solution. Aqueous 25% nitric acid is added and excess of 0.1 N silver nitrate solution is added, followed by dropwise addition of a concentrated solution of sodium acetate until precipitation is complete. The solution is then made up to volume, filtered, and the excess of silver nitrate back-titrated with 0.1 N ammonium thiocyanate solution. A micro Carius procedure for arsenic in which the sample is digested with fuming nitric acid to form quinquevalent arsenic, which is then determined by iodine titration, yields low results (Table 7, column I)[54,55]. When the digestion temperature and time were increased to 300°C for 10 and potassium chloride was added, all the solutions remained clear upon the addition of potassium iodide and the end-points became sharp[54]. The results (Table 7, column II) were in close agreement with theory.

Chloric acid has been recommended because it oxidizes organic matter smoothly and rapidly at 180°C, and is to be preferred to the more widely used sulphuric acid or sulphuric–nitric acid digestions or alkaline fusions[56]. Excess of chloric acid is easily removed by boiling to leave a perchloric acid solution of inorganic arsenic present in a

TABLE 7. Comparison of percentage of arsenic found by micro Carius method at different temperatures with and without potassium chloride

Sample	Theory	I: Digestion temp. 250°C, digestion time 8 h, KCl absent	II: Digestion temp. 300°C, digestion time 10 h, KCl present
As_2O_3	75.73	71.99	76.04
		72.01	75.66
			75.59
			75.66
$C_6H_5CH_2AsO_3H_2$	34.67	31.12	34.50
		31.63	34.49
$C_6H_5AsO_3H_2$	37.08	36.62	36.91
		34.06	36.88
		35.26	36.97
		35.21	
$HOC_6H_4AsO(OH)_2$	34.36	30.76	34.46
		31.70	34.42
$(C_6H_5)_3As$	24.46	Turbid	24.40
			24.46
			24.51
			24.49
$[(C_6H_5)_3AsCH_3]I$	16.72	Turbid	16.83
			16.72

higher valency state. The accuracy and reproducibility are good. Thus samples of *p*-arsanilic acid (theoretical arsenic content 34.5%) gave an average arsenic content of 34.6%, with a 99% confidence limit of 0.6%[57]. Chloride does not interfere in the procedure. Phosphorus interferes in the micro method by formation of a heteropoly blue similar to that formed by pentavalent arsenic. Phosphorus and arsenic, if present together, may be separated and determined after digestion of the sample with chloric acid[58].

3. *Magnesium fusion*[59–61]

The sample is heated in a sealed tube for 5 min with a mixture (3 + 1) of magnesium and magnesium oxide, which converts all the arsenic into magnesium arsenide. The tube is opened and the contents are decomposed by dilute sulphuric acid. Arsine is evolved and is absorbed in a 0.5% solution of silver diethyldithiocarbamate in pyridine. The colour produced has an absorption maxium at 560 nm and is proportional to concentration up to 20 μg of arsenic in 3 ml of solution. Alternatively, the arsine is oxidized by bromine and determined iodimetrically[61]. Methylated arsenicals have been determined by vapour generation atomic-absorption spectrometry[62].

B. Determination of Carbon

The determination of carbon and arsenic in organic compounds using magnesium fusion and the elemental analysis of organoarsines and organobromarsines have been reviewed[63,64]

C. Determination of Fluorine

A volumetric microdetermination based on oxygen flask combustion for 5–40% of fluorine in organic compounds containing arsenic has been described[65]. The sample (containing 0.2–0.6 mg of fluorine) is burnt in an oxygen combustion flask and the combustion products are absorbed in 5 ml of water. Hexamine and sufficient murexide–naphthol green B (C.I. Acid Green 1)–hexamine (1:3:100) are added and the flask contents are titrated with 0.005 M cerium (IV) sulphate to a green end-point. Methods have been reported for the determination of arsenic in organoarsenic compounds with halogen attached to the arsenic, organoarsenic compounds with halogen attached to the carbon, and in organoarsenic compounds with halogen in the anion[66].

D. Determination of Sulphur

Compounds containing arsenic which are unstable in oxygen or moist air can be stabilized by exposure to sulphur for 8–48 h *in vacuo*. The resulting compounds may be analysed by pyrolysis for sulphur and arsenic[67].

V. ORGANOBERYLLIUM COMPOUNDS

A. Determination of Carbon and Hydrogen

Carbon and hydrogen in dimethylberyllium and in beryllium hydride can be determined by combustion[68]. After weighing into a small tin capsule, the sample is ignited in a tube furnace at 600°C under a stream of pure helium to destroy any stable carbonates and then under a stream of pure oxygen at 1050°C. The resulting carbon dioxide and water are weighed off.

VI. ORGANOBISMUTH COMPOUNDS

A. Determination of Bismuth

Bismuth in pharmaceuticals such as bismuth tribromophenoxide, bismuth salicylate, and bismuth subgallate has been determined complexometrically[69]. The sample is shaken with dilute nitric acid for 5 min and then diethyl ether added with gentle swirling followed by water, and the solution is heated to 60°C. The solution is then cooled to 50°C, ignoring the precipitate, and a 1% trituration of methyl thymol blue in potassium nitrate is added and the solution titrated immediately with 0.1 M EDTA disodium salt to a pale red–violet colour. Then 10% aqueous ammonia is added and this solution titrated further until it becomes yellow.

VII. ORGANOBORON COMPOUNDS

A. Determination of Boron, Carbon, and Hydrogen

A combustion procedure has been used to determine carbon, hydrogen, and boron in organoboranes[70]. Samples containing 1–4 mg of boron are combusted by the standard Pregl-type procedure, using the Brinkman Heraeus microcombustion assembly. Carbon and hydrogen values are obtained in the normal fashion. The residue which remains behind in the boat is transferred to a beaker, dissolved in water, and the boric acid is titrated by the identical pH method, using 7.10 as the critical pH. The effects of adding oxidizing agents and of changes in heating temperature in the microdetermination by furnace methods of carbon and hydrogen in boroxins and other boron-containing compounds have been studied[71]. The best results were obtained with an oxygen flow-rate of 10 ml/min, a stationary furnace temperature of 900°C, and with the addition of about 100 mg of tungstic oxide to the sample, which was then heated to a final temperature of 1000°C. With these conditions, errors in the determination of carbon and hydrogen were reduced to ±0.28% and ±0.18%, respectively for the compounds studied. When organoboron compounds are oxidized by conventional methods low results are obtained for carbon and low or high results are obtained for nitrogen[72]. To reduce errors in the determination of organoboron compounds, oxidation conditions which are satisfactory for oxidizing methane should be used.

Boron has been determined by pyrolysing the organoboron compound in a stream of pure hydrogen containing methanol vapour, and burning the gaseous mixture at the end of a quartz tube (acting as a blowpipe) in a stream of oxygen[73]. The water-vapour formed contains the bulk of the boron and, after condensation, is collected in a flask containing a known volume of water that will absorb the boron. The Carius nitric acid oxidation procedure for the micro-determination of boron in organoboranes has been studied (Table 8)[74]. The standard deviation is ±0.33% absolute. A majority of the analytical results were slightly above the theoretical values, indicating a possible bias in the method. This may be due to leaching of boron from the borosilicate glass Carius tubes during the oxidation.

Boron and caron in alkyldecaboranes and related compounds have been determined by oxidation with alkaline potassium persulphate, followed by potentiometric titration of boric acid in the presence of mannitol[75]. Carbon contents are determined by a modification of the Van Slyke wet combustion technique. Although the method is limited to compounds which are non-volatile at room temperature it is rapid, does not require elaborate apparatus, and is usually accurate within 1.0 and 0.5% absolute for boron and carbon contents, respectively. The methods are readily applicable to the analysis of organoboron compounds containing nitrogen. The chief limitation of the

TABLE 8. Microdetermination of boron by the Carius method

Compound	Boron (%) Calculated	Found	No. of determinations
Dimethylamineborane, $(CH_3)_2NHBH_3$	18.36	18.66 ± 0.09	4
Sodium tetraborate, $Na_2B_4O_7$	21.50	21.69 ± 0.13	2
Decarborane, $B_{10}H_{14}$	88.46	89.49 ± 0.23	2
Isopropylamineborane, $C_3H_7NH_2BH_3$	14.83	15.24 ± 0.04	2
Trimethylamineborane, $(CH_3)_3NBH_3$	14.83	15.25 ± 0.02	2
Morpholineborane, $C_4H_9NOBH_3$	10.72	10.76 ± 0.02	2
Chlorodecaborane, $B_{10}H_{13}Cl$	69.02	69.74 ± 0.33	3
Bisacetonitriledecaborane, $B_{10}H_{12}(CH_3CN)_2$	53.46	53.17 ± 0.02	2

potassium persulphate oxidation method is that it cannot be used to analyse steam-volatile compounds. Thus, whereas decaborane and its alkyl derivatives dissolve rapidly during the oxidation, more highly alkylated derivatives of pentaborane, for example tetraethylpentaborane, are steam-volatile and remain unchanged. None of the compounds examined resisted oxidation by Van Slyke combustion fluid, but the method is limited to the analysis of compounds with a low vapour pressure at room temperature. A more complicated apparatus would be required to prevent loss of volatile material during preliminary evacuation of the system. Methods for determining boron in organoboron compounds based on oxidation with alkaline potassium persulphate have been compared with methods based on oxidation using (a) sodium peroxide, (b) alkaline hydrogen peroxide, and (c) trifluoroacetic acid[76,77]. Oxidation with sodium peroxide followed by titration with sodium hydroxide using mannitol as indicator gave an error not exceeding 1%. Oxidation with sodium peroxide or with alkaline hydrogen peroxide, followed by titration with barium hydroxide, gave errors as low as 0.1%.

Boron in borohydrides and organoboron compounds can be determined by oxidation with trifluoroperoxyacetic acid[78]. The sample is placed in a test-tube, methyl cyanide is added, a reflux condenser is fitted, and the tube is cooled in an ice-bath for a few minutes. Trifluoroperoxyacetic acid is cautiously added through the top of the condenser and, after any initial reaction has subsided, the tube is heated in a boiling-water bath for 5–60 min depending on the stability of the sample. When decomposition is complete the solution is washed into water and boiled gently for a few minutes before titrating the boric acid in the usual manner, after its conversion into mannitoboric acid.

Boron can be determined by combustion of micro samples of organoboron compounds[79]. The boric oxide formed by the combustion is quantitatively removed by refluxing water in the combustion tube and the resulting boric acid solution is titrated with standard sodium hydroxide with mannitol as a complexing agent with an initial pH of 8.4. Best results are obtained when the boron content of the solution is between 0.8 and 1.5 mg. Alternatively, the sample, covered with liquid paraffin, may be burnt in a Parr high-pressure oxygen bomb containing water[80]. An aliquot is brought to pH 4.8 and heated under reflux for 10 min, to remove carbon dioxide. While the solution is cooling a stream of pre-purified nitrogen is passed through by means of a filter-stick. The pH is adjusted to 5.15 and 10 ml of 10% aqueous mannitol solution are added, and the mixture is titrated with 0.05 N sodium hydroxide. The exact end-point (pH 7.5–8.0) is located by the method of differencing. A platinum-lined Parr bomb has also been used[82]. The sample is placed in the sample holder, water added, and the bomb fired in the inverted position. Combustion in an Inconel bomb at high temperature and pressure followed by separation of the products (boric oxide, carbon dioxide, and water) by conventional techniques has been applied successfully to several trialkylborates and -boranes[82]. For the carbon, hydrogen, and boron determinations, the precisions are ±0.11%, ±0.07%, and ±0.07 absolute, and the accuracies ±0.22%, ±0.12, and ±0.1% absolute, in the ranges 60–88%, 11–16%, and 3–11%, respectively. Since the method accurately determines all three elements it is not subject to the errors involved in only analysing two of the three elements and obtaining the concentration of the third element by a difference calculation. The method should be applicable to the analysis of all types of compounds containing boron, carbon, and hydrogen, if sufficient hydrogen is present to form water of hydration for all the boric oxide produced in the combustion reaction. It offers an easy and safe method of analysis for spontaneously flammable compounds such as boron hydrides and their hydrocarbon derivatives. Decomposition by fusion with sodium hydroxide in a nickel bomb at 800–850°C followed by a spectrophotometric determination of boron in the combustion residue may also be used[83]. Boron in the extract is determined spectrophotometrically at 415 nm with azomethine H[4-hydroxy-5-(salicylidene amino)-naphthalene 2,7-disulphonic acid]. This procedure has been applied to the determination of between 3 and 75% of boron in milligram samples of barenes (carboranes), dicarbaundecaboranes and their related compounds (containing, in addition to C, H, and B, appreciable amounts of N, Na, Si, P, Cl, Fe, Ni, Ge, As, Br, Sn, Cs, or Hg).

The oxygen flask combustion technique has been successfully applied to the determination of boron in organoboron compounds. The sample, together with sucrose, is burned in oxygen and the products are absorbed in water[84]. The solution is transferred with methanol to the cathode compartment of a coulometric titration cell and saturated sodium nitrate solution is added to both cathode and anode compartments. Mannitol is added, and the boric acid is titrated to the potentiometric end-point. If chlorine is present, a small amount of 30% hydrogen peroxide solution is added to the flask before combustion. Recoveries of boron are good except with very refractory materials such as boron carbide. Alternatively, the sample may be combusted over sodium hydroxide, then the solution is neutralized and the boron–mannitol complex is formed and determined by titration with standard 0.02 N sodium hydroxide[85]. The sample may with advantage be mixed with finely powdered potassium hydroxide before combustion[86].

An aspirating burner has been applied to the determination of boron in organoboron compounds[87,88]. The sample is placed in a flask containing sulphuric acid through which passes a rapid stream of nitrogen which leads into an oxy-hydrogen flame. When the sample has been decomposed and any water present driven off, methanol is added and the flask is heated to 70–80°C. On passage through the flame, methyl borate and other compounds are decomposed into boric oxide, water, and carbon dioxide. The boric oxide is absorbed in water and determined volumetrically. Recoveries of boron were within ±0.1 mg of the theoretical for H_3BO_3, $Na_2B_4O_7$, and sodium and potassium tetraphenylboron. Flame photometry may be used for the determination of boron in amounts down to 0.1% at 519.5 nm by volatilizing the compound in an organic solvent[89]. The accuracy and precision are of the order of 1–2% of the amount present. The method has been applied to the determination of boron in lubricating oils. Since boron emission in flame photometry depends not only on boron content, but also on the moleuclar structure of the sample, it is best to decompose the sample with nitric acid to boric oxide[90]. The coefficient of variation was 0.67%. Alternatively, the sample may be heated with a length of sodium wire in a nickel crucible at 400°C for 30 min[91]. The residue is then made up with water and the solution passed down a column of Amberlite IR-120 (H^+ form). The percolate is diluted with water and its boron content determined flame photometrically at 518 nm. All types of organoboron compounds could be determined by this method. No interference occurs from any nitrogen present in the sample. An absolute spectrographic method has been used for the determination of small amounts of organic boron in mixtures of terphenyls[92]. No preliminary treatment of the sample is required, and analyses are carried out using a large quartz spectrograph in a matrix graphite and terphenyls (9:1). In the spectrochemical analysis of methyl borate, the distillate was led from a micro-distillation apparatus directly through a bored electrode into the spark or arc gap[93]. Solutions of methyl borate were sprayed directly into the arc or spark gaps. Neutron absorption has been applied to the determination of boron in nitrogen-containing organoboron compounds[94]. Calibration graphs were prepared by plotting the number of slow neutrons passing through the sample against boron content. Down to 0.05%, boron could be determined with a relative error not exceeding ±8% in about 20 min.

B. Determination of Chlorine

No doubt oxygen flask combustion techniques[95], which are capable of determining organically bound boron, can be applied with suitable modifications, to the determination of halogens in organoboron compounds, as indeed can other types of combustion techniques described earlier. An accurate and reproducible method has been described for the determination of chlorine bound to boron in organoboron compounds containing boron—chlorine and carbon—chlorine bonding[95]. Analysis for chlorine bound to boron mixtures containing compounds which possess both chlorine bound to boron and to carbon necessitates a method which distinguishes each type of chlorine. Although procedures are available for the determination of chlorine in organoboron compounds (Parr bomb–sodium peroxide fusion, Volhard, and Carius methods), the values obtained give the total chlorine content. Aqueous hydrolysis of chloro-2-chlorovinylboranes, with subsequent titration of hydrochloric acid with either base or silver nitrate, fails to differentiate between chlorine bound to boron and to carbon, because these compounds break down in aqueous medium (above pH 3) to form hydrochloric acid, orthoboric acid, and acetylene. The hydrolysis of chlorine bound to carbon proceeds at a much slower rate than that for the chlorine bound to boron. Various attempts were made to stabilize the 2-chlorovinylboric acid produced upon hydrolysis of the carbon—chlorine bond[96]. When it was dissolved in benzene and titrated with a solution of potassium methylate in benzene containing a small amount of methanol, only chlorine bound to boron was determined. The titration was carried out to a thymol blue end-point. The analysis must be rapid since on standing some breakdown of the 2-chlorovinylboric acid occurred.

This procedure was found to be effective for chlorovinylboranes, but if sample mixtures contained chloroethylboranes, the method was inaccurate. Under these conditions all of the chlorine bound to boron in the chloroethylboranes did not dissociate. It is known[96] that chloroethylboranes can be hydrolysed quantitatively. The poor recovery of chlorine obtained when dealing with mixtures containing alkylboron chlorides by the benzene–potassium methylate titration technique is related to the low polarity of this solvent[95]

A polar solvent system enables the chlorine—carbon bonds to be stabilized. Accordingly, the compounds, or mixture of compounds containing both types of chlorine, are dissolved in a strong nitric acid–methanol solution. Hydrochloric acid formed through esterification of the chlorine bound to boron is titrated potentiometrically with standard silver nitrate. Chlorine bound to boron in the chloro-2-chlorovinylboranes and in the chloroethylboranes esterifies quantitatively; further the 2-chlorovinyldimethoxyborane is found to be very stable in this solvent system.

C. Determination of Nitrogen and Boron

Boron and nitrogen in aminoboranes may be determined by hydrolysing the aminoborane with acidified methanol to produce trimethyl borate, which is determined by conventional procedures using mannitol or glycerol[97,98]. Nitrogen is determined on a separate sample of the aminoborane by the Kjeldahl or the Dumas procedures. Alternatively, boron and nitrogen in aminoboranes may be determined by consecutive titrations[99]. Since difficulties are sometimes encountered in the determination of boron by procedures based on the use of mannitol or glycerol when the sample contains nitrogen, the aminoborane is first hydrolysed to produce a hydrolysate in which the amine fragment has not been degraded; consecutive titrations with a standard base are then carried out in aqueous solution using a pH meter, boron being determined by conventional titration in the presence of mannitol, and nitrogen by titration of the corresponding ammonium ion.

Boron, either alone or simultaneously with nitrogen and/or phosphorus in organoboranes, can be determined by digesting the sample with sulphuric acid in the presence of selenium powder and copper sulphate–potassium sulphate[100]. The boric acid produced is titrated with sodium hydroxide solution by the fixed-pH method. If nitrogen and phosphorus are to be determined simultaneously, a solution of hydrogen peroxide is added to the digestion reagents, and suitable aliquots are taken for measurement of the nitrogen by a micro-Kjeldahl method and of the phosphorus by a spectrophotometric method. A combustion technique has been described for the determination of carbon, hydrogen, and nitrogen in organoboron compounds[101]. By employing quartz combustion tubes and by heating to between 1000 and 1100°C in the unpacked section around the combustion boat, both carbon and hydrogen can be determined to within about $\pm 0.2\%$ and $\pm 4\%$, respectively. Determinations of nitrogen in acetanilide–BF_3 were improved by restricting the sample sizes to 7 mg or less or by employing a longer combustion tube with two successive packings.

D. Determination of Hydride and Active Hydrogen

Boranes and chlorinated boranes may be determined gas chromatographically by passing them, with argon as carrier gas, through a column of molecular sieve 5A moistened with water, then through a column of dry molecular sieve 5A to determine the amount of liberated hydrogen produced by hydrolysis of the B—H bonds in the sample[102]. The results are within 5% of the known values. If the original sample contains hydrogen as an impurity, it is frozen out in a cold trap at -78°C. Gaseous hydrogen is swept out of the trap with a current of argon, then the residiual boranes in the trap are allowed to warm up the room temperature and are collected in a gas burette. The sample is now ready for analysis by chemically active gas chromatography[103]. Pentaborane reacts with ethanol to produce hydrogen and triethyl borate quantitatively[104]. A very detailed study of the gas chromatographic technique for the determination of hydridic and active hydrogen in borane compounds has been reported[105]. This technique is based on the liberation of active or hydridic hydrogen in a microreaction cell incorporated in a gas chromatographic flow system and measurement of the hydrogen gas band in nitrogen gas carrier by thermal conductivity detection. The difference in thermal conductivity between hydrogen and nitrogen is detected with a Teflon-coated hot wire. The hydrogen gas is formed from the borane compounds by acid hydrolysis using 10% hydrochloric acid. Application of heat to the reaction cell accelerated the reaction rate and provided rapid and stoichiometric release of hydridic hydrogen from the compounds investigated. The main advantage of this method is that it can be applied to the analysis of a variety of borane compounds using a conventional commercially available gas chromatograph.

VIII. ORGANOCALCIUM COMPOUNDS

A. Determination of Carbon and Hydrogen

Carbon and hydrogen in alkali and alkaline earth metal compounds can be determined by igniting the sample at 900°C mixed with eight times its weight of tungstic oxide in oxygen.[106]

IX. ORGANOCHROMIUM COMPOUNDS

A. Determination of Chromium

Chromium in organochromium complexes used as additives in drilling fluids can be determined by evaporating a sample to dryness on a water-bath, mixing the residue with nitric acid and potassium chlorate solution, and again evaporating to dryness[107]. The residue is dissolved in water, the solution is boiled with 2.5 M sulphuric acid and silver nitrate solution, then gently boiled for 30 min with ammonium persulphate, sodium chloride solution is added, the mixture is filtered, and chromium in the filtrate is determined with diphenylcarbazide.

X. ORGANOCOBALT COMPOUNDS

A. Determination of Cobalt

Cobalt can be determined gravimetrically by the standard pyridine–thiocyanate method[108] or by X-fluorescence spectroscopy. [109]. For the latter only a minimum of sample preparation is required. Matrix effects and minor instrumental variables are compensated for by using a solution of cobalt octanoate or naphthenate in 2-ethylhexan-1-ol (containing 0.2% of cobalt) as internal standard. The lower limit of detection is about 10 ppm of cobalt, for concentrations of up to about 0.5% of cobalt, the precision (95% confidence level) being about 0.1%. The cobalt content of a solution of cyanocobalamin has been determined colorimetrically at 550 nm[110] by means of the nitroso-R salt reagent[111].

XI. ORGANOCOPPER COMPOUNDS

A. Determination of Copper

Copper in organocopper compounds may be determined by first oxidizing the sample by heating it with concentrated sulphuric acid in a Kjeldahl flask for 2–3.5 h[112]. Copper in the cooled and diluted residue is titrated with 5 mM EDTA (pH 4–4.5; disodium ethylbis(5-tetrazolylozo)acetate as indicator) or iodimetrically after the addition of potassium iodide solution. Alternatively, digestion may be followed by spectrophotometric determination using 1(2-pyridylazo)naphthol[113]. A polarographic method has been used for the determination of ionic copper in copper chlorophyllins[114]. In this method known amounts of dried copper sulphate are dissolved in 1.5 M aqueous ammonia solution and the solution is electrolysed at the dropping electrode.

XII. ORGANOGERMANIUM COMPOUNDS

A. Determination of Germanium

Germanium in organogermanium compounds can be determined by mixing the sample with a 100-fold amount of chromium(III) oxide and heating in a tube at 900°C under a current of oxygen[115,116]. Carbon dioxide and water are collected and weighed by conventional procedures to provide estimates for the carbon and hydrogen contents of the sample. The content of germanium is obtained from the change in weight of the ignition tube before and after the ignition. In a tube combustion procedure, which is applicable to the determination of germanium in volatile organometallic compounds oxygen is bubbled through a weighed portion of the sample until volatilization is complete[117]. The vapours are passed into a weighed silica tube and ignited in a plug of prepared asbestos. The silica tube is then disconnected and ignited to constant weight at 800°C. The germanium content of the sample can then be calculated from the residual weight of germanium oxide found in the silica tube. The method may be modified to accommodate samples that hydrolyse rapidly to non-volatile products[118]. It is accurate to ±0.5% and takes 3–4 h. In a bomb combustion procedure for the determination of organically bound germanium, the sample is fused with sodium hydroxide and sodium carbonate for 1.5 h in a sealed nickel bomb at 920–940°C[118]. After cooling, the residue is dissolved in water and ice-cold concentrated sulphuric acid added. The mixture is then heated almost to boiling point and diluted prior to determination of germanium and halogens.

A direct spectrographic procedure has been developed for the determination of germanium in organogermanium compounds[119]. A solution of the sample in cumene containing polymethylphenylsiloxane as internal standard is transferred to a fulgurator cooled to solid carbon dioxide–acetone temperature. Spectra are excited by a condensed spark discharge between the graphite rod in the fulgurator and graphite counterelectrode. X-ray fluorescence spectroscopy may also be used.[120]. The sample is dissolved in dioxane. Water-soluble species are dissolved in water and insoluble compounds are powdered and pressed into discs with $Na_2B_4O_7 \cdot 1OH_2$. Arsenic is added as an internal standard. An instrument with a lithium fluoride analysing crystal is suitable. Measurements are made of germanium and arsenic K radiation. The relative error for 0.15–0.30% of germanium was within ±3.3%. Atomic-emission spectrography has a detection limit for hydride of 0.4 μg[121].

B. Determination of Carbon and Hydrogen

Reproducible carbon and hydrogen analyses from tube combustions on organogermanium compounds either alone or mixed with tungstic oxide or other oxidants are difficult to obtain[122,123] unless a very slow combustion is used. Liquids are particularly difficult to analyse.

XIII. ORGANOIRON COMPOUNDS

A. Determination of Iron

Iron in ferrocene derivatives has been determined by first digesting 25–400 μg in a mixture of concentrated nitric acid and concentrated sulphuric acid in a sealed glass tube and heating to 300°C[124]. After the addition of hydroquinone and suitable buffering the iron is determined spectrophotometrically using 1,10-phenanthroline (Table 9).

In a polarographic method, the sample is decomposed with concentrated sulphuric acid in the presence of hydrogen peroxide and the unused hydrogen peroxide is destroyed by boiling[125]. The iron is then absorbed on KU-2 cationite, from which it is eluted with 4N hydrochloric acid. The pH of the eluate is adjusted to 9, the iron–catecholdisulphonic acid complex is formed, and the solution of this complex is analysed polarographically for iron. An absolute error of approximately ±0.5% is

TABLE 9. Determination of iron in ferrocene derivatives

Compound	Range of sample weight (μg)	Iron (%) Calc.	Iron (%) Found (mean)	No. of determinations	Standard deviation (%)	Range of errors (%)
Ferrocene	29.4–157.8	30.02	30.11	8	0.27	−0.34 to +0.43
1,1-Dibenzoyl-ferrocene	44.0–158.0	14.17	14.22	8	0.12	−0.15 to −0.23
Ferrocene 1,1-dicarboxylic acid	63.4–99.3	20.38	20.13	8	0.14	−0.45 to −0.10
Ferrocene 1,1′-dicarboxy-anilide	50.5–179.6	13.16	13.21	10	0.21	−0.30 to +0.40

claimed for this method. Iron can be determined spectrophotometrically following digestion with concentrated nitric acid and hydrogen peroxide[126]. Silicon and fluorine do not interfere in the procedure.

Simultaneous determination of iron and titanium in donor–acceptor complexes of the ferrocene bases can be effected by decomposing the sample with a mixture of nitric acid, anhydrous acetic acid, and aqueous bromine[127]. The iron is titrated complexometrically, with sulphosalicyclic acid as indicator, and the titanium is determined colorimetrically by extraction of its 8-hydroxyquinoline complex into chloroform. The standard deviation for each analysis is not greater than ±0.3%.

Ignition and mineralization techniques involve a risk of inflammation and sublimation of ferrocene derivatives. In a procedure which avoids these difficulties the sample is treated in a Kjeldhal flask with concentrated hydrochloric acid, followed by the addition of concentrated nitric acid[128]. After a few minutes the solution is heated to gentle boiling and, after 30 min, concentrated nitric acid is added and the solution boiled to expel nitrous fumes. After cooling, 110-volume hydrogen peroxide is added and the iron is determined spectrophotometrically using 1,10-phenanthroline. Iron in ferrocene and its organosilicon derivatives can be determined by dissolving the sample in carbon tetrachloride[129]. Hydrochloric acid and ammonium persulphate are added and the solution is stirred, whilst heating, until the blue colour of the solution changes to yellow. Boiling is continued until chlorine evolution ceases and then the solution is diluted with water. Potassium iodide solution is then added, the pH is adjusted to 4.5 and the solution titrated with standard EDTA, using sulphosalicylic acid as indicator, to a golden yellow colour change. In a volumetric microdetermination of iron in ferrocene and its derivatives the sample is shaken with acetone or anhydrous acetic acid and aqueous bromine is added[130]. The solution is then titrated to the colourless sulphosalicylic acid end-point with 0.01 M EDTA.

Iron has been determined spectrographically in ferrocene and its derivatives[131]. A cooled solution of the sample containing a known concentration of methylphenylpolysiloxane or of tributyl phosphate in cyclohexanone or kerosine is atomized into a spark between graphite electrodes. For standardization a similar solution was used containing known concentrations of pure ferrocene. The analytical line pairs used were either Fe 252.539 and Si 252.412 nm or Fe 249.318 and P 255.328 nm. For the determination of iron in haemoglobin the blood sample can be treated with concentrated sulphuric acid and aqueous potassium persulphate and the iron determined colorometrically by the thiocyanate procedure[132]. Alternatively, there is a simple reproducible method involving oxygen flask combustion which does not need internal standards, for the liquid scintillation counting of haemoglobin and haemin, labelled with carbon-14[133]. In a further method the blood sample, saturated potassium chlorate solution, and concentrated sulphuric acid are heated in boiling water[134]. After cooling, sodium tungstate solution is added and the mixture is centrifuged. Iron in the supernatant solution is determined using potassium thiocyanate. X-ray spectroscopy can be used to determine total blood iron[135]. Samples of serum are placed directly on to confined spots on paper, dried, and passed through the X-ray field. Results for total phosphorus in serum and total iron in whole blood showed no significant difference from those given by wet-washing procedures, except that the total iron is more precisely determined by wet-washing.

XIV. ORGANOLEAD COMPOUNDS

A. Determination of Lead

Lead may be determined by first decomposing the organolead compound with a mixture of 1:1 fuming sulphuric acid (25% SO_3) and fuming nitric acid (d = 1.52)[136]. The mixture is carefully heated until all of the sulphuric acid has been evaporated. The residue is dissolved in glacial acetic acid and 25% ammonia and diluted. After suitable buffering, lead is determined by titration with 0.05 M disodium EDTA to the Eriochrome Black T end-point. In a direct spectrographic procedure, a cumene solution of the sample and polymethylphenylsiloxane internal standard is poured into the inner vessel of a fulgurator, the outer vessel of which contains 50% aqueous monoethylene glycol cooled to $-70^\circ C$ in solid carbon dioxide–acetone[137]. Spectra are excited by a condensed spark discharge between the graphite rod in the fulgurator and a graphite counter electrode. Tetraethyllead has been analysed by β-ray back-scattering using ^{90}Sr as a source[138]. Good agreement with chemical analyses for lead content is reported by this method.

B. Determination of Organolead Compounds in Petroleum via the Determination of Lead

Atomic-absorption spectroscopy has been used to determine organolead compounds in petroleum[139–150]. The method is rapid, reproducible, and remarkably free from interferences by other elements present in the petroleum. The results obtained agree favourably with those obtained by X-ray fluorescence, wet chemical methods, and flame photometry. Lead absorbs strongly at 283.3 nm. At this wavelength the degree of absorption is so intense that the useful range of analysis is limited to between 0 and 70 ppm, so that sample dilution may be necessary. Tin, sodium, bismuth, copper, zinc, chromium, iron, and nickel do not interfere if an oxygen–hydrogen flame is used when present at about 1% of the concentration of lead in the sample[134]. There appeared to be no interference from sulphur, halogen or nitrogen compounds. Isoctane has also been used as the solvent[141]. A precision of about 1% for the determination of either tetraethyllead or tetramethyllead in petroleum using atomic-absorption spectroscopy is claimed. Satisfactory results for the determination of tetraethyllead in petroleum using an air–propane flame are obtained only if the standard solutions are prepared as dilutions of tetraethyllead since experiments with other lead compounds, such as lead nitrate or lead 8-hydroxyquinolate, have shown that the absorbance depends on the position of the absorption path in the air–propane flame. This is caused either by different 'burn-off' rates or easier atomization than tetraethyllead[142]. Consequently, the population of ground-state atoms is highest at the base of the flame when tetraethyllead is involved and measurements are best made at this point. Dif-

TABLE 10. Mean lead content values of selected petroleum samples

Lead content (ml/Imperial gallon)		
	Atomic-absorption spectroscopy	
Gravimetry[a] and/or polarography[b]	Isooctane	Acetone–isooctane[c]
2.15[d]	2.75	2.57
1.42	1.55	1.37
1.26	1.42	1.24
3.28	3.70	3.32

[a]Ref. 151.
[b]Ref. 152.
[c]1:1 mixture of acetone and isooctane used as a sample diluent instead of isooctane alone.
[d]The experimentally determined values for all three methods were found to vary by less than 0.025 ml/Imperial gallon from the mean values listed.

ficulties have been reported in the determination of lead in petroleum when using a high-efficiency burner of the total combustion type or a pre-mix burner[142,143]. The problem may be solved by using 1:1 v/v acetone–isooctane as the solvent for petroleum sample dilutions and the preparation of standard lead solutions rather than isooctane alone[143].

Table 10 compares results obtained on leaded petroleum samples by atomic-absorption spectroscopy[143] using isooctane alone and mixed isooctane–acetone as solvents with results obtained by various other established gravimetric[151] and polarographic[152] techniques. It is seen that much better agreement with these latter techniques is obtained when the mixed solvent is used in atomic-absorption spectroscopy. The problem seems to be due to the complex phenomena associated with the vaporization and burning of the lead solutions at the burner tip and in the flame itself[143], and the success of the mixed solvent may be due to its increased vapour pressure[140,144,145] brought about by its acetone content relative to that of raw petroleum. Alternatively, it may be merely that the presence of acetone in the mixed solvent increases the solubility of tetraethyllead. In spite of the number of attempts made[140–143,147] to exploit the obvious advantages of atomic-absorption spectroscopy in the direct determination of organolead compounds, in general these have not been successful. The main problem is that tetramethyllead and tetraethyllead give rise to different absorption coefficents for lead in the flame, necessitating very carefully matched standards. As the proportions of tetramethyllead and tetraethyllead may vary over a given batch of samples a method is required which is independent of the alkyl type ratio[147].

Lead alkyls may be efficiently extracted into aqueous iodine monochloride at room temperature with an extraction efficiency better than 96%[148]. This has been applied to the extraction of trace amounts of lead from a variety of petroleum products. The reagent reacts rapidly with lead alkyls, converting them to water-soluble dialkyl lead compounds. Inorganic lead compounds are also soluble in iodine monochloride, enabling the total lead content to be extracted from the gasoline. After extraction and suitable dilution, the aqueous iodine monochloride solution containing the lead can be sprayed directly into a lean air–acetylene flame, and calibration effected at 283.3 nm using aqueous solutions prepared from a lead salt. Table 11 compares results obtained

TABLE 11. Comparison of determination of lead in petroleum by iodide monochloride–atomic-absorption spectroscopic (AAS) and by X-ray fluorescence (XRF) methods

	Lead (g/Imperial gallon)	
Sample	AAS	XRF
Petrol A	2.80, 2.76	2.72
Petrol B	2.82, 2.86	2.82
Petrol C	2.76, 2.86	2.80

on petroleum samples by this method using a single water extraction and by X-ray fluorescence analysis.

Atomic-absorption spectroscopy has been applied to the determination of organolead compounds following their separation on a gas chromatographic column. Tetramethyllead, trimethylethyllead, dimethyldiethyllead, methyltriethyllead, and tetraethyllead have been separated on a column containing 60–80-mesh Chromosorb P coated with 1% potassium hydroxide operated at 85°C and lead in amounts down to 20 ng in the effluent determined by the absorption of the lead 283.3-nm emission line[149,150].

Procedures for the determination of tetraethyllead in petroleum by flame photometry have been described[153–156]. The sample is burned in a flame fed with oxygen and hydrogen at 293 and 14 nmHg pressure, respectively. The flame is measured with a monochromator at either 406 or 402 nm by means of a photocell previously calibrated with petroleum of known lead content. Emission spectrography using a medium-dispersion quartz spectrograph and a high-voltage spark discharge in conjunction with a rotating double-disc electrode has been used to determine lead[157]. The film of leaded petrol is transferred by contact onto spectrally pure carbon. The internal standard consists of a cobalt–pentanol complex. Standards are analysed by using the following line pairs: Pb 257.73–Co 276.42 nm; Pb 282.32–Co 301.76 or 276–42 nm; and Pb 287.33–Co 301.76 or 307.23 nm. The method permits the determination of 0.005–1% of lead with a relative error of ±3%.

X-ray fluorescence permits the determination of tetraethyllead in the concentration range 0.1–6 ml/Imperial gallon to be determined with a standard deviation of ±0.28 ml/Imperial gallon[158]. The error caused by sulphur in the sample is very small and that due to possible petrol additives such as phosphorus is negligible. The time required for one analysis is about 5–10 min. For samples of petroleum containing about 0.1 ml/l of tetraethyllead a molybdenum anti-cathode (50 kV, 13 mA), a curved crystal of lithium fluoride (r = 110 cm), and a scintillation counter to ensure maximum sensitivity have been used[159]. The lines $L\beta_1 + L\beta_2$ of lead and $K\beta$ (second order) of molybdenum are most suitable, but an internal standard must be used if other elements, e.g. 0.05–0.1% of sulphur, are at present. The standard deviation is 0.85%; the lower limit of determination is 25 ppm of lead. About 0.25% of bromine (as bromobenzene in tetralin) must be present as internal standard to compensate for the interference by sulphur and the fluorescence should be excited at 50 kV and 8 mA. The standard deviation is *ca*. 0.8%. Petroleum manufactured from different crude oils derived from different oil fields does not greatly effect the results in the determinations of tetraethyllead by X-ray fluorescence[160]. The organic bromine scavenger compound added to the petroleum with the tetraethyllead also has little effect, provided that the bromine to lead ratio in the sample is constant, which is the case for any particular additive composition. Good overall agreement in the determinations of lead in petrol between X-ray fluorescence using platinum metal as internal standard and a variety of chemical, gravimetric and X-ray absorption procedures has been reported[161]. Other references on the use of X-ray fluorescence for the determination of lead in petroleum include refs. 156 and 162–164.

Tetraethyllead in petroleum has been determined by X-ray absorption methods[165–169]. One method consists of measuring the absorption increment corresponding to the sublevel L_{111} of the tetraethyllead present in the petrol[165]. An apparatus of the General Electric type XRD3 was used with an anti-cathode tube of molybdenum (18 kV). The emitted radiation, rendered monochromatic by a crystal of sodium chloride, is passed through the cell containing the petroleum sample and its intensity is measured for various angular positions with a Geiger counter connected to a scaler. Constructional details of an X-ray absorptiometer and its application to the determination of tetraethyllead in petroleum over the range 0.0–1.5 ml/l with an accuracy of ±0.005 ml/l have been described[166]. Results obtained by the tritium Bremsstrahlung technique agree to within ±0.02% with those by the gravimetric procedure for concentrations of 0.5–2% of tetraethyllead in petroleum[169]. Absorption measurements were made on a sample dissolved in heptane. Equations were derived for calculating the content of tetraethyllead in the presence of dibromoethane and 1-chloronaphthalene.

Methods for the determination of lead anti-knock compounds in petroleum have been reviewed[151,170–173]. Results from ten cooperating laboratories indicate that the limits of reproducibility for compounds other than tetraethyllead are not as precise as those quoted in the ASTM Standard Method[151]. Consequently, wider limits are quoted in the revised standard for petroleum containing the more volatile anti-knock agents[151]. Critical comparisons have been given of flame photometric and complexometric methods[171] and of various polarographic, colorimetric, and gravimetric methods[172] for the determination of tetraethyllead.

C. Determination of Carbon and Hydrogen

The general opinion in the limited amount of published work[136,174,175] is that the presence of lead does not offer any serious difficulties in the determination of these elements in organolead compounds. About 4 mg of the sample may conveniently be burnt in a Heraeus furnace at 850°C and the combustion gases conducted in a stream of pure oxygen at 500°C over platinum gauze, silver permanganate, and silver gauze, respectively[136]. To avoid the risk of explosion the organolead sample is covered with about 20 mg of tungsten oxide.

D. Determination of Halogen

Organolead compounds containing ionic halogen can be titrated directly in ethanol or acetone solution with standardized silver nitrate[136]. If the halogen is covalently bonded, or if the ionic halide cannot be titrated directly, the sample can be completely decomposed by Parr bomb combustion in the presence of sodium peroxide. In the case of a bromine or iodine determination it is recommended that the decomposed sample is reduced with hydrazine in order to avoid losses through free halogen formation. It is also necessary to employ this reduction when determining chlorine in the presence of lead, otherwise some chlorine gas is evolved causing the chlorine content found to be low. The chlorine is formed probably through oxidation by tetravalent lead. This possibly accounts for the low chlorine recoveries reported[176].

XV. ORGANOLITHIUM COMPOUNDS

A. Determination of Lithium, Carbon, Hydrogen, and Oxygen

A microcombustion procedure has been described for the determination of lithium, carbon, and hydrogen in organolithium compounds[177]. The sample is intimately mixed with finely ground quartz in an empty tube and combusted in a stream of oxygen. Combustion in an empty tube prevents absorption of carbon dioxide by the lithium residue and permits carbon and hydrogen to be determined by standard methods. The amount of alkali metal can then be obtained from the weight increase of the ignition tube after the combustion. Other methods are described in ref. 178–181.

Very low levels of oxygen in butyllithium can be determined by using butyllithium as a source of tritons in the determination of oxygen by ^{18}F counting after activation according to the reaction $^{6}Li(n, \alpha)t$; $^{16}O(t, n)\ ^{18}F$ (ref. 182). The sample is mixed with butyllithium; if monomers are present, trimethylaluminium is added to overcome the polymerizing effect of butyllithium. The solvent or monomer is distilled in a high vacuum. The residue of butyllithium is irradiated with a neutron flux of $5 = 10^{1?}$ neutrons $cm^{-2}\,s^{-1}$. After addition of fluoride ion as carrier, the total fluorine is distilled as hydrofluosilicic acid; this is hydrolysed, and the fluoride ions are precipitated as lead chlorofluoride. The ^{18}F is determined by counting the positron-destroying radiation.

XVI. ORGANOMAGNESIUM COMPOUNDS

A. Determination of Carbon and Hydrogen

A combustion procedure which could probably be applied to the analysis of organomagnesium compounds has been described[178].

B. Determination of Alkyl Groups

The determination of alkyl groups in Grignard compounds is based on hydrolysis to produce a hydrocarbon which can then be determined gas volumetrically[183].

$$RMgX + H_2O \longrightarrow RH + Mg(OH)X \qquad (21)$$

The method, as used, is restricted to those RMgX compounds and their etherates which give a hydrocarbon that is gaseous at ordinary temperatures.

XVII. ORGANOMAGNANESE COMPOUNDS

A. Determination of Manganese

In one method for the determination of manganese in organic compounds, the sample is digested with nitric acid or aqua regia in a long-necked flask, the excess of acid is evaporated off, and dilute sulphuric acid containing hydrogen peroxide is added[184]. The solution of manganese(II) sulphate is then titrated with standard potassium permanganate. The error does not exceed $\pm 0.3\%$ (absolute).

In a further method, the sample is heated with potassium hydrogen sulphate and mercury(II) oxide moistened with 98% sulphuric acid[185]. The neutralized solution is titrated with 0.05 N potassium permanganate. This method was applied successfully to a wide range of cyclopentadienylmanganese tricarbonyl derivatives containing between 14 and 20% of manganese without interference by various elements present in the sample as major constituents, including chlorine, iodine, nitrogen, sulphur and mercury.

XVIII. ORGANOMERCURY COMPOUNDS

A. Determination of Mercury

Mercury in organic compounds can be determined by burning the sample at 750–800°C in a standard automatic combustion furnace[186,187]. The carrier gas (nitrogen mixed with a small amount of oxygen) sweeps the combustion products into a stationary furnace containing reduced copper gause and combustion catalyst (CuO, MnO_2, and Co_3O_4). Sulphur and halogens are absorbed in an auxiliary furnace in a tube containing MnO_2, Co_3O_4, and silver granules at 550°C. The absorption tube for the mercury is placed close to the last furnace so that its temperature remains between 40 and 100°C by radiation. Under these conditions mercury is quantitatively absorbed on silver granules and may be weighed. The results show a deviation of the mean from theoretical values of −0.06% to +0.08% with a standard deviation ranging from 0.10% to 0.17%. Tube combustion and wet oxidation methods for the determination of mercury in organic material are also available[188,189]. If halogens are present, loss of mercury as halide can be avoided if the sample is placed in a combustion tube filled with calcium oxide and burnt in a current of air at 370°C[190]. Mercury is then collected in a bubbler containing concentrated nitric acid, and determined by titration with 0.005 N potassium thiocyanate in the presence of hydrogen peroxide with iron alum as indicator.

In a further method the sample is heated in a quartz tube in a stream of pure dry nitrogen[191]. Nitrogen oxides, halogens, sulphur, phosphorus, and arsenic compounds are absorbed by a 6-cm layer of a decomposition product of potassium permanganate. The mercury is subsequently absorbed on silver sponge and weighed. The results quoted show an error of ±0.2%.

Organomercury compounds may be analysed by placing them between layers of calcium sulphide and heating in a slow stream of air[192]. The vapours are passed through successive layers of granular calcium oxide and silver pumice at 750–800°C and the mercury is absorbed in a tube containing gold leaf. The method is suitable for 1–20 mg of mercury. In a further method the sample is pyrolysed in a stream of hydrogen and burnt in an oxy-hydrogen flame, ensuring that a portion of the oxygen feeding the flame is bubbled through saturated bromine water[185]. The mercury(II) bromide so formed is collected in water. The excess of bromine in the condensate is removed by adding hydrazinium chloride in small portions until the colur is discharged. After stirring, 0.1 M EDTA (disodium salt) is added, followed by pyridine, and the mercury is determined by potentiometric titration with 0.01 M sodium diethyldithiocarbamate. The results by these last two methods range from 98 to 101% of the theoretical.

Several workers have described methods based on the Schöniger oxygen flask combustion technique for the determination of mercury in organomercury compounds[152,193–206]. The sample may be burned in a flask containing nitric acid in which the mercury is absorbed. Following adjustment of pH the mercury is titrated amperometrically using ethylenedinitriloacetic acid. The only commonly encountered interference comes from chloride ion, which stabilizes the mercury as mercury(I) chloride. It is then necessary to reflux the sample in the nitric acid to oxidize the mercury to the divalent form. For *o*(3-hydroxymercuri-2-methoxy-2-propylcarbamyl)phenoxyacetic acid, the 95% confidence interval is 15 parts per 1000 for both the micro and the semi-micro determinations. Alternatively, after oxygen flask combustion a simple visual titration method with sodium diethyldithiocarbamate may be used[196–198]. Chlorine and bromine, if present in the sample, can be titrated immediately after the mercury determination in the same combustion run. Iodine cannot however, be determined as it leads to indistinct end points and low recoveries (80–90%). Other procedures using an oxygen flask include (i) the use of 8 M nitric acid as an absorbent solution and determining mercury gravimetrically as $[Co(NH_3)_6][Hg(S_2O_3)_3]_3 \cdot IOH_2O$[199] and (ii) applying the sample to a strip of filter paper and, when burning it, using saturated aqueous bromine water as the absorbing liquid[200,201]. After combustion, excess of bromine is removed by aspiration. A measured volume of 0.005 M EDTA is then added and the buffered solution titrated with 0.01 M zinc chloride to the 1-(2-pyridylazo)-2-naphthol end-point in the presence of potassium iodide.

Several procedures based on digestion with mineral acids have been published for the determination of mercury in organic mercurials[202–210]. The sample may be digested by treatment with 75% sulphuric acid and then with potassium permanganate, heating until the odour of bromine has disappeared[202]. After the addition of acidic hydrogen peroxide the solution is boiled to destroy excess of peroxide and titrated with 0.1 N ammonium thiocyanate to the ferric alum end-point. The sample may also be digested with a mixture of sulphuric acid (60%) – nitric acid (70%) – perchloric acid (concentrated) (3:3:1) in an ignition tube for 30 min[203]. The mixture is washed into a beaker with water and excess of 0.005 M EDTA added. The pH is adjusted to 10 and excess of EDTA titrated with 0.005 M magnesium sulphate to the blue to purple Eriochrome Black T colour change. An error not exceeding ±0.4% is claimed for this method. In a further method the sample is wet oxidized with nitric acid, boiled until colourless, and mercury is determined by EDTA titration[204].

A procedure which determines total mercury, mercury(II) acetate, triacetoxymercuribenzene, and diacetoxymercuribenzene in crude phenylmercury(II) acetate involves heating the sample to fuming with concentrated sulphuric acid, diluting, and boiling with bromine water[205]. After making the solution alkaline with sodium hydroxide and aqueous ammonia, mercury(I) is titrated with 0.05 N sodium thioglycollate, with thiofluorescein as indicator. Mercury(II) acetate is determined in an aqueous extract by titration before and after decomposition with bromine water. The residue insoluble in hot 80% acetic acid is filtered off and weighed as triacetoxymercuribenzene; after dilution of the filtrate with water the precipitated diacetoxymercuribenzene is separated and the total mercury remaining in the clear solution is again determined.

A further digestion procedure for the determination of 2–20 mg amounts of mercury in organic matter depends on digestion of the sample with a mixture of nitric acid, sulphuric acid, and perchloric acids in a Kjeldahl flask fitted with a separating funnel, condenser, and receiver[206]. After digestion, a mixture of hydrochloric and hydrobromic acids is added to isolate the mercury in the residue, which is then extracted as the tetraiodo complex using ethyl acetate at pH 2–3 and finally determined colorimetrically using dithizone. If the sample does not contain antimony and tin, then the formation of the tetraiodide complex can be omitted. The recovery of mercury is between 87 and 100%. Mercury in plant protective substances such as phenylmercury(II) acetate, phenylmercury(II) chloride, and methoxyethylmercury(II) silicate may be determined by boiling the sample with an aqueous mixture of potassium iodide and iodine and concentrated sulphuric acid[207–209]. Excess of iodine is removed, the filtered residue is washed with boiling water, and enough EDTA (disodium salt) is added to complex contaminating metals, such as iron, aluminium, zinc, magnesium, manganese, nickel and cobalt, followed by the dropwise addition of concentrated aqueous ammonia to pH 7. The solution is boiled and treated with a boiling saturated solution of the reagent $[Cu(en)_2](NO_3)_2 \cdot 2H_2O$ (en = ethylenedimine) and mercury is determined gravimetrically as $[HgI_4][Cu(en)_2]$. The accuracy is about ±1%.

Precipitation of mercury with Reinecke salt has been used for the determination of mercury in organomercurial fungicides such as phenylmercury(II) acetate, nitrate, or borate, diphenylmercury(II) and methoxyethylmercury(II)silicate[211], and drugs[212].

A method for the identification and determination of mercury in *N*-organomercury compounds is based on the reaction of *N*-organomercury compounds with thiols, whereby *S*-aryl(alkyl)mercury compounds are quantitatively produced[213]. To determine *N*-organomercury compounds, the sample is treated with excess of an ethanolic sodium salt of 2-mercaptobenzothiazole. The precipitate is filtered off and excess of 2-mercaptobenzothiazole in the filtrate is titrated with 0.1 N iodine solution, with starch as indicator. The error of this determination is ±0.3%.

In an ignition method for the determination of mercury in phenylmercury(II) acetate and in mercury(II) acetate the sample, in a nickel crucible, is covered with layers of copper oxide, copper, iron, and calcium oxide and heated at 500–750°C[214]. The crucible is covered with a gold plate of known weight. The mercury evolved is retained by the gold plate and weighed. A photometric method for the determination of mercury in ethylmercury(II) chloride, phenylmercury(II) acetate, and phenylmercury(II) chloride uses copper diethyldithiocarbamate as the chromogenic reagent[215]. A very simple method for the determination of total mercury in organomercurials involves reduction of the sample with zinc amalgam in glacial acetic acid followed by dissolution of the filtered off amalgam in nitric acid and titration of the solution with standard ammonium thiocyanate[216]. A review has discussed the differential determination in organomercurials of phenylmercury(II) acetate and metallic mercury[217].

A semi-automated procedure has been developed for the determination of mercury in fish and animal tissue based on digestion with concentrated nitric and sulphuric acids at 58°C followed by flameless atomic-absorption spectroscopy of the treated extract[218–220]. Using Technicon AutoAnalyzer Equipment, samples can be analysed at the rate of 30 samples per hour with a recovery of 95% (standard deviation ± 3–8%). This digestion has been found to be satisfactory for all types of fish meat and other food products. Mercury in silicon-containing organomercurials can be determined by atomic-absorption spectroscopy[223]. Cold vapour atomic-absorption spectroscopy has been used for the determination of methylmercury in muscle of marine fish[224] and to determine low levels of organic mercury in natural waters after pre-concentration on a chelating resin[225]. Alkylmercury compounds in fish tissue have been analysed by using an atomic-absorption spectrometer as a specific gas chromatographic detector[226]. Volatile mercury compounds in air have been determined with a Coleman mercury analyser system[227]. Methylmercury compounds in fish have been analysed using a graphite furnace atomic-absorption spectrometer to analyse a dithizone-treated toluene extract of the sample[228]. A detection limit of 0.08 μg/g of mercury is claimed.

B. Determination of Carbon, Hydrogen, Sulphur, Halogens, and Oxygen

For the simultaneous micro-determination of carbon, hydrogen, and mercury in samples free from halogen the sample is decomposed in a stream of oxygen at 900–950°C[229,230]. The products of decomposition are burned at 600–650°C over Co_3O_4, and the mercury formed is collected on silver-impregnated pumice. The water and carbon dioxide are collected in anhydrite and Ascarite, respectively. The accuracy of the determination is within ±0.3% for carbon and hydrogen and within ±0.5% for mercury. In an alternative method for halogen-free samples the organomercury compound is burnt in a current of oxygen in a quartz tube containing the product of the thermal decomposition of silver permanganate heated at 60°C[231]. Carbon and hydrogen are determined with errors not greater than ±0.3%.

A rapid micro method for the determination of mercury in organomercury compounds that do not contain halogens involves burning them, absorbing the reaction products in concentrated nitric acid and titrating the mercury(II) ions formed with 0.1 N ammonium thiocyanate using ferric alum as indicator. In a combustion procedure that is suitable for halogen-containing compounds the sample is burnt in oxygen at 850–900°C and the combustion products are passed through a catalyst at 400–450°C[232]. The catalyst consists of the thermal decomposition products of potassium permanganate on asbestos. Mercury in the sample is oxidized and retained on the catalyst, as are halogens and their compounds. Carbon and hydrogen are determined gravimetrically by this procedure with an accuracy within ±0.25%.

Most methods reported for the determination of carbon and hydrogen in organic compounds containing mercury are based on slow combustion procedures in which, during the combustion, elemental mercury is retained temporarily by fillings of the Pregl universal type. Such fillings will, in subsequent determinations, pass on to the absorption train, resulting in high hydrogen values[233]. Various workers have overcome this effect by using gold wire in the beak-end of the combustion tube[234–237]. Another approach is to place a boat containing cerium(IV) oxide, litharge, silver dichromate, silver oxide and lead chromate immediately after the ceria–copper oxide–lead chromate combustion catalyst[238,239]. In a further method the exit tube and beak-end are packed with tightly coiled gold wire to remove the mercury[240,241]. It is necessary to regenerate the gold after 5–6 determinations. A 'rapid' empty-tube method[242] permits the analysis of mercury compounds to be carried out relatively rapidly and necessitates only one combustion apparatus for the determination of carbon, hydrogen, and mercury[233]. Methods based on the Dumas procedure for the micro or semi-micro determination in organomercury compounds of carbon and hydrogen[243], mercury and halogens[244], carbon, hydrogen, and mercury[245], sulphur[246], and halogens[247] have been described. These methods, especially those for sulphur[246] and halogens[244,247], are applicable to all solid and liquid organomercury compounds with a maximum error of ±0.3% and a maximum analysis time of 50 min. In all of the procedures the usual Dumas combustion apparatus is used, but because of the presence of mercury vapour the tube is packed either with a layer of the decomposition product of potassium permanganate or of silver permanganate as a combustion catalyst and absorbent for sulphur and halogens, or copper granules to decompose nitrogen oxides, or a layer of silver sponge to absorb mercury vapour, which causes errors in the conventional Dumas procedure. These procedures are described below.

1. Determination of carbon and hydrogen[243]

A layer of silver sponge is used to absorb the mercury in the cooler part of the combustion tube. Conditions recommended are an oxygen flow-rate of 15 ml/min, a layer of Co_3O_4 to catalyse the combustion, and silver in the hot zone to remove halogens and sulphur oxides. A 2.5-cm layer of silver sponge in a combustion tube of 11 mm diameter then suffices for 200 determinations.

2. Determination of carbon, hydrogen and mercury[245]

The combustion tube (heated at 550–600°C) is packed with a layer of the decomposition product of silver permanganate (prepared by heating $AgMnO_4$ at 90–95°C for 24 h), a layer of copper granules, and a further layer of the decomposition product. Mercury is trapped in an absorption tube packed with silver sponge and silver wool; carbon dioxide and water are absorbed in the usual manner.

3. Determination of mercury and halogens[244]

The combustion of the sample is carried out at 700–750°C in an atmosphere of nitrogen in the presence of granular decomposition products of potassium permanganate on a support of glass splinters. Mercury is absorbed in a layer of silver sponge (obtained by reduction of silver nitrate by acetaldehyde) and determined gravimetrically. The halogens are washed out of the combustion tube with water and titrated.

4. Determination of sulphur[246]

The sample is burnt in oxygen in the presence of Mn_2O_3; any $MnSO_4$ formed is extracted with boiling water and Mn^{2+} is determined complexometrically with EDTA. The error for determinations of sulphur in di(phenylthio)mercury varies from 0.11 to 0.09%.

In a micro-determination of carbon, hydrogen, mercury, and chlorine or bromine in organomercury compounds[249], the water and carbon dioxide produced by pyrolytic oxidation of the sample are absorbed in the conventional way and determined gravimetrically. Mercury is absorbed on fine-grain metallic bismuth and determined gravimetrically. To absorb halogens, a boat containing a product of the thermal decomposition of potassium permanganate is inserted in the combustion tube; the halogens are then determined by conventional procedures[250]. Rapid methods for the micro-determination of carbon, hydrogen, mercury, and halogen in a single combustion of the organomercury compound depend on igniting the substance in a stream of oxygen and gravimetric determination of the four elements[251]. Carbon, hydrogen, and halogen are determined by previously described procedures[252] and mercury is collected in the combustion tube on silver and determined by the increase in weight. The error for mercury is not greater than 0.7% absolute

C. Determination of Oxygen

A conventional combustion train may be used to determine oxygen in organomercury compounds[253,254]. The sample is decomposed at 1000°C in a stream of argon and passed through a combustion tube heated at 1120 ± 10°C. The tube is packed with a piece of platinum mesh, then asbestos, anthracene carbon black, and asbestos. Mercury vapour is absorbed on a layer of pumice (8 cm × 8 mm in diameter) at room temperature, and the carbon monoxide formed is oxidized to carbon dioxide over copper(II) oxide at 300°C and absorbed in Ascorite. The method has been used for about 80 determinations on samples containing up to 60% of mercury without replacement of the pumice. A single determination takes about 45 min and the error is less than ±0.3%.

XIX. ORGANONICKEL COMPOUNDS

A. Determination of Nickel

The quantitive determination of traces of nickel in oils has been discussed[255,256].

XX. ORGANOPALLADIUM COMPOUNDS

Palladium and chlorine in organopalladium complexes may be determined by decomposing the sample with sodium peroxide in an atmosphere of oxygen in a bomb, and then reducing the divalent palladium with sodium formate to palladium metal, which is determined gravimetrically[257]. The chloride ion in the filtrate is determined by potentiometric titration. The gravimetric determination of palladium with dimethylglyoxime[258] tends to give low results[259], so that gravimetric analyses based on 2-thiophene-*trans*-aldoxime[259] or precipitation of palladium as PdI_2[260] are preferred when only limited amounts of palladium are present.

XXI. ORGANOPHOSPHORUS COMPOUNDS

A. Determination of Phosphorus

As some types of organometallic compounds contain phosphorus in addition to a metal, the determination of phosphorus is discussed here. Various procedures have been described for the determination of organically bound phosphorus. The main procedures for sample decomposition which have been described, and which are discussed in the following sections, involve digestion with mixtures of sulphuric and perchloric acids, or with mixtures of nitric and perchloric acids, or with fuming nitric acid; fusion in a bomb with sodium peroxide and oxygen flask combustion have also been used extensively[261–263]. It is claimed that for mineralization, the open-tube wet-combustion method is the fastest, simplest, and most convenient for samples containing down to a few micrograms of phosphorus[262]. The sealed-tube method permits the analysis of both volatile compounds and aqueous solutions but takes longer, and large samples cannot be analysed. The flask combustion method works satisfactorily, but there is a slight tendency towards low results. The spectrophotometric methods investigated were the yellow molybdophosphoric acid and the molybdenum blue procedures. Measurement of the molybdophosphoric acid colour at 400 or 430 nm is the fastest, simplest, and most accurate of the colorimetric methods tested; 460 nm is preferable if a lower sensitivity is desired. Amyl acetate is an excellent extractant for separating molybdophosphoric acid completely from molybdosilicic acid and most other interfering substances. Molybdenum blue methods should be used when high sensitivities are required. The various sample digestion procedures that have been employed in the determination of phosphorus are discussed below.

1. *Digestion with mixtures of sulphuric acid, perchloric acid, and nitric acid*

Phosphorus at the microgram (30–500 μg) and milligram (3–5 mg) levels has been determined by a procedure involving preliminary digestion of the sample with a mixture of sulphuric acid and perchloric acid followed by spectrophotometric evaluation of the yellow molybdovanadophosphate complex at 430 nm[264,265]. Perchloric acid–sulphuric acid mixtures have been used for the digestion of organophosphorus compounds in colouring matters, plastics, insecticides, and pharamaceutical products[266]. Three different mineral acid systems and open and closed tube digestion have been compared[267–269]. The digestion reagents studied included nitric–sulphuric acids in a sealed tube, and either sulphuric–perchloric acid or 50% hydrogen peroxide in an open tube. The resulting phosphate was determined spectrophotometrically, either as molybdenum blue at 735 nm or as phosphovanadomolybdate at 315 nm[267]. It was concluded that: (i) The most sensitive, simple method available for the determination of microgram amounts of phosphorus involves measurement of the absorption of the phosphovanadomolybdate complex at 315 nm. Only one reagent is added; the complex forms rapidly and is stable for a considerable time; the sensitivity is 10–15 times that at 430 nm. On the other hand, measurement in the ultraviolet renders the method liable to interference from 'colourless' molecules or ions. (ii) For absorption in the visible region of the spectrum, the molybdenum blue method[270] is to be recommended. It is less sensitive than the phosphovanadomolybdate method, involves the addition of more reagents and requires 30 min for colour development. However, measurement at 735 nm should be liable to less interference than measurement at 315 nm. A hot flask technique[270] using sodium hypochlorite as absorbent has been applied successfully to liquid and solid compounds, including 'difficult' fluoro carbons. No blank was found, but it was essential to heat the absorbent after acidification, presumably to convert meta- into orthophosphate. Neither of the open-tube digestion procedures used (i.e. sulphuric–perchloric acid or 50% hydrogen peroxide) was completely successful for certain types of fluorophosphoroorganic compounds[267]. Sealed tube digestion with a mixture of nitric and sulphuric acids and hot flask combustion with sodium hypochlorite gave reasonably satisfactory results. The latter procedure was recommended with a phosphovanadomolybdate finish at 315 nm; although this is slightly less precise than the sealed tube digestion method and is more subject to interferences, it has the advantages of a shorter analysis time, ease of manipulation, and absence of reagent blank.

Other digestion methods that have been described include the following: (i) Digestion of organophosphorus compounds with mixtures of concentrated nitric and sulphuric acids in a Kjeldahl flask for 30 min followed by a molybdate finish gave a coefficient of variation of 0.68% in the determination of phosphorus in triphenylphosphine[272,273]. (ii) Good results were obtained for non-fluorinated compounds when phosphorus was determined gravimetrically as nitratopentamminecobaltidodecamolybdophosphate after wet oxidation with nitric and sulphuric acids[274]. (iii) Phosphorus in non-volatile organic phosphates has been determined by treatment at 215–230°C with a mixture (2:3 by volume) of concentrated nitric and perchloric acids (70% or by calcination with Eschta mixture[275]. After the former procedure, the phosphorus in the solution is determined volumetrically with silver nitrate in the presence of glycine, or gravimetrically as magnesium ammonium phosphate. After the calcination procedure, the gravimetric method is used. Results were reported for tritolyl phosphate, *OO*-diethyl-*o,p*-nitrophenylthiophosphate and lecithin. (iv) Mineralization with perchloric acid and nitric acid followed by phosphate determination by titration with lanthanum nitrate solution has been described[276]. (v) Mixtures of nitric and perchloric acids have been used for the decomposition of organophosphorus compounds followed by colorimetric determination of the phosphate produced as phosphovanadomolybdate[277]. (vi) Heating the sample for 2 h with concentrated sulphuric acid, iron(III) chloride and perchloric acid, followed by measuring the molar absorptivity of the molybdenum blue complex at 700 or 840 nm[278]. (vii) Digestion of the sample with fuming nitric acid in a sealed tube (Carius)[279]. (viii) Sealed tube digestion procedures using either concentrated sulphuric acid at 460°C[280] or fuming nitric acid[281] as the digestion reagent. (ix) Carius methods have been used for the determination of phosphorus in organic fluorine compounds[282]. Low recoveries have been reported for this technique and ascribed to the adsorption of phosphate on the walls of the tube[281]. An effective way of avoiding the deposition on the tube of a white insoluble material is to add a small amount of an alkali metal salt to the acid mixture in the decomposition tube[279]. Potassium chloride is the most effective salt. Arsenic, tungsten, tin, titanium, vanadium, and zirconium interfere in this procedure.

2. Sodium peroxide fusion

Following the finding that wet digestion with mixtures of concentrated nitric and sulphuric acids did not give reliable phosphorus determinations in fluorinated compounds, the applicability of fusion with sodium peroxide in a Parr bomb followed by determination of phosphate by Wilson's method[283] was examined for the determination of down to 2–3 mg of phosphorus in fluorinated organic compounds. Relatively large and variable blank values were obtained when a semi-micro (8.5-ml capacity) nickel bomb and the usual amounts[284] of reagents for semi-micro operation were used, i.e. 4 g of sodium peroxide, 200 mg of potassium nitrate, and the organic material made up to 200 mg with sucrose. These blank values were caused by silica picked up from the glass apparatus used during the weighing of the sodium peroxide and during the leaching of the bomb, and could be eliminated by using platinum apparatus for these operations, the bomb leachings being transferred to glass apparatus only after acidification. Tests carried out with a 3-ml micro-bomb with 25 mg of standard compounds, 25 mg of sucrose, 50 mg of potassium nitrate, and 1 g of sodium peroxide were satisfactory, as shown in Table 12.

TABLE 12. Phosphorus analysis of standard compounds by the micro-bomb fusion method

Compound	No. of determinations	Phosphorus calculated (%)	Phosphorus found (%)	Deviation from mean (%)
Triphenylphosphine	6	11.81	11.80	±0.14
Tri-*n*-butyl phosphate	4	11.63	11.69	±0.14
Tri-*m*-cresyl phosphate	5	8.41	8.42	±0.04

The presence of fluoride in the bomb leachings gave rise to positive errors in determinations of phosphorus, presumably by attack on the glass flask. This interference by fluoride was overcome by evaporating the bomb leachings, acidified with hydrochloric acid to dryness two or three times, or by adding boric acid. The following comments have been made about the method[271].

(a) Over 50 mg of material can be decomposed by using 1 g of sodium peroxide and 50 mg of potassium nitrate. When these amounts are used, no addition of sucrose is made.

(b) The use of platinum apparatus for weighing the sodium peroxide and in leaching the residue from the bomb assists materially in eliminating the blank values caused by the pick-up of silica. Easily measurable contamination by silica was found when the peroxide was weighed on a watch-glass. In the presence of fluoride, the leaching of the residue from the bomb, acidification, and evaporation must be carried out in platinum apparatus or very serious errors caused by high blank values will result. On the other hand, when platinum apparatus is used, the presence of fluorine assists in reducing the blank values by removing silica as silicon tetrafluoride during evaporation of the acidified leachings.

(c) Wilson's method for the determination of phosphate[283] has proved, with modification, to be suitable for the determination of 1–3 mg of phosphorus.

(d) Results should be within ±0.15% (absolute), and the majority of the results obtained on fluorinated materials are within ±0.5%.

(e) A single determination can be completed in about 2 h.

(f) It is thought that arsenic will interfere, although the method was not tested in the presence of arsenic. Silicon does not interfere if sufficient fluoride is present to remove silicon as the tetrafluoride or if the bomb leachings are evaporated to dryness twice after acidification with hydrochloric acid.

The determination of phosphorus in organophosphorus compounds by fusion with sodium peroxide in a Parr bomb has been studied by other workers[261,285]. Sodium peroxide fusion has been compared with two other procedures for the determination of phosphorus in glycerophosphates. The three methods involved are: (i) carbonization in a porcelain crucible in a gas flame (10 min), dissolution of the residue in concentrated nitric acid and evaporation; (ii) evaporation on a sand bath in perchloric acid–nitric acid mixture followed by dissolution in 10% nitric acid and evaporation; (iii) heating for 15 min in a Parr–Wurzschmitt bomb with sodium peroxide by a gas flame, and dissolution of the residue in water to decompose the excess of sodium peroxide. In each method phosphate was determined by precipitation with excess of bismuth nitrate in nitric acid and titration of the excess of bismuth with EDTA in the presence of catechol violet. The results were the same by all three methods, but method (ii) was of advantage in the presence of chloride ion which is eliminated as hydrochloric acid. The following ions do not interfere: NH_4^+, Li^+, Mg^{2+}, Ca^{2+}, Sr^{2+}, Ba^{2+}, Al^{3+}, Zn^{2+}, Ce^{3+}, Mn^{2+}, Co^{2+}, Ni^{2+}, Cd^{2+}, Cu^{2+}, Pb^{2+}, $UO_2{}^{2+}$, and Ag^+; however, Fe^{3+}, Ga^{2+}, In^{3+}, Zr^{4+}, Th^{4+}, Hg^{2+}, $SO_4{}^{2-}$, $AsO_4{}^{3-}$, $Cr_2O_7{}^{2-}$, and Cl^- should be absent. Method (ii) had the advantage of not being subject to interference by the presence of chlorine in the sample. Another group compared three methods of sample decomposition involving (i) fusion with sodium peroxide in a steel bomb in a burner flame, (ii) fusion with sodium peroxide in a calorimeter bomb, and (iii) heating with concentrated sulphuric acid–nitric acid mixture[287]. They recommend method (ii) for a variety of organophosphorus compounds, including polymers.

3. *Other digestion reagents*

A digestion reagent consisting of hydriodic acid, calcium hydroxide, water, phenol, and acetic acid has been described for the determination of microgram amounts of phosphorus in organic compounds[265]. During the removal of solvent and excess of reagent by volatilization and combustion, phosphorus is converted to orthophosphate. The molybdenum blue colour is developed. The procedure can be adapted to either ultramicro or trace analysis, and is applicable to the determination of organic phosphorus in a wide variety of solvents. The same group also described a rapid semimicro procedure which utilizes digestion in a sulphuric and perchloric acids, followed by formation of the phosphovanadomolybdate complex. Using this procedure, phosphorus was determined successfully in such compounds as 2,2-dichlorovinyldimethylphosphate, trimethyl phosphate, methyl parathion, *OOO*-tri-*p*-tolylphosphorothioate, *OOO*-triethylphosphorothioate, tricyclohexylphosphine oxide, and Phosdrin insecticide.

In addition to being applicable to these materials as an ultramicro method for phosphorus determination, the method was applicable as a trace method for phosphorus at the level of 1 ppm or less in various media, such as organic matter, water, lubricating oils, carbon tetrachloride, acetone, acetic acid, xylene, glycol, and mineral oil. Recovery of phosphorus from compounds dissolved in methanol, ethanol, and isopropanol, all of which might be expected to react with the hydriodic acid, was also quantitative. Apparently the reaction of hydriodic acid with the phosphate is much more rapid than with the alcohol solvent. The reagent reacted rapidly with most phosphate insecticides, requiring no digestion other than volatilization of solvent and excess of reagent on a steam bath and/or a hot-plate, a process which usually required about 10 min. However, some phosphorus-containing materials, including aryl esters of phosphates, phosphonamides, and dithiophosphate esters such as *OOS*-triethylphosphorodithioate, required 30 min or more of digestion on a steam bath.

Triphenylphosphine required an even longer digestion. Thiophosphate esters required no special treatment. Dithiophosphate esters, in aqueous or non-aqueous solution, were pre-treated with aqueous bromine solution. The excess of bromine was boiled off, reagent was added, and the procedure was continued as described. This allowed the determination of phosphorus without further digestion.

Persulphate[288] and hydrogen peroxide[289] oxidations and sodium carbonate–potassium nitrate fusion (1 g; 2:1) in a platinum crucible[290] have also been used for the determination of phosphorus in organic compounds. Sulphuric acid and potassium permanganate have been mentioned for the decomposition of organophosphorus compounds[291]. Following decomposition of various substituted phosphonic and phosphonothionic acids by this method, the determination of phosphate by the molybdenum blue method gave poor results. However, good results were obtained when phosphorus was determined in these compounds by a semi-micro method involving pyrolytic decomposition of the sample in a silica tube[292]. In a further method the organophosphorus sample was heated with potassium permanganate in a sealed glass tube at 400–500°C, thus oxidizing phosphorus to phosphate, which was then determined titrimetrically[293].

Decomposition with magnesium has been used for the microdetermination of phosphorus[294]. When organic compounds containing phosphorus are burned with metallic magnesium, the phosphorus is converted into magnesium phosphide, which can be decomposed in bromine water and oxidized to phosphoric acid, which can be determined photometrically as molybdophosphoric acid after having been extracted with ethyl acetate. The presence of nitrogen, sulphur, and halides does not cause interference. The method is suitable for 1.0–2.5 mg of sample, and the error is ±0.4%

4. Oxygen flask combustion

This technique has been extensively studied for the determination of phosphorus in organophosphorus compounds[287,295–307]. One procedure involves ignition over dilute nitric acid followed by reaction of the combustion products with magnesia mixture, which is filtered off and determined by reaction with ethylenediaminetetracetic acid to the Eriochrome Black T end-point (semi-micro method) or are reacted with molybdate reagent for a spectrophotometric finish (micro method)[296]. Other elements that form heteropoly acids with molybdates which are reducible to molybdenum blue (in the micro method) are silicon, arsenic, and germanium. Silicon, as silicate, does not interfere with the colorimetric method. Arsenic may be separated from the phosphate and determined quantitatively[297], but arsenic occurs rarely with phosphorus in organic compounds. Germanium is encountered infrequently in organic analysis. Complete transformation of phosphorus pentoxide to phosphoric acid requires boiling with dilute nitric acid for 10–15 min prior to application of the molbdate procedure. A further oxygen flask combustion uses 1 N sulphuric acid as the absorbent and a molybdate finish[298]. The analysis should be completed soon after the combustion, since phosphorus pentoxide may be lost from the solution (about 5% loss in 24 h, probably through adsorption by the glass).

A comparative study has been made of the methods available for the semi-micro determination of phosphorus in fabrics flame-proofed with organic phosphorus compounds[299]. In one method the products from oxygen flask combustions are absorbed in aqueous hydrogen peroxide solution[300]. After combustion of the sample over dilute hydrogen peroxide solution, the solution is boiled for 30 min and made slightly acidic. Eriochrome Black T solution and excess of 0.01 N lead nitrate are added, the excess being back-titrate with 0.01 N potassium dihydrogen phosphate.

In another very simple method the combustion products were absorbed in water and, after boiling, the resulting phosphoric acid was titrated with 0.1 N sodium hydroxide to the thymolphthalein end-point[301]. This method is obviously subject to interferences. A mixture of 0.4 N perchloric acid and 0.4 N nitric acid has been used as absorbent[277]. Phosphate can then be determined by the ammonium molybdovanadate method. In another procedure the combustion products were absorbed in 0.4 N sulphuric acid and the solution was boiled prior to adding ammonium persulphate[302]. An aliquot was diluted, treated with acidic $(NH_4)_2[MoO_4]$, bismuthyl carbonate, and ascorbic acid (as reducing agent), and the molybdenum blue measured at 710 nm. Sodium carbonate was used with silicon compounds to facilitate dissolution and boric acid was used with fluorine compounds to prevent etching of the glass. In the presence of arsenic and divalent nickel high and low results, respectively, were obtained.

Organophosphorus compounds can be burned over a solution of ammonia[303]. After boiling the resulting solution to remove excess of ammonia it is passed through a column of Amberlite IR-120 (H^+ form, 20–50 mesh). The percolate is acidified with nitric acid and pyridine, acetone, and a 0.1% dithizone solution in acetone added, and the solution is titrated with 0.01 M lead nitrate to a red colour. If fluoride is present, the absorber solution after removal of ammonia is acidified with nitric acid and evaporated to dryness; the residue is dissolved in water and then treated as before. Technicon AutoAnalyzer system has been adopted to the analysis of solutions of organophosphorus compounds decomposed by the oxygen flask technique[304]. This procedure is capable of determining less than 0.1 μg of phosphate per millilitre of test solution. A modification of the oxygen flask technique provides a flow of oxygen to the combustion bottle, the products from which pass into a Wickbold absorber[307]. The technique was applied successfully to the determination of phosphorus in solids and liquids (e.g. lubricating oil); recoveries from 92–106% were obtained for the range 10–1000 ppm of phosphorus.

5. *Miscellaneous methods*

The continuous band spectrum of phosphorus in ethanol solutions of organophosphorus compounds has been investigated[308]. For concentrations between 0.01 and 0.03 M the average error was 0.0006 M; sodium and calcium ions cause positive errors whereas nitrogen, iodine, sulphur, and chlorine do not interfere in amounts equivalent to that of phosphorus. Flame emission measurements at 540 nm were made on a variety of organophosphorus compounds. The standard deviation for tributyl phosphate in paraffin was ±0.88% at the 46% level and ±0.015% at the 1.7% level.

Three methods capable of detecting 0.2 μg of combined phosphorus are applicable to acids, esters, acyl halides, and anhydrides, together with their thio analogues[309]. They involve degradation by refluxing with concentrated sulphuric acid. Two of these methods then involve the use of *O*-dianisidine molybdate reagent and indicate phosphorus by the formation of a reddish brown precipitate. The third is based on the production of molybdenum blue, and hydrazine hydrate is used as the reducing agent. Methods of sampling and the application to air analysis were also discussed. Unstable organophosphorus compounds may be stabilized by exposure to sulphur for 8–48 h *in vacuo*, and the resulting compounds are analysed by pyrolysis[310].

B. Determination of Iron, Silicon, Titanium, Arsenic, and Phosphorus

To determine phosphorus and iron in organophosphorus compounds the sample may be fused with sodium peroxide in a bomb, the melt taken up in nitric acid and the phosphate titrated potentiometrically with standard lanthanum nitrate–ammonium chloride solution at pH 8[311]. In another aliquot of the solution, iron is determined photometrically with sulphosalicyclic acid in aqueous ammonia medium. The absolute error is less than 0.3% for each element.

Phosphorus and silicon occur together in certain types of organometallic compounds. They may be determined after decomposition by heating with potassium persulphate or hydrogen peroxide solution dissolved in concentrated sulphuric acid[312]. The silicic acid formed is determined separately[313]. The filtrate is used for the photometric determination of phosphorus as molybdenum blue. For samples containing 10–20% each of silicon and phosphorus, differences between the calculated and determined contents were about 0.2% in single determinations. When organic phosphorus silicon compounds are fused with potassium metal, phosphorus is reduced to potassium phosphide, which does not interfere with the amperometric titration of silicon; similarly, the potassium silicate formed does not interfere with the determination of phosphorus by amperometric titration with uranyl acetate solution[314]. Methods for the determination of free phosphorus, combined phosphorus, and silicon in reaction products of tetraalkoxysilanes with potassium halides are given in refs. 315 and 316. To determine titanium, phosphorus, and silicon in organic compounds that are difficult to decompose the sample may be fused with ammonium fluoride and potassium pyrosulphate and the cooled melt treated with concentrated sulphuric acid, followed by evaporation in an air-bath until fumes appear[316]. The residue is dissolved in 70% sulphuric acid and titanium, phosphorus, and silicon are determined by standard procedures.

Phosphorus and arsenic in organic compounds may be determined by first burning the sample by a modified oxygen flask method[281]. Phosphorus is determined by precipitation as quinoline molybdophosphate and titration with sodium hydroxide solution, or spectrophotometrically at 750nm as molybdenum blue, with iron(II) ammonium sulphate as the reductant. Arsenic is determined spectrophotometrically at 840 nm by a similar method, with hydrazine sulphate as reductant. The absolute accuracy is within ±0.5% for phosphorus by either method, and within ±1% for arsenic.

C. Determination of Carbon and Hydrogen

The determination of carbon and hydrogen in organic compounds containing phosphorus and sulphur is difficult, especially in compounds containing phosphorus[317]. This is due mainly to the formation of a phosphorus pentoxide–carbon film inside the combustion tube. The phosphorus pentoxide crystals surround the particle of carbon, making them thermoresistant, and a very high temperature (around 900–1000°C) is necessary to destroy this complex. The heaters of a standard Pregl combustion unit are capable only of temperatures around 700–800°C and it requires an extremely long time to decompose phosphorus pentoxide residue left inside the tube using such a technique. In order to provide a simple, fast, and economical method for the analysis of materials that are difficult to combust the Korbl method[318–321] is recommended[317]. This uses a packing of thermally decomposed silver permanganate without a buffer zone of asbestos. In order to effect complete combustion of phosphorus compounds, it is necessary to use a Fischer blast burner and to heat the sample vigorously for 5 min. The special heating element used by Korbl to prevent condensation of water in the capillary end of the absorption tube (Anhydrone) may be replaced by a simple steel hook connected to the stationary heater[317]. The total time required for one carbon and hydrogen determination is about 40 min. The principle of the method is that the organic material is burned at 600–700°C in an oxygen atmosphere. The conversion of carbon to carbon dioxide and of hydrogen to water is accelerated by passing the combustion products over decomposed silver permanganate. Sulphur and phosphorus oxides combine directly with silver wool and the contact mass of decomposed silver permanganate. Water produced from the combustion is absorbed in Anhydrone, and carbon dioxide is Ascarite tubes.

Carbon in organophosphorus compounds may be determined by wet combustion using a modified Van Slyke method[322,323]. For the micro-determination of carbon and hydrogen a standard combustion train with cobalt(III) oxide as the oxidizing agent may be used[324]. The combustion products from the sample are passed through finely divided silver supported on pumice to remove oxides of phosphorus, and the carbon dioxide and water are then determined by conventional gravimetry.

Other procedures involve combustion on pumice for the determination of carbon, hydrogen, and phosphorus[292], and for the simultaneous micro-determination of phosphorus, sulphur, carbon, and hydrogen in compounds containing these elements plus nitrogen pyrolysis of the sample followed by combustion with a large excess of hydrogen[325]. Phosphorus pentoxide produced in the combustion is absorbed by powdered quartz, which has been etched with caustic alkali. The method is claimed to be particularly suited to compounds with the C—P linkage.

D. Determination of Nitrogen

Nitrogen and phosphorus in organophosphorus compounds may be determined by first heating the sample with 70% perchloric acid[326]. In separate aliquots of the dilute solutions nitrogen is determined spectrophotometrically at 420 or 500 nm with Nessler reagent, and phosphorus is determined at 830 nm as molybdenum blue. This method has been applied to organic compounds and various natural products, such as egg and milk lipids, urine, meat extract, and some amino acids.

E. Determination of Oxygen

A novel approach to the carbon reduction method permits the direct determination of oxygen in organophosphorus compounds with an average recovery of 100.0% and a relative standard deviation of ±0.05%[327]. The method uses a carbon reduction bed contained in an induction-heated graphite pipe. The silica of the quartz reaction chamber is not directly exposed either to the corrosive vapours of the sample or to the hot reducing carbon of the graphite. This permits considerably more latitude in the operating temperatures of the carbon bed than was found necessary by earlier workers[270,328–331]. In a fluorination method for the determination of oxygen in organophosphorus compounds the sample is placed in a nickel vessel containing $BrF_2 \cdot SbF_6$, which is then evacuated and heated at 500°C to fluorinate the sample and convert the oxygen to oxygen gas[332]. The latter is determined by mass spectrometric analysis. This procedure is tedious and lengthy.

F. Determination of Halogens

In an ultramicro method for the determination of fluorine in 1–20 ±g samples of volatile organophosphorus compounds the sample is decomposed by combustion in an oxygen-filled flask and oxidized with a mixture of nitric and perchloric acids[333]. Fluorine is then determined photometrically with the zirconium–cyanine or the zirconium–norin complex. Another method depends upon the bleaching action of fluoride ions on the iron(III)–sulphosalicylic acid complex at pH 2.85–2.90[334]. The determination of fluorine and phosphorus in organic compounds using the Parr bomb technique followed by spectrophotometric determination has been discussed[335]. Phosphate present in the alkaline melt in the bomb after the fusion interferes in the determination of fluorine. Phosphate and fluoride are therefore separated by ion exchange. Other methods for determining fluorine involving precipitation of lead chlorofluoride[336,337] and the use of the thorium–alizarin lake complex[338–341]. For the latter, distillation of fluoride ions from the phosphate residue prior to titration is recommended because of the interference of orthophosphate with the fluorine determination[339].

Another method for total fluorine uses sodium ethylate in ethanol to form fluoride ion from phosphorus fluoridate while esterifying the phosphorus moiety[342]. This precludes the formation of orthophosphate or alkyl phosphonic acid during the conversion of fluoridate to fluoride. The fluoride is titrated using a thorium–alizarin method without prior distillation of fluoride from the phosphorus residue and in this sense is superior to the method describe above[338].

In a method for the determination of the halogen in phosphonitrile halides the sample is treated with pyridine and then with water[343]. Rapid hydrolysis occurs, and the halide is titrated with silver nitrate solution, preferably potentiometrically. Chlorine in 2-chloroethyl derivatives of phospho-organic acids may be determined by dissolving the substance in ethanediol, and boiling the solution under reflux with sodium hydroxide in ethanediol[344]. In ethanolic medium, the reaction is incomplete. After addition of aqueous nitric acid, the solution is cooled and treated with excess of standard silver nitrate, the excess of which is determined by back-titration with standard ammonium thiocyanate.

G. Determination of Sulphur

Sulphur in organophosphorus compounds may be determined by fusing the sample in a bomb with sodium peroxide[345,346]. The sulphate produced is titrated with 0.02–0.01 N barium chloride in the presence of one drop of 0.2% aqueous nitchromazo[347]. Procedures for overcoming phosphorus interference in the determination of sulphur have been discussed[348,349] and details have been given for the determination of organically bound sulphur and phosphorus by oxygen flask combustion[350]. An oxygen flask combustion method can be used for the determination of small amounts of fluorine or phosphorus or sulphur in substances of low volatility[333]. To determine sulphur, combustion products are oxidized with a nitric acid–perchloric acid mixture, and sulphur is reduced to hydrogen sulphide and titrated with cadmium chloride solution. It was not stated whether phosphorus interferes in the determination of sulphur by these procedures although it has been pointed out that methods employing the oxygen flask combustion and subsequent titration with a barium salt usually give high results because of the slightly soluble barium phosphate formed[351]. It is necessary, therefore, to eliminate the phosphate produced in the courseof the combustion before an accurate measurement of sulphur can be made. Phosphate ions may be masked with iron(III) ions, since the latter chelate more readily with phosphate than with sulphate ions in an acidic solution, and the excess of iron(III) ions can be back-titrated with EDTA[352]. Sulphate is titrated by a conventional procedure using standard barium chloride[353]. A combustion furnace procedure for the determination of carbon, hydrogen, sulphur, and phosphorus[325] and a rapid method for the determination of phosphorus—sulphur bonds in organophosphorus insecticides have been described[354].

XXII. ORGANOPLATINUM COMPOUNDS

Organoplatinum compounds may be broken down by gentle refluxing with a 50:50 mixture of concentrated hydrochloric and nitric acids followed by destruction of the remaining nitric acid by further boiling with hydrochloric acid[355]. A colorimetric analysis at 403 nm after treatment with tin(II) chloride solution enables platinum to be determined to within 1%[356].

XXIII. ORGANOPOTASSIUM COMPOUNDS

Three methods for the flame photometric determination of potassium in potassium tetraphenylborate have been described[357]. In the first the potassium tetraphenylborate is precipitated in aqueous solution, and the precipitate dissolved in acetone and then examined by flame photometry. In the second method the precipitate of potassium tetraphenylborate is heated for 20 min at 350°C prior to dissolving it in water for flame photometry. In the third method the potassium tetraphenylborate is converted to potassium chloride by boiling with an aqueous solution of mercury(II) chloride.

Carbon, hydrogen, and potassium in organic samples can be determined by mixing the sample intimately with finely ground quartz in an empty tube and combusting it in a stream of oxygen[177]. Combustion in an empty tube prevents absorption of carbon dioxide by the potassium residue and permits carbon and hydrogen to be determined by standard methods. The amount of potassium can then be obtained from the weight increase of the ignition tube after the combustion.

XXIV. ORGANOSELENIUM COMPOUNDS

A. Determination of Selenium

Various techniques have been employed for the determination of selenium in organoselenium compounds. These include combustion techniques, oxygen flask combustion, fusion with sodium peroxide, and digestion with acids. Selenium in organic compounds containing carbon, hydrogen, oxygen, and nitrogen can be determined by tube combustion of the sample in oxygen[358]. After the ignition, the oxygen intake is replaced with a Mariotte flask serving as an aspirator, the layer of sublimed selenium dioxide is treated with water, and the selenous acid produced is determined iodimetrically. Another method[359] involves igniting the sample in a stream of oxygen and collecting the selenium dioxide produced in an absorption funnel prior to colorimetric determination with 3,3′-diaminobenzidine[360]. Sulphur or halogens do not interfere in this procedure. Combustion of organoselenium compounds in a stream of oxygen over quartz wool in a quartz tube or in an oxygen-filled flask followed by iodimetric determination yields an accuracy of $\pm 2\%$[361,362].

Oxygen flask combustion has been used by several workers for the determination of selenium in organic compounds[361,363–369]. The sample may be wrapped in paper and the oxygen combustion conducted in a flask containing distilled water[361]. After the combustion is completed, the selenium may be determined iodimetrically[361,363,368,369] or with permanganate[364,365]. The latter method has been applied to samples containing up to 62% of organically bound selenium with an absolute systematic error of less than 0.12%. In a further procedure the sample is burnt in an oxygen-filled flask and the vapours are absorbed in water[366,367]. The selenite produced is converted to selenocyanate ion by adding potassium cyanide solution. Sodium tungstate is added, the solution neutralized and a slight excess of aqueous bromine added. Unconsumed bromine is destroyed with phenol solution and the cyanogen bromide produced is reacted with potassium iodide and the liberated iodine is titrated with 0.01 N sodium thiosulphate. The results obtained by this procedure for organoselenium compounds showed a mean overall error of $\pm 0.08\%$. The micro-determination of organic selenium has been carried out by fusion with sodium peroxide in a micro Parr bomb. Organic compounds containing selenium are readily decomposed by heating with sodium peroxide in a micro-bomb[370]. The product is dissolved in water, neutralized, and reacted with hydrazine, and the precipitate is filtered off in a fine-glass filter, dried at 110°C and weighed.

A colorimetric method has been described for the determination of selenium in organoselenium compounds following kjeldahl digestion[371]. Selenium is determined spectrophotometrically at 420 nm using chlorpromazine. Kjelkahl digestion with a mixture of concentrated sulphuric acid and potassium permanganate has been used as a preliminary to the iodimetric micro-determination of selenium in organic compounds[372]. Traces of selenium in organic matter can be determined using a combined spectrophotometric–isotope dilution method[373]. An earlier spectrophotometric method[360] was adapted to the micro-scale and the method improved by including an isotope-dilution procedure to compensate for the unavoidable loss of selenium. The sample (containing added ^{75}Se) is oxidized under reflux with a mixture of nitric, perchloric, and sulphuric acids and the selenium is then recovered as selenium tetrabromide by double distillation with hydrobromic acid and determined spectroscopically at 420 nm using 3,3-diaminobenzidine. X-ray emission can be used to determine down to 50 ppm of selenium in organic compounds[374]. The micro-determination of selenium in organic substances by chelatometry has been discussed[375].

B. Determination of Carbon and Hydrogen

Carbon and hydrogen in organic compounds containing selenium can be determined by combustion in oxygen in an empty tube, using finely ground quartz as a filter to retain the selenium dioxide produced[376]. An error not exceeding ±0.3% is claimed. Selenium in selenosemicarbazones can be determined by conversion to silver selenide and determination of silver by the Volhard method[377].

XXV. ORGANOTHALLIUM COMPOUNDS

A. Determination of Carbon, Hydrogen, and Thallium

Carbon and hydrogen in organic compounds containing thallium can be determined by a combustion method in which finely ground silica is used as a filler to prevent attack by thallium on the silica combustion tube and catalyst, and to prevent the formation of thallium compounds during the determination of carbon and hydrogen on some types of compounds[378]. The combustion tube contains a silver spiral, silver turnings, cobalt(III) oxide on corundum as catalyst and a platinum spiral; this zone of the furnace is heated at 680–700°C. The sample is covered with finely ground silica and ignited in a stream of oxygen at 1000°C and water and carbon dioxide are determined by standard procedures. In a further procedure in which thallium is also determined the sample is placed in a silica tube and covered with a layer of powdered silica[379]. The tube is then heated in a stream of oxygen and the pyrolysis products are passed over cobalt(III) oxide at 680–700°C. The water and carbon dioxide formed are absorbed in Anhydrone and Ascarite, respectively, in a Pregl apparatus and determined by weighing. Thallium is determined by weighing the residue (probably thallium silicate) in the silica tube. For compounds of the types tris(ethylenediamine)thallium nitrate and bipyridylthallium chloride, the error was within ±0.3% for carbon or hydrogen and within ±0.5 for thallium. Halogens do not interfere with the determination of carbon and hydrogen but interfere with that of thallium.

XXVI. ORGANOTIN COMPOUNDS

A. Determination of Tin

Numerous methods have been described for the determination of tin in organotin compounds[380–393]. These include gravimetric[381], volumetric[380,382,384], oxygen flask combustion[385], complexometric[383,386,387], photometric[380,388], X-ray fluorescence[393], X-ray spectrophotometric[389], spectrographic[390], and polarographic[394] methods. One volumetric procedure[382] determines tin in the presence of phosphorus. Another procedure involves a volumetric determination of tin after destruction of the organotin compound with a solution of bromine in chloroform[384]. A solution of bromine in carbon tetrachloride has also been used as a preliminary treatment for more volatile types of organotin compound[395]. Titrimetric methods[380,385] for the determination of tin are usually based on a final oxidation of tin(II) to tin(IV) and suffer from the disadvantages that all traces of the oxidizing agent used in the initial combustion must be removed and that an inert atmosphere must be maintained until the final titration is complete because of the ready oxidation of tin(II) by oxygen.

Samples may be digested by heating with concentrated nitric acid, then perchloric and hydrochloric acids[383], or nitric acid and sodium sulphate[386], prior to complexometric determination of tin. Organotin compounds can be decomposed by wet oxidation with a mixture of nitric and sulphuric acids, followed by ignition to tin(IV) oxide at 900°C, which is determined gravimetrically[396,397]. A successful procedure for very volatile compounds such as stannones is to aspirate the vapour of the sample by means of a current of nitrogen into a mixture of nitric and sulphuric acids and digestion mixture and continue by ignition to tin(IV) oxide as described above. A simple digestion method uses ammonium nitrate as oxidizing agent for the determination of tin, and also silicon and titanium in organometallic compounds[398]. An alternative digestion medium is sulphuric acid[399]. Oxygen flask combustion may be used for the micro-determination of tin[385]. Combustion in oxygen converts the organotin compound to a mixture of tin(IV) and tin(II) oxides. The combustion residue is warmed with freshly prepared chromium(II) sulphate solution, to dissolve the poorly soluble tin(IV) oxide. The tin(II) ions and unconsumed chromium(II) ions are oxidized by air, and the tin(IV) ions are then redued with sodium hypophosphite and titrated with standard potassium iodate solution.

Two procedures have been described for the determination of tin in volatile organotin hydrides[400,401]. One involves bubbling oxygen through the weighed sample until evaporation is complete[400]. The vapour is passed into a weighed silica tube and ignited in a plug of prepared asbestos. The silica tube is then removed and ignited to constant weight at 800°C to obtain the weight of tin(IV) oxide produced. With very volatile samples or samples of high boiling point it is necessary to control the temperature during the evaporation with a cold bath or an i.r. lamp. A modification of this procedure is available for handling samples that readily hydrolyse to produce non-volatile products. The procedure is accurate to within ±0.5% of the determined result and requires about 4 h per analysis.

Alkyltin compounds may be oxidized with sodium peroxide in a Parr bomb[402]. The product is boiled with water in the usual way, and tin is determined spectrophotometrically using cacotheline. For the spectrographic analysis of organotin compounds the sample is dissolved in cumen containing polymethylphenylsiloxane as a silicon internal standard[390]. The solution is passed into the inner vessel of an atomizer, the outer vessel of which is cooled. Spectra are excited by a condensed spark discharge between the graphite rod in the atomizer and graphite counter electrode. The reaction of methyllithium with triphenyltin hydride has been used as the basis for the determination in the latter compound[403]. Tetrabutyl- and tetraethyltin in factory air have been analysed by oxidation with a hydrogen peroxide–sulphuric acid mixture and colorimetric determination of total tin[404].

Residues of triphenyltin compounds used in crop protection can be determined by extraction of the material with dichloromethane, phase separation of triphenyltin compounds and their water-soluble decomposition products, and determination of tin after destruction of organic matter and distillation of the tin as the tetrabromide[405]. Triphenyltin, diphenyltin, and inorganic tin compounds have been analysed by conversion to triphenyltin hydroxide, which is extracted into chloroform and determined by conventional methods[406].

The determination of triphenyltin acetate has been described by various workers[407–414]. These include its determination in ruminants[409], plants[415], celery and apple[411], potato leaves[412], sugar beet leaves and in animals feeding thereon[414], and in milk[415]. Methods have been reported for determining microgram amounts of tin in animal and vegetable matter[415]. tin in foods[416], organotin fungicides used in potato blight control[417], organotin stabilizers in foods[418] and tricyclohexyltin hydroxides in fruits[419]. Work has been carried out on the determination of tin in aqueous leachates from organotin-containing antifouling paint compositions[420–423]. These include tributyltin compounds[420] and bis(tributyltin oxide)[421,422]. Tin analyses by atomic-absorption spectroscopy have been described, including the determination of butyltin compounds in textiles by graphite furnace atomic-absorption spectroscopy[424–426].

B. Determination of Carbon and Hydrogen

Methods have been described for the determination of carbon and hydrogen in organic compounds containing tin[427], arsenic, antimony, bismuth, and phosphorus[428] and complex compounds containing a tin halide[429].

C. Determination of Halogens

Methods have been described for the determination of halogen in organotin compounds[381,430,431] including long-chain alkyltin compounds[432]. These include the determination of halogens in alkyltin halides by high-frequency titration[381,433]. Trimethyl- triethyl-, and tributyltin halides and dimethyl-, diethyl-, dipropyl-, and dibutyltin halides all produce halide ions upon reaction with water, which can be determined by high-frequency titration with standard silver nitrate[433]. The error is claimed to be less than 3% in the 10–75% halogen range.

D. Determination of Nitrogen

Nitrogen can be determined in some types of organotin compounds by Kjeldahl digestion procedures[381].

E. Determination of Sulphur

Sulphur can be determined gravimetrically[381] or by a mercurimetric procedure[434] in which the sample is dissolved in toluene, methanol and aqueous sodium hydroxide and solid thiofluorescein–sodium sulphite is added. The solution is then titrated with standard *o*-hydroxymercuribenzoic acid to the disappearance of the blue colour. The accuracy is claimed to be within ±0.02% of sulphur. Cleavage with standard solutions of iodine has been used for the determination of Sn—S bonds[435]:

$$2R_3SnSR' + I_2 \longrightarrow 2R_3SnI + R'SSR' \qquad (22)$$

XXVII. ORGANOTITANIUM COMPOUNDS

A. Determination of Titanium

Digestion using hydrofluoric and nitric acids is suggested for the determination of titanium in titanium carbide–niobium alloys[436]. The determination of total and free carbon in titanium carbide, its cermet, and sintered cermet has been discussed[437]. A simple digestion procedure for the determination of titanium and tin in organometallic compounds involves dissolving the sample in a weighed platinum crucible by warming with sulphuric acid[398]. Crystalline ammonium nitrate is added to the cooled solution and the mixture shaken gently. The crucible contents are then heated until the weight of the metal oxide produced becomes constant. Titanium can then be determined by standard methods. Advantages claimed for this procedure include an absence of metal loss by volatilization.

In a wet combustion procedure for the analysis of organic compounds containing silicon, organic titanium, and phosphorus, the sample is heated with potassium persulphate and concentrated sulphuric acid for 2–4 h[438,439]. The filtrate is diluted with water and then excess of potassium hydroxide is added. This solution is passed through an ion-exchange column (KU-2, chloride form) which retains titanium. Phosphorus is determined photometrically as molybdate in the eluate. Titanium is eluted from the resin with 4 N hydrochloric acid and subsequently determined spectrophotometrically. The absolute error did not exceed 0.5% for determinations of titanium at the 10% level in the sample. In a further method for the analysis of difficult to decomposable organic compounds containing titanium, phosphorus, and silicon the sample is fused with sodium fluoride and potassium persulphate[440]. The cooled melt is then treated with concentrated sulphuric acid and evaporated to fuming in an air-bath. The residue is dissolved in 70% sulphuric acid and titanium, phosphorus, and silicon are determined by standard procedures.

Iron and titanium in donor-acceptor complexes of ferrocene bases may be determined by decomposing the sample with a mixture of nitric acid, anhydrous acetic acid and aqueous bromine[441]. Iron is titrated complexiometrically with sulphosalicyclic acid as indicator and titanium is determined colorimetrically by extraction of its 8-hydroxyquinoline complex into chloroform.

XXVIII. ORGANOZINC COMPOUNDS

A. Determination of Zinc

Zinc can be determined in organozinc compounds by a procedure involving complexometric titration with EDTA (disodium salt) in a suitably buffered medium[442]. The organozinc compound is diluted with an inert organic solvent and decomposed by the addition of dilute hydrochloric acid. Zinc can then be determined in the water extract by complexometric titration. Dialkylzinc preparations can be made by the reaction of an alkylaluminium compound with zinc chloride. Such preparations, even after distillation, usually contain residual amounts of organoaluminium compound. If the organozinc compound contains aluminium then this would be present in the aqueous extract and would interfere in the complexiometric determination of zinc. To overcome this the dilute hydrochloric acid extract obtained by decomposition of the organozinc sample is passed down a column of Amberlite IRA-400 ion-exchange resin. Percolation of the column with 2 N hydrochloric acid completely removes aluminium from the column, which can be determined in the eluate by complexometric titration. Subsequent percolation of the ion-exchange column with 0.2 N nitric acid then completely desorbs zinc, which can be collected separately and determined by complexometric titration[442]. Various procedures have been described for the determination of zinc in plants[443], organic material[444], fungicides such as zineb and ziram[445] and in zinc stearate[446]. None of these procedures is particularly relevant to the determination of zinc in pyrophoric organozinc compounds.

B. Determination of Halogens

The various procedures described in detail in Section II.C for the determination of halogens in organoaluminium compounds can also be applied to organozinc compounds.

C. Determination of Zinc-bound Alkoxide Groups

Alkoxide groups up to butoxide can be determined in organozinc compounds by hydrolysis with glacial acetic acid followed by spectrophotometric determination of the alcohol produced using cerium(IV) ammonium nitrate[447] as described in connection with the determination of alkoxide groups in organoaluminium compounds (Section II.G).

D. Determination of Lower Alkyl and Hydride Groups

The procedure for the determination of alkyl groups up to butyl and of hydride groups in organoaluminium compounds based on alcoholysis and hydrolysis of the sample described in Section II.D is applicable, without modification, to the determination of the same groups in organozinc compounds[448]. A rapid hydrolysis procedure for the analysis of zinc alkyls is also available[449].

XXIX. ORGANOZIRCONIUM COMPOUNDS

In one method for the determination of zirconium, the sample is fused with sodium carbonate, the cooled melt is dissolved in dilute hydrochloric acid, and sulphur (if present) is oxidized to sulphate with hydrogen peroxide[450]. Zirconium is determined by direct amperometric titration at a rotating platinum electrode with 0.002 M EDTA at an applied potential of +0.9 V, with hydrochloric acid as supporting electrolyte. The error is −0.5% (absolute). In a further method zirconium is determined in organic compounds, after fusion with sodium carbonate, by amperometric titration with EDTA using a graphite electrode impregnated with paraffin wax[451]. This electrode is preferred, because of its greater accuracy and applicability over a wider range of potentials, to the rotating platinum electrode for carrying out this determination. The $E_{1/2}$ value for the electro-oxidation of EDTA in hydrochloric acid medium at pH 3–4 is +1.2 V *vs*. SCE. The diffusion current of EDTA was measured at +1.3 V to find the end-points in the amperometric titration of zirconium.

XXX. REFERENCES

1. D. F. Hagen, D. G. Biechler, W. D. Leslie, and D. E. Jordan, *Anal. Chim. Acta*, **41**, 557 (1968).
2. T. R. Crompton, *Chemical Analysis of Organometallic Compounds,* Vol. V, Academic Press, London, New York, San Fransisco, 1977, Chapter 20, pp. 115–122.
3. E. Wannier and A. Ringbom, *Anal. Chem.*, **12**, 308 (1955).
4. F. Nydahl, *Talanta*, **4**, 1017 (1960).
5. G. S. Shvind'erman and E. N. Zavadovskaya, *Zavod. Lab.*, **31**, 32 (1965).

6. C. Hennart, *Chim. Anal.*, **43**, 283 (1961).
7. D. E. Jordan and W. D. Leslie, *Anal. Chim. Acta*, **50**, 161 (1970).
8. R. Belcher, B. Crossland and T. R. F. W. Fennell, *Talanta*, **17**, 639 (1970).
9. M. P. Strukova and T. V. Kirillova, *Zh. Anal. Khim.*, **21**, 1236 (1966).
10. M. F. Smith and R. A. Bayer, *Anal. Chem.*, **35**, 1098 (1963).
11. N. E. Gel'man and I. I. Bryushkova, *Zh. Anal. Khim.*, **19**, 369 (1964).
12. M. O. Korshun, *Anal. Abstr.*, **6**, 2625 (1959).
13. E. L. Head and J. Holley, *Anal. Chem.*, **28**, 1172 (1956).
14. T. R. Crompton, *Chemical Analysis of Organometallic Compounds,* Vol. V, Academic Press, London, New York, San Fransisco, 1977, Chapter 20, pp. 124–135.
15. E. Bonitz, *Chem. Ber.*, **88**, 742 (1955).
16. K. Ziegler, *Justus Liebigs Ann. Chem.*, **589**, 91 (1954).
17. T. R. Crompton and V. W. Reid, *Analyst (London),* **88**, 713 (1963).
18. T. R. Crompton, *Anal. Chem.*, **39**, 1464 (1967).
19. R. Dijkstra and E. A. M. Dahmen, *Z. Anal. Chem.*, **181**, 399 (1961).
20. Stauffer Chemical Company, Anderson Chemical Division, Exclusive Sales Agent, Weston, Michigan, USA, *Alkyls Bulletin, Triethylaluminium Analytical Methods*, 1959, T.68 5–4 and T.68 5–5.
21. B. J. Phillip, W. L. Mundry, and S. C. Watson, *Anal. Chem.*, **45**., 2298 (1973).
22. R. Z. Lioznova and M. L. Genusov, *Zavod. Lab.*, **26**, 945 (1960).
23. W. P. Neumann, *Justus Liebigs Ann. Chem.*, **629**, 23 (1960).
24. T. R. Crompton, *Chemical Analysis of Organometallic Compounds,* Vol. V, Academic Press, London, New York, San Fransisco, 1977, Chapter 20, pp. 186–199.
25. T. R. Crompton, *Chemical Analysis of Organometallic Compounds,* Vol. V, Academic Press, London, New York, San Fransisco, 1977, Chapter 20, pp. 206–213.
26. T. R. Crompton, *Analyst (London)*, **86**, 652 (1961).
27. A. I. Vogel, in *Elementary Practical Organic Chemistry, Part III, Quantitative Organic Analysis*, Longmans, Green & Co., London, 1958, p. 655 (Fig. XIV 3.3).
28. V. W. Reid and R. K. Truelove, *Analyst (London)*, **77**, 325 (1952).
29. V. W. Reid and D. G. Salmon, *Analyst (London)*, **80**, 704 (1955).
30. E. A. Bondarevskaya, S. V. Syavtsillo, and R. N. Potsepkina, *Zh. Anal. Khim.*, **14**, 501 (1959); *Chem. Abstr.*, **54**, 9611f (1959).
31. E. A. Bondarevskaya, S. V. Syavtsillo, and R. N. Potsepkina. *Tr. Kom. Anal. Khim. Akad. Nauk SSSR*, **13**, 178 (1963).
32. G. A. Belova, A. F. Kozlova, and T. N. Boltunova, *Prom. Sintet. Kauch.*, **4**, 19 (1967).
33. S. Mitev, N. Damyanov, and P. K. Ajasyan, *Zh. Anal. Khim.*, **28**, 821 (1973).
34. T. R. Crompton, *Chemical Analysis of Organometallic Compounds,* Vol. V, Academic Press, London, New York, San Fransisco, 1977, Chapter 20, pp. 222–232.
35. T. R. Crompton, *Chemical Analysis of Organometallic Compounds,* Vol. V, Academic Press, London, New York, San Fransisco, 1977, Chapter 20, pp. 232–235.
36. T. R. Crompton, *Chemical Analysis of Organometallic Compounds*, Vol. V, Academic Press, London, New York, San Fransisco, 1977, Chapter 20, pp. 235–238.
37. J. Jeník, *Chem. Listy*, **55**, 509 (1961).
38. M. Juraček and J. Jeník, *Sb. Věd. Prac. Vysoké Školy Chem. Technol. Pardubice*, 105 (1959).
39. Y. Kinoshita, *J. Pharm. Soc. Jap.*, **78**, 315 (1958).
40. H. Bieling and K. H. Thiele, *Z. Anal. Chem.*, **145**, 105 (1955).
41. A. P. Terent'ev, M. A. Volodina, and E. G. Fursova, *Dokl. Akad. Nauk SSSR*, **109**, 851 (1966).
42. M. Corner, *Analyst (London)*, **84**, 41 (1959).
43. R. Belcher, A. M. G. Macdonald, and T. S. West, *Talanta*, **1**, 408 (1958).
44. W. Merz, *Mikrochim. Acta*, 640 (1959).
45. K. Eder, *Mikrochim. Acta*, 471 (1960).
46. W. H. Guttermann, S. F. John, J. E. Barry, D. L. Jones, and E. D. Lisk, *J. Agric. Food Chem.*, **9**, 50 (1961).
47. R. Püschel and Z. Štefanac, *Mikrochim. Acta*, **6**, 1108 (1962).
48. Z. Štefanac, *Mikrochim. Acta*, **6**, 1115 (1962).
49. R. Belcher, A. M. G. Macdonald, S. E. Phang, and T. S. West, *J. Chem. Soc.*, 2044 (1965).

50. A. D. Wilson and D. T. Lewis, *Analyst (London)*, **88**, 510 (1963).
51. H. Kashiwagi, Y. Tukamoto, and M. Kan, *Ann. Rep. Takeda Res. Lab.*, **22**, 69 (1962).
52. G. Bähr, H. Bieling, and K. H. Thiele, *Z. Anal. Chem.*, **143**, 103 (1954).
53. R. Pietsch, *Z. Anal. Chem.*, **144**, 353 (1955).
54. C. DiPietro and W. A. Sassaman, *Anal. Chem.*, **36**, 2213 (1964).
55. A. Steyermark, in *Quantitative Organic Microanalysis*, 2nd ed., Academic Press, New York, London, 1961, p. 367.
56. M. M. Tuckerman, J. H. Hodecker, B. C. Southworth, and K. D. Fliescher, *Anal. Chim. Acta,* **21**, 463 (1959).
57. R. D. Strickland and C. M. Maloney, *Anal. Chem.*, **29**, 1870 (1957).
58. M. Jean, *Anal. Chim. Acta,* **14**, 172 (1956).
59. V. J. Neulenhoff, *Pharm. Weekbl.*, **100**, 409 (1968).
60. M. Juraček and J. Jeník, *Collect. Czech. Chem. Commun.*, **20**, 550 (1955).
61. M. Juraček and J. Jeník, *Collect. Czech. Chem. Commun.*, **21**, 890 (1956).
62. J. S. Edmonds and K. A. Francesconi, *Anal. Chem.*, **48**, 2019 (1976).
63. M. Juraček and J. Jeník, *Sb. Věd. Prac. Vysoké Skoly Chem. Technol. Pardubice*, 105 (1959).
64. L. Maier, D. Seyferth, F. G. A. Stove, and B. G. Rochow, *J. Am. Chem. Soc.*, **79**, 5884 (1957).
65. H. Trutnovsky, *Mikrochim. Acta*, **3**, 499 (1963).
66. G. Bähr, H. Bieling, and K. H. Thiele, *Z. Anal. Chem.*, **145**, 105 (1955).
67. A. P. Terent'ev, M. A. Volodina, and E. G. Fursova, *Dokl. Akad. Nauk SSSR*, **169**, 851 (1966).
68. E. L. Head and E. Holley, *Anal. Chem.*, **28**, 1172 (1956).
69. R. Schmitz, *Dtsch. Apoth.-Ztg.*, **100**, 693 (1960).
70. R. C. Rittner and R. Culmo, *Anal. Chem.*, **34**, 673 (1962).
71. S. Mizukami and T. Ieka, *Michrochem. J.*, **7**, 485 (1963).
72. G. Kainz and G. Chromy, *Mikrochim. Acta,* **16**, 1140 (1966).
73. F. Martin and A. Floret, *Chim. Anal.*, **41**, 181 (1959).
74. D. G. Shaheen and R. S. Braman, *Anal. Chem.*, **33**, 893 (1961).
75. I. Dunstan and J. V. Griffiths, *Anal. Chem.*, **33**, 1598 (1961).
76. S. Kato, K. Kimura and Y. Tsuzuki, *J. Chem. Soc. Jap. (Pure Chem. Sect.)*, **83**, 1039 (1962)
77. R. H. Pierson, *Anal. Chem.*, **34**, 1642 (1962).
78. R. D. Strahm and M. F. Hawthorne, *Anal. Chem.*, **32**, 530 (1960).
79. P. Arthur and W. P. Donahoo, *U.S. Atom. Energy Comm. Rep.*, CCC-1024-TR-221 1957.
80. J. A. Kuck and E. C. Grim, *Microchem. J.*, **3**, 35 (1959).
81. J. J. Bailey and D. G. Gehring, *Anal. Chem.,* **33**, 1760 (1961).
82. H. Allen and S. Tannenbaum, *Anal. Chem.*, **31**, 265 (1959).
83. T. M. Shanina, N. E. Gel'man, and V. S. Mikhailovskaya, *Zh. Anal. Khim.*, **22**, 732 (1967).
84. S. K. Yasuda and R. N. Togers, *Microchem. J.*, **4**, 155 (1960).
85. S. I. Obtemperanskaya and V. N. Likhosherstova, *Vestn. Mosk. Univ.*, **1**, 57 (1960).
86. M. Corner, *Analyst (London),* **84**, 41 (1959).
87. R. Wickbold, *Angew. Chem.*, **69**, 530 (1957).
88. R. Wickbold and F. Nagel, *Angew. Chem.*, **71**, 405 (1959).
89. B. E. Buell, *Anal. Chem.*, **30**, 1514 (1958).
90. T. Yoshizaki, *Anal. Chem.*, **35**, 2177 (1963).
91. A. Shah, A. A. Qadri, and R. Rehana, *Pak. J. Sci. Ind. Res.*, **8**, 282 (1965).
92. M. Giorgini and A. Lucchesi, *Ann. Chim. (Rome)*, **54**, 832 (1964).
93. L. Erdev, E. Gegus, and E. Kocsis, *Acta Chim. Acad. Sci. Hung.*, **7**, 343 (1955).
94. N. C. Malȳsheva, L. P. Starchik, I. S. Panidi, and Ya, M. Paushkin, *Zh. Anal. Khim.*, **18** 1367 (1963).
95. J. M. Thoburn, *Diss. Abstr.*, **15**, 3–4 (1955).
96. N. G. Nadeau, D. M. Oaks, and R. D. Buxton, *Anal. Chem.*, **33**, 341 (1961).
97. H. Nöth and H. Beyer, *Chem. Ber.*, **93**, 928 (1960).
98. G. E. Ryschkewitsch and E. R. Birnbaum, *Inorg. Chem.*, **4**, 575 (1965).
99. H. C. Kelly, *Anal. Chem.*, **40**, 240 (1968).
100. R. C. Rittner and R. Culmo, *Anal. Chem.*, **35**, 1268 (1963).

101. P. Arthur, R. Annino, and W. P. V. Donahoo, *Anal. Chem.*, **29**, 1852 (1957).
102. R. F. Putnam and H. W. Myers, *Anal. Chem.*, **34**, 486 (1962).
103. S. A. Greene and H. Pust, *Anal. Chem.*, **30**, 1039 (1958).
104. A. F. Zhigach, E. B. Kazakova, and R. A. Kigel, *Zh. Anal. Khim.*, **14**, 746 (1959).
105. I. Lysyj and R. C. Greenough, *Anal. Chem.*, **35**, 1657 (1963).
106. D. G. Newman and O. Tomlinson, *Mikrochim. Ichnoanal. Acta*, 1023 (1964).
107. E. A. Kalinovskaya and L. S. Sil'vestrova, *Zavod. Lab.*, **34**, 30 (1968).
108. O. Palade, *Zh. Anal. Khim.*, **21**, 377 (1966).
109. S. A. Bartkiewicz and E. A. Hammatt, *Anal. Chem.*, **36**, 833 (1964).
110. M. A. H. Sharif, *Pak. J. Sci. Ind. Res.*, **1**, 160 (1958).
111. O. Kidson and O. Eskew, *N.Z. J. Sci. Technol.*, **B21**, 178 (1940).
112. T. V. Reznitskaya and E. I. Burtsera, *Zh. Anal. Khim.*, **21**, 1132 (1966).
113. K. L. Cheng, *Microchem. J.*, **7**, 100 (1963).
114. W. L. Wuggatzer and J. E. Christian, *J. Am. Pharm. Assoc., Sci. Ed.*, **44**, 645 (1955).
115. T. Arány and A. Erden, *Magy. Kem. Lab.*, **20**, 164 (1965).
116. V. A. Klimova, M. O. Korshun, and E. G. Bereznitskaya, *Zhur. Anal. Khim.*, **11**, 223 (1956); *Anal. Abstr.*, **3**, 3663 (1956).
117. M. P. Brown and G. W. A. Fowles, *Anal. Chem.*, **30**, 1689 (1958).
118. V. A. Klimova and N. D. Vitalini, *Zh. Anal. Khim.*, **19**, 1254 (1963); *Anal. Abstr.*, **10**, 4716 (1964).
119. R. P. Kreshkov and E. A. Kucharev, *Zavod. Lab.*, **32**, 558 (1966).
120. M. Schülunz and A. Köster-Pflugmacher, *Z. Anal. Chem.*, **232**, 93 (1967).
121. R. S. Braman and M. A. Tompkins, *Anal. Chem.*, **50**, 1088 (1978).
122. H. Pieters and W. J. Buis, *Mickrochem. J.*, **8**, 383 (1964).
123. G. M. Van der Want, personal communication.
124. T. R. F. W. Fennell and J. R. Webb, *Talanta*, **9**, 795 (1962).
125. E. A. Terent'eva and T. M. Malolina, *Zh. Anal. Khim.*, **19**, 353 (1964).
126. H. M. Rosenberg and C. Riber, *Microchem. J.*, **6**, 103 (1962).
127. J. Jenik and F. Renger, *Collect. Czech. Chem. Commun., Engl. Ed.*, **29**, 2237 (1964).
128. J. Décombe and J. P. Ravoux, *Bull. Soc. Chim. Fr.*, **17**, 260 (1964).
129. L. V. Myshlyaeva, V. V. Krasnoshchekov, T. G. Shatunova, and I. V. Sedova, *Tr. Mosk. Khim. Tekhnol. Inst.*, **49**, 178 (1965); *Ref. Zh., Khim.*, **19GD**, Abstr. 18G254 (1966).
130. R. Renger and J. Jenik, *Sb. Věd. Prac. Vysoké Skoly Chem. Technol. Pardubice*, **1**, 55 (1963).
131. E. N. Kuchkarev, *Tr. Mosk. Khim. Tekhnol. Inst.*, **48**, 51 (1965); *Ref. Zh., Khim.*, **19GD**, Abstr. 11G230 (1966).
132. J. Fischl, *Clin. Chim. Acta*, **4**, 686 (1959).
133. D. G. Nathan, T. G. Gabuzda, and F. H. Gardner, *J. Lab. Clin. Med.*, **62**, 511 (1963).
134. J. Fine, *J. Clin. Pathol.*, **14**, 561 (1961).
135. S. Natelson and B. Shied, *Clin. Chem.*, **7**, 115 (1961).
136. L. C. Willemsens and G. J. M. Van der Kerk, in *Investigations in the Fields of Organolead Chemistry*, Institute for Organic Chemistry TNO, Utrecht, Published by the International Lead–Zinc Research Organization, New York, 1965, p. 87.
137. R. P. Kreshkov and E. A. Kucharev, *Zavod. Lab.*, **32**, 558 (1966).
138. F. B. Ashbel, A. M. Parshina, M. S. Goizman, L. T. Zhizhina, and K. M. Kuptsova, *Zavod. Lab.*, **31**, 1062 (1965).
139. G. R. Sirota and J. F. Uthe, *Anal. Chem.*, **49**, 823 (1977).
140. J. W. Robinson, *Anal. Chim. Acta*, **24**, 451 (1961).
141. D. J. Trent, *At. Absorpt. Newsl.*, **4**, 348 (1965).
142. R. M. Dagnall and T. S. West, *Talanta*, **11**, 1553 (1964).
143. H. W. Wilson, *Anal. Chem.*, **38**, 920 (1966).
144. A. C. Menzies, *Anal. Chem.*, **32**, 905 (1960).
145. C. S. Rann, in *The Element*, No. 9, Aztec Instruments, Westport, Conn., USA, 1965.
146. R. A. Mostyn and A. F. Cunningham, *J. Inst. Pet.*, **53**, 101 (1967).
147. E. J. Moore and J. R. Glass, *Microchem. J.*, **10**, 148 (1966).
148. K. Campbell and R. Moss, *J. Inst. Pet.*, **53**, 521,194 (1967).
149. P. R. Ballinger and I. M. Whittemore, *Proceedings of the American Chemical Society Division of Petroleum Chemistry, Atlantic City Meetings*, 13, 133 Sept. 1968 (1968).

150. B. Kolb, G. Kemnner, F. H. Schleser, and E. Wiedeking, *Z. Anal. Chem.*, **221**, 166 (1966).
151. American Society for Testing and Materials, *ASTM Standards on Petroleum Products and Lubricants*, Methods D526-56 (1956) and D526-61 (1961).
152. American Society for Testing and Materials, *ASTM Standards on Petroleum Products and Lubricants*, Methods D1269-53T (1963) and D1269-61 (1961).
153. W. Meine, *Erdöl Kohle,* **8**, 711 (1955).
154. W. Linné and H. D. Wülfken, *Erdöl Kohle,* **10**, 757 (1957).
155. J. Van Rysselberg and R. Leysen, *Nature (London)*, **189**, 478 (1961).
156. T. Okada, T. Ueda, and T. Kohzuma, *Bunko Kenkyu*, **4**, 30 (1956); *Chem. Abs.*, **53**, 65866 (1959).
157. P. Bunčák, *Ropa Uhlie*, **8**, 148 (1966).
158. F. W. Lamb, O. O. Niebylski, and E. W. Kiefer, *Anal. Chem.*, **27**, 129 (1955).
159. C. Campo, *Chim. Anal.*, **45**, 343 (1963).
160. L. Vajta and M. Moser, *Period. Polytech.*, **9**, 275 (1965).
161. E. L. Gunn, *Appl. Spectrosc.*, **19**, 99 (1965).
162. S. Szeiman, *Proc. VII Hungarian Symp. Emission Spectrogr., Pécs, Hungary, August 1964*, pp. 564–565; *Magy. Kém Foly.*, **70**, 511 (1964).
163. G. S. Smith, *Chem. Ind. (London)*, **22**, 907 (1963).
164. G. D. Christofferson and B. Y. Beach, *Colloq. Spectros. Int., 19th, Lyons*, **3**, 492 (1961).
165. A. Ferro and C. P. Galotto, *Ann. Chim. (Rome)*, **45**, 1234 (1955).
166. J. F. Brown and R. J. Weir, *J. Sci. Instrum.*, **33**, 222 (1956).
167. G. Calingaert, F. W. Lamb, H. L. Miller, and G. E. Noakes, *Anal. Chem.*, **22**, 1238 (1950).
168. H. K. Hughes and F. P. Hochgesang, *Anal. Chem.*, **22**, 1248 (1950).
169. M. Farkas and P. Fodor, *Magy. Kém Foly.*, **69**, 407 (1963).
170. American Society for Testing Materials, Materials Research and Standards Committee D2, **2**, 494 (1962).
171. G. Nagypataki and Z. Tamasi, *Magy. Kém. Lapja*, **17**, 140 (1962).
172. J. Krotký, *Paliva*, **36**, 124 (1956).
173. J. M. Lopez, *Colloq. Spectrosc. Int., 7th September 1958, University of Liège*; *Rev. Univ. Mines. Metall. Mec.*, **15**, 299 (1959).
174. L. C. Willemsens, in *Organolead Chemistry*, Institute for Organic Chemistry TNO, Utrecht, Published by the International Lead–Zinc Research Organization, New York, 1964, p. 69.
175. D, Colaitis and M. Lesbre, *Bull. Soc. Chim. Fr.*, 1069 (1952).
176. R. N. Meals, *J. Org. Chem.*, **9**, 211 (1944).
177. A. S. Zabrodina and U. P. Miroshino, *Vestn. Mosk. Univ.*, **2**, 195 (1957).
178. E. L. Head and C. E. Holley, *Anal. Chem.*, **28**, 1172 (1956).
179. J. B. Honeycutt (Ethyl Corp.), *US Pat.*, 3 059 036, Oct. 16th, 1962.
180. H. L. Johnson and R. A. Clark, *Anal. Chem.*, **19**, 869 (1947).
181. E. C. Juenge and D. Seyferth, *J. Org. Chem.*, **26**, 563 (1961).
182. H. Sinni and D. Aumann, *Makromol. Chem.*, **57**, 105 (1962).
183. H. Gilman, *Bull. Soc. Chim. Fr.*, 1356 (1963).
184. R. Reimschneider and K. Petzoldt, *Z. Anal. Chem.*, **176**, 401 (1960).
185. R. Reimschneider and K. Petzoldt, *Z. Anal. Chem.*, **193**, 193 (1963).
186. T. Mitsui, K. Yoshikawa, and Y. Sakai, *Microchem. J.*, **7**, 160 (1963).
187. T. Mitsui, O. Yamamoto and K. Yoshikawa, *Mikrochim. Acta,* 521 (1961); *Anal. Abstr.*, **9**, 174 (1962).
188. R. Belcher, D. Gibbons, and A. Sykes, *Mikrochem. Mirochim. Acta*, **40**, 76 (1952).
189. A. Sykes, *Mikrochim. Acta*, 1155 (1956).
190. T. Sudo and D. Shinoe, *Bunseki Kagaku*, **4**, 88 (1955).
191. V. Pechanec and J. Horáček, *Collect. Czech. Chem. Commun.*, **27**, 239 (1962).
192. F. Martin and A. Floret, *Bull. Soc. Chim. Fr.*, **4**, 610 (1960).
193. A. M. G. Macdonald, *Ind. Chem.*, **35**, 33 (1959).
194. A. M. G. Macdonald, *Analyst (London)*, **86**, 3 (1961).
195. B. C. Southworth, J. H. Hodecker, and K. D. Fleischer, *Anal. Chem.*, **30**, 1152 (1958).
196. P. Gouveneur and W. Hoedeman, *Anal. Chim. Acta*, **30**, 519 (1964).
197. R. Wickbold, *Z. Anal. Chem.*, **152**, 262 (1956).
198. H. Roth and W. Beck, in *Quantitative Organische Mikroanalyse* (Ed. F. Pregl and H. Roth), 7th ed., Springer-Verlag, Vienna, 1958, p. 184.

199. R. Donver, *Z. Chem.*, **5**, 466 (1965).
200. C. Vickers and J. V. Wilkinson, *J. Pharm. Pharmacol.*, **13**, 156T (1961).
201. C. A. Johnson and C. Vickers, *J. Pharm. Pharmacol.*, **11**, 218r (1959); *Anal. Abstr.*, **7**, 329 (1960).
202. K. Kámen, *Chem. Listy*, **47**, 1008 (1953).
203. K. Shigeo, *Microchem. J.*, **8**, 79 (1964).
204. H. Romanowski, *Chem. Anal. (Warsaw)*, **11**, 1027 (1966).
205. M. Wrónski, *Chem. Anal. (Warsaw)*, **5**, 601 (1960).
206. J. C. Merodio, *An. Asoc. Quim. Argent.*, **49**, 225 (1961).
207. D. Pirtea and I. Albescu, *Acad. R.P.R. Stud. Cercet. Chim.*, **7**, 137 (1959).
208. H. F. Walton and H. A. Smith, *Anal. Chem.*, **28**, 406 (1956).
209. G. Spacu and G. Suciu, *Z. Anal. Chem.*, **78**, 244 (1929).
210. L. Manger, *Kem. Ind.*, **14**, 317 (1965).
211. A. Dondrio and C. Molina, *Inf. Quim. Anal.*, **19**, 77 (1965).
212. T. Pelczar and J. Weyers, *Diss. Pharm. (Krakow)*, **13**, 243 (1961).
213. B. Hetnarski and K. Hetnarska, *Roczn. Chem.*, **34**, 457 (1960).
214. A. Kondo, *Bunseki Kagaku*, **10**, 658 (1961).
215. H. Takehara, T. Takeshita, and I. Hara, *Bunseki Kagaku*, **15**, 332 (1966).
216. U. H. Chambers, F. H. Cropper, and H. Crossley, *J. Sci. Food Agric.*, **7**, 17 (1956).
217. C. A. Johnson, *Mfg. Chem.*, **36**, 72 (1965).
218. F. A. J. Armstrong and J. F. Uthe, *Atom. Absorpt. Newsl.*, **10**, 101 (1971).
219. J. F. Uthe, F. A. J. Armstrong, and M. P. Stainton, *J. Fish Res. Board Can.*, **27**, 805 (1970).
220. J. F. Uthe, F. A. J. Armstrong, and K. C. Tam, *J. Assoc. Off. Anal. Chem.*, **54**, 866 (1971).
221. W. R. Hatch and W. L. Ott, *Anal. Chem.*, **40**, 2085 (1968).
222. P. D. Goulden and B. K. Afghan, *Technical Bulletin No. 27*, Inland Waters Branch, Department of Energy, Mines and Resources, Ottawa, Canada.
223. D. T. Burns, F. Glocking, V. B. Mahale, and W. J. Swindall, *Analyst (London)*, **103**, 985 (1978).
224. I. M. Davies, *Anal. Chim. Acta*, **102**, 189 (1978).
225. K. Minagania, Y. Takizawa, and I. Kifune, *Anal. Chim. Acta,* **115**, 103 (1980).
226. R. Bye, P. E. Paus, *Anal. Chim. Acta*, **107**, 169 (1979).
227. R. Dumarey, R. Heindryck, R. Dams, and J. Hoste, *Anal. Chim. Acta*, **107**, 159 (1979).
228. G. J. C. Shum, H. E. Freeman, and J. F. Uthe, *Anal. Chem.*, **51**, 414 (1979).
229. A. I. Lebedeva and K. S. Kramer, *Izv. Akad. Nauk SSSR, Otd. Khim. Nauk,* **7**, 1305 (1962).
230. M. O. Korshun and N. S. Sheveleva, *Zh. Anal. Khim.*, **7**, 104 (1952).
231. A. I. Lebedeva and E. F. Fedorova, *Zh. Anal. Khim.*, **16**, 87 (1961).
232. A. A. Abramyan and A. A. Kocharyan, *Izv. Akad. Nauk Arm. SSR, Khim. Nauk*, **20**, 515 (1967).
233. I. F. Holmes and A. Lander, *Analyst (London)*, **90**, 307 (1965).
234. A. Verdino, *Mikrochemie*, **6**, 5 (1928).
235. A. Verdino, *Mikrochemie*, **9**, 123 (1931).
236. F. Hernlet, *Mikrochemie, Pregl Festschrift*, 154 (1929).
237. M. Furter, *Mikrochemie*, **9**, 27 (1931).
238. G. Ingram, *J. Soc. Chem. Ind. (London)*, **61**, 112 (1942).
239. G. Ingram, *J. Soc. Chem. Ind. (London)*, **58**, 34 (1939).
240. L. Synek and M. Večeřa, *Chim. Listy*, **51**, 1551 (1957).
241. M. Večeřa, M. Vojtech, and L. Synek, *Collect,. Czech. Chem. Commun.*, **25**, 93 (1960).
242. R. Belcher and G. Ingram, *Anal. Chim. Acta*, **4**, 119 (1950).
243. V. Pechanec and J. Horáček, *Collect. Czech. Chem. Commun.*, **27**, 232 (1962).
244. V. Pechanec, *Collect. Czech. Chem. Commun.*, **27**, 2976 (1962).
245. V. Pechanec, *Collect. Czech. Chem. Commun.*, **27**, 2009 (1962).
246. V. Pechanec, *Collect. Czech. Chem. Commun.*, **27**, 1817 (1962).
247. V. Pechanec, *Collect. Czech. Chem. Commun.*, **27**, 1702 (1962); *Anal. Abstr.*, **10**, 1054 (1963).
248. V. Pechanec and J. Horáček, *Chim. Anal.*, **46**, 457 (1964).
249. A. A. Abramyan and R. A. Megroyan, *Arm. Khim. Zh.*, **21**, 115 (1968).

250. A. A. Abramyan and R. A. Megroyan, *Arm. Khim. Zh.*, **20**, 191 (1967); *Anal. Abstr.*, **16** 3049 (1969).
251. M. O. Korshun, N. S. Sheveleva, and N. E. Gel'man, *Zh. Anal. Khim.*, **15**, 99 (1960).
252. M. O. Korshun and N. E. Gel'man, *New Methods of Elementary Micro-Analysis*, State Chemistry Publications, Moscow, Leningrad, 1949.
253. A. I. Lebedeva and N. A. Nikolaeva, *Izv. Akad. Nauk SSSR, Ser. Khim.*, **10**, 1867 (1965).
254. A. I. Lebedeva and N. A. Nikolaeva, *Zh. Anal. Khim.*, **18**, 984 (1963).
255. G. Constantinides, G. Arich, and C. Lomi, *Chim. Ind. (Milan)*, **41**, 861 (1959).
256. C. W. Dwiggins and H. N. Dunning *Anal. Chem.*, **32**, 1137 (1960).
257. M. P. Strukova, I. I. Kashiricheva, and A. A. Lapshova, *Zhr. Anal. Khim.*, **22**, 1110 (1967).
258. L. J. Kanner, E. D. Slesin, and L. Gordon, *Talanta*, **7**, 288 (1961).
259. F. R. Hartley and J. L. Wagner, *J. Organomet. Chem.*, **42**, 477 (1972).
260. F. E. Beamish and J. Dale, *Ind. Eng. Chem., Anal. Ed.*, **10**, 697 (1938).
261. J. D. Burton and J. P. Riley, *Analyst (London)*, **80**, 391 (1955).
262. W. J. Kirsten and M. E. Carlsson, *Microchem. J.*, **4**, 3 (1960).
263. W. I. Stephen, *Ind. Chem.*, **37**, 86 (1961).
264. T. Salvage and J. P. Dixon, *Analyst (London)*, **90**, 24 (1965).
265. P. M. Saliman, *Anal. Chem.*, **36**, 112 (1964).
266. R. A. Apodacu, *Quim. Ind. (Bilbao)*, **9**, 167 (1962).
267. A. J. Christopher and T. R. F. W. Fennell, *Microchem. J.*, **12**, 593 (1967).
268. T. R. F. W. Fennell and J. R. Webb, *Microchem. J.*, **10**, 456 (1966).
269. T. R. F. W. Fennell, M. W. Roberts and J. R. Webb, *Analyst (London)*, **82**, 639 (1957).
270. H. Levine, J. J. Rowe, and F. S. Grimaldi, *Anal. Chem.*, **27**, 258 (1955).
271. W. J. Kirsten, *Mikrochem. J.*, **7**, 34 (1963).
272. R. A. Chalmers and D. A. Thomson, *Anal. Chim. Acta*, **18**, 575 (1958).
273. M. Kan and H. Kashiwagi, *Bunseki Kagaku*, **10**, 789 (1961).
274. R. Belcher and A. L. Godbert, *Analyst (London)*, **66**, 184 (1941).
275. I. Ubaldini and F. C. Maitan, *Chim. Ind. (Milan)*, **37**, 871 (1955).
276. J. Horáček, *Collect. Czech. Chem. Commun.*, **27**, 1811 (1962).
277. A. Nara, Y. Urushibata, and N. Oc, *Bunseki Kagaku*, **12**, 294 (1963).
278. M. Tanaka and S. Kanamori, *Anal. Chim. Acta*, **14**, 263 (1956).
279. C. DiPietro, R. E. Kramer, and W. A. Saisaman, *Anal. Chem.*, **34**, 586 (1962).
280. F. L. Schaeffer, J. Forg, and P. L. Kirk, *Anal. Chem.*, **25**, 343 (1953).
281. R. Belcher, A. M. G. Macdonald, S. E. Phang, and T. S. West, *J. Chem. Soc.*, 2044 (1965).
282. C. A. Rush, S. St. Cruickshank, and E. J. H. Rhodes, *Mikrochim. Acta*, 858 (1956).
283. H. N. Wilson, *Analyst (London)*, **76**, 65 (1951).
284. R. M. Lincoln, A. S. Carney, and E. C. Wagner, *Ind. Eng. Chem., Anal. Ed.*, **13**, 358 (1941).
285. A. Kondo, *Bunseki Kagaku*, **9**, 416 (1960).
286. R. Vasiliev and G. Anastasecu, *Rev. Chim. (Bucharest)*, **13**, 558 (1962).
287. F. S. Malyukova and A. D. Zaitseva, *Plast. Massý*, (1966); *Ref. Zh., Khim.*, **19 GD**, Abstr. 21G216 (1966).
288. F. F. Hoffman, L. C. Jones, O. E. Robbins, and F. F. Alsbert, *Anal. Chem.*, **30**, 1334 (1958).
289. R. May, *Anal. Chem.*, **31**, 308 (1959).
290. M. Maruyama and K. Hasegawa, *Ann. Rep. Takamine Lab.*, **13**, 173 (1961).
291. Ya A. Mandel'baum, A. F. Gropov, and A. L. Itskova, *Zh. Anal. Khim.*, **20**, 873 (1965).
292. M. O. Korshun, E. A. Terent'eva, and V. A. Klimova, *Zh. Anal. Khim.*, **9**, 275 (1954).
293. A. A. Abramyan, R. S. Sarkisyan, and M. A. Balyou, *Izv. Akad. Nauk Arm. SSR, Khim. Nauk*, **14**, 561 (1961).
294. M. Juraček and J. Jeník, *Chem. Listy*, **51**, 1312 (1957).
295. W. Schöniger, *Mikrochim. Acta*, 52 (1963).
296. F. D. Fleischer, B. C. Southworth, J. H. Hodécker, and M. M. Tuckerman, *Anal. Chem.*, **30**, 152 (1958).
297. M. Jean, *Anal. Chim. Acta*, **14**, 172 (1956).
298. L. E. Cohen and F. W. Czech, *Chemist-Analyst*, **47**, 86 (1958).
299. U. Bartels and H. Hayme, *Chem. Tech. (Berlin)*, **11**, 156 (1959).
300. E. Meier, *Mikrochim. Acta*, 70 (1961).
301. K. Zeigler, H. Till, and H. Schindbauer, *Mikrochim. Acta*, 1114 (1963).

302. H. Y. Yu and H. I. Sha, *Chem. Bull. (Peking)*, 557 (1965).
303. T. L. Hunter, *Anal. Chim. Acta*, **35**, 398 (1966).
304. D. E. Ott and F. A. Gunther, *Bull. Environ. Contam. Toxicol.*, **1**, 90 (1966).
305. N. E. Gel'man and T. M. Shanina, *Zh. Anal. Khim.*, **17**, 998 (1962).
306. J. E. Barney, J. G. Bergmann, and W. G. Tuskan, *Anal. Chem.* **31**, 1394 (1959).
307. L. L. Farley and R. A. Winkler, *Anal. Chem.*, **35**, 772 (1963).
308. D. W. Brite, *Anal. Chem.*, **27**, 1815 (1955).
309. C. M. Welch and P. W. West, *Anal. Chem.*, **29**, 874 (1957).
310. A. P. Terent'ev, M. A. Volodina, and E. G. Fursova, *Dokl. Akad. Nauk SSSR*, **169**, 851 (1966).
311. V. N. Kotova, *Zh. Anal. Khim.*, **22**, 1239 (1967).
312. B. M. Luskina, A. P. Terent'eva, and S. V. Syavtsillo, *Zh. Anal. Khim.*, **17**, 639 (1962).
313. A. P. Terent'ev, S. V. Sgavtsillo, and B. M. Luskina, *Zh. Anal. Khim.*, **16**, 83 (1961); *Anal. Abstr.*, **8**, 3309 (1961).
314. M. N. Chumachenko and V. P. Burlaka, *Izv. Akad Nauk SSSR, Otd. Khim. Nauk*, 5 (1963).
315. S. Ostrowski, R. Piekoś, and A. Radecki. *Chem. Anal. (Warsaw)*, **10**, 531 (1965).
316. E. A. Terent'eva and N. N. Smirnova, *Zavod. Lab.*, **32**, 924 (1966).
317. I. Lysyj and J. E. Zarembo, *Microchem. J.*, **2**, 245 (1958).
318. J. Körbl, *Chem. Listy,* **49**, 858, 862, 1532 (1955).
319. J. Körbl and K. Blabolil, *Chem. Listy*, **49**, 1664 (1955).
320. J. Körbl and P. Komers, *Chem. Listy*, **50**, 1120 (1956).
321. J. Körbl and R. Pribl, *Chem. Listy*, **50**, 232, 236 (1956).
322. J. Binkowski and B. Babrański, *Chem. Anal. (Warsaw)*, **9**, 515 (1964).
323. J. Binkowski and B. Babrański, *Chem. Anal. (Warsaw)*, **9**, 777 (1964).
324. J. Binkowski and M. Vecĕrä, *Mikrochim. Acta*, 842 (1965).
325. M. O. Korshun and E. A. Terent'eva, *Dokl. Akad. Nauk SSSR,* **100**, 707 (1955).
326. D. S. Galanos and V. M. Kapoulas, *Anal. Chim. Acta*, **34**, 360 (1966).
327. B. D. Holt, *Anal. Chem.*, **37**, 751 (1965).
328. R. D. Hinkel and R. Raymond, *Anal. Chem.*, **25**, 470 (1953).
329. W. H. Jones, *Anal. Chem.*, **25**, 1449 (1953).
330. I. J. Oita and H. S. Conway, *Anal. Chem.,* **26**, 600 (1954).
331. J. Unterzaucher, *Analyst (London)*, **77**, 584 (1952).
332. I. Sheft and J. J. Katz, *Anal. Chem.*, **29**, 1322 (1957).
333. G. Tölg, *Z. Anal. Chem.*, **194**, 20 (1963).
334. E. Debal, *Chim. Anal.*, **45**, 66 (1963).
335. C. Eger and J. Lipke, *Anal. Chim. Acta,* **20**, 548 (1959).
336. G. Stark, *Z. Anorg. Chem.*, **70**, 173 (1911).
337. J. I. Hoffman and G. E. F. Lundell, *J. Res. Nat. Bur. Stand.*, **3**, 581 (1929).
338. B. C. Saunders, in *Phosphorus and Fluorine*, Cambridge University Press, Cambridge, pp. 206–211, 1960.
339. W. M. Hoskins and C. A. Ferris, *Ind. Eng. Chem.*, Anal. Ed., **8**, 6 (1936).
340. W. D. Armstrong, *J. Am. Chem. Soc.*, **55**, 741 (1933).
341. H. H. Willard and O. B. Winter, *Ind. Eng. Chem., Anal. Ed.*, **5**, 7 (1933).
342. S. Sass, N. Beitsch, and C. U. Morgan, *Anal. Chem.*, **31**, 1970 (1959).
343. B. I. Stepanov and G. I. Migachev, *Zavod. Lab.*, **32**, 416 (1966).
344. E. L. Gefter, *Zavod. Lab.*, **29**, 419 (1963).
345. M. P. Strukova and T. V. Kirillova, *Zh. Anal. Khim.*, **21**, 1236 (1966); *Anal. Abstr.*, **15**, 3998 (1968).
346. V. V. Mikhailov and T. I. Tarasenko, *Zavod. Lab.*, **33**, 1380 (1967).
347. V. I. Kuznetsov and N. N. Basargin, *USSR Patent* 165566 (1964); *Chem. Abstr.*, 62, 4623a (1965); *Anal. Abstr.*, **13**, 4801 (1966).
348. J. S. Fritz and S. S. Yamamura, *Anal. Chem.*, **27**, 1461 (1955).
349. B. Jaselskis and S. F. Vas, *Anal. Chem.*, **36**, 1965 (1964).
350. A. M. G. Macdonald and W. I. Stephen, *J. Chem. Educ.*, **39**, 528 (1962).
351. R. B. Balodis, A. Comerford, and C. E. Childs, *Microchem. J.*, **12**, 606 (1967).
352. K. L. Cheng, *Anal. Chem.*, **33**, 783 (1961).
353. *Determination of Sulphate*, Technological Service Bulletin 6470-E-G-5, Arthur H. Thomas Co., 1958, California.
354. R. Levin, *Israel J. Chem.*, **3**, 11 (1966).

355. F. R. Hartley and L. M. Venanzl, unpublished results.
356. E. B. Sandell, *Colorimetric Metal Analysis*, Interscience, New York, 1959, p. 726.
357. M. G. Reed and A. D. Scott, *Anal. Chem.*, **33**, 773 (1961).
358. A. S. Zabrodina and M. R. Bagreeva, *Vestn. Mosk. Univ.*, 187 (1958).
359. N. Kunimine, H. Nakamaru, and M. Nakamaru, *J. Pharm. Soc. Jpn.*, **83**. 59 (1963).
360. K. L. Cheng, *Chemist Analyst*, **45**, 67 (1956); *Anal. Abstr.*, **4**, 1189 (1957).
361. Meier and N. Shaltiel, *Mikrochim. Acta*, 580 (1960).
362. Z. Štefanac and Z. Raković, *Mikrochim. Ichnoanal. Acta,* 81 (1965).
363. W. Ihn, G. Huse and P. Neuland, *Mikrochim. Acta*, 628 (1962).
364. Kh. Ya. Kuss and L. A. Lipp, *Zh. Anal. Khim.*, **21**, 1266 (1966).
365. S. Barabas and W. C. Cooper, *Anal Chem.*, **28**, 129 (1956); *Anal. Abstr.*, **3**, 1694 (1956).
366. L. Barcza, *Acta Chim. Hung.*, **47**, 137 (1966).
367. L. Barcza, *Acta Chim. Hung.*, **45**, 23 (1965); *Anal. Abstr.*, **13**, 6189 (1966).
368. A. S. Zabrodina and M. R. Bagreeva, *Vestn. Mosk. Univ.*, 187 (1958); *Anal. Abstr.*, **6**, 2197 (1959).
369. A. S. Zabrodina and A. P. Khlýstova, *Vestn. Mosk. Univ.*, 69 (1960).
370. A. Kondo, *Bunseki Kagaku,* **6**, 583 (1957).
371. D. Dingwall and W. D. Williams, *J. Pharm. Pharmacol.*, **13**, 12 (1961).
372. H. Bieling, W. Wagenknecht, and E. M. Arndt, *Z. Anal. Chem.*, **201**, 419 (1964).
373. W. J. Kelleher and M. J. Johnson, *Anal. Chem.*, **33**, 1429 (1961).
374. J. W. Shell, *J. Pharm. Sci.*, **51**, 731 (1962).
375. Z. Štefanac, *Bull. Sci. Yugosl.*, **8**, 136 (1963).
376. A. S. Zabrodina and S. Ya. Levina, *Vestn. Mosk. Univ.*, 181 (1957).
377. R. Huls and M. Renson, *Bull. Soc. Chim. Belg.*, **65**, 696 (1956).
378. A. I. Lebedeva, N. A. Nikalaeva, V. A. Orestova and E. V. Shikhman, *Izv. Akad. Nauk SSSR, Ser. Khim.*, **3**, 574 (1964).
379. A. I. Lebedeva, N. A. Nikalaeva, and E. V. Shikhman, *Zh. Anal. Khim.*, **20**, 832 (1965).
380. M. Farnsworth and J. Pekola, *Anal. Chem.*, **31**, 410 (1959).
381. D. Dunn and T. Norris, *Australia, Commonw. Dept. Supply, Defence Standard Lab. Rep.*, No. 269, 1964, 21 pp.
382. I. G. M. Campbell, G. W. A. Fowles, and L. A. Nixon, *J. Chem. Soc.*, 1398 (1964).
383. R. Geyer and H. J. Seidlitz, *Z. Chem.*, **4**, 468 (1964).
384. G. Tagliavini, *Studi Urbinati, Fac. Farm.*, **10**, 39 (1967).
385. R. Reverchon, *Chim. Anal. (Paris),* **47**, 70 (1965).
386. V. Chromý and J. Vřeštál, *Chem. Listy,* **60**, 1537 (1966).
387. L. V. Mÿshlaeva and T. G. Maksimova, *Zh. Anal. Khim.*, **23**, 1584 (1968).
388. S. Genda, K. Morikawa, and T. K. Kegaku, *To Kogyo (Osaka)*, **43**, 265 (1969).
389. F. Guenther, R. Geyer, and D. Stevenz, *Neue Hütte*, **14**, 563 (1969).
390. R. P. Kreshkov and E. A. Kucharev, *Zavod. Lab.*, **32**, 558 (1966).
391. P. Ochsenhein, *Kunststoffe*, **58**, 366 (1968).
392. J. C. Maire. *Ann. Chim. (Paris)*, **6**, 969 (1961).
393. C. Mohr and G. Z. Stock, *Anal. Chem.*, **221**, 1 (1966).
394. R. Geyer and H. T. Seidlitz, *Z. Chem.*, **7**, 114 (1967).
395. H. Gilman and W. B. King, *J. Am. Chem. Soc.*, **51**, 1213 (1929).
396. K. A. Kocheshkov, *Ber. Dtsch. Chem. Ges.*, **62**, 1659 (1926).
397. J. G. A. Luijten and G. J. M. Van der Kerk, in *Investigations in the Field of Organotin Chemistry*, Tin Research Institute, Greenford, Middlesex, 1966, p. 84.
398. S. Kohama, *Bull. Chem. Soc. Jap.*, **36**, 830 (1963).
399. S. Kohama, *Bull. Chem. Soc. Jap.*, **36**, 830 (1963).
400. M. P. Brown and G. W. A. Fowles, *Anal. Chem.*, **30**, 1689 (1958).
401. G. J. M. Van der Kerk, J. G. Noltes, and J. G. A. Luijten, *J. Appl. Chem.*, **7**, 366 (1957).
402. G. Fritz and H. Scheer, *Z. Anorg. Allg. Chem.*, **331**, 151 (1964).
403. H. Gilman and S. D. Rosenberg, *J. Am. Chem. Soc.*, **75**, 3592 (1953).
404. P. I. Seinorklin, *Hyg. Sanit.*, **31**, 270 (1966).
405. E. Kröller, *Dtsch. Pettenkofer Inst.*, **56**, 190 (1960).
406. R. Bock, S. Borback, and H. Oeser, *Angew. Chem.*, **70**, 272 (1958).
407. P. Nangniot and P. H. Martens, *Anal. Chim. Acta,* **24**, 276 (1961).
408. K. Hartel, *Agric. Vet. Chem.*, **3**, 19 (1962).

409. J. Herok and H. Götte, in *The Use of Radioisotopes in Animal Biology and the Medical Sciences*, Symposium, Mexico City, 21st Nov.–1st De
409. J. Herok and H. Götte, in *The Use of Radioisotopes in Animal Biology and the Medical Sciences, Symposium, Mexico City, 21st Nov.—1st Dec. 1961*, Vol. II, 4, Academic Press, London, New York, 1961, p. 177.
410. K. Bürger, *Z. Lebensm.-Unters.-Forsch.*, **114**, 1 (1961).
411. H. J. Hardon, A. F. H. Bessemer, and H. Brunink, *Dtsch. Lebensm.-Rundsch.*, **58**, 349 (1962).
412. G. A. Lloyd, C. Otaci, and F. T. Last, *J. Sci. Food Agric.*, **13**, 353 (1962).
413. J. Herok and H. Götte, *Int. J. Appl. Radiat. Isotop.*, **14**, 461 (1963).
414. J. Brüggemann, K. Barth, and K. H. Neisar, *Zentralbl. Veterinaermed., Reihe*, **11**, 4 (1964).
415. K. Bürger, *Z. Anal. Chem.*, **192**, 280 (1962).
416. J. Markland and F. C. Shenton, *Analyst (London)*, **82**, 43 (1957).
417. T. D. Holmes and I. F. Storey, *Plant Pathol.*, **11**, 139 (1962).
418. L. H. Adcock and W. G. Hope, *Analyst (London)*, **95**, 868 (1970).
419. H. B. Corbin, *J. Assoc. Off. Anal. Chem.*, **53**, 140 (1970).
420. P. Rivett, *J. Appl. Chem.*, **15**, 469 (1965).
421. L. Chromý and Z. Uhaez, *J. Oil Colour Chem. Assoc.*, **51**, 494 (1968).
422. R. Warchol, *J. Oil Colour Chem. Assoc.*, **53**, 121 (1970).
423. I. R. McCallum, *J. Oil Colour Chem. Assoc.*, **52**, 434 (1969).
424. P. N. Vijan and C. Y. Chan, *Anal. Chem.*, **48**, 1788 (1976).
425. V. F. Hodge, S. L. Seidel, and E. D. Goldberg, *Anal. Chem.*, **51**, 1256 (1979).
426. S. Kojima, *Analyst (London)*, **104**, 550 (1979).
427. D. Colaitis and M. Lesbre, *Bull. Soc. Chim. Fr.*, **19**, 1069 (1952).
428. F. C. Silbirt and W. R. Kirner, *Ind. Eng. Chem., Anal. Ed.*, **8**, 353 (1936).
429. U. S. Bazalitskaya and M. K. Dzhamletdinova, *Zavod. Lab.*, **33**, 427 (1967).
430. K. A. Kocheshkov, *Ber. Dtsch. Chem. Ges.*, **61**, 1659 (1928).
431. M. P. Strukova and I. I. Kashiricheva, *Zh. Anal. Khim.*, **24**, 1244 (1969).
432. G. J. M. Van der Kerk and J. G. A. Luijten, *J. Appl. Chem.*, **7**, 369 (1957).
433. H. Matsuda and S. Matsuda, *J. Chem. Soc. Jap., Ind. Chem. Sect.*, **64**, 539 (1961).
434. V. Chromý and L. Srp, *Chem. Listy*, **61**, 1509 (1967).
435. E. W. Abel and D. B. Brady, *J. Chem. Soc.*, 1192 (1965).
436. E. E. Kotlyar and T. N. Nazarchuk, *Zh. Anal. Khim.*, **16**, 631 (1961).
437. E. Niki and K. Masato, *Bunseki Kagaku*, **9**, 324 (1960).
438. B. M. Luskina, E. A. Terent'eva and N. A. Gradskova, *Zh. Anal. Khim.*, **20**, 990 (1965).
439. E. A. Terent'eva and M. V. Bernatskaya, *Zh. Anal. Khim.*, **21**, 870 (1966).
440. B. M. Luskina, E. A. Terent'eva, and N. A. Gradskova, *Zh. Anal. Khim.*, **19**, 1251 (1964); *Anal. Abstr.*, **13**, 1306 (1966).
441. J. Jenik and F. Renger, *Collect. Czech. Chem. Commun., Engl. Ed.*, **29**, 2237 (1964).
442. T. R. Crompton, *Chemical Analysis of Organometallic Compounds*, Vol. V, Academic Press, London, New York, San Fransisco, 1977, Chapter 21, pp. 378–386.
443. D. J. David, *Analyst (London)*, **83**, 655 (1958).
444. G. Westöö, *Analyst (London)*, **88**, 287 (1963).
445. H. J. Gudzinowicz and V. J. Suciano, *J. Assoc. Off. Anal. Chem.*, **49**, 1 (1966).
446. W. U. Malik, R. Hague and S. P. Verma, *Bull. Chem. Soc. Jpn.*, **36**, 746 (1963).
447. T. R. Crompton, *Analyst (London)*, **86**, 652 (1961).
448. T. R. Crompton, *Analyst (London)*, **91**, 374 (1966).
449. B. J. Phillip, W. L. Mundry, and S. C. Watson, *Anal. Chem.*, **45**, 2298 (1973).
450. E. A. Terent'eva and M. V. Bernatskaya, *Zh. Anal. Khim.*, **19**, 876 (1964).
451. E. A. Terent'eva and N. N. Smirnova, *Zavod, Lab.*, **32**, 924 (1966).

The Chemistry of the Metal–Carbon Bond
Edited by F. R. Hartley and S. Patai

CHAPTER **15**

Analysis of organometallic compounds: titration procedures

T. R. CROMPTON
'Beechcroft', Whittingham Lane, Goosnargh, Preston, Lancashire, UK

I. ORGANOALUMINIUM COMPOUNDS

A. Classical Titration Procedures

Aluminium in organoaluminium compounds of the type AlR_3 and AlR_2H can be determined with anhydrous ammonia[1]. A known weight of the organoaluminium sample is introduced into a nitrogen-filled reaction tube. An excess of ammonia is then passed through the sample:

$$AlR_3 + NH_3 \longrightarrow R_2AlNH_2 + RR \tag{1}$$

$$AlR_2H + NH_3 \longrightarrow R_2AlNH_2 + H_2 \tag{2}$$

Unreacted ammonia is swept away with a stream of dry nitrogen. Addition of ethanol to the reaction tube liberates ammonia, proportional in amount to the total AlR_3 plus AlR_2H content of the sample:

$$R_2AlNH_2 + C_2H_5OH \longrightarrow R_2AlOC_2H_5 + NH_3 \tag{3}$$

Finally, the liberated ammonia is swept into boric acid solution and determined by titration with standard acid. It is essential to use absolutely anhydrous nitrogen for purging. Any moisture in the nitrogen will decompose some of the dialkylaluminium amide derivative, causing loss of bound ammonia and consequent low analytical results:

$$R_2AlNH_2 + H_2O \longrightarrow R_2AlOH + NH_3 \tag{4}$$

Preliminary treatment of the nitrogen supply with a 10% solution of triisobutylaluminium dissolved in liquid paraffin is adequate for drying of nitrogen.

Compounds of the type AlR_2X, where X is OR, SR, or HN_2, do not react with ammonia and therefore do not interfere in the determination of active organoaluminium compounds. Higher molecular weight dialkylaluminium halides react with ammonia. The method can be used, therefore, to determine the concentration of [AlR_2(hal)] in mixtures containing [AlR(or)(hal)] and/or [AlR(OH)(hal)].

The ammonia method does not distinguish between the two types of organoaluminium compounds, i.e. AlR_3 and AlR_2H. It is possible, however, to determine these separately when both are present in mixtures. First the dialkylaluminium hydride content of the sample is determined by either the alcoholysis–hydrolysis procedure or the *N*-methylaniline method described in Section 14.II.D. Trialkylaluminium compound is then obtained, by difference, from this hydride determination and the results obtained by the ammonia method.

Iodometric methods for determining organoaluminium compounds have been reported[2,3]. Triethylaluminium reacts with iodine according to the following equation[2]:

$$Al(C_2H_5)_3 + 3I_2 \longrightarrow AlI_3 + 3C_2H_5I \qquad (5)$$

Dialkylaluminium chlorides and dialkylaluminium alkoxides consume, respectively, 2 and 1.25 mol of iodine per mole of organoaluminium compound[3]. This method is capable of analysing neat organoaluminium compounds or dilute hydrocarbon solutions thereof, containing alkyl groups up to octadecyl at concentrations down to a few millimoles per litre with an accuracy of ±3%. A suitable volume of the hydrocarbon solution of the organoaluminium compound is stirred with an excess of a solution of iodine dissolved in toluene. Alkyl groups in trialkylaluminium, dialkylaluminium chloride, and dialkylaluminium alkoxide compounds are completely iodinated within 20 min. Following the addition of dilute acetic acid, unreacted iodine is determined by titration with sodium thiosulphate solution. The concentration of organoaluminium compound is then calculated from the amount of iodine consumed in the determination. The presence of water in the iodine reagent causes interference by reacting with some of the alkyl groups. This is corrected by a suitable 'double titration' procedure.

Triisobutylaluminium has been determined by reaction with excess of standard mercaptan solution followed by amperometric titration of excess of mercaptan with standard silver nitrate in a methanolic ammonia/ammonium nitrate solution[4]. Acidic organoaluminium compounds have been titrated with basic titrants in the presence of acid–base indicators[5]. When a dichloroethane solution of methyl violet was added to toluene, benzene, or heptane solutions of various organoaluminium compounds, a colour change occurred from violet (alkaline) to yellow or green (acidic). Addition of bases such as butyl acetate, dimethylaniline, diethyl ether, or pyridine to this solution caused a reversion of colour to violet. Organoaluminium compounds containing alkoxide groups did not affect the colour of methyl violet. Other indicators such as crystal violet and gentian violet also gave colour changes in these circumstances.

Alkyl- and arylaluminium compounds behave as strong aprotic acids, giving titration curves with organic bases which are similar to those obtained in inorganic acid-based titrations. There are considerable differences in behaviour between various trialkylaluminium compounds and alkylaluminium halides[5]. The organoaluminium sample, diluted in xylene, can be introduced by a hypodermic syringe via the septum into a flask under argon (Figure 1) to remove the last traces of oxygen and moisture and to solubilize the indicator (methyl violet, basic fuchsin, phenazin, neutral red, or neutral violet)[6]. The solution is then titrated to the colour change with a standard base (0.2 M pyridine or isoquinoline in xylene) in a needle-tipped burette. Excess of titrant

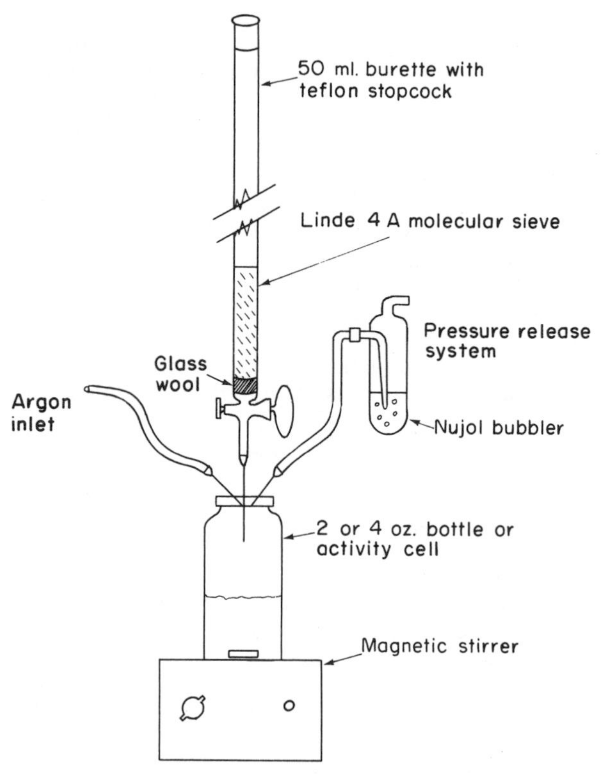

FIGURE 1. The Hagen and Leslie titration apparatus.

is added to the bottle and then the organoaluminium sample injected dropwise from a syringe until the indicator changes to the excess alkyl-colour. The solution is titrated to the end-point and the syringe reweighed to obtain the weight of the organoaluminium sample used between the two end-points.

Other types of triphenylmethane-type indicators can also function as reversible indicators[6]. For example, basic fuchsin is reversible, whereas methyl green does not display reversible complex behaviour. The triphenylmethane-type indicators which are reversible (methyl violet and other *p*-rosanilines) are unstable with respect to alkylaluminium hydrides and it appears that the hydride-sensitive part of the molecule is also responsible for its ability to act as an indicator. Reversible colour changes are observed with trialkylaluminium and ketones, such as anthrone, benzil, and Michler's ketone, and aldehydes such as *p*-dimethylaminobenzaldehyde. Coloured complexes can be formed with trialkylaluminium compounds by the use of compounds such as pyrazine, allowing the colorimetric determination of small amounts of the alkyl. Complexes of difunctional compounds containing azomethine linkages with dialkylaluminium hydrides are much more intensely coloured than those previously reported for monofunctional compounds such as isoquinoline and pyridine. *o*-Phenanthroline:α,α'-dipyridyl and 6,7-dimethyl-2,3-di(2-pyridyl)quinoxaline are capable of forming complexes with dialkylaluminium hydrides which are not readily displaced with a stronger base such as pyridine, whereas *m*-phenanthroline forms intensely coloured AlR_2H complexes which are readily decoloured by pyridine. Neutral red and neutral violet undergo several colour changes as the end-point is approached in the titration of trialkylaluminium compounds with pyridine, isoquinoline, or alcohols. Phenazine also gives this colour transition and it appears that with excess of alkylaluminium a green complex is formed which could be attributed to a quinoid-like compound. As the trialkylaluminium compound is displaced from the indicator with pyridine, the colour changes to red and finally to yellow. Dialkylaluminium hydrides destroy the ability of these indicators to give sharp colour changes at the end-point. These compounds never assume the green coloration but turn directly red upon addition of dialkylaluminium hydrides. Azine dyes in which one of the nitrogen atoms is pentavalent (amethyst violet, magdala red, and phenosafranine) do not function as reversible indicators, and this may be due to their greater basicity. Oxazines, pyronins, and sulphur compounds, such as methylene blue and dithizone, form highly coloured solutions with alkylaluminium compounds, but these appear to be too stable or are destroyed.

The titration procedure utilizing visual indicators is applicable to trialkyl samples that contain little or no hydride. Accurate determinations of dialkylaluminium hydrides are best obtained spectrophotometrically[6,7]. Total activity ($AlR_2H + AlR_3$) in samples containing too much dialkylaluminium hydride for the indicator titration is best obtained by a photometric titration technique.

A rapid, visual pyridine titrimetric method may be used to determine total trialkylaluminium plus diethylaluminium hydride reactivity as well as to differentiate between these moieties in mixed complex systems of aluminium alkyls[8]. Phenazine used as the indicator forms red to brown complexes with trialkylaluminium compounds and green to green–blue complexes with diethylaluminium hydride. The method is applicable to alkylaluminium compounds ranging from C_2 to at least C_{20} separately or in mixtures. The precision is ±1% (relative standard deviation) for both high and low concentrations of trialkylaluminium and/or dialkylaluminium hydride. This indicator is not used, however, for the titration of dialkylaluminium hydride compounds because of the possible formation of an Al—N bond with the dialkylaluminium hydride[8]. Phenazine forms reversible complexes with both trialkylaluminium and dialkylaluminium hydride. Pyridine is the most suitable base for forming complexes with trialkylaluminium and dialkylaluminium hydride. Pyridine displaces phenazine quantitatively from both trialkylaluminium and dialkylaluminium hydride to form more stable 1:1 complexes. Table 1 compares the results obtained by this procedure and the isoquinoline spectrophotometric procedure[6]. The results are generally in

TABLE 1. Comparison of results obtained by the described phenazine differential titration and the spectrophotometric isoquinoline procedures for ADAH[a] and total reactivity

	Mol-% ADAH[a]		Mol-% ATA[b]		Mol-% reactive	
Sample	Phenazine	Iso-quinoline	Phenazine	Iso-quinoline	Phenazine	Iso-quinoline
A	81.3 ± 0.85[c]	80.8	10.5	11.1	91.8 ± 0.92[c]	91.9
B	65.4	63.0	28.5	30.7	93.9	93.7
C	4.51 ± 0.05[d]	4.55	—	—	—	—
D	58.8	58.9	23.8	24.1	82.6	83.0
E	16.1	17.4	40.7	37.8	56.8	55.2
F	<0.3	—	>96.6	96.8	96.9	96.9

[a] ADAH = dialkylaluminium hydride.
[b] ATA = trialkylaluminium.
[c] Five or more replicate analyses.
[d] ATA = total reactive − ADAH.

excellent agreement. The greatest differences are observed for the trialkylaluminium values which are obtained, in each case, by difference. Of significance is the range of dialkylaluminium hydride determined by the method without alteration of the phenazine. The precision of the determination is slightly more than ±1% relative throughout the range of dialkylaluminium hydride values. The accuracy should approach the precision.

According to Jordan[8], simple complexation is involved in the reactions of both dialkylaluminium hydride and trialkylaluminium compounds with phenazine. However, a semiquinone has been identified by E.P.R. and N.M.R. spectroscopy, which suggests that reduction as well as complexation is involved in the reaction[10]. The diethylaluminium hydride is titrated after both the excess of triethylaluminium and the phenazine-complexed triethylaluminium, indicating its relatively weak Lewis acid nature. Phenazine has also been used as an indicator in the titration of organoaluminium compounds with nitrogen bases[11].

B. Conductometric Titration

Conductometric titration of various types of organoaluminium compounds with hydrocarbon solutions of either diethyl ether or isoquinoline solutions has been studied[12,13]. The sample, dissolved in anhydrous hexane, cyclohexane, or benzene, is titrated with a solution of diethylether or isoquinoline in the same solvent under dry, oxygen-free nitrogen. The curve obtained using diethyl ether as titrant shows a sharp conductivity maximum at about 2% before the molar ratio $N/M = 1.00$ (N = moles of diethyl ether or isoquinoline added, M = total moles of reactable compound in the sample) is obtained, i.e. at $N/M = 0.98$. Thus the weighed amount of triethylaluminium sample contains about 2 mol-% of a compound which is not titratable with diethyl ether. In the curve obtained by titrating triethylaluminium with isoquinoline two conductivity maxima occur. The first coincides with the single maximum obtained by titration with diethyl ether, i.e. $N/M = 0.98$, and corresponds to the complete titration of triethylaluminium in the sample to form a 1:1 triethylaluminium–isoquinoline complex. The minimum occurring between the two maxima obtained in the isoquinoline titration lies exactly at the molar ratio $N/M = 1.00$. It is at this point that the solution turns from yellow to red owing to the formation of some red 1:2 diethylaluminium hydride—isoquinoline complex. Thus, diethyl ether titrant determines only trialkylaluminium compounds and does not include the hydride. Conductometric titration with either diethyl ether or isoquinoline does not determine dialkylaluminium alkoxides.

Conductometric titration with isoquinoline of hydrocarbon solutions of pure dialkylaluminium hydrides such as dimethylaluminium hydride and diisobutylaluminium hydride has also been studied[12]. In both cases no increase in conductivity occurred for values of the molar ratio N/M up to 1.00, i.e. corresponding to the formation of a 1:1 dialkylaluminium—isoquinoline complex. Beyond this point, however, up to a ratio N/M of 2.00, the conductivity increased sharply and then either flattened out or started to decrease. At a value of N/M of 1.00 the test solution starts to become red and indeed this colour change may be used as a visual indication of the end-point. The intensity of the red colour increased during the titration of dialkylaluminium hydrides and reached a maximum value when N/M reached 2.00. This corresponds to the complete conversion of dialkylaluminium hydride to the red 1:2 dialkylhydride—isoquinoline complex. Based on this work, procedures have been devised for analysing mixtures of these two types of compounds (R_3Al and R_2HAl) in the presence of each other[12]. Diethylaluminium chloride can also be determined by conductometric titration with a cyclohexane solution of isoquinoline.

A procedure for the automatic recording of conductometric titration curves in the titration of organoaluminium compounds with a cyclohexane solution of isoquinoline uses an automatic burette powered by a synchronous electric motor[14]. The conductivity of the test solution is measured continuously by means of an ohmmeter and registered with a compensating recording apparatus equipped with a chart recorder. The operation of the burette and the recording apparatus are synchronously linked so that a known division on the recorder chart corresponds to a known volume of titrant added to the organoaluminium sample from the burette. The titration vessel containing the sample is equipped with silver electrodes and is suitably thermostated.

As previously mentioned, the isoquinoline titration procedure provides estimates of both the trialkylaluminium and the dialkylaluminium hydride contents of organoaluminium samples. The accuracy of the isoquinoline titration procedure has been obtained by comparing results on various neat, triethylaluminium–diethylaluminium hydride mixtures, that had also been analysed by the alcoholysis–hydrolysis procedure described in Section 14.II.D[15]. The isoquinoline method gives high hydride contents and low triethylaluminium contents for samples which contain less than 10% of diethylaluminium hydride. When the samples contain more than 10% of diethylaluminium hydride the isoquinoline method gives low hydride contents (and high triethylaluminium contents). In each analysis, however, the total determined isoquinoline consumption of the trialkylaluminium and the dialkylaluminium hydride constituents agree to within 2–8% of the value calculated from alcoholysis data. These results show that the isoquinoline titration procedure is unsuitable for the accurate analysis of trialkylaluminium–dialkylaluminium hydride mixtures as it neither distinguishes between different alkyl groups nor gives very accurate analyses of the individual compound types present. The more lengthy alcoholysis–hydrolysis procedures must always be used when an accurate analysis is required for these complex mixtures. The isoquinoline titration procedure does, however, give rapid and reasonably accurate estimates of the total AlR_3 plus AlR_2H organoaluminium content, even when the sample contains up to 60% of dialkylaluminium hydride. Also, the total AlR_3 plus AlR_2H organoaluminium content of hydrocarbon catalyst solutions, as dilute as 50–100 mmol/l, may be determined rapidly by the isoquinoline titration procedure.

A procedure for the analysis of mixtures of trialkylaluminium and dialkylaluminium compounds by conductometric titration with quinoline in light petroleum gives an error of about ±1%[16].

C. Potentiometric Titration

The potentiometric titration of organoaluminium compounds with a standard solution of isoquinoline uses the same titration apparatus and the general procedure described for conductometric titration[12]. Potential changes may be recorded between a bare platinum wire and a bare silver wire in a pH recording amplifier. In the potentiometric titration with isoquinoline of triethylaluminium containing a small amount of diethylaluminium hydride no change in potential occurs until the ratio N/M exceeds unity (where N is moles if isoquinoline added and M is total moles of triethylaluminium plus diethylaluminium hydride in the sample). Beyond this point (i.e. the formation of the 1:1 diethylaluminium hydride—isoquinoline complex) the potential starts to increase, reaching a maximum at a value of N/M of 2.00. Between values of N/M of 1.00 and 2.00 the solution becomes increasingly redder owing to the formation of the 1:2 diethylaluminium hydride—isoquinoline complex. Diethylaluminium chloride may also be titrated with isoquinoline. The automatic recording conductometric titration device is also applicable to the automatic potentiometric titration of organoaluminium compounds with isoquinoline[14].

A potentiometric titration procedure in which the sample is dissolved in cyclohexane and titrated potentiometrically with a standard cyclohexane solution of quinoline using platinum and silver electrodes in an inert gas atmosphere is claimed to give a more distinct end-point than the isoquinoline method[13]. The potential change occurring in the isoquinoline potentiometric titration is not always detectable or is inaccurate in the case of organoaluminium compounds containing aluminium hydride groups[16]. Superior potentiometric titration curves are obtained if the silver–platinum electrode system is replaced by an aluminium rod as titration electrode and an aluminium rod immersed in triethylaluminium solution as a reference electrode. The reference electrode is placed in contact with the organoaluminium sample solution by means of a porous disc. The establishment of the potential after addition of titrant is relatively slow[16]. Etherates of alkylaluminium compounds can be titrated potentiometrically with isoquinoline solution. These titrations are rather slow, however, and do not have the same high precision obtained with other types of organoaluminium compounds discussed[16].

The platinum–silver cell has been re-examined using a silver electrode having a large surface area, shielded and of special construction[17]. With this electrode it is possible to obtain differences of 250–400 mV in titrations with isoquinoline. Also, a stable potential is established immediately following every addition of reagent. α,β, and γ-methylpyridines, 2-ethylpyridine, 2,5-methylethylpyridine, 2,4,6-trimethylpyridine, and 2,6-lutidine are all excellent titrants, particularly the di- and tri-substituted pyridines, inasmuch as they can be easily purified by careful fractionation, and their solutions in benzene can be kept anhydrous and stable for long periods of time without change of titration. Moreover, these bases can be standardized easily and accurately by titration with perchloric acid in acetic solution using violet indicator.

Chlorodiethylaluminium and mixtures of tributylaluminium and dibutylaluminium hydride have been titrated potentiometrically with a 0.2 M solution of pyridine in anhydrous benzene using aluminium and silver amalgam electrodes in an atmosphere of dry oxygen-free nitrogen[18].

D. Amperometric Titration Procedures

Amperometric titration of mixtures of triethylaluminium and diethyliodoaluminium in octane can be performed in a cell equipped with two platinum electrodes[19]. Phenetole, diethyl, ether, tetrahydrofuran, and dioxan may be used as titrants. Titrations are carried out in an atmosphere of pure, dry argon. The sequential titration of diethyliodoaluminium and triethylaluminium with dibutyl ether, dioxane or tetrahydrofuran is possible without interference from any ethoxydiethylaluminium present. The experimental error of the method is $\pm 1\%$.

E. Activity by Dielectric Constant Titration

Trialkylaluminium compounds and diethylaluminium chloride, because of their tendency to accept a lone pair of electrons, form well-defined 'donor–acceptor' compounds which have a very high dipole moment[20]. Considerable evolution of heat also accompanies the formation of these compounds. In these donor–acceptor compounds the donor (e.g. a tertiary amine) supplies both bonding electrons to the organoaluminium acceptor molecule, thus disturbing the charge symmetry **(1)**. The donor becomes partially positively charged and the acceptor becomes partially negatively charged, and it is a consequence of this unsymmetrical charge distribution that such donor–acceptor complexes possess a considerable dipole moment, of the order of 4–6 Debye. Dialkylaluminium alkoxides, on the other hand, do not react with these donor molecules and have a low or zero dipole moment.

$$\begin{array}{c} \mathrm{R}\quad\mathrm{R} \\ | \quad\ \ | \\ \mathrm{R{-}\overset{\delta-}{Al}\leftarrow\overset{\delta+}{N}{-}R} \\ | \quad\ \ | \\ \mathrm{R}\quad\mathrm{R} \end{array}$$

(1)

A complexometric dielectric constant titration technique has been developed for the determination of organoaluminium compounds which form complexes of high dipole moment with suitable donor molecules[21]. The titration is carried out in an apparatus consisting of a thermostated titration vessel with magnetic stirrer, containing dry solvent and a known weight of the organoaluminium sample under pure nitrogen or argon. The dielectric constant measuring cell consists of a gold plated Teflon-lined immersion condenser of exactly 7 pF effective capacity used in conjuction with a Decameter. The donor titrant solution is added to the organoaluminium sample solution at a uniform rate by means of motor-driven piston burette synchronized with a suitable motor-driven chart recorder which automatically plots a curve of volume of donor solution added to the sample against dielectric constant. This procedure has been applied to the analysis of a mixture of diisobutylaluminium hydride (approx. 93 mol-%) and triisobutylaluminium (approx. 7 mol-%), using di-*n*-butyl ether as the titrant[21]. An inflection was obtained corresponding to 9.4 wt.-% of triisobutylaluminium in this sample. The slope of the curve then decreased on further addition of the ether.

F. Lumometric Titration Procedures

During an investigation in which a liquid scintillation counter was being used for the determination of radioactive carbon dioxide, it was found that the absorbing solution (a toluene solution of acetyldimethylbenzylammonium hydroxide) itself gave off measurable amounts of light[22]. The high background is due to luminescence associated with the quaternary ammonium compound. The phenomenon is not limited to solutions of quaternary ammonium bases but accompanies a wide variety of reactions. It may be used as a basis for aluminium analysis.

G. Thermometric Titration Procedures

A suitable apparatus for carrying out the thermometric titration of oxygen- and moisture-sensitive compounds has been described[21,23,24]. The essential components of the apparatus are a vacuum-jacketed titration flask from which air and moisture can be excluded, a constant delivery-rate syringe burette, a thermistor connected in a Wheatstone bridge circuit to detect temperature changes, and a strip-chart recorder to indicate bridge output as a function of titrant added (Figure 2). It is essential that air and moisture are excluded from the titration vessel and that the vessel has a low heat conductivity[24]. The syringe burette is calibrated in terms of millilitres of titrant per minute. The concentration of organoaluminium compound (mmol) is the product of this calibration factor, titrant concentration, the reciprocal of chart speed (in/min) and the distance in inches from start of titration to the intersection of lines drawn through straight segments of the curve just before and just after the inflection (see Figure 3). The various titrants used and the stoichiometries of their complexes with alkylaluminium compounds are shown in Table 2.

Triethylamine, isoquinoline, and 2,2′-bipyridyl (bipy) all give two-slope titration curves if both R_3Al- and R_2AlH-type compounds are present; typical curves obtained for the latter two titrants are shown in Figure 3. The amount of titrant consumed

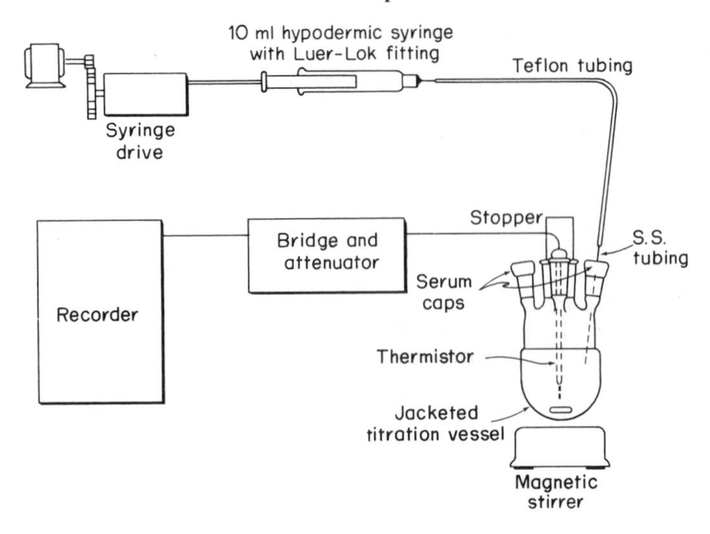

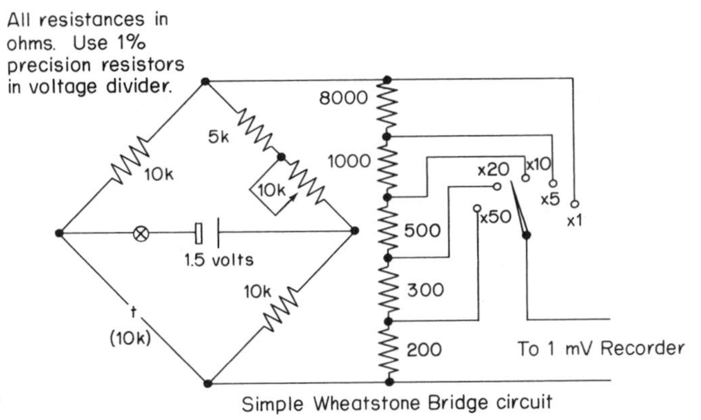

FIGURE 2. Titration assembly and bridge circuit. Thermometric titration of organometallic compounds.

between the first and second inflections is a direct measure (mole for mole) of the hydride content, but the stoichiometry at the first inflection point may vary (Table 3).

2,2′-Bipyridyl is unusual in that it forms both 1:2 and 1:1 complexes with triethylaluminium and other tri-*n*-alkylaluminium derivatives but only a 1:1 complex with triisoalkylaluminium derivatives. This is presumably because each nitrogen atom can form a complex with an aluminium atom in $(n\text{-R})_3\text{Al}$-type compounds, but that steric hindrance prevents a similar reaction with $(i\text{-R})_3\text{Al}$, i.e. addition of the first molecule blocks the entry of a second one. For the 2,2′-bipyridyl–R_2AlH reaction the 1:1 ratio (rather than 2:1 as with isoquinoline and benzalaniline) is understandable on the basis that the amide formed in the initial step dimerizes rather than reacting with a second molecule of reagent (equation 6). The reaction product is deep orange–red, similar to the corresponding isoquinoline compound. There is some indication (Figure 3, curve I) that diethylaluminium hydride reacts with a second molecule of 2,2′-bipyridyl but the energy of reaction is low.

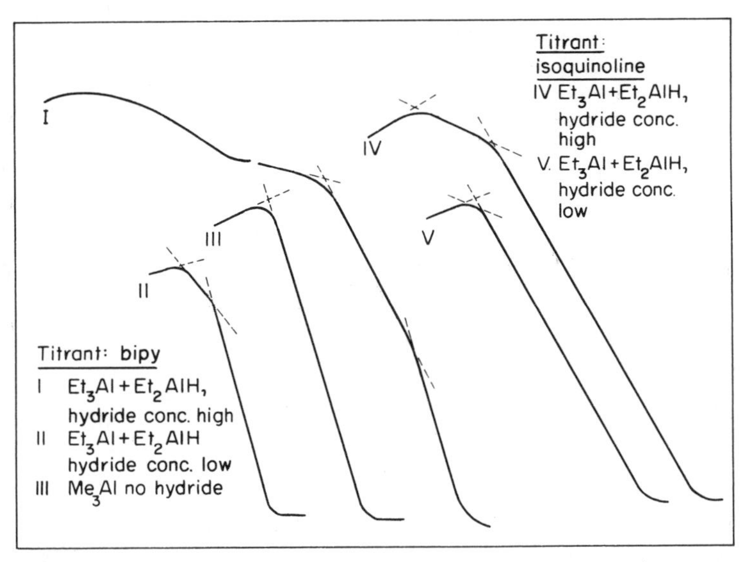

FIGURE 3. Typical thermometric titration curves obtained with 2,2′-bipyridyl (bipy) and isoquinoline.

$$C_5H_5N\cdot C_5H_5N + R_2AlH \longrightarrow (C_5H_6N)\text{—}AlR_2\text{—}NC_5H_5 \qquad (6)$$

Benzalaniline (*N*-benzilidineaniline) reacts similarly to isoquinoline but with triethylaluminium the curve shows considerable curvature, with a suggestion of an intermediate 1:2 complex. N-Methylaniline, which reacts selectively with

TABLE 2. Typical reactions of alkylaluminium compounds

	Reaction(s) as moles of titrant per mole of Al			
Titrant	R_3Al	R_2AlH	R_2AlCl[a]	Et_2AlOEt[a]
Triethylamine	1:1	1:1	1:1	None
Isoquinoline	1:1	1:1 and 2:1	1:1	None
2,2′-Bipyridyl	1:2[b] or 1:1	1:1	1:1	None
Di-*n*-Butyl ether	1:1	—[c]	1:1	None
t-Butyl alcohol	1:1	1:1	1:1[d]	None
Acetone, benzophenone	None[e]	1:1[e]	None[e]	None
Oxine	1:1, 2:1, and 3:1	1:1, 2:1, and 3:1	1:1, 2:1, and 3:1	1:1 and 2:1[f]

[a] RAlCl shows similar reactions to R_2AlCl. Et_2AlOEt was the only alkoxide studied in detail.
[b] Reaction is 1:2 with $(n\text{-}R)_3Al$ and 1:1 with $(i\text{-}R)_3Al$.
[c] Forms very weak 1:1 complex, not analytically significant.
[d] $EtAlCl_2$ reacts further to the 3:1 stage.
[e] In the presence of ether. Stoichiometry is different in the absence of ether.
[f] Slow, incomplete replacement of —OR group may occur after 2:1 reaction.

TABLE 3. Titration of trialkylaluminium and dialkylaluminium hydride with amines

	Moles of titrant per mole of Al, to first inflection point	
Titrant	R_3Al	R_2AlH
Triethylamine	1	0
Isoquinoline	1	1
2,2′-Bipyridyl	0.5 or 1	0

dialkylaluminium hydride at 30°C[25], reacts with triethylaluminium as well at room temperature and is unsuitable as a titrant.

Amine titrants generally can give, in a single titration, values of trialkylaluminium, dialkylaluminium hydride, and activity (sum of the two). However, this advantage is more apparent than real; for hydride, in particular, the precision and accuracy leave something to be desired. At the low hydride concentrations generally found in commercial triethylaluminium, isoquinoline gives a curve (Figure 3, curve V) in which the hydride segment is too small to measure accurately or is completely obscured by the normal slight rounding of the curve near the inflection points. 2,2′-Bipyridyl and triethylamine behave similarly.

At high hydride concentrations, 2,2′-bipyridyl and triethylamine give poorly defined first inflections (Figure 3, curve I), so that although the activity result is correct, the trialkylaluminium and dialkylaluminium hydride values cannot be determined very precisely. Isoquinoline gives well defined inflections with high-hydride samples but, although the result for activity (first inflection) is correct, the result for hydride tends to be low and variable. Results for trialkylaluminium are correspondingly high since, in this case, they are determined by difference. Presumably, the difficulty arises from slow reaction of the amide with a second molecule of isoquinoline; it is necessary to make potentiometric titrations with isoquinoline rather slowly in order to obtain a separate inflection for hydride[16]. These problems do not occur when a ketonic titrant is used. Alkylaluminium halides react like trialkylaluminium compounds with amine titrants forming 1:1 electron-sharing complexes.

Typical curves for oxygenated titrants are shown in Figure 4. Dialkyl ethers form electron-sharing complexes with trialkylaluminium compounds and alkylaluminium halides, giving 'normal' (type I) curves with a single, well defined inflection. Hydride does not interfere, although if it is present the post-inflection portion of the curve may show a slight rise, indicating the formation of weak ether—hydride complexes.

Ketones (in the absence of ethers) react readily with both R_3Al and R_2AlH compounds, although reaction with the latter is more rapid and energetic (see Figure 4, curve II, a typical commercial triethylaluminium sample containing a small amount of hydride impurity). Curve III shows the reaction of acetone with a mixture containing triethylaluminium and diethylaluminium hydride in about a 5:3 molar ratio.

Alcohols such as methanol, isopropanol, *t*-butyl alcohol, cyclohexanol, and 2-ethylhexan-1-ol react energetically with trialkylaluminium, dialkylaluminium hydrides, and aluminium halides[20,21]. In all cases the first stage (1:1) reaction is quantitative and the inflection (at room temperature) is sharp. With *t*-butyl alcohol the reaction essentially stops at this stage (but diethylaluminium chloride is an exception: see Figure 4, curve IV). Primary and secondary alcohols react further but in general the second stage reaction is slow and the third stage still slower. *t*-Butyl alcohol is perhaps the most useful of the alcoholic titrants because of its clear-cut, one-stage

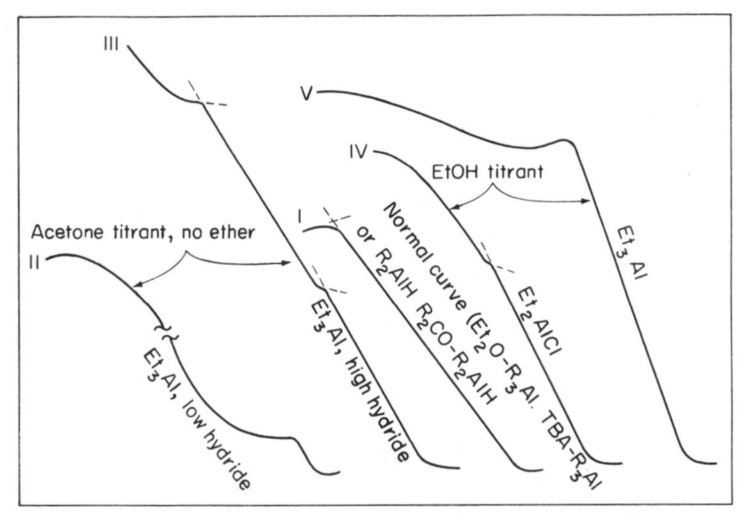

FIGURE 4. Typical titration curves obtained with oxygenated titrants.

reaction with many compounds. It is the only alcohol which gives satisfactory results if the clean-up technique is used. Other alcohols may be useful in special situations, e.g. where a second-stage inflection can be obtained.

In the thermometric titration procedure complete elimination of traces of active impurities from reagents and apparatus prior to analysis is difficult and time consuming. It is considerably simpler to remove them from the reaction system just before making the analysis, and the best reagent for this purpose is the sample itself. Table 4

TABLE 4. Typical analysis showing effect of clean-up procedure

Titrant[a]	Component determined	Determined value (mmol/g for sample portion)		
		Clean-up sample[b]	Second sample	Third sample
DNBE	$EtAlCl_2$	2.07	2.11	2.11
	R_3Al[c]	0.494	0.570	0.566
	Et_2AlCl	0.406	0.648	0.642
TBA	Me_3Al	0.609	0.665	0.679
	Activity[d]	0.645	0.680	0.683
IQ	Activity[c]	0.751	0.915	0.908
BZPH	Et_2AlH (Et_3Al present)	0.285	0.334	0.334
Bipy	Activity	0.580	0.622	0.628

[a]DNBE = di-*n*-butyl ether; TBA = *t*-butyl alcohol; IQ = isoquinoline; BZPH = benzophenone, bipy = 2,2′-bipyridyl.
[b]These results are not reportable, but give some idea of the amount of reactive impurities removed by clean-up. The effect is somewhat exaggerated because the clean-up sample is usually small.
[c]Commercial Et_3Al solution containing a small amount of Et_2AlH.
[d]Solution of $(i\text{-}Bu)_2AlH$ and $(i\text{-}Bu)_3Al$ in about a 6:1 molar ratio.

illustrates the effectiveness of the 'clean-up' procedure in obtaining precise results by thermometric titration.

II. ORGANOARSENIC COMPOUNDS

Potentiometric titration and colorimetric methods have been described for the determination of β-chlorovinyldichlorarsine (Lewisite)[26,27]. In the latter the organic arsenic is mineralized by refluxing with aqueous sodium hydrogen carbonate and then arsenic is oxidized to the pentavalent state by the addition of aqueous iodine, followed by conversion to molybdoarsenate and reduction to molybdenum blue by boiling under reflux with a sulphuric acid solution of ammonium vanadate and hydrazine sulphate. The method is sufficiently sensitive to determine down to 3 μg of Lewisite per millilitre of sample.

The pharmaceutical sodium methyl arsinate has been determined by non-aqueous titration with mercury(II) acetate[28].

III. ORGANOBORON COMPOUNDS

A. Classical Titration Procedures

A thiomercurimetric method has been described for the determination of potassium tetraphenylborate[29]. To the sample is added methanolic 0.05 N mercury(II) perchlorate and the solution is heated until dissolution is complete. Aqueous ammonia is added and the solution titrated with 0.05 N sodium mercaptoacetate in the presence of thiofluoroscein until a blue colour appears. In an alternative procedure, a known volume of 0.1 N mercury(II) acetate is added to the sample[30]. After warming and cooling and acidifying with nitric acid, the excess of mercury salt is titrated against aqueous standard ammonium thiocyanate. Another method[31] is based on reaction of the tetraphenylboron ion with the mercury(II)—EDTA complex, whereby EDTA is released in equivalent amount and is determined by titration in an acetate-buffered medium with standard zinc solution using 1-(2-pyridylazo)naphth-2-ol as indicator. Tetraphenylborate may be oxidized by chromium(VI) in concentrated sulphuric acid solution to carbon dioxide and water[32]. Excess of chromium(VI) is back-titrated with standard iron(II) solution in the presence of ferroin indicator.

The determination of dimethoxyborane in methylborate can be based on the reaction of hydridic hydrogen with iodine and subsequent titration of excess of iodine with sodium thiosulphate[33]. The method is directly applicable to dimethoxyborane in trimethyl borate solution in the dimethoxyborane concentration range 0–15%.

Pentaborane can be determined by its quantitative reaction with ethanol to produce hydrogen and triethyl borate, the latter being hydrolysed to boric acid, which is then converted to the mannitol complex and titrated with standard alkali[34]. Decaborane may be titrated as a monobasic acid using aqueous sodium hydroxide, although the results were consistently about 2% lower than theoretical[35]. The concentration of decaborane and diborane in air may be monitored by scrubbing the boranes out of the air into sodium hydrogen carbonate–potassium iodide electrolyte, then titrating with coulometrically generated iodine[36]. Diborane and decaborane in concentrations as low as 0.2 p.p.m. can be determined, but the various boranes cannot be differentiated. Materials that react with iodine, such as acetone and peroxides, interfere. Decaborane and tetraborane have been determined iodometrically[37]. They react instantaneously with a solution of iodine in methanol, evolving hydrogen. By back-titrating the excess of iodine with standard sodium thiosulphate solution, or measuring the volume of hydrogen evolved in a closed system, a quantitative determination of tetraborane or decaborane can be made. Since alkali titration methods for the determination of decaborane give high results owing to the presence of acidic impurities in the sample, an iodine titration has been recommended as a standard[38]. This may be based on the oxidation of decaborane with potassium iodate in glacial acetic acid, followed by an iodometric titration[39]. Although it is based on the reduction of 44 equivalents of oxygen per mole of decaborane, in actual practice the value is 3% lower than this; therefore, the reagents should be standardized with research-purity decaborane.

B. Potentiometric Titration

Various workers have described mercurimetric titration procedures for the determination of alkali metal tetraphenylborates[40–42]. Titrations may be carried out in 0.1 N sodium acetate, with standard mercury(II) nitrate or mercury(II) perchlorate solution as titrant, and the end-points detected potentiometrically with a mercury-coated platinum electrode and SCE joined by an agar bridge, or amperometrically with a dropping mercury electrode vs. SCE[40]. Potentiometric and amperometric titration procedures have been described for the determination of sodium tetraphenylborate using standard mercury(II) nitrate solution[41]. The potentiometric titration is reproducible in spite of the formation of intermediates (phenylboronic acid and diphenylmercury); the first break in the curve occurs after consumption of 3 equivalents of mercury(II) ion and the second after 8 equivalents. Boron in organic amine tetraphenylboron derivatives can be determined by potentiometric titration with standard silver nitrate solution[42]. The dry sample is dissolved in aqueous acetone (1:1) and the solution is buffered at pH 5. The resulting solution is then titrated with 0.06 N silver nitrate with a platinum indicator electrode and a shielded platinum reference electrode.

IV. ORGANOCALCIUM COMPOUNDS

An argentimetric method has been suggested for the determination of calcium acenaphthalene and similar types of compounds. This procedure involves reaction of the sample with dialkyl or diaryl sulphides and subsequent titration of the mercaptan formed with silver ion[43].

V. ORGANOCOBALT COMPOUNDS

Cyanobalamin has been analysed by decomposition with sulphuric acid and hydrogen peroxide, cobalt being precipitated as cobalt(III) hydroxide by addition of sodium hydroxide and hydrogen peroxide solution, and the dissolved precipitate being titrated iodometrically. The error in this method is not greater than ± 0.5%[44].

A rapid micro-determination of cobalt carbonyl anion $[Co(CO)_4]^-$ in organic solvents (glacial acetic acid–toluene) can be achieved by titration with methylene blue at 0°C under nitrogen[45]. The stoichiometric reaction of 2 mol of $[Co(CO)_4]^-$ with 1 mol of methylene blue takes place in the pH range 1.9–5.2. Cyanocobalamin has been titrated amperometrically with chromium(II) ion in an EDTA medium at pH 9.5[46]. The anodic polargraphic wave at $E_{1/2} = 0.3111$ V vs. SCE is probably due to the oxidation of the mercury of the electrode[47].

VI. ORGANOCOPPER COMPOUNDS

Copper in treated fabrics may be determined by titration with 8-hydroxyquinoline[48]. The sample is extracted with hydrochloric acid at 80°C and the solution is titrated potentiometrically with 0.1 N potassium bromate. Most cations and many organic substances can be tolerated, but antimony trioxide and certain non-ionic surfactants interfere.

VII. ORGANOGERMANIUM COMPOUNDS

Mercaptogermanes have been determined indirectly with iodine[49]. Alkoxygermanes have been determined by a modification of the Ziesel procedure[50]. Both GeOC and GeSC linkages can be determined by the perchloric acid-catalysed acetylation method[51]. The mechanism of the reaction of an alkoxy- or mercaptogermane with acetic anhydride is presumably the same as that proposed for the acetylation of SiOC[52,53] and SiSC linkages[54]. The acetylation method depends on the reaction of the alkoxy- or mercaptogermane with excess of acetic anhydride to form the corresponding acetate or thioacetate and acetoxygermane. The acetoxygermane and remaining acetic anhydride are rapidly hydrolysed to acetic acid, which is then titrated with standard base. The alkoxy or mercapto content is calculated from the difference in volume of base between the sample and blank titrations.

A further method for the assay of GeOC linkages is based on reaction with *in situ* generated hydrogen bromide[55]. This is based on an earlier procedure for the determination of epoxides and aziridines[56]. Under anhydrous conditions, standard perchloric acid is titrated into an acetic acid solution of tetraethylammonium bromide and the sample. The hydrogen bromide which is formed reacts with the alkoxygermane to form a bromide and the corresponding parent alcohol. The first excess of hydrogen bromide is detected by a blue BZL (Ciba 22062S) indicator end-point. This method allows the quantitative determination of alkoxygermanes in the presence of alkoxysilanes and alcohols.

Procedures based on acetylation and on reaction with hydrogen bromide for the determination of alkoxy and mercapto groups in organogermanium compounds, which by elemental analysis and gas chromatography were known to have a purity between 98.7 and 99.9%, have been compared[55]. Most alkoxy- and mercaptogermanes may be determined quantitatively by either of the above methods. Acetylation, however, is preferred for mercaptogermane determinations[55].

VIII. ORGANOIRON COMPOUNDS

A. Classical Titration Procedures

In determining the standard reduction potential of haemoglobin, dialysis affects the shape of the redox titration curves, thus leading to different end-point determinations[57]. At pH 7.9, the oxidation of ferrohaemoglobin by hexacyanoferrate(III) is not a simple reversible reaction, but appears to proceed by an irreversible stepwise mechanism.

The standardization of methods for the determination of carboxyhaemoglobin has been discussed[58]. Palladium chloride is reduced in an acidic medium by means of carbon monoxide to an equivalent amount of metallic palladium. The determination of excess of palladium chloride is carried out by the use of an indirect complexometric titration after addition of potassium tetracyanonickelate(II) ($K_2[Ni(CN)_4]$), the nickel ions liberated being titrated with EDTA (disodium salt) solution using murexide as indicator.

B. Potentiometric Titration

Aryl ferrocenes can be titrated potentiometrically with standard potassium dichromate solution and E^0 values may be calculated for the oxidation reaction[59]. The molecular weights of ferrocene derivatives may be obtained by potentiometric titration with potassium dichromate[60]. Ferrocene can be determined[61] by addition of an acetic acid solution of the sample to a solution of iron(III) chloride in 2 N hydrochloric acid and measurement of the oxidation potential of the resulting clear solution:

$$[Fe(C_5H_5)_2] + FeCl_3 \longrightarrow [Fe(C_5H_5)_2]^+ + FeCl_2 + Cl^- \quad (7)$$

The percentage purity or the equivalent weight of alkyl- and hydroxyalkylferrocenes can be determined by potentiometric titration with standard iron(III) chloride solution[62]. In addition to its rapidity, a particular advantage of this procedure is that ferrocene derivatives which have a carbonyl group adjacent to the cyclopentadiene ring do not titrate, so that it is possible to determine the amount of alkylferrocene in a mixture with alkyloxoferrocene. Potentiometric titration of alkyl- and hydroxyalkylferrocenes with 0.1 N potassium dichromate in acetic acid gave erratic results and over 100% recovery when the iron content of the sample indicated that the samples were less than 100% pure[62]. This is because dichromate, in strong acid, oxidizes some of the organic material, thereby causing high dichromate titrations. It may be corrected by carrying out the potentiometric titration with a methanolic solution of iron(III) chloride. A larger potentiometric break occurs when using methanol as solvent than with acetic acid. The method has been applied to the assay of a number of alkyl- and hydroxyalkylferrocenes and gives a standard deviation of 0.22%. Carbonyl-substituted ferrocenes do not titrate with iron(III) chloride.

IX. ORGANOLEAD COMPOUNDS

A. Classical Titration Procedures

A rapid procedure using complexone has been described for the determination of tetraethyllead in petrol[63,64]. The sample is pipetted into bromine solution in carbon tetrachloride until the colour persists. Methanol or ethanol is added and the solution is boiled, decolorized with a small excess of 1 N alcoholic potassium hydroxide, diluted with water, and boiled again. Lead is determined indirectly using EDTA. In a further procedure using bromine the petroleum sample is diluted with a high-boiling solvent then 30% bromine in carbon tetrachloride is added until the brown colour persists for 2 min[65]. This solution is shaken with 0.1 N nitric acid and the extract boiled to expel bromine fumes. Tartaric acid, Eriochrome Black T indicator, excess of 0.2 M magnesium chloride solution, and 0.1% potassium cyanide solution are added and the solution is buffered to pH 10 and titrated with 0.01 M EDTA.

Tetraethyllead may be extracted from petroleum into concentrated hydrochloric acid and the lead precipitated as lead sulphate[66]. The lead is then determined by complexometric titration with EDTA (disodium salt) using Eriochrome Black T as indicator. In a variant of this procedure, interferences caused by the presence in the petroleum of iron, dyes, and acid-extractable organic substances can be eliminated by oxidation of organic matter with sulphuric and nitric acids.

Leaded petroleum may be treated with concentrated hydrochloric acid and potassium perchlorate to extract lead into the acidic phase. This is determined by addition of excess of disodium EDTA, which is back-titrated to the Eriochrome Black

T end-point with standard zinc chloride solution[67]. Alternatively, tetraethyllead may be separated from the petroleum using the ASTM hot hydrochloric acid extraction procedure[68], and the lead ion titrated with disodium EDTA[69]. Serious interferences to the end-point in this titration due to iron, petroleum dyes, and organic compounds extracted from the petroleum are reported[70]. For example, 0.3 p.p.m. iron in the petroleum renders end-point detection impossible. These may be overcome with copper–PAN indicator[70]. However, fuels containing both tetraethyllead and (methylcyclopentadienyl)manganese tricarbonyl cannot be analysed for lead content by this method. The small amount of manganese extracted by the hot hydrochloric acid frequently leads to an error of several tenths of a millilitre of tetraethyllead per gallon[70].

A tentative DIN standard[71] has been issued for the determination of tetraethyllead in petrol based on decomposition with hydrochloric acid and complexometry. The method is not applicable to samples containing multivalent metals. The sample is boiled with hydrochloric acid to decompose the tetraethyllead, diluted with water, neutralized, and buffered to a pH value of between 10 and 11. A known excess of EDTA solution is added and the excess is back-titrated with standard zinc sulphate solution. From an examination of various methods for the determination of tetraethyllead in petrol it was concluded that direct complexometric titration of the lead chloride-containing hydrochloric acid extract in the presence of tartrate was the most accurate of the methods examined[72]. Following extraction of lead by the ASTM procedure[68], tartaric acid is added to the extract, which is then made alkaline with aqueous ammonia and the lead is determined indirectly using EDTA[73]. Interferences from copper, zinc, nickel, cobalt, cadmium, and manganese, but not from calcium or magnesium, are avoided by adding solid potassium cyanide after the aqueous ammonia.

Various oxidizing agents have been used to destroy organic matter in the hydrochloric acid extracts of petroleum prior to the complexometric determination of lead using EDTA[74–79]. The sample of petroleum may be treated with potassium chlorate and concentrated hydrochloric acid. Water is then added and the aqueous phase separated off. The organic matter is destroyed by heating with hydrogen peroxide or potassium chlorate and nitric acid[74]. A similar method extracts the lead by heating the petroleum sample under reflux with concentrated hydrochloric acid, evaporating the extract, and destroying the organic matter by heating with potassium chlorate and nitric acid[75]. A solution of chlorine in carbon tetrachloride has been used in the determination of organically bound lead in 'ethyl fluid' and in petroleum[76]. The sample is treated with a solution of chlorine in carbon tetrachloride and after a reaction period of 1 min, water is added, and the carbon tetrachloride removed by boiling. Excess of chlorine in the petroleum is then reduced with either 3% hydrogen peroxide, sodium thiosulphate, or sodium hydrogen sulphite and then lead is determined by addition of EDTA solution and back-titration with standard magnesium chloride solution.

A method based on titration with Karl Fischer reagent has been reported for the determination of PbOH groups[80]. An iodometric procedure for the determination of organolead compounds involves the reaction of the organolead compound with an excess of iodine. The unchanged iodine is then titrated with standard sodium thiosulphate using starch as indicator[81]. Alternatively, the sample is shaken with thiophen-free benzene, iodine solution is slowly added, and excess of iodine solution is then titrated with sodium thiosulphate[77]. In the argentimetric determination of tetraethyllead in antiknock mixtures excess of silver nitrate is added to the sample and the resulting metallic silver is filtered off, dissolved in nitric acid, and determined by the Volhard method[82]. In an organic solvent such as benzene and with a limited reaction time, tetraethyllead reacts quantitatively with iodine to form triethyllead iodide. This is the basis for the iodometric method for the determination of tetraethyllead[83]. With bromine or chlorine in an organic solvent, tetraethyllead reacts rapidly to form the corresponding diethyllead halide. In direct sunlight tetraethyllead is slowly converted to the triethyllead ion, then more slowly to the diethyllead ion, and finally to the simple lead ion. Under excessive heat tetraethyllead decomposes to form metallic lead and a variable mixture of hydrocarbon gases. Hexaethyllead reacts with silver nitrate to produce metallic silver, which can be separated, dissolved in nitric acid, and titrated by the Volhard method[84]. Tetraethyllead reacts similarly so that the method is not suitable for mixtures of these compounds.

Hexaethyldilead can also be determined by dissolving the sample in carbon tetrachloride, covering the solution with water containing starch indicator, and titrating in an atmosphere of nitrogen with 0.01 N iodine in potassium iodide solution. Two atoms of iodine react with one molecule of hexaethyldilead. Mixtures of tetraalkyllead, hexaethyldilead, and triethyllead chloride are analysed by titrating the hexaethyldilead with 0.01 N iodine as described above, separating the carbon tetrachloride layer, which now contains triethyllead iodide together with tetraethyllead and triethyllead chloride, and treating with silver nitrate which precipitates silver from tetraethyllead, silver iodide from triethyllead iodide, and silver chloride from triethyllead chloride. The precipitate is filtered off and dissolved in nitric acid and the silver is determined by Volhard's method. The carbon tetrachloride now contains triethyllead nitrate equivalent to the hexaethyldilead plus tetraethyllead plus triethyllead chloride originally present. This is converted to lead chloride by treatment with hydrogen chloride gas.

Triethyllead ions have been determined in aqueous solution using sodium tetraphenylborate. Precipitation of triethyllead ions from acetic acid solution of pH 4–5 by sodium tetraphenylborate is complete within 10 min. The sample solution is filtered and excess of sodium tetraphenylborate is titrated with 0.5% benzalkonium chloride solution[85].

B. Coulometric Titration

Tetraalkyllead compounds have been determined coulometrically using bromine and mercury(I) ion[86,87]. The titration of the tetramethyllead is carried out to the reaction

$$(CH_3)_4Pb + Br_2 \longrightarrow (CH_3)_3PbBr + CH_3Br \quad (8)$$

This reaction was studied in 0.5 M methanolic ammonium bromide as basal solution Hexaethyldilead may be determined in the presence of tetraethyllead, by coulometric iodination at constant current, with amperometric dead-stop end-point indication[88]. The sample, in an alcoholic solution containing iodide ions, is placed in the anodic compartment of an electrolysis cell where it undergoes the following reaction:

$$R_3M\text{-}MR_3 + I_2 \longrightarrow 2R_3MI \quad (9)$$

with the iodine electrolytically produced under constant current at the platinum anode. The end-point is observed by a rise in the indicator current, caused by excess of iodine, between a second pair of platinum electrodes sensitive to the I_3^-/I^- redox system[89]. An amperometric plot of indicator current *vs*. generation time can be obtained photographically in order to ensure an accurate determination. Iodine is employed in this method for the coulometric–amperometric titration of hexaethyldilead, because its rate of reaction is always greater than the rate of electrolytic generation of the iodine.

Following the discovery[84] that hexaethyldilead reacts quantitatively with silver ions:

$$(C_2H_5)_6Pb_2 + 2Ag^+ \longrightarrow 2[(C_2H_5)_3Pb]^+ + 2Ag \quad (10)$$

the possibility of titrating hexaethyldilead by means of electrolytically generated silver ion to an amperometric end-point[89] has been investigated[90]. The conditions whereby selective titrations of organolead species in the mixtures could be carried out by the utilization of quinoline as a complexing agent for silver have also been reported[90].

C. High-frequency Titration

High frequency titration with potassium permanganate has been used[91,92] for the determination of down to 0.01% of hexaethyldilead in tetraethyllead. The sample of tetraethyllead is diluted with acetone. Titration is carried out with a 0.05 N solution of potassium permanganate in acetone.

$$(C_2H_5)_3PbPb(C_2H_5)_3 + (O) \longrightarrow Pb_2(C_2H_5)_6O \quad (11)$$

X. ORGANOLITHIUM COMPOUNDS

A. Classical Titration Procedures

A double titration procedure for the analysis of alkyllithium compounds has been applied principally to *n*-butyllithium[93–95]. It cannot be used for the determination of methyllithium or phenyllithium owing to their low reactivity with the reagent. The total alkali (i.e. LiR + LiOR + LiOH + Li_2O) is first determined by hydrolysis of the sample solution under dry, oxygen-free nitrogen followed by titration with standard acid to the phenolphthalein end-point. To a further portion of the sample in dry diethyl ether is added benzyl chloride dissolved in diethyl ether, which reacts with the *n*-butyllithium:

$$n\text{-}C_4H_9Li + C_6H_5CH_2Cl \longrightarrow n\text{-}C_8H_{18} + C_6H_5CH_2C_4H_9\text{-}n + C_6H_5CH_2CH_2C_6H_5 \quad (12)$$

After reaction the solution is hydrolysed by the addition of water and finally titrated with standard acid. This titration is equivalent to the LiOR + LiOH + Li_2O content of the sample. The *n*-butyllithium content of the sample is then calculated from the difference between the two titrations. A four-fold excess of benzyl chloride with respect to the alkyllithium compound, a 5-min reaction time with the benzyl chloride and the use of not less than one volume of diethyl ether per volume of sample are recommended[96].

Low results are obtained when this double titration procedure employing benzyl chloride is used to determine methyllithium and aryllithium compounds[97–101]. In addition, the purity of the diethyl ether has a marked effect when the double titration procedure is used in analysing solutions for *n*-butyllithium content. Reasonably accurate results are obtained in *n*-heptane solutions. Purification of the ether by treatment with lithium aluminium hydride leads to more satisfactory results. The low results due to the fact that benzyl chloride does not react quantitatively with alkyllithium compounds to give non-basic products can be avoided by using alternative organic halides, particularly 1,2-dibromoethane for alkyllithium compounds and phenyllithium, and 1,1,2-tribromoethane or allyl bromide for butyllithium[97] and other alkyllithiums[102–104], as well as R_3MLi (M = Si, Ge, Sn)[105,106].

In Table 5 are shown typical titrations of *n*-butyllithium with both benzyl chloride and allyl bromide, being expressed as the percentage of the total basic content of the

TABLE 5. Double titrations of *n*-butyllithium. Solutions of *n*-butyllithium were analysed by double titration with 1,2-dibromoethane, allyl bromide, and benzyl chloride. The results are expressed as the percentage of the total basic content of the solutions which is due to C—Li

	Percentage of total base due to C—Li		
Preparation	$PhCH_2Cl$	$CH_2{=}CHCH_2Br$	$BrCH_2CH_2Br$
BuLi in Et_2O, run 1	88.1	93.6	
BuLi in Et_2O, run 2	90.1	95.1	
BuLi in Et_2O, run 3	76.3	81.2	
BuLi in hexane	94.7	98.6	
BuLi in hexane	97.1	99.2	
BuLi in hexane	97.0	99.1	99.1

solution which is due to RLi. The first three runs are for preparations in diethyl ether, and it may be noted that the allyl bromide values are consistently 5% above those obtained with benzyl chloride. This difference corresponds to the error previously estimated for the benzyl chloride method[99]. The high percentage found with the allyl bromide titrations is taken to indicate that the *n*-butyllithium in the solution is more completely consumed by allyl bromide than by benzyl chloride. 1,2-Dibromoethane has also been recommended as a reagent in the double titration of cyclopropyllithium[107].

Phenyllithium may be determined by pipetting inorganic halide-free 1-bromo-2-phenylethane into a flask and adding pure di-*n*-butyl ether[108]. The flask is flushed with nitrogen and an aliquot of phenyllithium dissolved in pure di-*n*-butyl ether added. The flask is left to react, then 1.5 M nitric acid is added, and halide ion is determined by a modified Volhard procedure[109].

The effect of lithium alkoxides on the determination of butyllithium (and butylsodium) compounds by the Gilman procedure was studied to establish whether any reaction between the halogen compound and any lithium alkoxide present had an effect on the accuracy of the lithium bound carbon determination[110]. Allyl bromide gave the most accurate results. When the Gilman double titration procedure is applied to tertiary lithium alkyls, then any tertiary lithium alkoxides present as an impurity in the sample react slowly with benzyl chloride or with 1,2-dibromoethane giving analyses for the alkyllithium component which are too high[111]. This may be overcome for the analysis of tertiary lithium alkyls through the observation that organic acids, including weak acids, can be titrated with sodium dimethyl sulphoxide in dimethyl sulphoxide solutions using diphenylmethane or triphenylmethane as indicator. The method involves the titration of a known amount of a standard organic acid with the organolithium solution of unknown titre in dimethyl sulphoxide–monoglyme–hydrocarbon solution with triphenylmethane as indicator. Benzoic acid is used as titrant because of the relative ease of observation of the yellow to red (alkyllithium) or green to red–brown (phenyllithium) end-point and because a monoglyme solution of this acid can be standardized by an aqueous base titration. The overall reactions occurring in the system are as follows:

$$RLi + C_6H_5CO_2H \longrightarrow C_6H_5CO_2Li + RH \quad (13)$$

and at the end-point when the standard acid is consumed:

$$RLi + (C_6H_5)_3CH \longrightarrow \underset{\text{(red)}}{RH + (C_6H_5)_3C^-Li^+} \quad (14)$$

The analysis of compounds of the type R_3ELi, where E = Si, Ge, or Sn, by a double titration procedure using allyl bromide and other organic halides as the reagent has been compared[105] with a method based on reaction of the R_3ELi compound with *n*-butyl bromide followed by titration of the released bromide ion by Volhard titration[112]. In general, better results were obtained using the allyl bromide double titration than with the *n*-butyl bromide Volhard analysis, triphenylgermanyllithium being an exception.

The present position regarding the applicability of the Gilman procedure to the assay of commercial alkyllithiums is that the ASTM has selected the Gilman benzyl chloride coupling procedure for the determination of *n*-butyllithium in hexane[113] and this can be taken as a measure of confidence in this procedure for this particular analysis. With experienced analysts, a reproducibility between two laboratories of about 0.2% is claimed. It has been pointed out[114–117] that, in spite of statements by earlier workers that the use of the benzyl chloride coupling reagent leads to low carbon-bound lithium values, it has frequently been possible to obtain as high as 99.2% of carbon-bound lithium (or 0.8% of non-carbon-bound lithium) on newly manufactured 15% hexane solutions of *n*-butyllithium. Obviously, avoidance of sample contamination by oxygen or moisture is a very important factor in obtaining these results. Thus, if benzyl chloride does give lower results for net assay (total base minus base left from Gilman coupling), the difference between 'actual' and assay values must be less than 0.8% of non-carbon-bound lithium. Results obtained by this procedure are therefore acceptably accurate. In a modification of the ASTM procedure used in Europe, a larger amount of benzyl chloride is used in the absence of ether. This method gives a result about 0.2% higher than that obtained by the ASTM assay. These comments apply strictly only to commercial organolithium preparations which contain a minimum amount of oxygenated impurities and do not necessarily apply to mixtures containing appreciable amounts of, for example, lithium alkoxides.

The ASTM method, using benzyl chloride, works well with phenyllithium provided that the coupling reaction is allowed to take place for at least 30 min. In most instances, allyl bromide can be substituted for benzyl chloride with no change in the analytical results.

Benzyl chloride does not work well with methyllithium or vinyllithium, but allyl bromide does react very readily in these cases, giving consistent analyses. Owing to the very limited solubility of methyllithium in diethyl ether, non-carbon lithium assays on solutions of 5% methyllithium in ether never exceed 0.03–0.05%, even if the true methoxide content of the sample is considerably higher than this.

Ethylene dibromide is recommended as a reagent for lithium alkoxides[117]. It reacts with lithium *t*-butoxide and lithium *n*-butoxide in hexane, and this affects the Gilman coupling correction. Any comparison of the reactivity of a coupling reagent with lithium alkoxides should be made in the presence of an alkyllithium compound as the lithium alkoxide is actually coordinated in the alkyllithium hexamers and should be more reactive in this mixed system towards the coupling than it would be in the absence of the alkyllithium compound.

A procedure has been described[118] for the determination of organolithium compounds based on cleavage of dialkyl or diaryl disulphides and subsequent titration of the lithium mercaptide formed with silver ion by the silver nitrate amperometric technique (equations 15 and 16)[119]. Only lithium metal has been found to complicate the cleavage reaction. Aromatic disulphides react rapidly and quantitatively with organolithium compounds in the presence or absence of ethers. A comparison of quantitative results obtained with tolyl disulphide and *n*-butyl disulphide in the presence and absence of ether showed good agreement. On the basis of the consistent results obtained in the presence or absence of ether with tolyl disulphide, this aromatic disulphide is recommended as the preferred reagent for the method.

$$R^1Li + RSSR \longrightarrow R^1SR + LiSR \quad (15)$$

$$LiSR + [Ag(NH_3)_2]^+ \longrightarrow RSAg + Li^+ + 2NH_3 \quad (16)$$

Oxygen, water, and alcohols and the products of their reaction with organolithium compounds do not interfere with the disulphide cleavage procedure except for the destruction of a stoichiometric amount of the organometallic compound. The reactions with water[120] and with alcohols[121,122] are the bases for published procedures for analysing organolithium compounds. The reaction product with oxygen has been used in a procedure for analysing dilithioaromatic compounds[123]. In the analysis of organolithium compounds, substances such as lithium metal, lithium hydride, lithium hydroxide, and lithium alkoxides should be considered as possible interfering ingredients[118]. Thus only lithium metal cleaved tolyl disulphide under the conditions of the analysis. Unreacted lithium metal can be readily determined[124]. A non-aqueous titration procedure has been described for the determination of *n*-, *sec*-, and *t*-butyllithium, based on titration with a standard solution of *sec*-butyl alcohol in xylene to the 1,10-phenanthroline or the 2,2′-biquinolyl colorimetric end-points[125]. Addition of a few milligrams of 2,2′-biquinolyl and about 5 ml of 1.5 M butyllithium in hexane to 20 ml of benzene produces a yellow–green or chartreuse-coloured solution. After titration with 1 M *sec*-butyl alcohol in xylene, the solution is clear and colourless; the disappearance of the green colour occurs sharply after addition of 1 mole-equivalent of titrant.

A titration method has been developed suitable for the determination of alkyllithium compounds in ether solutions at concentrations down to 10^{-3} M[126]. This analysis causes considerable difficulty because at room temperature and even at 0°C, the presence of ethers, such as 1,2-dimethoxyethane and tetrahydrofuran, as solvents causes rapid decomposition of *n*-butyllithium, as indicated by titration with a standard solution of *sec*-butyl alcohol. If, however, the ether was cooled to −78°C prior to the addition of *n*-butyllithium and kept at this temperature during the titration with *sec*-butyl alcohol then *n*-butyllithium decomposition was slow. Rapid, accurate, and precise analyses were obtained in this way by titration with a standard solution of *sec*-butyl alcohol using 2,2′-biquinolyl[125].

Vinyllithium can be determined by measurement of the amount of vinyltributyltin produced by reaction of the vinyllithium with tributyltin chloride[127] or the amount of tetraphenyllead obtained by reaction between tetravinyllead and phenyllithium[128,129]. Phenyllithium may be determined by measurement of the amount of tetraphenyltin produced in the transmetallation reaction between tetravinyltin and phenyllithium[129].

Organolithium compounds can be determined by iodination[130]. The organolithium compound is slowly added to an excess of a standardized diethyl ether solution of iodine and the unused iodine is back-titrated with standard sodium thiosulphate solution to the starch end-point. It is important to add the organolithium compound to an excess of the iodine solution, rather than the reverse. This ensures that interfering coupling reactions are minimized during iodination:

$$C_6H_5Li + I_2 \longrightarrow C_6H_5I + LiI \text{ (iodination)} \quad (17)$$

$$2C_6H_5Li + I_2 \longrightarrow C_6H_5{-}C_6H_5 + 2LiI \text{ (coupling)} \quad (18)$$

This method was applied successfully to the assay of phenyllithium solutions, giving results which agreed to within 3% of the theoretical result. The method was also shown to be applicable to the assay of butyllithium.

B. Potentiometric Titration

p-Phenylenedilithium has been determined by potentiometric titration with cerium(IV) nitrate[123]. This method, which determines even small amounts of this substance in the presence of monometallo-organics, involves the oxidation and hydrolysis of the dilithium compound to form hydroquinone, which is then titrated potentiometrically with standard cerium(IV) nitrate solution using a standard calomel electrode and a platinum reference electrode:

$$Li{-}C_6H_4{-}Li \xrightarrow{O_2} LiOO{-}C_6H_4{-}OOLi \xrightarrow{H_2O} LiO{-}C_6H_4{-}OLi \xrightarrow{H}$$

$$HO{-}C_6H_4{-}OH \xrightarrow{Ce^{4+}} O{=}C_6H_4{=}O + Ce^{3+} \quad (19)$$

As the method does not involve an acid–base titration, the presence of lithium hydroxide or other hydrolysis products does not interfere. Alkyllithiums and *m*-phenylenedilithiums give oxidation and hydrolysis products which cannot be oxidized by cerium(IV) ion; therefore, the method is selective for *o*- or *p*-phenylenedilithiums in the presence of other types of monometallo-organics.

n-Butyllithium in hydrocarbons has been determined by treatment with excess of vanadium pentoxide[98]. The reduced vanadium is then titrated potentiometrically with standard sulphatoceric acid solution. A comparison of results obtained by this method with those obtained by a method for determining total alkalinity, including butyllithium, lithium butoxide, lithium hydroxide, and other basic materials, is shown in Table 6 for solutions of *n*- and *t*-butyllithium in various solvents[98]. In all cases, the concentration of butyllithium as determined by the vanadium pentoxide method is less than the concentration of total base. This is to be expected, since any air oxidation of *n*-butyllithium results in the formation of lithium *n*-butoxide, which is soluble in solutions of *n*-butyllithium. However, in every case the difference between the two values is 4% or less, which indicates the presence of only a small amount of soluble base other than butyllithium. The analysis of similar solutions of *n*-butyllithium in *n*-heptane by the double titration method indicated that 4–5% of the total base present was non-butyllithium base[99]. This blank was fairly constant and it was concluded, probably wrongly, that it was inherent in the double titration method under the experimental conditions used.

Only *n*-, *sec*-, and *t*-butyllithium and ethyllithium solutions were assayed by the vanadium pentoxide method, but the procedure should be generally applicable to the determination of any alkyllithium compound in a hydrocarbon solvent. It cannot be

TABLE 6. Analysis of solutions of *n*- and *t*-butyllithium in various solvents

Compound	Solvent	BuLi by V_2O_5 method (M)	Total base (M)	Difference (%)
n-Butylithium	*n*-Haptane	1.67	1.70	1.8
	n-Heptane	2.74	2.80	2.1
	n-Heptane	1.23	1.24	0.8
	Cyclohexane	2.58	2.62	1.5
	Tolu-Sol	1.20	1.21	0.8
	Tolu-Sol	1.19	1.23	3.2
t-Butyllithium	*n*-Pentane	1.46	1.52	4.0

used for the determination of phenyllithium because, although phenyllithium rapidly reduces vanadium pentoxide, most solutions of phenyllithium contain lithium phenoxide because of air oxidation. On titration with sulphatoceric acid, the phenol is oxidized together with the reduced vanadium. This leads to high results for the phenylithium content.

C. Thermometric Titration

n- and *sec*-butyllithium in hydrocarbon solution can be determined by thermometric titration with a standard hydrocarbon solution of butyl alcohols (*n*-, *sec*-, *t*-)[24]:

$$\text{RLi} + \text{BuOH} \longrightarrow \text{LiOBu} + \text{RH} \qquad (20)$$

The reaction is stoichiometric; lithium butoxide, normally the major impurity, does not interfere. The simplicity of the method makes it more rapid and convenient than many of the alternative methods; the method is believed to be generally applicable to compounds containing lithium—carbon bonds. The apparatus is discussed in the section on the analysis of organoaluminium compounds[24,131]. A procedure for determining small amounts of *n*- or *sec*-butyl alcohol in the aqueous extract from hydrolysis of butyllithium provides an independent estimate of the accuracy of the impurity correction in the double titration method (butoxide is usually the major impurity)[24]. Because of the high energy of reaction and the ease with which large samples can be handled, the sensitivity is high; the detection limit for a 50-ml sample is estimated to be well below 0.01% of butyllithium. The sensitivity of the vanadium pentoxide method[98] appears to be comparable. Reasonable agreement is obtained between the butyl alcoholic thermometric titration method and the vanadium pentoxide method.

D. High-frequency Titration

High-frequency titration of various alkyllithiums as well as benzyl and phenyllithium compounds with a standard solution of acetone in benzene has a sensitivity of 0.01% of organolithium compound when a 50 ml sample is used[132]. Lithium alkoxides do not interfere. Acetone is used as a titrant because it is relatively easy to obtain pure and dry, and also because under ordinary conditions its reaction with lithium alkyls is rapid, complete, and irreversible.

$$\text{RLi} + (\text{CH}_3)_2\text{C}{=}\text{O} \longrightarrow (\text{CH}_3)_2\text{C(R)}{-}\text{O}^-\text{Li}^+ \qquad (21)$$

The results obtained by high-frequency titration generally agreed within 1% with results obtained by the vanadium pentoxide method[98], and for phenyllithium compounds the high-frequency titration results were in good agreement with those obtained by the double titration procedure[95] and gave a particularly good precision (0.5% agreement) when ethylene dibromide rather than the benzyl chloride reagent was employed in the latter procedure.

E. Lumometric Titration

A 2% toluene solution of *n*-butyllithium has been titrated lumometrically with air[133]. A very sharp increase in light intensity occurred at a point corresponding to one atom of oxygen for two atoms of lithium. Beyond this point no more oxygen was absorbed but light emission continued at a gradually decreasing rate for more than 24 h.

XI. ORGANOMAGNESIUM COMPOUNDS

A. Classical Titration Procedures

Grignard reagents cannot be determined by titration with a standard solution of iodine in diethyl ether until a pale iodine colour persists[134]:

$$RMgX + I_2 \longrightarrow RI + MgXI \qquad (22)$$

owing to the occurrence of a side-reaction simultaneously with the above reaction[135,136]:

$$2\ RMgX + I_2 \longrightarrow R{-}R + 2\ MgXI \qquad (23)$$

A satisfactory analysis can be obtained by adding an aliquot of the Grignard solution to an excess of standard iodine solution, that is in the reverse order of that above, and then titrating the excess of iodine with standard sodium thiosulphate solution[137,138]. The accuracy of the iodometric method has been checked by adding known amounts of water or methanol to a Grignard reagent and checking how much the titre ran back. It is claimed to be accurate to within 1%[139]. It has been suggested[140] that methods based on acid titration[135,137,141,142] on iodine titration[134], and on Volhard titration[143] give high results. Several alternative methods for the analysis of Grignard compounds which can also be applied to diaryl magnesium compounds are described below.

Titration with sulphuric acid[135,137,141,142]

Procedure. Add 1–2 ml of the organomagnesium sample to water. Boil for 10–15 min and add, after cooling, a known excess of 0.100 N sulphuric acid. Back-titrate excess of acid with 0.1 N sodium hydroxide using phenolphthalein as indicator.

Iodine titration[134]

Procedure. Add 1–2 ml of organomagnesium sample to 1 N iodine solution in dry diethyl ether. Let the reaction proceed for 10 min with shaking. Determine the unreacted amount of iodine by titration with 0.1 N sodium thiosulphate using starch as indicator.

Volhard halogen titration method[143]

Procedure. Add 1–2 ml of organomagnesium sample to water, boil until the solution is clear and adjust the pH to about 7 with sulphuric acid. Add a known excess of 0.1 N silver nitrate and heat the solution until the precipitate coagulates. Determine in the cooled solution the excess of unreacted silver nitrate with ammonium rhodanide using iron(III) sulphate as indicator.

Di-*sec*-butylmagnesium, *i*-butylmethylmagnesium, phenylmagnesium, and *n*-butylmagnesium chloride can be titrated directly under anhydrous conditions with *sec*-butyl alcohol using coloured indicators such as 1,10-phenanthroline or 2,2′-biquinolyl[144]. In a typical titration, 1,10-phenantholine is added to an ethereal solution containing approximately 0.1 M dialkylmagnesium to obtain a violet solution. Titration of the solution with standard 1 M *sec*-butyl alcohol in xylene causes no significant diminution in colour until two molecules of the titrant have been added per mole of magnesium compound, when the violet colour disappears sharply. End-points are sharper in ethereal solutions of organomagnesium compounds than in hydrocarbon solutions. Also, in hydrocarbon solution, precipitation of magnesium alkoxides may cause turbidity problems. Both diethyl and di-*n*-butyl ether were used as solvents for direct titration of butylmagnesium chloride. An analysis of a sample of this Grignard reagent, claimed to be about 2.8 M, showed it to be 2.73 M. Phenylmagnesium chloride was also analysed satisfactorily by direct titration.

An iodometric procedure for the determination of organoalkali metal compounds such as phenyllithium may also be applied to organomagnesium compounds[145]. A procedure for the iodometric determination of arylmagnesium compounds adds a measured volume of a chlorobenzene or anisole solution of the organomagnesium sample to a solution of iodine in benzene, toluene, or diethyl ether[146]. The iodine must be in three-fold excess with respect to the arylmagnesium compound. The solution is set aside for a few minutes at room temperature, then excess of iodine is titrated.

B. Amperometric Titration

A procedure for determining organomagnesium compounds is based on their cleavage of a dialkyl or diaryl disulphide[147]. The resulting thiol is titrated amperometrically in alcoholic medium with aqueous silver nitrate solution in the absence of air.

C. Potentiometric Titration

The organomagnesium sample is added to a 20–30% excess of a 1 N acetone solution in dry diethyl ether and then hydrolysed with methanol[140]. A solution of hydroxylamine formate in methanol is added and the free hydroxylamine formate is titrated potentiometrically with standard perchloric acid solution in dry dioxane. A procedure has also been described for overcoming interference by basic magnesium compounds in this method[140].

XII. ORGANOMERCURY COMPOUNDS

A. Classical Titration Procedures

Phenylmercury acetate in an aqueous acidic solution at pH 2–2.5 can be titrated with a carbon tetrachloride solution of copper diethyldithiocarbamate (1.5×10^{-4} M)[148]. The end-point is reached when the solvent layer becomes pale yellow–brown in colour. Non-aqueous titration with hydrochloric acid (0.1 N) in *n*-butanol medium has been used for the determination of phenylmercury acetate[149]. Thymol blue or diphenyl carbazide can be used as the indicator, except in the presence of basic compounds when only diphenyl carbazide is suitable. In an alternative non-aqueous titrimetric procedure the sample is dissolved in anhydrous acetic acid to acetolyse it and the reaction product is then titrated with 0.1 N perchloric acid dissolved in acetic acid to the *p*-naphtholbenzein or quinaldine red indicator end-point[150].

In a rapid volumetric procedure for the micro-determination of phenylmercury acetate, a weighed amount of phenylmercury acetate is dissolved in warm acetone and diluted with acetone[151]. To an aliquot is added sodium chloride and the solution is heated on a water-bath to evaporate the acetone. To the cooled solution is added nitric acid and ethanolic diphenyl carbazone, and the unconsumed sodium chloride is titrated with 0.02 N mercury(II) nitrate. The error of this method is less than $\pm 1\%$. In the case of fungicidal mixtures containing both phenylmercury acetate and organomercury halides, this method is combined with an argentimetric method[152], analysing separate sample aliquots by each method. In an alternative procedure for the determination of mercuriacetic acid in phenylmercury acetate, the sample is boiled with water and acidified with acetic acid[153]. Sodium chloride solution is added and the solution is diluted with water, then filtered. The filtrate is made alkaline with ammonia and the solution is titrated with sodium mercaptoacetate in the presence of thiofluorescein indicator.

Methoxyethylmercury chloride reacts with hydrochloric acid to produce mercury(II) chloride and ethyl methyl ether[154]:

$$CH_3OC_2H_4HgCl + HCl \longrightarrow HgCl_2 + CH_3OC_2H_5 \qquad (24)$$

To determine methoxyethylmercury chloride, the sample is dissolved in dilute hydrochloric acid and concentrated nitric acid. The solution is boiled, cooled, and diluted. An aliquot is mixed with aqueous ammonia, potassium cycanide, and aqueous potassium iodide and mercury is then determined by titration with 0.1 N silver nitrate.

Phenylmercury halides in technical fungicides may be analysed by dissolving the sample in dimethylformamide at room temperature and adding sodium hydroxide to make the solution blue (alkaline) to thymolphthalein[155,156]. Water is then added and the alkali titration is continued until the blue colour returns. This sequence of titrations is continued until the blue colour persists upon addition of water. Phenylmercury halides do not precipitate out under these conditions. Chloride resulting from the hydrolysis of phenylmercury chloride is then determined by argentimetric titration (0.05–0.1 N silver nitrate) to the potassium chromate end-point. The procedure has been applied to phenylmercury chloride and bromide and to the determination of phenylmercury halides in formulated fungicides containing powdered talc[156].

For the determination of ethylmercury chloride in technical products and in compounded products used as fungicides, the sample is dissolved in cold dimethylformamide and neutralized to thymolphthalein with 0.1 N sodium hydroxide to the pale blue end-point[157]. The chloride ion is titrated with 0.05 N silver nitrate in the presence of 5% potassium chromate.

Coulometric titration has been applied to the determination of various organomercury compounds. A 0.05 M solution of mercury(II) thioglycollate in acetate buffer (pH 5) can be used as generating solution with platinum and mercury electrodes. Oxygen is removed from the sample by nitrogen purging. Mercury(II) ions are generated, then the current is reversed and titration carried out with thioglycollic acid until a point of maximum potential inflection is reached (using silver amalgam and saturated calomel electrodes). The organomercury sample is added and thioglycollic acid generated to the same potential end-point as obtained previously. Samples that contain an Hg—C bond must first be heated with concentrated hydrochloric acid and brought to pH 5. The standard deviation was in the range 0.001–0.004 mg for 0.5-mg samples.

In a procedure for the determination of methylmercury salts in rat tissue and rat urine, the tissue is first homogenized with benzene and the extract digested with aqueous sodium sulphide and then oxidized with potassium permanganate[159]. Following decolorization with hydroxyammonium chloride, addition of urea and EDTA, and pH adjustment to 1.5, the solution is mixed with chloroform and titrated with standard dithizone solution until the colour of the chloroform layer is intermediate between the orange of the mercury complex and the green of the dithizone solution. Concentrations down to 1 ppm can be measured. Inorganic mercury does not interfere.

Phenylmercury compounds in paints and in fungicidal preparations have been determined by an iodometric method[160]. Microbiological assaying of mercurials in pharmaceutical products such as phenylmercury compounds in amounts down to 2–10 ppm have been discussed[161]. A method for the determination of mercury in biological tissue is based on electrolytic deposition of mercury from solution followed by titrimetric determination of mercury[162]. Organomercury compounds may be determined by formation of the *S*-organomercury derivative by reaction with excess of 2-mercaptobenzothiazide[163]. Excess of thiol is determined iodometrically.

B. Potentiometric Titration

A rapid volumetric determination of halogenated organomercury compounds is based on digestion of the sample with suitable solvents followed by argentimetric titration of the halogen, either potentiometrically or with potassium chromate as indicator[155]. Suitable solvents for use at room temperature are dimethylformamide and dioxane, and, when higher temperatures are needed, methanol and ethanol may be used.

Non-aqueous titrimetry has been used for the determination of organomercury compounds such as phenylmercury nitrate, phenylmercury acetate, mercury succinamide, Thiomersal, Nitromersol, and Meralluride based on acetolysis of the sample with glacial acetic acid followed by titration (potentiometric) with a standard solution of perchloric acid dissolved in anhydrous acetic acid to the *p*-naphthalbenzein or the quinaldine red end-point[150].

XIII. ORGANOPHOSPHORUS COMPOUNDS

A. Classical Titration Procedures

In a method for the determination of the phosphine group in hydrolysable phosphorus compounds in which phosphorus is linked to a less electronegative element such as diethyl(trimethylsilyl)phosphine, the sample is weighed and refluxed with 10% sodium hydroxide under nitrogen[164]. The outlet of the condenser is connected to absorption tubes containing aqueous mercury(II) chloride and the mixture is boiled under nitrogen. The precipitate formed in the first absorption tube is a compound of the type (chloromercuri)diethylphosphine; it is determined by adding potassium iodide and iodine solution and titrating the residual iodine. *OO*-Dialkyl hydrogen phosphorodithioates can be determined by conversion into the nickel salt, which is determined iodometrically without preliminary isolation[165]. Interference by hydrogen sulphide or thiols is eliminated by the formation of insoluble nickel sulphide or mercaptide, respectively. Sodium diethyl phosphorodithioate can be determined by iodometry and also photometrically as the bismuth complex[166]. The molecular weight of such salts can be obtained via the copper salt[167].

B. Potentiometric Titration

Trimethylsilyl dihydrogen phosphate [$(CH_3)_3SiOPO(OH)_2$] and its analogues can be titrated in non-aqueous medium with solutions of lithium, sodium, or potassium methoxide and tetraethylammonium hydroxide[168]. The end-point is determined potentiometrically with glass and saturated mercury(I) chloride electrodes, or visually in the presence of quinizarin, bromophenol blue, brilliant green, alkaline blue, or methyl red (the most accurate results are obtained with methyl red). Methyl cyanide, acetone, methyl ethyl ketone, methanol, ethanol, isopropyl alcohol, butanol, and benzyl alcohol can be used as titration media.

The nickel salt of diethylphosphorodithioate (dissolved in 0.01 M perchloric acid) can be titrated potentiometrically with 0.01 M iodine in potassium iodide solution or with 0.01 M silver perchlorate[169]. Platinum immersed in a solution saturated with an organic disulphide and containing diethyl phosphorodithioate is used as the indicator electrode. The relative error is ± 0.3% with either titrant. Similarly, the titration can be carried out with mercury(II) perchlorate solution.

XIV. ORGANOPOTASSIUM COMPOUNDS

Phenylisopropylpotassium has been determined by reaction with excess of *p*-ditolyl sulphide to produce a mercaptide which is determined by titration with standard silver nitrate solution[170].

XV. ORGANOSELENIUM COMPOUNDS

In a method for determining nitrobenzene selenyl bromides, thiocyanates, alkoxides and amides, and 2,4-dinitrobenzenselenyl, the substance is dissolved in ethyl acetate[171]. After addition of 96% ethanol and glacial acetic acid, standard sodium thiosulphate is added and the excess of sodium thiosulphate is back-titrated with standard iodine solution. When organoselenium compounds are dissolved in sulphuric acid, selenium is present as elemental selenium, 'dissolved' selenium, selenous acid, and organoselenium compounds[172,173]. A method for the separation and determination of elemental and 'dissolved' selenium and selenous acid is based on titration with sodium thiosulphate. An error of less than 1% is claimed for this method for the determination of total selenium in organic selenium compounds, and less than $\pm 0.3\%$ for the determination of 'dissolved selenium' and selenous acid.

XVI. ORGANOSODIUM COMPOUNDS

An iodometric method has been suggested for the assay of pentane solutions of amylsodium, although the absolute accuracy of the method is unknown[145]. A procedure for the high-frequency titration of cyclopentadienylsodium is based on titration of combined sodium with 0.5 N hydrochloric acid[174]. Indicators are excluded in this titration owing to the intense colour of the analysis solution. Also, potentiometric and conductometric methods are ruled out owing to electrode fouling.

The Gilman titration procedure can be used to determine butylsodium compounds containing various alkali metal alkoxides[175]. Allyl bromide is used to react with the organometallic compound for 1 min. Accurate assays can be obtained on organometallics in the presence of large amounts of alkoxides. Even in mixtures containing potassium *t*-butoxide at twice the concentration as butylsodium, accurate determination of the carbon-bound sodium was obtained.

XVII. ORGANOTIN COMPOUNDS

A. Classical Titration Procedures

A method for the determination of monobutyltin trichloride in technical dibutyltin dichloride and in dibutyltin oxide depends on the fact that both monobutyltin trichloride and dibutyltin dichloride form blue complexes with catechol violet; between pH 1.2 and 2.3, however, only the complex of monobutyltin trichloride is decomposed by EDTA, and this facilitates the determination of this substance in dibutyltindichloride[176]. The sample is dissolved in methanol, ethanolic catechol violet is added, and the resulting blue or green solution is titrated with 0.05 M EDTA to a reddish violet to red end-point. To determine butanestannonic acid in technical dibutyltin oxide, the sample is dissolved in warm methanolic hydrochloric acid, the solution is diluted with methanol, and catechol violet solution is added. Methanolic potassium hydroxide is added dropwise until the colour changes from red or green, and the monobutyltin trichloride produced is then determined as described above.

Salts of the di- and tri-basic di- and monoalkyl (or-aryl) compounds of tin can be detected by the deep blue complex formed with catechol violet; the tetraalkyl (or aryl) tin compounds and the salts of the monobasic compounds do not give the reaction[177]. The blue complexes are quantitatively destroyed by EDTA. Diphenyltin diacetate can be accurately determined in the presence of triphenyltin acetate by adding a solution of catechol violet to the test solution in methanol and titrating to a yellow colour with EDTA solution (disodium salt). Tin in PVC can be determined by complexometric titration with EDTA[178]. The sample is dissolved in hot tetrahydrofuran and treated with 50 ml of ethanol. The precipitated PVC is filtered off and washed with ethanol, and 0.1% catechol violet solution is added dropwise to the combined filtrate until the solution becomes blue. This blue solution is then titrated to a green end-point with 0.001 M EDTA. Complexometric titration using EDTA and back-titration with standard zinc acetate has been used for the determination of dibutyltin oxide and dibutyltin dichloride[179]. Analysis of the triphenyltin hydroxide—bis(triphenyltin) oxide system has been reported[180].

A Karl Fischer titration procedure can be used for the determination of and differentiation between trialkyl (aryl) organotin hydroxides and the corresponding oxides[181,182]. This method is based on the observation that in the determination of water by Karl Fischer reagent in silanols and silanediols consistently high water contents are obtained[183]. Investigation led to the conclusion that not only was the water content being determined, but also that the silanol was reacting quantitatively with the reagent. The Karl Fischer reagent is not only effective in the quantitative determination of triaryl- and trialkyltin hydroxides, but is also applicable to bis(triaryltin) oxides[181]. The following reactions are postulated:

$$(R_3Sn)_2O + I_2 + SO_2 + CH_3OH \longrightarrow 2R_3SnI + HSO_4CH_3 \quad (25)$$

$$R_3SnOH + I_2 + SO_2 + CH_3OH \longrightarrow R_3SnI + HSO_4CH_3 + HI \quad (26)$$

The R_3SnOH class of compounds consumes 1 mol of iodine for each tin atom, whereas with the $(R_3Sn)_2O$ type of compound the ratio is 0.5.

Organotin carboxylates have been determined by distillation with phosphoric acid and by titration in non-aqueous medium[184]. Allyltin compounds in the presence of propenyltin compounds can be determined by titration with a benzene solution of iodine[185]. Trace studies by the radioactivation method have been used for the semi-quantitative determination of residual tin compounds in laundered sanitized nylon cloth[186]. Non-aqueous titrimetry and atomic-absorption spectrometry have been used for the determination of organotin biocides in insect-proofed textile materials[187]. Triphenyltinlithium has been determined by a double titration procedure using allyl bromide[188].

B. Potentiometric Titration

Organotin compounds of the general type $R_{(4-n)}SnCl_n$ (R = methyl, ethyl, propyl, butyl, or phenyl; $n = 2$ or 3) have been titrated by potentiometric procedures using tetraphenylarsonium chloride or tetraethylammonium chloride or bromide in acetonitrile as titrants[189,190]. Mixtures of R_2Sn^{2+} and R_3Sn^+ compounds where R is methyl or ethyl have been determined in the 0.01–0.05 mM range by potentiometric titration with standard potassium hydroxide to determine total organotin ions, followed by amperometric titration of R_2Sn^{2+} on a second aliquot using 2 mM 8-hydroxyquinoline as titrant at pH 9.2 (aqueous ammonia–ammonium nitrate buffer)[191]. Amperometric titration was applied to the determination of dialkyltin perchlorates using as titrant either 8-hydroxyquinoline ($R = CH_3$, C_2H_5) or with hexacyanoferrate(II) ions ($R = C_4H_9$, C_6H_5). Amperometric titration is carried out to -1.4 V against a standard calomel electrode. The amperometric and potentiometric titration of R_3Sn^+ and R_2Sn^{2+} compounds has also been studied[192], as well as the potentiometric titration of methyltin chlorides and bromides[193].

C. Amperometric and Coulometric Titration

Amperometric titration with electrolytically generated iodine, bromine, or silver ion has been applied to the titration of hexaorganoditin compounds, such as hexamethylditin, hexaethylditin, hexapropylditin, hexabutylditin, hexaphenylditin, and trimethyltriphenylditin[194]. In early work the hexaalkylditin compound in an alcoholic solution containing bromide ions, X^-, was placed in the anodic compartment of an electrolysis cell, where it underwent the following reaction:

$$R_3M{-}MR_3 + X_2 \longrightarrow 2R_3MX \qquad (27)$$

owing to the X_2 (bromine) electrolytically produced under constant current at the platinum anode. The end-point was observed by a rise in the indicator current, caused by excess of halogen, between a second pair of platinum electrodes sensitive to the X_3^-/X^- redox system. An amperometric plot of indicator current against generation time was obtained photographically in order to ensure an accurate determination. In the determination of hexamethylditin, bromine had to be used because, unlike iodine, its rate of reaction with hexamethylditin was greater than the rate of electrolytic generation of bromine. For a definite rate of generation of bromine, the titration depends on an appropriate rate of reaction of hexamethylditin with the halogen. The rate of reaction can be modified not only by changing the temperature, but also by choosing a proper concentration of halide ion (X^-) according to the equilibrium

$$X_2 + X^- \rightleftharpoons X_3^- \qquad (28)$$

An attempt has been made to verify which of hexaethyl-, hexapropyl-, hexabutyl-, hexaphenylditins, and trimethyltriphenylditin could be titrated with bromide and which could be iodinated[194]. In addition, since the quantitative reaction

$$Et_6Pb_2 + 2Ag^+ \longrightarrow 2Et_3Pb^+ + 2Ag \qquad (29)$$

have been verified for hexadiethyllead[195], the possibility of titrating each of the above organotin compounds by means of electrolytically generated silver ion to an amperometric end-point[196] was explored[194]. Conditions were also studied whereby a selective titration of one species in the presence of another could be carried out by the utilization of a complexing agent (quinoline) for silver. In the iodination of hexaalkylditin compounds, even when the iodide concentration is greatly reduced, analytically correct results are obtained only for hexaethylditin. All of the ditin compounds could be determined with bromine except for hexaphenylditin. Even at elevated temperatures and with a bromide concentration of 0.01 M, the hexaphenyl compound gave unsatisfactory results.

The titration of hexaalkylditin compounds with silver ion has been verified for hexaphenylditin, hexamethylditin, and trimethyltriphenylditin[194]:

$$R_3Sn{-}SnR_3 + 2Ag^+ \longrightarrow 2R_3Sn^+ + 2Ag \qquad (30)$$

Known amounts of these substances dissolved in ethanol or alcohol–benzene mixtures were added to alcoholic silver nitrate. The precipitated metallic silver was then separated and titrated by the Volhard method. Sodium fluoride or sodium tetraphenylborate was added to the filtrate to precipitate the triphenyl and trimethyl ions. Coulometric–amperometric titrations with silver ions can be carried out for all the hexaalkylditin compounds.

Dibutyltin dichloride and dioctyltin dichloride can be determined by amperometric titration in weakly acidic medium with standard oxalic solution[197]. Coulometric titration has been used for the titration of dialkyltin perchlorates[198]. A coulometric method for the determination of hexamethylditin in tetramethyltin with amperometric indication of the end-point employs 0.5 M methanolic ammonium bromide solution as basal electrolyte[199].

XVIII. ORGANOZINC COMPOUNDS

A. Classical Titration Procedures

Iodometric methods as described for the determination of organoaluminium compounds are also applicable to the determination of organozinc compounds[200]. The procedure is capable of an accuracy of ±3%. The reactions involved when iodine reacts with various types of organoaluminium and organozinc compounds are as follows:

$$AlR_3 + 3I_2 \longrightarrow AlI_3 + 3RI \qquad (31)$$

$$AlR_2OR + I_2 \longrightarrow AlI_2OR + R{-}R \qquad (32)$$

$$AlR_2Cl + 2I_2 \longrightarrow AlI_2Cl + 2RI \qquad (33)$$

$$ZnR_2 + 2I_2 \longrightarrow ZnI_2 + 2RI \qquad (34)$$

In this method a suitable volume of the hydrocarbon solution of the sample is stirred with an excess of a solution of iodine dissolved in toluene. The alkyl groups are completely iodinated within 20 min. Following the addition of dilute acetic acid, unreacted iodine is determined by titration with sodium thiosulphate solution under conditions of vigorous stirring. The concentration of the dialkylzinc compound is then calculated from the amount of iodine consumed in the determination. The presence of water in the iodine reagent causes interferences by reacting with some of the alkyl groups. This is corrected by a suitable 'double titration' procedure.

For the iodometric determination of diethylzinc the titration vessel is first dried at 150°C and cooled[201]. Solid potassium iodide is introduced and the vessel put under a dry oxygen-free nitrogen purge. A measured volume of 0.4 N iodine is then introduced through the stopper, followed by the heptane-diluted diethylzinc sample. After 5 min, glacial acetic acid is added and excess of iodine is titrated with 0.2 N sodium solution to the starch indicator end-point. A reagent blank determination is run in parallel.

An unsuccessful attempt has been made to apply the amperometric silver nitrate titration procedure to the determination of diethylzinc[170]. In this procedure the sample is treated with dialkyl or diarly disulphide to produce a thiol, which is then titrated amperometrically with standard silver nitrate. Although the method was applied successfully to organolithium, diethylmagnesium, triisobutylaluminium and isopropylphenylpotassium, it was found that diethylzinc did not cleave the disulphide at a high enough rate to be of any practical use.

In a volumetric method for the determination of zinc in zinc dialkylphosphorodithioate the sample is ashed at 800–900°C and the ash is dissolved in hydrochloric acid containing concentrated nitric acid, then neutralized to methyl orange[202]. Zinc is determined complexometrically using standard EDTA. An alternative analysis of this substance is based on its precipitation as a silver salt or by oxidation to disulphate by means of iodine solution[203]. Zinc benzothiazolyl mercaptide can be determined iodometrically, either directly or after decomposition by acid.

B. Thermometric Titration

The thermometric titration of organozinc compounds[204] can be carried out in the same apparatus as used for organoaluminium compounds[23,24]. The following titrants were examined in detail[204]:

Compound	Concentration (M)	Solvent
o-Phenanthroline (phen)	0.8	Anisole
2,2′-Bipridyl (bipy)	1.0–1.3	Toluene
3-Hydroxyquinoline (oxine)	1.6–1.7	Anisole
Ethanol (absolute)	2.0	Toluene
Water	2.2	Dioxane

For titration, a weighed amount of 0.2–2.5 M diethylzinc in toluene was added by hypodermic syringe to 40–50 ml of toluene in a dry, nitrogen-purged titration flask. Figure 5 shows typical titration curves obtained by titrating diethylzinc with phen and bipy. Both compounds react exothermically, and the heats of reaction are about the same (10 ± 2 kcal/mol). However, bipy gives curves which are somewhat rounded, possibly indicating an unfavourable equilibrium; phen is a much better titrant, giving sharp, well defined breaks and considerably better precision.

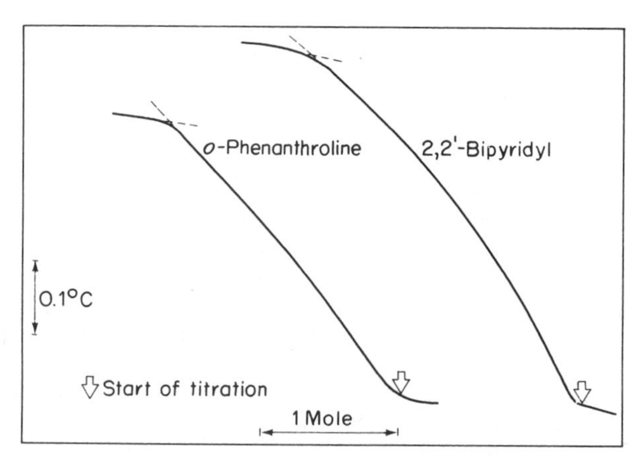

FIGURE 5. Thermometric titration of diethylzinc with *o*-phenanthroline and 2,2′-bipyridyl.

Figure 6 shows typical curves obtained in the titration of diethylzinc with oxine. The reaction is more exothermic than either the phen or the bipy reactions (33 ± 4 kcal/mol). Curve 1 in Figure 6 shows the type of curve obtained with relatively pure samples of diethylzinc. It shows three breaks. The first is fairly well defined and corresponds to the reaction of 1 mol of oxine per mole of diethylzinc. The second is poorly defined and sometimes is not observed; it is thought to be related to reaction of the Et—Zn bond in compounds of the Et—ZnO type, but agreement with this assumption is not good. In any event, this break is not useful analytically. The third break is sharply defined, usually with a slight characteristic peak at the inflection point; it represents the reaction of 2 mol of oxine per mole of Et_2Zn, EtZnOEt, or $Zn(OEt)_2$.

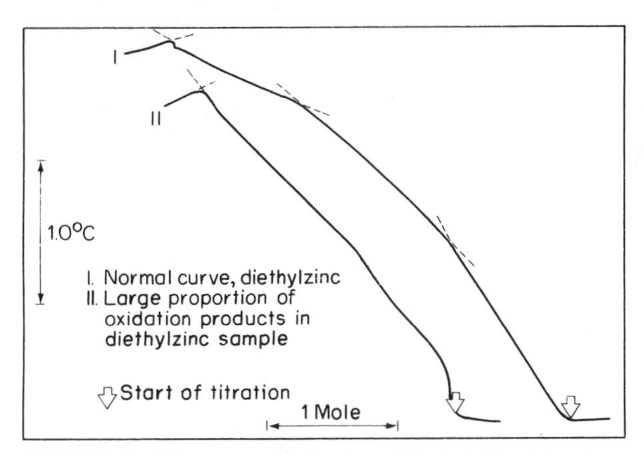

FIGURE 6. Thermometric titration of diethylzinc with oxine.

Reaction with EtZnOH proceeds past the 1:1 stage but is not stoichiometric. If the amount of oxidation products present in the diethylzinc sample is large, or if oxine is used to titrate a sample which was previously titrated with bipy or phen, then the curve obtained is similar to II in Figure 6. In this case only the final break is significant, and it is sharply defined. If much ethylzinc hydroxide is present in the sample, then the final break is considerably less sharp and may take the form of a smoothly rounded dome where no definite end-point can be located. The steep rise at the start of curve II (Figure 6) is abnormal; it is observed only when the oxine titration is made shortly after addition of ethanol or water to the diethylzinc sample. Similar segments of abnormally high slope are observed when such badly contaminated samples are titrated with phen or bipy.

There is little difference between results obtained by simple titration and those obtained using the 'clean-up' technique as described in the section on the thermometric titration of organoaluminium compounds. In general, it is considered that this technique is preferable to the simple titration because it is less vulnerable to contamination errors, and because comparison of results for the 'clean-up' sample and subsequent samples give some idea of the level of impurities present in the titration system and the effectiveness of the solvent purification and sample-vessel preparation techniques in use.

The thermometric titration of diethylzinc with gaseous oxygen from a motor-driven syringe burette has also been investigated to see whether quantitative indication of the reaction could be obtained using gas as titrant[204]. The oxidized solutions were subsequently titrated with oxine. Table 7 summarizes the results of these experiments.

These tests gave some indication that the reaction of diethylzinc with oxygen or reaction with ethanol gives equivalent products. Oxygen is not considered to be a practical titrant because of the large volume required, the need for slow addition, and the fact that the titration curve was not very well defined. The initial portion was normal, but a break was obtained at an O:Zn ratio of 0.65 rather than the expected 1.0. The temperature then remained approximately constant (i.e. heat production was balanced by heat loss) up to a second break at an O:Zn ratio of 2.06, after which the temperature decreased. The end product was assumed to be $Zn(OEt)_2$ rather than the monoperoxide, EtZnOOEt. There was no evidence of formation of the diperoxide reported elsewhere[9].

Thermometric titration with phen provides a simple, rapid, and precise method for

TABLE 7. Titration of diethylzinc with air or oxygen[a], then oxine

Test No.	Oxygen added as	Calculated composition			Found by oxine titration[b]		
		Et_2Zn	EtZnOEt	$Zn(OEt)_2$	Et_2Zn	EtZnOEt	Total Zn
1	—	—	—	—	0.435	0.040	0.478
2	Air, 0.020	0.418	0.060		0.436	0.044	0.480
3	O_2, 0.099	0.339	0.139	—	0.329	0.147	0.476
4	O_2, 491[c]	—	—	0.478	—	—	0.458

[a] All results in mmol/g.
[b] Et_2Zn from first oxine break, total Zn from final break, EtZnOEt by difference.
[c] Sample titrated with oxygen.

determining the net diethylzinc content of a solution which may also contain its oxidation or hydrolysis products. Titration with oxine gives a precise measure of the total zinc content in diethylzinc solutions which contain its oxidation products, but is not reliable if hydrolysis products are present. Under favourable circumstances oxine gives, in the same titration, a measure of both total zinc and diethylzinc.

XIX. REFERENCES

1. K. Ziegler and H. Gellert, *Justus Liebigs Ann. Chem.*, **629**, 20 (1960).
2. S. A. Bartkiewicz and W. J. Robinson, *Anal. Chim. Acta.*, **20**, 326 (1959).
3. T. R. Crompton, *Analyst (London)*, **91**, 374 (1966).
4. C. A. Uraneck, J. E. Burleigh, and J. W. Cleary, *Anal. Chem.*, **40**, 327 (1968).
5. Razuvaev and S. I. Graevsky, *Dokl. Akad. Nauk SSSR*, **128**, 309 (1959) (English translation, p. 747).
6. D. F. Hagen and W. D. Leslie, *Anal. Chem.*, **35**, 814 (1963).
7. J. H. Mitchen, *Anal. Chem.*, **33**, 1331 (1961).
8. D. E. Jordan, *Anal. Chem.*, **40**, 2150 (1968).
9. M. H. Abraham, *Chem. Ind. (London)*, 750 (1959).
10. G. W. Heunisch, *Anal. Chem.*, **44**, 741 (1972).
11. D. E. Jordan and W. D. Leslie, *Anal. Chim. Acta*, **50**, 161 (1970).
12. E. Bonitz, *Chem. Ber.*, **88**, 742 (1955).
13. A. I. Graevskii, S. SH. Shchegal, and Z. S. Smalian, *Dokl. Akad. Nauk SSSR*, **119**, 101 (1958).
14. E. Bonitz and W. Huber, *Z. Anal. Chem.*, **186**, 206 (1962).
15. T. R. Crompton, *Anal. Chem.*, **39**, 268 (1967).
16. M. Farina, M. Donati, and M. Ragazzini, *Ann. Chim. (Rome)*, **48**, 501 (1958).
17. L. Nebbia and B. Pagani, *Chim. Ind. (Milan)*, **44**, 383 (1962).
18. M. Uhniat and T. Zawada, *Chem. Anal. (Warsaw)*, **9**, 701 (1964).
19. V. P. Mardykin, E. I. Kvasyuk, and P. N. Gaponik, *Zh. Prikl. Khim. (Leningrad)*, **42**, 947 (1969).
20. E. G. Hoffman and W. Tornau, *Z. Anal. Chem.*, **188**, 321 (1962).
21. E. G. Hoffman and W. Tornau, *Z. Anal. Chem.*, **186**, 231 (1962).
22. M. Dimbat and G. A. Harlow, *Anal. Chem.*, **34**, 450 (1962).
23. W. L. Everson and E. M. Ramirez, *Anal. Chem.*, **37**, 806 (1965).
24. W. L. Everson, *Anal. Chem.*, **36**, 854 (1964).
25. W. P. Neumann, *Justus Liebigs Ann. Chem.*, **629**, 23 (1960).
26. P. Malatesta and A. Lorenzini, *Ric. Sci.*, **28**, 1874 (1958).
27. R. M. Fournier, *Mém. Poudres*, **40**, 385 (1958).
28. N. Z. Bruja, *Rev. Chim. (Bucharest)*, **17**, 359 (1966).
29. M. Wrónski, *Chem. Anal. (Warsaw)*, **8**, 299 (1963).
30. R. Montequi, A. Doadrio, and C. Serrano, *An. Rl. Soc. Esp. Fis. Quim. Ser. B*, **53**, 447 (1957).

31. H. Flaschka and F. Sadek, *Chem. Anal. (Warsaw)*, **47**, 30 (1958).
32. A. Schneer and H. Hartmann, *Magy. Kém Foly.*, **67**, 309 (1961).
33. A. J. Krol, L. B. Eddy and D. R. Mackey, *U.S. Atom. Energy Comm. Rep.*, CCC-1024-TR-239, 1957, 12 pp.
34. A. F. Zhigach, E. B. Kazakova and R. A. Kigel, *Zh. Anal. Chim.*, **14**, 746 (1959).
35. G. A. Guter and G. W. Schaeffer, *J. Am. Chem. Soc.*, **78**, 3346 (1956).
36. R. S. Braman, D. D. DeFord, T. N. Johnson, and L. J. Kuhns, *Anal. Chem.*, **32**, 1258 (1960).
37. A. E. Messner, *Anal. Chem.*, **30**, 547 (1958).
38. L. J. Kuhns, R. S. Braman, and J. E. Graham, *Anal. Chem.*, **34**, 1700 (1962).
39. M. I. Fauth and C. F. McNercy, *Anal. Chem.*, **32**, 91 (1960).
40. A. Heyrovský, *Z. Anal. Chem.*, **173**, 301 (1960).
41. A. Heyrovský, *Collect. Czech. Chem. Commun.*, **26**, 1305 (1961).
42. F. F. Crane, *Anal. Chim. Acta*, **16**, 370 (1957).
43. C. A. Uraneck, *Anal. Chem.*, **40**, 327 (1968).
44. O. Menyhárth, *Acta Chim. Hung.*, **41**, 195 (1964).
45. R. Iwanaga, *Bull. Chem. Soc. Jap.*, **35**, 247 (1962).
46. O. Boos, *Science.*, **117**, 603 (1953).
47. J. W. Collat and S. L. Tackett, *J. Electroanal. Chem.*, **4**, 59 (1962).
48. J. C. Chapin, *Am. Digest. Rep.*, **53**, 164 (1964).
49. H. H. Anderson, *J. Org. Chem.*, **21**, 869 (1956).
50. V. A. Klimova, K. S. Zabrodina, and N. L. Shitikova, *Izv. Akad. Nauk SSSR, Ser. Khim.*, **1**, 178 (1965).
51. J. S. Fritz and G. H. Schlenk, *Anal. Chem.*, **31**, 1808 (1959).
52. P. Dostal, J. Cermak, and B. Novotna, *Collect. Czech. Chem. Commun.*, **30**, 34 (1965).
53. J. A. Magnuson, *Anal. Chem.*, **35**, 1487 (1963).
54. A. Berger and J. A. Magnuson, *Anal. Chem.*, **36**, 1156 (1964).
55. J. A. Magnuson and E. W. Knaub, *Anal. Chem.*, **37**, 1607 (1965).
56. R. R. Jay, *Anal. Chem.*, **36**, 667 (1964).
57. K. Abel, *Biochim. Biophys. Acta*, **101**, 286 (1963).
58. A. Majerová and V. Porbuský, *Soudní Lék.*, **6**, 81 (1958).
59. J. G. Mason and M. Rosenblum, *J. Am. Chem. Soc.*, **82**, 4206 (1960).
60. M. Peterlik and K. Schögl, *Z. Anal. Chem.*, **195**, 113 (1963).
61. B. P. Nikol'shif, M. S. Zakhar'evskii and A. A. Pendin. *Zh. Anal. Khim.*, **19**, 1407 (1964).
62. D. M. Knight and R. C. Schlitt, *Anal. Chem.*, **37**, 470 (1965).
63. A. Grünwald, *Erdöl Köhle,* **6**, 550 (1953).
64. E. I. Uvarova and N. M. Vanyardina, *Zavod. Lab.*, **26**, 1097 (1960).
65. K. Katsumi, Y. Taguchi and S. Eguchi, *Bunseki Kagaku*, **12**, 435 (1963).
66. J. J. Russ and W. Reeder, *Anal. Chem.*, **29**, 1331 (1957).
67. K. H. Braun, *Chem. Tech. (Berlin)*, **10**, 159 (1958).
68. *ASTM Standards on Petroleum Products and Lubricants*, Methods D526-56 (1956) and D526-61 (1961), American Society for Testing and Materials, Washington D.C.
69. O. I. Milner and G. F. Shipman, *Anal. Chem.*, **26**, 1222 (1954).
70. M. Brandt and R. H. Van den Berg, *Anal. Chem.*, **31**, 1921 (1959).
71. DIN51781, *Erdöl Köhle*, **12**, 987 (1959).
72. A. Blumenthal, *Mitt. Geb. Lebensmitteluntes. Hyg.*, **51**, 159 (1960).
73. J. Mendes Cipriano, *Rev. Port. Quim.*, **5**, 129 (1963).
74. A. Fernández Perez, A. Peralonso, and F. G. Regalado, *Inf. Quím. Anal.*, **20**, 79 (1966).
75. J. Hurtado de Mendoza Riquelme, *Inf. Quím. Anal.*, **18**, 27 (1964).
76. L. G. Escolar and M. P. Castro, *Inf. Quím. Anal.*, **18**, 66 (1964).
77. R. Gelius and K. R. Pressner, *Chem. Tech. (Leipzig)*, **15**, 290 (1963).
78. V. S. Dimitrievskii, *Khim. Technol. Topliv Masel*, **3**, 59 (1958).
79. H. Saori, *Shoseki Giho*, **2**, 182 (1958).
80. H. Gilman and L. S. Miller, *J. Am. Chem. Soc.*, **73**, 2367 (1951).
81. A. F. Clifford and R. R. Olsen, *Anal. Chem.*, **32**, 544 (1960).
82. G. Tagliavini, *Chim. Ind. (Milan)*, **39**, 902 (1957).
83. L. Newman, J. F. Philip, and A. R. Jensen, *Ind. Eng. Chem., Anal. Ed.*, **19**, 451 (1947).
84. U. Belluco, G. Tagliavini, and R. Barbieri, *Ric. Sci.*, **30**, 1675 (1960).
85. S. Imura, K. Fukutaka, and Y. Takahiko, *Bunseki Kagaku*, **16**, 1351 (1967).

86. G. Pilloni and G. Plazzogna, *Ric. Sci., Parte 2, Sez. A*, **34**, 27 (1964).
87. G. Pilloni, *Farmaco, Ed. Prat.*, **22**, 666 (1967).
88. G. Tagliavini, U. Bulluco, and L. Ricoboni, *Ric. Sci., Parte 2, Sez. A*, **31**, 338 (1961).
89. H. L. Kies and G. Charlot, in *Modern Electroanalytical Methods*, Elsevier, Amsterdam, p. 14.
90. G. Tagliavini, *Anal. Chim. Acta*, **34**, 24 (1966).
91. A. L. Gol'dshteĭn, N. P. Lapisova, and I. M. Shtifman, *Zh. Anal. Khim.*, **17**, 143 (1962).
92. N. P. Lapisova and A. L. Gol'dshteĭn, *Zh. Anal. Khim.*, **16**, 508 (1961).
93. K. Ziegler, F. Croismann, H. Kleiner, and Schäfer, *Justus Liebigs, Ann. Chem.*, **31**, 473 (1929).
94. H. Gilman, P. D. Wilkinson, W. P. Fishel, and C. H. Meyers, *J. Am. Chem. Soc.*, **45**, 150 (1923).
96. Yu. N. Baryshnikov and G. I. Vesnovskaya, *Zh. Anal. Khim.*, **19**, 1128 (1964).
97. H. Gilman and F. K. Cartledge, *J. Organomet. Chem.*, **2**, 447 (1964).
98. P. F. Collins, C. W. Kamienski, D. L. Esmay, and R. B. Ellestad, *Anal. Chem.*, **33**, 468 (1961).
99. C. W. Kamienski and D. L. Esmay, *J. Org. Chem.*, **25**, 115 (1960).
100. W. L. Everson, *Anal. Chem.*, **36**, 854 (1964).
101. K. C. Eberly, *J. Org. Chem.*, **26**, 1309 (1961).
102. D. E. Applequist and A. H. Peterson, *J. Am. Chem. Soc.*, **83**, 862 (1961).
103. D. E. Applequist and D. F. O'Brian, *J. Am. Chem. Soc.*, **85**, 743 (1963).
104. H. J. S. Winkler, A. W. P. Jarvie, D. J. Peterson, and H. Gilman, *J. Am. Chem. Soc.*, **83**, 4089 (1961).
105. H. Gilman, F. K. Cartledge, and S. Y. Sim, *J. Organomet. Chem.*, **1**, 8 (1963).
106. C. Tamborski, F. E. Ford, and E. J. Soloski, *J. Org. Chem.*, **28**, 181 (1963).
107. H. Gilman, *Bull. Soc. Chim. Fr.*, 1356 (1963).
108. S. J. Crystal and R. S. Bly, *J. Am. Chem. Soc.*, **83**, 4027 (1961).
109. I. M. Kolthoff and E. B. Sandell, *Textbook of Quantitative Inorganic Analysis*, Macmillan, New York, 1949, p. 573.
110. R. R. Turner, A. G. Alterman, and T. C. Cheng, *Anal. Chem.*, **42**, 1835 (1970).
111. R. L. Eppley and J. A. Dixon, *J. Organomet. Chem.*, **8**, 176 (1967).
112. H. Gilman, R. A. Klein, and H. J. S. Winkler, *J. Org. Chem.*, **26**, 2474 (1961).
113. American Society for Testing and Materials, *Standard Method for Assay of n-Butyllithium Solutions,* ASTM Designation E233–68 (1968).
114. Foote Mineral Co., *Technical Data Bulletin,* No. T.D. 109 (December 1961).
115. R. A. Finnigan and H. W. Kutta, *J. Org. Chem.*, **30**, 4138 (1965).
116. H. W. Kutta, *MSc Thesis*, Ohio State University, 1964.
117. W. N. Smith, Foote Mineral Co., Exton, Pa., USA.
118. C. A. Uraneck, J. E. Burleigh, and J. W. Cleary, *Anal. Chem.*, **40**, 327 (1968).
119. I. M. Kolthoff and W. E. Harris, *Ind. Eng. Chem., Anal Ed.*, **18**, 161 (1946).
120. R. Adams, *Org. React.*, **6**, 353 (1951).
121. W. L. Everson, *Anal. Chem.*, **36**, 854 (1964).
122. S. C. Watson and J. F. Eastham, *J. Organomet. Chem.*, **9**, 165 (1967).
123. A. F. Clifford and R. R. Olsen, *Anal. Chem.*, **31**, 1860 (1959).
124. L. Kniel, *The Determination of Active Metal Content in Alkali Metal Dispersions*, CD-2817, Office of Rubber Reserve, USA, 1952.
125. S. C. Watson and J. F. Eastham, *J. Organomet. Chem.*, **9**, 165 (1967).
126. R. A. Ellison, R. Griffin, and F. N. Kotsonis, School of Pharmacy, University of Wisconsin, USA, personal communication, to be published in *J. Organomet. Chem.*
127. D. Seyferth and M. A. Weiner, *J. Am. Chem. Soc.*, **83**, 3583 (1961).
128. H. L. Johnson and R. A. Clark, *Anal. Chem.*, **19**, 869 (1947).
129. E. C. Juenge and D. Seyferth, *J. Org. Chem.*, **26**, 563 (1961).
130. A. F. Clifford and R. R. Olsen, *Anal. Chem.*, **32**, 544 (1960).
131. J. Jordan and T. G. Alleman, *Anal. Chem.*, **29**, 9 (1957).
132. S. C. Watson and J. F. Eastham, *Anal. Chem.*, **39**, 171 (1967).
133. M. Dimbat and G. R. Harslow, *Anal. Chem.*, **34**, 450 (1962).
134. Jolibois, *C. R. Acad. Sci.*, **155**, 213 (1912).
135. H. Gilman, P. D. Wilkinson, W. P. Fishel, and C. H. Meyers, *J. Am. Chem. Soc.*, **45**, 150 (1923).
136. D. Mitter, *J. Am. Chem. Soc.*, **41**, 287 (1919).

137. H. Gilman and C. H. Meyers, *Rec. Trav. Chim. Pays-Bas.*, **45**, 314 (1926).
138. J. Reich, *Bull. Soc. Chim. Fr.*, **33**, 1414 (1923).
139. S. Champtier and R. Kullman, *Bull. Soc. Chim. Fr.*, **693**, 155 (1949).
140. R. D'Hollander and M. Anteunis, *Bull. Soc. Chim. Belg.*, **72**, 77 (1963).
141. H. Gilman and C. H. Myers, *J. Am. Chem. Soc.*, **45**, 159 (1923).
142. H. Gilman, E. A. Zoellner and J. B. Dicky, *J. Am. Chem. Soc.*, **51**, 1576 (1929).
143. M. Kharash and D. Reinmuth, *Grignard Reactions of Non-metallic Substances*, Prentice Hall, New York, 1954.
144. J. F. Eastham and S. C. Watson, *J. Organomet. Chem.*, **9**, 165 (1967).
145. A. F. Clifford and R. R. Oben, *Anal. Chem.*, **32**, 544 (1960).
146. Yu. N. Barȳsknikov and A. A. Kvasou, *Zh. Anal. Khim.*, **19**, 117 (1964).
147. C. A. Unaneck, J. E. Burleigh, and J. W. Cleary, *Anal. Chem.*, **40**, 327 (1968).
148. N. Iritani, K. Ozawa, and H. Hoshida, *J. Pharm. Soc. Jpn.*, **80**, 1008 (1960).
149. K. K. Kundu and M. N. Das, *Sci. Cult.*, **23**, 660 (1958).
150. K. A. Connors and D. R. Swanson, *J. Pharm. Sci.*, **53**, 432 (1964).
151. I. Drăgusin, *Rev. Roum. Chim.*, **12**, 1235 (1967).
152. I. Drăgusin and N. Totir, *Stud. Cercet. Chim.*, **13**, 947 (1965).
153. W. Wrónski, *Chem. Anal. (Warsaw)*., **7**, 1011 (1962).
154. H. Lanbie, *Bull. Soc. Pharm. (Bordeaux)*, **96**, 65 (1957).
155. I. Drăgusin and N. Totir, *Rev. Chim. (Bucharest)*, **15**, 112 (1964).
156. I. Drăgusin and N. Totir, *Stud. Cercet. Chim.*, **13**, 947 (1965).
157. I. Drăgusin and A. German, *Rev. Chim. (Bucharest)*, **14**, 352 (1963).
158. F. H. Merkle and C. A. Discher, *J. Pharm. Sci.*, **51**, 117 (1962).
159. J. C. Gage, *Analyst (London)*, **86**, 457 (1961).
160. E. Hoffman and A. Saracz, *Z. Anal. Chem.*, **214**, 428 (1965).
161. D. V. Carter and G. Sykes, *Analyst (London)*, **83**, 536 (1958).
162. I. G. Druzhinin and P. S. Kislitsin, *Tr. Inst. Khim. Akad. Nauk Kirg. SSR*, **8**, 21 (1957).
163. B. Hetnarski and K. Hetnarska, *Bull. Acad. Polon. Sci., Ser. Chim. Geol. Geogr.*, **7**, 645 (1959).
164. G. Fritz and G. Pappenburg, *Z. Anorg. Chem.*, **331**, 147 (1964).
165. V. Bátora and Z. Veselá, *Sb. Pr. Vyzk. Ustavu Agrochem. Technol., Bratislava*, **1**, 85 (1961).
166. H. Bode and W. Arnswald, *Z. Anal. Chem.*, **185**, 99 (1961).
167. P. R. E. Lewkowitsch, *J. Inst. Pet.*, **48**, 217 (1962).
168. A. P. Kreshkov, V. A. Drozdov, and R. R. Tarasyants, *Plast. Massý.*, **4**, 57 (1963).
169. V. M. Shul'man, S. V. Larionov, and L. A. Padol'skaya, *Zh. Anal. Khim.*, **22**, 1165 (1967).
170. C. A. Uraneck, J. E. Burleigh, and J. W. Cleary, *Anal. Chem.*, **40**, 327 (1968).
171. O. Foss and S. R. Svendsen, *Acta Chem. Scand.*, **8**, 1351 (1954).
172. A. Kotarski, *Chem. Anal. (Warsaw)*, **10**, 541 (1965).
173. A. Kotarski, *Chem. Anal. (Warsaw)*, **10**, 321 (1965); *Anal. Abstr.*, **13**, 4805 (1966).
174. L. M. Shtifman, V. V. Lastovich, and L. G. Kuryakova, *Zvod. Lab.*, **29**, 546 (1963).
175. R. R. Turner, A. G. Alteron, and T. C. Cheng, *Anal. Chem.*, **42**, 1835 (1970).
176. J. Efer, D. Quaas, and W. Spichale, *Z. Chem.*, **5**, 390 (1965).
177. K. Bürger, *Z. Lebensm.-Unters.-Forsch.*, **114**, 1 (1961).
178. K. G. Bergner, U. Audt, and D. Mack, *Dsch. Lebensm.-Rundsch.*, **63**, 180 (1967).
179. R. Geyer and H. J. Seiditz, *Z. Chem.*, **4**, 468 (1964).
180. E. Friebe and H. Kelker, *F. Z. Anal. Chem.*, **192**, 267 (1963).
181. B. G. Kushlefsky and A. Ross, *Anal. Chem.*, **34**, 1666 (1962).
182. B. Kushlefsky, I. Simmons, and A. Ross, *Inorg. Chem.*, **2**, 187 (1963).
183. H. Gilman and L. S. Miller, *J. Am. Chem. Soc.*, **73**, 2367 (1951).
184. H. H. Anderson, *Anal. Chem.*, **34**, 1340 (1962).
185. W. P. Neumann and R. Sommer, *Justus Liebigs Ann. Chem.*, **701**, 28 (1967).
186. H. Owoki, H. Maeda, and N. Wada, *Nagoyashi Kogyo Kenkyusho Kenkyu Hakoku*, **24**, 92 (1960).
187. G. N. Freeland and R. M. Hoskinson, *Analyst (London)*, **95**, 579 (1970).
188. H. Gilman, F. K. Cartledge, and S. Y. Sim, *J. Organomet. Chem.*, **1**, 8 (1963).
189. G. Tagliavini and P. Zanella, *Anal. Chim. Acta*, **40**, 33 (1968).
190. V. Gutman and F. Mairinger, *Z. Anorg. Allg. Chem.*, **289**, 279 (1957).

191. G. Plazzogna and G. Pilloni, *Anal. Chim. Acta*, **37**, 260 (1967).
192. G. Pilloni, *Corsi. Semin. Chim.*, **9**, 98 (1968).
193. L. Magon, R. Portanova, A. Cassol, and G. Rizzardi, *Ric. Sci.*, **38**, 782 (1968).
194. G. Tagliavini, *Anal. Chim. Acta*, **34**, 24 (1966).
195. U. Belluco, G. Tagliavini, and R. Barbieri, *Ric. Sci.*, **30**, 1675 (1960).
196. H. L. Kies and G. Charlot, in *Modern Electrochemical Methods*, Elsevier, Amsterdam, 1958, p. 14.
197. L. Haasova and M. Pribyl, *Z. Anal. Chem.*, **249**, 35 (1970).
198. F. Magno and G. Pilloni, *Anal. Chim. Acta*, **41**, 413 (1968).
199. G. Tagliavini and G. Plazzogna, *Ric. Sci., Parte 2, Sez. A*, **2**, 356 (1962).
200. T. R. Crompton, *Analyst (London)*, **91**, 374 (1966).
201. K. Novák, *Chem. Prum.*, **12**, 551 (1962).
202. P. T. Makarov and K. F. Pershina, *Khim. Tekhnol. Topl. Masel*, **10**, 62 (1963).
203. O. Lorenz, E. Echte, and U. Kantsch, *Gummi*, **9**, 300 (1956).
204. W. L. Everson and E. M. Ramirez, *Anal Chem.*, **37**, 812 (1965).

The Chemistry of the Metal–Carbon Bond
Edited by F. R. Hartley and S. Patai

CHAPTER **16**

Analysis of organometallic compounds: spectroscopic methods

T. R. CROMPTON
'Beechcroft', Whittingham Lane, Goosnargh, Preston, Lancashire, UK

I. VISIBLE SPECTROSCOPY

A. Organoaluminium Compounds

The ultraviolet and visible spectra of isoquinoline alone (curve 1) and of mixtures of isoquinoline and diethylaluminium hydride (curves 2 and 3) are shown in Figure 1[1]. In Figure 2 are shown the spectra of isoquinoline alone (curve 1), mixtures of isoquinoline and diethylaluminium ethoxide (curve 2), and isoquinoline and triethylaluminium (curve 3). Comparison of curves 1 and 2 shows that the spectrum of diethylaluminium ethoxide remains unchanged in the presence of isoquinoline, suggesting that no reaction occurs between these two substances. The addition of triethylaluminium to isoquinoline, however, produces a distinct change in its spectrum, suggesting that complex formation occurs. Analysis of the difference between curves 1

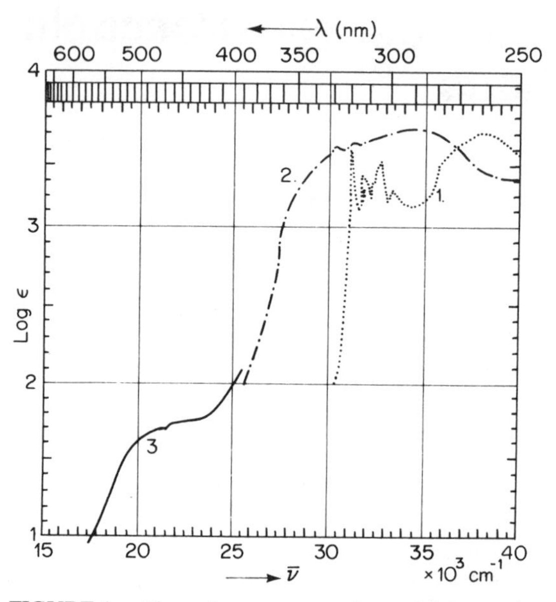

FIGURE 1. Absorption spectrum in cyclohexane (at 20°C) of isoquinoline alone (1); isoquinoline aluminium hydride, i.e. colourless 1:1 complex (2); and excess of isoquinoline plus diethylaluminium hydride, i.e. red 1:2 diethylaluminium hydride—isoquinoline complex (3)

1. N

2. N → $Al(C_2H_5)_2$ H

3. Red 1:2 diethylaluminium hydride—isoquinoline complex obtained with diethylaluminium hydride in the presence of 10 M excess of isoquinoline.

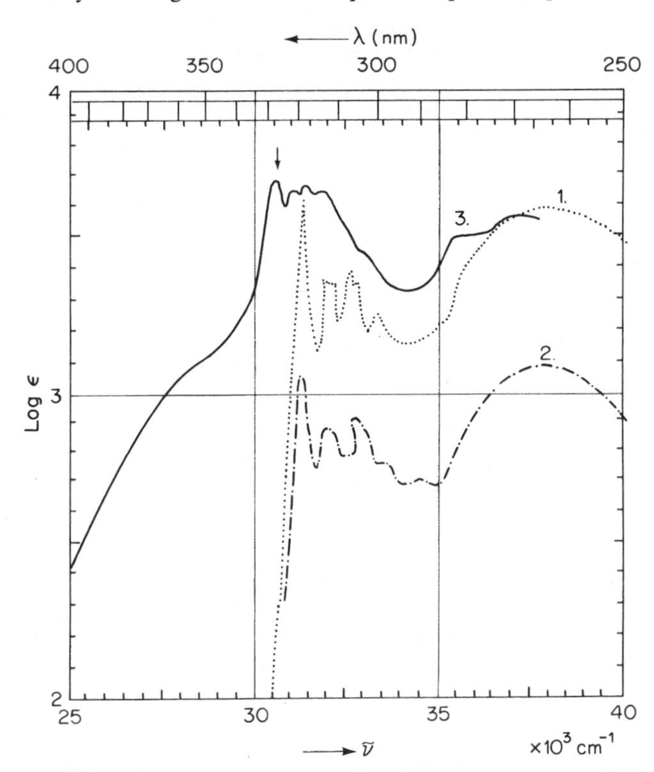

FIGURE 2. Absorption spectra in cyclohexane (at 20°C) of isoquinoline alone (1), isoquinoline plus diethylaluminium ethoxide (2), and isoquinoline plus triethylaluminium (3).

and 3 (Figure 2) shows that triethylaluminium and isoquinoline form a colourless 1:1 complex. Diethylaluminium hydride forms two complexes with isoquinoline. The 1:1 diethylaluminium hydride isoquinoline complex is colourless and absorbs only in the ultraviolet region of the spectrum (Figure 1, curve 2). The 1:2 diethylaluminium hydride—isoquinoline complex has an intense red coloration and this is shown in the absorption spectrum (see the strong absorption occurring above 400 nm in curve 3, Figure 1). Procedures have been devised based upon these observations for determining trialkylaluminium compounds and dialkylaluminium hydrides either singly or in the presence of each other[1].

The absorption of the triethylaluminium—isoquinoline complex occurring at about 328 nm (see arrow on curve 3, Figure 2) can be used for the colorimetric determination of trialkylaluminium compounds (isoquinoline itself does not absorb at this wavelength). The dialkylaluminium hydride—isoquinoline complex also absorbs at 328 nm and would, of course, interfere in this method of determination of trialkylaluminium compounds. Also, the absorption of the red 1:2 diethylaluminium hydride—isoquinoline complex occurring at about 500nm (see curve 3, Figure 1) can be used to determine dialkylaluminium hydrides without interference from any trialkylaluminium compounds present in the sample, as the colourless 1:1 trialkylaluminium—isoquinoline complex does not absorb at 500nm.

Dialkylaluminium hydrides can be used as visual indicators in the titration of organoaluminium compounds that form a 1:1 complex with isoquinoline. Thus, to determine the concentration of a trialkylaluminium compound in a solution, a small volume of diethylaluminium hydride is added (often the sample will contain a small amount of this as an impurity left in from the manufacture) and the solution is titrated with isoquinoline. When the trialkylaluminium compound and the dialkylaluminium hydride have both formed 1:1 complexes with isoquinoline then the solution suddenly becomes red owing to the formation of some 1:2 diethylaluminium hydride—isoquinoline complex. The volume of isoquinoline corresponding to the first appearance of the red colour can be taken as the end-point of the titration and is equivalent to the total trialkylaluminium plus dialkylaluminium hydride content of the sample. Every precaution must be taken to avoid contact of substances in the cell with air and moisture which would have a serious influence on the spectra obtained[1].

Work on the reaction occurring between 1 mol of dialkylaluminium hydride and 2 mols of isoquinoline to form strongly coloured 1:2 complexes was extended to form the basis for a spectrophotometric method for determining low concentrations of dialkylaluminium hydrides in trialkylaluminium compounds[2–4]. The reaction of dialkylaluminium hydrides with isoquinoline and with various other azomethines which form coloured 1:2 complexes has been studied[2]. It was concluded that isoquinoline and benzalaniline were the two most interesting azomethines. In Figure 3 (curves C and D) are shown the absorption curves in the range 350–550 nm obtained for a mixture of dialkylaluminium hydride with isoquinoline and a trialkylaluminium compound with isoquinoline, respectively. It can be seen that only the dialkylaluminium hydride—isoquinoline reaction product absorbs and this can be conveniently measured at a wavelength of 460 nm ($\log \varepsilon = 2.39$). At this wavelength the trialkylaluminium—isoquinoline reaction product exhibits no absorption. However, the isoquinoline colour does not correspond to any definite maximum in the spectrum. The absorption curve at 460 nm shows only a flat shoulder which rises sharply in the direction of the shorter wavelengths (Figure 3, curve C). A considerable improvement was obtained by the use of benzalaniline as reagent. It causes the formation of a colour with a broad absorption band at 450 nm ($\log \varepsilon = 2.54$) (Figure 3, curve A). The evaluation can also be made at 500 nm ($\log \varepsilon = 2.31$) and this has the advantage that any colour due to colloidal metals in the samples does not interfere in the analysis. It is essential to avoid contamination of the cell solution with moisture or oxygen during the spectrophotometric analysis. Very dry benzene is used as the solvent. Also, the benzalaniline reagent itself should contain some alkylaluminium, which completely dries this reagent. Flow-through glass spectrophotometer cells are employed in order to reduce sample contamination to zero by atmospheric water and oxygen.

In a further method for the spectrophotometric titration of isoquinoline at 460 nm with dialkylaluminium hydrides and trialkylaluminium compounds a sample cell fitted with a rubber serum cap is used[5]. Sample transfers are made using a hypodermic syringe which is weighed before and after transfer in order to obtain the weight of sample added to the spectrophotometer cell. In the analysis a portion of a benzene solution of isoquinoline is transferred to the nitrogen-filled spectrophotometer cell and a solution of diethylaluminium hydride in dry benzene is then added until a stable red colour is obtained having an absorbance of about 0.3 at 470 nm (i.e. the 1:2 diethylaluminium—isoquinoline complex is formed). To determine dialkylaluminium hydride in an unknown sample a weighed portion of benzene-diluted sample is added to the cell contents and the absorbance recorded. The increase in absorbance is due to dialkylaluminium hydride in the sample.

To determine the trialkylaluminium content of the same sample the addition of sample is continued until the absorbance passes through a maximum, when all the

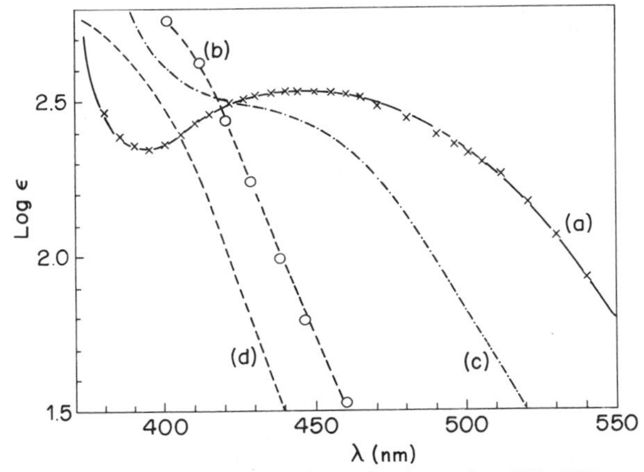

FIGURE 3. Absorption spectra in the region 350–550 nm. (a) Dialkylaluminium hydride—benzalaniline complex; (b) 1:1 trialkylaluminium—benzalaniline complex; (c) 1:2 dialkylaluminium hydride—isoquinoline complex; (d) 1:1 trialkylaluminium hydride—isoquinoline.

isoquinoline is bound as the red 1:2 dialkylaluminium hydride—isoquinoline complex and as the colourless 1:1 trialkylaluminium—isoquinoline complex. Thereafter, further sample addition destroys the red 1:2 complex in preference to the 1:1 complexes, decreasing the absorbance at 469 nm. Sample addition is continued until a convenient absorbance for measurement by the spectrometer is obtained (e.g. 1.5). Both the diethylaluminium hydride and triethylaluminium hydride is within ± 3% of the amount of dialkylaluminium hydride present[5].

In a further study[6] of the isoquinoline spectrophotometric method for the determination of dialkylaluminium hydrides and trialkylaluminium, the dialkylaluminium hydride reacts with isoquinoline to form red coloured 1:2 complexes which have an absorption maximum at 460 nm. Trialkylaluminium compounds form only a colourless

TABLE 1. Analysis of trialkylaluminium compounds. Comparison of the Wadelin[6] isoquinoline spectrophotometric titration method and the Ziegler and Gellert[7] 'ammonia method'

Compound	Activity (mmol/g)[a]	
	Isoquinoline method	Ammonia method
Tri-*n*-propylaluminium	6.00	6.02
	5.96	5.92
	6.58	6.57
	6.41	6.37
	6.30	6.34
	5.67	5.81
	5.47	5.81
Tri-iso-butylaluminium	4.85	4.80
Triisohexylaluminium	3.52	3.44

[a]Each result is the average of duplicate determinations.

1:1 complex with isoquinoline which has no absorption at 460 nm. Samples were analysed by this method and by the ammonia method[7]. The good agreement obtained by the two methods of analysis is demonstrated in Table 1.

In a further study of the isoquinoline spectrophotometric method for the determination of dialkylaluminium hydrides and trialkylaluminium, decomposition by air and moisture of organoaluminium samples in syringes was considerably reduced by using a syringe with a smooth-bore barrel and a machined Teflon plunger[8]. The syringes require no lubrication and it is claimed that organoaluminium samples can be stored in the syringe for several days without severe decomposition. N.m.r. spectroscopy has been applied to the elucidation of the structure of reaction products of alkylaluminium compounds with isoquinoline[9].

B Organoarsenic Compounds

Phenarsazine derivatives can be determined spectrophotometrically as the disodium salt of dinitrophenarsazinic acid at 520 nm[10]. The sample is dissolved in glacial acetic acid and oxidized and nitrated with an excess of nitric acid to form dinitrophenarsazinic acid. Addition of excess of sodium hydroxide yields a violet disodium salt suitable for photometric evaluation. From 1 to 8 μg/ml of phenarsazine can be determined by this method with an error of $\pm 4\%$. A spectrophotometric method has been described for the determination of 4-hydroxy-3-nitrophenylarsenic acid in animal feeds[11].

Two digestion methods have been compared for their effectiveness in releasing arsenic from three organoarsenicals introduced into wastewater samples[12]. The digestive methods utilized included a wet method employing hydrogen peroxide–sulphuric acid and ultraviolet photodecomposition. The organoarsenicals investigated were disodium methanearsonate, dimethylarsinic acid, and triphenylarsine oxide. All the digestive methods gave quantitative arsenic recoveries when applied to wastewater samples. The ultraviolet photodecomposition proved to be an effective digestive technique, requiring a 4 h irradiation to decompose a primary settled raw wastewater sample containing spiked amounts of the three organoarsenicals. Arsenic was determined in the digests by the silver diethyldithiocarbonate spectrophotometric method.

C. Organobismuth Compounds

In a spectrophotometric method for the assay of glycobiarsol tablets (bismuth glycolylarsanilate) the sample is allowed to react with aqueous EDTA disodium salt solution and the absorbance of the acidified solution is measured at 258 nm[13].

D. Organoboron Compounds

1. *Pentaborane*

Colorimetric methods based on the use of triphenyltetrazolium chloride have been described for the determination of pentaborane and for monitoring concentrations of pentaborane in air[14,15]. Instruments have been described for monitoring atmospheres: a portable field model and an automatic differential reflectance photometric analyser. Both methods depend on non-specific, highly sensitive reductions of the reagent by boron hydrides to give a red coloration. The field model can detect 0.1 ppm of decaborane and 0.5 ppm of pentaborane and the automatic instrument is capable of detecting 0.1 ppm of either compound.

An alternative excellent procedure for the determination of pentaborane in air samples is based on the formation of a coloured pyridine complex[16]. Using toluene as the solvent, Beer's law is followed through the concentration range 2–12 μg/ml at 400 nm. A trapping system is described which permits dynamic air sampling at rates as high as 15 l/min with over 95% efficiency. Using a 30-l air sample, the method is suitable for pentaborane concentrations as low as 0.1 ppm.

2. *Decaborane and other boranes*

Non-specific colorimetric methods have been described for the determination of decaborane based on colour formation with triphenyltetrazolium chloride[14,15], as well as colorimetric methods based on reaction with quinoline[17] and β-naphthoquinoline[18] in which red addition products with an absorption maxima at 490 nm are produced by addition of solutions of quinoline or β-naphthoquinoline in xylene medium to the sample. These methods are very efficient and accurate for dynamic air analysis, only one absorption bulb (containing 2% quinoline in xylene) being needed. Interference by diborane and pentaborane is negligible. A colorimetric procedure based on the use of 1,2-di(4-pyridyl)ethylene is claimed to have greater sensitivity, more rapid colour development and lower reagent blanks than the quinoline method[18]. This produces with decaborane a pink to red colour with an absorption maximum at 515 nm. The coloured complex formed between decaborane and benzo(*f*)quinoline is claimed to be more stable that than formed with quinoline and is more suited to spectrophotometric determination at 486 nm[20]. The absorption maximum of the decaborane—reagent complex occurs at 506 nm and of the diethylborane—reagent complex at 525 nm.

A colorimetric method using indigo carmine for the determination of decaborane, dimethylaminoborane, trimethylaminoborane, and pyridineborane is sensitive enough to detect 1–10 μg of boron[21]. The method does not distinguish between boron present as boron hydrides, boron acid, or boron oxides.

A method in which decaborane in water or cyclohexane is treated with an excess of N-diethylnicotinamide in water is claimed to be free from interference by boric acid, boron salts, diborane, and pentaborane[22]. After 90 min the orange–red colour intensity is measured at 435 nm and the decaborane concentration is derived from standards. In the colorimetric determination of boron hydrogen compounds with molybdophosphoric acid, addition of molybdophosphoric acid to decaborane, dimethylamineborane, or sodium borohydride produces a blue solution[23]. The colour intensity is directly related to the amount of boron present. A direct micro method for the determination of decaborane involves measurement of the strong u.v. absorption occurring at 265–270 nm of the solution in triethanolamine[18]. Beer's law is valid and from 1–25 $\leqslant$g/ml of boron can be determined accurately. The method is unaffected by the very slow hydrolysis of decaborane in the solvent and can be applied to the dynamic and static analysis of air and gases containing decaborane. The air or vapour is bubbled slowly through two glass bulbs containing aqueous triethanolamine the contents are afterwards combined and diluted and the u.v absorption is determined; the decaborane recovery is 98%. Static air samples are taken with a gas pipette (250 ml) into which 5 ml of the triethanolamine solution are introduced.

3. *Other organoboron compounds*

The photometric determination at 590 nm of diphenylborinic acid and its esters using diphenylcarbazide as reagent has been discussed[24]. In a colorimetric method for the determination of borinic acids in biological materials the frozen tissue sample is mixed with calcium hydroxide and concentrated sulphuric acid and rendered colourless by heating with hydrogen peroxide[25]. After complete destruction of hydrogen peroxide the determination is completed by reaction with 1,1-dianthrimide in concentrated sulphuric acid at 90°C and spectrophotometric evaluation at 620 nm. The error does not exceed 15%. An integrating monitor has been designed for determining low concentrations of gaseous boron hydrides in air[26]. The boron hydride vapours are quantitatively converted by burning into boron oxide, which is then determined colorimetrically with carmine at 585 nm. Tetraphenylborate ions have been determined spectrophotometrically, based on reaction with excess of standard rosaniline solution at pH 4.6. The precipitate is filtered off and the absorbance of the coloured filtrate is measured at 546 nm against water; the decrease in molar absorptivity of the filtrate is proportional to the concentration of tetraphenylborate ion.

E. Organocobalt Compounds

1. Cyanocobalamin

A general theory of partition has been developed and applied to the spectroscopic determination of cyanocobalamin and hydroxocobalamin[27]. In this method, the determination of the apparent partition coefficient is based on the spectrophotometric determination of total cobalamin in each phase of a benzyl alcohol and water partition at a wavelength (356 nm) at which both components have the same molar absorptivity. The total cobalamin concentration in each phase is given by

$$\frac{\varepsilon_{356\ \mathrm{nm}}}{0.0174}\ \mu\mathrm{g/ml}$$

and the apparent distribution coefficient for any particular mixture is given experimentally by

$$K = \frac{\varepsilon_{356\ \mathrm{nm}}(\mathrm{solvent})}{\varepsilon_{356\ \mathrm{nm}}(\mathrm{water})}$$

Good agreement is reported between determined and known concentrations of mixtures of cyanocobalamin and hydroxocobalamin. In a test for determining the purity of Cyanocobalamin Injection USP, the sample is extracted and after a working up procedure the cyanocobalamin determined spectrophotometrically at 361 nm[28]. A method for measuring the relative purity of cyanocobalamin has values of $\varepsilon_{1\ \mathrm{cm}}^{1\%}$ at 341 nm and 376 nm of 80.4 and 80.9 respectively the average ratio $\varepsilon_{341\,\mathrm{nm}}/\varepsilon_{376\,\mathrm{nm}}$ being 0.990[29]. The validity of this procedure was established by comparing results with the purity index found by a combination of tracer and spectrophotometric (361 nm) methods. The average difference between duplicate determinations of the absorbance ratio was 0.6%. Results obtained by this test agreed with those obtained by the USP XVI limit test for cyanocobalamin solids[30].

When exposed to light, cyanocobalamin in aqueous solution is converted into hydroxocobalamin, which has a lower molar absorptivity at 361 nm[31]. Errors in the spectrophotometric assay of cyanocobalamin may be avoided by avoiding exposure to light, or by measuring the absorbance at 356 nm, which is isosbestic for cyanocobalamin and hydroxocobalamin, using absolute ethanol as solvent. In a method for the chromatographic separation and spectrophotometric determination of cyano- and hydroxographic separation and spectrophotometric determination of cyano- and hydroxocobalamins in association with other pharmaceutical products, the two cobalamins are converted into the dicyano derivatives and extracted with butanol[32]. The two purified cobalamins are separated by paper chromatography and then measured spectrophotometrically at 367 nm (cyanocobalamin) and 580 nm (hydroxocobalamin). A method has been described for the determination of cyanocobalamin in injection liquids and in purified liver extracts[33].

A differential spectrophotometric method was developed for the determination of cyanocobalamin and hydroxocobalamin[34,35]. This is claimed to be much more precise than the direct spectrophotometry at 361 nm. To determine hydroxocobalamin in the presence of cyanocobalamin, the difference is measured between the absorbances at 349 nm in 0.01 N sodium hydroxide and in 0.01 N hydrochloric acid solution. To determine cyanocobalamin in the presence of hydroxocobalamin the ratio of the absorbances at 351 and 361 is measured in 0.01 N hydrochloric acid solution. A spectrophotometric method has been used for the determination of cyanocobalamin in concentrates and in supplements for compound feeding stuffs[36]. Cyanocobalamin is determined from absorbance measurements made at 360 and 535 nm[37]. A chromatographic separation and spectrophotometric determination can be used to determine cyanocobalamin in orange juice[38]. The cyanocobalamin is determined by measuring the absorbance at 530 nm.

A method suitable for the determination of cyanocobalamin in injections, tablets, and liver extracts depends on the colour produced with nitroso-R salt at controlled pH after oxidation of the sample with hydrogen peroxide[39]. The absorbance of the final solution is measured at 420 nm. Beer's law is obeyed for the range 100–600 μg of cyanocobalamin and the recovery is 100–103.8%.

An automated procedure has been developed for the determination of cyanocobalamin in pharmaceutical dosage forms[40]. Diluted samples are mixed with sulphuric acid and passed to a photolysis device. The hydrogen cyanide produced is distilled as it is formed, and trapped in sodium hydroxide, and this solution is mixed with sodium dihydrogen phosphate solution and chloramine T solution in an ice-bath, reacted with saturated aqueous 3-methyl-1-phenyl-2-pyrazolin-5-one, and the absorbance is measured at 620 nm. A spectrophotometric method for the determination of down to 0.25% of cyanocobalamin in hydroxocobalamin depends on photolysing the cyanocobalamin to hydrogen cyanide, which is then collected by diffusion and measured colorimetrically at 506 nm[41].

For the determination of cyanocobalamin in liver extracts and polyvitamin mixtures, the extracts are first extracted with benzyl alcohol, then treated with butanol to remove impurities that interfere in the spectrophotometric determination of cyanocobalamin[42,43]. The results obtained agree with those of the USP microbiological method; the method has the advantages of simplicity and speed. In a further method for the determination of cyanocobalamin in pharmaceutical products in the presence of liver extract and salts of cobalt, iron, and copper, the absorbance of a sample extract is measured at 570 nm[44]. The 3:4 complex formed by cyanocobalamin with Fast Navy 2R (CI Mordant Blue 9) has a maximum at 550 nm and an inflection at 620 nm; the latter point is used for the spectrophotometric determination of cobalt[45]. The presence of calcium, magnesium, manganese, lead, and zinc does not cause interference if EDTA is used as masking agent; interference from lead can also be suppressed by the use of tartaric acid, from manganese by the use of ascorbic acid, and from nickel by treatment with EDTA solution at pH 8.5 for 15 min. Further spectrophotometric methods have been described for the indirect determination of cyanocobalamin in cyanocobalamin preparations, liver extracts, and multivitamin preparations[46].

The determination of cobalt in aqueous solutions of cyanocobalamin can be based on the catalytic effect of cobalt on the fluorescence produced by hydrogen peroxide with luminol[47–49].

A counter-current method of analysis of cyanocobalamin has been developed[50], which can be applied to the determination of the purity of crystalline cyanocobalamin, oral-grade solids, and simple solutions. The presence or absence of pseudocobalamin and other non-cobalamin fractions may also be determined. The results agree well with those obtained by radioactive tracer methods. The assay of mixtures of cyanocobalamin and thiamine has been discussed[51].

2. *Hydroxocobalamin, cyanocobalamin and other cobalamins*

For the standardization of hydroxocobalamin the absorbance can be measured near 351 nm against the buffer solution ($\varepsilon^{1\%}_{1\,cm}$ for the pure dry substance is 190)[52]. For identification, the absorption spectrum from 250 to 600 nm is compared with that of a reference sample. The absorption spectra of cyano-, aquo-, and sulphitocobalamins in acidic, basic, and neutral solution, together with changes in the spectra resulting from ageing of the solution, have been studied[53]. The dissociation of cobalamins in acidic solution and the stability of the coordinate bond between cobalt and the nitrogen at position 3 of benzimidole are discussed together with the behaviour of sulphitocobalamin and its transformation to hydroxocobalamin.

3. *Miscellaneous*

The absorbance maxima of cobalt myristate and palmitate have been measured[54]. In pyridine, these were found to be 550 nm. A linear relationship existed between the absorbance and the concentration of the soap. This method is more rapid and precise than the gravimetric techniques normally used[54].

F. Organochromium Compounds

Rapid spectrophotometric methods for the determination of chromium stearate used as additives in lubricating oils have been devised[55]. When chromium stearate is heated with *o*- or *m*-toluidine at 180–200°C complexes are formed that have absorbance maxima at 480 and 540 nm, respectively. Beer's law is obeyed for concentrations of 1.2–10.7 mg/ml, so that the metal ion content of the soap can be determined even in dilute solution. The spectrophotometric behaviour at 425 nm of chromium(III) stearates, in amounts between 1 and 14 mmol/l, in non-aqueous solution has been studied[56,57]. The viscosity of non-aqeous solutions of chromium soaps can be used as a basis for determining these substances in the range 1–27 g/l[58].

G. Organocopper Compounds

The spectrophotometric determination of copper 8-hydroxyquinolinate in fabrics is subject to interference from some of the acid-soluble dyes used, but this can be overcome by examining a blank containing no 8-hydroxyquinolinate but otherwise treated and dyed in a similar manner[59]. A chromatographic method has also been described, but the spectrophotometric method is considered to be the most suitable.

For the determination of copper 8-hydroxyquinolinate in paint the copper complex is extracted by boiling the paint film with 0.5 M sulphuric acid and 8-hydroxyquinoline is then determined spectrophotometrically at 307, 317, or 355 nm[60,61]. To determine copper the film sample is treated with sodium sulphide solution and acidified. Barium sulphate is added to collect the precipitated copper sulphide, which is dissolved in nitric acid and the copper is determined photometrically with benzoin α-oxime or diphenylcarbazone. The determination of copper naphthenate is dealt with in both a British Standard[62] and in a paper[63].

H. Organoiron Compounds

1. *Haemoglobin in blood*

Many of the earlier methods for determination of haemoglobin in blood were based on direct visible spectrophotometry of the strongly coloured solution in a suitable medium. With dilutions of blood from 1:500 to 1:1500, haemoglobin can be determined by measurement of the absorption at 417 nm (Sorets method)[64]. Results obtained by this procedure were compared with those obtained by a gasometric method. The absorbance at 430 nm of haem chloride, obtained by dissolving blood in dilute hydrochloric acid, has been measured[65].

Critical examination of the alkaline haematin method[66] for determining haemoglobin and myoglobin in blood revealed some of the conditions affecting the behaviour of alkaline haematin[67]. The modified method is suitable as a comparative method of analysis for either blood haemoglobin or tissue myoglobin[67]. The determination is made spectrophotometrically at 380 nm. Standard solutions of haemin in alkali are not stable but show progressive fading, which is accelerated by exposure of the solution to light or heat, by agitation with air, or by the presence of traces of copper.

The cyanomethaemoglobin (hemiglobincyanide) method for the determination of haemoglobin in blood has been extensively discussed[68–77]. The haemoglobin is reacted under various conditions with potassium cyanide and the resulting cyanomethaemoglobin evaluated spectrophotometrically at a wavelength in the vicinity of 540 nm. A portion of blood or haemolysate is added by pipette to cyanide solution buffered at pH 7. The absorbance of this solution is measured at 540 nm between 1 and 4 h after mixing the reactants. The mean error ranges from 0.001 to 0.002 absorbance unit[74,75]. A wedge-shaped adjustable photometer cell that permits calibration of the instrument with a single cyanomethaemoglobin solution has been described[70]. A method in which total blood haemoglobin is converted into an azide complex of methaemoglobin is presented as an alternative to the cyanomethaemoglobin method[78]. Owing to the similarity of the absorption spectra of both coloured species, both procedures yield identical results. For the proposed method, a single reagent, containing potassium hexacyanoferrate(III) and sodium azide is required. A spectrophotometric iron(III) thiocyanate method is available for determining haemoglobin as iron[79]. A comparison of three methods for the assay of haemoglobin, determined as cyanomethaemoglobin, oxyhaemoglobin, and pyridine haemoglobin, indicated that the latter method gives the correct values, but the differences between the results are small enough to be negligible in routine work[80]. A comparison of results obtained in haemoglobin in blood determinations using seven commercial instruments concluded that for routine haemoglobinimetry in skilled and practised hands an absorptiometric method is suitable[81]. For survey work and in general practice, the American Optical Spenser haemoglobin meter provided an adequate and rapid determination. The British Standards Institution[82] specifies sealed glass cells containing suitable solutions for the photometric determination of haemoglobin. Haemoglobin in trout and carp has been determined using the acid haematin method and the cyanohaemoglobin method[83].

2. *Haemoglobin in plasma*

An ultraviolet spectrophotometric method for the determination of haemoglobin in plasma is based on the absorption of a solution of oxyhaemoglobin at 415 nm (i.e. in the Soret band) and also at 380 and 450 nm (to eliminate background absorption)[84]. This method is simple, each analysis taking only 5 min, and it can be applied to other media containing haemoglobin but not to icteric plasma or sera. The benzidine–

hydrogen peroxide method[85,86] involves reaction of the blood sample with these reagents in a suitable medium and photometric evaluation of the colour produced. The method is capable of determining 0.16–2.5 mg of haemoglobin per 100 ml of plasma and has been applied to the determination of haemoglobin in plasma and urine[87] and for the determination of small amounts of haemoglobin by the haemoglobin–haptoglobin–peroxidase reaction[88].

The cyanomethaemoglobin method has been used for the determination of haemoglobin in plasma[89,90]. Haemoglobin is converted to methaemoglobin by reaction with potassium cyanide. Both methaemoglobin and cyanomethaemoglobin obey Beer's law over the range 540–550 nm[89]. The molar absorptivity of cyanomethaemoglobin is 74.2 $l.\ mol^{-1}cm^{-1}$ at 420 nm and 7.3 $l.\ mol^{-1}\ cm^{-1}$ at 540 nm. The absorptions of the methaemoglobin and cyanomethaemoglobin solutions are measured at a specific wavelength and the difference gives a value which is proportional to the haemoglobin content of the plasma sample. The method can be used for the determination of low concentrations of haemoglobin, in the range 1–12 mg per 100 ml, in plasma and has a relative standard deviation of 1.09%. In applying it to plasma the turbidity of the plasma must be taken into account by making a correction or by adding a protein solubiliser[90].

3. *Haemoglobin in serum and urine*

Haemoglobin in serum has been determined by the benzidine–hydrogen peroxide photometric method[87]. The *o*-toluidine–hydrogen peroxide method has been used for both serum and urine[91,92]. Various spectrophotometric methods[92] have been applied to the determination of haemoglobin and bilirubin in serum and compared with a chemical method[93]. In general, good agreement was obtained. Direct and indirect bilirubin have essentially the same absorption spectra. Haemoglobin calculated from absorption equations showed little correlation at low levels with results obtained by the benzidine method.

4. *Oxyhaemoglobin in blood*

A simple spectrophotometric method for the analysis of uncontaminated blood for oxyhaemoglobin involves measurement of the absorbance at 660 nm against a saturated blood sample as blank[95]. A simply constructed cuvette can be used for the spectrophotometric determination of haemoglobin and oxyhaemoglobin by absorbance measurements at 660 and 850 nm[96]. Oxyhaemoglobin, haemoglobin, carboxyhaemoglobin, and methaemoglobin have been quantitatively determined spectrophotometrically in small blood samples[97]. Procedures have been described for the following determinations: (1) total haemoglobin, by conversion into cyanomethaemoglobin by treatment of the blood with potassium hexacyanoferrate(III) and sodium cyanide and measuring the absorbance at 540, 545, and 551 nm; (2) oxyhaemoglobin mixed with reduced haemoglobin by measuring the absorbance at 560 and 506 nm; (3) carboxyhaemoglobin mixed with oxyhaemoglobin by measuring the absorbance at 562 and 540 nm; (4) methaemoglobin mixed with oxyhaemoglobin in 0.1% aqueous ammonia solution by measuring the absorbance at 540 and 524 nm. The absorbance of oxyhaemoglobin at both the maximum (576 nm) and the minimum absorption (560 nm) has been measured[98]. Measurement of the absorbance of oxyhaemoglobin at the Soret band (412–415 nm) and how interference effects due to bilirubin at this wavelength may be overcome has been described[99].

5. *Carboxyhaemoglobin in blood*

Carboxyhaemoglobin mixed with oxyhaemoglobin can be determined by measuring the absorbance at 562 and 540 nm[97]. An earlier spectrophotometric method[100] for the determination of carboxyhaemoglobin at 555 nm has been modified[101]. A spectrophotometric method for the determination of carboxyhaemoglobin in blood, in which the absorption (A) of a 1:200 dilution of blood in 0.04% aqueous ammonia is read at 576, 560 and 541 nm, and the carboxyhaemoglobin calculated from the ratios of A_{541} to A_{560} and A_{576} to A_{560} with the aid of calibration graphs, sometimes gives erroneously high values[102]. This may be due to the formation of other haemoglobin derivatives (e.g. methaemoglobin or haematin), or to turbidity. The error due to turbidity can be corrected for by reading the absorbance at 700 and 660 nm (at which wavelengths the absorptions of oxyhaemoglobin and carboxyhaemoglobin are minimal), extrapolating the results to obtain blank values for 576, 560, and 541 nm, and subtracting these from the test readings.

For the determination of carboxyhaemoglobin in the presence of oxyhaemoglobin the absorbance is determined at 576, 560 and 541 nm[103–106]. The ratios A_{541}/A_{560} and A_{576}/A_{560} are calculated and the percentage of carboxyhaemoglobin is read off from a calibration graph. An alternative method[107,108] uses the difference in the Soret bands (absorption due to porphyrin nucleus) for oxyhaemoglobin and carboxyhaemoglobin; that for oxyhaemoglobin has a maximum at 420 nm, the latter being the greater[109]. In addition to Soret bands there are, in the case of oxyhaemoglobin, two bands with maxima at 544 and 578 nm and a maximum at 564 nm. These double peaks were used for the evaluation of the absolute haemoglobin concentration in the test solution[107]. The spectrum of carboxyhaemoglobin is similar to that of oxyhaemoglobin except that the peaks are in different positions. The difference spectrum shows a sharp peak at 422 nm, which represents the absorption of carboxyhaemoglobin. The absorptions at 414 and 426 nm correspond to zero carboxyhaemoglobin and are used as the base line.

A further method for determining carboxyhaemoglobin is based on the reduction of palladium(II) chloride to metallic palladium by the abstraction of carbon monoxide from the carboxyhaemoglobin:

$$PdCl_2 + CO + H_2O \longrightarrow CO_2 + 2HCl + Pd \quad (1)$$

Unchanged palladium(II) chloride can be determined colorimetrically after the addition of diethyl-p-nitrosoaniline[110]. Alternatively, the blood sample may be treated with dilute sulphuric acid and the volume of carbon monoxide measured[111–113], or the palladium complex may be determined by ultraviolet spectroscopy at 278 nm[114]. Determinations of carboxyhaemoglobin in the blood by the Van Slyke gasometric, photometric, and spectrophotometric methods have been compared[115]. All three methods give satisfactory and comparable results. The photometric method is recommended for convenience and simplicity. An account of the spectrophotometric determination at four wavelengths of carboxyhaemoglobin, and methaemoglobin makes special reference to the simultaneous determination of carboxyhaemoglobin and methaemoglobin in human blood[116].

6. *Methaemoglobin in blood*

Several groups have studied the determination of methaemoglobin in blood[97,116–121]. Methaemoglobin in mixtures with oxyhaemoglobin may be determined in a 0.1% aqueous ammonia medium by measuring the absorbance at 540 and 524 nm[97]. Methaemoglobin, which has a characteristic absorption peak at 630 nm, can be determined by first measuring the absorbance at 630 nm before and after addition of sodium cyanide. Another fraction of buffered haemolysed red cell solution is centrifuged, and the supernatent liquid treated with potassium hexacyanoferrate(III). Finally, the total haemoglobin content is determined by Crosby's method[122]. The concentration of methaemoglobin in the sample can then be calculated. In another method the absorbance of the whole blood haemolysate is measured at 578 and 525 nm[118]. The ratio of the absorbance at 578 nm to that at 525 nm corresponds to the content of methaemoglobin. The results compare well with those obtained by established methods, even at low concentrations of methaemoglobin. When a blood haemolysate is treated with sodium cyanide methaemoglobin is converted into cyanomethaemoglobin, and the reduction in the absorption of the solution at 630 nm is proportional to the amount of methaemoglobin in solution[119].

7. Iron carbonyl

Low concentrations of iron pentacarbonyl in commercial carbon monoxide can be determined by passing a gas sample through a train in which the iron pentacarbonyl is either condensed in a trap of solid carbon dioxide or absorbed in pure methanol[123]. After the sytem has been flushed with nitrogen, the concentration of iron pentacarbonyl is determined spectrophotometrically at 235 nm in a methanol solution. The error is about $\pm 1.3\%$ on samples containing not less than 0.04 mg of iron pentacarbonyl. To determine iron pentacarbonyl in air the sample (50 l) is drawn at 2–3 l/min through bubblers containing acidic iodine in potassium iodide, which traps 99% of the iron pentacarbonyl[124]. Iron is reduced to the iron(II) state with hydroxylammonium chloride solution and determined at 508 nm using 1,10-phenanthroline. For the determination of iron pentacarbonyl in amounts down to 0.01 ppm in town gas, the carbonyl is trapped in iodine monochloride solution and, after reduction, iron is determined by the ammonium thiocyanate procedure[125]. Iron pentacarbonyl in carbon monoxide and carbon dioxide has been determined by passing the gas sample through a silica tube containing a plug of silica-wool kept at 350°C[126]. The tube is allowed to cool while argon is flowing through it. The deposited iron is dissolved in hydrochloric acid and determined by conventional absorptiometric procedures.

I. Organolead Compounds

1. Determination of mono-, di, and trisubstituted organolead compounds

Dithizone has been used in the spectrophotometric determination of these compounds[127–131]. In one method the colour of the dithizone freed from the triethyllead dithizone complex by acidification is used as a measure of the triethyllead ion[131]. Interference from inorganic lead was eliminated by complexation with EDTA. In another method, a di- and trialkyllead chlorides in aqueous solution is converted into dithizonates, which are then extracted with chloroform and determined spectrophotometrically[128]. The absorption maxima are at 498–430 nm for dialkyllead dichlorides and trialkyllead chlorides (methyl or ethyl derivatives) and the proportions are obtained by differential analysis. If lead ions are present they must be complexed with EDTA to prevent interference. A rapid spectrophotometric method uses dithizone for the determination of triethyllead, diethyllead, and inorganic lead ions in tetraethyllead[129,132,133]. Since dithizone does not react with tetraethyllead the field of analysis is narrowed to three components. Each of the other three forms of lead has a distinctive chloroform-soluble dithizonate. Triethyllead ion forms a canary yellow complex, diethyllead ion an orange complex, and inorganic lead ion a red complex.

Absorption maxima are at wavelengths of 435, 487, and 520 nm, respectively. Measurement of absorbances of the mixed dithizonates at 424, 500, and 540 nm followed by the solution of three simultaneous equations enables the individual components to be determined. This method may not be applicable to the accurate determination of the individual diorganolead ions other than the ethyl or methyl homologues[129]. The absorption curve of diphenyllead dithizonate differs significantly from that of the diethyllead dithizonate.

Some workers prefer 4-(2-pyridylazo)resorcinol (PAR) to the less selective dithizone[134,135]. PAR reacts with compounds of the type R_2PbCl_2 (R = Me, Et, Ph) to form stable coloured complexes and forms no coloured complexes with R_3PbCl. The wavelengths of maximum absorption of $(C_2H_5)_2PbCl_2$ are unchanged at 514 nm in the pH range 5–10. The maximum colour intensity of the $(C_2H_5)_2Pb$—PAR complex develops immediately and is stable for at least 24 h. A plot of the absorbance at constant wavelength against pH indicated that the optimum pH is 9 for $(C_2H_5)_2PbCl_2$. Application of the method of continuous variations and the slope-ratio method show that only a 1:1 complex of PAR is formed with diethyllead dichloride. Beer's law is obeyed between 2×10^{-6} and 10^{-4} M $(C_2H_5)_2Pb$—PAR and the complex has a molar absorptivity coefficient of 41 000 l/mol. cm. For the diethyllead cation, spectrophotometric determination is possible provided that the molar ratio of reagent to cation exceeds 4. The minimum detectable amount of cation is about 0.7 ppm for 0.1 absorbance in a 19 mm cell. Because of the great solubility of PAR (monosodium salt) in water, the dilution factor may be as low as 2 and hence the sensitivity ($\varepsilon \times 10^{-3}$ divided by dilution factor) may reach 20. In the case of diethyllead dichloride, the complex is so stable that spectrophotometric titrations with standard PAR solution can be carried out. Compounds of the type R_2PbCl_2 readily decompose to lead chloride both in the solid state and in solution, and hence samples of the former often contain some of the latter compound. The absorbance of the $(C_2H_5)_2Pb$—PAR complex does not change on addition of EDTA[134,135], whereas the Pb—PAR complex[136] is quantitatively destroyed by a stoichiometric amount of EDTA. Accordingly, if an excess of PAR is added to a sample of diethyllead dichloride, any lead salt impurity can be determined by spectrophotometric titration with EDTA at 512 nm.

The stability constants of 1-(2-pyridylazo)-2-naphthol (PAN) with the diethyllead ion in aqueous 20% *v/v* dioxane are very high[137]. When dialkyllead ion is added to an aqueous dioxane solution of PAN, the yellow liquid changes to a red chelate which retains its colour intensity for at least 24 h. The diethyllead ion can be directly titrated spectrophotometrically with a standard PAN solution or determined as the $(C_2H_5)_2$—Pb(PAN)OH complex photometrically by measuring the absorbance of a chloroform extract at 555 nm which corresponds to the maximum of the uncharged complex[137]. Alkyllead ions form stable coordination compounds with 8-hydroxyquinoline[138,139], 2,2′-bipyridyl[139,140], *o*-phenanthroline[140–142], acetylacetone[142,143], and picolonic acid[142].

Triethyllead ions can be determined in blood and urine by extracting the triethyllead selectively into benzene from urine or deproteinized blood that has been almost saturated with sodium chloride[144]. After re-extraction into dilute nitric acid, the triethyllead is decomposed with sulphuric and nitric acid and the lead is determined by the dithizone method. The sensitivity of the method is 2 μg of lead and the precision is $\pm$6% but the extraction from blood samples is only 90% efficient (100% from urine).

2. Determination of tetrasubstituted organolead compounds

Tetraethyl and other tetraorganolead compounds may be determined by converting them to the ionic form by reaction with iodine and measuring the increase in ionic lead[129,132,133]. The precision of the method thus applied to tetraorganolead compounds is ±1.3% of the amount present. The method was used successfully in the determination of PbR_4 where R = Me, Et, *n*-Pr, *i*-Bu, *i*-Am, *n*-Am, $CH_2{=}CH$ and C_6H_5 individually, or in the presence of appreciable amounts of ionized lead.

In a spectrophotometric method capable of determining down to 10 ppm, the lead is extracted from the petroleum with a solution of potassium chlorate and sodium chloride in dilute nitric acid (Schwartz reagent) and then determined colorimetrically, as lead sulphide, by comparing the colour developed with that obtained by adding lead nitrate solution to sodium sulphide solution[133,145,146]. In an alternative method using dithizone, bromine is added to the sample to convert the tetraethyllead to lead bromide[133]. In a further method diisobutylene and bromine are added to produce hydrogen bromide gas, which then completes the decomposition of the organolead compound to inorganic lead ions. The lead is measured spectrophotometrically in the first procedure and by visual colour comparison in the second. The concentration range is 0.5–10 μg of lead and the accuracy and precision of both methods are within ±0.01 ppm. Spectrophotometric determination as lead sulphide has been used for the determination of tetraethyllead in petroleum[147].

For the determination of tetraethyllead in air, the sample is passed through an acidic solution of iodine monochloride to produce dialkyllead ions[148].

$$R_4Pb \longrightarrow PbR_3^+ \longrightarrow PbR_2^{2+} \longrightarrow Pb^{2+} \qquad (2)$$

Manual or automatic procedures can be used for the determination of the amount of tetraalkyllead collected. The manual method involves reaction of the dialkyllead ions with dithizone at high pH and matching the colour of the dialkyllead dithizonate with a standard disc. In the automatic procedure the dialkyllead is converted to the inorganic state before reaction with dithizone and colorimetric measurement as lead dithizonate (at 475 nm). The method measures lead-in-air concentrations down to 0.1 mg of lead per 10 m^3 of air, with sampling periods of at least 8 h.

Inorganic lead compounds may be present as dust in atmospheres that are monitored for tetraalkyllead vapour. Most of these compounds are soluble in the acidic solution of iodine monochloride and would, therefore, interfere in the subsequent determination of dialkyllead ions derived from tetraethyllead or tetramethyllead[148]. Filtration of the air under test, prior to contact with the iodine monochloride reagent, will avoid the posibility of this interference. The solids collected on the filter may also be analysed for lead content to give more complete results on the total toxic hazard of the atmosphere under test. At the 1–40 mg organic lead per 10 m^3 in air level, between 98 and 100% of tetraethyllead and of tetramethyllead is collected in a single scrubber by this technique[148]. A simple and rapid field method for the determination of organolead compounds in air utilizes as a scrubber a small disposable glass tube containing iodine crystals[149]. Lead is removed from the scrubber by acidic potassium iodide solution and is measured by a colorimetric dithizone method. Hydrogen sulphide does not interfere. Various other workers have reported on methods based on the formation of lead dithizonate for the determination of tetraethyllead in air[150,151]. Nephelometric determination as lead chromate has also been used[152,153].

Both solid iodine and solutions of iodine in methanol are effective in collecting tetraethyllead when using the Uni-jet air sampling equipment for sampling over a short period of time[154]. The volatility of both the iodine and the methanol would be an obvious disadvantage in applying these methods to air containing very low levels of organic lead owing to the prolonged sampling times that would be required in these circumstances. Reagent volatility is not a problem with iodine monochloride solution[148].

J. Organomanganese Compounds

Cyclopentadienylmanganese tricarbonyl and methylcyclopentadienylmanganese tricarbonyl anti-knock compounds have been determined in petroleum by a spectrophotometric procedure in which the petroleum is extracted first with bromine, then with sulphuric acid–nitric acid–water, and the extracted manganese is oxidised with potassium periodate[155]. The resulting permanganate is determined spectrophotometrically at 520 nm. A determination in the range of 0.03–0.3 g of manganese per US gallon has an average deviation of 0.002–0.004 g/gallon, and can be completed in about 2 h.

K. Organomercury Compounds

A method for the determination of diphenylmercury, either alone or in the presence of other phenylmercury compounds, has been described[156]. For the determination of diphenylmercury alone, it is extracted from a chloroform solution with 9 N hydrochloric acid. Phenylmercury chloride is formed quantitatively and is determined by reaction with dithizone and measurement of the colour produced at 629 nm. Diethylmercury is determined in the same way except that 12 N hydrochloric acid is used for the hydrolysis. For the determination of diphenyl- or diethylmercury in the presence of phenyl- or ethylmercury compounds, the latter are removed by extraction into acidified sodium thiosulphate, and the diphenyl- or diethylmercury is determined as above. Beer's law is obeyed over the ranges 1–30 and 90–120 μg with an error of less than 5%.

Phenylmercury compounds react with diphenylcarbazone in the same way as do inorganic mercury compounds, but are resistant to reduction by zinc dust. In a method for the determination of microgram amounts of some phenylmercury compounds and their separation from inorganic mercury salts, the sample solution is added to ethanolic diphenylcarbazone solution and the colour is measured against standards[157], which gives total mercury. The analysis is repeated after shaking the sample solution with zinc powder for 40 min, giving the phenylmercury content of the sample. Inorganic mercury is then obtained by difference. In a spectrophotometric method for the differential determination of phenylmercury acetate and inorganic mercury the mercury solution is shaken with hydrochloric acid and 2×10^{-4} M dithizone in carbon tetrachloride and the extract placed on an alumina column[158]. Successive elution with carbon tetrachloride, carbon tetrachloride–chloroform (19:1), and carbon tetrachloride–chloroform (1:1) gives a yellow eluate containing phenylmercury dithizonate and then an orange eluate containing mercury dithizonate. These eluates are stabilized with anhydrous acetic acid, and the absorptions are measured at 480 and 490 nm for phenylmercury and mercury dithizonates, respectively. Beer's law is obeyed for each compound up to 30 μg of mercury. The error in this procedure does not exceed 5%. Phenylmercury acetate forms a stable complex with diphenylcarbazone in alkaline media (maximum absorption in chloroform at 580 nm) that may be used for colorimetric analysis[159].

Various workers have described methods for the determination of mercury in tissues. Mercury in urine and kidney can be determined by forming a complex in aqueous trichloroacetic acid between potassium bromide and the mercury in certain organic mercurials[160]. After adjustment to pH 5.0 with formate buffer solution, the mercury in this form is extracted with dithizone solution in chloroform for spectrophotometric determination at 475 nm. The procedure can be used to detect as little as 1μg of mercury. Methods based on the use of dithizone can be used to determine phenylmercury acetate in urine, kidney, liver, muscle, spleen, and brain[161].

For the determination of phenylmercury compounds and total mercury in paints, phenylmercury compounds are converted into phenylmercury acetate by boiling with 1 N acetic acid; the solution is neutralized to pH 6 with sodium bicarbonate and phenylmercury acetate is extracted with benzene[162]. The colour obtained on the addition of 0.5% ethanolic diphenylcarbazone solution is compared with that given by the gradual addition of standard phenylmercury acetate to a reagent blank. The results are low because of absorption losses. Total mercury is determined by the Schöniger combustion technique, followed by reaction with ethanolic diphenylcarbazone; the sensitivity of this reaction can be reduced to a convenient level (about 5 μg of mercury) by the addition of potassium cyanide. Methods based on the use of dithizone have been described for the determination of mercury as methylmercury, ethylmercury, and phenylmercury in soil, turf, and grain samples[163,164]. This is preceded by wet oxidation of the organic material with dilute sulphuric acid and nitric acids in an apparatus in which the vapour from the digestion is condensed into a reservoir from which it can be collected or returned to the digestion flask as required[165]. The combined oxidized residue and condensate is diluted until the acid concentration is 1 N and nitrate is removed by addition of hydroxylammonium chloride with boiling. Any fat can be removed from the cooled solution with carbon tetrachloride and the liquid is then extracted with a solution of dithizone in carbon tetrachloride and mercury is determined spectrophotometrically at 485 nm using dithizone. Procedures for the determination of micro amounts of mercury in biological materials involving destruction of organic matter and the use of dithizone for mercury extraction have been reviewed[166].

The decomposition of organic fungicides in soil to mercury vapour and to methyl- or ethylmercury compounds has been studied and methods devised for the determination of these compounds in the vapours liberated from the soil sample[167]. The mixed vapours of mercury and organomercury compounds are passed successively through bubblers containing a carbonate–phosphate solution to absorb organic mercury and through an acidic potassium permanganate solution to absorb inorganic mercury vapour. In both cases the mercury in the scrubber solution is determined photometrically at 605 nm with dithizone. The method is capable of determining 10 μg or more of organic mercury per 1000 l of air in the presence of mercury vapour.

Alternative oxidizing agents to potassium permanganate have been examined for the absorption of metallic mercury vapour produced by the decomposition of organomercury fungicides in soil[167]. Acid potassium dichromate (1 of volume 18 N sulphuric acid plus 4 volumes of 5% potassium dichromate) is equally effective in trapping mercury vapour as acid permanganate. Mercury could be determined in the dichromate absorbers after first reducing the dichromate with excess of hydroxylammonium chloride. Using the technique of successive absorption of organic mercury in phosphate–carbonate solution followed by absorption of mercury vapour in dichromate solution (later reduced with excess of hydroxylammonium chloride), vapours produced by the decomposition in soil of phenyl- and alkylmercury compounds were collected and determined[168]. The air above the soil containing phenylmercury acetate contained mercury vapour and traces of phenylmercury acetate. Ethylmercury acetate produced about equal amounts of mercury vapour and an uncharacterized volatile ethylmercury compound, whereas methylmercury chloride and methylmercury dicyanamide both produced an uncharacterized methylmercury compound plus some mercury vapour. A method suitable for the determination in air samples or soil volatiles of amounts of methyl- and ethylmercury chlorides down to 1–5 μg in 50–100 ml of sample solution has been described[169]. An alternative method[170] is best used for the determination of amounts above 10 μg of methyl- and ethylmercury chloride in the carbonate–phosphate absorber solution previously mentioned[167]. Large aqueous sample volumes are not deleterious in this method as they are in the direct method of analysis mentioned above. A further method is suitable for the determination of below 30 μg of alkylmercury compounds in sample sizes of up to 100 ml of carbonate–phosphate absorber solution[171].

Dithizone procedures for the determination of phenylmercury chloride in fungicidal dust preparations and phenylmercury acetate in emulsifiable concentrates have been described[172,173]. For the determination of ethylmercury phosphate in emulsifiable concentrates the sample is dissolved in 4% acetic acid in aqueous methanol and 30% sodium chloride solution added[172,173]. An aliquot containing about 50 μg of mercury is shaken with chloroform and mercury is determined using dithizone at 478 nm. Calcium and magnesium stearates do not interfere in the procedure. Ethylmercury compounds may be separated from phenyl- and tolylmercury compounds by heating mixtures of these compounds with 1:3 hydrochloric acid and methanol[174,175]. This decomposes phenyl- and tolylmercury compounds in 30–60 min. Ethylmercury compounds remain unchanged and can be determined in a chloroform extract by the dithizone method described above.

In a rapid colorimetric method for the analysis of ethylmercury chloride the sample is dissolved in dimethylformamide and then mercury is determined spectrophotometrically using diphenylcarbazone[176]. Phenylmercury acetate has been determined spectrophotometrically in amounts down to 2 μg by evaluation at 550 nm of the coloured complex produced with Riehemann's purple, [2-(1,3-dioxoindan-2-yl)iminoindane-1,3-dione][177]. Analyses of phenylmercury acetate in fungicidal and herbicidal preparations obtained by ultraviolet spectroscopic methods have been compared with the thiocyanate method[178,179]. There is also found to be good agreement between the spectrophotometric method and the iodide–thiosulphate method[178,180]. A colorimetric method using dithizone has been used for the direct determination of 1–100 μg of methylmercury dicyanamide or methylmercury(II) chloride in fungicidal preparations[170]. Copper, cobalt, cadmium, iron, lead, nickel, silver, zinc, bismuth, and mercury(II) ions in amounts less than 1 mg do not interfere in this procedure. For the analysis of seed disinfectants based on phenylmercuricatechol the sample is shaken with 10% sodium hydroxide solution to produce catechol[181]. After preliminary work-up the catechol is determined spectrophotometrically at 560 nm using aqueous 4-aminophenazone and aqueous potassium hexacyanoferrate(III). The determination of methylmercury dicyanamide has been discussed[163].

L. Organonickel compounds

A sensitive, continuously recording instrument has been designed for detecting nickel carbonyl in the air in which a stream of air flows through a nozzle and impinges on a hot borosilicate glass disc, on which nickel carbonyl, if present, is deposited[182]. A collimated beam of plane-polarized light falls on the disc at the Brewsterian angle for borosilicate glass, its plane of polarization being perpendicular to the place of incidence. This arrangement results in extinction, so that no light is reflected from the disc until a deposit is formed. The intensity of the reflected light is then measured by a recording photometer calibrated in parts per million of nickel carbonyl. Concentrations in the range 0.05–4 ppm by volume can be measured and at a concentration of 1 ppm the accuracy is ±0.2 ppm. In another method for determining 0.002 mg or more of nickel tetracarbonyl in 1 litre of air, the sample is passed through 2 ml of a 1.5% solution of iodine in carbon tetrachloride[183]. After treatment with sodium sulphite the solution is analysed colorimetrically for nickel. Modifications to the method enable it to be extended to below 0.002 mg of nickel tetracarbonyl. In a further method for determining nickel tetracarbonyl in air, the air is dried by passage over calcium chloride, then passed through 0.05% ethanolic iodine in a special absorber cooled to

$-30°C$ in a bath of trichloroethylene–solid carbon dioxide[184]. The absorbent is then evaporated to dryness, the residue is dissolved in water, and nickel in the resulting solution is determined photometrically with dimethylglyoxime. The calibration graph is rectilinear for 17–64 μg of nickel per 50 ml of the aqueous solution. In yet another method for determining nickel tetracarbonyl in air, after passage through a filter to remove any solids, the sample is bubbled into 3% hydrochloric acid[185]. After neutralization, furildioxime solution in alcohol and chloroform is added and the colour is compared either visually or spectrophotometrically at 435 nm with standards.

In a spectrophotometric method for the determination of nickel and iron carbonyls in town gas the gas sample is passed through a sintered bubbler containing 1 M iodine monochloride in anhydrous acetic acid[125]. Nickel is determined spectrophotometrically using dimethylglyoxime. The detection limit of this method is 0.006 ppm of nickel tetracarbonyl. In a method for the determination of nickel tetracarbonyl in carbon monoxide and carbon dioxide, the gas sample is passed through a silica tube containing a plug of silica-wool kept at 350°C[126]. The tube is allowed to cool while argon is flowing through it. The deposited nickel is dissolved in hydrochloric acid and determined by conventional absorptiometric procedures.

The wavelength at which the absorbance maxima occur for pyridine solutions of nickel myristate and nickel palmitate is 550 nm[154]. The relationship between absorbance and concentration of nickel soap is linear.

M. Organophosphorus Compounds

The absorption in the visible region of nickel salts of some dialkyldithiophosphoric acids in benzene solution have been used for the colorimetric qualitative analysis of these compounds[186]. The method is applicable to the titration of crude industrial acids, especially for dimethyl- and diethyldithiophosphoric acids. It has an accuracy within 1%. In a method for the determination of sodium hydrogen *S*-(2-aminoethyl)phosphorothioate (sodium hydrogen cysteamine *S*-phosphate) the sample is added to a solution containing mercury(II) acetate in aqueous acetic acid, metol, and molybdate[187,188]. The absorbance is determined at 660 nm.

N. Organoselenium Compounds

In a method for the determination of small amounts of hydrogen selenide in air the sample is passed through an absorber containing hydrogen bromide solution with 18% of free bromine[189]. Unchanged bromine is destroyed with hydroxylammonium chloride. The colour is determined photometrically by the 3,3′-diaminobenzidine method[190]. Analytical aspects of organoselenium compounds have been reviewed[191].

O. Organothallium Compounds

Phenylthallium(III) dichloride forms with xylenol orange a complex having an absorption maximum at 576 nm, with a mean molar absorptivity of 3.69×10^4 l/mol. cm in the presence of 1–2% of methanol[192]. It can also be titrated with EDTA in aqueous methanol medium at pH 4.75 with 0.1% aqueous xylenol orange as indicator. The end-point is shown by a sudden colour change from red to lemon-yellow and corresponds to the formation of a 1:1 complex of phenylthallium dichloride and EDTA.

P. Organotin Compounds

A colorimetric method, based on the formation of a dithizone complex, can be used for the determination of diethyltin and triethyltin chloride and sulphate, either singly or as mixtures[193–195]. The absorption maximum for the diethyltin—dithizone complex, after being shaken with 10% trichloroacetic acid, is at 510 nm, whereas triethyltin does not react under these conditions. In the presence of borate buffer of pH 8.4, diethyltin and triethyltin compounds, with dithizone, give absorption maxima at 485 and 440 nm, respectively. At 510 nm the triethyltin—dithizone complex and dithizone have the same absorption. The separation and determination of diethyltin and triethyltin compounds can be based upon these findings and their distribution between chloroform and aqueous media. Interference from other metals is avoided by the use of EDTA.

The use of dithizone for the determination of trialkyltin compounds has the serious disadvantage that exposure to bright light causes a rapid change in the colour of the trialkyltin dithizone complex. This colour change is consistent with the conversion of the trialkyltin complex to the dialkyltin complex. Measurements of trialkyltin dithizone complexes must therefore be made in subdued light[196,197].

A spectrophotometric method[198] for the determination of down to 3 mg of dibutyltin dichloride in the presence of mono-, di-, and tetrabutyltin chlorides and several inorganic ions including zinc, manganese(II), iron(III), iron(II), lead, copper(II), copper(I), cadmium, tin(II) and tin(IV) depends on the fact that diphenylcarbazone produces a red colour with dibutyltin dichloride at pH 8.4[199]. Dibutyltin dichloride in the sample reacts with diphenylcarbazone at pH 1.8 to form a 3:1 complex with an absorption maximum at 510 nm. Butyltin trichloride, the only one of the butyltin compounds to interfere at this pH, is removed by extraction with EDTA. The success of this procedure in being able to determine only dibutyltin dichloride in the presence of a mixture of the various butyltin chlorides is due to the selective extraction of butyltin trichloride with EDTA. Dibutyltin dichloride together with butyltin trichloride is extracted in varying amounts at higher pH. Conceivably the tributyl tinchloride would not be extracted within this range and it could then be determined with dithizone[194].

4-(2-Pyridylazo)resorcinol (PAR)[200,201] and 1-(2-pyridylazo)-2-naphthol (PAN)[201,202] have been studied as reagents for the spectrophotometric determination of organotin ions. PAR reacts with compounds of the type R_2SnCl_2 ($R = CH_3$, C_2H_5, C_6H_5) to form coloured complexes, whereas compounds of the type R_3SnCl do not produce a colour using a 2×10^{-3} M solution of PAR as reagent[200]. The wavelength of maximum absorption (514 nm) of the 1:1 diethyltin dichloride—PAR complex is independent of pH in the pH range 3–8, with an optimum at pH 6. The colour develops immediately and is stable for 24 h. Reagent absorption is very small at this wavelength. Beer's law is obeyed in the range 2×10^{-6}–10^{-4} M of organotin ion, the molar absorptivity of the diethyltin dichloride—PAR complex being 42 500 l/mol. cm. A four-fold molar excess of reagent over organotin ion is necessary.

Dimethyl-, diethyl-, and di-n-butyltin dichlorides were examined spectrophotometrically using PAN as reagent[201]. In solutions of R_2Sn^{2+} there is considerable uncertainty regarding the ionic species which may be present. Potentiometric investigations of dimethyltin ion hydrolysis have shown the existence of the $(CH_3)_2Sn^{2+}$ ion alone up to a limiting pH of about 2[203–206]. These studies have been extended to other organotin ions in order to establish the pH values within which the ions do not undergo hydrolytic equilibria. Addition of dialkyltin ions to PAN solution in dioxane produces a colour change from yellow to red which is stable for at least 24 h. The wavelengths for maximum absorption and the corresponding molar absorptivities are listed in Table 2. The stability of the complexes of R_2Sn^{2+} with PAN decreases in the order

TABLE 2. Absorption data for PAN chelates, aqueous 20% dioxane

Compound	λ_{max} (nm)	$\varepsilon \times 10^{-3}$ (l/mol. cm)
$(CH_3)_2Sn^{2+}$	532	21.1
$(C_2H_5)_2Sn^{2+}$	538	22.6
$(C_4H_9)_2Sn^{2+}$	540	22.5
$(C_6H_5)_2Sn^{2+}$	542	22.0
$(C_2H_5)_2Pb^{2+}$	540	22.3

$C_6H_5 > C_4H_9 > C_2H_5 > CH_3$[201]. The high stability of these chelates agrees well with the established action of PAN as a tridenate ligand chelate system with five membered rings being formed[206–208]. Other compounds that form stable coordination compounds with alkyltin ions include 8-hydroxyquinoline[209–211], 2,2′-bipyridyl[209,211,212], phenanthroline[209,212,213], acetylacetone[213,214], picolinic acid[213], and Alizarin Red S[215]. The analytical chemistry of these complexes has not been extensively studied. Sodium dimethyldithiocarbamate can be used instead of dithizone for the determination of both types of organotin compounds. The alkyltin complexes with this compound show maximum absoption in the u.v. region, at about 280 nm, but unlike the dithizone derivative are not decomposed photochemically in daylight.

Two colorimetric methods have been described for the determination of organotin hydrides[216]. In one of these methods the organotin hydride reduces isatin in alcoholic medium in the presence of azobis(isobutyronitrile) to colourless dioxindole (hydroxyindolin-2-one). In the second method, ninhydrin in alcoholic medium is reduced by organotin hydrides, but not by other organotin compounds, to 2-hydroxyindan-1,3-dione, and a blue-violet colour is formed that is stable in the absence of air. The change in absorbance produced with either reagent can be used to determine organotin hydrides in concentrations down to 10^{-4} M. Other reducing agents must be absent. The formation of colourless dioxinole from the coloured isatin was utilized for the quantitative determination of organotin hydrides. The change in the absorbance of isatin solution is proportional to the concentration of added hydride. Organotin hydrides, such as tri-*n*-butyltin hydride, triphenyltin hydride, and diisobutyltin dihydride, give a blue–violet colour with ninhydrin[216]. The reaction can be carried out in various solvents in which both components are soluble, such as in alcohols, pyridine, acetone, dioxane, and chlorobenzene. The colour fades on exposure to air, but on careful exclusion of oxygen, and by use of alcohol, distilled under argon, as solvent, the absorbance does not change for at least 1 h. This colour reaction is not given by other types of organotin compounds such as oxides, hydroxides, or halides and is specific for organotin hydrides. The visible spectrum of the coloured reaction mixture in ethanol with tributyl or triphenyltin hydride shows a maximum absorption at 490 nm, and with diisobutyltin dihydride the maximum is at 525 nm. The absorption maxima are independent of the relative concentration of the reactants. The same maxima were obtained with a ten-fold excess of tributyltin hydride or a five-fold excess of ninhydrin. The absorbance in alcohol is proportional to the concentration of the hydride, thus permitting the quantitative determination of organotin hydrides.

Dibutyltin compounds have been used for a number of years as stabilizers for poly(vinyl chloride). They are generally present to the extent of 1–2% in the finished polymer. Other dialkyltin compounds, and in particular those of dioctylin, are equally effective as stabilizers whilst having no demonstrable mammalian toxicity. The organotin compound present in PVC can usually be determined by wet ashing a 5 g sample with sulphuric–nitric acids, reduction with aluminium or nickel, and titration of the tin(II) so formed with standard iodate solution. Precautions are necessary to prevent loss of tin by volatilization during the rapid evolution of hydrochloric acid in the early stages of the decomposition of the PVC. An alternative method for determining dialkyltin compounds involves extraction from the polymer with 1,2-dichloethane and spectrophotometric determination with dithizone at 490 nm[217]. In a method for differentiating between dibutyltin and dioctyltin compounds in PVC, the dibutyltin compounds are extracted from chloroform solution by 1 N sodium hydroxide, but those of dioctyltin are only slightly soluble, and can be readily detected in the chloroform solution with dithizone after extraction[217]. An aliquot of the dichloroethane solution form the tin determination is transferred to a separating funnel and tetrahydrofuran and 1 N-sodium hydroxide are added. The funnel is shaken and the two layers are allowed to separate. The organic layer is transferred to another separating funnel and shaken with trichloroacetic acid and dithizone solution. In the presence of a dibutyltin compound the dithizone remains unchanged while a dioctyltin compound produces a red coloration.

Dithizone forms coloured complexes with certain tin compounds and these complexes can be used for the determination of several dialkyltin and trialkyltin compounds[194]. When a solution of, for example, diethyltin dichloride is shaken with a chloroform solution of dithizone in the presence of 10% trichloroacetic acid, a red colour is produced, but under the same conditions neither the triethyltin nor the tetraethyltin compound reacts. In the presence of a borate buffer solution of pH 8.4, however, both tetraethyltin and triethyltin compounds gave a yellow colour, while the diethyltin derivative produces an orange colour.

A method for the estimation of dialkyltin stabilizers in aqueous extracts of PVC used in foodstuffs and drug packaging applications used 4-(2-pyridylazo)resorcinol–EDTA reagent[218]. The absorbance of the separated chloroform layer is measured at 518 nm and referred to a calibration graph prepared with 1–10 μM dibutyltin maleate. Various other aspects of the migratory tendencies of organotin stabilizers from PVC which have been studied include the analysis of aqueous and non-aqueous extracts of rigid and of plasticized PVC[219], and of tin in injection fluids packaged in PVC ampoules[220].

A colorimetric determination of dialkyltin compounds in fats and olive oil containing 2% of added oleic acid is suitable for testing the migration of substances from plastic packaging and has been used for determining dialkyltin stabilizers added to PVC film and extractable in small amounts by fats and olive oil under the conditions of accelerated tests[221]. To the sample of the used olive oil extractant is added alcoholic catechol violet solution. The absorbance of the clear upper layer is measured at 550 nm against a blank solution of similarly treated olive oil and is referred to a calibration graph prepared with standards. Triphenyltin residues in plant material can be determined by a spectrophotometric dithizone procedure[222,223]. Organotin compounds such as diethyltin dichloride, diethylbis(lauroyloxy)tin, dibutyltin dichloride, dibutyltin dilaurate, and diethyltin dicaprylate in air can be determined using diphenylcarbazone as a colorimetric reagent[224,225]. The air is drawn through butanol in an ice-cooled absorber. The absorbant solution is transferred to a Nessler tube and diluted with butanol. Dithizone solution is added and mixed and the colour is compared with similarly prepared standards containing 2–30 μg of the organotin compounds. A fluorometric method has been used for the determination of triphenyltin compounds in water[226].

Q. Organozinc Compounds

For the spectrophotometric determination of zinc diethyldithiocarbamate in rubbers as the copper complex, the finely divided rubber vulcanizate is extracted with benzene and the zinc compound separated from interfering substances on acetylated paper prior to spectrophotometric evaluation at 430 nm[227]. Zinc diethyldithiocarbamate cannot be determined in the presence of thiuram disulphide, which forms diethyldithiocarbamate during vulcanization and extraction. Other dialkyldithiocarbamates may also be determined by this method.

A procedure for the determination of zinc ethylenebisdithiocarbamate (zineb) residues on crops depends on the measurement of the absorbance at 434 nm of copper diethyldithiocarbamate in the presence of alkali[228].

II. ULTRAVIOLET SPECTROSCOPY

A. Organoboron Compounds

Decaborane can be determined using the absorption maximum at 272 nm in cyclohexane solution (molar absorptivity = 3000 l/mol. cm).[229]. This is a very useful method, but the samples analysed must be free from other materials which absorb at 272 nm.

B. Organolithium Compounds

The absorbance of diluted solutions of organolithium compounds calibrated by n.m.r. spectroscopy can directly measure the carbon-bound lithium content[230]. Suitable glass apparatus which connects the two physical methods such that adventitious contamination can be eliminated by pre-purging with sample within a closed system is desirable. The absorbance over the range 275–305 nm has been measured for a series of butyllithium concentrations from 6×10^{-3} to 3.1×10^{-2} mol/l. The maximum absorption shifted from 278 nm at the lowest concentration to 282 nm at the highest and changed monotonically with concentration. Beer's law was obeyed at 285 nm and a molar absorptivity of 91 l/mol. cm was calculated at this wavelength. Concentrations of approximately 10^{-3} mol/l in butyllithium can be satisfactorily determined to within 5% using the combined n.m.r.–u.v. technique. A check was made on the accuracy of the n.m.r. result by hydrolysing a sample of butyllithium–benzene with water and titrating against standard hydrochloric acid. The titration results were consistently 2–3% higher than the n.m.r. results. This is to be expected since titration gives the total lithium content, consisting of butyllithium, butoxide, and hydroxide.

C. Organomagnesium Compounds

Grignard reagents may be determined by breaking ampoules containing the Grignard solution in diethyl ether in a sealed glass tube containing the required amount of purified benzophenone, dissolved in the same ether[231,232]. After shaking, the sealed glass tube was opened and the reaction mixture treated with methanol and acetic acid, followed by the determination of the amount of unreacted benzophenone at 333 nm[233]. At the same time, in an aliquot of the methanol solution, the amount of total magnesium was determined titrimetrically using EDTA[234]. When applied to ethylmagnesium bromide uncontaminated by water or by oxygen, the benzophenone spectrophotometric method, the acid–base titration, and the gasometric method all agree within 1–3%[231]. Moreover, the large discrepancies obtained by others[233] in the determination of the concentration of ethylmagnesium bromide using the benzophenone photometric method and the acid–base titration method can only be ascribed to the way these others treated their Grignard solution[231].

When a Grignard reagent is allowed to react with 4,4′-bis(dimethylamino)benzophenone (Michler's ketone) a typical coloration occurs, providing a method for determining Grignard compounds[235]. After mixing with a 10% solution of Michler's ketone in dry chloroform the sample is hydrolysed with methanol–acetic acid–water (70:20:10). Then a solution of iodine in acetic acid is added and the concentration of unreacted Michler's ketone is determined at 615 nm.

D. Organomercury Compounds

The marked difference in absorption between diphenylmercury and phenylmercury compounds in the 226 nm region makes possible a determination of diphenylmercury in the presence of phenylmercury compounds[236,237]. Methods have been described for the determination of organomercury dusts and vapours in air. The sample of air is drawn through a furnace at 800°C in which the organomercury compounds are decomposed to metallic mercury, and finally through a u.v. spectrophotometer in which total mercury (original organo- and metallic mercury) is determined[238]. Acidified aqueous solutions of phenylmercury acetate exhibit absorption maxima at 250, 256, and 262 nm; a method suitable for the determination of 0.01–0.1 g of phenylmercury acetate with an accuracy of ±1% is based on measurements at 256 and 262 nm[178]. From detailed study of interference effects in this determination it was concluded that the addition of perchloric acid to the sample solution considerably reduces the error caused by impurities. Diphenylmercury and phenylmercury chloride do not interfere in the determination of phenylmercury acetate by this procedure.

E. Organotin Compounds

Some u.v. absorption bands for organotin compounds have been discussed[239]. It is now generally accepted that, in organodistannanes, there is intense absorption associated with the Sn—Sn bond which is not dependent upon the presence of aromatic groups joined to tin[240–242], although no observation of maxima in hexabutyldistannane was possible[242]. Similar absorption bands are observed with compounds in which tin is joined to other Group IVB metals. The intense absorptions recorded for some simple butyltin compounds are remarkable[239] and possible origins of these bands have been discussed[243]. Detailed analyses have been made of the u.v. absorption spectra of the phenyltin chlorides[244], of some vinyltin compounds[245], of compounds of the type $Me(SnMe_2)_nMe$[246], and also the corresponding ethyl compounds[247].

III. REFERENCES

1. E. Bonitz, *Chem. Ber.*, **88**, 742 (1955).
2. W. P. Neumann, *Angew. Chem.*, **69**, 730 (1957).
3. W. P. Neumann, *Justus Liebigs Ann. Chem.*, **629**, 23 (1960).
4. W. P. Neumann, *Dissertation*, University of Geissen, 1959.
5. J. H. Mitchen, *Anal. Chem.*, **33**, 1331 (1961).
6. C. N. Wadelin, *Talanta*, **10**, 97 (1962).
7. K. Ziegler and H. Gellert, *Justus Liebigs Ann. Chem.*, **629**, 20 (1960).
8. D. F. Hagen and W. D. Leslie, *Anal. Chem.*, **35**, 814 (1963).
9. G. W. Hennisch, *Anal. Chem.*, **46**, 1018 (1974).
10. G. Bruekner and M. Párkany, *Magy. Kem. Foly.*, **68**, 164 (1962).
11. J. W. Cavett, *J. Assoc. Off. Agric. Chem.*, **39**, 857 (1956).
12. C. E. Stringer and M. Attrep, *Anal. Chem.*, **51**, 731 (1979).
13. M. M. Auerbach and W. W. Haughtaling, *Drug Stand.*, **28**, 115 (1960).
14. W. H. Hill, L. J. Kuhns, J. M. Merrill, B. J. Palm, J. Seals, and U. Urquiza, *Am. Ind. Hyg. Assoc. J.*, **21**, 231 (1960).
15. L. J. Kuhns, R. H. Forsyth, and I. Masi, *Anal. Chem.*, **28**, 1750 (1956).

16. H. G. Offner, *Anal. Chem.*, **37**, 370 (1965).
17. W. H. Hill and M. S. Johnston, *Anal. Chem.*, **27**, 1300 (1955).
18. R. S. Braman, D. D. De Ford, T. N. Johnston, and L. J. Kuhns, *Anal Chem.*, **32**, 1258 (1960).
19. E. A. Pfitzer and J. M. Seals, *Am. Ind. Hyg. Assoc. J.*, **20**, 329 (1959).
20. R. S. Braman and T. N. Johnston, *Talanta*, **10**, 810 (1963).
21. W. H. Hill, J. M. Merrill, and R. H. Larsen, *U.S. Atomic Energy Comm.*, Rep. CCC-1024-TR-228, 1957, 13 pp.
22. D. L. Hill, E. I. Gipson, and J. F. Heacock, *Anal. Chem.*, **28**, 133 (1956).
23. W. H. Hill, J. M. Merrill, and B. J. Palm, *U.S. Atomic Energy Comm.*, Rep. CCC-1024-TR223, 1957, 10 pp.
24. D. Thierig and F. Umland, *Z. Anal. Chem.*, **215**, 24 (1966).
25. A. H. Soloway and J. R. Messer, *Anal. Chem.*, **36**, 433 (1964).
26. G. R. Friston, L. Bennett, and W. G. Bert, *Anal. Chem.*, **31**, 1696 (1959).
27. J. G. Heathcote and P. J. Duff, *Analyst (London)*, **79**, 727 (1954).
28. H. L. Newark and M. Leff, *Drug Stand.*, **25**, 177 (1957).
29. C. F. Bruening, W. L. Hall, and O. L. Kline, *J. Am. Pharm. Assoc. Sci. Ed.*, **47**, 15 (1958).
30. C. F. Bruening and O. L. Kline, *J. Pharm. Soc.*, **50**, 537 (1961).
31. J. Bayer, *Chimia*, **15**, 555 (1961).
32. M. Covello and O. Schettino, *Ann. Chim. (Rome)*, **52**, 1135 (1962).
33. H. Brink, *Pharm. Weekbl.*, **97**, 505 (1962).
34. J. Bayer, *Gyogyszereszet*, **8**, 21 (1964).
35. J. Bayer, *Pharmazie*, **19**, 602 (1964).
36. G. Asensi, *Quim. Ind. (Bilbao)*, **12**, 141 (1965).
37. D. Monnier, Y. Ghaliounglu, and R. Saba, *Anal. Chim. Acta*, **28**, 30 (1963); *Anal. Abstr.*, **10**, 5410 (1963).
38. T. Rahandraha, M. Chanez, and M. Sagot-Masson, *Ann. Pharm. Fr.*, **22**, 663 (1964).
39. R. K. Mitra, P. C. Bose, G. K. Ray, and B. Mukerji, *Indian J. Pharm.*, **24**, 152 (1962).
40. N. E. Dowd, A. M. Killard, H. J. Pazdera, and A. Ferrari, *Ann. N.Y. Acad. Sci.*, **130**, 558 (1965).
41. A. Sézerat, *Ann. Pharm. Fr.*, **22**, 159 (1964).
42. A. Dominquez, G. Oller, and M. Oller, *Galenica Acta*, **14**, 157 (1961).
43. G. O. Rudkin and R. J. Taylor, *Anal. Chem.*, **24**, 1155 (1952); *Anal. Abstr.*, **24**, 1155 (1952).
44. G. Parissakis and P. B. Issopoulos, *Pharm. Acta Helv.*, **40**, 653 (1965).
45. E. C. Abe, A. A. Raheem, and M. M. Dokhana, *Z. Anal. Chem.*, **189**, 389 (1962).
46. V. Brustier, A. Castaigne, E. de Montety, and A. Anselem, *Ann. Pharm. Fr.*, **22**, 373 (1964).
47. V. N. Danilova, *Ukr. Khim. Zh.*, **30**, 651 (1964).
48. A. K. Babko and N. M. Lukovskaya, *Zh. Anal. Khim.*, **17**, 50 (1962); *Anal. Abstr.*, **9**, 4078 (1962).
49. A. K. Babko and N. M. Lukovskaya, *Zavod. Lab.*, **29**, 4104 (1963); *Anal. Abstr.*, **11**, 1300 (1964).
50. W. J. Mader and R. G. Johl, *J. Am. Pharm. Assoc. Sci. Ed.*, **44**, 577 (1955).
51. J. Oslet-Conter, *Arch. Pharm. Chem.*, **68**, 529 (1961).
52. E. L. Smith, J. L. Martin, R. J. Gregory, and W. H. C. Shaw, *Analyst (London)*, **87**, 183 (1962).
53. G. Bellomonte and E. Cingolani, *R.C. Ist. Sup. Sanit.*, **26**, 1050 (1963).
54. W. U. Malik and R. Haque, *Z. Anal. Chem.*, **189**, 179 (1962).
55. W. U. Malik and S. I. Ahmed, *Indian J. Chem.*, **2**, 247 (1964).
56. W. U. Malik and R. Haque, *Z. Anal. Chem.*, **189**, 179 (1962); *Anal. Abstr.*, **10**, 1100 (1963).
57. W. U. Malik and S. I. Ahmed, *J. Am. Oil. Chem. Soc.*, **42**, 451 (1965).
58. W. U. Malik and S. I. Ahmed, *J. Am. Oil. Chem. Soc.*, **42**, 454 (1965).
59. A. G. Kempton, M. Greenberger, and A. M. Kaplan, *Am. Digest Rep.*, **51**, 19 (1952).
60. E. Hoffman, A. Saracz, and B. Z. Bursztyn, *Anal. Chem.*, **215**, 101 (1966).
61. E. Hoffman, A. Saracz, and Z. Bursztyn, *Z. Anal. Chem.*, **208**, 431 (1965); *Anal. Abstr.*, **13**, 3672 (1966).

62. British Standards Institution, BS 3769: 1964, 20 pp.
63. V. G. Hiatt, *J. Assoc. Off. Agric. Chem.*, **47**, 253 (1964).
64. T. N. Gladyshevskaya, Yas. Kvyatkovskaya, and B. A. Sobchuk, *Ukr. Biokhim Zh.*, **29**, 317 (1957); *Ref. Zh. Khim.*, Abstract No. 10, 852 (1958).
65. V. V. Popov and B. A. Sobshuk, *Lab. Delo*, **3**, 19 (1957).
66. O. O. Clegg, *Br. Med. J.*, **2**, 329 (1942).
67. J. F. Scaife, *Analyst (London)*, **80**, 562 (1955).
68. E. J. van Kampen and W. G. Zijlstra, *Clin. Chim. Acta*, **6**, 538 (1961).
69. E. J. van Kampen and W. G. Zijlstra, *Clin. Chim. Acta*, **6**, 538 (1961); *Anal. Abstr.*, **9**, 294 (1962).
70. E. J. van Kampen and W. G. Zijlstra, *Clin. Chim. Acta*, **7**, 147 (1962).
71. N. Ressler, N. A. Nelson, and I. M. Smith, *J. Lab. Clin. Med.*, **54**, 304 (1959).
72. W. G. Zijlstra and E. J. van Kampen, *Clin. Chim. Acta*, **7**, 96, (1962).
73. W. G. Zijlstra and E. J. van Kampen, *Clin. Chim. Acta*, **5**, 719 (1960).
74. W. Pilz, I. Johann, and E. Stelzl, *Z. Anal. Chem.*, **215**, 260 (1966).
75. W. Pilz, I. Johann, and E. Stelzl, *Z. Anal. Chem.*, **215**, 105 (1966); *Anal. Abstr.*, **14**, 1627 (1967).
76. O. Stadic, *J. Biol. Chem.*, **41**, 237 (1920).
77. L. Tentori, G. Vivaldi, and A. M. Salvati, *Clin. Chim. Acta*, **14**, 276 (1966).
78. G. Vanzetti and A. Nordeschi, *J. Lab. Clin. Med.*, **67**, 116 (1966).
79. H. V. Connerly and A. R. Briggs, *Clin. Chem.*, **8**, 151 (1962).
80. A. Kumlien, K. Paul, and S. Ljungberg, *J. Clin. Lab. Invest.*, **12**, 381 (1960).
81. P. C. Elwood and A. Jacobs, *Br. Med. J.*, **1**, 20 (1966).
82. British Standards Institution, BS 3420: 1961.
83. W. Steffens, *Naturwissenschaften.*, **49**, 109 (1962).
84. M. Harboe, *Scand. J. Clin. Lab. Invest.*, **11**, 66 (1959).
85. G. E. Hanks, M. Cassell, and H. Chaplin, *J. Lab. Clin. Med.*, **56**, 486 (1960).
86. G. V. Derviz and N. K. Byalko, *Lab. Delo*, **8**, 461 (1966).
87. T. R. Johnson, *J. Lab. Clin. Med.*, **53**, 495 (1959).
88. W. Pollmann, *Klin. Wochenschr.*, **44**, 789 (1966).
89. K. B. McCall, *Anal. Chem.*, **28**, 189 (1956).
90. A. Martin and M. Zade-Oppen, *Acta Soc. Med. Ups.*, **65**, 249 (1960).
91. A. F. Beau, *Am. J. Clin. Pathol.*, **38**, 111 (1962).
92. J. M. McKenzie, P. R. Fowler, and V. Fiorica, *Anal. Biochem.*, **16**, 139 (1966).
93. N. Chamori, R. J. Henry, and O. J. Golub, *Clin. Chim. Acta*, **6**, 1 (1961).
94. O. Malloy and O. Evelyn, *J. Biol. Chem.*, **119**, 481 (1937).
95. W. E. Huckabee, *J. Lab. Clin. Med.*, **46**, 486 (1955).
96. G. G. Nahas, *J. Appl. Physiol.*, **13**, 147 (1958).
97. W. G. Zijlstra, *Klin. Wochenschr.*, **34**, 384 (1956).
98. H. E. Refsum and S. L. Sveinssen, *Scand. J. Clin. Lab. Invest.*, **8**, 67 (1956).
99. R. G. Martinek, *J. Am. Med. Technol.*, **21**, 42 (1966).
100. O. Wolff, *Ann. Méd. Lég.*, 221 (1947).
101. T. P. Whitehead and S. Worthington, *Clin. Chim. Acta*, **6**, 356 (1961).
102. F. Vogt-Lorentzen, *Scand. J. Clin. Lab. Invest.*, **14**, 648 (1962).
103. L. Heilmeyer, in *Spectrometry in Medicine*, Adam Hilger, London, 1943.
104. H. Hartmann, *Ergeba Physiol.*, **39**, 413 (1937).
105. B. Steinmann, *Archw. Exp. Pathol. Pharm.*, **191**, 237 (1939).
106. P. L. Drabkin, in *Medical Physics*, Vol. II, Chicago Book Publishers, Chicago, 1950.
107. B. T. Commins and P. J. Lawther, *Br. J. Ind. Med.*, **22**, 139 (1965).
108. R. A. Knight, D. C. Ephraim, and F. E. Payne, *W. G. Pye Unicam Anal. News*, Vol. 3 (Dec. 1973).
109. R. Richterich, in *Clinical Chemistry*, Academic Press, New York, 1969.
110. A. A. Christman and E. L. Randall, *J. Biol. Chem.*, **102**, 595 (1933).
111. R. Wennesland, *Acta Physiol. Scand.*, **1**, 49 (1940).
112. C. H. Gray and H. Sandiford, *Analyst (London)*, **71**, 107 (1946).
113. M. T. Ryan, J. Nolan, and E. J. Conway, *Biochem. J.*, **42**, 94 (1948).
114. L. A. Williams, R. A. Linn, and B. Zak, *Am. J. Clin. Pathol.*, **34**, 334 (1960).
115. I. I. Datsenko, *Lab. Delo*, **9**, 8 (1963).

116. W. G. Zijlstra and C. J. Moller, *Clin. Chim. Acta*, **2**, 237 (1957).
117. G. E. Martin, J. I. Munn, and L. Biskup, *J. Assoc. Off. Agric. Chem.*, **43**, 743 (1960).
118. J. C. Kaplan, *Rev. Fr. Etud. Clin. Biol.*, **10**, 856 (1965).
119. G. V. Derviz, *Lab. Delo*, **9**, 527 (1966).
120. T. Leahy and R. Smith, *Clin. Chem.*, **6**, 148 (1960).
121. O. Evelyn and O. Malloy, *J. Biol. Chem.*, **126**, 655 (1960).
122. O. Crosby, *U.S. Armed Forces Med. J.*, **4**, 693 (1954).
123. J. Sendroy, H. A. Collison, and H. J. Mark, *Anal. Chem.*, **27**, 1641 (1955).
124. R. S. Brief, R. S. Ajemain, and R. G. Conter, *Am. Ind. Hyg. Assoc. J.*, **28**, 21 (1967).
125. A. B. Densham, P. A. A. Beale, and R. Palmer, *J. Appl. Chem.*, **13**, 576 (1963).
126. United Kingdom Atomic Energy Authority, Production Group Chemical Services Technical Department, Windscale, *UKAEA Rep.*, PG 483 (W), 1963, 9 pp.
127. H. Irving and J. J. Cox, *J. Chem. Soc.*, 1470 (1961).
128. R. Barbieri, G. Tagliavini, and U. Belluco, *Ric. Sci.*, **30**, 1963 (1960).
129. S. R. Henderson and L. J. Snyder, *Anal. Chem.*, **33**, 1172 (1961).
130. W. N. Aldridge and J. E. Cremer, *Analyst (London)*, **80**, 37 (1957).
131. J. E. Cremer, *Br. J. Ind. Med.*, **16**, 191 (1959).
132. S. R. Henderson and L. J. Snyder, *Anal. Chem.*, **31**, 2113 (1959).
133. M. E. Griffing, A. Rozek, L. J. Snyder, and S. R. Henderson, *Anal. Chem.*, **29**, 190 (1957).
134. G. Pilloni and G. Plazzogna, *Anal. Chim. Acta*, **35**, 325 (1966).
135. G. Pilloni, *Farmaco, Ed. Prat.*, **22**, 666 (1967).
136. R. M. Dagnall, T. S. West, and P. Young, *Talanta*, **12**, 583 (1965).
137. G. Pilloni, *Anal. Chim. Acta*, **37**, 497 (1967).
138. L. Roncucci, G. Faraglia, and R. Barbieri, *J. Organomet. Chem.*, **1**, 427 (1964).
139. D. Blake, G. E. Coates, and J. M. Tate, *J. Chem. Soc.*, 756 (1961).
140. D. L. Alleston and A. G. Davies, *J. Chem. Soc.*, 2050 (1962).
141. T. Tanaka, M. Komuna, Y. Kawasaki, and R. Okawara, *J. Organomet. Chem.*, **1**, 484 (1964).
142. M. Yasuda and R. S. Tobias, *Inorg. Chem.*, **2**, 207 (1963).
143. Y. Kawaskai, T. Tanaka, and R. Dkawara, *Bull. Chem. Soc. Jap.*, **27**, 903 (1964).
144. W. Bolanowska, *Chem. Anal. (Warsaw)*, **12**, 121 (1967).
145. V. A. Smith, W. E. Dilaney, W. J. Tanag, and J. C. Bailie, *Anal. Chem.*, **22**, 1230 (1950).
146. D. Loroue and G. Paul, *Fr. Tech. Pet. Bull. Assoc.*, **127**, (1963).
147. D. Loroue and G. Paul, *Rev. Inst. Fr. Pet. Ann. Combust. Liq.*, **17**, 830 (1962).
148. R. Moss and E. V. Browett, *Analyst (London)*, **91**, 428 (1966).
149. L. J. Snyder and S. R. Henderson, *Anal. Chem.*, **33**, 1175 (1961).
150. E. V. Gernet, *Nauchn. Roboty Khim. Lab. Gor'kovsk Nauchn. Issled. Inst. Gigieny Tr. Prof. Boleznei*, **6**, 5 (1957).
151. V. A. Khrustaleva, *Metody Opred. Vredn. Veshehestv Voz Dukhe, Moscow Sb.*, 105 (1961).
152. V. A. Khrustaleva, *Gig. Sanit.*, **25**, 57 (1960).
153. W. Dublinski and Z. Könsling, *Gaz. Woda Tech. Sanit.*, **34**, 258 (1960).
154. A. L. Linch, R. B. Davies, R. F. Stalzer, and W. F. Anzilotti, *Am. Ind. Hyg. Assoc. J.*, **25**, 81 (1964).
155. M. Pedinelli, *Chim. Ind. (Milan)*, **41**, 1180 (1959).
156. V. L. Miller and D. Polley, *Anal. Chem.*, **26**, 1333 (1954).
157. E. Hoffman, *Z. Anal. Chem.*, **174**, 48 (1960).
158. S. Ishikuro and K. Yokata, *Chem. Pharm. Bull. Jap.*, **11**, 939 (1963).
159. M. Iguchi, A. Nishiyama, and Y. Nagase, *J. Pharm. Soc. Jap.*, **80**, 1437 (1960).
160. E. J. Cafruny, *J. Lab. Clin. Med.*, **57**, 468 (1961).
161. V. L. Miller and D. Lillis, *Anal. Chem.*, **30**, 1705 (1958).
162. E. Hoffman, *Z. Anal. Chem.*, **182**, 193 (1961).
163. Y. Kimura and V. L. Miller, *Anal. Chim. Acta*, **27**, 325 (1962).
164. J. Story, K. Kratzer, and J. Prasilova, *Anal. Chim. Acta*, **100**, 643 (1978).
165. Metallic Impurities in Organic Matter Sub-Committee of Analytical Methods Committee, *Analyst (London)*, **90**, 515 (1965).
166. M. G. Ashley, *Analyst (London)*, **84**, (1959).
167. Y. Kimura and V. L. Miller, *Anal. Chem.*, **32**, 420 (1960).

168. Y. Kimura and V. L. Miller, *Agric. Food Chem.*, **12**, 253 (1964).
169. D. Polley and V. L. Miller, *Anal. Chem.*, **27**, 1162 (1955).
170. D. Polley and V. L. Miller, *J. Agric. Food Chem.*, **2**, 1030 (1954).
171. V. L. Miller and F. Swanberg, *Anal. Chem.*, **29**, 391 (1957).
172. T. Etchu, *Bull. Agric. Chem. Insp. Stn.*, **7**, 17 (1967).
173. T. Etchu, *Bull. Agric. Chem. Insp. Stn.*, **7**, 21 (1967).
174. T. Etchu, *Bull. Agric. Chem. Insp. Stn.*, **7**, 25 (1967).
175. D. Polley and V. L. Miller, *Anal. Chem.*, **24**, 1622 (1952).
176. M. Vancea and A. German, *Rev. Chim.*, **19**, 58 (1968).
177. K. Ozawa and S. Egoshira, *Bunseki Kagahu*, **11**, 506 (1962).
178. A. Eldridge and T. R. Sweet, *Anal. Chem.*, **28**, 1268 (1956).
179. I. M. Kolthoff and E. B. Sandell, *Text Book of Inorganic Quantitative Analysis*, 3rd ed., Macmillan, New York, 1952, p. 461.
180. G. Gran, *Sven. Papperstidn.*, **53**, 234 (1950).
181. G. Rentsch, *Z. Anal. Chem.*, **178**, 100 (1960).
182. J. E. McCarley, R. S. Saltzman, and R. H. Osborn, *Anal. Chem.*, **28**, 880 (1956).
183. V. S. Fikhtengol'ts and N. P. Kozlov, *Zavod. Lab.*, **23**, 917 (1957).
184. H. Suszuki and K. Oishi, *Bunseki Kagaku*, **12**, 1011 (1963).
185. R. S. Brief, F. S. Venable and R. S. Ajemian, *Am. Ind. Hyg. Assoc. J.*, **26**, 72 (1965).
186. M. Masoero and M. Perini, *Chim. Ind. (Milan)*, **37**, 945 (1955).
187. S. A. Kerfedt, *Acta Chem. Scand.*, **13**, 1479 (1959).
188. A. Gomori, *J. Lab. Clin. Med.*, **27**, 955 (1942).
189. M. Kawamura and K. Matsumoto, *Bunseki Kagaku*, **14**, 789 (1965).
190. K. L. Cheng, *Chemist Analyst*, **45**, 67 (1956); *Anal. Abstr.*, **4**, 1189 (1957).
191. A. J. Bawd, D. T. Burns, and A. G. Fogg, *Talanta*, **16**, 719 (1969).
192. B. L. Pepe and R. Barbieri-Rivarola, *Anal. Chem.*, **40**, 432 (1968).
193. W. N. Aldridge and J. E. Cremer, *Nature (London)*, **178**, 1306 (1956).
194. W. N. Aldridge and J. E. Cremer, *Analyst (London)*, **82**, 37 (1957).
195. E. B. Sandell, *Colorimetric Determination of Trace Metals*, Interscience, New York, London, 1944, p. 90.
196. R. Barbieri, G. Tagliavini, and U. Belluco, *Ric. Sci.*, **30**, 1963 (1960).
197. H. Irving and J. J. Cox, *J. Chem. Soc.*, 1470 (1961).
198. R. T. Skeel and C. E. Bricker, *Anal. Chem.*, **33**, 428 (1961).
199. G. L. Snable, *Senior Thesis*, Princeton University, 1959.
200. G. Pilloni and G. Plazzogna, *Anal. Chim. Acta*, **35**, 325 (1966).
201. G. Pilloni, *Anal. Chim. Acta*, **37**, 497 (1967).
202. G. Pilloni, *Farmaco, Ed. Prat.*, **22**, 666 (1967).
203. R. S. Tobias, I. Ogrins, and B. A. Nevett, *Inorg. Chem.*, **1**, 638 (1962).
204. R. S. Tobias and M. Yasuda, *J. Phys. Chem.*, **68**, 1820 (1964).
205. R. S. Tobias and M. Yasuda, *Can. J. Chem.*, **42**, 781 (1964).
206. R. S. Tobias, *Organomet. Chem. Rev.*, **1**, 93 (1966).
207. A. Corsino, I. May-Ling Yih, Q. Fernando, and H. Frieser, *Anal. Chem.*, **34**, 1090 (1962).
208. D. Betteridge, Q. Fernando, and H. Freiser, *Anal. Chem.*, **35**, 294 (1963).
209. T. Tanaka, M. Komura, Y. Kawasokai, and R. Okawara, *J. Organomet. Chem.*, **1**, 484 (1964).
210. L. Roncucci, G. Faraglia, and R. Barbieri, *J. Organomet. Chem.*, **1**, 427 (1964).
211. D. Blake, G. E. Coates, and J. M. Tate, *J. Chem. Soc.*, 756 (1961).
212. D. L. Alleston and A. G. Davies, *J. Chem. Soc.*, 2050 (1962).
213. M. Yasuda and R. S. Tobias, *Inorg. Chem.*, **2**, 207 (1963).
214. Y. Kawasaki, T. Tanaka, and R. Okawara, *Bull. Chem. Soc. Jap.*, **27**, 903 (1964).
215. A. Cassol and T. Magon, *J. Inorg. Nucl. Chem.*, **27**, 1297 (1965).
216. M. Frankel, D. Wagner, D. Gertner, and A. Zilha, *Israel J. Chem.*, **4**, 183 (1966).
217. A. H. Chapman, M. W. Duckworth, and J. W. Price, *Br. Plast.*, **32**, 78 (1959).
218. R. Sawyer, *Analyst (London)*, **92**, 569 (1967).
219. O. R. Klimmer and I. U. Nebel, *Arzneim.-Forsch.*, **10**, 44 (1960).
220. G. Gras and J. Castel, *Soc. Pharm. Montpellier*, **25**, 178 (1965).
221. J. H. Adamson, *Analyst (London)*, **87**, 597 (1962).
222. H. J. Hardon, H. Brunink, and E. W. Van der Pol, *Analyst (London)*, **85**, 847 (1960)

223. H. J. Hardon, A. F. H. Bessemer, and H. Brunik, *Dtsch. Lebensm.-Rundsch.*, **58**, 349 (1962).
224. P. I. Selivokhin, *Vestn. Tekhn. Ekon. Inform. Nauchno-Issled. Inst. Tekh.-Ekon. Issled. Gos. Kom. Khim. Prom. Gosplane, SSSR*, **8**, 27 (1964).
225. P. Selivokhin, *Gig. Sanit.*, **31**, 68 (1966).
226. S. J. Blunden and A. H. Chapman, *Analyst (London)*, **103**, 1266 (1978).
227. J. W. Zijp, *IRI Proc.*, **6**, 108 (1959).
228. M. C. Kerssen and P. Riepma, *Tijdschr. Plantenziekten*, **65**, 27 (1959).
229. G. C. Pimental, *Anal. Chem.*, **17**, 882 (1949).
230. J. R. Urwin and P. J. Reed, *J. Organomet. Chem.*, **15**, 1 (1968).
231. C. Blomberg, A. D. Vrengdenhil, and P. Vink, *Rec. Trav. Chim. Pays-Bas*, **83**, 662 (1964).
232. A. D. Vrengdenhil and C. Blomberg, *Rec. Trav. Chim. Pays-Bas*, **82**, 453 (1963).
233. D. Hollander and M. Anteunis, *Bull. Soc. Chim. Belg.*, **72**, 77 (1963).
234. H. Flaschka, *Mikrochemie*, **39**, 38 (1952).
235. O. Datta and O. Mitten, *J. Am. Chem. Soc.*, **41**, 287 (1919).
236. B. G. Gowenlock, *Anal. Abstr.*, **1**, 3020 (1954).
237. B. G. Gowenlock, *J. Chem. Soc.*, 1454 (1955).
238. G. A. Hamilton and A. D. Ruthven, *Lab. Pract.*, **15**, 995 (1966).
239. R. C. Poller, in *The Chemistry of Organotin Compounds*, Section 13, Logos Press, 1970.
240. H. M. J. C. Creemers and J. C. Noltes, *J. Organomet. Chem.*, **7**, 237 (1967).
241. D. N. Hague and R. H. Prince, *J. Chem. Soc.*, 4690 (1965).
242. W. Drenth, M. J. Janssen, G. J. M. van der Kerk, and J. A. Vliegenthart, *J. Organomet. Chem.*, **2**, 265 (1964).
243. C. W. N. Cumper, A. Meinikoff, and A. I. Vogel, *J. Chem. Soc. A*, 242 (1966).
244. V. S. Griffiths and G. A. W. Derwish, *J. Mol. Spectrose.*, **3**, 165 (1959).
245. V. A. Petukhov, V. F. Mironov, and A. L. Kravehenko, *Izv. Akad. Nauk SSSR, Ser. Khim.*, 156 (1966).
246. P. P. Shorygin, V. A. Petukhov, O. M. Nefedov, S. P. Kolesnikov, and V. I. Shiryaev, *Teor. Eksp. Khim. Akad. Nauk Ukr. SSR*, **2**, 190 (1966); *Chem. Abstr.*, **65**, 14660g (1966).
247. W. Drenth, J. G. Noltes, E. J. Bulten, and H. M. J. C. Creemers, *J. Organomet. Chem.*, **17**, 173 (1969).

The Chemistry of the Metal–Carbon Bond
Edited by F. R. Hartley and S. Patai

CHAPTER **17**

Analysis of organometallic compounds: polarographic techniques

T. R. CROMPTON
Beechcroft, Whittingham Lane, Goosnargh, Preston, Lancashire, U.K.

I. ORGANOARSENIC COMPOUNDS

Substituted diarsines ($R_2As—AsR_2$) can be determined polarographically[1], the $E_{1/2}$ of the anodic wave being independent of the nature of the substituent. The analysis must be carried out in the absence of oxygen, which oxidizes the As—As bond. Concentrations of diarsines down to about 10^{-4} M can be determined by this procedure. Various methods have been described for the determination of arsine in air and other gases. Various absorbing solutions have been used by different workers, including mixtures of potassium permanganate, concentrated sulphuric acid and bromine[2], a mixture of silver nitrate and silver diethyldithiocarbamate[3], and a solution of 1 N ammonium nitrate (9:1 *v*/*v*) in 95% ethanol[4]. The last solution is also a suitable medium for the polarographic determination of arsine. The method is sensitive enough to determine down to 5×10^{-4} M of arsine in gas mixtures. Phosphine interferes in this determination.

Differential pulse polarography has a detection limit of 10^{-8} M at pH 7.3 for phenylarsine oxide with a relative standard deviation of 1.7%[5]. Arsenic(III) and arsenic(V), monomethylarsonate, and dimethylarsinate have been determined by differential pulse polarography after separation by ion-exchange chromatography[6], detection limits for the latter two are 18 and 8 ppb, respectively. Pulse polarographic methods have been applied to aqueous and non-aqueous solutions of methylarsenic and dimethylarsenic acids at concentration levels down to 0.1 μg/ml[7]. Diphenylarsenic acid has been studied polarographically[8].

II. ORGANOBORON COMPOUNDS

A polarographic wave of $E_{1/2} = -1.55$ V *vs*. the mercury-pool electrode is obtained from a solution of potassium tetraphenylboron in dimethylformamide with tetrabutylammonium iodide as the supporting electrolyte[9]. The wave heights are proportional to concentration over the range 0.0002–0.0075 M, corresponding to 0.08–8.0 mg of potassium in the cell. Interference is caused by ammonium, rubidium, and caesium. The precision is within 3.0%.

III. ORGANOIRON COMPOUNDS

Various workers have described polarographic methods for the determination of ferrocene[10] and nitroferrocene[11]. A voltammetric method with a rotating platinum electrode was used for the determination of ferrocene in dimethylformamide medium[10]. Using 0.1 M sodium perchlorate in dimethylformamide as the base electrolyte, $E_{1/2} = +0.88$ V (reversible wave) relative to the silver–silver chloride saturated tetraethylammonium chloride electrode. Polarographic reduction of nitroferrocene in neutral of alkaline buffer solution is a diffusion-controlled, concentration-dependent, six-electron process.[11] The half-wave potential of the reduction moves 58 mV per pH unit. *In situ* electrochemical reduction of nitroferrocene within an electron spin resonance spectrometer produced an unstable radical. In aqueous buffer solution, nitroferrocene undergoes photochemical decomposition.

In a polarographic method for the determination of iron(III) dimethyldithiocarbamate (Ferbam), a freshly prepared acetone solution of Ferbam is diluted with a solution of disodium hydrogen phosphate and trisodium citrate and the solution is analysed by conventional polarography at 0.8–0.2 V *vs*. the saturated calomel electrode (SCE)[12]. Alternatively, a solution of sodium acetate and trisodium citrate is used to dilute the acetone solution of the sample and the mixture is analysed by cathode-ray polarography with a start potential of 0.5 V *vs*. the mercury pool. The limits of detection of Ferbam by the conventional and cathode-ray polarographic procedures are reported to be 2 and 0.02 μg/ml, respectively.

IV. ORGANOLEAD COMPOUNDS

In a procedure for the determination of lead in petroleum and lubricants, the lead is extracted with Schwartz reagent (potassium chlorate and sodium chloride in nitric acid) and then, after evaporation of the acidic solution, determined either polarographically in the presence of ammonium acetate and magenta or amperometrically by titrating with 0.01 M potassium dichromate in the presence of potassium nitrate[13]. By these methods increased accuracy and reduced time of operation are secured. With petroleum, agreement is obtained with standard methods within the specified limits of ±0.04 ml per imperial gallon. In a rapid polarographic method for the determination of tetraethyllead in petroleum the sample is dissolved in 2-ethoxyethanol (Cellulose)–hydrogen chloride solution, which simultaneously decomposes the tetraethyllead to lead chloride and extracts the latter[14]. If the 2-ethoxyethanol is cooled in an ice-bath during acidification with anhydrous hydrogen chloride, the residual current can be measured with great reliability.

In an alternative rapid polarographic method for the determination of tetraethyllead in petroleum at concentrations between 80 and 200 mg/l, the tetraethyllead is decomposed to lead chloride by the direct addition of concentrated hydrochloric acid to the petroleum[15]. The mixture is shaken, refluxed, then extracted with water, and 0.5% gelatin is added to the combined aqueous extracts. A portion of this solution is deoxygenated by nitrogen purging in the polariographic cell and the wave recorded between −0.2 and 0.7 V. A 0.02% solution of dithizone in chloroform at pH 8.5–9.0 in the presence of citrate to mask iron has been used to extract lead from petroleum prior to its determination by differential oscillopolarography[16]. The chloroform in the extract is volatilized, the brown residue is heated to fumes with 65% nitric acid – 70% perchloric acid (1:1), and the resulting white residue is dissolved rapidly and completely in the basal electrolyte (0.5 M ammonium tartrate – 0.1 M tartaric acid). The oscillopolarogram is recorded between −0.2 and 0.7 V. When thallium is present the basal electrolyte used to dissolve the white residue must be changed to EDTA–acetate buffer solution (1:1) so that the waves of lead and thallium are clearly separated. In a further procedure a slight excess of a 10% solution of bromine in carbon tetrachloride is added to the petroleum sample[17]. The precipitate of lead bromide obtained is washed with 50% ethanol, then dissolved in and evaporated with concentrated nitric acid until a white residue remains. This residue is dissolved in water, and the solution is titrated with 0.1 N potassium chromate by a conductometric or oscillometric techniche. Other polarographic procedures for the determination of lead in petroleum have been described[18–20].

A procedure for the analysis of tetraethyllead involves decomposition by bromination and solution in dilute nitric acid[21]. The lead is determined by anodic decomposition and an empirical correction factor of 0.86 is applied for the conversion of lead to lead dioxide. The solution is compared with a standard by placing the two solutions in two cells connected in parallel to an adjustable potential divider via calomel electrodes and twin electrodes. A null galvanometer between the two standard electrodes indicates any e.m.f. caused by a difference in concentration of lead ions between the two solutions. This method of comparison can be applied directly to a solution of the petroleum in 2-ethoxyethanol containing hydrogen chloride[14]. The method is rapid and accurate for normal concentrations of tetraethyllead in petroleums that are not rich in unsaturated hydrocarbons. An anodic stripping technique for the determination of lead in petroleum uses a solution of bromine in chloroform to decompose the tetraethyllead and then extraction of the lead ions into 0.1 M nitric acid[22]. After suitable working up, the solution is de-oxygenated with argon and submitted to anodic-stripping voltammetry in a modified cell with an SCE as reference electrode. Pre-electrolysis is carried out at −0.8 V for 1–5 min and lead is determined at −0.39 V. Capillary tubes in the cell are made water-repellent with paraffin wax. The detection limit is 8 parts of lead per 10^{13}, with a coefficient of variation on 7%.

Hexaethyldilead in tetraethyllead and in triethyllead chloride has also been determined polarographically. Because of the ease of hydrolysis of hexaethyldilead, it is necessary to conduct the titration in anhydrous ethanol[23]. Tetraethylammonium hydroxide is used as the base electrolyte and analysis is conducted in the absence of oxygen. Under these conditions hexaethyldilead has an $E_{1/2}$ value between 1.8 and 2.0 V. Between 0.5 and 10% of this substance could be determined in mixtures of tetraethyllead and triethyllead chloride with a mean error of 7%. An alternative polarographic method for the determination of hexaethyldilead and triethyllead chloride in tetraethyllead involves a direct polarographic measurment at $E_{1/2} = -0.24$ V *vs.* SCE of the concentration of hexaethyldilead, using 1:1 *v/v* benzene–methanol solvent with lithium chloride as supporting electrolyte medium[24]. There is no interference by tetraethyllead or other lead compounds. Concentrations of hexaethyldilead equivalent to the 0.1% level in tetraethyllead samples can be determined rapidly and accurately. The concentration of triethyllead chloride can be determined from the same polarogram, as it exhibits a separate polarographic wave at 0.98 V *vs.* SCE. Samples containing hexaethyldilead can be analysed simultaneously for triethyllead chloride[24]. Oxygen is a common contaminant and is reduced in the voltage range corresponding to the $E_{1/2}$ of triethyllead chloride. Care must be exercised to remove oxygen completely for this determination. The second wave for triethyllead chloride exhibited a maximum which could be suppressed by Triton X-100. No use was made of a suppressor in the measurements for triethyllead chloride as the pre-wave at −0.98 V could be measured without interference from the maximum.

The polarographic determination of tetraethyllead in air involves trapping the tetraethyllead from the air sample in fuming nitric acid[25]. Polarographic techniques have used for the analysis of leachates of antifouling paints containing triphenyllead acetate[26].

V. ORGANOMERCURY COMPOUNDS

In a polarographic method for determining phenylmercury halides in fungicidal preparations, a dimethylformamide extract of the sample is prepared and to the extract is added a solution of lithium monohydrate and 0.5% gelatin[27]. The mixture is diluted with lithium hydroxide and nitrogen is passed through for 25 min. The polarogram is then recorded at 25°C from −0.1 to −0.8 V; the error is ±1.5%. Polarography has also been used for the determination of down to 4 μg/ml of phenylmercury chloride[28]. The phenylmercury chloride is extracted into chloroform from acidified aqueous solution, the chloroform is removed, and the residue, dissolved in ethanol, is treated with 0.1 M potassium chloride–boric acid–sodium hydroxide buffer of pH 10 containing a small amount of Triton X-100. The polarogram is recorded between −0.4 and −1.6 V *vs.* SCE. Mercury(II) chloride is not extracted with chloroform but can be identified by re-extracting the aqueous sample solution with diethyl ether, evaporating the ether, dissolving the residue in dilute acid, and testing for mercury with thioacetamide.

In a polarographic procedure for the determination of ethylmercury chloride fungicide in mixtures with talc and mineral oil, fungicide is digested with dimethylformamide, a portion of the filtrate and a buffer solution (0.1 N boric acid–0.1 N sodium hydroxide, 1:3) are mixed, and nitrogen is passed through to sweep out dissolved oxygen[29]. The polarogram is recorded between −0.2 and −0.7 V at a sensitivity of 0.05. Ethylmercury chloride has an $E_{1/2}$ value of −0.435 V *vs.* SCE. Calibration graphs are prepared under identical conditions using recrystallized ethylmercury chloride as a reference standard.

Ethylmercury chloride and methoxyethylmercury chloride in mixtures have been determined polarographically in 0.1 M potassium nitrate containing Britton–Robinson buffer and 0.01% of gelatin[30]. Ethylmercury chloride gives a two-step wave ($E_{1/2}$ = 0.49 V and −1.6 V *vs*. SCE) at pH 1.9–11.8. The wave height is proportional to concentration up to 3×10^{-4} M, and is independent of pH. In the same solution methoxyethylmercury chloride gives one wave ($E_{1/2}$ = −0.49 V) at pH < 2.9, the height being proportional to concentration up to 2.5×10^{-4} M. At pH > 8, methoxyethylmercury chloride gives a two-step wave ($E_{1/2}$ = −0.49 V and −1.2 V), the height of the first wave being proportional to concentration, but almost half of that occurring at pH 2.9. Since the law of addition holds for their wave heights at −0.49 V, irrespective of the pH, a binary mixture can be analysed by measurment of the wave height at pH 2.9 and 10.0 using an empirical formula. The deviation is *ca*. 1.5%.

Polarography has been used for the determination of phenylmercury acetate. In neutral or slightly alkaline base electrolytes, phenylmercury acetate gives two waves, the $E_{1/2}$ value of the first wave being constant over a wide pH range whilst the $E_{1/2}$ of the second wave decreases with increase in both the pH and the concentration of phenylmercury acetate[31,32]. Polarographic determination of (a) ethylmercury chloride plus phenylmercury acetate in 0.2 M potassium nitrate base electrolyte at pH 10 and (b) ethylmercury chloride in 0.1 M tetramethylammonium bromide base electrolyte enables the phenylmercury acetate content of the sample to be obtained by difference[33]. In 0.1 M tetramethylammonium bromide solution of pH < 9 phenylmercury acetate exhibits a very low wave ($E_{1/2}$ = −0.27 V *vs*. SCE), with an almost constant wave height, and this does not cause interference in the determination of ethylmercury chloride ($E_{1/2}$ = −0.40 V, 0.4–1.6 mg per 10 ml). In 0.2 M potassium nitrate of pH about 10, both organomercury compounds give a reduction wave.

Polarographic and classical methods have been compared for the analysis of organomercury drugs[34,35]. A detailed study of the cathode-ray polarographic determination of merbromin showed that in Britton–Robinson buffer solution (pH 7.5)–potassium chloride solution merbromin exhibits reduction peaks at −0.28 V and −1.07 V *vs*. the silver–silver chloride anode, and the current at −1.07 V is directly proportional to concentration from 10 to 100 μg/ml of merbromin[34].

VI. ORGANOMANGANESE COMPOUNDS

In a method for the determination of cyclopentadienylmanganese tricarbonyl vapour in air, the carbonyl compound is absorbed in a fluidized bed of silica gel (0.4 mm), which permits complete absorption at high rates of flow of air (5–7 l/min)[36]. It is recovered by treatment with ethanol, of decomposed with nitric acid and the manganese is determined polarographically.

VII. ORGANONICKEL COMPOUNDS

A dropping mercury electrode polarographic procedure has been applied to the determination of nickel content of the pyridine complexes of nickel myristate and palmitate[37]. Lithium chloride or methyl hydrogen sulphate, each dissolved in benzene–methanol (1:1) or in potassium chloride in ethanediol, were used as supporting supporting electrolytes, and polarography was carried out in an atmosphere of purified nitrogen. The diffusion current varied directly with the concentration of the nickel soap in pyridine.

VIII. ORGANOCOBALT COMPOUNDS

A polarographic determination of cobalt carbonyl in the products of the Oxo process showed that dicobalt octacarbonyl gives a cathodic wave at −0.45 V *vs*. SCE in 1 M lithium chloride in isobutyl alcohol, which can be used for the determination of dicobalt octacarbonyl if the polarogram is recorded immediately after mixing the sample and the basal solution[38]. Dicobalt octacarbonyl disproportionates to $[Co(CO)_4]^-$ and solvated cobalt(II) ions in the basal medium and $[Co(CO)_4]^-$ gives an anodic wave, also at −0.45 V, but the solvated cobalt(II) ions give no wave. In this way, the concentration of dicobalt octacarbonyl (4×10^{-4} to 1.4×10^{-3} M) can be determined from the height of the anodic wave when disproportionation is complete. There is no interference from nickel tetracarbonyl, iron pentacarbonyl, or aldehydes and their oxidation products.

In a study of the polarographic behaviour of cobalt palmitate soaps in lithium chloride, methyl hydrogen sulphate, and potassium chloride base electrolytes polarograms were obtained for the pyridine complexes of the soaps in benzene-methanol (1:1) for lithium chloride and methyl hydrogen sulphate and ethanediol for potassium chloride base electrolytes[39]. Well defined waves were found, except for cobalt soaps in methyl hydrogen sulphate base electrolyte. The cobalt soaps were reducible at the dropping mercury electrode in the presence of these electrolytes. The diffusion current was a rectilinear function of concentration so that this method can be used for the determination of the metal content of soaps.

IX. ORGANOPHOSPHORUS COMPOUNDS

Alkyldithiophosphates have been determined by polarographic procedures. For compounds up to sodium dibutyldithiophosphates they may be polarographed *versus* the SCE with 0.1 N perchloric acid as base electrolyte, and a dropping mercury or a flowing junction platinum anode[40]. Electrolysis with micro-electrodes serves to isolate mercuric dialkyldithiophosphates as products from the mercury electrode and di(*OO*-dialkyldithiophosphoryl)disulphides from the platinum electrode, suggesting that the mercury takes part chemically in the reaction. The current *versus* concentration plot for the micro-electrode is linear up to 10^{-3} M. $E_{1/2}$ becomes increasingly more negative as the molecular weight of the sample increases. The electrocatalytic oxidation of dihydronicotinamide adenosine diphosphate with quinones and modified quinone electrodes has been reported[41] and a polarographic method for the determination of glyphosphate residues as their *N*-nitroso derivatives in natural waters has been described[42].

X. ORGANOSELENIUM COMPOUNDS

The polarography of 2-aminoethaneselenosulphuric acid and 2-aminoethanethiosulphuric acid has been studied under various conditions of pH, buffer composition, ionic strength, and temperature[43]. The former compound is reduced in two steps and the latter gives a single wave, all waves being irreversible at the dropping mercury electrode. Mechanisms for the electro-reduction of both compounds were given. The second wave of aminoethaneselenosulphuric acid is eliminated by the addition of a surface-active agent such as gelatine or polyacrylamide, which produces dithionate, which is inactive at the mercury electrode.

A further application of polarography to organoselenium compounds includes the determination of piazselenol (benzo-2,1,3-selenadiazole) and piazthiol (benzo-2,1,3-thiadiazole) in aqueous solutions[44]. The reduction of these compounds at the dropping mercury electrode in 0.1 M aqueous lithium perchlorate involves six electrons and yields *o*-phenylenediamine and hydrogen selenide and hydrogen sulphide, respectively.

XI. ORGANOTIN COMPOUNDS

Both a.c. and d.c. polarography have been used for the analysis of alkyltin chlorides. Tributyltin chloride in dibutyltin dichloride have been determined by a.c. polarography[45,46] in various base electrolytes, and also by d.c. polarography[47]. A commonly used base electrolyte in these methods is Britton–Robinson buffer at pH values between 9.3 and 10.3 containing also potassium chloride (0.5 M) and isopropanol (30%). Voltammetric methods for the determination of tributyltin chloride compounds at concentrations down to 0.5% (5×10^{-6} M) in dibutyltin dichloride have an average error of $\pm 5\%$[48]. The accuracy and sensitivity of a.c. polarography are claimed to be better than those obtained by d.c. polarography. By rectifying the alternating current using a phase-selective rectifier it is possible to suppress the much higher capacitive part in the a.c. polarography because of its different phase angle, leading to higher sensitivities in the reduction process. A detailed study of the a.c. polarographic capacity effects of organotin halides in alcoholic base electrolytes showed a capacity decrease caused by even low concentrations of the organotin compound[49].

A study of the polarographic behaviour of organotin compounds in strongly polar solvents and the electrode processes involved in the reduction of butyltin chlorides at the dropping mercury electrode established that butyltin trichloride, dibutyltin dichloride and tributyltin chloride are all completely hydrolysed at concentrations up to 1×10^{-4} M in aqueous solution[50,51]. Also, the reduction potentials of these organotin halides move towards more negative values in many electrolytes corresponding to decreasing polarity with an increasing degree of substitution on the tin. Parallel to an increase in polarizability, the potentials are shifted to more positive values when the chain lengths of the alkyltin derivatives increase, corresponding to a more facile reaction. The reduction of organotin compounds at the mercury electrode proceeds via two steps[51,52]. The first step involves an electron transfer of $n = 3$ for butyltin trichloride, $n = 2$ for dibutyltin dichloride, and $n = 1$ for tributyltin chloride. The second reduction wave is kinetic and strongly irreversible. Almost identical polarographic waves were obtained for compounds as different as tributyltin chloride and hexabutyldistannoxane.

Oscillographic polarography has been applied to butyltin trichloride, dibutyltin dichloride, dibutyltin diacetate, tributyltin chloride, tributyltin acetate, triphenyltin acetate and tetrabutyltin[53,54]. These substances could be determined at concentrations down to 0.005 mol-%.

The polarographic behaviour of some trialkyltin compounds of the type Et_3SnX, where X = F, Cl, Br, I, and R_3SnCl, where R = Pr, Bu, has been described[55,56]. The polarogram of, for example, triethyltin chloride shows three distinct waves, but the position and size of these waves depend on a number of variables, particularly the pH of the solution. Thus, in acidic solution up to pH 7 only the first wave is visible, whereas on increasing the pH above 7 the second and third waves appear. Further, the reduction potential of the first wave becomes progressively more electronegative with increasing pH, at the same time becoming smaller, finally disappearing at pH 12. In addition, a plot of the height of the first wave against the concentration of the trialkyltin compound is not linear, although the second and third waves give linear graphs. Unfortunately, the first wave is the best defined and it is not possible to make accurate measurements of the other waves owing to irregularities in their shape.

This work was extended to the determination of mixtures of dialkyltin and trialkyltin compounds by the examination of the polarographic behaviour of some dialkyltin compounds and by the use of a derivative circuit. With triethyltin hydroxide, a neutral solution containing isopropyl alcohol, potassium chloride, and gelatin gives three poorly defined waves which cannot be measured with any degree of accuracy. With a derivative circuit, however, the second and third waves give well defined peaks, the heights of which are directly proportional to the concentration of the organotin compound. This direct proportionality does not extend over a wide concentration range, but at concentrations between 0.2 and 0.6 mg/ml accurate and reproducible results can be obtained. With tributyltin compounds in hydrochloric acid solution a single wave is obtained which, again over a limited concentration range (0.1–0.4 mg/ml), is directly proportional to the concentration of the compound[55–57]. This wave is well defined using both the direct and the derivative circuit. Experiments with dibutyltin dichloride in the same medium show an irregular wave which is not proportional to the concentration, followed by a well defined wave similar to that produced by the tributyltin compound and separated from it by about 0.5 V. The height of the second wave is directly proportional to concentration over a limited range. It is thus possible to make a quantitative determination of a mixture of the dibutyltin and tributyltin compounds by either direct or derivative polarography, using the second wave of the dibutyltin compound and the single wave of the tributyltin compound.

Further work has been described on the application of polarography to various types of organotin chlorides[45,58–63], and to the determination of various particular compounds such as butyltin trichloride in dibutyltin dichloride and tributyltin chloride[64], trichlorethyltin[65,66], diethyldichlorotin[67], triethyltin halides[68], diethylchlorostannane[69], methyl-, ethyl-, and phenyltin trichlorides[70], triphenyltin fluoride[71], dialkyltin compounds[64], organotin(IV) halides[72], and dialkyldichlorotin compounds in water–methanol solutions[73]. The voltammetry of the aquodiethyltin(IV) cation–poly[diethyltin(II)] system has been discussed[74].

Detection limits of 1–100 μg are claimed for the polarography of trialkyl-substituted organotin compounds[75]. A polarographic method for determining triphenyltin acetate residues in vegetables can determine as little as 25 μg of triphenyltin acetate with a precision of $\pm 5\%$[76].

The oscillographic properties of various organotin acetates, including dibutyltin diacetate, tributyltin acetate and triphenyltin acetate, have been investigated[54]. Chronopotentiometry has also been applied to the determination of triphenyltin acetate at very low concentrations in plant material. In this method a hanging-drop electrode is used, at which the ions are reduced in a pre-electrolysis step at -0.7 V *vs.* the silver–silver chloride saturated potassium chloride electrode for 5 min; the potential is then increased gradually to -0.1 V, and the anodic diffusion current is registered at about -0.45 V. A peak height of about 0.3 A is obtained for a concentration of about 0.8 g/ml of tin.

Triphenyltin acetate has been determined polarographically in fresh leaves[77]. Tetraphenyltin has been determined in PVC by treating the sample with hydrogen peroxide solution and concentrated sulphuric acid[78]. When reaction ceases, the mixture is heated until it darkens, then concentrated hydrochloric acid is added and the mixture is boiled until polymer decomposition is complete. The cooled solution is diluted with 4 N ammonium chloride–10% hydrochloric acid (1:1). The solution is de-aerated and the tin(IV) polarogram starting at -0.2 V is recorded. Polarography has been applied to the determination of triethyltin hydroxide, triethyltin oxide and $R_3SnOCOC(Me){=}CH_2$ where R = Et or Bu[79], and amperometric and polarographic titration has been used to determine dialkyltin oxides in weakly acidic solution with standard oxalic acid solution.

XII. ORGANOZINC COMPOUNDS

The polarographic determination of zinc dialkyldithiophosphate in lubricating oils has been discussed[80].

XIII. REFERENCES

1. H. Matschiner and A. Tzschach, *Z. Chem.*, **5**, 144 (1965).
2. M. V. Alekseeva, *Inform. Metod. Materilý Gas Nauch. Issledovatel Sanit. Inst.*, **5**, 16 (1954); *Ref. Zh., Khim.*, 40,375 (1955).
3. H.M. Factory Inspectorate, Ministry of Labour, *Methods for the Detection of Toxic Substances in Air, No. 9, Arsine*, 2nd ed., H.M. Stationery Office, London, 1966.
4. V. Vašák, *Collect. Czech. Chem. Commun.*, **24**, 3500 (1959).
5. J. H. Lowry, R. B. Smart, and K. H. Mancy, *Anal. Chem.*, **50**, 1303 (1978).
6. F. T. Henry and T. M. Thorpe, *Anal. Chem.*, **52**, 80 (1980).
7. R. K. Elton and W. E. Geiger, *Anal. Chem.*, **50**, 712 (1978).
8. A. Watson, *Analyst (London)*, **103**, 332 (1978).
9. A. F. Findeis and T. De Vries, *Anal. Chem.*, **28**, 209 (1956).
10. J. B. Headridge, M. Ashraf, and H. L. H. Dodds, *J. Electroanal. Chem.*, **16**, 114 (1968).
11. A. M. Hartley and R. E. Visco, *Anal. Chem.*, **35**, 1871 (1963).
12. R. Engst, W. Schnaak, and H. Waggon, *Z. Anal. Chem.*, **222**, 388 (1966).
13. P. W. Swanson and P. H. Daniels, *J. Inst. Pet.*, **39**, 487 (1953).
14. W. Hubis and R. O. Clark, *Anal. Chem.*, **27**, 1009 (1955).
15. V. Sedivec and V. Flek, *Prac. Lek.*, **10**, 270 (1958).
16. P. Nangniot, *Chim. Anal.*, **47**, 592 (1965).
17. M. S. Jovanović, J. Tomić, Z. Masić, and M. Dragojević, *Chem. Anal. (Warsaw)*, **11**, 479 (1966).
18. *ASTM Proc.*, **52**, 365 (1952).
19. ASTM, *Standards on Petroleum Products and Lubricants*, Methods D1269-53T (1953) and D1269-61 (1961), American Society for Testing Materials.
20. I. A. Prashinskii and M. K. Frolova, *Tr. Vses. Nauchn.-Issled. Inst. Pererab Nefti Gaza Polycheniyu Iskusstv. Zhidkogo Topliva*, (6), 181 (1957).
21. J. Smelik, *J. Prakt. Chem.*, **5**, 9 (1954).
22. M. Roschig and H. Matschiner, *Chem. Tech. (Berlin)*, **19**, 103 (1967).
23. L. N. Vertyulina and I. A. Korschunov, *Khim. Nauka Prom.*, **4**, 136 (1959).
24. J. E. DeVries, A. Lauw-Zecha, and A. Pellecer, *Anal. Chem.*, **31**, 1995 (1959).
25. B. Skalická and M. Čejka, *Ropa Uhlie*, **16**, 246 (1966).
26. R. I. McCallum, *J. Oil Colour Chem. Assoc.*, **52**, 434 (1969).
27. I. Drăgusin and N. Totir, *Stud. Cercet. Chim.*, **13**, 955 (1965).
28. T. M. Hopes, *J. Assoc. Off. Anal. Chem.*, **49**, 840 (1966).
29. G. Rădulescu, I. Bădilescu, and A. Gilici, *Rev. Chim. (Bucharest)*, **15**, 164 (1964).
30. N. Shirota, M. Kotakersori, and H. Handa, *Ann. Rep. Takamine Lab.*, **9**, 198 (1957).
31. H. Sato, *Bunseki Kagaku*, **6**, 166 (1957).
32. H. Sato, *Japan Analyst*, **6**, 84 (1957); *Anal. Abstr.*, **5**, 705 (1958).
33. M. Kotakemori and H. Henda, *Ann. Rep. Takamine Lab.*, **8**, 231 (1956).
34. T. M. Hopes, *J. Assoc. Off. Agric. Chem.*, **48**, 585 (1965).
35. T. Medwick, in *Pharmaceutical Analysis* (Ed. T. Higuchi and E. Brockmann-Hanssen), Interscience, New York, 1961, Chapter XII.
36. M. S. Bȳkhovskaya, *Zavod. Lab.*, **29**, 667 (1963).
37. W. U. Malik and P. Hague, *Nature (London)*, **194**, 863 (1962).
38. R. Geyer and W. Gliem, *Z. Chem.*, **7**, 64 (1967).
39. W. U. Malik and R. Hague, *J. Am. Oil Chem. Soc.*, **41**, 411 (1964).
40. R. F. Makens, H. H. Vaughan, and R. R. Chelberg, *Anal. Chem.*, **27**, 1062 (1955).
41. D. C. Tse and T. Kuwana, *Anal. Chem.*, **50**, 1315 (1978).
42. J. O. Brønstad and H. O. Friestad, *Analyst (London)*, **101**, 820 (1976).
43. W. Stricks and R. G. Meuller, *Anal. Chem.*, **36**, 40 (1964).
44. V. Sh. Tsveniashvili, S. I. Zhdanov, and Z. V. Todres, *Z. Anal. Chem.*, **224**, 389 (1967)
45. H. Jehring and H. Mehner, *Z. Chem.*, **3**, 34 (1963).
46. H. Jehring and H. Mehner, *Z. Chem.*, **4**, 273 (1964).
47. H. Mehner and J. Jehring, *Z. Chem.*, **3**, 472 (1963).
48. H. Mehner, H. Jehring, and H. Kriegsmann, *3rd Analytical Conference, Budapest, 24–29 August, 1970.*
49. H. Jehring and H. Mehner, *Z. Anal. Chem.*, **1**, 136 (1967).
50. H. Mehner, H. Jehring, and H. Kriegsmann, *J. Organomet. Chem.*, **15**, 97 (1968).

51. H. Mehner, H. Jehring, and H. Kriegsmann, *J. Organomet. Chem.*, **15**, 107 (1968).
52. H. Jehring, H. Mehner, and H. Kriegsmann, *J. Organomet. Chem.*, **17**, 53 (1969).
53. J. Heyrovsky and R. Kalvoda, *Akademic Verlag, Berlin*, 1960.
54. R. Geyer and P. Rotermund, *Acta Chim. Acad. Hung.*, **59**, 201 (1969).
55. G. Costa, *Gazz. Chim. Ital.*, **80**, 42 (1950).
56. G. Costa, *Ann. Chim. (Rome)*, **41**, 207 (1951).
57. V. F. Toropova and M. K. Saikina, *Sb. Stat. Obshch. Khim. Akad. Nauk SSSR*, **1**, 210 (1953).
58. R. B. Allen, *Diss. Abstr.*, **20**, 897 (1959).
59. H. Jehring and H. Mehner, *Z. Chem.*, **3**, 472 (1963).
60. J. Lorberth and H. Nöth, *Chem. Ber.*, **98**, 969 (1965).
61. H. Kriegsmann and S. Pischtschan, *Z. Anorg. Allg. Chem.*, **308**, 212 (1961).
62. K. A. Kozeschkow, *Ber. Dtsch. Chem. Ges.*, **62**, 996 (1929).
63. D. Seyferth, *Naturwissenschaften*, 34 (1957).
64. V. A. Bork and P. I. Selivokhin, *Plast. Massy*, (10), 60 (1969).
65. M. Devaud, *C. R. Acad. Sci., Ser. C*, **262**, 702 (1966).
66. M. Devaud, P. Souchay, and M. Person, *J. Chim. Phys. Physicochim. Biol.*, **64**, 646 (1967).
67. M. Devaud, *C. R. Acad. Sci., Ser. C*, **263**, 1269 (1966).
68. M. Devaud, *J. Chim. Phys. Physicochim. Biol.*, **63**, 1335 (1966).
69. M. Devaud, *J. Chim. Phys. Physicochim. Biol.*, **64**, 791 (1967).
70. M. Devaud and P. Souchay, *J. Chim. Phys. Physicochim. Biol.*, **64**, 1778 (1967).
71. A. Vanachayangkul and M. D. Morris, *Anal. Lett.*, **1**, 885 (1968).
72. M. Devaud and S. Laviron, *Rev. Chim. Miner.*, **5**, 427 (1968).
73. V. N. Flerov and U. M. Tyurin, *Zh. Obshch. Khim.*, **38**, 1669 (1968).
74. M. D. Morris, *J. Electroanal. Chem. Interfacial Electrochem.*, **16**, 569 (1968).
75. T. L. Shkorbatova, O. A. Kochkin, L. D. Sirak, and T. V. Khavalit, *Zh. Anal. Khim.*, **26**, 1521 (1971).
76. A. Vogel and J. Deshusses, *Helv. Chim. Acta*, **47**, 181 (1964).
77. S. Gorbach and R. Bock, *Z. Anal. Chem.*, **163**, 429 (1958).
78. V. D. Bezuglyi, E. A. Preobrazhenskaya, and V. N. Dmitrieva, *Zh. Anal. Khim.*, **19**, 1033 (1964).
79. D. A. Kochkin, T. L. Shkorbatova, L. D. Pegusova, and N. A. Voronkov, *Zh. Obshch. Khim.*, **39**, 1777 (1969).
80. A. M. S. Alam, J. M. Martin, and P. Kapsa, *Anal. Chim. Acta*, **107**, 391 (1979).

The Chemistry of the Metal–Carbon Bond
Edited by F. R. Hartley and S. Patai

CHAPTER **18**

Analysis of organometallic compounds: gas chromatography

T. R. CROMPTON
'Beechcroft', Whittingham Lane, Goosnargh, Preston, Lancashire, UK

I. ORGANOALUMINIUM COMPOUNDS

In a direct gas chromatographic analysis of organoaluminium compounds the sample was purged with helium carrier gas on to a 1-m column of Chromosorb W containing 7.5% of paraffin wax mixed with triphenylamine (17:3) at a column temperature of 73–165°C[1]. A thermistor detector was used. Organoaluminium and organogallium compounds have been separated on a column containing silicone elastomer E301 on diatomaceous brick operated at 110°C[2]. Helium was used as the carrier gas and a katharometer as the detector. Aluminium, germanium, silicon, and titanium alkoxides have been separated on a 1-ft column of 1% Apiezon L on Chromosorb W packed in a PTFE tube[3]. The operating temperatures were in the range 60–150°C and a gas chromatograph with a dual column and dual thermal conductivity detector was used[3]. The best results were obtained with lightly loaded columns (1% liquid phase) using Apiezon L, silicone gum rubber SE-30, and silicone oil DC-200. A microwave emission detector[4,5] and a glow discharge tube[6] have been demonstrated to be useful for the detection of organoaluminium compounds[4,5].

II. ORGANOANTIMONY COMPOUNDS

A gas chromatographic method for the analysis of organoantimony compounds uses a special sample injector to avoid oxidation of the sample[7]. Separation was achieved on a 1-m column of Chromosorb W containing 7.5% of paraffin wax (m.p. 63–64°C)–triphenylamine (17:3), and using dry purified helium as the carrier gas and a thermistor detector. The column temperature ranged from 73 to 165°C, depending on the type of compound being determined. The gas chromatography–microwave plasma detector (GC–MPD) technique has been applied to the analysis of organoantimony compounds in environmental samples[8].

III. ORGANOARSENIC COMPOUNDS

Gas chromatography has been applied to the separation of eight substituted organoarsines and substituted organobromoarsines of the type RAsR′R″ where R is an alkyl or aryl group, R′ is an alkyl group (CF_3 or C_3F_7) and R″ is CF_3 or C_3F_7. ranging in molecular weight from 156 to 306[9,10]. There is an almost linear relationship between log(retention time) and either the boiling point or the molecular weight of each component of a homologous series. Chromatography was carried out on a column (6 ft × 0.25 in) of 5% of SE-30 silicone gum rubber on 80–100 mesh Chromosorb W; the column was operated at 290°C with argon at a flow-rate of 40 ml/min as the carrier gas; an argon ionization detector was used.

A modified Barber Coleman Model 10 argon ionization detector has been used for quantitative studies of organoarsenic and organobromoarsenic compound mixtures[10]. Using isothermal column operating conditions, the chromatograms for the separation of the arsenic derivatives investigated were obtained with a 6 ft × $\frac{1}{4}$ in o.d. stainless-steel column packed with 5% (w/w) dimethyl silicone polymer (General Electric SE-30 silicone gum rubber) as liquid stationary phase on 80–100 mesh Chromosorb W. Triphenylarsine (b.p. *ca.* 360°C) was eluted in 4.2 min using the following higher temperature operating conditions: column temperature, 290°C; flash heater, 340°C; detector temperature, 365°C; argon pressure, 30 lb/in^2; flow-rate, 40 ml/min. The same technique was also applied to the quantitative determination of substituted arsines. A mixture of dimethylbromoarsine (b.p. 128–130°C/720 mmHg) and trivinylarsine (b.p. 130°C) was resolved with a 6 ft × $\frac{1}{4}$ in o.d. copper column operated at 66°C and packed with 10% (w/w) of squalene on 35–80 mesh

Chromosorb W. With a helium inlet pressure of 25 lb/in^2 and a flow-rate of 350–355 ml/min, dimethylbromoarsine and trivinylarsine were eluted in 4.1 and 5.2 min, respectively.

In the determination of methylated arsenic species in natural waters, atomic-absorption spectrometry with electron-capture and/or flame-ionization detectors was used to achieve a detection limit of several nanograms of arsenic[11]. A commercial atomic-absorption spectrophotometer with a heated graphite tube furnace atomizer linked to a gas chromatograph has been used for the determination of trimethylarsine in respirant gases produced in microbiological reactions[12]. The gas chromatography–microwave plasma detector (GC–MPD) technique has been applied to the analysis of alkyl arsenic acids in environmental samples[13,14].

Methods involving reduction to produce hydrides followed by separation and detection by an emission-type detector have been investigated for the analysis of organoarsenic compounds[15]. The design of a glow discharge tube proposed earlier[16] as an element-specific detector for gas chromatography has been modified[17] to overcome its principal drawback, namely that it appeared to be subject to coating of the tube walls with decomposition products of the sample, thus attenuating the light signal as chromatographic peaks passed through the discharge. Although spectral background correction would be beneficial, it was not used in a simple helium glow discharge detector with a stable but inexpensive power supply that can detect various metals (Al, As, Cr, Cu) as well as P, Si, C, and S in gas chromatographic effluents[17]. An improved glow chamber design prevents degradation products from coating the observation window. The monochromator is provided with an internal beam splitter and a side-exit port. A moveable exit slit mounted on the latter permits background corrections to be made at the most suitable distance from the elemental line detected. Selectivity and versatility are greatly improved by this type of background detection.

Gas chromatography has been used[18] to determine arsine in hydrogen-rich mixtures. The arsine was detected on a column containing dioctylphthalate on polyoxyethyleneglycol as adsorbent with hydrogen as the carrier gas. The limit of detection as arsenious oxide was 0.001 mg. In addition, determinations of down to 1.5×10^{-3} g l of arsine in silane using a column (8 m × 5 mm i.d.) of alumina moistened with VKZL-94B silicone oil operated at 0°C or down to 4.2×10^{-4} g l of arsine in silane using a column (4 m × 5 mm i.d.) of diatomite brick treated with PFMS-4F silicone oil operated at 30°C have been developed[19]. Both procedures utilize dry nitrogen as carrier gas and a katharometer detector after passage of the dry gas issuing from the column through a furnace at 1000°C to decompose the arsine to hydrogen.

IV. ORGANOBERYLLIUM COMPOUNDS

Organoberyllium compounds may be analysed by gas chromatography on a 1-m column of Chromosorb W containing 7.5% paraffin wax (m.p. 63–64°C)–triphenylamine (17:3) employing dry helium as carrier gas and a thermistor detector[1]. Microwave emission detectors are useful for the detection of organoberyllium compounds[5].

V. ORGANOBORON COMPOUNDS

Gas chromatographic analysis of mixtures of boron alkyls ranging from triethylboron to tri-*n*-propylboron has been discussed[20,21]. A 1-m column packed with silicone oil on Celite at a carrier gas flow-rate of 100 ml/min at 80°C separated seven compounds in 13 min[20]. Another method uses a thermistor detector and a 1-m column of

Chromosorb W containing 7.5% paraffin wax (m.p. 63–64°C)–triphenylamine (17:3), dry pure hydrogen as the carrier gas and a column temperature between 73 and 165°C, depending on the type of compound[22]. Retention times of 1,3,5-trialkylborazoles have been calculated empirically for individual alkyl groups from the $\log t_R$ (logarithm of the individual retention times relative to mesitylene, $\log t_R = 2$) values of the symmetrical 1,3,5-derivatives by subtracting the $\log t_R$ of borazole (0.54) and dividing by 3[23]. The agreement between calculated and determined values was good. The columns used contained 13% Carbowax 400 at 100°C, with a flame-ionization detector and oxygen-free hydrogen as the carrier gas.

Gas chromatographic analysis of boron hydrides can be achieved using Celite coated with paraffin oil. (Octoils) or with tricresyl phosphate as the column packing[24]. Diborane, tetraborane, and pentaborane can be resolved without decomposition on the column. Extensive decomposition occurred, however, on the column in the case of dihydropentaborane. Of the three stationary phase liquids used to prepare the chromatographic columns, paraffin oil proved best for the separation of the boron hydrides themselves. The retention times for the boron hydrides were longest on the Celite–tricresyl phosphate column and shortest on the Celite–paraffin oil column. The peaks were well resolved on all columns in all cases, except where dihydropentaborane decomposition occurred. Mixtures of methyldiboranes can be almost completely resolved and determined on chromatographic columns of mineral oil on crushed firebrick, operating at 0°C[25]. A quantitative determination can be carried out by area measurement and is accurate to between 1 and 2% of the components present in a mixture.

A combination of gas chromatography with mass spectrometry has been studied for the analysis of mixtures of alkylboranes[20]. Separation of diborane, chloroboranes (B_2H_5Cl, $BHCL_2$, and BCl_3), and hydrogen and hydrogen chloride has been achieved on low-temperature (−78°C) columns containing powdered Teflon, silicone oil 703, Fluorolube GR 362, Kel F oil, liquid paraffin, or hexadecane[26]. It was apparent that to find a single partition column to resolve all of the components would be unrealistic. Thus, analytical requirements were successfully met through the development of several different columns with specific applications.

Figure 1 shows a gas chromatom obtained using a column containing 60–80 mesh Chromosorb coated with 20% (w/w) silicone oil at 0°C with a helium flow-rate of 1 ml/min. It is interesting that, although the carrier gas was helium, positive hydrogen peaks were invariably recorded and precise calibrations were obtained. Apparently the hydrogen segment leaving the 0°C column becomes warmer than the helium stream before it reaches the ambient temperature detector. Subsequently, the

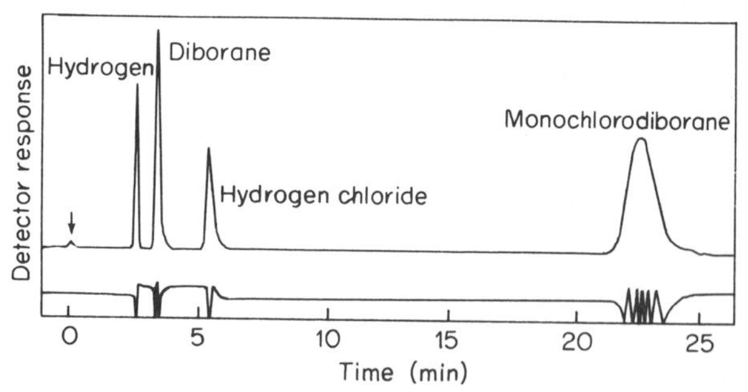

FIGURE 1. Chromatographic separation on a silicone oil column at 0°C.

temperature of the hydrogen becomes more important that its thermal conductivity. At the thermistor, then, less heat is conducted to the hydrogen segment than is lost to the cold helium flow, and a positive peak results. Once a constant helium flow is established, the temperature gradient between column exit and detector remains constant.

Another chromatographic separation medium for mixtures containing dichloroborane is based on the fact that this compound is more stable in the presence of boron trichloride[26]. If the partition liquid remains saturated with boron trichloride, dichloroborane and monochlorodiborane can be resolved even at 40°C. The $\frac{1}{4}$ in diameter column was 18 ft in length and filled with 60–80 mesh Chromosorb coated with 30% (w/w) *n*-hexadecane containing residual boron trichloride from a previous sample. The column was operated at 40°C with a helium flow-rate of 400 ml/min. Since boron trichloride was soon flushed from the column under these relatively drastic conditions, the column performance was found to be reproducible only when samples were introduced at regular intervals. Such a technique would, therefore, be more practical for use in a continuously operating plant stream analyser. Although the area of the unsymmetrical boron trichloride peak obtained in a separation at −78°C was used for quantitative determinations, more precise gas chromatographic methods were developed for mixtures in which hydrogen chloride and boron trichloride were the only components. Two different ambient temperature columns were used for the routine analysis of these mixtures. One column consisted of mineral oil (20% w/w) on Chromosorb and the other used Fluorolube (10% w/w) on Teflon. When using the 12 ft × $\frac{1}{4}$ in o.d. mineral oil column, which was operated at 25°C with a helium flow-rate of 230 ml/min, analyses were based on peak height measurements. The most precise measurements for boron trichloride were obtained by peak area determinations with the 4 ft × $\frac{1}{4}$ in o.d. Fluorolube column. If dichloroborane is present, however, it disproportionates during separation, and the result is an uneven line between the two peaks.

Pyrolysis followed by thermal conductivity cells has been used for the determination of deuterium in deuterated boron hydrides, their organic derivatives, and nitrogen compounds[27]. The volatile boron compounds were first passed over hot uranium metal (500–800°C), which pyrolysed various compounds with a recovery between 95 and 100% of hydrogen and deuterium. This gas mixture was then analysed using a thermal conductivity cell which had been previously calibrated against standard mixtures of hydrogen and deuterium.

Decaborane has been determined on a chromatograph with a 3 m × $\frac{3}{8}$ in i.d. column packed with 60–80 mesh Celite impregnated with 20% (w/w) Apiezon L[28,29]. It had a column efficiency of 12,000 theoretical plates; the retention time for decaborane relative to *n*-decane was 2.65 and to naphthalene 0.730. The helium flow-rate was 340 ml/min and the column and detector temperature was 150°C. Cyclohexane was used as a solvent for the decaborane. Alternative possible conditions include temperatures ranging from 90 to 220°C. Squalene, Apiezon L or M, silicone grease, and Fluorolube are suitable partitioning liquids. For decaborane samples the recommended conditions are a 0.5 m × $\frac{1}{4}$ in i.d. column packed with 100–120 mesh Celite impregnated with 20% (w/w) of squalene. The operating temperature should be about 140°C and the helium flow-rate 50 ml/min. Naphthalene is recommended as an internal standard.

VI. ORGANOCHROMIUM COMPOUNDS

A glow discharge tube and a microwave emission detector have been used as detectors for the gas chromatography of organochromium compounds[5,16,30].

VII. ORGANOCOPPER COMPOUNDS

A glow discharge tube and a microwave emission detector have been used to detect organocopper compounds separated on a gas chromatographic column[16,31].

VIII. ORGANOGALLIUM COMPOUNDS

The microwave emission detector has been demonstrated to be useful for the detection of organogallium compounds[31].

IX. ORGANOGERMANIUM COMPOUNDS

Alkylgermanium compounds exhibit very similar behaviour on a gas chromatographic column to those of silicon and may be separated on the same types of columns[33,39]. These include columns containing Apiezon L, SE-30, QF-1, XF 112, and *o*-nitrotoluene as the stationary phase with flame-ionization and thermal conductivity detectors[39]. Mixtures of organogermanium and organosilicon compounds can be separated[38] using fluorosilicone oil as the stationary phase on a column operated at 150°C using helium as the carrier gas.

Table 1 shows retention times determined for a number of alkyl compounds, compared with empirically calculated values[35]. Retention data in Table 1 (log t_R) are expressed as logarithms of retention times relative to mesitylene = 100. Estimates of the retention values of mixed alkylgermanes are made from observations on symmetrical tetraalkylgermanium compounds. The log t_R values of the latter were divided by 4, which gave the following constants representing the effect of single alkyl groups on the retention time of mixed alkyls: methyl 0.14, ethyl 0.45, *n*-propyl 0.69, and *n*-butyl 0.93. A constant of 0.14 was added to calculate the germane series.

Gas chromatography has been applied to the determination in germane of down to 10^{-4}–10^{-5}% of methane, ethane, and ethylene[40]. It is carried out at 40°C on a 5-m column containing porous glass using a flame-ionization detector. Silica furnaces heated to 400°C are placed before and after the column, to decompose the germane and prevent deposition of germanium dioxide in the detector. The error of the determination is claimed to be less than ±20%. The determination of dissolved oxygen

TABLE 1. Logarithm of retention times of germanes on squalane at 100°C

	Log t_R			Log t_R	
Alkyl group	Obs.	Calc.	Alkyl group	Obs.	Calc.
$Me_3(n\text{-}Bu)$	1.42	1.49	$Me(n\text{-}Bu)(n\text{-}Pr)_2$	2.61	2.58
$MeEt_3$	1.57	1.63	$MeEt(n\text{-}Bu)_2$	2.64	2.59
$Me_2(n\text{-}Pr)_2$	1.77	1.79	$Et(n\text{-}Pr)_3$	2.66	2.66
$Me_2Et(n\text{-}Bu)$	1.77	1.80	$(n\text{-}Pr)_4$	2.89	2.88
$MeEt_2(n\text{-}Pr)$	1.84	1.87	$Et(n\text{-}Pr)_2(n\text{-}Bu)$	2.91	2.89
Et_4	1.94	1.94	$Et_2(n\text{-}Bu)_2$	2.95	2.90
$Me_2(n\text{-}Pr)n\text{-}Bu$	2.04	2.04	$Me(n\text{-}Bu)_3$	3.14	3.07
$MeEt(n\text{-}Pr)_2$	2.09	2.10	$Pr_3(n\text{-}Bu)$	3.13	3.13
$Et_3(n\text{-}Pr)$	2.18	2.18	$Et(n\text{-}Pr)(n\text{-}Bu)_2$	3.16	3.14
$Me_2(n\text{-}Bu)_2$	2.31	2.28	$(n\text{-}Pr)_2(n\text{-}Bu)_2$	3.38	3.37
$Me(n\text{-}Pr_3)$	2.35	2.34	$Et(n\text{-}Bu)_3$	3.40	3.38
$Et_2(n\text{-}Pr)_2$	2.42	2.41	$(n\text{-}Pr)(n\text{-}Bu)_3$	3.61	3.62
$Et_3(n\text{-}Bu)$	2.45	2.42	$(n\text{-}Bu)_4$	3.85	3.86

nd nitrogen in germanium tetrachloride can be carried out at 60–90°C on a .5 m × 6 mm i.d. column containing 20% fluorosilicone oil 169 on firebrick (*ca.* .2 mm) using hydrogen as the carrier gas at a flow-rate of 125–150 ml/min[41].

X. ORGANOIRON COMPOUNDS

errocene derivatives have been completely separated by gas chromatography on an piezon L (2.5%) on Chromosorb W column (1.4 m × 4 mm i.d.) at 200°C[42]. The

TABLE 2. Retention times of ferrocene derivatives

Compound	Melting point (°C)	Molecular weight	Retention time (min)			
			125°C	150°C	175°C	200°C
Ferrocene	173–174	185.95	3.7	1.75	1.15	0.65
n-Butylferrocene	B.p. 84–86/0.2 mmHg	242.15	16.0	6.0	2.80	1.31
Ethylferrocene	B.p. 74–76/0.2 mmHg	213.97	6.7	2.64	1.48	0.83
Vinylferrocene	48–49	211.97	6.8	3.05	1.52	0.90
1,1′-Di-*n*-butylferrocene	–	298.22	73.0	20.2	7.7	2.90
Acetylferrocene	85.86	227.97	18.5	7.55	3.2	1.49
1,1′-Diacetylferrocene	122–124	269.99	77.0	26.2	9.7	3.70
Hydroxymethylferrocene	76–78	215.96	15.0	5.9	2.3	1.20
1,1′-Dihydroxymethylferrocene	85–86	245.97	76.0	20.5	9.2	3.40

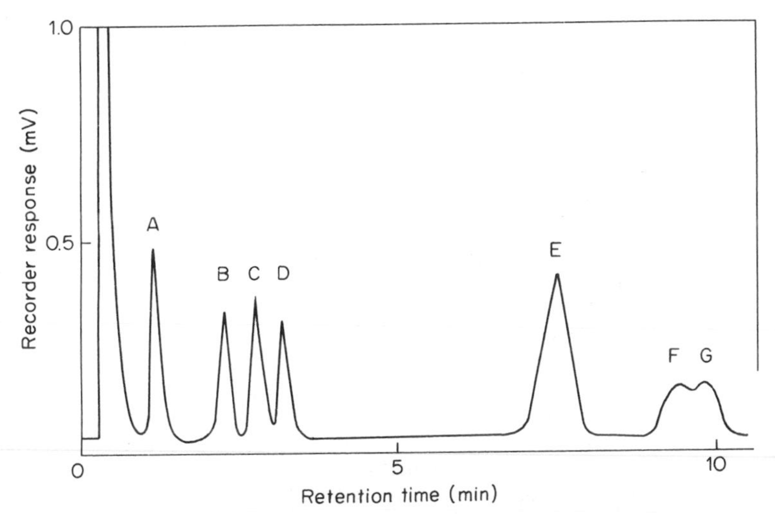

FIGURE 2. Separation of ferrocene (A), hydroxymethylferrocene (B), *n*-butylferrocene (D), 1,1′-di-*n*-butylferrocene (E), 1,1′-diacetylferrocene (F), and 1,1′-dihydroxymethylferrocene (G) by gas chromatography. Column of stainless, 5 ft × $\frac{1}{8}$ in o.d., containing 5% (w/w) SE-30 on 60–80 mesh Chromosorb W; flow-rate, 30 ml/min. Temperatures: column, 175°C; detector, 200°C; injection point, 195°C.

chromatograph was equipped with a thermal conductivity detector and helium at flow-rate of 50 ml/min was used as the carrier gas. Forty-one ferrocene and tw ruthenocene derivatives were separated on 2,2-dimethylpropane-1,3-diol adipat polyoxyethylene glycol adipate, polyoxyethylene glycol M-20, polyoxypropyler glycol, and Apiezon L (1.5% on Celite 545, 80–100 mesh) at 100–200°C[43] usir packed columns (1–1.2 m × 0.4 cm i.d.) in glass and stainless-steel tubes, and capillary column (45 m × 0.25 mm i.d.) and a β-ray detector. Best separations wer achieved on 2,2-dimethylpropane-1,3-diol adipate and polyoxyethylene glycol M-2 columns. Nitro-, dicyano-, diphenyl-, and diacylferrocenes could not be separate owing to their poor thermal stability at the column operating temperature. G chromatographic procedures utilizing SE-30 on 60–80 mesh Chromosorb W colum have been described for the separation of ferrocene and for butyl-, ethyl-, viny 1,1′-dibutyl-, acetyl-, 1,1′-diacetyl-, hydroxymethyl-, and 1,1′-b (hydroxymethyl)-ferrocenes[44]. The retention times of various other ferrocen derivatives at several temperatures are given in Table 2. Although all of the analyse were conducted under isothermal conditions, it is apparent that temperatur programming would be desirable in the separation and analysis of mixtures containin both volatile and relatively involatile ferrocene derivatives. The isothermal separatio of a seven-component mixture is shown in Figure 2.

Other workers have also discussed the gas chromatography of ferrocene and othe metallocenes. A microwave emission detector has been demonstrated to be useful fo the detection of organoiron compounds separated on a gas chromatographic column[3

XI. ORGANOLEAD COMPOUNDS

Almost all of the published work on the gas chromatography of organolea compounds is concerned with their analysis in petroleum solutions. The analysis of th lead alkyls has been investigated very extensively because of the widespread use o mixtures of tetramethyllead, trimethyllead, dimethyllead, methyltriethyllead, an tetramethyllead as hydrocarbon fuel additives. The analytical problem is complicate by the fact that the lead alkyls must be separated and analysed in the presence of complex mixture of hydrocarbons. The volatility of the lead compounds is such tha their peaks are superimposed upon hydrocarbon peaks. Selective detectors ar required.

An early approach to the problem used gas–liquid chromatographic columns t fractionate the lead alkyls and then determined the amount of lead in the fractio containing both lead and hydrocarbon compounds by a spectrophotometri method[46,47]. The lead alkyls were separated by a chromatographic column individually collected in iodine scrubbers, and measured by a dithizon spectrophotometric lead analysis procedure. Later modification employed a muc simpler chromatographic unit which consisted of a thick-walled aluminium tube whic served as a column[47]. A uniform temperature was maintained by means of electrica heating tape wrapped around the column, with control effected by a variabl transformer. The carrier gas flow-rate was controlled by the pressure regulator at th supply cylinder and was measured by a bubble flow meter. No detector elements wer necessary since the retention times were determined by calibration and remai unchanged under the fixed conditions of use. The column consisted of a 1 in o.d. aluminium cylinder 14 in long with a $\frac{3}{16}$ in hole bored full length through the centre The lower 12 in of the cavity was packed with 20% Apiezon M on 60–80 mesh wate washed Chromosorb W supported on glass-wool. Apart from innovations such a coupling spectrophotometric detection[46,47] or titrimetry[46] at the outlet end of the ga chromatographic column, conventional thermal conductivity or flame-ionization ethods of gas chromatographic analysis are not effective in the chromatography of ace amounts of tetraalkyl leads owing to the extreme complexity of the gasoline base ock, although the use of an ionization detector for the chromatography of traethyllead in petroleum has been discussed[48]. A selective electron-capture etector detects tetraethyllead with essentially no interference from hydrocarbons, hich have a much lower response factor[49]. A photoionization detector may be used measure the total amount of hydrocarbons[49]. The authors did not attempt to etermine the various lead alkyls separately. The chromatographic column was an 5 ft × 0.02 in i.d. stainless-steel tube coated with Apiezon L and operated at 90°C nd in 10 lb/in^2. Conventional packed columns are also satisfactory. The flash heater is ept below 100°C to prevent thermal decomposition of lead alkyls.

In addition to the methylethyllead alkyls, gasolines frequently contain ethylene ichloride and dibromide as scavengers. These compounds also give a high response in he electron-capture detector and frequently elute at the same time as one of the lead lkyls. This can be overcome by using a chemically active stationary phase, silver itrate in Carbowax 400, as a pre-column before the detector to remove the cavengers together with a silicone rubber on Chromosorb W analytical column[50]. Good separation of the five methylethyllead alkyls was obtained. However, the ensitivity of the electron-capture detector varied markedly with the applied voltage, which necessitated careful control of the operating conditions for quantitative analysis. nterchange of methyl and ethyl radicals between tetramethyllead and tetraethyllead ccurs on a column of 5% SE-30 silicone rubber on acid-washed Chromosorb W. Coating the Chromosorb with sodium hydroxide before the stationary liquid phase is pplied reduces interchange of radicals to an undetectable level. Slight interchange akes place when the silver nitrate packing is located at the column inlet. When this acking is located at the column exit, the lead alkyls are separated before contact with he silver nitrate and interchange is avoided. Excellent resolution of the five nethylethyllead alkyls was obtained with 37 V across the electron-capture detector. The maximum operating temperature for the tritium source is 200°C. The detector is naintained at 180°C to prevent condensation of sample and to obtain an increased esponse[51]. It is general practice to add a dilution gas to the column effluent to reduce he residence time of components in the detector and to maintain a desired flow-rate hrough the detector[52]. Increasing the dilution gas rate improves the stability but also lecreases the response. Optimum stability and sensitivity result from a dilution gas ate of 150 ml/min when the eluting gas rate is 100 ml/min.

The column packings remain effective for many analyses[50]. The silicone rubber packing continues to separate the lead alkyls and to avoid interchange after 4 months' use. The silver nitrate packing completely removes both scavengers for about 200 separations of 1 μl gasoline samples. The precision of the method was determined by six to ten analyses of petroleums containing known amounts of various lead alkyls. Standard deviations for the individual lead alkyls in terms of grams of lead per gallon are 0.01–0.02 when all five alkyls are present. The method was applied to the determination of individual lead alkyls in commercial petroleums. Results for total lead agreed well with those obtained by X-ray fluorescence.

1,2,3-Tris(2-cyanoethoxy)propane, a very polar liquid, when used as a packing material gave good resolution of the methylethyllead alkyls and retained the halogenated scavengers beyond the elution times for the lead alkyls so that these did not interfere[52]. Cell geometry, carrier gas flow-rate and electrometer voltage all effect the performance of the electron-capture detector so that frequent calibration is necessary for maximum accuracy. Relative response factors for the alkyllead compounds have been calculated for the flow conditions of this analysis. The detector response decreases with increasing molecular weight of the lead alkyl. The retention characteristics of the tetraalkylleads on Apiezon L, silicone SF-96, and 1,2,3-tris(2-cyanoethoxy)propane columns at two temperatures were reported. Several scavenger columns for the purification of the carrier gas and the removal of interfering peaks were discussed. An absolute accuracy of ±3% was obtained on standard samples made from weighed amounts of tetraethyl- and tetramethyllead.

Two types of electron-affinity detectors have been examined[53]. The first was of the parallel plate type equipped with a 100 mCi titanium tritide source, mounted on a Jarrell-Ash Universal 700 gas chromatograph, and the second was of the cylindrical type, consisting of a cylindrical 250 mCi titanium source cathode and a tubular inlet port anode supplied by Wilkins Instrument and Research Corporation (Acrograph Hi-Fi). The latter detector required several modifications to obtain adequate response and stability. The most important modification consisted of a heater to the detector chamber. The electron-affinity detector, while not directly sensitive to aliphatic hydrocarbons, can be blocked and rendered totally insensitive by condensation of a heavy and relatively involatile hydrocarbon film on the ionization source. A heater placed around the detector and maintained at a constant temperature of 150–180°C is necessary for long-term stability of this detector. A further modification consisted of introducing a voltage divider into the electrometer charging circuit, which permits the adjustment of the detector voltage to optimum conditions. The boiling points of the halogenated compounds added to leaded petroleum as scavengers overlaps the boiling range of three of the lead alkyls[53]. Thus the tetramethyllead peak is overlapped by one isomer of dibromoethane on the Apiezon L and silicone fluid SF-6 columns when the conditions of gas flow-rate, temperature and column length are adjusted to give tetraethyllead an elution time shorter than 90 min. Unfortunately, electron-affinity detectors are extremely sensitive to halogenated hydrocarbons.

A solution to the problem of halocarbon interference under isothermal conditions is to use 1,2,3-tris(2-cyanoethoxy)propane as the liquid phase. This material is an extremely polar liquid, which results in extremely long retention times for the halocarbons, e.g. 1,2-dibromoethane elutes after tetraethyllead. Higher molecular weight halocarbons are retained for such a time that the peaks are very broad and diffuse. The alkyllead compounds elute from this column in almost the same time as is required by the non-polar columns. A pre-column can be placed in the nitrogen carrier gas line before both chromatographs to obtain a high standing current at a low voltage (giving a wide linear detector range). Two columns were used, a 10 ft by $\frac{1}{2}$ in i.d. copper column packed with Linde 5A molecular sieve chilled in liquid nitrogen and a 10 ft by 0.25 in i.d. copper column packed with 10% silver nitrate on Chromosorb W. Both resulted in a 40% increase in standing current over the untreated nitrogen. The silver nitrate column was more convenient to use. It was concluded that the electron-affinity detector furnishes a simple and direct means for the analysis of the alkyllead isomers normally found in petroleum[53]. No essential difference due to cell geometry was noted between the circular and parallel plate detectors, with the exception that the circular plate detector was not as selective. The circular plate detector used by these workers, owing to its more radioactive source (250 mCi) proved to be 5 times more sensitive than the parallel plate detector (100 mCi). The smaller internal volume of the circular plate detector permitted a lower flow-rate of nitrogen to be used.

Another group[54] combined the use of 1,2,3-tris(2-cyanoethoxy)propane for the analytical column with a short scrubber column of silver nitrate on Carbowax 400 before the detector to remove the lead scavengers. They claim an analysis time for the five lead methylethyl alkyls of 10 min with an overall relative standard deviation of about 4% for each compound. Again, the sensitivities of compounds which have electron-affinity properties vary with the conditions of analysis such as column temerature, detector temperature, flow-rate, voltage applied across the cell, and the cleanness of the source, so that it is advisable to calibrate the instrument frequently with nown standards. The scrubber section is important. It absorbs the column material nd thereby maintains the full sensitivity of the detector. It also removes the halogeated lead scavengers by reacting with silver nitrate in the packing. Since these cavengers elute at approximately twice the retention time of tetraethyllead, the time f analysis may be considerably shortened by their removal. If the analysis of these ad scavengers is important, it may be included with the lead analysis by simply using scrubber without silver nitrate. The whole petroleum sample, including the alkyl ads and scavengers, is chromatographed at 60–70°C on a column (20 × 0.4 cm i.d.) f 10% of polyoxypropyleneglycol 400 on 30–60 mesh Chromosorb P using hydrogen s the carrier gas at a flow-rate of 40 ml/min. The eluate from this column is passed hrough a hydrogenator where the alkyl leads are catalytically converted over nickel at 40°C to ethane and methane. 1,2-Dichloroethane and 1,2-dibromoethane scavengers re similarly converted to ethane. The petroleum is almost unaffected. The hydrogeated eluate is then passed through a short column containing 3% liquid paraffin upported on charcoal and operated at 60–70°C, which retains almost permanently all naterials above propane. A flame-ionization detector placed at the end of the charcoal olumn detects the methane and/or ethane from the lead alkyl. In this way the etroleum 'background' can be separated from chemically produced ethane and methane y a specific chemical reaction followed by sorption. This method avoids completely he use of a specific detector.

This method has subsequently been modified in respect of the stationary phase to ermit the simultaneous determination of the lead alkyls and the scavengers[55]. The opper column (150 × 0.4 cm i.d.) used was packed with 20% of 1,2,3-ris(cyanoethoxy)propane on Chromosorb P (30–60 mesh) pre-coated with 1% of otassium hydroxide and operated at 80°C with hydrogen at a flow-rate of 40 ml/min s the carrier gas. Recoveries were between 98 and 102% for each tetralkyllead omponent in the range equivalent to 0.026–0.2 g/l of lead. Similar results were btained for the scavengers dibromo- and dichloroethane.

A major advance in lead alkyl analysis was made possible by the development of an lectron-capture detector capable of operating at high temperatures. The methods lescribed earlier in this section using electron-capture detection utilized tritium detecors with a safe upper operating temperature of 225°C. At this temperature highoiling components of the sample and column substrate can condense on the detector, giving an erratic response and necessitating frequent cleaning and calibration. A Ni^{63} lectron-capture detector operating at 300°C gives excellent long-term stability[56]. In a 3-week test period, a power failure occurred that permitted substrate to condense on the detector. Heating to 340°C restored the detector response to the original value. A 5 ft analytical column containing 20% 1,2,3-tris(2-cyanoethoxy)propane operated at 90°C separated the lead alkyls and scavengers in about 25 min. The lead alkyls could be determined in either petroleum or fuel oil with a sensitivity of 0.15 ppm for 1 μl of sample injected. The use of pre-concentration techniques[57] should make this technique readily applicable to atmospheric analysis.

It is claimed that using an electron-capture detector a complete analysis for tetra methyllead and tetraethyllead in petrol can be achieved on a column (3 m) of 10% of Apiezon L on Chromosorb W at 120°C with bromobenzene as internal standard[58]. To separate the mixed alkyls ethyltrimethyllead, diethyldimethyllead, triethylmethyllead, tetramethyllead, and tetraethyllead, it is necessary to combine the above analysis with use of a pre-column (5 cm) of 20% of Carbowax 400 saturated with silver nitrate on Chromosorb W impregnated with 8% of potassium hydroxide. The pre-column retains halogen-containing scavengers in the petrol, which would otherwise mask the ethy trimethyllead peak. As little as 0.002 g of tetramethyllead and tetraethyllead can b detected in 1 litre of petroleum in a 45 min analysis.

Basic work on the gas chromatography of tetraethyllead[59] and of tetramethyllead has been reported. Unfortunately, electron-capture detection, although sensitive, i not specific enough, nor is it a very easy method of detection to apply to organolea compounds. It requires extreme care, cleanliness, and rigid adherence to microchemi cal techniques. The alternative method consists of running the gasoline sampl through the gas chromatograph to separate the components, which are then intro duced, one by one, directly into the atomic-absorption burner. The atomic-absorptio spectrophotometer, which is set up for lead determination, records a peak absorptio for each lead compound as it passes from the chromatograph. This method is stanc ardized by using mixtures of known composition. Gasoline sample sizes are typicall 1μ litre; as little as 20 ng of lead as lead alkyl can be detected.

The sensitivities of the electron-capture, thermal conductivity, argon ionization an flame-ionization detectors in the chromatographic determination of organolead an aliphatic chlorine compounds in the atmosphere have been compared using a Wilkin HiFi Model 600 chromatograph with both hydrogen flame and electron-capture detec tors, a Beckmann Model GC-2A chromatograph with a thermal conductivity detector and a Research Specialities Model 600 chromatograph with an argon ionization detec tor (Table 3)[61]. All separations were made on 6 ft × $\frac{1}{8}$ or 0.25 in stainless-stee columns. It is apparent that the sensitivities of the thermal conductivity and the argo ionization detectors are independent of the molecular weight and the number o chlorine atoms in the chlorinated compounds, but the flame detector decreases slightl in sensitivity with increasing numbers of chlorine atoms. The electron-capture detecto has its greatest response to the chlorinated compounds at 10 V. With the electron capture detector the sensitivity is dependent in rather a complex manner on th

TABLE 3. Limits of detection for chlorinated aliphatic and lead alkyl compounds using gas chromatographs with various detectors

Compound	Thermal conductivity (μg × 10^{-1})	Argon ionization (μg × 10^{-3})	Flame ionization (μg × 10^{-3})	Electron capture (μg × 10^{-3})
Methyl chloride	1.2	2.0	3.0	8.5
Dichloromethane	4.2	5.0	1.3	8.6
Chloroform	6.0	4.3	20	0.08
Carbon tetrachloride	4.8	5.0	20	0.002
Ethyl chloride	1.4	6.0	1.6	11
1,2-Dichloroethane	3.4	4.1	13	13
1,1,1-Trichloroethane	2.6	5.2	6.0	0.03
1,1,2-Trichloroethane	2.8	4.0	8.6	0.07
1,1,2,2-Tetrachloroethane	5.0	8.0	16	0.008
1,2-Dichloropropane	0.9	5.5	8.8	23
1,2,3-Trichloropropane	2.8	3.8	4.0	0.07
Chloroethylene	0.2	1.9	2.2	2.3
cis-1,2-Dichloroethylene	4.0	6.5	2.6	13
trans-1,2-Dichloroethylene	2.2	3.5	2.5	8.4
Trichloroethylene	2.5	10	8.5	0.02
Tetrachloroethylene	3.2	5.3	21	0.003
Tetramethyllead	2.5	13	2.5	1.5
Tetraethyllead	3.3	33	6.6	3.3

molecular weight and the number of chlorine atoms in the chlorine compound. In the case of the two alkyl lead compounds examined no large gain in sensitivity of the electron-capture detector over the ionization detectors was realised, and the sensitivity gain was only a factor of 2. The sensitivity values given for tetramethyllead and tetraethyllead in air are about the same as reported elsewhere[62] for these compounds in gasoline. Here again, the greatest sensitivity was attained at 10 V, which is lower than others have found[50,62]. It was concluded that for the analysis of volatile chlorinated aliphatic hydrocarbons, the electron-capture detector is no more sensitive than the ionization detectors for compounds with one or two chlorine atoms[61]. For compounds with three of four chlorine atoms, the electron-capture detector is 100–1000 times more sensitive that the ionization detectors. For the two alkyllead compounds tested, the electron-capture detector gives little improvement in sensitivity but its discrimination towards the lead-substituted compounds as compared with unsubstituted hydrocarbons makes it a preferable detector for analysing mixtures of these two types of compounds. Other workers[63–65] have discussed the use of an electron-capture detector for the detection of organolead compounds.

To determine tetraethyllead in air the air may be passed through a sampling tube containing the material used for packing the chromatographic column[57]. When equilibrium conditions have been established, the sample is desorbed by flushing the tube heated at about 130°C with carrier gas, and injected directly into a glass column (1 m × 0.3 mm i.d.) packed with 10% of silicone rubber SE-52 on Chromosorb P (80–100 mesh), operated at 80°C with electron-capture detection using pure nitrogen at a flow-rate of 30 ml/min as the carrier gas. The method is sensitive to down to 0.1 ppm of tetraethyllead in the air.

A very elegant analytical technique for lead alkyls combines pressure programming with use of an atomic-absorption spectrophotometer as a specific detector to produce a rapid, precise, and sensitive analytical technique[66]. A 10 ft column packed with 20% 1,2,3-tris(2-cyanoethoxy)propane on 60–80 mesh Chromosorb P coated with 1% potassium hydroxide was operated at 85°C. Flow-rates were programmed from 10 to 200 ml/min. Analysis of the five lead alkyls was completed in less than 1.5 min. The amount of lead was determined by the absorption of the lead 283.3 nm emission line. The method could detect as little as 20 ng of lead as lead alkyl. The application of atomic-absorption spectrometry to the determination of lead alkyls separated by gas chromatography has also been discussed by others[67–72]. Detection limits for water, sediments, and fish of 0.5, 0.01, and 0.025 μg/g, respectively, are claimed[70]. When a silica furnace is used in the atomic-absorption unit the sensitivity limit for the detection of lead such as tetramethyllead in sediment systems and in the atmosphere can be enhanced by three orders of magnitude.

The gas chromatograph and atomic-absorption spectrophotometer were connected by means of stainless-steel tubing (2 mm o.d.) connected from the column outlet of the chromatograph to the silica furnace of the spectrometer. A four-way valve was installed between the carrier gas inlet and the column injection port so that a sample trap could be mounted, and the sample could be swept into the GC column by the carrier gas. The lead compounds separated by GC were introduced to the centre of the furnace through a side-arm. Hydrogen was introduced at the same point at a flow-rate of 1.35 ml/min; burning hydrogen improved the sensitivity. The furnace temperature was about 1000°C. The silica furnace was mounted on top of the atomic-absorption burner and aligned to the light path. The sample trap was a glass U-tube packed with 3% OV-1 on Chromosorb W, which was immersed in a dry-ice–methanol bath at *ca.* −70°C. A known amount of gaseous air sample was drawn through the trap by a peristaltic pump. After sampling, the trap was mounted to the four-way valve and heated to about 80–100°C by a beaker of hot water, and the adsorbed compounds were swept into the GC column. Liquid samples can be directly injected to the column through the injection port, without a sample trap.

A simple, rapid extraction procedure can extract the five tetraalkyllead compounds (Me_4Pb, Me_3EtPb, Me_2Et_2Pb, $MeEt_3Pb$, and Et_4Pb) into hexane or benzene from water, sediment, and fish samples. The extracted compounds are analysed in their authentic forms by a gas chromatographic atomic-adsorption spectrometry system[73]. Other forms of inorganic and organic lead do not interfere. The detection limits for water (200 ml), sediment (5 g), and fish (2 g) are 0.50 μg/l, 0.01 μg/g, and 0.025 μg/g respectively. Although this method would be applicable to the determination of tetraalkyllead compounds originating from automobile exhausts in water, fish, and sediment samples, the main interest was in the determination of organically bound lead produced by biological methylation of inorganic and organic lead compounds in the aquatic environment by microorganisms[74,75]. The extract was injected directly into the column injection port of the chromatograph. Ionic forms of lead such as lead(II), diethyllead dichloride, and trimethyllead acetate do not extract in the solvent phase. Tetraalkyllead compounds have high vapour pressures and are not stable in water[73]. In water containing 4.2 μg/l of tetramethyllead, the level decreased to 2.8 and 3.9 μg/l when stored at room temperature and at 4°C overnight, respectively. For this reason, water samples should not be filtered by suction but should be extracted with hexane immediately after collection. Another procedure[76] for determining tetraalkyllead compounds in fish samples employs vaccuum extraction of the tetraalkyllead into a cold trap under liquid nitrogen, followed by solvent extraction of the condensate for gas chromatographic determination.

Organic lead compounds have been trapped from street air and eluted directly into the flame of an atomic-absorption spectrometer, thus determining total organic lead[77]. In such a study it would be an advantage to employ furnace atomization[78] since organic lead exists in air at very low levels and the furnace can give a detection limit gain of up to 3 orders of magnitude. A gas chromatographic–atomic-absorption spectrophotometric technique for alkylleads in which a sample can be analysed in 5 min with a detection limit of 0.2 ppm of lead is suitable for determining trace lead in unleaded gasoline[79].

A special carbon furnace atomizer attached directly to a gas chromatographic column packed with 20% tricresyl phosphate on Chromosorb W operated at 100°C exhibits high sensitivity and eliminates many of the problems involved with interferences encountered with furnace atomization in the determination of organolead compounds in gasoline and air[80]. Many of the problems involved with commercial carbon atomizers persist when they are used in GC–AAS combination systems which are often unreliable. The commercial systems are certainly capable of performing analyses at very high levels of analytical sensitivity and precision, but the development of reliable quantitative procedures is much more difficult. The modified atomizer is left hot at all times when in use. The effluent from the gas chromatograph enters the base of the atomizer, where the gaseous sample is decomposed and atomization takes place. The atoms flow into the cross-piece, which is in the optical light path. The advantage of the process is that the peak of the solvent used is separate from the peak of the metal-bearing component on the gas chromatogram. The gas chromatograph separates the metal-bearing components from the rest of the material, which eliminates many of the problems encountered in the solvent evaporation step and other matrix effects. Decomposition is fairly rapid, although several seconds elapse from the time that the sample enters the carbon atomizer before the atoms reach the optical light path. This permits chemical decomposition to take place and virtually eliminates chemical interference, which is usually caused by varying rates of atomization from different compounds rather than by prevention of decomposition. Even if the rate varies, decomposition is virtually complete before the free atoms enter the light path.

Gas chromatography–mass spectrometry can be used to identify the separate organolead compounds in air[78,81]. The alkyl lead compounds are condensed from a 70-l air sample in a series of four traps at −72°C, separated by gas chromatography and determined at the 283.3 nm lead resonance line by atomic-absorption spectrometry with electrothermal atomization.

The hydrogen atmosphere flame-ionization detector (HAFID), introduced in 1972[82–86] and subsequently developed, is a sensitive and selective gas chromatographic detector for organometallic compounds. It has a selectivity for tetramethyllead over dodecane of 10^4 with a detection limit of 51 pg of tetraethyllead injected and is able to detect tetraethyllead in a gasoline diluted 1:100 with gasoline[85,86]. Minimum detectable amounts for certain metal-containing compounds extend to the low picogram and sub-picogram range with selectivities of 10^4 and 10^5 when compared to *n*-hydrocarbon responses. Because of the HAFIDs simple design and the high sensitivity and selectivity to organometallic compounds, an optimized design has been applied to the routine determination of organolead compounds in gasoline[87]. By a simple 1:10 dilution of a leaded gasoline, alkyllead compounds were detected with no interference from overlapping chromatographic peaks of hydrocarbons. Detection limits were calculated to be 7.3×10^{-12} g/l of lead.

Various workers[88,89] have examined the applicability of helium in microwave glow discharge detectors for the detection of organolead compounds leaving a gas chromatographic column. This uses the TM_{010} resonant cavity to sustain a plasma in helium at atmospheric pressure. The effluent from the gas chromatograph is split between a flame-ionization detector and a heated transfer line directing it to a small auxiliary oven containing a high-temperature valve. The valve allows the effluent to be directed either to a vent or to the plasma. Atomic emission at 283.3 nm from the lead entering the discharge is observed axially with an échelle grating spectrometer. The system allows for highly selective and sensitive detection of lead by monitoring an appropriate wavelength. A detection limit of 0.49 pg/l is claimed for this detector.

Another detector involves eluting the compounds from the gas chromatograph and directing them into a microwave discharge which is sustained in either argon or helium[90]. Observation of the optical emission spectrum resulting from the fragmentation and excitation of compounds entering the plasma affords sensitive element-selective detection. The applicability of a gas chromatograph coupled with a microwave plasma detector (GC–MPD) for the determination of tetraalkyllead species in the atmosphere has been discussed[88]. The tetraalkyllead species are collected by a cold trap. The volatile lead species are concentrated within an organic solvent, separated by a gas chromatographic column containing 3% OV-1 on Chromosorb W at 80°C, and determined by an MPD system which measures the emission intensity of the lead 405.78 nm spectral line. Previous workers have used an acidic solution of hydrochloric acid[47,48], activated charcoal[91,92], Apiezon L on a silanized universal support[93], the chromatographic support OV-1, and silicone rubber on Chromosorb P[94] to collect alkyllead compounds from the atmosphere. A cold trap containing SE-52 on Chromosorb P at −80°C has also been used for the collection at atmospheric alkyllead compounds[88].

XII. ORGANOLITHIUM COMPOUNDS

Hydrolysis coupled with identification of the gas produced by mass spectrometry[95] and by infrared analysis[96,97] has been used for the determination of organolithium compounds. A gas chromatographic method[98] for determining vinyl lithium in tetrahydrofuran and in diethyl ether is based on the reaction

$$CH_2{=}CH{-}Li + H_2O \longrightarrow CH_2{=}CH_2 + LiOH \qquad (1)$$

The observation that vinyllithium gave low recoveries in the double titration procedure, coupled with the observation that some samples contained small amount of lithium acetylide as an impurity, prompted a search for an alternative specific method of assay for vinyllithium[98]. This method involves preliminary hydrolysis of the sample followed by separation of the ethylene and acetylene produced on columns of 20% dimethyl phthalate on 60–80 mesh firebrick at 75°C or of 20% dimethyl sulphalone on 60–80 mesh firebrick at 35°C. There is good agreement between this gas chromatographic method and the vanadium pentoxide method[99]. In a gas chromatographic procedure for phenyllithium, a measured volume of phenyllithium solution is slowly transferred under nitrogen into an excess of a solution of iodine in diethyl ether[100]. The excess of iodine is removed from the ethereal phase by shaking with dilute sodium hydroxide solution and the concentration of iodobenzene in the ethereal phase is then determined by gas chromatography. Other organic compounds of lithium do not interfere. The accuracy was within 2%.

XIII. ORGANOMAGNESIUM COMPOUNDS

Gas chromatography has been used in the analysis of *n*-propylmagnesium bromide[101]. The organomagnesium compound is reacted with an excess of diisopropyl ketone to form a compound which at elevated temperatures produced propylene in almost quantitative yield. The propylene was determined gas chromatographically. A method for the analysis of arylmagnesium compounds, particularly *p-t*butylphenylmagnesium bromide, distinguishes between active Grignard reagent, RMgX, and hydrolysed Grignard reagent, Mg(OH)X[102]. Polypropylene glycol supported on firebrick was used as column packing and good agreement was obtained between this method and acid titration procedures. In a further gas chromatographic method for the analysis of Grignard reagents the sample is treated with a large excess of tributyltin chloride in tetrahydrofuran (reaction 2) and the magnesium salts formed (or their

$$\underset{}{CH_2{=}CHMgCl} + \underset{\text{excess}}{Bu_3SnCl} \xrightarrow{\text{thf}} Bu_3SnCH{=}CH_2 + Bu_3SnCl + MgCl_2 \quad (2)$$

tetrahydrofuran derivatives) are precipitated with hexane and filtered off[103]. The filtrate is heated to evaporate the solvent, and the residue, containing unconsumed butyltin compounds, is analysed chromatographically. This method distinguishes the vinylmagnesium from other titratable compounds resulting from hydrolysis, oxidation, or decomposition of the Grignard compounds. After a vinyl Grignard compound has been stored for some time, then hydrolysed with dilute acid, the gases produced, in addition to the expected ethylene, also contain considerable amounts of ethane, hydrogen, and C_4 and C_5 alkanes, which render gas volumetric procedures unsuitable for the assay of such samples[103].

A method, similar in principle to this last method, consists of coupling the reagent with an excess of dimethylphenylchlorosilane in ether and determining the amount of trimethylphenylsilane formed by gas chromatography using cumene as an internal standard[104]. In a further method, solutions of organomagnesium derivatives are analysed by hydrolysis with concentrated phosphoric acid in a micro-reactor and the hydrocarbons evolved are passed through a by-pass injector into a gas chromatographic apparatus (with an activated silica gel column, for C_1–C_3 hydrocarbons)[105]. Concentrations of alkyl groups are then calculated from the peak areas in the usual way.

XIV. ORGANOMANGANESE COMPOUNDS

A procedure for the analysis of manganese antiknock additives in gasoline uses a hydrogen atmosphere flame-ionization detector[106]. Detection limits are calculated to be 1.7×10^{-14} g/l of manganese. Methylcyclopentadienylmanganesetricarbonyl [$(CH_3C_5H_4)Mn(CO)_3$] in gasoline has been determined by gas chromatographic separation with interfaced specific manganese detection by means of d.c. argon plasma emission spectroscopy[107]. The procedure is rapid, free from interferences, specific, and requires little sample preparation. The use of cyclopentadienylmanganesetricarbonyl as an internal reference yields a relative standard deviation of 0.8–3.4%. The limit of detection is approximately 3 ng of manganese metal as the complex. The gas chromatograph was adapted for on-column injection on to a 6 ft × $\frac{1}{8}$ in o.d. stainless-steel column packed with 2% Dexsil 300 GC on Chromosorb 750, 100–120 mesh. The column effluent was split by an approximately 1:1 ratio between the flame-ionization detector of the gas chromatograph and a heated, thermal, and electrically insulated $\frac{1}{16}$ in o.d. stainless-steel transfer line to the d.c. plasma. Pre-heated argon sheath gas was required in addition to the argon supplied to sustain the plasma, in order to optimize spectral sensitivity. The column and injection port temperatures were set at 130 and 160°C, respectively, and the interface temperature was 170°C. The helium carrier gas flow-rate was 25 ml/min.

An alternative method for the determination of methylcyclopentadienylmanganese in amounts down to 1.7×10^{-14} g/l of manganese uses a hydrogen atmosphere flame-ionization detector (HAFID) modified from a commercial FID[87]. With the HAFID detector, the temperature was maintained at 250°C and the total hydrogen flow-rate was held at 1600 ml/min and for optimal response was doped with 34 ppm of silane by mixing pure hydrogen doped with 100 ppm of silane. Air at 120 ml/min was enriched with 150 ml/min of oxygen before entering the jet tip. With the FID detector, the detector temperature was maintained at 250°C. The flow-rates used were 30 ml/min for hydrogen and 240 ml/min for air. A further HAFID detector method uses a helium carrier gas flow-rate of 50 ml/min and an injection point temperature of 200°C[89]. The wavelength setting of the monochromator was optimized for manganese using a hollow-cathode lamp and a small mirror placed between the lens and the cavity. The wavelength setting was optimized by introducing small amounts of methylcyclopentadienylmanganese vapour into the plasma by connecting with a tee to a hydrocarbon solution of this compound.

XV. ORGANOMERCURY COMPOUNDS

In view of the comparatively high mercury content of fish found[108] in Swedish lakes and rivers[109–112], an extensive survey of the nature and the concentrations of mercury in fish from these waters has been made. Although several authors have described methods for the determination of organic mercury compounds, many either do not separate different compounds, such as methylmercury from phenylmercury compounds, or are designed for mercury contents higher than those usually encountered in foods. A combined gas chromatographic and thin-layer chromatographic method[109,110] identifies and determines methylmercury compounds in fish, animal foodstuffs, egg-yolk, meat, and liver, and a combination of gas chromatography and mass spectrometry has been used to identify and determine methylmercury compounds in fish[112]. To extract organically bound mercury from muscle tissue of fish the fish was homogenized with water and acidified with concentrated hydrochloric acid[109]. Organomercury(II) compounds were then extracted in one step with benzene. Methylmercury, either originally present or added to the fish, could be extracted, although with difficulty. From an aliquot of the benzene solution organomercury could be extracted with ammonium or sodium hydroxide solution, saturated with sodium sulphate, for elimination of lipids. The yields were low and variable, but could be improved by modifications to the sample working up procedure (not discussed here) prior to gas chromatography. An electron-capture detector was used together with a column containing Carbowax 1500 (polyethylene glycol, average mol. wt. 1500), 10% in Teflon 6, 35–60 mesh or on Chromosorb W, acid-washed DMCS, 60–80 mesh. Carbowax 20M has also been used. The nitrogen flow-rate was 65 ml/min, the temperature of the column was 130–145°C, and the temperature of the injector was 150–170°C. The peak of each sample solution was compared with the peak of a standard solution with about the same concentration of methylmercury. The purified benzene extracts of fish gave peaks with the same retention as methylmercury chloride.

A modification to this method renders it applicable to a wider range of foodstuffs (egg-yolk and egg-white, meat, and liver) by binding interfering thiols in the benzene extract of the sample to mercury(II) ions added in excess or by extracting the benzene extract with aqueous cysteine to form the cysteine—methylmercury complex[110,111]. Methylmercury compounds in fish can be identified and determined by combined gas chromatographic–mass spectrometric analysis and also by using a standard gas chromatograph with an electron-capture detector for detecting organic halogen compounds. Other methods for the determination of organomercury compounds in fish, biological materials, and water have used atomic-absorption, mass spectrometric, microwave emission and electron-capture detectors[113–120]. The separation of compounds of the type RHgBr, where R is methyl, ethyl, propyl, or *n*-butyl, has been achieved using hydrogen as the carrier gas and a column packed with Dow Corning silicone 550 and maintained at 190–220°C[121]. The sample (50 μl) was introduced into the column as a 10% solution in tetrahydrofuran.

Methods have been described for the determination of alkylmercury compounds in sediments[122], urine[123], hair[124], and blood[125]. All involve an initial extraction of methylmercury from the sample as a halide[126] with an organic solvent, followed by a clean-up prior to gas chromatography. A procedure based on reductive combustion in a flame-ionization detector of a conventional gas chromatograph followed by the cold vapour atomic-absorption detection of mercury vapour has been described for the determination of organomercurials in bacterial respirant gases[127,128]. Various workers have discussed clean-up procedures for removing fatty acids and amino acids from samples prior to gas chromatography for organomercury compounds. These would otherwise poison the column. The clean-up is achieved by adding to the organic phase a reagent, such as sodium sulphide[129], cysteine[130–132], sodium thiosulphate[115], or glutathione[133], which forms a strong water-soluble alkylmercury complex to extract the mercury complex into the aqueous phase. A halide is added to the aqueous phase, and the alkylmercury halides formed are re-extracted into an organic phase. Aliquots of this phase are finally injected into the gas chromatograph.

More recent work on the determination of alkylmercury compounds has centred on the applicability of the helium microwave glow discharge detector as a gas chromatographic detector[134–137]. An atmospheric pressure helium (or argon) plasma is used for the detection of diphenylmercury as this leads to high sensitivity and high optical resolution and selectivity[137]. A helium carrier gas flow-rate of 70 ml/min and an injection point temperature of 200°C are used. The wavelength setting of the monochromator was optimized for mercury using a hollow-cathode lamp and a small mirror placed between the lens and the cavity whilst introducing small amounts of dimethylmercury vapour into the plasma from a hydrocarbon solution. The detector response is significantly affected by the total flow-rate of helium through the discharge tube, remaining constant over the range 42–50 ml/min, then decreasing sharply with increasing flow-rate. The very large selectivity ratio (ratio of peak response per gram-atom of mercury to the peak response per gram-atom of carbon as dodecane) obtained for mercury results from a combination of two factors: (i) the high sensitivity observed for this element, and (ii) the favourable wavelength region employed with respect both to optical resolution of the monochromator and the minimal interference by molecular band emission from hydrocarbons. The detection limit, defined as the main flow-rate of element entering the plasma required to produce a signal-to-noise ratio of 2, was 1 pg/l with a selectivity 9.1×10^4 [137].

XVI. ORGANOMOLYBDENUM COMPOUNDS

The gas chromatography of molybdenum tricarbonyls has been discussed[138].

XVII. ORGANONICKEL COMPOUNDS

High-pressure gas chromatography above critical temperatures has been used to separate nickel actioporphyrin(II) and nickel mesoporphyrin(IX) dimethyl esters[139]. Dichlorodifluoromethane (critical temperature 112°C) at a starting pressure of 1830 psi was used to separate the two nickel porphyrins in 1-mg amounts on a column of polyoxyethylene glycol (33%) on Chromosorb W at 150–170°C. The porphyrins do not move on the column at a gas pressure of less than 600 psi.

XVIII. ORGANOPHOSPHORUS COMPOUNDS

The micropyrolytic gas chromatographic technique has been applied to the identification of organic radicals in organic phosphates and metal dialkylthiophosphates[140]. The compound is pyrolysed in the inlet system of a gas chromatograph, and the volatile flash pyrolysis products, generally olefins, are fractionated and collected individually for identification by mass or infrared spectrometry. The olefins are formed generally by the breaking of a carbon—oxygen bond and abstraction of hydrogen from a beta-carbon atom with no skeletal isomerization. Thus, the structures of the olefins are directly related to the structure of the alkyl groups initially present. Only when hydrogen is not available on a beta-carbon atom as in neopentyl radicals are olefins formed by carbon-skeletal rearrangement. Legate and Burnham determined the exact configuration of the alkyl radicals in several model organic phosphates and metal dialkyl thionothiophosphates and described a gas chromatographic inlet system suitable for pyrolysis or for conventional vaporization[140]. Statistical designs have been described for the optimization of the nitrogen–phosphorus gas chromatograph detector response[141].

XIX. ORGANORUTHENIUM COMPOUNDS

Ruthenocene derivatives have been separated on 2,2-dimethylpropane-1,3-diol adipate, polyoxyethylene glycol adipate, polyoxyethylene glycol M-20, polyoxypropylene glycol, and Apiezon L (1–5% on Celite 545, 80–100 mesh) at 100–200°C[43]. Packed columns (1–1.2 m × 0.4 cm i.d.) in glass and stainless-steel tubes, and a capillary column 45 m × 0.25 mm i.d.), were used, together with a β-ray ionization detector. The best results were obtained with the use of 2,2-dimethylpropane-1,2-diol and polyoxyethylene glycol M-20 columns.

XX. ORGANOSELENIUM COMPOUNDS

The gas chromatography of a range of organoselenium compounds including dialkyl diselenides, dialkyl selenides and ethyl selenocyanate has been studied using a hydrogen flame-ionization detector and an electron-capture detector[142]. Three column packings were examined: (i) 5 ft × $\frac{1}{8}$ in, 20% polymetaphenyl ether on 60–80 mesh Chromosorb W coated with hexamethyldisilazane; (ii) 10 ft × $\frac{1}{8}$ in, 20% Carbowax 20M on 60–80 mesh Chromosorb W treated with hexamethyldisilazene; and (iii) 10 ft × $\frac{1}{8}$ in, 20% silcone oil DC 550 on 60–80 mesh Chromosorb W coated with dimethyldichlorosilane. Nitrogen was used as the carrier gas at flow-rates between 20 and 30 ml/min. Using the hydrogen flame-ionization detector, the retention times of the alkylselenium compounds were determined on each of the three columns at column temperatures within the range 35–175°C. The injector temperature was set at 50–100°C higher than the column temperature. One per cent solutions of each selenium compound in carbon disulphide were used for the determinations, as carbon disulphide gives very little response with this detector system. For the preparative-scale gas chromatographic purification of organoselenium compounds a 5 ft × $\frac{1}{4}$ in column packed with silicone fluid (methyl) SF-96 on 60–80 mesh firebrick was used. Helium was the carrier gas at a flow-rate of 35–40 ml/min. Figure 3 shows the complete resolution of the seven alkylselenium compounds on the polymetaphenyl ether column at a column temperature of 150°C[142].

A procedure utilizing a commercial atomic-absorption spectrophotometer with a heated graphite tube furnace atomizer linked to a gas chromatograph has been used for the determination of dimethylselenium (and also trimethylarsenic and

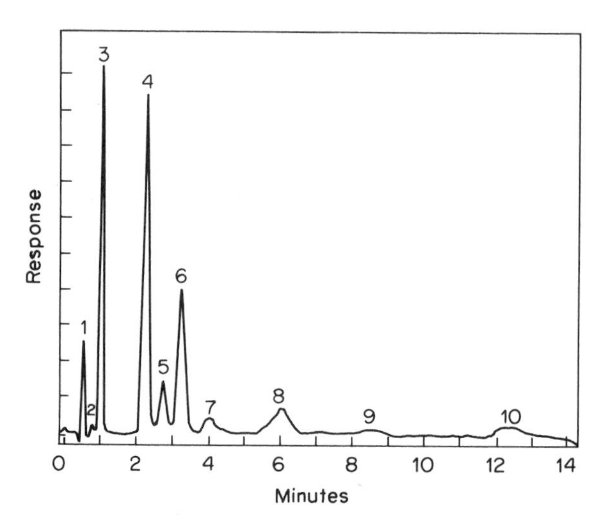

FIGURE 3. Separation of alkylselenium compounds on a poly–phenyl ether column, with a hydrogen flame-ionization detector. Column temperature, 150°C; injector temperature, 225°C; nitrogen carrier gas flow rate, 25 ml/min. 1% Solution of each compound in carbon disulphide, (1) Dimethyl selenide; (2) carbon disulphide; (3) diethyl selenide; (4) dipropyl selenide; (5) dimethyldiselenide; (6) ethyl selenocyanate; (7) methyl ethyl diselenide (?); (8) diethyl diselenide; (9) ethyl propyl diselenide (?); dipropyl diselenide.

etramethyltin) in respirant gases produced in microbiological reactions[143]. Atomic absorption and microwave plasmas have both been used as detectors[144,145]. Carbon diselenide can be determined gas chromatographically[146].

XXI. ORGANOTIN COMPOUNDS

The gas chromatography of organotin compounds has been extensively studied over the past 30 years[147]. The gas chromatography of σ- and π-bonded organotin compounds is difficult because of their instability towards oxygen and moisture and their thermal instability[148,149]. If a solid column support is insufficiently covered by the stationary liquid phase, e.g. 2–5%, adsorption on the exposed siliceous sites becomes significant with polar solutes, and tailing occurs. As a consequence, retention volumes are no longer directly proportional to the weight of solvent, and hence specific retention volumes can only be measured with columns containing a high proportion of stationary phase. Where organometallic solutes are involved, this adsorption becomes very important, and the band spreading is so extensive that squalane columns cannot be used for analysis of mixtures of such materials. Chemical instability gives rise to chemical change as the compound passes through the chromatographic column. This usually occurs through formation of bonds between the compound and reactive groups either on the column support (e.g. acid sites), or the stationary phase (e.g. hydroxy groups, as in the polyethylene glycol). This phenomena, termed trans-esterification[150], has been observed with organotin hydrides, chlorosilanes, and amino compounds such as hexamethyldisilazane. An early approach to pre-treatment of the support to remove this activity was to add small amounts of highly polar and involatile liquids to the support[150,151] or wash the support with acid and then alkali[152–154]. More recently there have been attempts to deposit solids such as silver on the support surface[155], but unfortunately this method cannot be used in the presence of thio compounds, e.g. silylthioethers. An alternative method is to treat the active sites of the support, which are presumed to be hydroxyl groups (—Si—O—H), and replace these by groups which should yield at least a weakly absorbing site. Both trimethylchlorosilane[156,157] and dimethyldichlorosilane[158,159] have been used successfully to reduce the activity. The surface reaction is presumed to be of the following type:

$$(CH_3)_3SiCl + \text{—}\overset{\overset{\large OH}{|}}{\underset{|}{Si}}\text{—} \longrightarrow \text{—}\overset{\overset{\large O-Si(CH_3)_3}{|}}{\underset{|}{Si}}\text{—} + HCl \qquad (3)$$

or:

$$\text{—}\overset{|}{\underset{\underset{\large OH}{|}}{Si}}\text{—O—}\overset{|}{\underset{\underset{\large OH}{|}}{Si}}\text{—} + (CH_3)_2SiCl_2 \longrightarrow \text{—}\overset{|}{\underset{\underset{\large O}{|}}{Si}}\text{—O—}\overset{|}{\underset{\underset{\large O}{|}}{Si}}\text{—} + 2HCl \qquad (4)$$

(in the product of (4) the two O atoms are bridged by $SiMe_2$)

When the hydroxyl groups are not adjacent, then a chlorosilyl ether group, $SiOSi(CH_3)_2Cl$, may be left, but this is not beneficial since it will be as reactive as the grouping replaced, because of the chlorine grouping. As an alternative to this treatment, hexamethyldisilazane has been used since it reacts quantitatively with hydroxyl groups[150,160,161]. It is now used to treat all the common solid supports[162]. Many advantages have been claimed for hexamethyldisilazane but it is expensive and gives a surface similar to the trimethylchlorosilane.

In the gas chromatography of organotin compounds, careful consideration must be given to the detector and its design since often when a compound is eluted from a column, decomposition occurs in the detector, invalidating the elution process[148]. When such decomposition occurs, the metal is deposited on the wires of filaments of the katharometer, or on the collector plates of a flame-ionization gauge, possibly causing the formation of tarry and finally carbonaceous deposits which foul the katharometer filaments[163]. The recommended treatment in such cases is regular flushing of the detector block with both polar and non-polar solvents. Although such treatment is beneficial, in the course of time the tarry deposits carbonize, leading to permanent changes in the katharometer resistance. The partial contacts of carbon deposits between helices of the coiled filament presumably are responsible for the increase in recorder base-line noise. When finely divided powder metal is deposited in the katharometer, especially on the filaments, a similar situation arises, but the bridge becomes permanently out of balance, since unfortunately no cleaning procedure can be used.

A similar situation is found with the flame-ionization detector, especially the conventional types where the collector electrode plate is vertically above the flame. A modified detector is needed and even when detection can occur, attention must be paid to saturation limits, since non-linearity of signal response and the inversion effect have been observed[164]. Trans-esterification of organotin compounds can be overcome by treatment of the supporting phase involving baking of Celite 545 (36–60 mesh B.S.S.) at 300°C for 5 h, acid and alkali washing, drying at 50°C, and treatment with trimethylchlorosilane[148]. After such treatment it is possible to chromatograph and separate the methylchlorosilanes and organotin hydrides. However, as indicated later, the choice of stationary phase is important for this type of compound. The gas–liquid chromatography of thermally unstable organometallic compounds has been effected[148] using separation techniques at normal temperatures (20–100°C) followed by combustion in a conventional micro-analytical furnace, absorption of water, and detection of the carbon dioxide with a Stuvé katharometer[165]. The metal oxide deposited in the furnace gradually poisons the copper oxide packing, and has to be replaced frequently. The 3 ft column contained 25% (w/v) di-2-ethylhexyl sebacate on Celite (36–60 mesh) at 56°C and the carrier gas was oxygen-free nitrogen at a flow-rate of 50 ml/min. Tin tetraalkyls and related compounds could be detected by a thermal conductivity cell, a modified flame-ionization gauge, and a commercial gas-density balance unit[166]. The latter unit has many advantages for such work, particularly that the sample is not subjected to a temperature greater than the column temperature, presumably a temperature at which the compound is stable. Results comparing these three detectors with tetramethyltin under the appropriate conditions are given below.

Thermal conductivity detector (Figure 4): Column, 6 ft, 25% (w/v) Apiezon M on treated Silocel (36–60 mesh); column temperature, 140°C; detector temperature, 150°C; carrier gas, hydrogen, 30.0 ml/min; recorder, 5 mV f.s.d. From Figure 4 it is obvious that the graph is linear for low sample volumes, but above 2.5 μl the thermal conductivity response is no longer linear[167]. The relative detector responses (relative to tetramethyltin = 100) are as follows: tetramethyltin, 100; *n*-hexane, 98; cyclohexane, 92; *n*-heptane, 87; and benzene, 116.

Flame-ionization detector: Column, 6 ft, Apiezon M on Silocel (36–60 mesh); column temperature, 140°C; bleed off, 98%; flow-rate of nitrogen through column, 7 ml/min; hydrogen flow-rate, 30 ml/min; air flow rate, 400 ml/min. The response results are shown in Figure 5, which indicates that the graph is again linear over the lower ranges. Figure 6 shows the saturation phenomena for a 4.0 μl sample size or above, compared with the shape of the elution peak for a 3.0 μl sample size.

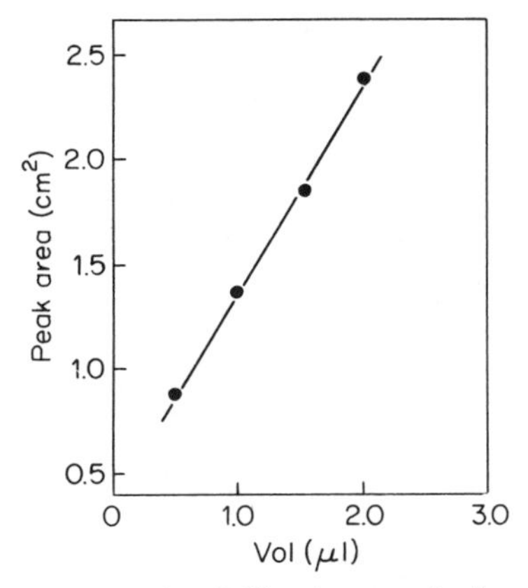

FIGURE 4. Calibration graph for tetramethyltin using a thermal conDuctivity detector.

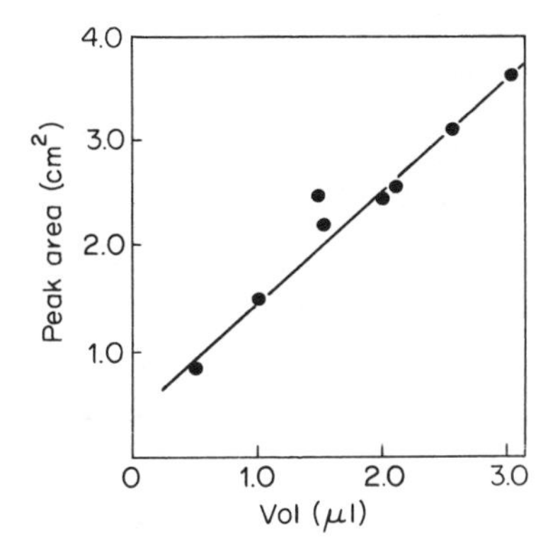

FIGURE 5. Calibration graph for tetramethyltin using a flame-ionization detector.

Gas-density balance detector: Column, 6 ft, 15% (w/w) silicone oil E301 on Celite 545 (36–60 mesh); column and balance temperature, 100°C; nitrogen flow-rate, 30 ml/min. Because of the non-linearity of signal response, the tetramethyltin was diluted with *n*-heptane in the ratios 1:1, 1:5, and 1:10.

The relative sensitivities are 1:10:1.700 for thermal conductivity, gas-density balance and flame-ionization detectors, respectively. The gas-density balance detector is excellent for the gas chromatography of alkyltin mono- and dihydrides without decomposition on the column[148]. It was possible to elute trimethyltin hydride through silicone E301, Apiezon L, and dinonyl phthalate phases, but when attempts to use a squalene column were made, decomposition occurred. A discussion[168] of the use of the flame-ionization detector in the gas chromatography of tetramethyltin, trimethylethyltin, dimethyldiethyltin, methyltriethyltin, and tetraethyltin included the usual sensitivity characteristics of the chromatography of these compounds and presented and correlated specific retention data and various thermodynamic

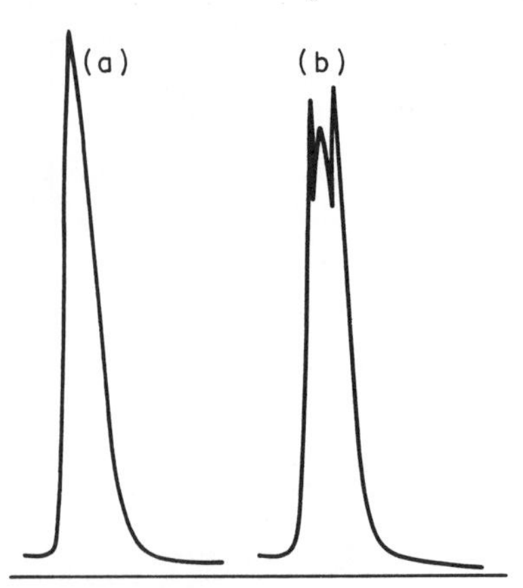

FIGURE 6. Elution pattern of tetramethyltin using a flame-ionization detector. (a) Normal response; (b) saturation response.

properties obtained on columns containing 15% silicone oil E301 (mol. wt. 700 000) on Celite[169] (treated by dry sieving to 36–60 mesh, washing with concentrated hydrochloric acid, methanol, and distilled water, and drying at 300°C).

The retention indices of 14 organotin compounds have been determined on columns of (i) 40% of Apiezon L on Celite, (ii) 20% of Carbowax 1500 on diatomaceous earth, and (iii) 20% of polyethylene glycol succinate polyester on diatomaceous earth (Table 4)[170,171]. The columns were all operated at 190°C, and before use were conditioned at

TABLE 4. Retention indices of organotin compounds

Compound	Apiezon L on Celite column	Carbowax 1500 on diatomaceous earth column	Polyethylene glycol succinate on diatomaceous earth column	B.p. (°C)	mol. wt.
Me_4Sn	630	676	692	77.4	178.8
Et_4Sn	1049	1074	1097	179.5	234.9
Pr_4Sn	1327	1347	1352	223.5	291.0
iPr_4Sn	1355	1364	1371		291.0
Bu_4Sn	1599	1606	1642	267.5	347.1
i-Bu_4Sn	1466	1515	1489		347.1
Me_3EtSn	728	745	744	108.2	192.7
Me_3PrSn	833	874	836	131.7	206.9
$Me_3(i$-Pr)Sn	794	849	818	123	206.9
$Me_3(t$-Bu)Sn	820	856	848	134	220.8
Me_3(Cyhex)Sn[a]	1208	1243	1300		247.0
Me_3ViSn[b]	703	800	800	99.5	190.7
Me_3Sn	1172	1498	1531	203	240.9
Vi_4Sn[b]	911	1153	1188	160	226.8
		Av. 1114	Av. 1122		

[a]Cyhex = cyclohexyl.
[b]Vi = vinyl.

200°C overnight and pre-treated with hexamethyldisilazane and dichlorodimethylsilane at room temperature. The nature of the compounds had little affect on the retention times of alkanes except on column (i), when the number of carbon atoms was less than 10. A correlation between retention index and boiling point existed for column (i), so that the retention index could be predicted from the boiling point with an average error of 1%. Distinct differences between saturated and unsaturated compounds were observed on the polar columns (ii) and (iii), and average values of retention index divided by molecular weight increased, with two exceptions, as the refractive index increased. A correlation of increasing retention index with increasing calculated molar refraction applied without exception to (i), and with three exceptions to (ii) and (iii).

The retention times, alone or in the presence of the organotin compounds, are given in Table 5. With Apiezon substantial deviations appeared below ten carbon atoms. the lack of linearity of the logarithm of the retention time on the polar columns is evident. Slight deviations appear above fourteen carbon atoms. A graph (Figure 7) of the retention index on Apiezon *versus* the total number of carbon atoms shows that the homologous series tetraethyl, tetrapropyl, tetrabutyl increases almost linearly, but that compounds not in this series deviate irregularly owing to boiling point and polarity differences. A plot (Figure 8) of the boiling points, where known, *versus* the total number of carbon atoms shows some degree of regularity. A plot (Figure 9) of the Apiezon retention index *versus* boiling point, where known, shows considerably more uniformity despite the variations in structure represented. From such a plot the retention index can be predicted from the boiling point with an average error of 1%, the poorest case being that of tetravinyltin, where the error is 6%.

Table 4 gives the retention data for the non-polar Apiezon column and the polar columns containing Carbowax 1500 and polyethylene glycol succinate on diatomaceous earth. As indicated by the retention times, the retention indices increase with the

TABLE 5. Retention times (minutes) of alkanes in the presence of organotin compounds. Column 1 gives the time for the alkanes alone, column 2 gives the time in the presence of the organotin. The columns were pre-conditioned at 200°C overnight. Each column was pre-treated before a series of runs with 20 μl of hexamethyldisilazane and 20 μl of dimethyldichlorosilane at room temperature. All runs were made at 190°C

	Apiezon L on Celite		Carbowax 1500 on diatomaceous earth		Polydiethyleneglycol succinate on diatomaceous earth	
Alkane	1	2	1	2	1	2
C_4						
C_5		0.10		1.75	1.75	1.78
C_6		0.24		1.81	1.81	
C_7		0.47	2.00	2.00		
C_8	0.99	0.87	2.30	2.30	2.11	
C_9	1.61	1.20	2.74	2.77	2.27	2.29
C_{10}	2.59	2.50	3.36	3.37	2.52	2.52
C_{11}	4.10	4.20	4.25	4.25	2.91	2.90
C_{12}	6.52	6.64	5.57	5.70	3.64	3.53
C_{13}	10.34	10.32	7.47	7.60	4.25	4.27
C_{14}	16.33	16.60	10.26	10.60	5.43	5.51
C_{15}	25.81	26.16	14.54	15.00	7.23	7.24
C_{16}	40.28	41.00	20.59	21.50	10.00	10.69
C_{17}		64.19		30.96		

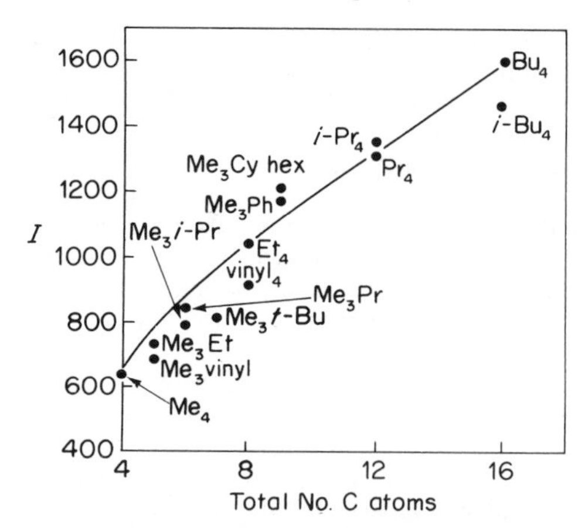

FIGURE 7. Retention indices (I) of organotin compounds on Apiezon L at 190°C versus total number of carbon atoms in the molecule.

polarity of the column, as expected. A plot (Figure 10) of the retention index on the Carbowax 1500 column *versus* boiling point shows distinct differences between the saturated and unsaturated compounds. The latter group included trimethylvinyltin, tetravinyltin, and trimethylphenyltin, all of which fall on a separate curve showing the substantial effect of their higher polarities in increasing their retention indices. A similar plot of the results on the polar polyethylene glycol succinate column shows the same effect.

An extensive number of retention times for the methylethyltin compounds as a function of temperature have been determined for a packing consisting of Sterchamol

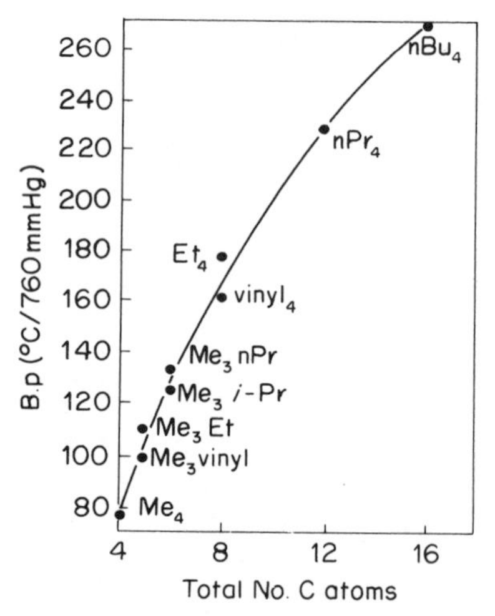

FIGURE 8. Boiling points of organotin compounds versus total number of carbon atoms in the molecule.

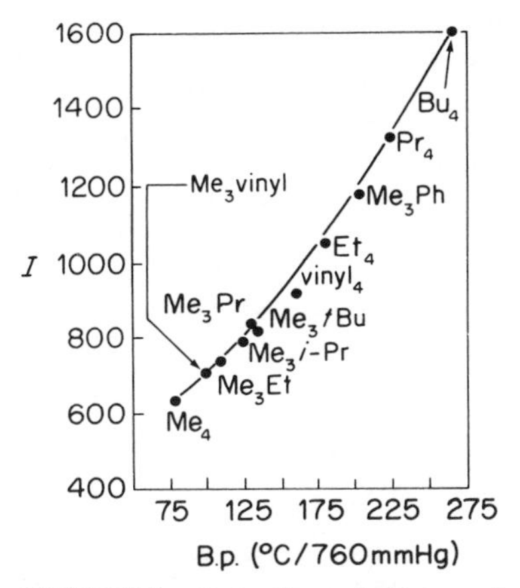

FIGURE 9. Retention indices of organotin compounds on Apiezon L at 190°C versus boiling point.

brick impregnated with 15% of high-vacuum stopcock grease (Zeiss No. 20) operated isothermally at 70–95°C and using argon as the carrier gas[172]. Plots of log (retention volume) *versus* number of chain carbon atoms are straight lines for $Sn(n\text{-}C_3H_7)_4$ through $Sn(n\text{-}C_4H_9)_4$ compounds[173]. Retention indices of several organotin compounds on columns containing Apiezon L, SE-30, QF-1, XF 112, and *o*-nitrotoluene as stationary phases and with flame-ionization and thermal conductivity detectors have been determined[174] in accordance with the recommendations of the Gas Chromatography Discussion Group Sub-Committee[175]. Some rules and generalizations of character relationships, based on ΔI values and retention index increments (δI), were established for successive homologous compounds.

A number of workers have reported aspects of the use of gas chromatography to determine tetraalkyltin compounds[143,176–182]. Specific mention has been made of the

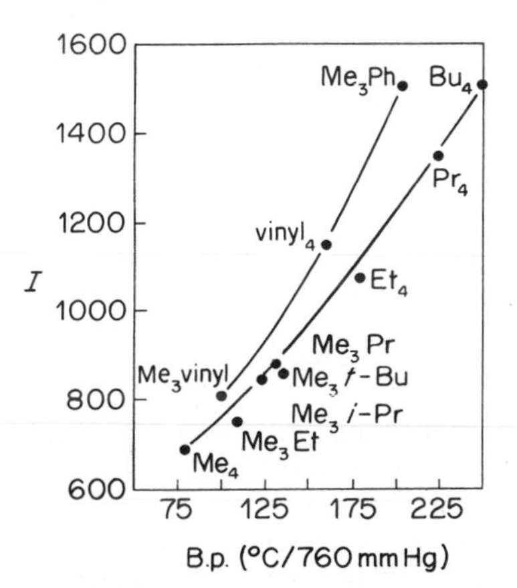

FIGURE 10. Retention indices of organotin compounds versus boiling point. Carbowax 1500 on diatomaceous earth column, 190°C.

use of Apiezon L columns[176], sample injection in the absence of air or water[180], flame-ionization[181,182], electron-capture[182], and atomic-absorption detectors[143]. Tetraalkyltin compounds have been separated and purified by preparative gas chromatography[183]. Diethyldimethyltin was separated from impurities (mainly ethyltrimethyltin and triethylmethyltin) by passing a 6 ml sample through a column (13 m × 25 mm i.d.) of Kieselguhr (particle size 0.4–0.5 mm) containing 18% of polyoxyethylene glycol 400. The column was operated at 105°C, and hydrogen was used as the carrier at a flow-rate of 64.5 l/h. The main fraction (retention time 29.5 min) contained only 10^{-3}% of impurity. An attempt to purify tetramethyltin by passage through a similar column was less successful.

$$R_3SnCl + RSnCl_3 \longrightarrow 2R_2SnCl_2 \quad (5)$$

$$R_4Sn + RSnCl_3 \longrightarrow R_3SnCl_2 + R_2SnCl_2 \quad (6)$$

$$R_4Sn + R_2SnCl_2 \longrightarrow 2R_3SnCl \quad (7)$$

When gas chromatography is used to separate tetraalkyltin compounds and alkyltin halides, problems due to redistribution (equations 5–7) may arise[184–186]. Thus, on a 2-m column packed with 18% of silicone oil OE 4011 on Sterchamol brick at 160–180°C, butyltin trichloride reacts with tetrabutyltin although mixtures containing tetrabutyltin, tributyltin chloride, and dibutyltin dichloride are stable[184]. Reaction 7 does not occur on Carbowax[185]. All four butyltin chlorides can be separated on 20% of GI 7100 FF on Sterchamol brick at 175°C, on 18% of OE 4007 D on Sterchamol brick at 178°C, or on OE 4011 supported on Sterchamol brick at 194°C[185,187]. Mass spectrometry provides a useful detector for identifying the species present[188]. A study of the relative retention volumes of the butyltin compounds on ten different liquid phases using a katharometer and hydrogen as the carrier gas clearly showed the effect of liquid phase polarity on the relative retention volumes[185]. In non-polar liquid phases the compounds are separated according to their molecular weights, whereas in polar liquid phases the relative retention volumes of compounds containing the Sn—Cl bond increase with increasing polarity of the liquid phase. In the case of Carbowax, tetra-*n*-butyltin is eluted as the first compound.

A hydrogen flame-ionization detector was used for the gas chromatographic analysis of butyl-, octyl-, and phenyltin halides, used as intermediates for the manufacture of stabilizers for plastics, fungicides for paints, and certain other biological and agriculteral chemicals[189]. For the butyltin halides, an injection temperature of 380–400°C was found to be sufficient, but for the octyltin and the phenyltin halides, a temperature of 400–425°C was necessary to give completely symmetrical peaks. At lower injection temperatures, the organotin halides appear to be insufficiently volatile for accurate quantitative work. The column used was a 16 cm U-shaped stainless-steel section of 4 mm i.d. packed with 5% (w/w) silicone oil (Midland Silicone MS 200) supported on Celite 545 prepared by the method of James and Martin[169], including a treatment with alkali[190]. Column temperatures of 110°C for butyltin halides, 180°C for phenyltin halides, and 210°C for octyltin halides were satisfactory. The chromatographs for phenyltins are substantially similar to those for the octyltins, with slightly less resolution of the early peaks. Little difference was observed between the retention times of alkyltin bromides and the corresponding chlorides, and the calibration graphs were also similar in all respects. There was no evidence for thermal decomposition and disproportionation of organotin halides during exposure to elevated temperatures whilst on the column as previously reported[191]. Disproportionation of mixtures of, for example, monooctyltin tribromide and trioctyltin monobromide does not occur during chromatography, as shown by the absence of a peak corresponding to dioctyltin dibromide on the chromatogram[191]. Butyltin bromides may be converted quantitatively to their more volatile butylmethyltin analogue with methyl Grignard reagent and then determined by gas–liquid chromatography with an estimated accuracy of about ±2%[192]. Direct chromatography of the alkyltin bromides appeared to produce severe sample decomposition. The method is stated to be applicable to a number of tetraalkyl- and/or tetraryltins and alkyl- or aryltin halides. A dual-column, linear-temperature programmed gas chromatograph equipped with a four-filament hot-wire thermal conductivity cell detector was used. The columns were 10 ft × $\frac{1}{4}$ in o.d. stainless-steel 304 tubing in a 4–5 in diameter coil with the ends bent parallel to the axis of the coil. The coils were packed with 15.5 ± 0.2 g of 20% (w/w) SE-30 silicone gum on 60–80 mesh Chromosorb W. It was found that in this method the Grignard reaction is indeed quantitative and that no loss of sample occurred during other steps in the sample preparation procedure. Nevertheless, large differences occur between the added and observed amounts of all of the components, excluding dibutyltin dibromide. This is due to the room temperature reaction of equimolar amounts of tin tetrabromide and tetrabutyltin to form butyltin tribromide and tributyltin bromide prior to reaction with the Grignard reagent. Dibutyltin dibromide is the only material that does not take part in this redistribution reaction. The data obtained for this material are then more indicative of the accuracy of the method and in this case 1% of the actual percentage. Other gas chromatographic methods for alkyltin halides are described in references 193–203; references 195–198 refer to the separation of organotin halides after total acetylation.

A method for the simultaneous determination of triphenyltin hydroxide and its possible degradation products tetraphenyltin, diphenyltin oxide, benzenestannoic acid, and inorganic tin in water, is rapid, sensitive to less than 0.01 μg/ml for most of the tin species, and exhibits no cross-interferences between the phenyltins[204]. The phenyltins are detected by electron-capture gas–liquid chromatography after conversion to their hydride derivatives, using lithium aluminium hydride, while organic tin is determined by a procedure which responds to tin(IV) oxide as well as aqueous tin(IV). Conversion of non-volatile hydroxyoxyphenylstannane ($PhSnO_2H$), oxodiphenylstannane (Ph_2SnO), and hydroxytriphenylstannane (Ph_3SnOH) to their hydrides by lithium aluminium hydride gives derivatives with excellent GLC properties, high response to electron-capture detection, and none of the attendent column stability problems encountered with other derivatives[204].

A method for the determination of nanogram amounts of methyltin compounds and inorganic tin in natural waters and human urine involves conversion by sodium borohydride at pH 6.5 to the corresponding volatile hydrides SnH_4, CH_3SnH_3, $(CH_3)SnH_2$, and $(CH_3)_3SnH$[205]. These hydrides are scrubbed from solution, cryogenically trapped in a U-tube, and separated upon warming. Detection limits are approximately 0.01 ng as tin when using a hydrogen-rich, hydrogen–air flame emission detector (SnH band) of a type having considerably lower detection limits than any previously reported[206–212]. Average tin recoveries ranged from 96 to 109% for seawater and from 83 to 108% for human urine samples for six samples, to which were added 0.4–1.6 ng of methyltin compounds and 3 ng of inorganic tin. Re-analysis of the analysed samples showed that all methyltin and inorganic tin is removed in one analysis procedure. The method was applied to rain and tap water, seashells, urine, fresh water, and saline and estuary water samples.

XXII. ORGANOZINC COMPOUNDS

Organozinc compounds have been gas chromatographed on a 1-m column of Chromosorb W, containing 7.5% of paraffin wax (m.p. 63–64°C)–triphenylamine (17:3), using dry purified helium as the carrier gas, a thermistor detector, and a

TABLE 6. Pyrolysis fragments of zinc dialkyldithiophosphates (wt.-% of volatile products)[215]

Olefinic fragment	Alkyl group				
	Iso-C_3	Iso-C_4	1-Pentyl	3-Pentyl	4-Methyl-2-pentyl
C_3''	100	3	2	—	—
C_4''-1/iso-C_4''	—	85	1	—	—
C_4''-2	—	12	—	—	—
C_5''-1	—	—	60	8	—
C_5''-2	—	—	36	92	—
3-MeC_4''-1	—	—	1	—	—
2-MeC_5''-1	—	—	—	—	3
2-MeC_5''-2	—	—	—	—	1
4-MeC_5''-1	—	—	—	—	34
4-MeC_5''-2	—	—	—	—	62

column temperature of 73–165°C, according to the type of organozinc compound being examined[213]. High-pressure gas chromatography (1000–1700 psi) has been applied to the separation of zinc(II) etioporphyrin from other metal chelates[214]. The column packed with 10% Epon 1001 on Chromosorb W, was operated at 145°C using dichlorodifluoromethane as the carrier gas. In a procedure for the identification of microgram amounts of zinc dialkylphosphorothioates in oils by pyrolysis–gas chromatography, the zinc compounds were separated from the oil by thin-layer chromatography on silica gel using a mixture of pyridine and acetic acid as developing solvent[215]. The separated zinc compounds are recovered from the silica gel adsorbent by extraction with methylene dichloride and then evaporating off the solvent and acetic acid on a water-bath in a stream of nitrogen. The extract is deposited on a platinum dish brazed to nichrome leads. Gas chromatographic separation of the pyrolysis products was achieved with a 6 ft × $\frac{1}{4}$ in i.d. column packed with 10% silicone oil on 60–80 mesh Embacel operated at room temperature. Five dialkyl zinc phosphorothioates with known alkyl groups were examined. The only significant decomposition products observed arose from simple cleavage of the carbon—oxygen bond (Table 6). Only very small amounts (less than 50%) of methane or C_2 hydrocarbons were observed in any of the cracking patterns, suggesting that the carbon—carbon bonds in the alkyl structure are more stable than the carbon—oxygen bonds, and also that further degradation to olefinic fragments occurred to a negligable extent. Pyrolysis of zinc diisopropylphosphorothioate led to the formation of propylene alone, thus phosphorothioate containing *n*- or iso-propyl groups could not be distinguished by this technique. The isobutyl compound, similarly, gives almost entirely isobutene. The cracking pattern given by zinc (4-methyl-2-pentyl)phosphorothiate is more informative. Having two beta-carbon atoms from which hydrogen may be removed, two olefins are likely to be formed, 4-methylpent-1-ene and 4-methylpent-2-ene.

In Table 7 the product proportions observed for the zinc salt[215] are compared with those for the lead salt[216]. The product distribution is almost identical in the two cases except for some 2-methylpent-1-ene in addition to other olefins in the case of zinc[215]. This may have been due to the presence of small amounts of the 4-methyl-2-pentyl and 2-methyl-1-pentyl structures. The preponderance of **2** over **1** in the products shows that the preferred direction of hydrogen abstraction is to give the olefin having the more centrally placed double bond. Pyrolysis of the 1-pentyl and 3-pentyl structures shows that in the former case, although the 1-ene is the main product, a considerable amount of double bond migration to the 2-position occurs. With the 3-pentyl

TABLE 7. Pyrolysis of zinc and lead dimethylpentyldithiophosphates (wt.-%)

Product	Zinc salt[215]	Lead salt[216]
4-MeC$_5$″-2	62	59
4-MeC$_5$″-1	34	35
2-MeC$_5$″-2	1	6
2-MeC$_5$″-1	3	—

compound, over 90% of the product appears as pent-2-ene, only 7.7% of pent-1-ene being observed.

$$>P{-}O{-}C(CH_3)H{-}CH_2{-}CHCH_3{-}CH_3 \longrightarrow CH_2{=}CH{-}CH_2{-}CHCH_3{-}CH_3 \; (\mathbf{1}) + CH_3{-}CH{=}CH{-}CHCH_3{-}CH_3 \; (\mathbf{2})$$

A micropyrolytic–gas chromatographic technique using a katharometer detector has been studied for the analysis of substrates such as the zinc salts of *OO′*-di-*n*-dodecylthionothiophosphate, *OO′*-di-neopentylthionothiophosphate, and *OO′*-dicyclohexylthionothiophosphate[216]. The compound is pyrolysed in the inlet system of a gas chromatograph and the volatile pyrolysis products, generally olefins, are separated and collected individually for identification, by mass or infrared spectrometry. The olefins are formed generally by the breaking of a carbon—oxygen bond and abstraction of hydrogen from a beta-carbon atom with no skeletal isomerization. The structure of the olefins produced is thus directly related to the structure of the alkyl groups in the original zinc compound. Only when hydrogen is not available on a beta-carbon atom, as in neopentyl radicals, are olefins formed by carbon skeletal rearrangement. The authors give examples of the determination of the exact configuration of the alkyl radicals in several model zinc dialkylthionothiophosphates[216].

XXIII. REFERENCES

1. P. Longi and R. Mazzocchi, *Chim. Ind. (Milan)*, **48**, 718 (1966).
2. G. V. Bortnikov, E. N. Vyankin, E. N. Gladyshev, and V. S. Andreevichev, *Zavodskaya Lab*., **35**, 1445 (1969).
3. L. M. Brown and K. S. Mazdiyasni, *Anal. Chem*., **41**, 1243 (1969).
4. R. M. Pagnall, T. S. West, and O. Whitehead, *Analyst (London)*, **98**, 647 (1973).
5. H. Kawaguchi, T. Sakamoto, and O. Mizuike, *Talanta*, **20**, 321 (1973).
6. C. Feldman and D. A. Batistoni, *Anal. Chem*., **49**, 2215 (1977).
7. P. Longi and R. Mazzocchi, *Chim. Ind. (Milan)*, **48**, 718 (1966).
8. Y. Talmi and V. E. Norvell, *Anal. Chem*., **47**, 1510 (1975).
9. B. J. Gudzinowicz and H. F. Martin, *Anal. Chem*., **34**, 648 (1962).
10. B. J. Gudzinowicz and J. L. Driscoll, *J. Gas Chromatogr*., **1**, 25 (1963).
11. M. O. Andreae, *Anal. Chem*., **49**, 820 (1977).
12. G. E. Pairris, W. R. Blair, and F. E. Brinckman, *Anal. Chem*., **49**, 378 (1977)
13. Y. Talmi and D. T. Bostick, *Anal. Chem*., **47**, 2145 (1975).
14. Y. Talmi and V. E. Norvell, *Anal. Chem*., **47**, 1510 (1975).

15. R. S. Braman, D. L. Johnson, C. C. Foreback, J. H. Ammons, and J. L. Bricker, *Anal. Chem*., **49**, 621 (1977).
16. R. S. Braman and A. Dyanko, *Anal. Chem*., **40**, 95 (1968).
17. C. Feldman and D. A. Batistoni, *Anal. Chem*., **49**, 2215 (1977).
18. M. Iguchi, A. Nishiyama, and Y. Nagese, *J. Pharm. Soc. Jap*., **80**, 1408 (1960).
19. A. D. Zorin, G. G. Devyatýkh, V. Ya Dudorov, and A. M. Amel'chenko, *Zh. Neorg. Khim*., **9**, 2526 (1964); *Ref. Zh., Khim*., 19GDE, 8G 141 (1965).
20. G. Schomburg, R. Köster, and D. Henneberg, *Z. Anal. Chem*., **170**, 285 (1959).
21. R. Koster and G. Bruno, *Justus Liebigs Ann. Chem*., **629**, 89 (1960).
22. P. Longi and R. Mazzocchi, *Chim. Ind. (Milan)*, **48**, 718 (1966).
23. J. A. Semlyen and C. S. G. Phillips, *J. Chromatogr*., **18**, 1 (1965).
24. J. J. Kaufman, J. E. Todd, and W. S. Koski, *Anal. Chem*., **29**, 1032 (1957).
25. G. R. Seely, J. P. Oliver, and D. M. Ritter, *Anal. Chem*., **31**, 1993 (1959).
26. H. W. Myers and R. F. Putman, *Anal. Chem*., **34**, 664 (1962).
27. W. S. Koski, P. C. Maybury, and J. J. Kaufman, *Anal. Chem*., **26**, 1992 (1954).
28. L. J. Kuhus, R. S. Braman, and J. E. Graham, *Anal. Chem*., **34**, 1700 (1962).
29. L. J. Kuhns and R. H. Forsyth, *Anal. Chem*., **28**, 1750 (1956).
30. F. A. Serravallo and T. S. Risky, *J. Chromatogr. Sci*., **12**, 585 (1974).
31. R. M. Dagnall, T. S. West, and P. Whitehead, *Analyst (London)*, **98**, 647 (1973).
32. E. W. Abel, G. Nickless, and F. H. Pollard, *Proc. Chem. Soc. (London)*, 288 (1960).
33. C. S. G. Phillips and P. L. Timms, *Anal. Chem*., **35**, 505 (1963).
34. F. H. Pollard, G. Nickless, and P. C. Uden, *J. Chromatogr*., **14**, 1 (1964).
35. J. A. Semlyen, G. R. Walker, R. E. Blofeld, and C. S. G. Phillips, *J. Chem. Soc*., 4948 (1964).
36. J. A. Semlyen and C. S. G. Phillips, *J. Chromatogr*., **18**, 1 (1965).
37. P. L. Timms, C. C. Simpson, and C. S. G. Phillips, *J. Chem. Soc*., 1467 (1964).
38. A. D. Snegova, L. K. Markov, and V. A. Ponomarenko, *Zh. Anal. Khim*., **19**, 610 (1964).
39. G. Garzo, J. Fekete, and M. Blazso, *Acta Chim. Acad. Sci Hung*., **51**, 359 (1967).
40. I. A. Frolov, *Tr. Khim. Tekhnol. (Gor'ki)*, **1**, (15), 107 (1966).
41. V. M. Gorbackev and G. V. Tret'yakov, *Zavod. Lab*., **32**, 796 (1966).
42. K. Tanikawa and K. Arakawa, *Chem. Pharm. Bul.*, **13**, 926 (1965).
43. A. N. Nesmeyanov, L. P. Yur'eva, N. S. Kochetkova, and S. V. Vitt, *Izv. Akad. Nauk SSSR, Ser. Khim*., **3**, 560 (1966).
44. O. E. Ayers, T. G. Smith, J. D. Burnett, and B. W. Pouder, *Anal. Chem*., **38**, 1606 (1966).
45. P. C. Uden, D. E. Henderson, F. F. DiSanzo, R. J. Lloyd, and T. Tetu, *Abstr. 174th National Meeting American Chemical Society, Chicago, Ill., August 1977.*
46. W. W. Parker, G. Z. Smith, and R. L. Hudson, *Anal. Chem*., **33**, 1170 (1961).
47. W. W. Parker and R. L. Hudson, *Anal. Chem*., **35**, 1334 (1963).
48. J. E. Lovelock and N. L. Gregory, in *Gas Chromatography International Symposium (1962)*, Vol. 3, 1962, p. 219. Rome.
49. J. E. Lovelock and A. Zlatkis, *Anal. Chem*., **33**, 1958 (1961).
50. H. J. Dawson, *Anal. Chem*., **35**, 542 (1963).
51. D. M. Coulson, *Pestic. Res. Bull*., **2**, 1 (1962).
52. J. E. Lovelock, *Anal. Chem*., **33**, 162 (1961).
53. E. M. Barrall and P. R. Ballinger, *J. Gas Chromatogr*., **1**, 7 (1963).
54. E. J. Bonelli and H. Hartmann, *Anal. Chem*., **35**, 1980 (1963).
55. N. L. Soulages, *Anal. Chem*., **39**, 1340 (1967).
56. L. E. Green, *Hewlett-Packard Facts Methods*, **8**, 4 (1967).
57. V. Cantuti and G. P. Cartoni, *J. Chromatogr*., **32**, 641 (1968).
58. K. Kramer, *Erdöl Kohle*, **19**, 182 (1966).
59. F. H. Pollard, G. Nickless, and P. C. Uden, *J. Chromatogr*., **14**, 1 (1964).
60. E. W. Abel, G. Nickless, and F. H. Pollard, *Proc. Chem. Soc. (London)*, 288 (1960).
61. E. A. Boettner and F. C. Dallos, *J. Gas Chromatogr*., **3**, 190 (1965).
62. E. J. Bonelli and H. Hartmann, *Wilkins Aerograph Res. Botes*, Autumn Issue (1963).
63. E. J. Bonelli and H. Hartmann, *Anal. Chem*., **35**, 1980 (1963).
64. H. J. Dawson, *Anal. Chem*., **35**, 542 (1963).
65. E. A. Boettner and F. C. Dallos, *J. Gas Chromatogr*., **3**, (6), 190 (1965).
66. P. R. Ballinger and I. M. Whittemore, *Proc. Am. Chem. Soc. Div. Pet. Chem., Atlantic City Meeting, Sept. 1968*, **13**, 133 (1968).

67. Y. K. Chan, P. T. S. Wong, and P. D. Goulden, *Anal. Chim. Acta*, **85**, 421 (1976).
68. B. Radzuik, Y. Thomassen, J. C. Van Loon, and Y. K. Chan, *Anal. Chim. Acta*, **105**, 255 (1979).
69. Y. K. Chan, P. T. S. Wong, G. A. Bengert, and O. Kramar, *Anal. Chem.*, **51**, 186 (1979).
70. Y. K. Chan, P. T. S. Wong, and H. Saitoh, *J. Chromatogr. Sci.*, **14**, 162 (1976).
71. D. A. Segar, *Anal. Lett.*, **7**, 89 (1974).
72. Y. K. Chan, P. T. S. Wong, and P. D. Goulden, *Anal. Chim. Acta*, **421**, 85 (1976).
73. Y. K. Chan, P. T. S. Wong, G. A. Bengent, and O. Kramer, *Anal. Chem.*, **51**, 51 (1979).
74. P. T. S. Wong, Y. K. Chan, and P. L. Luxon, *Nature (London)*, **253**, 263 (1975).
75. J. P. Dumas, L. Dazdernik, S. Belloncik, D. Bouchard, and G. Vaillancourt, *Proc. 12th Can. Symp. Water Pollut. Res.*, 91 (1977).
76. E. D. Mori and A. M. Beccoria, in *Proceedings of the International Experts Discussion on Lead Occurrence, Fate and Pollution in the Marine Environment, Rovinj, Yugoslavia*, 1977.
77. R. M. Harrison, R. Penny, and D. H. Slater, *Atmos. Environ.*, **8**, 1187 (1974).
78. B. Y. T. Radzuik, J. C. Van Loon, and Y. K. Chan, *Anal. Chim. Acta*, **255**, 105 (1979).
79. D. T. Coker, *Anal. Chem.*, **47**, 386 (1975).
80. J. W. Robinson, E. L. Kiesel, J. P. Goodhead, R. Bliss, and R. Marshall, *Anal. Chim. Acta*, **92**, 321 (1977).
81. A. Laveskag, *Second International Clean Air Congress Proceedings, American Chemical Society, Washington, D.C.*, 1970 p. 549.
82. W. A. Aue and H. H. Hill, *J. Chromatogr.*, **74**, 319 (1972).
83. W. A. Aue and H. H. Hill, *Anal. Chem.*, **45**, 729 (1973).
84. H. H. Hill and W. A. Aue, *J. Chromatogr. Sci.*, **12**, 541 (1974).
85. H. H. Hill and W. A. Aue, *J. Chromatogr.*, **122**, 515 (1976).
86. H. H. Hill, *PhD Thesis*, Dalhousie University, Halifax, N.S., Canada, 1975.
87. M. D. DePuis and H. H. Hill, *Anal. Chem.*, **51**, 292 (1979).
88. D. C. Reamer, W. H. Zoller, and T. C. O'Haver, *Anal. Chem.*, **50**, 1449 (1978).
89. B. D. Quimby, P. C. Uden, and R. M. Barnes, *Anal. Chem.*, **50**, 2112 (1978).
90. A. J. McCormack, S. C. Tong, and W. D. Cooke, *Anal. Chem.*, **37**, 1470 (1965).
91. L. J. Snyder and S. R. Henderson, *Anal. Chem.*, **33**, 1175 (1961).
92. L. J. Snyder, *Anal. Chem.*, **39**, 591 (1967).
93. R. H. Harrison, R. Perry, and D. H. Slater, *Atmos. Environ.*, **8**, 1187 (1974).
94. V. Cantuti and G. P. Cartoni, *J. Chromatogr.*, **32**, 641 (1968).
95. R. Waak and M. A. Doran, *J. Am. Chem. Soc.*, **85**, 165 (1963).
96. H. L. Johnson and R. A. Clack, *Anal. Chem.*, **19**, 869 (1947).
97. E. C. Juenge and D. Seyferth, *J. Org. Chem.*, **26**, 563 (1961).
98. W. S. Leonhardt, R. C. Morrison, and C. W. Kamienski, *Anal. Chem.*, **38**, 466 (1966).
99. P. F. Collins, C. W. Kamienski, D. L. Esmay, and R. B. Ellestad, *Anal. Chem.*, **33**, 468 (1961).
100. D. Bernstein, *Z. Anal. Chem.*, **182**, 321 (1961).
101. R. d'Hollander and M. Auteunis, *Bull. Soc. Chim. Belg.*, **72**, 77 (1963).
102. L. V. Guild, C. A. Hollingsworth, D. H. McDaniel, and H. J. H. Wotiz, *Anal. Chem.*, **33**, 1156 (1961).
103. A. Wowk and S. DiGiovanni, *Anal. Chem.*, **38**, 742 (1966).
104. H. O. House and W. L. Respers, *J. Organomet. Chem.*, **4**, 95 (1965).
105. M. A. Molinari, J. Lombardo, O. A. Lires, and G. J. Videla, *An. Assoc. Quím. Argentina*, **48**, 223 (1960).
106. M. D. DuPuis and H. H. Hill, *Anal. Chem.*, **51**, 292 (1979).
107. P. C. Uden, R. M. Barnes, and F. P. DiSanzo, *Anal. Chem.*, **50**, 852 (1978).
108. T. Westermark, *Kvicksilverfrageni Sverige*, **25**, 25 (1964).
109. G. Westhöö, *Acta, Chem. Scand.*, **20**, 2131 (1966).
110. G. Westhöö, *Acta Chem. Scand.*, **21**, 1790 (1967).
111. G. Westhöö, *Acta Chem. Scand.*, **22**, 2277 (1968).
112. G. Westhöö, B. Johansson, and R. Ryhage, *Acta Chem. Scand.*, **24**, 2349 (1970).
113. L. R. Kamps and B. McMahon, *J. Assoc. Off. Anal. Chem.*, **55**, 590 (1972).
114. M. L. Schafer, U. Rhea, J. T. Peeler, C. H. Hamilton, and J. E. Campbell, *J. Agric. Food Chem.*, **23**, 1079 (1975).
115. R. Bye and P. E. Paus, *Anal. Chim. Acta,* **107**, 169 (1979).
116. F. Frimmel and H. A. Winkler, *Wasser,* **45**, 285 (1975).

117. Y. Talmi and V. E. Norvell, *Anal. Chim. Acta*, **85**, 203 (1976).
118. C. J. Cappon and J. Crispin Smith, *Anal. Chem.*, **49**, 365 (1977).
119. L. Magous and T. W. Clarkson, *J. Assoc. Off. Anal. Chem.*, **55**, 966 (1972).
120. L. Fishbein, *Chromatogr. Rev.*, **13**, 83 (1970).
121. K. Broderson and U. Schlenker, *Z. Anal. Chem.*, **182**, 421 (1961).
122. J. E. Longbottom, R. C. Dreisman, and J. J. Lichtenbert, *J. Assoc. Off. Anal. Chem.*, **56**, 1297 (1973).
123. R. T. Ross and J. G. Gonzalez, *Bull. Environ. Contam. Toxicol.*, **10**, 187 (1973).
124. T. Giovanli-Jakubczak, M. R. Greenwood, J. C. Smith, and T. W. Clarkson, *Clin. Chem. (Winston-Salem, N.C.)*, **20**, 222 (1974).
125. R. Von. Burg, F. Farris, and J. C. Smith, *J. Chromatogr.*, **97**, 65 (1974).
126. J. A. Rodriguez-Valsques, *Talanta*, **25**, 299 (1978).
127. W. Blair, W. P. Iveson, and F. E. Brinkman, *Chromosphere*, **3**, 167 (1974).
128. J. E. Longbottom, *Anal. Chem.*, **44**, 111 (1972).
129. G. N. Turkel'taub, B. M. Luskina, and S. V. Syavtsillo, *Khim. Technol. Topl. Masel*, **12**, 58 (1967).
130. G. Westhöö, *Acta Chim. Scand.*, **20**, 2131 (1966).
131. G. Westhöö, *Acta Chim. Scand.*, **21**, 1790 (1967).
132. G. Westhöö, *Acta Chim. Scand.*, **22**, 2277 (1968).
133. K. Sumino and J. Kobe, *Med. Sci.*, **14**, 115 (1968).
134. C. A. Bach and D. J. Lisk, *Anal. Chem.*, **43**, 1950 (1971).
135. W. E. L. Grossman, J. Eug, and Y. C. Tong, *Anal. Chim. Acta*, **60**, 447 (1972).
136. Y. Talmi, *Anal. Chim. Acta*, **74**, 107 (1975).
137. B. D. Quimby, P. C. Uden, and R. M. Barnes, *Anal. Chem.*, **50**, 2112 (1978).
138. H. Veening, N. J. Graven, D. G. Clark, and B. R. Willeford, *Anal. Chem.*, **41**, 1655 (1969).
139. E. Klesper, A. H. Corwin, and D. A. Turner, *J. Org. Chem.*, **27**, 700 (1962).
140. C. E. Legate and H. D. Burnham, *Anal. Chem.*, **32**, 1042 (1960).
141. I. B. Rubin and C. K. Bayne, *Anal. Chem.*, **51**, 541 (1979).
142. C. S. Evans and C. M. Johnson, *J. Chromatogr.*, **21**, 202 (1966).
143. G. E. Parris, W. R. Blair, and F. E. Brinkman, *Anal. Chem.*, **49**, 378 (1977).
144. Y. K. Chan, P. T. S. Wong, and P. D. Goulden, *Anal. Chem.*, **47**, 2279 (1975).
145. Y. Talmi and W. W. Andren, *Anal. Chem.*, **46**, 2122 (1974).
146. J. R. Marquart, *Anal. Chem.*, **50**, 656 (1978).
147. R. K. Ingham, S. D. Rosenberg, and H. Gilman, *Chem. Rev.*, **60**, 459 (1960).
148. F. H. Pollard, G. Nickless, and D. J. Cooke, *J. Chromatogr.*, **13**, 48 (1964).
149. F. H. Pollard, G. Nickless, and D. J. Cooke, *J. Chromatogr.*, **17**, 472 (1965).
150. J. Bohemen, S. H. Langer, R. H. Perrett, and J. H. Purnell, *J. Chem. Soc.*, 2444 (1960).
151. F. F. Eggersten and S. Groennings, *Anal. Chem.*, **30**, 20 (1958).
152. A. James and A. J. P. Martin, *Biochem. J.*, **50**, 679 (1952).
153. A. Liberti, in *Gas Chromatography 1958* (Ed. D. H. Desty), Butterworths, London, 1958, p. 214.
154. C. H. Orr and J. E. Callen, *Ann. N.Y. Acad. Sci.*, **72**, 649 (1959).
155. E. C. Ormerod and R. P. W. Scott, *J. Chromatogr.*, **2**, 65 (1959).
156. A. V. Kiselev, D. H. Everett, and F. S. Stone, in *The Structure and Properties of Porous Materials*, Volume X, Colston Papers, Butterworths, London, 1958, p. 257.
157. H. W. Kohlschutter, P. Best, and G. Wirzung, *Z. Anorg. Allg. Chem.*, **285**, 336 (1956).
158. G. A. Haward and A. J. P. Martin, *Biochem. J.*, **46**, 532 (1950).
159. A. Kwantes and W. A. Rijnders, in *Gas Chromatography 1958* (Ed. D. H. Desty), Butterworths, London, 1958, p. 125.
160. S. H. Langer, S. Connell, and I. Wender, *J. Org. Chem.*, **23**, 50 (1958).
161. S. H. Langer, P. Pantages, and I. Wender, *Chem. Ind. (London)*, 1664 (1958).
162. R. H. Perrett and J. H. Purnell, *J. Chromatogr.*, **7**, 455 (1962).
163. R. G. Ackman, R. D. Burgher, J. C. Sipos, and P. H. Odense, *J. Chromatogr.*, **9**, 531 (1962).
164. J. Novák and J. Janák, *J. Chromatogr.*, **4**, 249 (1960).
165. J. Stuvé, in *Gas chromatography 1958* (Ed. D. H. Desty), Butterworths, London, 1958, p. 178.
166. A. J. P. Martin and A. T. James, *Biochem. J.*, **63**, 138 (1956).

167. G. R. Jamieson, *J. Chromatogr.*, **4**, 420 (1960).
168. F. H. Pollard, G. Nickless, and P. C. Uden, *J. Chromatogr.*, **14**, 1 (1964).
169. A. T. James and A. J. P. Martin, *Biochem. J.*, **50**, 679 (1952).
170. R. C. Putnam and H. Pu, *J. Gas Chromatogr.*, **3**, 160 (1965).
171. R. C. Putnam and H. Pu, *J. Gas Chromatogr.*, **3**, 289 (1965).
172. U. Prösch and H. J. Zoepfl, *Z. Chem.*, **3**, 97 (1963).
173. F. H. Pollard, G. Nickless, and P. C. Uden, *J. Chromatogr.*, **19**, 28 (1965).
174. G. Garzó, J. Fekete, and M. Blazó, *Acta, Chim. Hung.*, **51**, 359 (1967).
175. Gas Chromatography Discussion Group Sub Committee, *Gas Chromatography 1964*, Elsevier, Amsterdam, 1965.
176. E. A. Abel, G. Nickless, and F. H. Pollard, *Proc. Chem. Soc. (London)*, 288 (1960).
177. K. Höppner, U. Prösch, and H. J. Zoepfl, *Abh. Dtsch. Akad. Wiss. Berlin, Kl. Chem. Geol. Biol.*, 393 (1966).
178. D. J. Cooke, G. Nickless, and F. H. Pollard, *Chem. Ind. (London)*, 1493 (1963).
179. G. G. Devyatykh, V. A. Umilin, and U. N. Tsinovoi, *Tr. Khim. Khim. Tekhnol.*, 82 (1968).
180. P. Longi and R. Mazzocchi, *Chim. Ind. (Milan)*, **48**, 718 (1966).
181. K. Jergen and J. Figge, *J. Chromatogr.*, **109**, 89 (1975).
182. V. V. Brazhnikov and K. I. Sakodynskii, *J. Chromatogr.*, **66**, 361 (1972).
183. K. Höppner, U. Prösch, and H. Weigleb, *Z. Chem.*, **4**, 31 (1964).
184. H. Geissler and H. Kriegsmann, *Z. Chem.*, **4**, 354 (1964).
185. H. Geissler and H. Kriegsmann, Third Analytical Conference 24–29 August, Budapest, Hungary (1970).
186. H. Matsuda and A. Matsuda, *J. Chem. Soc. Jpn., Ind. Chem. Sect.*, **63**, 1960 (1960).
187. H. Giessler and H. Kriegsmann, *Z. Chem.*, **5**, 423 (1965).
188. V. A. Umilin and Yu. N. Tsinovoi, *Izv. Akad. Nauk SSSR*, *Ser. Khim.*, 1409 (1968).
189. B. L. Tonge, *J. Chromatogr.*, **19**, 182 (1965).
190. A. James, A. J. P. Martin, and G. Howard Smith, *Biochem. J.*, **52**, 238 (1925).
191. W. P. Neumann, *Angew. Chim., Int. Ed. Engl.*, **2**, 165 (1963).
192. R. D. Steinmeyer, A. F. Fentiman, and E. J. Kahler, *Anal. Chem.*, **37**, 520 (1965).
193. W. O. Gauer, J. N. Selber, and D. G. Crosby, *J. Agr. Food Chem.*, **22**, 252 (1974).
194. J. Franc, M. Wurst, and V. Moudrý, *Collect. Check. Chem. Commun.*, **26**, 1313 (1961).
195. W. Gerrard, E. F. Mooney, and R. G. Rees, *J. Chem. Soc.*, 740 (1964).
196. V. Chromý, A. Croagová, O. Pospichal, and K. Jurák, *Chem. Listy*, **60**, 1599 (1966).
197. W. P. Neumann and G. Burkhardt, *Justus Leibigs Ann. Chem.*, **663**, 11 (1963).
198. Y. Jitsu, N. Kudo, K. Sato, and T. Teshima, *Bunseki Kagaku*, **18**, 169 (1969).
199. R. A. Keller, *J. Chromatogr.*, **5**, 225 (1961).
200. H. D. Kaesz, F. L. Stafford, and F. G. A. Stone, *J. Am. Chem. Soc.*, **82**, 6232 (1960).
201. H. D. Kaesz, J. R. Phillips, and F. G. A. Stone, *J. Am. Chem. Soc.*, **82**, 6228 (1960).
202. B. M. Luskina and S. V. Syavtsillo, *Nov. Obl. Prom. Sanit. Khim.*, 186 (1969).
203. J. D. Smith, *Nature (London)*, **225**, 103 (1970).
204. C. J. Soderquist and D. G. Crosby, *Anal. Chem.*, **50**, 1435 (1978).
205. R. S. Braman and H. A. Tompkins, *Anal. Chem.*, **51**, 12 (1979).
206. H. D. Nelson, *Doctorial Dissertation*, University of Utrecht (1967).
207. R. S. Braman and A. Dynako, *Anal. Chem.*, **40**, 95 (1968).
208. R. Herman and C. T. J. Alkemade, *Chemical Analysis by Flame Photometry*, Interscience, New York and London, 1963.
209. W. R. S. Garton, *Proc. Phys. Soc.*, **64**, 591 (1951).
210. W. W. Watson and R. Simon, *Phys. Rev.*, **55**, 358 (1939).
211. R. M. Dagnall, K. C. Thompson, and T. S. West, *Analyst (London)*, **93**, 518 (1968).
212. W. A. Aue and H. H. Hill, *J. Chromatogr.*, **70**, 158 (1972).
213. P. Longi and R. Mazzocchi, *Chim. Ind. (Milan)*, **48**, 718 (1966).
214. N. M. Karayannis and A. H. Corwin, *J. Chromatogr.*, **47**, 247 (1970).
215. S. G. Perry, *J. Gas Chromatogr.*, **2**, 93 (1964).
216. C. E. Legate and H. D. Burnham, *Anal. Chem.*, **32**, 1042 (1960).

The Chemistry of the Metal–Carbon Bond
Edited by F. R. Hartley and S. Patai

CHAPTER **19**

Analysis of organometallic compounds: other chromatographic techniques

T. R. CROMPTON
'Beechcroft', Whittingham Lane, Goosnargh, Preston, Lancashire, UK

I. THIN-LAYER CHROMATOGRAPHY

A. Organoantimony Compounds

Thin-layer chromatography has been used to separate triphenylantimony from its phosphorus, arsenic, and bismuth analogues. R_F values have been reported for these compounds on alumina plates using light petroleum as developing solvent[1].

B. Organoarsenic Compounds

Thin-layer chromatography on alumina using light petroleum as the solvent has been used to separate $(C_6H_5)_3M$, where M is phosphorus, arsenic, antimony, or bismuth[1]. R_F values have been determined on silica gel G and neutral alumina for mixtures of labelled triphenylarsine, triphenyl phosphate, triphenyl phosphite, triphenylphosphine oxide, triphenylphosphine and tritolylphosphate using acetone–light petroleum, benzene–acetone (1:1 and 9:1) and chloroform–acetone (3:2 and 9:1) as solvents[2]. The chromatograms were automatically examined radiometrically. A range of aromatic compounds containing arsenic or phosphorus have been separated on silica gel, aluminium oxide, and magnesium silicate[3]. Twenty-four solvents were studied and R_F values and their standard deviations tabulated.

C. Organobismuth Compounds

Thin-layer chromatography on alumina has been applied to the separation of compounds of the type $(C_6H_5)M$, where M is bismuth, phosphorus, arsenic, or antimony[1]. Light petroleum is the recommended elution solvent.

D. Organoboron Compounds

Thin-layer chromatography has been used to separate some $\beta\beta\beta$-trichloroborazines[4]. 2,4,6-trichloroborazine and its 1,3,5-trimethyl and -triphenyl derivatives are separated from each other on 0.5 mm layers of microcrystalline cellulose with ethyl acetate or pyridine–ethyl acetate (1:1000) as solvent. For detection, the plates are sprayed with 6 N hydrochloric acid and then with a tincture of curcuma and heated on a hot-plate at 80°C. A red–brown spot on a yellow background indicates the presence of boron compounds. Pentaborane, decaborane, and some chlorinated boron compounds have been separated on silica gel G or binder-free alumina[5]. Between 3 and 50 μg of sample were applied, and the spots were detected with 1% potassium permanganate solution or iodine vapour, followed by spraying with a solution containing 5% each of silver nitrate and ethylenediamine.

In many cases, silver nitrate alone was sufficient. For pentaborane and decaborane and some halogenated compounds, a 10% solution of ethyl isonicotinate in hexane was a suitable reagent. R_F values in 10 solvents were listed for several boranes, and the chromatographic behaviour of the products of their reaction with some Lewis bases such as amines, hydrazine, trialkyphosphines, and dialkylsulphides was discussed. Difficulties were observed with low-boiling compounds or with those that were oxidized or hydrolysed readily. Hydrolysis occurred to some extent even in aprotic solvents, although in many instances the speed of the separation minimized the effect.

E. Organocobalt Compounds

For the thin-layer chromatographic separation of mixtures of cyanocobalamin and cobinamide the sample is applied to a thin layer of silica gel impregnated with thymol, and the chromatogram is developed with an aqueous solution containing 0.4% of pyridine, 3% of phenol, 0.01% of sodium cyanide, and 10% of acetic acid, saturated with thymol[6]. To determine cyanocobalamin the chromatogram is eluted with a 0.1% solution of sodium cyanide in 75% methanol, and the absorbance is measured.

Cyanocobalamin and hydroxocobalamin have been separated on thin films of silica gel[7]. Two methods of development were used, the first involving Merck Kieselgel G and elution with butanol–acetic acid –0.066 M potassium dihydrogen phosphate–methanol (4:2:4:1). In the second method the silica gel is suspended in 0.066 M potassium dihydrogen phosphate and development is effected with butanol–acetic acid–water–methanol (4:2:4:1). Identical results were obtained by both methods, but the latter method is more rapid, requiring only 5–6 h for development. If the spots are visible they are separately cut from the plate and extracted with 1% aqueous Tween 80 at 40–45°C, the extraction being repeated if necessary after centrifuging. The absorbance of the extract is measured, for cyanocobalamin at 361 nm and 548 nm. For hydroxocobalamin, the corresponding figures are 351 nm (165) and 527 nm (56). When the spots are not visible, duplicate plates are made, one of which is used for locating the cobalamin spots bioautographically, by incubation in contact with a plate of 3% agar in USP Difco assay medium, containing 0.02% of triphenyltetrazolium chloride, and inoculation with *Lactobacillus leichmannii* ATCC 7830. The corresponding zones are then removed from the second plate and eluted. The displacement of hydroxocobalamin relative to cyanocobalamin is 0.5 ± 0.05.

By using alumina G as adsorbent, and anhydrous acetic acid–water–methanol–chloroform–butanol (2:9:10:20:50) as solvent, cyanocobalamin was completely separated, but not hydroxocobalamin[8]. By using silica gel G as adsorbent, and anhydrous acetic acid–water–methanol–chloroform–butanol (9:11:5:10:25) as solvent, vitamin derivatives could be successfully separated, viz., hydroxocobalamin (R_F 0.05), cyanocobalamin (R_F 0.23), unknown substance (R_F 0.33), and thymidine (R_F 0.7). Spots were detected by bioautography on the plate with vitamin B_{12} assay agar medium and *Lactobacillus leichmannii* ATCC 7830 as test organism. The minimum amounts of cyanocobalamin detected on silica gel G and alumina G were 0.005 and 0.025×10^{-9} g, respectively. A thin-layer chromatographic–spectroscopic procedure for the determination of hydroxo- and cyanocobalamins associated with other medicaments in various pharmaceutical forms is based on extraction of the vitamins, purification by thin-layer chromatography, and spectrophotometric determination as dicyanocobalamin[9].

Methods for extracting and determining cobalamins have been reviewed under the headings biological, chemical, radiometric, and spectrophotometric[10]. A method particularly recommended that is applicable to all forms of cobalamin present in any type of extract or liver hydrolysate involves three double extractions wit *p*-chlorophenol solution with intermediate washings with water, followed by thre single extractions with *p*-chlorophenol solution. The extracts are combined an concentrated by several alternate extractions with decreasing volumes of water an *p*-chlorophenol solution, the alternate aqueous and *p*-chlorophenol extracts bein discarded. The organic concentrate so obtained is shaken with diethyl ether, then wit added water, and then with added ethyl oleate. After centrifuging, the organic phase i discarded, and the aqueous phase containing the cobalamins is washed with ether, the submitted to thin-layer chromatography on silica gel G, previously dried at 38°C b development with 50% aqueous ethanol for 3 h. Cyanocobalamin and coenzyme B are extracted individually with 50% ethanol. The hydroxocobalamin zone is treate with potassium cyanide solution, adjusted to pH 6 with acetic acid, and the resultin cyanocobalamin is extracted into 50% ethanol. Finally, cyanocobalamin and coenzym B_{12} are determined spectrophotometrically at 550 and 525 nm, respectively.

In a further method for separating hydroxocobalamin and cyanocobalamin b thin-layer chromatography, spots of aqueous solution adjusted to pH 8.5 with dilut aqueous ammonia were applied to layers of dry neutral alumina and developed wit isobutyl alcohol–isopropyl alcohol–water (6:4:5) (with addition of aqueous ammoni to pH 8.5)[11]. In this system, the R_F values of hydroxo- and cyanocobalamins were 0.3 and 0.46, respectively; 0.5 μg of each substance could be detected. For thin-laye chromatography of mixtures of cyanocobalamin (vitamin B_{12}), factor B_{12} pseudo-vitamin B_{12}, factor A, factor B, and factor VnB, a 5% solution of sodiu cyanide was added to an aqueous solution of the sample[12]. Spots of the solution wer applied to a layer (1 mm thick) of alumina and developed in isobuty alcohol–isopropyl alcohol–water (1:1:1). All the above substances were wel separated. For the analysis of a phenol solution of three B_{12} coenzymes, removal o inorganic ions was necessary for good results[13]. This was achieved by washing th phenol phase with water. Plates were coated with cellulose MN 300 or MN 300-CM and were developed with the lower layer of butan-2-ol-0.1 M acetate buffe (pH 3.5)–methanol (4:12:1).

F. Organocopper Compounds

Copper chlorophyllins in preserves have been determined by thin-laye chromatography[14]. For the extraction of chromopigments the sample is homogenize and acidified with concentrated formic acid, then extracted by stirring with isopropy alcohol–acetone–diethyl ether (5:3:2). The concentrates are developed on thin-layer of silica gel. For identifying copper chlorophyllins the chromatograms are develope with benzene–ethyl acetate–methanol (17:1:2); copper chlorophyllins remain at th origin as dark brown spots.

G. Organoiron Compounds

Thin-layer chromatography on Merck Kieselgel G has been applied to th separation of ferrocene and its derivatives[15]. Ferrocene compounds have bee separated by thin-layer chromatography on methylene chloride solutions on a plat coated with a 0.25 mm layer of Mallinckrodt Silic AR TLC–7G, a silicic acid gypsu absorbent, that has been activated for 1 h at 92°C[16]. The elution solvent wa benzene–acetone (30:1). Detection of ferrocene compounds was done visually an detection of benzenoid compounds by means of iodine vapour treatment. In mos cases the ferrocenyl analogue has a lower R_F than the phenyl compound. To ensur that no reaction had taken place on the adsorbent, a methylene chloride solution o errocenylphenylcarbinol was allowed to stand overnight in contact with Silic AR TLC-7G, then filtered and evaporated to dryness. The infrared spectrum proved to be dentical with the original, indicating that no reaction had taken place.

H. Organolead Compounds

The dithizonates of trialkyllead halide (yellow) and dialkyllead dihalide (red) compounds can be separated by thin-layer chromatography[17]. Mixtures containing much tributyllead acetate and a small proportion of dibutyllead diacetate are best chromatographed in acetone–water (1:1), and other mixtures can be separated in benzene. Following chromatography, the thin-layer plates are sprayed with a dithizone solution in chloroform (0.1%), after which yellow and/or red spots appear on a light green–blue background. Inorganic lead does not migrate and causes red spots at the baseline on the chromatogram.

I. Organomercury Compounds

Radio paper chromatography has been used to separate mixtures of mercury(II) bromide, phenylmercury bromide and *p*-bromophenylmercury bromide[18]. Various other aryl- and alkylmercury bromides were also separated. Various migration solvents were used, including *n*-butanol–ammonia, tetrahydrofuran–*n*-butanol, 1N ammonia (15:35:20), methyl acetate–*n*-butanol-2N ammonia (47:40:13), and $PO[N(CH_3)_2]_3$–*n*-butanol–2N ammonia (1:1:1).

J. Organophosphorus Compounds

Zinc *OO*-dialkylphosphorodithioates have been identified in mineral oil products by thin-layer chromatography[19]. The lubricant or additive concentrate is diluted with a paraffinic solvent. A portion of the solution is shaken with diethylamine to convert the additive into the ammonium salt, and an aliquot is applied to a plate of aluminium oxide G (activated at 120°C). The oil is removed by development with hexane, and the chromatogram is then developed with hexane–acetone–ethanol–triethylamine (20:30:1:1). The spots are made visible by spraying first with iodine azide and then with starch solution. Spots due to other compounds, particularly sulphur compounds, are visible before spraying. The R_F values increase with increasing size of the alkyl groups.

K. Organotellurium Compounds

Analytical aspects of organotellurium compounds have been discussed[20]. Mixtures of the *o*-, *m*-, and *p*-isomers of ditolyltellurium dissolved in light petroleum have been separated by thin-layer chromatography on alumina[21]. In the separation of organotellurium compounds containing phenyl, tolyl, or naphthyl radicals on alumina (activity III–IV, particle size 0.06–0.09 mm), the substances are revealed as yellow to red–brown spots by treating the chromatogram with iodine vapour[22]. Light petroleum is used as solvent to separate substances of the type R_2Te, where R is one of the organic radicals mentioned above, and ethyl acetate–methanol (1:1) and benzene–ethanol (9:1 or 4:1) are used to separate derivatives of the types R_2TeCl_2 and R_3TeCl.

L. Organotin Compounds

Thin-layer chromatography has been used to determine lubricants and organotin stabilizers in carbon tetrachloride extracts of PVC[23]. Radiometric methods have been used to study the migration of bis(2-ethylhexyl){di([1-^{14}C]octyl)stannylene}dithiodiacetate from rigid PVC into edible oils[24]. The extractability of dibutyl- and dioctyltin stabilizers from PVC into foodstuffs simulation liquids such as water, 3% acetic acid, 10% aqueous ethanol, and diethyl ether can be measured by extracting the organotin compound with chloroform from the simulation liquid and then applying the chloroform solution to a glass plate coated with a suspension of silica gel in water (1:23) that has been dried at 20°C and activated at 105°C just before use. The thin-layer plate is then developed with diisopropyl ether containing 1.5% of anhydrous acetic acid. The chromatogram is dried at 105°C and the separated organotin compounds are revealed by spraying with a 0.02% solution of dithizone in chloroform. The limit of detection is 10 μg of organotin compound. The procedure permits distinction between dibutyl- and dioctyltin compounds. Organotin stabilizers in PVC were identified using thin-layer chromatography[25]. The organotin compound is isolated from the polymer by shaking the diethyl ether extract with aqueous EDTA (disodium salt) followed by extraction with a chloroform solution of dithizone. Alternatively, the polymer film may be soaked in an oil (e.g. olive oil) and the organotin compounds absorbed from the oil on to activated silica gel. The mobile phase is either butanol–acetic acid (50–100:1) saturated with water or water–butanol–ethanol–acetic acid (20:10:10:1), and the spots are revealed with dithizone or diphenylcarbazone. Dibutyl-, dioctyl-, and dibenzyltin salts were separated and identified and mixtures of dialkyl- and trialkyltin compounds could be separated.

Traces of mono-, di-, and tervalent and uncharged tin compounds can be separated using silica gel as adsorbent and various solvents as the mobile phase[26]. The R_F values of the individual compounds are such that cations can be identified; the anionic part of the molecule has no effect. The recommended mobile phase for general use is a mixture (2:1) of isopropyl alcohol and ammonium carbonate solution (2 parts of 10% aqueous ammonium carbonate and 1 part of aqueous ammonia). For the separation of bivalent organotin compounds, isopropyl alcohol–1 N sodium acetate can be used. For monovalent compounds, butanol–2.5% aqueous ammonia is the most suitable. Spots are located by spraying the chromatogram with catechol violet solution and examining it in u.v. light. Phenyltin compounds dissolved in 96% ethanol are decomposed by u.v. light to give phenol, which can be detected with 4-aminophenazone solution. The method has been applied to the determination of the extractability of organotin stabilizers from poly(vinyl chloride) foil and the solubility of such stabilizers in fats and oils[26].

Thin-layer chromatography of organotin compounds on silica gel can be carried out with 1.5% acetic acid in diisopropyl ether as developing solvent[27]. Sulphur-containing compounds are identified by spraying the chromatogram with 2% ethanolic molybdophosphoric acid, and other tin compounds are located by examination under u.v. light and by treatment with 0.5% ethanolic catechol violet. In a procedure for the identification of organotin compounds commonly used as stabilizers in poly(vinyl chloride) packaging materials the stabilizers are extracted from the material with diethyl ether and separated on layers of silica gel G by radial development with 2,2,4-trimethylpentane–diisopropyl ether–acetic acid (80:3:8)[28]. The developed plates are exposed to bromine vapour then sprayed with a solution of Rhodamine B and catechol violet in acetone and examined under daylight and under u.v. irradiation (254 and 366 nm). Migration values (relative to dioctyltin diacetate) have been reported for 10 compounds. The detection limit is 1 μg.

Thin-layer chromatography has been applied to a wide variety of other types of organotin compounds, including bis(2-ethylhexyl)tin compounds[29], sulphur-containing organotin compounds[30], impurities in di-*n*-octyltin stabilizers[31], triphenyltin compounds[32], mixed allylphenyl and isobutylphenyltin compounds[33], mixed phenyltin chloride[34], and alkyltin compounds as their quercetin chelates[35]. Thus, dialkyltin compounds in amounts of 5–100 μg can be separated from other organotin species by thin-layer chromatography. They are determined by treatment with dithizone, elution, and photometric determination[30]. It is not usually easy to detect, for example on thin-layer chromatographic plates, tetraalkyltin compounds or trialkyl- or triaryltin halide compounds with the usual colorimetric reagents since complexes are either not formed or are too unstable. This difficulty can be overcome by exposing the plates to bromine vapour[36] so that tin—carbon bond cleavage occurs and the product has sufficient Lewis acidity to form a stable complex with the ragent.

M. Organozinc Compounds

A thin-layer chromatographic procedure for the determination of zineb in tobacco plants has a sensitivity of 10 μg of zineb per 100 g of tobacco[37].

II. PAPER CHROMATOGRAPHY

A. Organoarsenic Compounds

Six organoarsenic compounds (arsanilic acid, arsenosobenzene, arsphenamine, 4-hydroxy-3-nitrophenylarsonic acid, 4-nitrophenylarsonic acid, and *p*-ureidophenylarsonic acid) were separated by two-dimensional chromatography on sheets of Whatman No. 1 paper, with a solvent consisting of water and nitric acid diluted with methyl cyanide[38]. The compounds were located on the chromatogram and identified by their quenched or fluorescent areas in u.v. light. Final identification was made by spraying with ethanolic ammoniacal silver nitrate or ethanolic pyrogallol, with air drying between the sprayings. The identification limit is 1 μg for each compound.

B. Organobismuth Compounds

Basic bismuth gallate and free tribromophenol can be detected in small samples of bismuth tribromophenoxide by the simultaneous ascending chromatography of a neutral ethanolic suspension of one portion of the sample and an acidified ethanolic solution of another portion[39]. The two spots are placed side by side and developed with butanol saturated with water. After 2 h the chromatogram is treated with aqueous iron(II) ammonium sulphate solution at about halfway between the origin and the front and silver nitrate solution is applied close to the front. The presence of tribromophenol is shown by an orange fleck with silver nitrate on the neutral chromatogram and basic bismuth gallate by a blue–black spot with iron(II) ammonium sulphate solution on the acid chromatogram. An orange fleck on the acid chromatogram is due to the acid decomposition of the bismuth tribromophenoxide.

C. Organocobalt Compounds

For the determination of cobalamins, total cobalamins are determined spectroscopically in a pH 5.5 acetate buffer containing 25–30 μg (dry matter) of sample per millilitre[40]. Individual cobalamins are determined by applying a portion of the sample solution in a band to Whatman 3MM paper. Development is carried out for

18 h by the descending technique with water-saturated *sec*-butyl alcohol to which is added 1% of acetic acid and water to incipient turbidity. The individual cobalamin bands are eluted with measured volumes of 0.1% potassium cyanide solution (pH 6.0). Each individual substance is determined spectrophotometrically as cyanocobalamin. Hydroxo- and cyanocobalamins can be determined by converting them into the dicyano derivatives by treatment with potassium cyanide prior to paper chromatography[41]. The dicyano derivatives are concentrated by extraction with butanol of an aqueous solution of the cyanide-treated sample saturated with ammonium sulphate. Impurities are removed by paper chromatography, and the purified compounds are eluted from the spots with water. The absorbances are then measured spectrophotometrically at 367 and 570 nm, and the concentrations of the cobalamins are calculated with reference to calibration graphs.

Cyanocobalamin and its derivatives can be separated on paper with chloroform–phenol–butanol–water (12:2:5:20)[42,43]. Several R_F values were tabulated[42]. Cyanocobalamin can be extracted from the paper with water and determined absorptiometrically; the error was ±2%[43]. Hydroxycobalamin has been determined in parenteral injection solution by paper chromatography followed by spectroscopy[44]. Photodensitometric determinations of cyanocobalamin and hydroxocobalamin after separation on Whatman No. 1 paper, on thin layers of alumina or silica gel, or on Amberlite WA-2 paper show that the sensitivity decreases from 5 μg on Whatman paper to 15 μg on Amberlite paper[45]. Hydroxo- and cyanocobalamins give identical calibration graphs. The chromatogram is developed in a dark chamber at 20 ± 2°C in an atmosphere saturated with the solvent. The results for hydroxocobalamin are more reproducible than those for cyanocobalamin owing to losses of cyanide ion from cyanocobalamin during chromatography. The errors range from −5 to −8% for up to 30 μg of cyanocobalamin on paper, alumina, or silica gel, and from −1.8 to −2.4% on paper for 50–200 μg of cyanocobalamin. Results for the determination of cyanocobalamin by a paper chromatographic[46] and by a cup plate agar diffusion method[47] have been shown to be equally accurate.

D. Organocopper Compounds

In a paper chromatographic method for the quantitative determination of copper and zinc 8-hydroxyquinolinates, the 8-hydroxyquinolate, after extraction with 10% sulphuric acid and then with chloroform at pH 5–7, is chromatographed with butanol–hydrochloric acid–water (8:1:1) on Whatman No. 3 paper[48]. Two bands, one due to 8-hydroxyquinoline and Cu^{2+} ($R_F = 0.7$) and one to Cu^{2+} alone ($R_F = 0.4$), are formed. Both bands are cut out, and the copper is extracted with dilute acetic acid and titrated with EDTA, with 1-(2-pyridylazo)naphth-2-ol as indicator.

E. Organoiron Compounds

Paper chromatography of ferrocenes has been discussed[49,50]. In the paper chromatography of iron complexes of porphyrins the iron of uro-, copro-, haemato-, deutero-, meso, and protoporphyrins in the form of corresponding haemin, haematin, or haematin acetate are separated by reversed-phase paper chromatography with water–propanol–pyridine as the solvent system and silicone (Dow Corning No. 550) as the stationary phase[51]. A simplified lutidine–water system has been described for the determination of the number of carboxyl groups in haemins and free phosphyrins. Methods for the determination in plant extracts of iron chelates such as Fe—EDTA and Fe—1,2-diaminocyclohexanetetraacetic acid depend upon the removal of unwanted plant constituents and concentration of the chelates by passage through a column of cation-exchange resin or of activated charcoal, followed by evaporation and paper chromatography with phenol–water (4:1, w/v) in the presence of aqueous ammonia and potassium cyanide[52]. For the quantitative determination of free EDTA and 1,1-diaminocyclohexanetetraacetic acid, colorimetric methods depending on either the iron-(III)—salicylate complex or the copper—biscyclohexanoneoxalyl-dihydrazone complex are described. The iron chelates of free EDTA and 1,2-diaminocyclohexanetetraacetic acid may be determined by a modified 1,10-phenanthroline method; up to 200 μg of each can be determined.

Organolead Compounds

Mixtures of R_3PbCl, R_2PbCl_2, and R_4Pb (R = methyl, ethyl, or phenyl) can be separated by paper chromatography followed by conversion to the $[PbCl_4]^{2-}$ ion and spectrophotometric determination of the latter at 357 nm[53]. When applying this method to samples containing a single organolead compound, a suitably sized portion of the sample is mixed with 3 volumes of ethanol saturated with iodine. After 5 min the mixture is evaporated to dryness on a water-bath, the iodine is removed by heating on a hot-plate, and the residue is dissolved in 4 M potassium iodide and examined spectrophotometrically. Recoveries of lead range from 98 to 102.2% of theoretical. The solvents used for the di- and trimethyl- and di- and triethyllead chlorides are water and methanol, for the di-, tri-, and tetraphenyl compounds chloroform, acetone, or methanol, and for the tetramethyl and tetraethyl compounds *n*-heptane. Mixtures of di- and trimethyl- and di- and triethyllead chlorides with lead chloride containing 25–40 μg/ml of total lead can be separated chromatographically on paper[54]. The spots located by means of a reference strip, are cut out and extracted, and the extracts are treated as described above. The errors range from −2.0 to +1.9%. Treatment of the spots on the paper without previous elution was more rapid but less precise, the errors amounting to as much as 12.5%. Ionic lead present as an impurity can be determined by dissolving the sample in water, adding potassium iodide to a concentration of 4 M, and measuring the absorption at 357 nm, the organolead compounds having no absorption at this wavelength; lead ions are converted into $[PbI_4]^{2-}$.

A procedure for the paper chromatographic separation of trialkyllead chlorides and bromides uses two solvents: (i) benzene (6 vols.) and cyclohexane (3 vols.) saturated with water, the organic phase being filtered through a paper wetted with benzene and mixed with acetic acid (1 vol.); (ii) cyclohexane saturated with water[55]. Ascending development on 2043b paper (Schleicher and Schüll) is used. After irradiation with u.v. light the compounds are detected by spraying with ammonium sulphide solution. R_F values for several compounds in the two solvents were given. The water content of the solvents is critical; a lower water content leads to an increase in R_F, especially for slowly moving components. Better separations are obtained with trialkylead chlorides than with trialkyllead bromides.

Paper chromatography[56] has been applied successfully to the separation of mixtures of alkyllead compounds in petroleum. Tetramethyl- and tetraethyllead are detected by chromatography of their bromoderivatives (R_3PbBr). To the sample is added aqueous bromine, the mixture shaken, then concentrated aqueous ammonia is added in excess. The phases are left to separate, then a few drops of the aqueous phase are placed on Whatman No. 1 paper, and developed by descending chromatography for 12 h with the organic phase of butanol–concentrated aqueous ammonia (1:1). The air dried paper is sprayed with 0.1% dithizone in chloroform. The lead compounds appear as yellow–orange spots on an evanescent green background. The original compounds are identified from the R_F values of the derivatives, viz. tetramethyllead 0.6, ethyltrimethyllead 0.7, diethyldimethyllead 0.7 or 0.81, triethylmethyllead 0.81, and tetraethyllead 0.9. The mixed-alkyl compounds result from metathetical reactions that occur on storage when both tetramethyl- and tetraethyllead are present.

G. Organomercury Compounds

Several developing systems can be used for the paper chromatography of organomercury compounds if aqueous ammonia is one of the components[57]. The most suitable combination is butan-1-ol–95% ethanol28% ammonia (8:1:3). A chloroform solution of dithizone is the preferred spray reagent[58]. Alternatively, an aqueous sodium stannite solution can be used. The former is more sensitive but less selective than the latter. Both the ascending and the descending techniques were used. This procedure is useful for detecting relatively small amounts of organomercurials in the presence of large amounts of inorganic mercurials. Inorganic mercury salts did not migrate under these conditions. A study of the concentration limits showed that the least amount of mercurial which could be detected after chromatographic development contained from 1 to 1.5 μg of mercury. The maximum amount which could be chromatographed without tailing contained 20 μg of mercury.

In one procedure for the paper chromatography of organomercury compounds an aqueous solution is spotted on to a filter-paper and chromatographed with *n*-butanol saturated with 1 N aqueous ammonia solution[59,60]. The spots are detected with 1% diphenylcarbazone in ethanol. The R_F value increases with increasing number of carbon atoms in the organic radical (phenylmercury acetate 0.39, phenylmercury chloride 0.40, ethylmercury chloride 0.27, methoxyethylmercury chloride 0.18, methylmercury chloride 0.17; at 30°C); Hg_2^{2+} and Hg^{2+} remain at the baseline. Another method for the paper chromatographic separation of alkyl- and arylmercury compounds used as agricultural chemicals uses *n*-butanol–pyridine–1 N aqueous ammonia solution (35:34:31)[60]. The R_F values are as follows: *p*-tolylmercury acetate, 0.77; phenylmercury acetate, 0.68; di(ethylmercury phosphate, 0.62; 2-methoxyethylmercury chloride, 0.47; methylmercury chloride 0.41; and mercury(II) chloride, 0.06. The spots are revealed with a solution of 1% diphenylcarbazone in ethanol. Various methods based on paper chromatography, colour reactions, and u.v. spectrophotometry have been used for the detection of organomercury compounds[61]. In paper chromatography, butanol–glacial acetic acid–water (4:1:5) is a most useful solvent, and dithizone is a sensitive detection reagent. Both the Reinsch sublimation method and colour reaction with 2,2′-bipyridyl are suitable for qualitative analysis, but colour reactions with di-2-naphthylthiocarbazone and dithizone are unsuitable.

H. Organoselenium Compounds

Selenomethionine has been separated from methionine by paper chromatography[62]. Layers (0.25 mm thick) of silica gel G were used with isopropyl alcohol–butanol–water (1:3:1) as the solvent system. The spots were located with ninhydrin solution in butanol. The limit of detection was 0.2 μg of either compound. In a method for the identification of selenoamino acids by paper chromatography the sample containing selenomethionine and selenocystine is applied to the paper, and then exposed to the vapour of 15% hydrogen peroxide solution for 45–60 s; this oxidizes the selenium to selenoxide, but does not affect the sulphur of sulphur-containing amino acids[63]. After solvent development, the chromatogram is allowed to dry and is then lightly sprayed with starch–sodium iodide–hydrochloric acid solution. The selenium-containing zones give a purple colour in 45 s, and this colour changes to brown when the chromatogram is dried at 50°C. Sulphur-containing amino acids may be detected either by the ninhydrin reaction, after decolorization of the paper by exposure to ammonia vapour, or by a process similar to that described above, with hydrogen peroxide oxidation for 30 min to form the sulphoxides.

4. Organotin Compounds

Procedures for paper partition chromatography of compounds of the type R_nSnX_{4-n} using butanol–pyridine–water, butanol–ethanol–water, butanol–ammonia–water, and aqueous pyridine detect the organotin compounds by spraying with catechol violet[64,65]. Compounds of the type R_4Sn and R_3SnX are initally oxidized by u.v. irradiation before spraying with catechol violet. For the paper chromatography of diaryltin compounds the paper is impregnated with a 10% solution of olive oil in light petroleum[66]. Solutions of various concentrations of methanol in 1 N hydrochloric acid are used for development. The diaryltin spots are detected with a saturated solution of diphenylcarbazone in 50% aqueous methanol (purple spots). The minimum detectable amount of diphenyltin dichloride is 0.5 μg.

Butyltin compounds such as Bu_nSnX_{4-n} can be separated and identified by reversed-phase chromatography using liquid paraffin as the stationary phase[67]. Tetrabutyltin (R_F 0.0), tributyltin chloride ($R_F = 0.50$), dibutyltin dichloride (R_F 0.82), and butyltin trichloride (tailing) were separated by development with acetone–water–acetic acid (20:10:1), dichlorodioctyltin (R_F 0.69) and dibutyltin dichloride (R_F 0.95) with propanol–water–acetic acid (20:20:1), and butyltin trichloride and tin(IV) chloride with butanol saturated with 2 N hydrochloric acid in the absence of liquid paraffin.

An alternative reversed-phase paper chromatographic method for compounds of the type $(alkyl)_nSnX_{4-n}$ (where X = halogen or residue of monocarboxylic acid or dicarboxylic acid half-ester) uses 1-bromonaphthalene as the stationary phase with 50–70% acetic acid saturated with 1-bromonaphthalene as the developing solvent[68]. Spots are revealed by spraying with dithizone or catechol violet solution. Another method for the paper chromatography of the range of alkyltin chlorides uses Whatman No. 2 paper and a 30% solution of petroleum (boiling range 190–275°C) in benzene as the stationary phase and ethanol–acetic–water (20:1:14) as the mobile phase[69]. Spots are revealed by dithizone after 16–18 h at 20°C. Further methods are described in references 70, 71 and 72.

A method for the chromatographic identification of dibutyltin compounds in the presence of dioctyltin compounds in PVC subjects an acetone solution to chromatography on Whatman No. 4 paper (descending technique) with butanol–acetic acid–methanol–water (5:1:20:24)[73]. The air-dried chromatogram is sprayed with 0.1% butanolic diphenylcarbazone to produce red–violet spots. All of the dibutyltin compounds have an R_F value of 0.75 and the dioctyltin compounds remain at the origin. This procedure was also applied to PVC extracts prepared by extracting PVC with carbon tetrachloride, then converting the organotin compound into a chloride with concentrated hydrochloric acid[74]. Upon chromatographing as described above the dioctyltin stabilizers remained at the baseline, whereas the dibutyltin compounds migrated.

In a procedure for the identification of sulphur-containing organotin stabilizers such as dibutyltindithioacetate in the presence of sulphur-free organotin compounds, chromatograms of the stabilizers extracted with diethyl ether from the plastics are developed with hexane–acetic acid (12:1) and sprayed with a solution of dithizone in chloroform or ethanolic catechol violet[30]. The dioctyl, dibutyl, and trioctyl derivatives are clearly separated. The dialkyl compounds are revealed as red spots with dithizone reagent or as blue spots with catechol violet, and the trioctyl compounds as yellow spots with either reagent. The sulphur-containing and the sulphur-free compounds e.g. dioctyltin di(butyllthio)acetate and dioctyltin di(butyl maleate), have the same R_F values but can be distinguished from one another, as sulphur-containing compound form additional spots (blue with dithizone and yellow with catechol violet). Trioctyltin compounds can best be detected by conversion into the corresponding diocty compounds by exposing duplicate plates to bromine vapour for 10 min; the diocty compounds are then detected as indicated above. The minimum detectable amount by this method is 10 μg.

III. COLUMN CHROMATOGRAPHY

A. Organocobalt Compounds

In an ion-exchange chromatographic method for the determination of traces of cyanocobalamin, the solution is passed through a column of Amberlite CG-50 (100–150 mesh)[75]. The column is washed with 0.1 N hydrochloric acid and then the cyanocobalamin is eluted from the column with dioxane–0.25 N hydrochloric acid (3:2). To the elute is added 2% sodium chloride and concentrated nitric acid and the solution is evaporated to dryness, then redissolved in perchloric acid and evaporated to dryness again. The residue is dissolved in acetate buffer solution (pH 6) and nitroso-R salt solution, and then acidified. The absorption of this solution is measured at 520 nm against a reagent blank, and cyanocobalamin is determined by reference to a calibration graph or by means of an internal standard. Other vitamin B_{12} homologues interfere in this procedure.

In a procedure for the determination of hydroxo, aquo, and other cobalamins in injection solutions, the cobalamins are converted into cyanocobalamin by treatment with potassium cyanide at pH 7.5 (citrate buffer solution)[76]. The solution is adjusted to pH 4 with citric acid and applied to a column of Amberlite XE-97, which is then washed successively with water, citrate buffer solution, 0.1 N hydrochloric acid, 85% acetone, and 0.1 N hydrochloric acid to remove impurities. The red cyanocobalamin fraction is then eluted with dioxane into 0.1 N hydrochloric acid and converted into the dicyano compound by treatment with potassium cyanide. The absorption is read within 20–30 min at 578 nm, and compared with that of control and standard solutions.

Cyanocobalamin in a solution of the vitamin B complex has been separated and determined using ion-exchange chromatography on acidic Wafatit CP 300, sodium form[71]. The sample is placed on the column and then the various strengths of the resin are washed successively with water. This removes all the B vitamins other than cyanocobalamin. Cyanocobalamin is then eluted with acetone–water (1:1) containing 7.5% sodium chloride solution until a suitable volume has been collected. Finally, the absorption of this solution is measured at 361 nm against that of the solvent in the reference beam. Ion-exchange chromatography has been applied to the determination of cyanocobalamin in pharmaceutical syrups[78]. The sample at pH 4.5–5.1 is passed through a column of Amberlite XE-97 ion-exchange resin. The column is washed with 0.1 N hydrochloric acid, 80% acetone, and 0.1 N hydrochloric acid. Cyanocobalamin is then eluted from the column with dioxane–0.1 N hydrochloric acid (3:2) and the eluate collected and its absorption measured at 361 nm. Recoveries are claimed to be between 98 and 99.4% of the amount of cyanocobalamin added.

Neutral and basic cobalamins such as hydroxo- and cyanocobalamin and cobinamide have been separated on various ion-exchange papers under a variety of conditions[79]. R_F values have been listed for the vitamins of the B_{12} group on cellulose ion exchangers, alginic acid, and alginic acid–cellulose preparations using water, mixtures of water with various alcohols, and water with hydrogen cyanide as solvent[80].

B. Organoiron Compounds

Ferrocenes have been separated on columns[81]. Separation on a carboxymethylcellulose column was used in a simple and rapid method for the quantitative determination of haemoglobin A_2[82]. Packed red cells are lysed with 0.01 M phosphate buffer (pH 6.2) and diluted with buffer solution. A 5% potassium hexacyanoferrate(III) solution and 2% potassium cyanide are added. An aliquot is added to a column of carboxymethylcellulose and suspended in phosphate buffer solution (pH 6.2). The haemoglobin components other than A_2, are eluted first with 0.01 M phosphate buffer (pH 7.1) and then, when the red band reaches the bottom of the column, with buffer solution of pH 7.3. Haemoglobin A_2 is eluted with buffer solution of pH 7.7. The absorptions of the eluates are read at 540 nm. The behaviour of haemoglobins during chromatography is influenced by the pH and ionic concentration of the chromatographic developers, the state of equilibrium of the Amberlite IRC-50 resin with the developer, the temperature during equilibration and chromatography, the amount of haemoglobin on the chromatogram, and the oxidation state of the haem in the haemoglobin[83].

Two chromatographic procedures have been described for the determination of human haemoglobins A, B, C, and F on Amberlite IRC 50 (EX64) resin[84]. The results compare favourably with those of older methods. An investigation of the methods for determining foetal (Hb_f) and sickle-cell (Hb_b) haemoglobins in the presence of adult haemoglobin (Hb_a) led to the conclusion that for the determination of Hb_f the alkali denaturation method is not sufficient in the presence of Hb_a and Hb_b[85]. Amino acid composition provides a more definite answer to the presence of this protein. Only electrophoretic measurements can be used for the detection and determination of Hb_b. By the use of the four different methods it appeared that the alkali-resistant fraction, sometimes found in the blood of patients with sickle-cell anaemia, is not foetal haemoglobin[85].

The influence of gel structure and pore size of synthetic gels such as acrylamide gel on the resolution of various haemoglobins has been studied[86]. Human, horse, and dog haemoglobins have been separated on cation-exchange cellulose[87]. Chromatography of haemoglobins in the carboxy form is carried out on carboxymethycellulose eluted with 0.01 M sodium phosphate buffers to give a gradient of increasing pH. The eluted fractions are examined spectroscopically and, after concentration, by paper electrophoresis and sedimentation methods. The heterogeneous nature of adult, human, and horse carboxyhaemoglobins and the effects of temperature and of concentration of the urea solution on the chromatography were discussed. The chromatographic behaviour of normal haemoglobin and seven different abnormal human haemoglobins on the cation exchanger Amberlite IRC-50 has been studied[88,89], and the separation by electrophoresis, column chromatography, solubility, and alkaline denaturation of normal and abnormal human haemoglobins reviewed.[90].

Separation on Sephadex gels has been used for the determination of haemoglobin and myoglobin[91]. Separation was effected on Sephadex G-50 or G-75 (200–400 mesh). The columns were equilibrated at 4°C overnight with 0.05 M phosphate buffer (pH 7.4), which was also 0.05 M in sodium chloride. A mixture of the two proteins in buffer solution containing 25% of sucrose was applied to the column. The proteins were eluted with the buffer solution containing sodium chloride but no sucrose. The bands were located and determined by absorption measurements at 280, 410, and 577 nm. Recoveries exceeding 95% were usually obtained.

C. Organolead Compounds

High-performance liquid chromatography enables concentrations as low as 0.01 g of tetraethyllead as lead per imperial gallon in gasoline to be determined[92].

D. Organomercury Compounds

High-performance liquid chromatography has enabled dibenzo-18-crown-6 complexes of divalent mercury to be separated[93]. Methyl-, ethyl-, and phenylmercury cations have been determined in amounts down to 2 μg/g using liquid chromatography with differential pulse electrochemical detection[94].

E. Organophosphorus Compounds

Ethyl- and methylparathion at the parts per 10^9 level have been determined in run-off waters using high-performance chromatography[95]. The plasma chromatography of phosphorus esters has been investigated[96].

F. Organoselenium Compounds

Selenocystine and selenomethionine may be separated from their sulphur analogues and from leucine by chromatography on a column of sulphonated styrene–divinylbenzene copolymer resin; elution is carried out at 33°C with 0.2 N sodium citrate buffer, first at pH 3.28, changing to pH 4.25 after the passage of 300 ml through the column[97]. Selenocystine leaves the column in the eluate fraction between 405 and 415 ml, and selenomethionine between 450 and 500 ml.

IV. ELECTROPHORESIS

A. Organoiron Compounds

Several electrophoretic methods have been used for separating haemoglobin A from haemoglobin F. One uses starch hydrolysed in hydrochloric acid–acetone (1:100, v/v) at 38.5°C[98]. Electrophoresis is carried out at 20°C for 3 h under a potential gradient of 12 V/cm. A second method uses a layer of 2% agar in buffer solution (pH 5.7–5.9) supported on a glass plate covered with filter-paper[99]. Light green SF (CI Acid Green 5) is used for staining, followed by washing with 2% acetic acid until almost decolorized. The haemoglobins are well separated, haemoglobin A remaining near the point of application. Another method uses gel phase ion-exchange electrophoresis[100]. Carboxymethylcellulose gel is the supporting medium and electrophoresis is carried out in a sodium phosphate buffer (0.17–0.04 M sodium ion). The best overall separation is achieved with a sodium ion concentration of 0.07 M. A sensitive two-dimensional paper–agar electrophoretic method detects small amounts of haemoglobin A in the presence of haemoglobin F, and detects minor components in a haemoglobin solution[101]. Electrophoresis is carried out at pH 8.2 on paper. A strip cut along the centre of the paper in the direction of migration is inserted in a slit in the agar gel (pH 6.2), prepared in a plastic tray. Electrophoresis is then carried out at right-angles to the length of the paper strip. Haemoglobins A and F have been separated by the cation-exchange dextran gels SE-Sephadex C-50 and CM-Sephadex C-50[102,103]. The samples were converted into carboxyhaemoglobin and the haemoglobin was determined, after conversion into cyanomethaemoglobin, by measurement of the absorption at 420 or 540 nm. Both foetal and adult haemoglobin were eluted from each gel with 40 mN phosphate buffer (pH 6.0) containing an increasing concentration of sodium chloride.

In a rapid method for haemoglobin fractionation on cellulose acetate, the buffer solution (pH 8.6) is prepared from Tris, EDTA, and anhydrous boric acic[104,105]. A constant current of 0.4–0.5 mA per centimetre of strip width is applied for about

40 min, a clear separation of haemoglobins A, C, S, and A_2 being achieved. The spots are eluted into the buffer solution and the absorption of the eluate is measured at 400–410 nm. The factors influencing the chromatography and differentiation of similar haemoglobins have been discussed[83].

Various methods have been described for the determination of haptoglobin complexes. For zone electrophoresis *o*-dianisidine is the most satisfactory detection reagent[106]. Zones possessing peroxidase activity are stained brown–pink. To determine the haptoglobin group in sera containing low concentration of haptoglobin, haemoglobin is added to the sera in excess of the haemoglobin-binding capacity, and an electrophoretic separation is performed with a gel prepared by boiling starch with a phosphate buffer[107]. The free haemoglobin migrates towards the cathode, producing a clear separation of the haptoglobin groups, which migrate towards the anode. A solution of benzidine, acetic acid, and hydrogen peroxide is used to develop the electrophoretic pattern. Electrophoresis on acrylamide gel has been used to determine haptoglobin types. The three types of haptoglobin observed after separation by electrophoresis in starch or acrylamide gel can be distinguished readily by adding 10% normal adult haemoglobin solution to fresh, clear non-haemolysed serum to form the complexes, and staining the electropherogram with a benzidine reagent[108]. After electrophoresis, the gel is placed in the reagent and then washed with 15% acetic acid solution. In a specific separation of serum haptoglobin as a haemoglobin complex the serum is diluted five-fold with phosphate or acetate buffer solution and applied to a diethylaminoethylcellulose column, which has been previously washed with 0.2 M acetic acid, then with 0.2 M sodium hydroxide, and then with the buffer solution[109,110]. The column is washed with the buffer solution to elute the albumin and the α globulins, then washed again with the buffer solution containing a small amount of sodium chloride to elute other proteins. A calculated amount of horse haemoglobin dissolved in the buffer solution is added to saturate the haptoglobin, then the complexed haptoglobin–haemoglobin is eluted with buffer solution containing a higher concentration of sodium chloride than before. The complex is eluted rapidly and specifically.

In a discontinuous buffer system for human haemoglobins the electrode vessels are filled with barbitone buffer solution (pH 8.6) and the paper is immersed in a buffer solution (pH 9.1), prepared by dissolving tris(hydroxymethyl)aminomethane (Tris), disodium EDTA dihydrate, and boric acid in water[50,111]. The paper is placed in the electrophoresis chamber, buffered haemoglobin solution is applied, and the process is carried out in the usual way. The resolution of abnormal haemoglobins is superior to that in either buffer solution alone, and enables haemoglobin A_2 to be detected in small samples of haemolysate. The staining with Amido Black 10B (CI Acid Black 1) of patterns that on visual inspection show only haemoglobin A reveals the A_2 component if present. Agar gel[112], starch gel[113–118], paper[119,120], cellulose acetate[119–123], and acrylamide gel electrophoresis[124] have been applied to the determination of haemoglobin A_2. One group subjected the blood sample to electrophoresis on agar gel at 90–100 V and 50–50 mA[112]. After drying, the electropherograms are stained with Amido Black 10B (CI Acid Black 1) solution and measured in a densitometer at 500 nm. A good separation of haemoglobins A_1 and A_2 was achieved and up to three unidentified non-haemoglobin fractions were observed. Excellent resolution has been obtained in the starch-gel electrophoresis of normal and abnormal haemoglobins with relatively little trailing of components[116]. Separation occurs of haemoglobin A_2, two other pigmented fractions, and two non-pigmented protein fractions in normal red cell haemolysates, and the method can be used to distinguish between foetal haemoglobin and haemoglobin A when both are present.

Starch-gel electrophoresis of haemoglobin A_2 has been carried out by the vertical technique[125] with a gel made up in 0.025 M borate buffer (pH 8.6)[117,118]. After separation, the haemoglobin fractions are cut out and placed in 0.06 N aqueous ammonia and stored in a refrigerator for 3 days. The absorption is measured at 540 nm, and the haemoglobin A_2 content calculated.

Haemoglobin A_2 in blood can be determined quantitatively by paper electrophoresis[119]. The blood samples are collected with EDTA (dipotassium salt) as anticoagulant, haemolysed with water and toluene, and converted into carboxyhaemoglobin. Electrophoresis is carried out in vertical tanks with strips of Whatman No. 3 MM paper and tris(hydroxymethyl)aminomethane (Tris) buffer adjusted to pH 8.6 with boric acid. The treated samples are applied to the strips, which are then moistened with buffer on both sides to within 0.25 in of the point of application. After passing a current of 225 V and 2 mA per strip for 16 h, the strips are dried at 90–100°C, fixed in fresh 10% mercury(II) chloride solution and 10% glacial acetic acid in ethanol, redried, washed in water for 10 min, and then dried again. They are then stained in 1% bromophenol and 1% acetic acid in ethanol. The stained bands are cut out and eluted with 1.5% sodium carbonate solution in 50% methanol, and the absorbances of the eluates are read at 595 nm in a spectrophotometer. Another method uses electrophoresis of the sample prepared as a solution in 50% glycerol on paper for 3 h in barbitone buffer (pH 9.1) at 350 V[120]. After drying, the paper is stained with bromophenol blue and then examined visually or with a densitometer and automatic integrator to evaluate the separated haemoglobin A_2. A quantitative determination of haemoglobin A_2 by electrophoresis on cellulose acetate uses an electrophoretic cell solution of 0.05 M sodium carbonate, and a membrane buffer solution of Tris–EDTA solution adjusted to pH 8.8 with aqueous boric acid[123]. The strips are stained with Ponceau S (CI Acid Red 9), washed with 5% acetic acid, and the bands are eluted with 0.268 M aqueous ammonia. Acrylamide gel is superior to starch as it is easier to work with, gives more reproducible layers and faster separations, and affords fractions that can be evaluated by direct densitometry[124]. Haemolysates are diluted with an equal volume of Tris–borate–EDTA buffer solution and electrophoresis is carried out for 1 h at 300 V. The unstained gel patterns are scanned with a recording densitometer with a 500 nm interference filter. The normal range for haemoglobin A_2 found by this method was 1.4–4.4%.

A rapid method for the determination of haemoglobins uses microelectrophoresis on cellulose acetate[126]. The red-cell haemolysate is applied to a cellulose acetate membrane and subjected to electrophoresis for 1–1.5 h in a microelectrophoresis apparatus. The strips are then stained with 0.2% Ponceau S (CI Acid Red 9) solution in 3% trichloroacetic acid containing 5% of sulphosalicylic acid. After being washed with 5% acetic acid, the strips are cleared by immersion in methanol–acetic acid (7:3) and scanned in a photodensitometer. Cellulose acetate electrophoresis has also been used for haemoglobin fractionation[127]. The sample is subjected to electrophoresis on cellulose acetate with Tris–EDTA–boric acid buffer solution (pH 8.8). The electropherograms are stained with Ponceau S (CI Acid Red 9), rinsed in 5% acetic acid, cleared in acetic acid –95% ethanol (3:17), and then examined with a densitometer equipped with a blue filter.

Several groups have studied the application of starch gel electrophoresis to the separation of haemoglobins[125,127–130]. A simple inexpensive electrophoresis apparatus with water cooling is available[128]. Errors resulting from the instability of the haemoglobin solution can be reduced by adding a small amount of potassium cyanide. Reagents for the detection of haemoglobins separated by starch gel electrophoresis have been described. After completion of electrophoresis, the gel is cut into two portions, one of which is stained with benzidine and hydrogen peroxide to identify and locate the fractions. Zones containing the individual fractions are then cut out of the second portion of gel and dissolved in 10% sodium hydroxide solution. Each resulting solution is mixed with 1% benzidine solution in acetic acid and 1% (v/v) hydrogen peroxide solution and set aside, then 10% acetic acid is added. The mixture is again set aside and the absorption is measured at 515 nm.

A study of combined agar gel–paper electrophoresis for the resolution of haemoglobins showed that by carrying out the electrophoresis in highly purified agar gel with a buffer of decreased ionic strength and a high potential gradient, separation can be attained in 15 min[131,132]. For agar gel electrophoresis for the identification of haemoglobin types[133], blood samples may be prepared by centrifuging and addition of oxalic acid or heparinizing and addition of citric acid then removing the plasma and adding an equal volume of water[134]. The solution is shaken to haemolyse and the solution added to lead acetate barbitone solution at pH 8.6. After 30 min the solution is centrifuged at 2000 rev/min. Half saturated phosphate solution is then added and the solution is again centrifuged. It is claimed that this solution upon electrophoresis gives a sharp resolution of the constituent haemoglobins.

In an electrophoretic method for the separation of haemoglobin and cytochrome C, the sample components are separated on cellulose acetate at pH 8.6, and zones are revealed by staining with ethanolic Amido Black 10G (CI Acid Black 1)[135]. Similar procedures were used to separate haemoglobin from myoglobin.

B. Organolead Compounds

To separate 50 μg amounts of lead chloride, diethylleaddichloride and triethyllead chloride by paper electrophoresis, 2–3 M lithium chloride is used as the supporting electrolyte on a 5.5×40 cm strip of Whatman No. 1 paper, applying a potential of 135 V for 2 h[136].

C. Organothallium Compounds

Paper electrophoresis has been carried out on phenyl- and methylthallium compounds, using 10 V/cm for 1 h and aqueous sodium chloride containing 0.1 M hydrochloric and as the supporting electrolyte[137]. Unlabelled compounds were detected by spraying the paper with potassum rhodizonate and ^{204}Tl labelled compounds by autoradiography. Good separations were obtained between organothallium(III) chlorides and thallium(I) and thallium(III) chlorides.

V. REFERENCES

1. M. Vŏbetský, V. D. Nefedov, and E. N. Sinotova, *Zh. Obshch. Khim.*, **33**, 4023 (1963).
2. K. Berei, *J. Chromatogr.*, **20**, 406 (1965).
3. K. Berei and L. Vasána, *Magy. Kém. Foly.*, **73**, 313 (1967).
4. D. T. Haworth and A. F. Kardis, *J. Chromatogr.*, **27**, 302 (1967).
5. S. Heřmánek, J. Plešck, and V. Gregor, *Collect. Czech. Chem. Commun.*, **31**, 1281 (1966).
6. J. Huber, I. Rückbeil, and R. Kiessig, *Pharm. Zentralhalle Dtschl.*, **102**, 783 (1963).
7. L. Cima and R. Mantovan, *Farmaco, Ed. Prat.*, **17**, 473 (1962).
8. T. Ono, *Bitamin*, **30**, 280 (1964).
9. M. Covello and O. Schettino, *Farmaco, Ed. Prat.*, **19**, 38 (1964).
10. L. Cima, C. Levarato, and R. Mantovan, *Farmaco, Ed. Prat.*, **21**, 244 (1966).
11. Ya. G. Popova, K. Popov, and M. Ilieva, *J. Chromatogr.*, **24**, 263 (1966).
12. Ya. G. Popova, K. Popov, and M. Ilieva, *J. Chromatogr.*, **21**, 164 (1966).
13. T. Sasaki, *J. Chromatogr.*, **24**, 452 (1966).
14. C. Maglitto, L. Gianotti, and C. Maltarei, *Boll. Lab. Chim., Provinciali*, **15**, 354 (1964).

15. K. Schloegl, H. Pelousek, and A. Mohae, *Monatsh. Chem.*, **92**, 533 (1961).
16. R. E. Bozak and O. Fukuda, *J. Chromatogr.*, **26**, 501 (1967).
17. L. C. Willemsens and G. J. M. Van der Kerk, in *Investigations in the Field of Organolead Chemistry*, International Lead–Zinc Research Organization, New York, 1957, p. 84.
18. K. Broderson and U. Schlenker, *Z. Anal. Chem.*, **182**, 421 (1961).
19. L. Geldern, *Erdöl Kohle*, **18**, 545 (1965).
20. A. J. Bawd, D. T. Burns, and A. Fogg, *Talanta,* **16**, 719 (1969).
21. M. Vŏbetský, V. D. Nefedov, and E. N. Sinotova, *Zh. Obshch. Khim.*, **33**, 4023 (1963).
22. M. Vŏbetský, V. D. Nefedov, and E. N. Sinotova, *Zh. Obshch. Khim.*, **35**, 1684 (1965).
23. H. Huber and J. Wimmer, *Kunststoffe*, **58**, 786 (1968).
24. H. Seidler, H. Waggon, M. Haertig, and W. J. Uhde, *Nahrung*, **13**, 257 (1969).
25. M. Türler and O. Högl, *Mitt. Lebensmitt. Hyg. Bern*, **52**, 123 (1961).
26. K. Bürger, *Z. Anal. Chem.*, **192**, 280 (1962).
27. G. Neubert, *Z. Anal. Chem.*, **203**, 265 (1964).
28. D. Helberg, *Dtsch. Lebensm.-Rundsch.*, **62**, 178 (1966).
29. D. Helberg, *Dtsch. Lebensm.-Rundsch.*, **63**, 69 (1967).
30. R. F. Van der Heide, *Z. Lebensm.-Unters.-Forsch.*, **124**, 348 (1964).
31. K. Figge, *J. Chromatogr.*, **39**, 84 (1969).
32. Y. Jitsu, N. Kudo, and T. Sugujama, *Noyaku Seison Gijutsu*, **17**, 17 (1967).
33. D. Nraun and H. T. Heimes, *F. Z. Anal. Chem.*, **239**, 6 (1968).
34. V. D. Nefedov, V. E. Zhuravlev, N. G. Molchanova, and N. N. Kalinina, *Zh. Obshch. Khim.*, **38**, 1219 (1968).
35. H. Wieczorek, *Dtsch. Lebensm.-Rundsch.*, **65**, 74 (1969).
36. P. P. Otto, H. M. J. C. Creemers, and J. G. A. Luijten, *J. Labelled Comp.*, **2**, 339 (1966).
37. E. S. Kosmatyi, L. L. Bulic, and G. V. Gaurilova, *Fiziol. Biokhim. Kul't. Rast.*, **4**, 317 (1972).
38. L. C. Mitchell, *J. Assoc. Off. Agric. Chem.*, **42**, 684 (1959).
39. A. Castiglioni, *Z. Anal. Chem.*, **161**, 40 (1958).
40. C. Cardini, G. Cavina, E. Cingolani, A. Marioni, and C. Vicari, *Farmaco, Ed. Prat.*, **17**, 583 (1962).
41. M. Covello and O. Schettino, *Ann. Chim. (Rome)*, **52**, 1135 (1962).
42. J. Bayer, *J. Chromatogr.*, **8**, 123 (1962).
43. A. Sauciuc, L. Ionescu, and M. Albu-Budăi, *Revtă. Chim.*, **18**, 237 (1967).
44. J. L. Martin and W. H. C. Shaw, *Analyst (London)*, **88**, 292 (1963).
45. M. Covello and O. Schettino, *Farmaco, Ed. Prat.*, **20**, 581 (1965).
46. I. V. Konova, N. M. Neronov, N. D. Ierusalimskii, and A. I. Borisova, *Mikrobiologiya*, **28**, 490 (1959).
47. N. D. Ierusalimskii, I. V. Konova, and N. M. Neronova, *Mikrobiologiya*, **28**, 433 (1959).
48. T. D. Miles, A. C. Delasanta, and J. C. Barry, *Anal. Chem.*, **33**, 685 (1961).
49. A. N. DeBelder, E. J. Bourne, and J. B. Pridham, *Chem. Ind. (London)*, 996 (1959).
50. C. A. J. Goldberg, *Clin. Chem.*, **5**, 446 (1959).
51. T. C. Chu and E. J. H. Chu, *J. Biol. Chem.*, **212**, 1 (1955).
52. D. G. Hill-Cottingham and C. P. Lloyd-Jones, *J. Sci. Food Agric.*, **12**, 69 (1961).
53. R. Barbieri, U. Belluco, and G. Tagliavini, *Ric. Sci.*, **30**, 1671 (1960).
54. R. Barbieri, U. Belluco, and G. Tagliavini, *Ann. Chim. (Rome)*, **48**, 940 (1958).
55. H. Schafer, *Z. Anal. Chem.*, **180**, 15 (1961).
56. M. Pedinelli, *Chim. Ind. (Milan)*, **44**, 651 (1962).
57. J. N. Bartlett and G. W. Curtis, *Anal. Chem.*, **34**, 80 (1962).
58. V. L. Miller, D. Polley, and C. J. Gould, *Anal. Chem.*, **23**, 1286 (1951).
59. J. Kanazawa, K. Koyama, M. Aya, and R. Sato, *J. Agric. Chem. Soc. Jap.*, **31**, 872 (1957).
60. J. Kanazawa and R. Sato, *Bunseki Kagaku*, **8**, 322 (1959).
61. K. Sera, M. Kanda, A. Murakami, Y. Sera, Y. Kondo, and T. Yanagi, *Kumamato Med. J.*, **15**, 38 (1962).
62. K. R. Millar, *J. Chromatogr.*, **21**, 344 (1966).
63. J. Scala and H. H. Williams, *J. Chromatogr.*, **15**, 546 (1964).
64. D. J. Williams and J. W. Price, *Analyst (London)*, **85**, 579 (1960).
65. D. J. Williams and J. W. Price, *Analyst (London)*, **89**, 220 (1964).

66. O. A. Reutov, O. A. Putsÿna and M. F. Turchinskiĭ, *Dokl. Akad. Nauk SSSR*, **139**, 146 (1961).
67. Y. Tanaka and T. Morikawa, *Bunseki Kagaku*, **13**, 753 (1964).
68. J. Gasparič and A. Cee, *J. Chromatogr.*, **8**, 393 (1962).
69. J. Franc, M. Wurst, and V. Moudry, *Collect. Czech. Chem. Commun.*, **26**, 1313 (1961).
70. R. Barbieri, U. Belluco, and G. Tagliavini, *Ann. Chim. (Rome)*, **48**, 940 (1958).
71. A. Cassol, L. Magon, and R. Barbieri, *J. Chromatogr.*, **19**, 57 (1965).
72. A. Cassol and R. Barbieri, *Ann. Chim. (Rome)*, **55**, 606 (1965).
73. B. Visintin, A. Pepe, and S. A. Guiseppe, *Ann. Ist. Sup. Sanita*, **1**, 767 (1965).
74. K. Bürger, *Z. Anal. Chem.*, **192**, 280 (1962); *Anal. Abstr.*, **10**, 3742 (1963).
75. D. Monnier, Y. Ghaliounghi, and R. Salia, *Anal. Chim. Acta*, **28**, 30 (1963).
76. F. Gstirner and S. K. Baveja, *Arch. Pharm. (Weinheim)*, **298**, (2); *Mitt. Dtsch. Pharm. Ges.*, **35**, 29 (1965).
77. L. Klotz, *Pharm. Zentralhalle Dtsch.*, **104**, 393 (1965).
78. B. Petrangeli, *Boll. Chim. Farm.*, **105**, 770 (1966).
79. R. Hüttenrauch and L. Klotz, *J. Chromatogr.*, **12**, 464 (1963).
80. J. Pawelkiewicz, W. Walerych, W. Friedrich, and K. Bernhauer, *J. Chromatogr.*, **3**, 359 (1960).
81. M. Rosenblum and R. B. Woodword, *J. Am. Chem. Soc.*, **80**, 5443 (1958).
82. C. J. Muller and C. Pik, *Clin. Chim. Acta*, **7**, 92 (1962).
83. R. T. Jones and W. A. Schroeder, *J. Chromatogr.*, **10**, 421 (1963).
84. T. H. J. Huisman and H. K. Prins, *J. Lab. Clin. Med.*, **46**, 255 (1955).
85. P. C. Van der Schaaf and T. H. J. Huisman, *Rec. Trav. Chim. Pays-Bas*, **74**, 563 (1955).
86. S. Raymond and M. Nakamichi, *Anal. Biochem.*, **3**, 23 (1962).
87. F. J. Gulter, E. A. Peterson, and H. A. Sober, *Arch. Biochem. Biophys.*, **80**, 353 (1959).
88. T. H. J. Huisman and H. K. Prins, *Nature*, **175**, 903 (1955); *Anal. Abstr.*, **2**, 2498 (1955).
89. T. H. J. Huisman and H. K. Prins, *Clin. Chim. Acta,* **2**, 307 (1957).
90. H. K. Prins, *J. Chromatogr.*, **2**, 445 (1959).
91. E. Awad, B. Cameron, and L. Kotite, *Nature (London)*, **198**, 1201 (1963).
92. T. C. S. Ruo, M. L. Selucky, and O. P. Strauzz, *Anal. Chem.*, **49**, 1761 (1977).
93. A. Mangia, G. Parolari, E. Geatani, and C. F. Lamreci, *Anal. Chim. Acta*, **92**, 111 (1977).
94. W. A. MacCrehan and R. A. Durst, *Anal. Chem.*, **50**, 2108 (1978).
95. D. C. Paschal, R. Birknell, and D. Dresbach, *Anal. Chem.*, **49**, 1551 (1977).
96. J. M. Preston, F. W. Karasek, and S. H. Kim, *Anal. Chem.*, **49**, 1746 (1977).
97. J. L. Martin and L. M. Cummins, *Anal. Biochem.*, **15**, 530 (1966).
98. J. A. Owen and C. Got, *Clin. Chim. Acta*, **2**, 588 (1957).
99. F. Rappaport and M. Rabinovitz, *Clin. Chim. Acta*, **4**, 535 (1959).
100. E. R. Huehns and A. O. Jakubovic, *Nature (London)*, **186**, 729 (1960).
101. P. Fessás and A. Karaklis, *Clin. Chim. Acta*, **7**, 133 (1962).
102. A. M. M. Zade-Oppen, *Scand. J. Clin. Lab. Invest.*, **15**, 491 (1963).
103. A. M. M. Zade-Oppen, *Scand. J. Clin. Lab. Invest.*, **15**, 331 (1963).
104. R. C. Bartlett, *Clin. Chem.*, **9**, 325 (1963).
105. R. C. Bartlett, *Clin. Chem.*, **9**, 317 (1963); *Anal. Abstr.*, **11**, 713 (1964).
106. J. A. Owen, H. J. Silberman, and C. Got, *Nature (London)*, **182**, 1373 (1958).
107. C. B. Laurell, *Scand. J. Clin. Lab. Invest.*, **11**, 18 (1959).
108. T. G. Ferris, R. E. Easterling, and R. E. Budd, *Clin. Chim. Acta*, **8**, 792 (1963).
109. W. Dobryszycka, J. Moretti, and M. F. Jayle, *Bull. Soc. Chim. Biol.*, **45**, 301 (1963).
110. M. F. Jayle, *Bull. Soc. Chim. Biol.*, **33**, 876 (1951).
111. C. A. J. Goldberg, *Clin. Chem.*, **3**, 1 (1957).
112. V. J. Yakulis, P. Heller, A. M. Josephson, and L. Singer, *Am. J. Clin. Pathol.*, **34**, 28 (1960).
113. C. A. J. Goldberg, *Clin. Chem.*, **4**, 484 (1958).
114. C. A. J. Goldberg and A. C. Ross, *Clin. Chem.*, **6**, 254 (1960).
115. W. B. Gratzer and G. H. Beaven, *Clin. Chim. Acta*, **5**, 577 (1960).
116. R. L. Engle, A. Markev, J. H. Pert, and K. R. Woods, *Clin. Chim. Acta*, **6**, 136 (1961).
117. M. Aksay and S. Erdem, *Clin. Chim. Acta*, **12**, 696 (1965).
118. K. Aksay and S. Erdem, *Türk Tip Cemiy. Mecm.*, **31**, 593 (1965).

119. R. N. Ibbotson and B. A. Crompton, *J. Clin. Pathol.*, **14**, 164 (1961).
120. T. R. Johnson and O. N. Barrett, *J. Lab. Clin. Med.*, **57**, 961 (1961).
121. H. S. Friedmann, *Clin. Chim. Acta*, **7**, 100 (1962).
122. E. Afonso, *Clin. Chim. Acta*, **7**, 545 (1962).
123. A. S. Pinfield and D. O. Rodgerson, *Clin. Chem.*, **12**, 883 (1966).
124. T. G. Ferris, R. E. Easterling, and R. E. Budd, *Nature (London)*, **208**, 1103 (1965).
125. O. Smithies, *Biochem. J.*, **61**, 629 (1955).
126. J. L. Graham and B. W. Grunbaum, *Am. J. Clin. Pathol.*, **39**, 567 (1963).
127. R. O. Briere, T. Golias, and J. G. Batsakis, *Am. J. Clin. Pathol.*, **44**, 695 (1965).
128. H. R. Marti, *Experientia*, **17**, 235 (1961).
129. L. A. Lewis, *Clin. Chem.*, **12**, 596 (1966).
130. F. W. Sunderman, *Am. J. Clin. Pathol.*, **40**, 227 (1963).
131. B. Zak, F. Volini, J. Briski, and L. A. Williams, *Am. J. Clin. Pathol.*, **33**, 75 (1960).
132. B. Zak, E. M. Eggers, T. L. Jarkowski, and L. A. Williams, *J. Clin. Med.*, **54**, 288 (1959).
133. C. D. McDonald and T. H. J. Huisman, *Clin. Chim. Acta*, **8**, 639 (1963).
134. V. N. Tompkins, *Am. J. Clin. Pathol.*, **25**, 1430 (1955).
135. W. Leyko and R. Gondko, *Biochim. Biophys. Acta*, **77**, 500 (1963).
136. M. Guistiniani, G. Faraglia, and R. Barbieri, *J. Chromatogr.*, **15**, 207 (1964).
137. G. Foraglia, L. Roncucci, B. Fioroni, P. Lassandro, and R. Barbieri, *Ric. Sci. Riv.*, **37**, 986 (1967).

The Chemistry of the Metal–Carbon Bond
Edited by F. R. Hartley and S. Patai

CHAPTER **20**

Infrared and Raman spectroscopy of organometallic compounds

MICHAEL J. TAYLOR
Department of Chemistry, University of Auckland, Auckland 1, New Zealand

I. INTRODUCTION

Infrared and Raman spectroscopy and their chemical applications are well served by monographs, reviews, and collections of spectra[1–3]. The purpose of this chapter is to bring together those aspects particularly useful to the study of organometallic compounds. A further object is to describe the state of the art in experimental techniques and theoretical methods building on the available reviews (see Section II) and serving as a guide to future trends and developments.

Of previous accounts, the fullest is the compilation by Maslowsky[4], with comprehensive references up to 1975. Other treatments of organometallic spectra by Huggins and Kaesz[5], Adams[6], Downs[7], and Nakamoto[8,9] have an important place. More specialised accounts, which concentrate on particular groups of organometallic compounds, are cited in Section III.

Several different approaches have been adopted to the task of organizing organometallic spectroscopy for review. It is usual to examine metal–ligand vibrations in relation to symmetry and also to discuss the characteristic vibrations of ligands and the changes in these with bonding at the metal. This is the approach followed here.

II. THE TECHNIQUES OF VIBRATIONAL SPECTROSCOPY

A. General Aspects

Although infrared continues to be the more convenient fingerprint technique with which to guide preparative work, serious investigation of organometallic systems normally requires the joint application of i.r. and Raman methods. Technical developments in i.r. and Raman spectroscopy have been reviewed by Waters[10] and elsewhere[11–13]. Increasing use is being made of computational methods of data handling, spectrum averaging, band resolution, and solvent compensation, applied to both techniques. Infrared Fourier transform spectroscopy is finding wide application in both the mid- and far-i.r. regions[14–18]. Raman spectroscopy has a particular value where solution work can be used to distinguish the polarized character of totally symmetrical modes, and from a practical point of view it can be the more convenient technique for dealing with highly air- or moisture-sensitive materials whose samples can be examined in sealed all-glass systems. The method routinely requires only a small sample in a capillary tube (sealed if the sample is atmosphere-sensitive), and with suitable optics can handle microscopic amounts. The well known advantage of the Raman technique in its ability to handle aqueous solutions finds some application in organometallic chemistry. Some recent Raman studies of organometallic systems in liquid ammonia form part of a review of vibrational spectra of non-aqueous solutions[19].

The fact that Raman intensity is a near-linear function of concentration makes possible its quantitative use for analysis, or to study equilibria. Measurement of i.r. band intensities is not much applied to organometallic systems; exceptions are some studies of olefin and carbonyl complexes.

Far-i.r. spectroscopy (conveniently the range below 400 cm^{-1}) continues to be advanced by the application of both grating and Fourier transform spectrometers, preferably with computational accessories to allow spectrum accumulation and background compensation. Details of modern advances can be found in reviews already cited[10–18].

B. Spectroscopy at Non-ambient Temperatures and at High Pressure

The advantage of obtaining vibrational spectra at *low temperature* includes the improved resolution, especially in the region of low frequencies where lattice modes are often encountered or where information on low-energy deformations and torsional modes may be sought. The techniques and applications of low-temperature i.r. spectroscopy by conventional and Fourier transform methods have been discussed[20]. Complications due to change of phase and hence of crystal symmetry which are routinely taken into account in work on inorganic systems, especially single crystal studies, need to be watched for, however, and the collection of additional data at low temperature is seldom attempted unless with a particular aim in view, for example the determination of rotational barriers. Measurements above room temperature may be undertaken to follow dissociation and other equilibrium processes or to obtain Raman spectra of samples in the liquid phase, including depolarization ratios.

The additional information to be gained by studies of vibrational spectra under *high pressure* is beginning to be generally appreciated. Reviews of this area[21–23] have drawn attention to the pressure sensitivity of the skeletal modes of metal sandwich compounds as an aid to their assignment. Special cells[24] and procedures for obtaining i.r. spectra at high pressure have been described and have been used to study reactions of transition metal alkyls with carbon monoxide[25].

C. Vibrational Spectra of the Crystalline State

The principles which allow the i.r. and Raman spectra of crystalline solids to be predicted in terms of the space group of the crystal are well understood[26]. These find a limited use in the spectroscopy of organometallics in explaining site splitting and intermolecular factor group correlation effects. There is considerable scope for the extension of single crystal studies. Examples which have provided some unequivocal assignments include studies of $[Cp_2Ru]$[27], $[Cp_2Fe]$[28], and other cyclopentadienyls, $[Cp_2M]$ (M = Fe, Ru, Os, or Ni)[29], as well as [CpNiNO][30], $[(arene)Cr(CO)_3]$ (arene = benzene or 1,3-dimethylbenzene)[31], and the dimethylglyoximato complex *trans*-$[MeCo(py)(dmg)_2]$[32]. The examinations of solid-state organometallic spectra by Kettle and co-workers[33] concentrated particularly on the use of the ν(CO) modes. Main group compounds investigated in great detail by single-crystal Raman spectroscopy are $[Me_2SnF_2]$ and $[Me_2TlBr]$[34], $[Me_2Sn(acac)_2]$[35] and $[Me_2SnCl_2]\cdot 2DMSO$[36].

D. Matrix Isolation and the Spectra of Adsorbed Species and of Free Radicals

The techniques for obtaining i.r. and Raman spectra of *matrix-isolated molecules* are reaching a high level of development in a number of centres. Although many of the most thoroughly investigated systems are those of metal carbonyls, a range of organometallic species is also attracting attention. Advantages include the ability to trap otherwise unstable species, and the opportunity to observe sharp bands from isolated molecules. Surveys of this area[14,37–42] draw attention to studies of metal–

olefin and metal–aryl complexes, and a variety of alkyl-containing species. Examples are described in Section III.

The study of *adsorbed species* by i.r. spectroscopy[43,44] is well known though not widely applied. Although not at first sight a promising field for Raman scattering the situation may be changed if the vast enhancement in the effect which has been reported for some systems[45] proves to operate beyond a limited range of substrates and adsorbed species.

The vibrational spectroscopy of *free radicals*[46], which will undoubtedly contribute to mechanistic studies as well as to the understanding of structure and bonding, has yet to be applied to more than a few organometallic systems.

E. Resonance Raman Spectroscopy and Raman Optical Activity

The early promise of *resonance Raman spectroscopy* (RRS), the theory of which is well developed, is now being realised in areas including organometallic chemistry[47–50], and notable advances in both theory and practice of RRS are at the disposal of the well equipped spectroscopist. Excitation at a series of wavelengths, either by means of a dye-laser, or the various lines from blue to far-red offered by helium-cadmium, argon, helium–neon, and krypton lasers, makes it possible to obtain excitation profiles which offer a sensitive probe to the relationship between electronic structure and the vibrations of chromophoric groups.

Raman optical activity and the related topic of vibrational circular dichroism[51–53], which offer a probe for stereochemistry and magnetic structure in the investigation of organometallic complexes, have so far been applied mainly to organic molecules.

F. Coherent Anti-Stokes Raman Scattering

The CARS technique[54,55], which is capable of giving very high signal levels from a small volume of sample, has great scope for Raman studies of organometallic systems. Particular advantages are the ability to avoid the problem of fluorescence and the opportunity to make effective use of resonance enhancement when working near to an electronic absorption band.

G. Vibrational Analysis and Force Constant Calculations

The methods of vibrational analysis for the prediction of molecular spectra, the application of selection rules, and the further treatment of data, have been well described by a number of authors[56,56]. The need to use special point groups of higher symmetry than are required by the static model when dealing with non-rigid systems is illustrated in later sections. Reference is also made in Section III to some detailed studies of organometallic molecules by means of normal coordinate analysis. These illustrate the insight provided by force constants and the increasing use being made of the potential energy distribution to arrive at a better understanding of normal modes and their coupling in metal–ligand skeletal vibrations.

New theoretical advances have been described[58–61], as have experimental techniques such as isotope substitution (including metal isotopes)[62,63] which provide additional frequency data valuable for refined normal coordinate analysis (n.c.a) treatments. The subjects of torsional barriers and internal rotation in molecules[64,65] and the vibrations of ring compounds[66–68], of particular relevance to organometallic spectra, are treated elsewhere.

III. PROGRESS IN THE INTERPRETATION OF ORGANOMETALLIC SPECTRA

A. Vibrational Spectra of Main Group Organometallic Compounds

1. Organic ligand spectra

Vibrational spectra have been reported and analysed for neutral, covalent, *methyl* compounds of nearly every main group metal and metalloid, providing a basic knowledge of metal–carbon stretching and other skeletal frequencies and an appreciation of the ranges in which frequencies of the methyl ligand modes are located[4]. Six vibrations characterize the unit CH_3M. Taking account of the three-fold axis, these are as follows:

A_1 modes: $\nu(C{-}H)_s$, 2750–2950 cm^{-1}; $\delta(CH_3)_s$, 1100–1350 cm^{-1}; $\nu(M{-}C)$, variable with the nature of the M—C bond.

E modes: $\nu(C{-}H)_{as}$, 2810–3050 cm^{-1}; $\delta(CH_3)_{as}$, 1300–1475 cm^{-1}; $\rho(CH_3)$, 650–975 cm^{-1}.

The deuteromethyl frequency ranges are also well established[4]. The symmetric deformation is a useful characteristic mode in Raman spectra, while the rocking mode is often strong in the i.r. spectrum although weak in Raman.

In most compounds several methyl groups are bonded to the metal atom and so more than one peak can often be observed in the regions expected for modes of CH_3M. In addition there will be M—C stretching, CMC bending, and possibly CH_3 torsional modes for $[(CH_3)_nM]$ compounds (see the following sections). Pre-1975 data have been tabulated[4,9] for compounds with $[Me_nM]$ (n = 2–6), and for some complex ions $[Me_nM]^+$ and $[Me_nM]^-$ known to that date. The variation of skeletal frequencies with change of metal is thus well documented. Changes with increase in coordination number, oxidation state, or charge on the complex follow the established trends observed for other metal compounds, notably halides[6].

The *ethyl and higher alkyl* groups bonded to a metal in general exhibit bands due to CH_3, CH_2, and CH groups, and skeletal modes of increasing complexity. Some characteristic bands of ethyl, vinyl, and acetylenic compounds have been tabulated[4,9], but the difficulties caused by vibrational coupling and band overlapping are such that the interpretation of available data is often incomplete. The $\nu(M{-}C)$ frequencies in ethyl derivatives are often 20–50 cm^{-1} less than they are in the corresponding methyl compound. This is partly due to the increased mass of the ligand but the situation is complicated by coupling of M—C stretching with other modes, particularly M—C—C bending. Examples of normal coordinate analysis which help to clarify this situation are given in the sections which follow.

The *phenyl* group, when σ-bonded to a metal, exhibits bands characteristic of monosubstituted benzenes[69]. The $C_6H_5{-}M$ type of molecule has 30 fundamentals. Only six of these are appreciably sensitive to the change in metal[4,9], being the modes (in Whiffen's notation) q(A_1), 1050–1085; r(A_1), 620–700; y(B_1), 440–480; t(A_1), 200(bismuth)–420(aluminium); u(B_2), 160–330; and x(B_1), 145–240 cm^{-1}. The approximate forms of these modes are shown in Figure 1.

For all these organometallic systems the spectra of the $[R_nM]$ group are modified when other ligands are present in the compound or when the molecule takes part in complex formation. New spectroscopic data and improved assignments are continually being added. The following sections examine progress in the interpretation of the vibrational spectra of organometallic compounds and complexes, with particular emphasis on vibrations of the metal-to-carbon bond. Spectroscopic results for some examples of metal compounds with perhalogenated organic ligands are included.

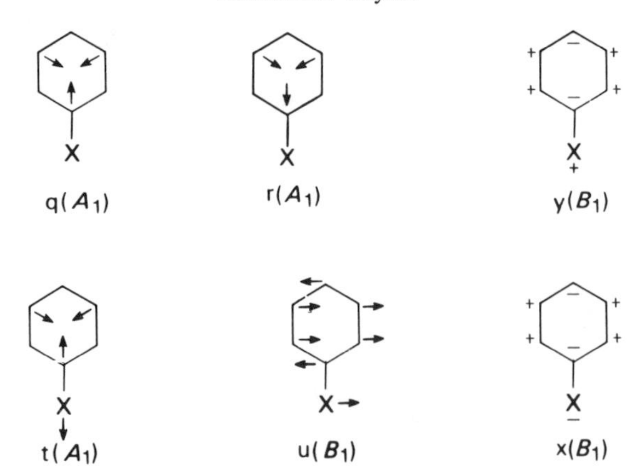

FIGURE 1. Symmetries of the X-sensitive modes of C_6H_5—X.

2. *Organometallic compounds of zinc, cadmium, mercury, and other Group I and II elements*

a. Alkyls of zinc, cadmium, and mercury: In a model study for the thorough understanding of metal alkyl spectra, Butler and Newbury[70] examined the i.r. and Raman spectra of $[(CH_3)_2M]$ (M = Zn, Cd, or Hg) in liquid, vapour, and solid states to derive secure assignments of ν(M—C) modes (Table 1) and to record changes in the fundamental frequencies and band contours through all three phases. Their analysis takes into account methyl group rotation and requires use of the double group $G_{36}^{\ddagger}$, as discussed elsewhere[71], rather than the D_{3d} (staggered) model of earlier work. The methyl groups rotate even in the solid state and variable-temperature Raman spectroscopy (200–15 K) reveals that a barrier to free rotation about the linear C—M—C axis gradually develops as the temperature falls. This work also examines the solid state spectra of the Group II dimethyls in terms of their known crystal space group, and reveals changes to lower symmetry at very low temperature. The stretching frequencies (and force constants) of the $[Me_2M]$ molecules are in the order Zn > Cd < Hg. In this series the assessment of bond stiffness from vibrational data is in line with thermodynamic bond strengths and with estimates of the chemical stability of the M—C bond from other sources.

Re-examination of the spectra of the methylmercury halides, $[CH_3HgX]$ and $[CD_3HgX]$ (X = Cl, Br or I)[72], using oriented single crystal techniques permits all vibrations to be accounted for in terms of the centrosymmetric space group $P_4/nmm(D_{4h}^7)$ and also allows the degree of intermolecular attraction (greatest for

TABLE 1. Skeletal vibrations (cm^{-1}) in $[Me_2M]$ molecules. Vibrations measured in the liquid state. Vapour phase values in parentheses

Parameter	$[Me_2Zn]$	$[Me_2Cd]$	$[Me_2Hg]$
$\nu(M—C)_s$	503	460	515
	(511)	(473)	(518)
$\nu(M—C)_{as}$	604	526	538
	(613)	(536)	(540)
δ(C—M—C)	157	140	160

TABLE 2. Hg—C stretching vibrations and force constants in [RHgX] molecules

Molecule	Parameter	X = Cl	X = Br	X = I
[MeHgX]	ν(Hg—C)(cm^{-1})	554	545	533
	f(Hg—C)(N/cm)	2.55	2.48	2.38
[EtHgX]	ν(Hg—C)(cm^{-1})	527	521	508
	f(Hg—C)(N/cm)	2.64	2.52	2.47

the most polar halide) to be assessed by observing the correlation field splitting between the i.r. and Raman-active modes of the crystal. A thorough study of the spectra and force fields of [EtHgX] (X = Cl, Br or I)[73] shows the Hg—C constants to be marginally higher than those for the corresponding methyl derivatives[74]. Table 2 compares the Hg—C stretching frequencies and force constants. Goggin *et al.*[72] noted that the *trans* interaction constant in the [MeHgX] molecules (about 1% of the primary bond-stretching force constant) is much smaller than the values reported for Au(I), Au(III), and Pt(II) systems, reflecting a difference in electronic structure and bonding.

In compounds [$CH_3HgC{\equiv}CR$] the Hg—CH_3 force constant remains near 2.50 N/cm, but that of the bond to the unsaturated ligand is larger: 2.83 N/cm where R = H[75] and 2.99 N/cm where R = CH_3[76]. In related work, the vibrational spectrum of [CH_3HgCN] has been assigned[77] and the effect of the *trans*-ligand on ν(Hg—C) in species [MeHgL] or $[MeHgL]^+$ (where L = CN, halide, or a sulphur or arsenic donor) has been compared[78] with coupling constants from n.m.r. spectra, and with the bond dissociation energy of [MeHgL] into Me• and LHg•, in an attempt to provide further insight into the electronic configuration of the Hg—C bond. However, the validity of such correlations is open to question[79].

I.r. and Raman spectra have provided the basis for some extended studies of analogues of dimethylmercury. Vibrational analysis of [CF_3HgCH_3][80] yields f(Hg—C) = 2.04 N/cm for the bond to CF_3 and 2.52 N/cm for the bond to CH_3. Compared with [$(CH_3)_2Hg$], wherein the Hg—C force constant is 2.38 N/cm[74] and [$(CF_3)_2Hg$] where it is 2.18 N/cm, it can be seen that in [CH_3HgCF_3] the bond Hg—CF_3 is weakened by the σ-donating methyl group. At the same time the electron-withdrawing CF_3 group strengthens the Hg—CH_3 bond to a value similar to that in [CH_3HgCl].

The desirability of using normal coordinate analysis to follow trends in metal–ligand bonding is further illustrated by studies of [$(CF_3)_2Hg$], [$(CCl_3)_2Hg$], and halide derivatives [CX_3HgX'] (X = F or Cl, X′ = F, Cl, or Br)[81]. The pseudohalide derivatives [CF_3HgY] (Y = N_3 or NCO)[82] have been examined by i.r. and Raman spectroscopy in conjunction with their crystal structural determination. The vibrational spectra of [$(CF_3)_2Hg$][83] in the gaseous state, in the molecular crystal, and in solution, are assignable to D_{3d} symmetry; use of double group $G_{36}^{\dagger}$ is not appropriate. In this molecule n.c.a. shows that the modes become extensively mixed. The following list shows the contributions to the potential energy of each of the totally symmetric modes from the coordinates of (1) C—F stretching, (2) Hg—C stretching, and (3) Hg—C—F bending:

$$\begin{aligned} A_{1g} \text{ modes of } (CF_3)_2Hg: \quad \nu_1 &= 1150\ cm^{-1}\ [58\%\ (1),\ 54\%\ (2),\ 20\%\ (3)] \\ \nu_2 &= 715\ cm^{-1}\ [44\%\ (1),\ 37\%\ (2),\ 6\%\ (3)] \\ \nu_3 &= 224\ cm^{-1}\ [76\%\ (3),\ 5\%\ (2)] \end{aligned}$$

Clearly, none of these frequencies can be compared directly with the symmetric Hg—CH_3 stretch, 503 cm^{-1}, of [$(CH_3)_2Hg$][70]. This same point will be recognized in making use of the 'Hg—C stretching' assignments advanced for the trimethylsilyl derivatives[84]:

$$[Hg\{CH(SiMe_3)_2\}_2][\nu(Hg—C)_s = 485\ cm^{-1};\ \nu(Hg—C)_{as} = 510\ cm^{-1}];$$

$$[Hg\{C(SiMe_3)_3\}_2][\nu(Hg—C)_s = 360\ cm^{-1};\ \nu(Hg—C)_{as} = 360\ cm^{-1}]$$

For comparison with the Group II dialkyls it is interesting to note a detailed study by i.r. and Raman spectroscopy of a silicon analogue, $[Hg(SiMe_3)_2]$[85], for which $\nu(Hg—Si)_s = 312$, $\nu(Hg—Si)_{as} = 318\ cm^{-1}$, and $f(Hg—Si) = 1.50$ N/cm, about 60% of the value of typical mercury–carbon force constants.

A different aspect of the vibrational spectroscopy of organomercury compounds involves the use of the methylmercury ion CH_3Hg^+ as a probe for complex formation. As a prelude to work on biological systems, the interaction of $MeHg^+$ with a variety of inorganic anions in aqueous solution has been studied by Raman spectroscopy and proton n.m.r. to determine the formation constants of the complexes[86]. In such complexes the characteristic methylmercury frequencies, $\nu(Hg—C)$ are in the range 535–580 cm^{-1} and the methyl group symmetric deformation, $\delta(CH_3)_s$, is 1180–1210 cm^{-1}. Similar Hg—C stretching frequencies, which assist in locating the bonding site of mercury, are readily observed in the i.r. or Raman spectra of $[MeHg(S_2CNEt_2)]$ and related complexes (510–528 cm^{-1})[87], and in MeHg derivatives of penicillamine (near 540 cm^{-1})[88]. Cationic complexes $[MeHgL]^+$, where L is pyridine or a pyridine derivative[89,90], display $\nu(Hg—C)$ near 565 cm^{-1}. The Raman spectrum has been used to investigate the solvation of the methylmercury halides, in comparison with HgX_2 molecules, in non-aqueous solvents[91] and in liquid ammonia[92].

In other methylmercury compounds the MeHg group takes the place of a proton on carbon. Examples studied by i.r. and Raman spectroscopy are $[MeHgCH_2CN]$ and $[(MeHg)_3CCN]$[93], and the acetylacetonato derivatives, $[MeHgCH(COCH_3)_2]$ and $[(MeHg)_2C(COCH_3)_2]$[94]. Skeletal Hg—C stretching frequencies in the range 450–500 cm^{-1} are shown by these compounds.

Organomercury ketenides, $[(CH_3COOHg)_2C{=}C{=}O]$ and $[(CH_3COOHgHg)_2$-$C{=}C{=}O]$, are of note since the latter appears to be the first organomercury(I) compound to be characterized[95]. Apart from acetate modes, the i.r. spectra show three bands: 2070s, $\nu(CCO)$; 620w, $\delta(CCO)$; and 276m cm^{-1}, tentatively assigned to $\nu(Hg—C)$. Sample fluorescence defeated an attempt to corroborate the presence of the Hg—Hg bond by Raman spectroscopy.

b. Phenyl derivatives: Diphenylmercury and its derivatives provide among the best understood examples of phenyl–metal spectra. Complete assignments of the 27 phenyl ligand modes in the range 3150–375 cm^{-1} are available for $[(C_6H_5)_2Hg]$ and $[C_6H_5HgL]$ (L = Cl, Br, I, CN, or OAc) and their perdeuterated analogues[96,97]. In $[(C_6H_5)_2Hg]$ only small differences are observed between the in-phase and out-of-phase modes of the two phenyl groups, showing that coupling between rings is slight. The t-mode (Section III.A.1), however, generates the two frequencies, $\nu(Hg—Ph)_s$ at 210 cm^{-1} and $\nu(Hg—Ph)_{as}$ at 252 cm^{-1}. The bending mode $\delta(Ph—Hg—Ph)$ has the frequency 62 cm^{-1}.

Assignments for the low frequency modes of phenylmercury halides are given in Table 3. The two A_1 frequencies involve coupled stretching of C—Hg and Hg—X bonds. Calculations of potential energy distribution[96] show that the band in the range 235–260 cm^{-1} (the lower of the two A_1 frequencies in [PhHgCl], but the higher in [PhHgBr] and [PhHgI]) has the predominant $\nu(Hg—C)$ character throughout. The corresponding mode has a frequency of 246 cm^{-1} in PhHgCN and 239 cm^{-1} in PhHgOAc[97].

c. Alkyls of lithium, beryllium, and magnesium: There has been some spectroscopic work on alkyllithium compounds, and an attempt has been made to delineate the

TABLE 3. Vibrations of [PhHgX] molecules below 375 cm^{-1}. Values are means of i.r. and Raman measurements (cm^{-1})

[PhHgCl]	[PhHgBr]	[PhHgI]	Symmetry	Major component
230	258	246	A_1	ν(Hg—Ph)
324	196	158	A_1	ν(Hg—X)
204	216	209	B_1	γ(Hg—X)
185	180	174	B_2	β(Hg—C)
99	97	89	B_1	γ(Hg—C)
78	59	52	B_2	β(Hg—C)

range of ν(Li—C) modes (*ca.* 400–500 cm^{-1}) in their complexes with donor molecules[98]. The stretching frequency ν(Li—C) is 450 cm^{-1} in Li_3BeMe_5 and bands at 335 and 405 cm^{-1} have been assigned to ν(Be—C) vibrations in this complex[99]. Group frequencies and ν(Be—C) assignments have been proposed for ethyl-[100] and butylberyllium[101] adducts.

The lightest alkylmagnesium derivatives, $[Me_2Mg]$ and $[Et_2Mg]$, are solids with infinite chain polymer structures. The spectra[102,103] at 300 and 90 K, including those of isotopic species, $[(CD_3)_2Mg]$, $[(CD_3CH_2)_2Mg]$, and $[(CH_3CD_2)_2Mg]$, can be interpreted on the basis of C_{2h} and C_{4h} factor group symmetry, respectively. Symmetric Mg—C stretching frequencies are assigned at 365 cm^{-1} in the $[Me_2Mg]$[102] and 340 cm^{-1} in $[Et_2Mg]$[103], and n.c.a. yields f(Mg—C) = 1.05 N/cm for the electron-deficient Mg—C bonds, less than half the value in dimethylzinc. Higher frequencies, ν(Mg—C) = 505–515 and 480–485 cm^{-1}, are given by $[MeMgX]\cdot 2Et_2O$ and $[EtMgX]\cdot 2Et_2O$ (X = Br or I), respectively[104,105].

The unstable compound $[Et_2Mg]\cdot Et_2O$ has been shown by i.r. and Raman spectroscopy[106] to have the structure of a centrosymmetric dimer with bridging ethyl groups, and in another aspect of this study[107] characteristic Raman bands have been used to calculate the thermodynamic quantities ΔG°, ΔH°, and ΔS° for the monomer–dimer equilibrium of diethylmagnesium in ether solution:

$$2[Et_2Mg]\cdot(Et_2O)_2 \rightleftharpoons [Et_2Mg\cdot Et_2O]_2 + 2Et_2O$$

The spectra of methylmagnesium alkoxides[108] and methyl-bridged complex anions $[Me_2Mg(Me)_2MgY_2]^{2-}$ (Y = Cl, N_3, or NCS)[109] have also been reported and provide structural contributions to understanding the nature and action of Grignard reagents.

3. *Organometallic compounds of the Group III elements, aluminium, gallium, indium, and thallium*

a. Alkyl derivatives: The Group III metals form complex anions $[Me_4M]^-$. Vibrational assignments[110] are supported by comparisons with $[Me_4M]$ and $[Me_4M]^+$ within the isoelectronic triads of Al, Si, P; Ga, Ge, As; and In, Sn, Sb. For all these tetrahedral species fundamental frequencies have been tabulated by Weidlein and co-workers for the T_d skeletal modes νMC_4 ($A_1 + F_2$) and δMC_4 ($E + F_2$), the methyl deformations $\delta_s CH_3$ ($A_1 + F_2$) and $\delta_{as} CH_3$ ($E + F_2$), and the rocking modes ρCH_3 ($E + F_2$). Valence force constants have then been derived for comparison with other bond properties which change progressively through Groups III, IV, and V. The i.r., Raman, and proton n.m.r. spectra have also been measured for the anions present in the compounds $[Me_4M^V]^+[Me_nM^{III}Cl_{4-n}]^-$, where M^V = As or Sb, M^{III} = Ga or In, and n = 1–3[111,112]. A useful observation is that the weighted means of the M—C stretching frequencies (in the range 510–590 cm^{-1} for gallium and 465–530 cm^{-1} for indium) show a linear correlation with the n.m.r. chemical shifts of the methyl protons.

Vibrational spectra of the Group III alkyls, [R_3M], have been reported in some detail. For the series, [R_3Ga], where R = Me, Et, *n-Pr*, *i*-Pr, *n*-Bu, or *i*-Bu, assignments have been based on the overall symmetry of the molecule and the local symmetry of the alkyl substituents[113]. The central GaC_3 group is planar in [Me_3Ga], [Et_3Ga], and [*i*-Pr_3Ga] and is probably so in the other derivatives. The skeletal frequencies of the trigonal MC_3 unit are assigned according to D_{3h} symmetry in Table 4.

TABLE 4. Skeletal vibrations (cm^{-1}) of [R_3Ga] molecules

Parameter	[Me_3Ga]	[Et_3Ga]	[*i*-Pr_3Ga]	
$\nu(Ga{-}C)_s$, A_1'	521	485	489	(Raman pol.)
$\nu(Ga{-}C)_{as}$, E'	570	537	525	(Raman dp.)
$\delta(GaC_3)$, E'	163	n.o.	n.o.	(Raman dp.)

For the higher alkyls, the Raman spectrum shows much broader bands in the region of metal–carbon stretching and this is ascribed to the simultaneous presence of several conformers, the *trans*-forms being the most stable ones in the crystalline state.

The series of propynyl derivatives [$Me_2MC{\equiv}CCH_3$] show the expected ligand frequencies, the skeletal modes of the Me_2M unit, and additional frequencies attributable to M—C stretching of the bond to the propynyl substituent at 360 cm^{-1} (M = Al), 325 cm^{-1} (Ga), and 283 cm^{-1} (In)[114]. Full analysis needs to take account of the fact that the molecules are dimeric, the π-system functioning as an electron donor to the vacant orbital of the metal. Tricyclopropylaluminium, although formally a saturated compound, gives rise to bridged systems of unusual stability. Physical methods, including i.r. spectroscopy[115], support a dimeric structure in solution and in the gas phase. Al—C stretching frequencies in the range 515–786 cm^{-1} can be identified by comparison of the spectra of [$(c\text{-}C_3H_5)_6Al_2$] with the spectra of [Me_6Al_2] and [Ph_6Al_2].

The halides [$RGaX_2$] (X = Me, Et, *n*-Pr, or *n*-Bu) are halogen-bridged dimers:

R X X
Ga Ga
X X R

The structures are centrosymmetric and the M—C stretching modes show small differences (*ca*. 10 cm^{-1}) between i.r. and Raman frequencies due to coupling across the molecule[116]. Taking the means of i.r. and Raman values, some ν(Ga—C) frequencies are given in Table 5. The frequencies suggest that the 'Ga—C stretching' is coupled to the metal–halogen stretching to a relatively small extent.

Spectroscopic data for the indium analogues [$RInX_2$] and for various dialkyls of gallium and indium, [R_2MX][117,118], reveal some differences in structural behaviour

TABLE 5. Ga—C stretching vibrations (cm^{-1}) in molecules [$RGaX_2$] where X = Cl, X = Br, X = I

[$MeGaX_2$]	603	595	580
[$EtGaX_2$]	568	562	549

from the systems just discussed. Freshly prepared samples of $[MeInI_2]$ are shown by the Raman spectrum[119] to have the usual haolgen-bridged structure, which then isomerizes to the ionic form, $[Me_2In]^+[InI_4]^-$. The linear cations $[Me_2In]^+$ and $[Me_2Tl]^+$, which are isoelectronic with $[Me_2Cd]$ and $[Me_2Hg]$, respectively, play a considerable part in the organometallic chemistry of the heavier Group III elements and are conveniently recognised by their i.r. and Raman spectra[4,9].

The Group III trialkyls are an important group of Lewis acids and their complexes with donor molecules are readily studied by means of vibrational spectroscopy. A full investigation of the molecules $[(CH_3)_3AlNH_3]$, $[(CD_3)_3AlNH_3]$, $[(CH_3)_3AlND_3]$, and $[(CD_3)_3AlND_3]$[120] provides insight into the vibrational coupling of the Al—C and Al—N bonds in stretching vibrations. According to this analysis '$\nu(Al—C)_s$' at 524 cm^{-1} in $[(CH_3)_3AlNH_3]$ is made up of 70% AlC_3 symmetric stretching and 20% Al—N stretching, whereas '$\nu(Al—N)$' at 451 cm^{-1} comprises 68% Al—N stretching and 22% AlC_3 stretching, the balance of the potential energy in each case coming from other modes. The doubly degenerate fundamental '$\nu(Al—C)_{as}$' at 597 cm^{-1} is predominantly of this coordinate (62%), with a substantial contribution from the NH_3 rocking mode (23%).

Vibrational spectra are widely used to characterize R_2M and RM derivatives of the Group III elements, especially by means of M—C stretching frequencies. Structural evidence and detailed assignments are available from the i.r. and Raman spectra for tetrameric complexes $[Me_2MCN]_4$, where M = Al, Ga, or In[121], and include a comparison with $[(Me_2MgCN)_4]^{4-}$ isoelectronic with the aluminium compound. Dimethylaluminium and -gallium azides are trimeric, $[Me_2MN_3]_3$, and form complex anions $[Me_2M(N_3)_2]^-$. I.r. and Raman spectra are known[122] for these and also for the complexes $[Me_3MN_3]^-$ and $[(Me_3M)_2N_3]^-$.

A dimethylindium complex, $[Me_2InNMe_2]_2$, with D_{2h} symmetry, provides an example of the trend to study the vibrational spectrum in conjunction with structure determination by X-ray crystallography. The frequencies $\nu(In—C)_s = 482$ and $\nu(In—C)_{as} = 510$ cm^{-1} were observed and a detailed assignment of other skeletal modes was made[123].

Thallium–carbon modes are assigned in the ranges $\nu(Tl—C)_s$ = 423–493 and $\nu(Tl—C)_{as}$ = 453–584 cm^{-1} in dialkylamides $[R_2TlNR'_2]$, where R = Me, Et, or *n*-Pr[124]. The tendency of thallium to form the $[Me_2Tl]^+$ ion has already been mentioned. Detection of its spectrum shows the azide $[Me_2TlN_3]$ and a diazomethane derivative $[(Me_2Tl)_2CN_2]$ to have predominantly ionic structures[125].

b. Phenyl derivatives: In extension of earlier work on phenyl derivatives of the Group III metals[4,9] vibrational spectroscopy has been used in the structural characterization of $[PhMX_2]$ and $[Ph_2MX]$ (M = Ga or In; X = Cl, Br, or I) and their dioxane adducts[126,127]. Whereas the gallium compounds[126] form dimeric molecules with bridging halogen atoms, those of indium[127] consist of a polymeric lattice. Assignments for the mode of predominant M—C stretching type are in the range 385–405 cm^{-1} for the phenylindium compounds. Surprisingly, the corresponding modes of the phenylgallium species occur in a lower range, 330–355 cm^{-1}. Other low-frequency skeletal modes were also assigned and comparisons made with Ph_3Ga, Ph_3In and some of their other derivatives[126,127].

4. *Organometallic compounds of the Group IV elements, silicon, germanium, tin, and lead*

Several reviews of the vibrational spectra of organic derivatives of the Group IV elements[128–131] are additional to the general reviews of organometallic spectroscopy cited in the Introduction. There is also a comprehensive guide to the pre-1970 literature[132].

a. Alkyl derivatives: The vibrational spectra of the Group IV tetramethyls have been recorded many times. Supplemented by measurements on the deutero-compounds there is enough information for n.c.a. calculations[133] which are not restricted to treating the methyl group as a point mass, and which show that the M—C constant decreases considerably from silicon to lead, while the C—H force constant increases slightly (Table 6). In the search for fuller data and an improved force field, i.r. spectra of matrix-isolated $[Me_4M]$ (M = Si, Ge, or Sn) and chlorides $[Me_{4-n}MCl_n]$ (n = 1–3)[134] have been recorded at 15 K to determine isotopic shifts for the M—C and M—Cl stretching modes as a prelude to further normal coordinate analyses.

TABLE 6. Force constants in $[Me_4M]$ molecules

Parameter	$[Me_4Si]$	$[Me_4Ge]$	$[Me_4Sn]$	$[Me_4Pb]$
f(M—C)(N/cm)	2.88	2.65	2.19	1.88
f(C—H)(N/cm)	4.61	4.68	4.77	4.80

Investigations of the series $[(CH_3)_3MCF_3]$ (M = Ge, Sn, or Pb) and the perdeuterated analogues allow comparison of the metal–carbon bond of methyl and perfluoromethyl ligands[135]. As in the case of mercury (Section III.A.2) it emerges that the $M—CF_3$ force constants are considerably weaker than the $M—CH_3$ constants, typically some 75% of the latter (Table 7).

TABLE 7. Force constants in $[Me_3MCF_3]$ molecules

Parameter	M = Ge	M = Sn	M = Pb
$f(M—CH_3)$(N/cm)	2.73–2.84	2.27	1.93
$f(M—CF_3)$(N/cm)	2.04–2.21	1.67	1.38

The higher alkyl derivatives $[R_4M]$ have attracted frequent attention, especially those of high symmetry[4,9]. Unusually low values of tin–carbon stretching frequencies are assigned to modes involving the bulky ligands of some tricyclohexyltin derivatives[136] for which $\nu(Sn—C)_s$ = 415–425 and $\nu(Sn—C)_{as}$ = 490–496 cm^{-1}. An example of vibrational analysis of an organometallic compound containing one of the higher alkyl groups is to be found in the study of the isotopic series of tertiary butyl compounds, $[t\text{-}Bu_3SnNH_2]$, $[t\text{-}Bu_3SnND_2]$ and $[t\text{-}Bu_3Sn^{15}NH_2]$ which has permitted a complete assignment and normal coordinate treatment of the C_3SnNH_2 group[137]. This analysis yields f(Sn—C) = 2.09 N/cm. Little coupling between cyclopropyl ligand modes occurs in $[(c\text{-}C_3H_5)_4Pb]$, whose i.r. and Raman spectra[138] resemble those of other metals. N.c.a. yields f(Pb—C) = 1.95 N/cm, a slightly larger value than in $[Me_4Pb]$.

The series $[Me_6M_2]$ has also been much investigated[4,9], a refinement being the interpretation of the spectra of hexamethyldisilane, $[(CH_3)_3SiSi(CH_3)_3]$, on the basis of double group $G^{\ddagger}_{36}$ symmetry to accommodate free rotation about the Si—Si bond[139]. For the molecule $[ClMe_2SnSnMe_2Cl]$, the most likely symmetry appears to be C_{2h}, corresponding to a staggered arrangement of the ligands[140]. However, the four Sn—C stretching frequencies, 520, 530, 543, and 547 cm^{-1}, are found in both i.r. and Raman spectra and the assignment is not finally established. A similar range, 510–560 cm^{-1}, spans the ν(Sn—C) modes of tetramethylditin complexes $[Me_4Sn_2(L)]$, where L is a bidentate ligand, O_2CR, S_2CNR_2, or $S_2P(OR)_2$[141,142]. A stretching mode of the tin–tin bond, ν(Sn—Sn), occurs in the region 190–210 cm^{-1} and the bridging acetate or other ligands are recognized in the spectra. The $Me_2SnSnMe_2$ structural unit is also present in the tin heterocycle

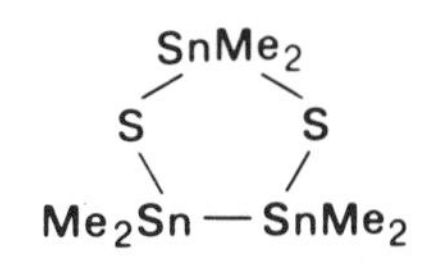

characterized[143] by Sn—S frequencies, 310–360 cm^{-1}, ν(Sn—Sn) = 190 cm^{-1}, and Sn—C stretches assigned to the Me_2Sn and Me_4Sn_2 units. In less detail the vibrational spectra of the stannacyclohexanes [$(CH_2)_5SnR_2$] (R = Me or Et)[144] can be assigned by considering the modes of the alkyl ligands, the fragment SnC_2, and the heterocycle $(CH_2)_5Sn$ in comparison with the corresponding silicon-containing ring system. The Me_2Pb group as part of some cyclic organolead compounds[145] gives $\nu(Pb—C)_s$ = 470–474 and $\nu(Pb—C)_{as}$ = 483–491 cm^{-1}, close to the values in [Me_4Pb].

Various branched-chain tin and lead organometallics have been characterized by their vibrational spectra[146]. In this category are the tetrastannanes [$RSn(SnMe_3)_3$], where R = Me, Et, *n*-Bu, *n*-pentyl, or phenyl[147], which in addition to the ν(Sn—C) modes show intense Raman bands at 106 ± 1 and 138 ± 2 cm^{-1} assigned to coupled ν(Sn—Sn) and δ(SnSnC) vibrations. Thorough studies of molecules of high symmetry being a cornerstone of vibrational spectroscopy, Müller *et al.*[148] performed a normal coordinate analysis of the regular tetrahedral species [$Me_4Sn_4S_6$]. This yields a valence force constant *f*(Sn—C) = 2.3 N/cm which, together with usual values for Sn—S and bending constants, suggests that bonds are approximately single and that π-bonding within the cage structure is negligible.

Organometallic derivatives of silicon, germanium, tin, and lead with unsaturated organic ligands have a special interest in the context of the metal–carbon bond because of the possibility of interaction between the π-electron system of the ligand and the unoccupied *d*-orbitals of the metal. Low values of ν(C≡C) in the compounds [$R_3MC≡CH$] (M = Si, Ge, or Sn) and higher i.r. band intensities than in the spectra of analogous compounds where the central atom is carbon have been attributed to an interaction of this kind[149].

In a series of studies of organoacetylenic compounds of Group IV and V elements, the vibrational spectra of [$Me_3M—C≡C—C≡C—MMe_3$] (M = C, Si, Ge, Sn, or Pb) have been examined and the assumption of D_{3d} symmetry found to be satisfactory. Detailed assignments for all these compounds have been tabulated[150].

Studies continue of many compounds of silicon, germanium, tin, and lead containing the groups RM, R_2M, and R_3M and ranges for M—C stretching frequencies, previously deduced[4,9], prove useful for their investigation. Typical examples include the organometallic amines [Me_3MNMe_2] (M = Si, Ge, or Sn)[151] and [R_3MNH_2], [$(R_3M)_2NH$], and [$(R_3M)_3N$] (M = Si, Ge or Sn; R = Me, Et, *n-Pr*, *i*-Pr, or *t*-Bu)[152]. In five-coordinate [$(R_3Sn(L_2)$] (R = Me or Et; L_2 = the anion of acetylacetone, benzylacetone or dibenzoylmethane)[153], ν(M—C) frequencies lie in the ranges 548–556 cm^{-1} (symmetric) and 574–589 cm^{-1} (asymmetric). In most R_3M derivatives the MC_3 unit is pyramidal and in R_2M derivatives it is usually bent. Simplest of the linear complexes are the $[Me_2M]^+$ cations found, for example, in the solutions of [Me_2Sn]O and [Me_2Pb]$(NO_3)_2$ in strong acids and identified by a single Raman band at 525 or 475 cm^{-1}, respectively[154].

Vibrational spectroscopy has been employed with n.m.r. and other techniques to characterize the bivalent organometallic derivatives of germanium, tin, and lead, [$M\{CH(SiMe_3)_2\}_2$], which are stabilized by the presence of the hexamethyldisilymethyl ligand[155]. Vibrational data including ν(Sn—C) assignments are conveniently used to identify the corresponding tin(IV) systems[156] and some analogous tin diamides[157,158].

The organometallic group R_3M, where M = Si, Ge, Sn, or Pb, is itself encountered as a ligand in compounds with other metals[159]. Spectra of such Group IV derivatives of the iron triad carbonyls have been reviewed[159] and other studies include those of nickel[160] and rhenium[161] complexes of the Me_3Sn ligand.

b. Phenyl derivatives: Vibrational assignments of the Group IV tetraphenyls, $[Ph_4M]$, have been substantiated in the case of tin by application of the metal isotope technique[162]. The i.r. spectra of isotopically pure $[Ph_4Sn]$, $[Ph_3SnI]$, and $[Ph_3SnOAc]$ labelled with ^{116}Sn or ^{124}Sn showed shifts of the order of 5 cm^{-1} in the metal-sensitive modes. The term 'tin–phenyl stretch' proves to have physical significance for the i.r. bands in the region 260–270 cm^{-1}, i.e. the asymmetric component of the t-mode. In triphenylvinyltin, $[Ph_3SnCH{=}CH_2]$, $\nu(Sn—Ph)_{as}$ is observed at 266 cm^{-1}, with $\nu(Sn—Ph)_s$ at 239 cm^{-1}[163] ν(Sn—C) of the vinyl group has a frequency of 526 cm^{-1}. Other phenyltin compounds, the benzoate $[Ph_3SnO_2CPh]$[164], complexes $[Ph_3Sn(NO_3)L]$[165], and some *N*-triphenylstannylcyanamides, $[Ph_3SnN(CN)R]$[166], exhibit $\nu(Sn—Ph)_{as}$ in the range 260–278 cm^{-1} and $\nu(Sn—Ph)_s$ 214–240 cm^{-1}. A substantially higher frequency for the Sn—Ph asymmetric stretching mode, near 320 cm^{-1}, has been advanced in tentative assignments of the i.r. spectra of the tin—tin bonded compounds, $[(Ph_3Sn)_3Sn(NO_3)_3]$ and $[Ph_3SnSnNO_3]$[167,168], the latter having the unusual feature of a tin(IV)—tin(II) bond which is confirmed by the crystal structure determination[167].

Solution measurements of the i.r. spectra of $[Ph_3SnCl]$ in the solvents benzene, acetone, and dimethyl sulphoxide[169] locate $\nu(Sn—Ph)_s$ at 240 ± 1 and $\nu(Sn—Ph)_{as}$ at 273 ± 2 cm^{-1}, the near constancy suggesting that solvent coordination plays little part. On the other hand, X-ray analysis has shown that dimethyl sulphoxide is coordinated to the seven-coordinate tin atom in $[Ph_2SnNO_3(DMSO)_3]^+$. The cationic nature probably explains the higher than usual Sn—Ph frequency of 285 cm^{-1} reported for this complex[170].

Fewer phenyl lead compounds have come under investigation[4,9]. Far-i.r. spectra of $[Ph_3PbX]$ and $[Ph_2PbX_2]$ (X = Cl, Br, I, or NCS) and various other complexes of Ph_2Pb with oxygen or nitrogen ligands[171,172] exhibit the t-modes $\nu(Pb—Ph)_s$ = 190–218 and $\nu(Pb—Ph)_{as}$ = 222—238 cm^{-1}. Other low-frequency X-sensitive modes are reported: 'u' in the range 160–185 and 'y' in the range 245–260 cm^{-1}.

The introduction of substituents in the benzene ring provides a familiar way of modifying the electronic environment in phenyl derivatives. From a study of the far-i.r. and Raman spectra of many substitued aryltrimethyltins, and the silicon and germanium analogues, Zuckerman and co-workers[173] have deduced that the frequencies which may be assigned to the aryl—tin stretching mode correlate well with the square root of the reciprocal mass of the substituted phenyl system, but not with electronic factors such as the Hammett σ-constant of the substituent attached to the benzene ring. Much useful information for the assignment of phenyl ring vibrations is compiled in the cited reference[173].

5. *Organometallic compounds of Group V and VI elements, arsenic, antimony, and bismuth; selenium and tellurium*

Elements of Group V, arsenic, antimony, and bismuth, and Group IV, selenium and tellurium, form a wide variety of organic derivatives which may involve the element in the group valence state or low-valent states, in some of which element-to-element covalent bonding is a structural feature[146].

a. Alkyl derivatives: Frequencies of methyl derivatives[174–176] are collected in Table

TABLE 8. Characteristic vibrations (cm^{-1}) of Group V methyl compounds

Compound	$\delta(CH_3)_s$	$\rho(CH_3)$	$\nu(M—C)_{as}$	$\nu(M—C)_s$	Reference
[Me_3As]	1250	880	583	568	174
[Me_4As_2]	1250,1235	889,824	578	564	175
[Me_5Sb]			516, 456	493, 414	176
[Me_3Sb]	1200	810	513	513	174
[Me_3Bi]	1160	780	460	460	174

8, which also illustrates the diversity in formula types. The vibrational spectra of trimethylantimony halides, [Me_3SbX_2], establish D_{3h} symmetry and normal coordinate analyses have been performed using several alternative force fields[177]. The symmetric stretching frequency of the equatorial SbC_3 unit and the Sb—C valence force constant, which decrease from fluoride to iodide, are given in Table 9. Also shown are the values of the $\nu(Sb—C)_s$ vibrations for the analogous ethyl and isopropyl compounds[178]. Structural analogues for which there are some vibrational data include [R_3MX_2] (M = As or Sb; R = Et, *i*-Pr, or Bz)[179]. Vibrational data for these compounds can be used to estimate the barrier height to intramolecular exchange of the ligand arrangements as demonstrated for a number of [MX_5] and [MX_3Y_2] species[180].

The tendency of antimony to attain a coordination number of 6 appears in the behaviour of [Me_2SbCl_3], which is shown by spectra to be monomeric in solution, with C_{2v} symmetry, but to crystallize as a chlorine-bridged dimer in which the methyl groups are in *trans*-positions (D_{2h} symmetry)[181]. Similarly, the compounds [Me_2SbCl_2Y] (Y = NCO or N_3) are dimeric and bridged through nitrogen of the pseudo-halide[182]. Bridging occurs through X in the complexes $[(Me_2SbCl_3)_2X]^-$ (X = F or CN)[183]. The ν(Sb—C) frequencies are in the range 510–582 cm^{-1}. The spectra of tetrahedral ions $[Me_4M]^+$ (M = As or Sb)[111,112,184] present in compounds [Me_4M][$Me_nM'Cl_{4-n}$] are mentioned in Section III.A.3.

In other Group V organometallic systems the As, Sb, or Bi atom is in the +3 oxidation state. Vibrational spectra of these compounds have been reviewed[185], and have featured in a number of studies of these molecules where they act as donor ligands coordinated to main group or transition metals. Examples whose spectra are additional to previous coverage[4,9] are the diazoalkyl derivatives [Me_2MCN_2R] and [$(Me_2M)_2CN_2$] (M = As, Sb, or Bi), and the related species [$Me_2AsCN_2MMe_2$] (M = Sb or Bi) and [$Me_2AsCN_2MMe_3$] (M = Ge, Sn, or Pb)[186]. The spectra of the latter are largely a superimposition of Me_2As and MMe_n bands, although the stretching modes of the CN_2 group can be identified and are sensitive to the nature of the metal M. There are also reports of the spectra of cyclopentadienyl derivatives [Me_2MCp] (M = As, Sb, or Bi)[187,188]. Assignments for the arsenic compound[187] are $\nu(As—C)_s$ = 572, $\nu(As—C)_{as}$ = 580 and ν(As—Cp) = 350 cm^{-1}.

Progress has been made in interpreting the spectra of perfluoromethyl derivatives of

TABLE 9. $\nu(Sb—C)_s$ vibrations (cm^{-1}) in organoantimony halides, R_3SbX_2, Values in parentheses are the force constants, *f*(Sb—C)(N/cm), in [Me_3SbX_2]

Compound	X = F	X = Cl	X = Br	X = I
[Me_3SbX_2]	546 (2.61)	538 (2.51)	526 (2.42)	508 (2.31)
[Et_3SbX_2]	513	492	482	475
[*i*-Pr_3SbX_2]	500	480	478	475

Group V elements, particularly $[(CF_3)_2ECH_3]$ (E = P, As, or Sb)[189], $[(CF_3)_2SbX]$ (X = H, Cl, Br, or I)[190], and $[(CF_3)_3Bi]$[191].

Vibrational spectra of organic derivatives of the Group VI metalloids are collected in Maslowsky's extensive survey[4]. The principles employed in interpreting these spectra are similar to those already illustrated. Reported ν(M—C) and ligand frequencies in compounds of selenium and tellurium fall within the established ranges. Spectra of the compounds $(MeSe)_3P$ and $(MeSe)_3As$[192] are richer in bands than expected for C_{3v} symmetry, which indicates the probable existence of conformational isomers. Strong i.r. bands, ν(Te—C), in the range 520–550 cm^{-1} have proved helpful in characterizing some tellurium-containing heterocycles[193,194].

There are many examples of molecules in which organic derivatives of the Group V and VI elements are coordinated to a main group or transition metal. A survey of diorganotellurides as ligands for transition metals[195] contains new spectroscopic data additional to the already large store (Section III.E).

b. Phenyl derivatives: A wide variety of phenyl derivatives of Group V and VI elements can be prepared, ranging in oxidation from +1, exemplified by Ph_2Te_2, to +5 in Ph_3SbX_2. Considerable progress has been made in applying i.r. and Raman spectra to their characterization and in deducing vibrational assignments. A list which encompasses spectroscopic work subsequent to the pre-1975 reviews[4,9] contains the following phenyl derivatives:

Group V: Ph_3SbX_2[196], $[Ph_2SbX_4]^-$[197], $[PhSbX_5]^-$[198], Ph_2SbOAc[199], Ph_3BiX_2[200], Ph_2BiX[201], $[Ph_2BiX_2]^-$, and $[Ph_2Bi(OAsPh_3)_2]^+$[202] (where X = F, Cl, Br, or in some cases a pseudohalide).

Group VI: Ph_2SeCl_2[203], $PhSeBr_3$[204], $PhSeMH_3$ (M = C, Si, or Ge)[205], Ph_2Se_2[206], Ph_2TeX_2 (X = Cl or Br)[207], Ph_2Te_2[206], PhTe(L)X (X = halogen, L = thiourea)[209], and $(C_6F_5)_2Te_2$[210].

For Ph_2TeX_2 (X = Cl or Br) and $(p\text{-}MeC_6H_4)_2TeBr_2$[207] definitive assignments of ν(Te—Ph) modes have been established by use of samples containing isotopically pure ^{126}Te or ^{130}Te. Values of ν(Te—Ph) stretching frequencies (Whiffen t-modes) from the above tellurium compounds are given in Table 10. Taken with the phenyl derivatives of other metals of Groups II–IV a consistent pattern emerges wherein the mass of M is the predominant factor in determining the frequencies of the skeletal M—Ph stretching modes and changes in oxidation state and coordination number have smaller effects.

TABLE 10. Low-frequency vibrations (cm^{-1}) of isotopically pure organotellurium compounds: i.r. values (supporting Raman data were also obtained)

Ph_2TeCl_2		Ph_2TeBr_2		
^{126}Te	^{130}Te	^{126}Te	^{130}Te	Assignment
286.5	284.5	160	160	$\nu(Te—X)_s$
265	263.5	186 br	185	$\nu(Te—X)_{as}$
271	270	272	270	$\nu(Te—C)_{as}(t)$
246	244.5	241.5	240	$\nu(Te—C)_s(t')$
251	251	258	258	Phenyl (u)
229	229			Phenyl (u′)
185	185	186 br	185	Phenyl (x)

B. Vibrational Spectra of Organo-transition Metal Compounds

In many transition metal organometallic compounds the M—C bond arises through the attachment of a coordinated ligand (Section III.C). The present section is concerned primarily with the spectroscopic investigation of alkyl, aryl, and acetyl complexes of the transition metals, for which early work was reviewed by Adams[6]. Most examples are provided by the later groups, especially Group VIII and the coinage metals, where some binary alkyl complexes are now well characterized.

1. *Binary alkyl complexes*

The complex anions $[Me_2Au]^-$ and $[Me_4Au]^-$ of gold and their deutero-analogues have been the subject of thorough i.r. and Raman studies[211,212]. The skeletal ν(M—C) frequencies of $[Me_2Au]^-$ and $[Me_4Au]^-$ differ little, being 526 and 530 cm^{-1}, respectively. The vibrations of linear $[Me_2Au]^-$ correlate smoothly with those of $[Me_2Hg]$, $[Me_2Tl]^+$, and $[Me_2Pb]^{2+}$. The trend of decreasing force constant from gold to lead is a reversal of the more usual variation with nuclear charge and serves to emphasize the stronger M—C bonds of the earlier members of this series. $[MeAu(Me_2PPh)]$ and $[Me_3Au(Me_2PPh)]$ show ν(Au—C) bands in the range 500–540 cm^{-1} [213]. The gold(III) methyl complex $[Me_2Au(PPh_3)_2]^+$ and some analogues exhibit intense ν(Au—C) Raman bands (516–542 cm^{-1}) which have been used to monitor reactions of these species[214]. Several other Me_2Au-containing complexes have been studied spectroscopically[215].

The methylplatinum complexes $[Me_4Pt]^{2-}$ and $[Me_6Pt]^{2-}$ and the phosphine derivatives $[Me_3Pt(PMe_3)]^+$, $[Me_4Pt(PMe_3)_2]$, and $[Me_2Pt(PMe_3)_2]$ are also well characterized by i.r. and Raman spectra[216,217]. Typically intense Pt—C stretching bands at 522 and 552 cm^{-1} are shown by the Raman spectrum of *cis*-$[Me_2Pt\{P(OMe)_3\}_2]$[218]. Other alkyl- and arylplatinum complexes are cited in the following section.

2. *Other alkyl and aryl complexes*

Early reviews of this area[3,5,6] have since been supplemented[4,219]. Stretching frequencies, ν(Ti—C) = 505–510 cm^{-1}, are definitely identified in $[MeTiCp(OR)_2]$ (R = Et or Pr), which also give a ν(Ti—Cp) band near 425 cm^{-1} [220]. In the complexes of $MeTiBr_3$ with donor ligands the ν(Ti—C) frequency drops to 450–470 cm^{-1} [221]. $[(C_6H_5CH_2)_3Ti]$ shows the frequencies of the benzyl ligand and an additional i.r. band at 540 cm^{-1} attributed to a Ti—C stretching mode[222] [compare 565 cm^{-1} in the titanium(IV) analogue]. The i.r. study[223] of $[(C_6H_5CH_2)_4M]$, where M = Ti, Zr, Hf, or W, also assigned W—C stretching modes $\nu(W—C)_s$ = 485 vw; $\nu(W—C)_{as}$ = 523 w, 535 m, 556 m.

In Groups V–VII, ν(V—C) = 547 cm^{-1} in $[MeVO(OR)_2]$[224] and ranges from 435 to 500 cm^{-1} in some methyltungsten derivatives[225]. A methylrhenium derivative $[Me_4ReO]$, for which Re—C stretching at 520 cm^{-1} is reported[226], probably has C_{4v} symmetry.

Group VIII provides a variety of organometal complexes for which there is vibrational information. I.r. has been used in conjunction with Mössbauer spectroscopy to compare the Fe—C and Fe—Si bonds in $[RFe(CO)_2Cp]$, where R represents a range of alkyl and silyl groups[227]. $[Me_2Co(PMe_3)_3]$[228], other methylcobalt derivatives[229], and dialkylcobalt complexes[230], as well as some palladium derivatives[231] provide examples which may be compared with the methylplatinum systems cited above. The compound palladium bis(cyanomethanide), $[Pd(CH_2CN)_2]$, and some derivatives $[R_2PdL_2]$ with amine and phosphine ligands, show ν(Pd—C) modes 580–612 cm^{-1} and ν(C≡N) 2180–2200 cm^{-1} [232].

Trends in Pt—C stretching frequencies in methylplatinum complexes, already well established, now include data for $[Me_2PtBr(H_2O)_3]^+$ [233] and $[Me_3PtX_3]^{2-}$ (X = Cl, Br, or NCS)[234]. In addition to the low-frequency information, the value of $\delta(CH_3)_s$ in the range 1225–1275 cm^{-1} is a useful characteristic parameter. Previous Raman and i.r. studies of the clusters $[Me_3PtX]_4$, mainly halides, reviewed by Hall[71] have been extended to the explosive perchlorate ($X = ClO_4$)[235] in which $\nu(Pt—C_3)_s$ = 565 vs (Raman); $\nu(Pt—C_3)_{as}$ = 650 vs (i.r.). The tetramer $[MePtSMe]_4$ shows bands ν(Pt—C) at 563 cm^{-1} and $\delta(CH_3)_s$ at 1255 cm^{-1} as polarized Raman features in chloroform solution[236]. Raman and i.r. spectra of the anions $[RPtCl_2(CO)]^-$ yield secure assignments for ν(Pt—C) at 570 cm^{-1} (R = Me), 540 cm^{-1} (Et, *n*-Pr, and *n*-Bu), and 510 cm^{-1} (*i*-Pr)[237]. When R = phenyl the Pt—Ph stretch (Whiffen t-mode) appears as a strong, polarized Raman band at 250 cm^{-1} and the Pt—C—C bend (Whiffen u-mode) gives a weak, depolarized band at 210 cm^{-1}.

Other vibrational analyses for aryl derivatives of the transition metals follow the lines of those of main group systems, and some useful assignments are available[4,9]. A further X-sensitive mode [still incorrectly described in some work as 'ν(M—C)'] is located in the 400–500 cm^{-1} region, for example at 448 cm^{-1} in $[PhAuCl_2]$ and in the range 460–500 cm^{-1} in other compounds $[RAuCl_2]$ (R = aryl)[238]. Systems containing the ligands C_6F_5[239] and C_6Cl_5[240,241] have also been studied, notably in divalent complexes of nickel, palladium, and platinum. Orthometallation of one of the phenyl rings of the Ph_3P ligand such as occurs in 5-coordinate iridium complexes $[MeIrClL_2]$[242] is accompanied by splitting of the 720 cm^{-1} band (the X-sensitive Whiffen r-mode), which helps to establish that the ring–metal bond is present. An i.r. band in this region, diagnostic of orthometallation, has been noted in other work on complexes of aromatic ligands[243].

3. *Ylid complexes*

No special features are expected in the vibrational spectra of ylid compounds, but ν(M—C) assignments should assist in structure determination. Characteristic i.r. frequencies of ylids, including ν(M—C) 510–544 cm^{-1}, have been reported by Tanaka and co-workers[244] who noted the existence of trends in the Pt—C stretching frequencies. I.r. bands in the range 495–550 cm^{-1}, assigned to ν(Au—C) modes, help to establish the structures of some ylid compounds of gold[245,246].

4. *Carbene, carbyne, and related complexes*

Carbene-type complexes in which ligands (—CH_2, —CR_2, and variants) are attached to a transition metal by a formal double bond should give rise to M—C stretching modes and internal vibrations, including ν(C—H) and/or ν(C—R) modes. Most reports have been concerned with the detection of i.r. bands in the latter category[247–264], for which data are summarized in Table 11. Various systems related to carbenes, and involving the formyl, thioformyl, and carbamoyl ligands, are also included. A Raman and i.r. study of isomeric forms of $[Re_2(CO)_9\{C(NHMe)Me\}]$[247] reveals low-frequency bands possibly associated with the carbene ligands, in addition to useful ν(NH) characteristics. The cumelene ligand (=C=C=C=O) analogous to carbene gives rise to a band (ν_1 of the ligand) at 2028 cm^{-1} in $[(C_3O)Cr(CO)_5]$[265].

In contrast to the absence of an obvious ν(M—C) band in carbene complexes[266], the triple bond frequency of carbynes is well established in several cases. The tungsten

TABLE 11. Characteristic frequencies of carbenes and related complexes

Ligand	Structure	Bond	Frequency (cm^{-1})	Source
Carbene (R = Me or Et)	$-C(OR)(CH_3)$	C—C	970	Re complex[247]
Carbene (R = Me, Et, *i*-Pr, *n*-Bu)	$-C(NR_2)(H)$	C⋯N	1535–1615	Rh complex[248]
		C—H	*ca.* 3050	
Carbene (R = *p*-tolyl)	$-C(NMeR)(H)$	C⋯N	1568–1575	Pt complexes[249]
Carbene	$-C(NMe_2)(H)$	C⋯N	1535–1640	Complexes of V, Cr, Mo, W, Mn, Re, Fe, Ru, Co[250]
Carbene (R = Me or Et)	$-C$ in ring $N(R)-CH_2-CH_2-N(R)$	C⋯N	1480–1540	Complexes of Fe, Ru, Co, Ni[251]
Carbene	$-C(NMe_2)(Cl)$	C⋯N	1510	Cr complex[252]

TABLE 11. (*continued*)

Ligand	Structure	Bond	Frequency (cm^{-1})	Source
Carbene	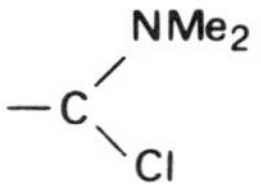 $-C(NMe_2)(Cl)$	C⋯N	1520–1578	Complexes of Cr, Mn, Rh[253]
		C—Cl	779–860	
Carbene (R = Me or Et)	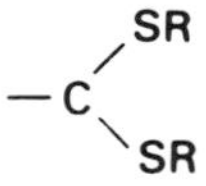 $-C(SR)(SR)$	C⋯S	840–998	Pt complex[254]
Carbene	$-C(S-CH_2)(S-CH_2)$ (ring)	C⋯S	860, 955	Os complex[255]
Carbene	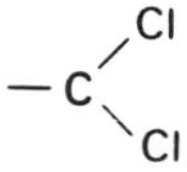 $-C(Cl)(Cl)$	^{12}C⋯Cl	872	Fe complex[256]
		^{13}C⋯Cl	841	
Carbenoid	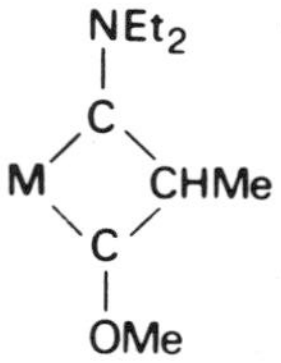 M–C(NEt_2)–CHMe–C(OMe)–M (ring)	C⋯N	1587	W complex[257]
		C⋯O	1322	

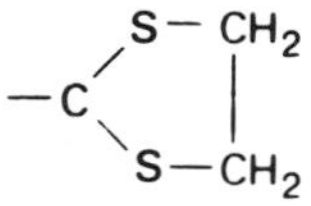

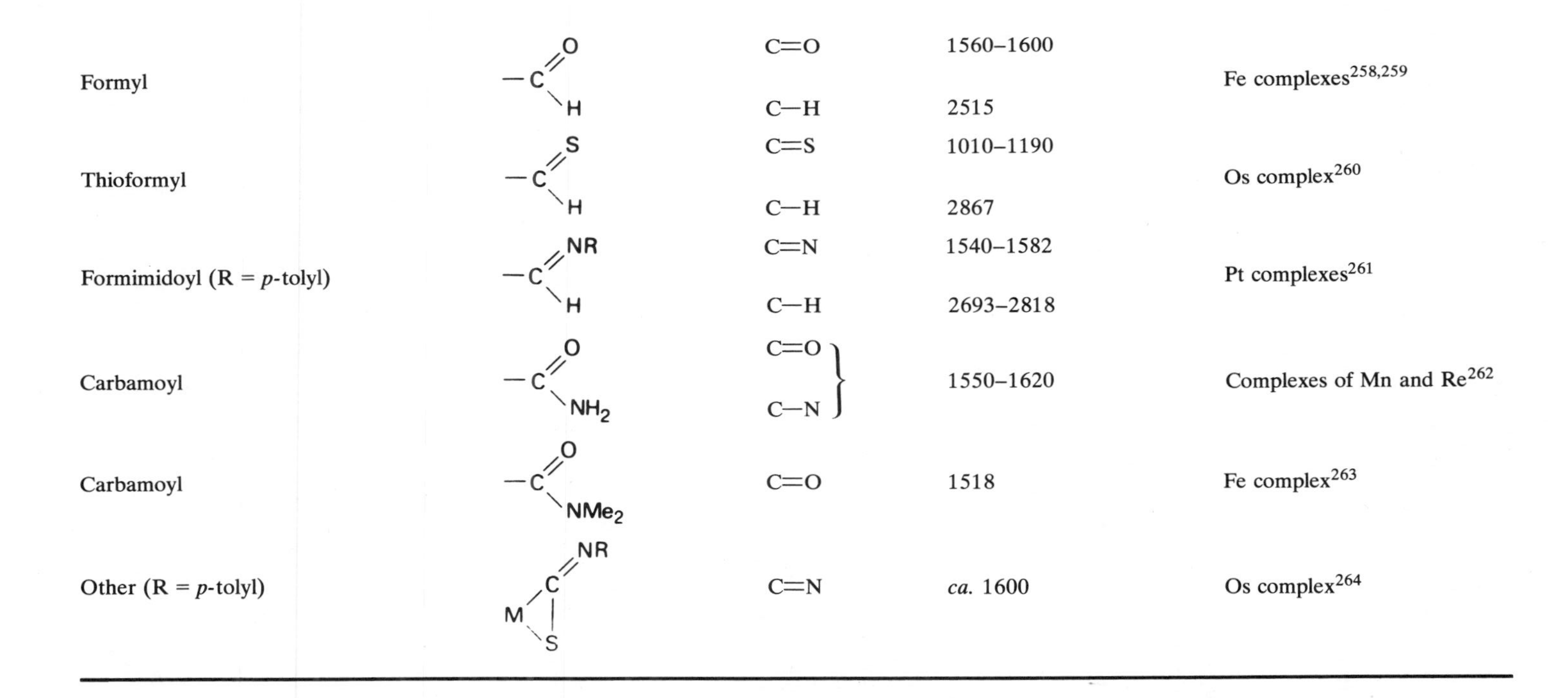

Formyl	—C(=O)H	C=O	1560–1600	Fe complexes[258,259]
		C—H	2515	
Thioformyl	—C(=S)H	C=S	1010–1190	Os complex[260]
		C—H	2867	
Formimidoyl (R = *p*-tolyl)	—C(=NR)H	C=N	1540–1582	Pt complexes[261]
		C—H	2693–2818	
Carbamoyl	—C(=O)NH_2	C=O } C—N	1550–1620	Complexes of Mn and Re[262]
Carbamoyl	—C(=O)NMe_2	C=O	1518	Fe complex[263]
Other (R = *p*-tolyl)	M(C(=NR))S	C=N	*ca.* 1600	Os complex[264]

carbyne complexes *trans*-$[X(CO)_4W{\equiv}CCH_3]$ (X = Cl, Br, or I)[267] exhibit the $\nu(W{\equiv}C)$ stretching band in the vicinity of 1315 cm^{-1} [compare $\nu(W{\equiv}N)$ at 1286 cm^{-1}]. Complete n.c.a. yields a suitably high force constant of 7.40 N/cm.

C. Vibrational Spectra of Organometallic Compounds with Carbon Donor Ligands

1. Reviews

The vibrational spectra of π-bonded organometal complexes were systematically treated by Davidson in 1972[268] and earlier by Fritz[269]. Davidson's important review tabulated the frequencies of all the common 2- to 8-electron donor ligands and discussed the best available assignments in their complexes with transition metals. These two surveys provide a sound basis for this part of the subject, and other accounts of the i.r. and Raman spectroscopy of metal π-complexes are available[4,6,9,270–272].

2. Olefin complexes

The early vibrational studies of olefin complexes, principally of Group VII and VIII transition metals, and also of main group metals, have been extensively reviewed[268,270,273–277].

Formation of the π-bond to metal characteristically lowers the C=C stretching frequency and renders this mode i.r. active, even for those olefins (including ethylene) which are centrosymmetric in the uncomplexed state. A convenient measure of the modification of the C=C bond, first demonstrated in complexes of platinum and silver[278], is the summed percentage lowering of bands due to the coupled $\nu(C{=}C)$, $\delta(CH_2)$, and/or $\delta(CH)$ vibrations. Also a useful indicator is a change in Raman intensities whereby the $\nu(C{=}C)$ band (near 1500 cm^{-1}) becomes progressively weaker relative to the $\delta(CH)$ band (near 1250 cm^{-1}) as the strength of the metal–olefin bond increases.

Direct observation of metal–ligand stretching frequencies, $\nu(M{-}C_2)$, assigned in the range 385–500 cm^{-1} for Pt–C_2H_4 and 270–290 cm^{-1} for Ag—C_2H_4 complexes[278], is becoming common. These spectroscopic criteria of the metal–olefin bond have been applied to iron[279,280], rhodium and iridium[281–283], copper, silver, and gold[284], as well as to many platinum–olefin complexes[285–288]. Vibrational spectra of olefin complexes [(ol)PtCl(acac)][289] reveal $\nu(Pt{-}C_2)_s$ near 400 cm^{-1}. The cationic complex $[(C_2H_4)PtCl(tmeda)]^+$ displays $\nu(Pt{-}C_2)_s = 368$ and $\nu(Pt{-}C_2)_{as} = 458$ cm^{-1} [290]. The value of $\nu(Pt{-}C_2)_s$ is lower than in neutral or anionic complexes because the positive charge on platinum lessens $d\pi{-}\pi^*$ interaction in the metal–olefin bond.

Among metal–olefin systems attractive for force constant calculation $[(C_2H_4)PtCl_3]^-$, the anion of Ziese's salt, continues to attract detailed attention[291]. Normal coordinate calculations for this complex and $[(C_2H_4)Fe(CO)_4]$[292] show that the original ethylene force field is very considerably perturbed by coordination. Assignments for the series $[(ol)Fe(CO)_4]$, where ol = C_2H_4, C_2HCl_3, $C_2H_2Cl_2$, and C_2Cl_4[293], which show $\nu(M{-}C_2)$ frequencies in the range 375–400 cm^{-1}, have provided the basis for force constant determinations and for calculating values of $\nu(C{=}C)$ which eliminate the effect of coupling to $\delta(CH_2)$ and $\delta(CH)$ modes. The decrease in $\nu(C{=}C)$ from the free olefin value thus calculated is 260–370 cm^{-1} and indicates C—C bond orders of 1.2–1.5.

Assignments for the metal complexes of more complicated olefinic ligands, including cycloocta-1,5-diene and various other dienes, have been reviewed[268,270,294]. Wertz and Moseley[295] in investigations of $[(cod)RhCl]_2$ and $[(cod)MCl_2]$, where M is Rh^I, Pd^{II} or Pt^{II}, have shown that the lowering of $\nu(C{=}C)$ and $\delta(CH)$ bands, which indicates the total σ- plus π-bonding between metals and olefin, follows the order Rh > Pt > Pd. On the basis of shifts in ν(M—L) frequencies in the range 400–580 cm^{-1}, they suggest a different order for the strength of π-interaction: Pt > Rh > Pd. The characteristic increase in the i.r. intensity of ν(C=C) upon coordination has been measured for the compounds [(nbd)M$(CO)_4$], where M = Cr, Mo, or W[296]. A study of the species formed in aqueous solution by the reaction of buta-1,3-diene with cobalt(III) cyanide complexes[297] provides a specialized example of the use of Raman spectroscopy in this field.

Other applications where vibrational spectroscopy contributes to the understanding of the M—C bond in olefin complexes occur in the fields of surface adsorption studies, including catalyst action, and metal atom—ligand reactions. Metal atom—olefin systems studied by i.r. spectroscopy include those of ethylene co-condensed in an argon matrix with tin[298], nickel[299], palladium[300], copper, silver, and gold[301,302]. Identifiable complexes are of the type [M$(C_2H_4)_n$], with n = 1, 2, or 3. Nickel complexes [Ni$(C_2F_4)_n$] and [Ni$(C_2H_3Cl)_n$], where n = 1, 2, or 3, have also been identified[303]. In related work, i.r. spectra have been obtained for ethylene chemisorbed on palladium and platinum[304,305], the interpretation of which suggests that π-bonded surface adducts co-exist with σ-bonded M—C_2H_4—M species. Developments in Raman spectroscopy are also expected to have an impact on this topic.

3. *Acetylides and acetylenic complexes*

The principal types of metal–acetylene derivatives, π-bonded complexes and acetylides, can usually be distinguished by vibrational spectroscopy. Characteristic of acetylides are intense ν(C≡C) bands only slightly different from the values in the free acetylene, e.g. in the range 2020–2040 cm^{-1} for Ni, Pd, and Pt complexes[306] (compare C_2H_2, 1974 cm^{-1}). These complexes also show metal–acetylene stretching frequencies, 510–600 cm^{-1}, in the order Ni < Pd < Pt attributable to increase of the σ-bond strength. Various complexes *cis*- and *trans*-[L_2Pt(C≡CR$)_2$] exhibit ν(Pt-acetylene) frequencies in the range 544–576 cm^{-1} as well as the expected one or two ν(C≡C) bands[307]. The tetrahedral anions [Mn(C≡CR$)_4]^{2-}$, where R = H, Me, or Ph[308], have been investigated and there are other spectral studies of acetylides of Re, Fe, Ru[309], Pd[310], Pt[311,312], Cu and Ag[313], and Hg[314].

Cluster compounds [As_4Cu_6(C≡CR$)_6$][315] and some dinuclear iron compounds[316] are examples where the —C≡CR ligand apparently acts as a 3-electron donor, forming σ- and π-bonds. These, and the series [Me_2M(C≡CMe)$]_2$, where M = Al, Ga, or In[317], have yielded insight into the characteristic spectra and have provided fairly complete vibrational assignments. The spectra of [$(C_2H_2)Co_2(CO)_9$] and related structures[276,318], also illustrate the participation of acetylene ligands in bridged and cluster compounds. Studies of [$(C_2R_2)Co_2(CO)_6$][319] show the effective use of Raman spectra to substantiate ligand assignments and to locate the intense, polarized bands of metal–alkyne stretching modes at 395, 380, and 275 cm^{-1} for C_2H_2, C_2Me_2, and $C_2(CF_3)_2$, respectively.

In some cases the spectra of complexes of π-bonded alkynes may fail to reveal any i.r. band for ν(C≡C). Alternatively, the frequency may be so much lower than that of the free alkyne as to lend support to a metallocyclopropene structure for the M(C_2R_2) unit. A case in point is the complex [$(C_2Ph_2)_2$Ti$(Cp)_2$CO] of known crystal structure[320] in which the ν(C≡C) frequency of PhC≡CPh, 2100 cm^{-1}, is replaced by a band at 1780 cm^{-1}. Shifts $\Delta\nu[(C{\equiv}C)_{free} - (C{\equiv}C)_{coord}]$ of 300–400 cm^{-1} are observed for a variety of π-bonded acetylene complexes of Mo[321], W[322], Fe[323,324], and Pt[325,326].

Another useful parameter from i.r. spectra is the integrated molar absorption coefficient of the ν(C≡C) band which has been shown, for complexes of the platinum group metals, to be proportional to the shift $\Delta\nu(C{\equiv}C)$ and to be useful in assessing the acceptor strength of these metals towards alkynes[327].

The strong metal–carbon bonding between zerovalent Ni, Pd, or Pt and acetylene has been investigated by i.r. studies of the interaction of acetylene with co-condensed metal atoms[328] and with metal films[329].

4. Allylic and allene complexes

In this category are complexes of the π-allyl ($CH_2{\cdots}CH{\cdots}CH_2$), π-methallyl, and π-crotyl groups for which ligand assignments are available in their complexes with Group VII and VIII transition metals[4,6,9,268]. Characteristically no C=C stretching band appears but instead three medium or strong bands 1375–1510 cm^{-1}, are observed and there are metal–ligand stretching vibrations in the range 280–570 cm^{-1}. A 1978 review[330] focuses on the frequencies of the metal–ligand modes (Figure 2) and shows that the out-of-plane allyl-tilting vibration can occupy any of three possible positions with respect to the two symmetrical vibrations. In the most studied systems the strengths of the metal–allyl bonds appear to vary thus: Ni $\approx$ Pd $<$ Pt $=$ Rh $<$ Ir.

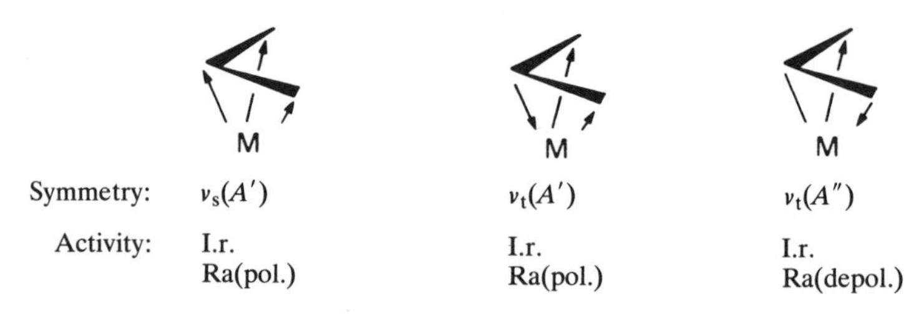

FIGURE 2. Metal—π—allyl stretching vibrations according to C_s symmetry.

Vibrational assignments for the η^3-allyl complexes $[(\eta^3\text{-}C_3H_5)Fe(CO)_3X]$, where X = Br, NO_3, or $[(\eta^3\text{-}C_3H_5)Fe(CO)_3]$[331], include Fe–allyl frequencies near 330 (A'), 360 (A'') and 400 (A') cm^{-1}. Other systems studied include $[(\eta^3\text{-allyl})Co(PF_3)_2\text{-}(PPh_3)]$[332] and derivatives of vanadium[333,334], zirconium, hafnium[335], and platinum[336]. These illustrate the differentiation between π-, σ-, and σ/π-bonding in allyl, methallyl, and crotyl complexes which can be accomplished by i.r. and Raman together with n.m.r. spectra.

There are few systems involving the allene molecule, $CH_2{=}C{=}CH_2$, as a donor ligand compared with the large number of allyl complexes. Coordination to a metal decreases the ν_{as} frequency of 1940 cm^{-1} by about 180–260 cm^{-1}, and this i.r. evidence for allene complexes of Pt, Rh, and Ir has been reviewed[337].

5. Benzene and related arene complexes

The typical π-complexes of benzene and other aromatic ligands involve the formal donation of six electrons to a low-oxidation-state metal centre. Best understood are the spectra of bisbenzenechromium and analogues $[(\eta^6\text{-}C_6H_6)M(CO)_3]$, where M = Cr, Mo, or W. For these compounds detailed assignments and normal coordinate calculations are available[31,338,339] and it has been shown that significant, although not large, changes occur in a number of the benzene force constants on complexation. A substantial Cr—(C_6H_6) stretching force constant of 3.74 N/cm is computed[339] and it has been shown that the lower wavenumber features, e.g. the A_1 band at 306 cm^{-1} of $[(\eta^6C_6H_6)Cr(CO)_3]$, involve heavily coupled vibrations. It is interesting to compare these studies with those of tricarbonyl(trimethylenemethane)iron $[\{\eta^6C(CH_2)_3\}Fe(CO)_3]$. This complex, with its η^6-coordinated hydrocarbon ligand, has been the subject of normal coordinate analyses[340,341] which yield a metal—ligand force constant of 2.83 N/cm comparable to those in more familiar η^6-arene complexes. Other i.r. studies include those to characterize an extensive range of η^6-arene tricarbonylchromium complexes[342–344]. The spectra of π-bonded complexes $[LCr(CO)_3]$, where L = thiophen[345] or a methyl-substituted pyridine[346], are also known.

Of considerable fundamental interest are studies of the i.r. spectra of C_6H_6, C_6D_6, and C_6H_5F co-condensed with Cr, Mn, Fe, Co, or Ni atoms which have shown that η^6-complexes, $(C_6H_6)M$ and/or $[(C_6H_6)_2M]$, are formed in all cases except with manganese[347]. It was deduced from the spectra that the relative strengths of the metal–arene bonds decrease in the order Cr > Fe > Co > Ni. Co-condensation of titanium atoms with benzene, toluene, or mesitylene yields complexes $[(arene)_2Ti]$ which show new i.r. bands near 400 cm^{-1} [348].

The interaction of main group metals with aromatic donor systems is less well understood. I.r. data are available for the π-arene complexes of zinc, cadmium, and mercury[349]. Dissociation of the adducts of $AlBr_3$ with aromatic hydrocarbons has been the subject of a variable-temperature Raman study[350].

6. *Cyclopentadiene complexes*

a. Structural types: Cyclopentadienyls are among the most numerous hydrocarbon complexes of metals. The various structural types can be distinguished by the vibrational spectra and symmetry rules for their interpretation, originally treated by Fritz[269], have been extended in more recent accounts[4,6,9,268].

(i) *Ionic complexes* are formed by the electropositive metals and give the spectrum of the $C_5H_5^-$ ion. I.r. active vibrations, whose frequencies change with the nature of the metal in predominantly ionic compounds[351], occur in the following ranges: ν(CH), 3020–3100; ν(CC), 1425–1500; β(CH), 1000–1010; and ρ(CH), 670–810 cm^{-1}.

(ii) *σ-bonded diene complexes (η^1 type)*: here cyclopentadiene functions as a *monohapto* ligand and these exhibit spectra similar to that of the hydrocarbon itself. Band assignments of $[(\eta^1\text{-}C_5H_5)_2Hg]$ and $[MeHg(\eta^1\text{-}C_5H_5)]$ are made accordingly[9]. The spectra of $[Me_3Sn(\eta^1\text{-}C_5H_5)]$ and $[(Me_3Sn)_2C_5H_4]$ which are of this type differ markedly from the spectrum of $[(\eta^5\text{-}C_5H_5)_2Sn]$ and allow a clear distinction to be made.

(iii) *Symmetrically σ- and π-bonded complexes (η^5 type)* contain the ligand in *pentahapto* coordination. Most are π-bonded and typically exhibit the following bands: ν(CH) 3000–3100; ν(CC) *ca.* 1100 and *ca.* 1400; δ(CH) *ca.* 1000; and π(CH) 750–900 cm^{-1}. At lower frequency are to be found the bands due to M—Cp stretching (ν_4, 200–350 cm^{-1}), ring-tilting (ν_{16}, *ca.* 400 cm^{-1}), and Cp—M—Cp bending (ν_{22}, 170–200 cm^{-1}). Examples of data for complexes $[Cp_2M]$ and $[CpM]^+$ have been collected by Nakamoto[9], who also listed typical M—Cp force constants.

(iv) *Cyclopentadienyl complexes of η^3 and other types* include cases where bonding is of an allylic or *trihapto*-type, and those where the Cp-ligand is associated with bridge- or polymer-links[352].

b. Spectroscopic characterization of cyclopentadienyl complexes: Frequencies are reasonably well established for all the above structural types and single-crystal vibrational spectra are available for ruthenocene[27], ferrocene[28], and several other cyclopentadienyl complexes[29,30] (Section II.C). Reassignments for some η^5-Cp ligand modes have been proposed[351]. Studies of $[Cp_2Be]$[353], [CpBeX] (X = Cl, Br or

TABLE 12. M—Cp vibrations (cm^{-1}) from Raman spectra

	$[Cp_2Mn]$	$[Cp_2V]$	$[Cp_2Cr]$	$[Cp_2Fe]$	$[Cp_2Ru]$	$[Cp_2Os]$
$\nu_4(A_{1g})$	203 m	258 m	273 s	303 s	325 s	349 vs

Me)[354,355], $[Cp_2Mg]$[356], $[Cp_3Sm]$[357], and $[CpAg(PPh_3)]$[358] have extended the spectroscopic data to metals outside the earlier compilations. For the series $[(\eta^5\text{-}C_5H_5)_2M]$, where M = Mn, Cr, V, Fe, Ru, or Os[359,360], whose spectra accord with D_{5d} symmetry the frequencies in Table 12 are attributed to ν_4, the totally symmetric M—Cp stretching mode. The frequency and also the Raman intensity increase through the series from the partially ionic manganese complex to the covalent species of the iron sub-group. Stretching force constants, f(M—Cp), of 3.04, 3.33 and 3.83 N/cm have Been obtained for $[Cp_2Fe]$, $[Cp_2Ru]$, and $[Cp_2Os]$, respectively[360] {compare f(Cr—Bz) = 3.74 N/cm for $[(\eta^6\text{-}C_6H_6)Cr(CO)_3]$[339]}. The Raman frequencies assigned to M—Cp stretching in $[Cp_2Co]$ (320 cm^{-1})[361] and $[Cp_2Ni]$ (245 cm^{-1})[362] show clearly the bond-weakening effect of the additional electron(s) compared with $[Cp_2Fe]$, where $\nu(Fe—Cp)_s$ = 478 cm^{-1}. Such changes have little effect on the characteristic cyclopentadiene modes, so that only slight ligand frequency shifts from $[Cp_2Fe]$ are observed for the protonated ferrocene cation, $[Cp_2FeH]^+$[363], or the ferricinium ion, $[Cp_2Fe]^+$[361,364].

The salts of the $[Cp_2Fe]^+$ ion have theoretical interest in that the Raman spectra include bands due to low-frequency electronic transitions[361,365,366]. These are predicted[367] to occur for various sandwich complexes of transition metals, and can be used to calculate the magnetic moment of the $[Cp_2M]$ system for small distortions from D_{5d} symmetry. In addition to the ferricinium salts, such Raman bands have also been observed for chromocene, $[Cp_2Cr]$[368,369], at 35 and 55 cm^{-1}. The lanthanide complexes $[Cp_3Ln]$ have been studied[370] to reveal typical bands of η^5-coordinated cyclopentadiene and low-frequency features, notably ν(Ln—Cp), a strong, polarized Raman band near 230 cm^{-1}, which varies little through the series of seven metals.

Coordinated cyclopentadiene is present in many metal complexes containing other polyatomic ligands. The i.r. spectra of $[Cp_2M(CO)_2]$, where M = Ti or Zr, exhibit $\nu(M—Cp)_{as}$ bands at 399 cm^{-1} (Ti) and 341 cm^{-1} (Zr) and $\nu(M—Cp)_s$ at 293 cm^{-1} (Ti)[371]. For the single cyclopentadienyl group in dithiocarbamates $[CpTi(S_2CNR_2)X_2]$ (X = Cl or Br), ν(Ti—Cp) occurs near 420 cm^{-1} and is shown to be coupled to ν(Ti—X) stretching by a shift of about 10 cm^{-1} on replacing Cl by Br[372]. Data for $[Cp_3C_5H_4Ti_2]$[373] illustrate the application of i.r. spectra to a complex with several Cp ligands. Detailed spectra of the complexes $[CpMn(CO)_2(CS)]$ with ν(Mn—Cp) = 342 cm^{-1} and $[CpMN(CO)(CS)_2]$ with ν(Mn—C) = 321 cm^{-1} have been reported[374]. A particularly thorough study was made of $[CpWR(CO)_3]$, where R = H, CH_3, Et, or σ-allyl[375], in which W—Cp stretching gives an intense Raman band at 328–333 cm^{-1} and the ring-tilting mode is located in the range 350–416 cm^{-1}. The Group IV anions $[CpM(CO)_3]^-$ show the ν(M—Cp) frequencies 319 (M = Cr) and 299 cm^{-1} (M = Mo or W) as strong Raman bands[376]. The vibrational spectra of compounds containing η^5-cyclopentadiene together with other ligands have been reviewed by Nakamoto[9]. There are newer studies of $[Cp_2M(NCSe)_2]$ (M = Ti, V or Cr)[377] with ν(M—Cp) in the range 282–306 cm^{-1}, and of some η^5-Cp complexes of Group VIII metals[378]. An investigation of $[(CH_3C_5H_4)Mn(CO)_3]$[379] includes the vibrational assignment of the methylcyclopentadienyl group and identification of metal–ligand modes.

Some examples of complexes of η^1- and η^3-bonded Cp ligands have already been mentioned. The Group III compounds $[CpMEt_2]$, where M = Al, Ga, or In[380], are monomeric and show cyclopentadiene acting as an allylic (η^3) ligand when attached to an unsaturated centre. Stretching vibrations, ν(M—Cp), polarized in the Raman spectrum, are observed at 345 (Al), 277 (Ga), and 255 cm^{-1} (In). Interestingly, the vibrational spectra of the adduct [$CpAlEt_2$]·OEt_2[380] show that here the ligand is η^5-coordinated and is an example of the centrally σ-bonded type. The spectrum of [$Cp_2SnFe(CO)_4$] from the reaction of [η^5-$Cp)_2Sn$] with [$Fe_2(CO)_9$] has been used to show the presence of an inequivalent pair of η^1-cyclopentadienyl ligands[381].

7. Complexes of other cyclic unsaturated ligands

Interpretation of the vibrational spectra of cyclic ligands, C_nH_n, owes much to the early systematization by Fritz[269]. In addition to benzene and the cyclopentadienyl ion, ligands in this category are cyclobutadiene, C_4H_4, the tropylium ion, $C_7H_7^+$, and the cyclooctatetraene ion, $C_8H_8^{2-}$. Complexes of these and some other cyclic unsaturated ligands are treated here.

Cyclobutadienylirontricarbonyl[382] displays ν(C_4H_4—Fe) stretching at 406 cm^{-1} observed in the Raman spectrum and confirmed by inelastic neutron scattering, and the spectra also reveal skeletal bending and ring-torsion modes. Recent data for the $C_8H_8^{2-}$ ion in its Na^+ and K^+ salts[383] have suggested some changes to earlier assignments[384] and there is a sound basis for the vibrational study of cot complexes. In recent work typical bands of η^8-C_8H_8 appear in the spectra of [(η^8-C_8H_8)Ti(C_9H_7)] (where C_9H_7 is the indenyl group)[385], [(η^8-$C_8H_8)_2$Ce][386], and [(η^8-$C_8H_8)_2$U][387]. The uranium derivative in a resonance Raman study exhibits a polarized band at 211 cm^{-1} assigned to metal–cot ligand stretching. In another uranium complex, [($Cp_3U)_2$-(C_8H_8)][388], the spectroscopic evidence points to bridging η^3-allylic coordination of the cot group. The derivative [($C_9H_7)_3$MX], where M = U or Th and X = Cl or Br[389], and the titanium compound above, provide examples of vibrational assignments of the indenyl ligand.

D. Carbonyls, Nitrosyls, Cyanides, and Related Complexes

The vibrational spectra of individual metal carbonyls lie outside the scope of this chapter; however, the value of ν(CO) and sometimes also ν(M—CO) observations in determining the structure of organometallic carbonyl complexes is well established and examples are given in Section III.C. The interpretation of numerous carbonyl spectra is thoroughly treated elsewhere[390–396]. Special topics, such as the spectra of matrix-isolated metal carbonyls[397] and of CO chemisorbed on metal surfaces, have also been reviewed[398].

Griffith[399] has collected vibrational data for cyanides, isocyanides, and nitrosyls, and cyanides have been the subject of further reviews[440–402] in which spectroscopic work was examined. Yaneff[403] has reviewed the area of thiocarbonyls and related complexes of the transition metals. There is a first report[404] of a homologous series of compounds (osmium complexes) in which the diagnostic ν(CE) frequency is observed for CO, CS, CSe, and CTe.

Spectra, usually infrared, are being used to characterize adducts of CO_2, CS_2, and related carbon-containing ligands; further information on this subject appears in Section III.B.4 dealing with carbene complexes.

E. Coordinated Inorganic Ligands and Other Structural Features of Organometallic Spectra

It is common for organometallic compounds to contain various other ligands, and the ν(M—L) and internal ligand frequencies of these are often very useful in inter-

TABLE 13. Coordinated ligands: a convenient organization for spectroscopic data purposes. This organization is similar to that used in *Specialist Periodical Reports*

1	Carbon donors
2(a)	Carbonyl, thiocarbonyl, and related complexes
2(b)	Cyanide and related complexes
3	Boron donors
4	Nitrogen donors
	(a) Molecular nitrogen, azido, and related complexes
	(b) Ammines
	(c) Amines and related ligands
	(d) Oximes
	(e) Ligands containing C=N groups
	(f) Nitriles and isonitriles
	(g) Nitrosyls and thionitrosyls
5	Phosphorus, arsenic, and antimony donors
6	Oxygen donors
	(a) Molecular oxygen, peroxo, and hydroxy complexes
	(b) Hydrates
	(c) Carbon dioxide and carbonato complexes
	(d) Carboxylato complexes
	(e) Acetylacetonates and related complexes
	(f) Keto, alkoxy, phenoxy, and ether ligands
	(g) O-bonded amides and ureas
	(h) Ligands containing N—O bonds
	(i) Ligands containing P—O, As—O, or Sb—O bonds
	(j) Ligands containing S—O, Se—O, or Te—O bonds
	(k) Ligands containing Cl—O, Br—O, or I—O bonds
7	Sulphur, selenium, and tellurium donors
8	Potentially ambident ligands
	(a) Cyanate and thiocyanate complexes and their iso-analogues
	(b) Ligands containing N and O donor atoms
	(c) Ligands containing N and S or Se donor atoms
	(d) Ligands containing O and S or Se donor atoms
9	Halogen and interhalogen donors

preting structure and bonding. Table 13 gives a convenient organization of ligands, similar to that in *Specialist Periodical Reports*[405], for the purpose of collecting such data. Useful reviews of spectra of common ligands include accounts of phosphines and their Group V analogues[406], other phosphorus compounds[407], thiocyanates and related complexes[408,409] organonitriles[410], alkoxides and dialkylamides[411], carboxylates[412], and trifluoroacetates[413].

The ranges of metal–hydrogen, –nitrogen, –oxygen, and –halogen frequencies are widely documented. Three volumes[414] have surveyed the spectroscopic literature prior to the commencement of *Specialist Periodical Reports*[405]. Broad treatments of mainly inorganic ligand systems are available in several well known books[6,9,415,416] and the spectroscopy of metal chelate compounds is the subject of a monograph[417]. The vibrational spectra of homo- and heteronuclear M—M bonded compounds have attracted a number of reviews[418–422] which examine this structural feature in both organometallic and purely inorganic systems.

Guidance in the interpretation of organometallic spectra will frequently be gained from the spectra of comparable organic molecules. Many collections of the i.r. and Raman spectra of organic compounds and discussions of their interpretation are available[1,2,69,423–428].

Monographs on the i.r. spectra of organosilicon[429,430], organogermanium[431], organophosphorus[432], and other Group V organometalloids[406] and organofluorine compounds[433] are a further adjunct to the elucidation of organometallic spectra. The vibrational spectra of other organic compounds which may be encountered as coordinated ligands, or which may have structural features in common with particular organometallic compounds, have been examined with particular emphasis on the functional group therein[434].

IV. REFERENCES

1. A. L. Smith, 'The Spectroscopic Literature' in *Applied Infrared Spectroscopy*, Chemical Analysis Monographs, Vol. 54, Wiley–Interscience, New York, 1979.
2. R. W. A. Oliver, 'Documentation of Molecular Spectroscopy', in *Introduction to Spectroscopic Methods in Organic Chemistry* (Ed. F. Scheinmann), Vol. 2, Pergamon Press, Oxford, 1973, p. 323.
3. M. Dub (Ed.), *Organometallic Compounds: Methods of Synthesis, Physical Constants and Chemical Reactions*, Springer-Verlag, New York, Vols. I–III, 1966–68, and 2nd Supplements, 1972–75.
4. E. Maslowsky, Jr., *Vibrational Spectra of Organometallic Compounds*, Wiley–Interscience, New York, 1977 (2133 references to 1975).
5. D. K. Huggins and H. D. Kaesz, *Prog. Solid State Chem.*, **1**, 417 (1964).
6. D. M. Adams, *Metal–Ligand and Related Vibrations*, Arnold, London, 1967.
7. A. J. Downs, 'Vibrational Spectroscopy', in *Spectroscopic Methods in Organometallic Chemistry* (Ed. W. O. George), Butterworths, London, 1970, p. 1.
8. K. Nakamoto, 'Infrared Spectroscopy', in *Characterisation of Organometallic Compounds*, Chemical Analysis Monographs, Vol. 26, Pt. 1 (Ed. M. Tsutsui), Wiley–Interscience, New York, 1969, p. 73.
9. K. Nakamoto, *Infrared and Raman Spectra of Inorganic and Coordination Compounds*, 3rd ed., Wiley–Interscience, New York, 1978, pp. 370–409.
10. D. N. Waters, *Molecular Spectroscopy*, Specialist Periodical Reports, Royal Society of Chemistry, London, Vol. 6, p. 46 (1979).
11. W. Kiefer, *Adv. Infrared Raman Spectrosc.*, **3**, 1 (1977).
12. P. J. Hendra, *Vibr. Spectra Struct.*, **2**, 135 (1975).
13. A. J. Barnes and W. J. Orville-Thomas (Ed.), *Vibrational Spectroscopy–Modern Trends*, Elsevier, Amsterdam, New York, 1977.
14. J. R. Ferraro and L. J. Basile (Ed.), *Fourier Transform Infrared Spectroscopy–Applications to Chemical Systems*, Academic Press, New York, Vol. 1, 1978, Vol. 2, 1979.
15. P. R. Griffiths, *Chemical Infrared Fourier Transform Spectroscopy*, Chemical Analysis Monographs, Vol. 43, Wiley–Interscience, New York, 1975.
16. J. Chamberlain, G. W. Chantry, and N. W. B. Stone, *Chem. Soc. Rev.*, **4**, 569 (1975).
17. R. Geick, *Topics Curr. Chem.*, **58**, 73 (1975).
18. E. Knozinger, *Angew. Chem.*, **15**, 25 (1976).
19. D. J. Gardiner, *Adv. Infrared Raman Spectrosc.*, **3**, 167 (1977).
20. J. E. Katon, J. L. Lauer, and M. E. Peterkin, in *Developments in Applied Spectroscopy* (Ed. E. L. Grove and A. J. Perkins), Vol. 9, Plenum, London, 1971, p. 3.
21. A. J. Melveger, J. W. Brasch, and E. R. Lippincott, *Vibr. Spectra Struct.*, **1**, 51 (1972).
22. D. M. Adams and S. J. Payne, *Annu. Rep. Chem. Soc.*, **69**, 3 (1972).
23. J. R. Ferraro, *Coord. Chem. Rev.*, **29**, 1 (1979).
24. T. A. Ballintine and C. D. Schmulbach, *J. Organomet. Chem.*, **164**, 381 (1979).
25. R. B King, A. D. King, M. Z. Iqbal, and C. C. Frazier, *J. Am. Chem. Soc.*, **100**, 1687 (1978).
26. D. M. Adams, *Coord. Chem. Rev.*, **10**, 183 (1973).
27. D. M. Adams and W. S. Fernando, *J. Chem. Soc., Dalton Trans.*, 2507 (1972).
28. I. J. Hyams, *Chem. Phys. Lett.*, **18**, 399 (1973).
29. Ya. M. Kimel'feld, E. M. Smirnova, and V. T. Aleksanyan, *J. Mol. Struct.*, **19**, 329 (1973).
30. A. Poletti, R. Cataliotti, and G. Paliani, *Spectrochim. Acta*, **29A**, 277 (1973).

31. H. J. Buttery, S. F. A. Kettle, and I. Paul. *J. Chem. Soc., Dalton Trans.*, 2293 (1974).
32. A. Bigotto, E. Zangrando, and L. Randaccio, *J. Chem. Soc., Dalton Trans.*, 96 (1976).
33. S. L. Barker, L. Harland, S. F. A. Kettle, and F. F. Stephens, *Inorg. Chim. Acta*, **31**, 217 and 223 (1978).
34. V. G. Ramos and R. S. Tobias, *Inorg. Chem.*, **11**, 2451 (1972).
35. V. G. Ramos and R. S. Tobias, *Spectrochim. Acta*, **29A**, 953 (1973).
36. V. G. Ramos and R. S. Tobias, *Spectrochim. Acta*, **30A**, 181 (1974).
37. H. E. Hallam (Ed.), *Vibrational Spectroscopy of Trapped Species*, Wiley, London, 1973.
38. J. J. Turner, M. Poliakoff, J. Burdett, and H. Dubost, *Adv. Infrared Raman Spectrosc.*, **2**, 1 (1976).
39. L. Andrews, *Vibr. Spectra Struct.*, **4**, 1 (1975).
40. M. Moskovitz and G. A. Ozin, *Vibr. Spectra Struc.*, **4**, 187 (1975).
41. B. M. Chadwick, *Molecular Spectroscopy*, Specialist Periodical Reports, Vol. 3, 1975, p. 281 and Vol. 6, p. 72, Chemical Society, London.
42. A. J. Downs and S. C. Peake, *Molecular Spectroscopy*, Specialist Periodical Reports, Vol. 1, Chemical Society, London, 1973, p. 523
43. R. P. Eischens, *Acc. Chem. Res.*, **5**, 74 (1972).
44. R. J. Kokes, *Acc. Chem. Res.*, **6**, 226 (1973).
45. W. F. Murphy (Ed.), *Proceedings of the VIIth International Conference on Raman Spectroscopy, Ottawa, 1980*. North Holland Publishing Co., Amsterdam, 1980, pp. 345–448.
46. R. E. Hester, *Adv. Infrared Raman Spectrosc.*, **4**, 1 (1978).
47. R. J. H. Clark and B. Stewart, *Struct. Bonding*, **36**, 1 (1979).
48. T. G. Spiro and P. Stein, *Annu. Rev. Phys. Chem.*, **28**, 501 (1977).
49. B. B. Johnson and W. L. Peticolas, *Annu. Rev. Phys. Chem.*, **27**, 465 (1976).
50. D. M. Kurtz Jr., D. F. Shriver, and I. M. Klotz, *Coord. Chem. Rev.*, **24**, 145 (1977).
51. L. D. Barron, *Adv. Infrared Raman Spectrosc.*, **4**, 271 (1978).
52. L. D. Barron, *Acc. Chem. Res.*, **13**, 90 (1980).
53. L. A. Nafie and M. Diem, *Acc. Chem. Res.*, **12**, 296 (1979).
54. J. W. Nibler, W. M. Shaub, J. R. McDonald, and A. B. Harvey, *Vibr. Spectra Struct.*, **6**, 173 (1977).
55. H. C. Anderson and B. S. Hudson, *Molecular Spectroscopy*, Specialist Periodical Reports, Vol. 5, Chemical Society, London, 1978, p. 142.
56. L. A. Woodward, *Introduction to the Theory of Molecular Vibrations and Vibrational Spectroscopy*, Oxford University Press, Oxford, 1972.
57. P. Gans, *Vibrating Molecules*, Chapman and Hall, London, 1971, and references cited therein.
58. C. Altona and D. H. Faber, *Topics Curr. Chem.*, **45**, 1 (1974).
59. J. L. Duncan, *Molecular Spectroscopy*, Specialist Periodical Reports, Vol. 3, Chemical Society, London, 1975, p. 104.
60. N. Mohan, S. J. Cyvin, and A. Muller, *Coord. Chem. Rev.*, **21**, 221 (1976).
61. M. Horák and A. Vítek, *Interpretation and Processing of Vibrational Spectra*, Wiley–Interscience, New York, 1978.
62. K. Nakamoto, *Angew. Chem.*, **11**, 666 (1972).
63. N. Mohan, A. Muller, and K. Nakamoto, *Adv. Infrared Raman Spectrosc.*, **1**, 173 (1975).
64. J. R. Durig, S. M. Craven, and W. C. Harris, *Vibr. Spectra Struct.*, **1**, 73 (1972).
65. W. J. Orville-Thomas (Ed.), *Internal Rotation in Molecules*, Wiley, Chichester, 1974.
66. J. F. Blanke and J. Overend, *Vibr, Spectra Struct.*, **7**, 270 (1978).
67. R. C. Lord and T. B. Malloy, Jr., *Topics Curr. Chem.*, **82**, 1 (1979).
68. T. C. Rounds and H. L. Strauss, *Vibr. Spectra Struct.*, **7**, 238 (1978).
69a. D. H. Whiffen, *J. Chem. Soc.*, 1350 (1956).
69b. D. H. Brown, A. Mohammed, and D. W. A. Sharp, *Spectrochim. Acta*, **21**, 659 (1965).
70. I. S. Butler and M. L. Newbury, *Spectrochim. Acta*, **33A**, 669 (1977).
71. J. R. Hall, in *Essays in Structural Chemistry* (Ed. A. J. Downs, D. A. Long, and L. A. K. Staveley), Macmillan, London, 1971, p. 433.
72. P. L. Goggin, G. Kemeny, and J. Mink, *J. Chem. Soc., Faraday Trans. II*, **72**, 1025 (1976).
73. J. Mink and P. L. Goggin, *J. Organomet. Chem.*, **185**, 129 (1980).
74. J. Mink and B. Gellai, *J. Organomet. Chem.*, **66**, 1 (1974).

75. Y. Imai, F. Watari, and K. Aida, *Spectrochim. Acta*, **36A**, 233 (1980).
76. Y. Imai and K. Aida, *Bull. Chem. Soc. Jap.*, **52**, 2875 (1979).
77. Y. Imai and K. Aida, *J. Inorg. Nucl. Chem.*, **41**, 963 (1979).
78. N. Iwasaki, *Bull. Chem. Soc. Jap.*, **49**, 2735 (1976).
79. J. Mink and P. L. Goggin, *J. Organomet. Chem.*, **156**, 317 (1978).
80. R. Eujen, *J. Mol. Struct.*, **53**, 1 (1979).
81. P. L. Goggin, R. J. Goodfellow, K. Kessler, and A. M. Prescott, *J. Chem. Soc. Dalton Trans.*, 328 (1978).
82. D. J. Brauer, H. Bürger, C. Pawelke, F. H. Flegler, and A. Haas, *J. Organomet. Chem.*, **160**, 389 (1978).
83. D. J. Brauer, H. Bürger, and R. Eujen, *J. Organomet. Chem.*, **135**, 281 (1977).
84. F. Glockling, N. S. Hosmane, V. B. Mahale, J. J. Swindall, L. Magos, and T. J. King, *J. Chem. Res. (M)*, 1201 (1977).
85. P. Bleckmann, M. Soliman, K. Reuter, and W. P. Neumann, *J. Organomet. Chem.*, **108**, C18 (1976).
86. D. L. Rabenstein, M. C. Towangeau, and C. A. Evans, *Can. J. Chem.*, **54**, 2517 (1976).
87. C. Chieh and L. P. C. Leung, *Can. J. Chem.*, **54**, 3077 (1976).
88. Y.-S. Wong, A. J. Carty, and C. Chieh, *J. Chem. Soc., Dalton Trans.*, 1157 and 1801 (1977).
89. A. J. Canty and A. Marker, *Inorg. Chem.*, **15**, 425 (1976).
90. K. H. Tan and M. J. Taylor, *Aust. J. Chem.*, **33**, 1753 (1980).
91. K. Sone, M. Aritaki, K. Hiraoka, and Y. Fukada, *Bull. Chem. Soc. Jap.*, **49**, 2015 (1976).
92. D. J. Gardiner, A. H. Haji, and B. Straughan, *J. Chem. Soc., Dalton Trans.*, 705 (1978).
93. F. Weller, *Z. Anorg. Chem.*, **415**, 233 (1975).
94. J. W. Macklin, *Spectrochim. Acta*, **32A**, 1459 (1976).
95. E. T. Blues, D. Bryce-Smith, and H. Karimpour, *J. Chem. Soc., Chem. Commun.*, 1043 (1979).
96. C. G. Barraclough, G. E. Berkovic, and G. B. Deacon, *Aust. J. Chem.*, **30**, 1905 (1977).
97. P. L. Goggin and D. M. McEwan, *J. Chem. Res. (S)*, 171 (1978).
98. V. M. Sergutin, V. N. Zgonnik, and K. K. Kalninsh, *J. Organomet. Chem.*, **170**, 151 (1979).
99. E. C. Ashby and H. S. Prasad, *Inorg. Chem.*, **14**, 2869 (1975).
100. N. Atam and K. Dehnicke, *Z. Anorg. Chem.*, **427**, 193 (1976).
101. J. Mounier, B. Mula, and A. Potier, *J. Organomet. Chem.*, **107**, 281 (1976).
102. J. Kress, D. Bougeard, and A. Novak, *Spectrochim. Acta*, **33A**, 161 (1977).
103. J. Kress, A. Novak, and J. Hervien, *J. Organomet. Chem.*, **121**, 7 (1976).
104. J. Kress and A. Novak, *J. Organomet. Chem.*, **99**, 23, and 199 (1975).
105. J. Kress and A. Novak, *J. Organomet. Chem.*, **86**, 281 (1975).
106. J. Kress and A. Novak, *J. Struct. Chem.*, **18**, 677 (1977).
107. J. Kress, *J. Organomet. Chem.*, **111**, 1 (1976).
108. E. C. Ashby, J. Nackashi, and C. E. Parris, *J. Am. Chem. Soc.*, **97**, 3162 (1975).
109. A. Klopsch and K. Dehnicke, *Chem. Ber.*, **108**, 420 (1975).
110. A. Tatzel, H. Schrem, and J. Weidlein, *Spectrochim. Acta*, **34A**, 549 (1978).
111. H. J. Widler, H.-D. Hausen, and J. Weidlein, *Z. Naturforsch.*, **30B**, 644 (1975).
112. H.-J. Widler, W. Schwartz, H.-D. Hausen, and J. Weidlein, *Z. Anorg. Chem.*, **435**, 179 (1977).
113. A. P. Kurbakova, L. A. Leites, V. T. Alexsanyan, L. M. Golubiushaya, E. N. Zerina, and V. I. Bregadze, *J. Struct. Chem.*, **15**, 961 (1974).
114. W. Fries, W. Schwartz, H.-D. Hausen, and J. Weidlein, *J. Organomet. Chem.*, **159**, 373 (1978).
115. D. A. Sanders, P. A. Scherr, and J. P. Oliver, *Inorg. Chem.*, **15**, 861 (1976).
116. M. Wilkinson and I. J. Worrall, *J. Organomet. Chem.*, **93**, 39 (1975).
117. M. J. S. Gynane, L. G. Waterworth, and I. J. Worrall, *J. Organomet. Chem.*, **43**, 257 (1972).
118. H.-U. Schwering, E. Jungk, and J. Weidlein, *J. Organomet. Chem.*, **91**, C4 (1975).
119. H. Schmidbauer and D. Koth, *Naturwissenschaften*, **63**, 482 (1976).
120. F. Watari, S. Shimizu, K. Aida, and E. Takayama, *Bull. Chem. Soc. Jap.*, **51**, 1602 (1978).
121. J. Müller, F. Schmock, A. Klopsch, and K. Dehnicke, *Chem. Ber.*, **108**, 664 (1975).

122. K. Dehnicke and N. Röder, *J. Organomet. Chem.*, **86**, 335 (1975).
123. K. Mertz, W. Schwartz, B. Eberwein, J. Weidlein, H. Hess, and H.-D. Hausen, *Z. Anorg. Chem.*, **429**, 99 (1977).
124. B. Walther, A. Zschunke, B. Adler, A. Kobe, and S. Bauer, *Z. Anorg. Chem.*, **427**, 137 (1976).
125. P. Krommes and J. Lorberth, *J. Organomet. Chem.*, **120**, 131 (1976).
126. S. B. Miller and T. B. Brill, *J. Organomet. Chem.*, **166**, 293 (1979).
127. S. B. Miller, B. L. Jelus, and T. B. Brill, *J. Organomet. Chem.*, **96**, 1 (1975).
128. T. Tanaka, *Organomet. Chem. Rev. A*, **5**, 1 (1970).
129. H. Bürger, *Organomet. Chem. Rev. A*, **3**, 425 (1968).
130. N. A. Chemaevskii, *Russ. Chem. Rev.*, **32**, 509 (1963).
131. J. R. Durig, A. D. Lopata, and P. Groner, *J. Chem. Phys.*, **66**, 1888 (1977), and previous work cited therein.
132. K. Licht and P. Reich, *Literature Data for I.r., Raman and N.m.r. Spectroscopy of Silicon, Germanium, Tin and Lead Compounds*, Deutscher Verlag, Berlin, 1971.
133. F. Watari, *Spectrochim. Acta*, **34A**, 1239 (1978).
134. J. D. Brown, D. Tevault, and K. Nakamoto, *J. Mol. Struct.*, **40**, 43 (1977).
135. R. Eugen and H. Bürger, *Spectrochim. Acta*, **35A**, 1135 (1979).
136. B. Y. K. Ho and J. J. Zuckerman, *J. Organomet. Chem.*, **96**, 41 (1975).
137. H.-J. Götze and G. Nergmann, *Z. Anal. Chem.*, **273**, 417 (1975).
138. L. Czuchajowski, J. Habdas, S. A. Kucharski, and K. Rogosz, *J. Organomet. Chem.*, **155**, 185 (1978).
139. F. Höffler, *Monatsh. Chem.*, **107**, 893 (1976).
140. B. Mathiasch, *Inorg. Nucl. Chem. Lett.*, **13**, 13 (1977).
141. T. Birchall and J. P. Johnson, *Can. J. Chem.*, **57**, 160 (1979).
142. B. Mathiasch and T. N. Mitchell, *J. Organomet. Chem.*, **185**, 351 (1980).
143. B. Mathiasch, *J. Organomet. Chem.*, **122**, 345 (1976).
144. G. Davidovics, M. Guiliano, J. Chouteau, R. Ouaki, and J. C. Maire, *Spectrochim. Acta*, **32A**, 30 (1976).
145. D. C. van Beelen, J. Wolters, and A. van der Gen, *J. Organomet. Chem.*, **145**, 359 (1978).
146. M. J. Taylor, *Metal-to-Metal Bonded States of the Main Group Elements*, Academic Press, London, New York, 1975.
147. T. N. Mitchell and M. El-Behairy, *J. Organomet. Chem.*, **141**, 43 (1977).
148. A. Müller, W. Nolte, S. J. Cyvin, B. N. Cyvin, and A. J. P. Alix, *Spectrochim. Acta*, **34A**, 383 (1978).
149. N. I. Shergina, B. A. Trofimov, and M. G. Voronkov, *Dokl. Phys. Chem.*, **225**, 1359 (1976).
150. J. Nakovich, Jr., S. D. Shook, F. A. Miller, D. R. Parnell, and R. E. Sacher, *Spectrochim. Acta*, **35A**, 495 (1979).
151. A. Marchand, M.-Th. Forel, M. Riviera-Baudet, *J. Organomet. Chem.*, **156**, 341 (1978).
152. H.-J. Götze and W. Garbe, *Spectrochim. Acta*, **35A**, 975 (1979).
153. G. M. Bancroft, B. W. Davies, N. C. Payne, and T. K. Sham, *J. Chem. Soc., Dalton Trans.*, 973 (1975).
154. Y. Kawasaki and M. Aritomi, *J. Organomet. Chem.*, **103**, 39 (1976).
155. P. J. Davidson, D. H. Harris, and M. F. Lappert, *J. Chem. Soc., Dalton Trans.*, 2268 (1976).
156. M. J. S. Gynane, M. F. Lappert, S. J. Miles, A. J. Carty, and N. J. Taylor, *J. Chem. Soc., Dalton Trans.*, 2009 (1977).
157. P. Foley and M. Zeldin, *Inorg. Chem.*, **14**, 2264 (1975).
158. J. D. Cotton, P. J. Davidson, and M. F. Lappert, *J. Chem. Soc., Dalton Trans.*, 2275 (1976).
159. A. Bonny, *Coord. Chem. Rev.*, **25**, 229 (1978), and references cited therein.
160. H. Schumann, L. Rösch, H. Neumann, and H.-J. Koth, *Chem. Ber.*, **108**, 1630 (1975).
161. R. A. Burnham and S. R. Stobart, *J. Chem. Soc., Dalton Trans.*, 1489 (1977).
162. N. S. Dance, W. R. McWhinnie, and R. C. Poller, *J. Chem. Soc., Dalton Trans.*, 2349 (1976).
163. V. Kunze and J. D. Koola, *Z. Naturforsch.*, **30B**, 91 (1975).

164. M. A. Mesubi, *Spectrochim. Acta*, **32A**, 1327 (1976).
165. M. Nardelli, C. Pelizzi, and G. Pelizzi, *J. Organomet. Chem.*, **125**, 161 (1977).
166. E. J. Kupchik and J. A. Feiccabrino, *J. Organomet. Chem.*, **93**, 325 (1975).
167. M. Nardelli, C. Pellizzi, G. Pellizzi, and P. Tarasconi, *Z. Anorg. Chem.*, **431**, 250 (1977).
168. C. Pellizzi, G. Pellizzi, and P. Tarasconi, *J. Chem. Soc., Dalton Trans.*, 1935 (1977).
169. G. Eng and C. R. Dillard, *Inorg. Chim. Acta*, **31**, 227 (1978).
170. L. Coghi, C. Pelizzi, and G. Pelizzi, *J. Organomet. Chem.*, **114**, 53 (1976).
171. R. Makhija and M. Onyszchuk, *J. Organomet. Chem.*, **135**, 261 (1977).
172. I. Wharf, M. Onyszchuk, J. M. Miller, and T. R. B. Jones, *J. Organomet. Chem.*, **190**, 417 (1980).
173. M. E. Bishop, C. D. Shaeffer, and J. J. Zuckerman, *Spectrochim. Acta*, **32A**, 1519 (1976).
174. P. Krommes and J. Lorberth, *J. Organomet. Chem.*, **93**, 339 (1975).
175. R. Goetze and H. Nöth, *Z. Naturforsch.*, **30B**, 343 (1975).
176. A. J. Downs, R. Schmutzler, and I. A. Steer, *J. Chem. Soc., Chem. Commun.*, 221 (1966).
177. B. A. Nevett and A. Perry, *J. Mol. Spectrosc.*, **66**, 331 (1977).
178. L. Verdonck and G. P. van der Kelen, *Spectrochim. Acta*, **31A**, 1707 (1975).
179. L. Verdonck and G. P. van der Kelen, *Spectrochim. Acta*, **35A**, 861 (1979); **33A**, 601 (1977).
180. R. R. Holmes, R. M. Deiters, and J. A. Golen, *Inorg. Chem.*, **8**, 2612 (1969).
181. K. Dehnicke and H. G. Nadler, *Chem. Ber.*, **109**, 3034 (1976).
182. H. G. Nadler and K. Dehnicke, *J. Organomet. Chem.*, **90**, 291 (1975).
183. K. Dehnicke and H. G. Nadler, *Z. Anorg. Chem.*, **418**, 229 (1975).
184. B. Eberwein and J. Weidlein, *Z. Anorg. Chem.*, **420**, 229 (1976).
185. E. Maslowsky, Jr., *J. Organomet. Chem.*, **70**, 153 (1974).
186. E. Glozbach and J. Lorberth, *J. Organomet. Chem.*, **132**, 359 (1977).
187. P. Krommes and J. Lorberth, *J. Organomet. Chem.*, **92**, 181 (1975).
188. P. Krommes and J. Lorberth, *J. Organomet. Chem.*, **88**, 329 (1975).
189. R. Demuth, J. Apel, and J. Grobe, *Spectrochim. Acta*, **34A**, 357 and 361 (1978).
190. P. Dehnert, R. Demuth, and J. Grobe, *Spectrochim. Acta*, **34A**, 857 (1978).
191. J. A. Morrison and R. J. Lagow, *Inorg. Chem.*, **16**, 1823 (1977).
192. J. W. Anderson, J. E. Drake, R. T. Hemmings, and D. L. Nelson, *Inorg. Nucl. Chem. Lett.*, **11**, 233 (1975).
193. T. N. Srivastava, R. C. Srivastava, and M. Singh, *J. Organomet. Chem.*, **157**, 405 (1978).
194. J. C. Dewan and J. Silver, *Inorg. Nucl. Chem. Lett.*, **12**, 647 (1976).
195. H. J. Gysling, H. R. Luss, and D. L. Smith, *Inorg. Chem.*, **18**, 2697 (1979).
196. B. A. Nevett and A. Perry, *Spectrochim. Acta*, **33A**, 755 (1977).
197. N. Bertazzi, *J. Organomet. Chem.*, **110**, 175 (1976).
198. N. Bertazzi, M. Airoldi, and L. Pellerito, *J. Organomet. Chem.*, **97**, 399 (1975).
199. S. P. Bone and D. B. Sowerby, *J. Organomet. Chem.*, **184**, 181 (1980).
200. R. G. Goel and H. S. Prasad, *Spectrochim. Acta*, **32A**, 569 (1976).
201. R. G. Goel and H. S. Prasad, *Spectrochim. Acta*, **35A**, 339 (1979).
202. T. Allman, R. G. Goel, and H. S. Prasad, *J. Organomet. Chem.*, **166**, 365 (1979).
203. E. R. Clark and M. A. Al-Turaihi, *J. Organomet. Chem.*, **96**, 251 (1975).
204. E. R. Clark and M. A. Al-Turaihi, *J. Organomet. Chem.*, **124**, 391 (1977).
205. J. E. Drake and R. T. Hemmings, *J. Chem. Soc., Dalton Trans.*, 1730 (1976).
206. E. R. Clark and M. A. Al-Turaihi, *J. Organomet. Chem.*, **134**, 181 (1977).
207. N. S. Dance and W. R. McWhinnie, *J. Chem. Soc., Dalton Trans.*, 43 (1975).
208. I. Davies and W. R. McWhinnie, *Inorg. Nucl. Chem. Lett.*, **12**, 763 (1976).
209. P. Klaeboe and O. Vikane, *Acta Chem. Scand.*, **A31**, 120 (1977).
210. N. S. Dance, W. R. McWhinnie, and C. H. W. Jones, *J. Organomet. Chem.*, **125**, 291 (1977).
211. G. W. Rice and R. S. Tobias, *Inorg. Chem.*, **14**, 2402 (1975).
212. G. W. Rice and R. S. Tobias, *Inorg. Chem.*, **15**, 489 (1976).
213. A. Johnson and R. J. Puddephatt, *J. Organomet. Chem.*, **85**, 115 (1975).
214. P. Gans, J. B. Gill, and M. Griffin, *J. Chem. Soc., Chem. Commun.*, 169 (1976).
215. G. C. Stocco, L. Pellerito, and N. Bertazzi, *Inorg. Chim. Acta*, **12**, 67 (1975).
216. G. W. Rice and R. S. Tobias, *J. Am. Chem. Soc.*, **99**, 2141 (1977).
217. G. W. Rice and R. S. Tobias, *J. Chem. Soc., Chem. Commun.*, 994 (1975).

218. B. Neruida and J. Lorberth, *J. Organomet. Chem.*, **111**, 241 (1976).
219. I. M. Cheremisina, *J. Struct. Chem.*, **19**, 286 (1978).
220. C. Blandy, R. Guerreiro, and D. Gervais, *J. Organomet. Chem.*, **128**, 415 (1977).
221. G. W. A. Fowles, D. A. Rice, and T. G. Sheenan, *J. Organomet. Chem.*, **135**, 321 (1977).
222. S. I. Beilin, S. B. Golstein, B. A. Dolgoplosk, L. Sh. Guzman, and E. L. Tinyakova, *J. Organomet. Chem.*, **142**, 145 (1977).
223. K.-H. Thiele, A. Russek, R. Opitz, B. Mohai, and W. Brüser, *Z. Anorg. Chem.*, **412**, 11 (1975).
224. A. Lachowitz and K.-H. Thiele, *Z. Anorg. Chem.*, **431**, 88 (1977).
225. C. Santini-Scampucci and J. G. Reiss, *J. Chem. Soc., Dalton Trans.*, 195 (1976).
226. K. Mertis, D. H. Williamson, and G. Wilkinson, *J. Chem. Soc., Dalton Trans.*, 607 (1975).
227. K. H. Pannell, C. C. Wu, and G. J. Long, *J. Organomet. Chem.*, **186**, 85 (1980).
228. H.-F. Klein and H. H. Karsch, *Chem. Ber.*, **109**, 1453 (1976).
229. H.-F. Klein and H. H. Karsch, *Chem. Ber.*, **108**, 944, and 956 (1975).
230. T. Ikariya and A. Yamamoto, *J. Organomet. Chem.*, **116**, 239 (1976).
231. T. Ito, H. Tsuchiya, and A. Yamamoto, *Bull. Chem. Soc. Jap.*, **50**, 1319 (1977).
232. G. Oehme, K.-C. Röber, and H. Pracejus, *J. Organomet. Chem.*, **105**, 127 (1976).
233. J. R. Hall and G. A. Swile, *J. Organomet. Chem.*, **139**, 403 (1977).
234. G. C. Stocco, F. S. Gattuso, N. Bertazzi, and L. Pellerito, *J. Coord. Chem.*, **5**, 55 (1976).
235. B. Neruda, E. Glozbach, and J. Lorberth, *J. Organomet. Chem.*, **131**, 317 (1977).
236. J. R. Hall, D. A. Hirons, and G. A. Swile, *J. Organomet. Chem.*, **174**, 355 (1979).
237. J. Browning, P. L. Goggin, R. J. Goodfellow, N. W. Hurst, L. G. Mallinson, and M. Murray, *J. Chem. Soc., Dalton Trans.*, 872 (1978).
238. P. W. J. de Graaf, J. Boersma, and G. J. M. van der Kerk, *J. Organomet. Chem.*, **105**, 399 (1976).
239. R. Uson, J. Fornies, and P. Espinet, *J. Organomet. Chem.*, **116**, 353 (1976).
240. J. M. Coronas and J. Sales, *J. Organomet. Chem.*, **94**, 107 (1975).
241. J. M. Coronas, C. Peruyero, and J. Sales, *J. Organomet. Chem.*, **128**, 291 (1977).
242. L. R. Smith and D. M. Blake, *J. Am. Chem. Soc.*, **99**, 3302 (1977).
243. M. Angoletta, L. Malatesta, P. L. Bellon, and G. Cagilo, *J. Organomet. Chem.*, **114**, 219 (1976).
244. H. Koezuka, G.-E. Matsubayashi, and T. Tanaka, *Inorg. Chem.*, **15**, 417 (1976).
245. H. Schmidbauer and R. Franke, *Chem. Ber.*, **108**, 1321 (1975).
246. H. Schmidbauer and R. Franke, *Inorg. Chim. Acta*, **13**, 79 and 85 (1975).
247. E. W. Post and K. L. Walters, *Inorg. Chim. Acta*, **26**, 29 (1978).
248. B. Cetinkaya, M. F. Lappert, G. M. McLaughlin, and K. Turner, *J. Chem. Soc., Dalton Trans.*, 1591 (1974).
249. D. F. Christian, H. C. Clark, and R. F. Stepaniak, *J. Organomet. Chem.*, **112**, 227 (1976).
250. A. J. Hartshorn, M. F. Lappert, and K. Turner, *J. Chem. Soc., Dalton Trans.*, 348 (1978).
251. M. F. Lappert and P. L. Pye, *J. Chem. Soc., Dalton Trans.*, 2172 (1977).
252. K. H. Dötz and C. G. Kreiter, *J. Organomet. Chem.*, **99**, 309 (1975).
253. A. J. Hartshorn, M. F. Lappert, and K. Turner, *J. Chem. Soc., Chem. Commun.*, 929 (1975).
254. D. H. Farrar, R. O. Harris, and A. Walker, *J. Organomet. Chem.*, **124**, 125 (1977).
255. T. J. Collins, K. R. Grundy, W. R. Roper, and S. F. Wong, *J. Organomet. Chem.*, **107**, C37 (1976).
256. D. Mansuy, M. Lange, J. C. Chottard, J. F. Bartoli, D. Chevrier, and R. Weiss, *Angew. Chem.*, **17**, 781 (1978).
257. W. Beck, H. Brix, and F. H. Köhler, *J. Organomet. Chem.*, **121**, 211 (1975).
258. C. P. Casey and S. M. Neumann, *J. Am. Chem. Soc.*, **98**, 5395 (1976).
259. S. R. Winter, G. W. Cornett, and E. A. Thompson, *J. Organomet. Chem.*, **133**, 339 (1977).
260. T. J. Collins and W. R. Roper, *J. Organomet. Chem.*, **159**, 73 (1978).
261. D. F. Christian, H. C. Clark, and R. F. Stepaniak, *J. Organomet. Chem.*, **112**, 209 (1976).
262. H. Behrens, R.-J. Lampe, P. Merbach, and M. Moll, *J. Organomet. Chem.*, **159**, 201 (1978).
263. W. Petz, *J. Organomet. Chem.*, **90**, 223 (1975).

264. K. R. Grundy and W. R. Roper, *J. Organomet. Chem.*, **113**, C45 (1976).
265. H. Berke and P. Harter, *Angew. Chem.*, **19**, 225 (1980).
266. M. F. Lappert, *J. Organomet. Chem.*, **100**, 139 (1975).
267. E. O. Fischer, N. Q. Dao, and W. R. Wagner, *Angew. Chem.*, **17**, 50 (1978).
268. G. Davidson, *Organomet. Chem. Rev. A*, **8**, 303 (1972).
269. H. P. Fritz, *Adv. Organomet. Chem.*, **1**, 239 (1964).
270. J. Yarwood (Ed.), *Spectroscopy and Structure of Molecular Complexes*, Plenum, London, 1973.
271. R. D. Johnson, *Adv. Inorg. Chem. Radiochem.*, **13**, 471 (1970).
272. E. O. Fischer and H. Werner (Ed.), *Metal π-Complexes*, Elsevier, Amsterdam, Vol. 1, 1966; Vol. 2, 1974.
273. L. D. Petit and D. S. Barnes, *Topics Curr. Chem.*, **28**, 87 (1972).
274. F. R. Hartley, *Chem. Rev.*, **69**, 799 (1969).
275. F. R. Hartley, *Angew. Chem.*, **11**, 596 (1972).
276. D. A. Duddell, in *Spectroscopy and Structure of Molecular Complexes* (Ed. J. Yarwood), Plenum, London, 1973, p. 427.
277. P. N. Gates and D. Steele, in *Spectroscopy and Structure of Molecular Complexes* (Ed. J. Yarwood), Plenum, London, 1973, p. 511.
278. D. B. Powell, J. E. V. Scott, and N. Sheppard, *Spectrochim. Acta*, **28A**, 327 (1972).
279. S. Sorriso and G. Cardaci, *J. Chem. Soc., Dalton Trans.*, 1041 (1975).
280. M. Bigorgne, *J. Organomet. Chem.*, **127**, 55 (1977).
281. H. L. M. van Gaal and F. L. A. van der Bekerom, *J. Organomet. Chem.*, **134**, 237 (1977).
282. A. C. Jesse, M. A. M. Meester, D. J. Stufkens, and K. Vrieze, *Inorg. Chim. Acta*, **26**, 129 (1978).
283. J. Howard and T. C. Waddington, *J. Chem. Soc., Faraday Trans. II*, **74**, 1275 (1978).
284. S. Komiya and J. K. Kochi, *J. Organomet. Chem.*, **135**, 65 (1977).
285. M. A. M. Meester, D. J. Stufkens, and K. Vrieze, *Inorg. Chim. Acta*, **16**, 191 (1976).
286. M. A. M. Meester, H. van Dam, D. J. Stufkens, and A. Oskam, *Inorg. Chim. Acta*, **20**, 155 (1976).
287. M. A. M. Meester, D. J. Stufkens, and K. Vrieze, *Inorg. Chim. Acta*, **21**, 251 (1977).
288. T. Iwayanagi and Y. Saito, *Inorg. Nucl. Chem. Lett.*, **11**, 459 (1975).
289. A. C. Jesse, D. J. Stufkens, and K. Vrieze, *Inorg. Chim. Acta*, **32**, 87 (1979).
290. L. Maresca, G. Natile, and G. Rizzardi, *Inorg. Chim. Acta*, **38**, 53 (1980).
291. R. Řeřicha, *Collect. Czech. Chem. Commun.*, **40**, 2577 (1975).
292. D. C. Andrews, G. Davidson, and D. A. Duce, *J. Organomet. Chem.*, **101**, 113 (1975).
293. M. Bigorgne, *J. Organomet. Chem.*, **127**, 55 (1977).
294. D. A. Edwards and R. Richards, *J. Organomet. Chem.*, **86**, 407 (1975).
295. D. W. Wertz and M. A. Moseley, *Inorg. Chem.*, **19**, 705 (1980).
296. D. J. Darensbourg, D. J. Tappan, and H. H. Nelson, *Inorg. Chem.*, **16**, 534 (1977).
297. H. J. Clase, A. J. Cleland, and M. J. Newlands, *J. Organomet. Chem.*, **93**, 231 (1975).
298 P. F. Meier, D. L. Perry, R. H. Hauge, and J. L. Margrave, *Inorg. Chem.*, **18**, 2051 (1979).
299. G. A. Ozin, W. J. Power, T. H. Upton, and W. A. Goddard, *J. Am. Chem. Soc.*, **100**, 4750 (1978).
300. H. Huber, G. A. Ozin, and W. J. Power, *Inorg. Chem.*, **16**, 979 (1977).
301. G. A. Ozin, H. Huber, and D. McIntosh, *Inorg. Chem.*, **16**, 3070 (1977).
302. D. McIntosh and G. A. Ozin, *J. Organomet. Chem.*, **121**, 127 (1976).
303. G. A. Ozin and W. J. Power, *Inorg. Chem.*, **17**, 2836 (1978).
304. Y. Soma, *J. Chem. Soc., Chem. Commun.*, 1004 (1976).
305. J. D. Prentis, A. Lesinnas, and N. Sheppard., *J. Chem. Soc., Chem. Commun.*, 76 (1976).
306. M. C. Barral, R. Jimenez, E. Royer, V. Moreno, and A. Santos, *Inorg. Chim. Acta*, **31**, 165 (1978).
307. K. Sonogashira, Y. Fujikura, T. Yatake, N. Toyoshima, S. Takahashi, and N. Hagihara, *J. Organomet. Chem.*, **145**, 101 (1978).
308. R. Nast and H. P. Müller, *Chem. Ber.*, **111**, 415 (1978).
309. O. M. Abu Salah and M. I. Bruce, *J. Chem. Soc., Dalton Trans.*, 2311 (1975).
310. R. Nast and V. Pank, *J. Organomet. Chem.*, **129**, 265 (1977).
311. R. Nast, J. Voss, and R. Kramolowsky, *Chem. Ber.*, **108**, 1511 (1975).

312. R. Nast and J. Moritz, *J. Organomet. Chem.*, **117**, 81 (1976).
313. V. T. Aleksanyan, I. A. Garbuzova, I. R. Gol'ding, and A. M. Sladkov, *Spectrochim. Acta* **31A**, 517 (1975).
314. M. C. Esquivel, A. S. Macías, and L. B. Reventós, *J. Inorg. Nucl. Chem.*, **39**, 1153 (1977).
315. R. W. M. ten Hoedt, G. van Koten, and J. G. Noltes, *J. Organomet. Chem.*, **133**, 113 (1977).
316. W. F. Smith, J. Yule, N. J. Taylor, H. N. Paik, and A. J. Carty, *Inorg. Chem.*, **16**, 1593 (1977).
317. W. Fries, W. Schwarz, H.-D. Hausen, and J. Weidlein, *J. Organomet. Chem.*, **159**, 373 (1978).
318. P. S. Santos, K. Kawai, and O. Sala, *Inorg. Chim. Acta*, **22**, 155 (1977).
319. S. F. A. Kettle and P. L. Stanghellini, *Inorg. Chem.*, **16**, 753 (1977).
320. G. Fachinetti, C. Floriani, F. Marchetti, and M. Mellini, *J. Chem. Soc., Dalton Trans.*, 1398 (1978).
321. E. A. Maatta, R. A. D. Wentworth, J. W. McDonald, and G. D. Watt, *J. Am. Chem. Soc.*, **100**, 1320 (1978).
322. H. G. Alt, *J. Organomet. Chem.*, **127**, 349 (1977).
323. R. Mathieu and R. Poilblanc, *J. Organomet. Chem.*, **142**, 351 (1977).
324. A. J. Carty, H. N. Paik, and G. J. Palenik, *Inorg. Chem.*, **16**, 300 (1977).
325. H. C. Clark and K. von Werner, *J. Organomet. Chem.*, **101**, 347 (1975).
326. D. Empsall, B. L. Shaw, and A. J. Stringer, *J. Chem. Soc., Dalton Trans.*, 185 (1976).
327. H. L. M. van Gaal, M. W. M. Graef, and A. van der Ent, *J. Organomet. Chem.*, **131**, 453 (1977).
328. V. T. Aleksanyan, G. M. Kyz'yats, and T. S. Kurtikyan, *Dokl. Chem.*, **216**, 331 (1974).
329. M. Ito and W. Suetaka, *Proc. Int. Vac. Congr., 7th*, 1043 (1977).
330. T. B. Chenskaya, L. A. Leites, and V. T. Aleksanyan, *J. Organomet. Chem.*, **148**, 85 (1978).
331. D. C. Andrews and G. Davidson, *J. Organomet. Chem.*, **124**, 181 (1977).
332. M. A. Cairns and J. F. Nixon, *J. Organomet. Chem.*, **87**, 109 (1975).
333. J. E. Ellis and R. A. Faltynek, *J. Organomet. Chem.*, **93**, 205 (1975).
334. M. Schneider and E. Weiss, *J. Organomet. Chem.*, **121**, 345 (1976).
335. E. G. Hoffmann, R. Kallweit, G. Schroth, K. Seevogel, W. Stemple, and G. Wilke, *J. Organomet. Chem.*, **97**, 183 (1975).
336. G. Carturan, A. Scrivanti, and V. Belluco, *Inorg. Chim. Acta*, **21**, 103 (1977).
337. F. L. Bowden and R. Giles, *Coord. Chem. Rev.*, **20**, 81 (1976).
338. D. M. Adams, R. E. Christopher, and D. C. Stevens, *Inorg. Chem.*, **14**, 1562 (1975).
339. E. M. Bisby, G. Davidson, and D. A. Duce, *J. Mol. Struct.*, **48**, 93 (1978).
340. D. C. Andrews, G. Davidson, and D. A. Duce, *J. Organomet. Chem.*, **97**, 95 (1975).
341. D. H. Finseth, C. Sourisseau, and F. A. Miller, *J. Phys. Chem.*, **80**, 1284 (1976).
342. E. W. Neuse, *J. Organomet. Chem.*, **99**, 287 (1975).
343. M. Bigorgne, O. Kahn, M. F. Koenig and A. Loutellier, *Spectrochim. Acta*, **31A**, 741 (1975).
344. F. van Meurs, J. M. A. Baas, and H. van Bekkum, *J. Organomet. Chem.*, **129**, 347 (1977).
345. B. V. Lokshin, E. B. Rusach, and Ya. D. Konovalov, *Izv. Adad. Nauk SSSR, Ser. Khim.*, 84 (1975).
346. H. G. Biedermann, K. Ofele, and J. Tajtelbaum, *Z. Naturforsch.*, **31B**, 321 (1976).
347. H. F. Efner, D. E. Tevault, W. B. Fox, and R. R. Smardzewski, *J. Organomet. Chem.*, **146**, 45 (1978).
348. M. T. Anthony, M. L. H. Green, and D. Young, *J. Chem. Soc., Dalton Trans.*, 1419 (1975).
349. L. C. Damude and P. A. W. Dean, *J. Organomet. Chem.*, **168**, 123 (1979).
350. K. Schumann and H.-H. Perkampus, *Spectrochim. Acta*, **33A**, 417 (1977).
351. V. T. Aleksanyan and B. V. Lokshin, *J. Organomet. Chem.*, **131**, 113 (1977).
352. H. Felkin and G. K. Turner, *J. Organomet. Chem.*, **129**, 429 (1977).
353. J. Lusztyk and K. B. Storowieyski, *J. Organomet. Chem.*, **170**, 293 (1979).
354. K. B. Starowieyski and J. Lusztyk, *J. Organomet. Chem.*, **133**, 281 (1977).
355. D. A. Drew and G. L. Morgan, *Inorg. Chem.*, **16**, 1704 (1977).

356. V. T. Aleksanyan, I. A. Garbuzova, V. V. Gavrilenko, and L. I. Zakharkin, *J. Organomet. Chem.*, **129**, 139 (1977).
357. S. Onaka and W. Furuichi, *J. Organomet. Chem.*, **173**, 77 (1979).
358. H. K. Hofstee, J. Boersma, and G. J. M. van der Kelen, *J. Organomet. Chem.*, **120**, 313 (1976).
359. V. T. Aleksanyan, B. V. Lokshin, G. K. Borisov, G. G. Devyatykh, A. S. Smirnov, R. V. Nazarova, J. A. Koningstein, and B. F. Gächter, *J. Organomet. Chem.*, **124**, 293 (1977).
360. B. V. Lokshin, V. T. Aleksanyan, and E. B. Rusach, *J. Organomet. Chem.*, **86**, 253 (1975).
361. B. F. Gächter, J. A. Koningstein, and V. T. Aleksanyan, *J. Chem. Phys.*, **62**, 4628 (1975).
362. K. Yokoyama, S. Kobinata, and S. Maeda, *Bull. Chem. Soc. Jap.*, **49**, 2182 (1976).
363. W. Siebert, W. Ruf, K.-J. Schaper, and T. Rank, *J. Organomet. Chem.*, **128**, 219 (1977).
364. E. F. Paulus and L. Schäfer, *J. Organomet. Chem.*, **144**, 205 (1978).
365. L. V. Haley, T. Parameswaran, J. A. Koningstein, and V. T. Aleksanyan, *Indian J. Pure Appl. Phys.*, **16**, 236 (1978).
366. D. M. Duggan and D. N. Hendrickson, *Inorg. Chem.*, **14**, 955 (1975).
367. B. F. Gächter, T. Parameswaran, and J. A. Koningstein, *J. Mol. Spectrosc.*, **54**, 215 (1975).
368. D. Rehder, *Z. Naturforsch.*, **31B**, 273 (1976).
369. T. Parameswaran, J. A. Koningstein, L. V. Haley, and V. T. Aleksanyan, *J. Chem. Phys.*, **68**, 1285 (1978).
370. V. T. Aleksanyan, G. K. Borisov, I. A. Gorbuzova, and G. G. Devyatykh, *J. Organomet. Chem.*, **131**, 251 (1977).
371. B. Demerseman, G. Bouquet, and M. Bigorgne, *J. Organomet. Chem.*, **132**, 223 (1977).
372. R. S. P. Coutts and P. C. Wailes, *J. Organomet. Chem.*, **84**, 47 (1975).
373. G. P. Pez, *J. Am. Chem. Soc.*, **98**, 8072 (1976).
374. G. G. Barna, I. S. Butler, and K. R. Plowman, *Can. J. Chem.*, **54**, 110 (1976).
375. G. Davidson and D. A. Duce, *J. Organomet. Chem.*, **120**, 229 (1976).
376. R. Feld, E. Hellner, A. Klopsch, and K. Dehnicke, *Z. Anorg. Chem.*, **442**, 173 (1978).
377. M. Morán and V. Fernández, *J. Organomet. Chem.*, **165**, 215 (1979).
378. P. Espinet, P. A. Bailey, P. Piraino, and P. M. Maitlis, *Inorg. Chem.*, **18**, 2706 (1979).
379. D. J. Parker, *Spectrochim. Acta*, **31A**, 1789 (1975).
380. J. Stadelhofer, J. Weidlein, P. Fischer, and A. Haaland, *J. Organomet. Chem.*, **116**, 55 (1976).
381. P. G. Harrison and J. A. Richards, *J. Organomet. Chem.*, **108**, 61, (1976).
382. J. Howard and T. C. Waddington, *Spectrochim. Acta*, **34A**, 445 (1978).
383. G. M. Kuzyants, A. O. Baronetskii, T. M. Chernyshova, and Z. V. Todres, *J. Organomet. Chem.*, **142**, 139 (1977).
384. H. P. Fritz and H. Keller, *Chem. Ber.*, **95**, 158 (1962).
385. J. Goffart and G. Duychaerts, *J. Organomet. Chem.*, **94**, 29 (1975).
386. A. Greco, S. Cesca, and G. Bertolini, *J. Organomet. Chem.*, **113**, 321 (1976).
387. R. F. Dallinger, P. Stein, and T. G. Spiro, *J. Am. Chem. Soc.*, **100**, 7865 (1978).
388. G. R. Sienel, A. W. Spiegl, and R. D. Fischer, *J. Organomet. Chem.*, **160**, 67 (1978).
389. J. Goffart, J. Fuger, B. Gilbert, L. Hocks, and G. Duychaerts, *Inorg. Nucl. Chem. Lett.*, **11**, 569 (1975).
390. P. S. Braterman, *Metal Carbonyl Spectra*, Academic Press, London, New York, 1975.
391. P. S. Braterman, *Struct. Bonding*, **10**, 57 (1972); **26**, 1 (1976).
392. J. J. Turner, in *Vibrational Spectroscopy—Modern Trends* (Ed. A. J. Barnes and W. J. Orville-Thomas), Elsevier, Amsterdam, New York, 1977, p.353.
393. S. F. A. Kettle, *Topics Curr. Chem.*, **71**, 111 (1977).
394. M. Bigorgne, *J. Organomet. Chem.*, **94**, 161 (1975).
395. S. F. A. Kettle and I. Paul, *Adv. Organomet. Chem.*, **10**, 199 (1972).
396. L. M. Haines and M. H. B. Stiddard, *Adv. Inorg. Chem. Radiochem.*, **12**, 53 (1969).
397. J. K. Burdett, *Coord. Chem. Rev.*, **27**, 1 (1978).
398. N. Sheppard and T. T. Nguyen, *Adv. Infrared Raman Spectrosc.*, **5**, 67 (1978).
399. W. P. Griffith, in *Comprehensive Inorganic Chemistry* (Ed. A. F. Trotman-Dickenson), Pergamon Press, Oxford, 1973, vol. 4, p. 105.

400. W. P. Griffith, *Coord. Chem. Rev.*, **17**, 177 (1975).
401. S. Jerome-Lerutte, *Struct. Bonding*, **10**, 153 (1972).
402. B. M. Chadwick and A. G. Sharpe, *Adv. Inorg. Chem. Radiochem.*, **8**, 84 (1966).
403. P. V. Yaneff, *Coord. Chem. Rev.*, **23**, 183 (1977).
404. G. R. Clark, K. Marsden, W. R. Roper, and L. J. Wright, *J. Am. Chem. Soc.*, **102**, 120 (1980).
405. *Spectroscopic Properties of Inorganic and Organometallic Compounds*, Specialis Periodical Reports, Chemical Society and Royal Society of Chemistry, London, 1968– continuing series.
406. E. Maslowsky, Jr., *J. Organomet. Chem.*, **70**, 153 (1974).
407. D. E. C. Corbridge, *Topics Phosphorus Chem.*, **6**, 235 (1969).
408. A. H. Norbury, *Adv. Inorg. Chem. Radiochem.*, **17**, 232 (1975).
409. R. A. Bailey, S. L. Kozak, T. W. Michelsen, and W. N. Mills, *Coord. Chem. Rev.*, **6**, 40 (1971).
410. B. N. Storhoff and H. C. Lewis, *Coord. Chem. Rev.*, **23**, 1 (1977).
411. D. C. Bradley, *Adv. Inorg. Chem. Radiochem.*, **15**, 259 (1972).
412. J. Catterick and P. Thornton, *Adv. Inorg. Chem. Radiochem.*, **20**, 291 (1972).
413. C. D. Gardner and B. Hughes, *Adv. Inorg. Chem. Radiochem.*, **17**, 2 (1975).
414. N. N. Greenwood, E. J. F. Ross and B. P. Straughan, *Index of Vibrational Spectra of Inorganic and Organometallic Compounds.* Butterworths, London, Vol. 1, (1935–60) 1972; Vol. 2 (1961–63), 1975; Vol. 3 (1964–66), 1977.
415. J. R. Ferraro, *Low Frequency Vibrations of Inorganic and Coordination Compounds*, Plenum, New York, 1971.
416. S. D. Ross, *Inorganic Infrared and Raman Spectroscopy*, McGraw-Hill, London, New York, 1972.
417. K. Nakamoto and P. J. McCarthy, *Spectroscopy and Structure of Metal Chelate Compounds*, Wiley, New York, 1968.
418. D. F. Shriver and C. B. Cooper, III, *Adv. Infrared Raman Spectrosc.*, **6**, 127 (1980).
419. E. Maslowsky, Jr., *Chem. Rev.*, **71**, 507 (1971).
420. M. J. Ware, in *Essays in Structural Chemistry* (Ed. A. J. Downs, D. A. Long, and L. A. K. Staveley), Macmillan, London, 1971, p. 404.
421. T. G. Spiro, *Prog. Inorg. Chem.*, **11**, 1 (1970).
422. K. L. Watters and W. M. Risen, Jr., *Inorg. Chim. Acta Rev.*, **3**, 129 (1969).
423. W. W. Simons (Ed.), *The Sadtler Handbook of I.r. Spectra*, Sadtler/Heyden, Philadelphia, London, 1978.
424. L. J. Bellamy, *The Infrared Spectra of Complex Moleucles*, 3rd ed., Chapman and Hall, London, 1975.
425. L. J. Bellamy, *Advances in Infrared Group Frequencies*, Chapman and Hall, London, 1975.
426. F. R. Dollish, W. G. Fateley, and F. F. Bentley, *Characteristic Raman Frequencies of Organic Compounds*, Wiley, New York, 1974.
427. G. Varsányi, *Assignments for Vibrational Spectra of Benzene Derivatives*, Hilger, London, 1974.
428. A. R. Katritsky (Ed.), *Physical Methods in Heterocyclic Chemistry*, Academic Press, New York, Vol. 2, 1963; Vol. 4, 1971.
429. A. L. Smith, *Spectrochim. Acta*, **16**, 87 (1960).
430. L. J. Bellamy, *The Infrared Spectra of Complex Molecules*, 3rd ed., Chapman and Hall, London, 1975, Chapter 20.
431. A. J. Cross and F. Glockling, *J. Organomet. Chem.*, **3**, 146 (1965).
432. L. C. Thomas, *Interpretation of the Infrared Spectra of Organophosphorus Compounds*,. Heyden, London, 1974.
433. J. K. Brown and K. J. Morgan, *Adv. Fluorine Chem.*, **4**, 253 (1965).
434. S. Patai (Ed.), *The Chemistry of the Functional Groups*, Wiley, London, New York.

The Chemistry of the Metal–Carbon Bond
Edited by F. R. Hartley and S. Patai

CHAPTER **21**

Multinuclear magnetic resonance methods in the study of organometallic compounds

JULIAN A. DAVIES
College of Arts and Sciences, Department of Chemistry, The University of Toledo, Toledo, Ohio 43606, USA

LIST OF SYMBOLS

A{X}	Nuclear magnetic resonance of isotope A with irradiation at or near the resonant frequency of X
Du	Term representing imbalance in the valence *d*-orbitals of a given atom
I	Nuclear spin quantum number
$^nJ(A, X)$	Coupling constant between nuclei A and X over *n* formal bonds
$^nK(A, X)$	Reduced coupling constant between nuclei A and X over *n* formal bonds [independent of $\gamma(A)$ and $\gamma(X)$]
N	Percentage natural abundance of a given isotope
Pu	Term representing imbalance in the valence *p*-orbitals of a given atom
Q	Nuclear quadrupole moment
R_X	Receptivity of nucleus X with respect to the proton
$\langle r^{-3}\rangle_{nx}$	Expectation value of the *n*th *x* orbital ($x = p, d$) of a given atom
T_1	Spin–lattice relaxation time
T_2	Spin–spin relaxation time
β	Bohr magneton
$\gamma(X)$	Gyromagnetic ratio of isotope X
$\delta(X)$	Chemical shift of isotope X in ppm (more positive shifts representing deshielding)
ΔE	Electronic excitation energy
Δ_R	Ring contribution to a chemical shift
η	Extent of Nuclear Overhauser Enhancement (NOE)
μ	Magnetic dipole moment
μ_0	Permeability of free space
μ_N	Nuclear magneton
$\Xi(X)$	Resonant frequency of isotope X, corrected to a field where the ^1H frequency of TMS is 100 MHz
σ	Shielding of a given nucleus
σ_d	Diamagnetic contribution to σ
σ_p	Paramagnetic contribution to σ
$\lvert\psi_{s(x)}(0)\rvert^2$	Magnitude of the valence *s*-orbital of X at the nucleus of X
ω	Linewidth of a signal at peak half-height

I. INTRODUCTION

A. N.m.r. in Organometallic Chemistry

Nuclear magnetic resonance (n.m.r.) techniques have been of great importance in the historical development of organometallic chemistry. Intense study of the ^1H n.m.r. spectra of organometallic compounds has led to an enormous increase in our knowledge of molecular structure in solution and molecular dynamics. Determinations of both rate and equilibrium constants by n.m.r. methods have also been of importance.

In recent years there have been two major developments in n.m.r. spectroscopy that have had particularly far-reaching effects. These developments concern the study of solids by high-resolution n.m.r.[1] and the study of nuclei other than the proton[2]. T date, solid-state high-resolution n.m.r. has only rarely been utilized in organometalli chemistry and studies of organometallic compounds using liquid crystal solvents[3] ar also relatively few. Accordingly, solid-state studies are considered in a section of thei own in this chapter and all other mention of n.m.r. refers to high-resolution measure ments of liquids or solutions. The other major development, n.m.r. measurements o nuclei other than the proton, has been of greater significance in organometalli chemistry. The advent of heteronuclear double resonance and more advanced multipl resonance techniques has been paralleled by the rapid development of very high fiel spectrometers with fully multinuclear capability. Reports of the n.m.r. of nuclei othe than the proton are now commonplace, although closer examination of the situatio shows that many areas are still in a state of infancy and that rapid development i likely. Undoubtedly, ^{1}H n.m.r. will always remain an important technique in the stud of organometallic compounds, but now it is not always the technique of choice and the future role of multinuclear studies in organometallic chemistry is becoming apparent.

In the chemistry of the metal—carbon bond, it is likely that the metal atom or the bonded carbon atom will be the centre of interest in a given reaction. In many cases, the nearest available proton for n.m.r. study may be two or three bonds away from the reaction centre and so may be in an environment which is little altered by the process under study. In order to move the n.m.r. 'probe' nearer to the region of interest in the molecule, it is necessary to examine the ^{13}C n.m.r. or the n.m.r. of the metal nucleus itself, if this is possible. The former method, most usually performed under conditions of broad-band proton decoupling, ^{13}C{^{1}H} n.m.r., has developed to an extent which almost rivals ^{1}H n.m.r. in popularity; for example, a recent research paper[4] routinely described the ^{13}C{^{1}H} n.m.r. of about 200 carbonyl complexes of nickel, chromium, and molybdenum. The alternative of examining the n.m.r. of the metal nucleus itself, can either be a simple matter or a research project in its own right, depending upon the metal nucleus in question. Thus, although almost all nuclei with $I = \frac{1}{2}$ have been investigated by n.m.r. methods, little of any chemical significance has been reported for several of them. Examination of Tables 1–3, which detail some important nuclear

TABLE 1. Nuclear properties of the $I = \frac{1}{2}$ non-metals[a]

Isotope	N (%)	Magnetic moment, μ/μ_N	Gyromagnetic ratio, γ(X) (10^7 rad T^{-1} s^{-1})	R_X[b]
^{1}H	99.985	4.8371	26.7510	1.000
^{3}H	—	5.1594	28.5335	—
^{3}He	1.3×10^{-4}	−3.6848	−20.378	5.75×10^{-7}
^{13}C	1.108	1.2162	6.7263	1.76×10^{-4}
^{15}N	0.37	−0.4901	−2.7107	3.85×10^{-6}
^{19}F	100	4.5506	25.1665	0.8328
^{29}Si	4.70	−0.9609	−5.3141	3.69×10^{-4}
^{31}P	100	1.9581	10.829	0.0663
^{77}Se	7.58	0.9223	5.101	5.26×10^{-4}
^{123}Te	0.87	—	−7.011	1.57×10^{-4}
^{125}Te	6.99	−1.528	−8.453	2.21×10^{-3}
^{129}Xe	26.44	−1.3380	−7.3995	5.60×10^{-3}

[a]Excluding radioisotopes (apart from ^{3}H). Data from reference 2.

[b]Receptivity relative to ^{1}H, given by $R_X = \left|\frac{\gamma(X)^3}{\gamma(^1H)^3}\right| \cdot \frac{N(X)I(X)\{I(X)+1\}}{N(^1H)I(^1H)\{I(^1H)+1\}}$; see reference 2.

TABLE 2. Nuclear properties of the $I = \frac{1}{2}$ main group metals[a]

Isotope	N (%)	Magnetic moment, μ/μ_N	Gyromagnetic ratio, $\gamma(X)$ (10^7 rad T^{-1} s^{-1})	R_X[b]
^{115}Sn	0.35	—	−8.7475	1.22×10^{-4}
^{117}Sn	7.61	—	−9.5301	3.44×10^{-3}
^{119}Sn	8.58	−1.8029	−9.9707	4.44×10^{-3}
^{203}Tl	29.50	—	15.288	5.51×10^{-2}
^{205}Tl	70.50	2.7914	15.438	0.1355
^{207}Pb	22.6	1.0120	5.5968	2.07×10^{-3}

[a]As [a] in Table 1.
[b]As [b] in Table 1.

TABLE 3. Nuclear properties of the $I = \frac{1}{2}$ transition metals[a]

Isotope	N (%)	Magnetic moment, μ/μ_N	Gyromagnetic ratio, $\gamma(X)$ (10^7 rad T^{-1} s^{-1})	R_X[b]
^{57}Fe	2.19	0.1563	0.8644	7.39×10^{-7}
^{89}Y	100	−0.2370	−1.3106	1.18×10^{-4}
^{103}Rh	100	−0.1522	−0.8420	3.12×10^{-5}
^{107}Ag	51.82	—	−1.0828	3.44×10^{-5}
^{109}Ag	48.18	−0.2251	−1.2449	4.86×10^{-5}
^{111}Cd	12.75	—	−5.6720	1.22×10^{-3}
^{113}Cd	12.26	−1.0728	−5.9330	1.34×10^{-3}
^{169}Tm	100	−0.400	−2.21	5.66×10^{-4}
^{171}Yb	14.31	0.8520	4.7117	7.81×10^{-4}
^{183}W	14.40	0.2013	1.1131	1.04×10^{-5}
^{187}Os	1.64	0.1114	0.6161	2.00×10^{-7}
^{195}Pt	33.8	1.0398	5.7505	3.36×10^{-3}
^{199}Hg	16.84	0.8623	4.7690	9.54×10^{-4}

[a]As in [a] Table 1. Transition metals taken to include the *f*-block.
[b]As in [b] Table 1.

properties of the $I = \frac{1}{2}$ elements, shows that of the 24 elements, 3 are main group metals and 11 are transition metals (including the *f*-block). Even more fortuitous for the organometallic chemist is the observation that 8 out of the remaining 10 non-metals are common donors as either neutral or negative supporting ligands in organometallic compounds. Of these 24 $I = \frac{1}{2}$ elements, neither thulium (^{169}Tm) nor ytterbium (^{171}Yb) has been studied, and reports of ^{3}He n.m.r. are as yet of little significance in organometallic chemistry. Perhaps most surprising is the scarcity of reports concerning ^{103}Rh n.m.r., of which no direct observation had been made when the standard text on multinuclear magnetic resonance[2] was written.

Thus, there are 14 metal nuclei with $I = \frac{1}{2}$ which have been, or potentially could be, investigated directly by n.m.r. To examine the n.m.r. of the metal centre in a compound not containing one of these elements, one has then to deal with the problems of obtaining n.m.r. spectra from quadrupolar ($I > \frac{1}{2}$) nuclei. Tables 4–6 detail some important nuclear properties of these elements. The overriding concern with n.m.r. of $I > \frac{1}{2}$ nuclei is the problem of line broadening associated with the quadrupolar moment (discussed in Section II.A.2) and this factor really dominates both natural abundance and receptivity as the major property that determines suitability for n.m.r. study.

TABLE 4. Nuclear properties of the $I > \frac{1}{2}$ non-metals[a]

Isotope	I	N (%)	Magnetic moment, μ/μ_N	Gyromagnetic ratio, $\gamma(X)$ (10^7 rad T^{-1} s^{-1})	Quadrupole moment[b], Q(10^{-28} m^2)	R_X^c
^{2}H	1	0.015	1.2125	4.1064	2.73×10^{-3}	1.45×10^{-6}
^{10}B	3	19.58	2.0793	2.8748	7.4×10^{-2}	3.89×10^{-3}
^{11}B	$\frac{3}{2}$	80.42	3.4702	8.5827	3.55×10^{-2}	0.133
^{14}N	1	99.63	0.5706	1.9324	1.6×10^{-2}	1.00×10^{-3}
^{17}O	$\frac{5}{2}$	0.037	−2.2398	−3.6266	-2.6×10^{-2}	1.08×10^{-5}
^{21}Ne	$\frac{3}{2}$	0.257	−0.8539	−2.1118	—	6.32×10^{-6}
^{33}S	$\frac{3}{2}$	0.76	0.8296	2.0517	-6.4×10^{-2}	1.71×10^{-5}
^{35}Cl	$\frac{3}{2}$	75.53	1.0598	2.6212	-7.89×10^{-2}	3.55×10^{-3}
^{37}Cl	$\frac{3}{2}$	24.47	0.8821	2.182	-6.21×10^{-2}	6.64×10^{-4}
^{73}Ge	$\frac{9}{2}$	7.76	−0.9693	−0.9332	−0.2	1.09×10^{-4}
^{75}As	$\frac{3}{2}$	100	1.8524	4.5816	0.3	2.51×10^{-2}
^{79}Br	$\frac{3}{2}$	50.54	2.7098	6.7021	0.33	3.97×10^{-2}
^{81}Br	$\frac{3}{2}$	49.46	2.9210	7.2245	0.28	4.87×10^{-2}
^{83}Kr	$\frac{9}{2}$	11.55	−1.069	−1.029	0.15	2.17×10^{-4}
^{121}Sb	$\frac{5}{2}$	57.25	3.9537	6.4016	−0.5	9.15×10^{-2}
^{123}Sb	$\frac{7}{2}$	42.75	2.8726	3.4668	−0.7	1.95×10^{-2}
^{127}I	$\frac{5}{2}$	100	3.3056	5.3521	−0.69	9.34×10^{-2}
^{131}Xe	$\frac{3}{2}$	21.18	0.8869	2.1935	−0.12	5.84×10^{-4}

[a] All n.m.r.-significant stable nuclei, including radioisotopes with half-life $>10^{10}$ years. Data from reference 2.
[b] Reported values of Q may involve significant errors, see reference 2.
[c] As [b] in Table 1.

TABLE 5. Nuclear properties of the $I > \frac{1}{2}$ main group metals[a]

Isotope	I	N (%)	Magnetic moment, μ/μ_N	Gyromagnetic ratio, $\gamma(X)$ (10^7 rad T^{-1} s^{-1})	Quadrupole moment[b], $Q(10^{-28}$ $m^2)$	R_x[c]
^{6}Li	1	7.42	1.1624	3.9366	-8×10^{-4}	6.31×10^{-4}
^{7}Li	$\frac{3}{2}$	92.58	4.2035	10.396	-4.5×10^{-2}	0.272
^{9}Be	$\frac{3}{2}$	100	−1.5200	−3.7594	5.2×10^{-2}	1.39×10^{-2}
^{23}Na	$\frac{3}{2}$	100	2.8610	7.0760	0.12	9.25×10^{-2}
^{25}Mg	$\frac{5}{2}$	10.13	−1.0110	−1.6370	0.22	2.71×10^{-4}
^{27}Al	$\frac{5}{2}$	100	4.3051	6.9706	0.149	0.206
^{39}K	$\frac{3}{2}$	93.1	0.5047	1.2484	5.5×10^{-2}	4.73×10^{-4}
^{41}K	$\frac{3}{2}$	6.88	0.2770	0.6852	6.7×10^{-2}	5.78×10^{-6}
^{43}Ca	$\frac{7}{2}$	0.145	−1.4914	−1.7999	0.2 ± 0.1	9.72×10^{-6}
^{69}Ga	$\frac{3}{2}$	60.4	2.596	6.421	0.178	4.18×10^{-2}
^{71}Ga	$\frac{3}{2}$	39.6	3.2984	8.1578	0.112	5.62×10^{-2}
^{85}Rb	$\frac{5}{2}$	72.15	1.5952	2.5829	0.247	7.58×10^{-3}
^{87}Rb	$\frac{3}{2}$	27.85	3.5391	8.7532	0.12	4.88×10^{-2}
^{87}Sr	$\frac{9}{2}$	7.02	−1.2043	−1.1594	0.36	1.89×10^{-4}
^{113}In	$\frac{9}{2}$	4.28	6.0761	5.8496	1.14	1.48×10^{-2}
^{115}In	$\frac{9}{2}$	95.72	6.0892	5.8622	1.16	0.332
^{113}Cs	$\frac{7}{2}$	100	2.9076	3.5089	-3×10^{-3}	4.74×10^{-2}
^{135}Ba	$\frac{3}{2}$	6.59	1.0745	2.6575	0.18	3.23×10^{-4}
^{137}Ba	$\frac{3}{2}$	11.32	1.2020	2.9729	0.28	7.77×10^{-4}
^{209}Bi	$\frac{9}{2}$	100	4.4652	4.2988	−0.4	0.137

[a] As [a] in Table 4.
[b] As [b] in Table 4.
[c] As [b] in Table 1.

TABLE 6. Nuclear properties of the $I > \frac{1}{2}$ transition metals[a]

Isotope	I	N (%)	Magnetic moment, μ/μ_N	Gyromagnetic ratio, $\gamma(X)$ (10^7 rad T^{-1} s^{-1})	Quadrupole moment[b], Q(10^{-28} m^2)	R_X[c]
^{45}Sc	$\frac{7}{2}$	100	5.3851	6.4989	−0.22	0.301
^{47}Ti	$\frac{5}{2}$	7.28	−0.9313	−1.5079	—	1.52×10^{-4}
^{49}Ti	$\frac{7}{2}$	5.51	−1.2498	−1.5083	—	2.07×10^{-4}
^{51}V	$\frac{7}{2}$	99.76	5.827	7.032	0.3	0.381
^{53}Cr	$\frac{3}{2}$	9.55	−0.6113	−1.5120	—	8.62×10^{-5}
^{55}Mn	$\frac{5}{2}$	100	4.075	6.598	0.55	0.175
^{59}Co	$\frac{7}{2}$	100	5.2344	6.3171	0.40	0.277
^{63}Cu	$\frac{3}{2}$	69.09	2.8668	7.0904	−0.16	6.43×10^{-2}
^{65}Cu	$\frac{3}{2}$	30.91	3.0711	7.5958	−0.15	3.54×10^{-2}
^{67}Zn	$\frac{5}{2}$	4.11	1.0330	1.6726	0.15	1.17×10^{-4}
^{93}Nb	$\frac{9}{2}$	100	6.7919	6.5387	−0.2	0.482
^{95}Mo	$\frac{5}{2}$	15.72	1.076	1.743	0.12	5.07×10^{-4}
^{97}Mo	$\frac{5}{2}$	9.46	−1.099	−1.780	1.1	3.25×10^{-4}
^{139}La	$\frac{7}{2}$	99.911	3.1312	3.7789	0.21	5.92×10^{-2}
^{181}Ta	$\frac{7}{2}$	99.988	2.653	3.202	3	3.60×10^{-2}
^{185}Re	$\frac{5}{2}$	37.07	3.7197	6.0227	2.8	4.94×10^{-2}
^{187}Re	$\frac{5}{2}$	62.93	3.7578	6.0844	2.6	8.64×10^{-2}
^{189}Os	$\frac{3}{2}$	16.1	0.8392	2.0756	0.8	3.76×10^{-4}
^{201}Hg	$\frac{3}{2}$	13.22	−0.7138	−1.7655	0.50	1.90×10^{-4}

[a] As [a] in Table 4. Transition metals taken to include the *f*-block.
[b] As [b] in Table 4.
[c] As [b] in Table 1.

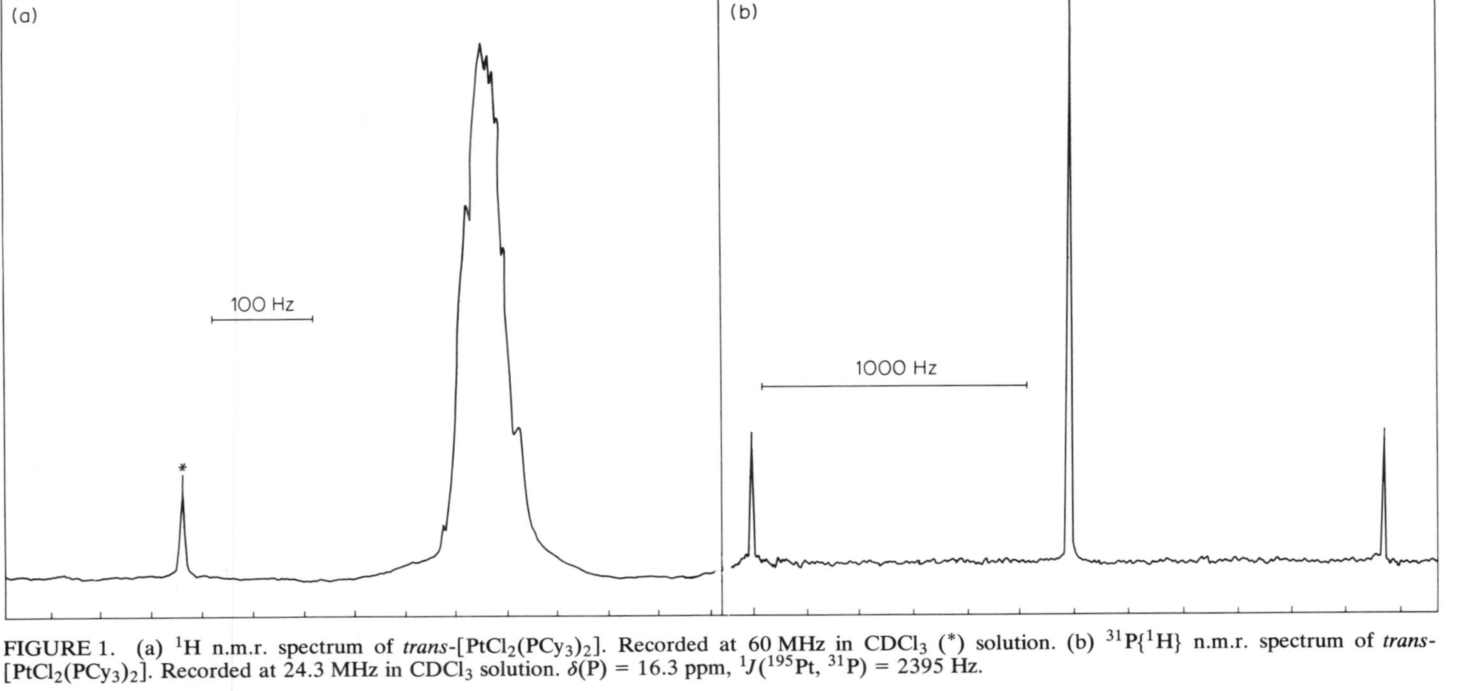

FIGURE 1. (a) ^{1}H n.m.r. spectrum of *trans*-$[PtCl_2(PCy_3)_2]$. Recorded at 60 MHz in $CDCl_3$ (*) solution. (b) $^{31}P\{^{1}H\}$ n.m.r. spectrum of *trans*-$[PtCl_2(PCy_3)_2]$. Recorded at 24.3 MHz in $CDCl_3$ solution. $\delta(P) = 16.3$ ppm, $^{1}J(^{195}Pt, ^{31}P) = 2395$ Hz.

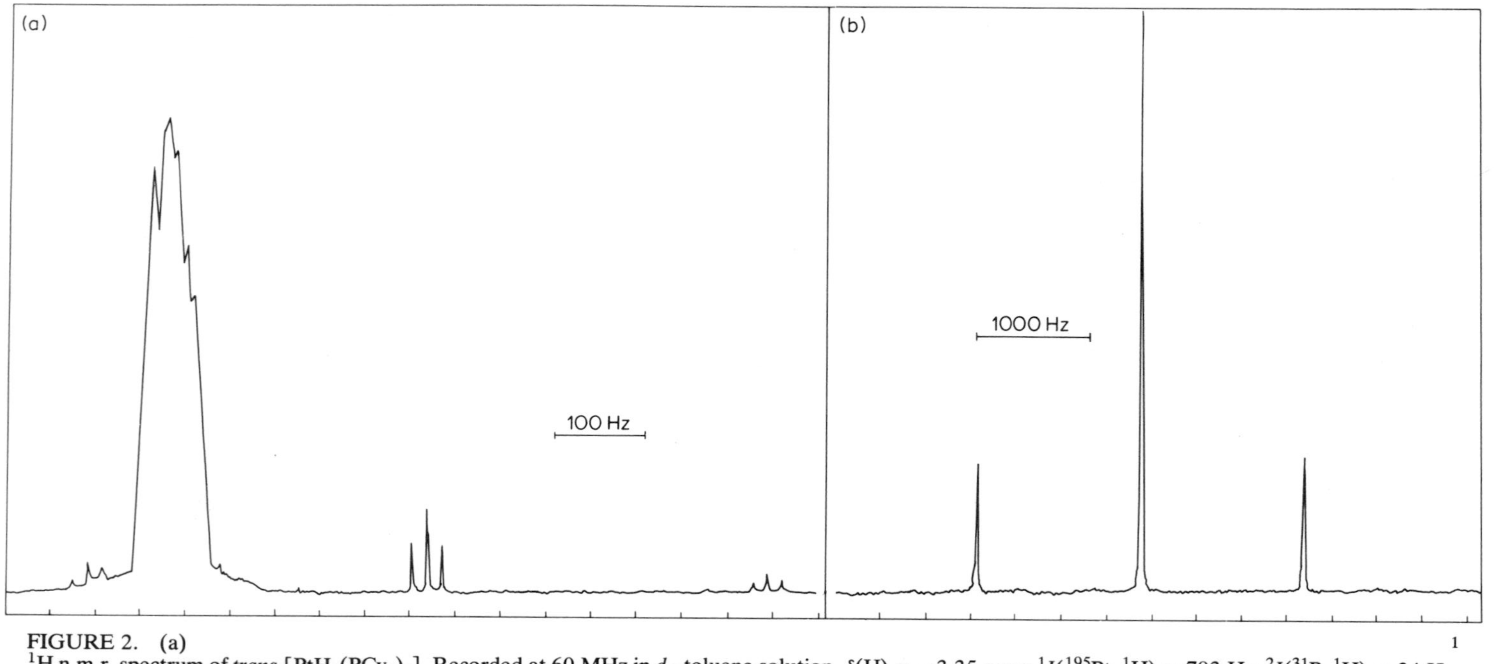

FIGURE 2. (a)
^{1}H n.m.r. spectrum of *trans*-$[PtH_2(PCy_3)_2]$. Recorded at 60 MHz in d_8-toluene solution. $\delta(H) = -3.35$ ppm; $^{1}J(^{195}Pt, ^{1}H) = 793$ Hz, $^{2}J(^{31}P, ^{1}H) = 34$ Hz.
(b) $^{31}P\{^{1}H\}$ n.m.r. spectrum of *trans*-$[PtH_2(PCy_3)_2]$. Recorded at 24.3 MHz in C_6H_6/C_6D_6 solution. $\delta(P) = 51.8$ ppm, $^{1}J(^{195}Pt, ^{31}P) = 2871$ Hz.

Depending upon the metal nucleus involved, it may well be feasible to study the n.m.r. of the two nuclei at the ends of a given metal—carbon bond. Further information still can be acquired by examining the n.m.r. of any donor atoms present as supporting ligands at the metal centre or any heteroatoms bonded to the carbon donor. The obvious extension of this approach is the study of organometallic compounds in which every atom has at least one n.m.r.-active isotope; structural characterization by this total-n.m.r. approach has proved to be highly successful and the expression 'as good as a crystal structure determination' justifiably applied.

A major advantage which arises from the ability to pick the most appropriate nucleus for study in a given situation is the simplicity of the spectra which can often result. For example, the ^{1}H n.m.r. spectrum of *trans*-$[PtCl_2(PCy_3)_2]$ (Figure 1a) is broadened and featureless owing to the complex coupling of the cyclohexyl protons to each other, ^{31}P, and ^{195}Pt and because of the very small chemical shift range of the various protons. The $^{31}P\{^{1}H\}$ n.m.r. spectrum (Figure 1b) is extremely simple, however, and allows the parameters $\delta(^{31}P)$ and $^{1}J(^{195}Pt, ^{31}P)$ to be obtained directly. Nonetheless, the power of ^{1}H n.m.r. spectroscopy as a tool in organometallic and coordination chemistry should not be underestimated; thus the ^{1}H n.m.r. spectrum of *trans*-$[PtH_2(PCy_3)_2]$ (Figure 2a) is definitive and allows the parameters associated with the hydridic centres, $\delta(^{1}H)$, $^{1}J(^{195}Pt, ^{1}H)$ and $^{2}J(^{31}P, ^{1}H)$ to be measured. The corresponding $^{31}P\{^{1}H\}$ n.m.r. spectrum (Figure 2b) is necessarily devoid of information concerning the hydridic centres and so only $\delta(^{31}P)$ and $^{1}J(^{195}Pt, ^{31}P)$ can be measured.

B. N.m.r. Literature

In this chapter, the application of multinuclear magnetic resonance to the study of organometallic and related coordination compounds is discussed, with particular emphasis on the detection and characterization (i.e. molecular structure determination) of organometallic compounds in solution. Solid-state studies are dealt with briefly in a section of their own. Theoretical considerations have been minimized and are presented only when necessary to understand the experimental results discussed. Many texts are available[5–8] which deal fully with the theory of the n.m.r. experiment and reference to these is made where appropriate. No attempt has been made to provide exhaustive lists of data and references on n.m.r. parameters of organometallic compounds; rather, a guide to the use of multinuclear magnetic resonance in organometallic chemistry and the literature surrounding the topic is presented to enable the scope and applicability of the method to be evaluated. Several review series are available which cover most aspects of n.m.r.[9–12], including the n.m.r. of inorganic and organometallic compounds[13]. The more recent attempts to survey the literature of n.m.r. in inorganic and organometallic chemistry demonstrate the vastness of the subject; thus a review of recent advances (up to late-1977) had nearly 1700 references[14].

The recent expansion in multinuclear magnetic resonance has led to the appearance of several general reviews dealing with the n.m.r. of nuclei other than the proton[15–22]. Reviews of the n.m.r. of the less-sensitive $I = \frac{1}{2}$ nuclei[23] and the less-common quadrupolar nuclei[24] have also appeared. The subject of multinuclear magnetic resonance gained a much needed impetus with the appearance of a text entitled *NMR and the Periodic Table*[2], now regarded as the standard work in this field. The single volume covers the n.m.r. of the less-common nuclei (i.e. excluding ^{1}H, ^{10}B, ^{11}B, ^{13}C, ^{14}N, ^{15}N, ^{19}F, and ^{31}P) in depth and the reader is referred to this work for details of n.m.r. studies which do not bear directly on organometallic chemistry, particularly work relating to nuclear relaxation, which as yet has little chemical significance in studies of metal—carbon compounds. The more common nuclei mentioned above have all been the subject of texts or review articles and these are detailed in the appropriate sections.

Mention is made of the various methods of direct and indirect observation of n.m.r spectra (see Sections III.A–III.C); spin–echo and two-dimensional spectroscopy are not included as few significant studies of organometallic compounds have been performed by these rather specialized techniques[25].

A text is available which deals with n.m.r. spectroscopy using liquid crystal solvents, including details of studies of organometallic compounds[3], and so this topic is not covered here.

II. MAGNETIC PROPERTIES OF NUCLEI

Almost all nuclei suffer from one or more disadvantages compared with the proton in their susceptibility to n.m.r. measurement. Low natural abundances, low gyromagnetic ratios and quadrupolar moments are frequently encountered when the n.m.r. of heteronuclei is measured. It is noteworthy, however, that although the low natural abundances of ^{13}C (1.108%) and ^{17}O (0.037%) cause problems in, for example, the study of transition metal carbonyls by ^{13}C and ^{17}O n.m.r., they are really a blessing as the ^{1}H n.m.r. of even a relatively simple organic molecule would otherwise be rendered incomprehensible by extreme coupling situations. The problem of low gyromagnetic ratios is exemplified by considering the ^{103}Rh nucleus; although $N = 100\%$ and $I = \frac{1}{2}$, the low value of $\gamma(^{103}Rh)$ and the correspondingly low resonant frequency had prevented any direct observation of its n.m.r. in solution until very recently. In fact, the resonant frequency of this nucleus is so low as to be below the range attainable by many of the earlier multinuclear high-field spectrometers. Nuclei with $I > \frac{1}{2}$ possess a quadrupolar moment which can adversely affect their relaxation behaviour in low symmetry environments (see Section II.A.2) and this can seriously affect the quality of the derived n.m.r. spectra. Fortunately, one of the most studied quadrupolar nuclei, ^{14}N ($I = 1$, $N = 99.63\%$) also has a congener with $I = \frac{1}{2}$, ^{15}N ($N = 0.37\%$). Accordingly, a choice between the problems of the nuclear quadrupole and the problems of low natural abundance can be made in a given situation. We are not so fortunate with many other quadrupolar nuclei as few such relationships exist (compare Tables 1–3 and 4–6).

The factors affecting nuclear relaxation processes, nuclear shielding, and coupling between magnetically inequivalent nuclei effectively define the final appearance of an n.m.r. spectrum. Relaxation processes tend to define signal shape and intensity whilst shielding and coupling effects relate to the number and positions of the observed signals. Some understanding of these effects, at least qualitatively, is necessary if the resulting spectra are to be meaningfully interpreted.

A. Relaxation Processes and Nuclear Overhauser Enhancement

As a molecule moves randomly in solution, fluctuations in the magnetic field at a given nucleus, X, can provide a pathway for the relaxation of a second nucleus, A, although the coupling energy averages to zero. Accordingly, irradiation at or near the resonant frequency of X (equivalent to proton decoupling where X = ^{1}H) can affect the observed intensity of the transition associated with the nucleus A. Interactions of this type, described as magnetic dipole–dipole interactions, can therefore allow some signal enhancement to be achieved in comparison with the observed spectrum of A without decoupling of X. The maximum possible enhancement, known as Nuclear

Overhauser Enhancement[26–29] (NOE), given that the relaxation of nucleus A occurs entirely by this route, is shown in equation 1,

$$\eta = \gamma(\mathrm{X})/2\gamma(\mathrm{A}) \tag{1}$$

where A = observed nucleus and X = decoupled nucleus. It therefore follows that the ratio of decoupled to undecoupled intensities will be $1 + \eta{:}1$ and thus, if $\gamma(\mathrm{A})$ is positive, the decoupling process will lead to signal enhancement. Some care is needed in considering such values as the Nuclear Overhauser Enhancement (NOE) has a value η, whereas the observable effect, known as the Nuclear Overhauser Effect (also abbreviated to NOE) has the value $1 + \eta$.

Unfortunately, not all nuclei have positive gyromagnetic ratios and in some cases the value of $\gamma(\mathrm{A})$ in equation 1 is negative; accordingly the signal 'enhancement' will also be negative. If other relaxation mechanisms (discussed below) detract from the magnetic dipole–dipole interactions then a case can be envisaged where partial negative enhancement occurs and nulls a signal to zero intensity. In $^{29}\mathrm{Si}\{^{1}\mathrm{H}\}$ n.m.r., resonances of reduced intensity, negative resonances, and completely nulled resonances have all been reported[30], depending upon the extent of mixing of the various relaxation processes.

There are five major relaxation processes which are generally operative, as follows.

1. Magnetic dipole–dipole interactions

This mechanism has been mentioned above in connection with the NOE. It should be emphasized that the two nuclear spins concerned need not be part of the same molecule for this mechanism to operate and both intramolecular and intermolecular interactions are possible. This mechanism is probably the most important process for relaxation of $I = \frac{1}{2}$ nuclei (except possibly at high temperature, see Section II.A.4) and an example which clearly demonstrates the necessity of bearing relaxation processes in mind when studying the n.m.r. spectra of organometallic compounds is the $^{13}\mathrm{C}\{^{1}\mathrm{H}\}$ n.m.r. spectra of the platinum(II) complexes **1** and **2**.

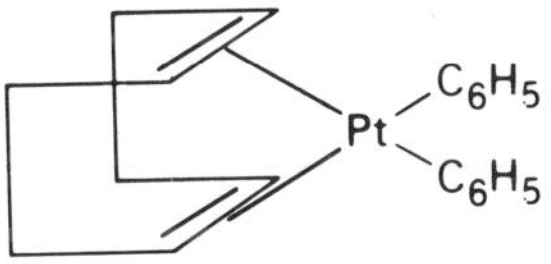

Pt C_6H_5 C_6H_5 **(1)**

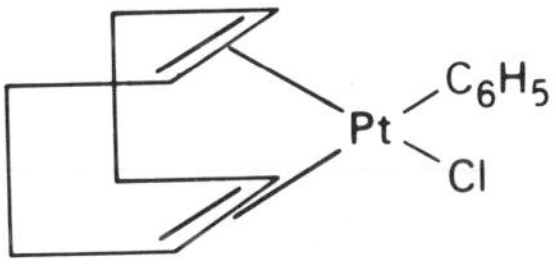

Pt C_6H_5 Cl **(2)**

As the $^{13}\mathrm{C}$ nucleus has a positive gyromagnetic ratio, proton decoupling would be expected to enhance the signal intensity in the spectrum, provided that the magnetic dipole–dipole interaction was a major relaxation route. In fact, resonances attributable to the olefinic and paraffinic carbon atoms of the cycloocta-1,5-diene ligand are readily observable, as are resonances for five of the six carbon atoms in each phenyl ring. The resonance associated with the phenyl carbon atom bonded directly to platinum is either not observed, or is very weak, depending upon the conditions[31,32]. The proposed explanation[31] for this observation is that the carbon atom bonded directly to platinum has no protons attached to it and so experiences no signal enhancement by NOE. In fact, it is entirely possible that the phenyl carbon bonded to platinum does experience a full NOE from the protons on the adjacent carbons but, as the rate of dipole–dipole relaxation is inversely proportional to the sixth power of the distance between the interacting nuclei, this unique carbon atom will possess a very long relaxation time and so will be difficult to observe under normal operating conditions because of spin-saturation problems.

Dipole–dipole interactions can also result from the presence of paramagnetic species in solution. Particularly important in this respect is the presence or absence of dissolved molecular oxygen. This trace paramagnetic is especially effective at enhancing relaxation of exposed nuclei, such as ^{1}H, but rather less so for nuclei such as ^{13}C and ^{31}P which tend to be 'buried' in the molecule. Long relaxation times of nuclei such as ^{13}C can often be reduced by the addition of a paramagnetic species capable of enhancing relaxation via intermolecular electron–nuclear interactions. A typical relaxation agent employed in organometallic chemistry is $[Cr(acac)_3]$. Of course, relaxation via interaction with a relaxation agent detracts from relaxation via dipole–dipole interactions with protons and so decreases the observed NOE. This is actually a useful fact as it can be employed to help remove the nulling effects previously mentioned for nuclei with negative gyromagnetic ratios.

The effects of added paramagnetics on n.m.r. spectra are actually extremely complex as shifts in resonance positions occur as well as relaxation enhancement. The former phenomena result largely from contact interactions and the species employed as shift reagents, notably rare earth ionic complexes, have a relaxation rate determined largely by the longitudinal relaxation of the unpaired electron and so are less effective at inducing dipole–dipole interactions. Accordingly, careful choice of the paramagnetics employed allows some division between relaxation agents and shift reagents to be made (see Sections V.B and V.C).

2. *Quadrupolar interactions*

Nuclei with $I > \frac{1}{2}$ have an electric quadrupole moment and so relaxation may be induced by fluctuations in the electric field at the nucleus caused by molecular motion in solution. Relaxation by this route is extremely efficient and thus is the preferred route for most quadrupolar nuclei. The symmetry of the electric field gradient at the nucleus relates to this effect and so in compounds where the ligand field is highly asymmetric, the central quadrupolar nucleus will relax predominantly via this interaction. In compounds where the symmetry about the nucleus is high (generally above tetrahedral) the electronic distribution is suitably symmetric to maintain the electric field in a pseudo-cylindrical geometry and in such cases quadrupolar interactions need not be the major relaxation route.

The ^{55}Mn nucleus has a relatively large quadrupolar moment ($Q = 0.55 \times 10^{-28}$ cm^{2}) and the effects of symmetry on the extent of quadrupolar relaxation, and hence the linewidth of the ^{55}Mn resonance, are clearly observed. Thus, the permanganate ion[33] (tetrahedral symmetry) has a linewidth at peak half-height of <10 Hz, the axially symmetric $[Mn_2(CO)_{10}]$ has a linewidth of 48 Hz[34], whilst the asymmetric field gradient produced in $[Mn_2(CO)_8(PPh_3)_2]$[34] results in a linewidth of 10.66 kHz. The problems associated with observing spin–spin coupling with such wide resonances can obviously be enormous.

3. *Scalar coupling interactions*

When two nuclei, A and B, are magnetically inequivalent then the spin–spin coupling between them can alter the field experienced by each nucleus. If the coupling between A and B is affected by chemical exchange or internal motion of the molecule then this effect can lead to relaxation, known as scalar coupling of the first kind. Alternatively, if nucleus B relaxes very much faster than nucleus A (say it is quadrupolar such as ^{11}B or ^{14}N) then the coupling between the two nuclei enables nucleus A to relax by this interaction, known as scalar coupling of the second kind. The importance of this type of relaxation is exemplified by the broad resonances observed in the ^{1}H n.m.r. of molecules containing quadrupolar nuclei. This line-broadening can be a problem if it is necessary to observe small couplings and can frequently be eliminated by heteronuclear double resonance techniques (see Section III.B).

Scalar coupling of the second kind may intuitively seem a reasonable relaxation route for ^{13}C in complexes such as transition metal cyanides ($[Fe(CN)_6]^{4-}$, $[Ni(CN)_4]^{2-}$, etc.) via the coupling interaction between ^{13}C and ^{14}N, but this point has proved to be controversial and is discussed further in Section VI.G.I.

4. Spin rotation interactions

The n.m.r. of $I = \frac{1}{2}$ nuclei in samples examined in the gas phase or as high-temperature (low-viscosity) liquids or solutions can demonstrate a decrease in the observed NOE from that expected if dipole–dipole interactions dominate the relaxation pathway. This is because the rapid rotation of molecules, or parts of molecules, results in fluctuating motion of the molecular magnetic moment which affects the field at the nucleus. At more ambient temperatures, freely rotating parts of molecules may also demonstrate this behaviour, which provides a path for relaxation, detracting from dipole–dipole interactions and so reducing the NOE.

In the $^{13}C\{^{1}H\}$ n.m.r. spectra of the complexes [PtMeCl(cod)] and $[PtMe_2(cod)]$ (i.e. the methyl analogues of **1** and **2**) the resonances of the methyl carbon atoms are of much reduced intensity compared with the resonances of the cycloocta-1,5-diene ligands. The explanation for this obviously cannot be the same as that proposed[31] for the phenylplatinum(II) complexes as here the carbon atoms bonded to platinum also have three protons bonded to them to facilitate dipole–dipole interactions and lead to an observable NOE. The explanation proposed[32] is that rapid rotation of the methyl groups about the Pt—C bond allows relaxation via spin rotation interactions, hence detracting from the magnetic dipole–dipole interaction and decreasing the observed NOE.

In general, dipole–dipole interactions dominate in the relaxation of a given heteronucleus in molecules where protons are also present unless a proton-containing group with a low barrier to rotation (such as a methyl group) is involved. In organometallic systems, spin rotation is most likely to be encountered in cases such as the ^{13}C n.m.r. of homoleptic metal carbonyls (such as $[Ni(CO)_4]$ and $[Fe(CO)_5]$) where the more favoured relaxation route is not available to the carbonyl carbon atoms.

5. Chemical shift anisotropy

Local variations in the magnetic field at the nucleus, which are sensitive to both temperature and the magnitude of the applied field, may open a new pathway for nuclear relaxation. The magnitude of this effect, the chemical shift anisotropy, is related to temperature and is proportional to the square of the applied field[35]. Accordingly, the effect is evident at low temperature and high applied field when the local variations in the field at the nucleus are most pronounced. Clear examples of this phenomenon are few, but a relevant example[36] is the ^{195}Pt n.m.r. spectrum of $[Pt_2(\mu\text{-}Cl)_2Cl_2(PBu^n_3)_2]$ which shows a clearly observable coupling, $^{2}J(^{195}Pt, ^{195}Pt) = 190$ Hz, when recorded under an applied field of 18.8 kG. Increasing the field to 58.7 kG results in an apparent loss of magnetic character of the ^{195}Pt nuclei and the two bond coupling collapses. This has been attributed to a change in relaxation mechanism as the applied field is increased until chemical shift anisotropy, characterized by long relaxation times, is dominant.

Care must be taken when comparing coupling situations for series of compounds where the spectra have been obtained at different applied fields. Thus, observable couplings, $^{n}J(^{195}Pt, ^{1}H)$, have been reported for some platinum(II) nucleoside and nucleotide complexes[37,38] (recorded on a 60-MHz instrument) whilst some very similar series of compounds (whose spectra were recorded on a 100-MHz instrument) were reported to show no such coupling[39–41]. Accordingly, great care must be taken when comparing coupling data from diverse sources, especially that involving heavy metal nuclei, such as ^{195}Pt or ^{207}Pb[36,42], where the effect is most often encountered in organometallic systems.

In summary, there are five major mechanisms which can contribute to nuclear relaxation. The extent to which each is involved defines both the extent of NOE which will be observed and the linewidth of the resonance. Some consideration of the mechanisms involved in nuclear relaxation can be a great asset in spectral analysis as the lack of certain resonances and expected couplings may (sometimes) be rationalized.

B. Nuclear Shielding and the Chemical Shift[43–45]

Before discussing the origins and variations in nuclear shielding and chemical shift in organometallic compounds, it is necessary to define exactly both sign convention and measurement standards.

1. *Sign conventions*[46]

The chemical shift (δ) of a resonance could be taken to imply either a change in resonant frequency or a change in nuclear shielding. The relationship between these parameters (the resonant frequency increases as the shielding decreases) makes a phrase such as '. . . the increases in chemical shift . . .' meaningless unless a sign convention is defined. The presently accepted convention is to refer chemical shifts, measured in parts per million (ppm), to changes in resonant frequency. As most modern spectrometers operate at fixed field and variable frequency this is obviously sensible as the expressions 'high field, low field' are effectively obsolete, and probably the best way to describe chemical shifts is by using a 'shielding, deshielding' nomenclature, although the expressions 'low frequency, high frequency' may also be acceptable in some circumstances.

Unfortunately, literature reports using both sign conventions are plentiful and frequently no attempt is made to define the convention at all. In this chapter, more positive values of the chemical shift represent deshielding and an attempt has been made to adapt all cited literature to this convention.

2. *Standards*

There is even less agreement concerning standards in n.m.r. circles than there is over the problem of sign convention. Currently, measurements for each nucleus are referenced to a given standard which is assigned zero chemical shift. Of course, not all systems lend themselves to the presence of an internal standard and so external standards are frequently used. The solvent medium employed for the standard can affect its chemical shift[47], as can the temperature, and so bulk susceptibility corrections should be made. In fact, such corrections are rarely applied and reports stating whether or not the correction has been made are extremely scarce. Unfortunately, as some standards are insoluble in certain solvents, some nuclei are referred to more than one shift standard [for example, tetramethylsilane (TMS) is insoluble in water]. Considering the choice of compound for a shift standard, several factors need to be evaluated, including solubility, stability towards decomposition, and extent of variation of chemical shift with temperature. Ideally, the standard should have a resonance which falls at the high shielding end of the range expected for that particular nucleus and accordingly the shifts in most other compounds will then be positive. Unfortunately, this is not always the case; for example, $[Me_4Sn]$ is the accepted standard for ^{119}Sn n.m.r. and the resonance of this standard falls about half way through the range of resonances known for tin compounds. Consequently, about half have positive sign and half have negative sign, a most undesirable situation as confusion and misrepresentation are bound to occur.

An alternative to using individual standards for each nucleus is to employ an absolute frequency scale whereby all resonances are referred to the proton resonance of TMS at 100 MHz[48]. As yet, this idea has not met with universal acceptance. In this chapter, the standards used for each nucleus are discussed in the appropriate sections; where more than one standard is currently being employed, the cited data are converted to the most suitable (in the author's opinion).

3. *Ranges of chemical shifts*

Considering diamagnetic compounds only, chemical shifts of resonances in ^{1}H n.m.r. spectra occur over a very small range, 0–15 ppm being typical for most compounds, with the exception of certain metal hydrides. In ^{205}Tl n.m.r., for example, the known shift range is *ca*. 34 000 ppm. This enormous difference can be understood by considering the factors affecting nuclear shielding and hence determining the chemical shift.

According to the model developed by Ramsey[49], the nuclear shielding, σ, consists of a 'diamagnetic' term (corresponding to free rotation of electrons about the nucleus) and a 'paramagnetic' term (corresponding to restriction of such rotation by other electrons and other nuclei in the molecule). Additionally, contributions due to ring currents, electric field effects, neighbouring group anisotropy, etc. may contribute to the total magnitude of σ.

Considering atoms in a molecular environment, then for small atoms, where there is little to affect the free rotation of electrons, σ_d is the dominant term. As larger atoms are considered, the role of the inner electrons in shielding the nucleus becomes less dominant and a number of factors contribute to an increase in σ_p (see below). Accordingly, for large atoms, σ becomes the sum of two numbers of opposite sign, one of which has a highly variable magnitude. With small atoms (such as ^{1}H, ^{2}H, ^{3}H), σ is dominated by one term (σ_d) with only a small contribution from the other, more variable, term (σ_p). In terms of shift ranges, we would thus expect small atoms to have a narrow range, whilst the range for large atoms will be much bigger, as the highly variable magnitude of σ_p determines the total shielding.

A number of factors contribute to determining the overall contribution of σ_p to σ; here we shall consider only the more important factors in determining the shifts of heavy nuclei in organometallic compounds. Fuller discussion is available elsewhere[44]. One of the approximations[50], made by the 'atom in a molecule' approach, for the paramagnetic term of a nucleus, A, is given in equation 2. In general terms, P_u and D_u are

$$\sigma_p(\mathrm{A}) = \frac{-2\mu_0\beta^2}{\pi(\Delta E)}\left[\langle r^{-3}\rangle_{\mathrm{np}}P_u + \langle r^{-3}\rangle_{\mathrm{nd}}D_u\right] \qquad (2)$$

factors representing imbalance of electrons in the p and d valence orbitals of atom A. The radial terms, $\langle r^{-3}\rangle$, represent the expectation values of the nth p and nth d orbitals and ΔE represents an effective excitation energy. In other words, the magnitude of σ_p is determined by a number of factors, which include the following:

(*i*) *Symmetry effects*. As σ_p is proportional to 'electron imbalance' in the p and d orbitals, increasing the symmetry about nucleus A will decrease the contribution of σ_p to σ and the shielding will increase. A frequent trend in the n.m.r. of the central metal atom in transition metal complexes is that the shieldings vary with geometry: square planar $<$ tetrahedral $<$ octahedral. Exceptions occur because of (*iii*) below.

(*ii*) *Electron density*. The $\langle r^{-3} \rangle$ term in equation 2 shows that σ_p will vary according to the fate of the valence electrons. If the valence electrons are removed (by delocalization, oxidation, etc.) then the remaining electrons are subject to an increased effective nuclear charge and so circulate closer to the nucleus. The magnitude of σ_p is thus increased and so deshielding results. Accordingly, the resonance of a central metal ion in a transition metal complex would be expected to vary with oxidation state and with the nephelauxetic effect of the surrounding ligands (see Section VI.E.5 for example).

(*iii*) *Excitation energy*. A small energy difference between occupied and unoccupied orbitals can lead to electric currents caused by transitions which enhance the paramagnetic term. With low-spin transition metal complexes, the ligand field splitting thus determines the size of this contribution. With main-group elements of the first row, the presence of lone-pairs of electrons, which can give rise to low energy $n \rightarrow \pi^*$ transitions, causes deshielding. Removing the lone-pair, say by donation to a Lewis acid, effectively increases ΔE in the denominator of equation 2 and hence decreases the magnitude of σ_p, resulting in increased screening.

The exceptions to (*i*) above occur when a low symmetry gives rise to a small excitation energy; accordingly, σ_p becomes large and deshielding occurs. The balance between these three factors is thus crucial in determining σ_p and hence the chemical shift of a heavy nucleus.

4. *Correlations between chemical shifts and physical properties*

Nuclear screening, and hence the chemical shift, of a nucleus is dependent upon its electronic environment (see Section II.B.3) and so correlations between the chemical shift of a nucleus and other physical properties of the compound, which relate to electronic structure, should be observable. In fact, the situation is extremely complex as the models used to describe nuclear screening (the independent electron model[51], the atom-dipole model[52], the average excitation energy method[50], etc.) are all approximations, to a greater or lesser extent, and a truly accurate model has not yet been found. Accordingly, while correlations with other spectroscopic data exist, these are frequently suitable only as rough guides and careful examination of the literature will usually produce one or two examples which do not fit the correlation. The subject of correlations between nuclear screening and other physical properties has recently been reviewed[44] with emphasis on inorganic compounds and so here such correlations will only be described in sections concerning particular nuclei where the relationship has some significance in organometallic chemistry.

C. Nuclear Spin–Spin coupling[53]

1. *Origin and variation in coupling constants*

Coupling between magnetic nuclei is another property which relates to the electronic structure of the molecule. Thus, one magnetic nucleus causes a perturbation in the valence electrons which in turn perturb a second nucleus, causing an indirect coupling to arise. Obviously, if coupling arises by perturbation of valence electrons in this manner then the two nuclei need not be directly bonded as the effect could be transmitted via a series of perturbations to a more distant magnetic nucleus within the molecule. Intuitively one would therefore expect the size of a coupling to relate to the size of the initial perturbation and the distance through which it is transmitted. In addition, the coupling will have a sign as it is a measure of an interaction; thus positive couplings are taken to imply that the interaction via perturbation of the valence electrons causes stabilization of antiparallel spins whilst a negative sign implies the converse.

Coupling constants have been treated theoretically by several valence-bond and molecular orbital approaches[54,55], and, according to Ramsey, arise via three major mechanisms of interaction. Briefly, these three mechanisms are as follows:

(*i*) Orbital–dipole interaction – whereby the nuclear moments interact with the orbital motion of the electrons.
(*ii*) Dipole–dipole interaction – whereby nuclear spin and electron spin interact.
(*iii*) Fermi contact interaction – whereby the nuclear spin causes an instantaneous correlation of electron spins.

It has been shown[56] that Fermi contact is the dominant mechanism for spin–spin coupling between the proton and any other nucleus. In terms of rationalizing coupling constants dominated by Fermi contact, the Pople–Santry expression[57] can be very useful. This expression (equation 3) follows from Ramsey's original work by an LCAO approximation, retaining only one-centre integrals.

$$^{1}J(\mathrm{A,X}) = -\gamma(\mathrm{A})\gamma(\mathrm{X})\cdot\frac{\hbar}{2\pi}\cdot\frac{256\pi^2}{9}\cdot\beta^2|\psi_{\mathrm{S(A)}}(0)|^2|\psi_{\mathrm{S(X)}}(0)|^2\sum_{i}^{\mathrm{OCC}}\sum_{j}^{\mathrm{UNOCC}}(^{3}\Delta E_{i\rightarrow j})^{-1}\cdot k \qquad (3)$$

where $|\psi_{\mathrm{S(A)}}(0)|^2$ = magnitude of the valence state s orbital of A at the nucleus of A; $k = c_{i\mathrm{S(A)}}c_{j\mathrm{S(A)}}c_{i\mathrm{S(X)}}c_{j\mathrm{S(X)}}$, where c is the coefficient of the S atomic orbitals of the coupled nuclei in both occupied (i) and unoccupied (j) molecular orbitals. Other symbols are defined in the list at the beginning of this chapter. The summation is carried out over all occupied and unoccupied orbitals and involves the triplet excitation energies, $^{3}\Delta E_{i\rightarrow j}$, and the coefficients, C_s, of the coupled atoms.

The Pople–Santry expression is particularly useful as it can be utilized to examine a number of possible relationships between the size of a coupling constant, in a series of related compounds, and a given variable within that series. Thus, by qualitative examination of equation 3, it is expected that the coupling constant, $^{1}J(\mathrm{A,X})$ will vary with:

(*i*) the products of the gyromagnetic ratios of A and X;
(*ii*) the magnitude of the valence state s orbital of A at the nucleus of A, for a series in which X is constant and A varies; and
(*iii*) the magnitude of the valence state s orbital of X at the nucleus of X, for a series in which A is constant and X varies.

The factors causing variation in points (*ii*) and (*iii*) (such as oxidation state, coordination number, electronegativity and/or *trans*-influence of supporting groups) give rise to a large number of possible variables to be explored. Rather than discussing examples for varying nuclei where the comparisons may not be obvious, this situation is discussed in some detail under ^{31}P n.m.r. studies (see Section VI.J.2) as metal—phosphorus compounds are particularly amenable to study in this respect. Variation in point (*i*) causes comparisons of the type $^{1}J(\mathrm{A,X})$ *vs*. $^{1}J(\mathrm{B,X})$ to be invalid as the difference may not result from any chemical implications, but may just reflect differences in the magnitudes of $\gamma(A)$ and $\gamma(B)$. Accordingly, a reduced coupling constant, ${}^nK(A, X)$, is defined (equation 4) to account for differences in gyromagnetic ratios between different nuclei.

$$ {}^nK(A, X) = \frac{2\pi}{h} \cdot \frac{{}^nJ(A, X)}{\gamma(A)\gamma(X)} \tag{4} $$

It is particularly important to stress that the treatment and rationalization of coupling constants by the Pople–Santry method is applicable only in cases where Fermi contact is the dominant interaction between the spins involved. In organometallic chemistry, the rationalization of spin–spin coupling between any two directly bonded nuclei by the Pople–Santry method is commonly encountered in the literature. In fact, it has only recently begun to be appreciated that, for nuclei apart from the proton, the other available mechanisms may make substantial, if not dominant, contributions to the interaction. In Section VI.J.2, the rationalization of one-bond metal—phosphorus couplings is examined and it can be clearly seen that the model predicts trends correctly, but not always accurately, and with more exotic couplings, such as those between two different metal nuclei, the extent of Fermi contact dominance is not known and treatments by the Pople–Santry and allied methods must be viewed with caution.

A final point to mention in introducing coupling constants is the question of sign. It was mentioned previously that coupling constants can be either positive or negative, but in fact the sign of a coupling constant cannot usually be derived simply from spectral analysis. Most coupling constants are cited without sign but this does not mean that they have been found to be positive. In this chapter, numerical values are given a sign if it has been determined and values quoted without sign may be either positive or negative.

2. *'Through-space' coupling*

This concept is invoked to explain coupling between nuclei which are separated by several bonds. Commonly, ${}^nJ({}^{19}F, {}^{19}F)$ and ${}^nJ({}^{19}F, {}^{1}H)$ values are explained by this phenomenon, attributed to an overlap of valence orbitals of the two nuclei. With ${}^nJ({}^{19}F, {}^{19}F)$ it seems likely that *p*-orbital overlap is involved in the transmission of the perturbation[58]. 'Through-space' couplings are known in several organometallic systems[59–61], although these usually require one or more nuclei with non-bonded electrons (such as ${}^{19}F$, ${}^{31}P$, and ${}^{199}Hg$) which can interact with a second nucleus to transmit the perturbation. Such contributions to the overall value of a coupling constant (i.e. 'through-space' contribution and through-bond contribution) can be useful in estimating the proximity of non-bonded groups in organometallic compounds, usually a difficult parameter to measure in solution. This topic is further discussed in Section VI.M.

3. *Analysis of spin systems*

Spectral analysis can often be a problem in multinuclear n.m.r. experiments, particularly where the observed nucleus is coupled to a heteronucleus which is not 100% abundant. For example, the complex $[Pt_2(\mu\text{-dppm})_2Cl_2]$, which has structure **3**, exhibits a complex ${}^{31}P\{{}^{1}H\}$ n.m.r. spectrum[62]. The complexity arises because ${}^{195}Pt$ is only 33.8% abundant and so there are three possible 'isotopomers' (isotopic isomers), one with no active platinum centres (43.8%), one with one active platinum (44.8%),

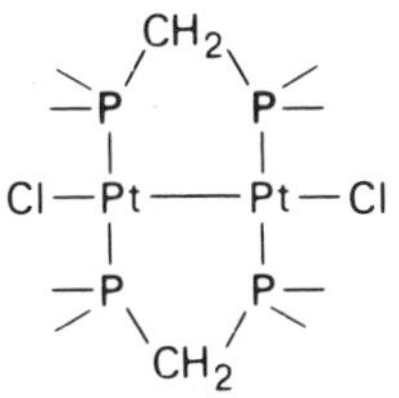

(3)

and one with two active platinums (11.4%). The last isotopomer represents the AA′A″A‴XX′ spin system[63] and as a consequence of the multiplicity caused by the complicated couplings and the low percentage abundance, this isotopomer is difficult to observe and fully characterize by $^{31}P\{^{1}H\}$ n.m.r. Perhaps unsurprisingly, computers play an ever increasing role in spectral analysis by iterative procedures. The increasing use of computers in n.m.r. procedures has been reviewed[64,65] and both texts and reviews are available on the analysis of complex spin systems[66,67]. The complexity of spin systems resulting from compounds containing magnetically inequivalent heteronuclei of $N < 100\%$ has led to some errors in structural assignment, particularly where spectra have been obtained by indirect observation (see Section III.C.1).

Selective population transfer (SPT) experiments coupled with difference-mode display allows an INDOR-type spectrum to be produced in Fourier transform n.m.r. Using these methods, separate spectral displays of the ^{31}P n.m.r. of three of the four possible isotopomers of $[Pt_2(\mu\text{-}S)(CO)(PPh_3)_3]$ have been described[68] (the minor, doubly active, isotopomer of 11.4% abundance was not observed). This method is experimentally simple to perform and will probably be applied to other heteronuclei in the future.

III. MEASUREMENT TECHNIQUES

A. Direct Observation Methods

1. Continuous sweep (and correlation spectroscopy)

The continuous sweep method of obtaining n.m.r. spectra is very familiar[69] from its extensive usage in ^{1}H n.m.r. spectroscopy of organic molecules. In this method, the exciting frequency is swept linearly through the appropriate range, at a rate which is sufficiently slow so as to avoid spin-saturation, and the nuclear magnetization measured. The method is very useful for high-abundance, high-sensitivity nuclei, such as ^{1}H and ^{19}F, and even less favourable nuclei, such as ^{31}P, provided that the compound under study contains a high proportion of the element and that it can be examined as a liquid or highly concentrated solution. Problems arise in organometallic chemistry, as in other areas, because of poor sensitivity, low natural abundance, unfavourable relaxation times, poor solubility, and complex coupling situations. Despite the inefficiency of this method (in which resonances are measured sequentially) it has two overwhelming advantages over all others. Firstly, it is technically a simple method to use, and secondly, it is an inexpensive facility to install and maintain. These two factors alone should ensure its continued usage in most branches of chemistry.

Correlation spectroscopy[70,71], which is a rapid sweep method employing cross-correlation to remove transient effects produced by fast scanning, has not been utilized to any significant extent in organometallic chemistry (possibly as the cross-correlation and subsequent Fourier transforms require sophisticated computing facilities) and so will not be discussed further.

2. *Pulsed excitation*

More efficient than continuous wave techniques are methods whereby all nuclei are excited simultaneously by a pulse of radiofrequency applied in the middle of the spectral range. When this pulse is completed, the nuclei relax in a manner characterized by T_2, producing a voltage change with time in the receiver coils of the spectrometer. This can be measured and is referred to as a free induction decay (FID) or time domain spectrum. A Fourier transform then relates the FID to the desired spectrum, in the frequency domain.

The advent of such spectrometers, commonly referred to as FT n.m.r. spectrometers, has been of enormous significance in many branches of chemistry. High signal-to-noise ratios can be achieved, enabling problems of sensitivity and abundance to be largely overcome and the multinuclear capability of the more sophisticated instruments, particularly the very high field spectrometers, has made multinuclear studies a reality. Several texts and review articles on Fourier transform n.m.r. are available[28,72–75].

3. *Other methods*

To examine very large spectral widths by pulsed techniques it is necessary to make efficient use of the exciting radiofrequency power, and this can be achieved by modulation with pseudo-random or white noise[76,77]. This technique is known as stochastic excitation and is likely to become extremely important as n.m.r. of heavy nuclei with large spectral ranges are further investigated.

Of importance in situations where complex homonuclear coupling arises is the tailored excitation method[78,79] which allows selective excitation of certain spectral regions. Currently the technique is fairly elaborate and as yet is really only of great interest to spectroscopists.

B. Double Resonance Techniques

A common application of double resonance in ^1H n.m.r. is in the irradiation of a multiplet arising from spin-coupling to other protons and so decoupling the nuclei. Of more interest are heteronuclear double resonance techniques whereby radiofrequency tickling at the resonant frequency of a nucleus X causes all couplings between X and a second nucleus A, the observed nucleus, to be removed[80]. Such a spectrum is referred to as an A{X} spectrum. The difference in chemical shift of the nuclei A and X is of importance as coupling between them can be re-established if the applied tickling frequency is of too high a power and the difference in shift is small[81]. In heteronuclear experiments this is unlikely to be a problem. Where nuclei with a range of shifts are to be decoupled, as in proton decoupling in multinuclear studies, a modulated frequency is applied in the centre of the range under study. The effects of proton decoupling on relaxation and hence NOE have previously been mentioned (see Section II.A).

The usefulness of double resonance techniques in n.m.r. studies should not be underestimated; apart from the simple heteronuclear proton decoupling method, which itself has made interpretation of heteronuclear n.m.r. spectra possible, several other useful experiments can be performed. Previously, the problems arising because of quadrupolar nuclei in molecules was mentioned in connection with line-broadening in ^1H n.m.r. spectra. By double resonance the line broadenings can be simply removed by decoupling the quadrupolar nucleus. Double resonance can also be used as an aid to assigning couplings in complex spectra, for example, $SnEt_2Cl_2$ has a ^1H n.m.r.

spectrum which is approximately first order and coupling to two of the three active tin isotopes can be observed[82] (^{115}Sn, $I = \frac{1}{2}$, is only 0.35% abundant). The peak heights of the satellites do not aid in determining which couplings are due to ^{117}Sn ($I = \frac{1}{2}$, $N = 7.61\%$) and which are due to ^{119}Sn ($I = \frac{1}{2}$, $N = 8.58\%$) as the abundances are so similar. The spectrum was recorded with ^{119}Sn decoupling, ^{1}H{^{119}Sn}, leaving only the satellites due to ^{117}Sn and so unambiguous assignment was possible[82]. Using this method of removing satellites, it is also possible to observe resonances which, in the undecoupled spectrum, may be 'buried' under a complex coupling pattern.

A less common application of double resonance n.m.r. in spectral assignment relates to the determination of molecular geometries by observing the effects of selective decoupling on the NOE of a resonance. Where relaxation of a nucleus A operates largely via magnetic dipole–dipole interactions with a nucleus X, then enhancement of intensity of the resonance due to A will occur when X is saturated. Accordingly, if A is coupled to two nuclei, X and X′, and selective decoupling of each in turn causes enhancement of the A resonance in one case, but not the other, then it may be possible to predict the geometric arrangement of the three nuclei, knowing which are coupled more strongly.

C. Indirect Methods of Observation

1. *Inter-nuclear double resonance (INDOR)*

In situations where an n.m.r.-active heteronucleus displays an observable coupling to a high-sensitivity nucleus (usually ^{1}H or ^{19}F), then it is possible to determine the chemical shift of the heteronucleus as well as both the magnitude and sign of the coupling involved. Thus, by monitoring the observed coupling in the ^{1}H (or ^{19}F) n.m.r. spectrum and slowly sweeping a secondary radiofrequency through the range expected for the resonant frequency of the heteronucleus, it is possible to simulate the n.m.r. spectrum of the heteronucleus by monitoring the extent of decoupling in the observed spectrum. As many heteronuclei have large ranges of resonant frequency it is often sufficient simply to measure the frequency of the applied secondary source which causes maximum decoupling, and this value will correspond to the chemical shift of the heteronucleus, the errors involved being less than the size of the coupling under observation. If greater accuracy is required, a series of tickling experiments can be performed and the INDOR spectrum plotted. Provided that certain conditions are met [especially that $\gamma(^{1}\text{H or }^{19}\text{F}) \gg \gamma(\text{heteronucleus})$], then the derived INDOR spectrum will closely resemble that obtained by direct observation.

The contribution made to organometallic chemistry by INDOR techniques is vast. Work described in the section concerning ^{119}Sn n.m.r. data (Section VI.H.2), for example, derives almost entirely from indirect observation by ^{1}H{^{119}Sn} n.m.r. Similarly, all early studies on ^{29}Si resulted from ^{1}H{^{29}Si} double resonance studies. The importance of INDOR as a technique in organometallic chemistry is exemplified by considering n.m.r. studies of rhodium compounds. Such compounds, which are greatly studied as homogeneous catalysts for a variety of organic transformations, could not be directly observed by ^{103}Rh n.m.r. until very recently because of the low gyromagnetic ratio of rhodium. Accordingly, INDOR was utilized and successful double resonance experiments including ^{1}H{^{103}Rh}[83], ^{13}C{^{103}Rh}[84], ^{19}F{^{103}Rh}[85], and ^{31}P{^{103}Rh}[86] have been reported.

Interest in INDOR techniques has declined since the advent of multinuclear spectrometers capable of direct observation of nuclei such as ^{29}Si and ^{119}Sn in a routine manner. Obviously, the major drawback of INDOR is that a clearly observable coupling to a high-sensitivity nucleus is essential or the method cannot be utilized.

Conversely, simple continuous wave spectrometers can be readily adapted for double resonance experiments and hence it is possible to observe the n.m.r. of heteronuclei, albeit indirectly, without the expense of establishing a multinuclear Fourier transform n.m.r. facility.

No direct pulse excitation equivalent of INDOR is available[75], although complex methods of achieving the same end are possible[28,68]. Unfortunately, this somewhat limits the sensitivity which can be achieved, even though INDOR relies on monitoring a high-sensitivity nucleus. Problems therefore arise in attempts to study complex spin systems of low abundance by indirect methods. For example, reaction of [PtMeCl(cod)] with the ligand $Ph_2PCH_2PPh_2$ has been reported[87] to yield an oligomeric product of the type $[\{PtMeCl(Ph_2PCH_2PPh_2)\}_n]$. The product was assigned the trimeric structure **4**, on the basis of ^{31}P n.m.r. data obtained by INDOR and molecular weight measurements. Re-examination of this system by direct observation[88,89] of the $^{31}P\{^1H\}$ n.m.r. of the product allowed its re-formulation as the cationic 'A-frame' organoplatinum(II) dimer $[Pt_2(\mu\text{-}Cl)(\mu\text{-}dppm)_2Me_2][Cl]$ (**5**). The product, **5**, is the same spin system as **3** and so the minor isotopomer also represents the AA′A″A‴XX′ spin system and the same problems in its detection apply.

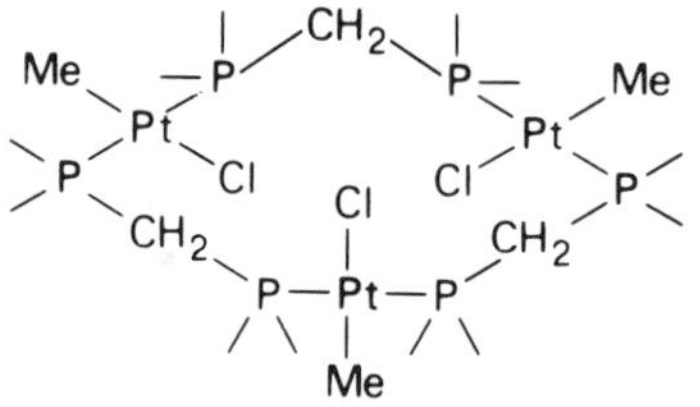

(**4**)

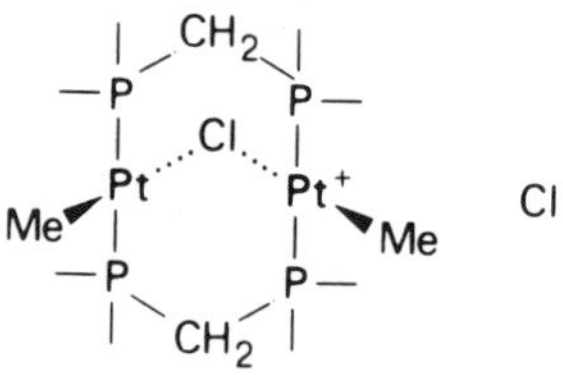

(**5**)

2. *Triple resonance*

An additional sophistication possible in indirect n.m.r. measurement is triple resonance. Thus, a high-sensitivity nucleus can be monitored with simultaneous irradiation at two different radiofrequencies. The power applied can be adjusted to allow total decoupling, spin-tickling or population transfer effects to be observed. Such triple resonance techniques, designated A{M, X}, have been successfully utilized in several diverse areas in organometallic and coordination chemistry. A straightforward example is the $^1H\{^{14}N, ^{59}Co\}$ n.m.r. spectra of a series of cobalt(III) complexes[90] which allowed observation of small proton–proton couplings by removal of the line-broadening effects of both the quadrupolar ^{14}N and ^{59}Co nuclei. In situations where INDOR is not suitable, triple resonance methods can sometimes be used; for example, tertiary phosphine complexes of tungsten have been successfully studied by INDOR, but in the corresponding trimethyl phosphite complexes, the value of $^4J(^{183}W, ^1H)$ is effectively zero and so $^1H\{^{31}P, ^{183}W\}$ experiments were performed[91,92]. Further mention of triple resonance experiments can be found in sections concerning the appropriate nuclei.

The topic of multiple resonance methods in n.m.r. has been reviewed many times[93–99].

IV. DYNAMIC SYSTEMS

Lineshapes of n.m.r. spectra are affected when two species are undergoing a dynamic process at a rate similar to the difference in chemical shift between the resonances associated with each species. Rates between 1 and 10^7 s^{-1} are thus of particular interest. This behaviour is very common with organometallic and coordination compounds, many of which are not 'well behaved' or static on the n.m.r. timescale. The processes which affect the n.m.r. spectra can be either intermolecular or intramolecular[100]. Intermolecular processes include dissociation, association, ionization, and other bond-making or bond-breaking reactions. Intramolecular processes refer to rearrangements of stereochemically non-rigid molecules, including processes where all the interconverting species are observably chemically and structurally equivalent. The latter class of stereochemically non-rigid molecules are referred to as fluxional.

Variable-temperature n.m.r., coupled with detailed lineshape analysis, can provide much information concerning dynamic processes. By varying the temperature of the sample under study it may be possible to reach a limit of slow exchange (low temperature) where the lifetime of the molecule is long compared with the dynamic process on the n.m.r. timescale. A limit of fast exchange (high temperature) may also be attained where the dynamic process is fast compared with the lifetime of the molecule on the n.m.r. timescale.

An initial problem is to determine whether the process is intermolecular or intramolecular. This *may* be possible if there are suitable spin–spin couplings to monitor in the n.m.r. spectrum. For example[101], a dihydride of the type **6** will exhibit a complex

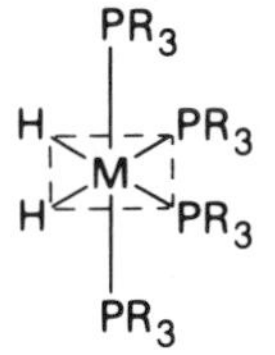

PR_3
H, H, M, PR_3, PR_3
PR_3

(**6**)

hydride resonance in the ^{1}H n.m.r. at the limit of slow exchange (an $AA'XX'Y_2$ spin system, A = H) whilst in the limit of fast exchange a quintet indicates coupling of the hydrides to four equivalent phosphorus nuclei. Thus no loss of $^2J(^{31}P, {}^1H)$ occurs in the dynamic process and it would appear reasonable to assume that the process was intramolecular. In fact another possibility exists; if dissociation of phosphine occurred and the species were prevented from drifting apart in solution, say by formation of a solvent lattice, then recombination could occur, after rearrangement, to give a new molecule in which spin coupling is maintained. If the lifetime of the dissociated species is short relative to the relaxation of the ^{31}P nucleus, then no loss of spin coupling will be observed in the spectrum. Such a process is perhaps simpler to envisage if the four PR_3 ligands in **6** are replaced by two chelating diphosphines.

An example of intermolecular exchange is the reversible dissociation of a ligand from a complex (equation 5).

$$ML_n \rightleftharpoons ML_{(n-1)} + L \qquad (5)$$

If M and a nucleus in L are spin–spin coupled, then the process may be monitored simply. In the limit of slow exchange, the undissociated complex can be detected, but in the immediate region broadening will occur until the limit of fast exchange is reached and a single peak is observed, representing an average of the chemical shifts of the dissociated species. A great many processes can be monitored in this fashion; of particular importance are the studies of metal phosphine complexes which act as homogeneous catalysts in organic transformations. An example[101] is the study of the complex $[RhH_2Cl(PPh_3)_3]$, formed according to equation 6. The ^{1}H and $^{31}P\{^1H\}$ n.m.r. spectra at two different temperatures are

$$[RhCl(PPh_3)_3] + H_2 \rightleftharpoons [RhH_2Cl(PPh_3)_3] \quad (6)$$

H, H, Rh, PPh_3 (2), Ph_3P (2), PPh_3 (1), Cl

(7)

shown in Figures 3a and 3b. The low-temperature ^1H n.m.r. spectrum shows two inequivalent hydrides, one with a large coupling $^2J(^{31}P, ^1H)$, establishing the geometry as **7**. The low temperature $^{31}P\{^1H\}$ n.m.r. spectrum shows the expected resonances for P_2 [doublet of doublets resulting from $^2J(^{31}P_1, ^{31}P_2)$ and $^1J(^{103}Rh, ^{31}P_2)$] and P_1 [doublet of triplets resulting from coupling to both P_2 nuclei, $^2J(^{31}P_1, ^{31}P_2)$ and to rhodium, $^1J(^{103}Rh, ^{31}P_1)$]. At high temperature, the ^1H n.m.r. spectrum is broadened, but otherwise uninformative, whilst the $^{31}P\{^1H\}$ n.m.r. spectrum shows that both P_2 nuclei are still coupled to rhodium, but that all couplings involving P_1 are lost [i.e. $^1J(^{103}Rh, ^{31}P_1)$, $^2J(^{31}P_1, ^{31}P_2)$]. This suggests that the phosphine *trans* to hydride (P_1) is dissociated to a considerable extent at 30°C and so an available coordination site at the metal centre is produced, a key step in the reaction of unsaturated organic molecules during homogeneous processes. There are many further examples of inter-

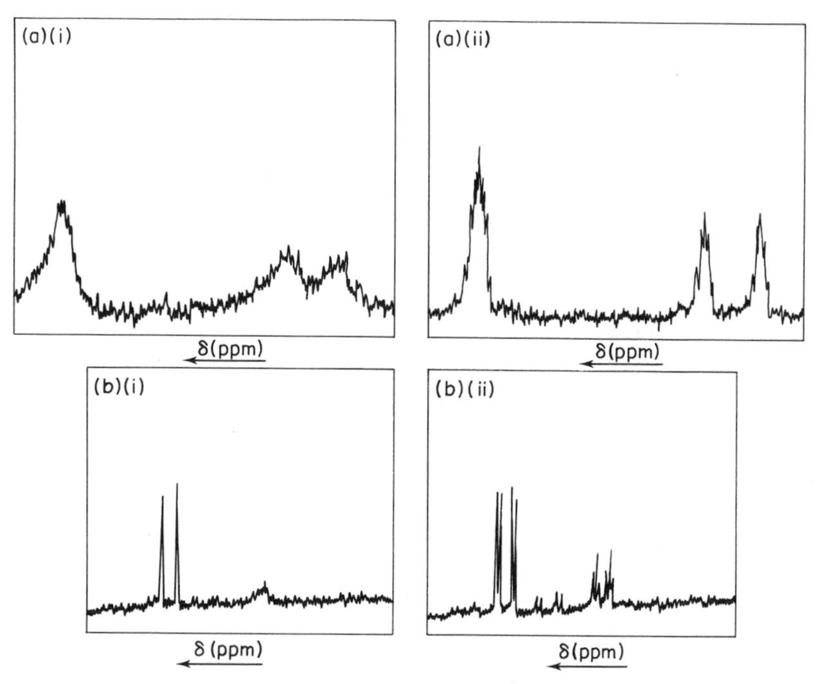

FIGURE 3. (a) ^1H n.m.r. spectra of $[RhH_2Cl(PPh_3)_3]$ at (i) +28°C and (ii) −25°C. ^1H n.m.r. spectra obtained at 90 MHz. (b) $^{31}P\{^1H\}$ n.m.r. spectra of $[RhH_2Cl(PPh_3)_3]$ at (i) +30°C and (ii) −25°C. Reproduced by permission of Academic Press from Jesson and Mutterties[101] in *Dynamic Nuclear Magnetic Resonance Spectroscopy*, edited by Jackman and Cotton, New York, 1975.

and intramolecular dynamic processes and an excellent review is available which describes these and the mathematical processes involved in lineshape analysis[101].

Fluxionality is of particular importance in the chemistry of metal complexes containing unsaturated carbon donors. The concept of fluxionality was first applied in 1956[102,103] when it was observed that the complex $[Fe(\eta^5\text{-}C_5H_5)(\eta^1\text{-}C_5H_5)(CO)_2]$ (believed to have η^1- and η^5-cyclopentadienyl groups on the basis of i.r. and chemical evidence) exhibited only two resonances (ratio 1:1) in its ^1H n.m.r. spectrum. Of course, an $\eta^5\text{-}C_5H_5$ would be expected to exhibit only one resonance, but an $\eta^1\text{-}C_5H_5$ should exhibit one resonance for the proton bonded to the Fe—C unit and a complex resonance for the four olefinic protons in the AA′BB′ spin system. The novel idea at that time was that the $\eta^1\text{-}C_5H_5$ unit was continually rotating, exchanging one Fe—C bond for another, at a rate that made all the protons equivalent on the n.m.r. time-scale. Later work[104] showed that this was correct and that the limit for slow exchange could be reached at -80°C. Complexes which exhibit this behaviour are commonly known as 'ring-whizzers'.

Since the initial discovery, many compounds, particularly cyclopentadienyl, cycloheptatrienyl, cyclopolyene, and allyl derivatives have been shown to exhibit fluxional behaviour. Fluxionality in complexes of this type has been the topic of many reviews[105–111].

Dynamic processes in metal—carbonyl and metal—hydride complexes, particularly polynuclear systems, have also been the subject of much investigation. In studies of carbonyl complexes, ^1H n.m.r. measurements[112] are of limited use unless other ligands (C_5H_5, CNR, etc.) can be substituted into the molecule under study. In this area, $^{13}C\{^1H\}$ n.m.r.[113–115] and, to a lesser extent, ^{17}O n.m.r.[116–118] have been of significance. These studies are further mentioned in the appropriate sections (Sections VI.G.3 and VI.L.1). With polynuclear hydrides[119], the current challenge is to synthesise non-fluxional complexes, as fluxionality is so common. A recent example[120] of such a system is the complex $[Rh_3H_3(\eta^5\text{-}C_5Me_5)_3O][PF_6]\cdot H_2O$, which has an oxygen atom capping a triangulo-rhodium cluster. ^1H n.m.r. studies showed that the complex was rigid up to at least 100°C, giving a minimum of $\Delta G^{\neq} = 109$ kJ mol^{-1} as the barrier to fluxionality. Highly rigid hydrido-complexes of this sort are uncommon.

Fluxionality in the solid-state was first observed[121,122] by wide-line ^1H n.m.r. in 1971 for the complexes $[Fe(C_8H_8)(CO)_3]$ and $[Fe_2(C_8H_8)(CO)_5]$. The few studies of solid-state samples of fluxional complexes by high-resolution n.m.r. methods are discussed in Section VII.

V. PARAMAGNETIC SYSTEMS

A. Paramagnetic Shifts and Spin Delocalization

The discussion so far presented has concentrated upon diamagnetic systems, with a brief mention of paramagnetic species as relaxation agents and shift reagents. Studies of the n.m.r. of paramagnetic compounds can be very informative about molecular electronic structure. In such systems, a contact interaction can occur between the nucleus under observation and the unpaired electron; such contact relies upon there being a finite electron density for the unpaired electron at the nucleus, such that contact can occur. Only electrons with *s* character have a finite electron density at the nucleus and so contact in these systems requires the unpaired electron to have some *s* character. The contact interaction between the nuclear spin and the spin of the unpaired electron gives rise to internal magnetic fields at the nucleus which add to, or subtract from, the applied field. Accordingly, the nucleus experiences a contact shift as the field is altered; the change in position of the resonance gives the magnitude of the shift, whilst the direction of the shift gives its sign. A further contribution to the shift can arise from dipolar coupling interactions between the unpaired electron and the observed nucleus, this contribution is referred to as the pseudocontact shift.

Among the first studies of paramagnetic compounds were reports of the ^1H n.m.r. spectra of the metallocenes $[M(Cp)_2]$ (M = Ni, V, Cr)[123,124]. Polycrystalline samples showed large resonance shifts in the positions of the cyclopentadienyl protons and this was explained in terms of a contact shift arising from delocalization of unpaired electron density over the cyclopentadienyl ring systems. The unpaired electron, formally from the metal atom, was thus seen to be involved in the interaction with the cyclopentadienyl ring. These observations laid the foundation for much research into the n.m.r. spectra of paramagnetic species with a view to gaining a better understanding of the electronic interactions between metal and ligand. The resonance observed for the cyclopentadienyl protons in $[Cp_2V]$ was attributed to an isotropic hyperfine contact and not a pseudocontact interaction. The basis for this assignment stems from the differences in spectra which may be observed between polycrystalline and solution samples. Thus, as contact shifts are determined by an electron spin–nuclear spin interaction, which is isotropic, solid-state and solution spectra are the same. Pseudocontact shifts, however, result from electron spin–orbit coupling, electron orbit–nuclear spin(dipolar)coupling, and electron spin–nuclear spin dipolar coupling[125–127]. The electronic g-tensor relating to these properties is anisotropic and so in the solid-state the electron spin will not generally be quantized in the direction of the applied field. In solution, however, free rotation makes such quantization possible; accordingly, the pseudocontact shifts in solution and in the solid-state will be different[127]. Applying these criteria to $[Cp_2V]$ proved that the shift originated via an isotropic hyperfine contact interaction[124].

Problems associated with obtaining satisfactory n.m.r. spectra of paramagnetic samples relate directly to the relaxation time of the observed nucleus. In non-quadrupolar species, relaxation is generally dominated by the nuclear spin–electron spin interaction. Fortunately, paramagnetic transition metals generally exhibit a much shorter electronic relaxation time than organic free radicals and so, in certain cases, the electronic relaxation time is sufficiently short that the resonances are not excessively broadened. The factors determining paramagnetic linewidths, and hence observability of contact shifts, have been the subject of a review[128].

There are three spin-transfer mechanisms normally invoked to explain spin delocalization in paramagnetic transition metal complexes; $L \rightarrow M$ σ-transfer, $L \rightarrow M$ π-transfer and $M \rightarrow L$ π-transfer[129]. Indirect spin-transfer processes may also be involved[130]. Certain qualitative observations[131] may allow distinction to be made between the various processes, particularly:

(*i*) σ-delocalization gives rise to high frequency shifts of aromatic protons. The magnitudes of the shifts attenuate according to the number of intervening bonds. Substituents on aromatic rings (particularly methyl groups) need not exhibit this behaviour;

(*ii*) σ-delocalization causes the ^{13}C hyperfine constants of carbon atoms in an aromatic ring to alternate in sign around the ring[130];

(*iii*) π-delocalization is often characterized[132] by opposite shifts for methyl and hydrogen substituents on an aromatic ring, at a given position.

It should be borne in mind, however, that (*iii*) is only diagnostic as exceptions are known. An example is the ^1H n.m.r. study[133] of the manganese isocyanide complexes $[Mn(CNR)_6][PF_6]$ and $[Mn(CNR)_6][PF_6]_2$. The former are Mn(I) and so are diamagnetic, but the latter contain Mn(II) and so paramagnetic shifts would be expected. The compounds studied (R = Ph, p-$CH_3C_6H_4$, Bz) exhibited shifts in the ^1H

n.m.r. spectra which alternated in direction for the *o*- and *m*-protons; this was interpreted in terms of π-spin delocalization [as criterion (*i*) is not obeyed, σ-delocalization can be discounted] and hence a $^2T_{2g}$ ground-state for the complexes.

Reviews on the n.m.r. of paramagnetic compounds are available which include discussion of paramagnetic coordination complexes and metallocene derivatives[130,134–137].

B. Relaxation Agents

Paramagnetic ions, particularly Cr(III) complexes, have previously been mentioned as suitable relaxation agents for enhancing the spin–lattice relaxation times of certain nuclei. It has been observed[138–140], however, that shifts in resonance position and increased linewidths can occur with the common relaxation agent [Cr(acac)$_3$]. Accordingly, care is needed, particularly in lineshape analysis of dynamic systems, when such agents are employed. Both [Cr(dppm)$_3$][140] and [Gd(dppm)$_3$][141] have been used as alternatives.

C. Shift Reagents

Lanthanide shift reagents[142–145] are commonly employed to render complex coupling patterns approximately first order by effectively 'spreading out' the spectrum. A more demanding application is in the determination of solution geometries of adducts.

The origins of the shifts[130] induced by lanthanide complexes are not straightforward. Ligand atoms adjacent to the rare earth ion are generally shifted by a contact interaction, whilst more distant atoms are shifted by a dipolar interaction. The exact contributions of each appear to vary across the lanthanide series[130]. In structure determination, several conditions must be met[146], particularly that the shifts are dipolar in origin, that only one geometric isomer of a single complex exists in equilibrium with the uncomplexed substrate, and that this isomer is magnetically axially symmetric with a known orientation of the axis with respect to the ligand. In cases where these conditions are met, as is the case with troponeiron tricarbonyl[147], conformations can be deduced. In other cases, such as in the interaction of [Eu(fod)$_3$] with [Fe(η^5-C_5H_5)(CO)$_2$(CN)][148], a simple 1:1 adduct is not formed and more complex equilibria exist and analysis becomes more difficult.

Chiral shift reagents[149,150] have been known for some years and the recent interest in chiral metal centres[151] has led them to be used in organometallic chemistry. A number of successful resolutions of optically active cyclopentadienyliron derivatives[152] were possible using chiral shift reagents.

VI. N.m.r. STUDIES OF ORGANOMETALLIC SYSTEMS

In this section, the study of organometallic compounds by n.m.r. methods is described with reference to the nucleus under study. Each nucleus for which relevant data have been obtained is mentioned, although the limitations of space restrict the scope of this section. The major limitations are as follows:

(*i*) Nuclei which have been thoroughly discussed in texts and review articles are discussed here only briefly. These are nuclei which are generally well known and will have been encountered by many readers. Few examples are given in these sections.

(*ii*) The more exotic nuclei are discussed in greater detail, where possible with tabulations of shift and coupling constant data and discussion of examples to allow these more unusual methods to be appreciated.
(*iii*) The theory of chemical shifts and coupling constants has briefly been mentioned (Sections II.B and II.C) and rather than engage in long discussion over each example, the correlations between theory and practice are considered in detail in only one section (Section VI.J).
(*iv*) When a particular compound is discussed, for example $[Fe(CO)_5]$ under ^{13}C n.m.r. studies, cross-reference to other sections (^{17}O and ^{57}Fe) is made, rather than repeating data in more than one section.
(*v*) Coupling constant data are usually considered under the section dealing with the heavier atom (e.g. $^{1}J(^{57}Fe, ^{13}C)$ under ^{57}Fe n.m.r. studies), unless it is particularly appropriate to consider the data in the earlier section.

With these restrictions in mind, it is intended that this section should provide sufficient data and examples for the various n.m.r. methods such that their utility and applicability to a given chemical problem in organometallic chemistry can be evaluated.

A. Proton N.m.r. Studies

The proton is a very familiar nucleus for n.m.r. study. The high sensitivity (Table 1) enables a great deal to be accomplished using conventional continuous wave measurement techniques. Relative to the widely accepted shift standard, TMS, nearly all resonances occur in the range 0–15 ppm. The major exception to this generality is the hydride resonance of many classes of transition metal hydrides and these are discussed below. The vast number of studies of organometallic compounds by ^{1}H n.m.r. precludes any general coverage here; application of ^{1}H n.m.r. spectroscopy to structural characterization and studies of dynamics can be found in other chapters. Attempts to review applications of ^{1}H n.m.r. spectroscopy in organometallic chemistry[153–155] are generally fairly outdated because of the rapid growth in this field.

One of the most fundamental uses of ^{1}H n.m.r. spectroscopy is to determine whether or not a specific organic group is coordinated to a metal centre and, if so, in what particular orientation. This problem is considered in the second section below, by way of examples.

For detailed lists of references concerning ^{1}H n.m.r. spectroscopy in organometallic chemistry, the reader is referred to the Specialist Periodical Report series[14]; appropriate reviews are also referenced in this series.

1. *Transition metal hydrides*

The hydridic proton of many classes of transition metal hydrides exhibits a characteristic resonance, more shielded than TMS, in a unique region of the spectrum. It should be remembered that the term 'hydride' results from an electron-counting formalism and does not imply any particular electron distribution in the metal–hydrogen bond; indeed, many hydrides are reasonably strong acids. The origin of the extreme shielding of hydridic protons arises from their proximity to the metal centre; valence-bond calculations for the molecule $CoH(CO)_4$[156] show that the shielding is a result of an 8% contribution from the hydrogen 1*s* electrons and a 92% contribution from the cobalt 4*s* and 4*p* electrons.

The highly definitive shift of a hydridic resonance is generally observed in most classes of transition metal hydrides. Certain exceptions do exist, however, a noteworthy example being the dimeric complex $[Pt_2(\mu\text{-}H)_2(Si\{OEt\}_3)_2(PMeBu^t_2)_2]$[157]. The ^1H n.m.r. spectrum of this compound at 75°C exhibits a single resonance at 2.95 ppm (ignoring resonances due to the organic groups) with a coupling to ^{195}Pt of 681 Hz. At −60°C the resonance is observed at 3.35 ppm, with two unequal couplings to ^{195}Pt of 456 and 894 Hz. These data imply that the molecule is undergoing a dynamic process at room temperature and above but, at −60°C, a static structure is observed with the hydride ligands having an uneven interaction with the two metal centres. In terms of the unusual shift position and the coupling data, the implied structure is suggested to involve an interaction of the platinum centre with a Si—H bond to yield a 3-centre, 2-electron bonding system, as shown in **8**. Interactions of this

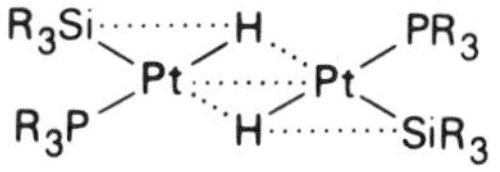

(8)

sort may be of relevance in the catalytic activation of Si—H (and perhaps even C—H) bonds by platinum metal complexes. Accordingly, hydride complexes may exhibit resonances in the ^1H n.m.r. spectrum that are either shielded or deshielded with respect to TMS. The important class of compounds which show no resonances at all, in either region, should also be borne in mind. An example related to **8** above, is the complex $[Pt_2(\mu\text{-}H)_2H_2(PCy_3)_2]$[158], known to contain both bridging and terminal hydride ligands from infrared and chemical evidence. The ^1H n.m.r. spectrum exhibits no resonances attributable to hydridic protons and the molecule is assumed to be dynamic, probably involved in a terminal-bridging hydride exchange process at all temperatures within the experimental range. As neither a limit of fast exchange nor a limit of slow exchange can be obtained, the spectra show only the intermediate region of broadening where the hydridic resonances are lost. The application of ^1H n.m.r. spectroscopy to the study of dynamic transition metal hydrides has been reviewed[119,159].

2. *Coordinated organic groups*

In determining whether or not an organic fragment is coordinated to a metal centre, the initial reaction is usually to look for a change in resonance position between the organic molecule itself (or facsimile of the organic fragment) and the supposed metal complex. The effects of coordination to a metal centre are not always straightforward; for example, formation of a chromium tricarbonyl π-complex with the organic ligand **9**[160] causes the aromatic protons, H_1–H_4, to be shielded by *ca*. 2 ppm. The proton H_5

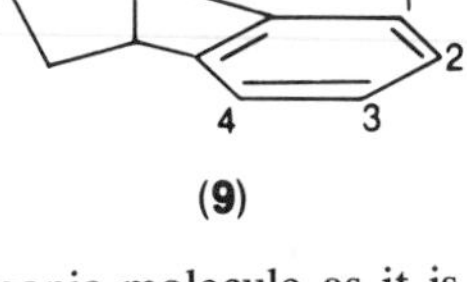

(9)

is very shielded in the free organic molecule as it is held in close proximity to the aromatic ring and so is affected by the ring currents. Upon coordination, the resonance position is not altered at all, implying that although coordination to chromium may be expected to perturb the ring current, this is evidently not the case. Accordingly, the proton taken to monitor complex formation must be selected carefully for, as in this case, the likely choice is not always the most useful.

The observation of spin–spin coupling between a proton in the organic molecule and the metal centre is, of course, indicative of complex formation, but necessarily requires the metal to have a reasonable abundance of an active isotope. Should this not be the case, it may be possible to incorporate a spin-probe into the molecule. Thus, inclusion of a PR_3 ligand in many organometallic transition metal complexes can facilitate reaction monitoring by ^1H n.m.r., by observation of indirect $^1H-^{31}P$ spin–spin coupling.

A less obvious monitoring system can result from the inclusion of chiral ligands in a complex[161]. For example[162], [$Au(PPh_3)((S)\text{-}2\text{-}Me_2NCHMeC_6H_4)$] exhibits a ^1H n.m.r. spectrum displaying a single resonance for the NMe_2 protons. Should there be a stable Au—N bond in this compound, then the nitrogen atom would be a prochiral centre (with no inversion possible) and, as the benzylic carbon atom is chiral, the two methyl groups at nitrogen would be diastereotopic and appear as two single resonances in the ^1H n.m.r. spectrum. As only one resonance is observed, inversion at nitrogen must render the methyl groups homotopic and hence no Au—N bonding can be occurring in this compound.

Having established that there is an interaction between a given organic fragment and the metal centre, ^1H n.m.r. spectroscopy can frequently be used to shed light on the geometry of the molecule. Examples of this abound in the literature and so only some relatively simple examples will be discussed here.

In the reaction between SbF_5 and **10** at $-20°C$, the product[163] is a bridged complex

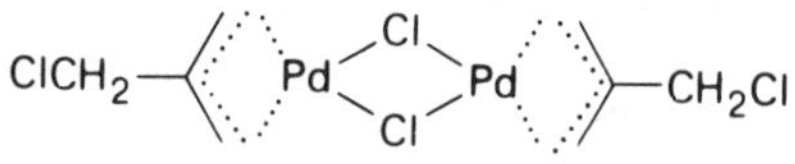

(10)

whose ^1H n.m.r. spectrum shows a single resonance at 5.70 ppm, indicating that halide abstraction, with the unusual η^4-trimethylenemethane complex, **11**, as a likely product,

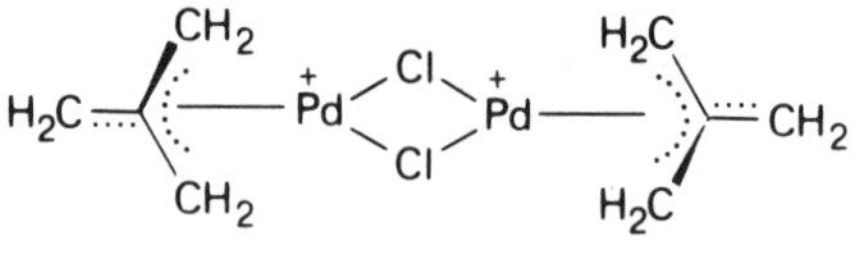

(11)

has occurred. Cooling the sample to below $-20°C$ resulted in considerable broadening of the resonance, but no low-temperature limit could be reached as the compound crystallized at $-80°C$. The broadening does suggest that a dynamic process is occurring and so the structure **11** is probably not a true representation of the molecule. More likely is the structure **12**, the positions C_1, C_2, and C_3 rapidly interconverting, involving the more common π-allyl ligand.

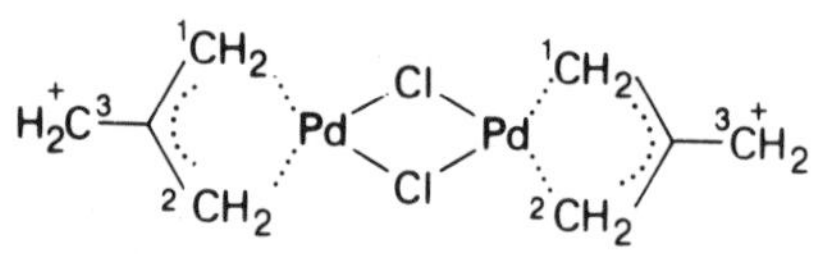

(12)

In the study of acetylene insertion into platinum—hydrogen bonds, it has been observed that a complex such as $[PtHCl(PEt_3)_2]$ can react with a disubstituted acetylene such as MeOOCC≡CCOOMe to yield four possible geometric isomers. Thus, the two phosphine ligands may be either *cis* or *trans* to each other and in both cases the insertion may lead to a *cis* or *trans* disposition of platinum and hydride groups about the resulting vinylic ligand. All four isomers have been prepared[164,165] and may be distinguished[165] from the magnitudes of the proton couplings to ^{31}P and ^{195}Pt (Table 7). ^{1}H n.m.r. studies of this type have led to a much clearer picture of the processes involved in the so-called insertion reaction.

A final example to show some of the uses of ^{1}H n.m.r. spectroscopy as a tool for molecular structure determination in solution comes from the chemistry of carbene complexes. The ^{1}H n.m.r. spectrum of the complex *cis*-$[PtCl_2(PMe_2Ph)(C\{OEt\}\{CH_2Ph\})]$ shows several interesting features[166]. The benzylic protons of the carbene moiety give rise to an AB pattern and the methyl resonances of the PMe_2Ph ligand are inequivalent. Additionally, the methylene protons of the ethoxy group are inequivalent, with a substantial difference in shift (*ca*. 0.6 ppm). These data imply that rotation

TABLE 7. $^{3}J(^{195}Pt, ^{1}H)$ and $^{4}J(^{31}P, ^{1}H)$ data of the four isomers of $[PtCl(PEt_3)_2(MeOOCC{=}C(H)COOMe)]$

Isomer[a]	$^{3}J(^{195}Pt, ^{1}H)$ (Hz)	$^{4}J(^{31}P, ^{1}H)$ (Hz)
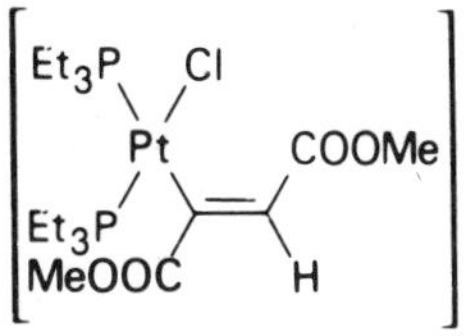 Et_3P, Cl, Pt, Et_3P, MeOOC, COOMe, H	80.5	16.6 1.1
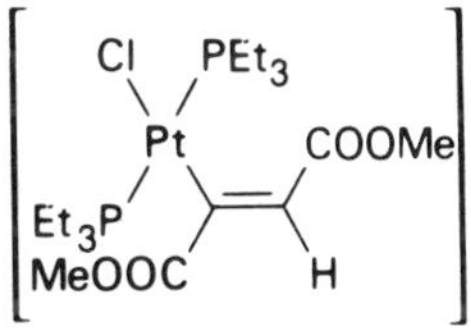 Cl, PEt_3, Pt, Et_3P, MeOOC, COOMe, H	138.0	1.6
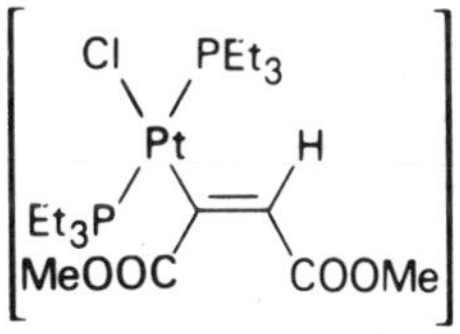 Cl, PEt_3, Pt, Et_3P, MeOOC, H, COOMe	102.2	1.5
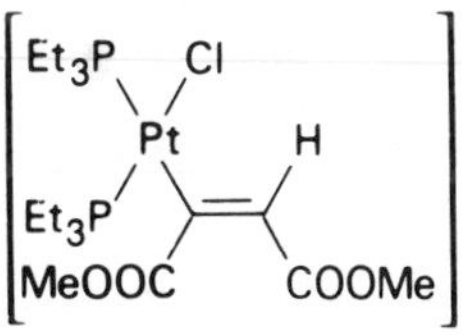 Et_3P, Cl, Pt, Et_3P, MeOOC, H, COOMe	60.5	10.0

[a]Data from reference 165.

of the carbene moiety about the Pt—C bond is restricted and that the methylene protons of the ethoxy group are in very different magnetic environments. The solid-state structure of this compound was determined[166] and showed that the orientation of the carbene group was such that a methylene proton of the ethoxy group was maintained in a position only 2.59(8) Å from the metal centre. This non-primary valence interaction is evidently maintained in solution, restricting rotation of the carbene moiety and resulting in different magnetic environments for each of the methylene protons.

B. Deuterium and Tritium N.m.r. Studies

Although deuterium is a low natural abundance isotope (Table 4), organic compounds enriched in ^{2}H are commonly available and routinely used as solvents in n.m.r. spectroscopy. The diverse applications of ^{2}H n.m.r. to chemical, biochemical, and physical problems[167–170] have been reviewed. The usefulness of ^{2}H n.m.r. in organometallic chemistry is currently rather limited as ^{2}H n.m.r. shows advantage over ^{1}H n.m.r. only in complex coupling situations where the low homonuclear coupling of deuterons gives rise to sharp, single resonances for each type of deuteron. Accordingly, ^{2}H n.m.r. has been applied to complex rearrangements of organic moieties in organometallic systems[171,172] where ^{1}H n.m.r. is unable to provide unambiguously interpretable spectra. Few other areas of interest have been examined so far.

Tritium is a radioactive isotope, principally of interest as it is more sensitive to n.m.r. detection than the proton (Table 1). The problems of handling radioisotopes disuade many workers from attempting to use ^{3}H n.m.r. in their particular field. The advantages again arise only when ^{1}H or ^{2}H n.m.r. produces overlapping resonances in highly complex molecules. Rearrangements of organic compounds over metal surfaces have been studied[173] by ^{3}H n.m.r. Reviews on ^{3}H n.m.r. are available[167,174].

C. Studies of Alkali Metal Nuclei

Lithium possesses two quadrupolar isotopes, ^{6}Li and ^{7}Li, the latter being more widely studied owing to its higher abundance, although the former has a smaller quadrupole moment; an attractive n.m.r. feature leading to pseudo-spin one half behaviour. The active nuclei of the remaining alkali metals (^{23}Na, ^{39}K, ^{41}K, ^{85}Rb, and ^{133}Cs; Fr being radioactive) are more typically quadrupolar in their n.m.r. behaviour (Table 5). The major areas of interest in alkali metal n.m.r. are as follows:

(*i*) Studies of aqueous solutions of ionic compounds[175]. The variation in linewidth associated with a change in symmetry during cation–anion interactions and variation in coordination sphere may be probed by alkali metal n.m.r.

(*ii*) Studies of non-aqueous solutions of ionic compounds[176]. Ion–ion and ion–solvent interactions cause substantial variations in chemical shift and linewidth. Effects of coordination number and solvent 'donicity' may thus be examined[177].

(*iii*) Studies of complex ions. Biochemical interactions of alkali metal ions with sugars. phosphate esters, nucleosides, etc., have been probed by these techniques[178,179]. Examination of cryptand and crown ether complexes of alkali metal ions[180] has allowed information concerning the complexing ability of these ions to be obtained.

Of interest in organometallic chemistry are the studies relating to organolithium compounds. Although organosodium and organopotassium compounds are also much studied systems, their alkali metal n.m.r. is as yet in a state of infancy.

The general topic of alkali metal n.m.r. has been reviewed[175,181].

1. Organolithium compounds

Lithium forms many compounds which are primarily covalent in character, thus differing considerably from the remaining alkali metals which tend to form ionic compounds. The elucidation of the structure of organolithium compounds, widely used as reagents in organic synthesis, has largely resulted from n.m.r. studies. For the ^{7}Li nucleus, the quadrupole coupling constant in organolithium compounds is extremely small[182] and quadrupolar relaxation relatively inefficient. Accordingly, very narrow resonances are obtained in the ^{7}Li n.m.r. of organolithium compounds[24]. This is particularly fortuitous as the chemical shift range is very small because the diamagnetic and paramagnetic contributions are of similar magnitude[183]. Cancellation of these parameters leaves a small shift range, governed by indirect contributions.

^{7}Li n.m.r. studies have done much towards the elucidation of the oligomeric structure of organolithium compounds and variation in molecularity with solvent, temperature and the nature of the organic group are known. The subject has been reviewed[175,178].

The relative merits of ^{6}Li and ^{7}Li n.m.r. in the study of *n*-PrLi have been reported[184]. The ^{6}Li nucleus has a long relaxation time, indicating that the quadrupolar mechanism is relatively inefficient, and the ^{6}Li n.m.r. spectrum of *n*-PrLi (96% enriched with ^{6}Li) consists of a resonance less than 1 Hz wide. The corresponding ^{7}Li n.m.r. spectrum shows a broad resonance. This broadening is frequently attributed to unresolved ^{7}Li–^{1}H coupling, but examination of a sample enriched to 90% with ^{13}C at C_1 shows no ^{7}Li–^{13}C coupling, indicating an exchange process. The line broadening is thus attributed to chemical exchange and to a quadrupolar contribution to ^{7}Li relaxation.

Studies[185] of the ^{7}Li n.m.r. of silyl- and germyllithium compounds have demonstrated that these compounds behave similarly to organolithium species with various dynamic processes, including inversion at Si (Ge or C), being evident. Lithium–carbon coupling has been investigated[186,187] and the sign shown to be positive for the methyllithium tetramer[186].

Further developments in alkali metal n.m.r. are likely to include the development of ^{6}Li as an alternative to ^{7}Li n.m.r. and the study of organosodium and organopotassium compounds by ^{23}Na and ^{39}K n.m.r. The development of a suitable reference for each technique should be forthcoming as none is currently well established.

D. Studies of Alkaline Earth Metal Nuclei

Alkaline earth metal n.m.r. is not a widely studied technique[24,188]. All these metals have quadrupolar nuclei with generally low receptivities (Table 5). Both beryllium and magnesium display the same covalent characteristics in their chemistry and are well known in organometallic systems. Despite the extensive chemistry of Grignard reagents and other organomagnesium compounds, ^{25}Mg n.m.r. does not appear to have been applied in these areas. Conversely, ^{9}Be n.m.r. has been utilized in the study of organoberyllium compounds. The weak quadrupole moment of ^{9}Be suggests that quadrupolar relaxation need not always be a dominant relaxation mechanism for this nucleus and accordingly it is the most suited to direct observation by n.m.r. spectroscopy.

1. Organoberyllium compounds

The few available data[189,190] on ^{9}Be n.m.r. of organoberyllium compounds suggest that shielding is related to the coordination number of Be (four coordinate > three coordinate) and to the electronegativity of the substituents (electronegative substituents cause deshielding). Coupling of ^{9}Be to ^{31}P has been observed[24].

Attempts to apply analogous methods to obtain ^{25}Mg n.m.r. of organomagnesium compounds have largely been unsuccessful[191].

E. Studies of Transition Metal Nuclei (Including Lanthanides and Actinides)

Interest in the n.m.r. of transition metal nuclei has developed in such a fashion that there is currently available a vast amount of data concerning nine or ten nuclei and virtually none for all the remainder. Historically, this awkward situation has arisen because of several factors, including:

(*i*) the quadrupolar nature of many transition metal nuclei (see Table 6);
(*ii*) the existence of paramagnetic oxidation states;
(*iii*) the ability of indirect methods to extract data only where coupling to a high sensitivity nucleus can be observed; and
(*iv*) a diversity of chemical interest, which concentrates only on certain elements.

A review on the n.m.r. of transition metal nuclei[192] dealing with inorganic and organometallic compounds and a more recent up-date are available[193]. More specific reviews concerning the n.m.r. of central metal ions in octahedral complexes[194] (mainly ^{59}Co, ^{93}Nb and ^{195}Pt) and metal ions as probes in the n.m.r. of biological molecules[195] are also available.

1. *Sc, Y, La, the lanthanides, and actinides*

Scandium exhibits a coordination chemistry reminiscent of aluminium. The unfortunate corollary of this, from an n.m.r. viewpoint, is that scandium compounds tend to be of low symmetry about the Sc(III) metal centre. As ^{45}Sc ($I = \frac{7}{2}$, N = 100%, Table 6) is quadrupolar, the linewidths of its resonances are dependent upon the field gradient (and hence the symmetry) at the nucleus. Only in the octahedral ion, $[ScF_6]^{3-}$, has coupling been observed, $^{1}J(^{45}Sc, ^{19}F)$ = 180 Hz[196], as large linewidths generally preclude this in less symmetrical species. The standard for shift measurements is $[Sc(ClO_4)_3]$(aq.), which displays a known shift dependence upon concentration[197,198]. No organoscandium compounds appear to have been studied, presumably because of the field gradient problem in compounds of lower symmetry or because of a lack of chemical interest in this field.

^{89}Y is an $I = \frac{1}{2}$ nucleus (Table 3) and suffers the opposite n.m.r. problems to ^{45}Sc; thus it has a low magnetic moment and exhibits long relaxation times, making spin-saturation a problem. Chemical shift measurements of a few inorganic salts have been made using special techniques to overcome this problem[199], but no organometallic yttrium compounds have been studied. Couplings to ^{89}Y are known from spectra of other nuclei; for example, $[Y(CH_2CMe_3)_3{\cdot}2thf]$ has been examined by ^{1}H n.m.r. spectroscopy[200] and shows coupling to methylene protons, $^{2}J(^{89}Y, ^{1}H)$ = 2.5 Hz. The scandium analogue is also known[200], but obviously no coupling could be observed. The molecules **13** have also been examined by ^{1}H n.m.r. spectroscopy[201]. The yttrium

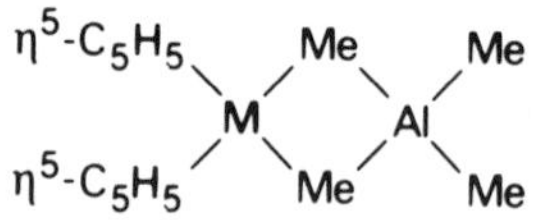

$(\eta^5\text{-}C_5H_5)_2M(\mu\text{-}Me)_2AlMe_2$

(13) M = Ti, Y

complex, whose X-ray structure is known[202], shows coupling to the bridging methyl protons, $^{2}J(^{89}Y, ^{1}H)$ = 5 Hz, at −45°C. At +40°C, collapse of the methyl resonances suggests a fluxional process, involving terminal-bridging methyl group exchange[201]. The ^{13}C n.m.r. spectrum of **13** (M = Y) shows $^{1}J(^{89}Y, ^{13}C)$ = 12.2 Hz[202]. The ethylaluminium analogue of **13** is similarly fluxional, but the static spectrum at −40°C shows coupling to the methylene protons of the bridging ethyl groups, $^{2}J(^{89}Y, ^{1}H)$ = 4 Hz[202]. Complexes of this sort are considered to be of importance as potential intermediates in Ziegler–Natta catalysis, the yttrium complex being especially suitable for study owing to its exceptionally high thermal stability[201].

^{139}La ($I = \frac{7}{2}$, N = 99.9%) is quadrupolar (Table 6) and the problems here parallel those for the ^{45}Sc nucleus. To date, ^{139}La n.m.r. has been of importance in two fields only. Firstly, in studies of ion pairing, the lowering of symmetry about the ^{139}La nucleus as cation–anion interactions occur causes linewidth changes which can be measured and interpreted in terms of inner-sphere complex formation. Secondly, in biochemistry, linewidth changes due to complex formation with biopolymers can be used as a structural probe. No reports on organolanthanum studies by ^{139}La n.m.r. appear to be available.

Although interest in the n.m.r. of the *f*-block element complexes[203,204], particularly organometallic complexes[205], is apparent from the availability of several reviews, no direct observation of metal nuclei in any organometallic compounds of the *f*-block elements appear to have been reported. No doubt this area will be further investigated in the future.

2. *Ti, Zr, Hf*

Titanium has two n.m.r.-active quadrupolar nuclei (Table 6) and both zirconium (^{91}Zr, $I = \frac{5}{2}$) and hafnium (^{177}Hf, N = 18.5%, $I = \frac{7}{2}$; ^{179}Hf, N = 13.75%, $I = \frac{9}{2}$) have suitable isotopes for study. Of these, only titanium has been investigated and parameters for halide salts reported[206]. No organometallic compounds have been examined, the quadrupole problem presumably deterring potential investigators.

3. *V, Nb, Ta*

^{51}V ($I = \frac{7}{2}$), ^{93}Nb ($I = \frac{9}{2}$), and ^{181}Ta ($I = \frac{7}{2}$) are all about 100% abundant (Table 6) and the ease of obtaining spectra of these nuclei relates directly to the magnitude of their quadrupole moments (Table 6). Both ^{51}V and ^{93}Nb have relatively small quadrupole moments, enabling resonances to be observed in most compounds, provided that the symmetry is not extremely low. ^{181}Ta, however, has a very large quadrupole moment (Table 6) and the only solution measurement is of the octahedral $[TaF_6]^-$ ion[207].

Many vanadium carbonyl derivatives[208–214], substituted with cyclopentadienyl or Group VB ligands, have been examined by ^{51}V n.m.r. spectroscopy. Chemical shift measurements are usually given relative to $VOCl_3$ (neat liquid), which resonates at the extreme de-shielding end of the shift range. Increasing shielding with decreasing oxidation state is observed, as expected in terms of electron density, and so most values are negative with respect to the reference. Various attempts have been made to correlate vanadium chemical shifts with ligand strength in complexes of the types $[V(C_5H_5)(CO)_3(PR_3)]$, $[V(C_5H_5)(CO)_2(PR_3)_2]$, $[V(CO)_5(PR_3)]^-$, and $[V(CO)_4(PR_3)_2]^-$. The term 'ligand strength' is generally fairly arbitrary, although correlations of $\delta(^{51}V)$ with ligand field splitting have been observed[192], which can be attributed to variations in ΔE in equation 2, altering the magnitude of σ_p. Similarly, variations in $\delta(^{51}V)$ with the electronegativity of the R group in PR_3 have been discussed[214], although these workers also observe a correlation[214] with Tolman's cone angle[215] for the phosphine; of course, to separate electronic and steric contributions in a single molecule is very difficult. Complexes of types $[V(CO)_4(LL)]^-$, $[V(C_5H_5)(CO)_2(LL)]$, and $[VH(CO)_4(LL)]$ also show a variation in $\delta(^{51}V)$ with the chelate-bite of the bidentate diphosphine, LL[214]. The most strained structures, with a four-membered ring, show minimum shielding of ^{51}V and maximum shielding of ^{31}P, indicative of a reduced V—P bond order (i.e. withdrawal of electron density by phosphorus). Vanadium(I) complexes exhibit[211,212] shifts of *ca.* -1250 ± 250 ppm, whilst vanadium(−I) complexes are[208,212] more shielded, *ca.* -1800 ± 150 ppm. The reported shift for $[V(CO)_6]$ of +5660 ppm[208] does not fit these general ranges and recent reviewers[192] have suggested that a contact or pseudocontact interaction with a paramagnetic decomposition product may be responsible for this enormous deshielding. The explanation is certainly credible and a re-investigation of the compound would appear to be warranted.

Coupling, $^1J(^{59}V, ^{31}P)$, is generally observed in the phosphine-substituted carbonyl complexes and the variations in magnitude appear to relate only to the nature of the phosphine[212] and thus are not diagnostically very informative[211].

^{93}Nb n.m.r. has been applied to a number of structural problems in inorganic niobium chemistry[192]. Only a narrow shift range is currently known (*ca.* 2300 ppm), but as only Nb(V) compounds have so far been studied, this can be expected to increase. Niobium has a limited organometallic chemistry and no organoniobium compounds have been examined by ^{93}Nb n.m.r. spectroscopy. Both $[NbCl_6]^-$ and $[NbF_6]^-$ have been used as shift standards[192]; the latter is preferred as, despite the coupling to fluorine, the fluoride resonates at the high shielding end of the shift range whereas the chloride resonates in the middle of the range.

As previously mentioned, no organotantalum compounds have been studied by ^{181}Ta n.m.r. spectroscopy.

4. *Cr, Mo, W*

^{53}Cr ($I = \frac{3}{2}$, $N = 9.55\%$) has a low sensitivity to n.m.r. detection and data relating to ^{53}Cr n.m.r. are scarce. The $[CrO_4]^{2-}$ ion has been examined[216], where symmetry constraints do not cause a large field gradient at the nucleus and $[Cr(CO)_6]$, where the ^{53}Cr nucleus is shielded by 1795 ppm relative to the chromate ion[217]. The higher shielding in the lower oxidation state is to be expected on electron density considerations. No other data relevant to organometallic systems appear to be available.

The two active molybdenum isotopes, ^{95}Mo ($I = \frac{5}{2}$, $N = 15.7\%$) and ^{97}Mo ($I = \frac{5}{2}$, $N = 9.5\%$), are quadrupolar with weak magnetic moments (Table 6). ^{95}Mo is preferred as it is slightly more sensitive and has a smaller quadrupolar moment. Few results from ^{95}Mo n.m.r. have yet been obtained. Both ^{95}Mo and ^{97}Mo resonances for $[MoO_4]^{2-}$ have been reported[218] and resonances at −1309 and −1856 ppm relative to $\delta(^{95}Mo)$ of the molybdate ion observed for $[Mo(CN)_8]^{4-}$ and $[Mo(CO)_6]$, respectively[219]. No other studies of relevance here appear to have been described. The decrease in shielding, Mo(0) > Mo(IV), is as expected (see above).

A few coupling constants involving molybdenum have been measured from the n.m.r. of the coupled nuclei. Particularly interesting are results obtained from the $^{31}P\{^1H\}$ n.m.r. of $[Mo(CO)_5L]$ (L = tertiary phosphine) where satellites assigned to coupling between phosphorus and both molybdenum isotopes were described[220]. These workers suggested that separate satellites could not be resolved because of the small differences between $^1J(^{95}Mo, ^{31}P)$ and $^1J(^{97}Mo, ^{31}P)$ in relation to the linewidth. The system has been re-investigated by $^{31}P\{^{95}Mo, ^1H\}$ triple resonance methods[221] and it appears that the satellites result entirely from coupling to ^{95}Mo. The explanation

TABLE 8. $\delta(^{17}O)$ and $^{1}J(^{183}W, ^{13}C)$ data for some tungsten carbonyl derivatives

Complex	$\delta(^{17}O)_{cis}$[a]	$\delta(^{17}O)_{trans}$[a]	$^{1}J(^{183}W, ^{13}C)_{cis}$ (Hz)	$^{1}J(^{183}W, ^{13}C)_{trans}$ (Hz)
$[W(CO)_6]$	356.8		124.5[b]	
$[W(CO)_5CPh_2]$	364.7	452.6	127.0	102.5[b]
$[W(CO)_5C(OMe)Ph]$	357.1	388.9	127.0	115.2
$[W(CO)_5C(NH_2)Ph]$	352.0	372.5	126.9	124.8
$[W(CO)_5\overline{CCBu^t{=}C}Bu^t]$	350.6	367.6	125.7	133.0[c]
$[W(CO)_5P(OMe)_3]$	353.7	359.0	125.1	139.1[d]
$[W(CO)_5PPh_3]$	353.6	$\geqslant$ 353.6[e]	—	—
$[W(CO)_5PBu^n_3]$	354.1	354.1	124.4	142.1[d]
$[W(CO)_4(Ph_2PCH_2CH_2PPh_2)]$	349.4	358.1	—	—
$[W(CO)_4I][NBu^n_4]$	349.0	349.0	127.0	175.8

[a] ppm relative to $H_2^{17}O$; data from reference 233.
[b] Data from reference 234.
[c] Data from reference 235.
[d] Data from reference 236.
[e] Shoulder on high-frequency side of *cis*-carbonyl resonance.

TABLE 9. $\delta(^{199}Hg)$ and $^{1}J(^{199}Hg, ^{183}W)$ data for some mercury–tungsten compounds, $[HgX\{W(CO)_3(C_5H_5)\}]$

X	$\delta(^{199}Hg)$ (ppm)[a]	$^{1}J(^{199}Hg, ^{183}W)$ (Hz)
Cl	−997	706
Br	−1200	690
I	−1529	630
SCN	−924	684

[a] Relative to 90% $HgMe_2$/10% C_6F_6 (internal lock). Data from reference 238.

proposed is that the large quadrupolar moment of ^{97}Mo causes rapid relaxation in isotopomers of ^{97}Mo which effectively decouples the phosphorus nucleus. In ^{95}Mo, the quadrupolar moment is substantially smaller and so relaxation is slower, enabling coupling to be observed.

Tungsten has one $I = \frac{1}{2}$ isotope, ^{183}W, which presents several problems relating to n.m.r. measurement. The isotope has a low sensitivity (Table 3), which hinders measurement by pulsed excitation, whilst the low gyromagnetic ratio leads to small values of $^{n}J(^{183}W, X)$, which restricts indirect observation, as the coupling must be clearly resolved before double resonance methods can be used. Nonetheless, $^{1}H\{^{183}W\}$ n.m.r. studies have been successfully performed for some tungsten phosphine complexes[91] in cases where $^{n}J(^{183}W, ^{1}H) > 1$ Hz. Even in cases where the coupling is less than this, the fact that $^{3}J(^{31}P, ^{1}H)$ and $^{1}J(^{183}W, ^{31}P)$ are both distinguishable allows $^{1}H\{^{31}P, ^{183}W\}$ triple resonance experiments to be performed[92] and so the data can be extracted by the indirect method.

Chemical shifts are measured relative to WF_6 and, apart from the usual variation with oxidation state, do not seem to be particularly sensitive probes for structural changes. Many carbonyl derivatives of types $[W(CO)_5L]$, $[W(CO)_4L_2]$, and $[W(CO)_3(C_5H_5)X]$ (L = tertiary phosphine, X = anionic group) have been studied[92,222] and a compilation of shifts is available[192]. The range of shifts for a given class of compound is small and hence correlations with other spectral parameters are not to be heavily relied upon.

As previously mentioned, couplings involving tungsten tend to be small and, although a substantial number of $^{1}J(^{183}W, ^{31}P)$ values have been deduced from the $^{31}P\{^{1}H\}$ n.m.r. of tungsten phosphine and phosphite complexes[223–229], few other couplings involving nuclei other than the proton are known. Couplings to fluorine in tungsten oxyfluorides[230,231] and to carbon in tungsten carbonyls have been reported[232] (Table 8), as have the values of $^{1}J(^{183}W, ^{119}Sn) = -150 \pm 5$ Hz by $^{1}H\{^{183}W\}$ double resonance of $[Me_3SnW(CO)_3(C_5H_5)]$[237] and $^{1}J(^{199}Hg, ^{183}W)$ (Table 9).

5. *Mn, Tc, Re*

^{55}Mn is a quadrupolar nucleus ($I = \frac{5}{2}$) of 100% abundance. It has a moderate quadrupole moment (Table 6) and accordingly it is to be expected that broad lines will result for complexes of low symmetry and that couplings involving ^{55}Mn will be difficult to determine unambiguously. These expectations are borne out in practice. Technetium has no stable isotopes and so is of limited interest in organometallic chemistry and will not be considered further here. Both ^{185}Re ($N = 37.07\%$) and ^{187}Re ($N = 62.93\%$) have $I = \frac{5}{2}$; ^{187}Re has the smaller quadrupole moment and is slightly more sensitive to n.m.r. detection (Table 6). Limited studies of rhenium n.m.r. have been described[192], but no data relevant to organometallic systems have been reported. Accordingly, only ^{55}Mn is of interest in this triad.

TABLE 10. $\delta(^{55}Mn)$ data for $[Mn(CO)_5X]$ complexes

X	$\delta(^{55}Mn)$ (ppm)[a]	X	$\delta(^{55}Mn)$ (ppm)[a]
H	−2630	CF_3	−1850
Me	−2265	I	−1485
CH_2F	−2130	Br	−1160
CHF_2	−1970	Cl	−1005

[a]Relative to $[K][MnO_4]_{(aq)}$; data from reference 34.

^{55}Mn chemical shifts are reported relative to $[K][MnO_4]_{(aq.)}$. The high oxidation state of manganese means that the resonance occurs at the low shielding end of the shift range and so most values are negative with respect to this reference. The high symmetry of the $[MnO_4]^-$ ion results in a narrow ^{55}Mn resonance and so, despite considerable solvent effects[192], the standard seems a good choice. Most compounds which have been studied involve low-valent manganese [Mn(I) and Mn(−I)] as the paramagnetic compounds of higher oxidation state are obviously not suitable.

Manganese chemical shifts have been measured for many complexes[33,34,239–244], largely substituted carbonyls, $[Mn(CO)_5L]$[34,240,242], $[MnH(CO)_{5-n}(PF_3)_n]$[239], and some dimeric carbonyls[34,241] (L = anionic ligand, n = 0–5). Data for some mononuclear $[Mn(CO)_5X]$ compounds are displayed in Table 10. The increase in shielding on moving from the $[X^- \text{----} {}^+Mn(CO)_5]$ types to the $[X^+ \text{----} {}^-Mn(CO)_5]$ types has been taken as a diagnostic of the σ-donor ability of the X ligand[240,242]. The increase in shielding reflects the increased electron density at the manganese nucleus.

It was previously mentioned that correlations between nuclear shielding and other spectroscopic parameters can be unreliable, a consequence of the number of variables in the shielding equations. Here, the ^{55}Mn shielding increases in the order Cl < Br < I (Table 10), exactly as predicted by the nephelauxetic series by 'cloud expansion' (i.e. variation in the radial term, $\langle r^{-3}\rangle$, in equation 2). This correlation is not always observed as changing from hard to soft halide donors usually causes ΔE to alter sufficiently that variations in $\langle r^{-3}\rangle$ are obscured. This variation is often described as normal or typical behaviour, whereas in reality it is not always observed with transition metal nuclei.

The variation in ^{55}Mn linewidth with symmetry provides a useful tool for measuring quadrupole coupling constants without resorting to NQR. The subject has been discussed in some detail[192].

Coupling data have yet to be extracted from ^{55}Mn n.m.r. spectra. The only coupling involving ^{55}Mn which appears to have been reported is $^1J(^{55}Mn, ^{17}O) \approx 30$ Hz[245,246] for the $[MnO_4]^-$ ion.

6. *Fe, Ru, Os*

The active nuclei in the iron triad must rank as one of the greatest disappointments in organometallic n.m.r. studies since, despite the very rich organometallic chemistry displayed by these elements, the difficulties encountered in observation of the n.m.r. of these nuclei almost precludes their use as structural probes.

^{57}Fe ($I = \frac{1}{2}$, N = 2.2%) has a poor sensitivity to n.m.r. detection (Table 3) and the long relaxation times associated with the $I = \frac{1}{2}$ nuclei cause saturation problems at high radiofrequency powers. Both active ruthenium nuclei, ^{99}Ru ($I = \frac{5}{2}$, N = 12.7%) and ^{101}Ru ($I = \frac{5}{2}$, N = 17.1%) are quadrupolar and no reports of their n.m.r. appear to be available. Osmium has both $I = \frac{1}{2}$ (^{187}Os) and $I = \frac{3}{2}$ (^{189}Os) nuclei; the former has a very low n.m.r. sensitivity and the latter is quadrupolar. Apart from resonances

TABLE 11. $\delta(^{57}Fe)$ data for some substituted ferrocenes

Compound[a]	$\delta(^{57}Fe)$ (ppm)[b]	Solvent
Fc—C(=O)CH$_3$	215.55	CS_2
Fc—CH(OH)CH$_3$	0.00	CS_2
Fc—CH$_2$CH$_3$	36.6	CS_2
Fc—H$^+$	−1098.85	$BF_3{\cdot}H_2O$

[a]Type: [(C_5H_4—R)Fe(C_5H_5)]; data from reference 248. Compounds enriched with up to 82% ^{57}Fe.

[b]Relative to $\delta(^{57}Fe)$ of ferrocene[c] = 0 (30°C, CS_2 solution).

[c]From $^{13}C\{^1H\}$ n.m.r., $^1J(^{57}Fe, ^{13}C) = 4.88 \pm 0.12$ Hz and $\delta(^{13}C) = 67.7$ ppm for ferrocene.

attributable to both isotopes in OsO_4[192], no other n.m.r. parameters have been measured. Only the ^{57}Fe nucleus has been studied in any depth for this triad and even here only a few significant results have been reported.

The ^{57}Fe resonance of [Fe(CO)$_5$] has been reported[247], but very sophisticated experimental techniques were necessary for its observation and the effort required would dissuade most workers from attempting similar experiments with other compounds. Substituted ferrocenes, enriched with ^{57}Fe (an expensive process), have also been examined[248,249], using multiple resonance methods. The data for these compounds are shown in Table 11. Double resonance studies of myoglobin enriched to 90% with ^{13}C and ^{57}Fe have also been described[250].

Apart from the data in Table 11, few coupling constants involving ^{57}Fe are known. Values of $^1J(^{57}Fe, ^{13}C) = 23.4$ Hz in [Fe(CO)$_5$] and $^1J(^{57}Fe, ^{31}P) \approx 26.5$ Hz in [Fe(CO)$_4$(PR$_3$)] complexes have been reported[251].

7. Co, Rh, Ir

^{59}Co ($I = \frac{7}{2}$, $N = 100\%$) is a sensitive nucleus for n.m.r. detection (Table 6); it has a moderate quadrupole moment and so linewidth variations with symmetry are to be expected. ^{103}Rh, despite being 100% abundant with $I = \frac{1}{2}$, is very difficult to observe directly by ^{103}Rh n.m.r. as it has a magnetic moment and a gyromagnetic ratio which are both of small magnitude (Table 3). Most measurements of ^{103}Rh n.m.r. parameters have been made indirectly. Iridium has two isotopes of $I = \frac{3}{2}$, ^{191}Ir and ^{193}Ir, both of which have large quadrupole moments. To date, no significant reports of iridium n.m.r. have appeared, and this is unlikely to be a useful area of study in organometallic systems as the linewidths will be prohibitively large in all but the most symmetric complexes.

Co(III) coordination complexes have been studied in depth by ^{59}Co n.m.r. [Co(II) being a paramagnetic oxidation state] and possibly more data on chemical shifts are available for Co(III) than for any other transition metal ion. Chemical shift data for cobalt complexes have recently been reviewed[192], with extensive tabulation of data for Co(III), and so the results, which do not bear directly on organometallic chemistry, will not be reconsidered here, although some results of relevance have appeared since the review article was published[252–259]. Of greater relevance to organometallic systems are data for low-valent cobalt complexes, largely Co(0) and Co(−I), for which rather fewer results are available.

Chemical shifts are generally measured relative to $[Co(CN)_6]^{3-}{}_{(aq.)}$, used as an external reference to prevent ligand exchange problems. The normal relationship between oxidation state and shielding appears to hold, i.e. Co(−I) > Co(0) > Co(III).

A number of cobalt(0) carbonyls and related cobalt(−I) derivatives have been examined[260–267], these compounds being of interest in catalytic hydroformylation and hydrosilylation reactions. The polynuclear carbonyls $[Co_2(CO)_8]$[260–262,267] and $[Co_4(CO)_{12}]$[261,262,264,267] have been examined by several workers; the former shows a single resonance at *ca.* −2100 ppm and the latter shows two resonances in the ratio 1:3 at *ca.* −400 and −1700 ppm, as expected. The data are approximate as there appears to be a substantial variation in shift with solvent. Some Co(−I) complexes, $[Co(NO)_2Br(PR_3)]$, have been studied[265] and the value of the cobalt shielding was found to correlate with the π-acceptor ability of the phosphine (i.e. good π-acceptors cause deshielding) as expected in terms of electron density at cobalt. Complexes of the type $[Co(CO)_4X]$ (X = SiR_3, GeR_3, H) have also been studied[261,263] and show the expected variation in $\delta(^{59}Co)$ with the electron withdrawing ability of X. The anion $[Co(PF_3)_4]^-$ gives rise to a cobalt resonance at −4220 ppm[261], comparison with a related Co(III) compound, $[Co(P\{OMe\}_3)_6]^{3+}$, $\delta(^{59}Co)$ = +304 ppm[258] demonstrates the considerable shielding experienced by low-valent cobalt nuclei.

As ^{59}Co is quadrupolar, few coupling constants involving this nucleus have been observed. For example, some Co(III) phosphite complexes have been reported[258,259], $[CoL_6]^{3+}$, where L = **14–18**, but $^1J(^{59}Co, ^{31}P)$ could only be obtained from the ^{59}Co n.m.r. of the complexes for L = **14** and **18**; quadrupolar broadening prevented observation of coupling for L = **15–17**. The values obtained were $^1J(^{59}Co, ^{31}P)$ = 443 Hz[259] for L = $P(OMe)_3$ and 412 Hz for L = **18**[259]. In the anion $[Co(PF_4)_4]^-$,

$P(OMe)_3$

(**14**)

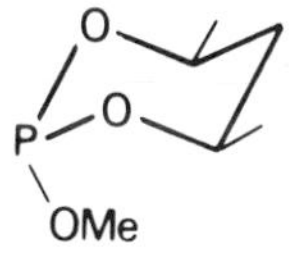

(**15**)

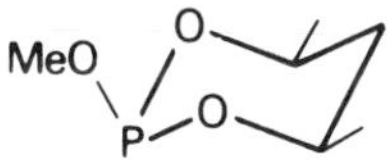

(**16**)

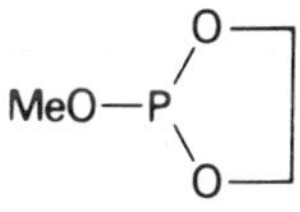

(**17**)

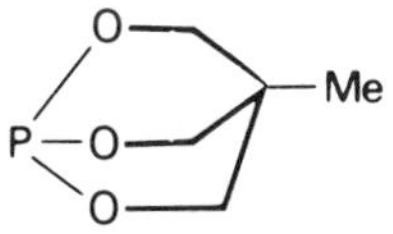

(**18**)

$^{1}J(^{59}Co, ^{31}P) = 1222$ Hz and $^{2}J(^{59}Co, ^{19}F) = 57$ Hz[261]. Coupling to carbon in $[Co(CO)_4]$, $^{1}J(^{59}Co, ^{13}C) = 287$ Hz[261] and $[Co(CN)_6]^{3-}$, $^{1}J(^{59}Co, ^{13}C) = 126$ Hz[268] has also been reported.

A variety of rhodium complexes ranging from simple mononuclear Rh(I)[85,86,269–271] and Rh(III)[85,86,269,272,273] species to complex carbonyl clusters[271], have been examined by double resonance methods. The choice of standard for ^{103}Rh n.m.r. shifts is currently rather arbitrary; *mer*-$[RhCl_3(SMe_2)_3]$ has been used[85,272], as this is especially suitable for $^{1}H\{^{103}Rh\}$ n.m.r. More recently, the symmetrical $[Rh_6C(CO)_{15}]^{2-}$ ion has been employed as a reference[274] in the direct observation of clusters where the resonant frequencies are corrected to the field strength where the protons of Me_4Si resonate at 200 MHz. A more acceptable approach is to use an essentially arbitrary value of 3.16 MHz as the resonant frequency of the rhodium reference corrected to the field strength where the protons of Me_4Si resonate at 100 MHz; i.e. $\Theta(Rh) = 3.16$ MHz. This arbitrary approach, relating resonance shifts to a given frequency, is used here to be consistent with the most recent articles dealing with ^{103}Rh n.m.r. data[192,193].

TABLE 12. $\delta(^{103}Rh)$ data for $[Rh(PPh_3)_3X]$ complexes

X	$\delta(^{103}Rh)$ (ppm)^a
Cl	−81
Br	−142
I	−268

^{a}Relative to $\Xi(^{103}Rh) = 3.16$ MHz; data obtained by $^{31}P\{^{103}Rh\}$ double resonance; reference 86.

Shift data[86] for the complexes $[Rh(PPh_3)_3X]$ (X = Cl, Br, I) are given in Table 12, and show a decrease in shielding, I > Br > Cl, as previously described for ^{55}Mn shielding. Variations in ΔE would usually cause the opposite trend to be observed, so here the effects of altering the halide ligand on $\langle r^{-3}\rangle$ must be dominant in determining the paramagnetic contribution. A large number of Rh(I) and Rh(III) complexes have been examined and their data have been discussed recently[192]. The most important recent development in ^{103}Rh n.m.r. is the direct observation of this nucleus in solution. A number of mononuclear complexes were initially studied[275] (Table 13), including some cyclopentadienyl derivatives which show no observable coupling to rhodium and

TABLE 13. ^{103}Rh n.m.r. parameters of mononuclear complexes obtained by direct observation

Complex	$\delta(^{103}Rh)$ (ppm)^a	$^{1}J(^{103}Rh, ^{31}P)$ (Hz)
$[Rh(PPh_3)_2(CO)Cl]$	−368	127 ± 4
$[C_5Me_5RhCl_2]_2$	+2303	—
$[C_5Me_5Rh(OCOMe)_2\cdot H_2O]_n$	+2644	—
$[C_5Me_5Rh(NO_3)_2]$	+2364	—
$[C_5Me_5Rh(PMe_2Ph)_2Cl][Cl]$	+425	104 ± 2
$[C_5Me_5Rh(PMe_2Ph)_2Cl][BPh_4]$	+415	143 ± 5

^{a}Relative to $\Xi(^{103}Rh) = 3.16$ MHz; data from reference 275.

TABLE 14. ^{103}Rh n.m.r. parameters of cluster complexes obtained by direct observation

Complex	$\delta(^{103}Rh)$ (ppm)[a]	$^{1}J(^{103}Rh, ^{31}P)$ (Hz)
$[Rh_6C(CO)_{15}]^{2-}$	−277.2	—
$[Rh_{17}S_2(CO)_{32}]^{3-}$ [b]	+879.3(1)	—
	−638.6(8)	—
	−1457.8(8)	—
$[Rh_9P(CO)_{21}]^{2-}$ [c]	−1423.1	46
	−1222.3	32
	−1051.8	36

[a]Relative to $\Xi(^{103}Rh)$ = 3.16 MHz; relative intensities in parentheses. Data from reference 274.
[b]See Figure 5.
[c]See Figure 6.

so could not be examined indirectly. The technique has since been utilized[274] to study some carbonyl clusters, $[Rh_9P(CO)_{21}]^{2-}$ and $[Rh_{17}S_2(CO)_{32}]^{3-}$. The data are shown in Table 14. The Rh_9 cluster has been characterized crystallographically[276] and the solid-state structure consists of eight rhodium atoms at the corners of a cubic antiprism with the ninth rhodium atom capping one square face. The phosphorus atom is encapsulated in the central cavity (see Figure 4). Initial $^{13}C\{^{1}H\}$ n.m.r. studies[276] showed that there was carbonyl fluxionality in the temperature range −40 to +45°C. The $^{31}P\{^{1}H\}$ n.m.r. spectrum[276] consisted of a ten-line multiplet at +40°C which increased in multiplicity down to −80°C. To investigate whether the rhodium skeleton was fluxional, as the ^{13}C and ^{31}P studies suggest, the ^{103}Rh n.m.r. spectrum was obtained[274]. The rhodium skeleton was indeed found to be non-rigid, as a static spectrum could be observed at −80°C consistent with the solid-state structure, whereas at +23°C only one doublet was observed. The shift and coupling constant of

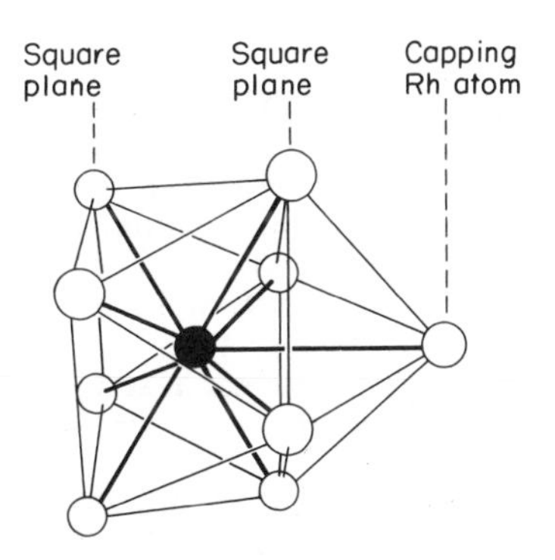

FIGURE 4. X-ray structure of $[Rh_9P(CO)_{21}]^{2-}$. Open spheres, Rh; closed spheres, P. Carbonyl ligands omitted. See Table 14 for ^{103}Rh n.m.r. data. Reprinted with permission from Vidal et al., *Inorg. Chem.*, **18**, 129 (1979). Copyright 1979 American Chemical Society.

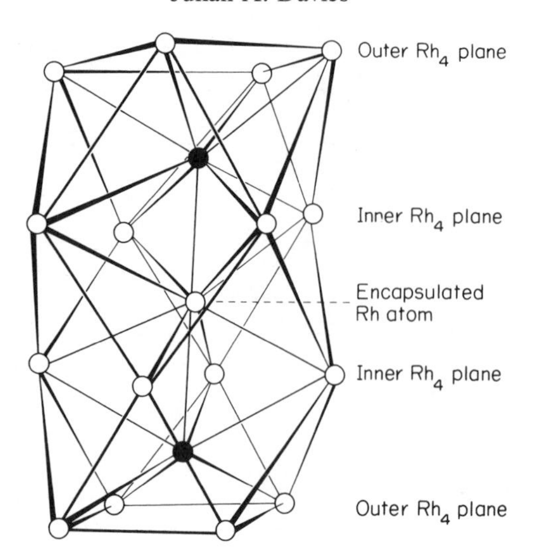

FIGURE 5. X-ray structure of $[Rh_{17}S_2(CO)_{32}]^{3-}$. Open spheres. Rh; closed spheres. S. Carbonyl ligands omitted. See Table 14 for ^{103}Rh n.m.r. data. Reprinted with permission from Vidal et al., *Inorg. Chem.*, **17**, 2574 (1978). Copyright 1978 American Chemical Society.

the doublet corresponded to a weighted mean of the three resonances in the static spectrum.

The $[Rh_{17}S_2(CO)_{32}]^{3-}$ ion gave a static spectrum[274]. The crystal structure of this compound is known[277] and consists of sixteen rhodium atoms at the corners of four square planes stacked horizontally with a 45° stagger between each layer. The S—Rh—S moiety lies along the C_4 axis of the cavity (Figure 5). The ^{103}Rh n.m.r. spectrum[274] shows resonances attributable to the two outer planes (8Rh), the two inner planes (8Rh) and the encapsulated atom (1Rh). The encapsulated atom is very deshielded, indicating a highly electropositive nature. The ^{13}C n.m.r. spectrum of this compound has also been reported[278], and exchange studies with ^{13}CO show that enrichment at all sites occurs at *ca*. 125°C whilst no enrichment occurs at lower temperatures (40°C). Only at high temperatures could carbonyl scrambling on the internal Rh_4 planes be observed. The ^{103}Rh n.m.r. results show[274] a static structure for the rhodium skeleton, but high-temperature studies are necessary for investigating the fluxionality observed in the ^{13}C n.m.r. work.

Another recent development in the study of carbonyl clusters is the use of selective $^{13}C\{^{103}Rh\}$ decoupling experiments[84]. The anion $[Rh_7(CO)_{16}]^{3-}$ has been investigated by this technique. The solid-state structure of this anion is known[279] and consists of a monocapped octahedron with seven terminal carbonyls (one per Rh), six edge-bridging and three face-bridging carbonyls (Figure 6). The $^{13}C\{^{103}Rh\}$ n.m.r.[84] demonstrated that one set of carbonyl groups moves around the basal triangle of rhodium atoms in the fluxional process and that fluxionality of the rhodium skeleton is not involved. Direct observation[274] of the ^{103}Rh n.m.r. of $[Rh_7(CO)_{16}]^{3-}$ confirms the rhodium chemical shifts derived from the $^{13}C\{^{103}Rh\}$ measurements[84] (Table 15) as three resonances in a 3:1:3 ratio were observed at similar shift values.

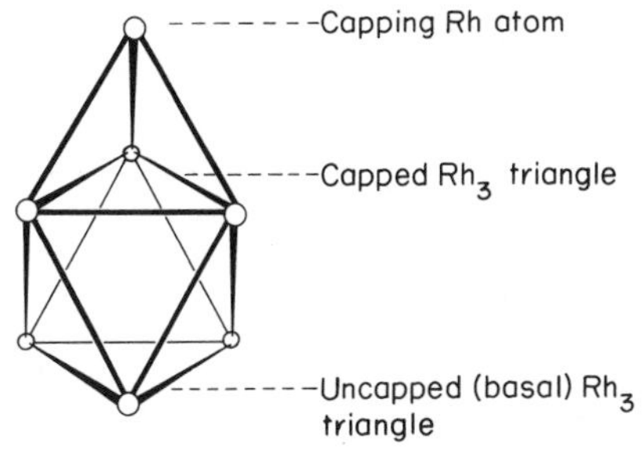

FIGURE 6. X-ray structure of $[Rh_7(CO)_{16}]^{3-}$, showing Rh skeleton only. Reproduced from reference 279 with permission. See Table 15 for $^{13}C\{^{103}Rh\}$ n.m.r. data.

TABLE 15. ^{103}Rh n.m.r. parameters of $[Rh_7(CO)_{16}]^{3-}$ obtained by $^{13}C\{^{103}Rh\}$ double resonance methods

Assignment[a]	$\delta(^{103}Rh)$ (ppm)[b]	Number of Rh—Rh bonds associated with atom
Capping Rh atom	+483	3
Capped Rh_3 triangle	−376	4
Uncapped (basal) Rh_3 triangle	+690	5

[a] See Figure 6.
[b] Relative to $\Xi(^{103}Rh) = 3.16$ MHz; data from reference 84.

As $\gamma(^{103}Rh)$ is of small magnitude, coupling constants involving the ^{103}Rh nucleus tend to be small in size. Couplings to ^{1}H[159], ^{13}C[232] and ^{31}P[280,281] have been discussed previously. Some telluride complexes have been described[83] and $^{1}J(^{125}Te, ^{103}Rh)$ measured; the data (Table 16) show the expected effects according to the *trans*-influence of the *trans*-ligand, but effects of changing the *trans*-halide are rather small and so not very diagnostically informative.

TABLE 16. ^{103}Rh n.m.r. parameters of *mer*-$[RhX_3(TeMe_2)_3]$ (X = Cl, Br, I) obtained by $^{1}H\{^{103}Rh\}$ double resonance methods

X	$\delta(^{103}Rh)$ (ppm)[a]	$^{1}J(^{125}Te, ^{103}Rh)$ (Hz)	
Cl	+3179	+71[b]	+94[c]
Br	+2567	+70	+93
I	+1352	+66	+69

[a] Relative to $\Xi(^{103}Rh) = 3.16$ MHz; data from reference 83.
[b] *trans* to $TeMe_2$.
[c] *trans* to X.

8. *Ni, Pd, Pt*

The ^{61}Ni nucleus ($I = \frac{3}{2}$, $N = 1.19\%$) has attracted very little attention from chemists, presumably the problems of low abundance and the nuclear quadrupole

TABLE 17. $\delta(^{61}Ni)$ and $^{1}J(^{61}Ni, ^{31}P)$ data for some $[Ni(CO)_3L]$ complexes

L	$\delta(^{61}Ni)$ (ppm)[a]	$^{1}J(^{61}Ni, ^{31}P)$ (Hz)
$PBu^t_2(SnMe_3)$	92.8	206
$PBu^t_2(GeMe_3)$	59.2	203
$PBu^t(GeMe_3)_2$	87.0	197
$PBu^t(SiMe_3)_2$	81.8	161
$PBu^t_2(SiMe_3)$	58.9	197
PBu^t_3	56.6	232

[a] Relative to $[Ni(CO)_4]$ in benzene. Data from reference 283.

dissuading many workers from attempting ^{61}Ni n.m.r. experiments. Despite reports to the contrary[24,192,193], the first solution measurement of an organometallic compound, $[Ni(CO)_4]$, by ^{61}Ni n.m.r. appears to have been performed as early as 1964[282], with no special problems of observing a resonance being mentioned. No other work appears to have been reported until very recently, when the ^{61}Ni n.m.r. of some $[Ni(CO)_3L]$ (L = tertiary phosphine) complexes were reported[283]. The values of $\delta(^{61}Ni)$ and $^{1}J(^{61}Ni, ^{31}P)$ obtained are shown in Table 17. The standard used for shift measurements was $[Ni(CO)_4]$ in benzene, rather an unpleasant choice in view of the toxicity and volatility of nickel carbonyl and the carcinogenicity of benzene, but necessary as the precedent for its measurement was available[282]. Undoubtedly this area will be one of rapid future growth in view of the extensive organometallic chemistry of nickel.

^{105}Pd ($I = \frac{5}{2}$, N = 22.23%) does not appear to have been the subject of chemically significant n.m.r. studies.

^{195}Pt ($I = \frac{1}{2}$, N = 33.8%) has been studied in some detail, initially by double resonance and now increasingly by direct observation.

Considerable solvent effects on the chemical shifts of both Pt(II) and Pt(IV) compounds have been noted[192] and accordingly the choice of shift standard is of considerable importance. Although a wide variety of compounds have been used as standards, none seem entirely satisfactory and again it seems best to employ an arbitrary frequency scale. Here, a value of $\Theta(^{195}Pt)$ = 21.4 MHz is used, in keeping with a previous article concerning ^{195}Pt n.m.r.[192]. A large number of coordination complexes of Pt(II) and, to a lesser extent, Pt(IV) have been examined by ^{195}Pt n.m.r. spectroscopy. As these studies have recently been discussed[192], only organoplatinum(II) compounds and the more recent results for related coordination compounds of significance in organometallic systems will be discussed here.

The shift range observed for ^{195}Pt resonances is very large, extending from +11 847 ppm for $[PtF_6]^{2-}$ [284] to −1528 ppm for $[PtI_6]^{2-}$ [285]. Clearly, as two Pt(IV) compounds occupy the extreme ends of the shift range, the magnitude of the ^{195}Pt chemical shift cannot be used to differentiate between oxidation states. There are, however, clearly observable trends in shift for related series of compounds. Some examples, useful in structural analysis, are as follows:

(*i*) Analogous substituted-phenylplatinum(II) complexes have similar shifts when substituted in *meta*- and *para*-positions as the shielding is not very sensitive to electron-withdrawing effects. Substitution in the *ortho*-position of the aryl ring causes deshielding, possibly a result of a sterically induced interaction between the *meta*-substituent and the platinum nucleus, resulting in an enhanced dipole–dipole interaction[286], or a result of a disrupted solvation sphere about the platinum nucleus.

TABLE 18. ^{195}Pt n.m.r. data for $[PtMe_2L_2]$ complexes, comparison of strain effects on $\delta(^{195}Pt)$

L_2	$\Delta\{\delta(^{195}Pt)\}$ (ppm)^a
dppm	+703.9
dppe	−9.2
dppp	+4.7
dppb	−58.6
2 PPh_2Me	0
$PBu^t_2(Bu^t_2PC_4H_8)$	+685

$^a\Delta\{\delta(^{195}Pt)\} = \delta(^{195}Pt)$ of complex $- \delta(^{195}Pt)$ of $[PtMe_2(PPh_2Me)_2]$ in ppm. Data from reference 287.

(*ii*) Platinum shieldings are very sensitive to bond angle strain[286] (see also ^{119}Sn and ^{207}Pb) and the formation of strained metallocycles is accompanied by considerable deshielding. An example is the series of compounds $[PtMe_2L_2]$ where $L_2 = Ph_2P\{CH_2\}_nPPh_2$ (n = 1–4) and L = $PMePh_2$. The shift data[287] (Table 18) for n = 2, 3, and 4 are all similar whereas the strained system, n = 1, is deshielded by *ca.* 700 ppm. This effect can be used to diagnose metallation in tertiary phosphine complexes; for example[287], in the metallated complex $[\overline{Pt(PBu^t_2CMe_2C}H_2)(PBu^t_3)Cl]$ the chemical shift is similar to that of $[PtMe_2(dppm)]$; data in Table 18 show that this value is typical of a strained platinacycle.

(*iii*) In related series of compounds, such as $[PtCl_3(PR_3)]^-$, changing the substituents on the donor atom has little effect on shielding (as in phenyl substitution in (*i*) above). For example the shieldings for R = Me, Ph, OMe lie within a 17 ppm range[192].

More useful in structural analysis is consideration of both chemical shift and coupling constant jointly. Some couplings for which data are available are as follows:

(*i*) $^nJ(^{195}Pt, ^1H)$. Values of the one-bond (n = 1) coupling are very well known; the coupling constant is usually obtained from the ^1H n.m.r. spectrum and consequently there are many reported values (e.g. ^1H n.m.r. of 123 platinum hydrides are mentioned in the 1976–77 annual reviews[14]). Collation of the available data[159,228] has shown that values of a *ca.* 1000 ± 300 Hz are typical, depending upon the *trans*-influence of the *trans*-ligand[288]. Couplings for n = 2[228] (e.g. Pt—CH_3) and n = 3[289] [e.g. Pt—$P(CH_3)_3$] have also been described.

(*ii*) $^nJ(^{195}Pt, ^{13}C)$. Values of one-bond couplings have been reviewed[232]. In studies[290] of the oxidative addition of XY (X = Y = Cl, Br, I; X = I, Y = CN) to $[Pt(CN)_4]^{2-}$ by ^{13}C and ^{195}Pt n.m.r. it was observed that $^1J(^{195}Pt, ^{13}C)$ in Pt(II) cyanides is *ca.* 1000 Hz whereas in Pt(IV) cyanides the value is typically *ca.* 850 Hz, the expected trend in terms of percentage *s*-orbital character in the theoretical hybrid orbitals of the Pt—C bonds (i.e. Pt(II) = dsp^2; 25% *s*-character; Pt(IV) = d^2sp^3; 16.6% *s*-character), assuming a simplified Fermi-contact model.

Multi-path platinum—carbon couplings have been investigated in some platinum—amine complexes[291]; the effects are probably general in most platinacycles. For example, coupling between Pt and C_1 in **19** is the sum of both a two-bond and three-bond coupling (i.e. Pt → N_1 → C_1 + Pt → N_2 → C_2 → C_1). The resulting coupling is very small (*ca.* 5 Hz) when compared with an analogous

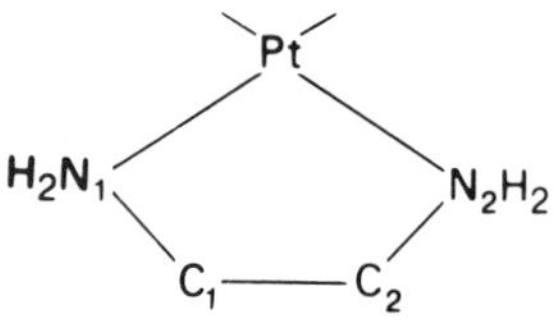

$$\text{Pt}(\text{H}_2\text{N}_1\text{C}_1\text{C}_2\text{N}_2\text{H}_2)$$

(19)

coupling in a bis(mondentate amine) complex (*ca*. 20 Hz) where no multipath coupling is possible. As three-bond couplings relate to the dihedral angle via a Karplus-type relationship[55], the ring strain or fixed bond angles result in an abnormally large magnitude for $^3J(^{195}\text{Pt}, ^{13}\text{C})$. Assuming the two and three-bond couplings to be of opposite sign, as expected, then the multipath effect reduces the coupling almost to zero as $|^2J(^{195}\text{Pt}, ^{13}\text{C})| \approx |^3J(^{195}\text{Pt}, ^{13}\text{C})|$.

Platinum chemical shifts and platinum—carbon coupling constants have been examined in studies of diastereomeric olefin complexes[292,293] and of platinum carbonyl clusters[294,295]. The major problem in spectral analysis with a carbonyl cluster of any size is the large number of possible isotopomers with ^{195}Pt (33.8%) and ^{13}C (1.108%). The ^{195}Pt and ^{13}C n.m.r. of the $[\text{Pt}_n(\text{CO})_{2n}]^{2-}$ anions (n = 3, 6, 9, 12, 15) have been reported[294,295] and the method of deducing isotopomer populations by pattern enumeration discussed[295].

The signs of $^1J(\text{Pt}, ^{13}\text{C})$ have been determined for $[\text{PtCl}_3(\text{C}_2\text{H}_4)]^-$ [296] and for some phosphine-substituted platinum carbonyls[297]. All values were positive.

(*iii*) $^nJ(^{195}\text{Pt}, ^{14}\text{N})$ and $^nJ(^{195}\text{Pt}, ^{15}\text{N})$. Coupling to ^{14}N may be resolved provided that quadrupolar relaxation is not too rapid. Values of $^1J(^{195}\text{Pt}, ^{14}\text{N})$ of *ca*. 300 Hz (substituted pyridine complexes)[298] and 200–500 Hz (*N*-thiocyanate and *N*-cyanate complexes)[299,300] have been reported. Studies of complexes enriched with ^{15}N by ^{15}N and ^{195}Pt n.m.r. generally give more definitive results. *S*- and *N*-coordination of thiocyanates can be differentiated[301] as $^1J(^{195}\text{Pt}, ^{15}\text{N})$ is *ca*. 550 Hz whereas $^3J(^{195}\text{Pt}, ^{15}\text{N})$ is *ca*. 15 Hz. Studies[302] of Schiff base complexes have shown the dependence of $^1J(^{195}\text{Pt}, ^{15}\text{N})$ on the *trans*-effect of the *trans*-ligand. Some interesting multinuclear studies on the reactions of amines by nucleophilic attack on platinum(II) olefin complexes have been described. Equation 7 describes the observed reaction[303].

$$(\text{C}_2\text{H}_4)\text{PtCl}_2(^{15}\text{NHMe}_2) + {}^{15}\text{NHMe}_2 \rightleftharpoons \text{Me}_2{}^{15}\overset{+}{\text{N}}(\text{H})\text{CH}_2\overset{-}{\text{C}}\text{H}_2\text{—PtCl}_2\text{—}^{15}\text{NHMe}_2 \quad (7)$$

(20) **(21)**

Complex **20** shows a coupling, $^1J(^{195}\text{Pt}, ^{15}\text{N})$ = 299 Hz, which collapses on addition of a trace of amine. Further addition causes the appearance of a new resonance, attributed to **21**, more shielded by 75.5 ppm, showing $^1J(^{195}\text{Pt}, ^{15}\text{N})$ = 107 Hz and $^3J(^{195}\text{Pt}, ^{15}\text{N})$ = 51 Hz. The large value of the three-bond coupling was interpreted in terms of a secondary Pt- - - -N interaction. The geometry of **20** appears to be critical, as further studies[304] of *cis*-complexes have shown a different reaction sequence (equation 8). The metallacycle shows the expected ^{13}C, ^{31}P and ^{195}Pt n.m.r. parameters, including $^1J(^{195}\text{Pt}, ^{15}\text{N})$ = 122 Hz. No three-bond coupling (as in **21**) was observed. The reaction has since been shown to occur with other neutral ligands[305].

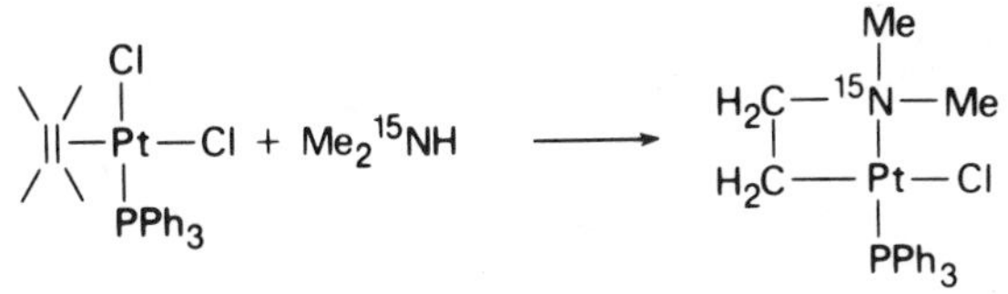

$$(C_2H_4)Pt(Cl)_2(PPh_3) + Me_2{}^{15}NH \longrightarrow [H_2C{-}{}^{15}N(Me)_2{-}CH_2{-}Pt(Cl)(PPh_3)] \qquad (8)$$

(*iv*) $^nJ(^{195}Pt, ^{19}F)$. Platinum—fluorine couplings have been used diagnostically in geometry determinations of fluoro-olefin complexes. $^2J(^{195}Pt, ^{19}F) = 279.5$ Hz in $[Pt(F_2C{=}CF_2)(PPh_3)_2]$[306]; this may be compared with the value of $^3J(^{195}Pt, ^{19}F)$ of +66.5 Hz in the analogous acetylene complex, $[Pt(F_3CC{\equiv}CCF_3)(PPh_3)_2]$[306]. A number of five-coordinate tris(pyrazolyl)borate fluoro-olefin complexes of platinum have been studied[61,307] and the magnitude of $^nJ(^{195}Pt, ^{19}F)$ found to vary with the orientation of the fluoro substituents with respect to the metal centre. Variation in $^2J(^{195}Pt, ^{19}F)$ in $Pt{-}CF_3$ complexes has been discussed[228]. See also Section VI.M.

(*v*) $^1J(^{195}Pt, ^{31}P)$. Variations in one-bond platinum—phosphorus couplings with the oxidation state of the metal, the nature of the *trans*-ligands and the bond angle strain at the metal centre are well known and very useful in structural assignments. Articles discussing $^1J(^{195}Pt, ^{31}P)$[228,280,281] are available (see also Section VI.J.2).

Less commonly observed couplings, $^1J(^{195}Pt, ^{29}Si)$[308], $^1J(^{195}Pt, ^{77}Se)$, $^1J(^{195}Pt, ^{125}Te)$[309–311], and $^1J(^{195}Pt, ^{119}Sn)$[312–314] have all been reported. Interestingly, the anion $[PtRh_5(CO)_{15}]$, whose crystal structure shows that the six metal atoms are arranged in an octahedron, shows both $^1J(^{195}Pt, ^{103}Rh) = 24.4$ Hz and $^2J(^{195}Pt, ^{103}Rh) = 73.2$ Hz[315]. The one-bond coupling to the four nearest rhodium atoms is much less than the two-bond coupling to the more distant single rhodium atom. This implies either a difference in sign between one and two-bond couplings or a strong interaction across the centre of the octahedron, or around the edges.

One of the most erratic coupling constants, with regard to magnitude, is the platinum—platinum coupling, $^1J(^{195}Pt, ^{195}Pt)$. In Section II.A.5 the problem of chemical shift anisotropy effectively collapsing this coupling[36] was discussed; the reported values of this coupling have been compiled[316] and show no correlation with the Pt—Pt bond order. The effect of the ligands *trans*- to the Pt—Pt moiety (whether formally bonded or not) seems to affect the coupling more than the bond order. For example, the formally Pt(I) dimer **3** has $^1J(^{195}Pt, ^{195}Pt) = 8197$ Hz whereas the Pt(I) complex **22** has $^1J(^{195}Pt, ^{195}Pt) = 188$ Hz. Both are estimated to have similar Pt—Pt distances[316]. As $[Pt_2(\mu\text{-}Cl)_2Cl_2(PBu^t_3)_2]$ has $^2J(^{195}Pt, ^{195}Pt) = 190$ Hz[36,317], where there is no formal metal—metal bond, this parameter must be utilized with caution.

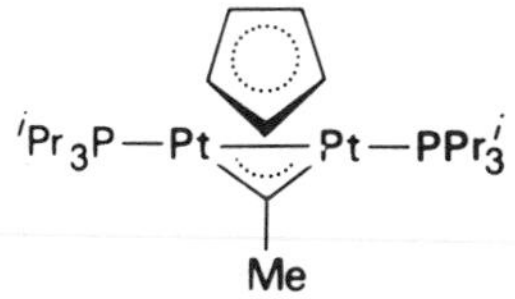

$$^iPr_3P{-}Pt(\mu\text{-}C_5H_5)(\mu\text{-}C_6H_4Me)Pt{-}PPr^i_3$$

(**22**)

9. *Cu, Ag, Au*

Both active copper nuclei are quadrupolar, but have relatively high receptivities. As Cu(II) is paramagnetic, only Cu(I) complexes are suitable for study. Very few reports of ^{63}Cu (the more sensitive isotope) n.m.r. have appeared, possibly indicating a lack of interest in Cu(I) chemistry, although the pronounced disproportionation to Cu(0) and Cu(II) is a deterrent. The $[Cu(P\{OMe\}_3)_4]^+$ ion has been studied by $^{31}P\{^{63}Cu, ^1H\}$ triple resonance[318] and the coupling, $^1J(^{63}Cu, ^{31}P) = 1210$ Hz, identified. The $[Cu(CN)_4]^-$ ion has also been examined[319,320]. Too few data exist to make meaningful comparisons of chemical shifts.

Silver has two $I = \frac{1}{2}$ nuclei, ^{107}Ag and ^{109}Ag, the latter being more sensitive to n.m.r. detection, despite its lower abundance, as its gyromagnetic ratio is larger in magnitude. The problem with ^{107}Ag n.m.r. is largely chemical as Ag(I) complexes tend to be highly labile and the formation of a number of species is possible upon dissolution of an Ag(I) salt in a coordinating solvent. Studies of this behaviour by ^{109}Ag n.m.r. have been discussed previously[192]. An example of relevance in organometallic chemistry is the effect of added arenes on the ^{107}Ag n.m.r. of AgX salts dissolved in coordinating solvents[321]. In weak donor solvents (such as methanol or dimethylformamide) addition of benzene or toluene causes deshielding of the ^{107}Ag nucleus by several hundred parts per million. In stronger donor solvents (such as acetonitrile or pyridine), addition of arenes has little effect. As silver(I) arene complexes are well known[322], these results have been interpreted in terms of π-arene complex formation in solution. These results support the well known empirical observation that silver(I) salts, such as $AgClO_4$, are soluble in aromatic hydrocarbons[323]. Similar solution studies of reactions of Ag(I) salts with S- and N-donor ligands have also been reported[324–326].

Couplings involving ^{107}Ag are uncommon owing to the lability of the complexes, although a few values of couplings to ^{1}H[327] and ^{19}F[328] are known. The system $[Ag(P\{OR\}_3)_n][X]$ ($n = 1, 2, 3, 4$) has been investigated by $^{31}P\{^{107}Ag, ^1H\}$ triple resonance[329], values of $^1J(^{107}Ag, ^{31}P)$ varying from *ca.* 340 Hz ($n = 4$, X = SCN) to *ca.* 1160 Hz ($n = 1$, $X = NO_3$). The corresponding phosphine system has been investigated by ^{31}P n.m.r.[330].

No chemically significant studies of ^{197}Au ($I = \frac{3}{2}$, $N = 100\%$) have been performed, a probable consequence of the relatively large quadrupolar moment, as there is significant interest in organogold systems[331].

10. Zn, Cd, Hg

^{67}Zn ($I = \frac{5}{2}$, $N = 4.11\%$) is a little studied nucleus. Reports of the ^{67}Zn n.m.r. of simple Zn(II) salts have appeared[192] and the linewidth is very sensitive to the symmetry of the field gradient at the nucleus, as expected for this quadrupolar species. No reports of organozinc compounds studied by ^{67}Zn n.m.r. have been described.

Both ^{111}Cd and ^{113}Cd are $I = \frac{1}{2}$ nuclei; the ^{113}Cd isotope is more sensitive to n.m.r. detection and so direct observations of this nucleus are usually made. Both Me_2Cd and $[CdSO_4]_{(aq.)}$ have been employed as shift standards, the former being suitable for double resonance work. A considerable solvent dependence of cadmium shifts can lead to appreciable differences in results in different solvents.

Alkylcadmium compounds have been studied by several workers[332–336]. The effect[334] of alkyl chain length and branching gives rise to an irregular increase in shielding, $CdMe_2 < Cd(CHMeEt)_2 < CdBu^n_2 < CdPr^n_2 < CdEt_2 < Cd(CHMe_2)_2$. These authors particularly note[334] that the substitution of a saturated carbon atom for a hydrogen atom in an alkyl chain has no regular effect on the central metal, but depends upon the nature of that metal. Some alkylcadmium alkoxides have also been investigated[335] and the results interpreted in terms of a tetramer–hexamer conversion.

Some results concerning the relaxation behaviour of organocadmium compounds have been described[333] and, although these will not be discussed in detail here, it is noteworthy that spin–rotation contributions are significant for Me_2Cd at high temperatures, as expected for a small organic group bonded to a heavy metal centre (see Section II.A.4).

As cadmium compounds tend to be labile in solution, few couplings involving ^{111}Cd or ^{113}Cd have been described, although the sign and magnitude of $^2J(^{113}Cd, ^1H)$ in R_2Cd are known[334,335] and couplings to ^{13}C[333] and ^{31}P[337] have been described.

Mercury has both $I = \frac{1}{2}$ (^{119}Hg) and $I = \frac{3}{2}$ (^{201}Hg) nuclei, the former being preferred for n.m.r. study as it is not quadrupolar. The accepted shift standard is Me_2Hg, which can be used as a neat liquid, so avoiding solvent effects which are known to be large for mercury shieldings[338]. The solvent effects on mercury shieldings are usually taken to imply an increase in coordination number, but recent reviewers[192] have pointed out that this is not always the case and the complex effects of fields produced by polar solvents must be considered. For example, the ^{199}Hg nucleus in the complex $[MeHg(NC_5H_5)][NO_3]$ is deshielded by substituting a methyl group for a proton in the 2-position of the pyridine ring and is further deshielded by a second substitution in the 6-position. This has been attributed to a disrupting effect on the solvent coordination sphere about the mercury nucleus by the methyl groups[339].

^{199}Hg chemical shifts of many simple organomercury compounds of the types R_2Hg and RHgX have been reported[340–348]; the differences in solvent and temperature make comparisons difficult although it is noteworthy[192] that the alkyl chain length affects $\delta(^{199}Hg)$ differently to $\delta(^{113}Cd)$. Studies of some cyclohexylmercury halides have shown[345] that axial and equatorial mercury halides can be distinguished by ^{199}Hg n.m.r., the equatorial groups being more shielded by *ca*. 90 ppm compared with axial groups. The explanation for this is not obvious, but the observation is useful for distinguishing different conformations of cycloalkylmercury derivatives.

^{199}Hg n.m.r. has been used to study anion-exchange reactions between mercury(II) cyanide and mercury(II) halides and the equilibria involved in these processes[349]. Mercury—phosphine complexes have also been examined[350] and some success achieved in differentiating between isomeric dimeric complexes.

A number of mercury–main group element complexes have been studied[351–354]; data for some mercury–silicon compounds are shown in Tables 19 and 20. The effect of increasing alkyl chain length on mercury results in shielding whereas increasing the alkyl chain length on silicon results in deshielding of the mercury nucleus. The

TABLE 19. ^{29}Si and ^{199}Hg n.m.r. data for trimethylsilylmercury compounds, $[RHg(SiMe_3)]^a$

R	$\delta(^{29}Si)$ (ppm)^b	$\delta(^{199}Hg)$ (ppm)^c	$^1J(^{199}Hg, ^{29}Si)$ (Hz)
Me	33.0	—	1367.0
Et	34.0	—	1213.0
n-Pr	35.2	166	1234.1
n-Bu	35.0	159	1225.9
i-Pr	34.0	51	1084.9
t-Bu	33.6	11	995.6
CH_2Cl	30.3	—	1137.0

^{a}Data from reference 351.
^{b}Relative to Me_4Si.
^{c}Data converted to Me_2Hg = 0 ppm; the original standard employed, $Hg(NO_3)_2$, was taken as having a chemical shift of −2361 ppm relative to Me_2Hg[355]. The solvent dependence of mercury chemical shifts may introduce errors into the tabulated data.

TABLE 20. ^{29}Si and ^{199}Hg n.m.r. data for bis(silyl)mercurials

Compound	$\delta(^{29}Si)$ (ppm)[a]	$\delta(^{199}Hg)$ (ppm)[b]	$^{1}J(^{199}Hg, ^{29}Si)$ (Hz)	Reference
$[H_3SiHgSiH_3]$	−10.3	+196.0	—	352
$[H_3Si^{(1)}HgSi^{(2)}Me_3]$	(1) −22.1	+327.1	—	352
	(2) +63.7			
$[Me_3SiHgSiMe_3]^c$	+64.0	+481.0	—	352
	+63.6	+566	989.6	351
$[Et_3SiHgSiEt_3]^c$	+35.1	+971	957.0	351

[a] Relative to Me_4Si.
[b] Relative to Me_2Hg.
[c] As [c] in Table 19.

former trend results from an electron density effect at the metal nucleus, whereas the latter may be due to an enhanced π-interaction causing a decrease in ΔE as the silicon becomes more electron-rich.

Some mercury—transition metal complexes have been studied[238]; the data in Table 21 show that no triad effect is observed in these compounds, probably because the small size of chromium leads to a decrease in covalency in the Hg—Cr bond and the resulting polarization shields the mercury nucleus. The halo-compounds (Table 9) show the decrease in shielding I > Br > Cl, typical of halomercury derivatives.

Coupling between ^{199}Hg and many other active nuclei have been reported. Coupling to ^{1}H and ^{19}F has been discussed previously[356]. Sign determinations have been performed for couplings to ^{13}C[357] and ^{31}P[358]. Examples of couplings to other hetero-nuclei are given in Tables 9, 19, 20, and 21.

Of the three nuclei in this triad, ^{199}Hg n.m.r. presents the most promise as a relatively simple and well behaved n.m.r. nucleus. A substantial data base now exists and more elaborate structural determinations by this technique should now be possible.

TABLE 21. ^{199}Hg n.m.r. data for some mercury–transition metal compounds

	$\delta(^{199}Hg)$ (ppm)[a]	
M	$Hg[M(CO)_3(C_5H_5)]_2$	$HgCl[M(CO)_3(C_5H_5)]$
Cr	−80	−542
Mo	+115	−617
W	-348^b	-997^c

[a] Relative to 90% $HgMe_2$/10% C_6F_6 (internal lock). Data from reference 238.
[b] $^{1}J(^{199}Hg, ^{183}W) = 151$ Hz.
[c] $^{1}J(^{199}Hg, ^{183}W) = 706$ Hz.

F. Studies of the Group IIIB Nuclei

1. Boron compounds

Boron has two n.m.r.-active isotopes, ^{10}B ($I = 3$, $N = 19.6\%$) and ^{11}B ($I = \frac{3}{2}$, $N = 80.4\%$), the latter being more favoured for n.m.r. study. Historically, boron n.m.r. has been of great importance in the development of the extensive chemistry associated with the boron hydrides and their derivatives[359–361]. Boranes, carboranes, and heteroatom boranes have all been studied in depth by ^{11}B n.m.r.[362,363]; a potential area of interest relevant to this chapter is the study of metallocarboranes[363]. Metallocarboranes containing either a metal—carbon σ-bond or a metal—boron σ-bond are known. A review of the chemistry of transition metal carboranes[364] has described ^{11}B n.m.r. as 'useless' for the structural characterization of such complexes. This rather pessimistic view arises from two inherent problems in ^{11}B n.m.r. Firstly, the multiplets arising from spin-coupling to ^{11}B (with $I = \frac{3}{2}$) in many compounds are highly complex owing to overlapping as the shift range for ^{11}B nuclei in a typical 'carborane-type' environment is relatively small. Secondly, many of the interesting transition metal carborane complexes are paramagnetic and hence give rise to broad resonances where small couplings (such as ^{11}B–^{1}H coupling) are not resolved. Nonetheless, this pessimistic attitude is not universal and a number of elegant structural determinations have been possible with favourable diamagnetic compounds[365,366]. Even with paramagnetic complexes the induced paramagnetic shifts may be of value diagnostically[367].

Also of interest are studies of main group organometallic compounds containing metal—boron σ-bonds. Several such compounds have been examined by double resonance methods in order to determine the sign of ^{1}J(M, ^{11}B) for comparison with ^{1}J(M, X) values for other elements. Some typical results are shown in Table 22, together with some values of $^{1}J(^{195}$Pt, ^{11}B) for which the magnitude, but not the sign, is known.

The widely accepted shift standard for ^{11}B n.m.r. is [$Et_2O \cdot BF_3$].

TABLE 22. ^{11}B n.m.r. data for some metal–boron compounds

Compound	$\delta(^{11}B)$ (ppm)^a	$^{1}J(M, ^{11}B)$ (Hz)	Reference
[$Me_3SnBCl(NMe_2)$]	$+44.4 \pm 0.3$	-1007 ± 10	237
[$Me_3SnB(NMe_2)_2$]	$+39.0 \pm 0.3$	-953 ± 10	237
[$Me_3PbB(NMe_2)_2$]	$+41.7 \pm 1$	-362 ± 2	368
[9,9-$(PMe_2Ph)_2$-6-9-$SPtB_8H_{10}$]	—	260^b 240^c	369

^{a}Relative to [$Et_2O \cdot BF_3$].
^{b}For $^{11}B_4$.
^{c}For $^{11}B_{8,10}$.

2. *Organoaluminium compounds*

^{27}Al is a quadrupolar nucleus and accordingly linewidths are sensitive to the field gradient, and hence the symmetry, about the metal centre. ^{27}Al n.m.r. has been reviewed[370,371] and data show that linewidths increase as the symmetry decreases, octahedral < tetrahedral < trigonal. Shifts correlate with the type of coordination[372] and so a combination of shift and linewidth measurements makes ^{27}Al n.m.r. suitable for studies of solvation[373] and metal binding in biological systems[372]. Of interest in organometallic chemistry are studies of aluminium hydrides[374–378], alkyls[375,379,380] and borohydrides[370,381–383]. The linewidth problem (e.g. for [Al_2Et_6], linewidth at half-height is *ca*. 1 kHz) normally prevents observation of coupling to ^{27}Al, although a few reports involving coupling to ^{1}H[370,376,377,381], ^{11}B[384], ^{13}C[385] and ^{31}P[370] have appeared.

The use of ^{27}Al n.m.r. in solving chemical problems concerning dimeric alanes, [Al_2R_6], or monomeric adducts, [$AlR_3 \cdot L$], is limited as dissociation and ligand-exchange problems add to the ever-present linewidth problem and really only comparative shift data are available[370,371].

3. Other systems

Gallium has two quadrupolar nuclei, ^{69}Ga and ^{71}Ga, the latter, less abundant, isotope being more suitable for n.m.r. study. The problems in ^{71}Ga n.m.r. parallel those for ^{27}Al n.m.r. and the limited number of studies reported[370,371] probably reflects a lack of chemical interest rather than inherent n.m.r. problems. A number of symmetrical species have been examined, but no compounds of particular interest in organometallic systems have been studied. A shift range of *ca.* 1200 ppm is now known[370,371] and coupling to ^{1}H[386] and ^{31}P[387] have been described.

Indium also has two quadrupolar nuclei, ^{113}In and ^{115}In, the former having poor receptivity and the latter being more sensitive. A few studies concerning the n.m.r. of In(III) salts have been reported and discussed[371], but no results of relevance in organometallic systems have been described.

In contrast to Ga and In, there are two $I = \frac{1}{2}$ nuclei of thallium, ^{203}Tl and ^{205}Tl, the latter being slightly more favourable for n.m.r. study. Both are sensitive nuclei and a large number of compounds, particularly organothallium derivatives, have been studied by ^{205}Tl n.m.r. Complexation of Tl(I) has been studied in depth in biological systems and in studies of solvation. A large number of alkyl- and arylthallium(III) compounds have been examined and a recent review[371] contains extensive tabulation of shift data and coupling constants. A number of generalizations concerning these data are possible:

(*i*) Thallium chemical shifts show a considerable concentration and solvent dependence, the shielding being related to the basicity of the solvent. For example, $[Me_2TlX]$ (X = ClO_4, NO_3) show a 200 ppm variation in $\delta(^{205}Tl)$ as the solvent is varied[388].

(*ii*) Thallium coupling constants show a solvent dependence. For example, variations in $^{1}J(^{205}Tl, ^{13}C)$ of *ca.* 2500–3000 Hz and $^{1}J(^{205}Tl, ^{1}H)$ of *ca.* 400–470 Hz in $[Me_2Tl]^+$ have been noted in different solvents[388].

(*iii*) Thallium chemical shifts can vary with ion pairing; thus $\delta(^{205}Tl)$ in $[Me_2TlX]$ (X = ClO_4, NO_3) show a dependence on the degree of ion association[388,389].

(*iv*) Thallium chemical shifts are sensitive to the degree of covalency in the Tl—C bond and changing the *p*-substituent in arylthallium complexes causes significant variation in $\delta(^{205}Tl)$[390].

It can clearly be seen that ^{205}Tl is of considerable importance in organometallic n.m.r. studies whereas ^{71}Ga and ^{113}In are unlikely to be of great interest for some time. The space devoted to ^{205}Tl in this chapter is certainly less than the number of reported studies merits, but this subject has recently been covered in intense detail[371], the authors of the excellent review being willing to supply lists of reference data upon request[371].

G. Carbon-13 N.m.r. Studies

^{13}C n.m.r. studies of organometallic compounds are now becoming routine in structural characterization and in studies of dynamic systems. The vast amount of background data available for organic compounds[391,392] is of considerable utility and various aspects of the application of ^{13}C n.m.r. to inorganic[393] and organometallic[394] chemistry have been described. The information obtained from ^{13}C n.m.r. spectra of organometallic compounds generally falls into three categories, as follows.

1. Relaxation behaviour

Relaxation times of the ^{13}C nuclei in organic compounds have been used as structural probes for many years[395]. The situation in organometallic chemistry is completely different, however; although many main-group organometallic compounds have been examined, these have generally been treated as organic molecules containing a heteroatom (which just happens to be a metal) and so from an organometallic viewpoint, few coherent series of data are available. Transition metal organometallic compounds have only recently been studied and as yet little data exist. Representative examples of the literature[396–403] concerning the relaxation behaviour of the ^{13}C nuclei in organometallic compounds of the transition metals show that the subject is still open to considerable debate and that much remains to be accomplished in this area.

In Section II.A.4, the possibility of the methyl carbon atoms in platinum(II) methyl complexes relaxing via spin–rotation interactions was mentioned. In relation to this, a detailed study of relaxation processes in some transition metal methyl complexes is noteworthy[401]. In this study, the relaxation of several methyl derivatives (of Os, Mo, Fe, Zr, Re, and Au) was examined and the results indicate that both ^{13}C–^{1}H dipolar interactions and internal spin–rotation are generally the major relaxation routes. The rate of molecular tumbling and the barrier to methyl group rotation appear to be the controlling factors in determining the relative contributions of each mechanism. Only when the metal ion is quadrupolar, as in [Re(CO)$_5$Me], does scalar relaxation of the second kind become important whilst chemical shift anisotropy appears to be negligible in all the cases studied.

A report[402] on the relaxation of cyano-carbon atoms in diamagnetic transition metal cyanides has suggested that both the spin–lattice and spin–spin relaxation times of such carbon atoms (ratio $T_1/T_2 = 10^2$–10^3) are typical of a mechanism involving scalar coupling of the second kind to ^{14}N. A generally observed trend, that T_1 increases with temperature, supports this conclusion. Further studies[403] on the ion $[Ni(CN)_4]^{2-}$ showed that T_1 decreases at higher applied fields, implying that, under such conditions, chemical shift anisotropy is an important relaxation route. The dipole–dipole interaction between ^{13}C and ^{14}N was also found to play an important role in spin–lattice relaxation. Scalar coupling to ^{14}N was found to be unimportant in spin–lattice relaxation, whilst making a significant contribution to spin–spin relaxation. The two reports are obviously at variance with each other, further examples being necessary for clarification.

In general, the relaxation of ^{13}C in organometallic compounds, particularly of the transition metals, is only poorly understood and further efforts in this area are necessary before a level comparable to organic systems can be reached.

2. *Chemical shift and coupling constant data*

The carbon-13 nucleus has a chemical shift range of *ca*. 700 ppm, making ^{13}C n.m.r. a useful technique for structural assignment. Tabulated data are available in reviews[393,394] for comparison purposes and will not be repeated here.

Spin–spin coupling between ^{13}C and other active nuclei, in conjuction with shift data, renders structural characterization of static molecules routine by ^{13}C n.m.r. Tabulated data for ^{13}C coupling constants are available[232] and coupling between ^{13}C and other first-row nuclei has been discussed in detail[404]. The theory of coupling constants involving ^{13}C has been described[405] and, at least with first-row elements, Fermi contact has been shown to be the dominant mechanism[406].

3. *Dynamic behaviour*

^{13}C n.m.r. has been instrumental in increasing our knowledge of dynamic systems involving organometallic compounds. Probably the major contribution of ^{13}C n.m.r. has been in the study of transition metal carbonyls. The known shift range of carbonyl carbon atoms is over 100 ppm, making ^{13}C n.m.r. a sensitive structural probe. The chemistry of dynamic transition metal carbonyls, and the use of ^{13}C n.m.r. in their study, has been reviewed many times[393,394,407–410]. The availability of carbon-13 labelled carbon monoxide can usually allow the preparation of enriched carbonyl complexes, so overcoming the problems of low natural abundance.

Possibly the most important compound so far studied in this class is $[Fe(CO)_5]$. Believed to have a trigonal-bipyramidal structure, only one carbonyl carbon resonance has been observed in the ^{13}C n.m.r. Attempts to find resonances attributable to both axial and equatorial carbonyls have failed, even using enriched samples and temperatures as low as $-170°C$[411–414]. Coupling, $^1J(^{57}Fe, ^{13}C)$, has been observed[415], so an intermolecular exchange of carbonyl ligands can be discounted and an intramolecular process is believed to occur which averages the carbonyl resonances. Results from ^{17}O n.m.r. are in agreement with this (see Section VI.L.I). Since the early studies of $[Fe(CO)_5]$, the use of ^{13}C n.m.r. in the study of dynamic processes has become well established and the reader is referred to the excellent reviews available[407–410] for further details.

Despite the wealth of data available from ^{13}C n.m.r. studies of organometallic compounds, there are several practical details which may cause experimental difficulties in obtaining satisfactory spectra. The long relaxation times (T_1) associated with quaternary carbon atoms and with carbonyl and cyano-carbon atoms can cause problems of saturation in pulsed n.m.r. experiments. The use of 'shiftless' relaxation agents, such as $[Cr(acac)_3]$, is almost obligatory in some experiments. Possible decomposition products of diamagnetic compounds may be paramagnetic species themselves, and accordingly any unusually large chemical shifts in systems where such decomposition is possible should be viewed with caution. Finally, in comparing literature data, it is noteworthy that although the accepted standard for chemical shifts is now TMS, early results may well be cited using benzene or carbon disulphide as the shift standard.

H. Studies of the Remaining Group IVB Nuclei

The remaining Group IVB nuclei present several diverse problems from an n.m.r. viewpoint. Silicon has one $I = \frac{1}{2}$ isotope, with a negative gyromagnetic ratio and germanium has one quadrupolar isotope, ^{73}Ge, with a negative gyromagnetic ratio. Tin has three $I = \frac{1}{2}$ isotopes, ^{115}Sn, ^{117}Sn, and ^{119}Sn, with negative gyromagnetic ratios, although ^{119}Sn is slightly more abundant and more sensitive than ^{117}Sn (^{115}Sn being of very low abundance) and so is the preferred nucleus for study. Lead has one $I = \frac{1}{2}$ isotope, ^{207}Pb, of reasonable abundance and with a positive gyromagnetic ratio.

1. *Organometallic Si and Ge systems*

^{29}Si n.m.r. studies have been performed by continuous wave, pulsed excitation, and double resonance methods. The major problem in obtaining ^{29}Si n.m.r. spectra arises because of the negative gyromagnetic ratio, which, as previously mentioned (see Section II.A.I) can lead to nulling of the resonances during decoupling. There are several ways to prevent this; by either adding to, or subtracting from, the observed NOE and thus causing the resonance to appear as either a negative or positive signal. Some applicable methods include:

(*i*) purging solutions with molecular oxygen (detracting from the NOE);
(*ii*) degassing solutions *in vacuo* or purging with nitrogen (i.e. removing dioxygen and so enhancing the NOE);
(*iii*) adding a 'shiftless' relaxation agent such as $[Cr(acac)_3]$ (see Section V.B);
(*iv*) applying gated decoupling techniques[416].

Assuming that the problem of nulling can be overcome by one of these methods, the only remaining problem is one of abundance ($N = 4.70\%$). Enrichment techniques to enhance the percentage of ^{29}Si in a molecule do not appear to be as common as they are with ^{13}C, and so it is necessary to use neat liquids or concentrated solutions for continuous wave observation. Indirect observation, where possible, and pulsed excitation methods are really necessary to observe more dilute solutions. This is frequently the case in organometallic systems.

There are several reviews dealing with ^{29}Si n.m.r.[417–422], some specifically with its application to the study of silicon polymers[419,421], to which reference should be made for background data, particularly that of ^{29}Si relaxation, which has been studied in depth in organic systems. The accepted standard for chemical shift measurement is Me_4Si (TMS).

^{73}Ge n.m.r. is restricted to high-symmetry molecules as the quadrupole adversely affects relaxation in situations where the field gradient at the nucleus deviates largely from cylindrical. Accordingly, only compounds of the type $[GeL_4]$ (L = halide, alkyl, alkoxy, thiolate) have been successfully examined. Unfortunately, whilst germyl ligands are of importance in organometallic systems, the symmetry at the germanium nucleus appears to be too low for observation of n.m.r. to be possible. A review discussing ^{73}Ge n.m.r. is available[421].

Although a vast number of organosilicon compounds have been studied by ^{29}Si n.m.r., surprisingly few organometallic or related compounds have been examined. A number of compounds containing metal—silicon bonds have been studied and these are of interest as analogues of metal—carbon compounds for comparison purposes.

^{29}Si n.m.r. studies of about twenty carbene and carbyne derivatives of Cr, Mo, and W have been reported[423]; representative data are shown in Tables 23 and 24. Comparison of these tables shows that the magnitude of $\delta(^{29}Si)$ is not diagnostic for these classes of compound. The data in Table 23 also show that no triad effect is observed here, as the deshielding reaches a minimum at molybdenum, instead of decreasing down the group as is usually observed (see Table 25 for an example of the triad effect on a nucleus one atom removed from the metal centre). This rather unusual trend has been explained[423] in terms of a metal–silicon π-interaction via hyperconjugation, facilitated by the availability of d-orbitals on silicon. The degree of shielding can be seen to relate to the electron-withdrawing ability of the carbene substituent by comparing —OMe and —NMe_2 derivatives (Table 23). In addition to carbene- and

TABLE 23. ^{29}Si n.m.r. data for some metal carbene complexes:

$$\left[(OC)_5M \cdots C \begin{matrix} \cdots R \\ \diagdown SiPh_3 \end{matrix}\right]$$

M[a]	R	$\delta(^{29}Si)$ (ppm)[b]	T (°C)
Cr	OMe	−24.2	−10
Mo	OMe	−30.3	−20
W[c]	OMe	−18.3	+32
Cr	NMe_2	−23.9	+1
Mo	NMe_2	−25.6	−20
W	NMe_2	−22.1	+32

[a] Data from reference 423.
[b] Relative to Me_4Si.
[c] $^2J(^{183}W, ^{29}Si) = 12.5$ Hz.

TABLE 24. ^{29}Si n.m.r. data for some metal carbyne complexes, $[Br(CO)_4W{\equiv}CSiR_3]$[a]

SiR_3	$\delta(^{29}Si)$ (ppm)[b]
$SiPh_3$	−23.4
$SiPh_2Me$	−17.8
$SiPhMe_2$	−10.6

[a] Data from reference 423.
[b] Relative to Me_4Si.

TABLE 25. The triad effect with $\delta(^{17}O)$ and $\delta(^{13}C)$ of some substituted carbonyls

$[M(CO)_4(NBD)]$	$\delta(^{17}O)_{cis}$[a]	$\delta(^{17}O)_{trans}$[a]	$\delta(^{13}C)_{cis}$[b]	$\delta(^{13}C)_{trans}$[b]
M = Cr	388.4	370.1	226.8	234.5
M = Mo	381.9	369.2	215	218
M = W	377.0	346.2	203.6	209.4

[a] ppm, relative to $H_2{}^{17}O$; data from reference 116.
[b] ppm, relative to $(^{13}CH_3)_4Si$; data from reference 116.

carbyne-substituted carbonyl complexes, the cyclopentadienyl derivative, **23**, was examined. The chemical shift, $\delta(^{29}Si) = -28.0$ ppm, is similar to the values for the Cr, Mo, and W carbene complexes in Table 23. Complexes with silicon directly bonded to a transition metal were studied by indirect methods[424] until recently, when studies by

$(C_5H_5)(CO)_2Mn{\equiv}C(OMe)(SiPh_3)$

(23)

selective population transfer[425] enabled simple observation of M—Si systems to be made. Among the first studies[424] were $^{19}F\{^{29}Si\}$ measurements of $[Co(SiF_3)(CO)_4]$; a value of −28.6 ppm, for the chemical shift of ^{29}Si was obtained. Comparing this value with −55.7 ppm, the ^{29}Si chemical shift of Me—SiF_3, shows how coordination of ^{29}Si to a transition metal causes deshielding of the silicon nucleus. This has been attributed to a low electronic excitation energy (ΔE in the denominator of equation 2, Section II.B.2) resulting from involvement of silicon *d*-orbitals in the bonding. These results correlate with ^{19}F n.m.r. studies of $[Co(CF_3)(CO)_4]$, where the fluorine nuclei are considerably deshielded. Similarly, direct observation[423] of the ^{29}Si n.m.r. of $[Fe(C_5H_5)(CO)_2(SiMe_3)]$ shows that the silicon chemical shift, +40.8 ppm, represents considerable deshielding in comparison with the organic analogue, H_3C—$SiMe_3$ [i.e. TMS, $\delta(^{29}Si) = 0$]. A further report[425] of the compound $[Fe(C_5H_5)(CO)_2(SiMe_3)]$ quotes a value of the ^{29}Si chemical shift as *ca.* +41.3 ppm (converted from a $Me_3SiOSiMe_3$ reference), which is within acceptable error of the earlier value of +40.8 ppm. A series of trimethylsilylmetal carbonyls were studied by the selective population transfer method previously mentioned and the chemical shift data are shown in Table 26. Moving across the first row of the transition series (Mn, Fe, Co) causes deshielding of the ^{29}Si nucleus, a probable result of an increased paramagnetic contribution. The data for the rhenium complex are included for completeness.

TABLE 26. ^{29}Si n.m.r. data for trimethylsilyl-metal carbonyl complexes

Complex	$\delta(^{29}Si)$ (ppm)[a]
$[Re(CO)_5(SiMe_3)]$	-14.1
$[Mn(CO)_5(SiMe_3)]$	$+17.8$
$[Fe(CO)_2(C_5H_5)(SiMe_3)]$	$+41.3$
$[Co(CO)_4(SiMe_3)]$	$+44.3$

[a] Relative to Me_4Si; data from reference 425.

^{29}Si n.m.r. studies of a number of cyclic metal systems have been reported[426,427], including the diiron complex **24**, which was studied by $^1H\{^{29}Si\}$ double resonance

$Fe(CO)_4$

Me_2Si

$Fe(CO)_4$

(**24**)

methods. The chemical shift, $\delta(^{29}Si) = +173.0$ ppm, represents extreme deshielding of the silicon nucleus by the close proximity of the two heavy nuclei. Double resonance methods have also been used[308] to examine the complex *trans*-$[PtCl(SiH_2Cl)(PEt_3)_2]$ ($\delta(^{29}Si) = -25.0$ ppm, $^1J(^{195}Pt, ^{29}Si) = -1600 \pm 100$ Hz, $^2J(^{31}P, ^{29}Si) = +18 \pm 3$ Hz), although in view of the extreme importance of the platinum metals in the activation of silicon compounds (hydrosilylation, silane alcoholysis and hydrolysis, etc.) the number of ^{29}Si n.m.r. studies is surprisingly few.

A large number of trimethylsilylmercury alkyls have been studied by ^{29}Si and ^{199}Hg n.m.r.[351]. Representative data are shown in Table 19. Studies of bis(silyl)mercurials have also been reported[351,352] and these data are shown in Table 20. The deshielding of the ^{29}Si nucleus by the close proximity of the ^{199}Hg atom has been ascribed to the so-called 'heavy atom effect'; an involvement of low-lying *d*-orbitals in the Hg—Si bond is postulated as an explanation for an increased paramagnetic contribution to σ, resulting in extensive deshielding[351]. Additionally, it has been observed that as $[HgR_2]$-type complexes are linear, or usually only slightly non-linear, an anisotropic contribution to σ is also possible[351]. The ^{199}Hg shieldings are discussed elsewhere (see Section VI.E.10).

Main-group organometallic compounds containing silicon have hardly been studied by ^{29}Si n.m.r. Data for the trimethyltin derivative $[Me_3SnSiMe_3]$ have been presented[327] [$\delta(^{29}Si) = 11.0 \pm 0.2$ ppm, $\delta(^{119}Sn) = -126.7$ ppm, $^1J(^{119}Sn, ^{29}Si) = +656 \pm 10$ Hz); substituting Ph_3Si— for Me_3Si— only slightly alters the tin–silicon coupling [$^1J(^{119}Sn, ^{29}Si) = +650$ Hz in $Ph_3SiSnMe_3$][428].

The potential of ^{29}Si n.m.r. in organometallic chemistry is indeed significant; further work, particularly on transition metal complexes active as catalysts in the activation of organosilicon compounds and on the chemistry of main-group organometallic compounds of silicon, will undoubtedly prove fruitful.

2. *Organotin compounds*

Double resonance techniques are ideal for the study of organotin compounds which display an observable coupling between tin and a high-sensitivity nucleus (such as the proton). As ^{119}Sn is the more abundant and more sensitive of the three active tin nuclei (Table 2), nearly all studies relate to ^{119}Sn n.m.r. The vast amount of data concerning tin shielding and coupling has been the subject of several comprehensive reviews[82,421,429–431]. More recently, pulsed excitation methods have enabled a large amount of ^{119}Sn n.m.r. data to be accumulated by direct observation; particularly noteworthy in this respect are several studies of platinum(II):tin(II) systems which gave insight into the role of tin(II) halides in promoting catalytic hydrogenation and hydroformylation by platinum(II) complexes. As the active tin nuclei have negative magnetogyric ratios, proton decoupling does not enhance the signal intensity and the problems discussed for the ^{29}Si nucleus apply. Nonetheless, proton decoupling is virtually a necessity in the direct observation of many organotin compounds, as the extreme tin–proton coupling situations could readily render the spectra uninterpretable. The accepted standard for chemical shift measurement is Me_4Sn, which has been discussed earlier (see Section II.B.2).

Tin chemical shift data are available for several hundred organotin compounds and some predominant trends can be distinguished as a function of structure. As this subject has been extensively reviewed only the most salient points will be discussed here.

a. Variation in coordination number. Increases in the coordination number of tin[432], from four to five or four to six, lead to increased shielding of the tin nucleus. Examples are shown in Table 27; clearly, the increase in coordination number causes shielding to the extent of 200–300 ppm. This effect is particularly important when comparing values of tin shieldings for compounds examined in different solvents[433], donor solvents frequently causing increased shielding. This effect can also be of importance for compounds which undergo auto-association; in these cases increasing the concentration (and hence auto-association) can cause shielding of the tin nucleus[434].

Various explanations have been proposed to account for increased shielding with higher coordination numbers. The possibility that π-bonding is involved, reducing d-electron imbalance at tin, has been mentioned[82], but this also involves a rehybridization of tin (from sp^3 to sp^3d in a five-coordinate compound) and so the electron imbalance of the p-orbitals will also be reduced. The presence of a donor ligand additionally increases the electron density at tin and so increases the diamagnetic term. Probably both of these factors contribute to the observed effect, which is diagnostically very useful.

b. Temperature effects. Related to the above effect is the temperature effect sometimes observed in the ^{119}Sn n.m.r. of compounds capable of auto-association. Increasing the temperature with compounds such as dialkyltin alkoxides[429], which are usually associated in solution, causes some dissociation to occur, resulting in an effectively decreased coordination number at tin and consequent deshielding of the tin nucleus.

TABLE 27. Effect of coordination number on $\delta(^{119}Sn)$ in simple organotin compounds

Compound	Coordination number	$\delta(^{119}Sn)$ (ppm)^a	Reference
$[Me_2SnBr_2]$	4	$+70^b$	435
$[Me_2SnBr_2(bipyr)]$	6	-245	432
$[Me_2SnCl_2]$	4	$+136.8^b$	436
$[Me_2SnCl_2(Me_2SO)_2]$	6	-84	432
$[Me_3SnCl]$	4	$+159^b$	436
$[Me_3SnCl(Me_2SO)]$	5	-86	432

^{a}Relative to Me_4Sn.
^{b}Benzene solution.

Compounds such as alkyltin trihalides, which are unassociated as pure liquids, display no temperature dependence of the chemical shift[437].

c. Multiple substitution effects. Substitution of an electronegative group for an alkyl group in $[R_{4-n}SnX_n]$ (where X is an electronegative group and n increases from 1 to 4) causes a decrease in shielding for the first substitution and then an increase in shielding for the subsequent substitution. The charactistic 'camel's hump' plot of ^{119}Sn shielding *vs*. n has been observed for many classes of compound[82] (X = OR, NR_2, halide, etc.). Organosilicon compounds show a similar effect[421].

Once again, π-bonding, a favourite bone of contention in tin chemistry[438], has been invoked in an attempt to explain this effect but, as ^{13}C shifts in substituted methanes also show this behaviour[439], π-effects are not a prerequisite in such shift anomalies. Possibly the p-electron imbalance contribution to σ_p and variations in σ_d, resulting from altering the electron density at tin by varying the number of electronegative ligands, may jointly contribute to this effect, but no really satisfactory explanation seems available as yet.

d. Single substitution effects. In a compound $[R_3SnX]$ the effect of varying the single X ligand is more predictable than varying the number of X ligands in $[R_{4-n}SnX_n]$. A linear correlation of $\delta(^{119}Sn)$ with the Pauling electronegativity of X has been observed[429], more electronegative groups causing deshielding[440]. Deshielding can be explained in terms of decreased electron density at tin (resulting in a decrease in the diamagnetic contribution), but the effects observed are so large that it seems likely that paramagnetic effects are also involved. Possibly, where the electronegativities of X and R are different, the p-electron imbalance is altered, more electronegative X groups increasing the electron imbalance, resulting in increased σ_p and so decreased total shielding.

The above four points cover the major causes of variations in tin shieldings; more minor effects relating to the inter-bond angles at tin[441] and the steric effects of bulky substituents[82] have also been noted.

Coupling constants involving tin[421] have been determined in many cases by observation of the coupled nucleus, or by double resonance. The latter method, of course, enables both sign and magnitude to be determined. As tin has a gyromagnetic ratio which is negative, coupling to a nucleus with a positive gyromagnetic ratio (^1H, ^{13}C, ^{19}F, etc.) gives rise to a negative reduced coupling constant if $J(^{119}Sn, X)$ is positive and *vice versa*. In the ^nX spectrum of an Sn—X compound it is usually possible to observe both $^nJ(^{119}Sn, X)$ and $^nJ(^{117}Sn, X)$; the low abundance of ^{115}Sn generally precludes observation of $^nJ(^{115}Sn, X)$. The ratios of couplings to ^{119}Sn and ^{117}Sn correspond to the quotient of their gyromagnetic ratios, as expected. The magnitudes of couplings involving tin, generally larger in magnitude than analogous couplings involving carbon, are of course dependent upon molecular structure; for example, $^2J(^{119}Sn, ^1H) = +96.9$ Hz in $Cl_3Sn—CH_3$ [442], but in $[Cl_3Sn—Pt(PPh_3)_2—H]$ the two-bond tin—hydrogen coupling is 1740 Hz[443]. The differences in molecular structure of these two compounds obviously preclude any meaningful comparison.

3. *Transition metal–tin systems*

Interest in metal—metal bonding between dissimilar species and the use of tin(II) halides in catalytic olefin activation by platinum metal halide complexes has led to substantial interest in the tin n.m.r. of transition metal:tin systems.

A series of trimethyltin derivatives of the chromium triad have been examined[444]; the shift data are presented in Table 28. Clearly, descending the triad causes shielding of the tin nucleus. The increase in polarizability in descending the triad undoubtedly contributes to this effect and is probably dominant with third row transition metals

TABLE 28. The triad effect with $\delta(^{119}Sn)$ in some transition metal:tin compounds

Compound	$\delta(^{119}Sn)$ (ppm)[a]
$[Me_3SnCr(CO)_3(\eta^5\text{-}C_5H_5)]$	+161
$[Me_3SnMo(CO)_3(\eta^5\text{-}C_5H_5)]$	+121
$[Me_3SnW(CO)_3(\eta^5\text{-}C_5H_5)]$	+43

[a] Relative to Me_4Sn; data from reference 444; benzene solution.

(the 'heavy atom effect'). With the first and second row transition metals, the extent of $d\pi$–$d\pi$ interaction is believed to give rise to low-energy excited states, causing lowering in ΔE, and contributing to deshielding[82]. Accordingly, an increase in shielding down the group is to be expected. With this particular set of compounds it was postulated[82] that the dipolar carbonyl groups may effect the field experienced by the tin nucleus and contribute to deshielding. There is slight evidence for this as the compounds $[Me_3SnMoCl(\eta^5\text{-}C_5H_5)_2]$ [$\delta(^{119}Sn) = +90 \pm 2$ ppm] and $[Me_3SnMo(CO)_3(\eta^5\text{-}C_5H_5)]$ [$\delta(^{119}Sn) = +121$ ppm][444] show that the carbonyl groups may cause some deshielding. There is very little conclusive evidence, however, that this is so[421].

There are several sets of compounds which demonstrate decreases in shielding as transition metals are coordinated to tin. Table 29 shows that increasing the number of Co or Mn atoms joined to a tin nucleus causes substantial deshielding. The same effect is evident in compounds **25** and **26**[444,445].

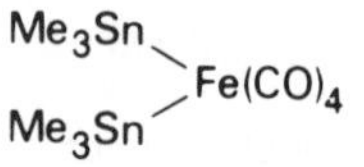

$(Me_3Sn)_2Fe(CO)_4$

$\delta(^{119}Sn) = +79$ ppm

(**25**)

$[Me_2SnFe(CO)_4]_2$ (cyclic: Me_2Sn and $SnMe_2$ bridged by two $Fe(CO)_4$ units)

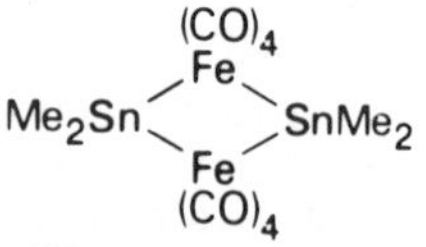

$\delta(^{119}Sn) = +257$ ppm

(**26**)

These observations demonstrate how important the paramagnetic contribution to the total shielding can be when π-interactions give rise to low-lying excited states. A small value of ΔE (equation 2, Section II.B.2) causes σ_p to dominate entirely and drastic deshielding can result.

TABLE 29. Multiple substitution effects of transition metals on $\delta(^{119}Sn)$ in transition metal:tin compounds

Compound	$\delta(^{119}Sn)$ (ppm)[a]
$[Me_3SnCo(CO)_4]$	+150.5[b]
$[Me_2Sn\{Co(CO)_4\}_2]$	+293 ± 1[b]
$[MeSn\{Co(CO)_4\}_3]$	+483 ± 1[b]
$[Me_3SnMn(CO)_5]$	+63 ± 1[b]
$[Me_2Sn\{Mn(CO)_5\}_2]$	+150 ± 1[b]
$[MeSn\{Mn(CO)_5\}_3]$	+284 ± 1[c]

[a] Relative to Me_4Sn; data from reference 444.
[b] Benzene solution.
[c] $CDCl_3$ solution.

Several platinum(II):tin(II) systems have been investigated by ^{119}Sn n.m.r.; the importance of these systems in catalytic hydrogenation[446] and hydroformylation[447] is well known and yet the role of the tin(II) salts, added to the systems to enhance activity, still remains something of a mystery.

The five-coordinate anion $[Pt(SnCl_3)_5]^{3-}$ is a particularly interesting sample to study; the crystal structure is known[448] and results of Mössbauer studies[449] suggest that the so-called 'red isomer' of $[PtCl_2(SnCl_3)_2]^{2-}$ [450] may in fact contain the five-coordinate anion. The ^{119}Sn n.m.r. of this anion[312], which has trigonal bipyrimidal geometry in the solid state[448], shows the presence of only one type of tin nucleus, $\delta(^{119}Sn) = -142$ ppm. The molecule must therefore be fluxional or of an extremely unusual geometry in solution. Coupling to platinum is maintained, $^1J(^{195}Pt, ^{119}Sn) = 16\,024$ Hz, and as coupling between different tin isotopes can be observed, $^2J(^{119}Sn, ^{117}Sn) = 6230$ Hz, it is possible to deduce that there are five tin nuclei coordinated to platinum from the satellite intensity ratios. Accordingly, fluxionality by an intramolecular process has been proposed[312]. The presence of different tin isotopes in a multi-tin species can generally be used to determine the molecularity, a useful factor in spectral analysis[451].

Complexes of the type $[Pt(PR_3)_2Cl_2]$ are generally believed to react with $SnCl_2{\cdot}2H_2O$ to yield insertion products containing the $SnCl_3^-$ ligand. A major problem in such systems is determining whether the $SnCl_3^-$ is present as a coordinated ligand or as a free anion. The system is well suited for multinuclear (^{31}P, ^{119}Sn, ^{195}Pt) n.m.r. studies and the results suggest a reaction sequence as in equation 9[313,443,452].

$$[PtCl_2(PR_3)_2] \xrightarrow{SnCl_2} \underset{\mathbf{(27)}}{[PtCl(SnCl_3)(PR_3)_2]} \xrightarrow{SnCl_2} \underset{\mathbf{(28)}}{[Pt(SnCl_3)_2(PR_3)_2]} \qquad (9)$$

The geometry of **27** depends on the nature of R, whilst complexes of type **28** all appear to be *trans*[312]. The complex **28** (R = Et) reacts with hydrogen to yield the hydride, *trans*-$[PtH(SnCl_3)(PR_3)_2]$ **(29)**[313]. Complexes such as **29** ($R_3 = Ph_2Bz$) react with activated acetylenes via insertion to yield the corresponding vinyl complexes[312].

These results are interesting from an n.m.r. viewpoint; geometry determinations for various complexes[313] of type **27** relate to magnitudes of coupling constants and the n.m.r. *trans*-influence (see Section VI.J.2). Also, complexes **28** contain two tin nuclei and so display a $^2J(^{119}Sn, ^{117}Sn)$ coupling[452]. The magnitude of $^2J(^{119}Sn, ^1H)$ in **29** has previously been mentioned[443], although not explained very satisfactorily. From the point of view of the catalytic system, it initially appeared that several subsequent stoichiometric steps had been defined. However, nearly all n.m.r. data were obtained at low temperatures as at higher temperatures broad, featureless spectra are generally observed[443], usually an indication that an exchange process is occurring. In fact, it is a general rule that any complex reactive enough to be part of a catalytic cycle is too short-lived to be observed. Actual conditions in catalytic reactions[446,447], of course, bear absolutely no resemblance to those used in n.m.r. experiments and so relating n.m.r. experiments to catalytic reactions must be done with the greatest caution. Solvent effects in n.m.r. experiments of this type have also been noted[453].

Other transition metal complexes of Ta, Rh and Ir containing tin ligands have been studied by ^{119}Sn n.m.r.[444,445], but the data are too few for meaningful comparisons to be made.

4. *Organolead compounds*

Several studies of organolead compounds have been reported[368,417,454–459]. Simple organolead halide compounds show much the same variation in lead chemical shift as

TABLE 30. ^{207}Pb n.m.r. data for some $[Me_3PbX]$ (X = Cl, Br, I) compounds

Compound	Solvent	$\delta(^{207}Pb)$ (ppm)[a]
$[Me_3PbI]$	CH_2Cl_2	$+203 \pm 0.3$
$[Me_3PbBr]$	CH_2Cl_2	$+367 \pm 4$
$[Me_3PbCl]$	CH_2Cl_2	$+432 \pm 1$
$[Me_3PbCl]$	$CDCl_3$	$+374.8$
$[Me_3PbCl]$	Me_2SO	$+258$
$[Me_3PbCl]$	C_5H_5N	$+216$
$[Me_3PbCl]$	HMPT	$+166$

[a] Relative to Me_4Pb; data from references 421 and 456.

analogous organotin compounds do for the ^{119}Sn nucleus. As a typical heavy metal nucleus, the polarizability of lead is reflected in chemical shift changes as the electron-withdrawing ability of the ligands is altered. Thus, the data in Table 30 show that the lead nucleus is deshielded as the electron-withdrawing ability of the halo-ligands in $[Me_3PbX]$ (X = Cl, Br, I) is increased. Solvent effects for the lead shielding of $[Me_3PbCl]$ (Table 30) suggest a variation in shielding with coordination number as described for ^{119}Sn. Similar variations in shielding have been noted for compounds with strained bond-angles at lead[456], although studies of spiroplumbanes are too few for the effects to be fully appreciated.

Some examples of lead coupling constants and chemical shifts in organolead compounds containing heteroatoms are shown in Tables 22 and 31. The large size of these coupling constants probably reflects the high value of the valence-state *s*-orbital expectation at the lead nucleus and the relatively high gyromagnetic ratio of lead. Diagnostically, variations in large couplings are obviously easier to detect than variations in smaller couplings; indeed, lead–proton couplings over six bonds have been detected; $^6J(^{207}Pb, ^1H) = 5.4$ Hz in $[Pb(p\text{-}CH_3C_6H_4)_4]$[460,461].

Surprisingly, although the variations in chemical shift and coupling constant are large and sensitive to structural variations in organolead compounds, few studies have employed ^{207}Pb n.m.r. in the elucidation of structural problems. This is particularly

TABLE 31. ^{207}Pb n.m.r. data for some lead–heteroatom compounds, $[Me_3PbX]$ (X ≠ H, halogen)[a]

X	$\delta(^{207}Pb)$ (ppm)[b]	δ(X) (ppm)	$^1J(^{207}Pb, X)$ (Hz)
$[PbMe_3]$	-281 ± 1	-281 ± 1	$+290 \pm 10$
$[SnMe_3]$	-323.8 ± 1.2	-57.0 ± 1.5[c]	-3570 ± 100
$[SnPh_3]$	-263.1 ± 0.2	-119.5 ± 1.5[c]	-2800 ± 50
[SeMe]	-196.5 ± 1	-1170 ± 100[d]	-61.9 ± 0.2
$[^{15}NMePh]$	$+226.5 \pm 2.5$	$+15.5 \pm 0.2$[e]	$+261 \pm 5$
$[PPh_2]$	$+40.5 \pm 4$	-35.9 ± 0.4[f]	-1335 ± 10
$[B(NMeCH_2)_2]$	-326.0 ± 2	$+41.7$[g]	-1330 ± 30

[a] Data from reference 368.
[b] Relative to $[Me_4Pb]$.
[c] Relative to $[Me_4Sn]$
[d] Relative to $[Me_2Se]$.
[e] Relative to $[Me_4NI]$, sample enriched with ^{15}N.
[f] Relative to 85% H_3PO_4.
[g] Relative to $[Et_2O{\cdot}BF_3]$.

notable in biochemical systems where the interactions of lead and inorganic lead compounds with biopolymers would make a number of interesting ^{207}Pb n.m.r. studies.

The standard employed for measuring lead chemical shifts is [Me_4Pb], which is relatively unaffected by solvent (unless coordination occurs) and temperature changes. Unfortunately, as with [Me_4Sn], the resonance of [Me_4Pb] falls in the middle of the shift range and so care with signs is necessary. Aspects of ^{207}Pb n.m.r. have previously been reviewed[417,421]. Chemical shift anisotropy can contribute to the relaxation of ^{207}Pb[42] and so care is needed in comparing data from different sources (see Section II.A.5).

I. Nitrogen N.m.r. Studies

Nitrogen is one of very few elements having both quadrupolar (^{14}N) and $I = \frac{1}{2}$ (^{15}N) isotopes available for n.m.r. study. ^{14}N has a high natural abundance ($N = 99.63\%$) and so is relatively simple to detect by n.m.r. methods (the receptivity is high and the gyromagnetic ratio is positive); however, as it is quadrupolar it is fairly unusual to resolve coupling to ^{14}N because of the linewidth problem. Accordingly, only chemical shifts and not coupling constants are usually obtained from ^{14}N n.m.r. spectra (there are, of course, many exceptions where the ^{14}N nucleus is in a highly symmetric environment, but these are few in organometallic systems).

^{15}N is the better nucleus to study as $I = \frac{1}{2}$ and so there are no problems of line broadening and spin-coupling can often be resolved. Unfortunately, this nucleus has a very low natural abundance ($N = 0.37\%$) and the gyromagnetic ratio is negative. Accordingly, spectra can usually be obtained only with high-field Fourier transform n.m.r. spectrometers or with samples artificially enriched with ^{15}N. Techniques are available for sensitivity enhancement in ^{15}N n.m.r. using population transfer methods[462].

Many texts and reviews are available which deal with various aspects of ^{14}N and ^{15}N n.m.r., including both theoretical and practical aspects as well as lists of reference data[463–468]. The application of ^{14}N n.m.r. to problems in inorganic chemistry has been discussed in some detail[468], and so only ^{15}N n.m.r. will be considered further here. An initial area of study in this field was the ^{15}N n.m.r. of amine and diamine complexes[469] of the transition metals, particularly platinum(II) amine complexes[470], which are known to exhibit chemotherapeutic anticarcinogenic activity. The parameter $^1J(^{195}Pt, ^{15}N)$ has been examined[471] as a function of the *trans*-ligand in some Schiff base complexes of type **30**. The variation in magnitude (290–500 Hz) serves as a sensitive

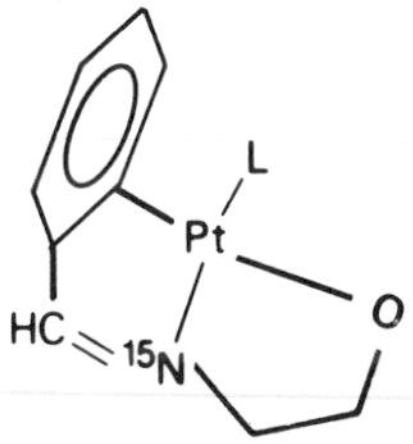

L
Pt
O
HC
^{15}N

(30)

measure of *trans*-influence and so is useful in structural assignments in much the same way as $^1J(M, ^{31}P)$ is used in ^{31}P n.m.r. studies (see Section IV.J.2). Coupling to ^{103}Rh has been exploited in a similar manner[472] and some ^{15}N n.m.r. studies of Rh(III) complexes have been performed at natural abundance, with no artificial isotope enhancement[473].

Among the more significant studies of organometallic and related coordination complexes are those of transition metal dinitrogen compounds, important as model compounds in studies of nitrogen fixation. A noteworthy example is the ^{15}N n.m.r study of the complex $[Ti(C_5Me_5)_2N_2]$[474] (prepared using doubly enriched nitrogen $^{15}N{\equiv}^{15}N$). The ^{15}N n.m.r. showed both a pair of doublets with $^1J(^{15}N, ^{15}N) = 7$ Hz and a singlet. The spectrum was interpreted in terms of an equilibrium (equation 10).

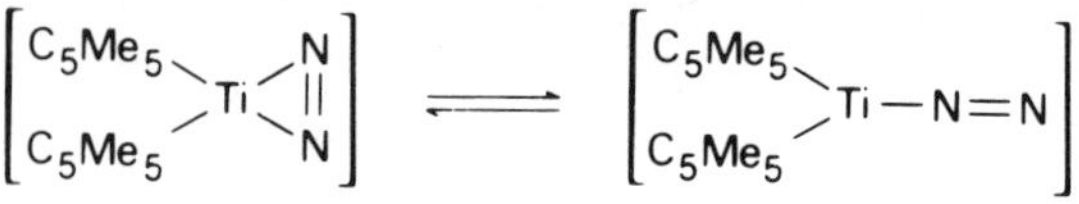

$$[(C_5Me_5)_2Ti(\eta^2\text{-}N{\equiv}N)] \rightleftharpoons [(C_5Me_5)_2Ti{-}N{=}N] \qquad (10)$$

(31) (32)

Complex **31** would be expected to show a single ^{15}N resonance as the side-on coordination of N_2 renders both ^{15}N nuclei equivalent, whereas complex **32** would be expected to show a doublet of doublets as the end-on coordinated N_2 ligand has two inequivalent ^{15}N nuclei, which will spin–spin couple.

The crystal structure determination[475,476] of this compound showed a solid-state structure equally in accord with the ^{15}N solution n.m.r. data and the spectroscopic data have now been reinterpreted[477] in terms of this single species, involving bridging (single resonance) and terminal (doublet of doublets) end-on coordinated N_2 ligands, **33**. Dinitrogen complexes of Zr[477], Mo, and W[478] have also been studied by ^{15}N n.m.r. and the technique shows considerable promise, especially when used in conjuction with other physical methods.

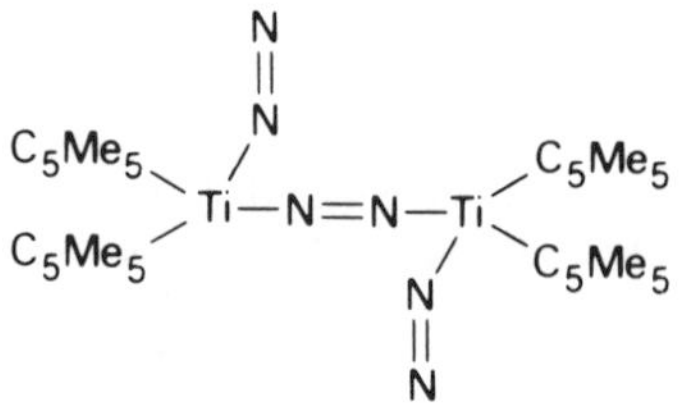

$$(C_5Me_5)_2Ti(N{=}N){-}N{=}N{-}Ti(N{=}N)(C_5Me_5)_2$$

(33)

There is some disagreement as to the most suitable shift reference for nitrogen n.m.r.; nitromethane is widely used, but results in most shifts being negative, whilst the $[NO_3]^-$ ion gives rise to problems of solvent, temperature, and concentration dependence. No ideal standard yet appears to have been found.

J. Phosphorus-31 N.m.r. Studies

^{31}P n.m.r. has a particularly important place in organometallic chemistry. The vast number of compounds containing phosphorus(III) ligands that have been synthesized have shown that tertiary phosphines and, to a lesser extent, phosphites, impart some almost unique characteristics to their compounds. Most significant are transition metal phosphine complexes, some of which act as very effective catalysts for a number of industrially important organic transformations. ^{31}P n.m.r. has contributed significantly to the development of tertiary phosphine chemistry; the ^{31}P nucleus has $I = \frac{1}{2}$ and $N = 100\%$ (Table 1) and a sensitivity that permits observation of ^{31}P n.m.r. by continuous-wave techniques in favourable cases (neat liquids, concentrated solutions, simple coupling situations, etc). Indirect and direct observation methods have previously been mentioned (Section III.C.1). A number of texts on ^{31}P n.m.r. are available[479–481] and a text[281] and review[280] on ^{31}P n.m.r. studies of transition metal—phosphorus(III) compounds.

In this section, the commonly measured parameters $\delta(^{31}P)$, $^{1}J(M, {}^{31}P)$, and $^{2}J(^{31}P, {}^{31}P)$ are discussed in the light of the theoretical descriptions of chemical shifts (Section II.B.3) and coupling constants (Section II.C.1). Chemical shifts are generally rather poorly understood theoretically and a suitable contribution of σ_d and σ_p to the observed shift can usually be rationalized for any given trend, provided much imagination is used. Coupling constants are most often discussed in terms of the Pople–Santry expression for Fermi contact. The expressions for the other two coupling mechanisms are rarely even mentioned, although the total dominance of Fermi contact is yet to be fully established for couplings between heavy atoms. Eaton and Lipscomb[360] have aptly described the rationalization of coupling constants by pseudo-theoretical treatments as 'orbital-waving' (being 'the quantum mechanical analogue of arm-waving rhetoric'). With this succinct phrase in mind, the following sections describe correlations to date.

It will quickly be seen that the theoretical expectations for variations in coupling constants and chemical shift are only poorly supported by practical results. Nonetheless, this rather unfortunate situation need not be a deterrent in organometallic and coordination chemistry as a large number of empirical relationships exist, enabling much to be accomplished in structural determination.

1. *Chemical shift variations*

Treatments of phosphorus chemical shifts assume the dominance of σ_p in the total shielding of the nucleus. Equation 2 (Section II.B.3) has been interpreted[280] in terms of σ- and π-bonding contributions as follows: the term P_u relates to electron imbalance in the valence *p*-orbitals and D_u to imbalance in valence *d*-orbitals. Accordingly, if σ-bonding relates to *s*- and *p*-electron involvement and π-bonding to *d*-electron involvement, these terms correspond to σ-bonding and π-bonding terms. In reality, this interpretation is open to question, the initial assumptions being constancy of the radial terms and the excitation energy. By carefully choosing a series of free tertiary phosphines in which π-bonding (between P and R in PR_3) is likely to be minimized, the constancy of the radial and excitation energy terms can be estimated by examining the variation in shielding with the extent of σ-bonding in the P—R bond. In other words, if all other facts are held constant, the shielding should be linearly related to parameters such as the electronegativity of R and the R—P—R bond angle which relate to the electron distribution in the phosphorus orbitals. Tables 32 and 33 show that correlations of $\delta(^{31}P)$ with the electronegativity of R and with the R—P—R bond angle are in fact poor. Thus, it appears that variations in ΔE or $\langle r^{-3}\rangle_{np}$ in equation 2 must be

TABLE 32. Values of $\delta(^{31}P)$ and the electronegativity of X in some PX_3 ligands

X	$\delta(^{31}P)$ (ppm)[a]	Electronegativity[b]
PF_3	127	4.0
PCl_3	170	3.0
PMe_3	−62	2.5[c]
PH_3	−240	2.1

[a]Data from references 215, 280, and 281.
[b]Data from reference 482.
[c]Electronegativity of carbon.

TABLE 33. Values of $\delta(^{31}P)$ and the cone-angle of some PR_3 ligands

Phosphine[a]	Cone-angle[b]	$\delta(^{31}P)$ (ppm)[c]
PEt_3	132°	−20
PBu^n_3	132°	−33
PBz_3	165°	−12
PCy_3	170°	+9
PBu^t_3	182°	+62

[a] Ligands presented are all trialkylphosphines, arylphosphines being excluded to minimize substituent electronic effects. Note that $\delta(^{31}P)$ and cone-angle do not correlate.
[b] Data from reference 215.
[c] Data from reference 483.

significant (alternatively, π-bonding between P and R may be considered to contribute). The major problem now encountered is how to consider yet another unknown factor, coordination to a metal centre. Here, π-interactions are likely to be significant in some (but not all) cases and the R—P—R bond angle not only increases upon donation of σ-electron density to the metal, but may be affected by steric interactions with other ligands. In some cases it may even be necessary to consider a (formal) rehybridization of phosphorus to account for the M—P bonding, which alters every variable in equation 2. Some correlations of phosphorus shielding with other parameters nevertheless do exist, despite the rather depressing account so far, and these are now considered.

Although for a series of different phosphorus(III) ligands there is no correlation between the shift of the free ligand and the shift of a given type of metal complex (see Table 34), some correlations do exist for $[ML_nX_m]^{p+}$ complexes where M, L, and X

TABLE 34. $\delta(^{31}P)$ for some free phosphorus(III) ligands and their Ni(0) complexes[a]

Ligand, L	$\delta(^{31}P)$ (ppm)	$\delta(^{31}P)$ of $[NiL_4]$ (ppm)	$\Delta\{\delta(^{31}P)\}$[b]
PCl_3	215	170	−45
PCl_2Ph	164	152	−12
PF_3	97	127	+30
$P(OEt)_3$	140	160	+20
PMe_3	−62	22.2	+82.2

[a] Data from references 280 and 281.
[b] Note that the coordination chemical shift may be either positive or negative.

are constant and n, m, and p vary. For example, in the complexes $[PdL_n]$ (Table 35), the ^{31}P nucleus is less shielded for $n = 3$ than for $n = 4$. It seems quite general that the ^{31}P shielding decreases in tertiary phosphine complexes as the coordination number increases. A variety of reasons are possible, including σ-bonding effects (variation in P_u), π-bonding effects (variation in D_u), and a decrease in ΔE. Probably all possible factors are involved.

Attempts to correlate shift changes with changes in oxidation state of the metal centre are generally less successful. Table 36 shows that no regular trend exists even in very closely related complexes. Changes in geometry at the metal centre often imply a change in the ligand *trans* to phosphorus also, thus two parameters vary simultane-

TABLE 35. $\delta(^{31}P)$ for some $[Pd(PR_3)_n]$ complexes

R	$\delta(^{31}P)$ of $[Pd(PR_3)_n]$ (ppm)[a]	
	$n = 3$	$n = 4$
Ph	22.6	18.4
Et	9.6	−1.5
n-Bu	−1.4	−7.9

[a]Data from reference 485. Note that shielding is consistently less for $n = 3$ than for $n = 4$.

TABLE 36. $\delta(^{31}P)$ for some $[PtCl_n(Bu^n_3P)_2]$ complexes ($n = 2, 4$)

Complex	$\delta(^{31}P)$ (ppm)[a]	Oxidation state of Pt
cis-$[PtCl_2(Bu^n_3P)_2]$	+1.2	2
cis-$[PtCl_4(Bu^n_3P)_2]$	+12.5	4
trans-$[PtCl_2(Bu^n_3P)_2]$	+5.1	2
trans-$[PtCl_4(Bu^n_3P)_2]$	0	4

[a]Data from reference 280. Note that shieldings vary Pt(IV) < Pt(II) for *trans*-complexes and Pt(II) < Pt(IV) for *cis*-complexes.

TABLE 37. Variation in $\delta(^{31}P)$ with the geometry of $[MCl_2(PR_3)_2]$ (M = Pd, Pt)

$[MCl_2(PPh_3)_2]$		
M	Geometry	$\delta(^{31}P)$ (ppm)[a]
Pt	*cis*	14.3
Pt	*trans*	19.8
Pd	*cis*	38.2
Pd	*trans*	28.5

[a]Data from references 486 and 487. Note that Pt(*trans*) is less shielded than Pt(*cis*) and that the converse is true for Pd.

ously and unsurprisingly the situation is far from straightforward. Table 37 shows that the *cis*- and *trans*-isomers of $[MCl_2(PR_3)_2]$ (M = Pd, Pt) show a different shift dependence on geometry. Generally, the effect of the *trans*-influence on $\delta(^{31}P)$ is not understood and cannot be used predictably with any confidence.

Most useful in characterization is the so-called ring contribution to the chemical shift which occurs in complexes containing a chelated phosphorus(III) ligand. The ^{31}P n.m.r. spectra of chelate complexes have been studied in much detail and a review discussing the ring contributions to ^{31}P n.m.r. parameters is available[484]. The ring contribution to the chemical shift is exemplified by considering the complex $[PtMe_2(PPh_2Me)_2]$ and its chelate analogues, $[PtMe_2\{Ph_2P(CH_2)_nPPh_2\}]$. The shift data (Table 38) show that the 4- and 5-membered rings ($n = 1$ and 2) result in a considerable shift relative to the parent complex, whilst the 6-membered ring ($n = 3$)

TABLE 38. Ring contribution to the chemical shift in some $[PtMe_2\{Ph_2P(CH_2)_nPPh_2\}]$ complexes

n	$\delta(^{31}P)$	Δ_R[a]
1	−40.0	−46.4
2	+45.4	+39.0
3	+3.2	−3.2

[a] Data from reference 484. $\Delta_R = \delta(^{31}P)$ of complex − $\delta(^{31}P)$ of $[PtMe_2(PPh_2Me)_2]$.

is scarcely affected. A recent review on this topic sensibly states that the theoretical aspects of the ring contribution are not clear[484]. Nonetheless, this contribution to the ^{31}P chemical shift appears to be general for transition metal complexes and can be particularly useful for studying the metallation of phosphorus(III) ligands, where a phosphorus—carbon chelate is formed (see also ^{195}Pt n.m.r., Section VI.E.8).

It can clearly be seen that, although a great many empirical relationships have been investigated for $\delta(^{31}P)$, the level of understanding for chemical shifts of even this, a greatly studied nucleus, is at an extremely primitive level. The utility of the data in solving chemical problems is undeniable, but the level of comprehension is very poor.

2. Variations in $^1J(M, ^{31}P)$

Understanding of the nature of the one-bond coupling between ^{31}P and a metal nucleus is, at the simplistic level, more acceptable. As previously mentioned, most correlations assume a dominance of Fermi contact in the coupling mechanism and so it is entirely possible that some data correlate with the theoretical expectation by coincidence as it is by no means proved that Fermi contact is a dominant interaction between coupled heavy nuclei. Assuming that equation 3 is valid, the magnitude of $^1J(M, ^{31}P)$ would be expected to vary as the metal is changed due to variations in the $|\psi_{s(m)}(0)|^2$ term. Indeed, this is so, and it has been shown that a plot of $|\psi_{s(m)}(0)|^2$ *vs*. the reduced coupling constant [hence eliminating dependence on $\gamma(M)$] is approximately linear[281]. Of course values of $^1J(M, ^{31}P)$ vary greatly for a given metal in different types of complex and so the errors involved in such a plot are large. It can clearly be seen that variations in $^1J(M, ^{31}P)$ for a given metal do not arise solely because of variation in the $|\psi_{s(m)}(0)|^2$ term by considering a complex such as [PtClPh(dppe)] where $^1J(^{195}Pt, ^{31}P) = 4192$ Hz for P *trans* to Cl and $^1J(^{195}Pt, ^{31}P) = 1638$ Hz for P *trans* to Ph[488]. The two very different coupling constants arise from interactions with a single ^{195}Pt nucleus, where the $|\psi_{s(m)}(0)|^2$ term must obviously be the same.

The corresponding term for phosphorus, $|\psi_{3s(p)}(0)|^2$, can also be seen to affect the magnitude of $^1J(M, ^{31}P)$. A plot of $^1J(^{183}W, ^{31}P)$ *vs*. the Sanderson electronegativity of R in $[W(CO)_5(PR_3)]$ complexes has been presented[281], which shows that the coupling constant increases with electronegativity in a roughly linear fashion. Further variation in $^1J(M, ^{31}P)$ is to be expected in terms of changes in the *s*-orbital bond order between M and P (the term k in equation 3). The problem here arises in considering the formal hybridization of the atoms involved, which may not give a true impression of the percentage *s*-character in the bond. The best known examples to illustrate this point are the complexes *trans*-$[PtCl_2(PBu^n_3)_2]$ with $^1J(^{195}Pt, ^{31}P) = 2392$ Hz and *trans*-$[PtCl_4(PBu^n_3)_2]$ with $^1J(^{195}Pt, ^{31}P) = 1474$ Hz. Assuming the *s*-character argument, then the ratio of Pt(II) (dsp^2) to Pt(IV) (d^2sp^3) couplings should be $\frac{1}{4} : \frac{1}{6}$. Accordingly the Pt(II) complex should have a coupling 1.5 times as large as the Pt(IV) complex, and indeed this is approximately the case. A recent text[281] points out, however, that

$[Pt(PEt_3)_4]$ and $[Pt(PEt_3)_4]^{2+}$ both have 25% *s*-character but $^1J(^{195}Pt, ^{31}P)$ is 3740 Hz in the former and 2342 Hz in the latter. The validity of the *s*-character argument is thus open to considerable doubt.

One of the most useful factors causing variation in $^1J(M, ^{31}P)$ is the *trans*-influence of the ligand *trans* to phosphorus in square-planar and octahedral metal complexes. Generally, high *trans*-influence groups give rise to low magnitudes of $^1J(M, ^{31}P)$ and *vice versa*. Considering either a σ- or π-dominance of the *trans*-influence requires that the effect operate indirectly, via the interaction with the metal centre. Indeed, the Fermi contact expression cannot account for π-effects and so we can circumvent this by assuming that the π-effects cause a synergic σ-effect in the *trans* M—P bond. Obviously this can hardly be called a satisfactory explanation at anything other than the most basic level, but nonetheless, the effect of *trans*-influence on $^1J(M, ^{31}P)$ is of the greatest importance in geometry determinations, especially those concerning σ-carbon ligands, which gave rise to very high *trans*-influence and hence a low value of $^1J(M, ^{31}P)$. The *trans*-influence has been discussed in detail[228] and its effects on ^{31}P n.m.r. spectra described[281]. Examples of geometry determinations which utilize the *trans*-influence argument are described below.

3. *Variations in* $^2J(^{31}P, ^{31}P)$

In ^{31}P n.m.r. spectra of complexes containing inequivalent phosphorus atoms, the two-bond phosphorus—phosphorus coupling is very important in determining the final appearance of the spectrum. Only in cases where the spectrum approximates to first order can the coupling be measured directly. Recently a qualitative molecular orbital model has been developed to explain variations in sign and magnitude of $^2J(^{31}P, ^{31}P)$[281]. This model again assumes the dominance of Fermi contact in the coupling mechanism and proposes that the magnitude of the triplet excitation energy, averaged as $^3\Delta E$, is dominant in determining the parameters of the coupling constant. The model assumes that excitation from one orbital to another of the same symmetry gives rise to a negative contribution to the coupling, whereas if two orbitals of opposite symmetry are involved, the contribution will be positive. By considering the relative energies of the metal orbitals in relation to the energies of the phosphorus orbitals contributing to the M—P bond, it is possible to show that couplings between *cis* and *trans* pairs of phosphorus nuclei will be of similar magnitude for complexes of first-row transition metals, but that second- and third-row transition metals will exhibit a smaller coupling for two *cis*-phosphorus nuclei than for two *trans*-phosphorus nuclei. The relative energies of the orbitals contributing to the M—P bond, and hence defining ΔE, similarly explain changes in $^2J(^{31}P, ^{31}P)$ with changing geometry and oxidation state of the metal centre. This model has been discussed in some detail previously[281].

Most useful in determining geometries is the observation that $^2J(^{31}P, ^{31}P)_{cis} \ll {}^2J(^{31}P, ^{31}P)_{trans}$ for second- and third-row transition metals. Some determinations of geometry utilizing this qualitative observation are discussed below.

4. *Geometry determinations*

Utilization of the parameters $\delta(^{31}P)$, $^1J(M, ^{31}P)$, $^2J(^{31}P, ^{31}P)$, and, less frequently, $^nJ(^{31}P, X)$ can often be an aid in structural assignment. Variations in combination of these parameters and occasionally just the magnitude of a single parameter can be indicative of a certain structure or geometry. Elegant examples abound in literature so only some relatively simple examples will be considered here.

In some cases $\delta(^{31}P)$ alone can be used in structural assignment. This is particularly evident in cases where chelation (in diphosphines, metallated phosphines, etc.) gives rise to a large ring contribution to the chemical shift. In addition, the difference in shift between free and coordinated ends of a potentially bidentate diphosphine may be indicative of monodentate coordination. The complex $[Fe(dppm)_3]$[484,489] illustrates both of these points. The observation of three resonances, $\delta(^{31}P_A) = 60.4$ ppm, $\delta(^{31}P_B) = 10.2$ ppm, and $\delta(^{31}P_C) = 49.8$ ppm, supports a structure **34**. Similarly, the

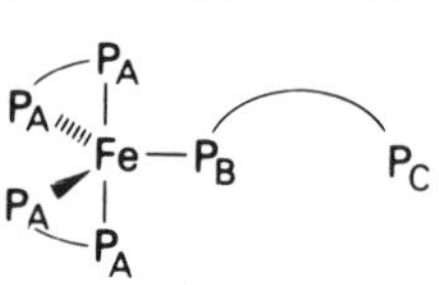

(34)

complex $[Fe_2(dppm)_5]$[484,489] is assigned the structure **35**, as one phosphorus resonance exhibits a large ring contribution to the shift, $\delta(^{31}P_B) = 60.4$ ppm, and the other does not, $\delta(^{31}P_A) = 7.2$ ppm. Another useful application of $\delta(^{31}P)$ to structural assignment, again related to the ring contribution to the shift, appears in the chemistry of phosphido-bridged complexes. Thus, in complex **36**, $\delta(^{31}P) = -127$ ppm, whereas in complex **37**, $\delta(^{31}P) = +155$ ppm (considering the phosphido resonances only). The enormous shift difference indicates a four-membered ring environment in **36** and a three-membered ring environment in **37**, thus supporting a metal—metal bonded structure in the latter[490,491].

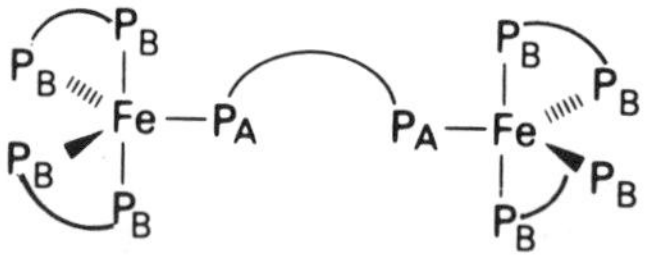

(35)

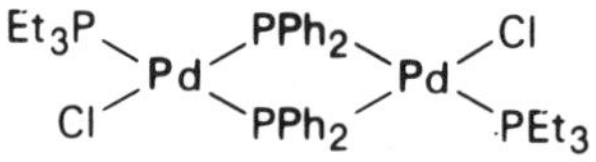

(36) 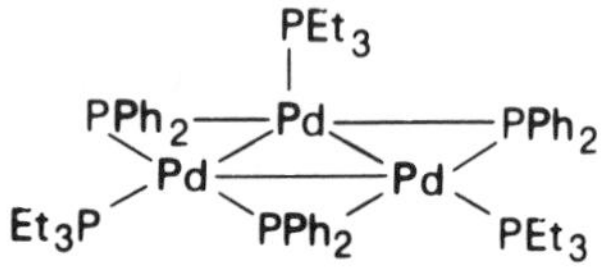 **(37)**

In a simple substitution reaction of [PtRCl(cod)] with the unsymmetrical ligand appe, the product may be of two possible geometries, **38** and **39**. The $^{31}P\{^1H\}$ n.m.r. spectra of the analogous [PtRCl(dppe)] complexes show that $^1J(^{195}Pt, ^{31}P) \approx 4200$ Hz for P *trans* to Cl and ≈ 1500 Hz for P *trans* to R. Accordingly, a value of $^1J(^{195}Pt, ^{31}P) \approx 4250$ Hz (4219 Hz for R = Ph, 4382 Hz for R = COPh) defines **38** as the sole product. Thus, the parameter $^1J(M, ^{31}P)$ defines the product in this case[448].

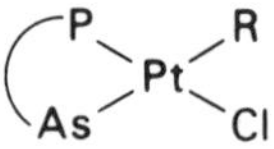

(38) 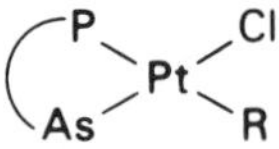 **(39)**

It is obviously more satisfactory to have two supporting parameters to identify a product. In the case of the complexes $[PtCl(PR_3)(appe)]^+$, two geometries, **40** and **41**, are possible.

The $^{31}P\{^1H\}$ n.m.r. spectra[488] of the products of reaction 11 indicate that both

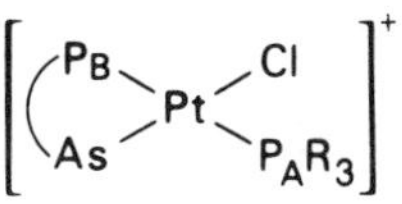

(40) 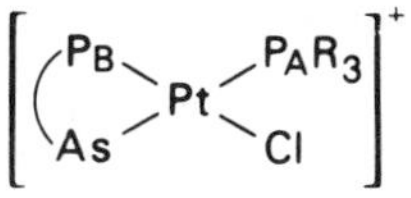 **(41)**

isomers are formed, with one isomer being more favoured. This is shown by $^{31}P\{^1H\}$ n.m.r. to be **41**. The P *trans* to P arrangement in **40** is supported by similar values of $^1J(^{195}Pt, ^{31}P_{A,B})$ (2313 *vs*. 2386 Hz), whereas in **41**, $^1J(^{195}Pt, ^{31}P_A)$ = 2845 Hz (*trans* to As) and $^1J(^{195}Pt, ^{31}P_B)$ = 4225 Hz (*trans* to Cl). The supporting evidence is the difference in values of $^2J(^{31}P, ^{31}P)$. In **40**, $^2J(^{31}P_A, ^{31}P_B)$ = 397 Hz (P *trans* to P), whereas in **41** $^2J(^{31}P_A, ^{31}P_B)$ = 14 Hz (P *cis* to P).

$$[PtCl_2(appe)] + AgClO_4 + PPh_3 \longrightarrow [PtCl(PPh_3)(appe)]^+ + AgCl \quad (11)$$

Coupling between ^{31}P and a heteroatom, other than the metal centre, can be very useful in product formulation. For example, the dimeric complex $[Pt_2(\mu\text{-}Cl)_2Cl_2(PEt_3)_2]$ is cleaved by Bu^{t}NC to yield a monomeric product of empirical formula $[PtCl_2(CNBu^t)(PEt_3)]$[488]. The $^{31}P\{^1H\}$ n.m.r. spectrum showed that both *cis*- and *trans*-isomers were present. The magnitudes of $^1J(^{195}Pt, ^{31}P)$ are indicative of this with values of 2822 Hz (P *trans* to C) and 3174 Hz (P *trans* to Cl) for the two isomers. The geometry is confirmed, however, as one resonance appears as a 1:1:1 triplet owing to coupling with ^{14}N, $^3J(^{31}P, ^{14}N)$ = 7 Hz, indicating a *trans* geometry, whilst the second resonance is a singlet (i.e. $^3J(^{31}P, ^{14}N) \approx 0$ Hz), indicating a *cis* geometry.

Occasionally a spectrum may contain more information regarding molecular structure than would initially be imagined. For example, the complex *trans*-$[PtH\{P(o\text{-tolyl})_3\}(PCy_3)_2]$ shows a resonance attributed to the PCy_3 group consisting of eight components[492]. This implies that the phosphorus nuclei of the PCy_3 groups couple not only to the *cis*-phosphorus but also to each other. Accordingly, it is likely that the molecule is distorted considerably from square-planar geometry in solution, hence rendering the PCy_3 groups inequivalent. The X-ray structure[492] shows that this is so in the solid state.

A final example in this section demonstrates how the multiplicity and magnitude of couplings to ^{31}P can be used to determine the numbers of ligands in a complex[488]. *trans*-$[PtH_2(PCy_3)_2]$ reacts with CO at −60°C in toluene solution to yield a single product observable in the $^{31}P\{^1H\}$ n.m.r. spectrum, with $\delta(^{31}P)$ = 20.2 ppm and $^1J(^{195}Pt, ^{31}P)$ = 3123 Hz. In order to determine the number of phosphorus ligands present, the reaction was repeated with 90% ^{13}CO and the $^{13}C\{^1H\}$ n.m.r. spectrum obtained. Here, a carbonyl resonance at $\delta(^{13}C)$ = 185.0 ppm was split into a triplet by two equivalent phosphorus nuclei, $^2J(^{31}P, ^{13}C)$ = 12 Hz, with ^{195}Pt satellites, $^1J(^{195}Pt, ^{13}C)$ = 1809 Hz. In order to determine the number of carbonyl ligands, the $^{31}P\{^1H\}$ n.m.r. spectrum of the ^{13}C-enriched sample was obtained. The phosphorus resonance appeared as a triplet [again, of course, with $^2J(^{31}P, ^{13}C)$ = 12 Hz], indicating that coupling to two equivalent carbonyl carbon atoms was present. Accordingly, these data establish[488] that two equivalent PCy_3 ligands and two equivalent ^{13}CO ligands are bound to platinum, defining the product as $[Pt(CO)_2(PCy_3)_2]$.

The enormous data base available for ^{31}P n.m.r. data of metal—phosphine and related complexes makes comparative structural assignment routine in many cases, although care in any theoretical interpretation of such results is obviously essential. Additionally, although the accepted standard for shift measurement is currently H_3PO_4 (the use of external P_4O_6 no longer being recommended by the American National Standards Committee), many arbitrary references are used and care in comparing literature data is necessary.

K. Studies of the Remaining Group VB Nuclei

^{75}As n.m.r. would seem to be a potentially useful technique in organometallic chemistry, in view of the very large number of transition metal complexes containing tertiary arsines as supporting ligands[493]. Unfortunately, as ^{75}As is quadrupolar (Table 4), the detectability of the spectra relate directly to the symmetry of the ^{75}As nucleus. To date, the majority of ^{75}As n.m.r. data are for tetrahedral or octahedral arsenic(V) derivatives (such as $[AsR_4]^+$ and $[AsF_6]^-$) and no transition metal, or other, low symmetry arsenic(III) derivatives have been observed. Studies of ^{75}As n.m.r. have been reviewed[24,494] and the standard used for chemical shift measurements is $[K][AsF_6]$ – a peculiar choice in view of the arsenic—fluorine coupling, $^1J(^{75}As-^{19}F) = 930$ Hz[495].

Antimony has two $I > \frac{1}{2}$ isotopes (^{121}Sb and ^{123}Sb; Table 4), although the slightly more sensitive ^{121}Sb nucleus is usually studied. The situation with regard to ^{121}Sb n.m.r. parallels that described for ^{75}As; thus, although it would be highly desirable to study the ^{121}Sb n.m.r. of transition metal stibine derivatives[493], this has not proved possible because of the low symmetry about the ^{121}Sb nucleus in such complexes which adversely affects it relaxation. Again, compounds of the types $[SbR_4]^+$ and $[SbF_6]^-$ have been observed, but attempts to observe Sb(III) derivatives have so far been unsuccessful[496]. The chemical shift standard is $[SbCl_6]^-$ (usually as the tetraethyl ammonium salt); reviews dealing with ^{121}Sb n.m.r. have appeared[24,494].

^{209}Bi ($I = \frac{9}{2}$; Table 4) has been little studied in chemical systems. The report[497] of a ^{209}Bi resonance for aqueous $Bi(NO_3)_3$ appears to be the only report of a solution study[494], indicating little interest in this element over the last 30 years.

L. Studies of Group VIB Nuclei

The lighter Group VIB elements are not particularly well suited for n.m.r. study. Oxygen has one quadrupolar isotope (^{17}O, $I = \frac{5}{2}$; Table 4) of low abundance, as does sulphur (^{33}S, $I = \frac{3}{2}$; Table 4). The heavier elements selenium and tellurium both have $I = \frac{1}{2}$ isotopes (^{77}Se, ^{123}Te, ^{125}Te; Table 1), the ^{125}Te isotope being the better suited of the two tellurium nuclei for n.m.r. study. No chemically significant studies of polonium have been performed.

Despite the problems of observing n.m.r. of low-abundance, quadrupolar nuclei with only average magnetic moments, much effort has gone into the study of ^{17}O n.m.r. The importance of oxygen in chemistry need not be stressed and this led to early work on the ^{17}O n.m.r. of organic and inorganic compounds. The early studies showed that resonances could usually be observed, provided that the oxygen nucleus was not in a completely unsymmetrical environment and that the molecule was sufficiently small that long molecular tumbling times do not become a problem. Several review articles are available, although the most recent is in Japanese[498–502].

1. ^{17}O n.m.r. of metal carbonyls

Among the first reports of ^{17}O n.m.r. were details of studies of several transition metal carbonyls[503]. Both $[Ni(CO)_4]$ (tetrahedral about Ni) and $[Mn_2(CO)_{10}]$ (octahedral about Mn) showed a single resonance, at 362 and 355 ppm, respectively. The nitrosyl derivatives $[Fe(CO)_2(NO)_2]$ and $[Co(CO)_3(NO)]$ also each give single resonances, at 418 and 377 ppm, respectively. No resonances attributable to the nitrosyl oxygens could be detected, presumably a result of quadrupolar broadening by the adjacent nitrogen atom, although signal averaging by dynamic oxygen exchange between nitrosyl and carbonyl groups cannot be ruled out. Possibly the most interesting sample studied was [$Fe(CO)_5$]. Again, only a single resonance could be observed, at 388 ppm, either as a pure liquid (30 to −20°C) or as an ether solution (30 to −63°C). This work thus substantiates the ^{13}C n.m.r. studies of [$Fe(CO)_5$], which also gave a single resonance (see Section VI.G.3); accordingly, no differentiation between a dynamic carbonyl averaging process and an unusual geometry with five equivalent carbonyls can be made. This work used the ^{17}O resonance of pure water as a reference and this has largely been accepted, although recent work[118] using the ^{17}O resonance of acetone as a reference has appeared. The former would appear to be the most satisfactory as the majority of organic compounds have positive shifts with respect to water. Shift changes with environment are known, however, and so it is possibly most satisfactory to use water as an external reference and apply the susceptibility correction.

Since the initial report appeared, studies of natural abundance ^{17}O n.m.r. of metal carbonyls have largely been concerned with the relationship of ^{13}C and ^{17}O parameters of a given carbonyl group. Unfortunately, although a very large and extensive data base exists for ^{13}C n.m.r. studies of metal carbonyls, this is not the case with ^{17}O n.m.r. and so the correlations proposed in the current literature may well need to be modified as further work increases our understanding of ^{17}O n.m.r.

A number of tungsten derivatives have been studied[233], of the type [$W(CO)_5L$] (L = carbene, phosphine, phosphite), and resonances attributable to carbonyl groups *cis* and *trans* to the L group were observed. The chemical shift data are shown in Table 8, together with the corresponding values of $^1J(^{183}W, ^{13}C)$ from $^{13}C\{^1H\}$ n.m.r. studies. It appears that the relationship $\delta(C^{17}O)_{trans} \geqslant \delta(C^{17}O)_{cis}$ is general. These workers[233] observe that deshielding of the oxygen nucleus is accompanied by a decrease in $^1J(^{183}W, ^{13}C)$, although inspection of Table 8 shows that the trend is not especially obvious. It has previously been mentioned that the value of $^1J(M, X)$ is related to the σ-donor/π-acceptor ability of the ligand *trans* to X (see Section VI.J.2) and, using these two correlations, a relationship between $\delta(^{17}O)$ and the σ-donor/π-acceptor ability of the *trans*-ligand, L, is proposed. A plot of $\delta(^{17}O)$ *vs*. $\delta(^{13}C)$ for each type of carbonyl group (*cis* or *trans*) was said to be approximately linear.

The latter point, the relationship between $\delta(^{17}O)$ and $\delta(^{13}C)$, has been the subject of some debate. One study[118] of a large group of substituted carbonyls (of Cr, Mo, W, Mn, Fe, and Co) showed a reasonably good correlation, whereas a second study[116] (of those metals previously mentioned together with Re, Rh, and Ni) showed a very poor correlation. A comparison[116] of ^{17}O and $^{13}C\{^1H\}$ n.m.r. data for substituted transition metal carbonyls raised the following points:

(*i*) electronegative substituents on M cause deshielding of ^{17}O but shielding of ^{13}C;
(*ii*) substitution of a carbonyl group by a Lewis base (PR_3, SbR_3, CNR, etc.) or by arenes causes shielding of ^{17}O and deshielding of ^{13}C in the remaining carbonyl groups;
(*iii*) $\delta(^{17}O)$ correlates poorly with the C=O stretching force constant whereas $\delta(^{13}C)$ shows a good correlation;
(*iv*) shielding of ^{17}O increases in the order cationic < neutral < anionic complex, the opposite of ^{13}C shielding; and
(*v*) both ^{17}O and ^{13}C exhibit triad effects; for example, Table 25 shows that descending the Cr, Mo, W triad in [$M(CO)_4(NBD)$] results in shielding of both ^{17}O and ^{13}C nuclei.

The opposite trends in ^{13}C and ^{17}O n.m.r. data have been rationalized by considering the effects of different *trans*-ligands on the C=O bond order and electron distribution. Thus, an electron-donating *trans*-ligand will enhance π-back-donation from M into the π^* orbitals of CO. This disrupts the CO bond order, causing the electron density on carbon to decrease and the electron density on oxygen to increase. This is

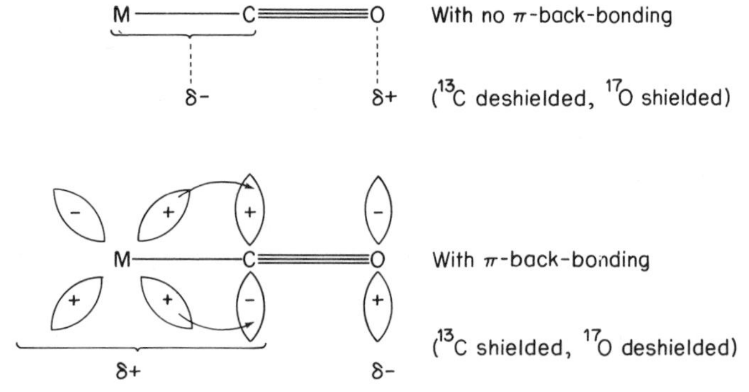

FIGURE 7. Schematic representation of the relationship between $\delta(^{13}C)$ and $\delta(^{17}O)$ in transition metal carbonyls.

shown schematically in Figure 7. The shielding of ^{13}C and ^{17}O will relate directly to the fate of these outer electrons (see Section II.B.3); as the *p*-orbitals on carbon contract with reducing electron density, so the contribution of σ_p to σ is increased and deshielding occurs. Conversely, the negative charge on oxygen increases, making the contribution of σ_p to σ less dominant and so shielding of the oxygen nucleus results. Accordingly, an inverse correlation between ^{17}O and ^{13}C shieldings should be observed although, as the role of the electrons in defining the contribution of σ_p to σ is very sensitive, any factors affecting the model described will have a dramatic effect on the shielding. Only strictly analogous systems should therefore be compared. Further, the points listed above do not appear to be entirely general; thus, comparison of substituted carbonyls of tungsten and chromium (Table 39) shows that point 2 is not general as the carbonyl groups in $[W(CO)_5P(OMe)_3]$ are both shielded (*cis*) and deshielded (*trans*) with respect to the parent compound, whereas in a similar chromium complex, $[Cr(CO)_5P(OPh)_3]$, both *cis*- and *trans*-carbonyl oxygens are shielded. In reality it is necessary to accumulate more data before valid statements concerning the relationship of ^{17}O and ^{13}C n.m.r. data of metal carbonyls can be made.

A number of derivatives, $[Cr(\eta^6\text{-}C_6H_{6-n}R_n)(CO)_2CX]$ (R = H, Cl, Me, OMe, NH_2, NMe_2, COOMe; X = O, S; n = 1–3) have also been studied[504] by natural abundance ^{17}O n.m.r. These compounds merit separate mention as a number of interesting 'through-space' interactions have been proposed. Molecular models show that the carbonyl groups may be close enough to the arene ring or, in the cases R = NH_2 or NMe_2, to the nitrogen lone-pair or methyl protons, for an interaction to occur. The total range of shieldings observed is 368.2–376.4 ppm and a possible electronic dominance by a 'through-space' interaction is proposed to account for the observed values.

TABLE 39. Effects of substitution on $\delta(^{17}O)$ in $[M(CO)_5L]$ complexes

Complex	$\delta(^{17}O)_{cis}$	$\delta(^{17}O)_{trans}$	$\Delta\delta(^{17}O)_{cis}$[a]	$\Delta\delta(^{17}O)_{trans}$[a]
$[W(CO)_6]$[b]	356.8		—	—
$[W(CO)_5P(OMe)_3]$[b]	353.7	359.0	+3.1	−2.2
$[Cr(CO)_6]$[c]	376.2		—	—
$[Cr(CO)_5P(OPh)_3]$[c]	373.6	368.4	+2.6	+7.8

[a] $\Delta\delta(^{17}O) = [\delta(^{17}O)_{\text{Unsubstituted}} - \delta(^{17}O)_{\text{Substituted}}]$ (ppm).
[b] Data from reference 233.
[c] Data from reference 116.

Complementary studies of the other nuclei (^{1}H and ^{13}C) are really necessary before this model can be rationalized.

Studies concerning several binuclear systems have also been reported. The complex [$Mn_2(CO)_{10}$], reported to show a single oxygen resonance in an early study[503] (at 355 ppm), has been re-examined[118] and resonances at *ca*. 368 and 387 ppm (corrected to $H_2{}^{17}O$ reference) observed. These were assigned to axial and equatorial carbonyls, respectively. Variable-temperature ^{17}O n.m.r. of [$Fe_2(CO)_6(Ph_2C_2COC_2Ph_2)$] showed that the three observable oxygen resonances broadened at +50°C as exchange began to occur[118]. Care is needed, however, with line-shape studies of ^{17}O n.m.r. spectra; for example, [$Co_2(CO)_6\{C_2(CH_2OH)_2\}$] gives a relatively narrow resonance at ambient temperature (ω = 100 Hz), but on cooling to −50°C the resonance broadens considerably (ω = 620 Hz). The effect is not related to fluxionality, however, as $^{13}C\{^{1}H\}$ n.m.r. shows that the molecule is still fluxional at −90°C. A possible explanation is that the molecular tumbling time is lengthened at low temperature and so the relaxation is adversely affected, resulting in a broad resonance[118]. The complex [$Co_2(CO)_6\{C_2(COOMe)_2\}$] also exhibits a very broad resonance, ω = 480 Hz, but this was explained by coincidental equivalence of the carbonyl and carboxyl oxygen shieldings. Linewidth measurements show that oxygen resonances of carbonyl groups are generally broadened by coordination to quadrupolar metals, but not as much as the ^{13}C resonances[505].

N.m.r. studies of metal carbonyls enriched with ^{17}O have been reported[505,506]. The series *cis*-[$Mo(C^{17}O)_4L_2$] (L = EPh_3; E = P, As, Sb; L = $PPhMe_2$) was studied[506], each compound showing two oxygen resonances. Again, the relationship $\delta(C^{17}O)_{trans} \geq \delta(C^{17}O)_{cis}$ appears to apply. The cluster [$Ru_3(C^{17}O)_{12}$] was examined and a single resonance observed. Possibly most interesting is the spectroscopic data concerning [$Fe(\eta^5\text{-}C_5H_5)(C^{17}O)_2]_2$; this complex showed a single oxygen resonance at 461.5 ppm at 55°C, indicating that a carbonyl-averaging process was occurring. Cooling to −46°C caused three resonances to appear, at 555.7, 359.7 and 458.8 ppm. The former two resonances were assigned to the *cis*-isomer, the more shielded resonance being broadened and consequently assigned to the bridging carbonyls. The latter resonance was assigned as an averaged signal for the *trans*-isomer, implying that the barrier to fluxionality is much lower for the *trans*-isomer than for the *cis*-isomer. Further evidence for these assignments comes from studies of the bis(triisopropylaluminium) adduct. ^{1}H n.m.r. studies have previously shown that coordination of aluminium to the bridging carbonyls slows the fluxional process to the extent that the molecule is static at +28°C. Here, the ^{17}O n.m.r. spectrum shows a single resonance at 371.2 ppm, attributable to a terminal carbonyl group in the *cis*-isomer. No resonance attributable to the bridging carbonyls of the *cis*-isomer could be observed, a probable consequence of quadrupolar broadening by the coordinated aluminium (^{27}Al, $I = \frac{5}{2}$, N = 100%).

The series [$Mn(CO)_5L$] (L = Cl, Br enriched with ^{17}O; L = CH_3 at natural abundance) have been studied[505] and each compound gives rise to two resonances in a 4:1 ratio, as expected. This is particularly interesting as the $^{13}C\{^{1}H\}$ n.m.r. spectrum of the methyl compound shows only a single broad resonance. This could be explained in terms of carbonyl exchange, or by quadrupolar broadening by ^{55}Mn causing two resonances to merge into one. The ^{17}O n.m.r. results demonstrate that the molecule is in fact static and so quadrupolar broadening must explain the $^{13}C\{^{1}H\}$ n.m.r. results.

2. *Other systems studied by ^{17}O n.m.r.*

Of the remaining ^{17}O n.m.r. studies of inorganic compounds, the majority have been concerned with transition metal oxy-anion complexes[500,501]. A few organometallic derivatives have been examined, such as $[Ti(\eta^5\text{-}C_5H_5)(Mo_5O_{18})][NBu^n_4]_3$[507], although the organic ligand is really coincidental and the oxy-anion is the centre of interest. Nonetheless, some very structural determinations[500,501,507,508] of oxy-anions have been performed by ^{17}O n.m.r. and the accumulated data will be of use in other fields.

Probably the most obvious use of ^{17}O n.m.r. is the study of reactions involving molecular oxygen. A number of transition metal complexes are known which react reversibly with dioxygen and these are important as analogues of biological oxygen carriers. ^{17}O n.m.r. has been applied to the study of such systems in a number of cases[500,509–512] with a singular lack of success. For example[509], a series of Vaska-type complexes, $[IrX(CO)(PR_3)_2]$ (X = Cl, I; R = Ph; X = Cl; R = C_7H_7), were reacted with $^{17}O_2$ and examined by ^{17}O n.m.r. No indication of the well known dioxygen complex was observed. The problem does not appear to be related to the field-gradient at the ^{17}O nucleus, as the complex $[IrI(CO)\{P(C_7H_7)_3\}_2(S^{17}O_4)]$ also could not be detected. In this sulphate complex no symmetry-determined field gradient problems would be expected and so the failure to observe oxygen resonances was interpreted in terms of a long molecular tumbling time for the relatively large unsymmetrical molecule. Such problems have only been observed elsewhere[118] at low temperatures, however.

3. ^{33}S, ^{77}Se, and ^{125}Te n.m.r. studies

^{33}S n.m.r. has been little studied[24,501]. The few available data indicate that there should be little problem in obtaining reasonably narrow resonances provided that the symmetry about the ^{33}S nucleus is not too low. To date, no organometallic compounds have been studied by ^{33}S n.m.r. and even studies of simple coordination compounds of charged or neutral sulphur donors do not appear to have been reported. Both $[Cs_2SO_4]_{(aq.)}$ and CS_2 have been employed as references and there is little to choose between them as $[Cs_2SO_4]$ shows little temperature or concentration dependence[513].

^{77}Se is more amenable to n.m.r. study and a great deal of data have been accumulated using both indirect and direct methods of observation. Reviews dealing with ^{77}Se n.m.r. have appeared[501,514,515]. The accepted standard for chemical shift measurements is Me_2Se, although a variety of other compounds [including selenophene, $H_2SeO_{3(aq.)}$ and $SeOCl_2$] have been used. The use of Me_2Se is particularly convenient in double resonance experiments.

The only complete series of organometallic compounds to be studied by ^{77}Se n.m.r. are the organotin selenides, $[Me_nSn(SeR)_{4-n}]$ (R = Me, Ph); the data obtained by double resonance techniques[516] are given in Table 40. The values of $\delta(^{77}Se)$ clearly show that the deshielding of the ^{77}Se nucleus increases as the electron-withdrawing

TABLE 40. Values of $\delta(^{119}Sn)$, $\delta(^{77}Se)$, and $^1J(^{119}Sn, ^{77}Se)$ in some organotin selenides, $[Me_nSn(SeR)_{4-n}]$

Compound	$\delta(^{119}Sn)$ (ppm)[a]	$\delta(^{77}Se)$ (ppm)[b]	$^1J(^{119}Sn, ^{77}Se)$ (Hz)
$[Me_3SnSeMe]$	+45.6	−276.7	1015 ± 10
$[Me_2Sn(SeMe)_2]$	+57.1	−237.0	1190 ± 10
$[MeSn(SeMe)_3]$	+14.8	−183.9	1340 ± 10
$[Sn(SeMe)_4]$	−80.5	−127.3	1520 ± 10
$[Me_3SnSePh]$	+55.0	+11.3	980 ± 10
$[Me_2Sn(SePh)_2]$	+54.1	+65.5	1190 ± 10
$[MeSn(SePh)_3$	−16.5	+124.0	1400 ± 15

[a]Relative to $Me_4\ ^{119}Sn = 0$; data from reference 516.
[b]Relative to $Me_4\ ^{77}Se = 0$.

ability of the attached groups increases. This has been attributed[516] to changes in σ_p arising from altered electron imbalance in the bonds to selenium. That the effect is greater for —SePh than for —SeMe may just reflect the greater electron-withdrawing ability of the —SePh group. The coupling between ^{77}Se and ^{119}Sn is also affected, electron-withdrawing groups causing the magnitude of $^1J(^{119}Sn, ^{77}Se)$ to increase, as expected in terms of equation 2.

A few transition metal complexes of selenium ligands have been examined by ^{77}Se n.m.r. Both *cis* [$\delta(^{77}Se) = +120$ ppm] and *trans* [$\delta(^{77}Se) = +135$ ppm] isomers of [$PtCl_2(SeMe_2)_2$] have been studied[311], as have a variety of dialkyldiselenocarbamate complexes[517] (including Ni, Pd, Pt, Zn, and Cd complexes). The complexes [$M(Se_2CNBu^i_2)_2$] (M = Ni, Pd, Pt) are of particular interest as the electronic spectra[518] show that the lowest energy transitions decrease in the order Pt > Pd > Ni and yet the shieldings of the selenium nuclei decrease in the order Pd > Ni > Pt. This has been interpreted in terms of a mesomeric effect which induces a shift of electron density from the —NR_2 group towards the metal, although the resulting order of shielding is not clear from this explanation. Variable-temperature ^{77}Se n.m.r. studies of [$Pt(Se_2CNBu^i_2)_2(PPh_3)$] have shown[519] that the molecule undergoes an intramolecular rearrangement in solution. No organometallic transition metal derivatives appear to have been studied by ^{77}Se n.m.r.

^{125}Te n.m.r.[501] is also reasonably promising; in many ways the technical problems are similar to those encountered in ^{77}Se n.m.r. and, once again, $^1H\{^{125}Te\}$ double-resonance techniques have been of the utmost importance. The only organometallic tellurium-containing compound which appears to have been reported[520,521] is [$(Me_3Sn)_2Te$] [$\delta(^{125}Te) = -1214$ ppm; $\delta(^{119}Sn) = -66.8$ ppm; $^1J(^{125}Te, ^{119}Sn) = -1385$ Hz]. From studies of analogous selenium- and tellurium-containing organic compounds, it seems that a plot of $\delta(^{125}Te)$ *vs*. $\delta(^{77}Se)$ is linear[520], enabling an approximate value of $\delta(^{125}Te)$ to be obtained when only the selenium analogue is available. It seems likely that such a relationship will only be true in simple sp^n hybridized compounds, where *p*-orbital electron imbalance will be similar in related Se and Te compounds. In transition metal derivatives, where *d*-orbital involvement must be considered, the relationship may not hold, so any extrapolation should be made with care.

M. Fluorine-19 N.m.r. Studies

The ^{19}F nucleus is unique among the halogens, being the only $I = \frac{1}{2}$ nucleus, and has been of considerable importance in organometallic n.m.r. studies. The very high receptivity to n.m.r. measurement of the ^{19}F nucleus enables analogies to be drawn with the proton. Thus, in suitable cases, continuous wave techniques may be applicable to its measurement, whilst the availability of suitable instrumentation enables $^{19}F\{X\}$ double resonance experiments to be performed, thus increasing the range of possibilities for the indirect observation of the heteronucleus (X).

There are introductory texts[522] and reviews[523,524] available which deal with most aspects of ^{19}F n.m.r. Particularly useful are compilations of chemical shifts[525] and coupling constants[526]. Reviews are available which deal specifically with the ^{19}F n.m.r. of fluoroalkyl and fluoroaryl complexes of the transition metals[527,528].

The problem of chemical shift standards is very important in ^{19}F n.m.r. This nucleus has a reported shift range of *ca*. 850 ppm and so various standards have been employed, according to the range of interest in a particular study. The widely accepted standard in current use is $CFCl_3$; conversion factors for the other common standards (about ten have been used fairly frequently in the past) are available[523].

As many reviews are available on ^{19}F n.m.r., only one particular aspect will be considered further here. The ^{19}F nucleus is almost unique in its ability to exhibit so-called 'through-space' coupling. This has been briefly mentioned in Section II.C.2 although no satisfactory theoretical explanation can be offered as yet. In organometallic systems, through-space coupling is occasionally encountered wit complexes of the heavy metals and some mechanism involving an overlap of valenc shell electrons often invoked as a possible explanation.

In the study of some substituted dialkylmercury(II) complexes, for example, it wa observed in the comparison of *o*-, *m*-, and *p*-[$(F_3CC_6H_4)_2Hg$] an -[$(F_3CC_6H_4)HgBr$][59] that the values of $^4J(^{199}Hg, ^{19}F)$ were large, in the rang 26–29 Hz, for the *o*-complexes. Owing to the proximity of the *o*-CF_3 groups to th mercury nucleus, a through-space interaction was proposed to account for this.

It has also been observed[529], in complexes of the type **42**, that $^2J(^{195}Pt, ^{19}F_3)$ corre lates well with the *trans*-influence of the ligand X (as estimated from other spectro scopic and chemical evidence), whereas $^3J(^{195}Pt, ^{19}F_2)$ and $^3J(^{195}Pt, ^{19}F_1)$ do not. A possible explanation has been proposed[529] which suggests that the magnitude o $^nJ(^{195}Pt, ^{19}F)$ relates to the degree of through-space interaction, determined largely b the orientation of the C—F bond relative to the square-plane occupied by the meta centre. Such observations, whatever the explanation, show the danger of correlatin physical parameters with the magnitude of a single n.m.r. parameter. Here, were othe data not available, a *trans*-influence series constructed on the basis of $^3J(^{195}Pt, ^{19}F_{1,2}$ values would be very misleading.

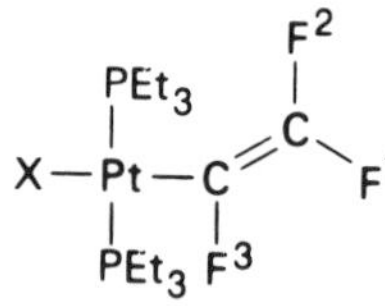

(**42**)

A further example[530] of through-space coupling is illustrated by the ^{19}F n.m.r spectrum of the product of the reaction between [$Pt(PPh_3)_4$] and hexafluoro bicyclo(2,2,0)hexa-2,5-diene, assigned the structure **43**. The resonance at −176. ppm is broad, probably because of unresolved $^{31}P—^{19}F$ coupling, and so is assigne to F^5 and F^6. The coupling to platinum, $^2J(^{195}Pt, ^{19}F_{5,6}) = 180 \pm 5$ Hz, was als observed. The resonance for F^1 and F^4 was not broadened, but also exhibited a substantial coupling, $^3J(^{195}Pt, ^{19}F_{1,4}) = 185 \pm 5$ Hz. Accordingly, the two- and three bond couplings are of the same magnitude, attributed to a through-space contribution to $^3J(^{195}Pt, ^{19}F)$. The assignments of these resonances relies largely on the broadening due to the unresolved $^{31}P—^{19}F$ coupling in this case.

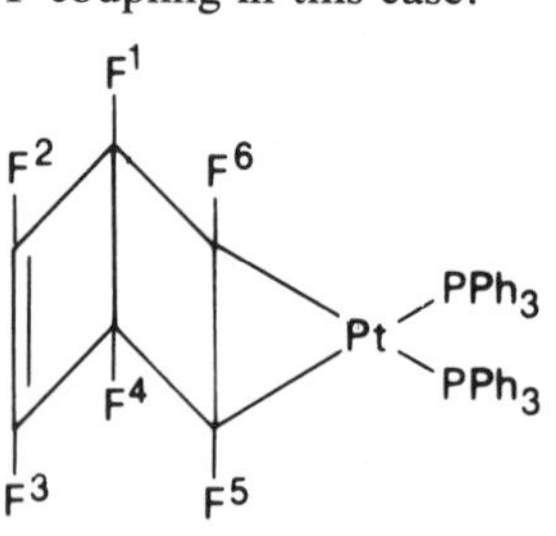

(**43**)

A final example of the phenomenon of through-space coupling involves the interaction of ^{19}F nuclei which are formally separated by several bonds. The complex **44**

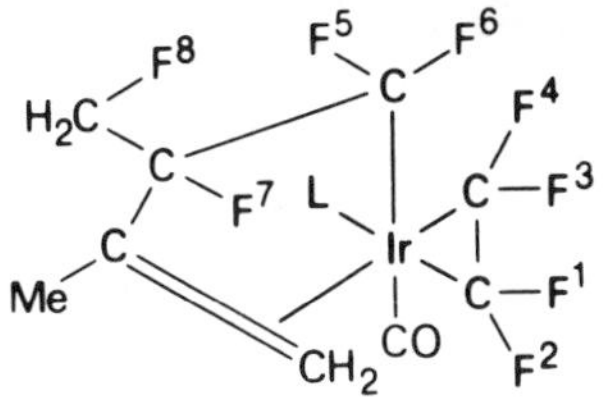

(**44**)

hows[531] strong coupling between F^1 and F^6 (53 Hz) and between F^4 and F^6 (51 Hz), espite separation by four formal bonds in each case. Additionally, where L = PPh_3, he value of $^3J(^{31}P, ^{19}F_5)$ is 67 Hz, also indicating a strong coupling over three formal onds. Molecular models show that the strongly coupled nuclei are maintained in ositions close to each other by the stereochemical constraints of the coordinated rganic group.

Through-space coupling is, of course, only one aspect of the ^{19}F n.m.r. studies of rganometallic compounds. The reader is referred to the available reviews for details f structural characterization and dynamic studies of other organometallic compounds.

. Studies of the Remaining Halogen Nuclei

Chlorine and bromine each have two stable $I > \frac{1}{2}$ isotopes (^{35}Cl, ^{37}Cl and ^{79}Br, ^{81}Br; able 4) and iodine has a single stable $I > \frac{1}{2}$ isotope (^{127}I; Table 4). Astatine is adioactive. Halogen n.m.r. has been the subject of a text[532] and recent review[533]. The major applications of halogen n.m.r. are in biochemistry[532,533] and no purposeful tudies of organometallic halogen-containing systems appear to have been reported. Mention of organometallic compounds in biological studies occurs occasionally, rganomercury(II) compounds being used as probes to investigate halide-binding in iopolymers. For example, the mercury-labelled hapten **45** was investigated[534] in tudies of the binding of the antidinitrophenyl antibody. In this system, the ^{35}Cl inewidth was monitored as a function of the concentration of **45** and it was observed hat the linewidth increases with concentration owing to rapid relaxation of $^{35}Cl^-$ by nteraction with the mercury(II) centre in the halide probe.

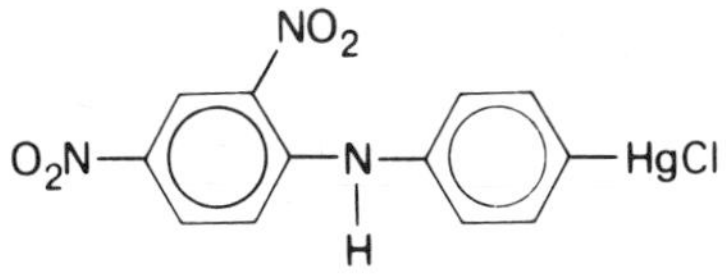

(**45**)

The major problem in obtaining n.m.r. spectra of halogen nuclei arises because of heir quadrupolar moments. Thus, in a covalently bound halogen compound (i.e. the ype we wish to study in organometallic systems), there is usually a very large field radient at the nucleus, caused by low symmetry of the halogen environment, and so he resonances are excessively broadened (see Section II.A.2). This problem appears o have deterred study, although the above example demonstrates that this effect can e put to good use in certain fields.

The standards for chemical shift ranges in spectroscopic studies of the halogens are queous solutions of the $[X]^-$ ions, extrapolated to infinite dilution.

O. Studies of the Noble Gas Nuclei

Of the five non-radioactive noble gases, there are two with isotopes of $I = \frac{1}{2}$ (^{3}H and ^{129}Xe; Table 1), two elements only with $I > \frac{1}{2}$ isotopes (^{21}Ne and ^{83}Kr; Table 4 and one, argon, with no stable isotope with non-zero spin. Xenon also has a quadrupo lar isotope (^{131}Xe; Table 4). Of these nuclei, only ^{129}Xe has been the subject of an chemically significant n.m.r. studies and this topic has recently been reviewed[535].

Of the reported work only studies of Xe(II) derivatives of Mo and W[536] are o marginal interest here, as no organometallic compounds containing xenon have bee reported.

VII. HIGH-RESOLUTION SOLID-STATE N.M.R. STUDIES[1]

The discussion so far presented in this chapter has concerned high-resolutio measurements of diamagnetic samples in solution, almost exclusively, for the simpl reason that such samples as these are of most interest to the majority of chemists. I the solid state, the ^{1}H n.m.r. spectra of most organic and organometallic compound consist of broad, featureless resonances, a result of direct dipole–dipole interaction between the nuclei. Such spectra (wide-line n.m.r.) are not informative from structural viewpoint, although in cases where motion in the solid allows som averaging of the dipole–dipole interactions, narrowing of the resonances can occur.

In the case of a nucleus such as ^{13}C[537], dipole–dipole interactions are likely to be smal in the solid state, a result of the low natural abundance of ^{13}C, which statisticall makes such interactions unlikely. Accordingly, there is a chance to obtai high-resolution n.m.r. spectra of 'dilute nuclei' in the solid state. Returning to the ^{13}C nucleus, the broad lines observed in a normal experiment arise largely from tw sources: firstly, carbon–proton dipolar interactions, and secondly, from ^{13}C chemica shift anisotropy. The former can be removed by high-power proton decoupling[538], th high power being neccessary as dipolar couplings up to several thousand hertz must b removed (compared with a solution ^{13}C{^{1}H} n.m.r. experiment where couplings ar rarely over 200 Hz). The chemical shift anisotropy can be removed by spinning th solid sample very fast (*ca.* 3000 Hz) at an angle, θ, of 54.7° to the field. This 'magi angle'[539] is chosen to make the term $(1-3\cos^2\theta)$ vanish from the expression relatin chemical shift anisotropy and field. A further sophistication is the technique o cross-polarization[540], which removes a dependence on the sometimes long ^{13}C relaxation time and relies on the shorter relaxation time of the proton.

A combination of these three techniques, cross-polarization (CP), very fast sampl spinning, and spinning at the 'magic angle' (MAS), enables high-resolution spectra to be obained[541]. The availability of such techniques enables several areas o organometallic chemistry to be explored which was not previously possible owing to constraints imposed by the solid state.

An example is the ^{13}C n.m.r. of solid $[Ru_3(COT)_2(CO)_4]$[541], which exhibits tw resonances for the carbonyl carbon atoms, but only a single resonance for the chemically inequivalent carbon atoms of the cycloctatetraene ligands. Since the MAS experiment produces isotropic values for the shift, this single resonance must be due to ring fluxionality in the solid state. Accordingly, this technique enables solid-stat fluxionality in organometallic compounds to be studied. Variable temperature work i now possible with the MAS experiment and the range of processes which may b examined is vast.

More recently, CP–MAS experiments designed to observe the ^{31}P nucleus hav been reported[542]. The spectra of the simple complexes *cis*-$[PtCl_2(PPh_2Me)_2]$ an *cis*-$[PtCl_2\{Ph_2P(CH_2)_2Si(OEt)_3\}_2]$ are shown in Figures 8a and 8b. The constraint

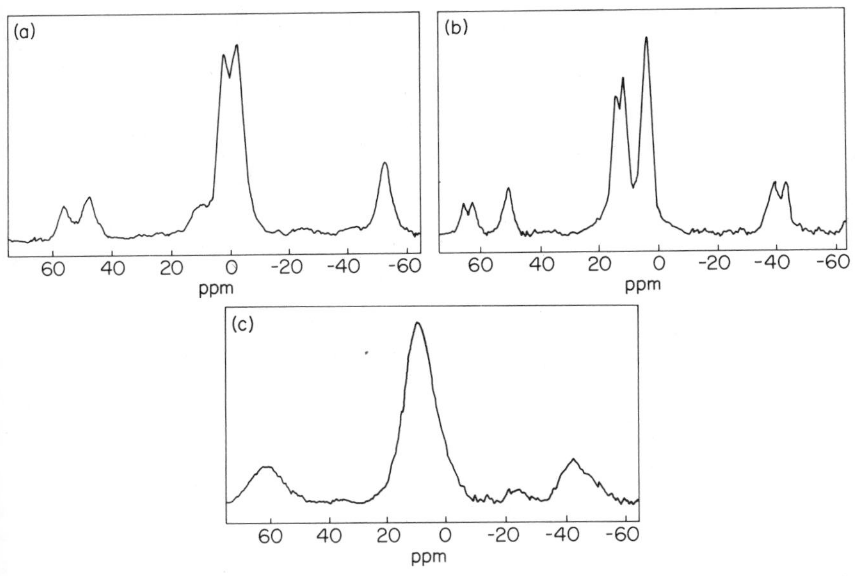

FIGURE 8. (a) Solid-state CP–MAS ^{31}P n.m.r. spectrum of $[PtCl_2(PPh_2Me)_2]$. Recorded at 36.442 MHz using matching and decoupling fields of approximately 10 G, spinning at 3 kHz. 3000 scans accumulated with a 1 s recycle time and a 1 ms single CP. The FID was transformed with 15 Hz line broadening. δ(P) = 1.6, −2.1 ppm; $^1J(^{195}Pt, ^{31}P)$ = 3466, 3759 Hz. (b) Solid-state CP–MAS ^{31}P n.m.r. spectrum of $[PtCl_2\{Ph_2P(CH_2)_2Si(OEt)_3\}_2]$. Details as for (a), except 6000 scans accumulated. δ(P) = 15.3, 12.9, 4.8 ppm; $^1J(^{195}Pt, ^{31}P)$ = 3740, 3711, and 3418 Hz, respectively. (c) Solid-state CP–MAS ^{31}P n.m.r. spectrum of $[PtCl_2\{Ph_2P(CH_2)_2Si(OEt)_3\}_2]$ supported on a glass surface. Details as for (a), except 14 000 scans accumulated, and FID transformed with 40 Hz line broadening. $\delta(^{31}P)$ = 10.45 ppm; $^1J(^{195}Pt, ^{31}P)$ = 3721 Hz.

imposed by the solid state cause the central resonance to appear as more than one component. Thus, crystallographic constraints may render the two phosphorus ligands inequivalent in a single molecule, or indeed there may be two distinct molecules in the crystal cell, either case resulting in the appearance of two resonances and a combination of these cases resulting in many possible resonances. These two spectra show that a considerable shift range is possible for a given nucleus as a result of solid-state constraints. Indeed, Figure 8c shows the spectrum of *cis*-$[PtCl_2\{Ph_2P(CH_2)_2Si(OEt)_3\}_2]$ after reaction with a glass surface, the width of the resonance (ω = 500 Hz) resulting from solid-state shift dispersion. This technique enables immobilized homogeneous transition metal catalysts to be studied[542], previously a difficult task owing to their solid nature and low-volume concentration of metal species.

Further development in this field, including studies of fluxionality and supported metal catalysts, is likely to be rapid and hindered only by the expense and sophistication of the apparatus required.

VIII. CONCLUSIONS

Although it has been possible to present only selected examples from the literature concerning multinuclear magnetic resonance studies of organometallic compounds, the scope and applicability of the subject can be seen to be vast. Future developmen are likely to include the direct observation of many more of the active nuclei in routine way and hence the development of a more substantial data base from whic the organometallic chemist can work in the elucidation of structural problems. Ther still exists a huge back log of problems which could never be solved by ^{1}H n.m.r. an hopefully, attention will be turned to these and the multinuclear approach applied. I this chapter many aspects of n.m.r. have been covered only briefly and some aspect not at all. Of these, multinuclear studies of paramagnetic compounds (in order t understand metal—ligand interactions a little better) and the development c high-resolution solid-state n.m.r. (and its application to fluxionality and immobilise organometallic compounds) are certainly going to be heard of in greater detail in th future.

In reading other chapters in these volumes, it will become clear that n.m.r. has bee and always will be, essential in studies of the metal—carbon bond.

IX. NOTES ADDED IN PROOF

The development of n.m.r. methods and their application to problems i organometallic chemistry continue to be reported at an ever increasing rate. Th following notes describe some of the more recent developments and applications. Th material is divided into the same sections as the main text, as set out in the table o contents (p. 813); sections for which no important new work has been detailed hav been omitted.

VI.B. Deuterium and Tritium Studies

^{2}H n.m.r. continues to be applied in specialized areas. The reactions of som transition metal (Re, Ru, Os, Rh, Ir) carbonyl clusters with $LiHBEt_3$ and $Li^{2}HBEt$ have been studied[543] by ^{1}H, ^{2}H, and ^{13}C n.m.r. and evidence presented suggesting th formation of new metal—formyl complexes and organic decomposition product (HCHO, CH_3OH). The importance of such homogeneous systems as models fo heterogeneous Fisher–Tropsch synthesis[544] makes this study particularly noteworthy The mechanism of reduction of organomercury(II) compounds by both $NaB^{2}H_4$ an sodium amalgam/$^{2}H_2O$/$NaOC^{2}H_3$ systems has been investigated by ^{2}H n.m.r. Th latter mode of reduction occurs stereospecifically and the results are interpreted i terms of the reaction mechanisms[545].

The selectivity of a range of heterogeneous and homogeneous catalysts in effectin hydrogen isotope exchange in organic molecules has been studied by ^{3}H n.m.r.[546]. Th complex $[RuCl_2(PPh_3)_3]$, for example, catalyses the specific exchange of ^{3}H for ^{1}H a the α-methylene position in the reaction of many primary alcohols with $^{3}H_2O$, a elevated temperatures.

VI.C. Studies of Alkali Metal Nuclei

While the n.m.r. of the alkali metals develops into an important technique in man areas of chemistry, few new results are of importance here. Further work on th mechanism of ^{7}Li relaxation has been reported[547] and both ^{7}Li and ^{27}Al n.m.r. studie of $[Li(AlH_4)]$ in ether solutions described[548]. Some aspects of the theory of ^{23}Na chemical shifts have also been described[549].

VI.E. Studies of the Transition Metal Nuclei (Including Lanthanides and Actinides)

1. *Sc, Y, La, the lanthanides, and actinides*

Still an untouched area in organometallic chemistry, ^{89}Y n.m.r. has been employed as a spin-relaxation probe for studying metal ion interactions with organic ligands (typically oxygen-donor ligands)[550] and ^{139}La n.m.r. continues to be of interest in studies of ionic complexation[551].

2. *Ti, Zr, Hf*

The first chemically significant study of ^{91}Zr n.m.r. has been reported by McGlinchey and co-workers[552]. Complexes of the type $[ZrCp_2X_2]$ (X = Cl, Br, I) and $[ZrCp_4]$ were examined and resonances observed for all the halide complexes (Table 41), but not the tetraorgano-complex. A temporary shift standard ($[ZrBr_2Cp_2]$ in thf solution) was employed until sufficient compounds have been examined to select a more convenient standard. The resonance due to $[ZrCp_4]$ was believed to be too broad to be observed.

No doubt further studies of ^{91}Zr n.m.r. will be forthcoming as no particular problems were observed in the initial study.

TABLE 41. ^{91}Zr n.m.r. data for some cyclopentadienyl complexes

Complex	$\delta(^{91}Zr)$ (ppm)[a]	ω (Hz)
$[ZrCp_2Cl_2]$	−121.9	276
$[ZrCp_2ClBr]$	−65.9	237
$[ZrCp_2Br_2]$	0.0	19
$[ZrCp_2I_2]$	126.0	134

[a]Relative to $[ZrCp_2Br_2]$ in thf solution; data from reference 552.

3. *V, Nb, Ta*

Further studies of ^{51}V n.m.r. of carbonyl and phosphine-substituted carbonyl complexes have been reported and correlations of shielding with variations in the contribution of σ_p to σ described[553]. The ^{51}V n.m.r. of $[V(PF_3)_6]^-$ has also been reported[554]. Interesting new work on the ^{93}Nb n.m.r. of some organoniobium and related complexes has been discussed. The complexes $[Nb(CO)_6]^-$ [554], $[NbCp(CO)_4]$, $[Nb_3Cp_3(CO)_7]$[555], and a series of phosphine-substituted $[NbCp(CO)_4]$ derivatives have been examined[556]. Various shift standards have been utilized, including $[NbCl_5]$ in diglyme[554], $[NbCl_5]$ in acetonitrile[556], and $[NbOCl_3]$ in acetonitrile[556]. The resonances are generally very broad in the less symmetrical complexes, being *ca*. 6000 Hz in the phosphine-substituted $[NbCp(CO)_4]$ derivatives[556].

4. *Cr, Mo, W*

Several reports of the ^{95}Mo n.m.r. of substituted $[Mo(CO)_6]$ derivatives have appeared[557–559], thus establishing a considerable data base in this area. A selection of data are presented in Table 42. The use of an alkaline (pH = 11) solution of 2 M

TABLE 42. ^{95}Mo n.m.r. data for some $[Mo(CO)_6]$ derivatives[a]

Compound[b]	Solvent	δ (ppm)	ω (Hz)	$^1J(^{95}Mo, ^{31}P)$ (Hz)
$[Mo(CO)_6]$	Thf	−1845.3	1	—
$[Mo(CO)_5(MeCN)]$	MeCN	−1439.7	39	—
$[Mo(CO)_4(MeCN)_2]$	MeCN	−1306.6	49	—
$[Mo(CO)_3(MeCN)_3]$	MeCN	−1113.8	10	—
$[Mo(CO)_5P(OPh)_3]$	CH_2Cl_2	−1819.1	36	234
$[Mo(CO)_5P(p\text{-}OMePh)_3]$	CH_2Cl_2	−1774.7	66	156
$[Mo(CO)_5(PPh_3)]$	CH_2Cl_2	−1742.7	54	139
$[Mo(CO)_5(PCy_3)]$	CH_2Cl_2	−1824.5	46	129
$[Mo(CO)_5(PBu^n_3)]$	CH_2Cl_2	−1842.8	16	129
$[Mo(CO)_5(PBu^t_3)]$	CH_2Cl_2	−1710.7	67	127
$[Mo(CO)_5(AsPh_3)]$	CH_2Cl_2	−1756.5	112	—
$[Mo(CO)_5(SbPh_3)]$	CH_2Cl_2	−1863.9	117	—
$[Mo(CO)_5(pip)]$	CH_2Cl_2	−1433.3	76	—
$[Et_4N][ClMo(CO)_5]$	CH_2Cl_2	−1512.9	108	—
$[Mo(CO)_4(dppe)]$	CH_2Cl_2	−1781.1	90	128
$[Mo(CO)_4(diars)]$	CH_2Cl_2	−1807.4	43	—
cis-$[Mo(CO)_4(P(OPh)_3)_2]$	CH_2Cl_2	−1753.7	36	250
cis-$[Mo(CO)_4(pip)[P(OPh)_3]]$	CH_2Cl_2	−1362.0	109	257
cis-$[Mo(CO)_4(PPh_3)_2]$	CH_2Cl_2	−1556.1	46	140
cis-$[Mo(CO)_4(MePPh_2)_2]$	CH_2Cl_2	−1637.1	57	133
cis-$[Mo(CO)_4(PBu^n_3)_2]$	CH_2Cl_2	−1741.7	93	123
cis-$[Mo(CO)_4(AsPh_3)_2]$	CH_2Cl_2	−1576.9	187	—
cis-$[Mo(CO)_4(SbPh_3)_2]$	CH_2Cl_2	−1807.0	247	—
cis-$[Mo(CO)_4(pip)_2]$	Dmf	−1092.6	94	—
trans-$[Mo(CO)_4(P(OPh)_3)_2]$	CH_2Cl_2	−1785.0	96	225
trans-$[Mo(CO)_4(MePPh_2)_2]$	$C_6H_5CH_3$	−1631.0	87	134
$[Mo(CO)_3(triphos)]$	CH_2Cl_2	−1759.5	43	129
$[Mo(CO)_3(MePPh_2)_3]$	CH_2Cl_2	−1427.1	7	126

[a]Relative to external aqueous alkaline 2 M $[K]_2[MoO_4]$; data from reference 557.
[b]diars = $o\text{-}C_6H_4(AsMe_2)_2$; pip = piperidine; triphos = bis(diphenylphosphino)phenylphosphine; dmf = dimethylformamide.

$[M]_2[MoO_4]$ (M = K, Na) is recommended as an external shift standard[557,559]. Coupling to ^{31}P is clearly observed (Table 42) in phosphine-substituted derivatives. This area will no doubt prove fruitful for chemically significant work now that background data have been presented.

5. *Mn, Tc, Re*

The ^{185}Re and ^{187}Re n.m.r. spectra of $[Re(CO)_6][Cl]\cdot HCl$ have been reported[560]. A single resonance at −3400 ppm relative to aqueous $NaReO_4$ was observed, with a linewidth at half-height of *ca.* 1.5 kHz. The same report[560] describes further ^{55}Mn data for $[Mn(CO)_5(PR_3)]^+$ complexes.

6. *Fe, Ru, Os*

^{57}Fe n.m.r. spectra of 35 diamagnetic iron complexes, largely $[Fe(CO)_3(diene)]$ compounds, have been reported[561]. Using Quadriga Fourier transform techniques[247], spectra were obtained for complexes at natural abundances of ^{57}Fe. The shift range so far observed covers *ca.* 3000 ppm and $[Fe(CO)_5]$ has been proposed as a secondary shift standard. The Quadriga Fourier transform method has proved to be especially useful ^{57}Fe n.m.r.[247,561–563] and in other areas, such as ^{103}Rh n.m.r. (see notes on Section VI.E.7, below), where the shift range is large and resonances can be very difficult to locate.

7. Co, Rh, Ir

^{13}C{^{103}Rh} n.m.r. studies[564] of the cluster $[Rh_{12}(CO)_{30}]^{2-}$ show that the solution structure at −72°C is the same as that in the solid state, as determined by X-ray crystallography[565]. Three ^{103}Rh resonances were monitored, at +168, −322 and −560 ppm (relative to $\Xi(^{103}Rh)$ = 3.16 MHz), although only two resonances were previously observed by direct observation of the ^{103}Rh nuclei[274]. The reason for this descrepancy is unknown.

Direct observation of the ^{103}Rh nucleus[566], utilizing Quadriga Fourier transform methods[247], has allowed relaxation studies of $[Rh(acac)_3]$ to be performed. Other studies[567], however, concerning relaxation measurements of $[Rh(acac)(CO)_2]$ and $[Rh(Bpz_4)(diene)]$ [Bpz_4 = tetrakis(1-pyrazole)borate; diene = duroquinone, cod, nbd) at high field, suggest some care is needed in the interpretation of such measurements. It appears that chemical shift anisotropy makes an important contribution to the relaxation mechanism at high field and that even at lower fields this contribution may still be significant. It was previously mentioned (Section II.A.5) that chemical shift anisotropy was most likely to be encountered as a problem in examining the n.m.r. of heavy metal nuclei, where substantial contributions to the observed linewidth (leading to problems in complex coupling situations) are to be expected. This seems to be the case for ^{103}Rh at high field and is similarly encountered with ^{199}Hg (see notes on Section VI.E.10, below) and ^{205}Tl (see notes on Section VI.F.3, below).

8. Ni, Pd, Pt

A paper discussing the applications of multinuclear magnetic resonance methods to the chemistry of platinum has appeared[568] and a number of papers discussing the ^{195}Pt n.m.r. spectra of $[Pt(PPh_3)_2(RCCR)]$[569], *cis*-$[PtCl_2(NH_3)_2]$ and its hydrolysis and oligomerization products[570], and ^{15}N-labelled diimine complexes[571] have been reported. Particularly interesting are reports of polynuclear complexes, including $[Pt_3(\mu\text{-}CO)_3(PR_3)_3]$[572] and $[PtMe_3X]_4$[573], as these involve the long-range couplings between active ^{195}Pt centres in the less abundant isotopomers.

At very high field[574] (9.4 T) the isotopomers of ^{35}Cl, ^{37}Cl, ^{79}Br, and ^{81}Br may be distinguished in chloro- and bromoplatinum(IV) complexes by ^{195}Pt n.m.r. In poorly resolved cases, the different isotope shifts thus cause a substantial line broadening. The linewidths are not field dependent (i.e. not dominated by chemical shift anisotropy) and relaxation by spin–rotation interactions is believed to occur. In the case of $[Pt(^{35}Cl)_n(^{37}Cl)_{6-n}]$ (n = 0–6), resonances for the five most abundant isotopomers have been clearly resolved. Undoubtedly, such high-field isotope shifts may cause problems of linewidth in more complex cases under conditions of poor digital resolution.

10. Zn, Cd, Hg

Many reports have appeared concerning ^{111}Cd, ^{113}Cd, and ^{199}Hg n.m.r. Particularly important are results[575,576] which show that the relaxation of ^{199}Hg is dominated by chemical shift anisotropy at high fields for species with symmetry less than tetrahedral. Chemical shift anisotropy measurements for Ph_2Hg yield a value of

6800 ± 680 ppm[575]. Also of interest are studies of [$HgX_2(PBu_3Se)_2$] (X = Cl, Br, I, SCN)[577] by ^{31}P, ^{77}Se, ^{199}Hg, ^{31}P {^{77}Se, ^{1}H}, and ^{31}P {^{199}Hg, ^{1}H} n.m.r. methods. A comparison of direct and indirect methods is presented, with useful practical details such as the time required for each type of measurement, instrumental sensitivity necessary, etc. By the indirect methods, $^{1}J(^{199}Hg, ^{77}Se)$ and $^{2}J(^{199}Hg, ^{31}P)$ were found to be negative in sign, relative to a negative value of $^{1}J(^{77}Se, ^{31}P)$.

VI.F. Studies of the Group IIIB Nuclei

3. *Other systems*

The ^{115}In n.m.r. spectra of some inorganic indium salts in non-aqueous solution have been reported[578]. Of particular interest is the complex of empirical formular [$MeInI_2$], known to exist as [$InMe_2$][InI_4] in the solid state, by X-ray crystallography[579]. Only a resonance for the [InI_4]$^-$ anion was observed in solution, indicating that the cation is of too low a symmetry (possibly because of solvation effects) to be detected. Clearly this technique is of limited structural use in organoindium chemistry. A standard of 0.5 M [$InCl_4$]$^-$ in CH_2Cl_2 solution is recommended[578].

Further studies on the effects of solvent[580] and temperature[581] on ^{205}Tl chemical shifts and coupling constants in [Me_2TlX] compounds have been reported. Significant relaxation measurements[582], at various applied fields, have been described for diorganothallium(III) derivatives and chemical shift anisotropy shown to be a dominant relaxation mechanism. The effects of this are clearly manifested in the ^{1}H n.m.r. of protons coupled to ^{205}Tl also.

VI.G. Carbon-13 N.m.r. Studies

A noteworthy ^{13}C n.m.r. experiment has been reported[583] whereby the cluster [$Rh_{12}(CO)_{30}$]$^{2-}$ was reacted with a gaseous CO–H_2 (2.1:1.0) mixture in a specially constructed high-pressure n.m.r. probe. For this experiment, a pressure of 850 bar was employed. Even under very high applied pressures of CO, the exchange of coordinated carbonyls for free CO is not fast at low temperature and so much structural information may be obtained. In the example under discussion, conversion to a cluster of lower nuclearity, [$Rh_5(CO)_{15}$]$^-$, occurs. The significance of this work is a clear demonstration that cluster degradation to mononuclear complexes does not occur under the type of conditions employed in catalytic organic synthesis, previously a disputed point in many areas of cluster chemistry. The report describes the probe design and hence further studies in related areas may be expected.

Papers describing studies of mononuclear metal carbonyls, with the emphasis on extraction of $^{2}J(^{13}C, ^{13}C)$ values from ^{13}C n.m.r. spectra have been reported[584, 585]. The utility in monitoring $^{2}J(^{13}C, ^{13}C)$ as a probe of inter- *versus* intramolecular carbonyl exchange in [$ML(CO)_5$] (M = Group VIII metal) complexes has also been detailed[585].

VI.H. Studies of the Remaining Group IVB Nuclei

1. *Organometallic Si and Ge systems*

^{73}Ge n.m.r. remains unapplied to organometallic systems, whereas ^{29}Si n.m.r. has now reached a level of real utility in organometallic chemistry. Many examples have been reported, a typical example[586] being the reformulation of (η^3-1-silapropenyl)tricarbonyliron complexes as η^2-(vinylsilane)tetracarbonyliron species after a study of the reactions of vinylsilanes with [$Fe_2(CO)_9$] by ^{29}Si n.m.r.

2. Organotin compounds

A ^{119}Sn n.m.r. study[587] of 7-coordinate organotin compounds confirms the general trend that an increase in coordination number results in an increased shielding of the ^{119}Sn nucleus. Thus, for [BuSn(oxalate)$_3$] (7-coordinate) $\delta = -561$, whereas for [BuSn(oxalate)$_2$Cl] (6-coordinate) $\delta = -395$.

The use of selective deuterium labelling in organotin compounds leads to observation of $^nJ(^{119}Sn, ^2H)$ in ^{119}Sn n.m.r. spectra[588]. The resonances for coupled nuclei thus appear as simple 1:1:1 triplets, which can be utilized as a structural probe in organotin chemistry.

3. Transition metal–tin systems

^{119}Sn n.m.r. studies of transition metal trichlorostannate complexes are now contributing significantly to the understanding of catalyst systems based on platinum metal halo-complexes with a tin(II) chloride co-catalyst. The $RhCl_3$–$SnCl_2$–$HCl_{(aq.)}$ system has been investigated[589] and parameters for $[Rh(SnCl_3)_nCl_{6-n}]^{3-}$ anions presented. An interesting redox reaction of $RhCl_3$ with $SnCl_2$ to yield a rhodium(I) derivative of the type $[Rh(SnCl_3)_5]^{4-}$ and tin(IV) species was also observed by ^{119}Sn n.m.r.[589]. Data for a series of compounds, $[Rh(SnCl_nBr_{(3-n)})(NBD)(PR_3)_2]$ have also been presented[590].

Most interesting is a study[591] by ^{119}Sn, ^{117}Sn, and ^{115}Sn n.m.r. of the complex $[Ru(SnCl_3)_5Cl]^-$, whose X-ray structure[591] shows an approximately octahedral array of ligands, with Cl^- occupying an equatorial sight. The spectra allowed observation of $^1J(^{117,119}Sn, ^{99}Ru)$, the first coupling to be observed for the quadrupolar ^{99}Ru nucleus. No couplings to ^{101}Ru were detected, presumably because of the larger quadrupole moment of this isotope. The study is particularly important in the light of the known catalytic activity of ruthenium:tin systems in reactions such as the trimerization of isobutene[592].

Studies of the reaction of *cis*-$[PtCl_2(CO)(PR_3)]$ complexes, including ^{13}CO-labelled analogues, with $SnCl_2{\cdot}2H_2O$ by ^{119}Sn n.m.r. and other methods (^{1}H, ^{13}C, ^{31}P, and ^{195}Pt n.m.r.)[593] have shown a remarkable solvent effect. Reactions in chloroform solution lead to a simple insertion of $SnCl_2$ *trans* to PR_3, whereas in acetone or acetonitrile solution a ligand rearrangement occurs. In the case of PPh_3, the products were identified by multinuclear magnetic resonance methods as *trans*-$[PtCl(CO)(PPh_3)_2]^+$, *trans*-$[PtCl(SnCl_3)_2(PPh_3)]^-$, *trans*-$[PtCl(SnCl_3)_2(CO)]^-$ and $[Pt(SnCl_3)_5]^{3-}$. The real utility of such solution studies is demonstrated by the fact that the only isolable product from this highly complex system is the simple compound *cis*-$[PtCl_2(PPh_3)_2]$. The system in acetone and acetonitrile solutions has previously been reported as an active and selective hydroformylation catalyst precursor[447].

4. Organolead compounds

While several studies of organolead compounds by ^{207}Pb n.m.r. have been reported, including a detailed study of vinyllead derivatives[594], most interesting is the observation[595] of ^{207}Pb—^{14}N couplings in the complexes $[(Me_3Pb)_2NSnMe_3]$ and $[Me_3PbN(SnMe_3)_2]$. Values of 170 ± 10 Hz were obtained, with the ^{207}Pb resonance appearing as an ill-defined triplet of 1:1:1 intensity (^{14}N, $I = 1$). Coupling of heavy metal nuclei to the ^{14}N nucleus are observed only rarely.

VI.I. Nitrogen N.m.r. Studies

The applications of nitrogen n.m.r. to problems in inorganic, organometallic, and bioinorganic chemistry have been reviewed[596]. ^{15}N n.m.r. continues to be applied to the study of molecular nitrogen activation by transition metal complexes. A series of related ligands ($^{15}N_2$, $^{15}N_2$|||||AlR_3, $^{15}N_2H$ and ^{15}NH) have been clearly distinguished by ^{15}N n.m.r. studies[597] of a wide range of transition metal (Mo, Re, W, Os, Rh, and Fe) complexes and the two forms of diazenido complex, **46** and **47**, clearly differentiated[598]. The ^{15}N—M resonance of **46** is shielded by *ca.* 300 ppm with respect to **47**, thus clearly allowing the geometries to be assigned[598].

M—N=N—R

(46)

M—N=N—R

(47)

An absolute ^{15}N n.m.r. shielding scale based on gas-phase studies of $^{15}NH_3$ has now been proposed[599].

VI.L. Studies of Group VIB Nuclei

3. ^{33}S, ^{77}Se and ^{125}Te n.m.r. studies

As predicted, ^{33}S n.m.r. studies of compounds relatively high in symmetry about the ^{33}S nucleus have been shown to produce resonances with linewidths small enough to permit substituent effects on the chemical shift to be monitored. Thus, although the ^{33}S resonance of sulphides could not be observed, the more symmetrical oxidation products, the sulphones, produce signals with small linewidths[600]. This n.m.r. technique may find application in the analysis of sulphur compounds in the petroleum field, but has yet to be applied to organometallic systems.

^{125}Te n.m.r. spectra of a series of tellurophenes have been reported[601] and the signs of couplings to ^{1}H and ^{13}C determined by SPT (selective population transfer) techniques[602]. An interesting application of ^{125}Te n.m.r. is in the study of the *cis/trans* isomerization of $[PtCl_2(TeR_2)_2]$ complexes, which has been described in detail[603]. The magnitudes of $^{1}J(^{195}Pt, ^{125}Te)$ are 900 Hz (*cis*) and 544 Hz (*trans*) for $R = CH_2CH_2C_6H_5$, thus clearly differentiating the isomers.

VII. High-resolution Solid-state N.m.r. Studies

A comprehensive and detailed review covering recent developments in high-resolution CP–MAS n.m.r. studies of solids is now available[604]. Further ^{13}C CP–MAS n.m.r. studies of organometallic compounds have been described, including $[Hg(CF_3)_2]$[605] and organothallium derivatives[606]. Although ^{29}Si CP–MAS n.m.r. spectroscopy has been applied to many areas of solid state chemistry, no organometallic systems incorporating silicon have been described[604]. Applications of ^{31}P CP–MAS n.m.r. continue to be reported, including studies[607] of several PPh_3 complexes of Rh(I), Cu(I) and some gold—phosphine clusters.

A series of asymmetric bidendate tertiary phosphines and their rhodium(I) complexes have been studied[608]. Such complexes are of interest in homogeneous asymmetric catalysis.

In the area of supported catalysts, further CP–MAS studies of Ni(II), Pd(II), and Pt(II) phosphine complexes and their analogues immobilized on silica and glass surfaces have been reported[609]. Using organic polymers as supports, CP–MAS ^{31}P n.m.r. has been employed[610] to study the incorporation of platinum(II) complexes by various routes and assignments of surface structures have been described.

X. REFERENCES

1. M. Mehring, *High Resolution NMR in Solids*, Academic Press, New York, 1976.
2. R. K. Harris and B. E. Mann, *NMR and the Periodic Table*, Academic Press, New York, 1978.
3. J. W. Emsley and J. C. Lindon, *NMR Spectroscopy Using Liquid Crystal Solvents*, Pergamon Press, Oxford, 1969.
4. G. M. Bodner, M. P. May, and L. E. McKinney, *Inorg. Chem.*, **19**, 1951 (1980).
5. J. A. Pople, W. G. Schneider, and H. J. Bernstein, *High Resolution Nuclear Magnetic Resonance*, McGraw-Hill, New York, 1959.
6. J. W. Emsley, J. Feeney, and L. H. Sutcliffe, *High Resolution Nuclear Magnetic Resonance Spectroscopy*, Vol. 1, Pergamon Press, Oxford, 1965.
7. R. M. Lyndon-Bell and R. K. Harris, *Nuclear Magnetic Resonance Spectroscopy*, Nelson, London, 1969.
8. E. D. Becker, *High Resolution NMR: Theory and Chemical Applications*, Academic Press, New York, 1969.
9. J. Emsley, J. Feeney, and L. H. Sutcliffe (Eds.), *Progress in NMR Spectroscopy*, Pergamon Press, Oxford, 1966–1981.
10. J. S. Waugh (Ed.), *Advances in Magnetic Resonance*, Academic Press, New York, 1965–1981.
11. E. F. Mooney and G. A. Webb (Eds.), *Annual Reports on NMR Spectroscopy*, Academic Press, London, 1971–1981.
12. R. J. Abraham and R. K. Harris (Eds.), *Specialist Periodical Reports on NMR*, Chemical Society and Royal Society of Chemistry, London, 1972–1981.
13. E. A. V. Ebsworth (Ed.), *Specialist Periodical Reports on Spectroscopic Properties of Inorganic and Organometallic Compounds*, Chemical Society and Royal Society of Chemistry, London, 1969–1981.
14. B. E. Mann, in *Specialist Periodical Reports on Spectroscopic Properties of Inorganic and Organometallic Compounds* (Ed. E. A. V. Ebsworth), Vol. II, Chemical Society, London, 1979, p. 1.
15. B. R. Garvey and R. J. Kurland, *NMR Paramagn. Mol.*, 555 (1973).
16. R. H. Cox, *Magn. Reson. Rev.*, **3**, 207 (1974).
17. G. A. Webb, *Magn. Reson. Rev.*, **3**, 327 (1974).
18. H. Zimmer and D. C. Lankin, *Method. Chim.*, **1(A)**, 351 (1974).
19. R. K. Harris, *Chem. Soc. Rev.*, **5**, 1 (1976).
20. Y. Arata, *Bunseki*, 445 (1979).
21. T. Axenrod and G. A. Webb (Eds.), *Nuclear Magnetic Resonance Spectroscopy of Nuclei Other than Protons*, Wiley–Interscience, New York, 1974.
22. I. K. O'Neill, *Proc. Anal. Div. Chem. Soc.*, **14**, 190 (1977).
23. R. R. Sharp, *Prog. Anal. Chem.*, **6**, 123 (1973).
24. F. W. Wehrli, *Annu. Rep. NMR Spectrosc.*, **9**, 125 (1979).
25. R. Freeman and H. D. W. Hill, in *Dynamic Nuclear Magnetic Resonance Spectroscopy* (Eds. L. M. Jackman and F. A. Cotton), Academic Press, New York, 1975, p. 131.
26. A. Abragam, *The Principles of Nuclear Magnetism*, Oxford University Press, London, 1961, Chapter 8.
27. T. C. Farrar, A. A. Maryott, and M. S. Malmsberg, *Annu. Rev. Phys. Chem.*, **23**, 193 (1972).
28. T. C. Farrar and E. D. Becker, *Pulse and Fourier Transform NMR*, Academic Press, New York, 1971, Chapter 4.
29. J. H. Noggle and R. E. Schirmer, *The Nuclear Overhaüser Effect,* Academic Press, New York, 1971.
30. G. C. Levy and J. D. Cargioli, in *Nuclear Magnetic Resonance Spectroscopy of Nuclei Other than Protons* (Eds. T. Axenrod and G. A. Webb), Wiley–Interscience, New York, 1974 p. 251.
31. M. H. Chisholm, H. C. Clark, L. E. Manzer, J. B. Stothers, and J. E. H. Ward, *J. Am. Chem. Soc.*, **97**, 721 (1975).
32. G. K. Anderson, H. C. Clark, and J. A. Davies, unpublished results.

33. D. Gudlin and H. Schneider, *J. Magn. Reson.*, **17**, 268 (1975).
34. F. Calderazzo, E. A. C. Lucken, and D. F. Williams, *J. Chem. Soc. A*, 154 (1967).
35. G. C. Levy, D. M. White, and F. A. L. Anet, *J. Magn. Reson.*, **6**, 453 (1972).
36. J.-Y. Lallemand, J. Soulié, and J.-C. Chottard, *Chem. Commun.*, 436 (1980).
37. P. C. Kong and T. Theophanides, *Inorg. Chem.*, **13**, 1167 (1974).
38. P. C. Kong and T. Theophanides, *Bioinorg. Chem.*, **5**, 51 (1975).
39. G. Y. H. Chu and R. S. Tobias, *J. Am. Chem. Soc.*, **98**, 2641 (1976).
40. R. E. Cramer and P. D. Dahlstrom, *J. Am. Chem. Soc.*, **101**, 3679 (1979).
41. G. V. Fazerkerley and K. R. Koch, *Inorg. Chim. Acta*, **36**, 13 (1979).
42. R. M. Hawk and R. R. Sharp, *J. Chem. Phys.*, **60**, 1522 (1974).
43. W. T. Raynes, *Nucl. Magn. Reson.*, **8**, 1 (1979).
44. J. Mason, *Adv. Inorg. Radiochem.*, **22**, 199 (1979).
45. D. W. Jones, *Nucl. Magn. Reson.*, **8**, 20 (1979).
46. R. K. Harris, *Educ. Chem.*, **14**, 44 (1977).
47. M. I. Foreman, *Nucl. Magn. Reson.*, **5**, 292 (1976).
48. S. Brownstein and J. Bornais, *J. Magn. Reson.*, **38**, 131 (1980).
49. N. F. Ramsey, *Phys. Rev.*, **78**, 689 (1950).
50. C. J. Jameson and H. S. Gutowsky, *J. Chem. Phys.*, **40**, 1714 (1964).
51. J. A. Pople, *J. Chem. Phys.*, **37**, 53 and 60 (1962).
52. W. H. Flygare and J. Goodisman, *J. Chem. Phys.*, **49**, 3122 (1968).
53. R. Grinter, in *Specialist Periodical Reports on NMR* (Eds. R. K. Harris and J. R. Abraham). Chemical Society, London, 1972–1981.
54. M. Barfield and D. M. Grant, *Adv. Magn. Reson.*, **1**, 149 (1965).
55. G. A. Webb, in *NMR and the Periodic Table* (Eds. R. K. Harris and B. E. Mann), Academic Press, New York, 1978, Chapter 3.
56. N. F. Ramsey, *Phys. Rev.*, **91**, 303 (1953).
57. J. A. Pople and D. P. Santry, *Mol. Phys.*, **8**, 1 (1964).
58. J. R. Llinas, E.-J. Vincent and G. Peiffer, *Bull. Soc. Chim. Fr.*, 3209 (1973).
59. W. McFarlane, *Chem. Commun.*, 609 (1971).
60. G. R. Miller, A. W. Yankowsky, and S. O. Grim, *J. Chem. Phys.*, **51**, 3185 (1969).
61. H. C. Clark and L. E. Manzer, *Chem. Commun.*, 870 (1973).
62. M. P. Brown, R. J. Puddephatt, M. Rashidi, and K. R. Seddon, *J. Chem. Soc., Dalton Trans.*, 951 (1977).
63. R. K. Harris, *Can. J. Chem.*, **42**, 2275 (1964).
64. P. Ansty and R. K. Harris, *Chem. Brit.*, **13**, 303 (1977).
65. C. Thibault and J. W. Cooper, *Magn. Reson. Rev.*, **5**, 101 (1980).
66. R. G. Jones, *Nucl. Magn. Reson.*, **3**, 181 (1974).
67. R. J. Abraham, *Analysis of High Resolution NMR Spectra*, Elsevier, Amsterdam, 1971.
68. A. L. Balch, personal communication.
69. B. I. Ionin and B. A. Ershov, *NMR Spectroscopy in Organic Chemistry*, Plenum Press, New York, 1970.
70. J. Dadok and R. F. Sprecher, *J. Magn. Reson.*, **13**, 243 (1974).
71. R. K. Gupta, J. A. Ferretti, and E. D. Becker, *J. Magn. Reson.*, **13**, 275 (1974).
72. K. Mullen and P. S. Pregosin, *Fourier Transform NMR Techniques: a Practical Approach*, Academic Press, New York, 1976.
73. D. Shaw, *Fourier Transform NMR Spectroscopy*, Elsevier, Amsterdam 1976.
74. D. Shaw, *Nucl. Magn. Reson.*, **5**, 188 (1976).
75. D. Shaw, in *NMR and the Periodic Table* (Eds. R. K. Harris and B. E. Mann), Academic Press, New York, 1978, Chapter 2.
76. R. Kaiser, *J. Magn. Reson.*, **3**, 28 (1970).
77. R. R. Ernst, *J. Magn. Reson.*, **3**, 10 (1970).
78. B. L. Tomlinson and H. D. W. Hill, *J. Chem. Phys.*, **59**, 1775 (1973).
79. R. Freeman, H. D. W. Hill, B. L. Tomlinson, and L. D. Hall, *J. Chem. Phys.*, **61**, 4466 (1974).
80. V. Royden, *Phys. Rev.*, **93**, 944 (1954).
81. T. M. Connor, D. W. Whiffen, and K. A. Lauchlan, *Mol. Phys.*, **13**, 221 (1977).
82. J. D. Kennedy and W. McFarlane, *Rev. Silicon, Germanium, Tin, Lead Compd.*, **1**, 235 (1973).

83. S. J. Anderson, J. R. Barnes, P. L. Goggin, and R. J. Goodfellow, *J. Chem. Res. (S)*, 286 (1978).
84. C. Brown, B. T. Heaton, L. Longhetti, D. O. Smith, P. Chini, and S. Martinego, *J. Organomet. Chem*., **169**, 309 (1979).
85. E. M. Hyde, J. D. Kennedy, B. L. Shaw, and W. McFarlane, *J. Chem. Soc., Dalton Trans*., 1571 (1977).
86. T. H. Brown and P. J. Green, *J. Am. Chem. Soc*., **92**, 2359 (1970).
87. T. G. Appleton, M. A. Bennett, and I. B. Tomkins, *J. Chem. Soc., Dalton Trans*., 439 (1977).
88. G. K. Anderson, H. C. Clark, and J. A. Davies, *J. Organomet. Chem*., **210**, 135 (1981).
89. R. J. Puddephatt, personal communication.
90. R. Bramley, A. E. Peppercorn, and M. J. Whittacker, *J. Magn. Reson*., **35**, 138 (1979).
91. W. McFarlane and D. S. Rycroft, *Chem. Commun*., 336 (1973).
92. H. C. E. McFarlane, W. McFarlane, and D. S. Rycroft, *J. Chem. Soc., Dalton Trans*., 1616 (1976).
93. W. McFarlane, *Chem. Brit*., **5**, 142 (1969).
94. W. McFarlane, in *Nuclear Magnetic Resonance Spectroscopy of Nuclei Other than Protons* (Eds. T. Axenrod and G. A. Webb), Wiley–Interscience, 1974, p. 31.
95. W. McFarlane and D. S. Rycroft, *Nucl. Magn. Reson*., **4**, 174 (1975).
96. W. McFarlane and D. S. Rycroft, *Nucl. Magn. Reson*., **6**, 67 (1977).
97. W. McFarlane and D. S. Rycroft, *Nucl. Magn. Reson*., **8**, 123 (1979).
98. W. McFarlane and D. S. Rycroft, *Annu. Rep. NMR Spectrosc*., **9**, 319 (1979).
99. D. E. Demco, *Stud. Cercet. Fiz*., **30**, 845 (1978).
100. R. D. W. Kemmitt and M. A. R. Smith, *Inorg. React. Mech*., **3**, 460 (1974).
101. J. P. Jesson and E. L. Muetterties, in *Dynamic Nuclear Resonance Spectroscopy* (Eds. L. M. Jackman and F. A. Cotton), Academic Press, New York, 1975, Chapter 8.
102. T. S. Piper and G. Wilkinson, *J. Inorg. Nucl. Chem*., **2**, 32 (1956).
103. T. S. Piper and G. Wilkinson, *J. Inorg. Nucl. Chem*., **3**, 104 (1956).
104. M. J. Bennett, F. A. Cotton, A. Davison, J. W. Faller, S. J. Lippard, and S. M. Morehouse, *J. Am. Chem. Soc*., **88**, 4371 (1966).
105. F. A. Cotton, in *Dynamic Nuclear Magnetic Resonance Spectroscopy* (Eds. L. M. Jackman and F. A. Cotton), Academic Press, New York, 1975, Chapter 10.
106. F. A. Cotton, *Acc. Chem. Res*., **1**, 257 (1968).
107. K. Vrieze and P. van Leeuwen, *Prog. Inorg. Chem*., **14**, 1 (1971).
108. K. Vrieze, in *Dynamic Nuclear Magnetic Resonance Spectroscopy* (Eds. L. M. Jackman and F. A. Cotton), Academic Press, New York, 1975, Chapter II.
109. N. M. Sergeyev, *Prog. Nucl. Magn. Reson. Spectrosc*., **9**, 71 (1975).
110. T. J. Marks, *Org. Chem. Iron*, **1**, 113 (1978).
111. J. W. Faller, *Adv. Organomet. Chem*., **16**, 211 (1977).
112. R. D. Adams and F. A. Cotton, in *Dynamic Nuclear Magnetic Resonance Spectroscopy* (Eds. L. M. Jackman and F. A. Cotton), Academic Press, New York, 1975, Chapter 12.
113. L. J. Todd and J. R. Wilkinson, *J. Organomet. Chem*., **77**, 1 (1974).
114. S. Aime and L. Milone, *Prog. NMR Sectrosc*., **11**, 183 (1977).
115. J. Evans, B. F. G. Johnson, J. Lewis, T. W. Matheson, and J. R. Norton, *J. Chem. Soc., Dalton Trans*., 626 (1978).
116. J. P. Hickey, I. M. Baibich, I. S. Butler, and L. J. Todd, *Spectrosc. Lett*., **11**, 671 (1978).
117. J. P. Hickey, J. R. Wilkinson, and L. J. Todd, *J. Organomet. Chem*., **179**, 159 (1979).
118. S. Aime, L. Milone, D. Osella, G. Hawkes, and E. W. Randall, *J. Organomet. Chem*., **178**, 171 (1979).
119. B. Pederson, *Hydrides Energy Storage, Proc. Int. Symp.* (Eds. A. F. Anderson and A. J. Maeland), Academic Press, New York, 1978, p. 83.
120. A. Nutton, P. M. Bailey, N. C. Braund, R. J. Goodfellow, R. S. Thompson, and P. M. Maitlis, *Chem. Commun*., 631 (1980).
121. A. J. Campbell, C. A. Fyfe, and E. Maslowky, *Chem. Commun*., 1032 (1971).
122. A. J. Campbell, C. A. Fyfe, and E. Maslowky, *J. Am. Chem. Soc*., **94**, 2690 (1972).
123. H. M. McConnell and C. H. Holm, *J. Chem. Phys*., **27**, 314 (1957).
124. H. M. McConnell and C. H. Holm, *J. Chem. Phys*., **28**, 749 (1958).
125. H. M. McConnell and D. S. Chesnut, *J. Chem. Phys*., **28**, 107 (1958).

126. N. Bloembergen and W. C. Dickinson, *Phys. Rev.*, **79**, 179 (1950).
127. H. M. McConnell and R. E. Robertson, *J. Chem. Phys.*, **29**, 1361 (1958).
128. T. J. Swift, in *NMR of Paramagnetic Molecules: Principles and Applications* (Eds. G. N. LaMar, W. de W. Horrocks and R. H. Holm), Academic Press, New York, 1973, p. 53.
129. G. N. LaMar, in *NMR of Paramagnetic Molecules: Principles and Applications* (Eds. G. N. LaMar, W. de W. Horrocks and R. H. Holm), Academic Press, New York, 1973, p. 85.
130. K. G. Orrell, *Annu, Rep. NMR Spectrosc.*, **9**, 1 (1979).
131. W. de W. Horrocks, in *NMR of Paramagnetic Molecules: Principles and Applications* (Eds. G. N. LaMar, W. de W. Horrocks and R. H. Holm), Academic Press, New York, 1973, p. 127.
132. J. A. Happe and R. L. Ward, *J. Chem. Phys.*, **39**, 1211 (1963).
133. P. Fanticci, L. Naldini, and F. Cariati, *J. Organomet. Chem.*, **64**, 109 (1974).
134. C. L. Honeybourne, *Nucl. Magn. Reson.*, **7**, 260 (1978).
135. G. A. Webb, *Annu. Rep. NMR Spectrosc.*, **6**, 1 (1975).
136. J. R. Wasson and D. K. Johnson, *Anal. Chem.*, **46**, 314R (1974).
137. G. A. Webb, *Annu. Rep. NMR Spectrosc.*, **3**, 211 (1970).
138. G. C. Levy and J. D. Cargioli, *J. Magn. Reson.*, **10**, 231 (1973).
139. G. C. Levy and U. Edlund, *J. Am. Chem. Soc.*, **97**, 4482 (1975).
140. G. C. Levy, U. Edlund, and J. G. Hexem, *J. Magn. Reson.*, **19**, 259 (1975).
141. J. W. Faller, M. A. Adams, and G. N. LaMar, *Tetrahedron Lett.*, 699 (1974).
142. R. E. Sievers (Ed.), *Nuclear Magnetic Resonance Shift Reagents*, Academic Press, New York, 1973.
143. J. Reuben, *Prog. Nucl. Magn. Reson. Spectrosc.*, **9**, 1 (1973).
144. A. F. Cockerill and G. L. O. Davies, *Chem. Rev.*, **73**, 533 (1973).
145. J. Reuben, *Handbook Phys. Chem. Rare Earths,* **4**, 483 (1979).
146. W. de W. Horrocks, in *NMR of Paramagnetic Molecules: Principles and Applications* (Eds. G. N. LaMar, W. de W. Horrocks and R. H. Holm), Academic Press, New York, 1973, p. 479.
147. B. F. G. Johnson, J. Lewis, and P. McArdle, *J. Chem. Soc., Dalton Trans.*, 1253 (1974).
148. R. Porter, T. J. Marks, and D. F. Shriver, *J. Am. Chem. Soc.*, **95**, 3548 (1973).
149. M. D. McCreary, D. W. Lewis, D. L. Wernick, and G. M. Whitesides, *J. Am. Chem. Soc.*, **96**, 1080 (1974).
150. G. R. Sullivan, *Top. Stereochem.*, **10**, 287 (1978).
151. H. Brunner, *Adv. Organomet. Chem.*, **18**, 152 (1980).
152. D. L. Reger, *Inorg. Chem.*, **14**, 660 (1975).
153. M. L. Maddox, S. L. Stafford, and H. D. Kaesz, *Adv. Organomet. Chem.*, **3**, 1 (1965).
154. P. S. Braterman, *J. Organomet. Chem.*, **123**, 75 (1976).
155. Y. Kawasaki, *Kakagu to Kogyo*, **52**, 247 (1978).
156. R. M. Stevens, C. W. Kern, and W. N. Lipscomb, *J. Chem. Phys.*, **37**, 279 (1962).
157. M. Ciriano, M. Green, J. A. K. Howard, J. Proud, J. L. Spencer, F. G. A. Stone, and C. A. Tsipis, *J. Chem. Soc., Dalton Trans.*, 801 (1978).
158. M. Green, J. A. K. Howard, J. Proud, J. L. Spencer, F. G. A. Stone, and C. A. Tsipis, *Chem. Commun.*, 671 (1976).
159. J. P. Jesson, in *Transition Metal Hydrides* (Ed. E. L. Muetterties), Marcel Dekker, New York, 1976, p. 75.
160. L. S. Keller, *Tetrahedron Lett.*, 2361 (1978).
161. W. B. Jennings, *Chem. Rev.*, **75**, 307 (1975).
162. G. Van Koten, C. A. Sharp, J. T. B. H. Jastrzebski, and J. G. Noltes, *J. Organomet. Chem.*, **186**, 427 (1980).
163. J. Lucas and P. A. Kramer, *J. Organomet. Chem.*, **31**, 111 (1971).
164. H. C. Clark and C. S. Wong, *J. Am. Chem. Soc.*, **99**, 703 (1977).
165. T. G. Attig, H. C. Clark, and C. S. Wong, *Can. J. Chem.*, **55**, 189 (1977).
166. G. K. Anderson, R. J. Cross, L. Manojlović-Muir, K. W. Muir, and R. A. Wales, *J. Chem. Soc., Dalton Trans.*, 684 (1979).
167. C. Brevard and J. P. Kintzinger, in *NMR and the Periodic Table* (Eds. R. K. Harris and B. E. Mann), Academic Press, New York, 1978, Chapter 5, p. 107.
168. Y. Fujise, *Kagaku*, **34**, 498 (1979).

169. H. H. Mantsch, H. Saito, and I. C. P. Smith, *Prog. Nucl. Magn. Reson. Spectrosc.*, **11**, 211 (1977).
170. H. Saito, *Farumashia*, **13**, 922 (1977).
171. L. S. Bresler, A. S. Khachaturov, and I. Ya. Podburnyi, *J. Organomet. Chem.*, **64**, 335 (1974).
172. D. L. Reger and E. C. Culbertson, *J. Am. Chem. Soc.*, **98**, 2789 (1976).
173. J. L. Garnett, C. A. Luckey, and P. G. Williams, *Aust. J. Chem.*, **33**, 1393 (1980).
174. J. A. Elvidge and J. R. Jones, *Isot. Org. Chem.*, **4**, 1 (1978).
175. B. Lindman and S. Forsen, in *NMR and the Periodic Table* (Eds. R. K. Harris and B. E. Mann), Academic Press, New York, 1978, Chapter 6, p. 129.
176. A. I. Popov, *Pure Appl. Chem.*, **51**, 101 (1979).
177. A. I. Popov, *Solute–Solvent Interact.*, **2**, 271 (1978).
178. P. Laszlo, *Angew. Chem.*, **90**, 271 (1978).
179. M. M. Civan and M. Shporer, *Biol. Magn. Reson.*, **1**, 1 (1978).
180. A. I. Popov, *Stereodyn. Molec. Symp., Proc. Symp.*, 197 (1979).
181. A. L. Van Geet, *Prog. Anal. Chem.*, **6**, 155 (1973).
182. E. A. C. Lucken, *J. Organomet. Chem.*, **4**, 252 (1965).
183. P. A. Scherr, R. J. Hogan, and J. P. Oliver, *J. Am. Chem. Soc.*, **96**, 6055 (1974).
184. G. Fraenkel, A. M. Fraenkel, M. J. Geckle, and F. Schloss, *J. Am. Chem. Soc.*, **101**, 4745 (1979).
185. J. B. Lambert and M. Urdaneta-Perez, *J. Am. Chem. Soc.*, **100**, 157 (1978).
186. W. McFarlane and D. S. Rycroft, *J. Organomet. Chem.*, **64**, 303 (1974).
187. T. Clark, J. Chandrasekhar, and P. von Ragué-Schleyer, *Chem. Commun.*, 627 (1980).
188. B. Lindman and S. Forsen, in *NMR and the Periodic Table* (Eds. R. K. Harris and B. E. Mann), Academic Press, New York, 1978, Chapter 7, p. 183.
189. J. J. Delpuech, A. Peguy, P. Rubini, and J. Steinmetz, *Nouv. J. Chim.*, **1**, 133 (1976).
190. F. W. Wehrli, unpublished results cited in reference 24.
191. F. Toma, M. Villemin, M. Ellenberger, and L. Brehamet, in *Magnetic Resonance and Related Phenomena* (Ed. I. Ursa), Publishing House of the Academy of the Socialist Republic of Romania, Bucharest, 1971, p. 317.
192. R. G. Kidd and R. J. Goodfellow, in *NMR and the Periodic Table* (Eds. R. K. Harris and B. E. Mann), Academic Press, New York, 1978, Chapter 8, p. 195.
193. R. G. Kidd, *Annu. Rep. NMR Spectrosc.*, **10A**, 1 (1980).
194. A. Yamasaki, *Kagaku Sosetsu*, **13**, 208 (1976).
195. S. J. Ferguson, *Tech. Top. Bioinorg. Chem.*, 305 (1975).
196. G. A. Melson, D. J. Olszanski, and E. T. Roach, *Chem. Commun.*, 229 (1974).
197. V. P. Tarasov and Y. A. Buslaev, *J. Magn. Reson.*, **25**, 197 (1977).
198. Y. A. Buslaev, S. P. Petrosyants, V. P. Tarasov, and V. I. Chagin, *Russ. J. Inorg. Chem.*, **19**, 975 (1974).
199. J. Kronenbiter and A. Schwenk, *J. Magn. Reson.*, **25**, 147 (1977).
200. M. F. Lappert and R. Pearce, *Chem. Commun.*, 126 (1973).
201. D. G. H. Ballard and R. Pearce, *Chem. Commun.*, 621 (1975).
202. J. Holton, M. F. Lappert, G. R. Scollary, D. G. H. Ballard, R. Pearce, J. L. Atwood, and W. E. Hunter, *Chem. Commun.*, 425 (1976).
203. R. D. Fischer, *NMR Paramagn. Molecules*, 521 (1973).
204. F. Y. Fradin, *Actinides: Electron. Struct. Related Prop.*, **1**, 181 (1974).
205. R. D. Fischer, *NATO Adv. Study Inst., Ser. C*, **44**, 337 (1978).
206. R. G. Kidd, R. W. Mathews, and H. G. Spinney, *J. Am. Chem. Soc.*, **94**, 6686 (1973).
207. L. C. Erich, A. C. Gossard, and R. L. Hartless, *J. Chem. Phys.*, **59**, 3911 (1973).
208. D. Rehder, *J. Organomet. Chem.*, **37**, 303 (1972).
209. D. Rehder and J. Schmidt, *J. Inorg. Nucl. Chem.*, **36**, 333 (1974).
210. D. Rehder, L. Dahlenburg and I. Muller, *J. Organomet. Chem.*, **122**, 53 (1976).
211. D. Rehder and W. L. Dorn, *Trans. Met. Chem.*, **1**, 74 and 233 (1976).
212. D. Rehder, *J. Magn. Reson.*, **25**, 177 (1977).
213. R. Talay and D. Rehder, *Chem. Ber.*, **111**, 1978 (1978).
214. D. Rehder, *J. Magn. Reson.*, **38**, 419 (1980).
215. C. A. Tolman, *Chem. Rev.*, **77**, 313 (1977).

216. Y. Egozy and A. Loewenstein, *J. Magn. Reson.*, **1**, 494 (1969).
217. B. W. Epperlein, H. Krüger, O. Lutz, A. Nolle, and W. Mayr, *Z. Naturforsch.*, **30A**, 1237 (1975).
218. R. R. Vold and R. L. Vold, *J. Magn. Reson.*, **19**, 365 (1975).
219. O. Lutz, A. Nolle, and P. Kroneck, *Z. Naturforsch.*, **31A**, 351 (1976).
220. D. S. Milbrath, J. G. Verkade, and R. J. Clark, *Inorg. Nucl. Chem. Lett.*, **12**, 921 (1976).
221. G. T. Andrews and W. McFarlane, *Inorg. Nucl. Chem. Lett.*, **14**, 215 (1978).
222. P. J. Green and H. T. Brown, *Inorg. Chem.*, **10**, 206 (1971).
223. S. O. Grim, D. A. Wheatland, and W. McFarlane, *J. Am. Chem. Soc.*, **89**, 5573 (1967).
224. S. O. Grim and D. A. Wheatland, *Inorg. Chem.*, **8**, 1716 (1969).
225. S. O. Grim, W. McFarlane, and D. A. Wheatland, *Inorg. Nucl. Chem. Lett.*, **2**, 49 (1966).
226. S. O. Grim and D. A. Wheatland, *Inorg. Nucl. Chem. Lett.*, **4**, 187 (1968).
227. S. O. Grim, P. R. McAllister, and R. M. Singer, *Chem. Commun.*, 38 (1969).
228. T. G. Appleton, H. C. Clark, and L. E. Manzer, *Coord. Chem. Rev.*, **10**, 335 (1973).
229. J. G. Verkade, *Coord. Chem. Rev.*, **9**, 1 (1972).
230. W. McFarlane, A. M. Noble, and J. M. Winfield, *J. Chem. Soc. A*, 948 (1971).
231. J. Banck and A. Schwenk, *Z. Phys.*, **20B**, 75 (1975).
232. B. E. Mann, *Adv. Organomet. Chem.*, **12**, 135 (1974).
233. Y. Kamada, T. Sugawara, and H. Iwamoto, *Chem. Commun.*, 291 (1979).
234. F. H. Kohler, H. J. Kalder, and E. O. Fischer, *J. Organomet. Chem.*, **113**, 11 (1976).
235. Y. Kamada and W. M. Jones, unpublished results cited in reference 233.
236. P. S. Braterman, D. W. Milne, E. W. Randall, and E. Rosenberg, *J. Chem. Soc., Dalton Trans.*, 1027 (1973).
237. J. D. Kennedy, W. McFarlane, G. S. Pyne, and B. Wrackmeyer, *J. Chem. Soc., Dalton Trans.*, 386 (1975).
238. M. J. Albright and J. P. Oliver, *J. Organomet. Chem.*, **172**, 99 (1979).
239. R. J. Miles, Jr., B. B. Garrett, and R. J. Clark, *Inorg. Chem.*, **8**, 2817 (1969).
240. S. Onaka, Y. Sasaki, and H. Sano, *Bull. Chem. Soc. Jap.*, **44**, 726 (1971).
241. E. S. Mooberry and R. K. Sheline, *J. Chem. Phys.*, **56**, 1852 (1972).
242. G. M. Bancroft, H. C. Clark, R. G. Kidd, A. T. Rake, and H. G. Spinney, *Inorg. Chem.*, **12**, 728 (1973).
243. J. P. Williams and A. Wojcicki, *Inorg. Chim. Acta*, **15**, L19 (1975).
244. T. Nakano, *Bull. Chem. Soc. Jap.*, **50**, 661 (1977).
245. O. Lutz, W. Nepple, and A. Nolle, *Z. Naturforsch*, **31A**, 1046 (1976).
246. M. Broze and Z. Luz, *J. Phys. Chem.*, **73**, 1600 (1969).
247. A. Schwenk, *J. Magn. Reson.*, **5**, 376 (1971).
248. A. A. Koridge, P. V. Petrovskii, S. P. Gubin, and E. I. Fedin, *J. Organomet. Chem.*, **93**, C26 (1975).
249. A. A. Koridge, N. M. Astakhova, P. V. Petrovskii, and A. I. Lutsenko, *Dolk. Akad. Nauk SSSR*, **242**, 117 (1978).
250. G. N. LaMar, D. B. Viscio, D. L. Budd, and K. Geronde, *Biochem. Biophys. Res. Commun.*, **82**, 19 (1978).
251. B. E. Mann, *Chem. Commun.*, 1173 (1971).
252. S. S. Dodwad and M. G. Datar, *Indian J. Chem.*, **17A**, 100 (1979).
253. N. Juranic, M. B. Celap, D. Vucelic, M. J. Malinar, and P. N. Radivojsa, *J. Magn. Reson.*, **35**, 319 (1979).
254. N. Juranic, M. B. Celap, D. Vucelic, M. J. Malinar, and P. N. Radivojsa, *J. Coord. Chem.*, **9**, 117 (1979).
255. K. D. Rose and R. G. Bryant, *Inorg. Chem.*, **18**, 1332 (1979).
256. D. M. Doddrell, D. T. Pegg, and M. R. Bendall, *Aust. J. Chem.*, **32**, 1 (1979).
257. D. M. Doddrell, M. R. Bendall, P. C. Healy, G. Smith, C. H. L. Kennard, C. L. Raston, and A. H. White, *Aust. J. Chem.*, **32**, 1219 (1979).
258. A. Yamasaki, T. Aoyama, S. Fujiwara, and K. Nakamura, *Bull. Chem. Soc. Jap.*, **51**, 643 (1978).
259. R. Wiess and J. G. Verkade, *Inorg. Chem.*, **18**, 529 (1979).
260. J. A. Seitchik, V. Jaccarino, and J. H. Wernick, *Phys. Rev.*, **138A**, 148 (1965).
261. E. A. C. Lucken, K. Noack, and D. F. Williams, *J. Chem. Soc. A*, 148 (1967).
262. H. Haas and R. K. Sheline, *J. Inorg. Nucl. Chem.*, **29**, 693 (1967).

263. H. W. Speiss and R. K. Sheline, *J. Chem. Phys.*, **53**, 3036 (1970).
264. E. S. Mooberry, M. Pupp, J. L. Slater, and R. K. Sheline, *J. Chem. Phys.*, **55**, 3655 (1971).
265. D. Rehder and J. Schmidt, *Z. Naturforsch*, **27B**, 625 (1972).
266. E. S. Mooberry and R. K. Sheline, *J. Chem. Phys.*, **56**, 1852 (1972).
267. M. A. Cohen, D. R. Kidd and T. L. Brown, *J. Am. Chem. Soc.*, **97**, 4408 (1975).
268. D. D. Traficante and J. A. Simms, *Rev. Sci. Instrum.*, **43**, 1122 (1972).
269. B. F. Taylor, *PhD Thesis,* University of Bristol, 1973, cited in reference 192.
270. R. J. Goodfellow, unpublished results cited in reference 192.
271. S. Martinego, B. T. Heaton, R. J. Goodfellow, and P. Chini, *Chem. Commun.*, 39 (1977).
272. H. C. E. McFarlane, W. McFarlane, and R. J. Wood, *Bull. Soc. Chim. Belg.*, **85**, 864 (1976).
273. S. J. Anderson, J. R. Barnes, P. L. Goggin, and R. J. Goodfellow, *J. Chem. Res. (S)*, 286 (1978).
274. O. A. Gansow, D. S. Gill, F. J. Bennis, J. R. Hutchinson, J. L. Vidal, and R. C. Schoening, *J. Am. Chem. Soc.*, **102**, 2449 (1980).
275. D. S. Gill, O. A. Gansow, F. J. Bennis, and K. C. Ott, *J. Magn, Reson.*, **35**, 459 (1979).
276. J. L. Vidal, W. E. Walker, R. L. Pruett, and R. C. Schoening, *Inorg. Chem.*, **18**, 129 (1979).
277. J. L. Vidal, R. A. Fiato, L. A. Crosby, and R. L. Pruett, *Inorg. Chem.*, **17**, 2574 (1978).
278. J. L. Vidal, R. C. Schoening, R. L. Pruett, and R. A. Fiato, *Inorg. Chem.*, **18**, 1821 (1979).
279. V. G. Albano, P. L. Bellon, and G. Ciani, *Chem. Commun.*, 1024 (1969).
280. J. F. Nixon and A. Pidock, *Annu. Rev. NMR Spectrosc.*, **2**, 345 (1969).
281. P. S. Pregosin, and R. W. Kunz, *NMR Basic Principles Prog.*, **16**, 1 (1979).
282. L. E. Drain, *Phys. Lett.*, **11**, 114 (1964).
283. H. Schumann, M. Meissmer and H. J. Kroth, *Z. Naturforsch. (B).*, **35**, 639 (1980).
284. P. L. Goggin and R. J. Goodfellow, unpublished results cited in reference 192.
285. P. S. Pregosin, unpublished results cited in reference 192.
286. J. D. Kennedy, W. McFarlane, R. J. Puddephatt, and P. J. Thompson, *J. Chem. Soc., Dalton Trans.*, 874 (1976).
287. S. Heitkamp, D. J. Stufkens, and K. Vreize, *J. Organomet. Chem.*, **169**, 107 (1979).
288. T. Miyamoto, *J. Organomet. Chem.*, **134**, 335 (1977).
289. P. L. Goggin, R. J. Goodfellow, J. R. Knight, M. G. Norton, and B. F. Taylor, *J. Chem. Soc., Dalton Trans.*, 2220 (1973).
290. C. Brown, B. T. Heaton, and J. Sabounchei, *J. Organomet. Chem.*, **142**, 413 (1977).
291. J. E. Sarneski, L. E. Erickson, and C. N. Reilley, *J. Magn. Reson.*, **37**, 155 (1980).
292. S. Shinoda, Y. Yamaguchi, and Y. Saito, *Inorg. Chem.*, **18**, 673 (1979).
293. P. S. Pregosin, S. N. Sze, P. Salvadori, and P. Lazzaroni, *Helv. Chim. Acta*, **60**, 2514 (1977).
294. C. Brown, B. T. Heaton, P. Chini, A. Fumagalli, and G. Longini, *Chem. Commun.*, 309 (1977).
295. C. Brown, B. T. Heaton, A. D. C. Towl, P. Chini, A. Fumagelli, and G. Longini, *J. Organomet. Chem.*, **181**, 233 (1979).
296. T. Iwayanagi and Y. Saito, *Chem. Lett.*, 1193 (1976).
297. G. K. Anderson, R. J. Cross, and D. S. Rycroft, *J. Chem. Res. (S)*, 240 (1980).
298. H. Motschi, S. M. Sze, and P. S. Pregosin, *Helv. Chim. Acta*, **62**, 2086 (1979).
299. S. J. Anderson and R. J. Goodfellew, *J. Chem. Soc., Dalton Trans.*, 1683 (1977).
300. S. J. Anderson, P. L. Goggin, and R. J. Goodfellow, *J. Chem. Soc., Dalton Trans.*, 1959 (1976).
301. P. S. Pregosin, H. Streit, and L. H. Venanzi, *Inorg. Chim. Acta*, **38**, 237 (1980).
302. H. Motschi and P. S. Pregosin, *Inorg. Chim. Acta*, **40**, 141 (1980).
303. I. M. Al-Najjar, M. Green, S. J. S. Kerrison, and P. J. Sadler, *J. Chem. Res. (S)*, 206 (1979).
304. I. M. Al-Najjar, M. Green, S. J. S. Kerrison, and P. J. Sadler, *Chem. Commun.*, 311 (1979).
305. I. M. Al-Najjar, M. Green, J. K. K. Sarhan, I. M. Ismail, and P. J. Sadler, *Inorg. Chim. Acta Lett.*, **44**, L187 (1980).
306. J. D. Kennedy, I. J. Colquhoun, W. McFarlane, and R. J. Puddephatt, *J. Organomet. Chem.*, **172**, 479 (1979).
307. L. E. Manzer, *PhD Thesis*, University of Western Ontario, 1973.
308. D. W. W. Anderson, E. A. V. Ebsworth, and D. W. H. Rankin, *J. Chem. Soc., Dalton Trans.*, 2370 (1973).
309. P. L. Goggin, R. J. Goodfellow, and S. R. Haddock, *Chem. Commun.*, 176 (1975).

310. W. McFarlane, *Chem. Commun.*, 755 (1968).
311. W. McFarlane and R. J. Wood, *J. Chem. Soc., Dalton Trans.*, 1397 (1972).
312. J. H. Nelson, V. Cooper, and R. W. Rudolph, *Inorg. Nucl. Chem. Lett.*, **16**, 263 and 587 (1980).
313. P. S. Pregosin and S. N. Sze, *Helv. Chim. Acta*, **61**, 1848 (1978).
314. K. H. O. Stargewski and P. S. Pregosin, *Angew. Chem.*, **92**, 323 (1980).
315. A. Fumagalli, S. Martinego, P. Chini, A. Albinati, S. Bruckner, and B. T. Heaton, *Chem. Commun.*, 195 (1978).
316. N. M. Boag, J. Browning, C. Crocker, P. L. Goggin, R. J. Goodfellow, M. Murray, and J. L. Spencer, *J. Chem. Res. (S)*, 228 (1978).
317. J. Soulie, J.-Y. Lallemand, and R. C. Rae, *Org. Magn. Reson.*, **12**, 67 (1979).
318. W. McFarlane and D. S. Rycroft, *J. Magn. Reson.*, **24**, 95 (1976).
319. P. D. Ellis, H. C. Walsh, and C. S. Peters, *J. Magn. Reson.*, **11**, 426 (1973).
320. Y. Yamamoto, H. Haraguchi and S. Fujiwara, *J. Phys. Chem.*, **74**, 4369 (1970).
321. A. K. Rahimi and A. I. Popov, *J. Magn. Reson.*, **36**, 351 (1979).
322. C. D. M. Beverwijk, G. J. M. Van der Kerk, A. J. Leusing, and J. G. Noltes, *Organomet. Chem. Rev.*, **A5**, 215 (1970).
323. A. E. Hill, *J. Am. Chem. Soc.*, **43**, 254 (1921).
324. P. M. Henrichs, S. Sheard, J. H. H. Ackermann, and G. E. Maciel, *J. Am. Chem. Soc.*, **101**, 3222 (1979).
325. P. M. Henrichs, J. H. H. Ackermann, and G. E. Maciel, *J. Am. Chem. Soc.*, **99**, 2544 (1977).
326. K. Jucker, W. Sahm, and A. Schwenk, *Z. Naturforsch*, **31A**, 1532 (1976).
327. H. Schmidbaur, J. Adlkofer, and W. Buchner, *Angew, Chem., Int. Ed. Engl.*, **12**, 415 (1973).
328. B. L. Dyatkin, B. I. Martynov, L. G. Martynova, N. G. Kizim, S. R. Sterlin, Z. A. Stumbrevichute, and L. A. Federov, *J. Organomet. Chem.*, **57**, 423 (1973).
329. I. J. Colquhoun and W. McFarlane, *Chem. Commun.*, 145 (1980).
330. E. L. Muetterties and C. W. Alegranti, *J. Am. Chem. Soc.*, **94**, 6386 (1972).
331. G. K. Anderson, *Adv. Organomet. Chem.*, in press.
332. G. E. Maciel and M. Borzo, *Chem. Commun.*, 394 (1973).
333. A. D. Cardin, P. D. Ellis, J. D. Odom, and J. W. Howard, *J. Am. Chem. Soc.*, **97**, 1672 (1975).
334. C. J. Turner and R. F. M. White, *J. Magn. Reson.*, **26**, 1 (1977).
335. J. D. Kennedy and W. McFarlane, *J. Chem. Soc., Perkin Trans. II*, 1187 (1977).
336. J. Jokisaari, K. Raisanen, L. Lajunen, A. Passoja, and P. Pyykko, *J. Magn. Reson.*, **31**, 121 (1978).
337. B. E. Mann, *Inorg. Nucl. Chem. Lett.*, **7**, 595 (1971).
338. P. Perringer, *Inorg. Chim. Acta*, **39**, 67 (1980).
339. A. J. Canty, P. Barron, and P. C. Healy, *J. Organomet. Chem.*, **179**, 447 (1979).
340. M. Borzo and G. E. Maciel, *J. Magn. Reson.*, **19**, 279 (1975).
341. M. A. Sens, N. K. Wilson, P. D. Ellis, and T. D. Osom, *J. Magn. Reson.*, **19**, 323 (1975).
342. A. J. Brown, O. W. Howarth, and P. J. Moore, *J. Chem. Soc., Dalton Trans.*, 1589 (1976).
343. H. Schmidbaur, O. Gasser, T. E. Fraser, and E. A. V. Ebsworth, *Chem. Commun.*, 334 (1977).
344. R. J. Goodfellow and S. R. Stobart, *J. Magn. Reson.*, **27**, 143 (1977).
345. P. E. Barron, D. Doddrell, and W. Kitching, *J. Organomet. Chem.*, **139**, 361 (1977).
346. P. L. Goggin, R. J. Goodfellow, and N. W. Hurst, *J. Chem. Soc., Dalton Trans.*, 561 (1978).
347. P. L. Goggin, R. J. Goodfellow, N. W. Hurst, L. G. Mallinson, and M. Murray, *J. Chem. Soc., Dalton Trans.*, 872 (1978).
348. J. L. Sudmeier, R. R. Birge, and T. G. Perkins, *J. Magn. Reson.*, **30**, 491 (1978).
349. P. Perringer, *Inorg. Nucl. Chem. Lett.*, **16**, 205 (1980).
350. P. L. Goggin, R. J. Goodfellow, D. M. McEwan, and K. Kessler, *Inorg. Chim. Acta Lett.*, **44**, LIII (1980).
351. T. N. Mitchell and H. C. Marsmann, *J. Organomet. Chem.*, **150**, 171 (1978).
352. S. Cradock, E. A. V. Ebsworth, N. S. Hosmane, and K. M. Mackay, *Angew. Chem., Int. Ed. Engl.*, **14**, 167 (1975).

353. Yu. A. Strelenko, Yu. K. Grishin, V. T. Bychkov, and Yu. A. Ustynyuk, *Zh. Obshch. Khim*., **49**, 1172 (1979).
354. M. F. Larin, D. V. Gendin, V. A. Pestunovich, O. A. Kruglaya, and N. S. Vyazankin, *Izv. Akad. Nauk SSSR, Ser. Khim*., 697 (1979).
355. G. E. Maciel and M. Borzo, *J. Magn. Reson*., **10**, 388 (1973).
356. V. S. Petrosyan and O. A. Reutov, *J. Organomet. Chem*., **76**, 123 (1974).
357. F. A. L. Anet and J. L. Sudmeier, *J. Magn. Reson*., **1**, 124 (1969).
358. P. L. Goggin, R. J. Goodfellow, D. M. McEwan, A. J. Griffiths, and K. Kessler, *J. Chem. Res. (S)*, 194 (1979).
359. E. L. Muetterties (Ed.), *Boron Hydride Chemistry*, Academic Press, New York, 1975.
360. G. R. Eaton and W. N. Lipscomb, *NMR Studies of Boron Hydrides and Related Compounds*, Benjamin, New York, 1969.
361. R. Schaeffer, in *Progress in Boron Chemistry* (Eds. H. Steinberg and A. L. McCloskey). Pergamon Press, Oxford, 1964, p. 417.
362. H. Nöth and B. Wrackmeyer, *NMR, Basic Principles and Prog*., **14**, 1 (1978).
363. L. J. Todd and A. R. Allen, *Prog. Nucl. Magn. Reson*., **3**, 87 (1979).
364. L. J. Todd, *Adv. Organomet. Chem*., **8**, 87 (1969).
365. G. E. Herberich, E. Brauer, J. Hengesbach, U. Kölle, G. Huttner, and H. Lorenz, *Chem. Ber*., **110**, 760 (1977).
366. W. Siebert and M. Bochmann, *Angew. Chem., Int. Ed. Engl*., **16**, 857 (1977).
367. R. V. Schultz, F. Sato, and L. J. Todd, *J. Organomet. Chem*., **125**, 115 (1977).
368. J. D. Kennedy, W. McFarlane, and B. Wrackmeyer, *Inorg. Chem*., **15**, 1299 (1976).
369. D. A. Thompson, T. K. Hilty, R. W. Rudolph, *J. Am. Chem. Soc*., **99**, 6774 (1977).
370. J. W. Akitt, *Annu. Rep. NMR Spectrosc*., **5A**, 465 (1972).
371. J. F. Hinton and R. W. Briggs, in *NMR and the Periodic Table* (Eds. R. K. Harris and B. E. Mann), Academic Press, New York, 1978, Chapter 9, p. 279.
372. M. Llinás and A. DeMarco, *J. Am. Chem. Soc*., **102**, 2226 (1980).
373. J. W. Akitt and R. H. Duncan, *J. Magn. Reson*., **25**, 391 (1977).
374. J. Huett, J. Durand, and Y. Infarnet, *Org. Magn. Reson*., **8**, 382 (1976).
375. H. Haraguchi and S. Fujiwara, *J. Phys. Chem*., **73**, 3467 (1969).
376. S. Hermanek, O. Kriz, J. Plesek, and T. Hanslik, *Chem. Ind. (London)*, 42 (1975).
377. R. G. Kidd and B. R. Truax, *Chem. Commun*., 160 (1969).
378. V. P. Tarasov and S. I. Bakum, *J. Magn. Reson*., **18**, 64 (1975).
379. H. E. Swift, C. P. Pool, and J. F. Itzel, *J. Phys. Chem*., **68**, 2509 (1964).
380. E. N. Dicarlo and H. E. Swift, *J. Phys. Chem*., **68**, 551 (1964).
381. P. H. Bird and M. G. H. Wallbridge, *J. Chem. Soc*., 3923 (1965).
382. P. R. Oddy and M. G. H. Wallbridge, *J. Chem. Soc., Dalton Trans*., 572 (1978).
383. V. V. Gavrilenko, M. I. Vinnikova, V. A. Antanovich, and L. I. Zakharin, *Izv. Akad. Nauk SSSR, Ser. Khim*, 1273 (1979).
384. P. C. Lauterbur, R. C. Hopkins, R. W. King, U. V. Ziebarth, and C. W. Heitsch, *Inorg. Chem*., **7**, 1025 (1968).
385. O. Yamamoto, *Chem. Lett*., **6**, 511 (1975).
386. V. P. Tarasov and S. I. Bakum, *J. Magn. Reson*., **18**, 64 (1975).
387. L. Rodehuesser, P. R. Rubini, and J. J. Delpuech, *Inorg. Chem*., **16**, 2837 (1977).
388. C. Schramm and J. I. Zink, *J. Magn. Reson*., **26**, 513 (1977).
389. R. W. Briggs, K. R. Metz, and J. F. Hinton, *J. Solution Chem*., **8**, 479 (1979).
390. J. I. Zink, C. Srivanavit, and J. J. Dechter, *Organomet. Polym. Symp*., 323 (1977).
391. G. C. Levy and G. L. Nelson, *Carbon-13 Nuclear Magnetic Resonance for Organic Chemists*, Wiley, New York, 1972.
392. J. B. Stothers, *Carbon-13 NMR Spectroscopy*, Academic Press, New York, 1972.
393. M. H. Chisholm and S. Godleski, *Prog. Inorg. Chem*., **20**, 299 (1976).
394. O. A. Gansow and W. D. Vernon, *Top. Carbon-13 NMR Spectrosc*., **2**, 270 (1976).
395. F. W. Wehrli, *Top. Carbon-13 NMR Spectrosc*., **2**, 243 (1976).
396. D. M. Doddrell, M. R. Bendall, A. J. O'Connor, and D. T. Pegg, *Aust. J. Chem*., **30**, 943 (1977).
397. G. C. Levy, *Tetrahedron Lett*., 3709 (1972).
398. H. W. Spiess, G. Grosescu, and U. Haeberlen, *Chem. Phys*., **6**, 226 (1974).

399. A. J. Brown, O. W. Howarth, and P. J. Moore, *J. Magn. Reson.*, **38**, 317 (1977).
400. R. F. Jordan, E. Tsang, and J. R. Norton, *J. Organomet. Chem.*, **149**, C53 (1978).
401. R. F. Jordan and J. R. Norton, *J. Am. Chem. Soc.*, **101**, 4853 (1979).
402. J. J. Pesek and W. R. Mason, *Inorg. Chem.*, **18**, 924 (1979).
403. R. E. Wasylichen, G. L. Neufield, and K. J. Friesen, *Inorg. Chem.*, **20**, 637 (1981).
404. R. E. Wasylichen, *Annu. Rep. NMR Spectrosc.*, **7**, 245 (1977).
405. P. D. Ellis and R. Ditchfield, *Top. Carbon-13 NMR Spectrosc.*, **2**, 434 (1976).
406. A. C. Blizzard and D. P. Santry, *J. Chem. Phys.*, **55**, 1950 (1971).
407. R. D. Adams and F. A. Cotton, in *Dynamic Nuclear Magnetic Resonance Spectroscopy* (Eds. L. M. Jackson and F. A. Cotton), Academic Press, New York, 1975, P. 489.
408. J. W. Faller, *Adv. Organomet. Chem.*, **16**, 211 (1977).
409. J. Evans, *Adv. Organomet. Chem.*, **16**, 319 (1977).
410. P. Chini, G. Longoni, and V. G. Albano, *Adv. Organomet. Chem.*, **14**, 285 (1975).
411. F. A. Cotton, A. Danti. J. S. Waugh, and R. W. Feesenden, *J. Chem. Phys.*, **29**, 1427 (1958).
412. R. Bramley, B. N. Figgis, and R. S. Nyholm, *Trans. Faraday Soc.*, **58**, 1893 (1962).
413. P. C. Lauterbur and J. B. Stothers, *Can. J. Chem.*, **42**, 1563 (1964).
414. A. Loewenstein, M. Schporer, and G. Navon, *J. Am. Chem. Soc.*, **85**, 2855 (1963).
415. B. E. Mann, *Chem. Commun.*, 1173 (1971).
416. G. C. Levy, J. D. Cargioli, P. C. Juliano, and T. D. Mitchell, *J. Am. Chem. Soc.*, **95**, 3445 (1973).
417. P. R. Wells, in *Determination of Organic Structures by Physical Methods* (Eds. F. C. Nachod and J. J. Zuckerman), Vol. 4, Academic Press, New York, 1971, p. 233.
418. H. C. Marsmann, *Chem. Ztg.*, **97**, 128 (1973).
419. R. K. Harris and B. J. Kimber, *Appl. Spectrosc. Rev.*, **10**, 117 (1975).
420. J. Schraml and J. M. Bellama, in *Determination of Organic Structures by Physical Methods* (Eds. F. C. Nachold and J. J. Zuckerman), Vol. 6, Academic Press, New York, 1976, p. 203.
421. R. K. Harris, J. D. Kennedy and W. McFarlane, in *NMR and the Periodic Table* (Eds. R. K. Harris and B. E. Mann), Academic Press, New York, 1978, Chapter 10, p. 309.
422. E. A. Williams and J. D. Cargioli, *Annu. Rep. NMR Spectrosc.*, **9**, 221 (1979).
423. F. H. Koehler, H. Hollfelder, and E. O. Fischer, *J. Organomet. Chem.*, **168**, 53 (1979).
424. R. B. Johannesen, F. E. Brinckmann, and T. D. Coyle, *J. Phys. Chem.*, **72**, 660 (1978).
425. S. Li, D. L. Johnson, J. A. Gladysz, and K. L. Servis, *J. Organomet. Chem.*, **166**, 317 (1979).
426. A. L. Bikovetz, O. V. Kuzmin, V. M. Volvin, and A. M. Krapivin, *Izv. Akad. Nauk SSSR, Ser. Khim.*, 2815 (1979).
427. A. L. Bikovetz, O. V. Kuzmin, V. M. Volvin, and A. M. Krapivin, *J. Organomet. Chem.*, **194**, C33 (1980).
428. H. Elser and H. Dreeskamp, *Ber. Bunsenges. Phys. Chem.*, **73**, 619 (1969).
429. P. J. Smith and L. Smith, *Inorg. Chim. Acta Rev.*, **7**, 11 (1973).
430. V. S. Petrosyan, *Prog. NMR Spectrosc.*, **11**, 115 (1977).
431. P. J. Smith, *Annu. Rep. NMR Spectrosc.*, **8**, 291 (1978).
432. P. G. Harrison, S. E. Ulrich, and J. J. Zuckerman, *J. Am. Chem. Soc.*, **93**, 5398 (1971).
433. J. J. Burke and P. C. Lauterbur, *J. Am. Chem. Soc.*, **83**, 326 (1961).
434. W. McFarlane and R. J. Wood, *J. Organomet. Chem.*, **40**, C17 (1972).
435. E. V. van der Berghe and G. P. van der Kelen, *J. Organomet. Chem.*, **26**, 207 (1971).
436. L. Smith, *PhD Thesis*, University of London, 1972, cited in reference 429.
437. A. G. Davies, L. Smith, and P. J. Smith, *J. Organomet. Chem.*, **39**, 279 (1972).
438. E. A. V. Ebsworth, in *Organometallic Compounds of the Group IV Elements* (Ed. A. G. McDiarmid), Vol. 1, Marcel Dekker, New York, 1968.
439. J. Mason, *J. Chem. Soc. (A)*, 1038 (1971).
440. B. K. Hunter and L. W. Reeves, *Can. J. Chem.*, **46**, 1399 (1968).
441. A. G. Davies, M. W. Tse, M. F. Ladd, J. D. Kennedy, W. McFarlane, and G. S. Pyne, *Chem. Commun.*, 791 (1978).
442. H. G. Kuivila, J. D. Kennedy, R. Y. Tien, I. J. Tyminski, F. L. Pelozar, and O. R. Khan, *J. Org. Chem.*, **36**, 2083 (1971).
443. K. A. Ostoja-Starzewski, H. Ruegger, and P. S. Pregosin, *Inorg. Chim. Acta*, **36**, L445 (1979).

444. D. H. Harris, M. F. Lappert, J. S. Poland, and W. McFarlane, *J. Chem. Soc., Dalton Trans.*, 311 (1975).
445. D. Harris, S. Keppie, M. F. Lappert, W. McFarlane, and J. Poland, unpublished results cited in reference 82.
446. H. Itatani and J. C. Bailar, Jr., *Ind. Eng. Chem. Prod. Res. Dev.*, **11**, 146 (1972).
447. H. C. Clark and J. A. Davies, *J. Organomet. Chem.*, **213**, 503 (1981).
448. R. D. Cramer, R. V. Lindsay, C. T. Prewitt, and U. G. Stolberg, *J. Am. Chem. Soc.*, **87**, 658 (1968).
449. R. V. Parish and P. J. Rowbottom, *J. Chem. Soc.*, 37 (1973).
450. J. F. Young, R. D. Gillard, and G. Wilkinson, *J. Chem. Soc.*, 5176 (1964).
451. R. W. Rudolph, W. L. Wilson, F. Parker, R. C. Taylor, and D. C. Young, *J. Am. Chem. Soc.*, **100**, 4629 (1978).
452. P. S. Pregosin, *Biennial Inorg. Chem. Symp., CIC and ACS, Guelph, Canada*, 8 (1980).
453. B. R. Koch, G. V. Fazakerley, and E. Dijkstra, *Inorg. Chim. Acta Lett.*, **45**, L51 (1980).
454. G. E. Maciel and J. L. Dallas, *J. Am. Chem. Soc.*, **95**, 3039 (1973).
455. M. J. Cooper, A. K. Holliday, P. H. Makin, R. J. Puddephatt, and P. J. Smith, *J. Organomet. Chem.*, **65**, 377 (1974).
456. J. D. Kennedy, W. McFarlane, and G. S. Pyne, *J. Chem. Soc., Dalton Trans.*, 2332 (1977).
457. T. N. Mitchell, J. Gmehling, and F. Huber, *J. Chem. Soc., Dalton Trans.*, 960 (1978).
458. R. H. Cox, *J. Magn. Reson.*, **33**, 61 (1979).
459. D. C. Van Beelan, H. O. van der Kooi, and J. Wolters, *J. Organomet. Chem.*, **179**, 37 (1979).
460. D. de Vos, *J. Organomet. Chem.*, **104**, 193 (1976).
461. W. Kitching, V. G. Kumar-Das, and P. R. Wells, *Chem. Commun.*, 1150 (1972).
462. G. A. Morris, *J. Am. Chem. Soc.*, **102**, 428 (1980).
463. M. Witanowski and G. A. Webb (Eds.), *Nitrogen NMR*, Plenum, London, 1973.
464. M. Witanowski, *Pure Appl. Chem.*, **37**, 225 (1974).
465. H. Saito, *Farumashia,* **13**, 922 (1977).
466. M. Witanowski, L. Stefaniak, and G. A. Webb, *Annu. Rep. NMR Spectrosc.*, **7**, 117 (1977).
467. G. C. Levy and R. L. Lichter, *Nitrogen-15 Nuclear Magnetic Resonance Spectroscopy*, Wiley, New York, 1979.
468. N. Logan, in *Nitrogen NMR* (Eds. M. Witanowski and G. A. Webb), Plenum, London, 1973, p. 318.
469. A. Yamasaki, Y. Miyakoshi, M. Fujita, Y. Yoshikawa, and H. Yamatera, *J. Inorg. Nucl. Chem.*, **41**, 473 (1979).
470. M. Alei Jr., P. J. Vergamini, and W. F. Wageman, *J. Am. Chem. Soc.*, **101**, 5415 (1979).
471. H. Motschi and P. S. Pregosin, *Inorg. Chim. Acta*, **40**, 141 (1980).
472. R. Meij, D. J. Stufkens, K. Vreize, W. van Gerresheim, and C. H. Stam, *J. Organomet. Chem.*, **164**, 353 (1979).
473. K. S. Bose and E. H. Abbott, *Inorg. Chem.*, **16**, 3190 (1977).
474. J. E. Bercaw, E. Rosenberg, and J. D. Roberts, *J. Am. Chem. Soc.*, **96**, 612 (1974).
475. R. D. Sanner, D. M. Duggan, T. C. McKenzie, R. E. Marsh, and J. E. Bercaw, *J. Am. Chem. Soc.*, **98**, 8351 (1976).
476. R. D. Sanner, D. M. Duggan, T. C. McKenzie, R. E. Marsh, and J. E. Bercaw, *J. Am. Chem. Soc.*, **98**, 8358 (1976).
477. J. N. Manriquez, D. R. McAlister, E. Rosenberg, A. M. Shiller, K. L. Williamson, S. I. Chan, and J. E. Bercaw, *J. Am. Chem. Soc.*, **100**, 3078 (1978).
478. J. Chatt, M. E. Fakley, R. L. Richards, J. Mason, and I. A. Stenhouse, *J. Chem. Res. (S)*, 44 (1979).
479. M. M. Crutchfield, C. H. Dungan, J. H. Letcher, V. Mark, and J. R. Van Wazer, *Top. Phosphorus Chem.*, **5**, 1 (1967).
480. G. Mavel, *Annu. Rep. NMR Spectrosc.*, **5B**, 1 (1973).
481. O. Stelzer, *Top. Phosphorus Chem.*, **9**, 1 (1977).
482. L. Pauling, *The Nature of the Chemical Bond*, Cornell University Press, Ithaca, New York, 1960.
483. G. M. Kossolapoff and L. Maier, *Organic Phosphorus Compounds*, Wiley, New York, 1972.
484. P. E. Garrou, *Chem. Rev.*, **81**, 291 (1981).

485. B. E. Mann and A. Musco, *J. Chem. Soc., Dalton Trans.*, 1673 (1975).
486. J. J. MacDougall and J. H. Nelson, *Inorg. Nucl. Chem. Lett.*, **15**, 315 (1979).
487. G. K. Anderson, unpublished results.
488. G. K. Anderson, H. C. Clark, and J. A. Davies, unpublished results.
489. C. A. Tolman, S. D. Ittel, A. D. English, and J. P. Jesson, *J. Am. Chem. Soc.*, **100**, 4080 (1978).
490. K. R. Dixon and A. D. Rattray, *Inorg. Chem.*, **17**, 1 (1978).
491. S. J. Cartwright, K. R. Dixon, and A. D. Rattray, *Inorg. Chem.*, **19**, 1120 (1980).
492. H. C. Clark, M. J. Dymarski, and J. D. Oliver, *J. Organomet. Chem.*, **154**, C40 (1978).
493. C. A. McAuliffe (Ed.), *Transition Metal Complexes of Phosphorus, Arsenic and Antimony Ligands*, Macmillan, London, 1973.
494. R. K. Harris, in *NMR and the Periodic Table* (Eds. R. K. Harris and B. E. Mann), Academic Press, London, 1978, Chapter 11, p. 378.
495. G. Baliman and P. S. Pregosin, *J. Magn. Reson.*, **26**, 283 (1977).
496. R. G. Kidd and R. W. Matthews, *J. Inorg. Nucl. Chem.*, **37**, 661 (1975).
497. W. G. Proctor and F. C. Yu, *Phys. Rev.*, **81**, 20 (1951).
498. B. L. Silver and Z. Luz, *Quart. Rev. Chem. Soc.*, **21**, 458 (1967).
499. H. Saito, *Farumashia*, **13**, 922 (1977).
500. W. G. Klemperer, *Angew. Chem.*, **90**, 258 (1979); *Angew. Chem., Int. Ed. Engl.*, **17**, 246 (1978).
501. C. Rodger, N. Sheppard, C. McFarlane, and W. McFarlane, *NMR and the Periodic Table* (Ed. R. K. Harris and B. E. Mann), Academic Press, New York, 1978, Chapter 12, p. 383.
502. T. Sugawara, Y. Kawada, and H. Iwamoto, *Kagaku no Ryoiki*, **34**, 207 (1980).
503. R. Bramley, B. N. Figgis, and R. S. Nyholm, *Trans. Faraday Soc.*, **58**, 1893 (1962).
504. D. Cozak, I. S. Butler, J. P. Hickey, and L. J. Todd, *J. Magn. Reson.*, **33**, 149 (1979).
505. R. L. Kump and L. J. Todd, *J. Organomet. Chem.*, **194**, C43 (1980).
506. R. L. Kump and L. J. Todd, *Chem. Commun.*, 292 (1980).
507. W. G. Klemperer and W. Shum, *Chem. Commun.*, 60 (1979).
508. K. F. Miller and R. A. D. Wentworth, *Inorg. Chem.*, **19**, 1818 (1980).
509. A. Lapidot and C. S. Irving, *J. Chem. Soc., Dalton Trans.*, 668 (1972).
510. C. S. Irving and A. Lapidot, *Nature New Biol.*, **230**, 224 (1974).
511. G. Pifat, S. Maricic, M. Petronivic, V. Kramer, J. Marsel, and K. Bonhard, *Croat. Chim. Acta*, **41**, 195 (1969).
512. S. Maricic, J. S. Leigh, Jr., and D. E. Sunko, *Nature (London)*, **214**, 462 (1967).
513. O. Lutz, A. Nolle, and A. Schwenk, *Z. Naturforsch.*, **28A**, 1370 (1973).
514. V. Svanholm, in *Organic Selenium Compounds: Their Chemistry and Biology* (Eds. D. L. Klayman and W. H. H. Gunther), Wiley, New York, 1973, p. 903.
515. M. Lardon, in *Organic Selenium Compounds: Their Chemistry and Biology* (Eds. D. L. Klayman and W. H. H. Gunther), Wiley, New York, 1973, p. 933.
516. J. D. Kennedy and W. McFarlane, *J. Chem. Soc., Dalton Trans.*, 2134 (1973).
517. W.-H. Pan and J. P. Fackler, Jr., *J. Am. Chem. Soc.*, **100**, 5783 (1978); see also errata, *J. Am. Chem. Soc.*, **100**, 8274 (1978).
518. K. A. Jensen, V. Krishnan, and C. K. Jørgensen, *Acta Chem. Scand.*, **24**, 743 (1970).
519. J. R. Fackler and W.-H. Pan, *J. Am. Chem. Soc.*, **101**, 1607 (1969).
520. H. C. E. McFarlane and W. McFarlane, *J. Chem. Soc., Dalton Trans.*, 2416 (1973).
521. J. D. Kennedy and W. McFarlane, *J. Organomet. Chem.*, **94**, 7 (1975).
522. E. F. Mooney, *An Introduction to ^{19}F NMR Spectroscopy*, Heyden, London, 1970.
523. L. Cavalli, *Annu. Rep. NMR Spectrosc.*, **6B**, 43 (1976).
524. K. Jones and E. F. Mooney, *Annu. Rep. NMR Spectrosc.*, **3**, 261 (1970).
525. C. H. Dungan and J. R. van Wazer, *Complication of Reported ^{19}F Chemical Shifts*, Wiley, New York, 1970.
526. J. W. Emsley, L. Phillips, and V. Wray, *Prog. NMR Spectrom.*, **10**, 83 (1976).
527. R. Fields, *Annu. Rep. NMR Spectrosc.*, **7**, 1 (1977).
528. P. M. Triechel and F. G. A. Stone, *Adv. Organomet. Chem.*, **1**, 143 (1964).
529. G. A. Rivett, *Diss. Abstr.*, **36B**, 2214 (1975).
530. B. L. Booth, R. N. Hazeldine, and N. I. Tucker, *J. Chem. Soc., Dalton Trans.*, 1439 (1975).
531. M. Green and S. H. Taylor, *J. Chem. Soc., Dalton Trans.*, 1128 (1975).

532. B. Lindman and S. Forsén, in *Chlorine, Bromine and Iodine NMR. Physico-Chemical and Biological Applications* (Ed. P. Diehl, E. Fluck, and R. Cosfield), Springer-Verlag, Berlin, 1976.
533. B. Lindman and S. Forsén, in *NMR and the Periodic Table* (Ed. R. K. Harris and B. E. Mann), Academic Press, London, 1978 Chapter 13, p. 421.
534. R. P. Haugland, L. Stryer, T. R. Stengle, and J. D. Baldeschwieler, *Biochemistry*, **6**, 498 (1967).
535. G. J. Schrobilgen, in *NMR and the Periodic Table* (Eds. R. K. Harris and B. E. Mann), Academic Press, London, 1978, Chapter 14, p. 439.
536. G. J. Schrobilgen, J. H. Holloway, P. Granger, and C. Brevard, *Inorg. Chem.*, **17**, 980 (1978).
537. J. Schaefer and E. O. Stejskal, *Top. Carbon-13 NMR Spectrosc.*, **3**, 283 (1979).
538. J. Schaefer and E. O. Stejskal, *J. Am. Chem. Soc.*, **98**, 1031 (1976).
539. E. R. Andrew, *Prog. Nucl. Magn. Reson. Spectrosc.*, **8**, 1 (1971).
540. A. Pines, M. G. Gibby, and J. S. Waugh, *J. Chem. Phys.*, **56**, 1776 (1972).
541. C. A. Fyfe, *Mol. Cryst. Liq. Cryst.*, **52**, 1 (1979).
542. L. Bemi, H. C. Clark, J. A. Davies, D. Drexler, C. A. Fyfe, and R. Wasylichen, *J. Organomet. Chem.*, **224**, C5 (1982).
543. R. C. Schoening, J. L. Vidal, and R. A. Fiato, *J. Organomet. Chem.*, **206**, C43 (1981).
544. E. L. Muetterties and J. Stein, *Chem. Rev.*, **79**, 479 (1979).
545. W. Kitching, A. R. Atkins, G. Wickham, and V. Alberts, *J. Org. Chem.*, **46**, 563 (1981).
546. J. M. A. Al-Rawi, J. A. Elvidge, J. R. Jones, R. B. Mane, and M. Saieed, *J. Chem. Res. (S)*, 298 (1980).
547. H. Weingartner, *J. Magn. Reson.*, **41**, 74 (1980).
548. H. Nöth, A. Rurländer, and P. Wolfgardt, *Z. Naturforsch.*, **36B**, 31 (1981).
549. A. Delville, C. Detellier, A. Gerstmans, and P. Laszlo, *J. Magn. Reson.*, **42**, 14 (1981).
550. G. C. Levy, P. L. Rinaldi, and J. T. Bailey, *J. Magn. Reson.*, **40**, 167 (1980).
551. C. A. M. Vijverberg, J. A. Peters, A. P. G. Kieboom, and H. van Bekkum, *Rec. Trav. Chim. Pays-Bas*, **99**, 287 (1980).
552. B. G. Sayer, N. Hao, G. Dénès, D. G. Bickley, and M. J. McGlinchey, *Inorg. Chim. Acta*, **48**, 53 (1981).
553. H. Schmidt and D. Rehder, *Transition Met. Chem.*, **5**, 214 (1980).
554. D. Rehder, H.-C. Bechthold, and K. Paulsen, *J. Magn. Reson.*, **40**, 305 (1980).
555. D. Rehder, unpublished work cited in reference 556.
556. H.-C. Bechthold and D. Rehder, *J. Organomet. Chem.*, **206**, 305 (1981).
557. E. C. Alyea, R. E. Lenkinski and A. Somogyvari, *Polyhedron*, submitted for publication.
558. S. Dysart, I. Geogii and B. E. Mann, *J. Organomet. Chem.*, **213**, C10 (1981).
559. A. F. Masters, R. T. C. Brownlee, M. J. O'Connor, A. G. Wedd, and J. D. Cotton, *J. Organomet. Chem.*, **195**, C17 (1980).
560. A. Kececi and D. Rehder, *Z. Naturforsch.*, **36B**, 20 (1981).
561. T. Jenny, W. von Philipsborn, J. Kronenbitter, and A. Schwenk, *J. Organomet. Chem.*, **205**, 211 (1981).
562. A. Schwenk, *Phys. Lett.*, **31A**, 513 (1970).
563. W. Sahm and A. Schwenk, *Z. Naturforsch.*, **29A**, 1763 (1974).
564. B. T. Heaton, S. Martinego, and P. Chini, *J. Organomet. Chem.*, **197**, C29 (1980).
565. V. G. Albano and P. L. Bellon, *J. Organomet. Chem.*, **19**, 405 (1969).
566. K.-D. Gruninger, A. Schwenk, and B. E. Mann, *J. Magn. Reson.*, **41**, 354 (1980).
567. M. Cocivera, G. Ferguson, R. E. Lenkinskii, P. Szczecinski, and F. J. Lalor, *J. Magn. Reson.*, submitted for publication.
568. P. S. Pregosin, *Chimia*, **35**, 43 (1981).
569. Y. Koie, S. Shinoda, and Y. Saito, *J. Chem. Soc., Dalton Trans.*, 1082 (1981).
570. C. J. Boreham, J. A. Broomhead, and D. P. Fairlie, *Aust. J. Chem.*, **34**, 659 (1981).
571. H. Vanderpoel, G. van Koten, D. M. Grove, P. S. Pregosin, and K. A. O. Starzewski, *Helv. Chim. Acta*, **64**, 1174 (1981).
572. A. Moor, P. S. Pregosin, and L. W. Venanzi, *Inorg. Chim. Acta*, **48**, 153 (1981).
573. T. G. Appleton and J. R. Hall, *Aust. J. Chem.*, **33**, 2387 (1980).
574. I. M. Ismail, S. J. S. Kerrison, and P. J. Sadler, *Chem. Commun.*, 1175 (1980).

575. D. G. Gillies, L. P. Blaauw, G. R. Hays, R. Huis, and A. D. H. Clague, *J. Magn. Reson.*, **42**, 420 (1981).
576. R. E. Lenkinskii, C. Rodger, and R. E. Wasylishen, *28th Canadian Spectrosc. Conf., Ottawa, Canada*, 1981.
577. I. J. Colquhoun and W. McFarlane, *J. Chem. Soc., Dalton Trans.*, 658, 1981.
578. B. R. McGarvey, C. O. Trudell, D. G. Tuck, and L. Victoriano, *Inorg. Chem.*, **19**, 3432 (1980).
579. J. S. Poland and D. G. Tuck, *J. Organomet. Chem.*, **42**, 315 (1972).
580. P. J. Burke, D. G. Gillies, and R. W. Matthews, *J. Chem. Res. (S)*, 124 (1981).
581. P. J. Burke, R. W. Matthews, I. D. Cresshull, and D. G. Gillies, *J. Chem. Soc., Dalton Trans.*, 132 (1981).
582. F. Brady, R. W. Matthews, M. J. Forster, and D. G. Gillies, *Inorg. Nucl. Chem. Lett.*, **17**, 155 (1981).
583. B. T. Heaton, J. Jonas, T. Eguchi, and G. A. Hoffman, *Chem. Commun.*, 331 (1981).
584. S. Aime and D. Osella, *Chem. Commun.*, 300 (1981).
585. D. J. Darensbourg, *J. Organomet. Chem.*, **209**, C37 (1981).
586. H. Sakuri, Y. Kaminyama, A. Mikoda, T. Kobayoshi, K. Sasaki, and Y. Nakadaira, *J. Organomet. Chem.*, **201**, C14 (1980).
587. J. Otera, T. Hinoishi, and R. Okawara, *J. Organomet. Chem.*, **202**, C93 (1980).
588. J. P. Quintard, M. D. Castaing, G. Dumartin, A. Rahm, and M. Pereyre, *Chem. Commun.*, 1004 (1980).
589. H. Moriyama, T. Aoki, S. Shinoda, and Y. Saito, *J. Chem. Soc., Dalton Trans.* 639 (1981).
590. M. Garralda, V. Garcia, M. Kretschmer, P. S. Pregosin, and H. Ruegger, *Helv. Chim. Acta*, **64**, 1150 (1981).
591. L. J. Farrugia, B. R. James, C. R. Lassigne, and E. J. Wells, *Inorg. Chim. Acta Lett.*, **53**, L261 (1981).
592. J. E. Hamlin and P. M. Maitlis, *J. Mol. Catal.*, **11**, 129 (1981).
593. G. K. Anderson, H. C. Clark, and J. A. Davies, unpublished results.
594. T. N. Mitchell and H. C. Marsmann, *Org. Magn. Reson.*, **15**, 263 (1981).
595. J. D. Kennedy, W. McFarlane, G. S. Pyne, and B. Wrackmeyer, *J. Organomet. Chem.*, **195**, 285 (1980).
596. J. Mason, *Chem. Rev.*, **81**, 205 (1981).
597. J. R. Dilworth, S. Donovan-Mtunzi, C. T. Kan, and R. L. Richards, *Inorg. Chim. Acta Lett.*, **53**, L161 (1981).
598. J. R. Dilworth, C. T. Kan, R. L. Richards, J. Mason, and I. A. Stenhouse, *J. Organomet. Chem.*, **201**, C24 (1980).
599. C. J. Jameson, A. K. Jameson, D. Oppusunggi, S. Wille, P. M. Burrel, and J. Mason, *J. Chem. Phys.*, **74**, 81 (1981).
600. R. Faure, E. J. Vincent, J. M. Ruiz, and L. Lena, *Org. Magn. Reson.*, **15**, 401 (1981).
601. N. Zumbulyadis and H. J. Gysling, *J. Organomet. Chem.*, **192**, 183 (1980).
602. M. L. Martin, M. Trierweiler, V. Galasso, F. Fringwelli, and A. Taticchi, *J. Magn. Reson.*, **42**, 155 (1981).
603. H. J. Gysling, N. Zumbulyadis, and J. A. Robertson, *J. Organomet. Chem.*, **209**, C41 (1981).
604. R. E. Wasylishen and C. A. Fyfe, *Annu. Rep. NMR Spectrosc.*, in press.
605. R. D. Kendrick, C. S. Yannoni, R. Aikman, and R. J. Lagow, *J. Magn. Reson.*, **37**, 555 (1980).
606. D. E. Wemmer, A. Pines and D. D. Whitehurst, *Phil. Trans. R. Soc. London, Ser. A*, **300**, 15 (1981).
607. J. W. Diesveld, E. M. Menger, H. T. Edzes, and W. S. Veeman, *J. Am. Chem. Soc.*, **102**, 7935 (1980).
608. G. E. Maciel, D. J. O'Donnell and R. Greaves, *Adv. Chem. Ser.*, in press.
609. L. Bemi, H. C. Clark, J. A. Davies, C. A. Fyfe, and R. E. Wasylischen, *J. Am. Chem. Soc.*, 104, 438 (1982).
610. H. C. Clark, J. A. Davies, C. A. Fyfe, P. Hayes, and R. E. Wasylishen, unreported observations.

The Chemistry of the Metal–Carbon Bond
Edited by F. R. Hartley and S. Patai

CHAPTER **22**

Mass spectrometry of organometallic compounds

T. R. SPALDING
Department of Chemistry, University College, Cork, Ireland

I. INTRODUCTION: COMPOUND IDENTIFICATION AND FRAGMENTATION BEHAVIOUR

A. Introduction

The initial steps in the study of the gas-phase ion chemistry of organometallics were taken long ago when several metal alkyls and carbonyls were used to investigate metal isotope ratios[1]. Subsequently the subject did not develop until well into the period when commercial mass spectrometers had become available. This was due to the misapprehension that metal-containing compounds would be particularly deleterious to mass spectrometer ion sources, focusing plates, etc. Such misgivings have proved almost entirely unfounded. Over the last 20 years the number of investigations has steadily increased, with a concomitant increase in the general understanding of the factors influencing organometallic spectra. However, a more fundamental understanding of the basic processes involved in the ionization and dissociation of these molecules and in particular ion structures is for the most part still lacking. A number of newer techniques are being developed to study these problems and we are at an exciting stage in the application of mass spectrometry to organometallic chemistry.

The purpose of this chapter is two-fold. First we discuss in general terms the appearance and basic features of the mass spectra of organometallics. This will be done initially according to the type of ion sources used, since the spectra depend largely on the ionization method employed. Electron impact ionization, which has been used for most of the studies to date, will probably continue as the most common method, but chemical ionization, desorption methods, and other ionization methods will gain in importance in the future. This first section will include results from studies using isotopically labelled compounds and metastable ion assignments. The development and application of inlet systems such as the use of the gas chromatographs in combined gas chromatography–mass spectrometry (GC–MS) and computers in data acquisition and handling will be noted but not treated as a primary concern. The second section will focus on investigations of ion structures and the energetics of ion decompositions leading to thermochemical information. Results from ion/molecule reaction studies will be discussed.

The previous literature on the mass spectrometry of organometallics has included two books[2,3] a number of general reviews[4–16], and reviews of a more specific nature[17-26]. Of these, the book by Litzow and Spalding[2], the *Specialist Periodical Reports* of the Chemical Society (now the Royal Society of Chemistry) (London)[12-16], and the monthly *Mass Spectrometry Bulletin*[27] are particularly useful as comprehensive reference sources. The *Specialist Periodical Reports* also cover most other aspects of mass spectrometry and offer excellent current reviews of the state of the art. On the more practical side the article by Miller and Wilson[11] discusses some of the problems encountered in obtaining spectra of organometallics and has sections on ion fragmentation, metastable ion techniques, and GC–MS. Specialized reviews have covered compounds of boron[23], main group IV[18,21,22,24], main group V[24,26], and transition metals[8,17,19,20,25].

B. Compound Identification, Assignment of Spectra, and Fragmentation Modes

1. Compound identification and assignment of spectra

The observation of ions in a mass spectrometer will depend on the machine's sensitivity and resolving power as well as on the sample. Sensitivity is usually not a problem. Spectra have been obtained with as little as a few nanograms of a compound. The resolving power (RP) depends on the design of the instruments' ion analysers. It is commonly defined as RP = $M/\delta M$, where two peaks of mass M and $M + \delta M$ of equal intensity are considered to be resolved if $M/\delta M$ is less than or equal to 0.1, i.e. 10%. This is called the 10% valley definition. Spectra are most commonly recorded in the low-resolution mode (RP = 1000–2000), but with double focusing spectrometers high resolution is obtainable (RP = 10 000 to 25 000+). With high resolution it is possible to measure the masses of ions to an accuracy of a few parts per million and consequently to determine their elemental composition. This extremely important function can also be achieved for resolved ions with double-beam instruments in the low-resolution mode (RP = 1 500+) by using one of the beams for a reference compound and computer acquisition of data from repeated scans of both sample and reference simultaneously. Mass measurement with computer-acquired data from single-beam instruments is usually less accurate (±5–10 ppm) than the other methods. (For further discussion of instrumentation see refs. 2, 11–16, 'organic' mass spectrometry textbooks such as refs. 28–31, and, for the application of computers in mass spectrometry, ref. 32.)

It is now common to record a low-resolution spectrum as the first step in the characterization of an organometallic compound. Since most metallic elements have distinctive polyisotopic patterns, a low-resolution spectrum usually offers immediate confirmation of the presence of a metallic element as well as information about the basic components in the molecular structure. (The only monoisotopic elements of the main groups and first three transition metal series are Na, Cs, Be, Al, P, As, Bi, Mn, Co, Nb, Rh, and Au.) Examples of the use of isotopic patterns in ion assignment are given in Figure 1. Germanium isotope ratios are compared with the parent molecular ions $M^{+\cdot}$ from [Ph_4Ge], Figure 1(a), and iron isotope ratios with M^+ ions from ferrocene, Figure 1(b). In both cases the respective patterns are very similar and easily recognized. However, they are not exactly superimposable since the presence of other isotopes, namely $^{12}C/^{13}C$ in these examples, has caused a shift in the basic metal patterns to higher mass/charge (m/z). Hence the most abundant iron isotope containing ion from ferrocene [$^{56}Fe^{12}C_{10}{}^1H_{10}$]$^+$ has a significant proportion (11%) as [$^{56}Fe^{13}C^{12}C_9{}^1H_{10}$]$^+$ at one mass unit higher. The proportions from combinations of polyisotopic elements can be calculated from elementary probability theory[2,28-31]. Computer programs are available to do this[33-36] and to calculate anisotopic spectra

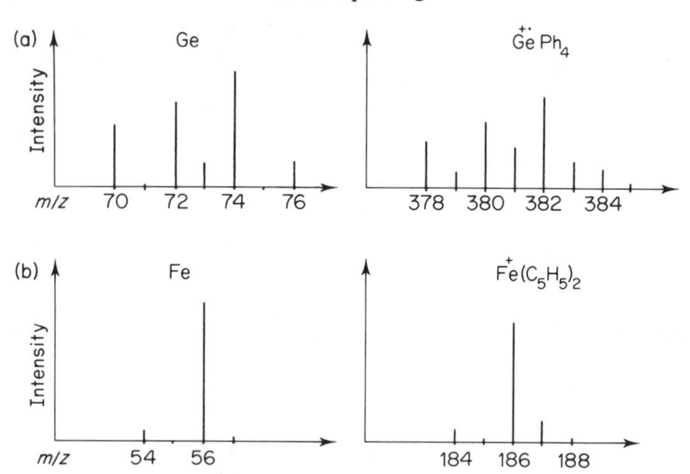

FIGURE 1. Isotope ratios and ion abundances from (a) germanium and $[GePh_4]$ and (b) iron and ferrocene.

from overlapping ions by deconvolution methods[35]. However, if losses of small masses (e.g. hydrogen) are suspected the results from deconvolution calculations should be carefully checked, preferably by high-resolution measurements.

The polyisotopic nature of metallic elements can make the calculation of spectra extremely arduous without computer assistance. For example, the compound $[Mo(SnMe_3)(CO)_3(\eta^5\text{-}C_5H_5)]$[37] produces M^+ ions covering 21 mass units from m/z 398$[^{92}Mo^{112}Sn^{16}O_3{}^{12}C_{11}{}^{1}H_{14}]^+$ to m/z 419$[^{100}Mo^{124}Sn^{16}O_3{}^{12}C_{10}{}^{1}H_{14}]^+$, and considerable overlap occurs in the parent molecular ion region between M^+, $[M - Me]^+$, $[M - 2Me]^+$, and $[M - CO]^+$.

Both polyisotopic and anisotopic spectra are commonly presented, either in bar-graph or table form. The most abundant ion (the base peak) is taken as the reference (100%).

The occurrence of a parent molecular ion and its subsequent fragmentation depend mainly on the sample and the method of ionisation (see below). Other factors to consider are the temperatures of the ion source and inlet system, and ion source and inlet system memory effects[2,11].

Temperature effects may be of two types. First there is the universal effect of decreasing the relative abundance of M^+ with increasing temperature[30,31]. The extent of this effect will depend on the number of atoms in the molecule, increasing with the number of atoms present. This is well known for organic compounds and can be quite dramatic; the M^+ ion of 2,2,3-trimethylpentane showed a five-fold decrease in abundance when the ion source temperature was increased from 175 to 225 °C[38]. Relatively little work has been reported with reference to organometallics, but the obvious implication is to record spectra at low ion source temperatures (i.e. 50–150 °C) when possible, if molecular weight information is required. The second temperature effect is that of thermal decomposition in the source or inlet systems. Organometallic compounds can be particularly susceptible and some care should be exercized in this respect[2,11]. Furthermore, thermally decomposed samples often act as catalysts in the decomposition of subsequent samples. This effect is most noticeable at heated metal surfaces. If a compound is suspected to be thermally unstable, spectra should be obtained with the lowest possible temperatures of the ion source and inlet system. The inlet system should preferably be glass and the path from the sample holder to the point of ionization should be as short as possible. Thus, a solid probe or direct all-glass inlet system is preferable to a gas chromatograph.

Spectra may be affected by traces of compounds retained in the ion source or inlet system. There are always residual amounts of water, N_2, O_2, and CO_2. Often these residues are used in counting the spectrum. However, if an organometallic compound is extremely moisture- or air-sensitive, unwanted reactions may occur. Hydrolysis is a common reaction and is evidenced by the appearance of a significant abundance of $[HX]^{+\cdot}$. In such cases the source and inlet system may be dried out with a volatile drying agent such as BCl_3 prior to inserting the sample. For the particularly susceptible triaryl derivatives of Ga and In, even the glass capillary sample holders had to be silanized by treatment with chlorotrimethylsilane before satisfactory spectra were obtained[39]. Another potential cause of confusion are halogenation reactions. Examples have been reported of halogen-containing ions appearing in spectra of halogen-free compounds, the halogen having been absorbed onto inlet and source surfaces from a previous sample[19]. Halogen exchange has also been observed, a notable case being the appearance of $[MeHgI]^{+\cdot}$ in the spectrum of MeHgCl[40]. Further details of the handling of moisture- and air-sensitive compounds are discussed in the review by Miller and Wilson[11].

2. *Fragmentation behaviour*

When analysing the fragmentation pattern of a compound it is necessary to adopt an empirical approach. This is because the processes involved in the ionization and dissociation of even simple molecules such as tetramethylsilane or ferrocene are too complex to treat rigorously by current versions of the quasi-equilibrium theory (QET) or other theories of mass spectrometry[41]. The lack of a suitable theoretical background has not inhibited the discussion of fragmentation mechanisms, nor has the almost total lack of knowledge of ion structures. On this point it is important to remember that most of the chemical formulae encountered in the mass spectrometry literature are not necessarily structurally significant. At best they may represent the most important form of an ion, and at worst they may be totally misleading. It is hoped that techniques currently being developed[41] such as photoelectron–photoion coincidence spectroscopy, ion photodissociation, collisional activation, and high-resolution emission spectroscopy will be applied increasingly to organometallics. At present, fragmentation paths are nearly always based on a general knowledge of molecular chemistry. This is not always a judicious approach, since in many cases ion fragmentation occurs with some degree of atomic rearrangement for which there are no known molecular analogies. However, in the absence of more detailed information one can only adopt a 'reasonable structure' approach.

In the spectra of organometallics the charge is generally assumed to be located mainly on the metal. This is consistent with the finding that in nearly all spectra most of the ion current is carried by metal-containing ions. For simple bond cleavages this can be rationalized in terms of Stevenson's rule[42], which states that the favoured process of dissociation is that producing the ion whose neutral species has the lowest ionization potential (I), i.e. mode 1a if $I(X_nM) < I(L)$. In general, metal-centred

$$X_nM^+ + L^\cdot \xleftarrow{(a)} X_nM{-}L \xrightarrow{(b)} L^+ + X_nM^\cdot \qquad (1)$$

$$X_nM^\cdot + {}^\cdot L \quad \text{(to } X_nM^+ + L^\cdot \text{ via } I(X_nM^\cdot)\text{; to } L^+ + X_nM^\cdot \text{ via } I(L^\cdot)\text{)}$$

species have lower I values than organic or non-metal centred species, and with simple cleavages the charge will be retained on the metal centred ion. Simple examples of this are shown in reactions 2 and 3.

$$Me^{\bullet} + Me_3Ge^+ \longleftarrow Me_4Ge^{+\bullet} \not\longrightarrow Me^+ + Me_3Ge^{\bullet} \quad (2)$$

$I(Me_3Ge^{\bullet}) = 7.1$ eV[43]; $I(Me^{\bullet}) = 9.82$ eV[44].

$$CO + Cr(CO)_5^+ \longleftarrow Cr(CO)_6^+ \not\longrightarrow CO^+ + Cr(CO)_5 \quad (3)$$

$I(Cr(CO)_5) = ?$, but $I(Cr(CO)_6) = 8.4$ eV[45] and $I(Cr) = 6.75$ eV[46]; $I(CO) = 14.0$ eV[45].

A corollary appears to exist for negative ions produced by dissociative electron capture processes; the favoured ion derives from the neutral with the higher electron affinity[47]. For more complex fragmentations, such as those involving rearrangements, rationalization with Stevenson's rule is less satisfactory because the neutral products are not always predictable. An example is the unexpected elimination of $(C_5H_5 + N)$, possibly as pyridine, from $[C_5H_5VNO]^+$ (reaction 4)[48]. Here the elimination of

$$C_5H_5V(NO)_2CO^+ \xrightarrow{-CO} \xrightarrow{-NO} C_5H_5VNO^+ \xrightarrow{-C_5H_5N} {}^+VO \quad (4)$$

molecules of CO then NO from $[C_5H_5V(NO)_2CO]^+$, typical of the fragmentation of metal carbonyls and nitrosyls, gives $[C_5H_5VNO]^+$, which must undergo considerable rearrangement to yield $[VO]^+$ in the next step. Although this process may be rationalized by Stevenson's rule it could not be predicted. Rearrangement reactions are common in mass spectrometry and their occurrence may be elucidated by the use of labelled atoms or groups. Using ^{13}CO in *cis*-$[M(CO)_4(^{13}CO)$piperidine$]$ (M=Cr, W), it was shown that initial loss of CO proceeded with complete scrambling between axial and equatorial sites[49].

A correlation has been suggested between the hard–soft acid–base character of the metallic element and the appearance of rearranged ions[50]. Thus in the spectra of $[(C_6X_5)_2Hg]$, with X = F or Cl, no Hg—X ions are produced but for X = Br, $[C_6Br_5HgBr]^{+\bullet}$, $[HgBr_2]^{+\bullet}$, and $[HgBr]^+$ are found. Similarly, from $[\{(C_6F_5)Me_2P\}AuX]$, P—Cl are favoured over P—I species.

Evidence for a fragmentation path is often sought from metastable peak analysis[41,51]. Metastable peaks are usually of low intensity, occurring at non-integral m/z values, and broadened compared with normal ion peaks. They are ordinarily observed from ions which are decomposing in the field-free regions between the ion source and magnet in single-focusing instruments, or between the electric and magnetic analysers in a double-focusing spectrometer. The centroid of a metastable peak occurs at an apparent m/z value $m^* = (m_2)^2/m_1$, where m_1 is decomposing to give m_2 and neutral species. The assignment of a metastable peak provides clear evidence that m_1 is a precursor of m_2. A number of methods have been devised to enhance the detection of metastable peaks[11,41,51]. Of these mass-analysed ion kinetic energy spectroscopy (MIKES) and linked scan techniques have proved extremely useful[11,51], since they provide direct evidence of fragmentation modes from selected ions. An extension of MIKES to the decomposition of metastable ions has led to the analysis of consecutive processes[52]. Most of the fragmentations discussed in this chapter are 'metastable-supported'. However, as noted in reaction 4, the form of the neutral species being lost is not always clear since only the mass is known. Furthermore, there is convincing evidence that in some fragmentations more than one neutral group is lost in each decomposition step[41].

II. ELECTRON IMPACT STUDIES

A. Introduction

Electron impact (EI) ionization has been used for most reported spectra (>99%). It will probably continue as the most important method on commercial machines because it is easy to operate and usually produces M^+ ions. Further, the electron beam ionizing energy can be continually altered between about 5 and 100 eV on most spectrometers, enabling I and appearance potential (A) measurements to be made. Typical ion source conditions are temperature *ca.* 100–150 °C, sample pressure in ionization area *ca.* 10^{-5} Torr, electron beam energy set at 70 eV. Under these conditions less than one sample molecule in several thousand is ionized. Both positive and negative ions are produced but the former are generally in at least a 1000-fold excess. Usually the positive ions are singly charged (reaction 5), but doubly and, rarely, triply charged ions are observed in low abundance from some compounds. Negative ion studies are feasible but not as common because of the lower intensities. In most compounds, impact by 70-eV electrons causes considerable fragmentation in the M^+ ions producing numerous fragment ions at lower m/z.

$$M + e^- \longrightarrow M^+ + 2e^- \qquad (5)$$

A discussion of EI spectra can be initiated according to whether the metallic element present is from a main group or is a transition metal.

B. Main Group Compounds[53]

Although compounds of almost all main group elements have been reported[2,12–16], most studies have concerned Group IV and V derivatives. Consequently, most of the examples given below contain elements from these groups. Several generalizations can be made about fragmentation behaviour:

1. For simple bond cleavages of type (1) the ion in higher abundance is the one whose radical has the lower I. In some cases the argument can be extended to make other deductions about the fragmentation processes. In the decomposition of the $[(C_6H_5)_2M]^+$ ion from Ph_3M (M = Sb or Bi) two alternative routes are observed[46] (reaction 6), depending on the relative I values of Sb (8.64 eV), Bi (7.29 eV), $C_{12}H_{10}$ as biphenyl (8.27 eV), and Ph (9.4 eV)[52]. These observations would seem to suggest that the hydrocarbon fragment is probably biphenyl. However, it must be remembered that there is no evidence about electronic states or structures of the ions and neutrals involved, so that the formation of $[Bi]^+$ by loss of two phenyl radicals could be an acceptable alternative mode.

$$C_{12}H_{10}^{+\bullet} \xleftarrow{-Sb} (C_6H_5)_2M^+ \xrightarrow[\text{or } 2Ph]{-C_{12}H_{10}} Bi^+ \qquad (6)$$

2. Generally, ions with even electron configurations are more favourable than odd-electron ions. This is directly analogous to organic mass spectrometry[28,30]. It follows that odd-electron ions will preferentially decompose by losing an odd-electron neutral to give an even-electron ion, and even-electron ions will decompose losing even-electron neutrals to other even-electron ions (reactions 7a and 7b).

$$Ph_4M^{+\bullet} \longrightarrow (C_6H_5)_3M^+ + Ph \qquad (7a)$$

$$(C_6H_5)_3M^+ \longrightarrow C_6H_5M^+ + C_{12}H_{10} \qquad (7b)$$

3. Reactions 7a and 2 are examples of very common fragmentation paths for $M^{+\bullet}$ ions from main group compounds R_nM and reflect the influence of the stable group oxidation state. The product $[R_{(n-1)}M]^+$ has the same overall electron configuration as the 'stable' derivative of the main group element to the left of M, e.g. $[Me_3Ge]^+$ is electronically equivalent to $[Me_3In]$, with M retaining the group oxidation state. It is common to observe an increase in stability of the (group−2) oxidation state ions as a group is descended. This is reflected in an increasing amount of the ion current being carried by ions such as $[R_{(n-3)}M]^+$, and is intimately related to point 4 below.
4. A decrease in bond energies E(M—C), E(M—H), etc., is generally observed as a group is descended. Thus there are fewer types of ions derived from $[Me_4Si]$ than $[Me_4Pb]$, since in tetramethylsilane Si—C, C—H, and Si—H bonds have similar strengths leading to Si—C and C—H cleavage and Si—H formation, but in tetramethyllead the Pb—C bond (and Pb—H) is much weaker than the C—H bond, favouring simple Pb—C cleavages without more complex fragmentation. From $[Me_4Si]$ twenty-five ions with an abundance >0.1% of the total ion current of metal-containing ions are formed[2]. Only fourteen are found from $[Me_4Pb]$.
5. Rearrangement ions are commonly found with compounds containing electronegative groups such as —OR, $—NR_2$, halogens, and fluorinated[54] or chlorinated hydrocarbons. A dramatic example is reaction 8 from $[(C_6F_5)_3B]$[11].

$$(C_6F_5)_3B^{+\bullet} \longrightarrow C_{18}F_{12}^{+\bullet} + BF_3 \tag{8}$$

1. Alkyl derivatives

Methyl compounds $[Me_nM]$ initially lose Me or, less commonly, H radicals from $M^{+\bullet}$ ions[2]. The latter mode is particularly noticeable with MeP^{III} and $MeAs^{III}$ compounds, where it probably yields $\overset{+}{M}{=}CH_2$-containing ions[55]. Further decomposition is mainly by loss of C_2H_4, C_2H_6, H_2, CH_4, and occasionally CH_2 species. As a group is descended simple M—Me cleavage becomes more important, yielding higher abundances of $[M—nMe]^+$ ions. Elimination of C_2H_4 or CH_2 groups suggests that rearrangements giving M—H bonds have occurred in several ions. Typical fragmentation patterns are summarized in equation 9 for Group IV compounds. Mode (a) is favoured for $[Me_4Si]$ and $[Me_4Ge]$ and mode (b) for $[Me_4Sn]$ and $[Me_4Pb]$.

$$Me_4\overset{+\bullet}{M} \xrightarrow{-Me} (CH_3)_3\overset{+}{M} \begin{cases} \xrightarrow{-C_2H_4} H_2\overset{+}{M}CH_3 \xrightarrow{-H_2} CH_3M^+ \\ \xrightarrow{-CH_2} H\overset{+}{M}(CH_3)_2 \xrightarrow{-C_2H_4} H_3M^+ \end{cases} \tag{9a}$$

$$(CH_3)_3\overset{+}{M} \xrightarrow{-Me} C_2H_6\overset{+\bullet}{M} \xrightarrow{-Me} CH_3M^{+\bullet} \xrightarrow{-Me} M^{+\bullet} \tag{9b}$$

Compounds of other alkyl groups produce spectra characterized by loss of R from odd-electron ions and of R—H from even-electron ions. Commonly, species such as C_2H_4, R + H, R_2, and H_2 are also eliminated. For phosphorus(III) compounds, removal of neutral $R—CH_2$ groups probably gives $^+P{=}CH_2$ species[55]. R_3M derivatives of P, As, and Sb produce abundant ions of the type $[R_2MH]^{+\bullet}$ and $[RMH_2]^{+\bullet}$. A relative abundance study of major ions from $[R_4Ge]$ (R = Me, Et, n-Pr, n-Bu, n-C_5H_{11}, and n-C_6H_{13})[56] showed that the abundance of $[R_3Ge]^+$ ions decreased as the alkyl group's chain length increased and the proportion of hydrocarbon ions increased

TABLE 1. Ion abundances (%) from $[R_4Ge]$ compounds

Ion	R					
	Me	Et	*n*-Pr	*n*-Bu	*n*-C_5H_{11}	*n*-C_6H_{13}
R_3Ge	71.1	26.6	20.4	14.7	9.1	11.6
Σ H/C ions / Σ Ge/H/C ions	3.5	3.0	7.4	18.0	28.3	30.5
$HGeR_2$	0.9	35.9	39.3	31.1	26.9	24.2

with increased chain length (Table 1). This is consistent with a decreasing D(Ge—R) bond strength and, more important, an increase in the number of decomposition modes available as the chain length increases, including C—C cleavage. The abundances of $[HGeR_2]^+$ indicates that Ge—H formation is dependent on chain length, being negligible for R = Me and most significant for R = *n*-Pr. It has been suggested that loss of C_2H_4 from Et_4M compounds specifically involves the β-H atom, but no convincing evidence has been put forward to justify this claim. Furthermore, a study of labelled Bu^n_3B with deuterium labels at α-,β-, and γ-carbon atoms and ^{13}C at the α-carbon atom has shown that the fragmentation of $[M - Bu]^+$ occurs with very extensive H/D and $^{12}C/^{13}C$ scrambling[57]. The mechanism was suggested to involve reversible reactions of protonated boracyclopropane- and cyclopropane-based ions which decomposed by loss of C_nH_{2n+2}, C_nH_{2n+1}, and C_nH_{2n} species, including CH_2. Some of the reactions are given in Scheme 1.

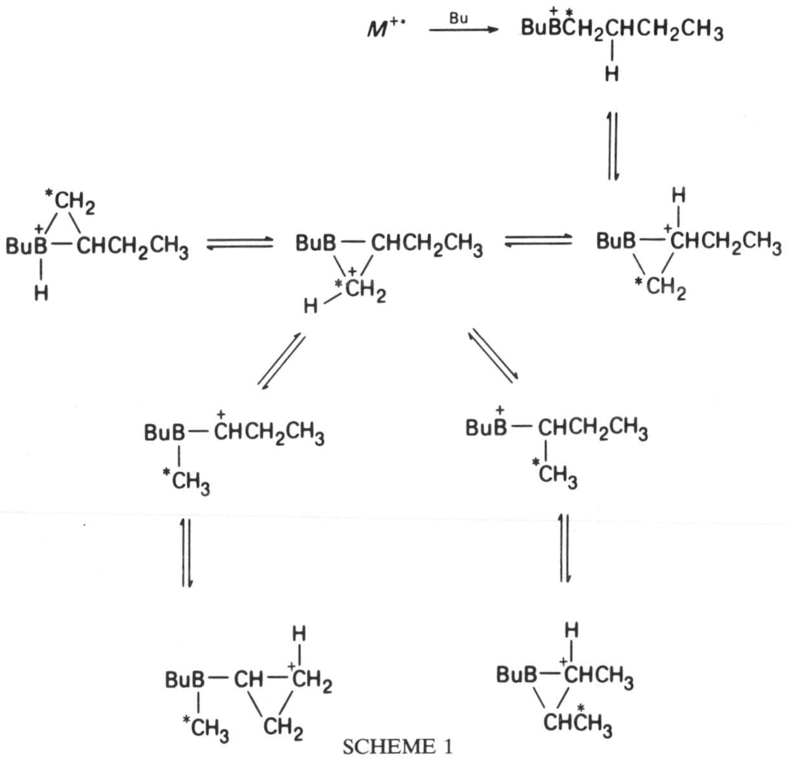

SCHEME 1

Relatively few simple compounds containing alkenyl or alkynyl groups have been reported. Fragmentation of Group IV tetravinyls[58] and tetraallyls[59] were similar in that the germanium and tin compounds tended to lose R, (R—H), and R_2 neutral species whereas the silicon derivatives exhibited more complex spectra with a wider variety of hydrocarbon neutral species lost. This is consistent with the bond strength arguments used previously to discuss other $[R_4M]$ species.

Compounds containing cycloalkyl or cycloalkenyl groups show unusual fragmentation behaviour in that olefin loss is often the most important path even from odd-

$$\text{spiro-}(C_4H_8)_2\dot{Sn}^{+} \xrightarrow{-C_2H_4} \xrightarrow{-C_2H_4} C_4H_8\dot{Sn}^{+} \quad (10)$$

electron ions (reaction 10)[60]. Analysis of the spectrum of **1** showed that the olefin lost was exclusively CH_2CD_2[61]. Similar reactions occur with cycloalkenyl derivatives. From **2** both $M^{+\cdot}$ and $[M - Me]^+$ lost $C_4H_4R_2$ (possibly as 2,3-di-R-buta-1,3-diene)[62].

H2 D2 R Ge D2 R H2

(1)

R Me Ge R Me

(2)

The spectra of $[(C_5H_5)_3M]$ (M = As, Sb) show major fragmentation modes which involve loss of C_5H_5 then CH_3, C_2H_4, and another C_5H_5 group[63]. Elimination of H_2 to give ions with fused ring structures, a common feature in the spectra of diphenyl compounds, was not important from $[(C_5H_5)_2M]^+$.

2. *Aryl derivatives*

Phenyl compounds $[Ph_nM]$ of main groups III–V fragmented predominantly with initial loss of a C_6H_5 radical. Thereafter removal of C_6H_6, $C_{12}H_{10}$, C_2H_2, and H_2 molecules provided the common decomposition modes. Loss of H was also observed in the spectra of B[64], Ga, In[39], Si, Ge, P, and As compounds[2], but only with Ph_3P was it an important pathway. Deuterium-labelled derivatives of Ph_3P showed evidence for ring fusion in several of the ions observed, including $[M - H]^+$ [65,66] (Scheme 2). Such ions are suspected to be a feature of the spectra of related As and Sb compounds but appear to be less important in compounds of main group IV[65]. Another noticeable feature in the spectra of main group V compounds of both types R_3M and R_5M (M = P, As, Sb)[67] and derivatives of Be[68], Hg[69], and B[64] was the loss of main group element containing neutrals. Both beryllium and boron compounds lost MH from $M^{+\cdot}$. Diarylmercury derivatives lose Hg from $M^{+\cdot}$ and $[ArHg]^+$ ions. The spectrum of Ph_2Be was generally unlike those discussed above. It did not show any major ions due to C_6H_5, C_6H_6 or $C_{12}H_{10}$ losses from $M^{+\cdot}$ but rather eliminated H, H_2, C_2H_2, and BeH groups.

Data on other series of $[Ar_nM]$ compounds are sparse. Tolyl derivatives of Ga, In[39], Ge[70], and main group V elements[71] have been reported. The *m*- and *p*-isomers gave almost identical spectra. Typically, C_7H_7, C_7H_8, and $C_{14}H_{14}$ groups are lost. Also, Me, CH_4, H, H_2, and central element containing species are lost from the derivatives of the lighter elements. The $[(o\text{-tolyl})_3P]$ spectrum differed considerably from those of the other main group V compounds in that $[M - Me]^+$ was the base peak, presumably

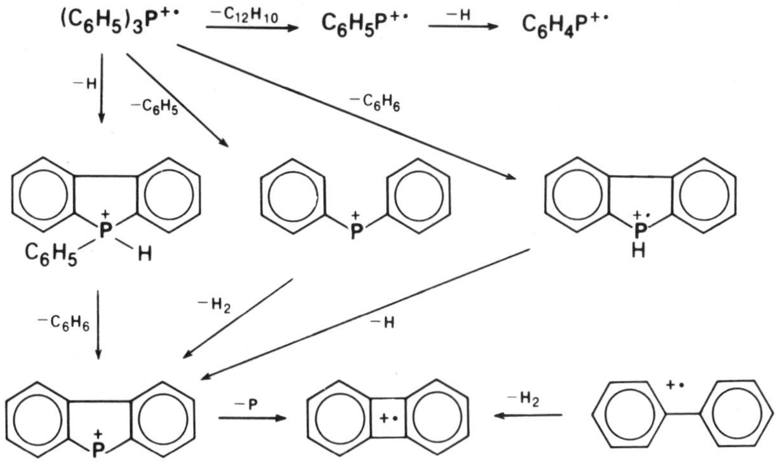

$$(C_6H_5)_3P^{+\bullet} \xrightarrow{-C_{12}H_{10}} C_6H_5P^{+\bullet} \xrightarrow{-H} C_6H_4P^{+\bullet}$$

SCHEME 2

with a fused ring structure analogous to $[M - H]^+$ from Ph_3P. The related ion from $[(o\text{-tolyl})_4Ge]$ was not observed but was found in *m*- and *p*-derivatives.

From the preceding discussion it can be seen as a general rule that aryl groups more complex than Ph can be expected to provide more fragmentation routes and consequently more complicated spectra.

3. *Compounds with alkyl and aryl groups*

Compounds with a variety of alkyl or aryl hydrocarbon groups attached generally produce spectra in which the fragmentation of the groups occurs by the same routes as in the symmetrical compounds $[R_nM]$ or $[Ar_nM]$. Alkyl—M bonds cleave more readily than aryl—M bonds. This, together with the possibility of greater delocalization of charge in aryl—M containing ions and the lower I valves of Ar-substituted compared with R-substituted species, tends to produce Ar-containing ions as the most abundant species[2,4]. Although this is the general trend it is sometimes complicated by interactions between hydrocarbon groups. Thus $[Me_2PhM]$ (M = P, As, Sb) lost CH_2 as well as Me from $M^{+\bullet}$, and C_7H_7 was eliminated from $[C_6H_5As(CH_3)_2]^{+\bullet}$[72]. In a releated reaction, the $[C_7H_7Ge(CH_3)_2]^+$ ion from dibenzyldimethylgermane lost C_8H_{10} to give $[CH_3Ge]^+$[70].

4. *Halocarbon derivatives*

Halocarbon derivatives are notorious for producing rearranged ions and neutrals with M—X bonds. This is especially so for fluorocarbon or chlorocarbon compounds of B, Si, Ge, and P[2,11,73] (equation 11).

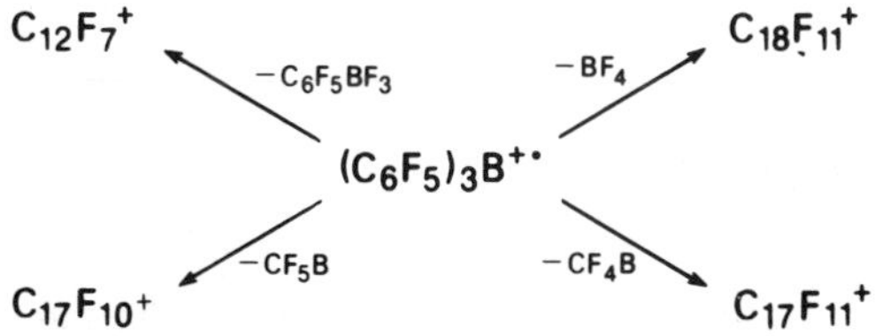

$$C_{12}F_7^+ \xleftarrow{-C_6F_5BF_3} (C_6F_5)_3B^{+\bullet} \xrightarrow{-BF_4} C_{18}F_{11}^+$$
$$C_{17}F_{10}^+ \xleftarrow{-CF_5B} (C_6F_5)_3B^{+\bullet} \xrightarrow{-CF_4B} C_{17}F_{11}^+ \qquad (11)$$

Other fragmentations of C_6F_5 derivatives involve the loss of F,C_6F_5, and $C_{12}F_{10}$. Trifluoromethyl compounds commonly lose F and CF_n(n = 2–4) groups. Chlorocarbon derivatives generally produce fewer M—X containing species. Two examples are given in equations 12[74] and 13[75].

$$Me_3\overset{+\bullet}{Si}CCl_3 \xrightarrow{-Cl} (CH_3)_3\overset{+}{Si}CCl_2 \xrightarrow{-C_3H_6} CH_3\overset{+}{Si}Cl_2 \quad (12)$$

$$(p\text{-}ClC_6H_4C{\equiv}C)_2Hg^{+\bullet} \xrightarrow{-HgCl_2} C_{16}H_8^{+\bullet} \quad (13)$$

5. Compounds with M—H bonds

Compounds containing M—hydride bonds often show significant abundances of ions produced by loss of (H + R) species (R = H, alkyl, or aryl), as well as loss of H and fragmentations associated with R groups. Some examples are given from main group IV compounds in equations 14[70] and 15[76].

$$(PhCH_2)_2GeD^+ \longrightarrow C_7H_7Ge^+ + C_7H_7D \quad (14)$$

$$PhGeD^{+\bullet} \longrightarrow Ge^+ + C_6H_5D \quad (15)$$

6. Compounds with M—N bonds

One of the simplest N-containing organometallics studied is $[(Me_3Si)_2NH]$[77]. Initial loss of Me was followed by elimination of CH_4, C_2H_4, and $C_2H_6SiH_2$ molecules as well as the N-containing species NH_3, HCN, and SiNH. The fragmentation of $[(Me_3Si)_2NMe]$[77] and bis(trimethylsilyl)- and bis(trimethylgermyl)carbodiimide[78] also proceeded with loss of M—N containing species. The spectra of $[Ph(CH_2)_nNHSiMe_3]$ (n = 1,2) and related compounds[79] showed an abundant $[CH_2{=}NHSiMe_3]^+$ ion which lost $CH_2{=}NH$ to give $[Me_3Si]^+$. Derivatives with M—NR_2 bonds generally show loss of NR_2 and (NR—H) neutrals (equation 16)[80].

$$EtP(NMe_2)_2^{+\bullet} \xrightarrow{-Et} {}^+P[NC_2H_6]_2 \xrightarrow{-CH_2=NMe} H\overset{+}{P}(NC_2H_6) \quad (16)$$

Heterocyclic derivatives which have been reported include borazines and related compounds, silazanes, and phosphonitrilic derivatives[2]. The ring systems in these compounds are resilient to fragmentation and the major ions are usually produced by simple cleavage of peripheral groups.

7. Compounds with M—O or M—S bonds

Most studies have concerned oxygen containing groups. Elimination of H_2O is a characteristic mode in the later stages of fragmentation of R_nM—OH compounds such as Me_3SiOH and $R_2P(O)OH$[2]. For compounds of the $R_nP(X)$ type (R > Me) elimination of (R—H) and rearrangement to P(XH)-containing ions (X = O or S) is an important decomposition mode[2].

A formidable number of M—OR containing compounds have been studied, especially with $OSiMe_3$ or P—OR groups. Commonly occurring fragmentations result from the cleavage of M—OR, MO—R, and $MOCH_2$—R^1 bonds, and olefin elimination from alkyl R groups (if R > Me) giving H-rearranged ions with M—OH bonds. If methoxy groups are present elimination of CH_2O provides the characteristic mode of decomposition. Aryloxy derivatives generally show fewer rearranged ions. Some typical initial fragmentations from alkyl derivatives $R^1(OR^2)_2P(O)$ are given in equation

17. All fragment ions undergo further loss of olefin groups giving other (OH)—P containing ions.

$$
\begin{array}{l}
R^1\overset{+\bullet}{P}(O)(OR^2)_2 \xrightarrow{-\text{olefin}} \overset{+\bullet}{P}(OH)(OR^2)_2 \\
R^1\overset{+\bullet}{P}(O)(OR^2)_2 \xrightarrow{-R^1} \overset{+}{P}(O)(OR^2)_2 \\
R^1\overset{+\bullet}{P}(O)(OR^2)_2 \xrightarrow{-\text{olefin}} R^1\overset{+\bullet}{P}(O)(OR^2)(OH) \\
R^1\overset{+\bullet}{P}(O)(OR^2)_2 \xrightarrow{-Me} R^1\overset{+\bullet}{P}(O)(OR^2)(OX) \\
R^1\overset{+\bullet}{P}(O)(OR^2)_2 \xrightarrow{-OR^2} R^1\overset{+\bullet}{P}(O)(OR^2) \\
R^1\overset{+\bullet}{P}(O)(OR^2)_2 \xrightarrow{-(R^2-H_2)} R^1\overset{+\bullet}{P}(OR^2)(OH)_2
\end{array}
\qquad (17)
$$

The spectra of the $[(Me_3Si)_2X]$ derivatives (X = O, S, Se, and Te) are reported to show a characteristic fragmentation pattern (equation 18) with the loss of Me_2SiX species[81]. The telluride was unusual in that the loss of Te from $M^{+\bullet}$ was observed.

$$M^{+\bullet} \xrightarrow{-Me} [M - Me]^+ \xrightarrow{-Me_2SiX} Me_3Si^+ \qquad (18)$$

The spectra of a large number of Me_3SiO derivatives of alcohols[79,82–84] and acids[83,85] have been discussed in detail with the aid of deuterium labelling. The more interesting fragmentations involved rearrangements such as those in equations 19–22. Whilst

$$p\text{-}XC_6H_4CH_2OSiMe_3\text{—}SiMe_3^{+\bullet} \xrightarrow{-Me} \xrightarrow{-CH_2O} XC_6H_4\overset{+}{Si}Me_2 \qquad (19)$$

$$p\text{-}XC_6H_4O_2CSiMe_3^{+\bullet} \xrightarrow{-Me} \xrightarrow{-CO_2} XC_6H_4\overset{+}{Si}Me_2 \qquad (20)$$

$$RO(CH_2)_nCO_2SiMe_3^{+\bullet} \xrightarrow{-Me} \xrightarrow{-(CH_2)_nCO_2} RO\overset{+}{Si}Me_2 \qquad (21)$$

$$RO(CH_2)_nOSiMe_3^{+\bullet} \xrightarrow{-Me} \xrightarrow{-O(CH_2)_n} RO\overset{+}{Si}Me_2 \qquad (22)$$

$Ph(CH_2)_nCO_2SiMe_3$ showed rearranged $[C_6H_5SiMe_2]^+$ ions[85], acyloxytriphenyl compounds $[(Ph_3M)CO_2R]$ (M = C, Si, Ge; R = Me, Ph) exhibited the rearranged ions $[(C_6H_5)_2MR]^+$ in reasonable abundance only with M = Si or Ge (equation 23)[86]. Such ions generally become progressively less important as a group is descended, reflecting the decreasing M—O bond strengths. Thus, $[Ph_3SnCO_2Me]$ fragments mainly by simple cleavages of Sn—Ph and Sn—OC(O)Me bonds[87].

$$Ph_3MCO_2R^{+\bullet} \xrightarrow{-C_6H_5} (C_6H_5)_2\overset{+}{M}\text{—}O\text{—}C(=O)R \xrightarrow{-CO_2} (C_6H_5)_2MR^+ \qquad (23)$$

Ketonic derivatives $[Me_nPh_{3-n}SiC(O)Ar]$ exhibited several ions formed by migration of the Ar group on to Si[88]. Alkyl ketone compounds $[Me_3Si(CH_2)_3C(O)R]$ preferred to fragment by loss of C_2H_4 then Me[89].

A number of interesting fragmentation paths have been reported from the phosphorus compounds $[R_3P{=}X]$ (X = O or S) and $[(RX)_nR_{3-n}P{=}X]^1$, etc. Eliminations of H, C_6H_5, and C_6H_5X from $M^{+\bullet}$ of $[Ph_3PX]$ (X = O or S) were observed[65].

Data from deuterium-labelled compounds showed the probable formation of fused ring structured ions similar to those from Ph_3P[65]. Similar ring fusions were suggested in the spectra of diarylphosphinic acids and esters (equation 24)[90].

$$Ph_2\overset{+\bullet}{P}(O)Me \xrightarrow{-H} [\text{dibenzophospholium, } P^+(OMe)(OH)] \xrightarrow{-MeOH} [\text{dibenzophospholium } P^+{=}O] \xrightarrow{-PO} [\text{biphenylene}]^{+\bullet} \quad (24)$$

Rearrangements involving group migration from O to S (equation 25) were observed from $[(PhO)_2(RO)P{=}S]$ (R = Me, Et), and provided several unexpected ions such as $[M - SH]^+$, $[M - SMe]^+$, and $[M - SPh]^+$ as well as those expected from loss of OPh and R groups[91].

$$PhO{-}\overset{+\bullet}{P}(OMe)(=O){-}SPh \longleftrightarrow PhO{-}\overset{+}{P}(OMe)(OPh){=}S \longleftrightarrow PhO{-}\overset{+}{P}(=O)(OPh){-}SMe \quad (25)$$

Heterocyclic derivatives of many kinds have been reported, particularly of boron[23], silicon, phosphorus, and arsenic[2,41]. Although they are too diverse to allow a detailed discussion, it may be noted that elimination of MO-containing neutrals is a common mode of fragmentation, (equations 26[92], 27[93], and 28[94]).

$$[\text{cyclic } C_2H_4O_2BPh]^{+\bullet} \xrightarrow{-CH_2BO_2} {}^{+}C_7H_7 \xleftarrow{-BO} [M-OCH_2]^+ \quad (26)$$

$$[(Me_2SiO)_3]^{+\bullet} \xrightarrow{-Me} \begin{cases} \xrightarrow{-C_2H_6SiO} C_3H_9\overset{+}{Si}_2O_2 \\ \xrightarrow{-C_2H_6} C_3H_9\overset{+}{Si}_3O_3 \xrightarrow{-SiO} C_3H_9\overset{+}{Si}_2O_2 \end{cases} \quad (27)$$

$$[\text{phenoxarsine, As(=O)OH}]^{+\bullet} \xrightarrow{-AsO_2H} [\text{dibenzofuran}]^{+\bullet} \quad (28)$$

8. *Compounds with M—halogen bonds*

Compounds containing M—halogen bonds exhibit ions due to loss of halogen or elimination of a neutral RX group (R = H, alkyl, aryl). Molecular eliminations are favoured for X = F and Cl and to a lesser extent Br or I, whereas simple bond cleavages occur more commonly from M—Br and M—I compounds. Examples from main group IV are given in equations 29 and 30. Another feature in the spectra of

$$Me_3\overset{+\cdot}{Ge}Cl \xrightarrow{-Me} (CH_3)_2\overset{+}{Ge}Cl \xrightarrow{-HCl} C_2H_5Ge^+ \qquad (CH_3)_2\overset{+}{Ge}Cl \xrightarrow{-MeCl} CH_3Ge^+ \tag{29}$$

$$Ph_3\overset{+\cdot}{Sn}X \xrightarrow{-Ph} (C_6H_5)_2\overset{+}{Sn}X \xrightarrow{-C_6H_5X} C_6H_5Sn^+ \tag{30}$$

X = F, Cl, Br, or I

many M—halogen compounds is the elimination of MX-containing species (equations 31[95], 32[60], and 33[94]).

$$(MeAlCl_2)_2^{+\cdot} \xrightarrow{AlCl_3} CH_3\overset{+}{Al}Cl \tag{31}$$

$$^{+}C_{18}H_{15} \xleftarrow{-SnCl} Ph_3\overset{+\cdot}{M}Cl \xrightarrow[M\,=\,Ge,\ Sn]{-C_6H_5MCl} {}^{\cdot+}C_{12}H_{10} \tag{32}$$

$$\text{(phenoxarsine radical cation, As–Cl)} \xrightarrow{-AsCl} \text{(dibenzofuran)}^{+\cdot} \tag{33}$$

9. *Compounds with M—M bonds*

Compounds containing bonds between main group elements supporting different R groups show characteristic rearrangement ions, involving redistribution of R groups (equations 34 and 35[4,96]).

$$Ph_3SiSiEt_3^{+\cdot} \xrightarrow{-C_6H_5SiEt_2} (C_6H_5)_2\overset{+}{Si}Et \tag{34}$$

$$Ph_3SnGeMe_3^{+\cdot} \longrightarrow Ph_n\overset{+}{Ge}Me_{3-n} + Ph_n\overset{+}{Sn}Me_{3-n} \quad (n = 0\text{–}3) \tag{35}$$

However, with main group IV compounds most of the ion current is usually carried by ions derived from an initial simple M—M bond cleavage. The spectrum of diphenyl diselenide is an interesting case, giving abundant ions due to simple Se—Se cleavages, $[C_6H_5Se]^+$, Ph rearrangement, $[(C_6H_5)_2Se]^{+\cdot}$, and Se elimination from $M^{+\cdot}$ [97].

C. Transition Metal Compounds

Here the distinction between fragmentation of even- or odd-electron ions becomes blurred by the fact that transition metals have accessible *d*-orbitals which are capable of accepting electrons. Thus, although most transition metal compounds and ions are apparently odd-electron species they may show a dual character (equation 36).

$$[L_nM^NX]^{+\bullet} \longleftrightarrow [L_nM^{N-1}X]^+ \qquad (36)$$

The concept of variable valency has often been invoked to explain fragmentation effects in the spectra of transition metal inorganic and coordination compounds, but this approach is generally of limited use in organometallics since the bonding is usually more complex than the basically 'simple' σ systems found in most inorganic and coordination compounds. Furthermore, metal—ligand rearrangements are more common and more extensive in organometallics. Again, information about structural and electronic states in gas-phase ions is extremely limited.

As with main group compounds, most of the ion current is carried by metal-containing ions and it is usually assumed that the charge is located mainly at the metal. Subsequently the weakened M^+ decomposes, leaving the charge still predominantly at the metal centre. The reactions which occur in the spectra of transition metal compounds can involve simple M—L bond cleavages, ML^1—X bond cleavages within the ligands, or reactions giving rearranged ions containing new M—X bonds.

1. *Metal—Carbon σ-bonded compounds*

Simple methyl derivatives have been reported for a number of metals. Hexamethyltungsten produced no appreciable M^+, only abundant $[WMe_n]^+$ ions (n = 0–5)[98]. Hexamethylrhenium gave M^+ and the ions $[ReMe_n]^+$ (n = 0–5)[99]. The difference in behaviour can be simply understood in terms of the formal oxidation states of W^{VI} (d^0) and Re^{VI} (d^1). The fragmentation by Me loss is similar to that found with the heavier main group element methyls (equation 9b) and presumably reflects the weak M—Me bonding. However, the choice of the R group also has an effect since from $[Cr(t\text{-Bu})_4]$ $[M - Bu]^+$ appeared at the highest m/z while [Cr(neopentyl)$_4$] produced M^+ [100]. Neither μ-methylene complex $[(\mu\text{-}CH_2)\{M(CO)_n(\eta^5\text{-}C_5H_5)\}_2]$ (M = Mn or Rh, n = 2 or 1) showed fragmentation by the direct loss of the CH_2 bridge[101]. Rather, elimination of four CO groups occurred initially from the manganese compound, followed by loss of Mn, MnCH, and $MnCH_2$ groups. The rhodium derivative fragmented differently, losing one CO molecule then CH_2O, CH_2CO, $RhCH_2CO$, or the second CO group. Rearranged ions $[MnCH_2(C_5H_5)_2]^+$ and $[Rh(C_5H_5)_2]^+$ constituted the base peaks; both can be formulated as having 18 electron rule structures. The related rhenium vinylidene complex $[(\mu\text{-}C{=}CHPh)\{Re(CO)_2(\eta^5\text{-}C_5H_5)\}_2]$ fragments by losing four CO molecules consecutively then H_2 and C_2H_2 molecules[102]. These latter decomposition paths are also typical of acetylene and benzene derivatives, and are observed from the $[M - 2CO]^+$ ion of $[Re(CO)_2(C = CHPh)(\eta^5\text{-}C_5H_5)]$ but the manganese analogue prefers to lose the C_2HPh ligand.

Relatively few aryl—M compounds have been reported.

Carbene[103] and carbyne[104] compounds have received some detailed attention. Carbenes $[Cr(CO)_5(CXY)]$ (X = Me or Ph; Y = OR, NHR, NR_2, or SR) show major decomposition paths from the base peak ions $[Cr\ CXY]^+$ dependent on X and Y (equations 37–40).

$$(Y = OMe, NMe_2)Cr^+ \xleftarrow{-Y} {}^+CrY \xleftarrow{-C_2H_3} {}^+CrC(Me)Y \xrightarrow{-Me} \qquad (37)$$

$$^+CrY \xrightarrow{-Ph} (Y = SPh)\overset{+}{Cr}S \qquad\qquad ^+CrY \xrightarrow{-H} {}^+CrNAr\ (Y = NHAr)$$

$$Cr^+ \xleftarrow{-Me} {}^+CrMe \xleftarrow{-(CY-Me)} \xleftarrow{-Me} {}^+CrC(Me)Y \xrightarrow{-CMeY} Cr^+ \quad (38)$$

$$(Y = OMe, NMe_2)$$

$${}^+CrCNAr \xleftarrow{-H} \xleftarrow{-Me} {}^+CrC(Me)NHAr \xrightarrow{-MeCN} \xrightarrow{-H} {}^+CrAr \quad (39)$$

$${}^+CrC(Me)NHAr \xrightarrow{-Ar} \xrightarrow{-Me} \xrightarrow{-HCN} Cr^+$$

$${}^+CrC(Me)SPh \xrightarrow{H_2S} {}^+CrC_2HPh \xrightarrow{H} {}^+CrC_2Ph \quad (40)$$

For X > Me, olefin elimination commonly occurred; thus the ion $[CrCMe(OH)]^+$ was produced from $[CrC(Me)OEt]^+$.

Carbynes $[MX(CO)_4CR]$ (M = Cr or W; X = halogen; R = Me or C_6H_4Y) tended to lose the CR group from $[M - 4CO]^+$ when M = Cr but with W, degradation of the ligand often occurred (equation 41).

$$\xleftarrow{-C_2H_2} \xleftarrow{-C_2H_2} \xleftarrow{-H} Br\overset{+}{W}{\equiv}C{-}C_6H_4{-}Me \quad \begin{matrix} \xrightarrow{-C_3H_4} \xrightarrow{-C_2H_2} \\ \xrightarrow{-C_2H_2} \xrightarrow{-C_2H_2} \end{matrix} \quad (41)$$

As with the main group compounds, fluorocarbon derivatives are typified by M—F containing ions and elimination of neutral MF_n species. An example is $[Fe(C_3F_7)(I)(CO)_4]$, which first loses the carbonyl molecules then decomposes by loss of I, F, and C_3F_7 groups[105]. From $[(C_3F_7)Fe]^+$ elimination of FeF_2 gave $[C_3F_5]^+$. The ready formation of M—F containing ions is illustrated by the appearance of abundant $[MnF]^+$ ions from $[Mn(CH_2C_6H_4F\text{-}p)(CO)_5]$ and $[Mn(CH_2C_6F_5)(CO)_5]$[106]. Loss of HF may also be important if a hydrocarbon group is present[73]. A number of similar rearrangements have been reported from chlorocarbon compounds of transition metals.

2. *Olefin complexes*

Relatively few complexes have been reported with simple olefins. Cyclopentadienyldiethylenerhodium shows predominant elimination of successive C_2H_4 molecules (equation 42)[107].

$$C_5H_5\overset{+}{Rh}(C_2H_4)_2 \xrightarrow{-C_2H_4} \xrightarrow{-C_2H_4} C_5H_5Rh^+ \xrightarrow{-C_5H_5} Rh^+ \quad (42)$$

In contrast, a number of studies of more complicated polyolefins and cycloolefins have indicated very complex modes of fragmentation often involving rearrangement with other hydrocarbon ligands when present. The fragmentation of $[Fe(diene)(PF_3)_3]$ complexes depended on whether the diene was cyclic or not[108]. Typical routes are shown in equations 43 and 44 from $[M - 3PF_3]^+$.

The loss of a hydrogen molecule has been suggested to convert the olefin from a $2n$ to a $2n + 2$ electron donor (n = number of olefin to metal 2-electron donor bonds).

$$\text{(2-methylbutadiene)}Fe^{+} \xrightarrow{-C_2H_2} {}^{+}FeC_3H_6; \quad \rightarrow {}^{+}FeC_2H_2; \quad \xrightarrow{-\text{olefin}} {}^{+}FeC_3H_4 \qquad (43)$$

$$\text{(cycloocta-1,5-diene)}Fe^{+} \xrightarrow{-H_2} \xrightarrow{-H_2} \xrightarrow{-C_2H_4} {}^{+}FeC_6H_4 \qquad (44)$$

Deuterium labelling in [(cyclohexadiene)$Fe(CO)_3$] showed hydrogen loss to be highly, but not totally, stereospecific[109]. From the *endo* D_2 olefin, D_2 was lost in preference to H_2 in a ratio of 9:1.

Sometimes very complex interactions occur between different hydrocarbon groups attached to a metal. One can consider as an example of this the cyclopentadienyl cycloocta-1,5-diene complexes of cobalt and rhodium (Scheme 3)[107,110]. Paths for which metastable peaks were found for Co or for Rh are marked. Initial losses of H_2, C_2H_6, C_4H_6, C_4H_6, or C_8H_{12} are followed by more complex eliminations of C_5H_6 and C_6H_6 molecules from rearranged ions.

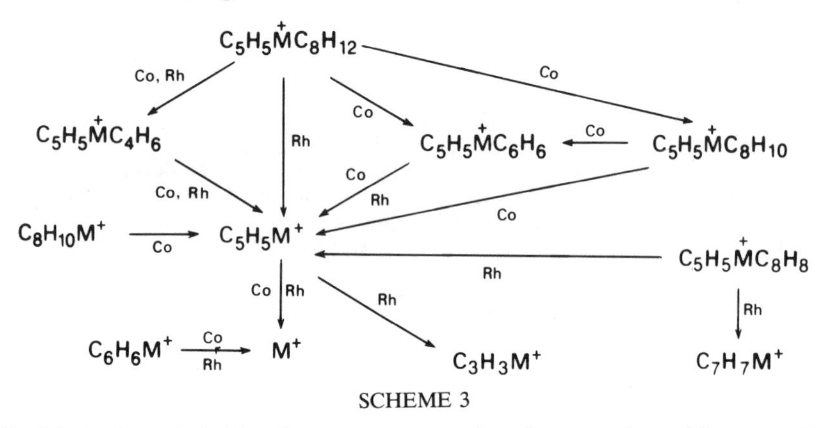

SCHEME 3

Cyclobutadiene derivatives have been reported to show a variety of decomposition modes. Whereas $[Co(\eta^4\text{-}C_4H_4)(\eta^5\text{-}C_5H_5)]$ loses one then another C_2H_2 molecule[111], $[Fe(CO)_3(\eta^4\text{-}C_4H_4)]$ loses C_4H_4 only[112]. Successive losses of two C_2Ph_2 molecules were reported from some cobalt and iron compounds of C_4Ph_4[113]. However, $[Mo(CO)_2(\eta^4\text{-}C_4Ph_4)_2]$ loses both carbonyls and three C_2Ph_2 molecules before $[MoC_2Ph_2]^+$ undergoes consecutive CPh loss[114].

Some related [(cyclopentadienone)$M(CO)_3$] complexes of iron and ruthenium show elimination of four CO molecules to give $[MC_4R_4]^+$ ions, probably with similar properties to the ions derived from cyclobutadiene complexes[115,116]. Decarbonylation of the ligand also occurs in a wide variety of substituted dieneirontricarbonyl complexes (equation 45)[117].

$$R^1\text{–CH=CH–CH=CH–C(=O)}\dot{R}^2\ \cdot Fe(CO)_3 \xrightarrow{-(1-3)\ CO} \xrightarrow{-CO} R^1\text{–CH=CH–CH=CH–}R^2\cdot Fe^{+} \qquad (45)$$

Substantially different structures are suspected to arise for the $[FeOC_6H_6]^+$ ion formed from **3** or **4**[118]. From **3** the ion decomposes by loss of CO or HCO, but from **4** C_6H_5 or FeOH is lost.

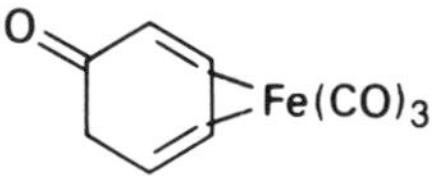

(**3**) 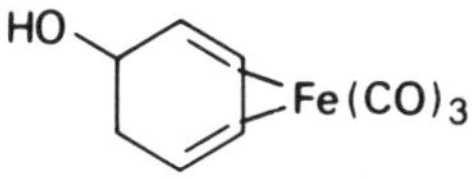 (**4**)

3. *Acetylene complexes*

Complexes of acetylenes $[Co_2(CO)_6(C_2R_2)]$ (R = H, CF_3, CH_2Cl, Ph, CO_2Me) showed a marked reluctance to fragment by cleavage of the acetylene ligand[119]. Of the more interesting decompositions, those shown in equations 46 and 47 were noteworthy.

$$Co_2(CO)_6C_2Ph_2^+ \xrightarrow{-(1\text{–}6)CO} Co_2C_2Ph_2^+ \xrightarrow{-Co} CoC_2Ph_2^+ \xrightarrow{-Co} C_2Ph_2^+ \quad (46)$$

$$Co_2(CO)_6C_2(CF_3)_2^+ \xrightarrow{-(1\text{–}6)CO} Co_2C_4F_6^+ \xrightarrow{-CoF_2} CoC_4F_4^+ \quad (47)$$

4. *Allyl complexes*

Allyl derivatives $[RhCl(allyl)_2]_2$ typically lose the allyl group, H_2 and (allyl X) (X = halogen)[120]. Propene is eliminated from the base peak $[Rh(C_3H_5)_2]^+$ and this requires a transfer of H from one group to the other. Methallyl derivatives of Rh and Fe[121] have been postulated to undergo rearrangement to butadiene type ion structures. Complexes of rhodium (**5**) with formally 16-electron structures were reported to lose up to three H_2 molecules and (R + H) groups[122]. Related 18-electron species with a five electron cycloolefin donor only lost up to two H_2 molecules and (R + 3H) groups.

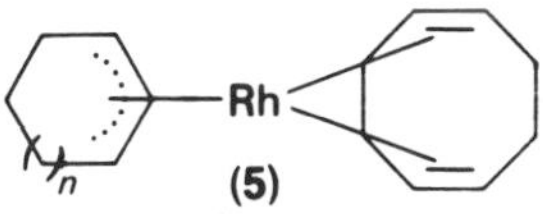

(**5**)
(n = 1 − 3)

5. *Cyclopentadienyl complexes*

Cyclopentadienyl derivatives of virtually all transition metals have been studied[20]. As a group, ferrocenes have been examined most extensively and a number of interestingly rearrangements have been reported involving substituents on the C_5 ring. Fragmentation of a C_5H_5 group attached to a metal is predominantly by loss of C_5H_5 radicals or C_2H_2 molecules. Less importantly, H, C_3H_3, and CH_3 radicals and C_3H_4 molecules are lost. Occasionally a metal atom is expelled to produce a hydrocarbon ion (equation 48)[123].

$$C_5H_5FeC_5H_4^+ \xrightarrow{-Fe} C_{10}H_9^+ \quad (48)$$

The incorporation of a side chain group, i.e. $\{C_5H_4(C_nH_{2n+1})\}$ ($n > 2$) produces $[C_7H_7Fe]^+$ and $[C_7H_7]^+$ions[20,124,125]. The transfer of groups from side-chain on to Fe has also been a commonly discussed decomposition mode[20,123]. Thus compounds **6** produce $[C_5H_5FeOH]^+$ions[126], whilst for 2-ferrocenylethyl derivatives (**7**)[127] the

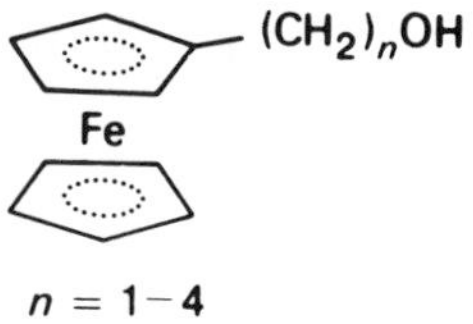

$(CH_2)_nOH$ Fe $n = 1-4$ (**6**)

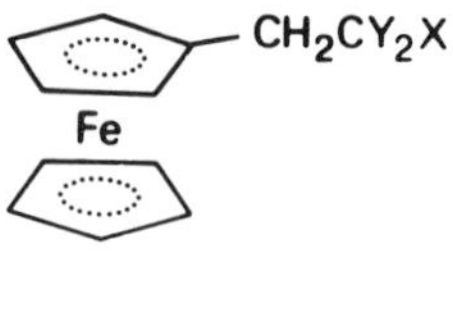

CH_2CY_2X Fe (**7**)

common elimination of CY_2X gave $[C_5H_5FeC_6H_6]^+$, an 18-electron species which loses C_6H_6. The importance of rearranged ions is confirmed by a study of deuterium labelled methylferrocenes where $[FeC_nH_n]^+$ (n = 5, 6 or 11) and $[FeC_5H_6]^+$ were produced with complete H/D randomization[128].

Steric effects in various *endo* and *exo* isomers of **8** have been discussed[129]. The stereoselective elimination of C_6H_5R occurred with transfer to Fe of the *endo*-H- from *exo*-R-substituted derivatives. The loss of C_5H_6 from M^+ was also stereoselective from this isomer.

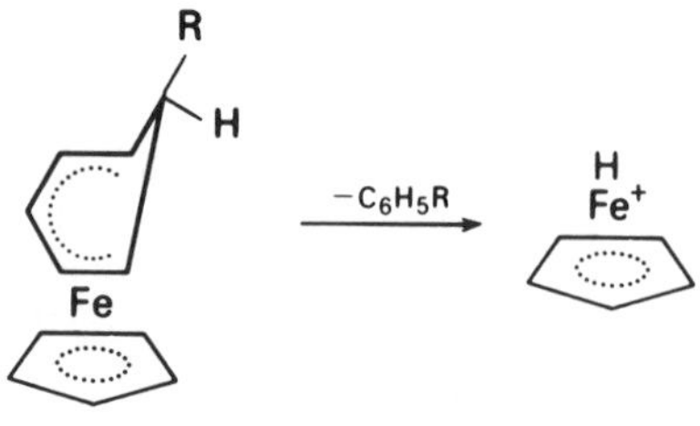

$$\text{(8)} \xrightarrow{-C_6H_5R} \text{H–Fe}^+(C_5H_5) \tag{49}$$

6. *Arene complexes*

Complexes with arene ligands are generally characterized by ions due to loss of the arene ligand, those ions formed by partial decomposition of the ligand being much less abundant. Doubly charged ions were of significance from $[Cr(C_6H_6)_2]$ and a most unusual decomposition was observed (equation 50)[130,131].

$$(C_6H_6)_2Cr^{2+} \longrightarrow C_6H_6Cr^+ + C_6H_6^{+\cdot} \tag{50}$$

A series of substituted benzene—$Cr(CO)_3$ complexes showed rearranged $[CrX]^+$ ions (X = F, I, OR, NH_2) and $[C_6H_5Cr]^+$ [132]. A study of styrene and $[Cr(CO)_3(\eta^6\text{-}C_6H_5CH{:}CH_2)]$ showed the randomization of H in the $[CrC_8H_8]^+$ ion was different to that occurring in $[C_8H_8]^{+\cdot}$ from styrene. This is presumably due to the influence of the metal[133]. Isomeric substituted benzyl— and cycloheptatriene—$Cr(CO)_3$ complexes decomposed differently and therefore did not reflect the closely similar fragmentations found from the benzyl and cycloheptatriene ligands[134].

7. *Complexes with η^7-C_7 or η^8-C_8 hydrocarbon ligands*

For complexes containing C_7, C_8, etc., ligands the most typical modes of decomposition are by loss of a complete ligand, C_2H_2 or H_2[2]. Rearrangement ions formed by interactions with other ligands are often produced (equation 51)[130].

$$C_{10}H_{10}V^+ \xleftarrow{-C_2H_2} C_5H_5VC_7H_7^+ \xrightarrow{-C_6H_6} C_6H_6V^+ \xrightarrow{-C_6H_6} V^+ \quad (51)$$

8. *Organic ligands containing atoms from main groups III–VI*

Apart from hydrocarbon ligands, other common groups have elements from main groups III–VI at bonding sites. We shall briefly consider some which contain a group III–VI atom in the organic ligand, but recommend that details of metal to N, P, S, etc., bonds are sought in the literature[2,12–16].

(*a*) Compounds containing B atoms have included $[Cr(CO)_3\{\eta^6\text{-PhB}(CH_2)_5\}]$[135] and $[Cr(\eta^6\text{-}C_5H_5BR)_2]$ (R = Me or Ph)[136]. The former lost hydrocarbon fragments, $C_6H_5BH_2$, and $C_3H_3BCH_2$ from $[M - 3CO]^+$. The latter eliminated H_2 and RH molecules, and from R = Me, $MeBCH_2$.

(*b*) Compounds with main group IV ligands MX_3 bonded to transition metals are characterized by cleavages of the transition metal—M bonds, production of ions typical of main group species MX_3, and sometimes rearranged ions such as $[C_5H_5Sn]^+$ which have been observed from R_3Sn compounds of cyclopentadienylmolybdenum,-tungsten and -iron[2].

Transfer of X(X = F, Cl, Br, or Ph) groups to the transition metal centre has also been reported, as in the spectrum of $[Mn(SnBr_2Ph)(CO)_5]$ which gave $[BrMn]^+$ and $[C_6H_5Mn]^+$ ions[137]. Mention may also be made here of several isocyanide-containing complexes which have been studied[2]. Typical are $[Cr(CO)_5CNR]$ (R = alkyl, aryl) which produced $[CrCNR]^+$, $[CrCNH]^+$, and $[CrCN]^+$ ions[138], and $[Fe(CO)_2$(olefin)(CNL)] complexes[139].

(*c*) Numerous complexes with main group V containing ligands have been studied. The formation of M—O containing ions from $[V(CO)(NO)_2(\eta^5\text{-}C_5H_5)]$ mentioned previously (equation 4)[48] is also observed with other nitrosyl complexes such as those shown in equations 52[48] and 53[140].

$$Cr\{C(OMe)Ph\}NO(C_5H_5)^{+\cdot} \xrightarrow{-Me} \xrightarrow{-CO} \xrightarrow{-C_6H_5N} {}^+CrOC_5H_5 \quad (52)$$

$$Cr_2(NH_2)(NO)_3(C_5H_5)_2^+ \longrightarrow \overset{+}{Cr_2}O(C_5H_5)_2 \quad\text{and}\quad \longrightarrow \overset{+}{Cr_2}O(NH_2) \quad (53)$$

Elimination of HCN from pyridine-containing organic ligands is a common process (equations 54[141] and 55[142]).

$$^+\overline{M(CO)_3CH_2C_5H_4N}(C_5H_5) \xrightarrow{-(1-3)CO} \xrightarrow{-HCN} (C_5H_5)_2M^+ \quad (54)$$

M = Mo, W

$$^+M(CO)_4(R)N{=}CHC_5H_5N) \xrightarrow{-(1-4)CO} \xrightarrow{-R} \xrightarrow{-HCN} C_5H_5NM^+ \xrightarrow{-C_5H_5N} M^+ \quad (55)$$

Azaferrocene $[Fe(\eta^5\text{-}C_5H_5)(\eta^5\text{-}C_4H_4N)]$ also showed loss of HCN from M^+ as well as C_2H_2, C_4H_4N, and C_2H_2N molecules[143].

Organometallic compounds containing (PX_3) ligands (X = alkyl, aryl, halogen, OR, etc.) tend to fragment by loss of PX_3 and cleavage of P—X or O—R bonds[2]. There have been some examples of interactions with other groups in the molecule such as those shown in equations 56[144], 57[145] and 58[146].

$$PtMe_2(PPh_3)_2^+ \begin{cases} \longrightarrow {}^+PtMe_2(PPh_3)(PPh_2) \\ \longrightarrow {}^+PMe_2Ph_2 \end{cases} \tag{56}$$

$$C_5H_5MnPX_3^+ \begin{cases} \longrightarrow C_5H_5PX^+ \longrightarrow C_5H_4P^+ \\ \longrightarrow C_5H_5MnX^+ \longrightarrow MnX^+ \end{cases} \tag{57}$$

$$C_5H_5CoP(OR)_3^+ \begin{cases} \longrightarrow {}^+CoOR \\ \longrightarrow C_5H_5CoOR^+ \end{cases} \tag{58}$$

(*d*) Sulphur-containing compounds generally produce abundant $[MS]^+$ ions and often show elimination of MS_n and H_2S molecules. Examples are found in the spectra of the norbornadiene iron complex containing two SMe bridging ligands (equation 59[147] and cyclopentadienyl iron compounds (equation 60)[148].

$$C_7H_8\overset{+}{Fe}_2(CO)_4(SMe)_2 \xrightarrow{-(1-4)CO} \xrightarrow{-2Me} C_7H_8\overset{+}{Fe}_2S_2 \longrightarrow \overset{+}{Fe}_2S_2 \tag{59}$$

$$C_7H_8\overset{+}{Fe}_2S_2 \longrightarrow C_5H_5\overset{+}{Fe}_2S_2 \longrightarrow \overset{+}{Fe}_2S_2$$

$$C_5H_5FeSCH_3^+ \xrightarrow{-H_2S} C_6H_6Fe^+ \tag{60}$$

Complexes with more complicated ligands (**9**)[149], often produce MS-containing ions (equation 61).

$$C_5H_5(CO)_2\overset{+}{M}(S{\cdots}CR^1{\cdots}NR^2) \xrightarrow{-2CO} \xrightarrow{-CR^1{=}NR^2} C_5H_5\overset{+}{M}S \tag{61}$$

(**9**)

9. Cluster compounds

Cluster compounds are characterized by the appearance of ions containing the core atoms. From carbonyl clusters $[Co_3(CO)_9CX]$[150], $[Ru_3H_3(CO)_9CX]$[151], or $[Ru_6C(CO)_{16}]$[152], the ions $[Co_3CX]^+$, $[Ru_3CX]^+$, and $[Ru_6C]^+$ are observed as well as the expected CO-containing ions. The further fragmentation of $[Co_3CX]^+$ occurs by cleavage of C—X and, unusually, the loss of a Co atom. A cyclopentadienylnickel compound $[Ni_4H_3(C_5H_5)_4]$[153] and related cobalt complex $[Co_4H_4(C_5H_5)_4]$[154] eliminate H atoms from M^+ ions.

D. Negative Ion Spectra by Electron Impact

1. Main group compounds

These studies have been relatively few but a number of compounds of main group elements Hg, Si, Sn, and P have been reported. At 30 eV the base peak from $[Me_4Si]$ was $[(CH_3)_3SiCH_2]^{-\cdot}$ with $[(CH_3)_2SiCH]^-$, $[CH_3SiCH_2]^-$, and $[CHSi]^-$ also reasonably abundant[155]. From $[Me_4Sn]$, $[(CH_3)_3Sn]^-$ was the most abundant ion with $[CH_3SnCH_2]^-$, $[(CH_3)_2SnCH_2]^{-\cdot}$, and $[C_2H_4Sn]^{-\cdot}$ in order of significance[155]. Spectra of alkenylsilanes $[R^1Me_2SiCH_2CH{=}CHCH_2R^2]$ contain $M^{-\cdot}$ and ions due to cleavage of Si—C, Si—H, $H_2C—R^2$, or $CH_2—CH{=}CHCH_2R^2$ bonds[156]. Prominent in the negative ion spectrum of $[Si(C{\equiv}CCF_3)_4]$ were $[M - CF_3]^-$, $[C_{12}F_9]^-$ and $[C_6F_4]^-$[157]. Both $[Me_6Si_2]$ and $[(Me_2Si)_6]$ show $M^{-\cdot}$ but $(Me_3Si)_2O$ does not[158]. Trimethylsilyl derivatives of ethers (**10**, **12**) and esters (**11**, **13**) have been studied[159]. Basic fragmentation leads to $[M - Me_3Si]^-$ and $[M - Me_3Si - CO_2]^-$ ions, respectively. Migrations of $SiMe_3$ groups were common in *ortho*-substituted compounds and abundant M^{2-} ions were features of several spectra.

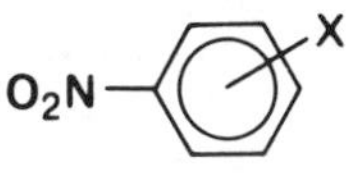

(**10**) X = $OSiMe_3$
(**11**) X = O_2CSiMe_3
(**12**) X = $OSiMe_3$ 
(**13**) X = O_2CSiMe_3

Deuterium-labelling experiments showed that loss of C_6H_5 and $C_{12}H_{10}$ from $[Ph_4Si]^{-\cdot}$ proceeded without scrambling but elimination of H_2 from the base peak $[(C_6H_5)_2Si]^-$ occurred with complete randomization[62]. The base peaks in the spectra of related main group IV and V compounds were $[M - Ph]^-$ [66,160]. Of the group V derivatives only Sb and Bi gave $M^{-\cdot}$ ions. It was noticeable that $[Ph_3Bi]$ produced the greatest variety of ions, the reverse of the positive ion spectra[66]. This complementary effect has also been noticed with some transition metal compounds. Spectra of PBu^n_3, Ph_2PCl, $P(OEt)_3$ were reported to show abundant $[M - H]^-$ ions and ions due to P—X cleavage[161].

Carbonyl-stabilized phosphoranes $Ph_3PCR^1C(O)R^2$ decomposed by initial loss of Ph or Ph_3P[162]. Further fragmentation of $[M - Ph]^-$ gave ions of formula $(C_6H_5)_2PO$ and $(C_6H_4)_2PO$, which have direct counterparts in the positive ion spectra (equation 62).

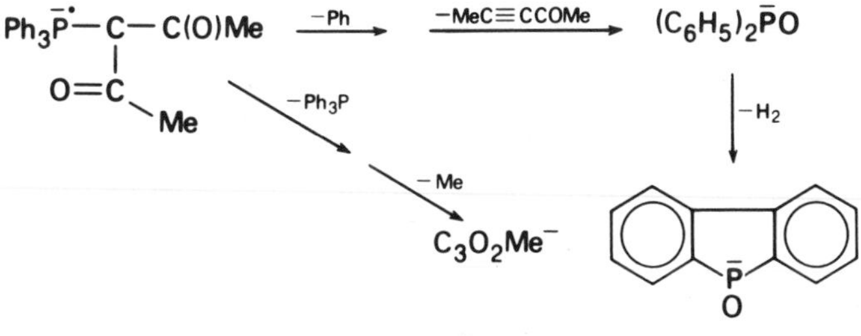

$$Ph_3\bar{P}^{\cdot}-C(-C(O)Me)(-C(=O)Me) \xrightarrow{-Ph} \xrightarrow{-MeC\equiv CCOMe} (C_6H_5)_2\bar{P}O \xrightarrow{-H_2} (C_6H_4)_2\bar{P}O; \quad \xrightarrow{-Ph_3P} \xrightarrow{-Me} C_3O_2Me^- \tag{62}$$

Pentafluorophenylmercury derivatives $[(C_6F_5)HgX]$ ($X = C_6F_5$, halogen) showed $M^{-\cdot}$ ions[163]. An unusual reaction (X = hal) leading to ion-pair production was reported from these compounds (equation 63), and also from bistrichlorovinylmercury (equation 64)[164].

$$[C_6F_5HgX] \longrightarrow C_6F_5X^- + Hg^+ \quad (63)$$

$$[(Cl_2C{=}CCl)_2Hg] \longrightarrow C_2Cl_4Hg^- + C_2Cl_2^+ \quad (64)$$

2. *Transition metal compounds*

Transition metal compounds containing cyclopentadienyl and arene ligands have been studied. These include $[M(CO)_n(\eta^5\text{-}C_5H_5)]$ compounds of V, Mn, and Co[165,166] related $[M(CO)_n(\eta^5\text{-}C_5H_5)]_2$ (Cr, Mo, Fe, and Ni) derivatives[166], and $[Cr(CO)_3(arene)]$ derivatives[167]. Carbonyl losses predominated in the fragmentation paths. Symmetrical cleavage of the dimers was a feature of their spectra. Most spectra showed abundant $[M - CO]^-$ ions which are often base peaks and formally isoelectronic at M with M^+. Abundant M^- ions were observed from $[MCl_2(\eta^5\text{-}C_5H_5)_2]$ (M = Ti, Zr, Hf) and major fragment ions arose through loss of Cl or C_5H_5 radicals[168].

Numerous studies of metal carbonyls have been reported[2,169–171], including tetranuclear carbonyl clusters of cobalt, rhodium, and iridium[172]. The electronic formality between $[M - CO]^-$ and M^+ in the spectra of carbonyl derivatives has been the subject of some discussion[173]. The appearance of $[M - CO]^-$ ions as base peaks in the spectra of most η^5-cyclopentadienyl metal carbonyls has already been noted[166,167]; however, from $[Co(CO)_2(\eta^5\text{-}C_5H_5)]$ the base peak was M^- [173]. This was rationalized by postulating a change from $\eta^5\text{-}C_5H_5$– to $\eta^3\text{-}C_5H_5$–M interaction. Further, η^4-diene iron tricarbonyls gave abundant M^-, which were suggested to contain η^3-allylic bonded species. Similar arguments were used in interpreting the spectra of η^4-cycloheptatriene—iron, and η^6-cycloheptatriene—chromium, —molybdenum, and —tungsten derivatives[173].

III. OTHER IONIZATION TECHNIQUES

A. Chemical Ionization

Chemical ionization (CI) occurs in an ion source by ion/molecule reactions between ions of a reagent gas, commonly CH_4 or $i\text{-}C_4H_{10}$, and sample molecules M (equation 65)[174]. Usually the ratio of reagent gas to sample is of the order of 1000:1 and the

$$CH_4 \longrightarrow CH_4^{+\bullet} \xrightarrow{CH_4} CH_5^+ + CH_3 \quad (48\%)$$

$$CH_4 \longrightarrow CH_3^+ + H \xrightarrow{CH_4} C_2H_5^+ + H_2 \quad (40\%)$$

then

$$CH_5^+ + M \longrightarrow MH^+ + CH_4 \quad (65)$$

or

$$C_2H_5^+ + M \longrightarrow MH^+ + C_2H_4$$

$$C_2H_5^+ + M \longrightarrow [M + C_2H_5]^+$$

reagent gas pressure is *ca*. 1 Torr. Since the sample-containing ions are even-electron species and, more important, the energy transferred in the ionization process is much lower than by electron impact (EI), the proportion of M-containing ions is greater and consequently the amount of fragmentation is less. Often quasimolecular ions such as

$[M + H]^+$, $[M + C_2H_5]^+$, and $[M - H]^+$ are the base peaks in a spectrum. Selection of the reagent gas is all-important in determining the spectrum obtained. Isobutane has become a popular choice since the reagent ion is >90% $[C_4H_9]^+$, it is a weaker Brönsted acid than $[CH_5]^+$, and transfers a proton with much less energy. Other gases have been NH_3, NO, H_2O, and Me_4Si for positive-ion CI, and CH_3ONO, N_2O mixed with H_2, CH_4, or N_2, and CH_2Cl_2 for negative-ion CI[174]. The application of inert gases (He, Ar, Xe) or N_2 as reagent gases in CI sources, causing odd-electron reagent gas ions to transfer charge to a sample molecule, has been described. This is usually termed charge-exchange CI. The molecular ion formed is odd-electron, as in EI, but the energy of the process is more controlled and hence fragmentation is not as extensive.

Many advantages are to be found with CI over EI, especially in the determination of molecular weights and the simplification of fragmentation patterns. However, some slight disadvantages also exist, including excessive ion source contamination from the high pressure of reagent gas and poor reproductility since CI spectra are often very temperature dependent.

The use of CI for organometallics was first reported in 1971, since when its adoption has been rather slow. We shall discuss the results obtained for main group compounds first, then those from transition metal derivatives. As a reagent gas, Me_4Si produces $[(CH_3)_3Si]^+$ and $[(CH_3)_7Si_2]^+$. These have been subjected to reaction with a variety of compounds to give $[M + SiMe_3]^+$ ions[175,176]. With methane as reagent gas, Me_4Si gave $[(CH_3)_3Si]^+$ ions predominantly[177]. In contrast, Me_nSiH_{4-n} (n = O–3) produced $[M - H]^+$ [177].

CI spectra of dialkylmercury compounds and organomercury acetates have been reported[178,179]. The fragmentation of the acetates depended on R (equation 66). Path (b) was favoured most for R = Ph and less for R = Et or heptyl[179].

$$\left[\begin{array}{c}\text{OMe}\\|\\\text{RCHCH}_2\text{HgOAc}\end{array}\right] \xrightarrow{(a)} \begin{array}{c}\text{H}\overset{+}{\text{O}}\text{Me}\\|\\\text{RCHCH}_2\text{HgOAc}\end{array} \xrightarrow[-\text{RCHCH}_2]{-\text{MeOH}} \text{AcOHg}^+$$

$$\left[\begin{array}{c}\text{OMe}\\|\\\text{RCHCH}_2\text{HgOAc}\end{array}\right] \xrightarrow{(b)} \begin{array}{c}\text{OMe}\\|\\\text{RCHCH}_2\text{Hg}\overset{+}{\text{O}}\text{Ac}\\|\\\text{H}\end{array} \xrightarrow[-\text{Hg}]{-\text{HOAc}} \begin{array}{c}\text{OMe}\\|\\\text{RCH}\overset{+}{\text{C}}\text{H}_2\end{array} \qquad (66)$$

Both Ph_3Ga and Ph_3In showed $[(C_6H_5)_2M]^+$ as the only significant ion in CI spectra[39]. It was suggested that this ion was formed by elimination of C_6H_6 after initial protonation at a phenyl group since neither $[M + H]^+$ nor $[M - H]^+$ ions were observed. Some butyltin compounds $[Bu_nSnX_{4-n}]$ are reported to produce three-coordinate ions by simple Sn—Bu or Sn—X cleavages and no ions due to C_4H_8 loss[180]. However, in the more complex γ-OH substituted butyltins, $[Bu_2(X)Sn(CH_2)_2CH(OH)Me]$, the 1,3-deoxystannylation reaction giving $[M—H_2O]^+$ was significant and more important than the corresponding 1,4-deoxystannylation from δ-OH compounds[181]. A correlation was found between the electron-donating ability of X and stability of $[M - H_2O]^+$ from $[Me_3Sn(CH_2)_2CH(OH)C_6H_4X]$. The isobutane CI spectra of some benzyltin compounds have been studied[182]. Evidence from p-substituted-benzyl compounds suggested that the reagent ion $[C_4H_9]^+$ react predominantly at the p-position (equation 67).

Alkyldiphenylphosphine oxides produced as major ions $[(C_6H_5)_2POH]^{+\bullet}$

$$Me_3SnCH_2Ph \xrightarrow{C_4H_9^+} Me_3Sn{-}CH_2{-}C_6H_5^+(H)(C_4H_9) \xrightarrow{-C_{11}H_{16}} Me_3Sn^+ \quad (67)$$

(with H transfer from the β-C atom of the alkyl groups), $[(C_6H_5)_2POCH_2]^{+\cdot}$ $[(C_6H_5)_2P(OH)CH_2]^{+\cdot}$ and $[M - R]^+$ [183]. The CI of some organophosphonates has been reported[184].

Other compounds reported have included MeAs derivatives[185], RB derivatives of boronate esters[186,187], and an Me_2Al—carborane[188].

The [(diene)$Fe(CO)_3$] complexes **14–16** were the first transition metal organometallics to be studied[189,190]. All produced abundant $[M + H]^+$ ions whereas under EI no M^+ ions were found. Likewise, quasimolecular ions were observed from cyclooctatetraene—and cyclobutadiene—$Fe(CO)_3$ complexes. A later study of [(η^4-olefin)$Fe(CO)_3$] compounds was reported using hydrogen, methane, isobutane, and ammonia as reagent gases[191]. With the olefins cyclobutadiene, buta-1,3,-diene, penta-1,3-diene, and cyclohexa-1,3-diene, protonation of M was observed with all reagent gases except ammonia. Thus these organoiron compounds have proton affinities lower than that of ammonia (205 kcal/mol). Base peaks in the spectra were usually $[M + H]^+$ or $[M + H - CO]^+$ and fragmentation occurred mainly by CO and H_2 loss. With the polyene ligands cycloheptatriene, cycloheptatrienone, and cyclooctatetraene, protonation occurred with all of the reagent gases used. It was suggested that the difference in the behaviour of the diene and polyene complexes could be due to the possible protonation of an uncoordinated C=C bond in the latter.

X
(CO)$_3$Fe
(14) X = CH_2
(15) X = O

CHO
Fe(CO)$_3$
(16)

Cyclopentadienyl compounds which have been studied with methane reagent gas include $[M(C_5H_5)_2]$ (M = Fe, Ru, Os, Co, Ni) and $[MCl_2(C_5H_5)_2]$ (M = Ti, Zr, Hf)[190]. The metallocenes afforded only three prominent ions, M^+, $[M + H]^+$, and $[M + C_2H_5]^+$, whereas the dichloro derivatives produced M^+, $[M + H]^+$, and $[M - Cl]^+$ as the base peak ion.

[(Arene)$Cr(CO)_3$] complexes (C_6H_5X; X = H, F, Cl, Me, CO_2Me) with methane reagent gas produced abundant M^+ and $[M + H]^+$ ions and weak fragment ions by loss of CO, H, and arene groups[192]. No Cr—F ions were observed in the spectrum of the fluorocarbon compound, in contrast to the EI spectrum. Protonation reactions were observed for [$Cr(CO)_3$(η^6-arene)] complexes (arene = C_6H_6, toluene, methylbenzoate, and acetophenone) and η^6-cycloheptatriene tricarbonyl derivatives of Mo and W using hydrogen, methane, isobutane, and ammonia as reagent gases[193]. Ions derived from ammonia failed to protonate [$Cr(CO)_3(\eta^6$-$C_7H_8)$], whereas those from H_2, CH_4, and C_4H_{10} succeeded. Relatively high abundances of M^+ ions were observed particularly with methane as the reagent gas. A charge exchange reaction (equation 68) was postulated for their formation, which is in agreement with the known ionization potentials for the species involved. A similar explanation was suggested for the formation of M^+ ions from metallocene derivatives[190]. A number of studies of metal carbonyls have been reported, including [$M(CO)_6$][190].

$$[Cr(CO)_3(\eta^6\text{-arene})] + C_2H_5^+ \longrightarrow [Cr(CO)_3\text{arene}]^+ + {}^{\bullet}C_2H_5 \qquad (68)$$

B. Field ionization (FI), Field Desorption (FD) and Related Techniques

In FI, vaporized molecules are ionized by a very high electric field gradient (*ca.* 2 V/Å) between electrodes. This is a 'softer' ionization than EI and consequently M^+ ions are more probable. Relatively few studies of organometallics have been reported. They include substituted phenylphosphines[194], bis(arene)—chromium iodides[195] and biphenyl derivatives [$(\mu$-η^6-$C_6H_5)_2\{Cr(\eta^6$-$C_6H_6)\}_2$][196].

Field desorption (FD) usually involves deposition of the sample from solution on to a specially prepared emitter electrode[41]. The sample is desorbed under the influence of a high field gradient under normal mass spectrometer vacuum conditions to produce molecular or quasimolecular ions and relatively few fragment ions. This technique has particular relevance to thermally labile or ionic compounds. It is not as simple to apply as EI or even CI, and there can be problems with reproducibility and interpretation of spectra. However, it has proved successful for several compounds which do not give satisfactory EI spectra, including [C_4Li_4][197], R_4PBr[198], and ionic transition metal derivatives such as [$Fe(\eta^5$-$C_5H_5)$ (η^6-arene)]PF_6[199], [$Mn(CO)_3(\eta^5$-$Me_2SC_5H_4)$]PF_6[200], [$Fe(CO)_3(\eta^5$-olefin)]BF_4, [$Co(\eta^5$-$C_5H_5)(\eta^5$-$C_8H_{11})$]BF_4[199], and [$Fe\{P(OMe)_3\}_3(\eta^5$-olefin)]BF_4[201], as well as neutral complexes such as [$M(CO)_3(\eta^6$-$C_7H_8)$](M = Cr, Mo, W), and [$Ru(X)_2(L)(\eta^6$-$C_6H_6)$][202] and the cluster compounds [$Rh_3(CO)_3(\eta^5$-$C_5H_5)_3$][203], [$Ir_7(CO)_{12}(C_8H_{12})(C_8H_{11})(C_8H_{10})$][204], [$Pt_n(CO)_{n+1}(PPh_3)_4$]($n$ = 4, 5)[205], and polynuclear carbonyls[206].

A related 'soft' ionization technique is the attachment of ions to a neutral compound preceding desorption. This has not yet been reported for organometallics but both $[M_n + Na]^+$ and $[M + I]^-$ ions were observed from NaI in glycerol (M) using electrohydrodynamic ionization[207]. Cationization has also been observed in conjunction with FD[208] and CI[209]. Several other modifications exist of the general ionization techniques described above, including thermal desorption of samples combined with EI[210] or CI[211]. Added to these the thermal evaporation of the cations from phosphonium salts has been reported[212].

C. Other Techniques

Recently developed alternative methods of ionization as yet untried for organometallics include the use of radionuclides and lasers[41]. Photoionization sources have been developed for some time but relatively few studies of organometallic spectra have been reported[41]. Rather, the interest in these sources has been in the determination of accurate I and A data.

Newer methods of sample introduction such as combined liquid chromatography–mass spectrometry (LC–MS)[41] and 'in-beam' sample introduction[213] to obtain CI spectra will be of future interest to chemists studying organometallics.

IV. ION STRUCTURES AND ENERGETICS

A. Introduction

Throughout this chapter reference has been made to the lack of structural informa-tion for gas-phase ions. Fortunately, a number of recently developed techniques are beginning to alter this situation significantly. Most of these techniques are only just being applied to organometallics. Of particular importance are metastable ion studies including collisional activation[41,51], photoelectron–photoion coincidence experi-ments[41,214], and photodissociation studies[41,215]. These, together with the measurement of I and A data and results from ion cyclotron resonance (ICR) spectroscopy[41,216], are the subjects of this section.

B. Metastable Ion Studies[51,217]

Intensity ratios of metastable ions from competing fragmentations can be used to characterize ion structures (or mixtures of structures)[41]. It is generally accepted that if intensity ratios of precursor and m* ions are different by a factor of greater than 5, this constitutes evidence for precursor ions having different structures (or mixtures of structures). Thus, the spectra of the *o*-, *m*-, and *p*-derivatives of $(tolyl)_3P$ all showed an ion $[C_{13}H_{10}P]^+$ which fragmented by loss of H, H_2, and PH neutrals[71]. The ratios of the intensities of the metastable peaks for these fragmentations were very similar and it was suggested that the structures, or mixture of structures, of $[C_{13}H_{10}P]^+$ were the same in each case. However, it should be noted that, although a similar intensity ratio may be taken as suggesting identical structures, it is not a proof since the energy distribution in the ions may be substantially different even if the ion structures are the same. Hence evidence from metastable ion intensity ratios of ion similarities should be taken as a guide and not as a rule.

Developments in instrumentation have led to an increase in the study of meta-stable ions[41,51]. A technique for scanning both magnetic (B) and electric (E) fields simultaneously on double-focusing spectrometers provides more specific and sim-plified fragmentation patterns and information about fragmentation paths[41,218]. Meta-stable decompositions occurring in the first field-free region (i.e. between source and analysers) can be observed. Scanning at a constant B/E ratio at a constant m/z reveals the daughter ions arising from the chosen precursor. Scanning at constant B^2/E reveals the precursor ions of chosen daughter ions. In a study of $[Cr(CO)_3(\eta^6\text{-}PhCH_3)]$ the molecular ion was shown to fragment by CO losses exclusively (equation 69)[219]. Subsequently $[M - 3CO]^+$ decomposed by four different routes.

$$C_7H_8Cr(CO)_3^+ \rightarrow C_7H_8CrCO^+,\quad C_7H_8Cr(CO)_3^+ \rightarrow C_7H_8Cr^+,\quad C_7H_8Cr(CO)_3^+ \rightarrow C_7H_8Cr(CO)_2^+ \rightarrow C_7H_8Cr^+ \tag{69}$$

$$C_7H_8Cr^+ \rightarrow {}^{+\cdot}C_7H_8,\quad {}^{+}C_7H_7,\quad {}^{+}CrH,\quad {}^{+}Cr$$

Another technique of future importance is the study of metastable ions produced in collision induced decomposition or collisional activation spectroscopy[41]. Activation of ions by collision with neutral molecules of a rare gas such as helium in a field-free region of a mass spectrometer leads to the addition of internal energy and the fragmentation of the ions. For a particular ion this produces a characteristic collisonal activation (CA) spectrum which is related to the structure of the precursor ion. It appears that the internal energy of the precursor ion has a negligible effect on such a spectrum and comparison of CA spectra for ions of the same formulae derived from different precursors therefore gives information about structural similarities. This technique has not yet been applied to organometallics.

C. Photoelectron–Photoion Coincidence Spectroscopy[41,214]

By this technique ions can be formed in selected internal energy states and their subsequent decompositions or reactions with other species can be studied. Mostly small inorganic or organic molecules have been investigated and a number of interesting effects have already been observed, including isolated state behaviour in the fragmentation of molecular ions such as $[C_2F_6]^{+\cdot}$ [220]. This constitutes a breakdown of a basic assumption in quasi-equilibrium theory (QET) of mass spectra that electronic energy is randomized on ionization and will provoke a reassessment of the theory. A study of the unimolecular decay of ions from $[Me_2Hg]$ has been reported to show deviations from QET predictions at high initial internal energies[221]. In particular the formation of $[CH_3Hg]^+$ from $[Me_2Hg]$ may involve incomplete statistical energy partitioning in the reaction (equation 70). At about 5 eV above the threshold of this reaction a new fragmentation mode emerges in which it seems possible that an electronically excited methyl radical is involved.

$$Hg(CH_3)_2^{+\cdot} \longrightarrow {}^+HgCH_3 + CH_3 \tag{70}$$

D. Ion Photodissociation Studies[41,215]

These concern the impact of photons on ions in ICR traps or in beams[215], and can be represented by reaction 71. Most work has used ICR ion traps. By investigating

$$AB^+ \xrightarrow{h\nu} A^+ + B \tag{71}$$

reaction 71 as a function of photon wavelength it is possible to obtain a photodissociation spectrum. Such studies can be interpreted to yield spectroscopic, thermodynamic, kinetic, and dynamic data on the trapped ions and their reactions. Virtually all of the photodissociation work reported has involved organic molecules. An example is the study $[C_7H_8]^{+\cdot}$ ions produced from toluene, cycloheptatriene, and norbornadiene[222]. This showed that, contrary to previous ideas based on fragmentation patterns, the ions' structures were not common. The photodissociation spectra were different in each case and demonstrated that little or no interconversion of structures was occurring within the time scale of a few seconds.

With this technique the thermochemistry of the interactions between $[Li]^+$ and derivatives of benzene and ferrocene have been investigated[223]. A series of $D(M - Li)^+$ ionic bond dissociation energies were obtained for ground- and excited-state ions.

An initial study of $[Ni(NO)(\eta^5\text{-}C_5H_5)]$ determined $D(C_5H_5Ni - NO)^+$ as $<43 \pm 2$ kcal/mol and identified two low-lying excited states of the $[Ni(NO)(C_5H_5)]^+$ ion[224].

$$C_5H_5NiNO^+ \xrightarrow{h\nu} C_5H_5Ni^+ + NO \quad (72)$$

The reaction studied was reaction 72. The value for the ionic bond dissociation energy compares well with 45.9 ± 1 kcal/mol determined using photoionization.

E. Ionization and Appearance Potential Measurements[41,217]

In principle, the measurement of I or A data should be achieved relatively simply by following an ion's ionization efficiency curve to the onset potential, OP (Figure 2a). In practice there are many pitfalls associated with these measurements. Both the sensitivity of the spectrometer and the signal-to-noise (S/N) ratio of the ion current will play significant parts in the determination of the onset potential (Figure 2b). Other com-

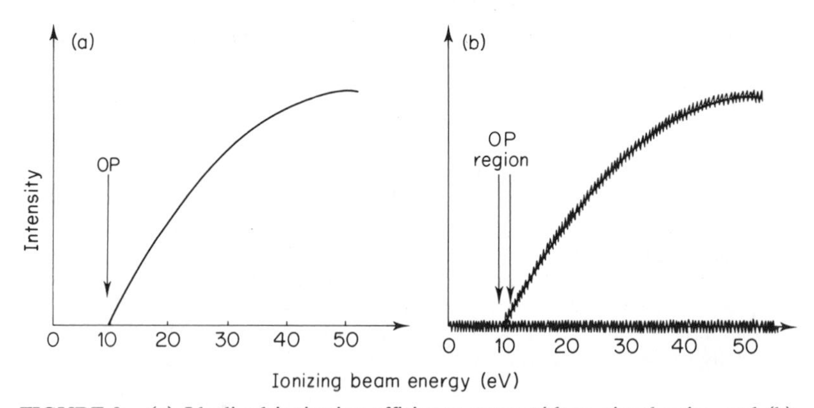

FIGURE 2. (a) Idealized ionization efficiency curve without signal noise, and (b) more typical curve with signal noise.

plicating factors will include the possibility of the formation of the ion with excess energy and, if a conventional EI source is used, the inhomogeneous ionizing beam. Concerning sensitivity, measurements of I or A are most successful at high spectrometer sensitivity with abundant ions at a high S/N ratio. The most accurate results are usually for I's rather than A's, but of course photo-electron spectroscopic I values are preferable to mass spectrometrically determined values. It has been commonly accepted that an I measured with EI ionization refers to the vertical I_v rather than the adiabatic I_a because ionization by EI is a Franck–Condon controlled process. While it is clear that reported EI data are usually higher than the corresponding I_a values determined by photoelectron spectroscopy, it should be remembered that there is a small but finite chance of producing ions with internal energies lower than that corresponding to I_v, even some with internal energies corresponding to the threshold notwithstanding the Franck–Condon restrictions. The importance of these ions will depend on each molecule studied but it may be noted that differences in I_a and I_v are commonly of the order 0.1–1 eV in organometallic compounds[46]. Further, there is usually the complicating effect of an inhomogeneous electron beam. Hence I and especially A data should be interpreted cautiously, taking the values as upper limits unless there is other evidence to substantiate them.

In a fragmentation reaction (equation 73), the $A(X^+)$ value is related to the standard enthalpies of formation of the species involved by equation 74, where E is a conglomerate term representing the inclusion of any excess energy in the formation of

$$M + e^- \longrightarrow X^+ + Y + 2e^- \quad (73)$$

$$A(X^+) = \Delta H_f^\circ(X^+) + \Delta H_f^\circ(Y) - \Delta H_f^\circ(M) + E \quad (74)$$

$[X]^+$ or Y at the threshold for the production of $[X]^+$. It is possible that E would contain contributions from (a) the kinetic shift, (b) the competitive shift, (c) a thermal shift, and (d) an energy of activation for the reverse reaction, i.e. recombination. The first of these arises because extra energy may have to be imparted in order that $[X]^+$ may be formed with a rate constant $>10^6$ s^{-1} so that a sufficient number of ions are ejected from the ion source to be subsequently detected. The kinetic shift is the excess energy above the minimum $A(X^+)$ necessary to bring about decomposition with this rate constant. If competing reactions are occurring at the threshold then the intensity of $[X]^+$ may increase very slowly with electron beam energy. A competitive shift similar to the kinetic shift may be experienced requiring an extra amount of energy before $[X]^+$ is detected. The thermal shift arises from the fact that molecules in the ion source possess thermal energy and therefore less than the threshold $A(X^+)$ will be required for the reaction. This effect increases as molecular size increases and would counter the kinetic and competitive shifts. An activation energy ($\neq 0$) for the recombination reaction of reaction 73 may mean the products $[X]^+$ and Y are formed with excess internal or kinetic energy. The size and partitioning of these will depend on the reaction. With the possibility of any of the above excess energy components being present it is surprising that in most studies the excess energy term E is assumed to be zero. However, this drastic assumption appears to be valid within the precision of the measurements for a large number of 'simple' fragmentations. By 'simple' we mean to exclude reactions producing rearrangement ions. Simple cleavages are generally synonymous with the absence of abundant metastable peaks and significant competing reactions.

Very few investigations of the excess terms have been made with organometallics. The fragmentation reaction 75 (M = Cr, Mo, W)[225] has been reported and a study of the CO molecules eliminated showed them to be predominantly in the first vibrationally excited state with an energy of 0.27 eV. However, when the precision of the measurements is considered, typically ±0.05–0.1 eV for I and ±0.1–0.3 eV for A, it is doubtful if such effects would be observable in most cases.

$$^+M(CO)_n \longrightarrow {}^+M(CO)_{n-1} + CO^* \quad (75)$$

The inhomogeneity of ionizing beams, generally of the order of 0.1 eV at low beam energies, is another source of inaccuracy. Several laboratories have designed monochromated photon or electron beams and results obtained with them are to be preferred. An alternative approach has been to remove the effects of the normal energy spread by Fourier transformation or deconvolution techniques applied to the ionisation efficiency curve. Some success has been achieved with these and also empirical ionization efficiency curve analysis by the critical slope curve matching (CSCM) method[226]. After comparison with a number of commonly used methods including the semi-logarithmic plot, energy compensation and extrapolated voltage difference, second derivative, and critical slope methods, it was concluded that the CSCM and a deconvolution method (the double energy distribution differences method) were preferred to give accurate values for A (to within 0.05 eV)[227]. The empirical CSCM method is relatively easy to apply and can be used with inferior S/N levels.

Unfortunately, most measurements of I and A prior to 1974 were obtained by the less reliable empirical methods. Even now results using these early methods are still being published, but it is hoped that this will soon stop. Although trends in such results are probably noteworthy, their accuracy will remain suspect unless good agreement is

TABLE 2. Ionization and appearance potential measurements

Main group derivatives

Compound	Reference	Compound	Reference
$(RLi)_n$	229		
R_2Be (R = alkyl, aryl)	230, 68	R_2Zn	233
$(C_5H_5)_2Be$	231		234[a]
$(C_5H_5)_2Mg$	232	Me_2Cd	234[a]
		R_2Hg	235–237
			234[a]
Ph_3M (M = B, Al, Ga, In)	64	RBF_2	238
$(Me_3Al)_2$	233		
Me_4M (M = Si)	43, 228, 239–241, 242[a], 243[a]	Containing Ge—halogen	244, 258
		Containing Sn—halogen	247
(M = Ge)	43, 235, 244	Me_nSn—R (R = alkyl, aryl)	43, 246, 259
(M = Sn)	43, 240, 245–247	(R = heterocyclic group)	260
(M = Pb)	43, 240	Me_3Pb—Bu^t	43
Containing Si—H	228, 239, 241, 248–251	$Me_nMR_{(4-n)}$ (M = several of Si, Ge, Sn, or Pb)	
Si—halogen	228, 239, 241, 243, 254	(R = alkyl or aryl)	261–263
Si—OR	239, 252–254	(R = SMe)	264
Si—NR_2	239	$Ph_nMX_{(4-n)}$ (M = Si)	249, 256
		(M = Sn)	60, 245, 265
Me_nSi—R (R = alkyl, aryl)	228, 239, 248–250 255, 256, 257[a]	Containing M—M′ bonds	43, 239, 241, 246, 249, 256, 265–267
R_3M (M = P or As)	145, 268, 269, 270[a]		
$R_3M{=}X$	271		
R_2Te	272		
Selenophanes	273		

Transition metal derivatives

Compound	Reference
Olefin complexes:	
[Mn(CO)$_2$olefin(η^5-C_5H_5)]	145
Carbene complexes:	
[Cr(CO)$_5$CXY]	103, 274–276
Acetylene complexes:	
[Co_2(CO)$_6$$C_2R^1R^2$]	277
Arene complexes:	
[M(Arene)$_n$]	130, 131, 286, 301–305
[M(CO)$_x$(Arene)$_y$]	134, 286, 289, 305–307
Complexes with C_7 or C_8 rings:	
[M(C_7 ring)$_y$]	130, 300, 308
[M(CO)$_x$(C_7 ring)$_y$]	300, 309
[U(η^8-C_8H_8)$_2$]	310
Cyclopentadienyl complexes:	
[M(η^5-C_5H_5)$_n$]	63, 232, 278–287
[M(CO)$_x$(η^5-C_5H_5)$_y$] and related (CS) complexes	145, 288–293
[M(L)$_x$(η^5-C_5H_5)$_y$]	
(L = isocyanide)	145
(L = PX_3 or AsX_3)	145, 294–296
(L = SX_2)	145, 297
(L = boron-containing group)	298
[M(H)$_x$(η^5-C_5H_5)$_n$]	285
[M(Hal)$_x$(η^5-C_5H_5)$_n$]	63, 298, 299
[M(η^5-C_5H_5)$_x$(arene)$_y$]	130, 300
[Fe(η^5-C_4H_4N)(η^5-C_5H_5)]	301

Compounds with main group IV—transition metal bonds

Compound	Reference
X_3M—M′(L)$_n$	
—Ta	311
—Cr, Mo, W	37, 311
—Mn, Re	311–314
—Fe	314–316
—Co	311, 312, 317–319

[a] Indicates a photoionization study.

established with accepted thermochemical data from other methods. A good example of the use of mass spectrometrically determined data combined with thermochemical data is the EI determination of bond dissociation energies and enthalpies of formation of organosilicon compounds containing Si—Me, Si—Et, Si—Si, Si—H, and Si—Cl bonds[228]. This study used a monochromated electron beam. Among the compounds studied the methylsilanes, $[Me_nSi\ H_{4-n}]$, were of particular interest since few if any reliable enthalpies of formation have been obtained for such compounds from calorimetric measurements. Combined with thermochemical data, the A values from 47 reactions gave consistent results. The enthalpies of formation were calculated and bond additivitvy schemes formulated for the calculation of enthalpies of formation of silicon compounds containing Si—alkyl, Si—Si, Si—H, and Si—Cl bonds. It was found that for any compound the Si—C, Si—Si, and Si—H bond dissociation energies were almost independent, having very small interaction corrections. For the Si—Cl bond there was greater variation with values from 116 kcal/mol in Me_3Si—Cl to 104 kcal/mol in Cl_3Si—Cl.

Table 2 lists organometallic compounds for which I and A data have been reported. A number of trends are noteworthy from these studies. Generally, the bond dissociation energies in ions $D(X_nM{-}R)^+$ are less than in molecules $D(X_nM{-}R)$ for main group organometallics but this is not necessarily so for transition metal compounds. Some examples are given in Table 3 for M—Me, M—C_5H_5 and M—CO bonds.

$$D(X_nM{-}R)^+ = A(X_nM)^+ - I(X_nMR) \tag{76}$$

Ionic bond dissociation energies were calculated from equation 76 assuming that no excess energy was present. A decrease in the values from both ions and molecules with

TABLE 3. Bond strengths in ions and molecules (kcal/mol)

Main group IV Me_4M[43]

Compound	$D(Me_3M{-}Me)$ ±5	$D(Me_3M{-}Me)^{+\bullet}$ ±5
Me_4Si	91	39
Me_4Ge	82	19
Me_4Sn	73	19
Me_4Pb	49	12

Transition metal cyclopentadienyl compounds $[M(C_5H_5)_2]$

Compound(M)	$E(M{-}C_5H_5)$[4]	$\bar{D}(M{-}C_5H_5)^+$ [285]
Fe	75	83.5
Co	79	97.5
Ni	71	74.5

Transition metal carbonyls $[M(CO)_6]$ and related $[M(CO)_5CS]$ compounds[225]

Compound (M)	$\bar{D}(M{-}CO)$	$\bar{D}(M{-}CO)^+$
Cr	25.6	25.4
Mo	36.2	37.2
W	42.7	45.4
	$D(CO)_5(M{-}CO)^+$	$D(CO)_5(M{-}CS)^+$
Cr	33.0	52.5
Mo	36.0	64.2
W	39.2	74.8

increasing atomic number of M in main group compounds is noted. The smaller differences between and ionic and molecular values for transition metal compounds reflects the fact that ionization from the predominantly transition metal-centred highest occupied molecular orbital (h.o.m.o.) in these compounds will have less effect on metal—ligand bonding than ionization from a h.o.m.o. which is more M—L bonding in character such as is found in most main group compounds[46].

The relative strengths of several metal—ligand bonds have been reported. In $[Mn(CO)_2(L)(\eta^5\text{-}C_5H_5)]$ (L = CO or CS) complexes the Mn—CO bond was weaker than the Mn—CS bond[292,293]. This was related to the fact that CS is a better σ-donor/π-acceptor than CO and CS is primarily a σ-donor. The effect on the Fe—Si bond in $[Fe(SiMe_3)(CO)(L)(\eta^5\text{-}C_5H_5)]$ of replacing L = CO by PPh_3 was studied and it was found that D(Fe—Si) increased with Ph_3P substitution[315]. Changing X from Me or Ph to Cl in $[Mn(SnX_3)(CO)_5]$ and $[Fe(SnX_3)(CO)_2(\eta\text{-}C_5H_5)]$ increased the M—Sn bond strength[310]. This agreed with Mössbauer evidence and the observed shortening of M—Sn bond lengths.

The assignment of ion structures on the basis of I and A data has often been attempted in organic mass spectrometry but has been less frequent with organometallics. A case in point is the formation of $[C_7H_8Cr]^+$ ions in the spectra of $[Cr(CO)_3(\eta^6\text{-}PhCH_3)]$ and $[Cr(CO)_3(\eta^6\text{-cycloheptatriene})]$[309]. In this study it was shown that the energies for the formation of $[C_7H_8Cr]^+$ from either molecule were very nearly identical. It was then argued that since the fragmentation behaviour of this ion is significantly different in each case, the ion did not have a common structure. This conclusion was reinforced by a study of the corresponding $[C_8H_{10}Cr]^+$ ions from $[Cr(CO)_3(\eta^6\text{-}PhEt)]$ and $[Cr(CO)_3(\eta^6\text{-7-}exo\text{-methylcycloheptatriene})]$.

F. Ion/Molecule Reaction Studies[41,216]

Generally these have been carried out with (a) normal, commercial mass spectrometers operating at higher than usual pressures, (b) ion cyclotron resonance (ICR) spectrometers or more recently (c) ion beam and molecular beam methods[230,321], and (d) flowing afterglow techniques[322,323]. ICR[216] has proved most useful to date for studying reactions between organometallic ions and M molecules, and between organometallic ions and potential ligand molecules. ICR techniques enable precursors to be identified, reaction rates to be measured and, in suitable cases, thermochemical properties to be calculated.

1. *Reactions in normal mass spectrometers*

A number of reports of mononuclear transition metal organometallics giving dimetallic ions in normal mass spectrometers have been published. Arene—chromium derivatives[324,325] $[Cr(CO)_3(arene)]$ gave ions of types $[Cr_2(arene)_2CO_n]^+$ by reaction 77, and even trinuclear ions $[Cr_3(arene)_2(CO)_n]^+ (n = 1\text{–}6)$ for arene = C_6H_6.

$$[Cr(CO)_3arene]^+ + [Cr(CO)_3arene] \longrightarrow [Cr_2(arene)_2(CO)_n]^+ (n = 0\text{–}3) \quad (77)$$

Displacement of CO ligands was studied in $[Mn(CO)_3(\eta^5\text{-}C_5H_5)]$[326]. Displacing ligands included MF_3(M = P, As, Sb) and SF_4 (equation 78). A similar study was made of $[Ni(NO)(\eta^5\text{-}C_5H_5)]$, with NO displaced by numerous groups including amines or olefins[327]. Other cyclopentadienylmetal carbonyls which have been studied include compounds of V, Mn, and Co[134,328].

$$Mn(CO)_3C_5H_5^+ + L \longrightarrow Mn(CO)LC_5H_5^+ + 2CO \quad (78)$$

Ferrocene is reported to show $[Fe_2(C_5H_5)_3]^+$ ions[329]. Reactions between ions derived from $[M(\eta^5\text{-}C_5H_5)(\eta^6\text{-}C_6H_6)]$ (M = Cr, Mn) or $[M(\eta^5\text{-}C_5H_5)(\eta^7\text{-}C_7H_7)]$(M = V, Cr) and various ligands including H_2O, NH_3, phosphines, acetone, benzene, and olefins have been discussed[330]. Other work in this area has concerned ferrocene derivatives[331], carbene complexes[332], and exchange of PF_3 for other ligands in $[Fe(PF_3)_5]$[333].

2. *Ion cyclotron resonance techniques*

ICR techniques have been applied to numerous reactions of organometallics, including those in Table 4.

TABLE 4. Studies of ion/molecule reactions

Ions derived from	Neutral	Reference
Li	Various organic species	334, 335
CH_4, Bu^iH	Mg	336
Me_2Hg	C_2H_4	337
	Allene	338
	MeI	339
	Alkenes	340
Me_3B	Me_3B	341
SF_6[a]	R_3B and R_2BF	342
$MeBC_5H_6$	Various species	343
Al	RX	344
SiH_4	CH_4	345
	C_2H_4	346
	C_2H_2	347
	C_6H_6	348
	CF_4	349
CH_4	Methyl silanes	350
$MeSiH_3$	$MeSiH_3$	351
Me_4Si	Me_4Si	175, 176
	Alcohols	352
	Ketones, esters, carboxylic acids	353
	Ethers	354
$(Allyl)_2Me_2Si$	Alcohols, ethers	355
(Various anions[a])	(Me_4Si, Me_3SiPh, other silanes)	323
Me_3SiNMe_2, other amine derivatives	Amines	356
$(Me_nH_{(3-n)}Si)_2O$	Various species	357
SF_6[a]	$Me_nSiF_{(4-n)}$	358
$Me_nSiF_{(4-n)}$	$Me_nSiF_{(4-n)}$	359
Me_3SiCl	Amines	360
MCl_4	MeF	361
Et_4Pb	Et_4Pb	362
Me_3P	Me_3P	363
$Me_nPH_{(3-n)}$	$Me_3PH_{(3-n)}$	364
$Me_3P(X)$	$Me_3P(X)$	
X = CH_2		365, 366
X = O, NH, NMe		366
Phosphiran	Phosphiran	367
$(Me_2P)_2$	$(Me_2P)_2$	368
Me_3As	Me_3As	369
Fe	Alkanes	370
Co	Alkanes	321
Fe, Co, Ni	Alkyl halides, alcohols	335

TABLE 4. *continued*

Ions derived from	Neutral	Reference
Fe or Co	MeI	339
Fe	Phenyl halides	371
$TiCl_4$	Small olefins	372
	Chlorohydrocarbons	373
Cu	Alkyl chlorides	374
	Methyl cyanide	375
$Fe(\eta^5-C_5H_5)_2$	$Fe(\eta^5-C_5H_5)_2$	376
	Arenes	377
HL, ligands	$M(\eta^5-C_5H_5)_2$ (M = Fe or Ni)	378
$Ni(NO)(\eta^5-C_5H_5)$	σ- and π-donor ligands	379
	Aldehydes	380
$M(\eta^5-C_5H_5)X_n$	Methyl halides	381
M = Mn; $X_n = (CO)_3$	NH_3	382
M = Co; $X_n = (CO)_2$[a]	$Co(CO)_2(\eta^5-C_5H_5)$	383
$[Fe(CO)_3(CH_2)(\eta^5-C_5H_5)]$	Various species	384
$[Mn(CH_3)(CO)_5]$	Various species	385, 386
$Fe(CO)_5$	Various ligands	387
$Co(CO)_3(NO)$	Various ligands	388

[a] Signifies a negative ion study.

a. Ion reactivities. Ion/molecule reactions can be extremely complex, as in the case of $Me_3P{=}O$ where 68 independent reactions between M and $M^{+\cdot}$ or fragment (product) ions occurred[366]. The most intense product ion was $[(CH_3)_3POH]^+$, itself formed

$$M^+ + M \longrightarrow (CH_3)_3\overset{+}{P}OH + POC_3H_8 \quad (79)$$

$$[M - Me]^+ + M \longrightarrow (CH_3)_3\overset{+}{P}OH + POC_2H_5 \quad (80)$$

$$H_2PO^+ + M \longrightarrow (CH_3)_3\overset{+}{P}OH + HPO \quad (81)$$

by three separate reactions (79–81). This ion undergoes seven ion/M reactions (82). Fortunately, in other cases the reactions observed are more specific, permitting a more detailed study of intermediates in fragmentation and allowing some insight into possible mechanisms.

$$[M + (CH_3)_3\overset{+}{P}OH] \xrightarrow{} {}^+M_2H \quad (82)$$

(other arrows from $[M + (CH_3)_3\overset{+}{P}OH]$ labelled: $-CH_3OH-C_2H_6$; $-(CH_3)_2POH$; $-C_2H_6-CH_4$; $-CH_3OH$, $-2\,H_2$; $-CH_3OH$, $-H_2$; $-CH_3OH$)

A number of $[Me_3Si]^+$—ether systems have been investigated[354]. The trimethylsiliconium ion is the predominant ion derived from Me_4Si. Reactions with ethers produce a 1:1 adduct ion $[Me_3SiORR^1]^+$ which generally decomposes be elimination of a neutral (R—H) group for R > Me. Experiments with deuterium-labelled species suggested that this elimination proceeded by β-hydrogen transfer to oxygen through a

$$Me_3Si-\overset{+}{O}(Et)(H\cdots CH_2-CH_2) \xrightarrow{-C_2H_4} Me_3Si-\overset{+}{O}(Et)H \qquad (83)$$

four-membered transition state, as in reaction 83. The hydrogen transfer appeared to be involved in the rate-determining step. For $R \neq R^1$ the (R—H) eliminated primarily is from whichever R is larger. Cyclic ethers $O(CH_2)_n$ formed adducts which, except for $n = 2$, decomposed by loss of $(CH_2)_{n-1}$ groups (equation 84). Formation of $[Me_3SiO(CH_2)_2]^+$ as a product ion was a minor feature from $n = 4$ and 5.

$$Me_3Si-\overset{+}{O}\langle CH_2(CH_2)_{n-2}CH_2\rangle \xrightarrow{(CH_2)_{n-1}} Me_3Si\overset{+}{O}CH_2 \qquad (84)$$

A number of interesting studies involving transition metal centred ions have been initiated. These include reactions between olefins and ions derived from $TiCl_4$[372] and reactions between alkyl halides or alcohols and metal ions[335]. In the former study it was shown that $[TiCl_4]^{+\cdot}$ was unreactive but $[TiCl_n]^+$ ($n < 4$) ions underwent a variety of reactions, typical reactions are 85 and 86. Both $[Ti]^+$ and $[TiCl]^+$ complexed with

$$Cl\overset{+}{Ti}C_6H_6 \xleftarrow{-3\,H_2} {}^{+}TiCl \xrightarrow{-H_2} Cl\overset{+}{Ti}(\text{diene}) \qquad (85)$$

$$Cl_3\overset{+}{Ti}C_2H_4 \xleftarrow{-C_4H_8} {}^{+}TiCl_3 \xrightarrow{-HCl} Cl_2\overset{+}{Ti}(\text{allyl}) \qquad (86)$$

olefins and then eliminated H_2 molecules whereas elimination of HCl was a common mode from the complexes with $[TiCl_2]^+$ and $[TiCl_3]^+$ ions. Complexes of $[TiCl_n]^+$ ($n = 1$–3) and larger olefins with five or more carbon atoms lost smaller olefinic fragments. Reactions of M^+ ions (M = Fe, Co, Ni), produced by electron impact on metal carbonyls, with alkyl halides or alcohols (RX) were postulated to involve an intermediate ion $[RMX]^+$[335]. This suggestion presumes a gas-phase oxidative addition and was supported by $[MX]^+$ and $[MR]^+$ ions in some cases and reactions such as 87, which are easiest to rationalize in terms of such intermediates.

$$[CH_3FeOH]^+ + CD_3OH \longrightarrow CD_3OH\overset{+}{Fe}OH + CH_3 \qquad (87)$$

Evidence was found for the participation of M—H bonds in fragmentation by dehydrohalogenation (X = halogen) and dehydration (X = OH). This has been a commonly postulated feature in the fragmentations of transition metal organometallics, and is well established in solution reactions[389]. The reaction between deuteriated ethyl iodide and $[Fe]^+$ showed HI and DI elimination occurring in the expected statistical ratio (2:3) for randomization via a β-hydrogen shift (equation 88). It was noticeable that $[Fe]^+$ and $[Co]^+$ tended to form $[MR]^+$ and $[M\text{—olefin}]^+$ ions in preference to $[MX]^+$ and $[MXH]^+$. The opposite tendency was found for $[Ni]^+$. Several reactions of $[M(CO)_n]^+$ ions were discussed, including CO displacement giving, e.g., $[M(CO)_{n-1}RX]^+$ and $[M(CO)_{n-1}R]^+$. The results suggested that

$$
\begin{array}{c}
Fe^+ + CD_3CH_2I \\
\downarrow \\
CD_3CH_2\overset{+}{Fe}I \rightleftharpoons (D)(I)\overset{+}{Fe}-(CH_2{=}CD_2) \xrightarrow{DI} \\
\Downarrow \\
\xleftarrow{HI} (H)(I)\overset{+}{Fe}-(CHD{=}CD_2) \rightleftharpoons CH_2DCD_2\overset{+}{Fe}I
\end{array}
\qquad (88)
$$

M—alcohol bonds were stronger than M—alkyl halide bonds. The MR-containing ions tended to undergo proton transfer from R. The importance of this was proportional to the acidity of $[R]^+$.

A number of studies of CO displacement or ligand addition to an organometallic ion have been reported (Table 4). With ions derived from $[Fe(CO)_5]$ the ability to displace CO and the number of carbonyls displaced depended on the basicity of L and the value of y in the ions $[Fe(CO)_y]^+$ [387]. Ligands with higher proton affinity (PA) than CO displaced it, including NH_3, H_2O, Me_2O, NO, C_6H_6, and MeF, but not HCl. More CO groups were displaced by ligands with higher proton affinities such that up to three H_2O molecules could be introduced but only one MeF molecule. Ions with lower values of y showed more extensive substitution.

b. Thermochemical studies[216]. The use of ICR techniques in determining gas-phase acidities or basicities as well as bond dissociation energy and enthalpy of formation data is increasing. Proton affinity may be defined as the negative of the enthalpy change of reaction 89.

$$M + H^+ \rightleftharpoons MH^+ \qquad (89)$$

Relative acidities/basicities may be deduced from a study of reactions of type 90. Since gas-phase ion/molecule reactions are assumed to proceed with negligible activation energy, the observation of a reaction 90 can be interpreted as showing that the

$$M_1^+ + M_2H \longrightarrow M_1H^+ + M_2 \qquad (90)$$

proton affinity of M_1H is greater than that of M_2H. Thus, the unknown proton affinity of a molecule M_1 may be established by 'bracketing' reactions of type 90 using compounds M_2H with known proton affinities. The accuracy of the determination will depend on the available calibrated compounds. A sufficiently large number of these are usually available for proton affinities to be determined to within ±4–5 kcal/mol. The gas-phase basicity may also be calculated. It is defined as the negative of the free energy change for the protonation reaction 89. Similarly hydrogen affinity (HA) may be defined (equation 91) and the relationship between equations 89 and 91 expressed as equation 92, where adiabatic I values are used for H and M. Proton affinities for a

$$MH^+ \rightleftharpoons M^+ + H \qquad (91)$$

$$PA(M) - HA(M^+) = I(H) - I(M) \qquad (92)$$

number of organometallics have been reported (Table 5), including methylphosphines, Me_nPH_{3-n}, and ferrocene. A study of the reactions between ferrocene and $[LH]^+$ ions (L = NH_3, $MeNH_2$, PH_3, H_2S, and Pr^i_2O) established the proton affinity of ferrocene to be 213 ± 4 kcal/mol, just greater than that of ammonia (205 kcal/mol)[378].

TABLE 5. Proton affinities (PA) of organometallics

Type	Compound	PA (kcal/mol)	Reference
Main group compounds	$Me_3Si(CH_2)_nNMe_2$:		
	$n = 1$	230.6	356
	$n = 2$	231.1	356
	$n = 3$	231.1	356
	$Bu^tMe_2SiNMe_2$	229.1	356
	Me_2SiCH_2	22.7	360
	PH_3	191.1	391
	$MePH_2$	204.8	392
	Me_2PH	217.1	392
	Me_3P	225.7	392
	Et_3P	231.7	391
	AsH_3	183.6	393
	Me_3As	213.7	369
	$\overline{CH_2CH_2PH}$	194.8	367
Transition metal compounds	$[Fe(\eta^5\text{-}C_5H_5)_2]$	213	378
	$[Ni(\eta^5\text{-}C_5H_5)_2]$	219	378
	$[Mn(CO)_3(\eta^5\text{-}C_5H_4Me)]$	197	382
	$[Mn(Me)(CO)_5]$	188	385
	$[Re(Me)(CO)_5]$	<191	385
	$[Fe(CO)_5]$	204	387

It is worth noting that attempts to measure acid/base strengths are sometimes thwarted by unwanted alternative reactions. Thus in a study of silylamines, transamination reactions upset the protonation equilibria 93[356].

$$\begin{aligned} Me_3SiNMe_2 + \text{piperidine-}H^+ &\rightleftharpoons Me_3SiNMe_2H^+ + \text{piperidine} \\ Me_3SiNMe_2H^+ + \text{piperidine} &\longrightarrow Me_3Si\ \text{piperidine}^+ + Me_2NH \end{aligned} \tag{93}$$

ICR methods have been used to obtain bond strength and enthalpy of formation data. An interesting example is the determination of ΔH_f^0 of 1,1-dimethylsilaethylene[360]. Trimethylsilylcations, produced by EI on Me_3SiCl, were subjected to reaction with a variety of bases, B, some of which deprotonated $[Me_3Si]^+$ (reaction 94). Of the bases used, piperidine (proton affinity 225.4 kcal/mol) failed to

$$Me_3SiCl \xrightarrow[-Cl^{\cdot}]{e} Me_2\overset{+}{Si}{-}CH_3 + B \longrightarrow Me_2Si{=}CH_2 + BH^+ \tag{94}$$

deprotonate $[Me_3Si]^+$ but $EtPr^iNH$ (proton affinity 226.9 kcal/mol) did. The average of these proton affinities was taken as that of $Me_2Si{=}CH_2$ and combined with the enthalpy of formation of $[Me_3Si]^+$ and $[H]^+$ yielded a value for ΔH_f^0 $(Me_2Si{=}CH_2) = 20.5$ kcal/mol with an estimated error ± 2 kcal/mol. This value was consistent with those obtained previously by other methods[390]. Finally, the Si=C π-bond energy was estimated using known thermochemical data to be 34 kcal/mol, significantly less than that of the C=C π-bond (60–65 kcal/mol in ethylene).

A study of the formation of CH_4 from protonation of $[Mn(Me)(CO)_5]$ has led to some information about possible intermediates[385]. Protonation of $[Mn(Me)(CO)_5]$ yields two products depending on the proton affinity (PA) of the base B (equations 95

$$[Mn(Me)(CO)_5] + BH^+ \longrightarrow \overset{+}{Mn}(CO)_5 + CH_4 + B \quad PA_B \leqslant 203 \text{ kcal/mol} \qquad (95)$$

$$[Mn(Me)(CO)_5] + BH^+ \longrightarrow \overset{+}{Mn}(Me)H(CO)_5 + B \quad PA_B \leqslant 188 \text{ kcal/mol} \qquad (96)$$

and 96). It is a surprising observation that $[Mn(Me)(H)(CO)_5]^+$ is formed only with bases with proton affinities substantially lower than those producing abundant $[Mn(CO)_5]^+$ ions. A simple interpretation of methane formation via the conjugate acid intermediate $[Mn(Me)(H)(CO)_5]^+$ would therefore appear untenable, especially as the further products from sufficiently highly exothermic reactions of type 97 are not $[Mn(CO)_5]^+$ and methane, but $[Mn(Me)(H)(CO)_4]^+$ and CO.

$$[Mn(Me)(H)(CO)_5]^{+*} \longrightarrow [Mn(Me)(H)(CO)_4]^+ + CO \qquad (97)$$

It was suggested that reactions 95 and 96 were not competitive in the sense of showing some common or readily interconverted intermediate. Instead, it was proposed that methane was eliminated with virtually no activation barrier by protonation of the Mn—Me bond. A second protonation site accessible with stronger donors led to an Mn—H bonded species which eliminated CO in preference to methane. The calculated value of $D[Mn(Me)(CO)_5{-}H]^+$ was 67 ± 3 kcal/mol, comparable to those for other first-row M—H bond strengths. Similar conclusions were reached with the rhenium analogue.

The chemistry of $[MnCH_2]^+$ has been studied[386]. Generation of this ion was via $[Mn]^+$ formed by EI on $[Mn_2(CO)_{10}]$, then reactions 98, 99, and 101.

$$Mn^+ + N_2O \longrightarrow MnO^+ + N_2 \xrightarrow{C_2H_4} MnCH_2^+ + CH_2O \qquad (98)$$

$$Mn^+ + \triangle \longrightarrow MnCH_2^+ + C_2H_4 \qquad (99)$$

$$Mn^+ + \overset{O}{\triangle} \longrightarrow MnO^+ + C_2H_4 \qquad (100)$$

$$Mn^+ + \overset{O}{\triangle} \longrightarrow MnCH_2^+ + CH_2O \qquad (101)$$

Reaction 99 established a lower limit for $D(Mn{-}CH_2)^+$ of 92 kcal/mol, and a further reaction (102), an upper limit of 100 kcal/mol. The manganese carbene ion underwent metathetical reactions with olefins CD_2CD_2 and $(CD_3)_2CCD_2$ giving only $[MnCD_2]^+$ and $[MnC(CD_3)_2]^+$, respectively (equation 103).

$$MnCH_2^+ + CD_2CD_2 \longrightarrow Mn^+ + C_3H_2D_4 \qquad (102)$$

$$MnCH_2^+ + CR^1_2{=}CR^2_2 \longrightarrow \begin{matrix} ^+Mn{-}CR^1_2 \\ | \quad\quad | \\ H_2C{-}CR^2_2 \end{matrix} \longrightarrow {}^+MnCR^1_2 + H_2C{=}CR^2_2 \qquad (103)$$

Analysis of the reactions of $[Mn(CH_2)(CO)_5]^+$ with olefins and of A data lead to an estimated $D(Mn(CO)_5{-}CH_2)^+ = 77 \pm 5$ kcal/mol, significantly lower than $D(Mn{-}CH_2)^+$. From these observations it was concluded that a strong $M{=}CR_2$ bond was desirable from the point of view that it would reduce the tendency for olefin homologation reactions (equation 102) but encourage metathesis (reaction 103).

It is obvious from a consideration of the examples given above that ICR and other ion/molecule techniques have a significant future in the study of organometallic chemistry.

V. REFERENCES

1. F. W. Aston, *Mass Spectra and Isotopes*, 2nd ed., Edward Arnold, London, 1942.
2. M. R. Litzow and T. R. Spalding, *Mass Spectrometry of Inorganic and Organometallic Compounds*, Elsevier, Amsterdam, 1973.
3. J. Charalambous (Ed.), *Mass Spectrometry of Metal-containing Compounds*, Butterworths, London, 1975.
4. D. B. Chambers, F. Glockling, and J. R. C. Light, *Q. Rev. Chem. Soc.*, **22**, 317 (1968).
5. M. I. Bruce, *Adv. Organomet. Chem.*, **6**, 273 (1968).
6. R. W. Kiser, in Characterisation of Organometallic Compounds, Part I, (Ed. M. Tsutsui), Wiley, New York, 1969.
7. T. R. Spalding, in *Spectroscopic Methods in Organometallic Chemistry* (Ed. W. O. George), Butterworths, London, 1970.
8. J. Müller, *Angew. Chem., Int. Ed. Engl.*, **11**, 653 (1972).
9. J. G. Dillard, *Chem. Rev.*, **73**, 589 (1973).
10. A. Maccoll (Ed.), *Mass Spectrometry, MTP International Review of Science, Physical Chemistry Series 2*, Vol. 5, Butterworths, London, 1975.
11. J. M. Miller and G. L. Wilson, *Adv. Inorg. Chem. Radiochem.*, **18**, 229 (1976).
12. M. I. Bruce, in *Mass Spectrometry* (Ed. D. H. Williams), Vol. 1, Specialist Periodical Reports, The Chemical Society, London, 1971, Chapter 5.
13. M. I. Bruce, in *Mass Spectrometry* (Ed. D. H. Williams), Vol. 2, Specialist Periodical Reports, The Chemical Society, London, 1973, Chapter 5.
14. T. R. Spalding, in *Mass Spectrometry* (Ed. R. A. W. Johnstone), Vol. 3, Specialist Periodical Reports, The Chemical Society, London, 1975, Chapter 5.
15. T. R. Spalding, in *Mass Spectrometry* (Ed. R. A. W. Johnstone), Vol. 4, Specialist Periodical Reports, The Chemical Society, London, 1977, Chapter 12.
16. T. R. Spalding, in *Mass Spectrometry* (Ed. R. A. W. Johnstone), Vol. 5, Specialist Periodical Reports, The Chemical Society, London, 1979, Chapter 13.
17. J. Lewis and B. F. G. Johnson, *Acc. Chem. Res.*, **1**, 245 (1968).
18. F. Glockling, *The Chemistry of Germanium*, Academic Press, New York, 1969.
19. R. B. King, *Fortschr. Chem. Forsch.*, **14**, 92 (1970).
20. M. Cais and M. S. Lupin, *Adv. Organomet. Chem.*, **8**, 211 (1970).
21. M. Lesbre, P. Mazerolles, and J. Satge, *The Organometallic Compounds of Germanium*, Wiley, New York, 1971.
22. V. Yu. Orlov, *Russ. Chem. Rev.*, **42**, 529 (1973).
23. R. H. Cragg and A. F. Weston, *J. Organomet. Chem.*, **67**, 161 (1974).
24. T. Nishiwaki, *Heterocycles*, **2**, 473 (1974).
25. P. E. Gaivoronskii and N. V. Larin, *Russ. Chem. Rev.*, **46**, 466 (1974).
26. I. Granoth, in *Topics in Phosphorus Chemistry* (Ed. J. Griffith and M. Grayson), Vol. 8, Wiley, New York, 1976.
27. *Mass Spectrometry Bulletin*, Mass Spectrometry Data Centre, UKCIS, The University, Nottingham.
28. J. H. Beynon, R. A. Saunders, and A. E. Williams, *The Mass Spectra of Organic Molecules*, Elsevier, Amsterdam, 1968.
29. J. Roboz, *Introduction to Mass Spectrometry, Instrumentation and Techniques*, Wiley–Interscience, New York, 1968.
30. D. H. Williams and I. Howe, *Principles of Organic Mass Spectrometry*, McGraw-Hill, New York, 1972.
31. B. J. Millard, *Quantitative Mass Spectrometry*, Heyden, London, 1979.
32. J. R. Chapman, *Computers in Mass Spectrometry*, Academic Press, London, 1978.
33. L. R. Crawford, *Int. J. Mass Spectrom. Ion Phys.*, **10**, 279 (1972).
34. E. McLaughlin and R. W. Rozett, *J. Organomet. Chem.*, **52**, 261 (1973).

35. G. F. Lanthier, J. M. Miller, S. C. Cohen, and A. G. Massey, *Org. Mass Spectrom.*, **8**, 235 (1974).
36. E. McLaughlin, L. M. Hall, and R. W. Rozett, *J. Phys. Chem.*, **77**, 2984 (1973); but see N. N. Greenwood, J. D. Kennedy, and D. Taylorson, *J. Phys. Chem.*, **82**, 623 (1978).
37. D. J. Cardin, S. A. Keppie, M. F. Lappert, M. R. Litzow, and T. R. Spalding, *J. Chem. Soc. A*, 2262 (1971).
38. J. H. Beynon, *Mass Spectrometry and its Application to Organic Chemistry*, Elsevier, Amsterdam, 1960, p. 429.
39. S. B. Miller, B. L. Jelus, J. H. Smith, B. M. Munson, and T. B. Brill, *J. Organomet. Chem.*, **170**, 9 (1979).
40. B. Johanson, R. Ryhage, and G. Westoo, *Acta Chem. Scand.*, **24**, 2349 (1970).
41. (a) *Mass Spectrometry* (Ed. D. H. Williams), Vol. 1, Specialist Periodical Reports, The Chemical Society, London, 1971; (b) Vol. 2, 1973; (c) Vol. 3 (Ed. R. A. W. Johnstone), 1975; (d) Vol. 4, 1977; (e) Vol. 5, 1979.
42. D. P. Stevenson, *Discuss. Faraday Soc.*, **10**, 35 (1951).
43. M. F. Lappert, J. B. Pedley, J. Simpson, and T. R. Spalding, *J. Organomet. Chem.*, **29**, 195 (1971).
44. T. Koenig, T. Balle, and W. Snell, *J. Am. Chem. Soc.*, **97**, 662 (1975).
45. R. E. Ballard, *Photoelectron Spectroscopy and Molecular Orbital Theory*, Adam Hilger, Bristol, 1978.
46. C. E. Moore, *Atomic Energy Levels*, Vol. 3, Circular 467, National Bureau of Standards, US Government Printing Office, Washington, D.C., 1958.
47. S. Pignataro, S. Torroni, G. Innota, and A. Foffani, *Gazz. Chim. Ital.*, **104**, 97 (1974).
48. J. Müller, *J. Organomet. Chem.*, **23**, C38 (1970).
49. R. Davis and D. J. Darensbourg, *J. Organomet. Chem.*, **161**, C11 (1978).
50. J. M. Miller, T. R. B. Jones, and G. B. Deacon, *Inorg. Chim. Acta*, **32**, L75 (1979).
51. R. G. Cooks, J. A. Beynon, R. M. Caprioli, and G. R. Lester, *Metastable Ions*, Elsevier, Amsterdam, 1973.
52. J. M. Miller and G. L. Wilson, *Int. J. Mass Spectrom. Ion Phys.*, **12**, 225 (1973).
53. T. R. Spalding, in reference 2, Part 2, Introduction, p. 79.
54. J. R. Majer and C. R. Patric, *Trans. Faraday Soc.*, **58**, 17 (1962).
55. R. G. Kostyanovsky and V. G. Pekhanov, *Org. Mass Spectrom.*, **6**, 1183 (1972).
56. J. J. de Ridder and G. Dijkstra, *Recl. Trav. Chim. Pays-Bas*, **86**, 1325 (1976).
57. D. J. Pasto and P. E. Timony, *Org. Mass Spectrom.*, **10**, 222 (1975).
58. F. Glockling, M. A. Lyle, and S. R. Stobart, *J. Chem. Soc., Dalton Trans.*, 2537 (1974).
59. M. Fishwick and M. G. H. Wallbridge, *J. Chem. Soc. A*, 57 (1971).
60. (a) D. B. Chambers, F. Glockling, and M. Weston, *J. Chem. Soc. A*, 1759 (1967); (b) K. C. Williams, *J. Organomet. Chem.*, **19**, 210 (1969).
61. A. M. Duffield, C. Djerassi, P. Mazerolles, J. Dubac, and G. Manuel, *J. Organomet. Chem.*, **12**, 123 (1968).
62. P. Mazerolles, J. Dubac, and M. Lesbre, *J. Organomet. Chem.*, **12**, 143 (1968).
63. J. Müller, *Chem. Ber.*, **102**, 152 (1969).
64. F. Glockling and J. G. Irwin, *J. Chem. Soc., Dalton Trans.*, 1424 (1973).
65. D. H. Williams, R. S. Ward, and R. G. Cooks, *J. Am. Chem. Soc.*, **90**, 966 (1968).
66. J. H. Bowie and B. Nussey, *Org. Mass Spectrom.*, **3**, 933 (1970).
67. D. Hellwinkel, C. Wünsche, and M. Bach, *Phosphorus*, **2**, 167, (1973).
68. F. Glockling, R. J. Morrison, and J. W. Wilson, *J. Chem. Soc., Dalton Trans.*, 95 (1973).
69. S. W. Brewer, T. E. Fear, P. H. Lindsay, and F. G. Thorpe, *J. Chem. Soc. C*, 3519 (1971).
70. F. Gockling and J. R. C. Light, *J. Chem. Soc. A*, 717 (1968).
71. T. R. Spalding, *Org. Mass Spectrom.*, **11**, 1019 (1976).
72. K. Henrick, M. Mickiewicz, N. Roberts, E. Shewchuk, and S. B. Wild, *Aust. J. Chem.*, **28**, 1473 (1975).
73. T. R. B. Jones, J. M. Miller, S. A. Gardner, and M. D. Rausch, *Can. J. Chem.*, **57**, 335 (1979).
74. D. R. Dimmel, C. A. Wilkie, and F. Ramon, *J. Org. Chem.*, **37**, 2665 (1972).
75. N. A. Bell and S. W. Breuer, *J. Chem. Soc., Perkin Trans. II*, 717 (1974).
76. J. Laane, *J. Am. Chem. Soc.*, **89**, 1144 (1967).

77. J. Tamas, K. Ujszazy, T. Szekely, and G. Bujtas, *Acta Chim. Acad. Sci. Hung.*, **62**, 335 (1969).
78. J. E. Drake, B. M. Glavincevskii, H. E. Henderson, and C. Wong, *Can. J. Chem.*, **57**, 1162 (1979).
79. J. Diekman, J. B. Thomson, and C. Djerassi, *J. Org. Chem.*, **32**, 3904 (1967).
80. R. G. Kostyanovskii, I. A. Nuretdinov, N. P. Grehkin, and I. I. Chervin, *Bull. Acad. Sci. USSR (Eng. transl.)*, **11**, 2429 (1969).
81. J. E. Drake, B. M. Glavincevskii, and C. Wong, *J. Inorg. Nucl. Chem.*, **42**, 175 (1980).
82. A. G. Starkey, R. A. Friedel, and S. H. Langer, Anal. Chem., **29**, 770 (1957).
83. G. H. Draffen, R. N. Stillwell, and J. A. McCloskey, *Org. Mass. Spectrom*, **1**, 669 (1968).
84. J. Diekman, J. B. Thomson, and C. Djerassi, *J. Org. Chem.*, **33**, 2271 (1968).
85. J. Diekman, J. B. Thomson, and C. Djerassi, *J. Org. Chem.*, **34**, 3147 (1969).
86. A. G. Brook, A. G. Harrison, and P. F. Jones, *Can. J. Chem.*, **46**, 2862 (1968).
87. S. Safe and O. Hutzinger, *Mass Spectrometry of Pesticides and Pollutants*, CRC Press, Cleveland, Ohio, 1973, Chapter 20.
88. B. Y. K. Ho, E. M. Dexheimer, L. Spialter, and L. D. Smithson, *Org. Mass Spectrom.*, **14**, 185 (1979).
89. G. Sonnek, K. G. Baumgarten, and D. Habisch, *J. Prakt. Chem.*, **322**, 94 (1980).
90. P. Haake, M. J. Frearson, and C. E. Diebert, *J. Org. Chem.*, **34**, 788 (1969).
91. R. G. Cooks and A. F. Gerrard, *J. Chem. Soc. B*, 1327 (1968).
92. R. H. Cragg, G. Lawson, and J. F. J. Todd, *J. Chem. Soc., Dalton Trans.*, 878 (1972).
93. V. Yu. Orlov, *J. Gen. Chem. USSR*, **37**, 2188 (1967).
94. J. C. Tou and C. S. Wang, *Org. Mass Spectrom.*, **3**, 287 (1970).
95. J. Tanaka and S. R. Smith, *Inorg. Chem.*, **8**, 265 (1969).
96. D. B. Chambers and F. Glockling, *J. Chem. Soc. A*, 735 (1968).
97. A. Benedetti, C. Preti, G. Tosi, and P. Zannini, *J. Chem. Soc., Dalton Trans.*, 1467 (1980).
98. A. J. Shortland and G. Wilkinson, *J. Chem. Soc., Dalton Trans.*, 872 (1973).
99. K. Mertis and G. Wilkinson, *J. Chem. Soc., Dalton Trans.*, 1488 (1976).
100. W. Kruse, *J. Organomet. Chem.*, **42**, C39 (1972).
101. K. K. Meyer and W. A. Herrmann, *J. Organomet. Chem.*, **182**, 361 (1979).
102. V. F. Sizoi, Yu. S. Nekrasov, Yu. N. Sukharev, N. E. Kolobova, O. M. Khitrova, N. S. Obezyuk, and A. B. Antonova, *J. Organomet. Chem.*, **162**, 171 (1978).
103. J. Müller and J. A. Connor, *Chem. Ber.*, **102**, 1148 (1969).
104. W. Kalbfus, E. O. Fischer, and J. W. Buckler, *J. Organomet. Chem.*, **129**, 79 (1977).
105. M. I. Bruce, *Org. Mass Spectrom.*, **2**, 63 (1969).
106. J. D. Hawthorne, M. J. Mays, and R. N. F. Simpson, *J. Organomet. Chem.*, **12**, 407 (1968).
107. R. B. King, *J. Organomet. Chem.*, **14**, P19 (1968).
108. T. Krück, L. Knoll, and J. Laufenberg, *Chem. Ber.*, **106**, 697 (1973).
109. T. H. Whitesides and R. W. Arhart, *Tetrahedron Lett.*, 297 (1972).
110. K. Yasufuku and H. Yamazaki, *Org. Mass Spectrom.*, **3**, 23 (1970).
111. M. Rosenblum, B. North, D. Wells, and W. P. Giering, *J. Am. Chem. Soc.*, **94**, 1239 (1972).
112. G. F. Emerson, L. Watts, and R. Petit, *J. Am. Chem. Soc.*, **87**, 131 (1965).
113. R. B. King and A. Efraty, *Org. Mass Spectrom.*, **3**, 1233 (1970).
114. A. Efraty, J. A. Potenza, L. Zyonty, M. H. A. Huang, and B. Toby, *J. Organomet. Chem.*, **145**, 315 (1978).
115. M. I. Bruce and J. R. Knight, *J. Organomet. Chem.*, **12**, 411 (1968).
116. M. M. Bursey, F. E. Tibbetts, and W. F. Little, *J. Am. Chem. Soc.*, **92**, 1087 (1970).
117. N. Maoz, A. Mendelbaum, and M. Cais, *Tetrahedron Lett.*, 2087 (1965).
118. J. Calleja, R. Davis, and I. A. O. Ojo, *Org. Mass Spectrom.*, **12**, 109 (1977).
119. O. Gambino, G. A. Vaglio, R. P. Ferrari, M. Valle, and G. Cetini, *Org. Mass Spectrom.*, **6**, 723 (1972).
120. M. S. Lupin and M. Cais, *J. Chem. Soc. A*, 3098 (1969).
121. A. N. Nesmeyanov, Y. S. Nekrasov, N. P. Avakyan, and I. I. Kritskaya, *J. Organomet. Chem.*, **33**, 375 (1971).
122. J. Müller, H. O. Stuhler, and W. Goll, *Chem. Ber.*, **108**, 1074 (1975).
123. A. Mendelbaum and M. Cais, *Tetrahedron Lett.*, 3847 (1964).
124. B. V. Zhuk, G. A. Domrachev, N. M. Semenov, E. I. Mysov, R. B. Materikova, and N. S. Kochetkova, *J. Organomet. Chem.*, **184**, 231 (1980).

125. D. T. Roberts, W. F. Little, and M. M. Bursey, *J. Am. Chem. Soc.*, **90**, 973 (1968).
126. H. Egger, *Monatsh. Chem.*, **97**, 602 (1966).
127. C. C. Lee, S. C. Chen, and R. G. Sutherland, *Can. J. Chem.*, **53**, 232 (1975).
128. D. V. Zagorevskii, Yu. S. Nekrasov, and D. A. Lemenovskii, *J. Organomet. Chem.*, **146**, 279 (1978).
129. C. C. Lee, R. G. Sutherland, and B. J. Thompson, *Tetrahedron Lett.*, 2625 (1972); A. N. Nesmeyanov, Yu. S. Nekrasov, N. I. Vasyukova, L. S. Kotova, and N. A. Vol'kenau, *J. Organomet. Chem.*, **96**, 265 (1975).
130. J. Müller and P. Göser, *J. Organomet. Chem.*, **12**, 163 (1968).
131. G. E. Herberich and J. Müller, *J. Organomet. Chem.*, **16**, 111 (1969).
132. J. Müller and P. Göser, *Chem. Ber.*, **102**, 3314 (1969).
133. M. M. Bursey, F. E. Tibbetts, W. F. Little, M. D. Rausch, and G. A. Moser, *Tetrahedron Lett.*, 1649 (1970).
134. J. Müller and K. Fenderl, *Chem. Ber.*, **103**, 3128 (1970).
135. R. Goetze and H. Nöth, *J. Organomet. Chem.*, **145**, 151 (1978).
136. G. E. Herberich and W. Koch, *Chem. Ber.*, **110**, 816 (1977).
137. M. J. Mays and R. N. F. Simpson, *J. Chem. Soc. A*, 1936 (1969).
138. J. A. Connor, E. M. Jones, G. K. McEwen, M. K. Lloyd, and J. A. McCleverty, *J. Chem. Soc., Dalton Trans.*, 1246 (1972).
139. P. Merbach, P. Würstl, H.-J. Seibold, and W. Popp, *J. Organomet. Chem.*, **191**, 205 (1980).
140. B. W. S. Kolthammer and P. Legzdins, *J. Chem. Soc., Dalton Trans.*, 31 (1978).
141. R. B. King, *Org. Mass Spectrom.*, **2**, 287 (1969).
142. H. Brunner and W. A. Herrmann, *J. Organomet. Chem.*, **57**, 183 (1973).
143. R. B. King, *Appl. Spectrosc.*, **23**, 148 (1969).
144. F. Glockling, T. McBreide, and R. J. I. Pollock, *Inorg. Chim. Acta*, **8**, 81 (1974).
145. J. Müller and M. Herberhold, *J. Organomet. Chem.*, **13**, 399 (1968); J. Müller and K. Fernderl, *J. Organomet. Chem.*, **19**, 123 (1969).
146. V. Harder, J. Müller, and H. Warner, *Helv. Chim. Acta*, **54**, 1 (1971).
147. E. Kostiner, M. L. N. Reddy, D. S. Urch, and A. J. Massey, *J. Organomet. Chem.*, **15**, 383 (1968).
148. R. B. King, *J. Am. Chem. Soc.*, **90**, 1429 (1968).
149. H. Brunner, K. K. Mayer, and J. Wachter, *Chem. Ber.*, **110**, 730 (1977).
150. R. B. King, *Org. Mass Spectrom.*, **2**, 257 (1969).
151. C. R. Eady, B. F. G. Johnson, and J. Lewis, *J. Chem. Soc., Dalton Trans.*, 477 (1977).
152. B. F. G. Johnson, R. D. Johnston, and J. Lewis, *J. Chem. Soc. A,* 2865 (1968).
153. J. Müller, H. Dorner, G. Huttner, and H. Lorenz, *Angew. Chem., Int. Ed. Engl.*, **12**, 1005 (1973).
154. J. Müller and H. Dorner, *Angew. Chem., Int. Ed. Engl.*, **12**, 842 (1973).
155. R. G. Kostyanovski, *Tetrahedron Lett.*, 2271 (1968); *Izv. Akad. Nauk SSSR, Ser. Khim.*, 2784 (1967).
156. I. I. Furlei, V. P. Yur'ev, V. I. Khvostenko, G. A. Tolstikov, and S. R. Rafikov, *Proc. Acad. Sci. USSR (Phys. Chem. Sect.)*, **222**, 616 (1975).
157. B. C. Pant and R. E. Sacher, *Inorg. Nucl. Chem. Lett.*, **5**, 549 (1969).
158. R. S. Gohlke, *J. Am. Chem. Soc.*, **90**, 2713 (1968).
159. H. H. Bowie and B. A. Hale, *Org. Mass Spectrom.*, **11**, 1105 (1976).
160. J. H. Bowie and B. Nussey, *J. Chem. Soc. D*, 17, (1970).
161. I. I. Furlei, U. M. Dzhemilev, V. I. Khvostenko, and G. A. Tolstikov, *Bull. Acad. Sci. USSR*, **25**, 1991 (1977).
162. R. G. Alexander, D. B. Bigley, and J. F. J. Todd, *J. Chem. Soc., Chem. Commun.*, 553 (1972); R. G. Alexander, D. B. Bigley, and J. F. J. Todd, *Org. Mass Spectrom.*, **7**, 963 (1973).
163. S. C. Cohen and E. C. Tifft, *J. Chem. Soc., Chem. Commun.*, 226 (1970).
164. S. C. Cohen, *Inorg. Nucl. Chem. Lett.*, **6**, 757 (1970).
165. R. E. Winters and R. W. Kiser, *J. Organomet. Chem.*, **4**, 190 (1965).
166. M. R. Blake, J. L. Garnett, I. K. Gregor, and S. B. Wild, *Org. Mass Spectrom.*, **13**, 20 (1978).
167. M. R. Blake, I. W. Fraser, J. L. Garnett, I. K. Gregor, and R. Levot, *J. Chem. Soc., Chem. Commun.*, 1004 (1974).

168. J. G. Dillard, *Inorg. Chem.*, **8**, 2148 (1969).
169. R. E. Sullivan, M. S. Lupin, and R. W. Kiser, *J. Chem. Soc., Chem. Commun.*, 655 (1969).
170. R. N. Compton and J. A. D. Stockdale, *Int. J. Mass Spectrom. Ion Phys.*, **22**, 47 (1976).
171. A. Sh. Sultanov, U. M. Dzhemilev, M. S. Mifatkhov, V. I. Khovstenko, and G. A. Tolstikov, *J. Gen. Chem. USSR*, **47**, 1229 (1977).
172. P. M. Lausarot, G. A. Vaglio, and M. Valle, *Inorg. Chim. Acta*, **35**, 227 (1979).
173. M. R. Blake, J. L. Garnett, I. K. Gregor, and S. B. Wild, *J. Organomet. Chem.*, **178**, C37 (1979).
174. K. R. Jennings, in *Gas Phase Ion Chemistry* (Ed. M. J. Bowers), Vol. 2, Academic Press, New York, 1979, Chapter 12.
175. T. J. Odiorne, P. Vouros, and D. J. Harvey, *J. Phys. Chem.*, **76**, 3217 (1972).
176. T. J. Odiorne, D. J. Harvey, and P. Vouros, *J. Org. Chem.*, **38**, 4274 (1973).
177. J. R. Krause and P. Potzinger, *Int. J. Mass Spectrom. Ion Phys.*, **18**, 303 (1975).
178. L. Simonotti and C. Trombini, *Inorg. Chim. Acta*, **21**, L27 (1977).
179. R. H. Fish and R. L. Holmstead, *Tetrahedron Lett.*, 2497 (1976).
180. R. H. Fish, R. L. Holmstead, and J. F. Casida, *Tetrahedron Lett.*, 1303 (1974).
181. R. H. Fish, R. L. Holmstead, and H. G. Kuivila, *J. Organomet. Chem.*, **137**, 175 (1977); R. H. Fish, E. C. Kimmel, and J. F. Casida, *J. Organomet. Chem.*, **118**, 41 (1976).
182. R. H. Fish, R. L. Holmstead, M. Gielen, and B. De Poorter, *J. Org. Chem.*, **43**, 4969 (1978).
183. S. D. Goff, B. L. Jelus, and E. E. Schweizer, *Org. Mass Spectrom.*, **12**, 33 (1977).
184. S. Sass and J. L. Fisher, *Org. Mass Spectrom.*, **14**, 257 (1979).
185. F. Kober, *Chem. Ztg.*, **101**, 532 (1977); J. Kaufmann, F. Kober, and L. Zimmer, *Z. Chem.*, **16**, 280 (1976).
186. M. Claeys and D. Van Haver, *Org. Mass Spectrom.*, **12**, 531 (1977).
187. S. J. Gaskell and C. J. W. Brooks, *Org. Mass Spectrom.*, **12**, 651 (1977).
188. C. P. Magee, L. G. Sneddon, D. C. Beer, and R. N. Gaines, *J. Organomet. Chem.*, **86**, 159 (1975).
189. G. T. Rodeheaver, and G. C. Farrant, and D. J. Hunt, *J. Organomet. Chem.*, **30**, C22 (1971).
190. D. F. Hunt, J. W. Russell, and R. L. Torian, *J. Organomet. Chem.*, **43**, 175 (1972).
191. M. R. Blake, J. L. Garnett, I. K. Gregor, and D. Nelson, *J. Organomet. Chem.*, **188**, 203 (1980).
192. W. P. Anderson, N. Hsu, C. W. Stanger, and B. Munson, *J. Organomet. Chem.*, **69**, 249 (1974).
193. M. R. Blake, J. L. Garnett, I. K. Gregor, and D. Nelson, *J. Organomet. Chem.*, **193**, 219 (1980).
194. L. Horner and U. M. Duda, *Phosphorus*, **5**, 135 (1975).
195. I. L. Agafonov and V. I. Faerman, *J. Anal. Chem. USSR*, **30**, 338 (1975).
196. C. Elschenbrioch and J. Heck, *Angew. Chem., Int. Ed. Engl.*, **16**, 479 (1977).
197. G. Rauscher, T. Clark, D. Poppinger, and P. von R. Schleyer, *Angew. Chem., Int. Ed. Engl.*, **17**, 276 (1978).
198. H. J. Veith, *Org. Mass Spectrom.*, **13**, 280 (1978).
199. D. E. Games, A. H. Jackson, L. A. P. Kane-Maguire, and K. Taylor, *J. Organomet. Chem.*, **88**, 345 (1975).
200. V. I. Zdanovitch, N. E. Kolobova, N. I. Vasynkova, Yu. S. Nekrasov, G. A. Panosyan, P. V. Petrovskii, and A. Zh. Khakaeva, *J. Organomet. Chem.*, **148**, 63 (1978).
201. C. N. McEwen and S. D. Ittel, *Org. Mass Spectrom.*, **15**, 35 (1980).
202. D. E. Games, J. L. Gower, M. Gower, and L. A. P. Kane-Maguire, *J. Organomet. Chem.*, **193**, 229 (1980).
203. R. J. Lawson and J. R. Shapley, *J. Am. Chem. Soc.*, **98**, 7433 (1976).
204. C. G. Pierpoint, G. F. Stuntz, and J. R. Shapley, *J. Am. Chem. Soc.*, **100**, 617 (1978).
205. Th. Würminghausen, H. J. Reinecke, and P. Braunstein, *Org. Mass Spectrom.*, **15**, 38 (1980).
206. D. G. Tuck, G. W. Wood, and S. Zhandire, *Can. J. Chem.*, **58**, 833 (1980).
207. B. P. Stimpson, D. S. Simons, and C. A. Evans, *J. Phys. Chem.*, **82**, 660 (1978).
208. T. L. Youngless, M. M. Bursey, and C. E. Rechsteiner, *Anal. Chem.*, **50**, 1951 (1978).
209. R. V. Hodges and J. L. Beauchamp, *Anal. Chem.*, **48**, 825 (1976).

210. B. Soltmann, C. C. Sweeley, and J. F. Holland, *Anal. Chem*., **49**, 1164 (1977).
211. D. F. Hunt, J. Shabanowitz, F. K. Botz, and D. Brent, *Anal, Chem*., **49**, 1160 (1977).
212. R. Stoll and F. W. Rollgen, *J. Chem. Soc., Chem. Commun*., 789 (1980).
213. G. Hanson and B. Munson, *Anal. Chem*., **50**, 1130 (1978).
214. T. Baer, in *Gas Phase Ion Chemistry* (Ed. M. T. Bowers), Vol. 1, Academic Press, New York, 1979, Chapter 5.
215. R. C. Dunbar, in *Gas Phase Ion Chemistry* (Ed. M. T. Bowers), Vol. 2, Academic Press, New York, 1979, Chapter 14.
216. J. L. Beauchamp, *Annu. Rev. Phys. Chem*., **22**, 527 (1971).
217. J. H. Beynon and J. R. Gilbert, in *Gas Phase Ion Chemistry* (Ed. M. T. Bowers), Vol. 2, Academic Press, New York, 1979, Chapter 13.
218. R. K. Boyd and J. H. Beynon, *Org. Mass Spectrom*., **12**, 163 (1977).
219. R. Davis, M. L. Webb, D. S. Millington, and V. Parr, *Org. Mass Spectrom*., **14**, 289 (1979).
220. I. G. Simm, C. J. Danby, and J. H. D. Eland, *Int. J. Mass Spectrom. Ion Phys*., **14**, 285 (1974).
221. C. S. T. Cant, C. J. Danby, and J. H. D. Eland, *J. Chem. Soc., Faraday Trans. II*, **71**, 1015 (1975).
222. R. C. Dunbar and E. W. Fu, *J. Am. Chem. Soc*., **95**, 2716 (1973); R. C. Dunbar, *J. Am. Chem. Soc*., **95**, 472 (1973).
223. B. S. Freiser and J. L. Beauchamp, *J. Am. Chem. Soc*., **99**, 3214 (1977).
224. R. C. Burnier and B. S. Freiser, *Inorg. Chem*., **18**, 906 (1979).
225. G. D. Michels, G. D. Flesch, and H. J. Svec, *Inorg. Chem*., **19**, 479 (1980).
226. R. A. W. Johnstone and B. N. McMaster, *J. Chem. Soc., Chem. Commun*., 730 (1973); R. A. W. Johnstone and B. N. McMaster, *Adv. Mass Spectrom*., **6**, 451 (1974).
227. J. L. Occolowitz, B. J. Cerimele, and P. Brown, *Org. Mass Spectrom*., **8**, 61 (1974).
228. P. Potzinger, A. Ritter, and J. Krause, *Z. Naturforsch*., **30A**, 347 (1975).
229. J. Berkowitz, D. A. Bafus, and T. L. Brown, *J. Phys. Chem*., **65**, 1380 (1961); G. E. Hartwell and T. L. Brown, *Inorg. Chem*., **5**, 1257 (1966).
230. D. B. Chambers, G. E. Coates, and F. Glockling, *Discuss. Faraday Soc*., **47**, 157 (1969); D. B. Chambers, G. E. Coates, and F. Glockling, *J. Chem. Soc. A*, 741 (1970).
231. D. A. Drews and G. L. Morgan, *Inorg. Chem*., **16**, 1705 (1977).
232. L. Friedman, A. P. Irsa, and G. Wilkinson, *J. Am. Chem., Soc*., **77**, 3689 (1955).
233. R. E. Winters and R. W. Kiser, *J. Organomet. Chem*., **10**, 7 (1967).
234. G. Distefano and V. H. Dibeler, *Int. J. Mass Spectrom. Ion Phys*., **4**, 59 (1970).
235. B. G. Hobrock and R. W. Kiser, *J. Phys. Chem*., **66**, 155 (1962).
236. B. G. Gowenlock, R. M. Haynes, and J. R. Majer, *Trans. Faraday Soc*., **58**, 1905 (1962).
237. F. Glockling, I. G. Irwin, R. J. Morrison, and J. J. Sweeney, *Inorg. Chim. Acta*, **19**, 267 (1976).
238. W. C. Steele, L. D. Nichols, and F. G. A. Stone, *J. Am. Chem. Soc*., **84**, 1154 (1962).
239. G. G. Hess, F. W. Lampe, and L. H. Sommer, *J. Am. Chem. Soc*., **86**, 3174 (1964); G. G. Hess, F. W. Lampe, and L. H. Sommer, *J. Am. Chem. Soc*., **87**, 5329 (1965).
240. B. G. Hobrock and R. W. Kiser, *J. Phys. Chem*., **65**, 2186 (1961).
241. S. J. Band, I. M. T. Davidson, and C. A. Lambert, *J. Chem. Soc. A*, 2068 (1968); S. J. Band, I. M. T. Davidson, C. A. Lambert, and T. L. Stephenson, *Chem. Commun*., 723 (1967).
242. G. Distefano, *Inorg. Chem*., **9**, 1919 (1970).
243. M. K. Murphy and J. L. Beauchamp, *J. Am. Chem. Soc*., **99**, 2085 (1977).
244. J. Tamas, G. Czira, A. K. Maltsev, and O. M. Nefedov, *J. Organomet. Chem*., **40**, 311 (1972).
245. J. L. Occolowitz, *Tetrahedron Lett*., **43**, 5291 (1966).
246. A. L. Yergey and F. W. Lampe, *J. Organomet. Chem*., **15**, 399 (1968); A. L. Yergey and F. W. Lampe, *J. Am. Chem. Soc*., **87**, 4204 (1965).
247. T. R. Spalding, *J. Organomet. Chem*., **56**, C65 (1973).
248. J. A. Connor, G. Finney, G. H. Leigh, R. N. Hazeldine, P. J. Robinson, R. D. Sedgwick, and R. F. Simmons, *Chem. Commun*., 178 (1966).
249. J. M. Gaidis, P. R. Briggs, and T. W. Shannon, *J. Phys. Chem*., **75**, 974 (1971).
250. G. Dube and V. Chvalovsky, *Collect. Czech. Chem. Commun*., **39**, 2621 (1974); G. Dube and V. Chvalovsky, *Collect. Czech. Chem. Commun*., **35**, 2641 (1970).

251. G. Innorta, L. Szepes, and J. Borossay, *Acta Chim. Acad. Sci. Hung.*, **89**, 23 (1976).
252. C. Koppel, H. Schwarz, and F. Bohlmann, *Org. Mass Spectrom.*, **9**, 567 (1974).
253. M. K. Murphy and J. L. Beauchamp, *J. Am. Chem. Soc.*, **99**, 4992 (1977).
254. E. Gey and G. Dube, *Int. J. Mass Spectrom. Ion Phys.*, **22**, 103 (1976).
255. H. Bock and H. Seidel, *J. Organomet. Chem.*, **13**, 87 (1968).
256. H. Bock and H. Alt, *J. Am. Chem. Soc.*, **92**, 1569 (1970).
257. L. E. Gusel'nikov, V. K. Potapov, E. A. Volnina, V. Yu. Orlov, V. M. Vdovin, and N. S. Nametkin, *Proc. Acad. Sci. USSR (Phys. Chem. Sect.)*, **229**, 753 (1976).
258. J. Tamas, K. Ujszaszy, A. K. Maltsev, and O. M. Nefedov, *J. Organomet. Chem.*, **87**, 275 (1975).
259. F. W. Lampe and A. Niehaus, *J. Chem. Phys.*, **49**, 2949 (1968).
260. J. K. Terlouw, W. Heerma, P. C. M. Frintrop, G. Dijkstra, and H. A. Meinma, *J. Organomet. Chem.*, **64**, 205 (1974).
261. P. E. Rakita, M. K. Hoffman, M. N. Andrew, and M. M. Bursey, *J. Organomet. Chem.*, **49**, 213 (1973).
262. D. I. Maclean and R. E. Sacher, *J. Organomet. Chem.*, **74**, 197 (1974).
263. G. Innota, S. Torroni, G. Distefano, D. Pietropaolo, and A. Ricci, *Org. Mass Spectrom.*, **12**, 766 (1977).
264. G. Distefano, A. Ricci, R. Danieli, A. Foffani, G. Innorta, and S. Torroni, *J. Organomet. Chem.*, **65**, 205 (1974).
265. D. B. Chambers and F. Glockling, *Inorg. Chim. Acta*, **4**, 150 (1970).
266. C. G. Pitt, M. M. Bursey, and P. F. Rogerson, *J. Am. Chem. Soc.*, **92**, 519 (1970).
267. J. J. de Ridder and G. Dijkstra, *Org. Mass Spectrom.*, **1**, 647 (1968).
268. G. Distefano, G. Innorta, S. Pignataro, and A. Foffani, *J. Organomet. Chem.*, **14**, 165 (1968).
269. G. M. Bogolyubov, N. N. Grishin, and A. A. Petrov, *Zh. Obshch. Khim.*, **39**, 2244 (1969); G. M. Bogolyubov, N. N. Grishin, and A. A. Petrov, *Zh. Obshch. Khim.*, **41**, 1710 (1971).
270. A. N. Smirov, L. A. Yagodina, V. M. Orlov, A. I. Bokanov, and B. I. Stepanov, *J. Gen. Chem. USSR*, **46**, 435 (1976).
271. G. M. Bogolyubov, N. N. Grishin, and A. A. Petrov, *Zh. Obshch. Khim.*, **41**, 811 (1971).
272. N. S. Dance, W. R. McWhinnie, and C. H. W. Jones, *J. Organomet. Chem.*, **125**, 291 (1977).
273. B. Cederlung, R. Lantz, A.-B. Hörnfeldt, O. Thorstad, and K. Undheim, *Acta Chem. Scand.*, **31B**, 198 (1977).
274. E. O. Fischer, C. G. Kreiter, H. J. Kollmeier, J. Müller, and R. D. Fischer, *J. Organomet. Chem.*, **28**, 237 (1971).
275. E. O. Fischer, H. J. Kollmeier, C. G. Kreiter, J. Muller, and R. D. Fischer, *J. Organomet. Chem.*, **22**, C39 (1970).
276. E. O. Fischer, M. Leupold, C. G. Kreiter, and J. Müller, *Chem. Ber.*, **105**, 150 (1972).
277. O. Gabmino, G. A. Vaglio, R. P. Ferrari, and M. Valle, *J. Organomet. Chem.*, **75**, 89 (1974).
278. P. Schissel, D. J. McAdoo, E. Hedaya, and D. W. McNeil, *J. Chem. Phys.*, **49**, 5061 (1968).
279. J. L. Thomas and R. G. Hayes, *J. Organomet. Chem.*, **23**, 487 (1970).
280. G. G. Devyatykh, S. G. Krasnova, G. K. Bonsov, N. V. Larin, and P. E. Gaivoronskii, *Dokl. Akad. Nauk SSSR*, **193**, 1069 (1970).
281. G. D. Flesch, G. A. Junk, and H. J. Svec, *J. Chem. Soc., Dalton Trans.*, 1102 (1972).
282. G. G. Davyatykh, P. E. Gaivoronskii, N. V. Larin, G. K. Borisov, S. G. Krasnova, and L. F. Zyugina, *Russ. J. Inorg. Chem.*, **19**, 496 (1974).
283. L. R. Crisler and W. G. Eggerman, *J. Inorg. Nucl. Chem.*, **36**, 1424 (1974).
284. G. M. Begun and R. N. Compton, *J. Chem. Phys.*, **58**, 2271 (1973).
285. J. Müller and L. D'Or, *J. Organomet. Chem.*, **10**, 313 (1967).
286. S. Pignataro and F. P. Lossing, *J. Organomet. Chem.*, **10**, 531 (1967).
287. A. Foffani, S. Pignataro, G. Distefano, and G. Innorta, *J. Organomet. Chem.*, **7**, 473 (1967).
288. R. E. Winters and R. W. Kiser, *J. Organomet. Chem.*, **4**, 190 (1965).
289. J. Müller, *J. Organomet. Chem.*, **18**, 321 (1969).
290. J. R. Krause and D. R. Bidinosti, *Can. J. Chem.*, **53**, 628 (1975).
291. A. Efraty, M. H. A. Huang, and C. A. Weston, *Inorg. Chem.*, **16**, 79 (1977).

292. A. Efraty, M. H. A. Huang, and C. A. Weston, *Inorg. Chem.*, **14**, 2796 (1975).
293. A. Efraty, R. Arneri, and M. H. A. Huang, *J. Am. Chem. Soc.*, **98**, 639 (1976).
294. E. O. Fischer, W. Bathelt, M. Herberhold, and J. Müller, *Angew. Chem.*, **80**, 625 (1968).
295. E. O. Fischer, W. Bathelt, and J. Müller, *Chem. Ber.*, **103**, 1815 (1970).
296. J. Müller, K. Fernderl, and B. Mertschenk, *Chem. Ber.*, **104**, 700 (1971).
297. G. Distefano, G. Innorta, S. Pignataro, and A. Foffani, *Int. J. Mass Spectrom. Ion Phys.*, **7**, 383 (1971).
298. G. E. Herberich, G. Greiss, H. F. Heil, and J. Müller, *Chem. Commun.*, 1328 (1971).
299. J. G. Dillard and R. W. Kiser, *J. Organomet. Chem.*, **16**, 265 (1969).
300. J. Müller and B. Mertshenk, *J. Organomet. Chem.*, **34**, 165 (1972).
301. R. Cataliotti, A. Foffani, and S. Pignataro, *Inorg. Chem.*, **9**, 2594 (1970).
302. G. G. Devyatykh, N. V. Larin, and P. E. Gaivoronskii, *Zh. Obshch. Khim.*, **39**, 1823 (1969).
303. G. G. Devyatykh, P. E. Gaivoronskii, and N. V. Larin, *J. Gen. Chem. USSR*, **43**, 1121 (1973).
304. N. V. Lavin, G. G. Devyatykh, and P. E. Gaivoronskii, *Russ. J. Inorg. Chem.*, **17**, 841 (1972).
305. G. Innorta, S. Pignataro, and G. Natile, *J. Organomet. Chem.*, **65**, 391 (1974).
306. J. R. Gilbert, W. P. Leach, and J. R. Miller, *J. Organomet. Chem.*, **49**, 219 (1973).
307. G. Simonneaux, G. Jaouen, R. Dabard, and P. Guernot, *J. Organomet. Chem.*, **132**, 321 (1977).
308. J. Müller and B. Mertschenk, *J. Organomet. Chem.*, **34**, C41 (1972).
309. R. Davis, I. A. Ojo, and M. L. Webb, *Org. Mass Spectrom.*, **13**, 547 (1978).
310. R. G. Bedford, *J. Phys. Chem.*, **81**, 1284 (1977).
311. D. H. Harris and T. R. Spalding, *Inorg. Chim. Acta,* **39**, 189 (1980).
312. R. A. Burnham and S. R. Stobart, *J. Chem. Soc., Dalton Trans.*, 1269 (1973); *J. Organomet. Chem.*, **86**, C45 (1975); R. A. Burnham and S. R. Stobart, *J. Chem. Soc., Dalton Trans.*, 1489 (1977).
313. F. E. Saalfeld, M. V. McDowell, J. J. De Corpo, A. D. Berry, and A. G. MacDairmid, *Inorg. Chem.*, **12**, 48 (1973).
314. H.C. Clarke and A. T. Rake, *J. Organomet. Chem.*, **82**, 159 (1974).
315. G. Innorta, A. Foffani, and S. Torroni, *Inorg. Chim. Acta*, **19**, 263 (1976).
316. T. R. Spalding, *J. Organomet. Chem.*, **149**, 371 (1978).
317. F. E. Saalfield, M. V. McDowell, S. K. Gondal, and A. G. MacDairmid, *Inorg. Chem.*, **7**, 1465 (1968).
318. F. E. Saalfeld, M. V. McDowell and A. G. MacDairmid, *J. Am. Chem. Soc.*, **92**, 2324 (1970).
319. F. E. Saalfeld, M. V. McDowell, A. G. MacDairmid, and R. E. Highsmith, *Int. J. Mass Spectrom. Ion Phys.*, **9**, 197 (1972).
320. W. R. Gentry, in *Gas Phase Ion Chemistry* (Ed. M. T. Bowers), Vol. 2, Academic Press, New York, 1979, Chapter 15.
321. P. B. Armentrout and J. L. Beauchamp, *J. Am. Chem. Soc.*, **102**, 1736 (1980).
322. D. Smith and N. G. Adams, in *Gas Phase Ion Chemistry* (Ed. M. T. Bowers), Vol. 1, Academic Press, New York, 1979, Chapter 1.
323. C. H. DePay, V. M. Bierbaum, L. A. Flippin, J. T. Grabowski, G. K. King, R. J. Schmitt, and S. A. Sullivan, *J. Am. Chem. Soc.*, **102**, 5012 (1980).
324. J. Müller and K. Fenderl, *Chem. Ber.*, **104**, 2199 (1971).
325. J. R. Gilbert, W. P. Leach, and J. R. Miller, *J. Organomet. Chem.*, **30**, C41 (1971); J. R. Gilbert, W. P. Leach, and J. R. Miller, *J. Organomet. Chem.*, **56**, 295 (1973).
326. J. Müller and K. Fenderl, *Chem. Ber.*, **104**, 2207 (1971).
327. J. Müller and W. Goll, *Chem. Ber.*, **106**, 1129 (1973).
328. J. Müller and W. Goll, *Chem. Ber.*, **107**, 2084 (1974).
329. S. M. Schildcrout, *J. Am. Chem. Soc.*, **95**, 3846 (1973).
330. J. Müller, W. Holzinger, and W. Kalbfus, *J. Organomet. Chem.*, **97**, 213 (1975).
331. D. V. Zagorevskii, Yu. S. Nekrasov, and G. A. Nurgalieva, *J. Organomet. Chem.*, **194**, 77 (1980).
332. J. Müller and W. Goll, *J. Organomet. Chem.*, **69**, C23 (1974).
333. R. C. Dougherty, M. A. Krevalis, and R. J. Clark, *J. Am. Chem. Soc.*, **101**, 2642 (1979).

334. R. D. Wieting, R. H. Staley, and J. L. Beauchamp, *J. Am. Chem. Soc.*, **97**, 924 (1975); R. L. Woodin and J. L. Beauchamp, *J. Am. Chem. Soc.*, **100**, 501 (1978).
335. J. Allison and D. P. Ridge, *J. Am. Chem. Soc.*, **101**, 4998 (1979).
336. P. L. Po and R. F. Porter, *J. Am. Chem. Soc.*, **99**, 4923 (1977).
337. R. D. Back, J. Ganglhofer, and L. Kevan, *J. Am. Chem. Soc.*, **94**, 6860 (1972).
338. R. D. Back, J. Patane, and L. Kevan, *J. Org. Chem.*, **37**, 257 (1975).
339. J. Allison and D. P. Ridge, *J. Am. Chem. Soc.*, **98**, 7445 (1976); J. Allison and D. P. Ridge, *J. Organomet. Chem.*, **99**, C11 (1975).
340. R. D. Back, A. J. Weibel, J. Patane, and L. Kevan, *J. Am. Chem. Soc.*, **98**, 6237 (1976).
341. M. K. Murphy and J. L. Beauchamp, *J. Am. Chem. Soc.*, **98**, 1433 (1976).
342. M. K. Murphy and J. L. Beauchamp, *Inorg. Chem.*, **16**, 2437 (1977).
343. S. A. Sullivan, H. Sandford, J. L. Beauchamp, and A. J. Ashe, *J. Am. Chem. Soc.*, **100**, 3737 (1978).
344. R. H. Hodges, P. B. Armentrout, and J. L. Beauchamp, *Int. J. Mass Spectrom. Ion Phys.*, **29**, 375 (1979).
345. G. W. Stewart, J. M. S. Henis, and P. P. Gaspar, *J. Chem. Phys.*, **57**, 1990 (1972); J. M. S. Henis, G. W. Stewart, and P. P. Gaspar, *J. Chem. Phys.*, **61**, 4860 (1974).
346. T. M. Mayer and F. W. Lampe, *J. Phys. Chem.*, **78**, 2433 (1974); W. N. Allen and F. W. Lampe, *J. Am. Chem. Soc.*, **99**, 6816 (1977).
347. T. M. H. Cheng, T. Y. Yu, and F. W. Lampe, *J. Phys. Chem.*, **78**, 2645 (1974).
348. W. N. Allen and F. W. Lampe, *J. Chem. Phys.*, **65**, 3378 (1976).
349. J. R. Krause and F. W. Lampe, *J. Am. Chem. Soc.*, **98**, 7826 (1976); J. R. Krause and F. W. Lampe, *J. Phys. Chem.*, **81**, 281 (1977).
350. G. W. Goodloe, E. R. Austin, and F. W. Lampe, *J. Am. Chem. Soc.*, **101**, 3472 (1979).
351. T. M. Mayer and F. W. Lampe, *J. Phys. Chem.*, **78**, 2422 (1974); T. M. Mayer and F. W. Lampe, *J. Phys. Chem.*, **78**, 2429 (1974).
352. I. A. Blair, G. Phillipou, and J. H. Bowie, *Aust. J. Chem.*, **32**, 59 (1979).
353. I. A. Blair and J. H. Bowie, *Aust. J. Chem.*, **32**, 1389 (1979).
354. V. C. Trenerry, J. H. Bowie, and I. A. Blair, *J. Chem. Soc., Perkin Trans. II*, 1640 (1979).
355. I. A. Blair, V. C. Trenerry, and J. H. Bowie, *Org. Mass Spectrom.*, **15**, 15 (1980).
356. K. H. Shea, R. Gobeille, and J. F. Wolf, *J. Organomet. Chem.*, **156**, 323 (1978).
357. C. G. Pitt, M. M. Bursey, and D. A. Chatfield, *J. Chem. Soc., Perkin Trans. II*, 434 (1976).
358. J. G. Dillard, *Inorg. Chem.*, **13**, 1491 (1974); M. K. Murphy and J. L. Beauchamp, *J. Am. Chem. Soc.*, **99**, 4992 (1977).
359. M. K. Murphy and J. L. Beauchamp, *J. Am. Chem. Soc.*, **98**, 5781 (1976).
360. W. J. Pietro, S. K. Pollack, and W. J. Hehre, *J. Am. Chem. Soc.*, **101**, 7126 (1979).
361. R. Kinser, J. Allison, T. G. Dietz, M. de Angelis, and D. P. Ridge, *J. Am. Chem. Soc.*, **100**, 2706 (1978).
362. R. C. Dunbar, J. F. Ennever, and J. P. Fackler, *Inorg. Chem.*, **12**, 2735 (1973).
363. K.-P. Wanczek and Z. C. Profous, *Int. J. Mass Spectrom. Ion Phys.*, **17**, 23 (1975).
364. K.-P. Wanczek, *Z. Naturforsch.*, **30A**, 329 (1975); R. H. Staley and J. L. Beauchamp, *J. Am. Chem. Soc.*, **96**, 6252 (1974).
365. O. R. Hartmann, K.-P. Wanczek, and H. Hartmann, *Z. Naturforsch.*, **31A**, 630 (1976).
366. O.-R. Hartmann, K.-P. Wanczek, and H. Hartmann, in *Ion Cyclotron Resonance Spectrometry* (Ed. H. Hartmann and K.-P. Wanczek), Springer-Verlag, Berlin, 1978, p. 283.
367. Z. C. Profous, K.-P. Wanczek, and H. Hartmann, *Z. Naturforsch.*, **30A**, 1470 (1975).
368. K.-P. Wanczek, *Z. Naturforsch.*, **31A**, 414 (1976).
369. R. V. Hodges and J. L. Beauchamp, *Inorg. Chem.*, **14**, 2887 (1975).
370. J. Allison, R. B. Freas, and D. P. Ridge, *J. Am. Chem. Soc.*, **101**, 1332 (1979).
371. T. G. Dietz, D. S. Chatellier, and D. P. Ridge, *J. Am. Chem. Soc.*, **100**, 4905 (1978).
372. J. Allison and D. P. Ridge, *J. Am. Chem. Soc.*, **99**, 35 (1977).
373. J. S. Uppal and R. H. Staley, *J. Am. Chem. Soc.*, **102**, 4144 (1980).
374. R. W. Jones and R. H. Staley, *J. Am. Chem. Soc.*, **102**, 3794 (1980).
375. R. C. Burnier, T. J. Carlin, W. D. Reents, R. B. Cody, R. K. Lengel, and B. S. Freiser, *J. Am. Chem. Soc.*, **101**, 7127 (1979).
376. M. S. Foster and J. L. Beauchamp, *J. Am. Chem. Soc.*, **97**, 4814 (1975).
377. P. H. Hemberger and R. C. Dunbar, *Inorg. Chem.*, **16**, 1246 (1977).

378. R. R. Cordermann and J. L. Beauchamp, *Inorg. Chem.*, **15**, 665 (1976); M. S. Foster and J. L. Beauchamp, *J. Am. Chem. Soc.*, **97**, 4814 (1975).
379. R. R. Cordermann and J. L. Beauchamp, *J. Am. Chem. Soc.*, **98**, 3998 (1976).
380. R. R. Cordermann and J. L. Beauchamp, *J. Am. Chem. Soc.*, **98**, 5702 (1976).
381. R. R. Cordermann and J. L. Beauchamp, *Inorg. Chem.*, **17**, 68 (1978).
382. J. Fernando, G. Faigle, A. M. da C. Ferreira, S. E. Galembeck, and J. M. Riveros, *J. Chem. Soc., Chem. Commun.*, 126 (1978).
383. R. R. Cordermann and J. L. Beauchamp, *Inorg. Chem.*, **16**, 3135 (1977).
384. A. E. Stevens and J. L. Beauchamp, *J. Am. Chem. Soc.*, **100**, 2584 (1978).
385. A. E. Stevens and J. L. Beauchamp, *J. Am. Chem. Soc.*, **101**, 245 (1979).
386. A. E. Stevens and J. L. Beauchamp, *J. Am. Chem. Soc.*, **101**, 6449 (1979).
387. M. S. Foster and J. L. Beauchamp, *J. Am. Chem. Soc.*, **97**, 4808 (1975).
388. G. H. Weddle, J. Allison, and D. P. Ridge, *J. Am. Chem. Soc.*, **99**, 105 (1977).
389. W. Lamanna and M. Brookhart, *J. Am. Chem. Soc.*, **102**, 3490 (1980), and references cited therein.
390. I. M. T. Davidson, N. A. Ostah, D. Seyferth, and D. P. Duncan, *J. Organomet. Chem.*, **187**, 297 (1980); L. E. Gusel'nikov and N. S. Nametkin, *J. Organomet. Chem.*, **169**, 155 (1979).
391. D. H. Aue and M. T. Bowers, in *Gas Phase Ion Chemistry* (Ed. M. T. Bowers), Vol. 2, Academic Press, New York, 1979, Chapter 9.
392. R. H. Staley and J. L. Beauchamp, *J. Am. Chem. Soc.*, **96**, 1604 (1974).
393. J. F. Wolf, R. H. Staley, I. Koppel, I. Taagepera, R. T. McIver, J. L. Beauchamp, and R. W. Taft, *J. Am. Chem. Soc.*, **99**, 5417 (1977).

Author Index

This author index is designed to enable the reader to locate an author's name and work with the aid of the reference numbers appearing in the text. The page numbers are printed in normal type in ascending numerical order, followed by the reference numbers in parentheses. The numbers in *italics* refer to the pages on which the references are actually listed.

Subject Index